THE M.V.M. CAPPELLIN GLASSWORKS AND THE YOUNG CARLO SCARPA 1925–1931

THE M.V.M. CAPPELLIN GLASSWORKS AND THE YOUNG CARLO SCARPA 1925-1931

edited by
Marino Barovier and Carla Sonego

LE STANZE DEL VETRO

fondazione GIORGIO CINI onlus

SKIRA

The M.V.M. Cappellin Glassworks and the Young Carlo Scarpa 1925–1931

Exhibition organized by
LE STANZE DEL VETRO

Island of San Giorgio Maggiore
Venice

LE STANZE DEL VETRO

A joint initiative of
Fondazione Giorgio Cini
and Pentagram Stiftung

Scientific committee
Luca Massimo Barbero, *Chairman*
Marino Barovier
Rosa Barovier Mentasti
David Landau
Jean-Luc Olivié
Nico Stringa
Giorgio Vigna

Coordination management
Fondazione Giorgio Cini
and Pentagram Stiftung
Maria Novella Benzoni

Exhibition space design
Selldorf Architects
F. Cattaruzza and F. Millosevich
Architetti Associati

Lighting design
Alessandro Diaz de Santillana
FontanaArte

Fondazione Giorgio Cini

Chairman
Giovanni Bazoli

Secretary General
Pasquale Gagliardi

Facility coordination and management
Maria Novella Benzoni
Erica Galvan

Technical department
Massimo Altieri
Adriano Longhin
Gloria Pasqualetto

Marketing and communication department
Emilio Quintè
Serena Concone
Giovanna Pesaro

Press office
CASADOROFUNGHER
Comunicazione

Management
Michele Ballarin

Institute of Art History

Director
Luca Massimo Barbero

Coordination management
Simone Guerriero

Secretariat
Rossella Patrizio

Glass Study Center

Marzia Scalon
Sabina Tutone

Pentagram Stiftung

Chairman
Marie-Rose Kahane

Head of cultural projects
Francesca Nisii

Exhibition curated by
Marino Barovier

Exhibition project
Marino Barovier

Head of conservation
Luisa Mensi

Registrar
Laura Corazzol

Graphic design of communication materials
bruno - Andrea Codolo and Giacomo Covacich

Communication
CASADOROFUNGHER
Comunicazione

Social media
Silvia Casalini Martinelli

Organisational coordinator
Fiammetta Calzavara

Educational
Artsystem
Anna Fornezza

Secretariat
Chiara Brescacin

Video
Gianluigi Calderone
Antonio Pintus

Exhibition set-up and lighting
OTT ART prodotti per l'arte
Giacomo Andrea Doria
Andrea Riato

Transports
Apice - Venice
Alice Zanon

Insurance
Assicurazioni Generali
Marine & Aviation JLT – Divisione Fine-Art
Alberto Magni
Zilkens Fine Art, Insurance Broker GmbH

Reception
Artsystem
Valentina Stella

Security
Iniziative Venete

Technical sponsor
Artsystem

Catalogue edited by
Marino Barovier *and* Carla Sonego

Scientific research and coordination
Carla Sonego

Catalogue graphic project
Tapiro Camplani+Pescolderung

Coordination of editorial design and layout
Enrico Fiorese
Francesca Tiengo

Photographs
Enrico Fiorese

Revision of the texts
Anna Albano

Translations from Italian into English
Scarlet Sweetman
Mícheál Ua Nunáin

First published in Italy in 2018 by
Skira editore S.p.A.
Palazzo Casati Stampa
via Torino 61
20123 Milano
Italy
www.skira.net

Printed and bound in Italy. First edition

ISBN: 978-88-572-3925-5

Distributed in USA, Canada, Central & South America by ARTBOOK | D.A.P. 75, Broad Street Suite 630, New York, NY 10004, USA.
Distributed elsewhere in the world by Thames and Hudson Ltd., 181A High Holborn, London WC1V 7QX, United Kingdom.

Photo credits
Museum Kunstpalast, Düsseldorf, Glasmuseum Hentrich, LVR-ZMB, Stefan Arendt, p. 109, no. 5669; p. 134, no. 5718; p. 209, no. 5952; p. 296, no. 5970
BIB 706146, J. Stanley Brothers Collection, The Rakow Research Library, The Corning Museum of Glass, Corning, New York, p. 422

© Christie's Images / Bridgeman Images, p. 359, no. 3871
© Comune di Milano, tutti i diritti riservati, Raccolte d'Arte Applicata, Castello Sforzesco, Milano, p. 483
© La Triennale di Milano - Archivio Fotografico, pp. 38, 44, 134, 135, 170, 467, 475, 481
© Philadelphia Museum of Art, p. 453, nos. 4709, 4710, 4712; p. 454, nos. 4711, 4713, 4714
© The Cleveland Museum of Art, p. 461, n.i.
© The Corning Museum of Glass, p. 249, no. 5856; p. 285, no. 5952; p. 350, no. 5970; p. 376, no. 6005
© The Metropolitan Museum of Art, Foto Scala Firenze, p. 382, no. 5536; p. 350, no. 5928
© The Newark Museum, p. 164, no. 5700
© The Steinberg Foundation, p. 385, n.i; p. 355, no. 1281

Servizi Fondazione Pentagram is at the disposal of the entitled parties as regards all unidentified iconographic sources

We would like to thank the museums, foundations, public and private archives, collectors, scholars and all those who have contributed to this project:

Giuliana Altea
Massimo Altieri
Guido Antonello
Selldorf Architects
Marco Arosio
Riccardo Avanzi
Luca Massimo Barbero
Alberto Barigazzi
Marina Barovier
Rosa Barovier Mentasti
Claudio Bazzichetto
Carl Bellavia
Gabriella Belli
Cristina Beltrami
Sara Benetti
Fabio Benzi
Maria Novella Benzoni
Ketty Bertolaso
Giorgina Bertolino
Marco Berton
Robert Beyer
Maurizio Bianchini
Giancarlo Bisazza
Christina and Bruno Bischofberger
Elisa Bissacco
Sara Blumberg
Angelo Boscolo
Anna Bozano
Chiara Brescacin
Warren Bunn
Massimo Busetto
Renzo Busetto
Gianluigi Calderone
Fiammetta Calzavara
Davide Cammi
Enrico Camponi
Carlo Capovilla
Alessandra Cappellin
Consuelo Cappellin
Luigi Cappellin
Chiara Carraro
Leonardo Carrer
Elena Casadoro
Silvia Casalini Martinelli
Fabrizio Cattaruzza
Enrico Cattelan
Emma Cavazzini
Giancarlo Chimento
Giulia Chimento
Renata Codello
Andrea Codolo
Serena Concone
Laura Corazzol
Serena Corvi Mora
Aud and Paolo Cuniberti
Francesco Curato
Gianni De Carlo
Irene De Guttry
Laura de Santillana
Barbara De Zorzi
Franco Deboni
Alessandro Degnoni
Marco Della Chiesa
Alba Di Lieto
Alessandro Diaz de Santillana
Manuela Divari
Giacomo Andrea Doria
Roberto Dulio
Daniela Ferretti
Gerard Figliola
Luigi Fiore
Enrico Fiorese
Anna Fornezza
Anna Maria Foscolo Cappellin
Ferruccio Franzoia
Francesca Fungher
Marica Gallina
Erica Galvan
Roberto Gasparotto
Anna Gastel
Mario Gemin
Luciano Gemin
John and Jean Geresi
Anthony Germano
Sara Gerometta
Roberta Giacoi
Giorgio Grassi Damiani
Guido Grassi Damiani
Silvia Grassi Damiani
Simone Guerriero
Marc Heiremans
Lutz H. Holz
Paolo Kahlberg
Elizabeth Kessenides
Martin Kline
Richard Kossmann
Orietta Lanzarini
Reto Lardelli
Christiane Leister
Enrico Lepschy
Howard Lockwood
Anna Lombardi
Carlo Longagnani
Silvia Ludi-Sokalski
Carla Macola Bonsembiante
Maria Paola Maino
Lucia Mannini
Cen Massobrio Maciotta
Bruna Mateotti Moretti
Luisa Mensi
Francesco Millosevich
Marta Moretti
Manlio Morterra
Giampiero Mughini
Giordana Naccari
Francesca Nisii
Jim Oliveira
Jean-Luc Olivié
Gloria Pasqualetto
Rossella Patrizio
Nerina and Giancarlo Pellegrini
Luigi Pescolderung
Stefano Piantini
Antonio Pintus
Paola Placentino
Stefano Poli
Doreno Pontiggia
Emilio Quintè
Rita Ragazzi
Elvia Redaelli
George Renand
Roberta Rinaldi Moretti
Sofia Rinaldi
Marco Rocca
Pierre Rosenberg
Francesca Rossi
Kathleen and David Roth
Giovanni Salmistrari
Sheri Sandler
Marzia Scalon
Simona Scarchilli
Susanne Schneeweiss
Fred Schroeder and Allison Whiting
Giuseppe Scognamiglio
Annabelle Selldorf
Monica Selva
Alberto Serra
Anna Somers-Cocks
Anita Sonego

Nancy Olnick and Giorgio Spanu
Chiara Squarcina
Michele Stavagna
Valentina Stella
Chris Stines
Carolin Strawson
Nico Stringa
Elisa Sturiale
Scarlet Sweetman
Manuel Tarlà
Christoph Thun-Hohenstein
Francesca Tiengo
Francesco Toniatti Giacometti
Graziella Tonnarelli
Maurizio Torcellan
Isotta Tortorelli
Filippo Trevisanello
Giuliana Tucci
Mario Tucci
Sabina Tutone
Mícheál Ua Nunáin
Francesca and Massimo Valsecchi
Rosanna Vascon
Teresa and Giorgio Veronese
Davide Vianello
Massimo Vitta Zelman
Lambert von Grasern
Dedo von Kerssenbrock-Krosigk
Anna Wagner
Verena Wasmuth
Karol White
Karol and Richard Wright
Nicoletta Zanetti
Cesare Zanini
Tommaso Zanini
Alice Zanon
Francesco Zanon
Paolo Zanon
Alma Zevi
Patricia and Ranier Zietz

Fitzwilliam Museum, Cambridge
Fondazione Boschi Di Stefano
Fondazione Fiera di Milano
Fondazione Giorgio Cini Onlus, Venice
Fondazione Musei Civici di Venice:
Museo del Vetro, Murano
La Biennale di Venezia, Venice
MAXXI Museo nazionale delle arti del XXI secolo, Rome. Collezione MAXXI, Architettura. Archivio Carlo Scarpa
Metropolitan Museum of Art, New York
Musei del Castello Sforzesco, Milan
Museo di Castelvecchio, Verona
Museo Rizzarda, Feltre
Museum Kunstpalast, Düsseldorf
Newark Museum, New Jersey
Philadelphia Museum of Art, Philadelphia, PA
Stedelijk Museum, Amsterdam
The Cleveland Museum of Art, Cleveland, OH
The Corning Museum of Glass and The Rakow Library, Corning, NY
The Steinberg Foundation, Vaduz

Archivi della Triennale di Milano, Milan
Archivi Centro Studi del Vetro, Fondazione Giorgio Cini, Venice
Archivio Carlo Scarpa - Museo di Castelvecchio, Verona
Archivio Howard J. Lockwood, Fort Lee, NJ
Archivio Marc Heiremans, Brussels
Archivio Marco Arosio, Milan
Archivio Marino Barovier, Venice
Archivio Museo del Vetro, Venice
Archivio Storico Venezia, Camera di Commercio, Venezia, Rovigo, Delta Lagunare
Archivio Storico Venini, Murano

Auctioneers:
Cambi
Christie's
Il Ponte
Phillips
Quittenbaum
Ripley
Sotheby's
Wright

We would also like to thank all those who prefer to remain anonymous

Marino Barovier *also wishes to express his thanks to the authors of the essays in the catalogue. He is particularly grateful to* Marie-Rose Kahane, David Landau *and* Pasquale Gagliardi

PRODIGALITY AND REFINEMENT

The exhibition documented by this catalogue illustrates the essential role played by Giacomo Cappellin in the renaissance of Murano art glass and in the launching of the 'Venini era'. This, together with other exhibitions presented in 'Le Stanze del Vetro' leads one to reflect on the notion of the 'authorship' or 'paternity' of a work and to pose this problem. When admiring the products of the Maestri Vetrai Muranesi Cappellin & C. we ask ourselves if it is possible — and if so how — to identify and weigh up the respective roles of Cappellin, the tenacious, inspired and spendthrift entrepreneur (who encouraged the glassblower — as Marino Barovier reminds us in his essay — to 'waste' all the time necessary to try out a technical innovation), the personal ability of the glassblower, the pictorial sensitivity of the artistic director, the creative contribution of the young Carlo Scarpa (a gifted student of architecture in the Accademia di Belle Arti of Venice, whose name is not even mentioned in the various exhibitions as a designer of glass pieces but only of the layout of the events).

And yet, looking at these objects, we can only be struck by a kind of 'fingerprint', a trademark which fully reflects the artistic personality of the entrepreneur, an extraordinary combination of formal refinement and executive precision, 'an example not so much rare as unique of an artist who expresses himself exclusively through interpreters whose thought and will he moulds just as his workers mould the liquid sand with their breath' (Bernardi, quoted by Barovier). On the altar of the excellence of the results, the prodigal Cappellin was prepared to sacrifice all the time and money needed. This cultural approach created a financially unsustainable situation: the company closed after only a few years. Cappellin passed like a meteor through the 1920s but the traces of his whirlwind passage are deep and visible: he brought Murano glass to the United States and to all the important European exhibitions and demonstrated the importance of experimentation and the revolutionary value of excellence, contributing in a decisive way to the renewal process of the decorative arts in Italy. He sensed the creative potential of the young Carlo Scarpa: introducing him to the world of art glass, he initiated a fruitful dialogue with him, transmitted his love for glass to him and progressively led him to establish himself both as a designer of glass pieces and a planner of interiors and furnishing. This is probably the biggest gift that he has left to us: without Cappellin, Carlo Scarpa would never have said, as is recalled by Carla Sonego in her essay, 'After the Academy I found work in a Murano glassworks. I learned to work with a marvellous material [...]. I know glass and I know what can be done with it'.

I hope that this exhibition will have the success it deserves and would like to express — also in the name of the Chairman of the Management Committee of the Fondazione Cini — my most sincere gratitude to all those who have contributed to bringing it about, in particular Marino Barovier, who has curated it with such attention and passion. Special recognition is due once again to David Landau, Marie-Rose Kahane and all the colleagues of the Pentagram Stiftung, the loyal partner of the Fondazione Giorgio Cini in the project 'Le Stanze del Vetro', which Xavier Salomon has described in Apollo as being 'the most successful cultural institution created in Italy in the past decade'.

Pasquale Gagliardi
Secretary-General of the Fondazione Giorgio Cini
Venice, 9 July 2018

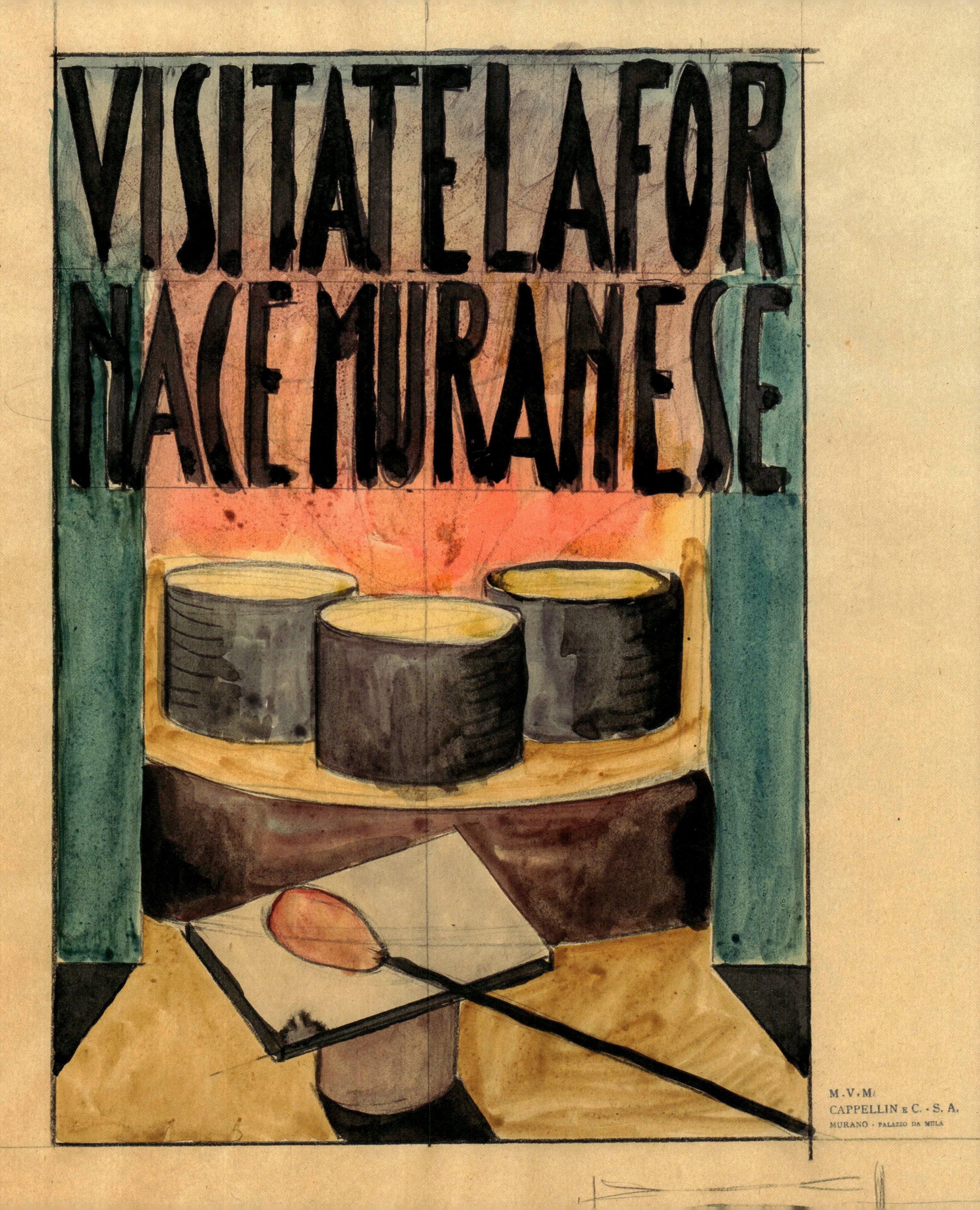

Carlo Scarpa, study for poster, c. 1928, CSV

CONTENTS

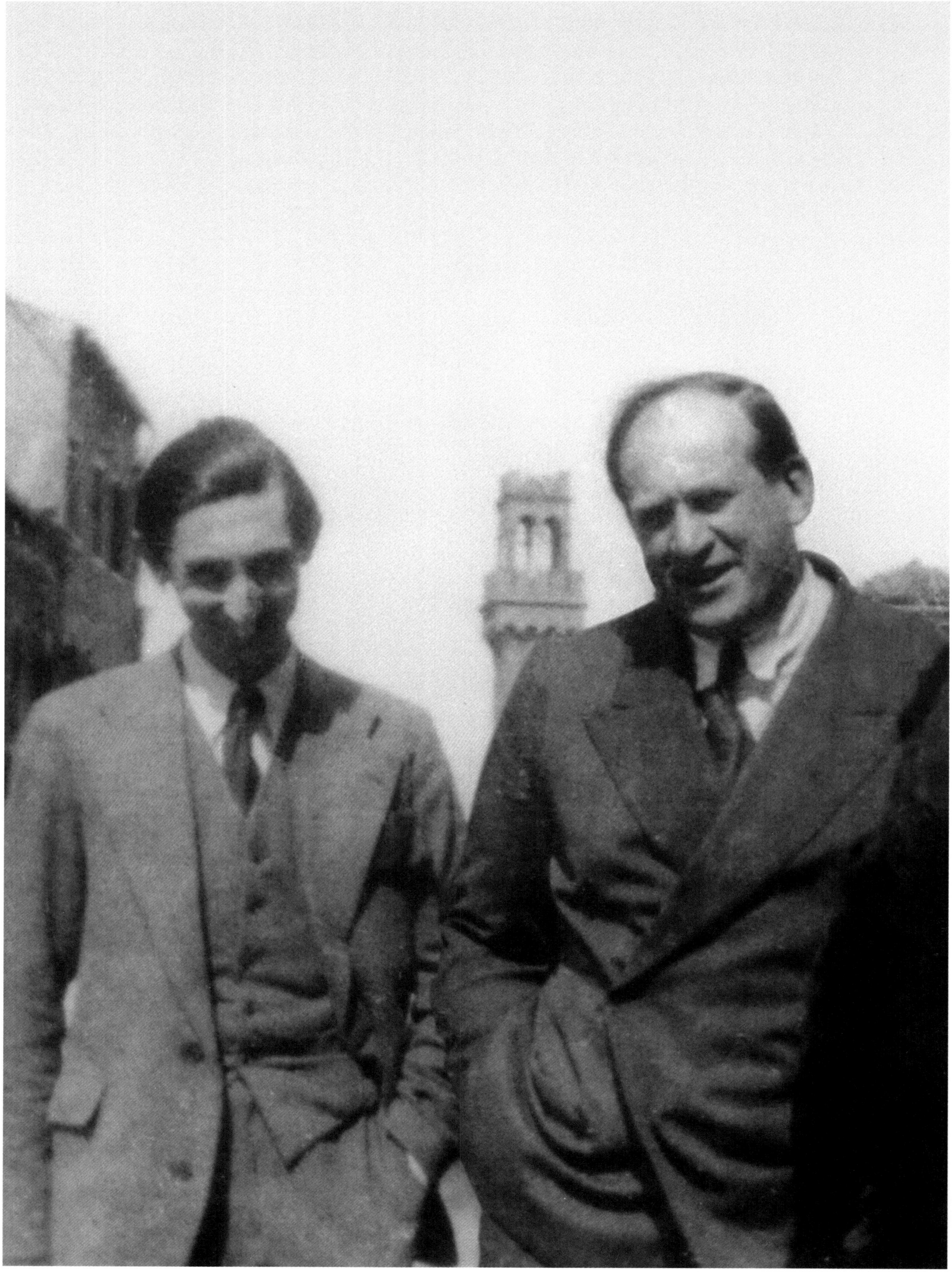

PREFACE

Marino Barovier

After his departure from the company V.S.M. Cappellin Venini e C., Giacomo Cappellin (1887-1968) founded the Maestri Vetrai Muranesi Cappellin & C. in June 1925, a new glassworks company which affirmed itself as one of the best through the modernity and quality of its production, constituting an ideal rival for that of his ex-partner, Venini.
Animated by the enthusiasm and passion of its founder who, with his strong personality, guided the production, from the end of 1926 the furnace also benefited from the contribution of the young Venetian architect, Carlo Scarpa (1906-1978).
Having come into contact with M.V.M. in 1925 on the occasion of the restoration of Palazzo da Mula, headquarters of the firm, where he had supervised the works, Scarpa began to work with M.V.M. as a designer during the period of the artistic direction of Vittorio Zecchin (1925-26) but, very soon, also as a creator of glass pieces.
In some cases he acted as an interpreter for Cappellin himself and progressively acquired a certain autonomy in the planning of models characterised by the clean lines of the design and the use of geometric shapes.
M.V.M. Cappellin & C., often in advance of the times, was able to propose new series, the fruit of research on glass techniques and forms, producing artefacts of great technical perfection aimed at an élite market.
Ancient working techniques, such as the *filigrana a reticello* and the *decoro fenicio* were also revisited and, taking up the challenge of opaque glass, compositions with a notable chromatic impact were achieved: from the *lattimo aurato* and *pasta vitrea* pieces to objects in brilliantly coloured cased glass, where the contribution of Scarpa, who worked with the glassworks up to its closure (1931), is evident.
With these glass pieces M.V.M. Cappellin & C. was repeatedly present at the major decorative art exhibitions in Italy and abroad and achieved a significant success. In the extraordinary production exhibited, there were also some stained-glass pieces produced after cartoons by Scarpa and important artists such as Casorati, Sironi etc., works which bore further witness to the excellent level achieved by the furnace.
In January 1932, however, the company went bankrupt, possibly because of imprudent management aggravated by the economic crash of 1929.
So came to an end a period marked by the passionate direction of Giacomo Cappellin and the fertile creativity of Carlo Scarpa, leading to very considerable results with an important impact also on subsequent Murano production.
Reconstructing the history of the glassworks required long, demanding and wide-ranging research

<
Carlo Scarpa and Giacomo Cappellin, 1928-1930, private archive

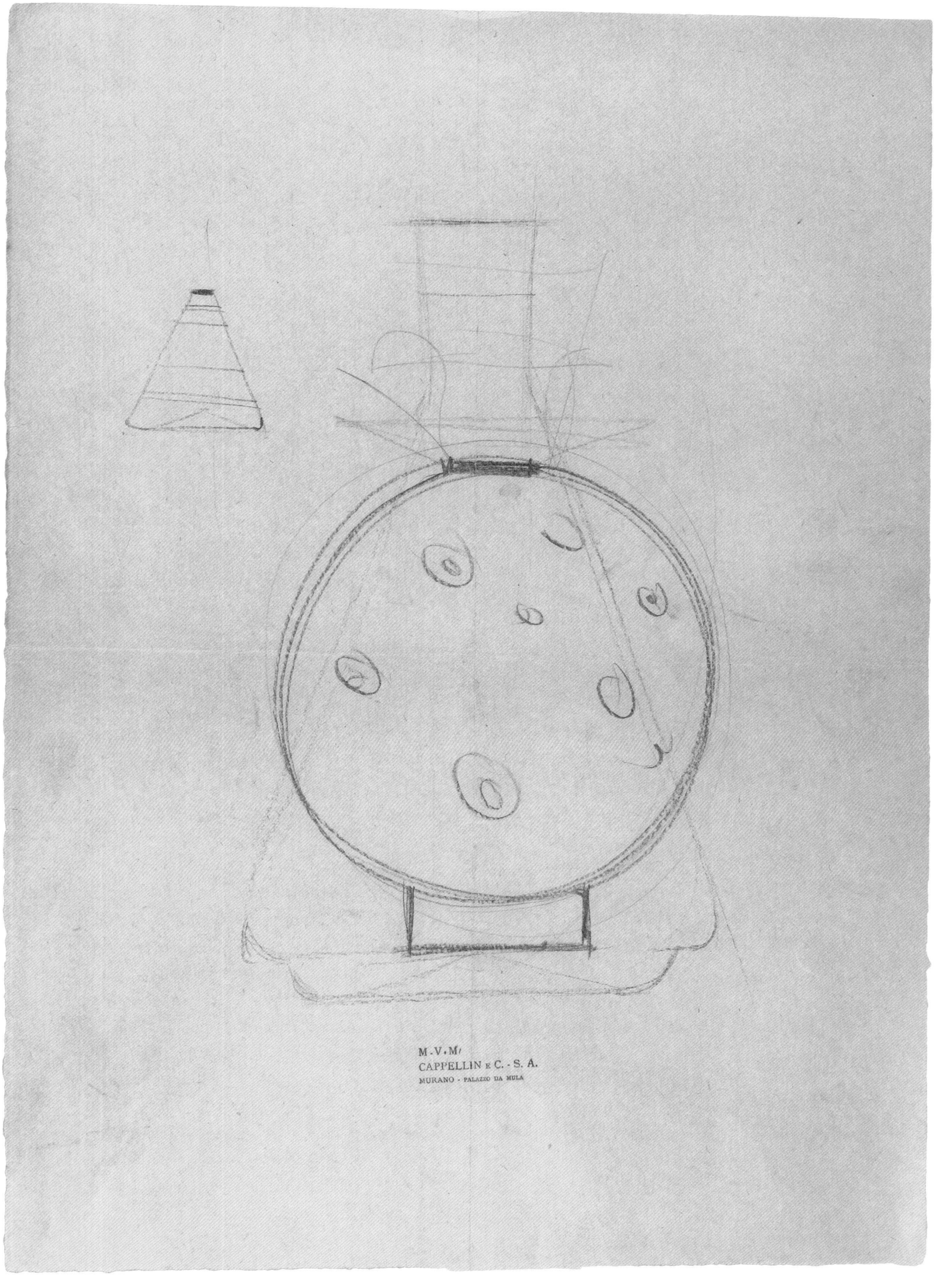
M·V·M
CAPPELLIN e C. - S. A.
MURANO - PALAZZO DA MULA

in public and private archives because, after the failure of the firm, all its material, including the documentation, was first acquired by the Pauly glassworks (1933) and subsequently split up, sold and dispersed in Italy and abroad.
Up to now, in order to recognise the glass pieces produced by M.V.M. Cappellin & C., only partial and incomplete documentation had been available. Thanks to this research and the re-discovery of many drawings and period photographs, it has been possible to reconstruct the catalogue of the glassworks in its entirety, identifying their production areas (vases, figurines, plants, animals, lighting, stained glass etc.), a production in many cases conditioned by the presence of the young Scarpa.
The consultation of the archive material kept in the Carlo Moretti and Giovanni Moretti Fund, Centro Studi del Vetro, Giorgio Cini Foundation, Venice, in the Carlo Scarpa Archive, Museo di Castelvecchio, Verona, in the Museo del Vetro Archive, Murano and in numerous private archives, has been of fundamental importance.
Recourse was also had to other sources (catalogues, magazines, daily newspapers of the time etc.), and such information was collated with the objects made available by museums, public and private institutions and Italian and foreign collectors.

<
Carlo Scarpa, study for vases, 1929-30, Museo del Vetro archive, Murano

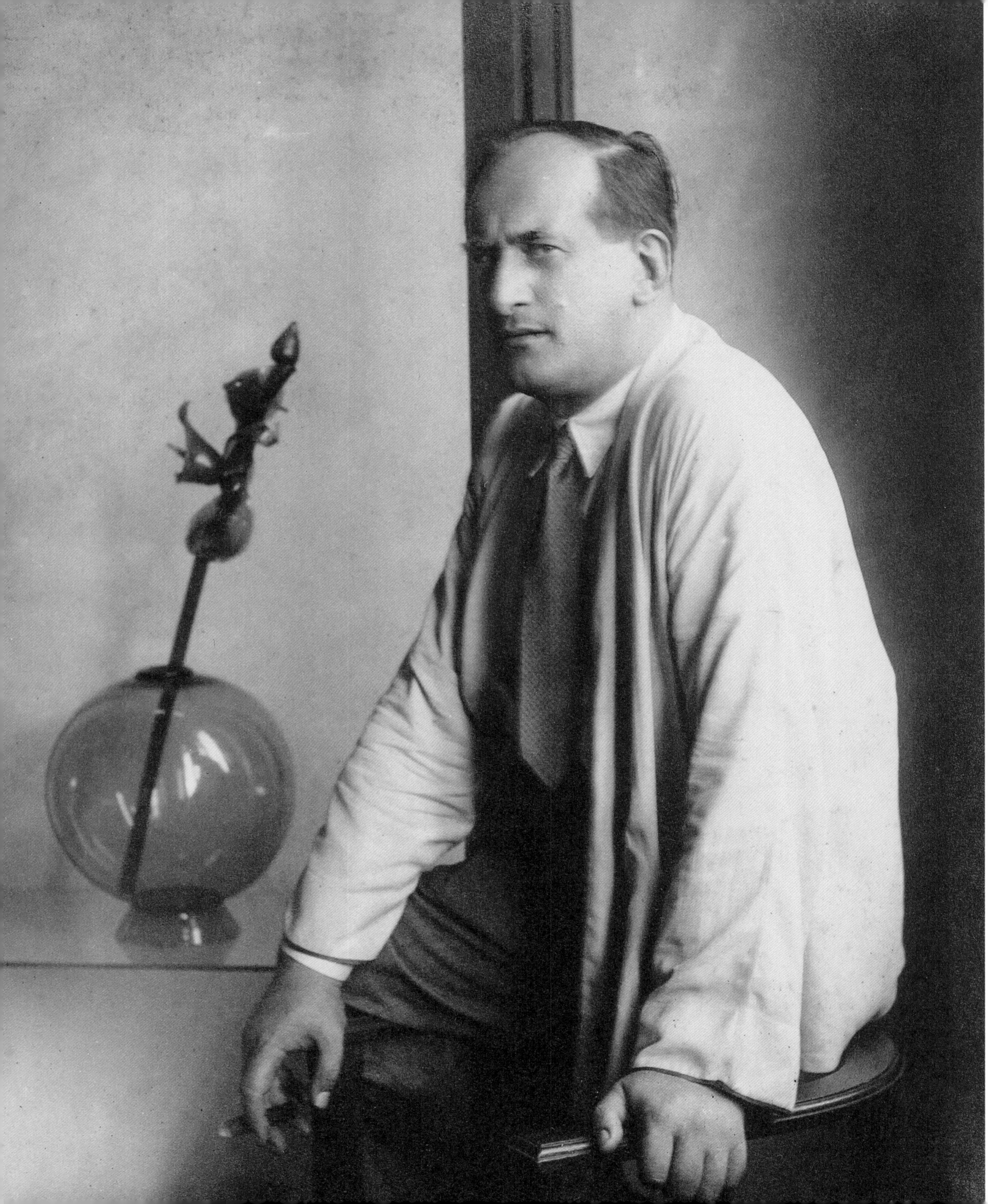

'EL TEMPO XE MIO, PROVA!'

Marino Barovier

In documents concerning the history of Murano glass in the 20th century, among the architects of the renaissance of Murano glass the name of Paolo Venini is associated with that of Giacomo Cappellin, who was Venini's first partner and then his ideal adversary in a rivalry that in the 1920s aimed at restoring this extraordinary manufacturing to its preeminent position. [1]

Gio Ponti, in a wide-ranging article dedicated to Cappellin's activity published in *Domus* in December 1928, maintained that 'if one had to tell the story of the present thriving renaissance of the decorative arts in Italy, a place of importance would have to be given to the Murano people, not only for the entirely modern, renewed glory of their glass pieces but also because of their great influence on taste and the encouragement that the felicitous example of the Murano glass-makers had provided in orienting the development of the Italian decorative arts. For all of this, great merit must be given, together with some few others known to all [...], to Giacomo Cappellin, from whose furnaces came the beautiful glass pieces that decorate these pages; and it is also acknowledged by all that this merit is associated with the very beginnings of the modern glory of Murano glass, with the appearance of the first exquisite nude glass pieces, with the efforts and initial success which has never since abandoned this most Italian of arts.'[1]

Born in Venice on 29 May 1887, Giacomo Cappellin, known as 'Bapi', graduated from the technical school and, having first worked for a lawyer and then as a proofreader at the *Giornale di Venezia*, opened a little shop of Venetian 'glass and pearls' near the San Moisè bridge[2] around 1915, moving a short time later to Campo San Moisè. Here, as well as *objets d'art* and antiques, he offered handcrafted pieces by the master glassmaker Andrea Rioda 'who made glass pieces of a contemporary kind and copies of the Old Masters for him'.[3]

During the First World War he was employed as a sergeant in the Red Cross and sent to various places, including Milan where, as early as September 1916, he was accorded honourable discharge. Integrating into the environment of the Lombard capital city, he transferred his activity to Via Montenapoleone, 'with the few glass pieces remaining to me', recalled Cappellin, '[and] I made up for the lack of merchandise with antique (or old) glass pieces in which I had become specialised'.[4]

He acquired a certain notoriety and in 1921, thanks to the goodwill of some financial backers, such as the Venetian Luigi Ceresa, Cappellin involved the young Milanese lawyer, Paolo Venini, who frequented his shop, in the project of founding a new glasswork firm in the spirit of the age. The glassworks would come into being following the take-over of the Andrea Rioda furnace, in difficulty at the time, and would have the painter Vittorio Zecchin as artistic director.

As is known, due to the early death of Rioda, the launching of the Vetri Soffiati Muranesi Cappellin Venini & C. was not free of difficulty, but thanks to the enthusiasm of its two founding members and the

<
[1] Portrait of Giacomo Cappellin, c. 1929, G. De Carlo collection

availability and cooperation of Zecchin, who initially followed the direction given to him by Cappellin, the enterprise was fruitful and signalled a turning-point in the history of Murano.[5]

Writing in 1954 to Agnoldomenico Pica, Cappellin recalls in his letter that: 'When Paolo Venini and I went to Murano, he a new boy in town, we had to fight and fight hard. The factories, perhaps less than ten (I'm talking about blown glass) were in a state of crisis. On our first visit to Paris we had to sweat to get ourselves received, I don't say to sell, I mean to be received by the buyers who didn't want to hear any talk of Murano and the miserable level of production to which it had fallen. It was our courage, our struggle, it was the Milan shop, it was that of the *piazzetta* [dei Leoncini in Venice] that had the courage, the daring to show the new production of "forms and colours". There were those then who looked at us, in fact who appreciated us. And this was our ambition: every Saturday to exhibit the production of the week in the windows of the *piazzetta*. Such was our joy and our love that we did not notice that indiscreet eyes and — allow me to say it — thieves, passed, saw, copied. It was known now that the "Cappellin-Venini" was successful. Following this the first participation in Monza with a corner all to ourselves. It was a revelation.'[6]

The 1st Biennale of Monza in 1923 sealed the success of the glassworks, which, alone in the panorama of Murano in those years, proposed highly refined transparent glass pieces of a sober design, mostly of a classical inspiration and with notably delicate colours, being able to count on a highly qualified workforce. At the same time the firm had built up a significant commercial network with shops in Venice and Rome, in addition to the Cappellin shop in Milan, and with representatives in various European cities. Such results were possible because of the work both of Vittorio Zecchin and of the two partners, of whom, as the French critic Robert Linzler observed in the pages of *Art et Décoration*, 'Cappellin is the inspirer, the creator, the technician. Venini, with his acute mind, wide culture, sure knowledge of the men and things of his time, is the administrator, the man of action, in this partnership'.[7]

Subsequently, despite tensions in the company occurring in 1924 that resulted in Luigi Ceresa leaving the firm and Enrico Galvani — the Pordenone ceramics industrialist — joining it, the glassworks participated in the *Exposition internationale des Arts Décoratifs et industriels modernes*, opening on 28 April 1925 in Paris, where the pieces designed by Zecchin were awarded the *Grand Prix*.

In the meantime, however, the differences between the partners had become insurmountable so that, shortly after the opening of the Paris exhibition, Giacomo Cappellin ended his partnership with Paolo Venini and left the company (14 May) to found the Maestri Vetrai Muranesi Cappellin & C. (23 June), with the financial support of the Venetian lawyer Raffaello Levi and that of Galvani. In this new 'enterprise', he was further followed by Vittorio Zecchin, and the expert and old master glassmaker Giovanni Seguso (Nane Patàre), together with other master glassmakers who were counted also among the partners.[8] [2]

As production and exhibition headquarters, the choice fell on the historic and imposing Palazzo da Mula with its annexes, which had been used on several occasions by Murano glassmakers in the course of the centuries. The building was acquired by the financial backers and renovated at the expense of the company, in a costly operation which was probably a great burden on the financial management of the firm, the complex affairs of which came to an end less than seven years later, on 12 January 1932, with the bankruptcy of the company.[9]

Notwithstanding the fact that the glassworks had always distinguished itself over the years for the extraordinary quality of its production, the somewhat imprudent management, aggravated by the effects of the crash of 1929, led to its closure.

Initially, the M.V.M. Cappellin catalogue was entrusted to Vittorio Zecchin who, reconfirmed in his role of artistic director 'in collaboration' with Cappellin,[10] proposed new blown-glass pieces in essential lines which were presented alongside those designed by the same artist for V. S. M. Cappellin Venini e C.

After the break-up between Cappellin and Venini, in fact, the 'old' Zecchin models continued to be produced both by M.V.M. Cappellin & C., sometimes with slight variations, and by V.S.M. Venini e C., the

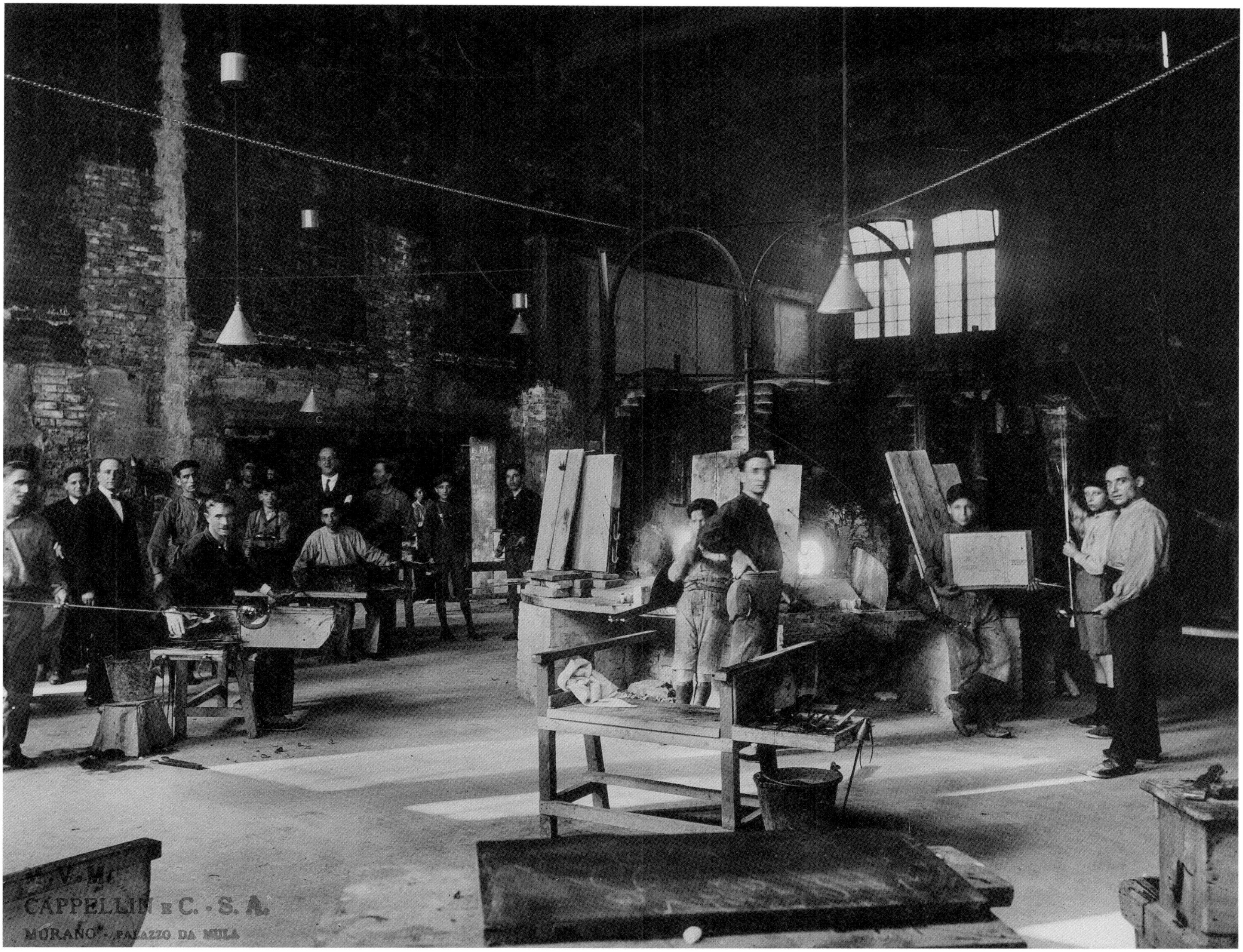

[2] Inside view the M.V.M. Cappellin & C. furnace, Murano, c. 1926, Fondo Carlo Moretti and Giovanni Moretti, Centro Studi del Vetro, Giorgio Cini Foundation, Venice. On the left in front of the entrance the upright figure of Giacomo Cappellin can be seen, dressed in a suit and tie

glassworks re-established by Paolo Venini with other partners (among whom was Napoleone Martinuzzi) and a new workforce, having the former Cappellin Venini furnace and shops still available to it. The new Zecchin production was exhibited in Venice in 1926, at the 15th *Biennale internazionale d'arte* and at the 17th exhibition of the Ca' Pesaro artists and enjoyed a critical success, the painter being once again appreciated for his glass objects but also his highly coloured panels in Murano glass, which opened a new prospect in the Cappellin repertoire that would be developed in the following years.[11]
However, the collaboration with Vittorio Zecchin came to an end on 30 October of that year when the artist left the company just before *Salon d'automne* of Paris, an event at which M.V.M. Cappellin & C. 'made its début' with a pavilion of considerable dimensions (a '*boutique de verres*'), designed by Mme Laroche-Garrus. [3]
Among the blown-glass pieces exhibited some very large ones were particularly noticeable, characterised by a truncated-cone-shaped foot and geometrical forms, very different from those designed by Vittorio Zecchin and traceable back to the presence in the glassworks of Carlo Scarpa, the young and talented student of architecture at the Accademia di Belle Arti of Venice. [5] Scarpa, who in November 1926 would obtain the title of Professor of Architectural Design, came into contact with the Cappellin firm in 1925 on the occasion of the renovation of Palazzo da Mula, where he had supervised the works, and began to collaborate with the furnace, first as draftsman but very soon also as a designer of glass pieces.
Exhibited in Paris with other pieces was the spherical vase with the truncated-cone-shaped foot, which was included in the catalogue in different sizes and which, accompanied by a glass flower, became the symbol of the furnace. With regard to this model, Gio Ponti, in the December 1928 article already referred to, had noted its 'bold geometric nature', an example of those 'more explicitly determined forms' which characterised the recent production of the Cappellin firm.[12]
From 1927, in fact, the glass pieces with classical lines were progressively accompanied by models of a more essential design which often arose out of an original reinterpretation of the work of Vittorio Zecchin in agreement with and at the suggestion of Giacomo Cappellin, a man of considerable artistic sensitivity and great 'communicative powers'. These 'powers' had been discussed at length by the critic Marziano Bernardi, co-director of the *Teatro di Torino* that had been restored following the design of Gigi Chessa in 1925 and had included the use of lamps and chandeliers by M.V.M. Cappellin. In the catalogue of the exhibition *Exposition d'artistes italiens contemporains* held in Geneva in 1927, Bernardi had pointed out how 'Giacomo Cappellin, even if as a designer he does not know how to hold a pencil in his hand, succeeds in giving to the hundreds and hundreds of iridescent glass pieces that come out of his glorious Murano furnace the exact imprint of his most original artistic personality. An example [...], not so much rare as unique, of an artist who expresses himself exclusively through interpreters whose thought and will he moulds just as his workers mould the liquid sand with their breath.'[13]
In Scarpa, due to his notable graphic ability and his creativity, Cappellin found an ideal interpreter who — as shown also by the development of the glassworks catalogue — in his dialogue with the director of the furnace and the master glassmakers progressively acquired more technical knowledge and independence.
Despite this, perhaps for commercial reasons, in accordance with a practice which had sometimes applied even to Zecchin, Scarpa's name did not appear in the various exhibitions as a designer of glass pieces but only as an exhibition designer (3rd Biennale of Monza, 1927) or as a creator of stained glass (4th Triennale of Monza, 1930). On the other hand, on more than one occasion, for example in 1927 at the 3rd Biennale of Monza and 93rd *Esposizione di Belle arti* of Rome, the artistic management in the respective catalogues was attributed to Giacomo Cappellin.
Contrary to this, on the same occasions and in the same catalogues, the names of the executants of the glass pieces were also given, perhaps Cappellin's homage to the glass-blower who had continued to work with him.

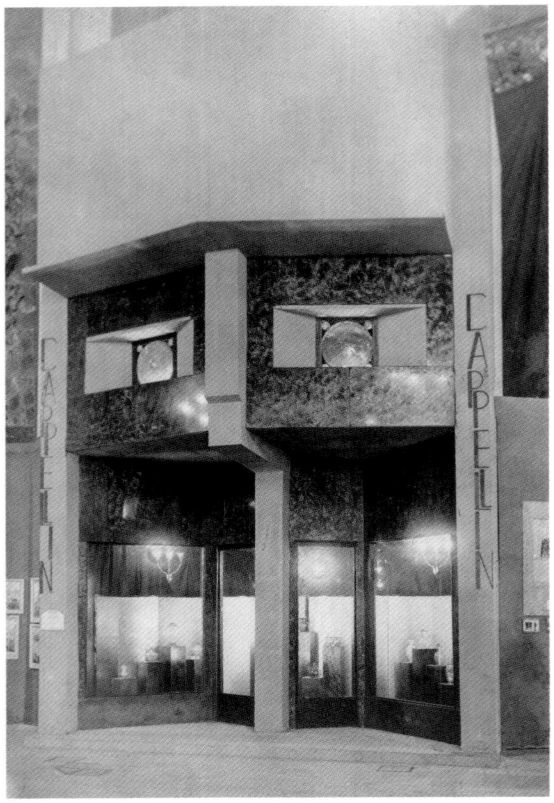

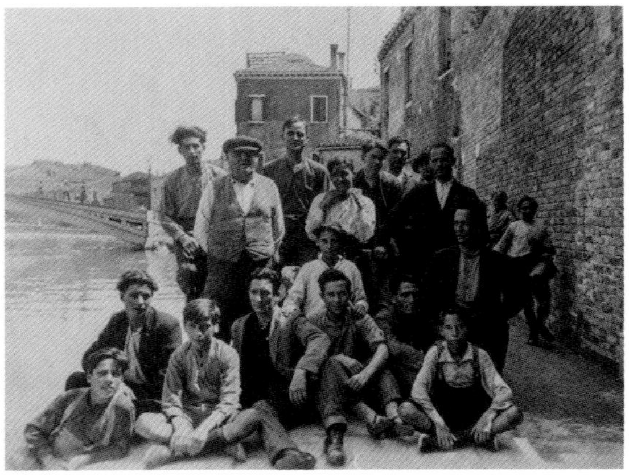

[3] M.V.M. Cappellin and C. Pavilion at the *Salon d'automne* of Paris, 1926, private collection

[4] Group photograph of M.V.M. Cappellin and C. workers in the vicinity of Palazzo da Mula, Murano, 1925-26, collection of the architect Luigi Cappellin

>
[5] Portrait of Carlo Scarpa at Palazzo da Mula, Murano, c. 1927, private collection

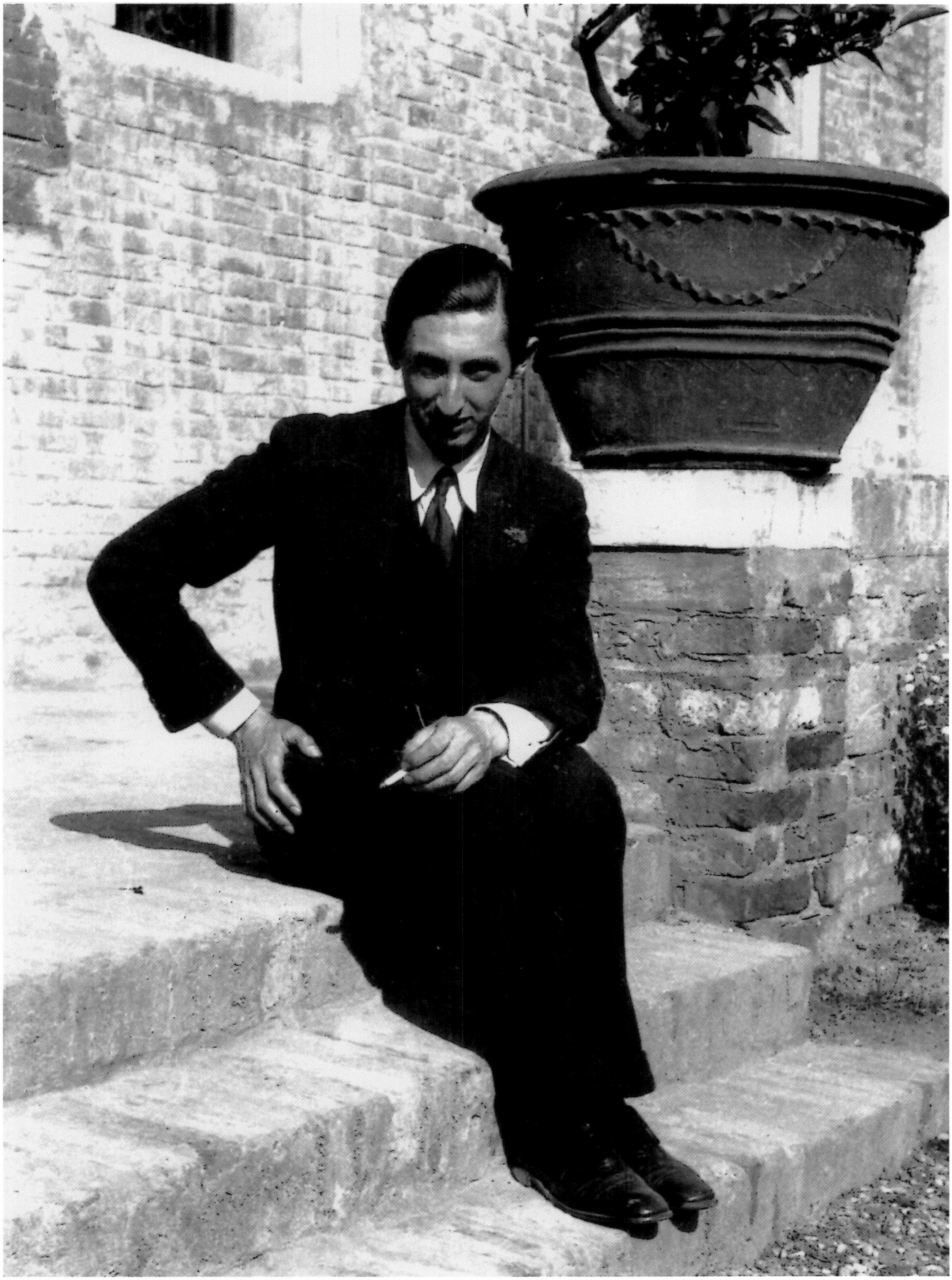

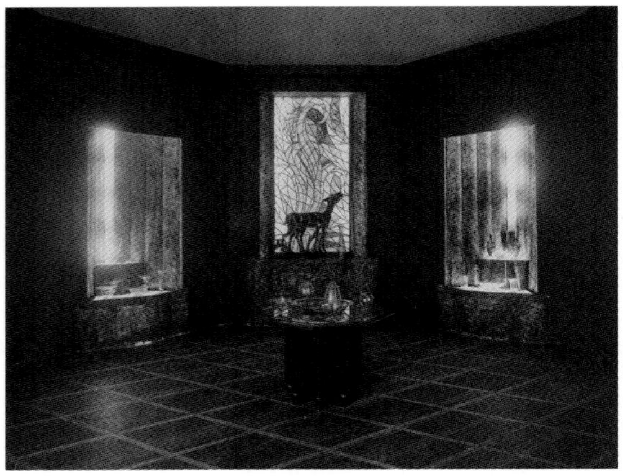

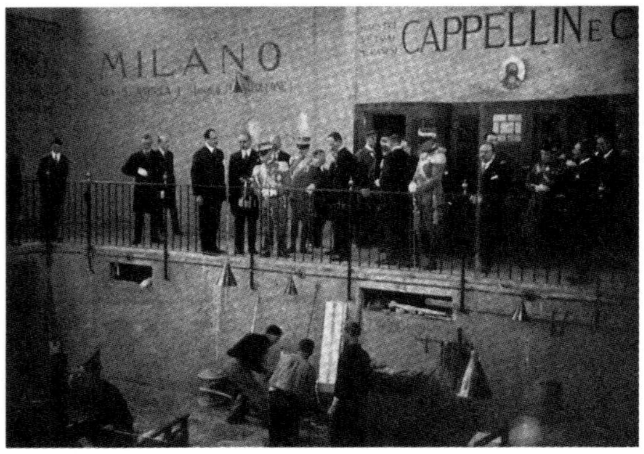

[6] M.V.M. Cappellin and C. room at the 3rd Biennale of Monza, 1927, Carlo Scarpa Archives, Museo di Castelvecchio, Verona

[7] Visit of the King to the M.V.M. Cappellin and C. demonstration furnace, Milan Trade Fair, 1928, Marino Barovier Archives

>
[8] Carlo Scarpa (?), Models for exhibition at the *Salon d'automne* of Paris in 1928, Museo del Vetro Archives, Murano

While demanding great expertise of them in their execution, to the point of smashing into smithereens with an ebony stick the glass pieces 'that he did not consider worthy of exhibition',[14] Cappellin had been able to establish a privileged relationship with his master glassmakers. [4] In a letter of 1954 to Agnoldomenico Pica regretting the sad end of M.V.M., he wrote that 'the glassmakers who are now in their forties or fifties, [...] those, in secret, I know, loved him [Cappellin] and would have liked to have him near them to suggest a restriction or a *pizzegada* or a movement that he — unfortunately — did not know how to and could not make [...]. That same Cappellin who when one of them had a piece fall from the blow-pipe would pat him on the back and say *wait a while, don't get annoyed, later, tomorrow you'll try again*. That same Cappellin who when he gave him [the glassmaker] a piece to do that he considered too difficult for him saying *I'm not able I'll just be wasting time*, would put his hand on his shoulder and say: *The time is mine, just try!* This was Cappellin and this is why the great old Ori [Vittorio Zanetti, glassmaking technician] liked him so much.'[15]

Many of these 'tries' were to be seen in 1927 at the 2nd Biennale of Monza, in which M.V.M. Cappellin & C. participated with a great effort, whether by making lamps and objects 'on commission', in particular for the 'Torino group', or by presenting its own production in various areas of the exhibition, and also in a small personal room. [6]

Here, together with two stained glass pieces executed after a cartoon by Ugo Ortona and Felice Casorati, there were many examples of the new production, starting with 'simple' transparent glass pieces with their intense colours. Among these was the spherical vase with the truncated-cone-shaped foot, a feature to be found also in various other models exhibited. Other transparent pieces were characterised by their 'palm-leaf motif' decoration in red *pasta vitrea* applied in relief on the surface of the vase or partially incorporated into the wall. The decoration was obtained by applying to the glass, during the working process, three or more horizontal threads which were 'combed' downwards in the middle with an appropriate instrument, so as to modify their rectilinear progression and obtain a design suggesting a palm leaf. The process seems to have originated in the ancient technique of *decoro fenicio* that Zecchin had already adopted for four models designed for the Cappellin Venini firm[16] and which would be taken up again subsequently to create an extraordinary series of opaque glass objects.

Another ancient Murano technique, *filigrana a reticello*, of 16th-century origin, was at the basis of other artefacts exhibited at Monza, characterised by a glass technique with crossed canes obtained through a complex work process. The technique was used in the variant with *filigrane lattimo* to execute the 'rim' — united using *incalmo* — of classical-style bowls and vases in transparent glass and, in the variant with polychrome canes, to give form to artefacts of exceptional chromatic impact, which gave rise to differing opinions among the commentators.

Alessandro De Stefani noted for example that: 'The latest novelty in the area of Murano glass, [...] designed and executed with the patience of the Venice lagoon, is the multicoloured plates where the glass is veined geometrically by lines of different colour, so that it seems to rival Scottish tartan, a marvel that would have astonished even the most imaginative glassmakers of the past.'[17] Roberto Papini, never tender with his criticisms of the entire Cappellin production, too variegated in his opinion, pointed out, in different terms, that 'the factory [...] tries to imitate *murrine* in certain coloured plates with squares that look like the Scottish shawls of our grandmothers'.[18] On the other hand these *filigrane* pieces were chosen by Gio Ponti, together with the *filigrane lattimo* of the V.S.M. Venini e C. firm, designed by Martinuzzi, to honour the best Murano glassmakers in the launch number of *Domus*, in January 1928, where he underlined the 'modern and personal decorative intelligence of these works'.[19]

In 1928 the glassworks again confirmed its exhibition 'policy' by participating in numerous events, from those of a purely artistic kind, like the Biennale of Venice, to those of a more commercial nature, like the Fiera [Trade Fair] of Milan, dedicated that year to the tenth anniversary of the victory in World War One. In this event, marked by the attempt on the life of King Vittorio Emanuele III, who had arrived in the Lombard capital for the opening of the exhibition, M.V.M. also participated with a demonstration

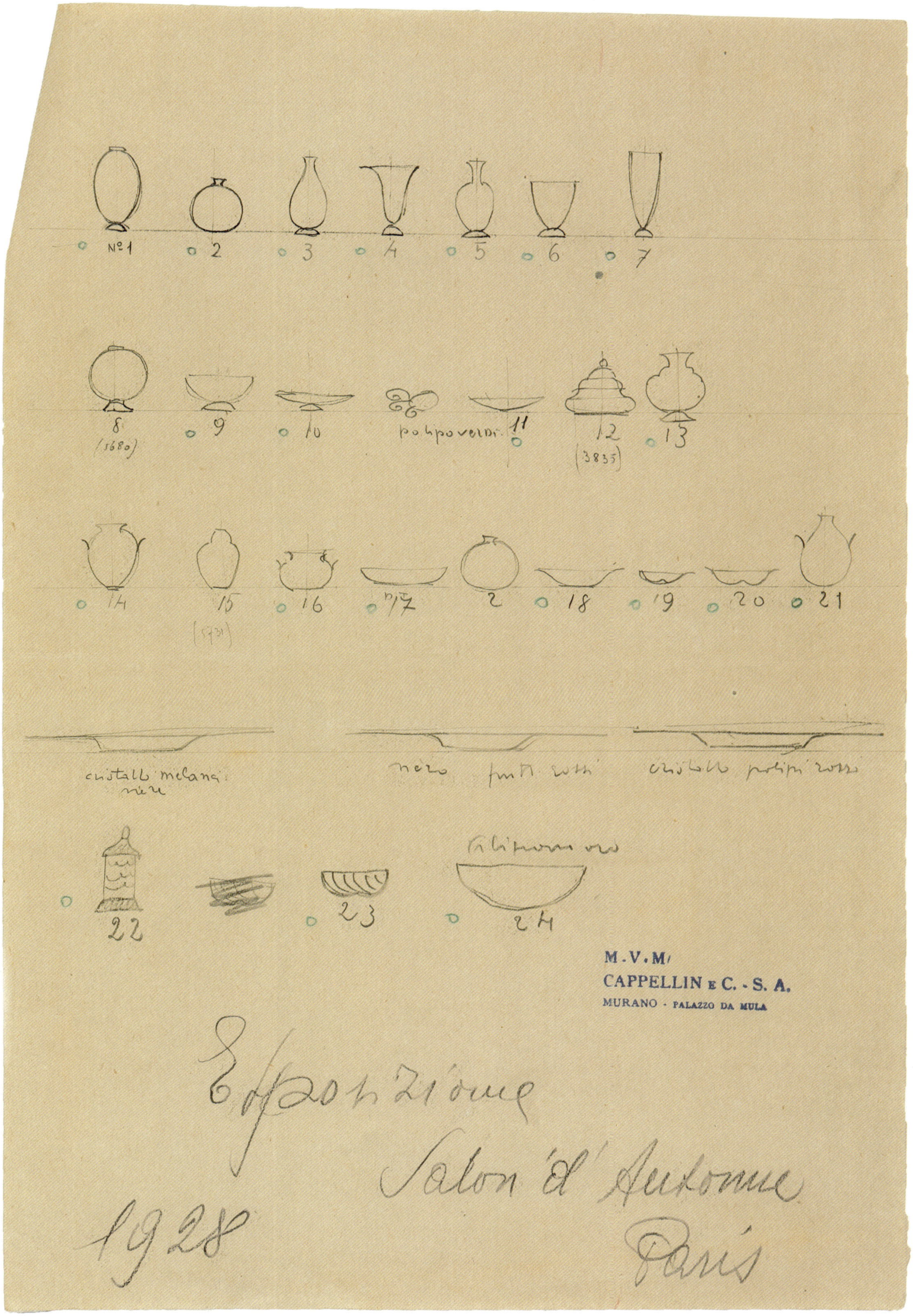
M.V.M.
CAPPELLIN E C. - S. A.
MURANO - PALAZZO DA MULA
Esposizione
Salon d'Automne
Paris
1928

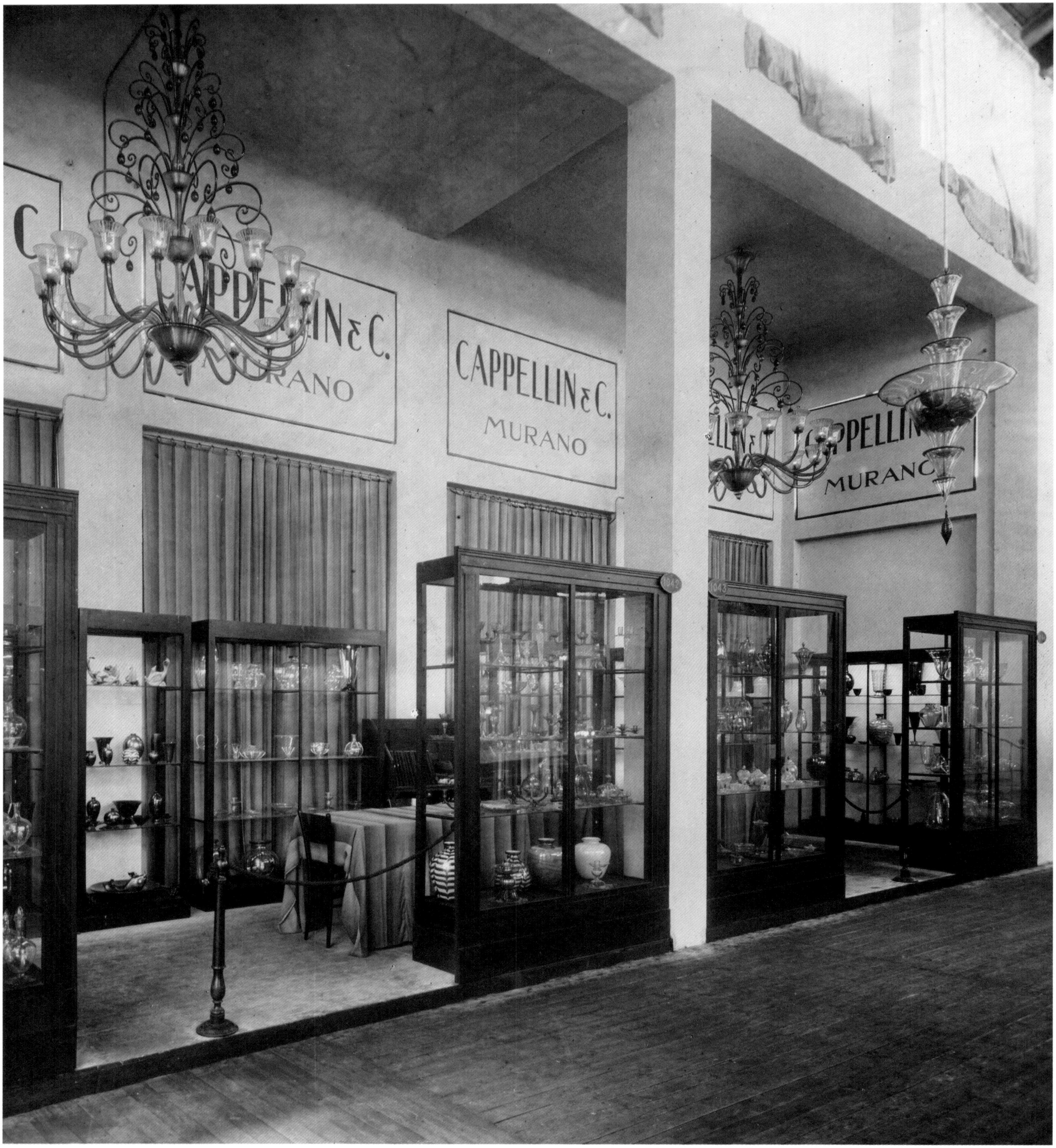
CAPPELLIN & C.
MURANO

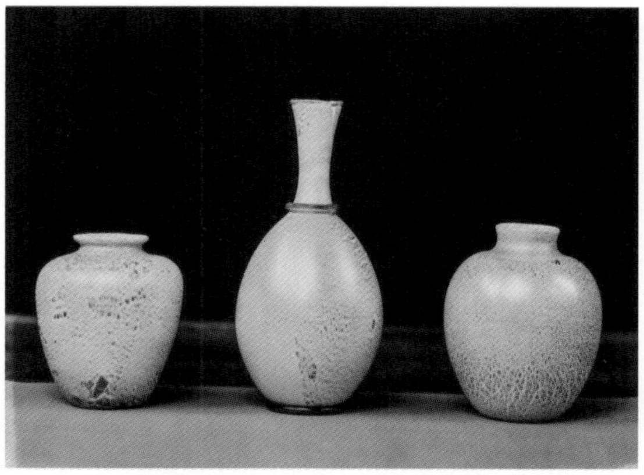

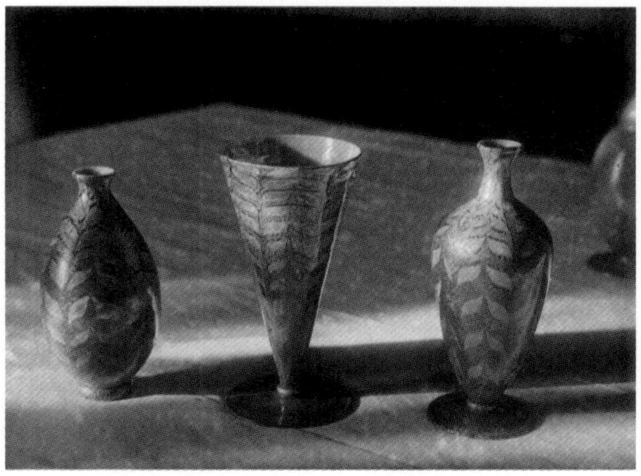

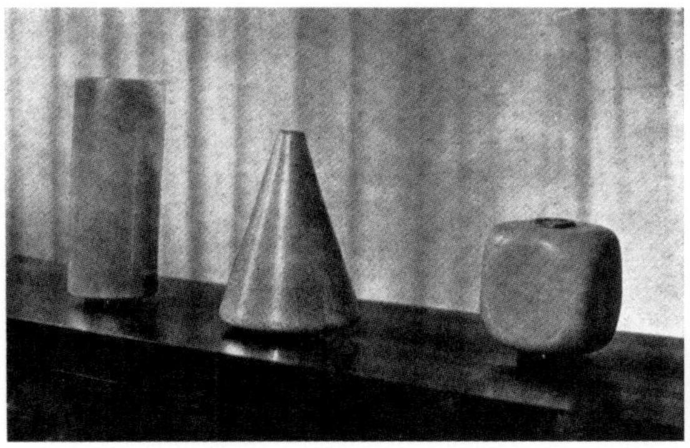

<

[9] M.V. M. Cappellin and C. Pavilion at the Milan Trade Fair in 1929, detail, Carlo Scarpa Archives, Museo di Castelvecchio, Verona

[10] Vases in *lattimo* glass with applied gold leaf, c. 1929, Marino Barovier Archives

[11] Vases in glass with *decoro fenicio*, c. 1929, Carlo Scarpa Archives, Museo di Castelvecchio, Verona

[12] 'The new 'geometric' vases in coloured opaque glass' in *Tentoonstelling van glas, aardewerk en kant, van moderne italiaansche kunstnijverheid. Mostra di vetri, ceramiche e merletti d'arte moderna italiana*, Amsterdam 1931, p. 26

>>

[13] Cappellin glass pieces at the *Tentoonstelling van glas, aardewerk en kant, van moderne italiaansche kunstnijverheid. Mostra di vetri, ceramiche e merletti d'arte moderna italiana*, Amsterdam 1931, detail, Howard Lockwood collection

furnace which caught the public's eye. [7] In general, with regard to the production exhibited, this was presumably pieces in transparent glass with ever more geometric lines (cf., for example, Register of the Works below, transparent pieces nos. 5825, 5827). However, as Gio Ponti remarked in December 1928, 'in the most recent Cappellin works the material is also transformed. This can be seen in the milky and precious glass pieces which, at the *Salon d'automne*, once again showed the superior virtues of our glassmakers.'[20] [8]

They were highly elegant objects such as the 'three superb examples' published in April 1929 in *Domus*, which constantly documented the novelties coming from the glassworks, particularly appreciating the *lattimo* pieces because of the refinement of the material, the 'knowledgeable workmanship', the 'simple and pure design'.

The finishings with which these generally small vases were proposed were in fact extraordinary: from the simple iridescence associated with foot and/or side handles in *pasta vitrea*, to the application of gold leaf combined, in some cases, with the coral-red *pasta vitrea* of the finishings at the mouth and/or foot. [10] Other *lattimo* pieces, characterised in general by a disk-shaped foot, were decorated with variously coloured undulating bands applied to the body. If sometimes these models presented somewhat affected details, in the majority of cases they were characterised by great rigour of form, whether of classical or accentuated geometric inspiration, the latter evident in the recurrence of spherical or ovoid forms.

This was an initial response to the demands of a changing taste, which had also seen the V.S.M. Venini e C. glassworks propose new pieces, such as the *pulegosi* of Napoleone Martinuzzi, exhibited first at the Biennale of Venice in 1928 and then at the 4th Triennale of Monza in 1930.

M.V.M. Cappellin, however, under the management of Giacomo Cappellin and with the input of Scarpa, chose another road, aiming at experimentation and research on the material and, between 1928 and 1930, creating a new series of elegant artefacts with varying opaque glass techniques. Accordingly, side by side with the 'traditional' transparent blown glass pieces, there was a diversified production, from the *lattimo* series to which were progressively added *decoro fenicio* pieces, with their characteristic bichrome texture 'with festoons', often embellished by gold leaf, [11] and the highly coloured *paste vitree* finished by strong iridisation or by applied gold leaf or oxidised silver.

The *pasta vitrea* was in fact adopted several times to produce the Cappellin opaque glass pieces, whether as a 'base' for some vases with *decoro fenicio* or monochrome objects with intense colouration and notable material consistency, these characteristics being due to the particular chemical composition of the *pasta*, which required the use of microcrystals of opacifier and a great quantity of pigments. It is worthwhile emphasizing that M.V.M. Cappellin was among the few furnaces to use this because of the difficulties of the execution techniques it required, due to the long cooling times which made the outcome difficult to manage.

Examples of these productions, characterised by 'elementary and simple forms' which 'also make the qualities [...] of the material more attractive', were documented by *Domus* in January 1930. Already in the course of 1929, however, coloured opaque glass pieces had been exhibited at certain exhibitions in which the glassworks had participated in Italy and abroad (Leipzig, Paris etc.). [9]

In the meantime, also in 1929, financial problems had become apparent, which led to significant corporate changes, such as the master glassmakers giving up their shares and the entry of a new partner who contributed capital.

Presumably this allowed the glassworks to prepare for its participation in the 4th Triennale of Monza in 1930 where, in a space designed by Scarpa, an extraordinary collection of glass pieces was exhibited, for the most part in opaque glass, which abundantly illustrated the variety and remarkably high quality attained by the glassworks in this field. Published several times in the specialised magazines, and not only in these, this production was welcomed unanimously by the critics, receiving the appreciation of, among others, Carlo Alberto Felice who noted how 'these new Cappellin works are among

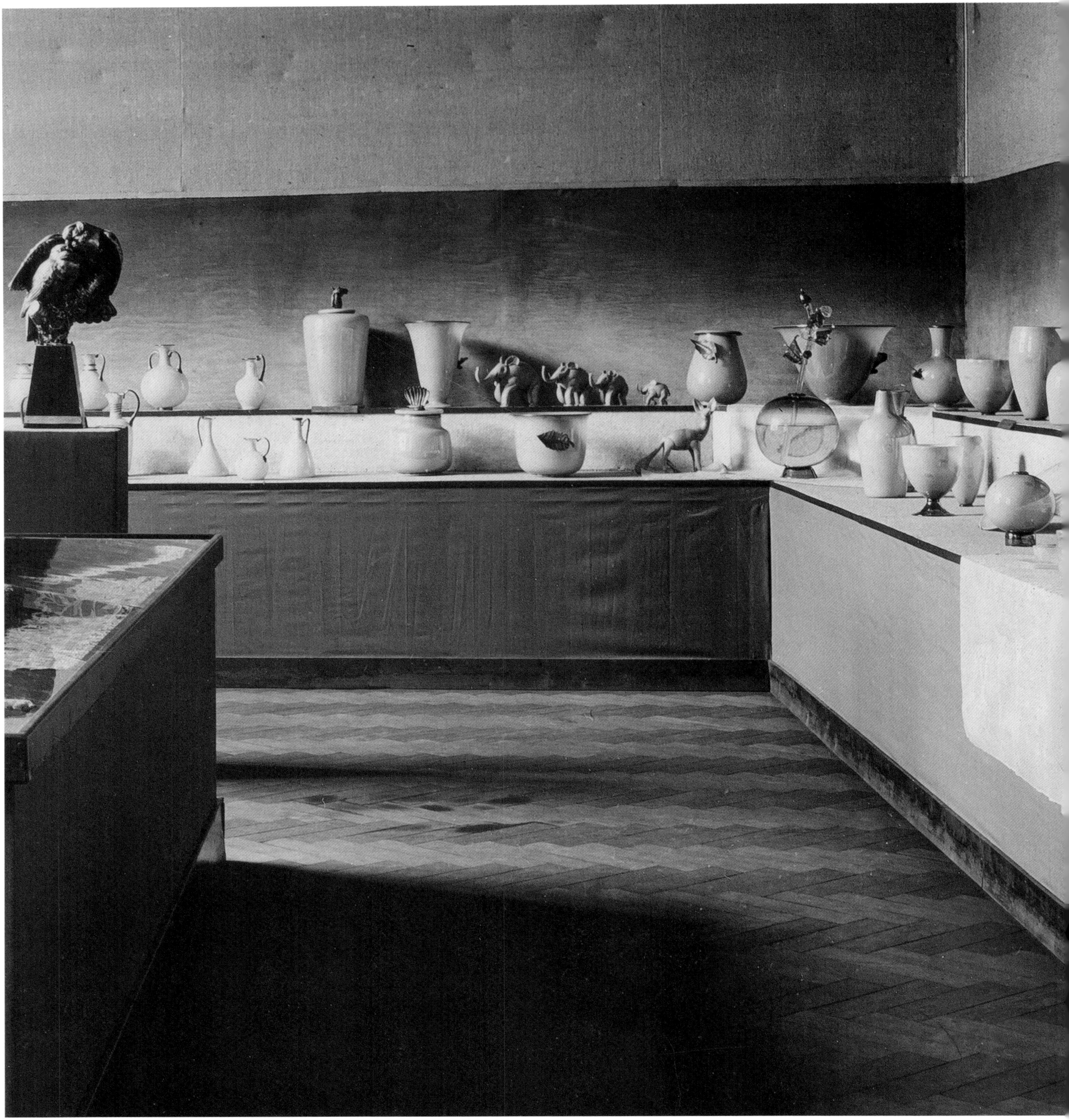

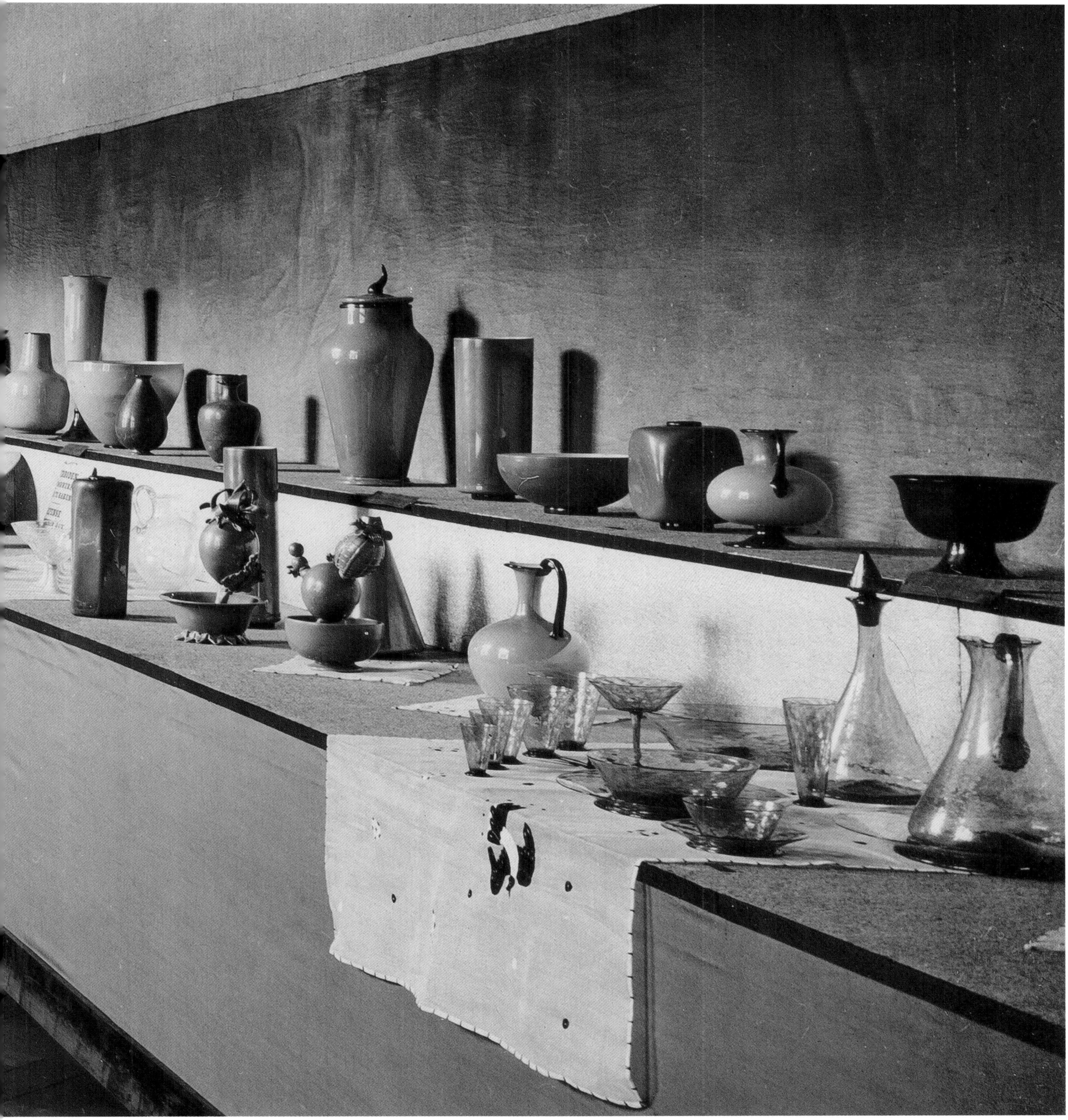

the most appropriate ornaments for a modern house, in perfect stylistic accord with the furniture and fittings that are fashionable today.'[21]

'Sober geometric forms and ones inspired directly by totally classic models', argued the magazine *Domus*, 'harmonise in the Cappellin presentation in their artistic tenor, in the same perfection of execution, in the high quality of the material'.[22]

Among the glass pieces with 'sober modern forms' attributable to Scarpa, as shown by surviving studies in his hand, a particular series stands out with elementary geometric forms such as the cone, the cylinder and the sphere, which were proposed in 'lucid, compact, vivid colourations'. [12] With regard to the 'totally classical models', executed in general on the basis of Zecchin's 'old' designs for transparent blown glass pieces, these were transformed with the use in some cases of *lattimo* glass, in others, of iridised black glass with the application of gold leaf.

Alongside these there were also new interpretations of the *filigrana a reticello*, whether in *lattimo* or in black glass, and some pieces, transparent and opaque, elegantly embellished by applied decorations (shells, leaves or animal heads).

In addition to vases and bowls, fun animals were also presented (fish, octopus and owl) and amusing figurines in glass, an example of that production already under way from the late 1920s to satisfy market demands, in particular those of the American market, for which a series of figurines inspired by jazz bands had been produced in 1929.

In the Monza room, finally, some stained-glass pieces after cartoons by Scarpa were exhibited, which together with those executed after the cartoons by Sironi and placed in the Graphic Arts Gallery, documented one of the distinctive production areas of the furnace.

With similar works (after cartoons by Enrico Del Debbio and Mario Romano) the Cappellin firm also participated, early in 1931, at the 1st Quadriennale d'arte of Rome, while for the Mostra del Giardino italiano held in the same year in Florence it produced the monumental centrepiece in crystal glass, 'composition inspired by the Italian Garden, designed and executed in a lively modern spirit but with love — in the use of the material — of the purest Murano tradition' that — said Giuseppe Dell'Oro — 'obtained a success as sensational as it was merited.'[23]

In the meantime a major exhibition had been inaugurated in Amsterdam dedicated to the decorative arts of Venice in which the art of glassmaking played an important part and, with it, the production of the Cappellin firm. [13] Together with some stained-glass, the furnace had brought its most recent works, among which the elegant vases with vertical bands and the refined series of glass pieces with *millefiori* murrine. Particular attention, noted Ugo Nebbia, the curator of the exhibition, was merited however by the 'new tonalities of its *paste vitree* [...] yellow, jade, grey, ennobled by distinction of form and by a tasteful spirit of simplification, sufficient to save it from the dangers to which sometimes its imagination exposed it in its search for originality and a place of pre-eminence among the glassmakers.'[24]

This extraordinary exhibition at which, as the catalogue pointed out, one could 'understand what the present aspirations of Murano glass [were]', was however the Cappellin firm's last.[25] Despite all Giacomo Cappellin's last-ditch efforts, the worsening of the financial situation obliged him to close down his business in early 1932, so ending a season, sometimes difficult but always the expression of his enthusiasm, passion and love for the material of glass, a season that he had shared with a promising young man, Carlo Scarpa, who, initially as a simple interpreter, was able to mark with his personality this excellent production.

1 Dir. [G. Ponti], 'Murano: oggi fatto d'arte', in *Domus,* December 1928, p. 59. The reference to 'some few others that we know' is to Paolo Venini and the V.S.M. Venini & C. firm, works by which *Domus* had already published in the course of 1928.

2 G. Cappellin, *appunti autobiografici,* manuscript, undated [1965], private archive, hereinafter referred to as AP. For the biographical details concerning Giacomo Cappellin, documents stored in this archive were consulted.

3 *Lettera di G. Cappellin ad A. Gasparetto*, 2 February 1960, draft, AP.

4 Cf. note 2.

5 Cf. M. Barovier, C. Sonego (edited by), *Vittorio Zecchin: i vetri trasparenti di Cappellin e Venini*, Milan 2017.

6 *Letter of G. Cappellin to A. Pica*, 5 July 1954, draft, AP.

7 L. Linzler, 'Les verreries de Cappellin et Venini', in *Art et Décoration*, March 1923, translated into Italian and published as 'I vetri soffiati muranesi di Cappellin e Venini', in *Arte pura e decorativa*, June 1923, p. 14.

8 Archive of the Chamber of Commerce, Venice, company register (hereinafter ACCVeR), Company 5432, *Repertorio 40571, Costituzione di società, 23 giugno 1925*. In the articles of association the partners are Giacomo Cappellin (manager), the Murano industrialist Gino Camozzo (manager), with four shares, the Venice lawyer Raffaele Levi, with eight, and, with one share each, Vittorio Zecchin and the master glassmakers: Giovanni Seguso, Diego Barovier, Attilio Moratto, Raffaele Ferro, Romano Zanetti, Malvino Pavanello, Antonio Toso. In the following November also Enrico Galvani formally joined the company. In the articles of association it is specified that 'sig. G. Camozzo is responsible for the accounting and administrative management [...]; sig. G. Cappellin is responsible for the artistic and industrial management, the sales branch, the supervision of shops, exhibition agencies and the signing of contracts'.

9 ACCVeR, Company 5432, G. Campagna, *Relazione del curatore fallimentare, 10 febbraio 1932*. On 28 January 1933 the company name and all the assets of M.V.M. Cappellin were acquired by Gaetano Ceschina, proprietor of the Pauly glassworks. Giacomo Cappellin was arrested on the charge of fraudulent bankruptcy, was found not guilty and moved to France up to the end of the Second World War, continuing to devote himself to glass and in particular the production of perfume bottles.

10 'The limited partner, Vittorio Zecchin, has the function of artistic director in cooperation with the manager Giacomo Cappellin and is free to occupy himself with pure and decorative art on his own account also, and to exhibit in his own name, except for glasswork production', *Costituzione di società..., op. cit.*

11 Cf. the article by Lucia Mannini in the present volume.

12 Dir. [G. Ponti], 'Murano...', *op. cit.*

13 M. Bernardi, 'Giacomo Cappellin (Maestri Vetrai Muranesi)', in *Exposition d'artistes italiens contemporains,* Turin 1927, n.p.n.

14 The anecdote recalled by Scarpa is given in L. Gemin, 'Una testimonianza', in M. Barovier, *Carlo Scarpa i vetri di un architetto*, Milan 1997, p. 19.

15 [...Quel Cappellin che quando a qualcuno di loro cadeva dal pontello (canna da soffio) un pezzo gli batteva la spalla e gli diceva *aspetta un poco, non arrabbiarti, dopo, domani proverai di nuovo*. Quel tal Cappellin che quando dava loro da fare un pezzo che reputava troppo difficile per lui e diceva *non sono capace perdo tempo* gli buttava la mano sulla spalla e gli diceva: *Il tempo è mio, prova!...*], *Lettera di G. Cappellin ad A. Pica..., op. cit.*

16 Cf. M. Barovier, C. Sonego (edited by), *Vittorio Zecchin..., op. cit.*, p. 213.

17 A. De Stefani, 'Il trionfo del vetro a Monza', in *Il lavoro fascista*, 7 June 1927

18 R. Papini, 'Le arti a Monza nel 1927. I Gli italiani', in *Emporium*, 391, 1927, p. 24.

19 Dir. [G. Ponti], 'Filigrane di Murano', in *Domus*, January 1928, p. 29.

20 Dir. [G. Ponti], 'Murano...', *op. cit.*

21 C. A. Felice, 'I vetri alla Triennale di Monza', *in Dedalo,* 1930-31, II, p. 322

22 'Alla Triennale di Monza', in *Domus*, September 1930, p. 32.

23 Gida, 'Giardini veneti e veneziani alla mostra di Palazzo Vecchio', in *Rivista di Venezia*, June 1931, p. 212.

24 U. Nebbia, 'Una mostra d'arte decorativa italiana ad Amsterdam', in *Emporium*, 439, 1931, p. 56.

25 'Cappellin & C. – Murano', in *Tentoonstelling van glas, aardewerk en kant, van moderne italiaansche kunstnijverheid. Mostra di vetri, ceramiche e merletti d'arte moderna italiana*, Amsterdam 1931, p. 19.

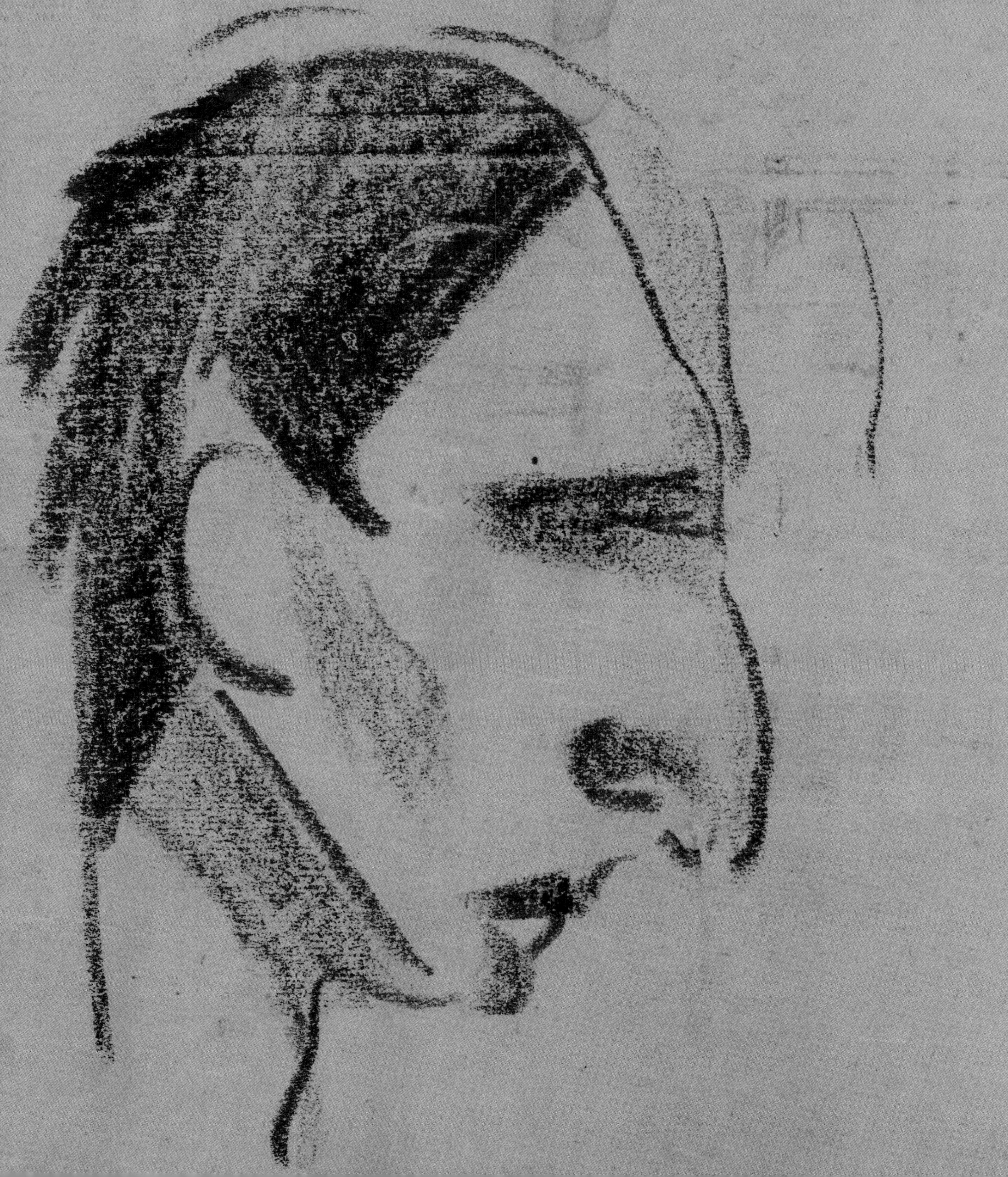

CARLO SCARPA, FROM ARCHITECTURE TO GLASS

Carla Sonego

When he arrived at the Cappellin firm in 1925 to oversee the Palazzo da Mula building works at Murano, Carlo Scarpa was attending the last year of a specialisation course in architecture at the Accademia di Belle Arti of Venice.[1] [1]

At that time attendance at such courses was an established practice for anyone who did not choose the School of Engineering or the Higher School of Architecture of Rome, opened in 1919, but nevertheless wished to devote himself to this discipline. Also in 1925, however, from an 'offshoot of the local Academy, an experimental two-year course in Architecture came into being, the first nucleus of the *Scuola Superiore di Architettura* of Venice, which was officially opened in December 1926. Scarpa, however, chose to complete the 'traditional' academic pathway and attended a supplementary year to achieve, in November 1926, the qualification of Professor in Architectural Design with First Class Honours.

Although in Venice for some time an effort was being made to provide future architects with a minimal scientific culture not provided for in the normal study structure, the Academy had up to that point been above all a school for artists, with a small number of teachers, themselves artists. Scarpa, in a rare interview recalled: '...I thought of the Academy as an art school where one became an artist. I thought: perhaps I'll become a painter or an artist'.[2]

At the Academy Scarpa, who had a natural inclination for drawing, developed great ease in the use of representational techniques and learned to measure and plot. Drawing from life, the copying of architectural features, orders and mouldings from plaster or from tables with 19th-century reliefs and exercises in perspective were the basis of his scholastic journey.

Many of these lessons were taught by the Ancona architect, Guido Cirilli, who, as Professor, was in fact the 'principal protagonist' in the special Architecture course.[3] With Cirilli, favourite pupil of Giuseppe Sacconi, the creator of the *Altare della Patria* of Rome, Scarpa had a special relationship, also because of the latter's exceptional ability in drawing. An expert draftsman himself, Cirilli preferred 'to show in concrete terms by the line traced rather than by the word how architecture should be done [...] since', recalled Samonà, 'he had always thought that the line was the language of the architect. The pupils [among whom Scarpa] owe to him this knowledge, this ability to penetrate the problems of the material, its plasticity, to express themselves effectively with colour and in *chiaroscuro* in order to bring out

<
[1] Carlo Scarpa, *Self-portrait*, c. 1927, MAXXI Architecture Collection. Carlo Scarpa Archives

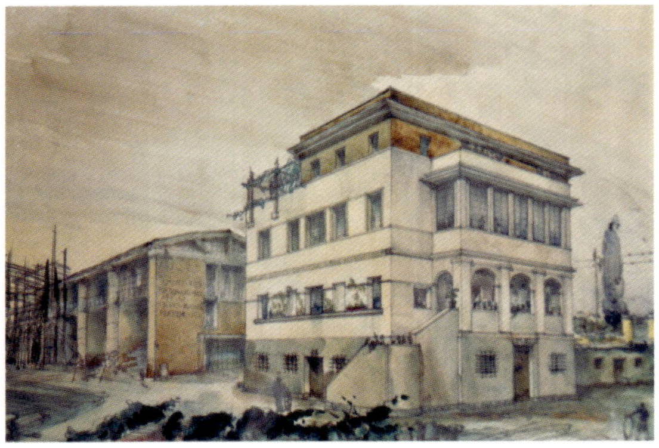

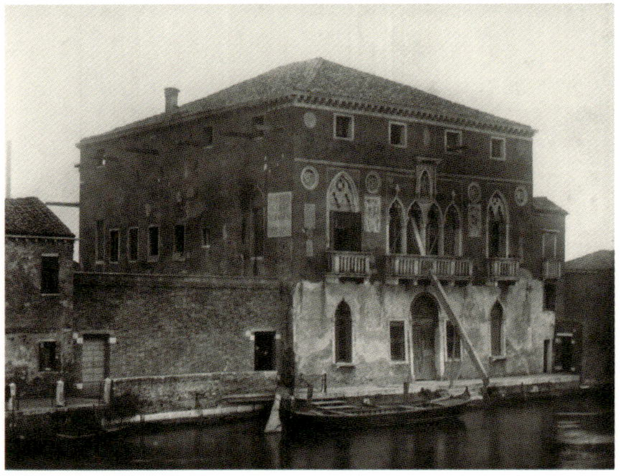

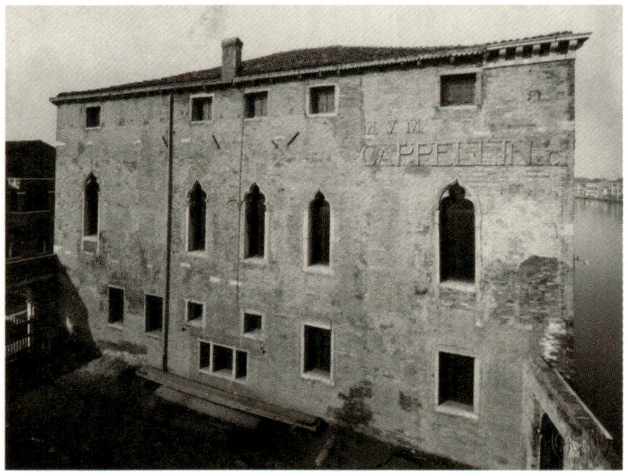

a detail in a clearer way'.[4] Cirilli, who saw himself as an 'artist' and 'builder', did not limit himself in his teaching to a theoretical approach but, backed by his professional experience as a designer, restorer and 'perfect cutter of marble', did not fail to deal with 'living techniques, materials, walls, proportions',[5] emphasizing, among other things, the colour aspect of the various materials.

With regard to this period, Scarpa often said that: 'I work somewhat in a traditional way, as I learned at school. I have found that none of what I learned at school was to be thrown away. I don't know: accuracy, working with clarity, not botching, for example'.[6]

The esteem that Cirilli had for his young pupil was confirmed also in late 1926 when, having finished his studies, Scarpa was called upon by the Ancona architect to join his studio and to be an assistant at the Scuola Superiore di Architettura [the Higher School of Architecture] of Venice, so embarking in an academic career.[7] In December of the same year, moreover, Cirilli backed Scarpa's attempt to obtain the professional title of architect, making use of the ratification procedure for lecturers in architectural drawing in the context of the law governing the profession.[8] In his reference Cirilli wrote: 'If architecture is not a pure and simple derivative of formulas but rather the expression of a mind sensitive to form, balanced relationships, colour etc. [...], then Scarpa is truly an architect — and with an intelligence which allows him, at the necessary time, to deal with the shortcomings that only long experience makes it possible to remedy.'[9]

In the curriculum vitae drawn up for the occasion, the activities carried out by Scarpa in the preceding years were documented. From the end of 1922 he had worked at the studio of the Venice architect, Vincenzo Rinaldo, first as designer and then, between 1923 and 1924, supervising the sites of religious buildings in the Venetian hinterland. Among his most recent jobs, however, was his collaboration with Franco Pizzuto, a fellow student with whom, around 1926, he designed and constructed two buildings of an industrial nature and others of a residential kind on behalf of the building company of Angelo Velo, located in Fontaniva (Padua). In the residential projects of the said Velo and for the little house designed for Father Aldo Martinati, for which graphic evidence remains,[10] the inspiration drawn from the Mitteleuropean school of architecture is evident, particularly the work of Otto Wagner and Josef Hoffmann, who were beginning to be known through foreign journals which were discreetly starting to circulate and of which Cirilli, even if he did not encourage his pupils to consult them, had some copies in his studio.[11] [2]

For the period 1925-26, Scarpa reported in his curriculum vitae that he had directed and supervised, for 'the company Maestri Vetrai Muranesi Cappellin & C.', the construction sites of 'various buildings (1200 mq), among which a furnace building, some offices and warehouses. At the same time and for the same company [...] the works for the restoration of Palazzo da Mula, in Murano, Fondamenta Vivarini'.[12] [3]

For the headquarters of the new glassworks, formally established on 23 June 1925, Giacomo Cappellin and his partners had chosen the ancient and imposing mansion with its numerous outbuildings.[13] This remarkable edifice, the origins of which are considered to go back to the twelfth century, had been redeveloped in the course of the fifteenth century and, to an even greater extent, in the sixteenth, becoming the summer residence of the Venetian family, Da Mula.

Since the eighteenth century, the building had on several occasions become the premises of art glassworks: first with the Fontanellas and Ferraris for the production of enamel canes for glass beads and in the nineteenth century for the production of glassware with Domenico Bussolin (1838), known for his white and coloured filigree, and above all in 1866 with the important activities of Antonio Salviati. Finally, in the first decades of the twentieth century, Luciano Barbon produced multicoloured glass canes there. It was probably also for its historic and symbolic value that Cappellin & C. chose this complex as its own manufacturing, commercial and exhibition headquarters, a location which, as Giacomo Cappellin wrote in a letter to Ugo Ojetti as early as April 1925, 'Must become a necessary destination for anyone going to Murano'.[14]

[2] Carlo Scarpa, plan for Velo House at Fontaniva (PD) in association with F. Pizzuto, 1926, private archive

[3] Palazzo da Mula in Murano during restoration work, 1925-26, Marino Barovier Archive

[4] Palazzo da Mula after restoration. East side, c. 1928, Carlo Scarpa Archive, Museo di Castelvecchio, Verona

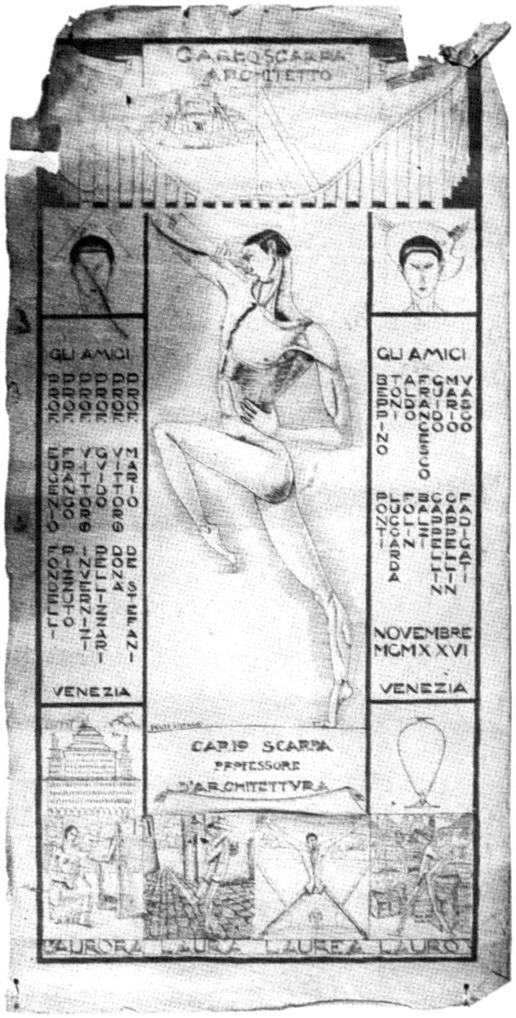

[5] Pino Ponti, 'Degree papyrus' for Carlo Scarpa's diploma, November 1926, in *Les Cahiers de la recherche architecturale*, *Carlo Scarpa*, 19, 1986

The significance of this choice was underlined in a small volume (*Il Palazzo da Mula di Murano)*, written by the historian Giuseppe Morazzoni and published, not for general sale, by Maestri Vetrai Muranesi Cappellin & C., in 1926. It produced echoes in certain articles appearing in local and national publications and dedicated to the new glassworks and its headquarters.[15] Among other things, these articles duly reported that one of the Cappellin master glassmakers was none other than Giovanni Seguso, known as Nane Patàre, who, having already worked at the Palazzo da Mula with Salviati in 1866, represented 'the link between the new and the old masters', guaranteeing the continuity of an important tradition.

The conversion works on the buildings, some of which were in a dilapidated state of repair, were entrusted to the largest building company in Venice, Pasqualin & Vienna.[16] The direction of the works, as documented by the said curriculum vitae, was undertaken by the young Scarpa, who had come into contact with the glassworks presumably because of his friendship with the nephews of Giacomo Cappellin and also as a result of his relationship and collaboration with Guido Cirilli.

In the 'restoration' of the Palazzo, in accordance with the wishes of the 'Partners', 'the main preoccupation was to follow the original layout of the buildings and to restore the glory of the original decorations, refraining from ornamental or reconstruction works for fear of becoming arbitrary.'[17] The intervention therefore consisted of structural reinforcement measures and the re-establishing of blocked openings, among which the partially blinded three-mullioned window at the back. Architectural and decorative elements no longer visible or in part rebuilt were brought to light and the polychrome decoration of some of the rooms was restored.

The internal space was organised to accommodate the permanent exhibition of M.V.M. Cappellin, in which the furnishing harmonised with the 'elegant and sober Fortuny draperies'.[18] At the back, between the mansion and the furnaces, the later accretions were eliminated and a 'grassy courtyard' obtained, overlooked by a Byzantine arch of the twelfth century. It is not known which choices are to be attributed to Scarpa; however, a document in his own hand vouches for the design of the large business sign in capital letters with the name of the company, which was placed on the right side of the east face of the Palazzo da Mula, so as not to interfere with the structure and decorations of the principal façade.[19] The inscription, an evocation of Renaissance epigraphy, was produced in metal so that it stood out from the bare texture of the wall and was visible from the canal and the Ponte Vivarini. [4]

The disposition of Palazzo da Mula and its outbuildings, on the one hand allowed Scarpa to experiment with the complexity of works on such a multifarious structure, while on the other he was introduced to the world of glass, which significantly marked the years of his training.

Presumably, before the end of his studies, Scarpa began to take part in the life of the company also as a designer, as would appear from the presence of the *Veronese* vase on his 'degree papyrus', produced by Pino Ponti on the occasion of Scarpa being conferred with the diploma of lecturer in architectural design in November 1926.[20] [5]

This vase, a kind of icon in 'modern' production, was designed by the Murano painter Vittorio Zecchin who, having already been artistic director of the V.S.M. Cappellin Venini & C, continued to play this role in M.V.M Cappellin & C.[21]

During this period therefore Scarpa was able to become familiar with the work of the Murano artist who, however, left the glassworks in October 1926.

From an analysis of the M.V.M. catalogue it is difficult to establish with certainty the authorship of the glass pieces produced between 1926 and 1927 as, together with the models by Zecchin, which continued to be produced after his departure, there are others in a classical style which took up the same themes.[22] Other objects again, however, where the involvement of Scarpa is probable, are characterised by extremely geometrical forms and by the recurrent use of a typical truncated-cone-shaped foot, as in the case of the spherical vase which became the symbol of the glassworks and was exhibited in November 1926 at the *Salon d'automne* of Paris. [6]

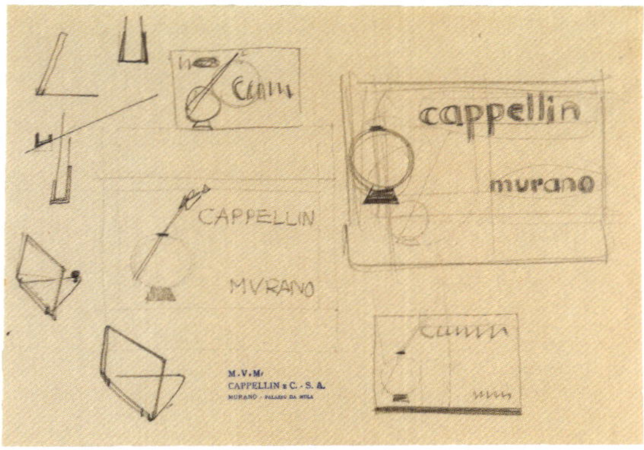

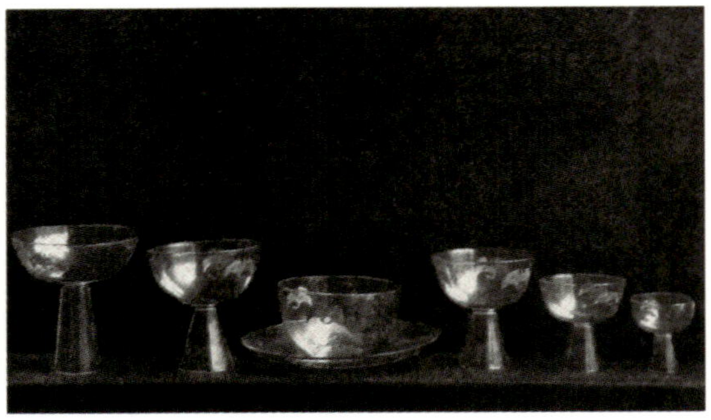

The presence in his library of an edition of '*Art et Décoration*', documenting the participation of the Cappellin company at the Paris exhibition,[23] is significant in this regard.

Even if the truncated-cone foot is to be found (once only) in the designs of Zecchin for a dinner service [7], exhibited at the 1st Biennale of Monza in 1923 and decorated with a 'golden heraldic heron',[24] its constant recurrence in the new Cappellin production, combined with essential geometrical shapes, would suggest another hand, identifiable with that of the young 'architect' who was looking with a new eye at the work of Zecchin.

This was perhaps an initial answer to the observations made by Vincenzo Costantini in the pages of *Le arti plastiche* in July 1926 who, while not sparing praise of the production of the glassworks, had commented regarding the 'forms': '...it would not be a bad idea if the Cappellin company added to its essential merits those of an inventiveness less inclined to repose and more open to novelty. But the Cappellin company' — continued Costantini — 'in its great fervour nourished by fundamental experiments will certainly find the elements necessary to an unflagging creative genius'.[25]

At the M.V.M., one of the best glassworks on the island, Scarpa was able to count on workers who had 'such a confident relationship with glass as to be able to shape it to the most bizarre of fantasies' and who were capable of achieving extraordinary artefacts characterised in particular by 'colour in all its variety of shades and quality'.[26] Giacomo Cappellin, defined by Costantini as the 'helmsman [...] who guides the crew', had in fact directed production to very high qualitative levels, even if sometimes to the detriment of the financial aspect.[27]

'Enamoured of the material' of glass, as he described himself,[28] and a man of refined taste, Cappellin frequented circles and personalities in the world of culture and art, especially that of Milan linked to the Famiglia Artistica and the Galleria Pesaro and that of Turin and the circle of Riccardo Gualino. Particularly committed to the activity of the glassworks, intervening also in the designing of the models, he was a convinced supporter of the importance of participation in decorative art exhibitions in Italy and abroad, in which he did not fail to take part.

As an 'artistic consultant' of M.V.M.,[29] Scarpa was involved in different areas of the life of the furnace and established a fruitful dialogue with Cappellin, based on great esteem, which progressively led him to affirm his own personality as a designer both of glass pieces and of interiors and layouts.

A preferred interpreter of Cappellin's — to whom he sometimes offered his drawing ability to give shape to the latter's ideas — in 1927 Scarpa was entrusted with the re-arrangement of the new shop on the Lungarno in Florence, the opening of which formed part of the project of establishing a wide-ranging commercial network.[30]

Of the shop, inaugurated in December 1927, there remain two preliminary studies and some photographic evidence of the completed work.[31] [8–11] In the proposal illustrated by the studies, Scarpa makes self-confident use of elements derived from historic styles, giving shape to the space with pairs of columns on pedestals which, repeating themselves at the centre of the area in a subordinate rhythm, define a passageway and side spaces where, on transparent shelves, the glass objects were placed.

The front of the shop was designed with a certain monumental character, particularly in the careful design of the three windows, each surmounted by a tympanum, a solution owing much to his academic studies.

The implementation of the exterior however, while retaining some motifs present in the study design, simplified the architectural elements and placed emphasis on the shop sign with the name of the firm standing out over the three windows and again executed in capital letters made of metal. With regard to the interior, also greatly simplified compared with the initial proposal, this was arranged in an *enfilade* of spaces linked together by arched openings, among which the principal arch was notable for the repetition of niches with glass shelves on which to display the artefacts. In general, the shop was characterised by the essential lines of the wooden furnishings, of Mitteleuropean inspiration, and by

[6] Carlo Scarpa, studies for the Cappellin company visiting card with the spherical vase, c. 1929, Museo del Vetro Murano Archives

[7] Vittorio Zecchin, 'Glass dinner service decorated with gold', [c. 1923], detail in M. Marangoni, *Le arti del fuoco: ceramica, vetri, vetrate*, Milan 1927, plate 112

>
[8-9] Carlo Scarpa, studies for the M.V.M. Cappellin & C. shop in Florence, 1927, MAXXI Architecture Collection. Carlo Scarpa Archive

>
[10-11]
M.V.M. Cappellin & C. shop in Florence: exterior and interior, 1928-29, Arch. Luigi Cappellin collection

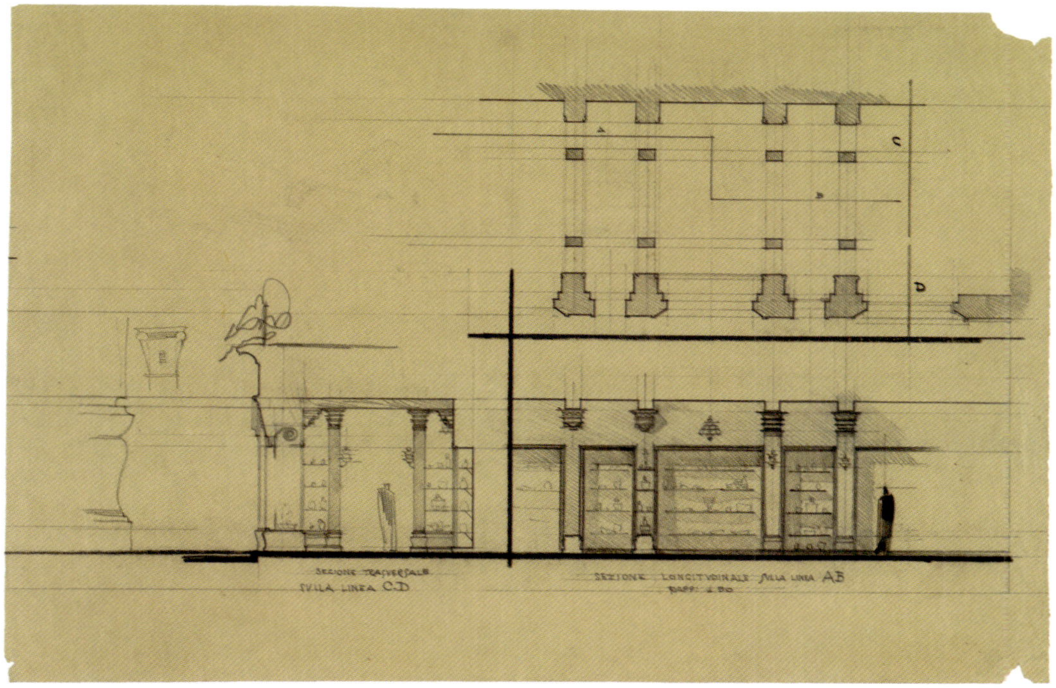
SEZIONE TRASVERSALE
SULLA LINEA C.D
SEZIONE LONGITUDINALE SULLA LINEA AB

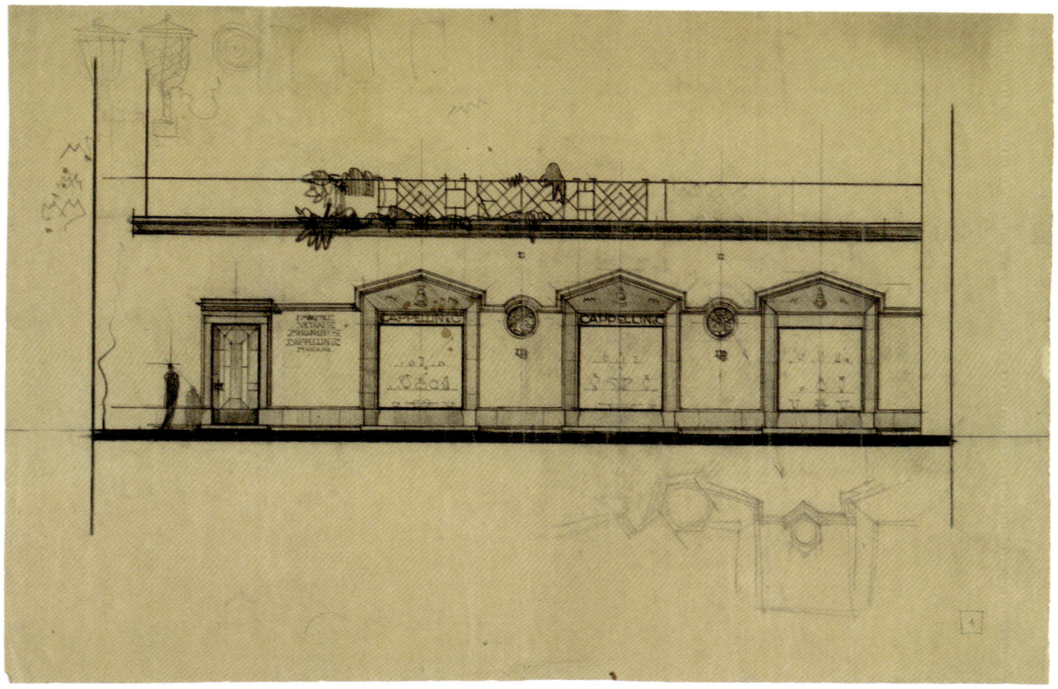

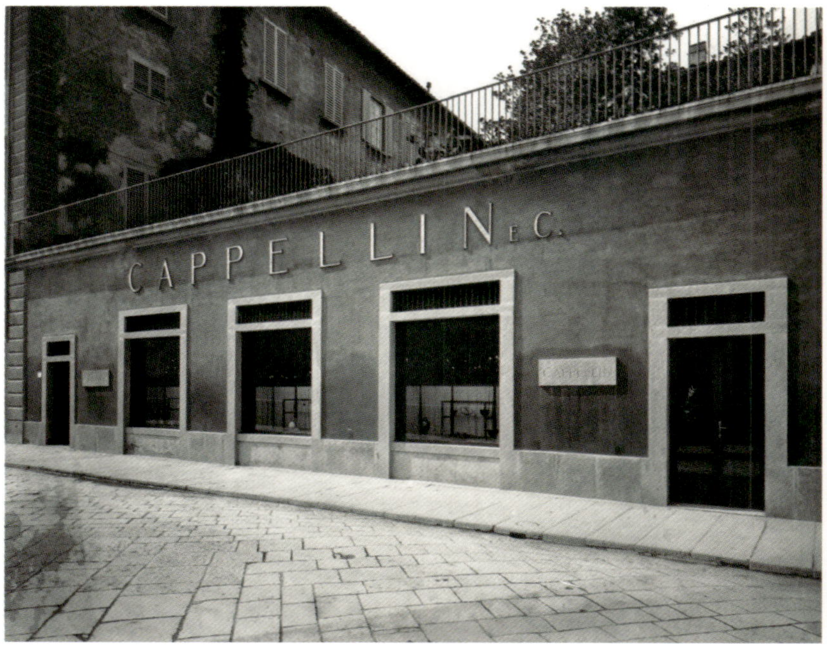
CAPPELLIN E C.

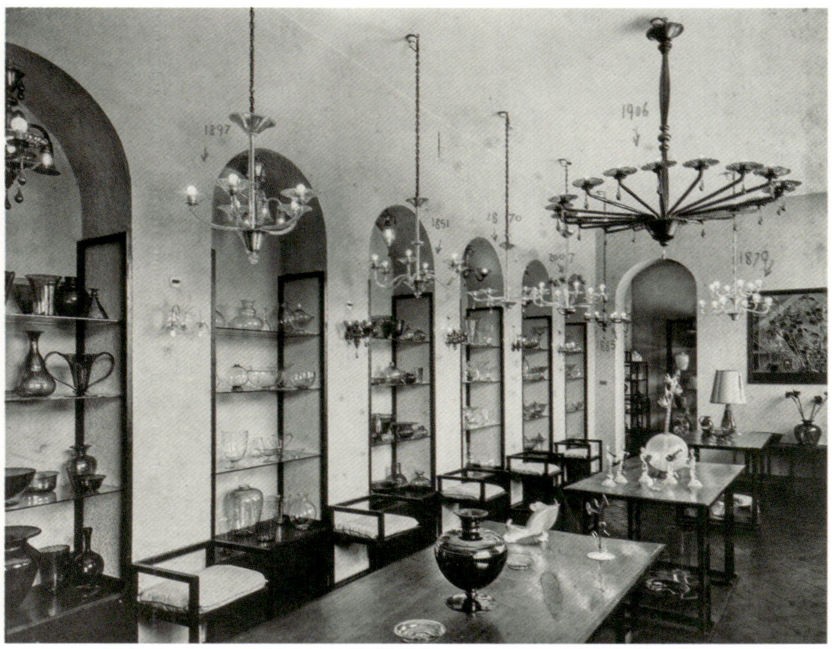

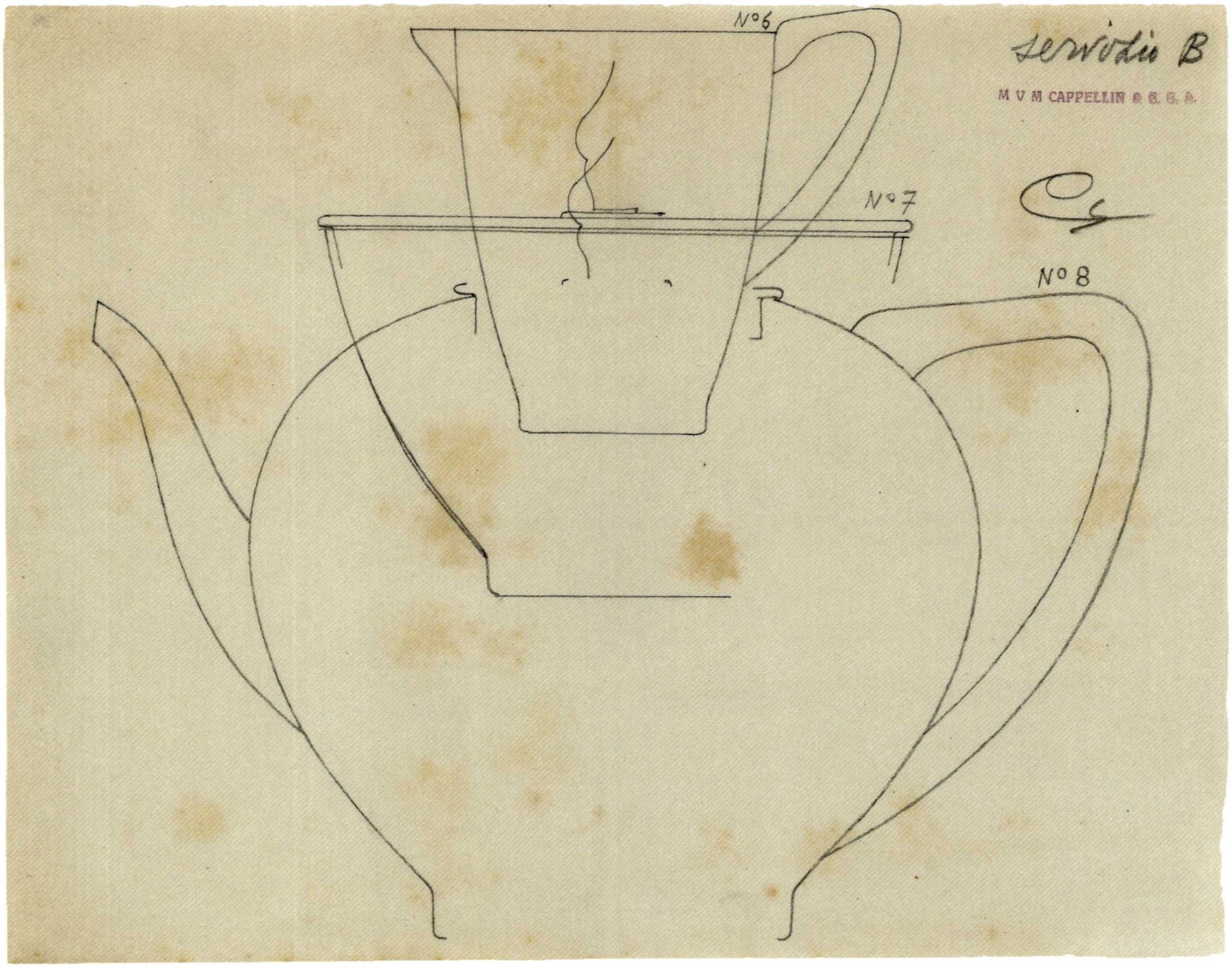
servizio B
M V M CAPPELLIN & C. S. A.
N°6
N°7
N°8

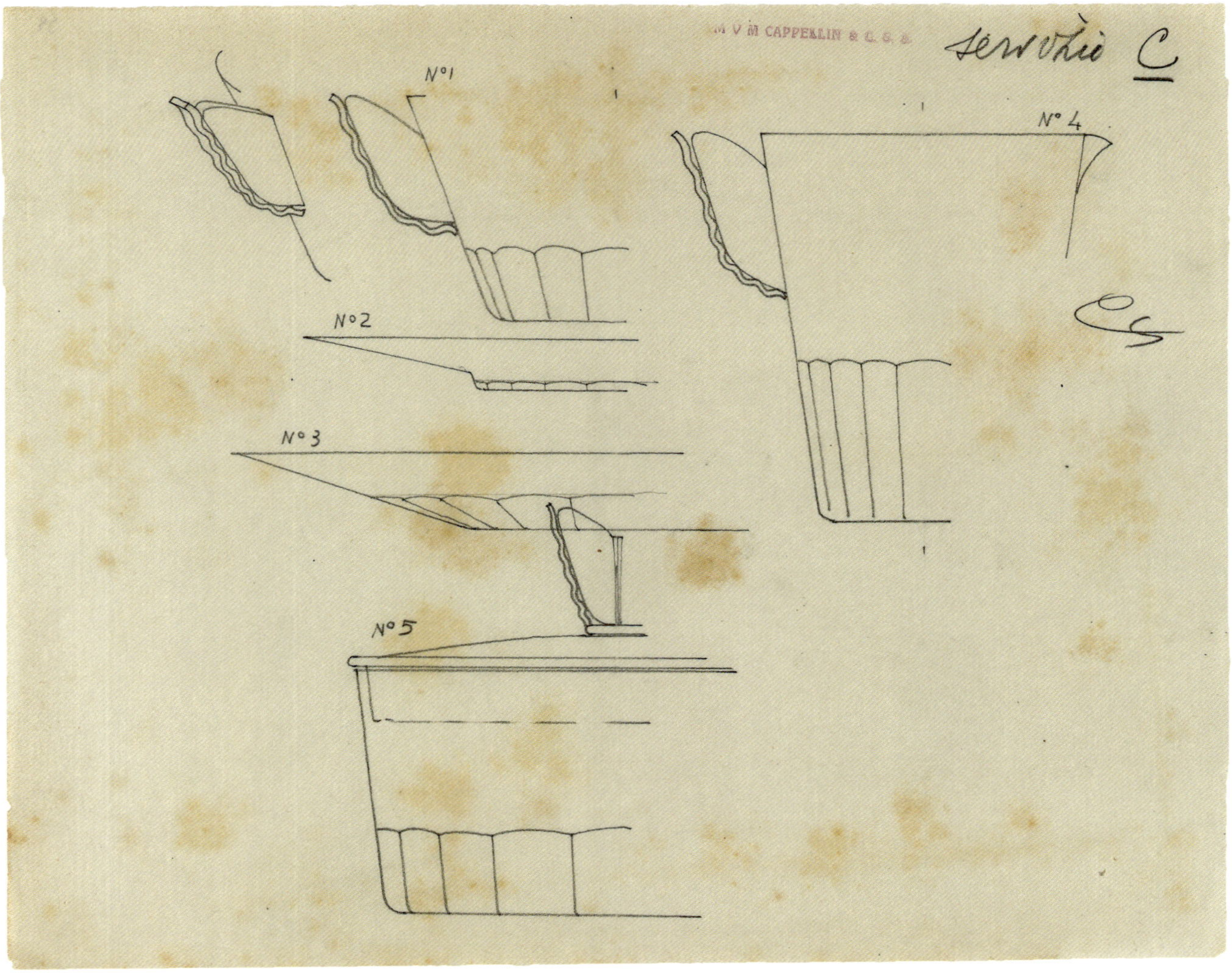
M V M CAPPELLIN & C. S. A.
servizio C
N°1
N°2
N°3
N°4
N°5

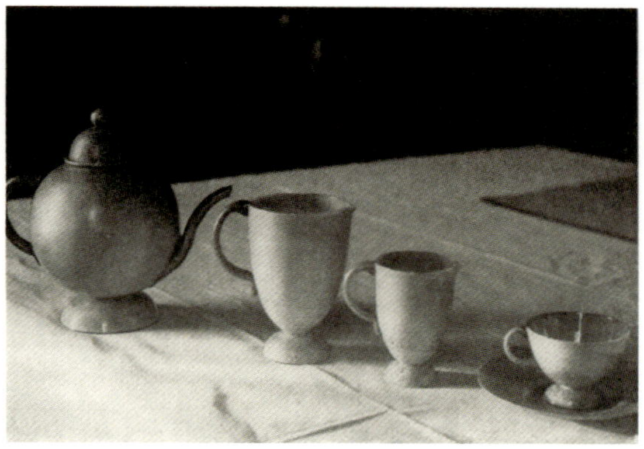

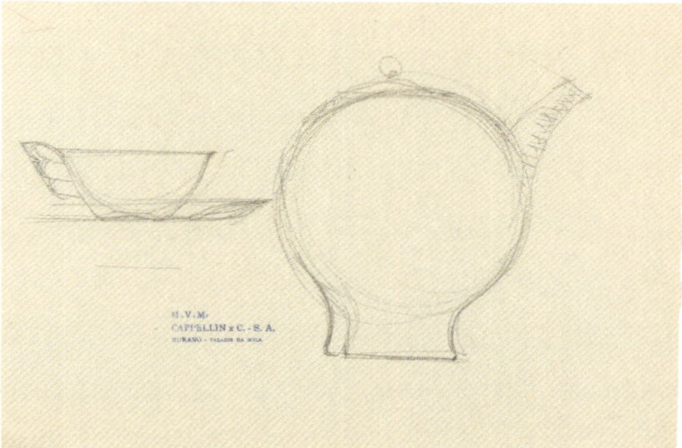

<
[12] Carlo Scarpa, designs for a tea service, solution 'B', c. 1929, Carlo Scarpa Archives, Museo di Castelvecchio, Verona

<
[13] Carlo Scarpa, designs for a tea service, solution 'C', c. 1929, Carlo Scarpa Archives, Museo di Castelvecchio, Verona

[14] 'Tea service in *lattimo* glass – yellow and gold', in *Mostra 'Artisti contemporanei': Galleria Pesaro, Milan*, Milan 1929, no page number

[15] Carlo Scarpa, studies for tea service, c. 1929, AMuVe

the architectural structure designed to emphasize the quality of the Cappellin production. This seems to have been the intention which, perhaps together with demands of a financial kind, guided Scarpa in the development of the project, resulting in his abandoning a solution which was certainly more erudite and structured and which might have placed the architecture in the foreground instead of the glass pieces themselves.

In Florence Scarpa also designed the polychrome stained glass door with two leaves which closed one of the arched openings.[32] The design of such pieces, in which the M.V.M. distinguished itself, allowed him to tackle the figurative arts directly, in which he cultivated a growing interest in those years, whether through his reading or by devoting himself personally to painting.[33] The study and execution of the 'Murano' stained-glass pieces was an occasion for him to broaden his artistic experience and explore the light-colour relationship through the material of glass.

Also in 1927, Scarpa was responsible for the layout of the Cappellin exhibition at the 3rd Mostra Internazionale delle Arti Decorative of Monza, where it was possible for him to have a direct appreciation of some of the most recent research in the field of architecture and the decorative arts (for example those exhibited in the 'Germanic section'), and presumably to come into contact with some of the protagonists in these areas, thanks to the circles frequented by Cappellin.[34]

For the M.V.M. exhibition he prepared a structured display system, transforming the room into an octagonal space, in the outer walls of which niches varying in depth, breadth and height were made to accommodate the glass pieces on display.

In this space, with its Art Déco-type solution, together with objects attributable to the intervention of Cappellin, some of the first works by Scarpa were shown, among which, in addition to the blown-glass pieces with truncated-cone-shaped foot, were also objects in netted glass or with crossed canes with the characteristic multi-coloured glass technique.

In general, as was possible to appreciate also on the occasion of the other exhibitions in which the glassworks participated in the same year, in Rome and Venice,[35] the new production was distinguished by its elegant polychrome nature, a choice ascribable to the young architect. Up to this time, the glass pieces were distinguished by their monochrome elegance several times noted by the commentators; the new ones, however, were mostly proposed in bichrome versions, whether by varying colours of the same tone (for example purple and amethyst) or by juxtaposing 'contrasting' colours, such as yellow and green, white (crystal glass) and purple etc. In some cases also, recourse was had to finishings in coral-red *pasta vitrea* combined with green or *pagliesco*-glass surfaces, as in the case of the series with a 'palm-leaf motif' decoration.

The work on the colour of the material led in the following years to new and extraordinary outcomes, as a result also of the greater familiarity acquired by Scarpa with the production techniques due to his increasingly assiduous attendance at the furnace. The experiments, presumably encouraged by Cappellin, were accordingly directed towards opaque glass based on the *lattimo* glass that Zecchin and also Cappellin had used intermittently in the early stages of their association with V.S.M. Cappellin Venini & C., as was seen at the *Salon d'automne* of Paris in 1922.[36]

The new productions, which were shown at the same Paris exhibition in 1928,[37] were, however, very distant from these examples, both in the consistency of the material and in the design. Many of these objects were embellished for example with applied gold leaf which, during processing with the effect of heat and blowing, disintegrated, almost becoming powder, so giving the vase a characteristic flecked texture resembling to some extent that of certain ceramics such as those of Robert Lallemand exhibited in 1926 at the *Salon d'automne* and published in *Art et Décoration*.[38] The cross-influences coming from the ceramic world are also evident in certain *lattimo* pieces decorated by horizontal polychrome bands.

In the same way it is possible to conjecture that the repertory of forms, characterised by significant simplification in design and often of oriental inspiration, was partly borrowed from this world. If it is

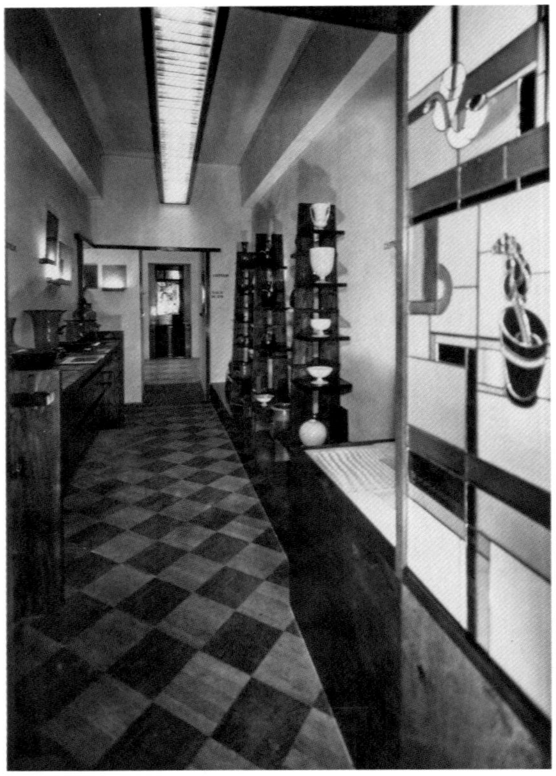

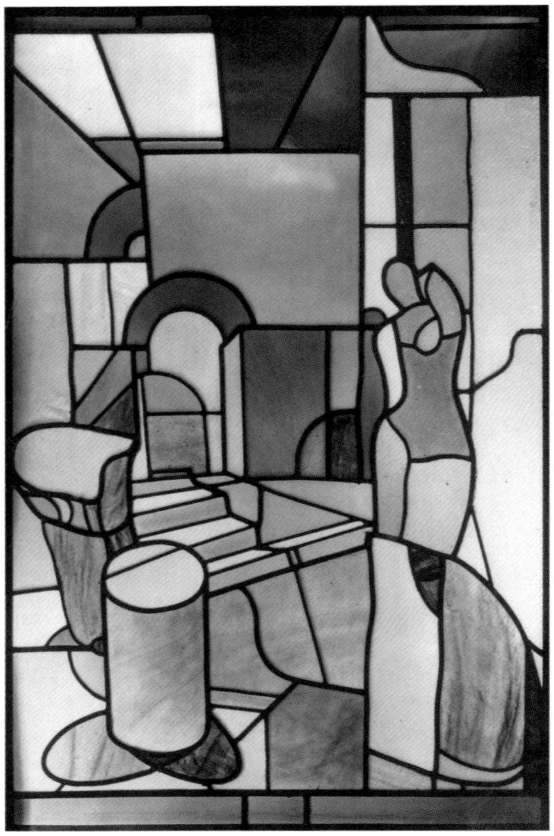

[16] Cappellin Room at the 4th Triennale of Monza, 1930

[17] Stained-glass after cartoon by Carlo Scarpa, 4th Triennale di Monza, 1930, © La Triennale di Milano - Archivio fotografico

true that in the glass museum of Murano there were certain eighteenth century *chinoiseries* in *lattimo* imitating porcelain, it is also true that in those years, as shown in particular by the German exhibition at the 3rd Biennale di Monza in 1927, much of contemporary ceramic production was marked by this source of inspiration.[39]

A case in point is the tea service in gold *lattimo* exhibited by Cappellin in the Galleria Pesaro of Milan at the exhibition *Artisti italiani contemporanei* in 1929.[40] [14] It was one of three variants (Variant A) that Scarpa, as documented by some graphic evidence, [12-13, 15] had designed for the Deutsche Werkstätten of Munich, which marketed M.V.M. products from c. 1928.[41]

In the various alternatives the inspiration of oriental art and ceramic production is discernible: while Variant C is based on certain somewhat finicky details (design of the handle of the cup and jug) combined with the Murano technique of the *meza stampaura* for the decoration of the base, Variants A and B are more linear. In particular the teapot in Variant A, produced in gold *lattimo*, has a markedly geometrical shape and seems to refer back, particularly in the design of the handle and spout, to the ceramic artefacts of the Bauhaus school and, above all, to those of Otto Linding.[42]

However, these glass objects, as shown also by the extraordinary production of highly coloured opaque pieces (with a *decoro fenicio*, in *pasta vitrea* etc.), the fruit of further experiments on the material and the colour realised between 1928 and 1930, did not aim at imitating the consistency of ceramics but, as Carlo Alberto Felice emphasized on the occasion of the 4th Triennale di Monza in 1930, 'even if they do not follow the classic understanding of glass [...] they do not aspire to be or appear to be anything other than glass, which in them reveals others of its innate qualities.'[43]

The Cappellin firm had in fact exhibited at the Monza event [16] — which allowed for much discussion of opaque glass — a significant collection of its recent production characterised, for the greater part, by glass techniques with a great expressive value which, to paraphrase Felice, showed the 'infinite [...] possibilities and resources of this exciting material'.[44]

Prominent among them was the series of 'geometric' glass pieces (so described in the firm's catalogue) which Scarpa designed with an eye once again on the avant-garde production beyond the Alps. In addition to the vases some examples of *bibelots* (figurines and animals) were also exhibited, a type much in fashion at the time, which the architect redesigned in a personal key. Remarkable in this respect is the model of a winged creature (owl?), seen at Monza, which refers in the design to that of an antique Chinese bronze work reproduced in the volume *Chinese Art*, published in 1927, and which he presumably had been able to consult.[45]

Also on the occasion of the Triennale of 1930 the layout of the Cappellin exhibition was entrusted to Scarpa,[46] who organised the space — a long narrow passageway — in a rhythmic succession of carefully designed vertical display units.

As a visual 'finishing line' in this ideal journey through the Cappellin glass, there was a stained glass piece, after a cartoon by the architect, which appeared repeatedly in the publications covering the event.[47] [17] The construction of this work and the presence in the foreground of certain elements, such as the stylised female figure (a mannequin) and the shadowy geometric solids, are a reference to the imaginative world of metaphysical painting, on which, as shown also by *La statua* (oil on cardboard executed in the same period), Scarpa was reflecting.[48]

As though to reinforce this pathway, along the entire length of the room, in a position below the ceiling, a lighting structure had been placed supported by a metal frame.

This was the implementation of a system that the Cappellin firm had patented on 25 May 1929 and that provided for the construction of a backlit diffuser screen, obtained by juxtaposing colourless or coloured transparent glass canes 'appropriately embedded at their extremities'.[49] In the lighting sector also, the Cappellin firm was showing itself to be in the avant-garde and, by investing in experimentation, was pointing to new ways of executing lighting appliances based on modular elements, which would be developed by others in the following years.[50]

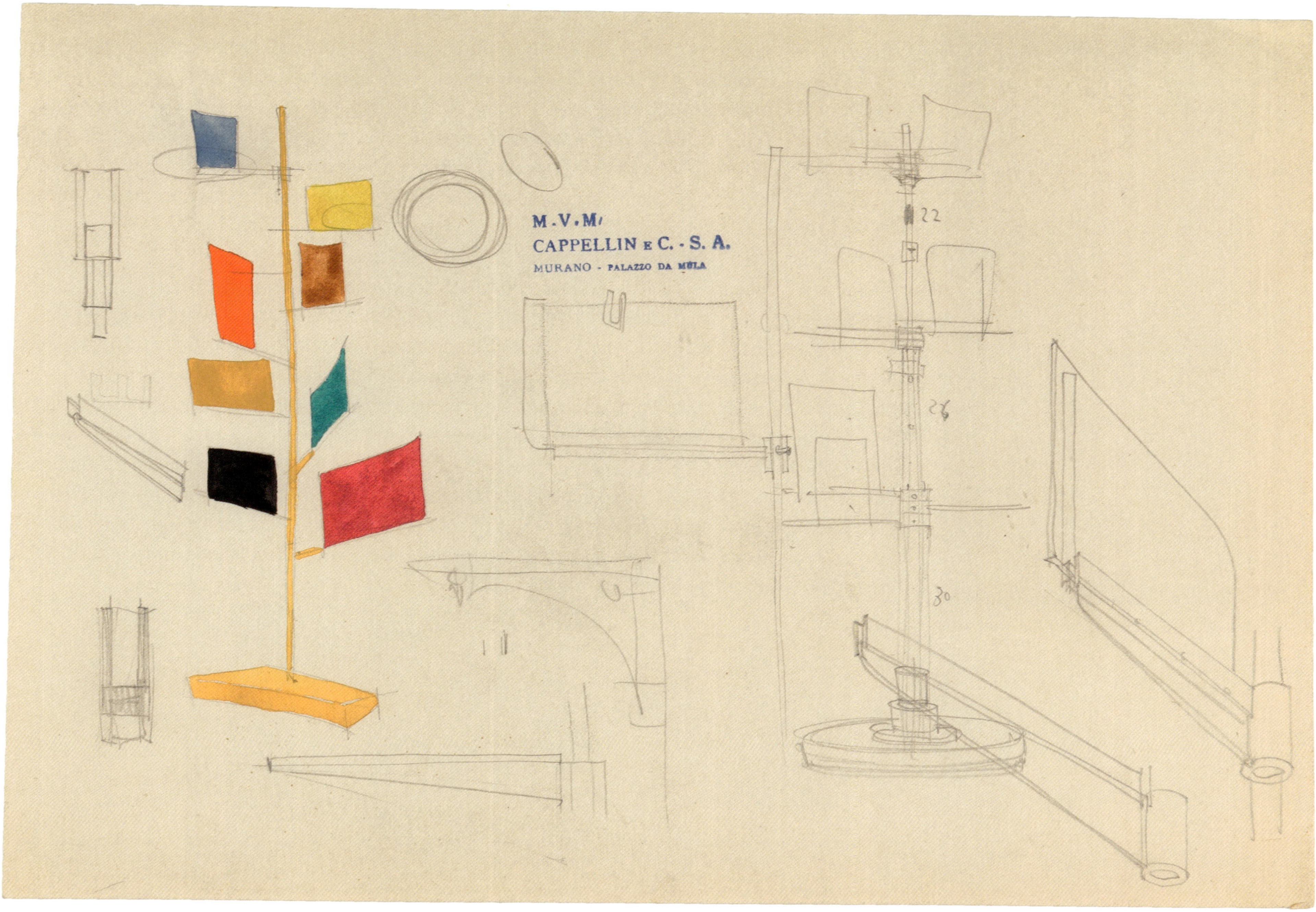

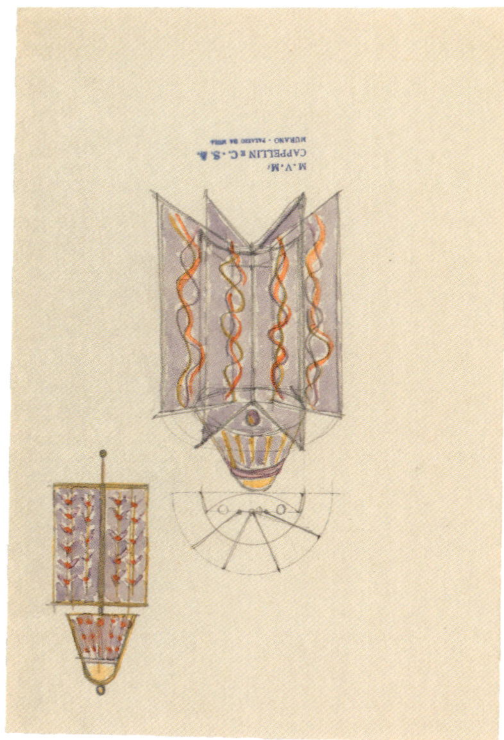

[18] Studies for floor lamp (or decorative structure?) with sheets of coloured glass, 1929-30, AMuVe

[19] Carlo Scarpa, studies for lamp with glass sheets, 1929-30, AMuVe

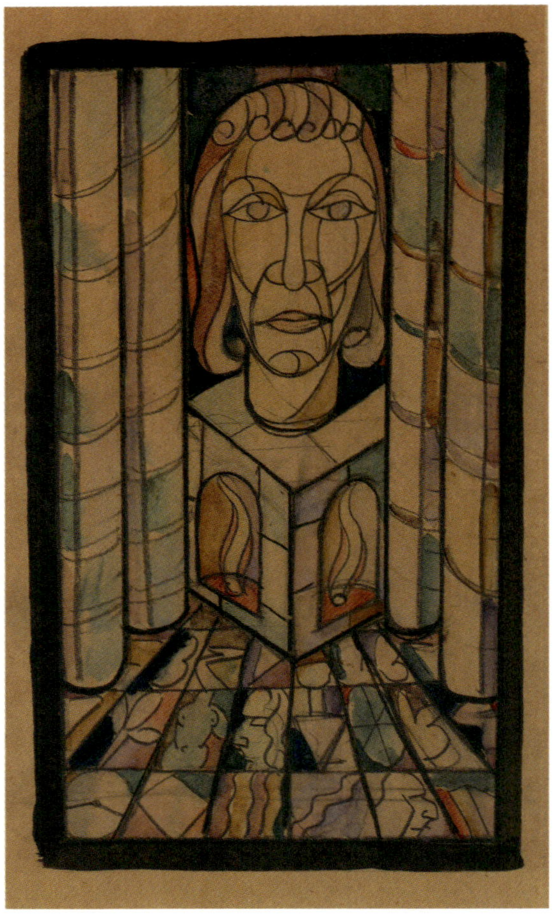

Other solutions to be seen at Monza provided for the use of polychrome sheets; with different glass techniques, which acted as a diffuser screen mounted on the wall. As shown by a considerable number of drawings executed in his hand, Scarpa was also involved in the designing of lighting projects.[51] For example, he repeatedly studied ways of using glass sheets, whether for the construction of suspension lamps or wall or ceiling lamps, or also standard lamps. Of significance in this regard is a study drawing for a floor lamp (or decorative structure?) with mobile screens fixed to a central shaft, where Scarpa uses, with great self-confidence and powers of abstraction, sheets of exceptional chromatic richness. [18-19]

The exhibition of 1930 was the last occasion on which Scarpa was present at the Triennale together with Cappellin, since the glassworks closed because of bankruptcy in January 1932.

A sketch, perhaps an ironic proposal for a stained glass piece, sums up the probable relationship between the young architect and Giacomo Cappellin who, believing in him, introduced him into the world of art glass, so enabling him to express his creative genius.[52] [20]

At the centre of the composition, over a cube with flames (a sort of altar or allusion to the fire of the furnace), there is the representation of a head drawn in the 'classical' way, resembling the style of Wildt, in which the features of Cappellin can be recognised, like a divinity or *dominus* of the glassworks. On the tiles of the floor seen in perspective, among the different elements of the composition, appear two male profiles, probably self-portraits of Scarpa, who in 1978 still recalled that 'After the Academy I found work in a Murano glassworks. I learned to work with a marvellous material [...]. I know glass and I know what can be done with it'.[53]

[20] Carlo Scarpa, study for stained glass (?), c. 1928, Carlo Scarpa Archive, Museo di Castelvecchio, Verona

1 *Documenti per l'iscrizione all'albo degli architetti*, published in *Documents pour l'inscription à l'ordre des architectes de la province de Venise* in *Les cahiers de la recherche architecturale Carlo Scarpa*, 19, single-subject number, 1986, p. 48. With regard to Scarpa's school training, cf. C. Sonego, 'Carlo Scarpa: gli anni '20', in K. W. Foster, P. Marini, *Studi su Carlo Scarpa 2000-2002*, Venice 2004, pp. 27-41; C. Sonego, 'La cultura tecnica nella formazione di Scarpa', in W. Tegethoff, V. Zanchettin (edited by), *Carlo Scarpa. Struttura e forme*, Venice 2007, pp. 83-97.

2 B. Radice (edited by), 'Colloquio con Carlo Scarpa, un architetto a regola d'arte', in *Modo*, 16, 1979, p. 19

3 With regard to Guido Cirilli cf. C. Sonego, *Carlo Scarpa: gli anni...*, *op. cit.*, pp. 27-46; A.G. Cassani, G. Zucconi (edited by), *Guido Cirilli architetto dell'Accademia*, Padua 2014.

4 G. Samonà, 'Relazione per l'inaugurazione dell'anno accademico', in *Annuario I.U.A.V. 1945/46*, p. 6.

5 *Op. cit.* In the area of restoration work, understood as being in continuity with the architectural design and aimed at restoring the original state, particular mention should be made of Cirilli's experience as a superintendent of monuments in Venezia Giulia, between 1918 and 1922.

6 M. Cascavilla, G. Favero (edited by), *Un'ora con Carlo Scarpa*, rai 1972, transcribed in O. Lanzarini (edited by), *Fermare la complessità. La videobiografia.* 'Un'ora con Carlo Scarpa', *1972*, in J. Gudelj, P. Nicolin (edited by), *Costruire il dispositivo storico. Tra fonti e strumenti*, Milan 2006, pp. 247-273.

7 M. Dominguez (edited by), 'Intervista a Carlo Scarpa', in F. Dal Co, G. Mazzariol (edited by), *Carlo Scarpa Opera completa*, Milan 1984, p. 297. Scarpa was Cirilli's assistant from the academic year 1926/27 to 1928/29.

8 Article 10, Act of 24 June 1923 no. 1395. In addition P. Nicoloso, *Gli architetti di Mussolini. Scuole e sindacato, architetti e massoni, professori e politici negli anni del regime*, Milan 1999, pp. 70-74. Scarpa's application was not accepted, however. In accordance with the account given by Nicoloso, the Scarpa dossier probably continued until May 1929, perhaps preventing him from taking other steps to obtain accreditation.

9 Translation from the French by the present writer of the text published in *Documents pour l'inscription...*, *op. cit.*, p. 50.

10 *Ibid.*, p. 53.

11 Cf. C. Sonego, *La cultura tecnica...*, *op. cit.*, p. 91.

12 Cf. note 9.

13 With regard to Giacomo Cappellin and the M.V.M., see the essay by Marino Barovier in the present volume.

14 *Lettera di Giacomo Cappellin ad Ugo Ojetti*, 13 April 1925 in Galleria Nazionale d'Arte Moderna, Rome, Fondo Ojetti, cass. 393. The other headquarters identified were a 'furnace at san Baseggio and the property of Mazzega in rio dei Vetrai', G. Cappellin, *Fallimento Soc. an. Cappellin & C. Murano. Appunti per la memoria difensiva*, March 1932, typewritten, in Private Archive.

15 'Il ripristino di Palazzo da Mula a Murano', in *La Gazzetta di Venezia*, 5 October 1926; C. [V. Costantini], 'Maestri vetrai muranesi "Cappellin"', in *Le arti plastiche*, 16 July 1926; V. Costantini, 'Murano e le sue celebri vetrerie', in *Le vie d'italia e dell'America latina*, June 1927, pp. 704-707.

16 G. Cappellin, *Fallimento Soc. an. Cappellin & C...*, *op. cit.* Regarding Pasqualin & Vienna cf. M. Carraro, *Venice 1913-1947: architetti, istituzioni, città*, degree thesis, IUAV, 1998-99, p. 52.

17 G. Morazzoni, *Il palazzo da Mula..., op. cit.*, p. 13. The photographic documentation consulted regarding the restoration is in the Marino Barovier archive, Venice (hereinafter AMB).

18 *Il ripristino..., op. cit.*

19 The design is kept in the archives of the Museo del Vetro, Murano (hereinafter referred to as AMuVe).

20 *Les cahiers de la recherche architecturale..., op. cit*, p. 51.

21 Cf. M. Barovier, C. Sonego (edited by), *Vittorio Zecchin. I vetri trasparenti di Cappellin e Venini*, Milan 2017.

22 Cf. *1925-1931. Catalogue of models* in the present volume.

23 Specifically the article by P. Du Colombier, 'Le Salon d'automne', in *Art et Décoration*, November, 1926, p. 187. Cf. C. Sonego, 'Carlo Scarpa alla Maestri Vetrai Muranesi Cappellin & C. 1925-31' in M. Barovier, *Carlo Scarpa. I vetri di un architetto*, Milan 1997, p. 33.

24 R. Papini, 'La Mostra delle arti decorative a Monza II. Le arti della terra', in *Emporium*, 343, July, 1923, p. 10. The service, present in the V.S.M. Cappellin Venini e C. catalogue under the no. 120, appears in G. Marangoni, *Le arti del fuoco: ceramica, vetri, vetrate*, Turin 1927, plate 112.

25 C. [V. Costantini], *Maestri Vetrai Muranesi..., op. cit.*

26 V. Costantini, *Murano e le sue celebri vetrerie..., op. cit.*, p. 705.

27 In an article that appeared in the magazine of the Fiera (Trade Fair) of Milan and was based on Cappellin's own notes (cf. *Lettera di G. Cappellin a Pietro Camperio, 10 gennaio 1929,* in AMuVe), with regard to the work of the master glassmakers, the following point is made: 'it is not the daily quantity of objects that a master can blow that counts, but rather the quality and kind of work carried out. Every artistic bowl or vase must be turned to perfection without humps or inequalities, otherwise it must inexorably be destroyed', 'Il vetro di Murano', in *La Fiera di Milano,* March 1929, p. 20.

28 R. Papini, 'Le arti a Monza nel 1927. I Gli italiani', in *Emporium*, 391, 1927, p. 24.

29 C. Scarpa, *Curriculum vitae*, typewritten, 1951, p. 1, also G. Mazzariol (edited by), *Curriculum vitae di Carlo Scarpa*, typewritten, 1961, p. 1, MAXXI Museo nazionale delle arti del XXI secolo (National Museum of the Arts of the 21st Century), Rome. Collezione MAXXI Architettura. Carlo Scarpa Archives (hereinafter referred to as ACS). Both curricula vitae were drawn up for the accreditation examinations as a university lecturer.

30 Cf. C. Sonego, *Carlo Scarpa alla Maestri..., op. cit.*, p. 36. After the separation from his ex-partner Paolo Venini, in 1925, Giacomo Cappellin was left without a commercial network which, however, as early as 1928 included presences and distributors in Murano, Venice, Florence, Rome, Palermo, Turin, Genoa, Bergamo, Brescia. Outside Italy there were retailers in London, Paris, Berlin, Munich, Dresden, Majola (Switzerland), Stockholm, M.V.M. Cappellin & C. Murano, advertising folder for the Fiera (Trade Fair) of Milan 1928, AMB.

31 The designs for the Cappellin shop in Florence are two: one an elevation study, the other with a plan and two sections, ACS.

32 Cf. C. Sonego, 'Le vetrate di Carlo Scarpa', in M. Barovier, *Carlo Scarpa..., op. cit.*, p. 43. With regard to this and other stained-glass pieces of M.V.M., see also the essay by Lucia Mannini in the present volume.

33 Cf. *C. Sonego, Carlo Scarpa: gli anni '20..., op. cit.*, note 157, pp. 81-82.

34 *Catalogo ufficiale della III mostra Internazionale delle Arti Decorative*, Milan 1927, p. 30.

35 Cf. *XCIII Esposizione di Belle Arti della Società amatori e cultori di belle arti. Catalogue*, Rome 1927; *Catalogue of the XVIII Mostra dell'Opera Bevilacqua La Masa*, Venice 1927.

36 M. Barovier, C. Sonego (edited by), *Vittorio Zecchin..., op. cit.*, p. 38.

37 Dir. [G. Ponti], 'Murano: oggi fatto d'arte', in *Domus,* December 1928, p. 62. According to the catalogue of the *Esposizione di belle arti di Roma* of 1927, already on display at this exhibition were '4 lattimo glass pieces with varied and glittering decorations', for which however no other documentation has been found, *XCIII Esposizione di Belle Arti..., op. cit.*

38 P. Du Colombier, *Le Salon d'automne..., op. cit.*, p. 186.

39 Cf. R. Papini, 'Le arti a Monza nel 1927. II Gli stranieri', in *Emporium*, 392, 1927, pp. 71-79.

40 *Mostra Artisti italiani contemporanei*, *Galleria Pesaro, Milano,* Milan 1929, no page number.

41 The designs for this series are kept in the Carlo Scarpa Archive, Museo di Castelvecchio, Verona (hereinafter ASCVr), while some studies are to be found in AMuVe.

42 Cf., among others, M. De Michelis, A. Kohlmeyer (edited by), *Bauhaus 1919-1933*, Milan 1996, pp. 106-109.

43 C.A. Felice, 'I vetri alla Triennale di Monza', in *Dedalo*, II, 1930-31, p. 322.

44 *Ibid*, p. 312.

45 R.L. Hobson, *Chinese art*, London, 1927. The volume came into Scarpa's library some years after 1937, cf. J.K.M. Pierconti, 'Appunti per uno scenario di cultura cinese nei vetri e nell'opera di Carlo Scarpa', in M. Barovier (edited by), *Carlo Scarpa Venini 1932-47*, Milan 2012 p. 85.

46 The exhibition catalogue makes no reference to it; however the fact is reported in C. Scarpa, *Curriculum..., op. cit.,* p. 1 e G. Mazzariol (edited by), *Curriculum..., op. cit.,* p. 8. In 1930 Carlo Scarpa was also involved in the renovation of the new Cappellin facilities, in San Gregorio Venice, were a special fornace was set-up for demostration purpose.

47 C.A. Felice, *I vetri...*, cit., p. 90. C.A. Felice, *Arte decorativa 1930 all'esposizione di Monza*, Milan 1930, plate 89; E. Serretta, 'Una giornata alla Triennale di Monza', in *L'illustrazione italiana*, 22 June 1930, p.1106.

48 C. Sonego, *Le vetrate..., op. cit.*, p. 43. For the pictorial work see G. Scarpa, *le pitture..., op. cit.*, p. 153.

49 Patent n. 279918, 'Process for producing decorative motifs in glass and industrial products obtained using the said process', in the Central Archives of the State, Rome, Ministry of Industry and Commerce and Craftsmanship, Italian Office for patents and trademarks, Invention series. (hereinafter ACS MICA). The patent provided for the use of pierced solid glass canes or for coupling them to produce lighting systems or decorative motifs as was the case with the base of the centrepiece exhibited in 1931 at the Mostra del Giardino Italiano in Florence. This system was used in the lighting proposal for the Grand Hotel Thermes of Acqui Terme, inaugurated on 5 October 1930 (cf., F. Mangone, G. Belli, M.G. Tampieri, *Architettura e paesaggi della villeggiatura in Italia tra Otto e Novecento*, Milan 2015, p. 218), to which bear witness certain graphics produced by Carlo Scarpa, now kept in ACSVr. The Cappellin firm also took out a patent for 'innovation in the system of electric connections for chandeliers and lamps with moveable arms', no. 279142, 6 May 1929, ACS MICA.

50 C. Sonego, 'Carlo Scarpa nell'ufficio tecnico della Venini', in M. Barovier (edited by), *Carlo Scarpa Venini 1932-47..., op. cit.*, pp. 45-66.

51 Most of the lighting designs are to be found in AMuVe. Others are in The Wolfsonian-Florida International University, Miami Beach Florida, The Mitchell Wolfson jr. Collection, Miami, in Archivio Arti Applicate Italiane, Rome, in the Cammi collection and private collection. The designs concern different types of chandelier: with arms, suspension, wall etc.

52 The design, datable between 1928 and 1930, is to be found in ACSVr.

53 M. Dominguez (edited by), *Intervista..., op. cit.*, p. 297.

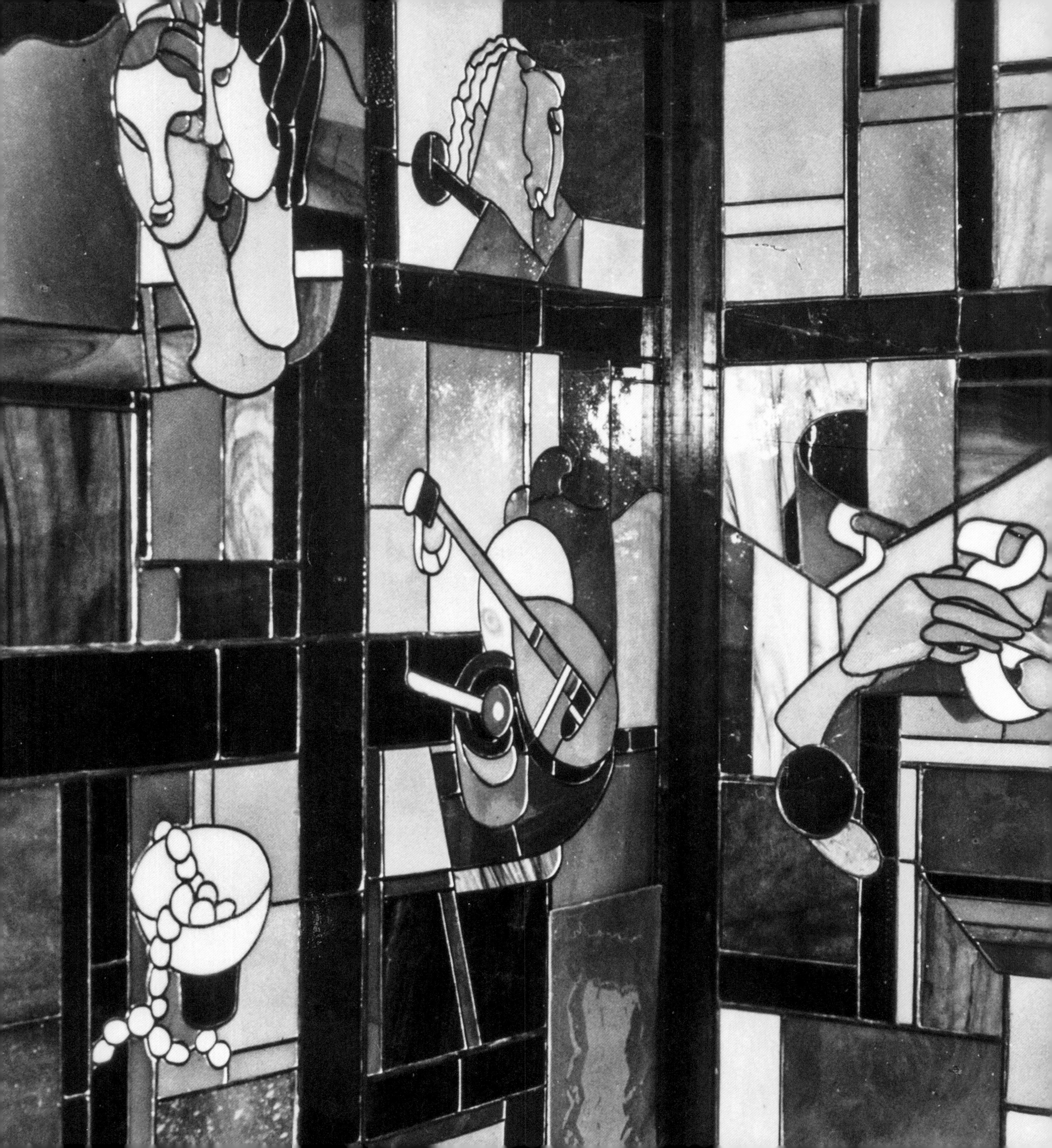

FROM THE NOTTE STELLATA TO THE STRADA DI NOTTE: PROJECTS AND ACHIEVEMENTS OF THE CAPPELLIN GLASSWORKS

Lucia Mannini

The twentieth century 'marks a renaissance in the art of the stained-glass window in Murano, particularly because of the work of Cappellin, who dedicated himself to it with the passion of an artist, with the broad understanding of the technician, with a secure vision of the moment. [...] The Veneto stained-glass school does not perhaps have the industrial development of the Tuscan, still less that of the Lombard school; one must, however, acknowledge that its contribution is very impressive'.[1]

This testimony, even if it has no particular critical value in itself, is a good summary of the role played by Giacomo Cappellin in twentieth century Italian stained glass, and that of Veneto in particular. In comparison with the vast and highly qualified production which had developed in Lazio, Tuscany and Lombardy, where there were many firms that had made their mark in the first three decades of the century, the modern stained-glass work of Venice was represented exclusively by that of the M.V.M. Cappellin firm which, between 1926 and 1931, was able to establish itself above all in the production and use of artisan glass, as against the industrial kinds that the others adopted, importing them from abroad.
Given the few works produced in Venice in the early twentieth century,[2] it is no surprise that the story of Venetian stained glass should have started in Milan, in via Montenapoleone, to which Giacomo Cappellin had moved his own antiquarian shop after World War one. 'The gallery soon became a place of encounter between interesting people, belonging to the world of artistic production, such as the stained glass artist Pietro Chiesa, and to the more cultivated and up-to-date elements of the bourgeoisie', such as Paolo Venini, whom Cappellin involved in the founding of a 'glass workshop of purely modern inspiration', and which owed its catalogue of models to Vittorio Zecchin.[3] Although the Venice glassworks Vetri Soffiati Muranesi Cappellin, Venini & C. did not produce stained glass, Vittorio Zecchin, its artistic director from 1921, designed cartoons which were executed in Milan by Pietro Chiesa. If, as is probable, Chiesa was responsible for the execution of the *Notturno* by Zecchin, a stained-glass piece exhibited at the Biennale of Venice in 1922,[4] the same Milan workshop produced the stained-glass pieces exhibited at the Venice exhibition in 1924[5] and also those displayed in the refectory at the Monza exhibition in 1925.[6] In that year, following the separation between Giacomo Cappellin and Paolo Venini, the association between Pietro Chiesa and Vittorio Zecchin also came to an end, the former consolidating his ties with Venini, the latter becoming artistic director of the new Cappellin glassworks. In this new glass workshop — M.V.M. Cappellin as it was called — the stained-glass sector was introduced as one of the prime elements in the activity.
Drawing on his experience with Pietro Chiesa, Cappellin supplied glass sheets produced in his Murano furnaces, while Vittorio Zecchin designed the works chosen for the first official appearance of the

<
[1] Carlo Scarpa, stained-glass piece exposed at the 4th Triennale, Monza 1930, detail

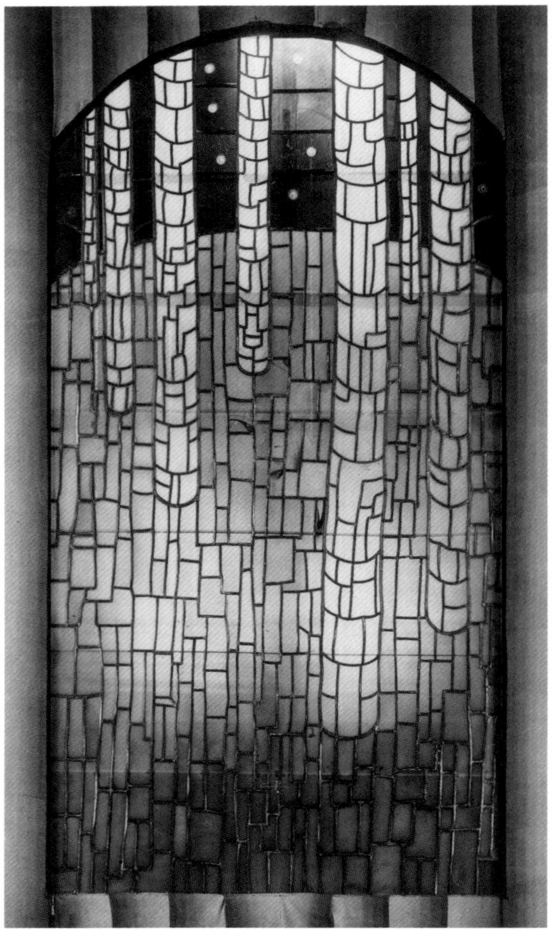

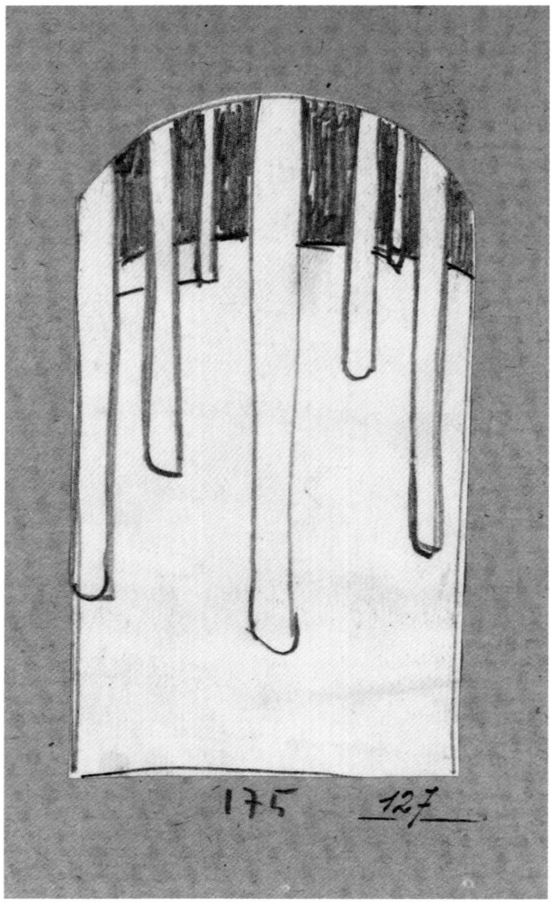

[2] *Betulle e notte stellata*, M.V.M. Cappellin stained-glass piece after a design by Felice Casorati, c. 1927. Period photograph, ©La Triennale di Milano-Archivio Fotografico

[3] Felice Casorati, album design. Turin, Galleria Civica d'Arte Moderna e Contemporanea

'stained-glass sector', which took place in Venice in 1926 and was decidedly 'Venetian' in its themes: *Bragozzi chioggiotti* and *Navi veneziane*.[7] In complete continuity with the previous stained-glass pieces — and in line with the tendency taking hold also in Italy for modern stained glass — production was carried out using a pure mosaic of coloured blown glass sheets, without recourse to *grisaille* painting; it was therefore an 'intarsia of pure brilliant notes',[8] with results that to some appeared 'something of an orgy of colour'[9] and 'too intense' in colouration.[10]

The vivid colourations of the *pâte de verre* certainly had the effect of drawing attention to a production that had no equal in Venice, in support of which recourse was had to a very modern technique and to themes strongly rooted in local culture, to that enchanted and fabulous collective imagination steeped in folklore which had been so dear to the creative world of Zecchin for over a decade.

From then on stained glass would never be absent from Cappellin exhibition events, rather it would be a characteristic element of them, but the departure of Zecchin in October of 1926 entailed the necessity of finding creators capable of producing cartoons suitable for treatment by that specific technique, which required both powers of graphic synthesis and knowledge of the materials. Among the number of items exhibited at the Biennale of Monza in 1927, uncertain in kind and lacking a defined formal orientation,[11] the stained-glass work *Betulle e notte stellata*, designed by Felice Casorati [2] constituted a sign of continuity. The choice of the nocturnal scene and dense forest was consistent with the matrix of Casorati's training, with that interest in the Vienna Secession and Klimt which, from the second decade of the twentieth century, he had shared with the artists of Ca' Pesaro and Vittorio Zecchin.[12] As any documentation that might throw light on the relationships between Casorati and the glassworks and on the genesis of the work — as the mark of his first visual idea — has been lost, we may instance a small, undated drawing in which there is a clear definition of the compositional layout and the relationship between the shapes of the seven slender trees and the scalloped format of the image.[13] [3] The translation into stained glass, however, led to a quite different outcome. From the only known photograph, one has the impression in fact that any illusion of prospective depth is sacrificed in favour of a tight and almost abstract synthesis, especially in the lower part where the lead connections frame a dense network of tesserae. The black-and-white image does not permit considerations about the tonalities adopted in the glass but in this respect it must be said that the judgement delivered by Roberto Papini was decidedly severe, as he defined the Cappellin stained-glass pieces as 'quite unsuccessful': 'not only are the tones far from being happy in their relationships but when, in a blue pane of glass, holes are made to represent stars, then the level is that of puppet theatres and the cribs of country churches', he wrote.[14] The fulminations of Papini, who was openly hostile to Cappellin, were therefore directed in particular against the stained-glass work designed by Casorati: the expedient of the holes made in the glass intended to correspond to the stars and to produce an original illusionistic effect could not fail to enrage a supporter, such as was Papini, of the correspondence of material and subject (and a stained-glass piece with holes could only be seen as an enormous contradiction of the function) while the formal solution was far from the critic's declared inclination in favour of 'twentieth century purism'.

The idiom of the stained glass *I cerbiatti* was more easily accessible and the subject accorded well with the *déco* climate of the 1920s, if in an older naturalistic vein, synthetically interpreted thanks to the graphic touch acquired by Ugo Ortona from the experience of illustrator and the example of Duilio Cambellotti.[15] Together with the Casorati stained glass, that of *I cerbiatti* was put on display at the exhibition of Monza of 1927 in a room entirely reserved for the Cappellin firm, Giacomo Cappellin and Carlo Scarpa being responsible for the order and layout.[16]

In the glass works displayed in Monza in this room, Papini had seen a 'notable decline' in Cappellin standards, which in his opinion had lost 'any character of personality' and above all suffered from the lack of an artist to guide production, since, after Zecchin's departure, Cappellin himself had taken on the task of replacing him.[17] However, Carlo Scarpa was in fact becoming such an artist, ever more in-

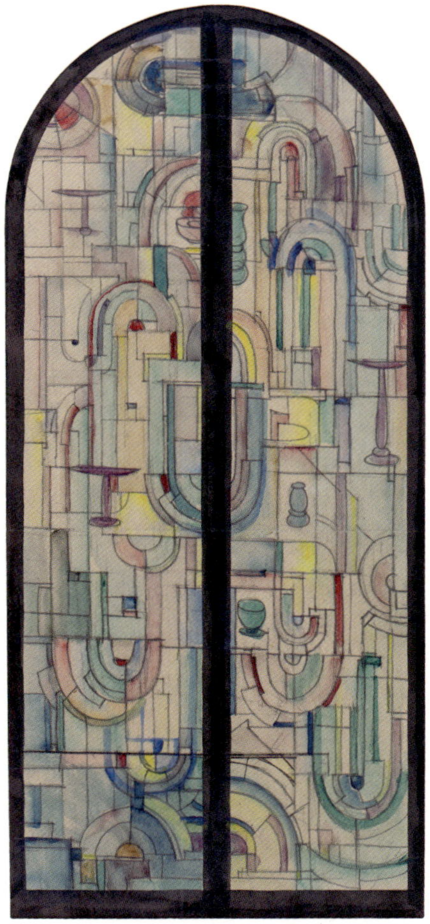

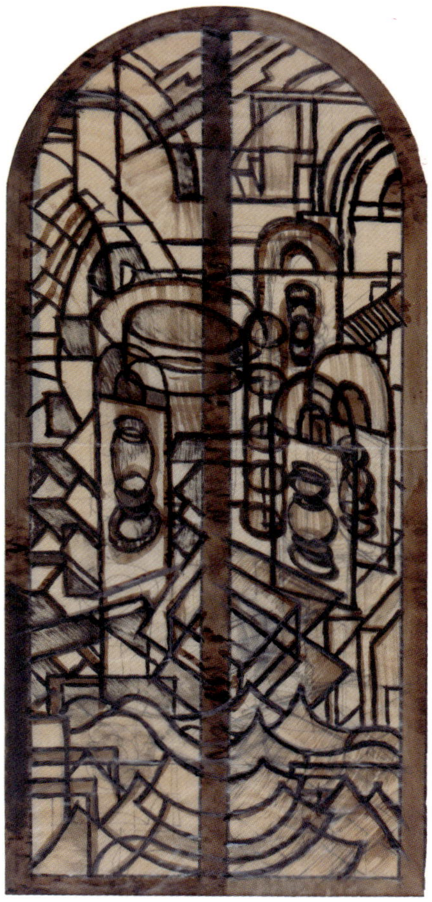

[4] Carlo Scarpa, project for stained-glass doors, 1927-28, Fondo C. Moretti e G. Moretti, Centro Studi del Vetro, Fondazione Giorgio Cini, Venice

[5] Carlo Scarpa, project for stained-glass doors, 1927-28, Fondo C. Moretti e G. Moretti, Centro Studi del Vetro, Fondazione Giorgio Cini, Venice, detail

volved as he was in the design of glass and stained-glass pieces, but also in the creation and layout of display spaces, prepared on the occasion of exhibitions or for the shops that the firm was opening in Italy and abroad.

A substantial set of drawings attributed to Scarpa is evidence of the search for formal solutions that the architect was carrying out in those years and which found in stained glass a preferential field of experimentation, an opportunity to deepen understanding of the expressive power of colour and light.[18]

Some of these designs can be traced to the Florence shop, of which Scarpa had also designed the furnishings. Opened in Lungarno Guicciardini in December 1927, it is known thanks to five photographs portraying both the exterior and interior, dated approximately 1928-29, from which one can understand how the stained glass, in addition to fulfilling a function of 'furnishing', also played a role in promoting the glass sheets, the objects produced and the activity of the furnace.[19] If in the iconography of stained glass the representation of themes linked to the furnace or the pieces produced was not at all unusual — and to go no further than Florence we can cite the example of the Quentin firm[20] — the range of solutions tried out by Scarpa was often oriented towards translating the caption or narrative content into geometrical motifs on the verge of abstraction, certainly owing much to the vast visual culture that he was developing and his interest in contemporary European artistic movements, from Cubism to Le Corbusier to De Stijl, and possibly also his knowledge of the more recent German, Dutch or French works in stained glass.[21]

There is an interpretation of De Stijl solutions to be seen in the designs for the stained glass of an internal door of the Florence shop — later executed, as the photographs show — in which pure clean geometrical forms springing from the initials M.V.M are deployed, notably asymmetrical in the two leaves of the door, creating an aerial, practically three-dimensional architecture;[22] the research on light and colour by Robert and Sonia Delaunay is what guides the projects for stained glass doors or stained glass with transom lights in which, from the forms of the vases, different planes interpenetrate, or in which concentric circles intersect in different chromatic fields;[23] [4] studied compositional cadences and calculated visual rhythms are predominant when the repetition of a vase — especially the spherical vase with the truncated-cone-shaped base — or of the mouth of the furnace takes place in a succession of planes until it becomes an abstract motif.[24] This is the case, for example, of the design which has on the left edge the sketch of a display connected with the Florence shop.[25] [5]

If not all the drawings by Scarpa that have come down to us were intended for the Florence shop,[26] neither were all the stained glass pieces of the Florence shop from the hand of Scarpa, as is witnessed by the *Canestro di frutta* by Thayaht, which can be recognised in a photograph, framed and hanging on the back wall of a room.[27] The bowl full of fruit, very different from the repertoire of abstract decorative motifs designed at the time by Thayaht, is to be compared with the artist's pictorial work, the simplifying spirit of which can certainly be seen in the geometrical symmetries of the background which mitigate the naturalistic definition of the subject. This stained glass piece is moreover the only tangible evidence of the artist's association with the Cappellin firm 'for the new premises of the firm in Florence and elsewhere' as a monograph of 1932[28] informs us. The word 'elsewhere' is probably to be associated with other designs of Thayaht's for stained glass with allusions to objects in glass and the processing thereof, probably intended to be produced by Cappellin.[29] Although no documentation about the relationships between Cappellin and Thayaht has emerged, that particularly close ties did exist in those years is evidenced by the photographs published in 1929 with Cappellin glass pieces side-by-side with works by Thayaht and Ram, some of which are indicated as having been displayed at the exhibition of Barcelona in that year.[30]

At this Barcelona exhibition, the Cappellin firm won a prize for a luminous ceiling which was not stained glass — as were those of Pietro Chiesa with lictor s fasces and flags — but the demonstration of the application of its glass sheets in an illumination system conceived by Piero Portaluppi, re-exploring

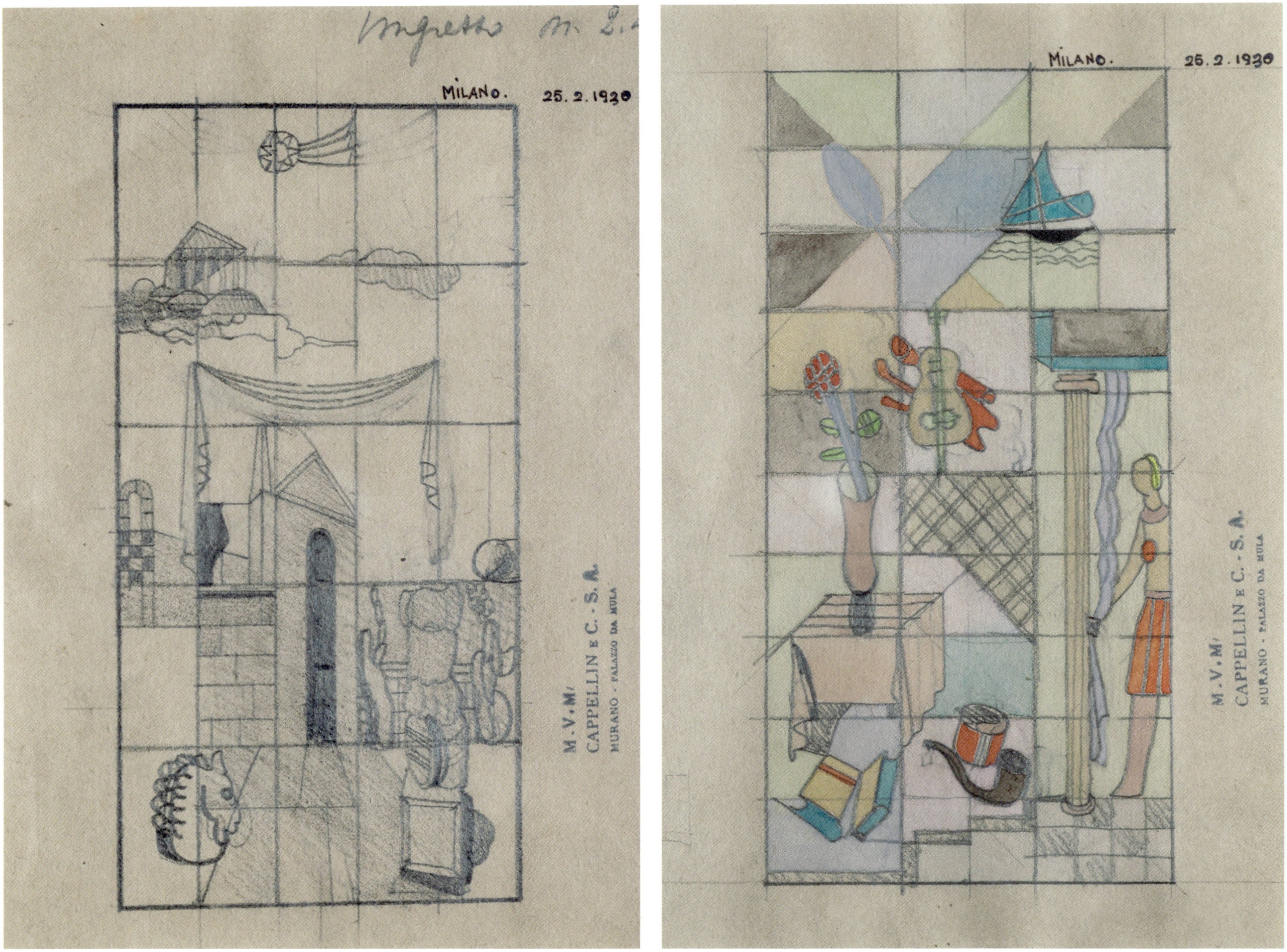

[6] Carlo Scarpa, two projects for stained-glass pieces, 1930, Fondo C. e G. Moretti, Centro Studi del Vetro, Fondazione Giorgio Cini, Venice

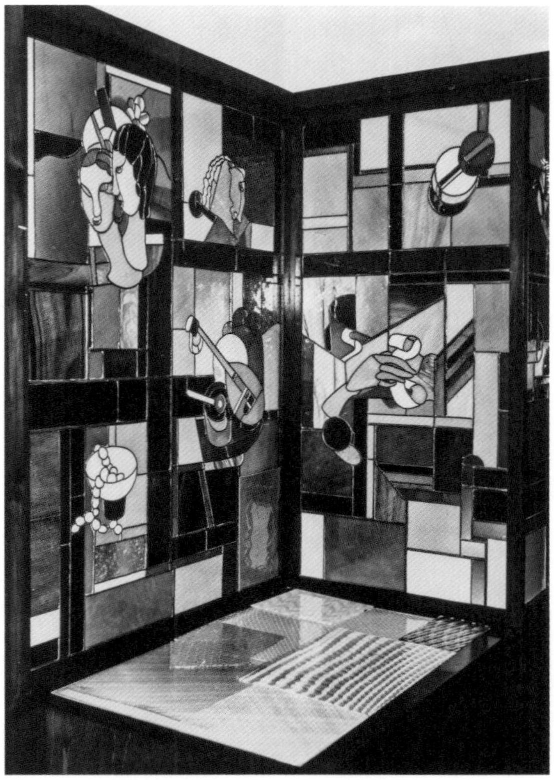

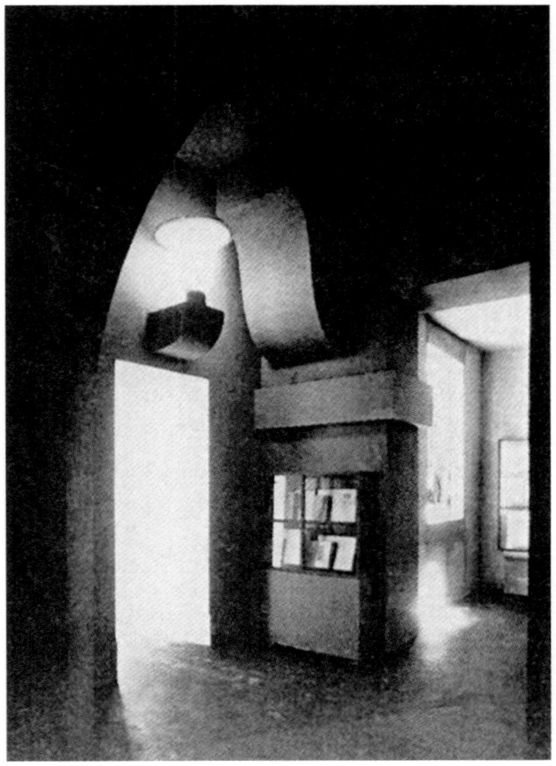

[7] Carlo Scarpa, stained-glass piece in the M.V.M. Cappellin room at the 4th Triennale, Monza 1930

[8] Giovanni Muzio and Mario Sironi, layout of the display of the Arti grafiche at the Monza exhibition of 1930

the idea of a suspended labyrinth of some years before.[31] The episode is a reminder, however, of how a new aesthetic philosophy of illumination was coming to the fore also in Italy and how at the same time the idea was gaining ground that stained glass could participate in this, leaving aside its primary role as a diaphragm of natural light and gaining that of a furnishing element exalted by artificial light, as was being shown by the luminous walls and ceilings realised by the major creators of the time, such as Louis Barillet, one of the protagonists in the renewal of French stained glass.

That a certain amount of attention was being paid to French and international examples is shown by the project of a Cappellin shop in Paris — that we know from some designs dated 1930 but are not sure whether it was executed — in which Scarpa undertook the challenge of great walls of glass — probably Murano glass — which defined the space, subdivided into regular rectangles or interrupted by geometrical inserts.[32]

There was a similar geometrical articulation in the structure of the stained-glass pieces produced for the exhibition of Monza in 1930. [1, 7] At the Triennale of Monza, in fact, to enter the room where the Cappellin firm was exhibiting its most recent productions, one passed beside a structure made up of a horizontal shelf (used, as a photograph shows, to place objects on) and a tripartite vertical wall: while sheets of glass of varying texture were fused in the shelf, in the walls there was retro-illuminated stained glass designed — like the rest of the display — by Carlo Scarpa.[33]

If on the technical side the 'panes striped with stained glass' were counted among the 'new shoots' springing from 'old Murano trunk',[34] in the design there was appreciation of the 'composite fantasies'[35] of the still lives that animated the geometrical network: the pipe with the tobacco pouch and the mandolin with the drum (taken from Braque and from Cubism) and the little vase with the succulent plant (very fashionable at the time as a synonym of modern furnishings), the hand with banderol and a horse's head (like an archaeological find), two female heads and a sphere projecting its shadow, and finally the inevitable reference to an object produced by the firm in the small, geometrically shaped bowl, produced in white cased glass with a black base, like the one actually on display in the room.

The idea of a geometrical mosaic of glass sheets in different colours and with various surfaces interrupted by still lives became a decorative cypher of the time, as is witnessed by two drawings for the stained glass of an apartment with the indication 'Milano 25.2.1930'[36] [6] where there is a recurrence of motifs dear to Scarpa, references to the classical tradition or modern figurative culture: the horse's head with the curling mane like a fragment of an ancient statue, the guitar, the pipe coupled with the tobacco pouch and, in one of the two drawings, the customary allusion to the production of glass objects, easily recognisable in the pink cased-glass vase with the black foot containing a glass flower. In one of the two drawings there is also a female figure, between Carrà (the mannequin with the pleated skirt) and Léger (the spherical breast).

A female figure was also the subject of the stained glass piece which at the Monza exhibition of 1930 closed the prospect of the corridor, which one would have described rather as a mannequin, slotted into a 'metaphysical' space made up of trapped solids, a recurring subject in Scarpa's research, an indication of dimensional relationships or an unexpected appearance between the architectonic lines. Similarly situated between metaphysics and enigmatic architectures was the stained glass designed by Mario Sironi for the Galleria delle Arti Grafiche at the Monza exhibition, also a Cappellin production, which, unsurprisingly, Carlo Alberto Felice published beside the female figure by Scarpa.[37] Unlike Scarpa's stained glass piece, the design of which appears entirely based on the synthesis of the glass tesserae, the, 'quite excellent' 'coherent cartoon[38] of Mario Sironi did not renounce reference to painting, not only because the compositional and iconographic elements — figures, architectonic motifs and solids in strong perspective — derived directly from contemporary pictures but also because their translation into stained glass required the help of painting on the glass — in the definition of the faces and hands and to give volume to the architectures. [8] The large-scale stained glass by Sironi, the only pictorial 'presence' in the Galleria delle Arti Grafiche, constituted an indispensable element in a deco-

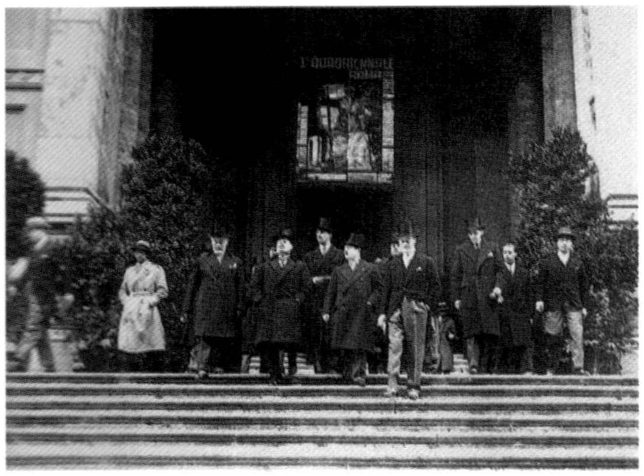

rative and complex whole, orchestrated according to a principle of harmony between the arts (painting and sculpture) directed by architecture, but above all it contributed to reaffirming the importance accorded by Sironi and Giovanni Muzio to the role of light, exalted in the calculated use of colours, in the strong contrasts and in the use of the most diverse materials: in the luminousness of the white walls, 'dazzling' on the black linoleum, in the capitals and luminous glass vases (also produced by Cappellin) and in the other smaller stained-glass pieces, the simple geometric elements of which stood out 'in vivid red on a light background'.[39]

The theme of light — which in those years was debated in its aesthetic and technological aspects with respect to the resources offered by electricity — was also crucial in the layout of the exhibition hall in Rome, which was totally renovated in accordance with the directions of Enrico Del Debbio on the occasion of the first Quadriennale, held in 1931. This renovation was carried out in order to create a context in which architecture could 'usefully combine with the sister arts to produce a more logical presentation' of the works.[40] There were very few transformations externally but a conspicuous sign of stylistic and technological modernity was to be seen in the luminous stained glass in which the Cappellin firm had reproduced the manifesto of the exhibition designed by the architect Mario Romano.[41] [9] 'A large luminous American-style advertisement over the front will discharge the task of advising even the blind that modern art exists', ironically noted the commentator of 'Il Lavoro Fascista';[42] a 'luminous advertisement' such as those that the Cappellin firm, starting in 1926, had proposed to Depero and Campari[43] which was part of the growing enthusiasm for 'luminous notices', 'beautiful in their entirely new but sure beauty', to quote Marinetti.[44]

With regard to the interior of the Quadriennale, none of the reporters or critics failed to note how the stylistic transformation went hand in hand with technological innovation, marked by an architecture that was moulded by light. The insertion of stained-glass pieces — all of them by Cappellin[45] — in some spaces of the exhibition was part of this. One reads that when entering the rotunda designed by Pietro Aschieri, for example, 'all contrasts are modulated with equal intensity by the light, which, entirely distributed in an indirect way, filters in having been mellowed by the broad veils and, at the entrance, by two large, harmonious stained glass pieces, beautiful in their inventiveness and colour, designed by the architect Mario Romano and executed by the Cappellin firm of Murano'.[46] [10] The colouration of the two stained-glass pieces must have caused a particular contrast with the uniform white of the surfaces and the 'invention' must have had a powerful effect in the monumental union of pure geometries and figurative parts.[47] These geometries, in fact, assumed the architectonic forms of the chimney stacks of industrial plants or of square dwellings (between Giotto and Carrà), or created three-dimensional spaces between which still lives had infiltrated, rendered in a tight, dry stylisation: fragments of classical sculptures (perhaps even a memory of the *Colossus* of Constantine) or of mannequin sculptures, references ranging from contemporary Italian painting (Sironi's prism and sphere or Carrà's compositions with geometric solids, graceless sculptured heads or stray everyday objects) to international painting (the pipe now understood as the 'initials' of modernity) and finishing with the little vase with the cactus plant (indispensable element of furnishing in any house that aspired to fashion), conspicuously follow the example, in a tighter and more concise form, of the repertoire of objects with which Scarpa had animated the geometrical rigour of the stained-glass piece designed for the Triennale di Monza in 1930. Less 'floating' than the Monza still lives, defined by firmly anchored space solids, these compositions give rise to questions about the little explored personality of Mario Romano: while we cannot say in what terms and to what extent he availed himself of the guidance of Scarpa, he certainly followed the same 'metaphysical muse'.

Much simpler, but close to these two stained-glass pieces in spatial distribution and compositional modalities, was the composition designed by Del Debbio for a room on the first floor,[48] while the design of the stained-glass piece mounted in the bookshop on the ground floor was completely different. Having crossed the reception room one entered a large transversal room where the publishing house,

[9] The entrance to the Quadriennale of Rome in 1931 with the M.V.M. Cappellin stained glass after a design by Mario Romano

>
[10] *La pittura*; *La scultura*, Cappellin stained-glass pieces after a design by Mario Romano, in *Domus*, April 1931, pp. 68-69

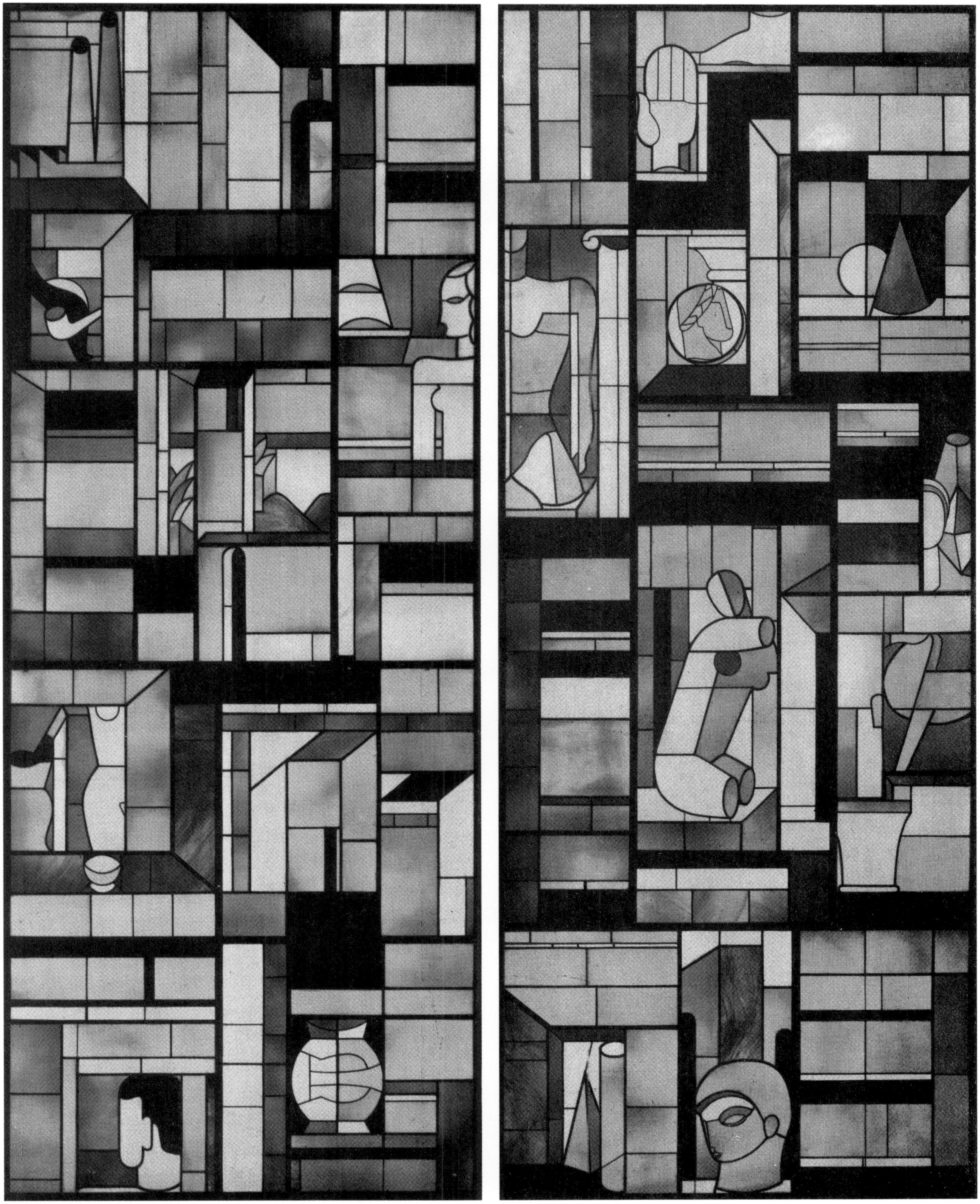

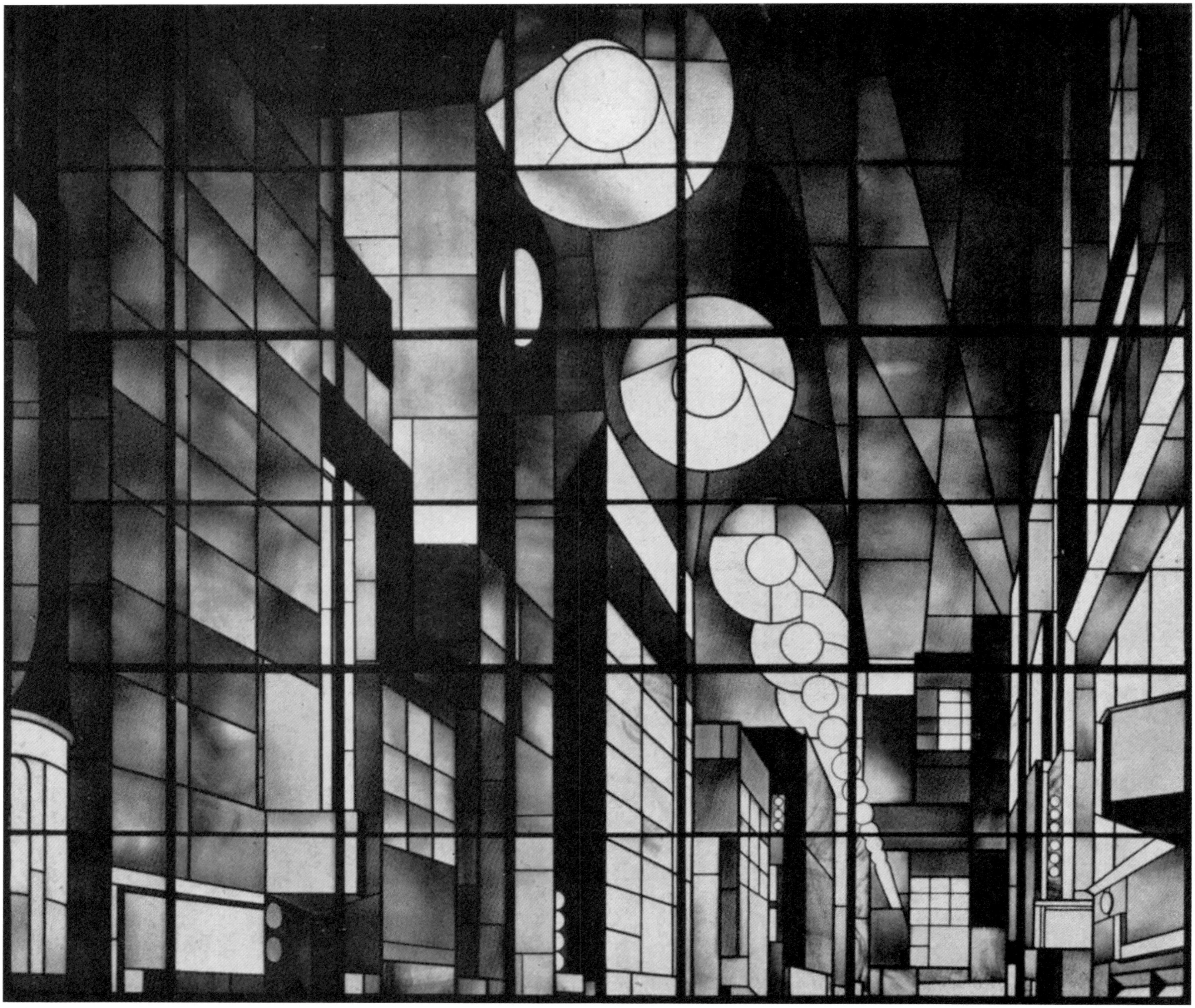

[11] *La strada*, M.V.M. Cappellin stained-glass piece after a design by Mario Romano, in *La casa bella*, March 1931, p.16

Bestetti and Tuminelli, had prepared an art library, 'laid out with great elegance by Del Debbio and adorned by a most original lateral stained-glass piece', *La strada*, produced by Cappellin after a design of Mario Romano's, as the sources inform us.[49] [11]

While *Domus* reproduced only the stained glass with 'metaphysical' still lives, tracing these back to the perspective frames designed by the favourite of *Domus*, Pietro Chiesa, in 1926,[50] *La casa bella* chose to reproduce only *La strada*, starting with a brief note on the architectures of the Quadriennale, which referred exclusively to the contributions of Pietro Aschieri, interpreted as 'a sign of the times that were now ready for a modern Italian architecture and as a decided affirmation of the new style', 'in the very heart of the neoclassical movement, and in a predominantly official place.'[51] In this clear attempt to number the contribution of Aschieri among the expressions of Rationalism — an operation supported by a part of the press of the time[52] —, *La casa bella* (perhaps Persico himself) chose to reproduce the very 'modern' stained-glass pieces — which, however, were by Romano in an architecture by Del Debbio — as an element qualifying the transformation. In this piece, in fact, the profiles of the buildings were certainly Rationalist, with iron and glass surfaces defining the street of a modern metropolis, an ideal reconstruction of a stock of images leafed through in magazines or an exemplary project image. The design of the stained glass, presented to us by a Mario Romano very distant from the Romano of the metaphysical still lives on display at the same exhibition, was also a reponse to the new aesthetic of the artificial light illuminating the nights of modern cities and it again calls for comparisons with contemporary projects, such as Piero Bottoni's cover for the magazine *Illuminotecnica* of May-June 1929. As one can understand from the films of the Istituto Luce,[53] the artificial lighting behind the stained glass made the nocturnal effect particularly evocative — in particular in the large globe-shaped lamps (similar to the bulbs used by Del Debbio in the Quadriennale winter garden) produced using paler glass sheets — a nocturnal effect marking a profound cultural and formal distance from those of Zecchin and Casorati. In *La strada di notte*, there was furthermore a surprising contrast: not only were the artisan sheets in Venice glass made to allude to the new industrial glass materials then gaining ground in the construction area, but there was a celebration of that 'glass revolution' (to use the definition of Pier Maria Bardi) which was helping to bring about the crisis of figured stained glass. The exorbitant costs invested by Cappellin in stained glass — despite important commissions obtained[54] — was probably a factor in the failure of the firm in January 1932.

Notwithstanding the efforts made for these 'enormous stained-glass works based on an ingenious, brilliant and modern technique destined for the best of fortunes', these 'best of fortunes' would never materialize.[55]

1 L. Arimattei, 'L'arte fenestriera veneta', in *L'industria del vetro e della ceramica*, 1935, pp. 14-15.

2 There are reports of stained-glass pieces by Giuseppe Maffioli displayed at the Biennale of Venice in 1903 and 1905 and stained glass designed by Wolf Ferrari for the exhibition of Ca' Pesaro in 1912; for the stained glass of Andrea Rioda, whose glassworks were taken over by Venini and Cappellin, see *Le industrie artistiche italiane. Vetri e vetrate*, Rome 1912, p. 34, fig.15.

3 Testimony of Giacomo Cappellin in 1960 reported in R. Barovier Mentasti, *Venini a Murano e nel mondo*, in A. Bettagno (edited by), *Gli artisti di Venini, per una storia del vetro d'arte veneziano*, exhibition catalogue, Milan 1996, p. 13.

4 See *Catalogo della XII esposizione d'arte della città di Venezia*, exhibition catalogue, Milan, 1922, p. 70, in which the execution is not indicated.

5 See *XIVa Esposizione internazionale d'arte della città di Venezia*, exhibition catalogue, Venice 1924, p. 137, stained-glass pieces *Barca corsara* and *Gli uccelli della tempesta*.

6 See *Seconda mostra internazionale delle arti decorative alla Villa Reale di Monza*, exhibition catalogue, Milan 1925, pp. 116-117.

7 See *XV esposizione internazionale d'arte della città di Venezia*, exhibition catalogue, Venice, 1926, p. 126. The stained-glass piece *Bragozzi chioggiotti* is reproduced in the aforesaid catalogue (p. 175) and in U. Nebbia, *La XV esposizione d'arte internazionale di Venezia*, Bergamo, 1926, p. 187. The two stained-glass works were exhibited also in Rome at the maritime art exhibition (*I Mostra Nazionale d'Arte Marinara. Promossa dalla Lega Navale Italiana. Roma MCMXXVI-MCMXXVII*, exhibition catalogue, Rome, 1926, p. 48). The stained glass piece *Navi veneziane* is reproduced in the booklet published by the Cappellin firm on the Palazzo da Mula with the purpose of presenting the image of the new firm (*Il Palazzo da Mula di Murano. Note di G. Morazzoni*, out-of-print editions edited by M.V.M. Cappellin and C., Milan 1926, p. 36). The sketches for the two stained-glass works are in the Zecchin Ramani collection in Trieste. For Zecchin's stained-glass pieces see also M. Barovier and C. Sonego (edited by), *Vittorio Zecchin. I vetri trasparenti per Cappellin e Venini*, exhibition catalogue, Milan 2017 and L. Mannini, 'La "colorata anima del Vittoriale". Aggiunte e precisazioni sulle vetrate di Pietro Chiesa', in *Quaderni del Vittoriale*, no. 9, 2013, pp. 105-119.

8 U. Nebbia, *La XV esposizione ...*, cit., p. 186.

9 *Enciclopedia delle moderne arti decorative italiane diretta da Guido Marangoni. III. Arti del fuoco: ceramica, vetri, vetrate*, Milan 1927, p. 109.

10 E. Pansini, *XV Biennale internazionale d'arte della città di Venezia*, Naples 1928, p. 58.

11 In addition to the two in the dedicated room, of which we will have more to say, there were two Cappellin stained-glass pieces in the oratory of the Ligurian section: *La Crocifissione* and *San Sebastiano* after cartoons by Paolo S. Rodocanachi (see *Catalogo ufficiale della III mostra ..., op. cit.*, p. 35).

12 The most direct comparison is with the snowfall and night scenes realised in different techniques between 1913 and 1914.

13 Kept at the Galleria Civica d'Arte Moderna e Contemporanea of Turin, from album FD 107, no. 127, published in R. Maggio Serra (edited by), *Arte moderna a Torino. 200 opere acquisite per la Galleria civica d'Arte Moderna*, Turin 1986, p. 138.

14 R. Papini, *Le arti a Monza nel 1927. Gli italiani*, in 'Emporium', LXVI, 1927, p. 24. As pointed out by Giorgina Bertolino on the occasion of the conference *La vetreria Cappellin* (27 February 2018, Giorgio Cini Foundation, Venice), this expedient is to be compared with the glass panes to be etched produced in the second decade of the 20th century in which Casorati made holes which were to correspond to the constellations.

15 See *The Works, Stained glass*. Ugo Ortona's stained glass aroused more interest than that of Casorati and was reproduced, for example, in G. Marangoni, *La III mostra internazionale delle arti decorative nella villa Reale di Monza MCMXXVII. Notizie, rilievi, risultati*, Bergamo, [1927], plate 5; P. Torriano, 'La III mostra d'arte decorativa a Monza. II', in *L'Illustrazione Italiana*, 9 October 1927, p. 290; G. Marangoni, 'La vetrata', in *La casa bella*, April 1928, p. 14. There is a sketch of the stained-glass work at the Centro Studi del Vetro, Fondo Carlo Moretti e Giovanni Moretti (hereinafter referred to as CSV). It is possible that the Cappellin stained glass exhibited in Rome in April 1927 (see *XCIII Esposizione di Belle Arti della Società amatori e cultori di belle arti. Catalogo*, Rome, 1927, p. 79) should be identified as that designed by Ortona: the original destination of the piece for the Roman exhibition would explain the relationship of the Venice glassworks with Ortona, an artist with ties to the artistic world of Rome.

16 *Catalogo ufficiale della III mostra ..., op. cit.*, p. 30.

17 R. Papini. *Le arti a Monza nel 1927..., op. cit.*, p. 24. It would seem relevant to point out here that in the volume by Roberto Papini published in 1930 as a collection of the architecture and decorative arts of Europe, the Cappellin firm is not mentioned.

18 The designs for the Cappellin stained glass pieces are subdivided between: Venice, CSV; Venice, Museo del Vetro; Venice, De Carlo private archive; Verona, Carlo Scarpa Archive at the Museum of Castelvecchio; Treviso, Carlo Scarpa Archive for the MAXXI Architettura collections at the Archivio di Stato. Not all the designs for Cappellin stained glass present in these archives are by Scarpa, but it is probable that he carried out a role of verification and control on the projects of other artists. With regard to Scarpa stained glass, see in particular L. Baldin, 'Alcuni disegni inediti di Carlo Scarpa', in *Venezia Arti*, no. 6, 1992, pp. 138-140; C. Sonego, 'Vetrate di Carlo Scarpa', in M. Barovier (edited by), *Carlo Scarpa. I vetri di un architetto*, exhibition catalogue, Milan 1997, pp. 43-46; C. Sonego, 'Carlo Scarpa: Gli anni '20', in K. W. Forster and P. Marini (edited by), *Studi su Carlo Scarpa 2000-2002*, Venice 2004, pp. 54-104.

19 See C. Sonego, 'Carlo Scarpa alla Maestri Vetrai Muranesi Cappellin & C 1925-1931', in M. Barovier (edited by), *Carlo Scarpa op. cit.*, p. 36 (note). The shop closed in June 1932 after the firm went bankrupt, as is confirmed by the low-price sales advertisements appearing from that month in *L'illustrazione toscana*.

20 Stained glass pieces showing glassmaking activities are reproduced in *La vetrata moderna. Glass windows and decorated glass art, edited by Manifattura Felix Quentin di Firenze*, advertising catalogue, Florence [1929], plates 8 and 9.

21 For Italian stained glass inspired by international models, see L. Mannini, 'Italian Stained Glass in the 1920s and 1930s: Standing the Test of Modernity' in L. De Girolami Cheney (edited by), *Radiance and Symbolism in Modern Stained Glass: European and American Innovations and aesthetic interrelations in material culture*, Cambridge 2016, pp. 315-342.

22 Design no. 12.; Carlo Scarpa Archive at the Museo di Castelvecchio, no. 295.

23 See *The Works, Stained glass*.

24 See *The Works, Stained glass*; Museo del Vetro, designs nos. 577, 579.

25 CSV.

26 Some designs could have been intended for the Turin shop. See Museo del Vetro nos. 913, 914.

27 See *The Works, Stained glass*. Regarding the stained-glass work and the donation thereof to the Musei Civici Veneziani (Inventory of the Museo Correr, *Inv. Vetri e smalti classe VI, 944*), see, *Recenti doni al Museo Correr*, in *Bollettino dei Musei Civici Veneziani*, no. 1, 1969, p. 29. A preparatory drawing for the stained glass is with the CSV.

28 *Ernesto Thayaht*, presentations by A. Maraini, F. T. Marinetti, F. Bellonzi, Florence, 1932, p. 29. Comparisons can be made with pictures such as *Spiralfiori* (1924) or *Frutta* (1928).

29 Published in F. Antonacci, D. Lapiccirella e C. Cerutti (edited by), *Thayaht e Ram. Dal Futurismo al Novecento*, exhibition catalogue, Rome, 2004, Sheet 9. See also drawings in M. Pratesi (edited by) *Futurismo e bon ton: i fratelli Thayaht e Ram*, exhibition catalogue, Florence, 2005, pp. 84-85. Some designs for stained glass kept at the CSV could be attributed to Thayaht. Thayaht could have been the author of the stained glass piece with flowers that can be seen mounted as a window in a photograph of the Florence shop, where the motif in the upper part reminds one of the dynamic lines which characterise many of the artist's projects, from fashion illustrations to decorative panels. My thanks in this respect to Riccardo Michahelles and the Thayaht and Ram Archive for their support and precious assistance.

30 P. Bottoni, 'Ernesto Thayaht decoratore', in *Rassegna di architettura*, August 1929, pp. 310-313.

31 See *La partecipazione italiana all'Esposizione Internazionale di Barcellona. 1929, a cura del commissario generale del Governo*, Milan 1930, p. 108

32 The drawing is published in F. Dal Co, G. Mazzariol (edited by), *Carlo Scarpa. Opera completa*, Milan 1984, no. 13.

33 See R. Targetti (edited by), *Catalogo ufficiale della IV Esposizione Triennale Internazionale delle Arti Decorative ed Industriali Moderne. Villa Reale Monza*, exhibition catalogue, 2nd ed. Milan 1930, p. 236. Scarpa is not named in the catalogue but the preparation of the Cappellin room is indicated by him in his typewritten *curriculum vitae* in 1951 and mentioned in the subsequent monographs dedicated to the artist.

34 R. Papini, 'Bilancio dell'esposizione di Monza 1930', in *L'Illustrazione Italiana,* 1930, no. 42, II (26 October), p. 640.

35 C. A. Felice, 'I vetri alla Triennale di Monza', in *Dedalo*, 1930-1931, XI, II, p. 326.

36 CSV. It is not known whether these stained-glass pieces were produced.

37 See *The Works, Stained glass*. C. A. Felice, *Arte decorativa 1930 all'esposizione di Monza*, Milano 1930, plate 89. See also 'Una giornata alla Triennale di Monza', in *L'Illustrazione Italiana,* 1930, I (22 June), no. 25, pp. 1103-1106: pp. 1104, 1106. Sironi's stained glass was donated by the Cappellin firm in 1930 to the Museo del Castello Sforzesco (see G. Mori, 'La collezione di vetri moderni del Castello Sforzesco acquisiti negli anni Venti e Trenta', in R. Barovier Mentasti, M. Chirico (edited by), *Tra creatività e progettazione. Il vetro italiano a Milano, 1906-1968*, exhibition catalogue, Milan 1998, pp. 62- 67, 176). For an extensive bibliography on the layout of the room, see S. Tosini Pizzetti (edited by), *Sironi metafisico. L'atelier della meraviglia*, exhibition catalogue, Milan 2007, pp. 167-169, 205-213; preceded by the sheet G. Ginex, 'La Gallerie delle Arti Grafiche alla IV Triennale di Monza', in A. Sironi (edited by), *Sironi. La grande decorazione*, Milan 2004, pp. 192-197; and by the investigations *Mario Sironi. Il mito dell'architettura*, Milan 1990.

38 C. A. Felice, I vetri alla Triennale di Monza, in *Dedalo*, 1930-1931, XI, II, p. 326.

39 A. Pica, 'La mostra grafica a Monza', in *La casa bella*, May 1930, p. 21, 24 (in the caption).

40 In 'Eclettismo e innovazione. L'arte italiana negli intendimenti politici e nella pratica espositiva. 1931-1935: le prime Quadriennali romane', in G. Bonasegale (edited by), *Catalogo generale Galleria Comunale d'arte moderna di Roma*, Rome 1995, p. 62.

41 See C. Salaris, *La Quadriennale. Storia della rassegna d'arte italiana dagli anni Trenta a oggi*, Rome-Venice 2004, p. 20.

42 Anonymous, 'Come viene allestita la sede della prossima Quadriennale a Roma', in *Il Lavoro Fascista*, 12 November 1930.

43 See the letter that Cappellin sent to Depero on 9 August 1926 kept at the MART, Archivio del Novecento, Fondo Depero, Dep. 3.1.6.4: 'In conversation with the Comm. Davide Campari we thought we saw the possibility of applying our stained glass pieces to his company advertisements and better if they were drawn from your cartoons. Have you any sketch that would work for this purpose. My company is prepared to make one gratis so that the Comm. Campari can see the effect and whether or not one can begin this branch of activity. Remembering that during your visit to the furnace you expressed an analogous idea, I do not doubt that you will wish to write something to me in this regard'.

44 F. T. Marinetti, 'Gli avvisi luminosi. Lettera aperta a S.E. Mussolini', in *L'Impero*, 12 February 1927.

45 In addition to the stained-glass works which we will discuss, I would point out that a Cappellin design (Museo del Vetro, no 202) is very close to the skylight executed in the winter garden.

46 P. Torriano, 'La prima Quadriennale nazionale d'arte a Roma', in *L'Illustrazione italiana*, 1st semester (11 January 1931), p. 50.

47 The dimensions were impressive, each measuring 2.45 metres at the base and 6.07 metres in height (see design Archivio Scarpa Treviso).

48 See *The Works, Stained glass*. See *Gazzetta di Venezia*, 4 January 1931 and *Roma. Alla Prima Quadriennale d'Arte Nazionale tutte le tendenze dell'arte italiana vi sono rappresentate,* Giornale Luce A0713 01/1931, Archivio Luce.

49 'La prima Quadriennale nazionale d'arte a Roma', in *L'Illustrazione italiana*, 11 January 1931, pp. 50-53: p. 50. The stained-glass was positioned on the short side of the room: see *Roma. Alla Prima Quadriennale..., op. cit.* and *I Quadriennale d'arte nazionale*, documentario, 1931, Archivio Luce M005501.

50 'Arti applicate alla Quadriennale romana', in *Domus*, April 1931, pp. 68-69: 'the design, by Mario Romano, resounds with a synthetic stylisation in full and graduated tones, without brush-stroke light-and-shade effects, a composition which was dear also to another grand master of stained glass, Pietro Chiesa', referring to the still lives presented by the latter at the Biennale of Venice in 1926.

51 'Architetture alla Quadriennale', in *La casa bella*, March 1931, pp. 16-17. It was reproduced also in G. Dottori, 'Prima Quadriennale d'arte Nazionale', in *Oggi e Domani*, 5 January 1931; F. Geraci, 'I veneti alla I Quadriennale romana d'arte', in *Le tre Venezie*, February 1931, pp. 77-81; A. Lancellotti, *La prima Quadriennale d'arte romana*, Rome, [1931], p. 236.

52 See C. F. Carli, *Tra Déco e razionale. Allestimenti di qualità*, in *XIV Quadriennale di Roma. Retrospettive 1931/1948*, exhibition catalogue, Milan 2005, pp. 49-58.

53 See previous notes.

54 The Cappellin firm was commissioned in 1931 with the execution of *La Carta del Lavoro* by Sironi, replaced in the work later by Pietro Chiesa. See M. Margozzi, 'La Carta del Lavoro', in *Sironi. La grande decorazione..., op. cit.*, pp. 198-217.

55 *Gazzetta di Venezia*, 4 January 1931. From the documentation traced down and the history of the few stained-glass pieces identified, it emerges that they mostly remained unsold, therefore representing an unprofitable financial investment. Cappellin in subsequent years would attempt to open a glasswork firm that would also produce stained glass, but would not succeed, as is shown by certain letters sent to Fernanda Ojetti in the years 1945-1946 (at the Galleria Nazionale d'Arte Moderna, Roma, Fondo Ojetti).

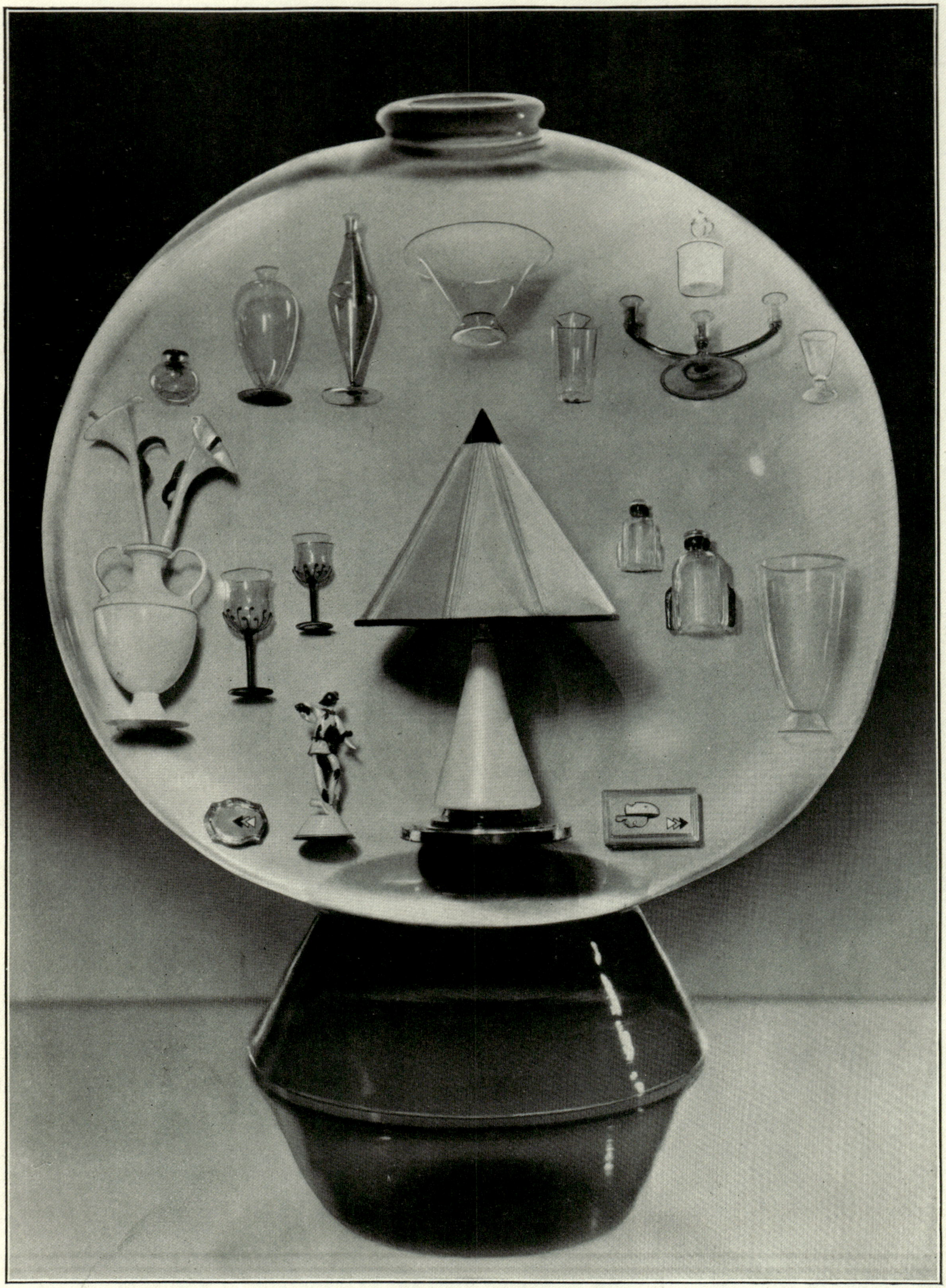

CAPPELLIN GLASSWARE IN THE UNITED STATES

Howard J. Lockwood

On November 16th, 1932, Marshall Field's, a Chicago department store, announced a sale of 5000 pieces of Cappellin glassware. This sale is a testament to the vision of Giacomo Cappellin, a vision that saw the United States as a large, potential market and a vital instrument to the growth and success of MVM Cappellin. Cappellin Glassware in the United States is the culmination/combination of the dreams of Giacomo Cappellin. In the United States his pride and joy, the Palazzo da Mula, was held in high esteem. His glassware appeared in Museum exhibitions and was even acquired for some permanent collections. Finally, his glass was available to the consumer throughout the United States.

Giacomo Cappellin always saw the world outside Murano as his main market. In the early years of Cappellin Venini, the company exhibited at the 1922 Paris Salon for which they received much publicity. By 1924, Cappellin Venini had dealers in Rome, Milan, Paris, London, Amsterdam, and Buenos Aires. On April 29th, 1925, the *Exposition internationale des Arts décoratifs et industriels modernes* opened in Paris to worldwide acclaim. Cappellin Venini exhibited at the *Exposition* where they won two Grand prizes and several honorable mentions. Since the partners had separated just before the opening, V.S.M. Venini received most of the publicity, but several sources did use the name Cappellin Venini.[1]

After the separation from Paolo Venini, Giacomo Cappellin opened his own glasshouse on June 23th, 1925, in the Palazzo da Mula, a fifteenth century Palazzo on Murano. Vittorio Zecchin was his designer and they started producing glass from the old Cappellin Venini catalogue interspersed with new designs. Cappellin also started to build a network of retail outlets in Milan, Florence, Paris, Rome, and Buenos Aires, Argentina.[2] The United States was not in his early sights but Paolo Venini would change that viewpoint.

On January 19th, 1926 the *Exhibition of Modern Italian Art* opened at The Grand Central Galleries in New York City and after a month-long display travelled to the major museums of Boston, Chicago, San Francisco and Washington. Though it was an exhibition of paintings and sculpture, the exhibition included Venini vases from the old Cappellin Venini catalogue designed by Vittorio Zecchin. In an article about the exhibition, the *International Studio* illustrated some of the Venini vases and reported 'the glass includes a number of pieces by Paolo Venini of Venice whose work recalls the classic period of Venetian glass.'[3] Since it was organized around the time of the separation from Venini, Cappellin would have known about this exhibition and seen the review. If he had seen the actual catalogue of the exhibition, he would have found a full-page advertisement for Venini glass.[4] The United States was definitely a market that Giacomo Cappellin wanted, and with increasing expenses due to the renovation of the Palazzo da Mula, he needed it.

The United States market was a vastly different environment from the European market. The first major obstacle to trade was the distance and time involved in getting items from Venice to the ports of

<
[1] 'Behold the smartest Gifts of the present and the future foretold in the mystic crystal of the past', 'Cappellin Glassware inc.', advertisement, in *Arts and decoration*, December 1931, p. 1, detail

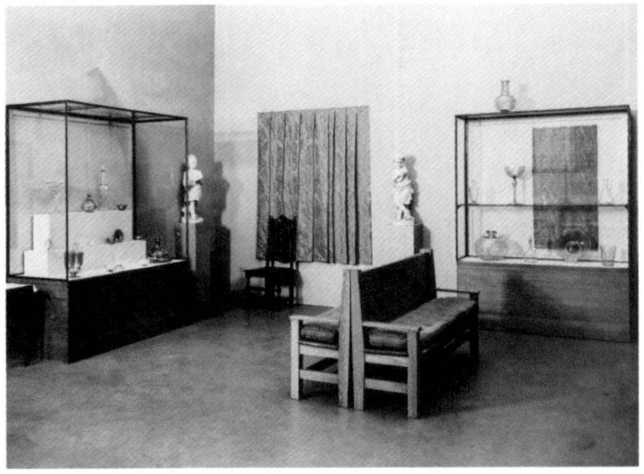

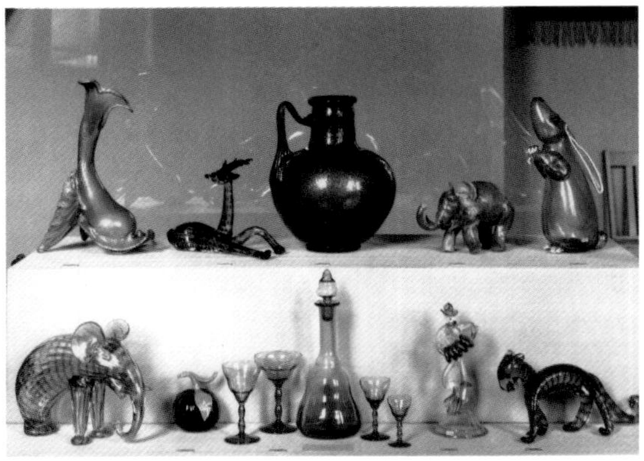

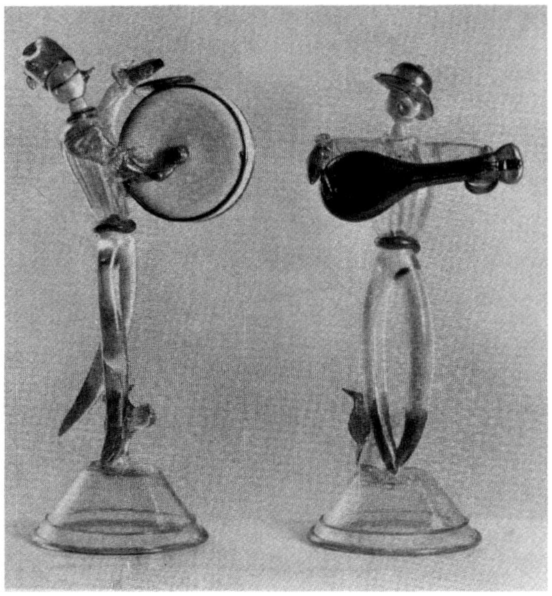

[2] A photograph of the exhibition *Five Hundred New Objects, 1928,* featuring Cappellin glassware. Courtesy The Newark Museum of Art

[3] A cabinet from a display of Italian Glass at the *International Exhibition of Glass and Rugs*. Top row left to right: Cappellin *Fish*, Barovier *Giraffe*, Venini *Pulegoso Ewer*, Cappellin *Elephant* and a Venini *Rabbit*. Bottom Row: Barovier *Elephant*, Cappellin *Fruit*, Cappellin table set (probably), Cappellin *Figure* and a Barovier *Tiger*. Courtesy the Cincinnati Museum

[4] An illustration of two Cappellin figures (no. 244) from the unpaginated exhibition catalogue *International Exhibition of Glass and Rugs*. This illustration was also used for publicity and reviews

>
[5] Cappellin advertisement in the literary magazine *The New Yorker*, 19 October 1929

>
[6] An advertisement in the trade magazine *Gift and Art Shop*, March, 1930, p. 48

Boston or New York. In addition there was the actual size of the United States. It took 20 hours by train to travel the 1541 kilometers between New York and Chicago. To get to California could take an additional three days. Adding to the complexity, every city in the United States had at least one major department store staffed with its own buyers plus smaller, specialized gift stores. Some of the major department stores would send its buyers to Europe to purchase inventory for sale in their store. For the majority of glass houses in Murano, this was their only way to reach the American market. They had to hope and wait for the various buyers to come to them. The other stores would send their buyers to either New York City or Chicago to survey the latest trends and replenish their inventory.

In 1927 one of the largest department stores, New York's R.H. Macy's, planned *An International Exposition of Art in Industry.* The idea was to bring to New York the best of modern design and create a true museum-type experience, including lectures and demonstrations. In a period of six months, Macy's buyers visited 300 different companies in six countries and chose 5000 pieces for their exhibition. On May 14th, 1928, the exhibition opened to great acclaim. Among the glass displayed throughout the store were thirty-eight examples of Venini, seven examples of Balsamo Stella, and in the first documented United States appearance, seven examples of M.V.M. Cappellin. A review in the trade journal *The Crockery and Glass Journal* reported that the 'blown glass of the Murano craftsmen, Cappelini [sic] and Venini, emphasized that Italy is still so conscious of its class traditions that even the moderns of that country definitely retain their ties with the past.'[5] Though the exhibition lasted only two weeks, it was estimated that 250,000 people visited it.[6]

In early 1928, John Cotton Dana, the director of the Newark Museum of Art in Newark, NJ, visited the Cappellin store in Rome and acquired 40 pieces of glass for the museum from G. Guarnati. These are the first known Cappellin pieces to enter a collection of an American museum and possibly the first in the world. The pieces were exhibited in the Newark Museum's *Five Hundred New Objects* exhibition that opened in November 1928.[7] [2]

In 1929 The American Federation of Arts organized a travelling exhibition, the *International Exhibition of Contemporary Glass and Rugs.* It opened at New York's Metropolitan Museum of Art on November 4th, 1929 and then travelled to major museums in Boston, Dayton, Pittsburgh. Chicago, St. Louis, Cincinnati, Baltimore, and Philadelphia. Every major European glasshouse from Orrefors to Marinot was included in the exhibition of 709 objects. Balsamo Stella, Venini, Vetreria Artistica Barovier and M.V.M. Cappellin were the chosen Italian participants. [3] Items included in the exhibition could be purchased through the museum. Additionally, local stores and galleries in each city stocked different examples of the various companies' work. The exhibition catalogue even listed MacAllister & Alvord as the American agent for Cappellin.[8]

Cappellin was well represented in the exhibition, with 24 pieces of glass, 'ashtrays, examples of table sets, fish, group of jazz musicians, fruit, group of bowls and vases; blown glass.'[9] Two of the six jazz musicians were illustrated in the exhibition catalogue [4] and in various newspaper and museum bulletins. The *Chicago Daily Tribune* wrote, 'You will be startled and thrilled by the grace of a group of glass musicians, blown glass with steadying touches of dark blue and instruments of reddish brown, one of the loveliest and most exotic, most decorative achievements in glass that has ever been exhibited in America.'[10] Joseph Down wrote in the *Philadelphia Museum Bulletin* '...a pair of blown glass musicians mounted on bell shaped vases gives the movement and spirit of these gay entertainers. The essential forms are there, the quickly cooling glass allows no fine details, these are left to the imagination.'[11] The show toured through 1930. In 1931 the Metropolitan Museum of Art purchased three pieces of Cappellin, though it is not known if they were pieces from the exhibition.

Due in large part to Paolo Venini's success at the *Exhibition of Modern Italian Art*, the visiting American department store buyers to the Palazzo da Mula, and the upcoming American Federation show, Cappellin realized he could not afford to ignore the large potential United States market. He had two options. He could wait for the buyers to come to Murano and then compete with every factory for the

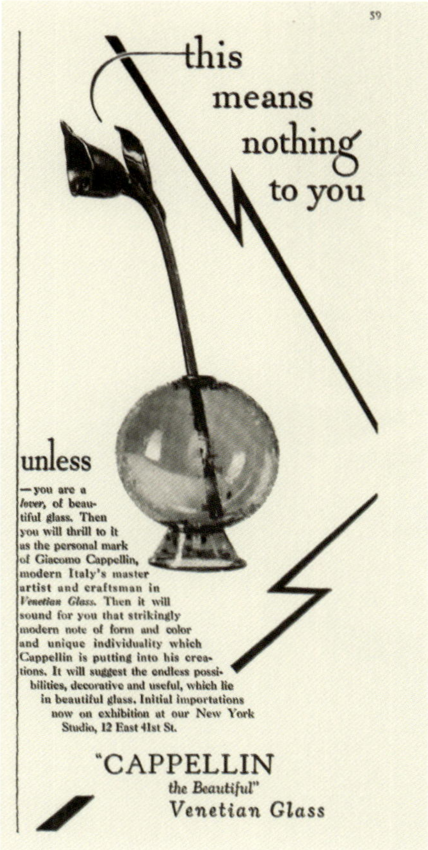

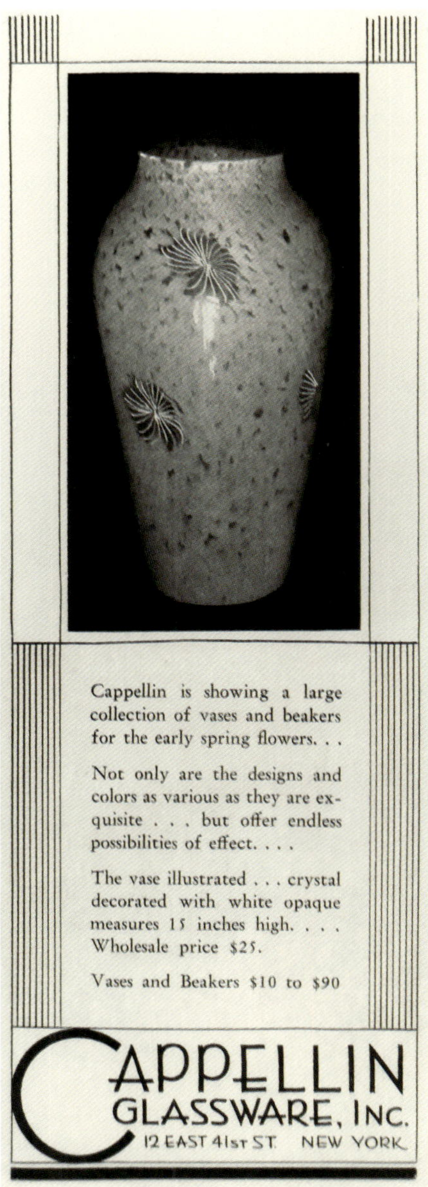

buyer's business or go directly to the buyers. Like Venini, Cappellin decided to approach the American market from America itself. Around 1927, through agents, MacAllister and Alvord, a wholesale/retail store was opened in New York City with the goal of meeting the department store buyers who did not travel to Europe.

Not much is known about MacAllister and Alvord. They were listed first as furniture upholsterers and then as designers on New York's East 41st Street. On August 17th, 1927, it was announced in the *New York Times* that 'the interior decorators MacAllister and Alvord'[12] leased half of the fifteenth floor at 16 East 52nd street. This was to become the new headquarters of Cappellin Glassware and a piece of Murano in New York City. On June 11th, 1929, Cappellin Glassware was incorporated with three shareholders: Paul MacAllister, William Horrocks and Buell Alvord. Giacomo Cappellin is not listed in the incorporation papers and neither MacAllister, Horrocks nor Alvord appear on any ship manifests. It is unknown how they became connected with Giacomo Cappellin.

Because there are no photographs or documents from Cappellin Glassware, advertisements offer the only view of Cappellin in the United States. The first series of advertisements were geared towards the end consumer and were found in consumer magazines like *Town and Country*, *Home and Field*, and *The New Yorker* magazine. [5]

In order to entice the many gift shop and department store buyers Cappellin advertised in the specialized magazines *Spur,* the *Crockery and Glass Journal,* and *The Gift and Art Shop* and *Arts and decoration*. [6] In most instances, utilitarian ware was featured.

Giacomo Cappellin's pride and joy, The Palazzo da Mula, became a focus in many advertisements and promotions. The April 1930 issue of *International Studio* described the New York showroom: 'On the fifteenth of March the American agents for Giacomo Cappellin opened their new studios at 16 East 52 Street. Here, in a metropolitan penthouse, there has been reproduced a room from the magnificent Byzantine Palazzo da Mula, the thirteenth century Murano headquarters of the House of Cappellin. Against the dull golds and blues of the Mariano Fortuny hangings, in racks of fifteenth century hand-forged ironwork and on long tables of Tuscan walnut the glass is displayed, while set against the subdued draperies and suspended from the beams of the delicately polychromed ceiling are brackets and chandeliers in which present day simplicity of line has curbed the elaborations of an earlier day.'[13]

The *Crockery and Glass Journal* also had an extensive article about an actual visit to the Palazzo da Mula. It described the factory and showroom in detail. The last paragraph highlighted the New York showroom: 'What was our amazement when we chanced to visit Cappellin's studio atop one of New York's skyscrapers...this glamorous scene repeated. It is as though we are in the Palazzo da Mula on the bank of the grand canal in Murano. About the rooms are tables set with glass.'[14] [7]

Cappellin Glassware worked diligently to acquire a larger United States audience for their glass. They had two outside agents; Edna Young Scott of Detroit, MI, and W.J. Saylor in Los Angeles, who represented Cappellin throughout the United States. To entice the department store buyers and give the sales agents a selling tool, a photographic catalogue of both utilitarian and decorative glassware was developed. The catalogue, as found, consisted of a hard covered, cloth bound presentation-styled book that allowed for individual photo pages to be inserted into the catalogue as needed. Each photo page consisted of two photographs of a group of five to ten objects in each group. [8] The opposite side of the book had a holder for the relevant typed price list. The range of illustrated items is interesting. There are many different table service items and seven different figurines, with the large musician figures offered at $35 each. Three different large plates are illustrated, all with the same model number, but one is transparent glass, one appears to be opaque (black?) glass plate and one is a *reticello* plate. There are several Carlo Scarpa designs illustrated and one photo included a vase with applied butterflies and a vase with applied birds.[15] [8] The vase with birds could accompany the birdcage that was illustrated in the *Architectural Review*.[16] [11] This is not a complete production catalogue and does not include many well known pieces, including the Scarpa 'signature' vase no. 5673 [5]. It is not known

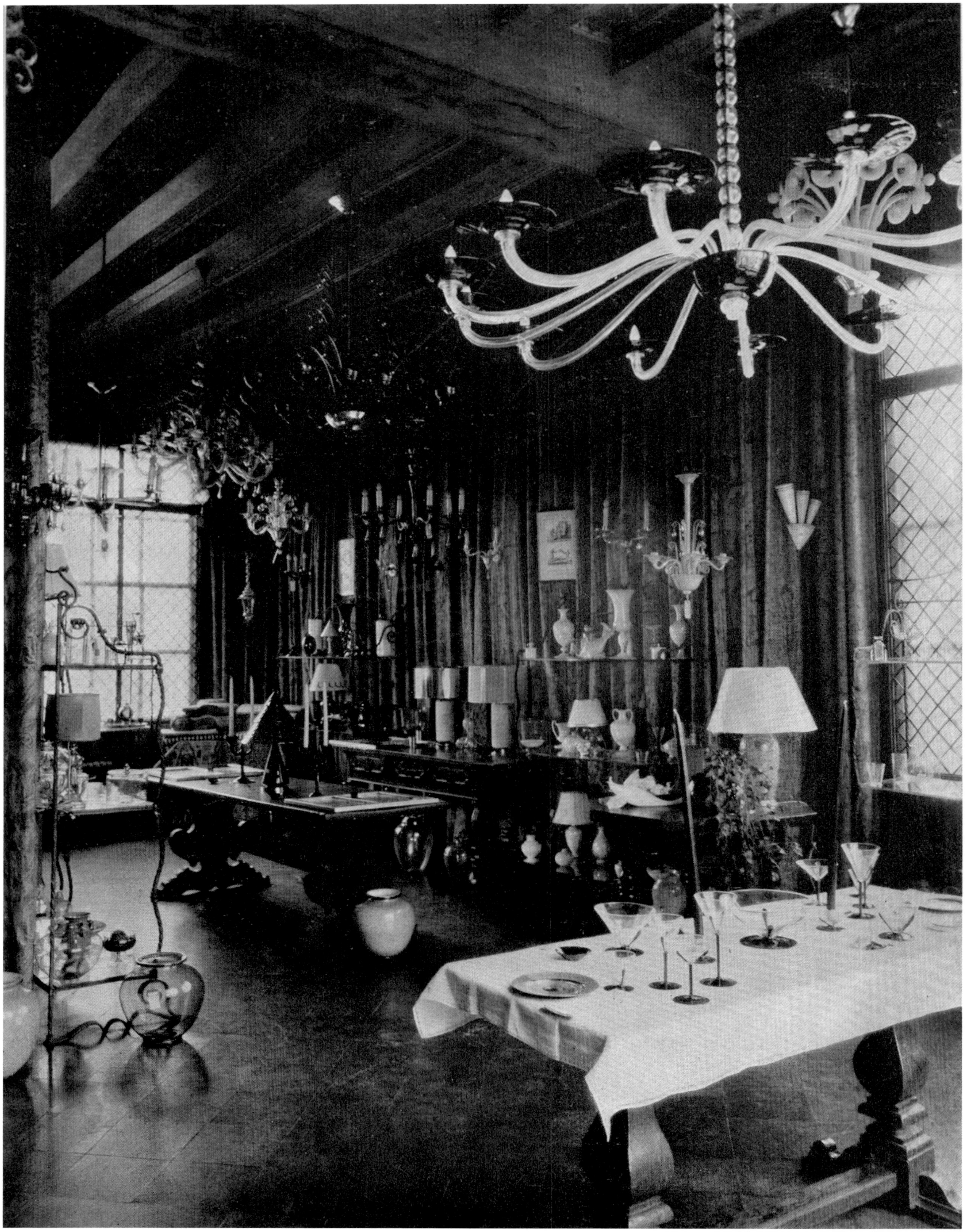

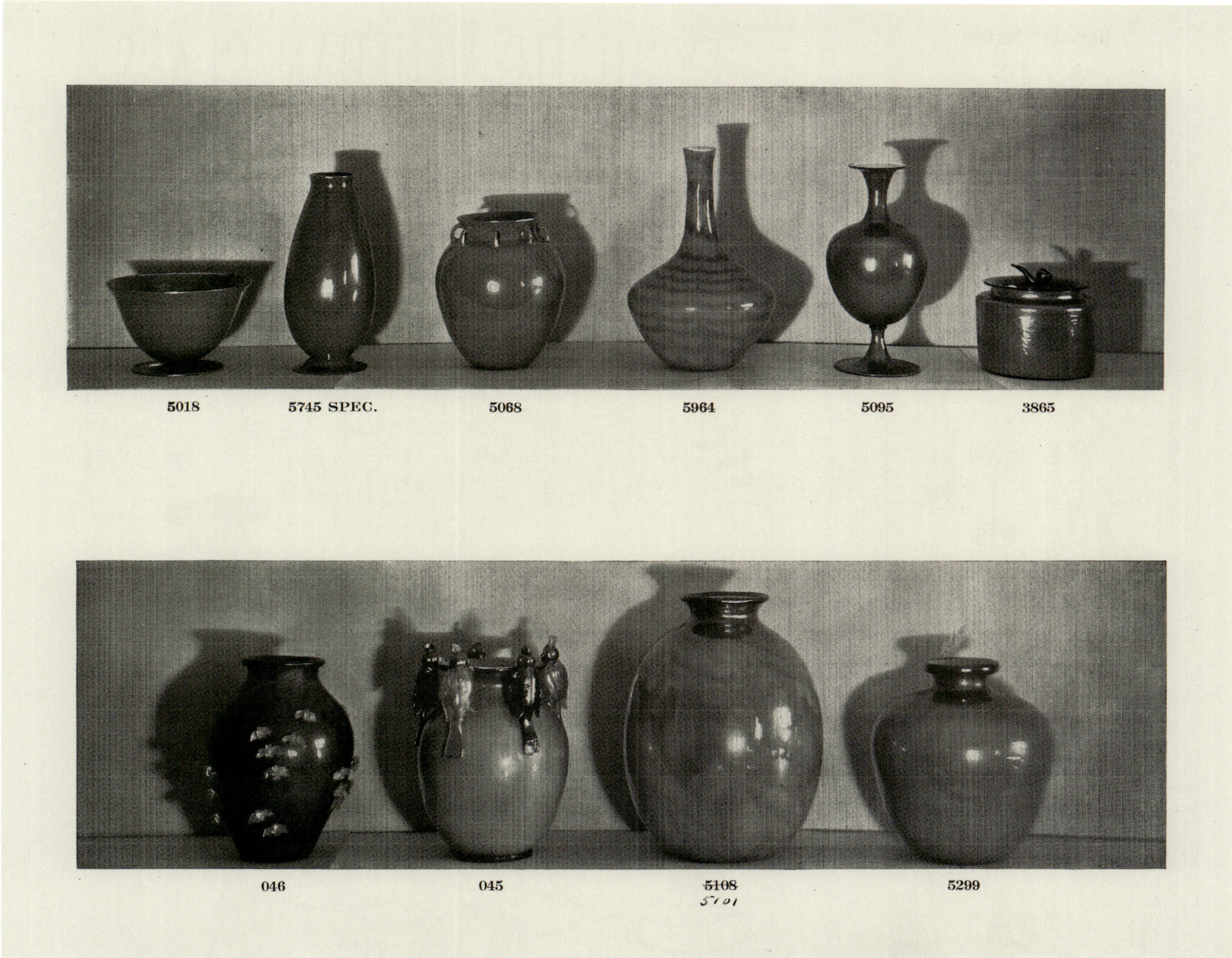

<
[7] 'Cappellin Glassware inc.' showroom, New York in 'Cappellin Glassware inc.', advertistment in *Arts and decoration*, November 1931, p. 17, detail

[8] *Cappellin Glassware Catalog*, undated, Howard Lockwood collection

BONWIT TELLER

SALE CAPPELLIN VENETIAN GLASS

½ OFF

AND MORE

NOW 2^{50} to 45^{00}

Formerly 5^{00} to 95^{00}

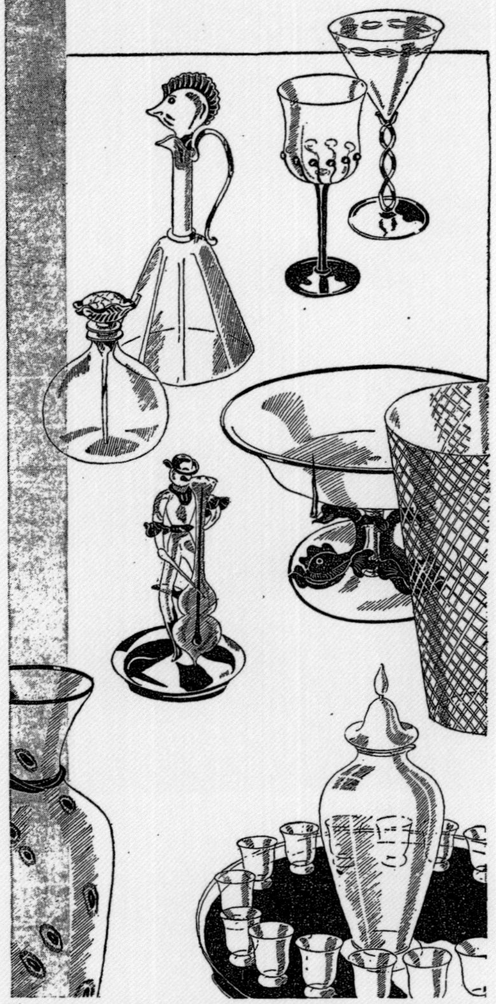

HANDMADE AT THE PALAZZO DI SECULO, ON THE ISLAND OF MURANO, NEAR VENICE. EACH PIECE SIGNED BY Cappellin

No other glassmaker in the world creates pieces like those which come from Cappellin, hailed by connoisseurs as the greatest master in all Italy.

For nine generations, the Cappellins have fashioned glass famous for its exquisite beauty. Only in comparatively recent years has this lovely glass been offered to the public, for originally the Cappellins fashioned their masterpieces only for the Doges and other dignitaries of the Republic of Venice.

Very thin, yet surprisingly strong this rare glass has a clear, bell-like ring which even the layman's unrrained ear recognizes as being decidedly different. Unique effects of color or design can be found in practically any article of glassware you may be seeking. Main Floor

BONWIT TELLER

FIFTH AVENUE AT 56TH STREET

EXQUISITE VENETIAN GLASS

It's new. It's different. Unlike any other Venetian Glass.

Hand made by the famous Cappellin glassmakers.

We bought 5,000 pieces.

We're featuring it in a sensational selling, beginning this morning.

25c 50c $1 $2 $3 Up to $10

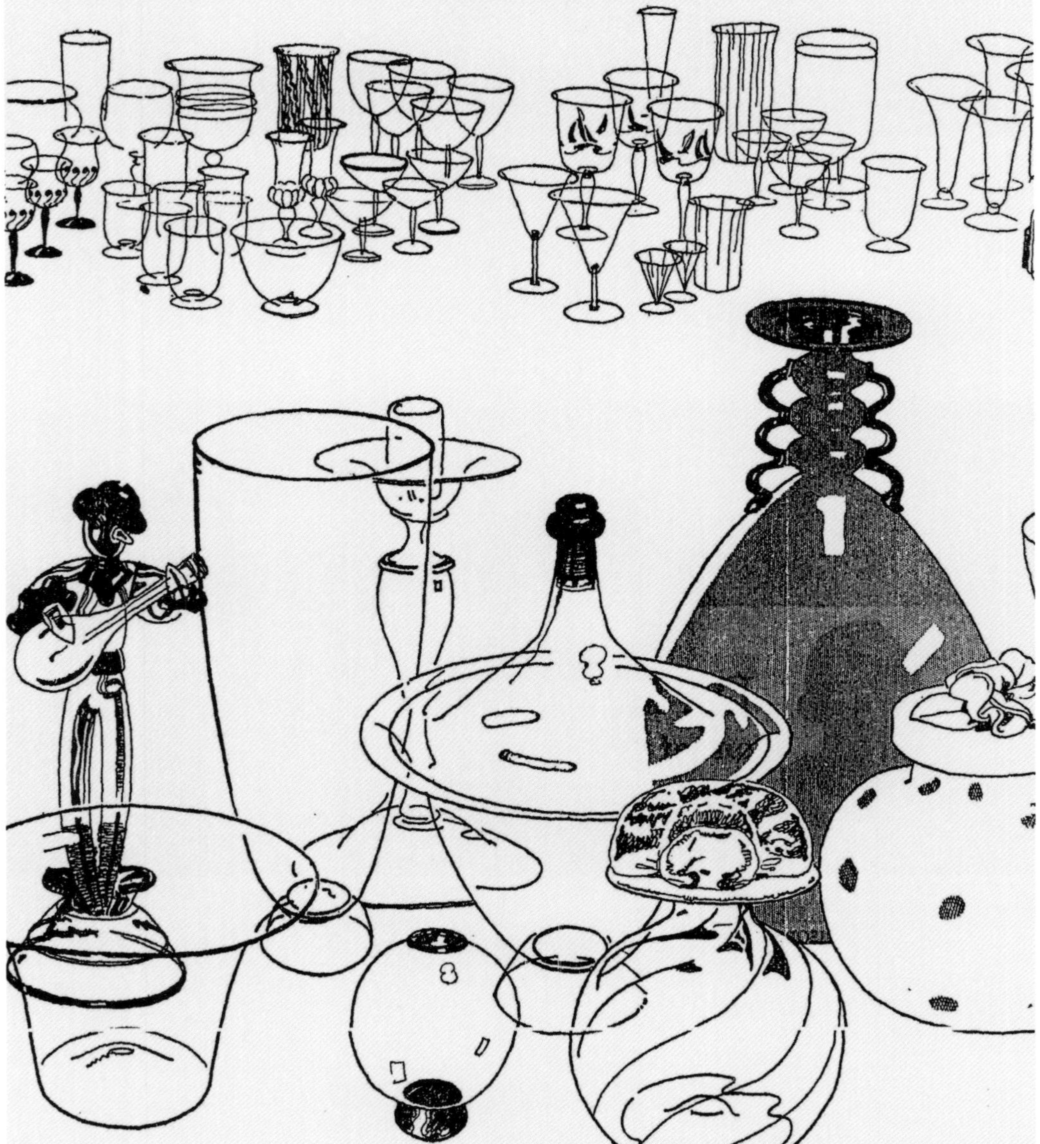

Imagine Venetian stemware for as little as 25c ... imagine Venetian bottles as low as $1 ... large, lovely flower vases as low as $2.50! The sheer beauty of this glass lies in its utter simplicity ... absence of over-decoration ... its delicate colorings and original shapes. New "swirly" effects ... new twists. Every piece compellingly low priced. Conveniently grouped on tables according to price. Second Floor, Wabash.

MARSHALL FIELD & COMPANY

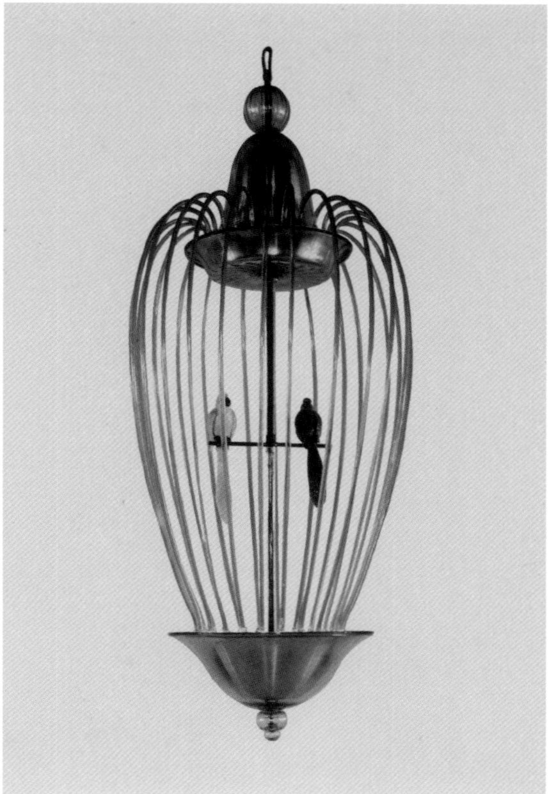

if this copy is complete since the catalogue was designed to be flexible. It is possible that the salesperson would leave the catalogue with important buyers, or leave the buyer with individual pages. Though it can be seen that prior to the beginning of the Depression, MVM Cappellin was in dire financial straits, the world-wide financial crisis hastened the inevitable closing. On April 6th, 1932, three months after the actual bankruptcy, the New York department store Bonwit Teller's advertised a 50% off sale of Cappellin glass. [9] On November 16th, 1932, Marshall Fields advertised a closeout sale of 5000 pieces of Cappellin glass for prices of 25 cents to $10.00. 5000 pieces of glass! America truly was a huge market! [10]

<
[9] Bonwit Teller', advertisement, in *New York Times*, 6 April 1932

<
[10] An advertisement for the Chicago department store Marshall Field's advertising a 'Sell-Out sale of 5000 pieces of Cappellin glass.' in *Chicago Daily Tribune*, November 11, 1932

[11] MVM Cappellin birdcage in *The Architectural Review*, February 1930.

1 The 'Diplome de Medaille' certificate awarded to Raffaele Ferro by the jury for his glass, names 'Collaboration de la maison Cappellin-Venini et Cie'. Giulio Lorenzetti in *L'Italia all'Esposizione internazionale delle arti decorative ed industriali moderne. Parigi MCMXXV*, Roma, 1926, listed the display of the glass as 'Società Vetri Soffiati Muranesi Cappellin Venini e C. (ora: Venini e C.)' while he lists the chandelier as 'Lampadario della Ditta Venini & C'.

2 'La Missione argentina a Murano: Il ricevimento in Municipio', in *La Gazzetta di Venezia,* 9 July 1925, p. 5. The Secretary of the National Commission of Fine Arts in Argentina, Francesco Armellini, was an ex-Venetian and well aware of the Muranese glass industry. On 8 July 1925, the Argentinian Trade Mission Minister, S. E. Perez, visited various Murano factories to discuss Argentina as a prospective market. M.V.M. Cappellin was definitely one of the factories visited because Cappellin opened a gallery in Buenos Aires at 1508 Calle Charcas.

3 'Here and Everywhere', in *International Studio*, March 1926, p. 92.

4 *Exhibition of Modern Italian Art,* New York 1926, p. 69. Though the advertisement featured many forms of Venini vases, it did not name Venini as the maker, instead featured the name Benello Brothers, the agent for Venini.

5 'When Art goes to Work for Trade', in *Crockery and Glass Journal*, June 1928, p.29.

6 Advertisement in *New York Times*, May 1928.

7 In 1955, 28 of the original pieces were deaccessioned from the collection and were used for flowers or other functions.

8 'Maestri Vetrai Muranesi Cappellin & C.', in *Catalogue. International Exhibition. Contemporary Glass and Rugs,* Portland, 1929. Unpaginated.

9 *Ibid.*

10 E. Jewett, 'Glass Exhibition at Art Institute Opens with Beautiful Colors', in *Chicago Daily Tribune,* 20 March 1930, p. 35.

11 J. Downs, 'An Exhibition of Glass and Rugs', in *The Bulletin of the Pennsylvania Museum*, no. 131, January 1930, p. 13.

12 'Beauty Laboratory on Downtown Lease', in *New York Times*, 17 August 1927, p. 40.

13 'Seen in the Galleries' in *International Studio,* April 1930, p. 10.

14 C. Rager, 'Venetian Glass: Made in an Old Castle', in *Crockery & Glass Journal,* 1 March 1930, p. 64.

15 *Cappellin Glassware*, an undated photographic catalog published in the United States.

16 Y. Maraini, 'The Modern Italians', in *The Architectural Review,* February 1930.

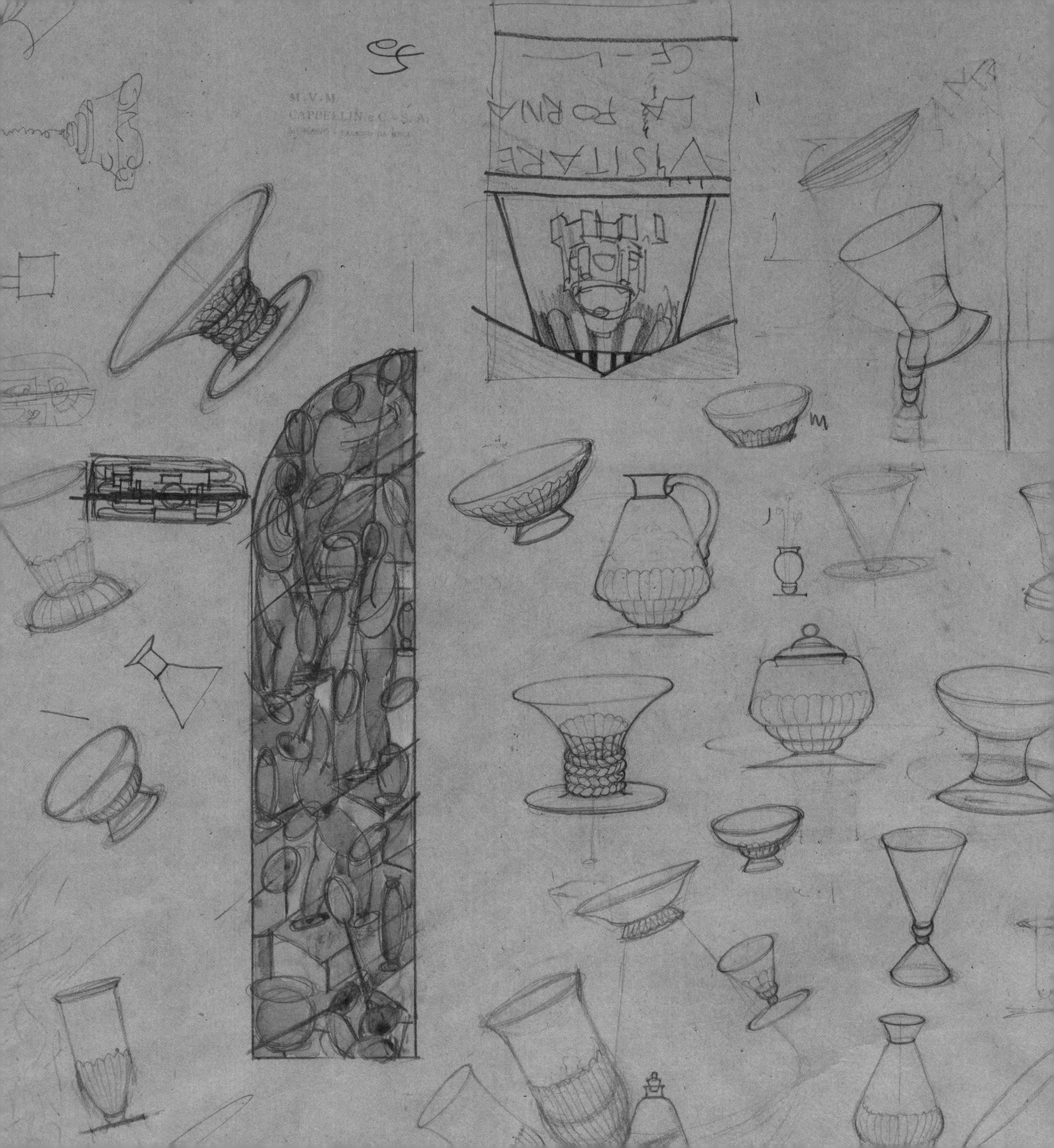
M.V.M.
CAPPELLIN e C. S.A.
VISITARE ROMA

THE DOCUMENTS ON THE CAPPELLIN GLASSWORKS IN THE CARLO SCARPA ARCHIVES IN VERONA

Alba Di Lieto, Ketty Bertolaso

The relationship between the art of glass and the production of the Murano glassworks was examined in depth in 1960 by Licisco Magagnato, Director of the Verona museums, together with Astone Gasparetto and of course Carlo Scarpa, on the occasion of the *Vetri di Murano 1860-1960* exhibition, organised in the Palazzo della Gran Guardia in Verona in the context of a wide-ranging research project on the so-called minor arts in Venice.[1] In the exhibition the most important furnaces and the glass pieces that best represented the change of taste and the stylistic revolution occurring between the nineteenth and twentieth centuries were shown. The catalogue identified the artistic directors of the glassworks and the often high-profile artists who collaborated with them, such as for example Napoleone Martinuzzi, Tomaso Buzzi, Vittorio Zecchin, Carlo Scarpa, Ettore Sottsass jr., and concluded with 'the most avant-garde works in this area' by Vinicic Vianello.

These were the same years in which Magagnato started his collaboration with Carlo Scarpa, calling on the master for the preparation of the exhibition *Da Altichiero a Pisanello* of 1958, which in fact would turn out to be the first stage in the complex re-organisation of the Castelvecchio Museum of Verona. Another of Magagnato's achievements was the felicitous and far-sighted idea of acquiring the designs for the restoration at the end of the works in 1973 directly from Professor Scarpa: these would constitute the core of the 'Carlo Scarpa Archives' of Verona.[2] To the initial *corpus* another 200 drawings were added, which Angelo Rudella and Angelo Aldrighetti, two invaluable collaborators of the museum, had collected and kept over the period of the building works.[3]

Since then the archives have grown in size and the management of the Musei d'Arte has taken many initiatives in collecting, identifying, photographing, assessing and cataloguing every single sheet relating to some of Scarpa's projects.[4] A significant increase was made possible thanks to the financial resources the Veneto Region made available with Regional Act No. 2 Article 41/2002. The systematic campaign to search for the graphic works of the architect found in the collections of private and work associates, conducted with dedication by Aldo Businaro, made it possible to acquire a portion of the Cappellin archives — lodged since 2004 at the museum — archives partly included in the collection of the Moretti brothers and in that of the Museo Vetrario di Murano and partly remaining with the Cappellin family heirs.[5] The collection owned by the Veneto Region and preserved in Verona consists of graphic documentation amounting to 295 pages belonging for the greater part to the catalogue of the Maestri Vetrai Muranesi Cappellin & C., active between 1925 and 1931. The drawings represent various types of numbered glass pieces: vases, jugs, bombonières, bowls, dinner services, ornaments (figurine, animals, succulent plants and flowers), [2] small boxes, perfume bottles and ashtrays. Twelve pages marked with stamps and inscriptions are related to specific orders of the glassworks, while 68 pages probably come from another glassworks, still to be identified, which also contain models similar

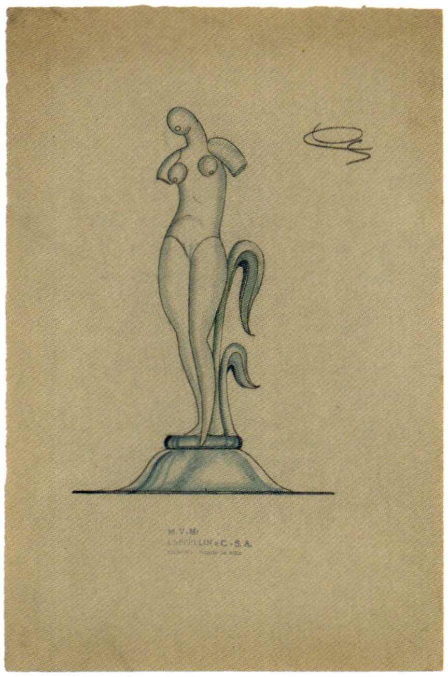

<

[1] Carlo Scarpa, studies for vases, glazed shutters in Murano glass and sign, c. 1928, graphite, tinted water colours on cardboard, 750 x 457 mm, detail. Carlo Scarpa Archives, Museo di Castelvecchio, Verona (RC288)

[2] Carlo Scarpa, study for female nude to be produced in glass, 1929-30, graphite, tinted water colours on cardboard, 375 x 250 mm, detail. Carlo Scarpa Archives, Museo di Castelvecchio, Verona (RC262)

to those of Cappellin.[6] Accompanying all of these there are also 78 period photographs relating to the Vetri Soffiati Muranesi Cappellin Venini & C., a company which was active between 1921 and 1925, the winding up of which led to the founding of M.V.M. Cappellin.

The drawings for the pieces were produced mainly by designers internal to the glassworks; only a few — more precisely 18[7] — can be attributed to the Muranese painter Vittorio Zecchin, artistic director of the glassworks from 1925 to 1926, while the most considerable part must be attributed to the young Carlo Scarpa.

Scarpa was called upon by Giacomo Cappellin for the restoration of the Gothic Palazzo da Mula, head office of the company, and only later became involved also as a designer of glass pieces. In this context he distinguished himself as the creator of some of the most innovative forms of glassware, characterised initially by a particular truncated-cone-shaped foot.[8]

Following a targeted and wide-ranging marketing strategy in Italy and abroad[9], the Cappellin company opened a large shop in 1927 in Florence on Lungarno Guicciardini which transmitted an uncompromisingly modern image, due particularly to the style of Carlo Scarpa. His contribution can be seen even from the outside in the shop sign, in applied metal block letters recalling classical epigraphy.[10]

The internal decoration must also be attributed to the architect, as can be seen from the photograph and the sketch for the stained glass door of the shop.[11] On the two glass door-leaves forming a round arch the abbreviation M.V.M. can be made out and the mouth of a furnace in a style close to Cubism.

Among the 14 studies for stained glass present in the collection, 6 can be attributed to Carlo Scarpa, [1, 3] including the study produced for the *4th Triennale delle Arti Decorative di Monza* of 1930 which contains clear references to the Metaphysic movement. In the collection there are projects for stained glass by other artists, such as that by Mario Romano for the *Quadriennale d'Arte di Roma* of 1931 and that of Ernesto Michahelles, a.k.a. Thayaht, for the Cappellin shop in Florence.[12]

Another group of 11 drawings concern the production of an ashtray with a small fish projected in different variants and intended to determine the financial costs of implementation. This was a piece for the Genoa Cappellin shop of Raffaele Profumo and Arnaldo Fuselli, who from 1928 to 1930 sold elegant pieces produced by the glassworks.[13] [4]

Finally, there are 12 drawings pertaining to lighting, such as wall lamps and chandeliers designed for the interiors of hotels like the recently identified Grand Hotel Thermes of Acqui Terme.[14] The master's hand can be recognised in the artistic quality of these sheets: in the sure graphite stroke, the watercolouring, the mastery of perspective and the characteristic geometric design of the floor. [5]

Cross-verification between the archive documents and the catalogue of the glassworks has added further elements to our knowledge of the activity of the young Carlo Scarpa during his association with the Cappellin glassworks.

The exceptional presence of an archive dedicated to Carlo Scarpa and kept in the museum of Castelvecchio, itself restructured by the architect, made it possible during the management of Paola Marini (1993-2015) to develop research regarding design, architecture and contemporary art.[15]

In this context, attention should be drawn to the exhibition dedicated to Vinicio Vianello, *Il design del vetro,* for which Toni Follina, the nephew of the artist and a famous architect himself, re-used the show-cases designed by Carlo Scarpa for the aforementioned *Vetri di Murano 1860-1960* exhibition, in which the glass pieces of Vianello[16] had been shown.

Glass attracted renewed attention in Castelvecchio in 2014 with *Acquaria*, a permanent installation by Giorgio Vigna,[17] an artist profoundly attached to research on this material, who inserted quite naturally into the architecture of the fountain by Carlo Scarpa stones made of blown glass; solidified bubbles that come to surface in the rippling of the water and which, with studied artifice, blend, merge and emit a new light, establishing a point of departure for new explorations in this extraordinary material.

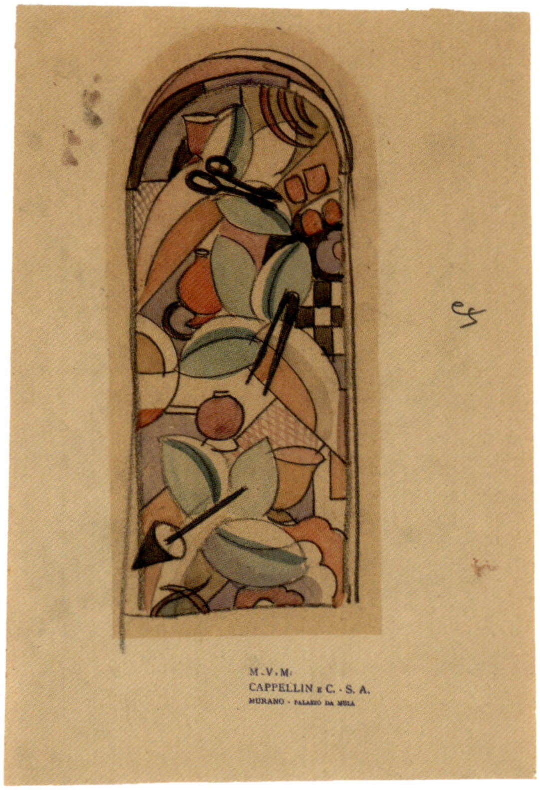

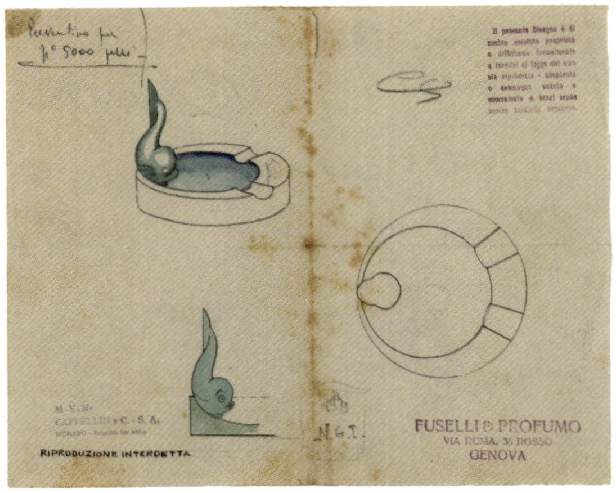

[3] Carlo Scarpa, study for stained glass in Murano glass, c. 1928, graphite, charcoal, tinted water colours on cardboard, 305 x 210 mm, detail. Carlo Scarpa Archives, Museo di Castelvecchio, Verona (RC294)

[4] Carlo Scarpa, *Preparatory sketch of ashtray with fish,* 1928-31, graphite, tinted water colours on cardboard, 315 x 215 mm. Carlo Scarpa Archives, Museo di Castelvecchio, Verona (RC025)

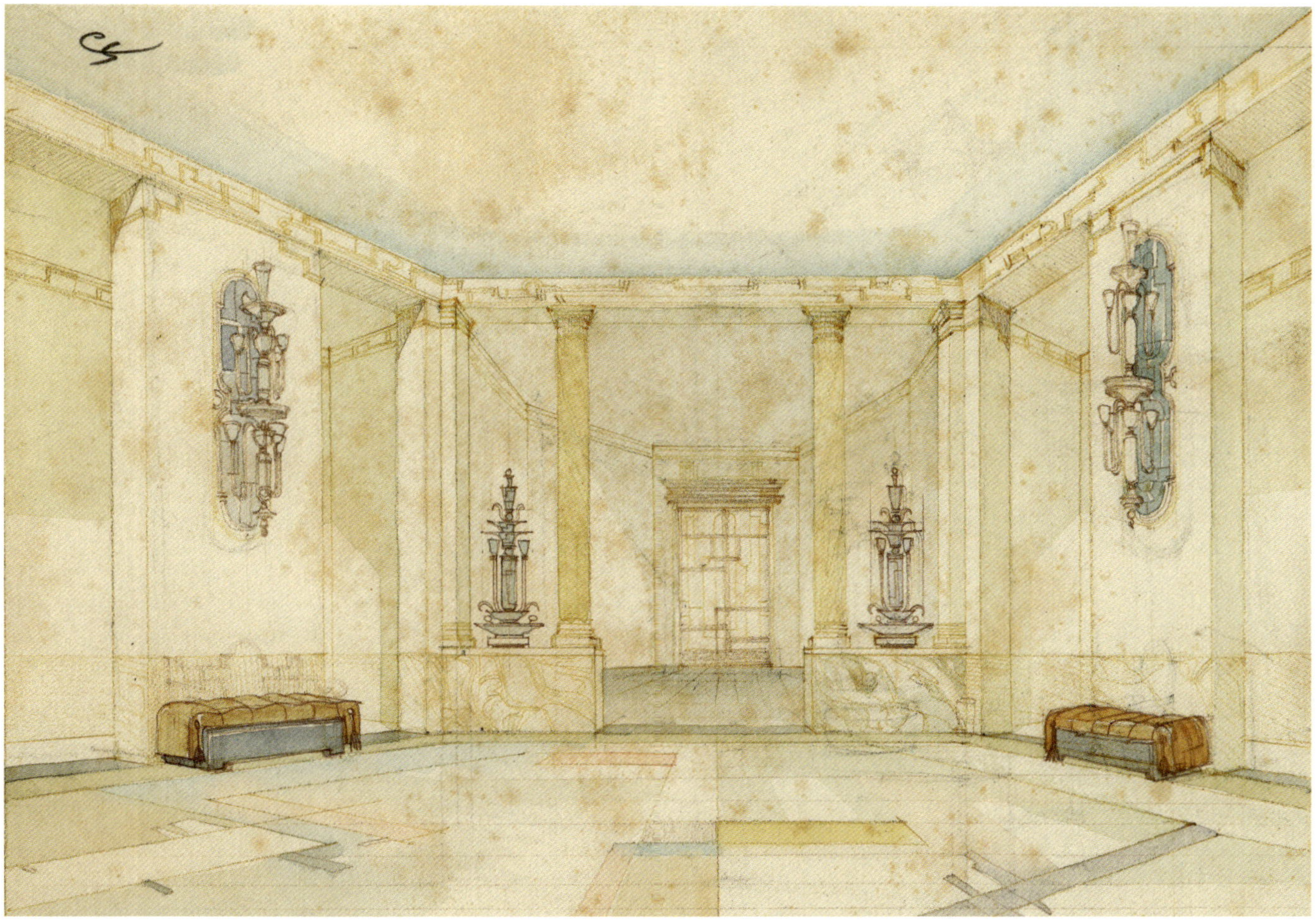

[5] Carlo Scarpa, study for lighting of the Hall of the Grand Hotel Thermes of Acqui Terme, c. 1929, graphite, tinted water colours and coloured crayon on cardboard, 262 x 321 mm. Carlo Scarpa Archives, Museo di Castelvecchio, Verona (RC276)

The authors would like to thank Marino Barovier, Elisa Festa, Marzia Guastella, Anna Pasti, Carla Sonego and Giorgio Vigna.

1 A. Gasparetto (edited by), *Vetri di Murano 1860-1960*, Verona 1960.

2 L. Magagnato, 'Storia e genesi dell'intervento', in Id., *Carlo Scarpa a Castelvecchio*, Milan 1982, p. 33, note 2. In the note Magagnato summarised the expenses for the reorganisation of the museum and the remuneration for services provided by Prof. Carlo Scarpa, including the purchase of the designs.

3 P. Marini, 'L'Archivio di Carlo Scarpa a Castelvecchio', in M. Guccione (edited by), *Documentare il contemporaneo. Archivi e Musei di architettura, Atti della giornata di studio MAXXI Museo nazionale delle arti del XXI secolo*, Rome 2009.

4 The Italian state purchased the entire Carlo Scarpa archive from his son Tobia Scarpa, archives consisting of more than 12,000 graphic designs, 11,454 drawings and sketches, 12,200 photographs, 8,000 cassettes and videos, for the MAXXI collections, at present kept at the Centro Carlo Scarpa in the Archivio di Stato of Treviso. At the Museo di Castelvecchio of Verona there are some collections of Carlo Scarpa designs that can be consulted on-line at the site www.archiviocarloscarpa.it. An overview of the activities carried out by the management of the Musei d'Arte between 2003 and 2013 can be found in: P. Marini, A. Di Lieto and K. Bertolaso (edited by), 'Castelvecchio sottotraccia', appendix to *Architettiverona*, no. 2, 21st year, May-August 2013.

5 L. Baldini, 'Alcuni disegni inediti di Carlo Scarpa', in *VeneziaArti*, 6, 1992, pp. 138-140.

6 C. Sonego, 'I disegni per i vetri', in *Castelvecchio Sottotraccia*, appendix to *ArchitettiVerona*, no. 2, 21st year, May-August 2013.

7 The project of cataloguing, digitising and putting on-line on the site www.archiviocarloscarpa.it of the graphic and photographic collection, was carried out under the management of Paola Marini and curated by Alba Di Lieto. The identification of the designs and the bibliography were entrusted to Carla Sonego; the computerised cataloguing was carried out by Ketty Bertolaso with the collaboration of Eva Di Lorenzo.

8 'As generally happened at the time in the world of glasswork, the name of the author of the pieces was often sacrificed to that of the furnace or its proprietor. Scarpa, taking up the research initiated by Zecchin, first designed light transparent blown-glass pieces, sometimes also re-presenting the work of the Murano painter in a personal version. Scarpa's hand is, however, recognisable in the characteristic recurrence of elementary geometric forms and the presence, particularly in the first models, of a typical truncated-cone-shaped foot not to be found in the work of Zecchin', quotation from C. Sonego, 'Carlo Scarpa: gli anni '20', in K.W. Forster, P. Marini (edited by), *Studi su Carlo Scarpa 2000-2002*, Venice 2004, pp. 54-55.

9 In particular the international side of the Cappellin strategy has been highlighted by Cristina Beltrami, in her presentation 'Giacomo Cappellin e Parigi' of 27 February 2018 on the occasion of the symposium *La Vetreria Cappellin* (*Fondazione G. Cini/Le stanze del vetro*, Venice), the same subject as that of a previous essay: C. Beltrami, 'Giacomo Cappellin: gli anni di Parigi', in *Saggi e Memorie*, XIII, 2017, pp. 108-122.

10 I. Abbondandolo, *Carlo Scarpa e la forma delle parole*, Venice 2011, p. 13. About the shop in Florence, see the contribuition of C. Sonego in this catalogue.

11 See *Works*, *Stained glass* in this catalogue.

12 The design was shown in the exhibition: *1927. Il ritorno in Italia.* See L. Mannini, 'La Firenze 'industriosa' degli anni venti', in S. Ricci, C. Sisi (edited by), *1927. Il ritorno in Italia. Salvatore Ferragamo e la cultura visiva del Novecento*, Florence 2017, p. 306. See *Works*, *Stained glass* in this catalogue.

13 M. Fochessati, G. Franzone, 'La trama dell'arte. Arte e design nella produzione MITA', in *Collana di studi della Fondazione Conservatorio Fieschi*, Genoa 2016, p. 15.

14 See the contribution of C. Sonego in this catalogue. See also C. Sonego, 'Carlo Scarpa nell'ufficio tecnico della Venini', in M. Barovier, C. Sonego (edited by), *Paolo Venini e la sua fornace*, Milan 2016, pp. 45-47.

15 F. Rapisarda, 'Ospiti in museo', in *Castelvecchio Sottotraccia*, appendix to *Architettiverona*, no. 2, 21st year, May-August 2013.

16 A. Bassi, P. Marini, A. Di Lieto (edited by), *Vinicio Vianello. Il design del vetro*, Venice 2007.

17 G. Vigna, *Stati Naturali*, Milan 2013.

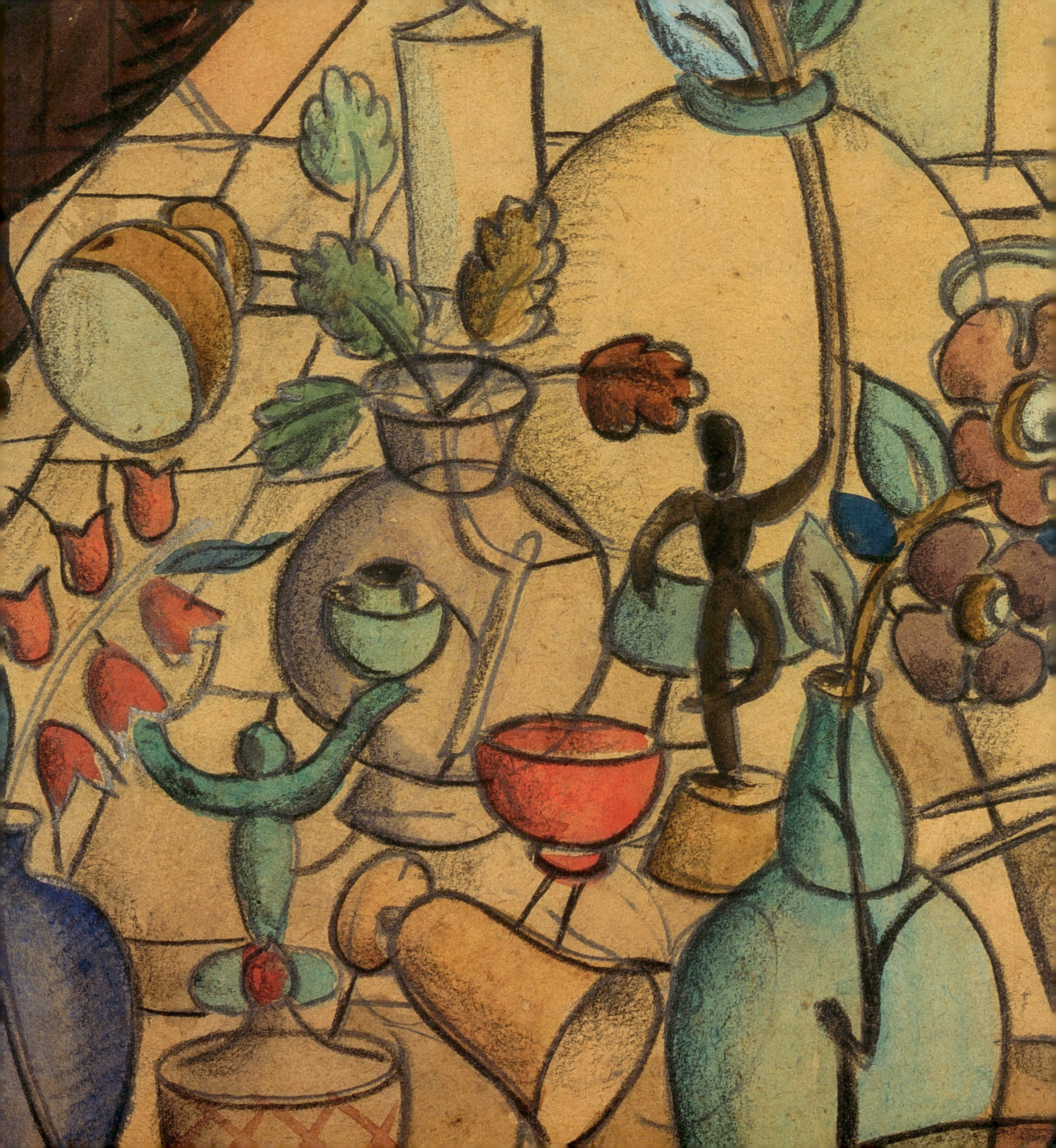

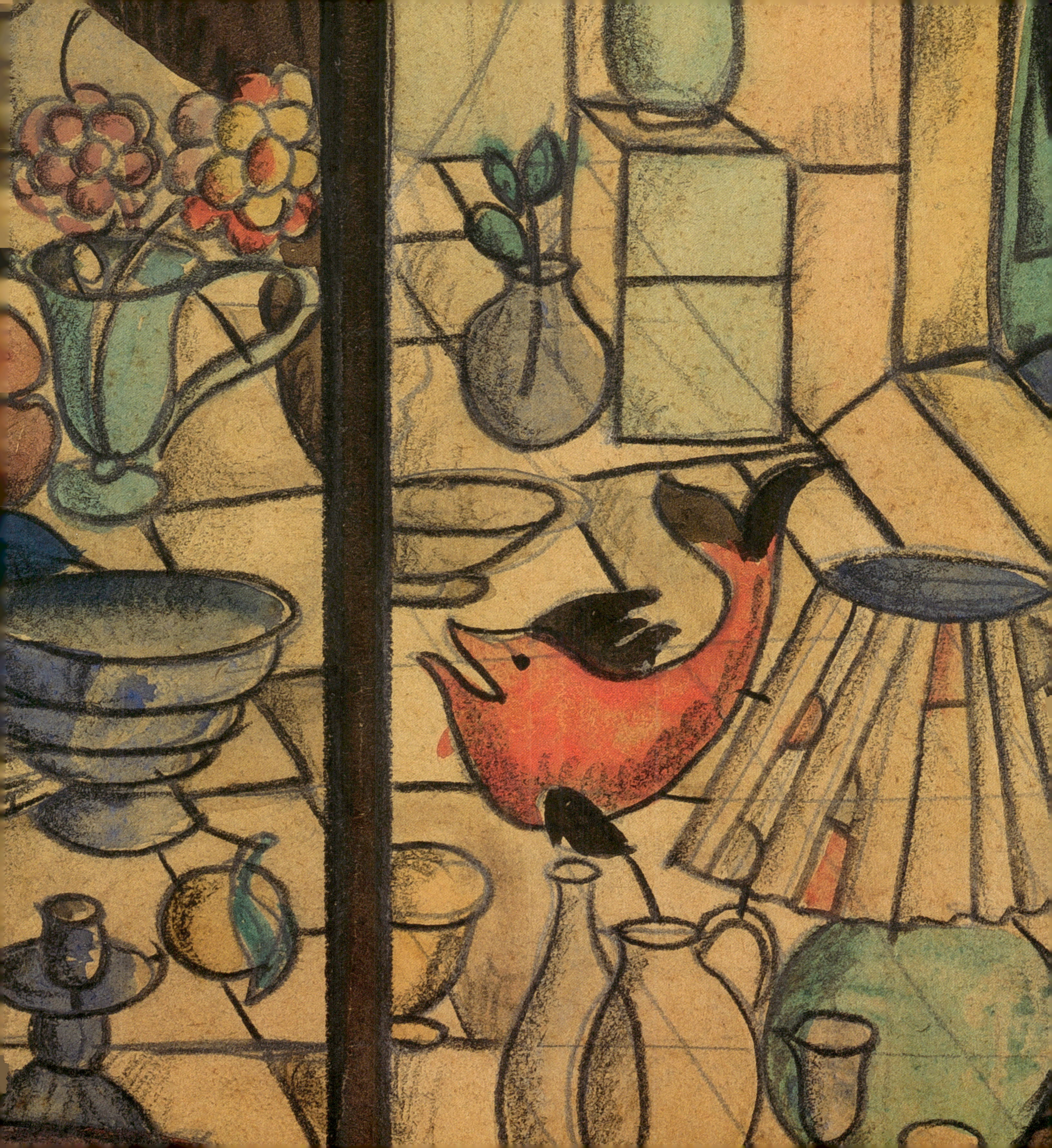

THE WORKS

Marino Barovier and *Carla Sonego*

The catalogue of the works, presented below, consists of a careful selection of the glass pieces that M.V.M. Cappellin & C. produced from October 1926 (after the departure from the company of Vittorio Zecchin, artistic director between 1925 and 1926) up to the final closure of the firm (1931).
The glass pieces are subdivided by type (grouping together models produced using the same glass technique) and by category of object, observing the chronological order as far as possible.
In each chapter explanatory entries have been drawn up for the objects that have been traced and those documented by period photographic reproductions.
The entries are accompanied by a brief description in which the original terminology is mostly used, whether for the execution technique or the colours.
The dates given in each entry refer to the period of presentation of the series and whithin it to that of the new models added progressively.
To date the artefacts cross-checking was carried out based on the presence of the glass pieces at different exhibitions, the year in which they were published in the sector-specific magazines, the dates of the period photographs and the factory documentation (price list, lists of glass pieces to be shown at exhibitions etc.).
The different models are illustrated by a silhouette of the piece, where possible by a working or catalogue drawing and by an example of the actual object or detail taken from a period photograph.
Where there is no catalogue for the model and no corresponding number because the available documentation did not make identification possible, the entry is accompanied by a grey square with the initials n.i. (not identified).
The silhouettes are reproduced maintaining correct proportions with the real measurements but this criterion was not followed for the working and catalogue drawings, to facilitate reading.
Given the heterogeneous nature of the archiving references for the catalogue and working drawings and the studies, for each of these the origin has been indicated.
For the bibliography, please refer to the general bibliography.

<<
Carlo Scarpa, study for stained glass, *c. 1928, detail, G. De Carlo Collection*

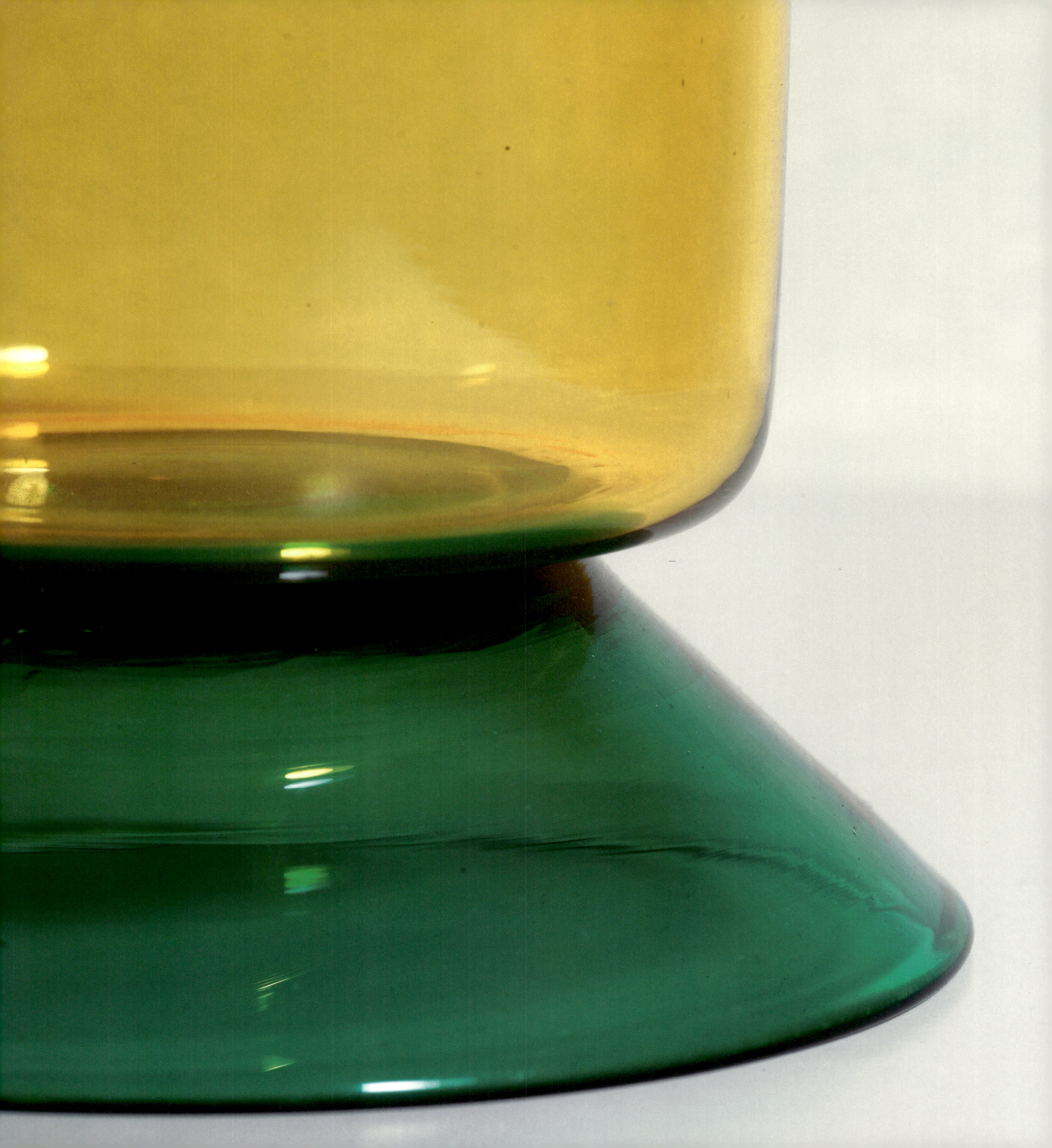

Following the departure from M.V.M. Cappellin & C. of Vittorio Zecchin, where he had worked as Artistic Director from 1925 to October 1926, the glassworks proposed new transparent blown-glass pieces, which in some cases followed the example of those designed by Zecchin himself. In addition to these, there were also items marked by a more essential design, the result of the use of geometrical forms and a truncated-cone-shaped foot, attributable to the presence of the young Carlo Scarpa at the furnace. An outstanding example of this is the spherical vase with truncated-cone-shaped foot that was exhibited at the *Salon d'automne* in Paris and which, accompanied by a glass flower, became the symbol of the Cappellin firm. New models in transparent glass appeared particularly in the period up to late 1928 approximately, from which time the more modern-style production was directed towards opaque glass. Between 1930 and 1931, however, another small blown-glass series was realised with softer lines and applied decoration. Some transparent glass pieces were executed in a monochrome as elegant as it was 'classical', with preference given to purple and Savoy blue. A greater number, however, was in a two-colour version, whether varying the tones of the same colour (e.g. purple and amethyst) or associating 'contrasting' colours, such as yellow and green, white (crystal glass) and purple etc. Moreover, in some cases there were finishings in coral-red *pasta vitrea* combined with green or *pagliesco*-glass surfaces. All the transparent pieces had an iridised finish.

TRANSPARENT GLASS 1926-31

Exhibitions

1926, Paris, *Salon d'automne*
1927, Monza, *III Mostra Internazionale delle Arti Decorative*
1927, Rome, *XCIII Esposizione di Belle Arti della Società amatori e cultori di belle arti*
1927, Rome, *II Mostra d'arte marinara*
1927, Venice, *XVIII Mostra dell'Opera Bevilacqua La Masa*
1927, Geneva, *Exposition d'artistes italiens contemporains*
1927, Paris, *Salon d'automne*
1927, Tripoli, Fiera campionaria [Trade Fair]
1928, Milan, Fiera campionaria [Trade Fair]
1928, Turin, *Torino 1928*
1928, Venice, *XVI Biennale d'Arte Internazionale*
1928, New York, *Macy's International Exposition of Art in Industry*
1928, Newark (NJ), Newark Museum, *Five Hundred New Objects*
1928, Paris, *Art au Foyer 1928*
1929, Milan, Fiera campionaria [Trade Fair]
1929, Paris, *Salon des Artistes dècorateurs*
1930, Monza, *IV Esposizione Triennale Internazionale delle Arti Decorative ed Industriali Moderne*
1931, Venice, *XX Esposizione dell'opera Bevilacqua La Masa*
1931, Amsterdam, *Tentoonstelling van glas, aardewerk en kant, van moderne italiaansche kunstnijverheid. Mostra di vetri, ceramiche e merletti d'arte moderna italiana*

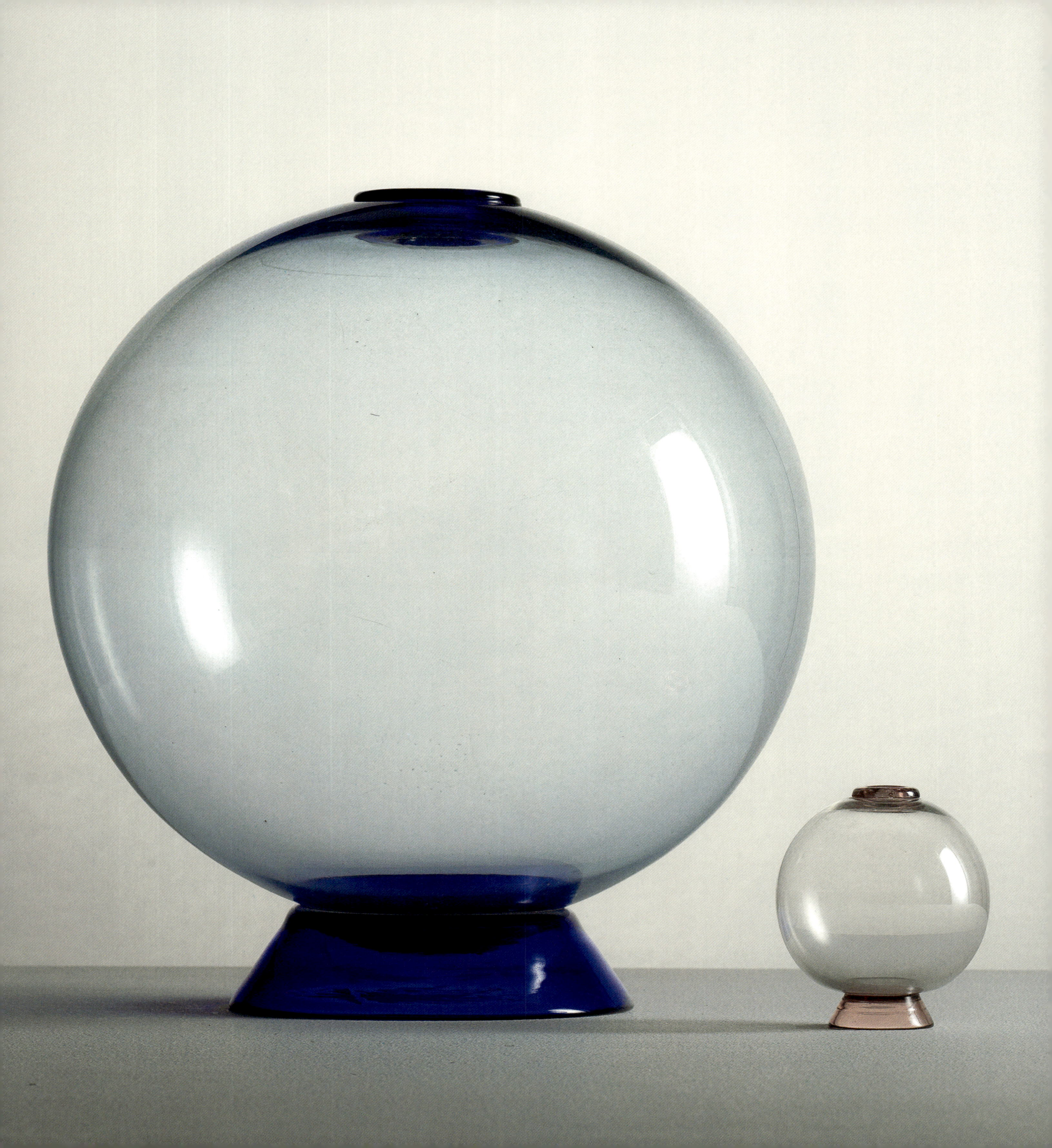

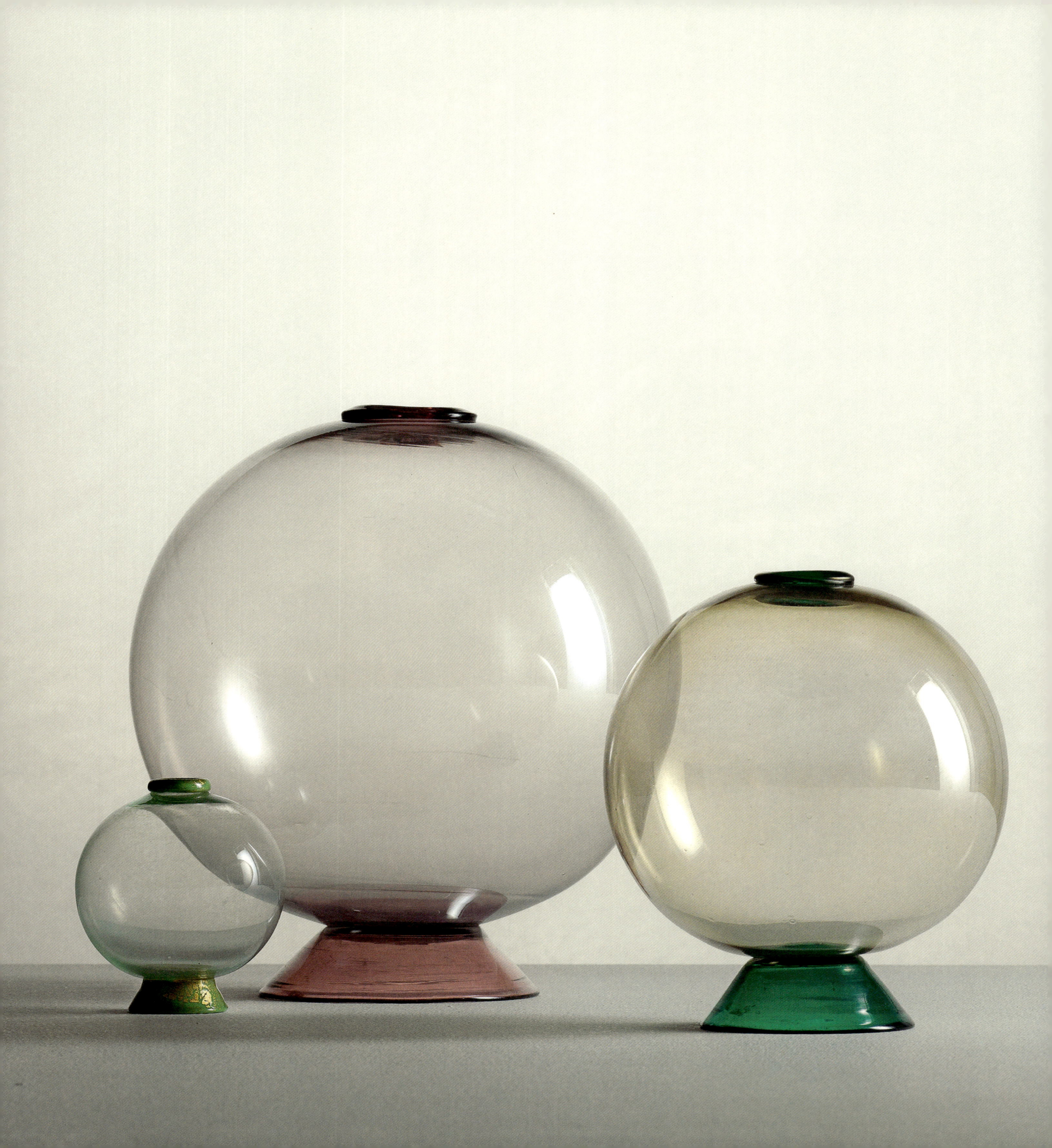

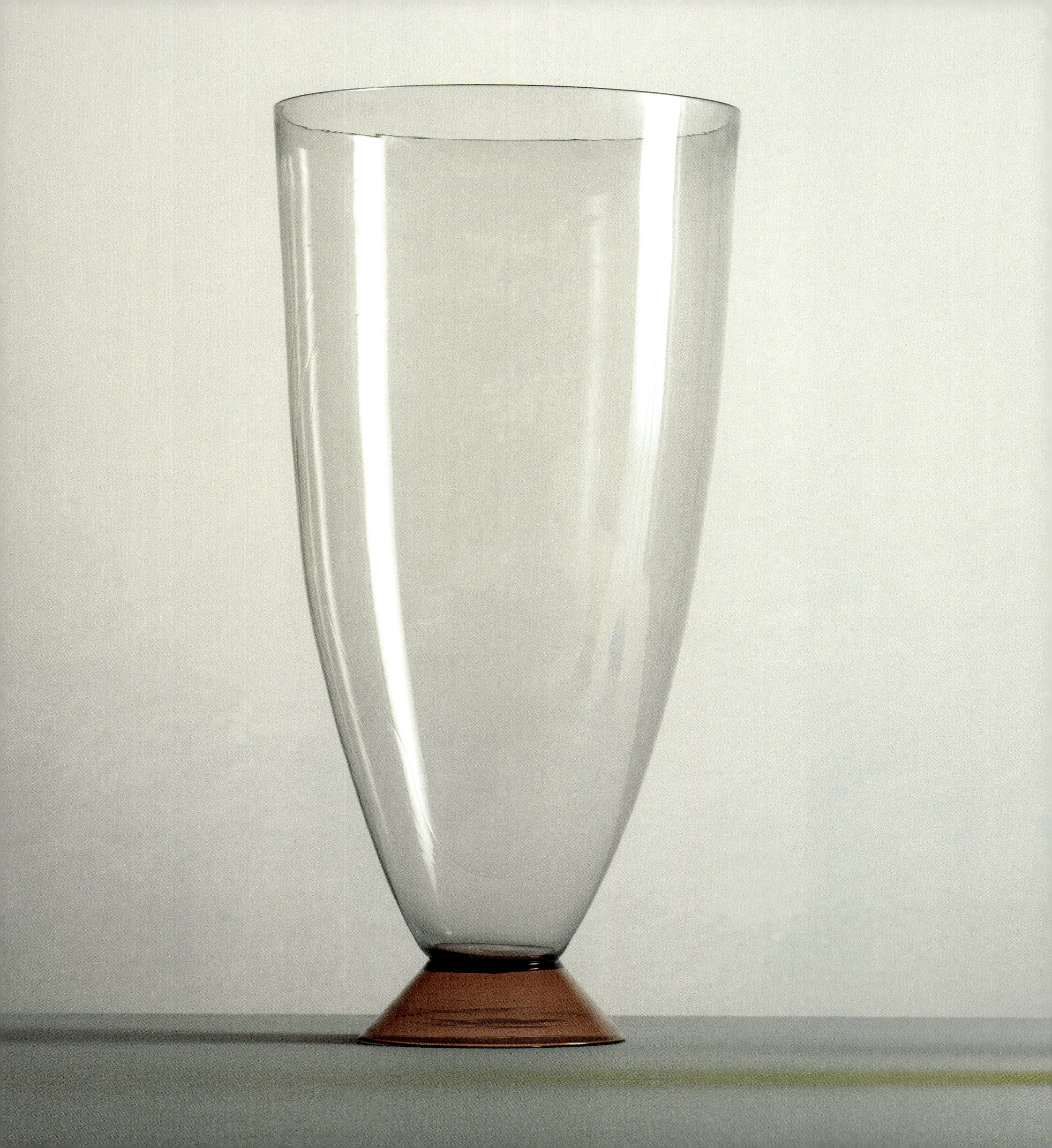

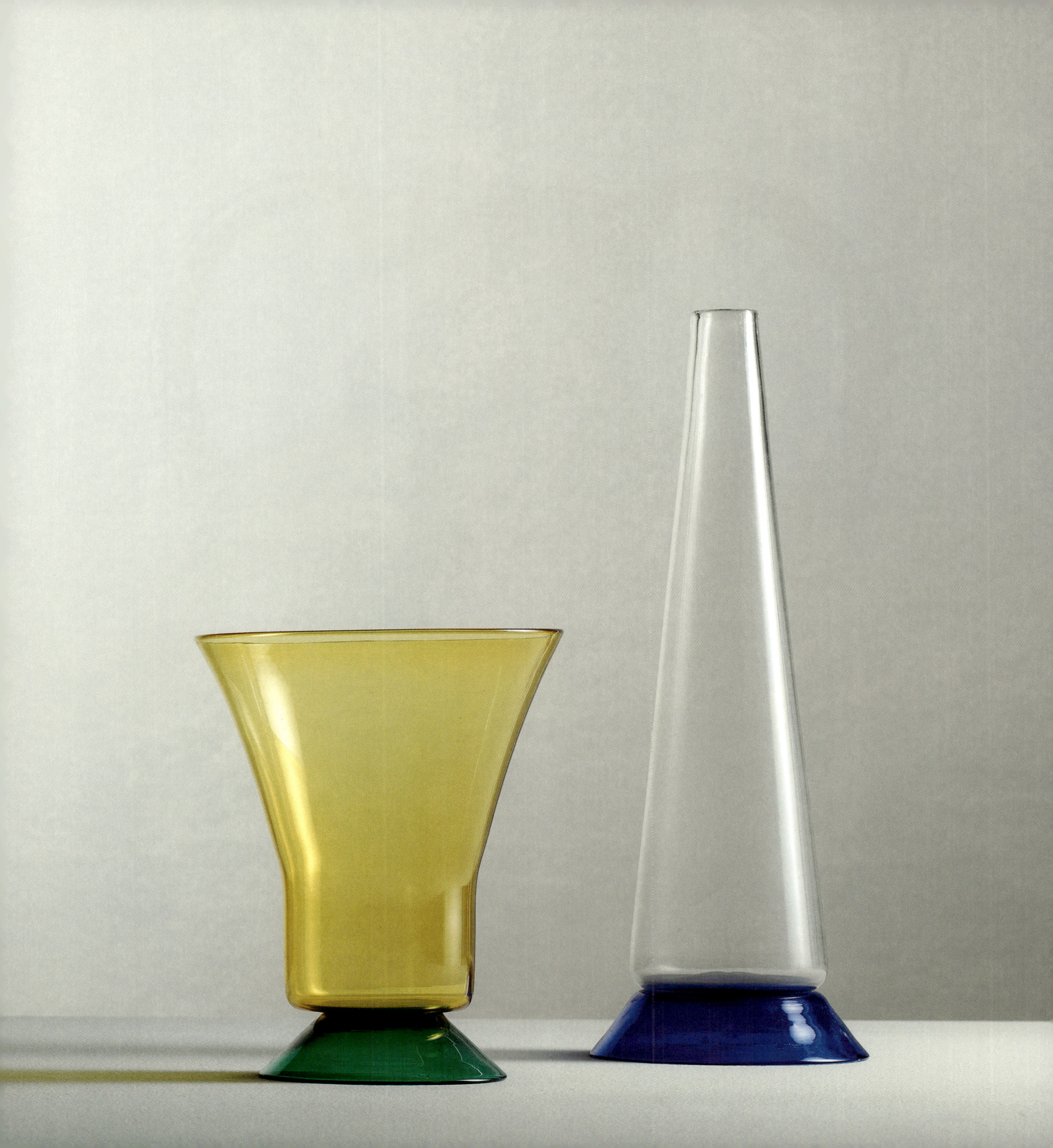

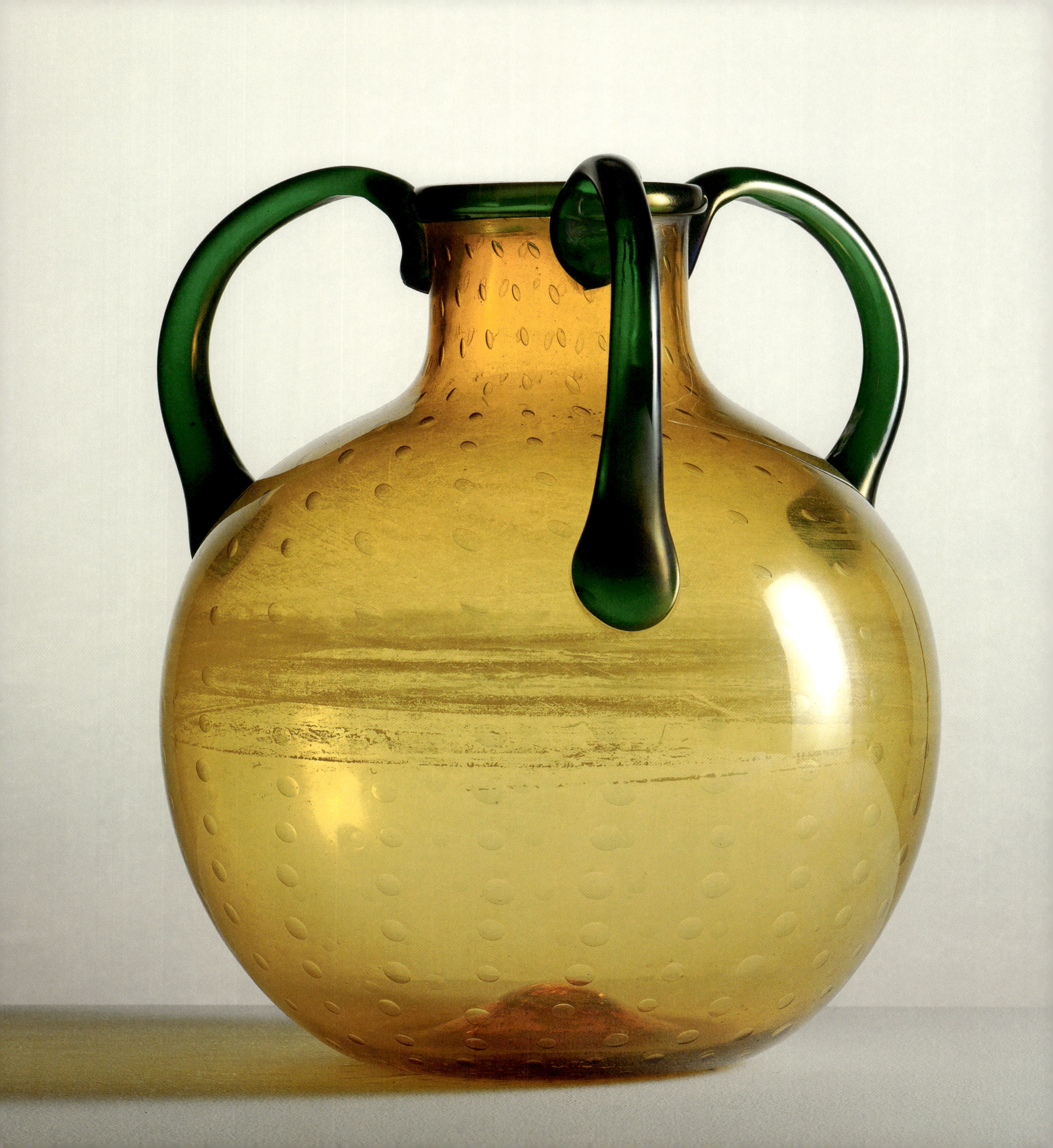

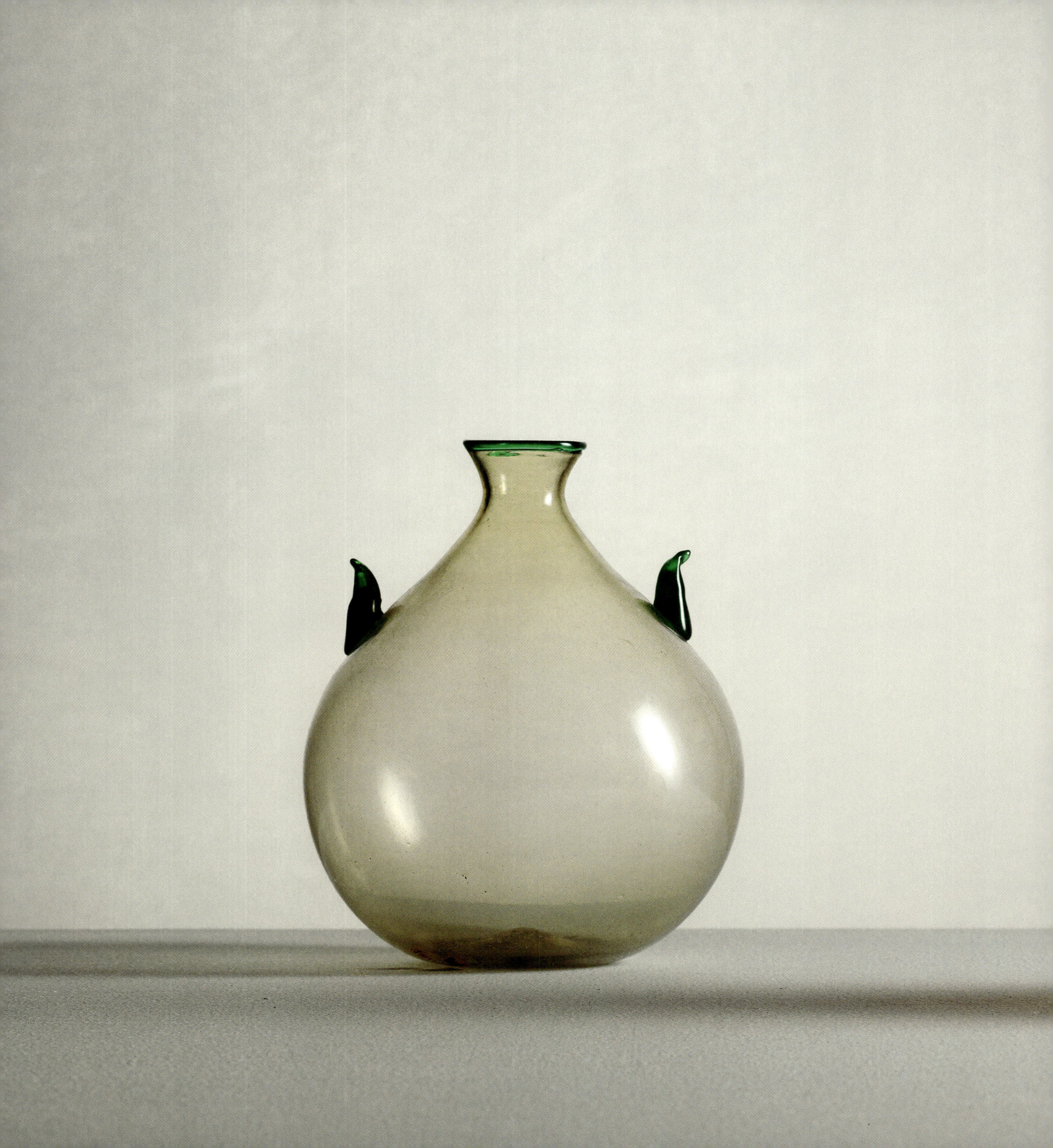

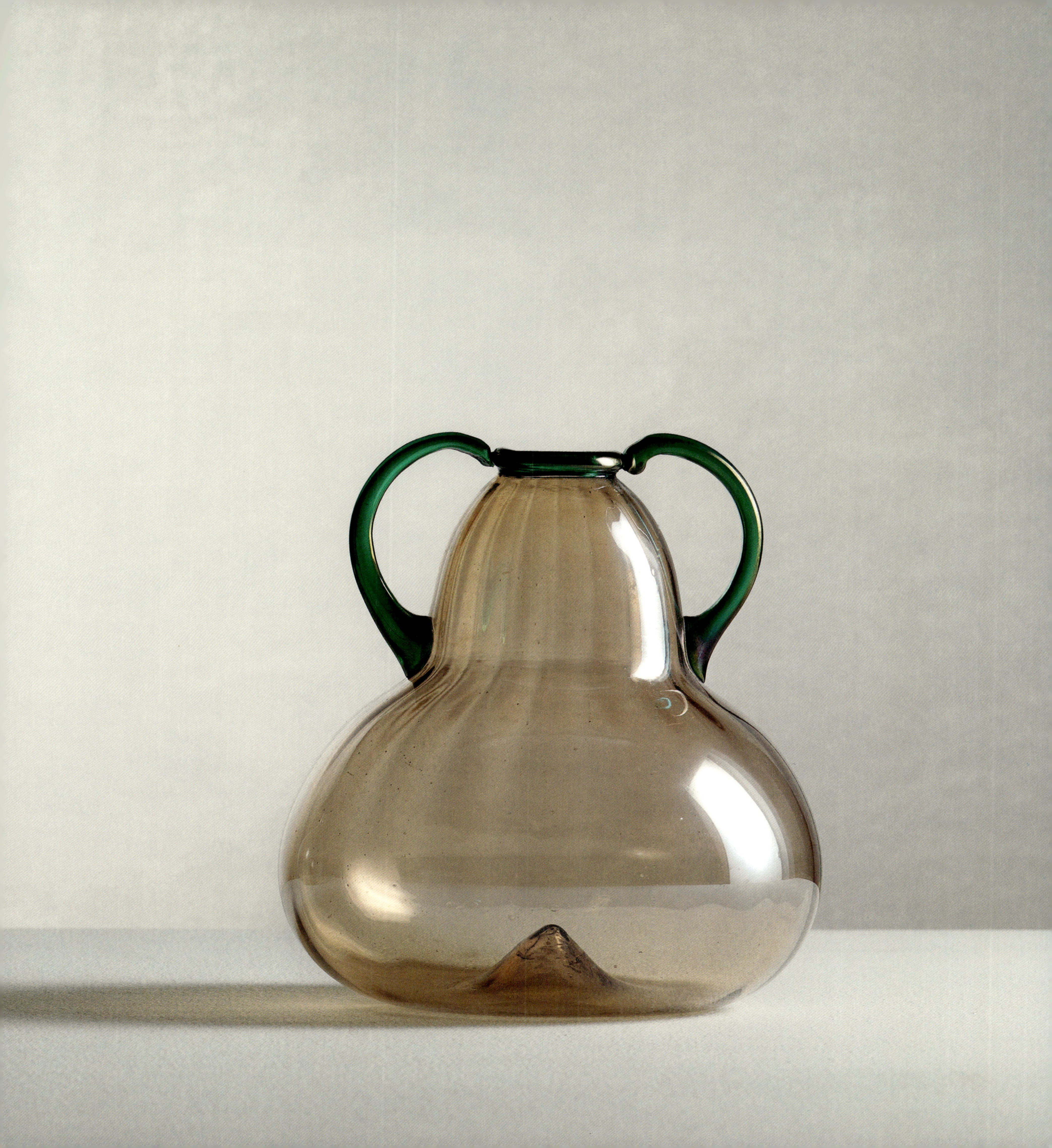

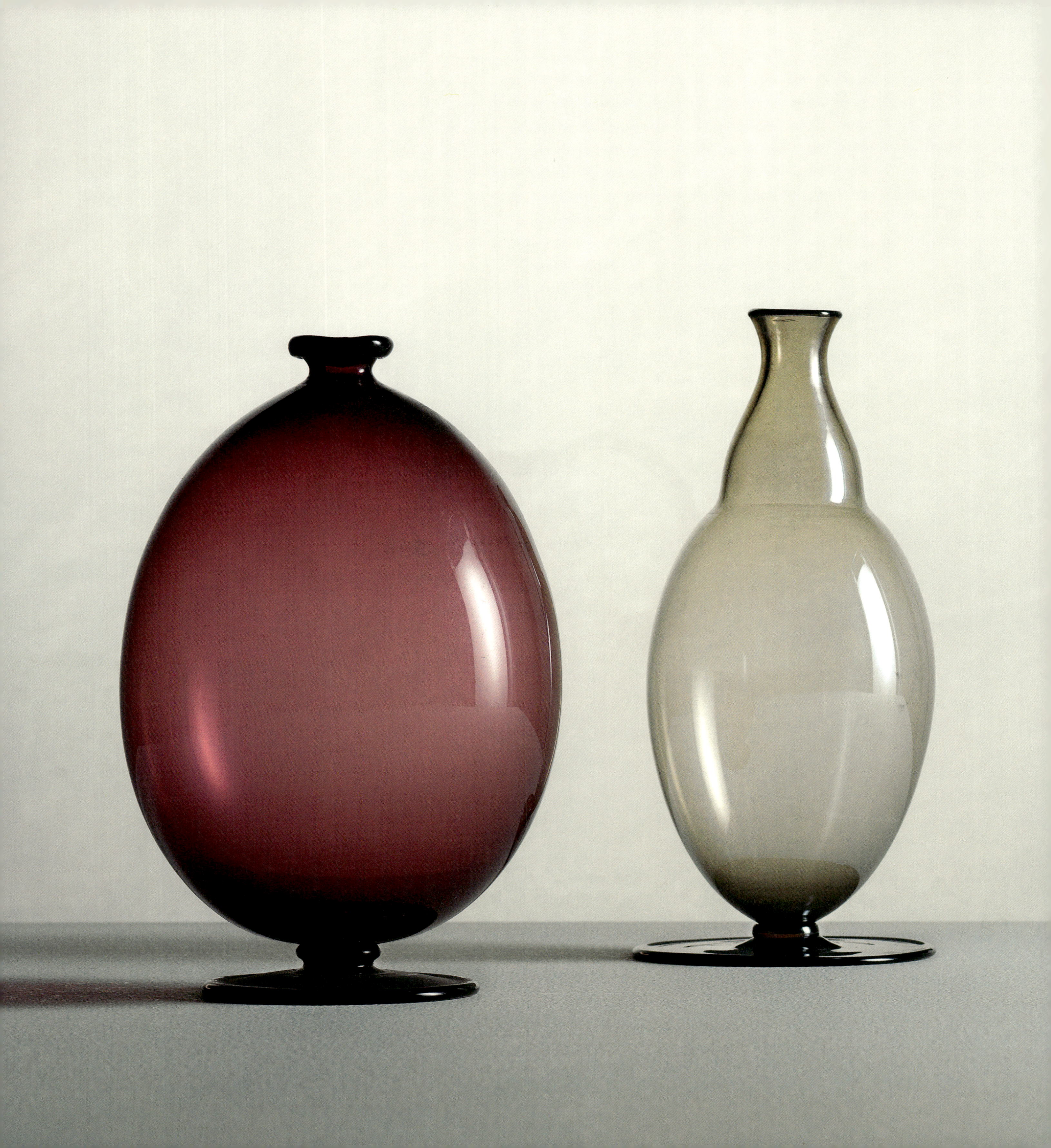

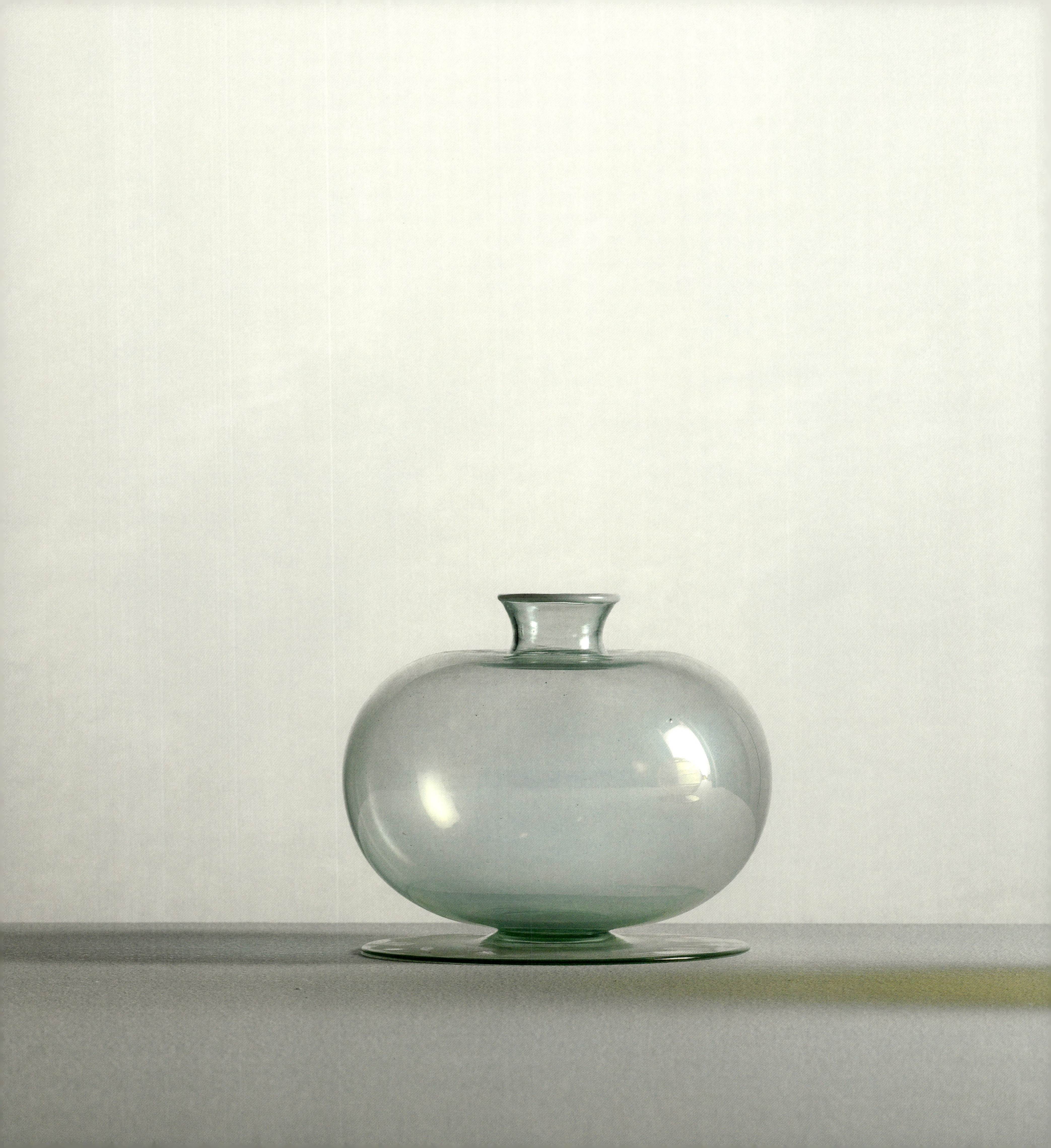

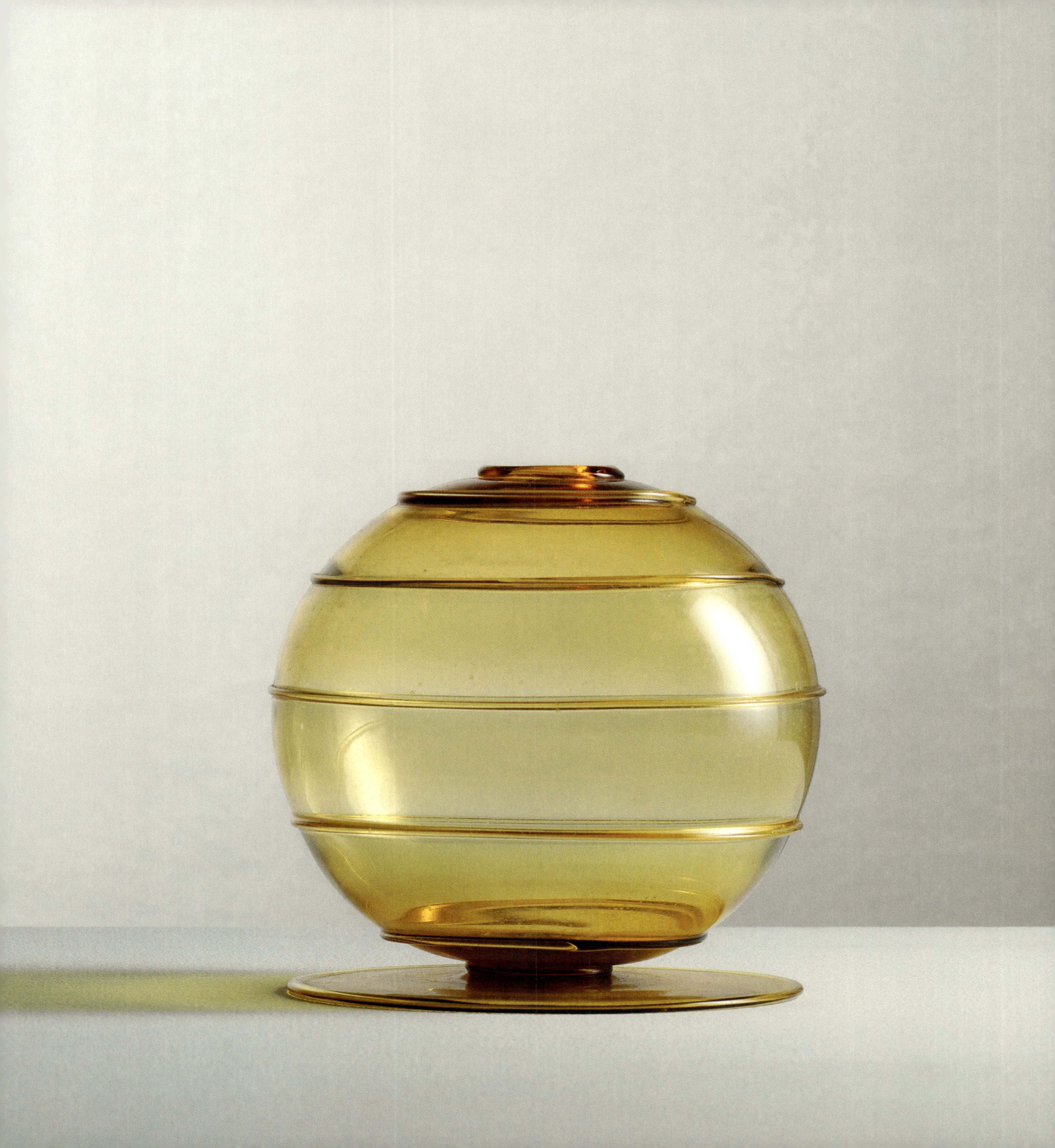

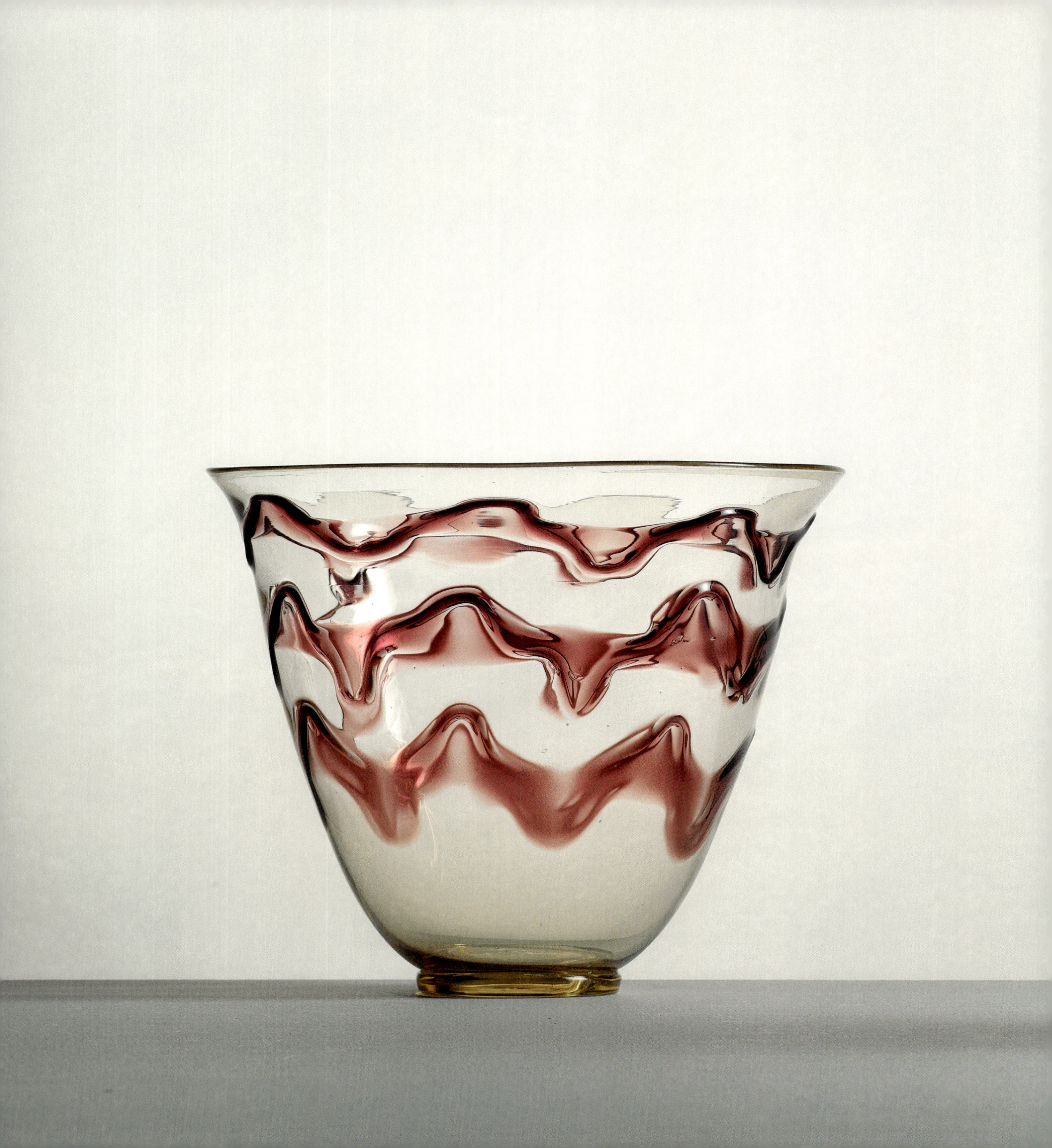

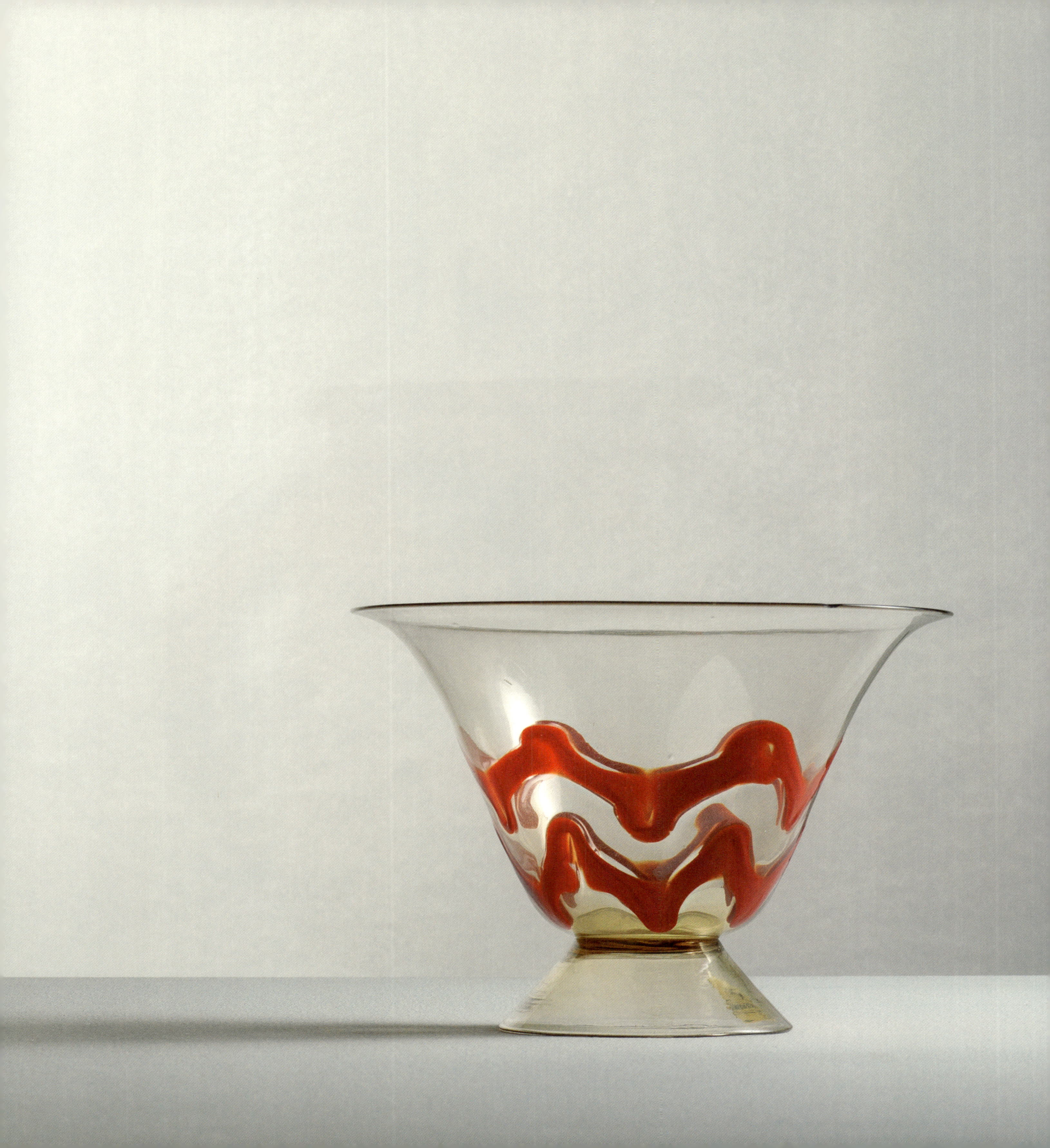

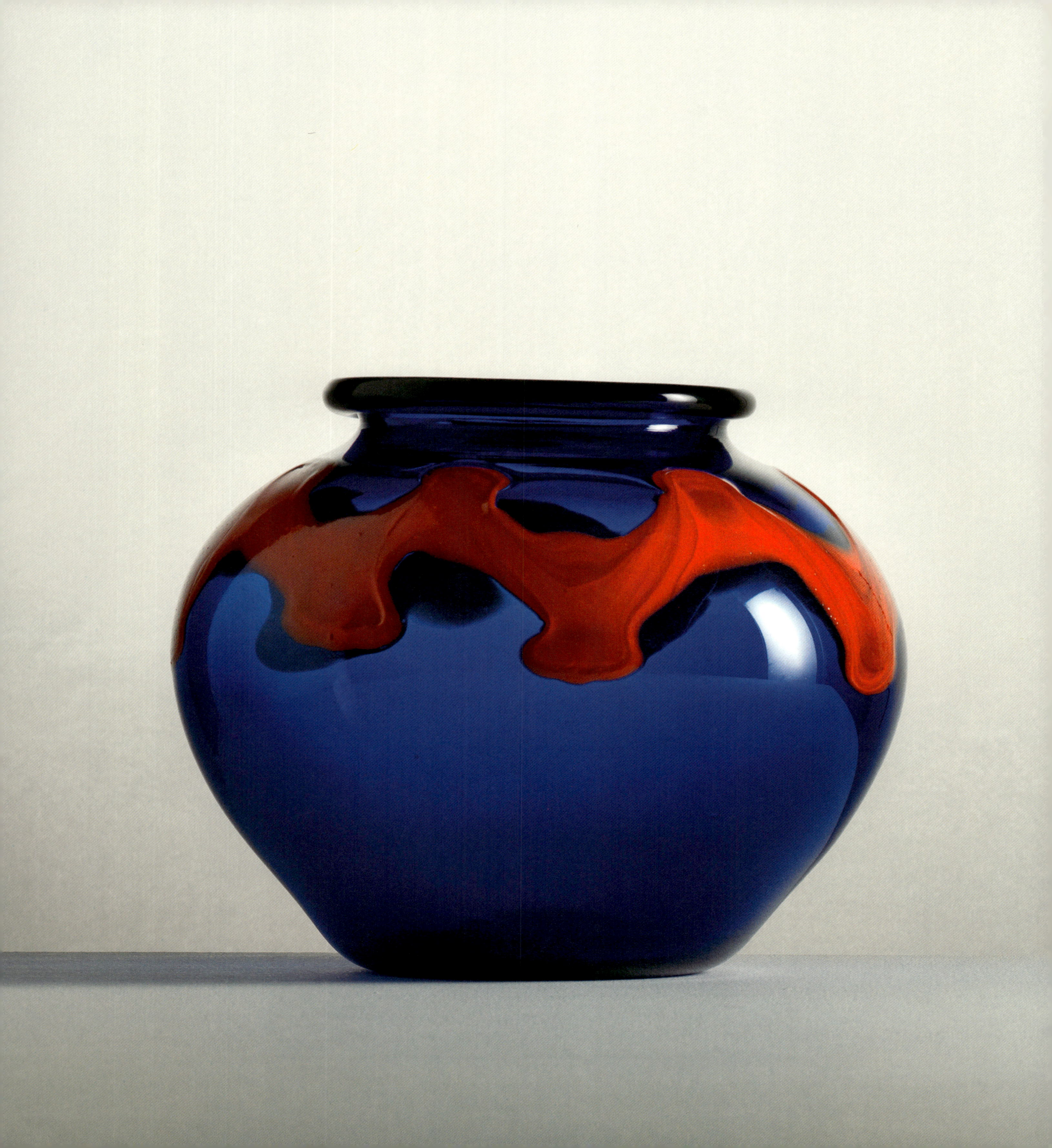

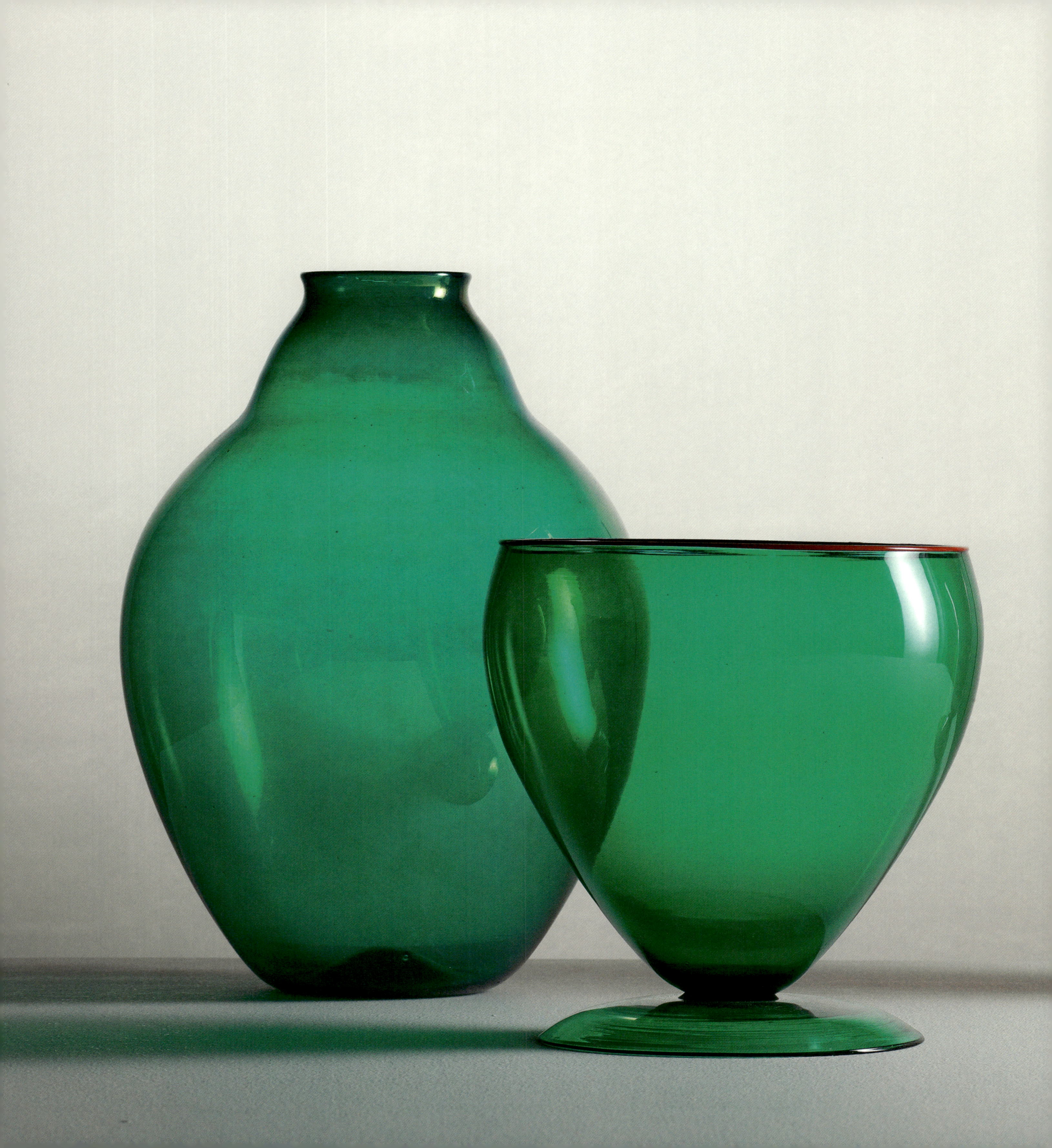

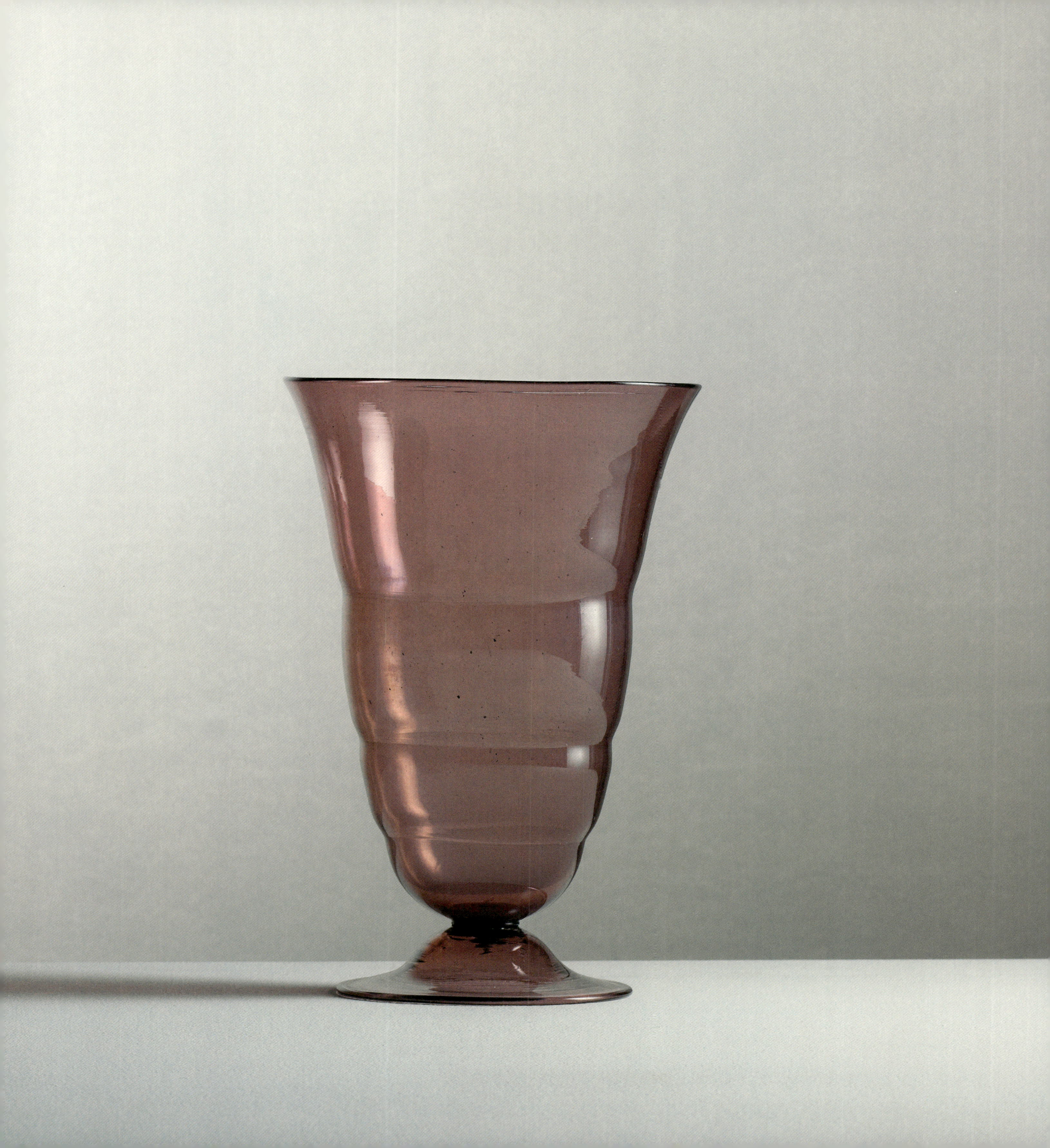

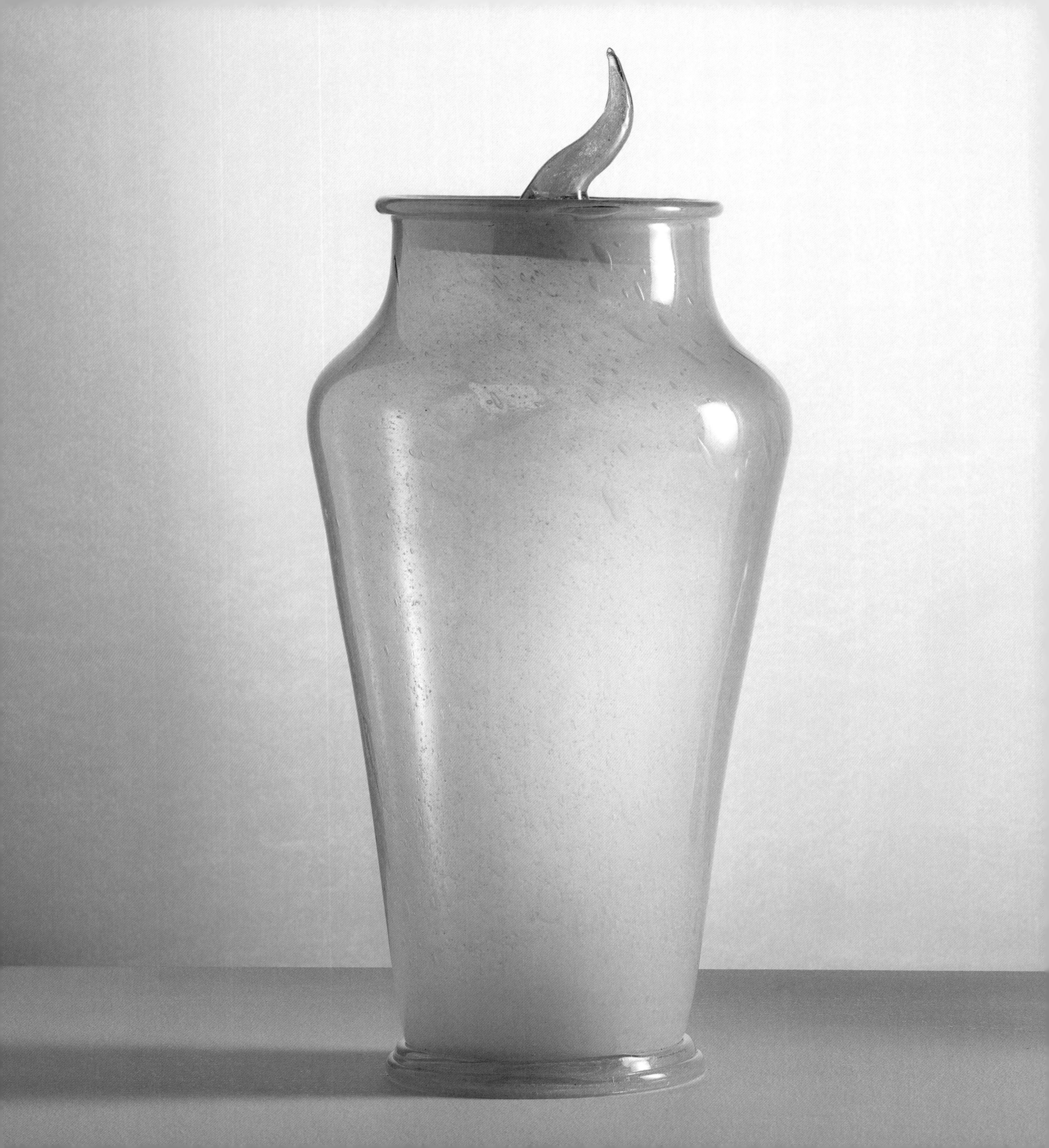

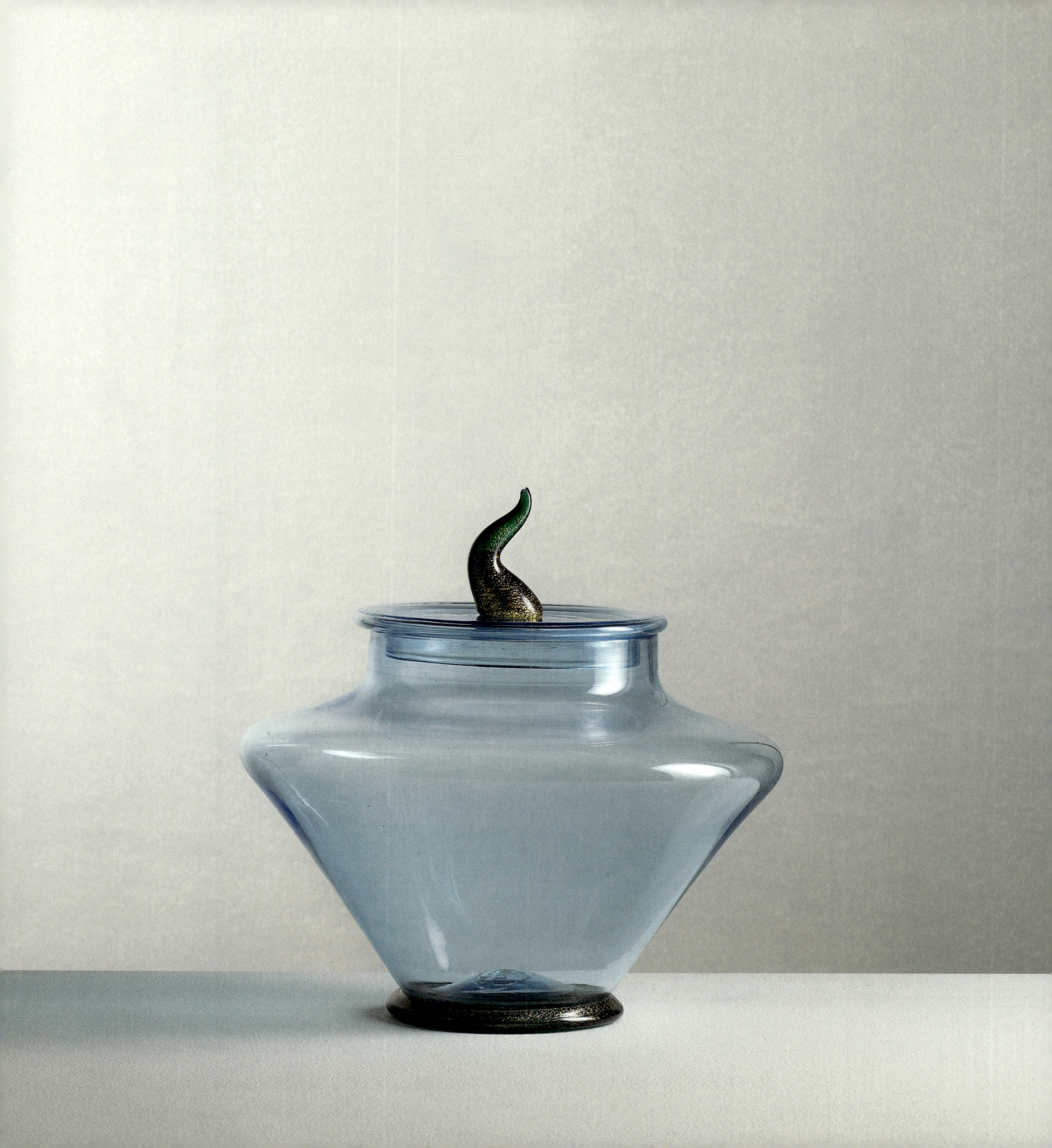

>
'Ball-shaped vase with flowers in white and purple Murano glass' in Domus, *December 1928, p. 59*

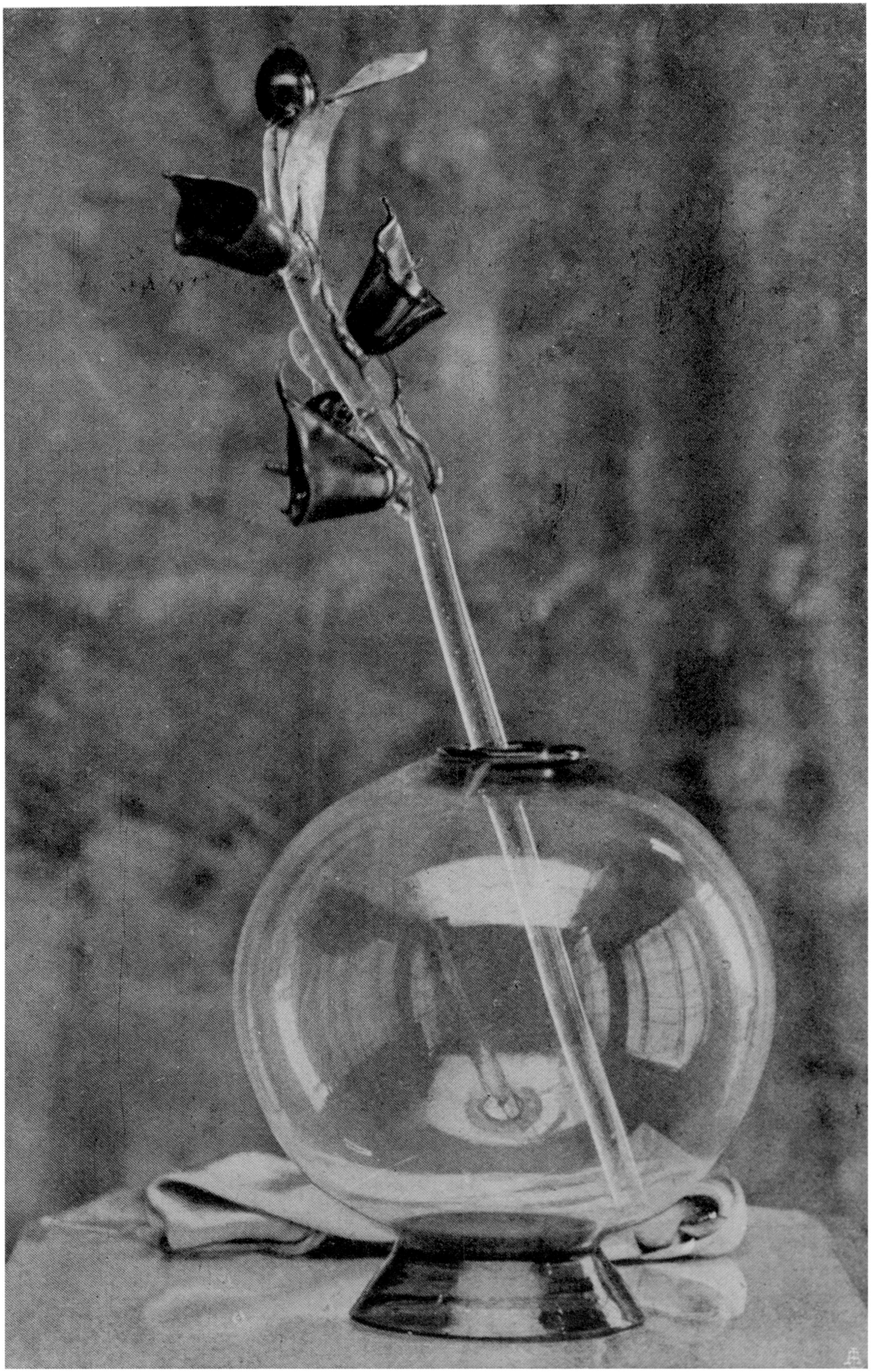

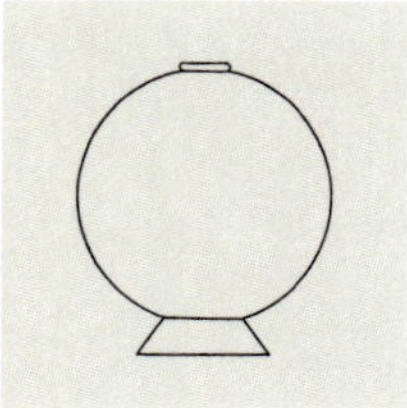

5673

Spherical vase in light-blue transparent glass with truncated-cone-shaped foot and mouth in Savoy-blue transparent glass. Surface with iridised finish. The model was exhibited in 1926 at the Salon d'automne of Paris and, accompanied by a glass flower, became the symbol of the glassworks. The same model was included in the catalogue in a smaller size (no. 5680, h 17 cm), which enjoyed a greater circulation.

The spherical vase with truncated-cone-shaped foot was proposed with various glass techniques (e.g. *filigrana a reticello* [net filigree], *pasta vitrea*, *lattimo*, with 'palm-leaf motif' decoration, cased etc.).
1926
32 cm high
M.V.M. Cappellin Murano *acid stamp*
Bischofberger collection, Switzerland

exhibitions
1926, Paris, Salon d'Automne
1927, Monza, 3rd Biennale
1927, Venice, 18th Mostra dell'Opera Bevilacqua La Masa
1928, Paris, Art au Foyer
1928, Paris, Salon d'automne (?)
1931, Amsterdam, Exhibition of glass, ceramics and lace

>
Page of notebook with the models exhibited at the Salon d'automne, Paris, 1926, CSV

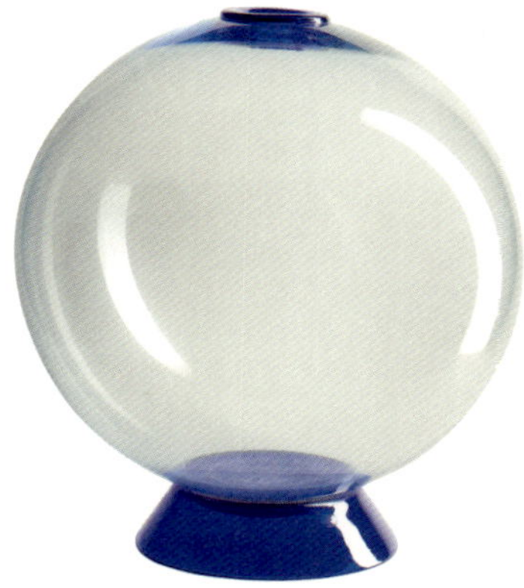

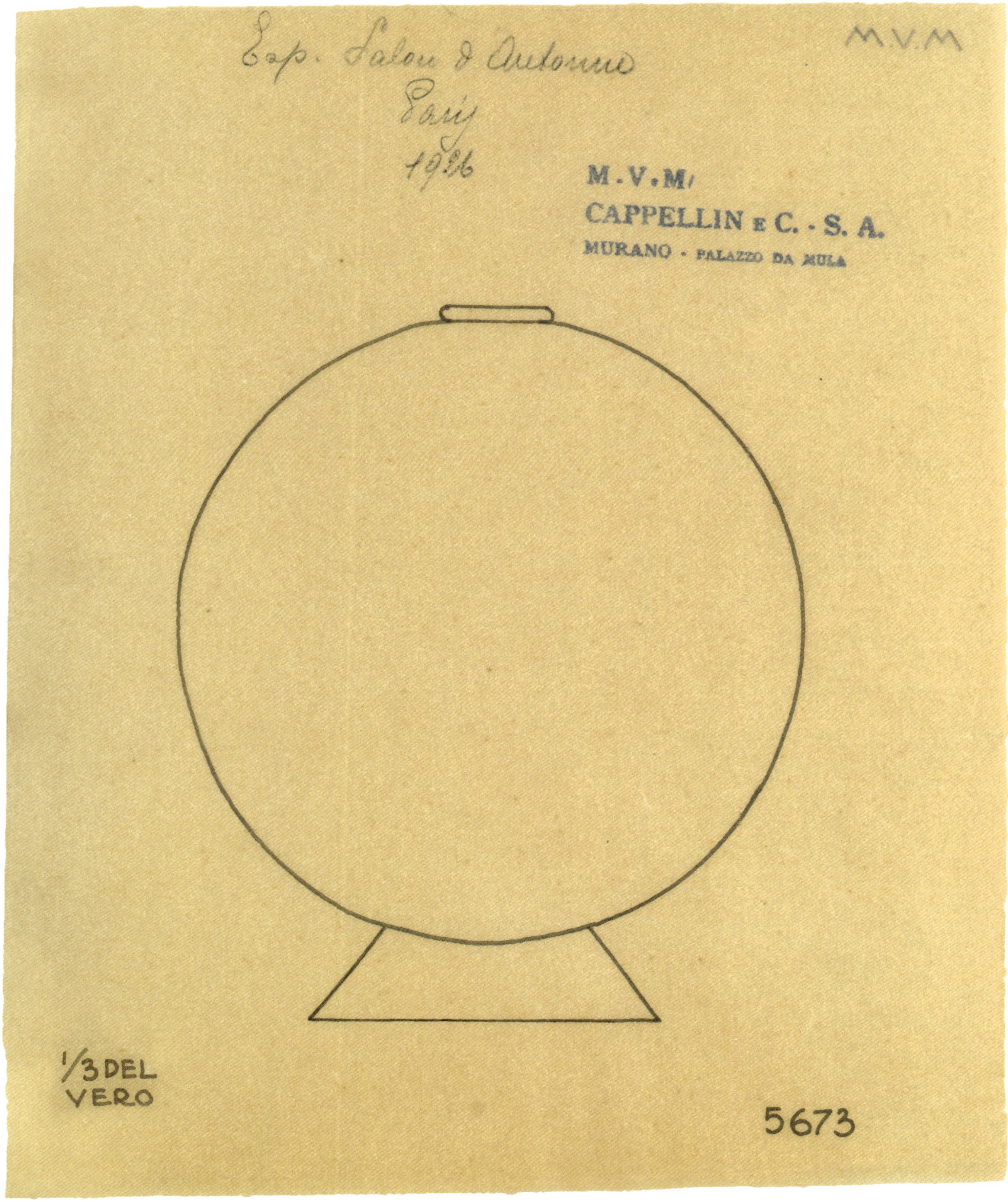

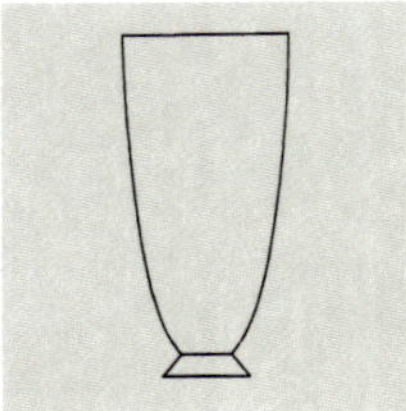

5714

Large vase in crystal glass with truncated-cone-shaped foot in amethyst transparent glass. Surface with iridised finish. A similar example in 'purple' transparent glass was exhibited in 1927 at the 3rd Biennale of Monza.
1927
40 cm high
Private collection

exhibitions
1927, Monza, 3rd Biennale

›
5714, catalogue drawing, AMB

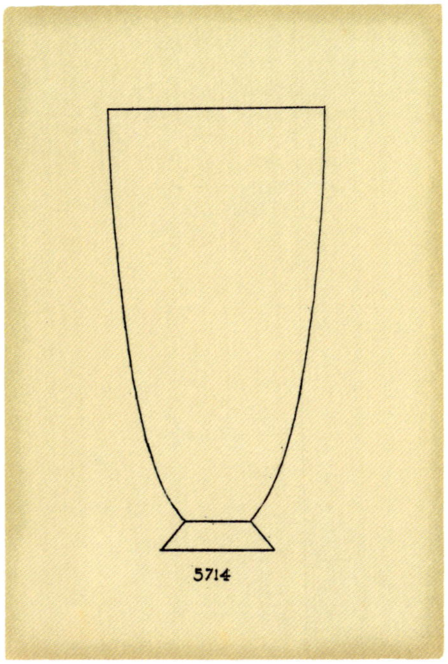

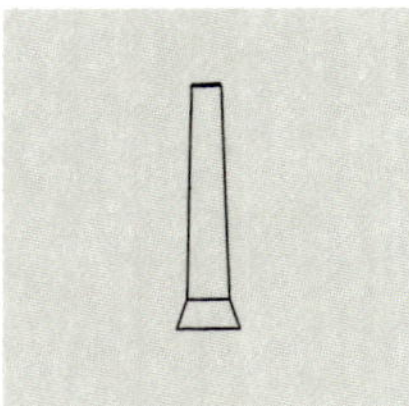

5827

Tall, narrow, truncated-cone-shaped vase, in crystal glass. Truncated-cone-shaped foot in Savoy-blue transparent glass. Surface with iridised finish. The piece is characterised by the small cross-section of the mouth and body designed to contain a single flower.
c. 1928
32 cm high
C.V.M. *acid stamp*
Ferraris-Rocca collection, Venice

›
5827, catalogue drawing, AMB

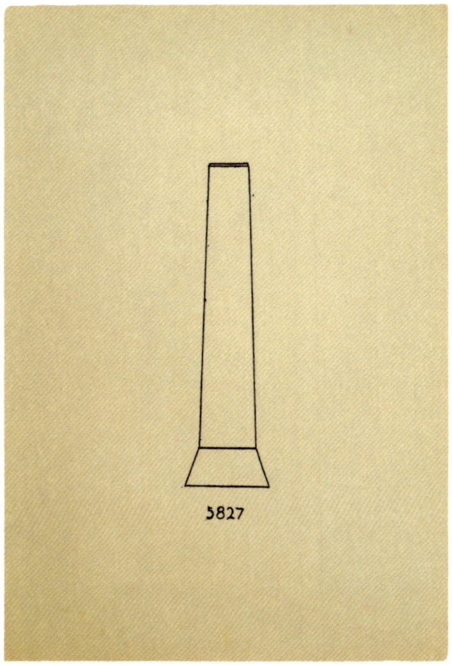

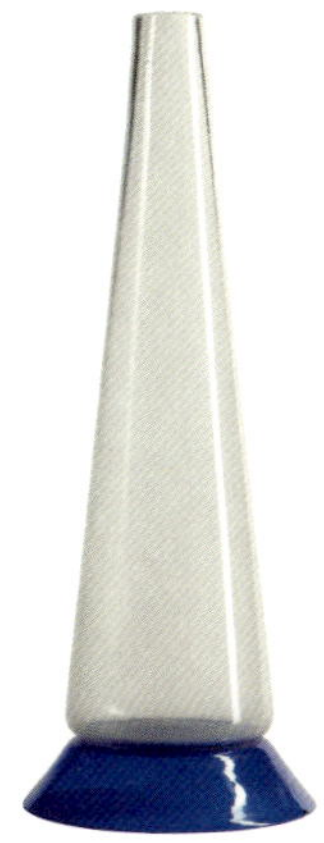

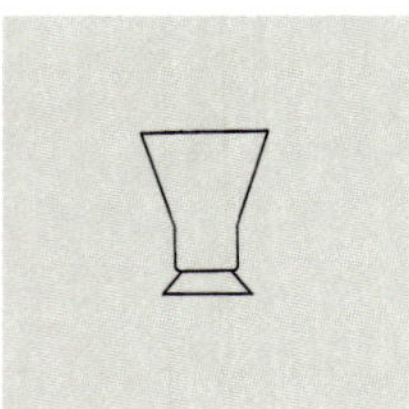

5825

Vase in yellow transparent glass with truncated-cone-shaped foot in green transparent glass. Surface with iridised finish.
c. 1928
18 cm high
M.V.M. Cappellin Murano *acid stamp*
Private collection

›
5825, catalogue drawing, AMB

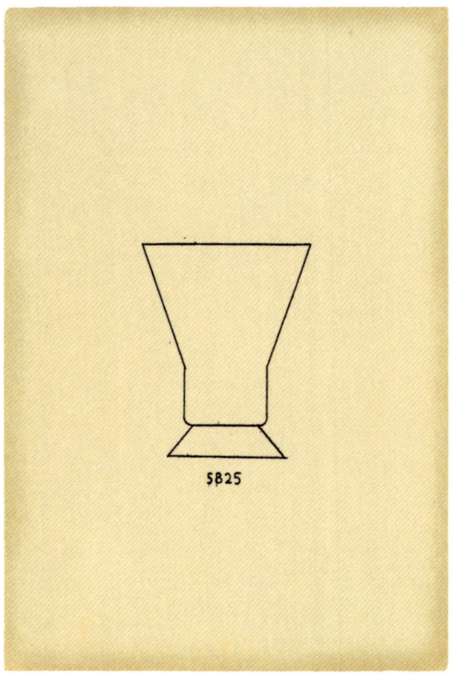

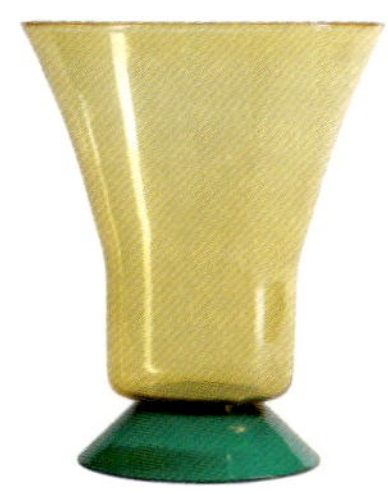

>
Vases, bowls and wine glasses in transparent glass exhibited in the Cappellin shop in Florence. Period photograph, 1928–29, AMB. From left, on the low shelf, models nos 5814 e 5815 can be seen

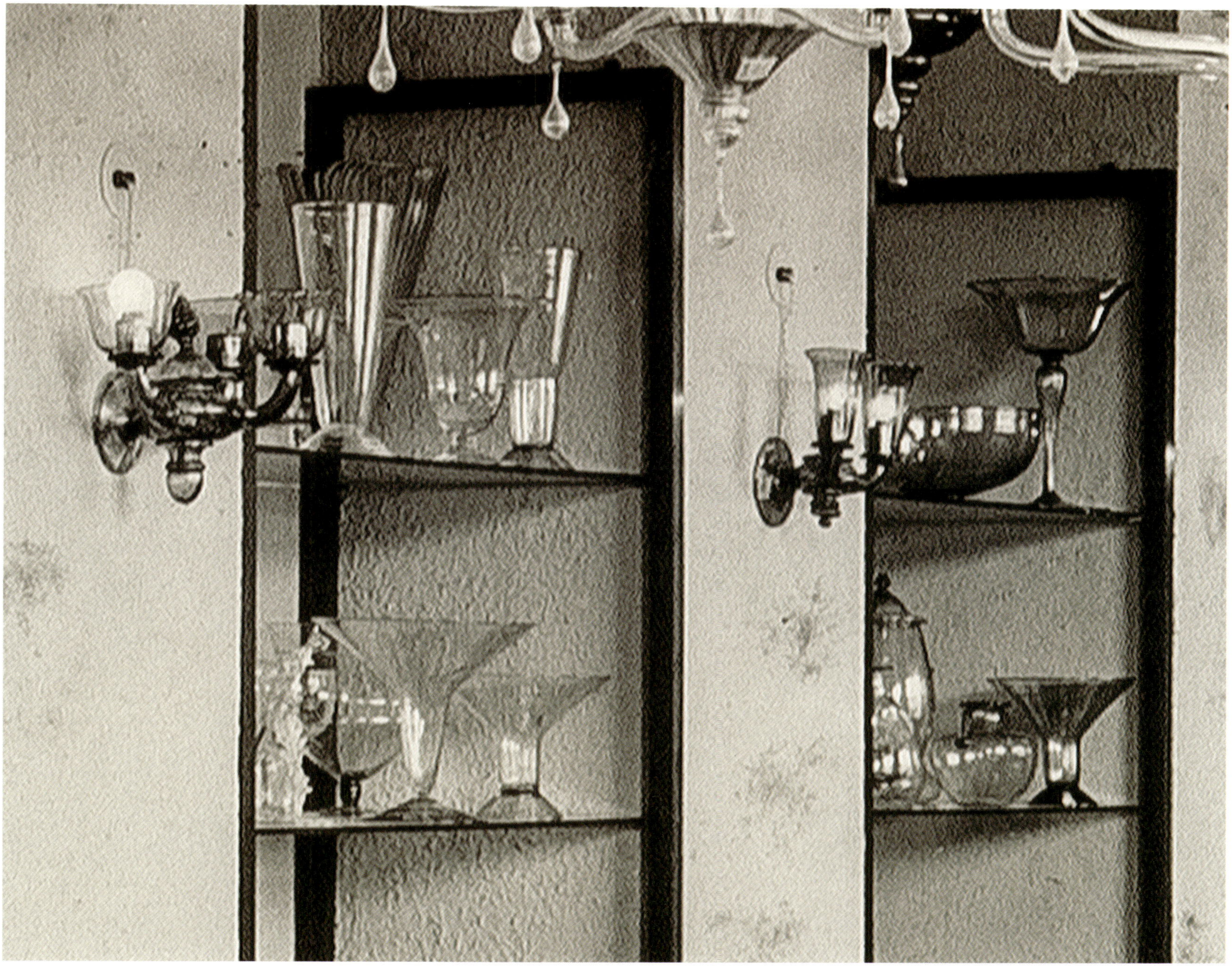

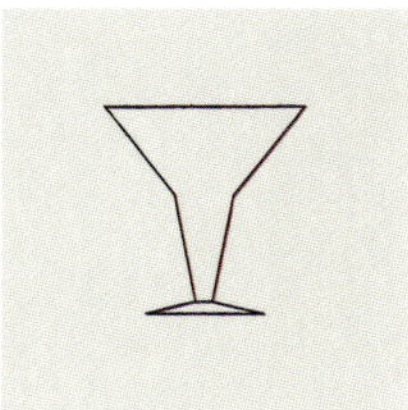

5814

Catalogue drawing for a vase in transparent glass with low truncated-cone-shaped foot and very flared neck, AMB.
c. 1928
24 cm high

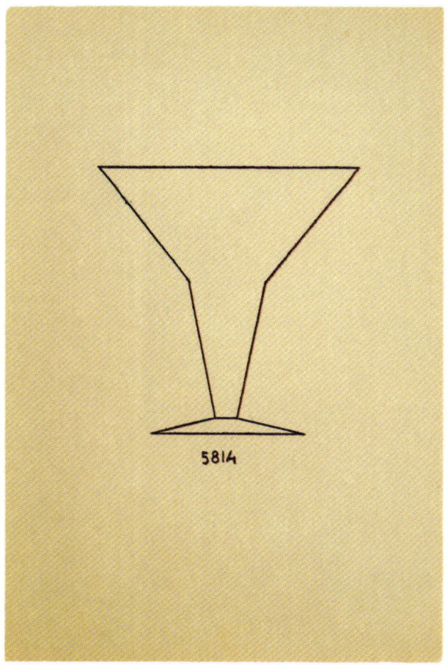

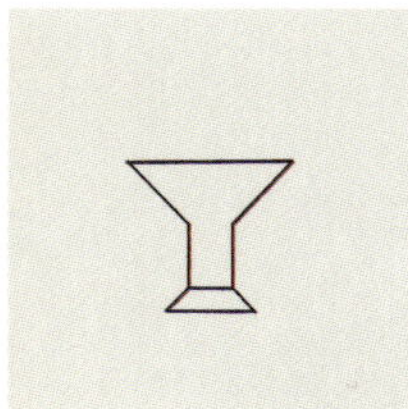

5815

Catalogue drawing for a vase in transparent glass with truncated-cone-shaped foot and very flared neck, AMB.
c 1928
17.5 cm high

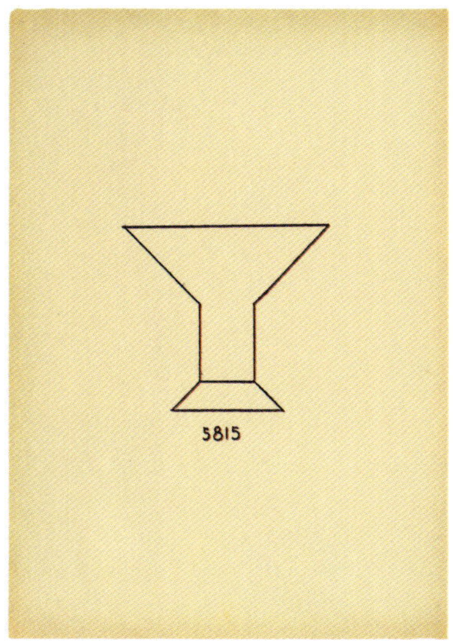

>
Transparent glass pieces in shades of blue at the 'Mostra di Cappellin e Rizzarda' 93rd Esposizione di belle arti della Società amatori e cultori di belle arti, Rome, 1927, CSV

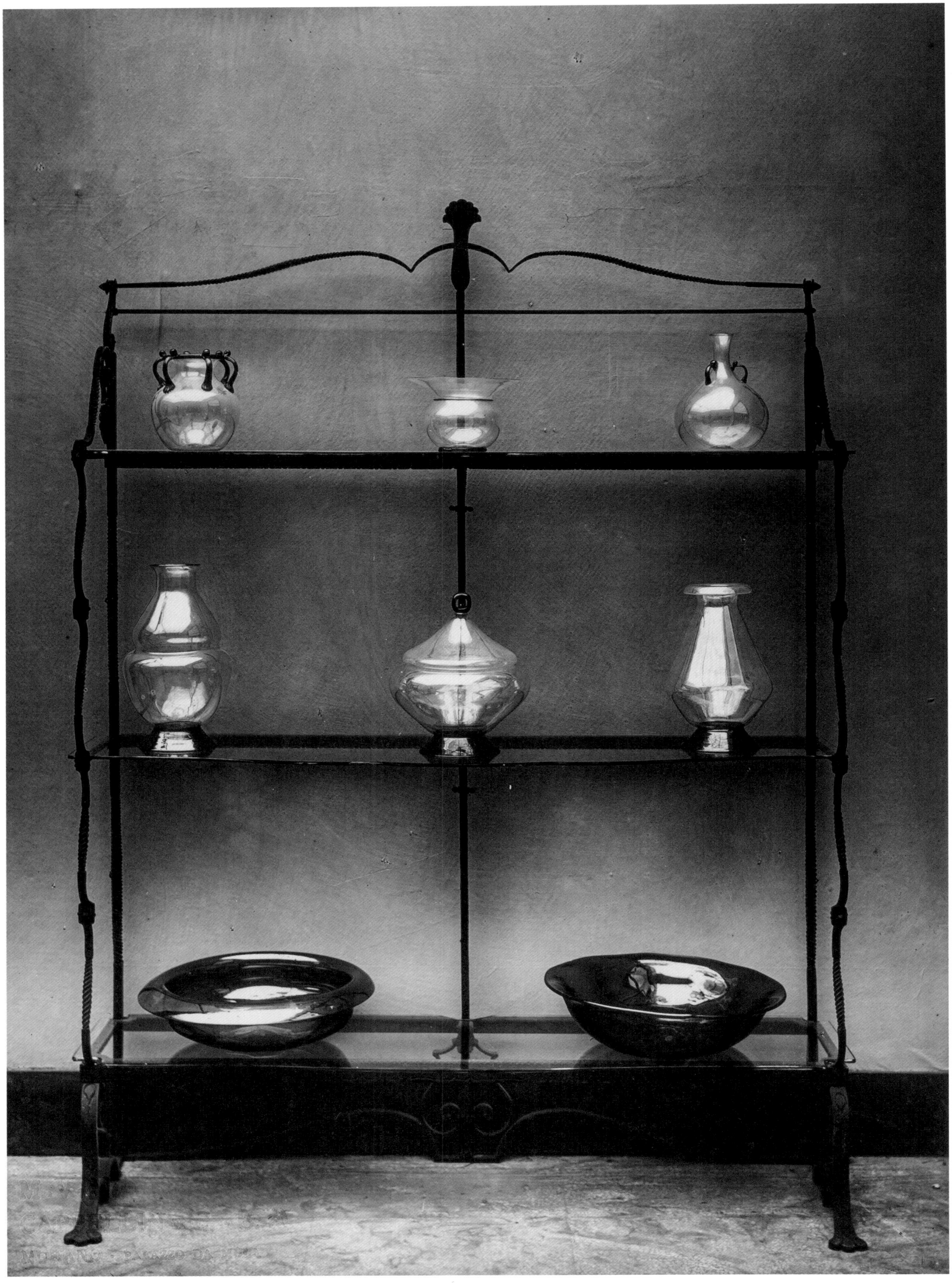

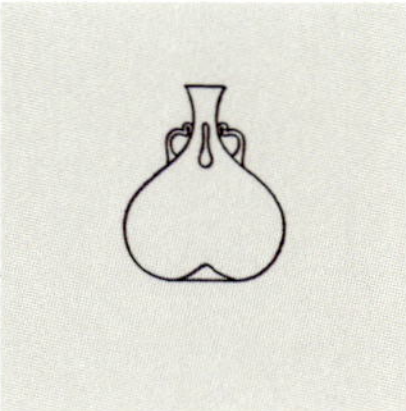

5677

Vase in crystal glass decorated on the neck with '3 handles' in coral-red and gold *pasta vitrea*. Strongly iridised surface. A similar example was shown in 1926 at the Salon d'automne of Paris and the following year at the 93rd Esposizione di belle arti of Rome. On the latter occasion the vase was exhibited in 'iridised light-blue dark-blue dec[oration]'.
1926
23 cm high
Arch. Luigi Cappellin collection, Venice

exhibitions
1926, Paris, Salon d'automne
1927, Rome, 93rd Esposizione di belle arti

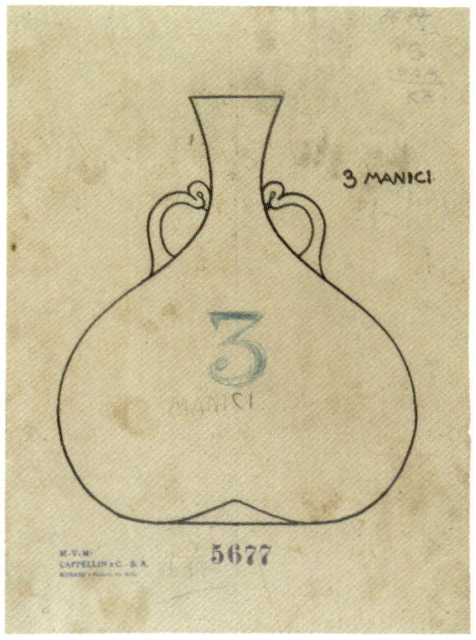

›
5677, working drawing, CSV

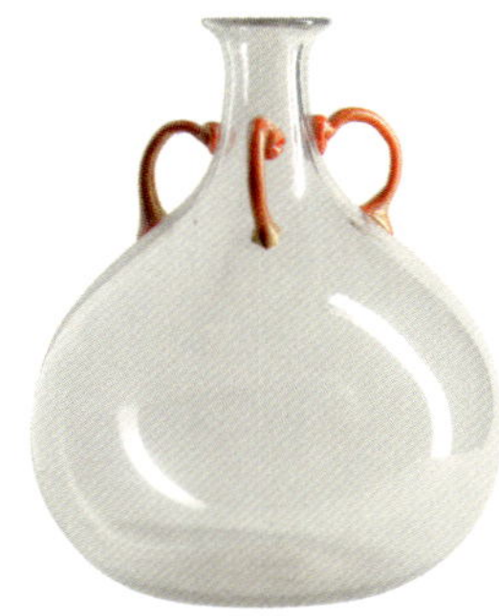

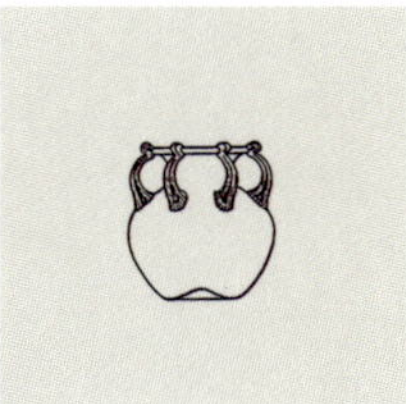

5667

Catalogue drawing for one of the vases presented in 1926 at the Salon d'automne of Paris and the following year at the 93rd Esposizione di belle arti of Rome, AMB. On the latter occasion the vase was exhibited in 'iridised light-blue dark-blue dec[oration]'.
1926
18.5 cm high

exhibitions
1926, Paris, Salon d'automne
1927, Rome, 93rd Esposizione di belle arti

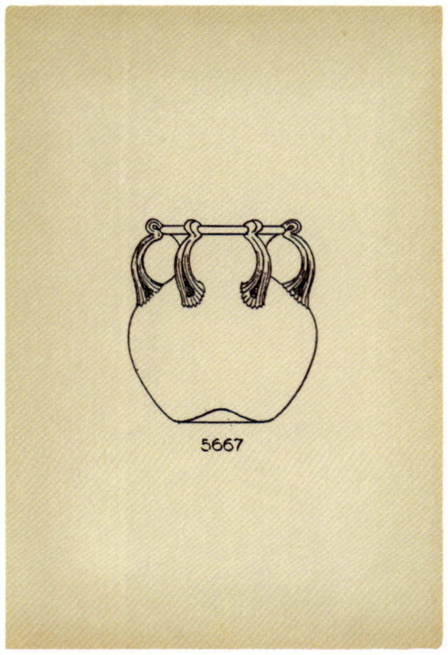

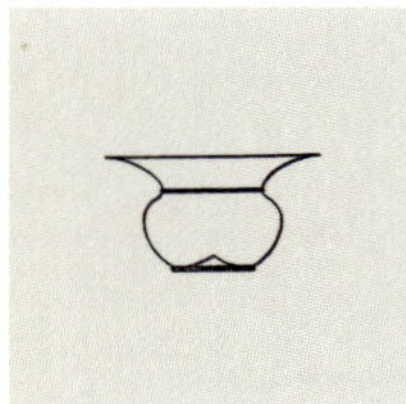

5670

Catalogue drawing for one of the vases displayed in 1926 at the Salon d'automne of Paris and the following year at the 93rd Esposizione di belle arti of Rome in 1927, AMB. On the latter occasion the vase was exhibited in 'iridised light-blue dark-blue dec[oration]'. The model was also produced in filigree.
1926
13.5 cm high

exhibitions
1926, Paris, Salon d'automne
1927, Rome, 93rd Esposizione di belle arti

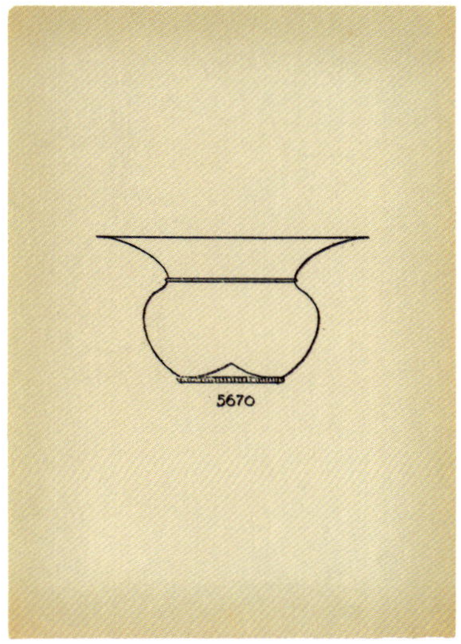

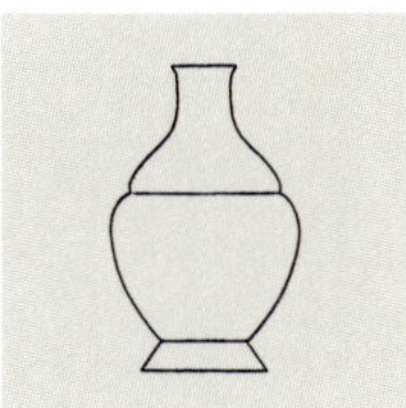

5685

Working drawing for one of the vases with truncated-cone-shaped foot presented in 1927 at the 93rd Esposizione di belle arti of Rome, CSV. On that occasion the vase was exhibited in 'iridised crystal dark-blue dec[oration]'.
1927
36 cm high

exhibitions
1927, Rome, 93rd Esposizione di belle arti

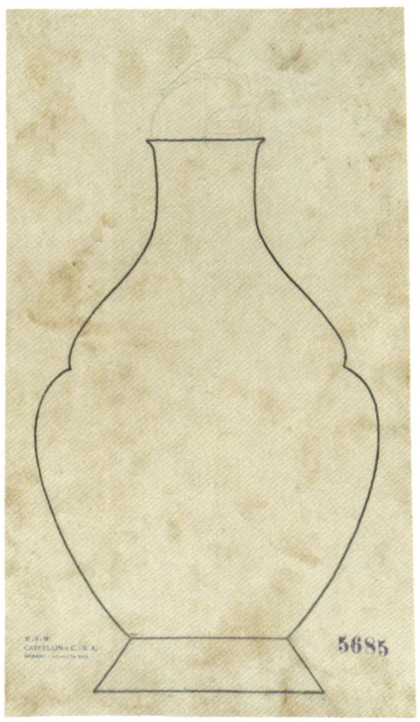

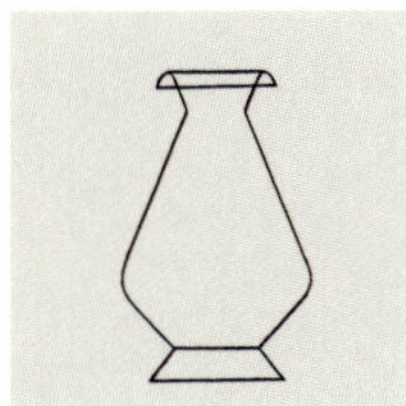

5690

Working drawing for one of the vases with truncated-cone-shaped foot presented in 1927 at the 93rd Esposizione di belle arti of Rome, CSV. On that occasion the vase was exhibited in 'iridised crystal dark-blue dec[oration]'.
1927
36 cm high

exhibitions
1927, Rome, 93rd Esposizione di belle arti

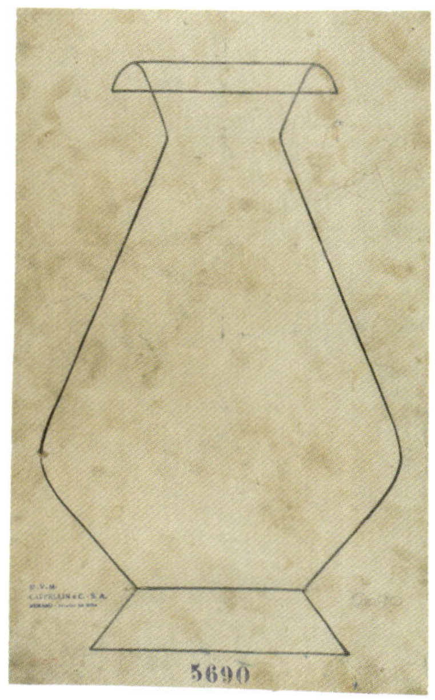

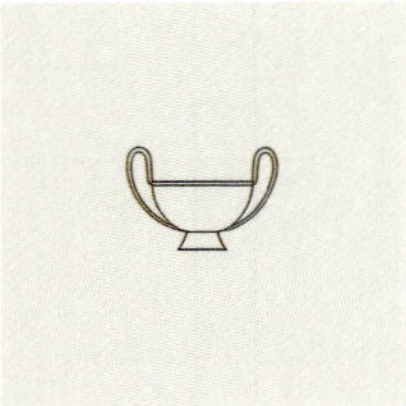

5693

Bowl with truncated-cone-shaped foot and side handles in amethyst transparent glass. Surface with iridised finish. A similar example was presented in 1927 at the 93rd Esposizione di belle arti of Rome. On that occasion the bowl was exhibited in 'pagliesco glass dec[oration] [handles and finishings] in green'. The design of the bowl refers back to that of the Etruscan *kantharoi*.
1927
14 cm high; 23 cm ø
M.V.M. Cappellin Murano *acid stamp*
Arch. Luigi Cappellin collection, Venice

exhibitions
1927, Rome, 93rd Esposizione di belle arti

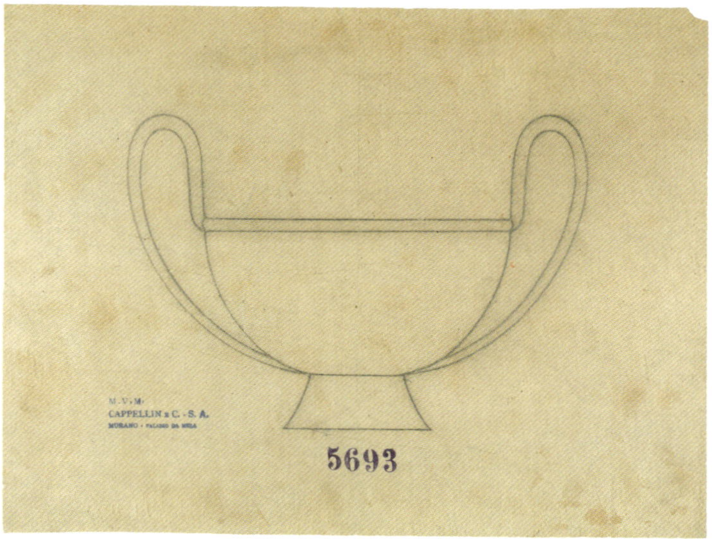

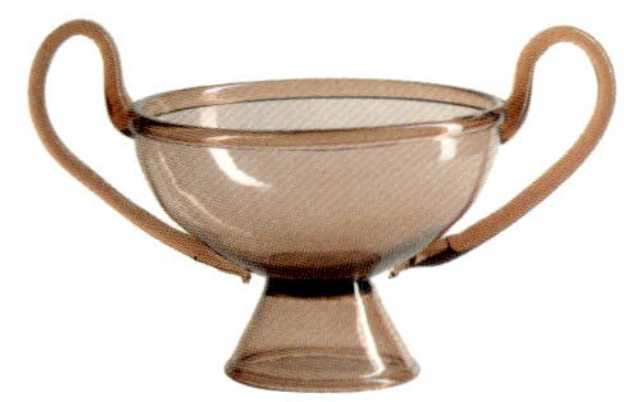

>
5693, working drawing, CSV

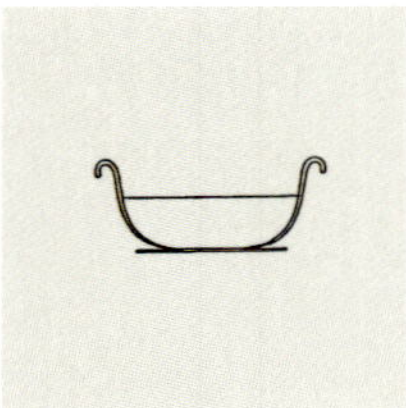

5695

Working drawing for one of the bowls presented in 1927 at the 93rd Esposizione di belle arti of Rome, CSV. On that occasion the bowl was exhibited in 'pagliesco glass dec[oration] [handles and finishings] in green'.
1927
11 cm high; 27 cm ø

exhibitions
1927, Rome, 93rd Esposizione di belle arti

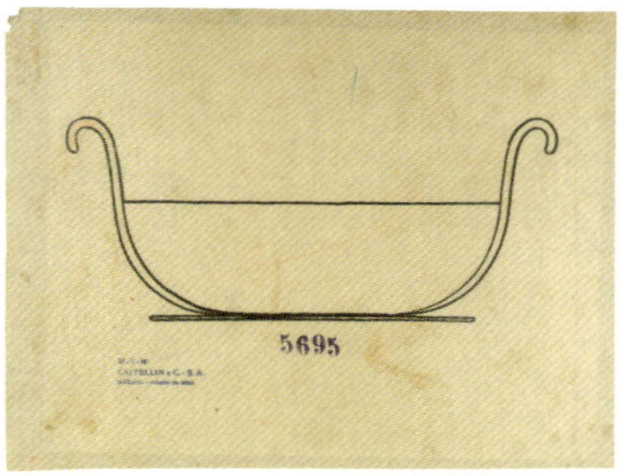

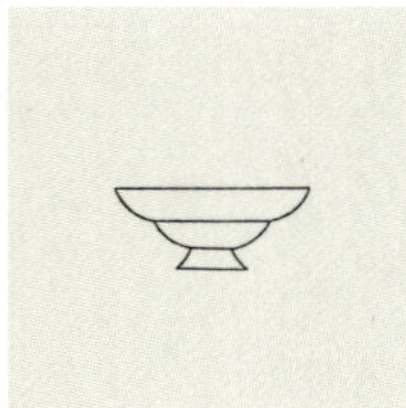

5692

Catalogue drawing for one of the bowls presented in 1927 at the 93rd Esposizione di belle arti of Rome, AMB. On that occasion the bowl, with truncated-cone-shaped foot produced no. 5692 'speciale' (c. h 15 cm), was exhibited in 'pagliesco' glass with 'green' foot.
1927
9 cm high

exhibitions
1927, Rome, 93rd Esposizione di belle arti

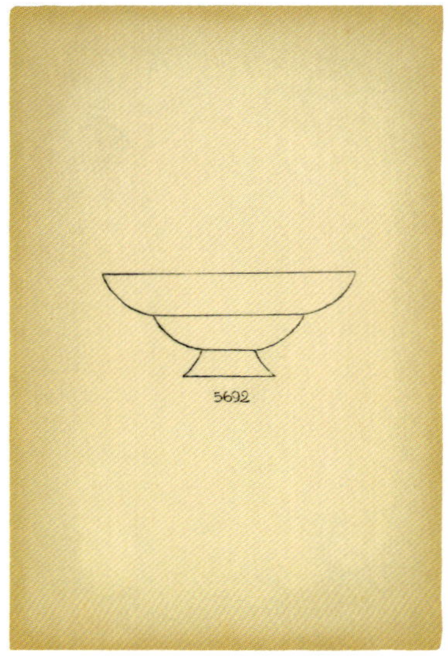

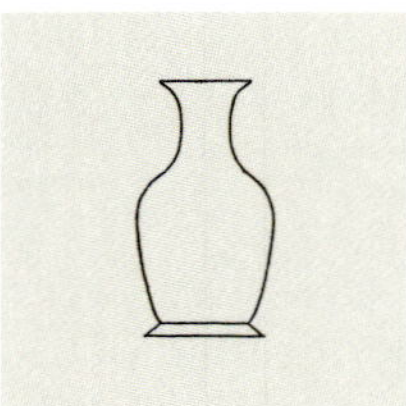

5687

Working drawing for one of the vases with truncated-cone-shaped foot, presented in 1927 at the 93rd Esposizione di belle arti of Rome, CSV. On that occasion the vase was exhibited in '*pagliesco*' with 'green' foot.
1927
30 cm high

exhibitions
1927, Rome, 93rd Esposizione di belle arti

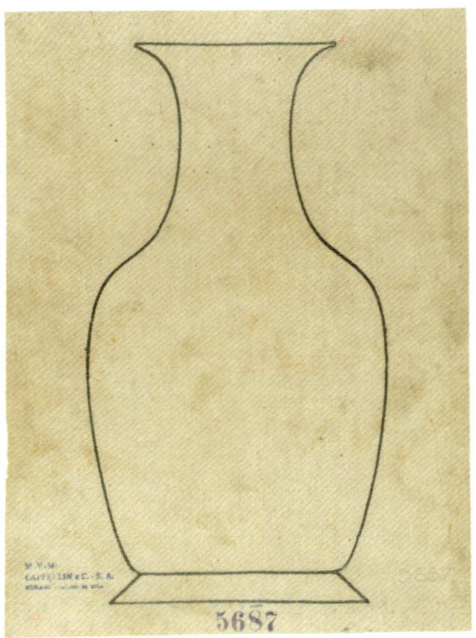

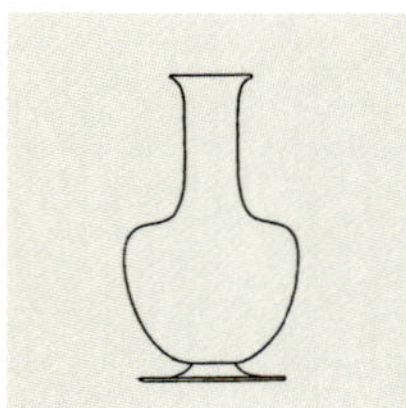

5689

Catalogue drawing for a bottle presented in 1927 at the 93rd Esposizione di belle arti of Rome, AMB. On that occasion the bottle, with a low flared foot, was exhibited in 'pagliesco' glass, with foot and thread around the mouth in 'green'.
1927
36 cm high

exhibitions
1927, Rome, 93rd Esposizione di belle arti

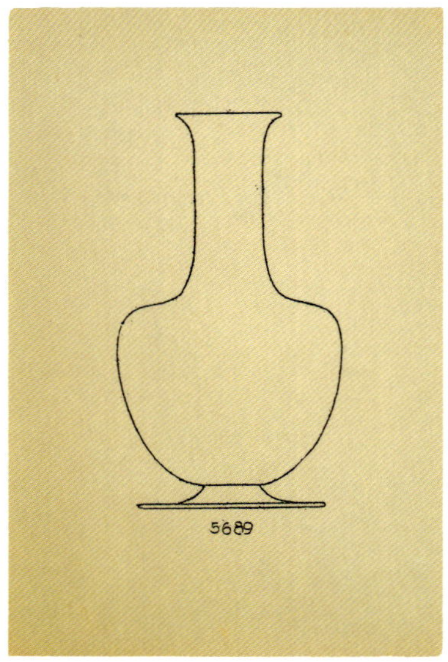

>
Transparent yellow and green glass pieces at the 'Mostra di Cappellin e Rizzarda', 93rd Esposizione di belle arti della Società amatori e cultori di belle arti, Rome 1927, CSV

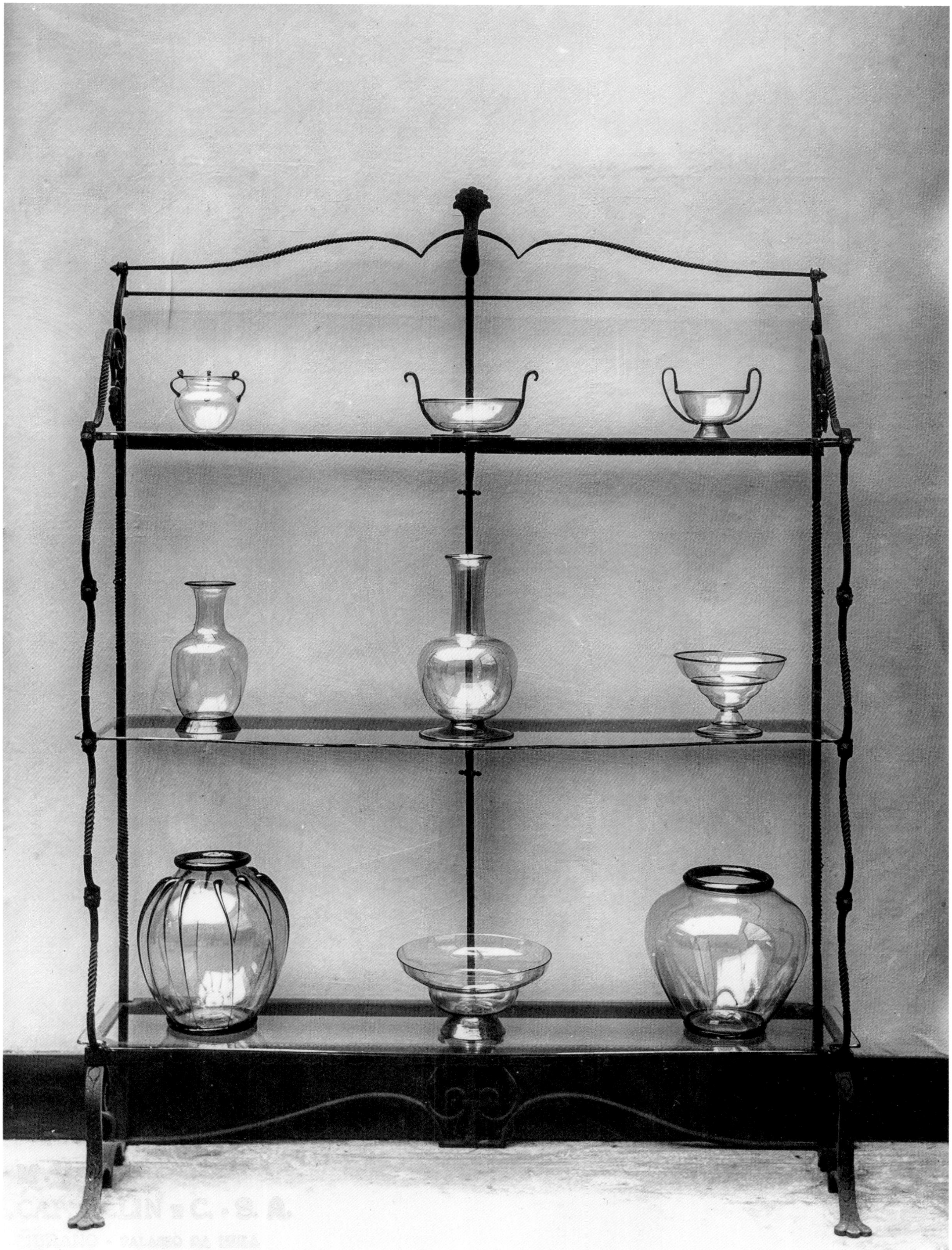

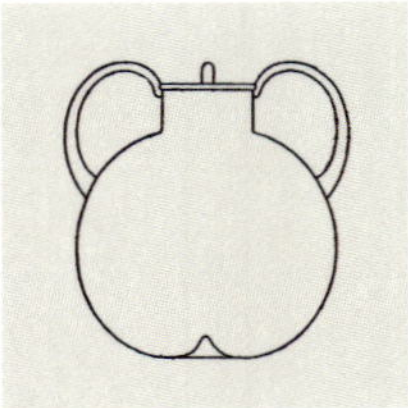

5757

Vase in yellow transparent glass decorated with incorporated bubbles, with three applied handles and thread around the mouth in green transparent glass. Strongly iridised surface. A similar example in the same colour was exhibited in 1927 at the 3rd Biennale of Monza.
1927
36 cm high
M.V.M. Cappellin Murano *acid stamp*
Private collection, Venice

exhibitions
1927, Monza, 3rd Biennale

>
5757, catalogue drawing, AMB

>
'Venezianer gläser der fa. Cappellin', in Das schöne Heim, *November 1930, p. 50*

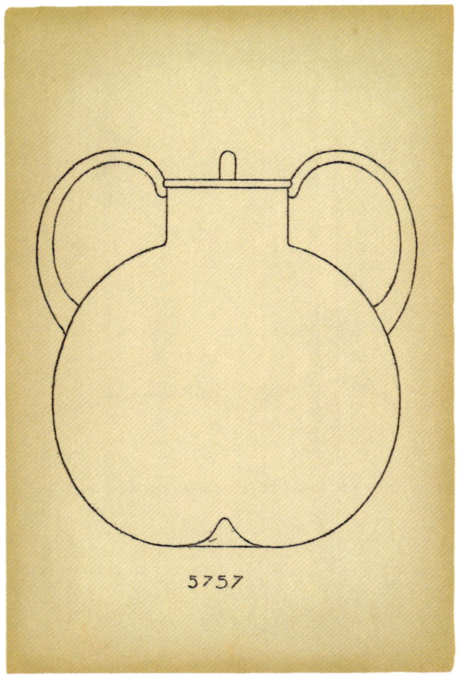

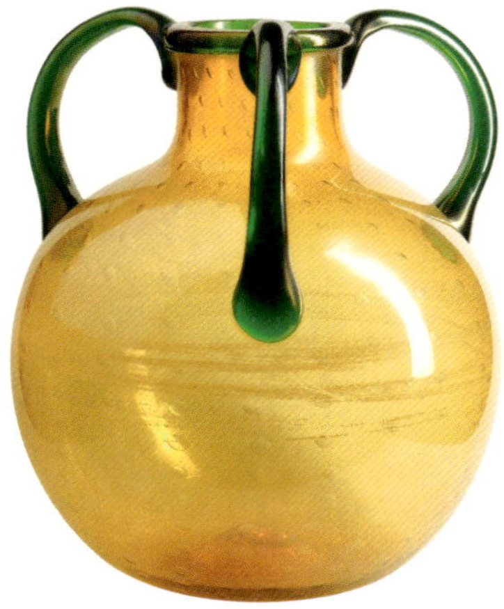

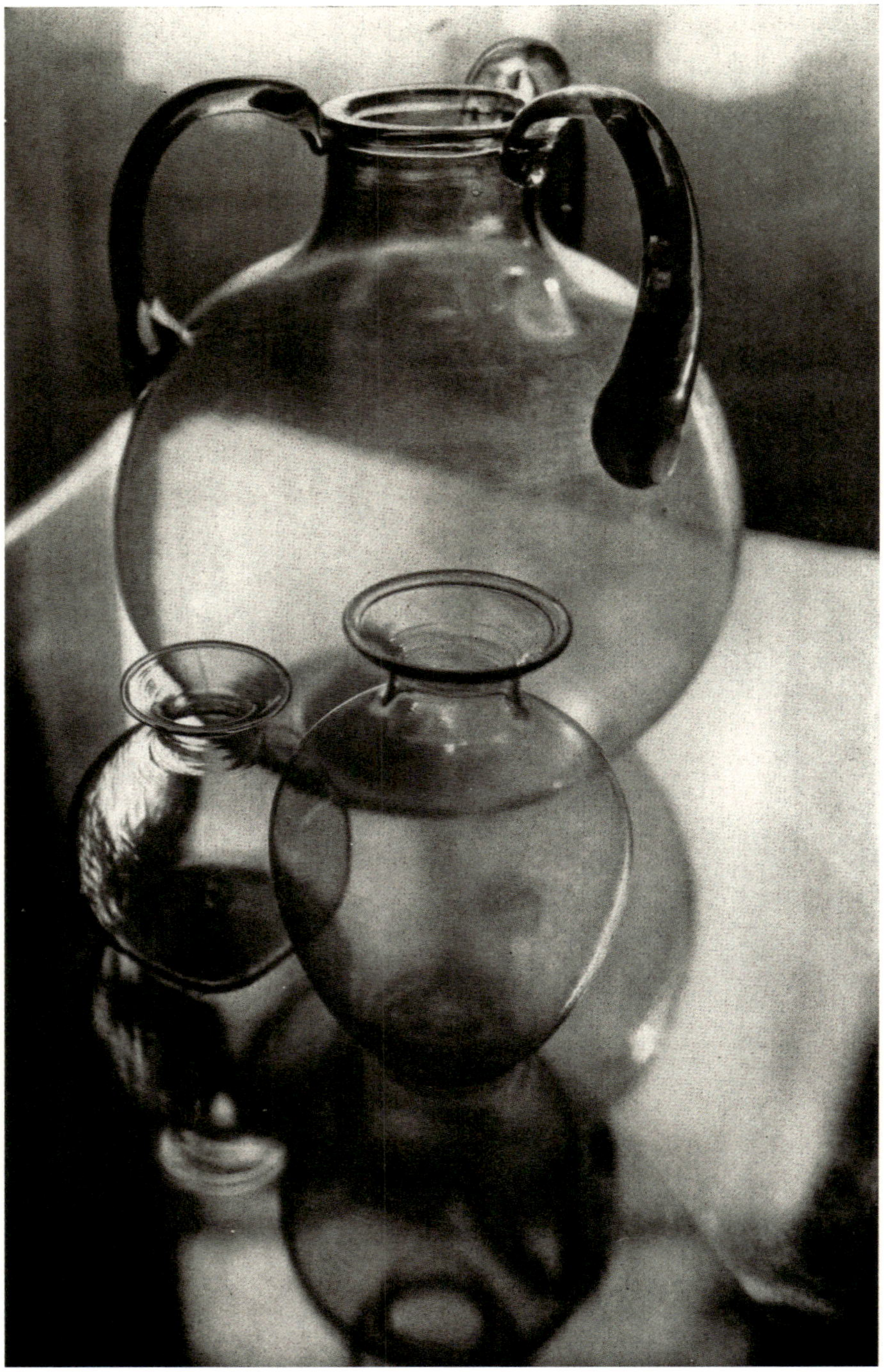

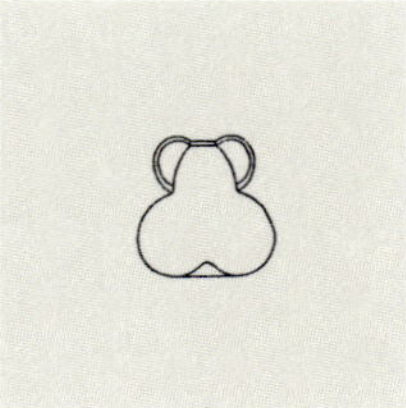

5741

Vase, slightly ribbed, in *pagliesco* transparent glass with side handles and thread around the mouth in green glass. Surface with iridised finish.
1927
18 cm high
Elizabeth Kessenides collection, New York

>
5741, working drawing, CSV

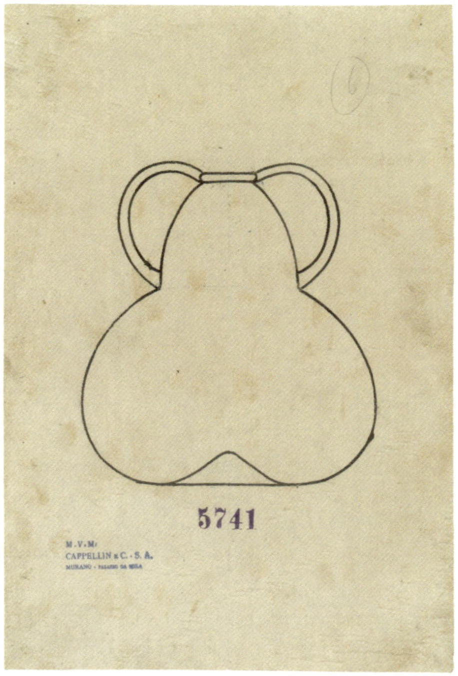

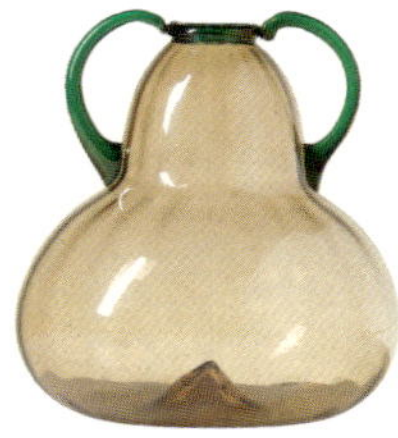

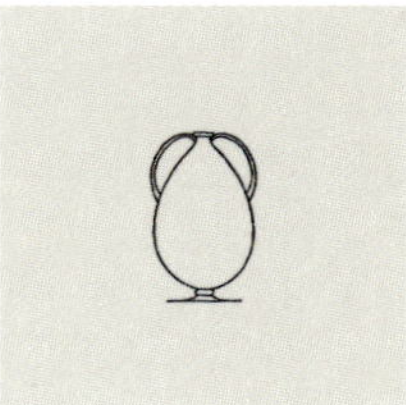

5743

Vase in amethyst transparent glass with disk-shaped foot, flattened knot and applied lateral handles. Surface with iridised finish.
1927
20 cm high
Private collection, Feltre

>
5743, working drawing, CSV

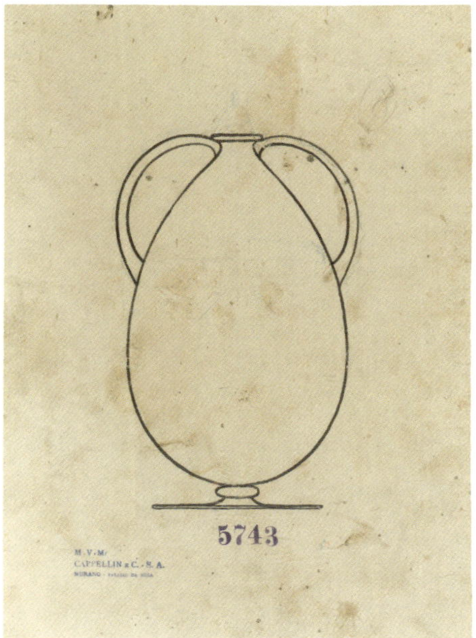

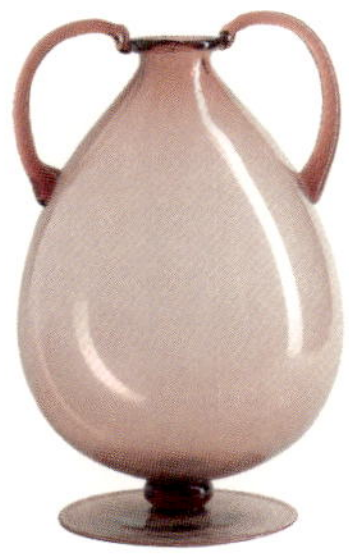

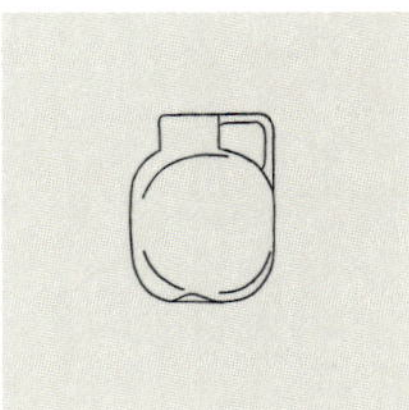

5878

Jug, body with triangular cross-section, in amethyst transparent glass. Thread around the mouth and squared lateral handle with veining, in ruby transparent glass. Surface with iridised finish.
c. 1928
42.5 cm high
M.V.M. Cappellin Murano *acid stamp*
Private collection

>
5878, catalogue drawing, AMB

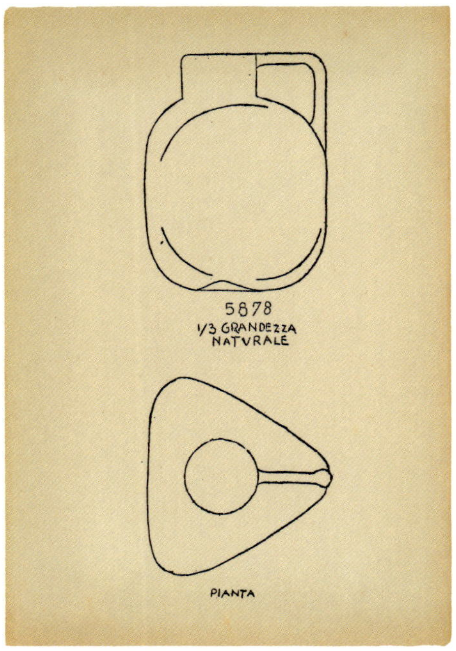

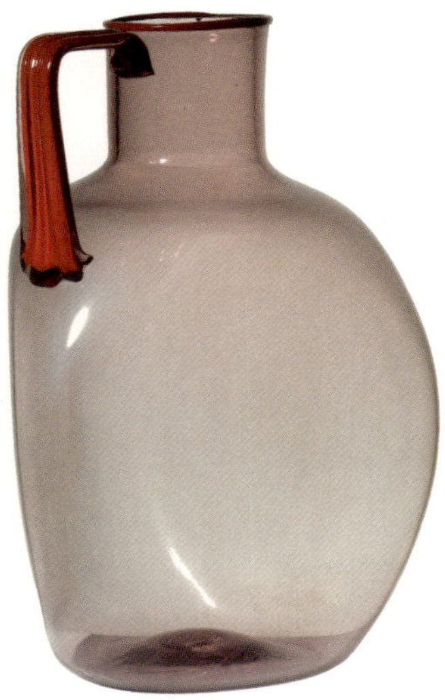

>
Cappellin glass pieces at the 18th Mostra dell'Opera Bevilacqua La Masa, Venice, 1927, CSV

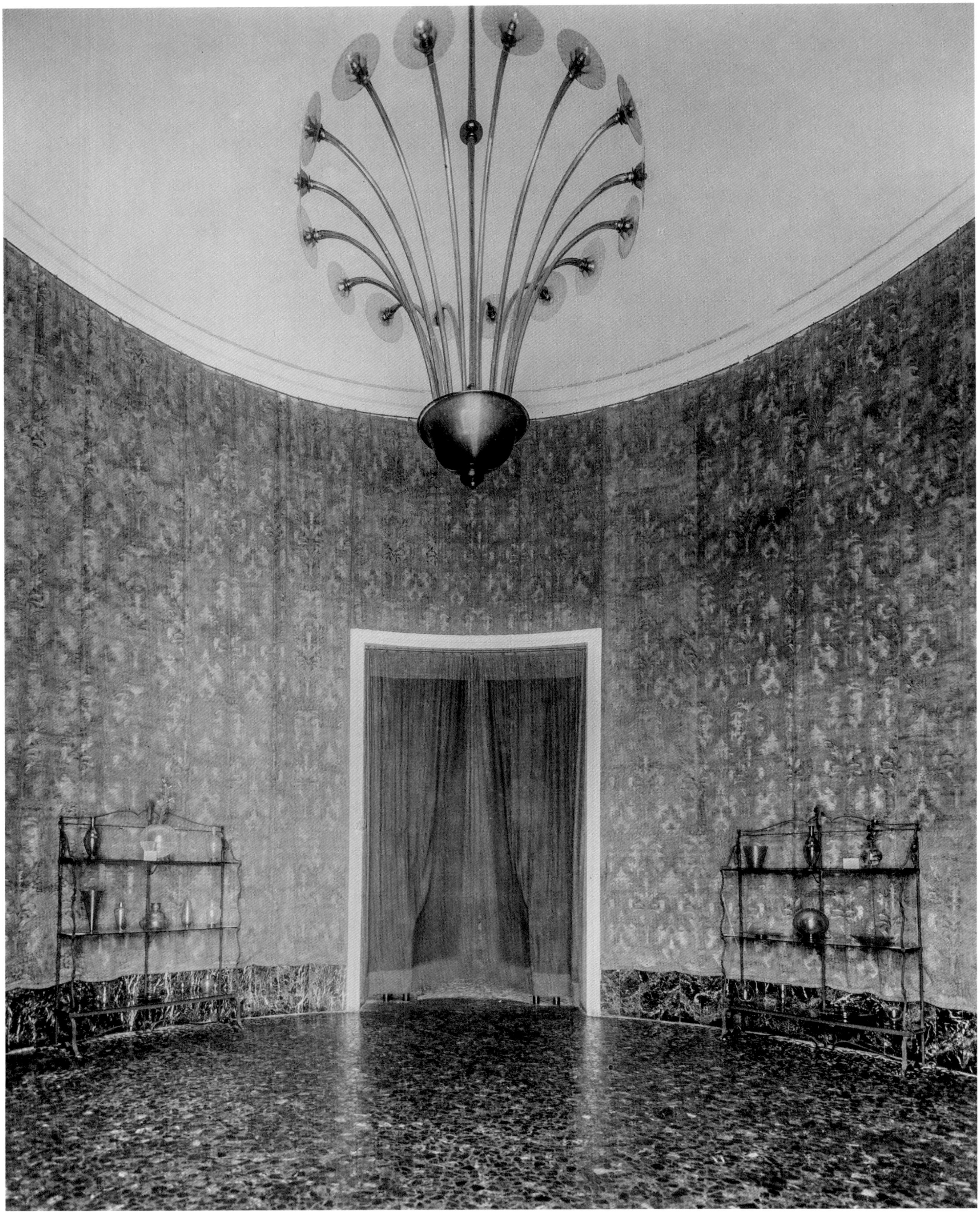

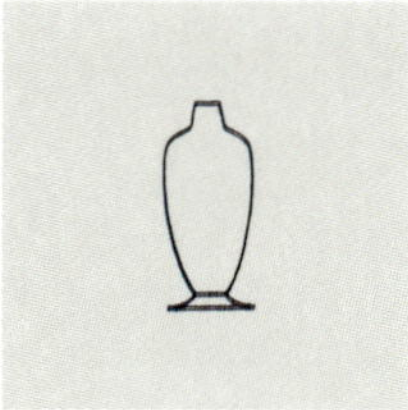

5732

Vase in Savoy-blue transparent glass with flared foot. Surface with iridised finish. A similar example in 'iridised purple' glass was exhibited in 1927, in Venice, at the 18th Mostra dell'Opera Bevilacqua La Masa, while a version in 'green glass [with] coral-red rim' was shown, in the same year, at the 3rd Biennale of Monza.
1927
25 cm high
M.V.M. Cappellin Murano *acid stamp*
Private collection

exhibitions
1927, Monza, 3rd Biennale
1927, Venice, 18th Mostra dell'Opera Bevilacqua La Masa
1964, Venice, 23rd Biennale

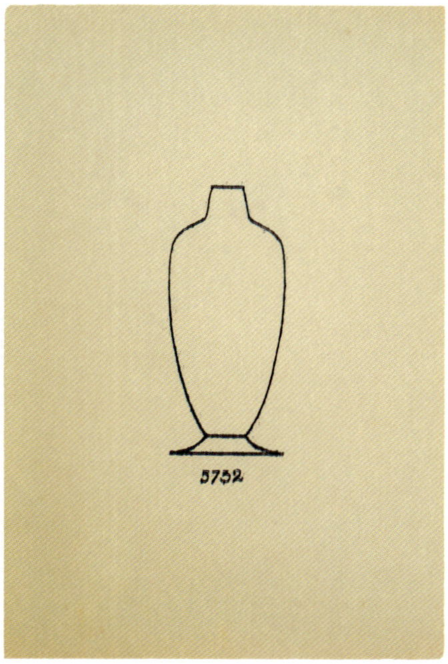

>
5732, catalogue drawing, AMB

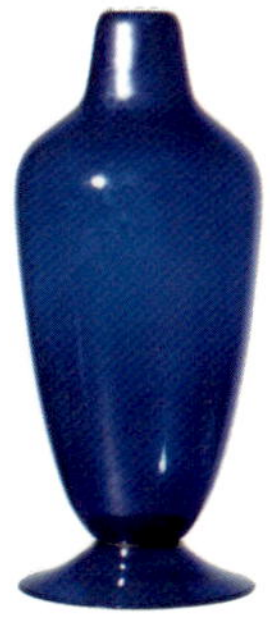

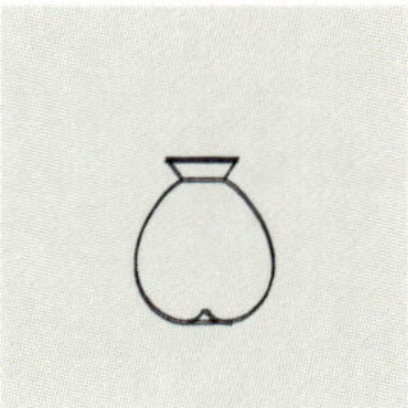

5761

Pear-shaped vase with flared neck in amethyst transparent glass. Strongly iridised finish. A similar example in 'green glass [with] coral-red thread' was exhibited in 1927, in Venice, at the 18th Mostra dell'Opera Bevilacqua La Masa.
1927
19,5 cm high
M.V.M. Cappellin Murano *acid stamp*
Rob Beyer collection, New York

exhibitions
1927, Venice, 18th Mostra dell'Opera Bevilacqua La Masa

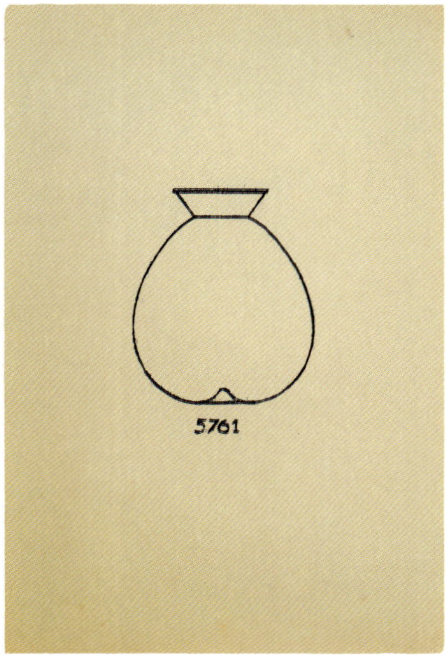

>
5761, catalogue drawing, AMB

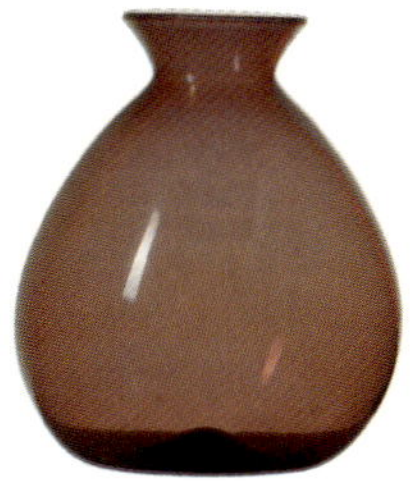

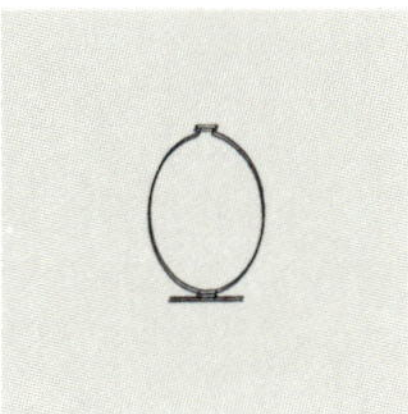

5726

Ovoid vase in purple transparent glass with disk-shaped foot. Strongly iridised finish. A similar example in 'green with coral-red rim' was exhibited in 1927 at the 3rd Biennale of Monza.
1927
19,5 cm high
M.V.M. Cappellin Murano *acid stamp*
The Lockwood collection, Fort Lee, NJ

exhibitions
1927, Monza, 3rd Biennale

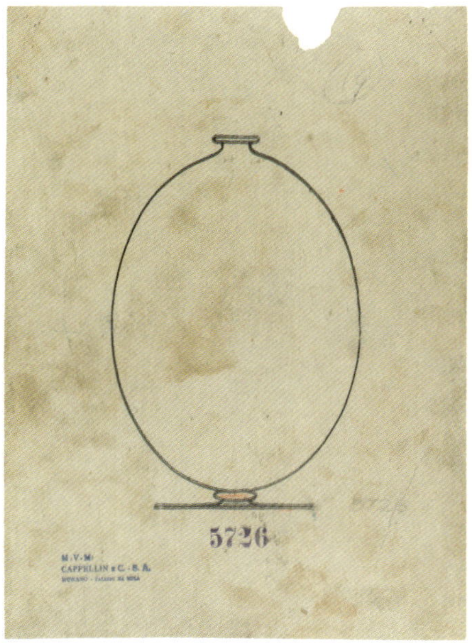

>
5726, working drawing, CSV

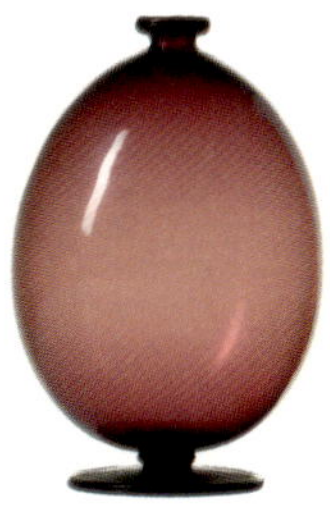

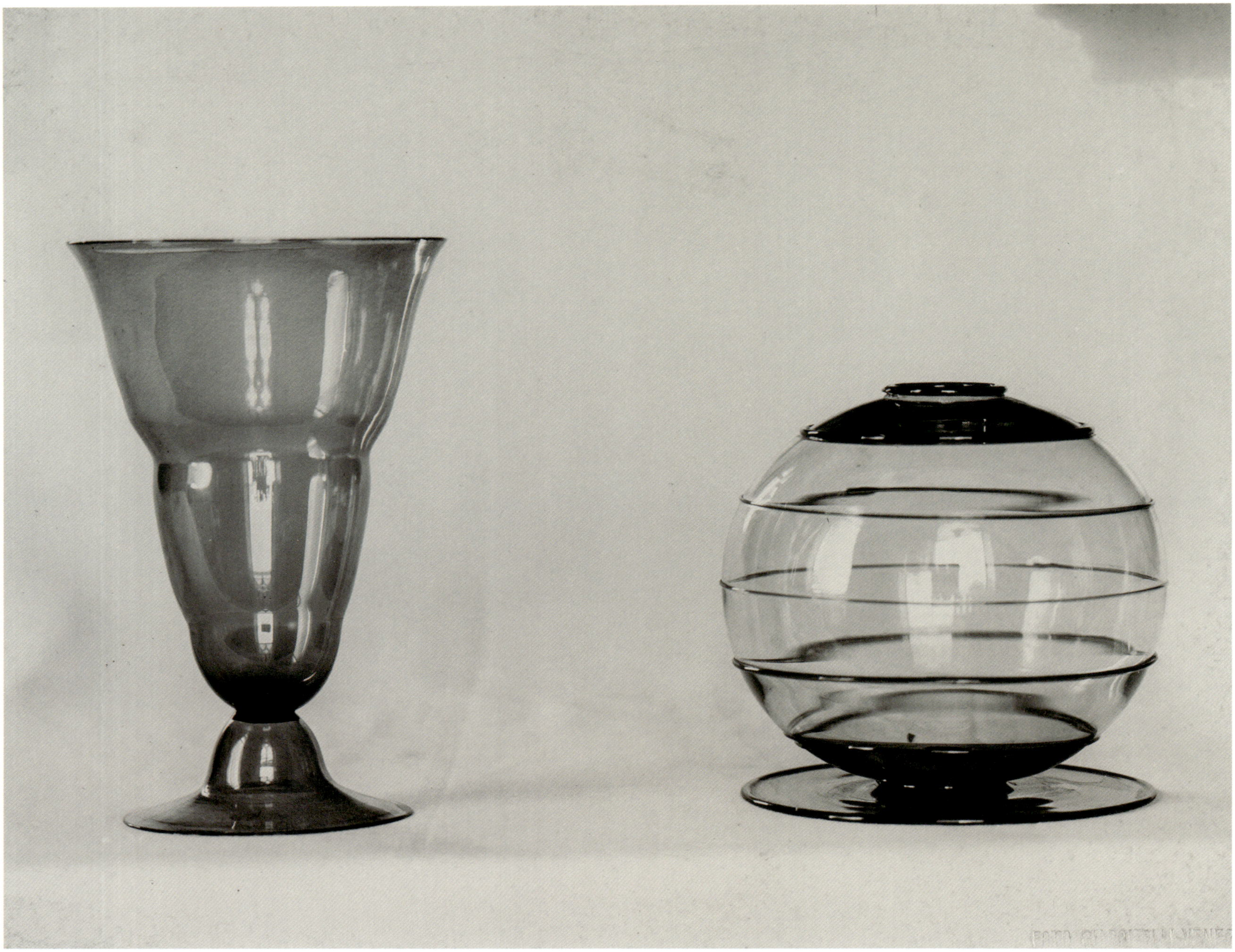

>
Vases no. 5735 and no. 5822 in transparent glass. Period photograph, c. 1928, CSV

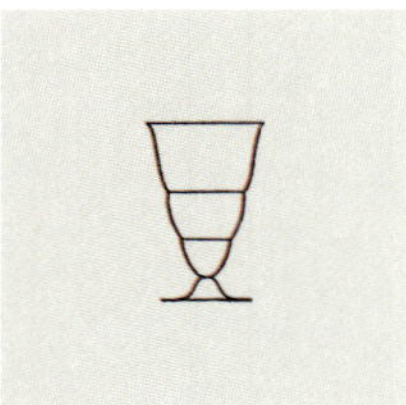

5735

Vase with bell-shaped foot in amethyst transparent glass with three slight constrictions on the body. Surface with iridised finish. The piece was produced in a variant no. 5735 with three constrictions (h 21 cm). An example of no. 5735 in 'green glass [with] coral-red rim' was exhibited in 1927 in Venice, at the 18th Mostra dell'Opera Bevilacqua La Masa, and in Monza, at the 3rd Biennale.
1927
28 cm high
Private collection, Treviso

exhibitions
1927, Monza, 3rd Biennale
1927, Venice, 18th Mostra dell'Opera Bevilacqua La Masa

>
5735, working drawing, CSV

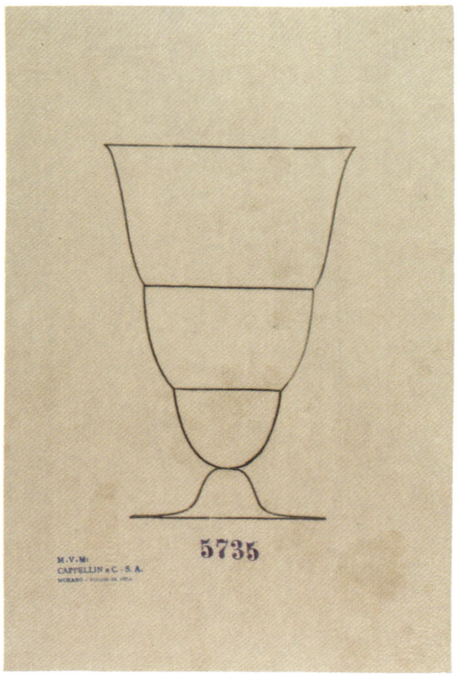

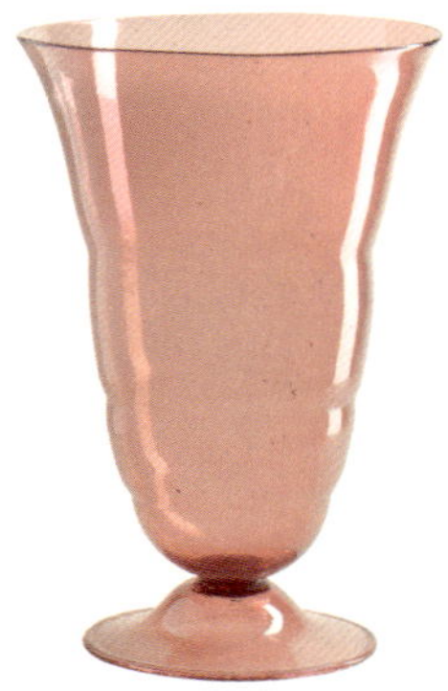

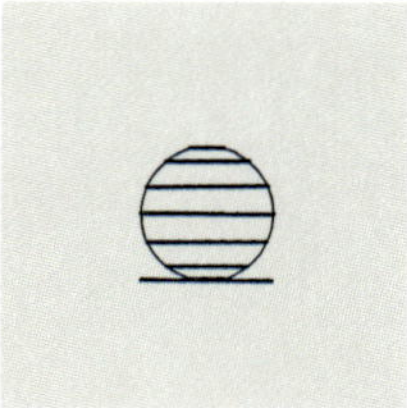

5822

Spherical vase with disk-shaped foot in transparent yellow glass, decorated on the body with five horizontal threads. Surface with iridised finish.
c. 1928
16 cm high
M.V.M. Cappellin Murano *acid stamp*
Private collection

>
5822, catalogue drawing, AMB

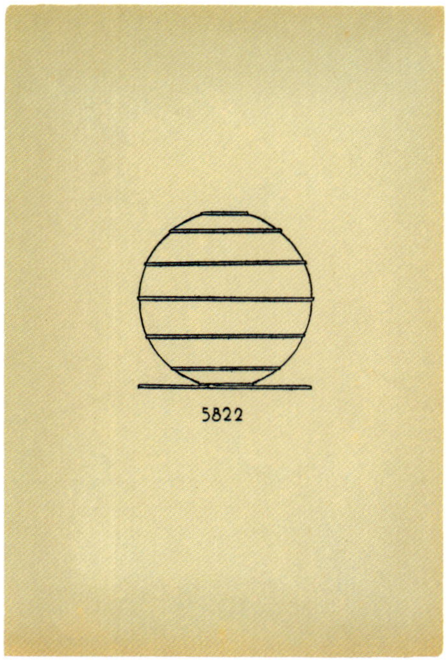

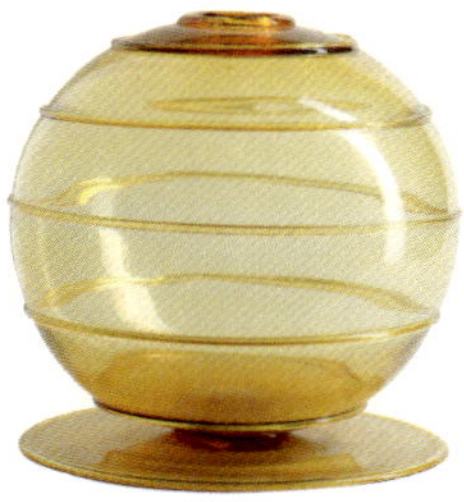

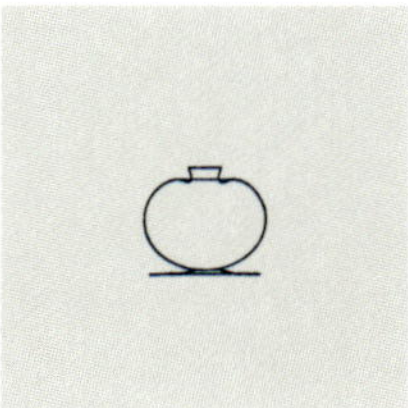

5733

Vase in light-green transparent glass with flattened globular body and disk-shaped foot. Thread around the mouth in *lattimo* glass. Surface with iridised finish. A similar example in 'green glass [with] coral-red rim' was exhibited in 1927 at the 3rd Biennale of Monza.
1927
12 cm high
M.V.M. Cappellin Murano *acid stamp*
The Lockwood collection, Fort Lee, NJ

exhibitions
1927, Monza, 3nd Biennale

>
5733, working drawing, CSV

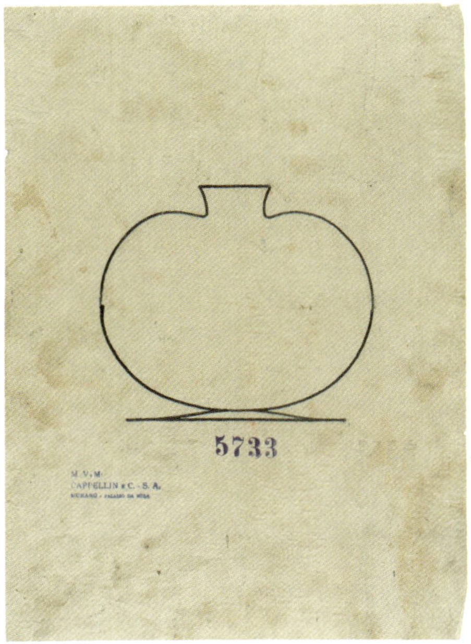

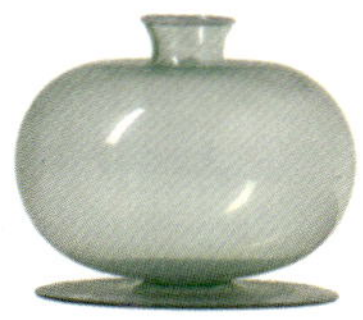

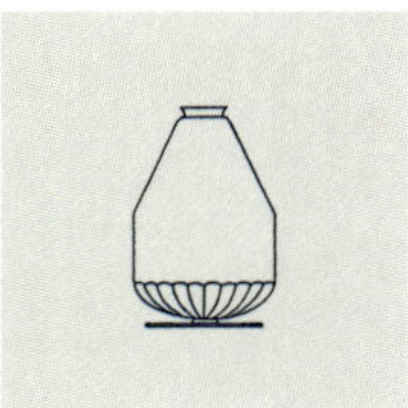

5861

Vase in crystal glass with disk-shaped foot and base decorated with *meza stampaura* (half moulding). Surface with iridised finish.
c. 1928
21 cm high
C.V.M. *acid stamp*
Private collection, Feltre

>
5861, catalogue drawing, AMB

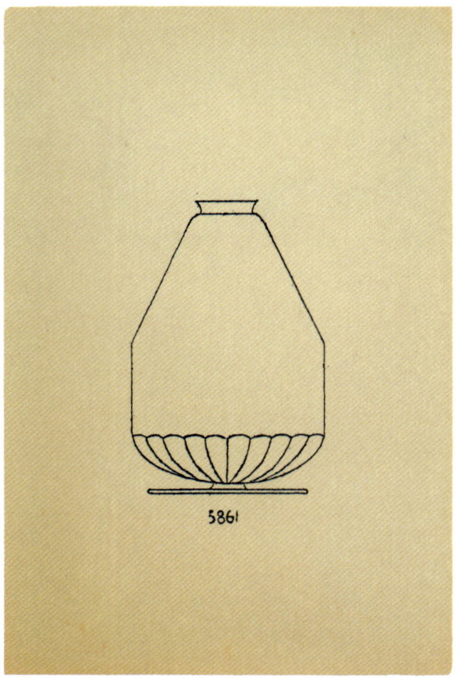

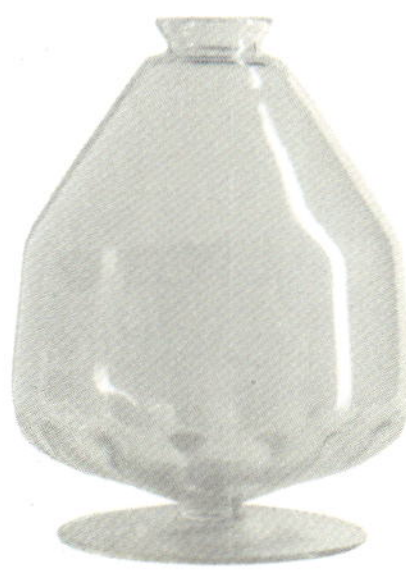

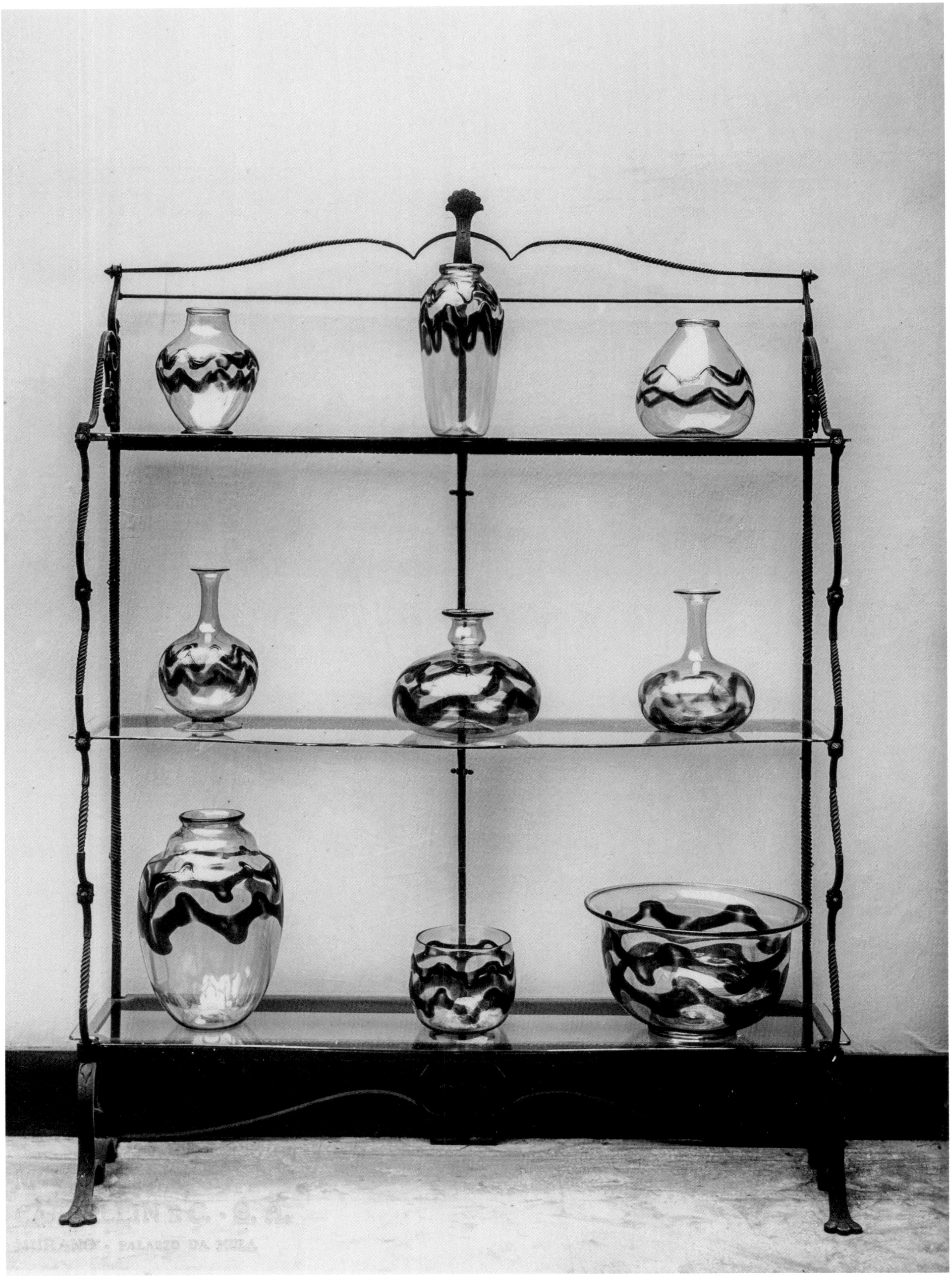

>
Transparent amethyst glass pieces with zig-zag purple bands at the 'Mostra di Cappellin e Rizzarda', 93rd Esposizione di belle arti, Rome 1927, CSV. Most of the pieces exhibited are a revisiting, with zig-zag decoration, of models designed earlier by Vittorio Zecchin (1921–25)

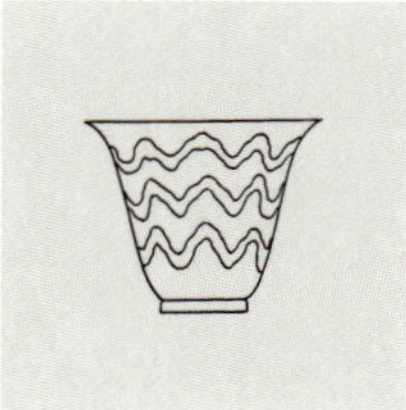

5773

Vase in transparent *pagliesco* glass decorated with three zig-zag bands in purple glass. Ring-shaped foot with wound thread in *pagliesco* glass. Surface with iridised finish.
1927
21 cm high
M.V.M. Cappellin Murano *acid stamp*
Private collection, Treviso

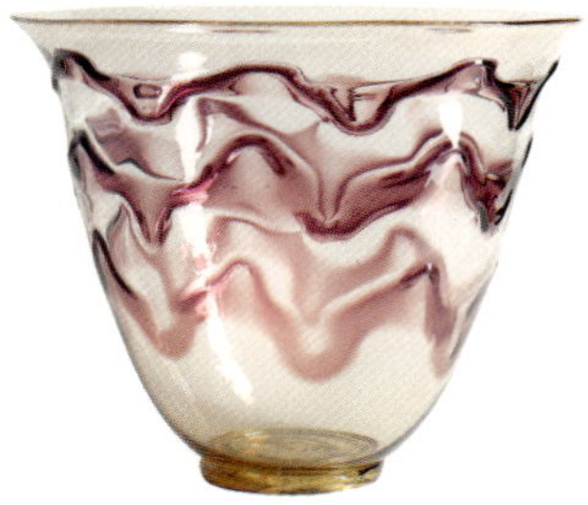

n.i.

Vase with truncated-cone-shaped foot in transparent *pagliesco* glass decorated with two zig-zag bands in coral-red *pasta vitrea*.
c. 1927
15 cm high
C.V.M. *acid stamp*; Venezia Murano old label
Private collection

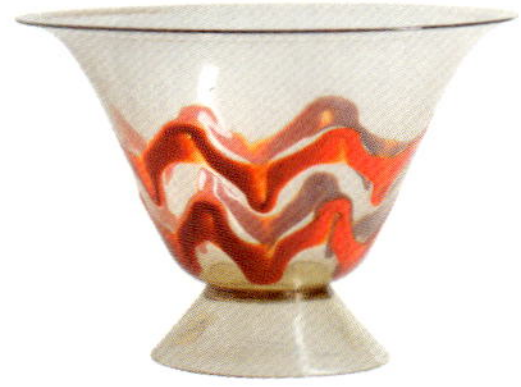

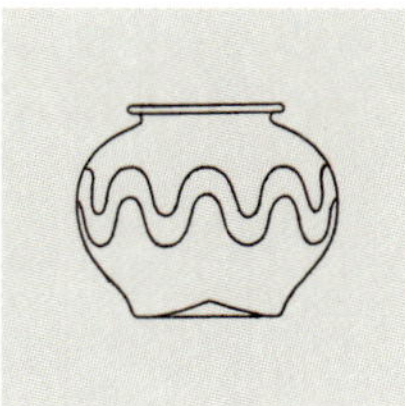

5315

Vase in transparent Savoy-blue glass decorated with a zig-zag band in coral-red *pasta vitrea*. Surface with iridised finish. The piece was produced after a model in the catalogue of the glassworks designed by Vittorio Zecchin (1921–25).
c. 1927
23 cm high
Mughini collection, Rome

>
5315, catalogue drawing, CSV

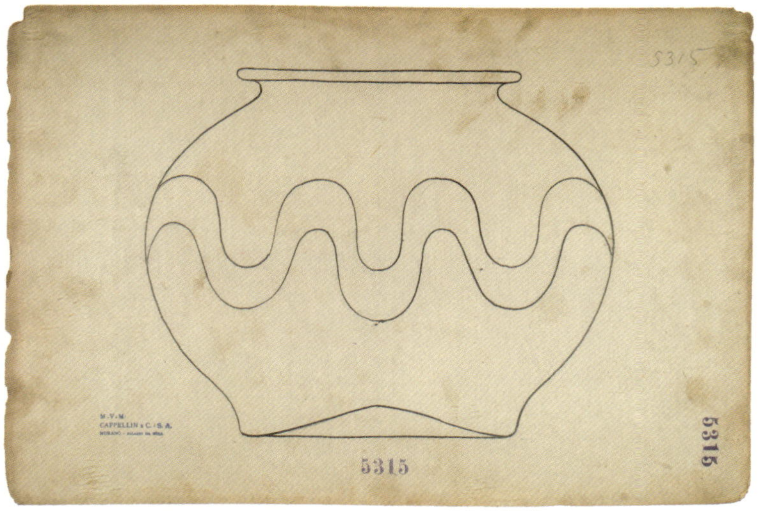

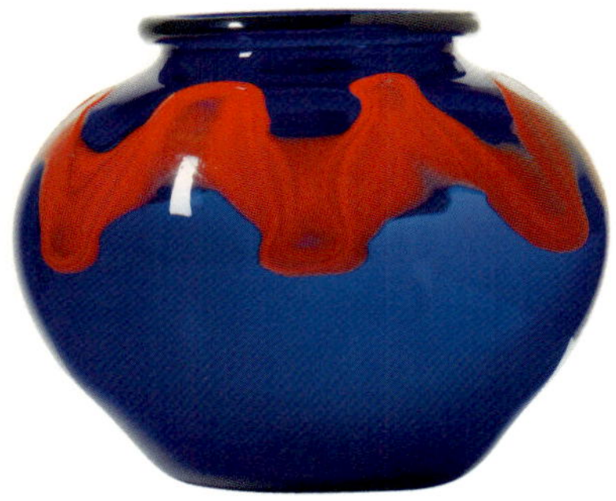

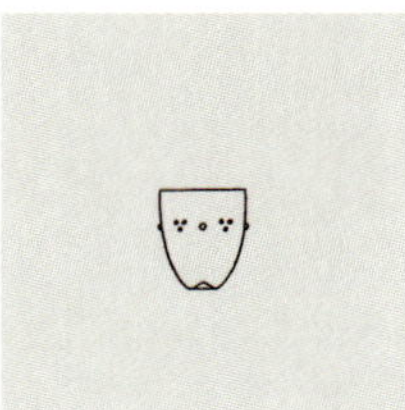

5669

Vase in transparent yellow glass, decorated with a band of yellow pearls applied on the wall of the piece in groups of three, alternating with single pearls. Surface with iridised finish. A similar example was exhibited in 1926 at the Salon d'automne of Paris.
1926
11.5 cm high
M.V.M. Cappellin Murano *acid stamp*
Stiftung Museum Kunstpalast collection, Düsseldorf-Glasmuseum Hentrich

exhibitions
1926, Paris, Salon d'automne

>
Page of notebook with the models exhibited at the Salon d'automne, Paris, 1926, CSV

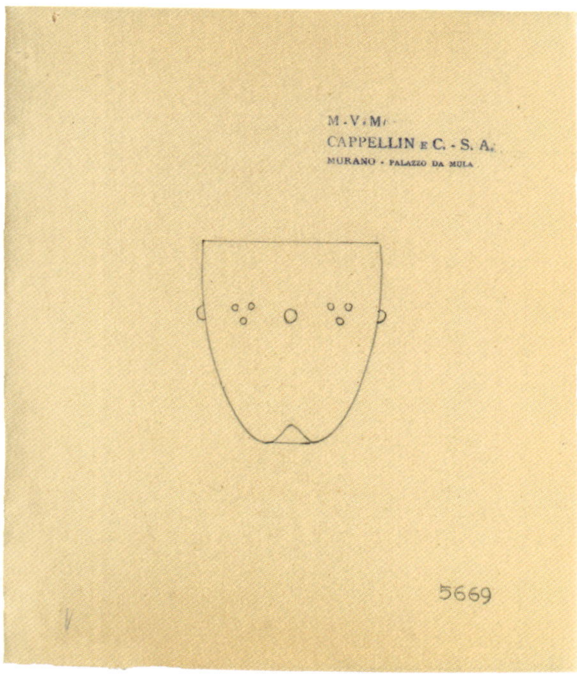

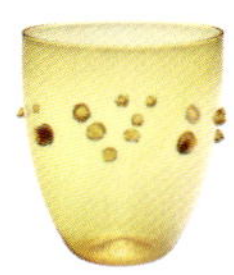

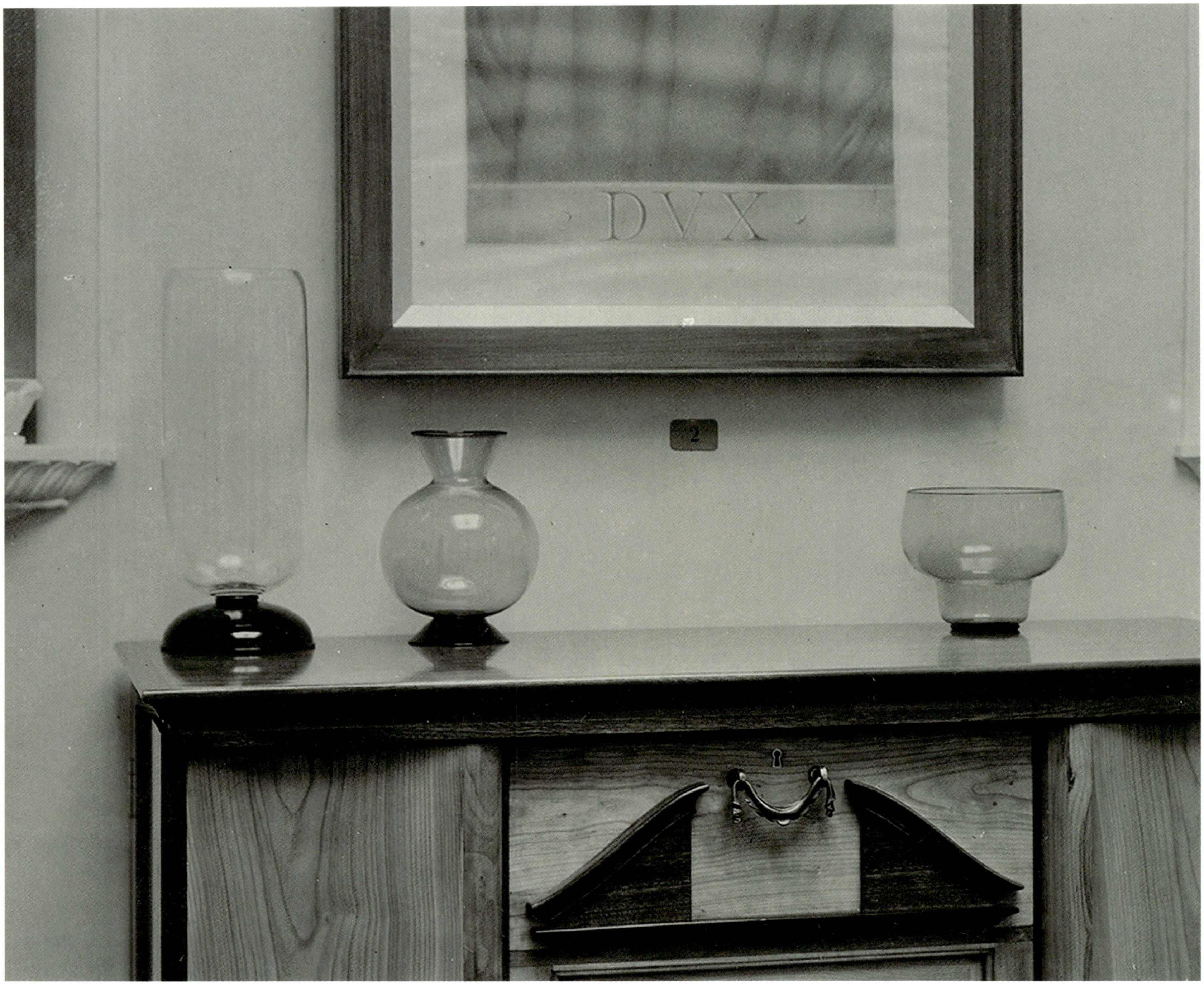

>
Cappellin glass pieces in room 32, at the 16th Biennale of Venice, 1928, ASAC, detail. The architect Duilio Torres and Guido Valconi were responsible for the layout of the room

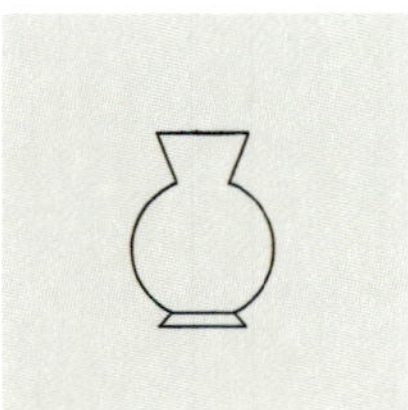

5837

Catalogue drawing for one of the vases in transparent glass (after a design by Gigi Chessa?) exhibited in room 32, at the 16th Biennale of Venice (1928), AMB.
1928
23 cm high

exhibitions
1928, Venice, 16th Biennale d'arte

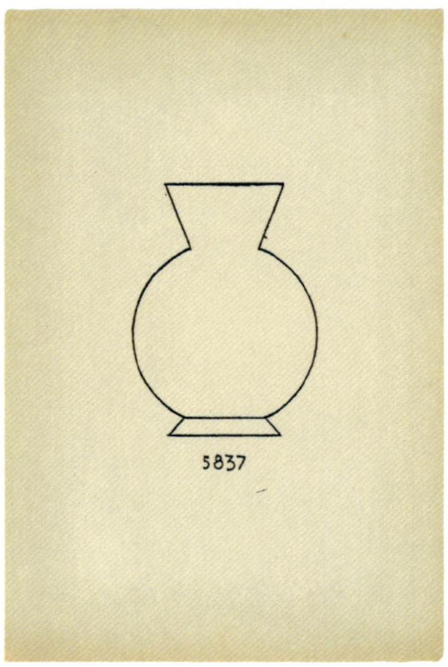

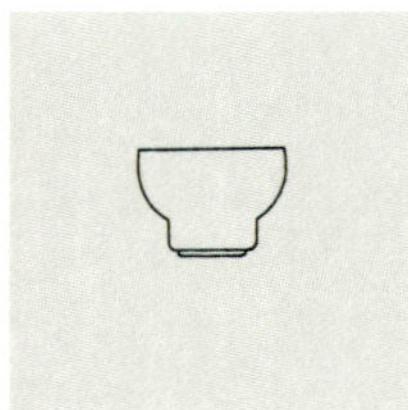

5841

Catalogue drawing for one of the vases in transparent glass (after a design by Gigi Chessa?) exhibited in room 32, at the 16th Biennale of Venice (1928), AMB.
1928
12.5 cm high

exhibitions
1928, Venice, 16th Biennale d'arte

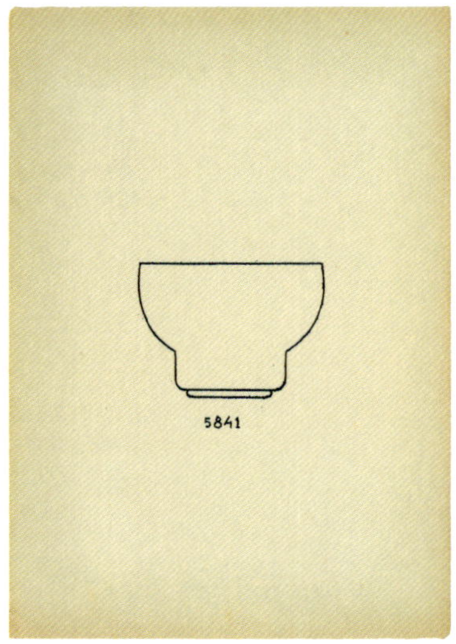

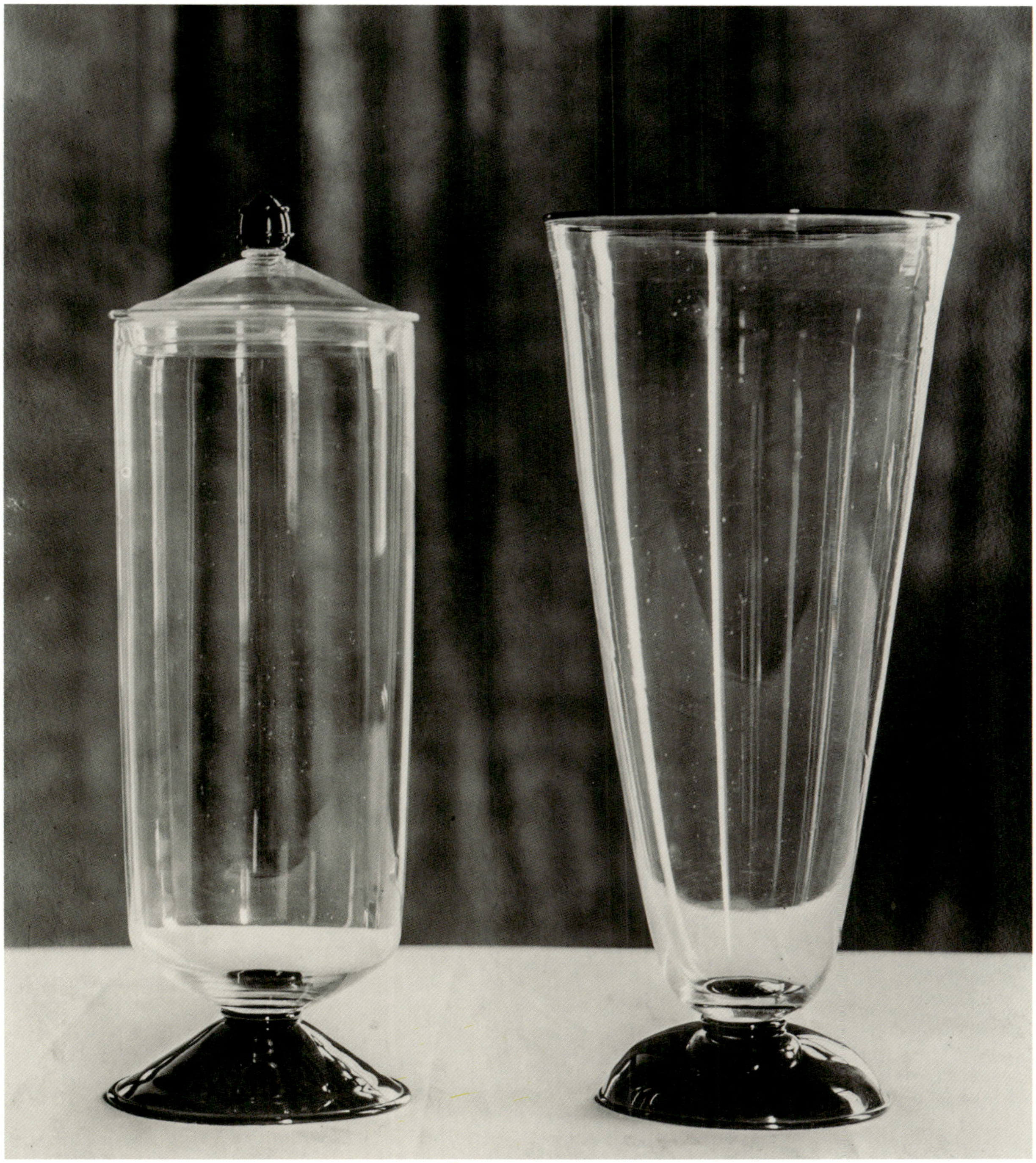

>
Compote bowl no. 3845 and vase no. 5854, after a design by Gigi Chessa at the 16th Biennale of Venice, 1928, CSV, detail

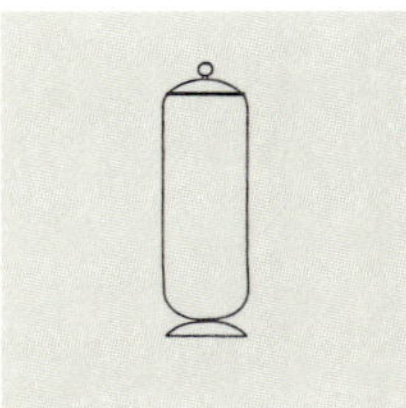

3846

Catalogue drawing for a tall container (after a design by Gigi Chessa?) displayed in room 32, at the 16th Biennale of Venice (1928), AMB. On this occasion the container was exhibited without a lid.
1928
48 cm high

exhibitions
1928, Venice, 16th Biennale d'arte

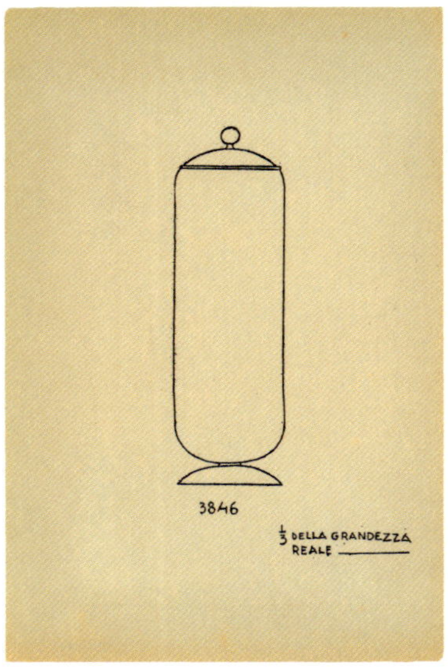

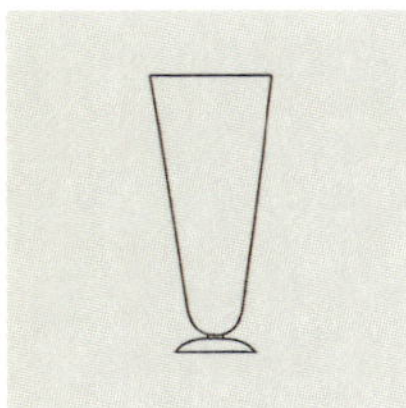

5854

Catalogue drawing for one of the transparent glass pieces, after a design by Gigi Chessa, exhibited in room 34, the room of *Bianco e Nero* (White and Black), at the 16th Biennale of Venice (1928), AMB. The vase, with a skullcap foot, was on display on a piece of furniture together with the compote bowl with truncated-cone-shaped foot no. 3845.
1928
50 cm high

exhibitions
1928, Venice, 16th Biennale d'arte

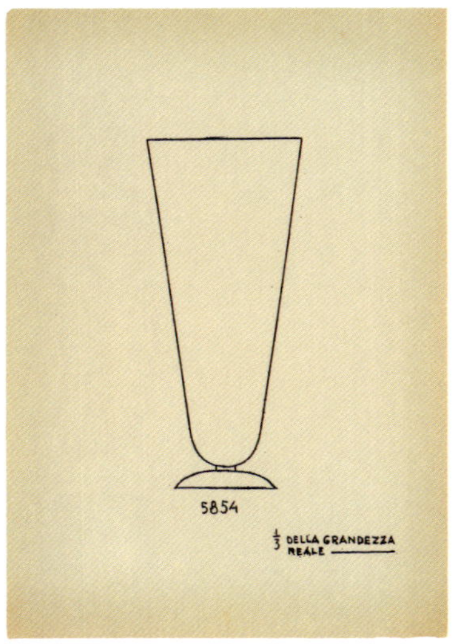

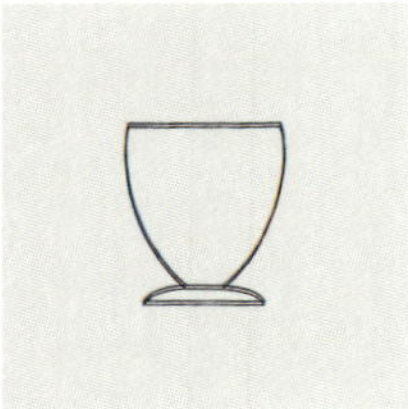

5821

Vase in green transparent glass with low skullcap foot. Coral-red thread around the mouth. Surface with iridised finish. Model no. 5821 was also used for the production of opaque glass pieces as could be seen on the occasion of the 4th Triennale of Monza (1930).
c. 1928
20 cm high
M.V.M. Cappellin Murano *acid stamp*
Dr. Richard Kossmann collection, New Jersey

>
5821, catalogue drawing, AMB

>
Cappellin glass pieces in the living-room designed by the architect Ottavio Cabiati, 4th Triennale of Monza, Furniture Gallery, 1930

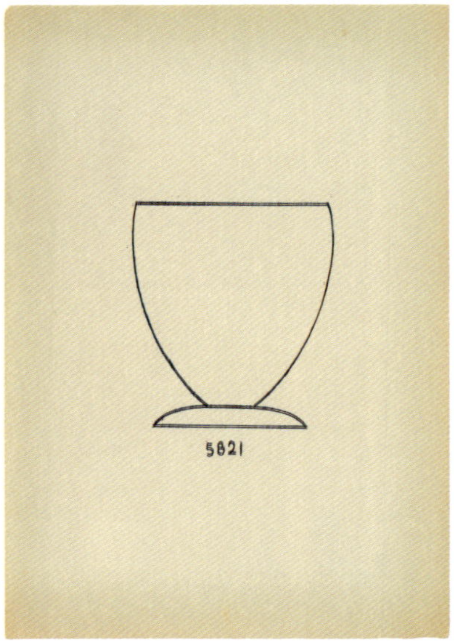

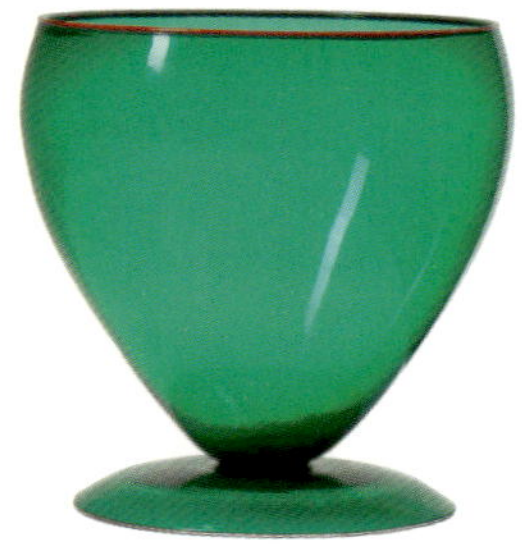

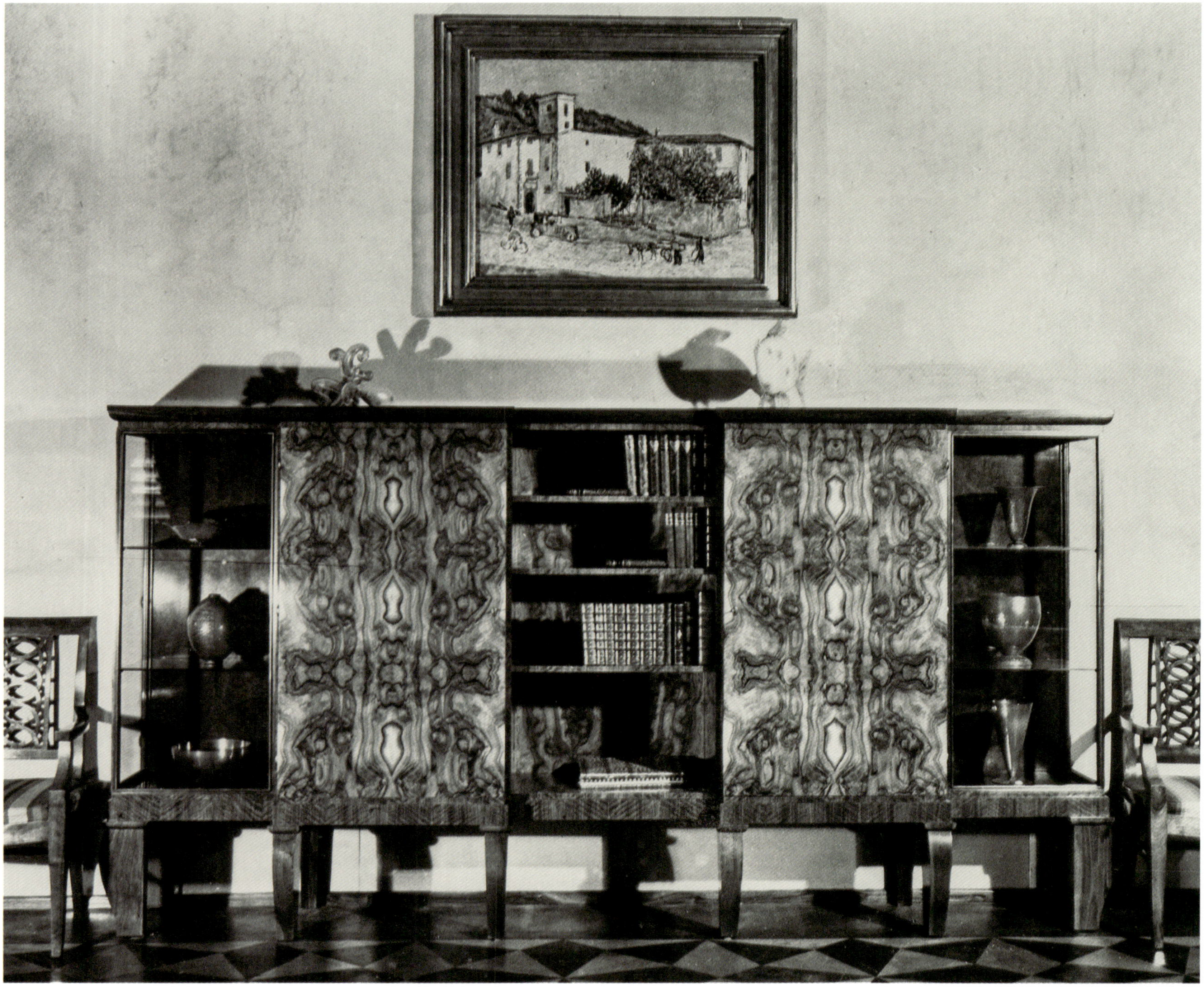

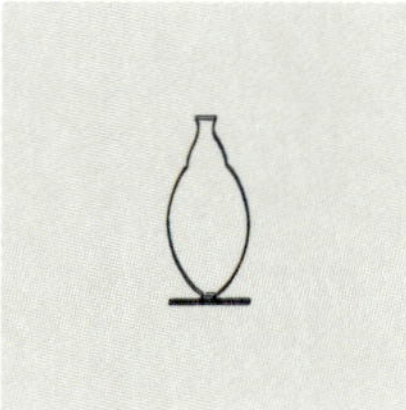

5921

Vase in smoke-coloured transparent glass with disk-shaped foot and constriction at the base of the neck. Surface with iridised finish. Model no. 5921 was also produced with other glass techniques (e.g. *lattimo aurato*, vertical canes).
1928–29
21 cm high
M.V.M. Cappellin Murano *acid stamp*
Private collection, Venice

>
5921, catalogue drawing, AMB

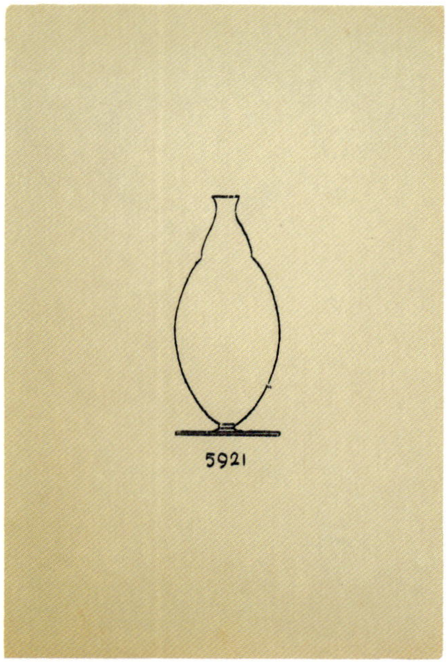

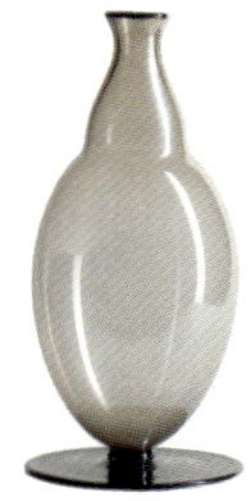

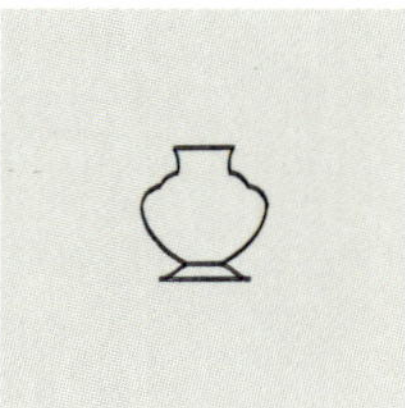

5938

Vase in smoke-coloured glass with a double constriction at the base of the neck. Flared foot and thread around the mouth in black glass. Strongly iridised finish. No. 5938 was also produced with other glass techniques (e.g. *lattimo aurato*, *decoro fenicio*, *pasta vitrea*, silvered black).
1928–29
17 cm high
M.V.M. Cappellin Murano *acid stamp*
Private collection

>
5938, catalogue drawing, AMB

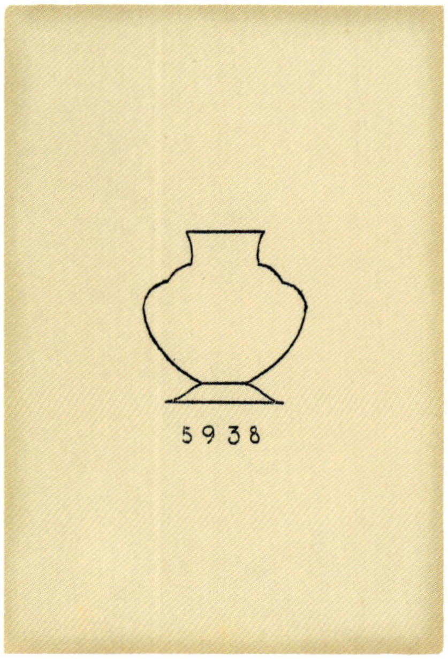

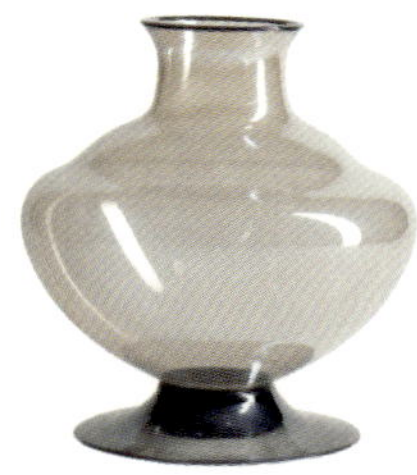

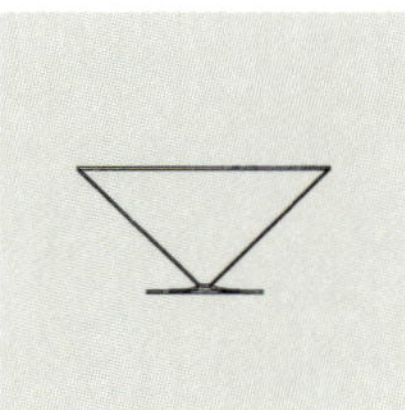

6000

Upturned-cone-shaped vase in *pagliesco* transparent glass with disk-shaped foot. Surface with iridised finish. No. 6000 was also produced with other glass techniques (e.g. *decoro fenicio*, cased).
c. 1930
15 cm high; 28 cm ø
M.V.M. Cappellin Murano *acid stamp*
Private collection

>
6000–6001, catalogue drawings, AMB

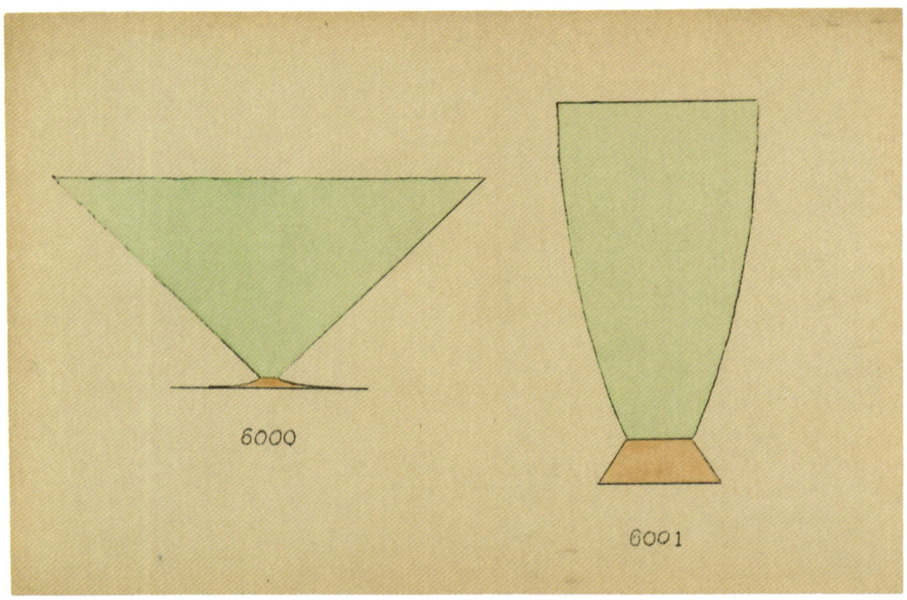

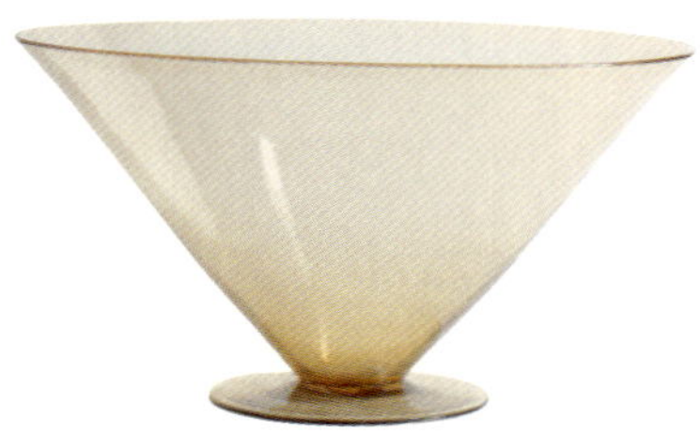

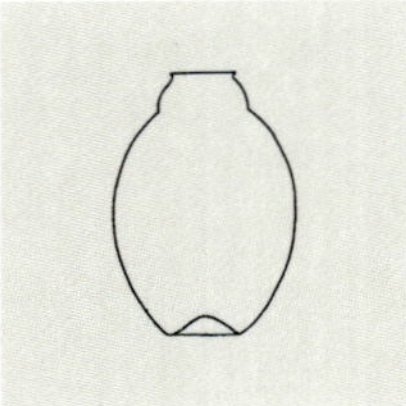

5887

Vase in green transparent glass with two slight constrictions at the neck. Strongly iridised surface. A similar example finished with applied gold or silver leaf was exhibited in 1930 at the 4thTriennale of Monza, in the sitting-room designed by the architect Ottavio Cabiati.
c. 1928
34 cm high
M.V.M. Cappellin Murano *acid stamp*
Private collection, Feltre

>
5887, catalogue drawing, AMB

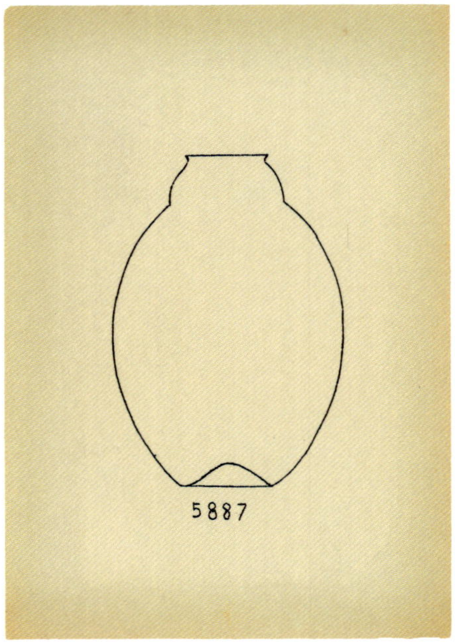

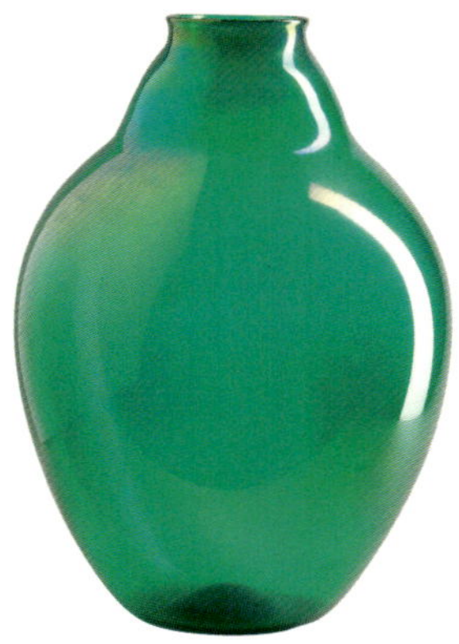

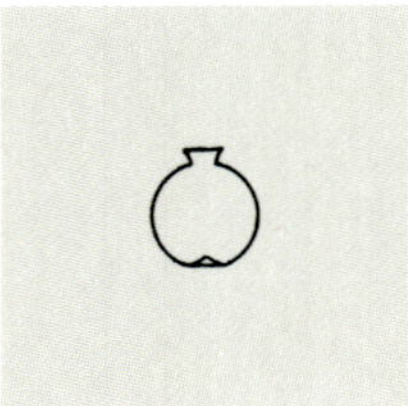

5933

Vase in Savoy-blue transparent glass with applied gold leaf. Surface with iridised finish.
1928–29
15 cm high
The Olnick Spanu Collection, New York

>
5933, catalogue drawing, AMH

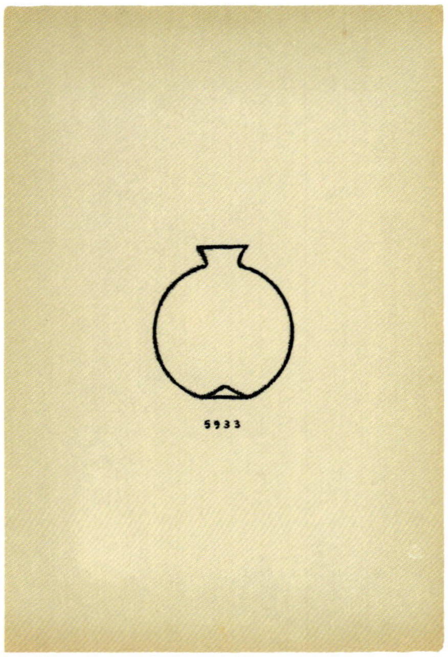

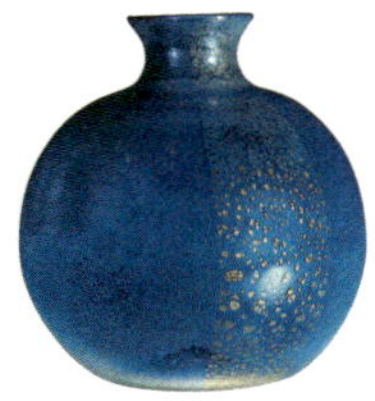

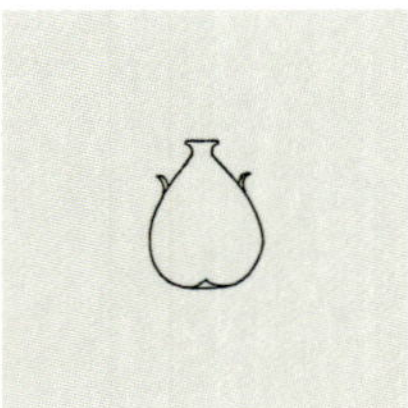

5941

Pear-shaped vase in *pagliesco* transparent glass with small lateral handles and thread around the mouth in green transparent glass. Surface with iridised finish. Model no. 5941 was also produced with other glass techniques (e.g. *pasta vitrea*, silvered or iridised black).
1928–29
17 cm high
M.V.M. Cappellin old label
Private collection

>
5941, catalogue drawing, AMB

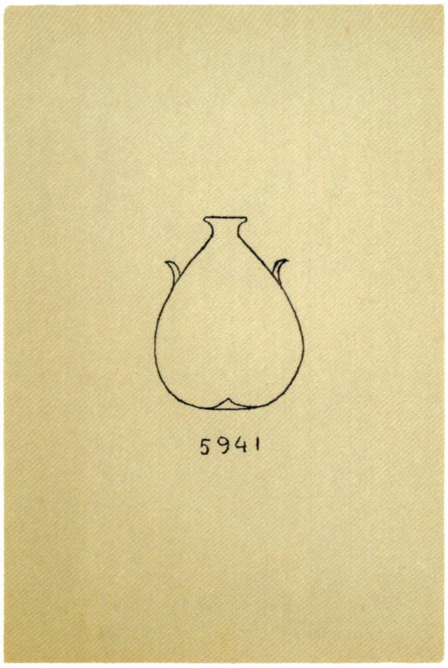

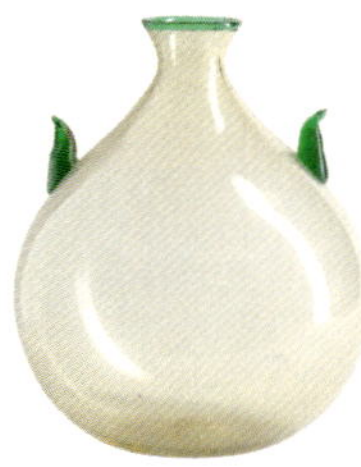

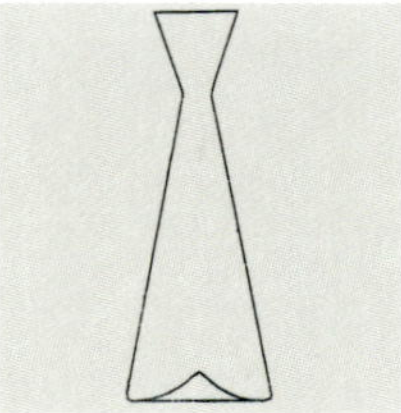

5762

Large vase in light-blue transparent glass with incorporated bubbles. Surface with iridised finish. A similar example in 'yellow glass [with] green thread around the mouth' was exhibited in 1927 at the 3rd Biennale of Monza.
1927
48 cm high
C.V.M. *acid stamp*; Venezia Murano Company old label
Private collection

exhibitions
1927, Monza, 3nd Biennale

>
2, drawing for model to be exhibited at the 3rd Biennale of Monza, c. 1927, AMuVe

>
Vases in trasparent glass. Period photograph, c. 1929, ACSVr

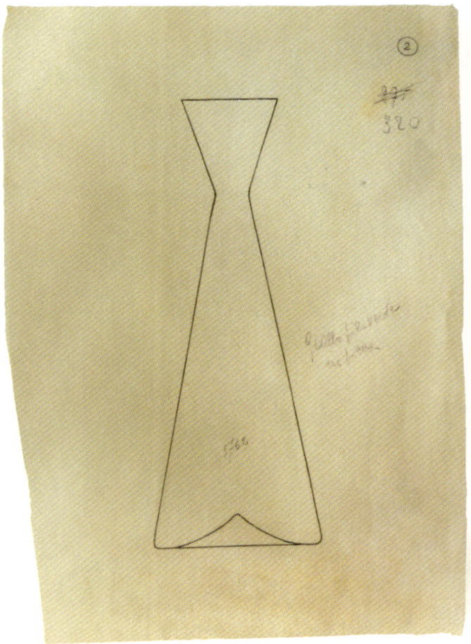

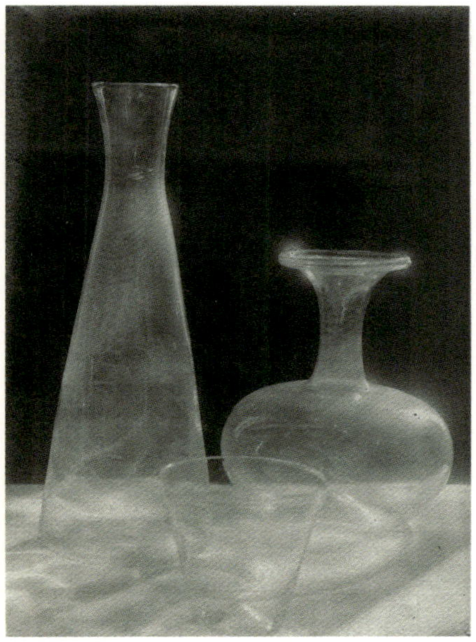

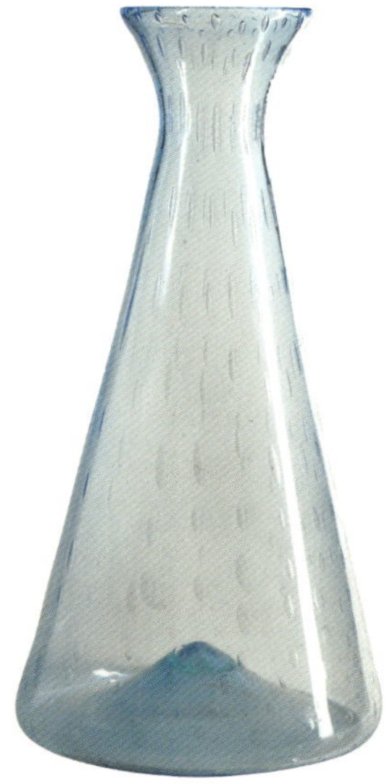

---

n.i.

Bowl in yellow transparent glass, with incorporated bubbles, tall foot and thread around the mouth in dark coral-red and gold *pasta vitrea*. Surface with iridised finish.
1930–31
17 cm high; 20 cm ø
M.V.M. Cappellin Murano *acid stamp*
The Olnick Spanu Collection, New York

>
Carlo Scarpa, study for bowl with tall foot and body in glass with bubbles, 1929–30, AMuVe

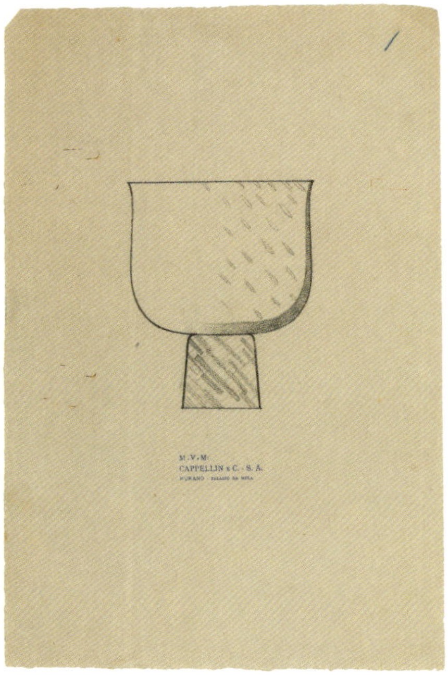

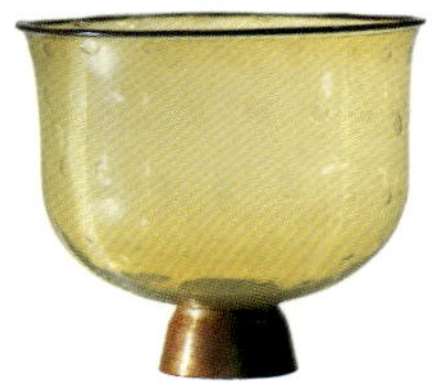

---

n.i.

Vase in yellow transparent glass with incorporated bubbles, having a tall foot and a thread around the mouth in dark coral-red and gold *pasta vitrea*. Surface with iridised finish.
1930–31
23 cm high
Venezia Murano Company old label
Private collection

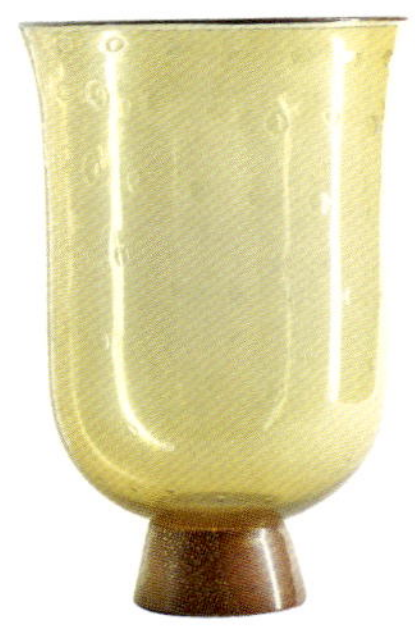

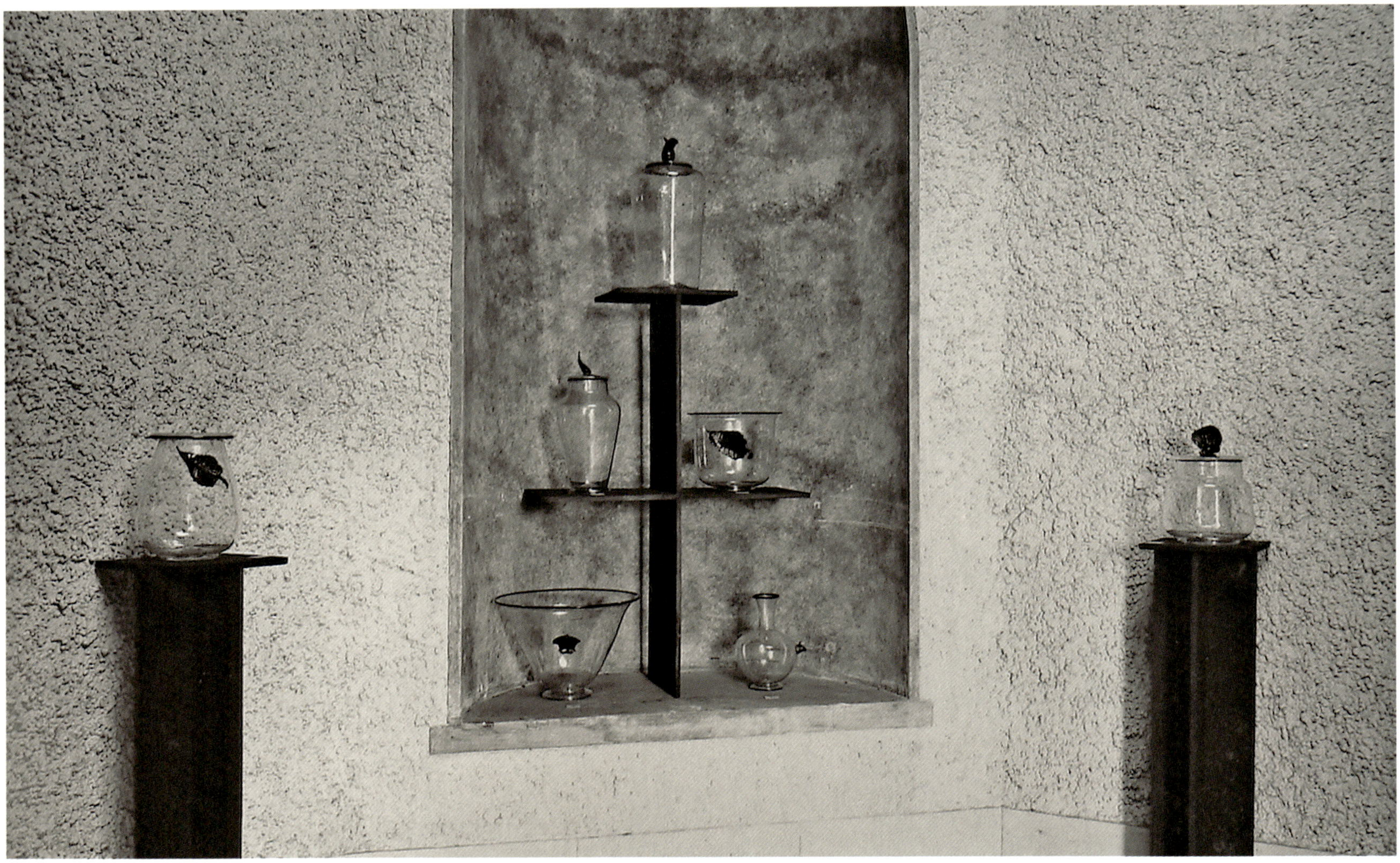

>
Cappellin glass pieces at the 22nd Esposizione dell'Opera Bevilacqua La Masa at the Lido of Venice, 1931

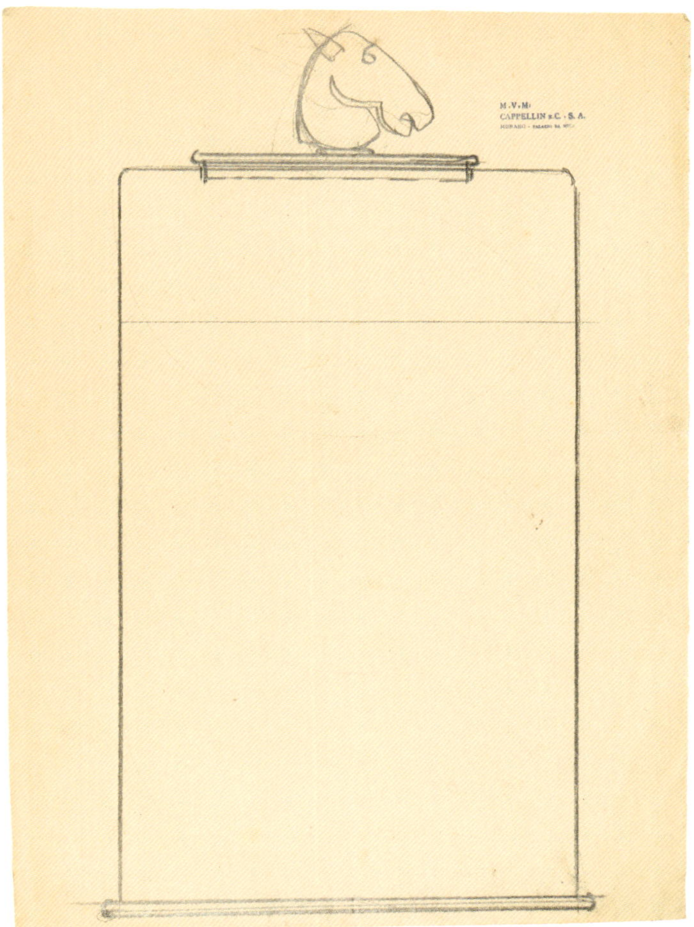

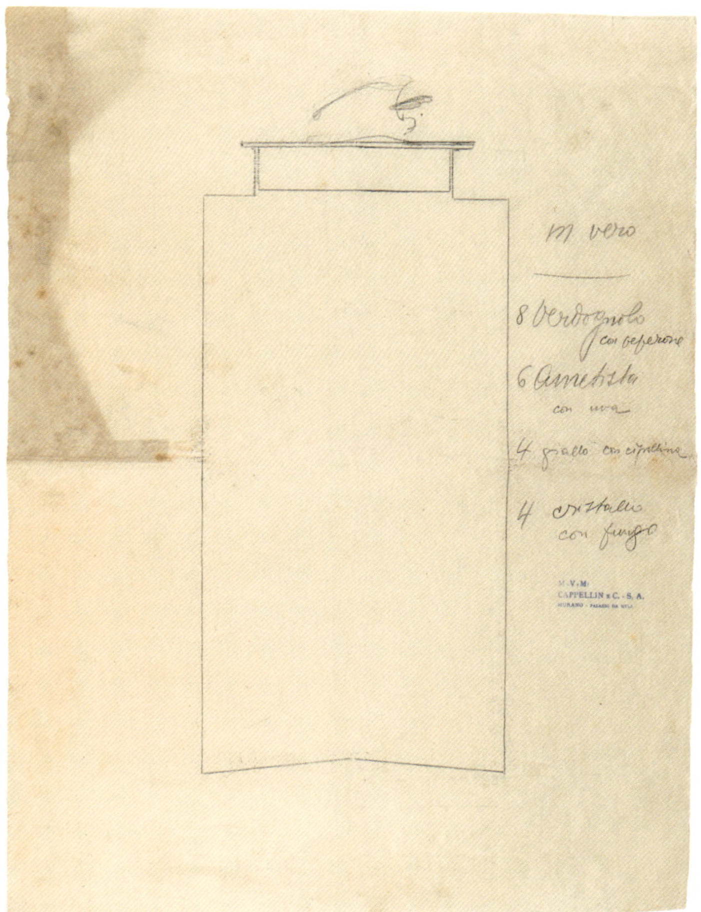

>
Carlo Scarpa, study for compote bowl decorated on lid with horse's head, c. 1930, AMuVe

>
Carlo Scarpa, study for compote bowl decorated on lid with piece of fruit, c. 1930, AMuVe. As indicated in the margin notes of the drawing, four variants were planned: one in 'greenish glass with sweet pepper', one 'amethyst with grapes', one 'yellow with baby onion' and one 'crystal with mushroom'

n.i.

Compote bowl in transparent light-blue glass with small foot and horn applied to the lid in transparent green glass with applied gold leaf. Surface with iridised finish.
1930–31
25 cm high
Pauly old label
Kathleen and David Roth collection, USA

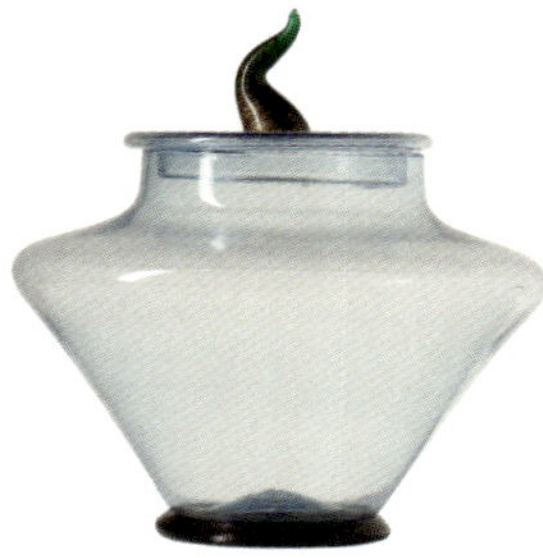

n.i.

Large compote bowl in white *pulegoso* glass with lid decorated by an applied horn. Surface with iridised finish. A similar example in transparent glass with foot and horn in *pasta vitrea* was exhibited in 1931 at the 22nd Esposizione dell'Opera Bevilacqua La Masa of Venice.
1930–31
47 cm high
M.V.M. Cappellin Murano *acid stamp*
Bischofberger collection, Switzerland

exhibitions
1931, Venice, 22nd Esposizione dell'Opera Bevilacqua La Masa

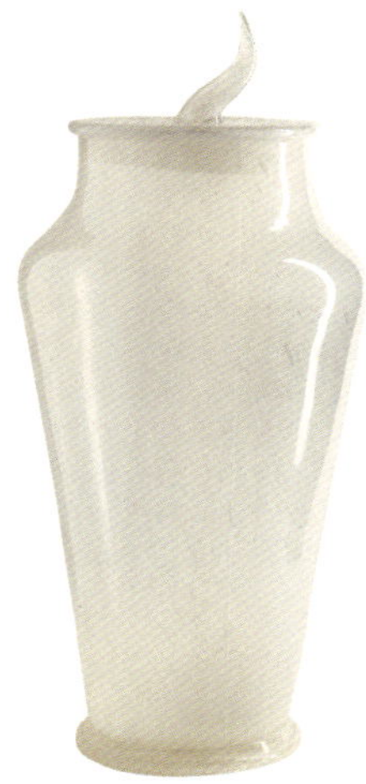

n.i.

Large compote bowl in transparent light-blue glass with incorporated bubbles. Foot and shell applied to the lid in green transparent glass with applied gold leaf. Surface with iridised finish.
1930–31
53 cm high
Private collection

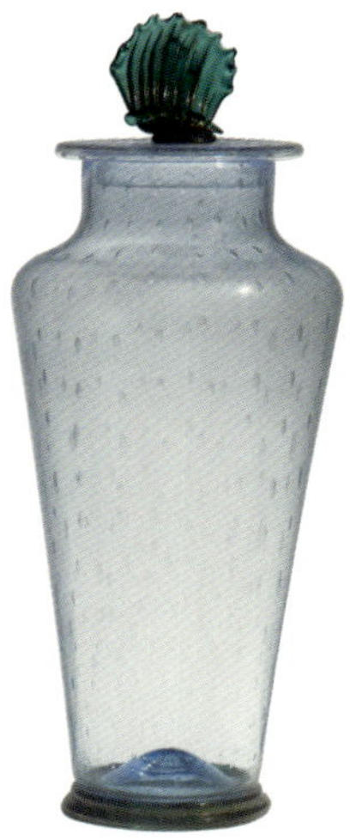

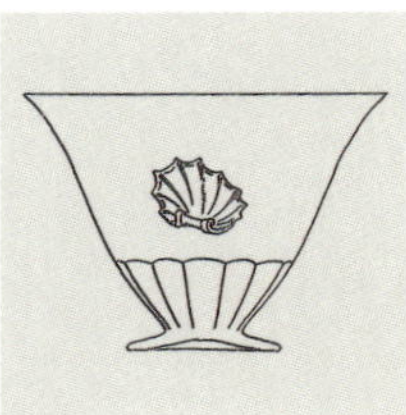

6022

Vase in light-green *pulegoso* glass. Base decorated with *meza stampaura* (half moulding) and a shell applied. A similar example in transparent glass with a thread around the mouth and shell in *pasta vitrea* was exhibited in 1931 at the 22nd Mostra dell'Opera Bevilacqua La Masa of Venice.
1930–31
33 cm high; 48 cm ø
Private collection

exhibitions
1931, Venice, 22nd Esposizione dell'Opera Bevilacqua La Masa

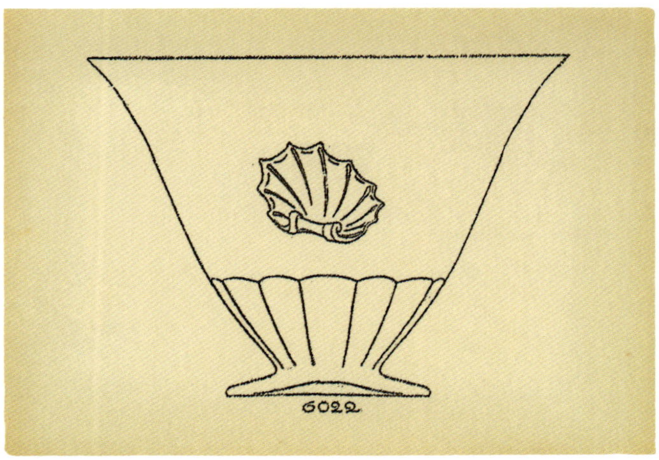

>
6022, catalogue drawing, private collection

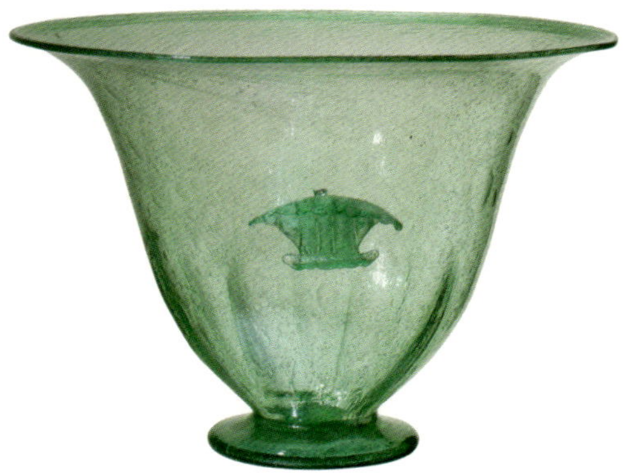

018

Perfume bottle, with disk-shaped foot, in transparent amethyst glass, decorated with a couple of supports in amethyst and dark amethyst applied to the base of the body. Surface with iridised finish. The model, equipped with a circular two-coloured stopper, is present in the American catalogue of the Cappellin Glassware in four different sizes (nos. 018–020 and special one). Three examples of this series were published in *Vogue*, November 1931, p. 82
1928–29
16 cm high
M.V.M. Cappellin Murano *acid stamp*
Private collection

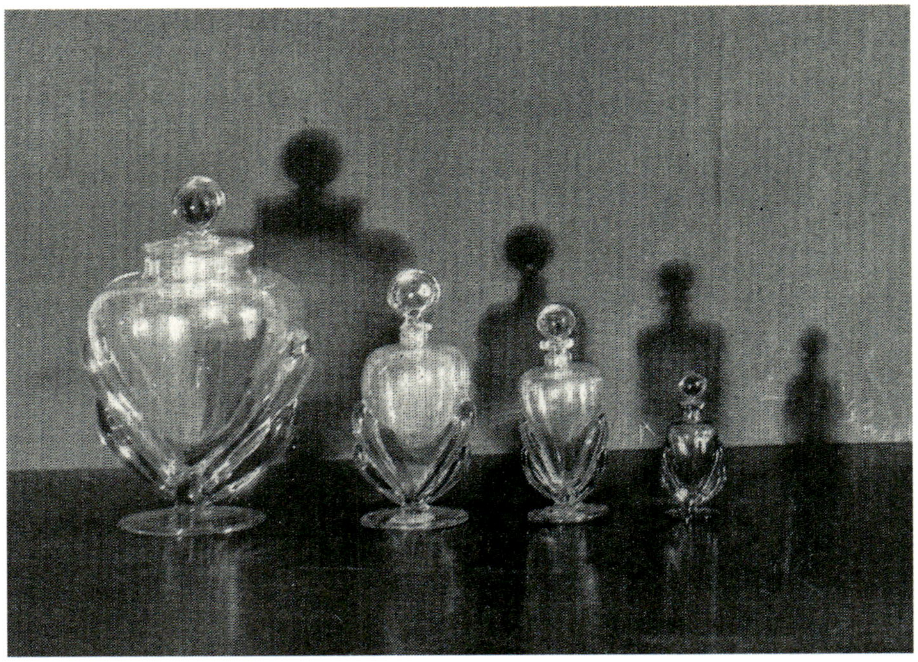

>
Perfume bottles in Cappellin Glassware*, catalogue, undated, detail, AHL*

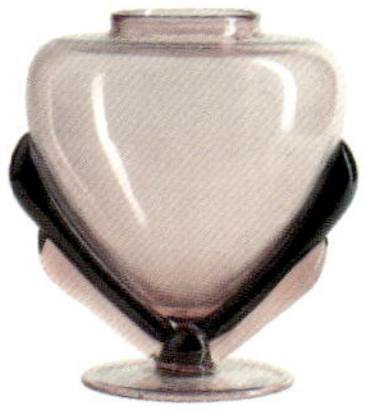

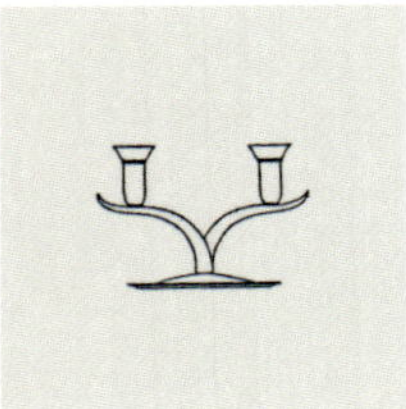

2692

Candelabrum with two arms and disk-shaped foot in light-green transparent glass. Thread around the foot and finishings of the candle holders in green glass. Surface with iridised finish. The model was also proposed with a single arm (no. 3001).
c. 1927
19 cm high
M.V.M. Cappellin Murano *acid stamp*
Private collection

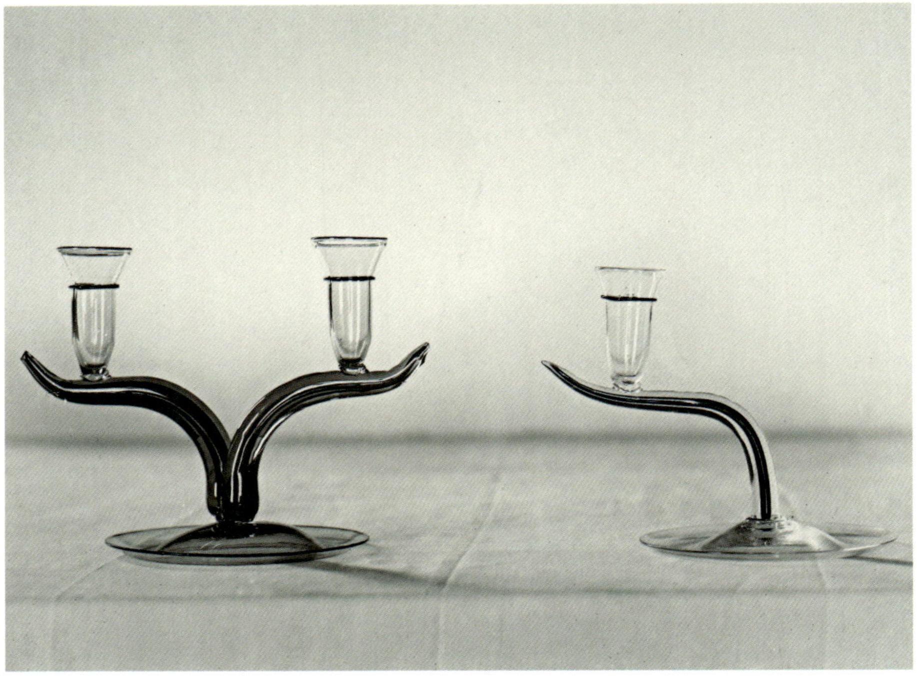

>
Candlestick no. 2692 and no. 3001 in transparent glass. Period photograph, c. 1927, ACSVr

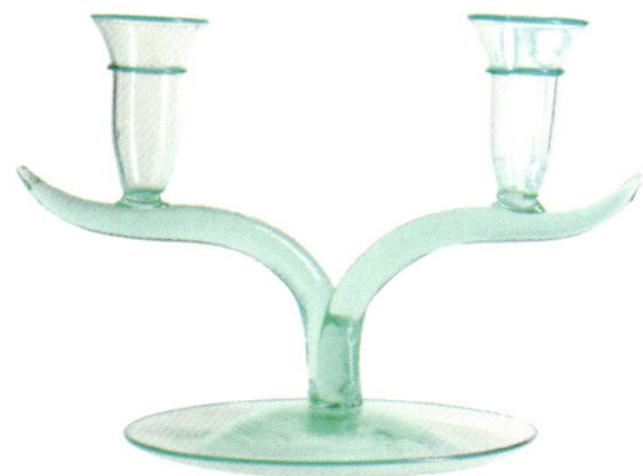

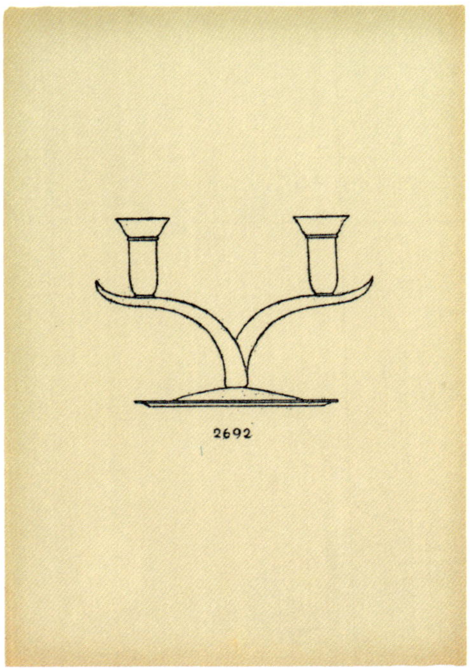

>
2692, catalogue drawing, AMB

n.i.

Candelabrum with two arms and three candle holders, disk-shaped foot, in light-blue transparent glass. Finishings around the foot and the candle holders and applied leaves in coral-red *pasta vitrea*. Surface with iridised finish.
1928–29
12 cm high
M.V.M. Cappellin Murano *acid stamp*
Private collection, Treviso

›
Candelabrum, goblet, plate and elements for table centrepiece with a floral motif. Period photograph, c. 1929, CSV

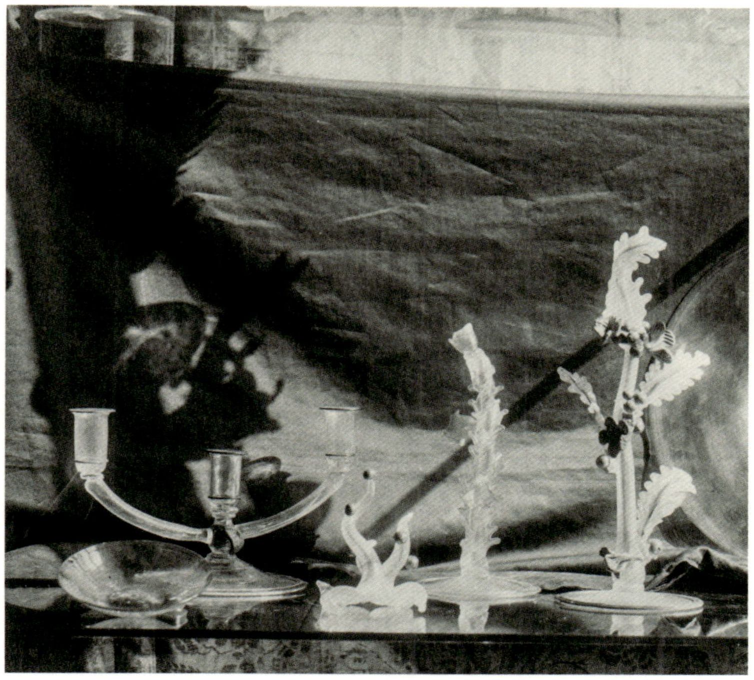

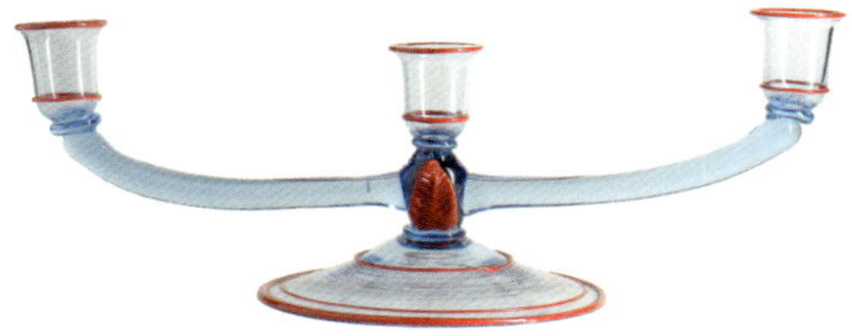

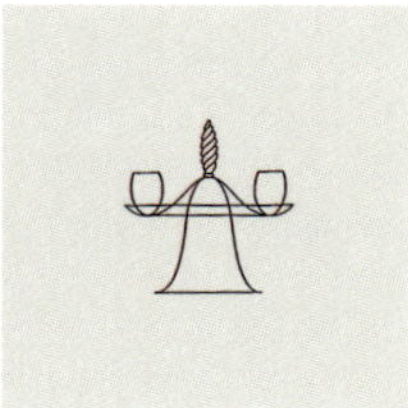

2707

Candelabrum, with a structure of superimposed bells, in light-blue transparent glass and surface with iridised finish.
1927–28
21 cm high
M.V.M. Cappellin Murano *acid stamp*
Fondazione Musei Civici di Venezia, Museo del Vetro collection

›
2707, working drawing, CSV

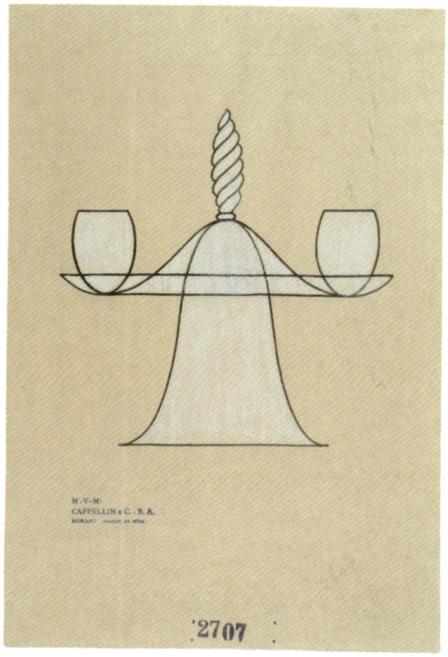

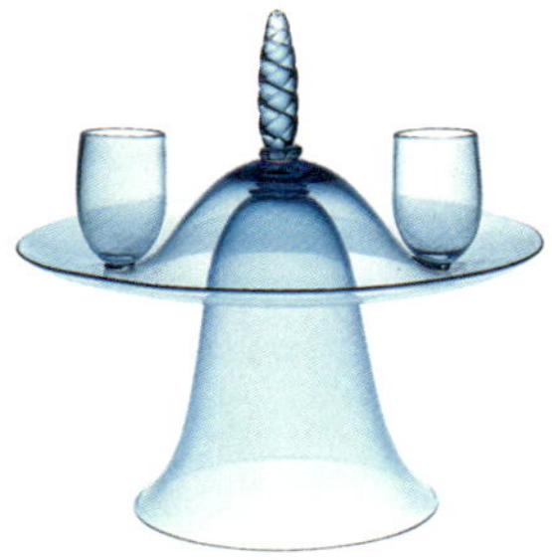

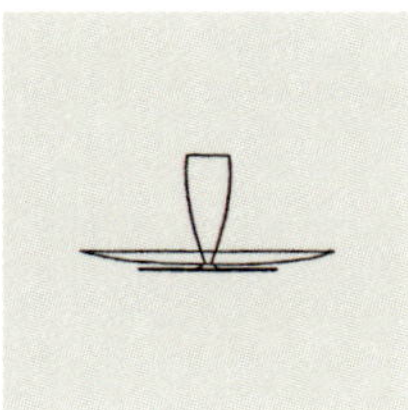

4708

Fruit dish in light-green transparent glass with disk-shaped foot surmounted by a plate with a small vase at the centre. Thread around the mouth in *lattimo*. Surface with iridised finish.
1926–27
14 cm high; 27 cm ø
M.V.M. Cappellin Murano *acid stamp*
Private collection, Feltre

›
4708, catalogue drawing, AMB

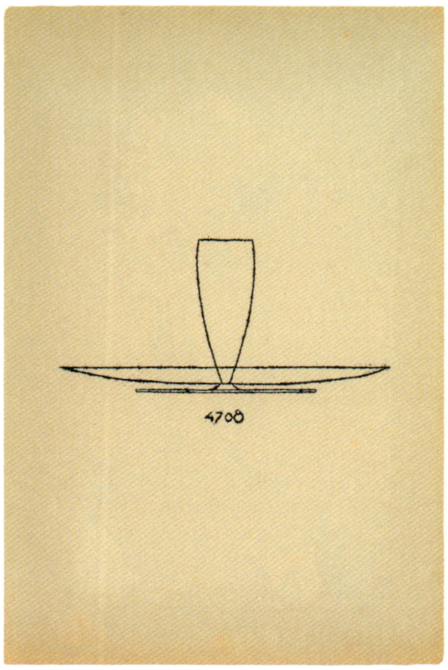

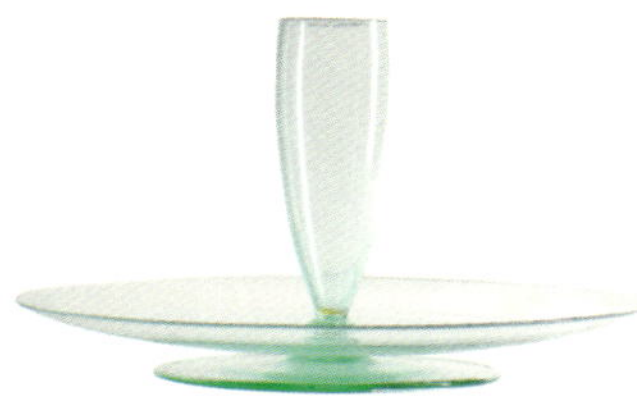

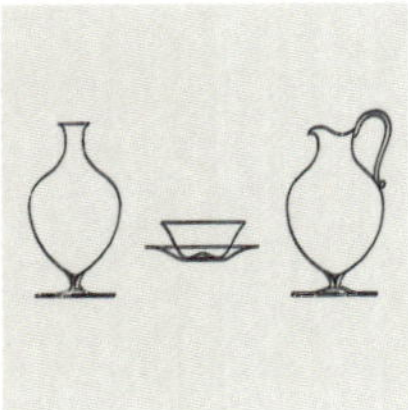

59

Glasses in crystal glass with baluster-shaped stem with a dark-blue nucleus belonging to dinner service no. 59.
c. 1927
13 cm high
M.V.M. Cappellin Murano *acid stamp*
The Lockwood collection, Fort Lee, NJ

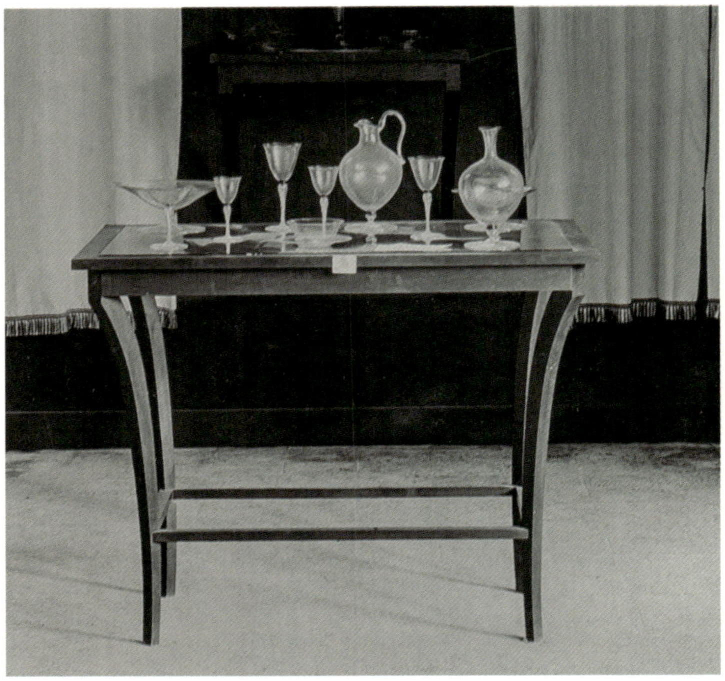

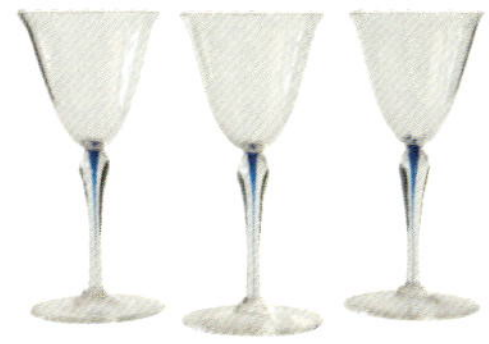

›
Dinner service no. 59 at an exhibition of Cappellin glass. Period photograph, c. 1927, CSV, detail

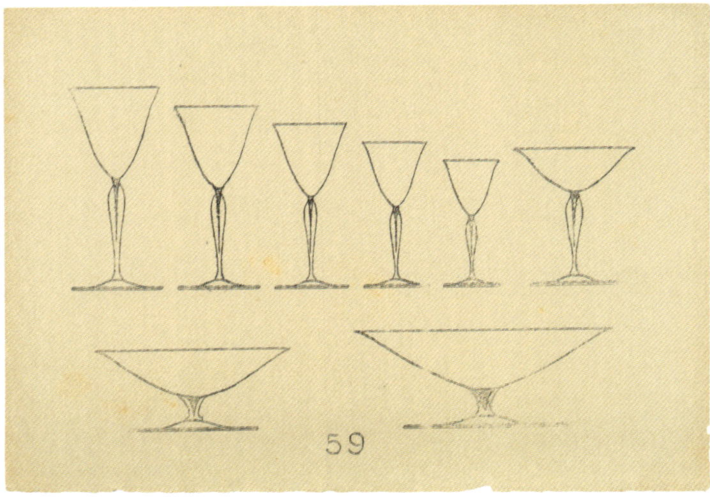

›
59, catalogue drawing, AMB

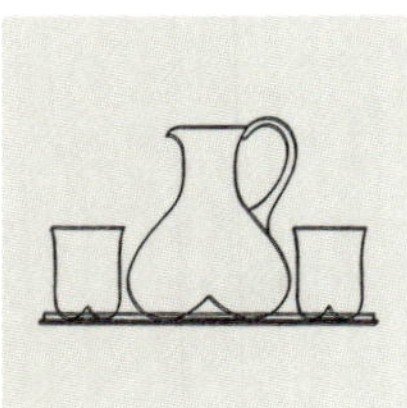

261

Jug and glasses in crystal glass with incorporated bubbles belonging to the drinks service no. 261. Thread around the mouth in coral-red *pasta vitrea*. Surface with iridised finish.
1927–28
16; 24 cm high
M.V.M. Cappellin Murano *acid stamp*
The Lockwood collection, Fort Lee, NJ

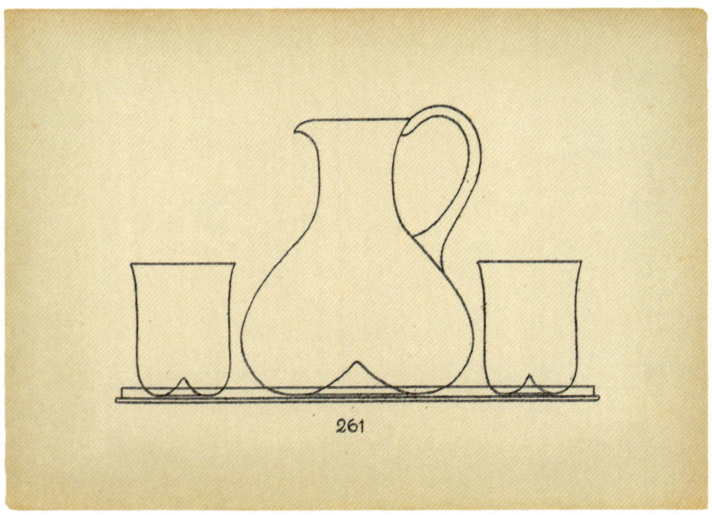

›
261, catalogue drawing, AMB

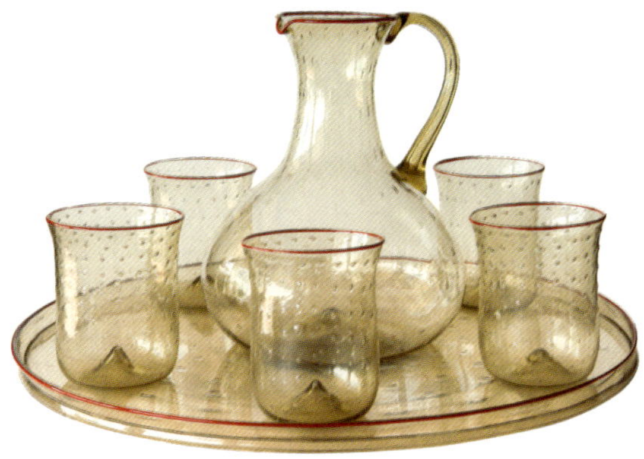

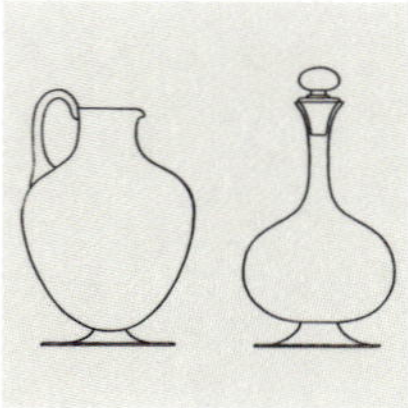

9

Dinner service in transparent glass and in opaque glass (*lattimo*?) exhibited in 1931 at the Amsterdam Exhibition of glass, ceramics and lace. The service, described in the exhibition catalogue as 'a new service in two colours transparent and opaque', represents a variant with foot and stem in opaque glass of no. 9, already present in the catalogue and designed by Vittorio Zecchin (1921–25).
1930–31
27; 24 cm high

exhibition
1931, Amsterdam Exhibition of glass, ceramics and lace

›
9, catalogue drawing, CSV

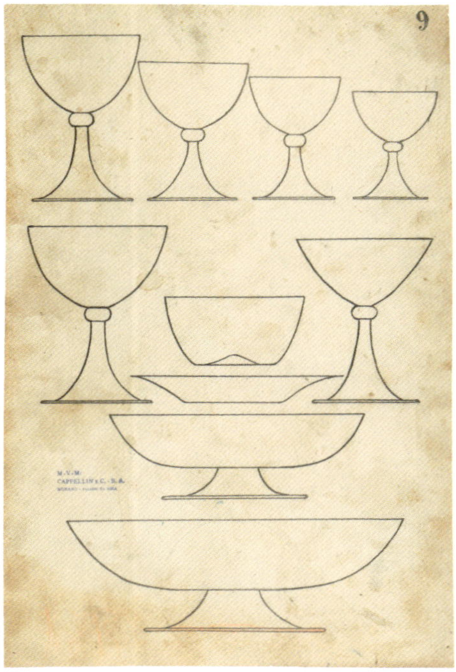

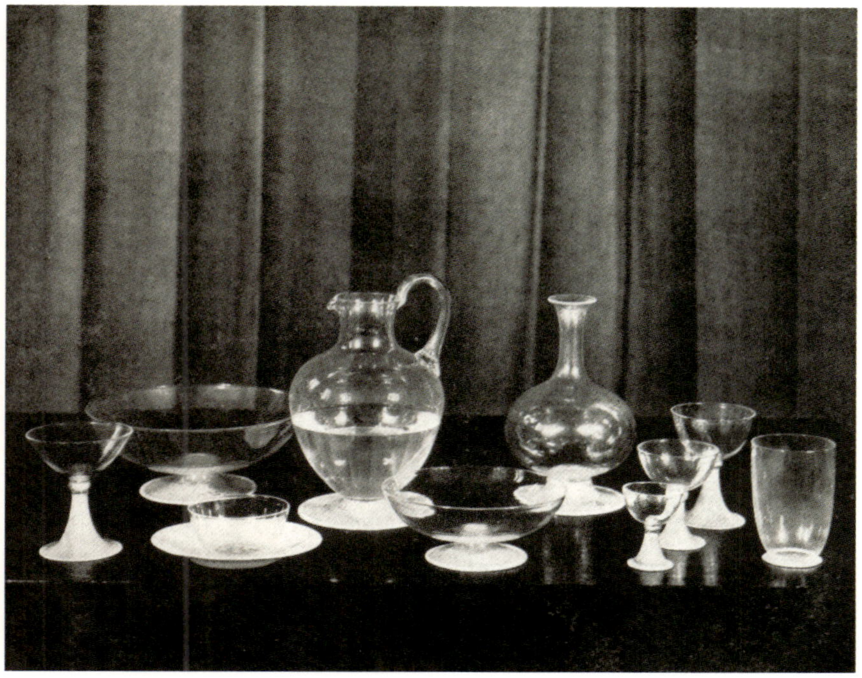

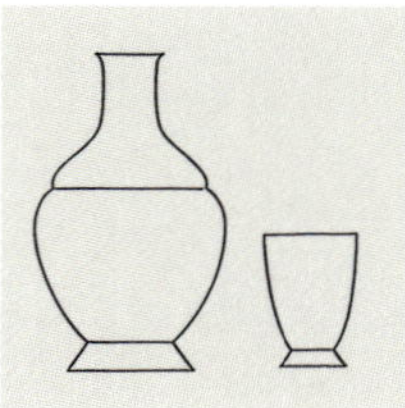

262

Glass with truncated-cone-shaped foot in light-green transparent glass. Foot in green transparent glass with applied gold leaf. The glass is accompanied by a small plate in light-green transparent glass. Both pieces have an iridised finish.
The truncated-cone-shaped foot also characterises the drinks service, no. 262, to be produced in transparent glass.
c. 1927
15.5 cm high
M.V.M. Cappellin Murano *acid stamp*
Private collection, Feltre

›
262, working drawing, CSV

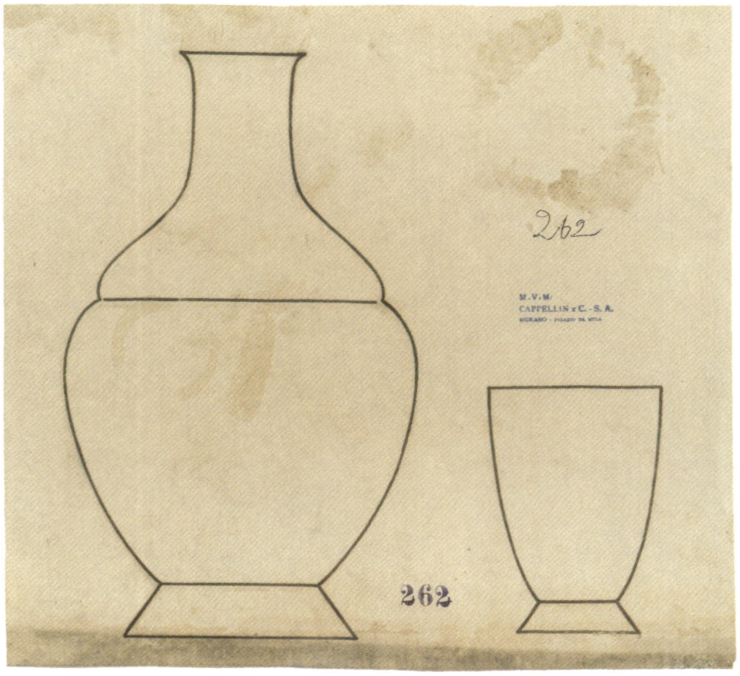

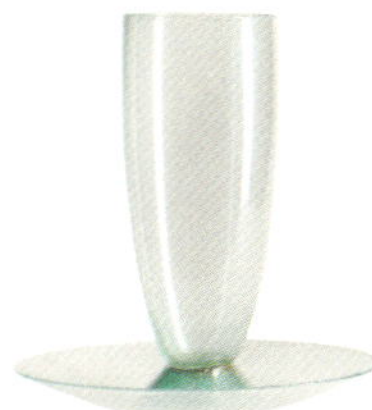

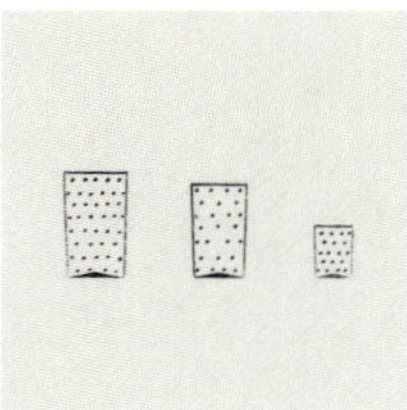

781

Tumblers in crystal glass with irregular bubbles partially incorporated in the wall of the piece. The presence of the irregular bubbles in the wall of the piece also characterises the spirits service, no. 781.
c. 1930
7.5; 8; 9.5; 10 cm high
Private collection, Venice

›
781, catalogue drawing, AMA

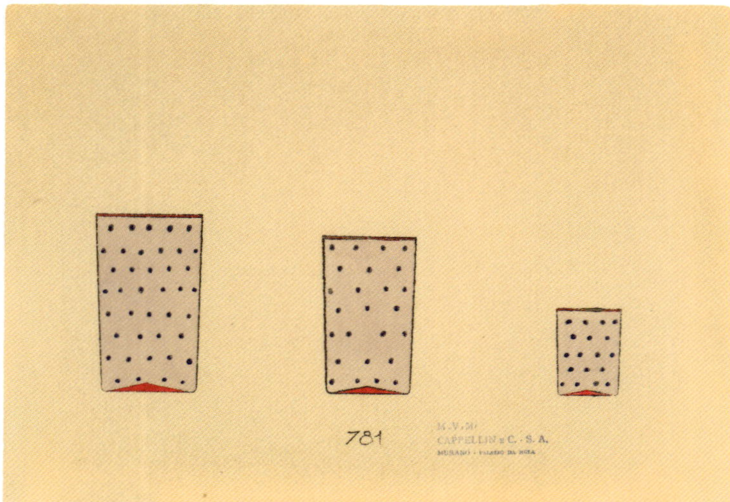

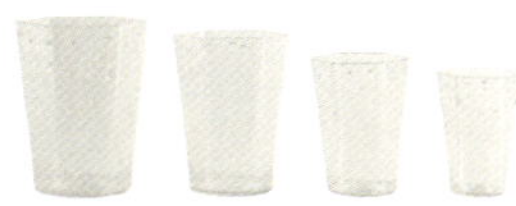

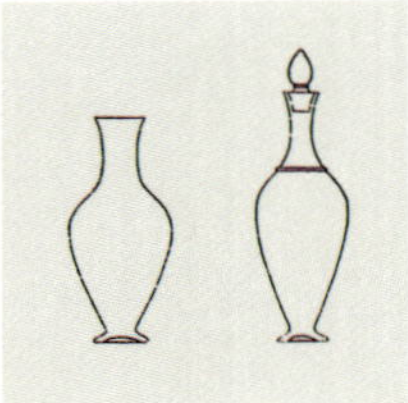

67

Catalogue drawing for water and wine bottle of dinner service no. 67, AMA. The service is characterised by elements with foot worked from the body of the piece.
1929
26; 34 cm high

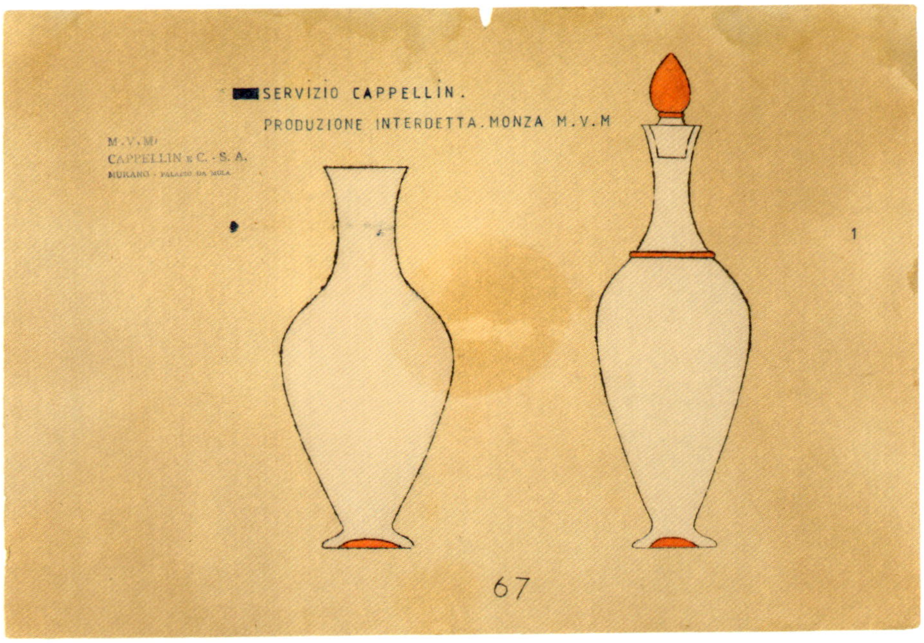

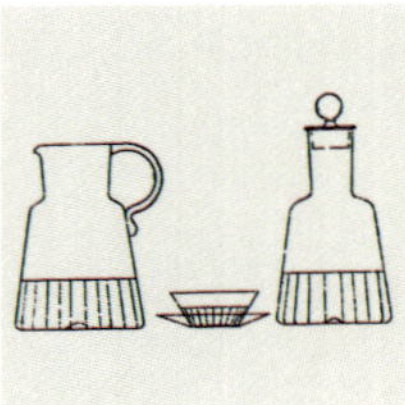

68

Catalogue drawing for jug, wine carafe and bowl of dinner service no. 68 in transparent glass with elements decorated at the base with *meza stampaura* (half moulding), AMA.
1929–30
26; 32 cm high

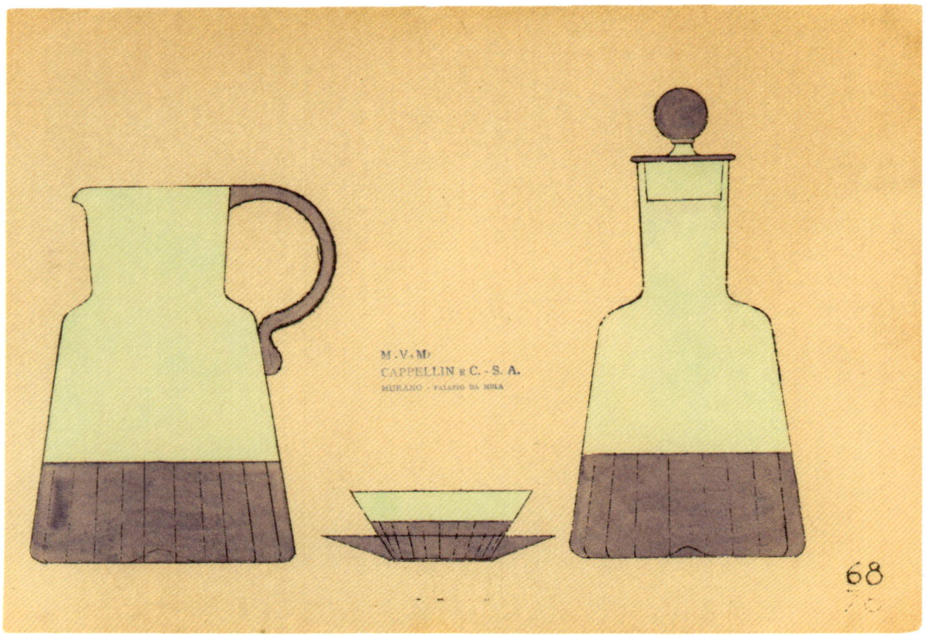

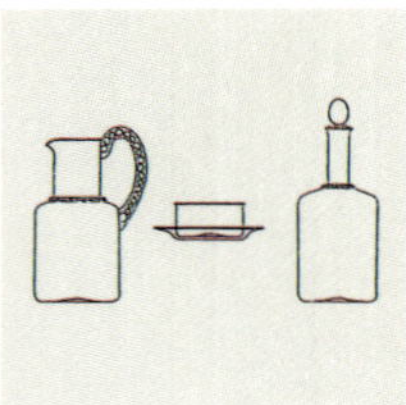

69

Catalogue drawing for jug, wine carafe and bowl of dinner service no. 69 in transparent glass with finishings in coral-red *pasta vitrea*, neck and lateral handle with *filigrana a retortoli*, AMA.
1929–30
28; 25 cm high

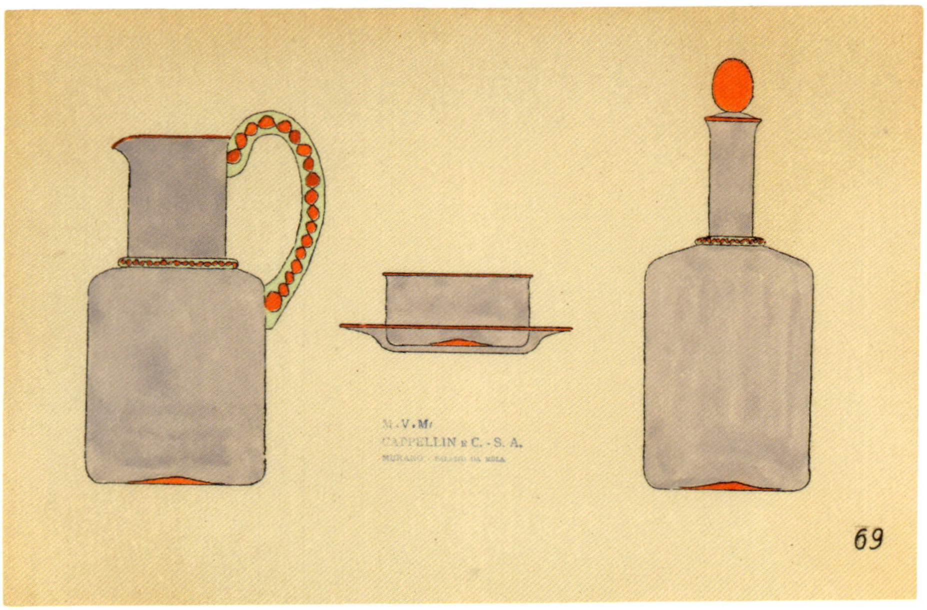

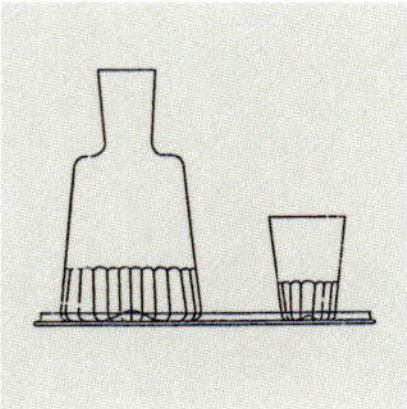

266

Catalogue drawing for the drinks service no. 266 in transparent glass with decorative elements at the base with *meza stampaura* (half moulding), AMA.
1929–30
30; 13 cm high

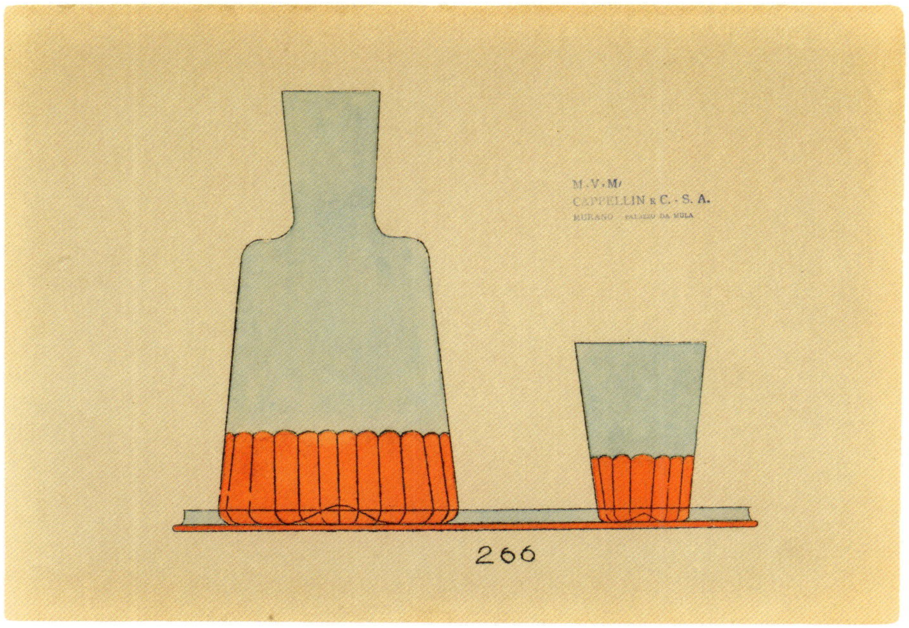

>
Carlo Scarpa, studies for lighting and furnishing, 1929–30, AMuVe. At centre a bottle and a drinking glass decorated at the base with meza stampaura.

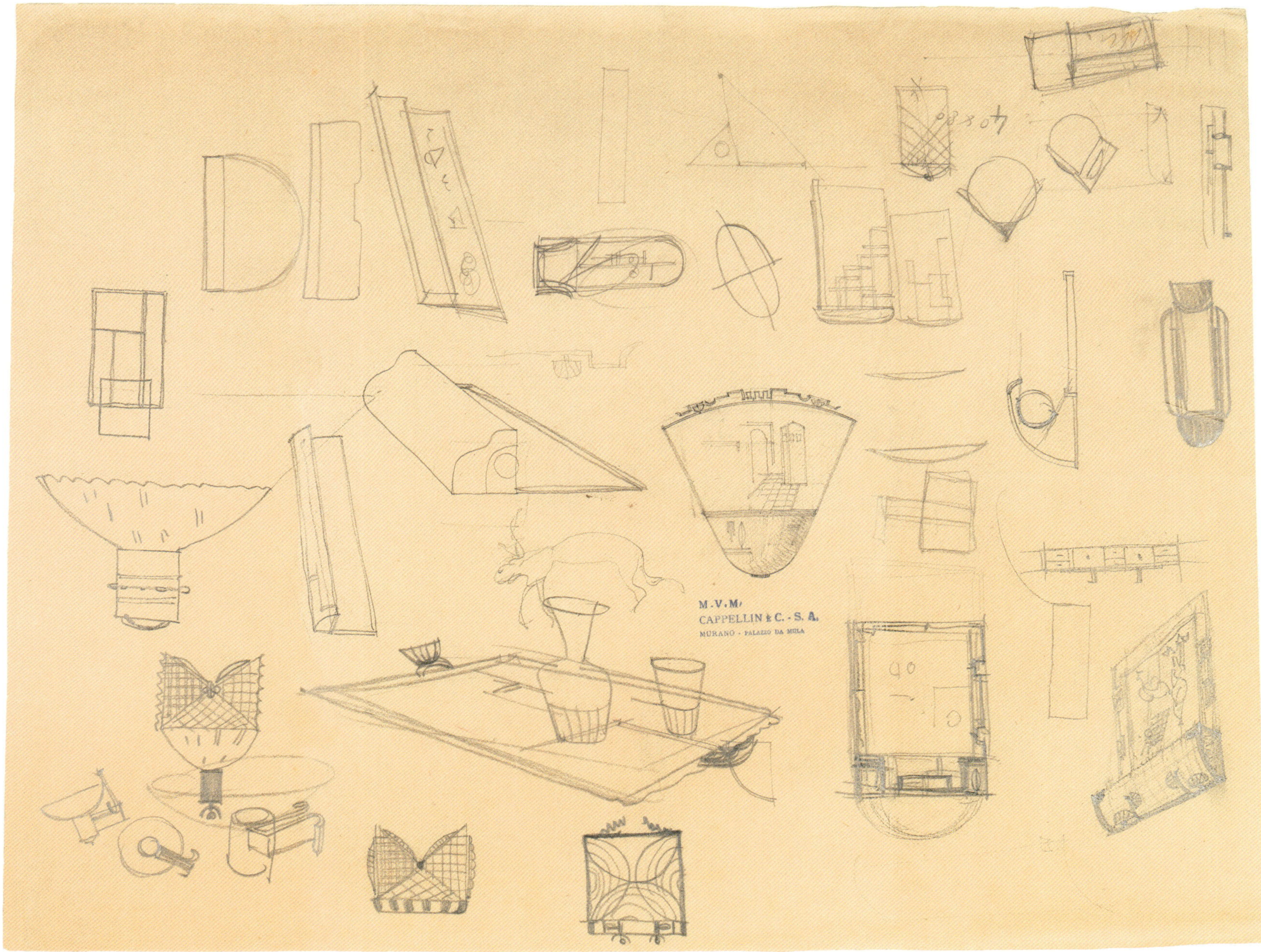

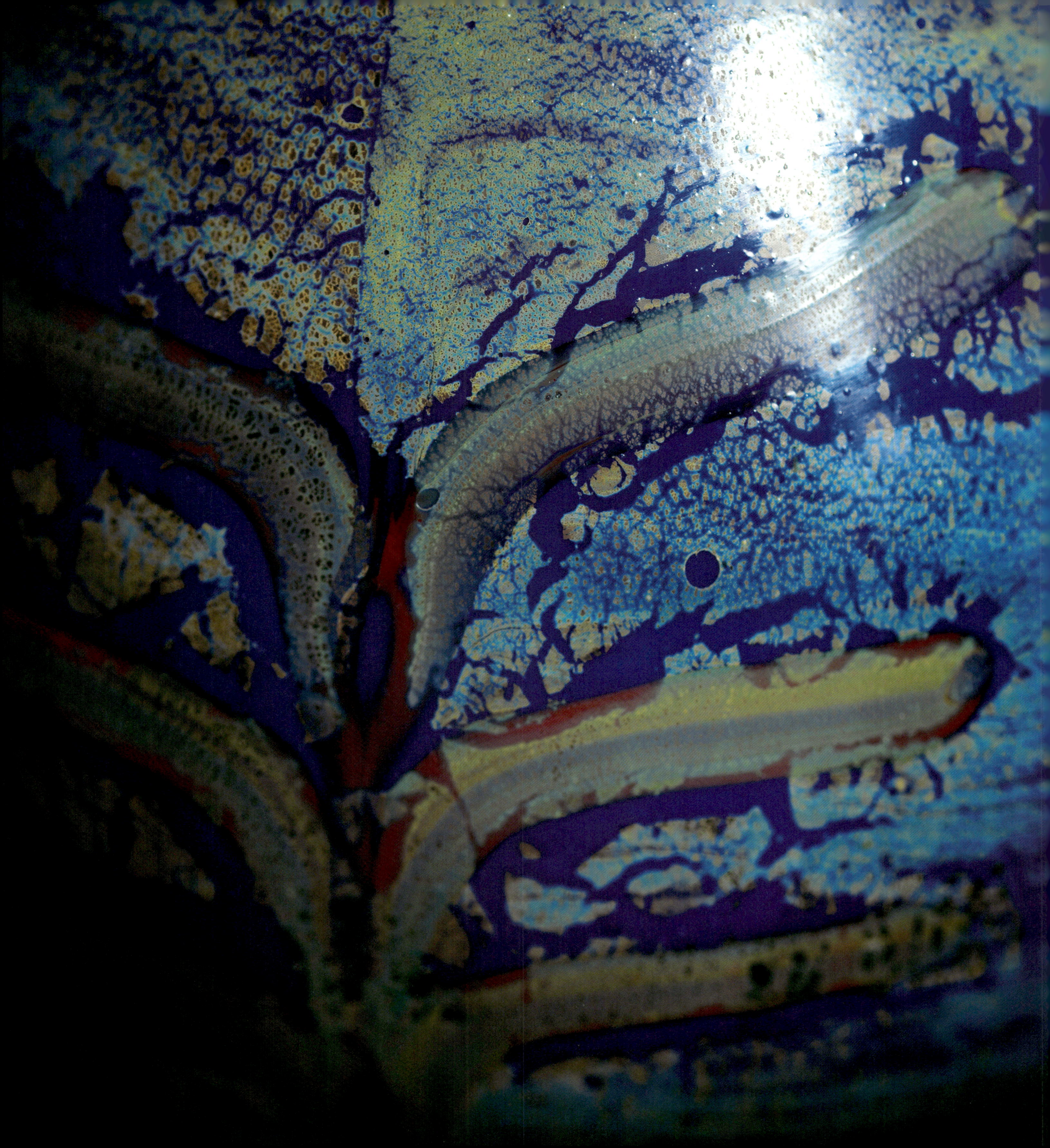

This is a small series of transparent glass pieces which are characterised by an *a palmette* (palm-leaf motif) ornamentation, applied as a relief on the surface of the vase or partially incorporated into its wall. The technique for executing the ornamentation is similar to that of the *decoro fenicio*, and was obtained by applying to the glass, during production, three or more horizontal threads which were 'combed' downwards in the middle with an appropriate instrument, so changing the rectilinear pattern and obtaining a design suggesting a palm leaf. The series includes a certain number of large vases, together with bowls and vases of smaller size and a fairly simplified design. Exhibited at the 3rd Biennale of Monza in 1927, it was proposed in *pagliesco* transparent glass with ornamentation and finishing in coral-red *pasta vitrea*, but other variants are also documented (e.g. with dark-blue decoration and finishing). Around 1929, with the increasing dominance of opaque glass, some models were produced in *pasta vitrea* and/or in black glass.

PALM-LEAF MOTIF 1927-29

Exhibitions

1927, Monza, *III Mostra Internazionale delle Arti Decorative*
1927, Venice, *XVIII Mostra dell'Opera Bevilacqua La Masa*

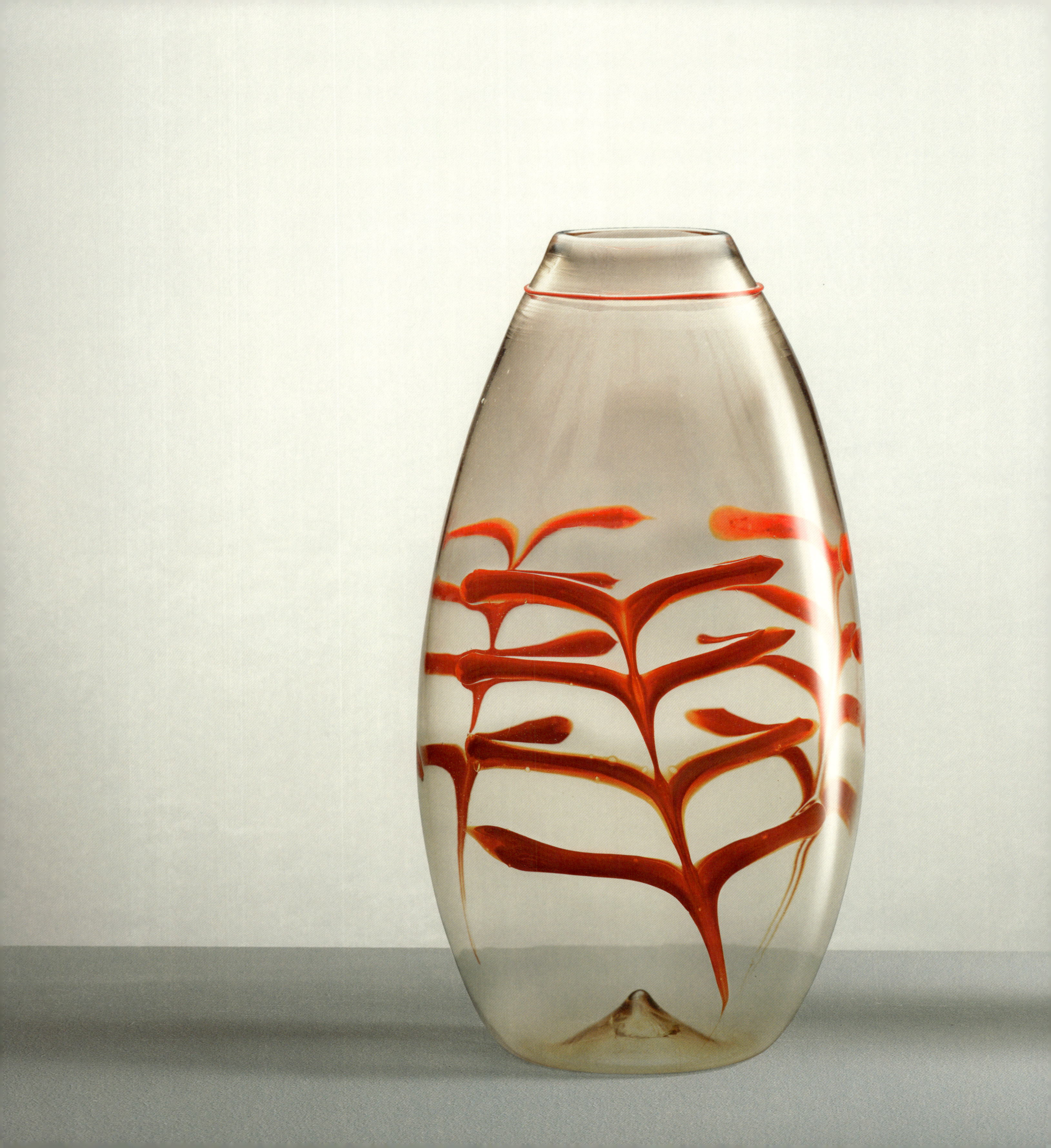

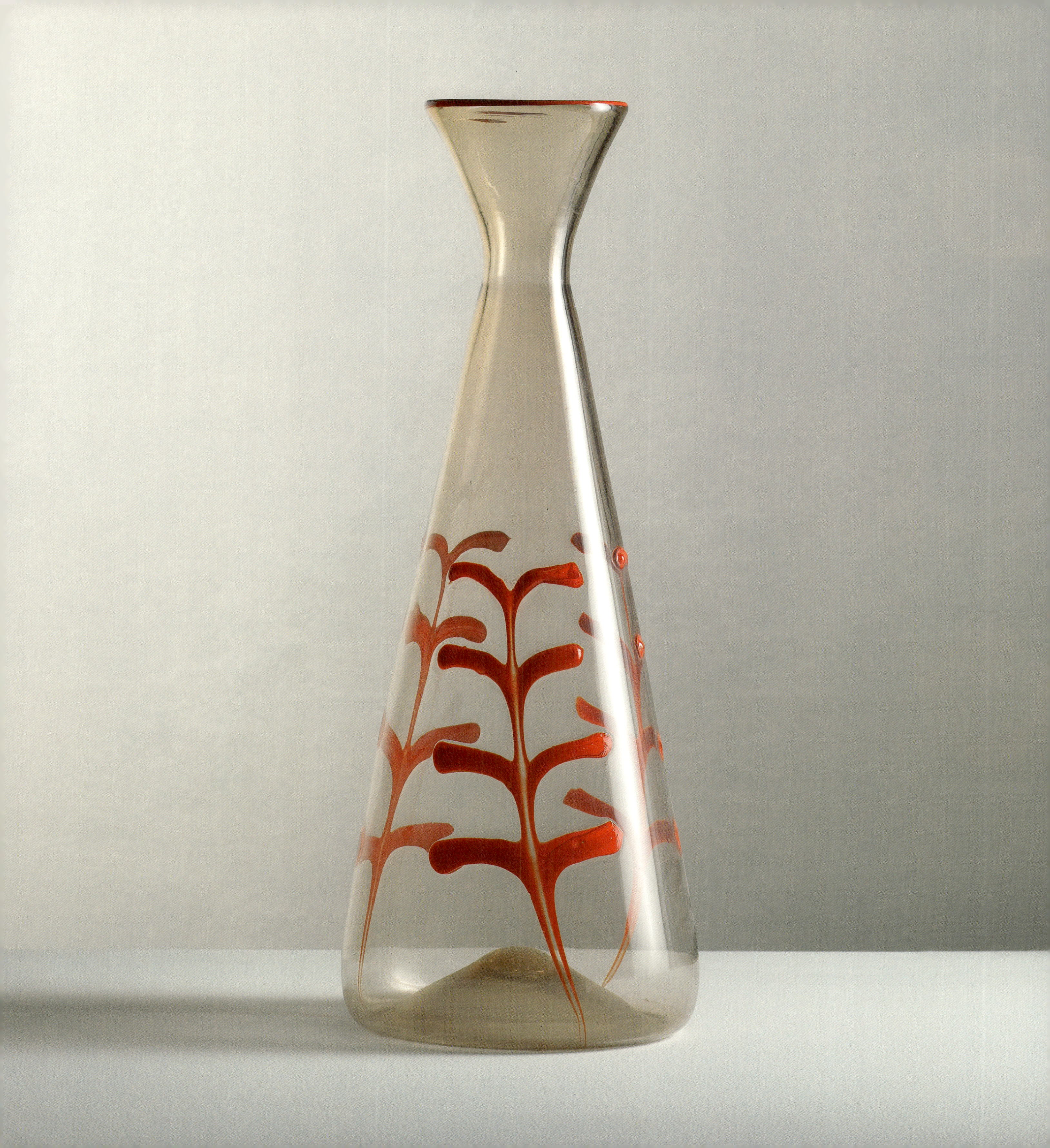

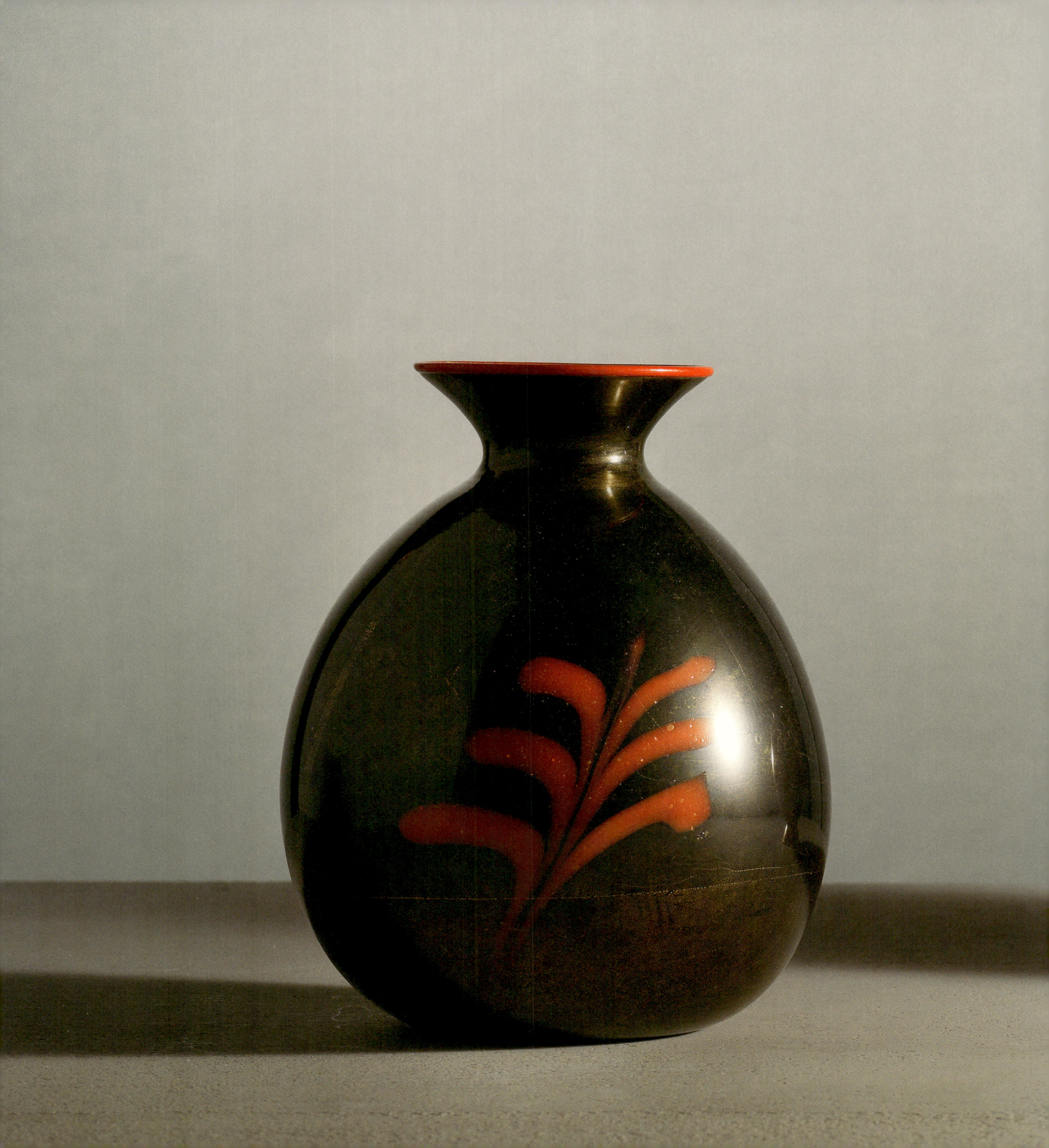

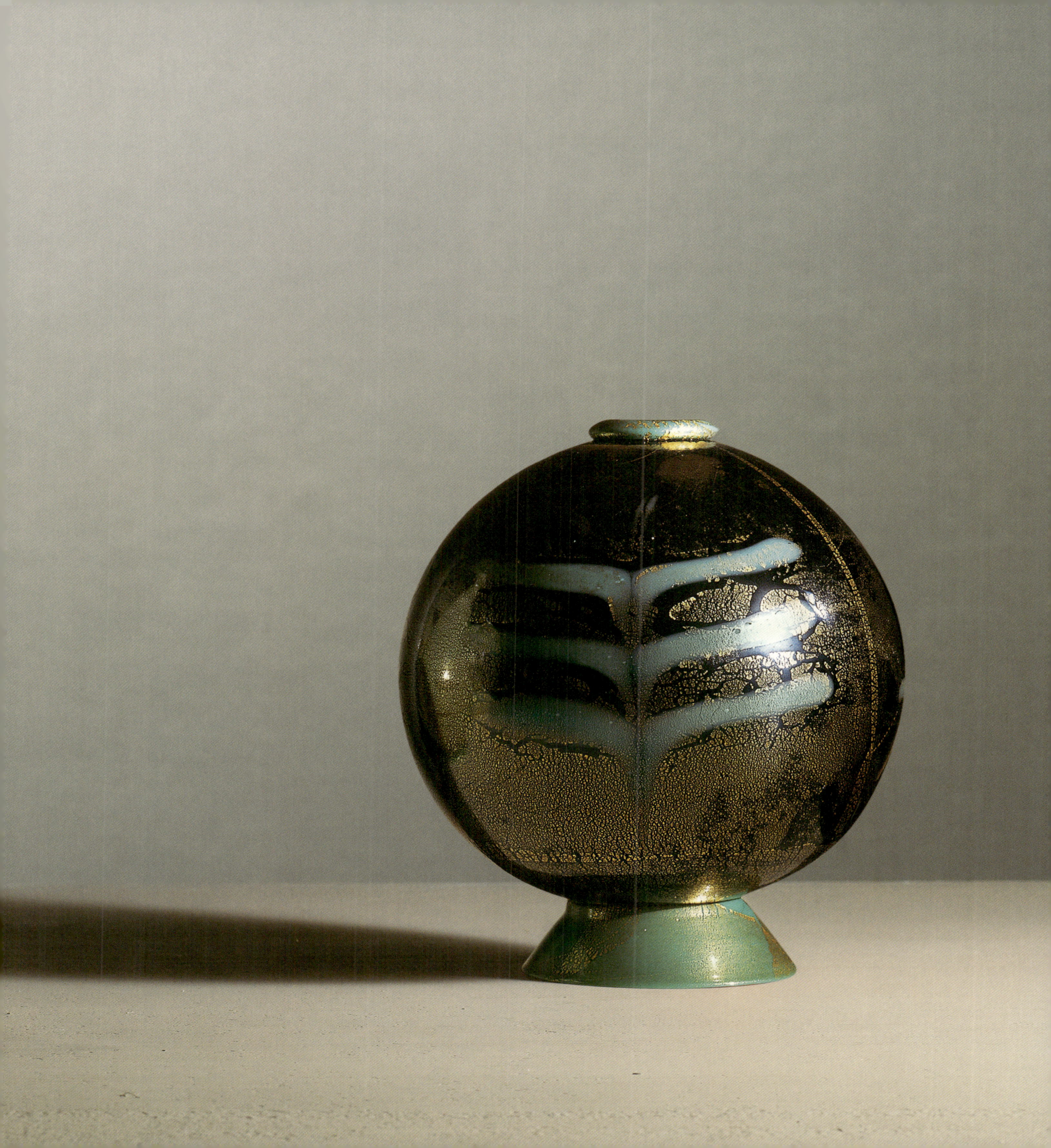

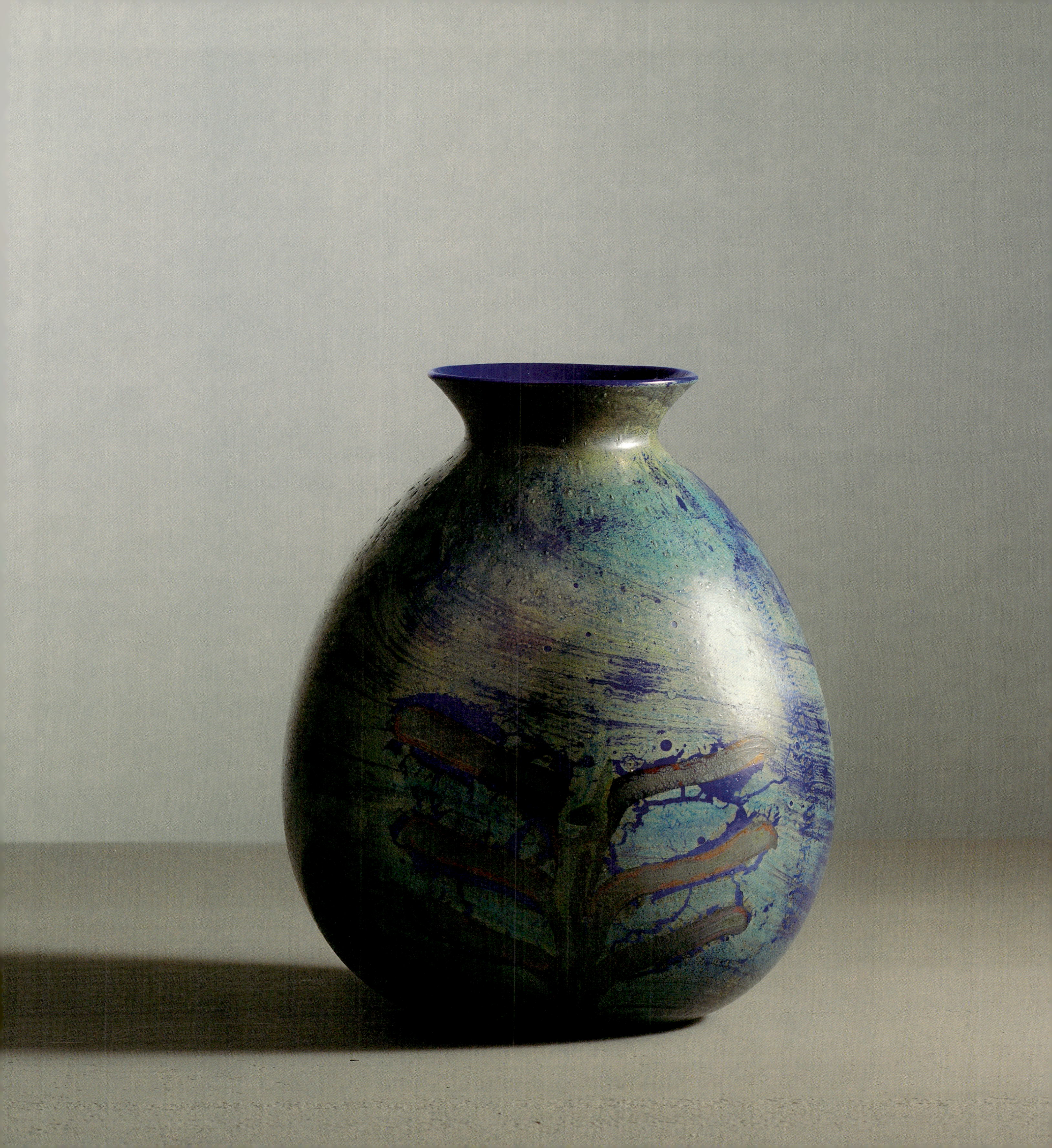

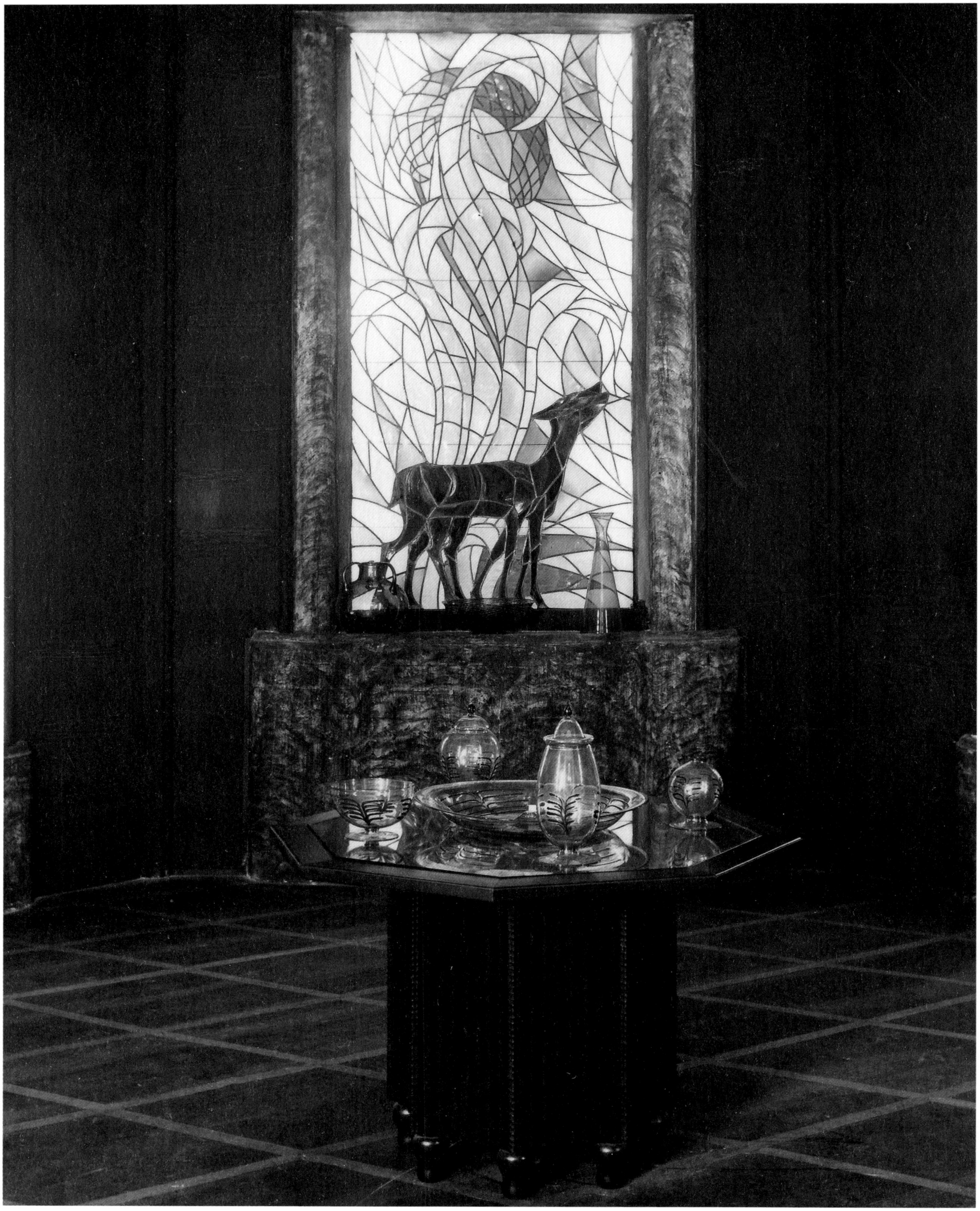

›
M.V.M. Cappellin room at the 3rd Biennale of Monza, 1927, detail, ACSVr. In the foreground: glass pieces with 'palmette' decoration

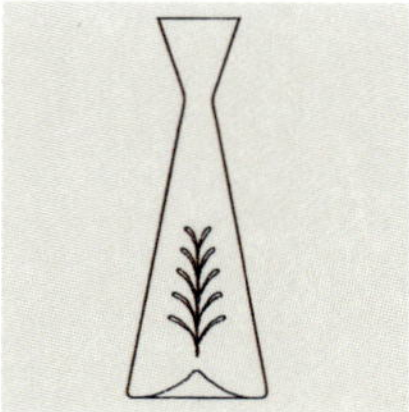

5717

Large vase in *pagliesco* glass with flared neck, decorated with three palm-leaf motifs applied on the body of the vase. Palm-leaf motifs and thread around the mouth in coral-red glass. Surface with iridised finish. The series with palm-leaf motif decorations was exhibited in 1927 at the 3rd Biennale of Monza. On this occasion a similar vase was also exhibited, without decoration (no. 5762), in transparent yellow glass and mouth with a green thread.
1927
46 cm high
M.V.M. Cappellin Murano *acid stamp*
Private collection

>
5717, catalogue drawing, AMB

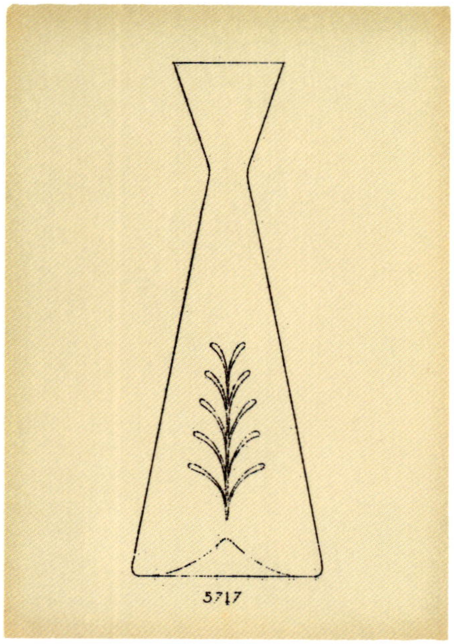

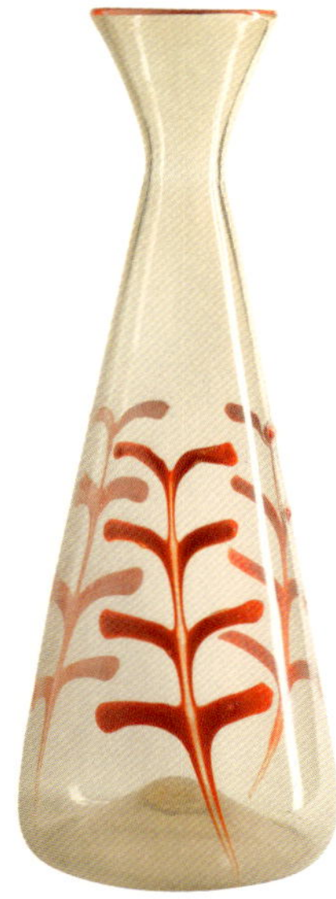

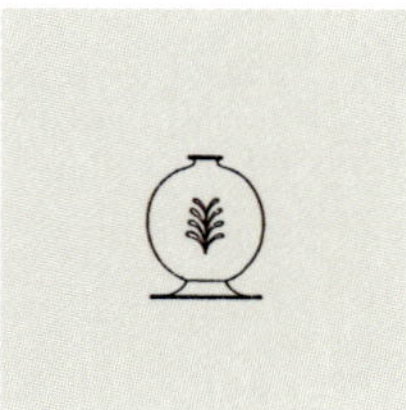

5723

Catalogue drawing for a spherical vase in transparent glass with applied palm-leaf motifs decoration, 1927-1931, AMB. An example in 'pagl[iesco] coral-red dec[oration]' glass was exhibited in 1927 at the 3rd Biennale of Monza. The model was also included in the catalogue without the decoration (no. 5759).
1927
17 cm high

exhibitions
1927, Monza, 3rd Biennale

>
43, design for a model with a palm-leaf motifs decoration to be exhibited at the 3rd Biennale of Monza, c. 1927, AMuVe

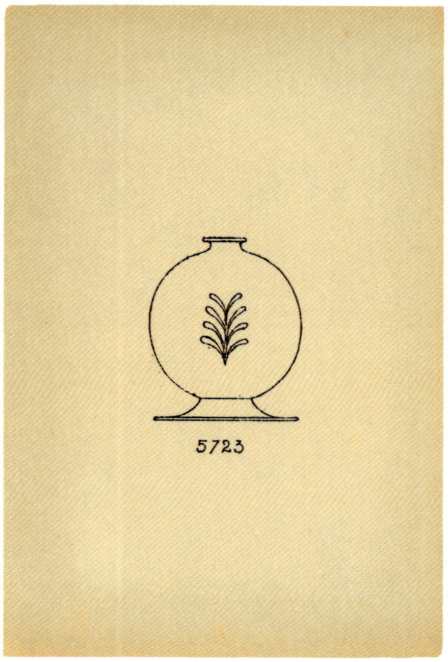

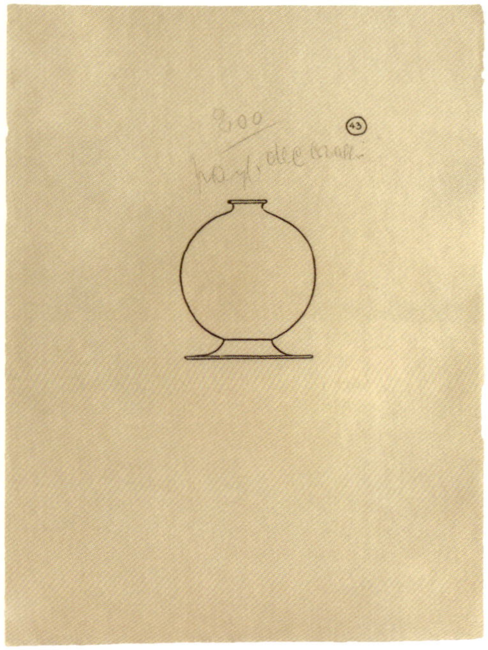

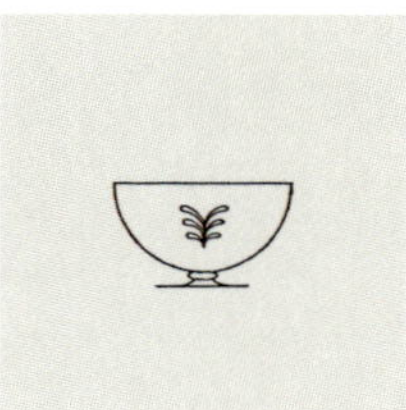

5724

Working drawing for a bowl in transparent glass decorated with 3 applied palm-leaf motifs, CSV. Two examples in '*pagliesco* coral-red dec[oration]' were exhibited in 1927 at the 3rd Biennale of Monza. The model also appeared in the catalogue without decoration (no. 5760).
1927
12 cm high; 20 cm ø

exhibitions
1927, Monza, 3rd Biennale

>
Bowl no. 5724 in a showcase of the Deutsche Werkstätten. Period photograph, c. 1929, detail, ACSVr

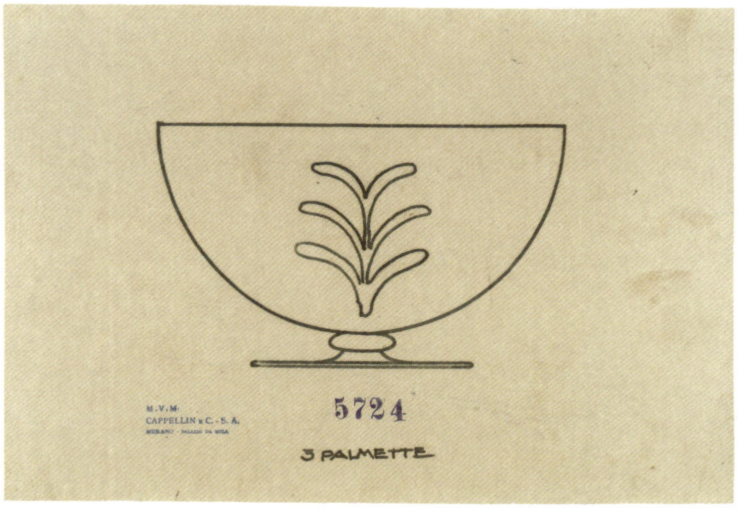

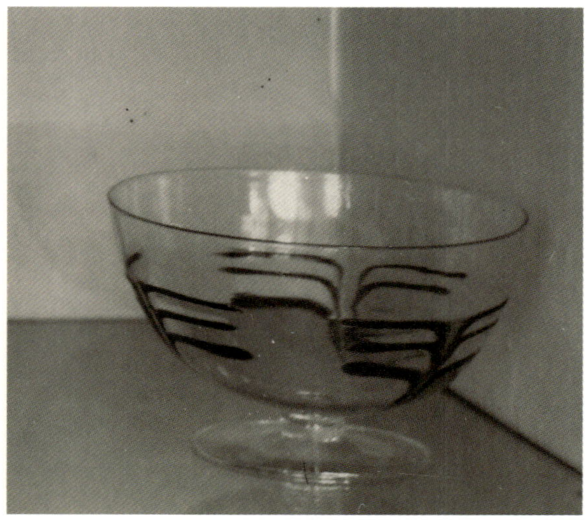

>
Glass pieces exhibited at the 3rd Biennale of Monza, 1927, detail, AFT

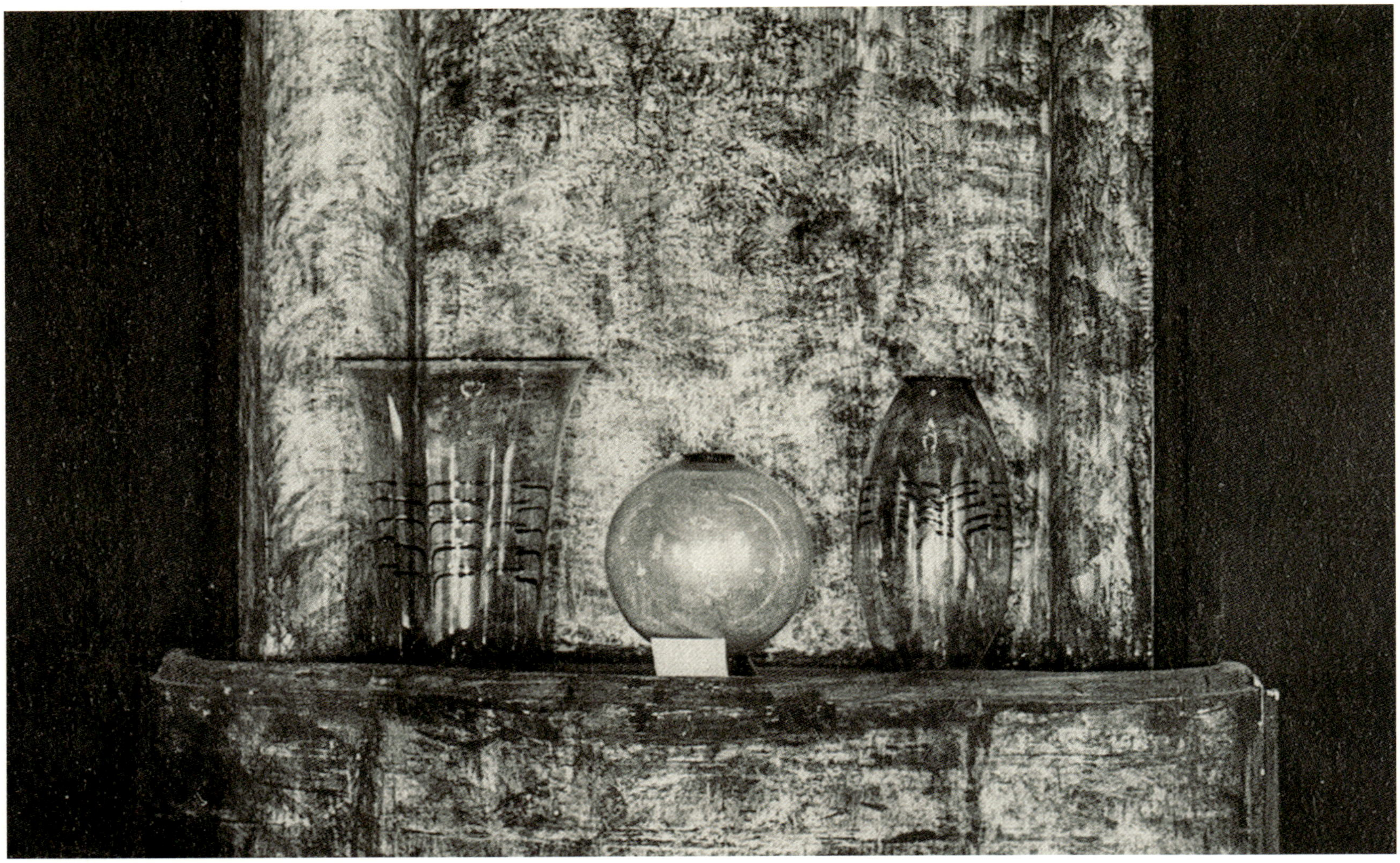

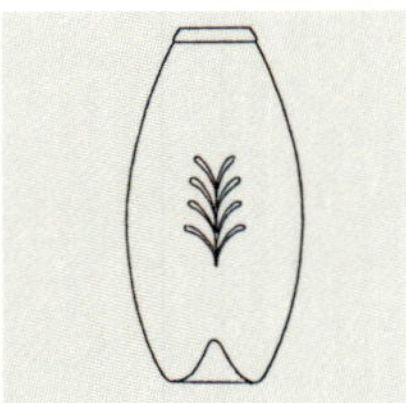

5718

Large vase in *pagliesco* glass decorated with three palm-leaf motifs applied on the body of the vase. Palm-leaf motifs and collar in coral-red glass. Surface with iridised finish. A similar example was exhibited in 1927 at the 3rd Biennale of Monza. The model also appeared in the catalogue without decoration (no. 5763).
1927
43 cm high
Stiftung Museum Kunstpalast collection, Düsseldorf-Glasmuseum Hentrich

exhibitions
1927, Monza, 3rd Biennale

>
5718, catalogue drawing, AMB

>
44, design for a model with palm-leaf motif decoration to be exhibited at the 3rd Biennale of Monza, c. 1927, AMuVe

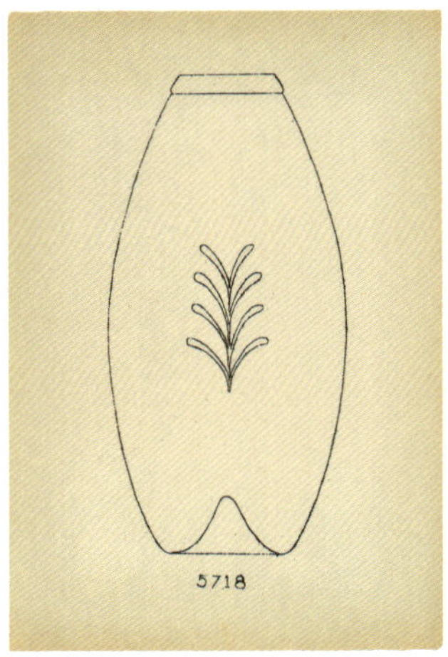

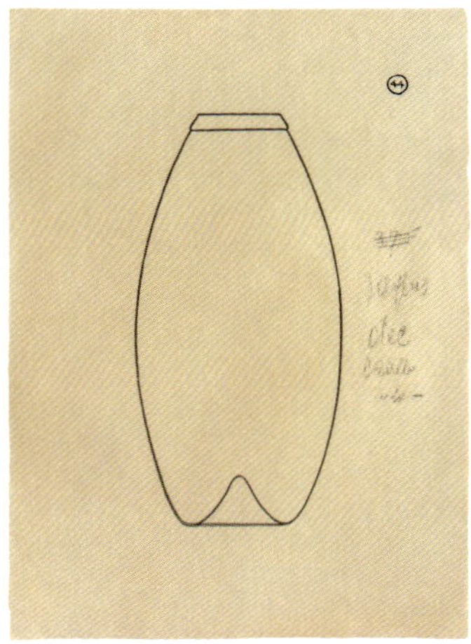

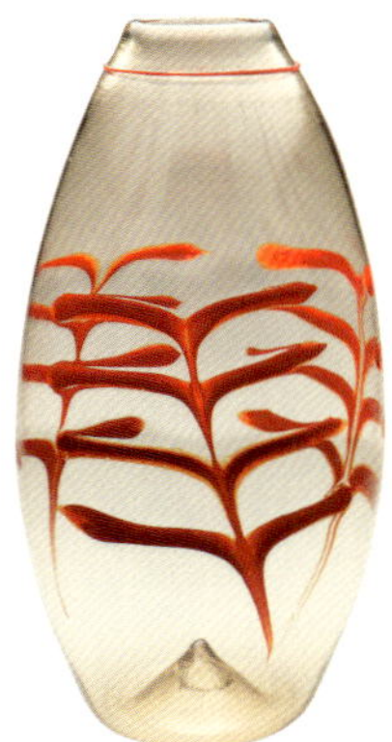

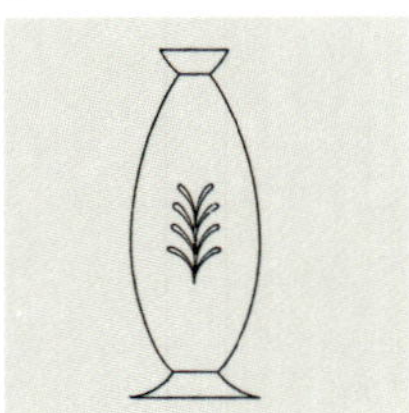

5719

Catalogue drawing for a large vase, with a flattened cross-section, in transparent glass with an applied palm-leaf motif decoration, AMB. Flared foot and neck. Two examples of this model, in '*pagl[iesco]* coral-red dec[oration]' glass were exhibited in 1927 at the 3rd Biennale of Monza. The model also appeared in the catalogue without decoration (no. 5764).
1927
48 cm high

exhibitions
1927, Monza, 3rd Biennale

>
46, design for a model with palm-leaf motif decoration to be exhibited at the 3rd Biennale of Monza, c. 1927, AMuVe

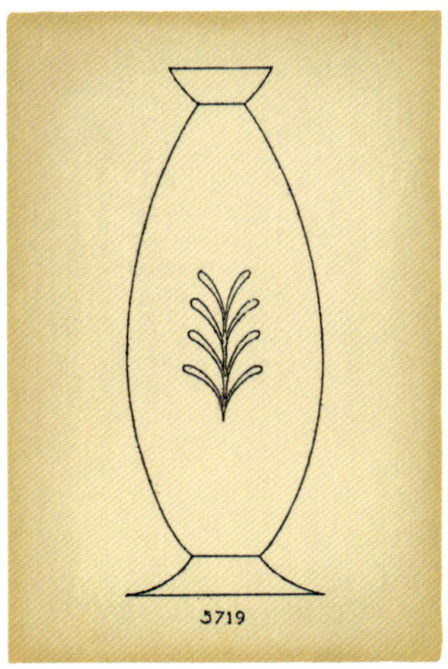

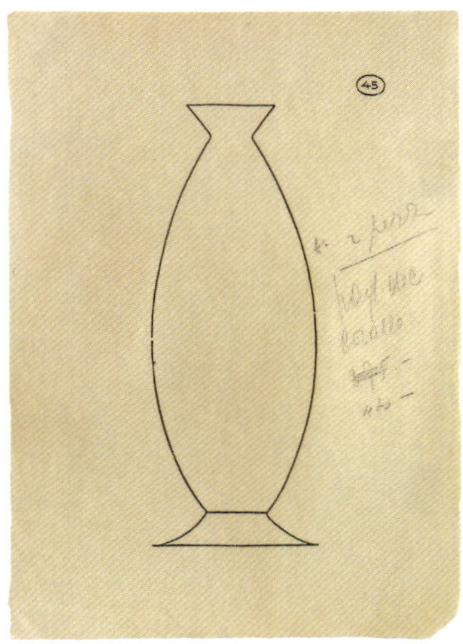

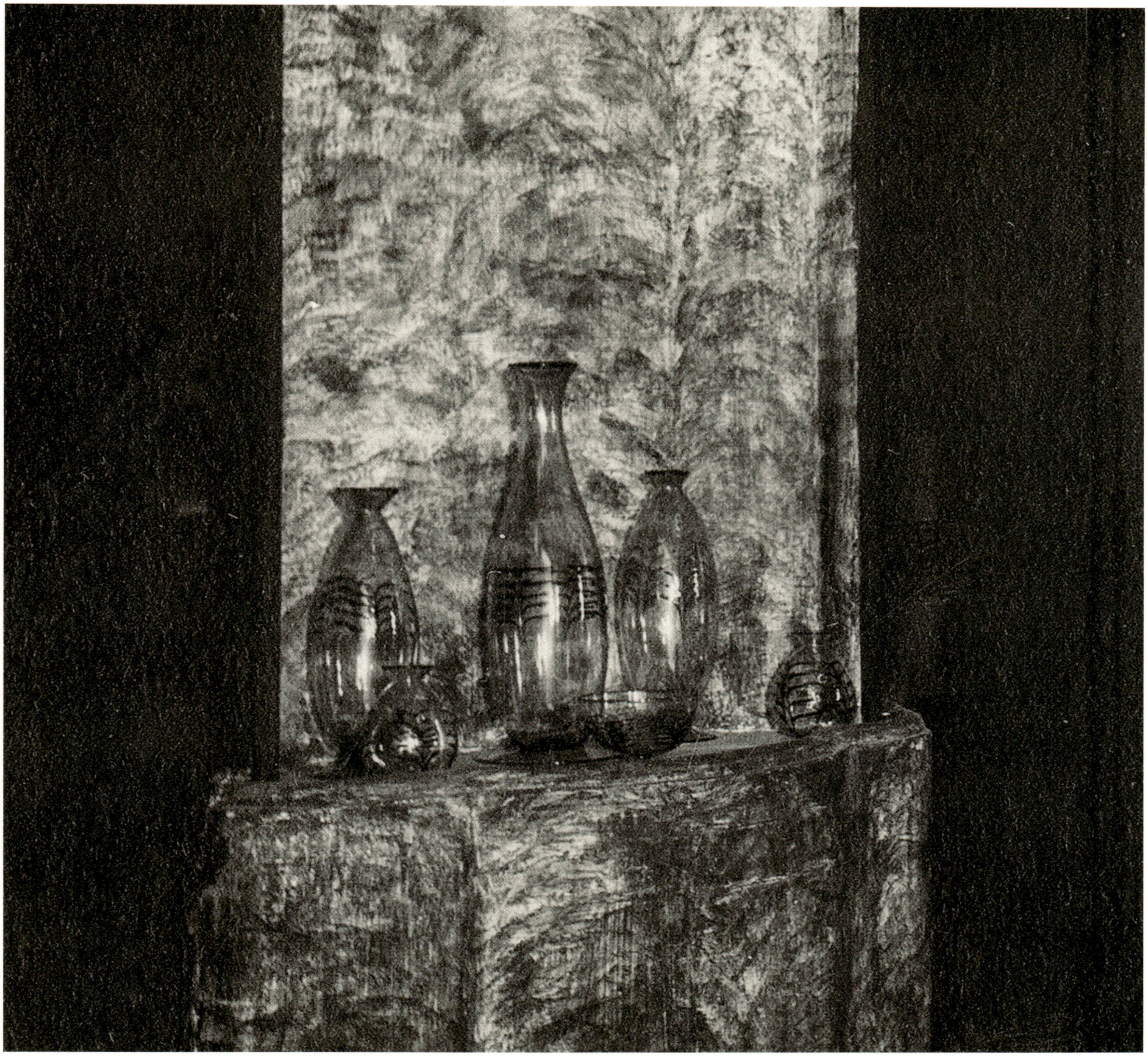

›
Glass pieces exhibited at the 3rd Biennale of Monza, 1927, detail, AFT

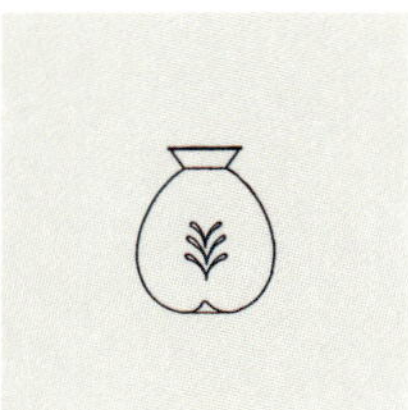

5725

Working drawing for a vase with a flared neck and applied palm-leaf motif decoration, CSV. Two examples of this model, in '*pagl[iesco*] coral-red dec[oration]' glass, were exhibited in 1927 at the 3rd Biennale of Monza. The model also appeared in the catalogue without decoration (no. 5761) and was subsequently produced also in opaque glass.
1927
19 cm high

exhibitions
1927, Monza, 3rd Biennale

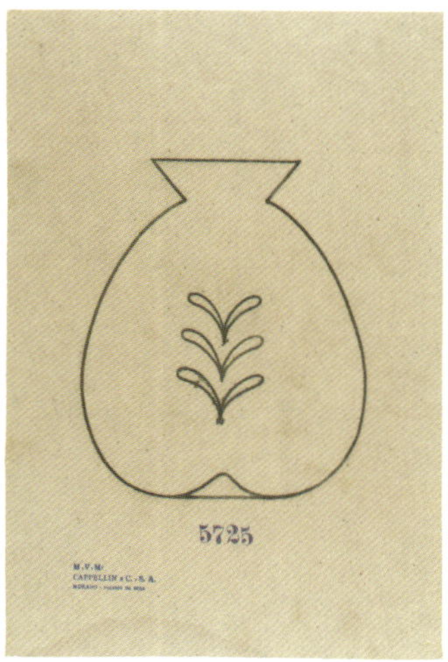

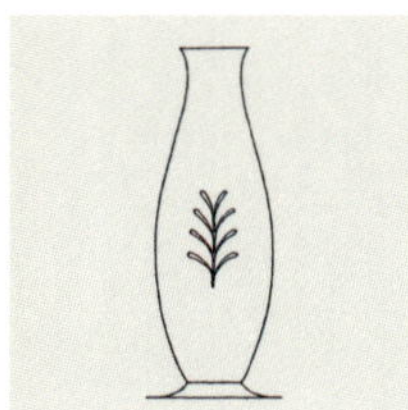

5720

Catalogue drawing for a large vase with a flattened cross-section in transparent glass with applied palm-leaf motif decoration, AMB. Small flared foot. A similar example, in '*pagliesco* coral-red dec[oration]' glass, was exhibited in 1927 at the 3rd Biennale of Monza. The model also appeared in the catalogue without decoration (no. 5765).
1927
61,5 cm high

exhibitions
1927, Monza, 3rd Biennale

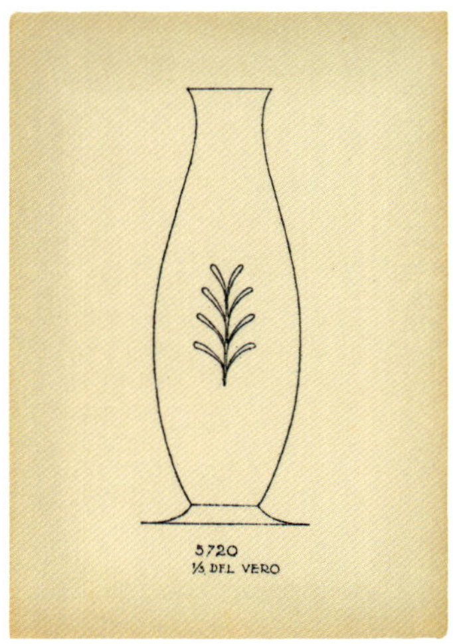

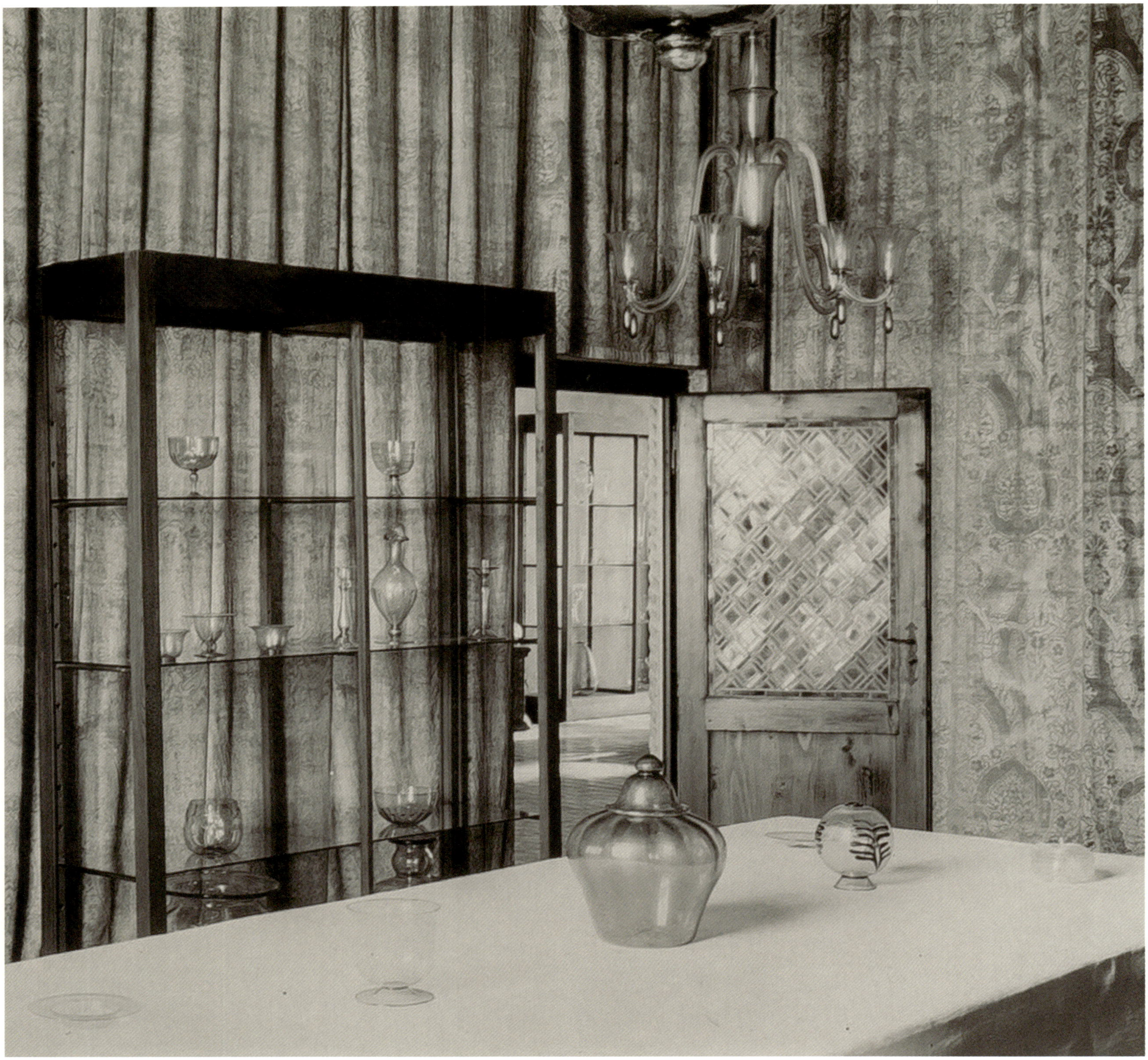

> *Exhibition room at the Palazzo Da Mula, Murano. Period photograph, c. 1928, detail, CSV*

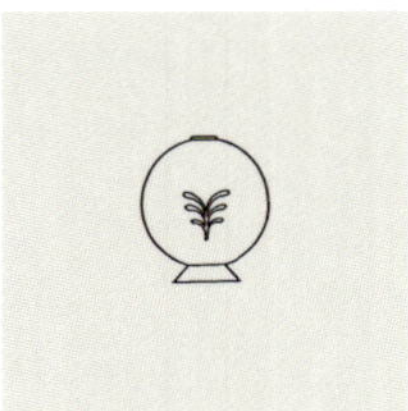

5755

Spherical vase with truncated-cone-shaped foot in crystal glass decorated with three palm-leaf motifs applied to the body of the vase. Palm-leaf motifs and finishings in dark-blue glass. Surface with iridised finish. A similar example in '*pagliesco* [and] coral-red' was exhibited in 1927, in Venice, at the 28th Mostra dell'Opera Bevilacqua La Masa. The model was subsequently also produced in opaque glass.
1927
18 cm high
M.V.M. Cappellin Murano *acid stamp*
Private collection

exhibitions
1927, Venice, 28th Mostra dell'Opera Bevilacqua La Masa

> *5755, catalogue drawing, AMB*

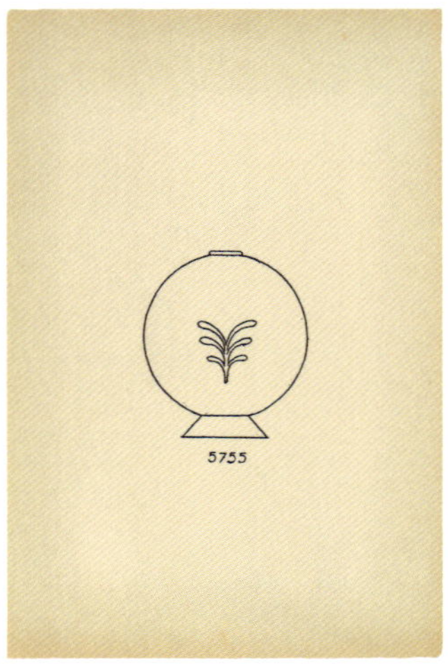

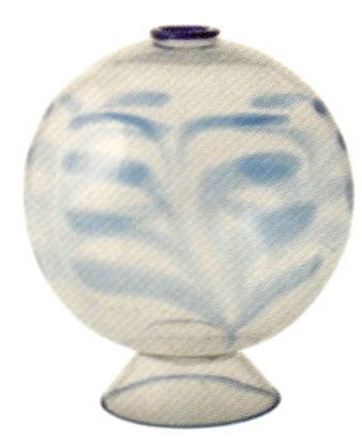

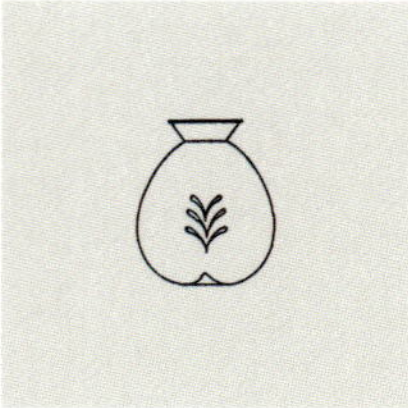

5725

Vase in black glass with flared neck decorated with three applied palm-leaf motifs, in coral-red glass. Surface finished with applied gold leaf. Finishing at the mouth in coral-red glass.
c. 1929
20 cm high
Private collection

>
5725, catalogue drawing, AMB

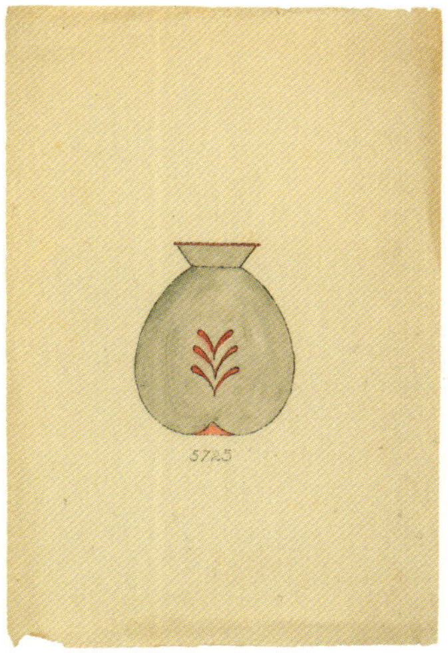

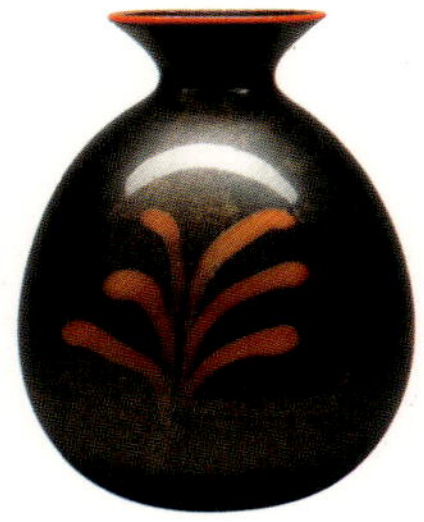

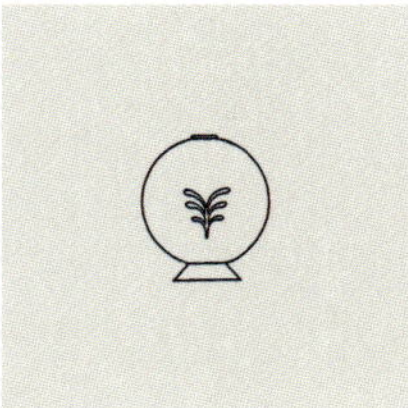

5755

Spherical vase in black glass decorated with palm-leaf motifs applied in green *pasta vitrea*. Truncated-cone-shaped foot and mouth in green *pasta vitrea*. Surface finished with applied gold leaf.
c. 1929
16 cm high
M.V.M. Cappellin Murano *acid stamp*
The Olnick Spanu Collection, New York

>
5755, catalogue drawing, AMB

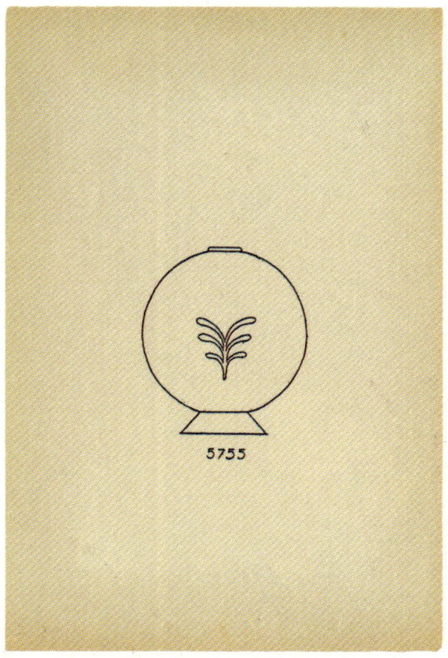

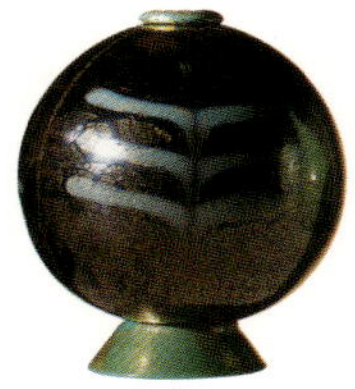

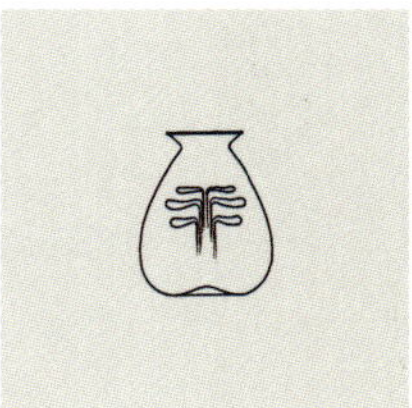

5962

Vase in dark-blue *pasta vitrea* decorated with three applied palm-leaf motifs in coral-red glass. Surface finished with applied oxidised silver leaf.
c. 1929
19 cm high
M.V.M. Cappellin Murano *acid stamp*
Private collection, Venice

>
5962, catalogue drawing, AMB

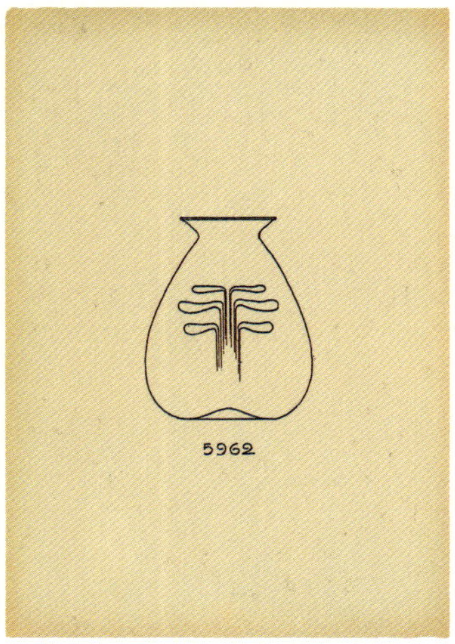

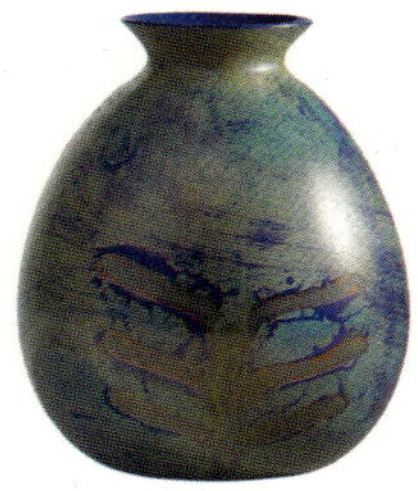

---

n.i.

Spherical vase in black glass decorated with applied palm-leaf motifs in coral-red glass. Surface finished with applied gold leaf.
c. 1929
21 cm high
Private collection

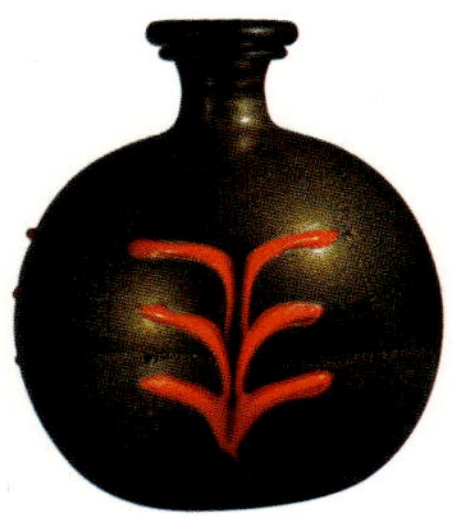

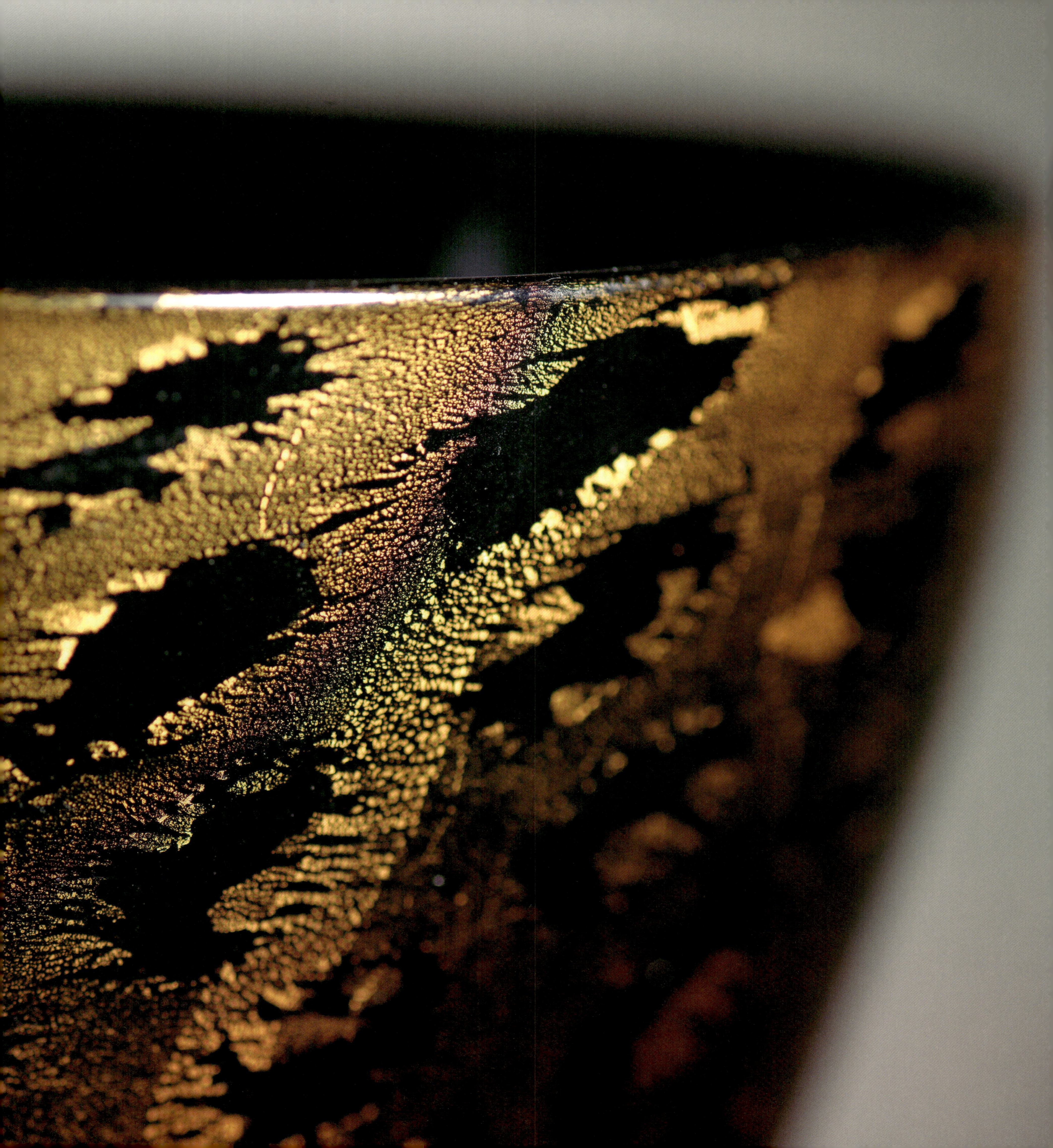

Among the prized glass techniques used by Cappellin to characterise its production is that which in Murano is called '*oro spacà*' (*oro spaccato* – broken gold), with which a number of artefacts were produced towards the end of the 1920s, using the *balloton* technique and the application of gold leaf. The processing included the application of gold leaf on the '*péa*' (globe in the shape of a pear constituting the initial state of the object being made) which is then blown in a special open metallic mould, called a *balloton,* which contains little pyramid-shaped points with a square base. These, pushed into the wall of the semi-finished piece, provide the surface with a unique decorative motif, generally in relief, amplified by the use of gold which, at the end of processing, appears to form a characteristic honeycomb network. This procedure was used to produce several models, whether in transparent or opaque glass, the gold leaf sometimes being replaced by silver leaf.

BALLOTON c. 1929

Exhibitions

1931, Amsterdam, *Tentoonstelling van glas, aardewerk en kant, van moderne italiaansche kunstnijverheid. Mostra di vetri, ceramiche e merletti d'arte moderna italiana*

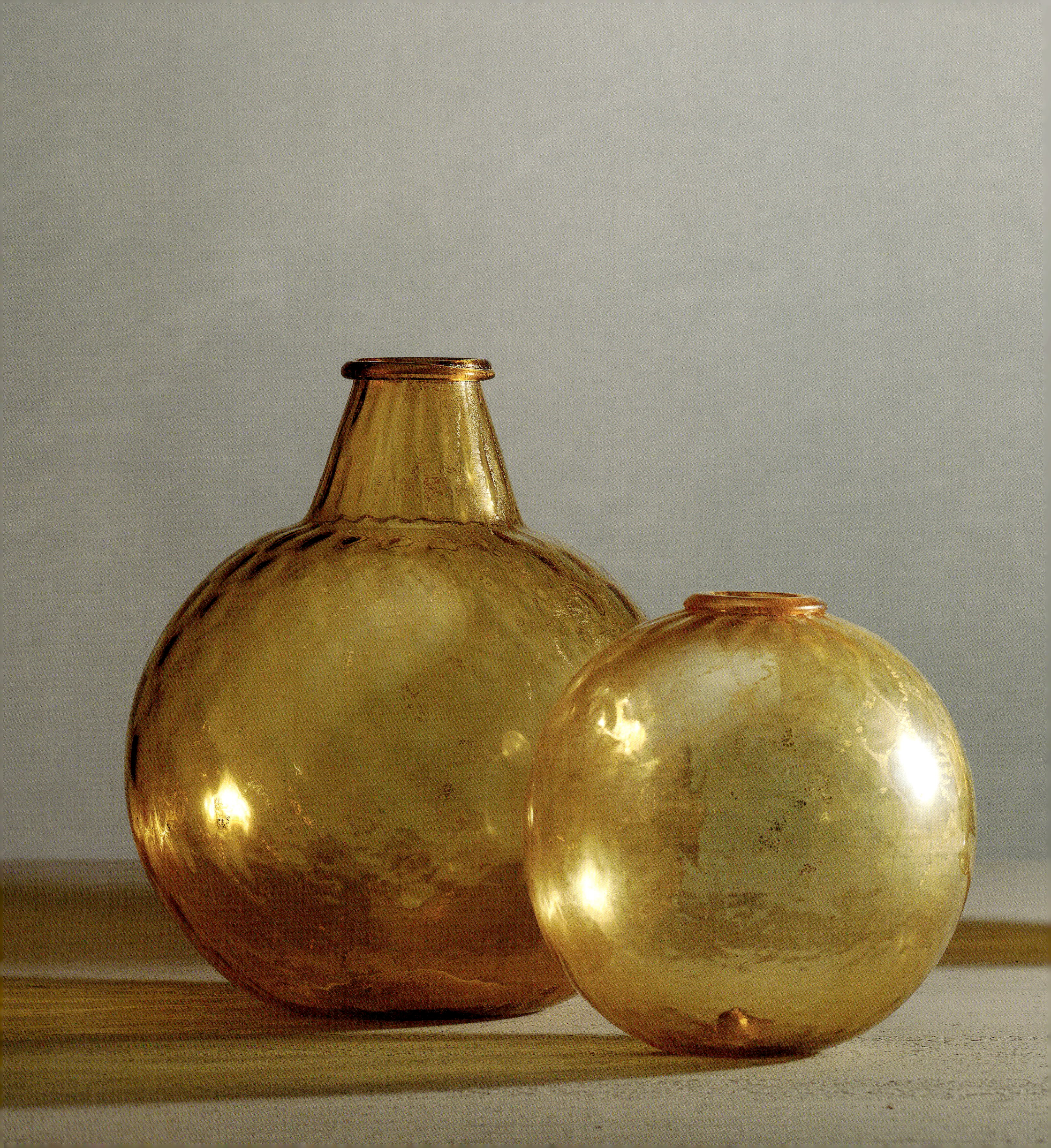

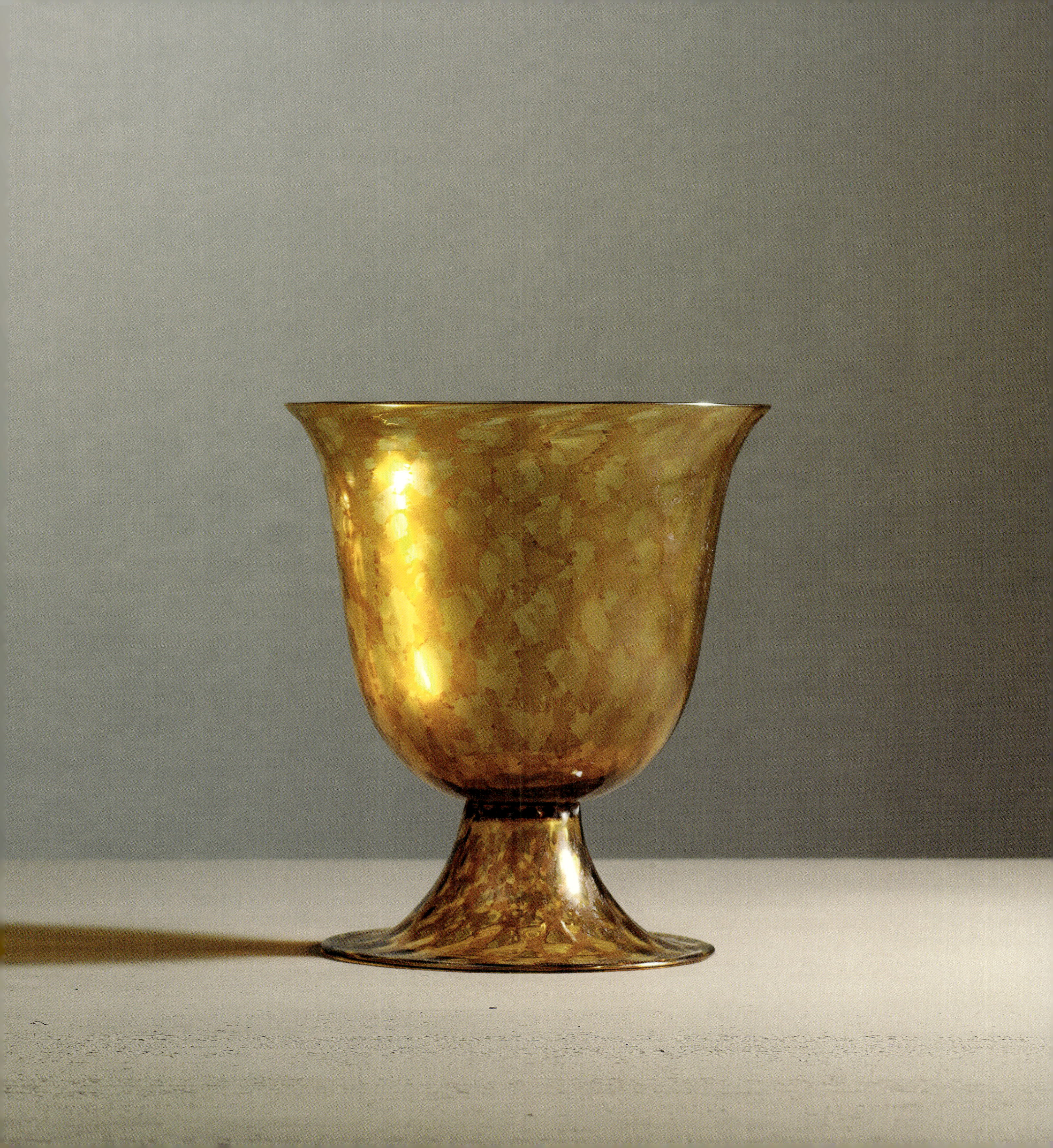

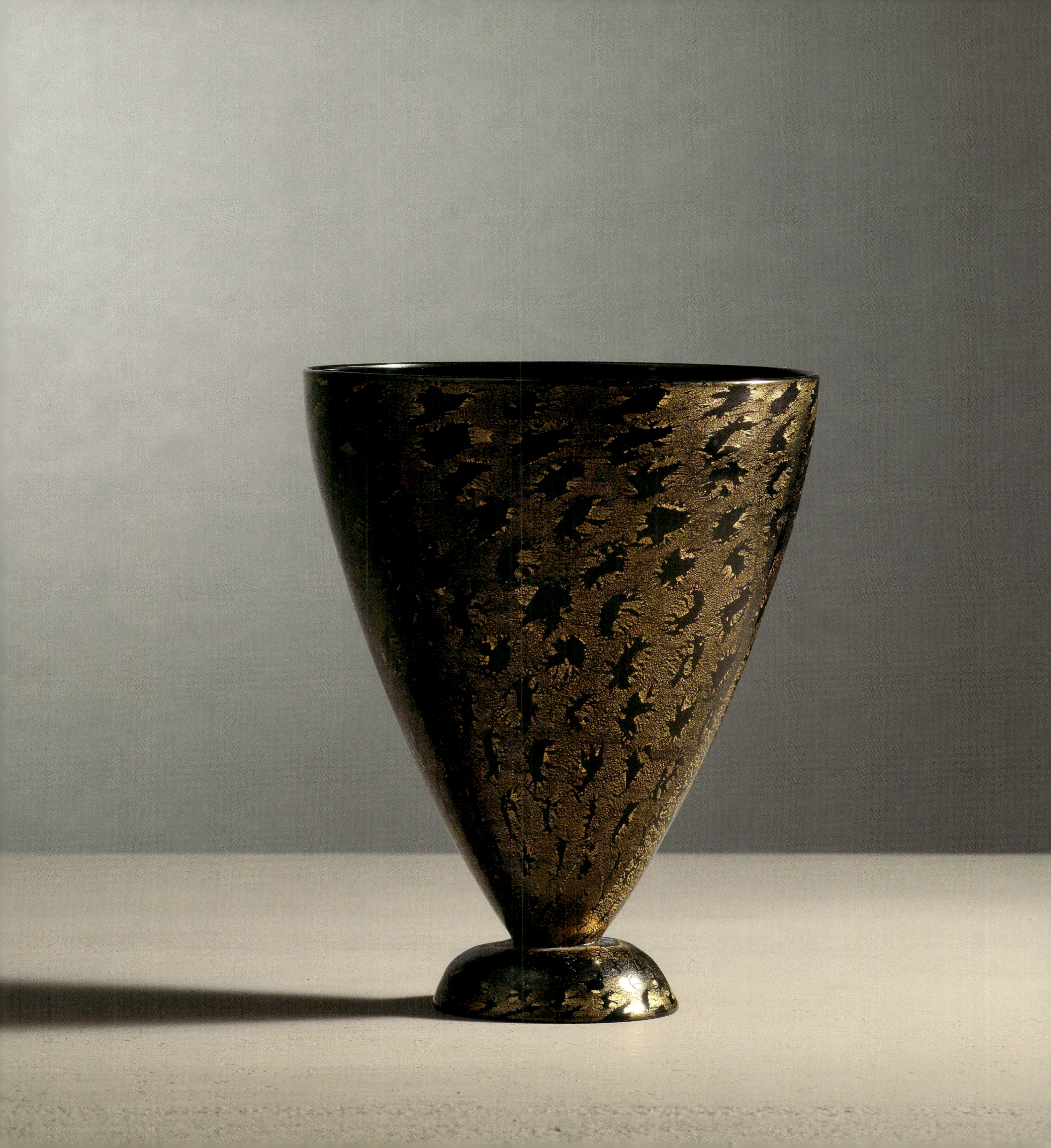

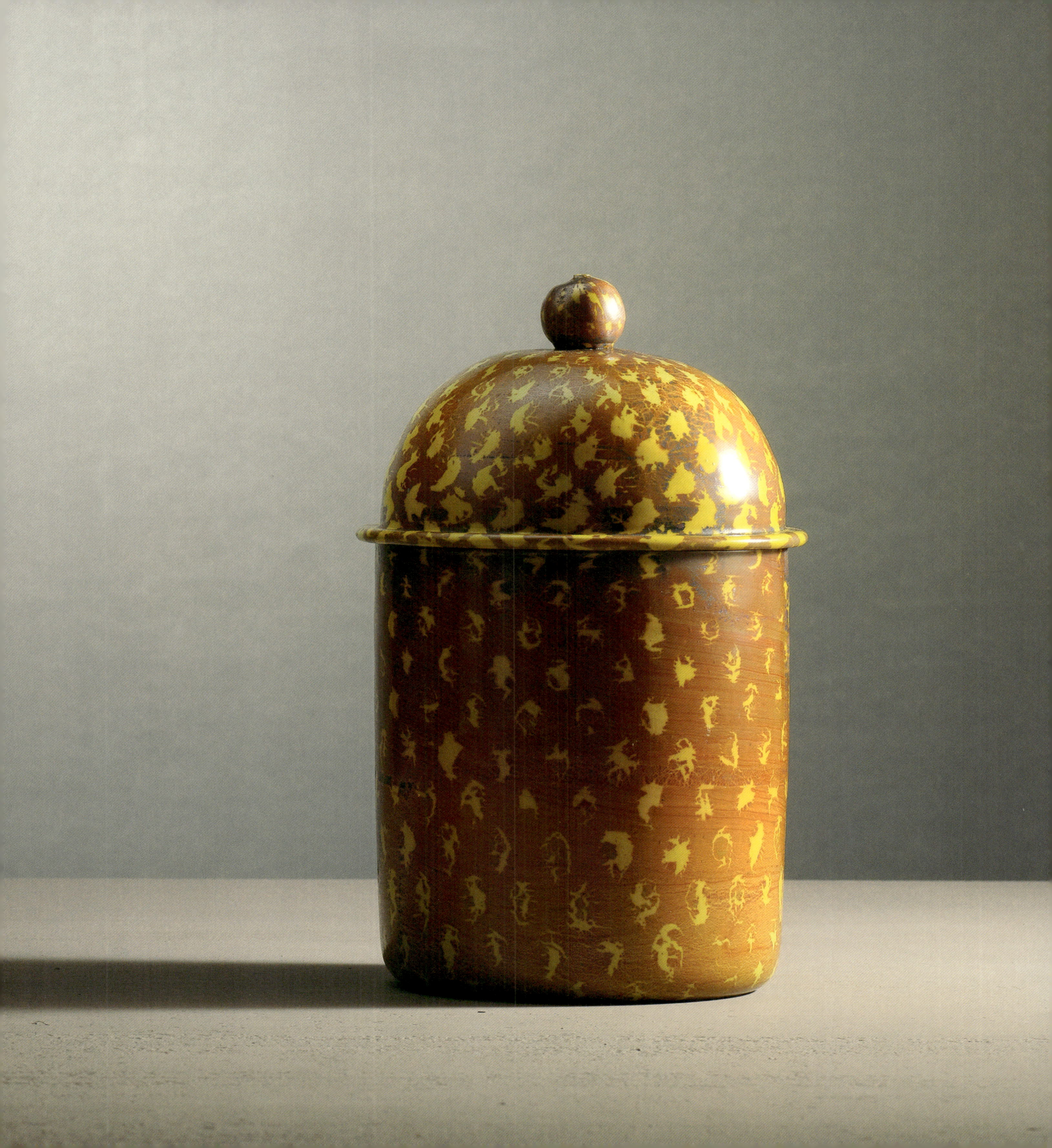

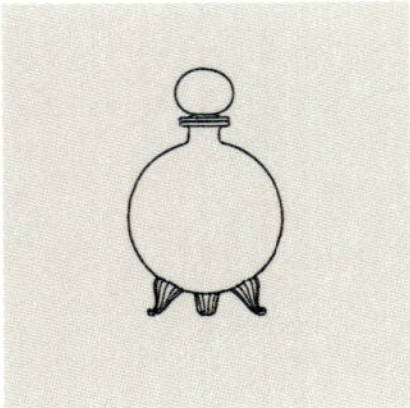

1285

Bottle in transparent glass with *balloton* decoration and applied gold leaf, documented by a period photograph, c. 1929, CSV. Spherical body supported by three globular feet. The piece was produced as a variant of no. 1285.
c. 1929
29 cm high

>
1285, working drawing, CSV

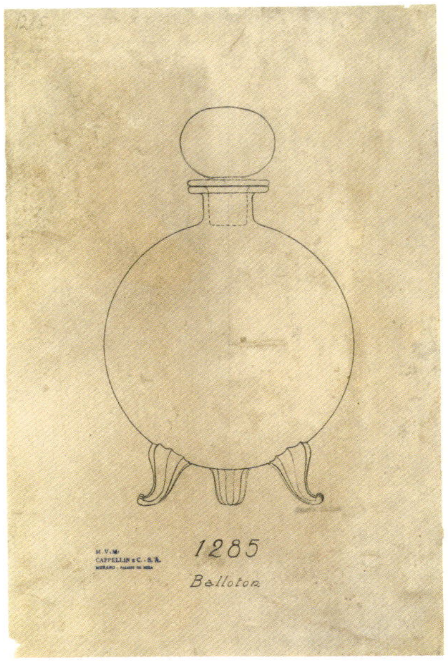

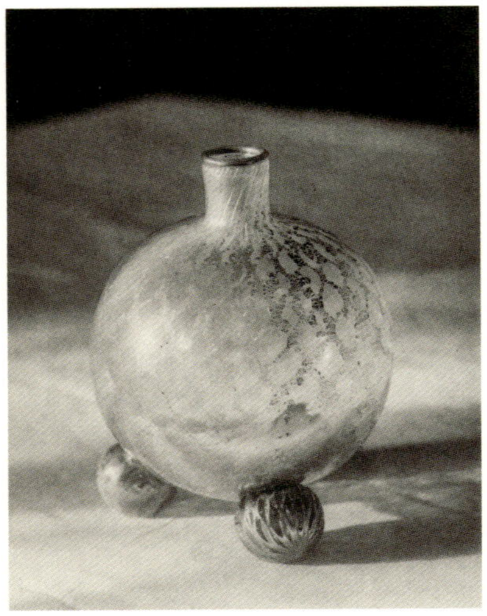

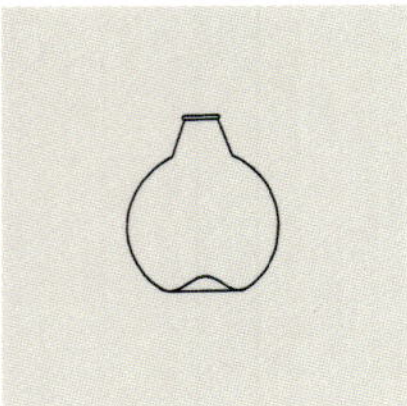

5858

Vase in yellow transparent glass with *balloton* decoration obtained with a mould. Surface finished with applied gold leaf. Model no. 5858 was also produced with other glass techniques (e.g. smooth transparent, *pasta vitrea*).
c. 1929
21 cm high
M.V.M. Cappellin Murano *acid stamp*
John and Jean Geresi collection, USA

>
5858, catalogue drawing, AMB

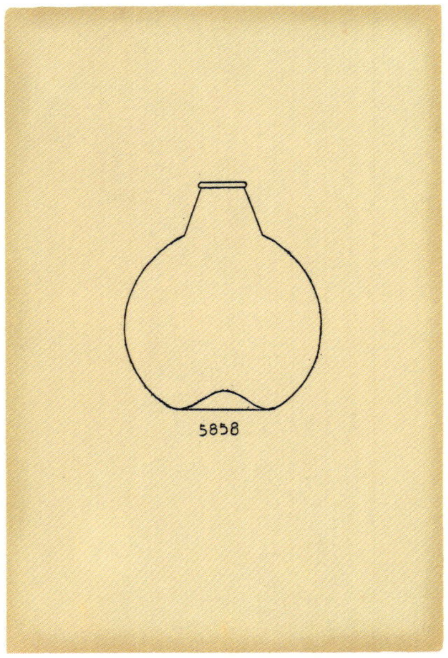

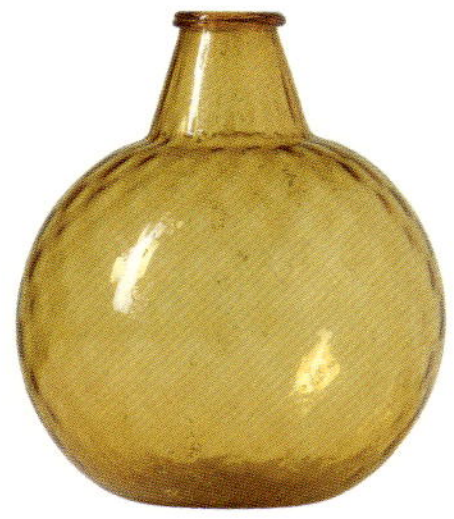

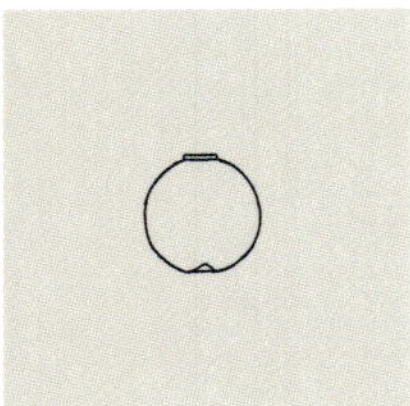

5934

Spherical vase in yellow transparent glass with *balloton* decoration obtained with a specific mould. Surface finished with applied gold leaf. Model no. 5934 was also produced with other glass techniques (e.g. *lattimo aurato*, *decoro fenicio*, *filgrana a reticello*).
c. 1929
13 cm high
Private collection

>
5934, catalogue drawing, AMB

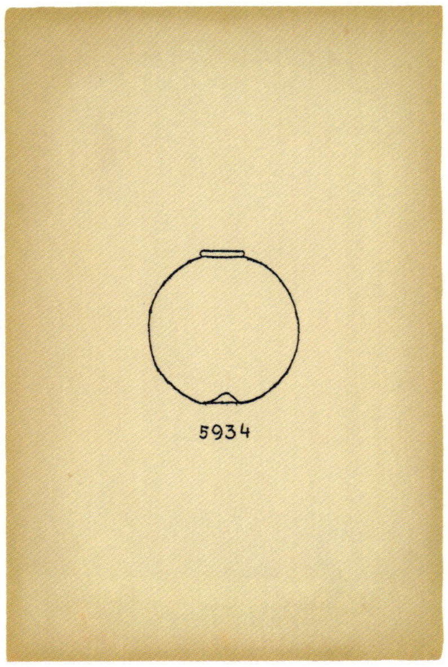

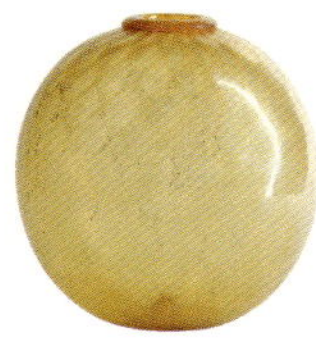

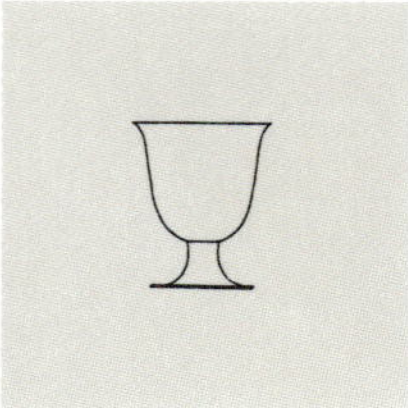

5286

Goblet in yellow transparent glass with *balloton* decoration obtained with a mould. Surface finished with applied gold leaf. The goblet was produced after a model in the catalogue of the glassworks designed by Vittorio Zecchin (1921–23).
c. 1929
19 cm high
M.V.M. Cappellin Murano *acid stamp*
Martin Kline collection, NY

>
5286, catalogue drawing, AMB

>
Octopus in opaque glass and goblet in transparent glass with balloton *decoration and applied gold leaf. Period photograph 1930–31, detail, CSV*

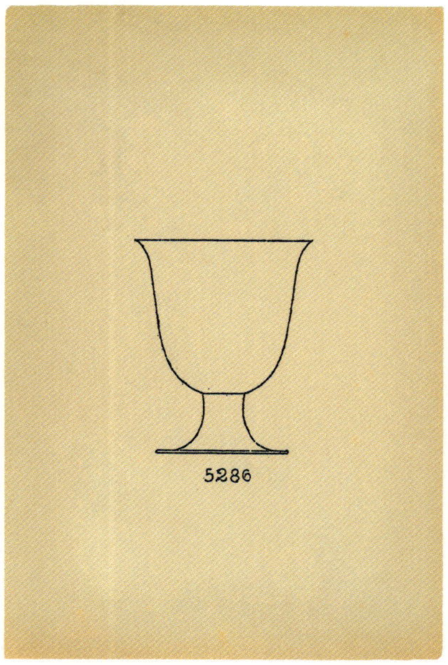

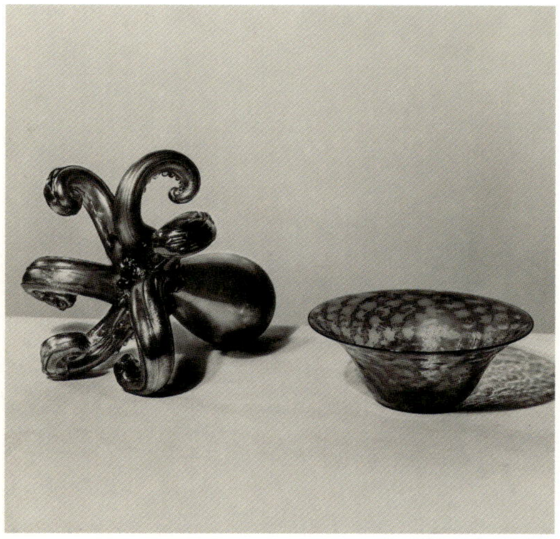

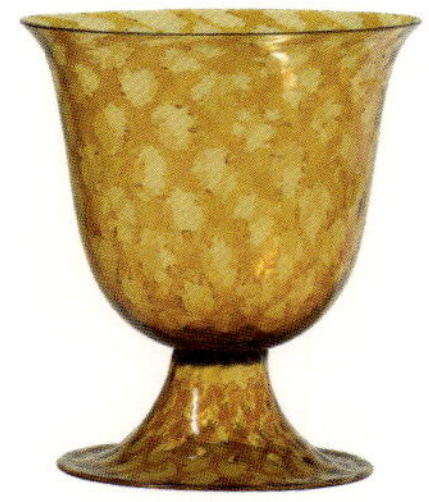

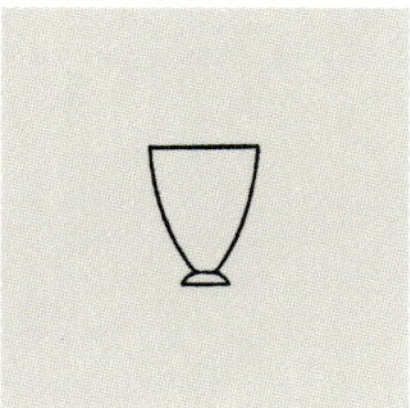

5937

Vase in black glass with *bal-loton* decoration and applied gold leaf. Model no. 5937 was also produced with other glass techniques (e.g. *decoro fenicio*, iridised black etc.).
c. 1929
17 cm high
Francesca and Massimo Valsecchi collection, Palazzo Butera

>
5937, catalogue drawing, AMH

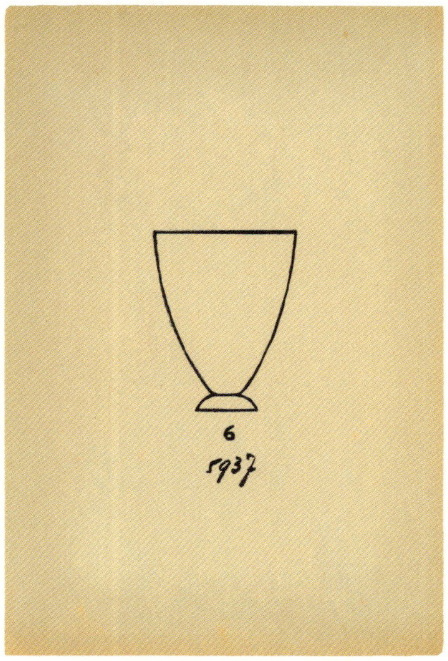

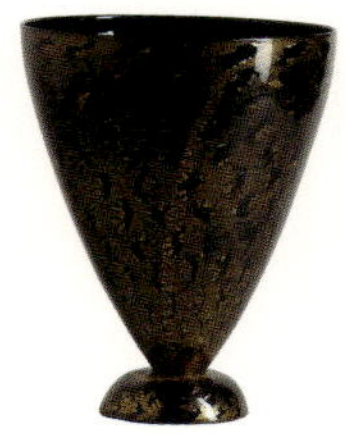

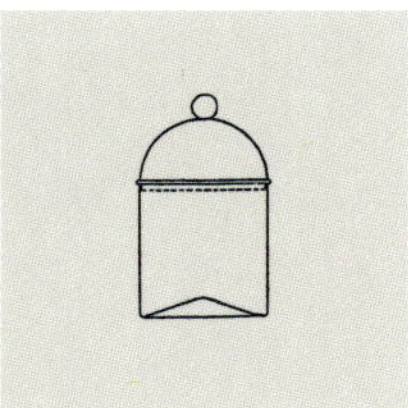

3848

Compote bowl in yellow opaque glass with applied oxidised silver leaf and characteristic diamond-shaped decoration obtained with the *balloton* technique. Cylindrical body and skullcap lid with spherical knob. Model no. 3848 was also produced with other glass tech-niques (e.g. transparent, *filigra-na a reticello*).
c. 1929
29 cm high
Anna Bozano collection, Genoa

>
3848, catalogue drawing, AMB

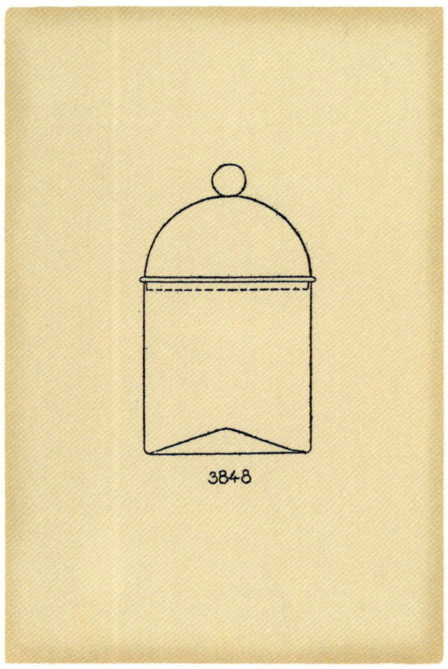

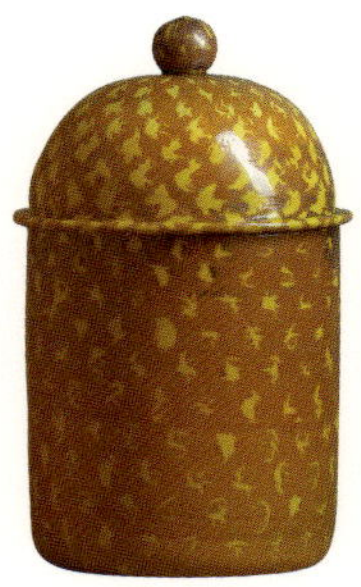

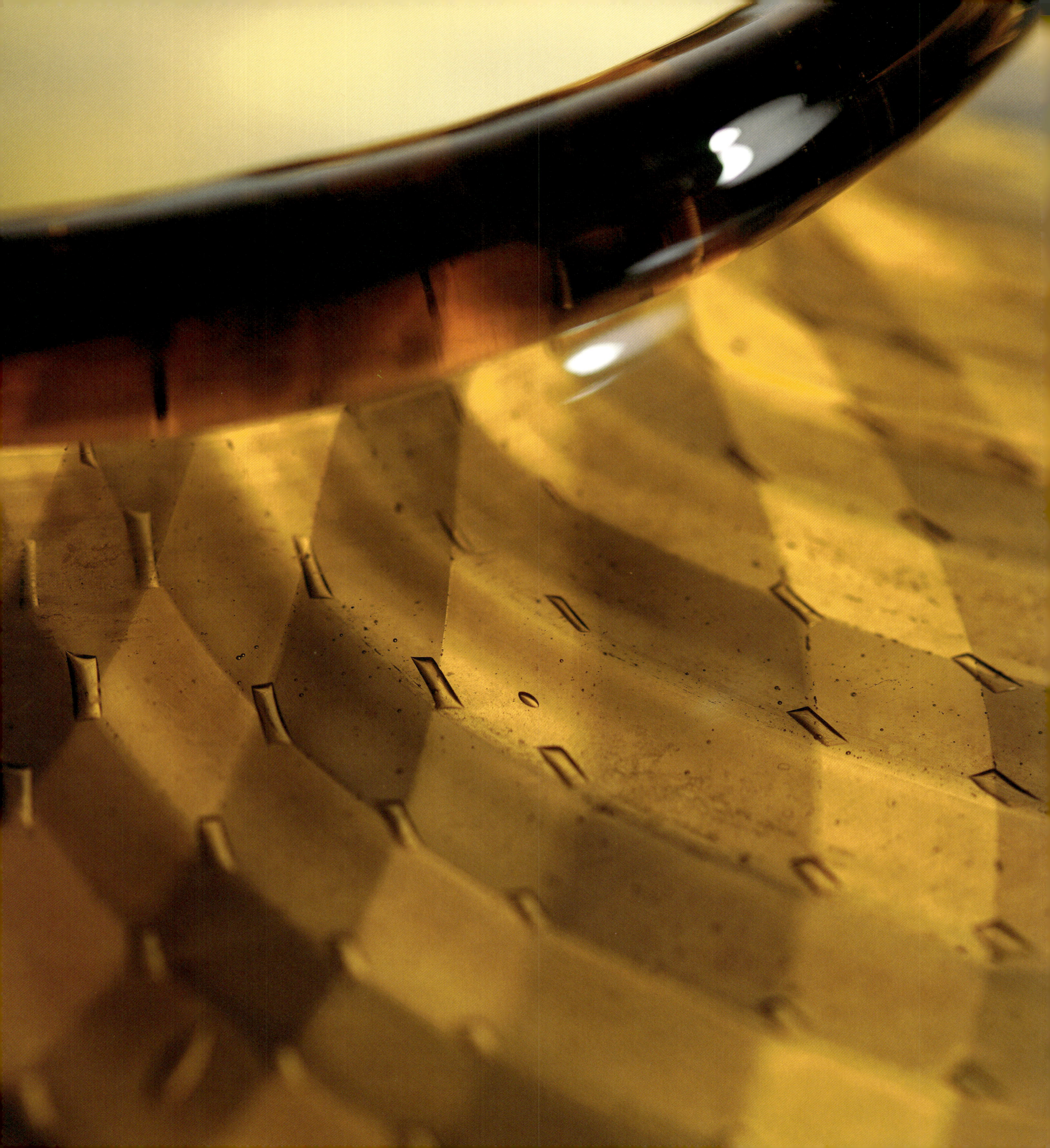

Among the novelties of 1927 was the series of *filigrana a reticello* (netted filigree) glass pieces using the characteristic glass technique of crossed canes. The technique, going back to the 16th century and enjoying a wide revival in the 19th, required the use of transparent glass canes containing an opaque white (*lattimo*) or coloured rectilinear thread inside. These canes are placed one beside the other on a metallic plate that is introduced into the furnace at a temperature of about 700 °C, so obtaining a unified glass material when the heat has fused the canes. The glass material is then collected and closed to form a cylinder which, before being blown, is rotated in such a way that the threads take on a spiral pattern. To obtain the netted filigree, two cylinders are prepared with opposite spirals which, blown and appropriately opened, are inserted one inside the other, welding the walls together. In this way a characteristic diamond design is achieved, at the centre of whose lozenges a small bubble of air remains trapped. The series was generally proposed combining *pagliesco* glass with coral-red or black or dark-blue filigree. One variant is constituted by artefacts where the filigree was replaced or associated with monochrome canes in bright colours, the patterns of which, as documented by a series of drawings, were carefully studied by Carlo Scarpa. He was also involved with the 'twisted canes' series, which subsequently was also proposed in opaque glass.

FILIGRANE 1927-30

Exhibitions

1927, Monza, *III Mostra Internazionale delle Arti Decorative*
1927, Rome, *XCIII Esposizione di Belle Arti della Società amatori e cultori di belle arti*
1927, Venice, *XVIII Mostra dell'Opera Bevilacqua La Masa*
1928, Newark (NJ), Newark Museum, *Five Hundred New Objects*
1929, Leipzig, Trade Fair
1929-30, New York, *Macy's International Exposition of Art in Industry*
1930, Monza, *IV Esposizione Triennale Internazionale delle Arti Decorative ed Industriali Moderne*
1931, Amsterdam, *Tentoonstelling van glas, aardewerk en kant, van moderne italiaansche kunstnijverheid. Mostra di vetri, ceramiche e merletti d'arte moderna italiana*

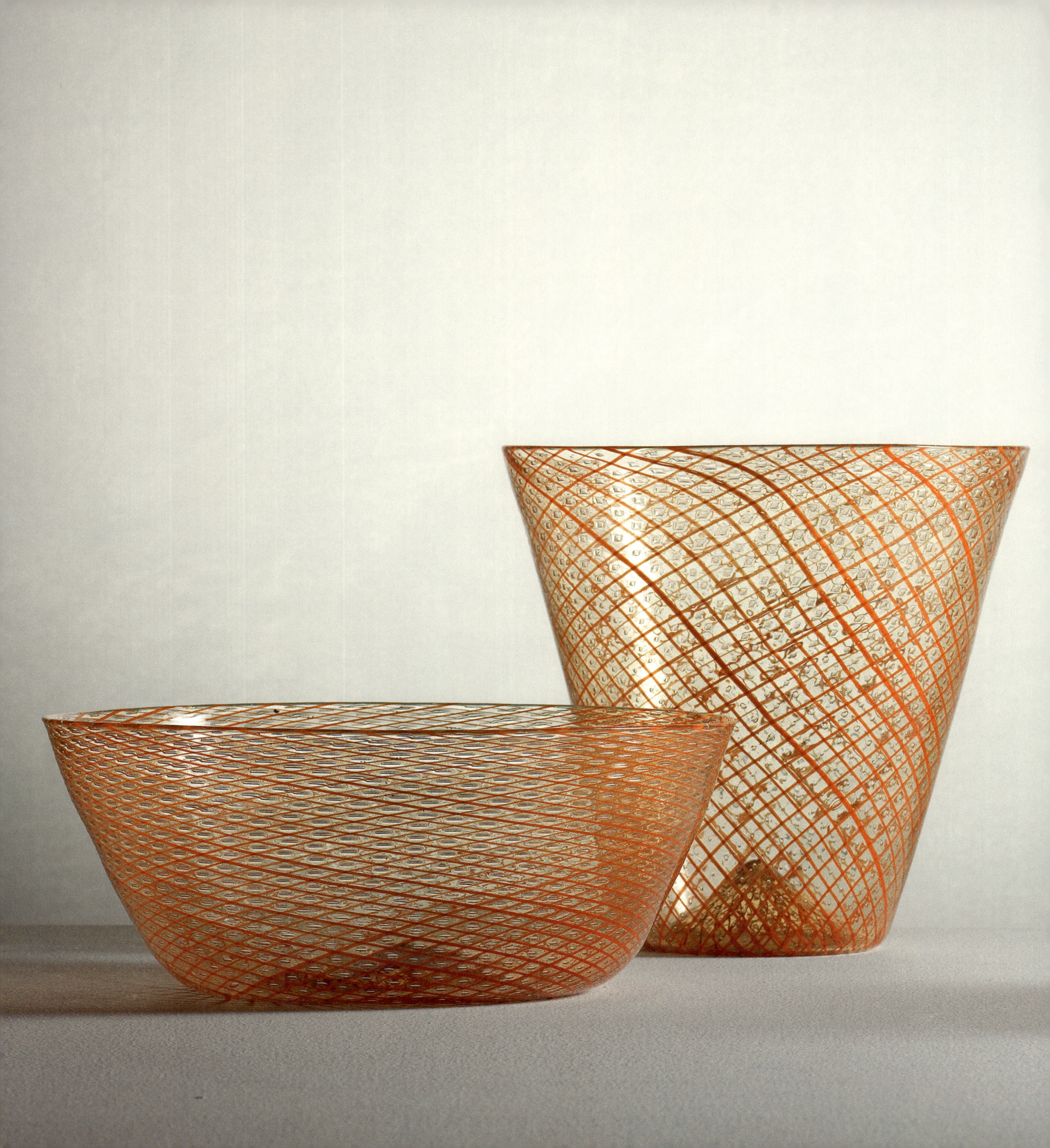

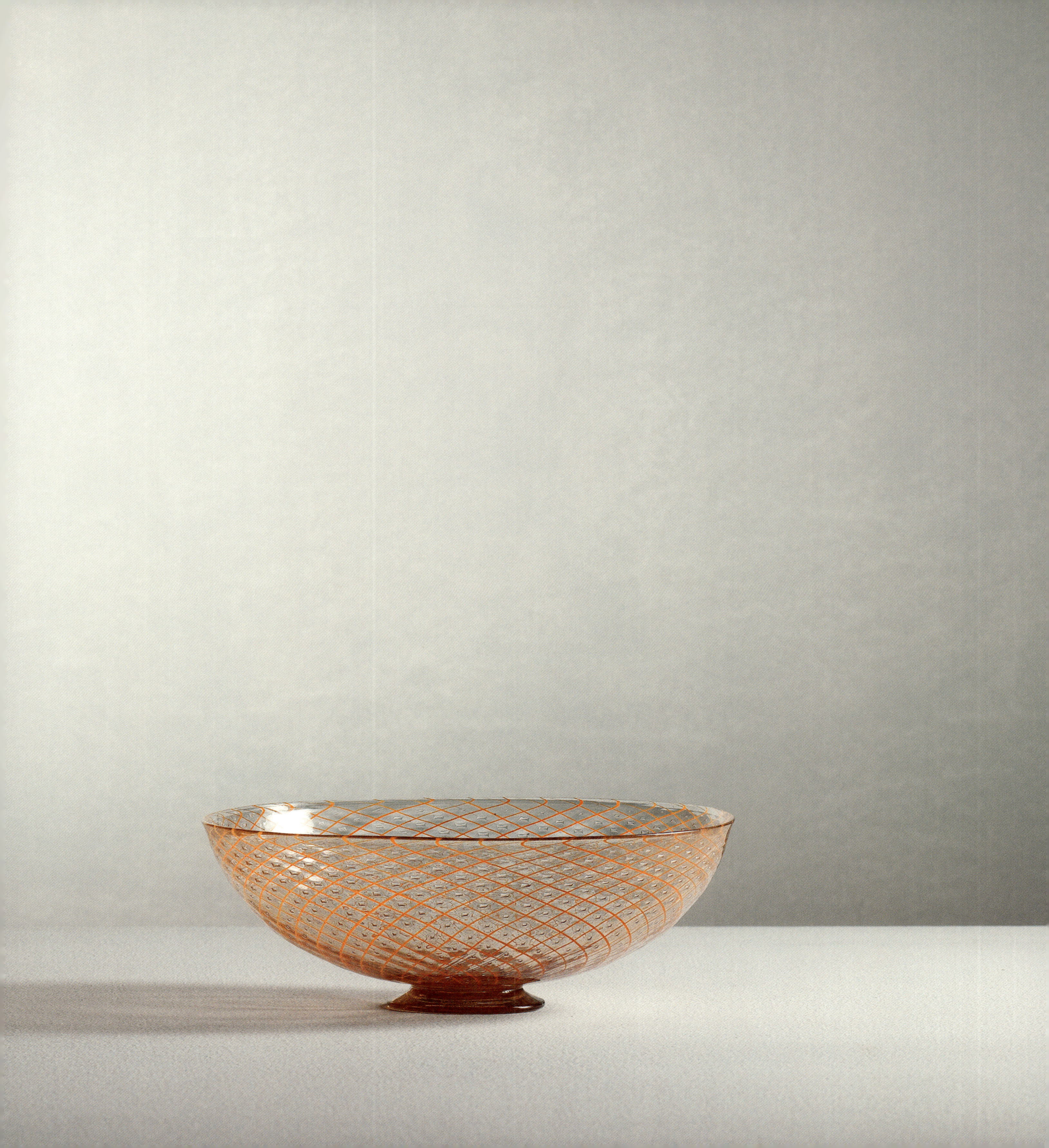

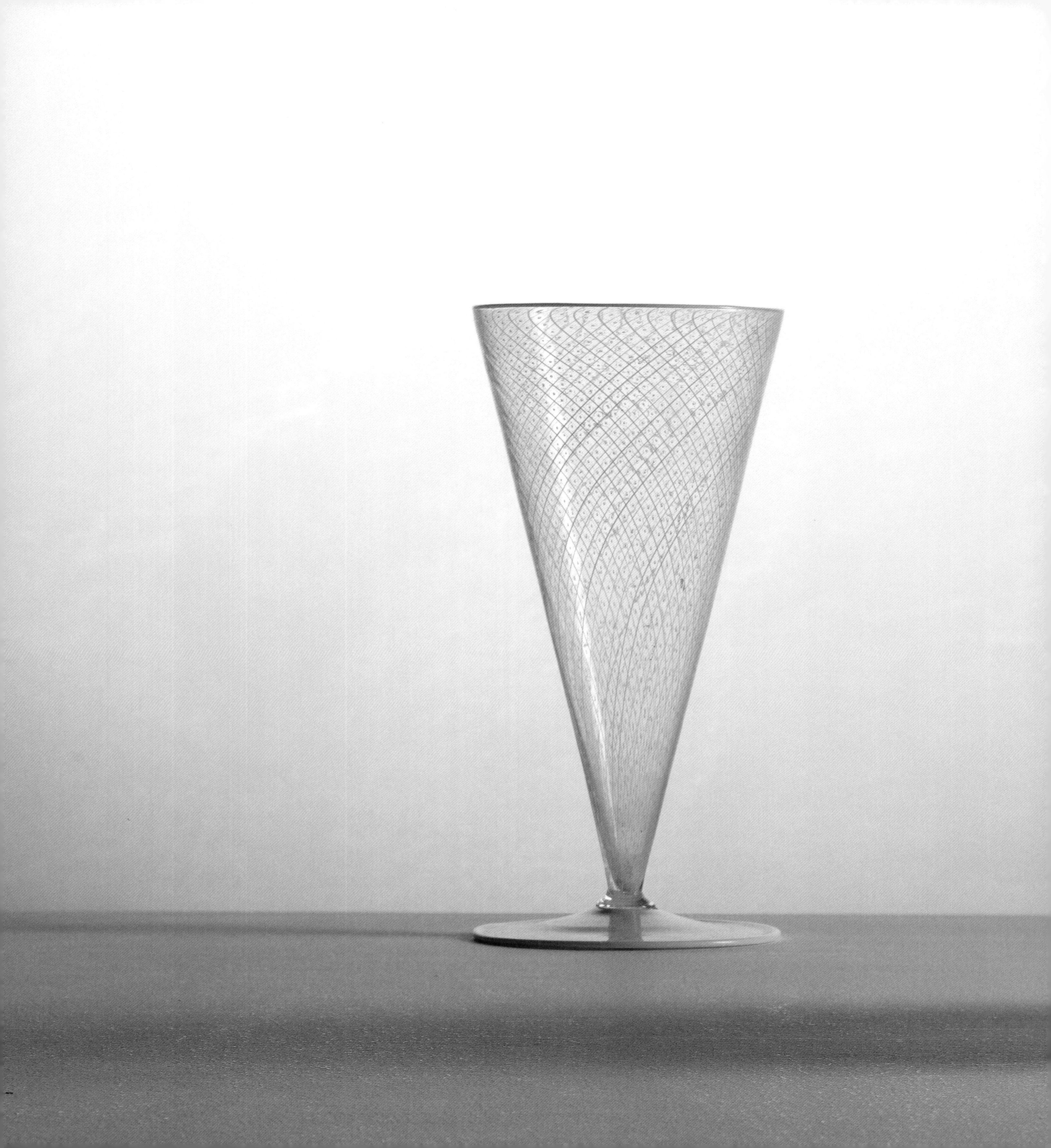

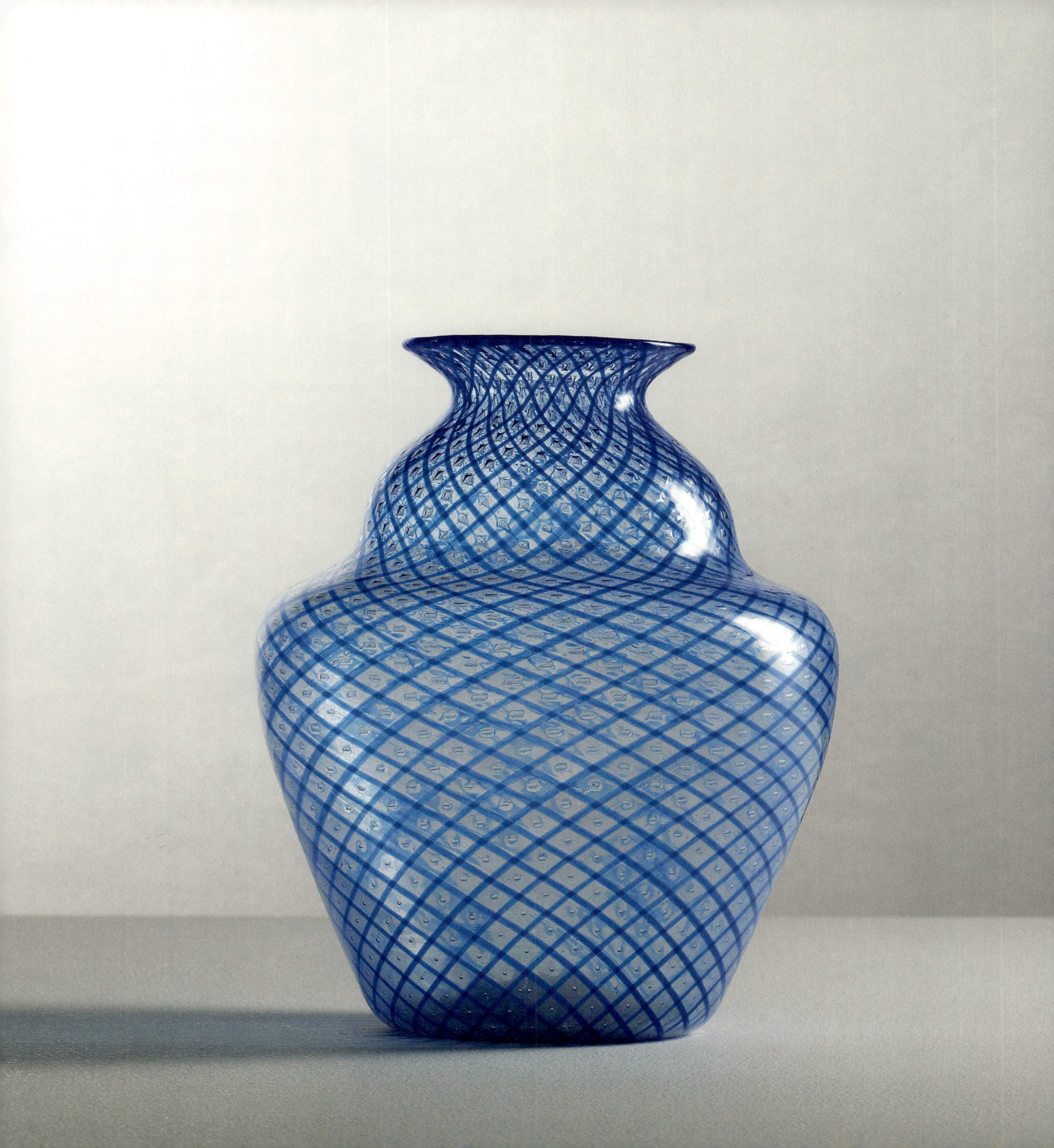

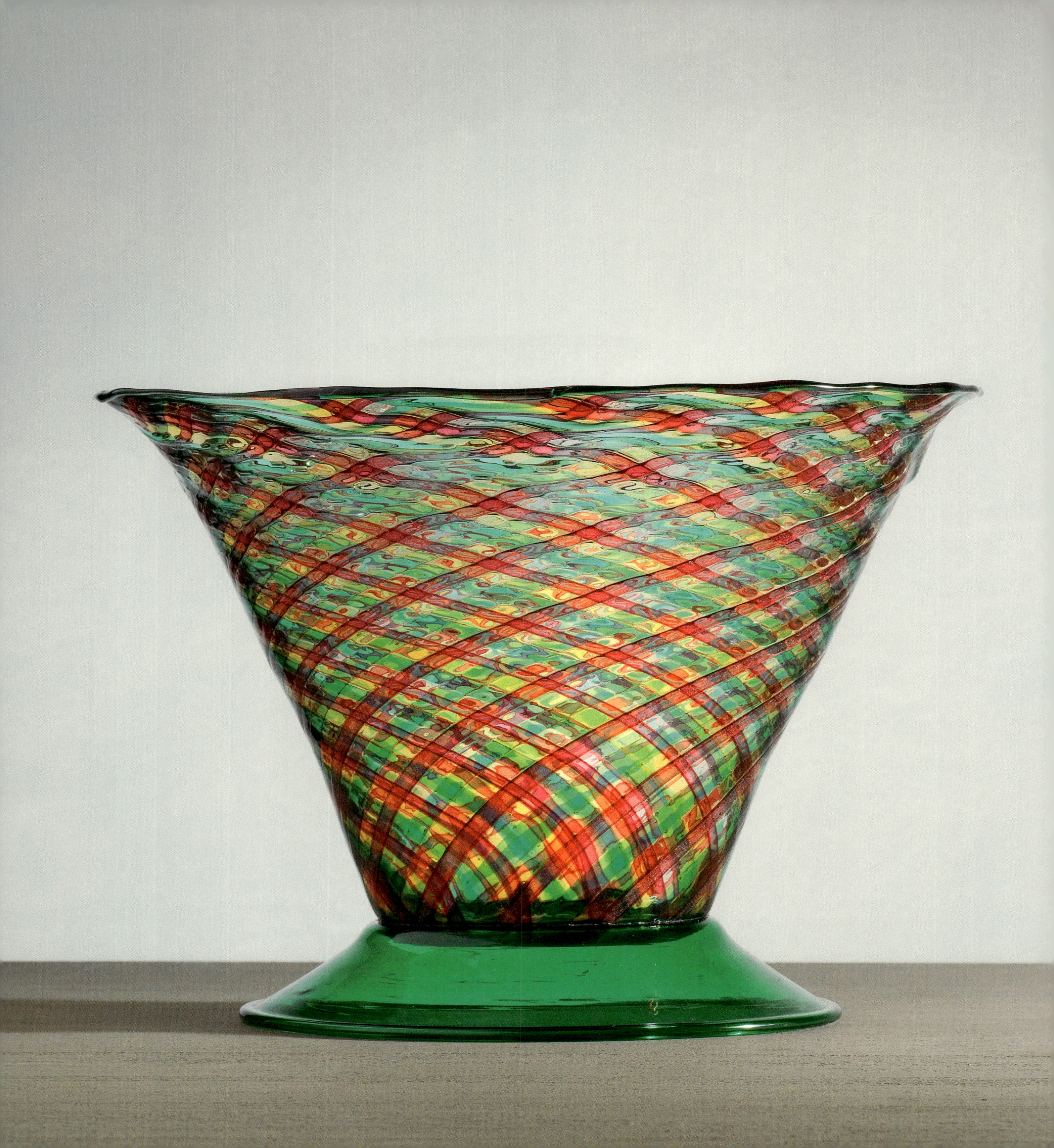

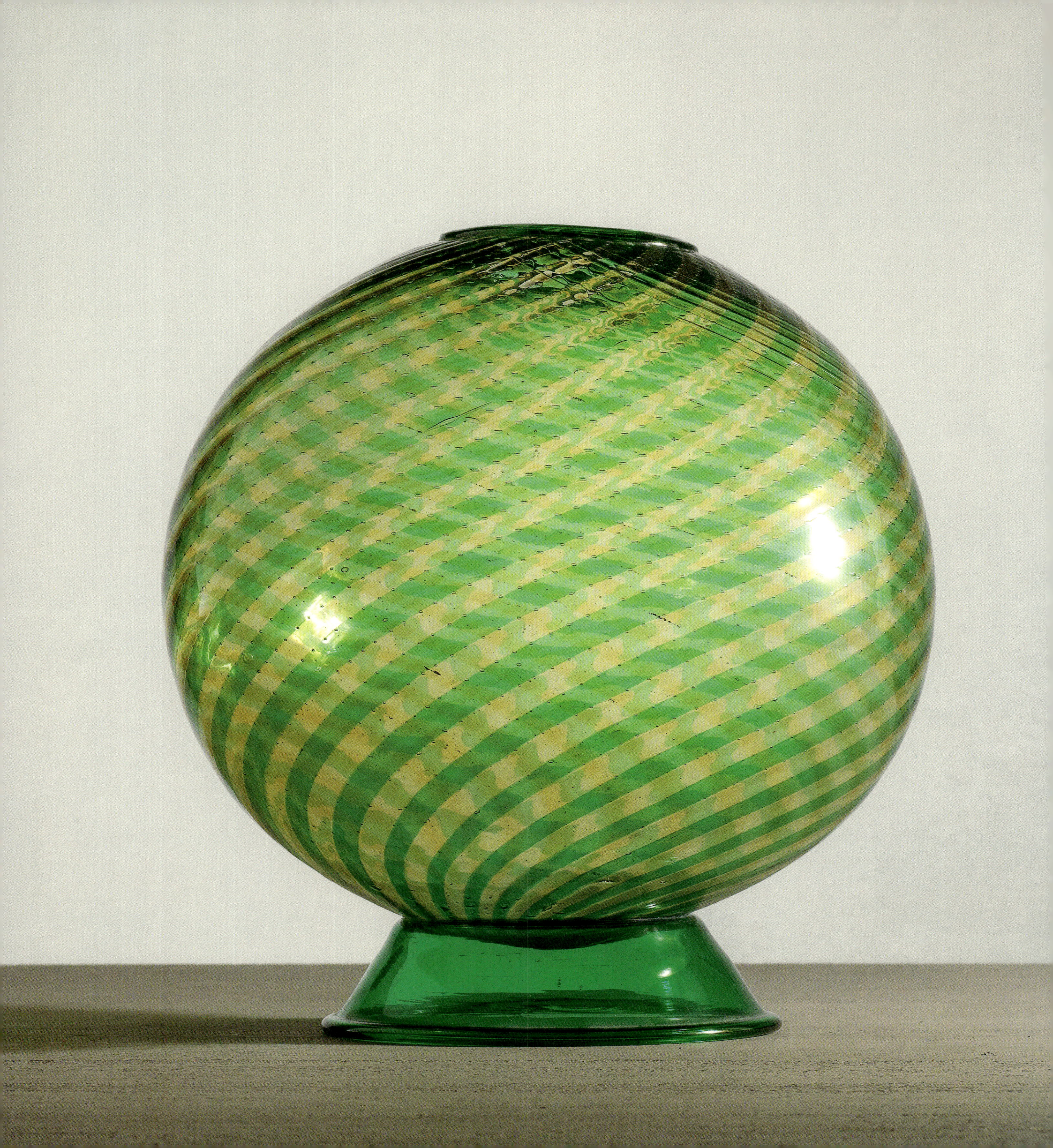

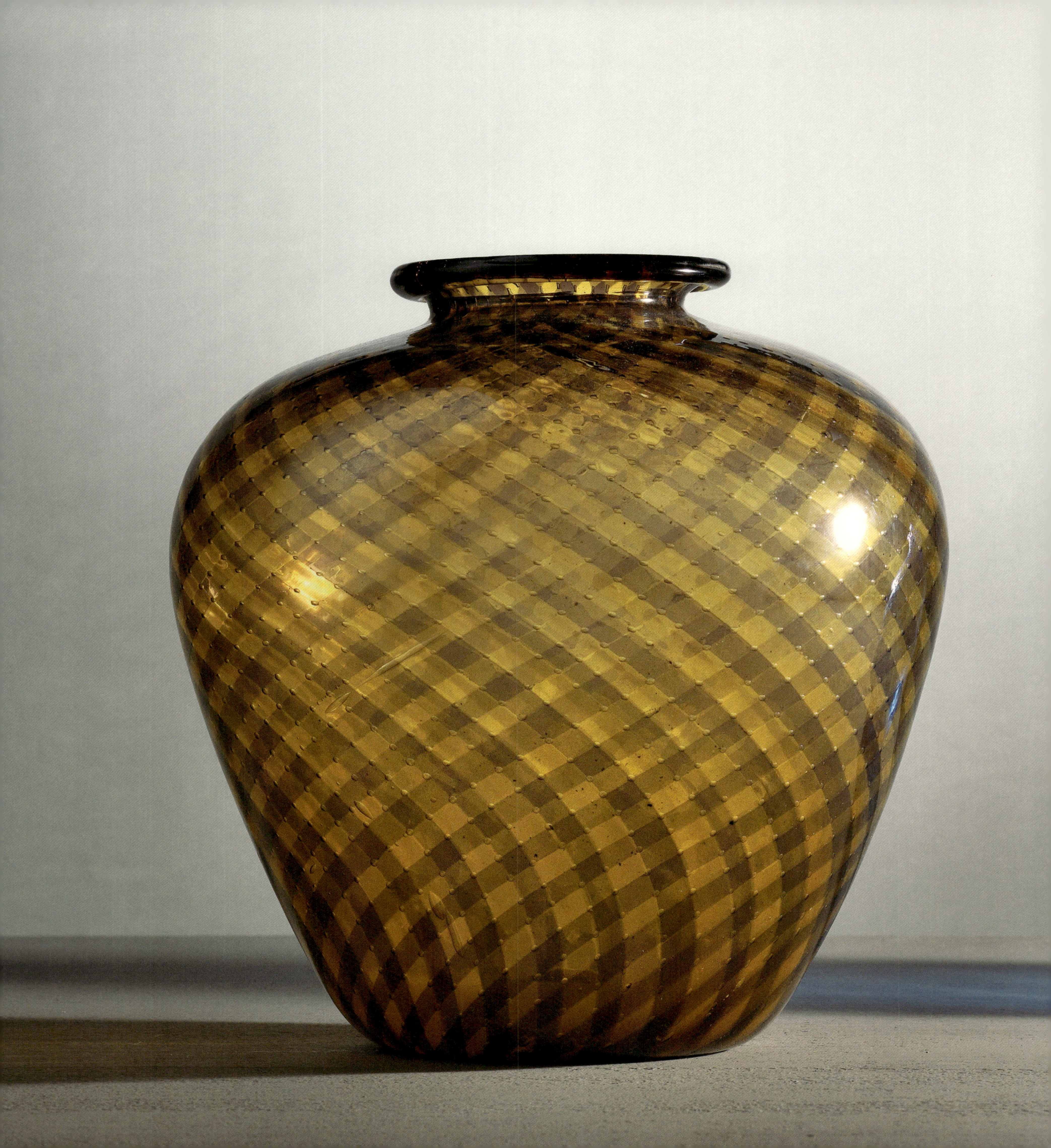

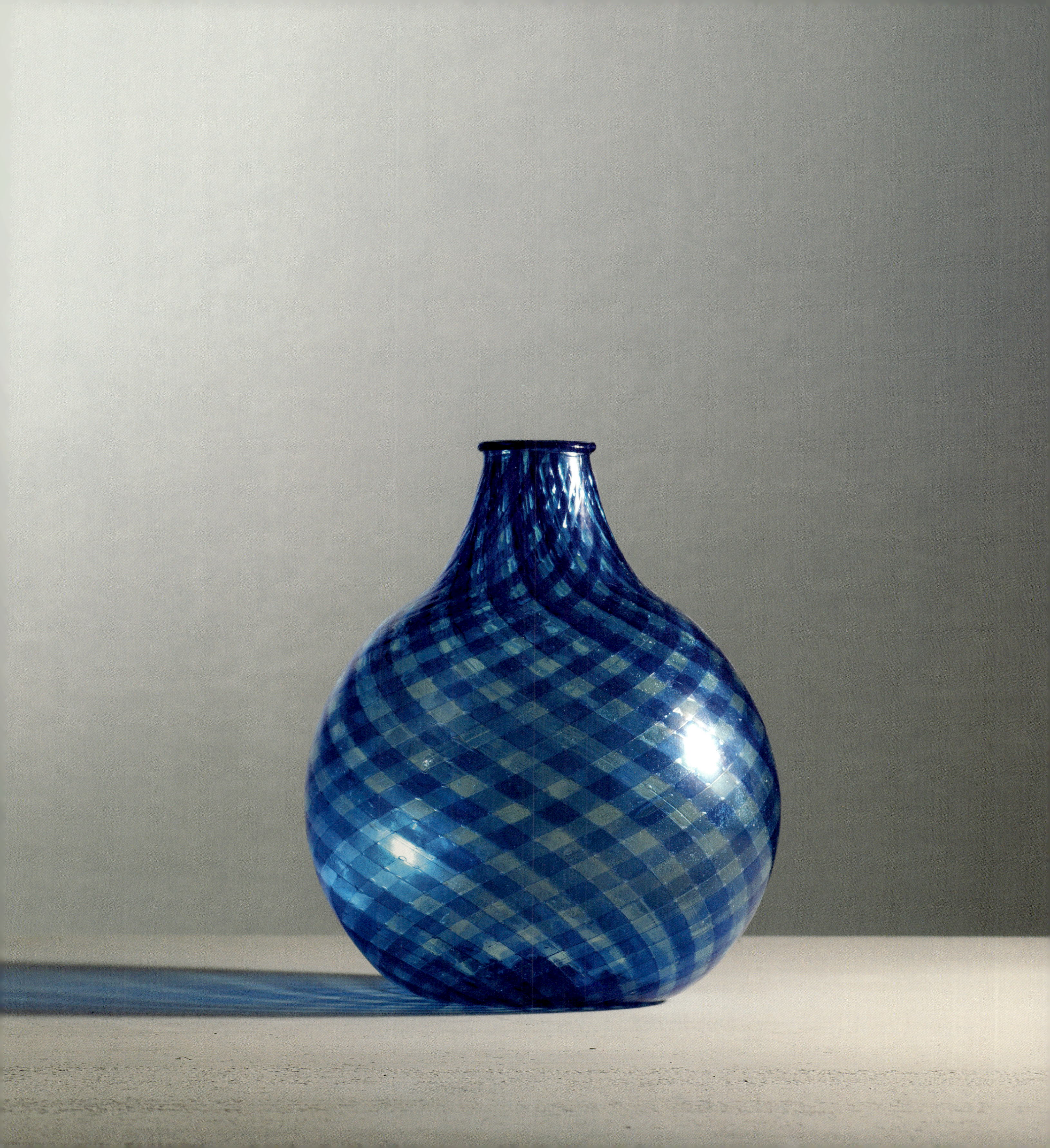

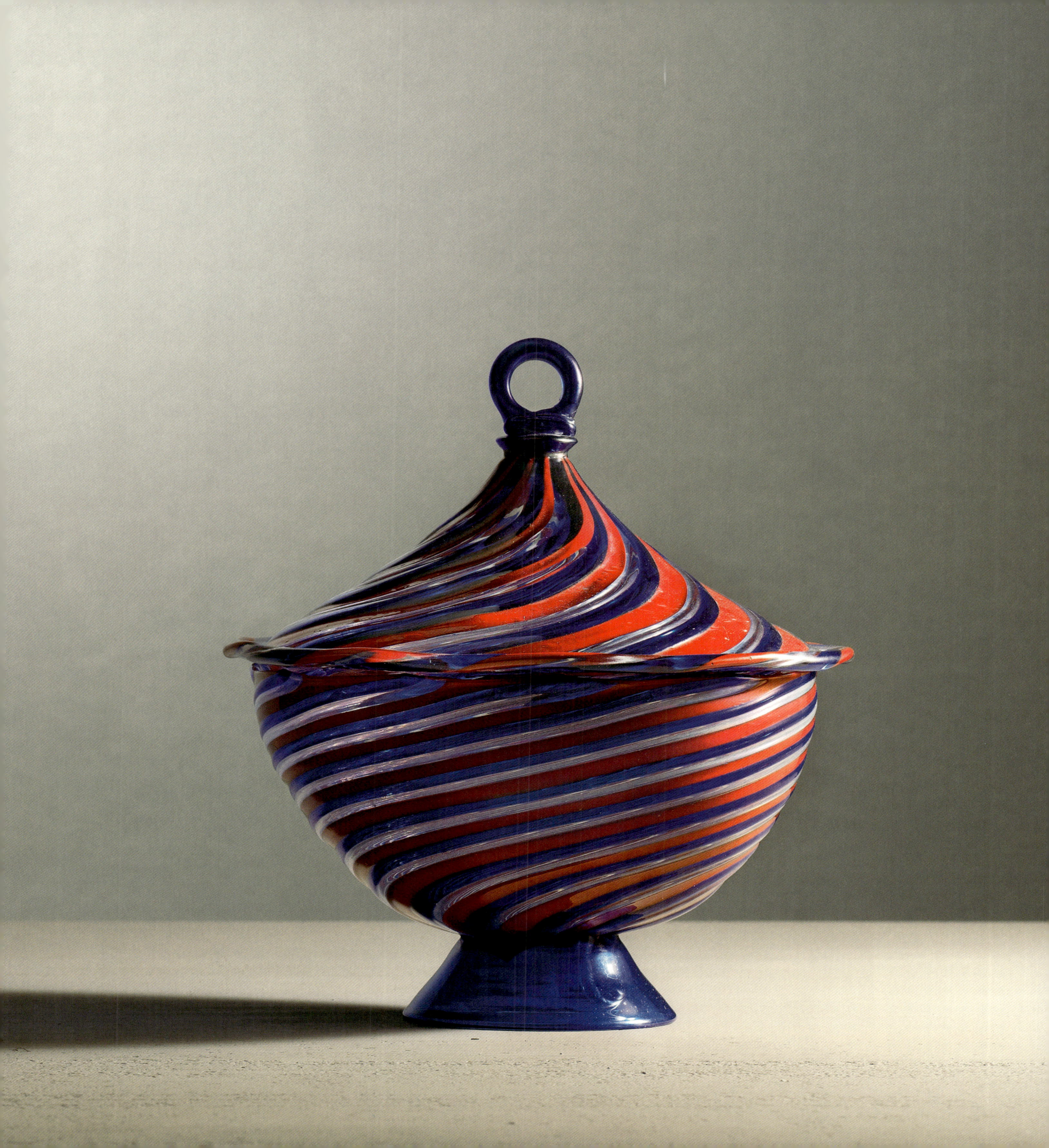

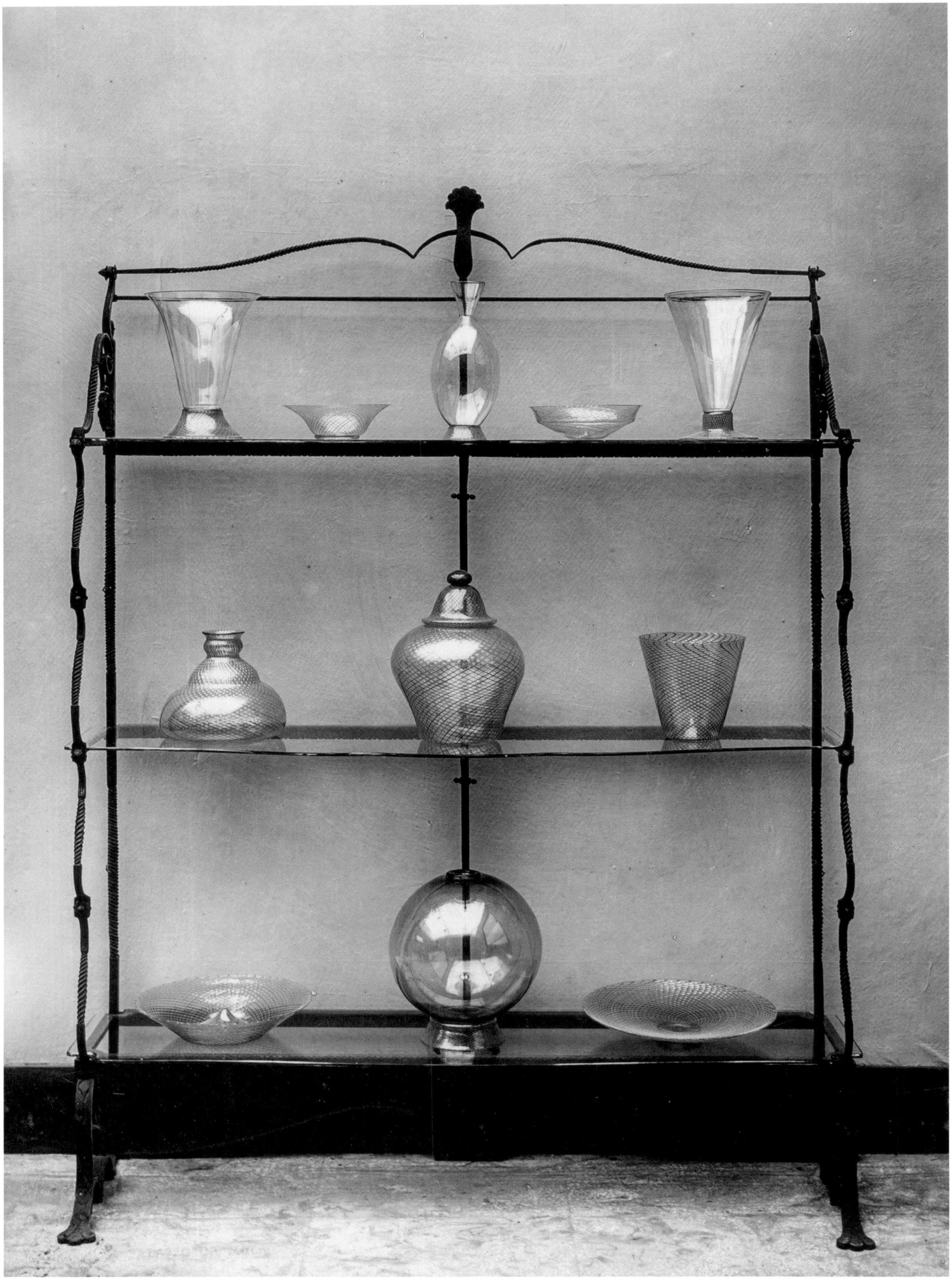

>
Glass pieces in filigrana a reticello at the 'Mostra di Cappellin e Rizzarda', 93rd Esposizione di belle arti della Società amatori e cultori di belle arti, Rome 1927, CSV

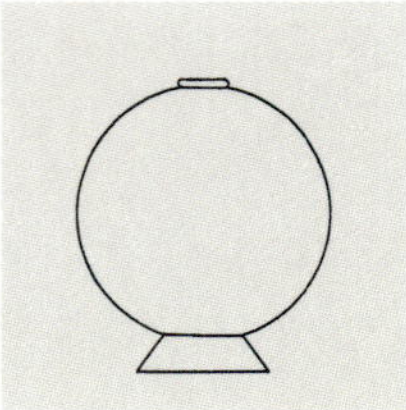

5673

A spherical vase in red *filigrana a reticello* (net filigree). Mouth and truncated-cone-shaped foot in red transparent glass.
A similar example in *pagliesco filigrana corallo* (coral-red *pagliesco* filigree) glass was exhibited in 1927 in Venice, at the 18th Mostra dell'Opera Bevilacqua La Masa. In the same year the model was shown in Rome, at the 93rd Esposizione di belle arti, in the variant with the body in '*pagliesco* glass and foot in *filigrana corallo*' (coral-red filigree). Model no. 5673, symbol of the Cappellin firm, was also produced in other glass techniques (e.g. transparent, *reticello* with canes, opaque).
1927
33 cm high
Private collection

exhibitions
1927, Rome 93rd Esposizione di belle arti
1927, Venice, 18th Mostra dell'Opera Bevilacqua La Masa

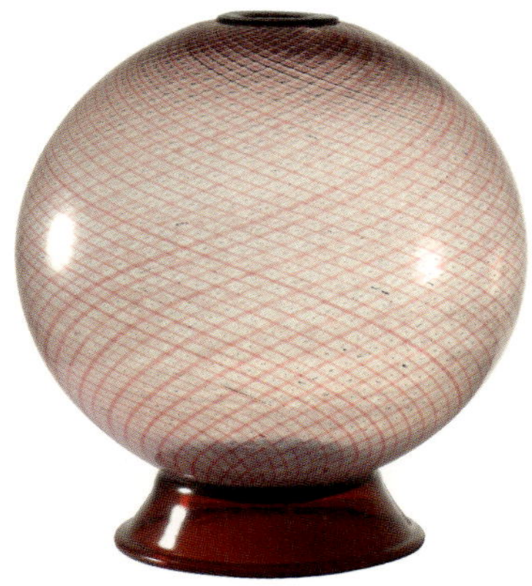

>
5673, working drawing, CSV

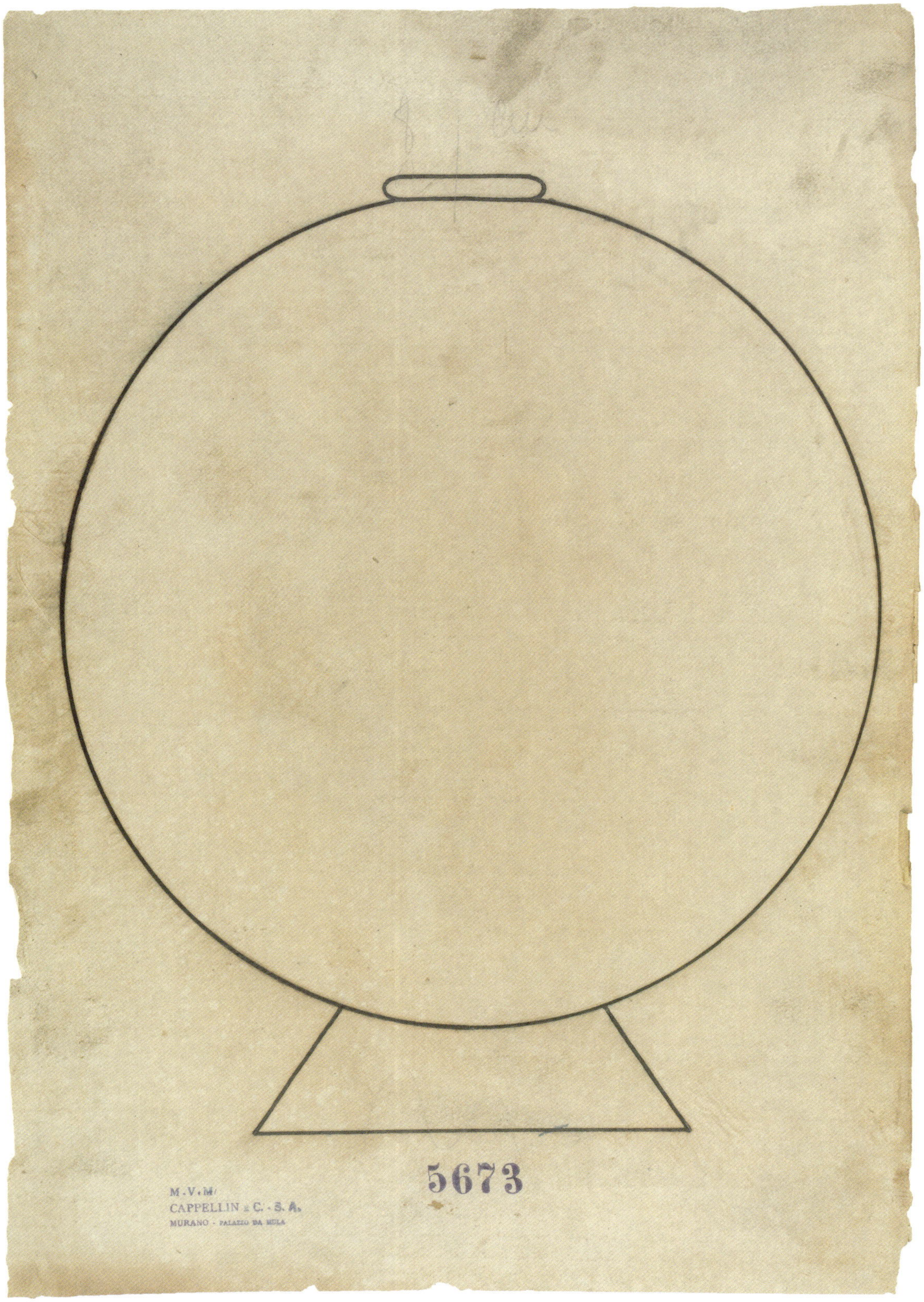

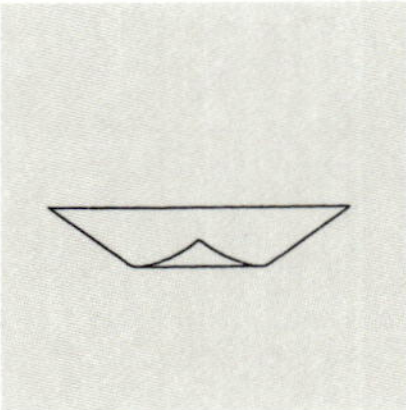

5700

Bowl in *pagliesco* and coral-red *filigrana a reticello* (net filigree). A similar example, in the same colours, was exhibited in 1927 in Rome, at the 93rd Esposizione di belle arti, and in Venice, at the 18th Mostra dell'Opera Bevilacqua La Masa. The bowl was acquired by the then Director of the Newark Museum of Art (NJ, USA) in the Cappellin shop in Rome in 1928.
1927
3.5 cm high; 14 cm ø
Newark Museum collection, New Jersey

exhibitions
1927, Rome, 93rd Esposizione di belle arti
1927, Venice, 18th Mostra dell'Opera Bevilacqua La Masa
1928, Newark (USA), *Five Hundred New Objects*

>
5700, working drawing, CSV

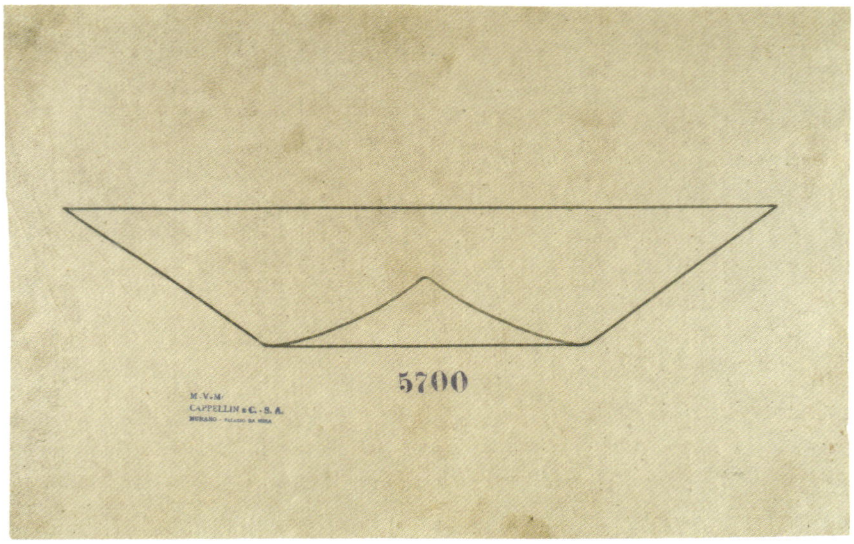

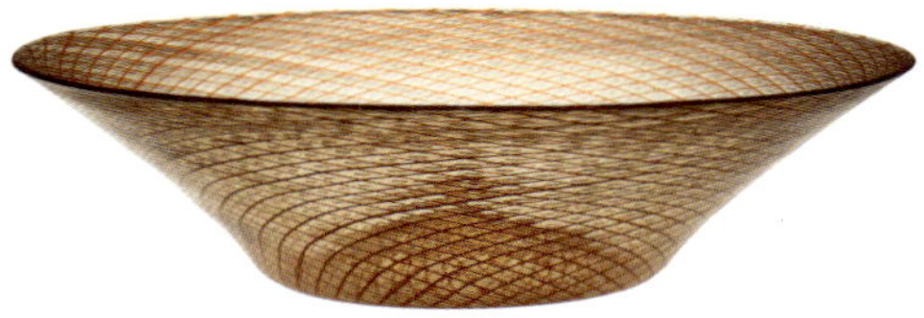

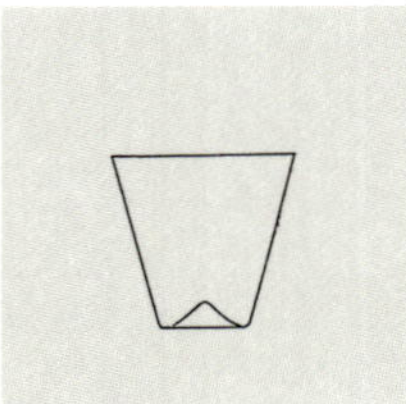

5699

Vase in *pagliesco* and coral-red *filigrana a reticello* (net filigree). A similar example, in the same colours, was exhibited in 1927 at the 93rd Esposizione di belle arti of Rome. In the same year, the model was also shown in Venice, at the 18th Mostra dell'Opera Bevilacqua La Masa, in the variant *'pagliesco filigrana rossa'* (red *pagliesco* filigree).
1927
21 cm high
M.V.M. Cappellin Murano *acid stamp*
Bischofberger collection, Switzerland

exhibitions
1927, Rome, 93rd Esposizione di belle arti
1927, Venice, 18th Mostra dell'Opera Bevilacqua La Masa

>
5699, working drawing, CSV

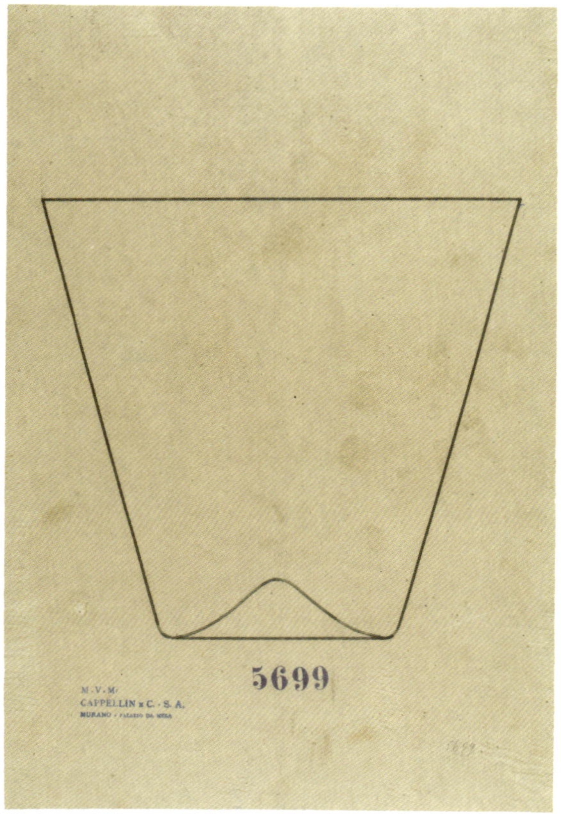

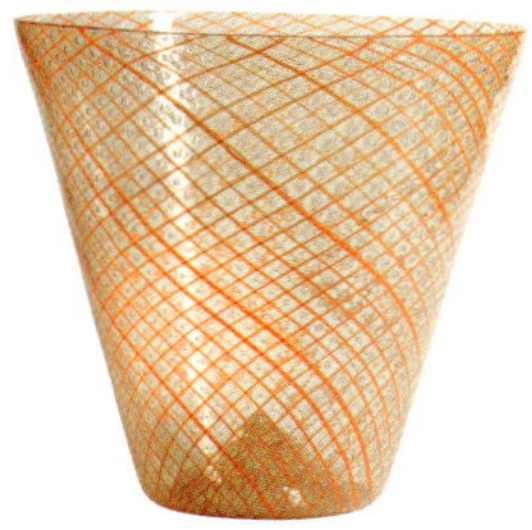

---

n.i.

Bowl in *pagliesco* and coral-red *filigrana a reticello* (net filigree).
1927
11 cm high; 27 cm ø
M.V.M. Cappellin Murano *acid stamp*
Bischofberger collection, Switzerland

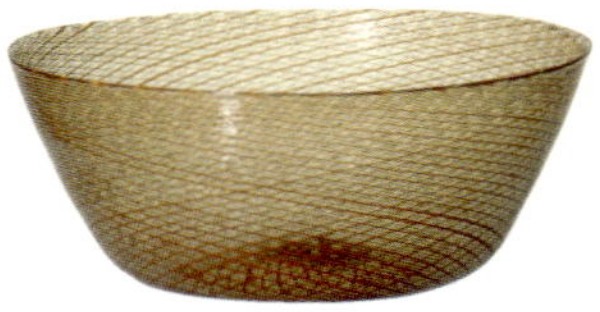

---

n.i.

Bowl in amethyst *filigrana a reticello* (net filigree). Foot in transparent glass.
1927
12 cm high; 22 cm ø
M.V.M. Cappellin Murano *acid stamp*
Private collection, Udine

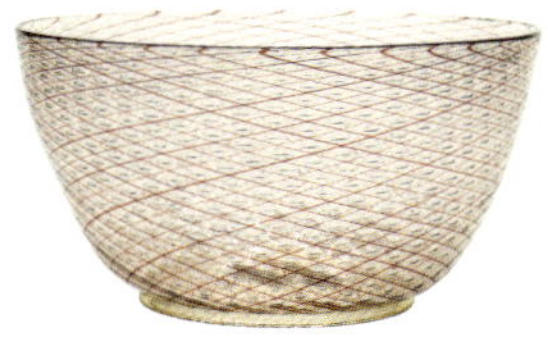

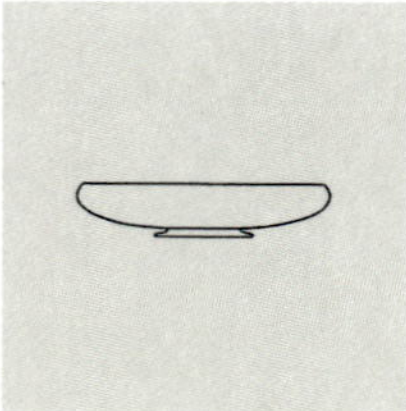

5623

Bowl in *pagliesco* and coral-red *filigrana a reticello* (net filigree). Small foot in *pagliesco* transparent glass.
1927
30 cm ø
Private collection, Venice

›
5623, catalogue drawing, AMB

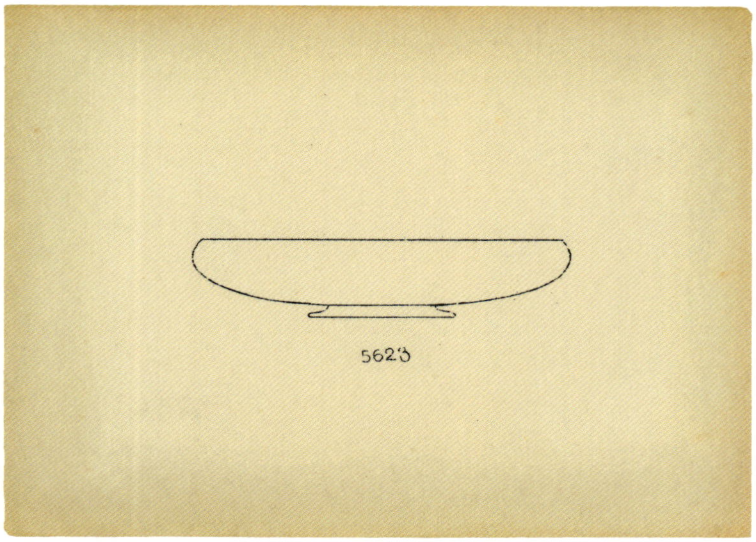

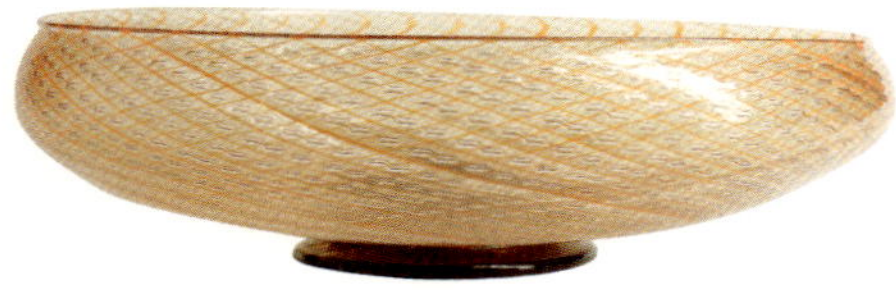

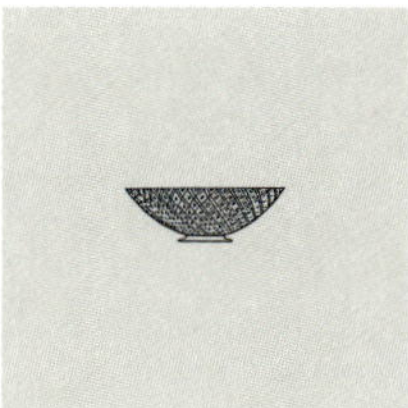

1792

Bowl in *pagliesco* and coral-red *filigrana a reticello* (net filigree). Small flared foot in *pagliesco* transparent glass.
1927
7 cm high; 20 cm ø
M.V.M. Cappellin Murano *acid stamp*
Private collection, Udine

›
1789–1792, working drawings, CSV

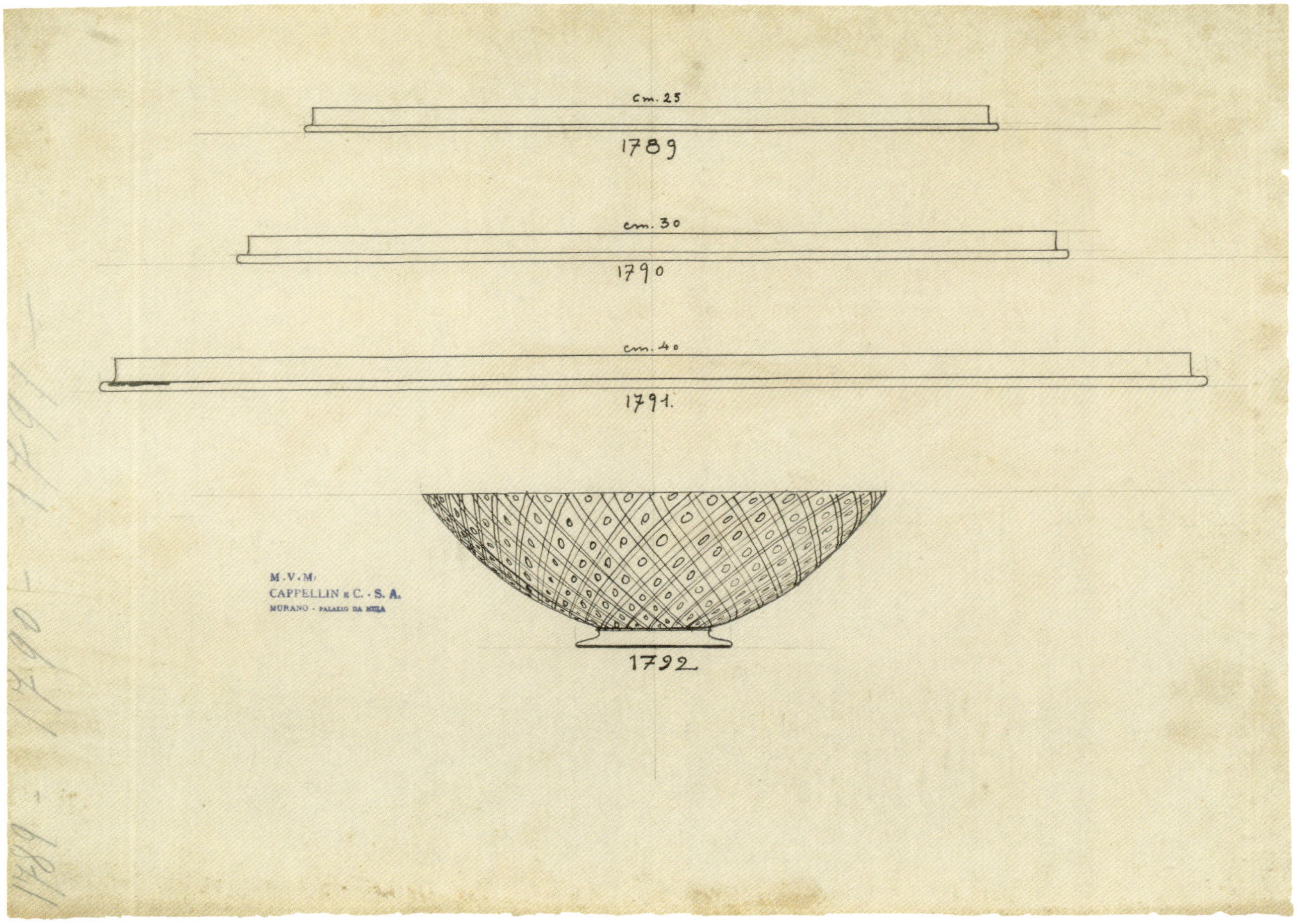

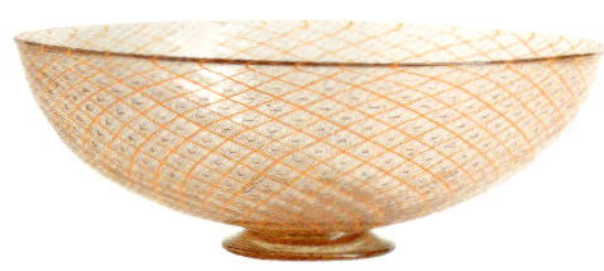

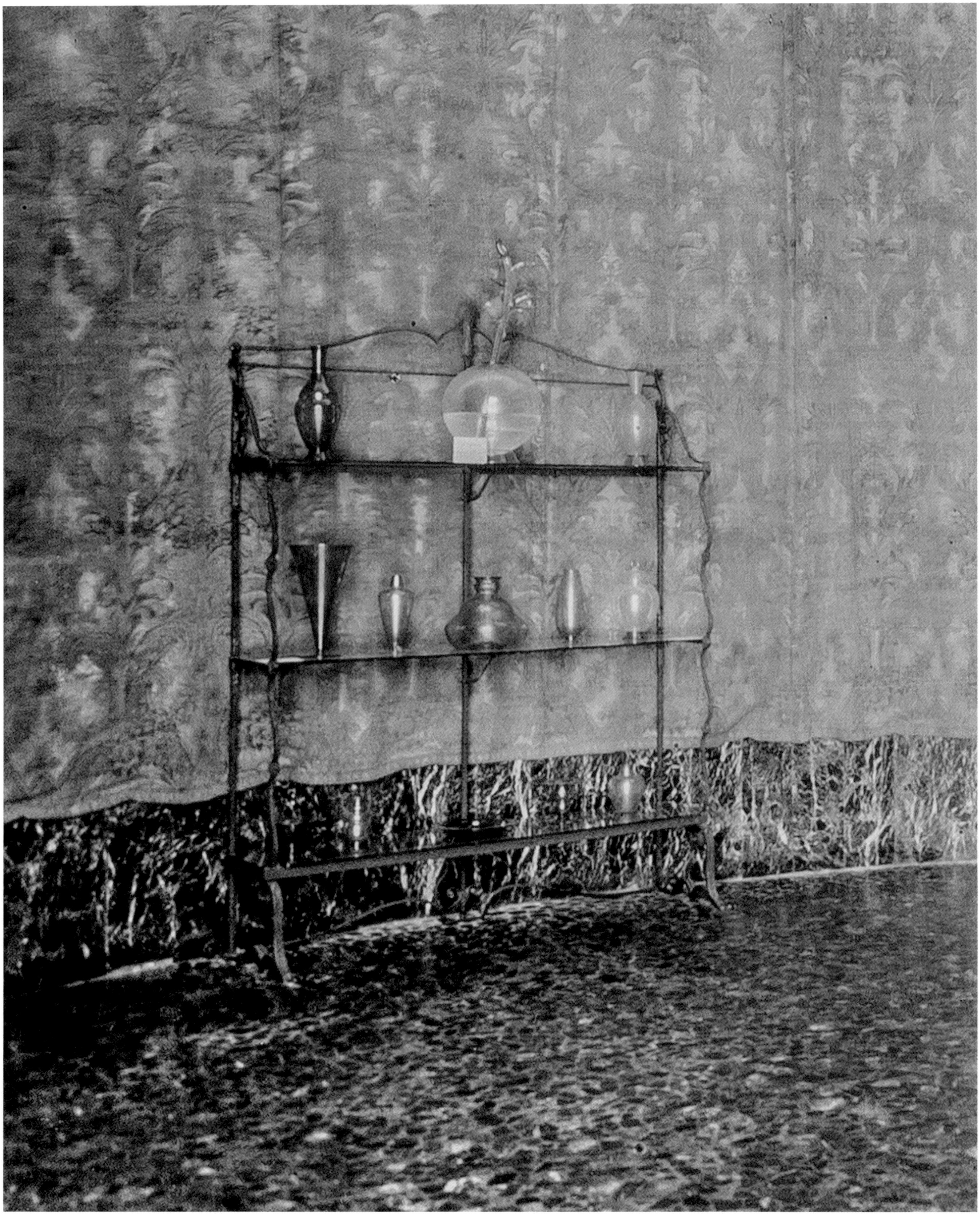

>
M.V.M. Cappellin glass pieces at the 28th Mostra dell'Opera Bevilacqua La Masa, Venice 1927, detail CSV

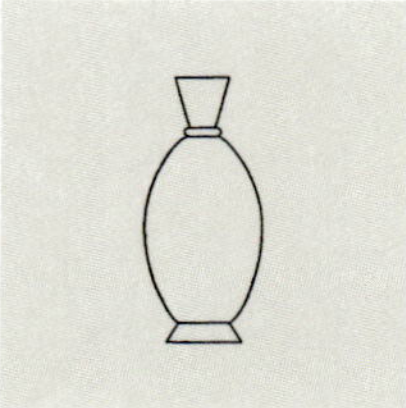

5675

Working drawing for one of the vases in *filigrana a reticello* (net filigree) exhibited in 1927 at the 93rd Esposizione di belle arti of Rome, CSV. On that occasion the vase was presented in *pagliesco* glass 'without a rope motif and [with] a coral-red filigree foot'. Also in 1927, model no. 5675 was exhibited at the 18th Mostra dell'Opera Bevilacqua La Masa in the variant with amethyst transparent glass and a violet foot and collar.
1927
31 cm high

exhibitions
1927, Rome, 93rd Esposizione di belle arti
1927, Venice, 18th Mostra dell'Opera Bevilacqua La Masa (transparent)

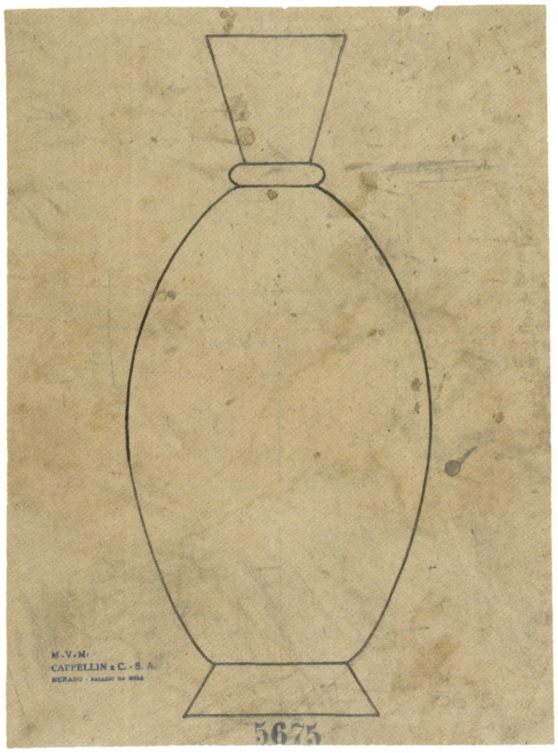

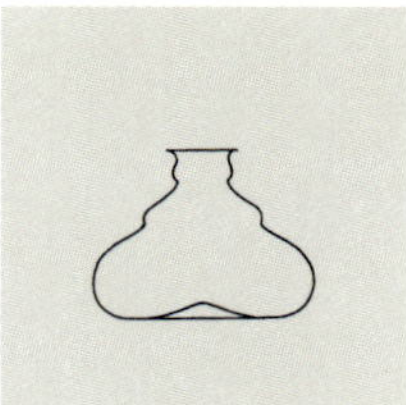

5682

Working drawing for one of the vases in *filigrana a reticello* (net filigree) exhibited at the 93rd Esposizione di belle arti of Rome in 1927, CSV. On that occasion the vase was presented in the *pagliesco* version with *filigrana nera* (black filigree). Also in 1927, model no. 5682 was shown at the 18th Mostra dell'Opera Bevilacqua La Masa of Venice in the '*pagliesco* coral-red filigree' variant.
1927
20 cm high

exhibitions
1927, Rome, 93rd Esposizione di belle arti
1927, Venice, 18th Mostra dell'Opera Bevilacqua La Masa

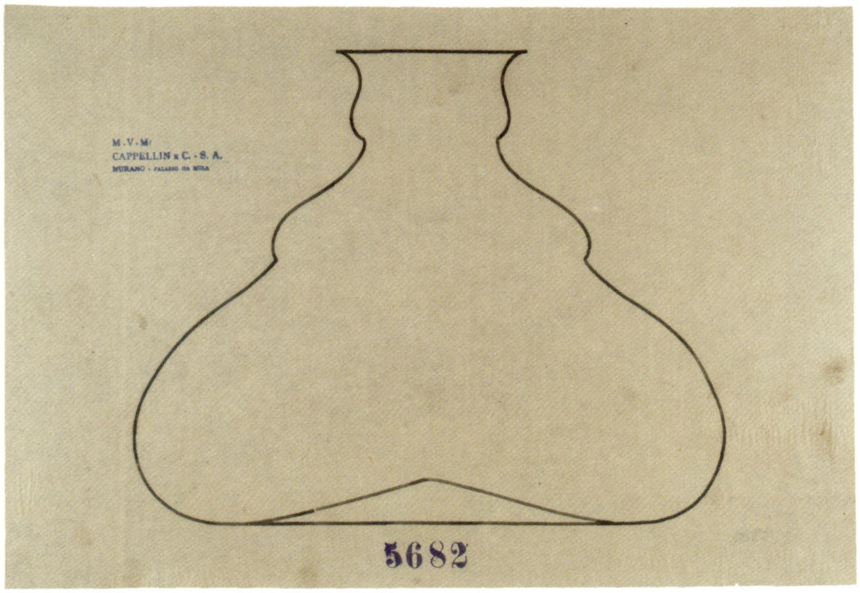

n.i.

Vase in Savoy-blue *filigrana a reticello* (net filigree) with a double constriction at the base of the neck.
1927
24 cm high
M.V.M. Cappellin Murano *acid stamp*
Zanfirico di di Vizia M. collection, Turin

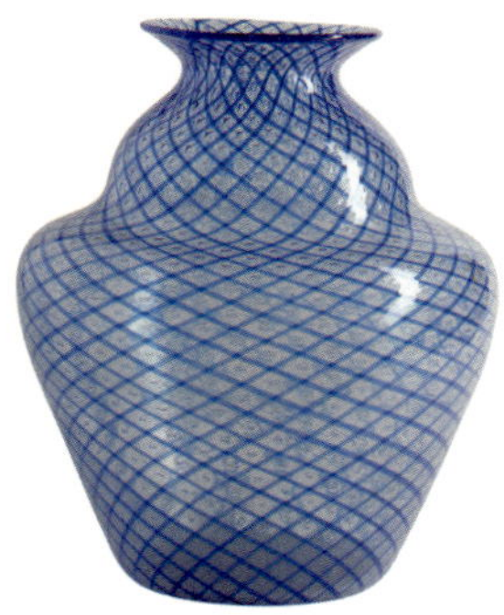

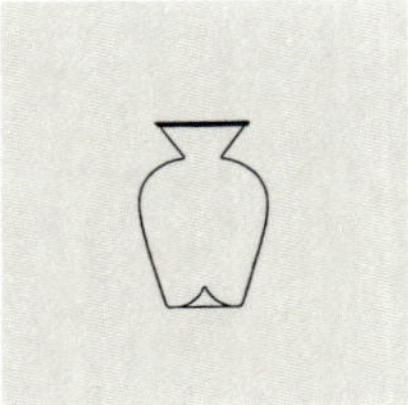

5662

Drawing for one of the vases in *filigrana a reticello* (net filigree) presented at an exhibition of Cappellin glass pieces around 1927, AMB. The model, which was exhibited together with the compote no. 3816, was produced also in other glass techniques (e.g. transparent).
c. 1927
22 cm high

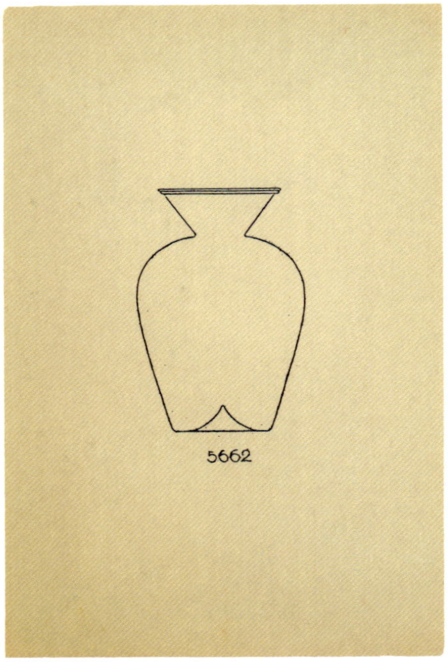

»
Exhibition of M.V.M. Cappellin glass pieces, c. 1927. Period photograph, CSV

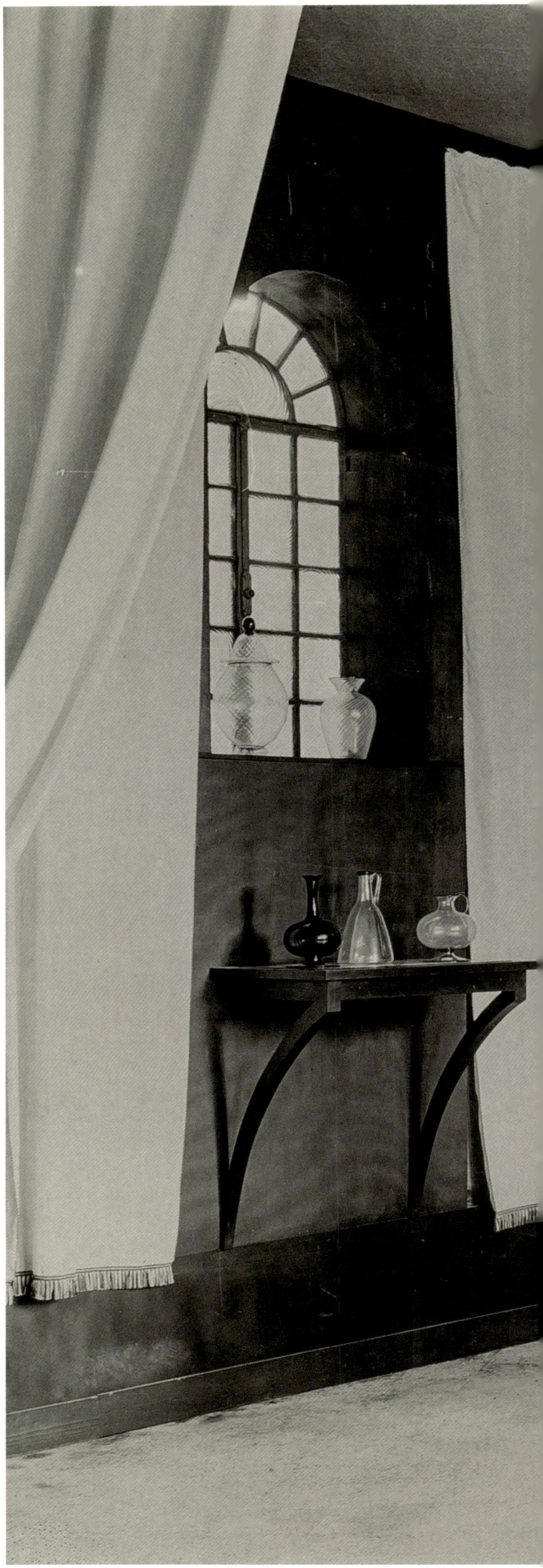

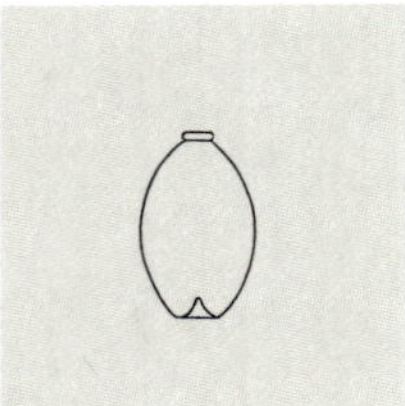

5663

Drawing for one of the vases in *filigrana a reticello* (net filigree) presented at an exhibition of Cappellin glass pieces around 1927, AMB. The model was exhibited together with the compote bowl no. 3814.
1927
22 cm high

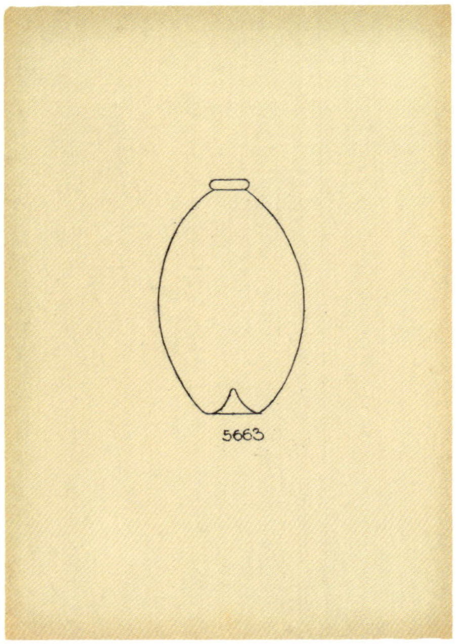

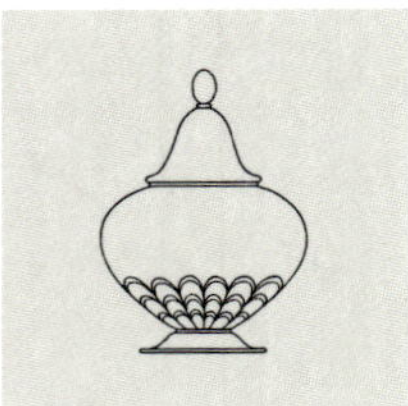

3815

Working drawing for one of the compote bowls in *filigrana a reticello* (net filigree) presented at an exhibition of Cappellin glass pieces around 1927, CSV. On that occasion the compote bowl was exhibited together with vase no. 5660. As the working drawing documents, it was also planned to produce no. 3815 with a *decoro fenicio* at the base.
1927
33 cm high

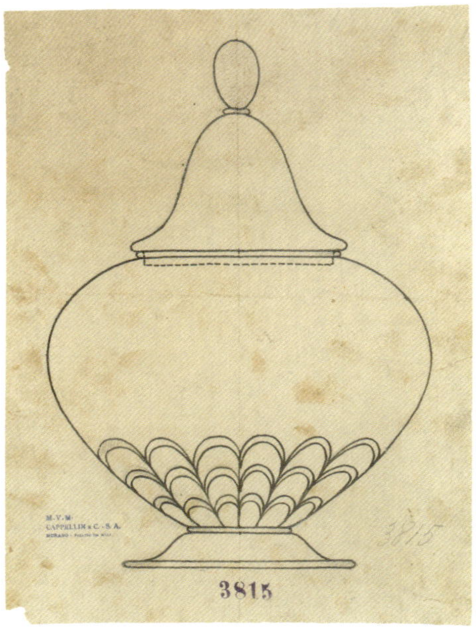

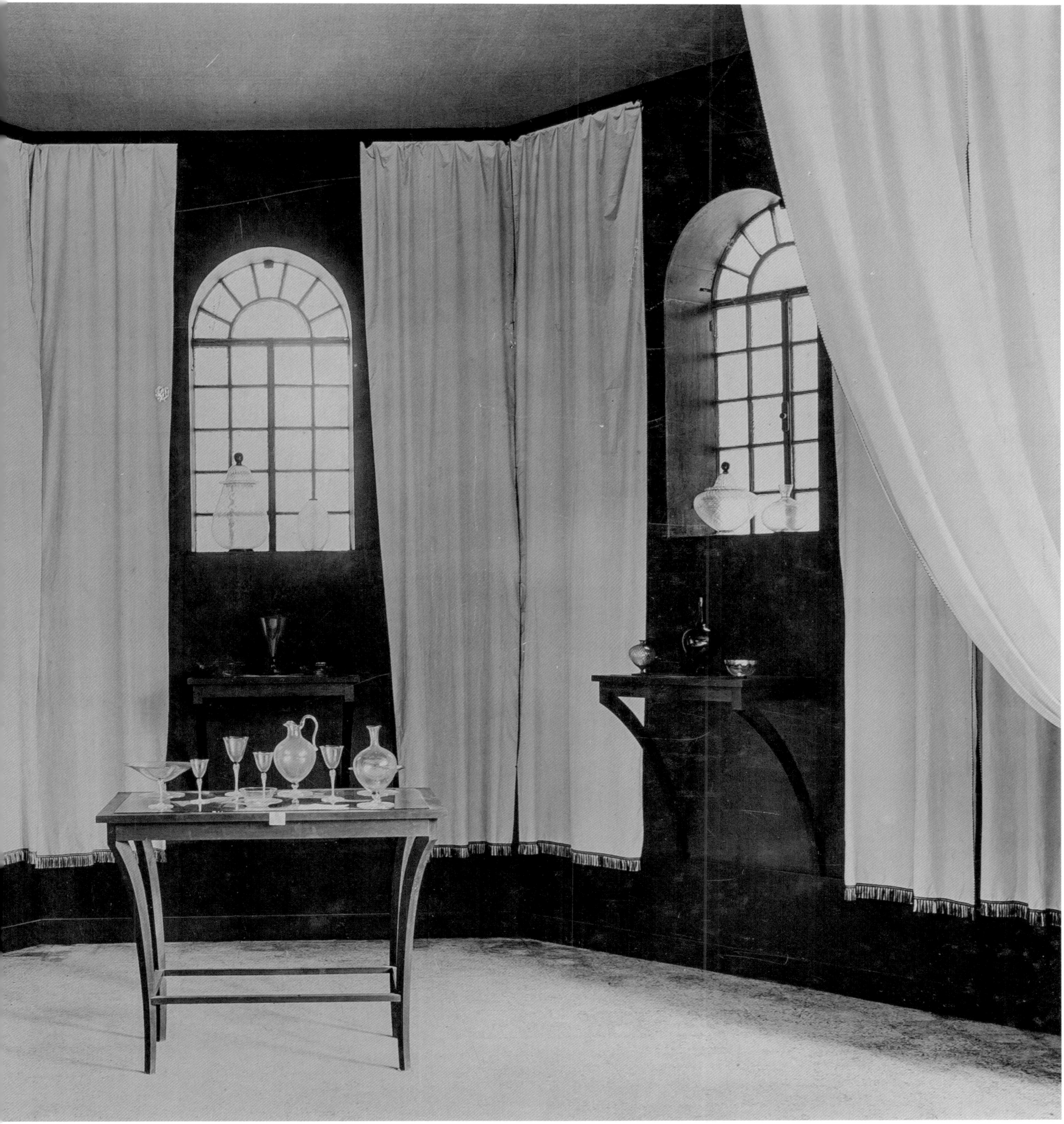

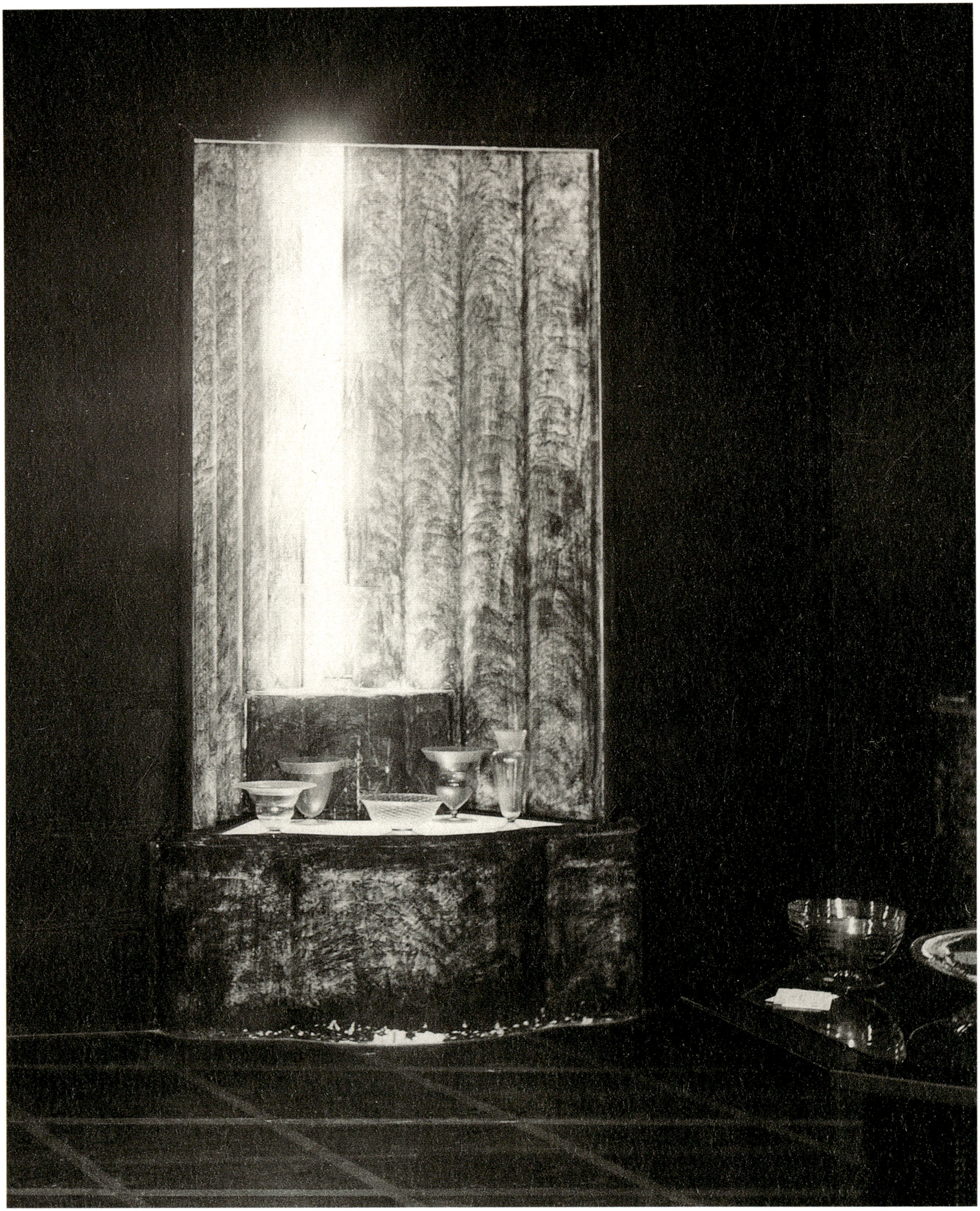

>
Glass pieces in filigrana a reticello exhibited at the 3rd Biennale of Monza in the M.V.M. Cappellin & C. room, 1927, detail, AFT. On the upper shelf vase no. 5749 can be seen. This was produced in 'green pagliesco-filigrana' with foot in green transparent glass

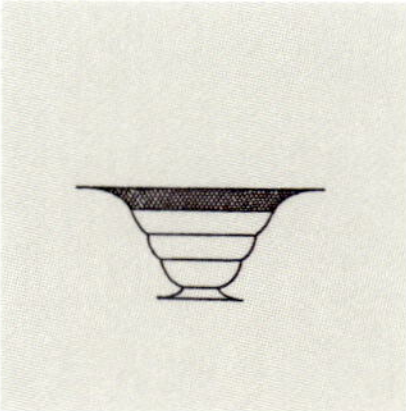

5710

Drawing for one of the bowls in *filigrana a reticello* (net filigree) presented at the 3rd Biennale of Monza in 1927, AMuVe. On that occasion model no. 2 (corresponding to the catalogue number 5710) was proposed in light-blue transparent glass with a 'rim [of] white filigree'.
1927
14 cm high; 30 cm ø

exhibitions
1927, Monza, 3rd Biennale

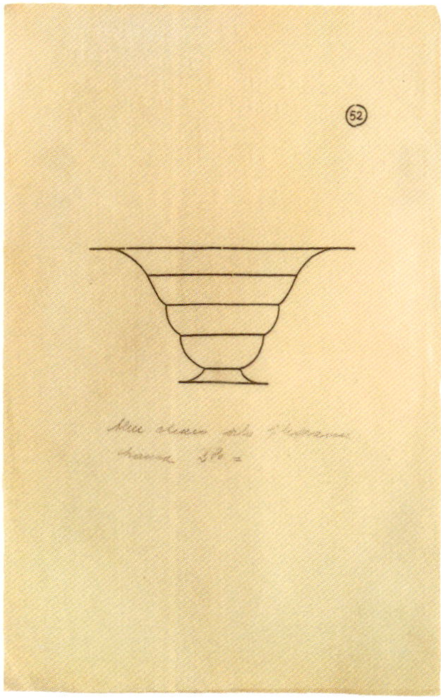

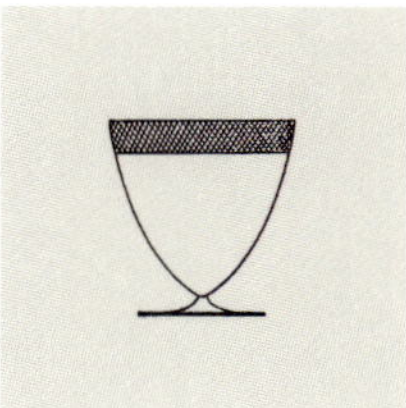

5712

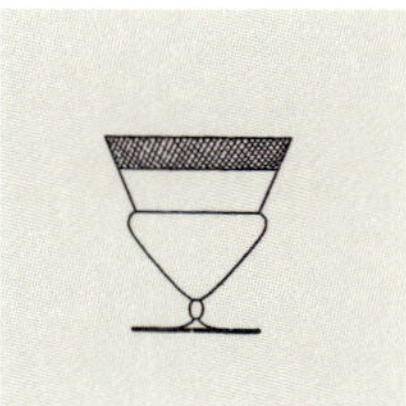

5711

Drawings for vases in *filigrana a reticello* (net filigree) exhibited at the 3rd Biennale of Monza in 1927, AMuVe. On that occasion nos. 53 and 54 (corresponding to catalogue nos. 5712 and 5711) were presented in dark-blue transparent glass with 'light-blue neck [and] *lattimo* filigree'.
1927
23 cm high

exhibitions
1927, Monza, 3rd Biennale

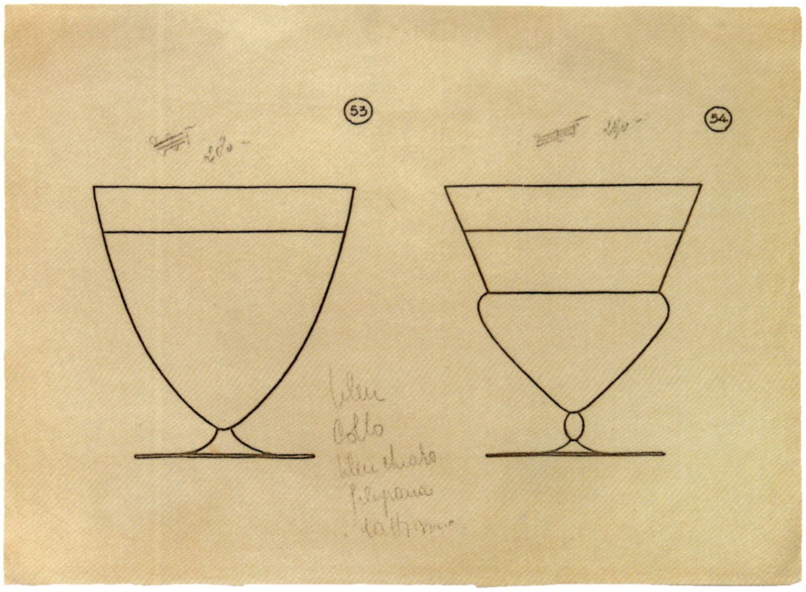

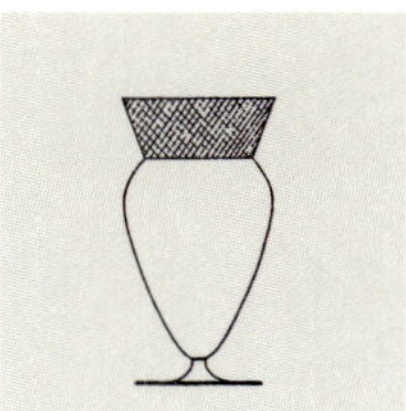

5752

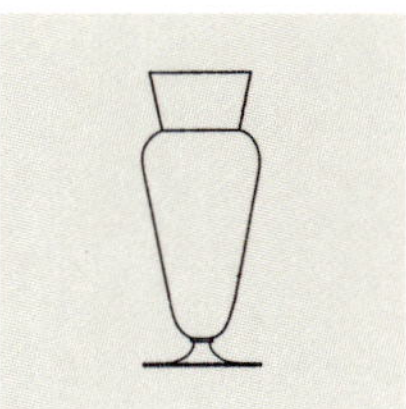

5716

Drawings for vases in *filigrana a reticello* (net filigree) to be exhibited at the 3rd Biennale of Monza in 1927, AMuVe. On that occasion, of nos. 39 and 40 (corresponding to catalogue nos. 5752 and 5716) only the latter was exhibited, which was presented in transparent blue with 'light-blue neck [and] *lattimo* filigree' and also in the variant "crystal violet rim".
1927
33.4 cm high

exhibitions
1927, Monza, 3rd Biennale

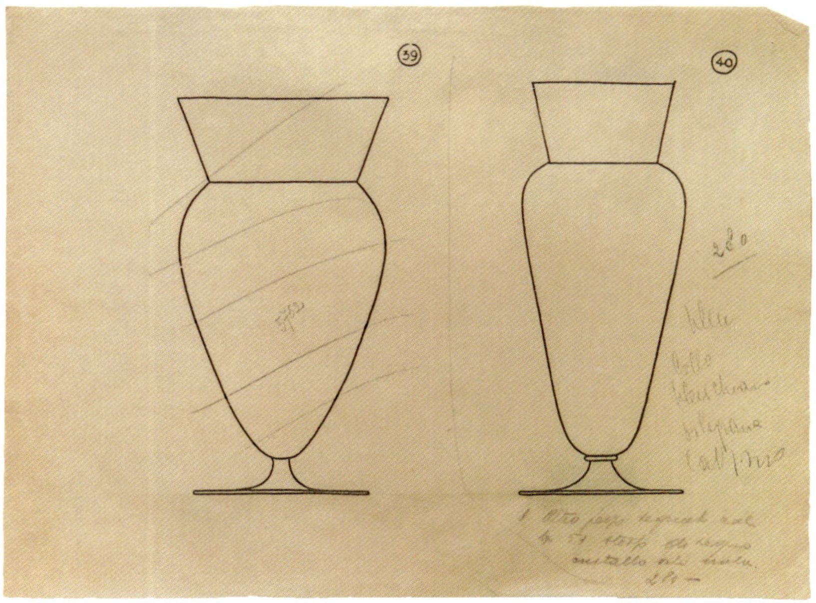

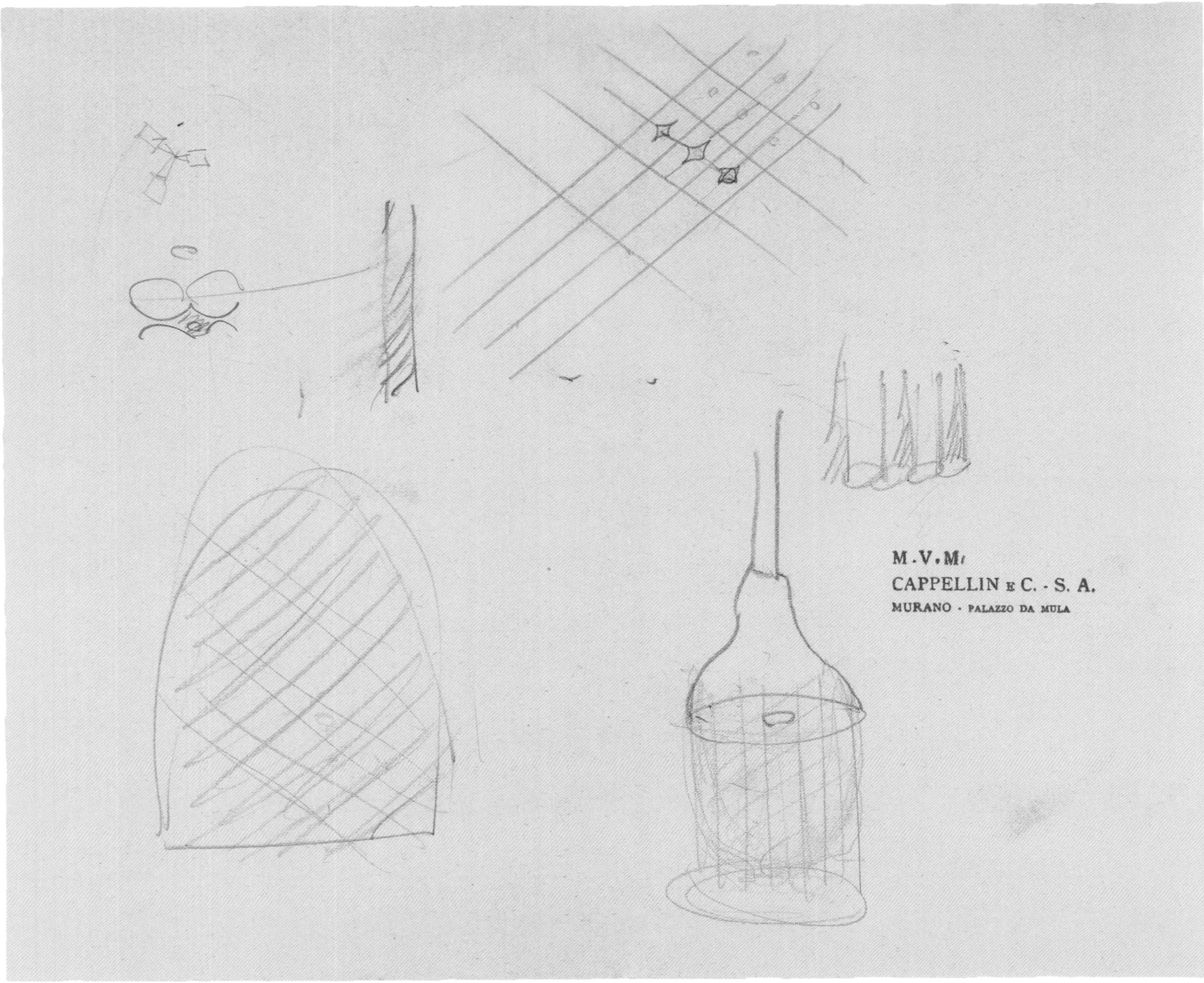

>
Carlo Scarpa, explanatory drawings for the filigrana a reticello *technique, c. 1927, AmuVe*

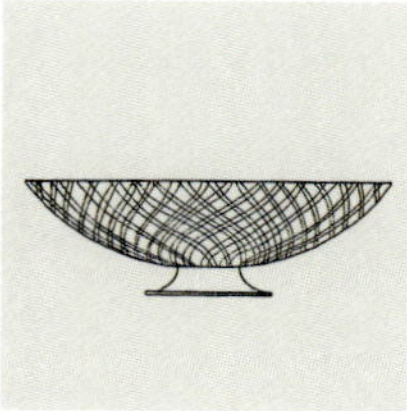

5999

Large bowl in white (*lattimo*) *filigrana a reticello* (net filigree). Flared foot in glass with a silvered mirror finish.
c. 1930
19 cm high; 47 cm ø
Private collection, Treviso

>
5999, catalogue drawing, AMA

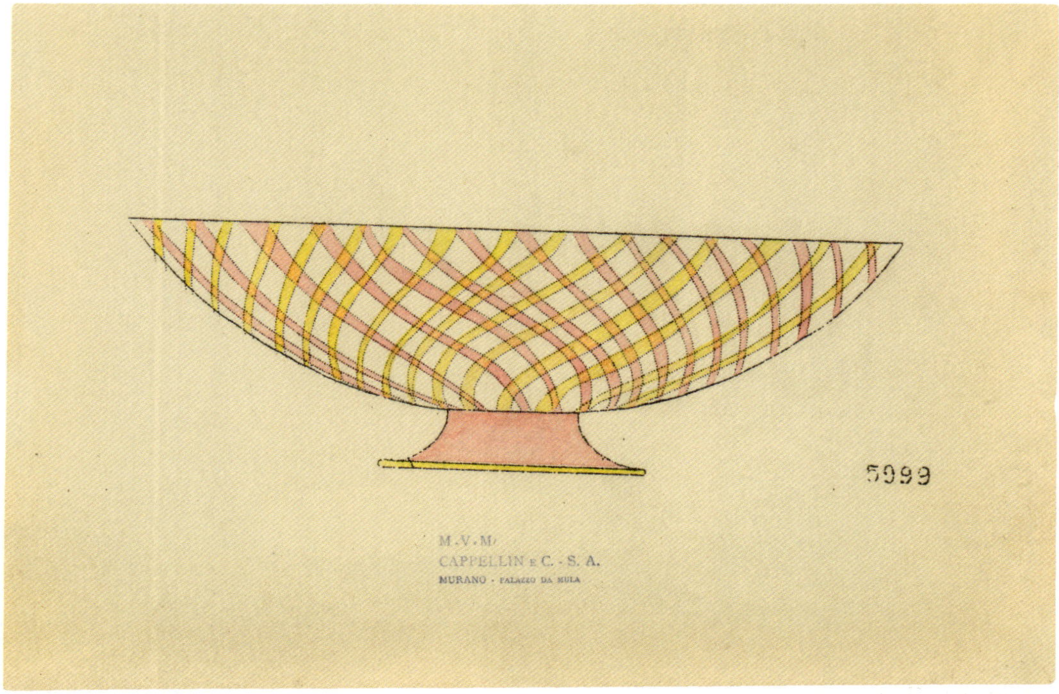

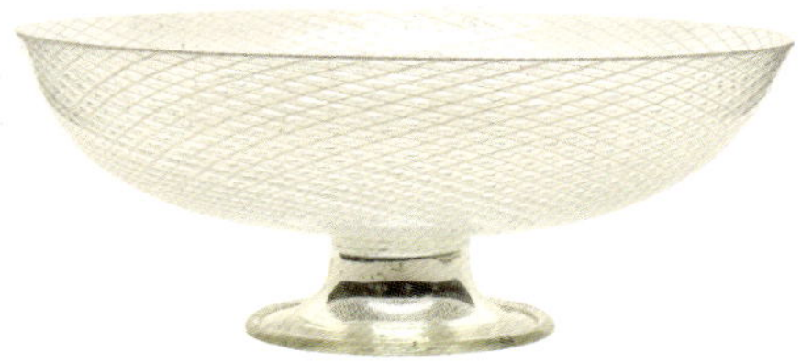

---

n.i.

Bowl in *filigrana a reticello* with pea-green and *avventurina* threads. Rim turned outwards.
1927–31
32 cm ø
Private collection

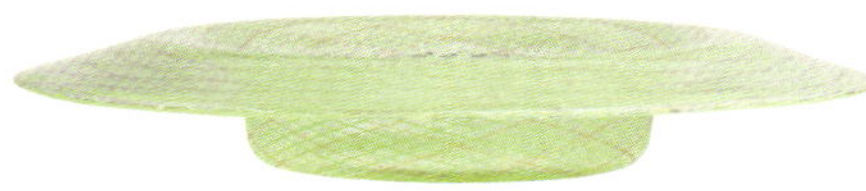

---

n.i.

Vase in *filigrana a reticello* (net filigree) with *lattimo* and yellow threads. Foot drawn from the body of the vase.
1927–31
20 cm high
M.V.M. Cappellin Murano *acid stamp*
Private collection

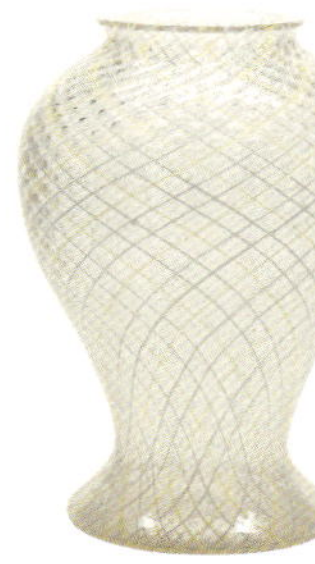

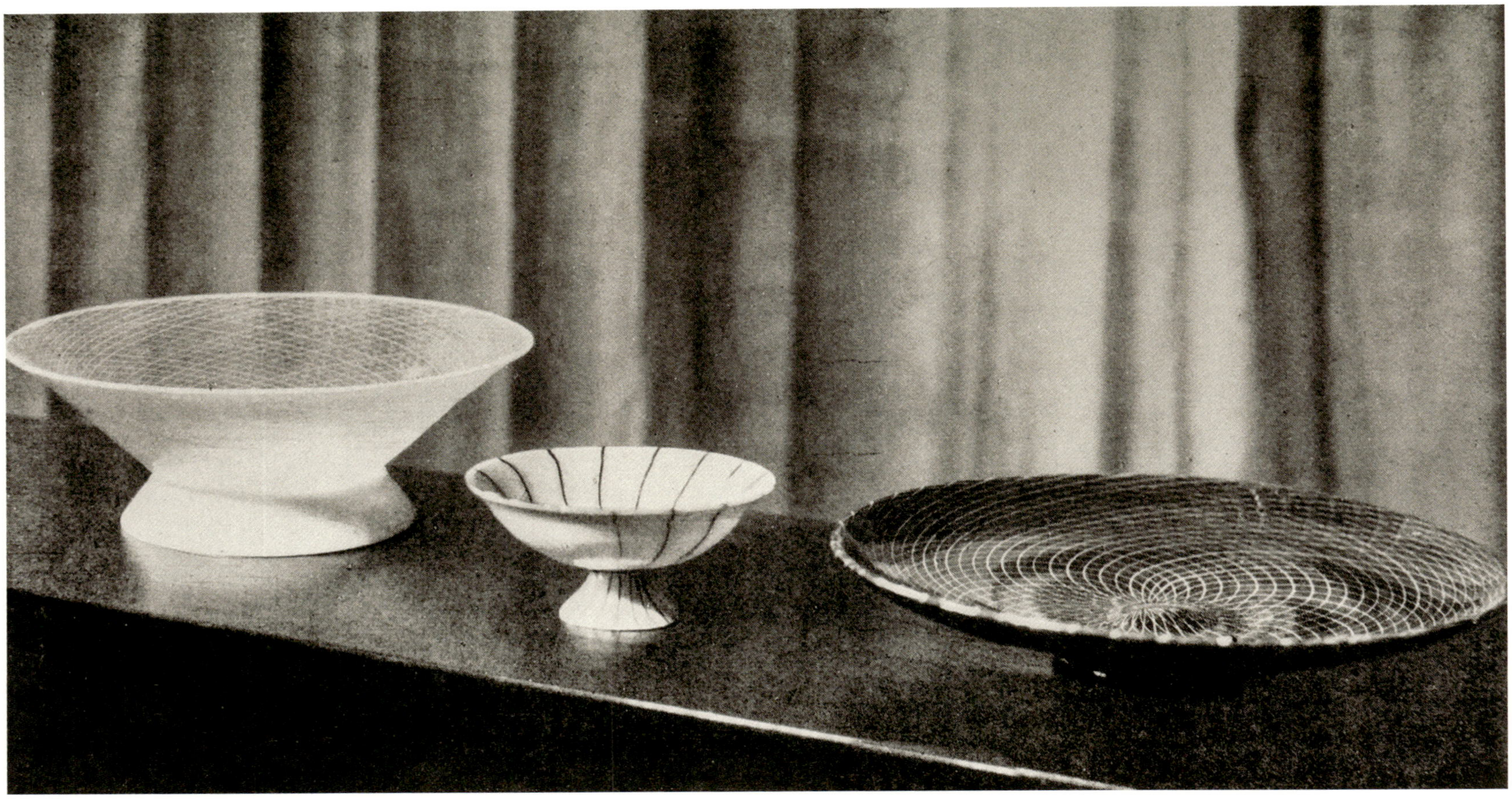

>
A reticello *glass pieces in* Tentoonstelling van glas, aardewerk en kant, van moderne italiaansche kunstnijverheid. Mostra di Vetri, Ceramiche e Merletti d'Arte Moderna Italiana, *Amsterdam 1931, p. 21*

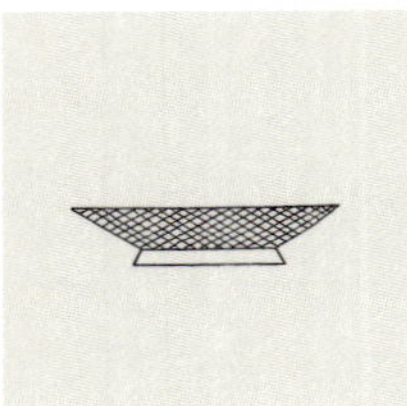

6010

Catalogue drawing for one of the bowls in *filigrana a reticello* (net filigree) presented at the 4th Triennale of Monza in 1930 and at the Amsterdam glass, ceramics and lace exhibition in 1931, AMB. As documented by period photographs, the bowl was produced in *lattimo* filigree with a *lattimo* foot in the shape of a truncated cone.
1930
7 cm high; 32 cm ø

exhibitions
1930, Monza, 4th Triennale
1931, Amsterdam, Exhibition of glass, ceramics and lace

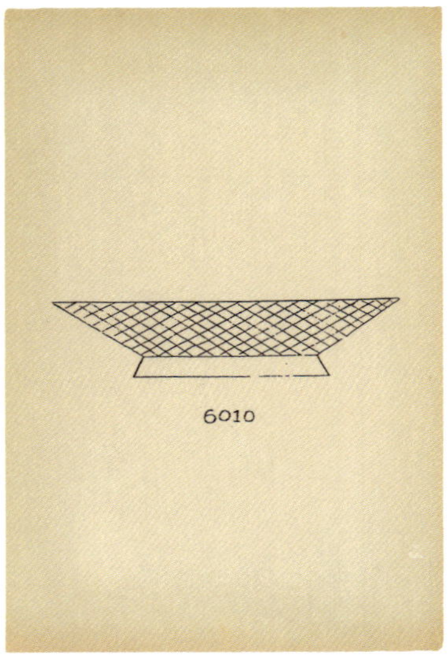

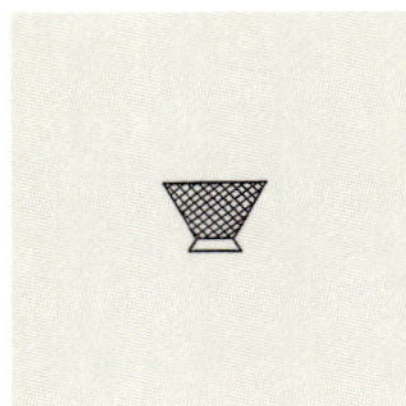

6025

Working drawing for one of the bowls in *filigrana a reticello* (net filigree) presented at the 4th Triennale of Monza in 1930, CSV. As documented by period pictures, the bowl was produced in *lattimo* filigree with a *lattimo* foot in the shape of a truncated cone.
1930
9 cm high; 12 cm ø

exhibitions
1930, Monza, 4th Triennale

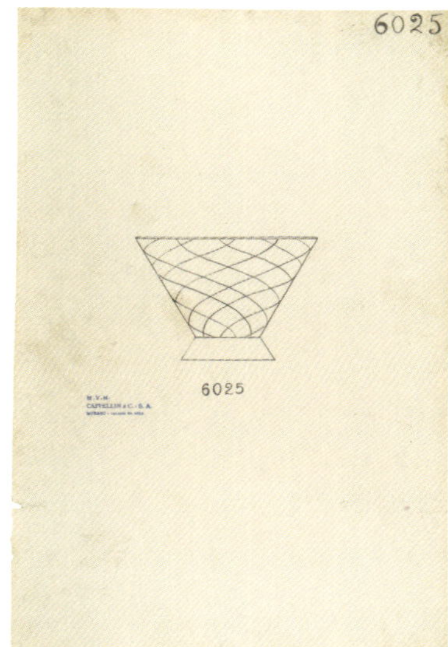

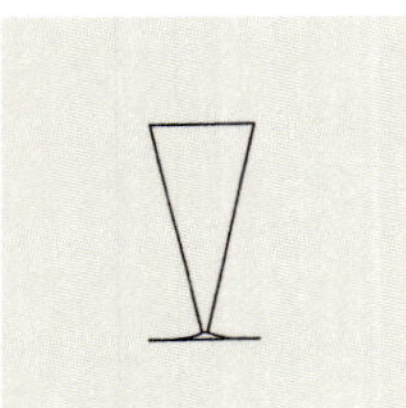

6008

Cone-shaped vase in white (*lattimo*) *filigrana a reticello* (net filigree). Rim and disk-shaped foot in *lattimo* glass.
c. 1930
1930
25 cm high
Private collection

>
6008, catalogue drawing, AMB

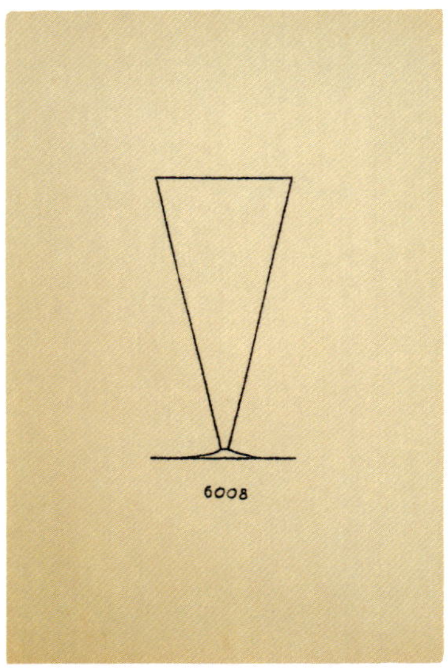

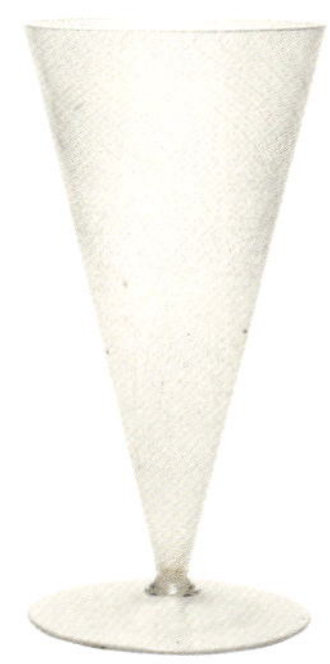

>
Cappellin glass pieces at the 4th Triennale of Monza, 1930, detail

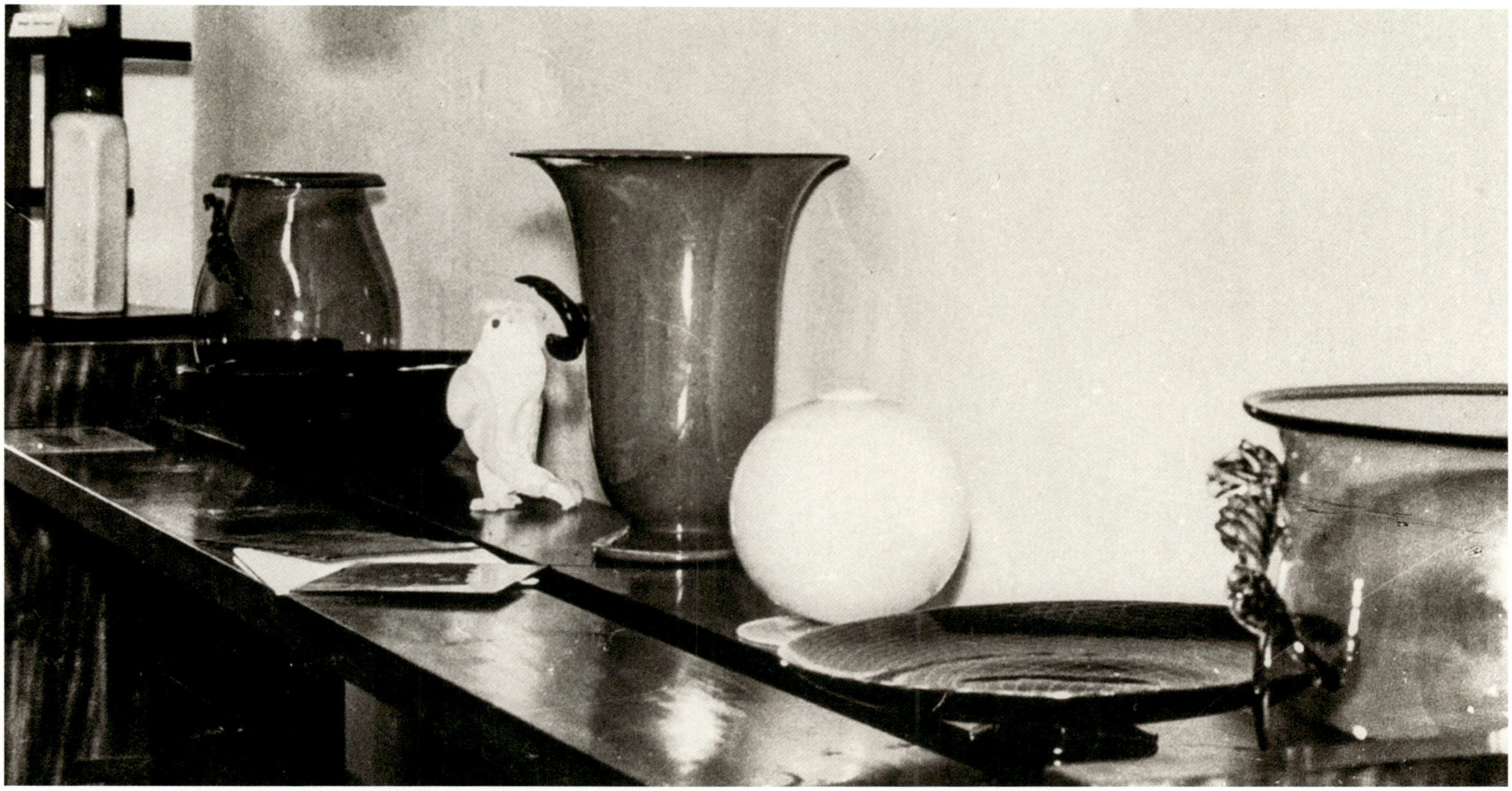

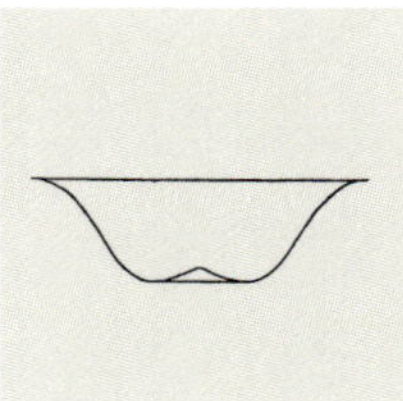

1785

Bowl in black and white (*lattimo*) *filigrana a reticello* (net filigree). Strongly iridised surface. Because of the particular process used in producing the piece, the characteristic pattern, white, with lozenges, is visible only inside the bowl. A similar example was exhibited at the 4th Triennale of Monza in 1930.
1930
11 cm high; 38 cm ø
M.V.M. Cappellin Murano *acid stamp*
Private collection, Venice

exhibitions
1930, Monza, 4th Triennale

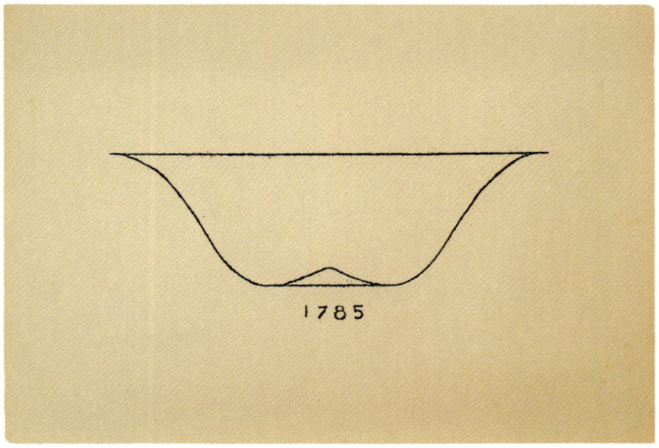

>
1785, catalogue drawing, AMB

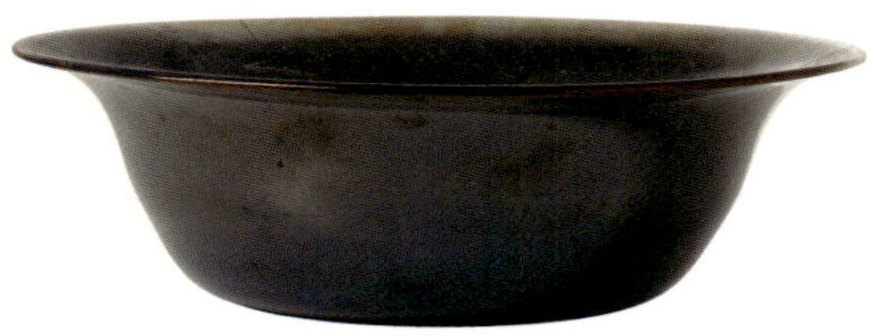

n.i.

Bowl in black and white (*lattimo*) *filigrana a reticello* (net filigree). Because of the particular process used in producing the piece, the characteristic pattern, white, with lozenges, is visible only inside the bowl. The piece is one of a small series in black and white which was exhibited in 1930 at the 4th Triennale of Monza and the following year at the Amsterdam Exhibition of glass, ceramics and lace.
1930
28 cm ø
M.V.M. Cappellin Murano *acid stamp*
Private collection

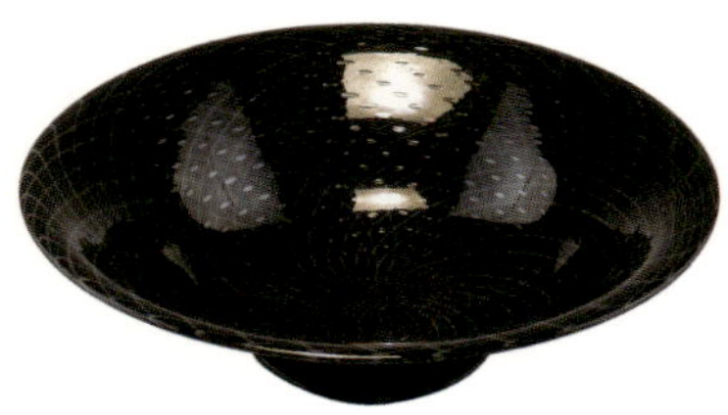

>
Bowl in filigrana a reticello *with polychrome canes and vase with 'twisted canes', in* Domus, *1, January 1928, p. 29.*

>
Carlo Scarpa, studies for bowl and vase and for polychrome 'filigree and rotated canes', c. 1927, AMuVe

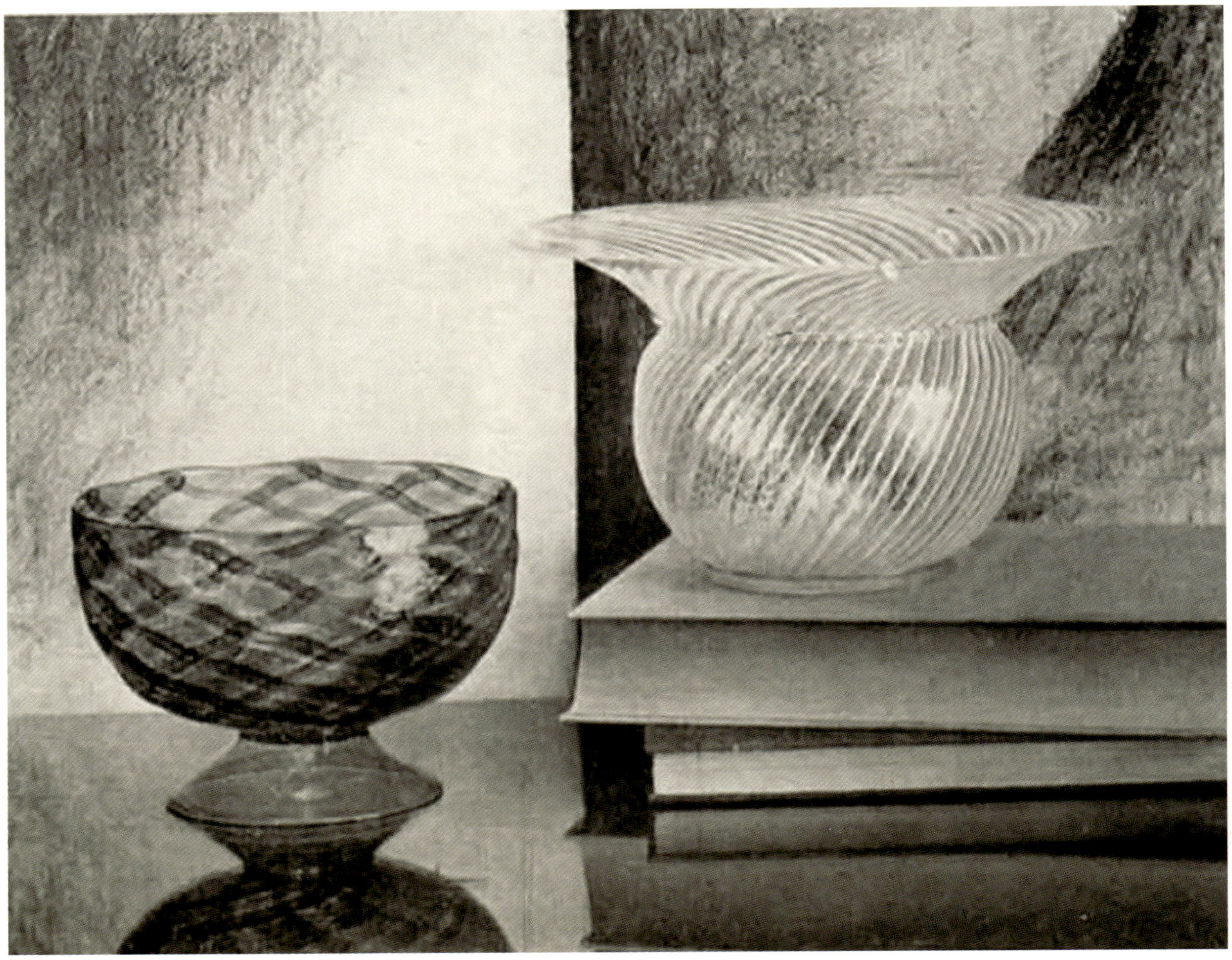

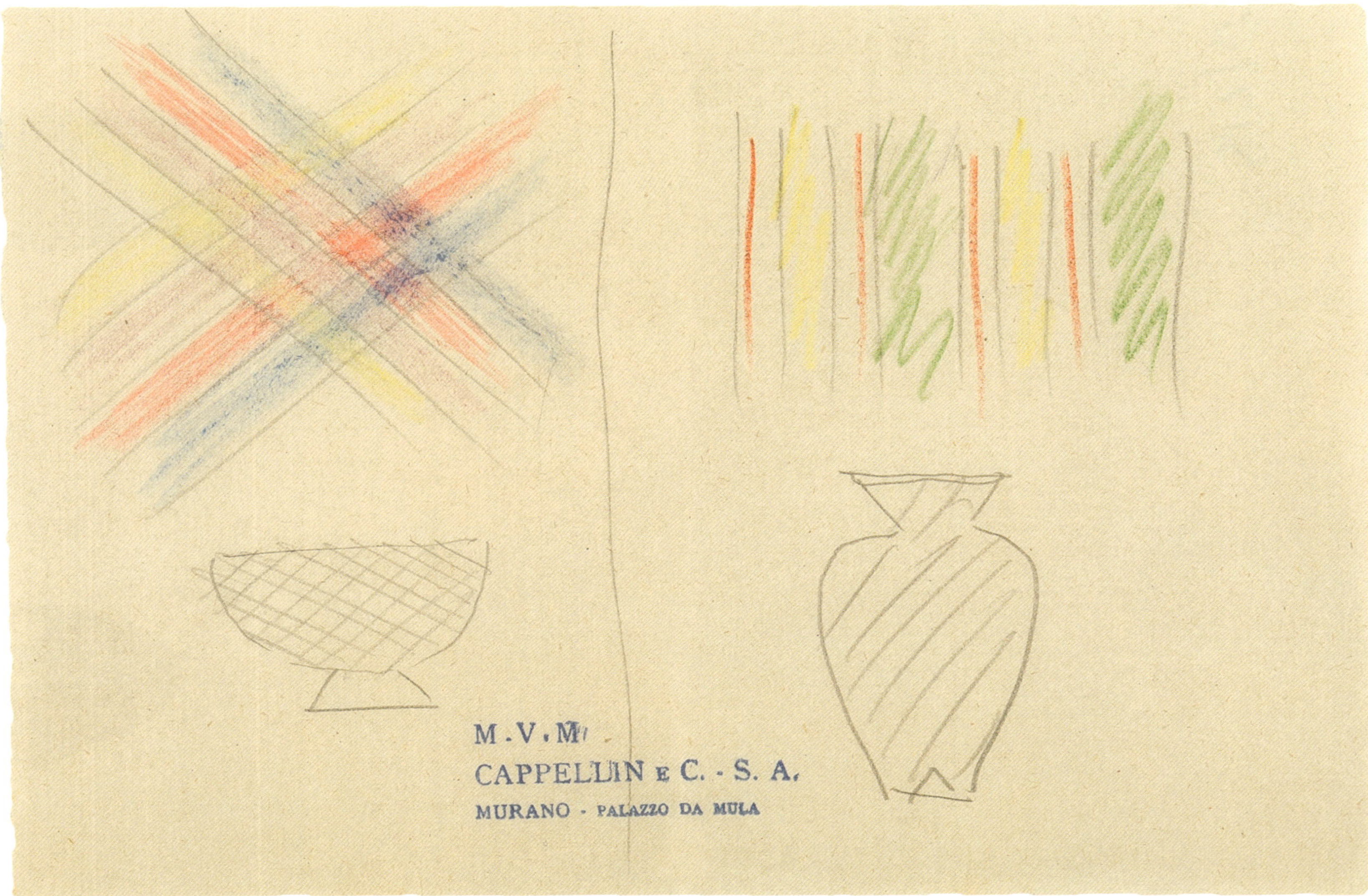

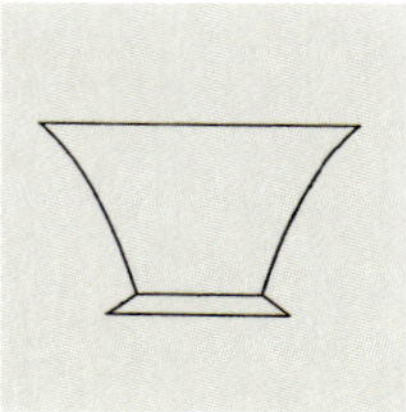

5702

Vase in *filigrana a reticello* (net filigree) with ruby, green and yellow canes. Foot in green transparent glass. A similar example in 'dark-blue' glass '[with] 'red canes' was exhibited at the 3rd Biennale of Monza in 1927 (no. 70 corresponding to catalogue number 5702).
c. 1927
28 cm high; 38 cm ø
M.V.M. Cappellin label
Private collection

exhibitions
1927, Monza, 3rd Biennale

>
70, drawing for a model to be exhibited at the 3rd Biennale of Monza, AMuVe

>
Carlo Scarpa, studies for 'turned filigrees and canes'. Above: 'twisted canes: yellow / Savoy-blue / ruby / green'. Below: 'filigree canes: yellow / light violet / ruby / Savoy-blue', c. 1927, AMuVe

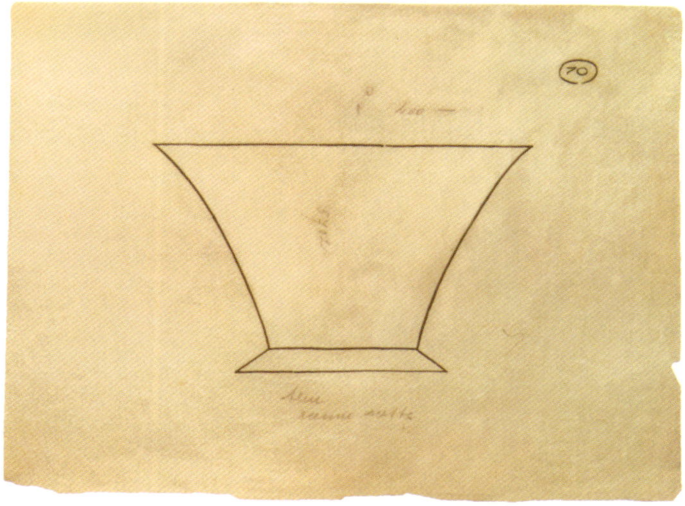

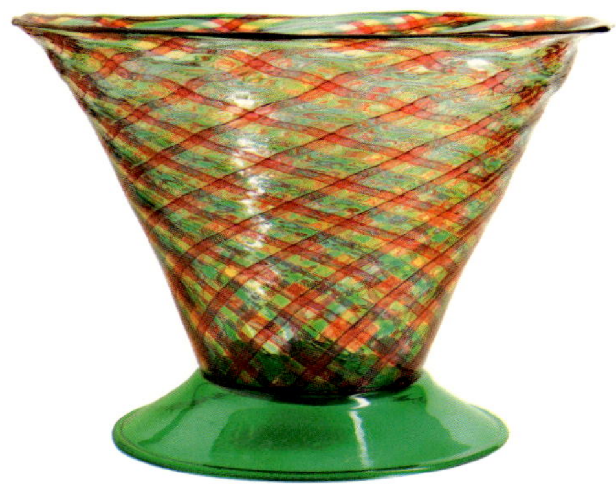

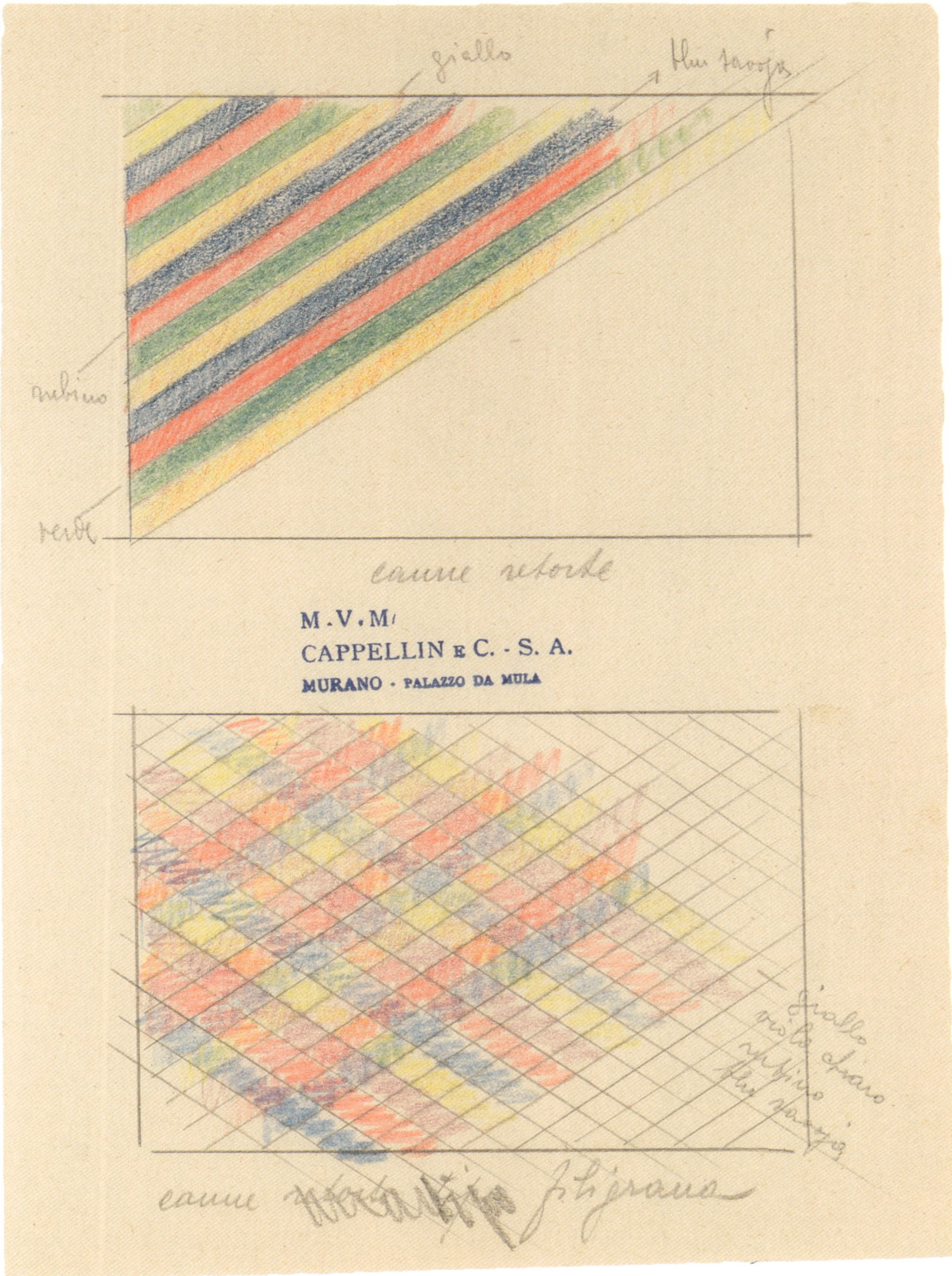

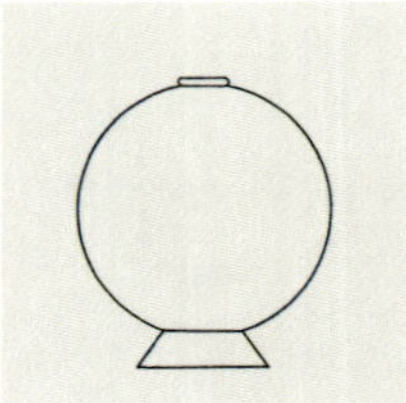

5673

Spherical vase in *filigrana a reticello* (net filigree) with green, yellow and white (colourless) canes. Foot in the shape of a truncated cone in green transparent glass. Model no. 5673, symbol of the Cappellin firm, was also produced in other glass techniques (e.g. transparent, *reticello* with threads, opaque).
c. 27
31 cm high
M. V. M. Cappellin label
Private collection, Udine

>
Vase no. 5673 with glass flower. Period photograph 1927–30, CSV

>
5673, working drawing, CSV

>
Carlo Scarpa, studies for 'turned filigrees and canes'. Above: 'twisted canes: pagliesco with coral-red / yellow / green thread': Below: "canes: crystal with coral-red / green/ [crystal canes coral-red thread] / thread', AMuVe

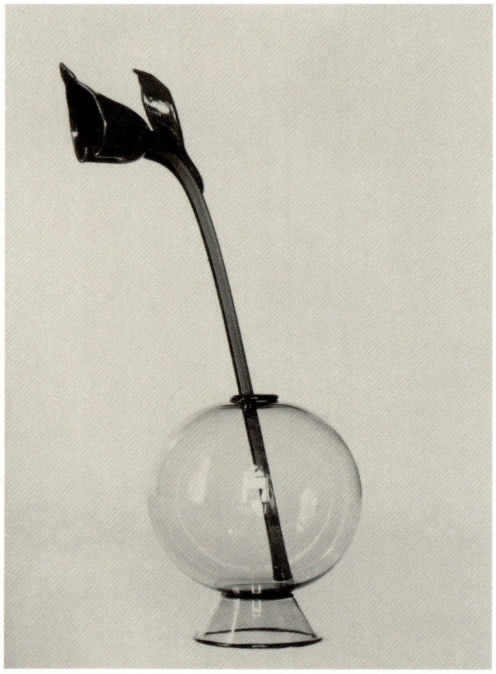

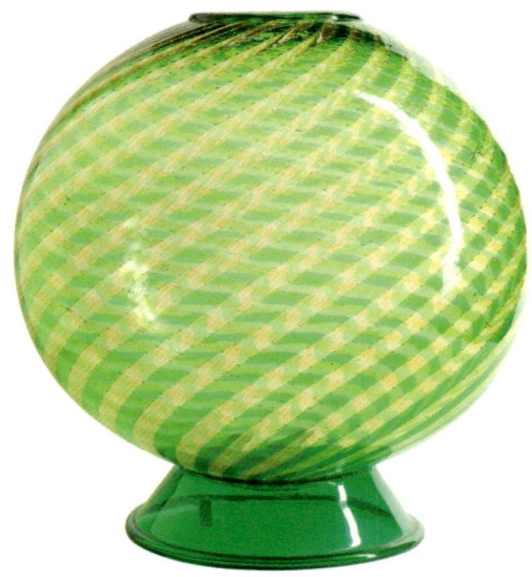

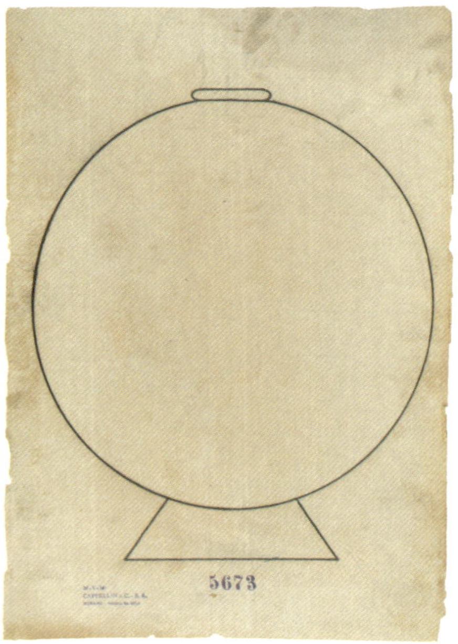

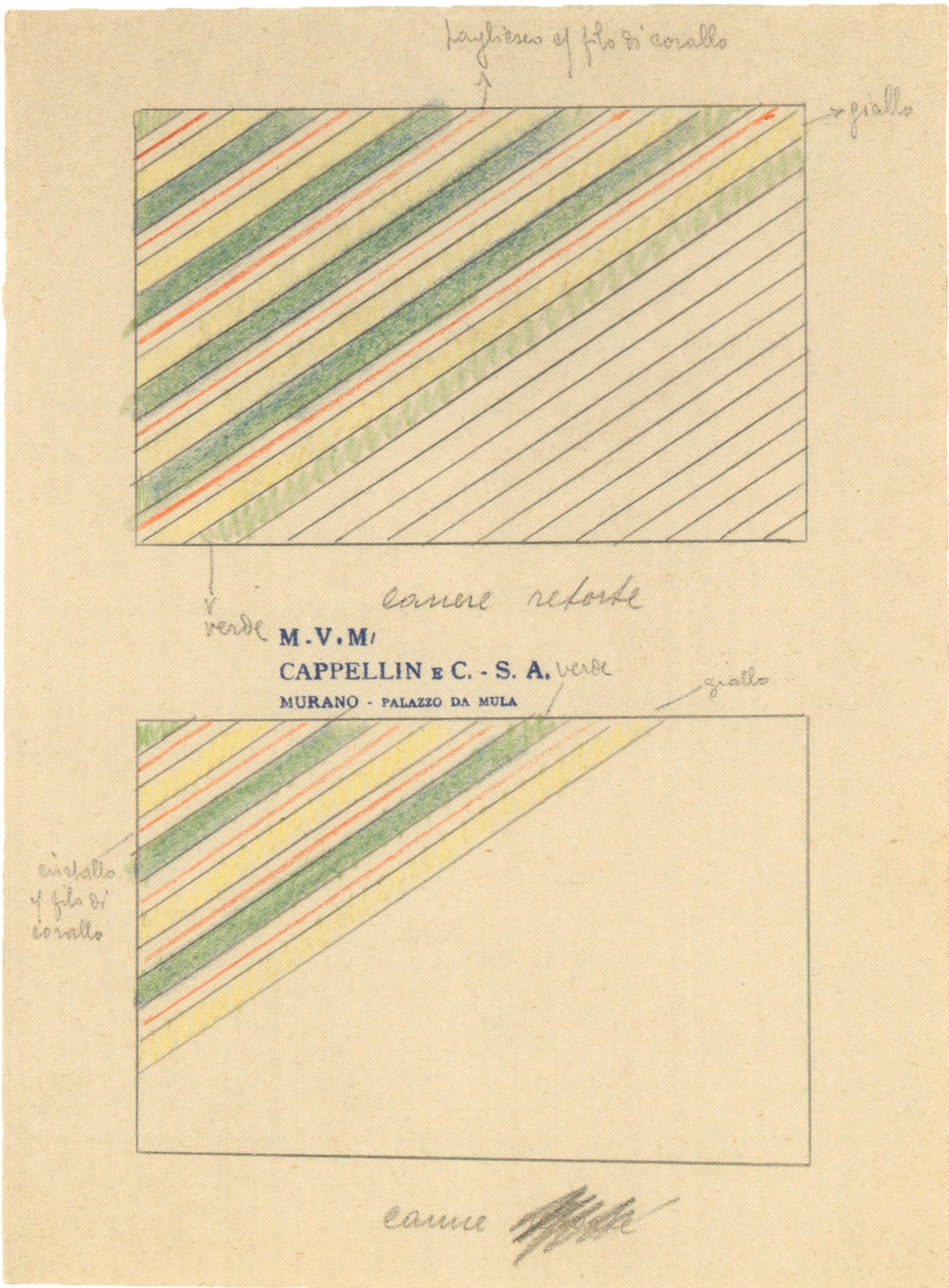

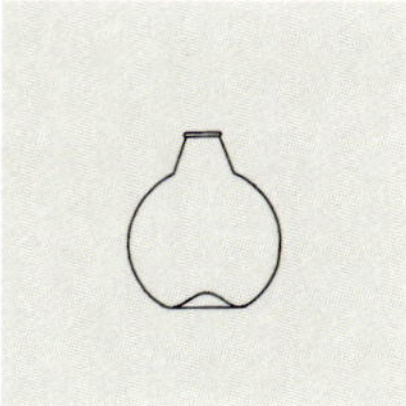

5858

Vase in *filigrana a reticello* (net filigree) with light-blue and Savoy-blue canes. Mouth in Savoy-blue glass. Model no. 5858 was also produced in other glass techniques (e.g. transparent *balloton*, *pasta vitrea*).
c. 1928
21 cm high
M.V.M. Cappellin Murano *acid stamp*
Private collection

>
5858, catalogue drawing, AMB

>
Carlo Scarpa, studies for 'turned filigrees and canes'. Above and in the centre: twisted canes: 'yellow/ amethyst" and "light blue / pagliesco'. Below filigrana a reticello (net filigree) with canes: violet, amethyst and dark blue, AMuVe

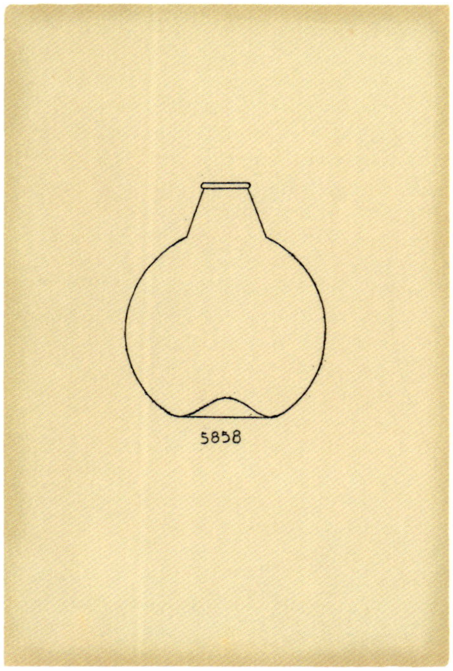

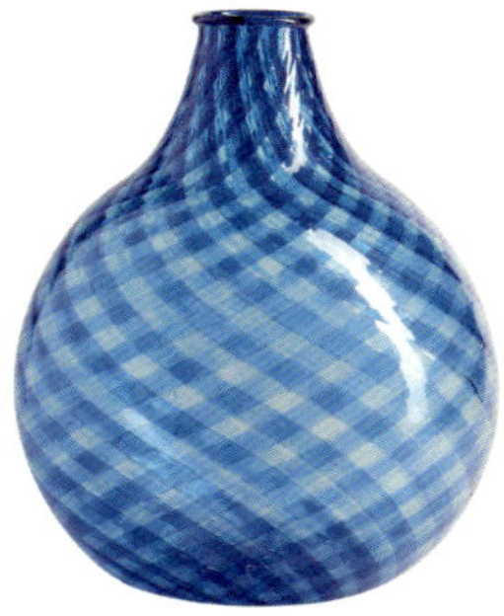

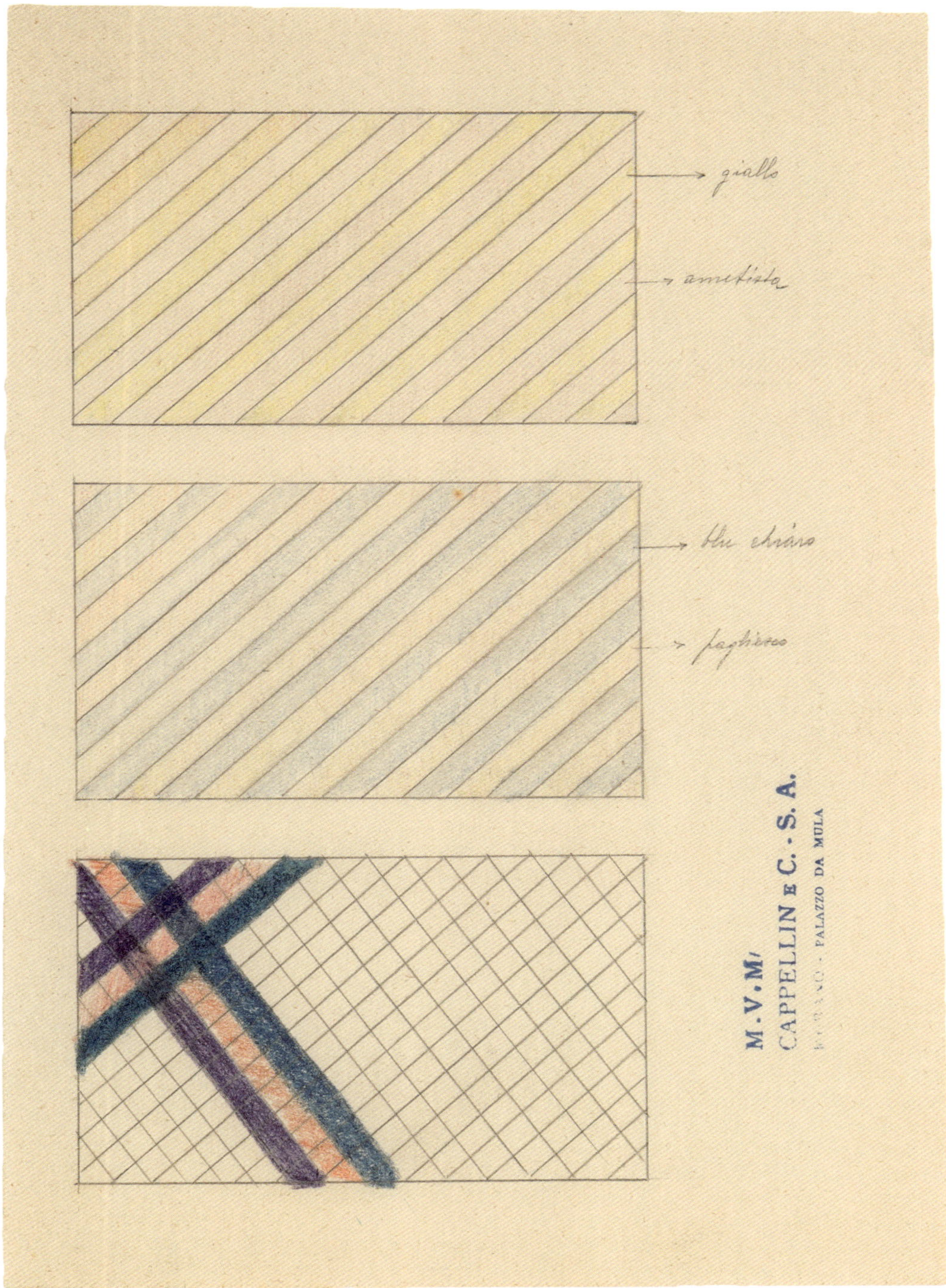

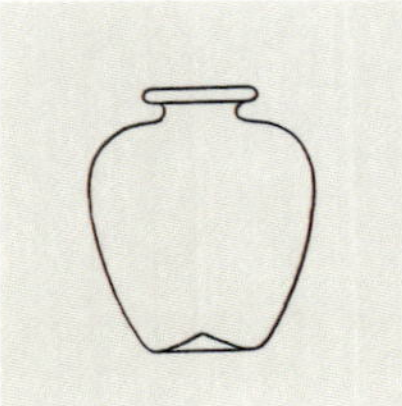

5299

Vase in *filigrana a reticello* (net filigree) with *pagliesco* and amethyst canes. The vase was produced following a model in the catalogue of the glassworks designed by Vittorio Zecchin (1921–23).
c. 1927
28 cm high
M.V.M. Cappellin Murano *acid stamp*
Private collection, Venice

›
5299, catalogue drawing, AMB

›
1793–1795, working drawings, CSV

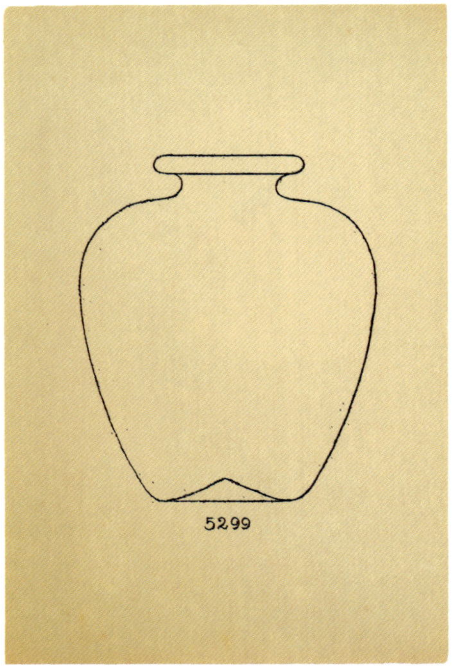

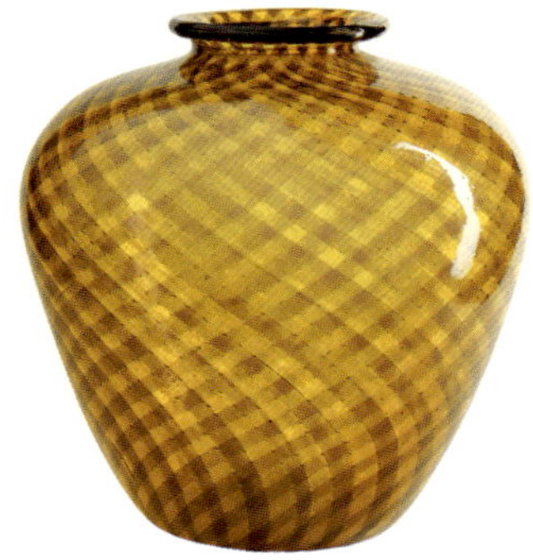

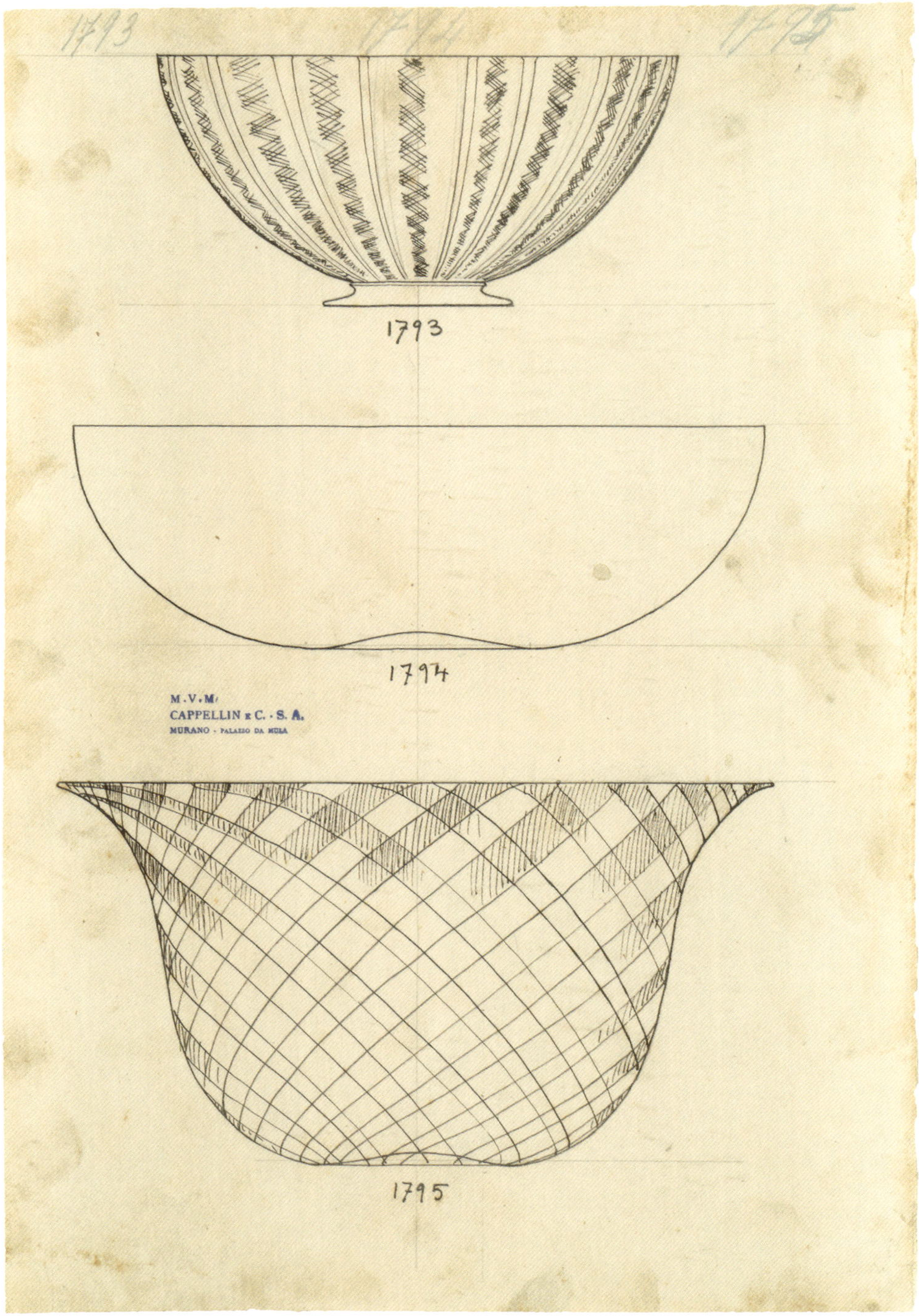

>
Vase and compote bowl in filigrana a reticello. *Period photograph, 1927–30, CSV*

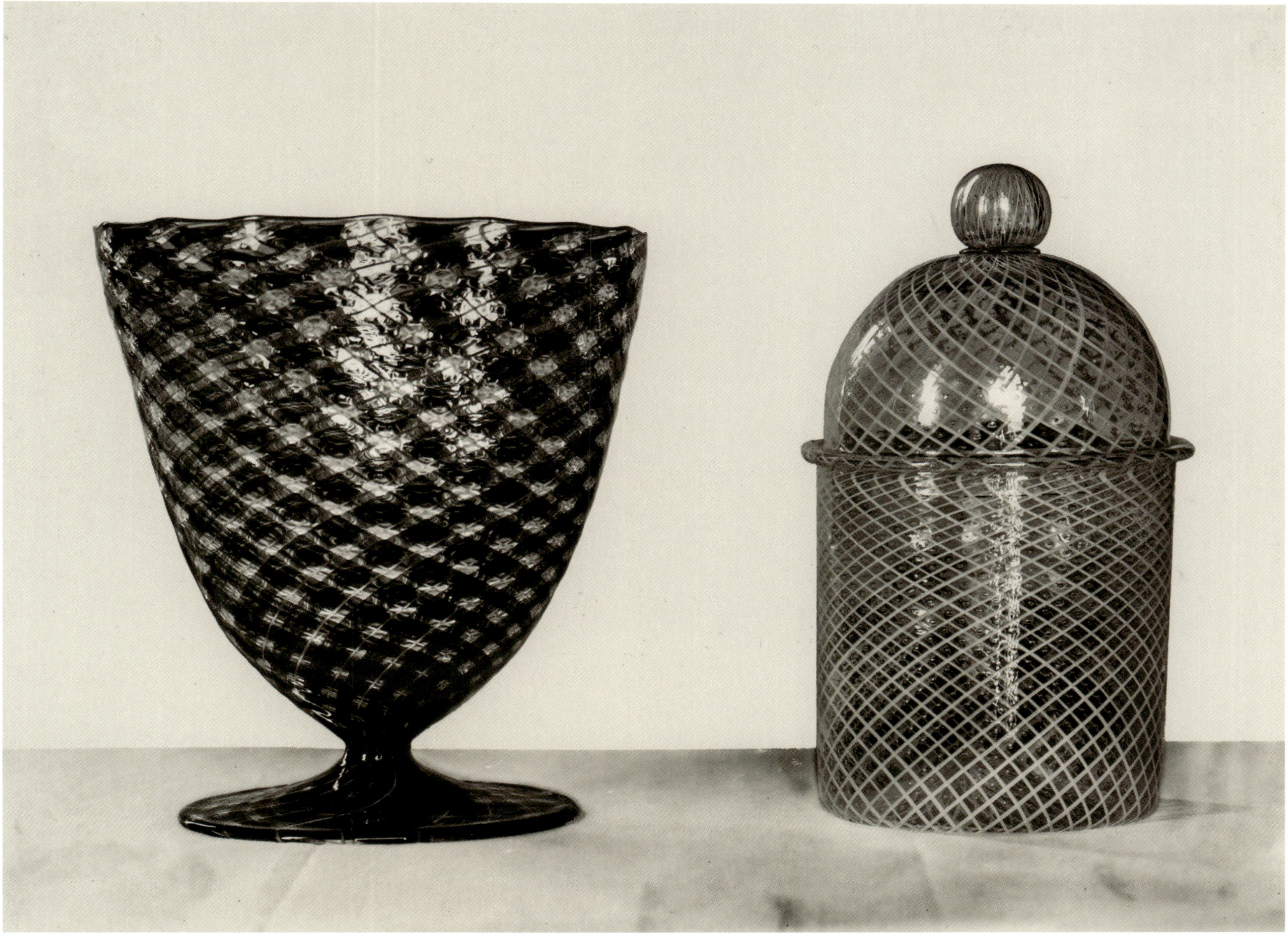

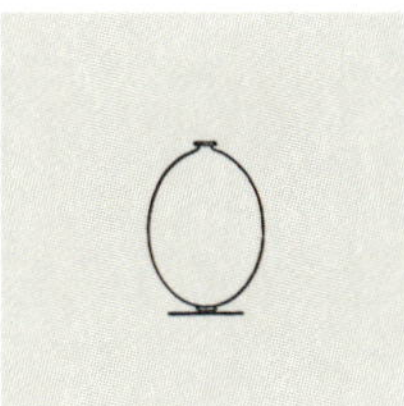

5726

Ovoid vase in *filigrana a reticello* (net filigree) with *pagliesco* and amethyst canes. Disk-shaped foot. Model no. 5726 was also produced in other glass techniques (e.g. transparent).
c. 1927
21.5 cm high
M.V.M. Cappellin Murano *acid stamp*
Private collection

>
5726, working drawing, CSV

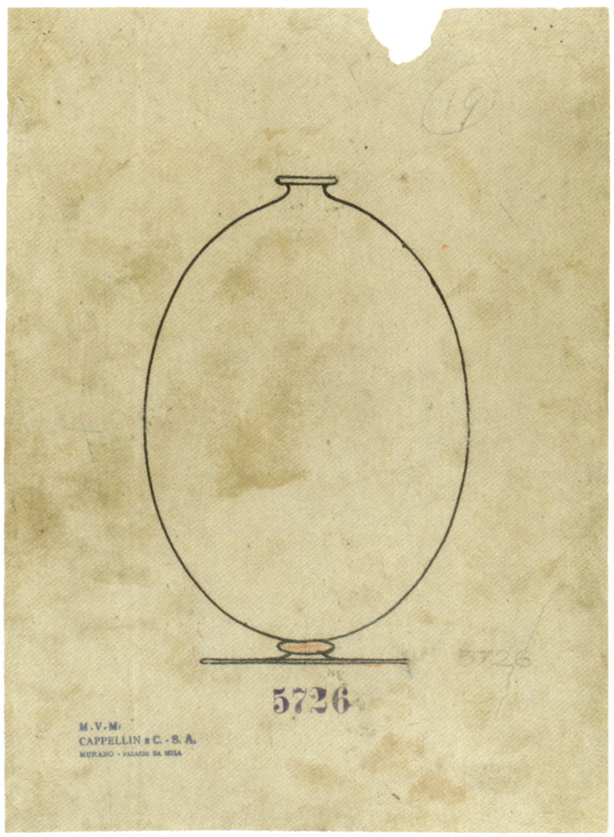

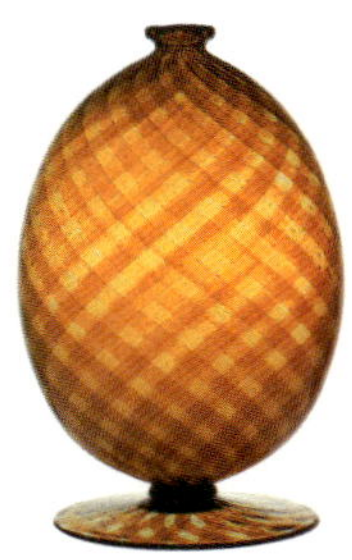

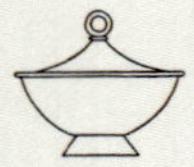

3820

Compote bowl in glass with Savoy blue, coral-red and crystal twisted bands. Truncated-cone-shaped foot and ring handle on the cover in transparent blue glass. The piece was produced in a variant of no. 3820, initially designed in transparent glass and proposed in this version at the Salon d'automne of Paris in 1926.
1927
23 cm high
M.V.M. Cappellin Murano *acid stamp*
Gina Nanni collection

>
3820, working drawing, CSV

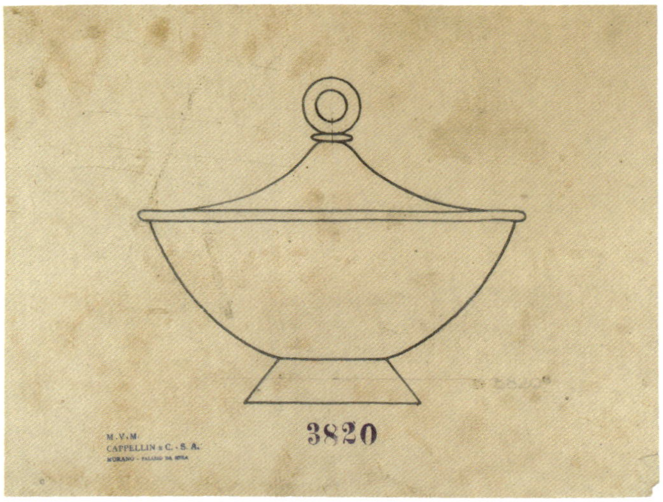

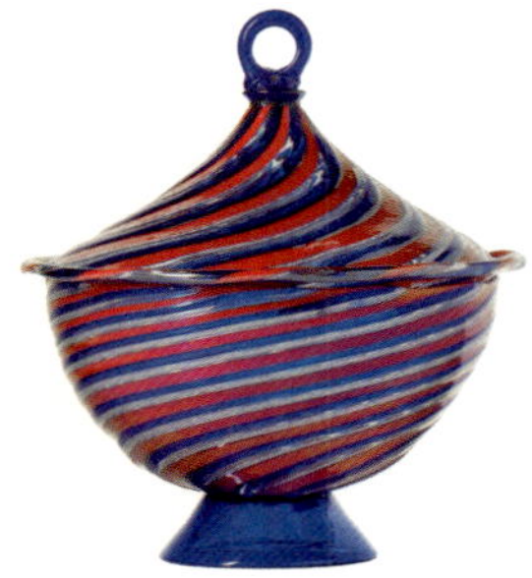

n.i.

Bowl in glass with twisted *lattimo* and light-green canes. Small *lattimo* and gold foot.
c. 1929
14 cm high; 25 cm ø
Venezia Murano Company *acid stamp*
Ferraris-Rocca collection, Venice

>
Carlo Scarpa, studies for 'filigrees and turned canes'. Above: 'light blue / Savoy-blue'; in the centre: 'ruby red / yellow'; Below: 'dark-blue / Savoy blue / light blue', c. 1927, AMuVe

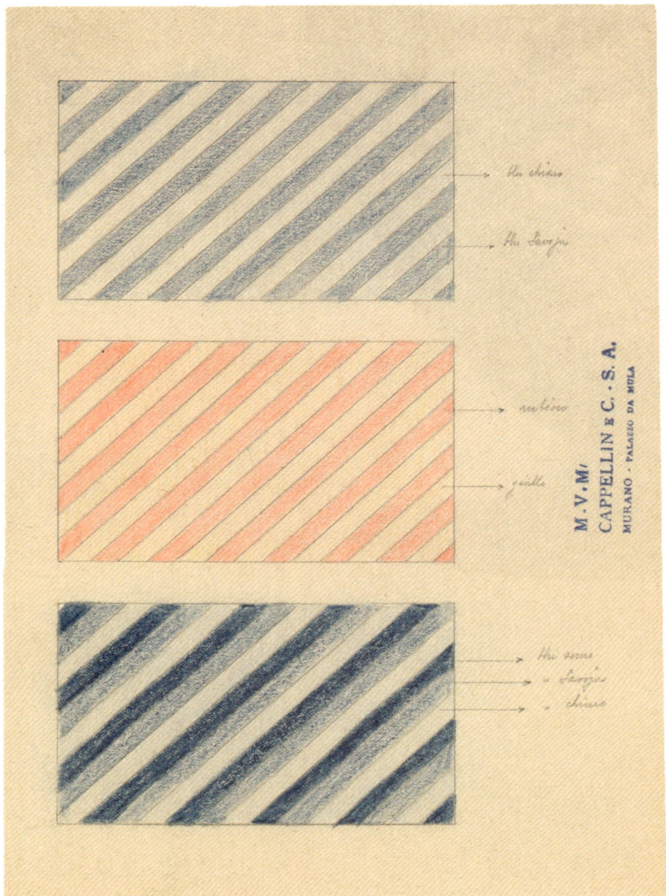

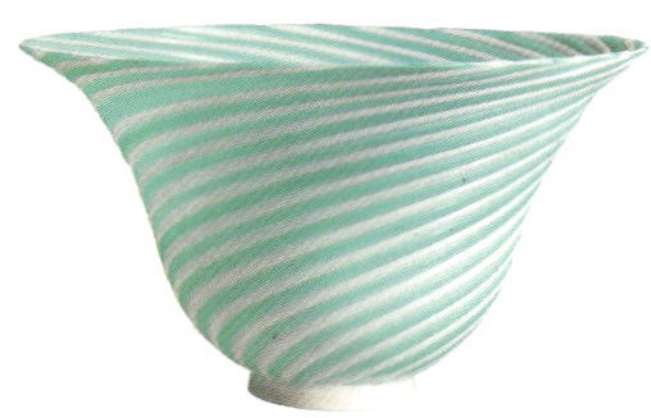

n.i.

Glass bowl with twisted *lattimo* and amethyst canes. Flared foot in glass with a silvered mirror finish.
1929–30
10 cm high; 18 cm ø
Private collection, Udine

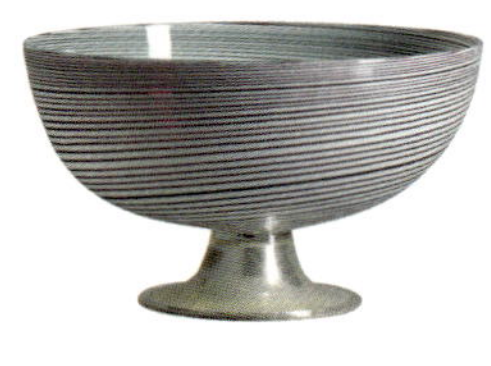

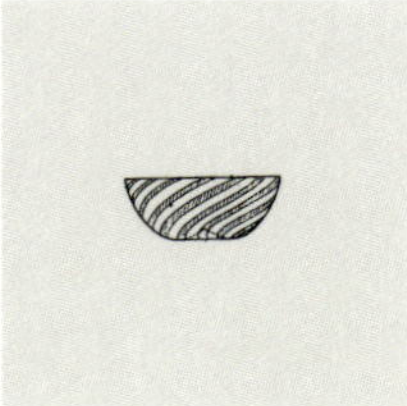

913

Bowl in glass with twisted canes in two colours, documented by a period photograph c. 1929, detail CSV.
c. 1929
8 cm high; 19 cm ø

›
913, catalogue drawing, ACSVr

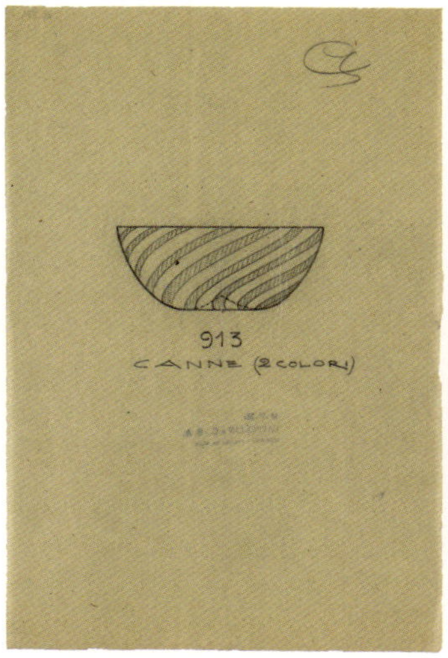

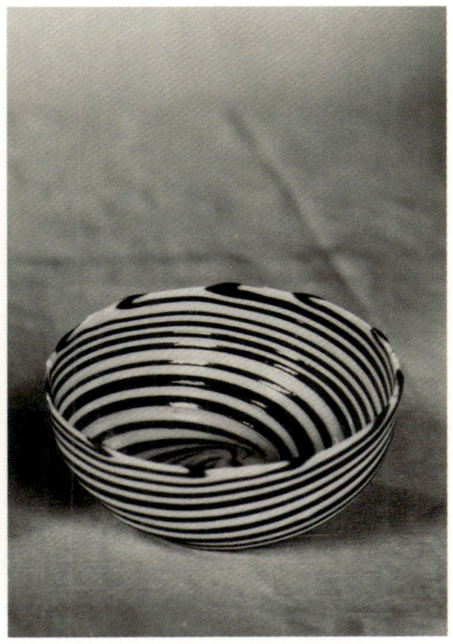

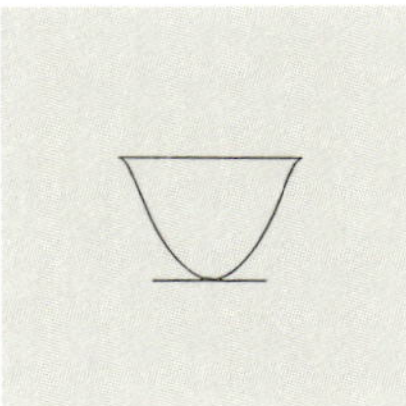

5018

n.i.

Bowl with disk-shaped foot in glass with characteristic decoration obtained with the appropriate mould, the vitreous thread having been wound around the wall of the vessel. Period photograph, c. 1929, CSV. The piece was produced after a model included in the catalogue of the glassworks and designed by Vittorio Zecchin (1921–25). In the same illustration there is an opaque glass plate obtained with the same technique.
c. 1929
14 cm high; 21 cm ø

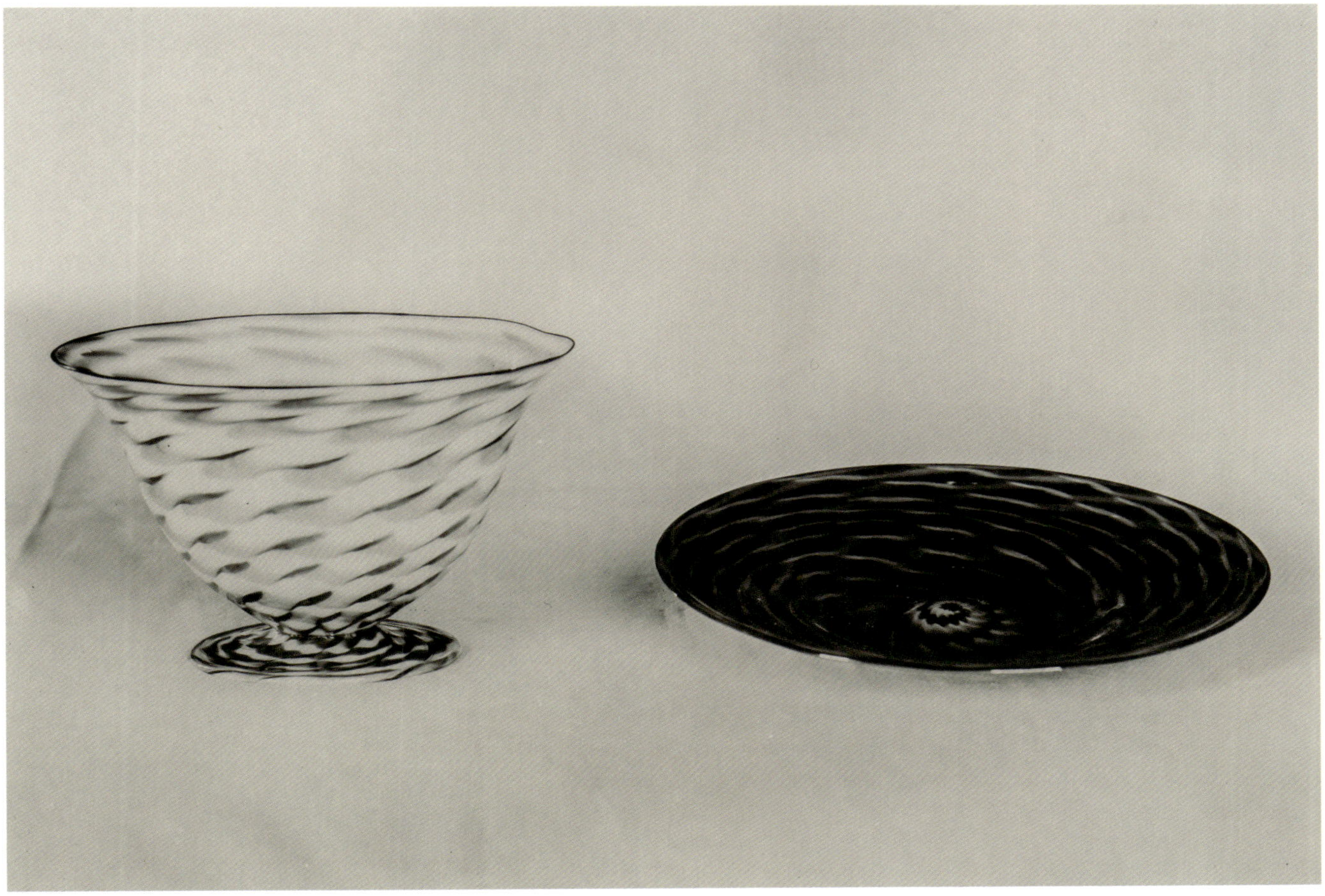

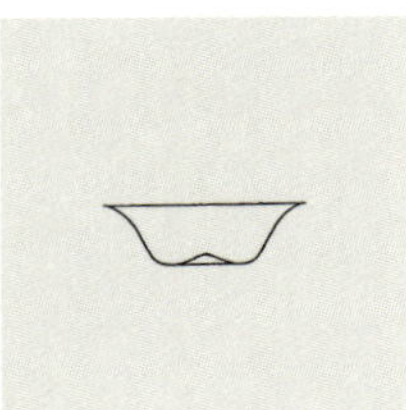

5864

Bowl in black glass with the inside decorated by a thread wound in a spiral, segmented by means of a mould. Strongly iridised surface.
c. 1929–30
5 cm high; 21.5 cm ø
M.V.M. Cappellin Murano *acid stamp*
The Olnick Spanu Collection, New York

›
5864, catalogue drawing, AMB

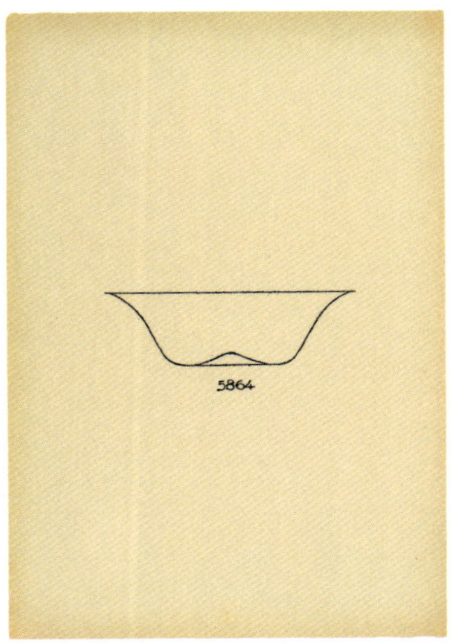

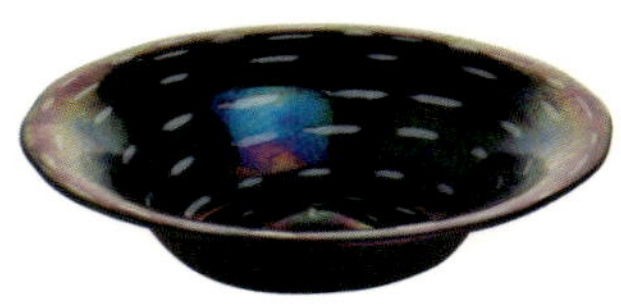

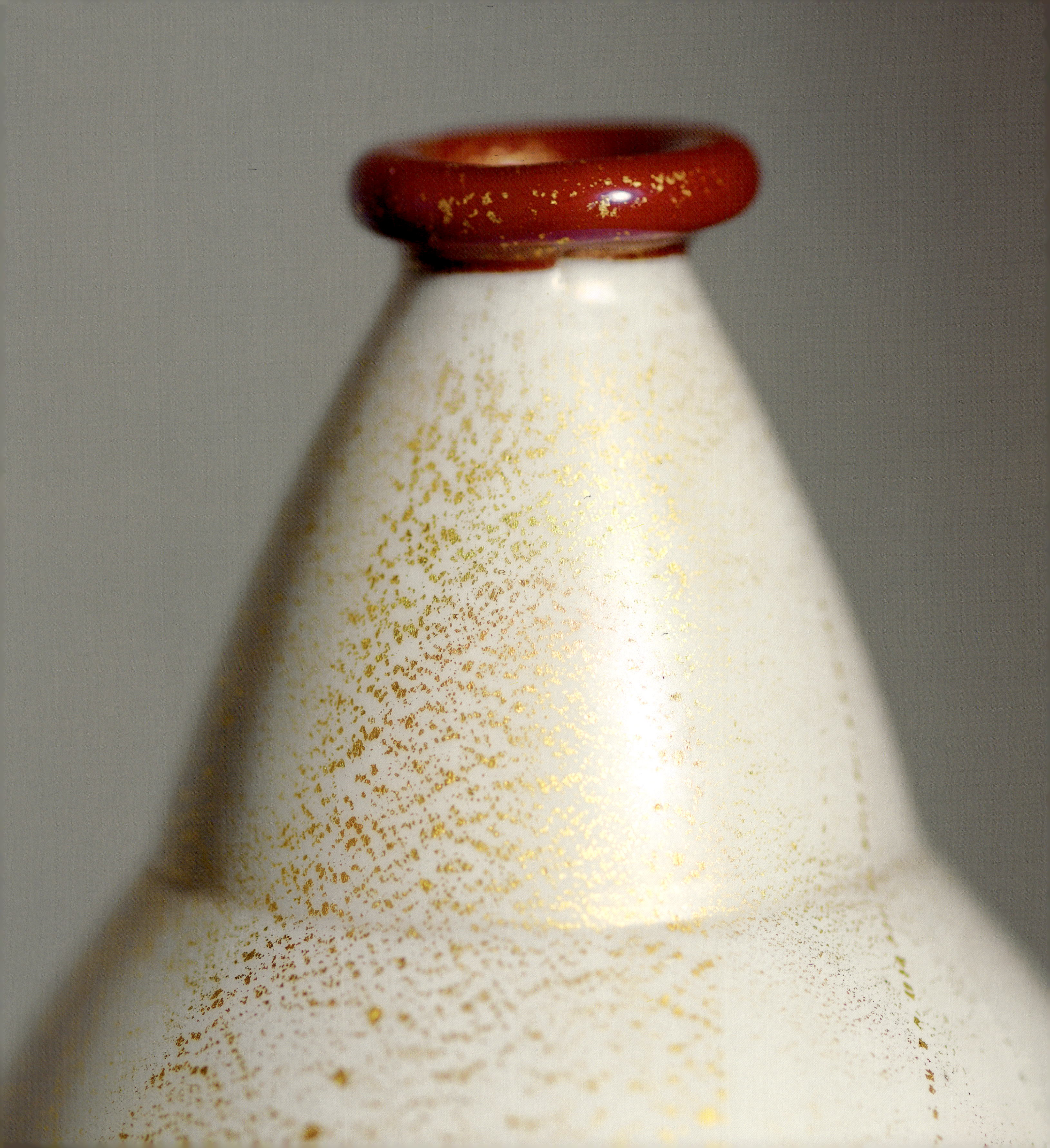

Lattimo glass pieces are white and opaque and are obtained using tin dioxide. The Cappellin firm used this type of material starting from 1928, creating precious, extraordinary items, generally of small dimensions, which were proposed with various finishings: from simple iridisation associated with a foot and/or little side handles in *pasta vitrea*, to the application of gold leaf combined, in some cases, with the coral-red *pasta vitrea* at the mouth and/or foot. Other *lattimo* pieces, often characterised by a disk-shaped foot, were decorated with undulating bands of varying colours applied on the body. Sometimes these models contained somewhat fussy details but in the majority of cases they were marked by very rigorous forms, some of classical inspiration, others influenced by oriental art, and others again characterised by the accentuated geometric nature of their design.

LATTIMI 1928-30

Exhibitions

1928, Paris, *Salon d'automne*
1929, Leipzig, Trade Fair
1929, Milan, *Mostra Artisti italiani contemporanei*
1929, Paris, *Salon d'automne*
1930, Milan, Trade Fair
1930, Monza, 4th *Esposizione Triennale Internazionale delle Arti Decorative ed Industriali Moderne*
1931, Amsterdam, *Tentoonstelling van glas, aardewerk en kant, van moderne italiaansche kunstnijverheid. Mostra di vetri, ceramiche e merletti d'arte moderna italiana*

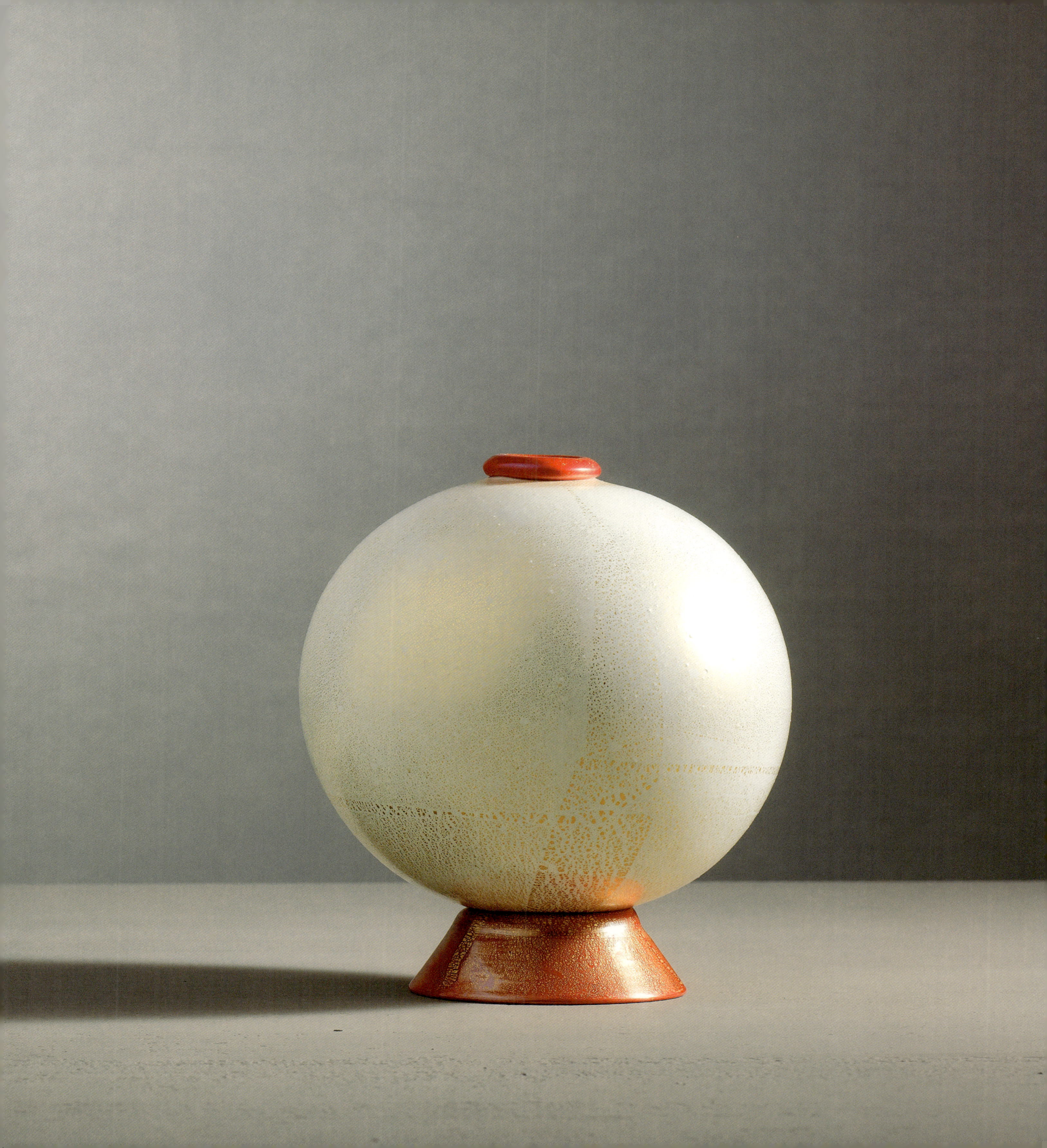

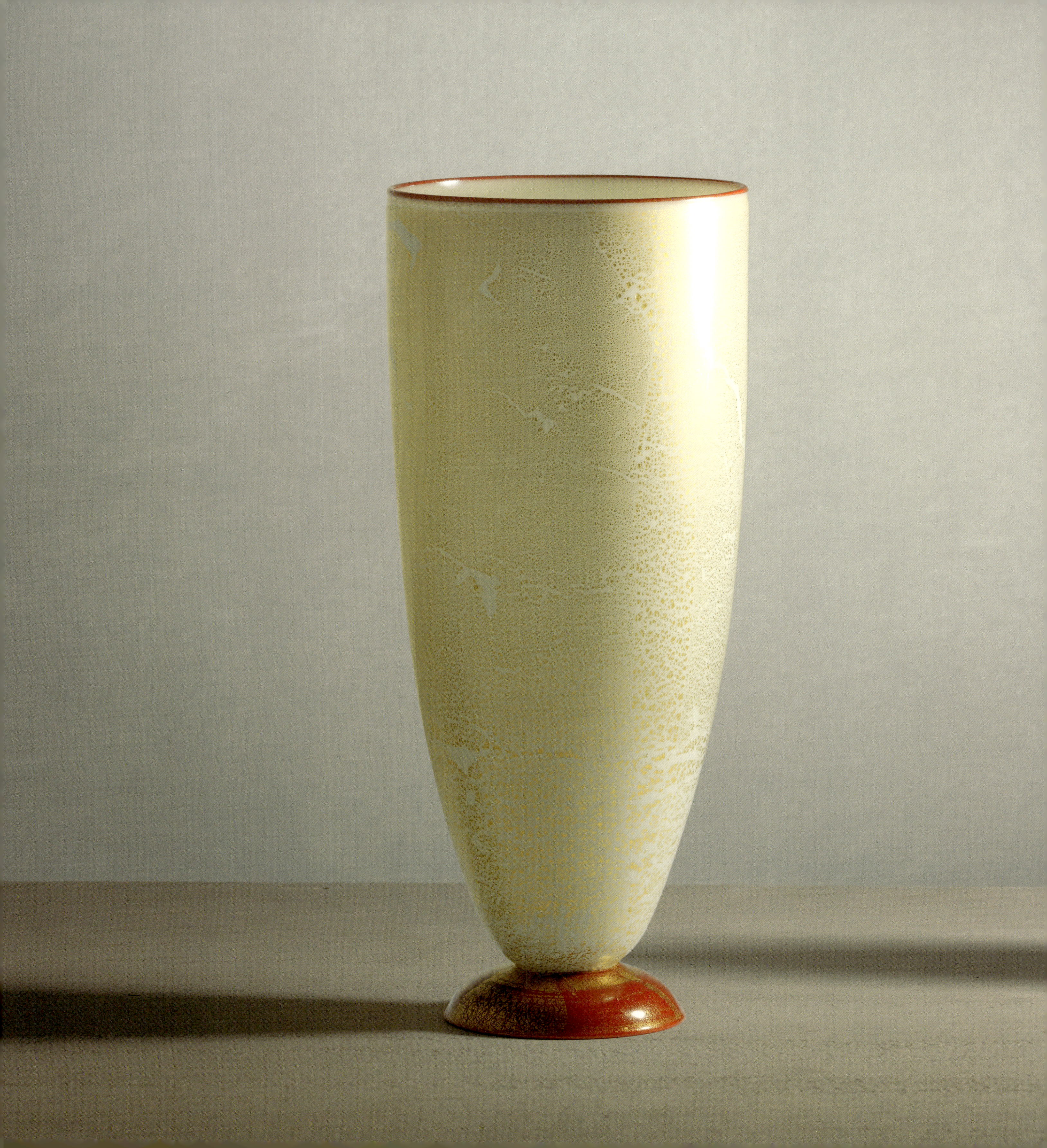

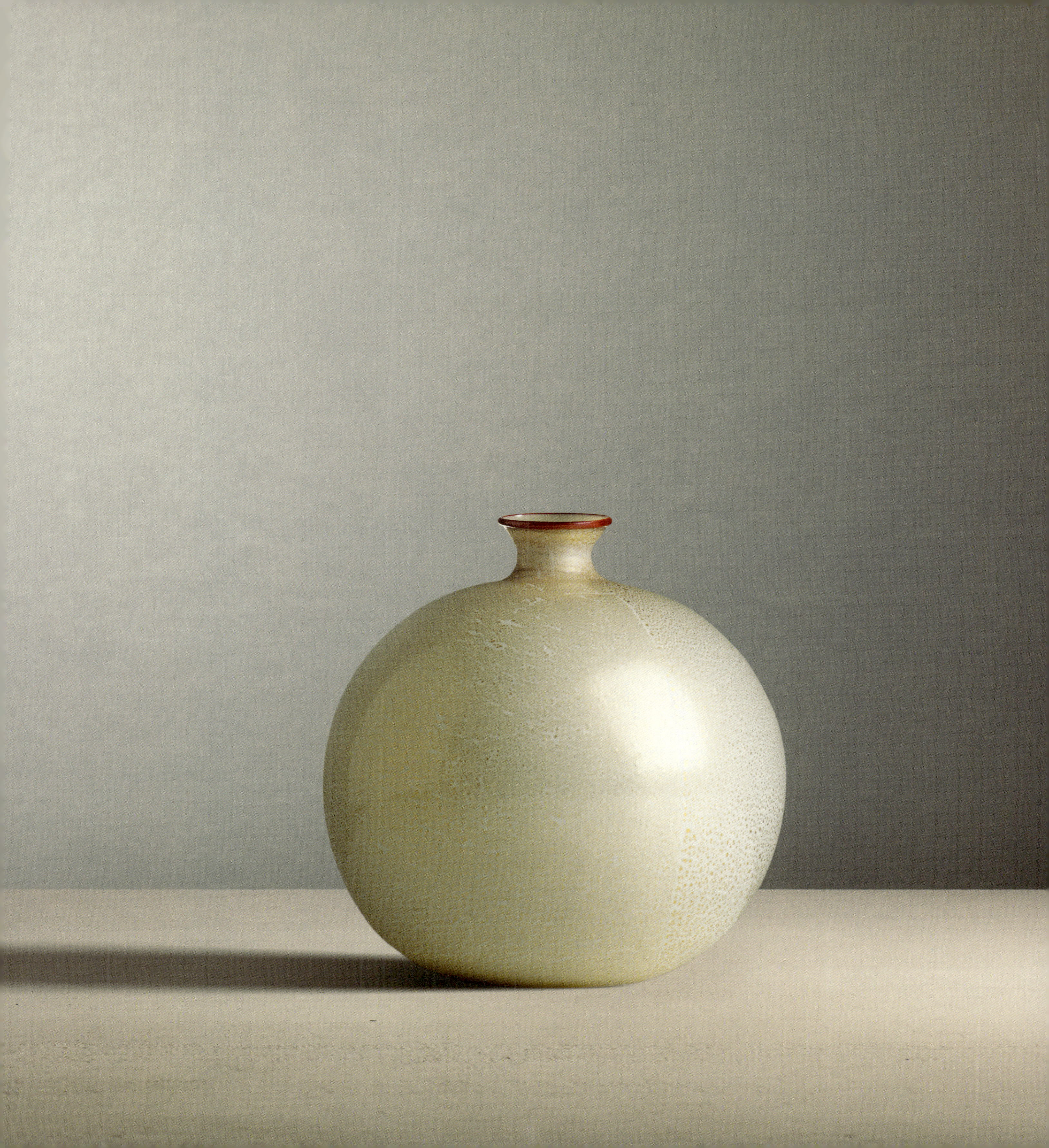

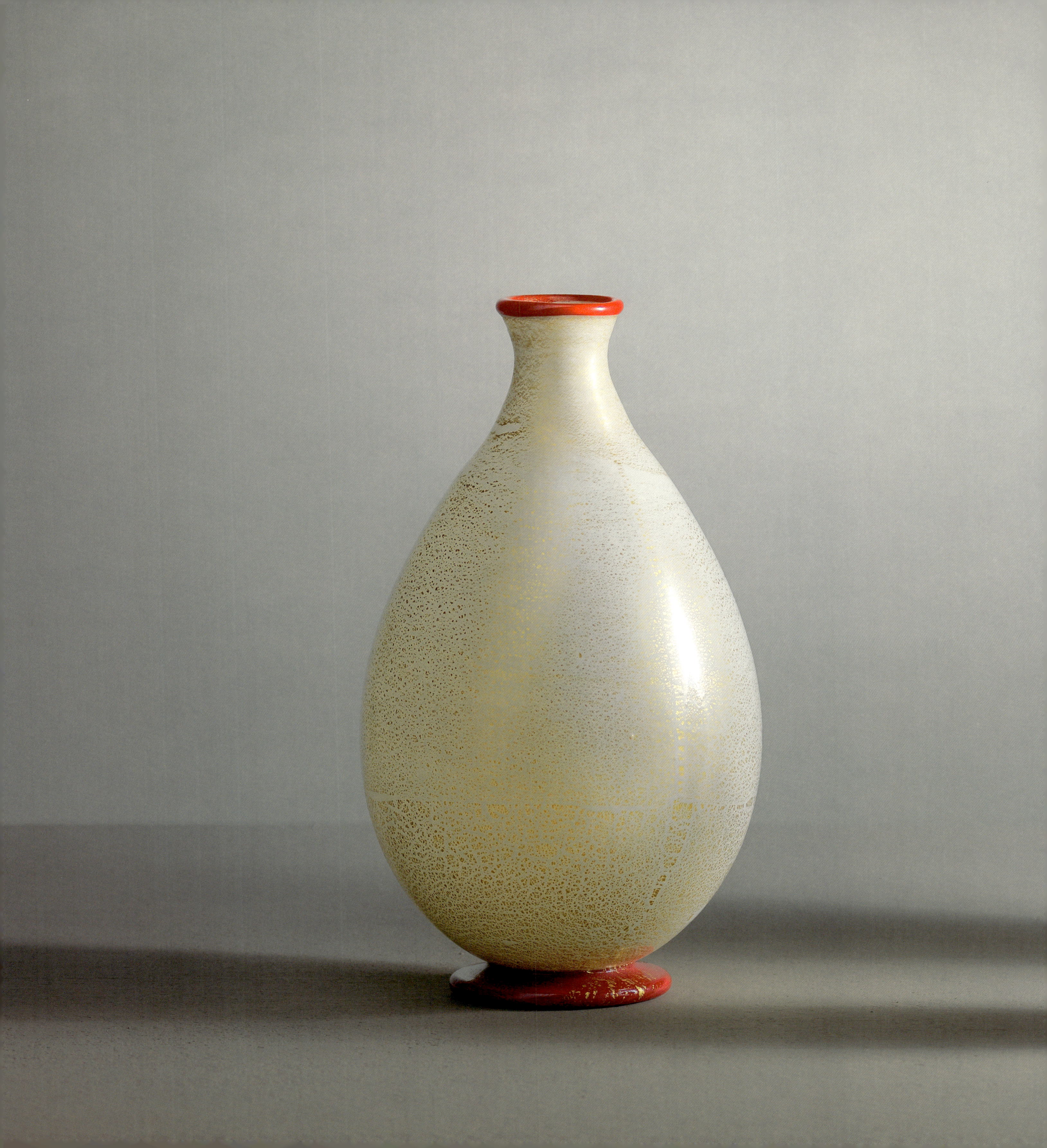

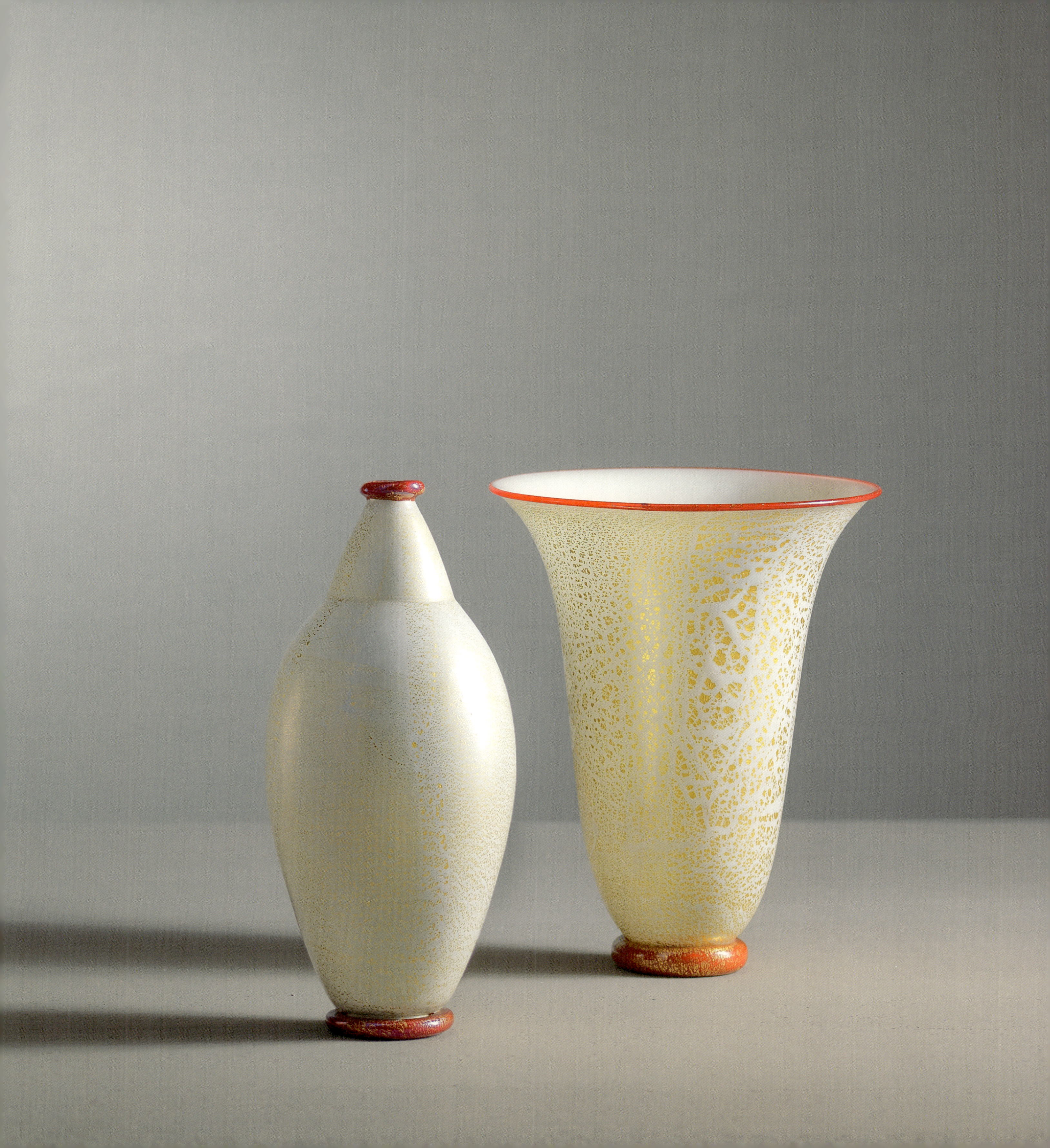

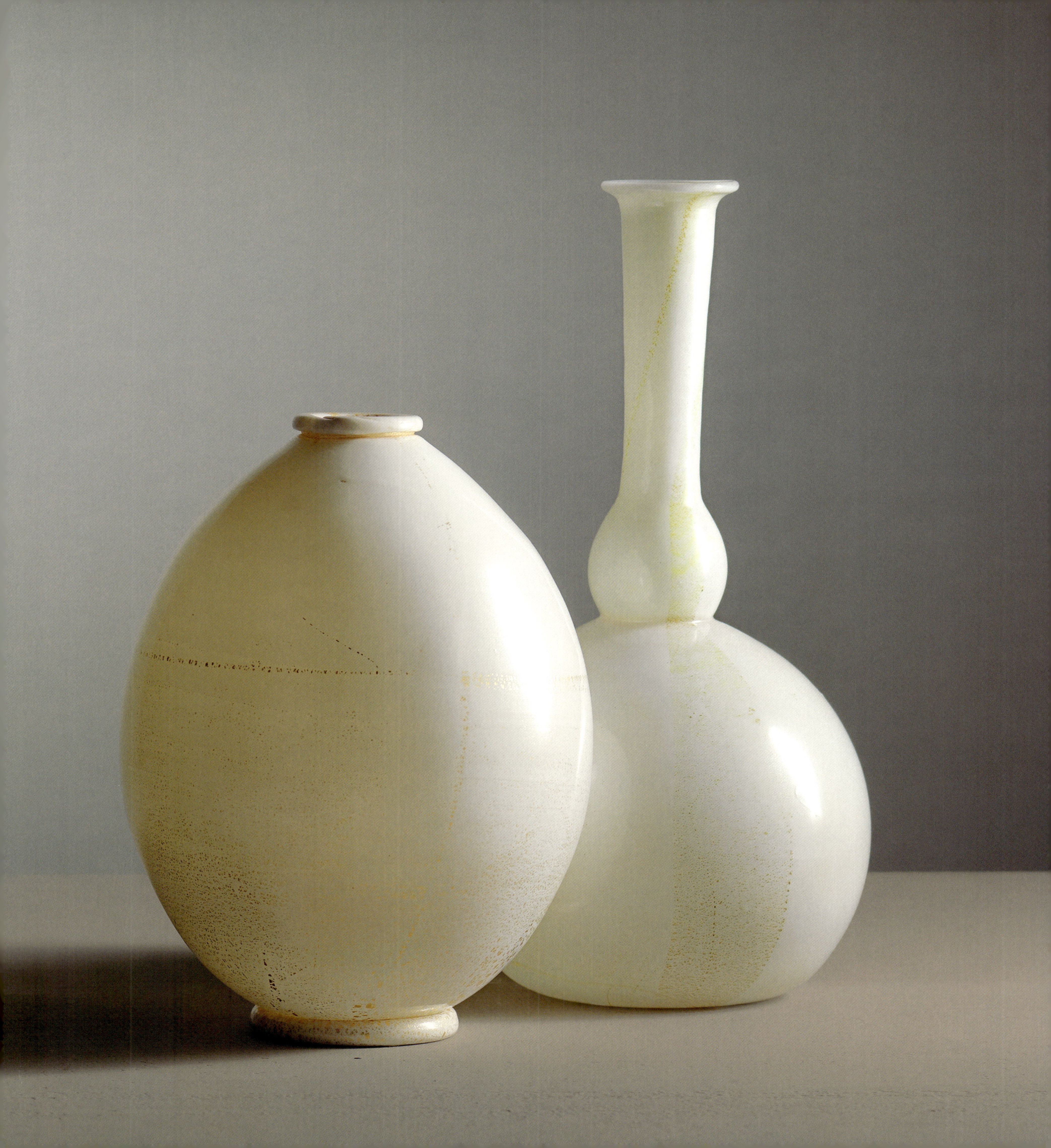

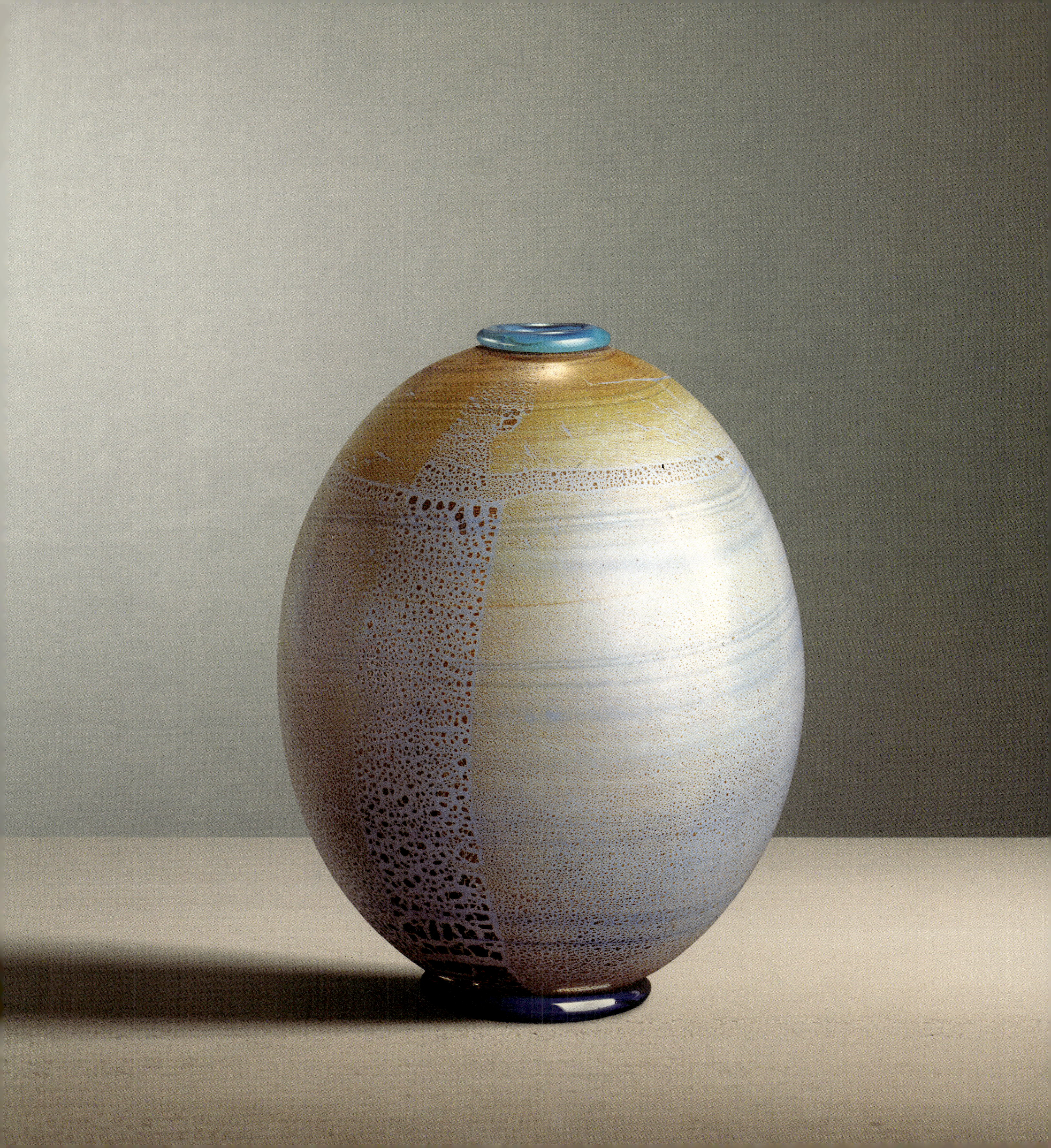

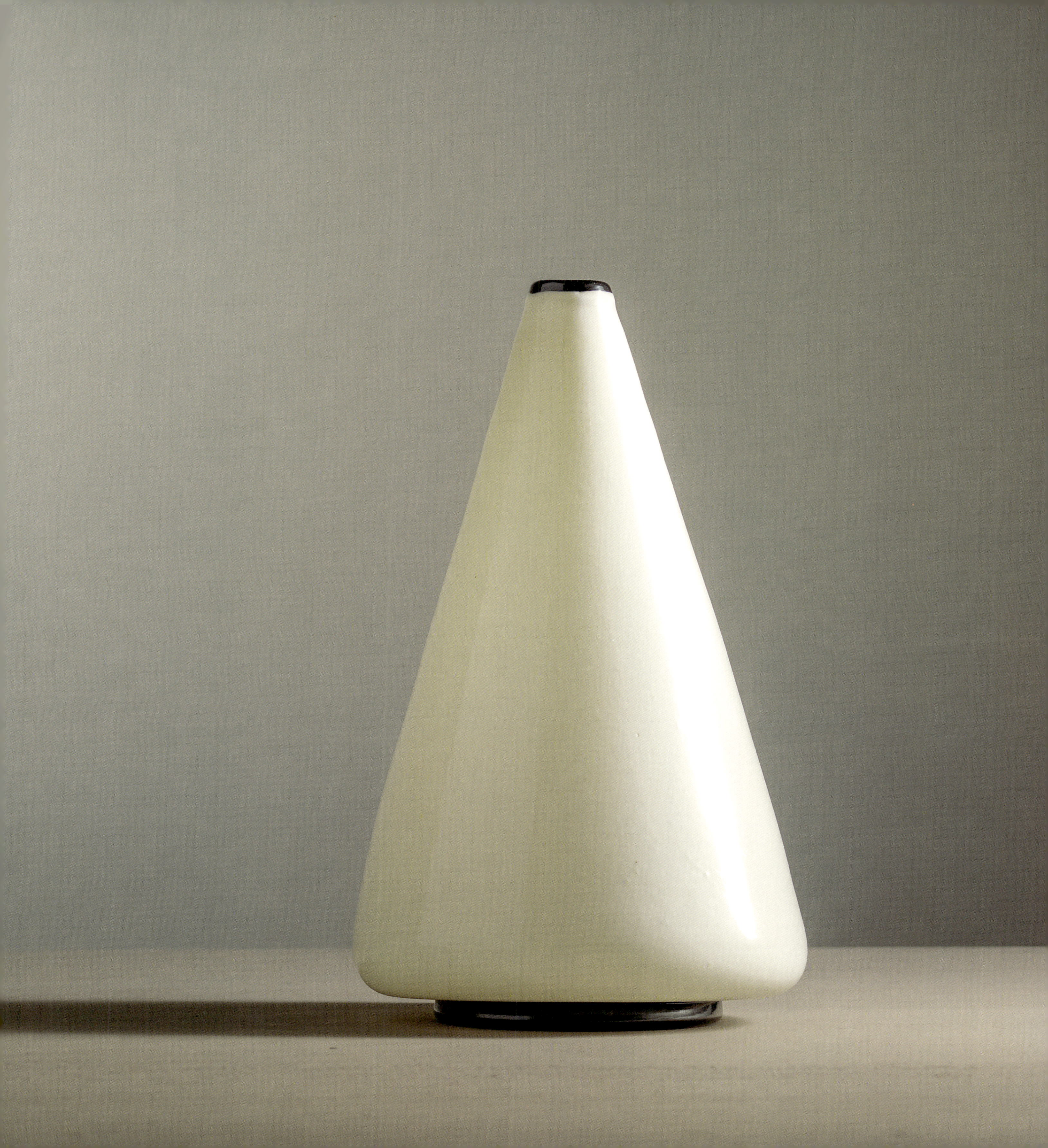

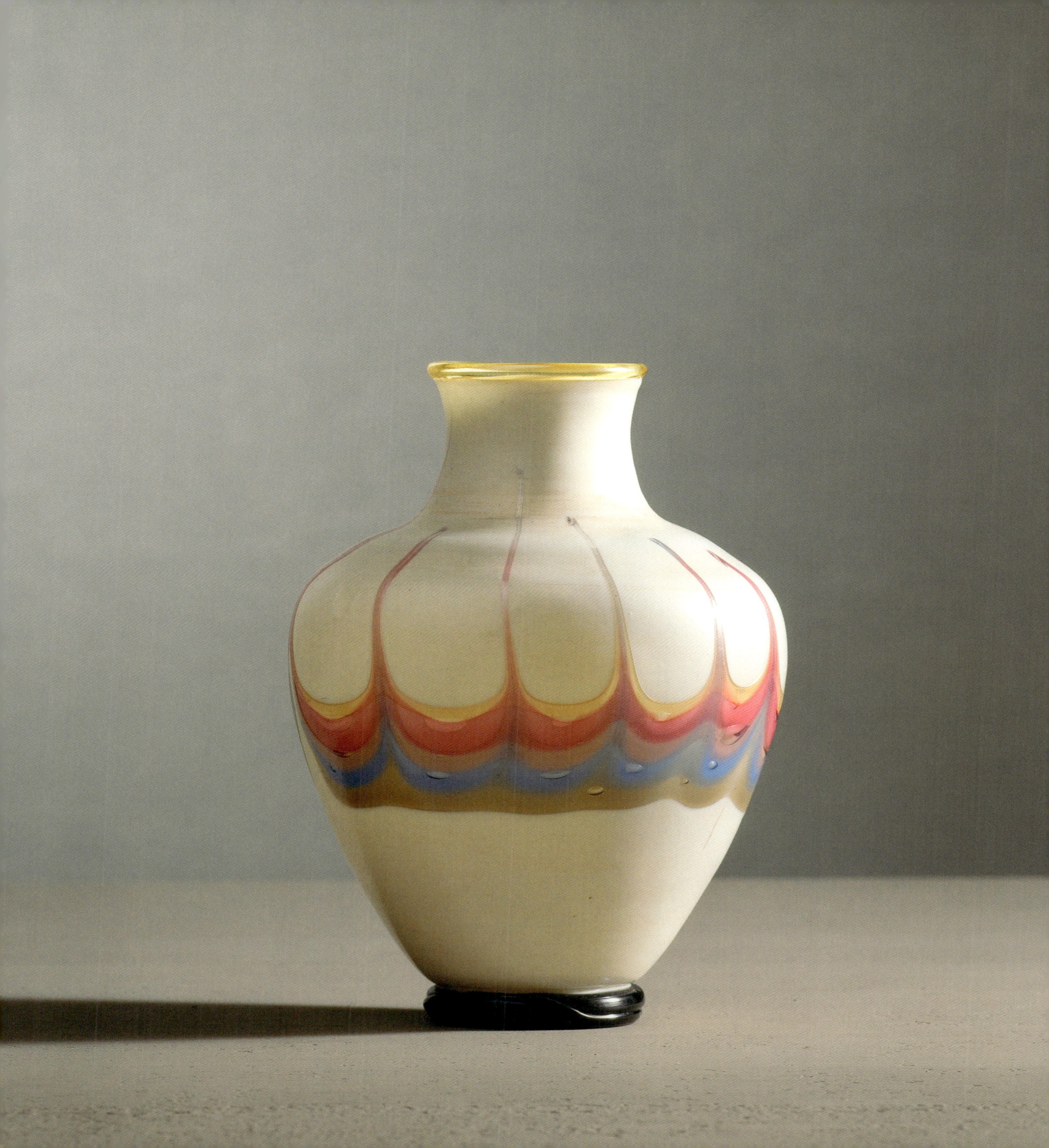

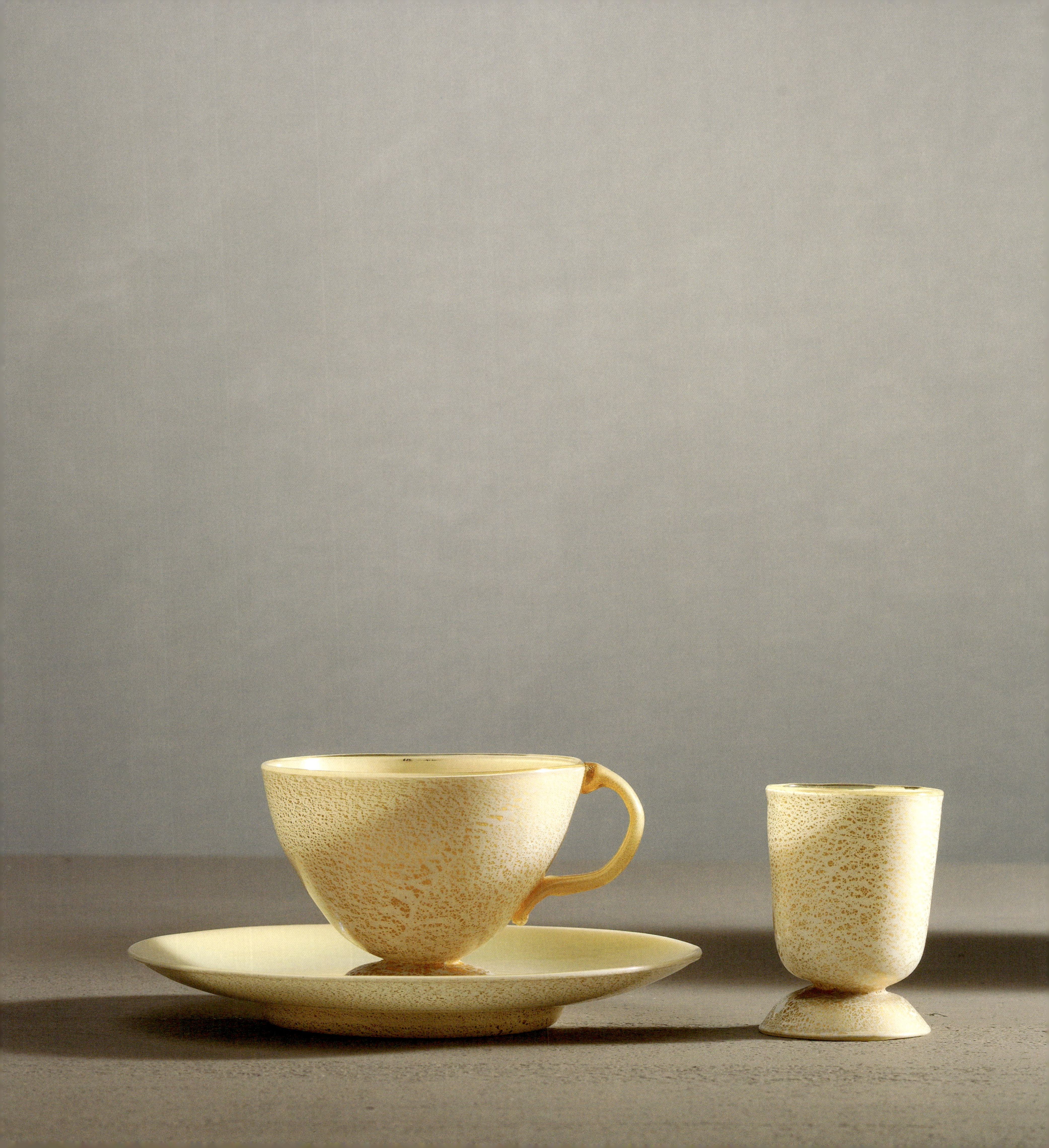

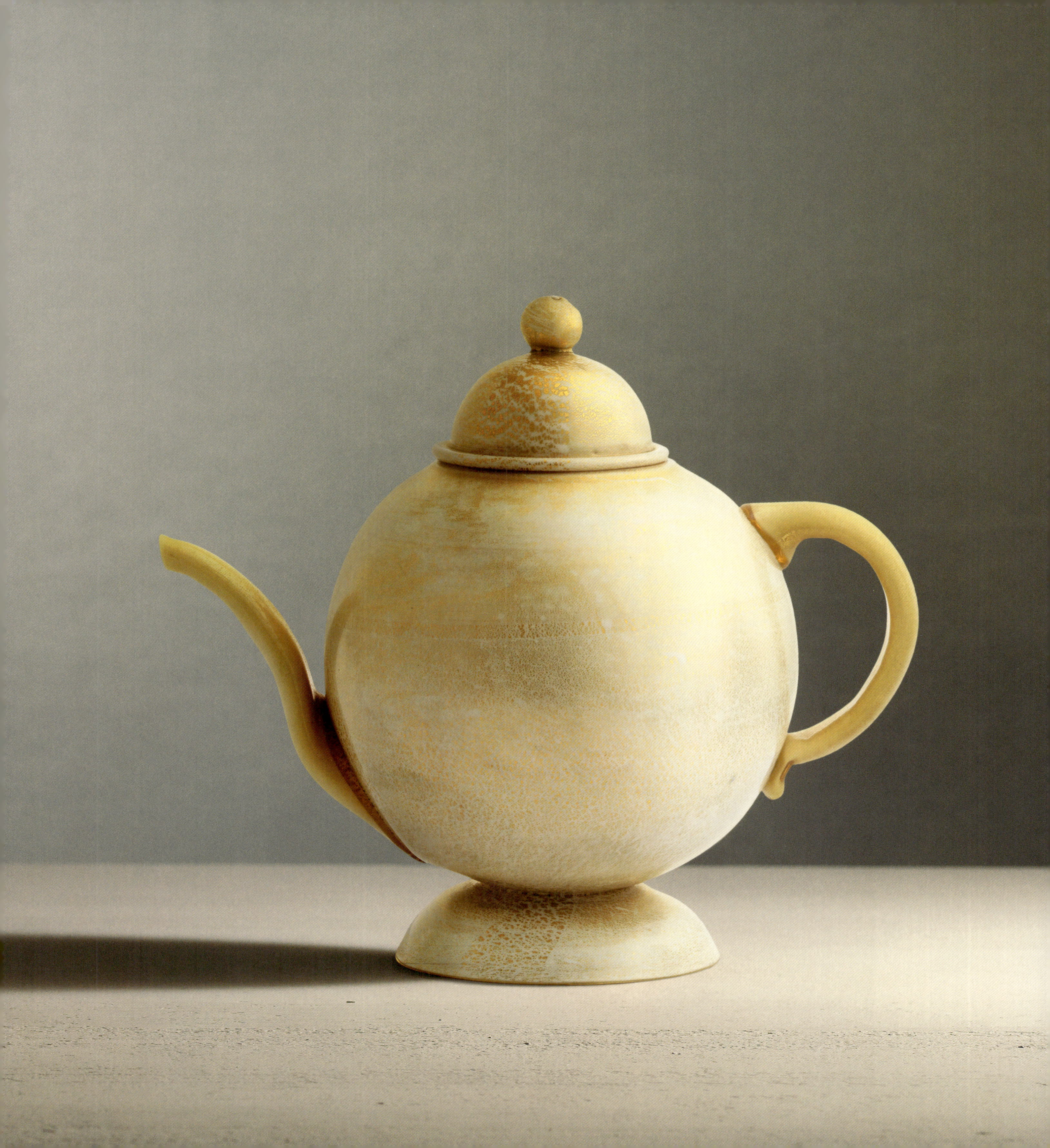

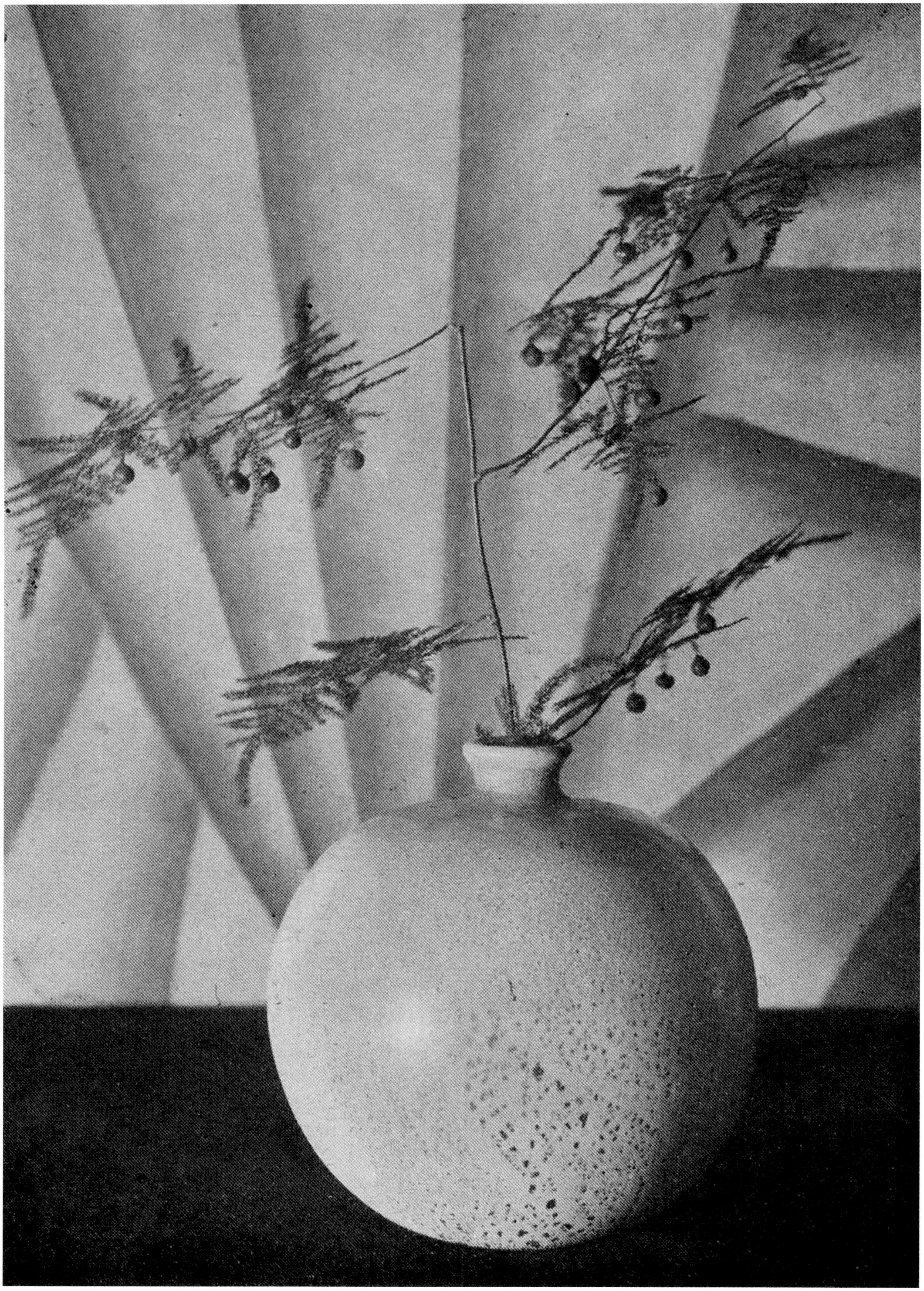

>
Lattimo *vase with applied gold leaf, in* Domus, *April 1929, p. 35*

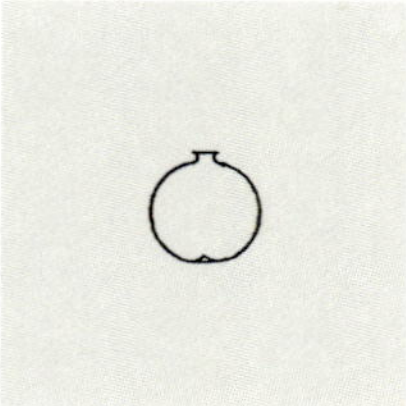

5932

Spherical vase in *lattimo* glass with applied gold leaf. Thread around the mouth in coral-red *pasta vitrea*. In 1928 an example of this model, probably in *lattimo aurato*, was exhibited at the Salon d'automne of Paris. The following year the model was published in the April number of *Domus* as a 'superb' example of the recent production of the glassworks. No. 5932 was also produced in other glass techniques (e.g. *pasta vitrea*, *decoro fenicio*).
c. 1928
13 cm high
The Olnick Spanu Collection, New York

exhibitions
1928, Paris, Salon d'automne (?)
1929, Leipzig, Trade Fair

>
5932, catalogue drawing, AMB

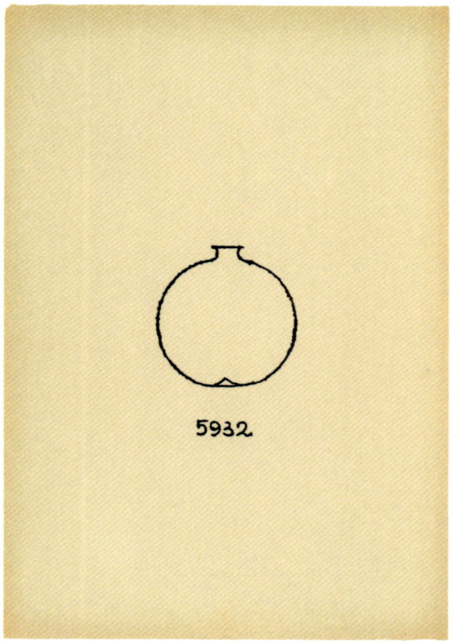

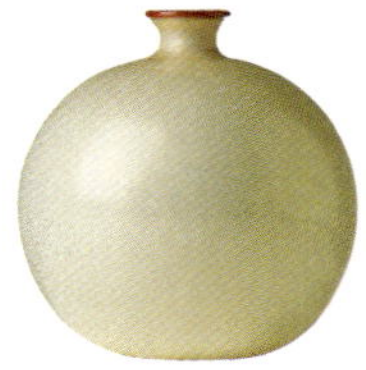

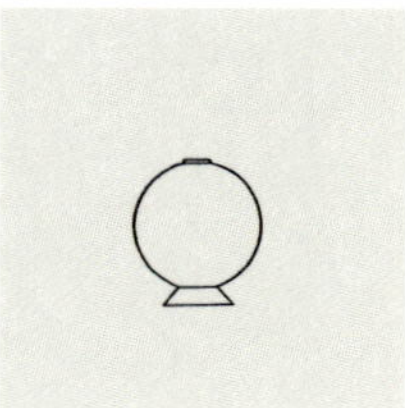

5680

Spherical vase in *lattimo* glass with applied gold leaf. Foot in the shape of a truncated cone and mouth in coral-red and gold *pasta vitrea*. Model no. 5680 was also produced in other glass techniques (e.g. transparent, *pasta vitrea*).
1928–29
16.5 cm high
M.V.M. Cappellin Murano *acid stamp*
Private collection

>
5680, working drawing, CSV

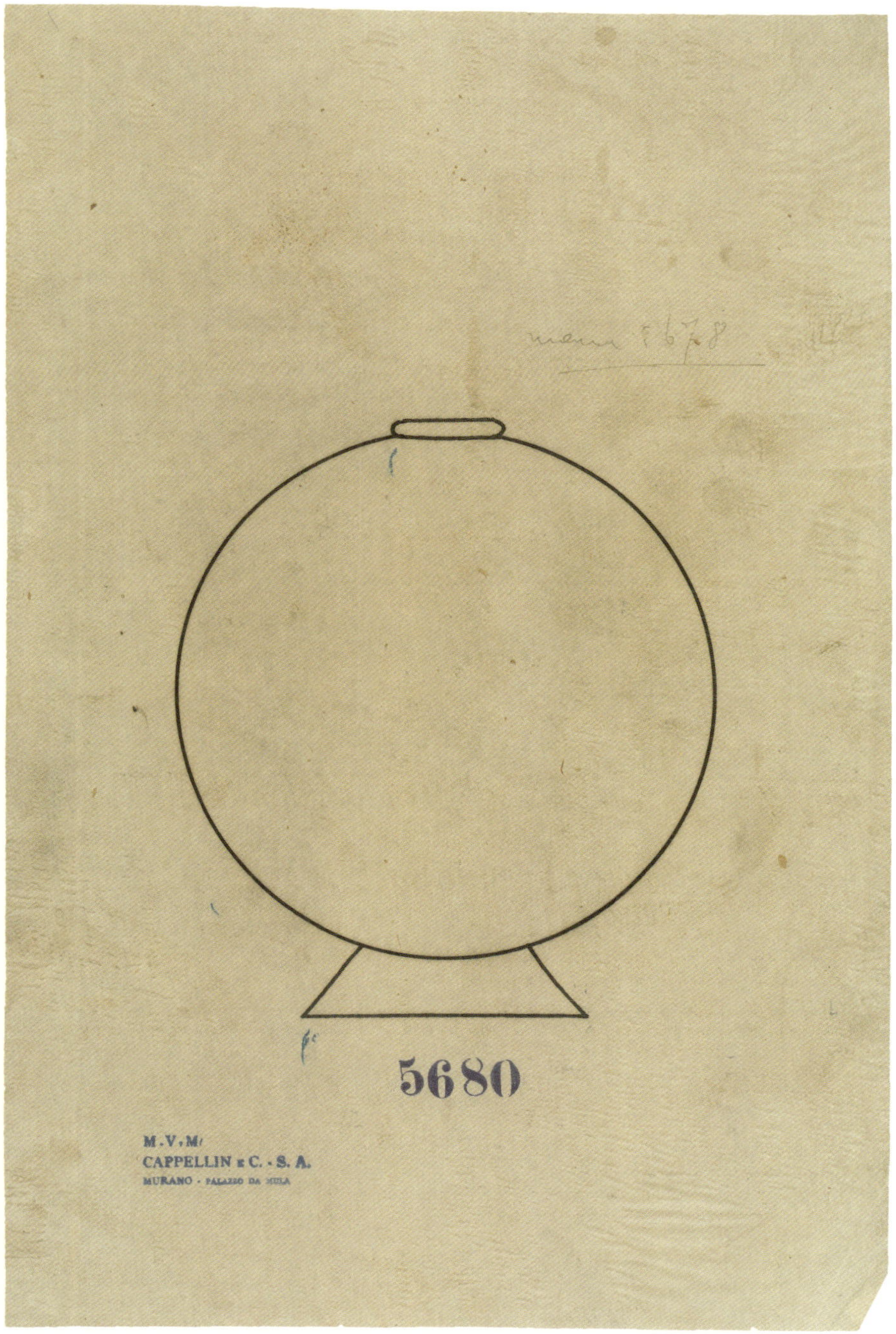

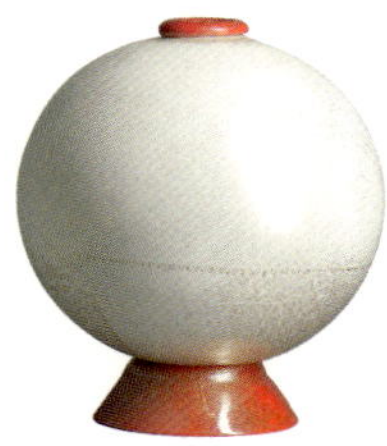

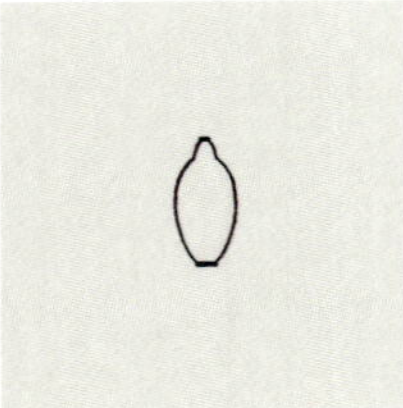

5920

Vase, with a constriction at the bottom of the neck, in *lattimo* glass with applied gold leaf. Foot and mouth in coral-red and gold *pasta vitrea*. A similar example was exhibited in 1929 at the Leipzig Trade Fair and the following year at the 4th Triennale of Milan. Model no. 5920 was also produced in other glass techniques (e.g. *pasta vitrea*).
1928–29
16 cm high
Private collection, Feltre

exhibitions
1929, Leipzig, Trade Fair
1930, Monza, 4th Triennale

>
5920, catalogue drawing, AMB

>
Cappellin glass items on a piece of dining-room furniture (architect Gino Maggioni) at the 4th Triennale of Milan, interior design gallery, 1930

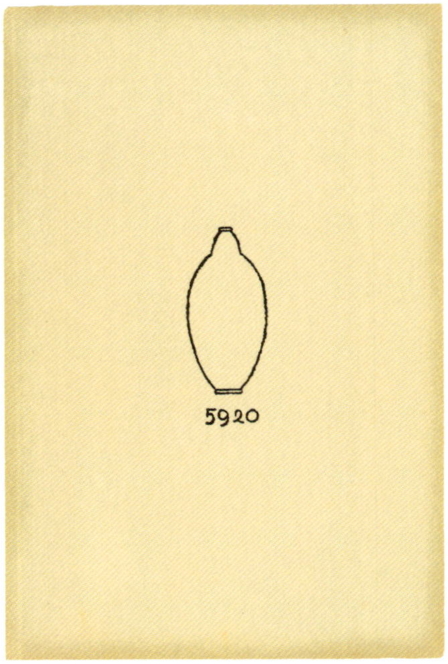

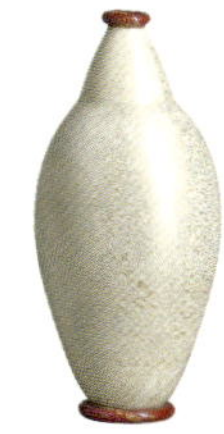

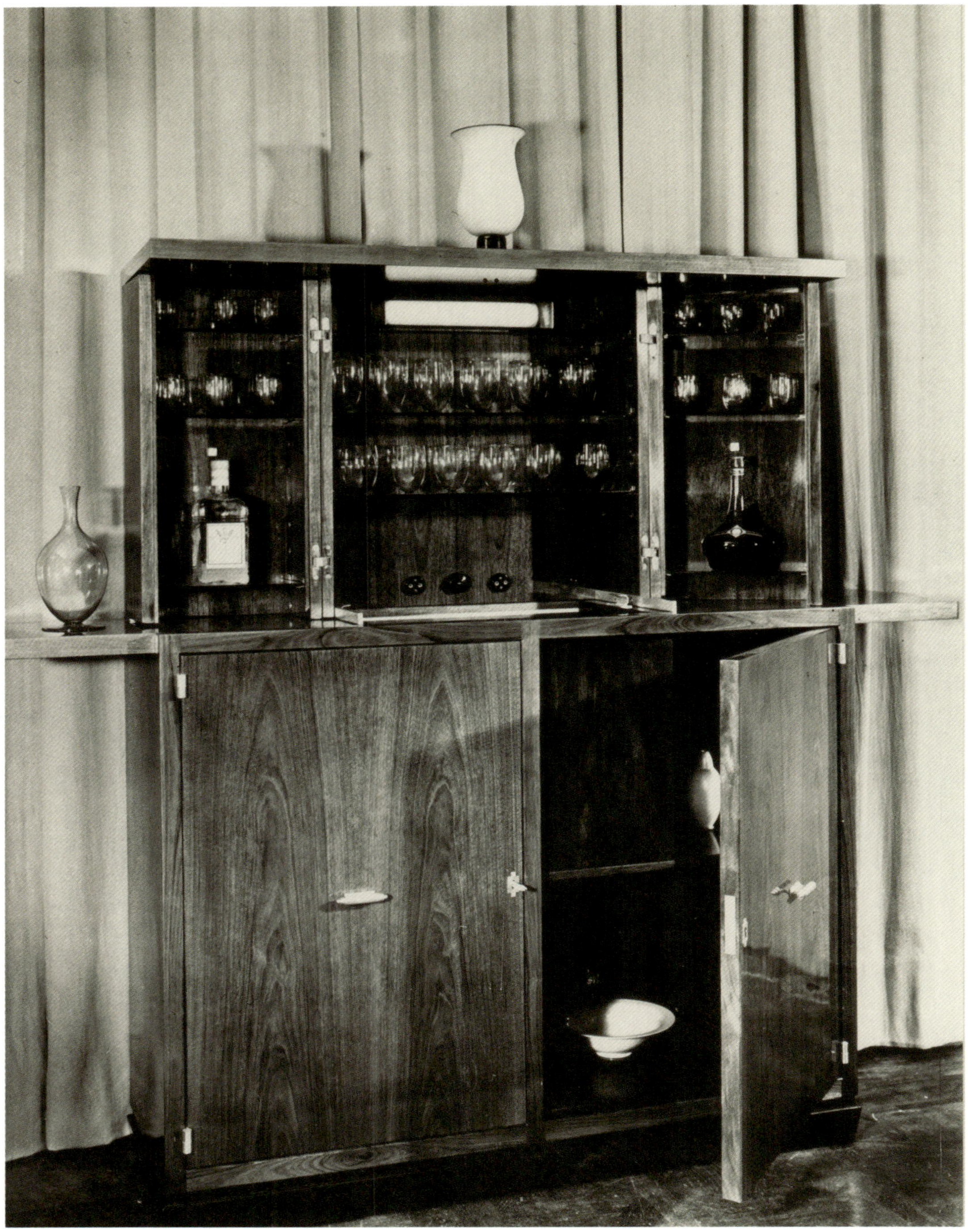

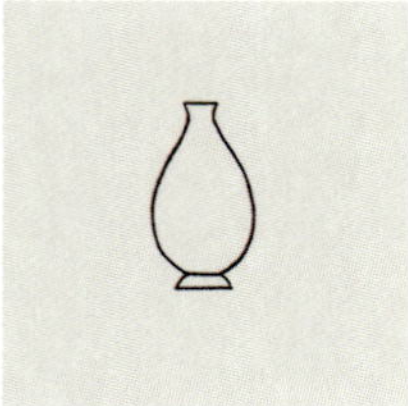

5946

Vase in *lattimo* glass with applied gold leaf. Foot and thread around mouth in coral-red and gold *pasta vitrea*. In 1928 an example of this model, probably in *lattimo aurato*, was exhibited at the Salon d'automne of Paris. Model no. 5946 was also produced in other glass techniques (e.g. *pasta vitrea*, *decoro fenicio*, iridised black).
c. 1928
21 cm high
Private collection

exhibitions
1928, Paris, Salon d'automne (?)

>
5946, catalogue drawing, AMB

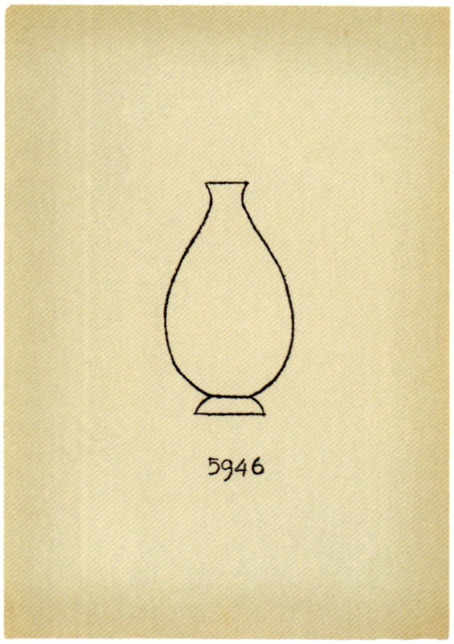

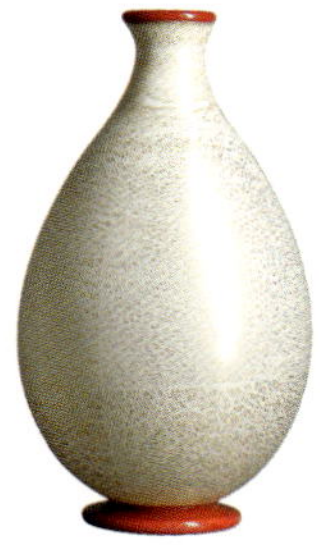

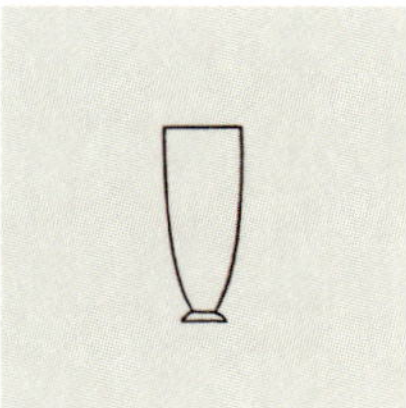

5948

Vase in *lattimo* glass with applied gold leaf. Skull-cap foot and thread around mouth in coral-red and gold *pasta vitrea*. Model no. 5948 was also produced in other glass techniques (e.g. *pasta vitrea*, cased glass).
c. 1928
23 cm high
M.V.M. Cappellin Murano *acid stamp*
Private collection

exhibitions
1928, Paris, Salon d'automne (?)

>
5948, catalogue drawing, AMB

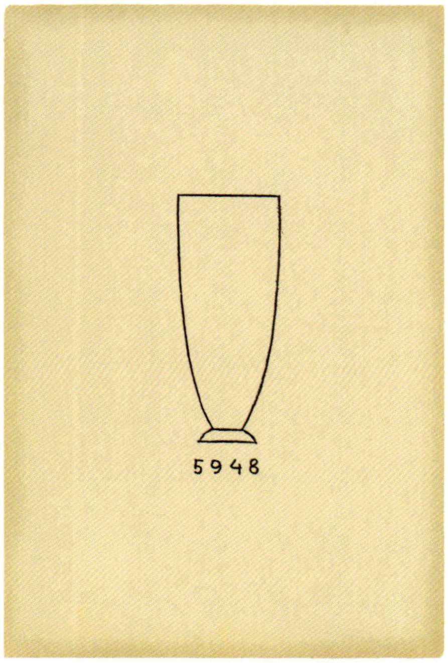

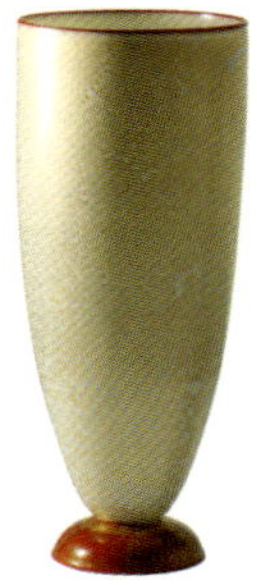

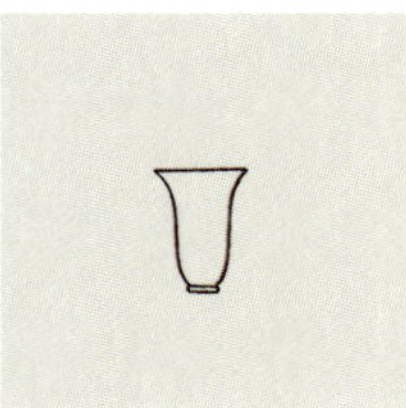

5924

Vase in *lattimo* glass with applied gold leaf. Foot and thread around mouth in coral-red and gold *pasta vitrea*. A similar example was exhibited at the Leipzig Trade Fair in 1929. Model no. 5924 was also produced in other glass techniques (e.g. *pasta vitrea*, *decoro fenicio*).
1928–29
14.5 cm high
M.V.M. Cappellin, Murano label
Private collection

exhibitions
1929, Leipzig, Trade Fair

>
5924, catalogue drawing, AMB

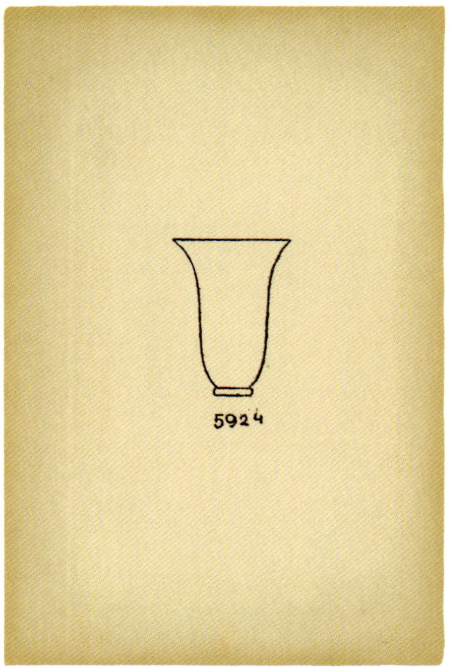

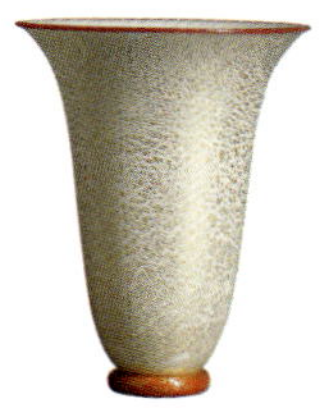

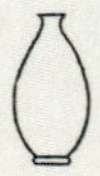

5925

Vase in *lattimo* and gold glass with applied gold-leaf decoration depicting a concertina player and a stylised landscape. Foot and thread around the mouth in coral-red and gold *pasta vitrea*. This is one of the rare examples of artefacts with figured decorations produced by the Cappellin firm.
1929 ca.
18 cm high
Private collection

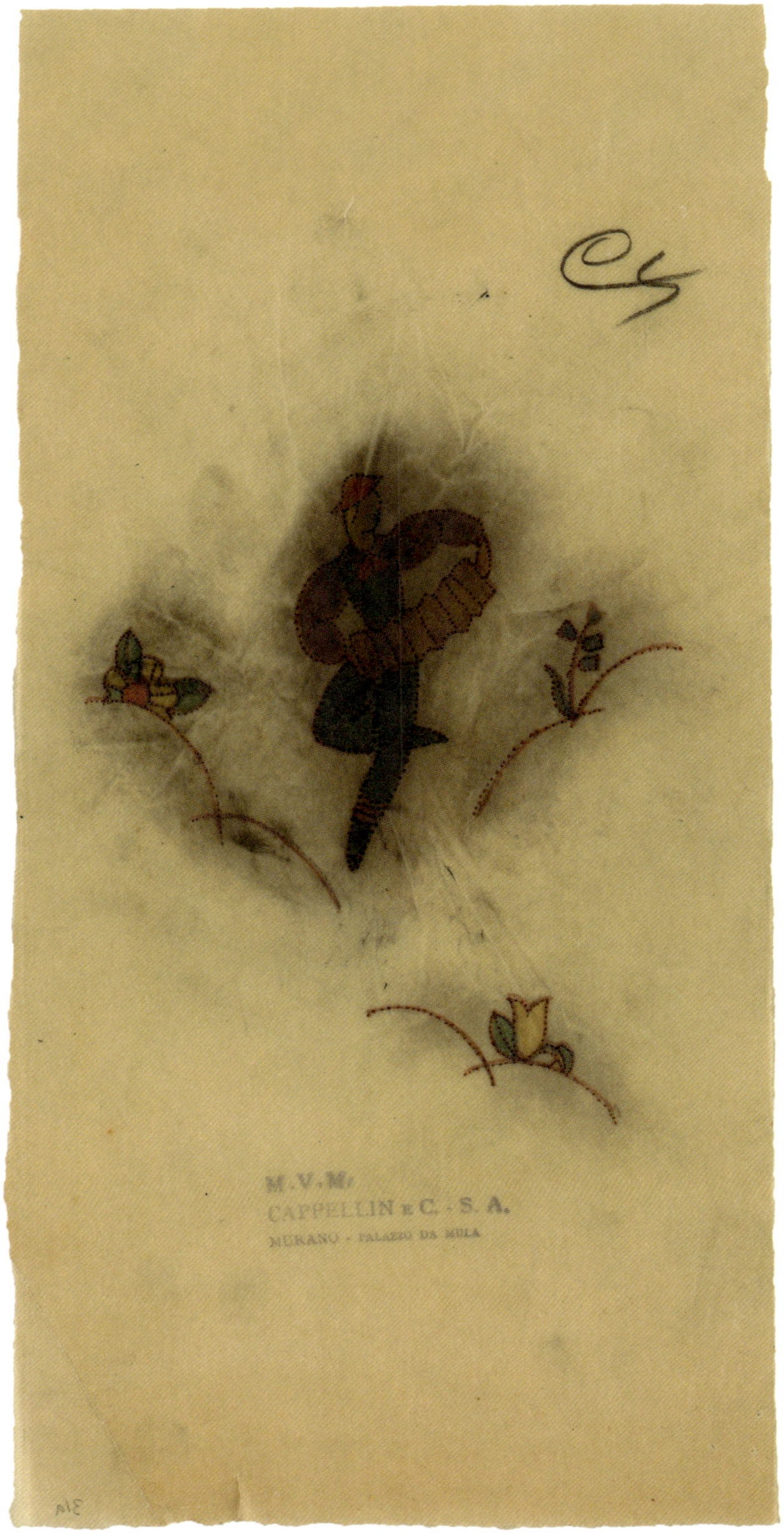

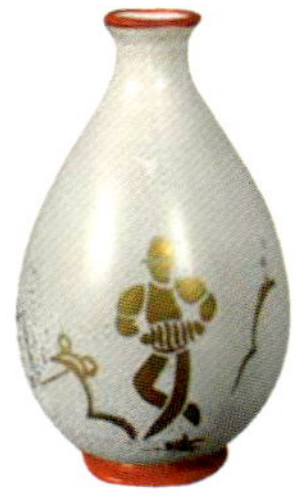

\>
Working drawing for decoration with concertina player to be produced by the pouncing technique, 1928–29, ACSVr

n.i.

Vase in *lattimo* glass with applied gold leaf. The model is characterised by a spherical body and a long neck with two constrictions.
1928–29
30 cm high
M.V.M. Cappellin Murano *acid stamp*
Private collection

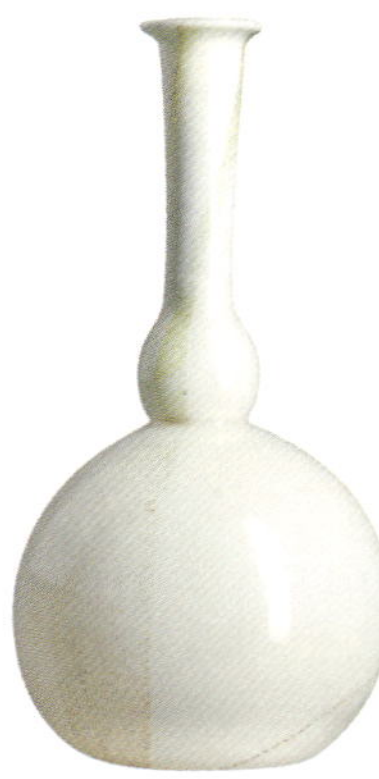

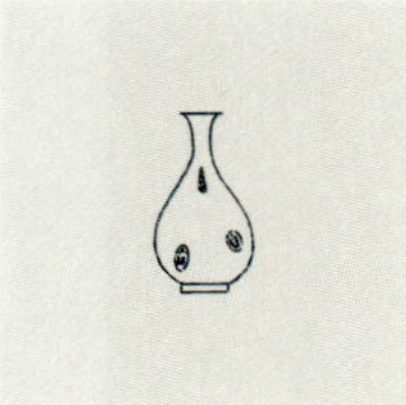

6014

Vase in *lattimo* glass with applied gold leaf. The piece is a variant in *lattimo* and gold of no. 6014 designed with murrine in the wall of the piece.
c. 1930
21 cm high
Private collection

>
6014, catalogue drawing, AMA

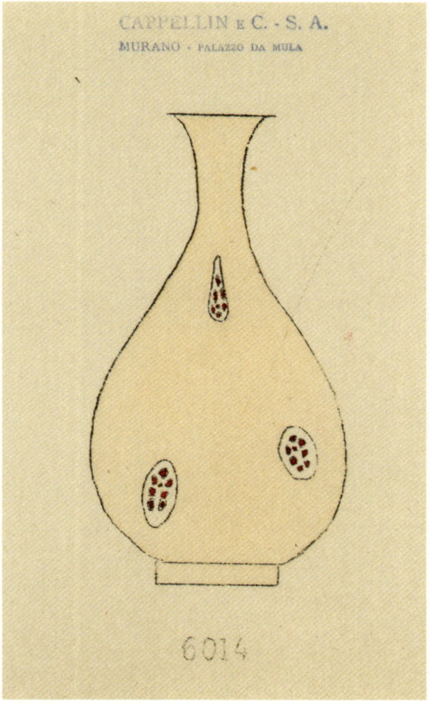

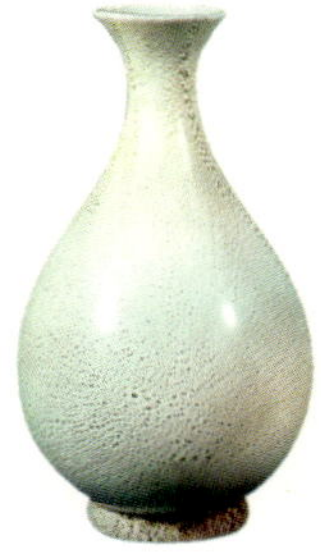

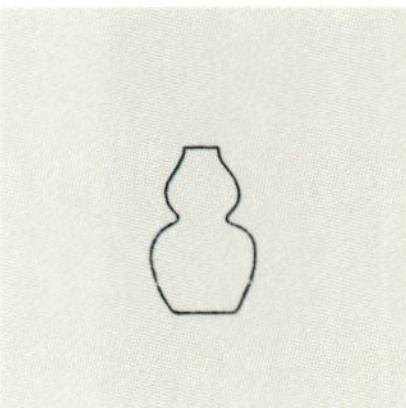

6015

Vase in *lattimo* glass with applied gold leaf. The design of the vase, characterised by a deep constriction in the body, refers back to similar models found in Chinese porcelain.
c. 1930
22 cm high
The Olnick Spanu Collection, New York

>
6015, catalogue drawing, AMA

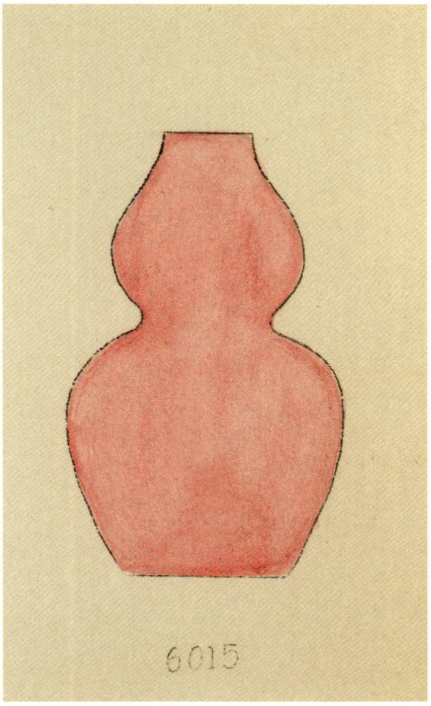

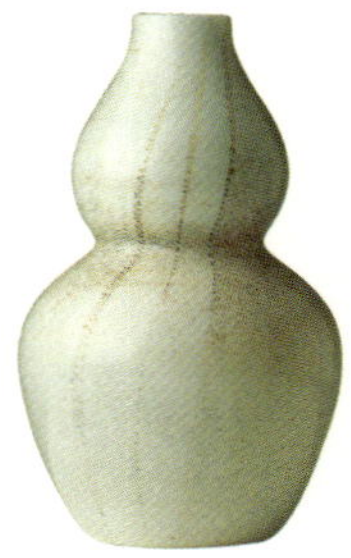

---

n.i.

Vase with flattened cross-section in *lattimo* glass with applied gold leaf. Small flared foot. The design of the model is a reference to oriental art.
c. 1930
30 cm high
Private collection, Venice

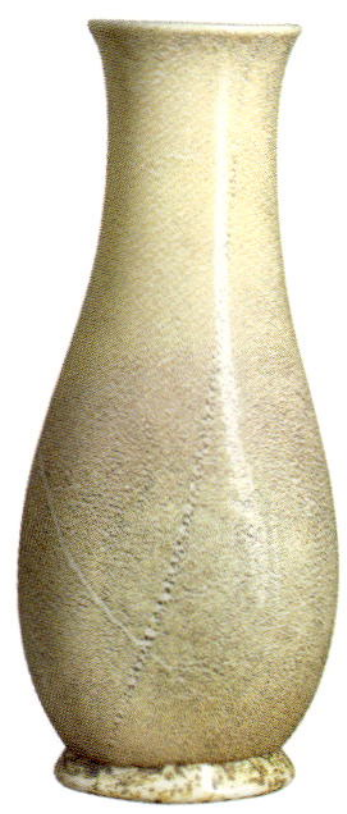

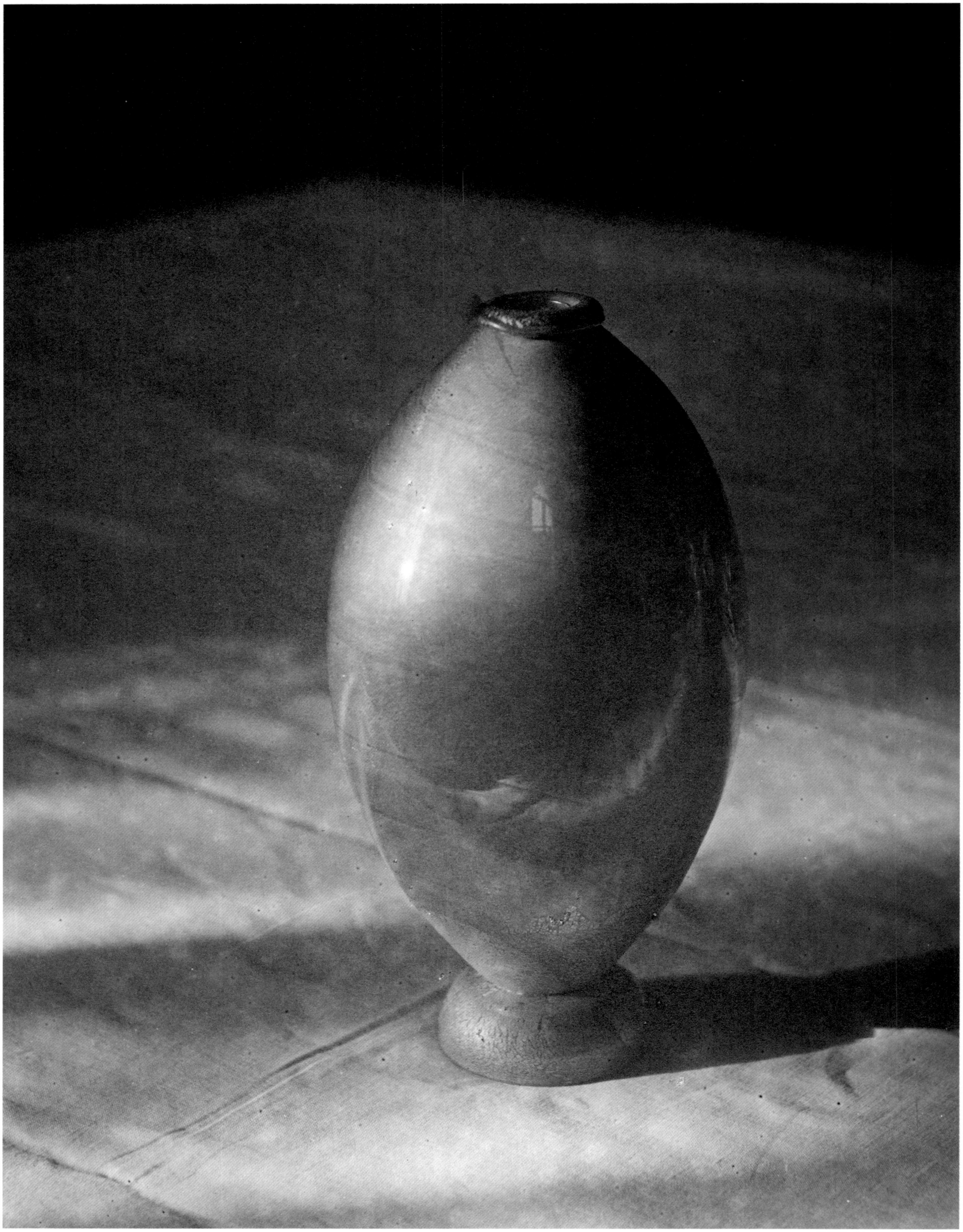

>
Vase in lattimo *glass with applied gold leaf. Period photograph, c. 1929, ACSV*

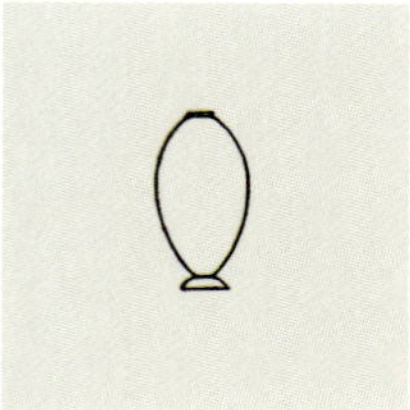

5935

Ovoid vase with skull-cap foot in *lattimo* glass with applied gold leaf. Model no. 5935 was also produced in other glass techniques (e.g. *decoro fenicio*, *pasta vitrea*, cased glass, iridised black).
c. 1928
22 cm high
Private collection

exhibitions
1928, Paris, Salon d'automne (?)

>
5935, catalogue drawing, AMH

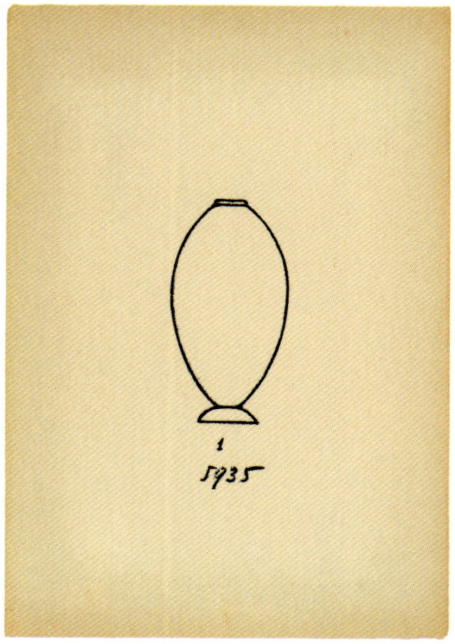

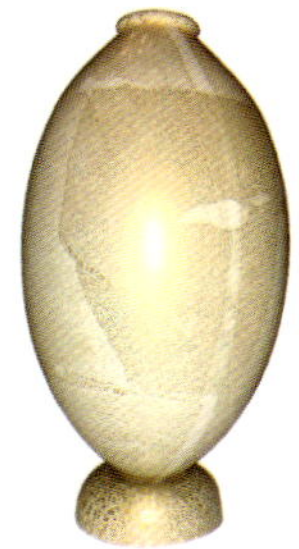

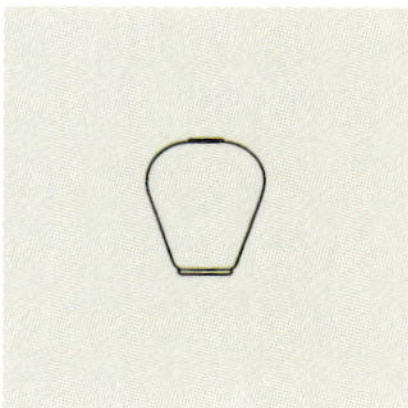

5856

Vase in *lattimo* glass with applied gold leaf. Small foot and mouth in *lattimo* and gold. Model no. 5956 was also produced in other glass techniques (e.g. *decoro fenicio*, *pasta vitrea*, silvery black).
1928–29
19 cm high
Private collection

>
5856, catalogue drawing, AMB

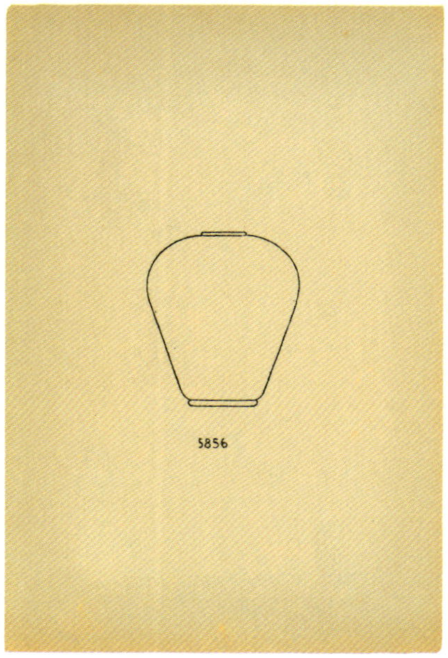

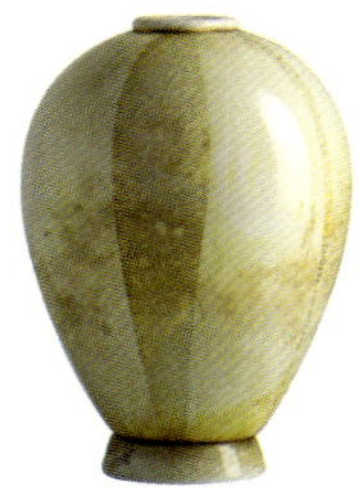

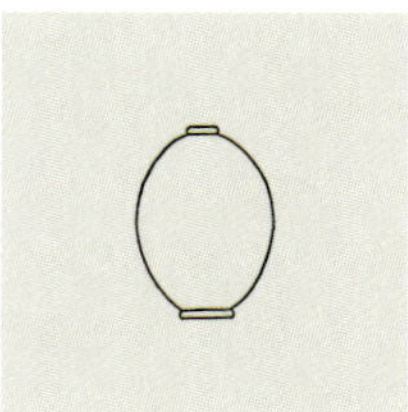

5952

Ovoid vase in light-blue *lattimo* cased glass with applied gold leaf. Ring-shaped foot and mouth in light-blue transparent glass. Surface with iridised finish. Model no. 5952 was also produced in other glass techniques (e.g. *decoro fenicio*, *pasta vitrea*).
1928–29
21.5 cm high
Stiftung Museum Kunstpalast collection, Düsseldorf-Glasmuseum Hentrich

>
5952, catalogue drawing, AMB

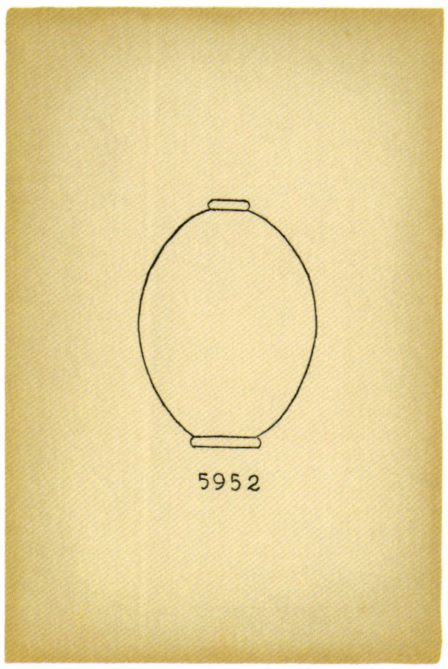

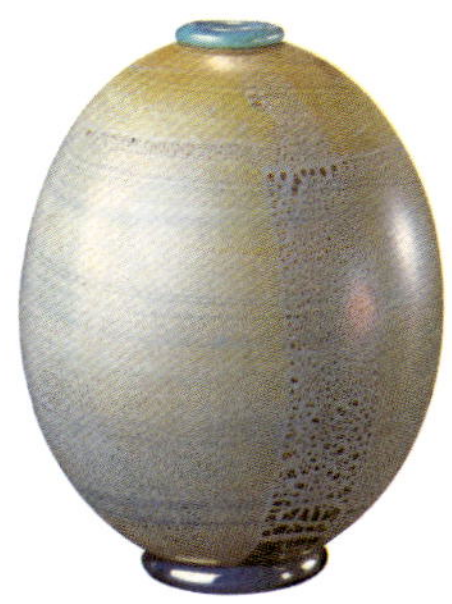

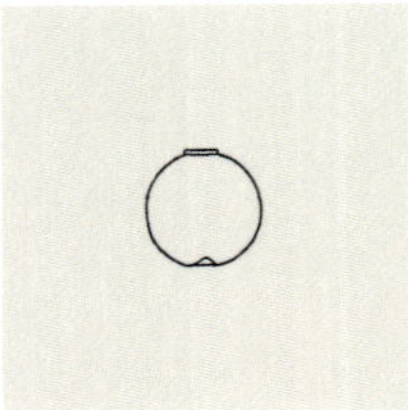

5934

Spherical vase in milky glass with applied gold leaf. Model no. 5934 was also produced in other glass techniques (e.g. transparent *balloton*, *filigrana a reticello*, *decoro fenicio*, vertical canes, silver black).
1928–29
13 cm high
M.V.M. Cappellin Murano *acid stamp*
Private collection, Treviso

>
5934, catalogue drawing, AMB

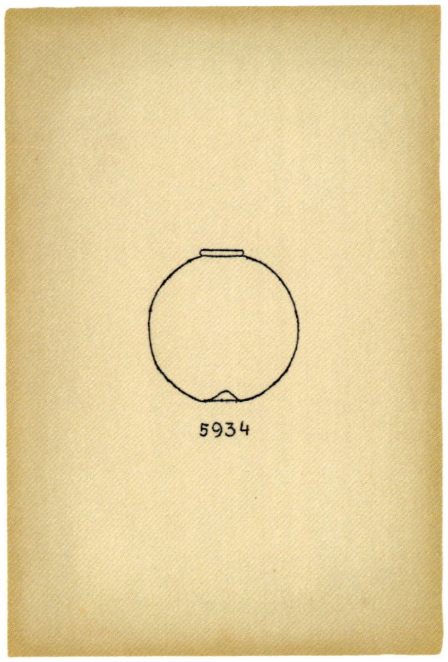

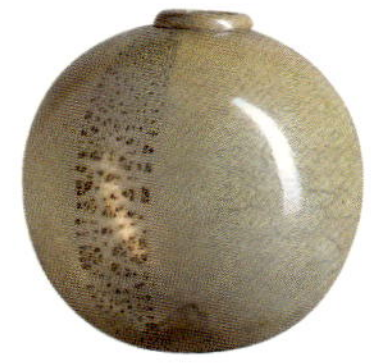

1275

Small spherical vase in *lattimo* glass with applied oxidised gold leaf. Model no. 1275 was also produced in other glass techniques (e.g. *pasta vitrea*).
1929–30
5 cm high
Private collection

>
1275, catalogue drawing, AMA

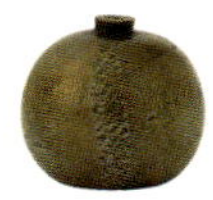

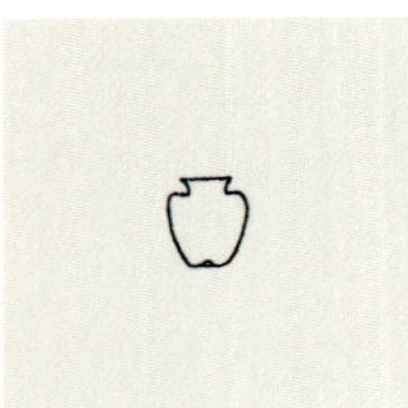

5927

Vase in *lattimo* glass with applied gold leaf. A similar example was published in *Domus* in October 1929.
1928–29
10 cm high
Private collection

>
5927, catalogue drawing, AMB

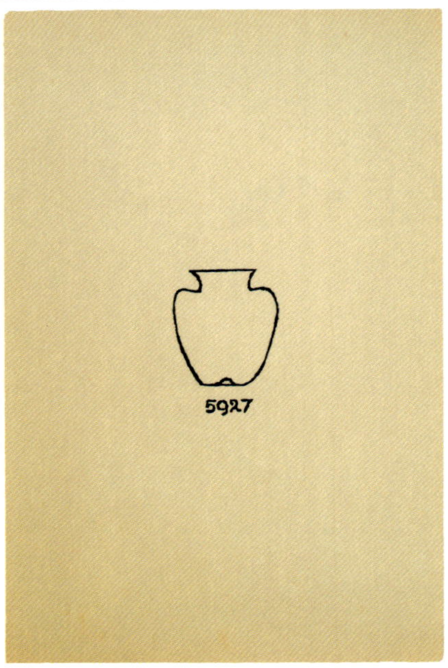

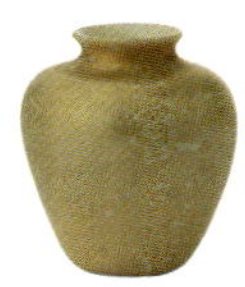

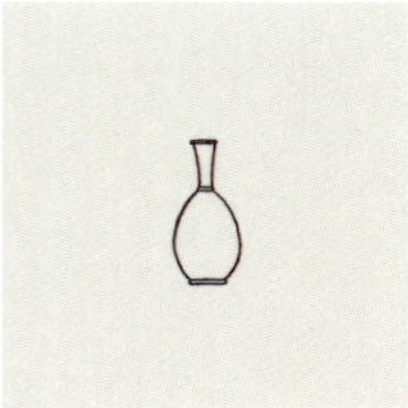

5918

Catalogue drawing for one of the vases in *lattimo* glass with applied gold leaf documented in a period photograph, AMB. The illustration was published in *Domus* in October 1929. An example of this model, in *lattimo* and gold, with a coral-red collar, was exhibited in 1929 at the Leipzig Trade Fair. On this occasion, no. 5918 was proposed also in the variant in pea-green with gold *pasta vitrea*.
1928–29
17 cm high

exhibitions
1929, Leipzig, Trade Fair

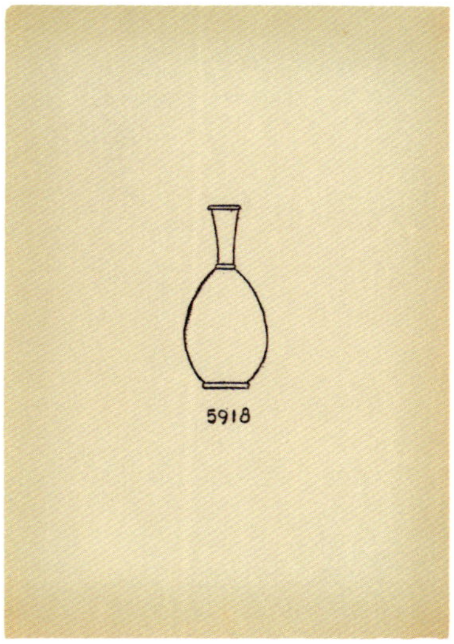

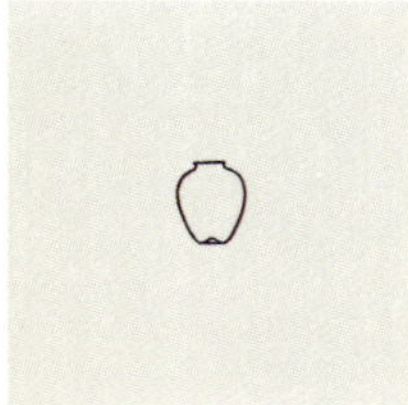

5929

Catalogue drawing for one of the vases in *lattimo* glass with applied gold leaf documented in a period photograph, AMB. An example of this model, in *lattimo* and gold, was exhibited in 1929 at the Leipzig Trade Fair. On that occasion, model no. 5929 was proposed also in the 'gold-sprayed iridised pea-green' variant or in the 'silver-sprayed iridised black' variant.
1928–29
10 cm high

exhibition
1929, Leipzig, Trade Fair

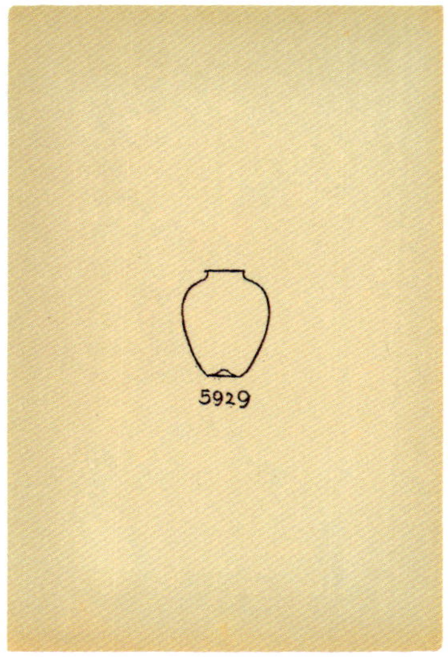

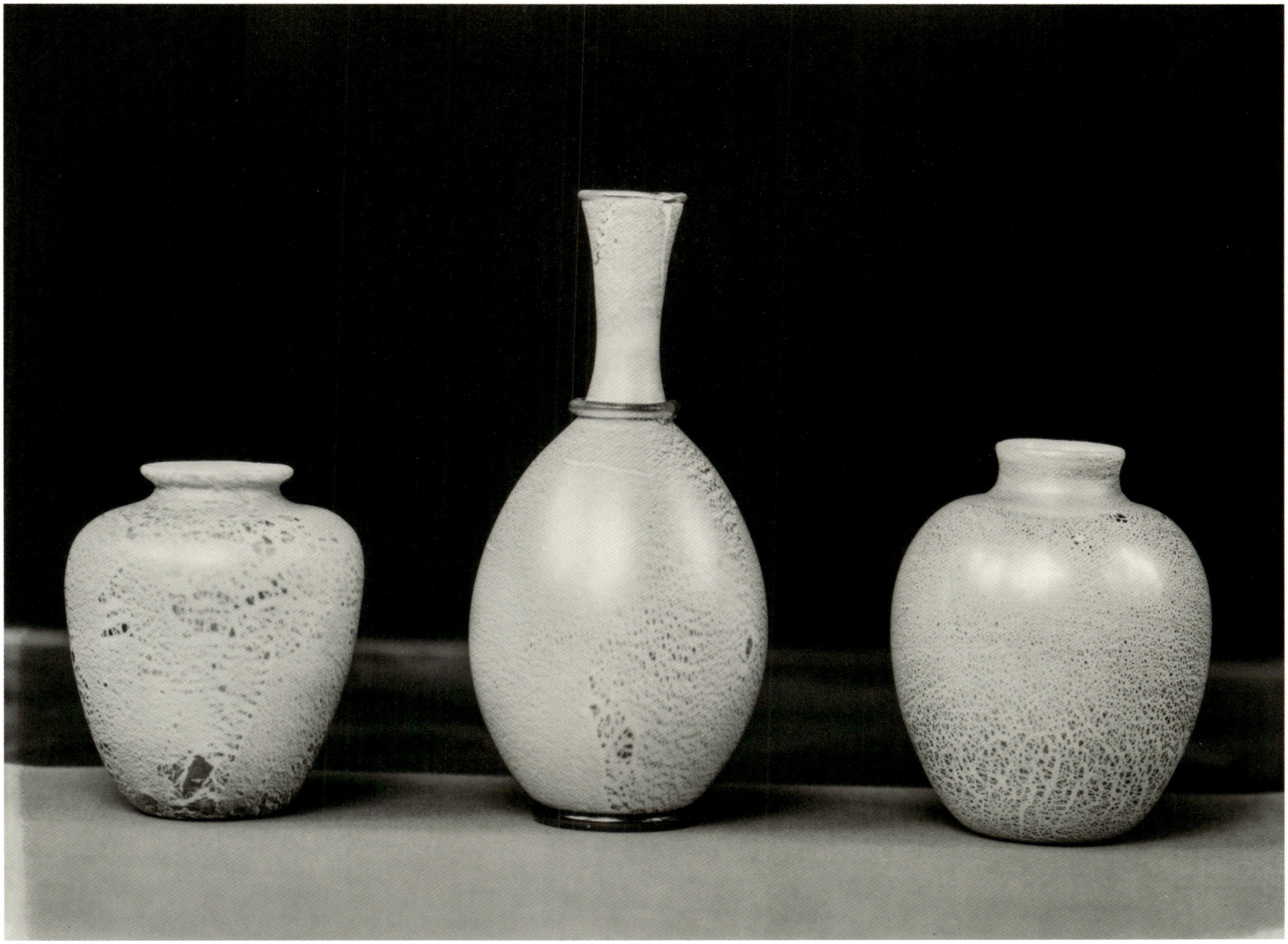

›
Vases in lattimo *glass with applied gold leaf. Period photograph, c. 1929, AMB*

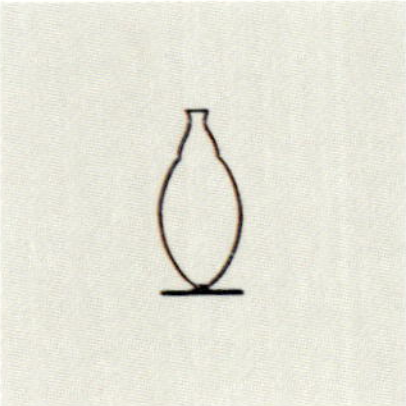

5921

Vase in *lattimo* glass with applied gold leaf. The piece is characterised by its disk-shaped foot and the constriction at the base of the neck. Model no. 5921 was also produced in other glass techniques (e.g. transparent, vertical canes).
1928–29
23 cm high
Private collection, Montebelluna

>
5921, catalogue drawing, AMB

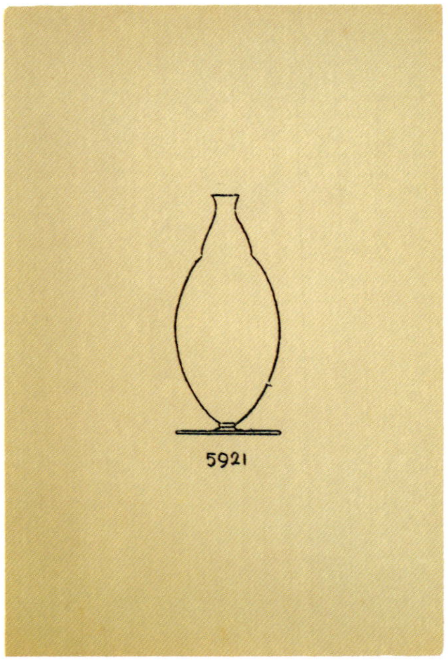

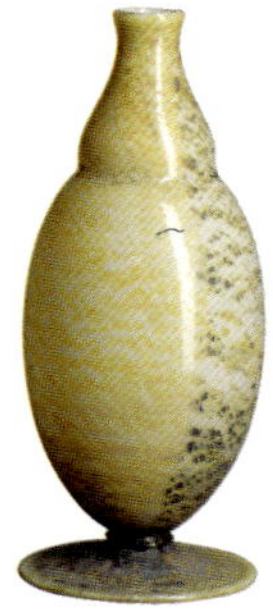

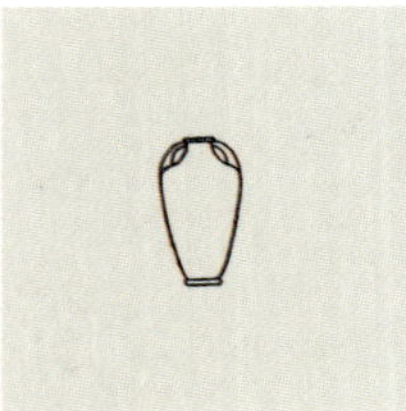

5913

Vase in *lattimo* glass with applied gold leaf. The piece is characterised by the squared shoulders, side handles and ring-shaped foot. Model no. 5913 was also produced in a variant in *lattimo* glass with side handles and finishings in coloured glass as well as in other glass techniques (e.g. silvered *pasta vitrea*).
1928–29
17 cm high
Private collection, Udine

>
5913, catalogue drawing, AMB

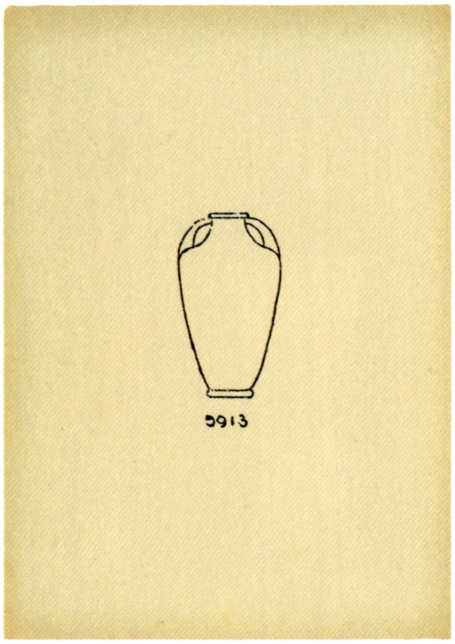

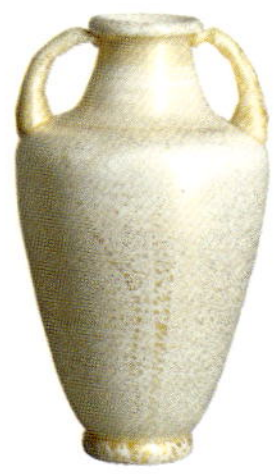

n.i.

Ovoid vase in *lattimo* glass with applied gold leaf. The piece, with small flared foot, is characterised by two small handles on the sides of the neck. Because of these, the vase is attributed to the production of the Cappellin glassworks.
1929–30
24 cm high
Private collection

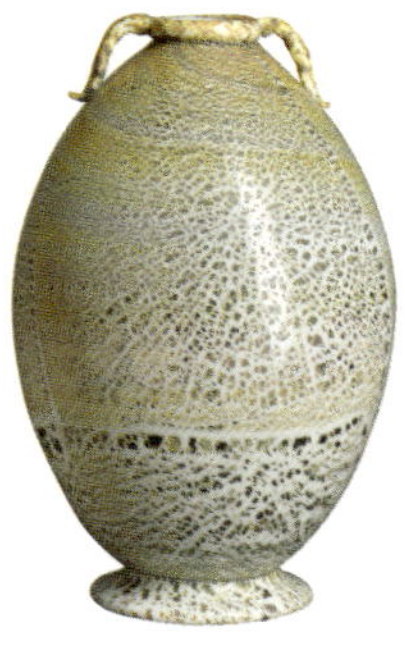

>
Vases in lattimo *glass with finishings in coloured glass. Period photograph, c. 1929, AMB*

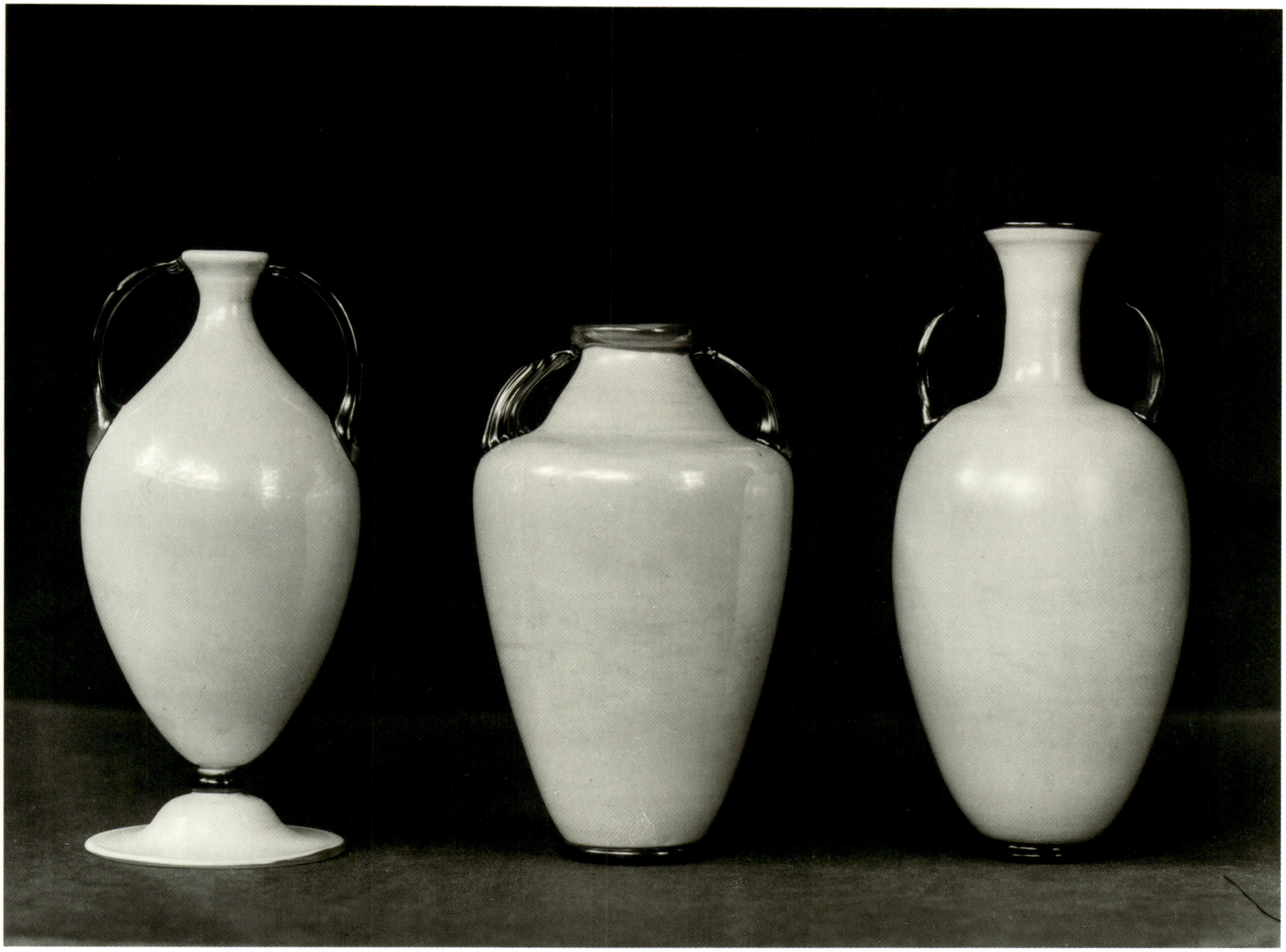

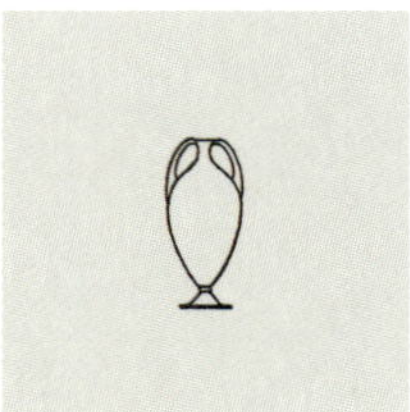

5898

Catalogue drawing for one of the vases in *lattimo* glass with finishings in coloured glass documented by a period photograph, AMB.
1928–29
20 cm high

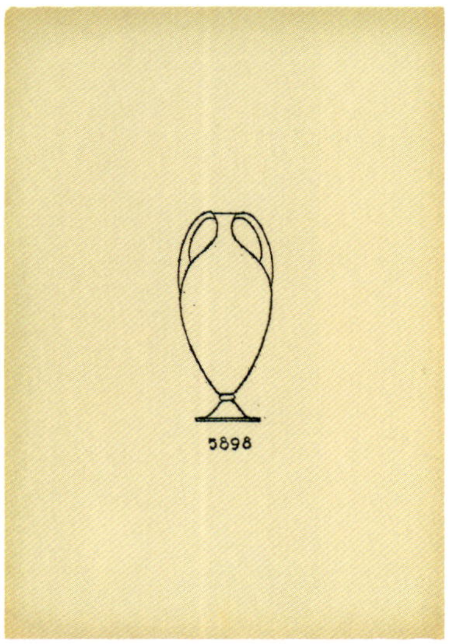

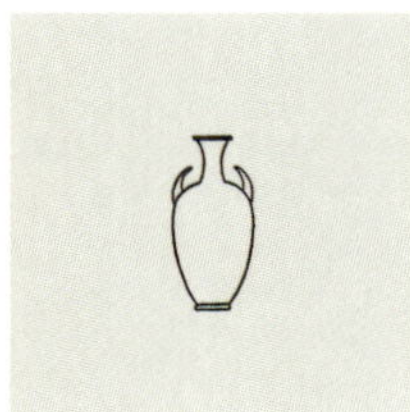

5897

Catalogue drawing for one of the vases in *lattimo* glass with finishings in coloured glass documented by a period photograph, AMB.
1928–29
20 cm high

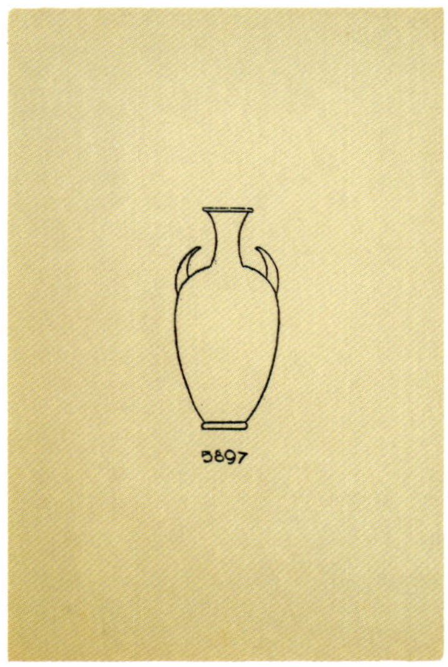

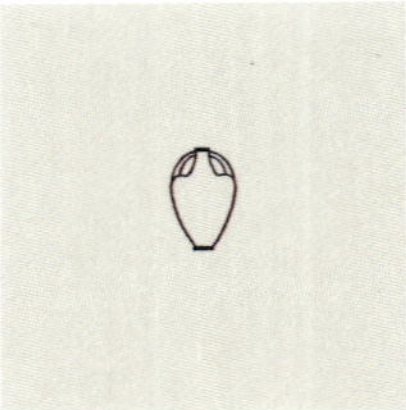

5907

Catalogue drawing for one of the vases in *lattimo* glass with applied gold leaf and finishings (handles, foot, mouth) in coloured glass, documented by a period photograph, AMH. An example of this model was exhibited in 1929 at the Leipzig Trade Fair in the variant 'iridised lattimo grey dec[oration]'.
1928–29
12 cm high

exhibitions
1929, Leipzig, Trade Fair

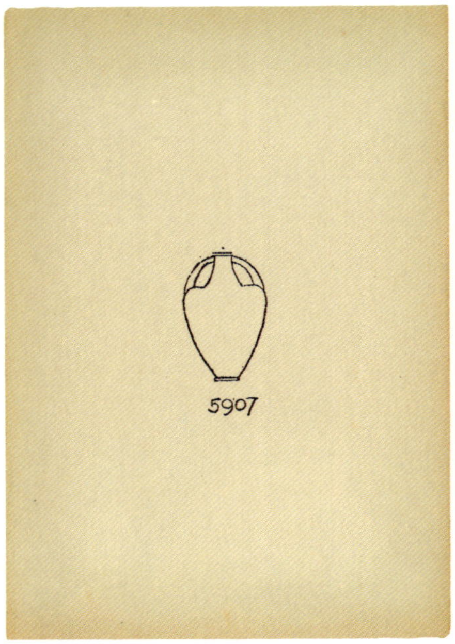

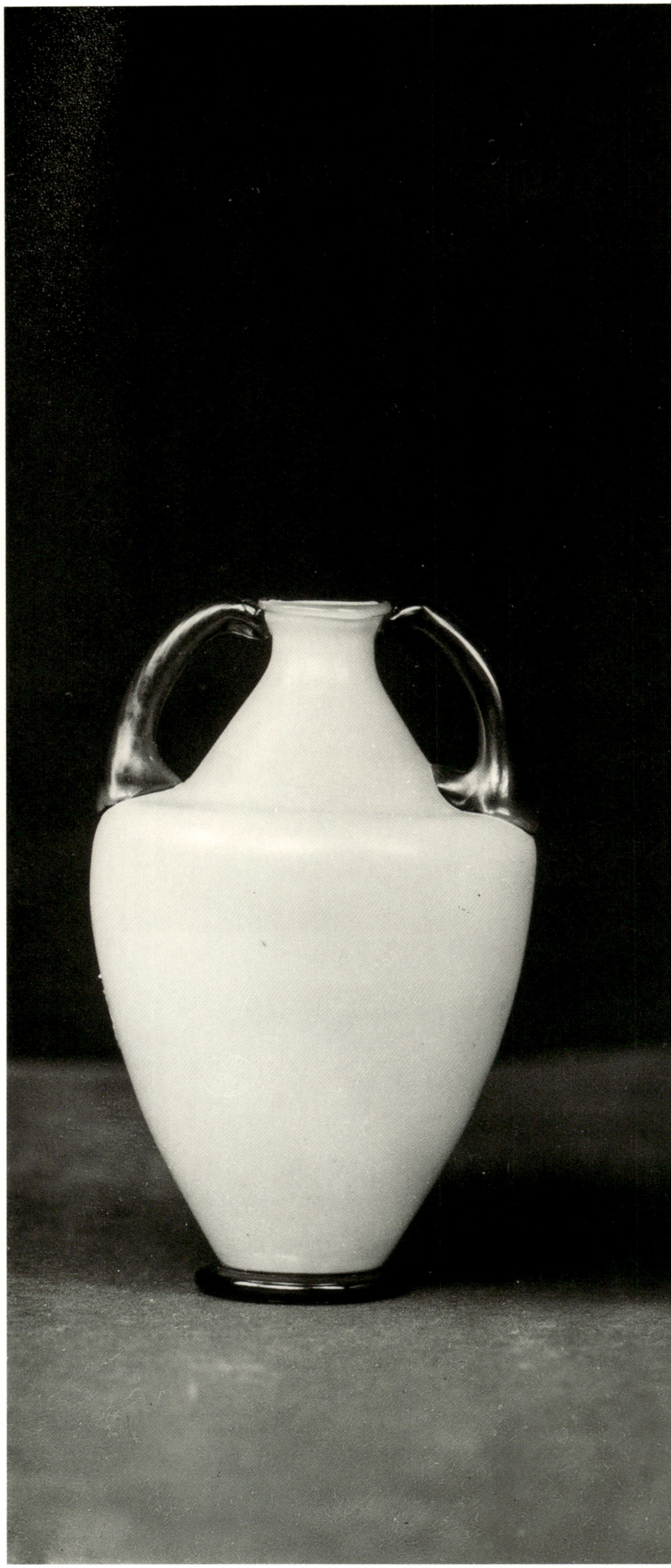

››
Vases in lattimo *glass with finishings in coloured glass (grey?). Period photograph, c. 1929, AMB*

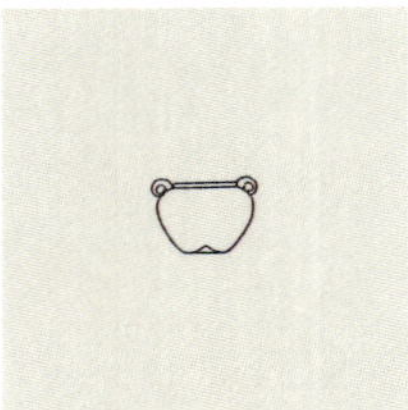

5903

Catalogue drawing for one of the vases in *lattimo* glass with applied gold leaf and finishings (handles, foot, mouth) in coloured glass, documented by a period photograph, AMH. An example of this model was exhibited in 1929 at the Leipzig Trade Fair in the variant 'iridised lattimo grey dec[oration]'.
1928–29
9 cm high

exhibitions
1929, Leipzig, Trade Fair

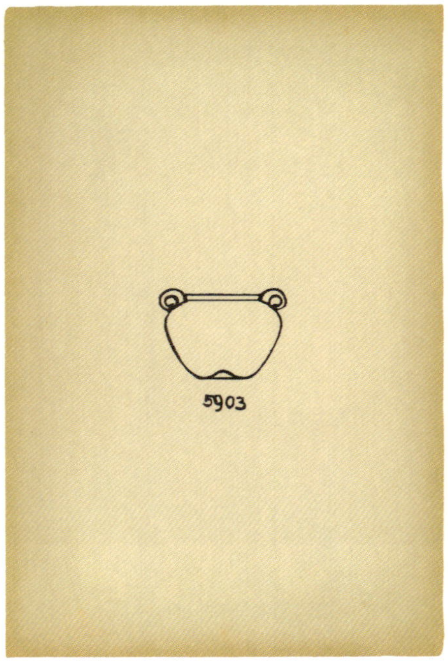

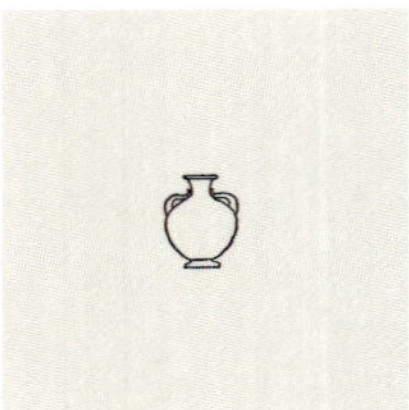

5909

Catalogue drawing for one of the vases in *lattimo* glass with applied gold leaf and finishings (handles, foot, mouth) in coloured glass, documented by a period photograph, AMB. An example of this model was exhibited in 1929 at the Leipzig Trade Fair in the variant 'iridised lattimo grey dec[oration]'.
1928–29
11 cm high

exhibitions
1929, Leipzig, Trade Fair

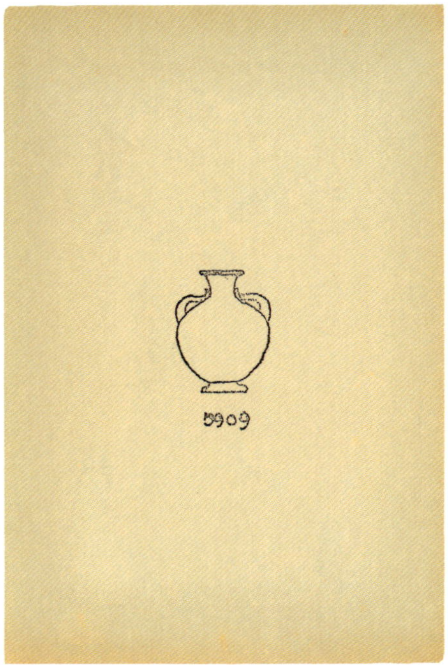

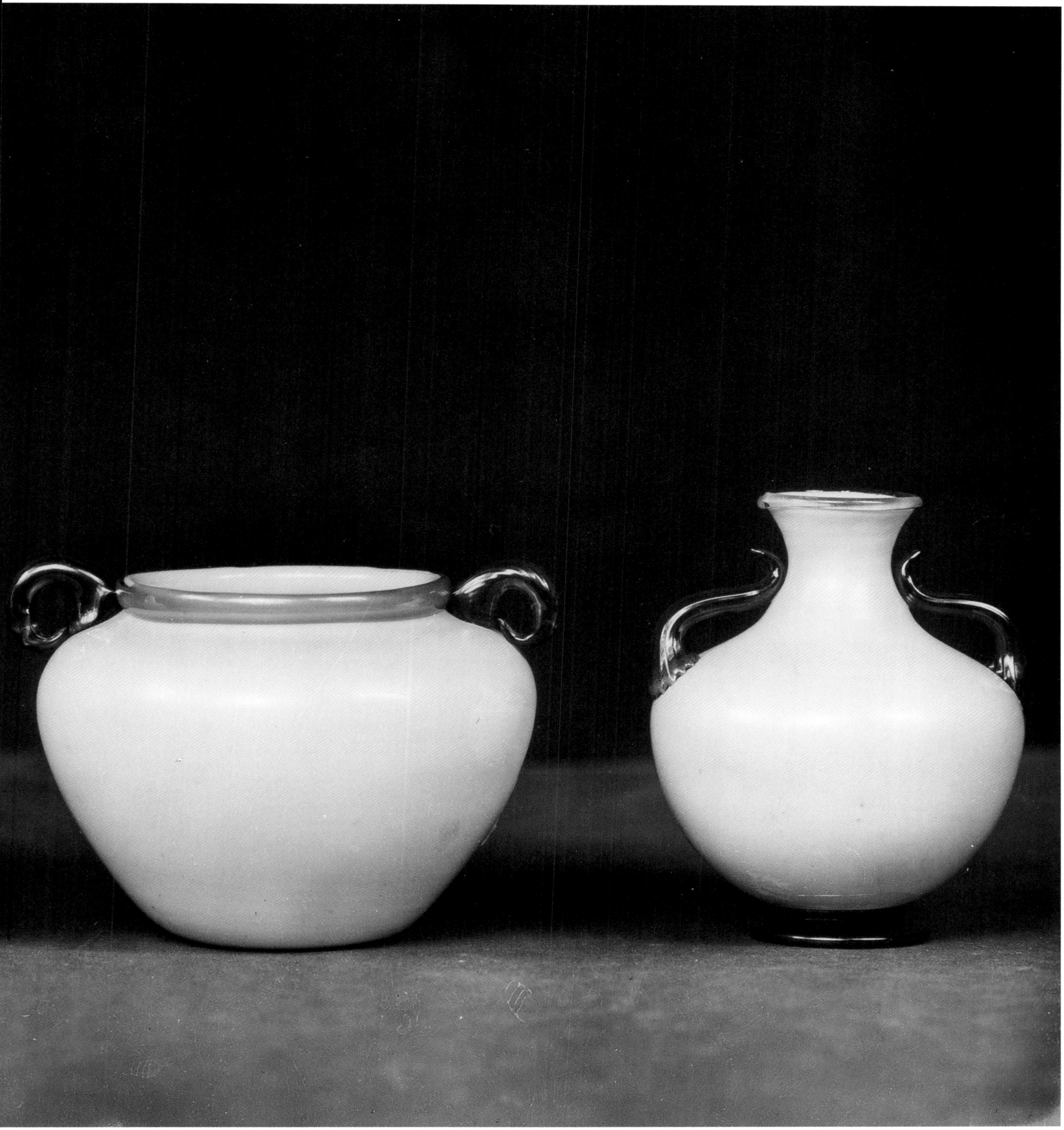

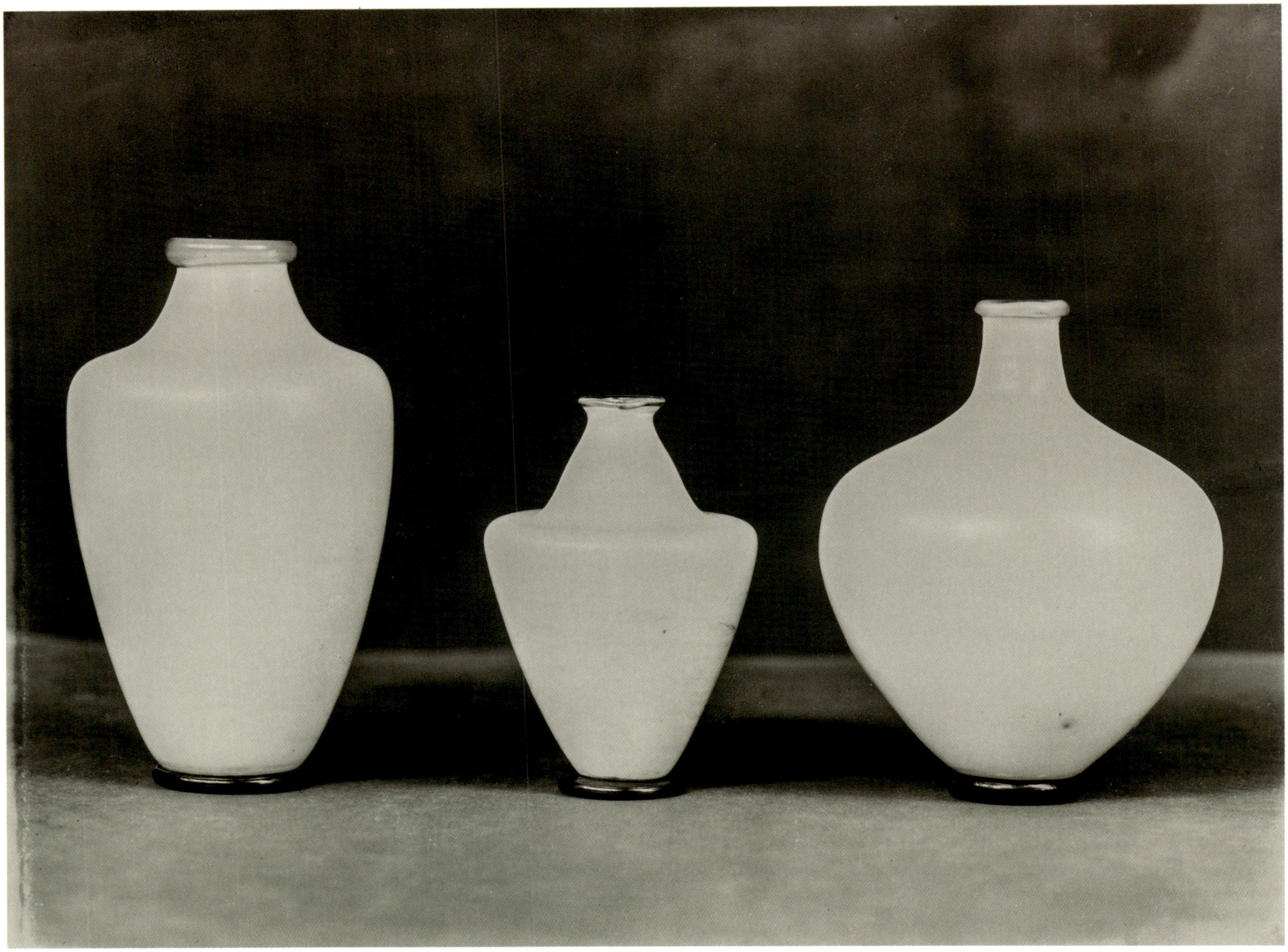

>
Vases in lattimo *glass with finishings in coloured glass. Period photograph, c. 1929, AMB*

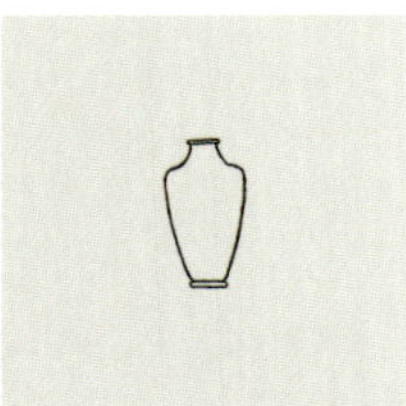

5914

Catalogue drawing for one of the vases in *lattimo* glass with applied gold leaf and finishings (foot and mouth) in coloured glass, documented by a period photograph, AMB. An example of this model was exhibited in 1929 at the Leipzig Trade Fair in the variant 'iridised lattimo grey dec[oration]'.
1928–29
18 cm high

exhibitions
1929, Leipzig, Trade Fair

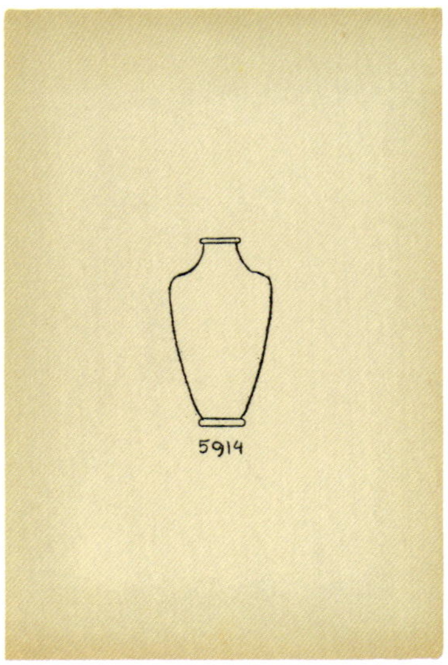

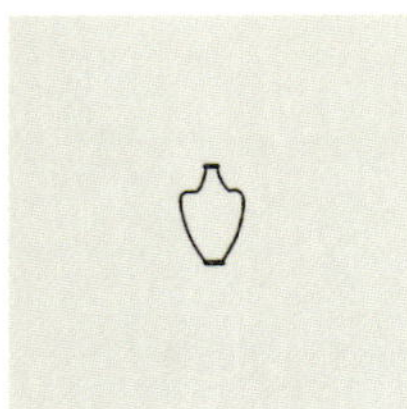

5908

Catalogue drawing for one of the vases in *lattimo* glass with applied gold leaf and finishings (foot and mouth) in coloured glass, documented by a period photograph, AMB. An example of this model was exhibited in 1929 at the Leipzig Trade Fair in the variant 'iridised lattimo grey dec[oration]'.
1928–29
12 cm high

exhibitions
1929, Leipzig, Trade Fair

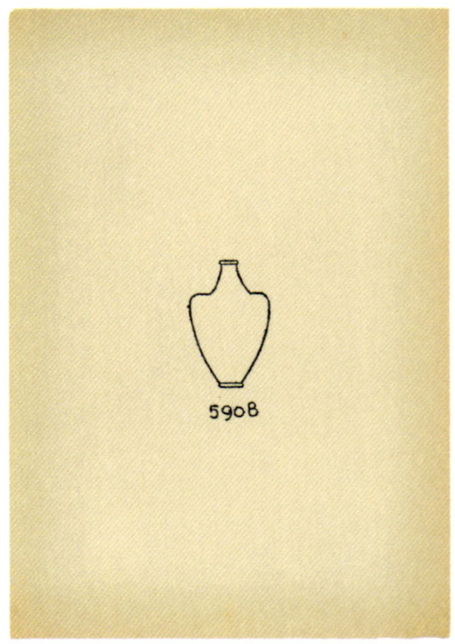

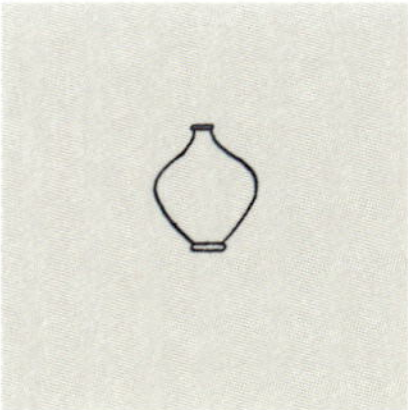

5912

Catalogue drawing for one of the vases in *lattimo* glass with applied gold leaf and finishings (foot and mouth) in coloured glass, documented by a period photograph, AMH. An example of this model was exhibited in 1929 at the Leipzig Trade Fair in the variant 'iridised lattimo grey dec[oration]'.
1928–29
15 cm high

exhibitions
1929, Leipzig, Trade Fair

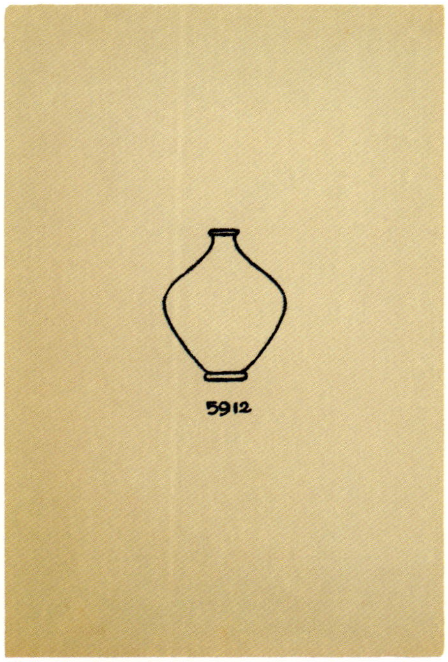

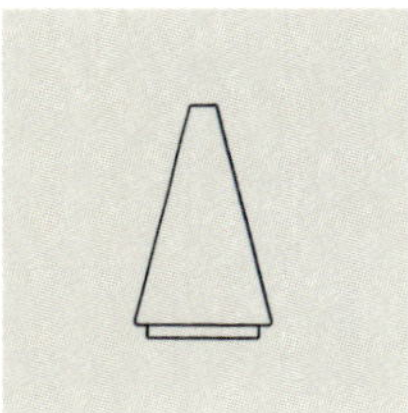

5971

Vase in the shape of a truncated cone in *lattimo* glass with foot and thread around the mouth in black glass. Surface with iridised finish. Model no. 5971 was also produced in other glass techniques (e.g. *pasta vitrea*, iridised black, cased glass).
1930
25.5 cm high
M.V.M. Cappellin Murano *acid stamp*
Cappellin collection, Venice

>
5971, working drawing, CSV

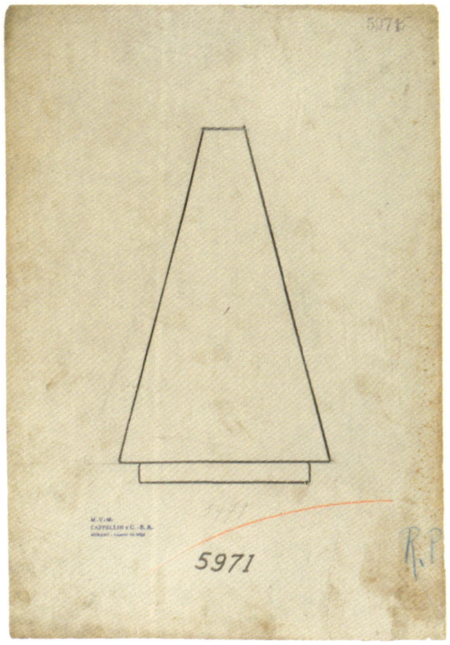

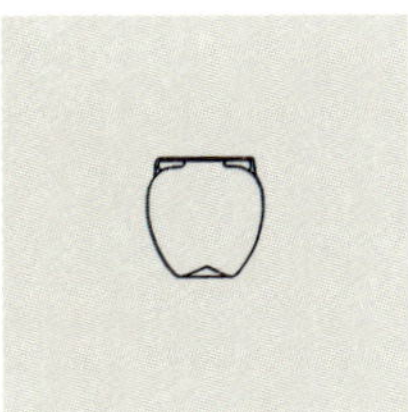

5267

Vase in *lattimo* glass with side handles in black glass. The vase was produced after a model in the catalogue of the glassworks designed by Vittorio Zecchin (1921–25).
1928–30
14 cm high
M.V.M. Cappellin Murano *acid stamp*
The Olnick Spanu Collection, New York

>
5267, catalogue drawing, AMB

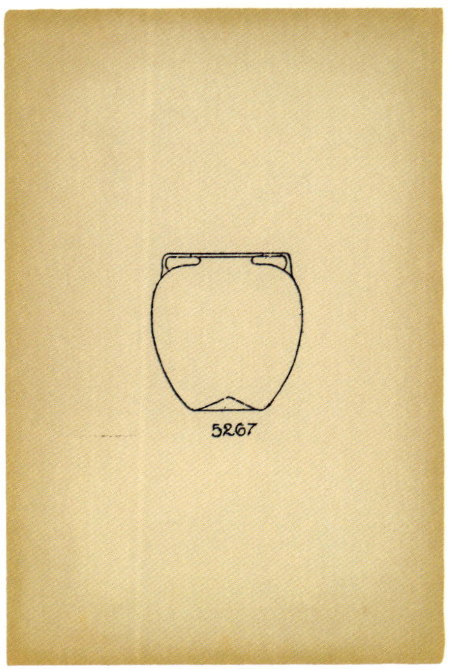

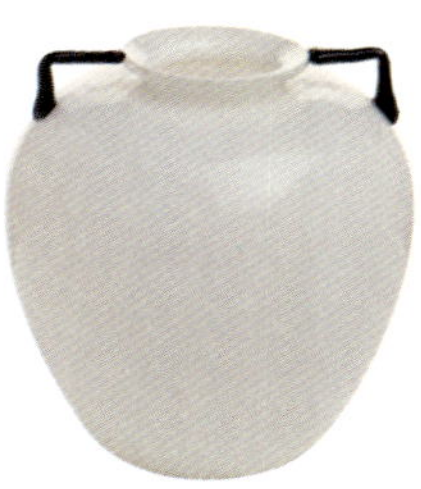

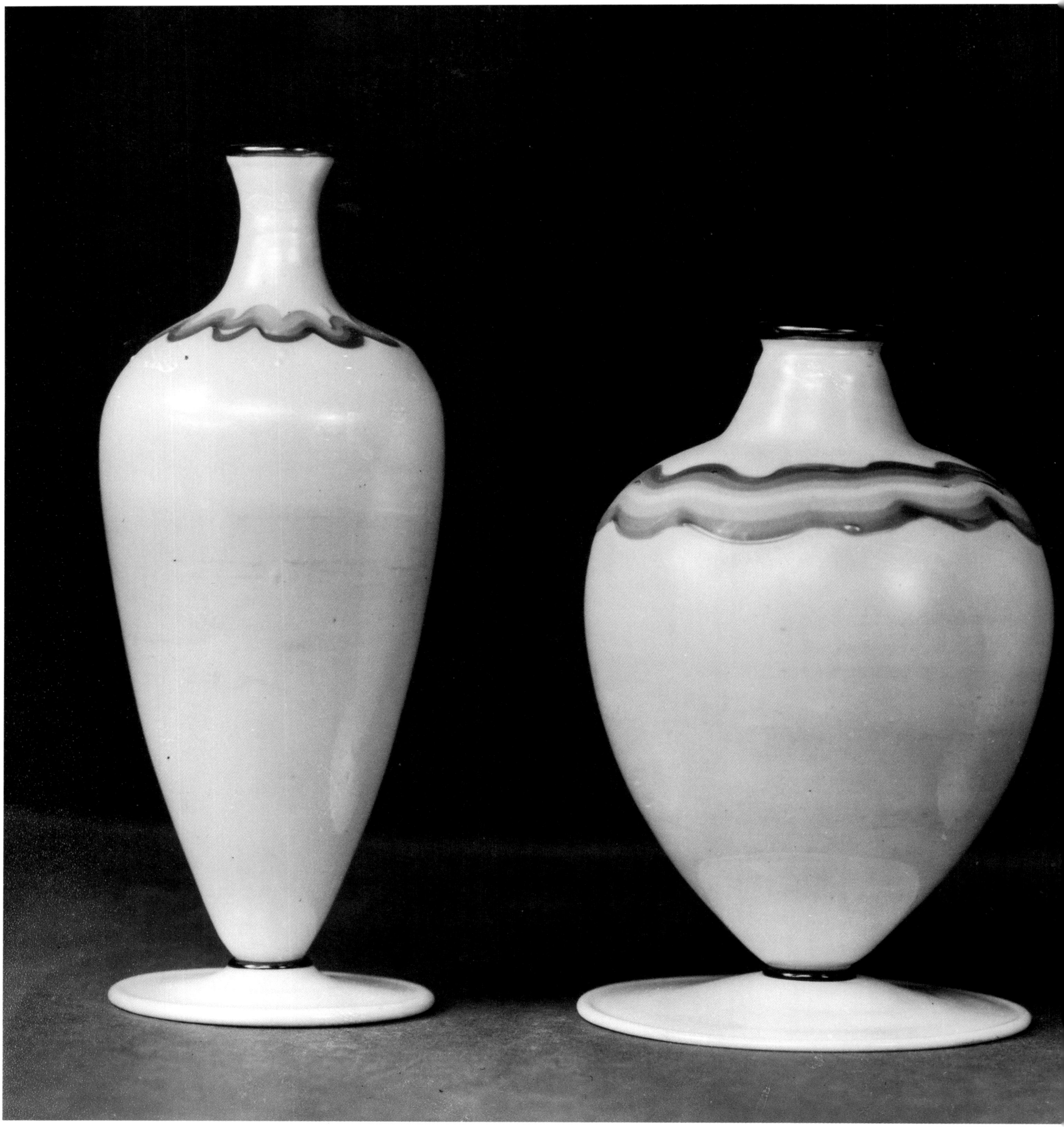

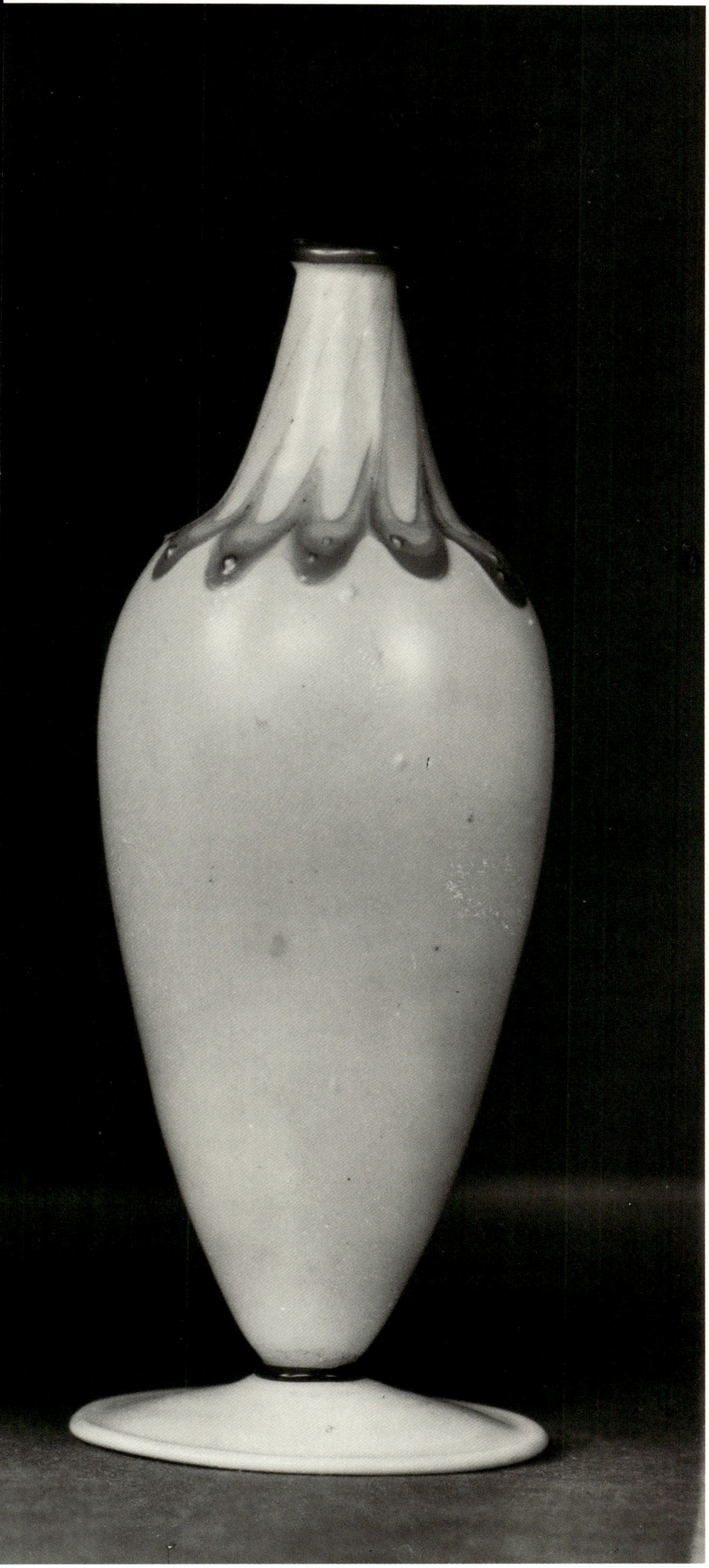

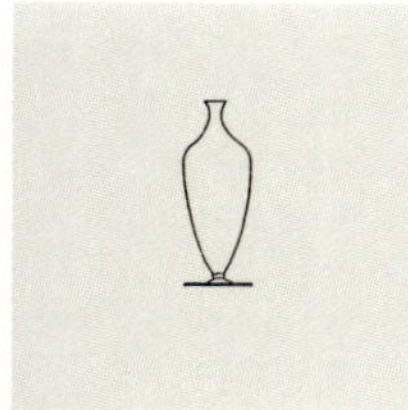

5894

Catalogue drawing for one of the vases in *lattimo* glass, decorated with undulating polychrome bands, published in *Domus* in October 1929, AMB.
1928–29
22 cm high

<< Vases in lattimo *glass with polychrome undulating bands applied at the base of the neck. Period photograph, c. 1929, AMB*

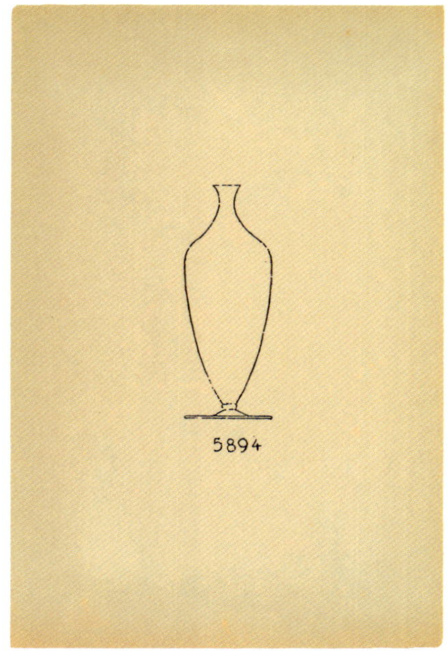

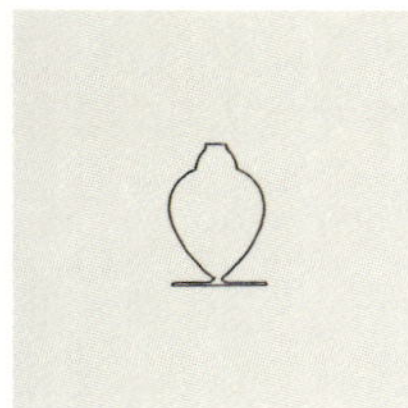

5895

Catalogue drawing for one of the vases in *lattimo* glass, decorated with undulating polychrome bands, published in *Domus* in October 1929, AMB. Model no. 5895 was also produced in other glass techniques (e.g. *pasta vitrea*).
1928–29
17 cm high

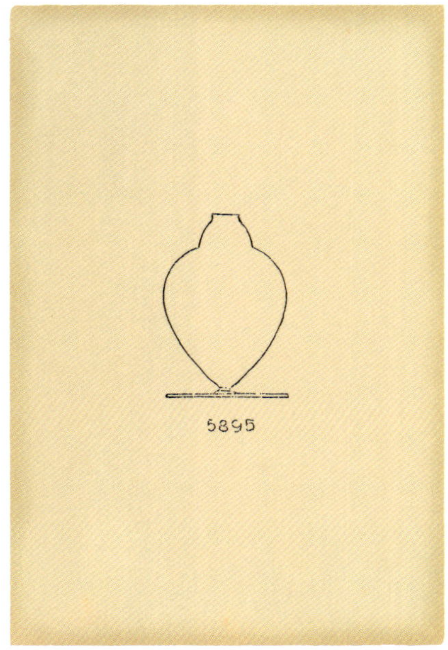

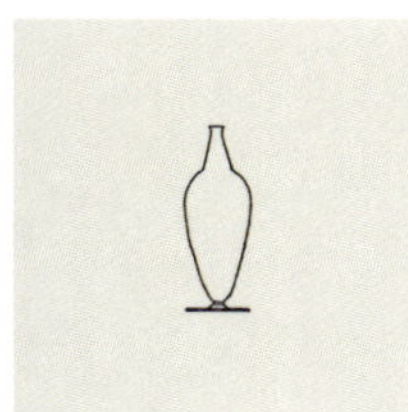

5923

Catalogue drawing for one of the vases in *lattimo* glass, decorated with undulating polychrome bands, published in *Domus* in October 1929, AMB. Model no. 5823 was also produced in other glass techniques (e.g. *decoro fenicio, pasta vitrea,* iridised black).
1928–29
22 cm high

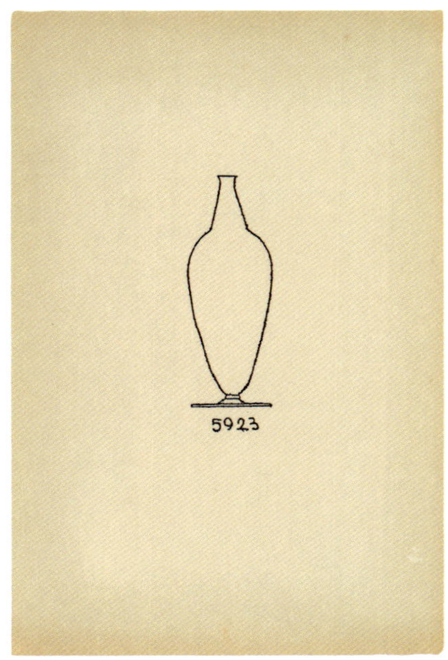

>
Vases in lattimo *glass in* Domus, *April 1929, p. 35*

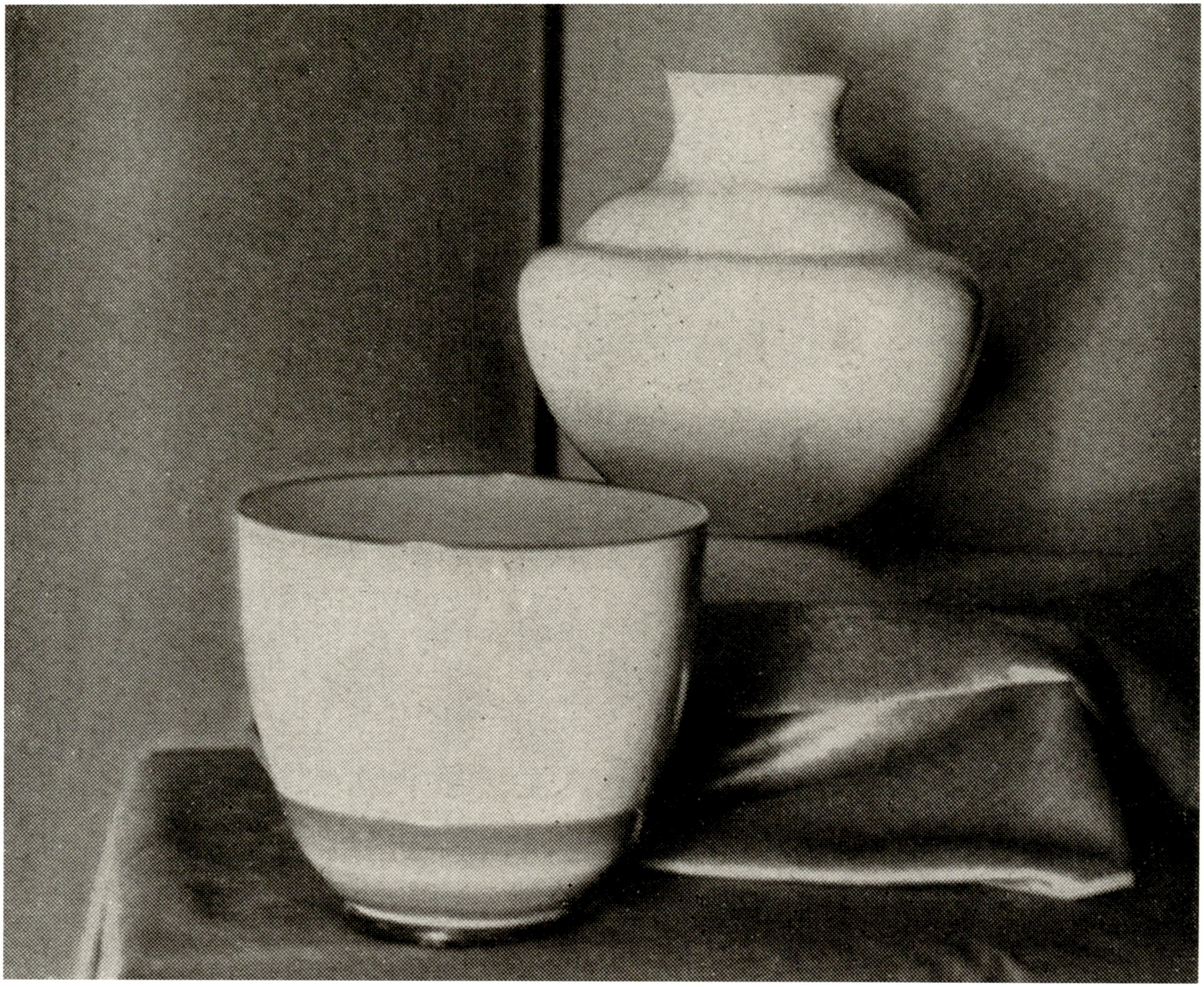

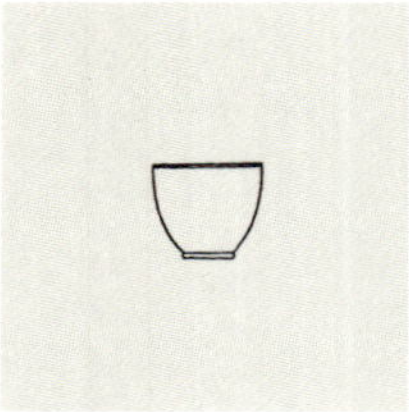

5931

Catalogue drawing for one of the vases in *lattimo* glass published in *Domus* in April 1929, AMB. This piece is characterised by its horizontal polychrome bands on the lower part of the body.
1928–29
11 cm high

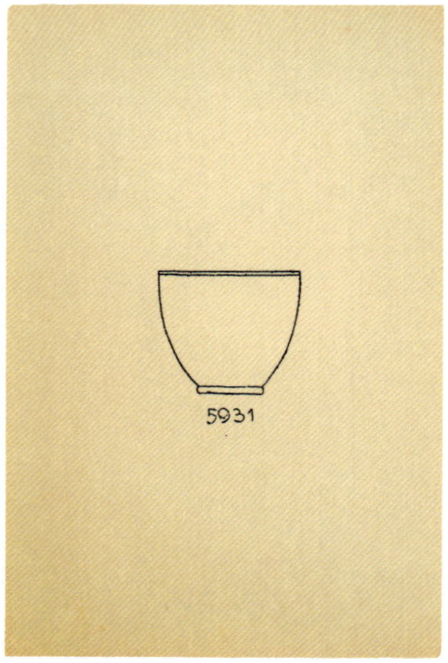

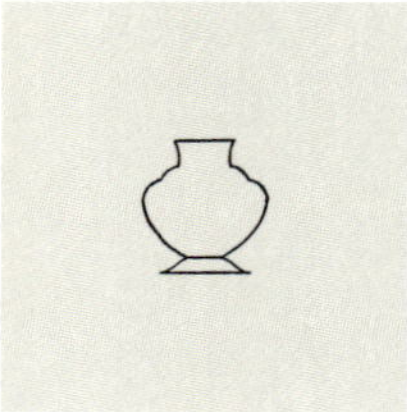

5938

Catalogue drawing for one of the vases in *lattimo* glass published in *Domus* in April 1929, AMB. An example of this model, presumably in iridised *lattimo*, was exhibited at the Salon d'automne of Paris in 1928, while the variant in 'iridised lattimo sprayed with gold and coral-red dec[oration]' was on display the following year at the Leipzig Trade Fair. Model no. 5938 was widely used in the production of the glassworks and was also executed in other glass techniques (e.g. transparent, *decoro fenicio*, silvered *pasta vitrea*, silvered black).
c. 1928
16 cm high

exhibitions
1928, Paris, Salon d'automne (?)
1929, Leipzig, Trade Fair

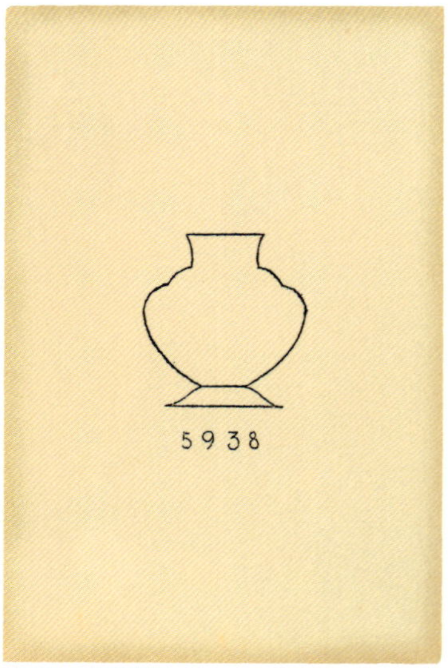

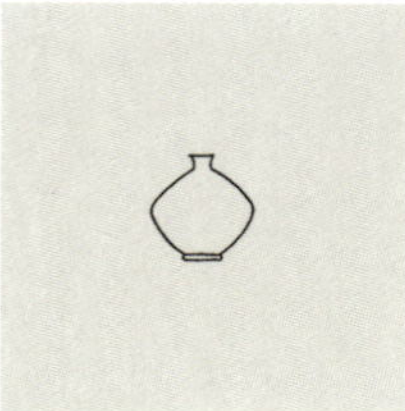

5919

Vase in *lattimo* glass decorated on the body with a polychrome undulating band, in *Domus*, October 1929, p. 31.
1928–29
13 cm high

>
5919, catalogue drawing, AMB

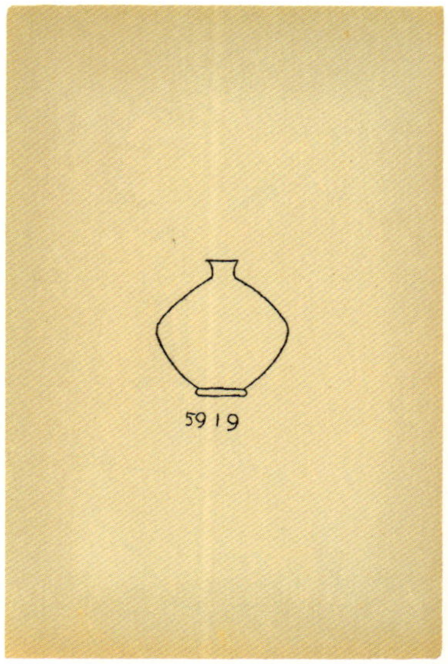

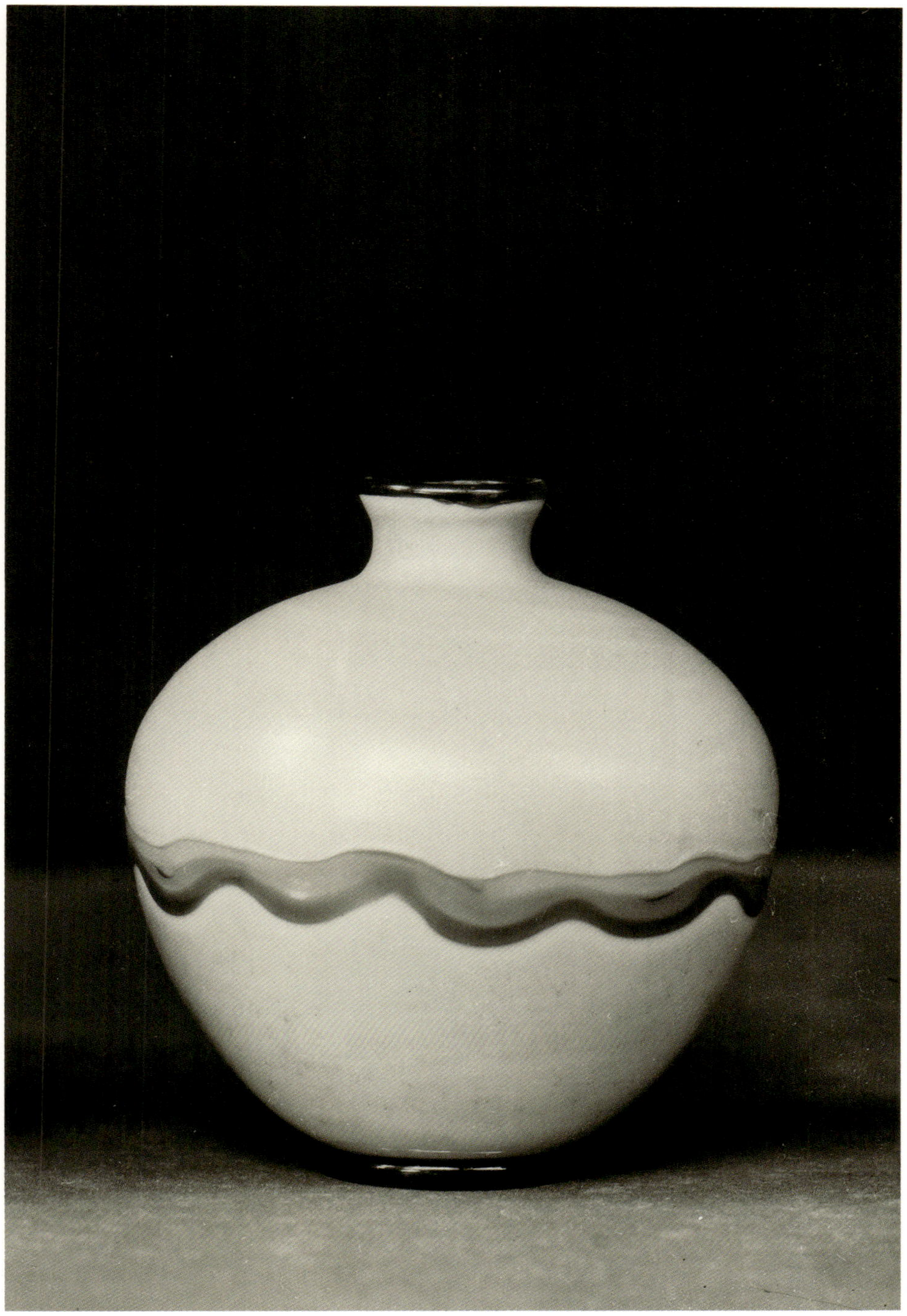

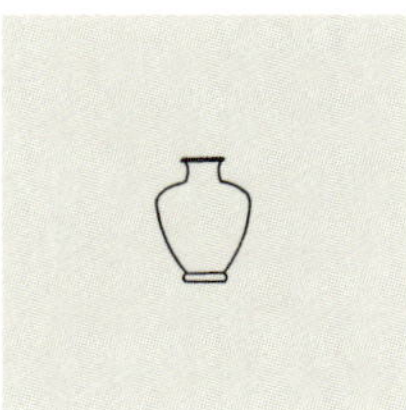

5917

Vase in *lattimo* glass decorated on the body with a grey and light-blue, *pagliesco* and amethyst undulating band. Foot in black glass and mouth in *pagliesco* glass. Surface with iridised finish.
c. 1929
14 cm high
Sara Blumberg and Jim Oliveira collection, New York

>
5917, catalogue drawing, AMB

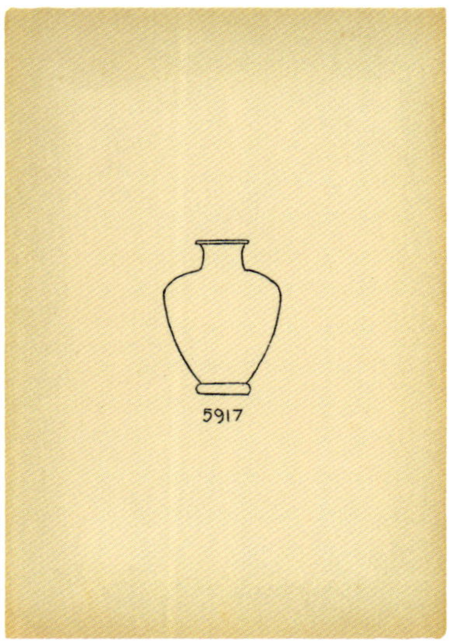

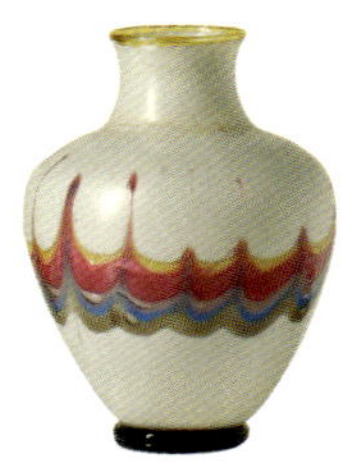

>
Vases in cased latti-mo *glass, with chain ornamentation. Period photograph, c. 1929, CSV*

>>
Service no. 651 ('breakfast set') in lattimo *cased* pa-gliesco *glass, slightly ribbed in* Cappellin Glassware, *catalogue, undated, AHL, detail*

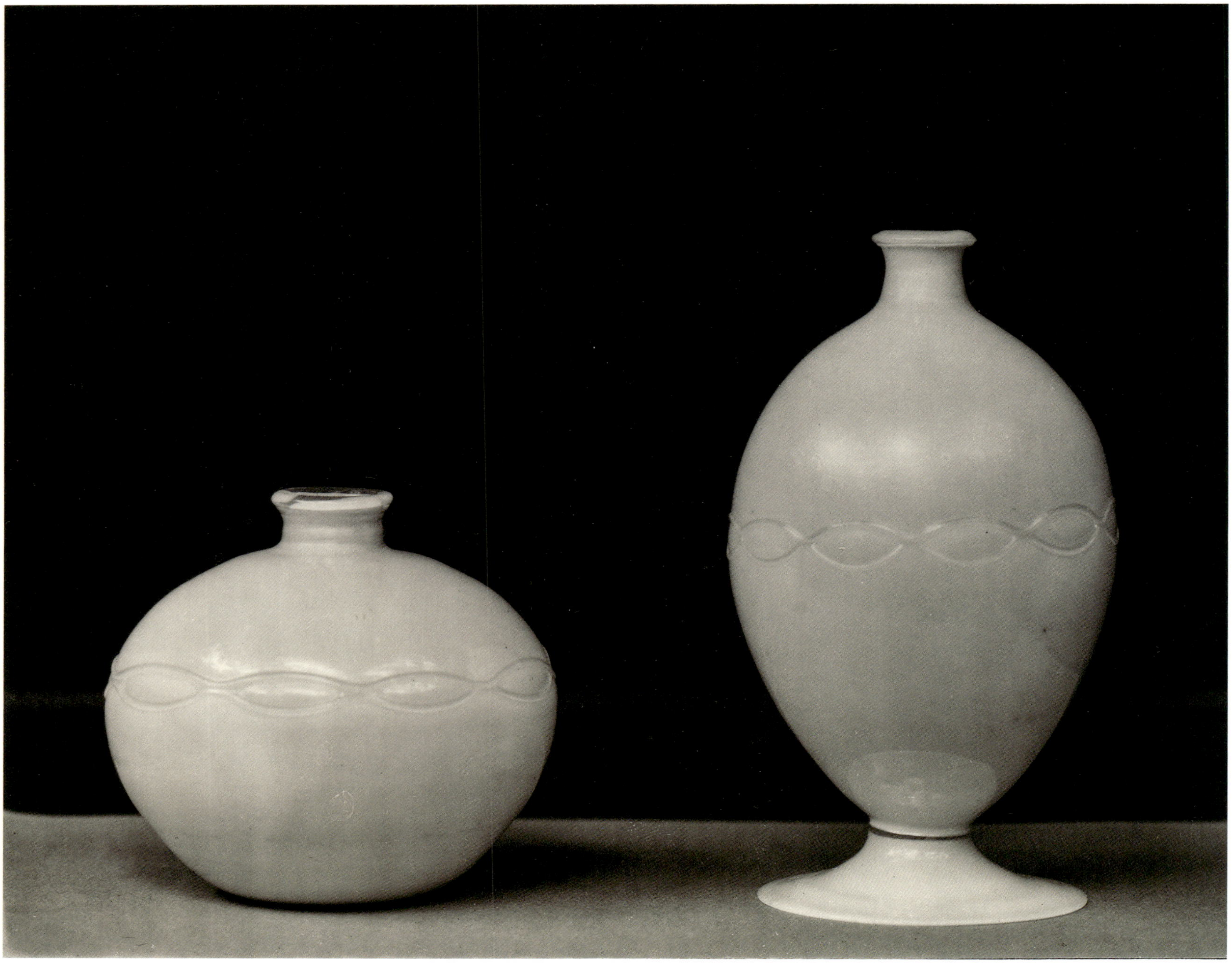

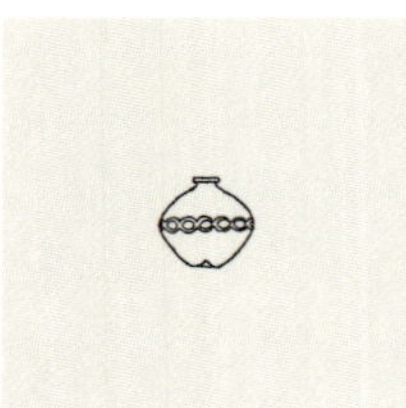

5945

Catalogue drawing for one of the vases in *lattimo* cased glass with a chain decoration documented by a period photograph, AMB. Three examples of this model were exhibited at the Leipzig Trade Fair in 1929. On that occasion the model was presented in 'yellow iridised lattimo', 'light-blue iridised lattimo' and 'grey iridised lattimo'.
c. 1929
10 cm high

exhibitions
1929, Leipzig, Trade Fair

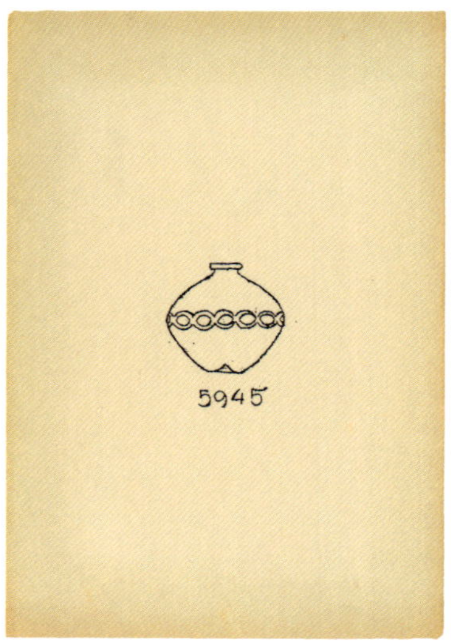

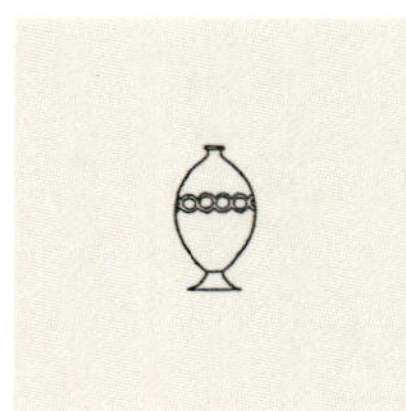

5944

Catalogue drawing for one of the vases in *lattimo* cased glass with a chain decoration documented by a period photograph, AMB. Two examples of this model were exhibited at the Leipzig Trade Fair in 1929. On this occasion the model was presented in 'yellow iridised lattimo' and 'light-blue iridised lattimo'. To these was added the *decoro fenicio* variant in *lattimo* and amethyst.
c. 1929
17 cm high

exhibitions
1929, Leipzig, Trade Fair

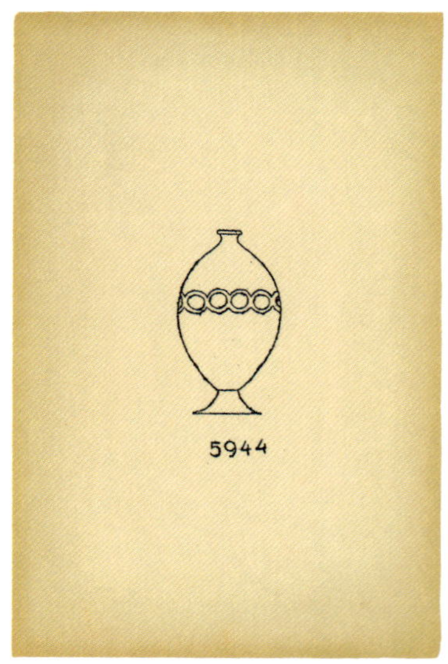

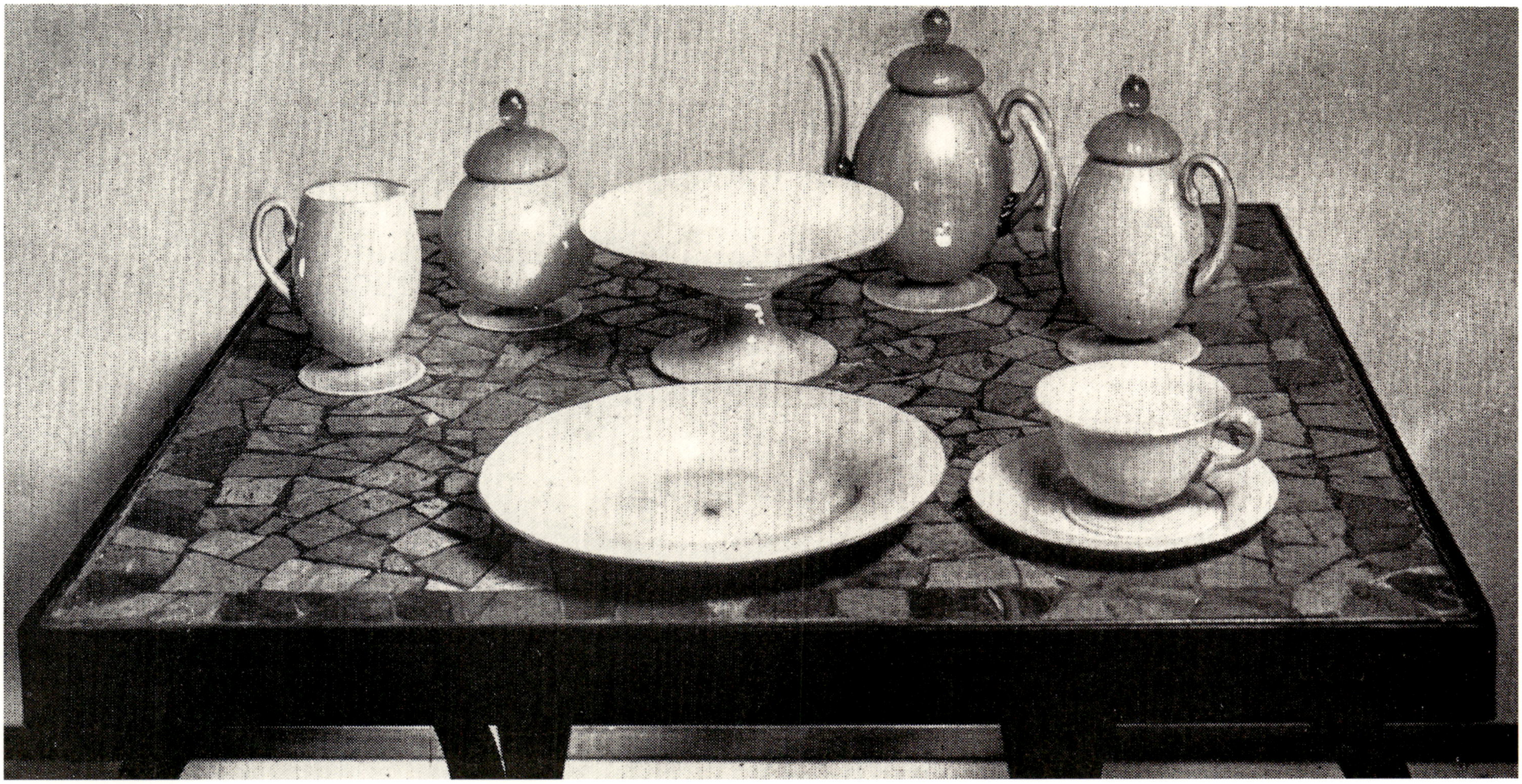

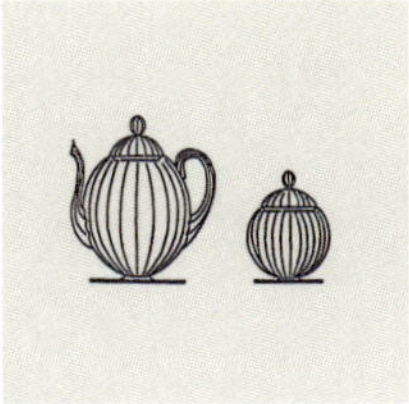

651

Catalogue drawing for service no. 651 ('breakfast set') in *lattimo* cased *pagliesco* glass, slightly ribbed, documented in the American photographic catalogue of Cappellin Glassware, undated, AMB.
c. 1929
20; 13 cm high

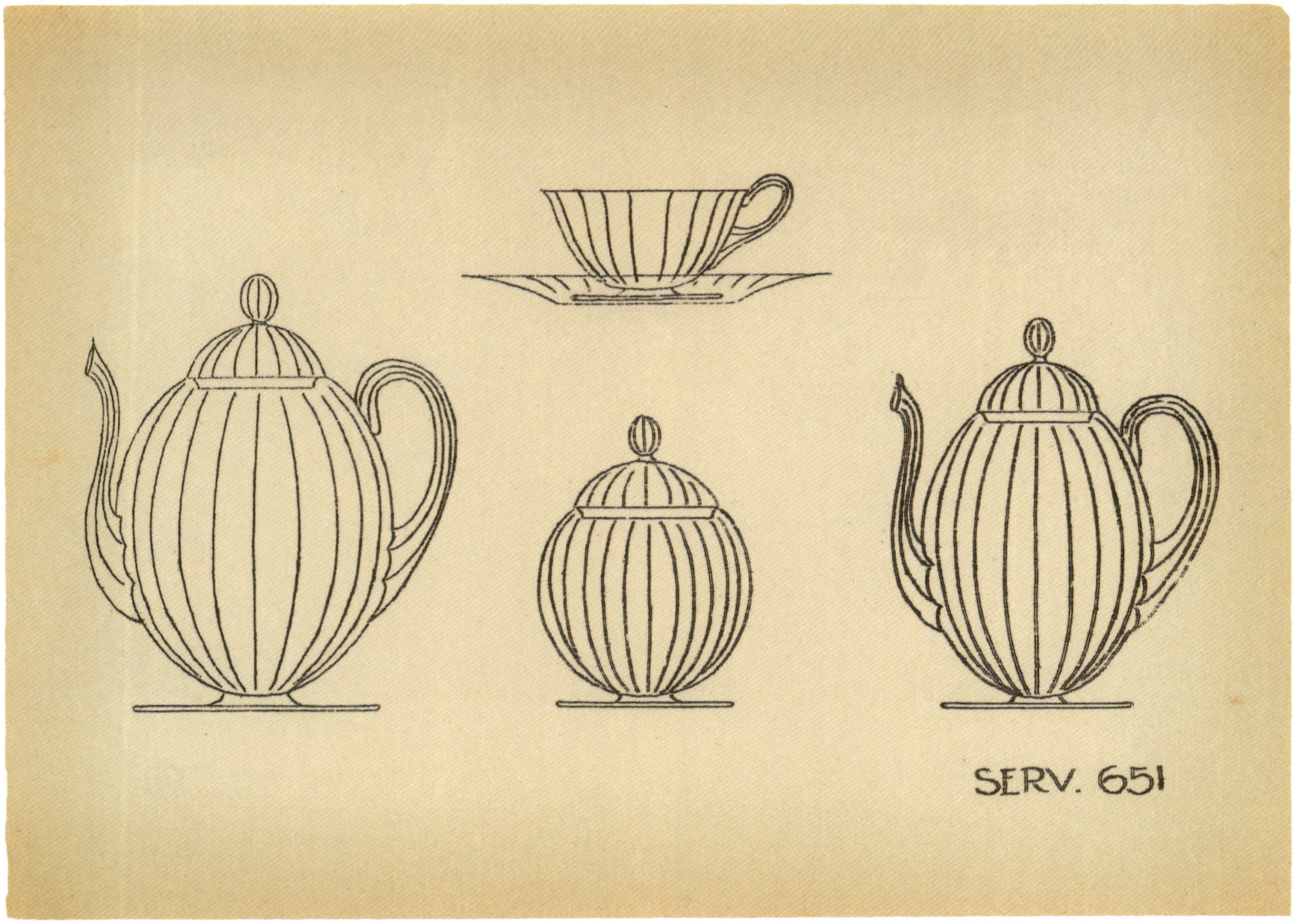

A

Sugar bowl, cup and plate and tea pot, in *lattimo* glass with applied gold leaf belonging to the 'A' tea service. The service was exhibited at the *Mostra Artisti italiani contemporanei* [Exhibition of Contemporary Italian Artists] held at the Galleria Pesaro of Milan in 1929.
c. 1929
7; 7; 23 cm high
C.V.M. *acid stamp* (cup and plate)*;* Venezia Murano Company *acid stamp* (tea pot)
The Lockwood collection, Fort Lee, NJ; private collection, Udine; private collection

exhibition
1929, Milan, *Mostra Artisti italiani contemporanei* [Exhibition of Contemporary Italian Artists]
1931, Amsterdam Exhibition of glass, ceramics and lace

>
Carlo Scarpa, design for the tea pot of the 'A' Service, Cappellin of Munich, c. 1929., ACSVr

>
Carlo Scarpa, design for various elements of the 'A' Service, c. 1929, ACSVr

Both designs were attached to an order of the Cappellin commercial firm in Munich, presumably intended for the Deutsche Werkstätten, to which three different tea service proposals were sent.

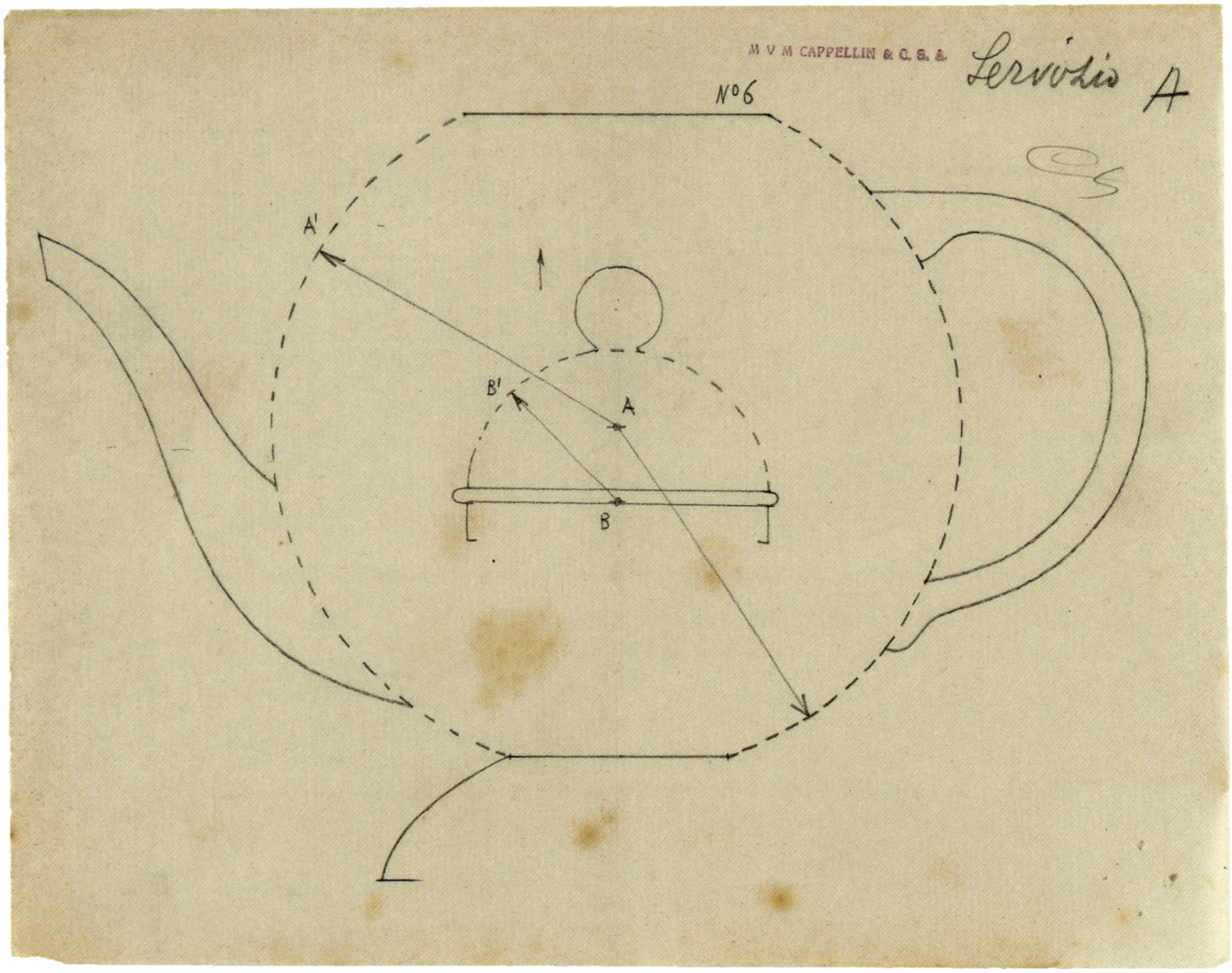

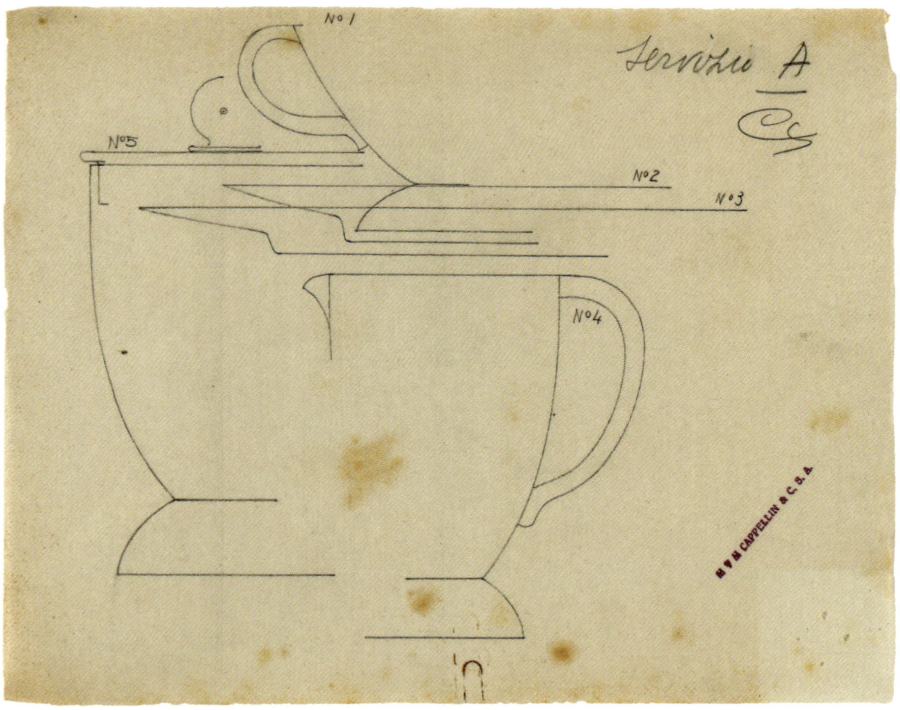

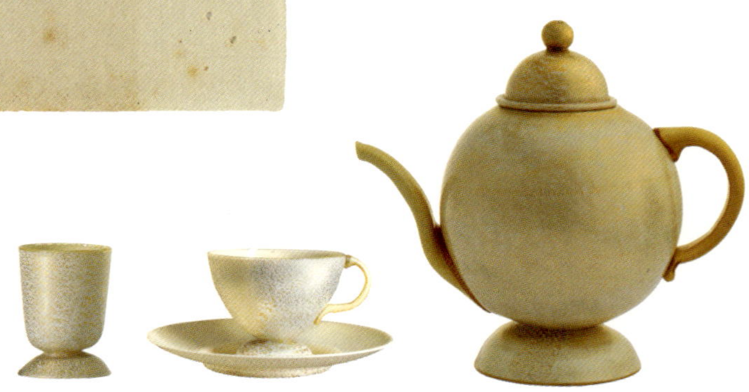

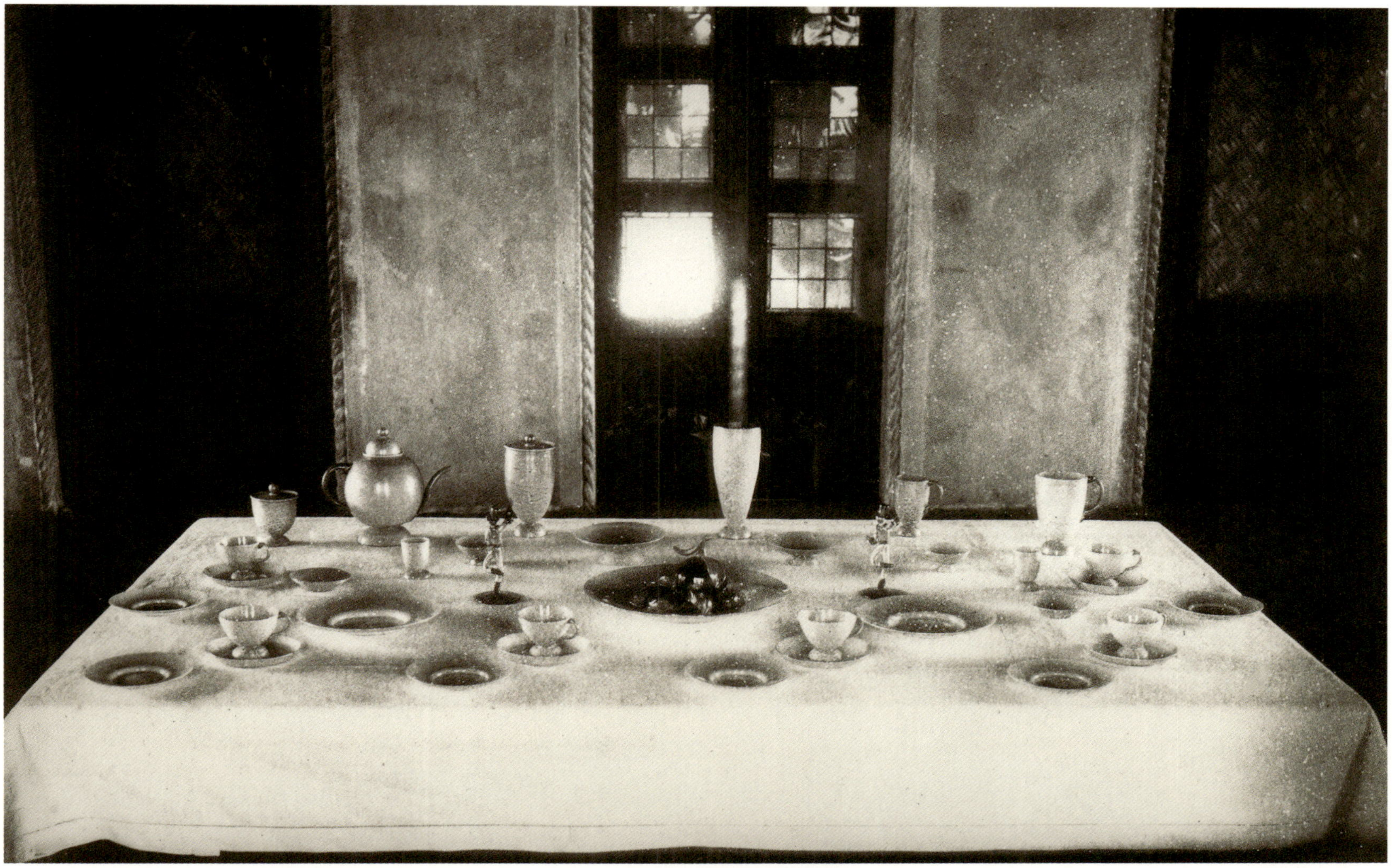

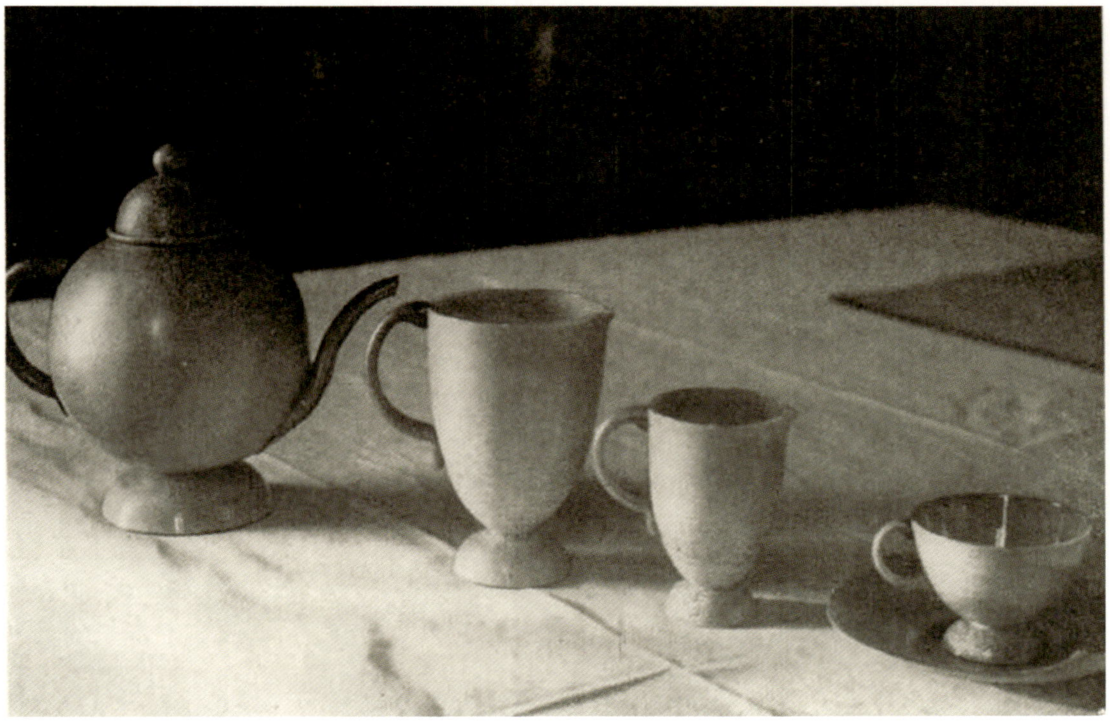

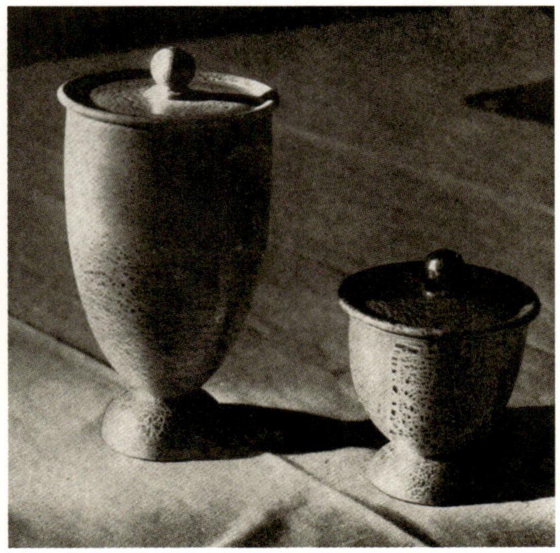

>
Table prepared with dinner service in lattimo *aurato glass Period Photograph, 1929 ca., ACSVr*

>
'Tea service in yellow and gold lattimo *glass' in* Mostra Artisti italiani contemporanei *[Exhibition of Contemporary Italian Artists], Milan 1929. The illustration was also published in* Domus, *January 1930*

>
Sugar bowl in aureate lattimo *glass, in the* Tentoonstelling van glas, aardewerk en kant, van moderne italiaansche kunstnijverheid. Mostra di vetri, ceramiche e merletti d'arte moderna italiana, *Venice 1931, p. 21*

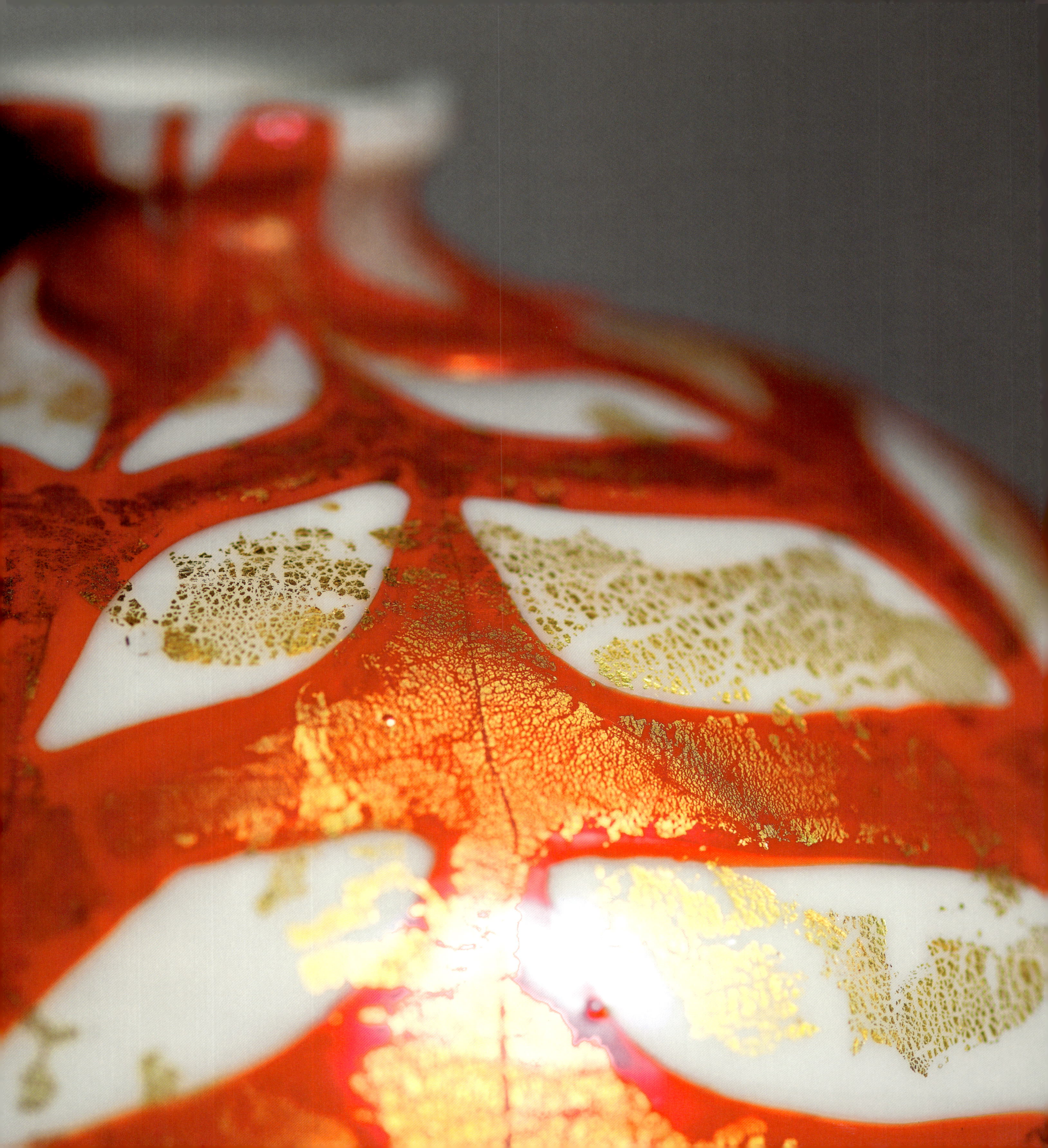

The pieces with *decoro fenicio*, an ancient glass-making technique referred to in Scarpa's notes as 'graffito', is distinguished by polychrome glass with a typical festoon decoration. The glass is obtained by the application under heat around the wall of the object of a glass thread wound in a spiral, which is subsequently 'combed' with a specific hooked instrument. For this series, colour combinations with a strong impact were chosen (e.g. coral-red – *lattimo*), often embellished by applications of gold leaf, but also more delicate associations with the use of lighter colours or playing with the shades of white (*lattimo*) and *pagliesco* or of grey or amethyst. A particular variant is constituted by twisted *decoro fenicio* with its evocative glass composition which is obtained by rotating the already decorated semi-finished product on its axis in the blowing phase. Generally proposed in opaque glass, the series of pieces with *decoro fenicio* also includes examples in transparent glass.

DECORO FENICIO 1928-29

Exhibitions

1929, Leipzig, Trade Fair
1929, Milan, Trade Fair
1930, Milan, Trade Fair

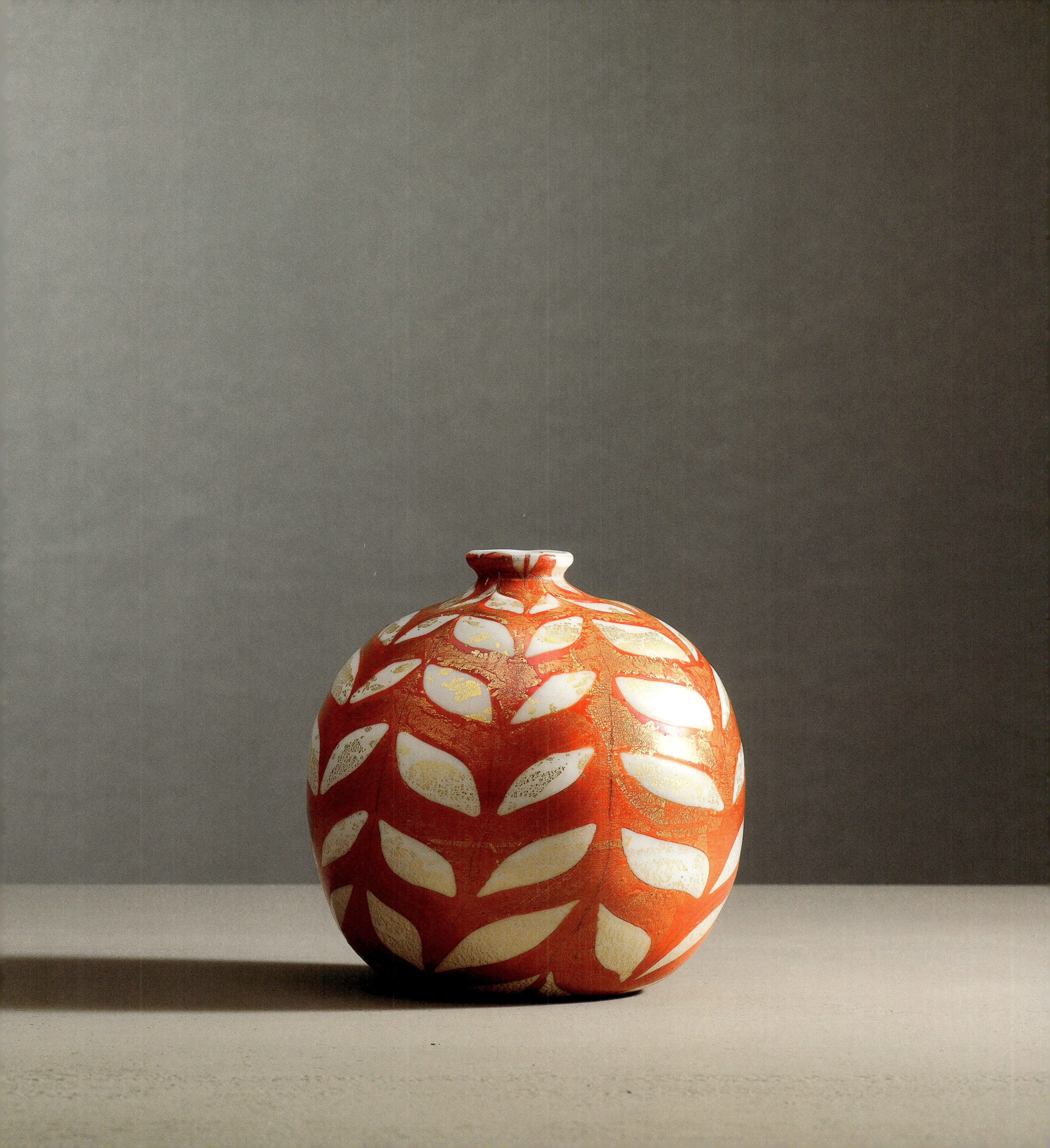

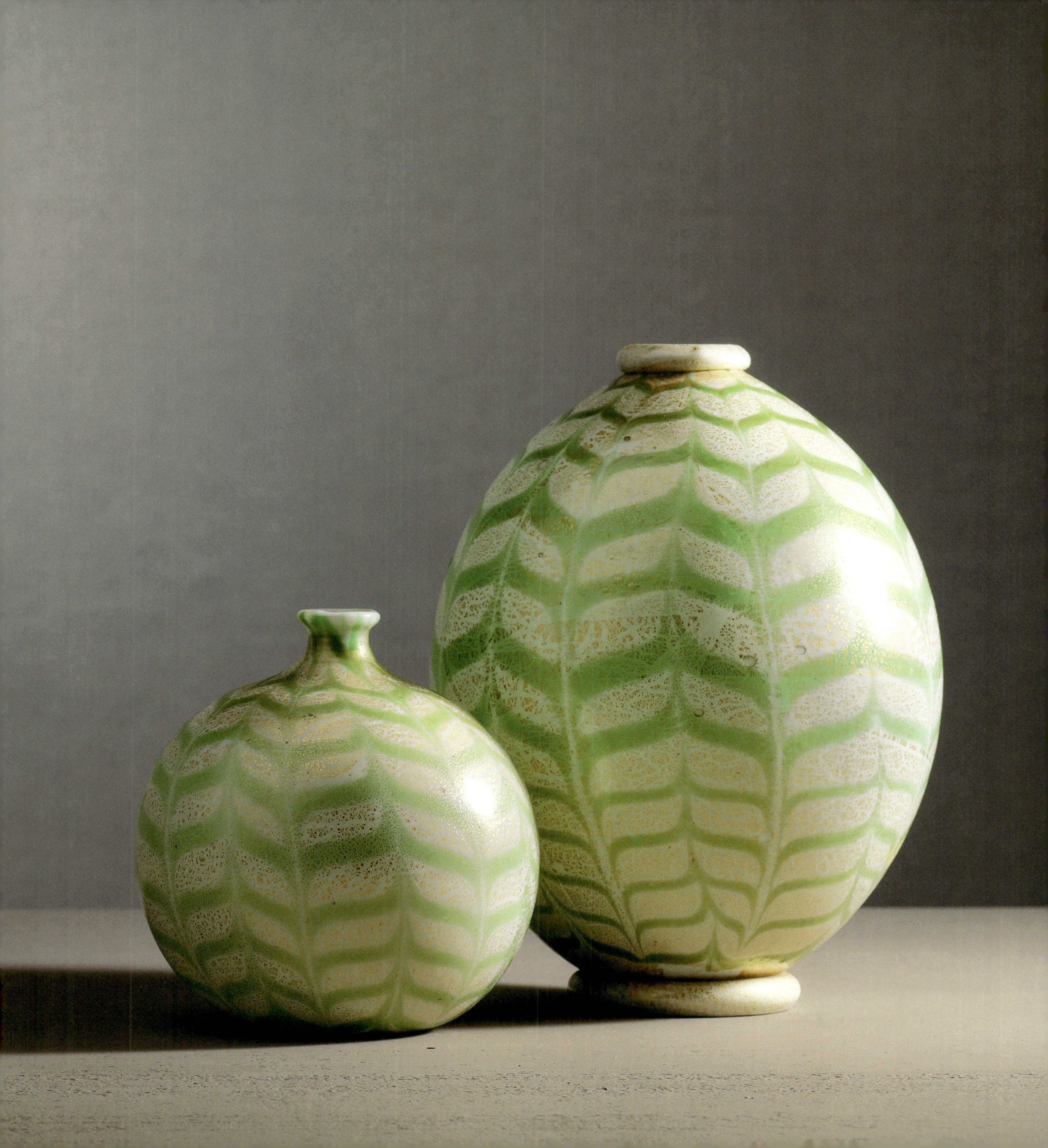

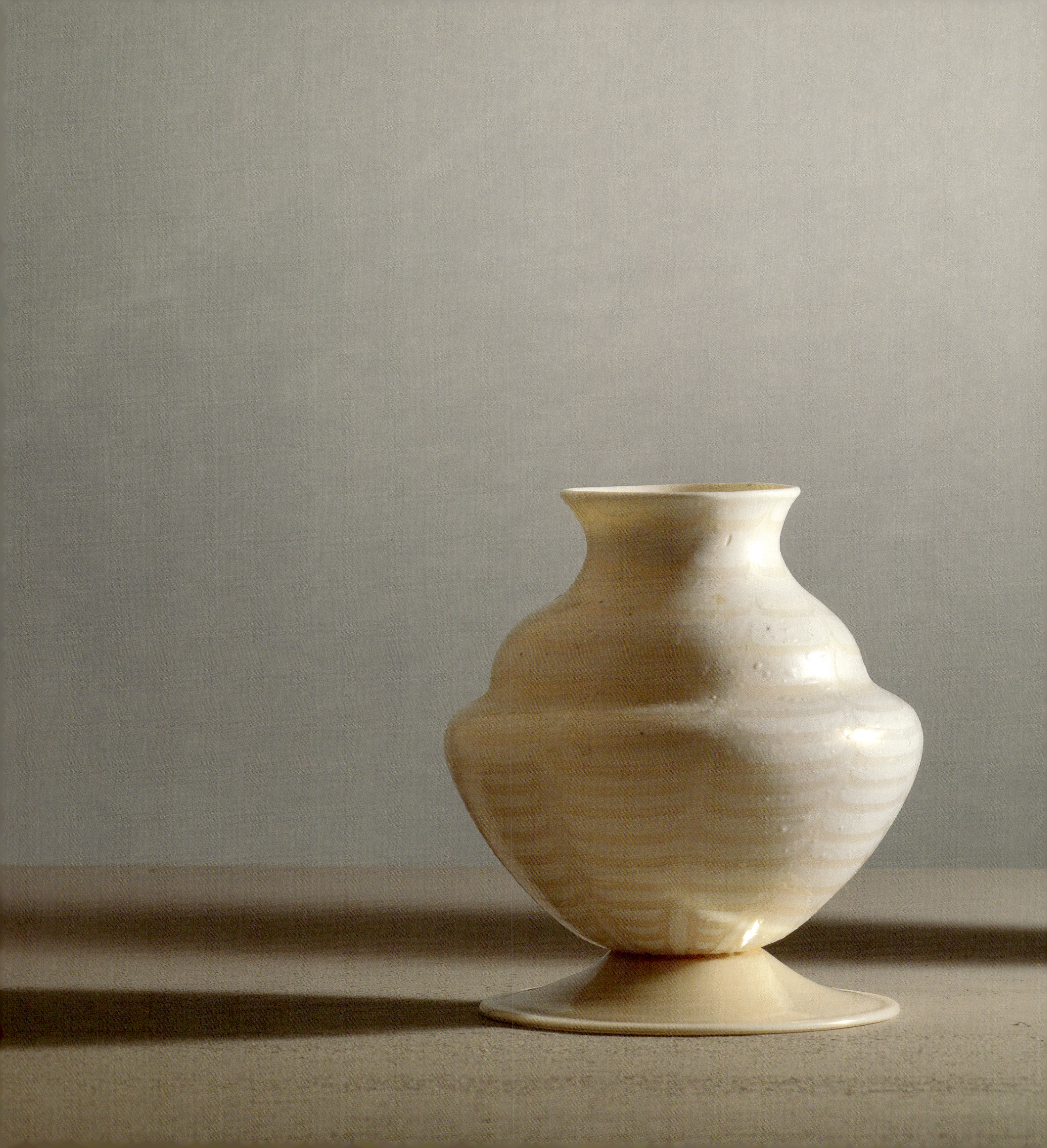

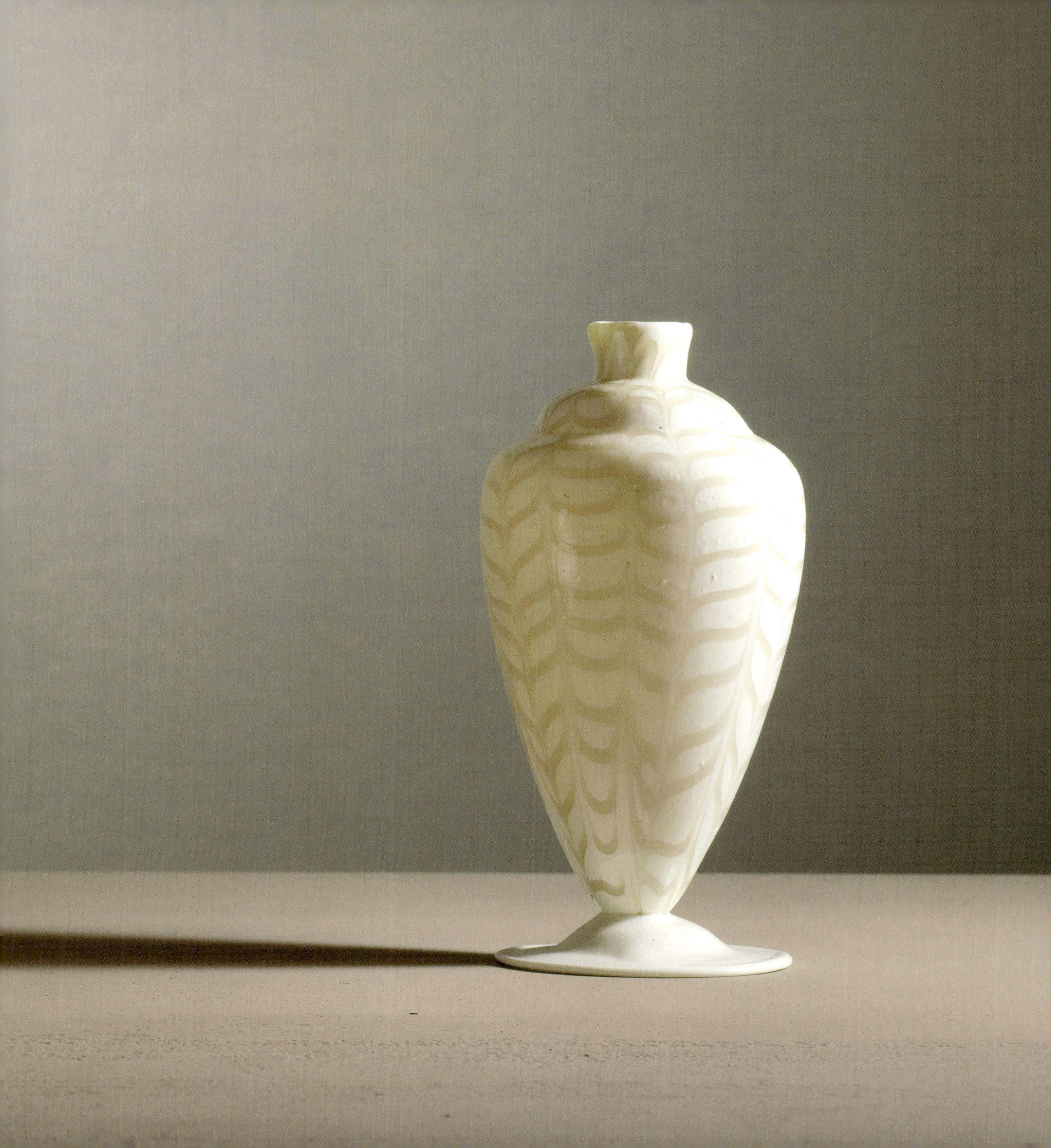

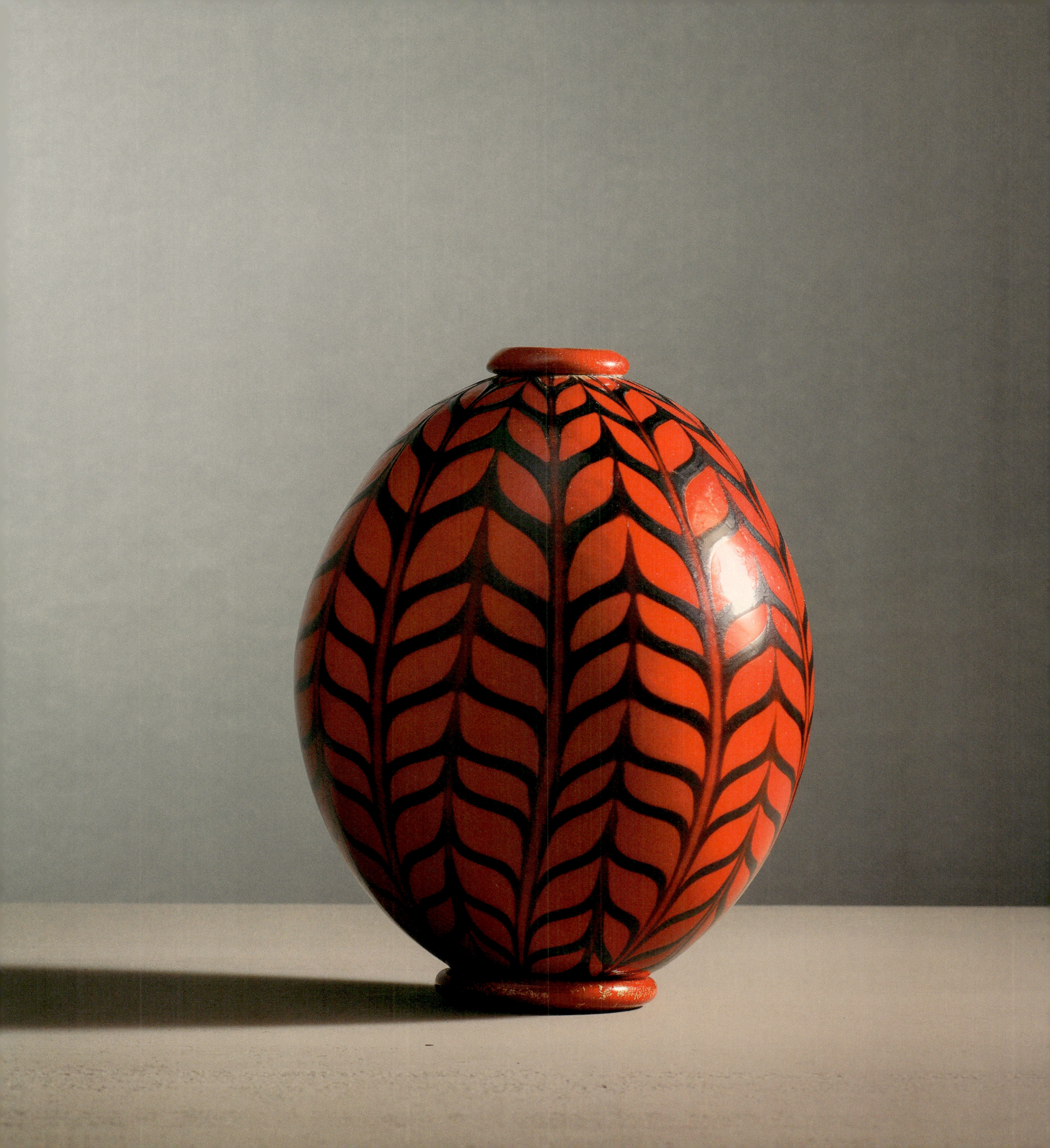

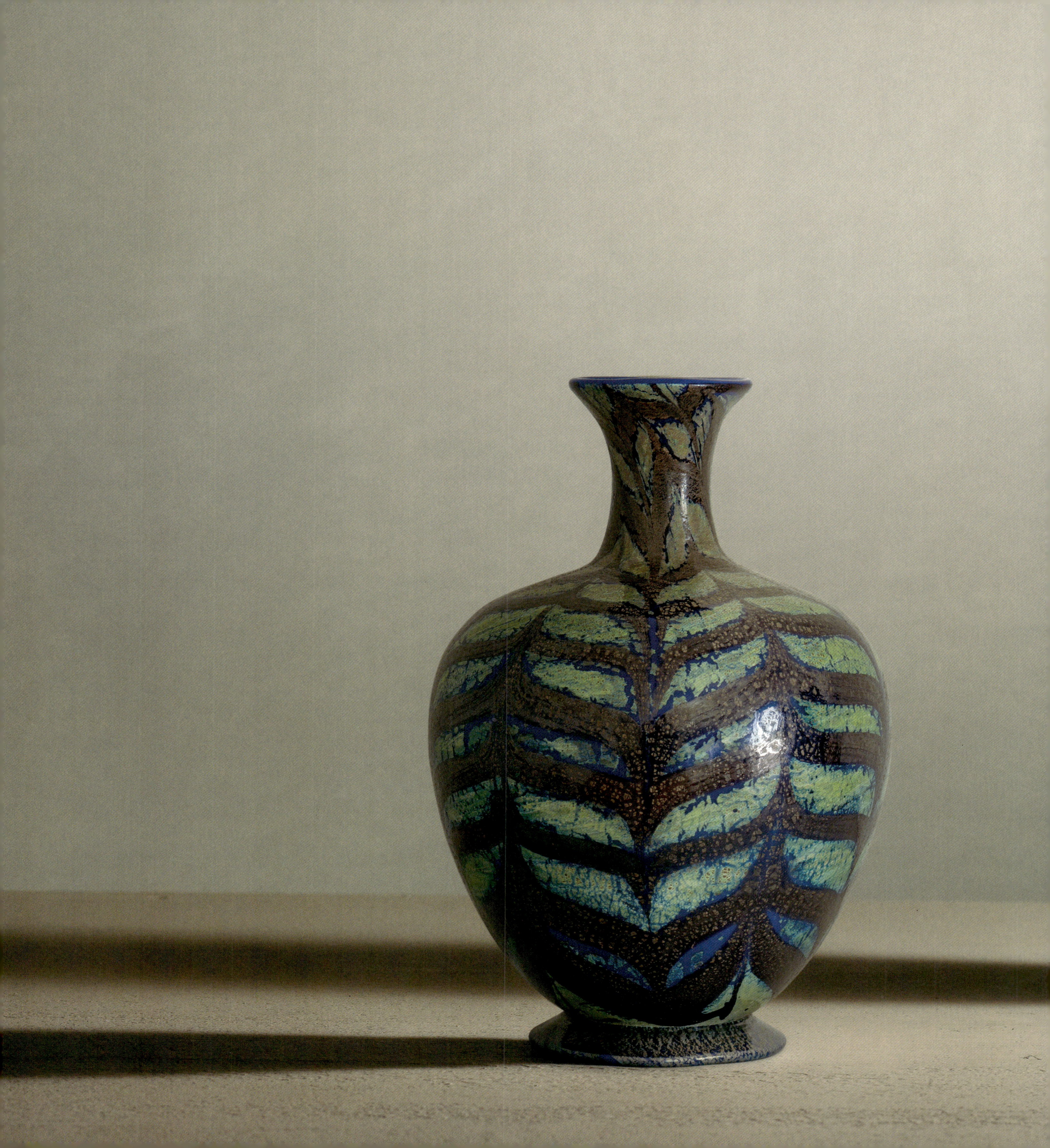

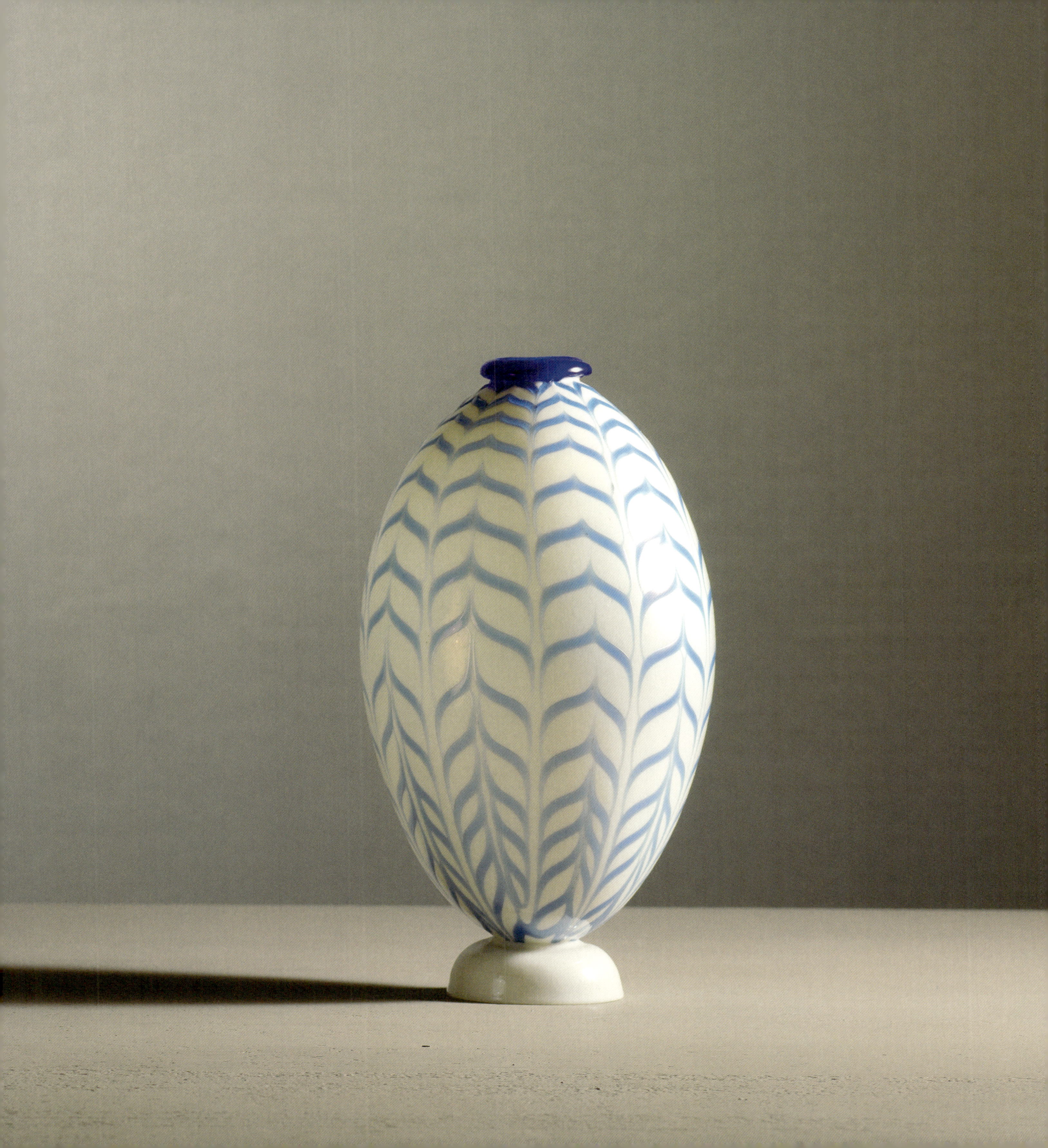

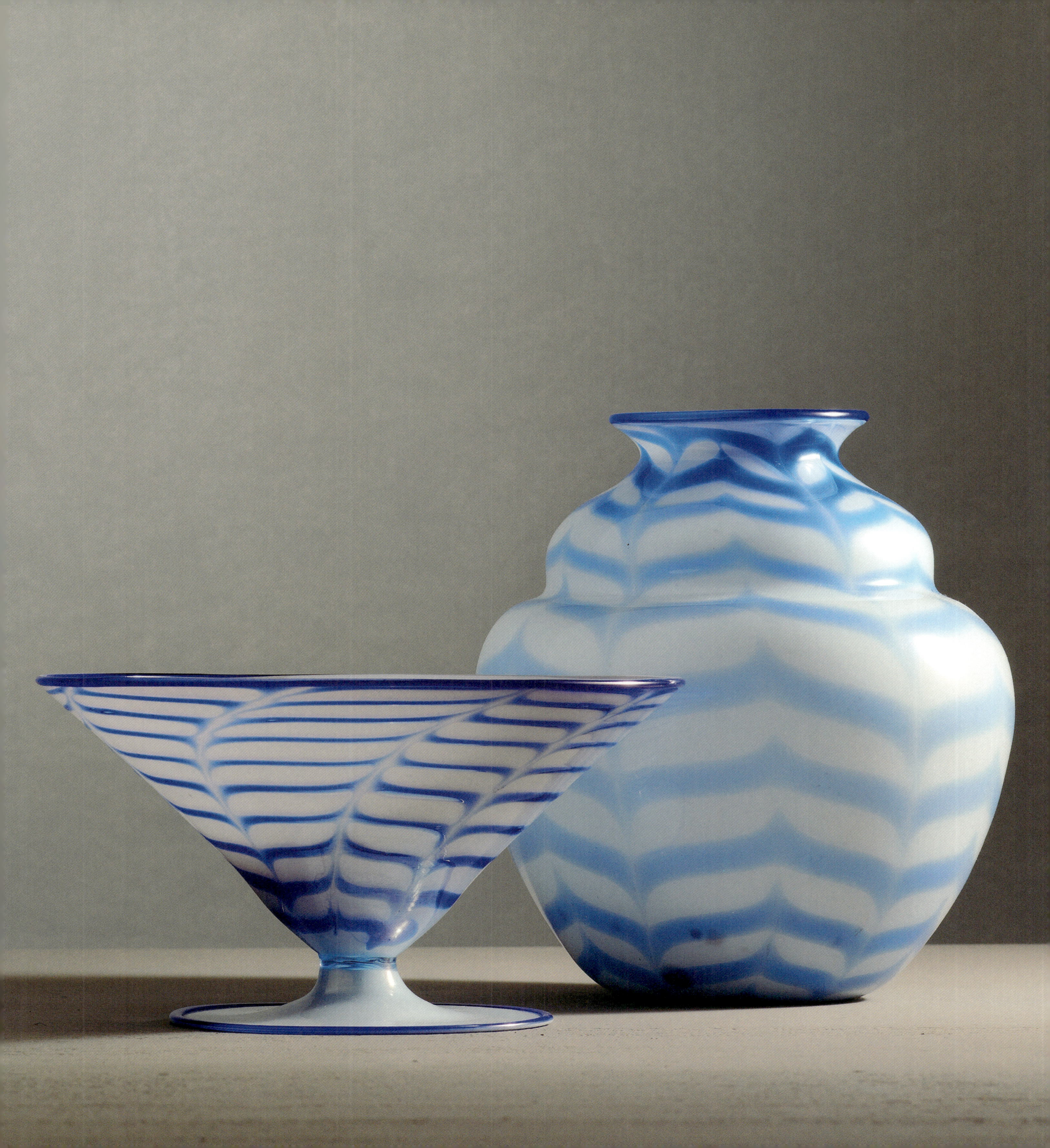

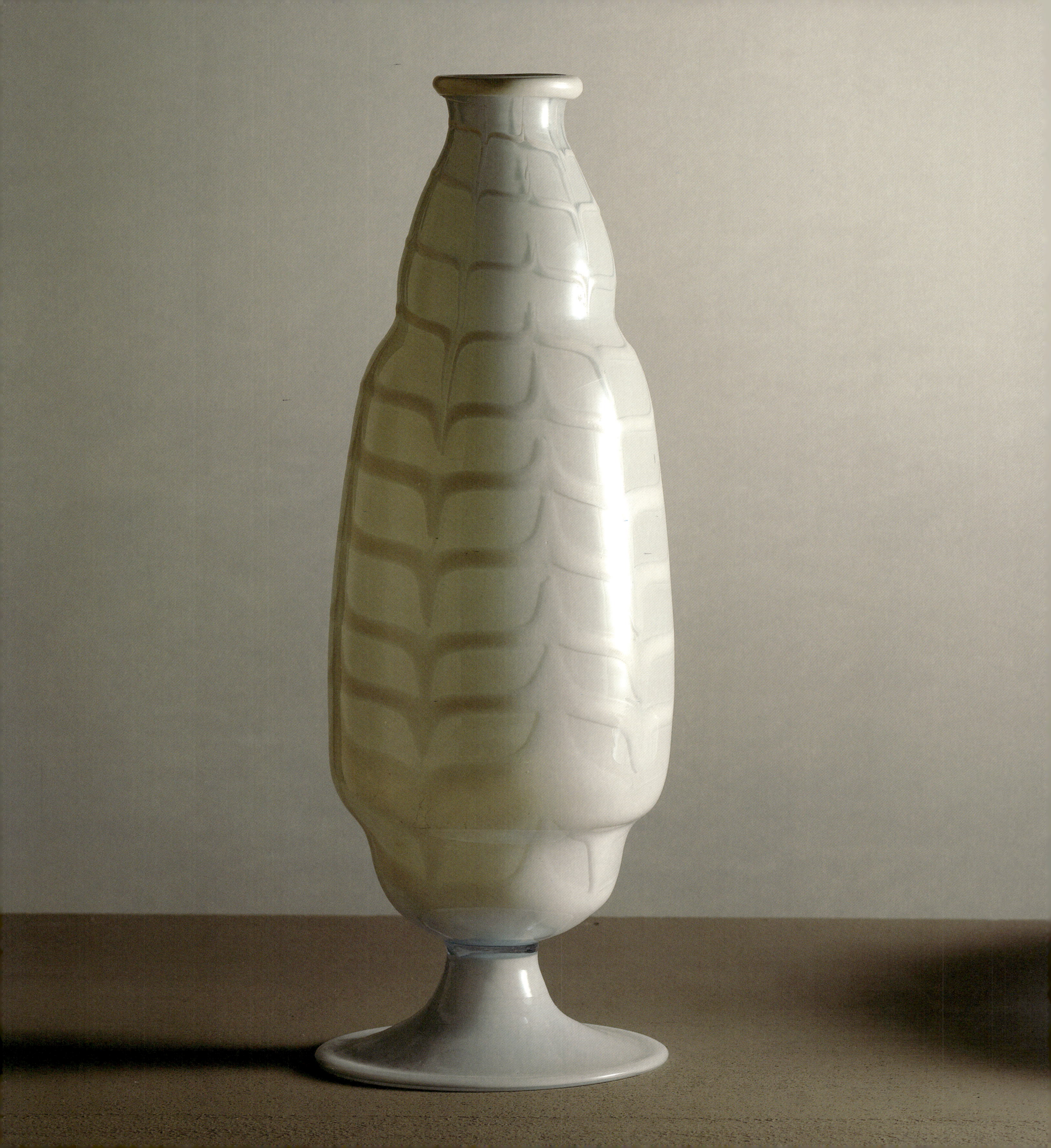

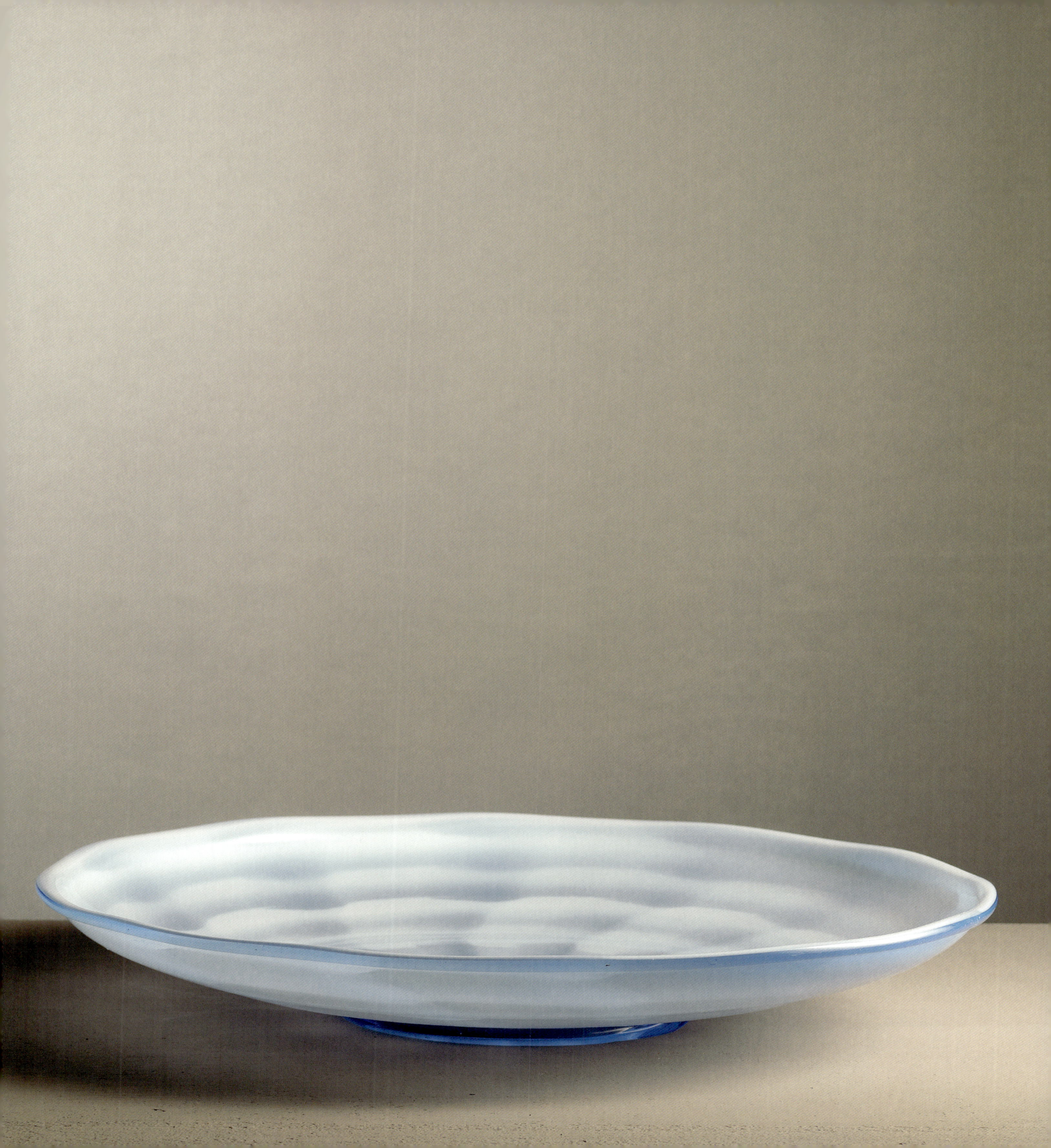

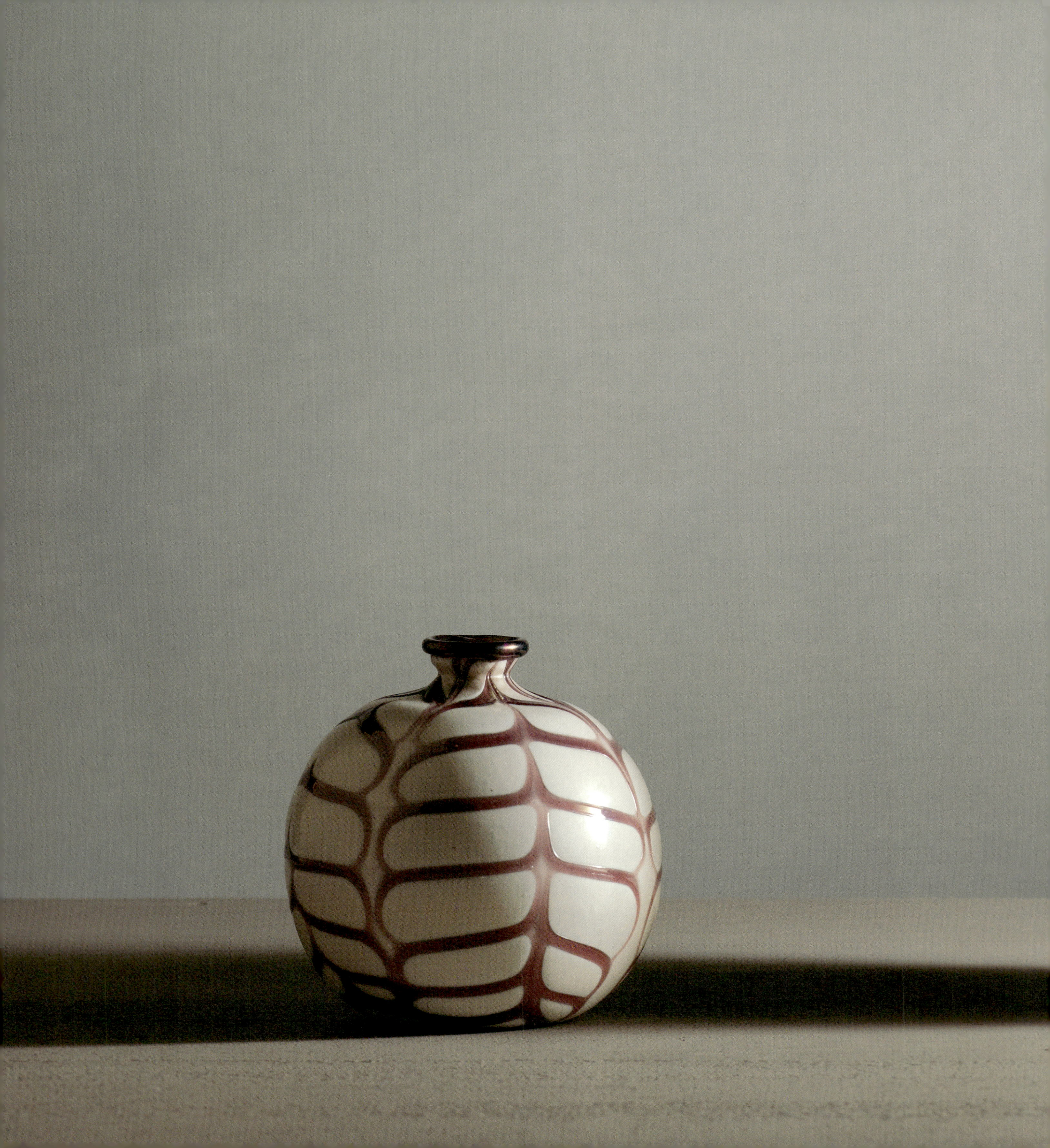

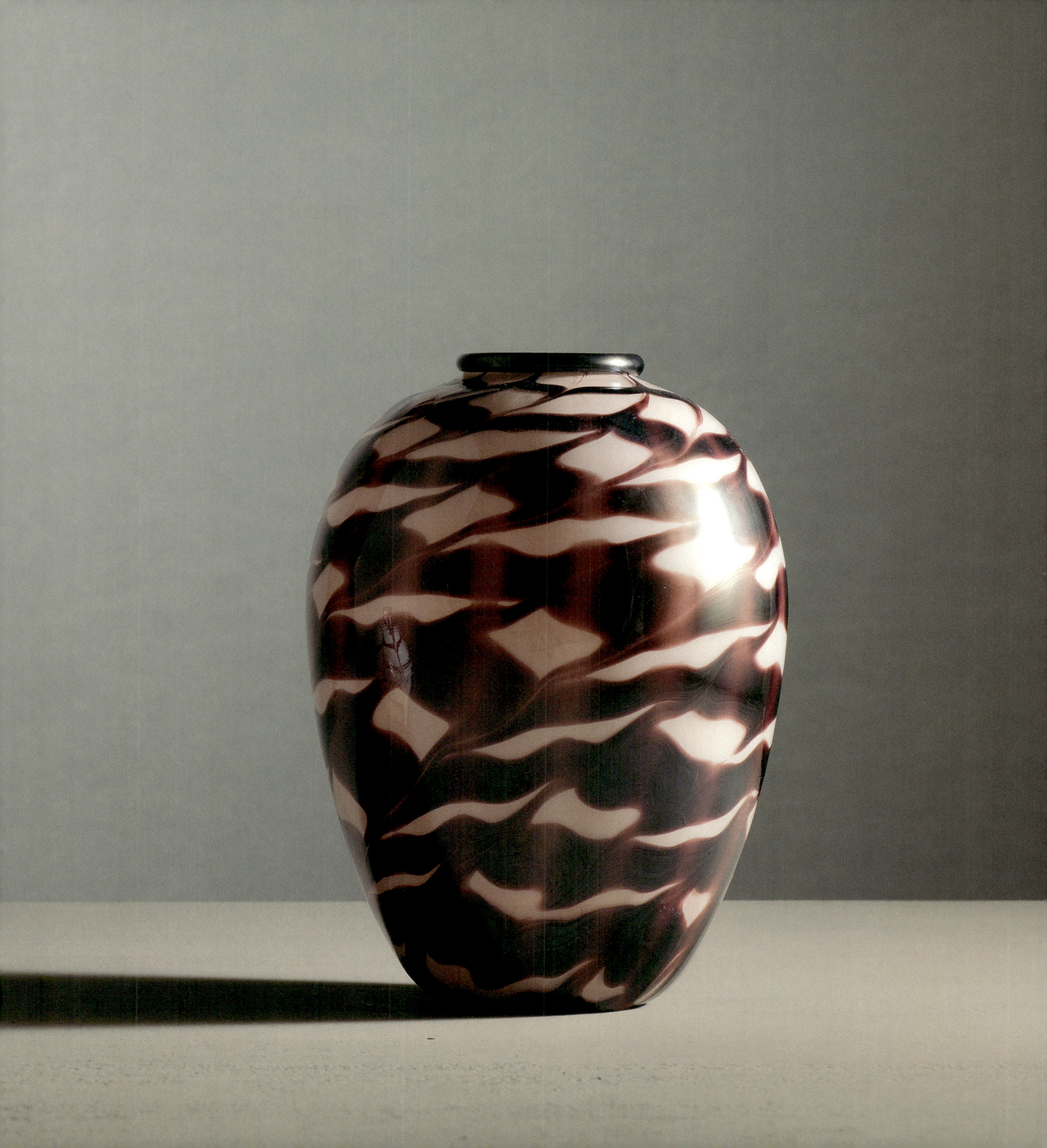

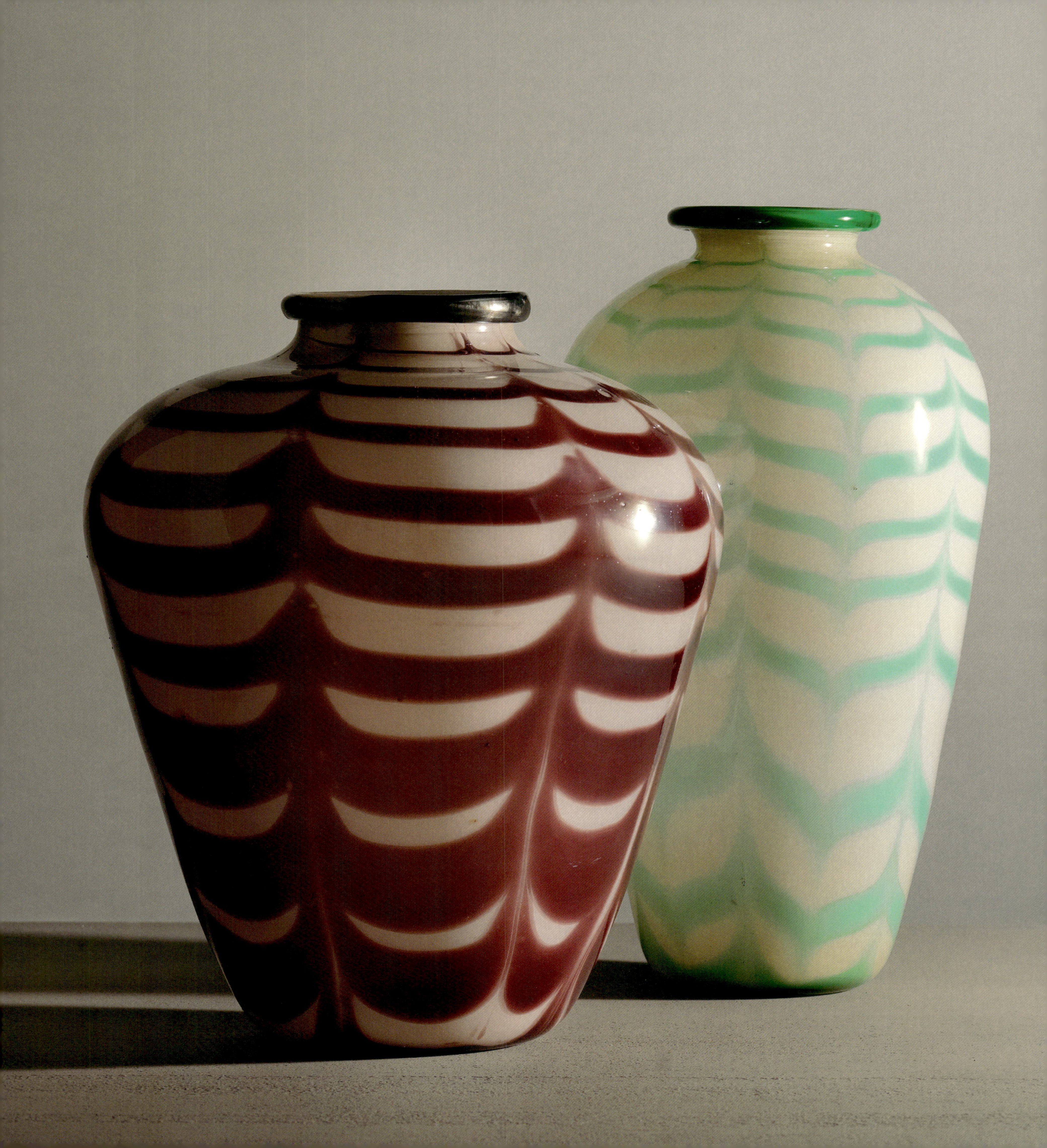

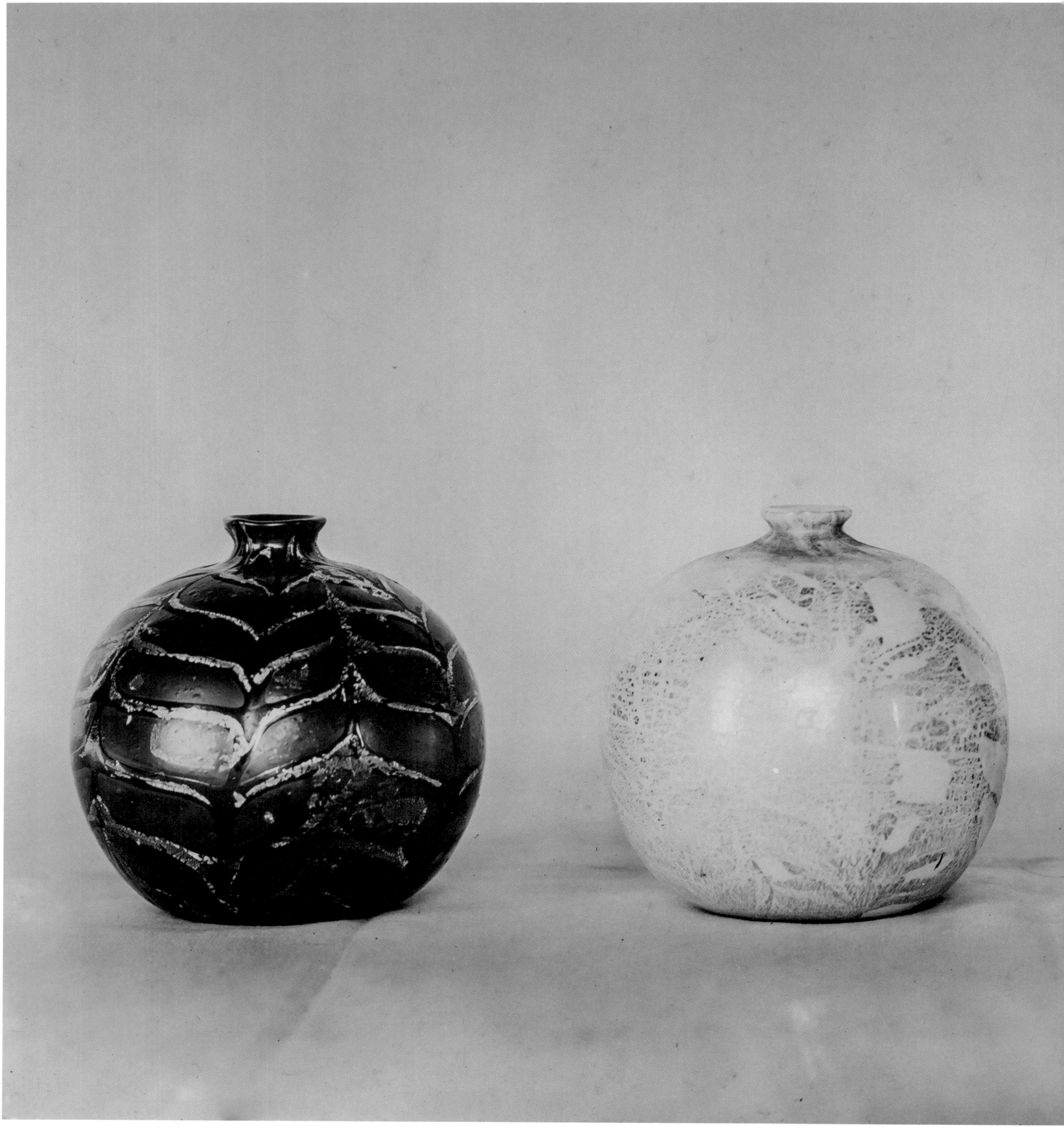

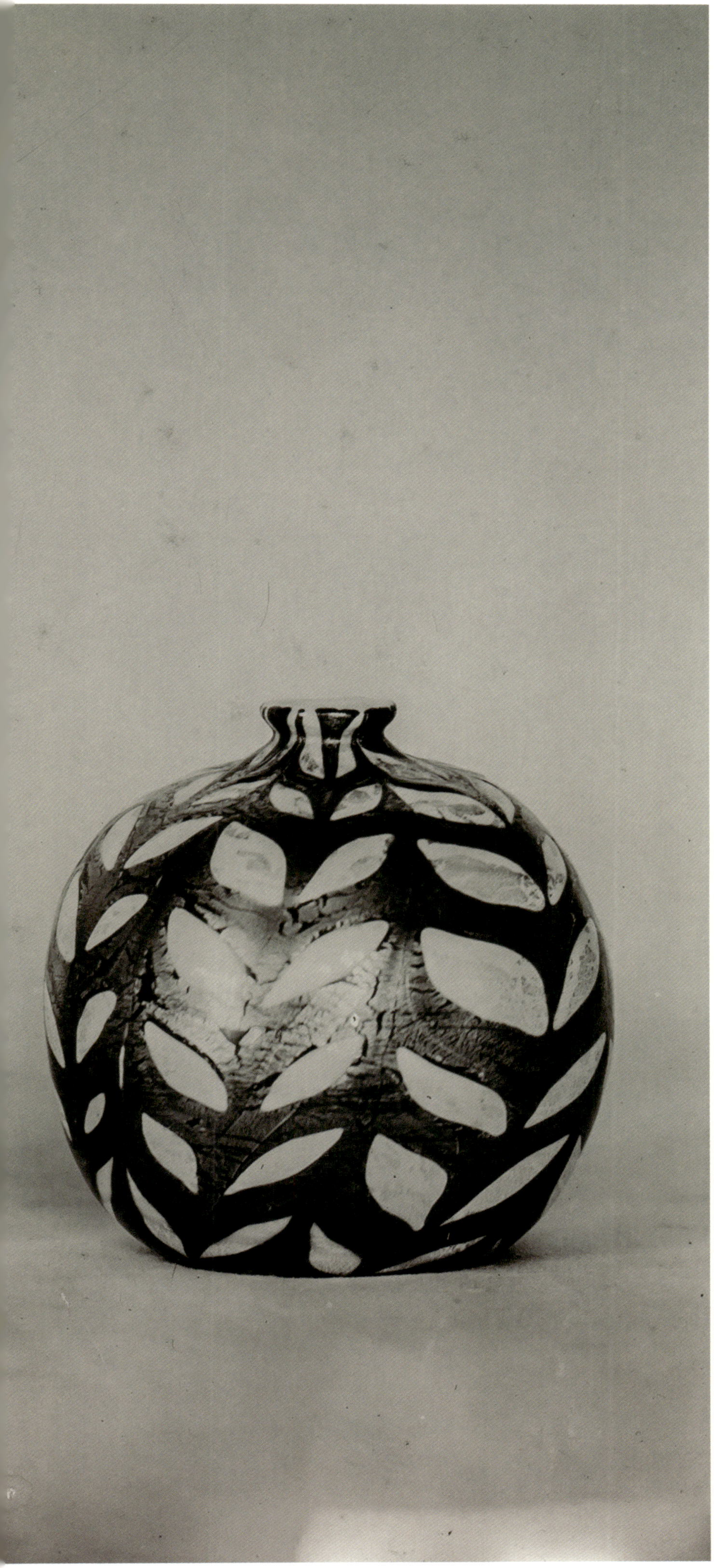

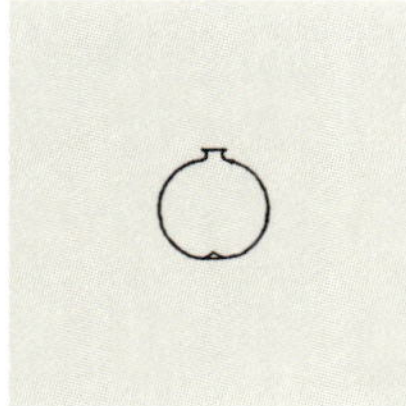

5932

Spherical glass in *lattimo* glass with coral-red *decoro fenicio* and applied gold leaf. A similar example with the same colours was exhibited in 1929 at the Leipzig Trade Fair. Model no. 5932 was also produced in other glass techniques (e.g. *lattimo aurato*, *pasta vitrea*).
1928–29
13 cm high
M.V.M. Cappellin Murano *acid stamp*
Private collection

exhibitions
1929, Milano Fair
1929, Leipzig, Trade Fair

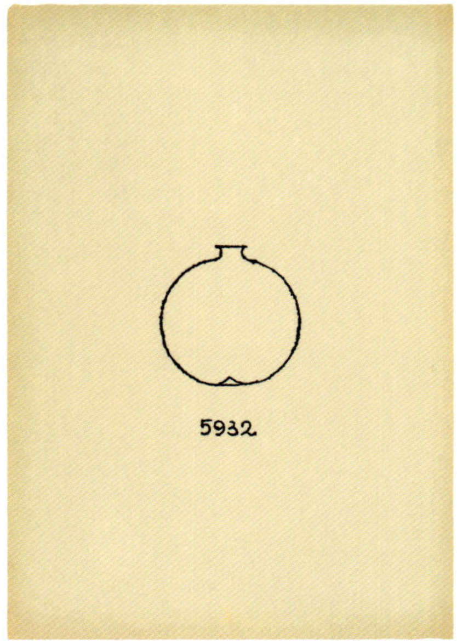

> 5932, catalogue drawing, AMB

<< Vases no. 5932 in glass with a fenicio decoration. Period photograph, c. 1929, CSV

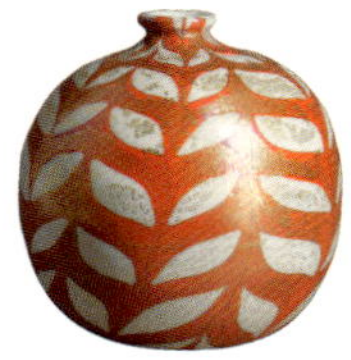

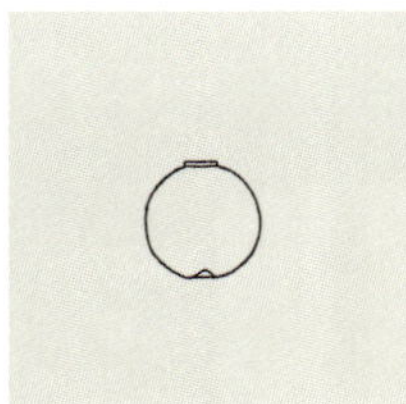

5934

Spherical vase in *lattimo* glass with coral-red *decoro fenicio* and applied gold leaf. Mouth in coral-red *pasta vitrea*. Model no. 5934 was also produced in other glass techniques (e.g. *lattimo aurato*, *filigrana a reticello*, *balloton*, with canes).
1928–29
13 cm high
The Olnick Spanu Collection, New York

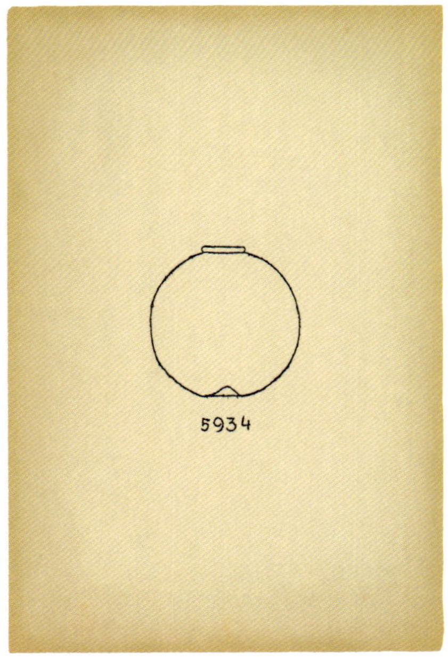

> 5934, catalogue drawing, AMB

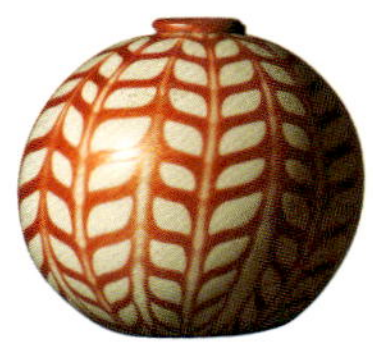

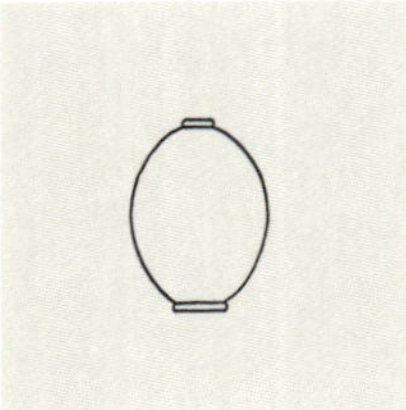

5952

Ovoid vase in coral-red *pasta vitrea* with amethyst *decoro fenicio*. Foot and mouth in coral-red and gold *pasta vitrea*. Model no. 5952 was also produced in other glass techniques (e.g. *lattimo aurato*, *pasta vitrea*).
1928–29
21.5 cm high
Bischofberger collection, Switzerland

>
5952, catalogue drawing, AMB

>
Interior of an apartment on Lungotevere, Rome, architect Giuseppe Capponi, in Domus, *January 1931, p. 27*

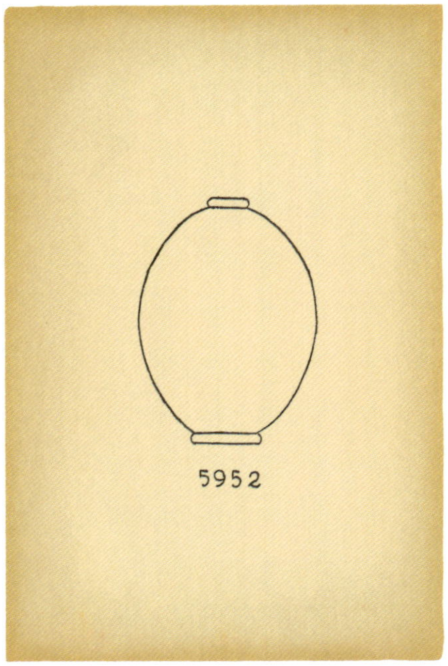

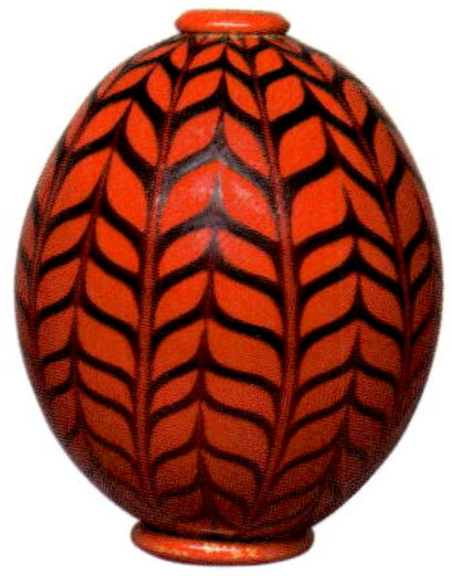

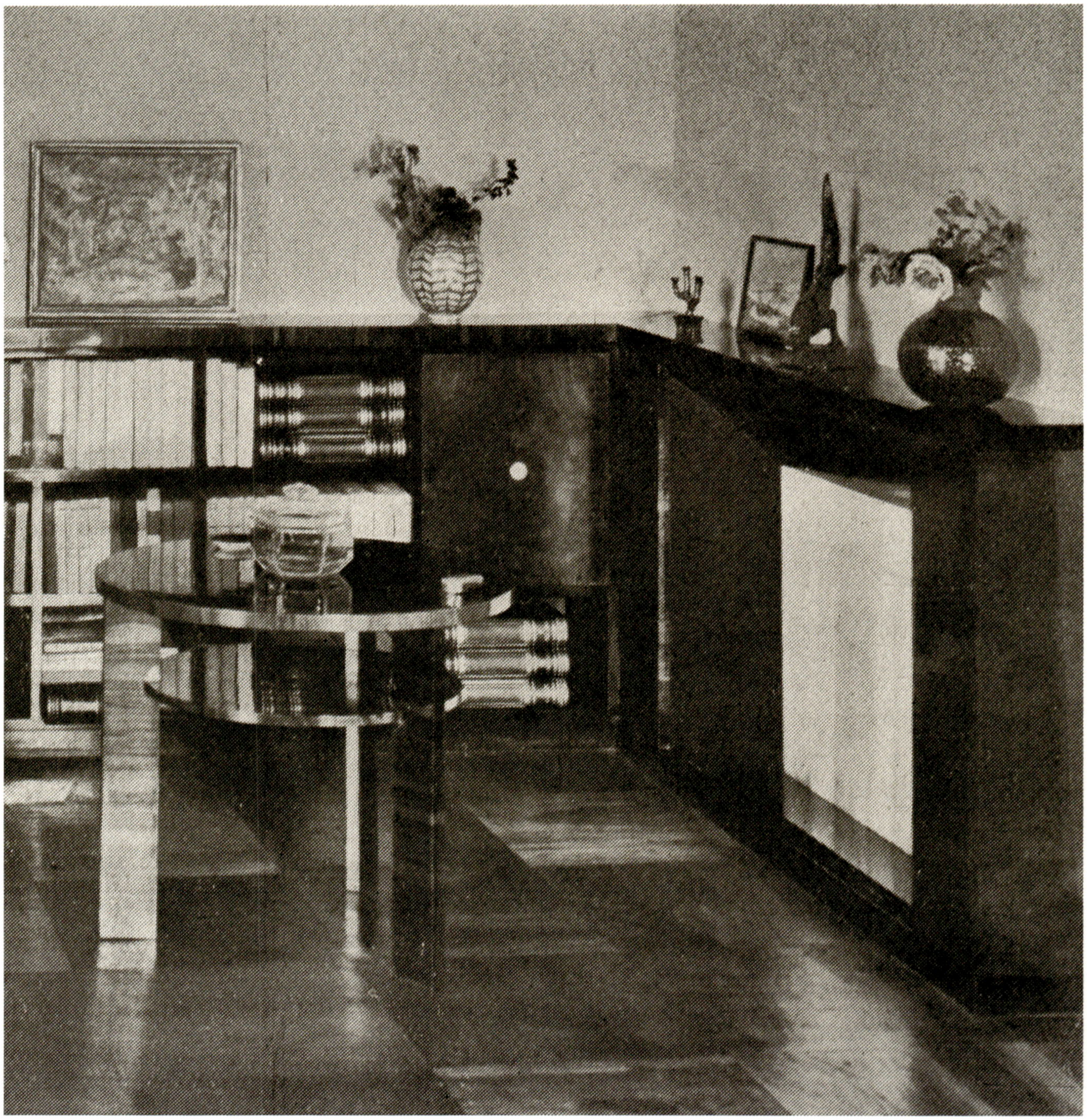

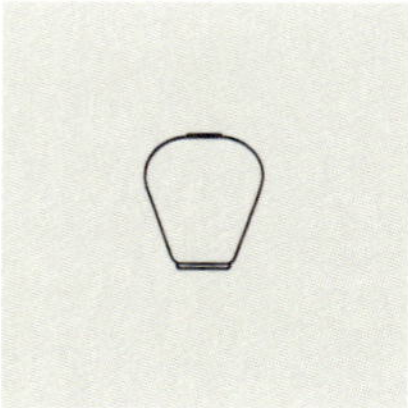

5856

Vase in Savoy-blue glass paste with coral-red *decoro fenicio* and applied gold leaf. Foot and mouth in coral-red *pasta vitrea*. Surface with iridised finish. Model no. 5856 was also produced in other glass techniques (e.g. *lattimo aurato*, *pasta vitrea*).
1928–29
16 cm high
Corning Museum of Glass collection, Corning, NY, USA

>
5856, catalogue drawing, AMB

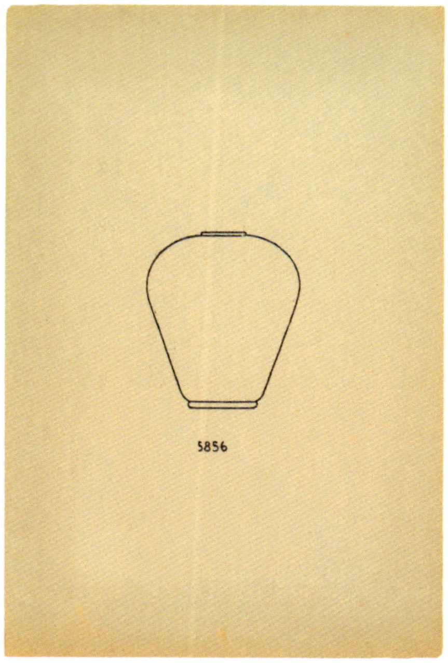

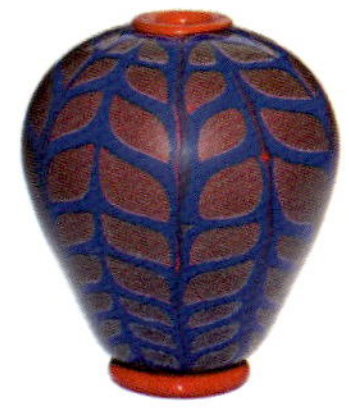

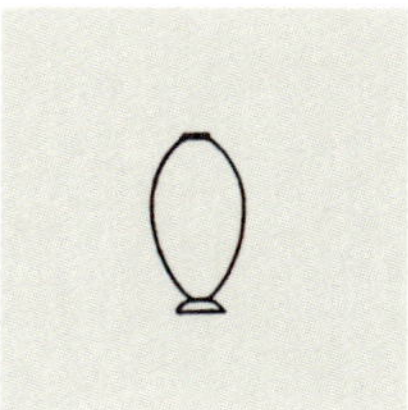

5935

Ovoid vase in fumé cased *lattimo* glass with *lattimo decoro fenicio*. Skull-cap foot in *lattimo* and mouth in fumé trasparent glass. Surface with iridised finish. Model no. 5935 was also produced in other glass techniques (e.g. *lattimo aurato*, cased glass).
1928–29
23 cm high
Private collection, Treviso

>
5935, catalogue drawing, AMH

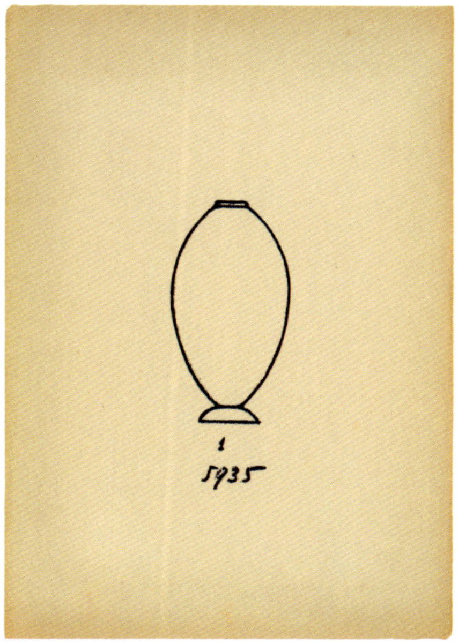

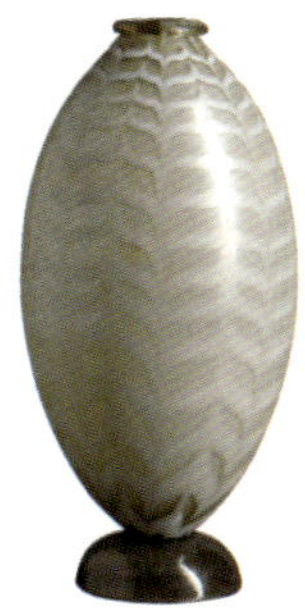

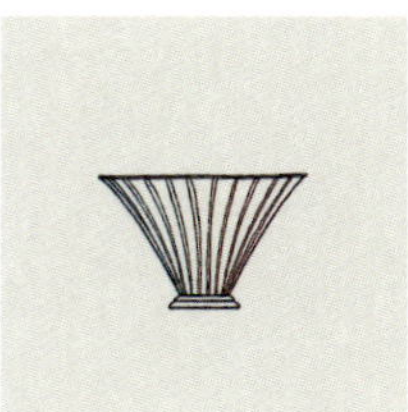

5990

Vase in coral-red *pasta vitrea* with *lattimo decoro fenicio* and applied gold leaf. Gradated foot with double-wound thread, in coral-red and gold *pasta vitrea*. The piece was produced after a model of the vertical bands series.
1930–31
17 cm high; 27.5 cm ø
Private collection

>
5990, catalogue drawing, AMB

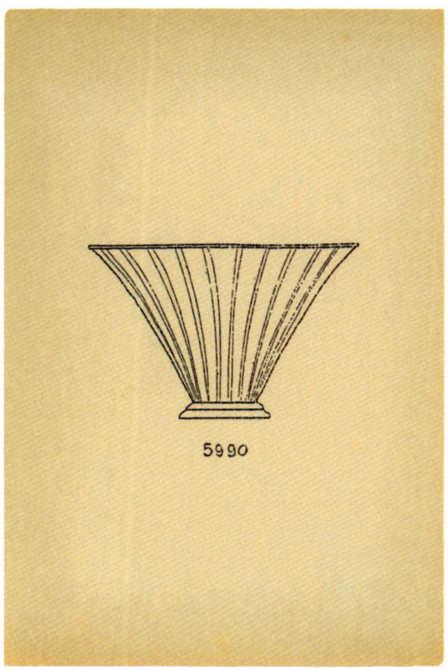

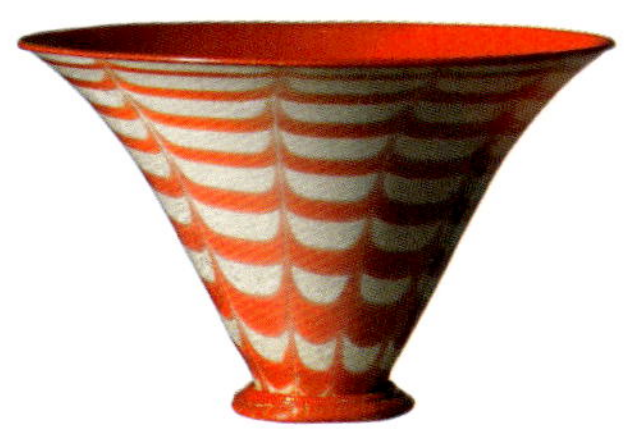

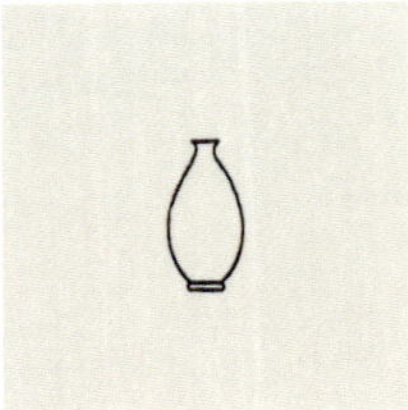

5925

Catalogue drawing for one of the vases in *decoro fenicio* glass documented by a period photograph, AMB. Model no. 5925 was also produced in other glass techniques (e.g. *lattimo aurato*).
1928–29
18 cm high

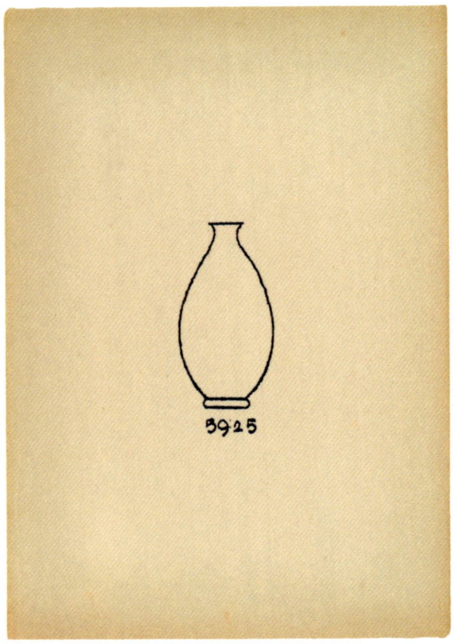

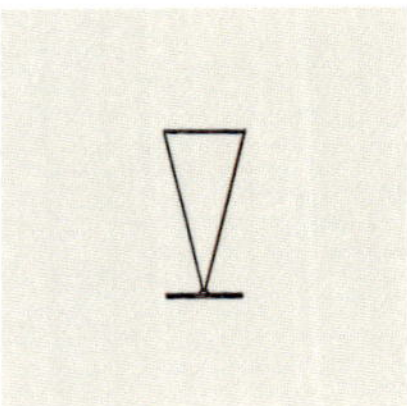

5930

Catalogue drawing for one of the vases in *decoro fenicio* glass documented by a period photograph, AMB. Model no. 5930 was also produced in other glass techniques (e.g. *pasta vitrea*).
1928–29
20 cm high

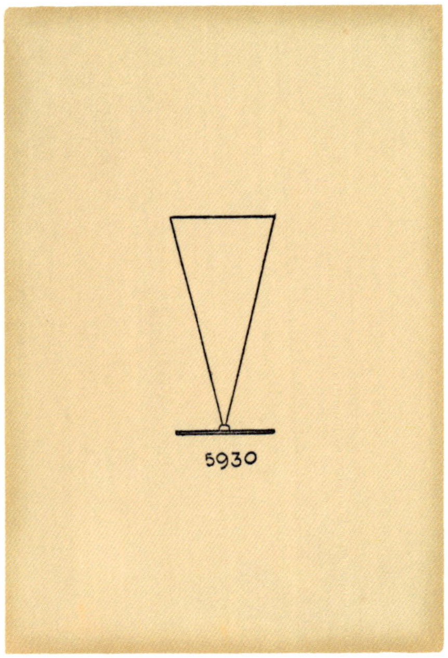

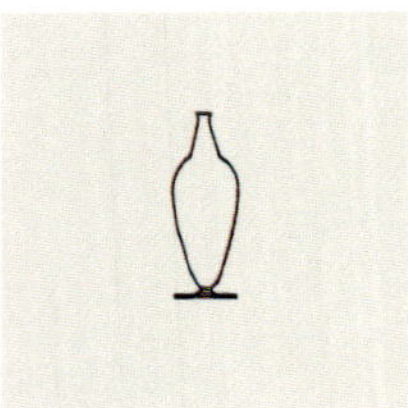

5923

Catalogue drawing for one of the vases in *decoro fenicio* glass documented by a period photograph, AMB. Model no. 5923 was also produced in other glass techniques (e.g. iridised black, *lattimo*).
1928–29
22 cm high

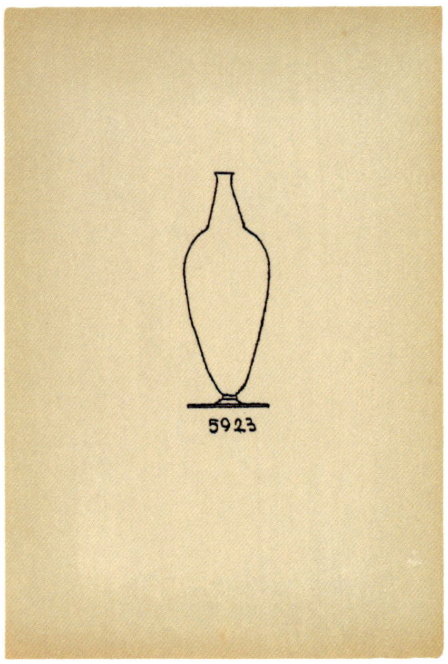

>
Vases in glass with a fenicio decoration. Period photograph, c. 1929, ACSVr

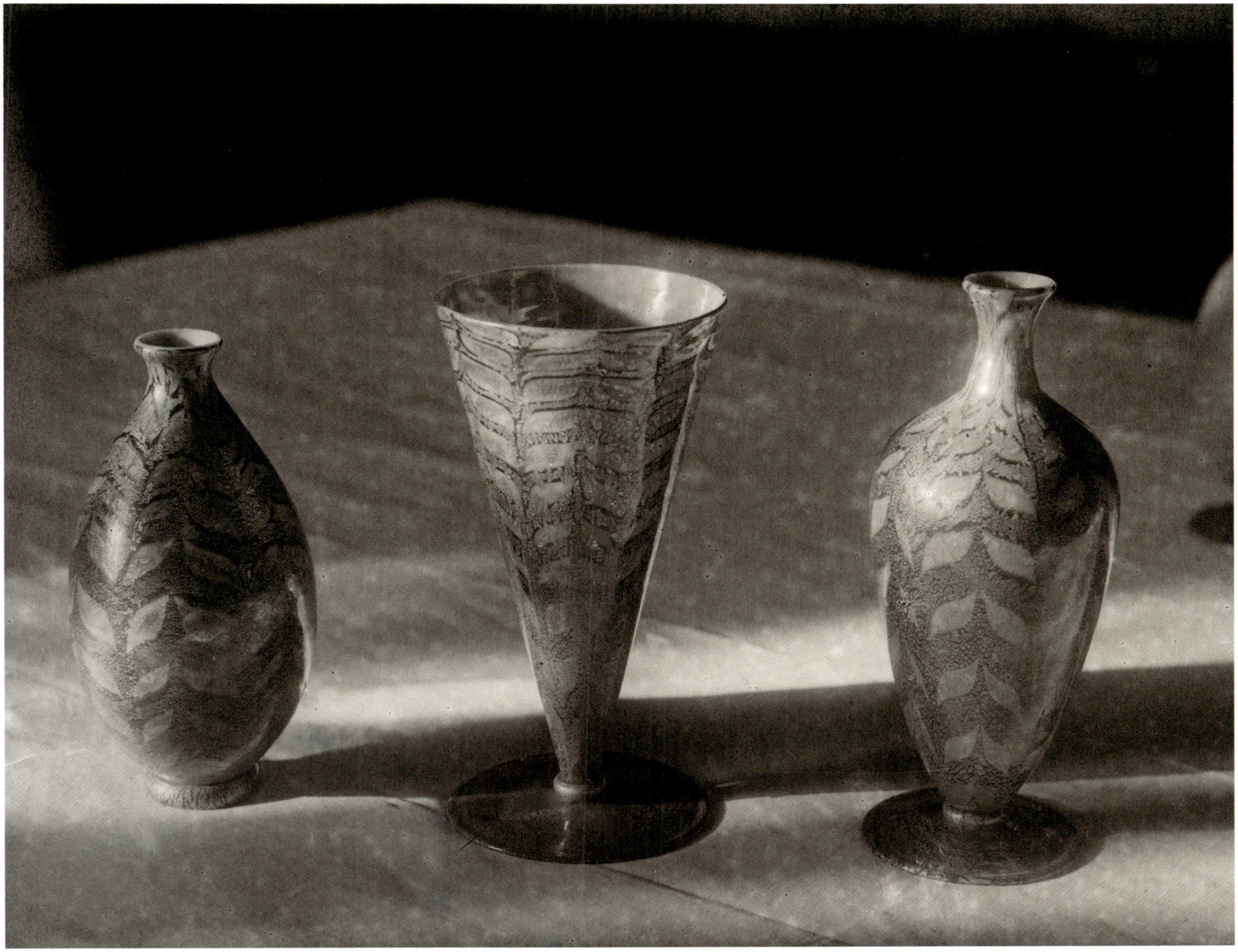

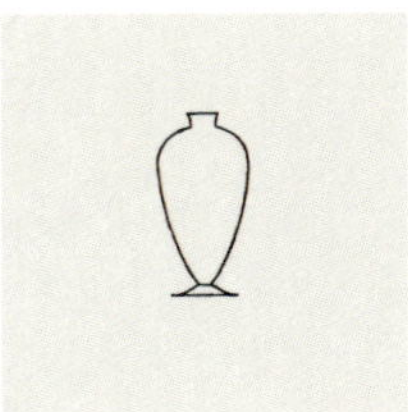

5870

Vase in black glass with light-amethyst *decoro fenicio* and applied gold leaf. Disk-shaped foot in black and gold. Model no. 5870 was also produced in other glass techniques (e.g. transparent, *pasta vitrea*, iridised black, *lattimo*).
1928–29
21 cm high
M.V.M. Cappellin Murano *acid stamp*
Private collection

>
5870, catalogue drawing, AMB

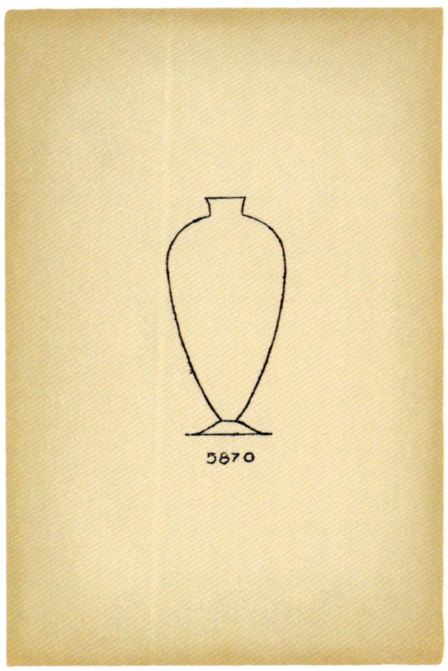

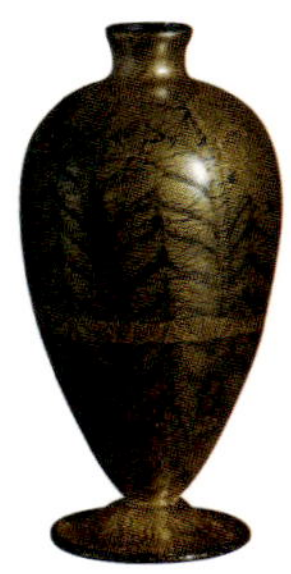

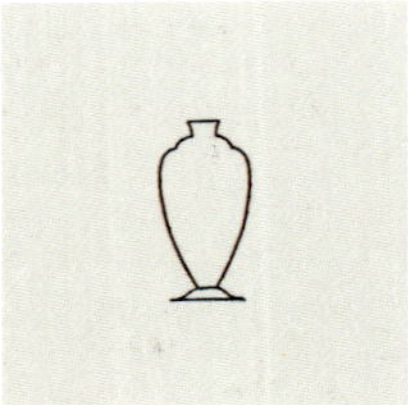

5936

Vase in *lattimo* glass with *pagliesco decoro fenicio*. Disk-shaped foot in *lattimo* glass. Surface with iridised finish. The model is characterised by a double constriction at the base of the neck. No. 5936 was also produced in other glass techniques (e.g. *pasta vitrea*, *lattimo aurato*).
1928–29
21 cm high
M.V.M. Cappellin Murano *acid stamp*
Private collection

>
5936, catalogue drawing, AMB

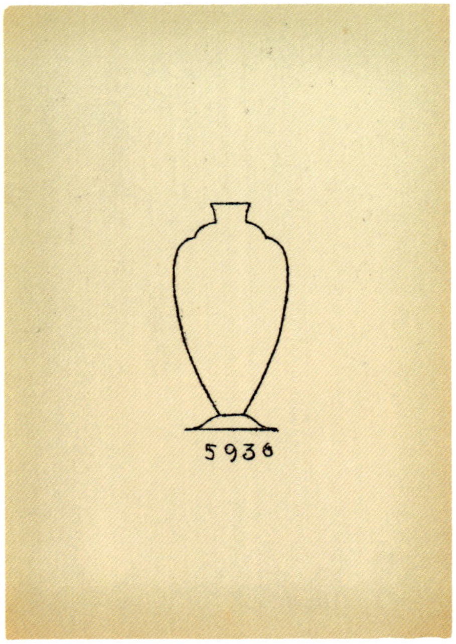

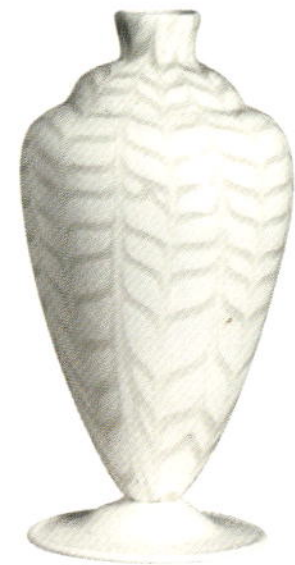

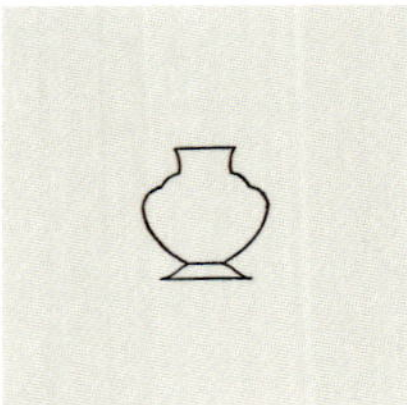

5938

Vase in *lattimo* cased and *pagliesco* glass with light-amethyst *decoro fenicio*. Disk-shaped foot in *lattimo* cased and *pagliesco* glass. Surface with iridised finish. The model is characterised by a double constriction at the base of the neck. No. 5938 was also produced in other glass techniques (e.g. transparent, *pasta vitrea*, silvered black).
1928–29
15.5 cm high
Francesca and Massimo Valsecchi collection, Palazzo Butera

>
5938, catalogue drawing, AMB

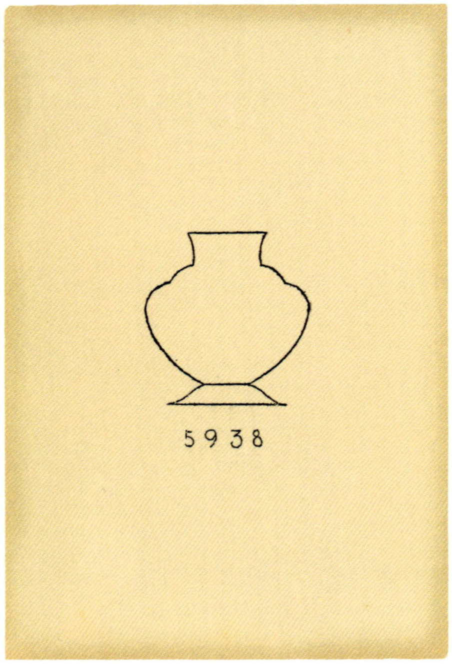

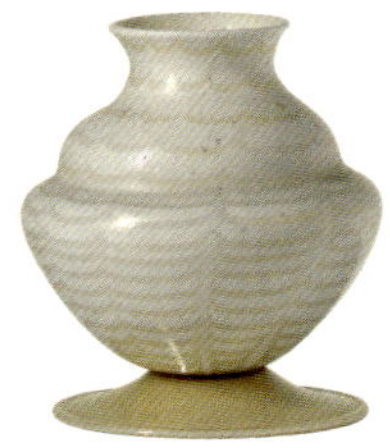

n.i.

Vase in *lattimo* glass with grey *decoro fenicio*. Disk-shaped foot and mouth in grey cased glass. Surface with iridised finish. The model is characterised by the constrictions at the extremities of the body.
1928–29
46 cm high
M.V.M. Cappellin Murano *acid stamp*
Private collection

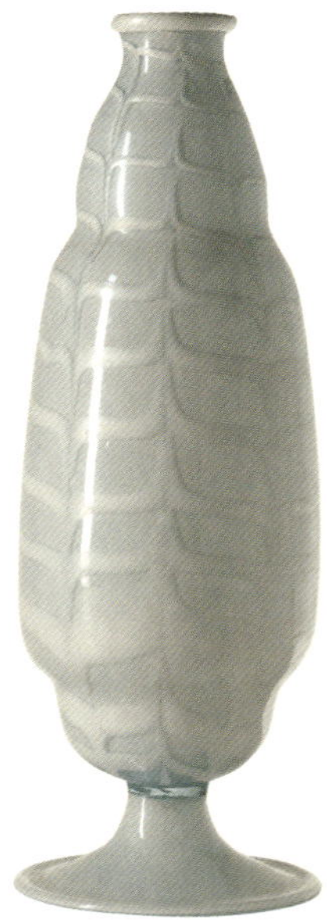

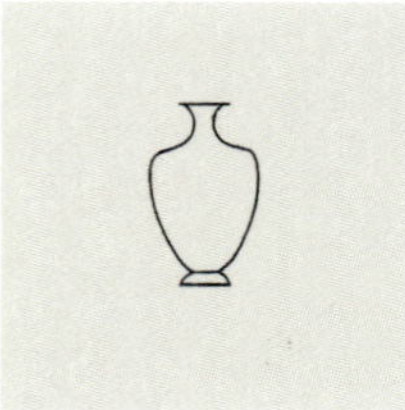

5947

Vase in *lattimo* glass with light-blue *decoro fenicio*. Foot in *lattimo* cased and light-blue glass. Surface with iridised finish. Model no. 5947 was also produced in other glass techniques (e.g. *pasta vitrea*, iridised black).
1928–29
20 cm high
M.V.M. Cappellin Murano *acid stamp*
Fred Schroeder and Allison Whiting collection, New York

>
5947, catalogue drawing, AMB

>
'From our collection of contemporary Lighting Fixtures. This Lamp of white opaque glass', *Cappellin Glassware advertising insert in* Gift and Art Shop, *January 1931, detail*

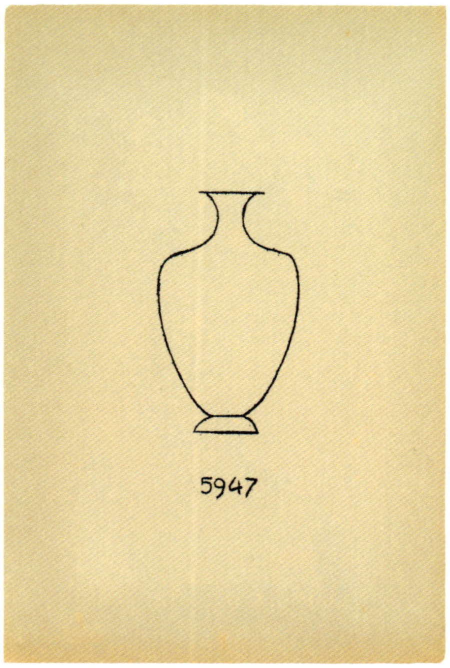

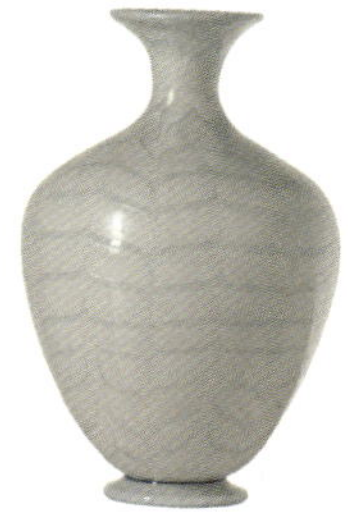

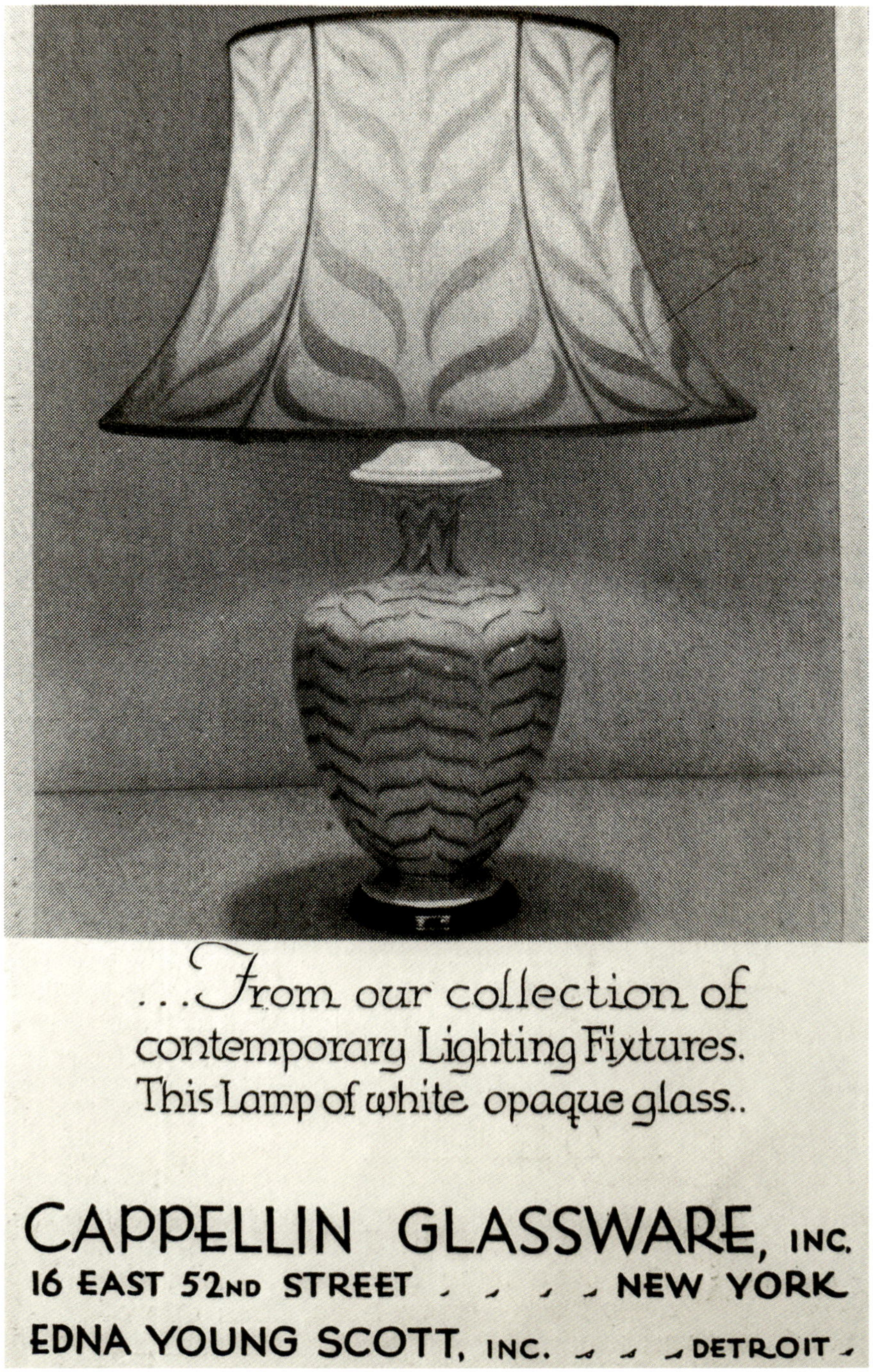

>
Vases in glass with decoro fenicio. *Period photograph, c. 1929, AMB*

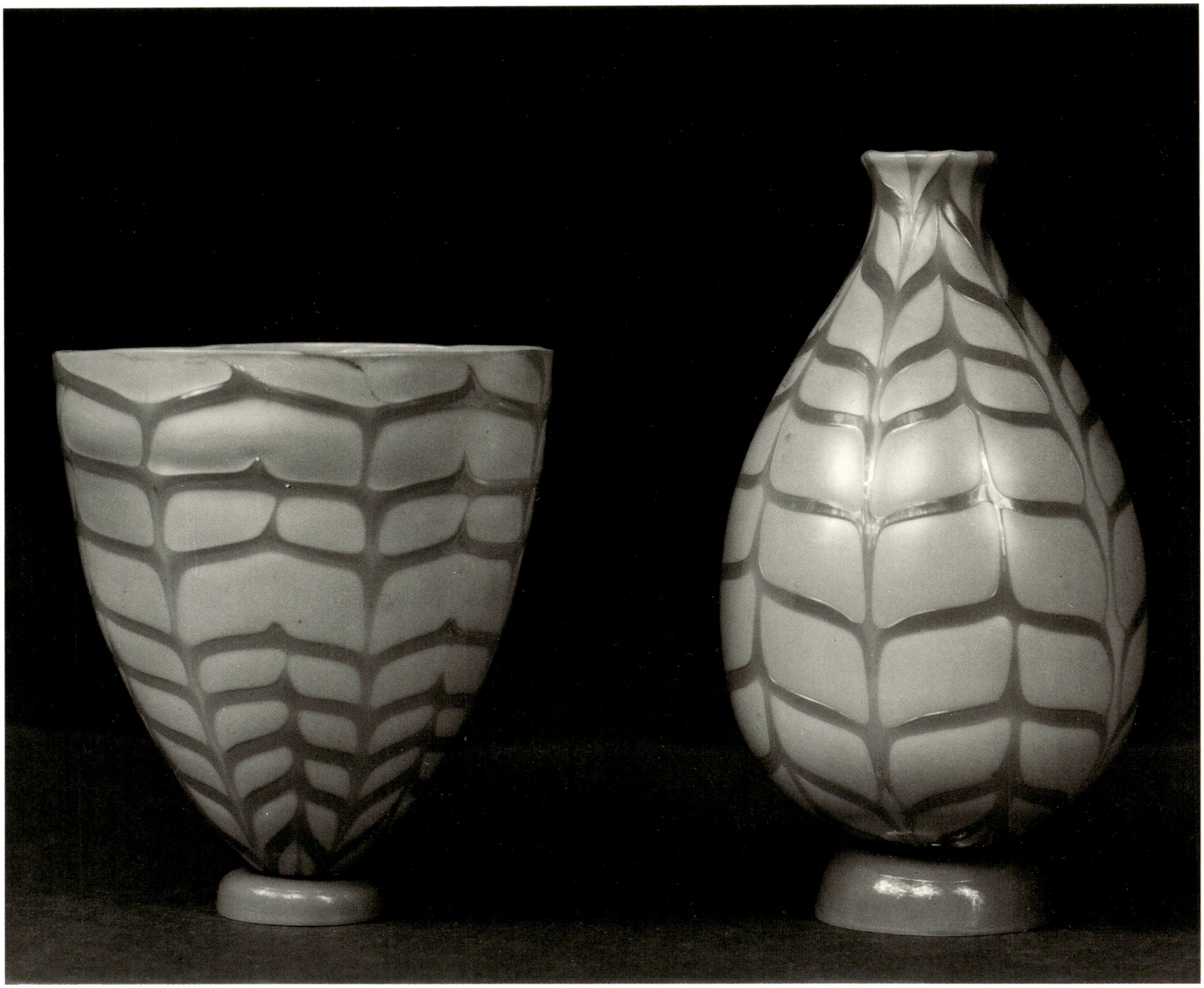

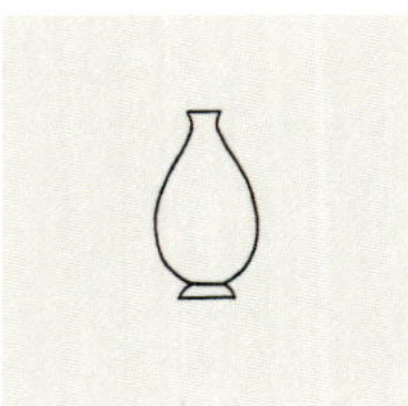

5946

Vase in *lattimo* glass with *decoro fenicio* fumé. Skull-cap foot in *lattimo* glass. Surface with iridised finish. A similar example was exhibited in 1929 at the Leipzig Trade Fair. Model no. 5946 was also produced in other glass techniques (e.g. *lattimo aurato*, iridised black).
1928–29
22 cm high
Private collection

exhibitions
1929, Leipzig, Trade Fair

>
5946, catalogue drawing, AMB

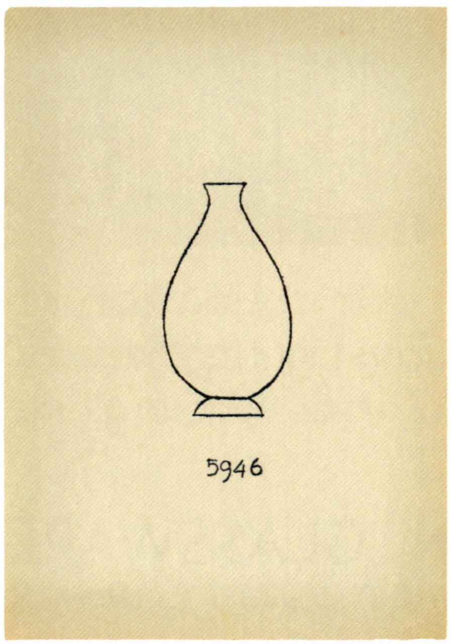

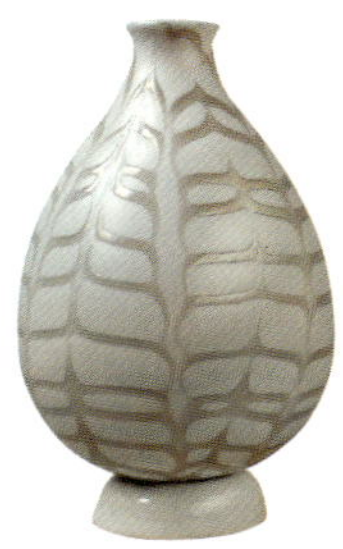

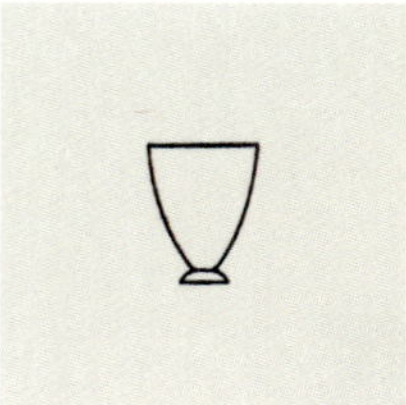

5937

Catalogue drawing for one of the vases in *decoro fenicio* glass documented by a period photograph, AMH. An example in *lattimo* with amethyst *decoro fenicio* was exhibited in 1929 at the Leipzig Trade Fair. Model no. 5937 was also produced in other glass techniques (e.g. *balloton*, iridised black).
1928–29
17 cm high

exhibitions
1929, Leipzig, Trade Fair

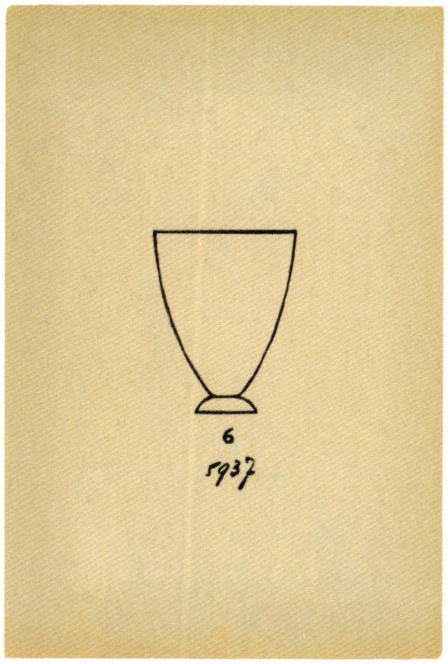

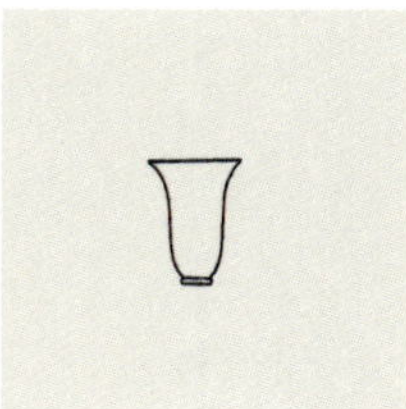

5924

Vase in *lattimo* glass with *pagliesco decoro fenicio*. Foot in *lattimo* cased and *pagliesco* glass. Surface with iridised finish. Model no. 5924 was also produced in other glass techniques (e.g. *lattimo aurato*, *pasta vitrea*).
1928–29
15.5 cm high
M.V.M. Cappellin Murano *acid stamp*
Private collection

›
5924, catalogue drawing, AMB

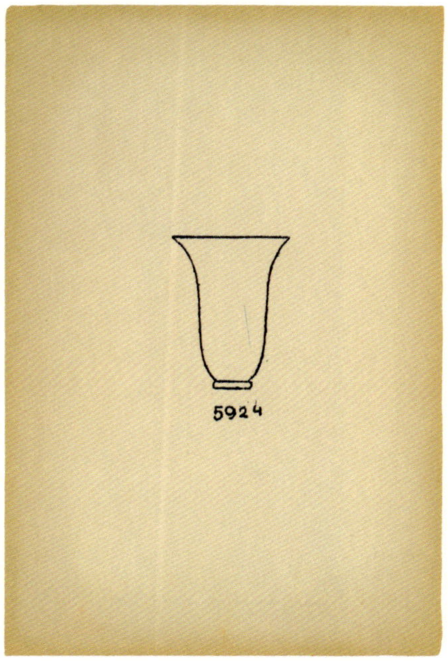

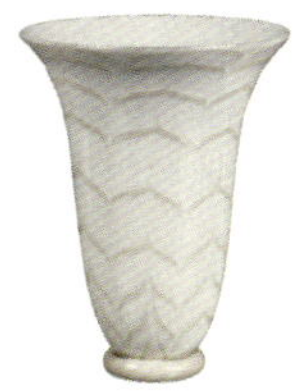

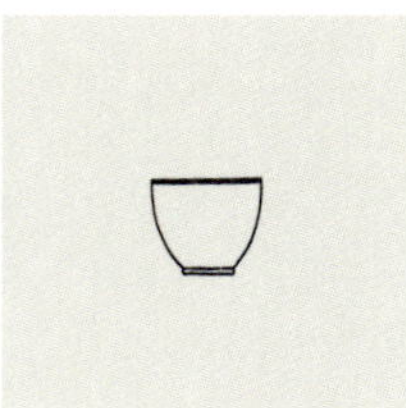

5931

Vase in *lattimo* glass with crystal and gold *decoro fenicio*. Foot in *lattimo* glass. Surface with iridised finish. Model no. 5931 was also produced in other glass techniques (e.g. *lattimo*).
1928–29
12 cm high
Private collection

›
5931, catalogue drawing, AMB

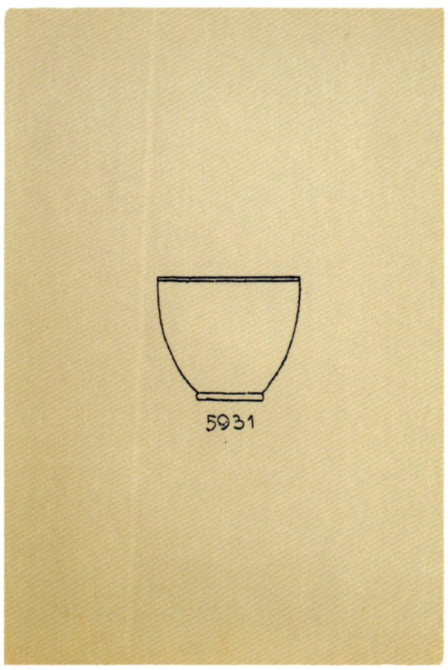

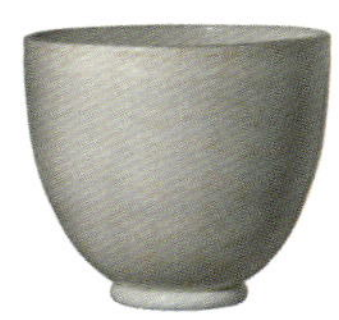

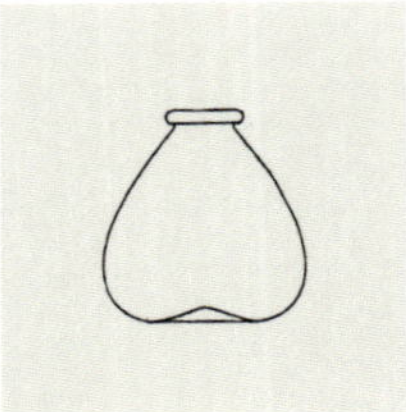

5183

Vase in *lattimo* glass with light-blue *decoro fenicio*. Mouth in light-blue glass. Surface with iridised finish. The vase was produced after a model present in the catalogue of the glassworks and designed by Vittorio Zecchin (1921–25). A similar example was exhibited in 1929 at the Milano Fair.
1928–29
26 cm high
M.V.M. Cappellin Murano *acid stamp*
The Olnick Spanu Collection, New York

exhibitions
1929, Milano Fair
1930, Milan, Fiera

>
5183, working drawing, CSV

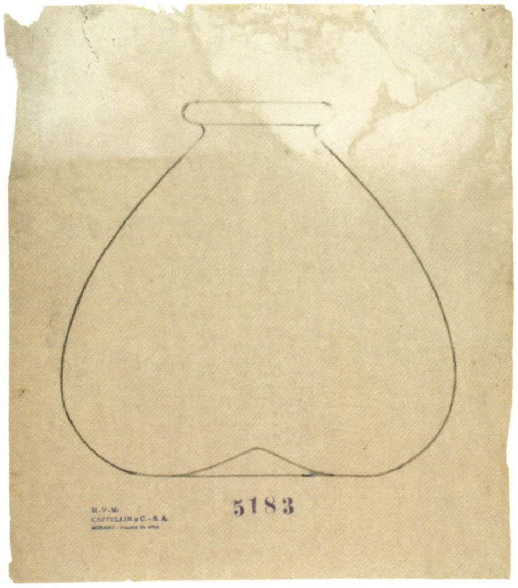

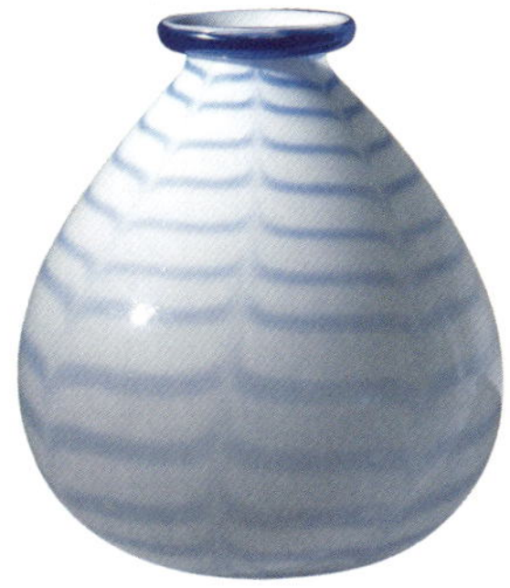

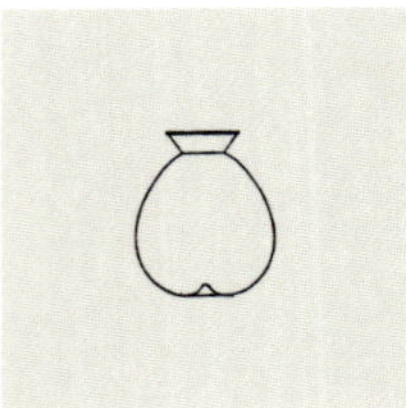

5761

Vase in *lattimo* glass with amethyst *decoro fenicio*. Thread around the mouth in amethyst glass. Surface with iridised finish. Model no. 5761 was also produced in other glass techniques (e.g. transparent).
1928–29
19 cm high
M.V.M. Cappellin Murano *acid stamp*
The Olnick Spanu Collection, New York

>
5761, catalogue drawing, AMB

>
Vas no. 5761 in glass with decoro fenicio *in the American catalogue of Cappellin Glassware, undated, detail, AHL*

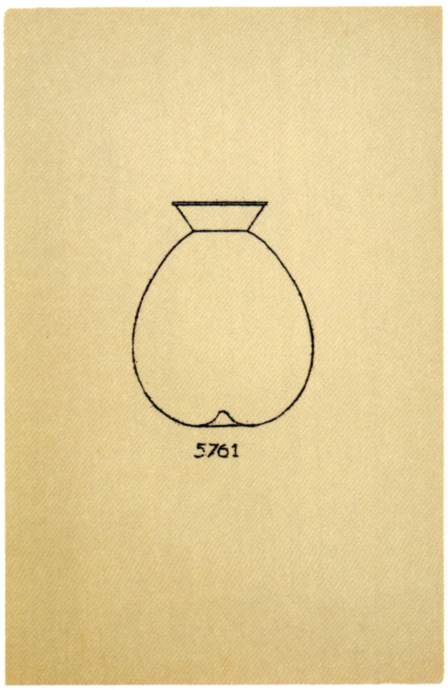

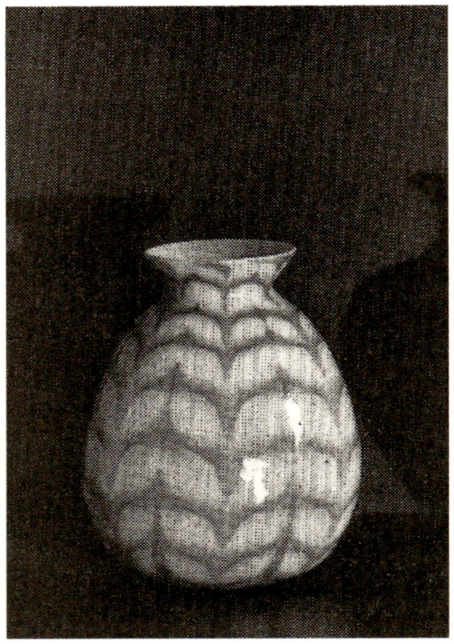

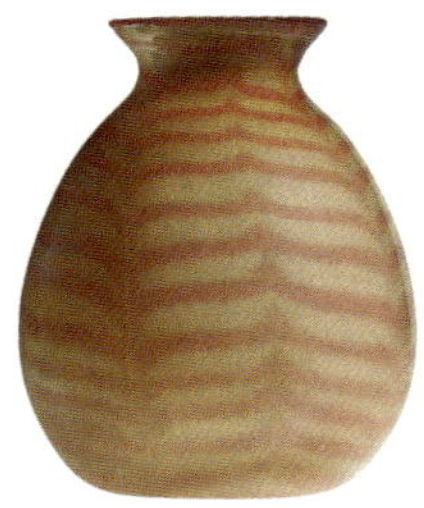

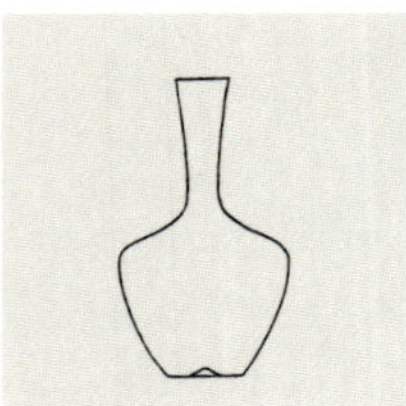

5964

Vase no. 5964 in *decoro fenicio* glass documented in the American photographic catalogue of the Cappellin Glassware, undated, detail, AHL.
c. 1929
35 cm high

>
5964, catalogue drawing, AMA

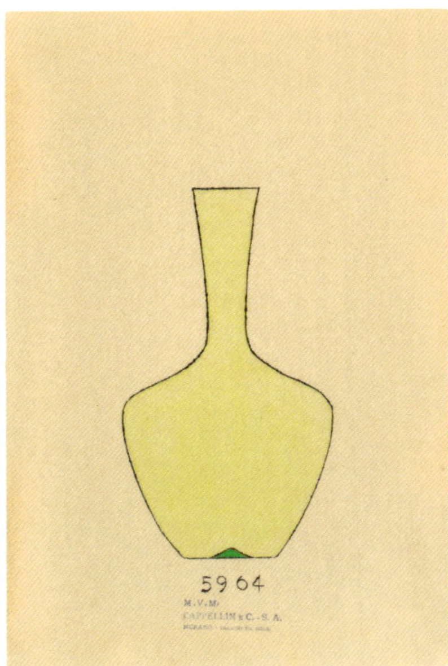

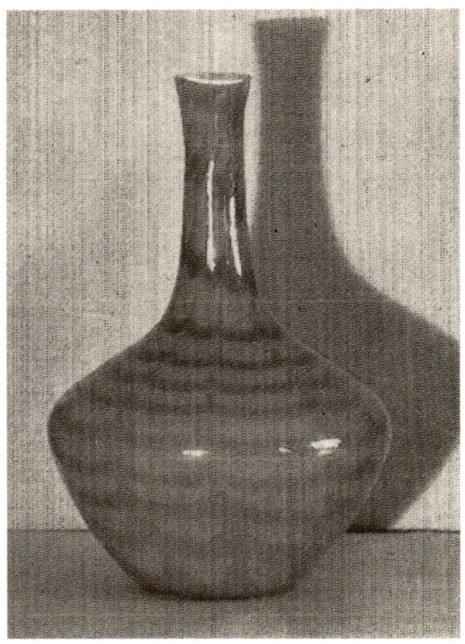

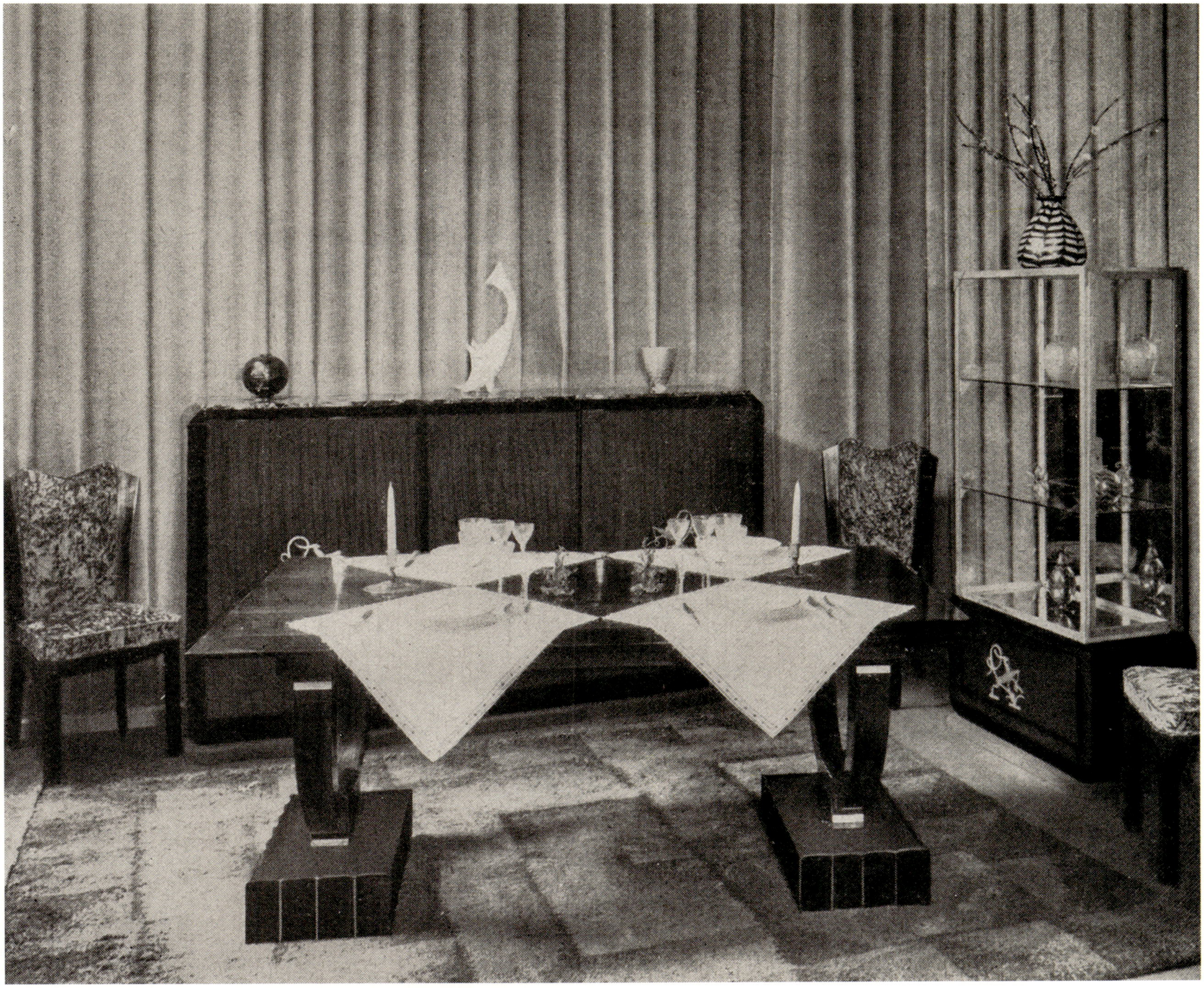

›
Cappellin glass pieces at the stand of the 'Salone del mobilio s.a.' of the Trade Fair of Milan in 1930 in La casa bella, *April 1930, p. 69*

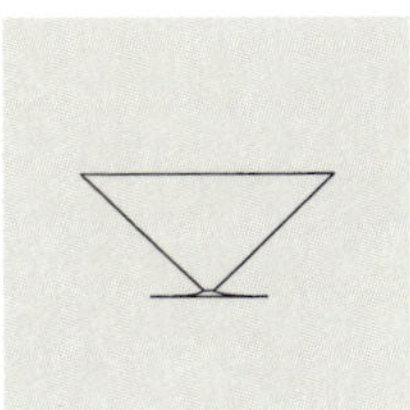

6000

Bowl in *lattimo* glass with Savoy-blue *decoro fenicio*. Dish-shaped foot in *lattimo*. Thread around the foot and mouth in dark-blue glass. Surface with iridised finish. Model no. 6000 was also produced in other glass techniques (e.g. transparent, cased glass).
c. 1930
13 cm high; 24 cm ø
Private collection, Montebelluna

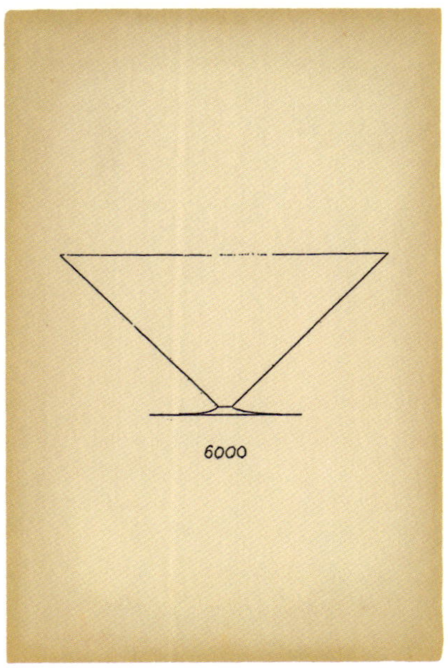

›
6000, catalogue drawing, AMB

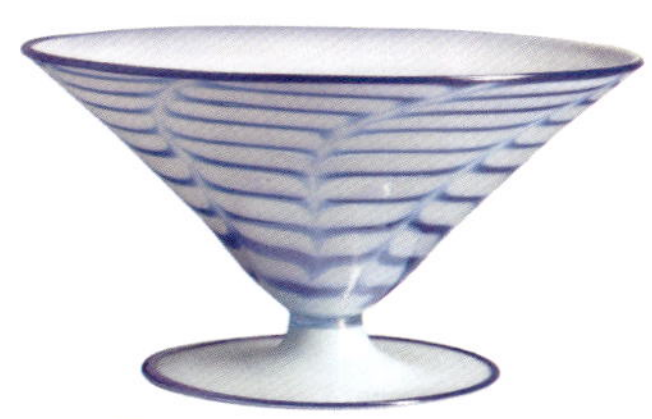

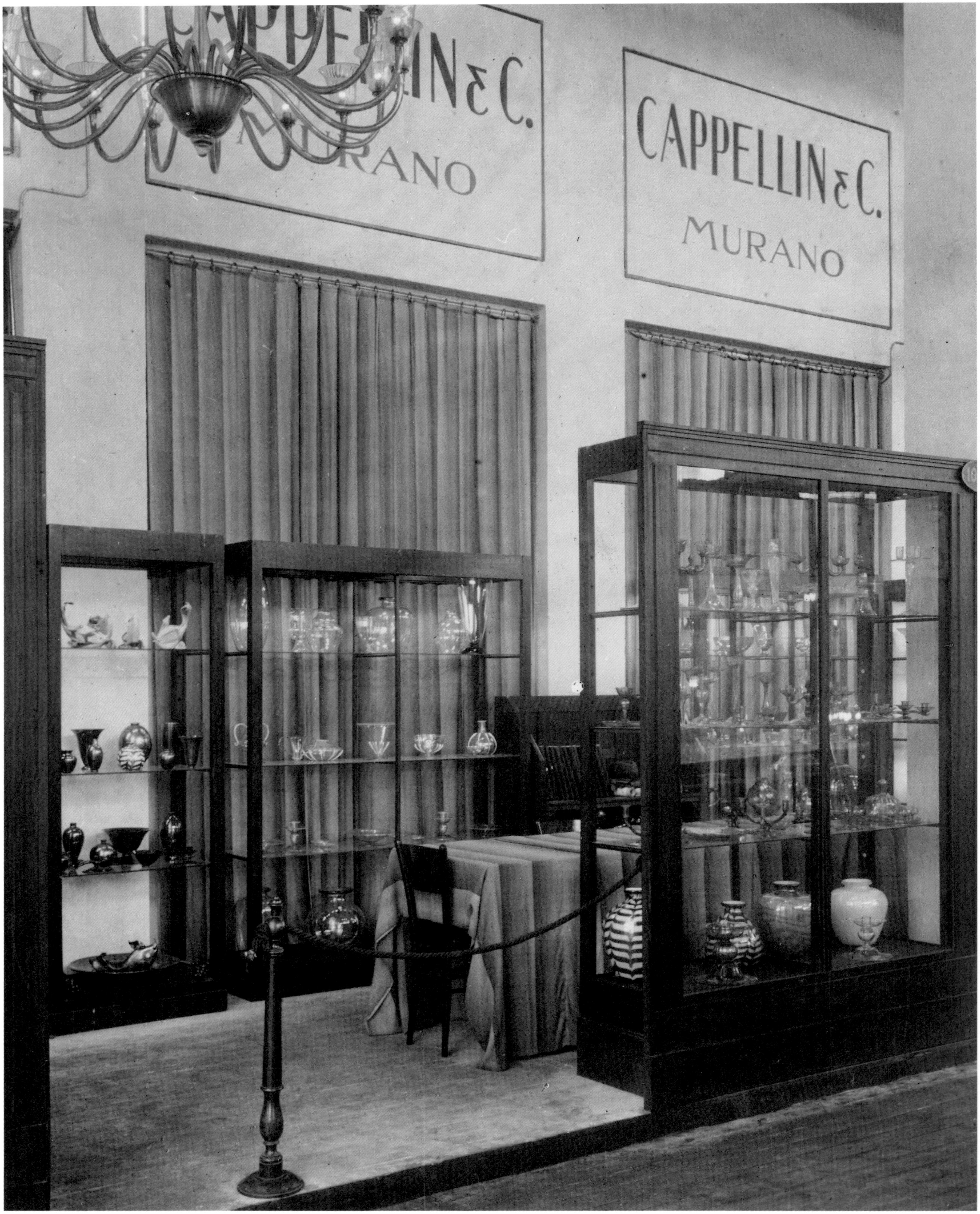

>
Cappellin Pavilion at the Trade Fair of Milan, 1929, detail, ACSVr. Bottom right are some vases in glass with decoro fenicio

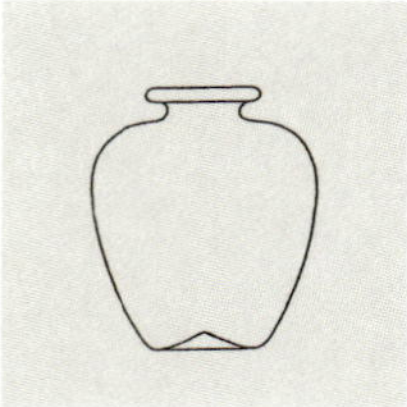

5299

Vase in *lattimo* cased glass with *decoro fenicio* in dark-amethyst glass. Rim of the mouth in dark-amethyst glass. Two similar examples, one in the same colours, the other in different shades of dark blue, were exhibited in 1929 at the Leipzig Trade Fair. The vase was produced after a model present in the catalogue of the glassworks and designed by Vittorio Zecchin (1921–23).
1928–29
36 cm high
Private collection

exhibitions
1929, Leipzig, Trade Fair
1929, Milano Fair

›
5299, catalogue drawing, AMB

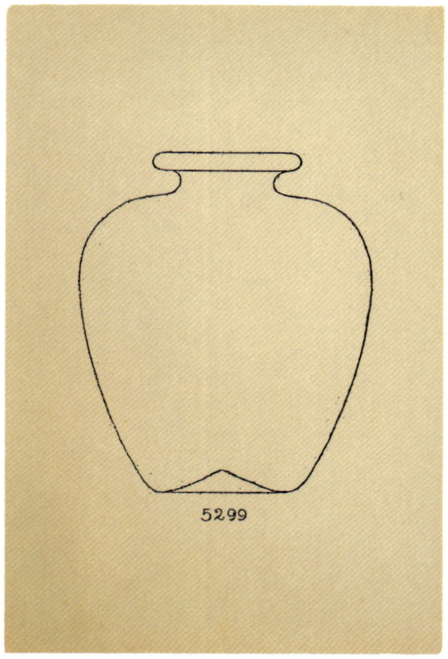

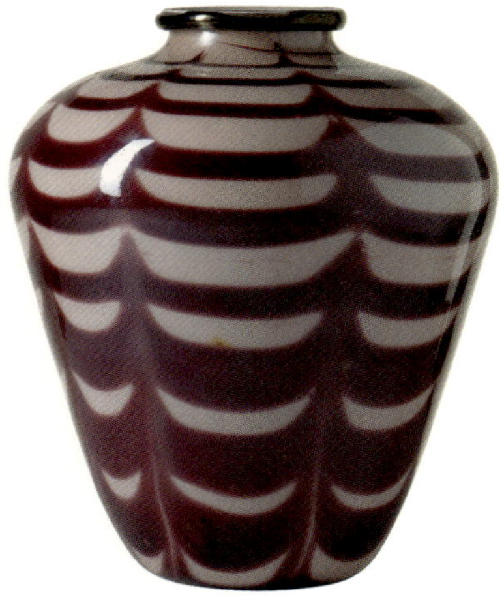

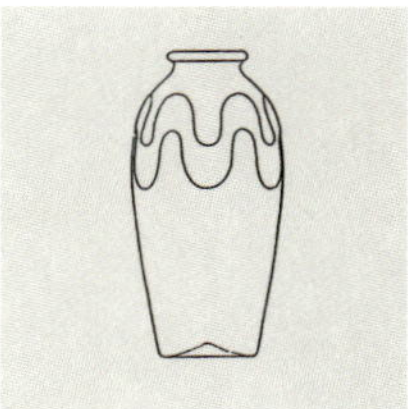

5316

Vase in *lattimo* glass with *decoro fenicio* in turquoise glass. Rim of the mouth in green glass. Surface with iridised finish. The vase was produced after a model in the catalogue of the glassworks designed by Vittorio Zecchin (1921–25).
1928–29
40 cm high
Private collection, Treviso

›
5316, catalogue drawing, AMB

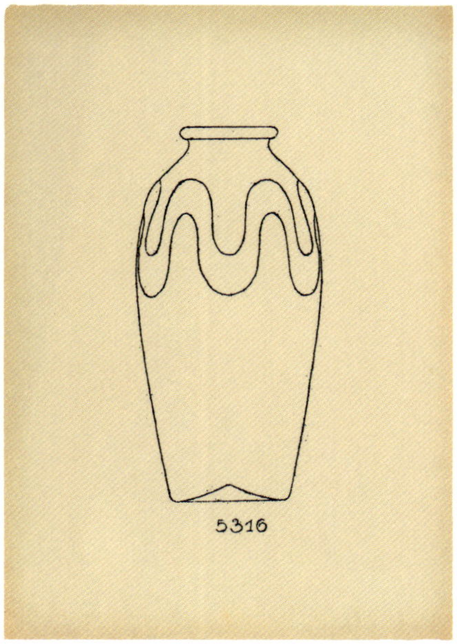

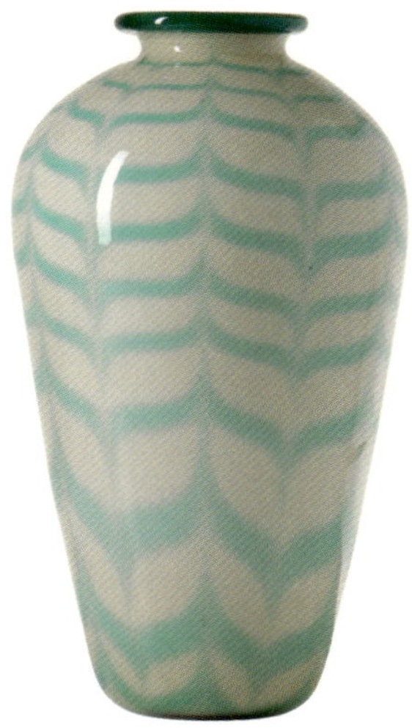

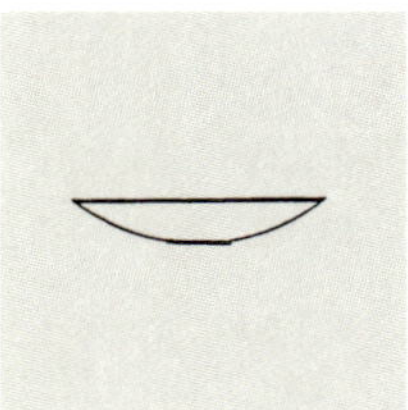

1783

Plate in *lattimo* glass with *decoro fenicio* in light-blue glass. Ringed foot with wound thread, and thread around the mouth in light-blue glass. Surface with iridised finish. The piece is a larger variant of no. 1783 (Ø 30 cm).
1928–29
47 cm ø
Private collection, Venice

›
1783, catalogue drawing, AMB

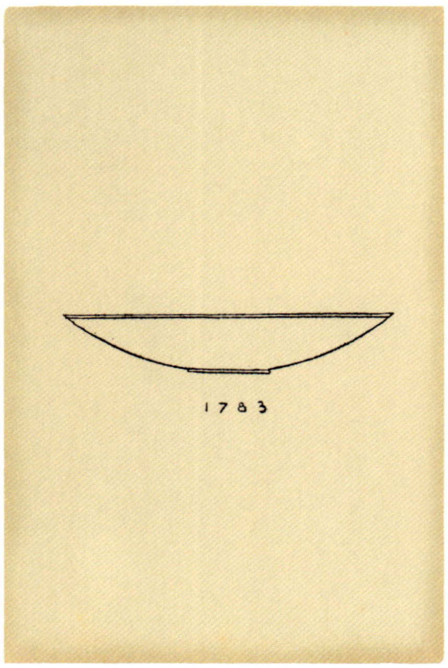

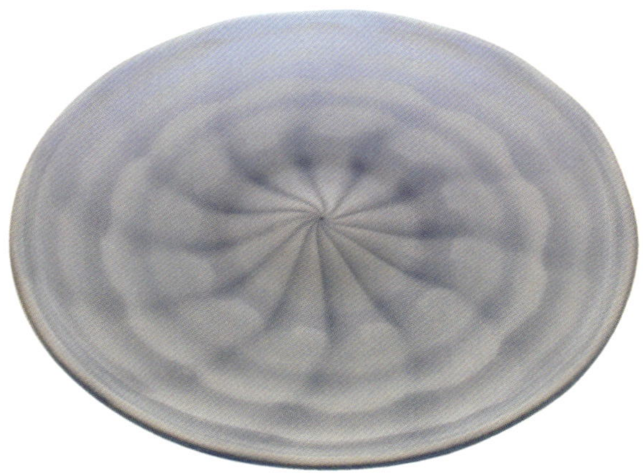

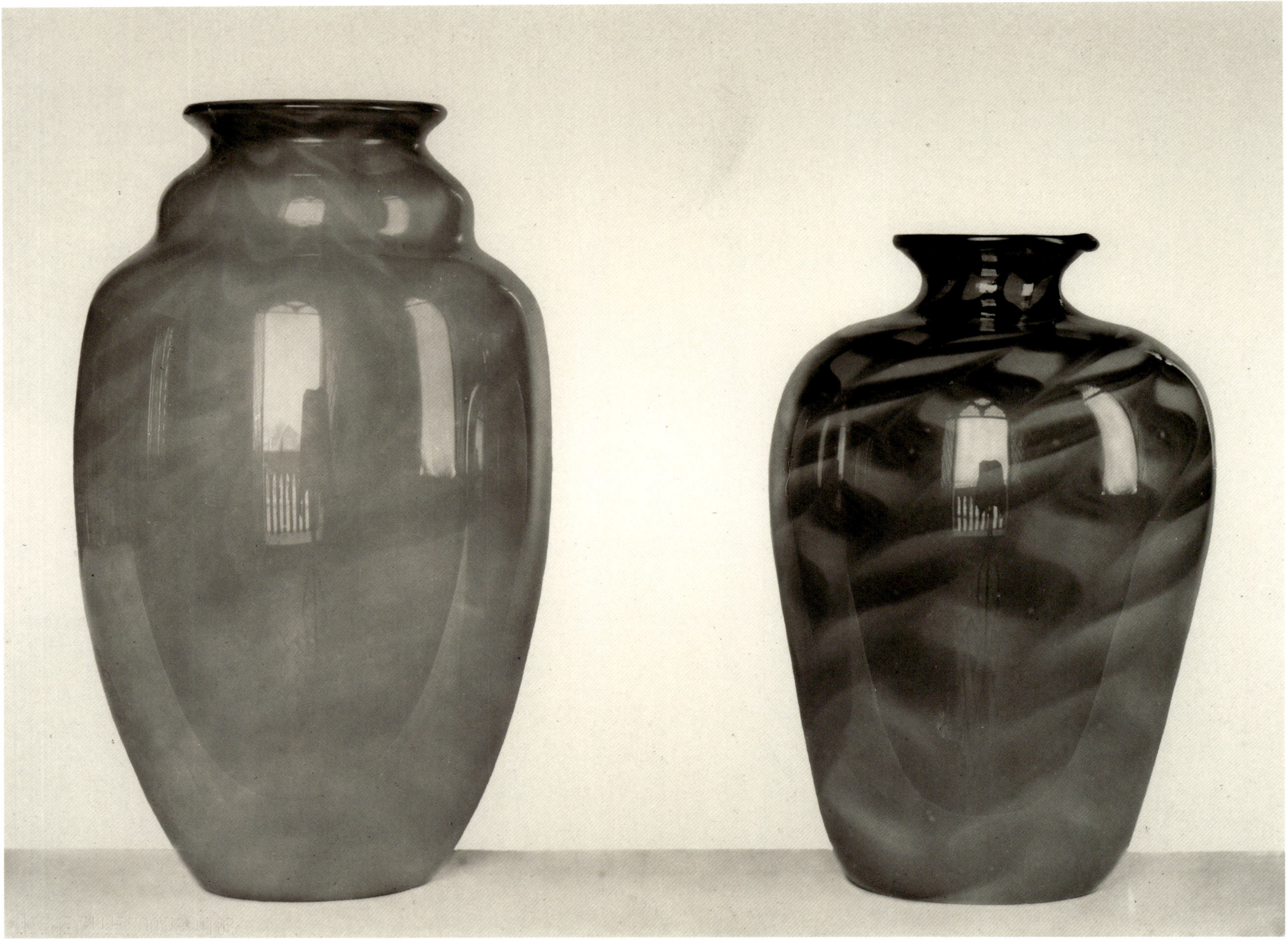

>
Vases in glass with a fenicio ritorto decoration. Period photograph, c. 1929, ACSVr

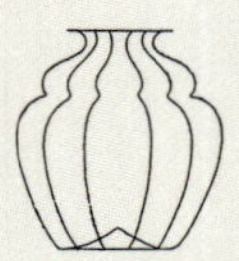

5411

Vase in *lattimo* glass with *decoro fenicio* in light-blue glass. Thread around the mouth in light-blue glass. Surface with iridised finish. The vase was produced after a model included in the catalogue of the glassworks and designed by Vittorio Zecchin (1925–26).
1928–29
23 cm high
Private collection

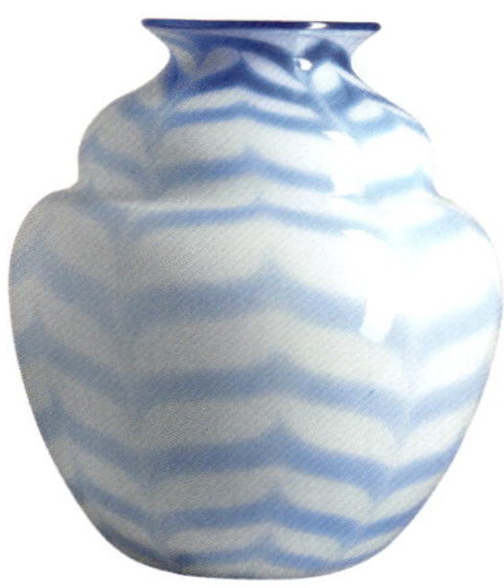

n.i.

Vase in *lattimo* glass with twisted *decoro fenicio* in dark-amethyst glass. Mouth in dark-amethyst glass. Surface with iridised finish. Some vases with twisted *decoro fenicio* were exhibited in 1929 at the Milano Fair
1928–29
21 cm high
C.V.M. *acid stamp*
Bischofberger collection, Switzerland

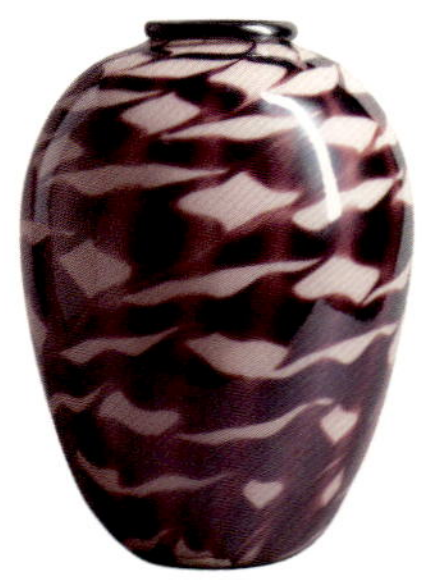

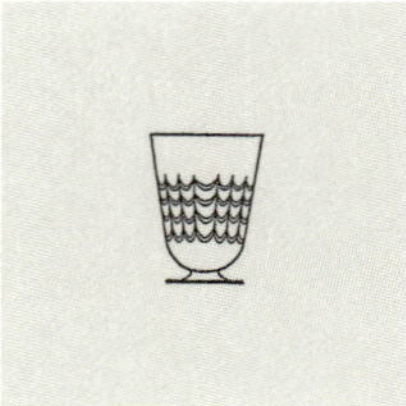

5591

Vase in black glass with disk-shaped foot and central band with green *decoro fenicio*. Thread around the mouth in green glass. Surface with iridised finish. The vase was produced after a model included in the catalogue of the glassworks and designed by Vittorio Zecchin (1925–26).
1928–29
17 cm high
M.V.M. Cappellin Murano *acid stamp*
Private collection

›
5591, working drawing, CSV

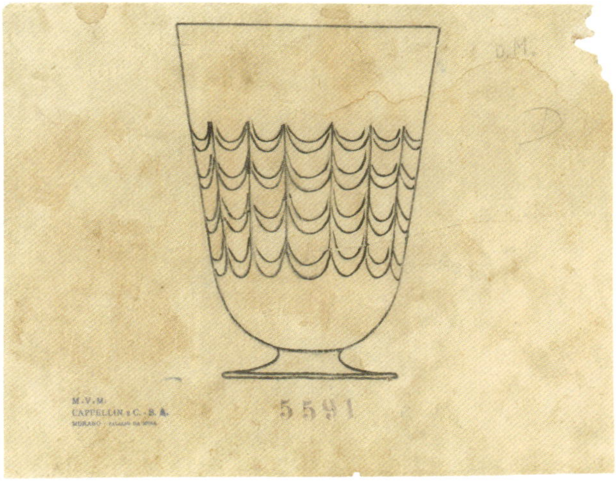

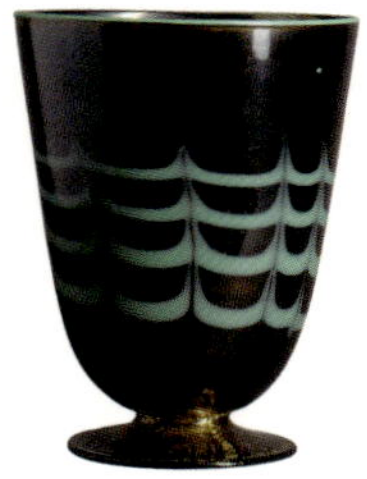

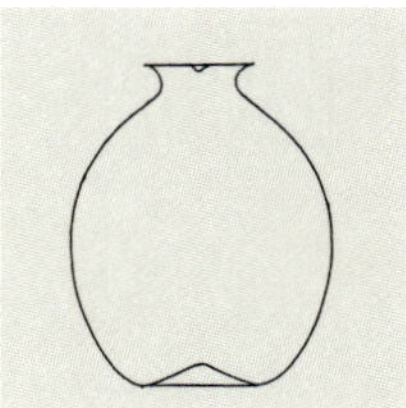

5101

Vase in crystal glass with amethyst *decoro fenicio*. Surface with iridised finish. A similar example in *lattimo* cased and *pagliesco* glass with amethyst *decoro fenicio* was exhibited in 1929 at the Leipzig Trade Fair. The vase was produced in a variant without a lip of model no. 5101 present in the catalogue of the glassworks and designed by Vittorio Zecchin (1921–23).
1928–29
36 cm high
M.V.M. Cappellin Murano *acid stamp*
Private collection

exhibition
1929, Leipzig, Trade Fair

›
5101, catalogue drawing, AMB

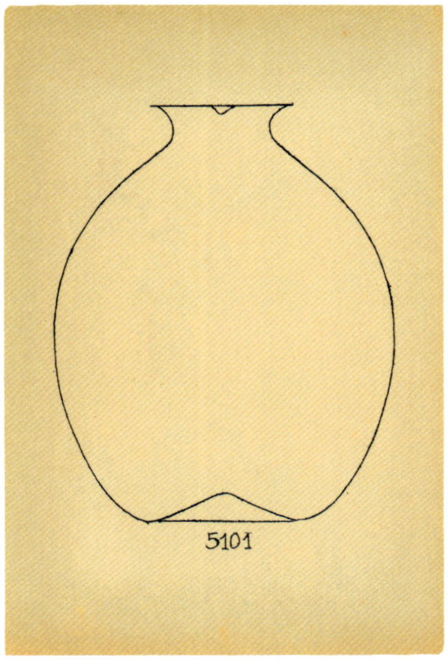

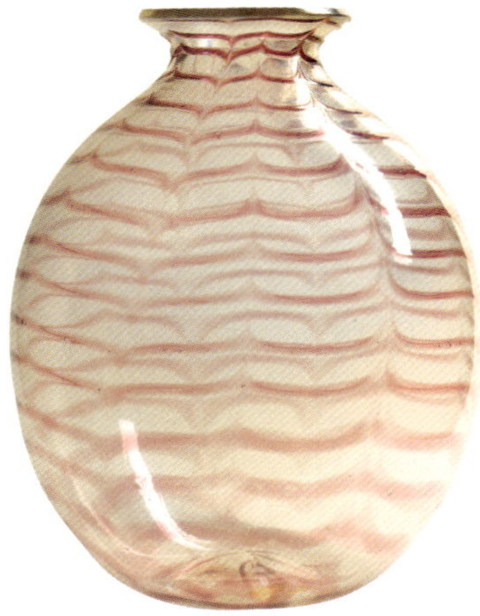

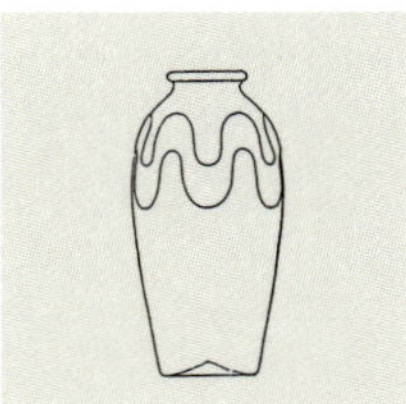

5316

Vase in crystal glass with amethyst *decoro fenicio*. Mouth in amethyst glass. Surface with iridised finish. The vase was produced after a model present in the catalogue of the glassworks and designed by Vittorio Zecchin (1921–25).
1928–29
33 cm high
Dr. Richard Kossmann collection, New Jersey

›
5316, catalogue drawing, AMB

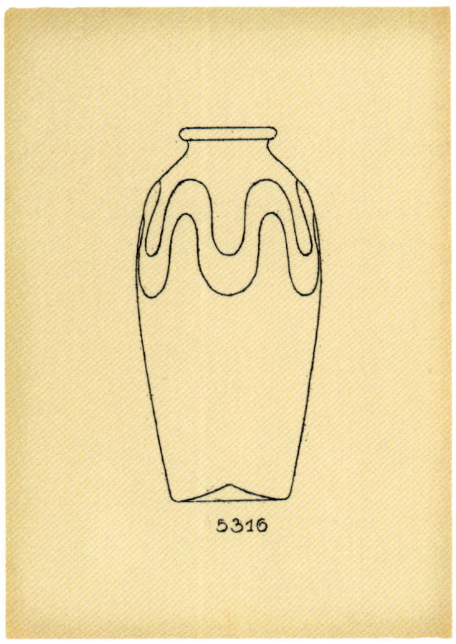

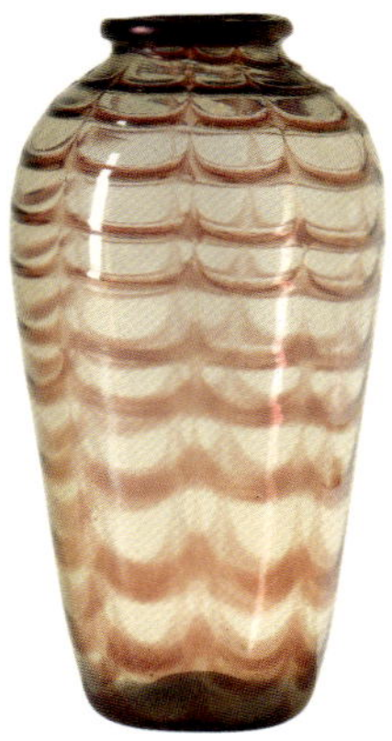

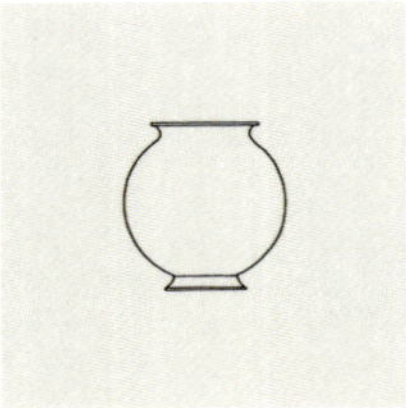

6446

Spherical, ribbed vase in crystal glass with dark-blue *decoro fenicio*. Low truncated-cone-shaped foot in crystal glass. Thread at the foot and mouth in dark-blue glass. Surface with iridised finish.
1930–31
21 cm high
Private collection, New York

›
6446, working drawing, CSV

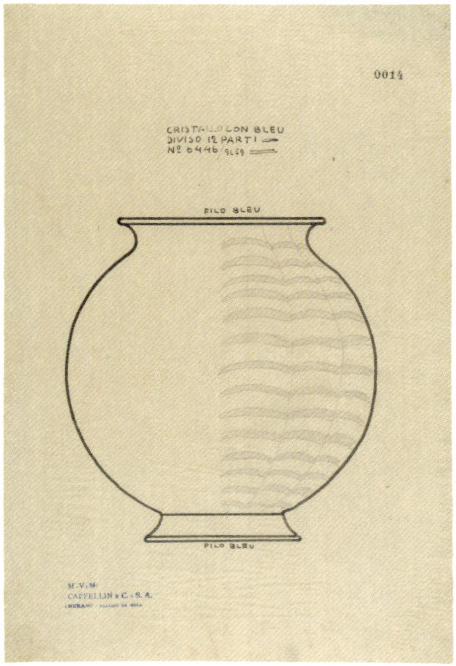

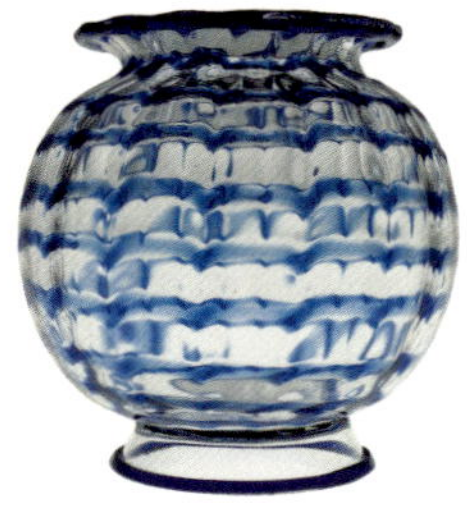

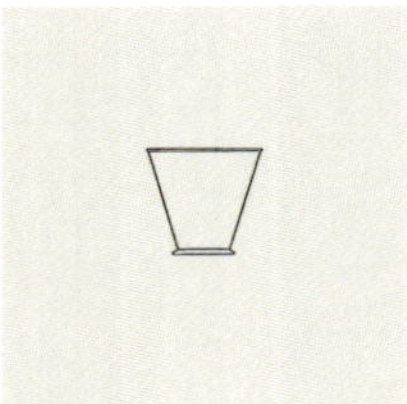

6448

Working drawing for a vase produced in crystal glass with twisted *decoro fenicio* in dark blue glass, CSV. Small flared foot in crystal glass.
1930–31
15 cm high

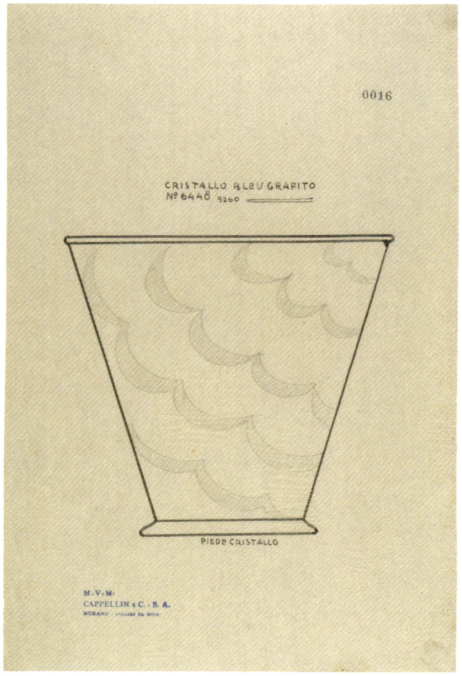

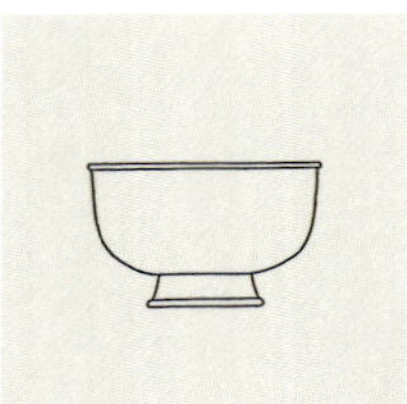

6445

Spherical, ribbed vase in crystal glass with dark-blue *decoro fenicio*. Low truncated-cone-shaped foot in crystal glass. Thread at the foot and mouth in dark-blue glass. Surface with iridised finish.
1930–31
17 cm high; 28 cm ø

›
6446, working drawing, CSV

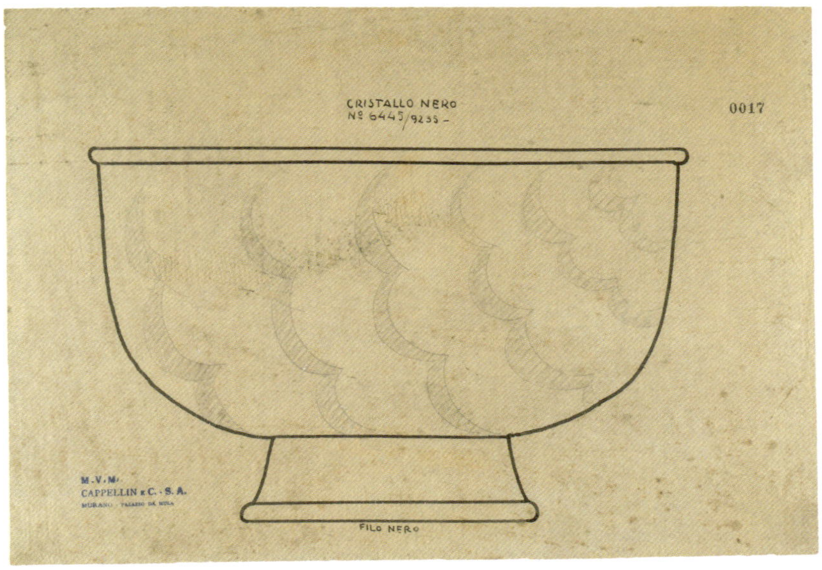

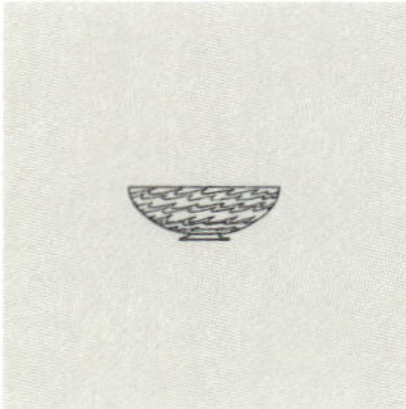

1787

Bowl in *pagliesco* glass with strongly twisted amethyst *decoro fenicio*. Small foot and brim in dark-amethyst glass. Surface with iridised finish. The piece is a variant with a thread around the mouth of no. 1787.
1930–31
8 cm high; 23 cm ø
Private collection, Venice

›
1787–1788, working drawings, CSV

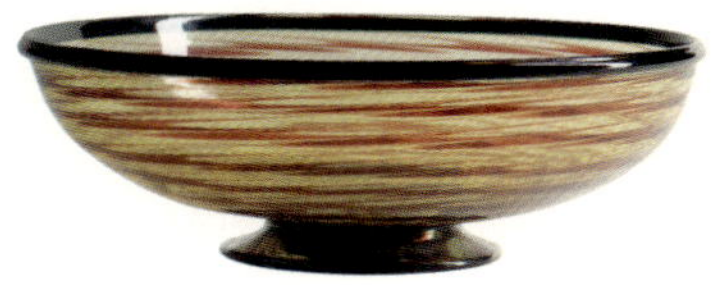

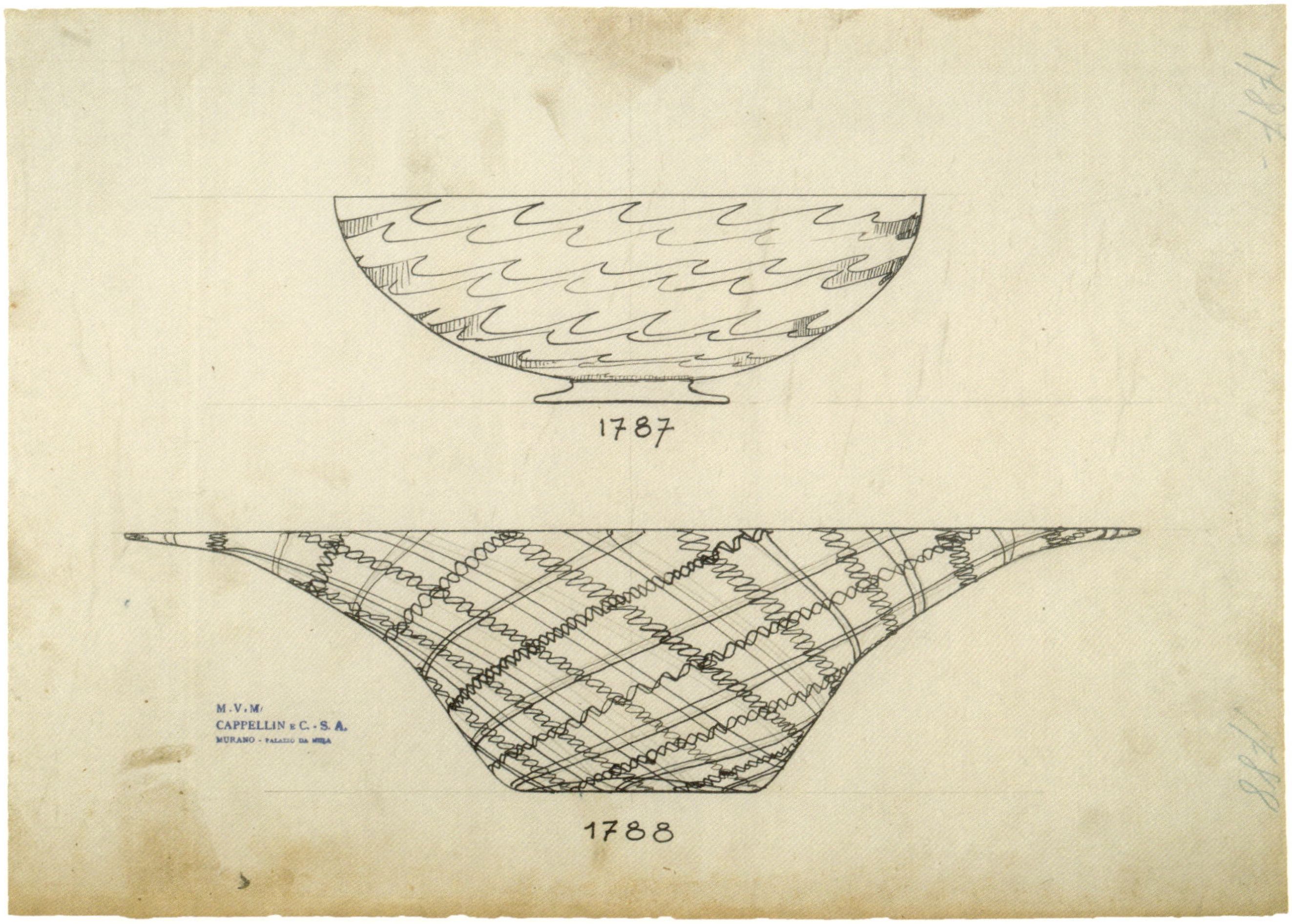

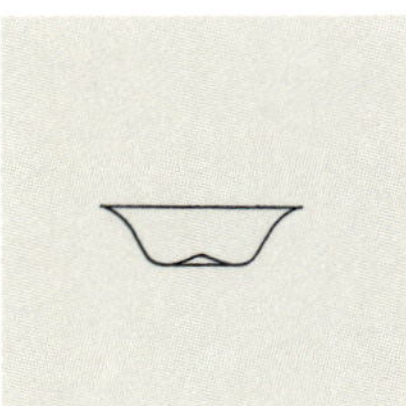

5864

Bowl in transparent *pagliesco* glass with variegated amethyst decoration obtained using the technique of strongly twisted *decoro fenicio*. Surface with iridised finish.
1930–31
9.5 cm high; 32.5 cm ø
M.V.M. Cappellin Murano *acid stamp*
Private collection, Treviso

›
5864, catalogue drawing, AMB

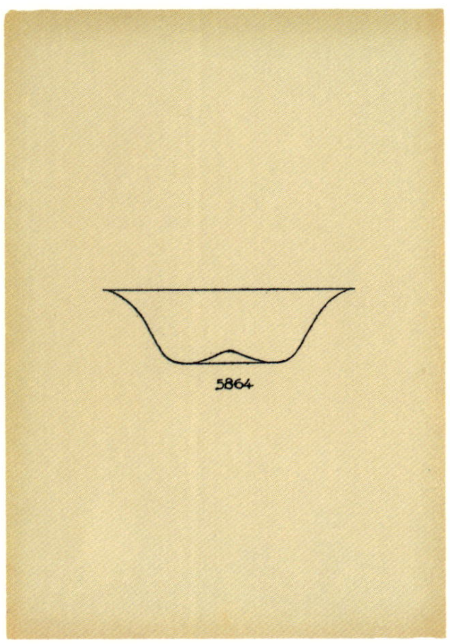

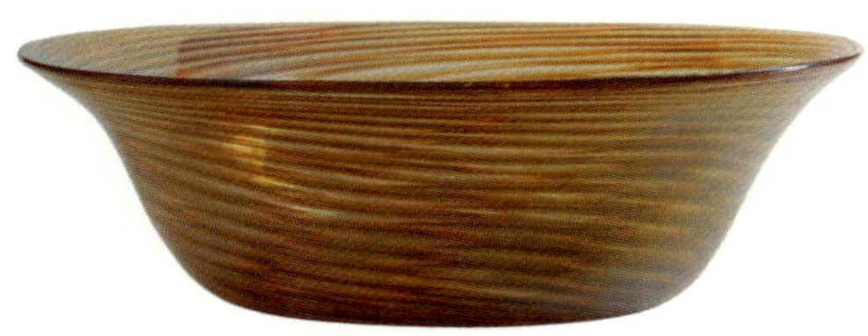

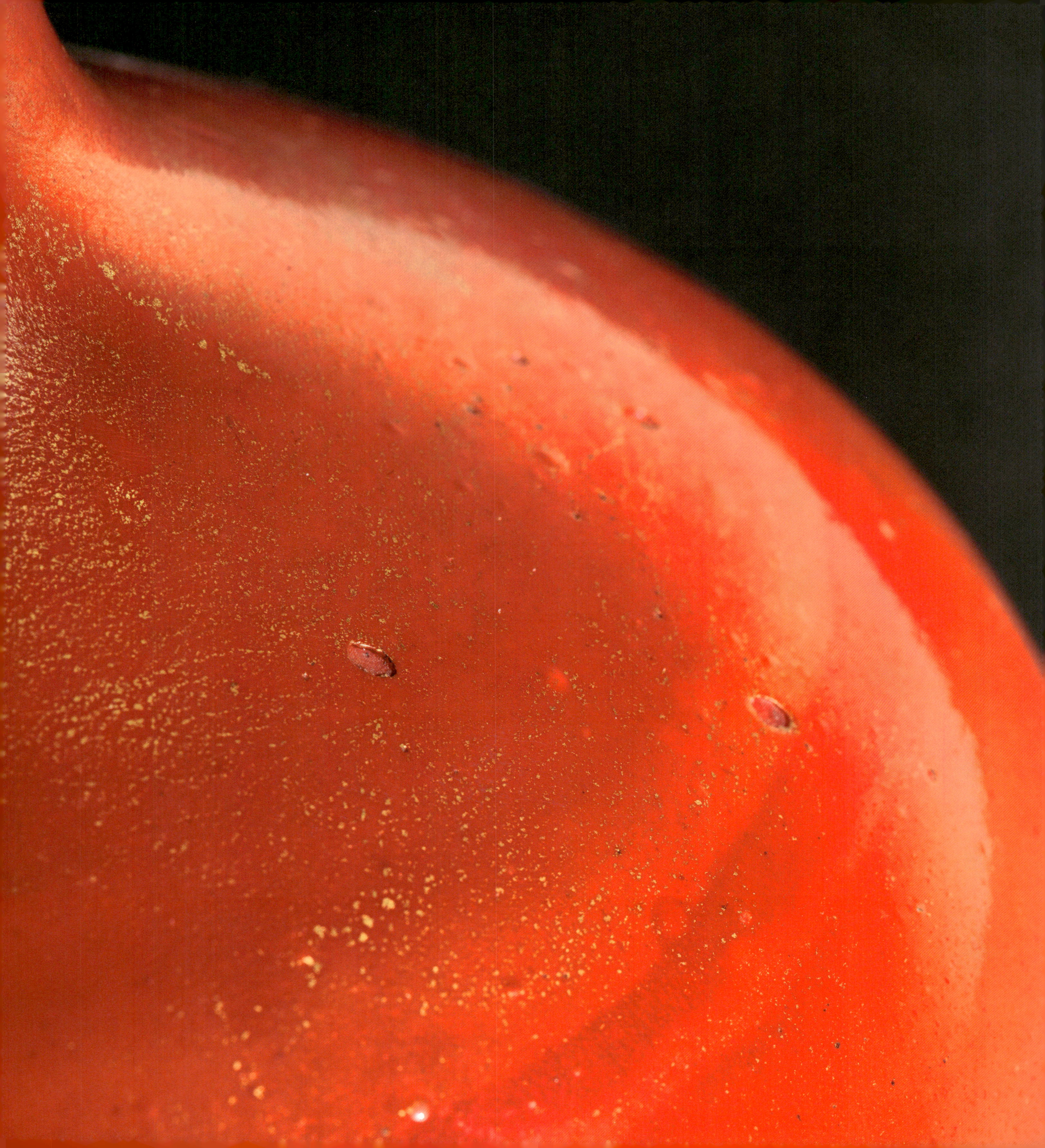

These are opaque glass pieces which owe their particular colour and material consistency to the notable quantity of opacifiers and pigments added to the transparent glass base. This type of material was repeatedly adopted by the Cappellin firm to produce highly refined artefacts, mostly of somewhat small dimensions, which were characterised by their intense colours (red, dark blue, green, yellow), often embellished by applications of gold leaf or oxidised silver leaf. The series is marked by the accentuated stylisation of the forms, clearly to be seen in the 'geometric' pieces exhibited with great success in 1930 at the 4th Triennale of Monza and in 1931 at the Exhibition of modern Italian art glass, ceramics and lace held in Amsterdam. This production is particular in the general context of Murano glass also because M.V.M. Cappellin was one of the few furnaces to use *pasta vitrea* given the technical difficulties it entailed, because of the long cooling times which made the final outcome difficult to control.

PASTA VITREA 1929-31

Exhibitions

1929, Leipzig, Trade Fair
1929, Milan, Trade Fair
1930, Monza, 4th *Esposizione Triennale Internazionale delle Arti Decorative ed Industriali Moderne*
1931, Amsterdam, *Tentoonstelling van glas, aardewerk en kant, van moderne italiaansche kunstnijverheid. Mostra di vetri, ceramiche e merletti d'arte moderna italiana*

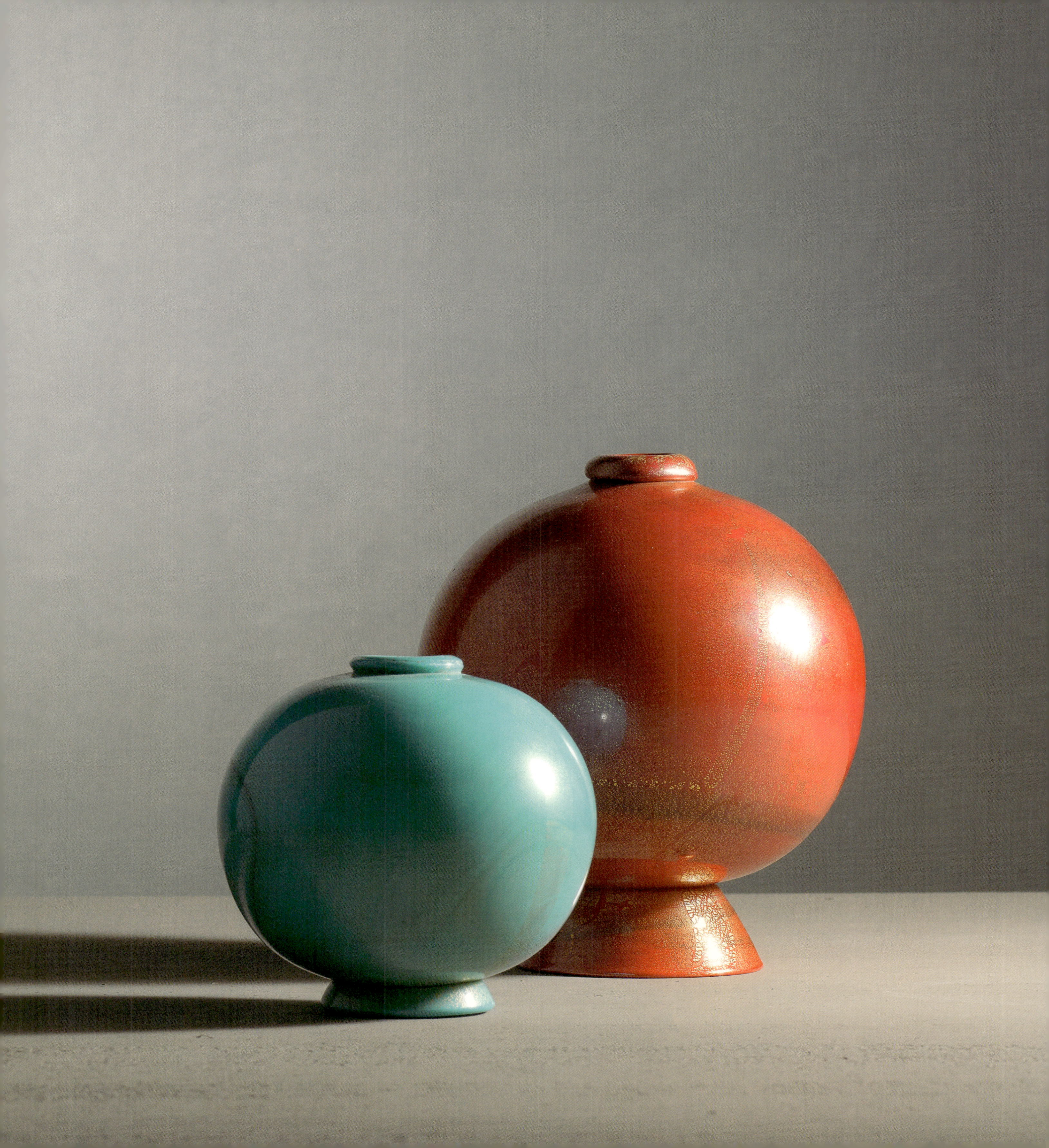

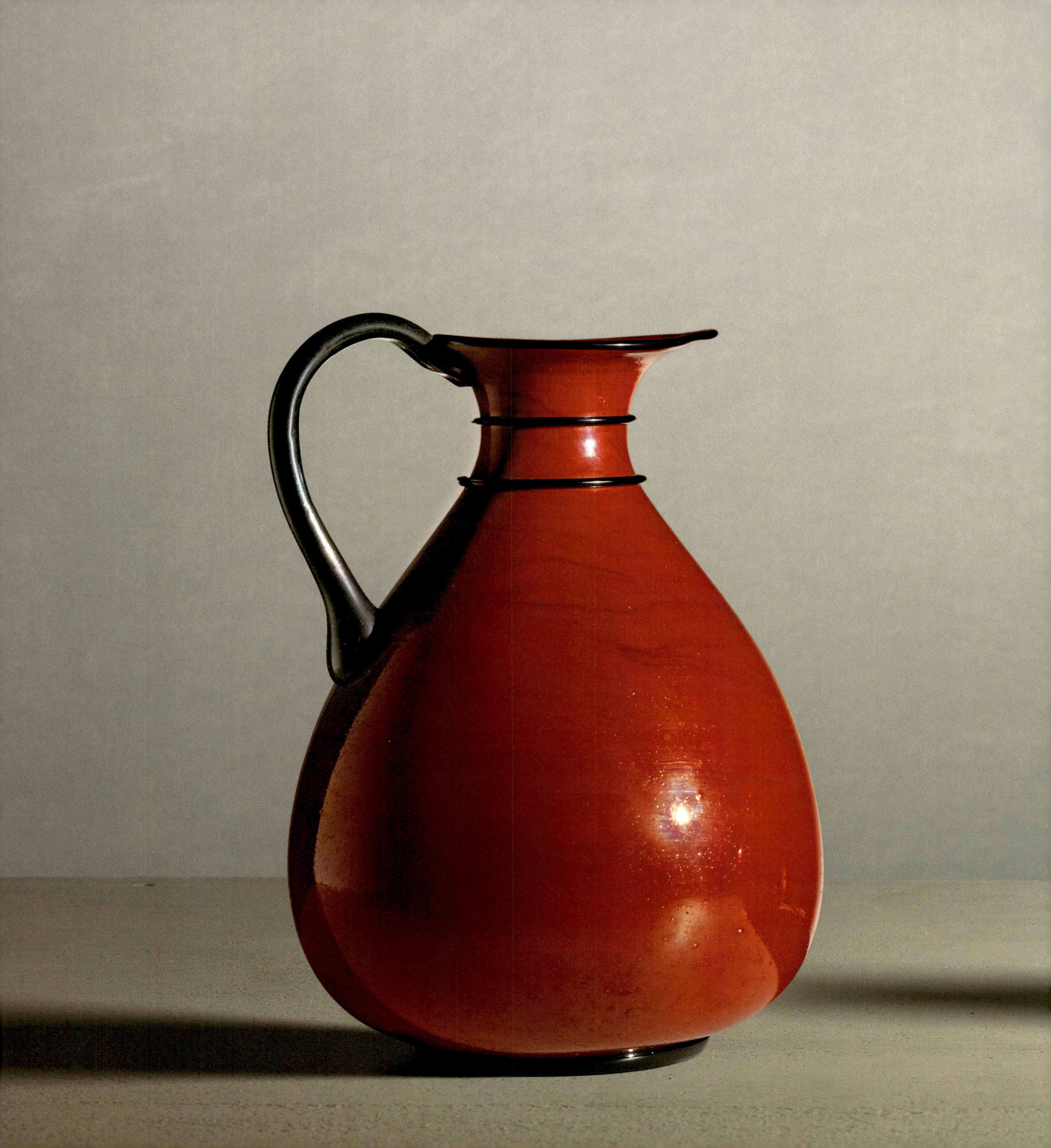

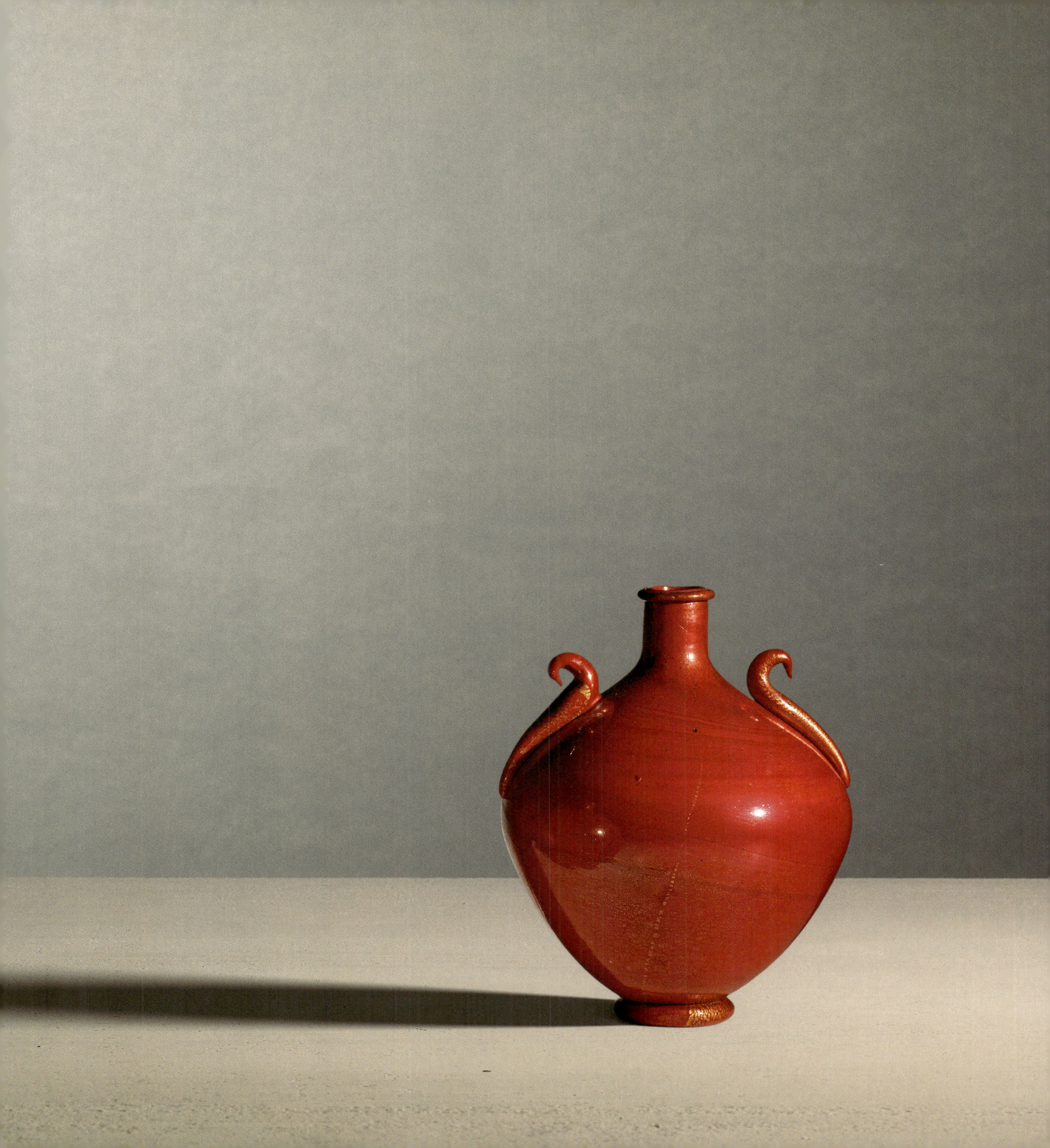

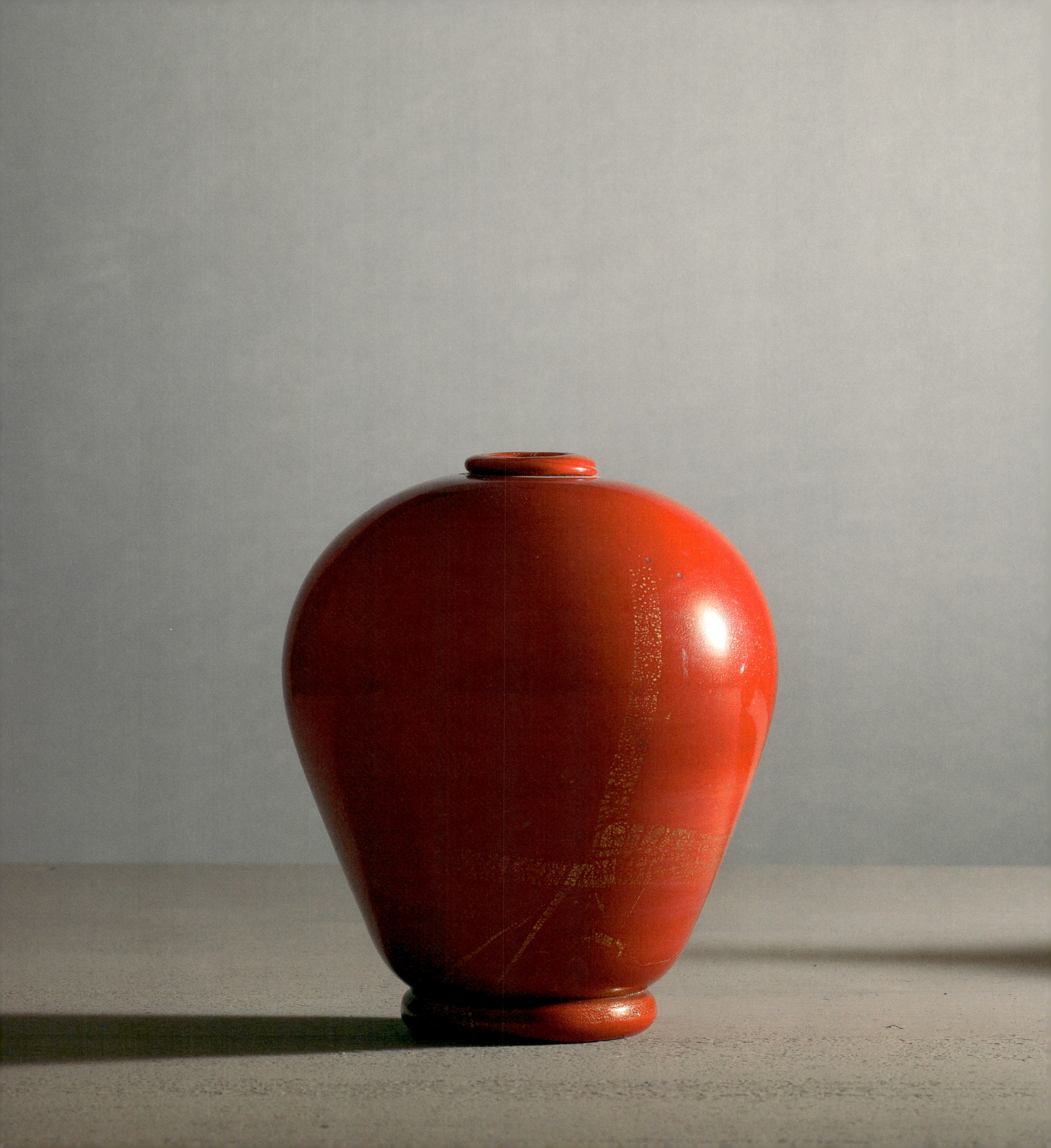

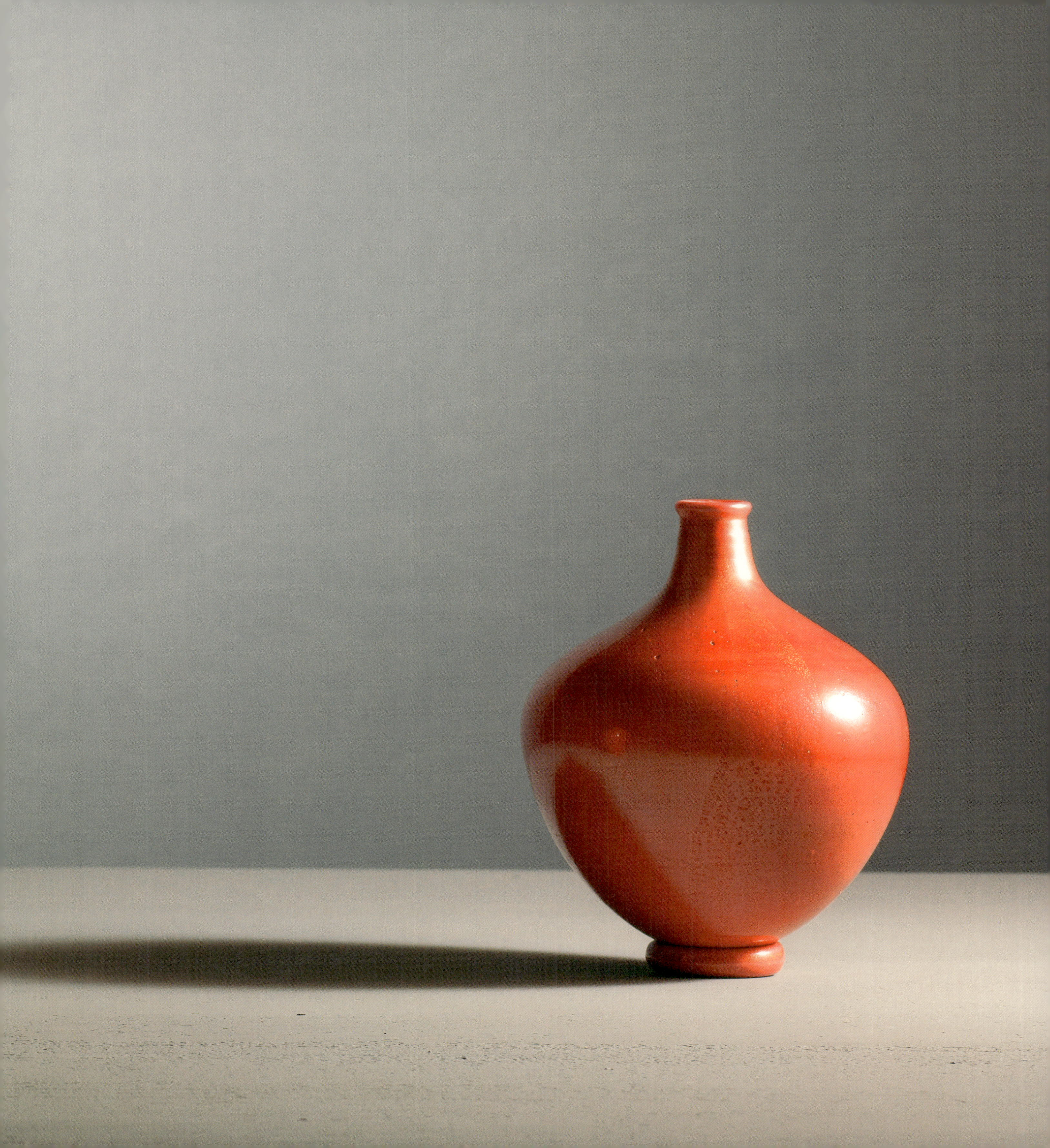

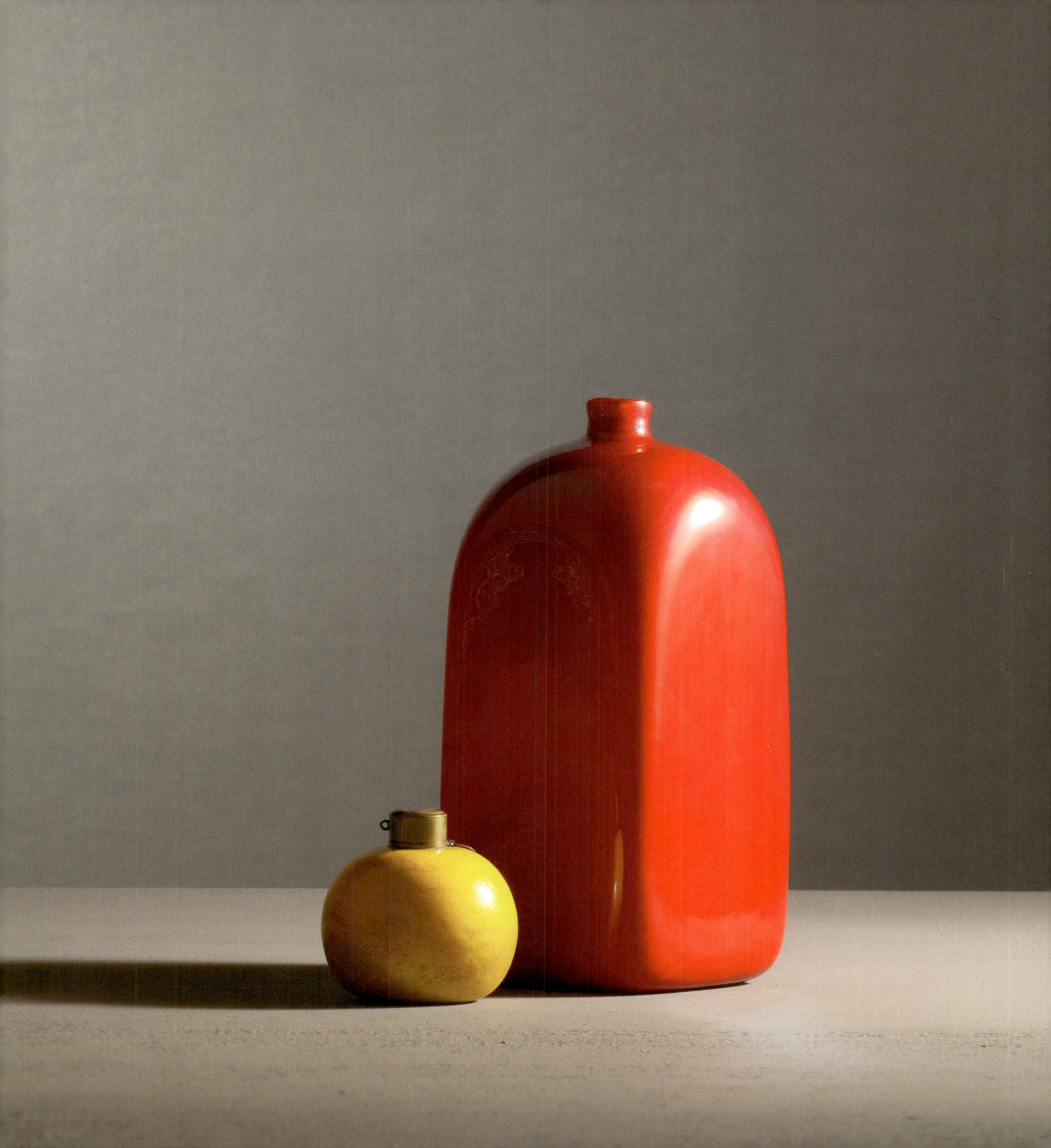

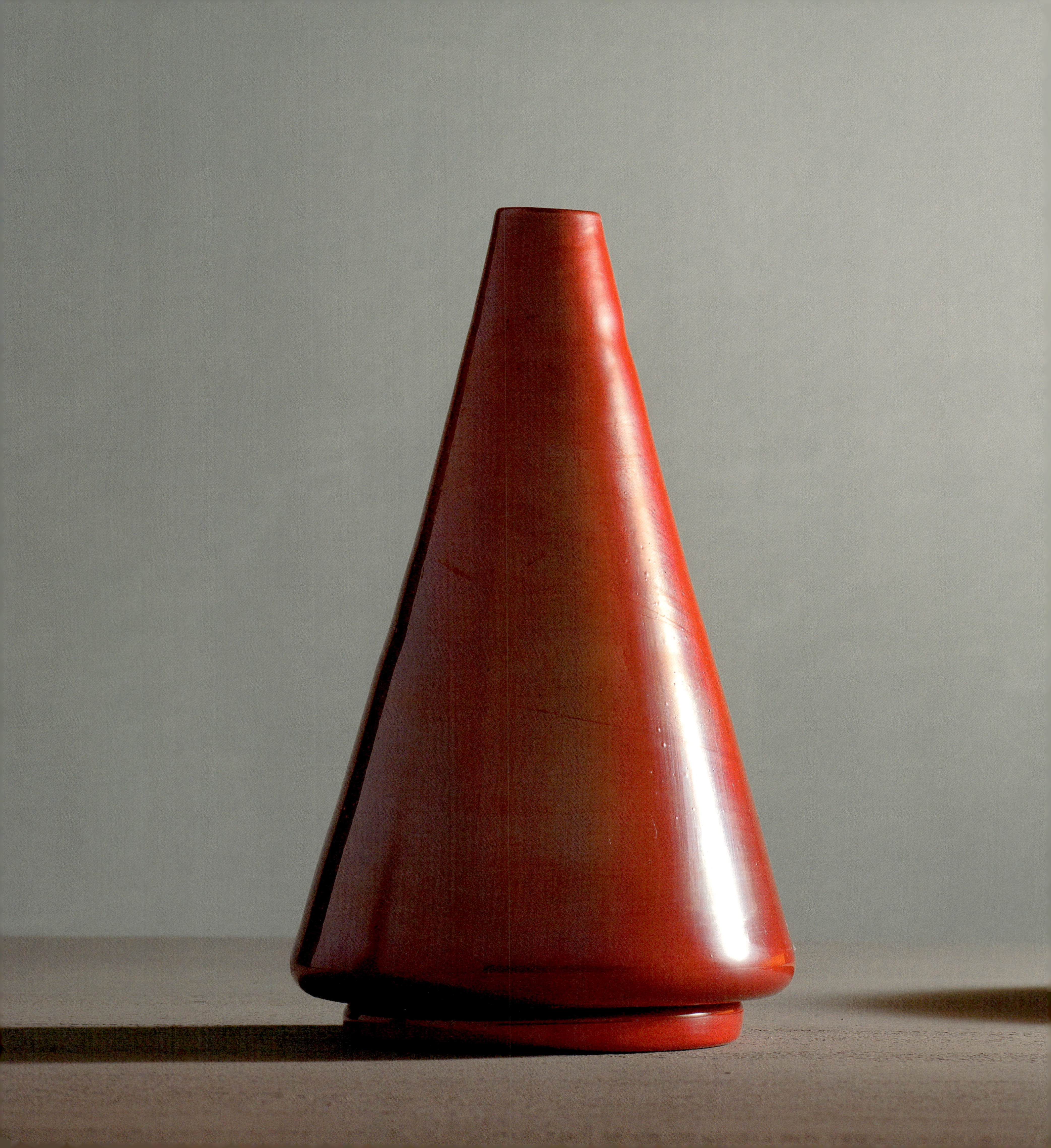

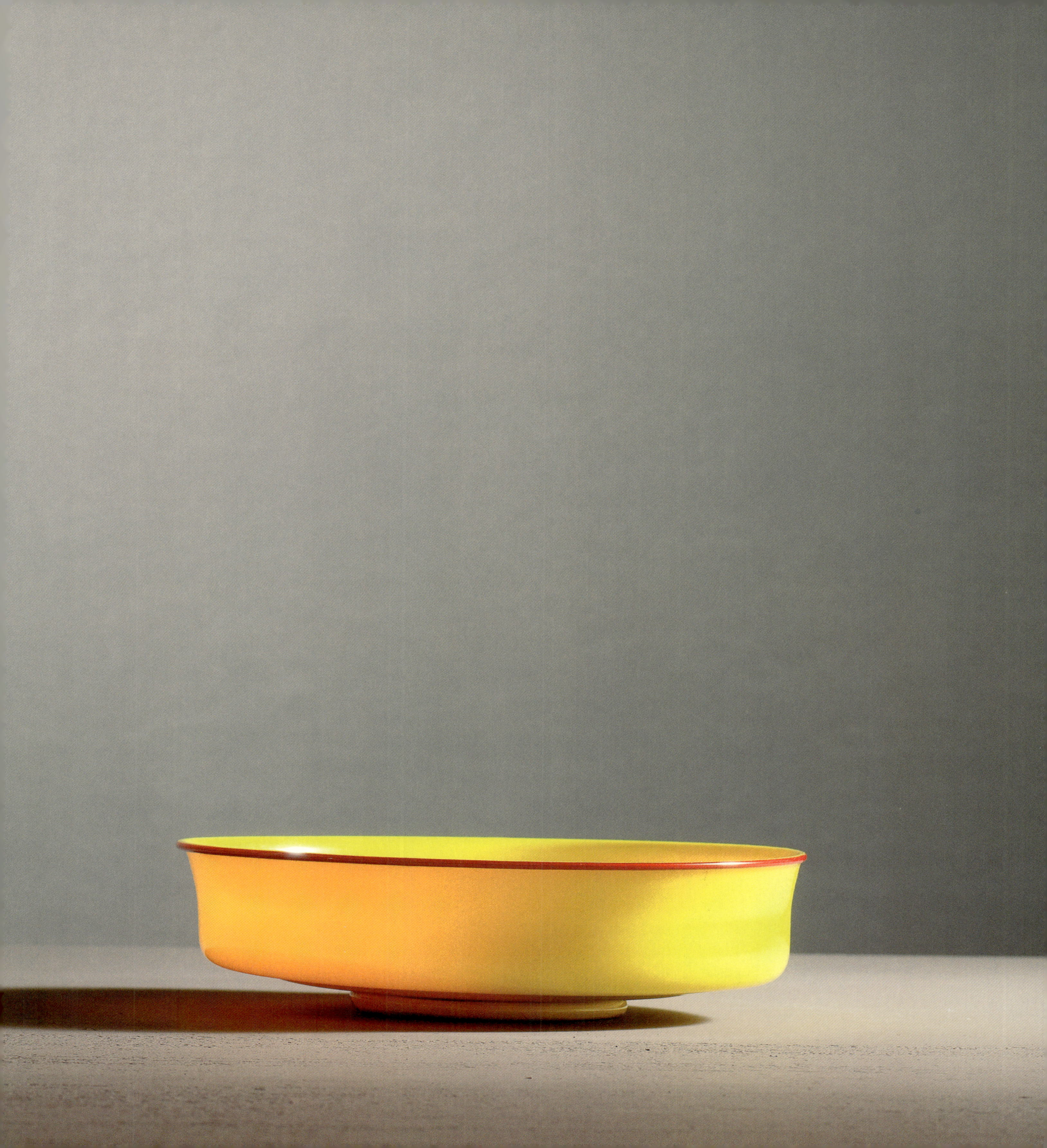

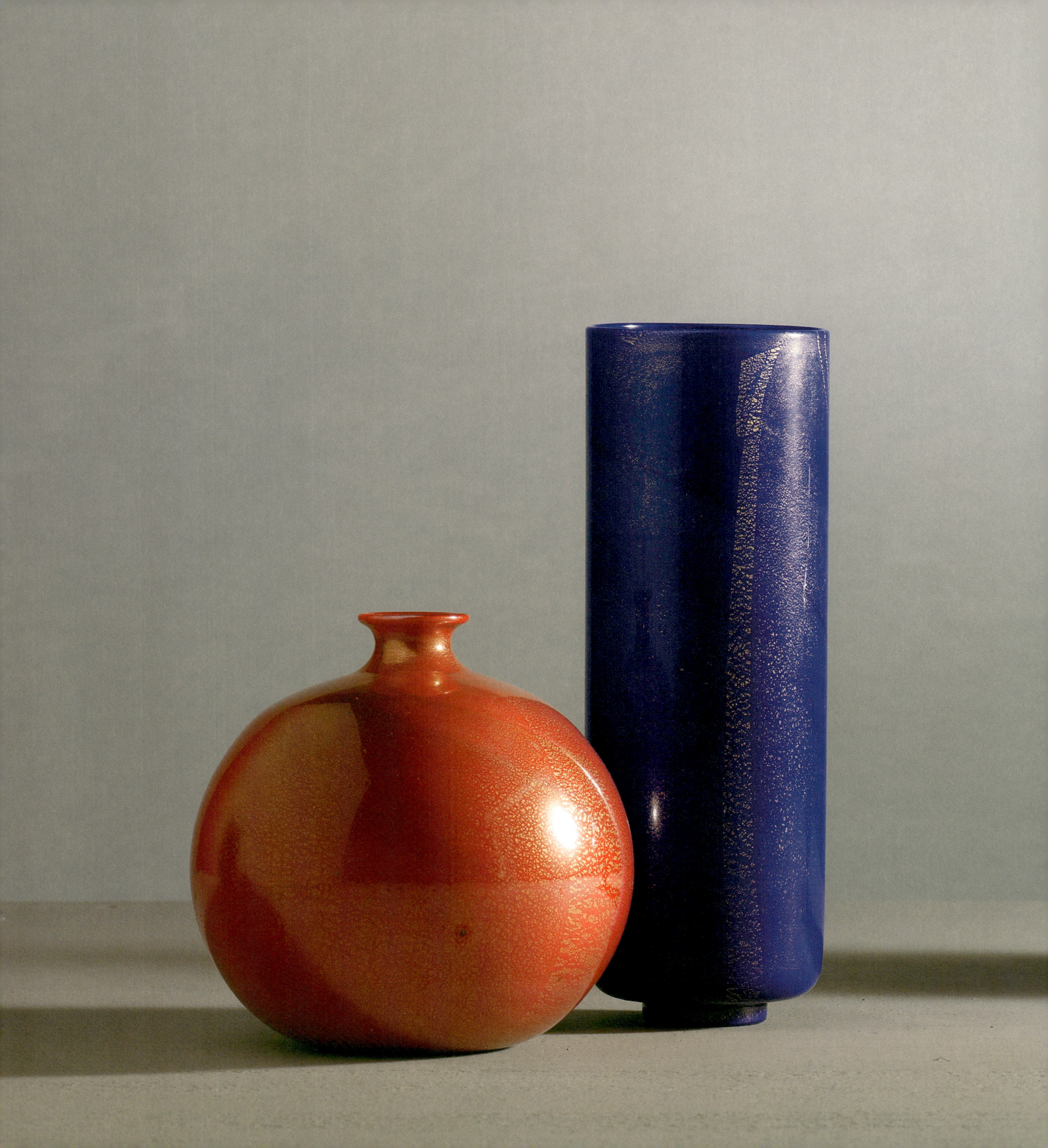

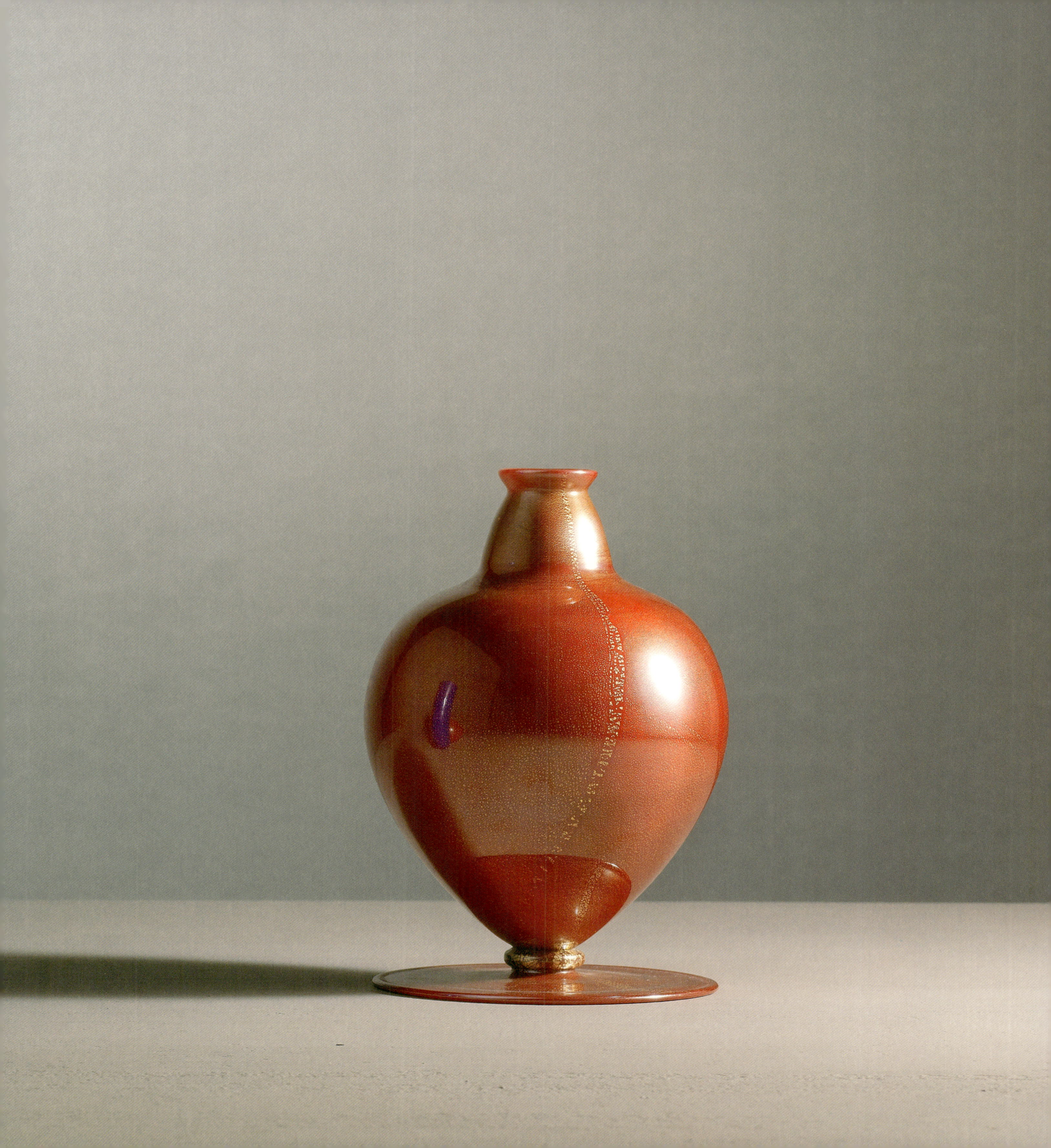

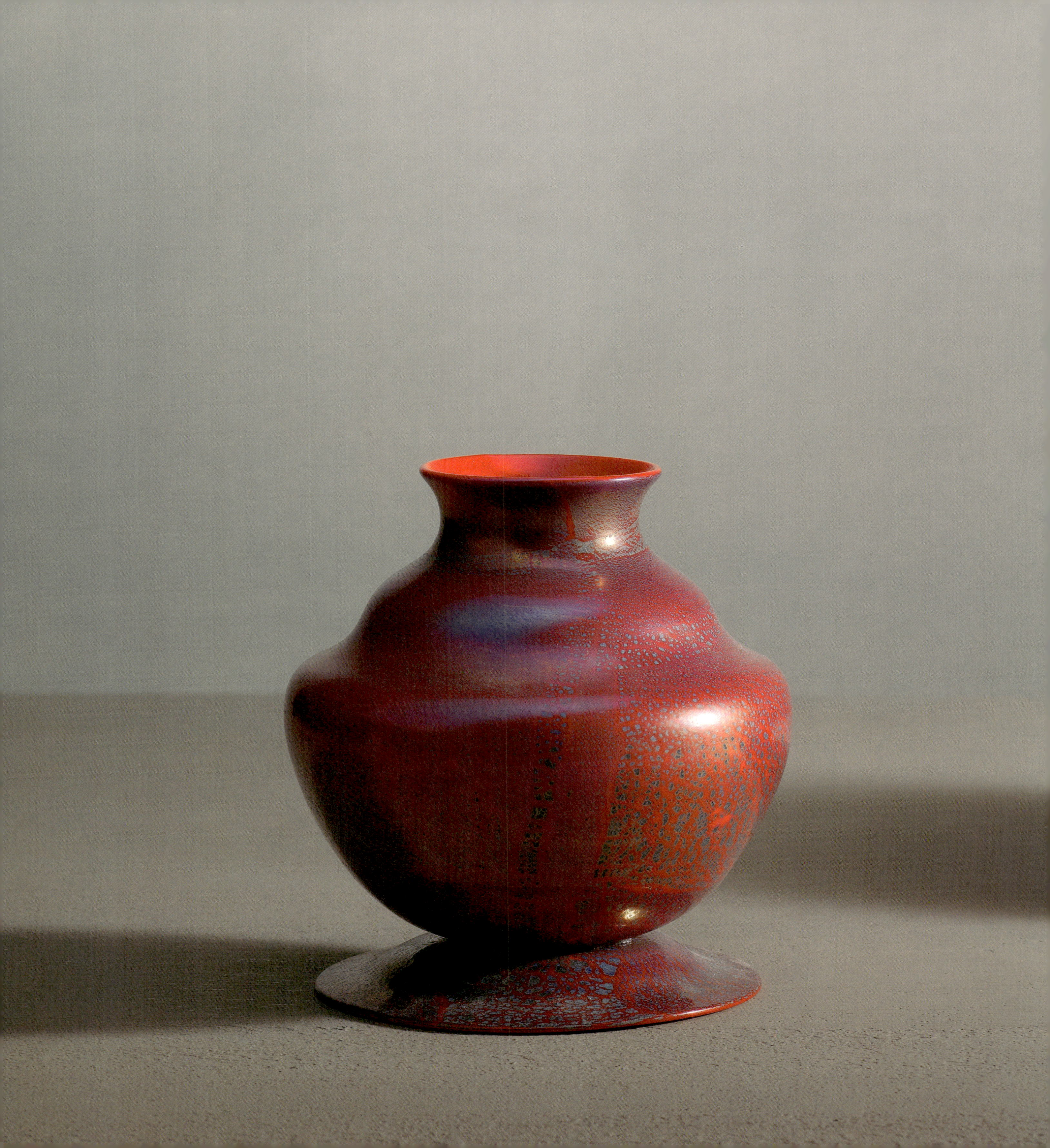

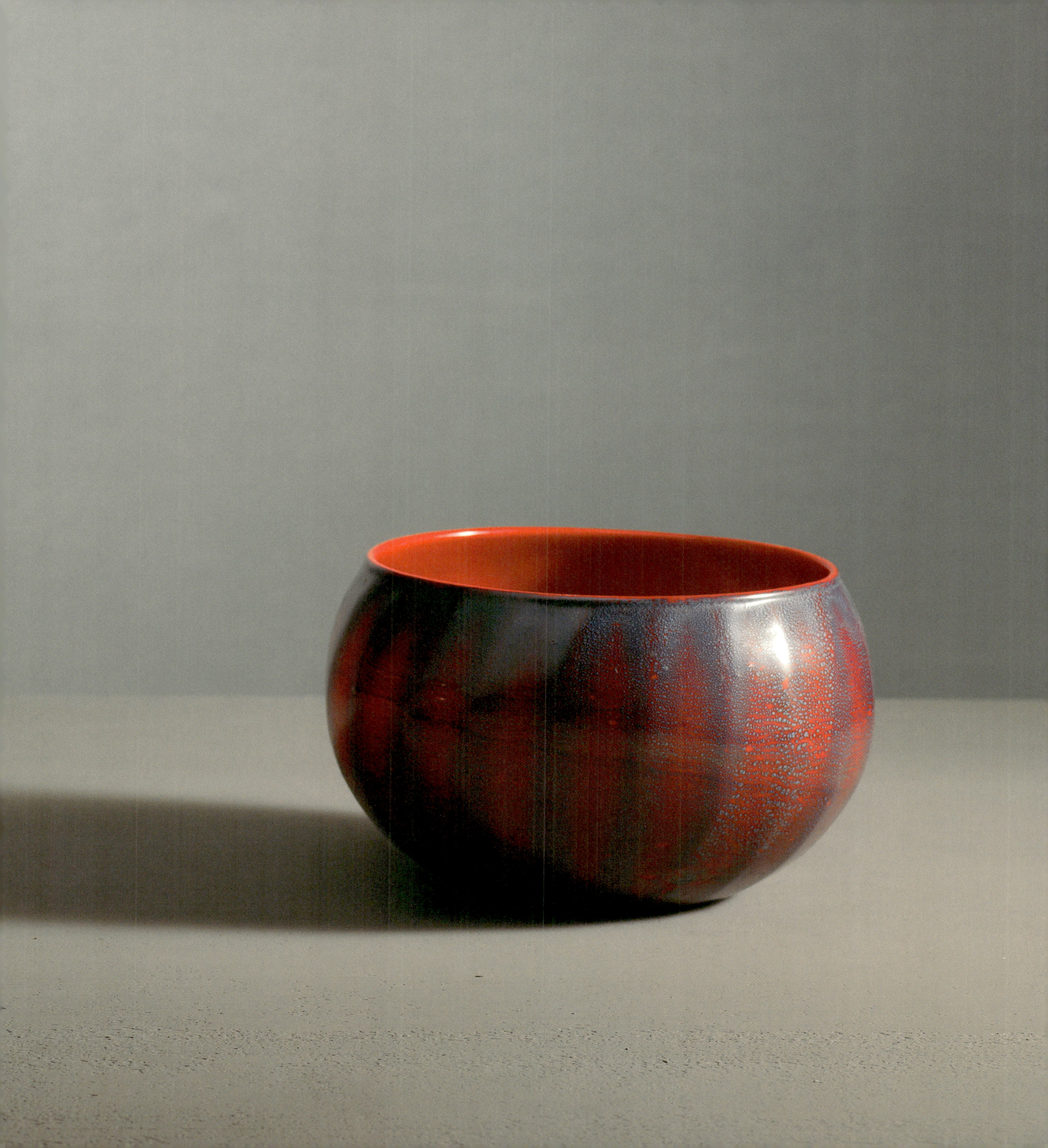

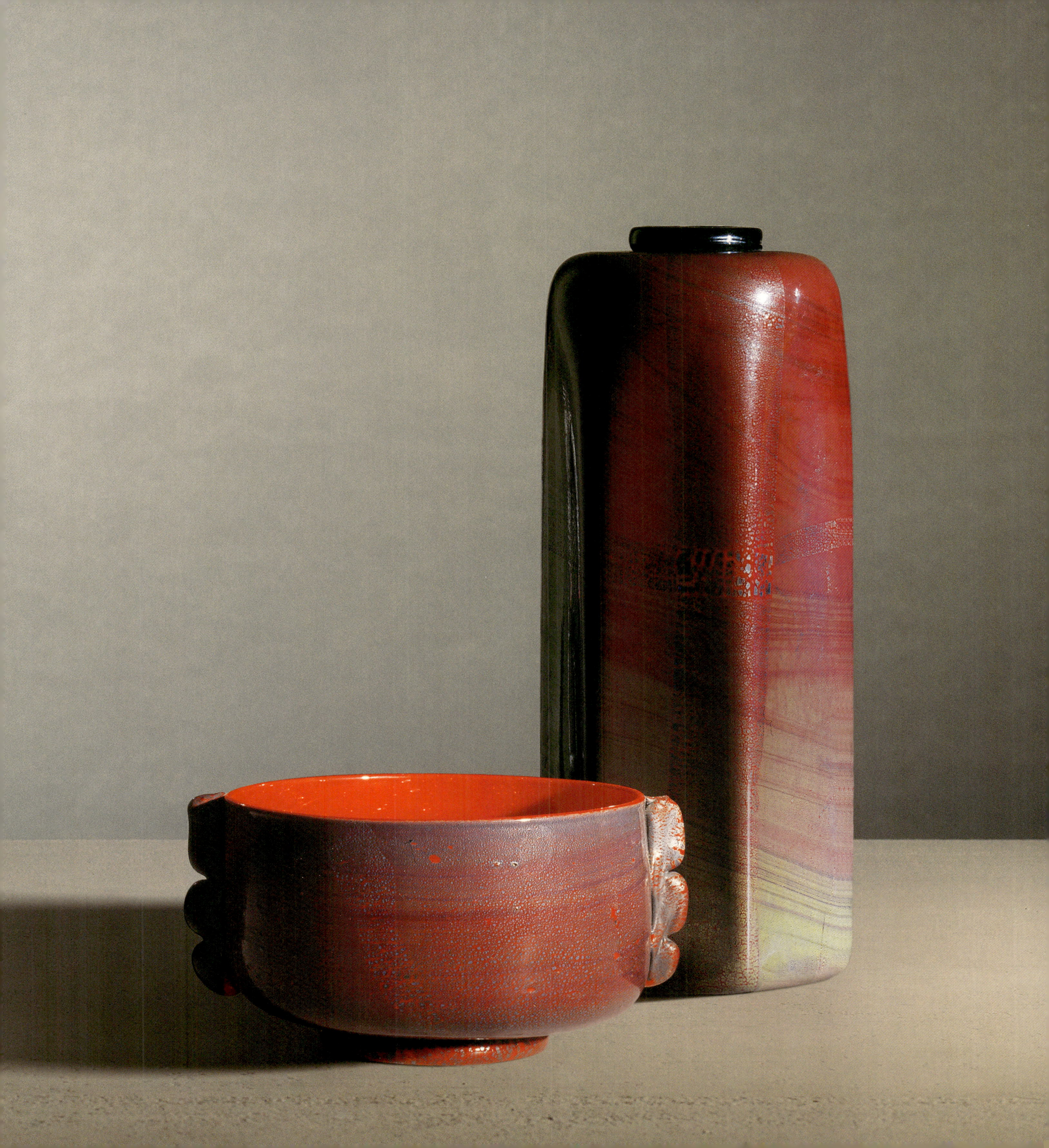

>
Bowls and plate in pasta vitrea. *Period photograph, c. 1929, AMB*

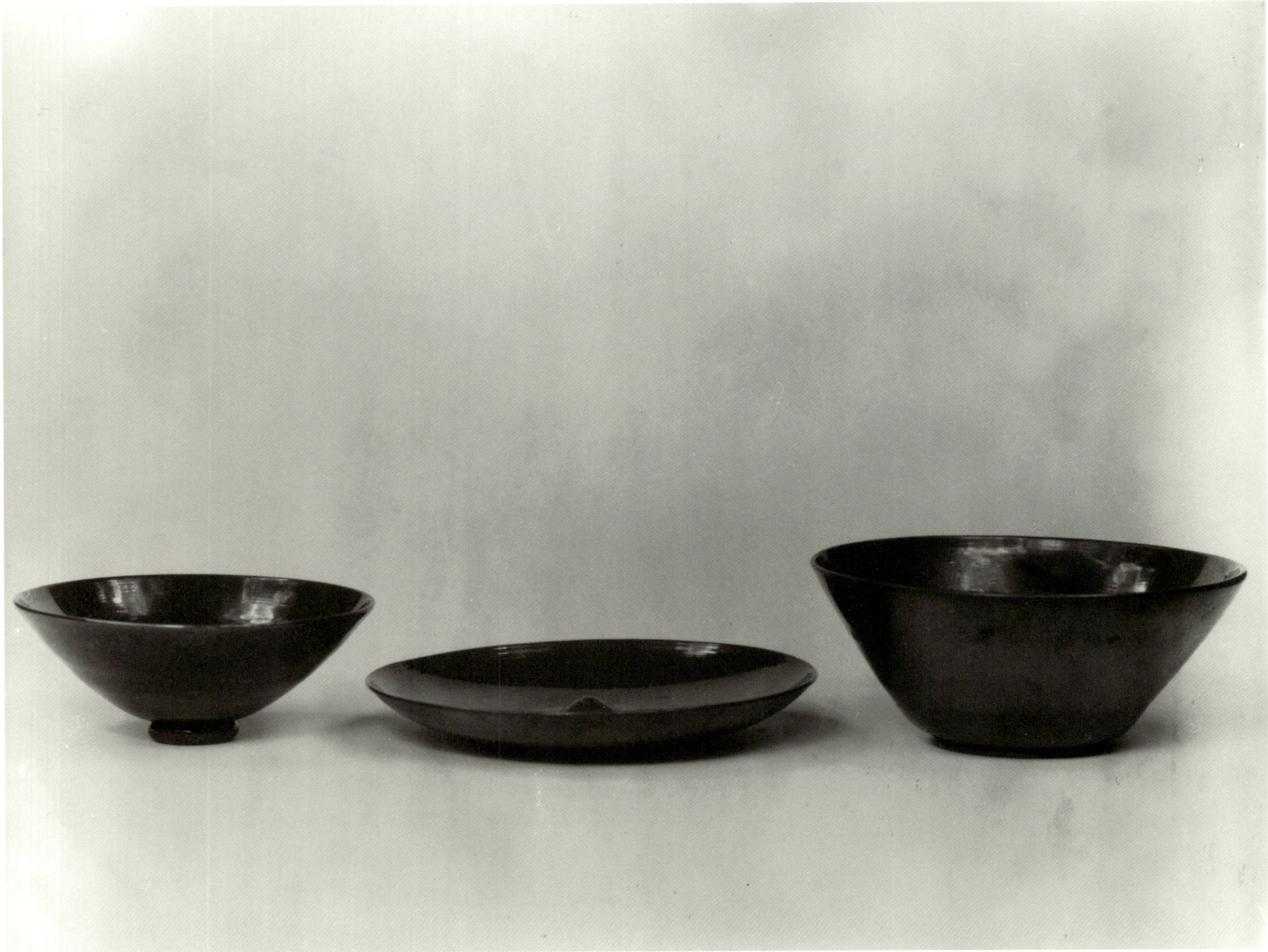

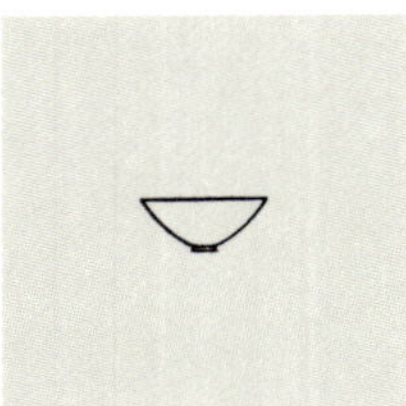

5928

Catalogue drawing for a small bowl in *pasta vitrea* documented by a period photograph, AMB. An example of this model produced in 'pea-green iridised gold-sprayed' *pasta vitrea* was exhibited in 1929 at the Leipzig Trade Fair. On that occasion the variant in 'lattimo iridised gold-sprayed and coral-red decoration' was also exhibited. Model no. 5928 was also produced in cased glass.
c. 1929
6 cm high; 16 cm ø

exhibitions
1929, Milano Fair
1929, Leipzig, Trade Fair

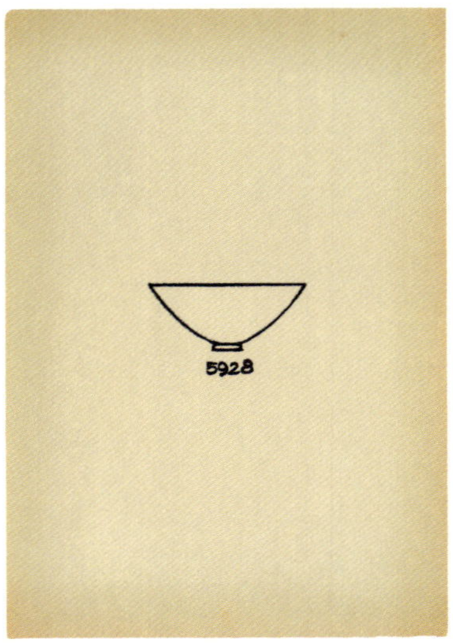

n.i.

Bowl in coral-red *pasta vitrea* with iridised finish.
c. 1929
19 cm ø
Private collection

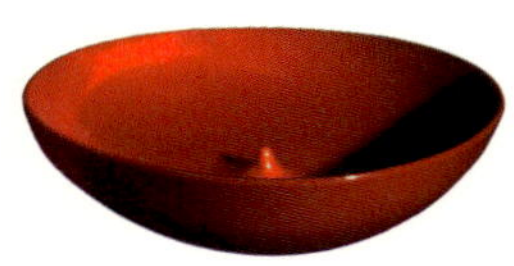

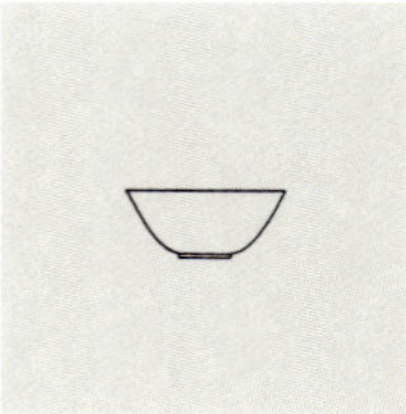

5813

Bowl in coral-red *pasta vitrea* with iridised finish. Small ringed-shaped foot wound round with thread.
c. 1929
8 cm high; 20 cm ø
Biennale 1964 old label
Private collection

exhibitions
1964, Venice, 32nd Biennale d'arte

›
5813, catalogue drawing, AMB

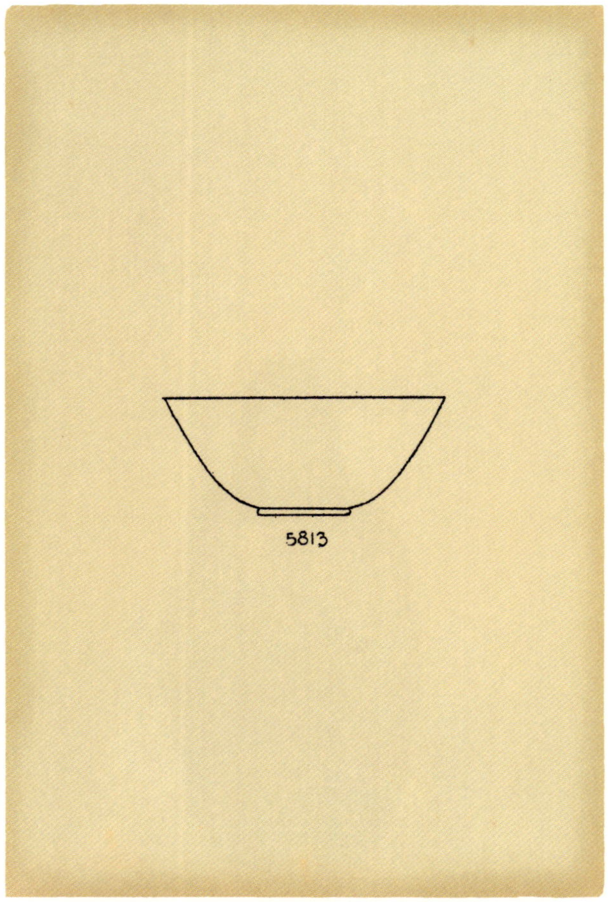

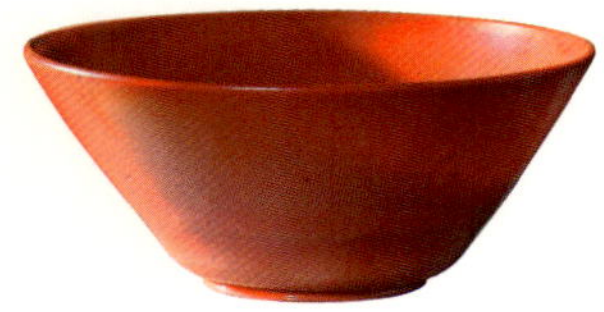

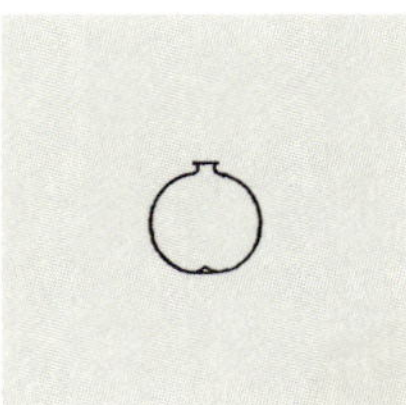

5932

Spherical vase in coral-red *pasta vitrea* with applied gold leaf and iridised finish. A similar example was exhibited in 1929 at the Milano Fair. Model no. 5932 was also produced in other glass techniques (e.g. *lattimo aurato*, *decoro fenicio*).
c. 1929
13 cm high
Fondazione Musei Civici di Venezia, Museo del Vetro collection

exhibitions
1929, Milano Fair
1964, Venice, 32nd Art Biennale

›
5932, catalogue drawing, AMB

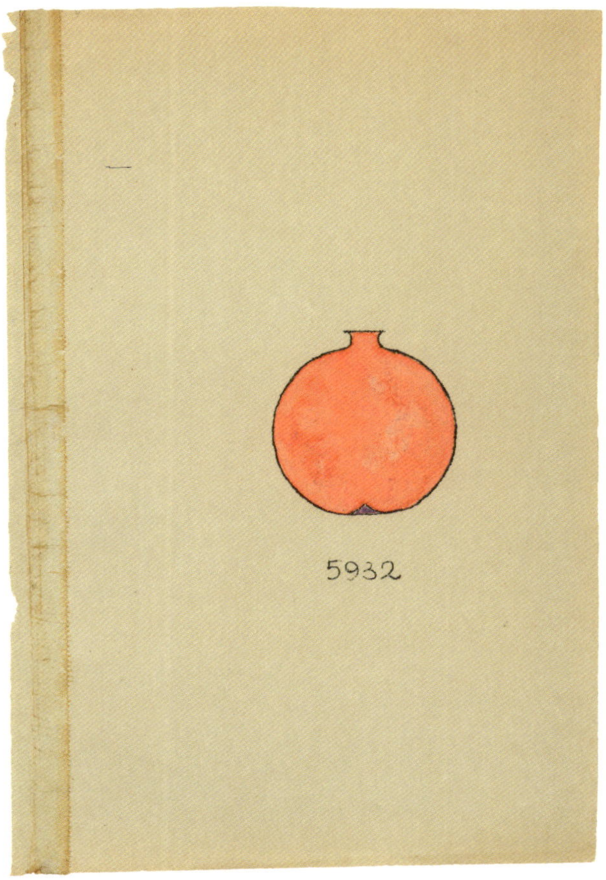

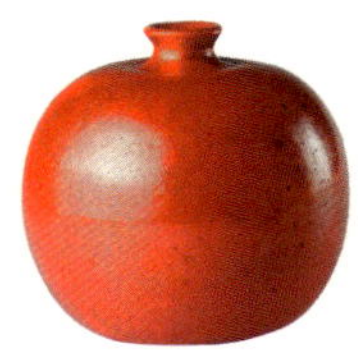

>
Vases with foot in pasta vitrea. *Period photograph, c. 1929, AMB*

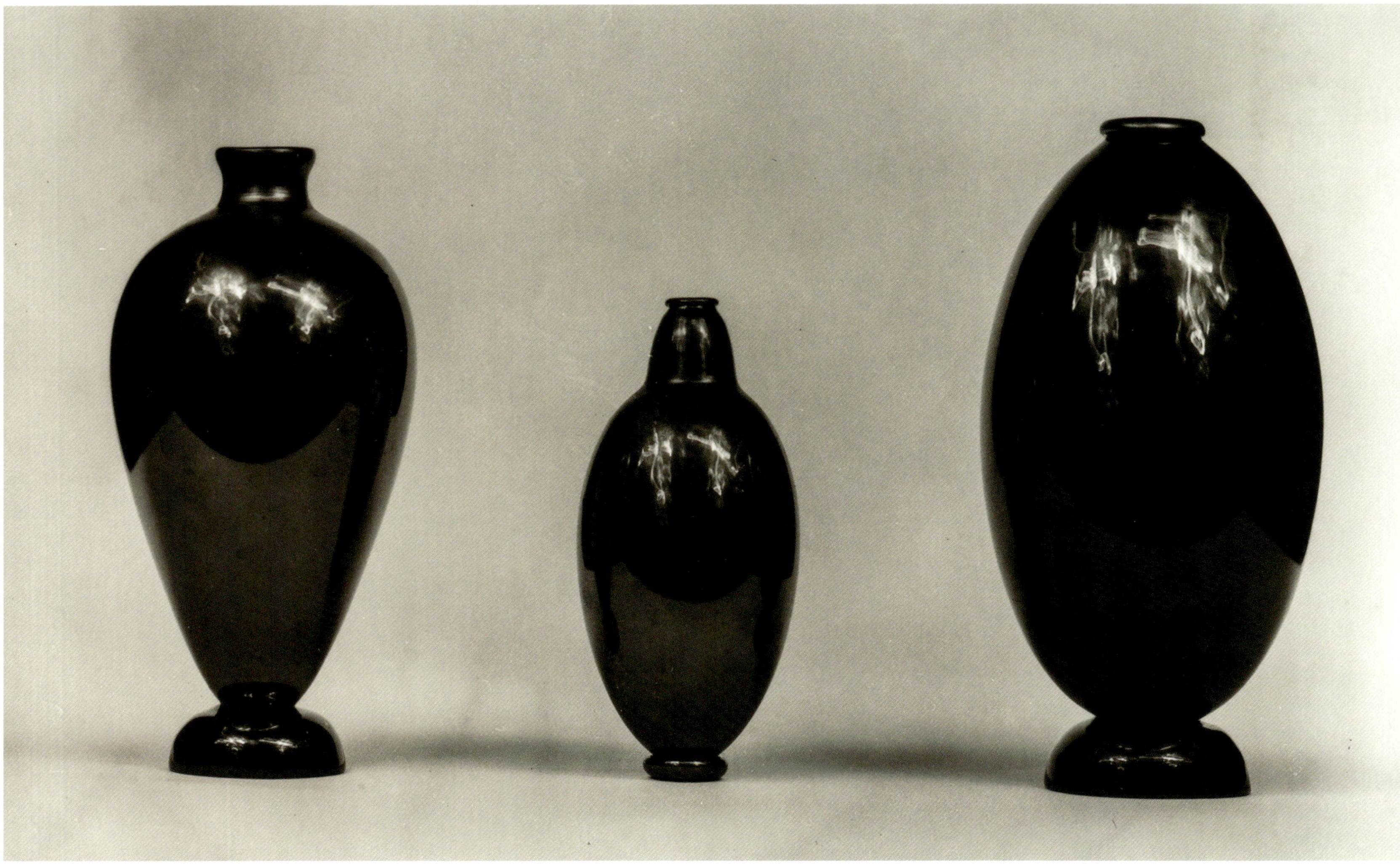

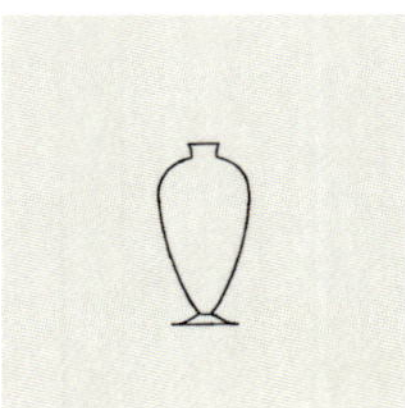

5870

Vase with skull-cap foot in coral-red *pasta vitrea* with iridised finish. A similar example was exhibited in 1929 at the Milano Fair. Model no. 5870 was also produced in other glass techniques (e.g. transparent, *decoro fenicio*, iridised black, *lattimo*).
c. 1929
21 cm high
M.V.M. Cappellin Murano *acid stamp*
The Olnick Spanu Collection, New York

exhibitions
1929, Milano Fair

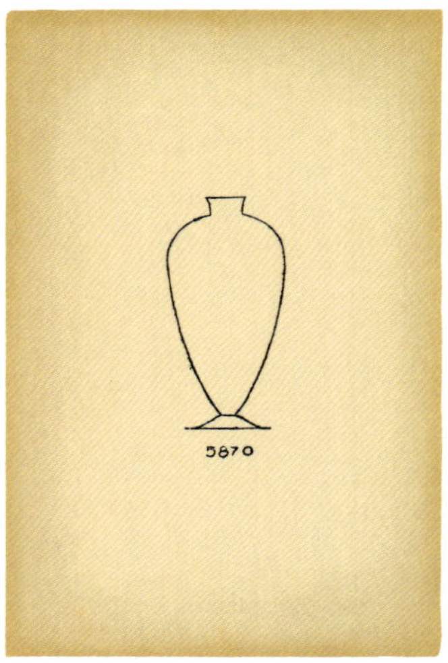

>
5870, catalogue drawing, AMB

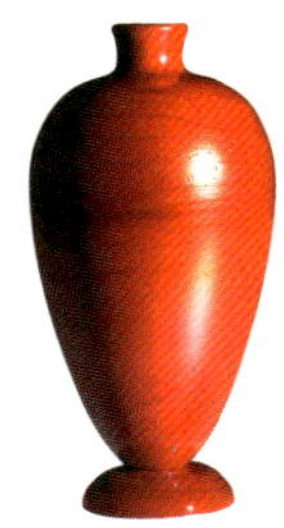

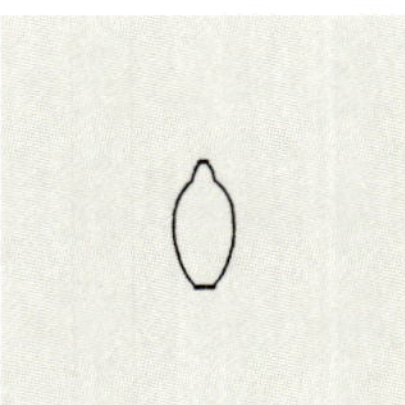

5920

Catalogue drawing for one of the vases in *pasta vitrea* documented by a period photograph, AMB. Two examples of no. 5920 produced in *pasta vitrea*, one 'gold-sprayed iridised pea-green', the other in 'iridised coral-red' were exhibited in 1929 at the Leipzig Trade Fair. Model no. 5920 was also produced in other glass techniques (e.g. *lattimo aurato*).
c. 1929
15 cm high

exhibitions
1929, Milan, Trade Fair
1929, Leipzig, Trade Fair

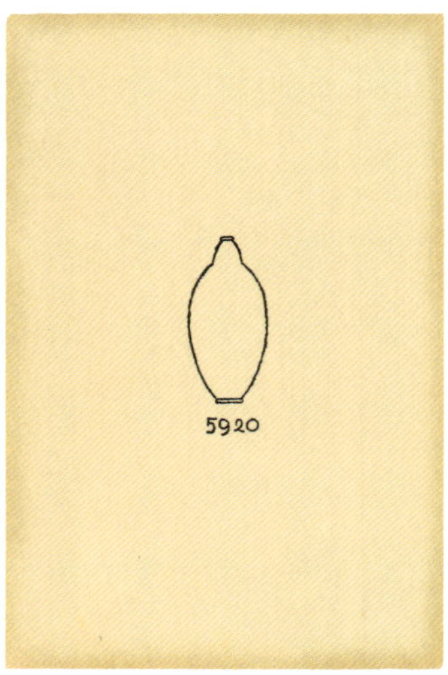

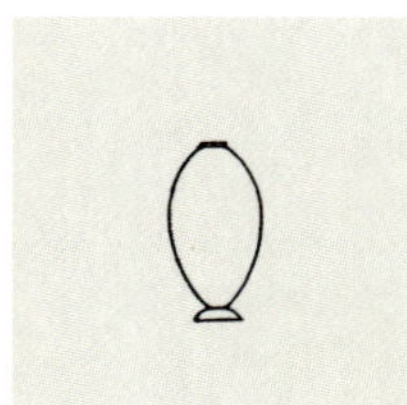

5935

Catalogue drawing for one of the vases in *pasta vitrea* documented by a period photograph, AMH. Two examples of no. 5935 produced in *pasta vitrea*, one 'gold-sprayed iridised coral-red', the other in 'iridised coral-red' were exhibited in 1929 at the Leipzig Trade Fair. Model no. 5935 was also produced with other glass techniques (e.g. *lattimo aurato*, *decoro fenicio*, cased, iridised black).
c. 1929
22 cm high

exhibitions
1929, Leipzig, Trade Fair
1929, Milan, Trade Fair

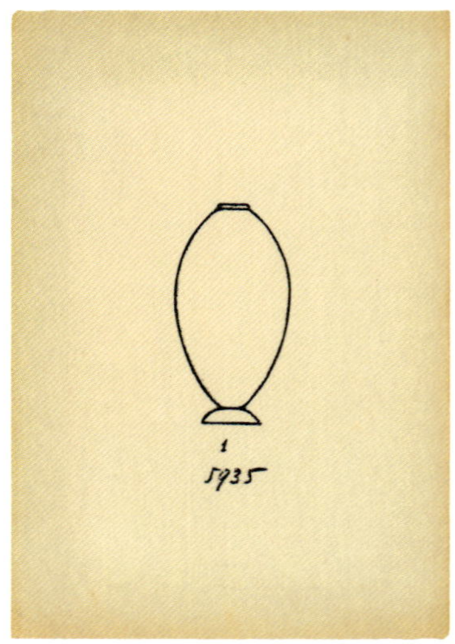

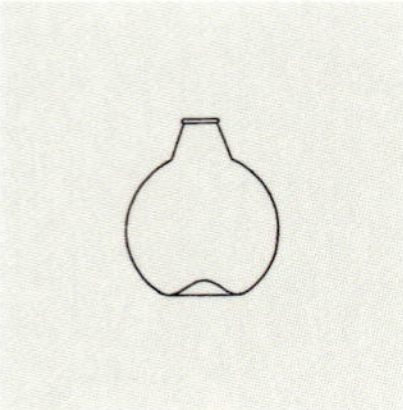

5858

Pear-shaped vase in *pasta vitrea* with applied gold leaf documented by a period photograph, c. 1929, AMB. The piece, which is depicted in the photograph together with vase no. 5952, is distinguished also by the presence of characteristic rectangular depressions on the body and neck, obtained by using a mould. An example of no. 5858 produced in 'coral-red iridised gold-sprayed' *pasta vitrea* was exhibited in 1929 at the Leipzig Trade Fair.
The same model was also produced in other glass techniques (e.g. smooth transparent, *balloton* transparent, *filigrana a reticello*, silvered black).
c. 1929
21 cm high

exhibition
1929, Leipzig, Trade Fair

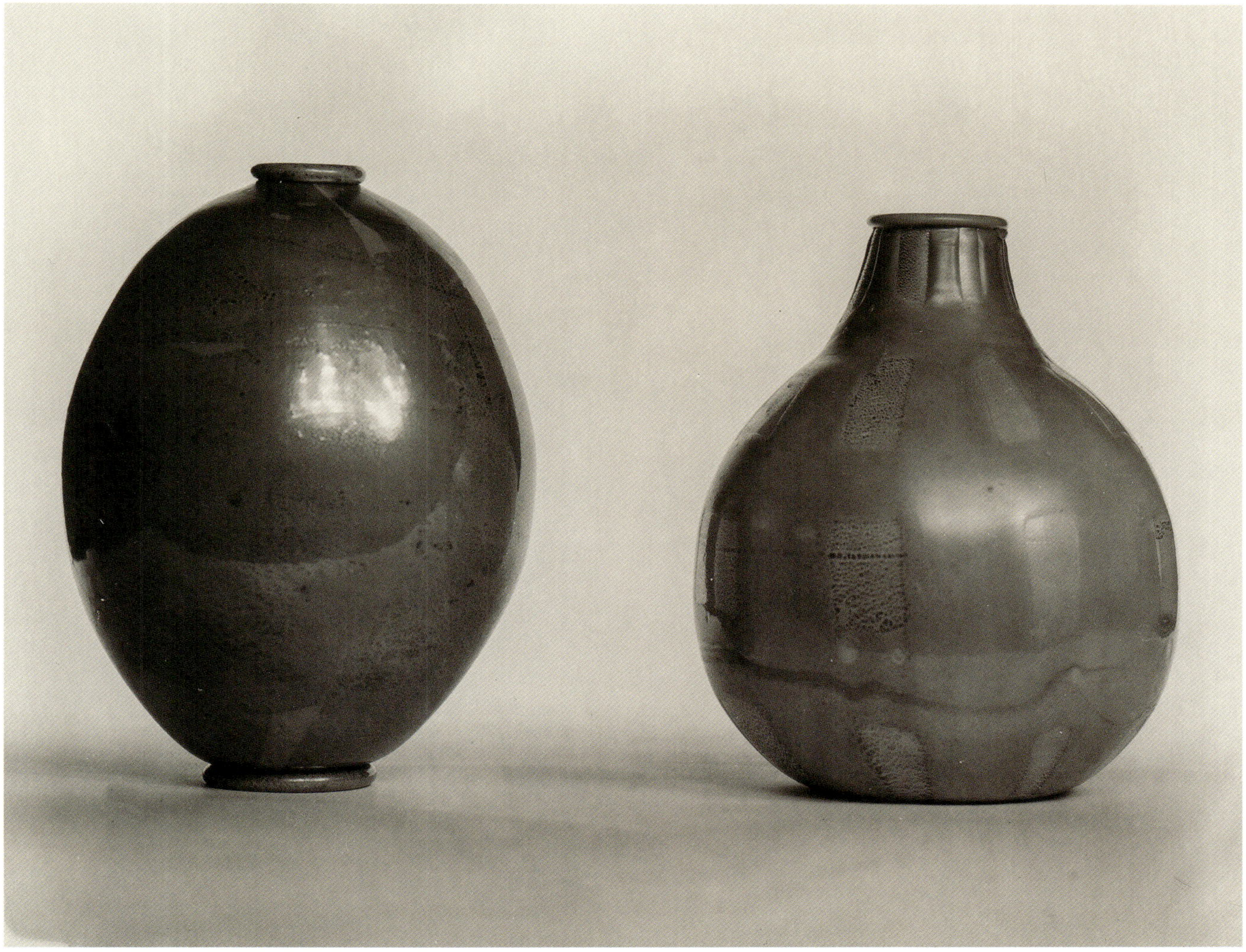

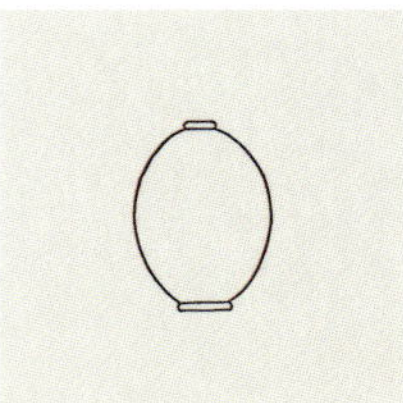

5952

Ovoid vase in coral-red *pasta vitrea* with applied silver leaf. An example in 'pea-green iridised gold-sprayed' *pasta vitrea* was exhibited in 1929 at the Leipzig Trade Fair.
c. 1929
21 cm high
Corning Museum of Glass collection, Corning, NY, USA

exhibitions
1929, Leipzig, Trade Fair

>
5952, catalogue drawing, AMB

>
'Casa T. in Florence', in Domus, *October 1933, p. 528, detail*

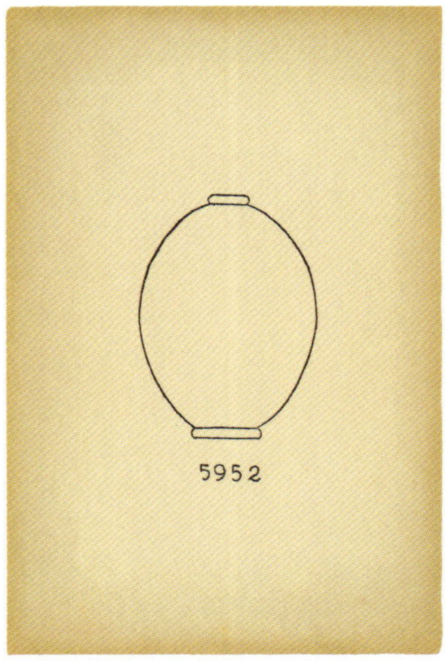

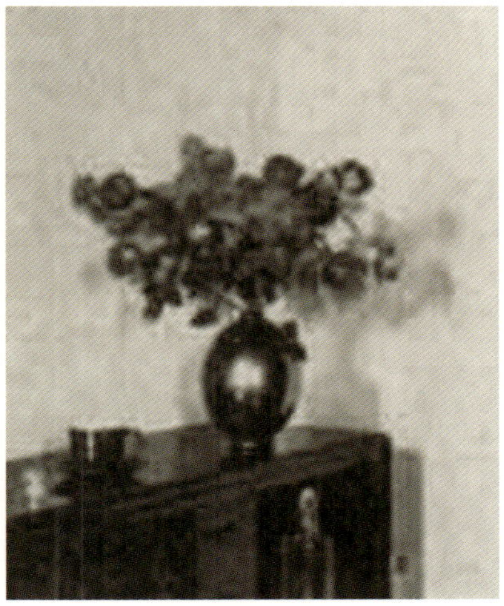

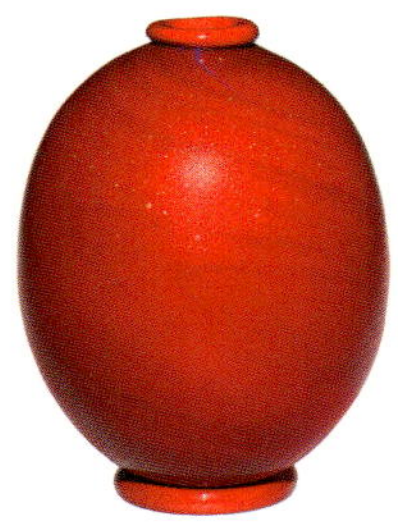

n.i.

Vase in green *pasta vitrea* with iridised finish. Small foot wound with thread.
1929-31
21 cm high
M.V.M. Cappellin Murano *acid stamp*
The Olnick Spanu Collection, New York

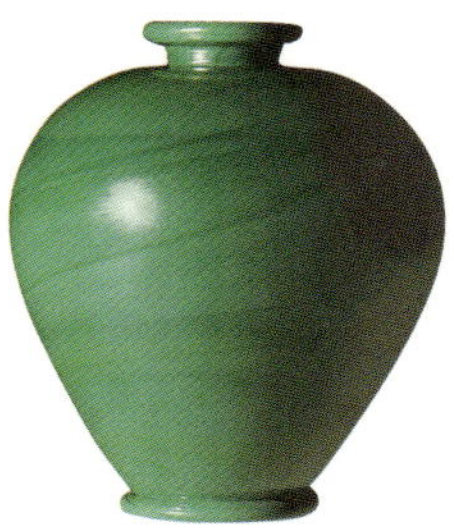

n.i.

Truncated-cone-shaped vase in green *pasta vitrea*, slightly ribbed, iridised finish. Small flared foot.
1929–31
24 cm high
Private collection

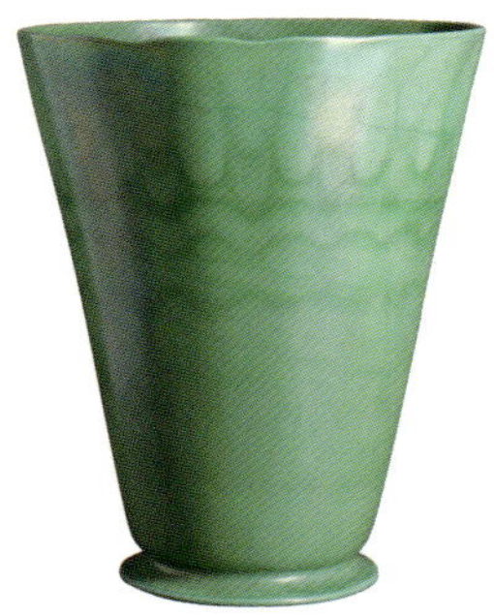

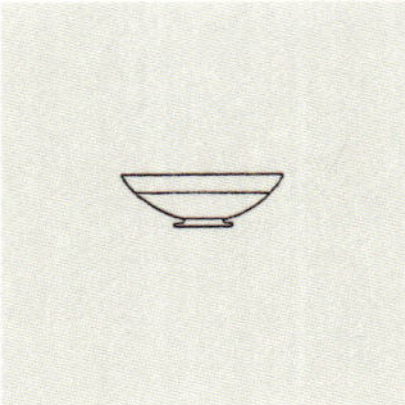

5982

Catalogue drawing for a bowl to be produced in black glass and coral-red *pasta vitrea* joined using the 'incalmo' technique, AMA. An illustration of the bowl was published in *Domus* in January 1930. Model no. 5982 was also produced in other glass techniques (e.g. cased glass).
c. 1929
6 cm high; 19 cm ø

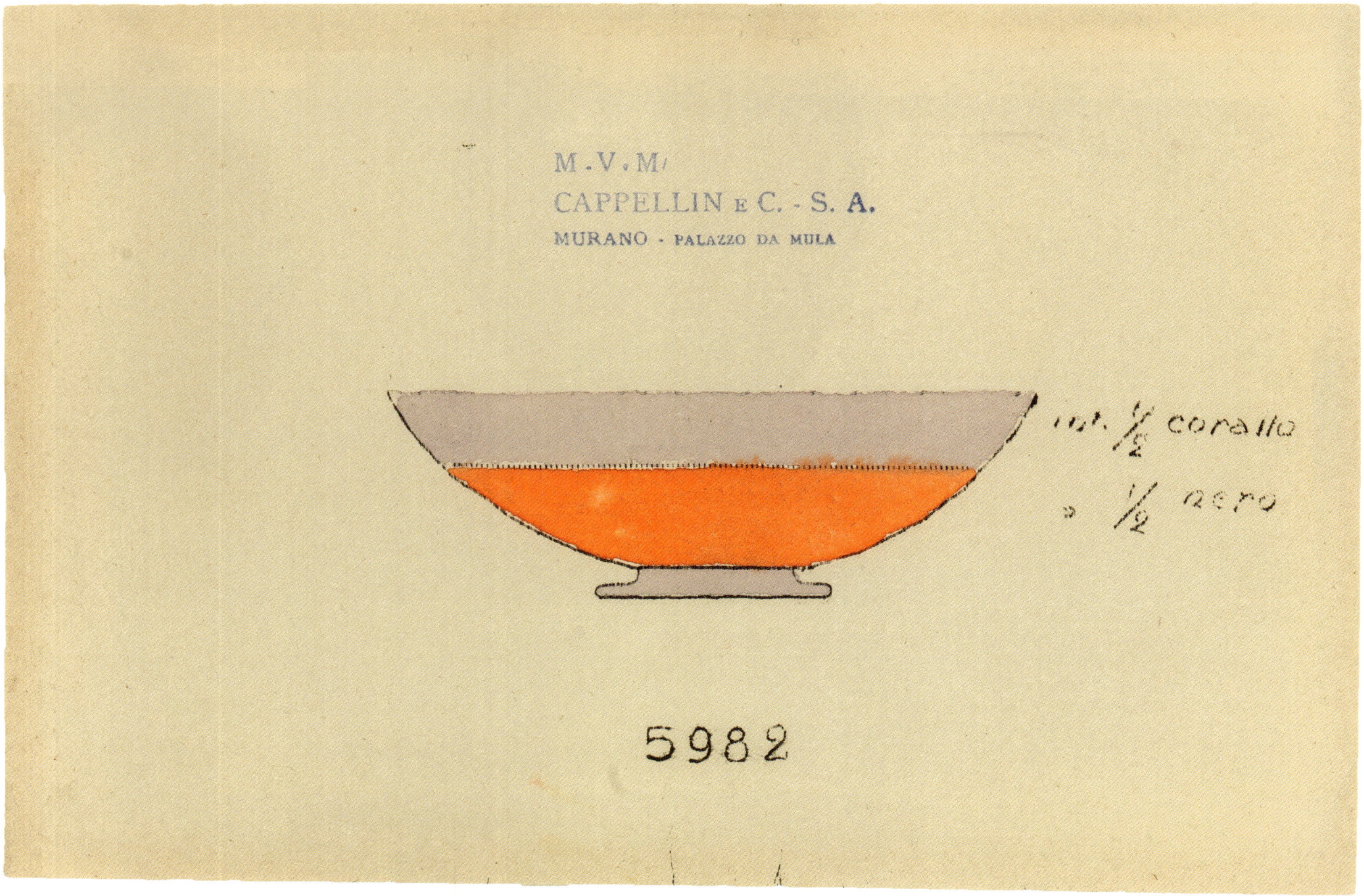

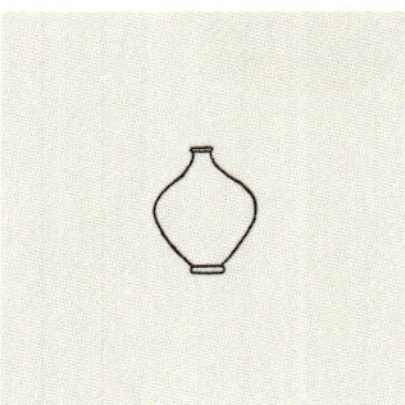

5912

Vase in coral-red *pasta vitrea* with applied gold leaf, iridised finish. Ring-shaped foot wound with thread. Model no. 5912 was also produced in other glass techniques (e.g. *lattimo*).
An illustration of the bowl was published in *Domus* in January 1930.
c. 1929
15 cm high
Private collection, Treviso

›
5912, catalogue drawing, AMB

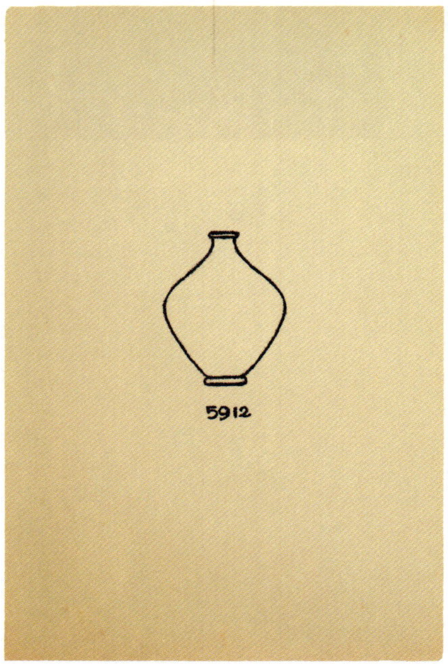

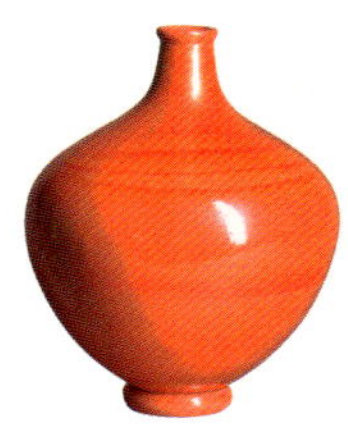

>
Vase and bowl in opaque glass, in Domus, January 1930, p. 38

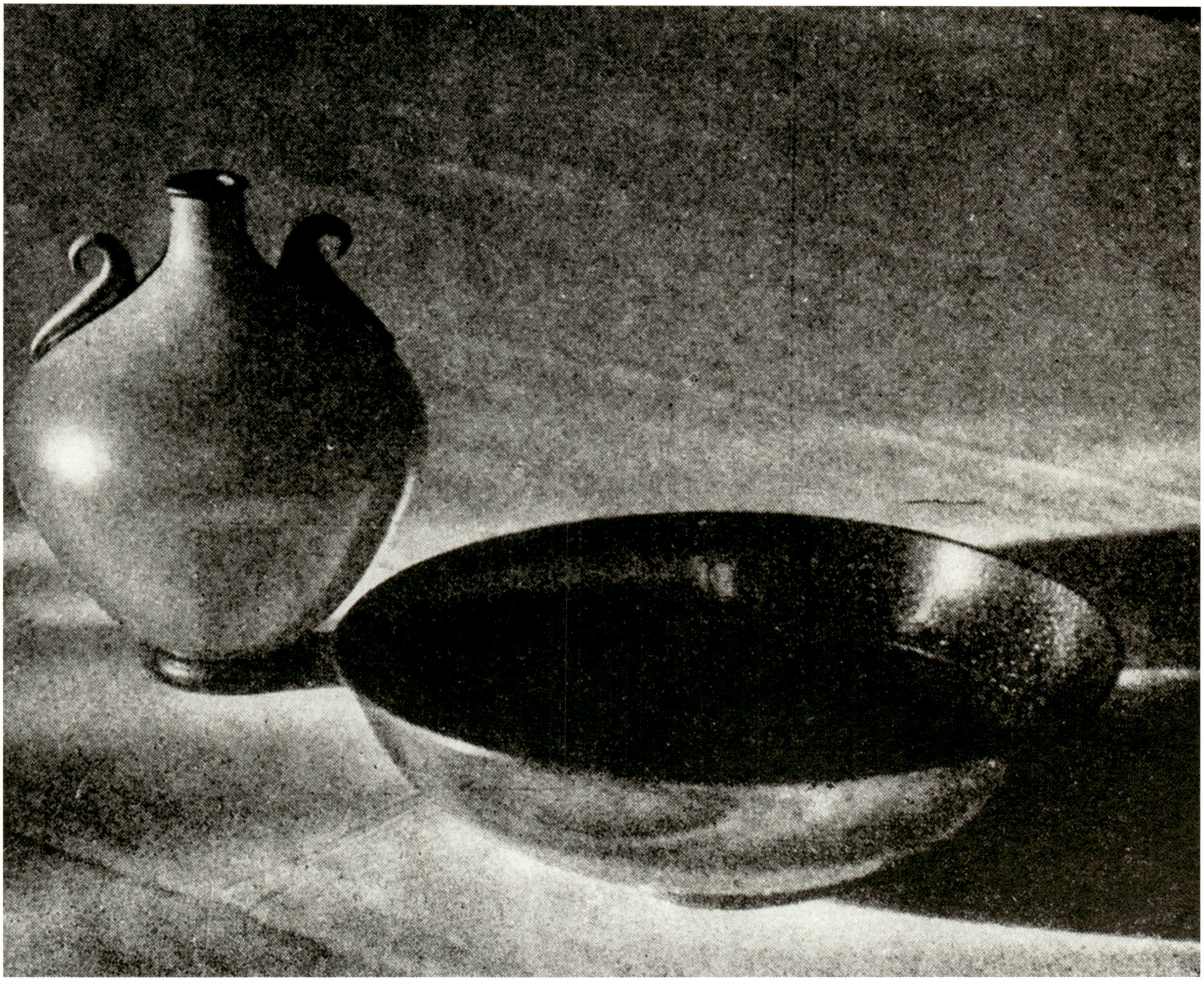

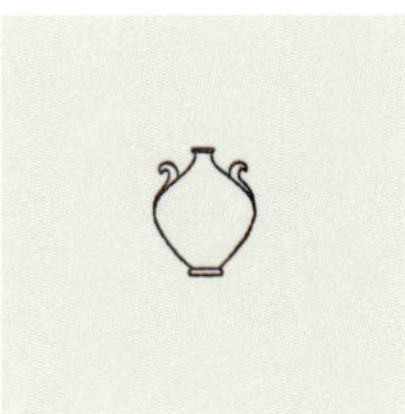

5911

Vase in coral-red and gold *pasta vitrea* with curled side handles and ring-shaped foot wound with thread. Surface with iridised finish.
c. 1929
15 cm high
The Olnick Spanu Collection, New York

exhibitions
1931, Amsterdam, Exhibition of glass, ceramics and lace

>
5911, catalogue drawing, AMB

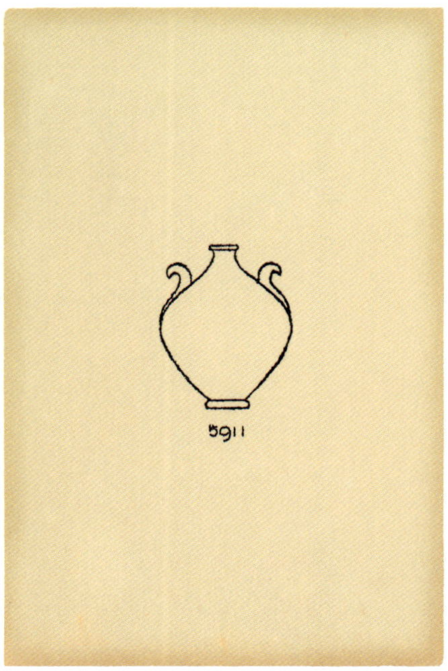

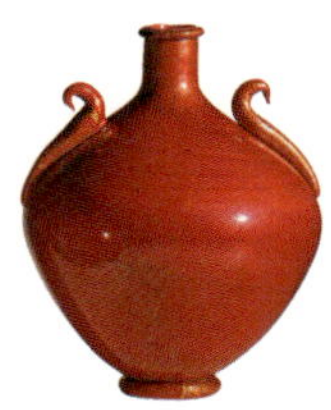

>
Vase and bowls in pasta vitrea *with side handles. Period photograph, c. 1929, AMB*

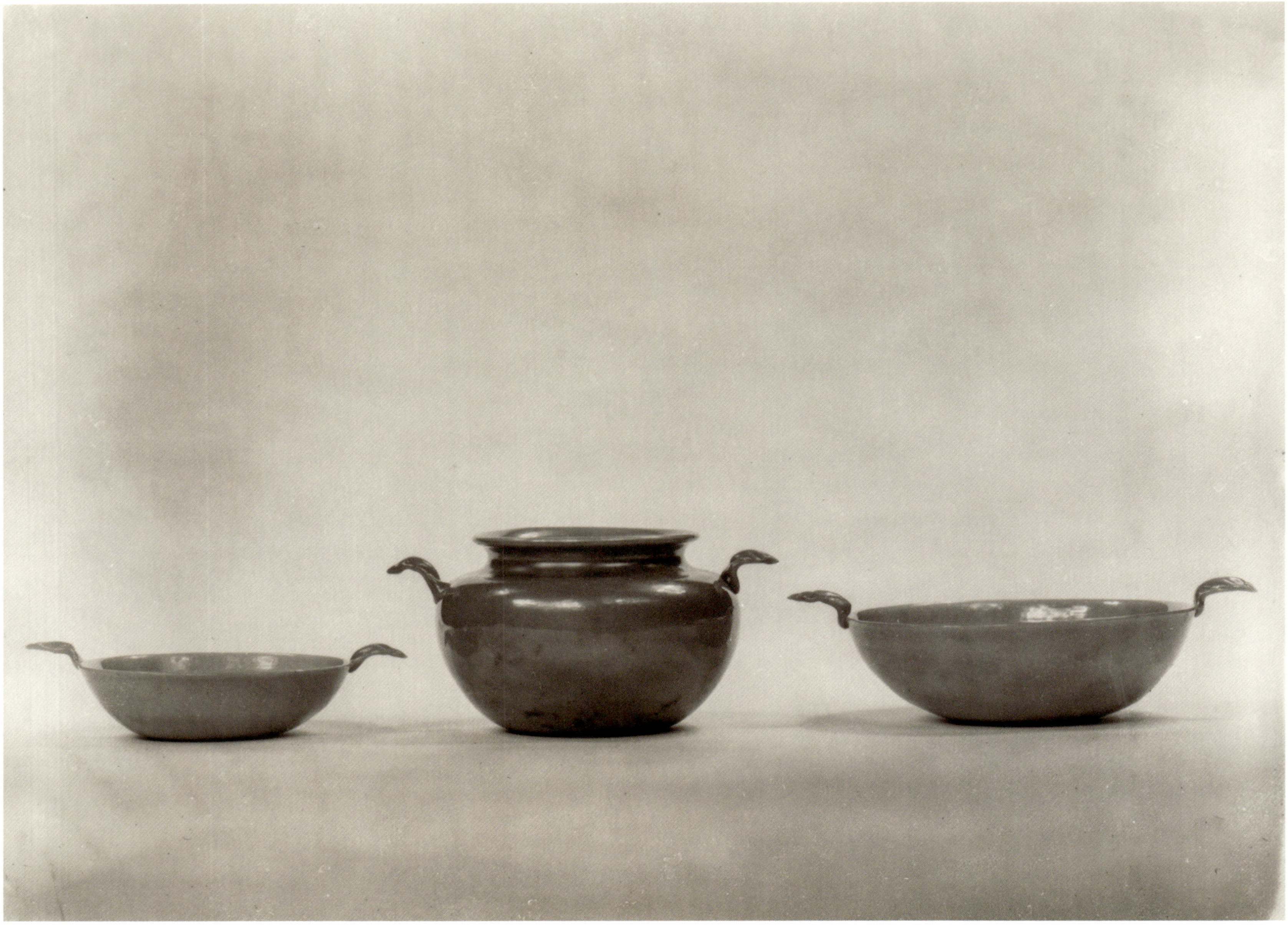

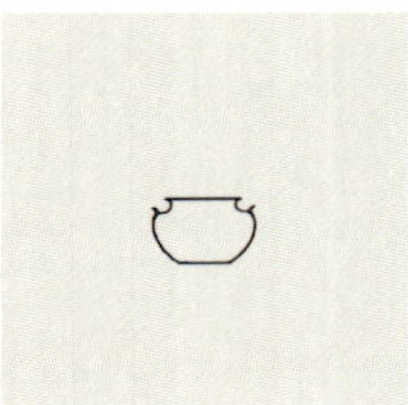

5939

Catalogue drawing for a small vase in *pasta vitrea* with side handles documented by a period photograph, AMB. The model was exhibited in 1929 at the Leipzig Trade Fair in the variant 'gold-sprayed iridised lattimo and coral-red decoration'.
c. 1929
8 cm high

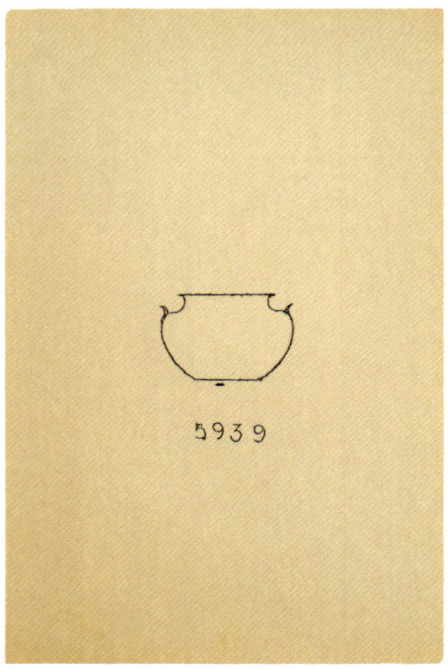

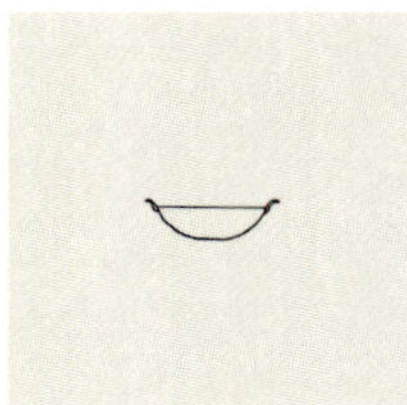

5940

Working drawing for a small vase in *pasta vitrea* with side handles documented by a period photograph, CSV. An example of this model produced in 'coral-red iridised gold-sprayed' *pasta vitrea* was exhibited in 1929 at the Leipzig Trade Fair.
c. 1929
16 cm ø

exhibitions
1929, Leipzig, Trade Fair

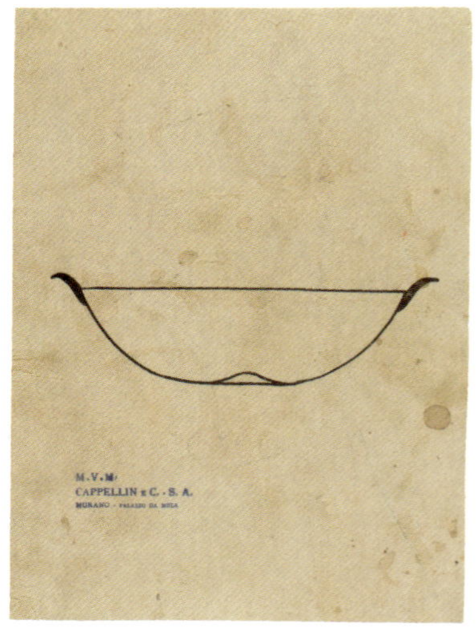

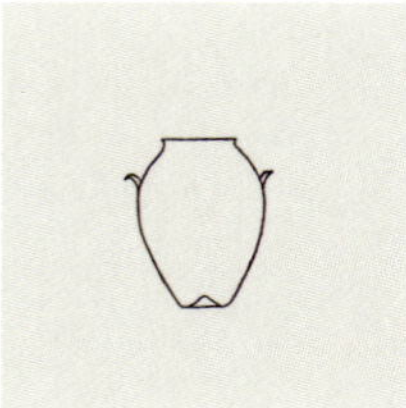

5942

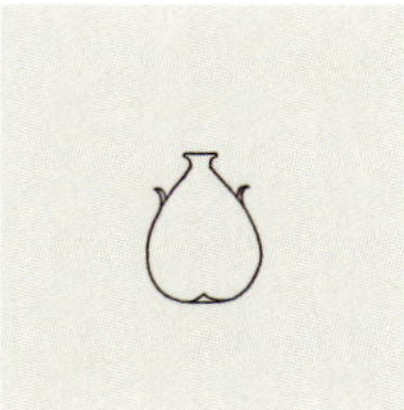

5941

Vases in *pasta vitrea* with side handles documented by a period photograph, c. 1929, AMB. Model no. 5941 was also produced in other glass techniques (e.g. transparent, iridised black, silvered black).
c. 1929
20; 18 cm high

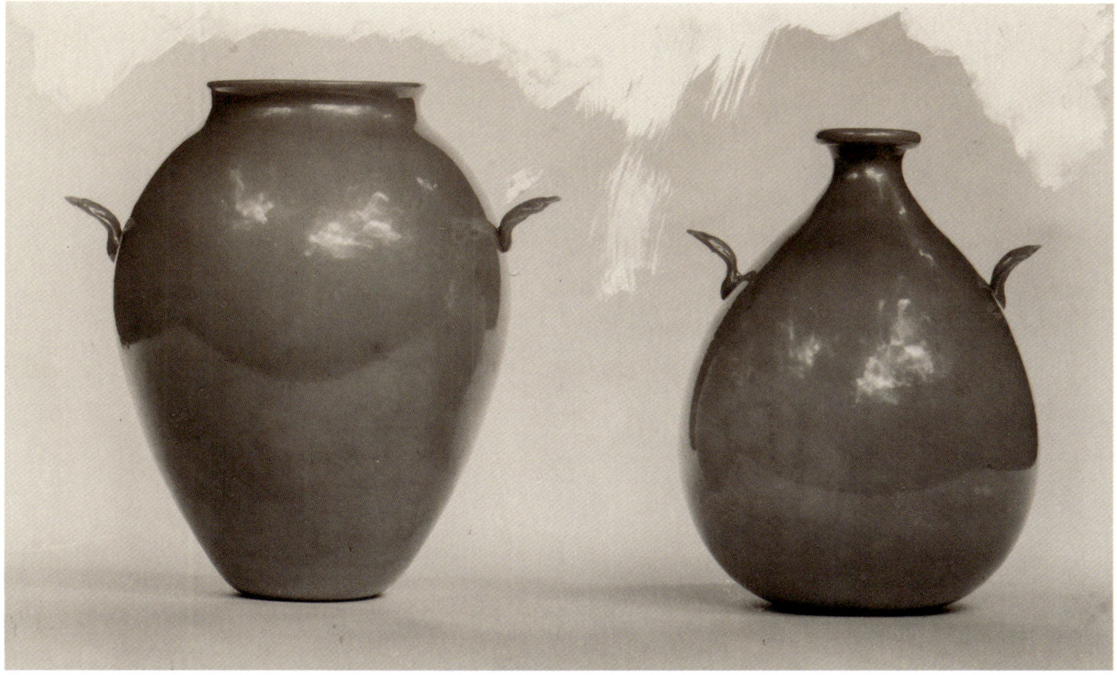

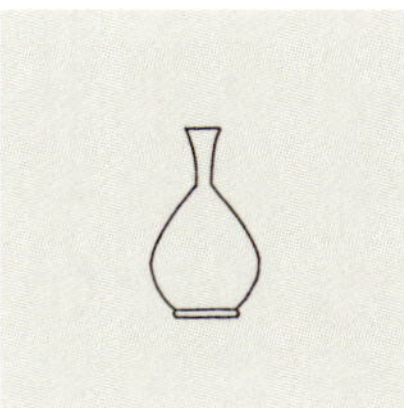

5916

Vase in coral-red and gold *pasta vitrea* with long flared neck. Ring-shaped foot wound with thread. Surface with iridised finish. An example of no. 5916 produced in 'pea-green iridised gold-sprayed' *pasta vitrea* was exhibited in 1929 at the Leipzig Trade Fair. On that occasion the variant in 'iridised gold-sprayed lattimo and coral-red decoration' was also shown.
c. 1929
22 cm high
Lutz H. Holz collection

exhibitions
1929, Leipzig, Trade Fair

>
5916, catalogue drawing, AMB

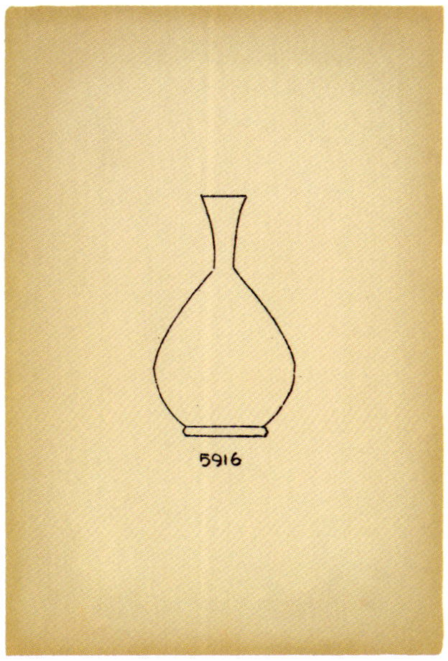

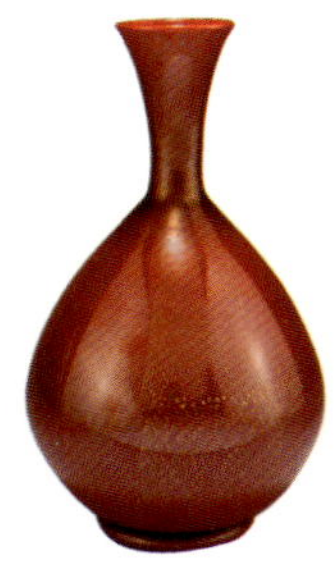

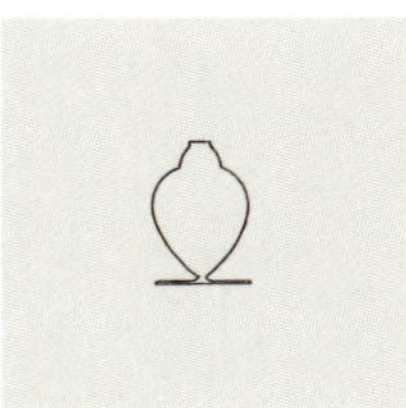

5895

Vase with disk-shaped foot in coral-red and gold *pasta vitrea*, iridised finish. Model no. 5895 was also produced in other glass techniques (e.g. *lattimo*).
c. 1929
17.2 cm high
Martin Kline collection, NY

>
5895, catalogue drawing, AMB

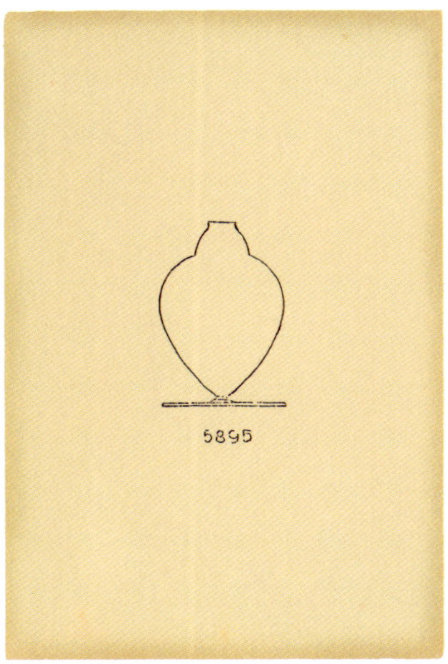

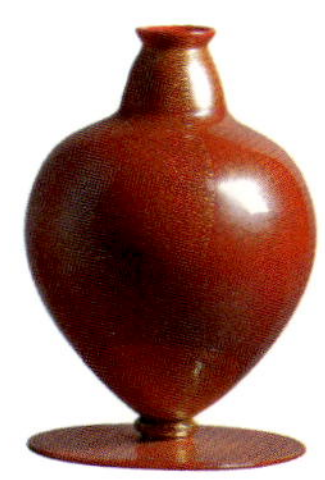

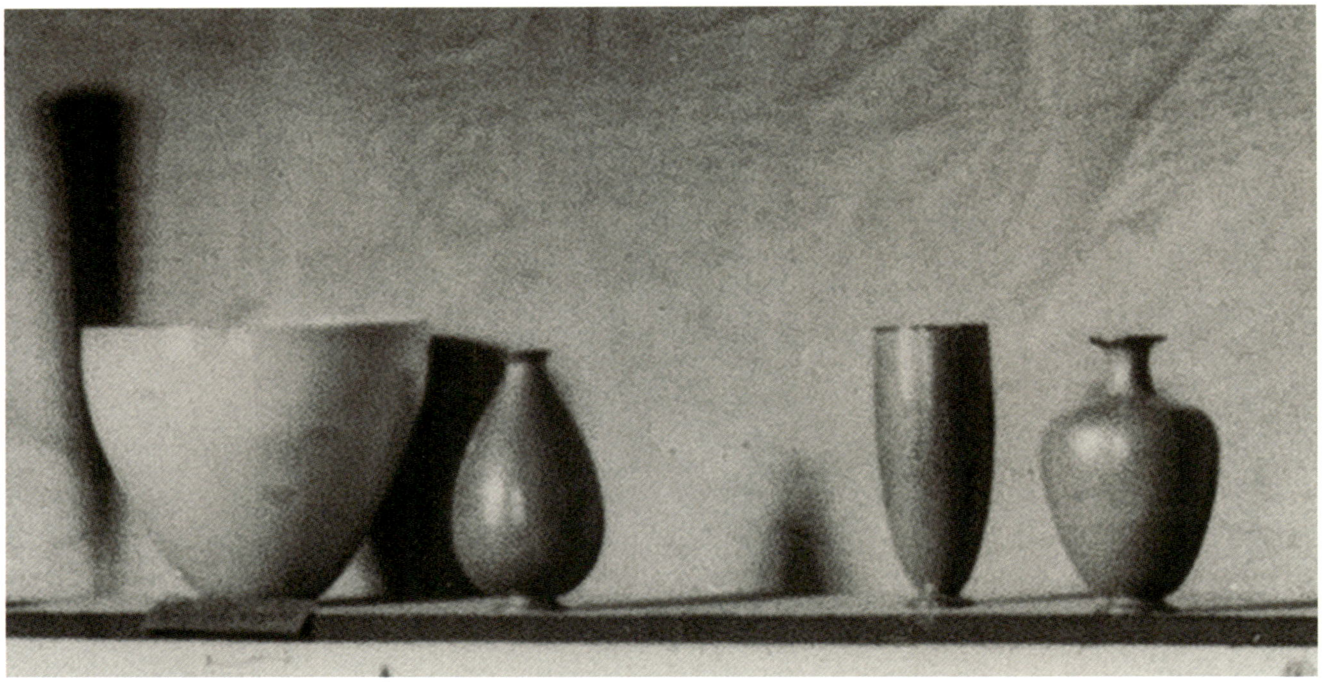

›
Cappellin glass pieces exhibited at the Amsterdam Exhibition of glass, ceramics and lace, 1931, AHL, detail

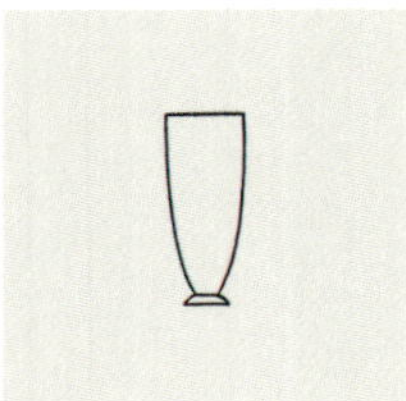

5948

Vase with skull-cap foot in coral-red *pasta vitrea* with applied gold leaf. A similar example was exhibited in 1929 at the Milano Fair and in 1931 at the Amsterdam Exhibition of glass, ceramics and lace. Model no. 5948 was also produced in other glass techniques (e.g. *lattimo aurato*, cased glass).
c. 1929
22 cm high
Private collection

exhibitions
1929, Milano Fair
1931, Amsterdam, Exhibition of glass, ceramics and lace

›
5948, catalogue drawing, AMB

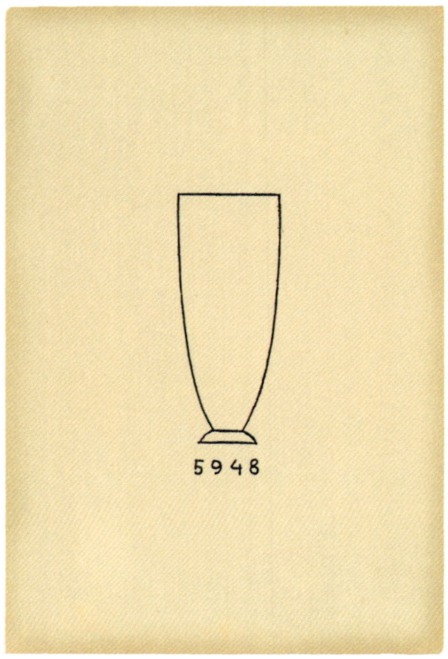

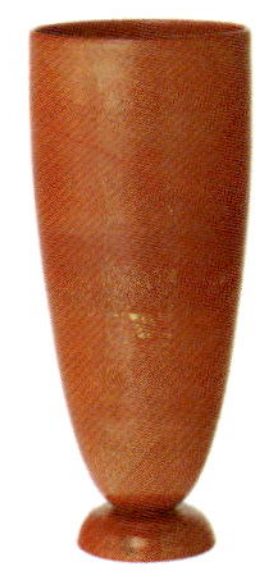

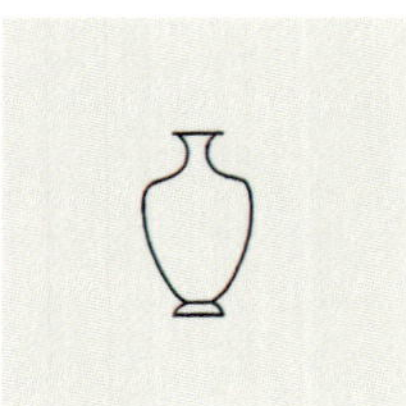

5947

Vase in coral-red *pasta vitrea*, with small foot, iridised finish. A similar example was on display in 1931 at the Amsterdam Exhibition of glass, ceramics and lace. Model no. 5947 was also produced in other glass techniques (e.g. *decoro fenicio*, iridised black).
c. 1929
20 cm high
Galleria d'arte moderna Carlo Rizzarda collection, Feltre

exhibitions
1931, Amsterdam, Exhibition of glass, ceramics and lace

›
5947, catalogue drawing, AMB

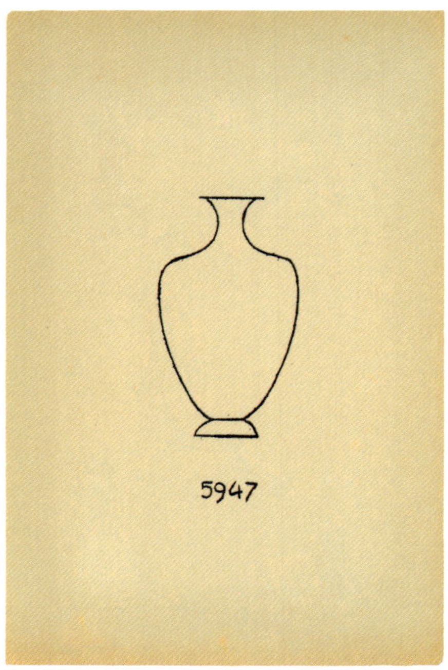

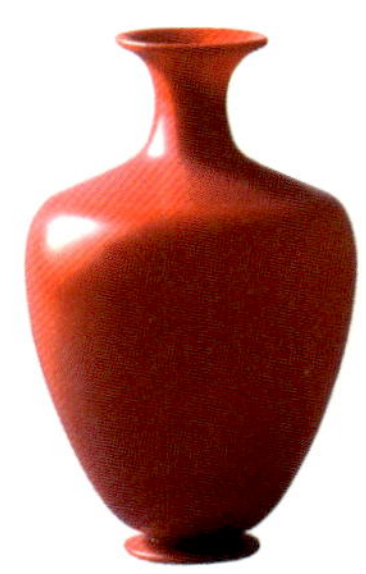

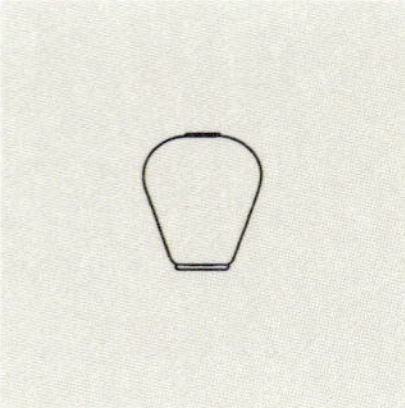

5856

Vase in green *pasta vitrea* with applied gold leaf. Thread around the mouth and ringed-shaped foot wound with thread. Two examples of no. 5856 produced one in 'pea-green iridised gold-sprayed' *pasta vitrea*, the other in 'iridised coral-red' were exhibited in 1929 at the Leipzig Trade Fair. On that occasion the variant in 'iridised silver-sprayed black' was also exhibited. Model no. 5856 was also produced in other glass techniques, such as *decoro fenicio* and *lattimo aurato*.
c. 1929
16.5 cm high
M.V.M. Cappellin Murano *acid stamp*
The Olnick Spanu Collection, New York

exhibitions
1929, Leipzig, Trade Fair

>
Vases nos 5858 and 5856 in pasta vitrea. *Period photograph, c. 1929, AMB*

>
5856, catalogue drawing, AMB

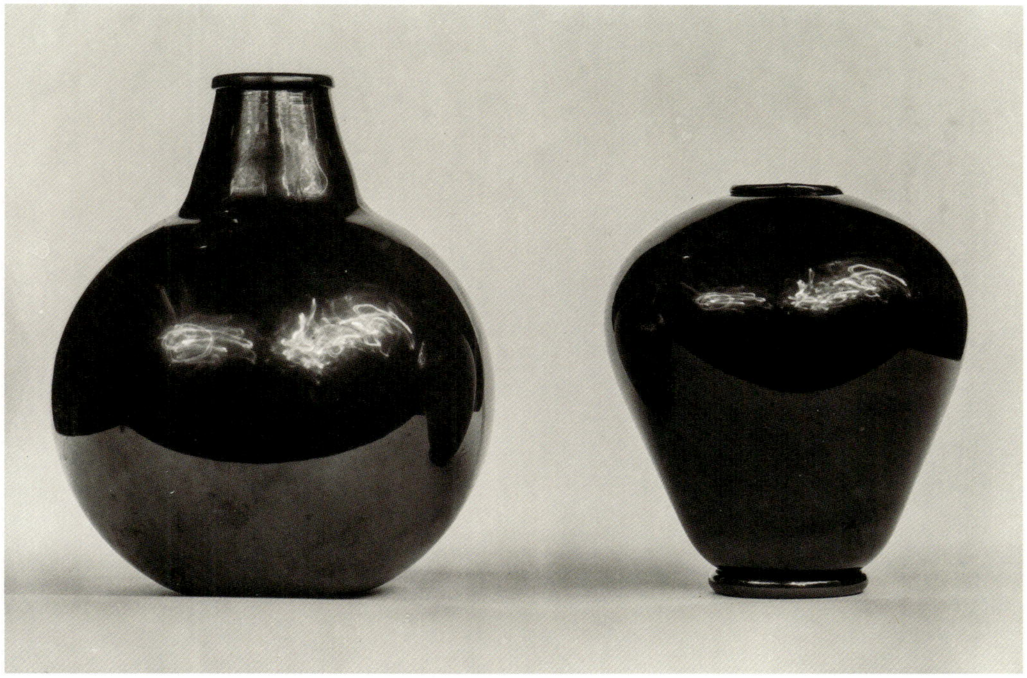

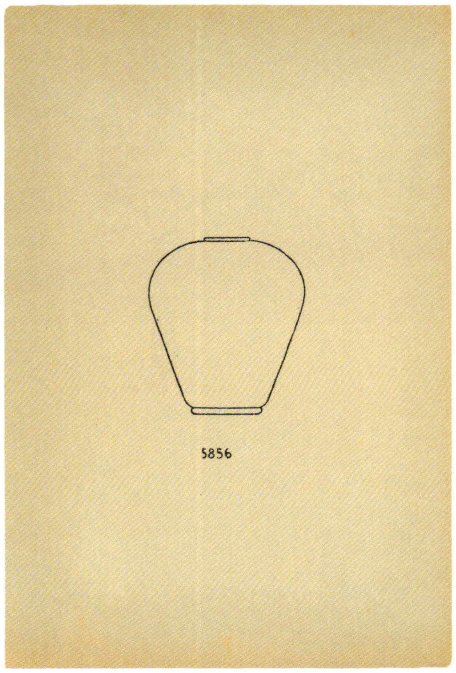

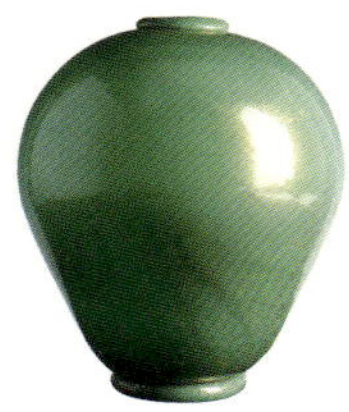

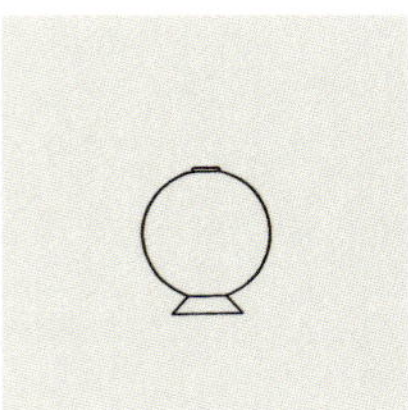

5680

Spherical vase with truncated-cone-shaped foot in coral-red *pasta vitrea* with applied gold leaf. Model no. 5680 was also produced in other glass techniques (e.g. transparent, *lattimo aurato*, cased glass).
c. 1929
18 cm high
M.V.M. Cappellin Murano *acid stamp*
Private collection, Montebelluna

>
5680, catalogue drawing, AMB

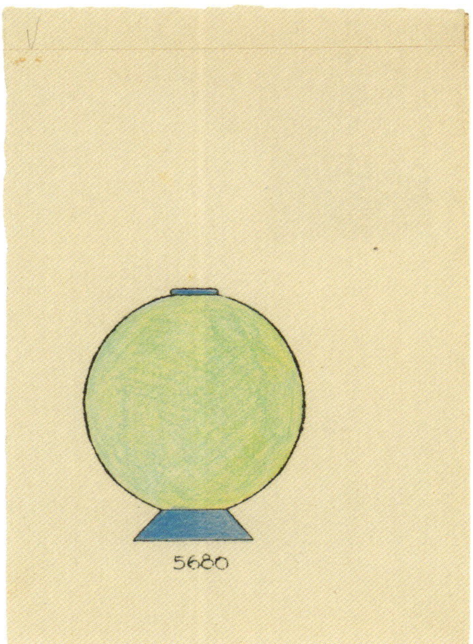

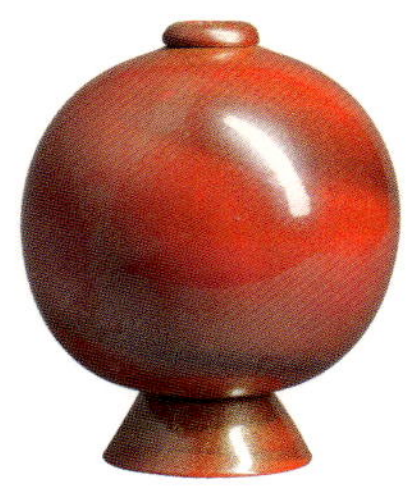

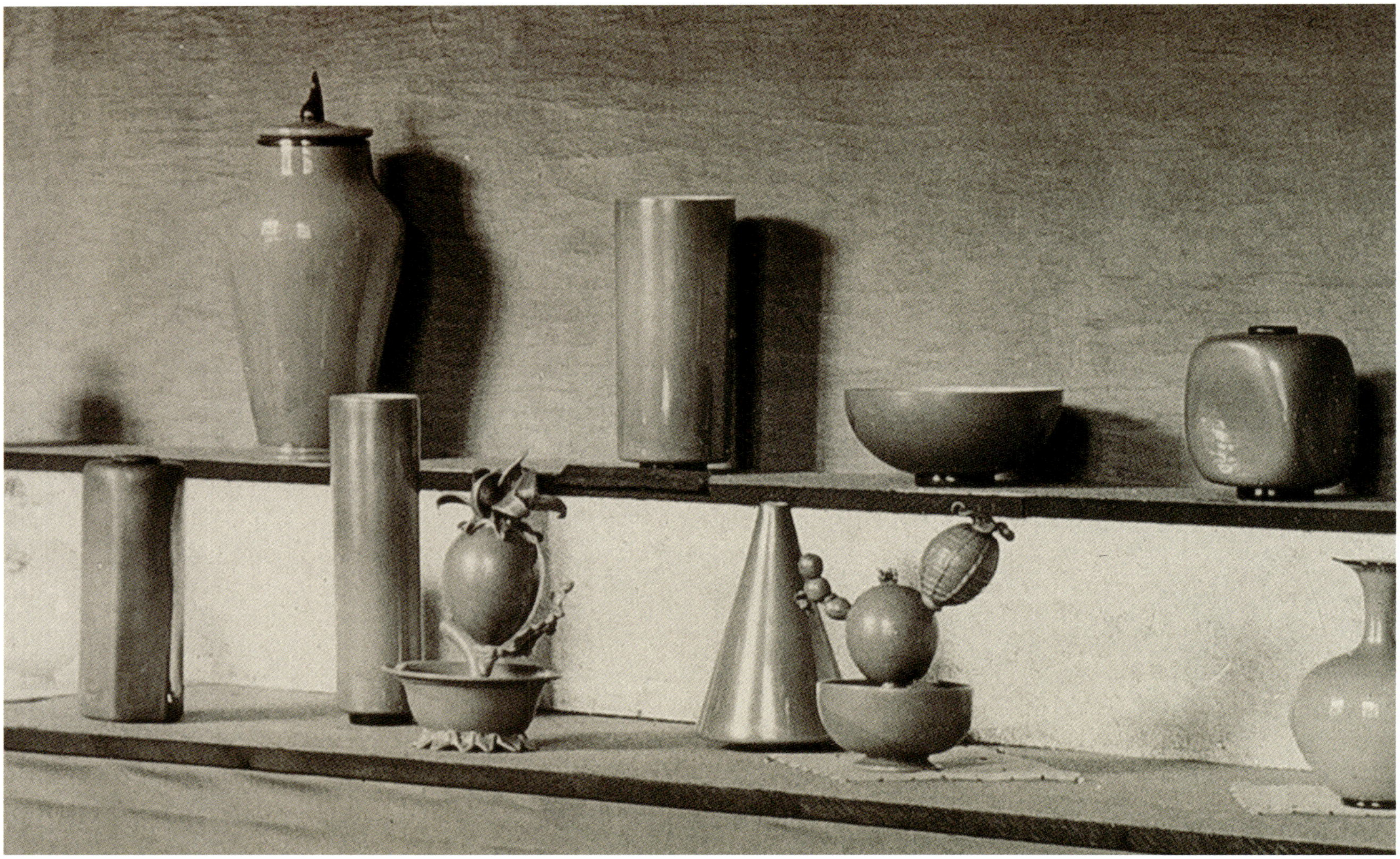

>
Cappellin glass pieces exhibited at the Amsterdam Exhibition of glass, ceramics and lace, 1931, AHL, detail

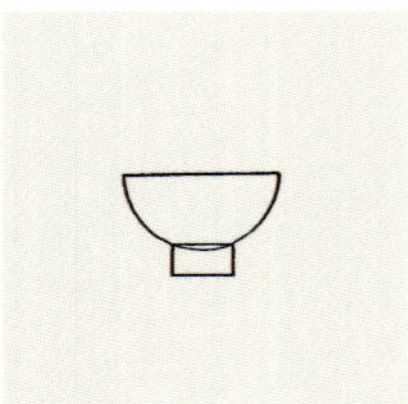

5995

Catalogue drawing for a bowl with cylindrical foot to be produced in coral-red *pasta vitrea* and in black glass (foot and thread around the mouth), AMA. On the same page of the catalogue, model no. 5996 is also shown, with circular *murrine* decorations.
c. 1930
12 cm high; 18 cm ø

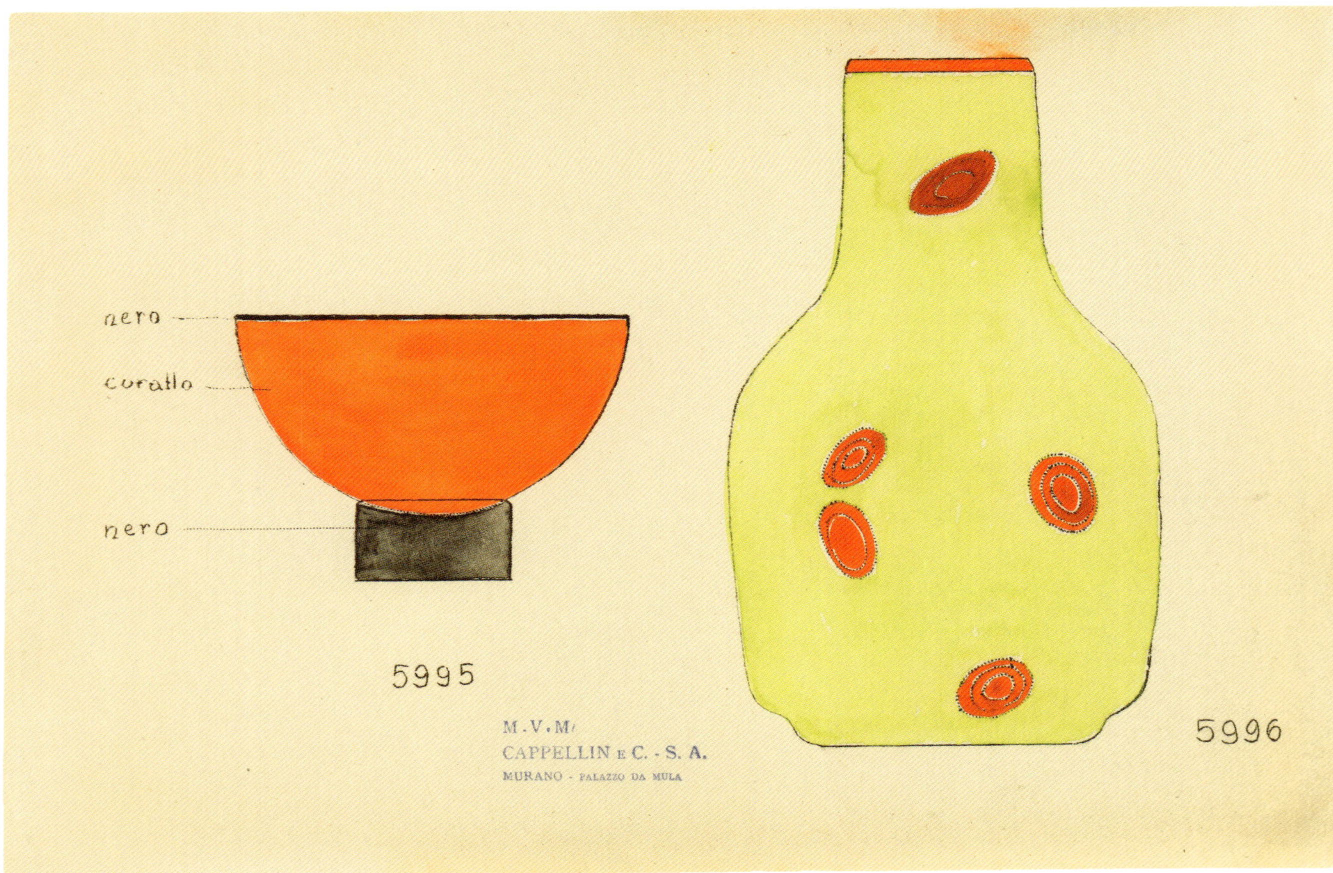

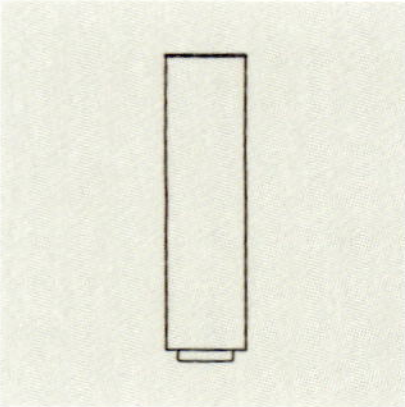

5973

Cylindrical vase with small foot in Savoy-blue *pasta vitrea* with applied gold leaf. The vase is a variant in a smaller size of no. 5973 (h 36 cm). An example of no. 5973, produced in *pasta vitrea*, was shown in 1931 at the Amsterdam Exhibition of glass, ceramics and lace. Model no. 5973 was also produced in other glass techniques (e.g. cased glass).
c. 1930
22 cm high
Private collection, Treviso

exhibitions
1931, Amsterdam, Exhibition of glass, ceramics and lace

>
5973, catalogue drawing, AMA

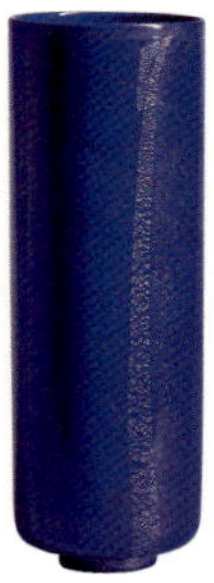

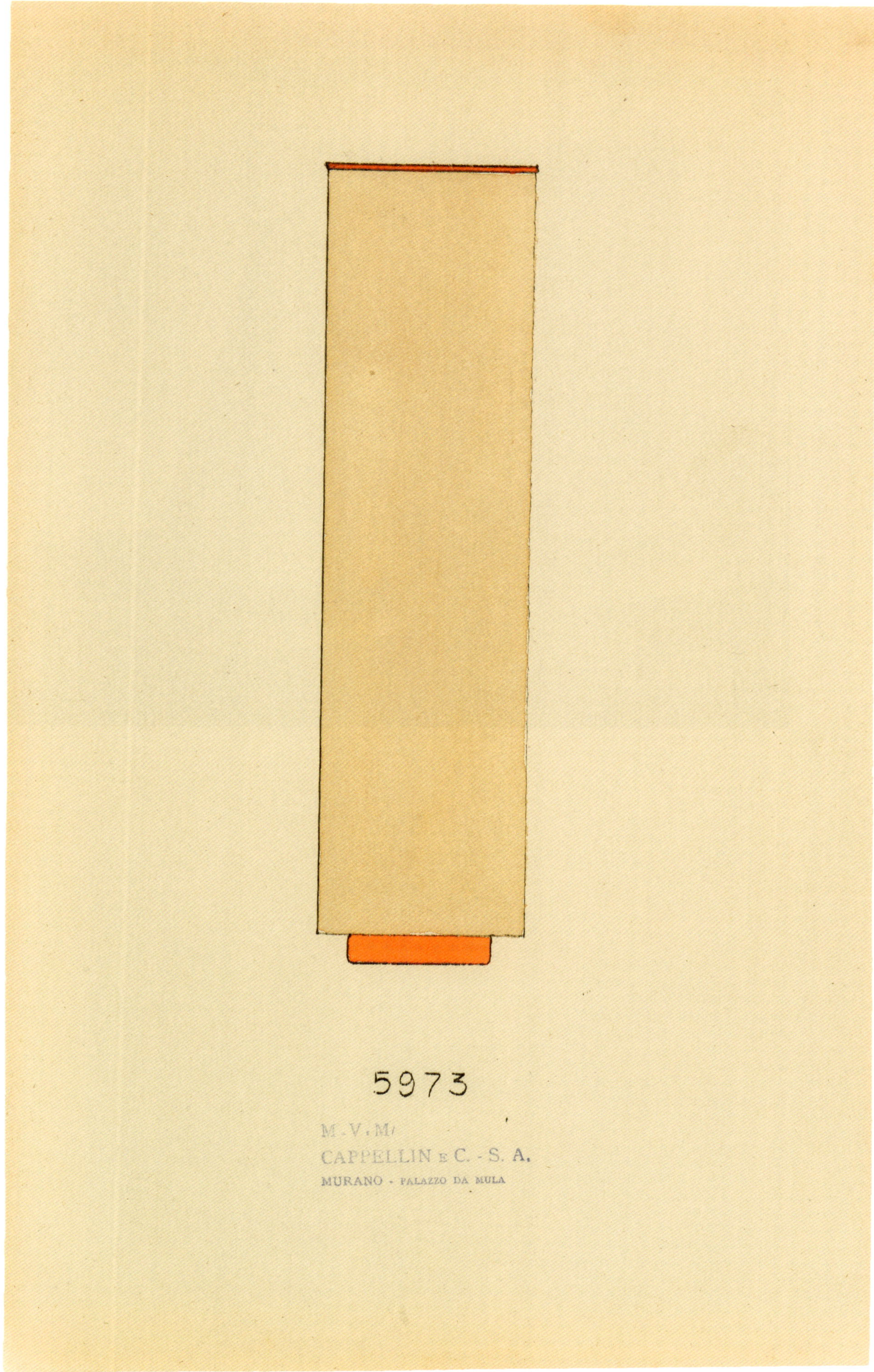

>
'The new "geometric" vases in coloured opaque glass', in the Tentoonstelling van glas, aardewerk en kant, van moderne italiaansche kunstnijverheid. Mostra di Vetri, Ceramiche e Merletti d'Arte Moderna Italiana, *Amsterdam 1931, p. 26*

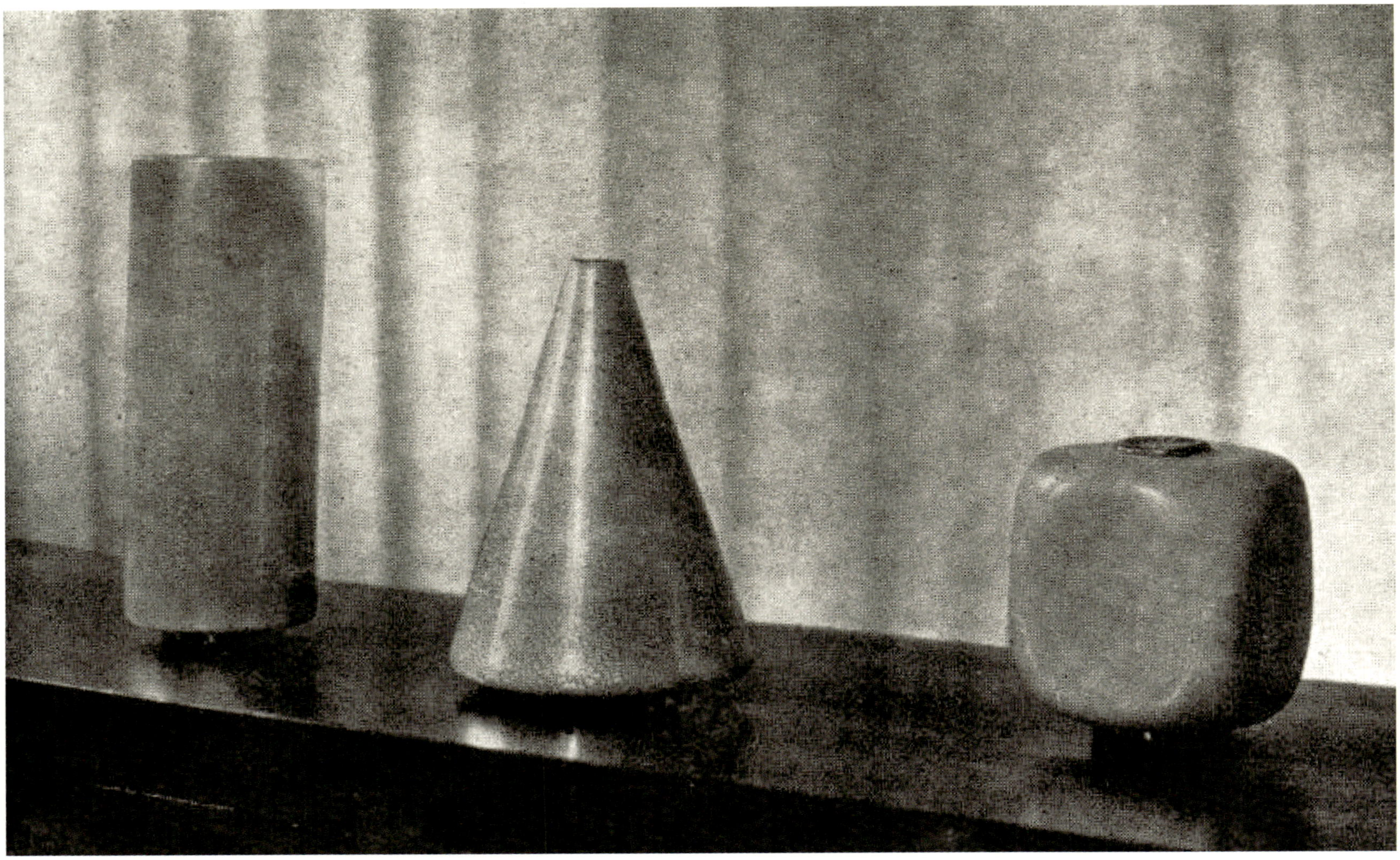

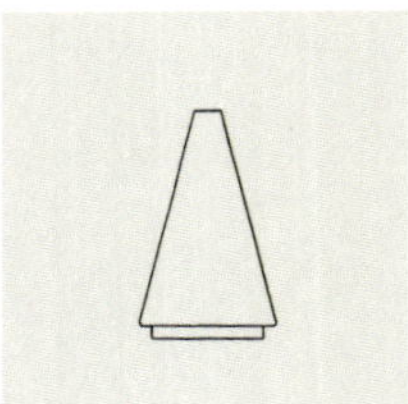

5971

Truncated-cone-shaped vase with foot in strongly iridised coral-red *pasta vitrea*. An example of no. 5971, produced in *pasta vitrea*, was displayed in 1931 at the Amsterdam Exhibition of glass, ceramics and lace. Model no. 5971 was also produced in other glass techniques (e.g. *lattimo*, cased glass).
c. 1930
27 cm high
M.V.M. Cappellin Murano *acid stamp*
Francesca and Massimo Valsecchi collection, Palazzo Butera

exhibitions
1931, Amsterdam, Exhibition of glass, ceramics and lace

>
5971, catalogue drawing, AMA

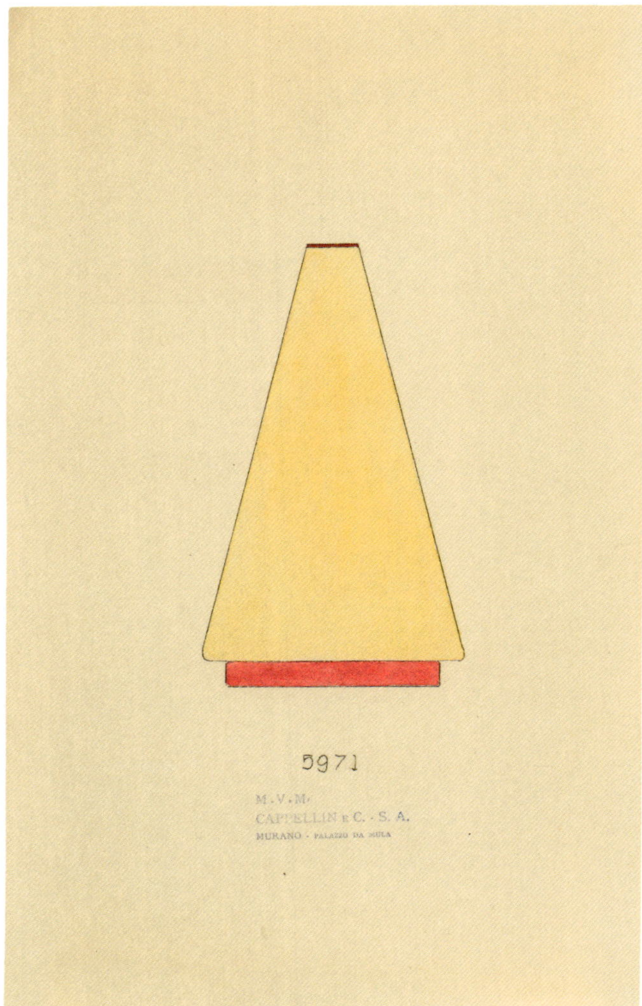

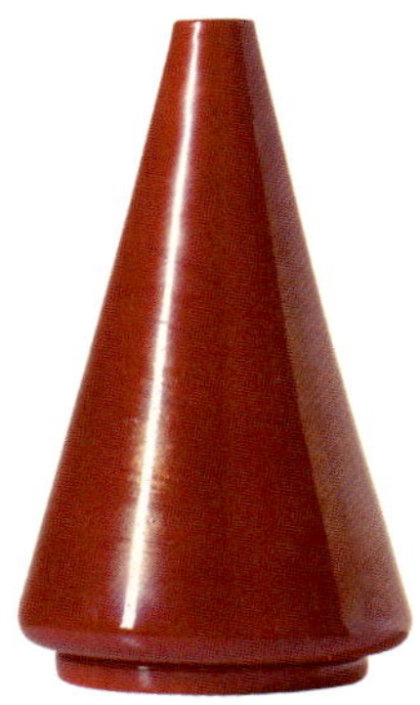

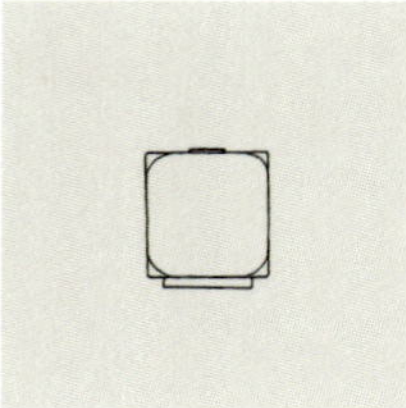

5972

Catalogue drawing for one of the vases in *pasta vitrea* exhibited in 1931 at the Amsterdam Exhibition of glass, ceramics and lace, AMA. The same model was also used to produce cased glass pieces.
c. 1930
16 cm high

exhibitions
1931, Amsterdam, Exhibition of glass, ceramics and lace

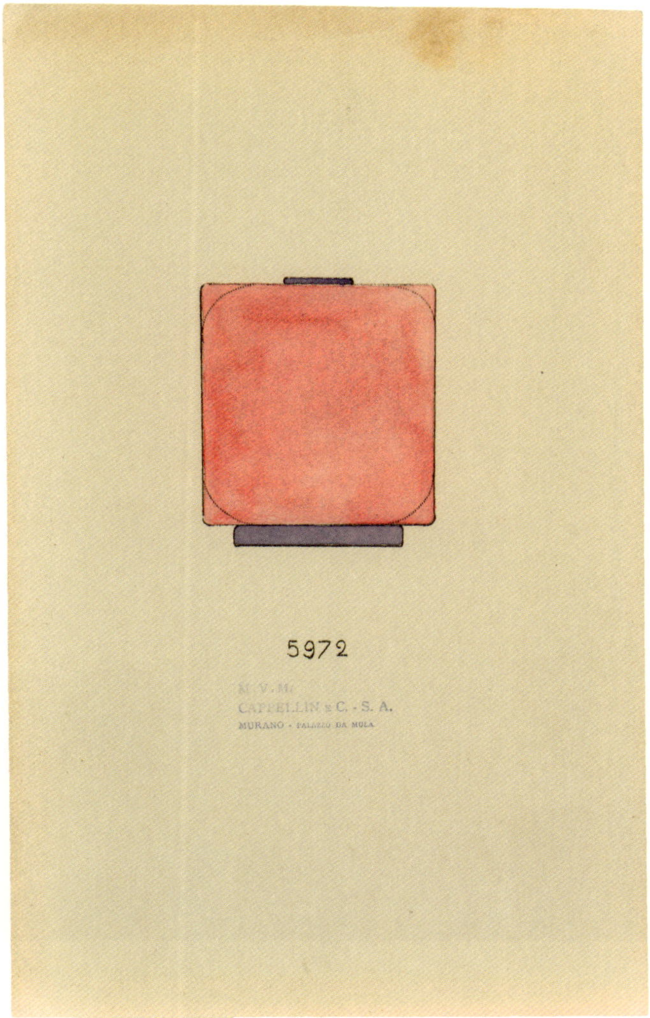

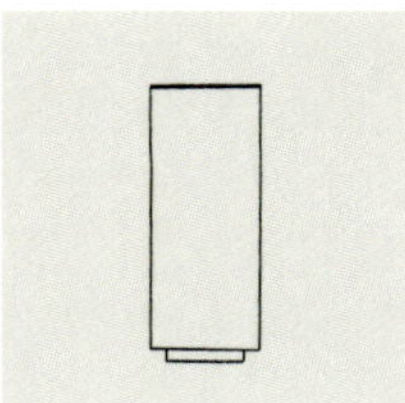

5974

Catalogue drawing of one of the vases in *pasta vitrea* exhibited in 1931 at the Amsterdam Exhibition of glass, ceramics and lace, AMA. The same model was also used for the production of cased glass pieces.
c. 1930
33 cm high

exhibitions
1931, Amsterdam, Exhibition of glass, ceramics and lace

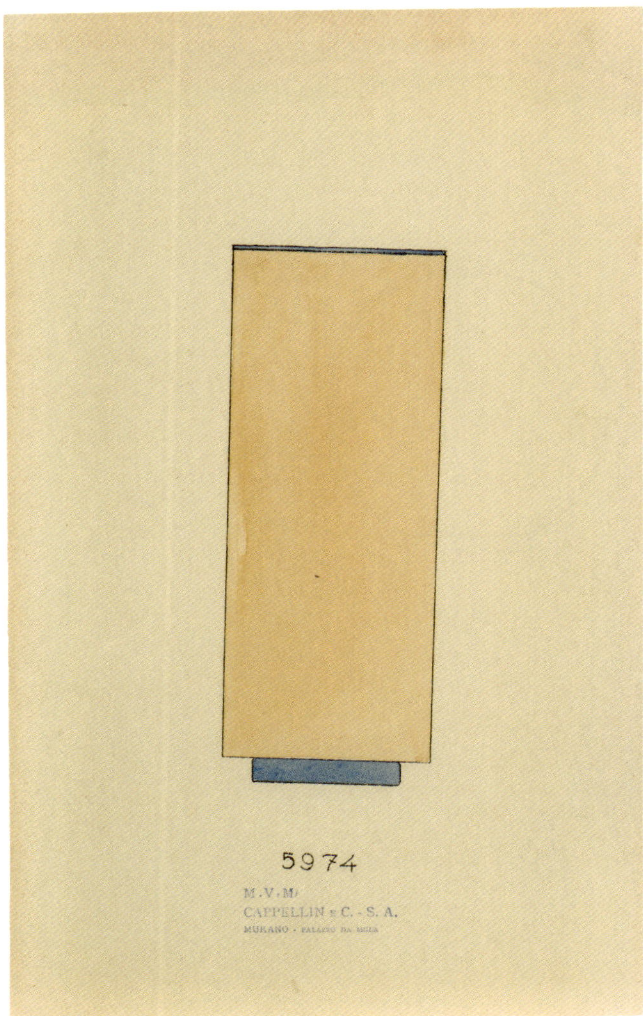

>
'The new "geometric" vases in coloured opaque glass', in the Tentoonstelling van glas, aardewerk en kant, van moderne italiaansche kunstnijverheid. Mostra di Vetri, Ceramiche e Merletti d'Arte Moderna Italiana, *Amsterdam 1931, p. 26*

>>
'Bowl and sculptures in polychrome glass', in the Tentoonstelling van glas, aardewerk en kant, van moderne italiaansche kunstnijverheid. Mostra di Vetri, Ceramiche e Merletti d'Arte Moderna Italiana, *Amsterdam 1931, p. 26*

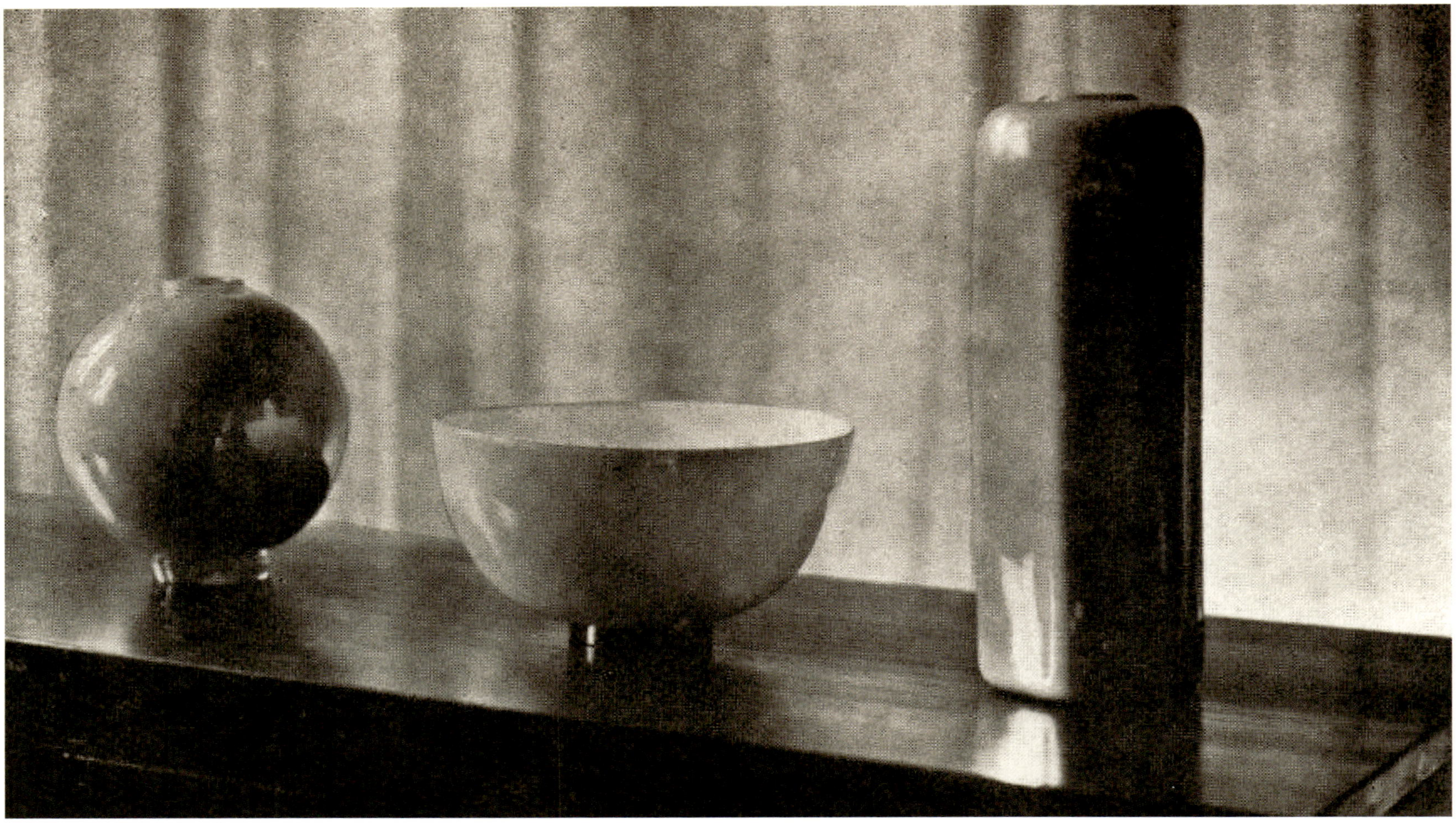

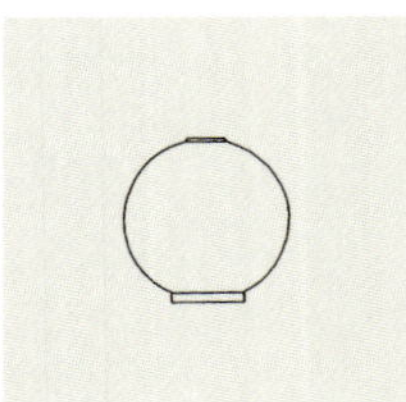

5970

Spherical vase in coral-red *pasta vitrea* with applied gold leaf. Foot and thread around the mouth in coral-red *pasta vitrea*. Model no. 5970 was also produced in other glass techniques (e.g. cased glass).
c. 1930
19 cm high
Stiftung Museum Kunstpalast collection, Düsseldorf-Glasmuseum Hentrich

exhibitions
1931, Amsterdam, Exhibition of glass, ceramics and lace

>
5970, catalogue drawing, AMB

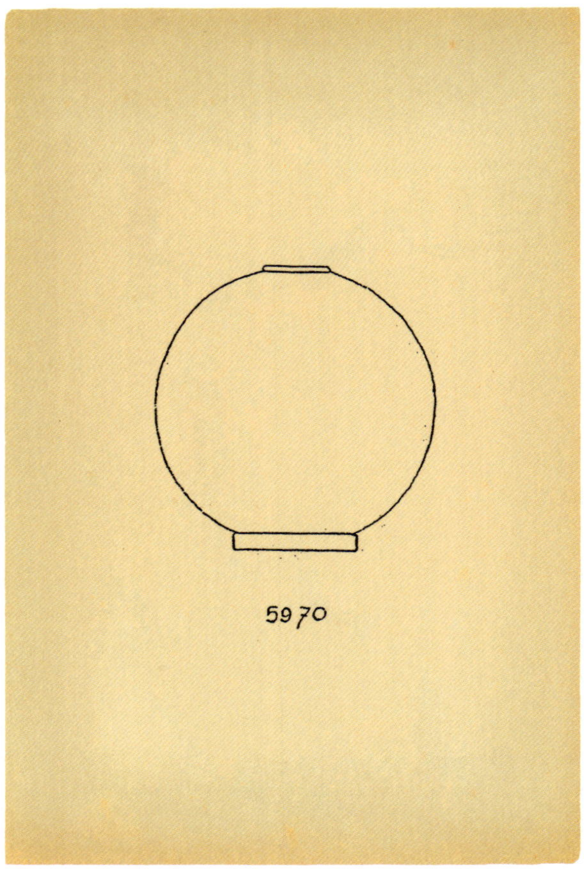

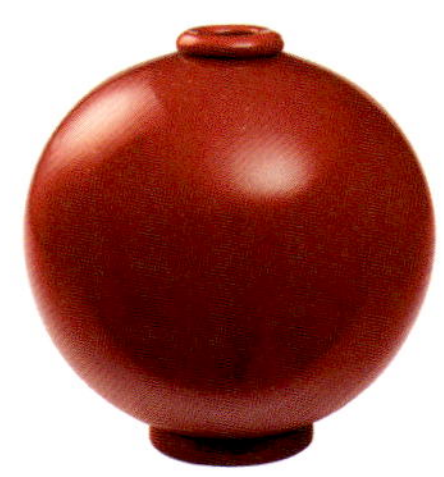

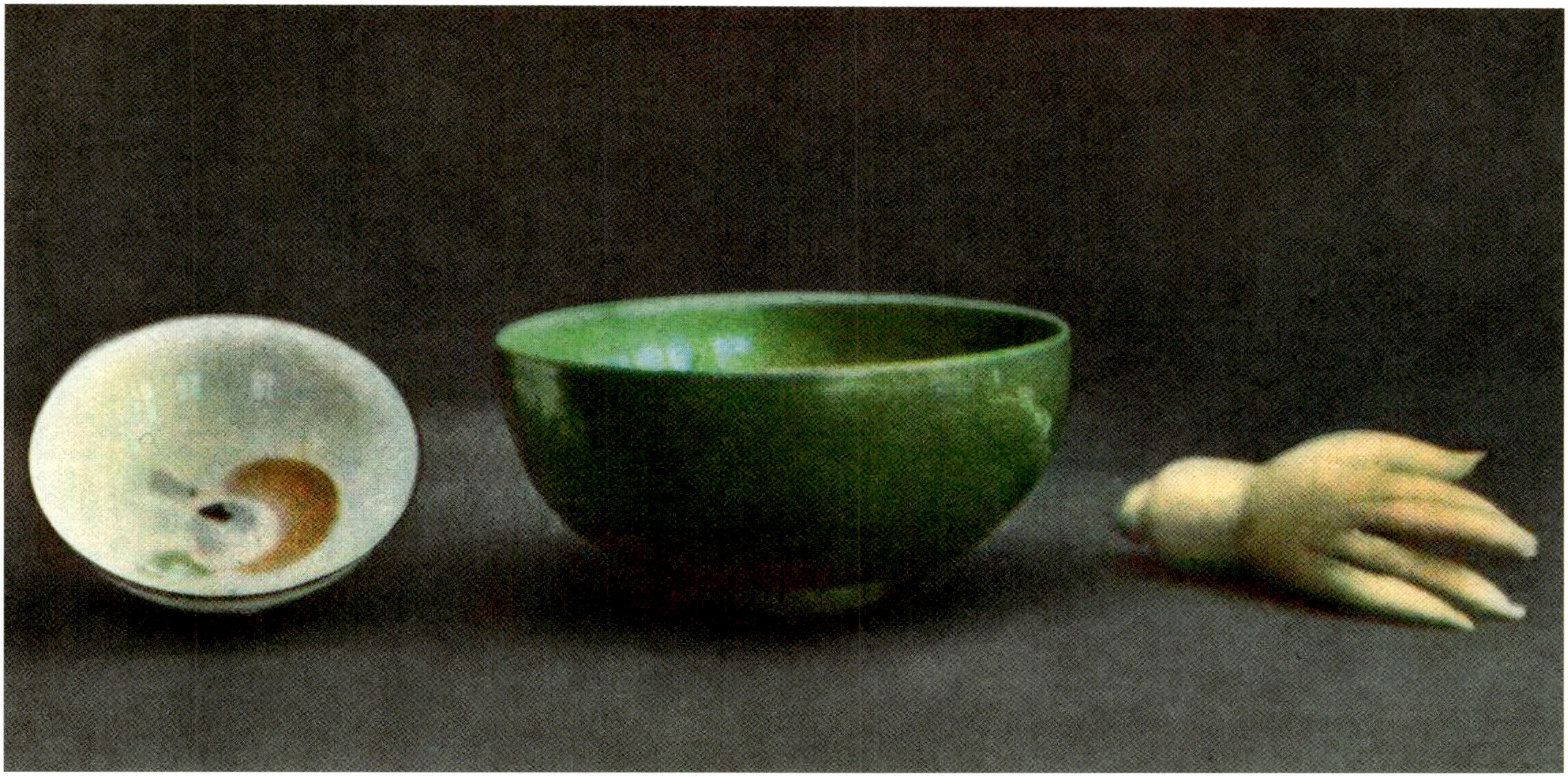

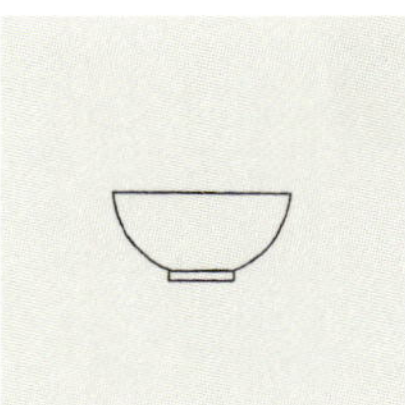

5994

Bowl in Savoy-blue *pasta vitrea* with applied gold leaf. Foot and thread around the mouth in dark-blue *pasta vitrea*.
c. 1930
24 cm ø
C.V.M. *acid stamp*
Private collection

>
5992–5994, catalogue drawing, AMA

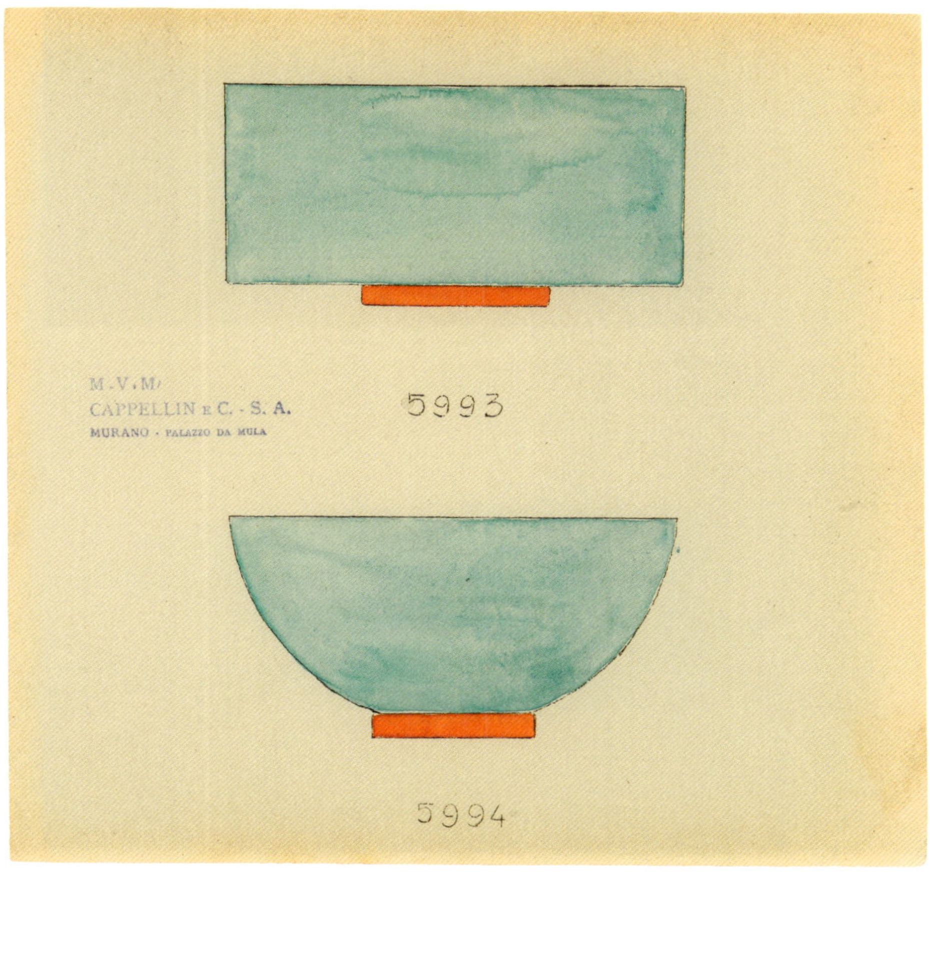

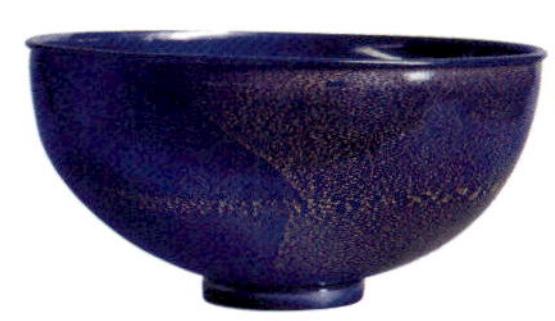

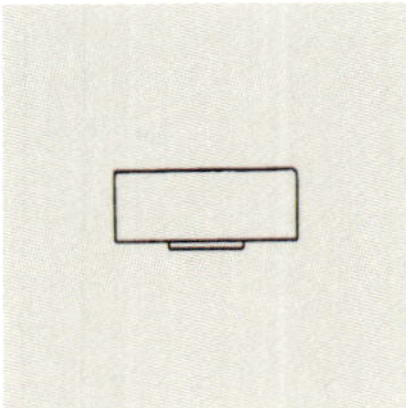

5993

Cylindrical bowl in yellow *pasta vitrea*. Foot and thread around the mouth in coral-red *pasta vitrea*. The series of 'geometric' glass pieces was presented in 1930 at the 4th Triennale of Monza.
c. 1930
12 cm high; 22 cm ø
M.V.M. Cappellin Murano *acid stamp*
Private collection

>
'Geometric' glass pieces at the 4th Triennale of Monza, 1930

>
5993, catalogue drawing, AMB

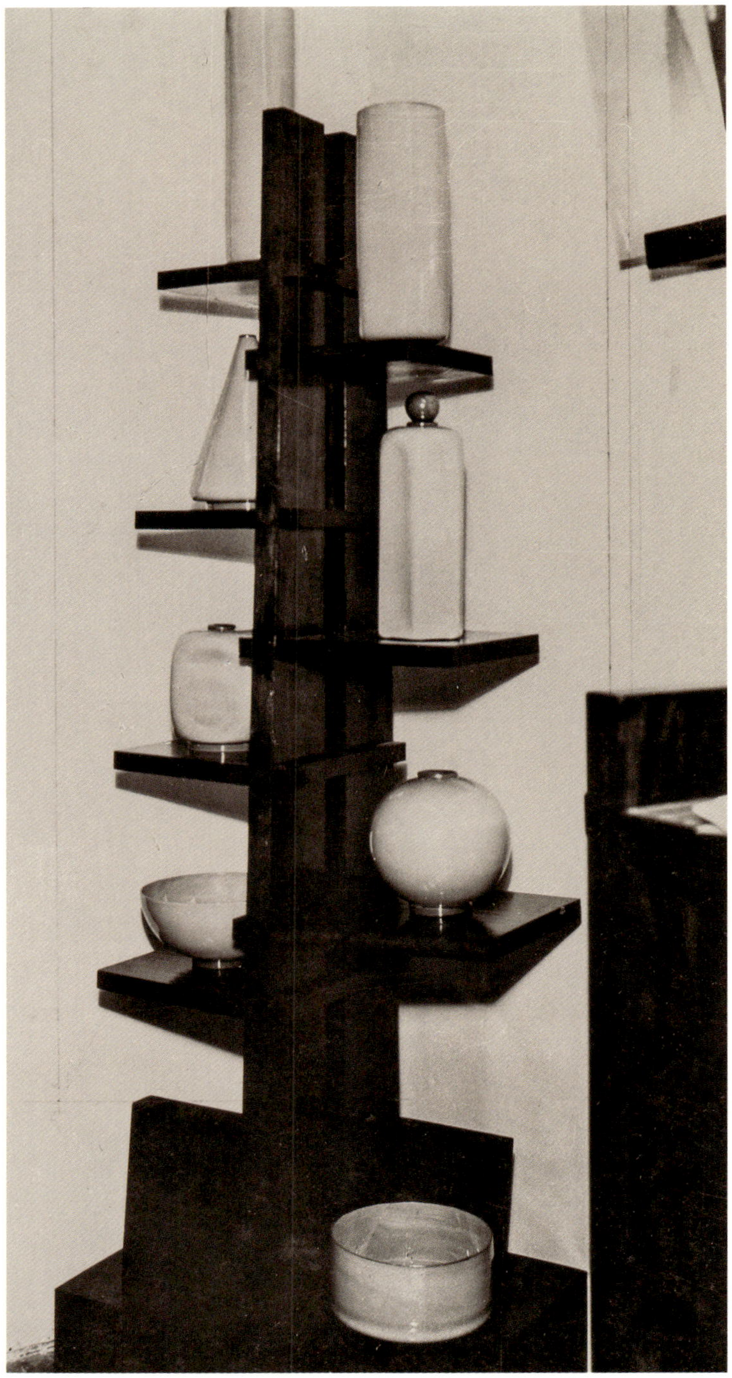

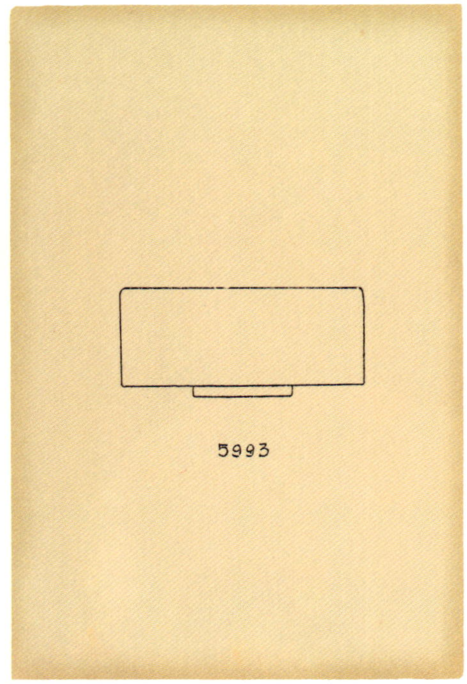

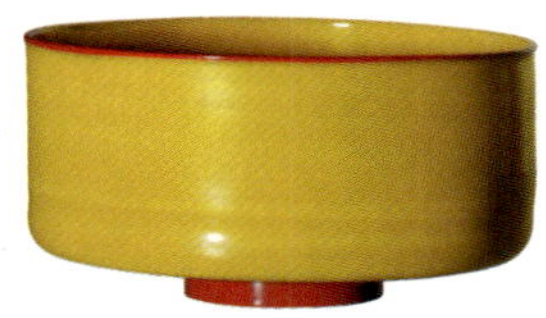

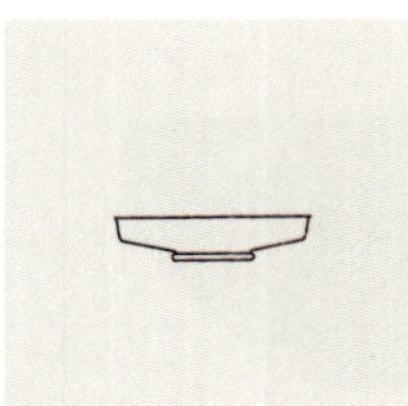

5283

Bowl in yellow *pasta vitrea*. Foot and thread around the mouth in coral-red *pasta vitrea*. The bowl was produced after a model in the catalogue of the glassworks designed by Vittorio Zecchin (1921–25).
1930–31
6 cm high; 23 cm ø
M.V.M. Cappellin Murano *acid stamp*
John and Jean Geresi collection, USA

>
5283, catalogue drawing, AMB

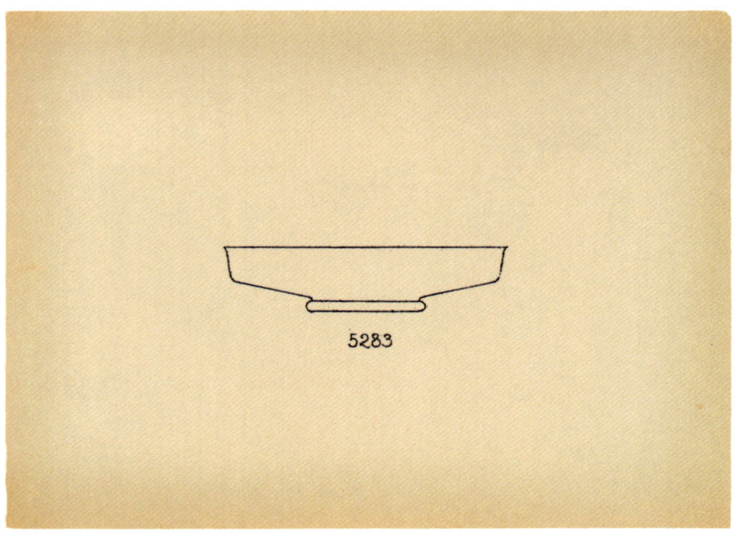

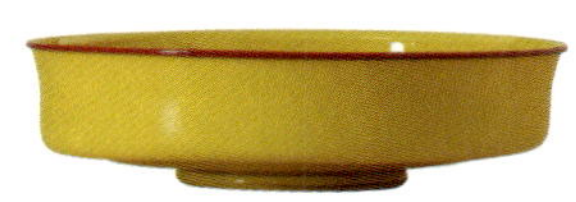

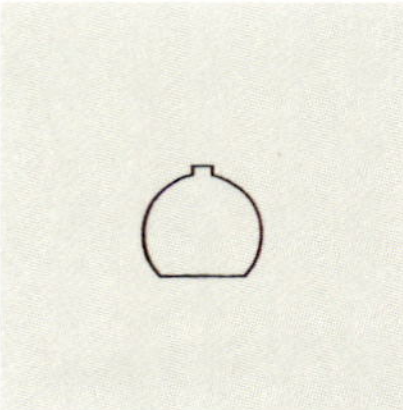

1271

Spherical perfume bottle in crystal-cased coral-red *pasta vitrea* with cylindrical metallic stopper. The stopper carries the brand-name 'Nicky', with the name of the fashion designer, Nicky Chini, among whose creations was also a series of perfumes. The model was produced in various dimensions (no. 1272, h 10 cm; no. 1273, h 8 cm; 1275, h 5.5 cm), in various glass techniques (e.g. iridised black) and was also proposed without a stopper.
1929–30
12 cm high
Private collection

›
1271, catalogue drawing, AMA

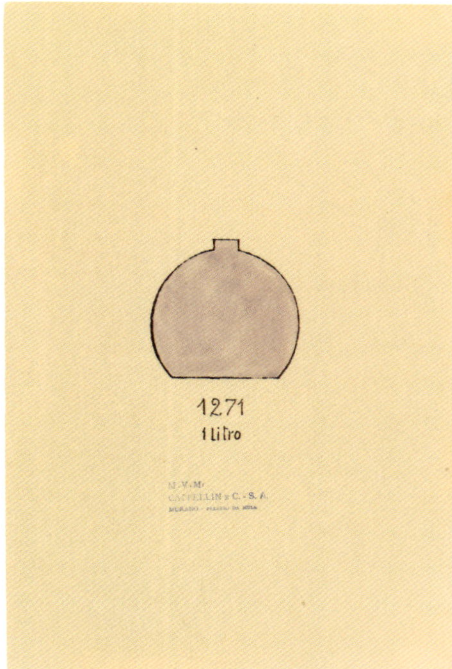

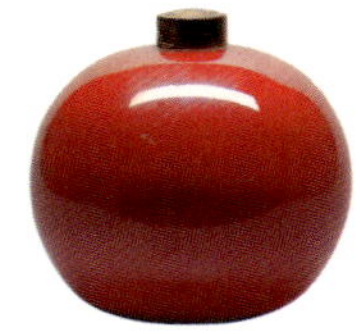

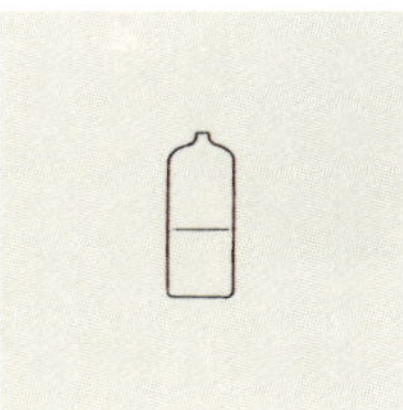

1276

Vase in coral-red in crystal-cased coral-red *pasta vitrea* with iridised finish. The model was proposed in various dimensions and used also as a perfume bottle with a cylindrical metallic stopper for the Eau de Cologne 'Cuoio bulgaro' of 'Nicky Chini' of Milan.
1929–30
19 cm high
Private collection, Treviso

›
1276–1278, catalogue drawings, AMA

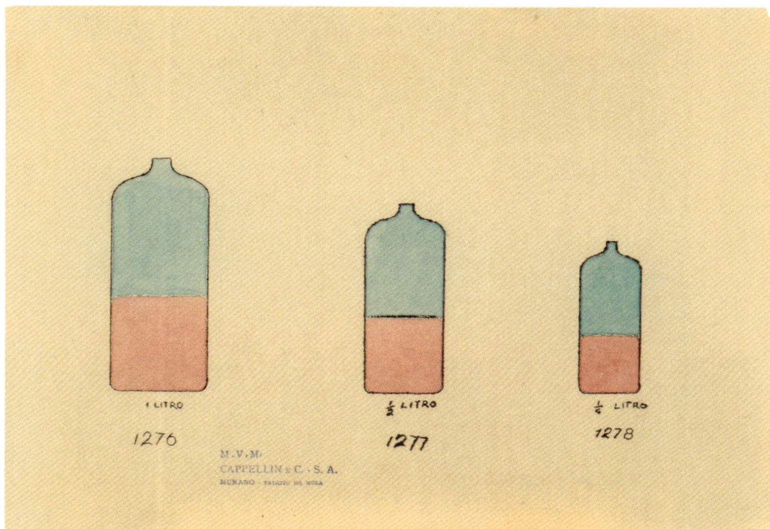

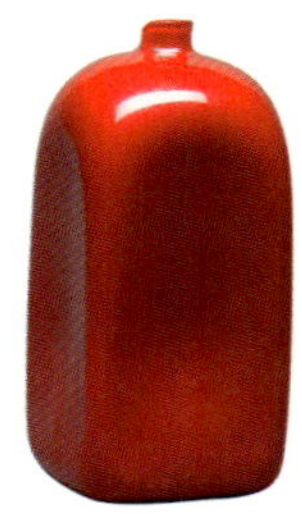

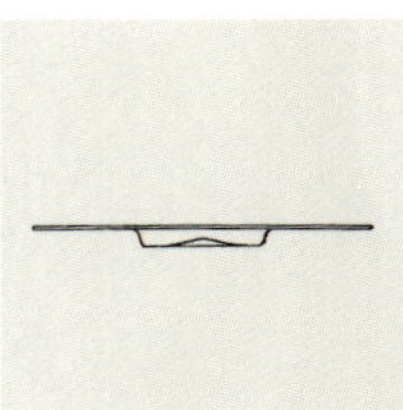

1755

Plate in coral-red crystal-cased *pasta vitrea*. The plate was produced after a model in the catalogue of the glassworks designed by Vittorio Zecchin (1921–25).
c. 1929
23 cm ø
M.V.M. Cappellin Murano *acid stamp*
Private collection

›
1755, catalogue drawing, AMB

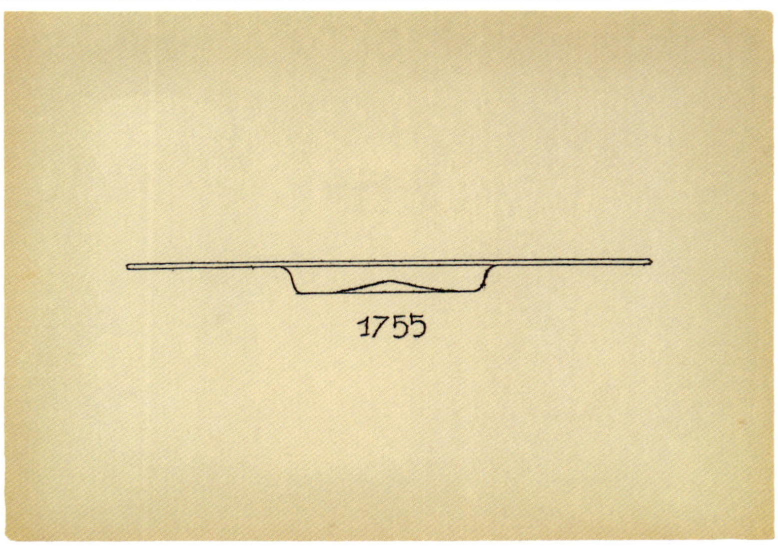

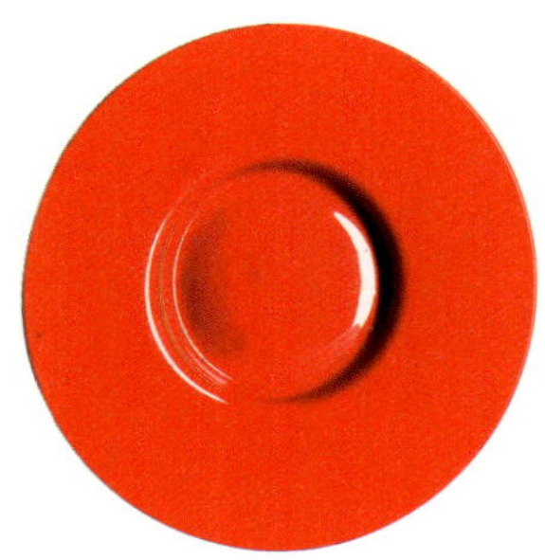

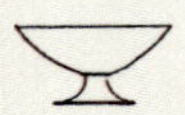

1786

Bowl with flared foot in 'malachite'-coloured *pasta vitrea*
1929–31
19 cm ø
Private collection, Treviso

›
1786, working drawing, CSV

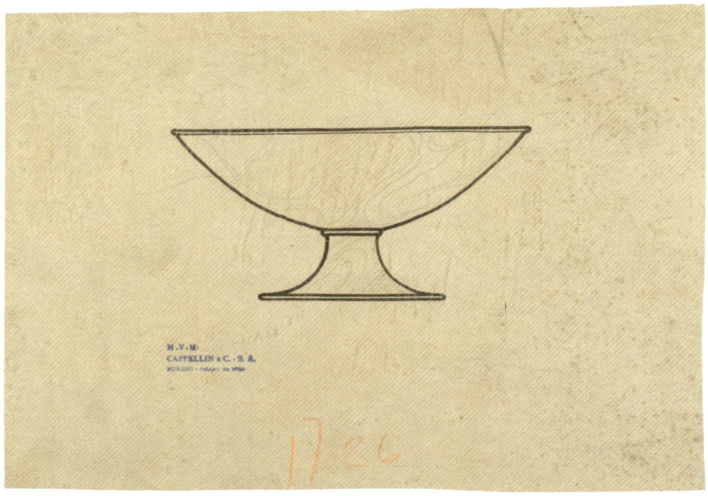

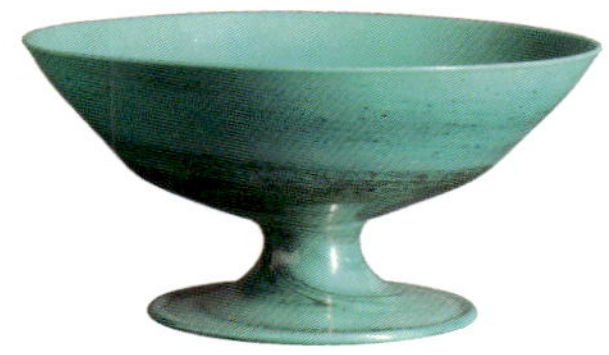

n.i.

Spherical vase with low truncated-cone-shaped foot in turquoise *pasta vitrea* with applied gold leaf.
1929–31
11 cm high
Private collection, Treviso

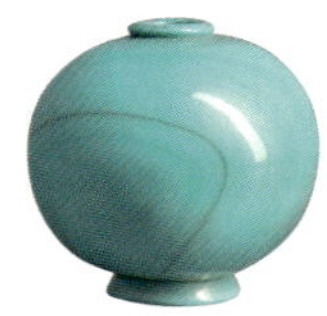

n.i.

Large bottle in coral-red *pasta vitrea* with applied gold leaf, decorated at the base with two applied threads. Surface with iridised finish.
c. 1930
40 cm high
M.V.M. Cappellin Murano *acid stamp*
Private collection

›
Bottle in pasta vitrea *for area of private-bar use (architect Serafini), furniture gallery, at the 4th Triennale of Monza, 1930, detail*

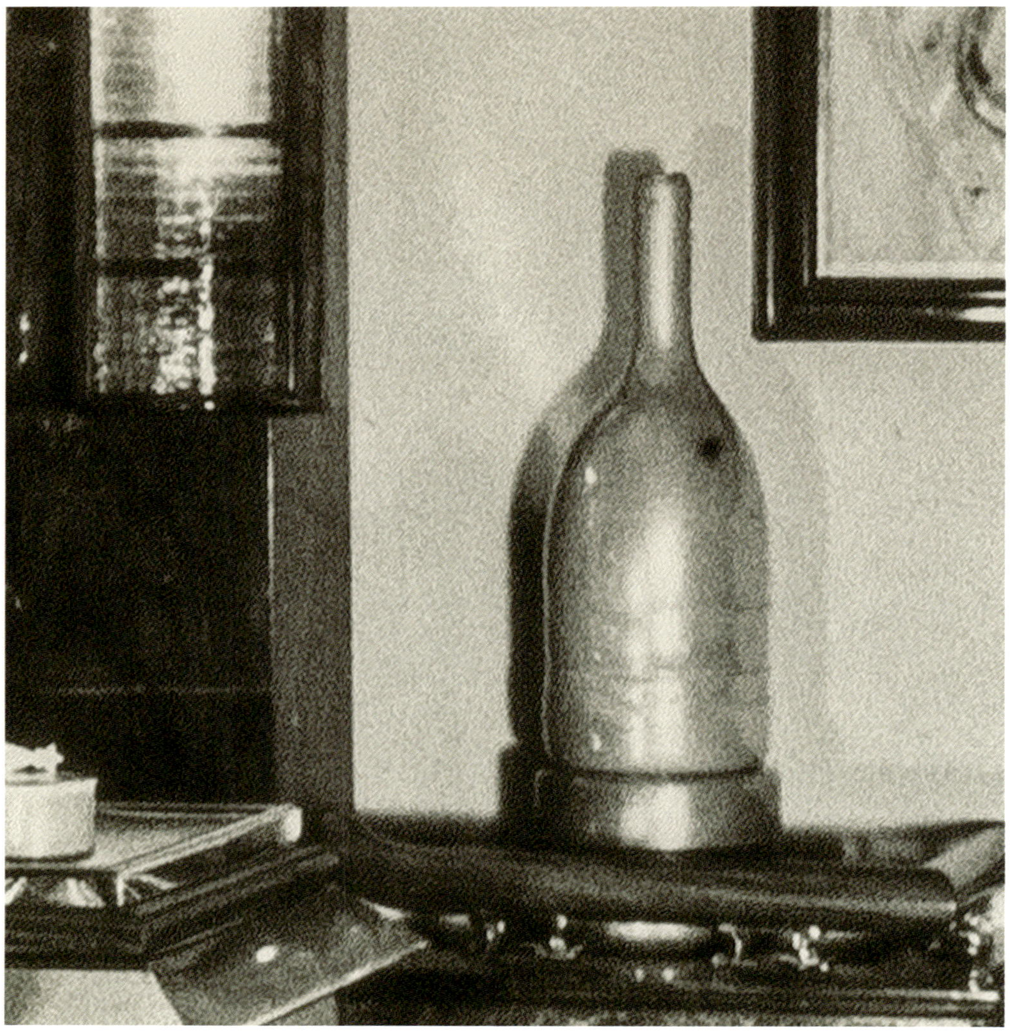

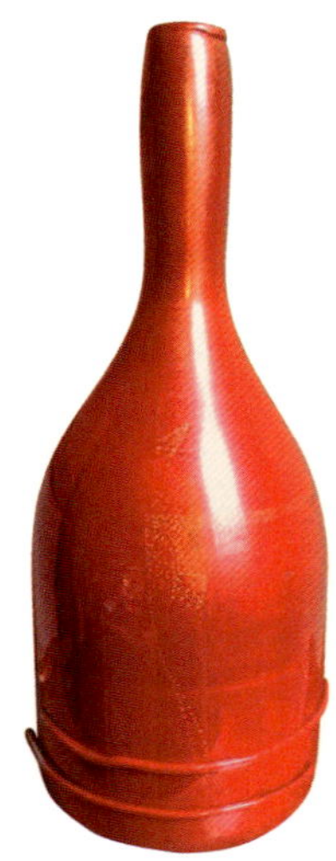

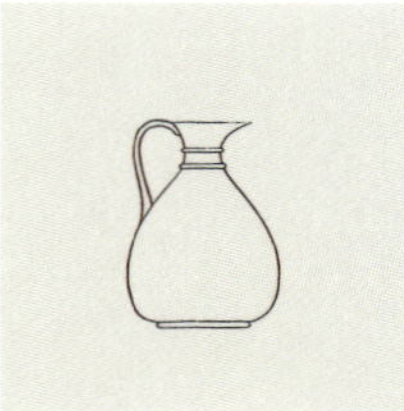

5536

Jug in coral-red *pasta vitrea* with foot and finishings in black glass. Strongly iridised surface. The jug was produced after a model in the catalogue of the glassworks designed by Vittorio Zecchin (1925–26).
1930–31
24,5 cm high
M.V.M. Cappellin Murano *acid stamp*
The Olnick Spanu Collection, New York

›
5536, working drawing, CSV

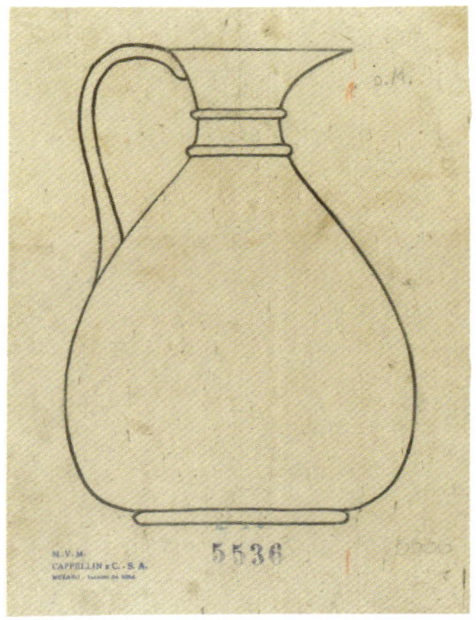

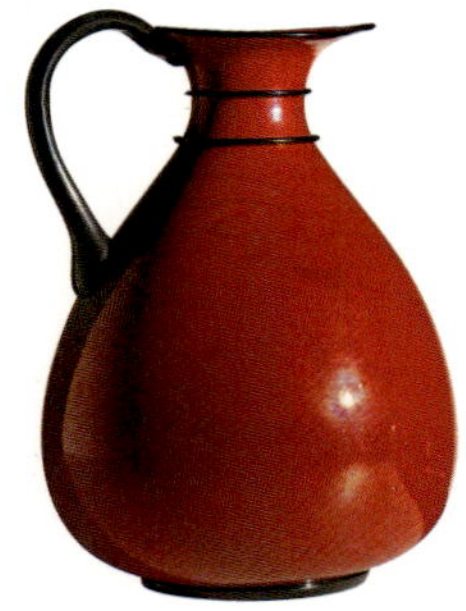

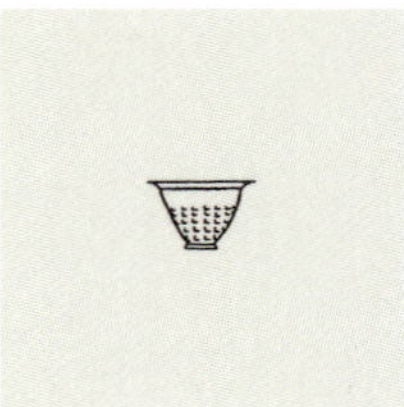

5954

Working drawing for one of the bowls in *pasta vitrea* with relief decorations documented by a period photograph, CSV. This model, together with other examples of this series, was exhibited at the Milano Fair.
1929
8 cm high

exhibitions
1929, Milano Fair

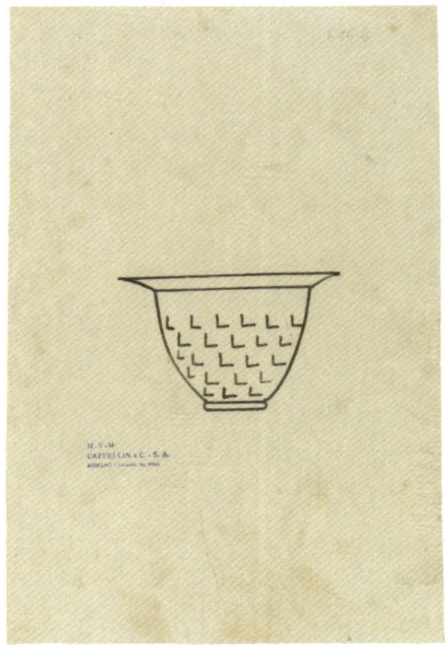

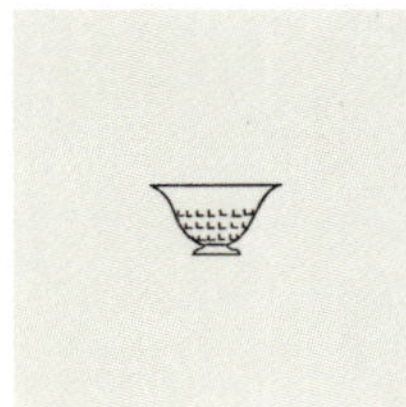

5956

Catalogue drawing for one of the bowls in *pasta vitrea* with relief decorations documented by a period photograph, AMB. This model, together with other examples of this series, was exhibited in 1929 at the Milano Fair.
1929
8 cm high; 14 cm ø

exhibition
1929, Milano Fair

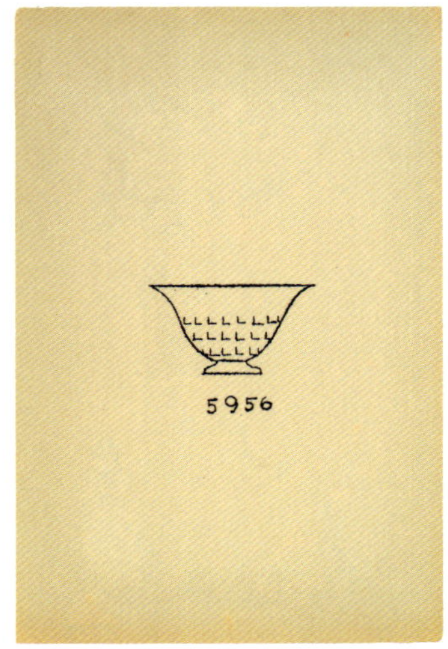

›
Bowls in pasta vitrea *with characteristic relief decoration obtained with a mould. Period photograph, c. 1929, ACSVr*

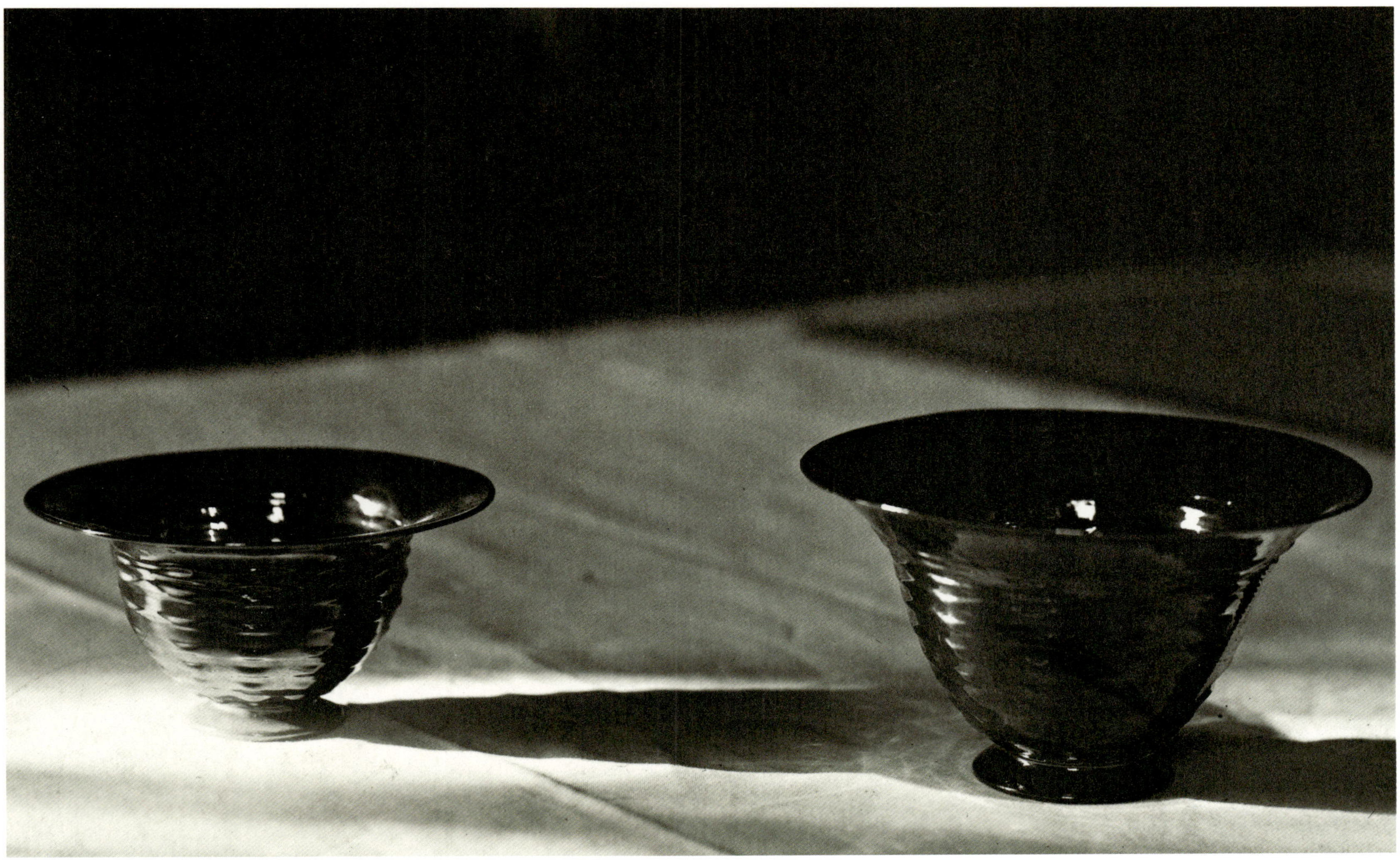

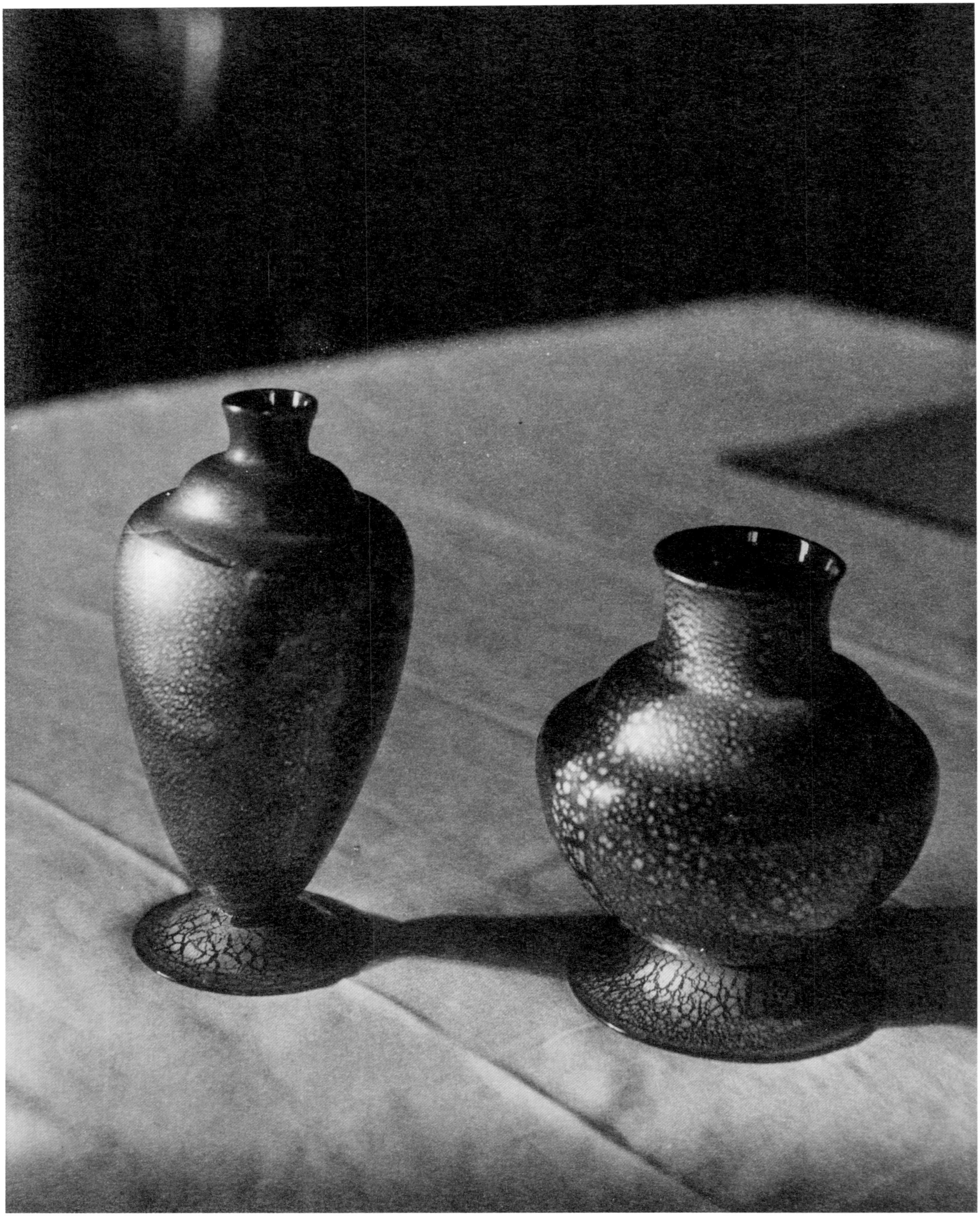

>
Vases no. 5936 and no. 5938 in pasta vitrea *with applied oxidised silver leaf. Period photograph, c. 1929, CSV*

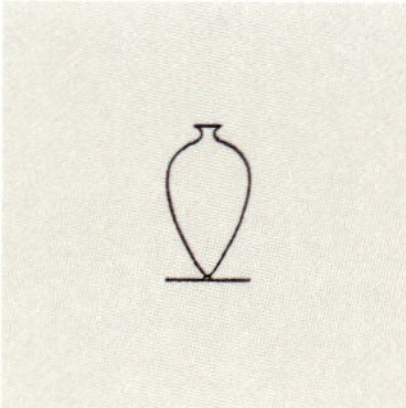

5092

Vase with disk-shaped foot in coral-red *pasta vitrea* with applied oxidised silver leaf. The vase was produced after a model in the catalogue of the glassworks designed by Vittorio Zecchin (1921–22).
c. 1929
21.5 cm high
Private collection

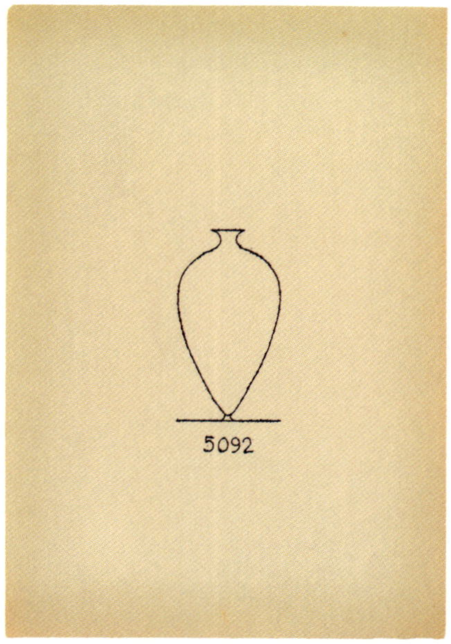

>
5092, catalogue drawing, AMB

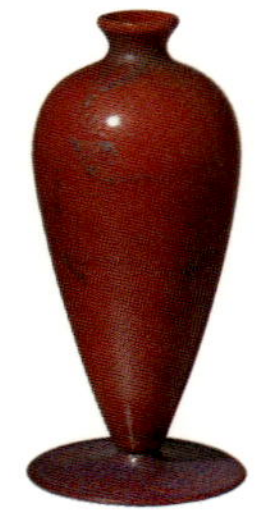

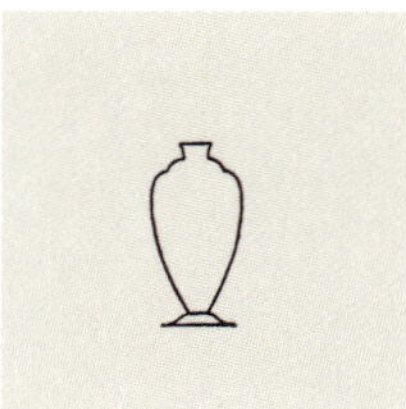

5936

Vase in pea-green *pasta vitrea* with applied oxidised silver leaf. The model is characterised by a double constriction at the base of the neck. An example of no. 5936 executed in 'coral-red iridised gold-sprayed' *pasta vitrea* was exhibited in 1929 at the Leipzig Trade Fair. On that occasion the 'iridised gold-sprayed lattimo with coral-red decorations' variant was exhibited. Model no. 5936 was also produced in other glass techniques, such as *decoro fenicio* and transparent *balloton*.
c. 1929
22.5 cm high
Private collection

exhibitions
1929, Leipzig, Trade Fair

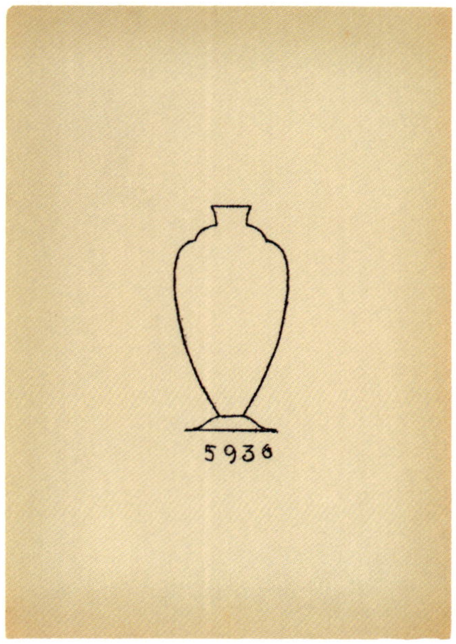

>
5936, catalogue drawing, AMB

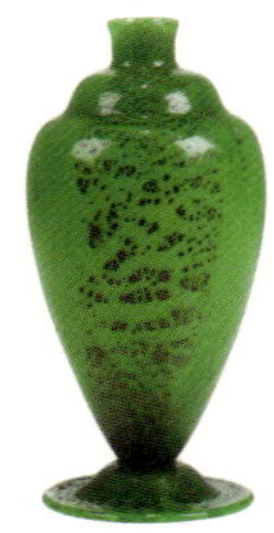

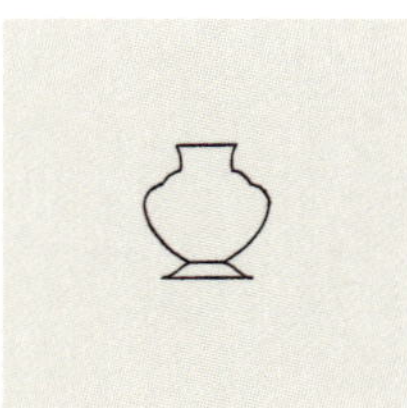

5938

Vase with disk-shaped foot in coral-red *pasta vitrea* with applied oxidised silver leaf. The model is characterised by a double constriction at the base of the neck. No. 5938 was also produced in other glass techniques (e.g. transparent, *lattimo aurato*, *decoro fenicio*, silvered black).
c. 1929
15 cm high
M.V.M. Cappellin Murano *acid stamp*
The Lockwood collection, Fort Lee, NJ

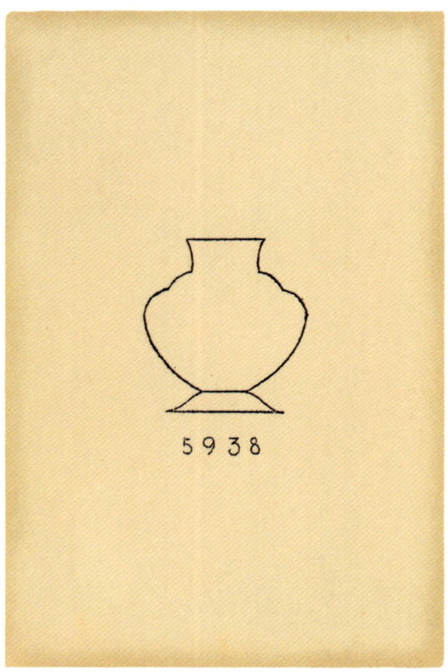

>
5938, catalogue drawing, AMB

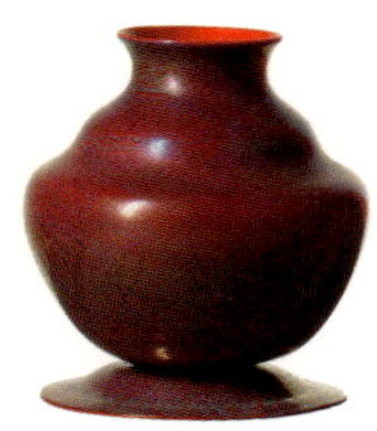

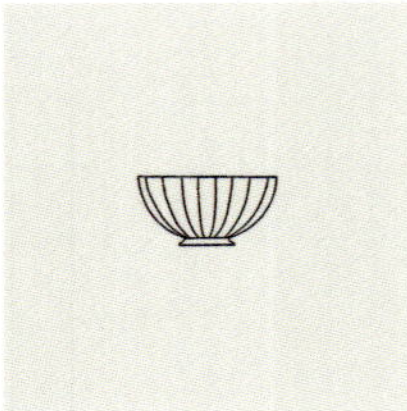

565

Ribbed bowl in pea-green *pasta vitrea* with applied oxidised silver leaf.
c. 1930
16.5 cm ø
Giuseppe Scognamiglio collection, Naples

›
565, catalogue drawing, AMA

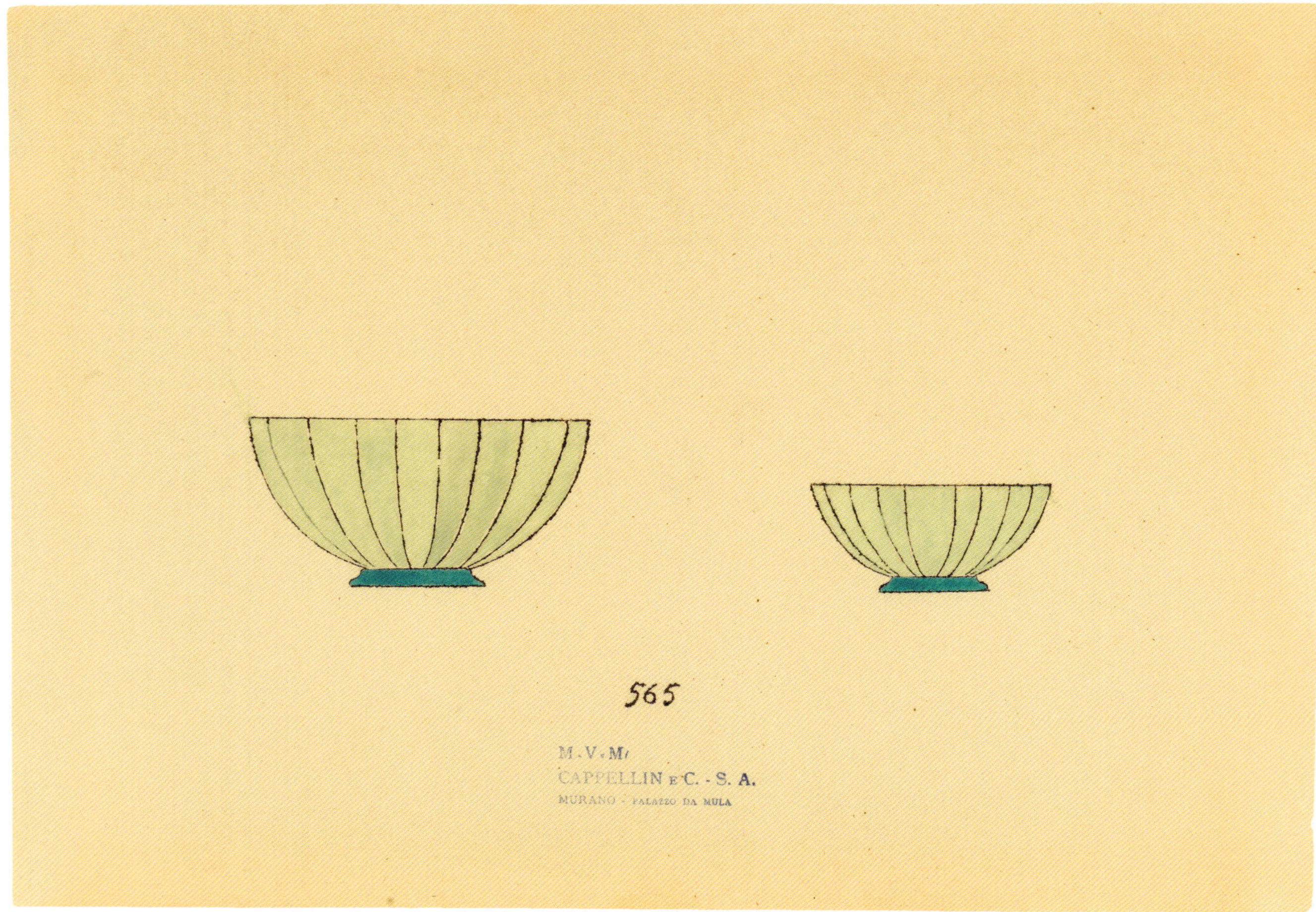

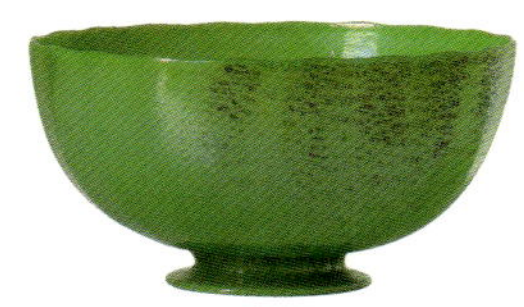

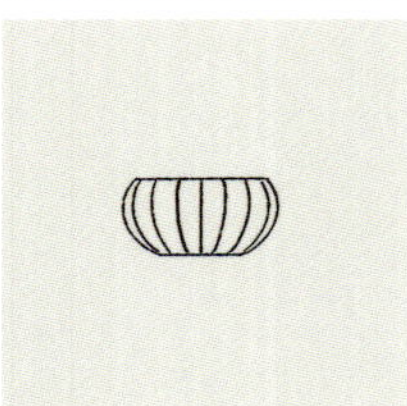

5058

Ribbed bowl in coral-red *pasta vitrea* with applied oxidised silver leaf. The bowl was produced after a model in the catalogue of the glassworks designed by Vittorio Zecchin (1921–22).
c. 1930
10 cm high
Private collection, Turin

›
5058, catalogue drawing, AMB

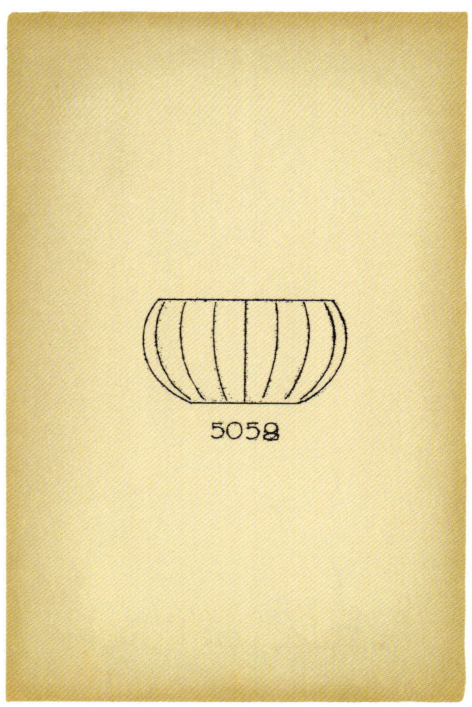

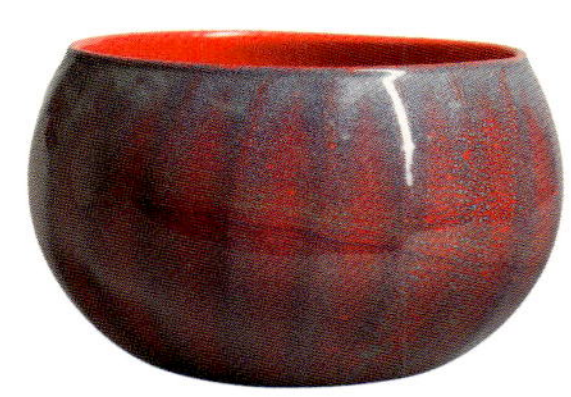

n.i.

Bowl in coral-red *pasta vitrea* with applied oxidised silver leaf. Small applied foot and lateral mixtilinear handles. Surface with iridised finish.
c. 1930
9 cm high; 13 cm ø
C.V.M. *acid stamp*
The Olnick Spanu Collection, New York

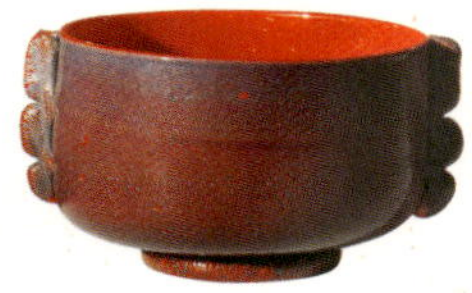

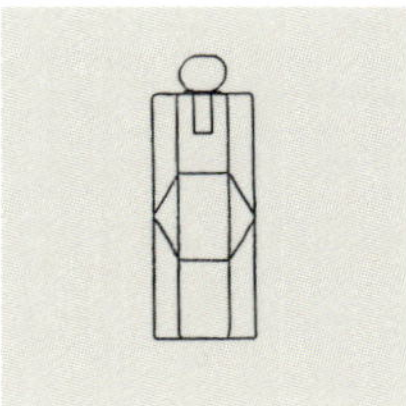

5992

Vase with a hexagonal cross-section in coral-red *pasta vitrea* with applied oxidised silver leaf. Mouth in black glass. Surface with iridised finish. Model no. 5992 was also produced in other glass techniques (e.g. silvered black, cased glass).
c. 1930
28.5 cm high
The Olnick Spanu Collection, New York

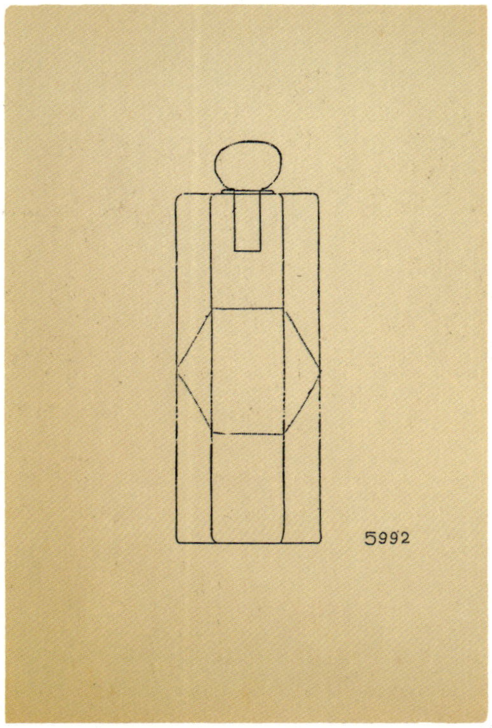

> *5992, catalogue drawing, AMB*

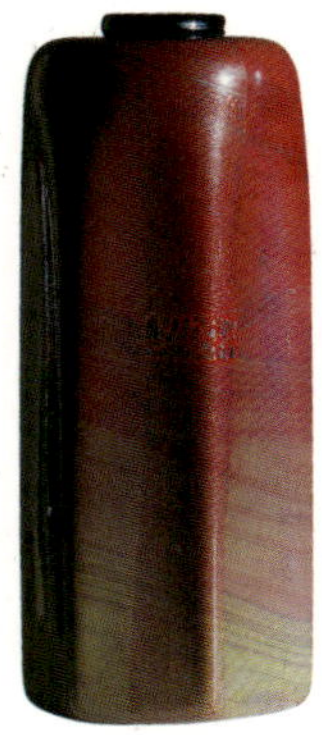

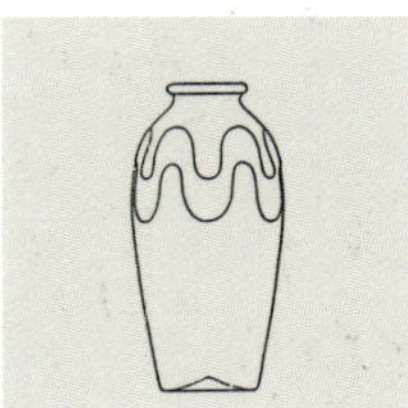

5316

Vase in dark-amber *pasta vitrea* decorated with an undulating band of the same colour included in the wall of the piece. Strongly iridised surface. The vase was produced after a model in the catalogue of the glassworks designed by Vittorio Zecchin (1921–25).
c. 1930
35 cm high
Venezia Murano Company *acid stamp*
Private collection

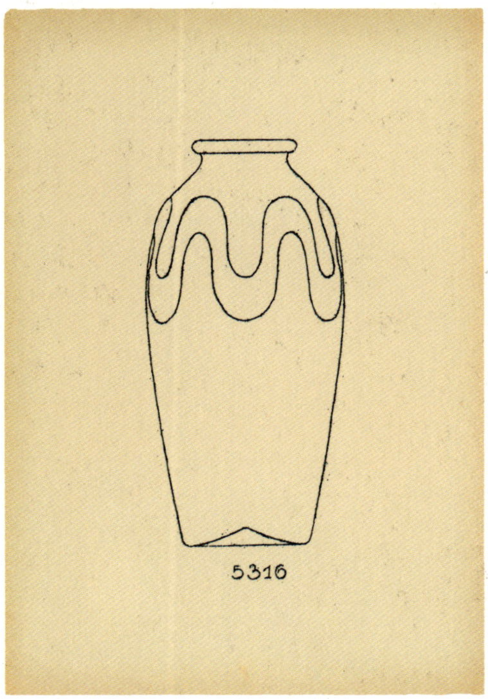

> *5316, catalogue drawing, AMB*

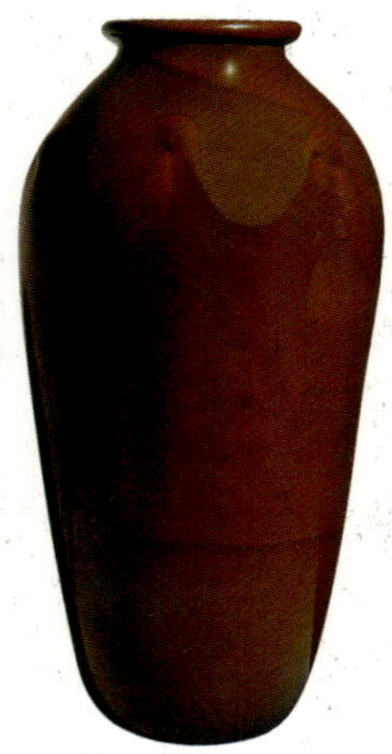

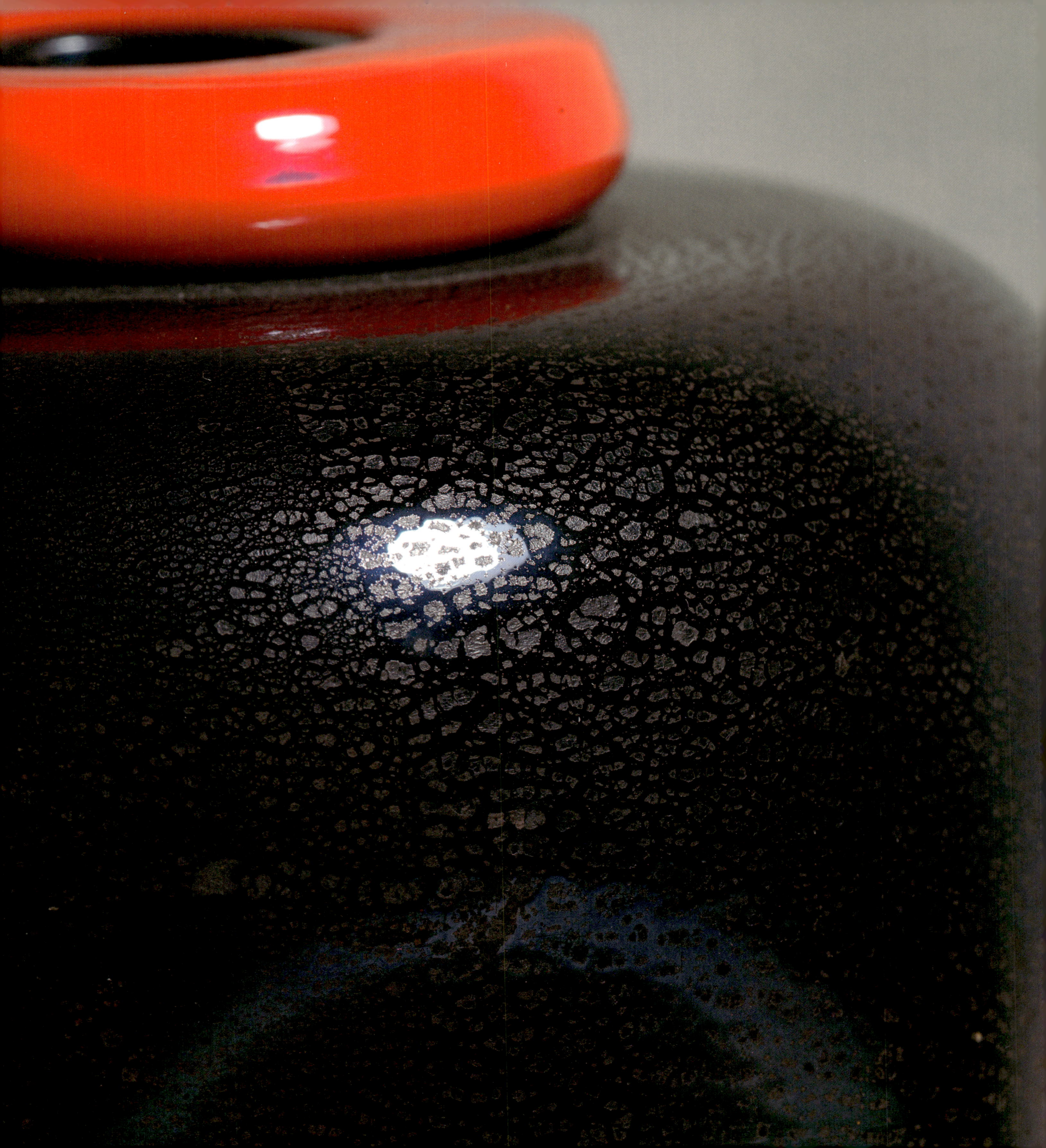

Among the opaque pieces the Cappellin firm proposed there was also a varied series of artefacts produced in black glass with different finishings. Some of these are characterised by pronounced iridisation obtained by exposing the object to tin or titanium vapours which, having been deposited on the wall, form a thin iridescent metallic patina. Others are decorated with applied oxidised silver leaf, which creates evocative effects in the glass material. Others again are marked by the refined coral-red *pasta vitrea*, used in details such as the thread at the mouth or foot or for the foot itself. For this last series the recent models designed by Carlo Scarpa were used in some cases, in others the old transparent glass models designed by Zecchin were proposed again, having been transformed to achieve new effects through the use of opaque glass.

IRIDISED BLACK, SILVERED BLACK 1929-31

Exhibitions

1929, Leipzig, Trade Fair
1929, Milan, Trade Fair
1930, Monza, 4th *Esposizione Triennale Internazionale delle Arti Decorative ed Industriali Moderne*

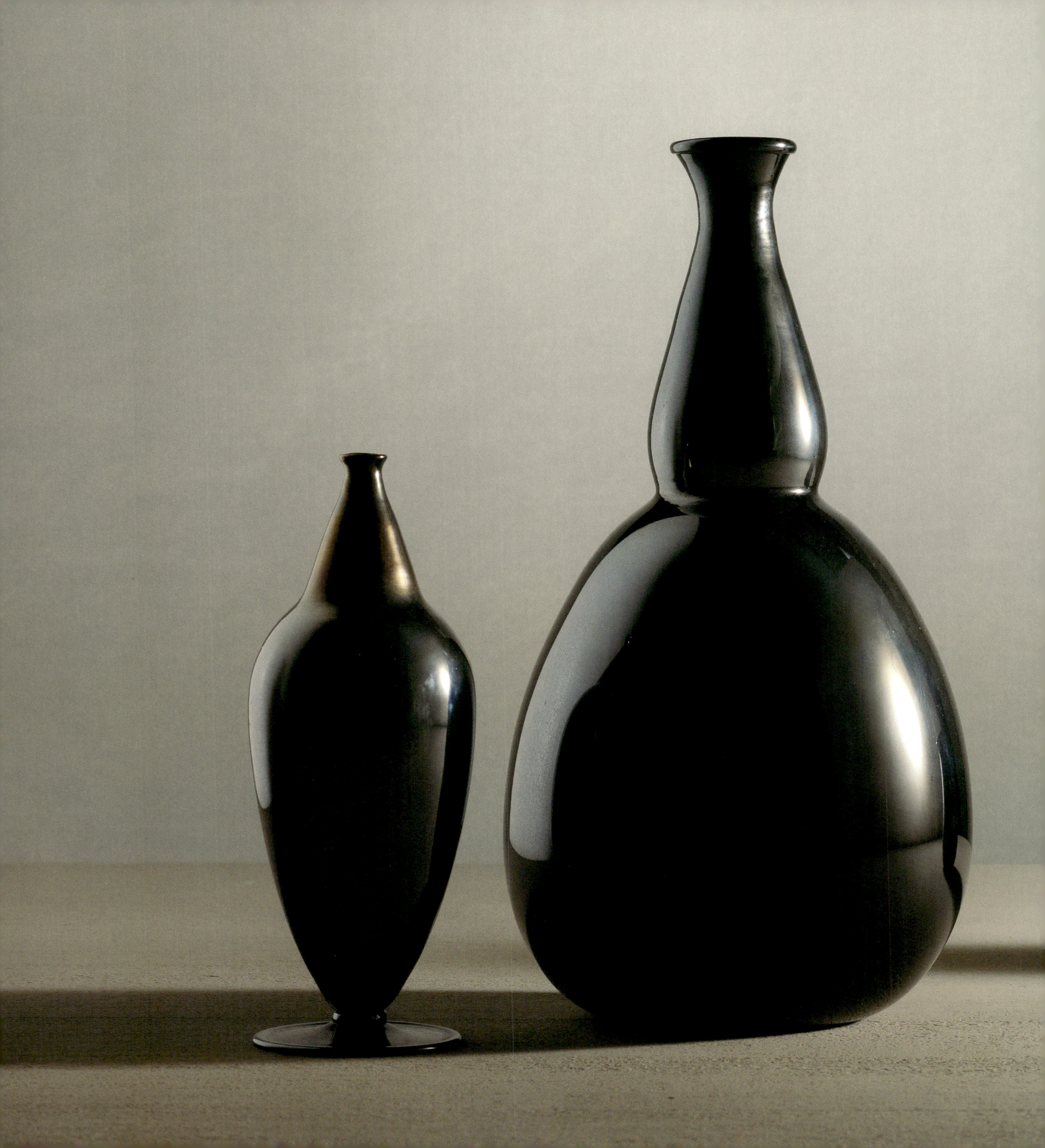

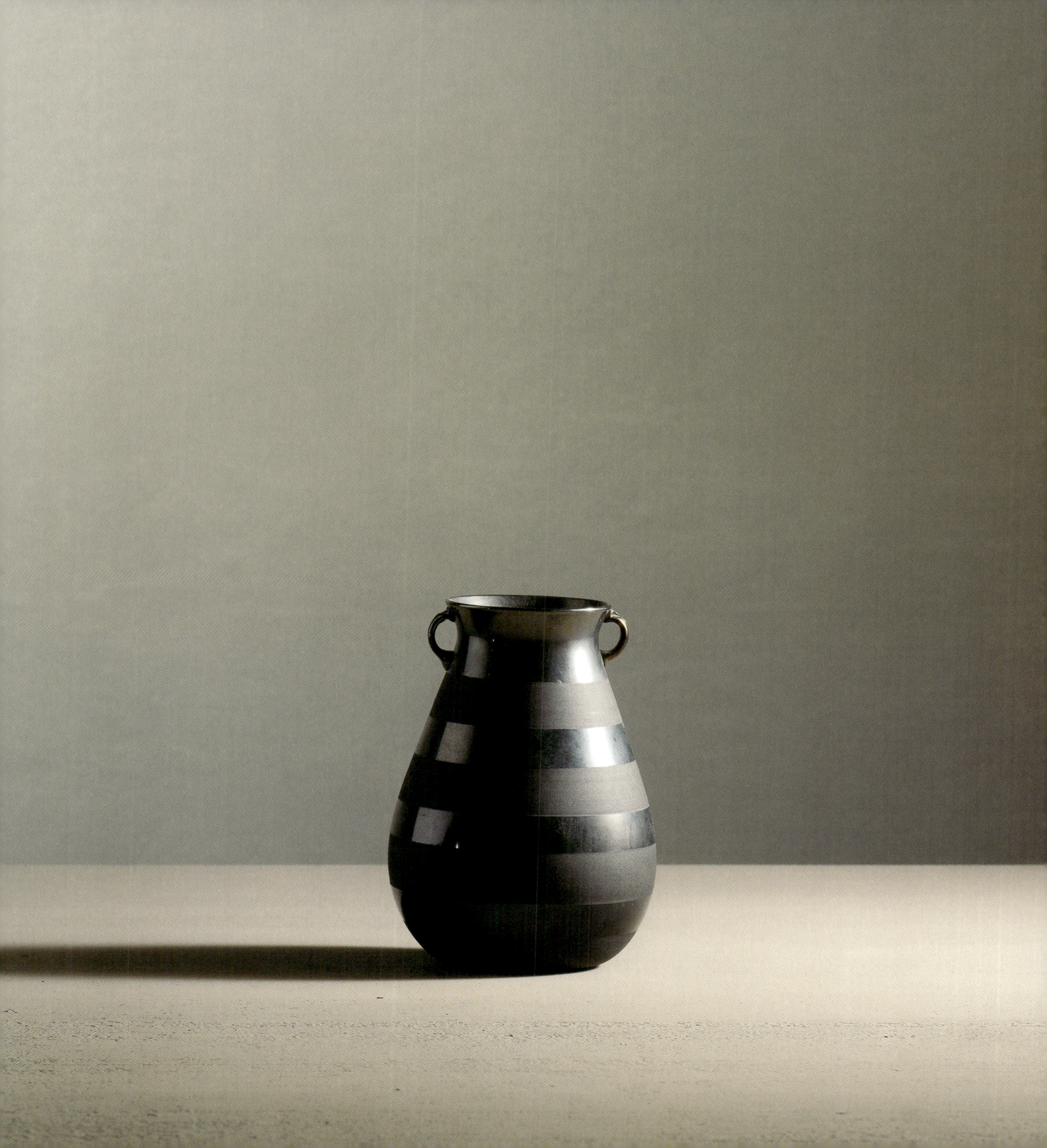

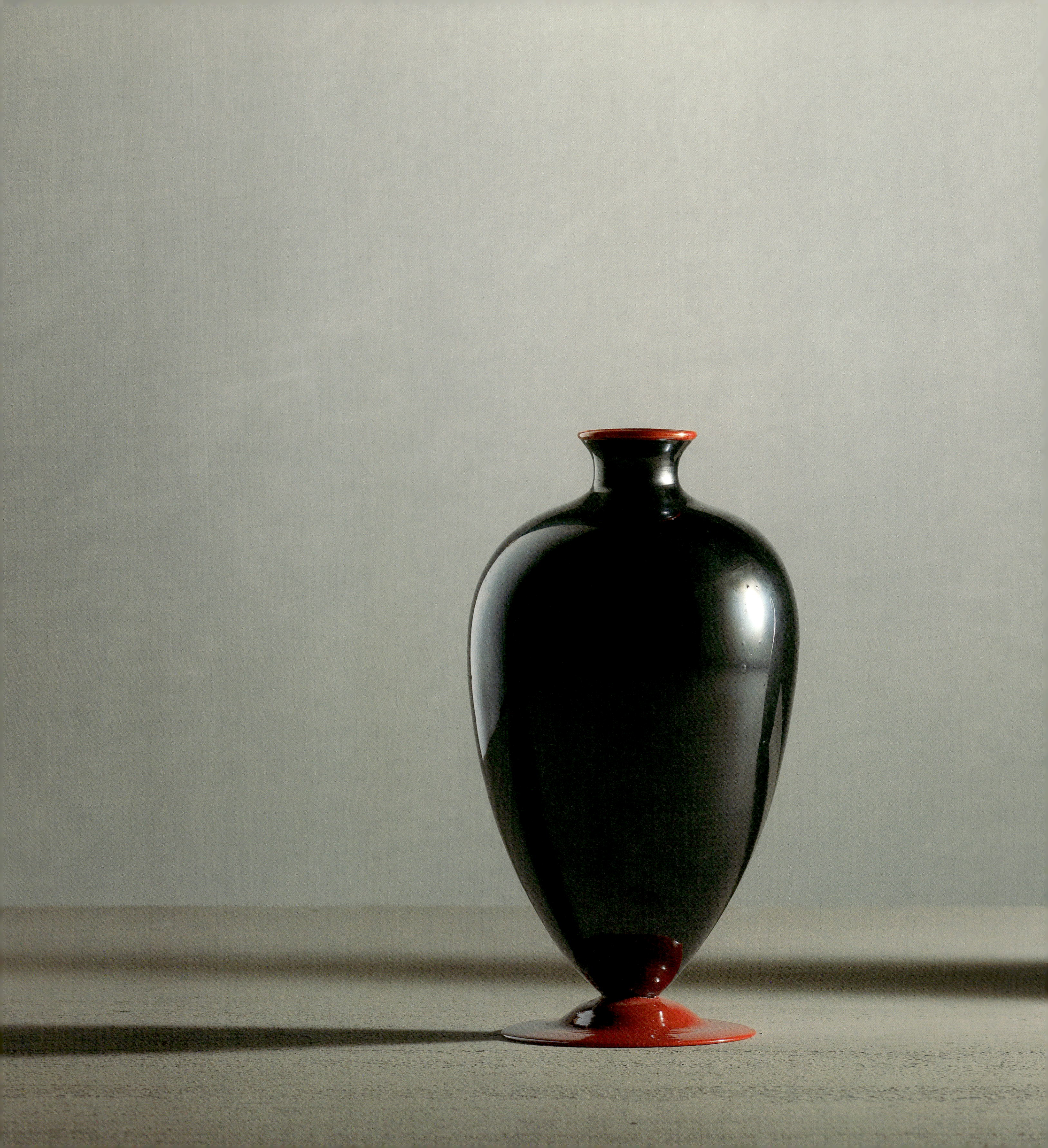

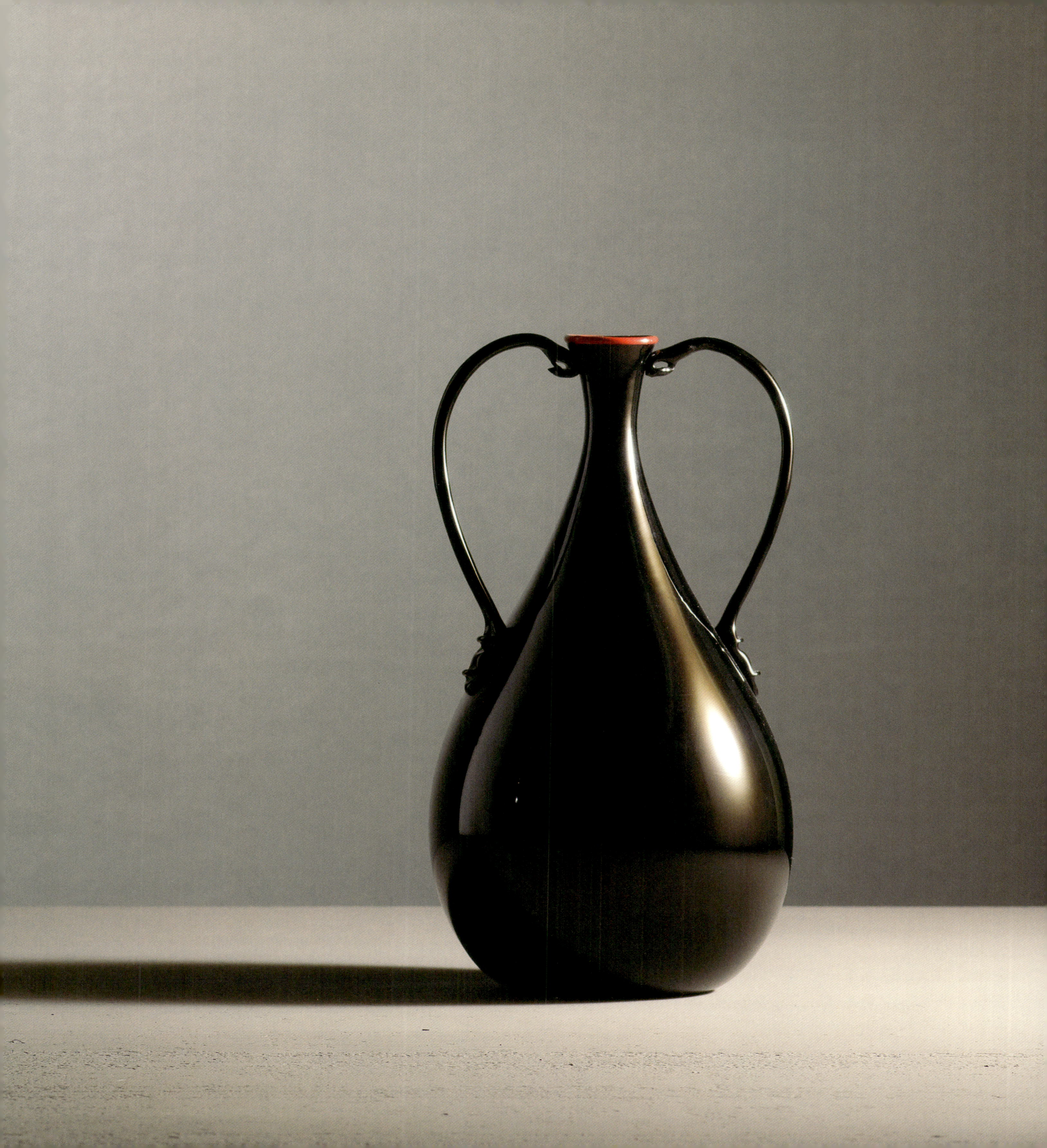

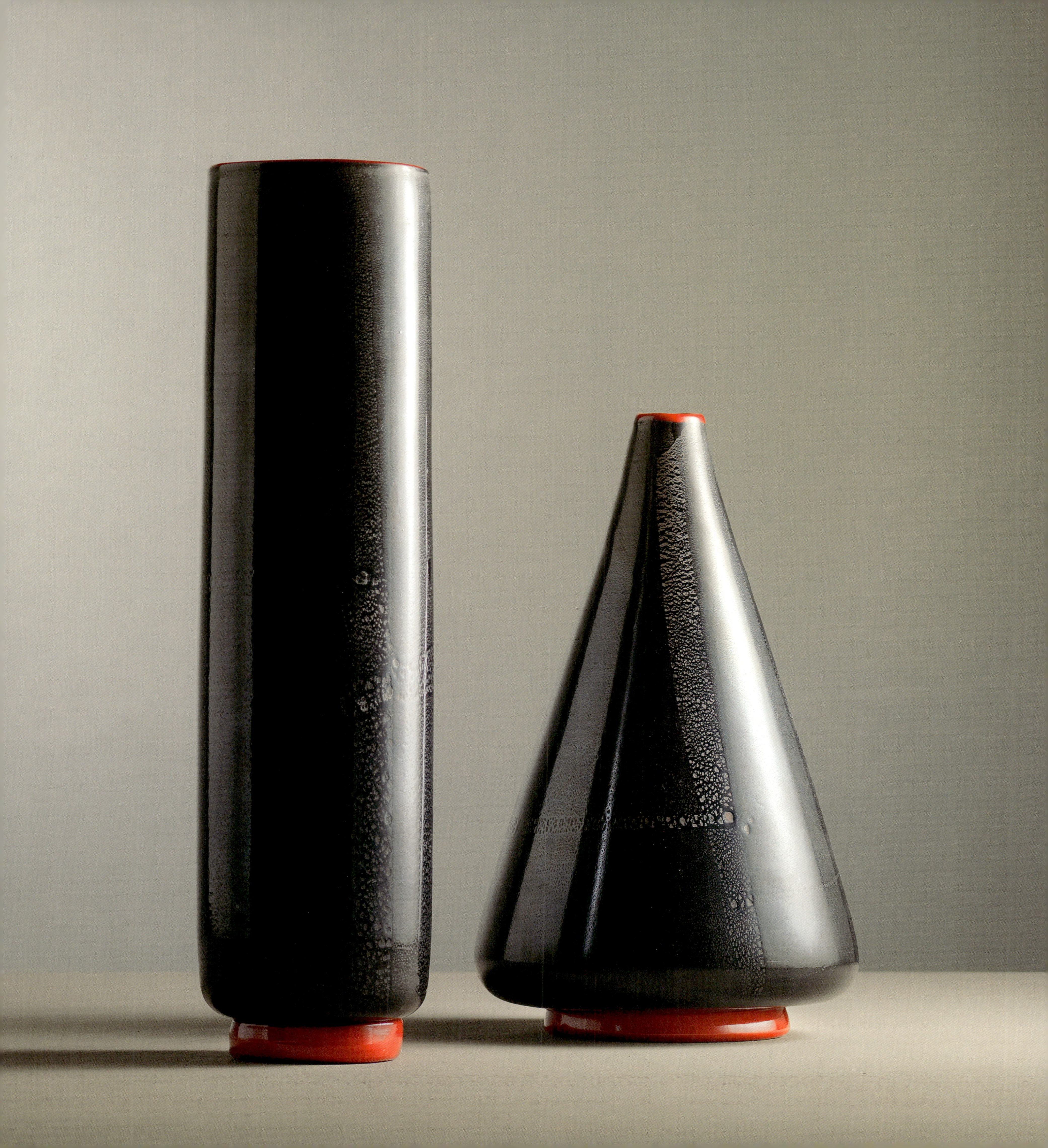

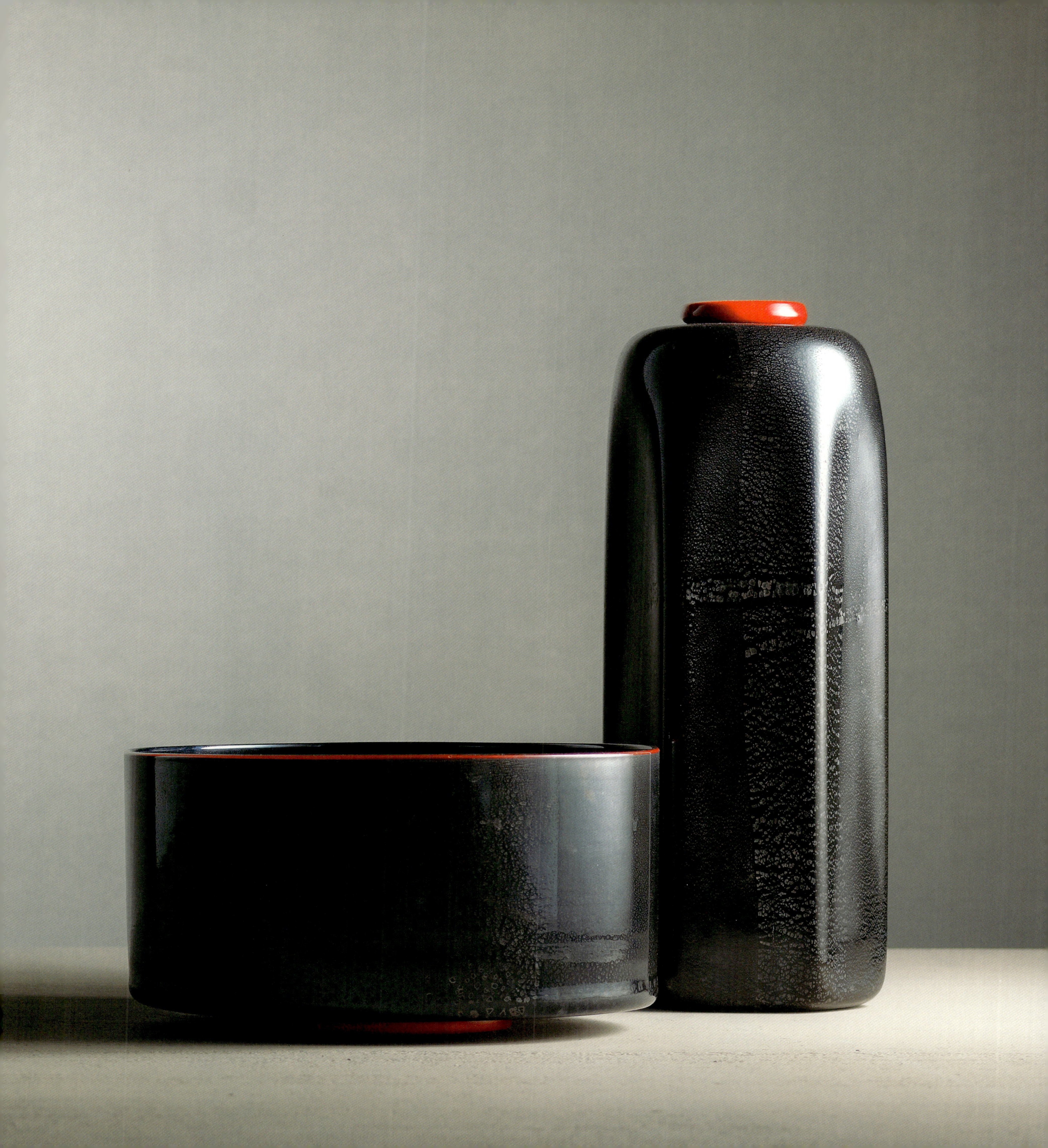

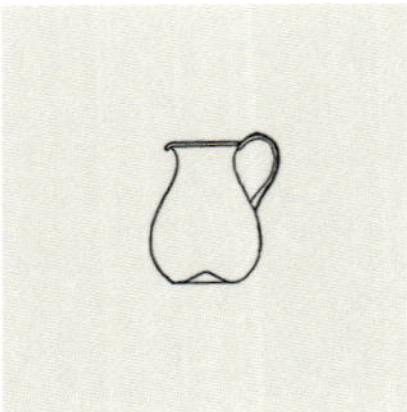

5963

Catalogue drawing for a jug of the 'black and silver' series, exhibited in 1930 at the 4th Triennale of Monza, AMB. Most of them were produced after models already designed by Vittorio Zecchin (nos. 5536, 5573, 5389, 5564), AMB.
1929–30
18 cm high

exhibitions
1930, Monza, 4th Triennale

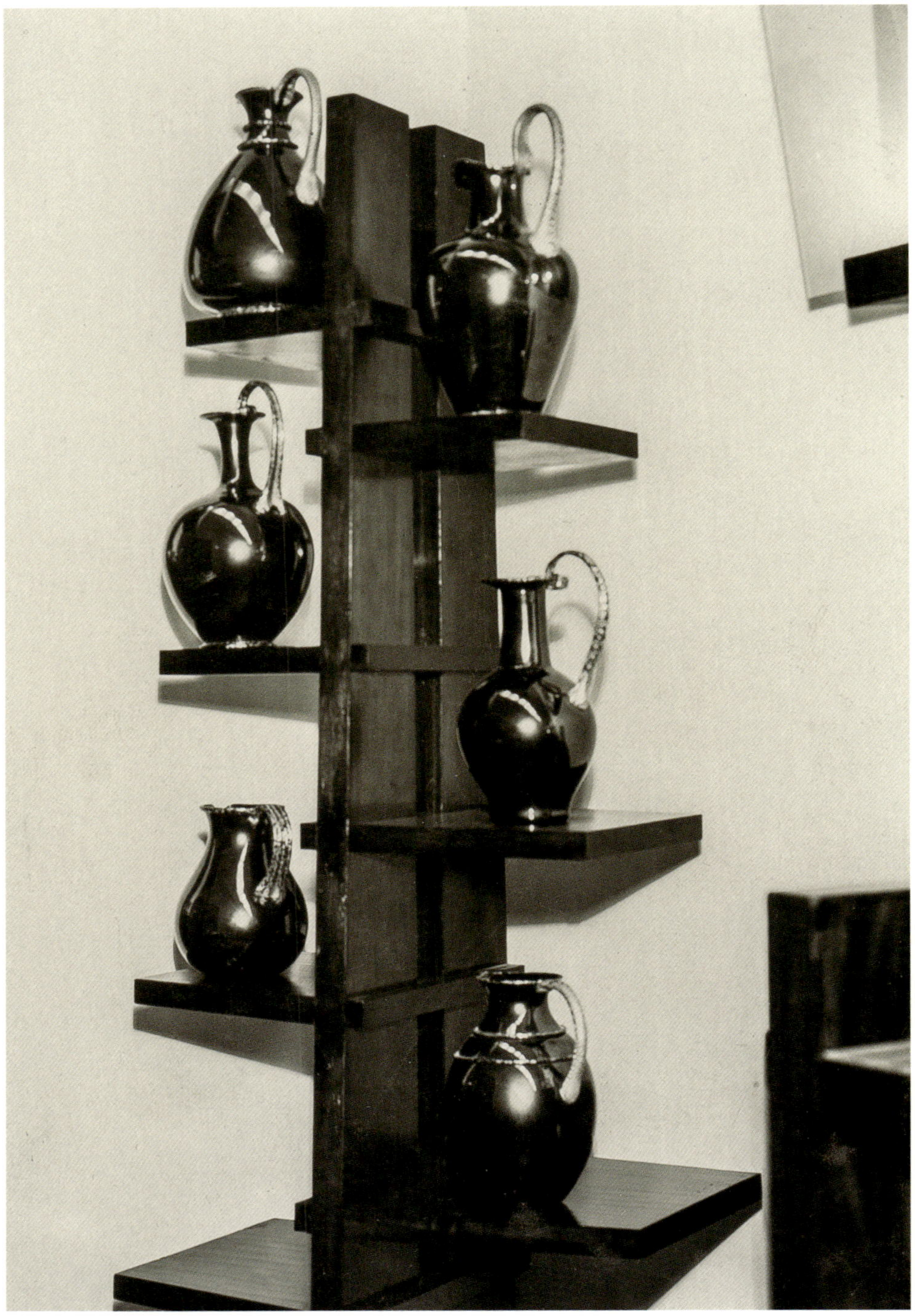

>
'Series of "black and silver" of Cappellin', in Domus, *July 1930, p. 41. The illustration was published also by* Emporium *to document the presence of the glassworks at the 4th Triennale of Monza*

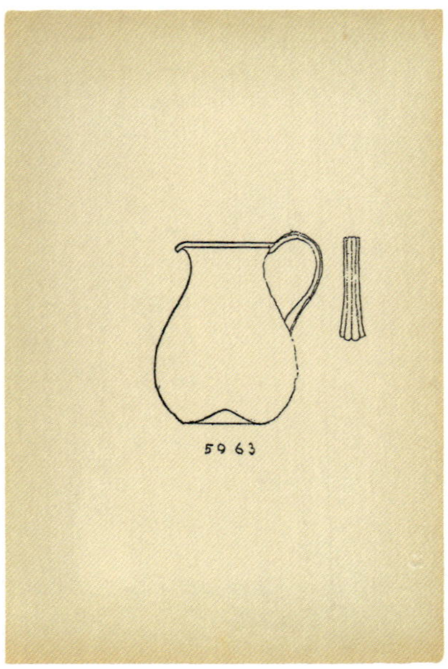

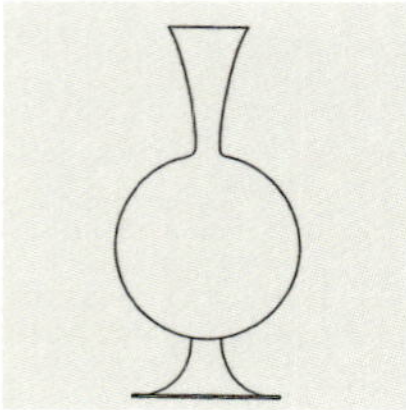

5678

Flask-shaped vase with a flattened cross-section in strongly iridised black glass. The design of the piece is inspired by a Vittorio Zecchin catalogue model (no. 5093), the lines of which have been simplified.
c. 1929
38 cm high
Private collection

>
5678, catalogue drawing, AMB

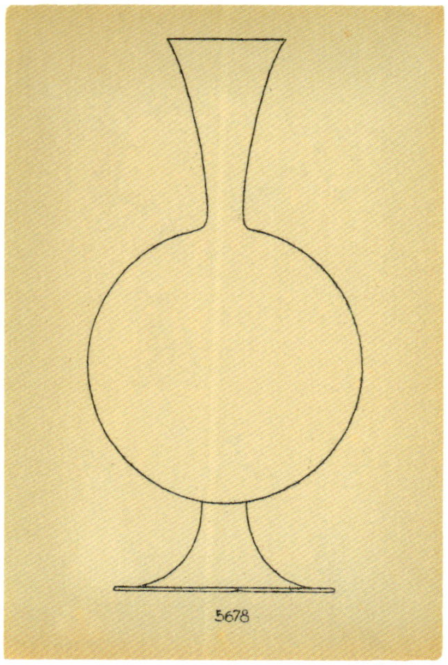

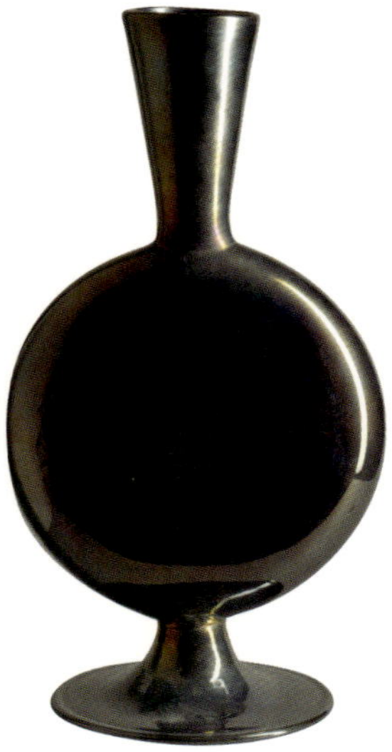

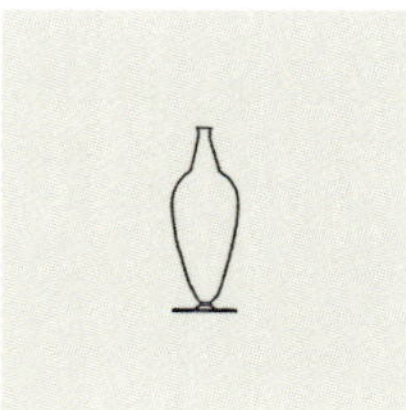

5923

Vase, with disk-shaped foot, in strongly iridised glass.
c. 1929
21.5 cm high
M.V.M. Cappellin Murano *acid stamp*
Private collection

>
5923, catalogue drawing, AMB

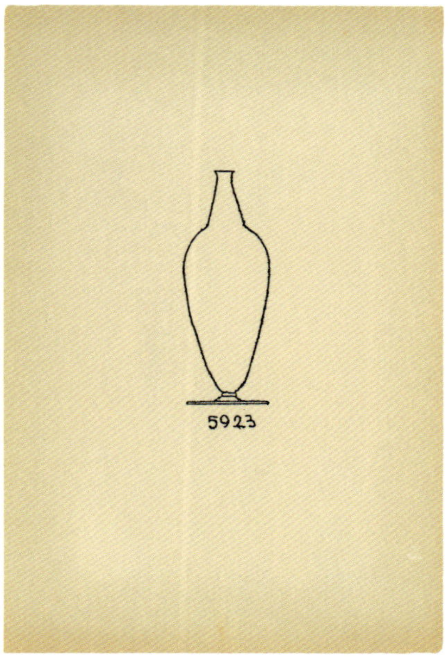

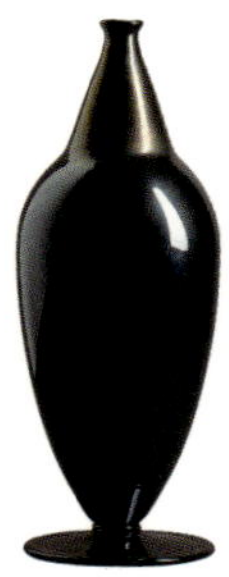

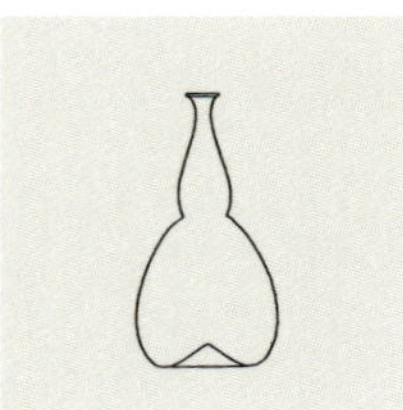

5275

Vase in strongly iridised black glass with a deep constriction in the body. The vase was produced after a model in the catalogue of the glassworks designed by Vittorio Zecchin (1921–25).
c. 1929
32 cm high
M.V.M. Cappellin Murano *acid stamp*
Bischofberger collection, Switzerland

>
5275, catalogue drawing, AMB

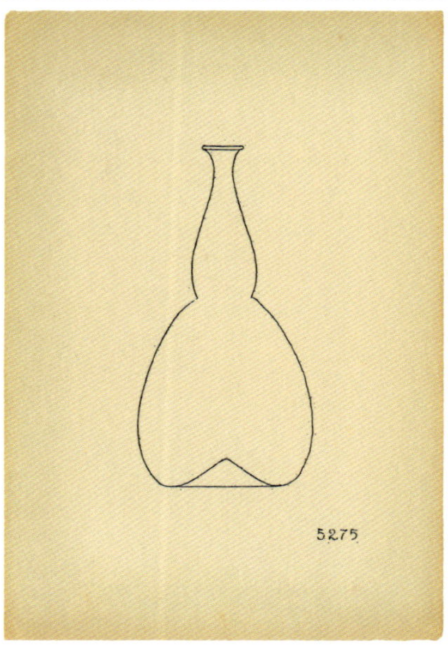

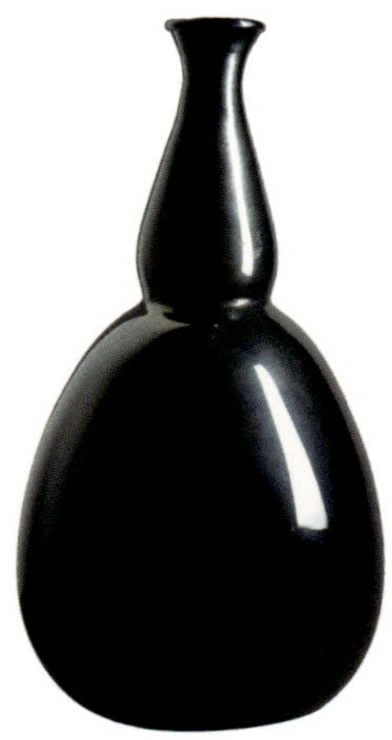

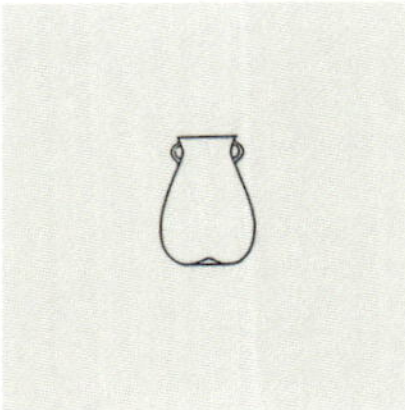

5262

Vase in strongly iridised black glass with small handles on the sides and body decorated with satin-finish horizontal bands. The vase was produced after a model in the catalogue of the glassworks designed by Vittorio Zecchin (1921–23).
1930–31
15 cm high
M.V.M. Cappellin Murano *acid stamp*
The Lockwood collection, Fort Lee, NJ

>
5262, catalogue drawing, AMB

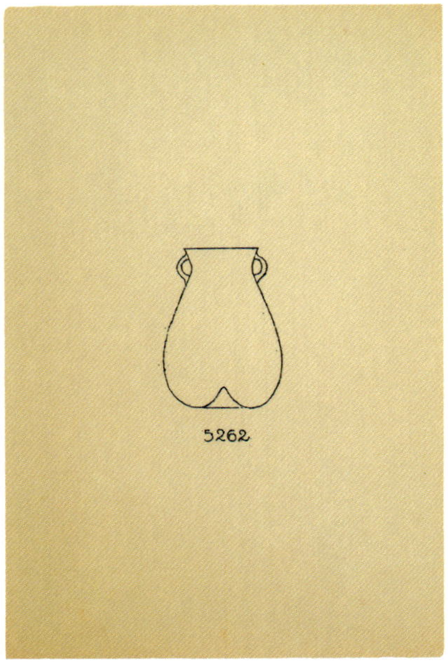

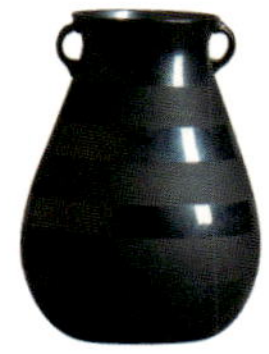

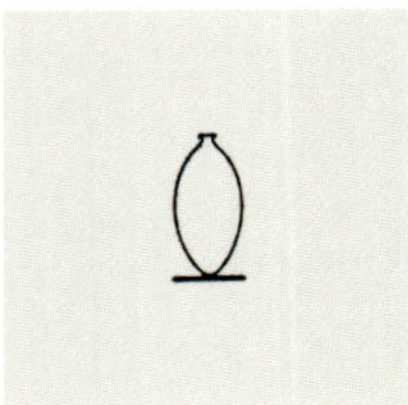

5922

Ovoid vase with disk-shaped foot in black glass with applied oxidised silver leaf. Model no. 5922 was also produced in other glass techniques (e.g. polychrome vertical canes).
c. 1929
18.5 cm high
M.V.M. Cappellin Murano *acid stamp*; old label
Private collection

>
5922, catalogue drawing, AMH

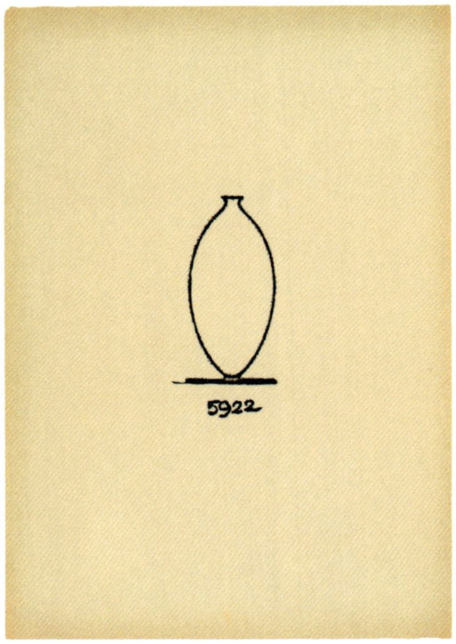

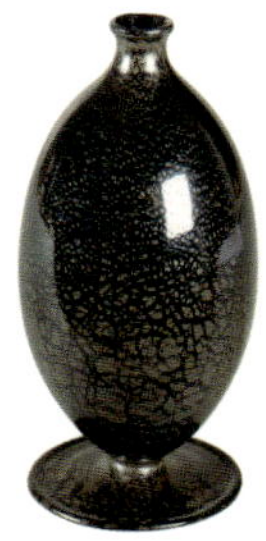

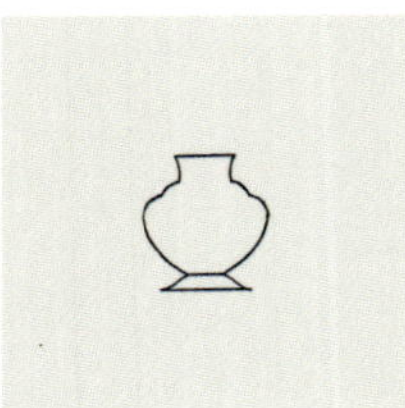

5938

Vase with disk-shaped foot in black glass with applied oxidised silver leaf. The model is characterised by a double constriction at the base of the neck. No. 5938 was also produced in other glass techniques (e.g. transparent, *lattimo aurato*, *decoro fenicio*, *pasta vitrea*).
c. 1929
16 cm high
M.V.M. Cappellin Murano *acid stamp*
Private collection, Udine

>
5938, catalogue drawing, AMB

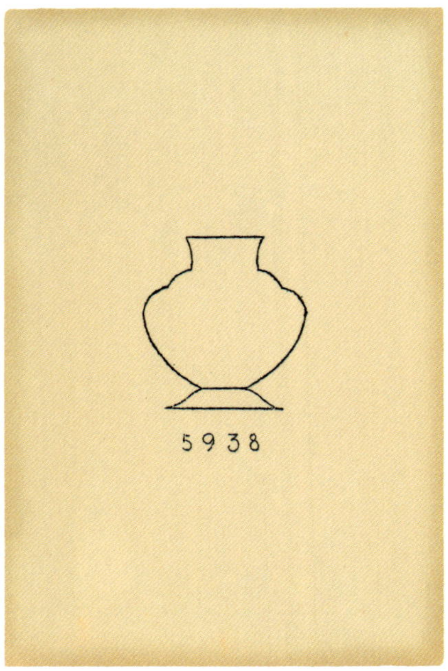

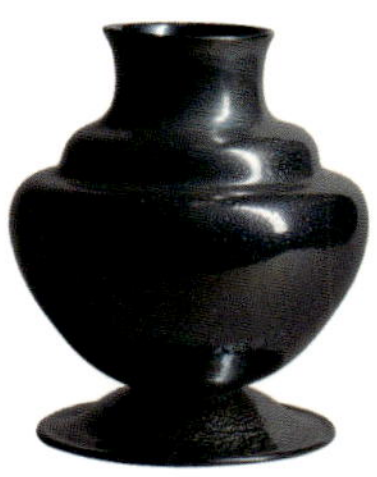

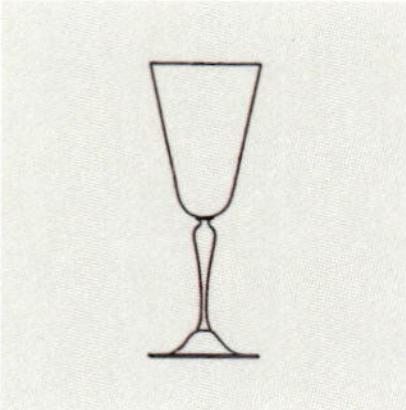

5436

Large goblet in black glass with disk-shaped foot and baluster-shaped stem. Thread around the mouth, around the foot and '*avolio*' (junction) in coral-red *pasta vitrea*. Surface with iridised finish. The goblet was produced after a model in the catalogue of the glassworks designed by Vittorio Zecchin (1921–23).
c. 1929
32.5 cm high;
Bischofberger collection, Switzerland

exhibitions
1929, Milano Fair

>
5219, catalogue drawing, AMH

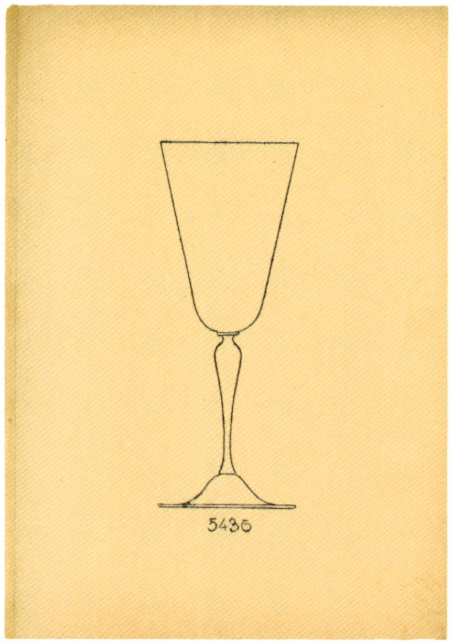

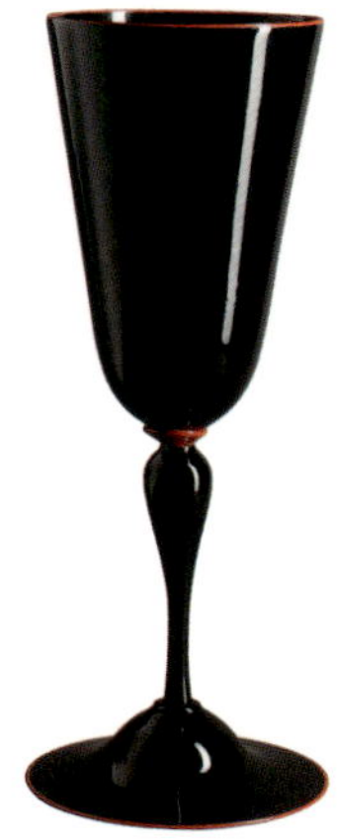

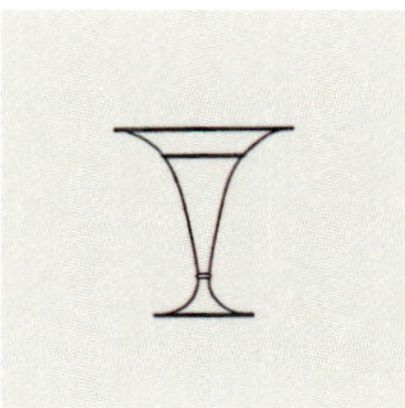

5561

Vase with flared foot and trumpet-shaped body in black glass. Finishings in coral-red *pasta vitrea*. Surface with iridised finish. The vase was produced after a model in the catalogue of the glassworks designed by Vittorio Zecchin (1925–26).
c. 1929
24.5 cm high;
Bischofberger collection, Switzerland

>
5561, working drawing, CSV

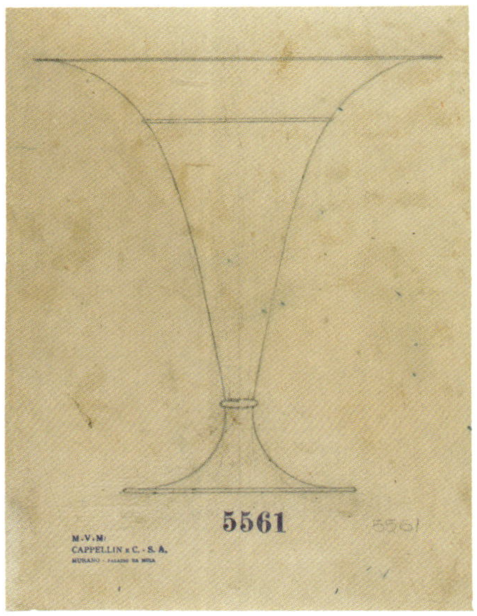

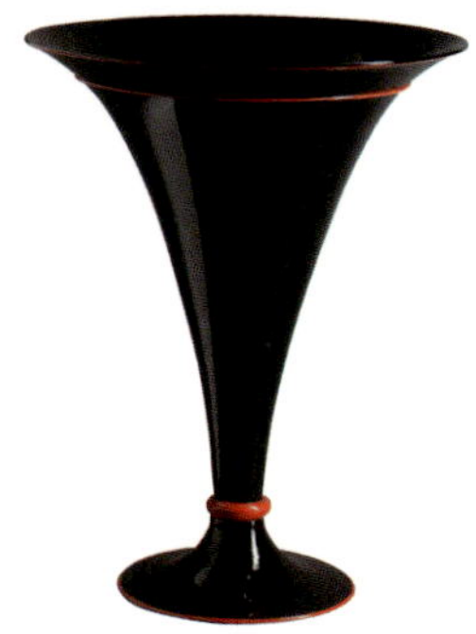

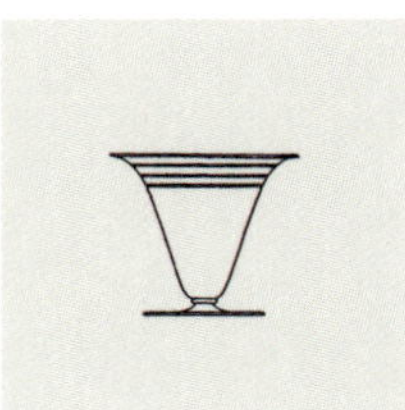

5585

Vase with disk-shaped foot in black glass. Finishings in coral-red *pasta vitrea*. Surface with iridised finish. The vase was produced after a model in the catalogue of the glassworks designed by Vittorio Zecchin (1925–26).
c. 1929
21.5 cm high;
Bischofberger collection, Switzerland

>
5585, working drawing, CSV

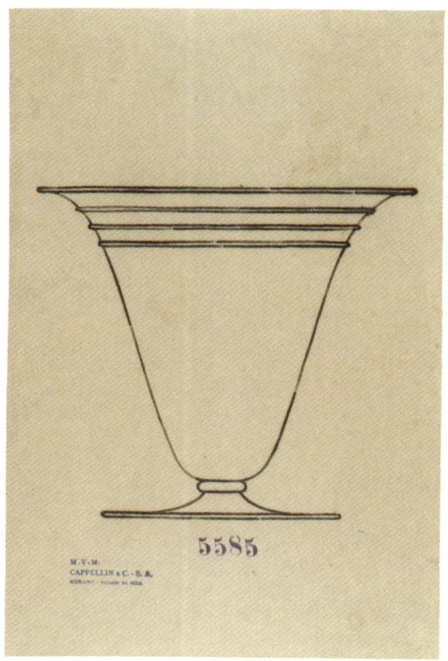

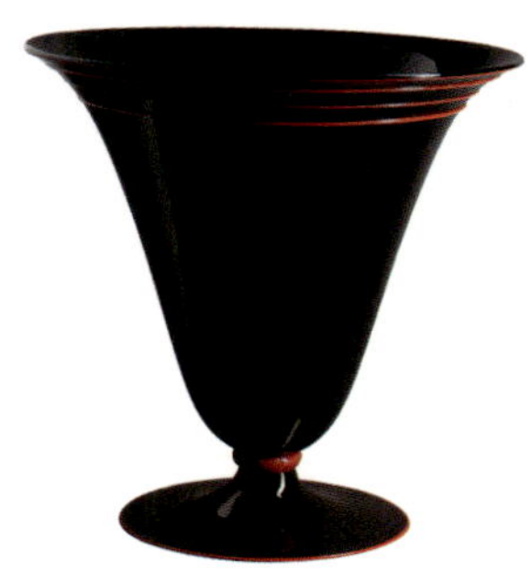

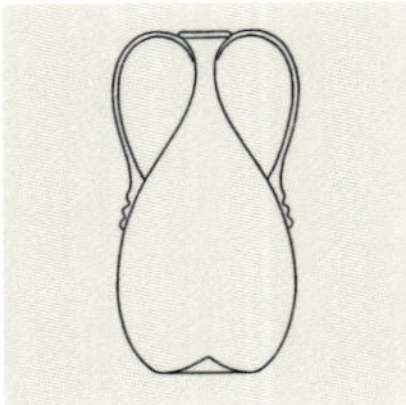

5217

Vase in black glass with widely extending handles decorated with *morise*, in black glass. Thread around the mouth in coral-red *pasta vitrea*. Surface with iridised finish. The vase was produced in a small-scale version of the model no. 5217 (h 40 cm) designed by Vittorio Zecchin (1921–22).
c. 1929
24.5 cm high
Bischofberger collection, Switzerland

>
5217, catalogue drawing, AMB

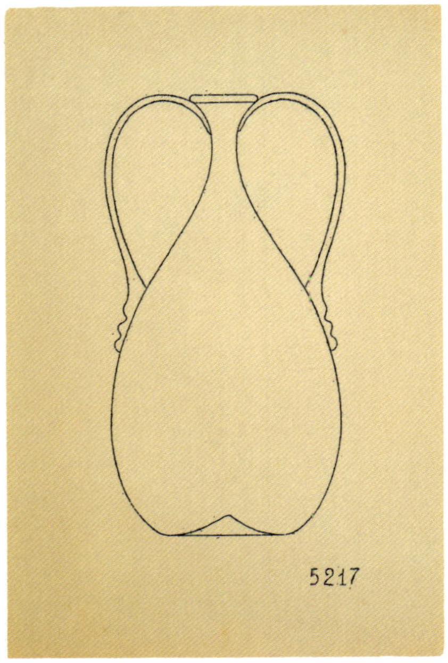

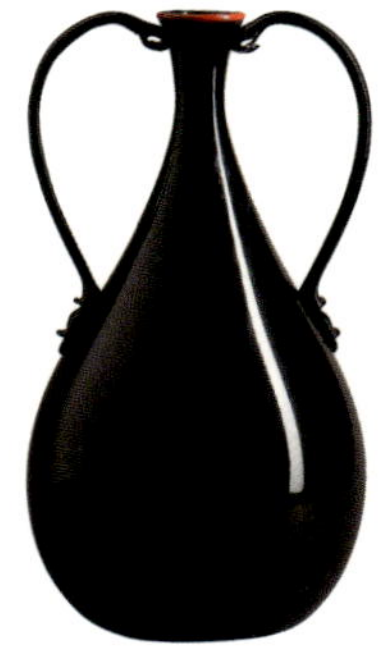

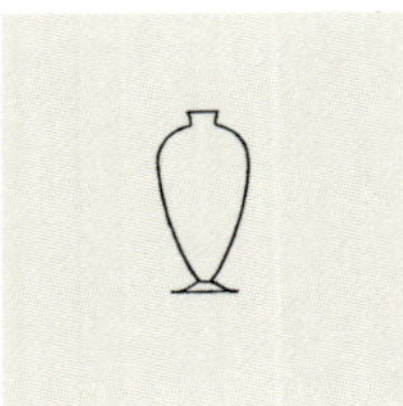

5870

Vase in black glass with disk-shaped foot and thread around the mouth in coral-red *pasta vitrea*. Surface strongly iridised. Model no. 5870 was also produced in other glass techniques (e.g. transparent, *pasta vitrea*, *decoro fenicio*, *lattimo*).
c. 1929
20.5 cm high
M.V.M. Cappellin Murano *acid stamp*
The Lockwood collection, Fort Lee, NJ

>
5870, catalogue drawing, AMB

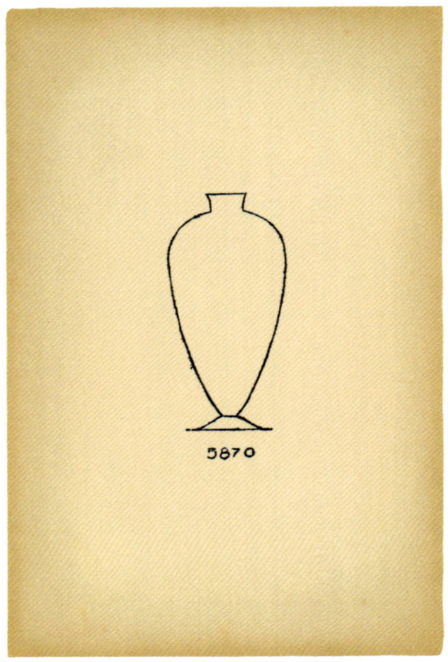

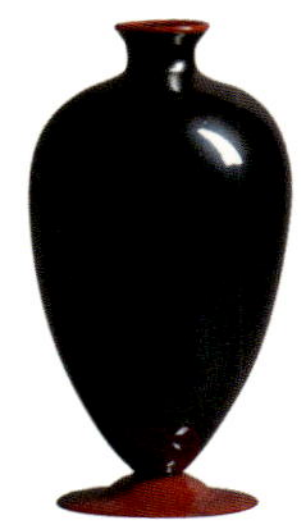

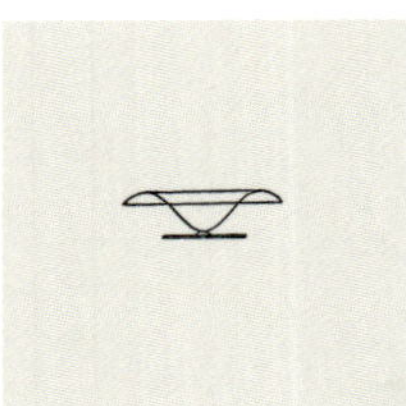

5877

Vase with disk-shaped foot and rim folded outwards in black glass. Finishings in red *pasta vitrea*. Surface with iridised finish. The design of the piece was inspired by a model by Vittorio Zecchin present in the catalogue (no. 5044), the proportions of which were here revised.
c. 1929
7 cm high; 28 cm ø
M.V.M. Cappellin Murano *acid stamp*
The Olnick Spanu Collection, New York

>
5877, working drawing, CSV

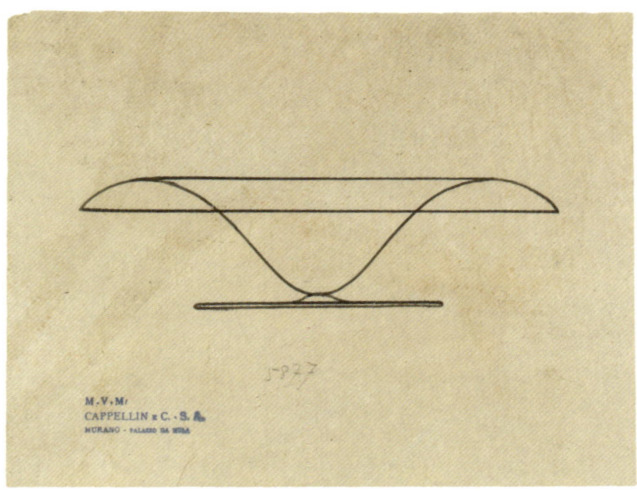

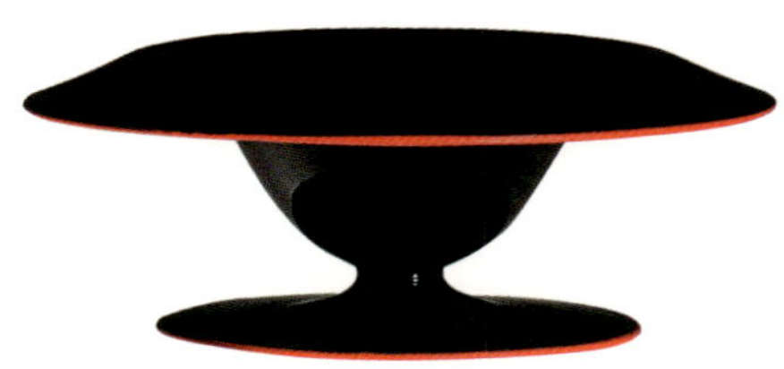

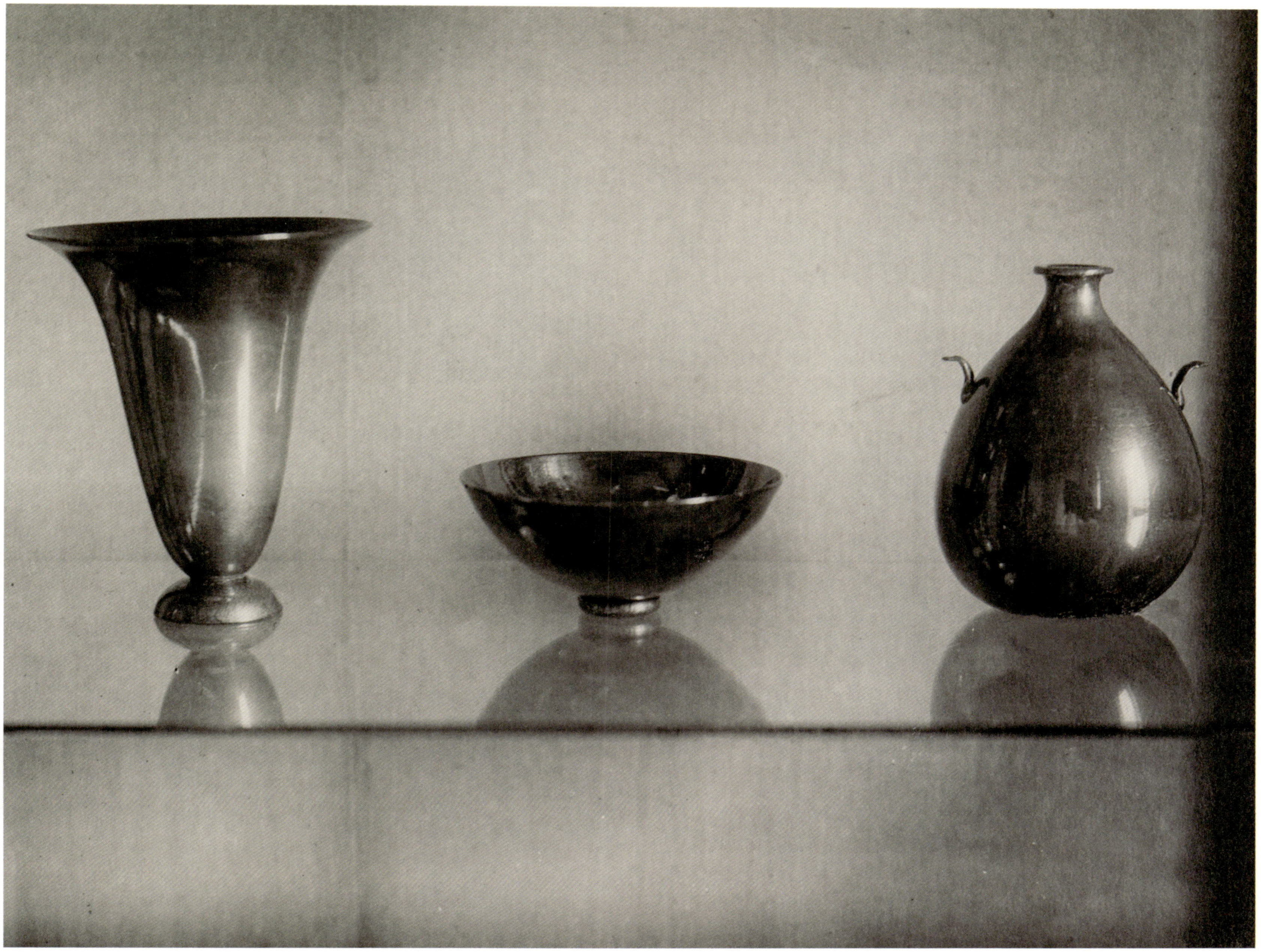

›
Vases in opaque glass. Period photograph, c. 1929, CSV

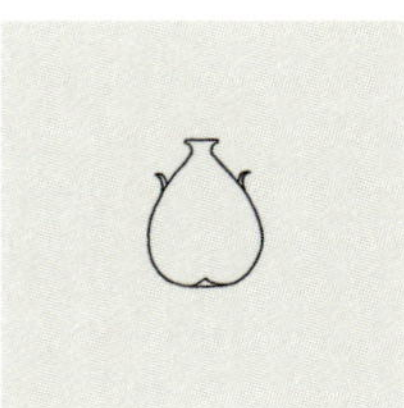

5941

Pear-shaped vase with small lateral handles in black glass and a thread around the mouth in coral-red *pasta vitrea*. Surface with iridised finish. Model no. 5941 was produced also in other glass techniques (e.g. transparent, *pasta vitrea*, silver black).
c. 1929
19.5 cm high
Private collection

›
5941, catalogue drawing, AMB

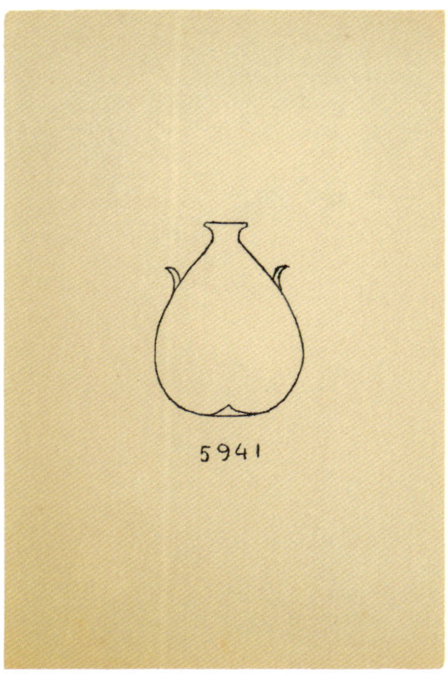

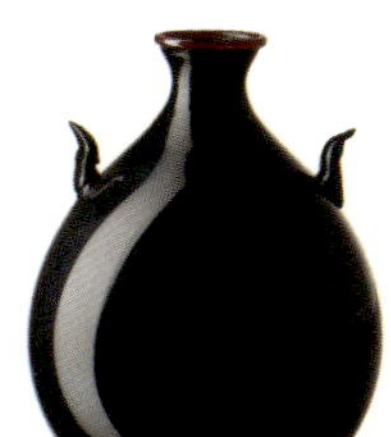

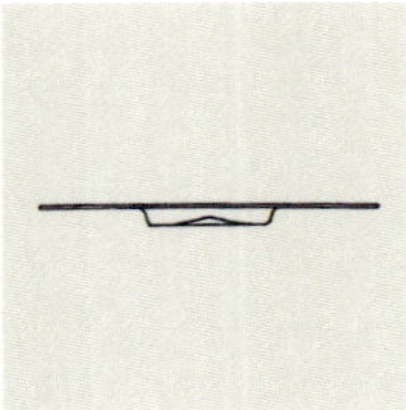

1755

Plate in black glass with rim in red *pasta vitrea*. Surface with iridised finish. The plate was produced after a model in the catalogue of the glassworks designed by Vittorio Zecchin (1921–25).
c. 1929
22 cm ø
M.V.M. Cappellin Murano *acid stamp*
The Lockwood collection, Fort Lee, NJ

›
1755, catalogue drawing, AMB

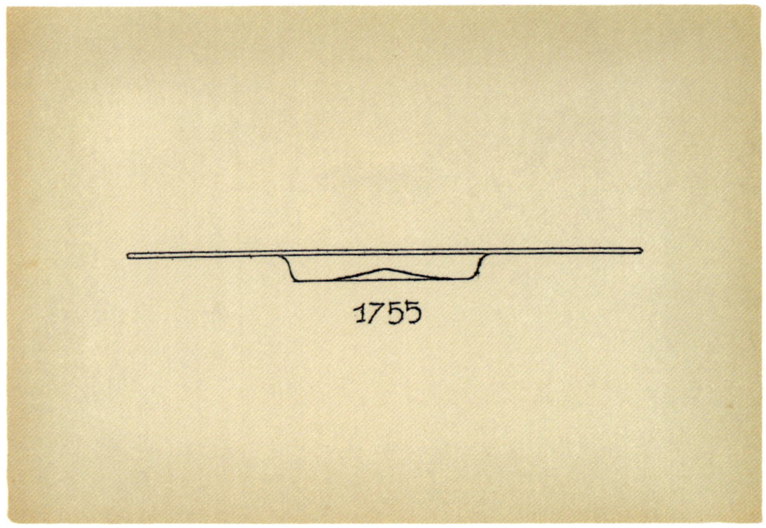

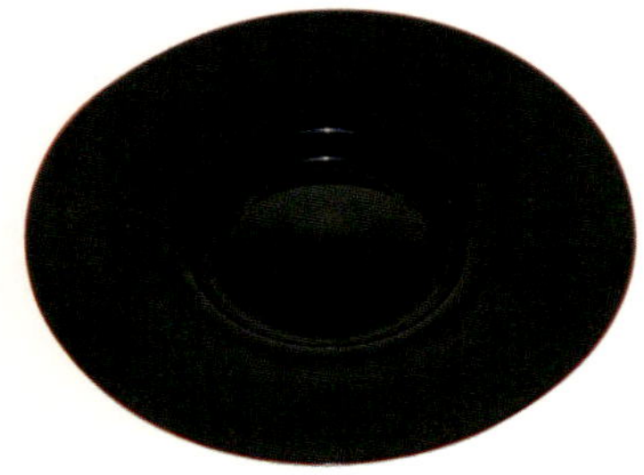

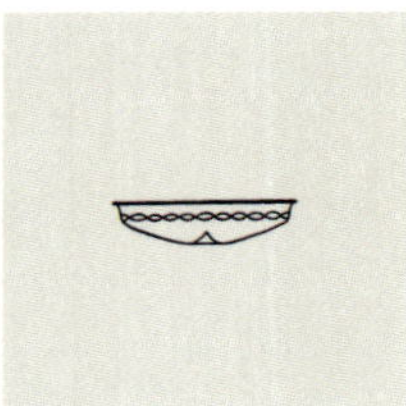

5659

Bowl in black glass with chain decoration in coral-red *pasta vitrea*. Surface finished with applied gold leaf and iridisation.
c. 1929
5 cm high; 21.5 cm ø
Private collection

›
5659, working drawing, CSV

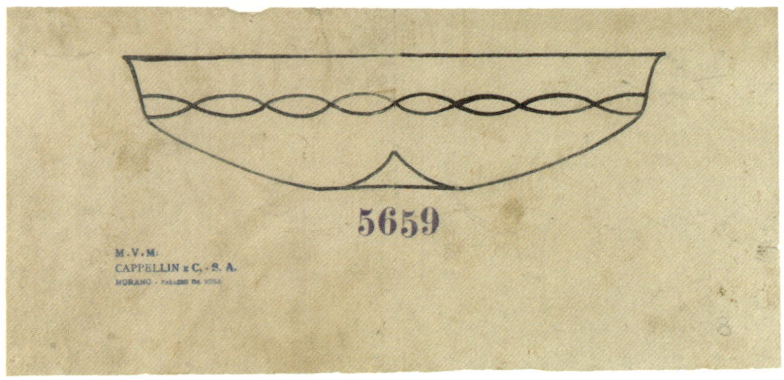

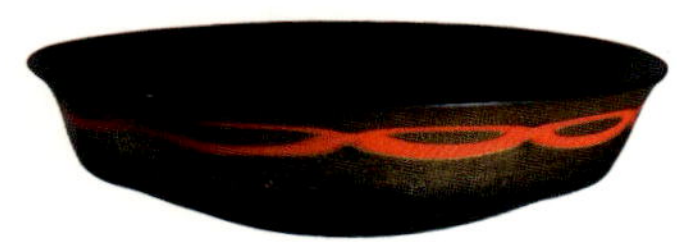

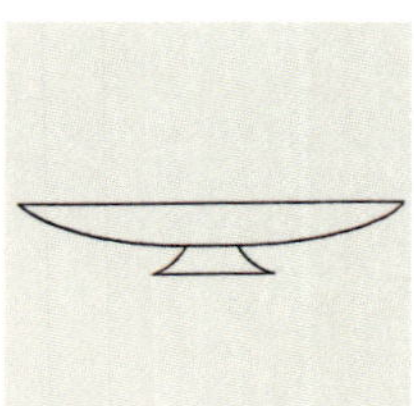

5705

Cake stand in black glass with foot and thread around the mouth in coral-red *pasta vitrea*. Surface with iridised finish. The piece was produced in a small-scale variant of no. 5705 (Ø 52 cm).
c. 1929
10 cm high; 46.5 cm ø
C.V.M. *acid stamp*
Private collection

›
5705, catalogue drawing, AMB

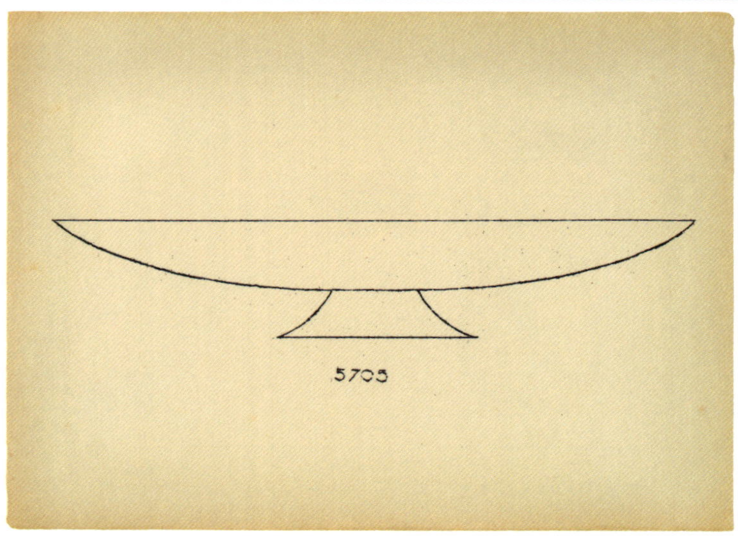

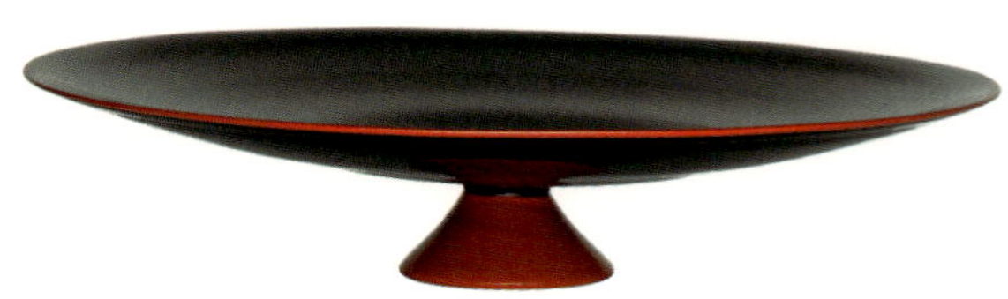

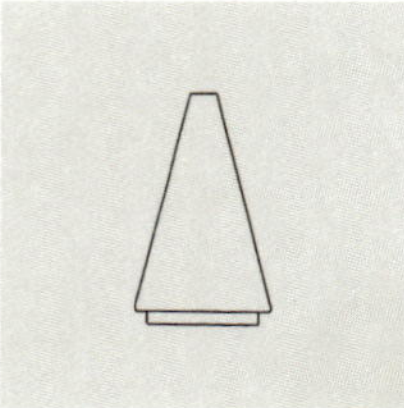

5971

Truncated-cone-shaped vase in black glass with applied oxidised silver leaf. Foot and thread around the mouth in coral-red *pasta vitrea*. Surface with iridised finish. Model no. 5971 was also produced in other glass techniques (e.g. *pasta vitrea*, *lattimo*, cased glass).
c. 1930
26 cm high
M.V.M. Cappellin Murano *acid stamp*
Francesca and Massimo Valsecchi collection, Palazzo Butera

exhibitions
1931, Amsterdam, Exhibition of glass, ceramics and lace

>
5971, working drawing, CSV

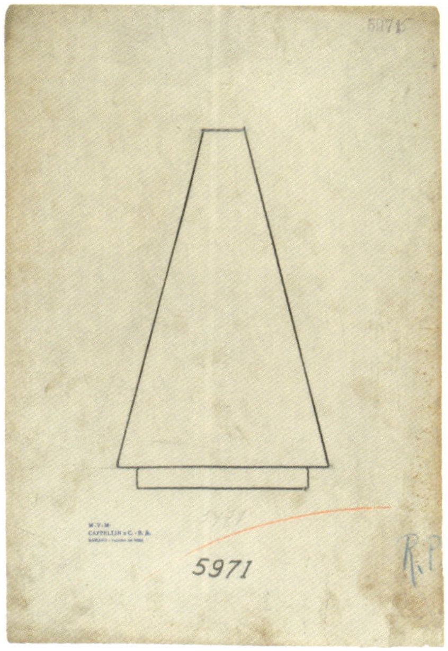

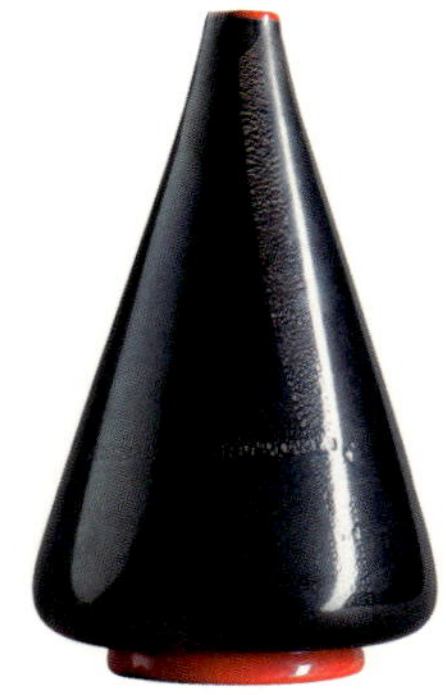

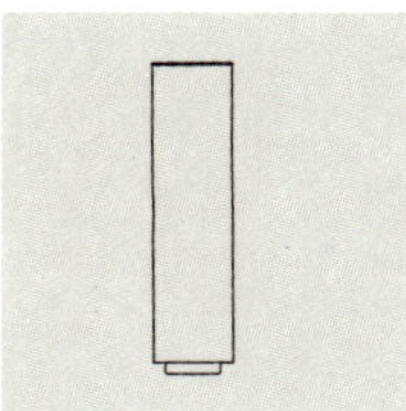

5973

Cylindrical vase in black glass with applied oxidised silver leaf. Foot and thread around the mouth in coral-red *pasta vitrea*. Surface with iridised finish. Model no. 5973 was produced in other glass techniques (e.g. *pasta vitrea*, cased glass).
c. 1930
35.5 cm high
Francesca and Massimo Valsecchi collection, Palazzo Butera

>
5973, working drawing, CSV

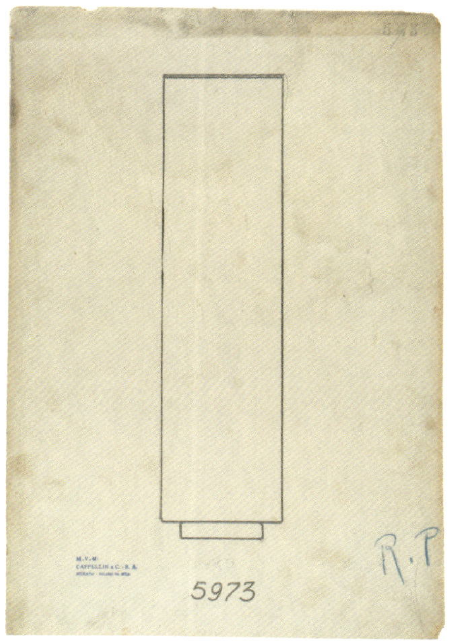

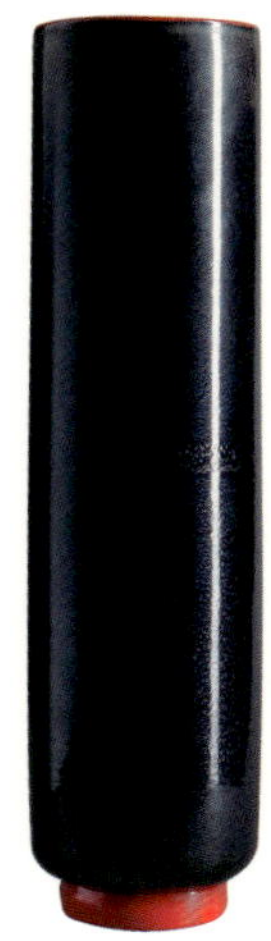

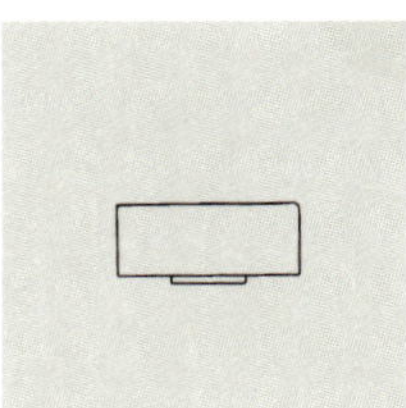

5993

Cylindrical bowl in black glass with applied oxidised silver leaf. Foot and thread around the mouth in coral-red *pasta vitrea*. Surface with iridised finish. Model no. 5993 was also produced in other glass techniques (e.g. *pasta vitrea*, cased glass).
c. 1930
11 cm high; 21.5 cm ø
M.V.M. Cappellin Murano *acid stamp*
Francesca and Massimo Valsecchi collection, Palazzo Butera

>
5993, catalogue drawing, AMB

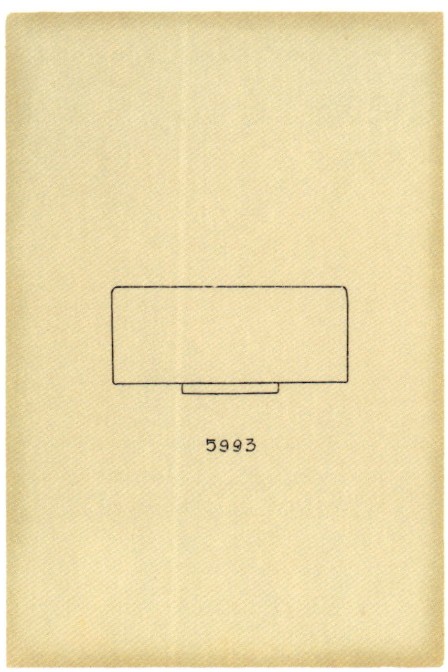

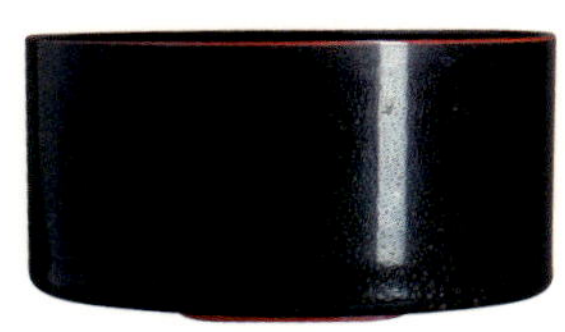

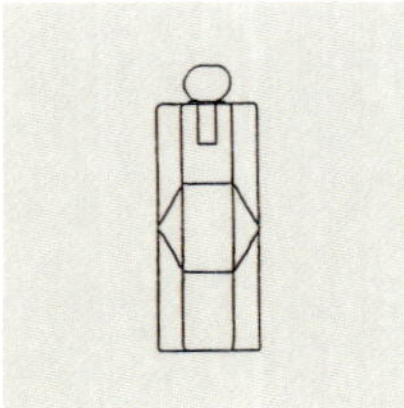

5992

Vase having a hexagonal cross-section in black glass with applied oxidised silver leaf. Mouth in coral-red *pasta vitrea*. Surface with iridised finish. The model no. 5992 was also produced in other glass techniques (e.g. *pasta vitrea*, cased glass).
c. 1930
29 cm high
Bischofberger collection, Switzerland

>
5992, catalogue drawing, AMB

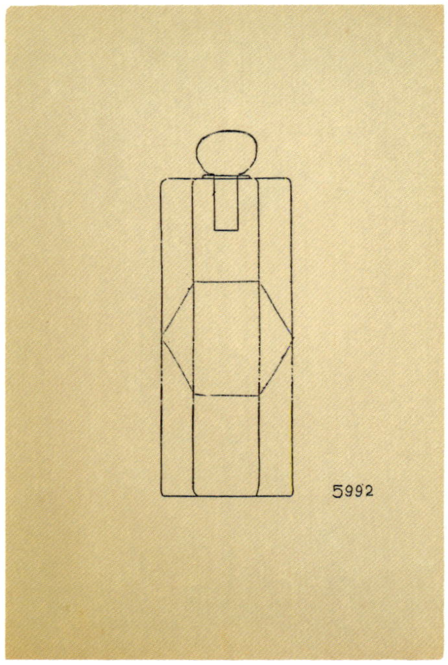

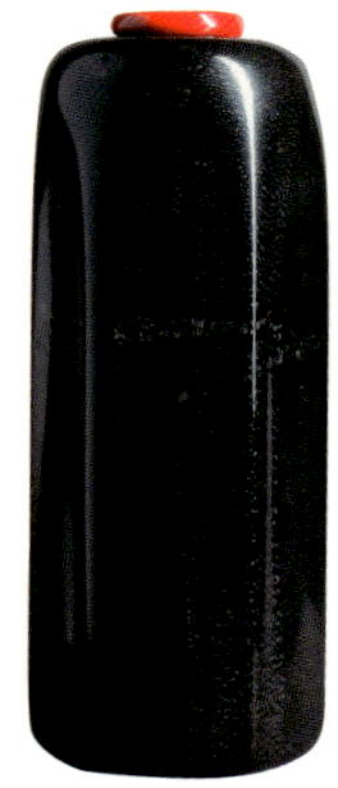

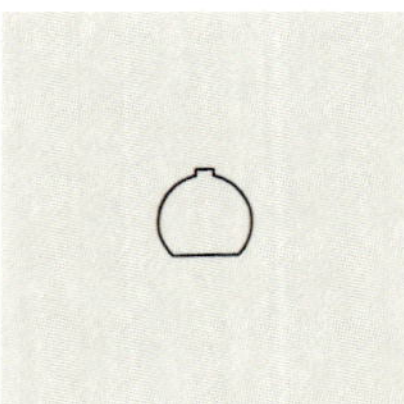

1272

Spherical vase in black glass having surface with iridised finish. The model was proposed in four different sizes and also as a perfume bottle.
1929–30
11 cm high
Biennale 1964 old label
Private collection

Exhibitions
1964, Venice, 32nd Biennale d'arte

>
1272, catalogue drawing, AMA

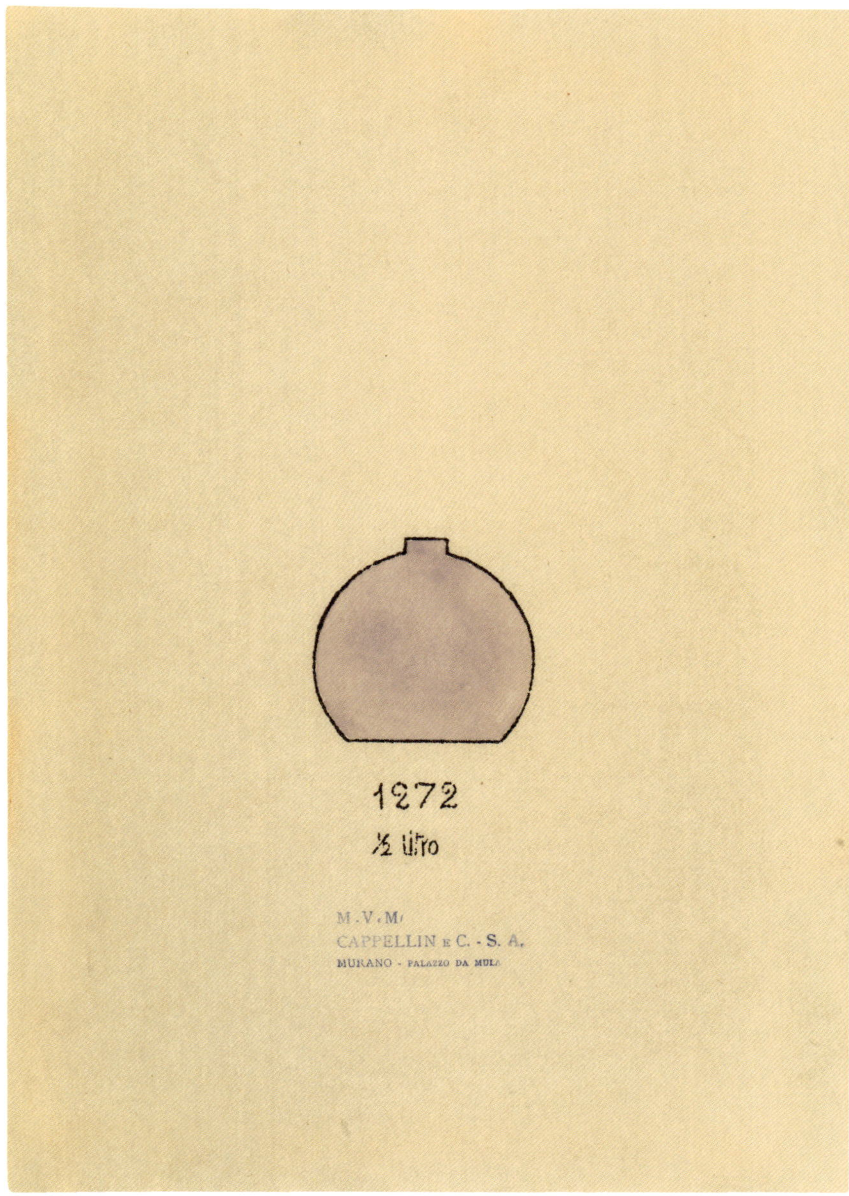

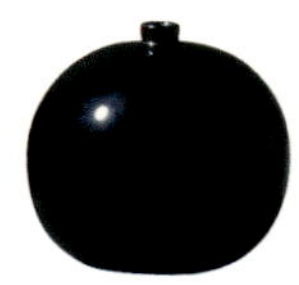

>
Vases no. 1272 and no. 1273. Period photograph, 1929–30 CSV, detail

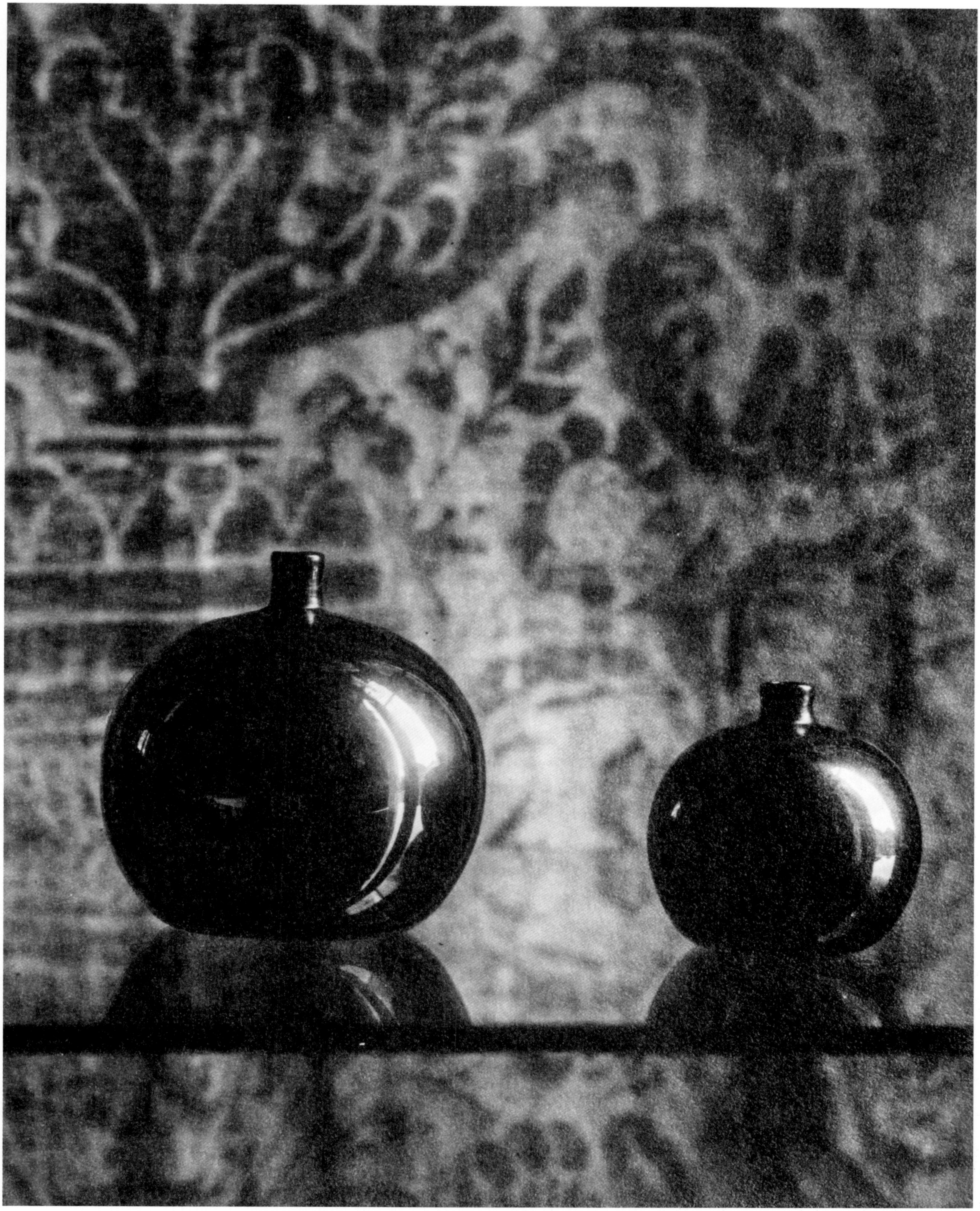

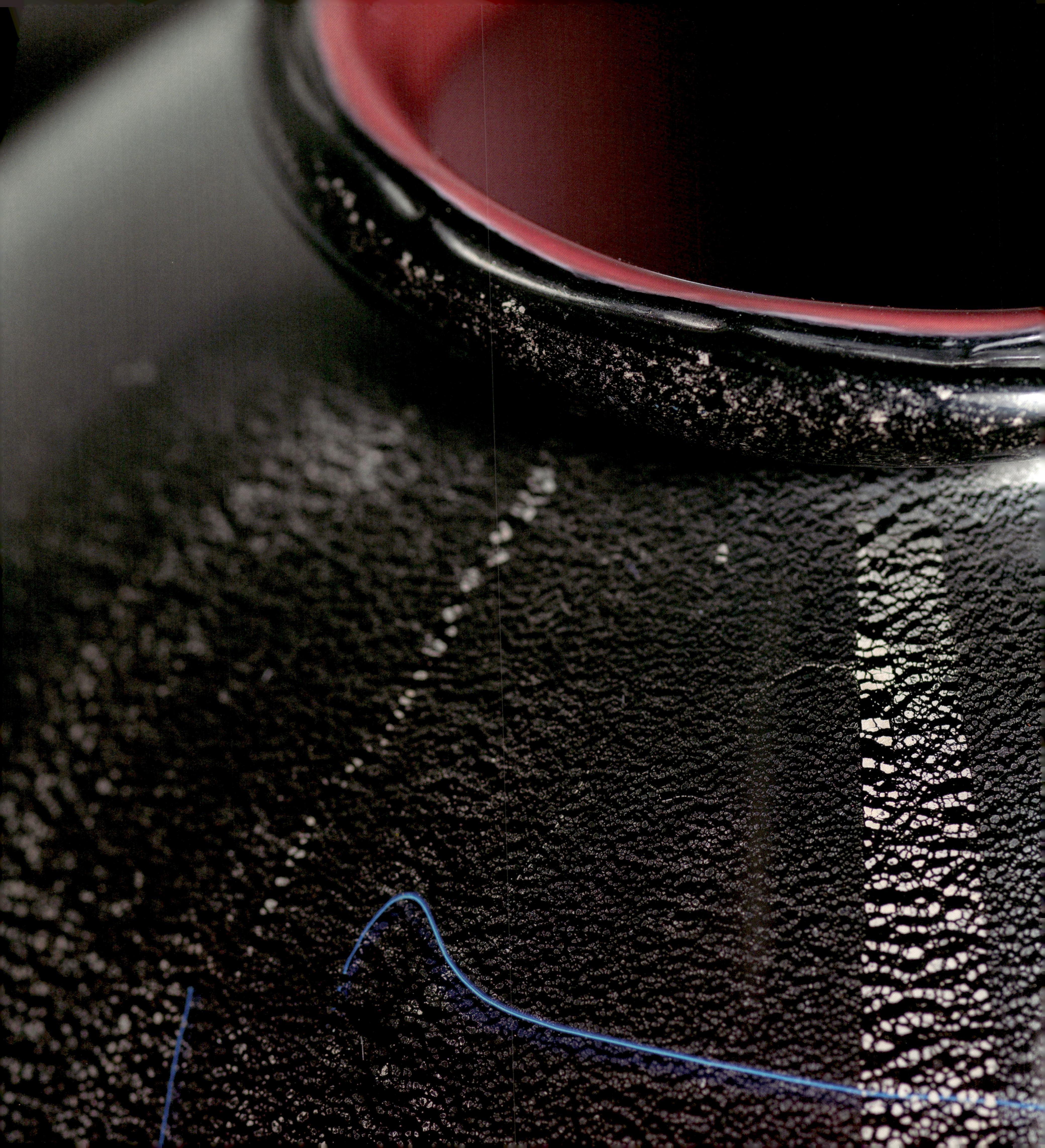

These are opaque glass pieces generally produced by superimposing one or more layers of coloured glass or *pasta vitrea* on a *lattimo* base. Some variants were executed by superimposing several layers of colour or *pasta vitrea* divided by a layer of *lattimo*, in such a way that the inside of the vase was different from the outside (e.g. green on the inside, coral-red on the outside). The technique was increasingly preferred to that of *pasta vitrea* because it provided greater guarantees for the outcome of the working process. Extraordinary examples of this series — also characterised by the recourse to geometric and essential forms — were exhibited at the 4th Triennale of Monza in 1930, where they were appreciated also for their 'lucid, compact, vivid colours' (C.A. Felice, *Arti decorative 1930,* Milan, 1930, p. 38). In Monza in 1930 and in Amsterdam in 1931 (*Mostra di vetri, ceramiche e merletti d'arte moderna italiana*), elegant cased-glass pieces with softer lines and applied decorations (shells, leaves, animal heads) were also shown.

CASED GLASS 1929-31

Exhibitions

1930, Monza, 4th *Esposizione Triennale Internazionale delle Arti Decorative ed Industriali Moderne*
1931, Amsterdam, *Tentoonstelling van glas, aardewerk en kant, van moderne italiaansche kunstnijverheid. Mostra di vetri, ceramiche e merletti d'arte moderna italiana*

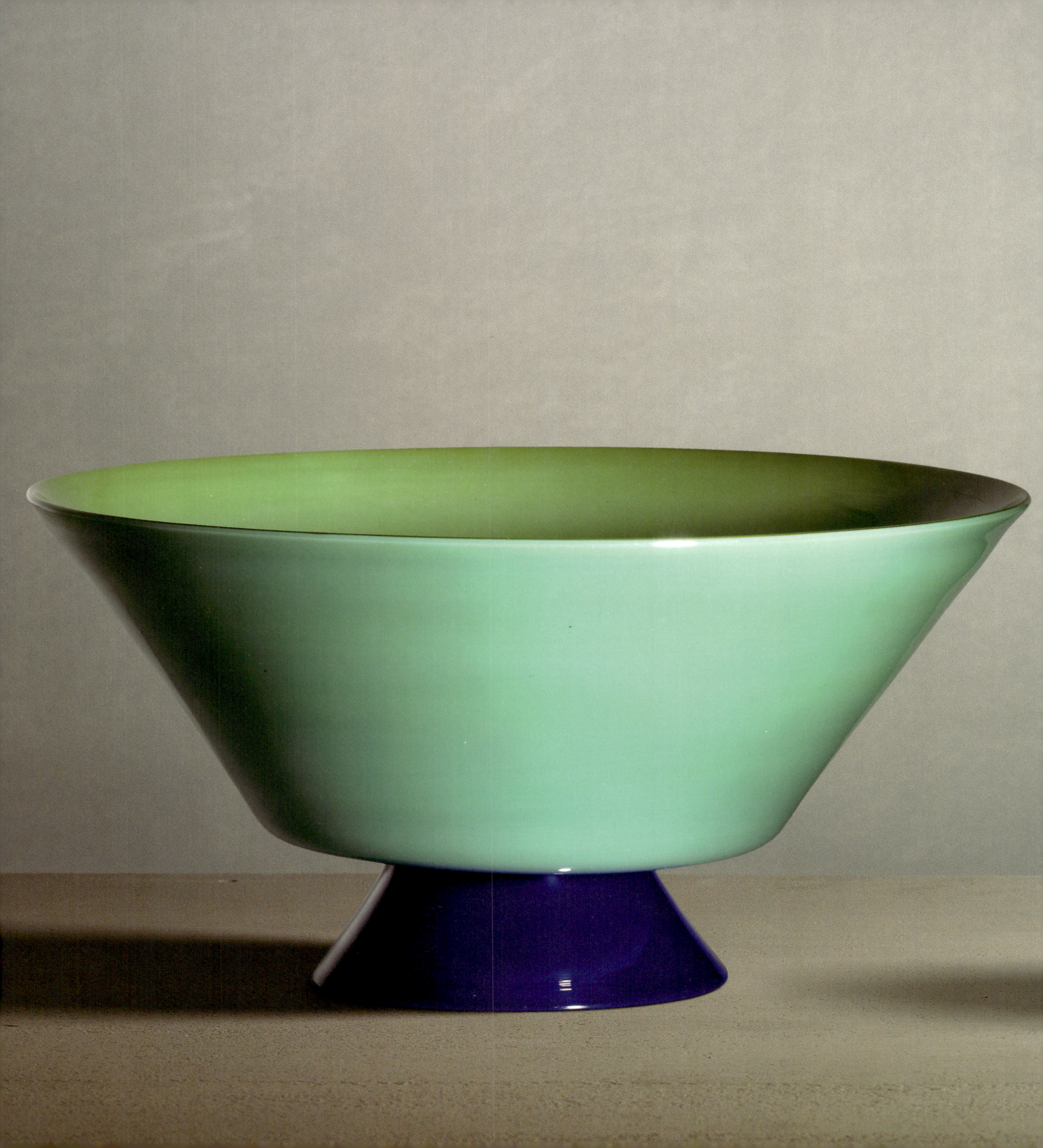

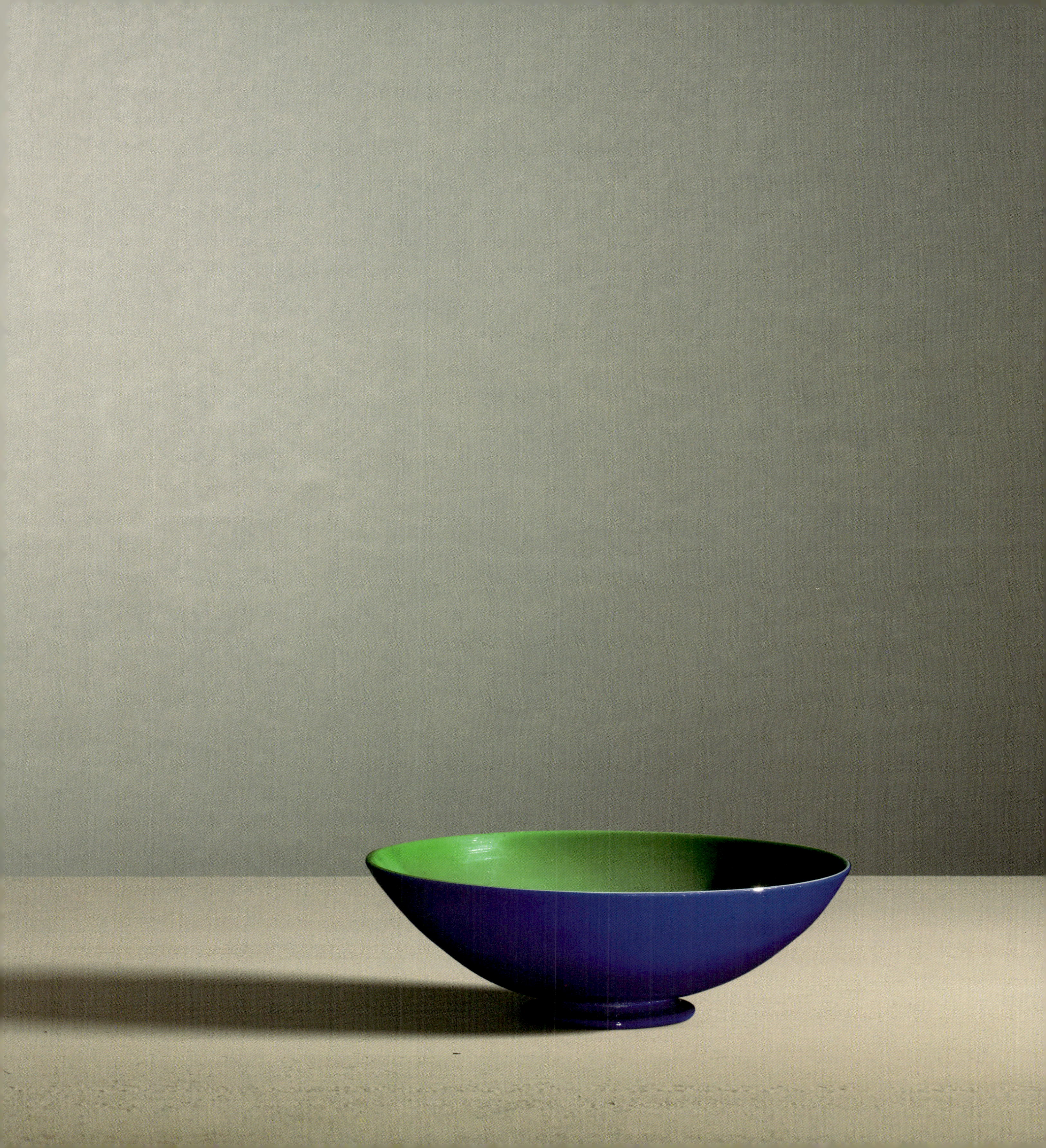

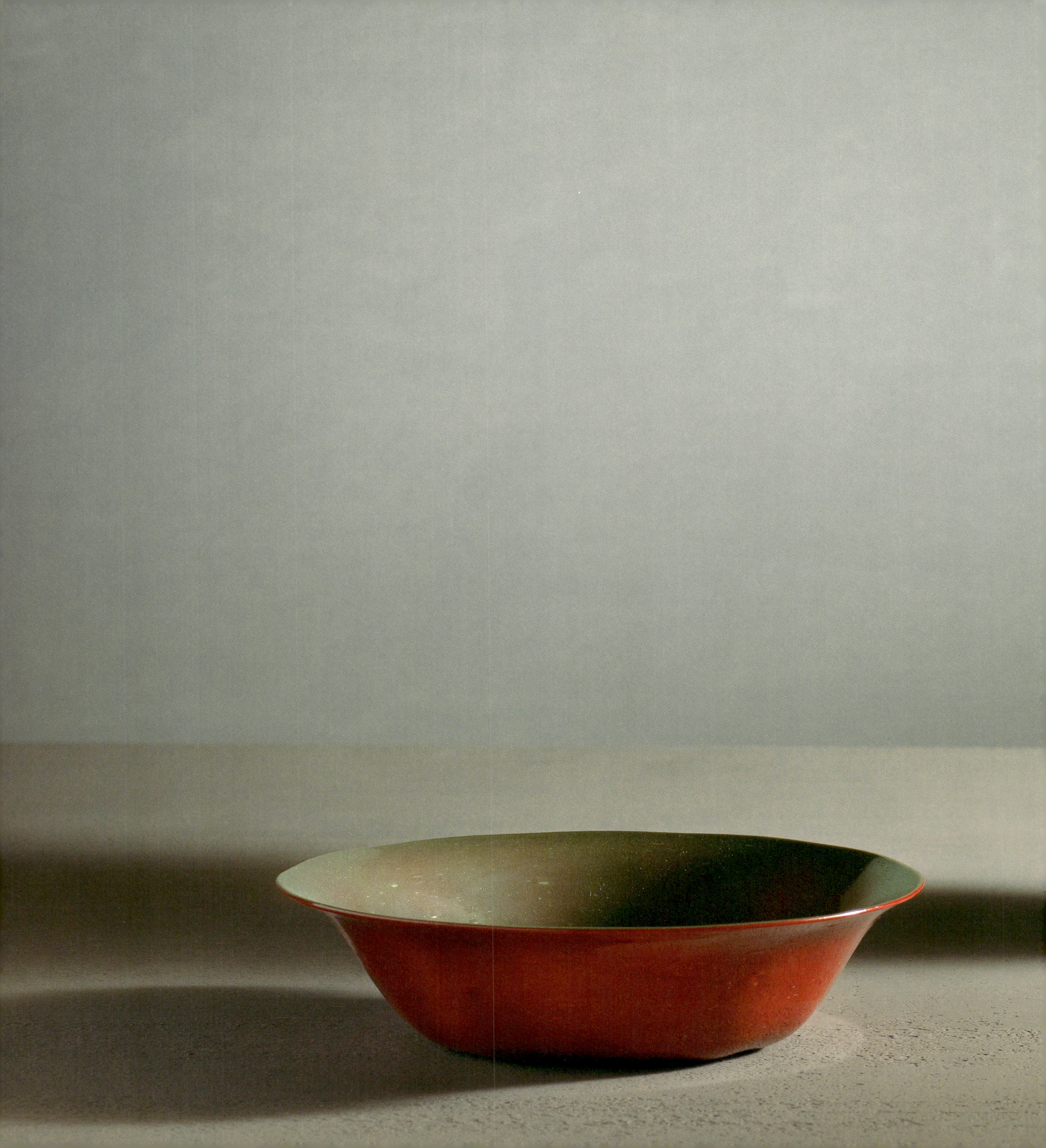

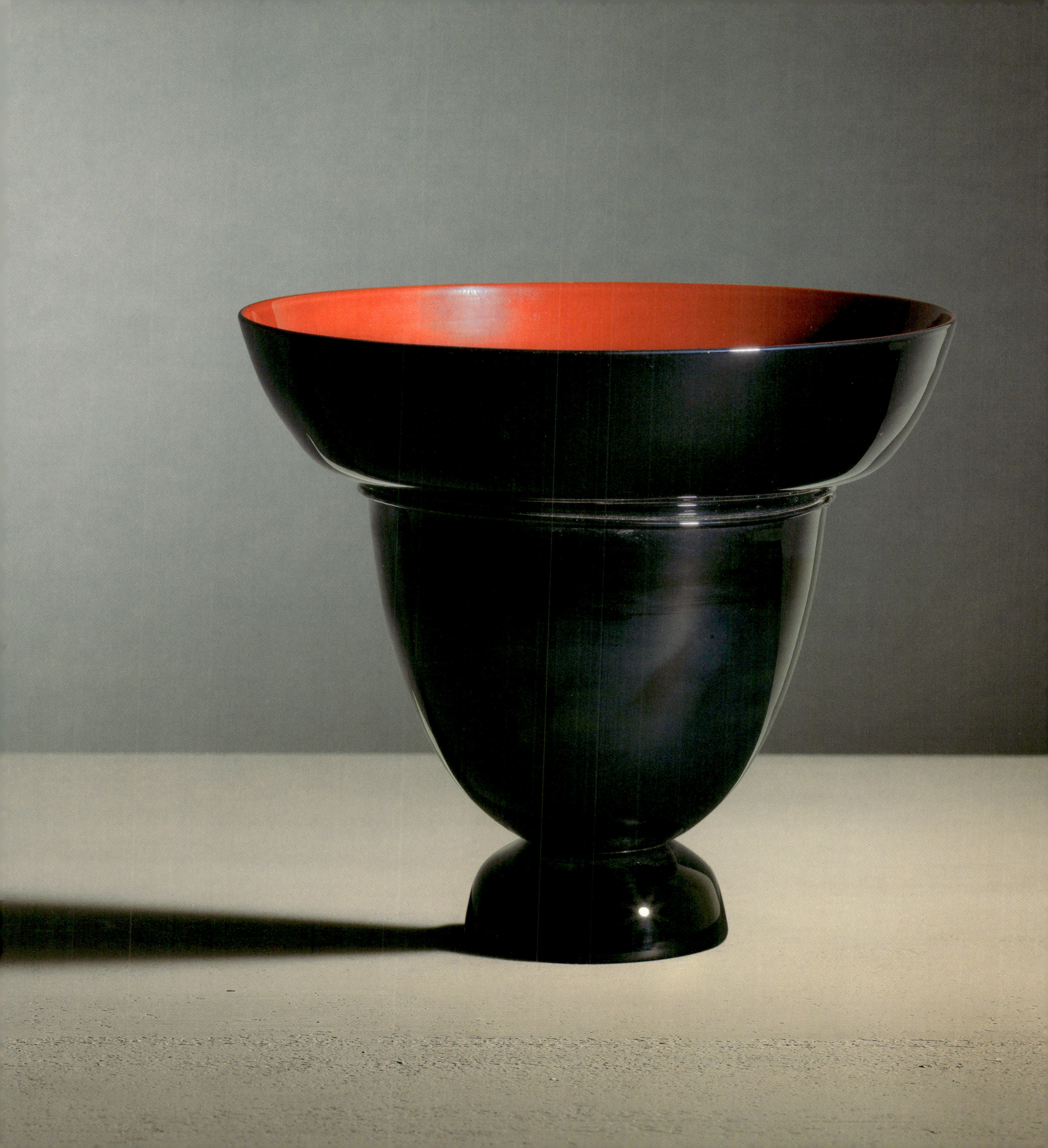

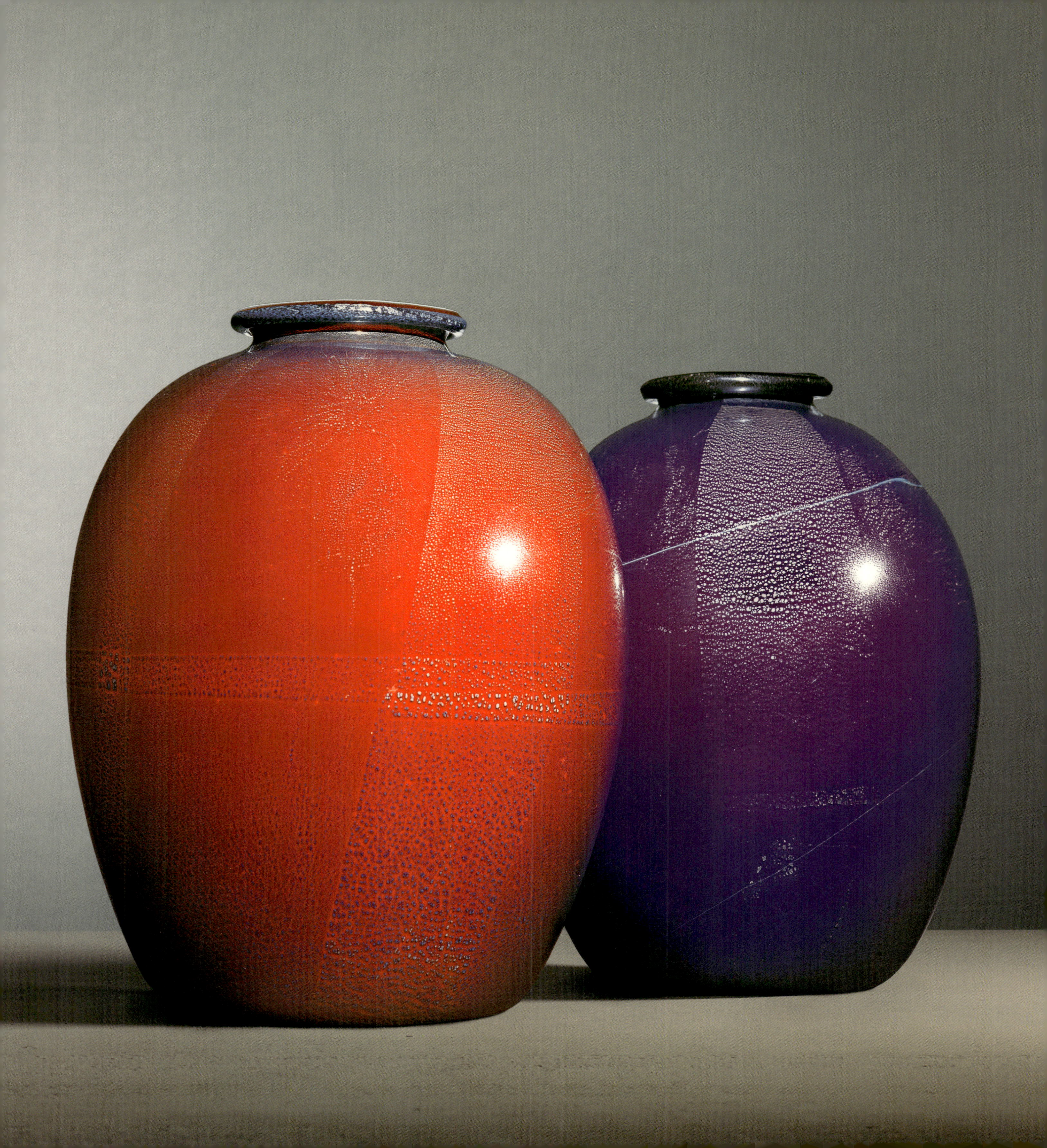

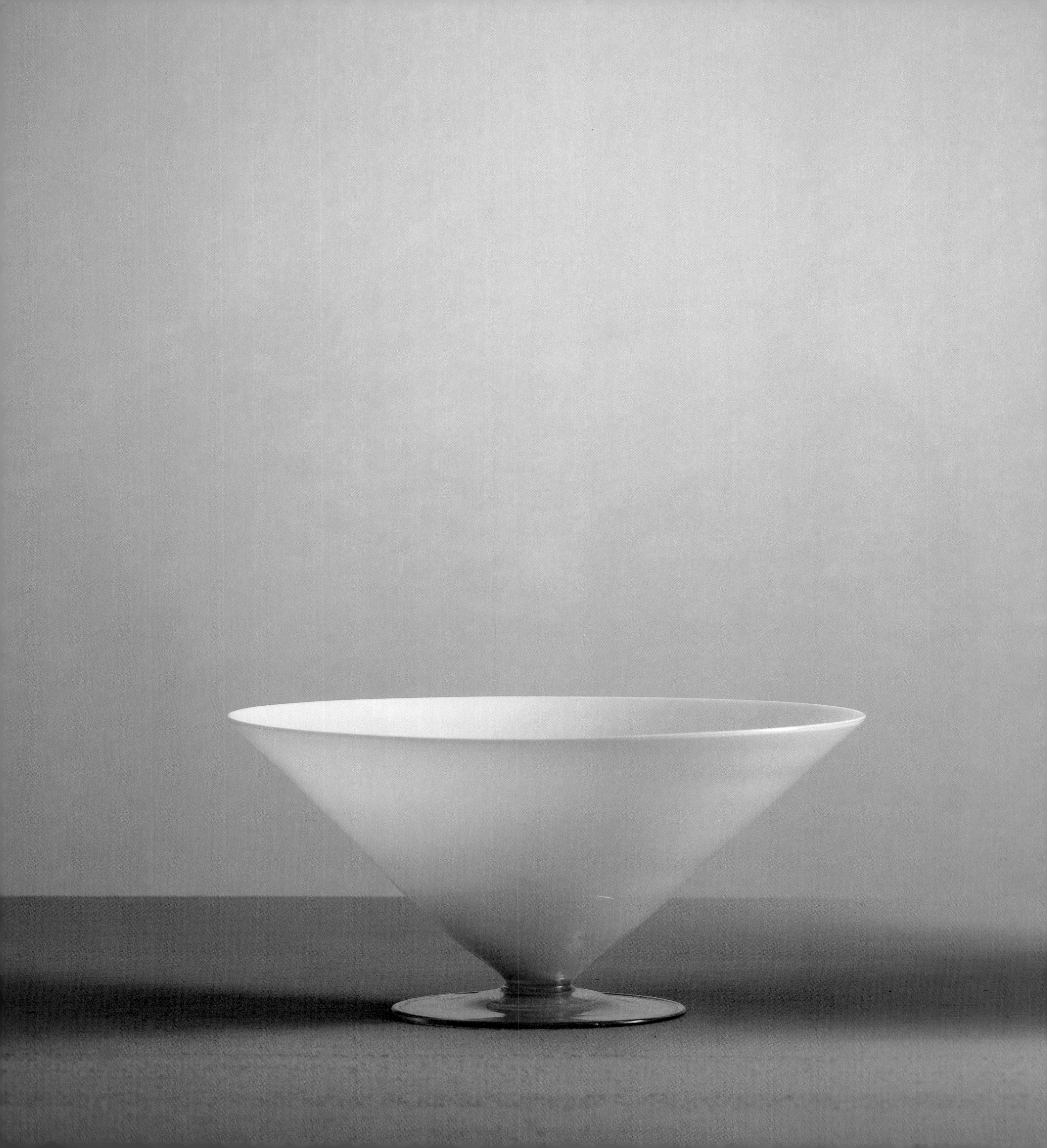

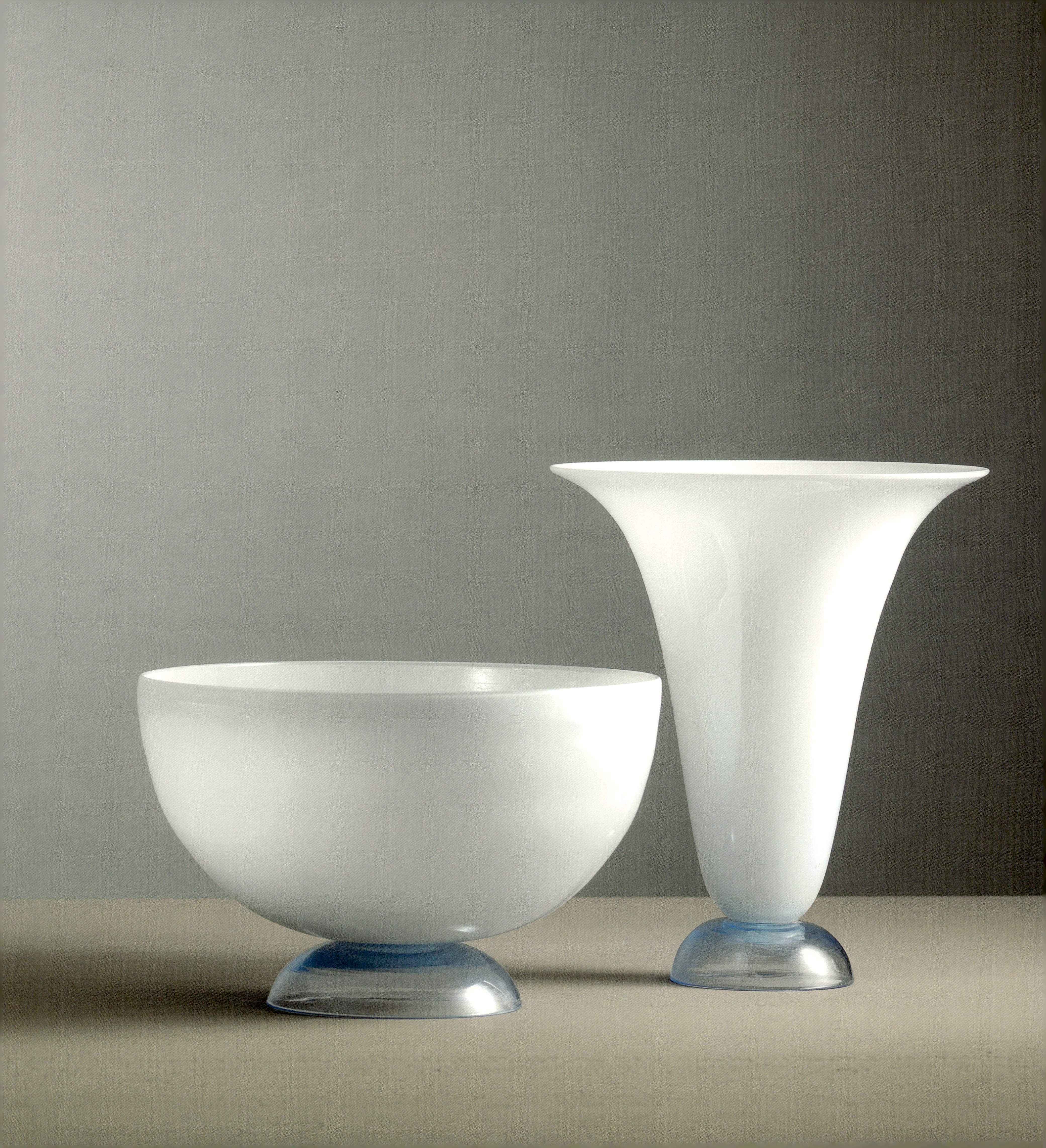

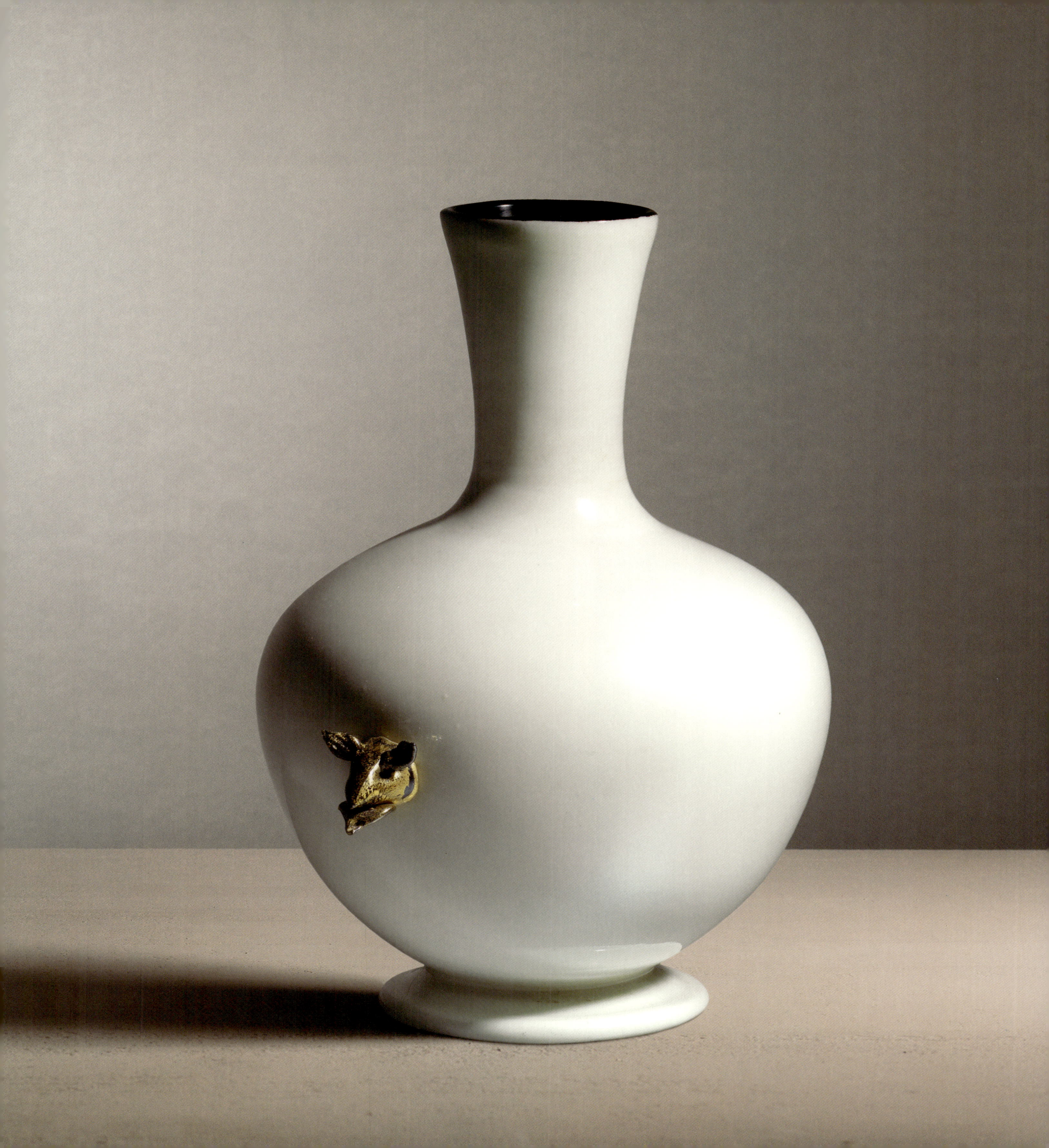

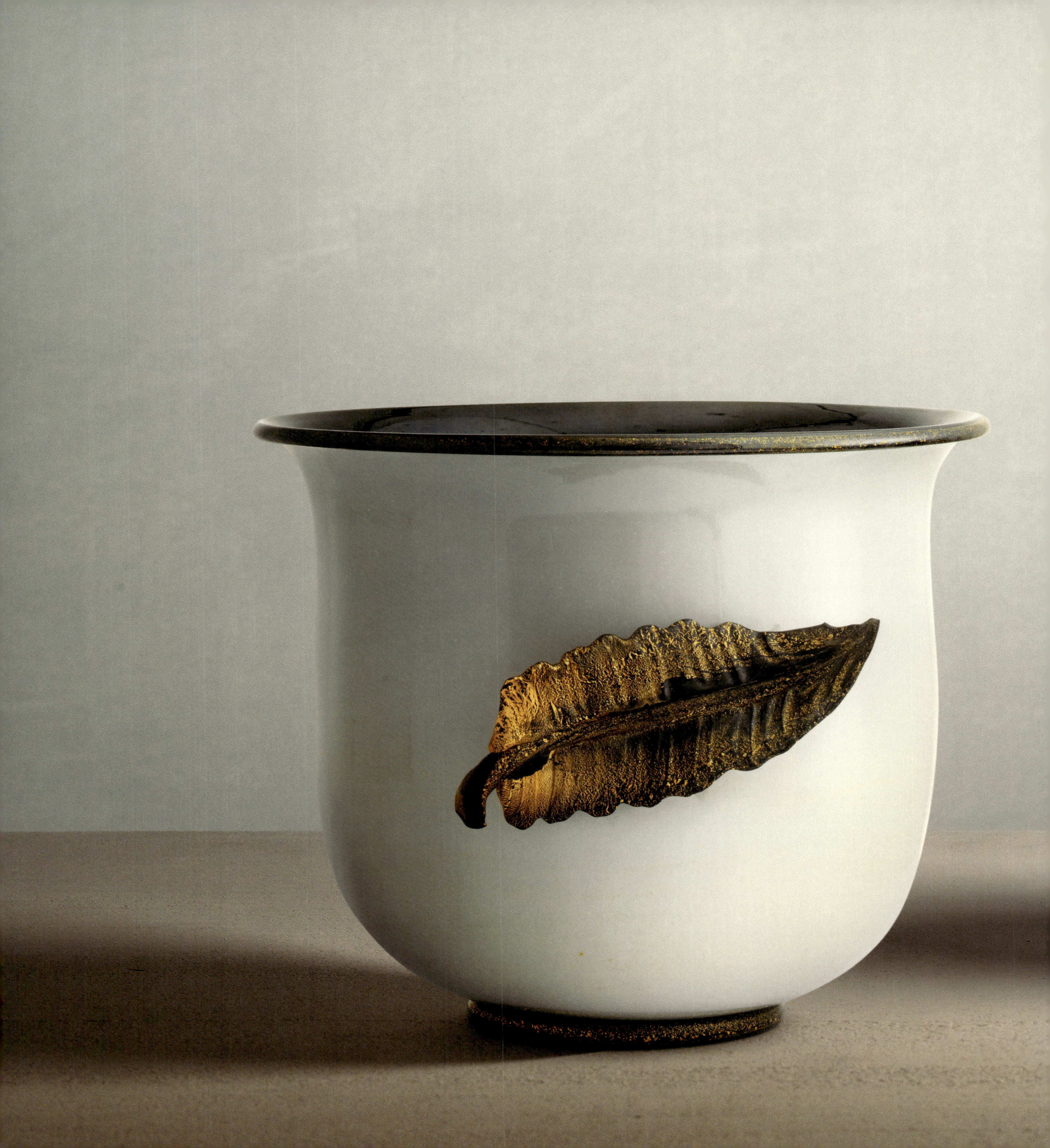

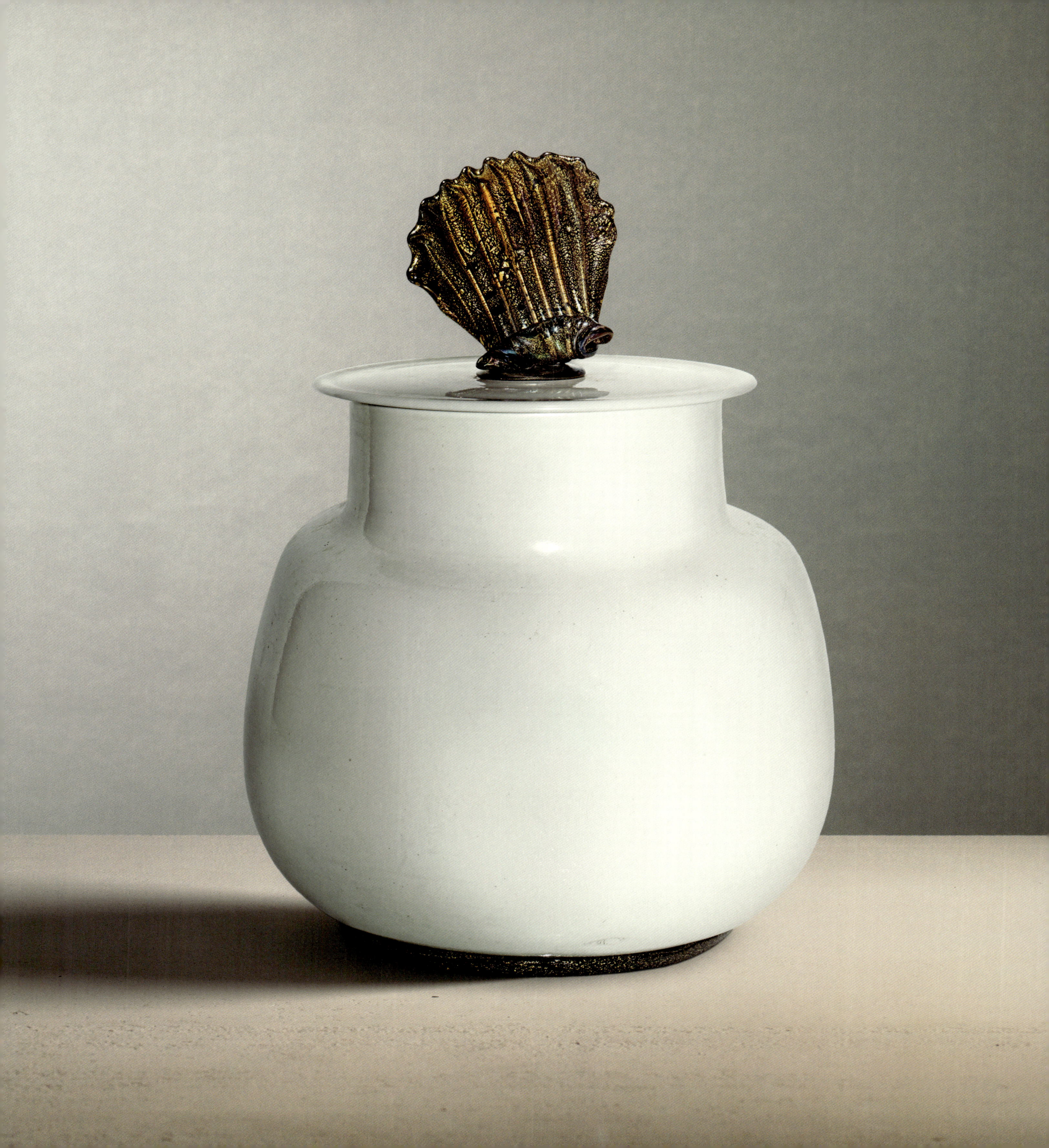

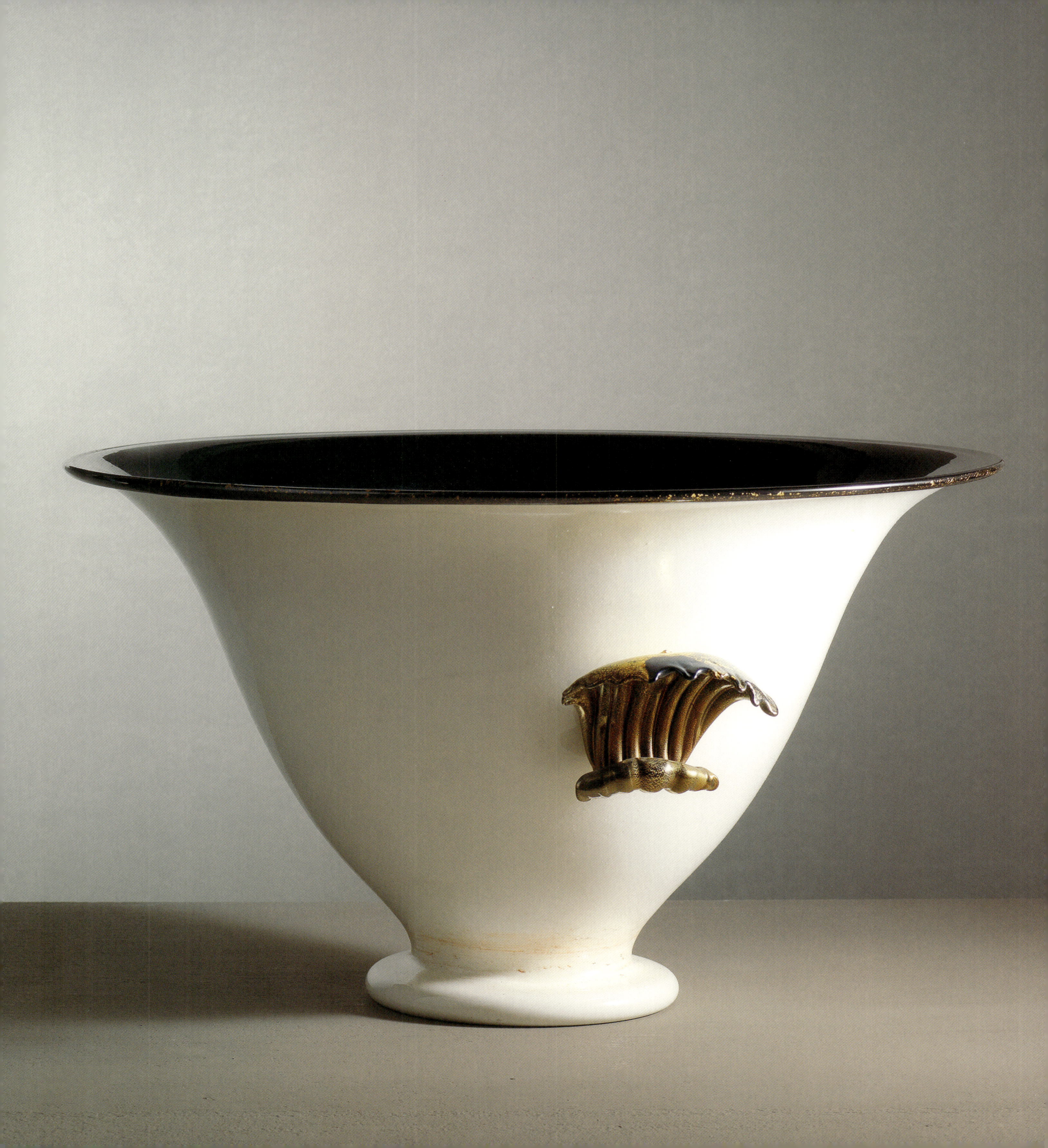

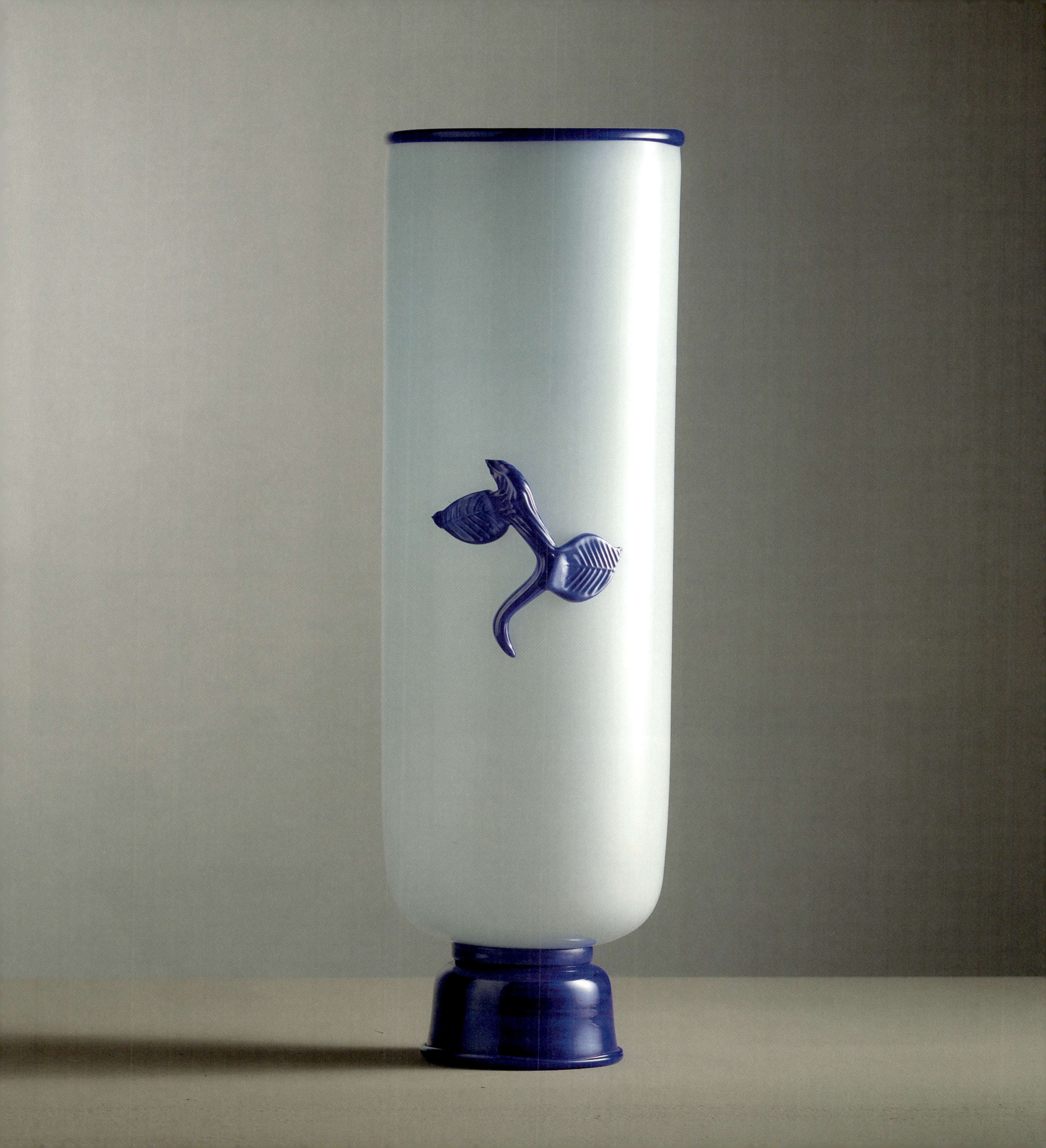

›
Bowls with truncated-cone-shaped foot in cased glass. Period photograph, c. 1929, CSV

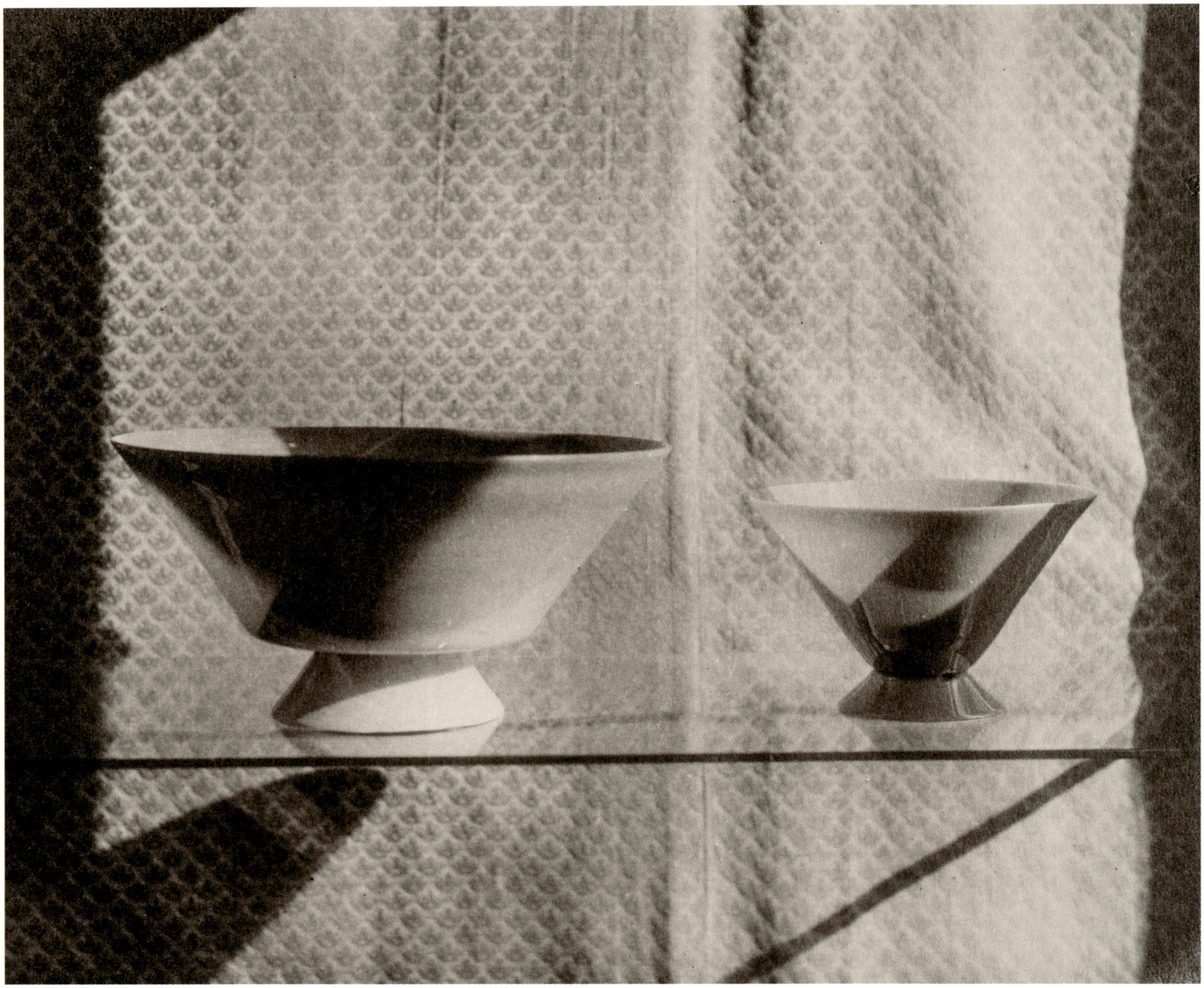

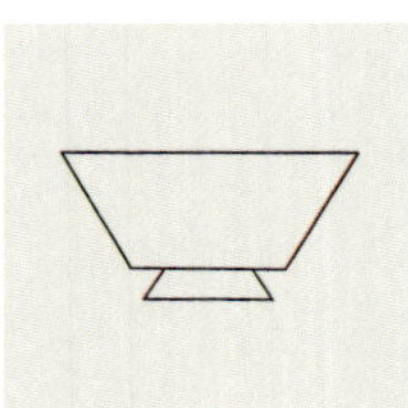

5817

Bowl in cased green glass with truncated-cone-shaped foot in Savoy-blue *pasta vitrea*. Surface with iridised finish.
c. 1929
19 cm high; 36 cm ø
M.V.M. Cappellin Murano *acid stamp*
Private collection

›
5817, catalogue drawing, AMB

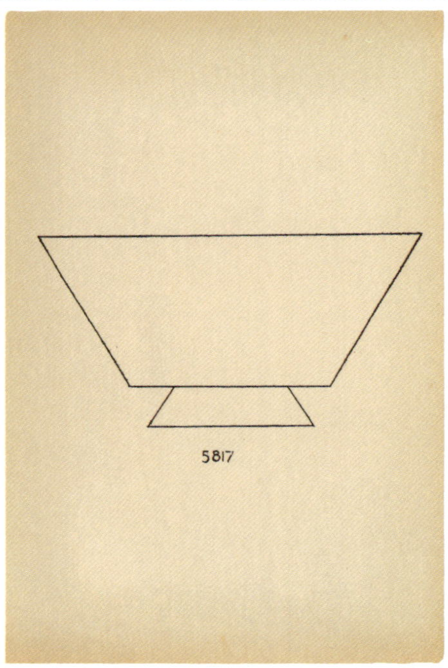

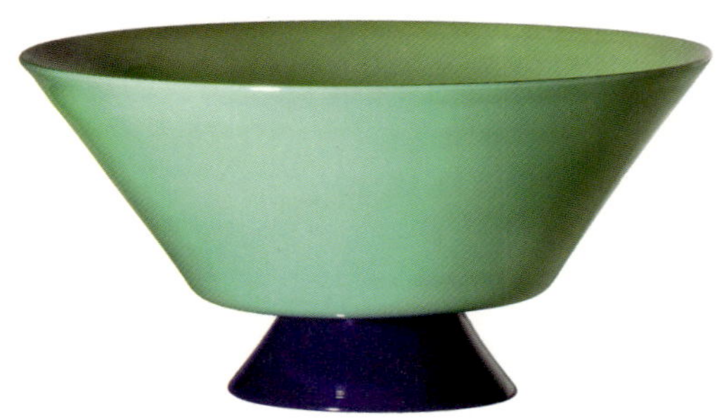

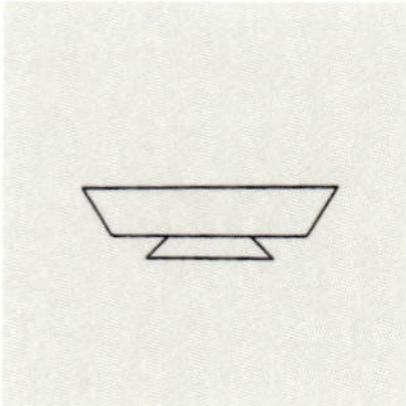

5820

Catalogue drawing for one of the vases with truncated-cone-shaped foot produced in cased glass documented by a period photograph, AMB.
c. 1929
8.5 cm high; 30 cm ø

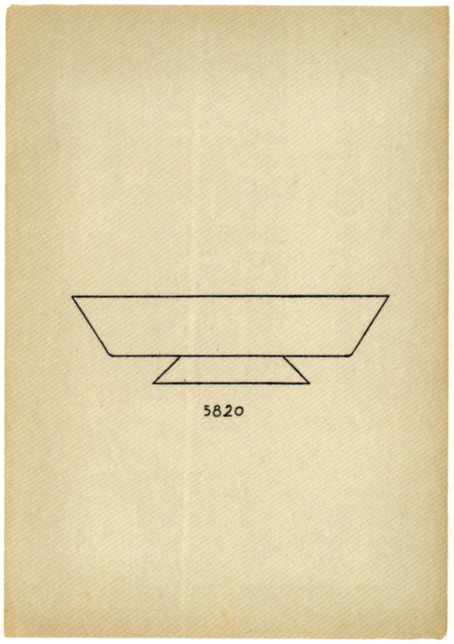

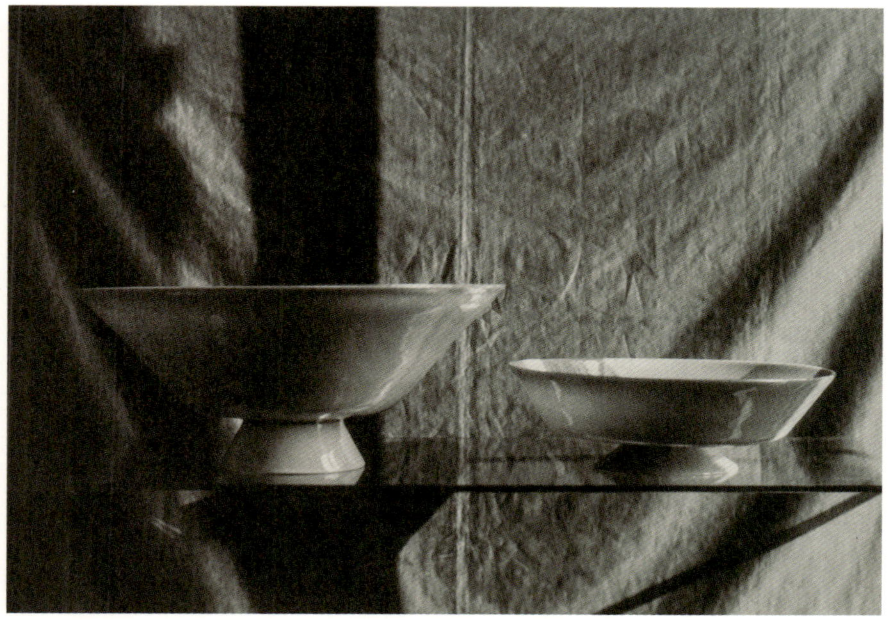

>
Bowls with truncated-cone-shaped foot in cased glass, c. 1929, CSV

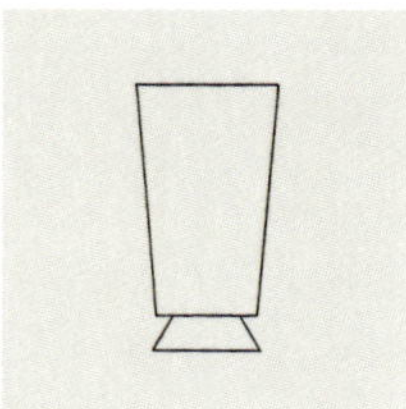

5819

Catalogue drawing for one of the vases with truncated-cone-shaped foot produced in cased glass documented by a period photograph, AMB.
c. 1929
31 cm high

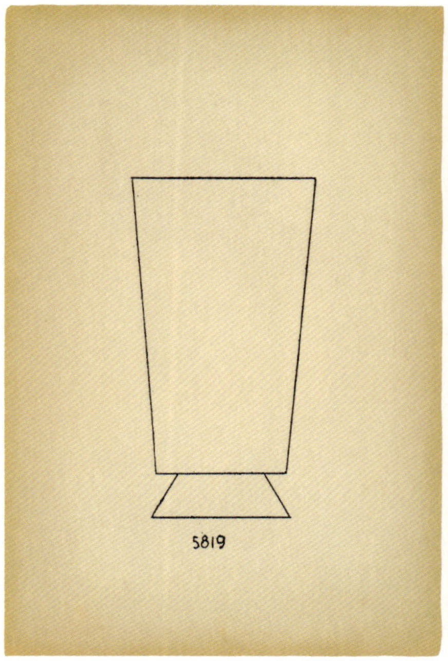

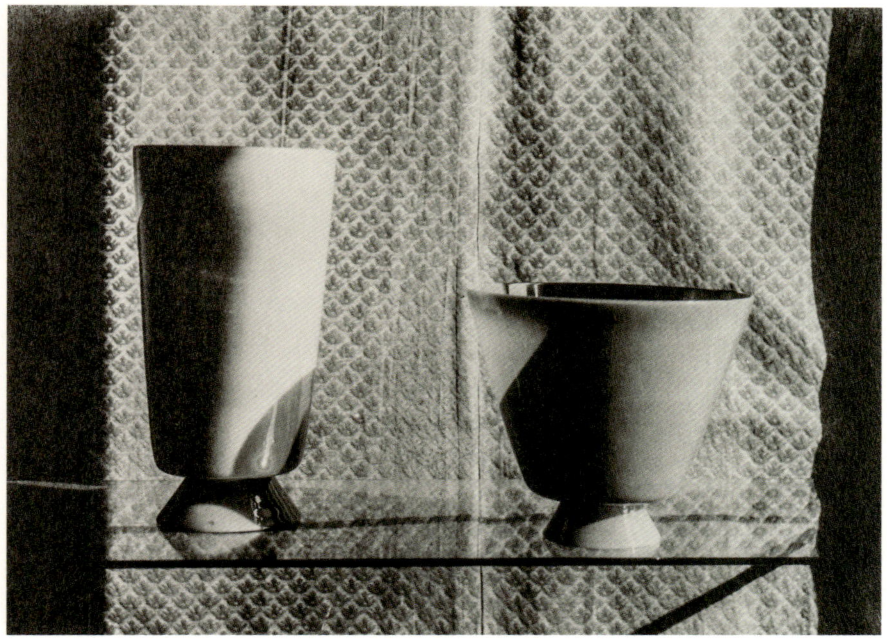

>
Vases with truncated-cone-shaped foot in cased glass. Period photograph, c. 1929, CSV

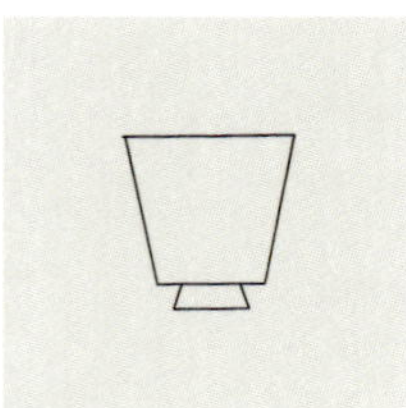

5818

Catalogue drawing for one of the vases with truncated-cone-shaped foot produced in cased glass documented by a period photograph, AMB.
c. 1929
20 cm high

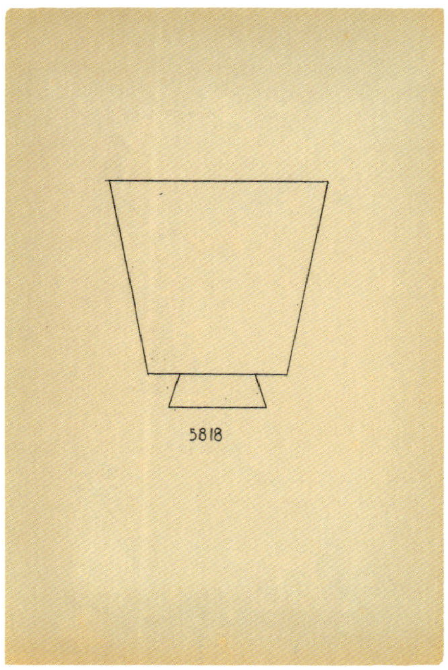

> *Vases in cased glass exhibited at the 4th Triennale of Monza, 1930, detail*

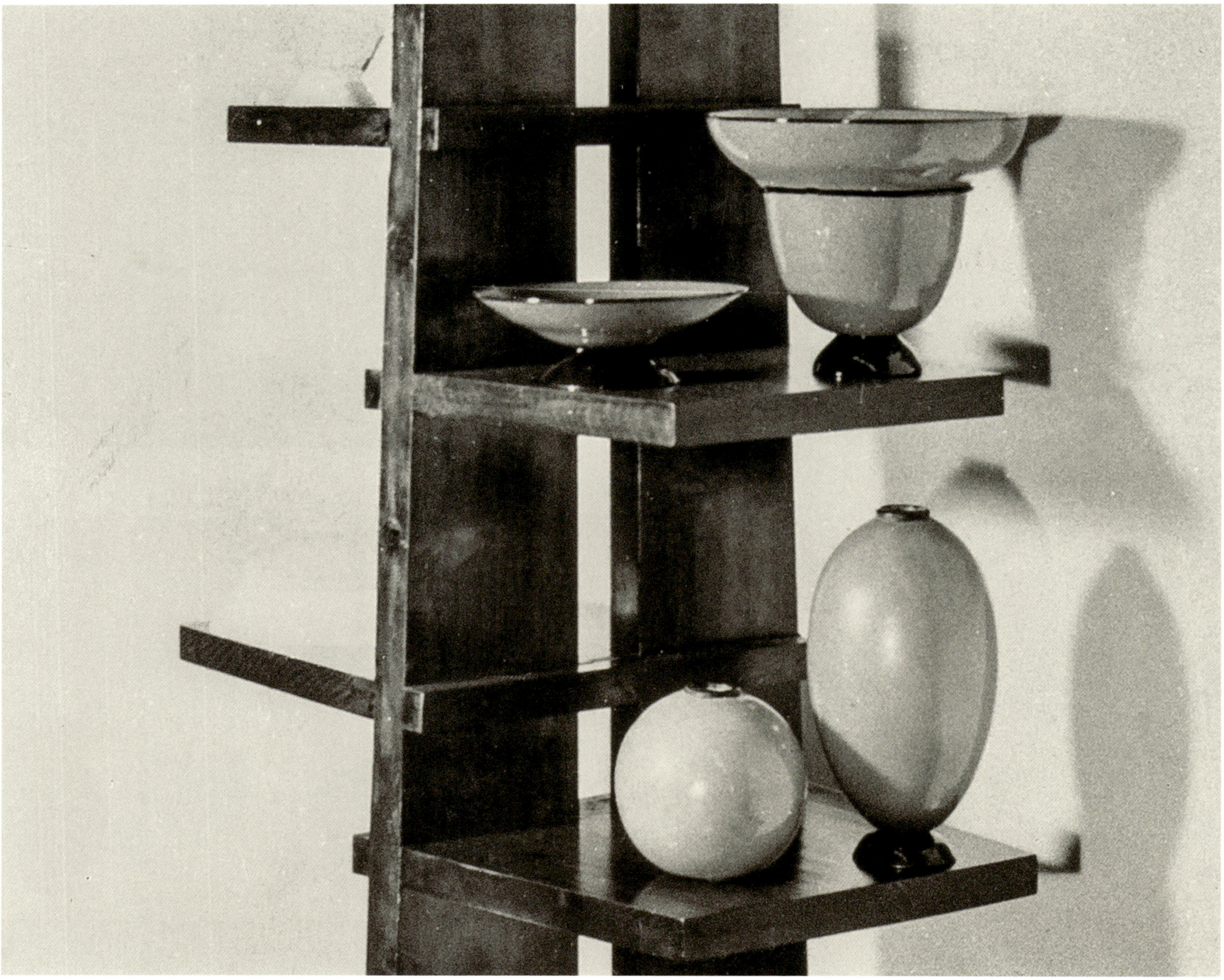

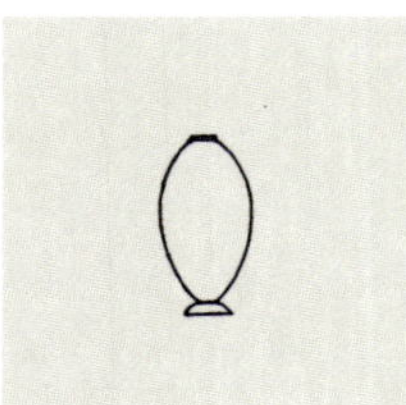

5935

Catalogue drawing for one of the vases in cased glass with skullcap foot that were exhibited in 1930 at the 4th Triennale of Monza, AMH. Model no. 5935 was also produced with other glass techniques (e.g. *decoro fenicio*, *lattimo aurato*, *pasta vitrea*, iridised black).
c. 1930
22 cm high

exhibitions
1930, Monza, 4th Triennale

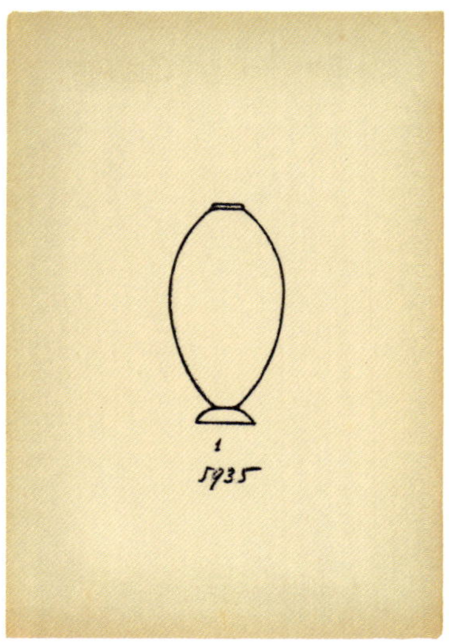

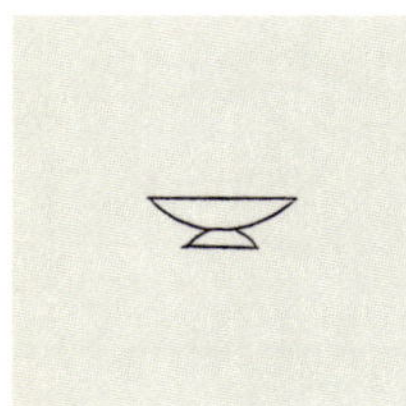

5951

Catalogue drawing for one of the vases in cased glass with skullcap foot that were exhibited in 1930 at the 4th Triennale of Monza, AMH. Model no. 5951 was also produced with other glass techniques (e.g. iridised black).
c. 1930
6 cm high; 17 cm ø

exhibitions
1930, Monza, 4th Triennale

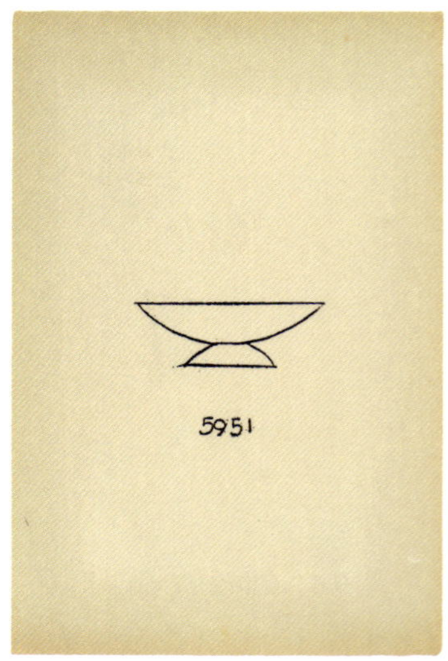

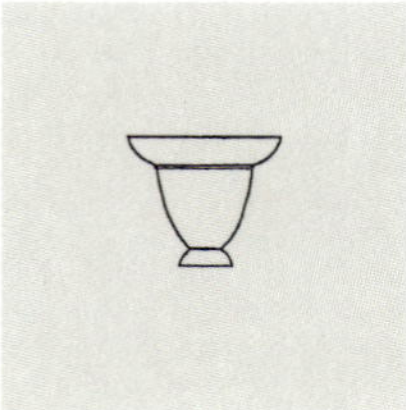

5960

Vase in black cased glass with inside surface in coral-red glass. Surface with iridised finish. The model is characterised by the skullcap foot and outspread rim. An example of no. 5960, produced in cased glass, with foot and body in contrasting colours, was exhibited in 1930 at the 4th Triennale of Monza.
c. 1930
17 cm high
Private collection

exhibitions
1930, Monza, 4th Triennale

>
5960, catalogue drawing, AMB

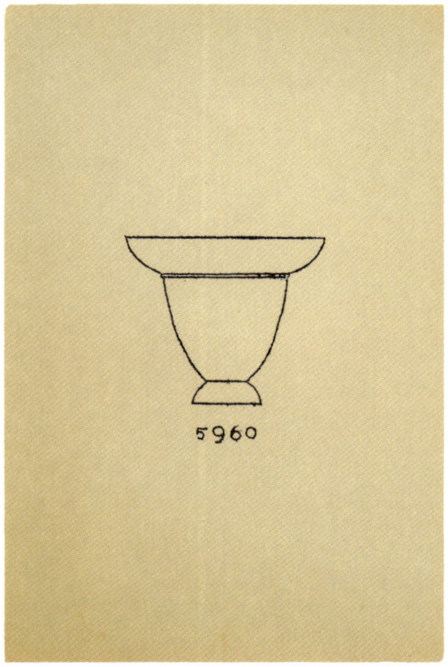

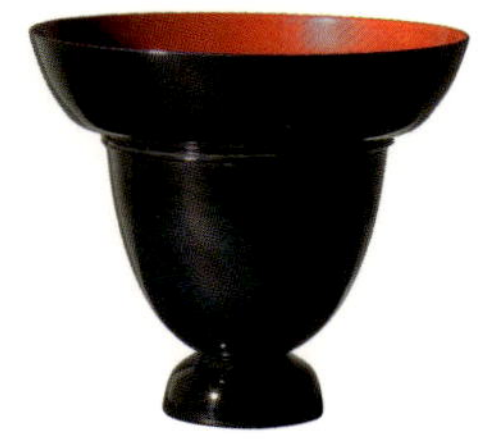

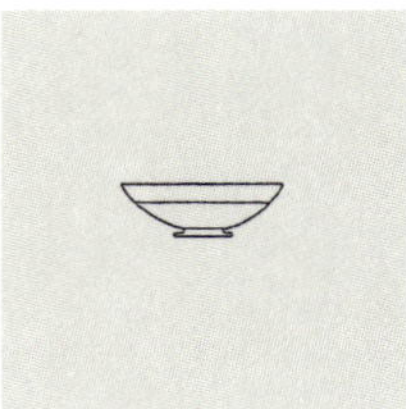

5982

Bowl in Savoy-blue cased glass with internal surface in pea-green. Surface with iridised finish. Model no. 5982 was also produced with other glass techniques (e.g. *pasta vitrea*).
1929–31
6 cm high; 17 cm ø
M.V.M. Cappellin Murano *acid stamp*
Private collection

>
5960, working drawing, CSV

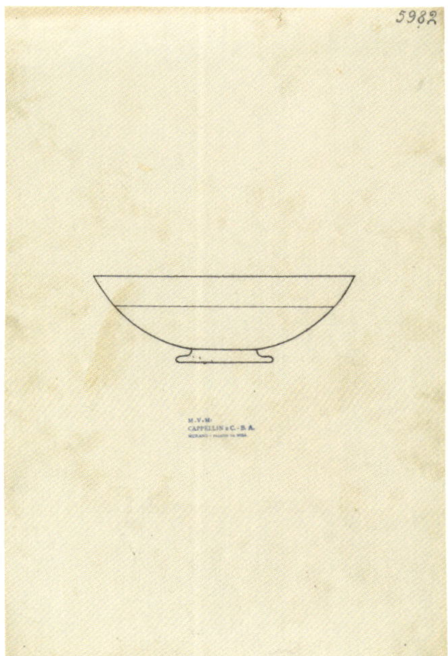

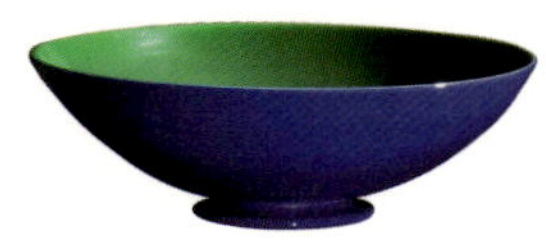

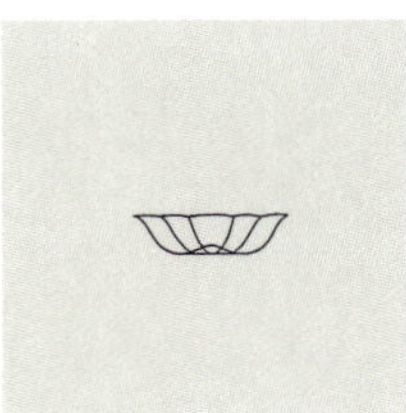

5593

Ribbed bowl in coral-red *pasta vitrea* with internal surface in pea-green. Surface with iridised finish. The bowl was produced after a model in the catalogue of the glassworks designed by Vittorio Zecchin (1925–26).
1929–31
17 cm ø
M.V.M. Cappellin Murano *acid stamp*
Private collection, Udine

>
5593, catalogue drawing, AMB

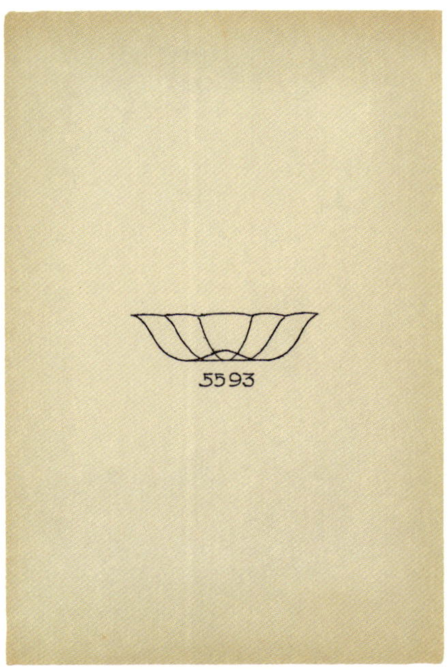

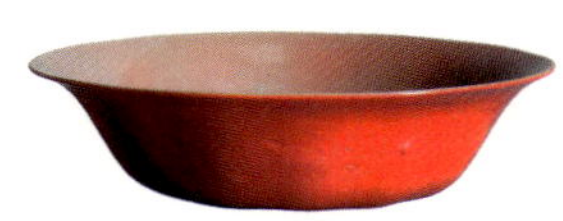

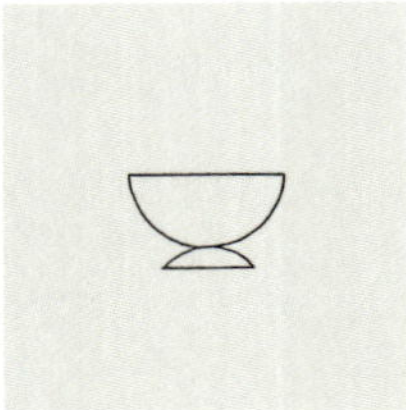

5950

Bowl in cased *lattimo* and light-blue glass with skullcap foot in light-blue transparent glass. Surface with iridised finish.
An example of n. 5950, in cased glass, with foot and body in contrasting opaque colours, was shown in 1930 at the 4th Triennale of Monza. Model no. 5950 was also produced with other glass techniques (e.g. iridised black).
1930–31
12 cm high; 19,5 cm ø
C.V.M. *acid stamp*
Private collection

exhibitions
1930, Monza, 4th Triennale

›
5950, catalogue drawing, AMH

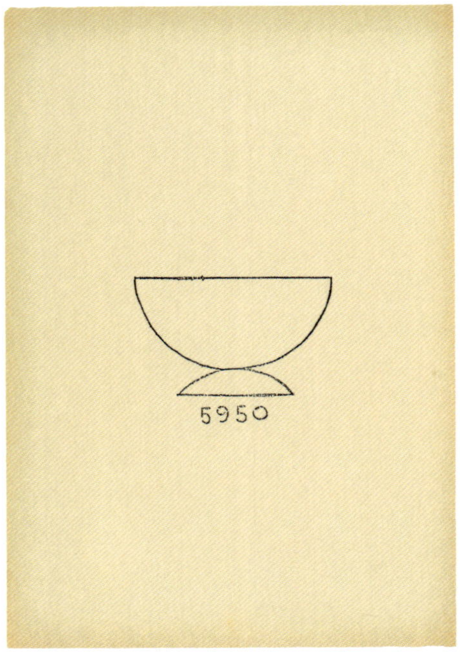

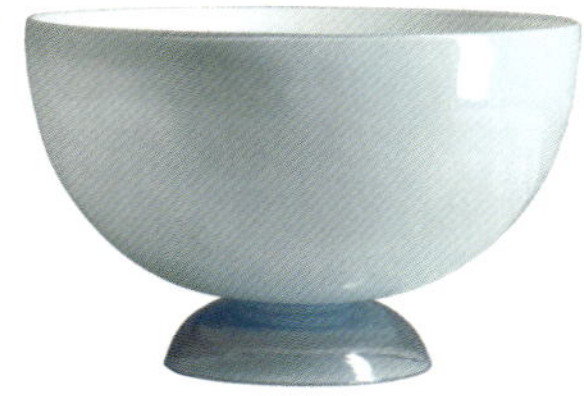

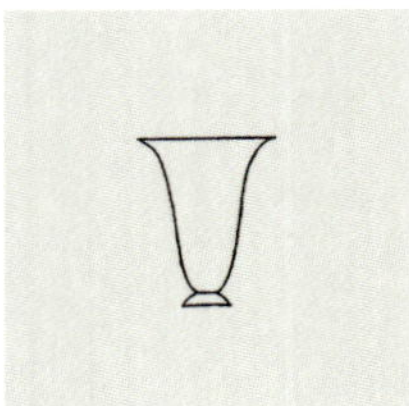

5949

Vase in cased *lattimo* and light-blue glass with skullcap foot in light-blue transparent glass. Surface with iridised finish. Model no. 5949 was also produced with other glass techniques (e.g. *pasta vitrea*, iridised black, transparent *balloton*). In the American catalogue of the Cappellin Glassware there is also a model in opaque glass with enamel decorations.
1930–31
20 cm high
Private collection

exhibitions
1930, Monza, 4th Triennale

›
5949, catalogue drawing, AMB

›
Vases in iridised opaque glass. Period photograph, c. 1929, CSV

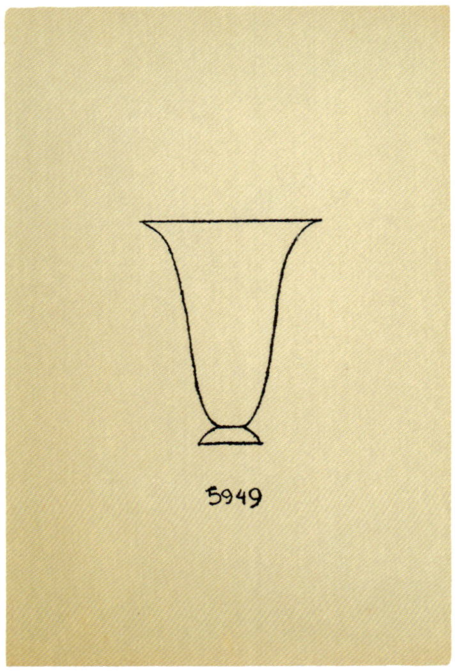

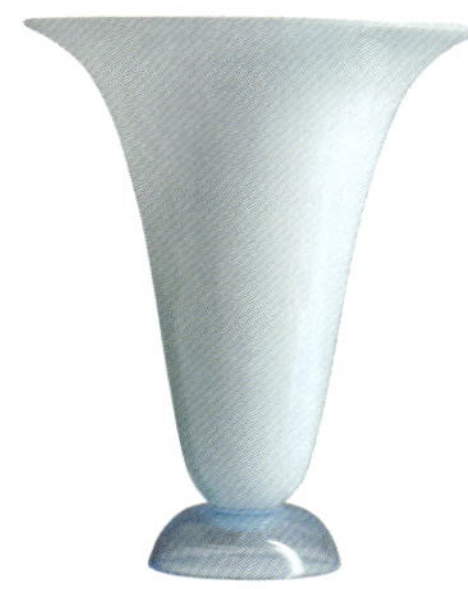

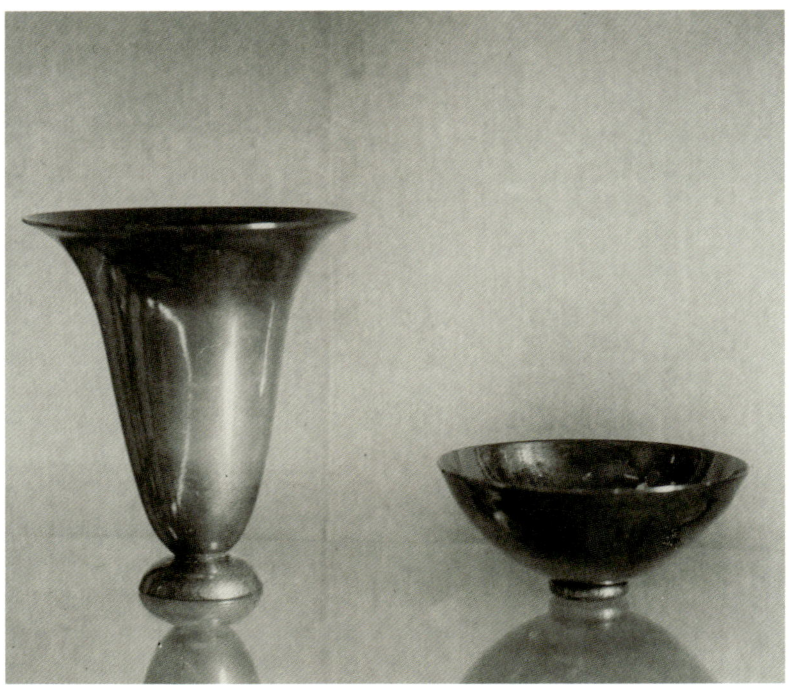

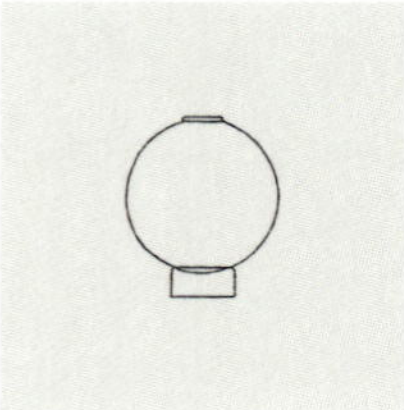

5988

Spherical vase in cased *lattimo* and *pagliesco* glass with a cylindrical foot and thread around the mouth in transparent *pagliesco* glass. Surface with iridised finish. A similar example was exhibited in 1931 at the Amsterdam Exhibition of glass, ceramics and lace.
1930–31
25 cm high
Private collection, Venice

exhibitions
1931, Amsterdam, Exhibition of glass, ceramics and lace

>
5988, catalogue drawing, AMB

>
Cappellin glass pieces at the Amsterdam Exhibition of glass, ceramics and lace, 1931, AHL, detail

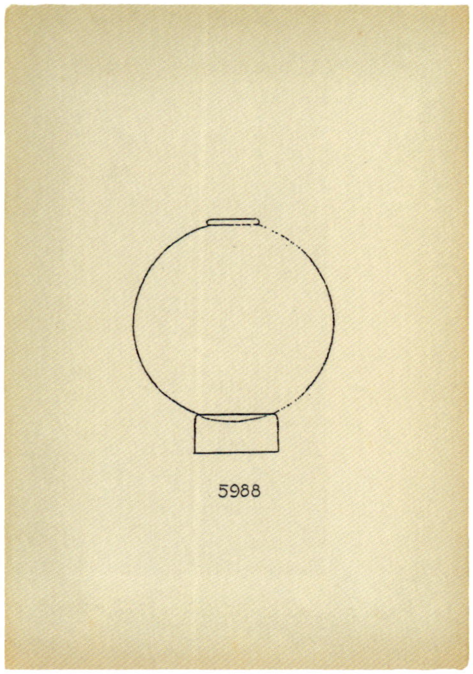

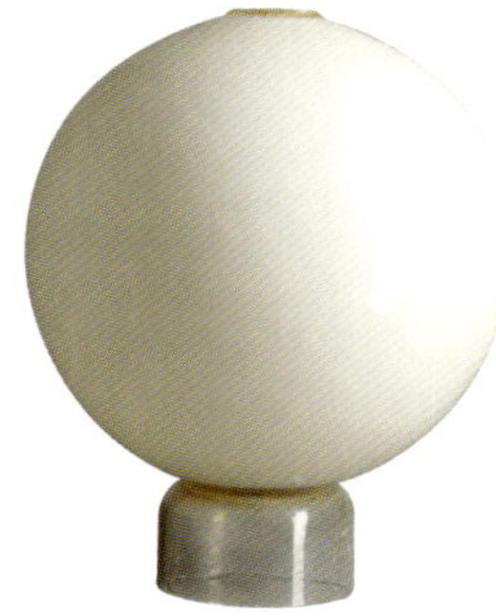

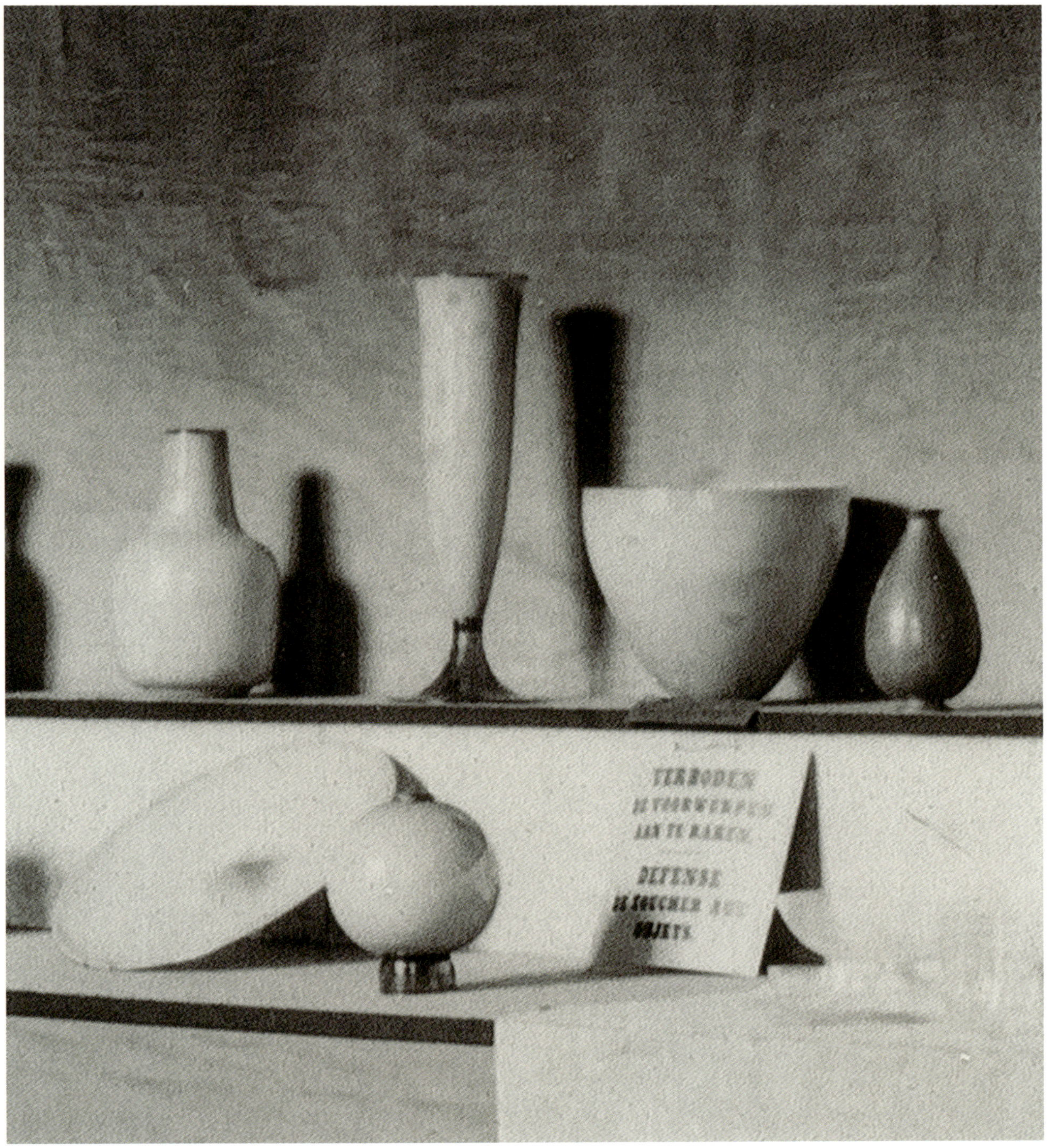

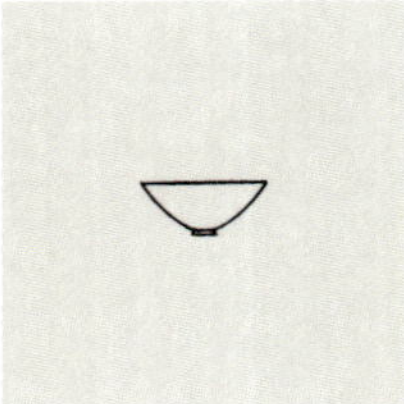

5928

Bowl in cased *lattimo* and light-blue glass. Small foot and thread around the mouth in light-blue glass. Strongly iridised finish. The bowl became part of the collection of the Metropolitan Museum of Art, NY, in 1931. Model 5928 was also produced with other glass techniques (e.g. *pasta vitrea*, *lattimo aurato*).
1929–30
7 cm high; 17 cm ø
The Metropolitan Museum of Art collection, NY

>
5928, catalogue drawing, AMH

>
Bowl no. 5928 in Cappellin Glassware, *catalogue, undated, AHL, detail*

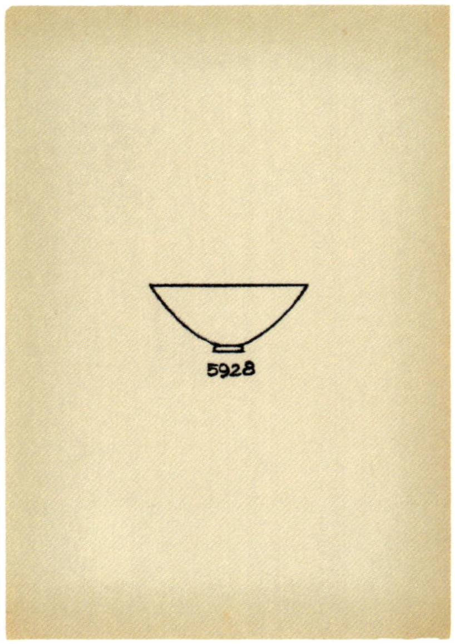

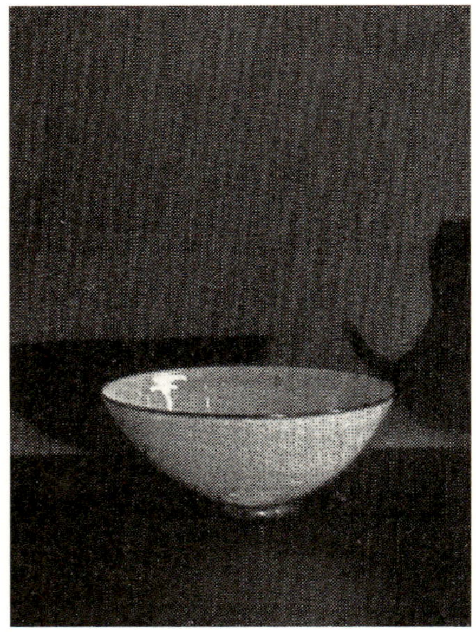

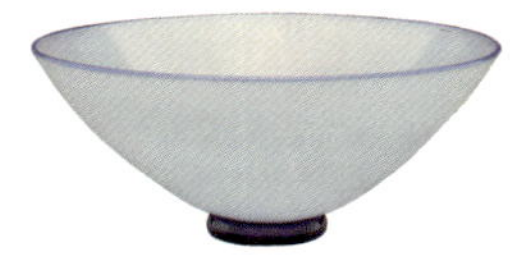

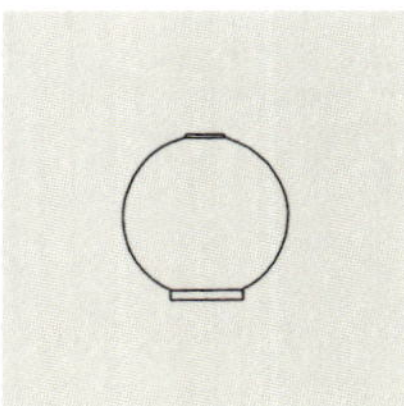

5970

Spherical vase in Savoy-blue cased and crystal glass with finishing in Savoy-blue transparent glass. Strongly iridised surface. An example in cased glass was exhibited in 1930 at the 4th Triennale of Monza as part of the 'geometric' series of glass pieces. It was also produced with other glass techniques (e.g. *pasta vitrea*).
c. 1930
19 cm high
Corning Museum of Glass collection, Corning, NY, USA

exhibitions
1930, Monza, 4th Triennale
1931, Amsterdam, Exhibition of glass, ceramics and lace

>
5970, catalogue drawing, AMB

>
Vases no. 5970 in cased glass at the 4thTriennale of Monza, 1930

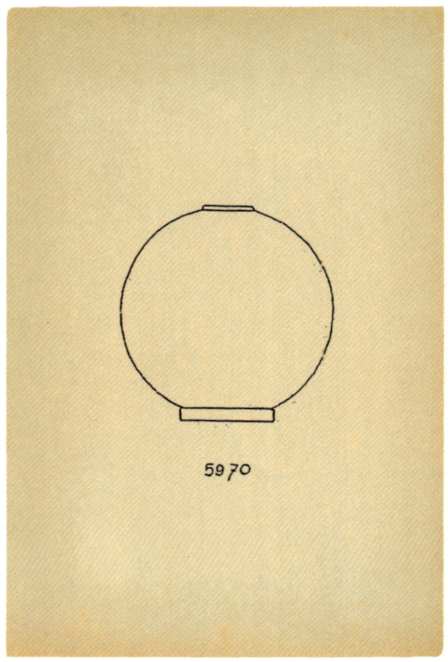

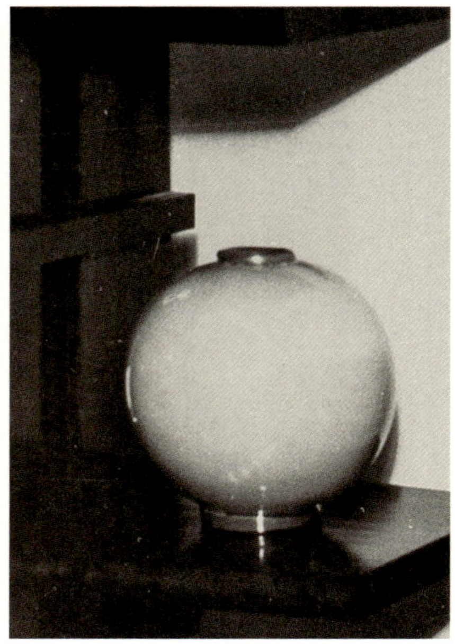

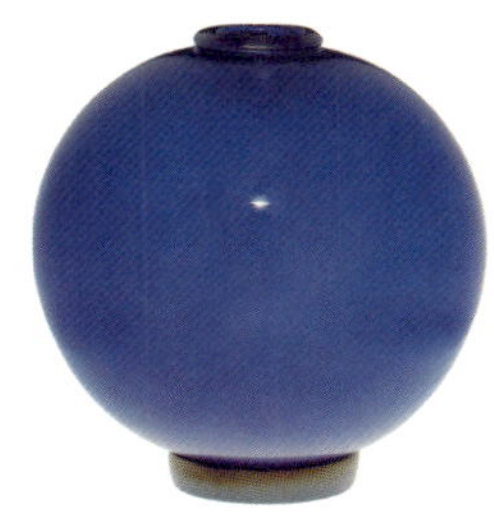

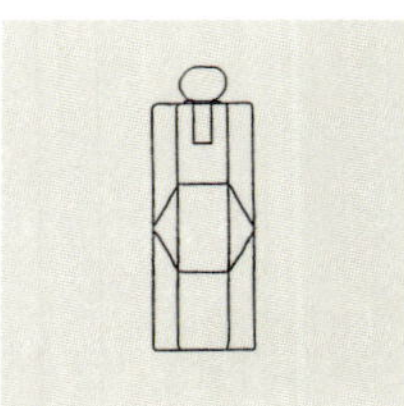

5992

Jar with hexagonal cross-section in yellow cased glass with mouth and stopper in green glass. An example of no. 5992 in cased glass was exhibited in 1930 at the 4th Trienale of Monza as part of the 'geometric' series of glass pieces. Model no. 5992 was also produced with other glass techniques (e.g. *pasta vitrea*, silvered black).
c. 1930
34 cm high
M.V.M. Cappellin Murano *acid stamp*
Private collection

exhibition
1930, Monza, 4th Triennale

>
5992, catalogue drawing, AMB

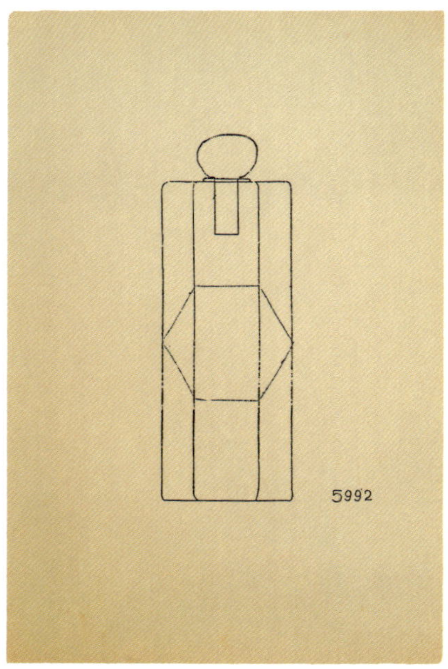

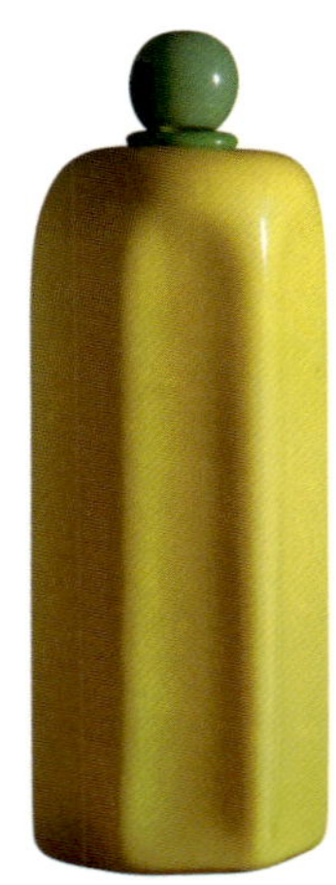

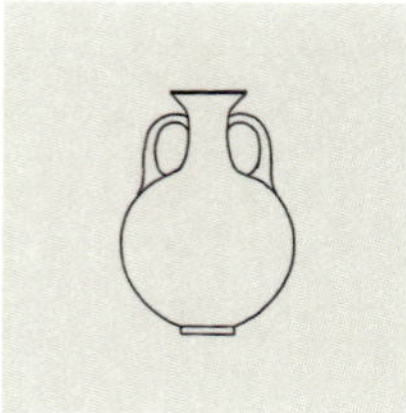

6018

Amphora in cased *lattimo* and *pagliesco* glass with finishings in black glass. A similar example was exhibited in 1931 at the Amsterdam Exhibition of glass, ceramics and lace. The piece was presented together with a series of jugs in clear-coloured cased glass with handles and finishings in black glass. Most of these were produced after models previously designed by Vittorio Zecchin
1930–31
28 cm high
Private collection

exhibitions
1931, Amsterdam, Exhibition of glass, ceramics and lace

>
6018, catalogue drawing, AMB

>
Glasses exhibited at the Amsterdam Exhibition of glass, ceramics and lace, 1931, AHL, detail

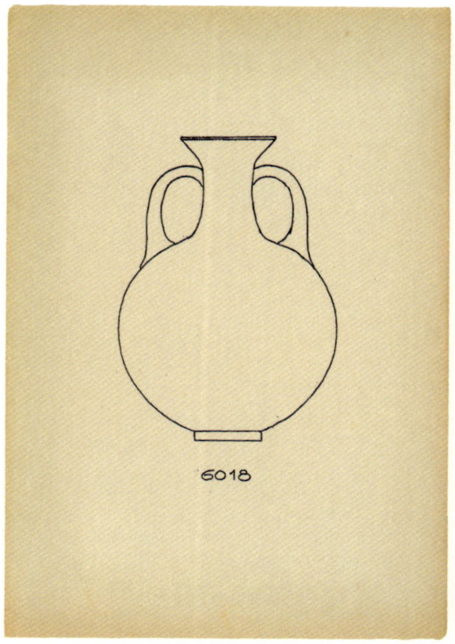

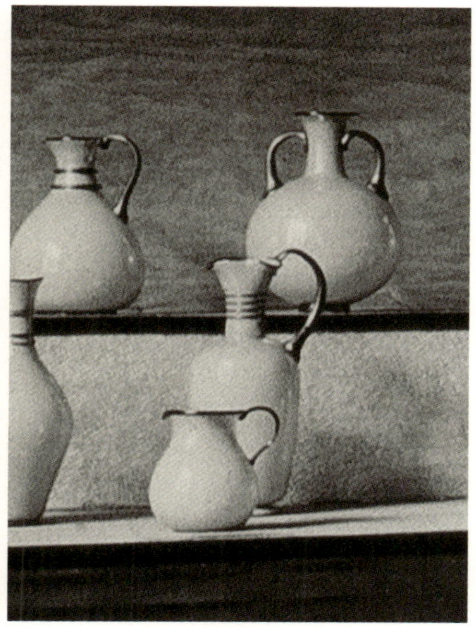

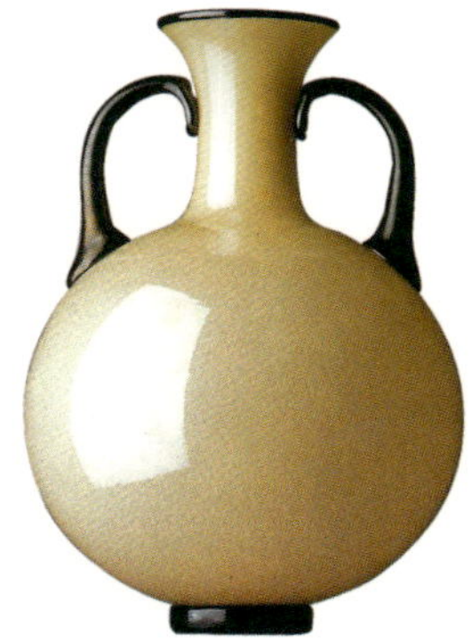

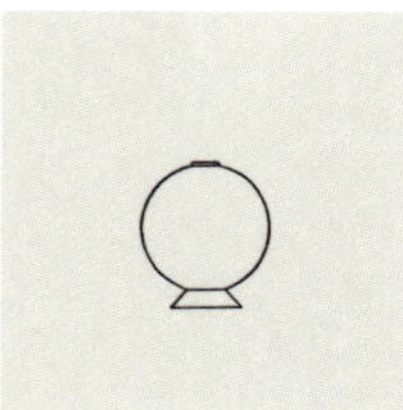

5680

Spherical vase with truncated-cone-shaped foot in coral-red and crystal cased glass. Surface with iridised finish. Model no. 5680 was also produced with other glass techniques (e.g. transparent, *lattimo aurato*, *pasta vitrea*).
1929–31
17 cm high
The Olnick Spanu Collection, New York

>
5680, working drawing, CSV

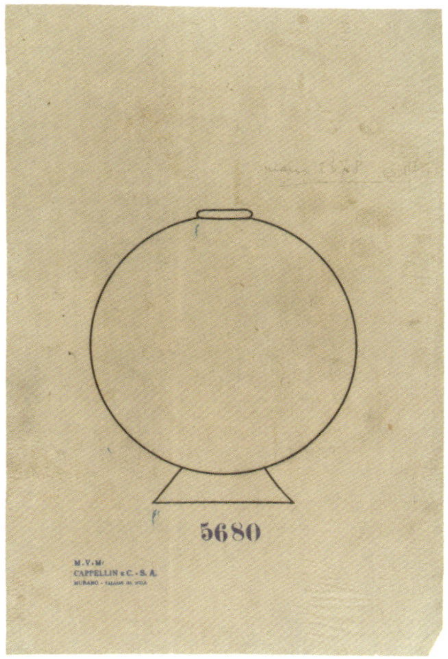

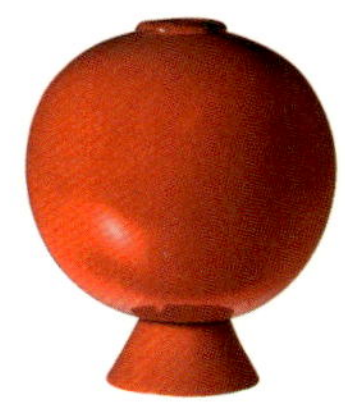

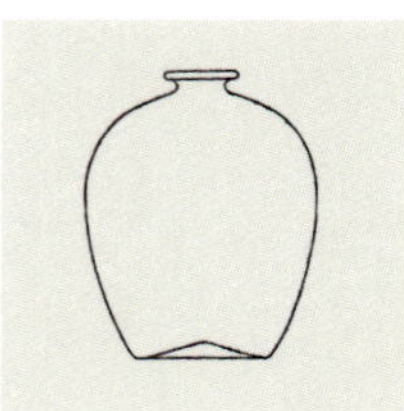

5087

Vase in cased *lattimo* and coral-red glass with applied oxidised silver leaf. The vase was produced after a model in the catalogue of the glassworks designed by Vittorio Zecchin (1921–25).
1929–31
33 cm high
The Olnick Spanu Collection, New York

>
5087, catalogue drawing, AMB

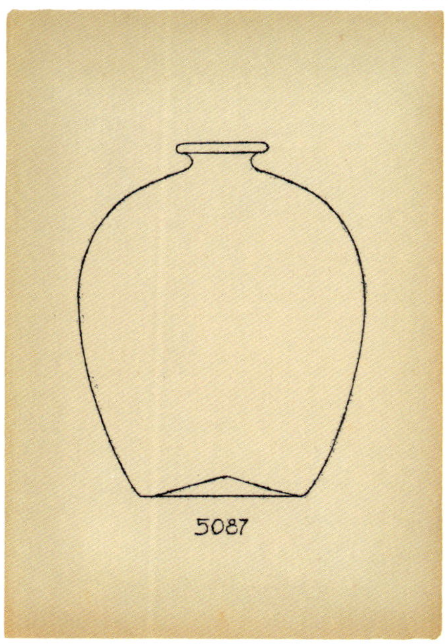

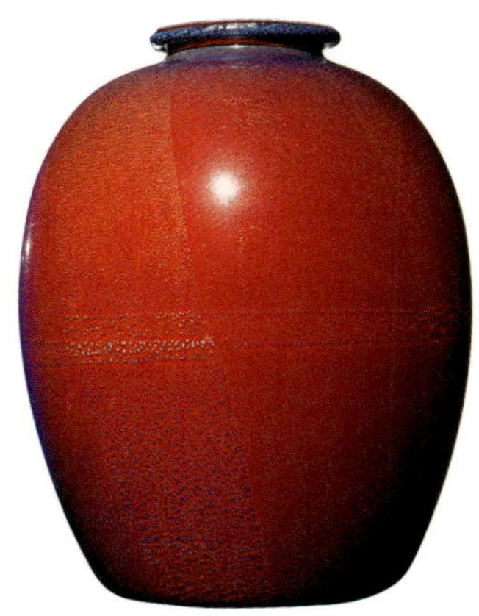

>
'The new vases in coloured opaque glass', in the Tentoonstelling van glas, aardewerk en kant, van moderne italiaansche kunstnijverheid. Mostra di Vetri, Ceramiche e Merletti d'Arte Moderna Italiana, *Amsterdam 1931, p. 26*

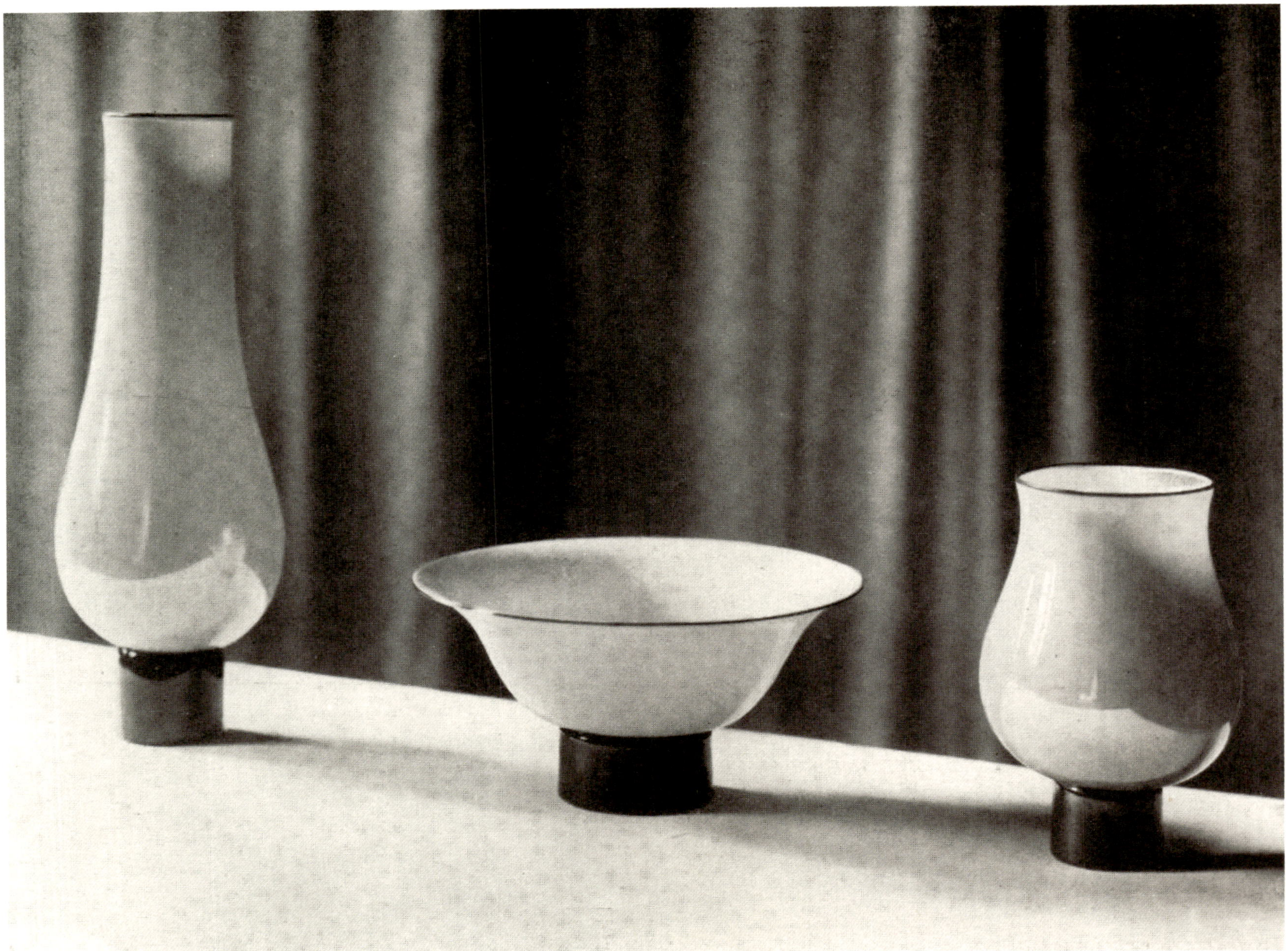

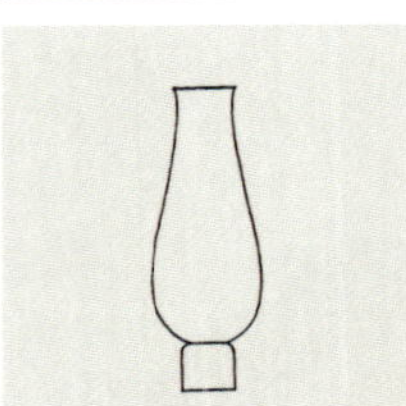

6031

Catalogue drawing for one of the vases in cased glass with cylindrical foot exhibited in 1930 at the 4th Triennale of Monza, AMH. On this occasion five different models of the series were shown.
1930
36 cm high

exhibitions
1930, Monza, 4th Triennale

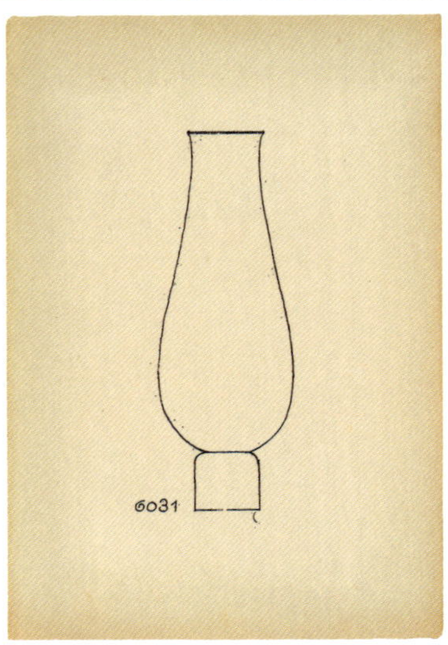

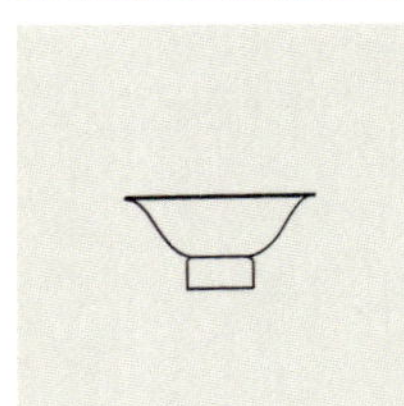

6034

Catalogue drawing for one of the bowls in cased glass with cylindrical foot, exhibited in 1930 at the 4th Triennale of Monza.
1930
11 cm high; 22 cm ø

exhibitions
1930, Monza, 4th Triennale

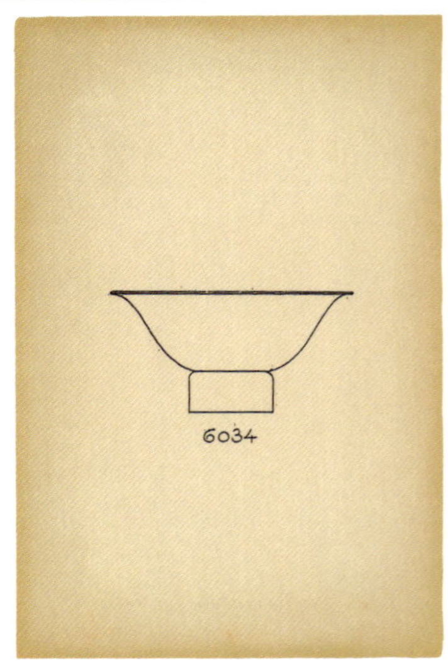

n.i.

Vases in cased glass with cylindrical foot exhibited in the little room occupied by M.V.M. Cappellin at the 4th Triennale of Monza, 1930, detail. The catalogue number (6033 or 6035?) for the bowl on the right (h c. 15 cm) is not known.
1930
15 cm high

exhibitions
1930, Monza, 4th Triennale

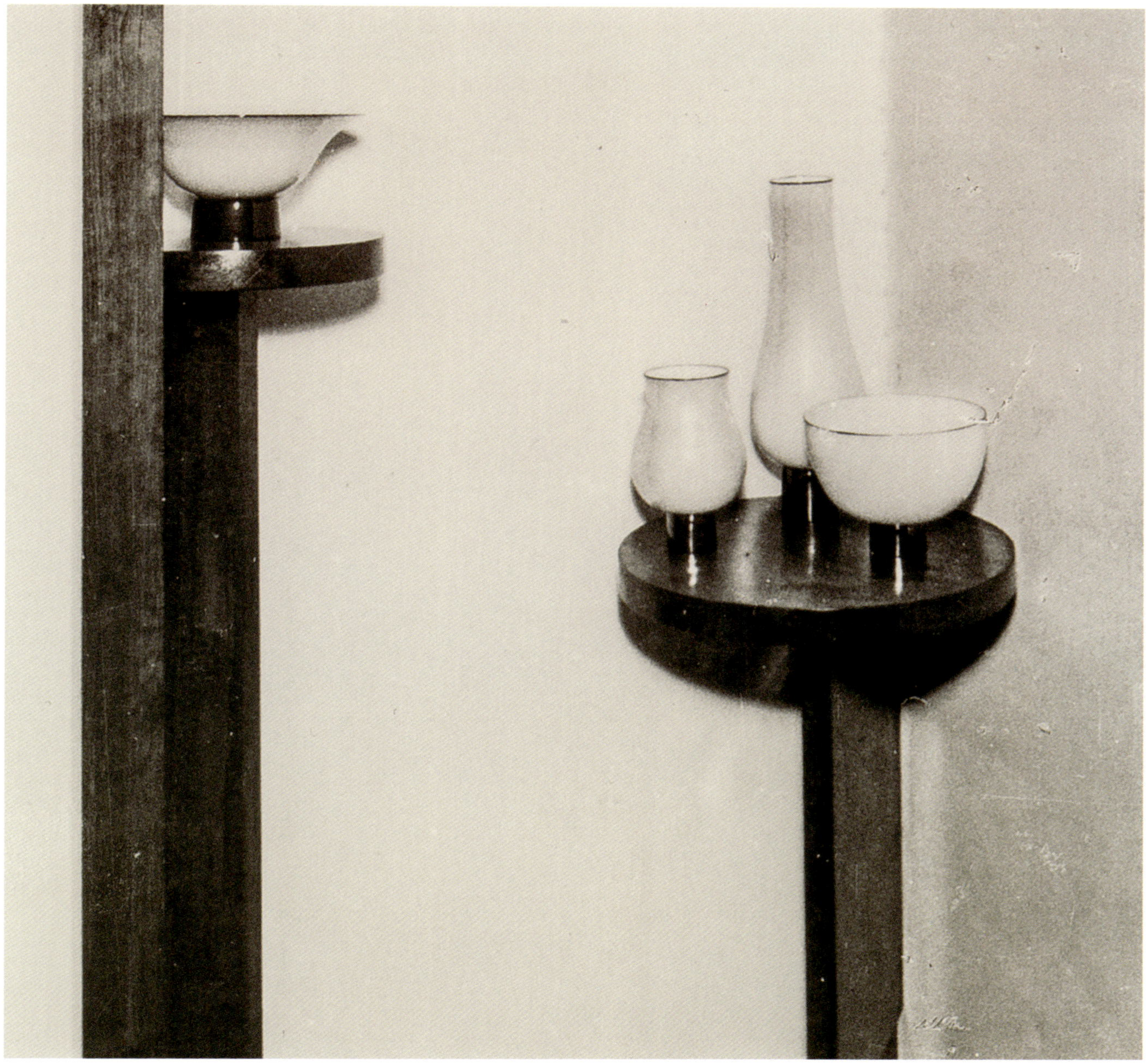

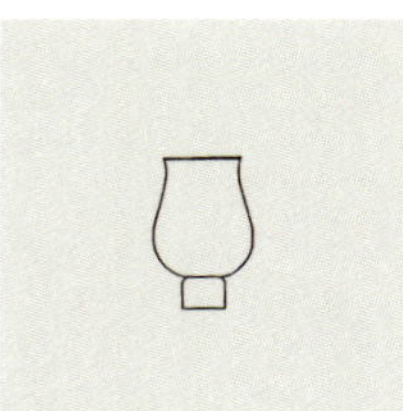

6032

Catalogue drawing for one of the vases in cased glass and with cylindrical foot, exhibited in 1930 at the 4th Triennale of Monza, AMH.
1930
18 cm high

exhibitions
1930, Monza, 4th Triennale

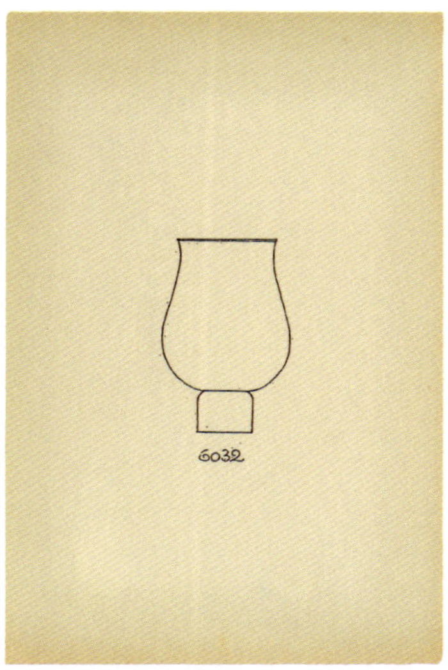

n.i.

Vase in cased glass and with cylindrical foot exhibited at the 4th Triennale of Monza, Furnishing Gallery, in the living room designed by the architect Gino Maggioni, 1930, detail. The catalogue number (6033 or 6035?) of the vase (h c. 22 cm) is not known.
1930
22 cm high

exhibitions
1930, Monza, 4th Triennale

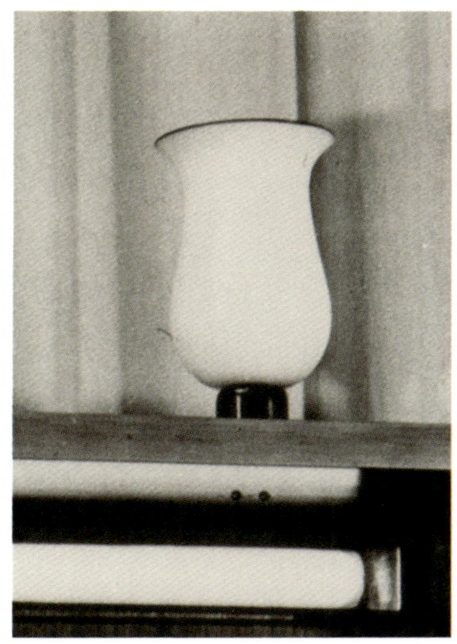

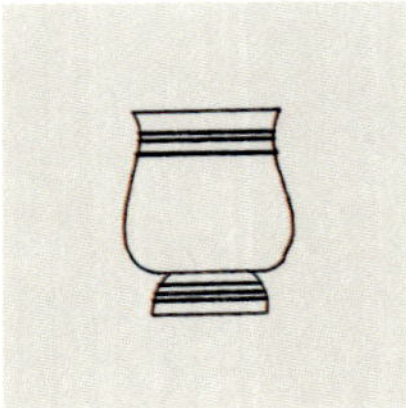

6037

Vase with skullcap foot in light amethyst cased glass. Three applied threads in light-amethyst glass on the foot and near the mouth. Surface with iridised finish. The piece is part of a series of five pieces (nos. 6037–6041), among the last to be documented in the Cappellin catalogue, which are characterised by a skullcap foot, proposed in different heights, and a decoration of three applied threads on the foot and near the mouth.
1930–31
23.5 cm high
M.V.M. Cappellin Murano *acid stamp*
Private collection, Treviso

>
6037, working drawing, CSV

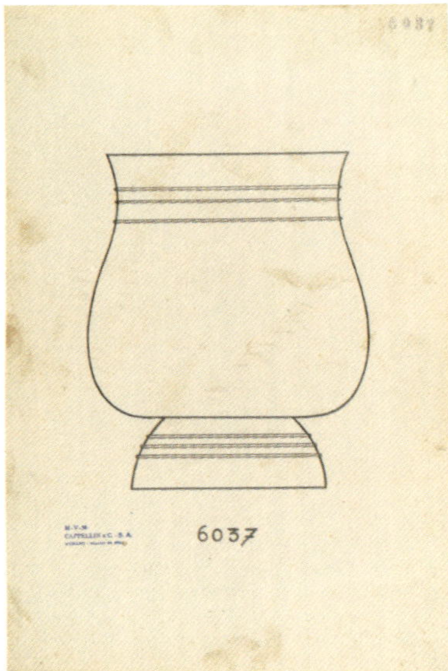

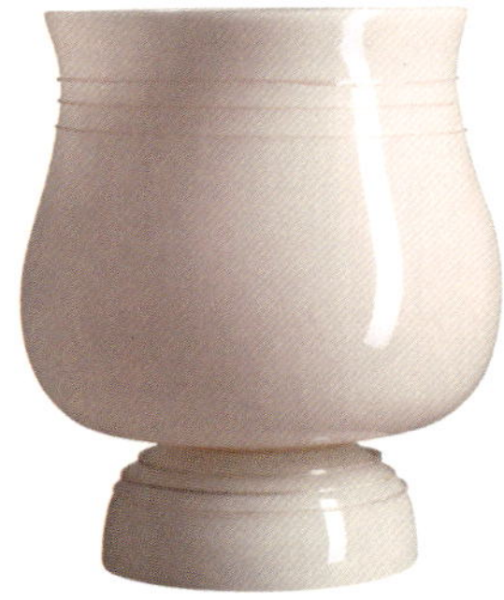

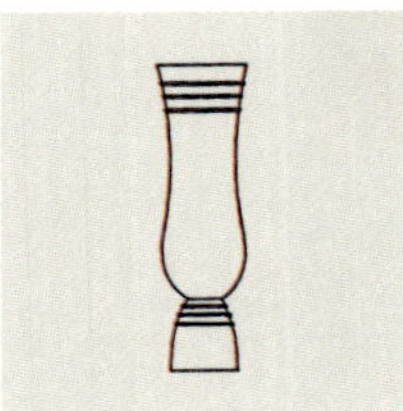

6040

Working drawing for one of the vases with skullcap foot and decoration with three applied threads on the foot and close to the mouth, CSV. The model is characterised by the height of the foot and the elongated shape of the body.
1930–31
36 cm high

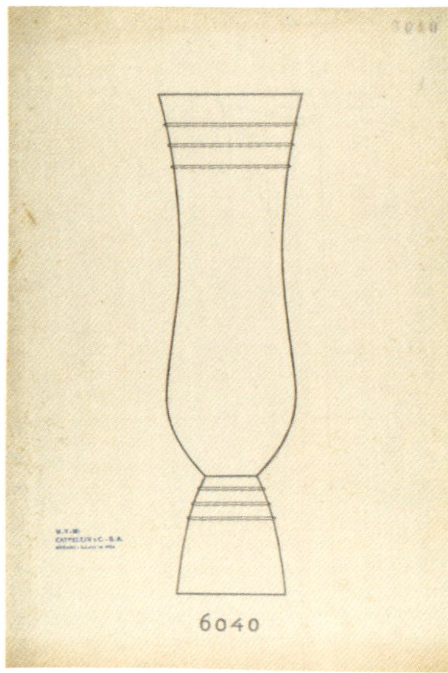

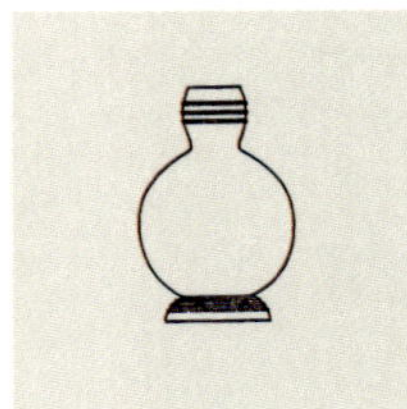

6039

Working drawing for one of the vases with skullcap foot and decoration with three applied threads on the foot and close to the mouth, CSV. The model is characterised by its spherical body and a constriction at the base of the neck.
1930–31
27 cm high

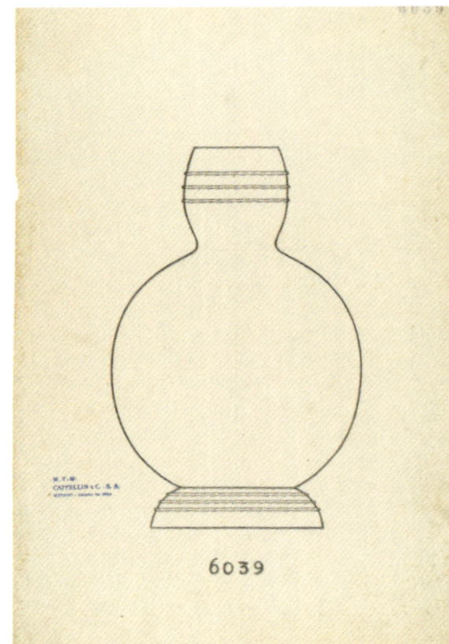

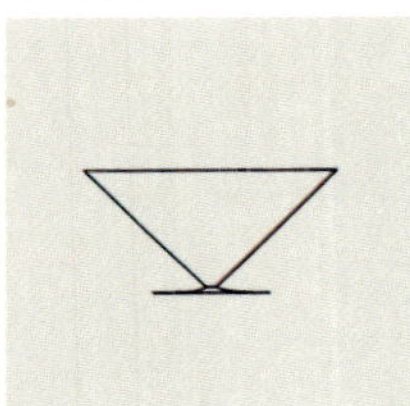

6000

Bowl in shape of upturned cone in cased *lattimo* and light-blue glass. Disk-shaped foot in transparent light-blue glass. Surface with iridised finish. Model no. 6000 was also produced with other glass techniques (e.g. transparent, *decoro fenicio*).
1930–31
23 cm ø
C.V.M. *acid stamp*
Private collection

>
6000, catalogue drawing, AMB

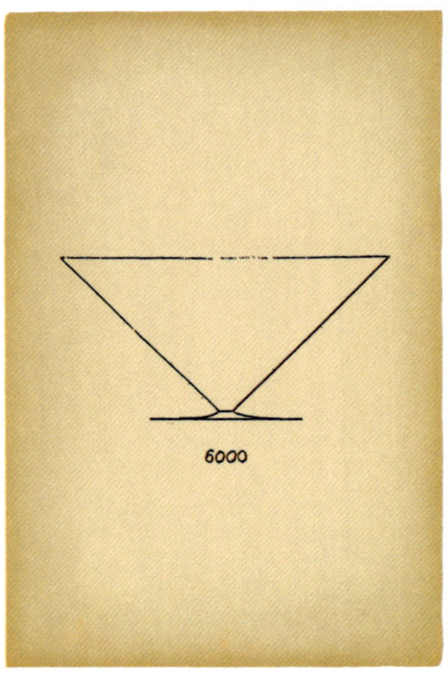

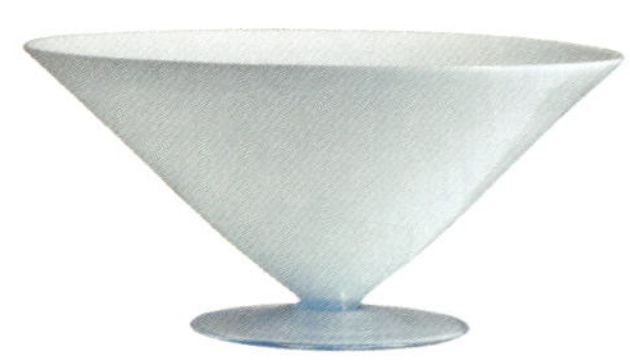

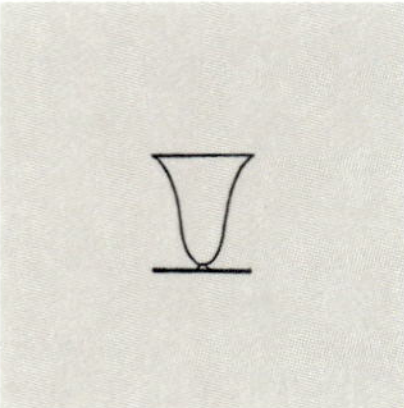

5875

Vase in cased *lattimo* and crystal glass with disk-shaped foot in crystal glass. Surface with iridised finish.
1930–31
18 cm high
M.V.M. Cappellin Murano *acid stamp*
Private collection, Feltre

>
5875, working drawing, CSV

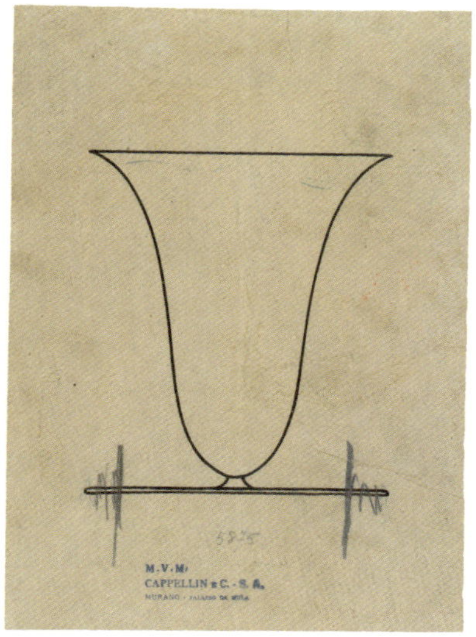

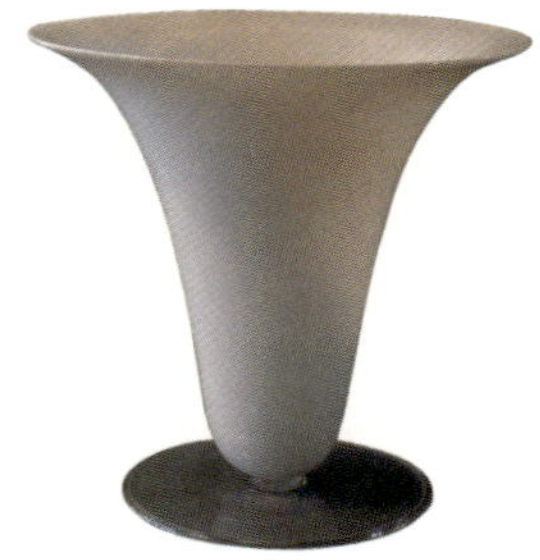

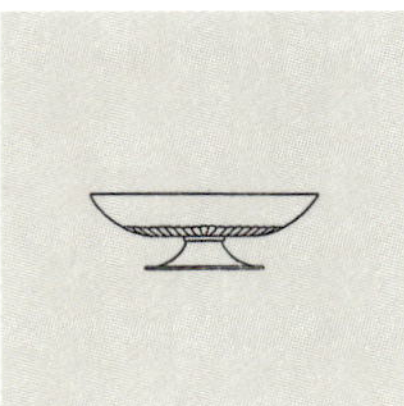

5514

Bowl with flared foot in cased *lattimo* glass on the outside and turquoise on the inside. '*Meza stampaura*' (half moulding) decoration at the base of the bowl. Surface with iridised finish.
The bowl was produced after a model in the catalogue of the glassworks designed by Vittorio Zecchin (1925–26).
1929–31
9 cm high; 24 cm ø
M.V.M. Cappellin Murano *acid stamp*
Private collection

>
5514, catalogue drawing, AMB

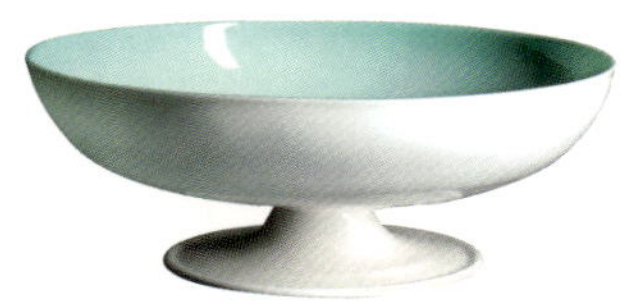

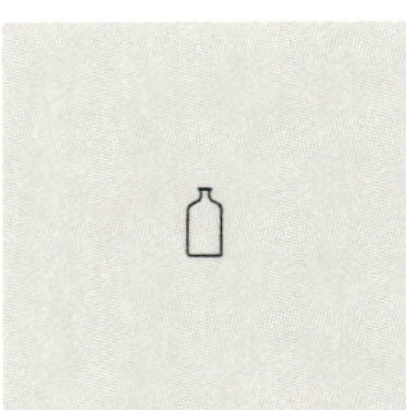

1281

Bottle in cased *lattimo* and green glass with thread around the mouth in coral-red *pasta vitrea*. Surface with iridised finish.
The model was produced as a variant of no. 1281 (h 8 cm).
1929–31
14 cm high
M.V.M. Cappellin Murano *acid stamp*
The Steinberg Foundation collection / The Corning Museum of Glass, Corning, NYW

>
1281, catalogue drawing, AMA

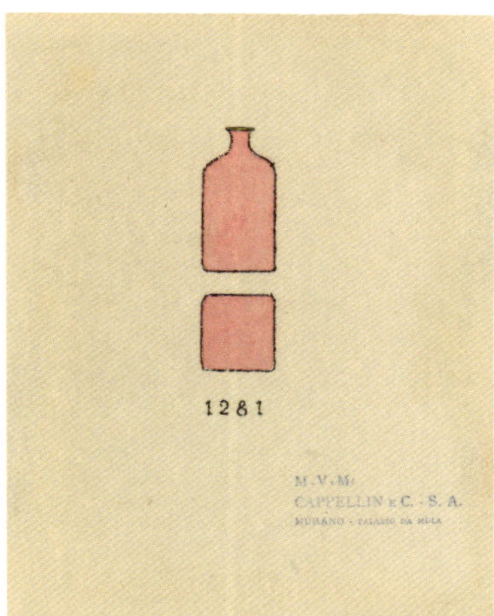

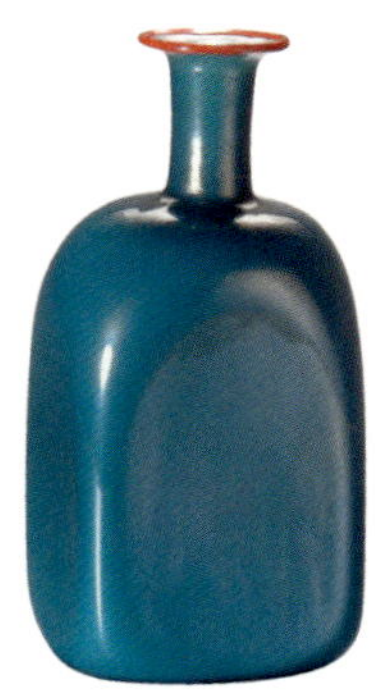

>
Cappellin glass pieces at the Amsterdam Exhibition of glass, ceramics and lace, 1931, AHL, detail

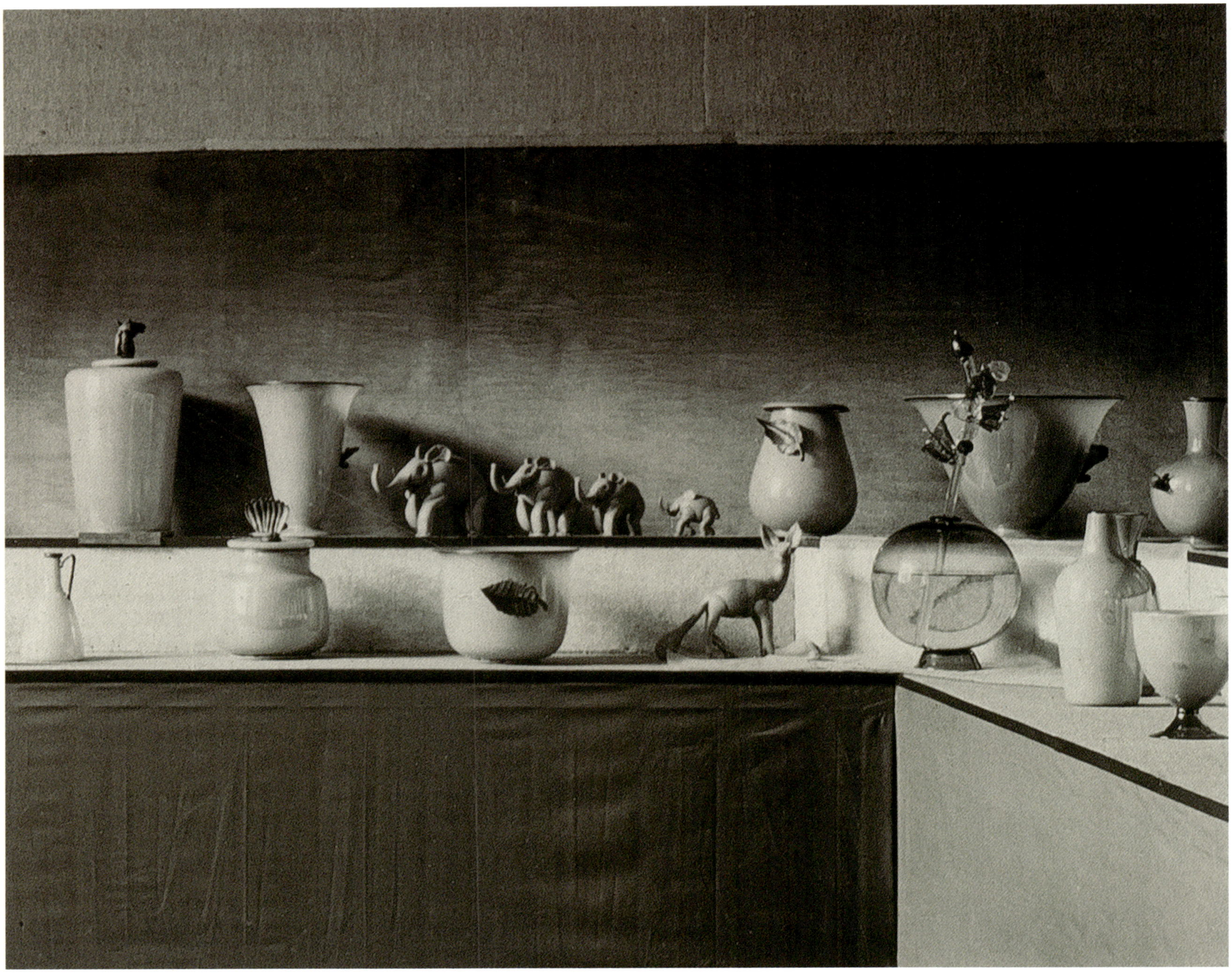

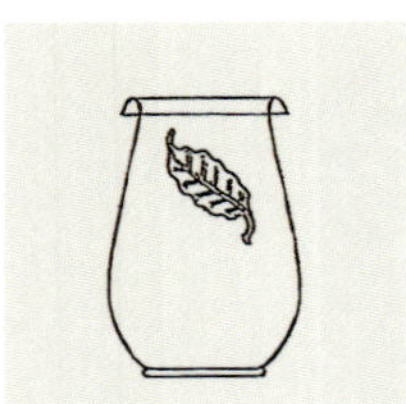

6024

Working drawing for a vase in glass decorated with an applied leaf on the body, CSV. An example of this model, produced in cased glass was exhibited in 1931 at the Amsterdam Exhibition of glass, ceramics and lace. no. 6024 was also proposed in transparent glass.
1930
32 cm high

exhibitions
1931, Amsterdam, Exhibition of glass, ceramics and lace

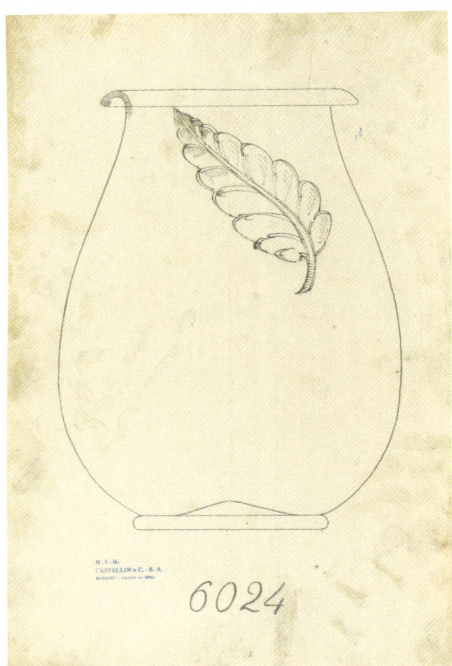

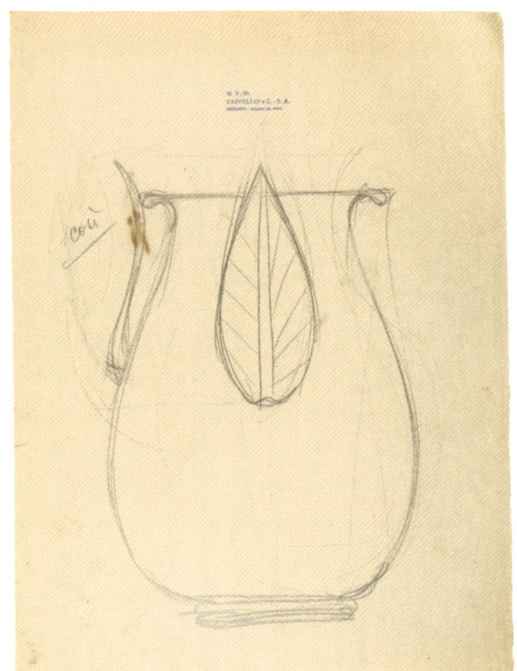

>
Carlo Scarpa, study for vase decorated with applied leaf, c. 1930, AMuVe

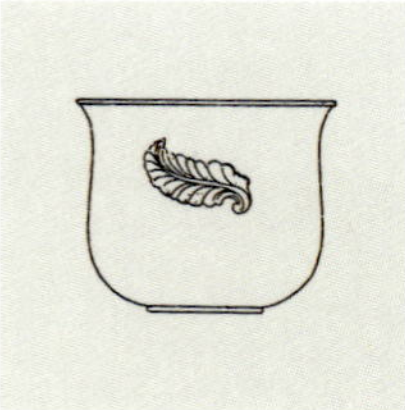

6027

Vase in cased *lattimo* glass with black and gold glass on the inside. Foot and applied leaf on the body in black and gold glass. An example of no. 6027, produced in cased glass was exhibited in 1930 in Monza at the 4th Triennale and the following year at the Amsterdam Exhibition of glass, ceramics and lace. The model was also produced in transparent glass.
1930
26 cm high
Private collection

exhibitions
1930, Monza, 4th Triennale
1931, Amsterdam, Exhibition of glass, ceramics and lace

>
6027, catalogue drawing, AMA

>
Carlo Scarpa, study for a vase decorated on the body with an applied leaf, c. 1930, MuVe

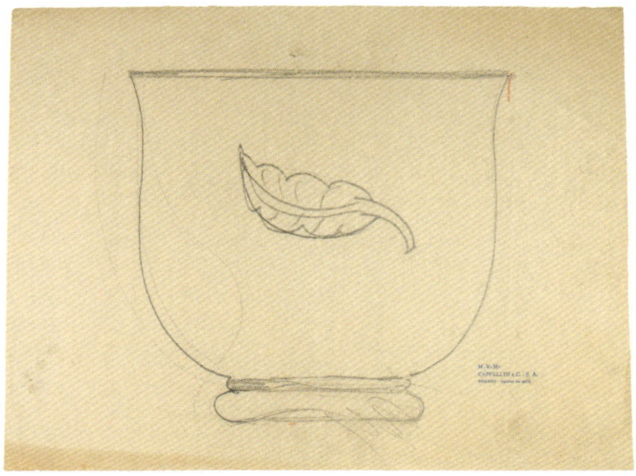

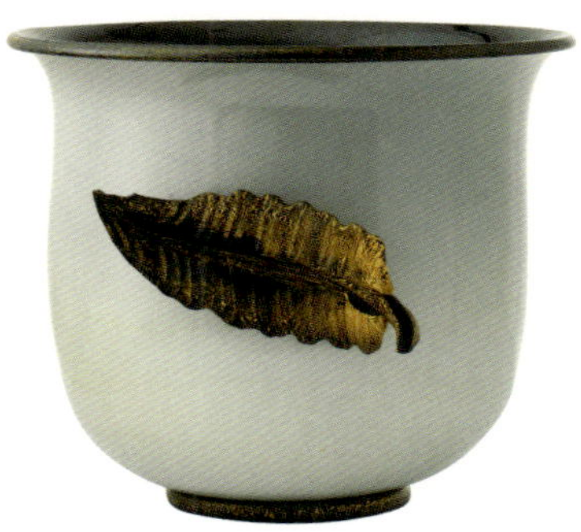

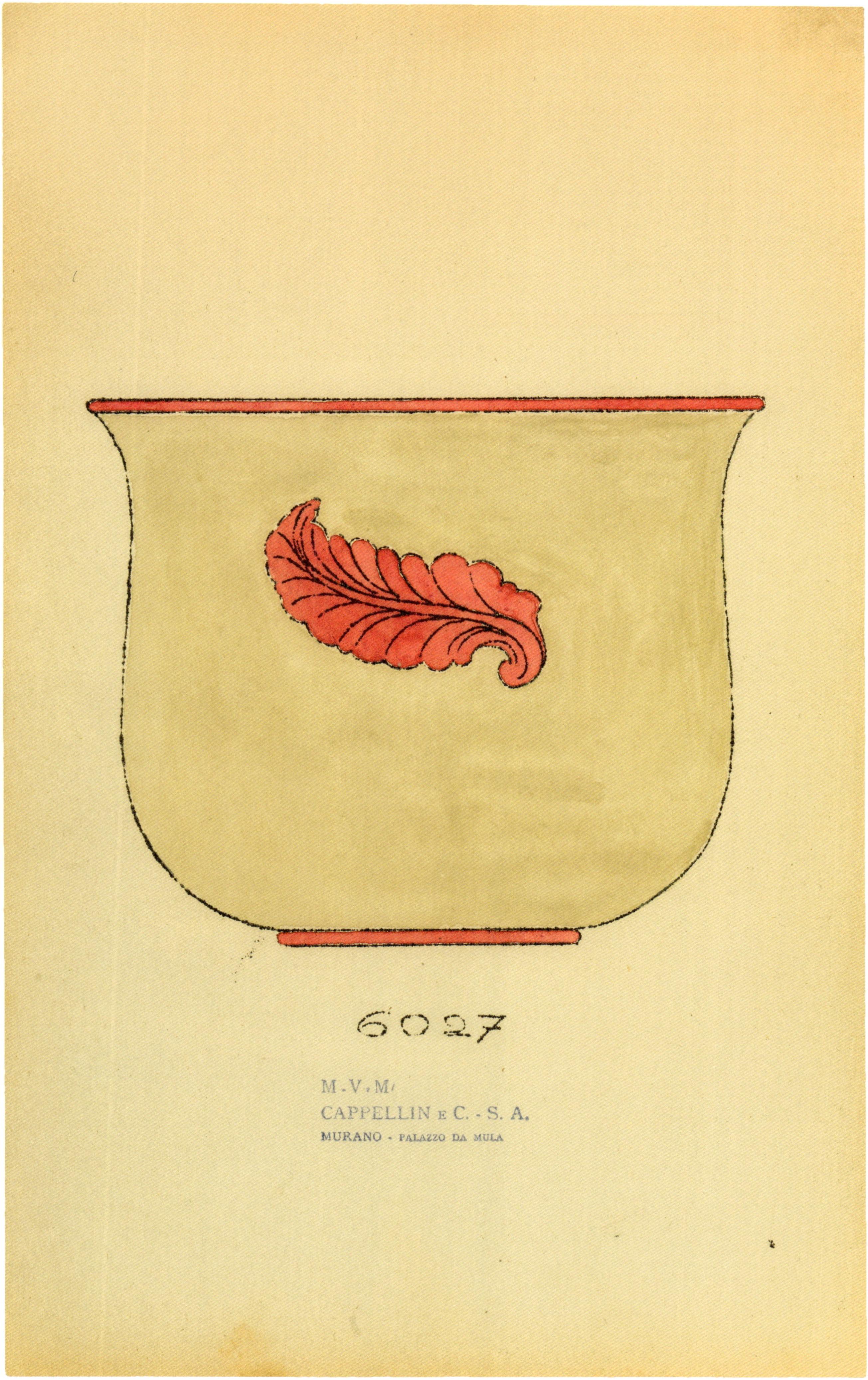

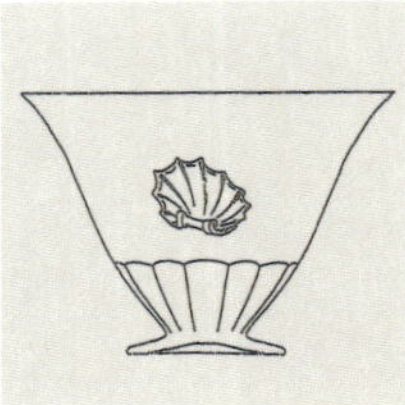

6022

Vase with foot worked out of the body of the piece, in *lattimo* cased and amethyst glass with applied gold leaf. Thread around the mouth and shell applied to the body in coral-red *pasta vitrea.* The piece was produced in a variant without *meza stampaura* (half moulding) of no. 6022. A similar example was exhibited in 1931 at the Amsterdam Exhibition of glass, ceramics and lace. The model was also produced in transparent glass.
1930
37.3 cm high
The Olnick Spanu Collection, New York

exhibitions
1930, Monza, 4th Triennale
1931, Amsterdam, Exhibition of glass, ceramics and lace

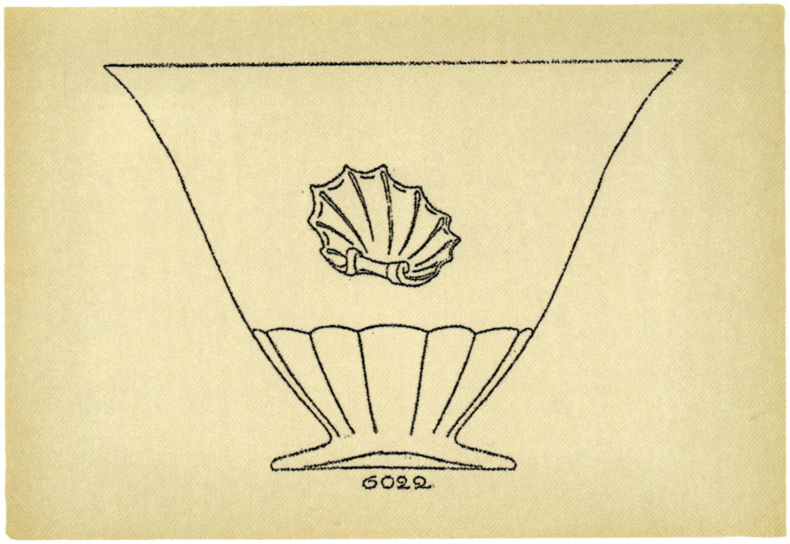

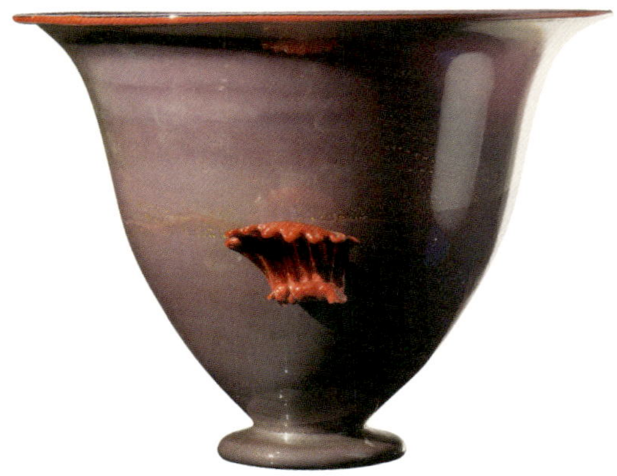

>
6022, catalogue drawing, private collection

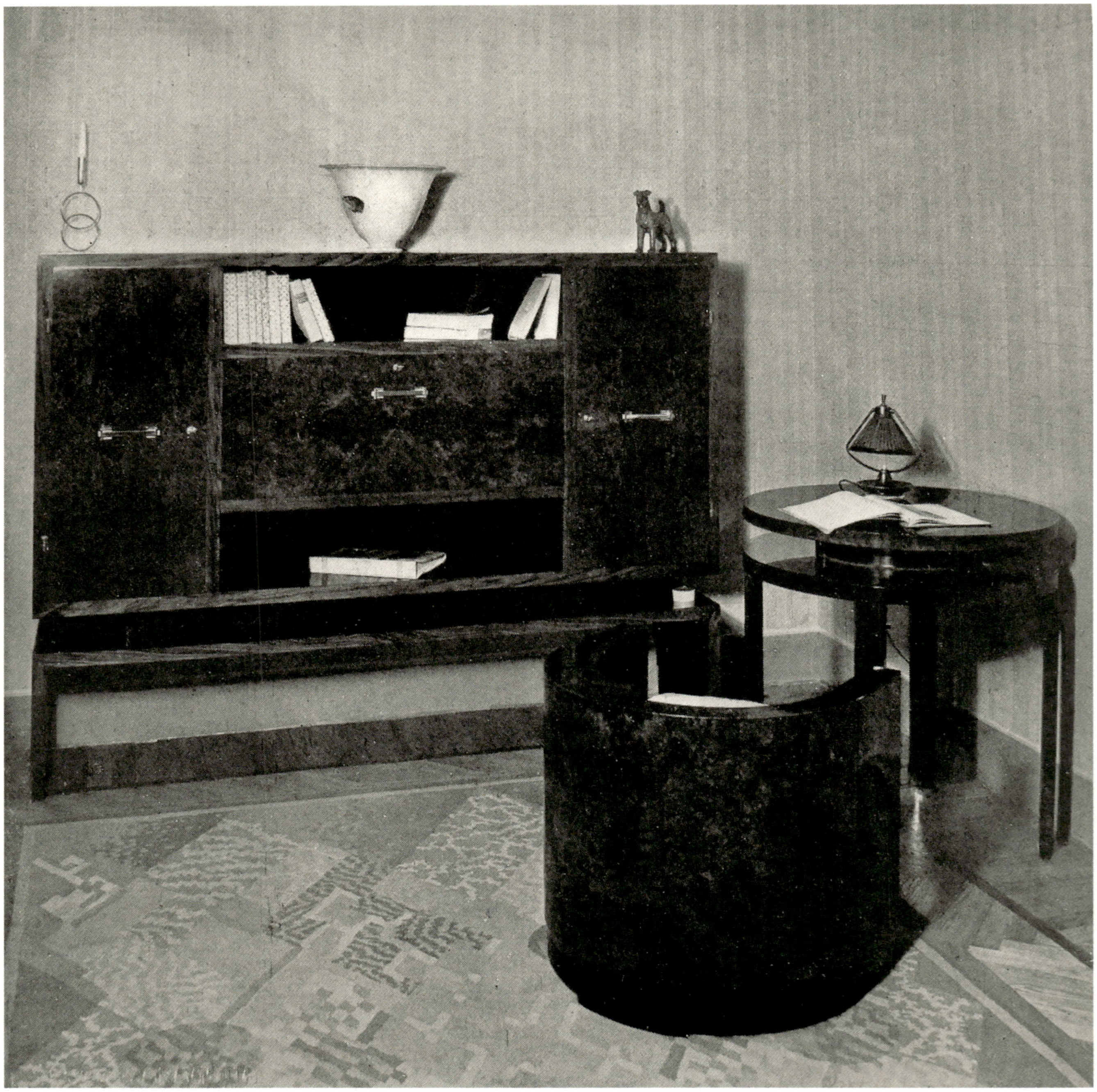

>
Vase no. 6022 in a sitting room designed by Michele Merigi, Milano, in Domus, July 1932, p. 424

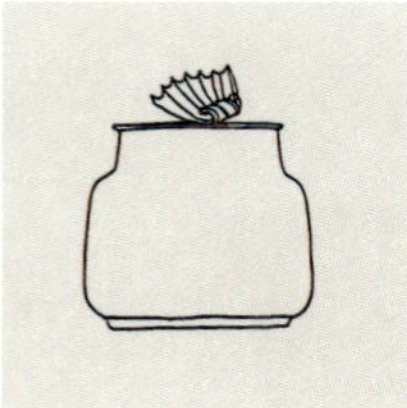

3871

Compote bowl in *lattimo* glass with black glass on the inside with foot and applied shell on the lid in black and gold glass. An example of no. 3871 was exhibited in 1931 at the Amsterdam Exhibition of glass, ceramics and lace. The model was also produced in transparent glass.
1930
37 cm high
Private collection

exhibitions
1931, Amsterdam, Exhibition of glass, ceramics and lace

>
Carlo Scarpa, study for a compote bowl, lid decorated with an applied shell, c. 1930, MuVe

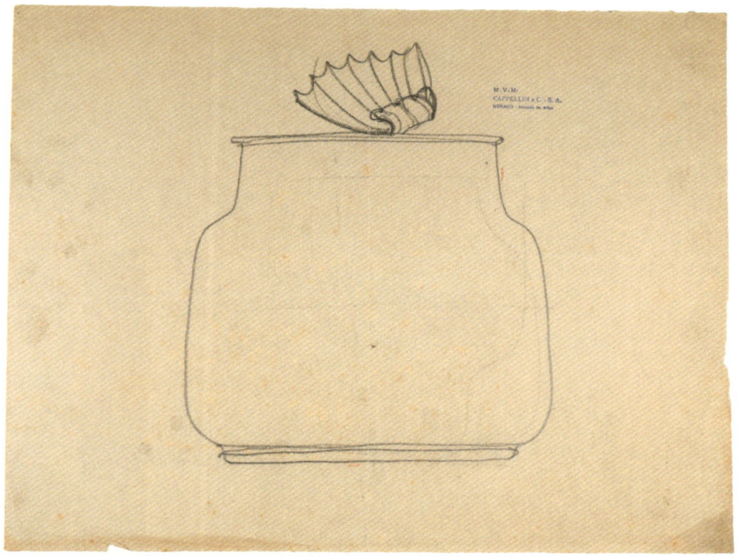

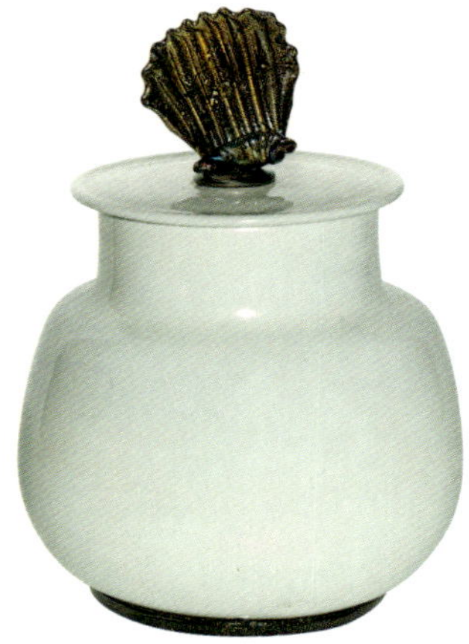

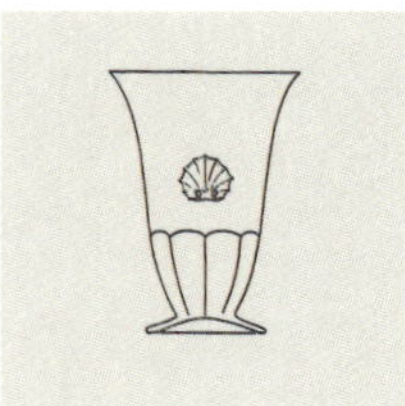

6021

Vase in cased glass with foot worked out of the body and applied shell, exhibited at the 4th Triennale of Monza (1930). The piece was produced in a variant without *meza stampaura* (half moulding) of no. 6021. A similar example was also presented in 1931 at the Amsterdam Exhibition of glass, ceramics and lace.
1930
36 cm high

exhibitions
1930, Monza, 4th Triennale
1931, Amsterdam, Exhibition of glass, ceramics and lace

>
6021, working drawing, CSV

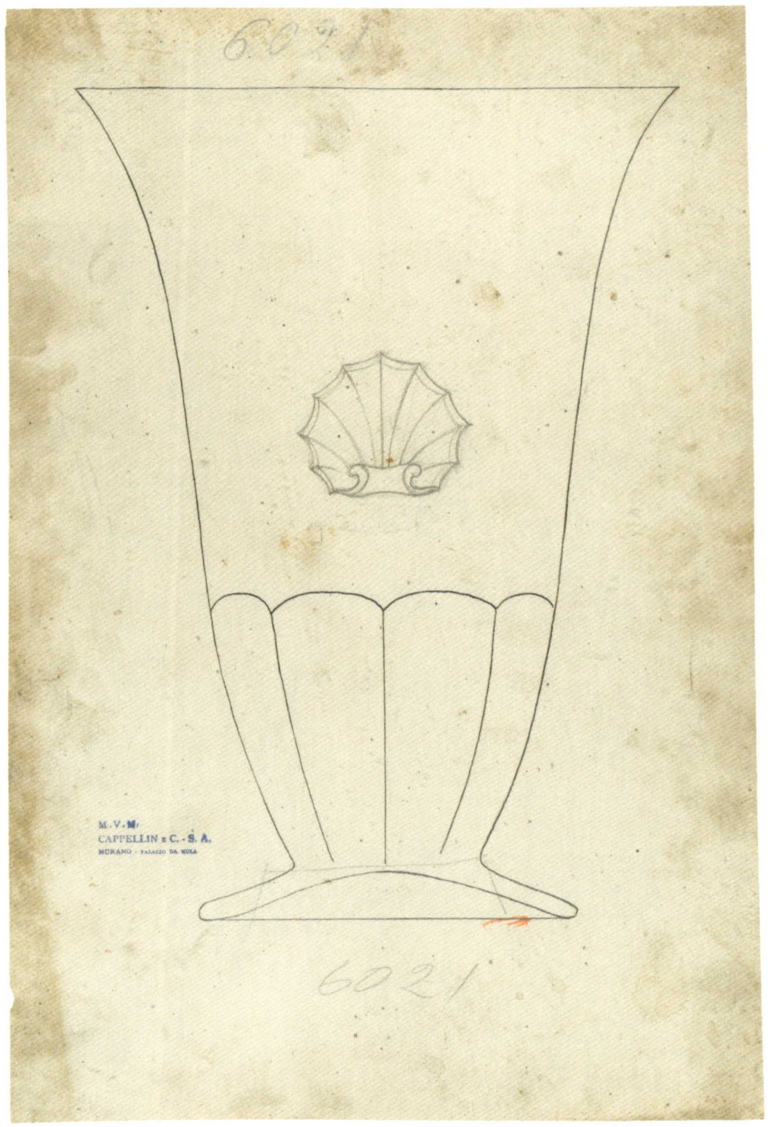

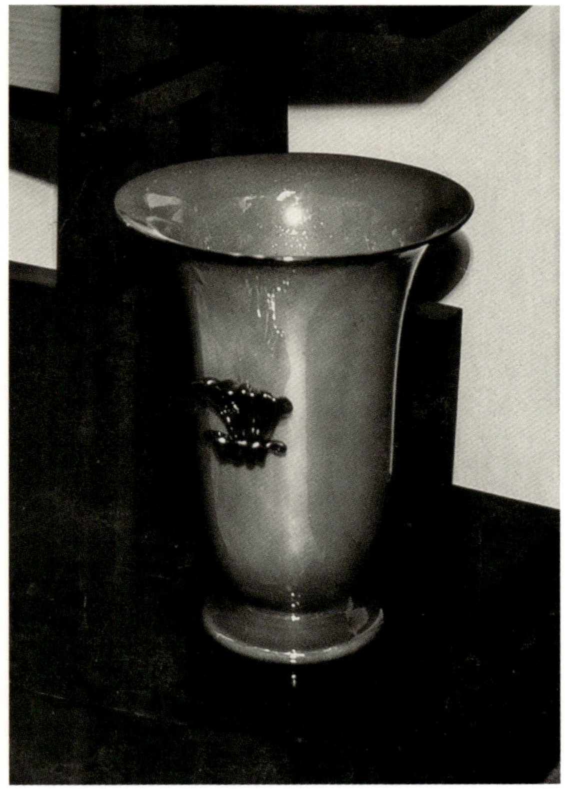

n.i.

Vase in *lattimo* glass with black glass on the inside, decorated with an animal head in black and gold applied on the body. Foot worked out of the body of the piece. A similar example was exhibited in 1930 at the 4th Triennale of Monza and in 1931 at the Amsterdam Exhibition of glass, ceramics and lace. The model (no. 6016?) was also proposed in transparent glass. No. 6017 also belongs to the same series with an applied lion's head on the body.
1930
32 cm high
Private collection

exhibitions
1930, Monza, 4th Triennale
1931, Amsterdam, Exhibition of glass, ceramics and lace

>
6017, catalogue drawing, AMA

>
Vase decorated with an animal head at the 4th Triennale of Monza, 1930, detail

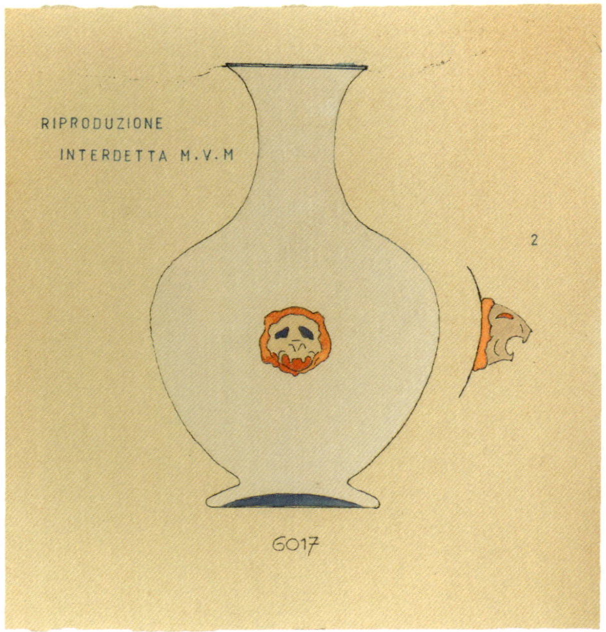

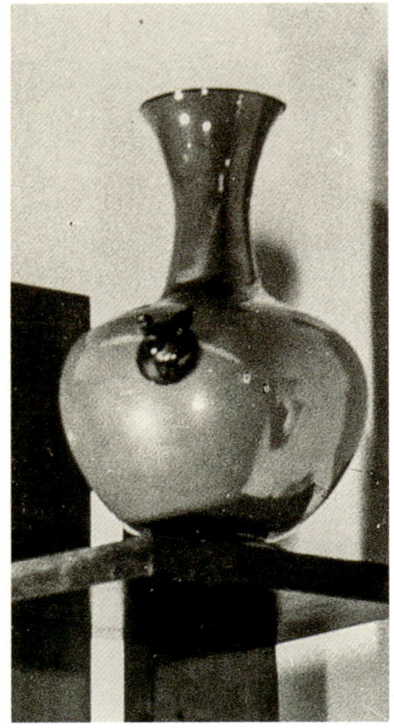

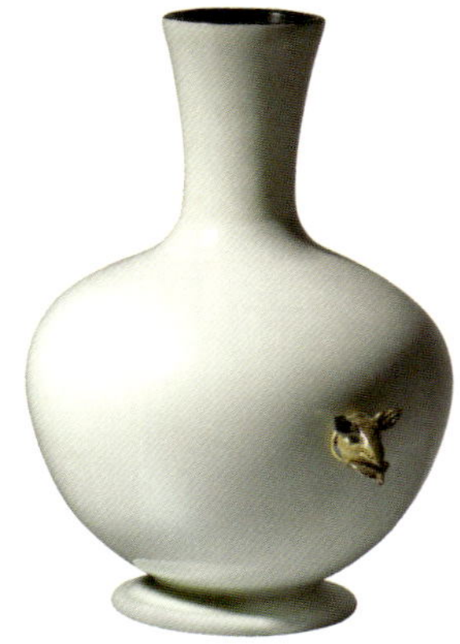

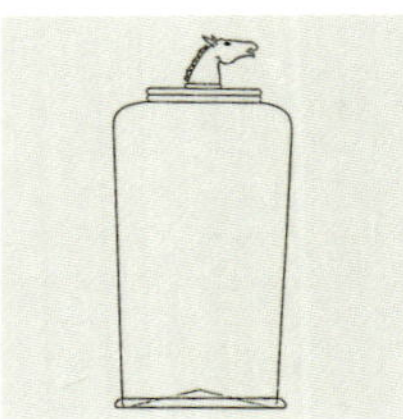

3872

Top left: compote bowl in cased glass with lid decorated with a horse's head, exhibited at the Amsterdam Exhibition of glass, ceramics and lace in 1931, AHL, detail. No. 6024 was also produced in transparent glass.
1930–31
52 cm high

exhibitions
1931, Amsterdam, Exhibition of glass, ceramics and lace

>
3872, catalogue drawing, AMA

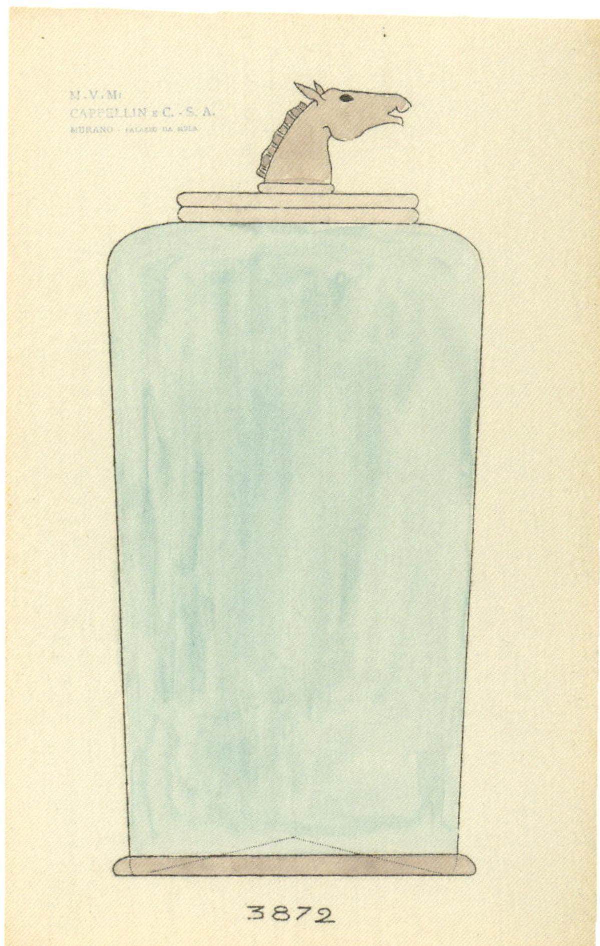

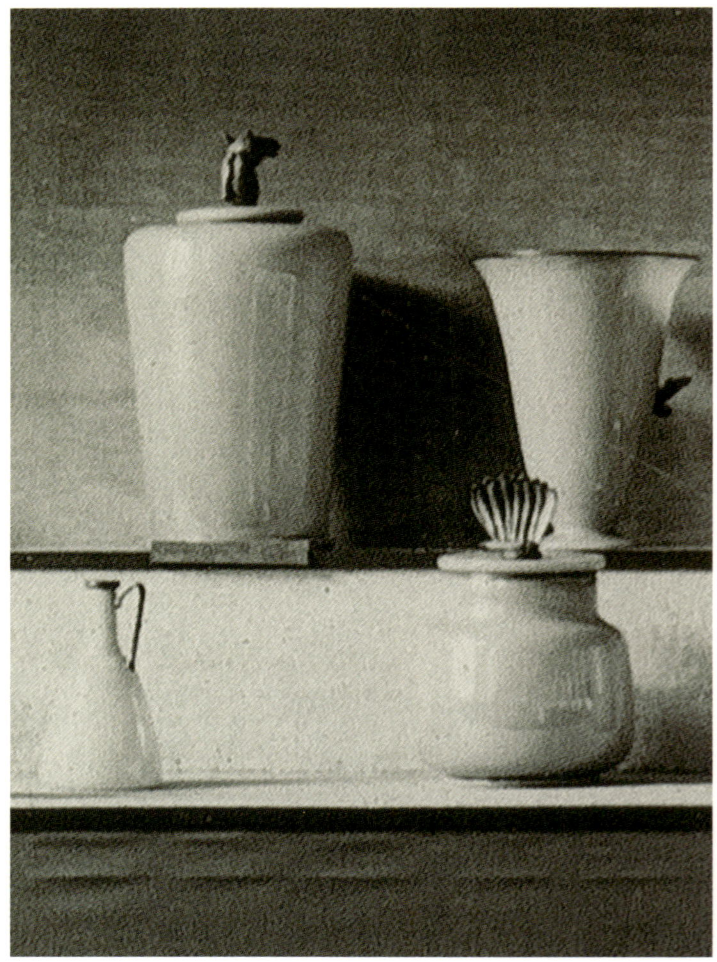

n.i.

Cylindrical vase in *lattimo* cased crystal glass with tall foot, thread around the mouth and branch with applied leaves in Savoy-blue *pasta vitrea*.
1930–31
47 cm high
Bischofberger collection, Switzerland

>
Carlo Scarpa, study for vase with applied tall foot decorated with branches and leaves, c. 1930, MuVe

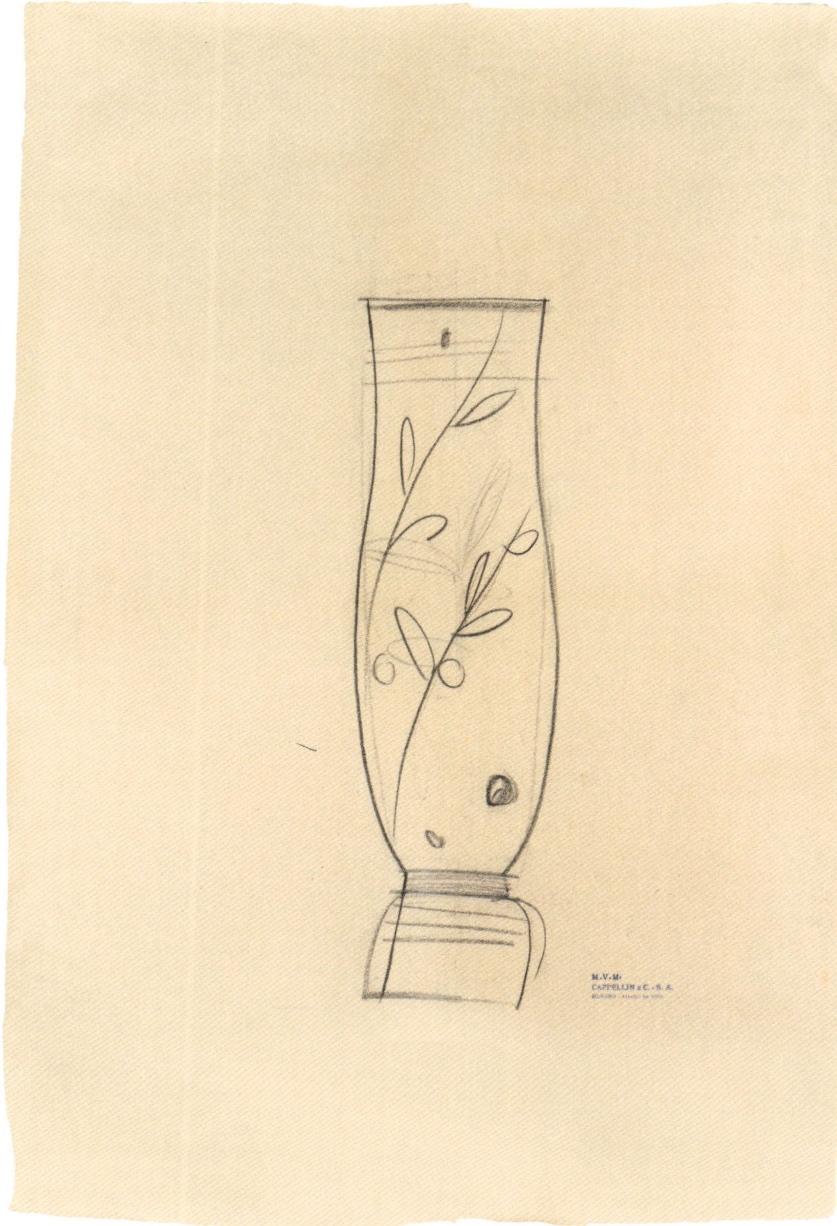

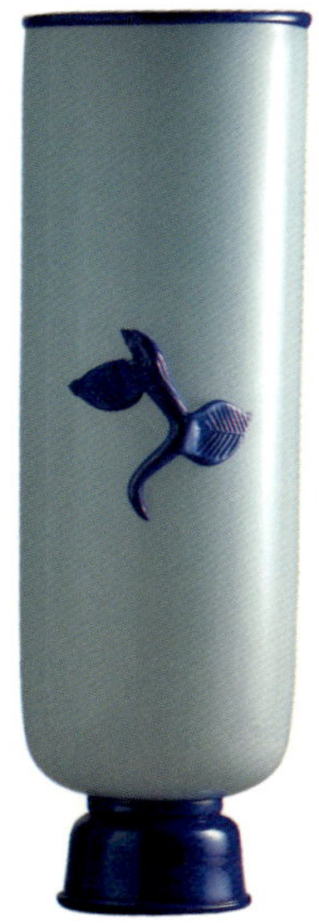

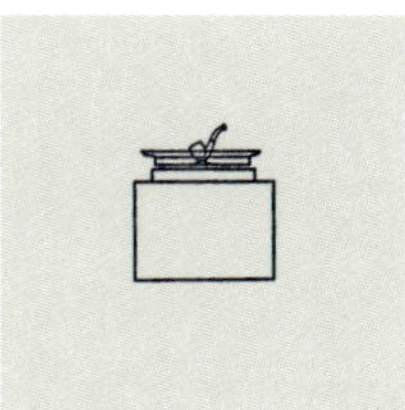

3865

Compote bowl no. 3865 in opaque glass documented in the American photograph catalogue of Cappellin Glassware, undated, detail, AHL. The model is characterised by the lid which is decorated with an applied pipe. A similar example was exhibited in 1930 at the 4th Triennale of Monza, in the furnishings Gallery in the bar prepared by the architect Serafini.
1930
18 cm high

exhibitions
1930, Monza, 4th Triennale

>
3865, catalogue drawing, AMB

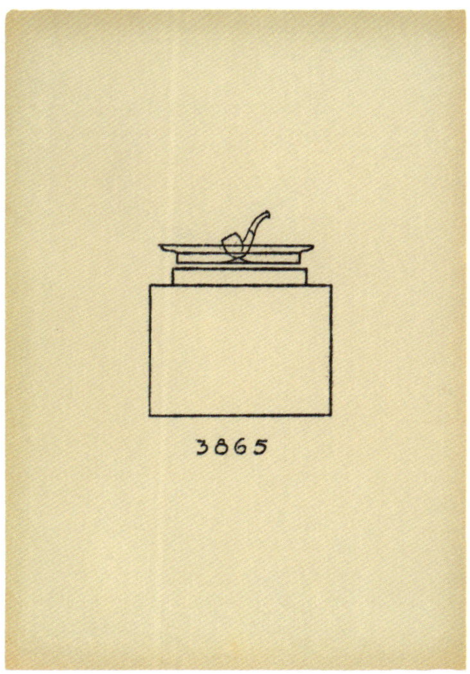

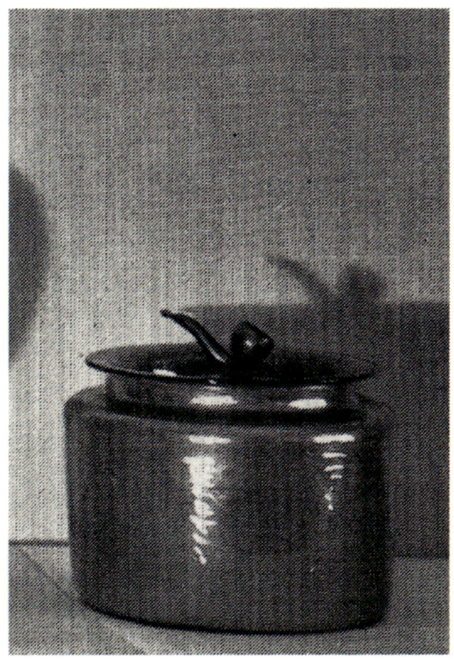

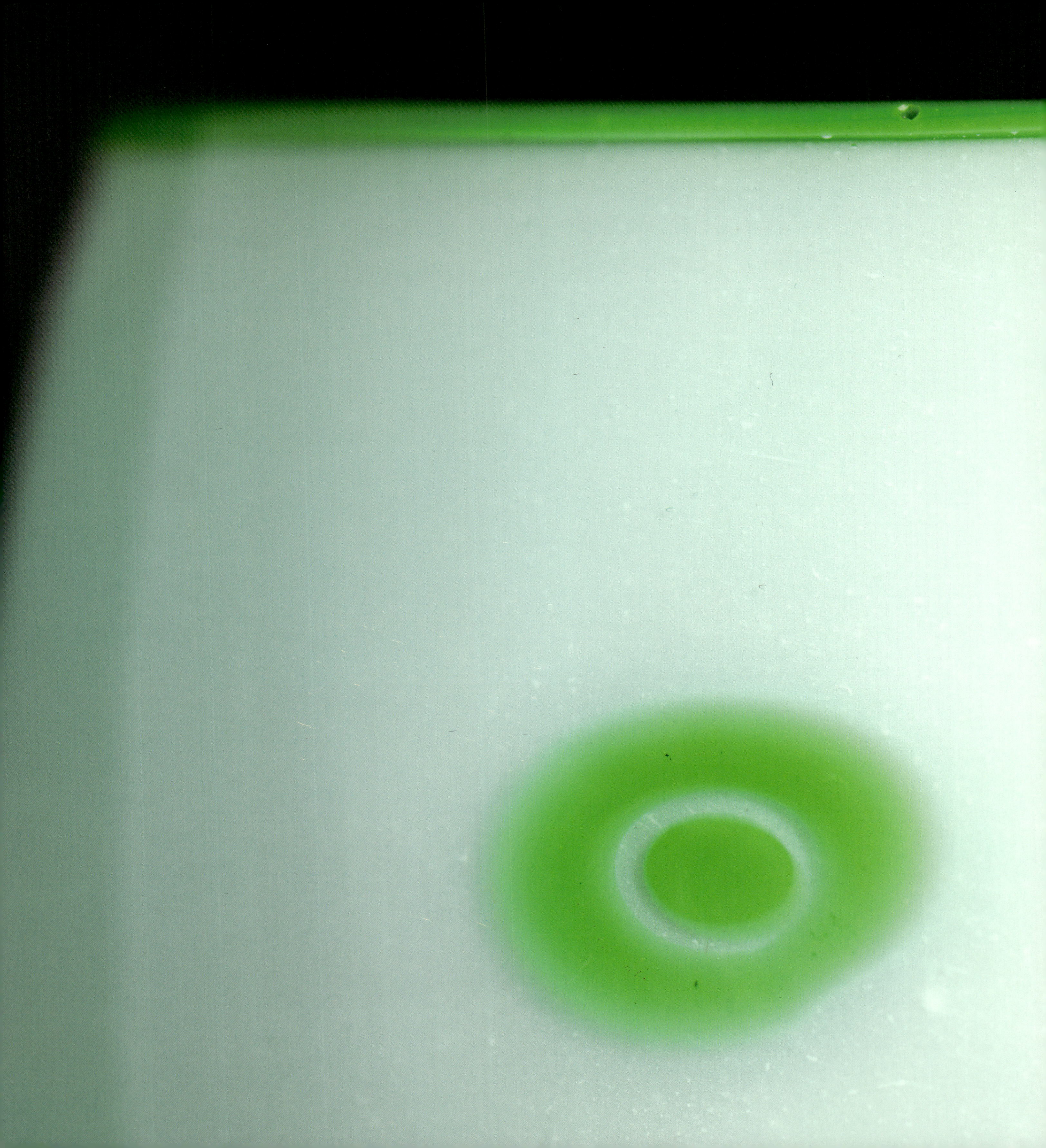

In Monza in 1930 (4th Triennale) and Amsterdam in 1931 (*Mostra di vetri, ceramiche e merletti d'arte moderna italiana*) among the most recent novelties in the Cappellin production there were also highly refined glass pieces decorated by *murrine*. This was a series of objects in *lattimo* glass with slight 'casing' of coloured glass in light shades (*pagliesco*, light blue, light green etc.) which are characterised by the presence of delicate *murrine* in concentric circles, produced in the same colour, irregularly distributed and included in the wall of the piece. The greater part of these vases and bowls were finished at the mouth and/or at the foot with transparent glass in the same shade. In the final production of the glassworks, before it closed down, the theme was further developed using new *murrine* with an abstract design and new shapes. In a presumably experimental way, around 1929, the *murrine* were also used in producing the glass composition of certain objects (of which a vase and a bowl survive) characterised by a floral and star-like design.

MURRINE 1929-31

Exhibitions

1930, Monza, 4th *Esposizione Triennale Internazionale delle Arti Decorative ed Industriali Moderne*
1931, Amsterdam, *Tentoonstelling van glas, aardewerk en kant, van moderne italiaansche kunstnijverheid. Mostra di vetri, ceramiche e merletti d'arte moderna italiana*

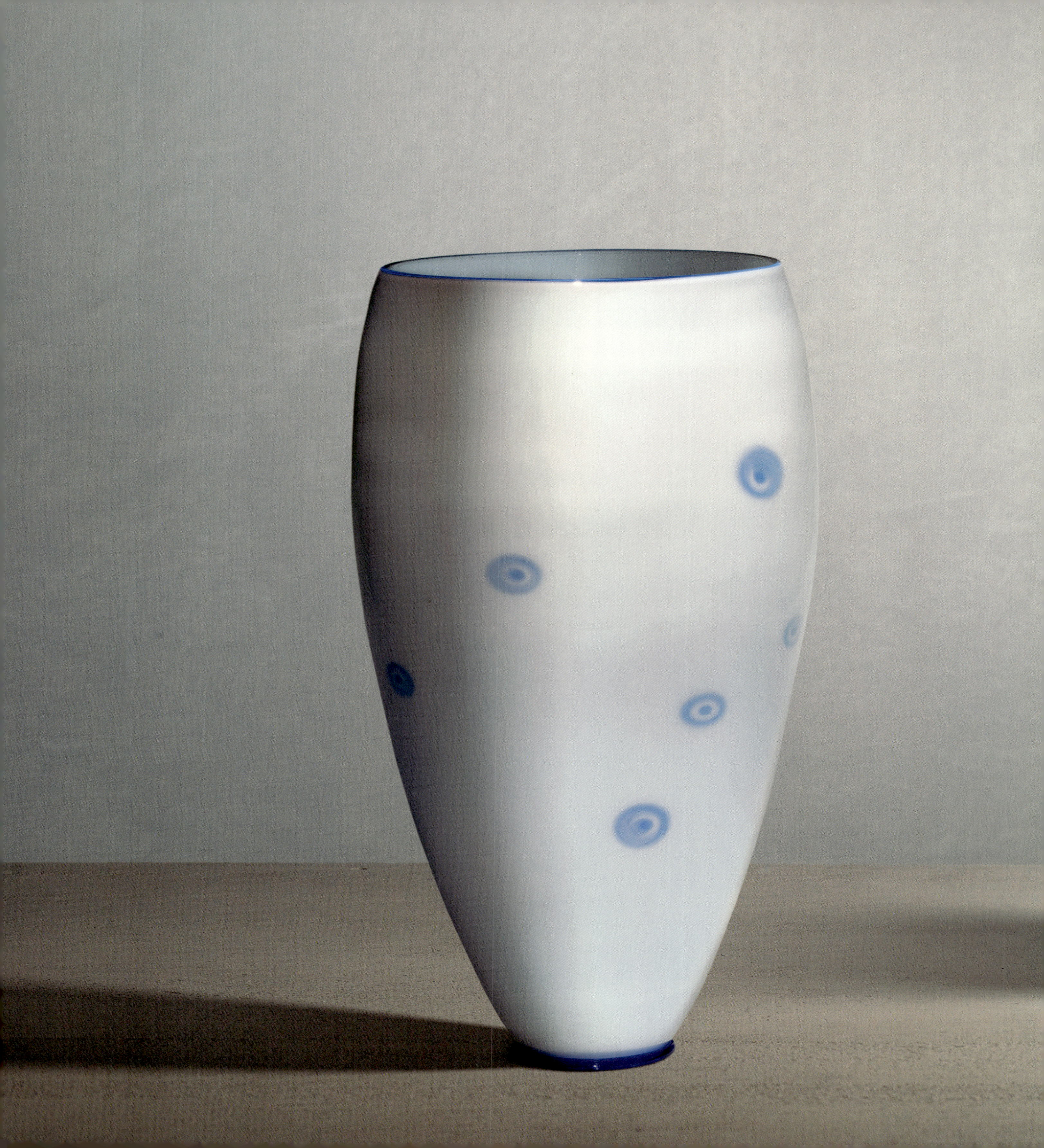

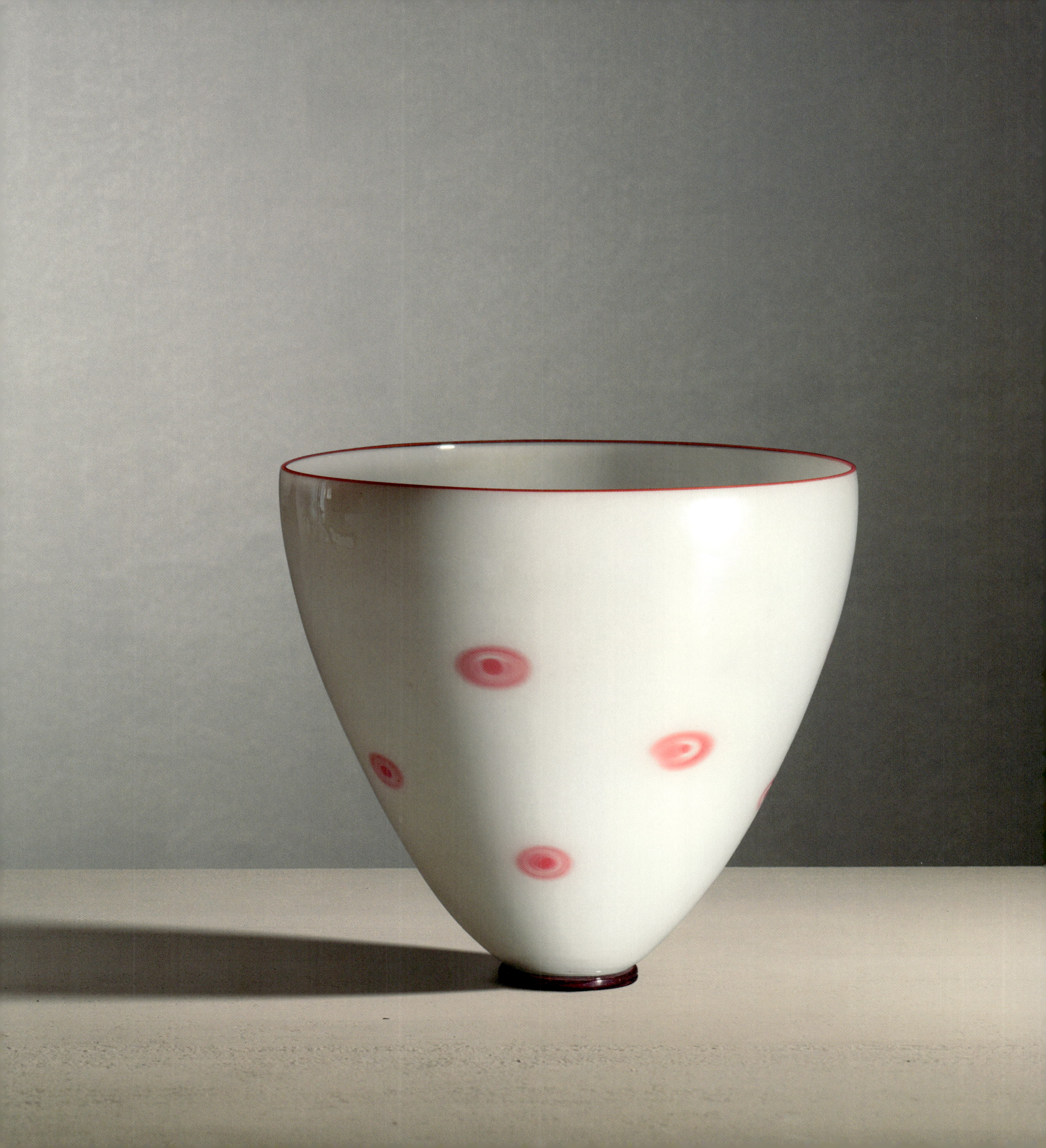

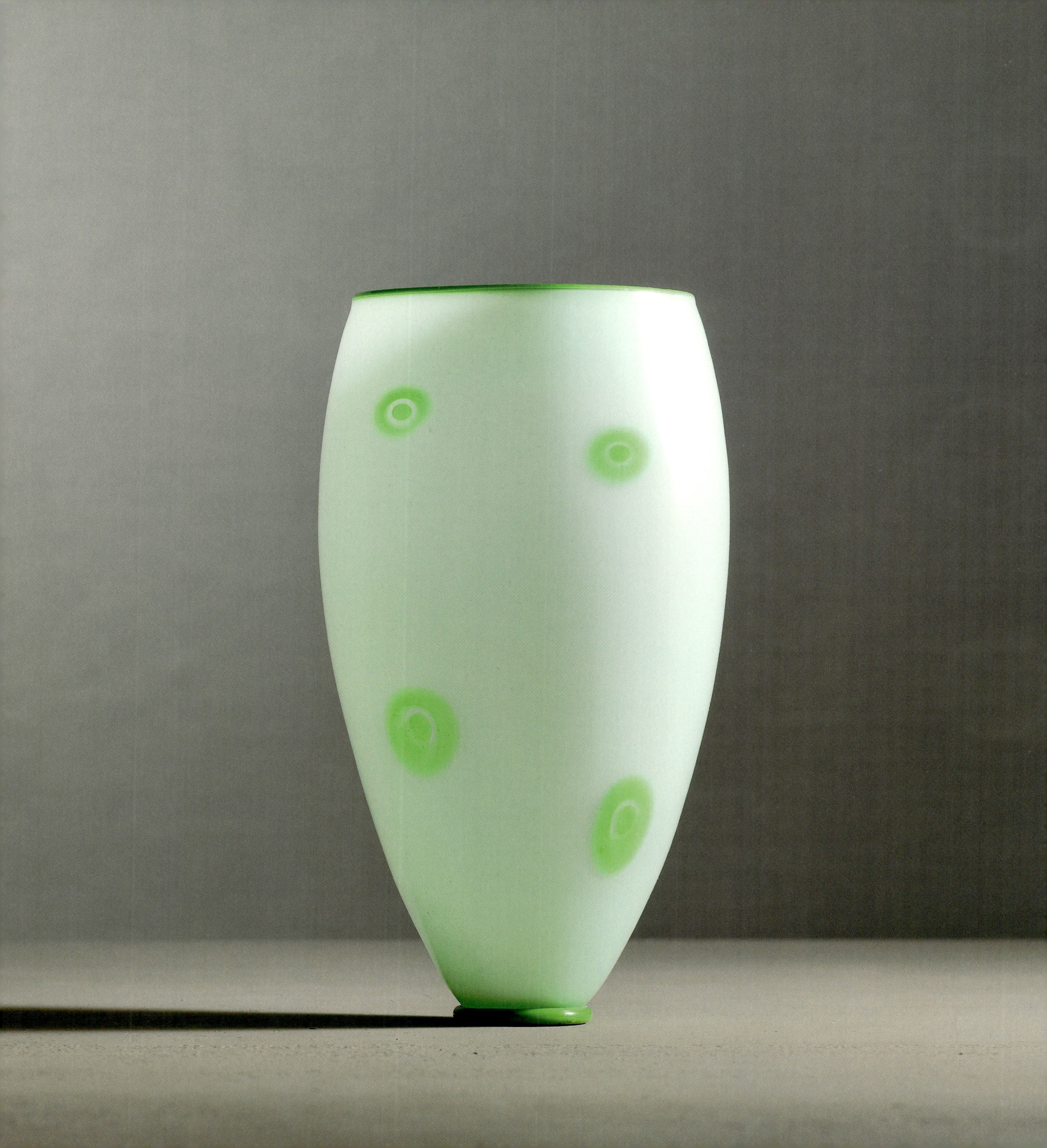

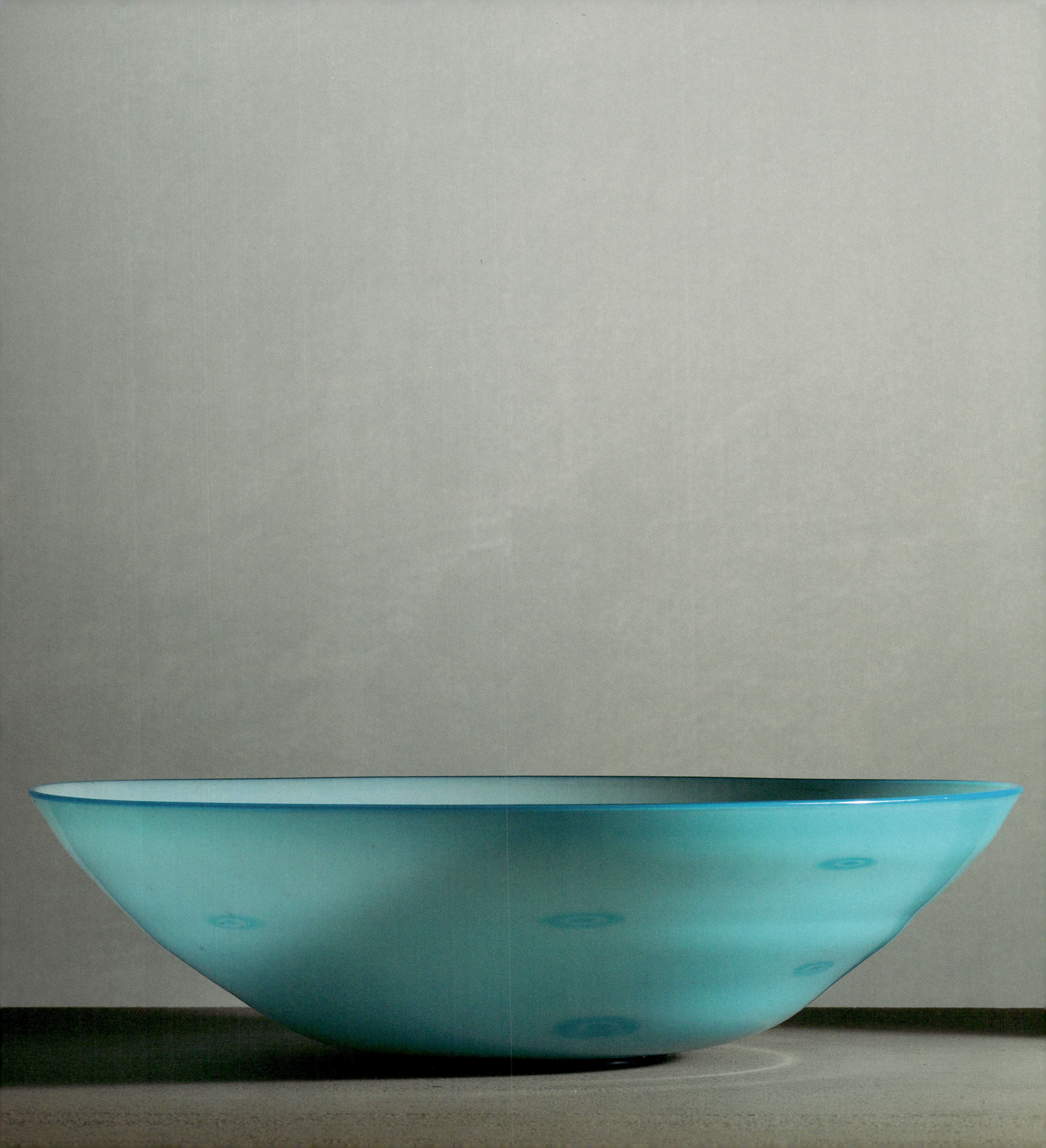

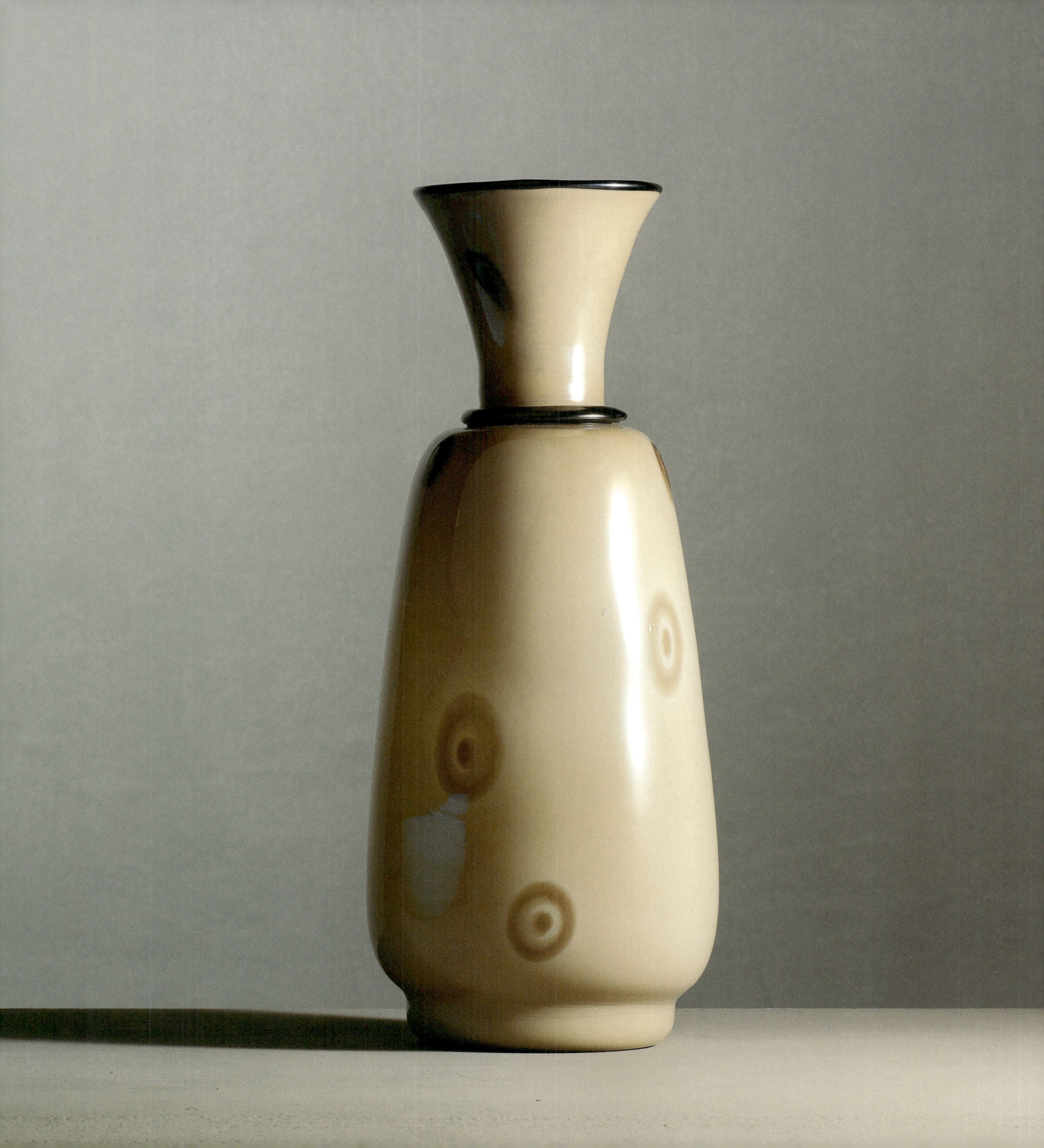

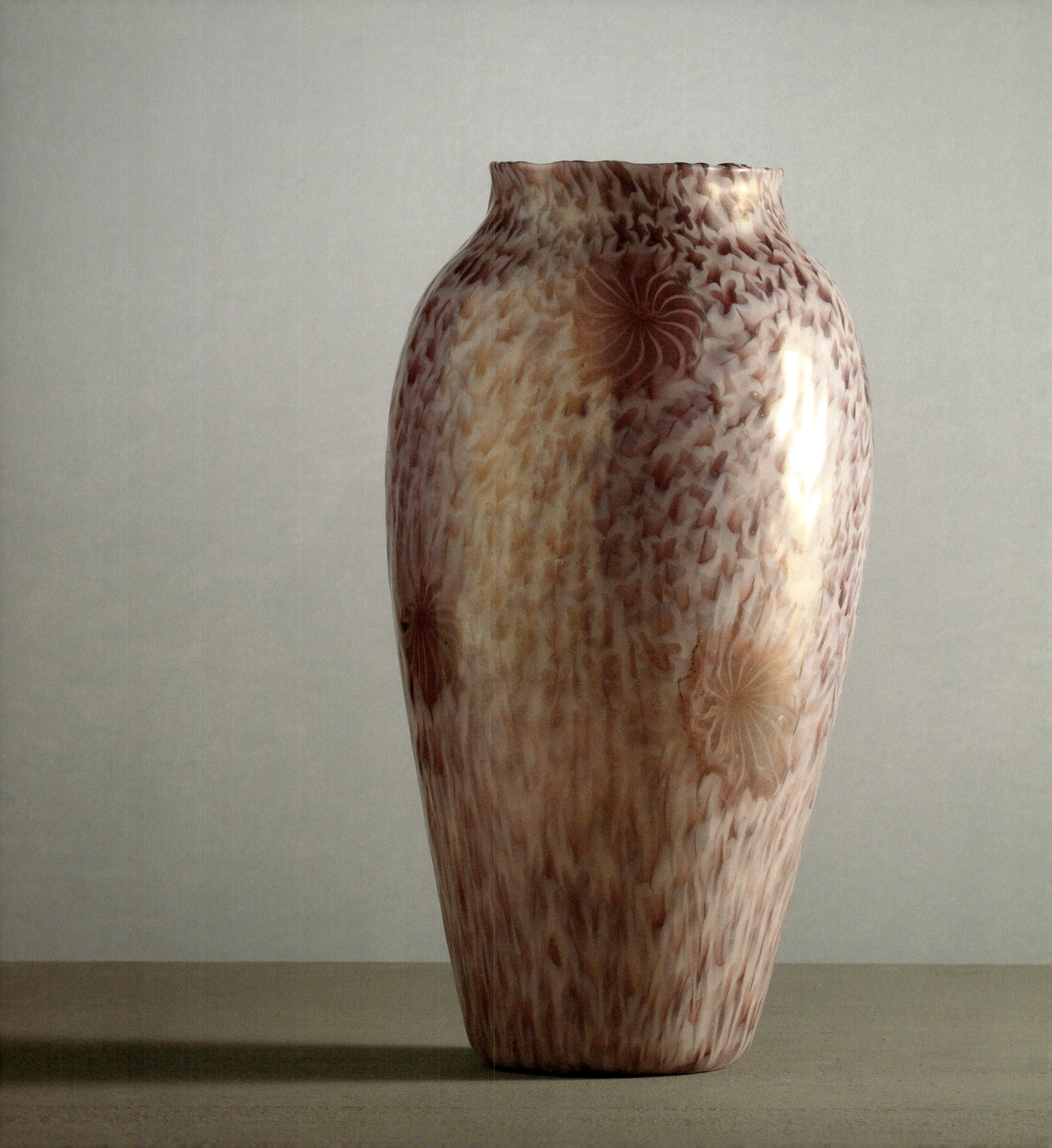

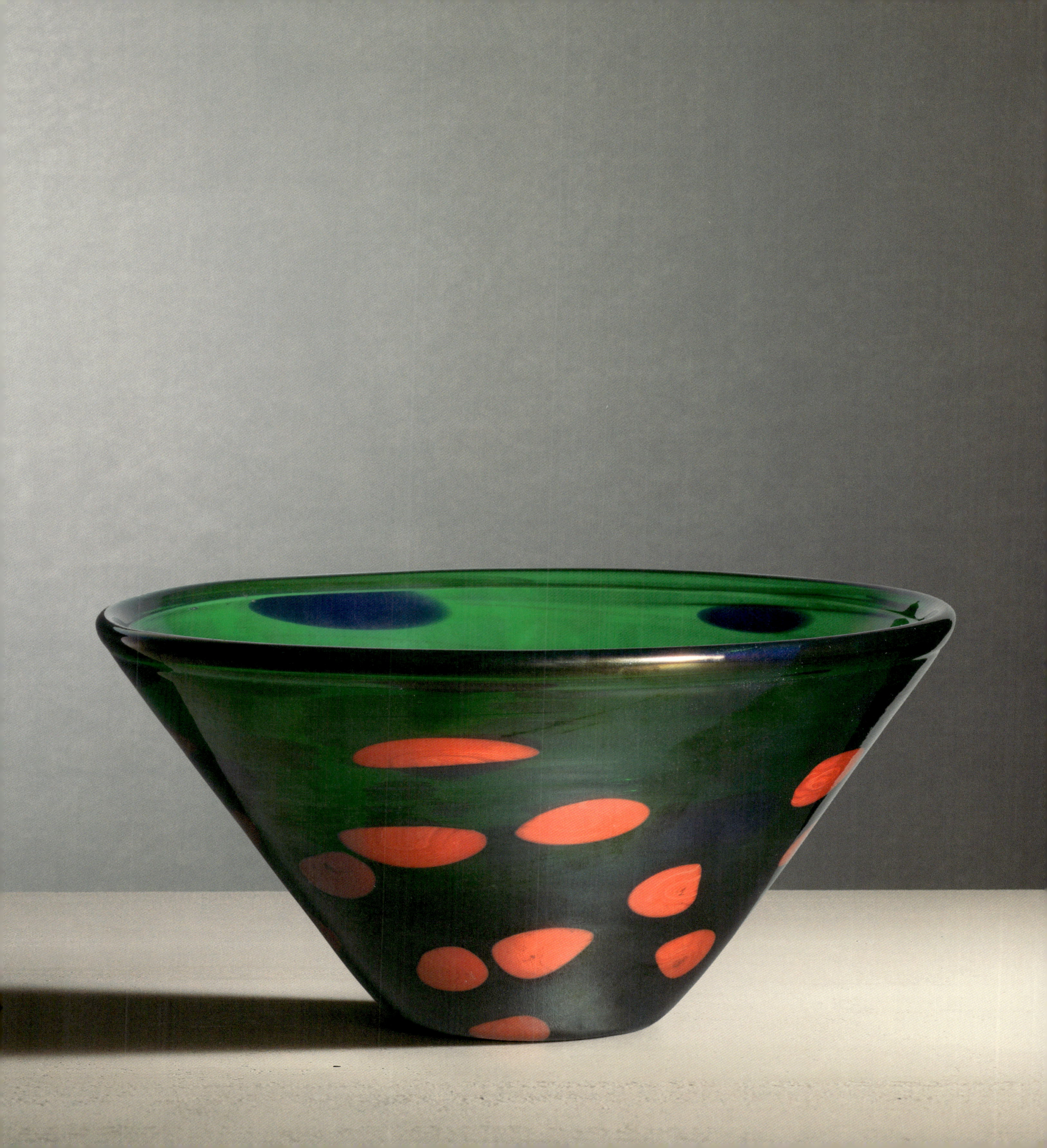

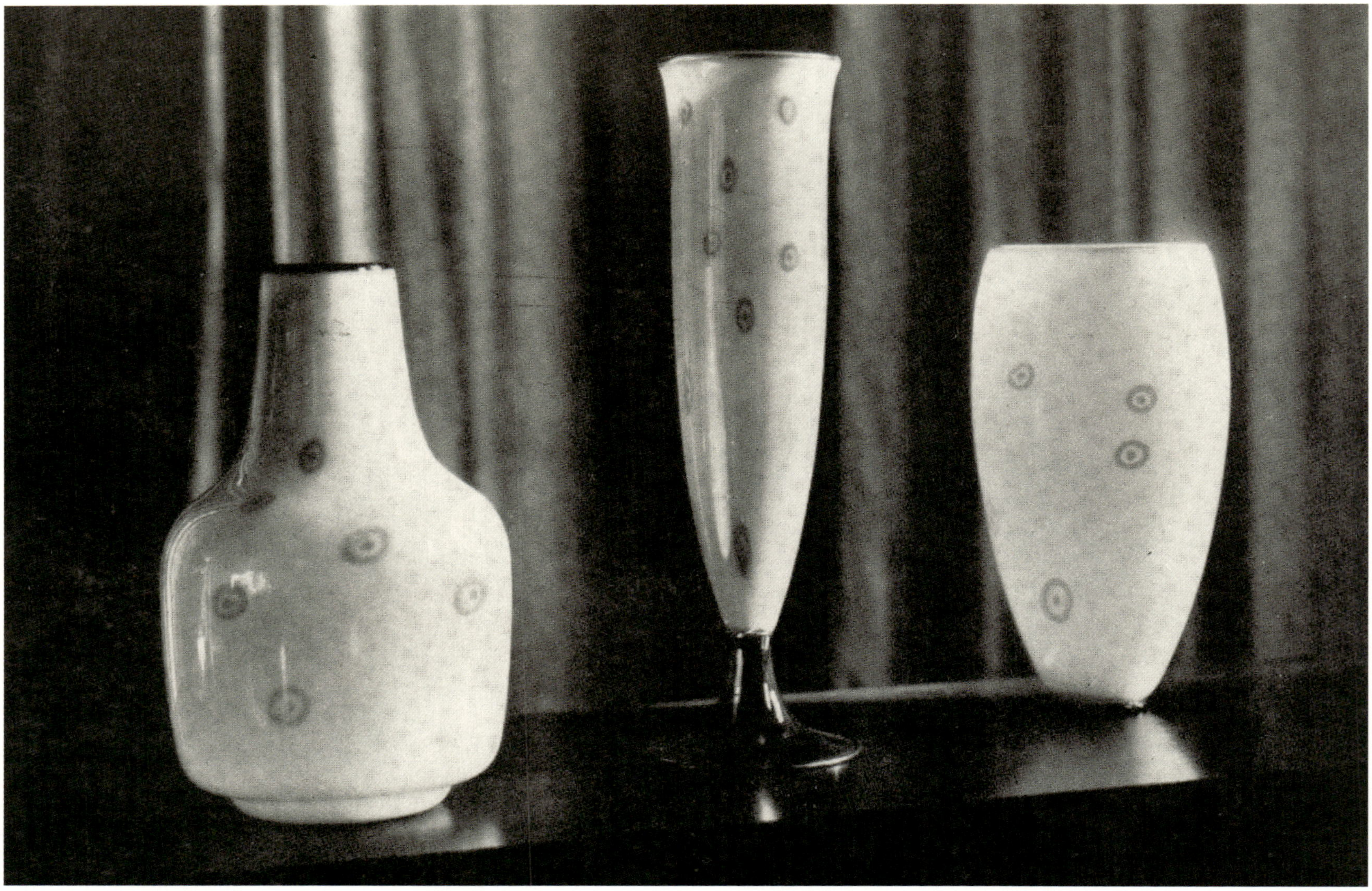

'The new vases in opaque glass with touches of colour by Giacomo Cappellin', in Domus, *May 1931, p. 26*

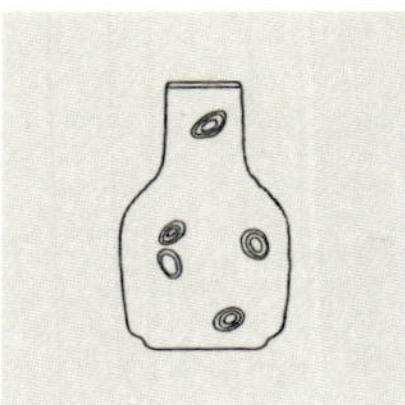

5996

Vase in *lattimo* and fumé cased glass decorated with amethyst and crystal concentric circles of murrine, included in the wall of the piece. Dark amethyst glass thread around the mouth. Surface with iridised finish. A similar example was exhibited in 1931 at the Amsterdam Exhibition of glass, ceramics and lace.
c. 1930
32 cm high
M.V.M. Cappellin Murano *acid stamp*
Private collection

exhibitions
1931, Amsterdam, Exhibition of glass, ceramics and lace

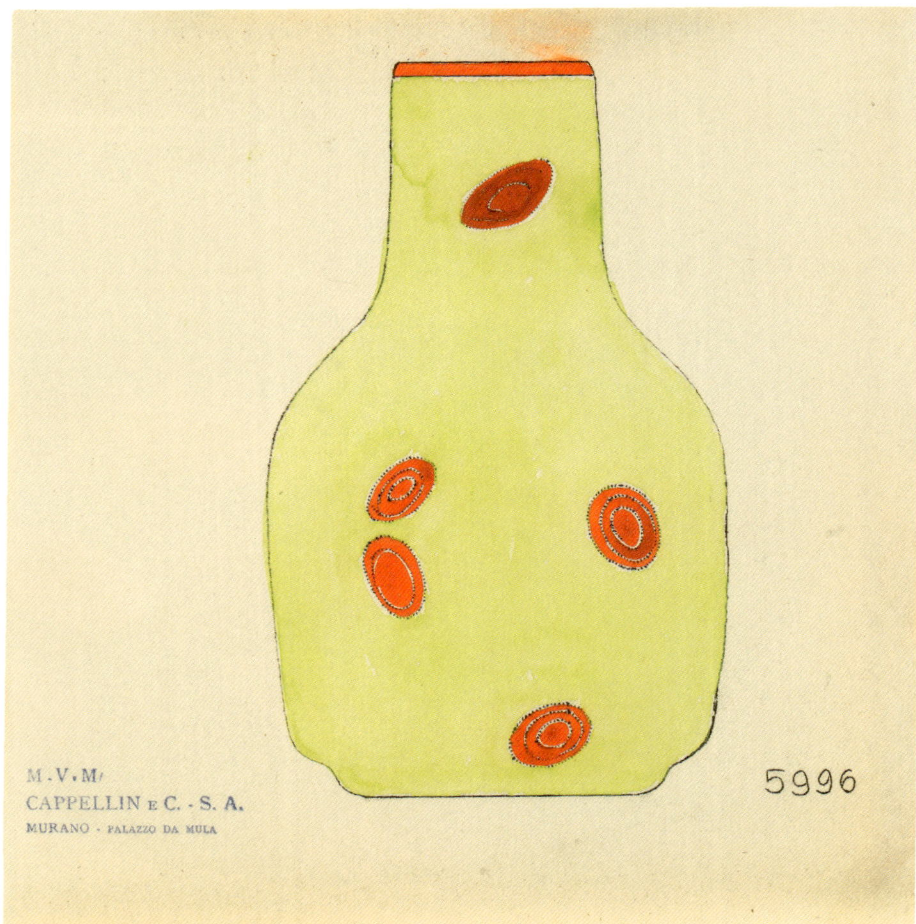

5996, catalogue drawing, AMA

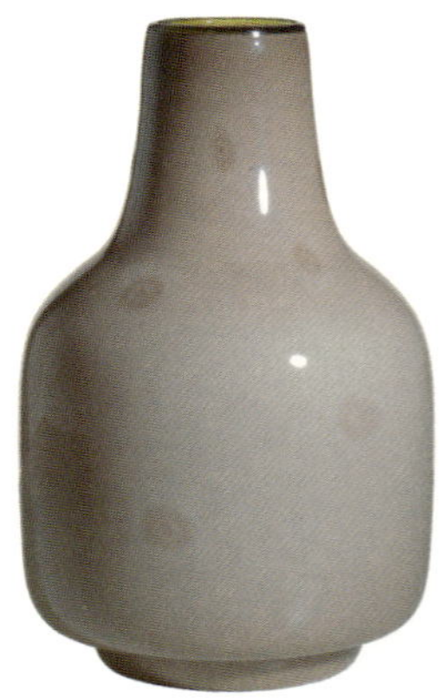

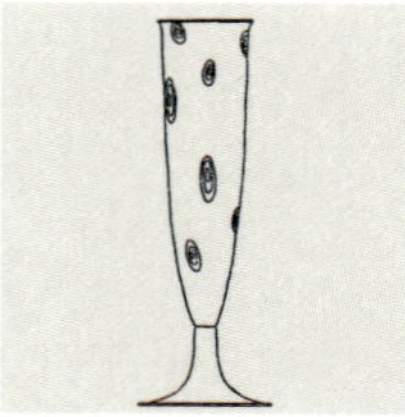

5987

Catalogue drawing for a large goblet-shaped vase, of the *millefiori* series, which was displayed in 1931 at the Amsterdam Exhibition of glass, ceramics and lace, AMA. As documented by period photographs, on this occasion the flared foot was in transparent glass.
c. 1930
45 cm high

exhibition
1931, Amsterdam, Exhibition of glass, ceramics and lace

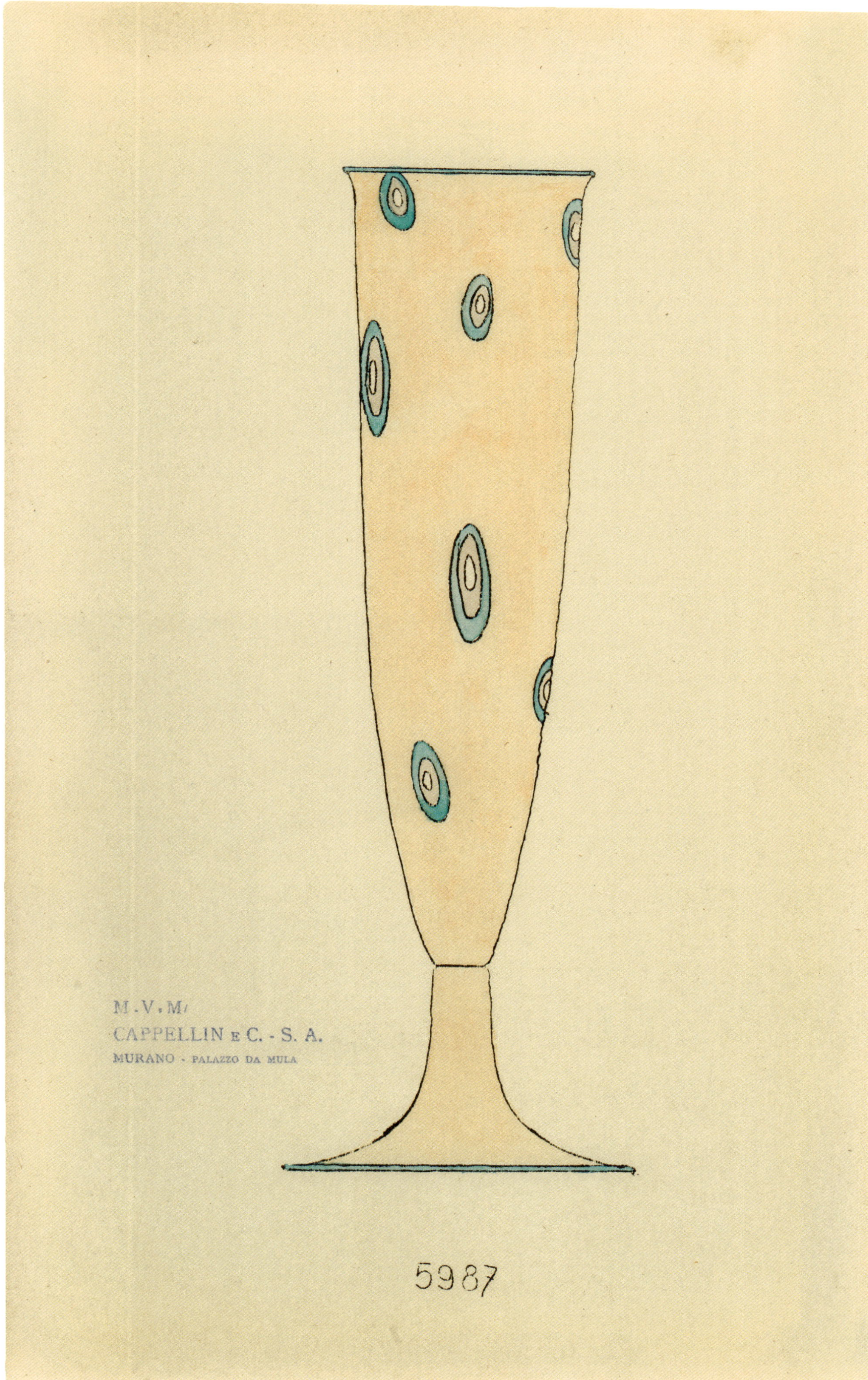

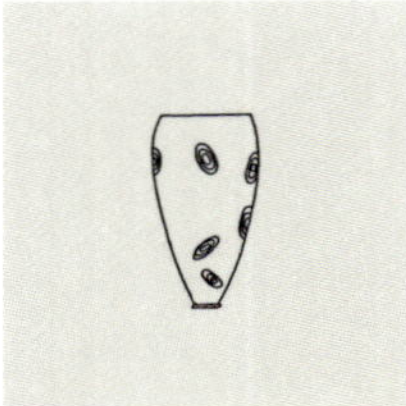
6004

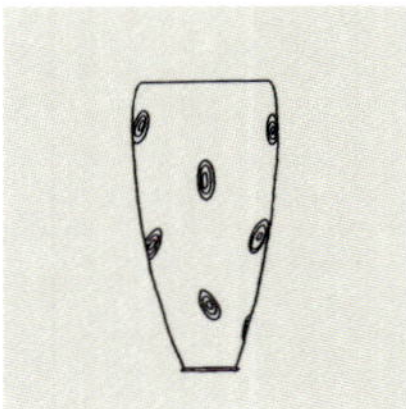
6005

Vase in *lattimo* cased and light green glass decorated with concentric circles of green and crystal murrine, included in the wall of the piece. Foot and thread around the mouth in green glass. Surface with iridised finish.
c. 1930
22.5 cm high
M.V.M. Cappellin Murano *acid stamp*
Private collection, Udine

exhibitions
1931, Amsterdam, Exhibition of glass, ceramics and lace

Vase in *lattimo* cased and light-blue glass decorated with concentric circles of blue and crystal murrine included in the wall of the piece. Foot and thread around the mouth in blue glass. Surface with iridised finish. A similar example was exhibited in 1930 at the 4th Triennale of Monza and the following year at the Amsterdam Exhibition of glass, ceramics and lace.
c. 1930
32.5 cm high
M.V.M. Cappellin Murano *acid stamp*
The Steinberg Foundation collection/ The Corning Museum of Glass, Corning, NY

exhibitions
1930, Monza, 4th Triennale
1931, Amsterdam, Exhibition of glass, ceramics and lace

>
6004–6005, catalogue drawing, AMA

>
Vase no. 6005 in room 78: 'area for private bar use' (architect Serafini), Furnishings Gallery at the 4th Triennale of Monza, 1930, detail

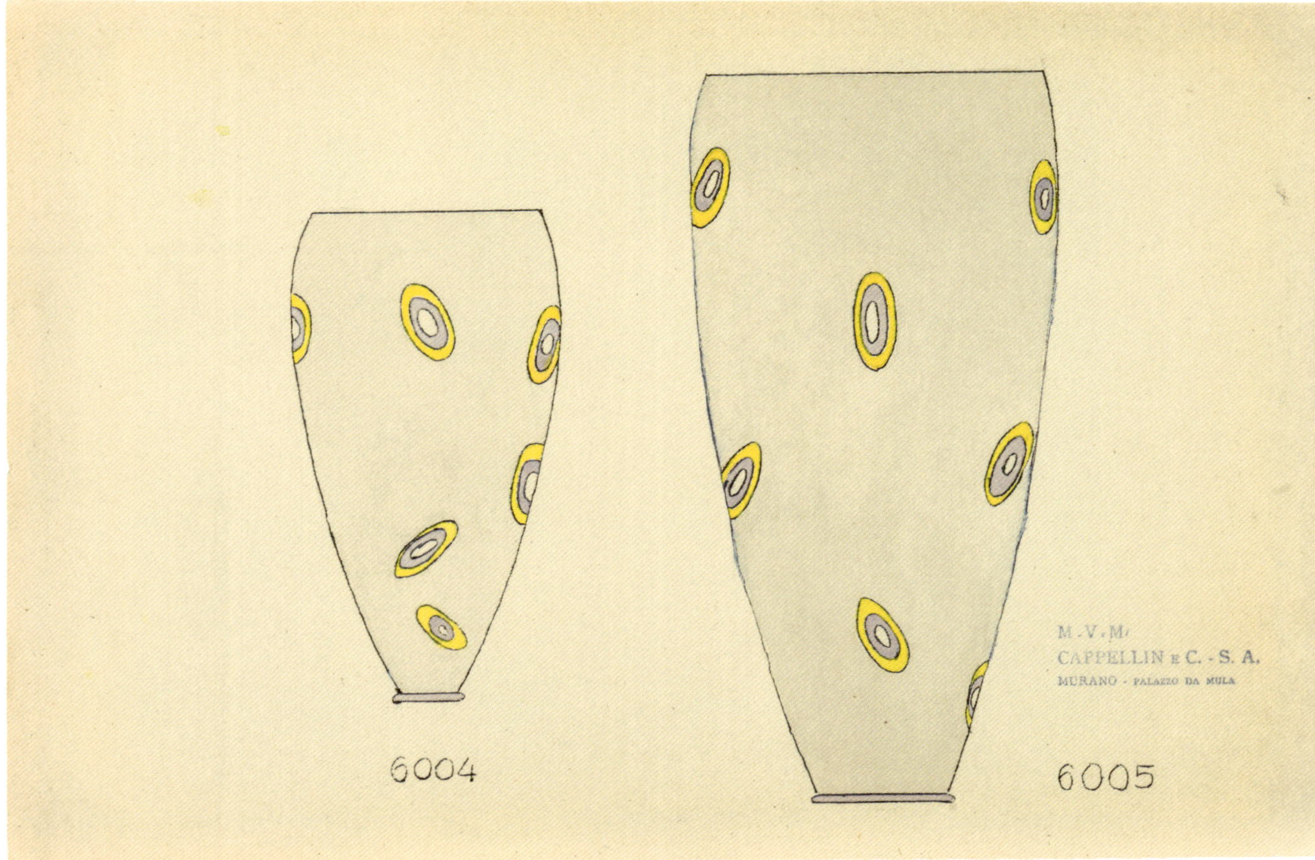

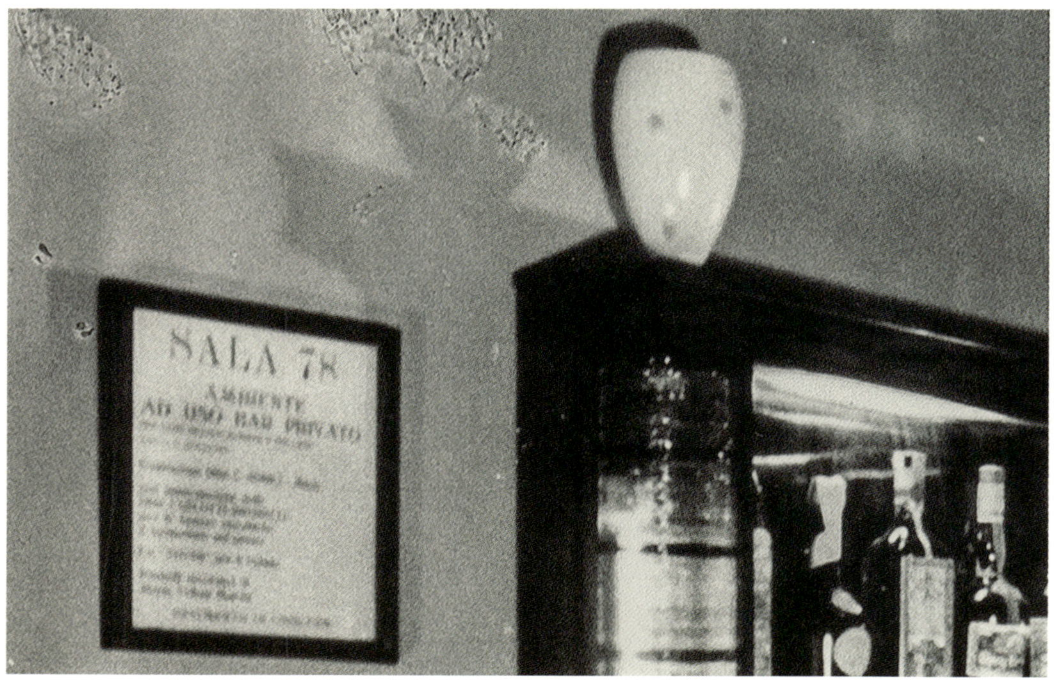

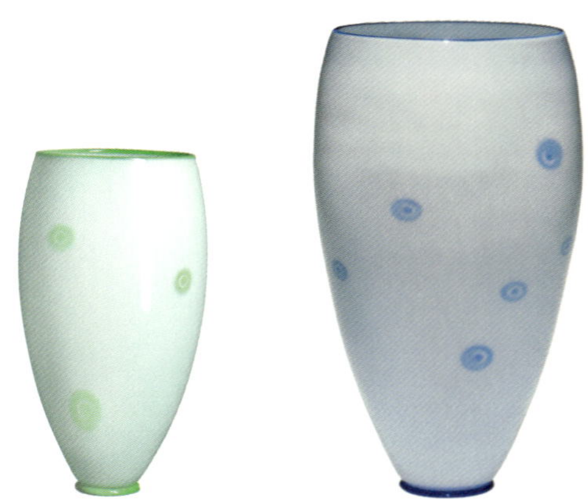

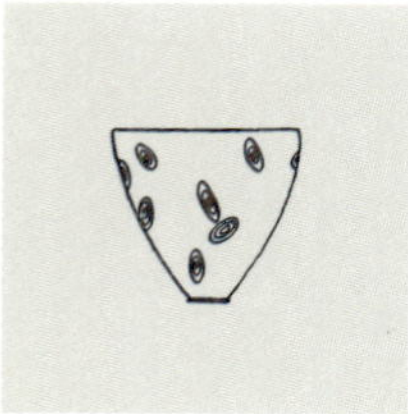
6006

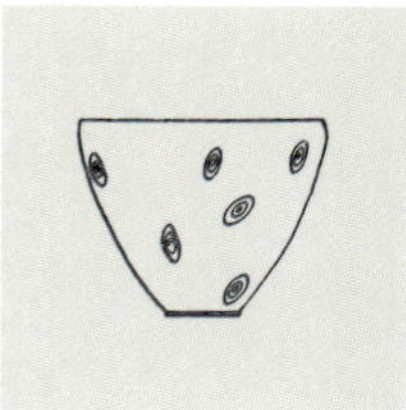
6007

Vase in *lattimo* cased and crystal glass decorated with concentric circles of red and crystal murrine included in the wall of the piece. Foot and thread around the mouth in red glass. Surface with iridised finish. The model was also proposed in larger dimensions, no. 6007 (h 23 cm). Both versions were on display in 1931 at the Amsterdam Exhibition of glass, ceramics and lace.
c. 1930
20 cm high
Private collection

exhibitions
1931, Amsterdam, Exhibition of glass, ceramics and lace

>
6006–6007, catalogue drawings, AMA

>
Vases of the millefiori series at the Amsterdam Exhibition of glass, ceramics and lace, 1931, detail, AHL

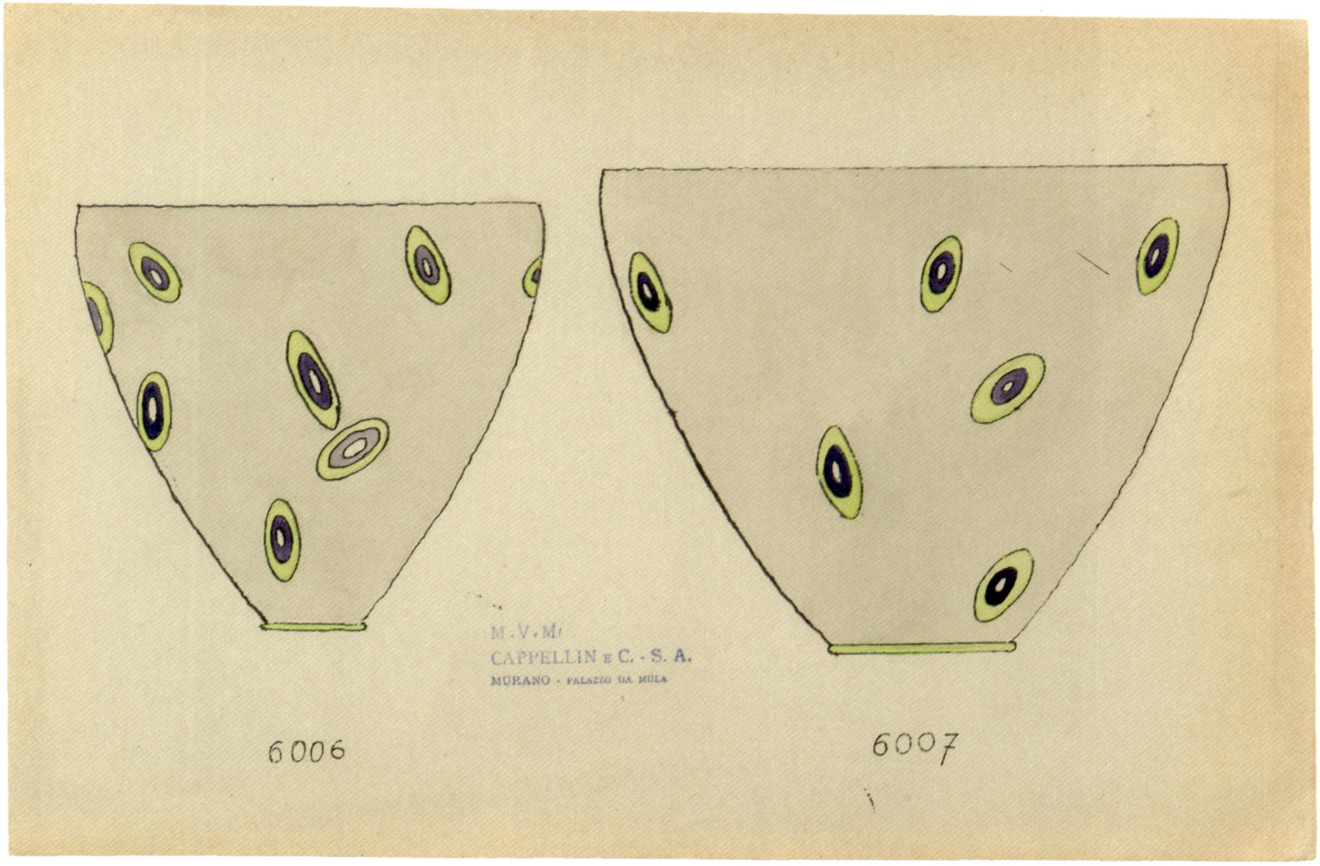

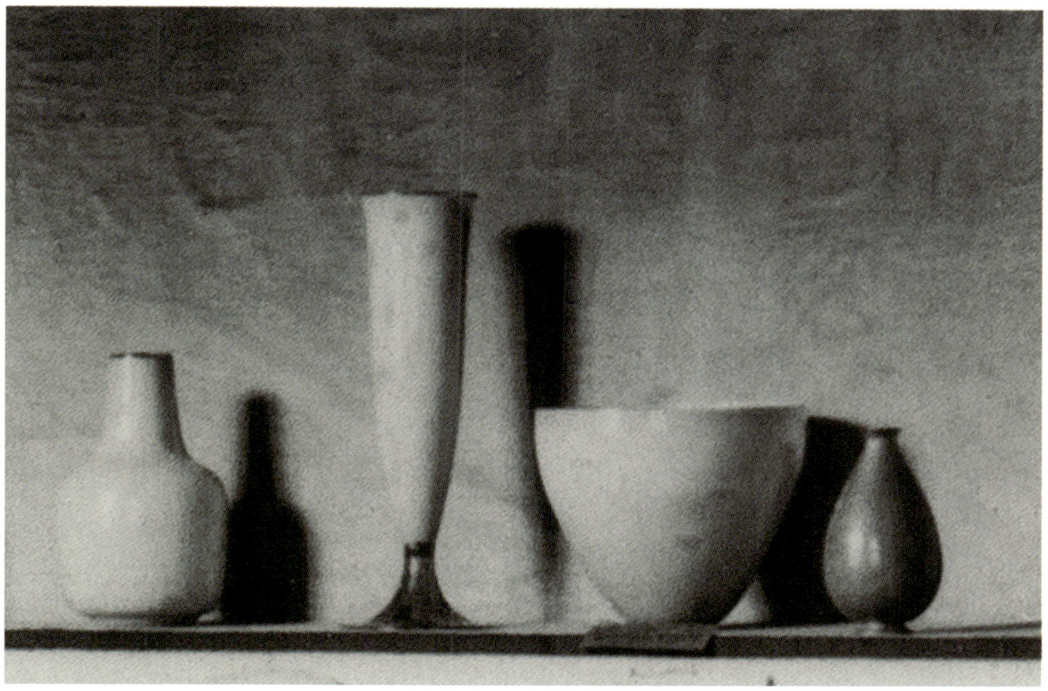

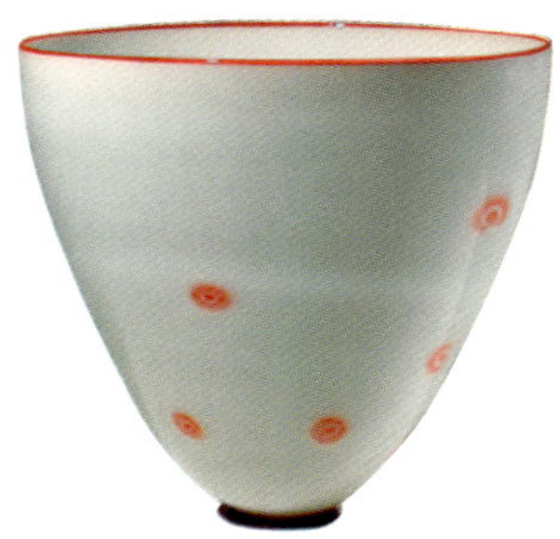

> Cappellin glass pieces at the Amsterdam Exhibition of glass, ceramics and lace, 1931, detail, AHL

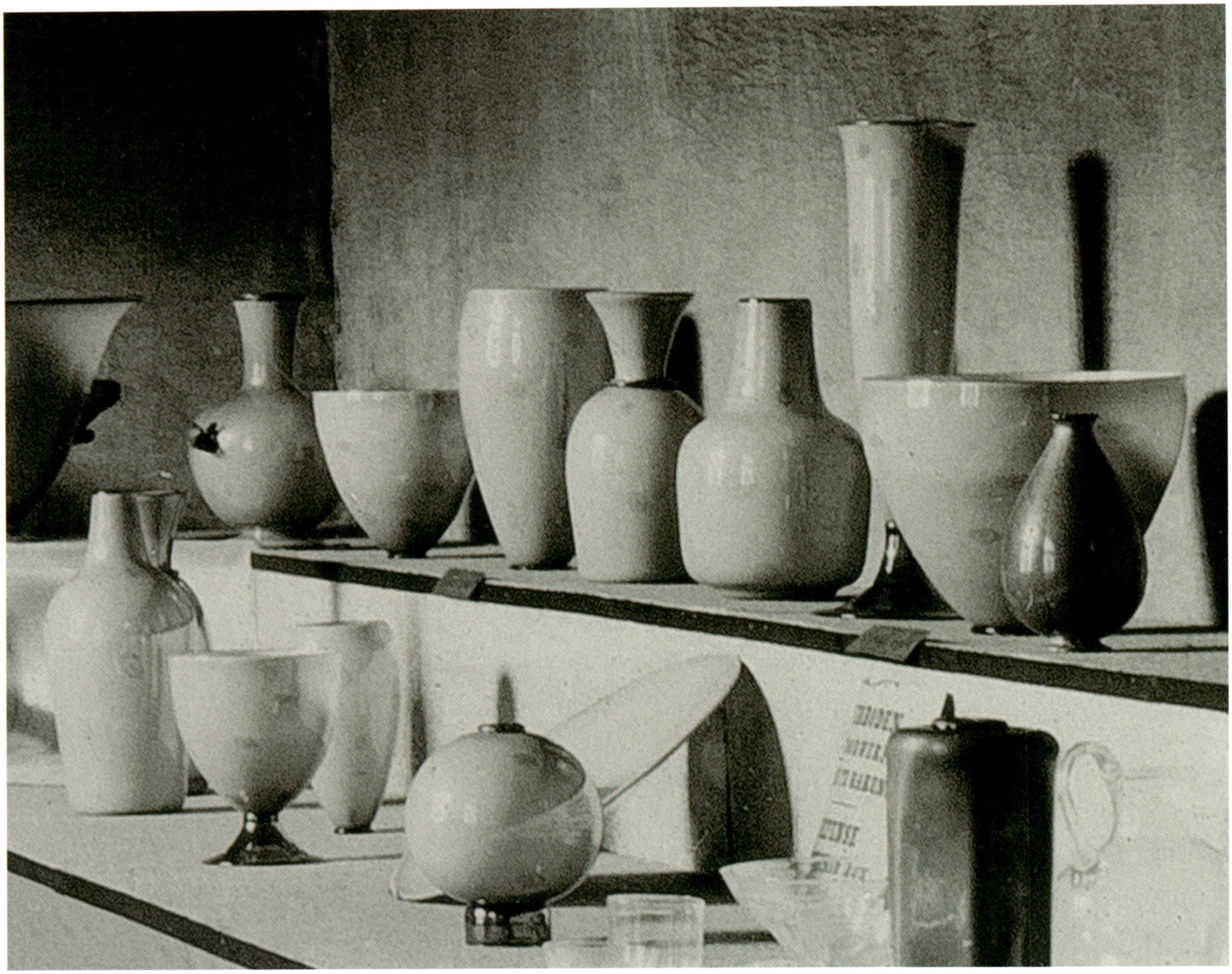

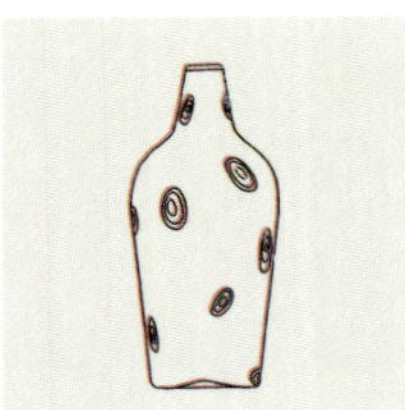

5983

Bottle in *lattimo* cased and light-blue glass decorated with concentric circles of red and crystal murrine included in the wall of the piece. Foot and thread around the mouth in red glass. Surface with iridised finish. A similar example was exhibited in 1931 at the Amsterdam Exhibition of glass, ceramics and lace.
c. 1930
36 cm high
M.V.M. Cappellin Murano *acid stamp*
Private collection

exhibitions
1931, Amsterdam, Exhibition of glass, ceramics and lace.

> *5983, catalogue drawing, AMB*

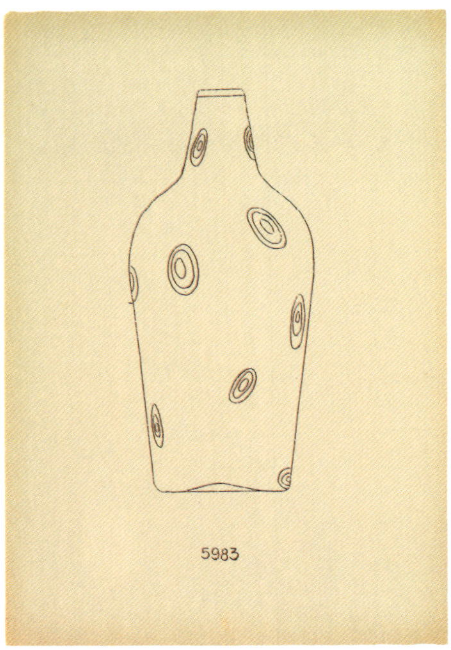

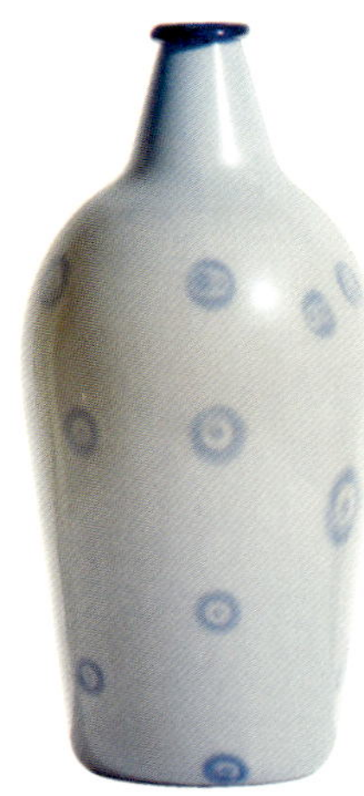

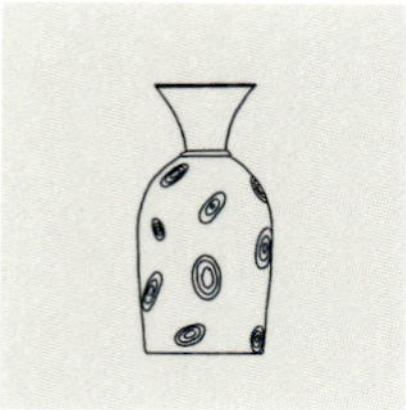

5984

Catalogue drawing for a large vase with flared neck of the *millefiori* series, which was displayed in 1931 at the Amsterdam Exhibition of glass, ceramics and lace, AMA.
c. 1930
32 cm high

exhibitions
1931, Amsterdam, Exhibition of glass, ceramics and lace

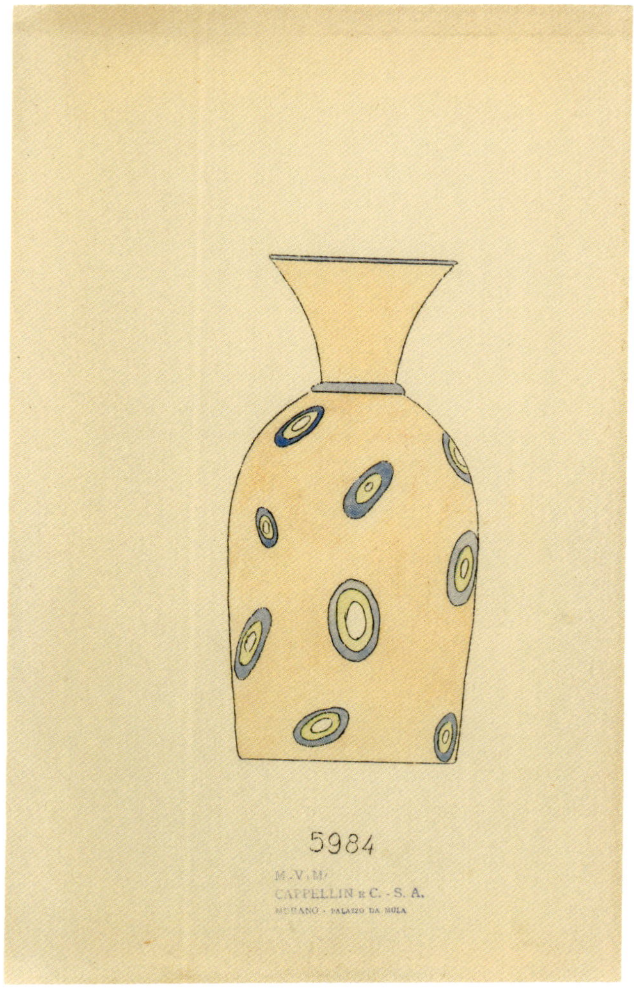

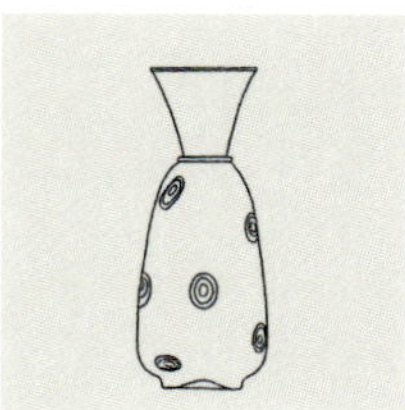

5985

Vase in *lattimo* and fumé cased glass decorated with concentric circles of amethyst and crystal murrine included in the wall of the piece. Collar at the base of the neck and thread around the mouth in black glass. Surface with iridised finish.
c. 1930
39 cm high
M.V.M. Cappellin Murano *acid stamp*
Private collection, Treviso

>
5985, catalogue drawing, AMA

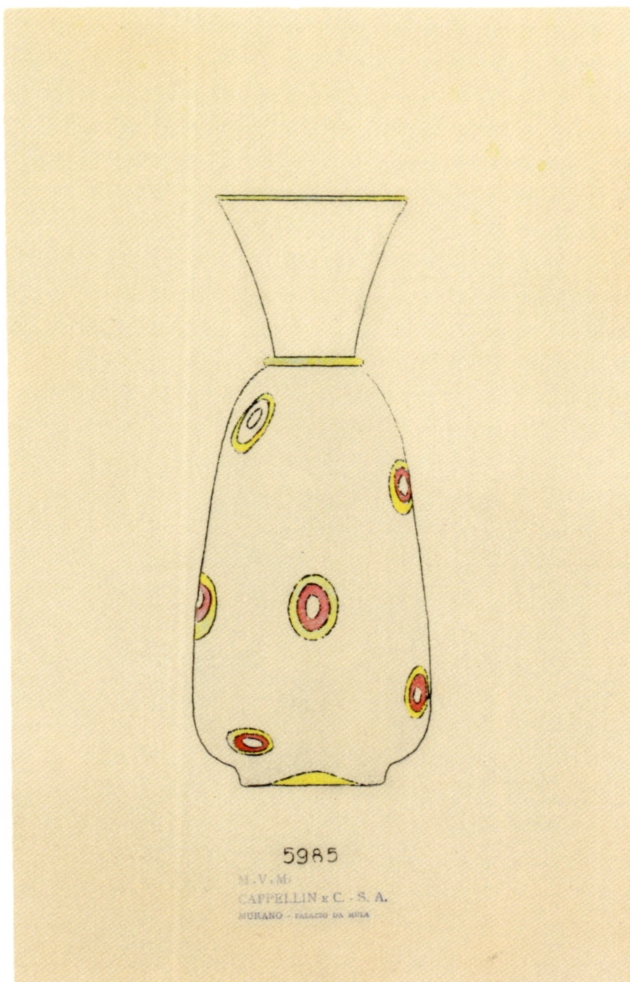

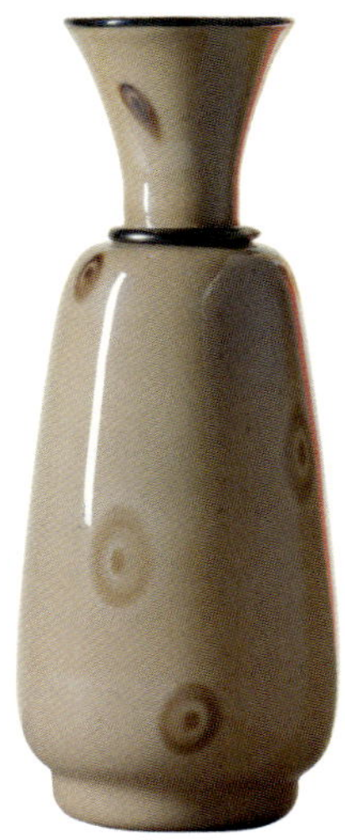

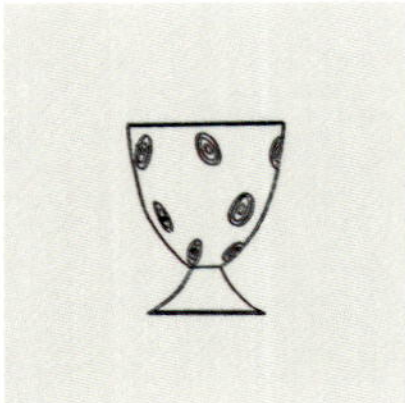

5986

Catalogue drawing for a vase with flared foot, of the *millefiori* series, displayed in 1931 at the Amsterdam Exhibition of glass, ceramics and lace, AMA. As documented by period photographs, on this occasion the foot was in transparent glass.
c. 1930
23 cm high

exhibitions
1931, Amsterdam, Exhibition of glass, ceramics and lace

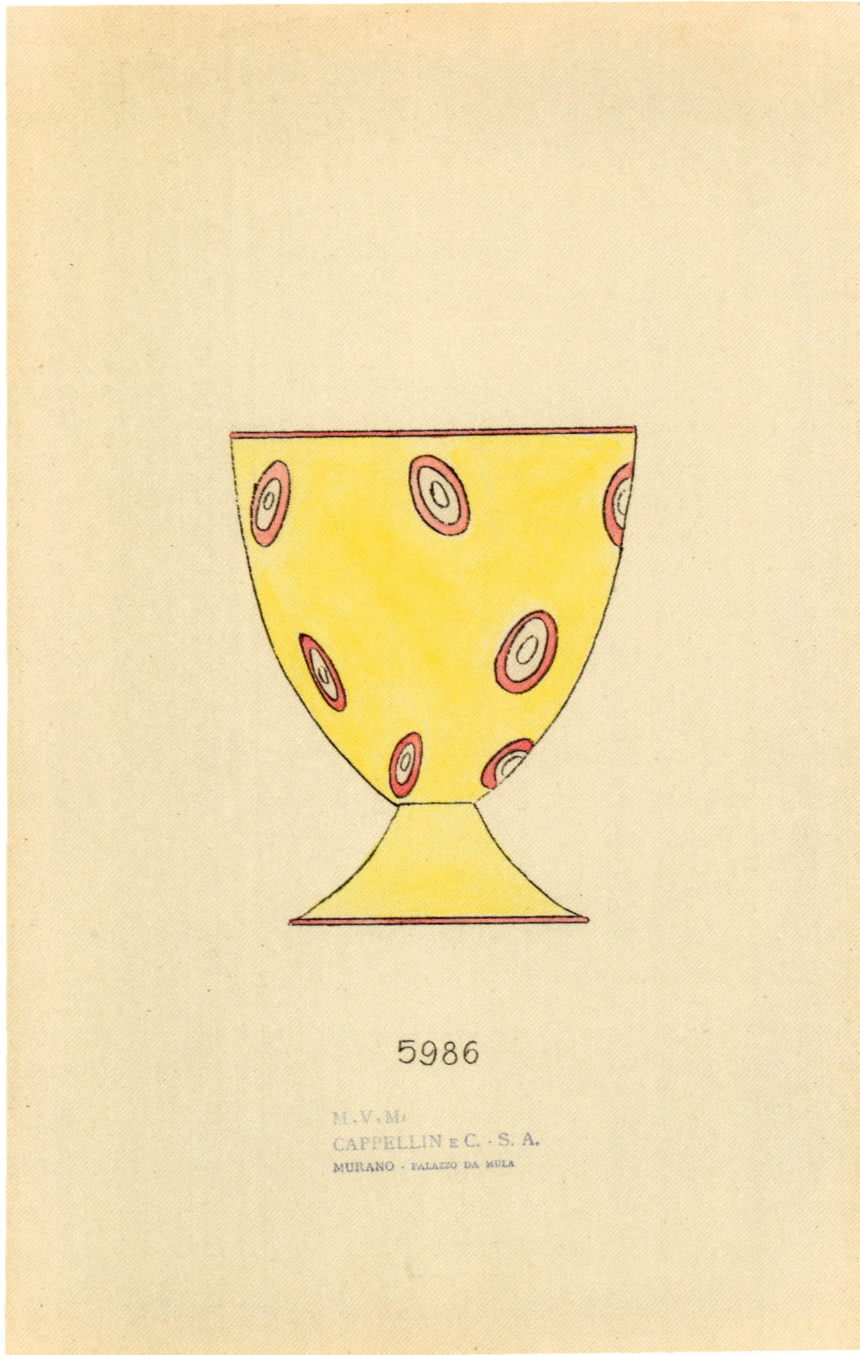

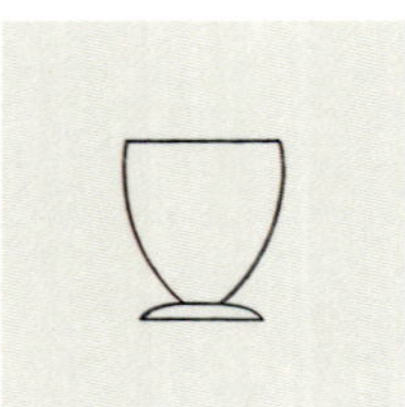

5821

Vase with skull-cap foot in *lattimo* cased and dark-blue glass decorated with concentric circles of dark-blue and crystal murrine included in the wall of the piece. Surface with iridised finish. Model no. 5821 was also produced in other glass techniques (e.g. transparent).
c. 1930
21 cm high
M.V.M. Cappellin Murano *acid stamp*
The Olnick Spanu Collection, New York

>
5821, catalogue drawing, AMB

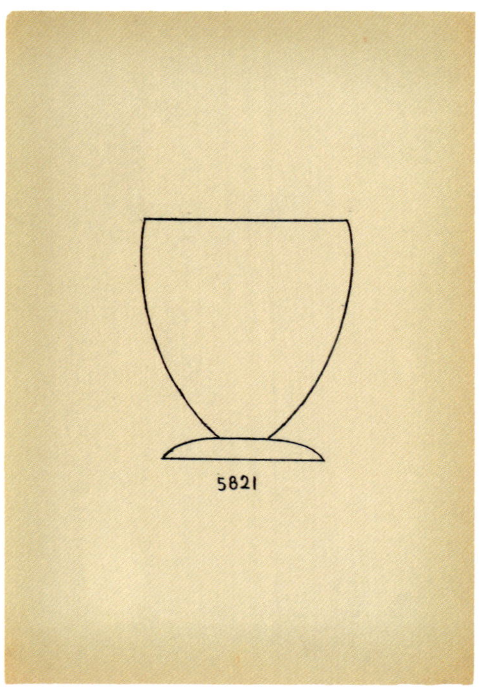

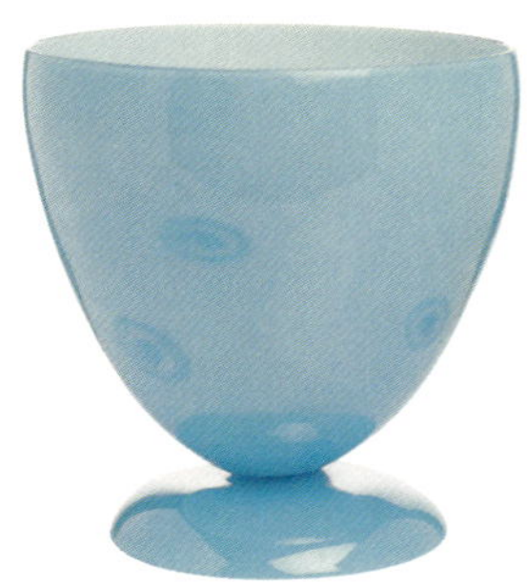

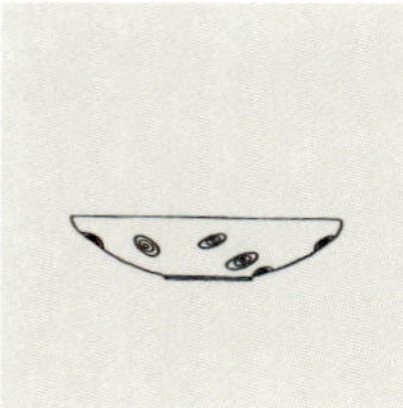

6002

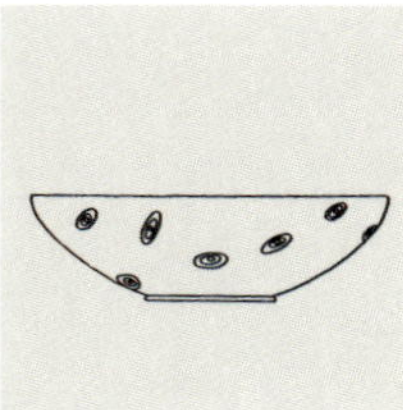

6003

Large bowl in *lattimo* cased and light-blue glass decorated with concentric circles of dark-blue and crystal murrine included in the wall of the piece. Foot and thread around the mouth in dark-blue glass. Surface with iridised finish. The model was also proposed in the catalogue in smaller dimensions (no. 6002, Ø 34 cm).
c. 1930
12 cm high; 43.5 cm ø
M.V.M. Cappellin Murano *acid stamp*
Private collection, Treviso

>
6002–6003, catalogue drawings, AMA

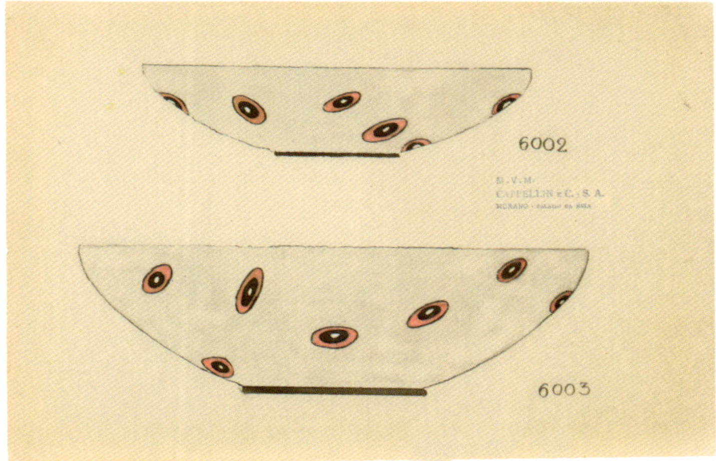

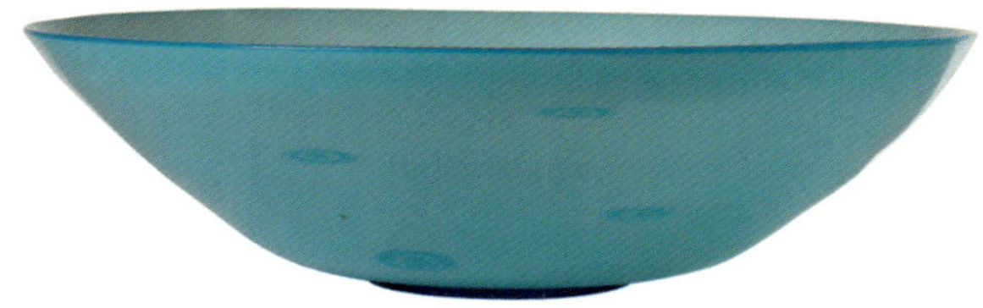

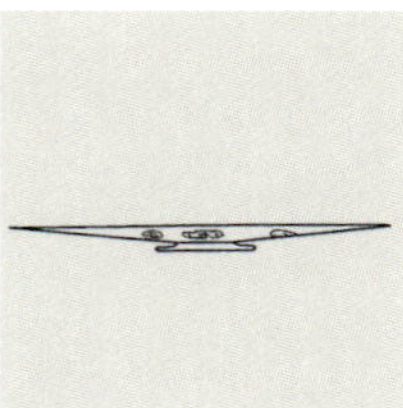

5998

Large cake stand in *lattimo* cased and *pagliesco* glass decorated inside with concentric circles of amethyst and crystal murrine included in the wall of the piece. Surface with iridised finish. Flared foot in *lattimo* cased and *pagliesco* glass.
c. 1930
47 cm ø
M.V.M. Cappellin Murano *acid stamp*
Museo Venini collection

exhibitions
1931, Amsterdam, Exhibition of glass, ceramics and lace

>
5998, catalogue drawing, AMB

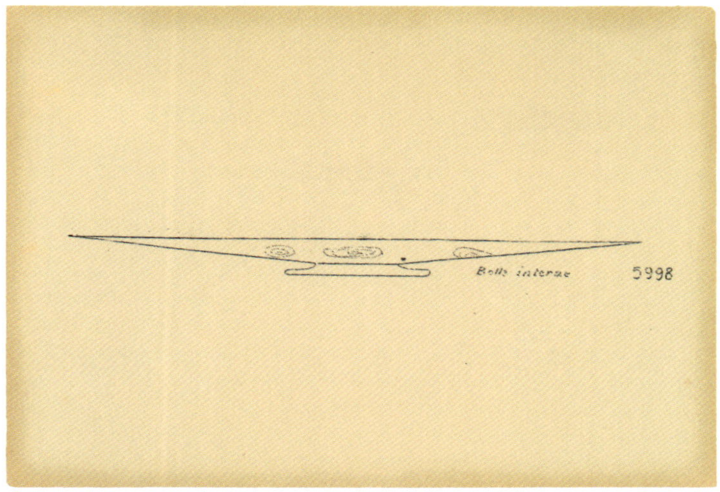

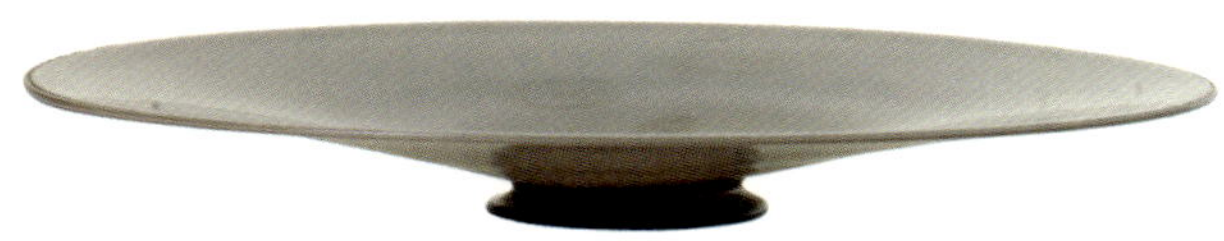

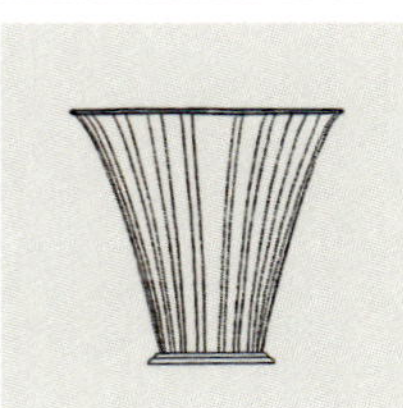

5989

Vase in *lattimo* cased and light-amethyst glass decorated with concentric circles of amethyst and crystal murrine included in the wall of the piece. Foot and thread around the mouth in dark-amethyst glass. Surface with iridised finish. The piece was produced after a model from the vertical band series.
c. 1930
29 cm high
Private collection

>
5989, catalogue drawing, AMB

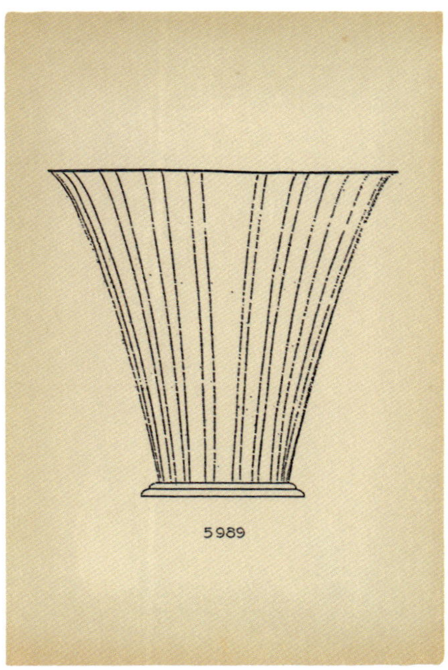

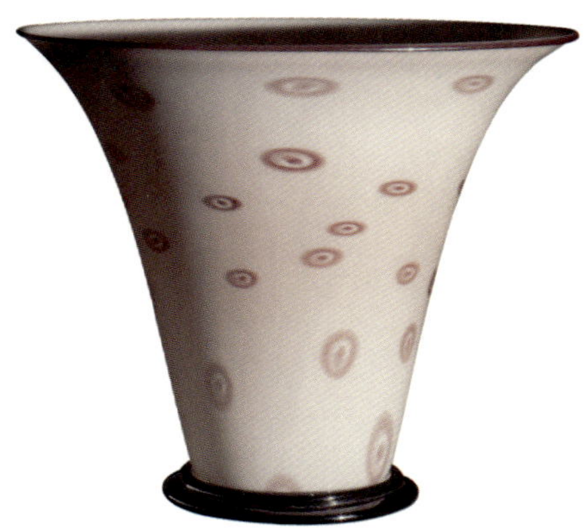

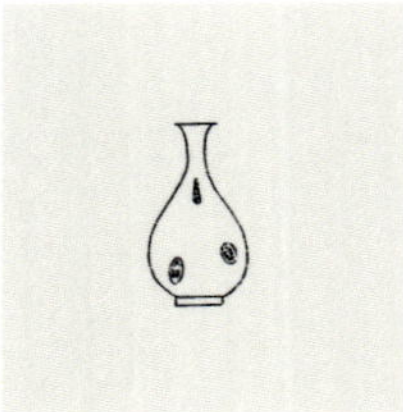

6014

Catalogue drawing for a vase in opaque glass decorated with murrine having a characteristic dotted design, AMA. The drawing refers back to similar models in Chinese porcelain. The model is presented in the catalogue together with no. 6013, of a similar inspiration.
c. 1930
21 cm high

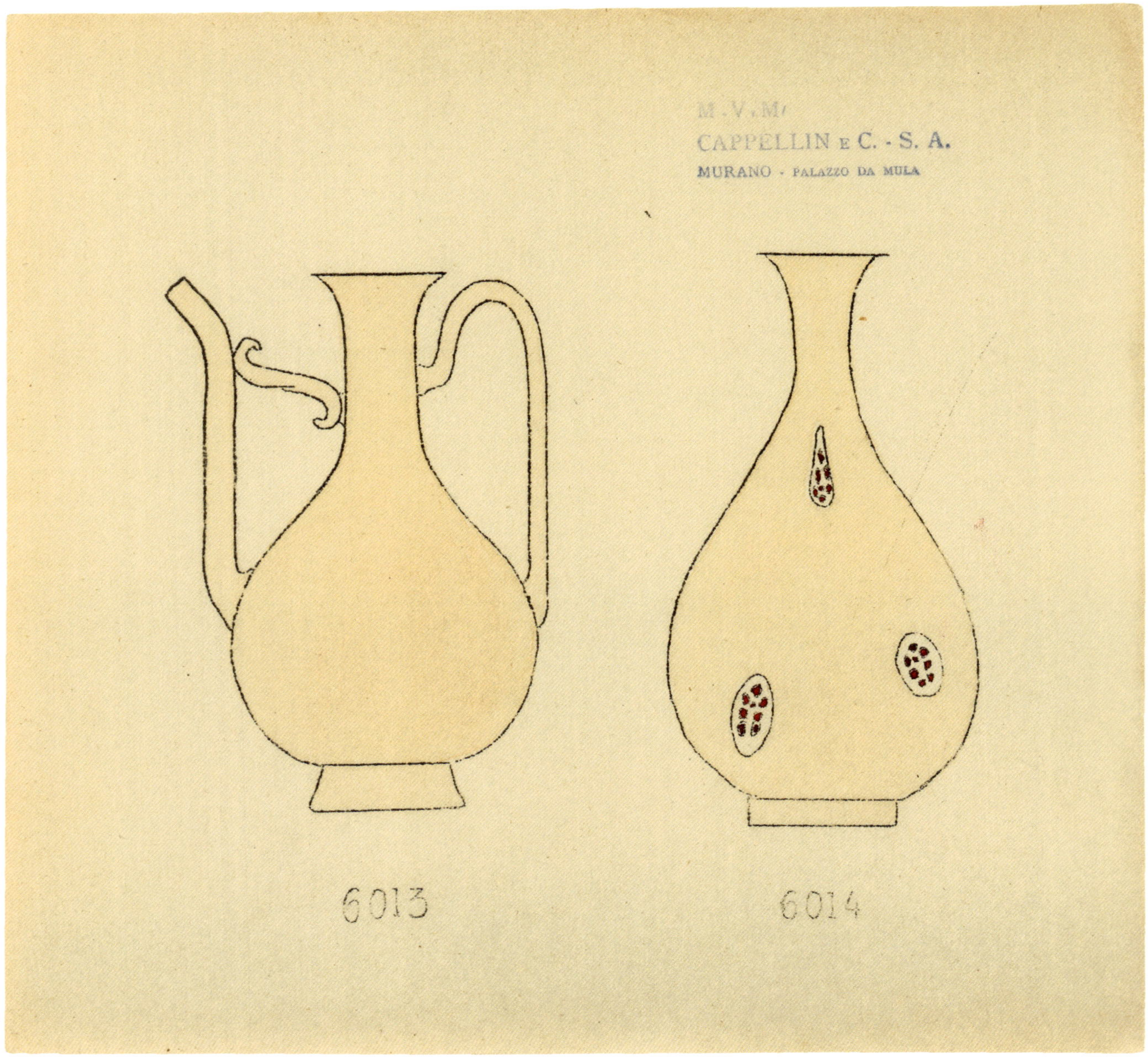

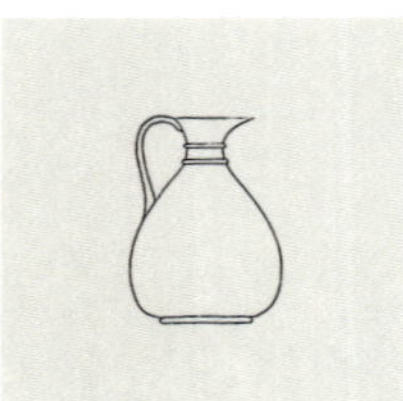

5536

Jug in *lattimo* cased and crystal glass with amethyst *mezzaluna* murrine included in the wall of the piece. Handle, foot, collar and thread around the mouth in dark amethyst glass. Surface with iridised finish. The piece was produced after a model in the catalogue of the glassworks and was designed by Vittorio Zecchin (1925–26). The jug became part of the collection of the Metropolitan Museum of Art, NY, in 1931.
c. 1930
25.5 cm high
The Metropolitan Museum of Art collection, NY

›
5536, working drawing, CSV

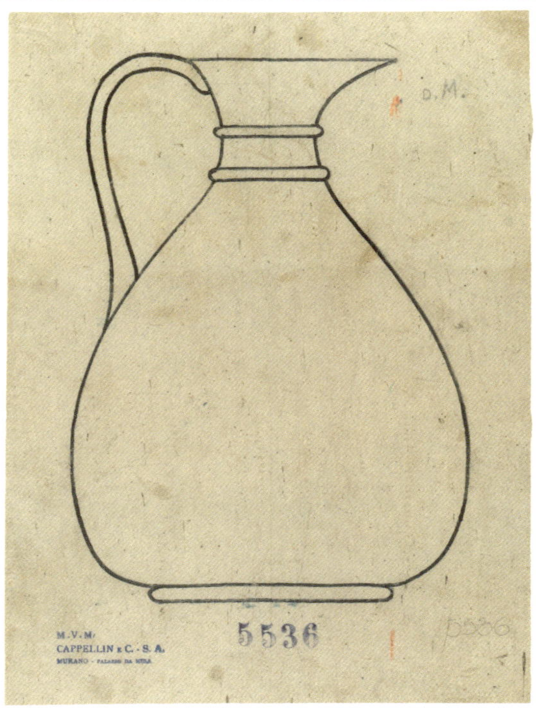

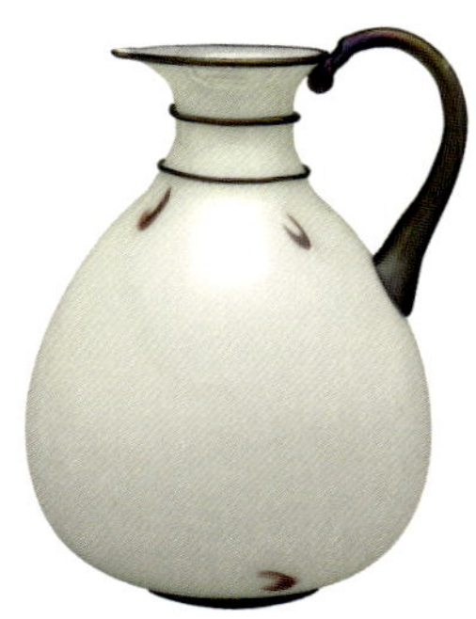

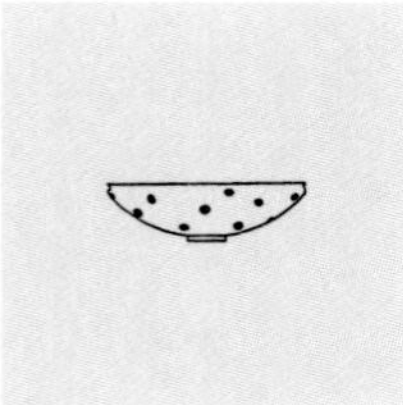

6042

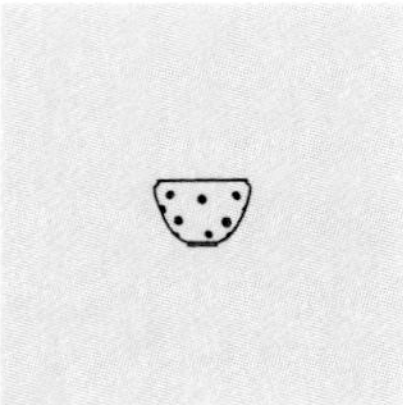

6043

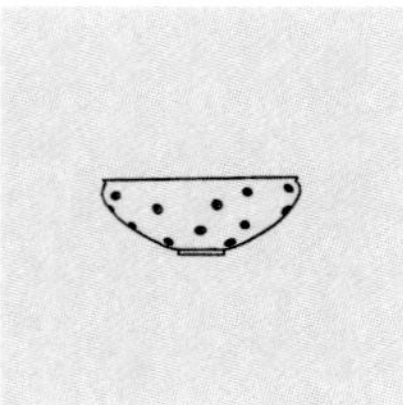

6044

Working drawing for three bowls of different shapes and dimensions in opaque glass, with small foot, decorated by circular murrine with a square nucleus, CSV.
1930–31
23; 12; 24 cm ø

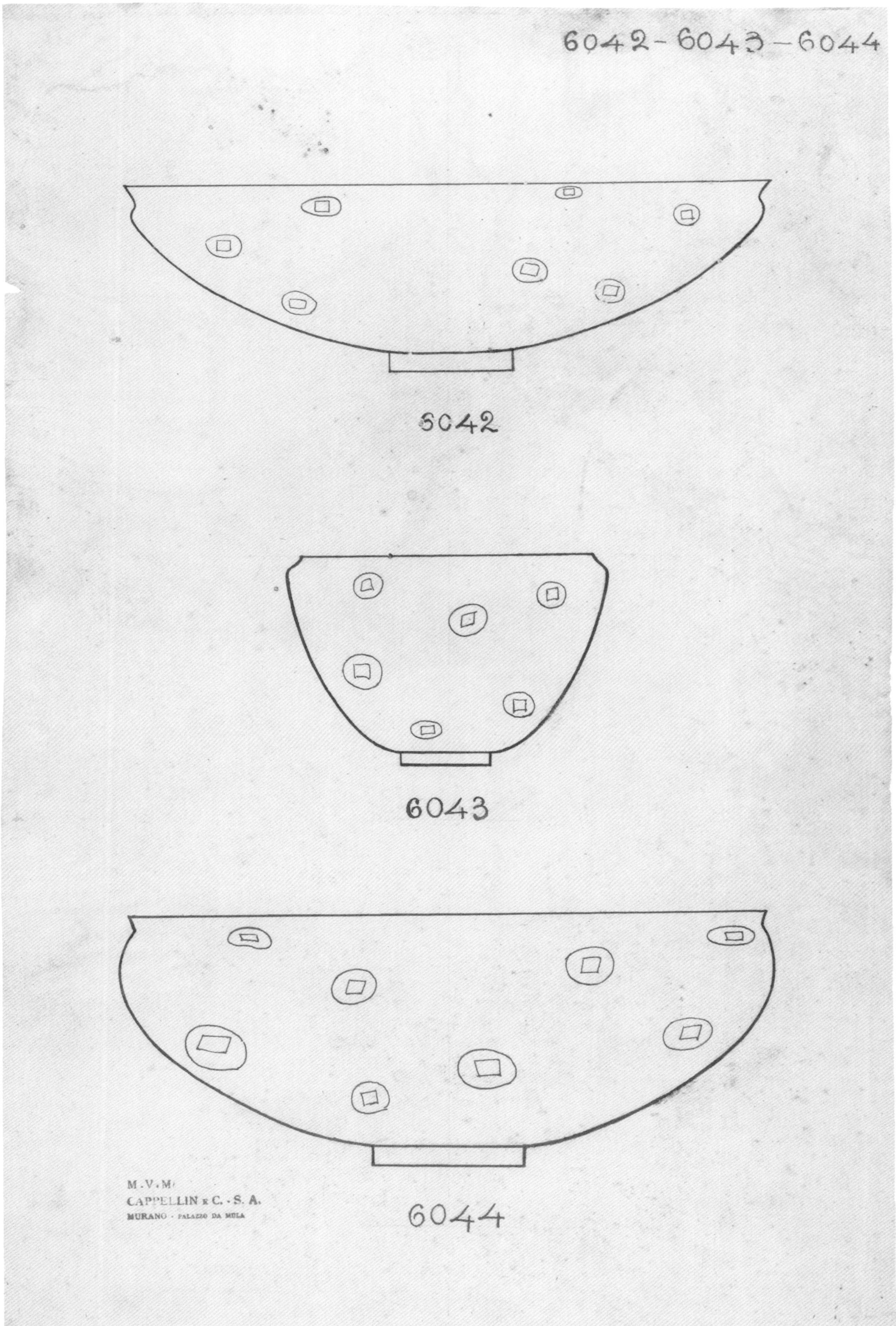

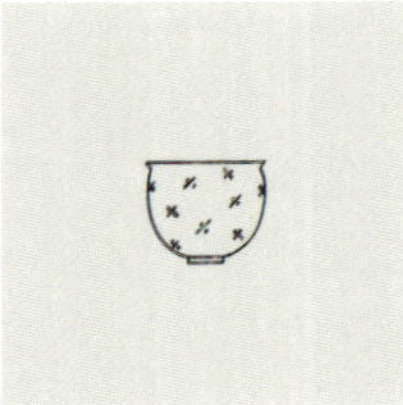

6030

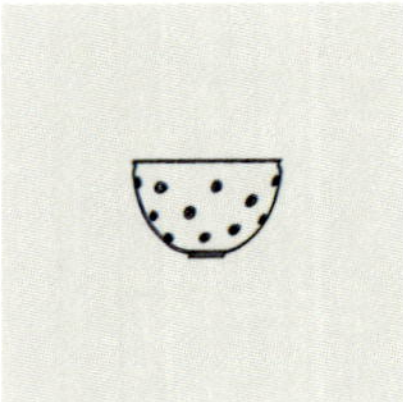

6045

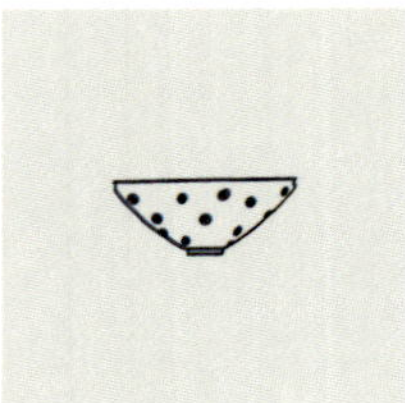

6046

Working drawing for three bowls of different shapes and dimensions and with small foot decorated as follows: no. 6030 with murrine and two dots and a dividing line, nos. 6045 and 6046 with circular murrine with a square nucleus, CSV.
1930–31
17; 18; 22 cm ø

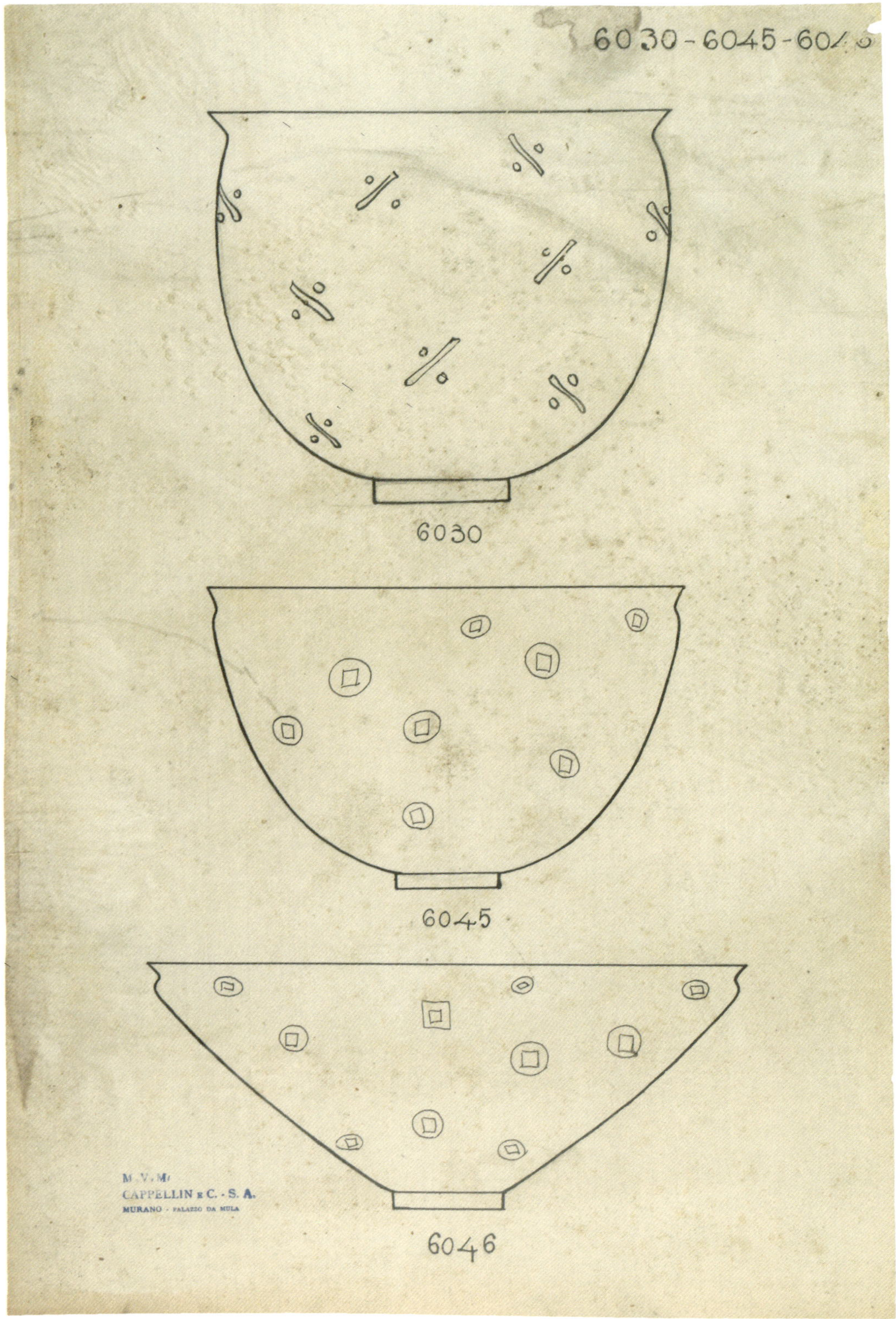

---

n.i.

Vase in transparent yellow glass decorated with pairs of *lattimo* canes included in the wall of the piece. Surface with iridised finish. Probably one of the last works of the Cappellin firm.
1930–31
18 cm high
C.V.M. *acid stamp*
Private collection

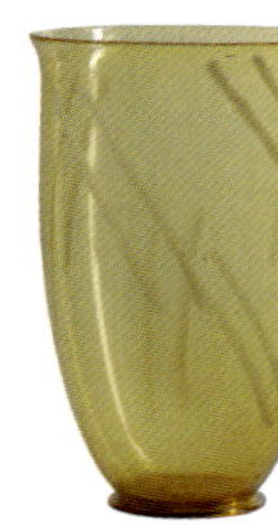

n.i.

Bowl in *lattimo* cased and blue glass decorated with sections of dark-amethyst canes included in the wall of the piece. Small foot and thread around the mouth in dark-amethyst glass. Surface with iridised finish. Probably one of the last works of the Cappellin firm.
1930–31
20.3 cm ø
The Steinberg Foundation collection

>
Carlo Scarpa, study for bowl in lattimo glass with foot in yellow cased glass, body decorated with green and yellow murrine, 1930–31, AMA

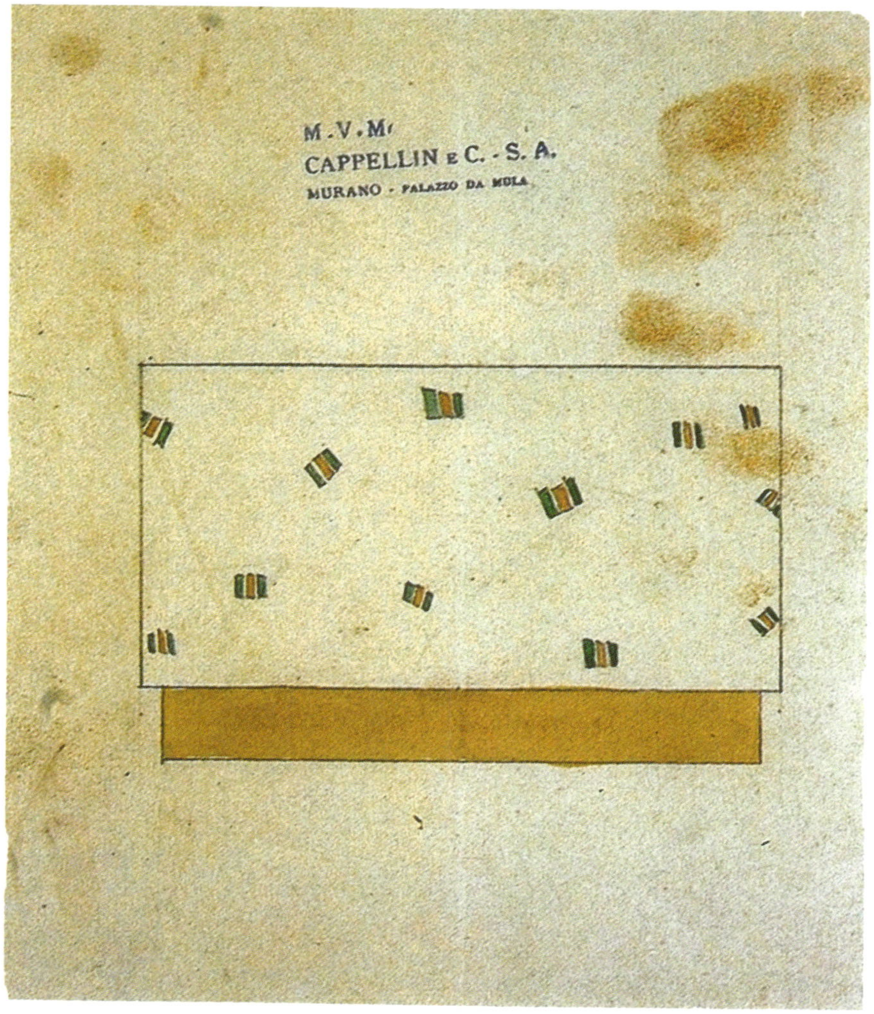

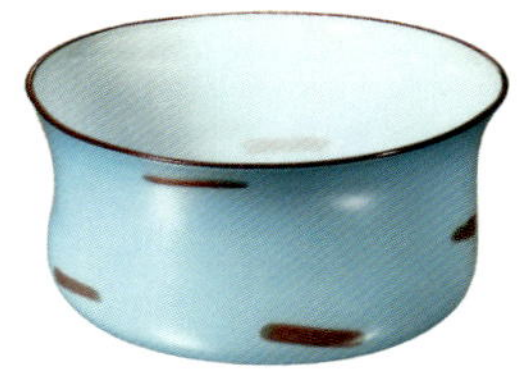

n.i.

Vase in *lattimo* cased and blue glass decorated with sections of dark-amethyst canes included in the wall of the piece. Small foot and thread around the mouth in ruby *pasta vitrea*. Surface with iridised finish. Probably one of the last works of the Cappellin firm.
1930–31
30 cm high
Private collection

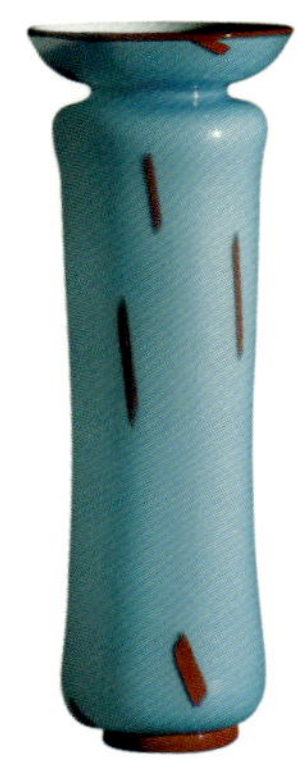

n.i.

Vase in *lattimo* cased and light-blue glass decorated with segments of dark-amethyst canes included in the wall of the piece. Small foot and thread around the mouth in ruby *pasta vitrea*. Surface with iridised finish. The piece is characterised by its elongated shape and slightly flared mouth. It is probably one of the last works of the Cappellin firm.
1930–31
32 cm high
Pauly collection

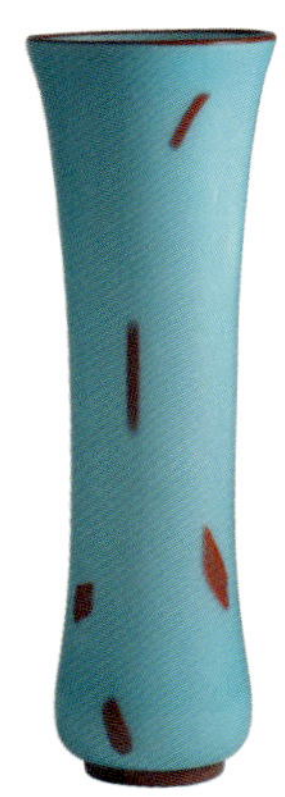

Bathroom jar in *lattimo* glass decorated with circular and star-like murrine in *avventurina*. Cover decorated with an applied flower in *lattimo* cased and amethyst glass.
c. 1930
15 cm high
The Olnick Spanu Collection, New York

n.i.

>
Working drawing for a bathroom jar with applied flower, CSV

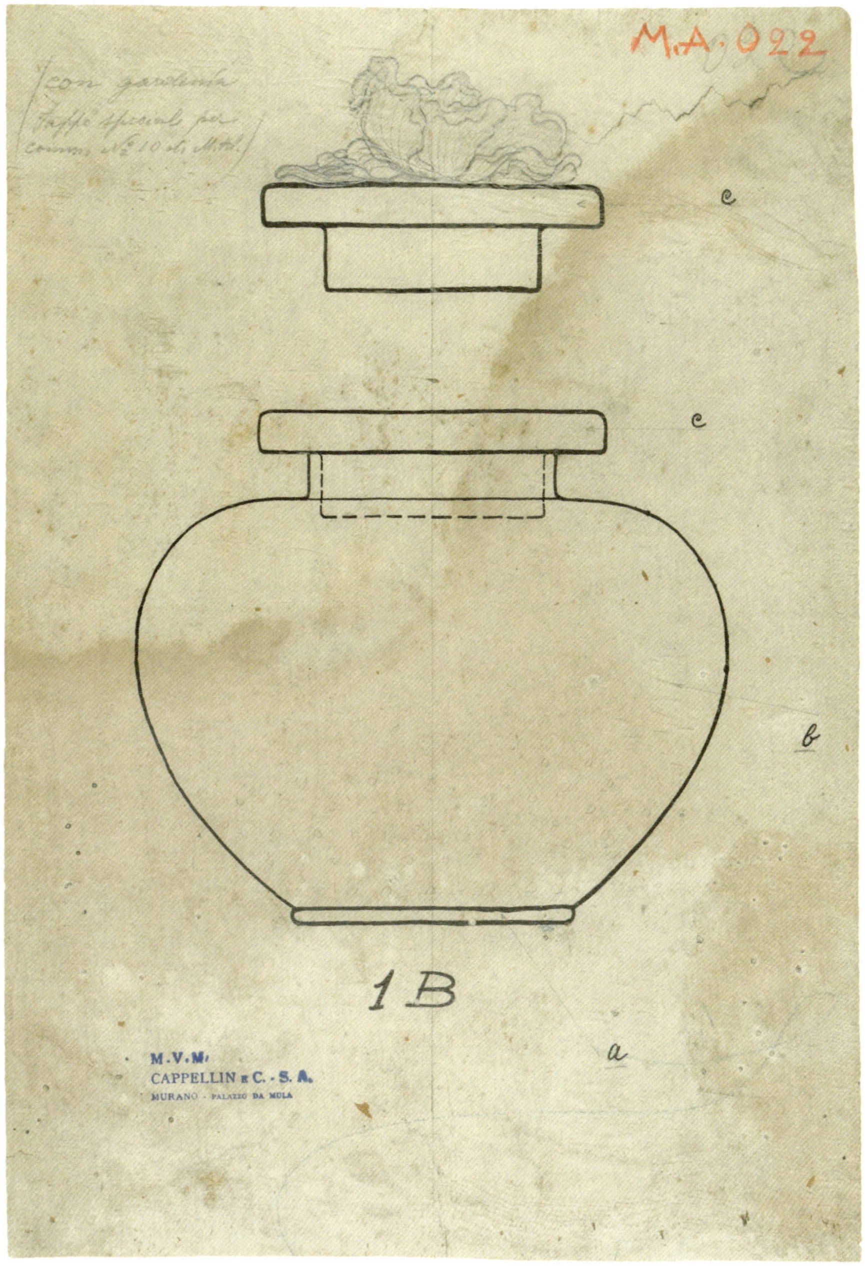

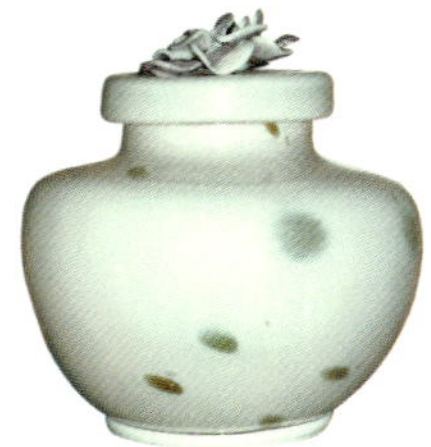

Bowl in green transparent glass decorated inside with blue murrine and outside with coral-red murrine included in the wall of the piece. Strongly iridised surface. Probably one of the last works of the Cappellin firm.
1930–31
10 cm high; 20 cm ø
C.V.M. *acid stamp*
Private collection

n.i.

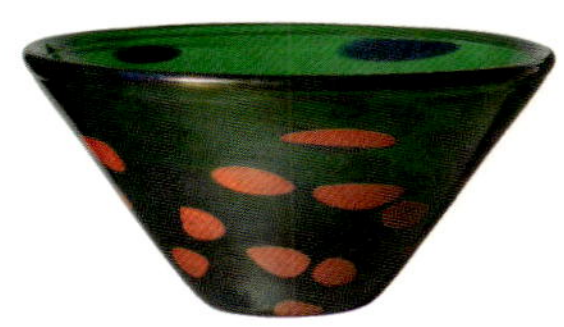

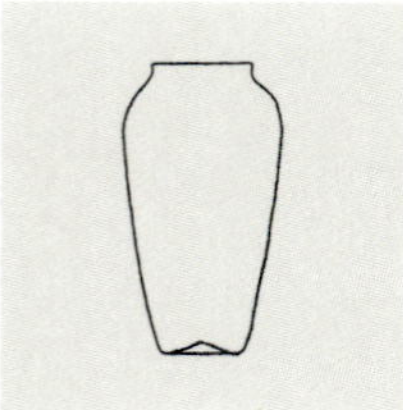

5943

Vase in glass with *lattimo* and amethyst quadrilobate floral murrine. Large star-like murrine included in the glass. Surface with iridised finish. A similar example ('crystal decorated with white opaque') was shown in an advertising insert of Cappellin Glassware, published in the American magazine *The Gift and Art Shop*, in March 1930, p. 48.
1928–29
36 cm high
M.V.M. Cappellin Murano *acid stamp*
Martin Kline collection, NY

>
5943, catalogue drawing, ACSVr

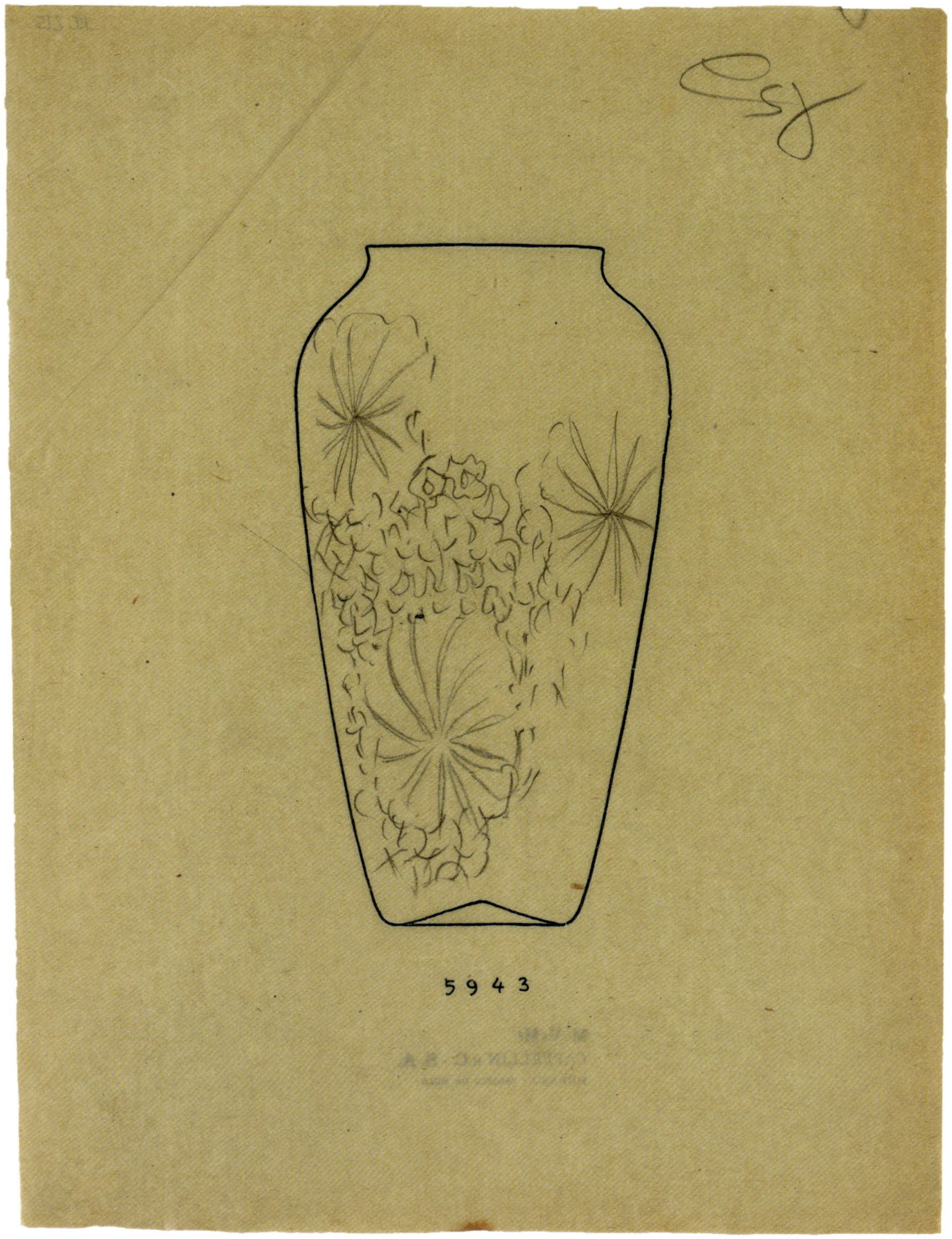

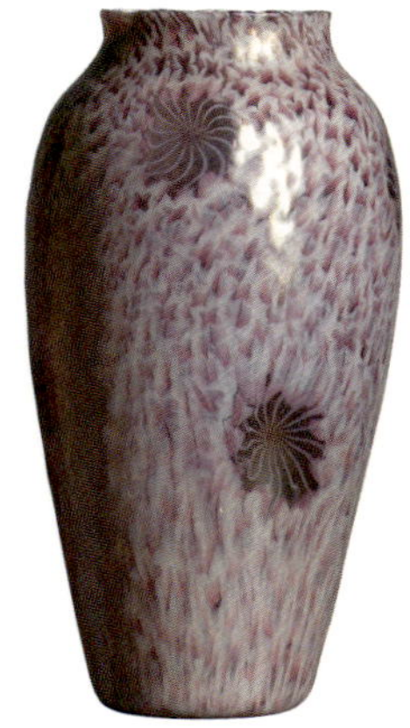

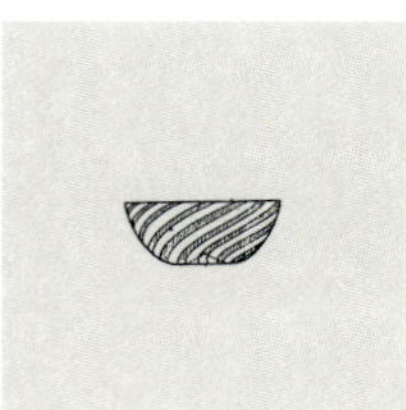

913

Bowl in glass with murrine, with star-like murrine, documented by a period photograph, c. 1929, detail, CSV. The piece was produced after model no. 913, present in the catalogue in glass with bands.
c. 1929
8 cm high; 21 cm ø

>
913, catalogue drawing, ACSVr

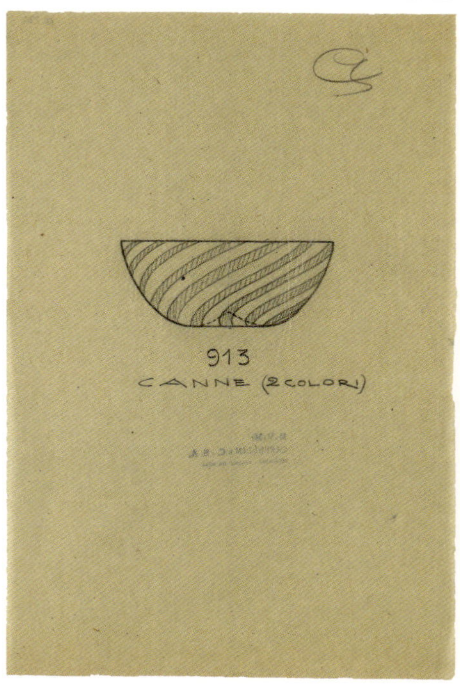

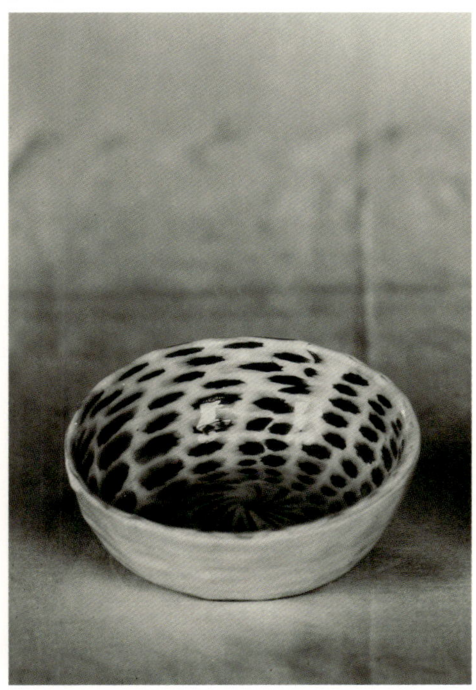

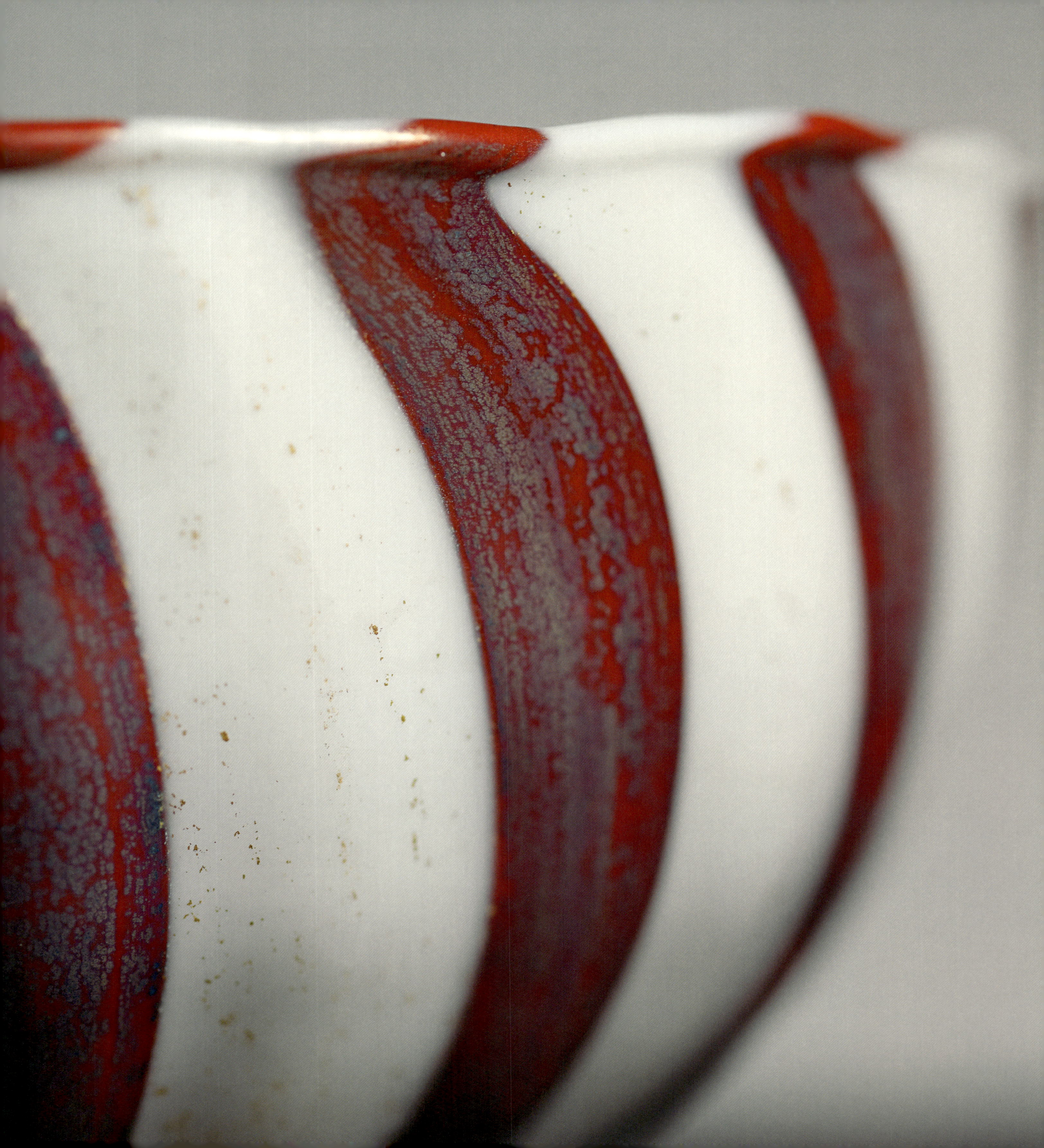

The first examples of the vertical-band 'series' (1928-29) are the artefacts, documented by period photographs, produced using a glass technique consisting of the union under heat of polychrome transparent glass canes arranged in alternating colours to form a precise decorative motif. Artefacts were also produced in opaque glass and executed in two different ways: one that required the use of coloured canes on the wall of the object, generally in *lattimo*, the other where the same wall was obtained by juxtaposing and fusing under heat polychrome canes of contrasting colours. In the first case the *lattimo* was combined with canes in delicate shades, while in the second the juxtapositions were more striking, such as red and yellow, but more particularly *lattimo* and dark amethyst. In this technique very linear vases with a small foot were generally produced, sometimes built up in height, playing also on the differences in flaring of the object.

VERTICAL BANDS 1928-31

Exhibitions

1930, Monza, 4th *Esposizione Triennale Internazionale delle Arti Decorative ed Industriali Moderne*
1931, Amsterdam, *Tentoonstelling van glas, aardewerk en kant, van moderne italiaansche kunstnijverheid. Mostra di vetri, ceramiche e merletti d'arte moderna italiana*

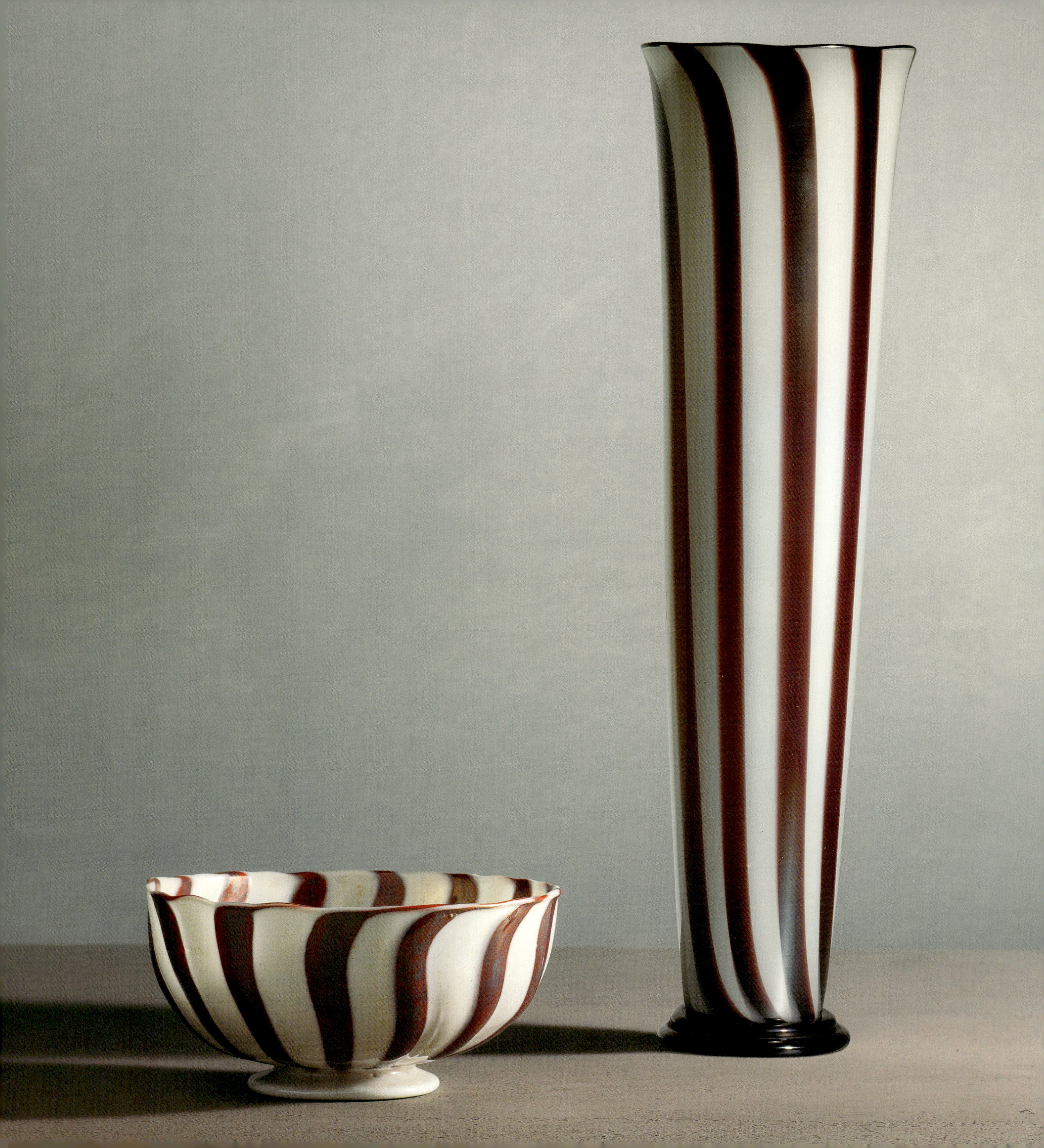

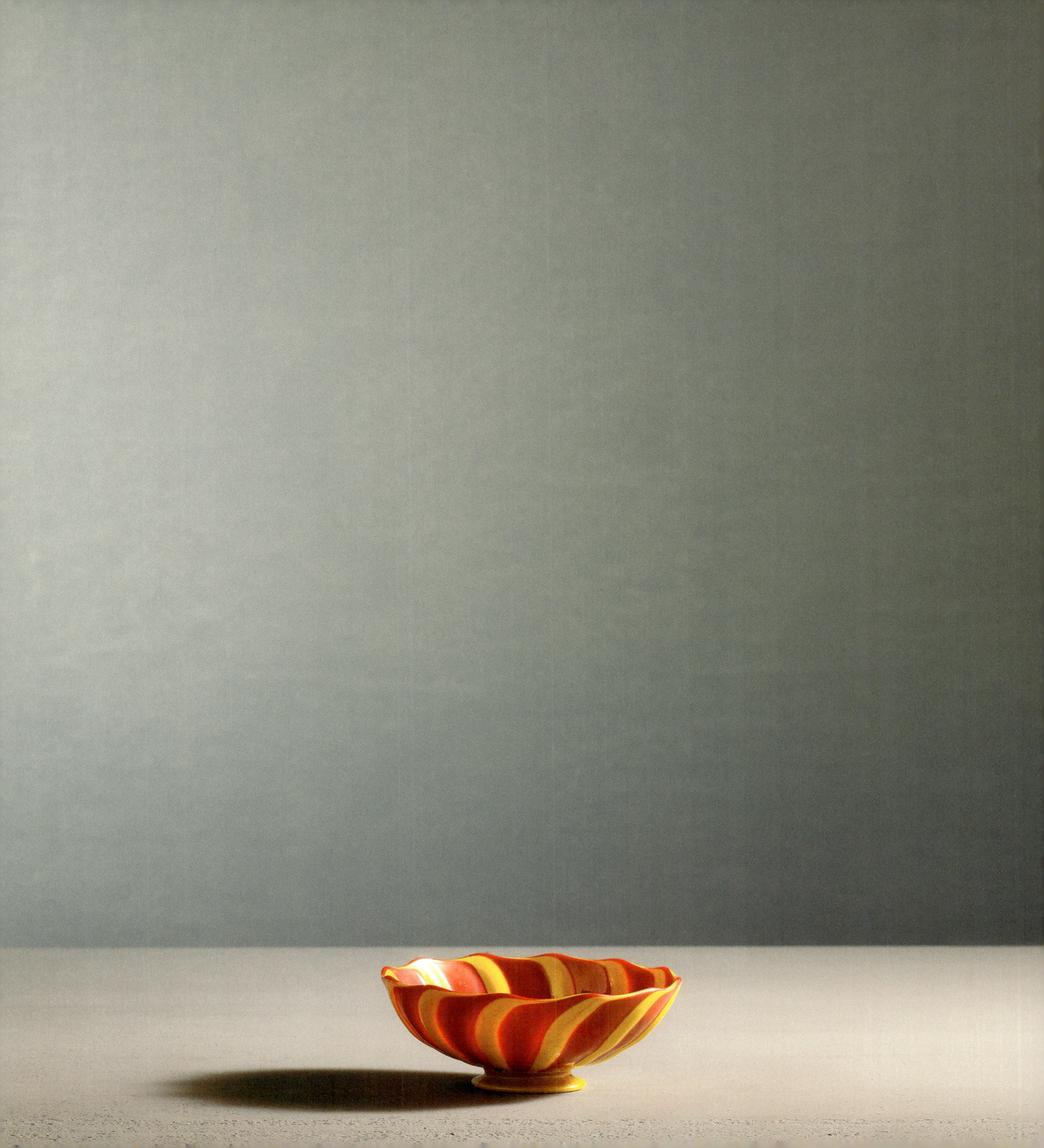

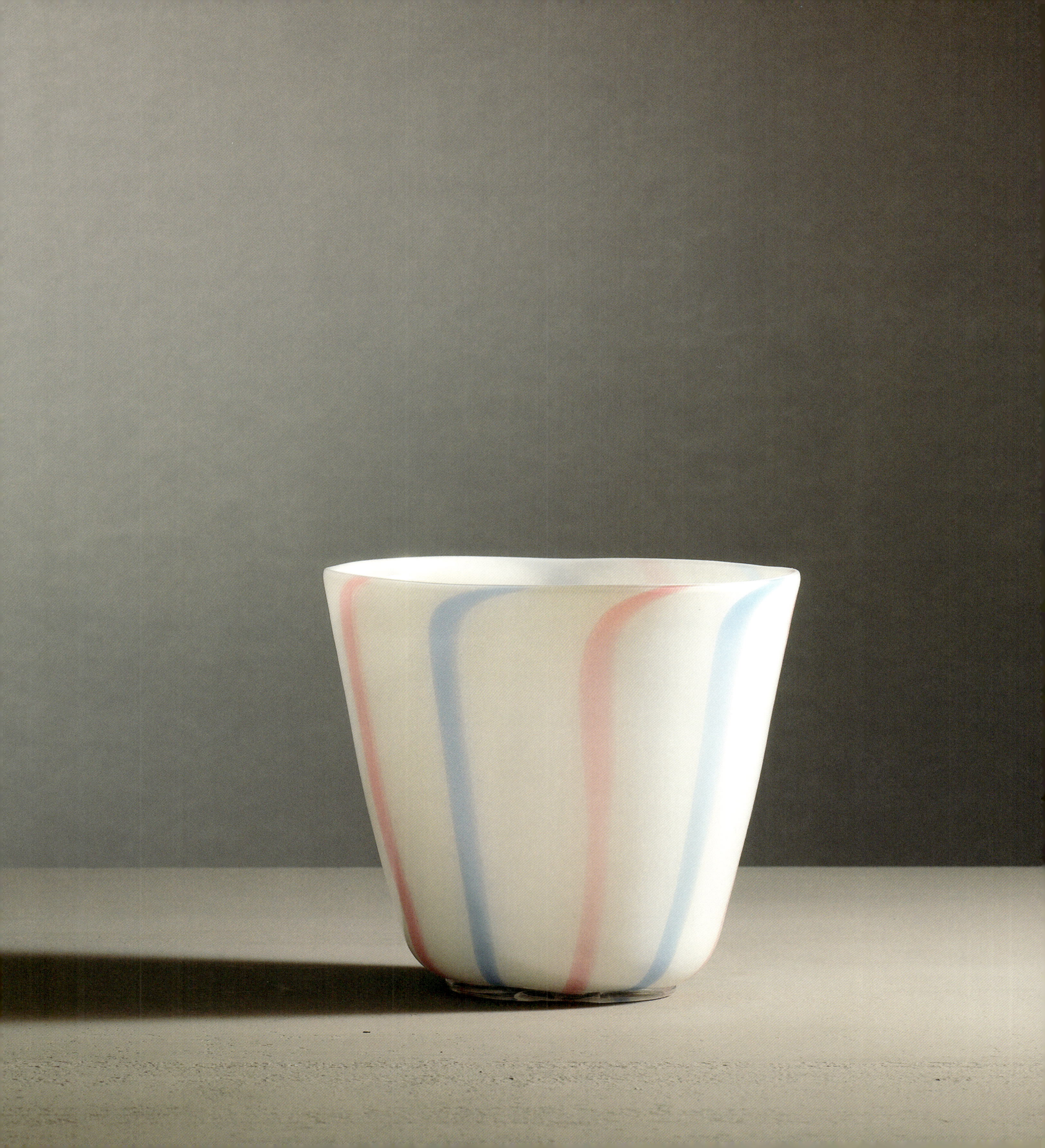

>
Ribbed vases in glass with polychrome vertical canes.
Period photograph, 1928–29, CSV

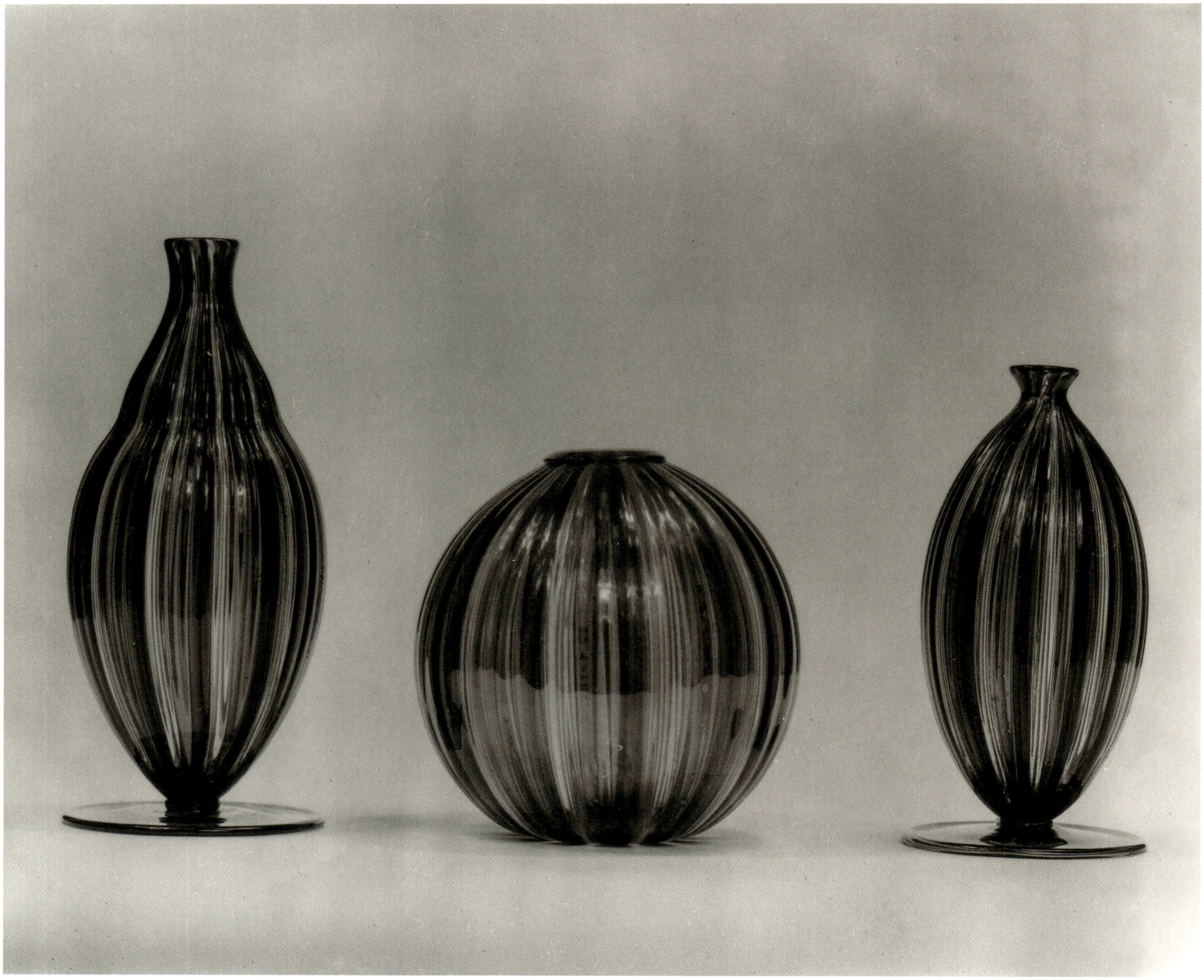

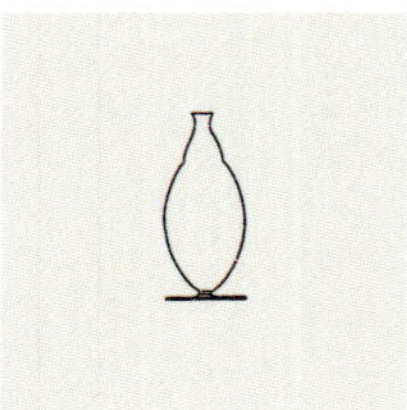

5921

Catalogue drawing for one of the ribbed vases in glass with polychrome vertical bands documented by a period photograph, AMB. Model no. 5921 was produced also in other glass techniques (e.g. transparent, *aurato*).
1928–29
22 cm high

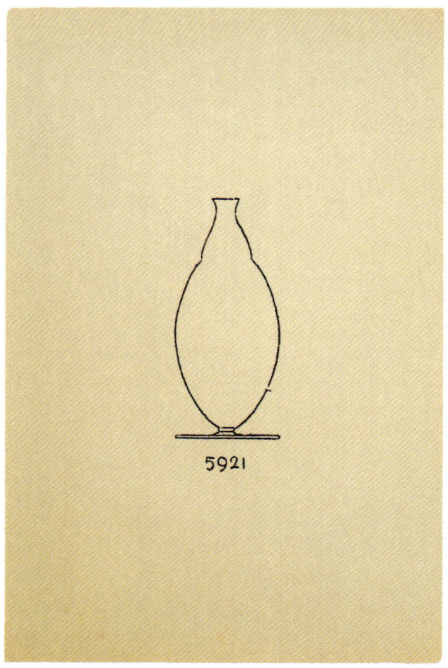

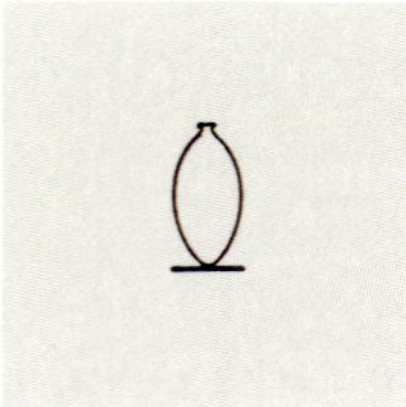

5922

Catalogue drawing for one of the ribbed vases in glass with polychrome vertical bands documented by a period photograph, AMB. Model no. 5922 was produced also in other glass techniques (e.g. silvered black).
1928–29
18 cm high

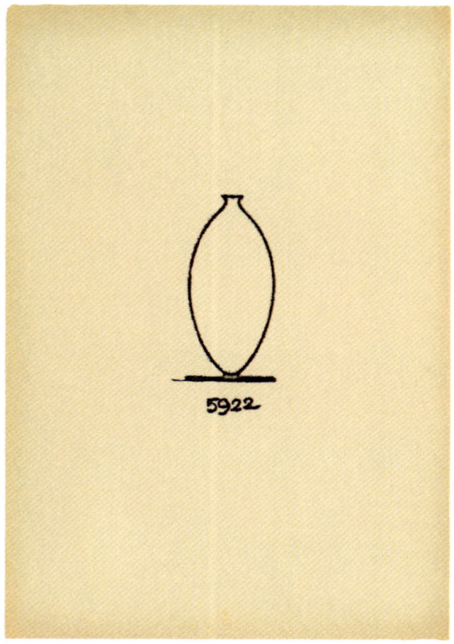

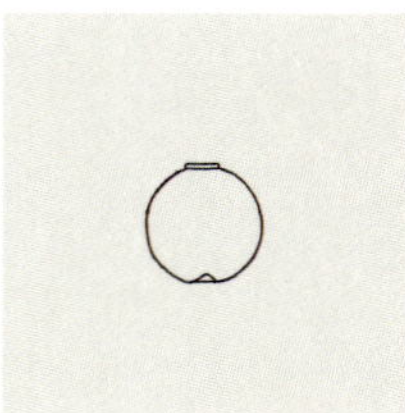

5934

Catalogue drawing for one of the ribbed vases in glass with polychrome vertical bands documented by a period photograph, AMB. Model no. 5934 was produced also in other glass techniques (e.g. *filigrana a reticello* (netted filigree), transparent *balloton*, *aurato*, *decoro fenicio*).
1928–29
14 cm high

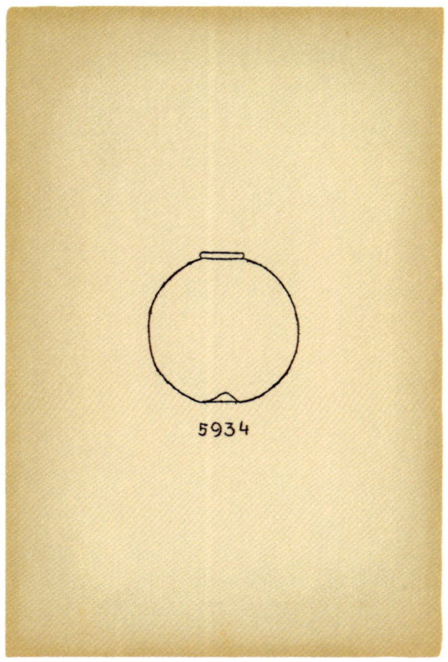

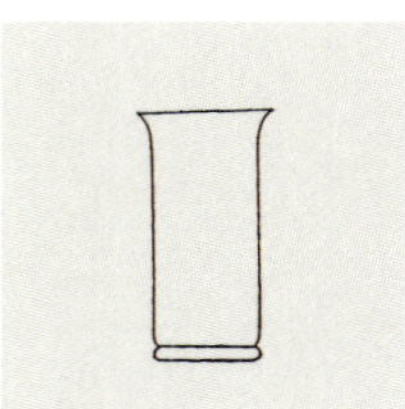

5881

Vase no. 5881 in glass with canes and polychrome vertical bands documented in the American photographic catalogue of Cappellin Glassware, undated, detail, AHL. Foot in opaque glass.
1928–29
29 cm high

>
5881, working drawing, CSV

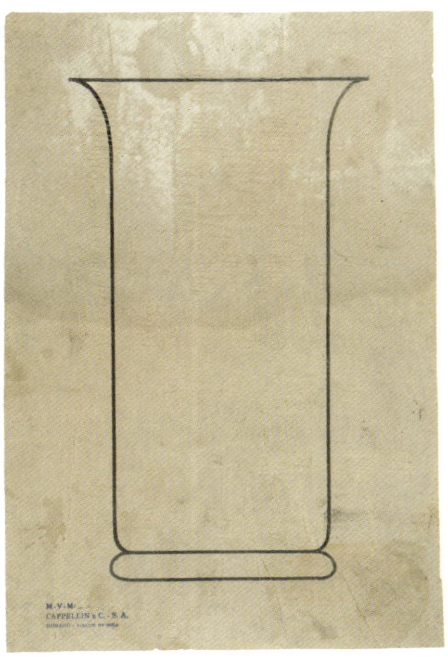

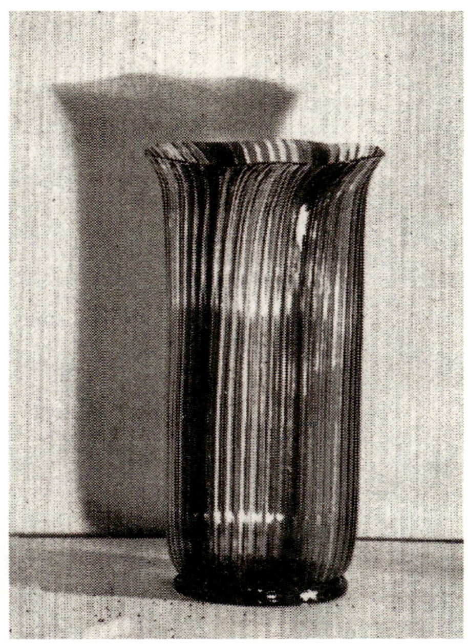

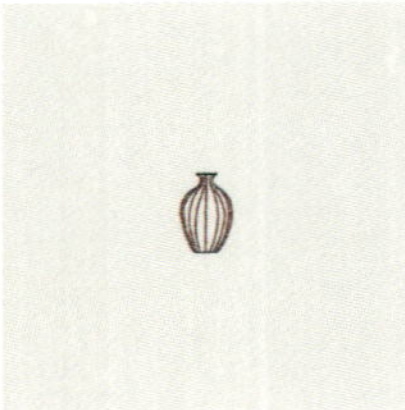

1280

1282

Catalogue drawings for bottles in monochrome glass and glass with polychrome bands, AMA. The two models in glass with bands were illustrated in the American photographic catalogue of Cappellin Glassware.
1929–30
9; 8 cm high

>
Bottles nos. 1280 and 1282, Cappellin Glassware, catalogue, undated, detail, AHL

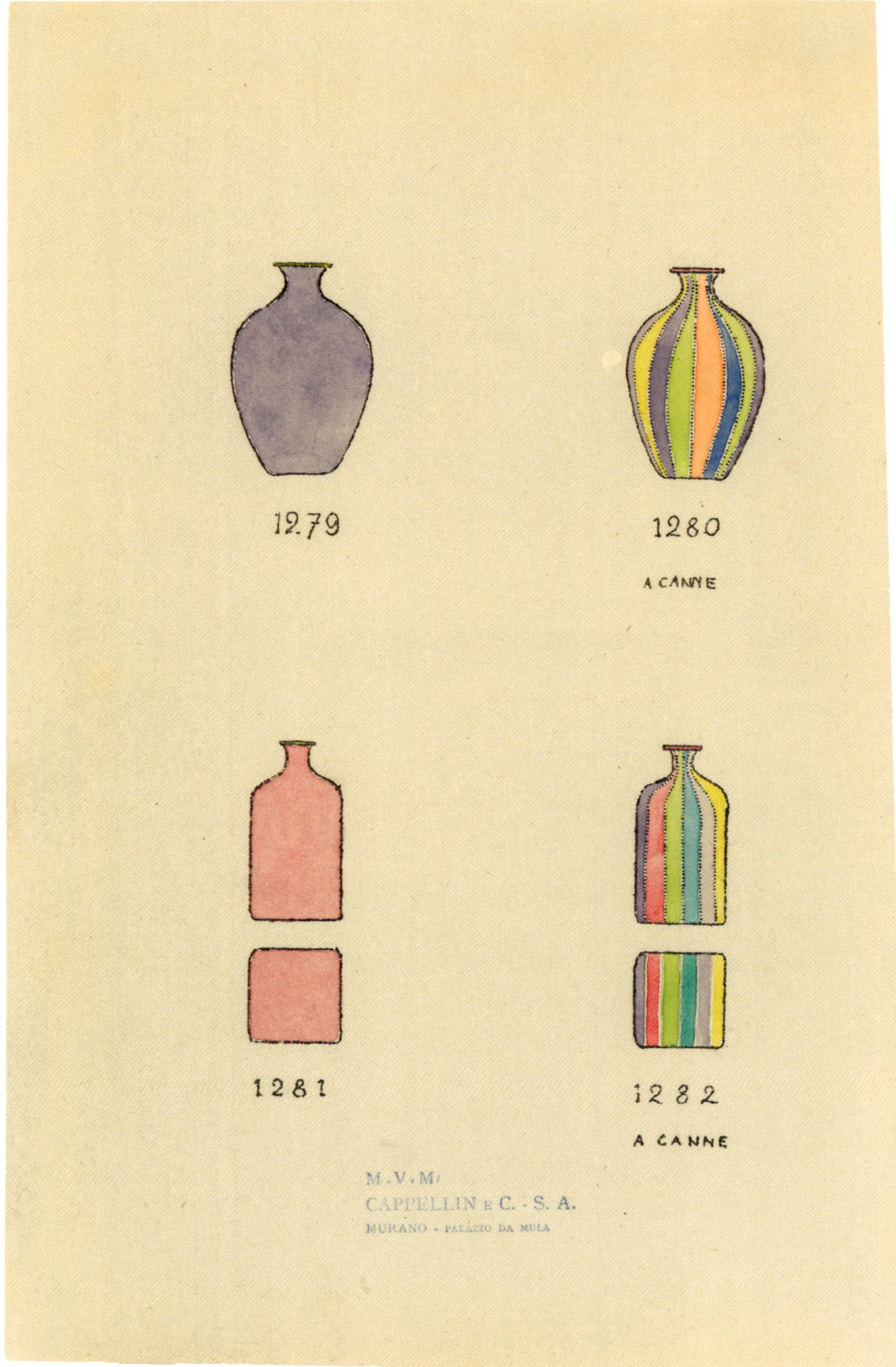

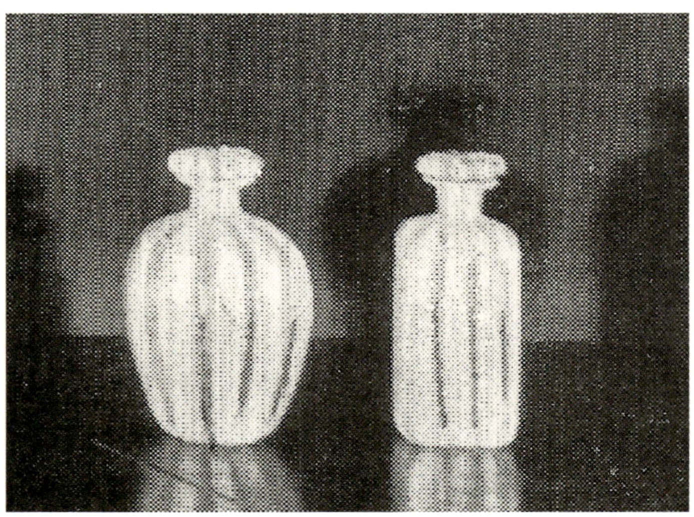

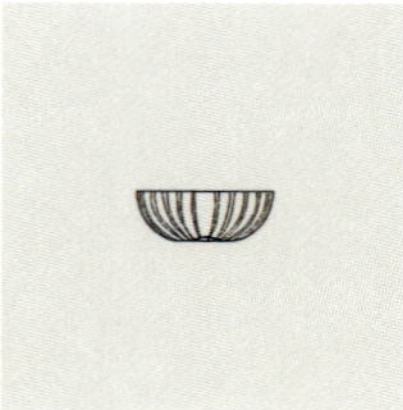

5980

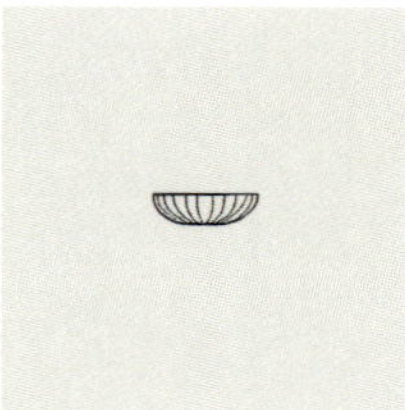

5981

Catalogue drawing for bowls no. 5980 and no. 5981 in glass with vertical polychrome bands, AMA. An example of model no. 5980 was published among the *Nuovi vetri di Cappellin* [New Cappellin glass pieces] in *Domus* in January 1930.
c. 1929.
16; 12 cm ø

>
Glass bowl with vertical bands and glass bowl with decoro fenicio. *Period photograph, c. 1929, ACSVr*

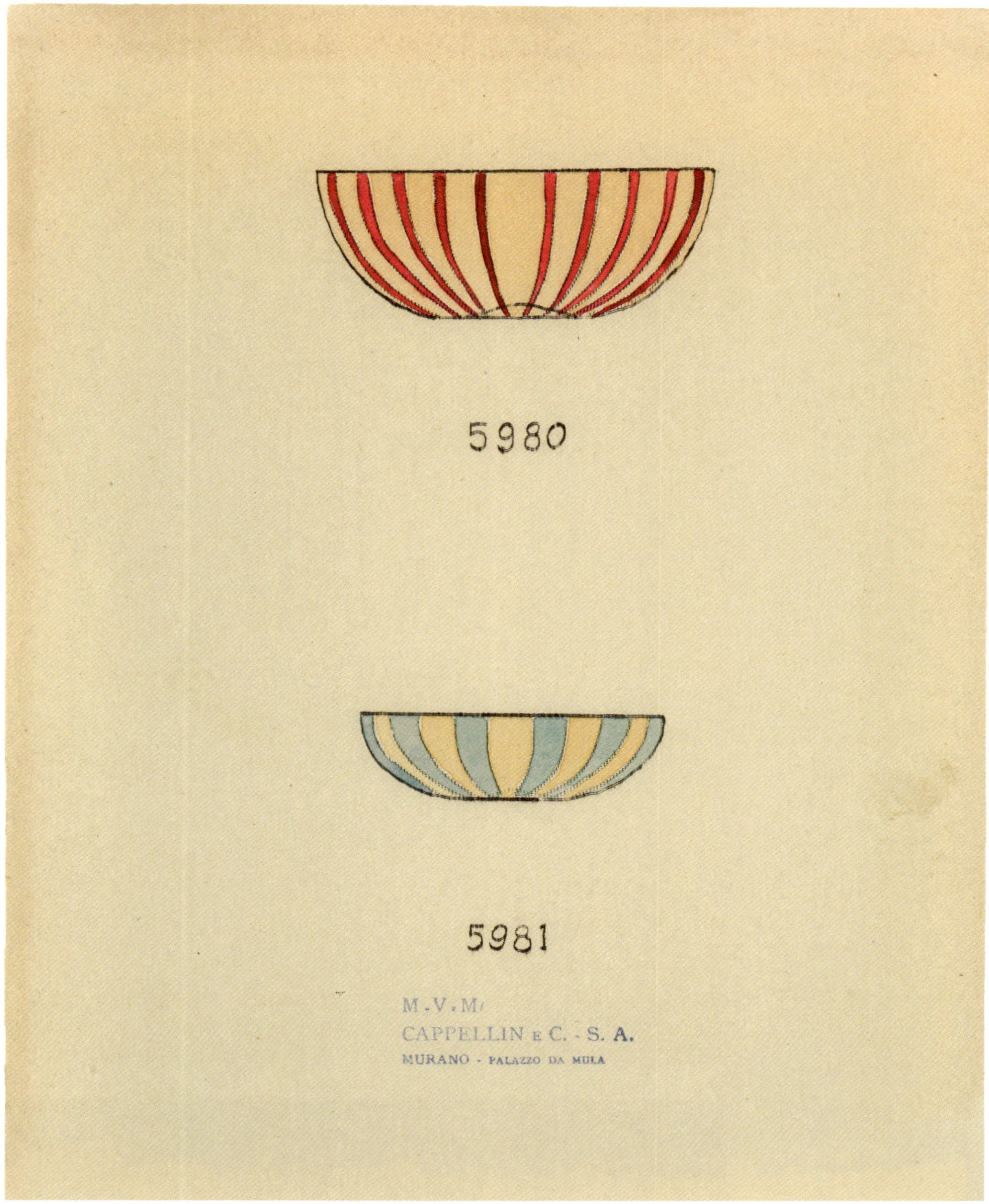

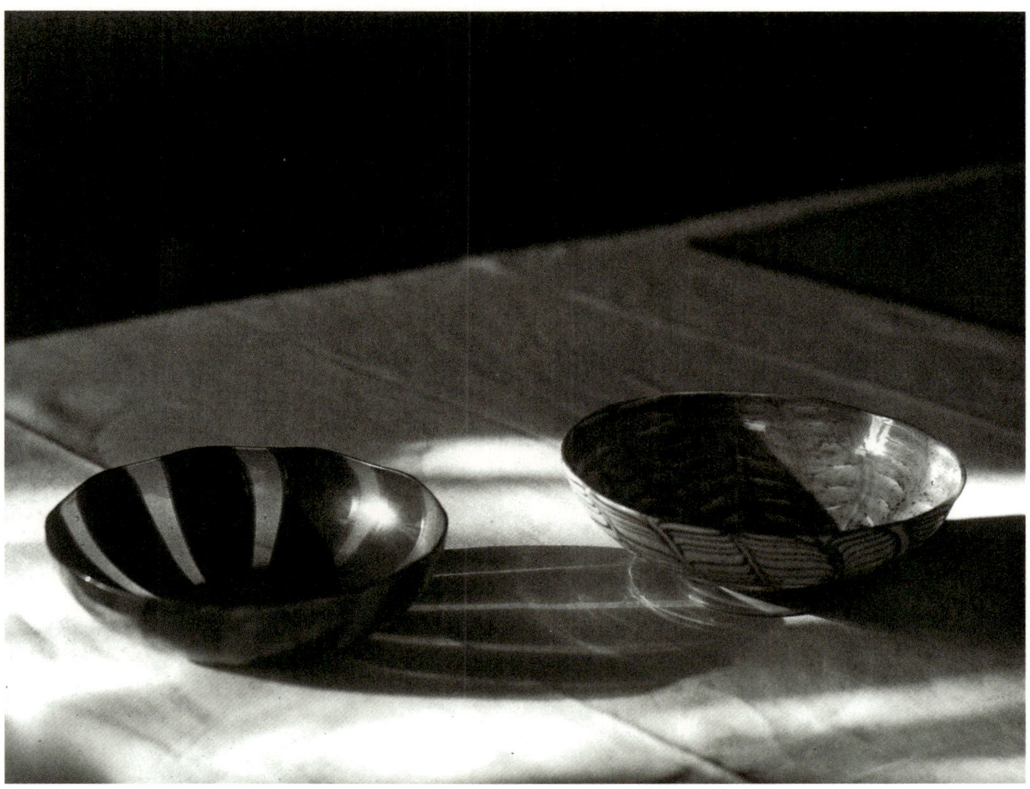

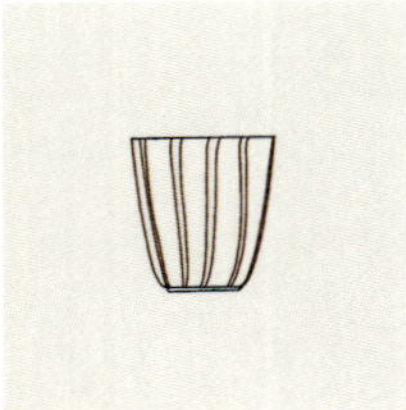

5968

Vase in *lattimo* cased crystal glass decorated with light-blue and light amethyst vertical canes included in the wall of the piece. Small foot in crystal glass. Surface with an iridised finish. An example of no. 5968 was published among the *Nuovi vetri di Cappellin* in *Domus* in January 1930.
c. 1929
15 cm high
M.V.M. Cappellin Murano *acid stamp*
Bischofberger collection, Switzerland

>
5968, catalogue drawing, AMB

>
Vase in decoro fenicio *glass and vertical bands. Period photograph, c. 1929, ACSVr*

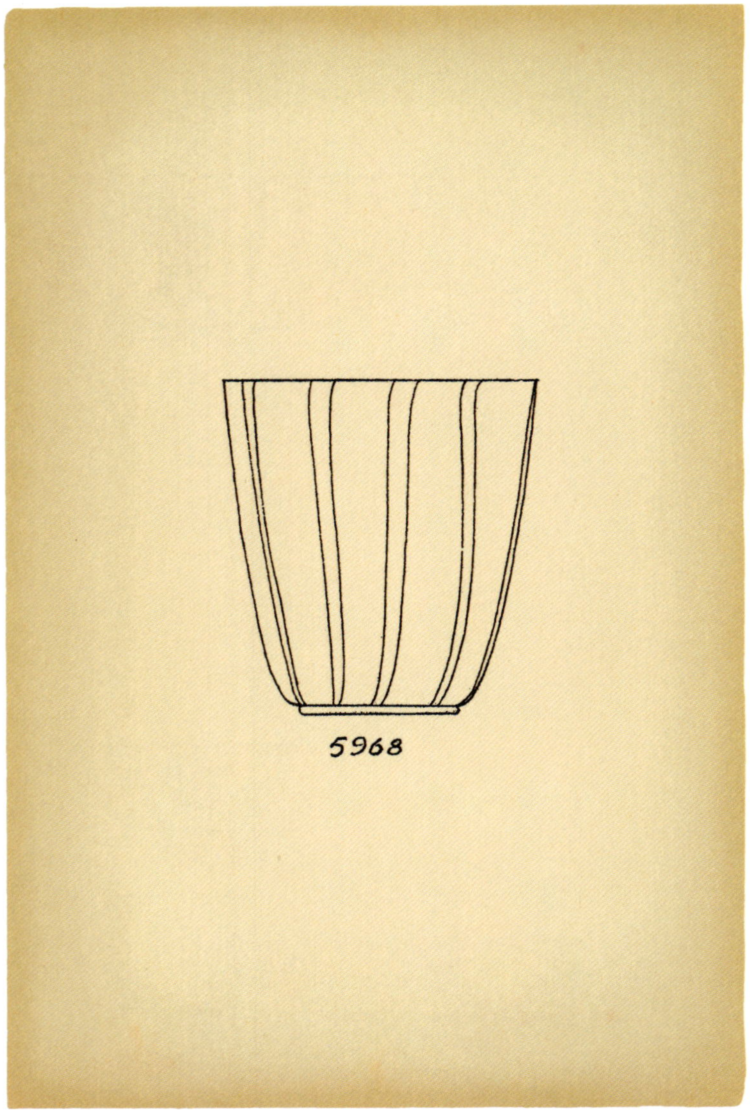

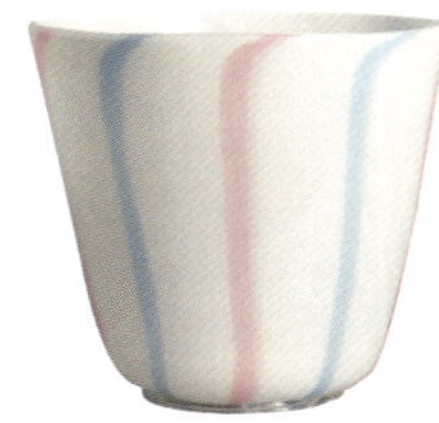

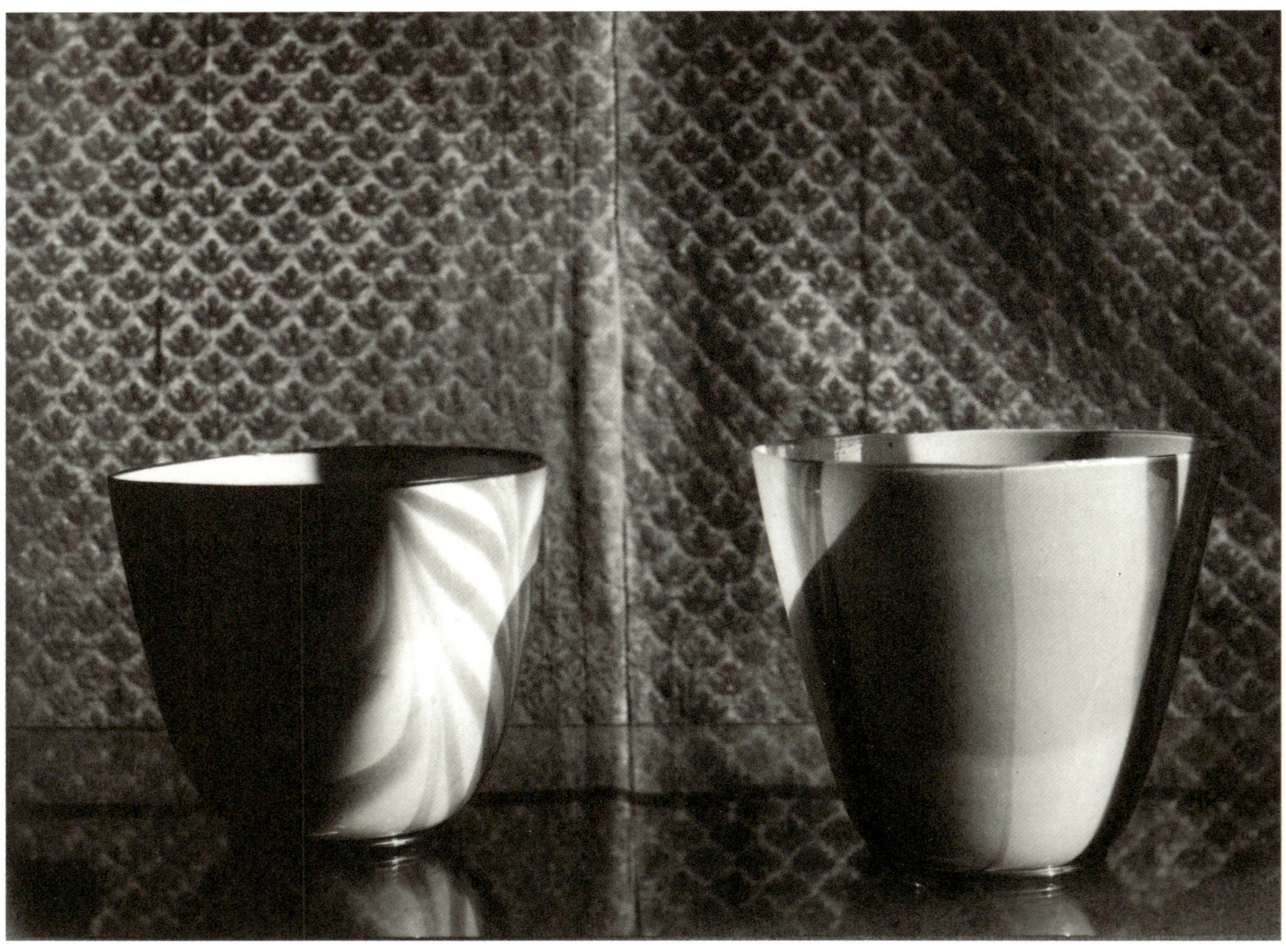

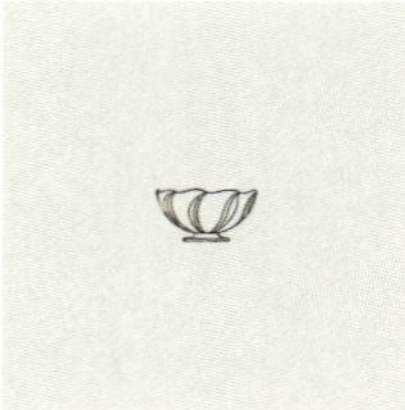

917

Glass bowl with coral-red and opaque yellow vertical bands. Small flared foot in opaque yellow. Surface with an iridised finish.
c. 1930.
12 cm ø
Private collection

>
917, catalogue drawing, AMA

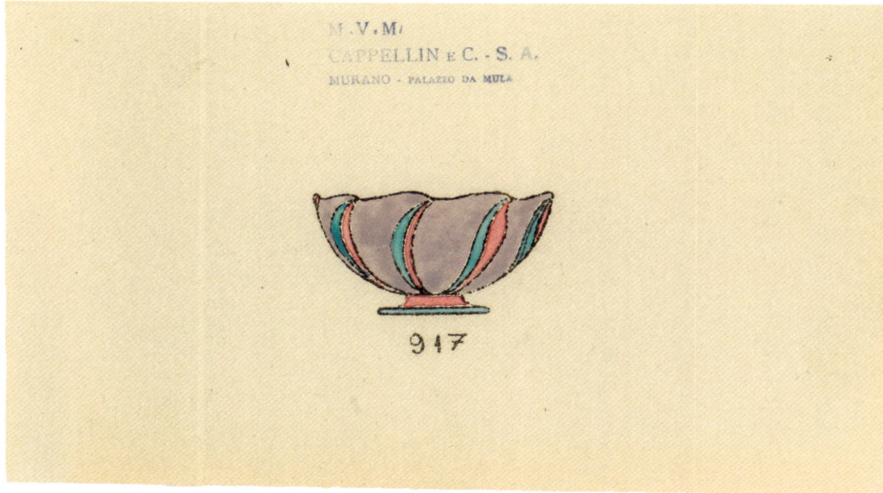

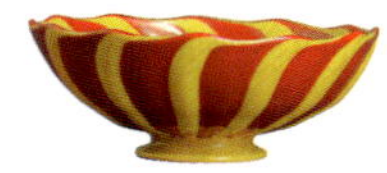

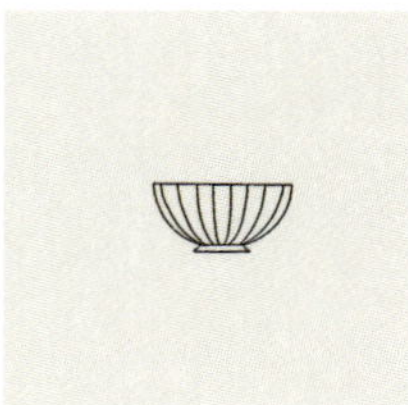

565

Glass bowl with *lattimo* bands and coral-red bands with applied oxidised silver leaf. Small flared foot in *lattimo*. Surface with an iridised finish.
c. 1930.
16 cm ø
M.V.M. Cappellin label
Private collection, Treviso

>
565, catalogue drawing, AMA

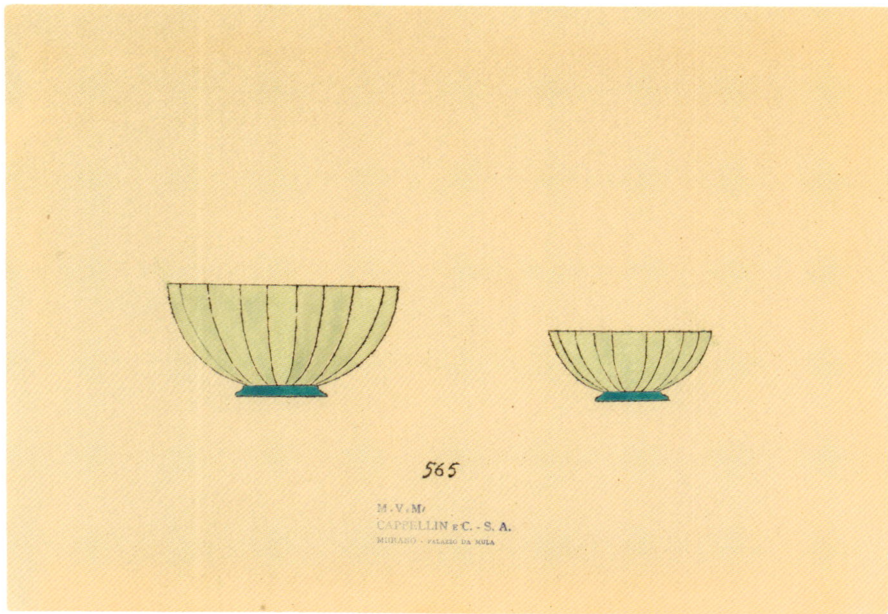

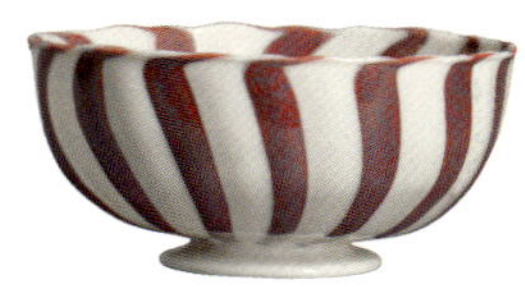

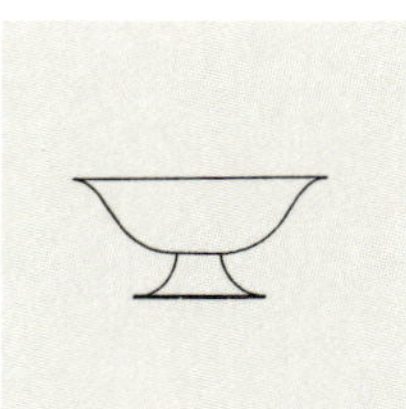

5709

Bowl with vertical bands and coloured glass exhibited in 1931 at the Amsterdam Exhibition of glass, ceramics and lace and documented by a period photograph (*Tentoonstelling van glas, aardewerk en kant, van moderne italiaansche kunstnijverheid. Mostra di vetri, ceramiche e merletti d'arte moderna italiana*, Venice 1931, p. 21). The bowl was produced in a smaller-sized variant of no. 5709 (h 7 cm).
c. 1930
14 cm high; 30 cm ø

>
5709, catalogue drawing, AMB

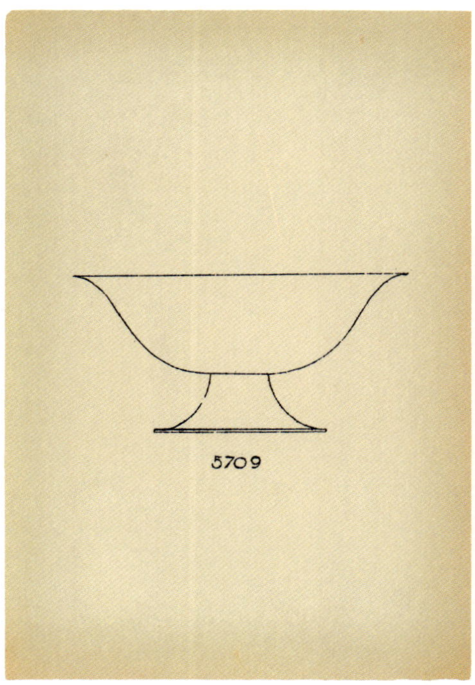

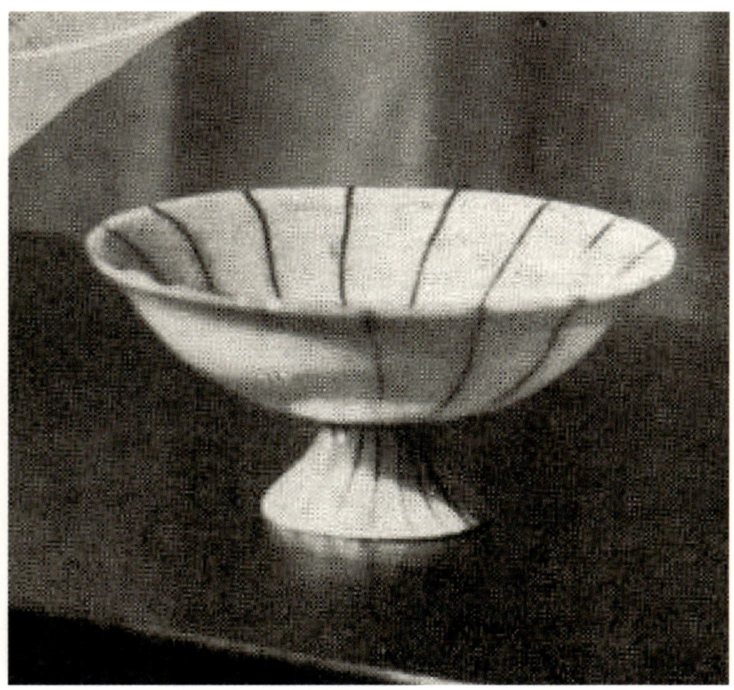

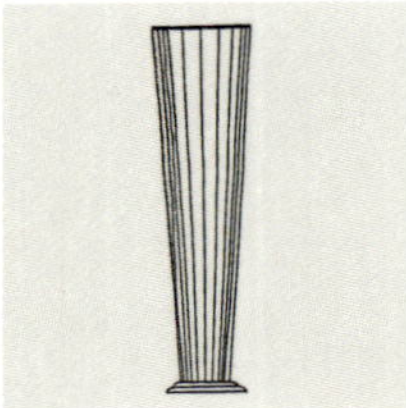
5977

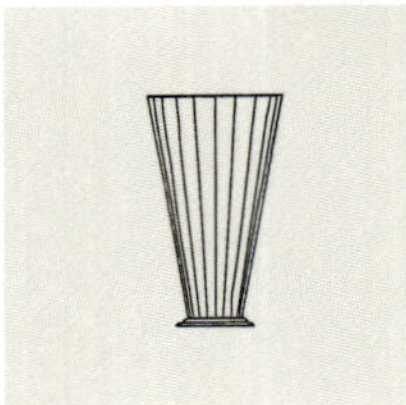
5978

Large glass vase with dark-amethyst vertical bands. Gradated foot with double-wound thread and a dark-amethyst thread around the mouth. Surface with iridised finish. The model was also proposed in other dimensions (no. 5978, h 27 cm). Both models (5977 and 5978), presumably produced in a monochrome version or with colours of the same shade were exhibited in 1931 at the Amsterdam Exhibition of glass, ceramics and lace.
c. 1930
43 cm high
M.V.M. Cappellin Murano *acid stamp*
Private collection

exhibitions
1931, Amsterdam Exhibition of glass, ceramics and lace

>
5977–5978, catalogue drawing, AMA

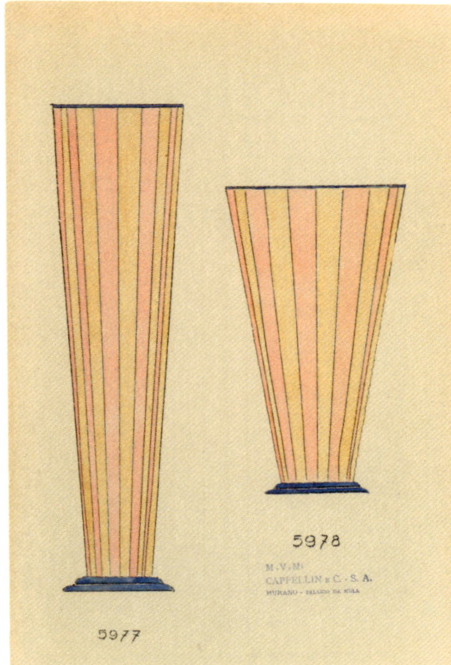

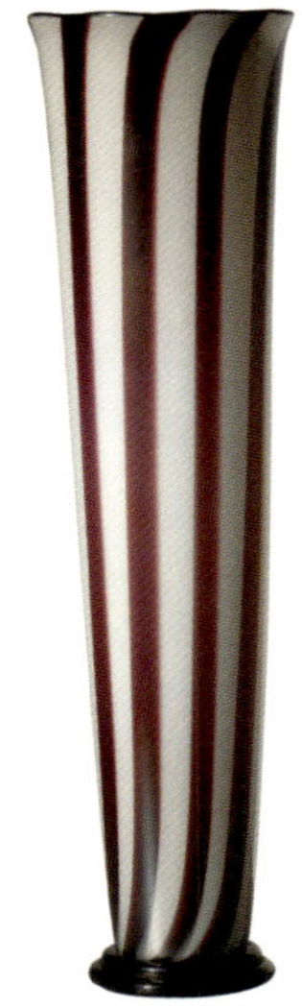

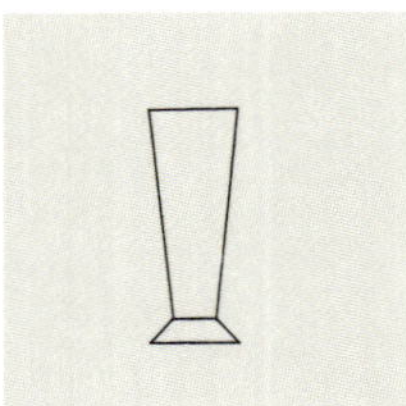
5829

Lattimo glass vase decorated with dark-amethyst vertical bands included in the wall of the piece. Truncated-cone-shaped foot and dark-amethyst thread around the mouth. Surface with iridised finish. A variant of the piece was produced in dimensions larger than no. 5829 (h 28 cm), a model initially designed in transparent glass.
c. 1930
38 cm high
M.V.M. Cappellin Murano *acid stamp*
Private collection

>
5829, catalogue drawing, AMB

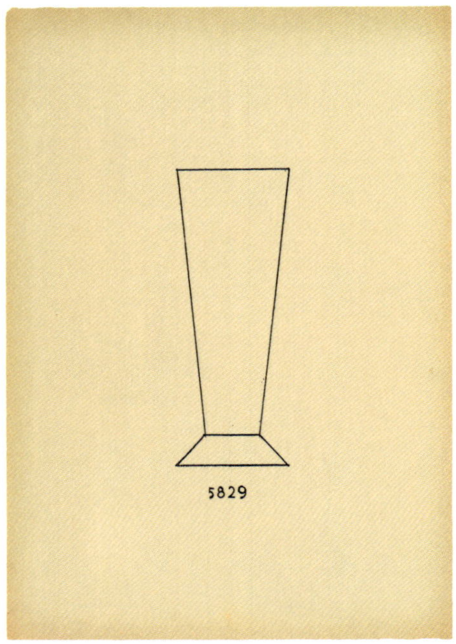

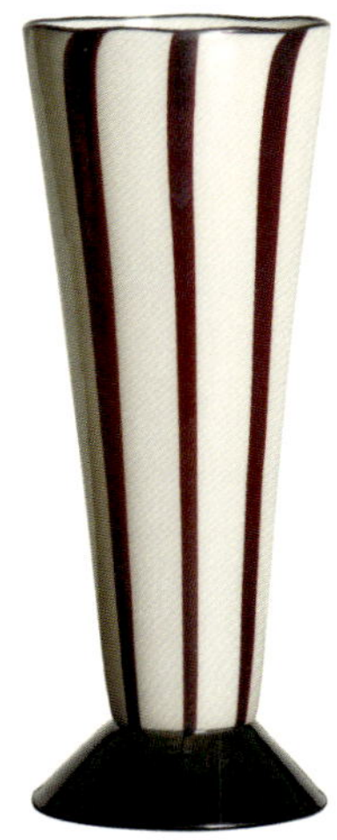

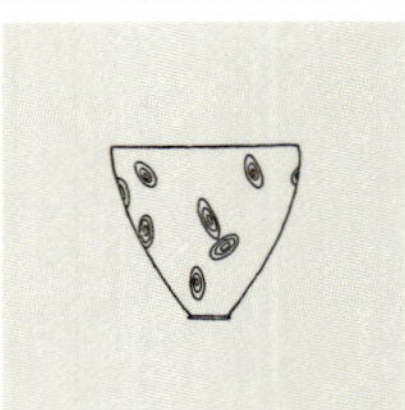
6006

Lattimo glass vase with applied silver leaf decorated with dark-amethyst vertical bands included in the wall of the piece. Small foot in *lattimo* and oxidised silver. Surface with iridised finish. The piece was produced in a variant no. 6006 (h 20 cm).
c. 1930
17 cm high
Private collection, Montebelluna

>
C. Scarpa (?), study for a bowl with polychrome vertical bands, 1929-30, AmuVe

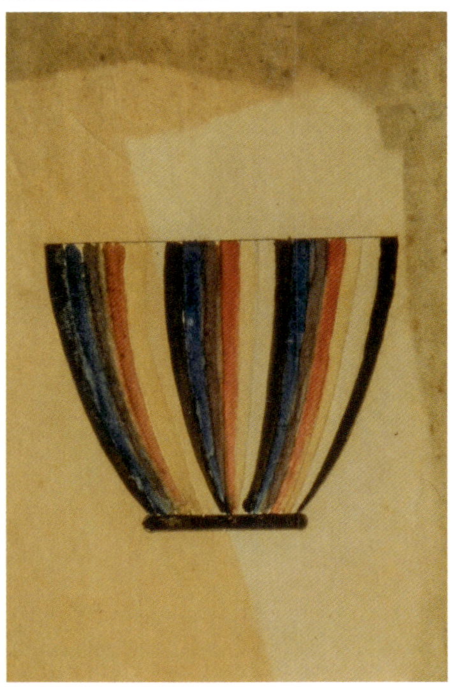

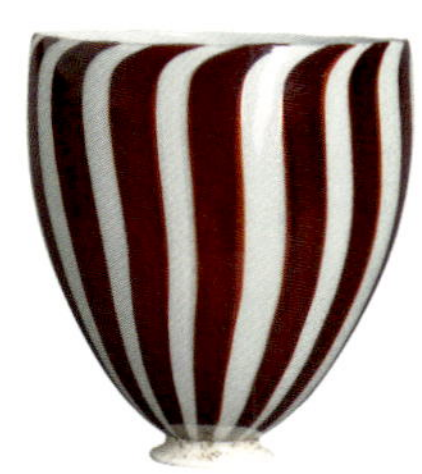

>
Strongly iridised glass pieces with vertical bands at the Amsterdam Exhibition of glass, ceramics and lace, 1931, AHL, detail

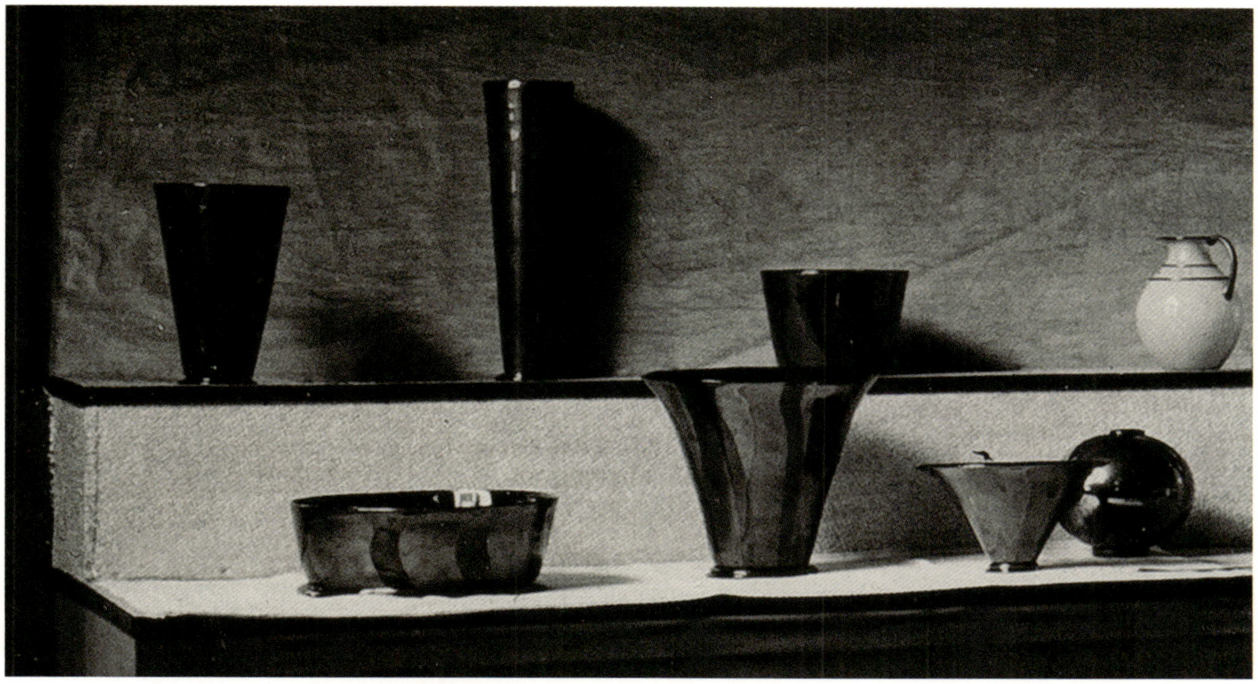

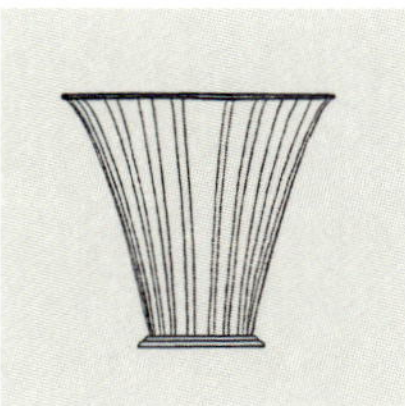

5989

Catalogue drawing for one of the vases with vertical bands, and gradated foot wound round with a thread, exhibited in 1931 at the Amsterdam Exhibition of glass, ceramics and lace, AMB.
c. 1930
30 cm high

exhibitions
1931, Amsterdam, Exhibition of glass, ceramics and lace

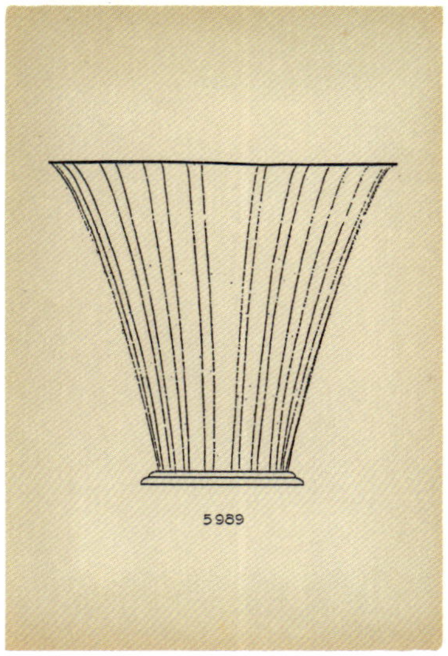

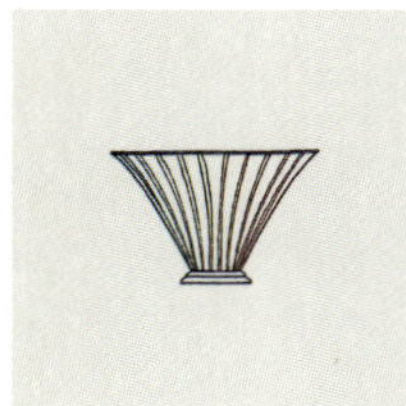

5990

Catalogue drawing for one of the vases with vertical bands and gradated foot wound round with a thread, exhibited in 1931 at the Amsterdam Exhibition of glass, ceramics and lace, AMB.
c. 1930
16 cm high

exhibitions
1931, Amsterdam, Exhibition of glass, ceramics and lace

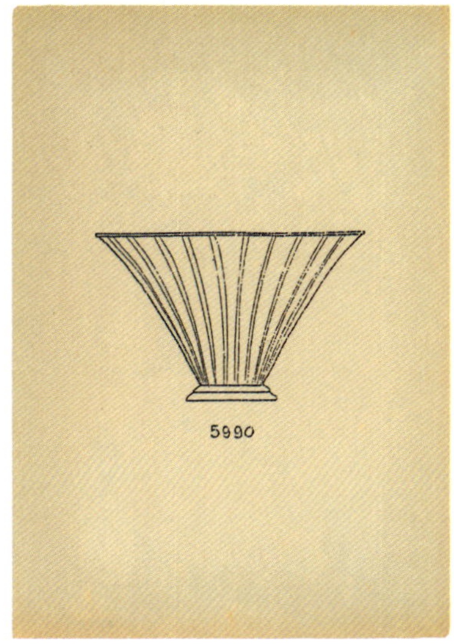

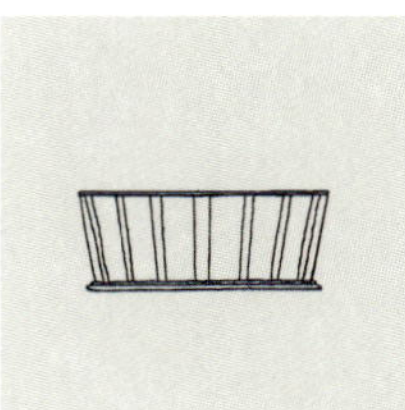

5991

Catalogue drawing for one of the vases with vertical bands and gradated foot wound round with a thread, exhibited in 1931 at the Amsterdam Exhibition of glass, ceramics and lace, AMB. An example of this model produced in glass with vertical polychrome bands was on display in 1930 at the 4th Triennale of Monza, in the Furnishing Gallery, in a room prepared by the architect Serafini and in the Borsani of Varedo dining-room.
c. 1930
11 cm high

exhibitions
1930, Monza 4th Triennale
1931, Amsterdam, Exhibition of glass, ceramics and lace

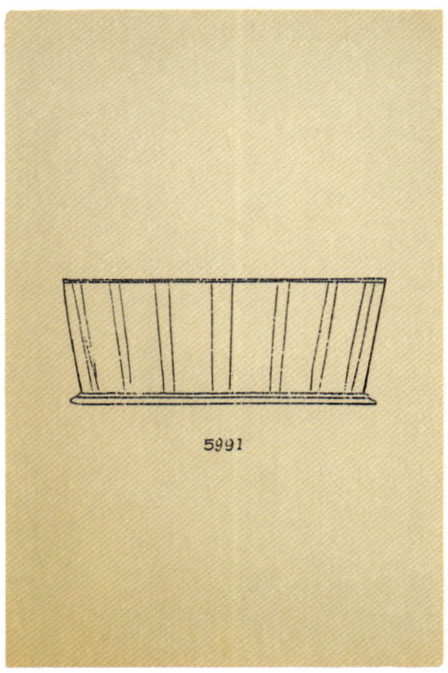

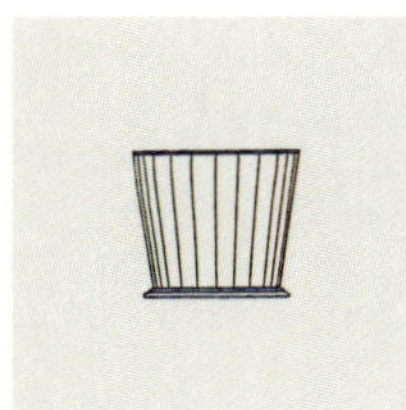

5979

Catalogue drawing for one of the vases with vertical bands and gradated foot wound round with a thread, exhibited in 1931 at the Amsterdam Exhibition of glass, ceramics and lace, AMA.
c. 1930
17 cm high

exhibitions
1931, Amsterdam, Exhibition of glass, ceramics and lace

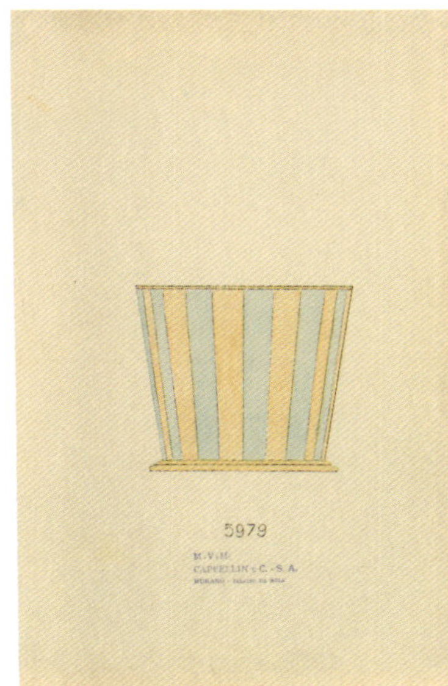

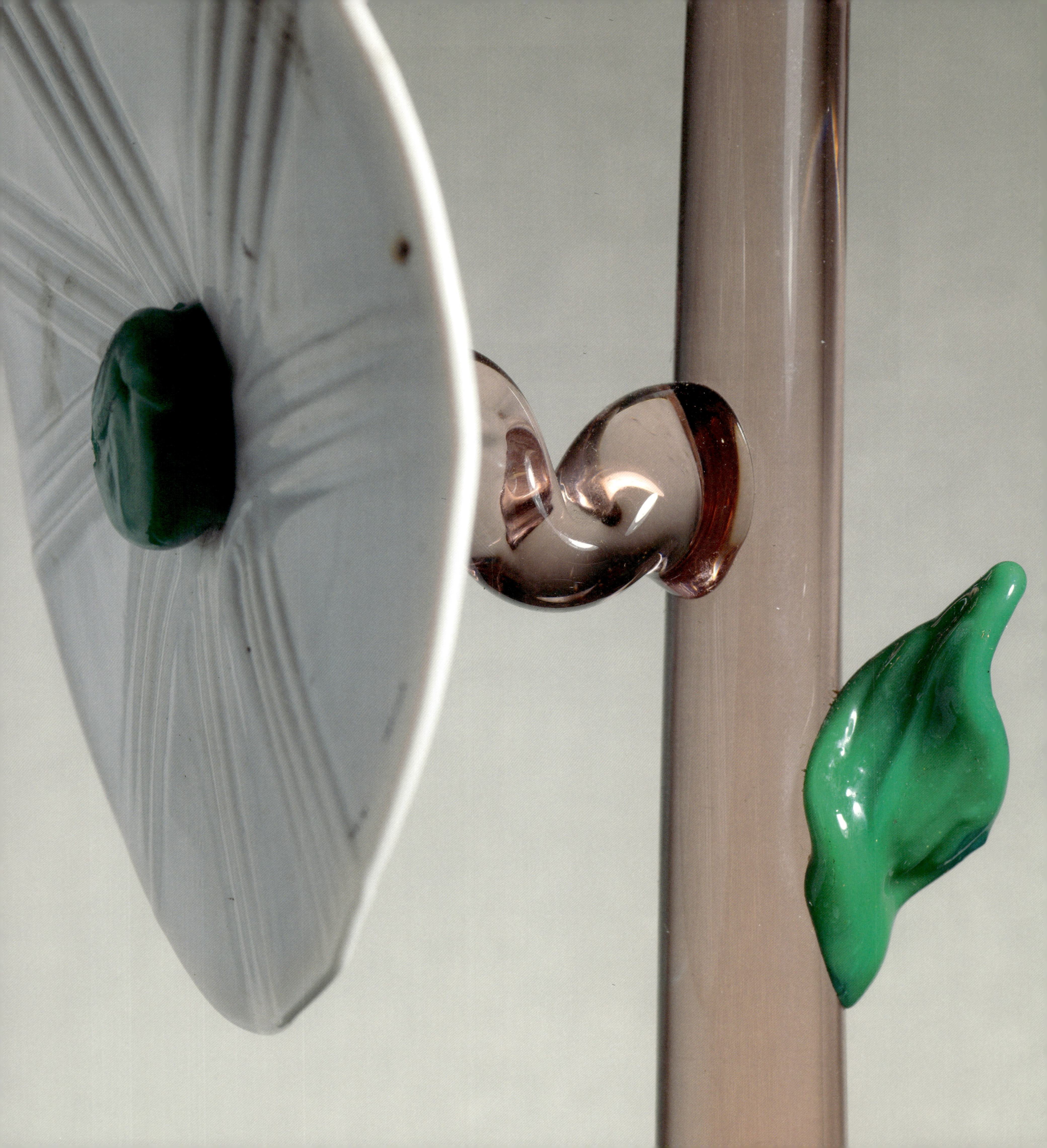

In the catalogue of the glassworks different models are also documented for the traditional glass flowers, either entirely in transparent glass or with a corolla in opaque glass. It should be noted that a glass flower combined with a spherical vase with a truncated-cone-shaped foot is to be found in the emblem of the glassworks. Generally, these artefacts are designed to represent in a more or less realistic way certain kinds of flower (chrysanthemum, rose, calla lily etc.), the shape of which has been simplified in some cases. The reference to the vegetable kingdom can be found again in some elements for a centrepiece that the Cappellin firm proposed in response to a fashion of the time. Finally, a singular response to market demands is the small series of succulent and aquatic plants that were designed with a certain irony and playfulness.

FLOWERS AND PLANTS 1927-30

Exhibitions

1927, Venice, 18th *Mostra dell'Opera Bevilacqua La Masa*
1929, Leipzig, Trade Fair
1931, Amsterdam, *Tentoonstelling van glas, aardewerk en kant, van moderne italiaansche kunstnijverheid. Mostra di vetri, ceramiche e merletti d'arte moderna italiana*
1931, Florence, Italian garden exhibition

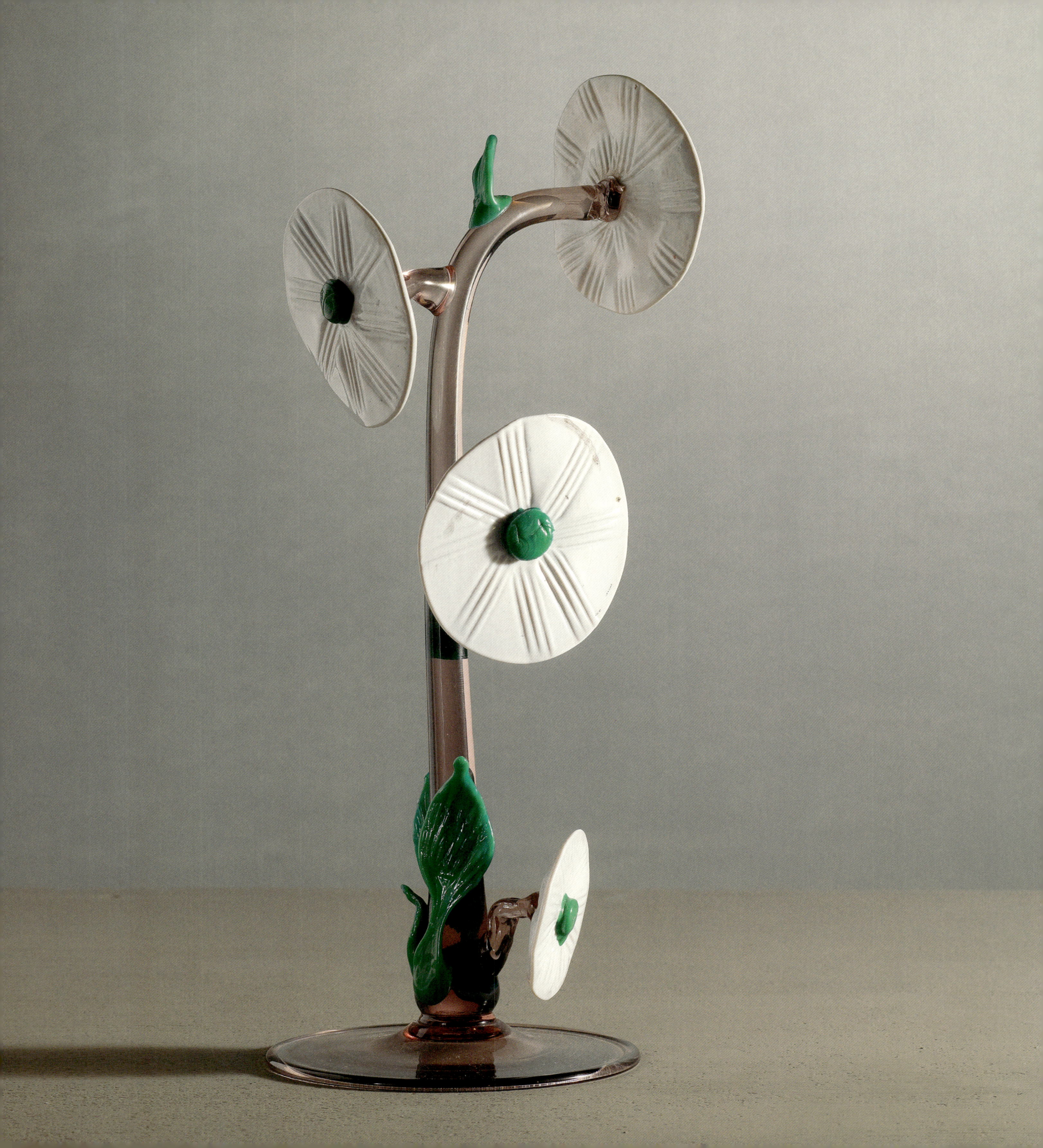

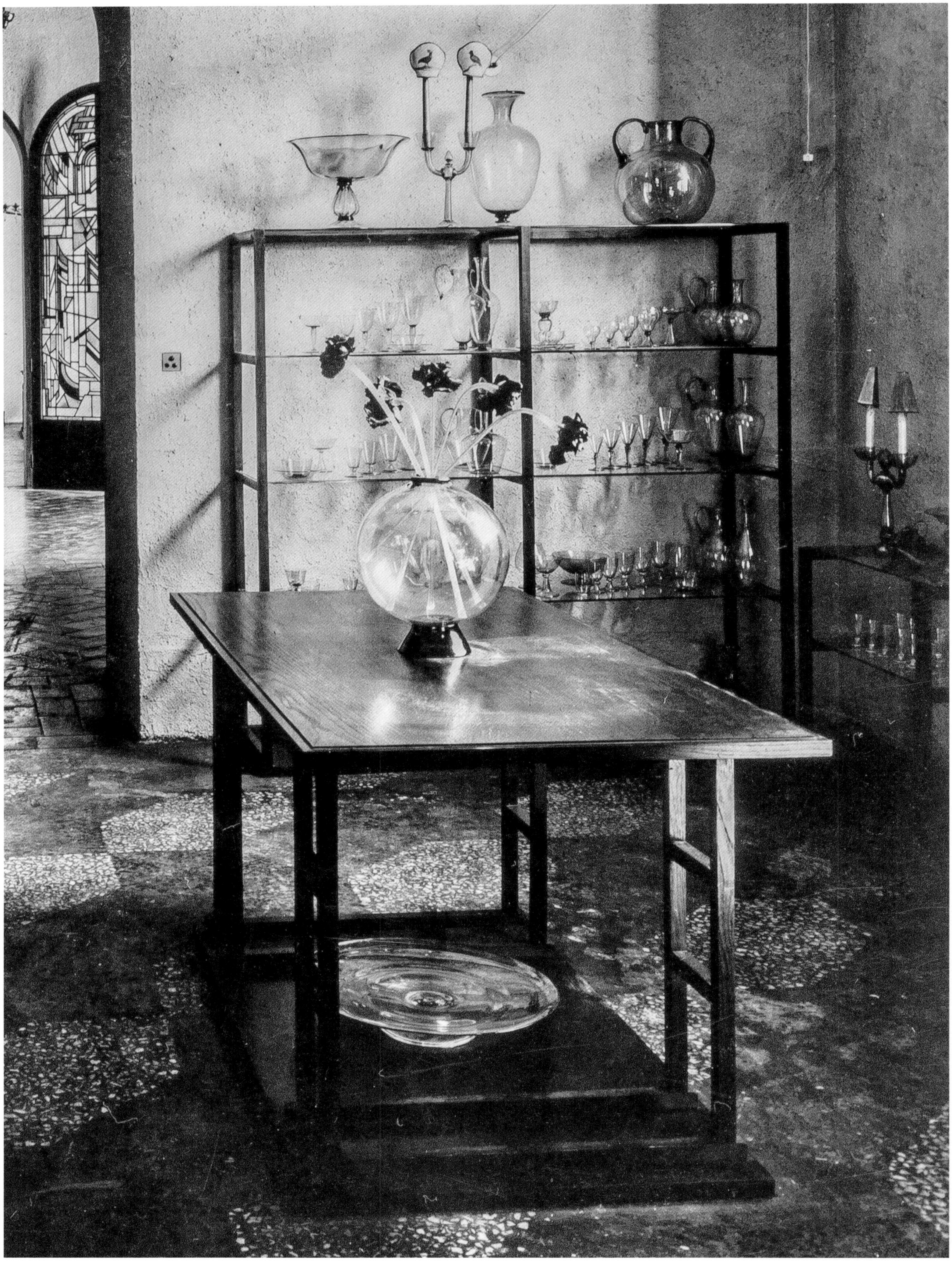

>
Vase no. 5673 with flowers no. 3187 in the Cappellin shop in Florence. Period photograph, 1928-29, detail, CSV

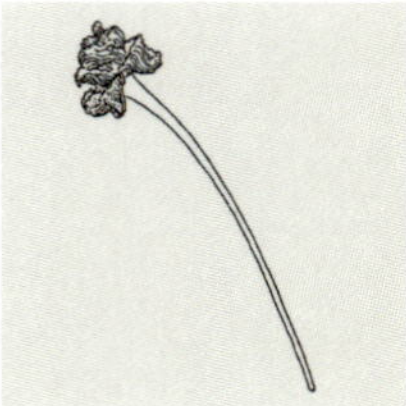

3187

Catalogue drawing for flower (carnation type) to be produced with stem in transparent glass and petals in opaque polychrome glass, AMuVe. As witnessed by factory documents ('Price List' and drawings for models to be exhibited at the 3rd Biennale of Monza Exhibition), an initial carnation ('Carnation of Venice", n.i.) was exhibited in 1927 at 3rd Biennale of Monza Exhibition.
c. 1929
50 cm wide

>
Vase in transparent glass no. 5299 with flowers in polychrome glass. Period photograph, c. 1928, CSV

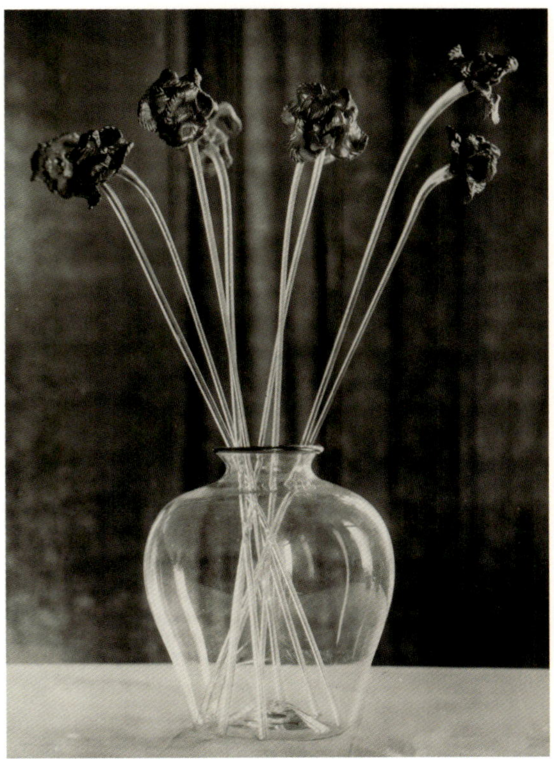

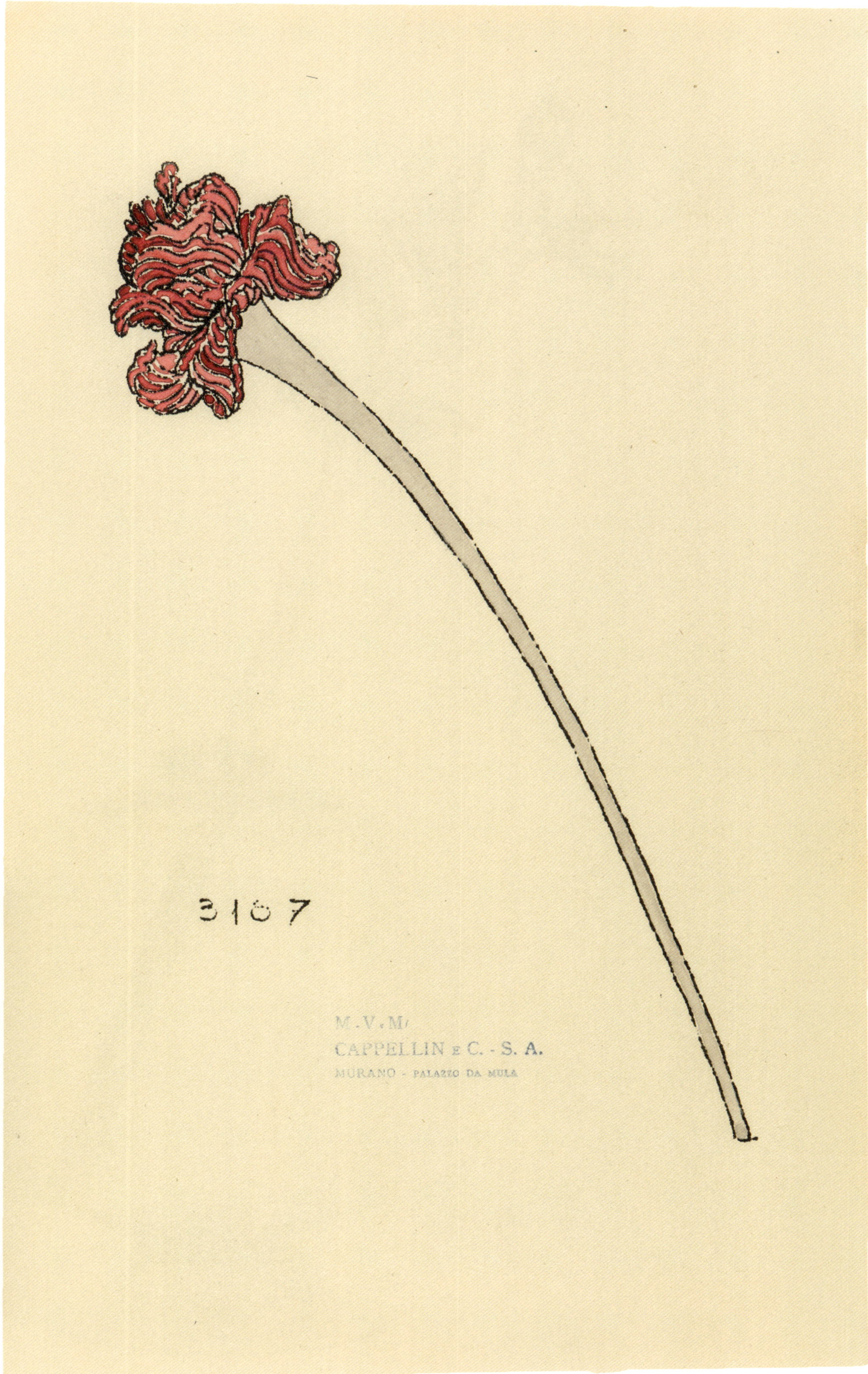

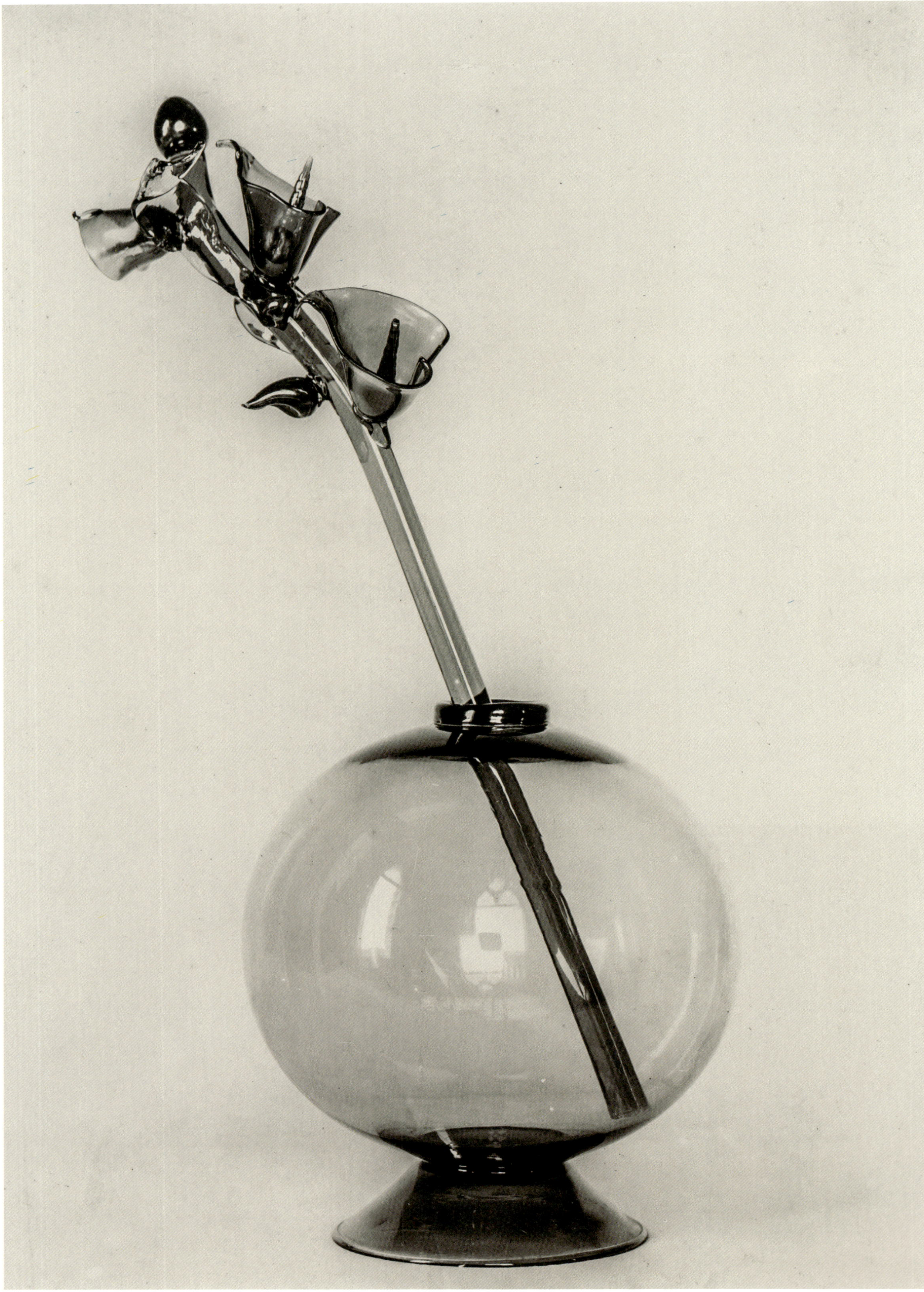

>
Vase no. 5673 with flowering branch. Period photograph, c. 1927, CSV

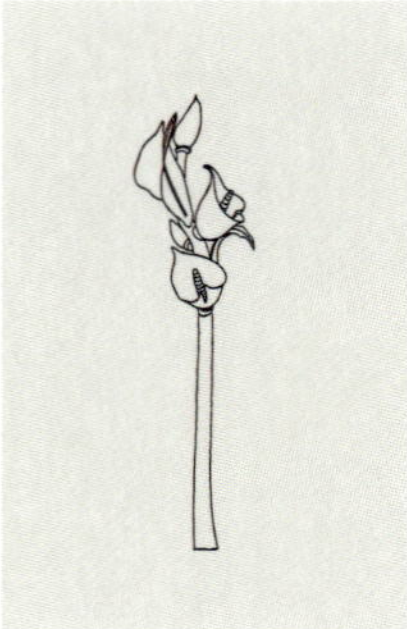

3170

Catalogue drawing for branch with calla lily flowers to be produced in polychrome glass, AMB. A 'white and purple' example, together with a spherical vase with truncated-cone-shaped foot, was published in *Domus* in December 1928, p. 59.
c. 1927
80 cm high

exhibitions
1927, Venice, 18th Mostra dell'Opera Bevilacqua La Masa
1931, Amsterdam, Exhibition of glass, ceramics and lace

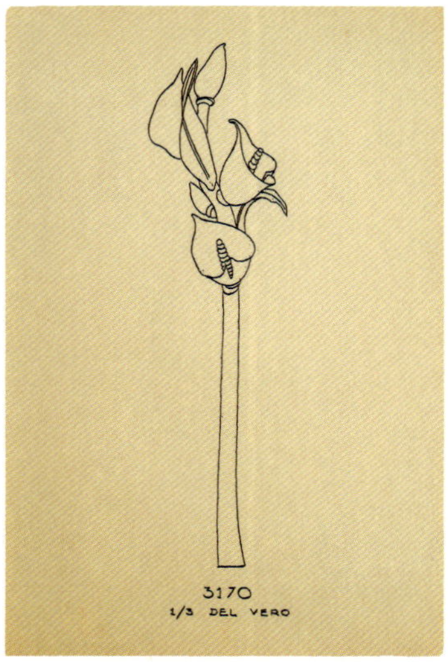

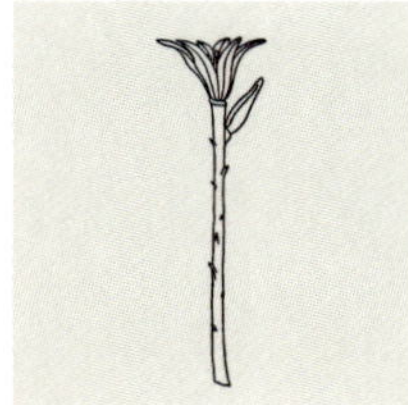

3168

Catalogue drawing for branch with leaf and flowers of the lily type, to be produced in polychrome glass, AMB.
c. 1927
41 cm high

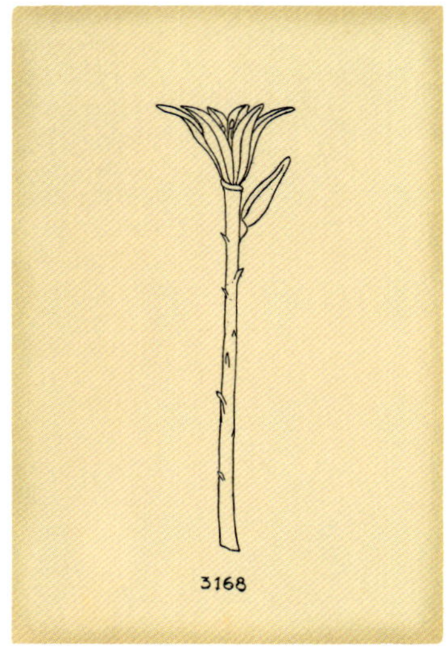

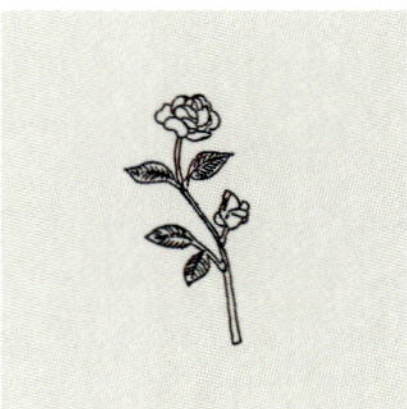

3179

Catalogue drawing for flowering branch with rose, bud and leaves, to be produced in glass, AMB
30 cm high
c. 1929

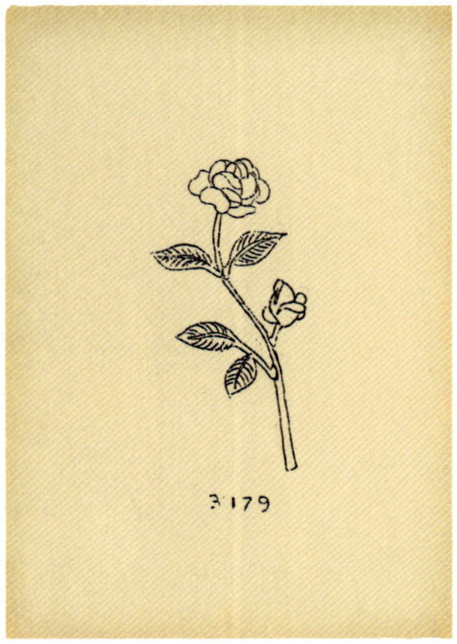

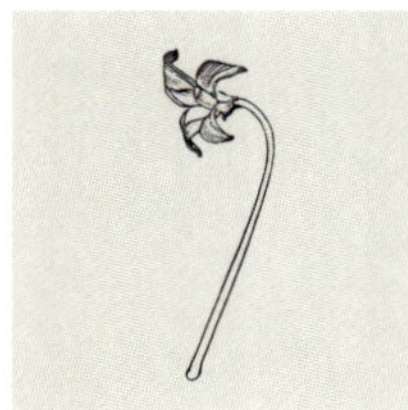

3191

Working drawing for *Orchid* to be produced with stem and petals in *rigadin* glass (with veining), CSV.
1929–30
45 cm high

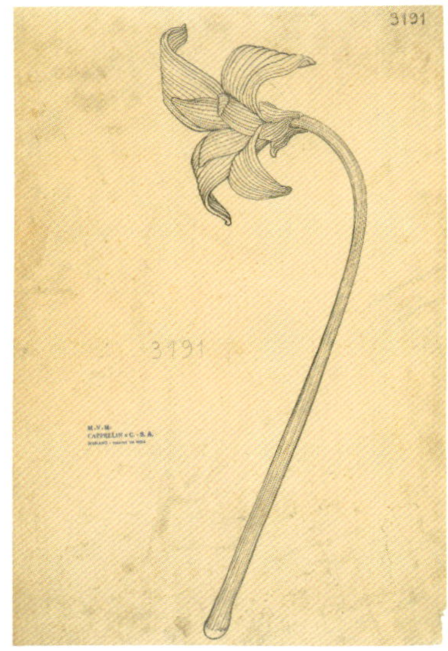

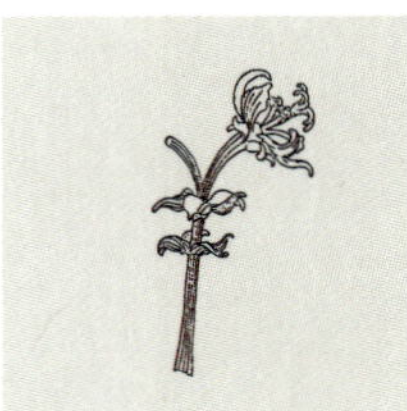

3188

Catalogue drawing for flowering branch to be produced in polychrome glass; petals, stem and leaves in *rigadin* (with veining), AMuVe.
1929–30
36 cm high

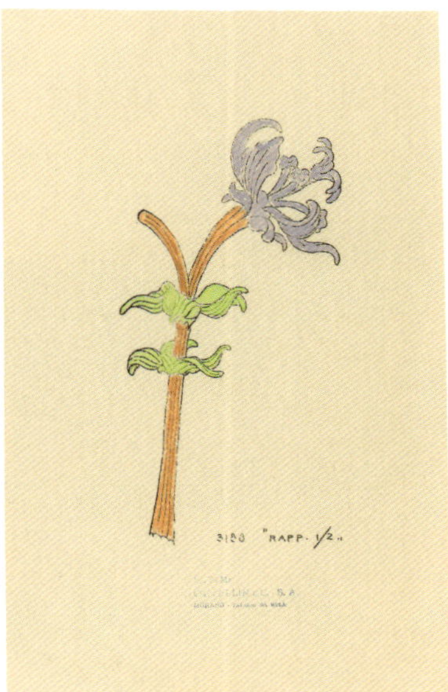

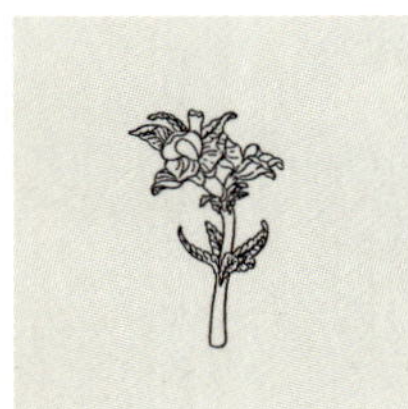

3189

Catalogue drawing for flowering branch with leaves and double inflorescence to be produced in polychrome glass, AMuVe.
1929–30
28 cm high

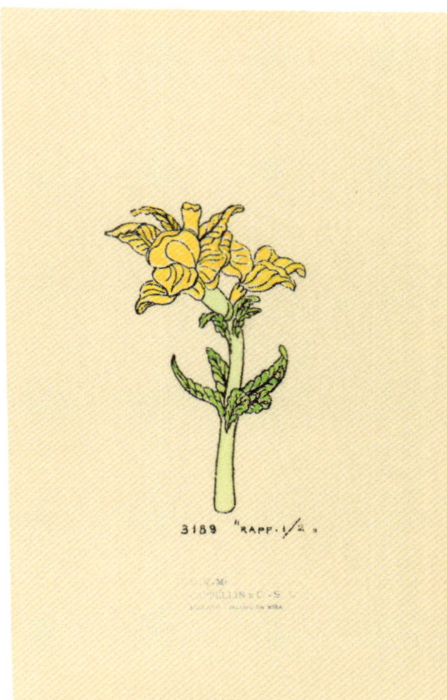

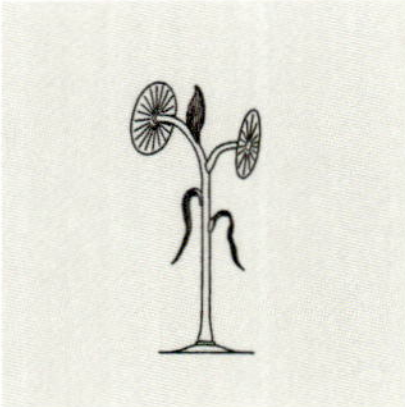

3166

Flowering plant on disk-shaped foot in amethyst transparent glass. Corollas in circular form in *lattimo* glass, leaves and finishing in green *pasta vitrea*. The piece is a variant with four flowers of no. 3166, which was also produced in monochrome transparent glass. The model was also proposed in the catalogue with more flowers and greater height (no. 3165, h 42 cm).
c. 1927
30 cm high
Private collection

>
3166, catalogue drawing, AMB

>
Interior of the Cappellin shop in Florence. Period photograph, 1928-29, detail, CSV. Model no. 3166 can be seen in the foreground

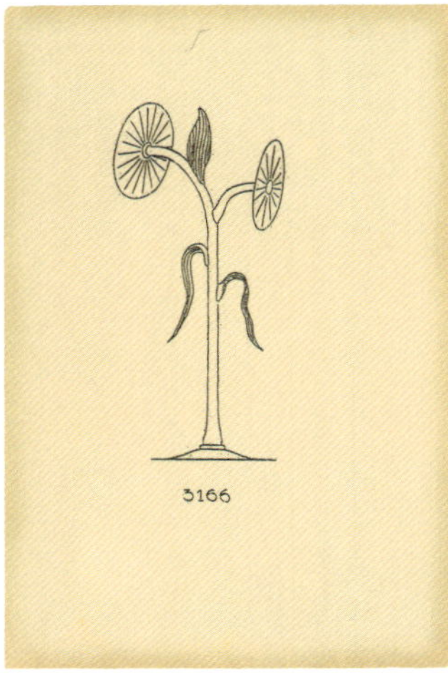

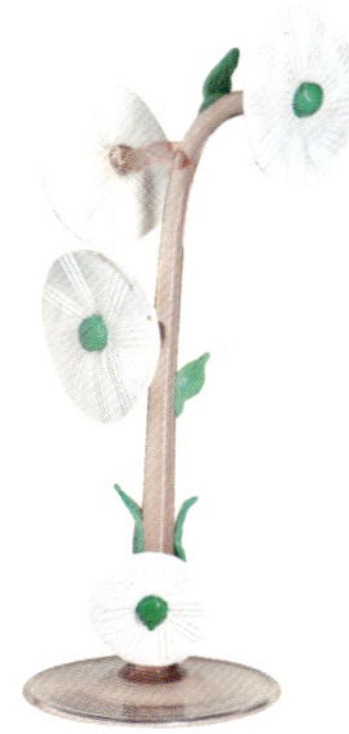

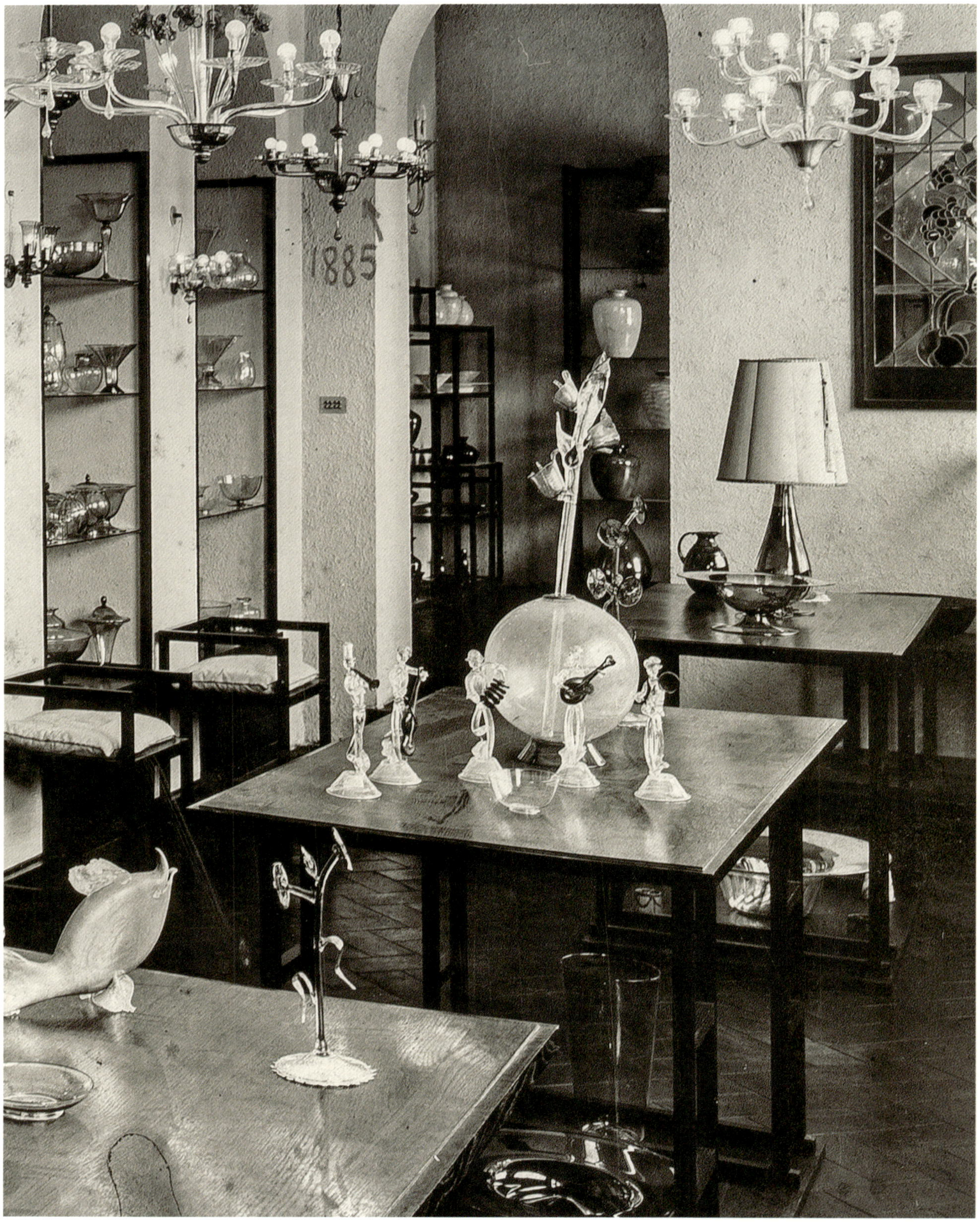

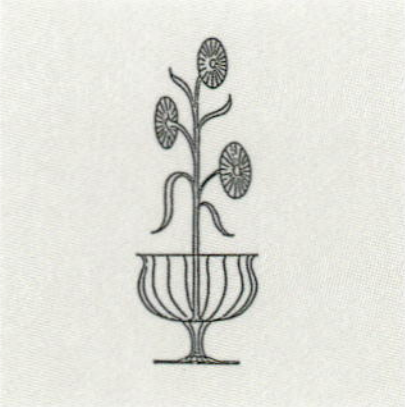

5688

Ribbed goblet on stem and disk-shaped foot, in *pagliesco* glass with flowering plant. Plant with two flowers in circular form and leaves in *pagliesco* glass. Surface with iridised finish. The piece is a variant in a reduced size of the model of the vase with plant no. 5688 (h 38.5 cm and three corollas).
c. 1927
29 cm high
M.V.M. Cappellin Murano *acid stamp*
Private collection, Treviso

>
5688, catalogue drawing, AMB

>
Window of the Cappellin shop in Milan. Period photograph, c. 1928, CSV

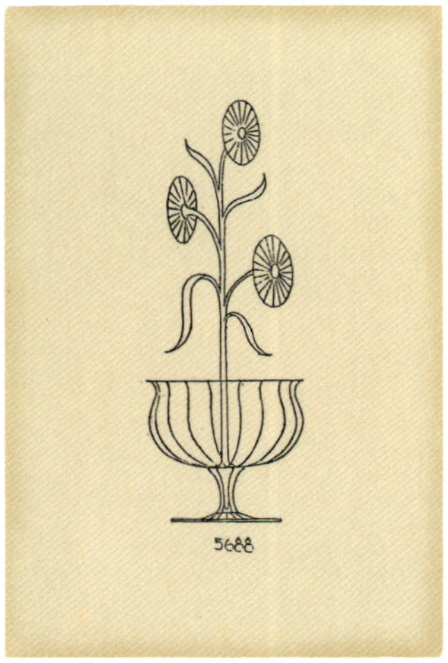

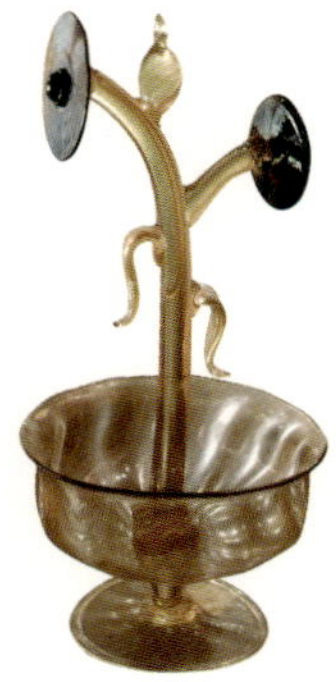

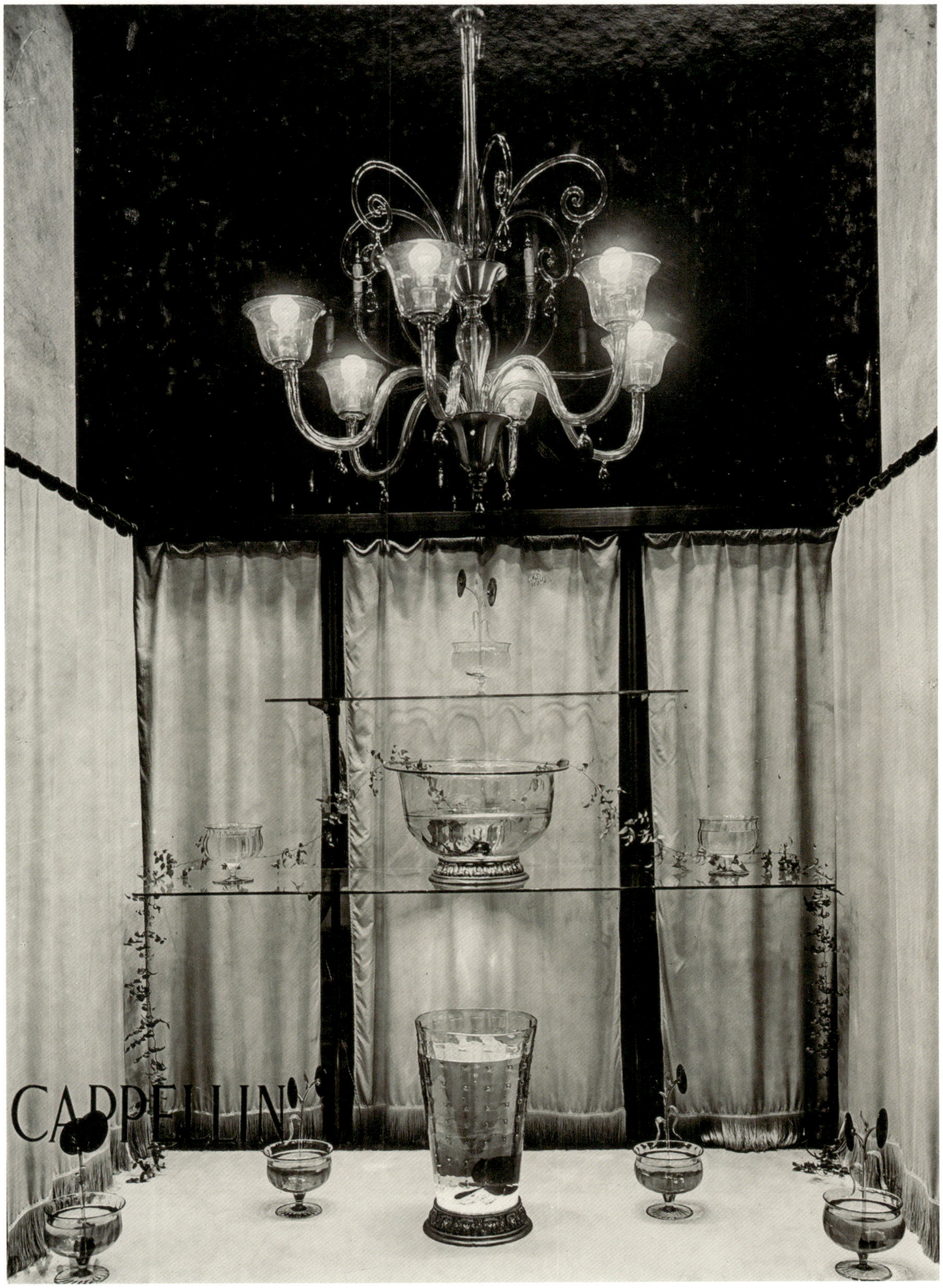

>
Candlestick, bowl, plate and elements for centrepiece with floral motif. Period photograph, c. 1929, CSV

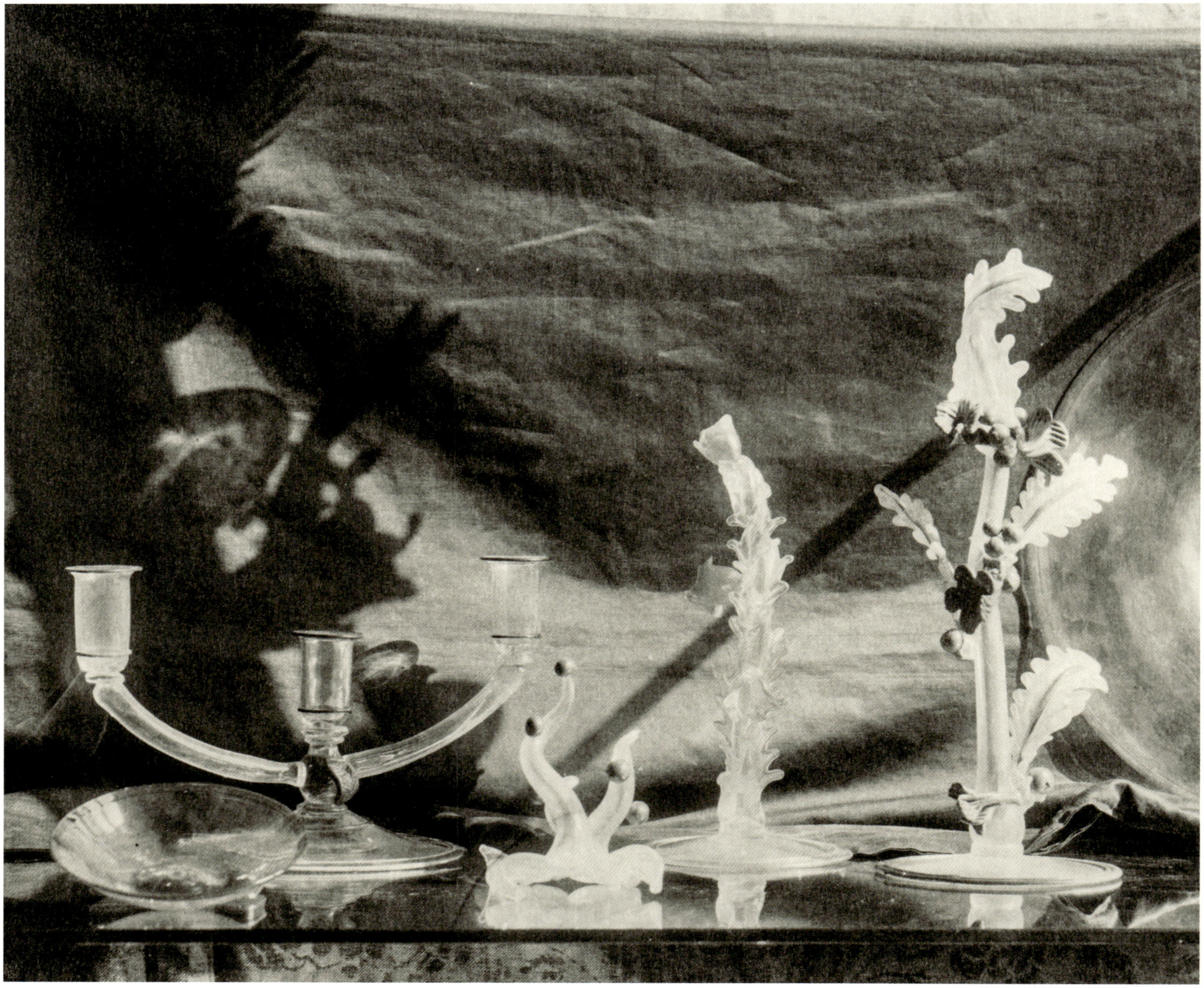

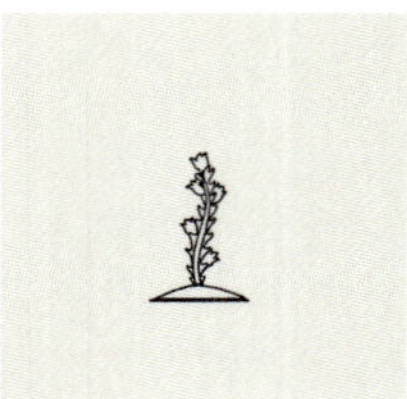

3172

Flowering branch on disk-shaped foot in crystal glass with finishings in ruby-coloured glass. Jagged stem with ruby nucleus and four small flowers in ruby-coloured glass. Surface with iridised finish. Some similar examples in 'iridised crystal [with] coral-red dec[oration]' were exhibited in 1929 at the Leipzig Trade Fair as part of a centrepiece.
c. 1929
18 cm high
Private collection

exhibitions
1929, Leipzig, Trade Fair

>
3172, catalogue drawing, AMB

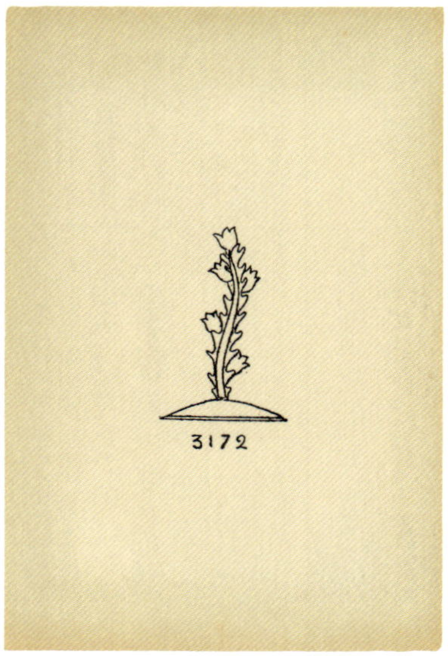

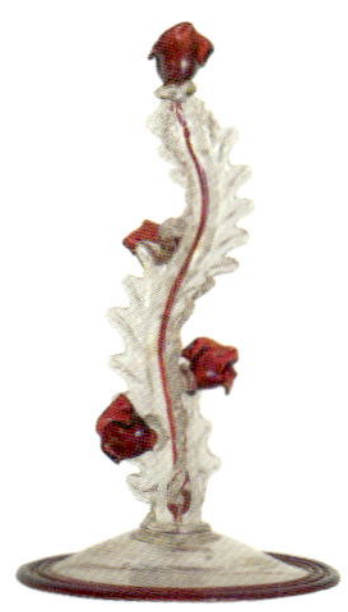

n.i.

Decorative flower in cased *lattimo* and amethyst glass with leaves and short stem in crystal glass. Strongly iridised surface. It is probably a prototype, without a catalogue number, produced in the final period of the glasswork's activities.
1930–31
9 cm high; 20 cm ø
Cappellin collection, Venice

>
Carlo Scarpa, study for decorative flower, 1930–31, ACSVr

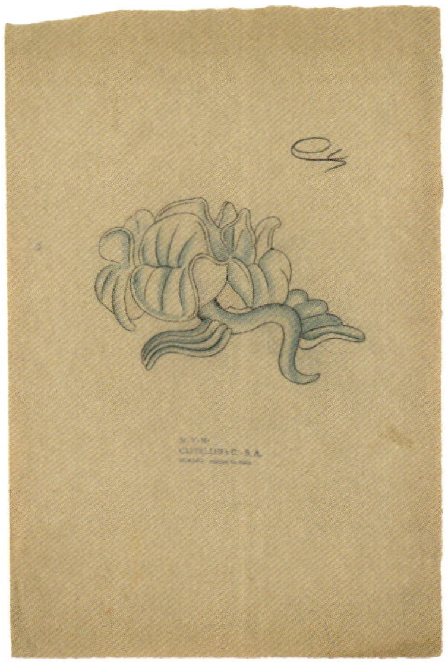

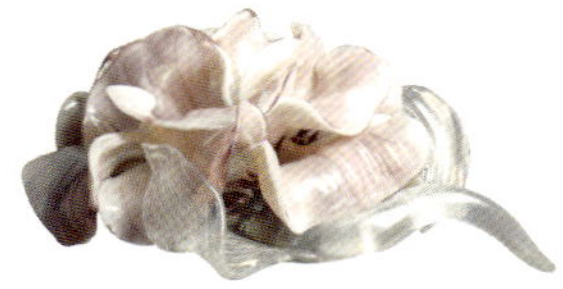

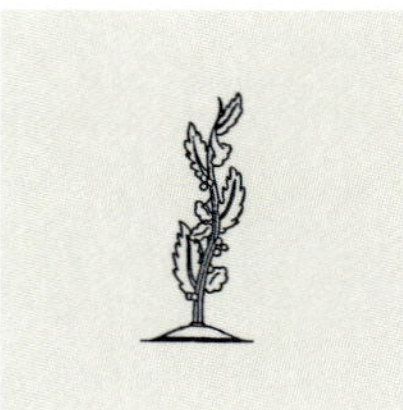

3171

Catalogue drawing for a flowering branch on disk-shaped foot to be produced in glass, AMB. A version of the model is documented in transparent glass with finishings in *pasta vitrea*.
c. 1929
29 cm high

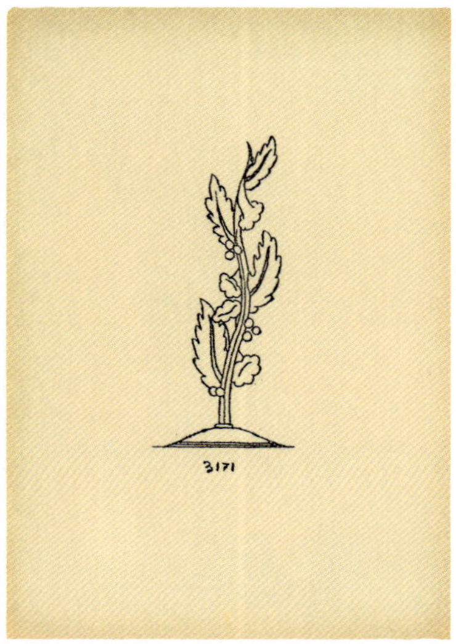

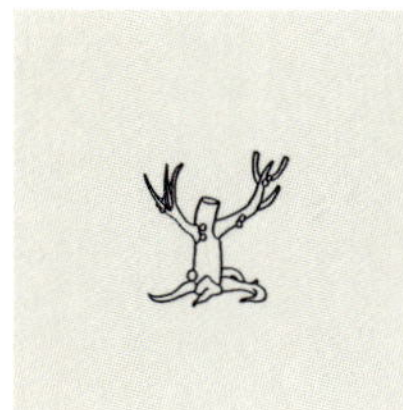

3182

Catalogue drawing for a branching aquatic plant with little globular fruit to be produced in glass, AMB.
c. 1929
17 cm high

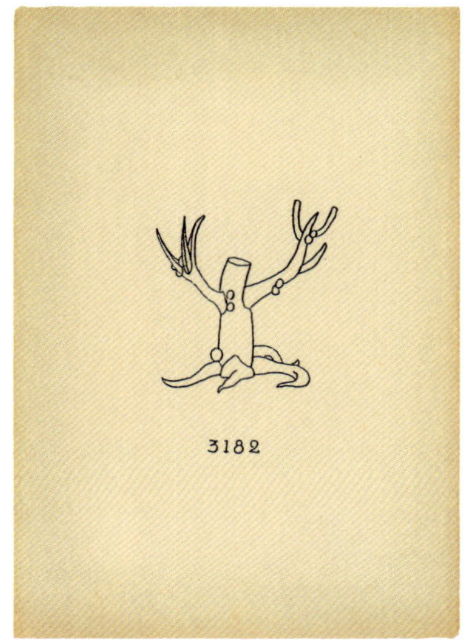

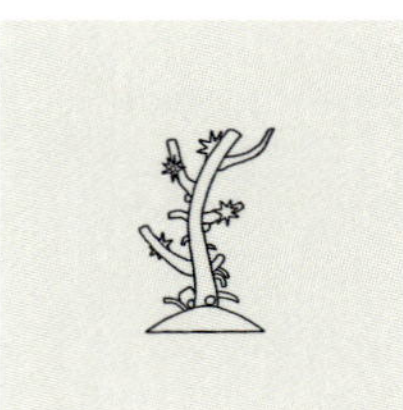

3183

Catalogue drawing for a branching aquatic plant with sinuous stem and flowers, to be produced in glass, AMB. Skullcap foot.
c. 1929
24 cm high

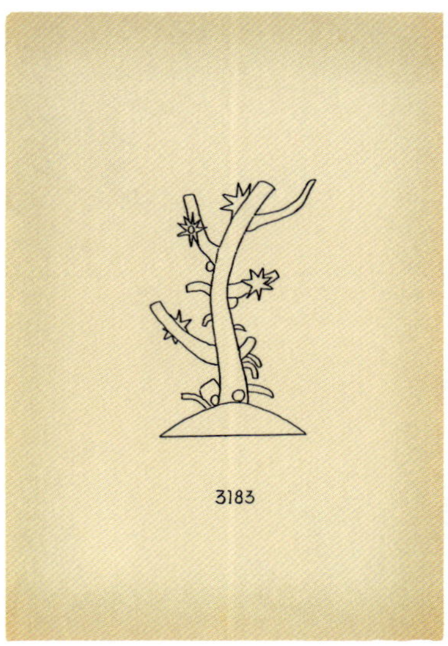

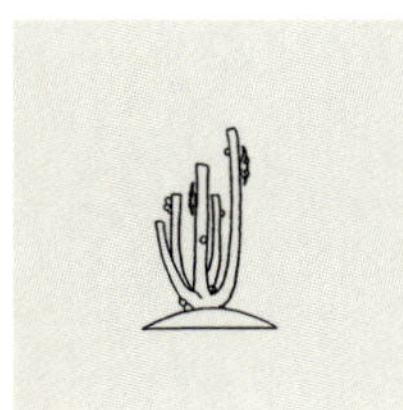

3184

Catalogue drawing for a branching aquatic plant with flowers, to be produced in glass, AMB. Skullcap foot.
c. 1929
23 cm high

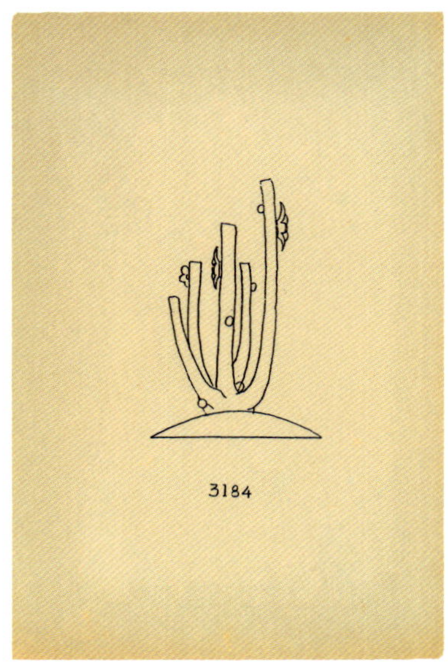

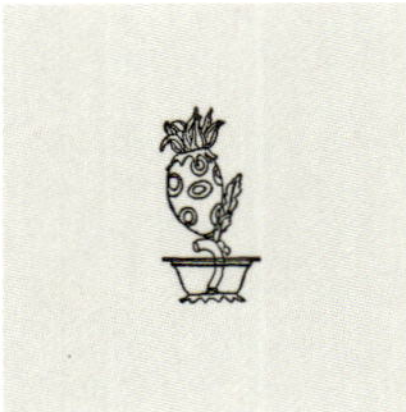

3181

Catalogue design for a succulent plant with vase to be produced in glass, AMB. As the design would seem to indicate, the body of the plant was planned with a *millefiori* murrine decoration. A similar example, produced in *pasta vitrea* without decoration, was exhibited in 1931 at the Amsterdam Exhibition of glass, ceramics and lace.
c. 1929
22 cm high

exhibitions
1931, Amsterdam, Exhibition of glass, ceramics and lace

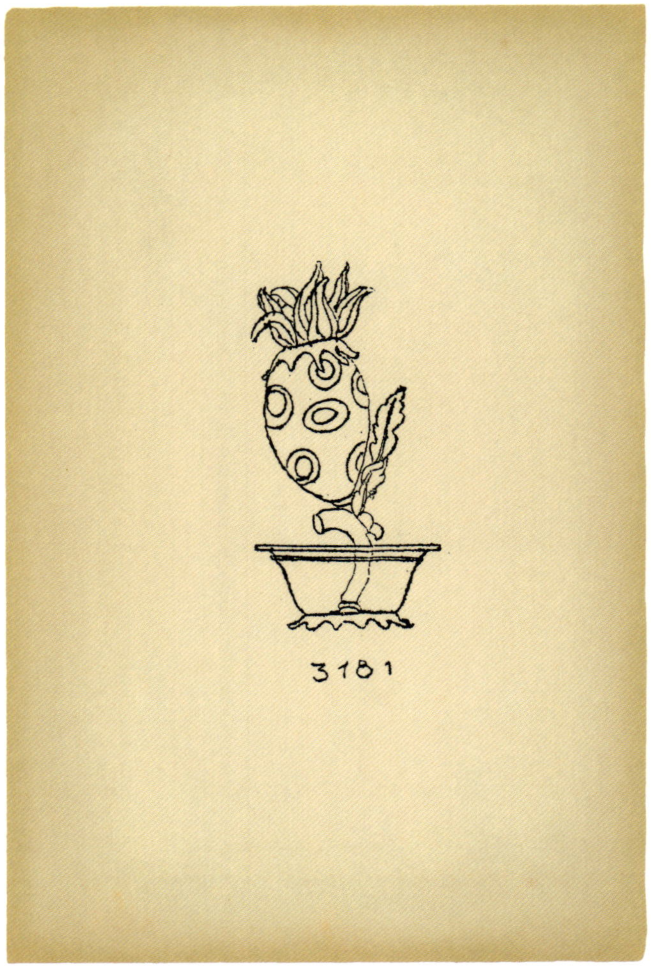

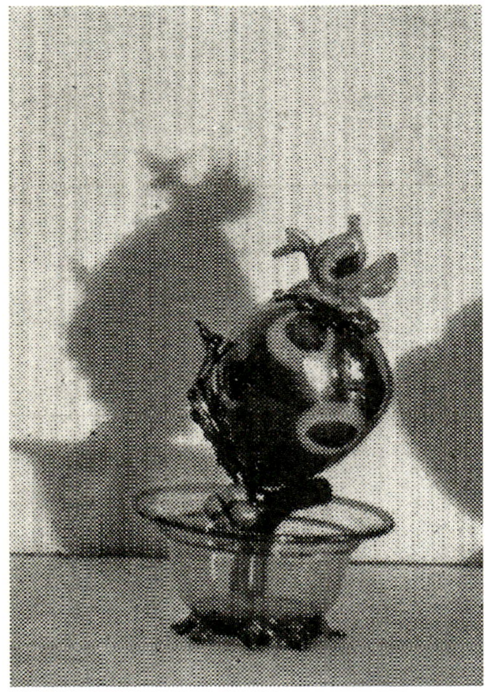

>
Succulent plant no. 3181 in Cappellin Glassware, catalogue, undated, detail, AHL

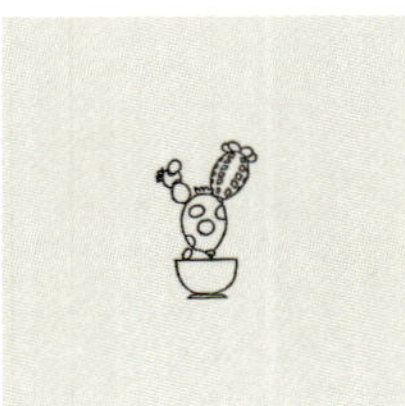

3180

Catalogue drawing for a succulent plant with vase to be produced in opaque glass, 1929–31, AMB. As the drawing would seem to indicate, the body of the plant was planned with irregular spots and segmented canes. A similar example, produced in *pasta vitrea*, with a crossed-thread decoration, was exhibited in 1931 at the Amsterdam Exhibition of glass, ceramics and lace.
c. 1929
18 cm high

exhibitions
1931, Amsterdam, Exhibition of glass, ceramics and lace

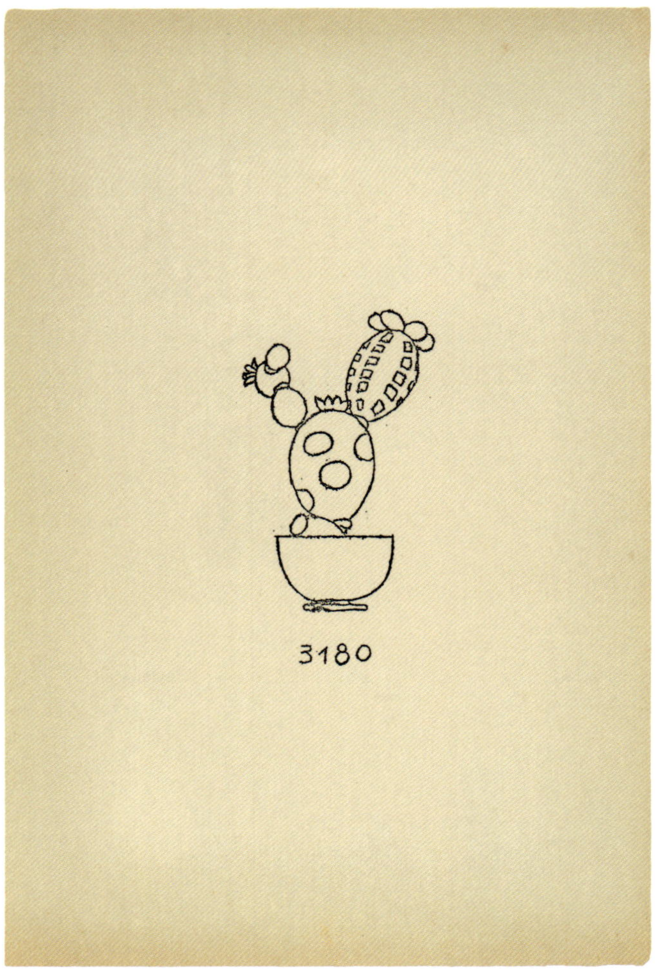

>
Succulent plants and vases in pasta vitrea *exhibited at the Amsterdam Exhibition of glass, ceramics and lace in 1931, detail, AHL*

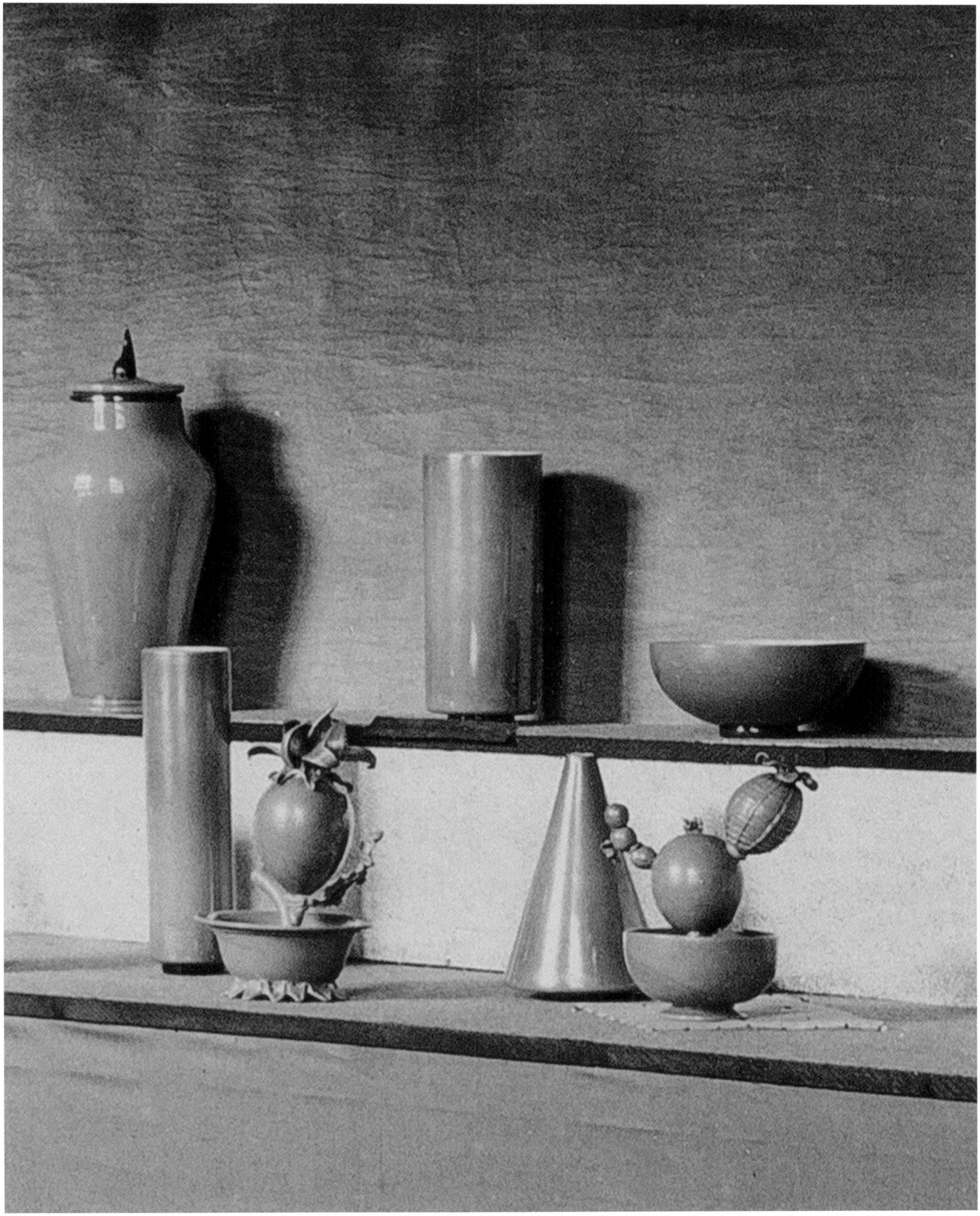

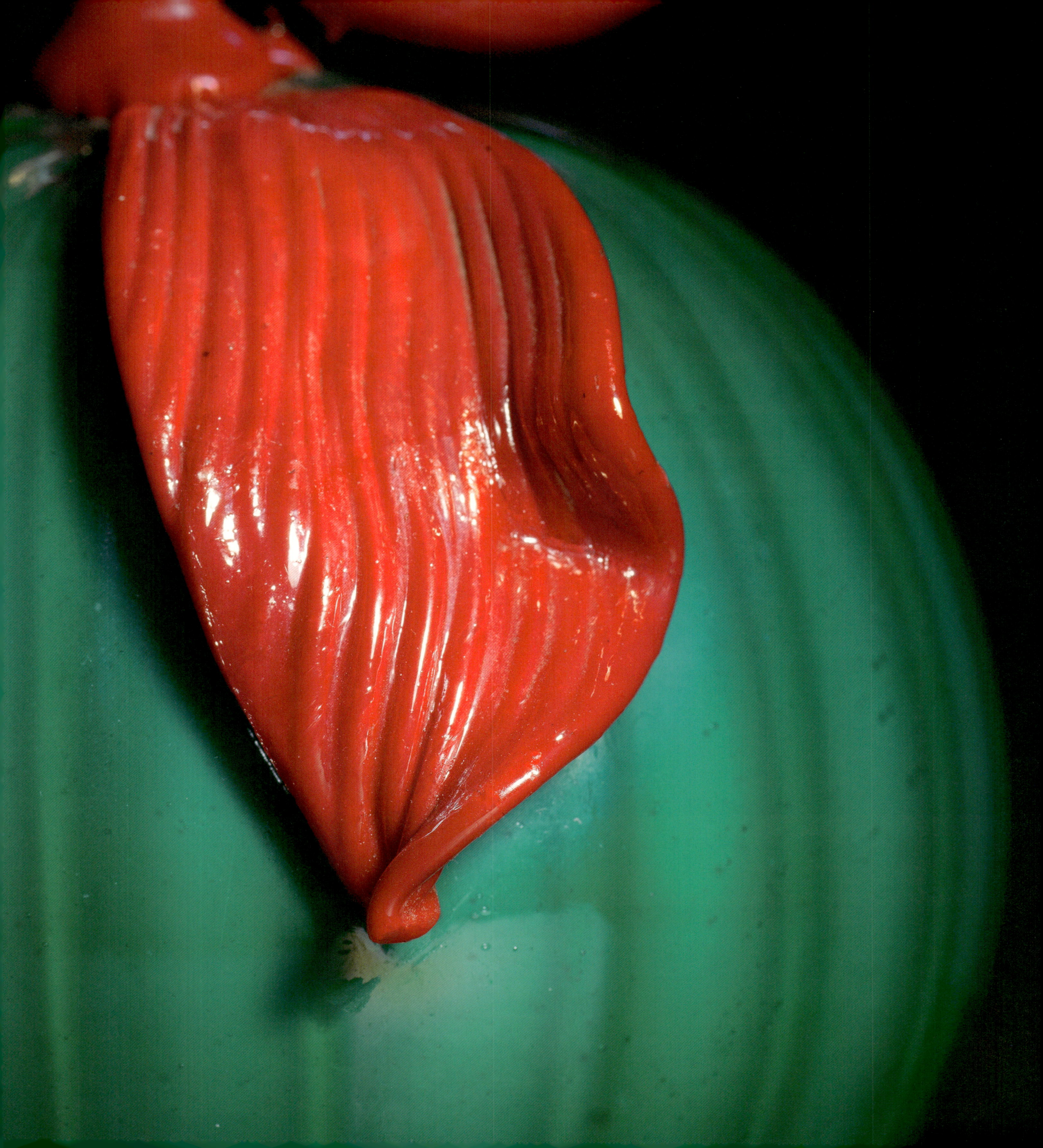

Among the products of the late 1920s there were also new models of fruit which differed from those designed earlier by Zecchin in 1922 for V.S.M. Cappellin Venini e C. by a greater freedom in the design and the material in which they were executed (*lattimo* glass and polychrome opaque glass). In various cases their unrealistically 'large' dimensions contributed to the originality of the pieces, designed with a certain playfulness and often finished with ribbing. As could also be appreciated on the occasion of the American travelling exhibition *Contemporary glass and rugs* (1929-30), these pieces were quite often characterised by a self-confident use of intense colours, allowing for unusual associations of colour, for example black and red.

FRUIT 1928-30

Exhibitions

1928, Paris, *Salon d'automne*
1929-30, USA *International Exhibition of Contemporary Glass and Rugs*, travelling exhibition
1930, Monza, 4th *Esposizione Triennale Internazionale delle Arti Decorative ed Industriali Moderne*
1931, Amsterdam, *Tentoonstelling van glas, aardewerk en kant, van moderne italiaansche kunstnijverheid. Mostra di vetri, ceramiche e merletti d'arte moderna italiana*

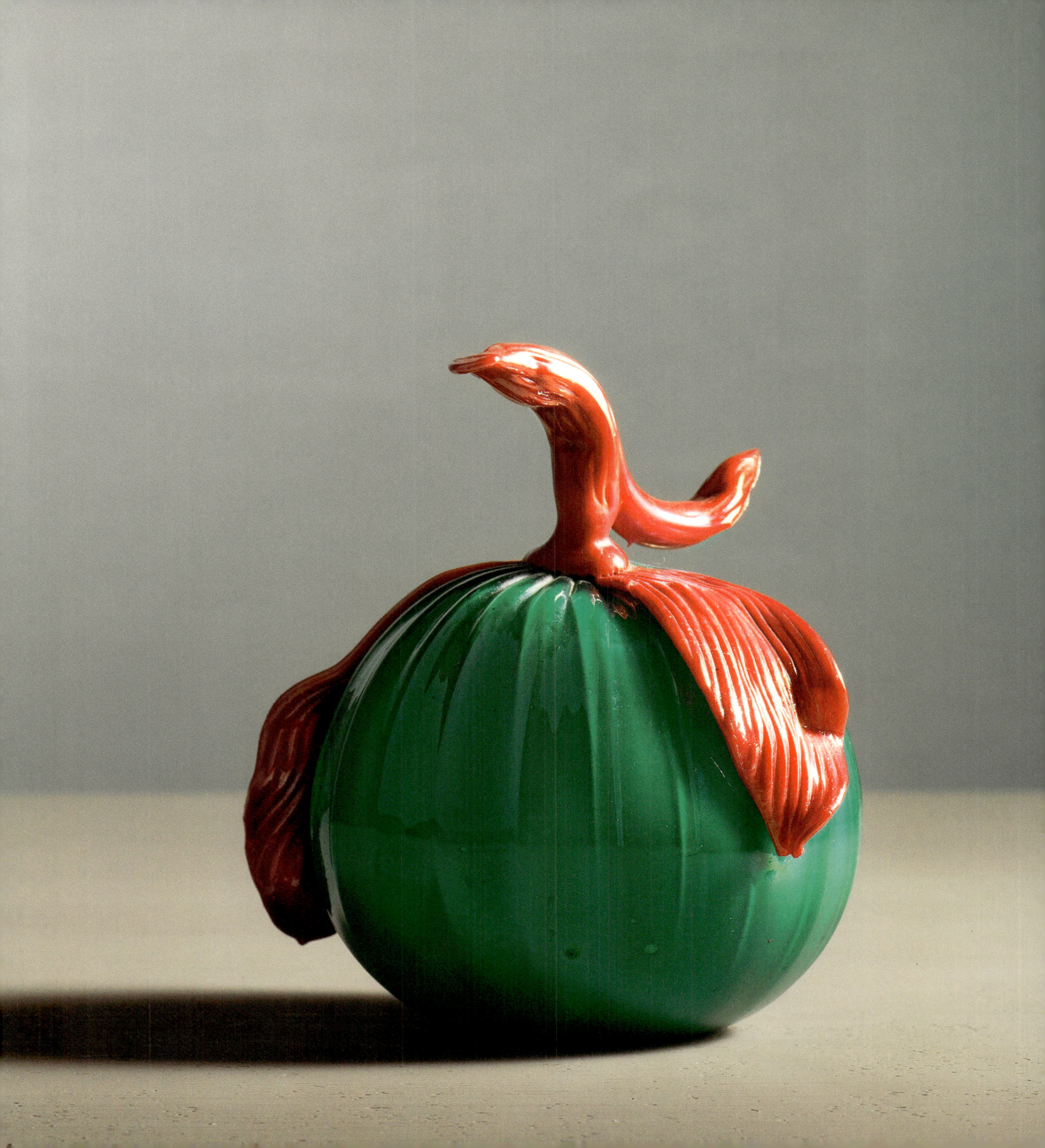

>
'Sculptures in polychrome glass', in Tentoonstelling van glas, aardewerk en kant, van moderne italiaansche kunstnijverheid. Mostra di Vetri, Ceramiche e Merletti d'Arte Moderna Italiana, *Amsterdam 1931, p. 23, detail*

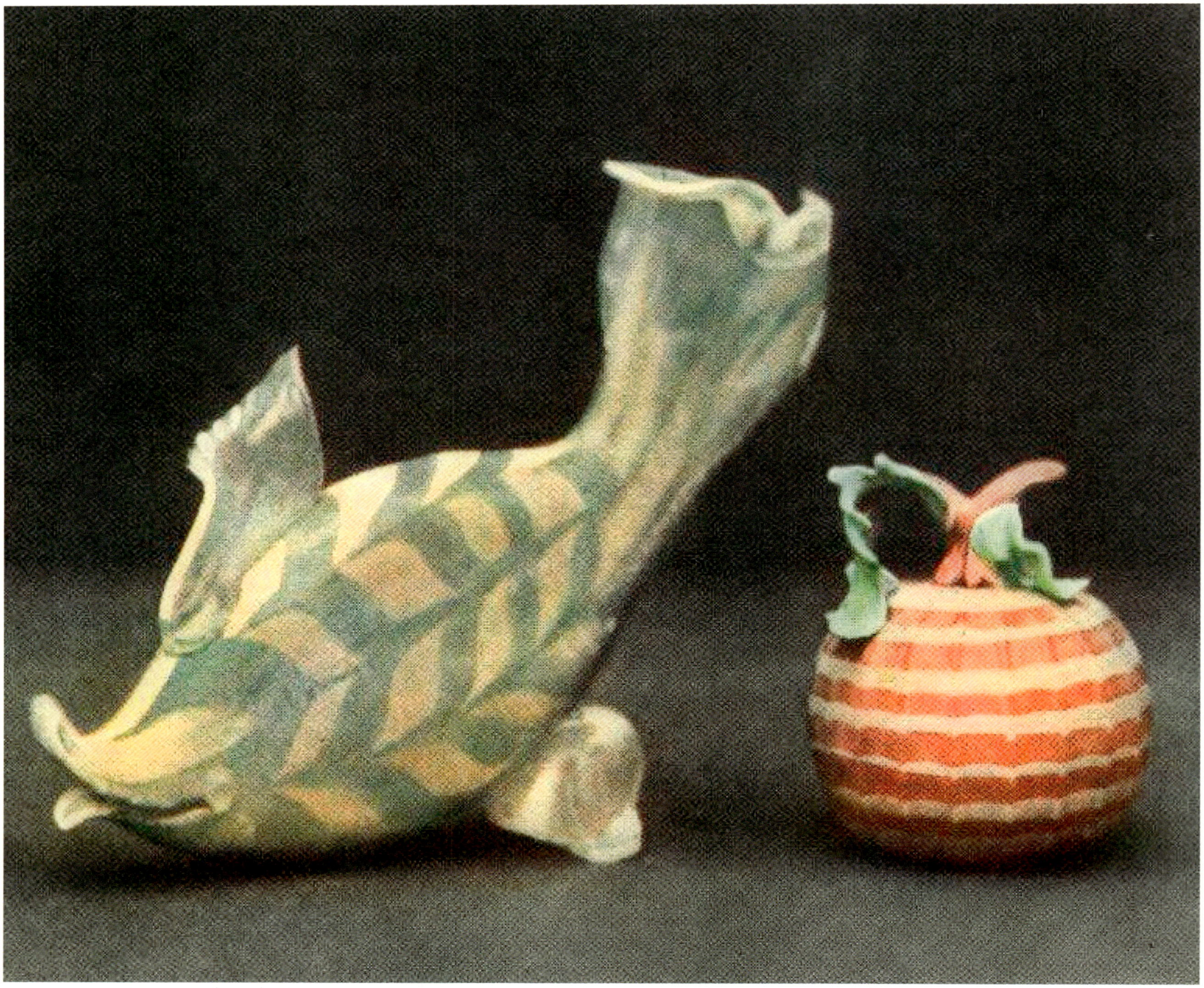

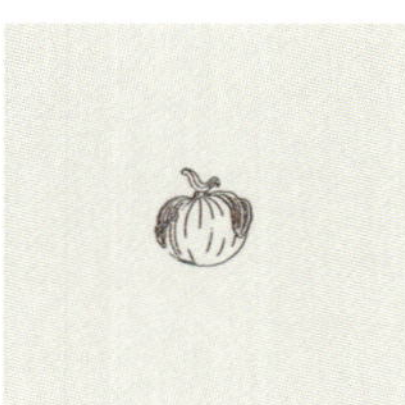

4732

Catalogue drawing for an apple with stem and two leaves to be produced in polychrome glass or in *lattimo* glass, 1928–30, AMuVe. A variant in *lattimo* glass with a red band wound in a spiral was published in the catalogue for the Amsterdam glass, ceramics and lace exhibition (1931).
1928
11 cm high

exhibitions
1931, Amsterdam, Exhibition of glass, ceramics and lace

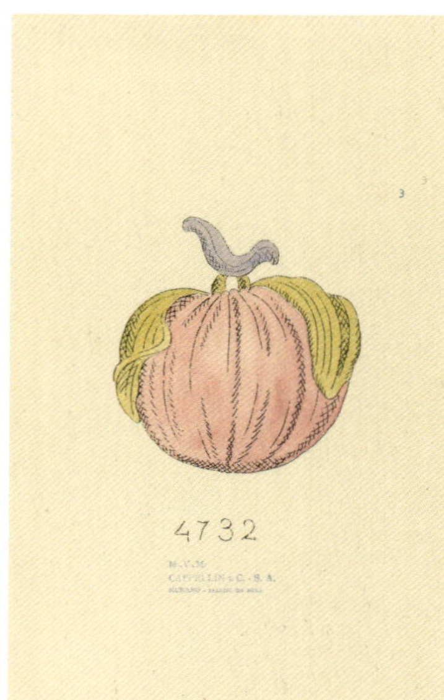

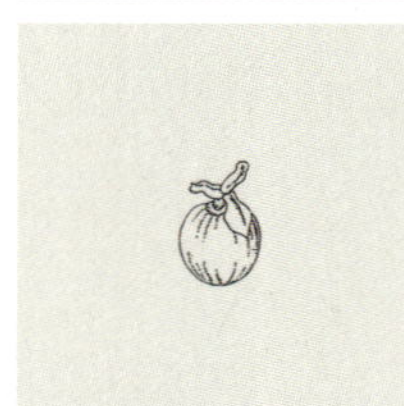

4727

Catalogue drawing for an orange with stem and leaf to be produced in polychrome glass or *lattimo* glass, AMuVe.
1928
14 cm high

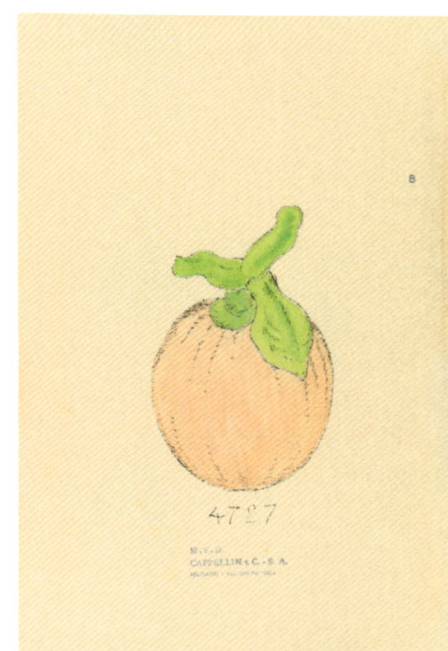

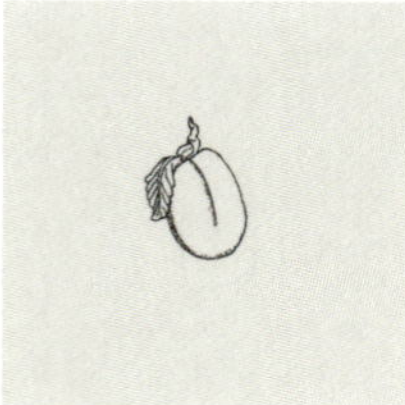

4724

Catalogue drawing for a plum with stem and leaf, to be produced in polychrome glass or *lattimo* glass, AMuVe.
1928
15.5 cm high

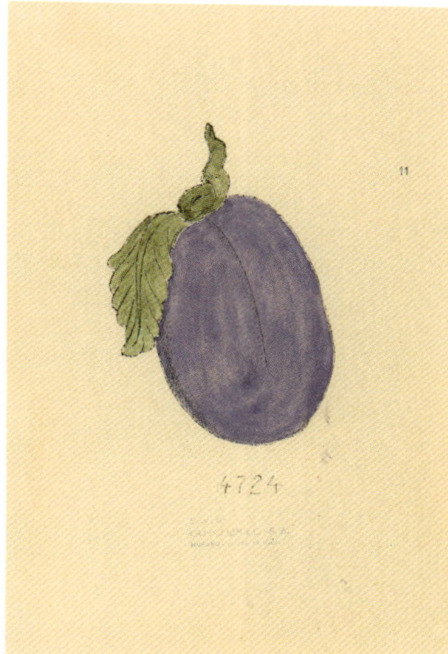

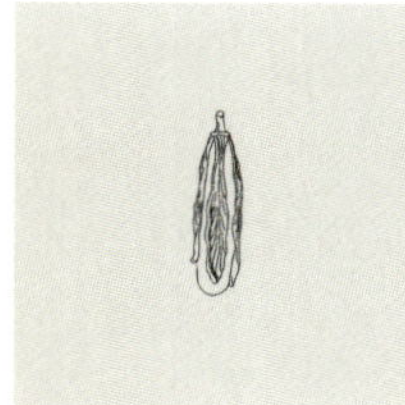

4730

Catalogue drawing for an aubergine (?) with stem and three long leaves to be produced in polychrome glass or *lattimo* glass, AMuVe.
1928
21 cm high

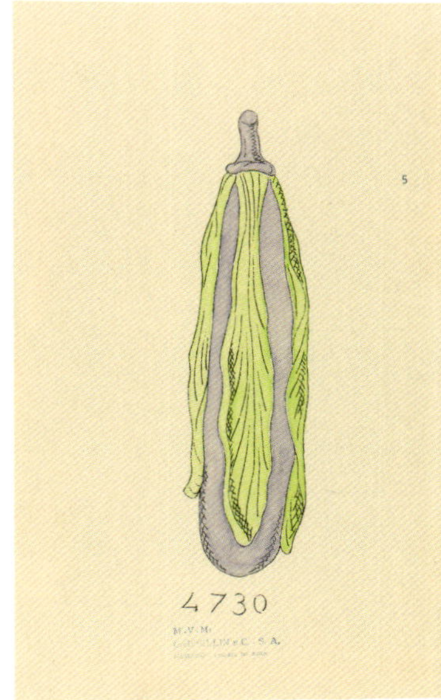

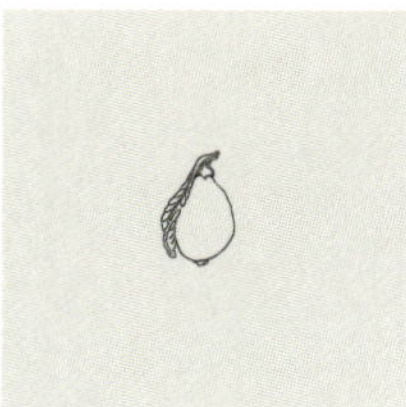

4734

Catalogue drawing for a lemon with stem and leaf, to be produced in polychrome glass or *lattimo* glass, AMuVe.
1928
13 cm high

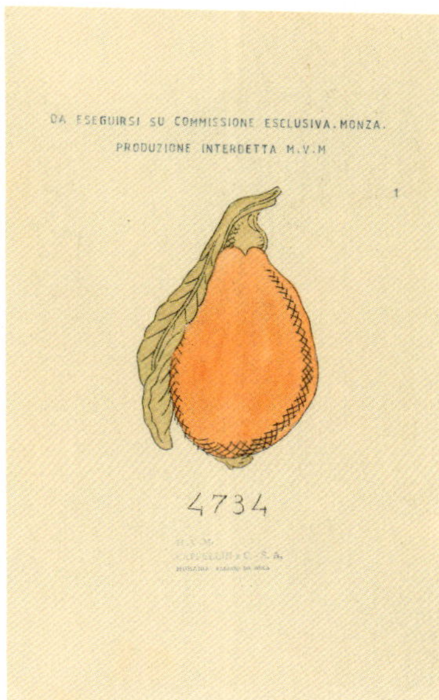

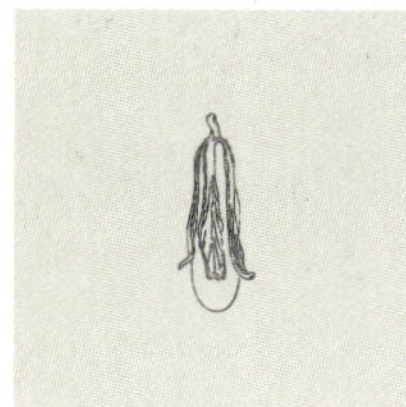

4733

Catalogue drawing for a courgette with stem and three long leaves to be produced in polychrome glass or *lattimo* glass, AMuVe.
1928
22 cm high

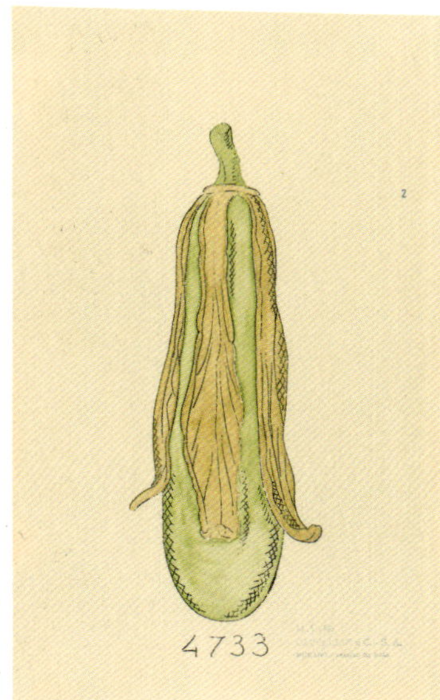

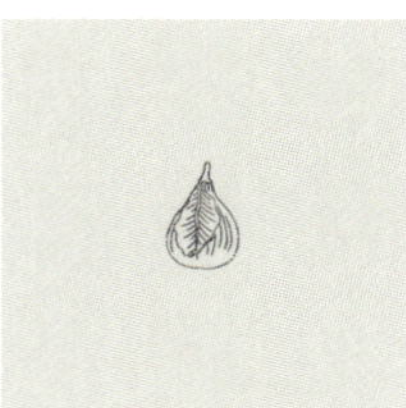

4731

Catalogue drawing for a fig with stem and leaf to be produced in polychrome glass or *lattimo* glass, AMuVe.
1928
14.5 cm high

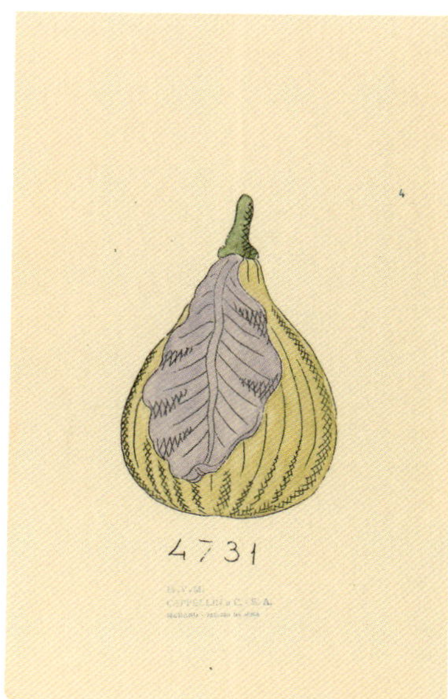

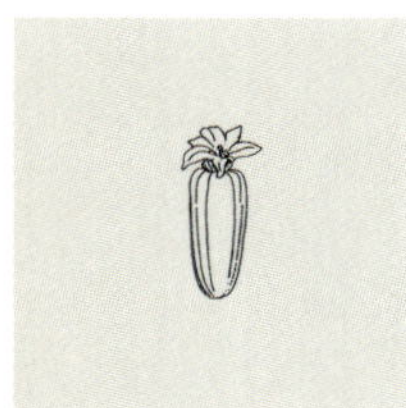

4725

Catalogue drawing for an oblong ribbed fruit with leaves arranged in a crown, to be produced in polychrome glass or *lattimo* glass, AMuVe.
1928
19 cm high

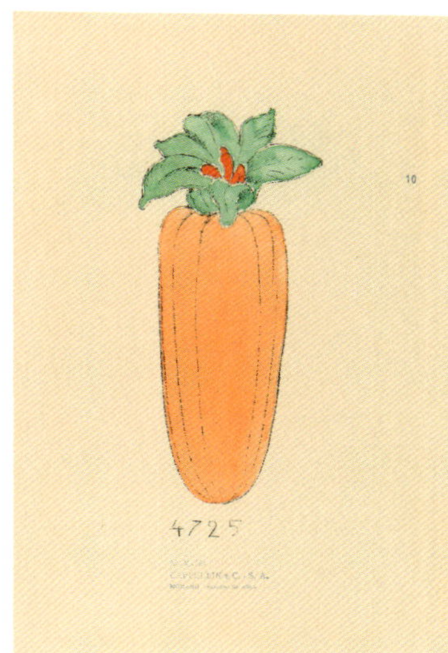

>
Cappellin fruit and vases in Arts and Decoration, *November 1930, p. 1*

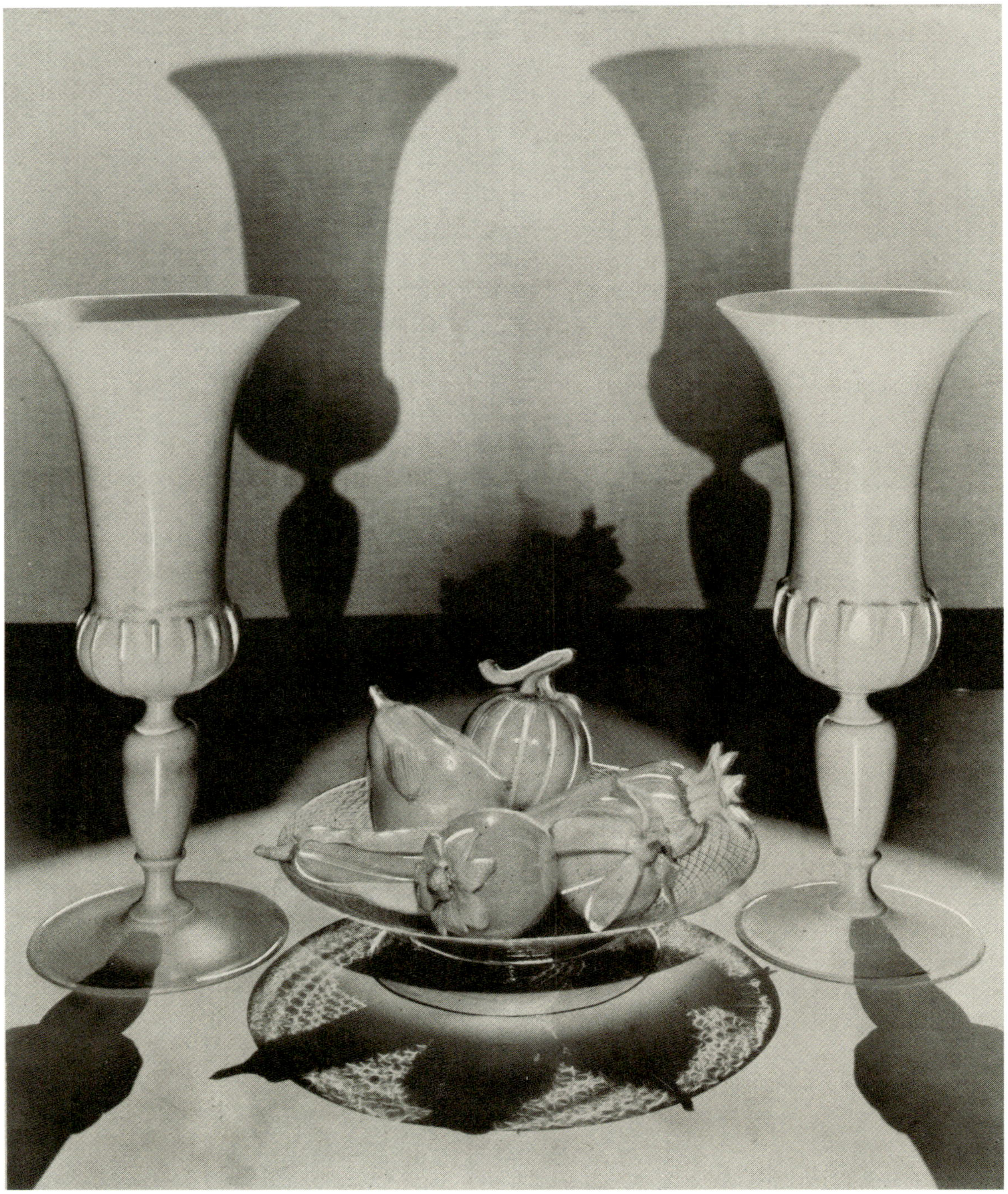

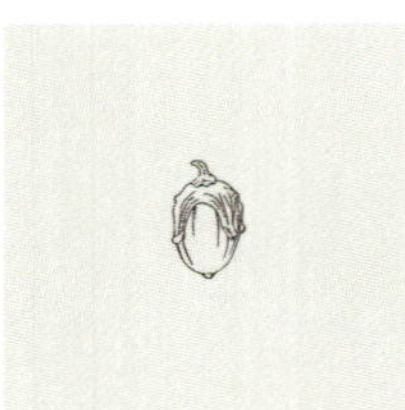

4726

Fruit in *lattimo* glass with veins, stem and four leaves. Strongly iridised surface. The model was also produced in polychrome opaque glass.
1928
13 cm high
Chris Stines collection, Alabama

>
4726, catalogue drawing, 1928–30, AMuVe

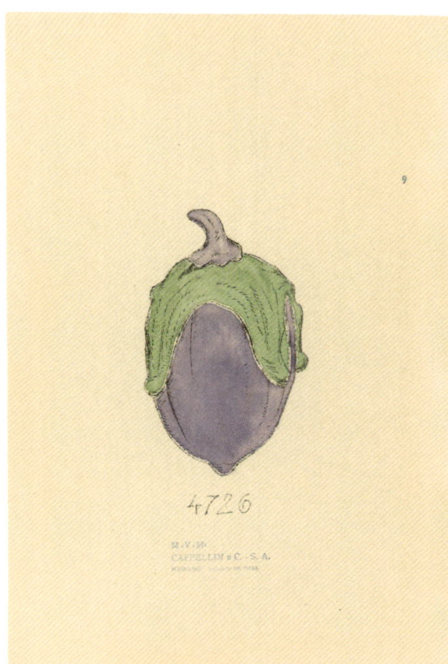

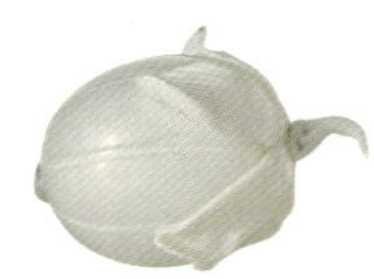

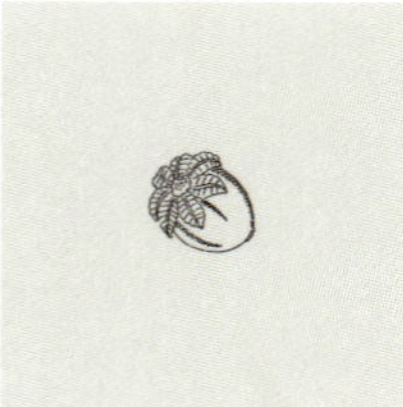

4728

Fruit in *lattimo* glass with six small leaves and inflorescence at the top. Strongly iridised surface. The model was also produced in polychrome opaque glass.
1928
10 cm high
Chris Stines collection, Alabama

>
4728, catalogue drawing, AMuVe

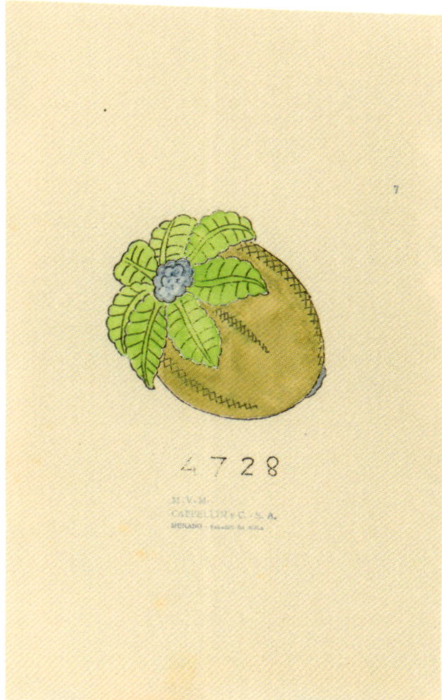

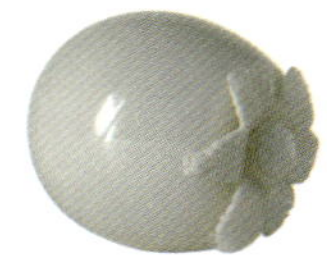

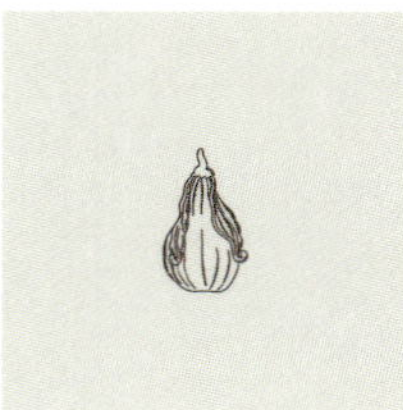

4729

Pear in *lattimo* glass with veins, stem and two leaves. Strongly iridised surface. The model was also produced in polychrome opaque glass.
1928
15 cm high
Chris Stines collection, Alabama

>
4726, catalogue drawing, AMuVe

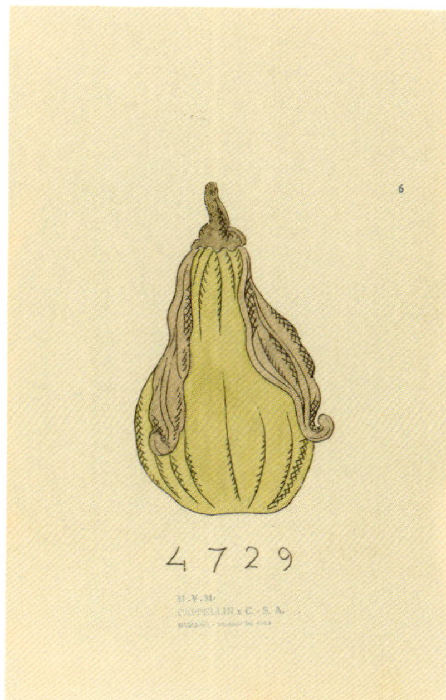

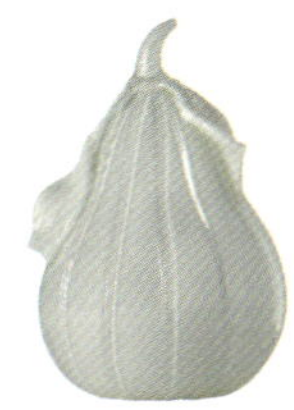

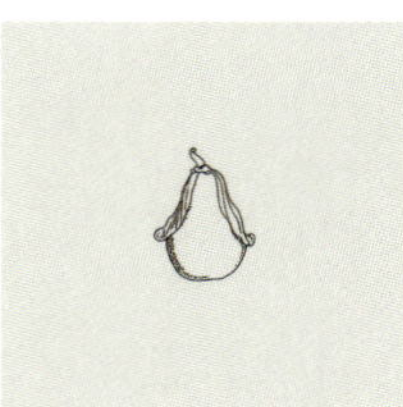

4723

Catalogue drawing for pear with stem and two leaves to be produced in polychrome glass or in *lattimo* glass, AMuVe.
1928
14.5 cm high

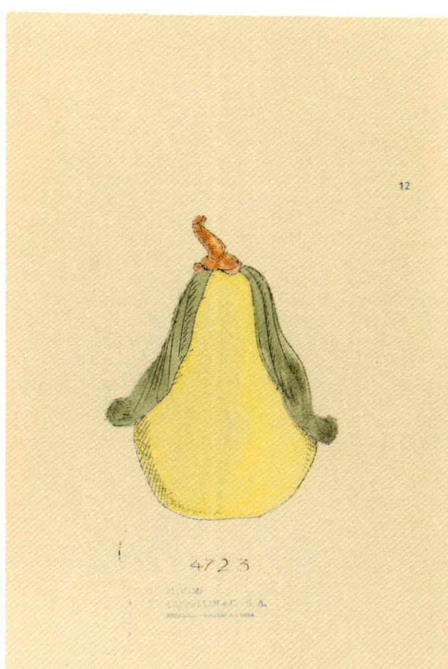

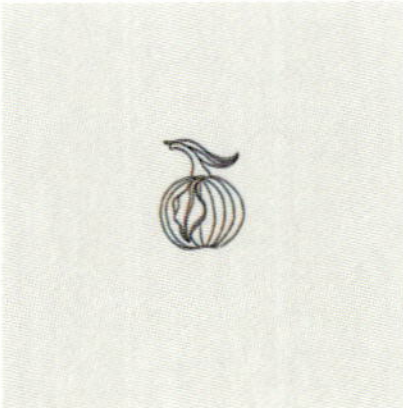

4742

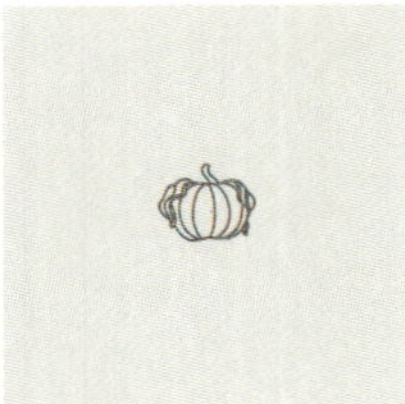

4743

Ribbed ornamental pumpkin in black glass with stem and leaves in red *pasta vitrea*. Surface with iridised finish. The piece is a variant on a larger scale of model no. 4742 (h. 12.5 cm). An example of no. 4742 in red and black glass was shown at the travelling American exhibition, Contemporary Glass and Rugs (1929–30). This model is in the catalogue along with another ribbed pumpkin with a more flattened shape and with two leaves (h cm 9).
c. 1929
20 cm high
Galleria d'arte moderna Carlo Rizzarda collection, Feltre

exhibitions
1929–30, USA, Contemporary Glass and Rugs

>
4742–43, catalogue drawings, AMuVe

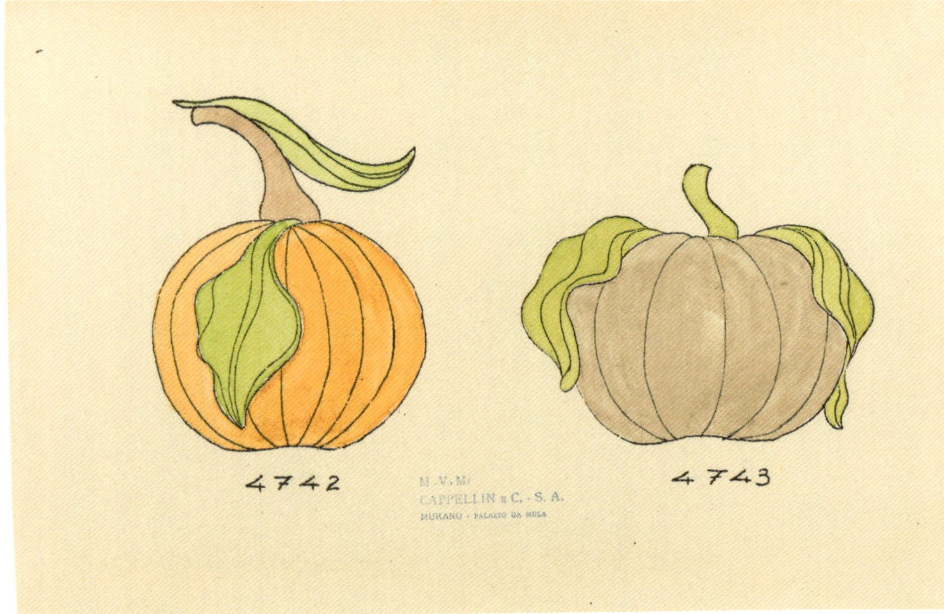

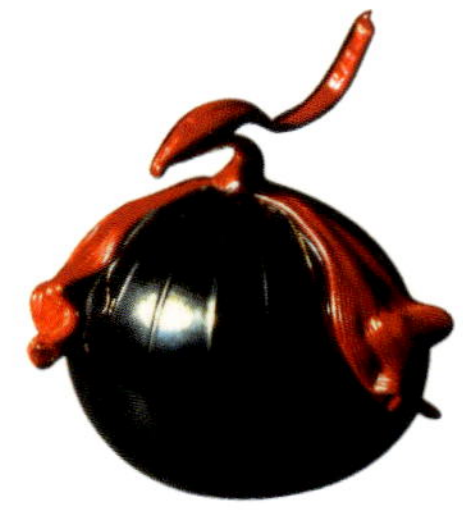

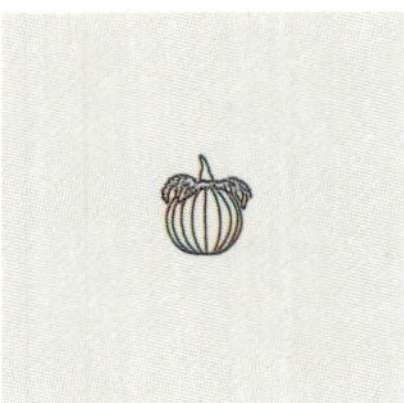

4750

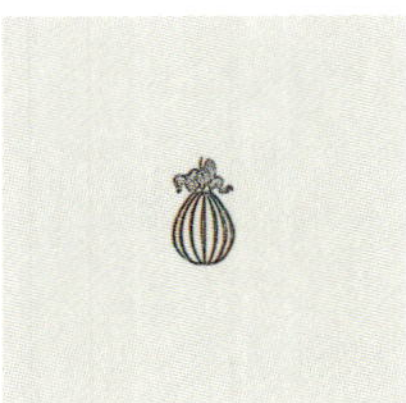

4751

Catalogue drawings for 'large fruit' models to be produced in polychrome or *lattimo* (?) glass: no. 4750, round, ribbed, with stem and four small leaves; no. 4751, pear-shaped, ribbed with superimposed leaves in *rigadin* glass, AMuVe.
1929–30
11.5; 15.5 cm high

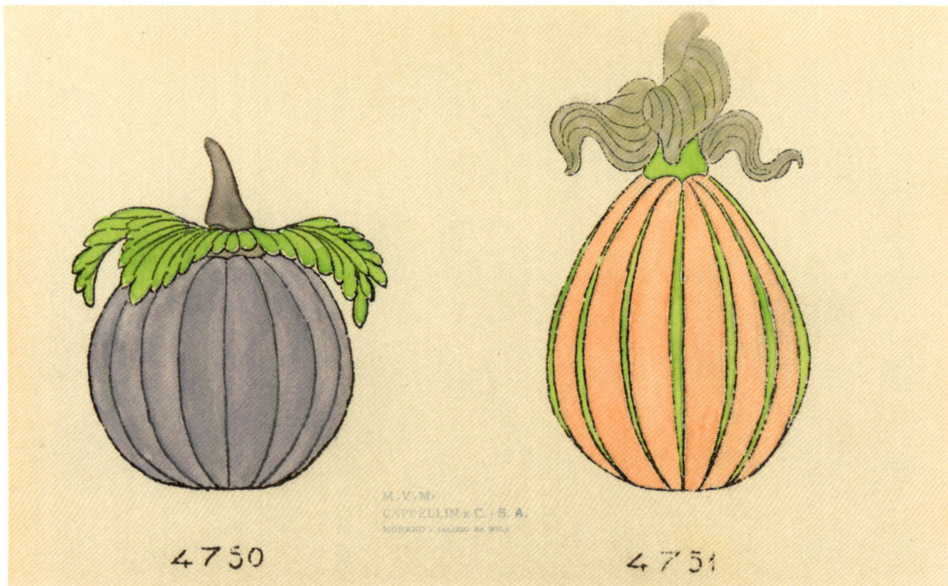

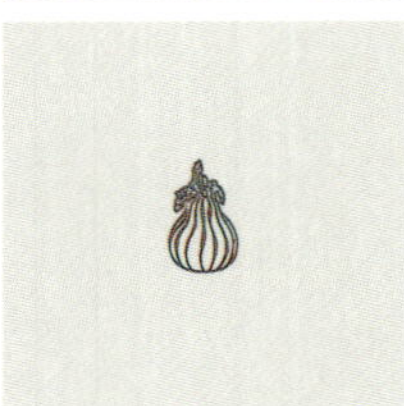

4752

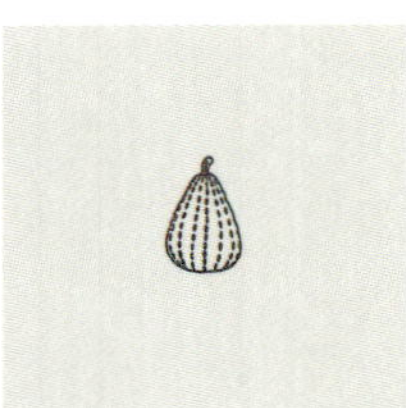

4753

Catalogue drawings for 'large fruit' models to be produced in polychrome or *lattimo* (?) glass: no. 4752, pear-shaped, ribbed with three small leaves; no. 4753, pear-shaped, with stem and segmented cane decoration (with veining), AMuVe.
1929–30
13.5; 14 cm high

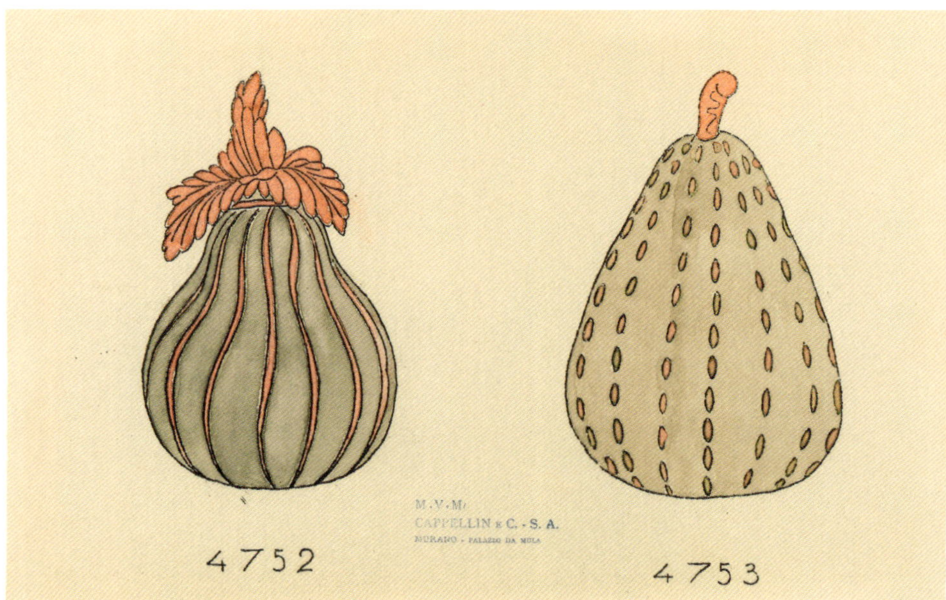

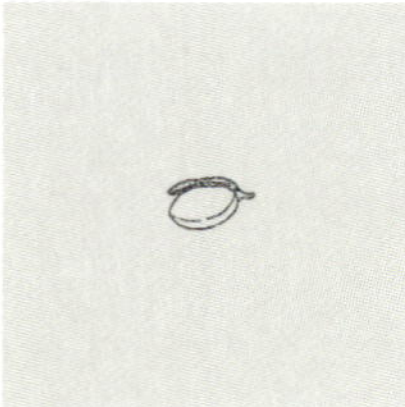

4754

Catalogue drawings for the 'small fruit' model, no. 4754, and the 'large fruit' model no. 4756 (apricot?), to be produced in polychrome or in *lattimo* glass (?), AMuVe.
1929–30
8 cm wide

>
Plate with fruit no. 4754 in pasta vitrea. Period photograph, 1929–30, CSV

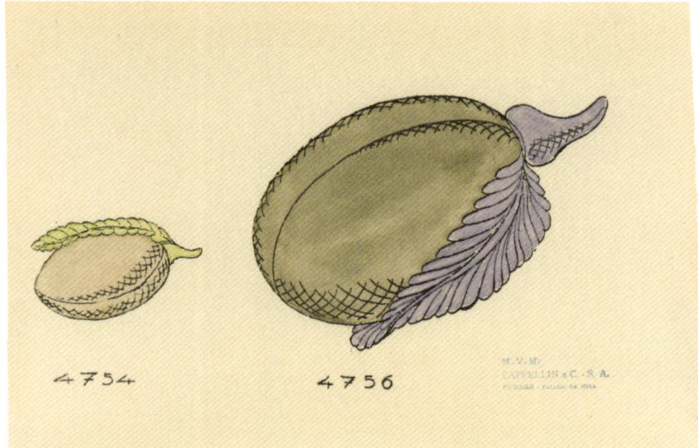

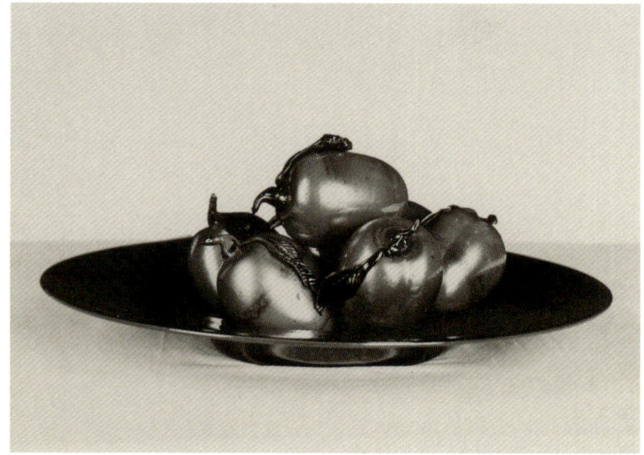

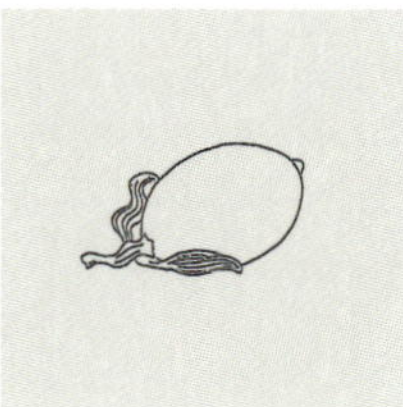

4748

Catalogue drawing for a 'large citron', with stem and two leaves, to be produced in polychrome or *lattimo* glass (?), AMuVe.
1929–30
30 cm wide

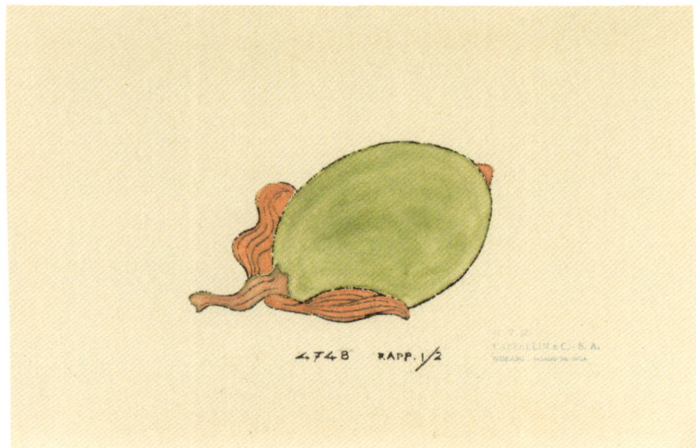

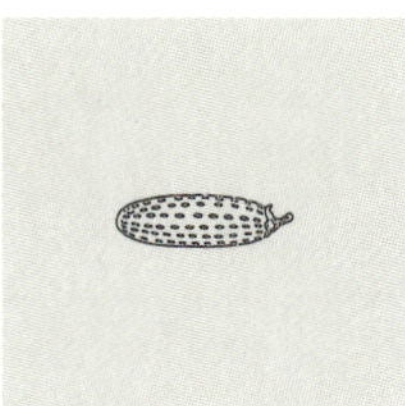

4749

Catalogue drawing the 'large fruit' model (cucumber?) no. 4749, with stalk and segmented cane decoration, to be produced in polychrome or in *lattimo* glass (?), AMuVe.
22 cm wide

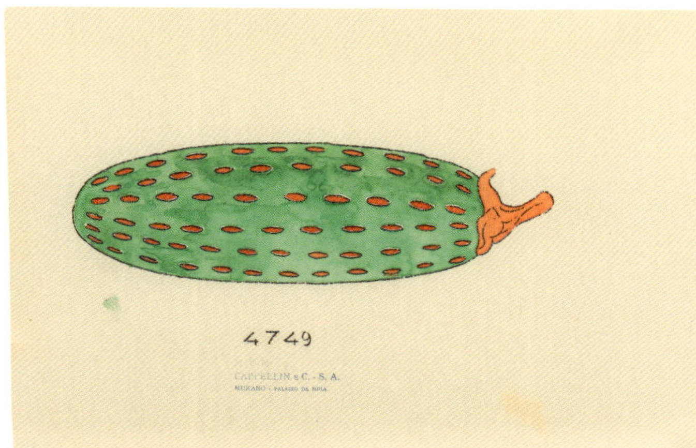

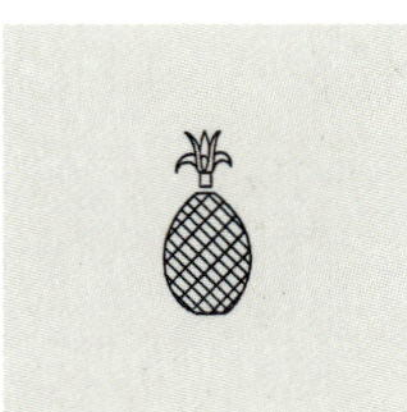

1274

Pineapple in black with a *balloton* decoration. Stopper with superimposed leaves in *rigadin* glass (with veining). Strongly iridised surface. The model is a variant on no. 1274 (bottle with stopper in the shape of a pineapple).
1929–30
35 cm high
Anna Gastel collection, Milan

>
1274, catalogue drawing, AMB

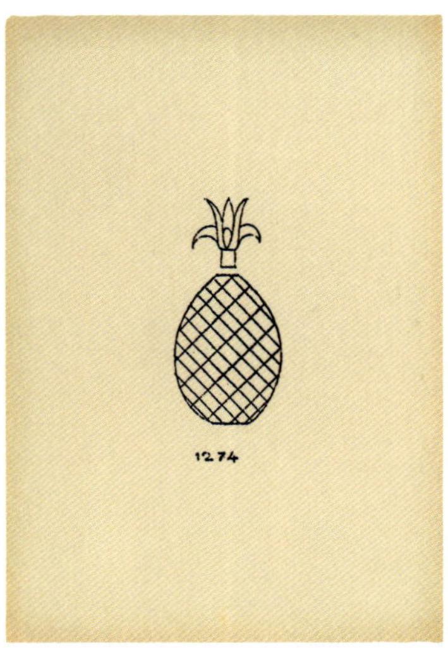

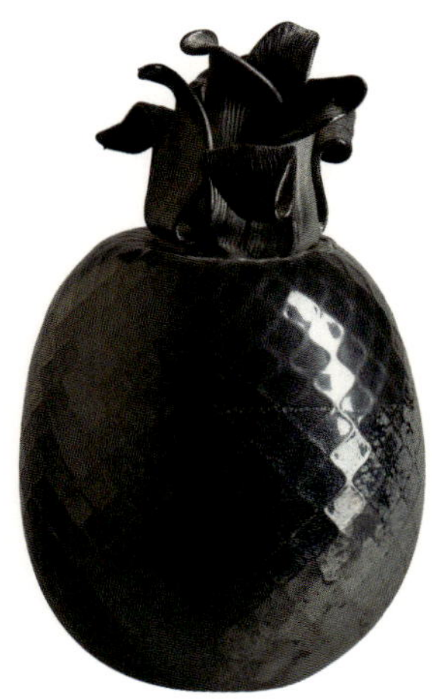

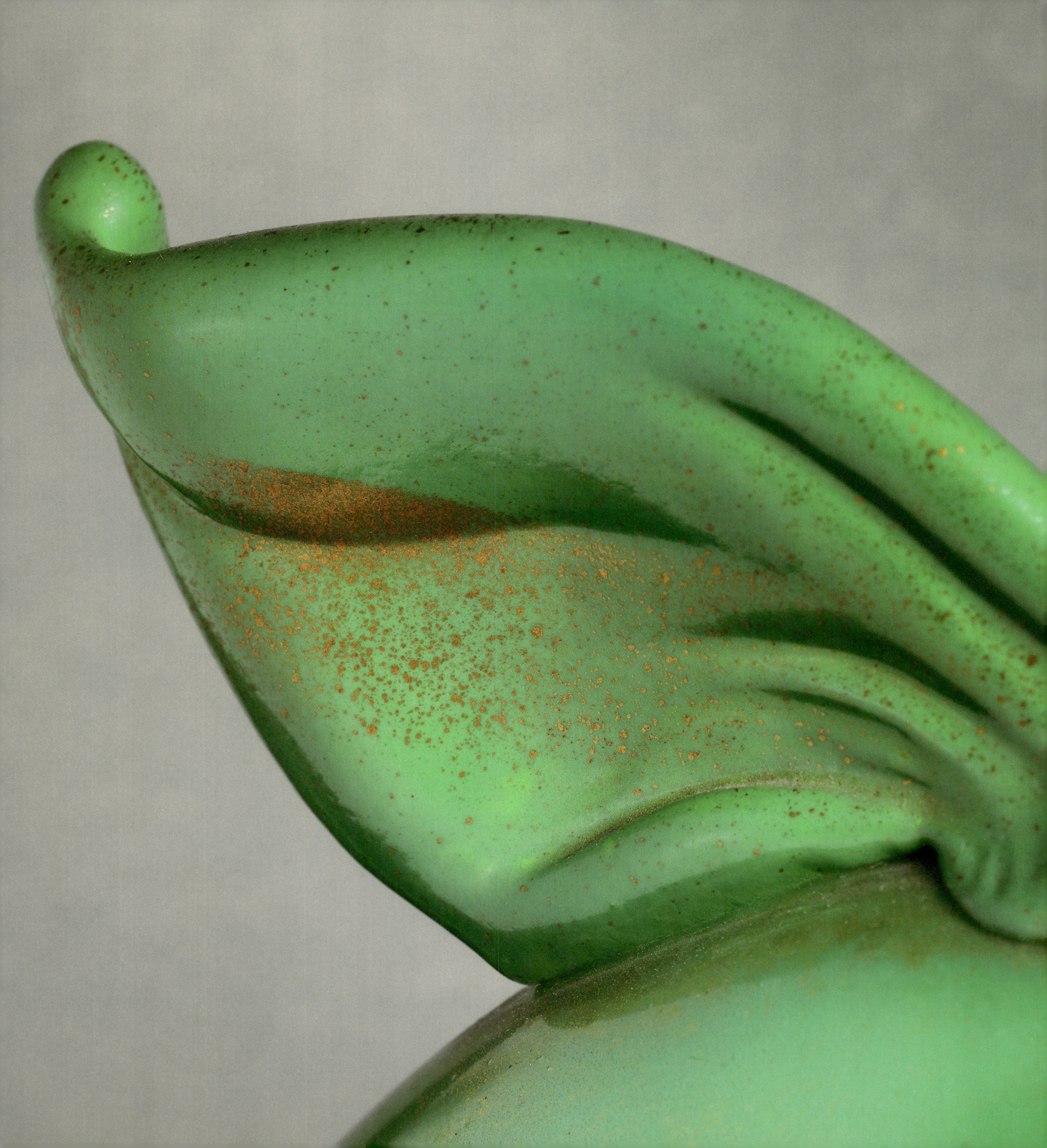

Like V.S.M. Venini e C. and the Vetreria Artistica Barovier, the Cappellin firm also included in its repertoire some models of animals, treating the then fashionable theme in a personal way. These artefacts, due to the contribution of Carlo Scarpa, were marked by a light and playful design and the refined quality of the material in which they were executed. A small number of animals were produced in *pulegoso* glass (fish and elephants), but the majority were in *pasta vitrea*, often refined by gold or silver leaf, in *decoro fenicio* glass or glass decorated with circular *murrine*. Fish were a recurrent subject, proposed in differing variants of form, dimension and material. There was no lack of subjects taken from the repertoire of aquatic fauna, such as the octopus, or of an exotic kind, like the tiger.

ANIMALS 1928-31

Exhibitions

1929, Paris, *Salon d'automne*
1929-30, USA, *International Exhibition of Contemporary Glass and Rugs*, travelling exhibition
1930, Milan, Trade Fair
1930, Monza, 4th *Esposizione Triennale Internazionale delle Arti Decorative ed Industriali Moderne*
1931, Amsterdam, *Tentoonstelling van glas, aardewerk en kant, van moderne italiaansche kunstnijverheid. Mostra di vetri, ceramiche e merletti d'arte moderna italiana*

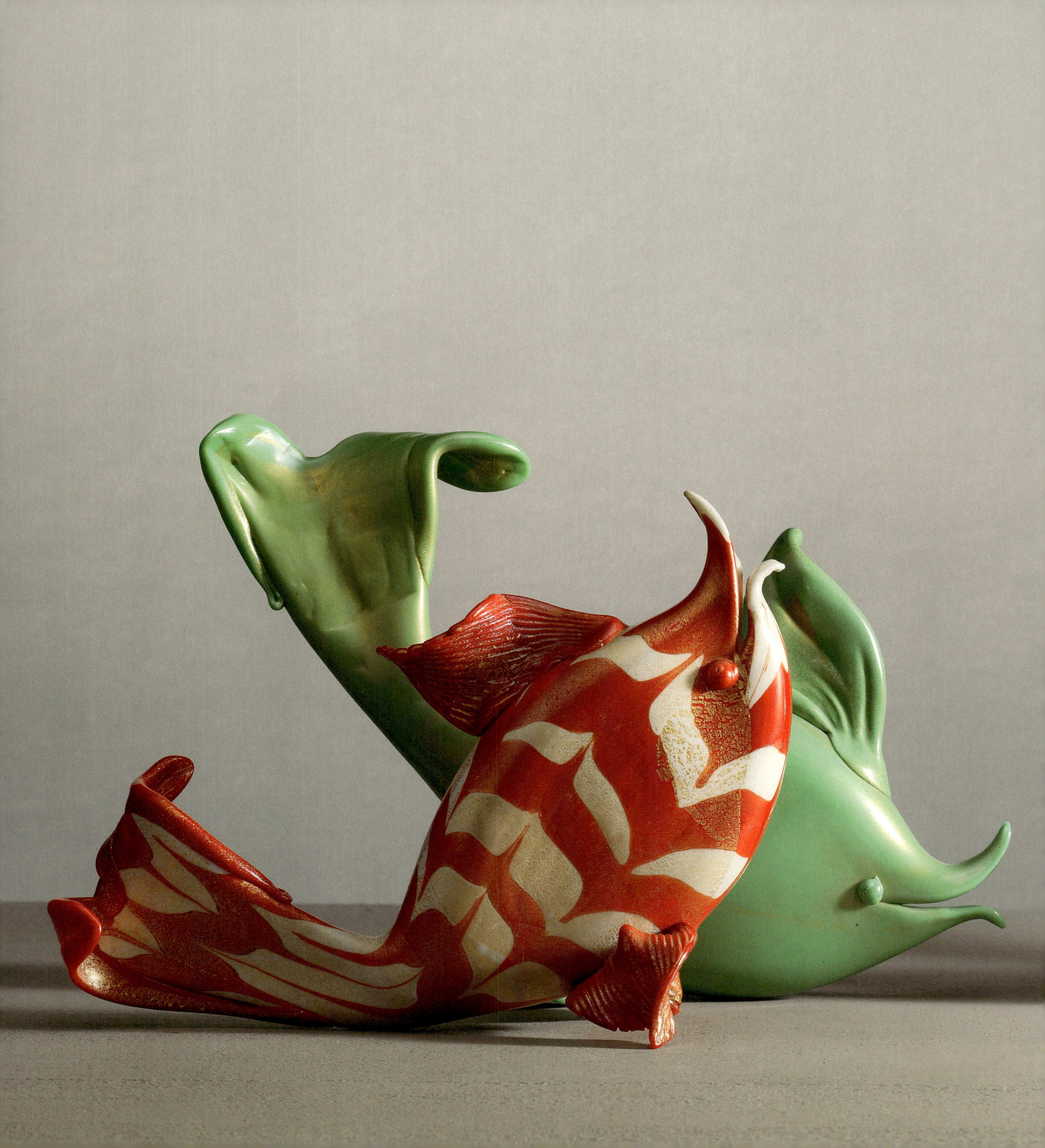

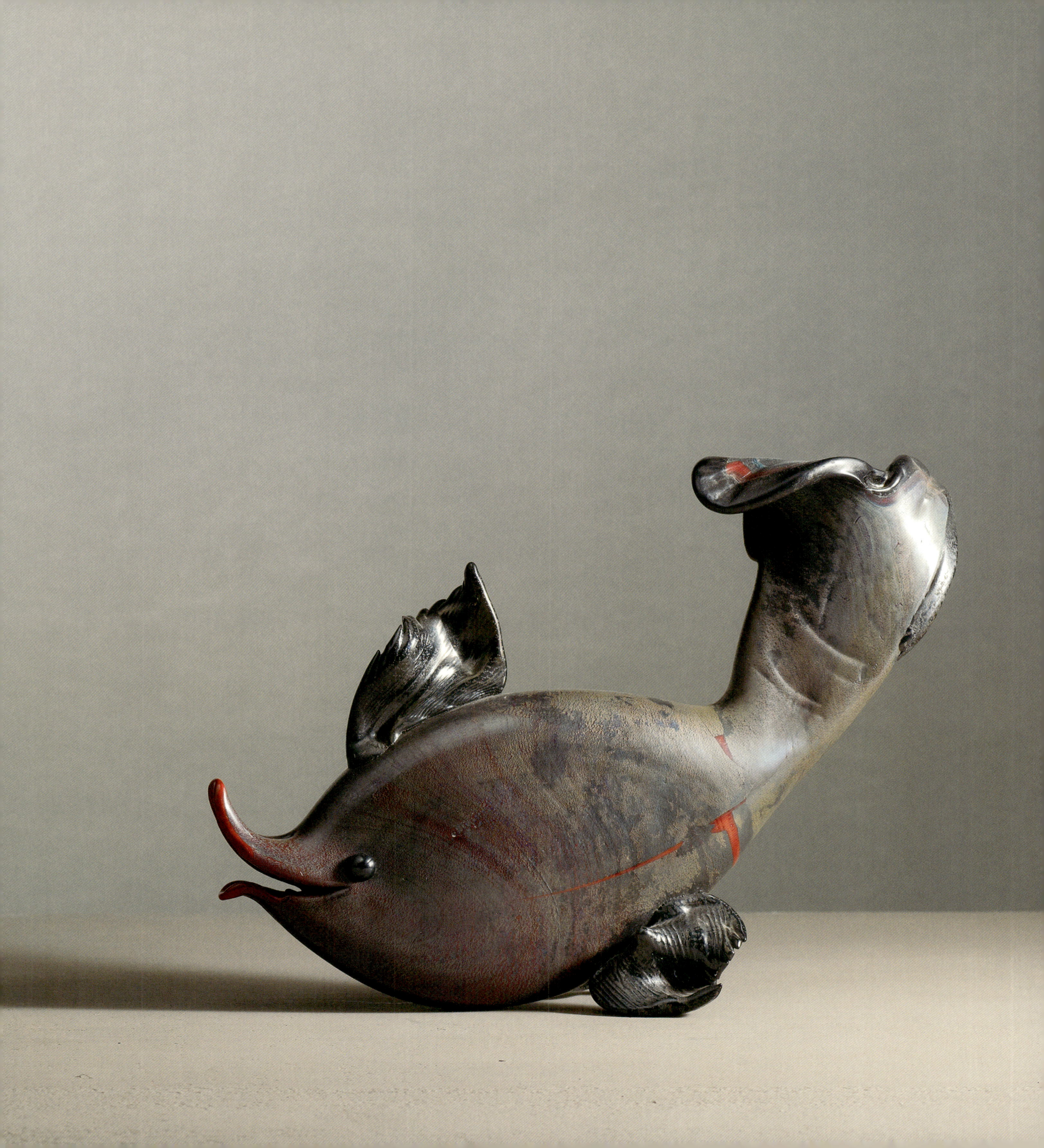

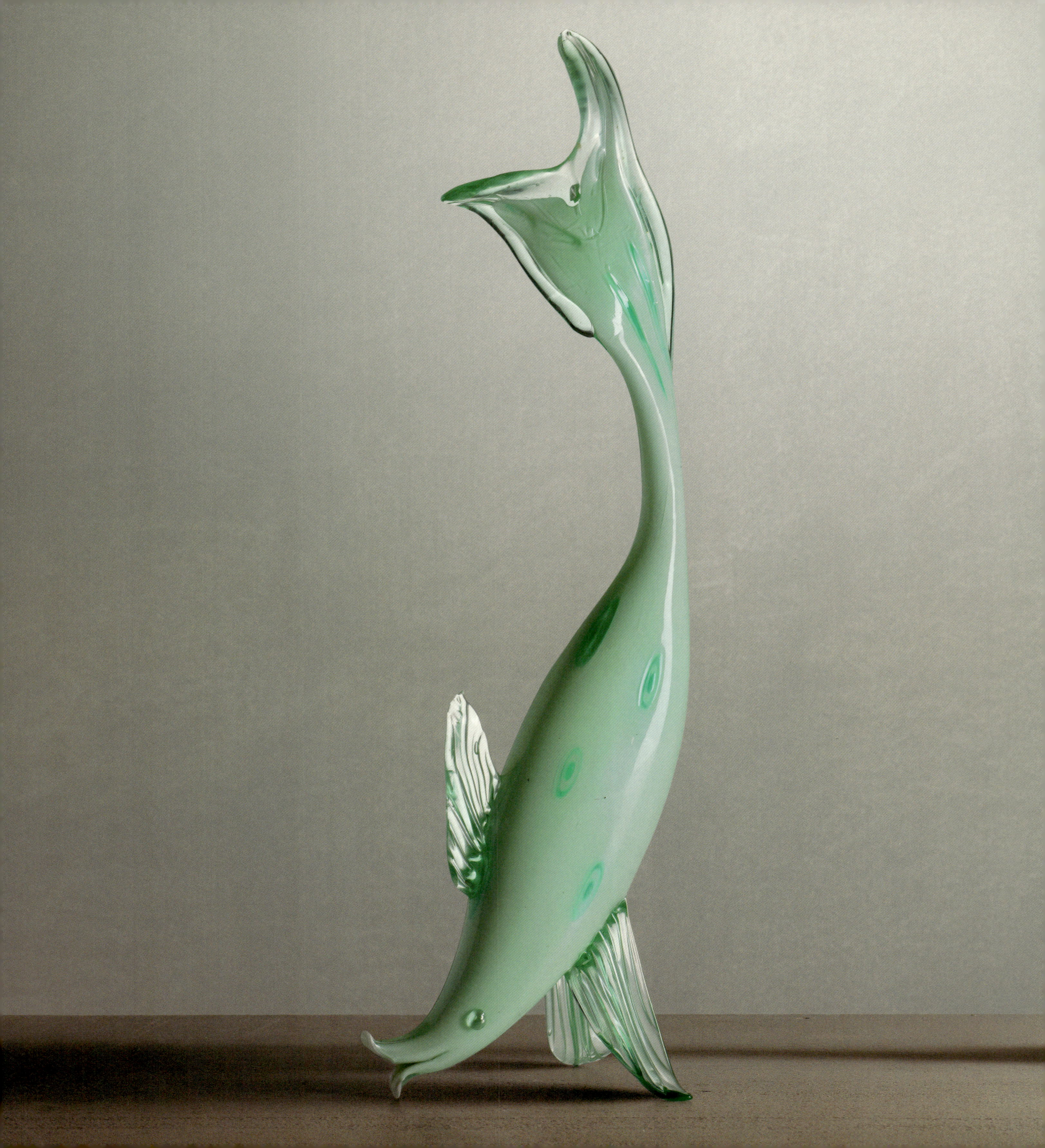

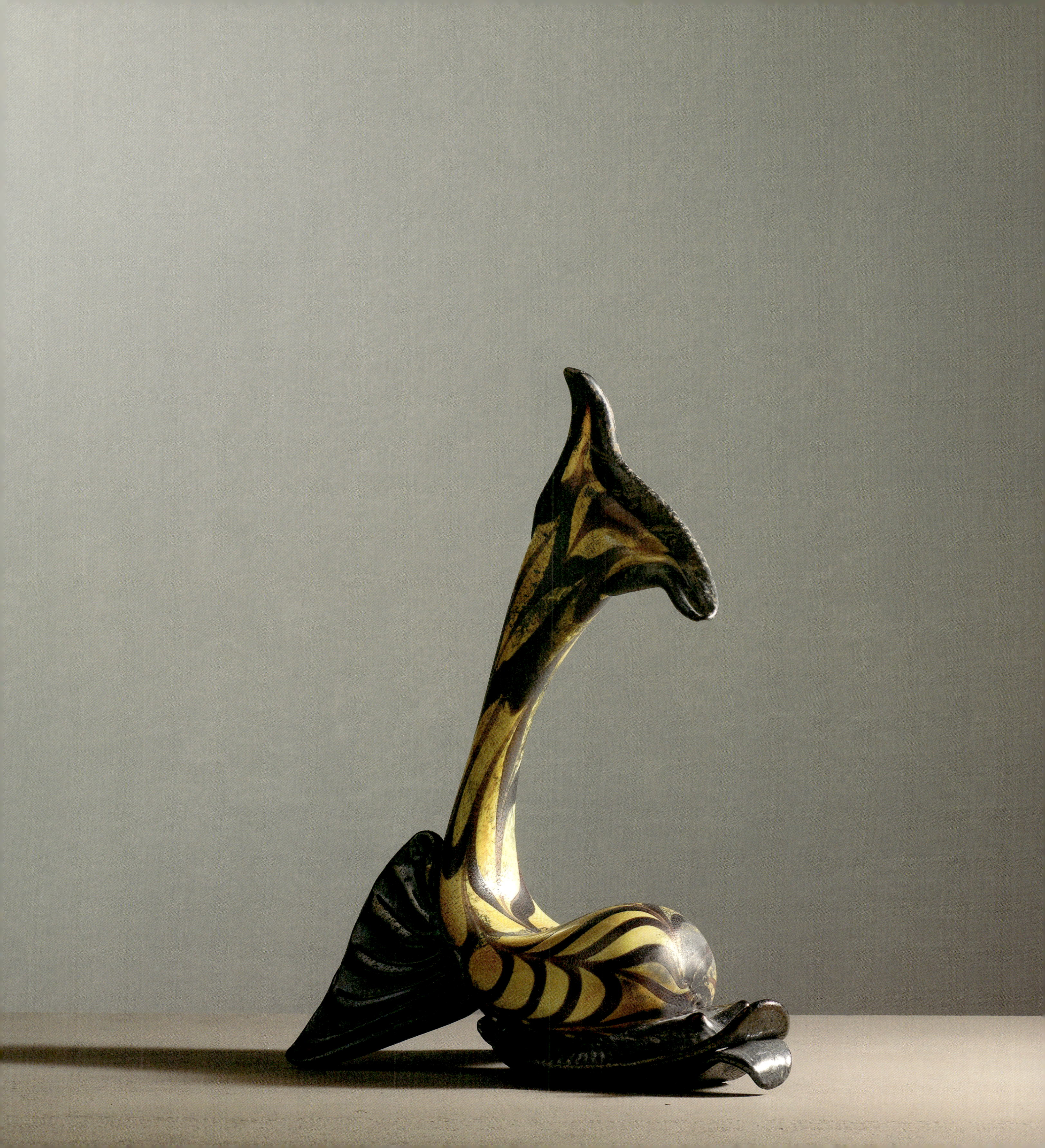

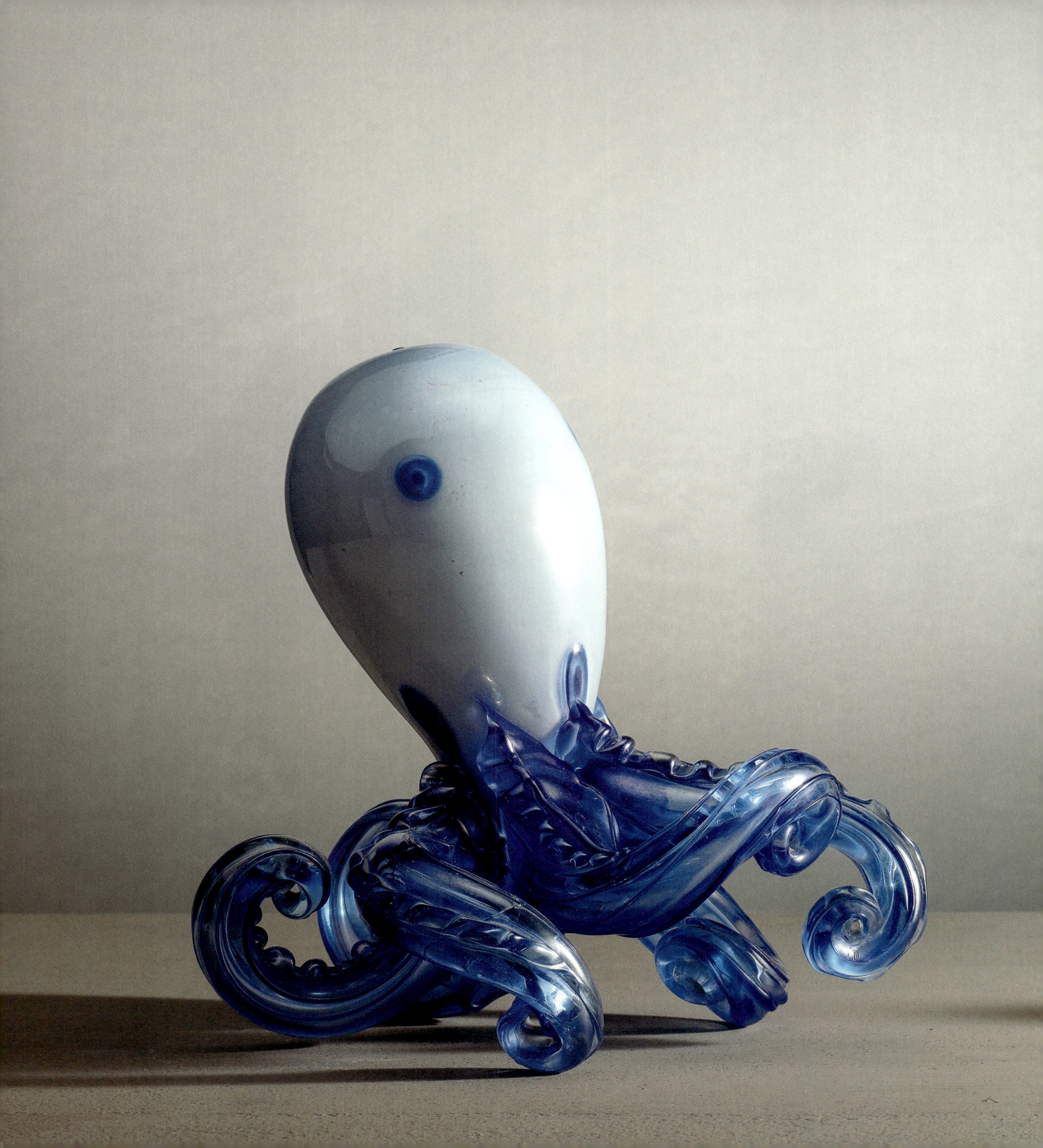

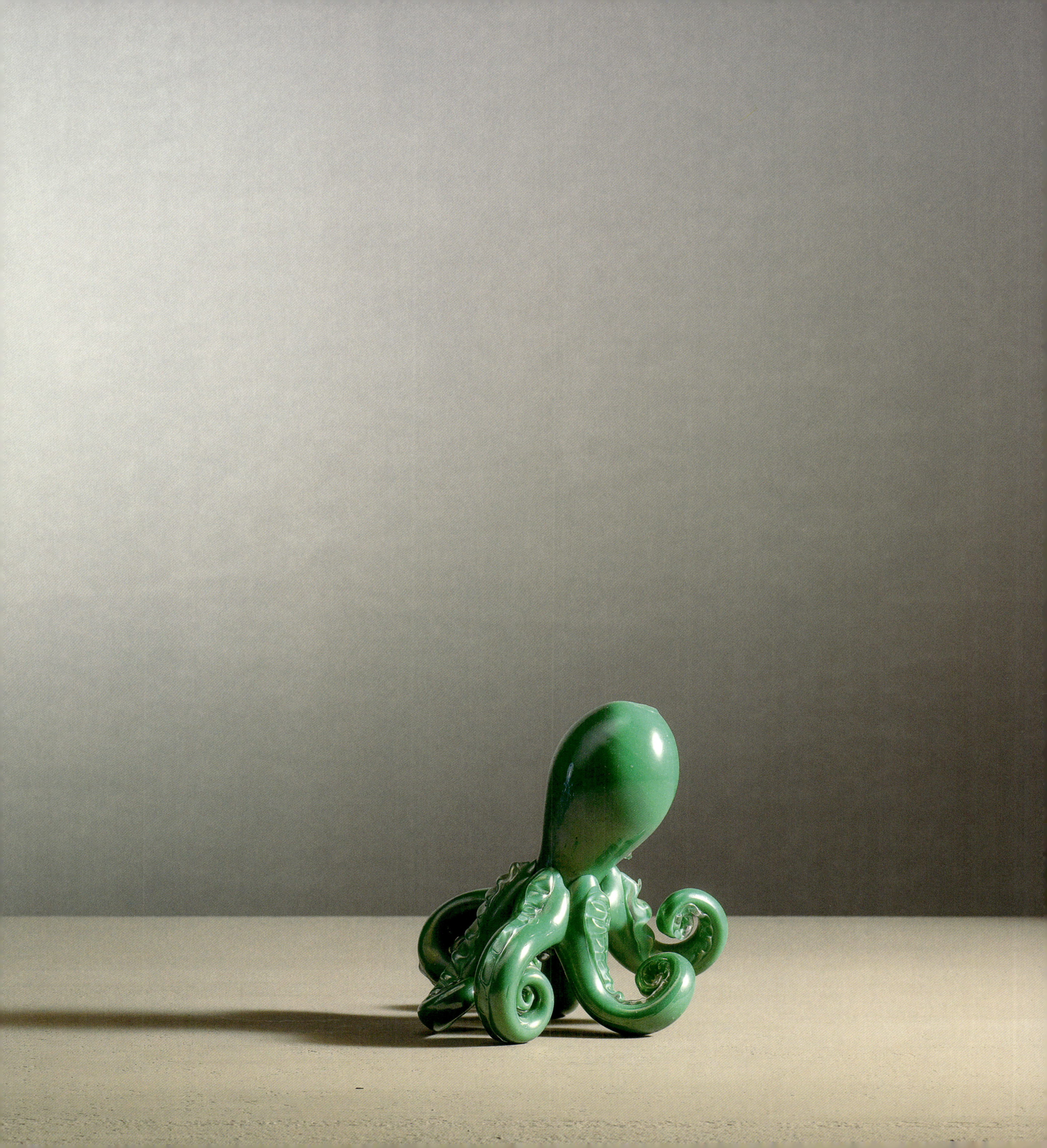

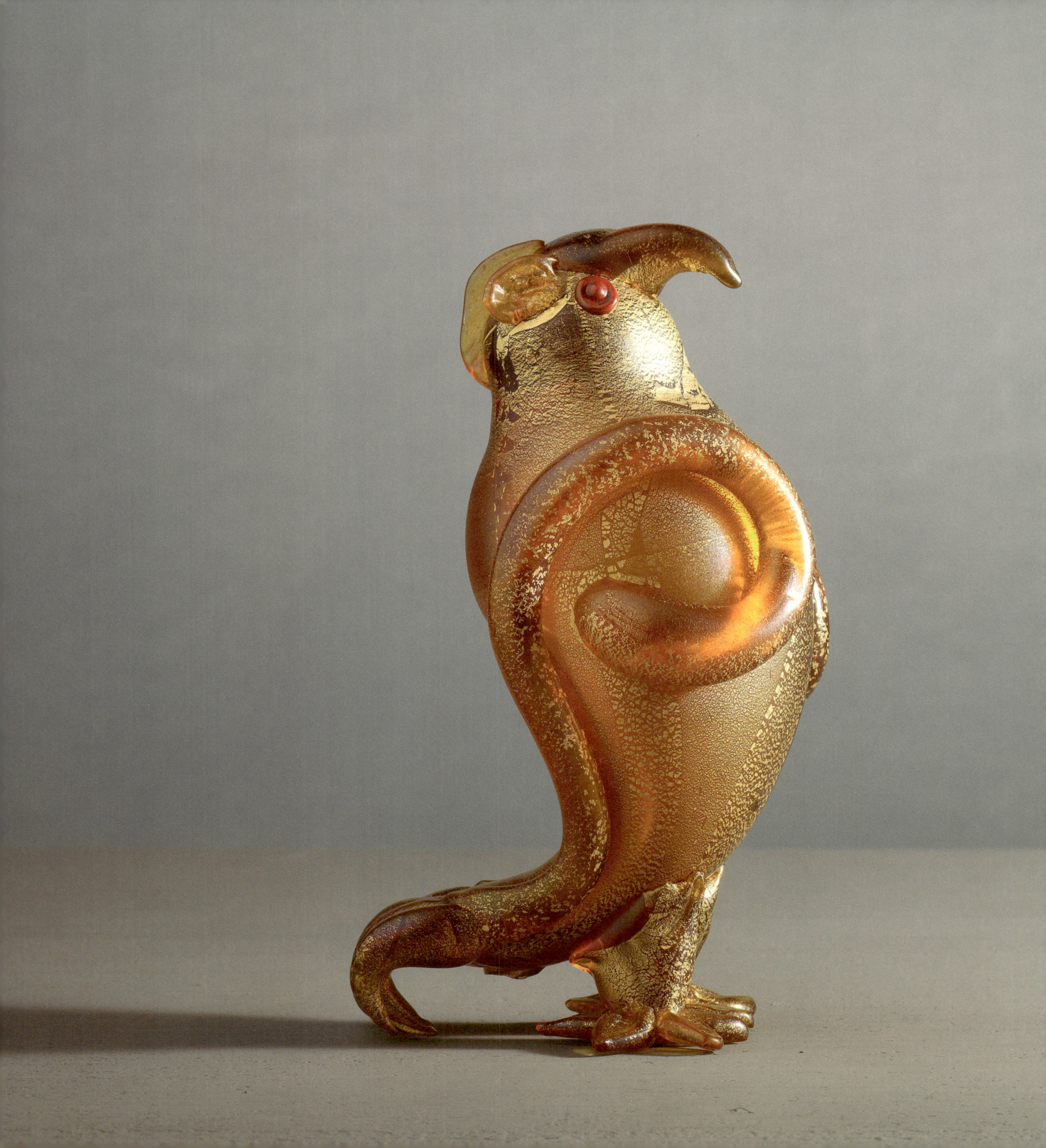

>
Fishes in pulegoso *glass with fins and finishing in coloured transparent glass. Period photograph, c. 1929, CSV. These are probably the first examples of fish produced by the Cappellin firm*

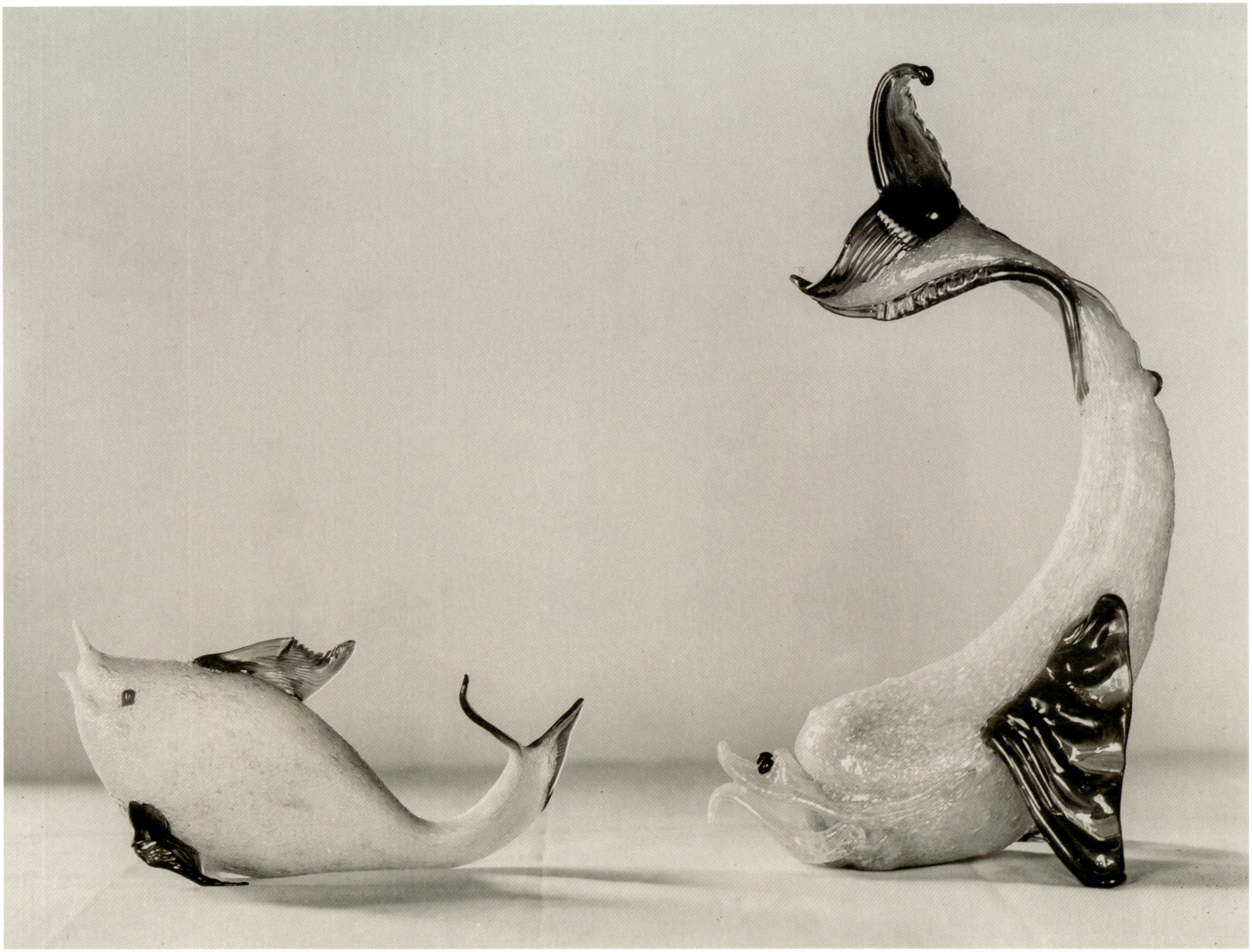

Fish in *pulegoso* glass with fins and finishings in coloured transparent glass. Period photograph, c. 1929, CSV
1928–29
30 cm high

n.i.

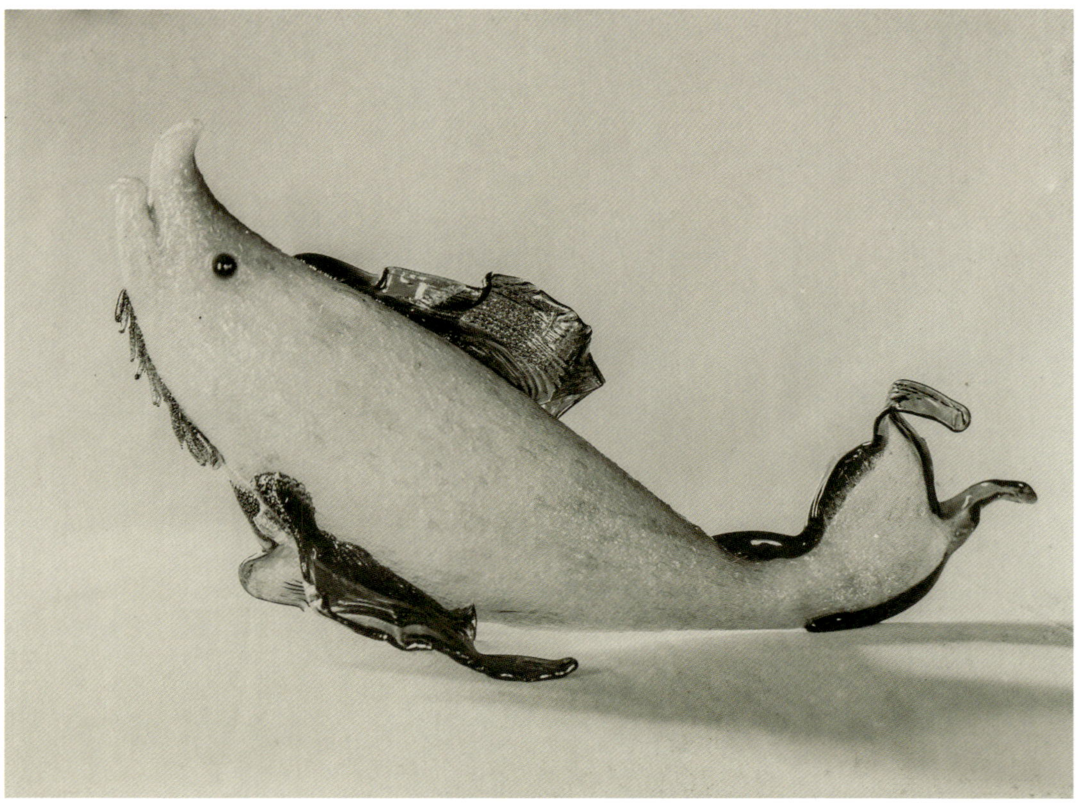

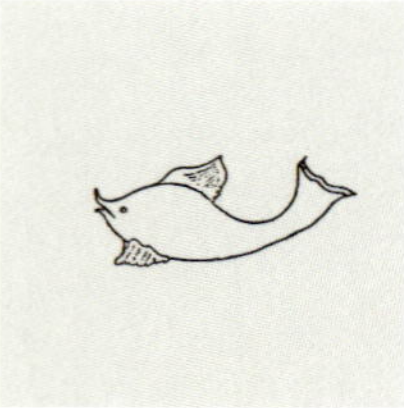

4722

Fish in *lattimo* glass with an amethyst *decoro fenicio*, finished with applied gold leaf. Fins and finishings in amethyst and gold transparent glass. The model was produced in two versions which differ in the position of the lower fins: one, more like the catalogue drawing, with the fins fixed in a forward position in front of the belly, the other with the fins more towards the rear (4722v). The latter variant, documented also by *Domus* (April 1929), enjoyed greater distribution. Examples of both versions were exhibited in 1929 at the Milano Fair.
1928–29
31 cm wide
Carl Bellavia collection

exhibitions
1929, Barcelona, International Exhibition

>
4722, catalogue drawing, AMB

>
Fish in pasta vitrea *and fish in* pulegoso *glass with fins and finishings in glass of contrasting colours. Period, photograph, c. 1929, CSV*

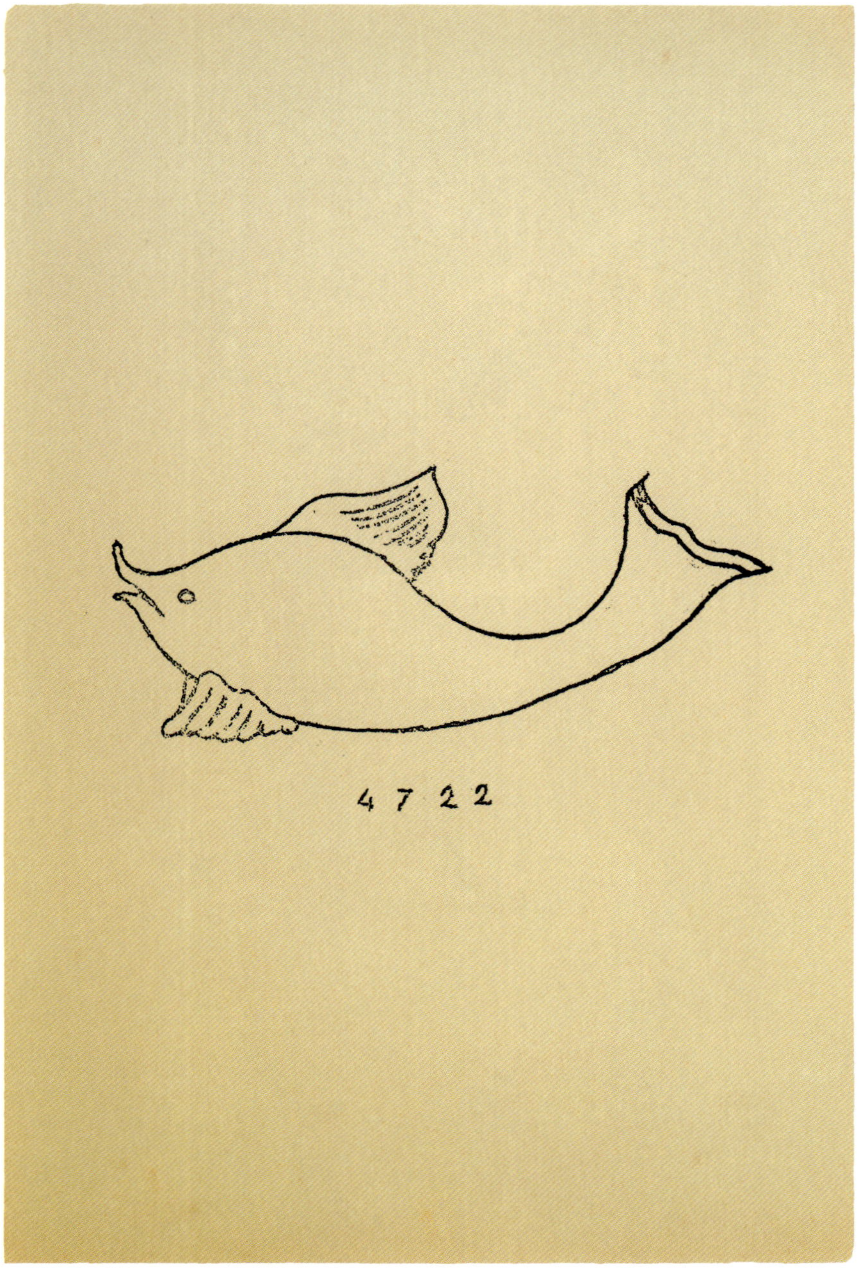

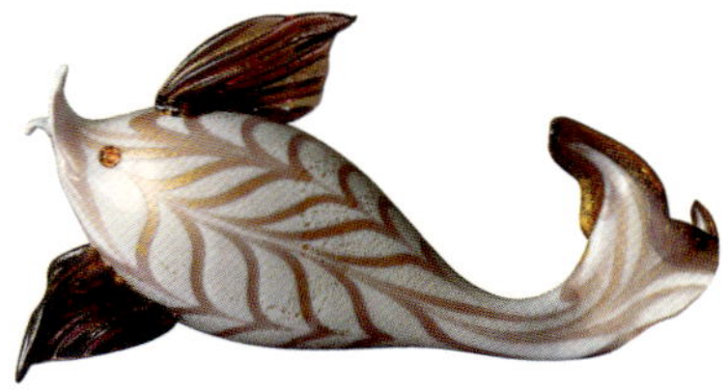

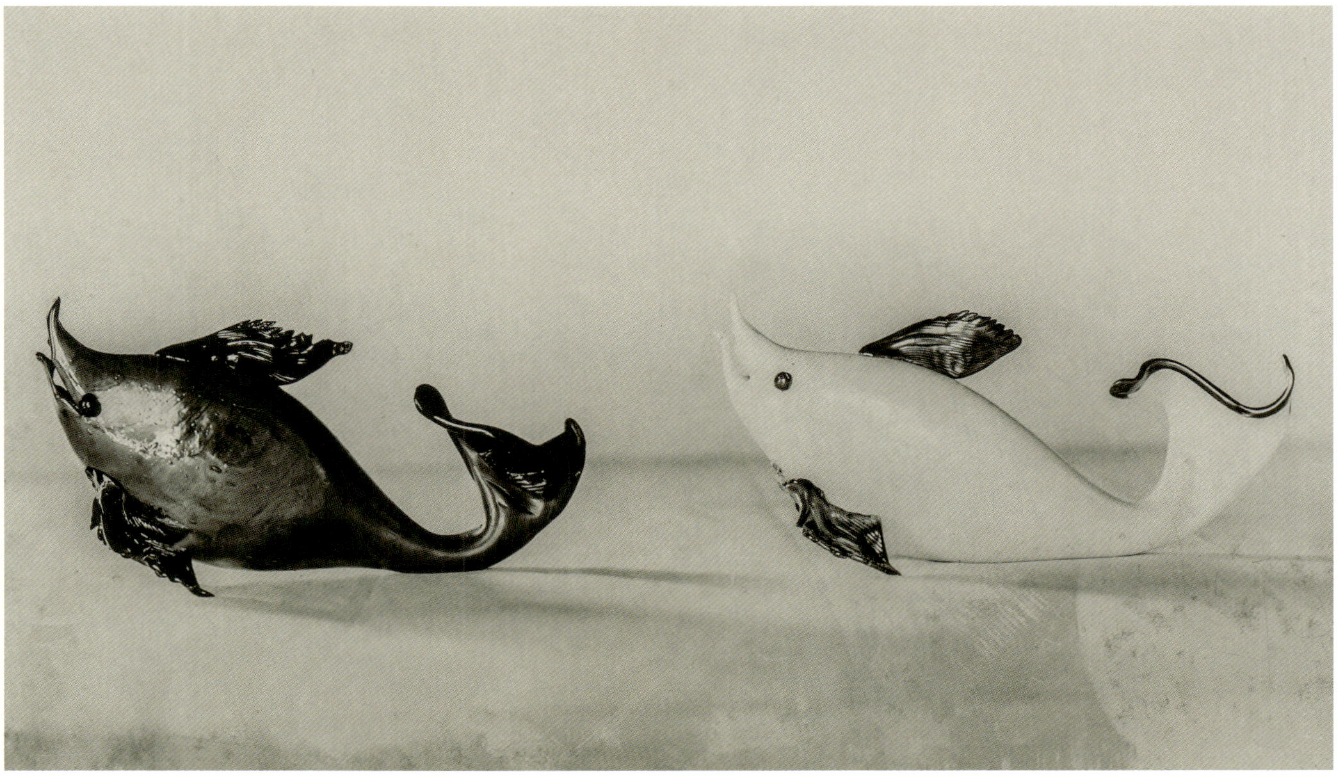

4722v

Fish in pea-green *pasta vitrea* with applied gold leaf. The model was also produced in other glass techniques (*decoro fenicio*, *pulegoso*, black and silver etc.).
1928–29
35 cm wide
Private collection

exhibitions
1929, Barcelona, International Exhibition
1930, Milan, Fiera
1931, Amsterdam, Exhibition of glass, ceramics and lace

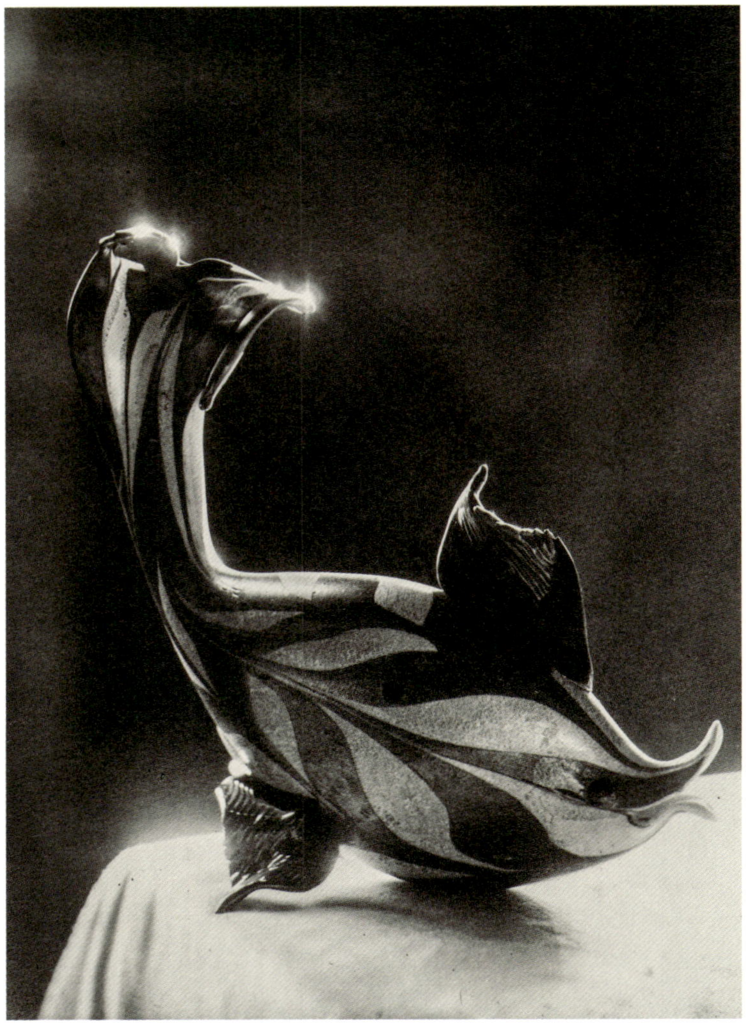

>
Fish in decoro fenicio *glass. Period photograph, c. 1929, CSV*

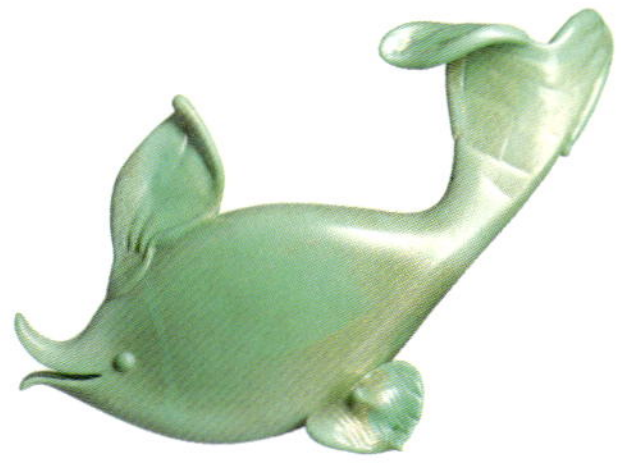

n.i.

Fish in *lattimo* glass with dark amethyst *decoro fenicio*. Fins, mouth and finishings in black glass. Strongly iridised surface. A similar example in opaque monochrome glass with applied gold leaf was exhibited in the travelling American exhibition, Contemporary Glass and Rugs (1929–30). The model was also produced in *pulegoso* glass.
1928–29
36 cm high
Private collection

exhibitions
1929–30, USA, Contemporary Glass and Rugs

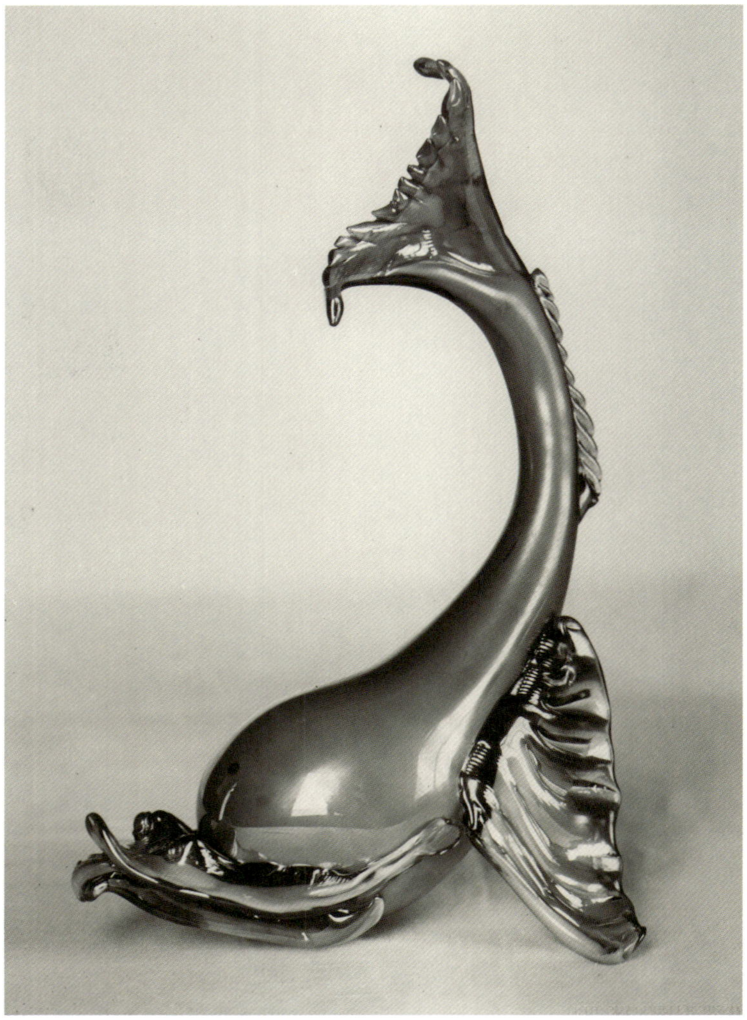

>
Fish in opaque glass. Period photograph, c. 1929, ACSVr

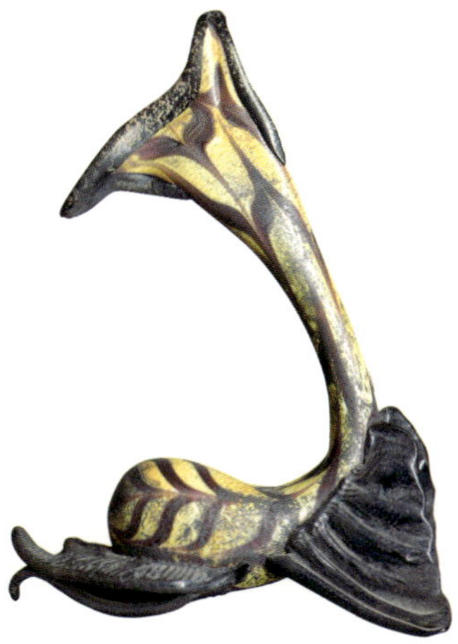

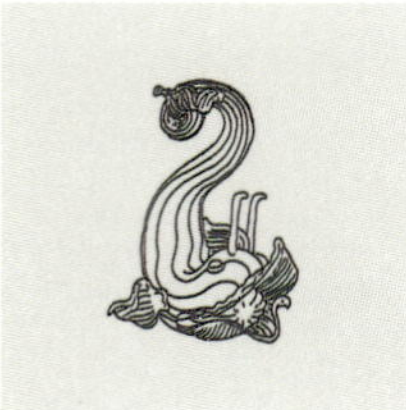

4758

Catalogue drawing for a fish ("dolfin") with antennae, ribbed body and open mouth, AMuVe. As documented by a period photograph published in the 1931 Amsterdam exhibition catalogue, the animal was produced with a body and tail in opaque glass and mouth, fins and finishings in transparent glass.
c. 1930
32 cm high

exhibitions
1931, Amsterdam, Exhibition of glass, ceramics and lace

>
Fish and fruit in glass. Period photograph c. 1930, CSV. The image was published in Tentoonstelling van glas, aardewerk en kant, van moderne italiaansche kunstnijverheid. Mostra di vetri, ceramiche e merletti d'arte moderna italiana, *Venice 1931, p. 20*

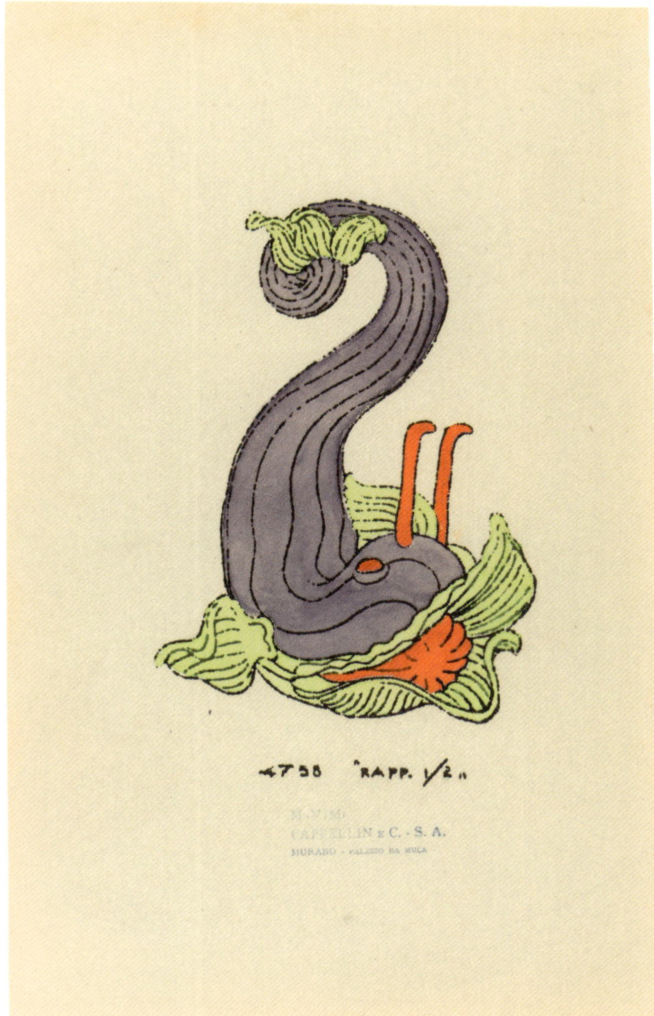

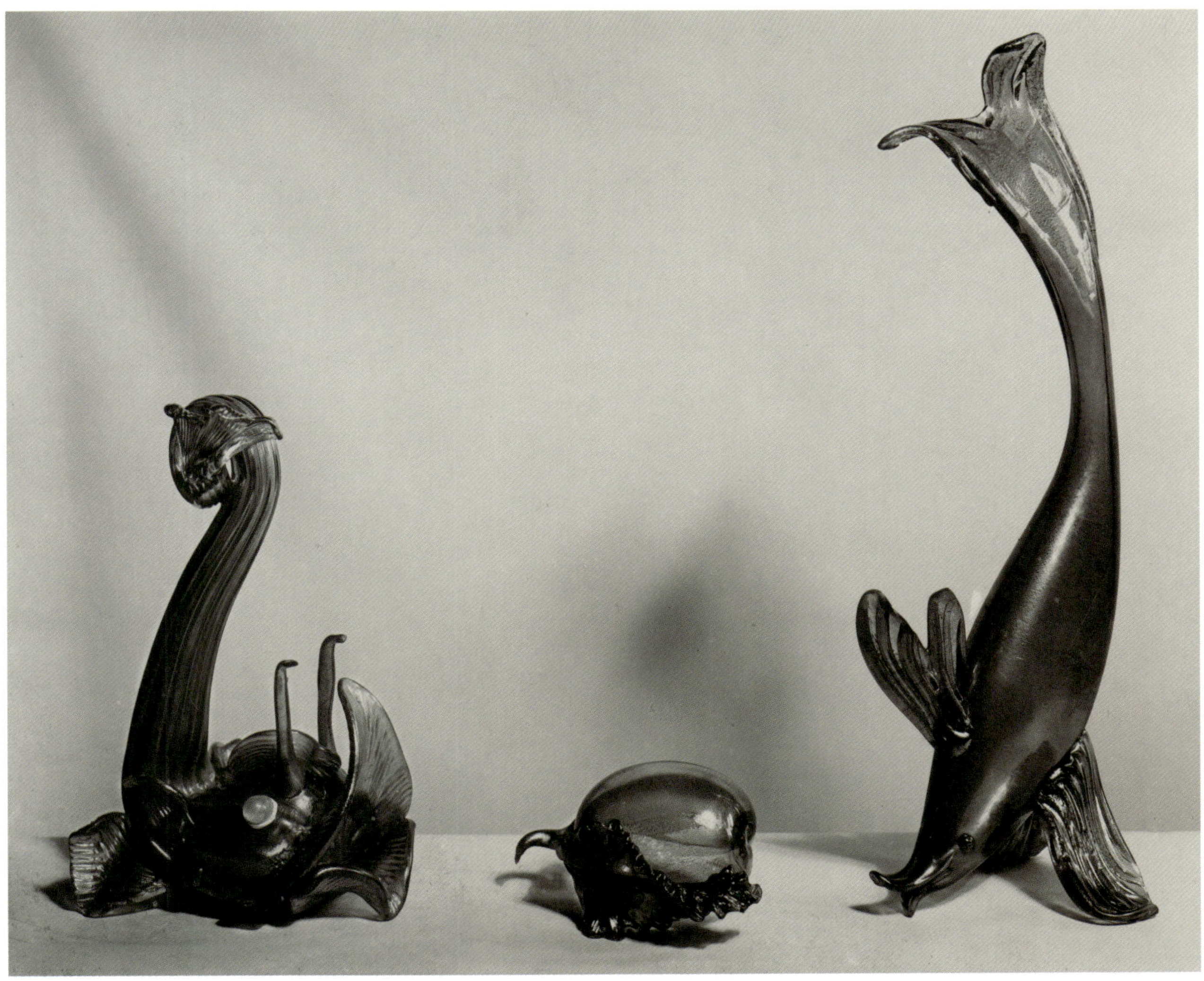

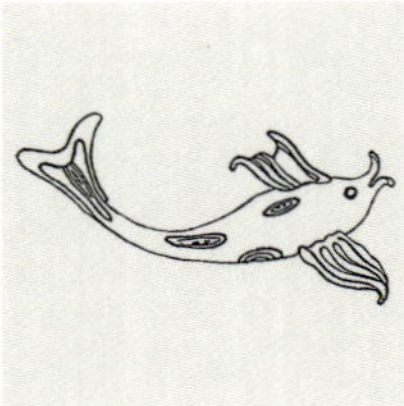

4745

Large fish in cased *lattimo* and green glass with green *millefiori* ("pearls") murrine. Fins and finishings in green transparent glass. Surface with iridised finish. A similar example, with a different colour variant, was exhibited in 1930 at the 4th Triennale of Monza. In the following year the model, produced in opaque glass, appeared in the catalogue of the Amsterdam Exhibition of glass, ceramics and lace. A variant of this model in with *decoro fenicio* was also offered: this was included in the catalogue of the glassworks under the number 4746.
1929–30
57 cm high
Private collection

exhibitions
1930, Monza, 4th Triennale
1931, Amsterdam, Exhibition of glass, ceramics and lace

>
4745, catalogue drawing, AMuVe

>
4746, catalogue drawing, AMuVe

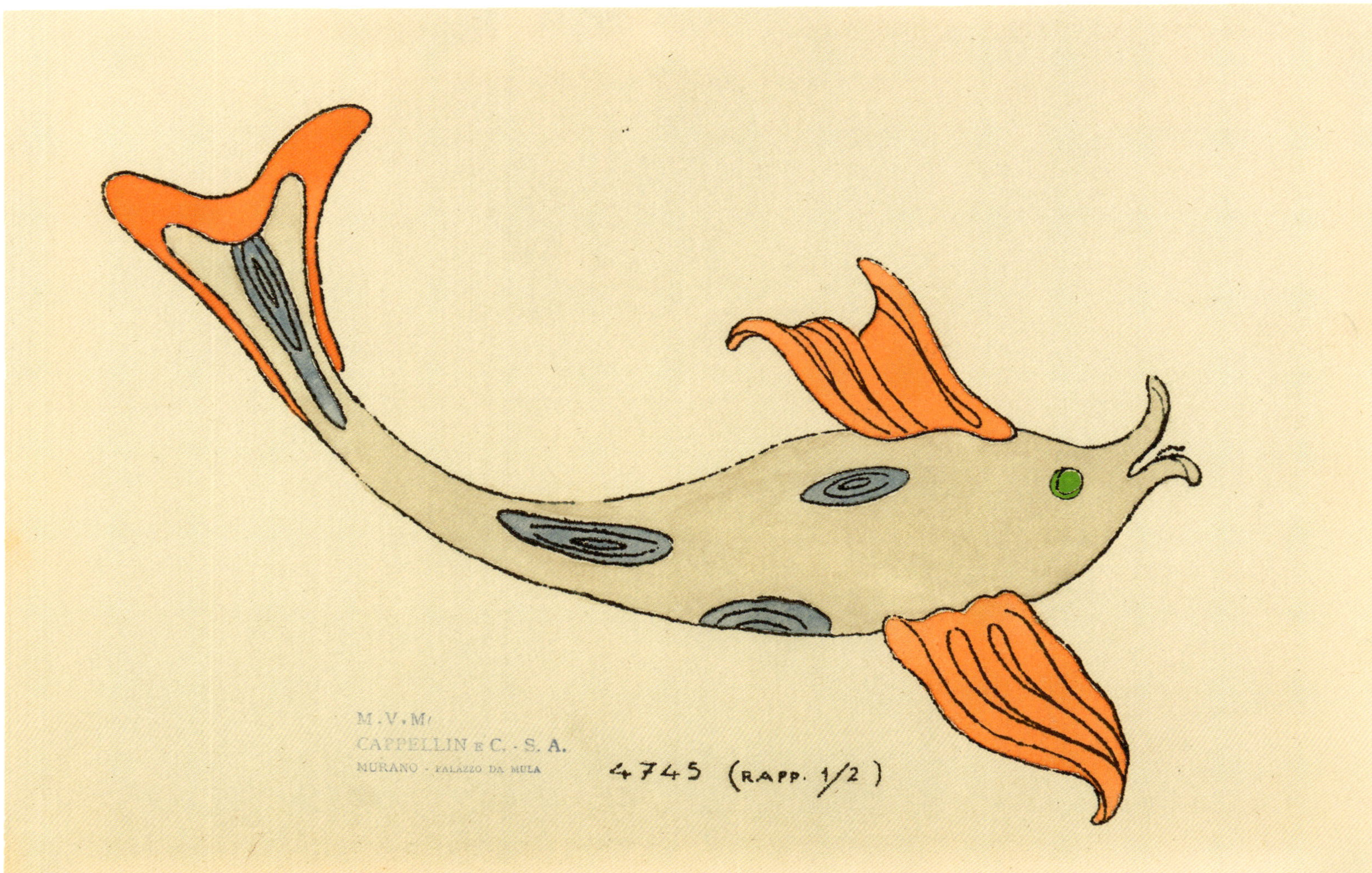

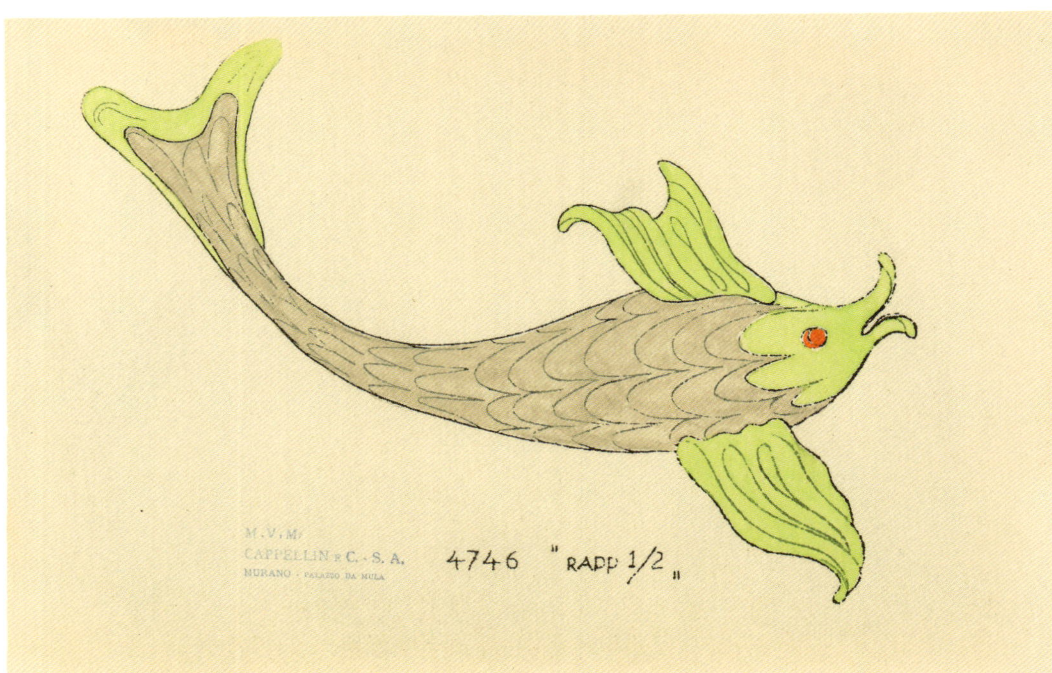

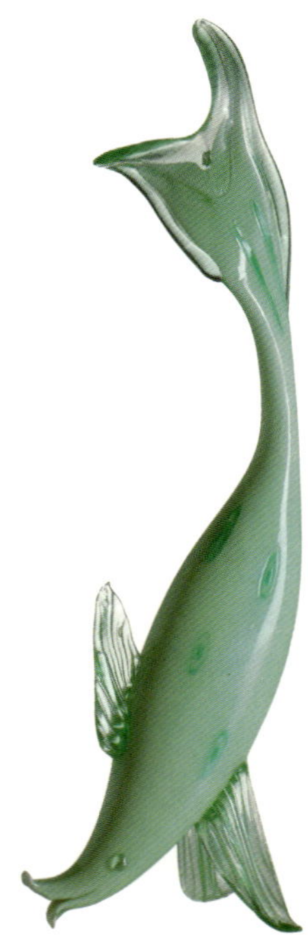

>
Cappellin glass pieces at the 4th Triennale of Monza, 1930

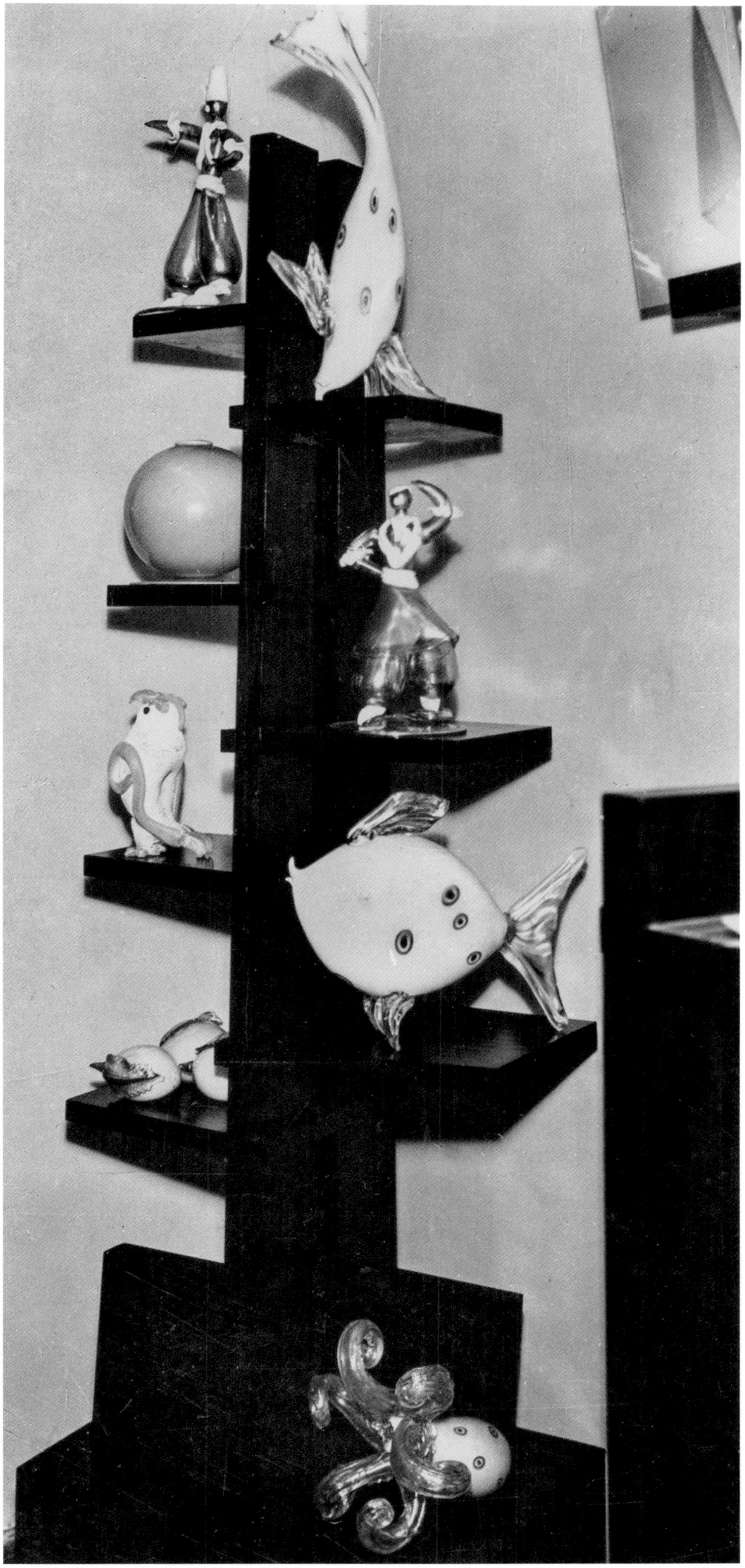

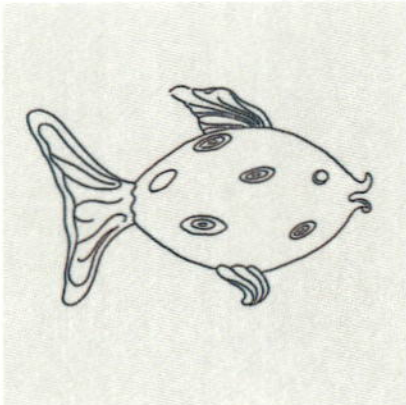

4755

Fish in cased *lattimo* and green glass decorated with circular *murrine* in green and crystal concentric circles included in the wall of the piece. Fins and finishings in transparent green glass. Surface with iridised finish. The fish is a variant in a smaller size of no. 4755 (39 cm wide). Some examples of no. 4755 were exhibited in 1930 at the 4th Triennale of Monza.
c. 1930
30 cm wide
Private collection, Venice

exhibitions
1930, Monza, 4th Triennale

>
4755, catalogue drawing, AMuVe

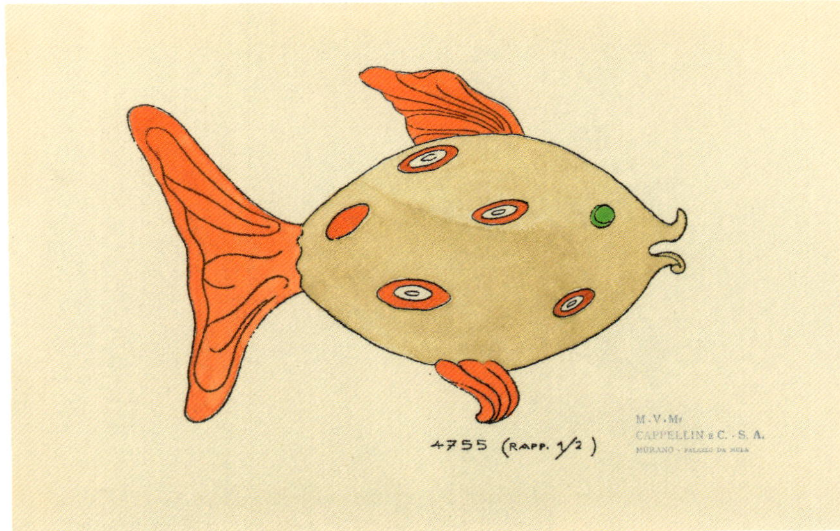

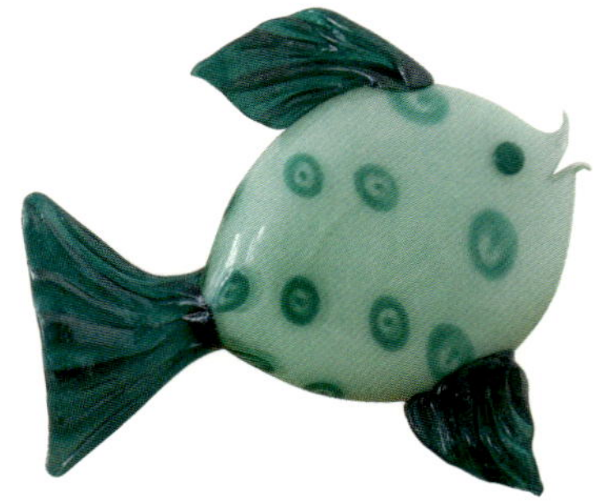

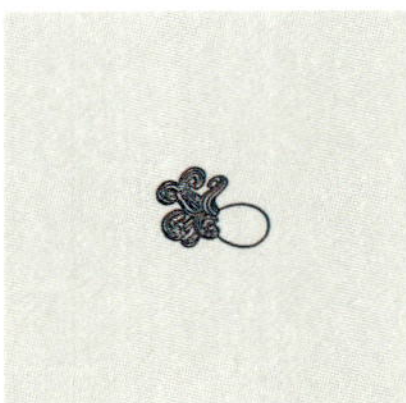

4747

'Octopus' in green *pasta vitrea* with crystal finishings. As witnessed by factory documents ('Price List' and list of pieces 'Salon d'Automne Paris 1928' Exhibition), an initial octopus model (no. 4740 *bis*) was exhibited in 1928 at the Salon d'automne of Paris, in red and green.
c. 1929–30
11.5 cm high
Bischofberger collection, Switzerland

exhibitions
1928, Paris, Salon d'automne
1931, Amsterdam, Exhibition of glass, ceramics and lace

>
4747, catalogue drawing, AMuVe

>
Octopuses in opaque glass. Period photograph, c. 1929, CSV

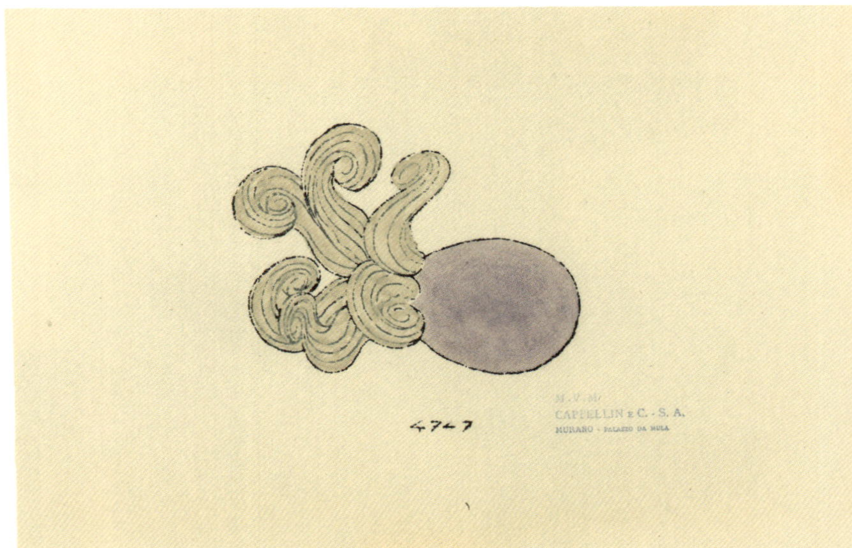

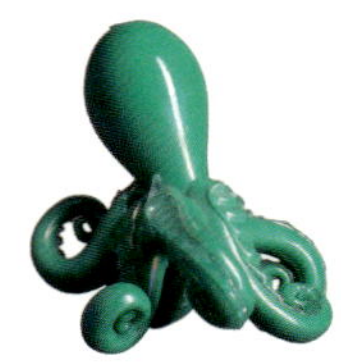

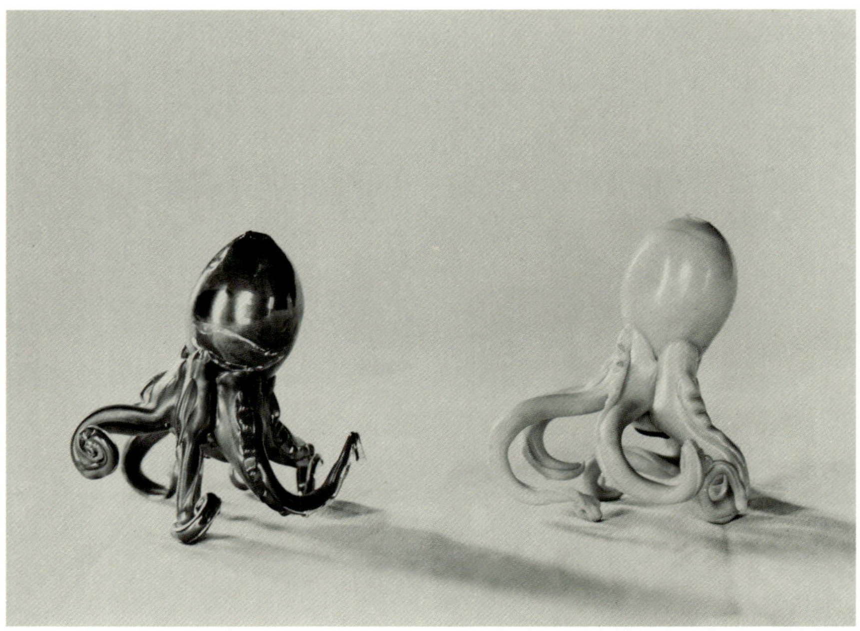

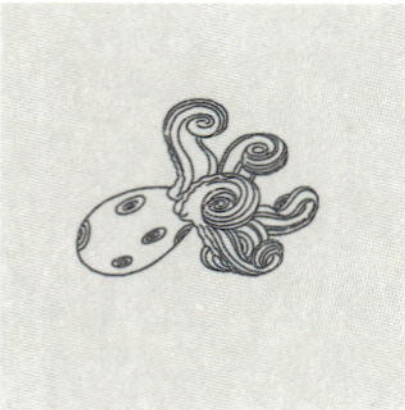

4744

'Large Octopus' in cased *lattimo* and green glass with green *millefiori* ("pearls") murrine and ribbed tentacles in green transparent glass. Strongly iridised surface. Some similar examples were exhibited in 1930 at the 4th Triennale of Monza.
1929–30
24 cm high
Fondazione Musei Civici di Venezia, Museo del Vetro collection

exhibitions
1930, Monza, 4th Triennale

>
4744, catalogue drawing, AMuVe

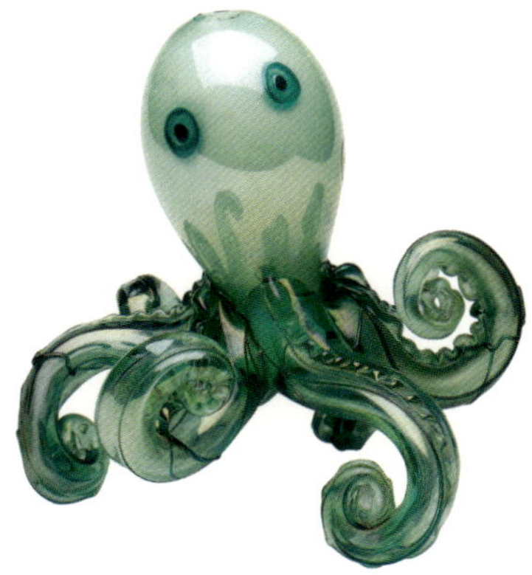

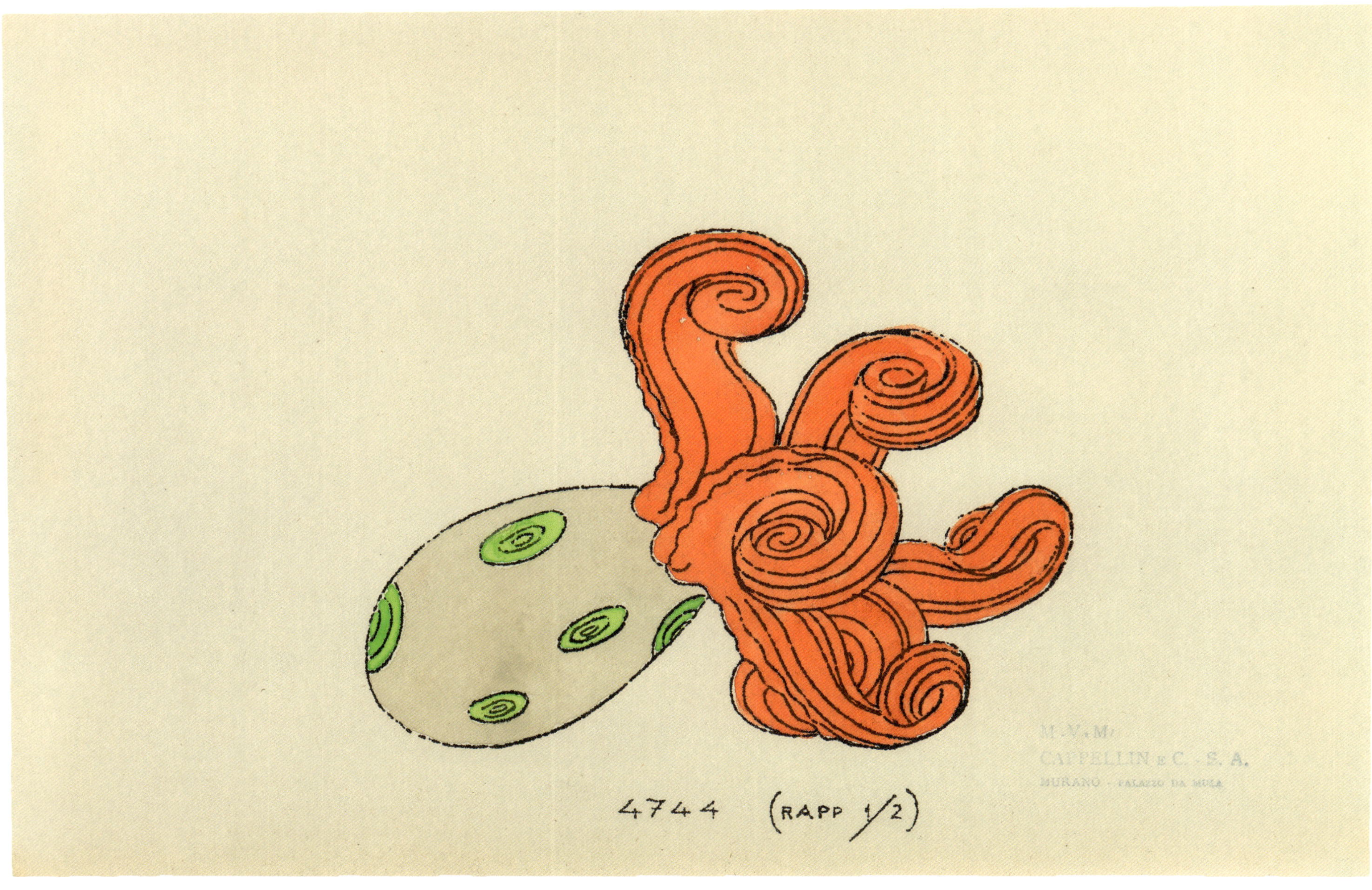

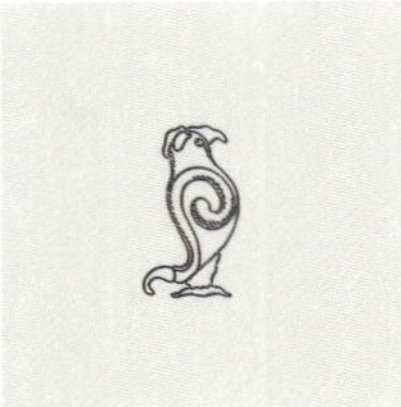

4757

Owl in yellow transparent glass with applied gold leaf. Strongly iridised surface. Eyes in red *pasta vitrea*. A similar example appears in the catalogue for the Amsterdam glass, ceramics and lace exhibition of 1931. In 1930, the model was exhibited at the 4th Triennale of Monza in the *lattimo* glass variant with finishings in coloured opaque glass. The design of the piece was inspired by ancient Chinese bronzes.
c. 1930
20 cm high
Private collection, Turin

exhibitions
1930, Monza, 4th Triennale
1931, Amsterdam, Exhibition of glass, ceramics and lace

>
'Bowl and sculptural pieces in blown polychrome glass', in Tentoonstelling van glas, aardewerk en kant, van moderne italiaansche kunstnijverheid. Mostra di vetri, ceramiche e merletti d'arte moderna italiana, *Venice 1931, p. 20*

>>
4757, catalogue drawing, AMuVe

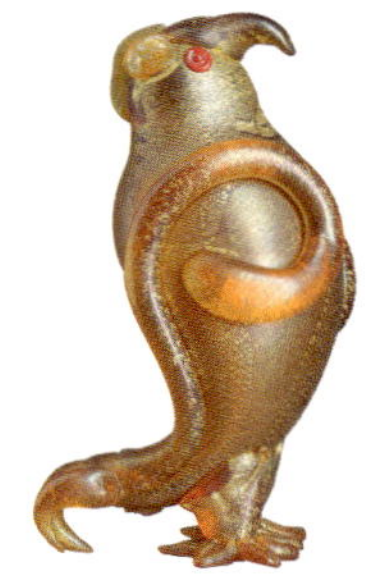

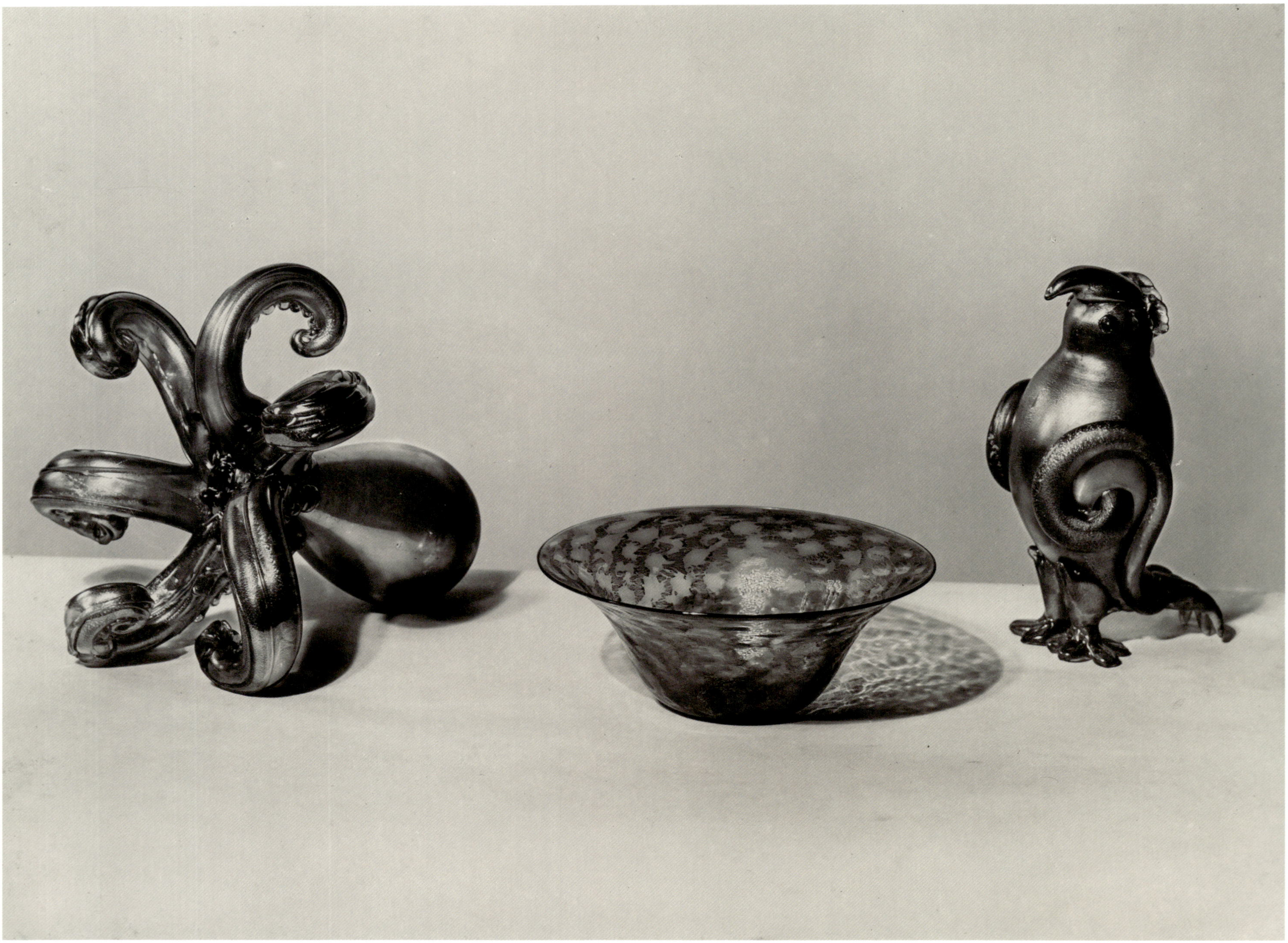

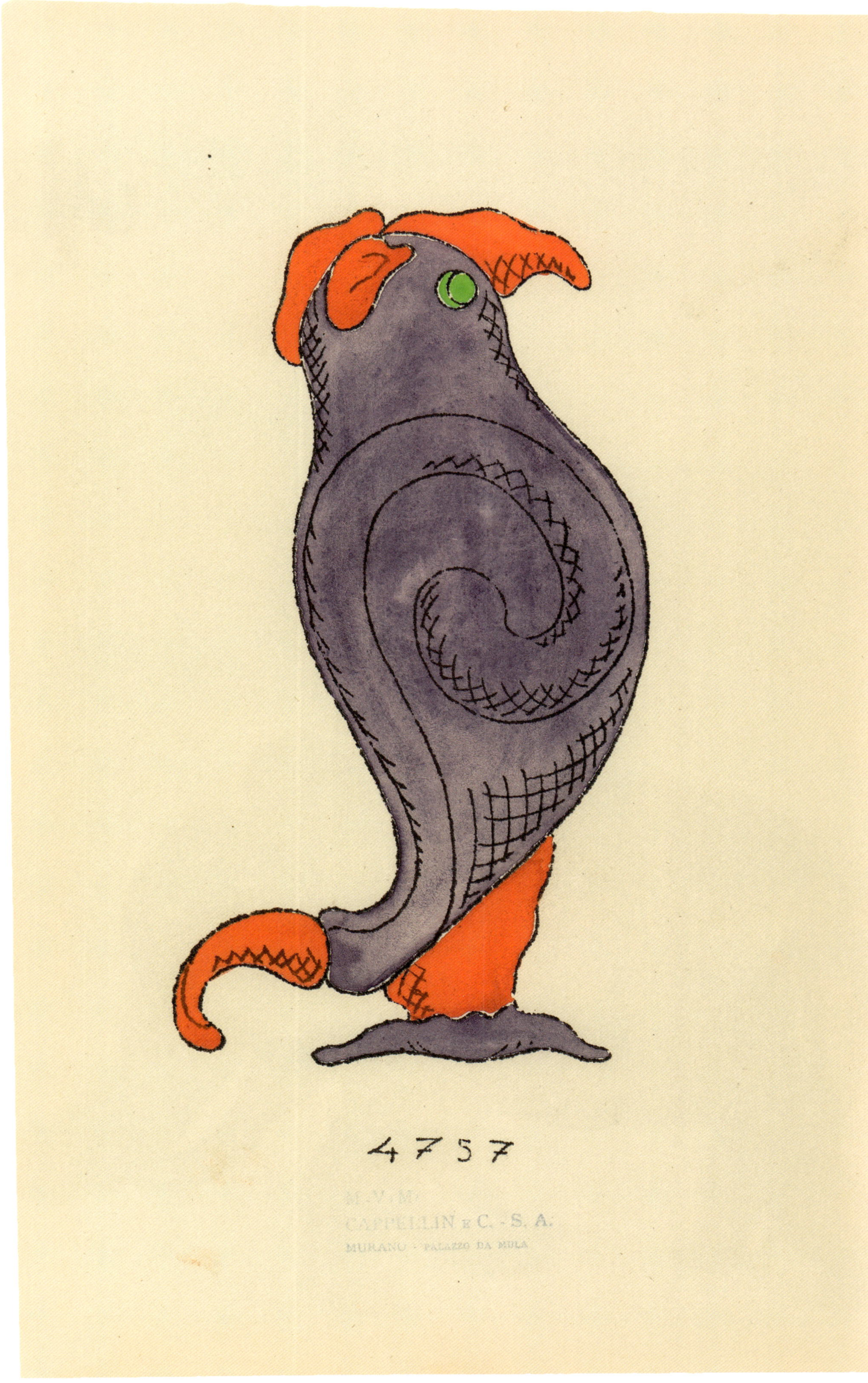
4757
M.V.M.
CAPPELLIN E C. - S. A.
MURANO - PALAZZO DA MULA

4774

Elephant in grey *pulegoso* glass with tusks in red *pasta vitrea* and eyes with black and yellow murrina. A similar example was exhibited in the travelling American exhibition, Contemporary Glass and Rugs (1929–30). Four *pasta vitrea* elephants, of different dimensions, were exhibited in 1931 at the Amsterdam Exhibition of glass, ceramics and lace (see no. 4770).
c. 1929
17 cm high
Private collection

exhibitions
1929–30, USA, Contemporary Glass and Rugs
1931, Amsterdam, Exhibition of glass, ceramics and lace

>
Elephant in pulegoso *glass. Period photograph, c. 1929, CSV*

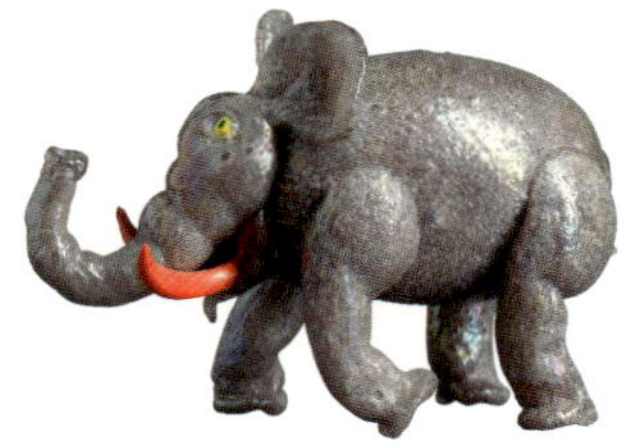

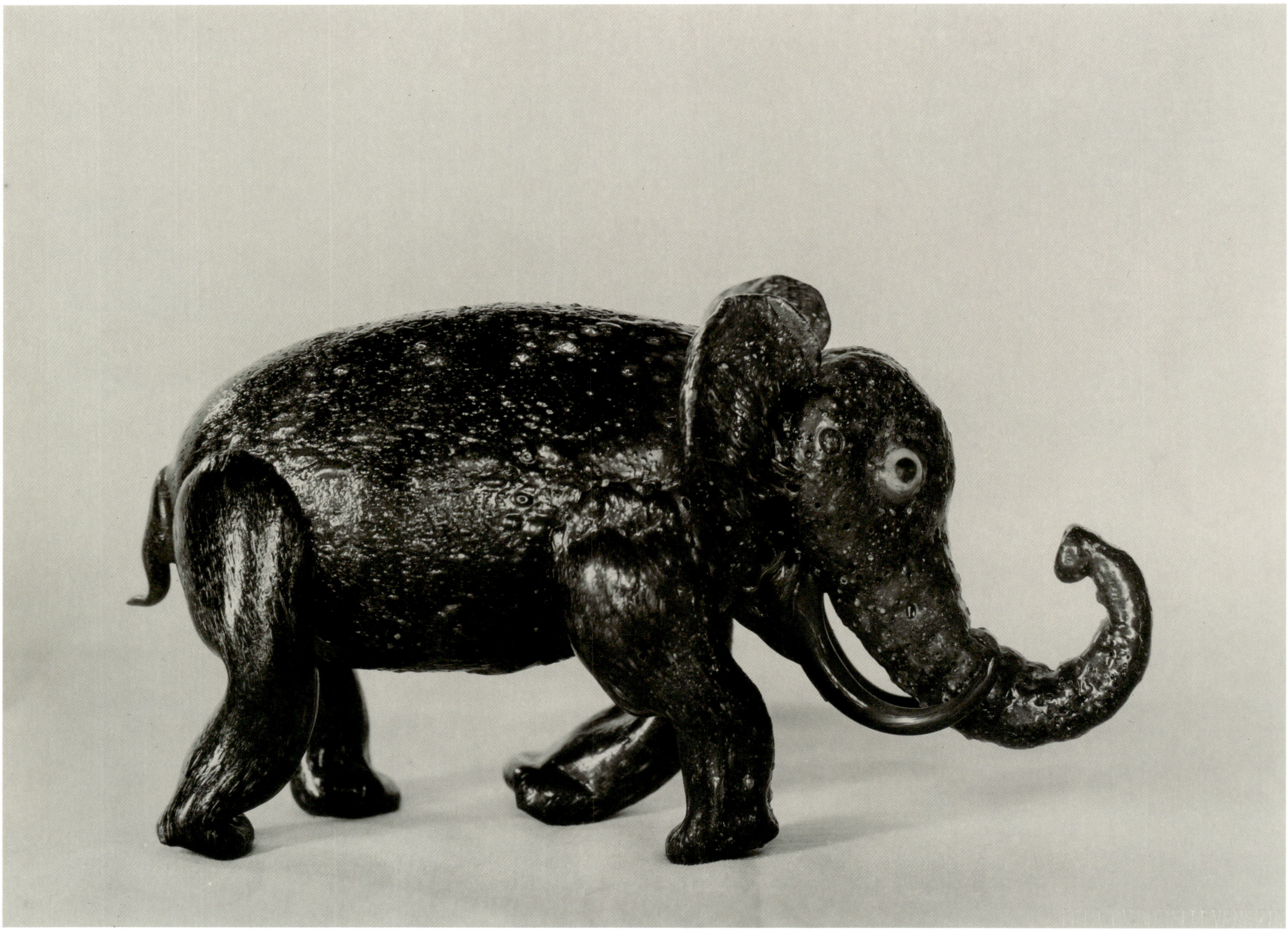

4770

Fox in opaque glass exhibited at Amsterdam Exhibition of glass, ceramics and lace in 1931. Period photograph, detail, AHL.
c. 1929
30 cm high

exhibitions
1931, Amsterdam, Exhibition of glass, ceramics and lace

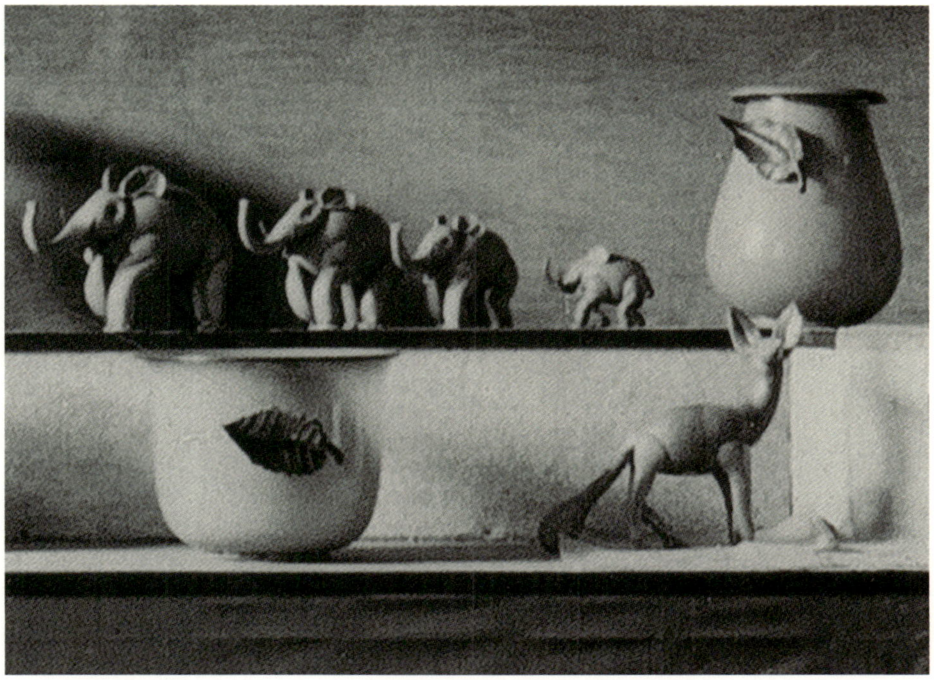

4471

Tiger in opaque glass. Period photograph, c. 1929, CSV. The image appeared in the magazine *Poligono* in December 1929 and in *Domus* in January 1930, where the object was described as "bibelots [sic] in milky and *aurato* glass".
c. 1929
20 cm wide

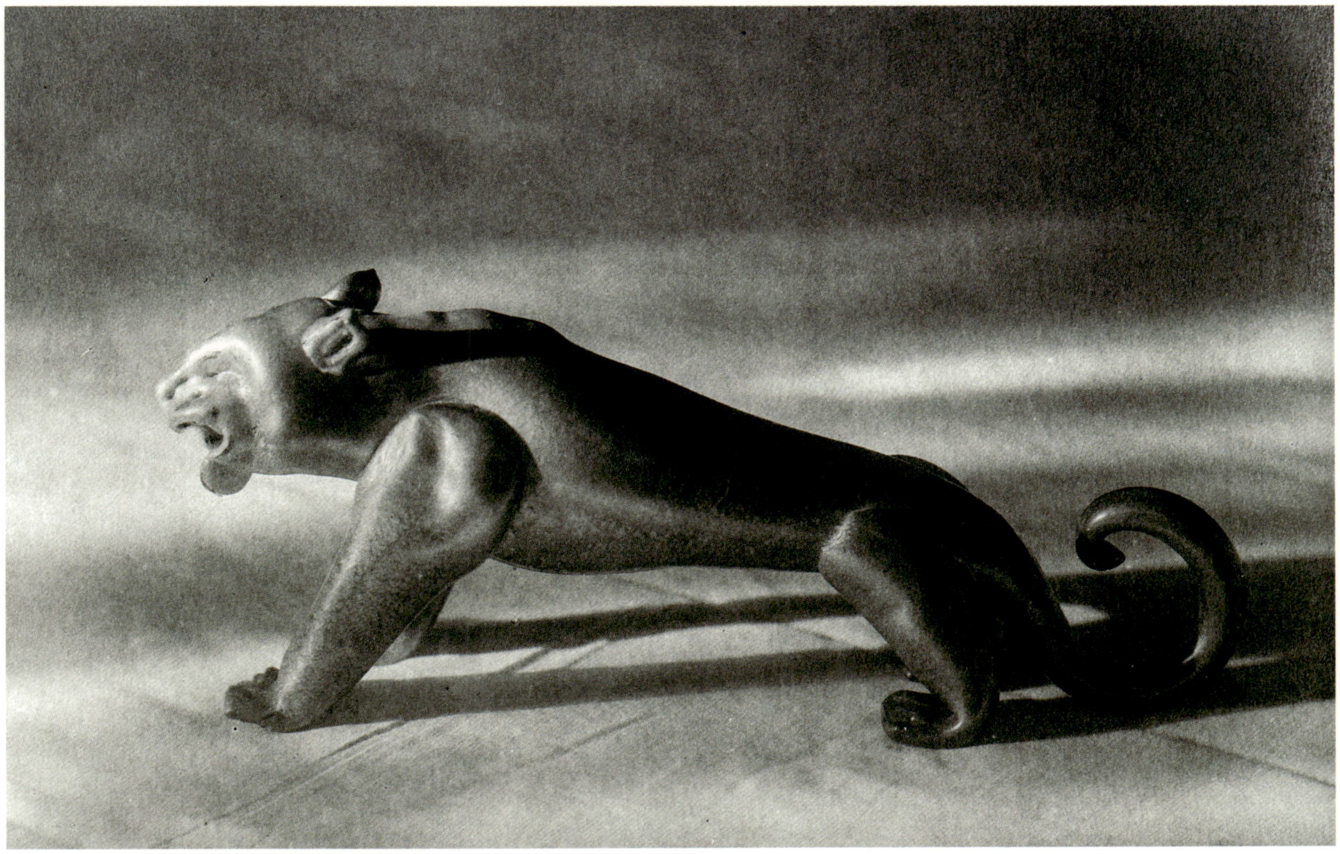

n.i.

Rabbit in light-blue transparent glass and squirrel in violet transparent glass, both with funnel-shaped tails. Strongly iridised surface. These are two non-catalogue items, the design of which refers back to the animal-shaped oil lamps of the 17th century, with particular reference to an artefact in the shape of a rabbit kept in the Museum of Capodimonte in Naples.
1928–29
9; 15 cm high
M. V. M. Cappellin label
Anna Gastel collection, Milan

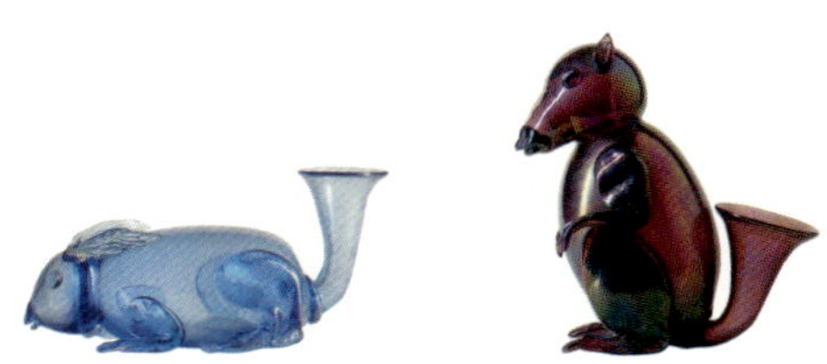

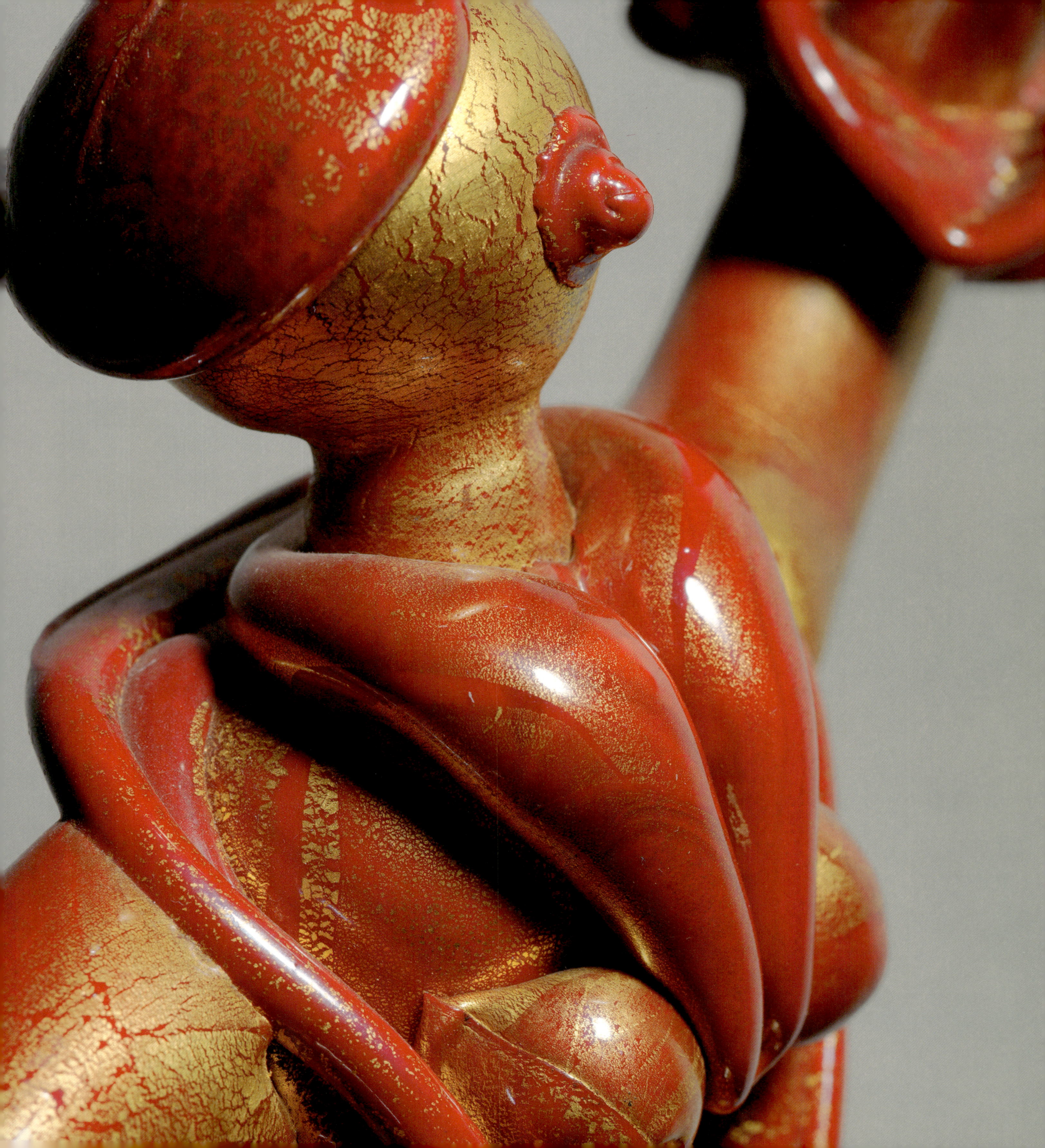

On the occasion of the American travelling exhibition *Contemporary Glass and Rugs,* in 1929, a series of musicians presented as a jazz band was shown, and met with particular success with the public. Beginning with this subject, the repertoire was extended to include a series of dancers, male and female, which were also used to decorate ashtrays. These artefacts were usually proposed in transparent glass but there are documented variants in *lattimo* with finishings in polychrome glass, among them a dancing *Harlequin*. Other new models belonging to the series of the *Turks* were exhibited in 1930 at the 4th Triennale of Monza.

FIGURINES 1929-30

Exhibitions

1929, Leipzig, Trade Fair
1929-30, USA, *International Exhibition of Contemporary Glass and Rugs*, travelling exhibition
1930, Monza, 4th *Esposizione Triennale Internazionale delle Arti Decorative ed Industriali Moderne*

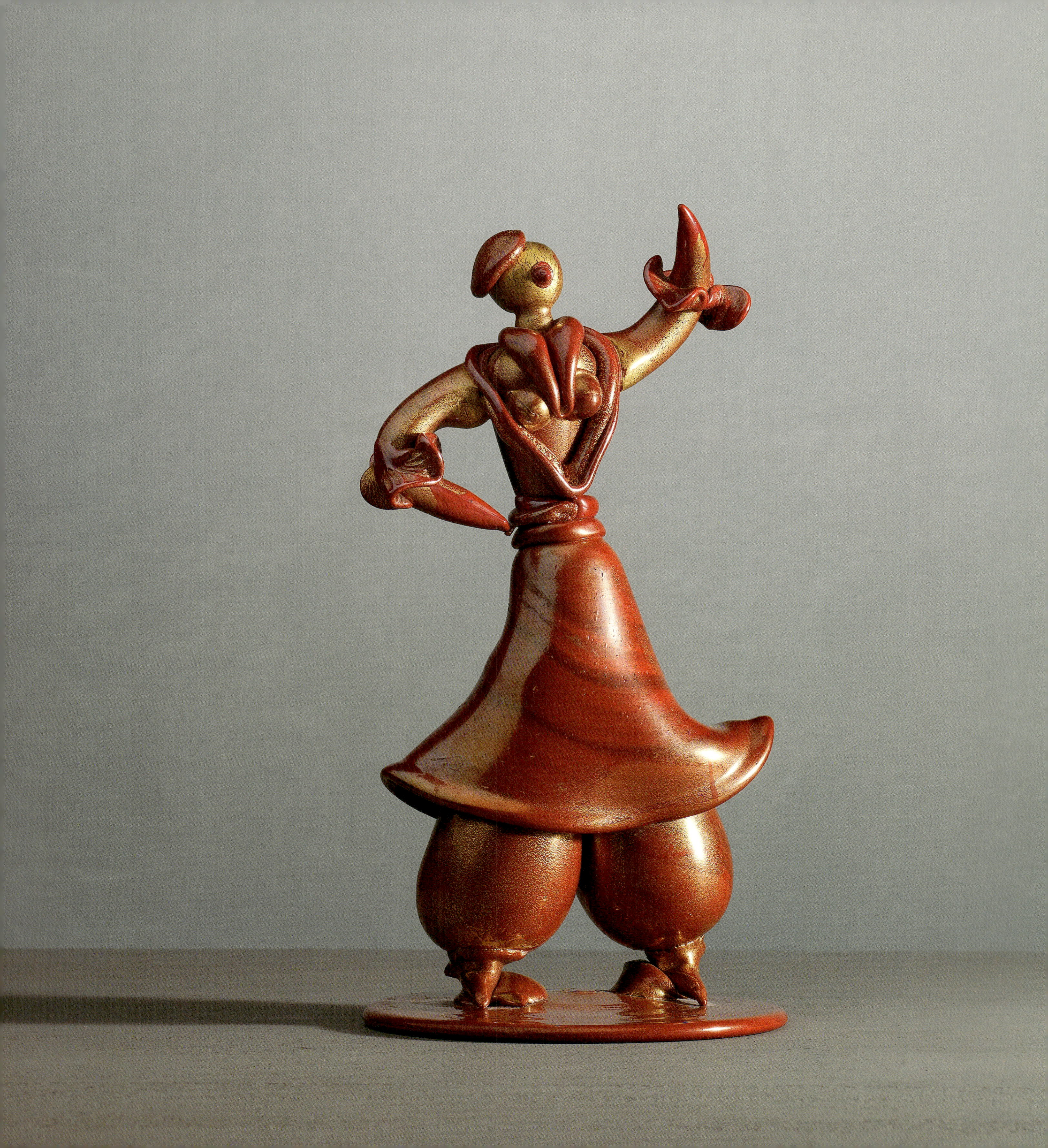

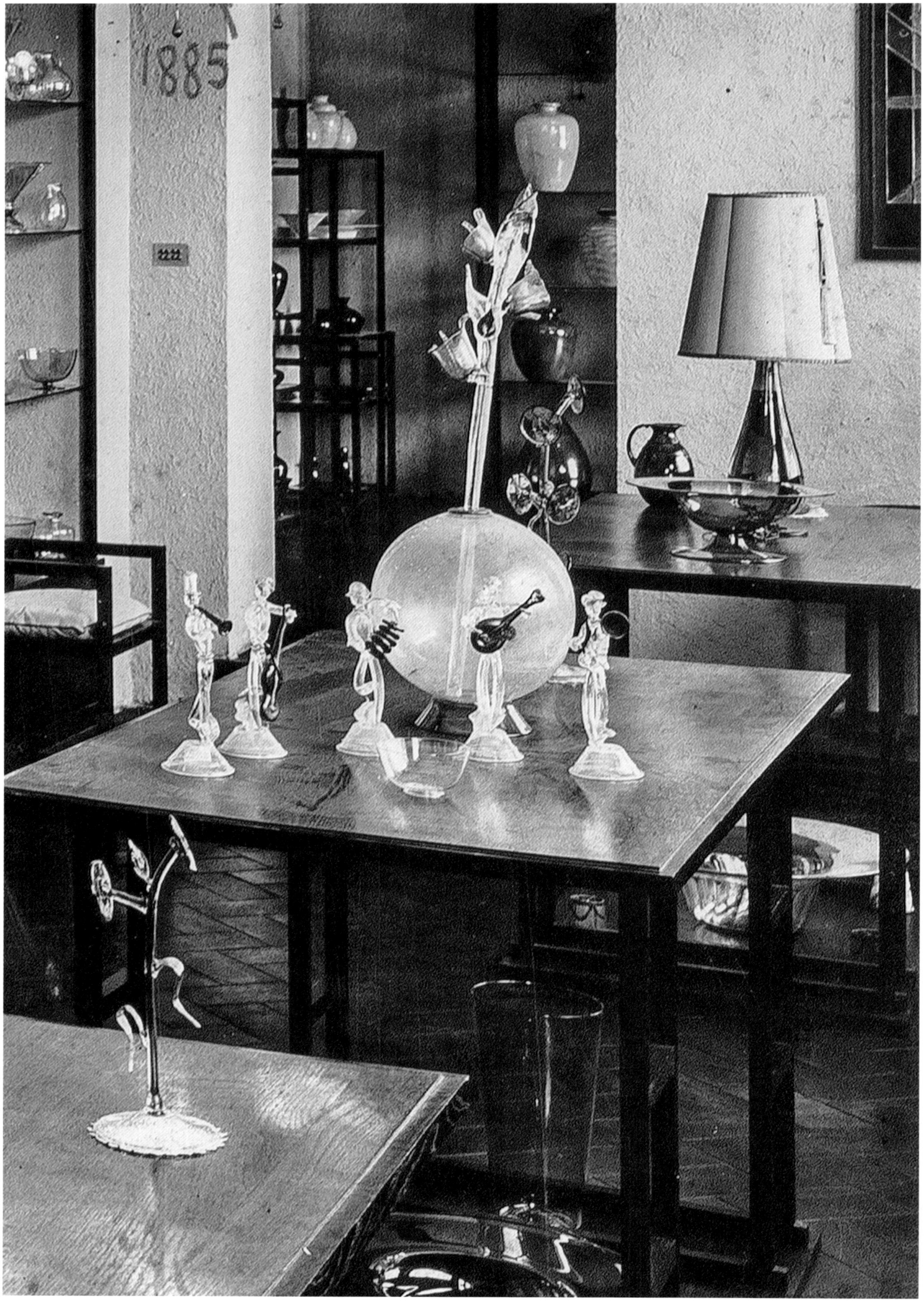

> Group of musicians in glass exhibited in the Cappellin shop in Florence. Period photograph, 1928–29, AMB

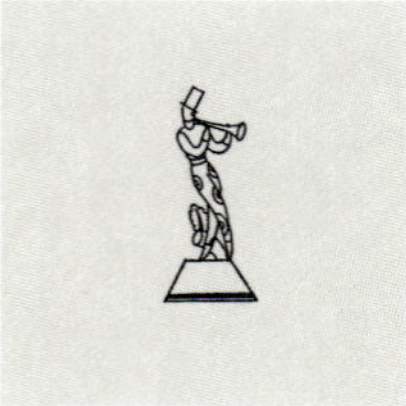

4710

Cornet Player on truncated-cone-shaped base with branch and leaf, in *pagliesco* glass and in black glass. Strongly iridised surface. Model no. 4710 is part of a series of six 'instrumentalists' (4709–4714) that were presented as members of a jazz band and exhibited at the travelling American exhibition, *Contemporary Glass and Rugs*. This model was used also to decorate an ashtray (no. 4760).
c. 1929
26.4 cm high
Philadelphia Museum of Art: Gift of Mrs. Rodolphe Meyer de Schauensee, 1977, 196-3e

exhibitions
1929–30, USA, *Contemporary Glass and Rugs*

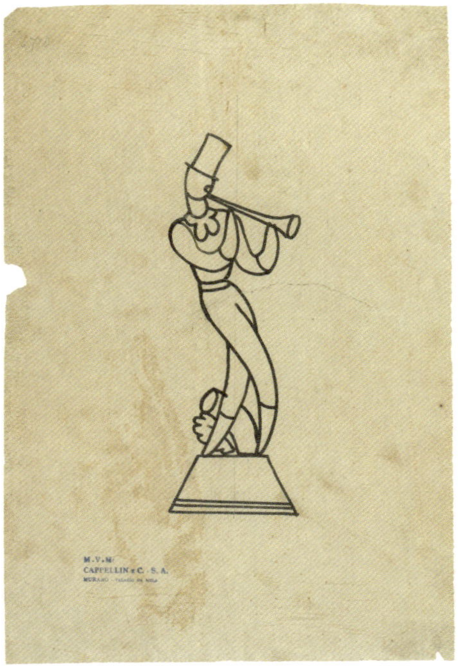

>
4710, working drawing, CSV

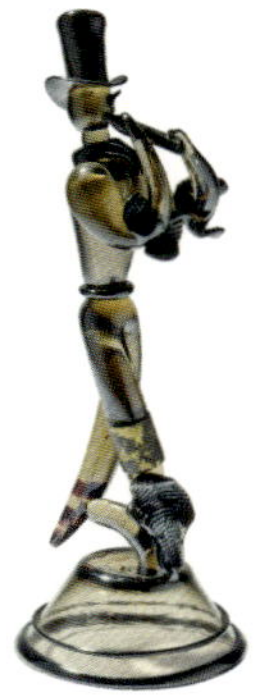

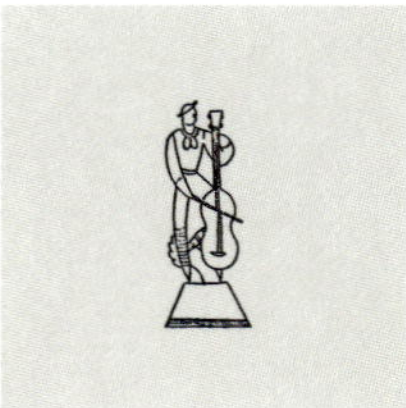

4712

Violone Player on truncated-cone-shaped base with branch and leaf, in *pagliesco* glass and in black glass. Strongly iridised surface. Model no. 4712 is part of a series of six 'instrumentalists' (4709–4714) that were presented as members of a jazz band and exhibited at the travelling American exhibition, *Contemporary Glass and Rugs*. This model was used also to decorate an ashtray (no. 4762)
c. 1929
23 cm high
Philadelphia Museum of Art: Gift of Mrs. Rodolphe Meyer de Schauensee, 1977, 196-3a

exhibitions
1929–30, USA, *Contemporary Glass and Rugs*

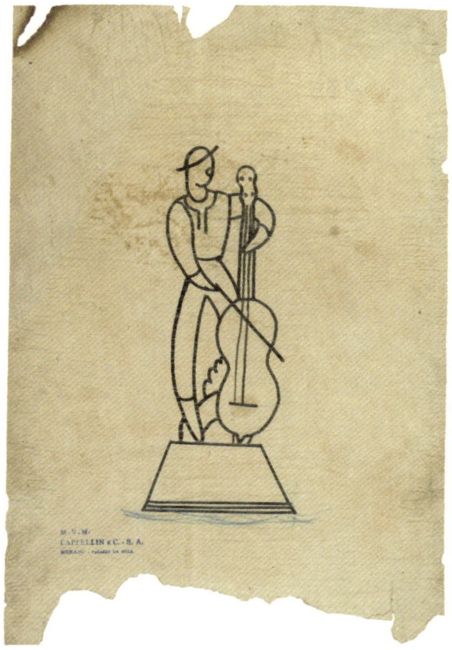

>
4712, working drawing, CSV

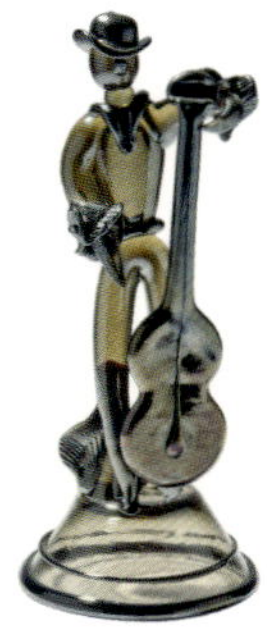

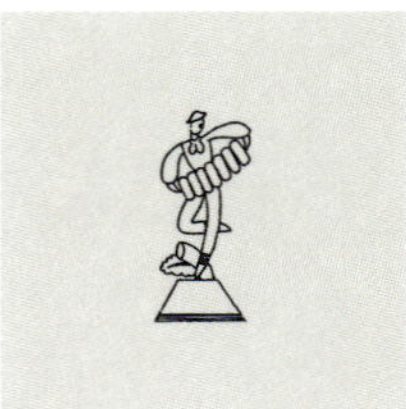

4709

Concertina Player on truncated-cone-shaped base with branch and leaf, in *pagliesco* glass and in black glass. Strongly iridised surface. Model no. 4709 is part of a series of six 'instrumentalists' (4709–4714) that were presented as members of a jazz band and exhibited at the travelling American exhibition, *Contemporary Glass and Rugs*. This model was used also to decorate an ashtray (no. 4759).
c. 1929
23.5 cm high
Philadelphia Museum of Art: Gift of Mrs. Rodolphe Meyer de Schauensee, 1977, 196-3b

exhibitions
1929–30, USA, *Contemporary Glass and Rugs*

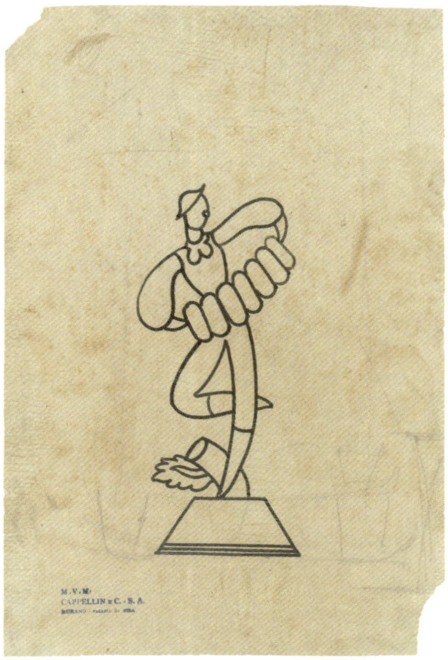

>
4709, working drawing, CSV

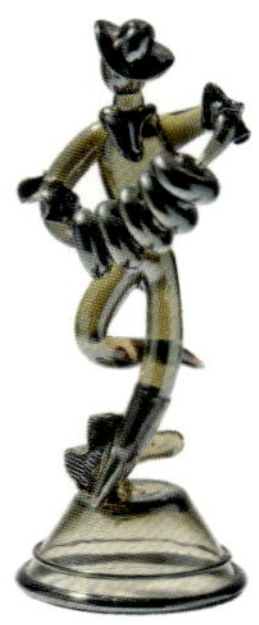

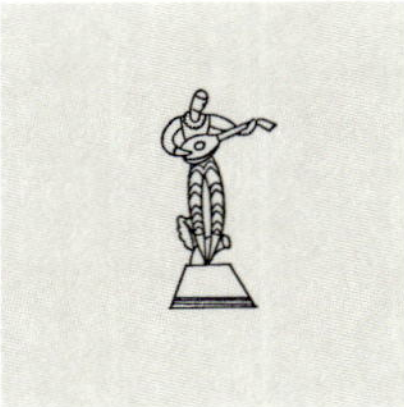

4713

Mandola Player on truncated-cone-shaped base with branch and leaf, in *pagliesco* glass and in black glass. Strongly iridised surface. Model no. 4713 is part of a series of six 'instrumentalists' (4709–4714) that were presented as members of a jazz band and exhibited at the travelling American exhibition, *Contemporary Glass and Rugs*. This model was used also to decorate an ashtray (no. 4763) and as the base of a lamp.
c. 1929
24 cm high
Philadelphia Museum of Art: Gift of Mrs. Rodolphe Meyer de Schauensee, 1977, 196-3d

exhibitions
1929–30, USA, *Contemporary Glass and Rugs*

>
4713, working drawing, CSV

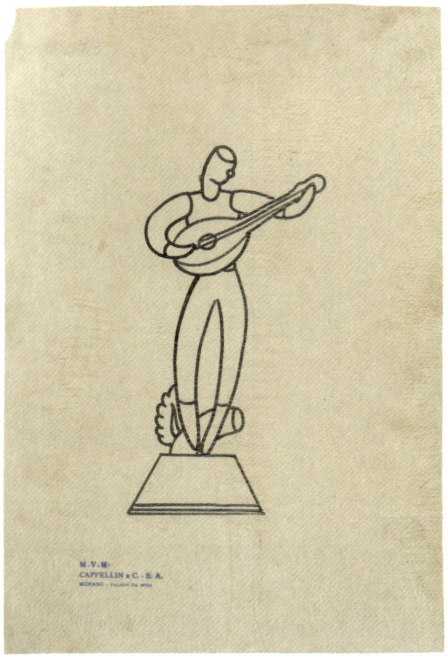

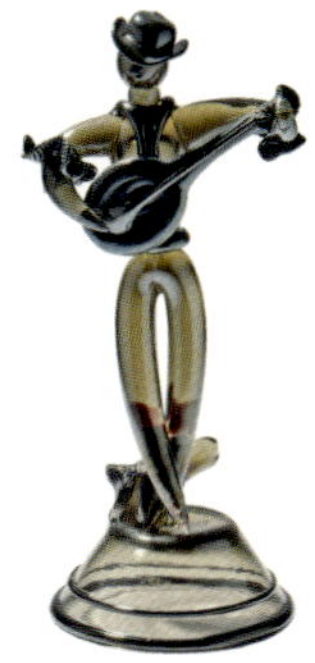

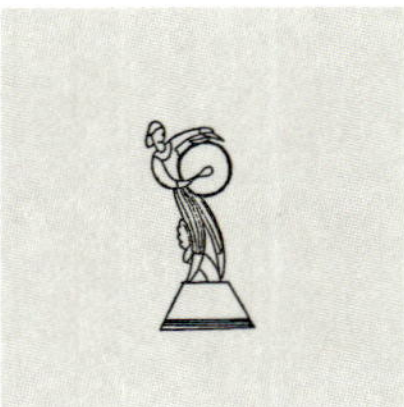

4714

Bass Drum Player on truncated-cone-shaped base with branch and leaf, in *pagliesco* glass and in black glass. Strongly iridised surface. Model no. 4714 is part of a series of six 'instrumentalists' (4709–4714) that were presented as members of a jazz band and exhibited at the travelling American exhibition, *Contemporary Glass and Rugs*. This model was used also to decorate an ashtray (no. 4764)
c. 1929
23.5 cm high
Philadelphia Museum of Art: Gift of Mrs. Rodolphe Meyer de Schauensee, 1977, 196-3f

exhibitions
1929–30, USA, *Contemporary Glass and Rugs*

>
4714, working drawing, CSV

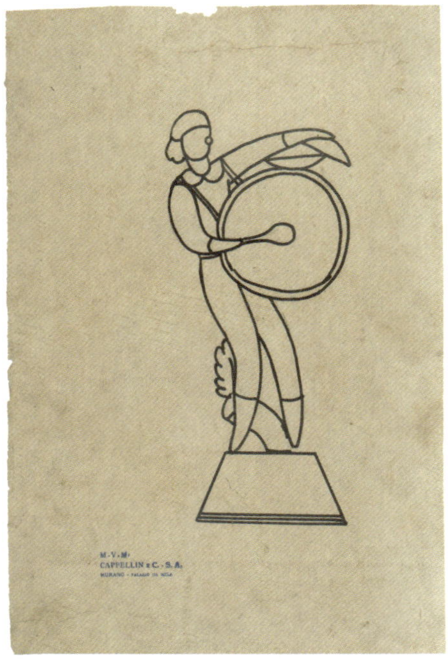

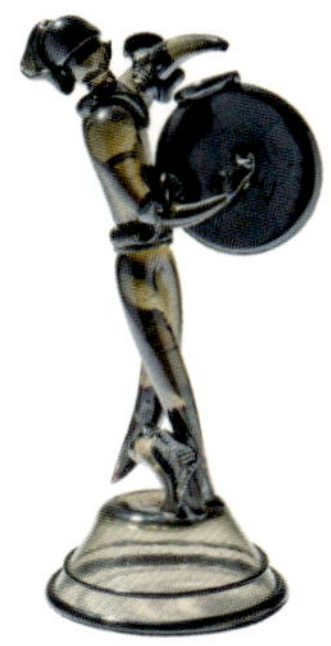

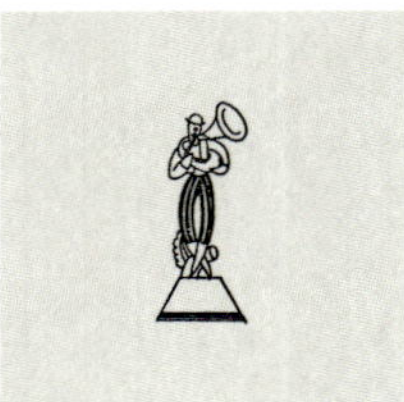

4711

Trombone Player on truncated-cone-shaped base with branch and leaf, in *pagliesco* glass and in black glass. Strongly iridised surface. Model no. 4711 is part of a series of six 'instrumentalists' (4709–4714) that were presented as members of a jazz band and exhibited at the travelling American exhibition, *Contemporary Glass and Rugs*. This model was used also to decorate an ashtray (no. 4761)
c. 1929
23.5 cm high
Philadelphia Museum of Art: Gift of Mrs. Rodolphe Meyer de Schauensee, 1977, 196-3c

exhibitions
1929–30, USA, *Contemporary Glass and Rugs*

>
4711, working drawing, CSV

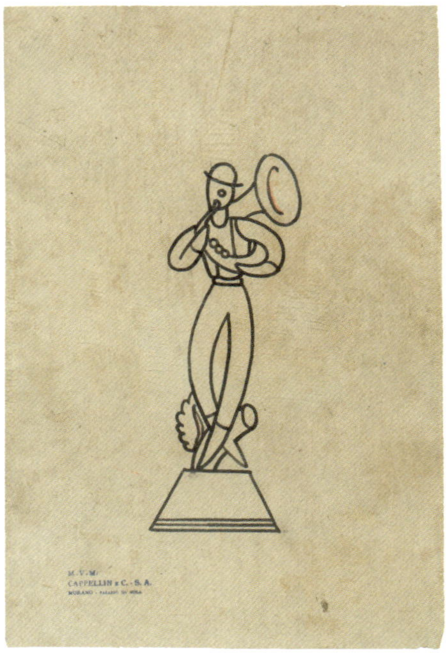

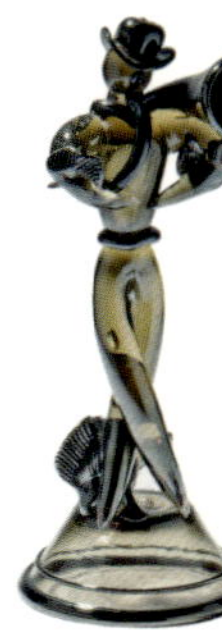

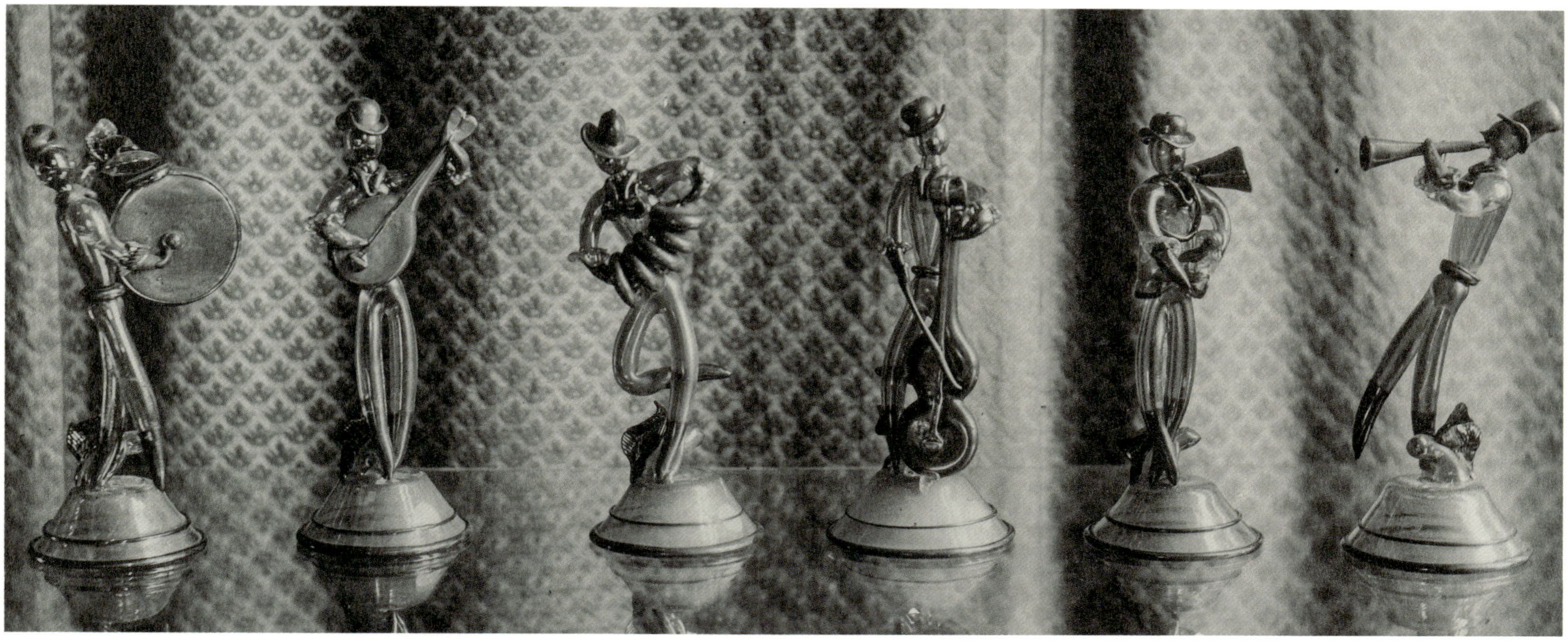

>
Jazz Band *in transparent glass. Period photograph, c. 1929, CSV*

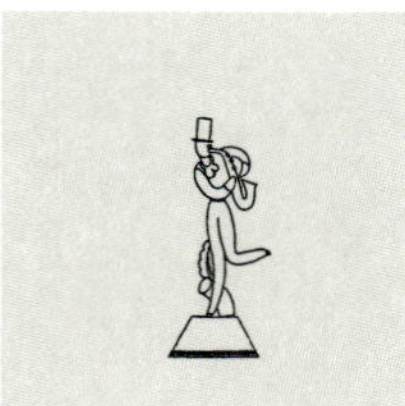

4717

Working drawing for *Saxophone Player* on truncated-cone-shaped base, with branch and leaf, to be produced in glass, CSV. The model is part of a series of figurines representing instrumentalists designed for the American market. This model was also used to decorate an ashtray (n. 4766).
c. 1929
28 cm high

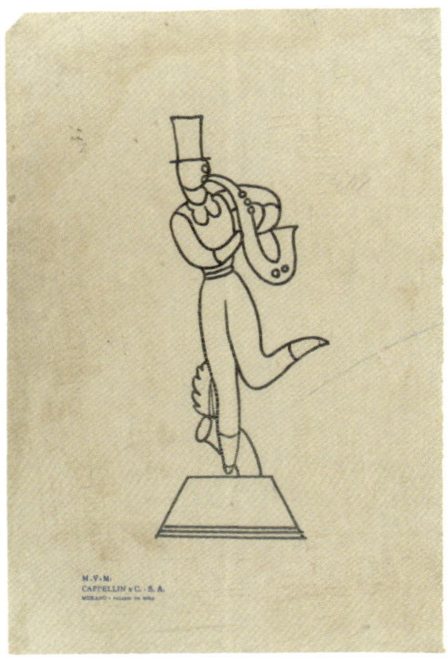

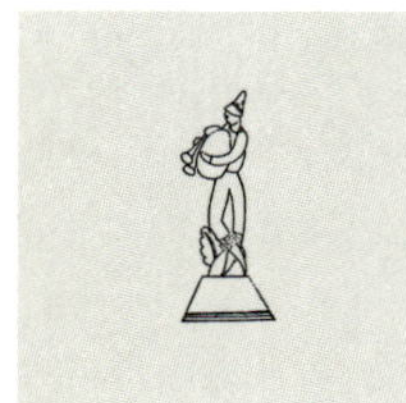

4718

Catalogue drawing for *Bagpipe Player* on truncated-cone-shaped base, with branch and leaf, to be produced in glass, AMB. The model is part of a series of figurines representing instrumentalists designed for the American market. Presumably this model did not have much success as already in 1930 the number '4718' was used to label the figurine of a *Turk*.
c. 1929
27 cm high

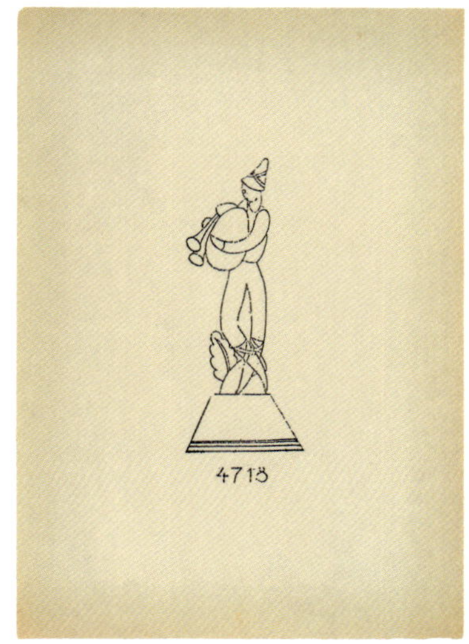

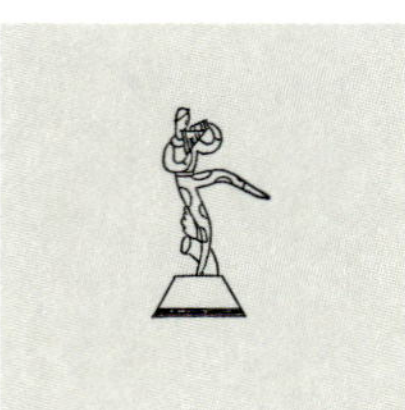

4715

Catalogue drawing for a *Bagpipe Player* on a truncated-cone-shaped base with branch and leaf, to be produced in glass, AMB. The model is part of a series of figurines representing instrumentalists designed for the American market. Presumably this model did not have much success, as already in 1930 the number '4715' was used to label the figurine of a *Turk*.
c. 1929
25 cm high

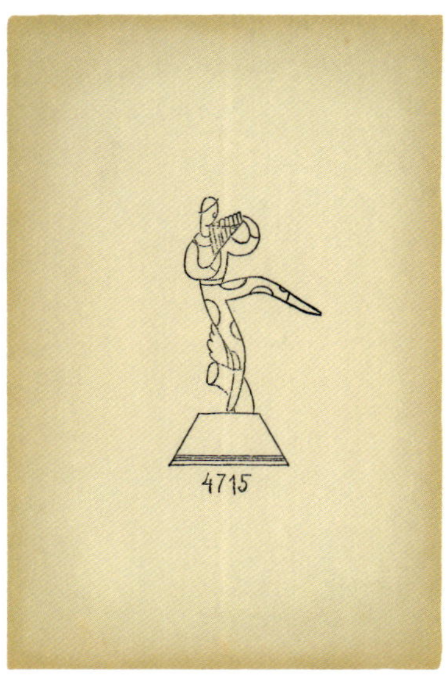

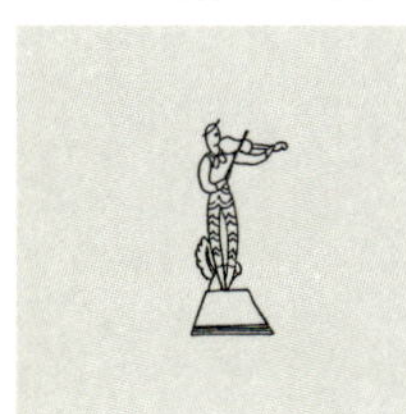

4716

Catalogue drawing for *Violin Player* on truncated-cone-shaped base, with branch and leaf, to be produced in glass, AMB. The model was part of a series of figurines of instrumentalists designed for the American market. This model was used also to decorate an ashtray (n. 4765).
c. 1929
25 cm high

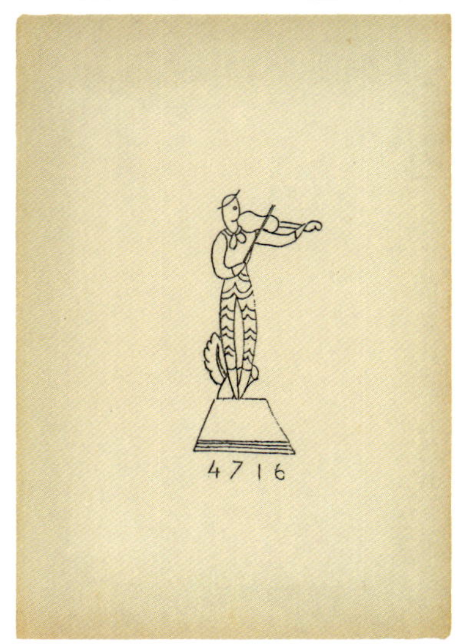

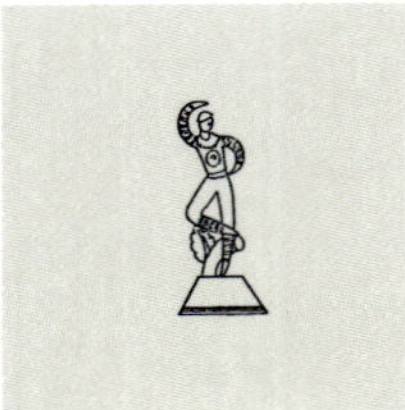

4719

Catalogue drawing for a *Male Dancer with Stripes* on a truncated-cone-shaped base, with branch and leaf, to be produced in glass, AMB. The model was also used to decorate an ashtray (no. 4767).
c. 1929
25 cm high

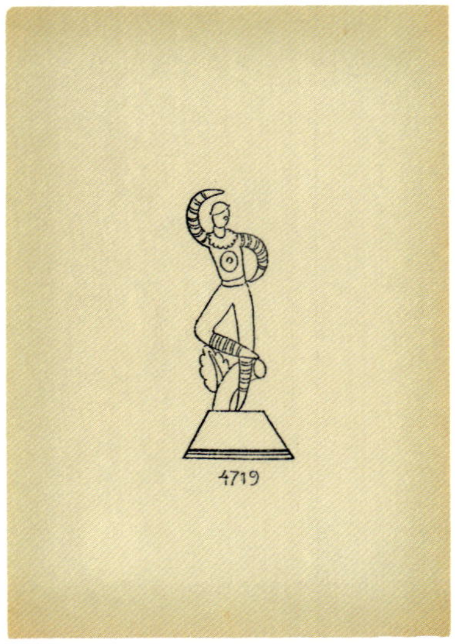

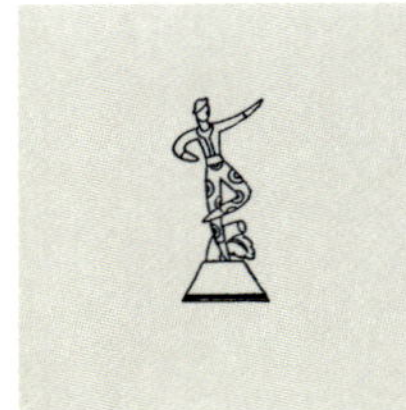

4720

Catalogue drawing for a *Male Dancer with Pearls* (*murrine*) on truncated-cone-shaped base, with branch and leaf, to be produced in glass, AMB. The model was also used to decorate an ashtray (no. 4768).
c. 1929
24 cm high

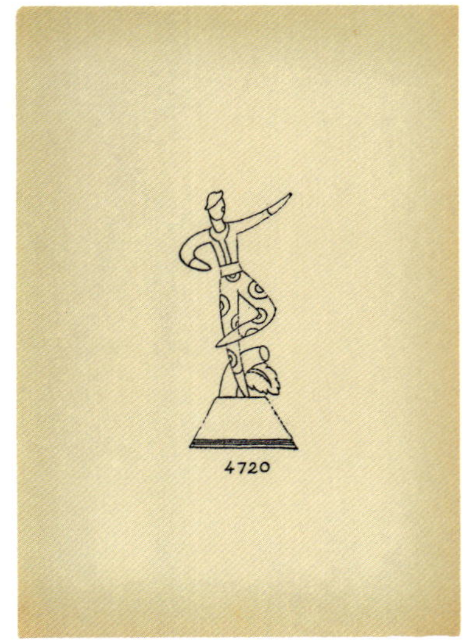

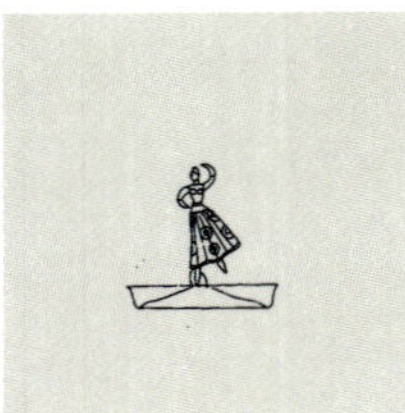

4769

Catalogue drawing for ashtrays no. 4770 and no. 4769 decorated one with a 'harlequin' figure and the other with a 'female dancer' with a dress decorated with murrine, AMuVe. Both the figures were also proposed on a truncated-cone-shaped base (no. 4735, no. 4721).
c. 1929
13 cm high

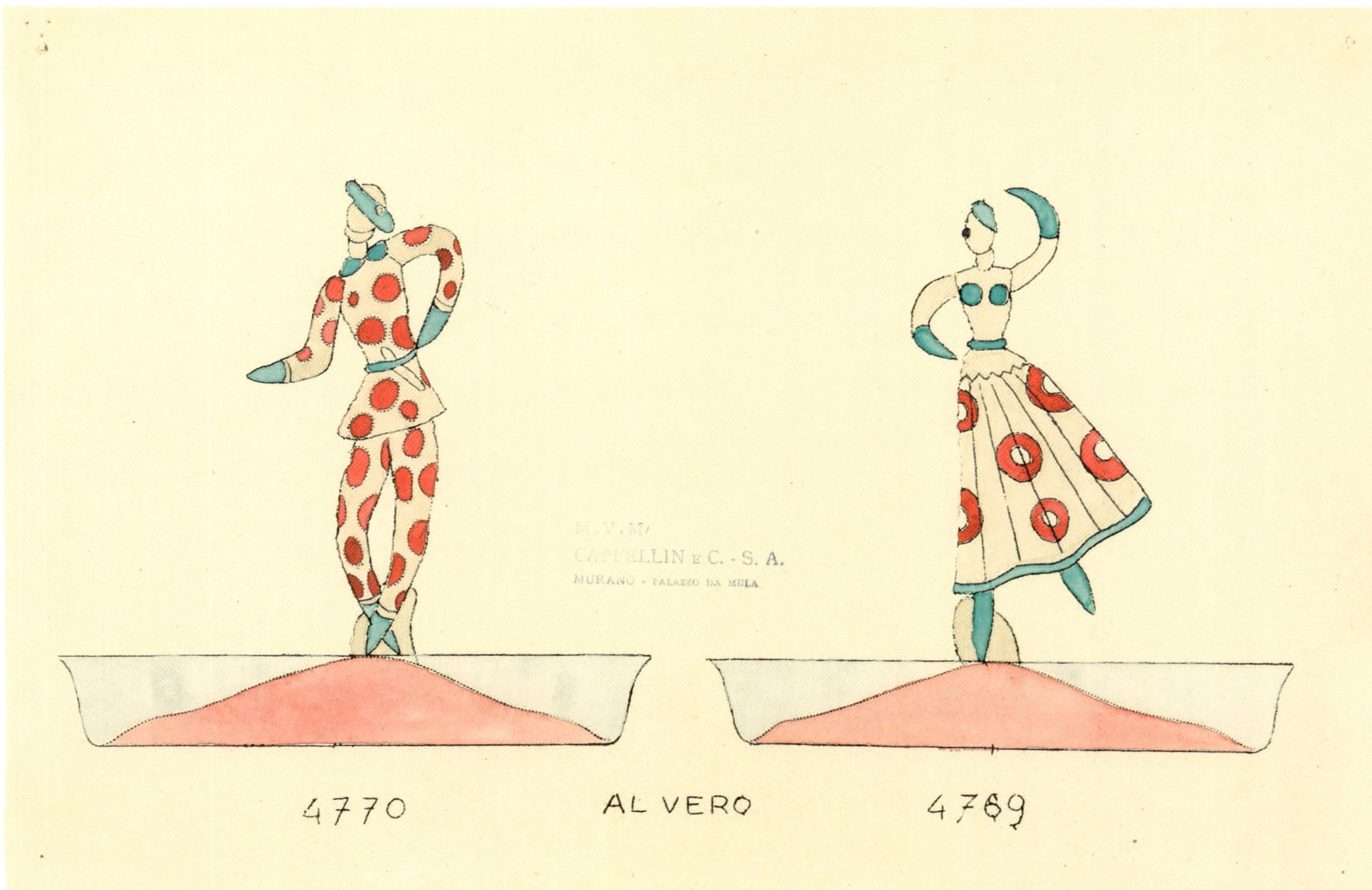

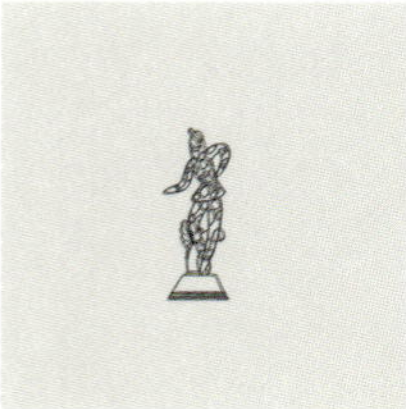

4735

Carlo Scarpa, study for *Arlecchino* on a truncated-cone-shaped base, with branch and leaf, to be produced in *lattimo* glass with 'crystal / ruby / amethyst / dark green / black / [...] yellow', decoration, ACSVr. The same model was also used to decorate an ashtray (no. 4770).
c. 1929
20 cm high

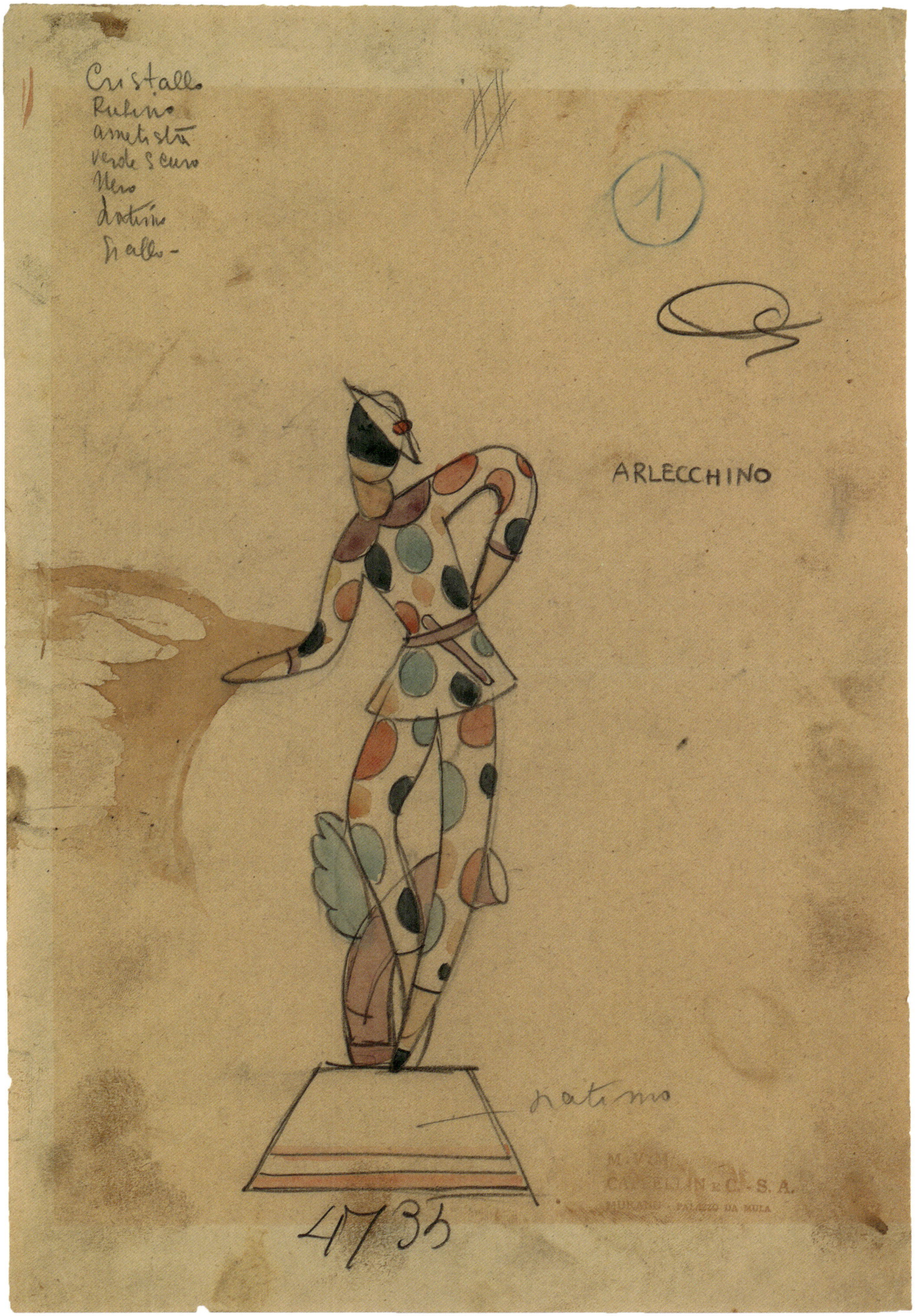

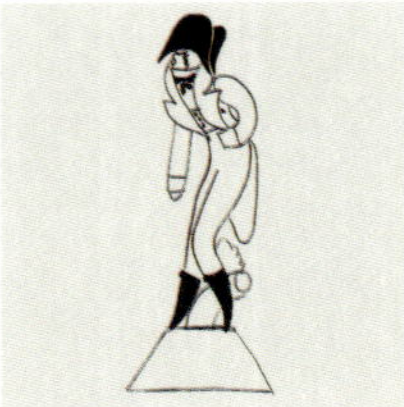

1810

Carlo Scarpa, *Incroyable*, study for male figurine in period dress (French end of the 18th century), on truncated-cone-shaped base with branch and leaf, c. 1929, ACSVr.
c. 1929
45 cm high

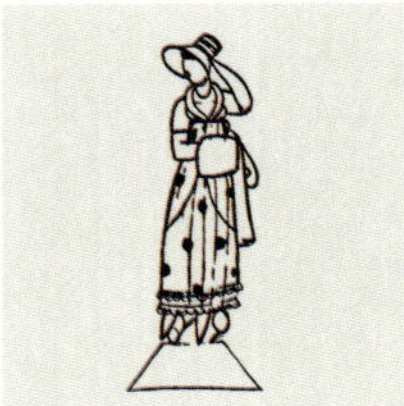

1810

Carlo Scarpa, *Merveilleuse*, study for female figurine in period dress (French end of the 18th century), on truncated-cone-shaped base with branch and leaf, c. 1929, ACSVr.
c. 1929
45 cm high

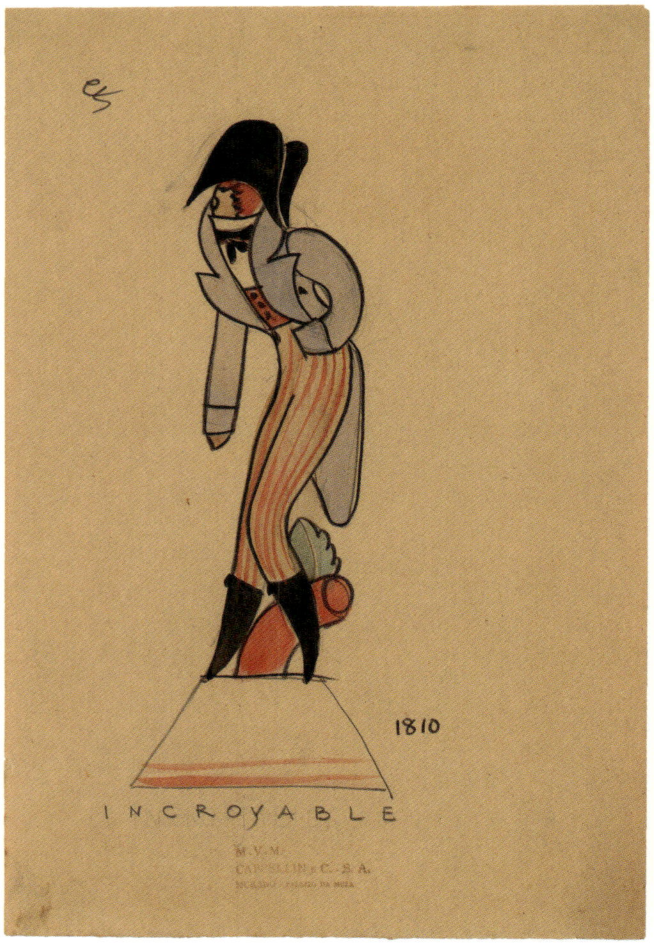

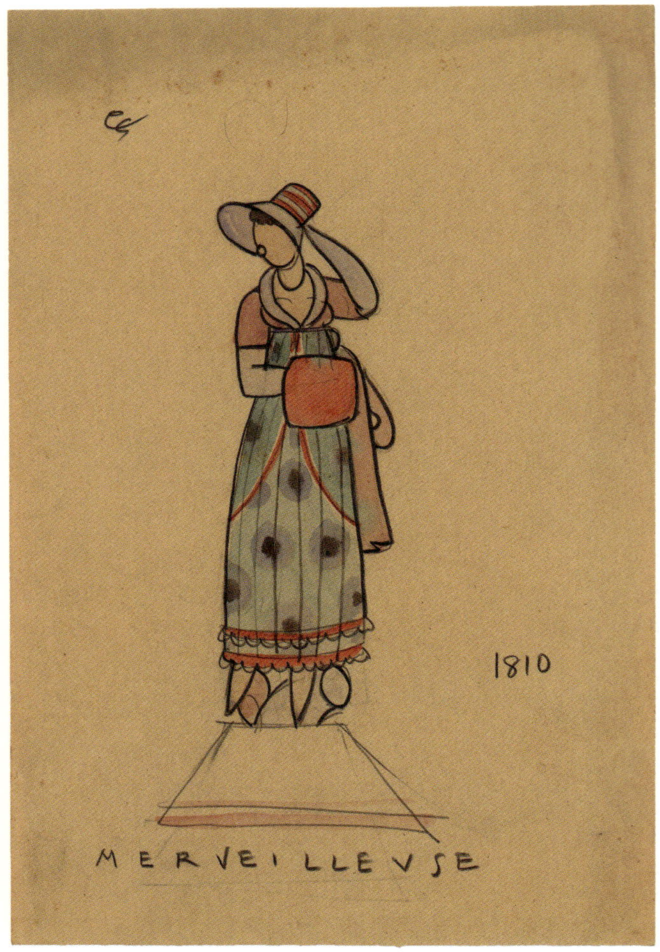

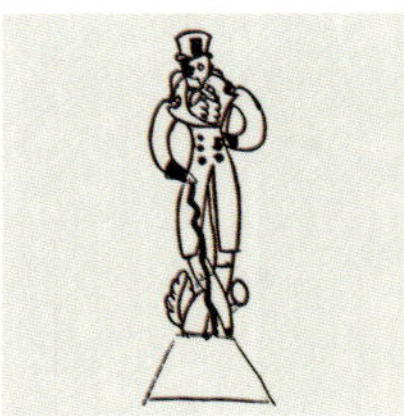

1802

Carlo Scarpa, study for male figurine in period dress on a truncated-cone-shaped base with branch and leaf, c. 1929, ACSVr.
c. 1929
48 cm high

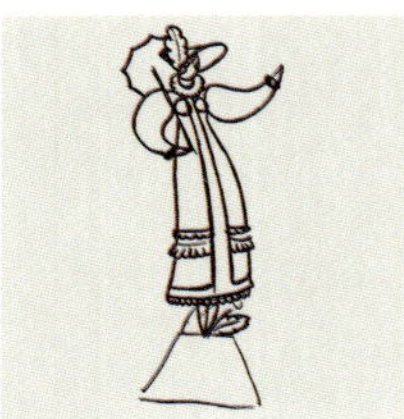

1814

Carlo Scarpa, study for female figurine in period dress, on truncated-cone-shaped base with branch and leaf, c. 1929, ACSVr.
c. 1929
52 cm high

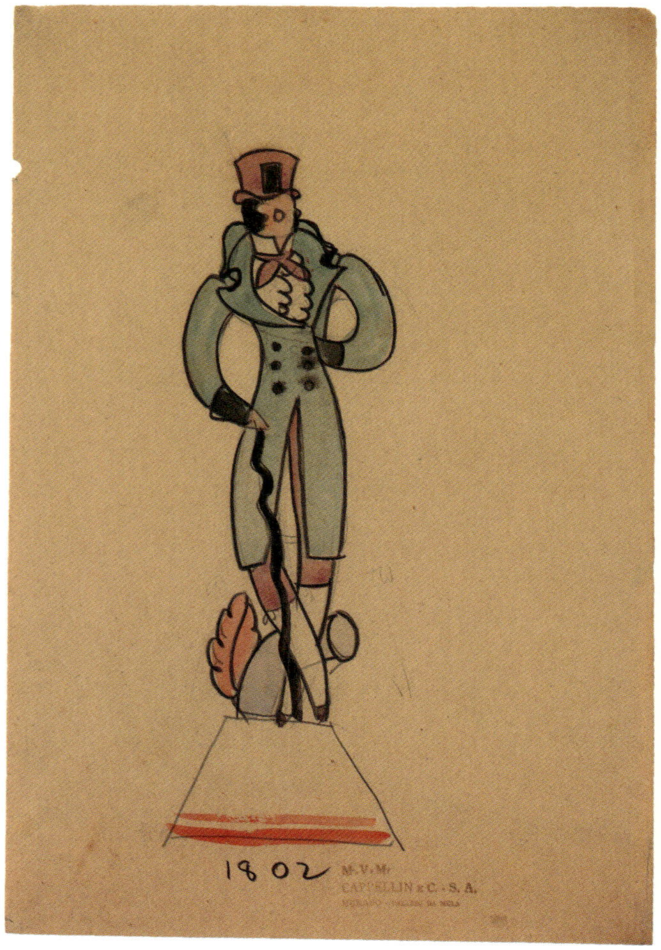

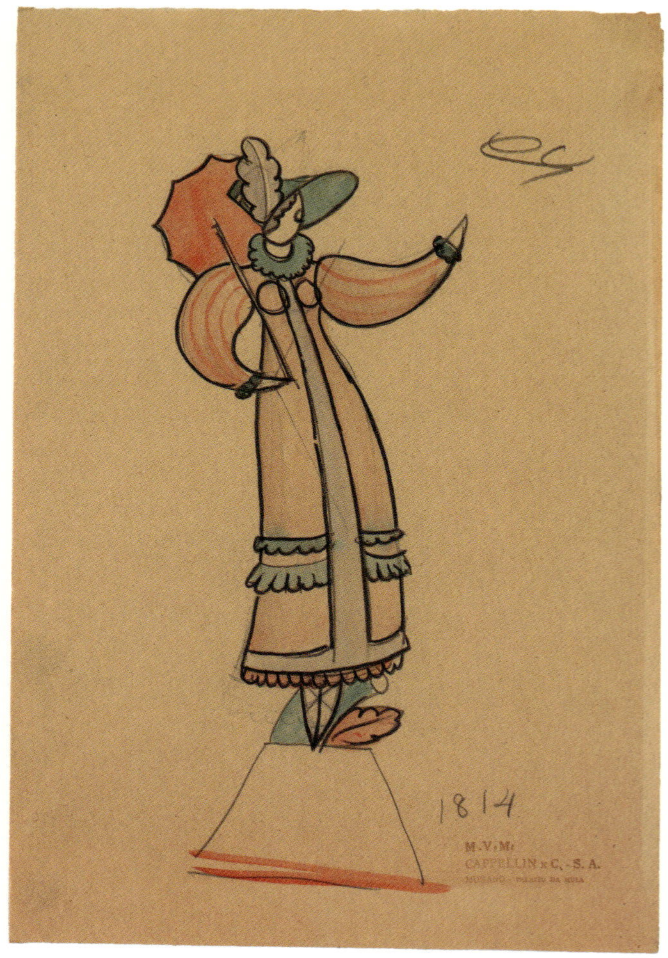

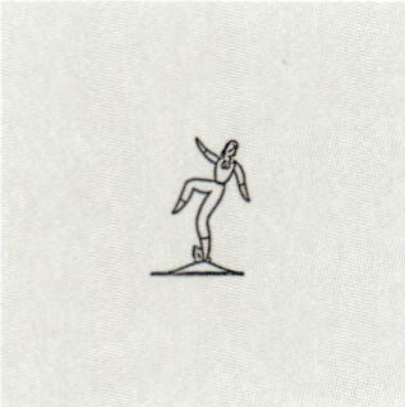

4740

On the right: *Male Dancer* on circular base with leaf, with kerchief tied at the neck. Period photograph, c. 1929, CSV. Model no. 4740 is depicted together with *Male Dancer* no. 4741. Both models were also used to decorate ashtrays (no. 4774 and no. 4776).
c. 1929
12 cm high

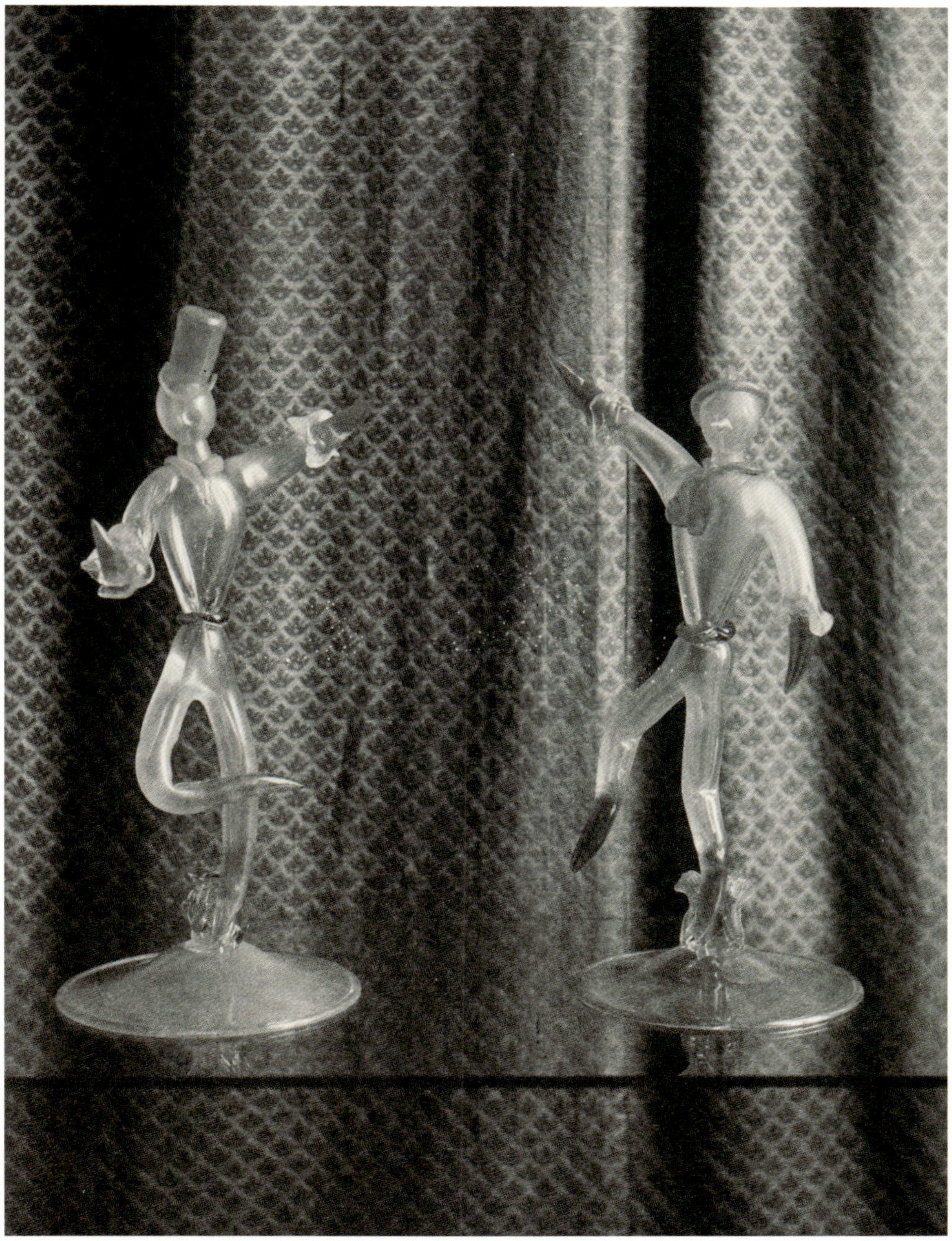

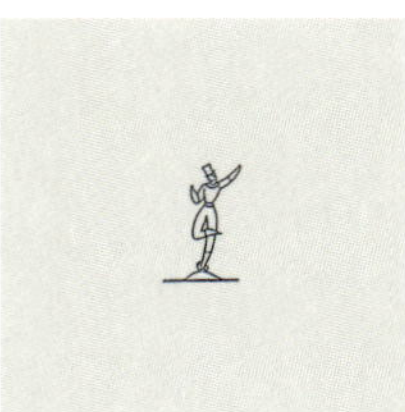

4741

Male Dancer in crystal glass with finishings in *lattimo* glass. The piece is a variant of model no. 4741 (h 12 cm) with one leg bent.
c. 1929
20 cm high
The Lockwood collection,
Fort Lee, NJ

>
4741, catalogue drawing, CSV

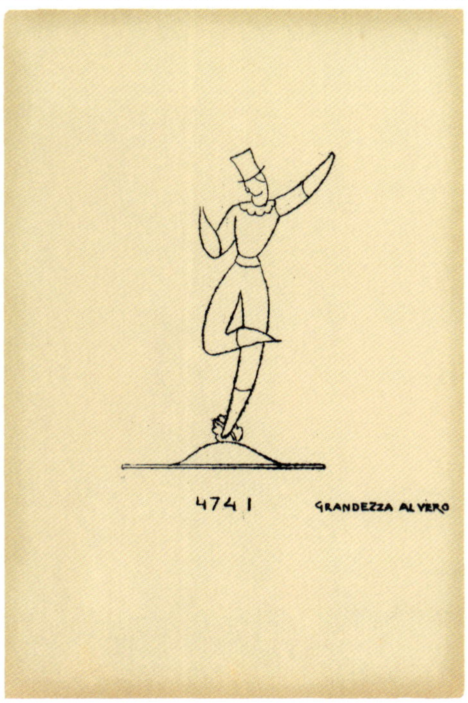

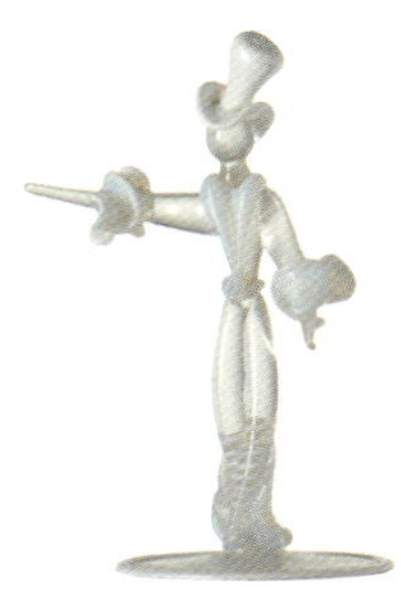

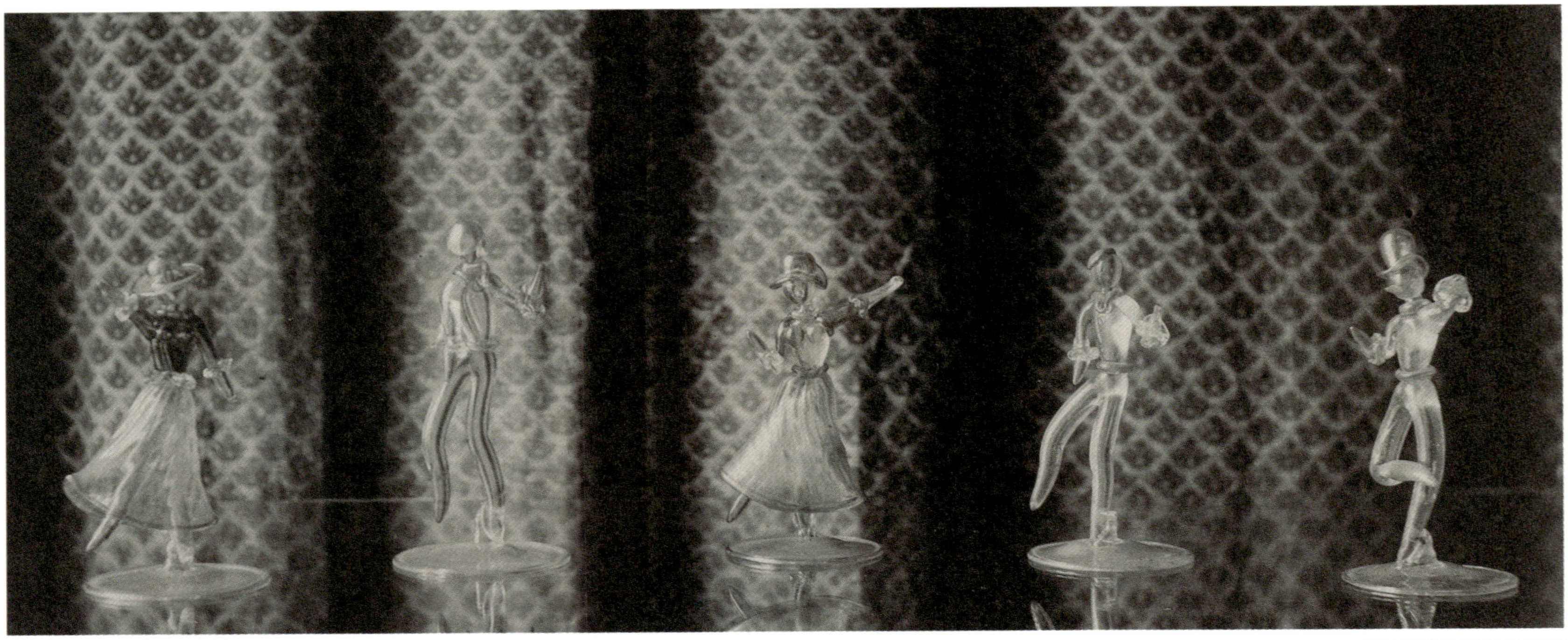

>
Group of dancers.
Period photograph,
c. 1929, CSV

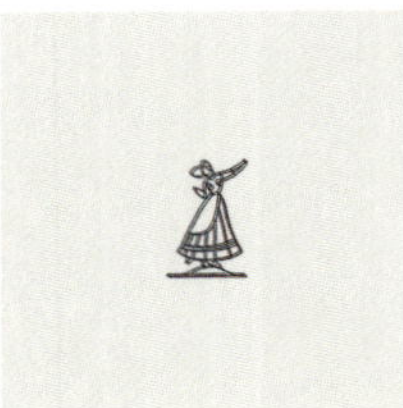

4736

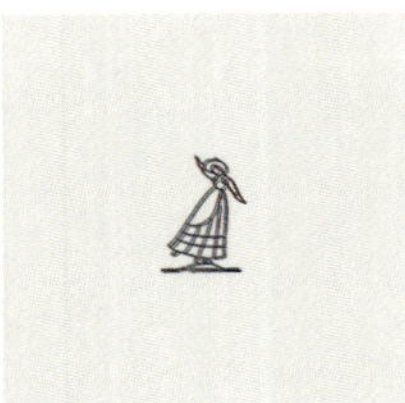

4739

Female Dancer on a circular base with leaf, in light-amethyst transparent glass with amethyst finishings. Strongly iridised surface. The model was also proposed with a different style of hat and a different position of the arms (no. 4739). Both models were also used to decorate ashtrays (no. 4771 and no. 4775).
c. 1929
16 cm high
M.V.M. Cappellin Murano *acid stamp*
Private collection

>
4736, catalogue drawing, CSV

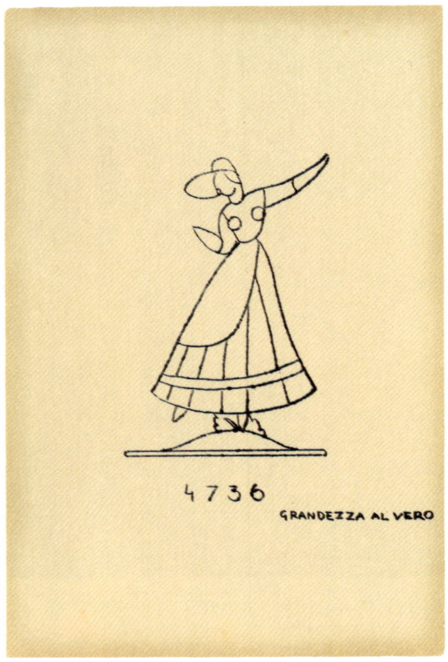

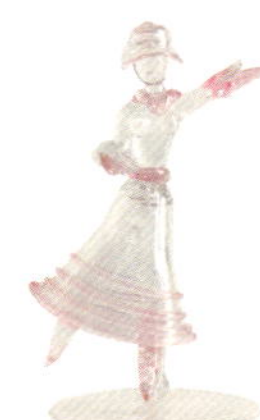

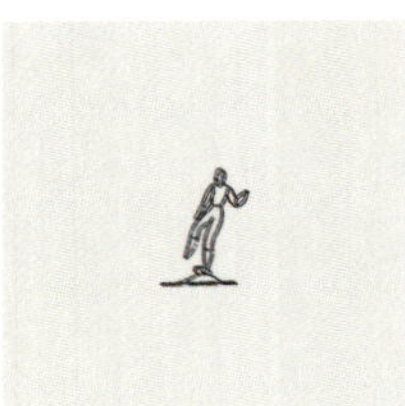

4737

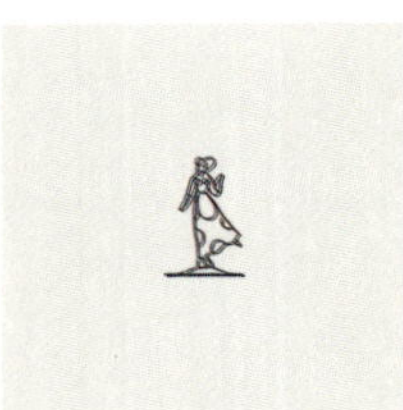

4738

Male Dancer and *Female Dancer* in polychrome glass on circular base with leaf in the Cappellin Glassware catalogue, undated, period photograph, detail, AHL. Both models were also used to decorate ashtrays (no. 4773 and no. 4772).
c. 1929
13; 14 cm high

>
4738, catalogue drawing, CSV

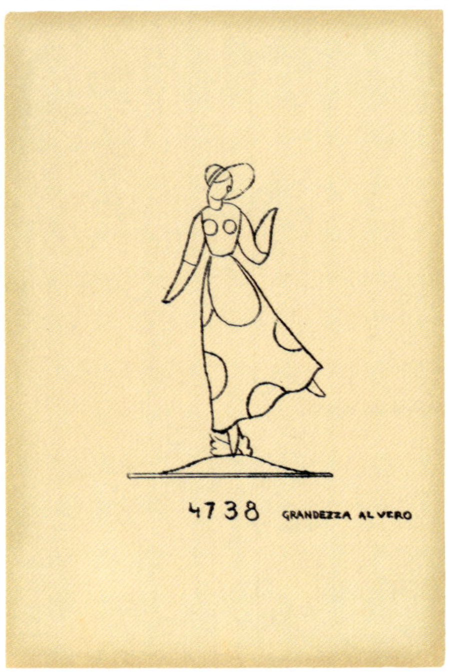

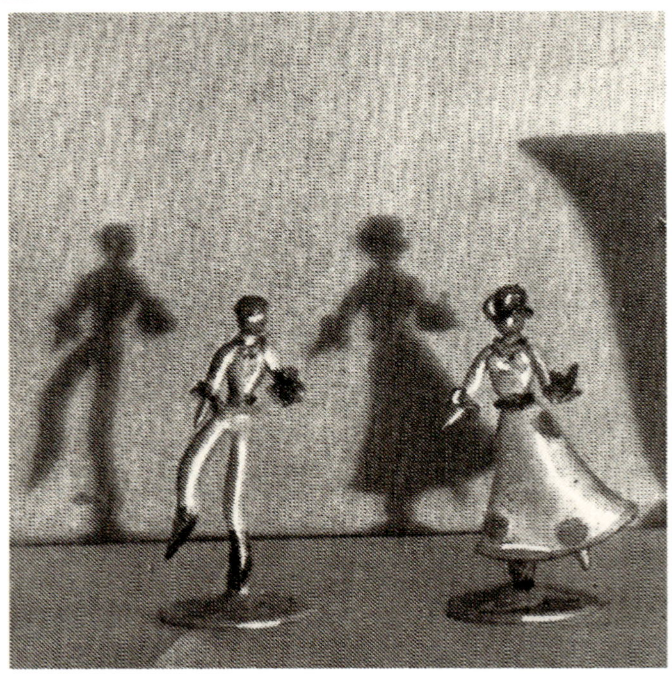

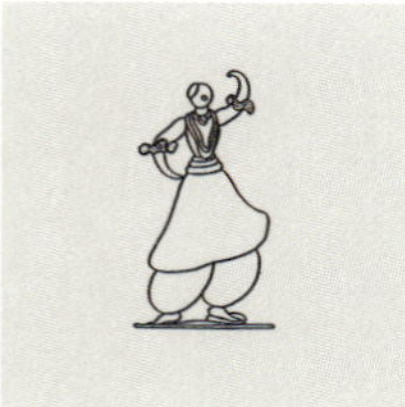

4715

Turk on circular base in coral-red *pasta vitrea* with applied gold leaf. Surface with iridised finish. An example of this model in coloured glass (transparent?) with finishings in *lattimo* was exhibited in 1930 at the 4th Triennale of Monza.
1930
32 cm high
M.V.M. Cappellin Murano *acid stamp*
Private collection, Feltre

exhibitions
1930, Monza, 4th Triennale

>
4715, catalogue drawing, private collection, Feltre

>
Turk *at the 4th Triennale of Monza, 1930, detail*

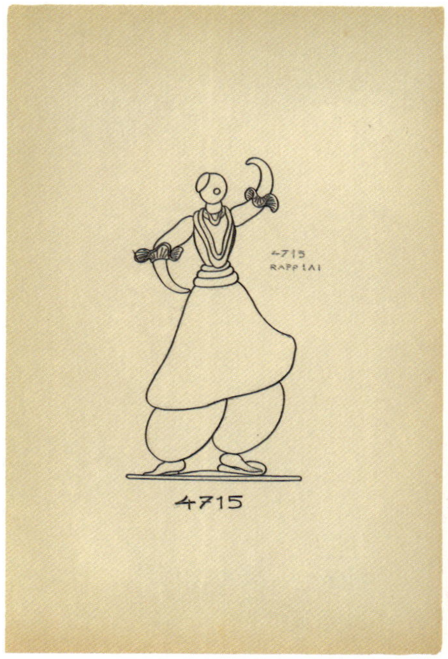

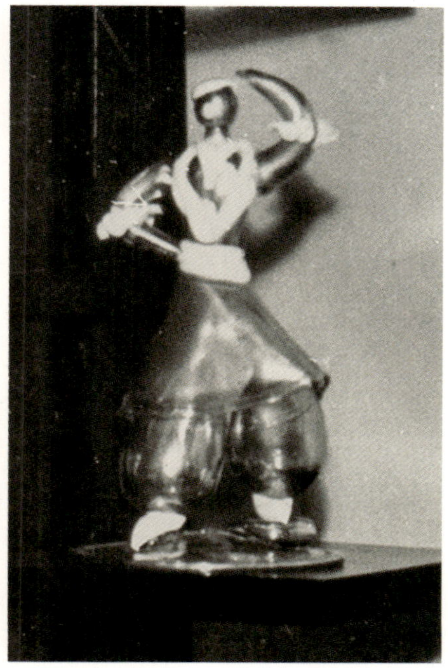

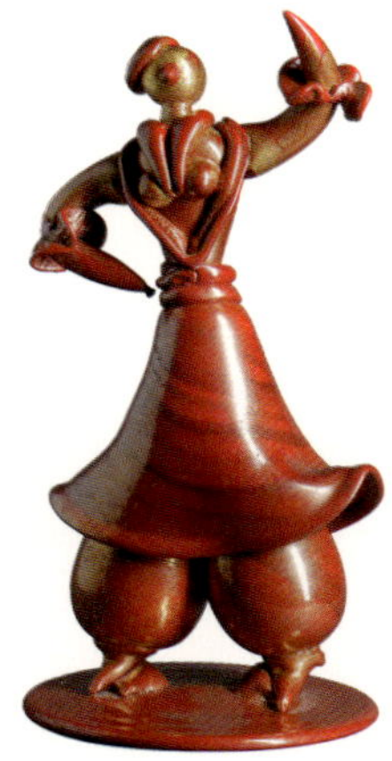

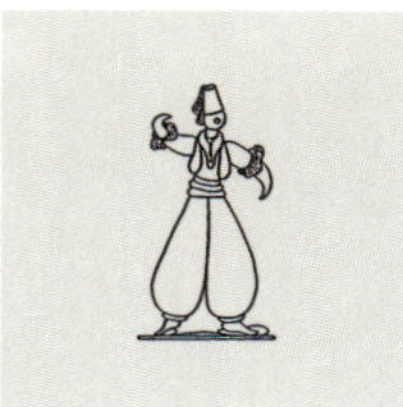

4718

Catalogue drawing for a *Turk with Fez* on a circular base, ACSVr. An example of this model in coloured glass (transparent?) with finishings in *lattimo* was exhibited in 1930 at the 4th Triennale of Monza.
1930
30 cm high

exhibitions
1930, Monza, 4th Triennale

>
4718, catalogue drawing, ACSVr

>
Turk *at the 4th Triennale of Monza, 1930, detail*

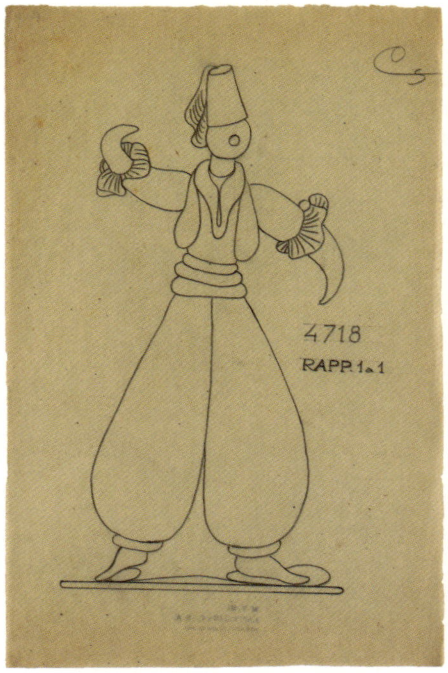

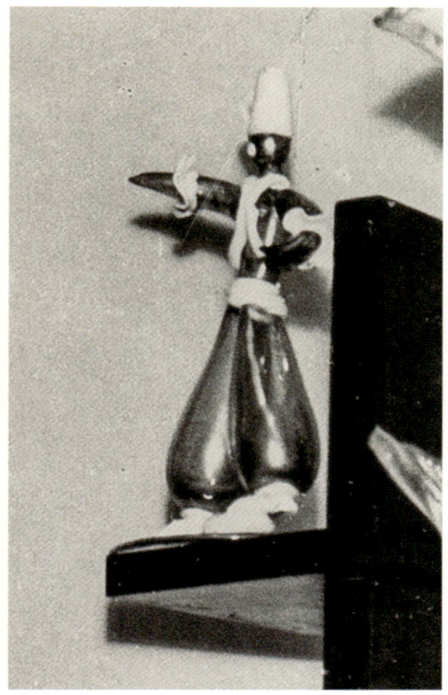

---

n.i.

Turk with Fez in crystal glass and in crystal and gold. Circular base in transparent light-blue glass with a piece of fruit in transparent light-blue and blue and gold glass. Surface with iridised finish.
c. 1930
24 cm high
The Cleveland Museum of Art, The Mary Spedding Milliken Memorial Collection, Gift of William Mathewson Milliken, Ohio

>
Study for figurine couple with inflorescence on a circular base (no. 4806), 1929–30, ACSVr

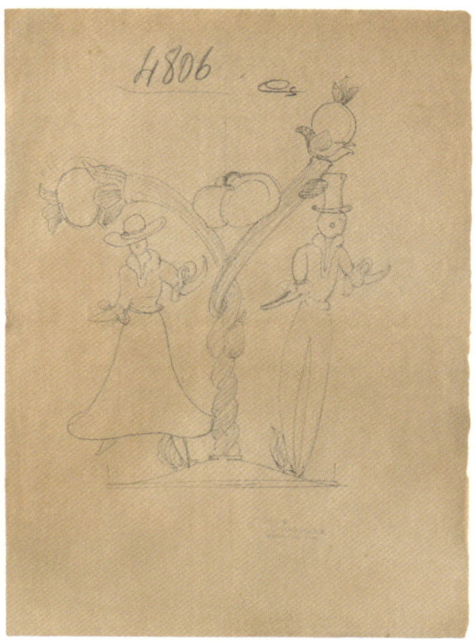

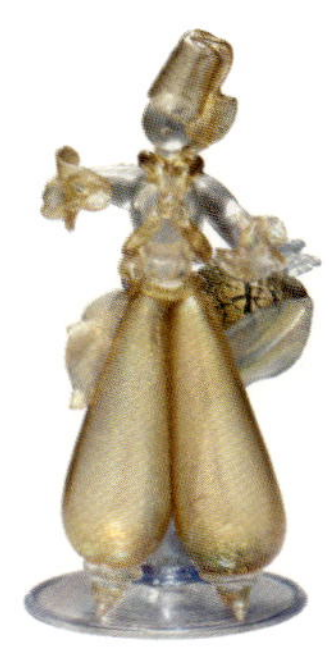

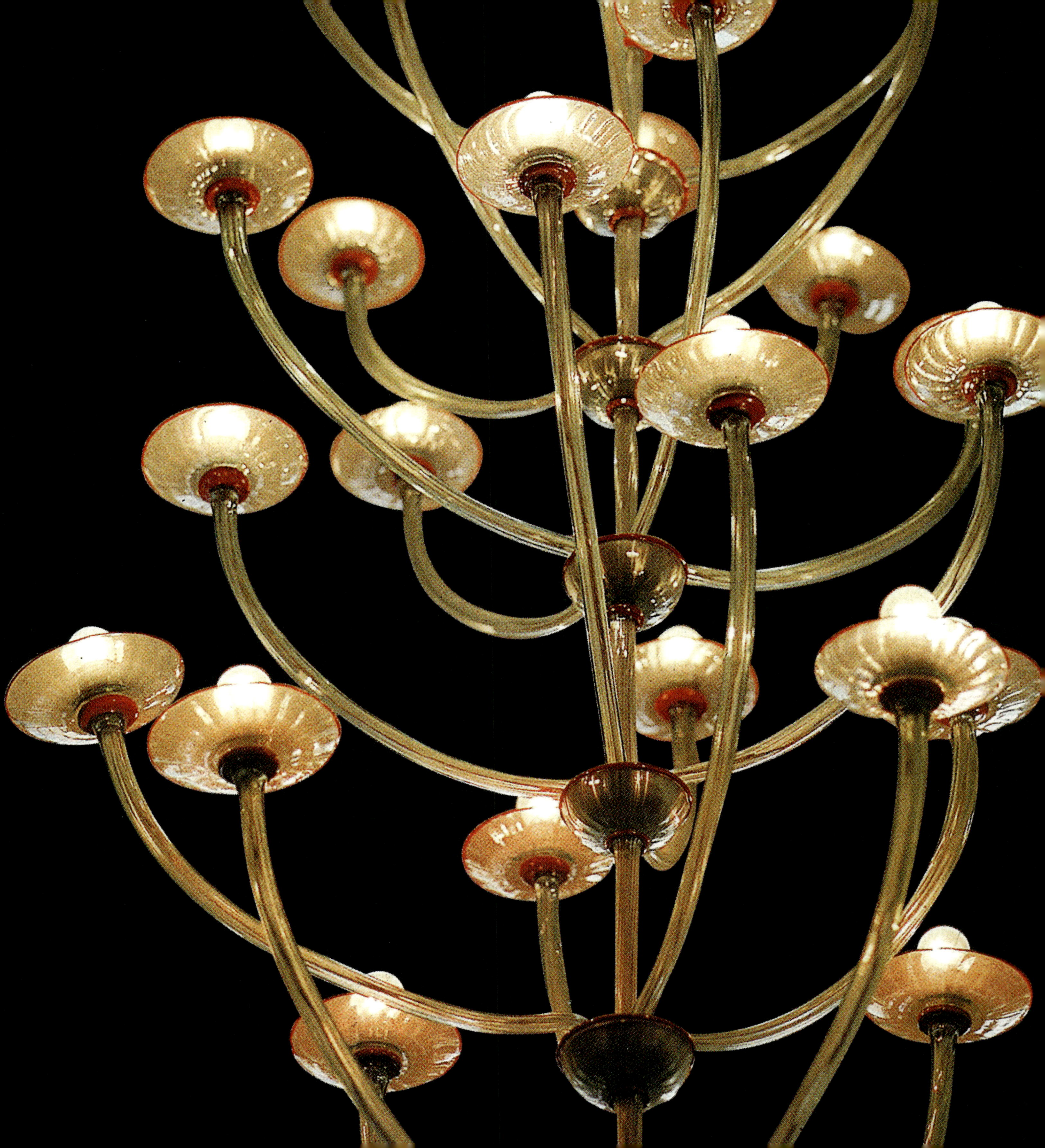

During all its activities, the Cappellin firm paid particular attention to the lighting sector and continued to excel in it, even after the departure of Vittorio Zecchin in October 1926, because of continuing research developed with the contribution of Carlo Scarpa. In this area too, a progressive simplification of design could be seen, accompanied by the study of new types of lighting appliance, in a response to the changed demands in taste. A careful selection of these appliances, intended for both domestic and formal environments, is proposed here by way of example.

LIGHTING 1927-31

n.i.

Chandelier with 12 lights in light-amethyst transparent glass with arms in *rigadin* (with veining) and ribbed cups with a flat border. A similar example was presented in 1927 at the 18th Mostra dell'Opera Bevilacqua La Masa of Venice and the following year at the exhibition of Turin 1928, in the Enapi pavilion. The model is described as a '20th-Century-style chandelier' in *La Fiera di Milano*, March 1929, p. 20.
1927
215 cm high
M.V.M. Cappellin Murano *acid stamp*
Private collection

exhibitions
1927, Venice, 18th Mostra dell'Opera Bevilacqua La Masa
1928, Turin, Torino Exhibition 1928

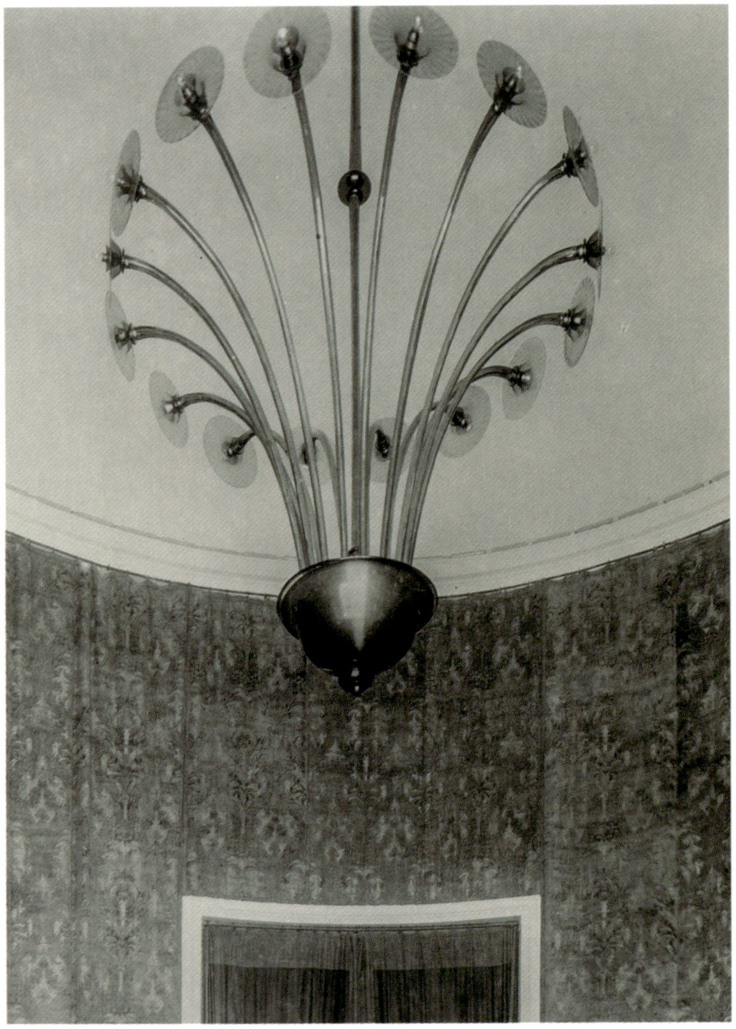

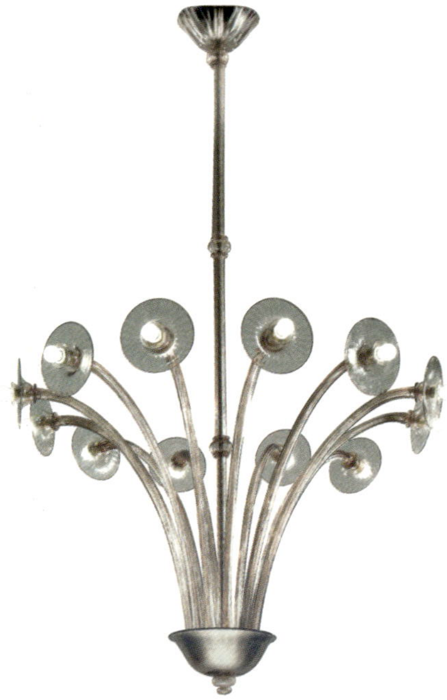

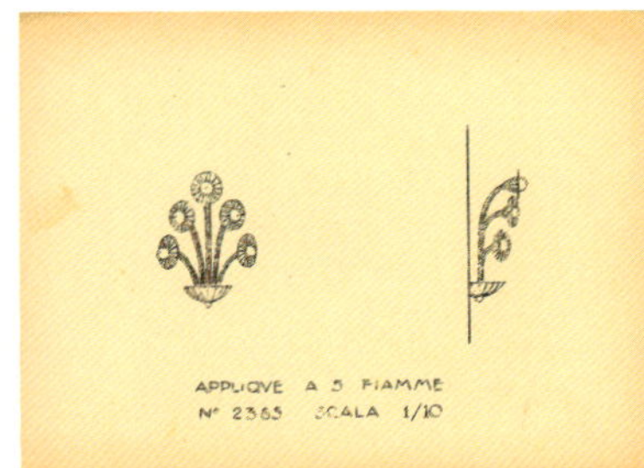

>
Chandelier with 12 lights at the 18th Mostra dell'Opera Bevilacqua La Masa of Venice, 1931, detail, CSV

>
2385, catalogue drawing, AMH

n.i.

Chandelier with 24 lights in transparent glass with tapering diffuser bodies in an exhibition room in the Palazzo da Mula, Murano. Period photograph c. 1929, CSV. A similar example was exhibited in 1927 at the 3rd Biennale of Monza.
1927
260 cm high

exhibitions
1927, Monza, 3rd Biennale

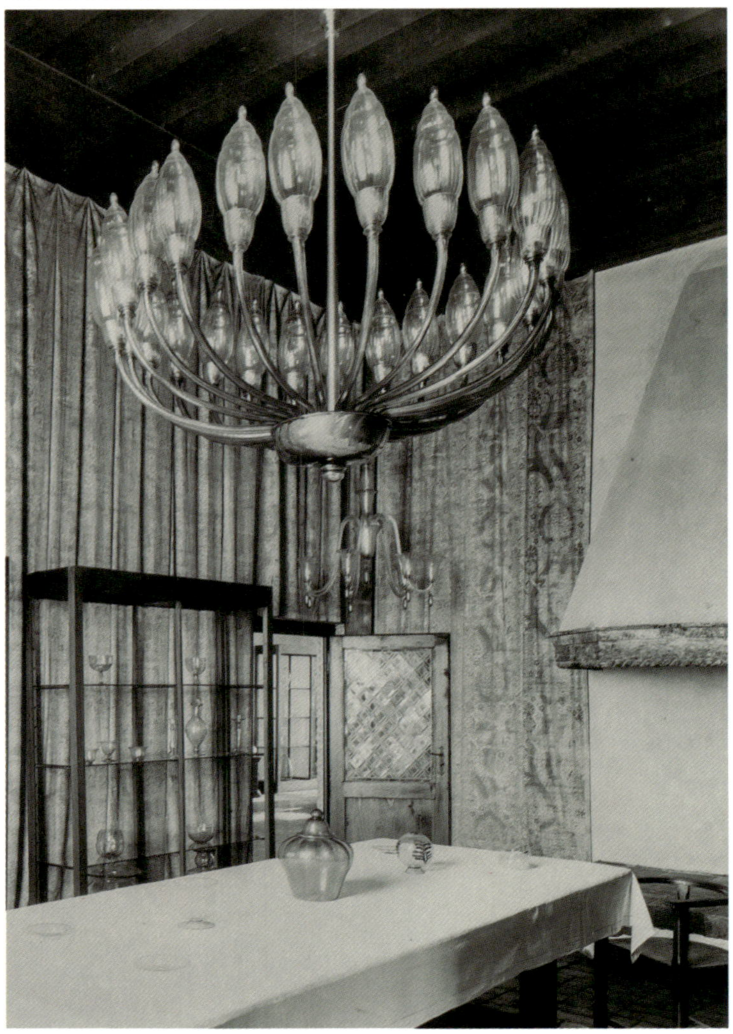

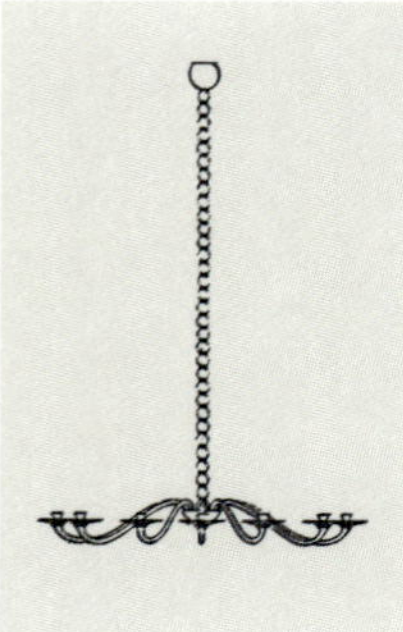

1950

Chandelier with 12 lights in crystal glass and finishings in dark-blue glass. Rectilinear rod with superimposed spheres alternating with rings in dark-blue glass. Strongly iridised surface.
1927–28
230 cm high
Private collection

>
1950, catalogue drawing, AMH

>
Chandelier no. 1950 in an exhibition room of the Palazzo da Mula, Murano. Period photograph, c. 1929, CSV

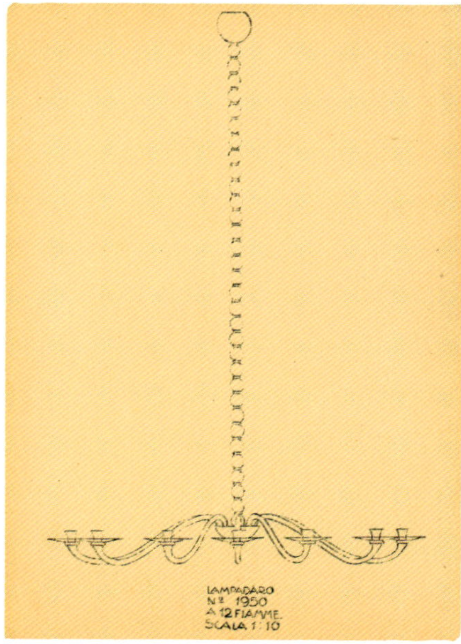

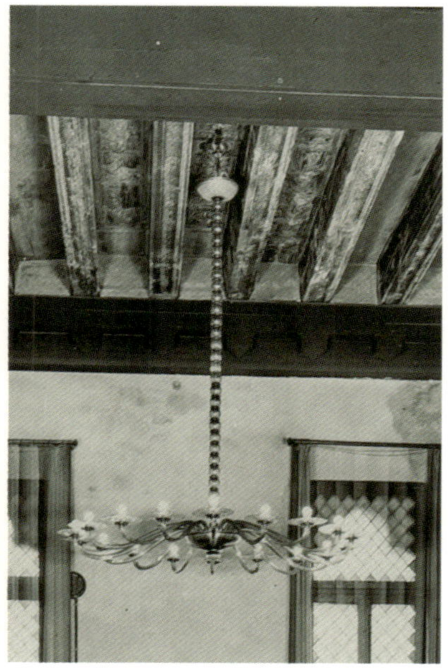

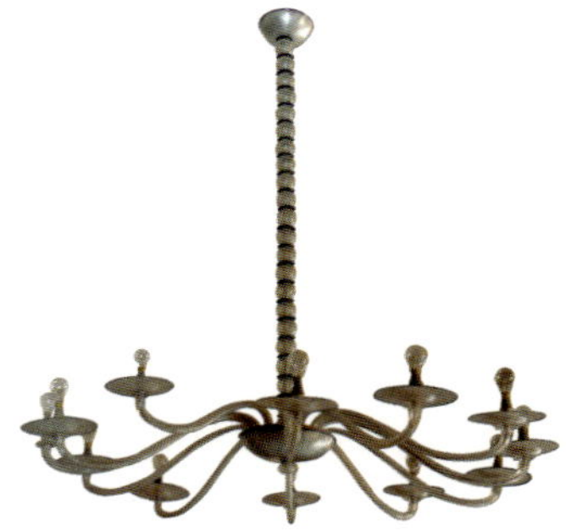

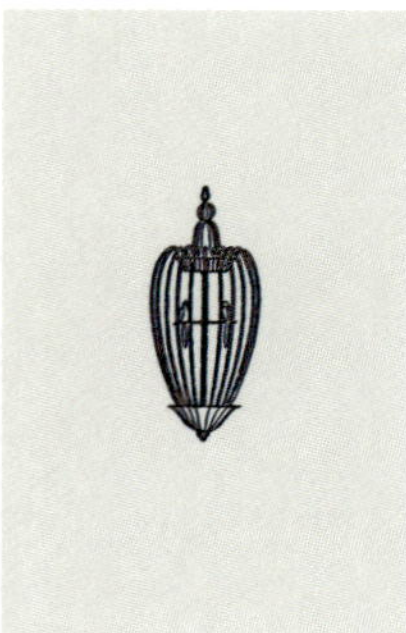

2695

Cage in green transparent glass with a couple of parrots in green glass and finishings in coral-red *pasta vitrea*. The light source is hidden at the base of the cage. The model was also proposed in a larger size (no. 2693, h 205 cm).
1928–29
75 cm high
Private collection

>
2695, catalogue drawing, AMB

>
Carlo Scarpa (?), preparatory drawing for model no. 2695, 1928–29, G. De Carlo collection

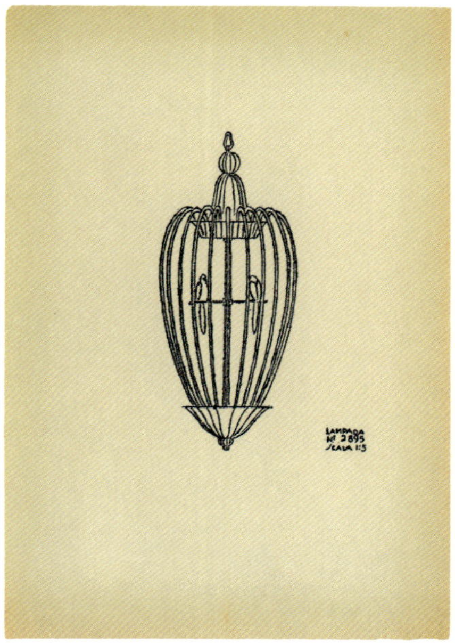

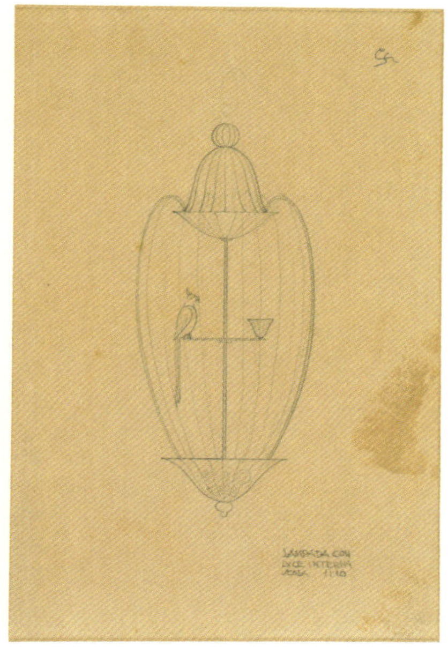

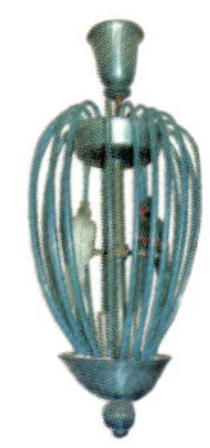

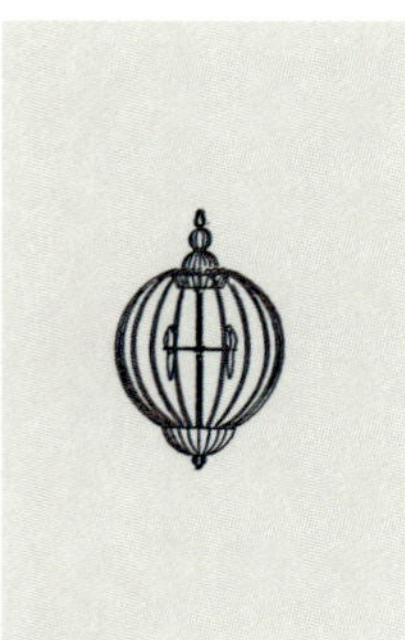

2696

Spherical cage with a couple of parrots in green glass and light source inserted at the base. The model was also proposed in a larger size (no. 2964, h 150 cm).
1928–29
75 cm high
Private collection

>
2696, catalogue drawing, AMB

>
Carlo Scarpa, studies for illuminated cage with parrots, 1928–29, G. De Carlo collection

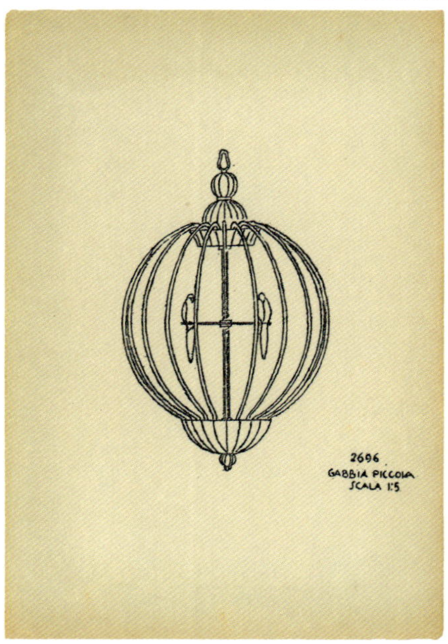

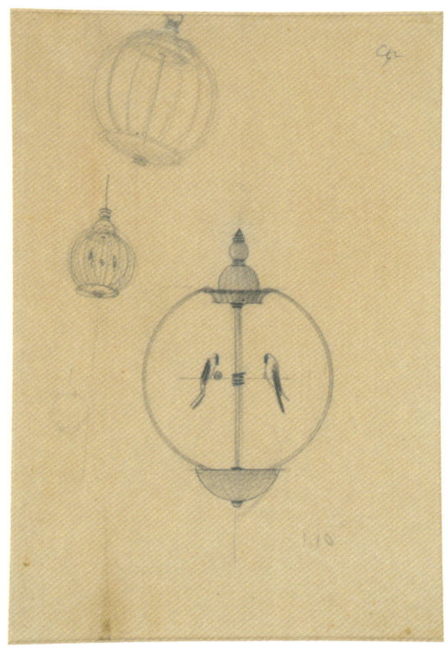

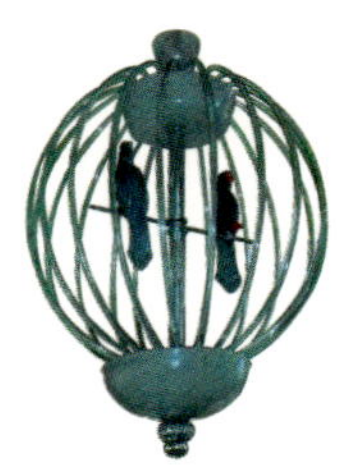

n.i.

Chandelier with three lights in transparent glass with upturned-cone-shaped cups, period photograph, c. 1928, private collection. The model was described as a 'Cubist-style chandelier' in *La Fiera di Milano*, March 1929, p. 21.
c. 1928
c. 150 cm high

exhibitions
1928, Milan, Trade Fair (?)

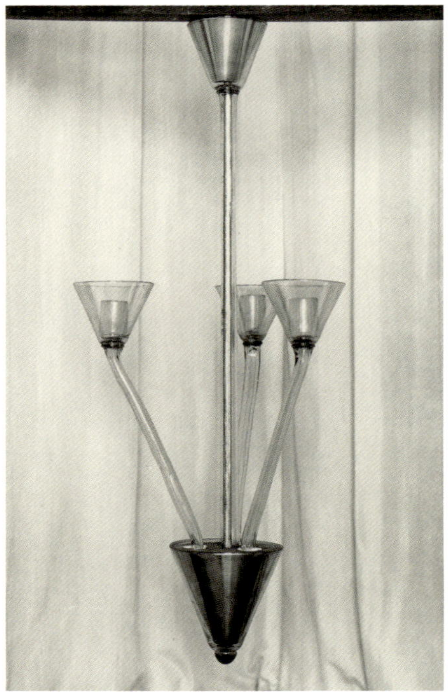

n.i.

Chandelier in transparent glass with upturned-cone-shaped elements, in various sizes, stacked one over the other, period photograph 1929, ACSVr, detail. The chandelier with light ribbing was exhibited at the Trade Fair of Milan in 1929.
1928–29
150 cm high

exhibitions
1929, Milan, Trade Fair

>
Carlo Scarpa, study for chandelier, 1928–29, AAAI

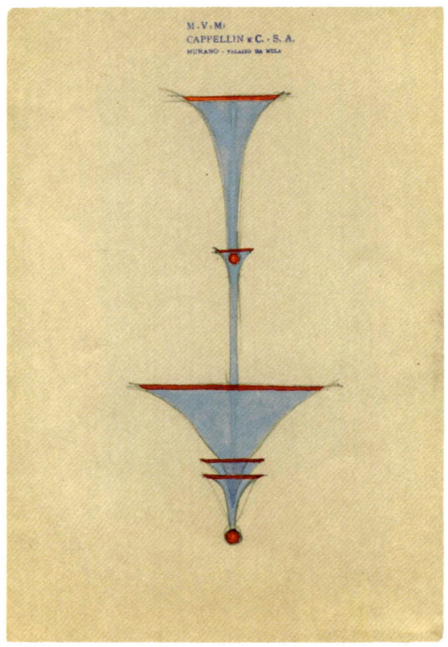

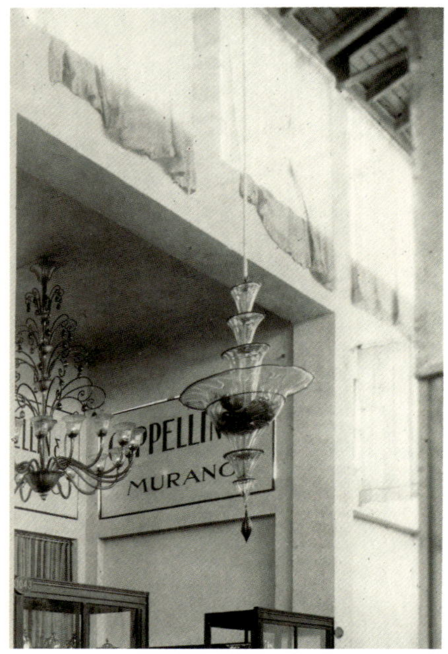

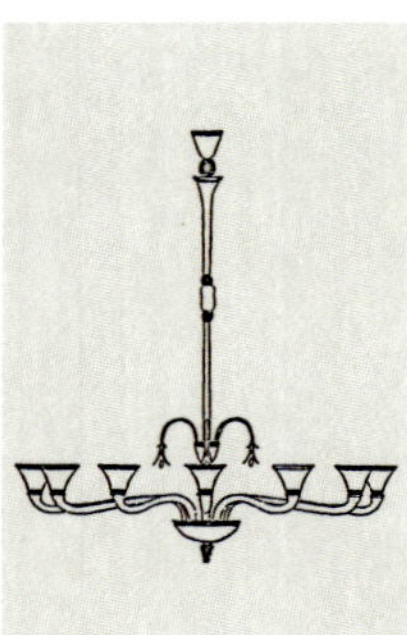

2013

Catalogue drawing for a chandelier with 12 lights, finishings in *pasta vitrea* and hanging decorations at the base of the rod, AMuVe.
1930
150 cm high

>
2013, catalogue drawing, AMuVe

n.i.

Large chandelier with 24 lights, structured in six levels, in transparent *pagliesco* glass with finishings in coral-red *pasta vitrea*. A similar example was exhibited in the Lighting Gallery at the 4th Triennale of Monza,1930.
1930
300 cm high
Private collection

exhibitions
1930, Monza, 4th Triennale

>
Chandelier with 24 lights at the 4th Triennale of Monza, 1930, AFT

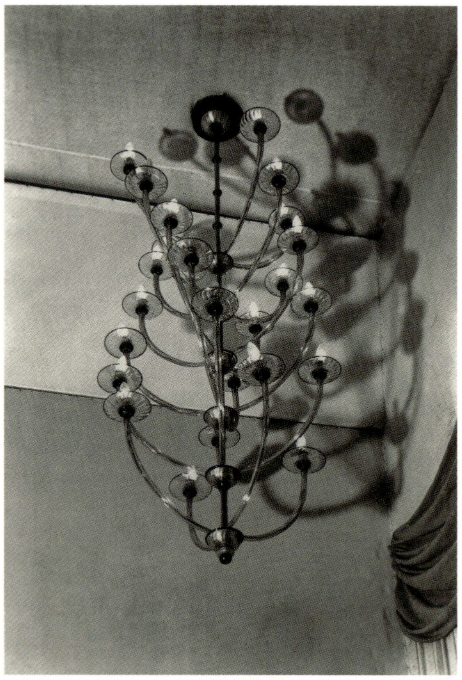

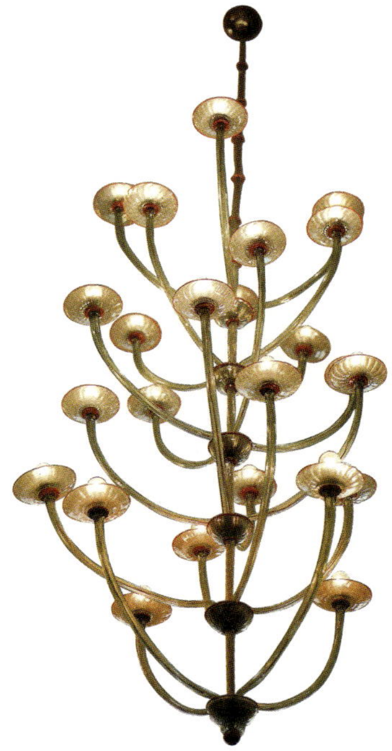

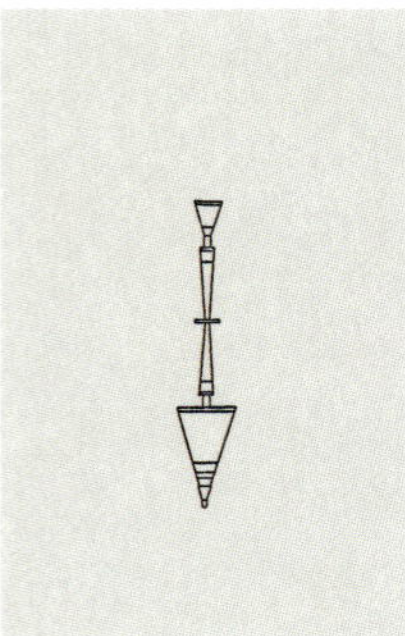

2723

Upturned-cone-shaped hanging lamp in white opal glass with finishings in green transparent glass. The model appears in the catalogue together with no. 2723 with a cylindrical light-diffuser body.
c. 1930
140 cm high
Casa Museo Boschi Di Stefano collection, Comune di Milano

>
2723–2724, catalogue drawing, AMuVe

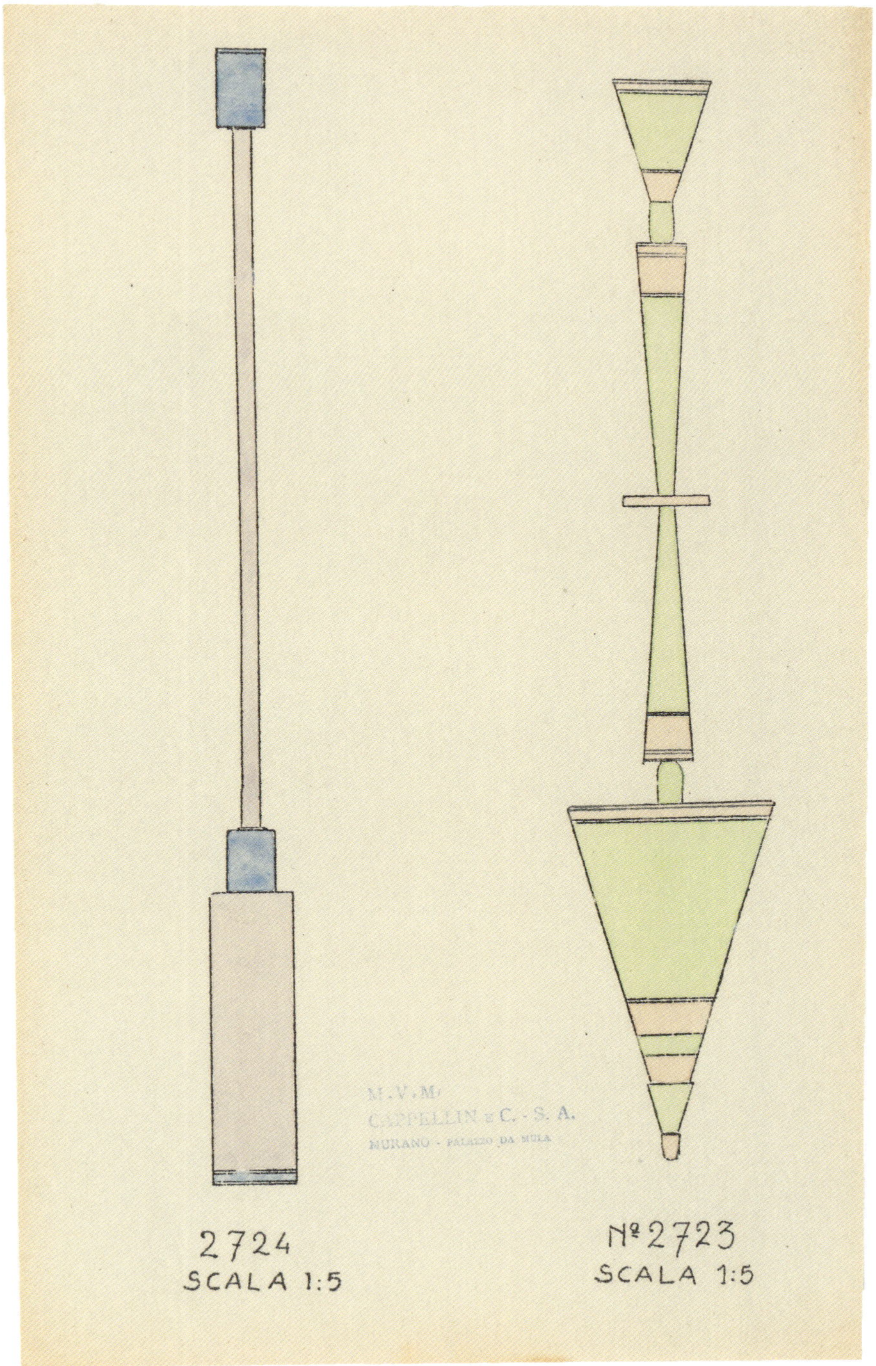

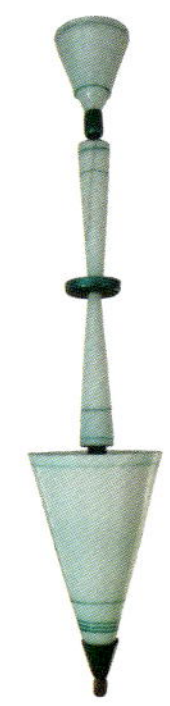

2725 -2726

Catalogue drawings for hanging lamps with geometrically shaped diffuser body, AMuVe.
1930
150; 160 cm high

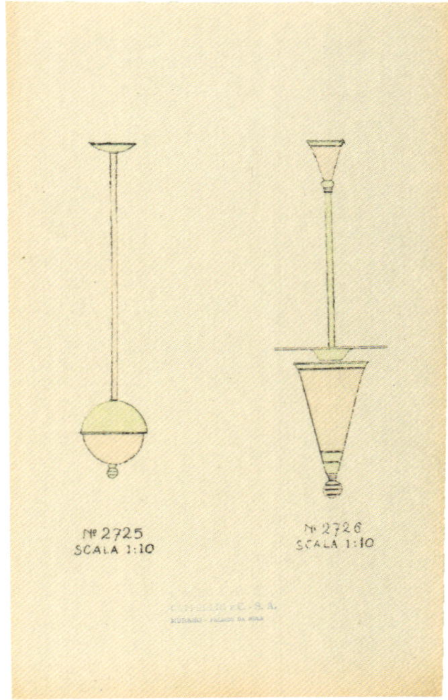

>
Carlo Scarpa, studies for hanging lamps, 1929–30, ED Gallery, Piacenza. On the left two studies can be seen for a lamp with double-half-spherical diffuser body similar to no. 2725

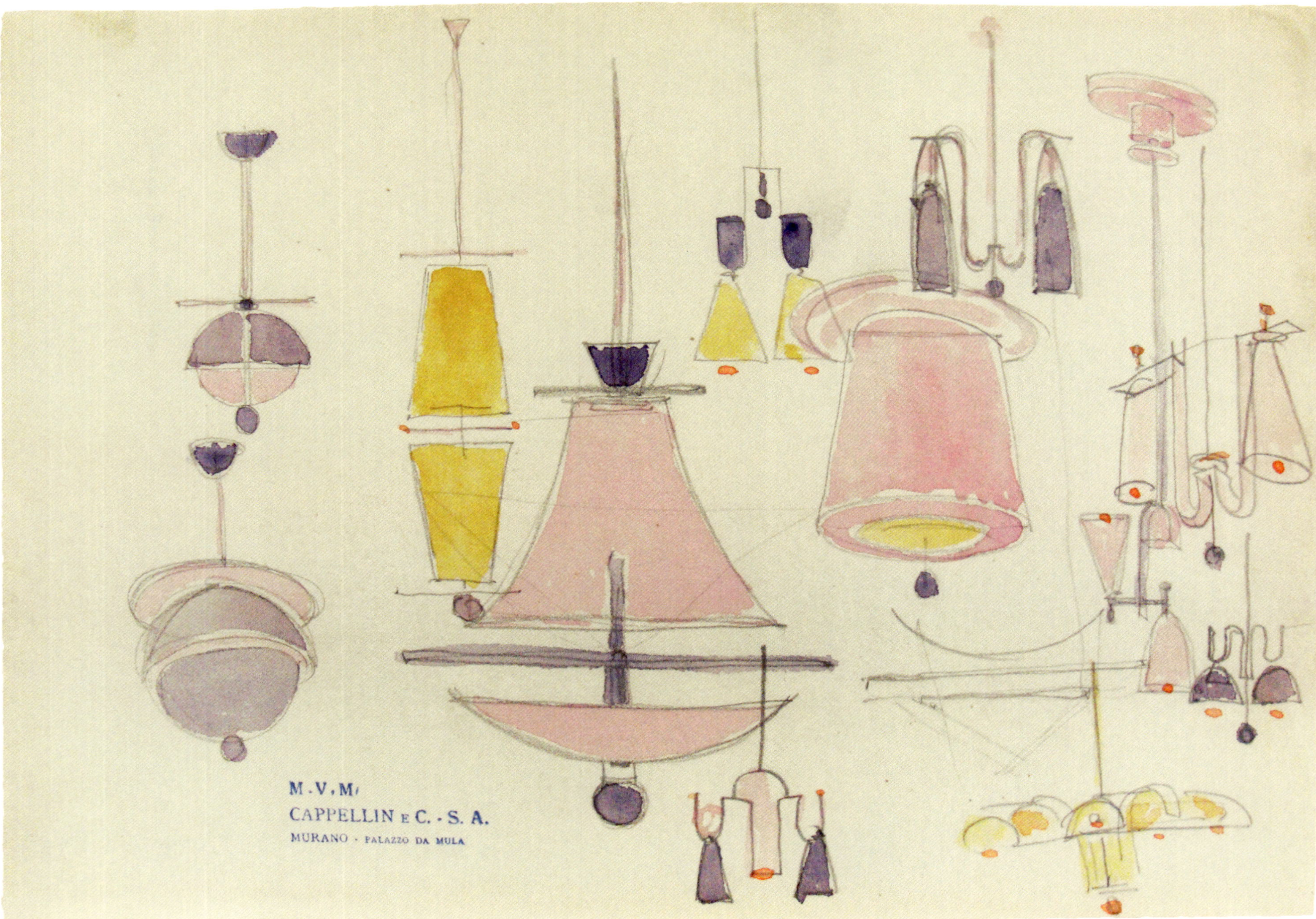

2731-2732

Catalogue drawings for hanging lamps, AMuVe. no. 2732 is characterised by the diffuser body with two counterpoised flat skullcaps.
1930
45; 70 cm high

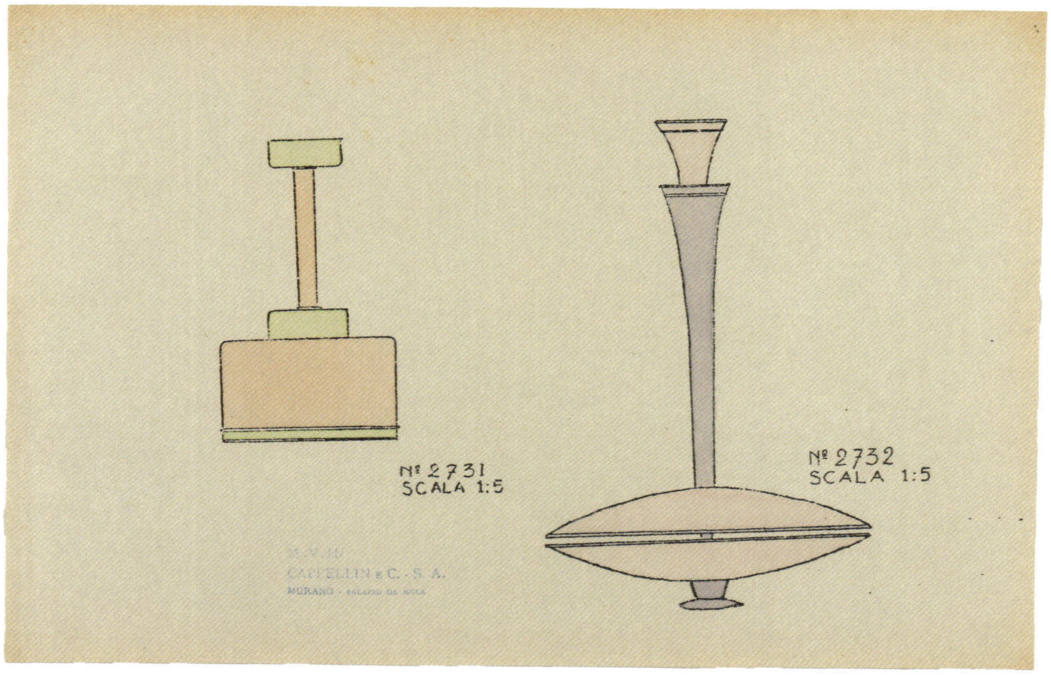

>
Carlo Scarpa, studies for wall lamps, 1929–30, AMuVe

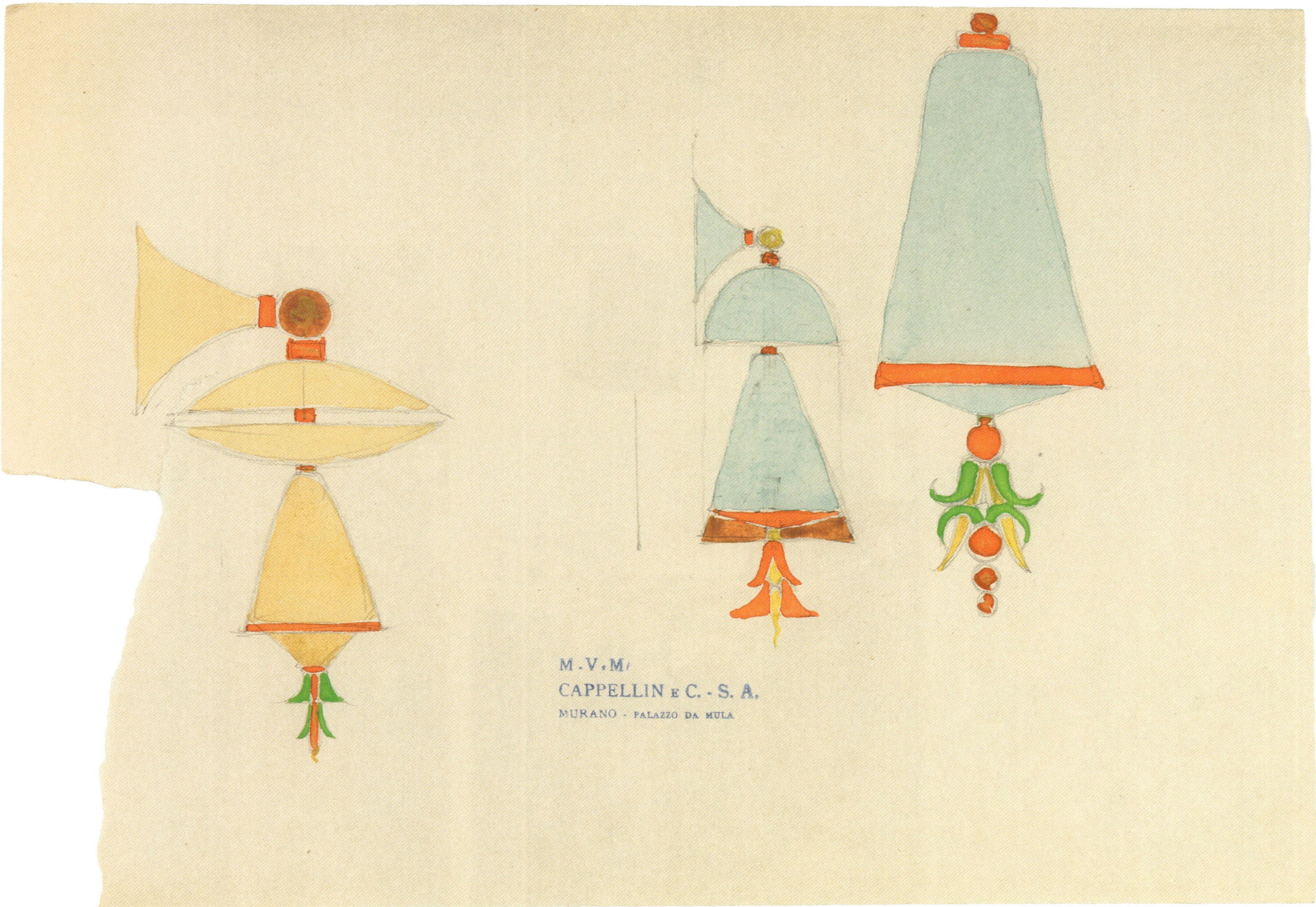

2427-2429

Catalogue drawings for wall lamps, AMuVe. The design and decorative motifs of these models refer to series of glass pieces presented at the 4th Triennale of Monza in 1930. In particular no. 2428 is characterised by the applied animal head, no. 2429 by the geometrical shape of the globe and no. 2427 by the glass technique decorated with murrine in concentric circles.
1930
60; 30; 60 cm high

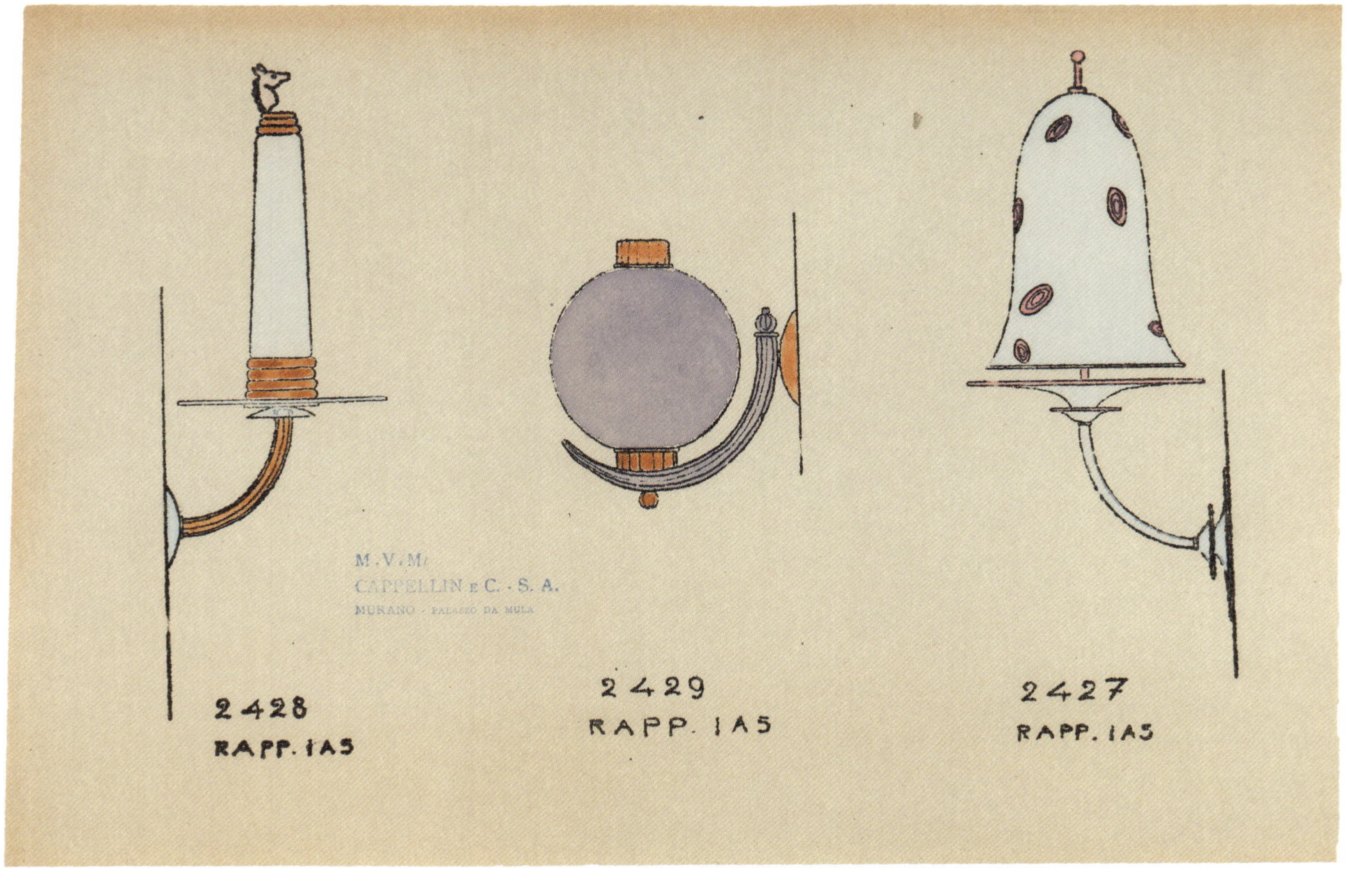

n.i.

Wall lamp with screen diffuser produced with sheet glass at the 4th Triennale of Monza in the M.V.M. Cappellin room, 1930, detail.

Lighting structure with diffuser composed of closely juxtaposed canes exhibited at the 4th Triennale of Monza in the M.V.M. Cappellin room, 1930, detail.

exhibitions
1930, Monza, 4th Triennale

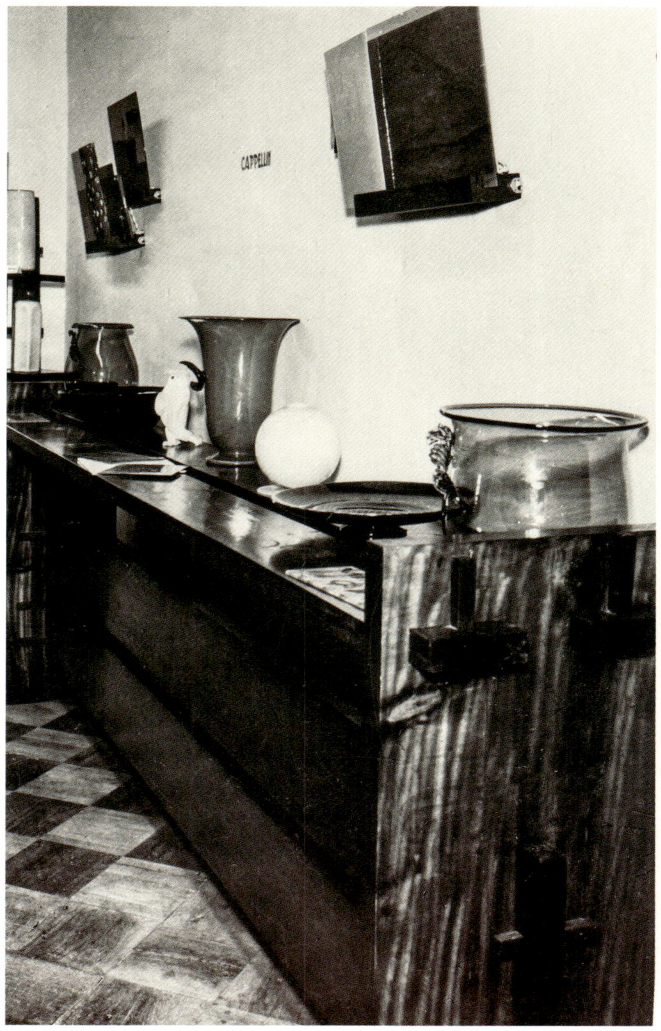

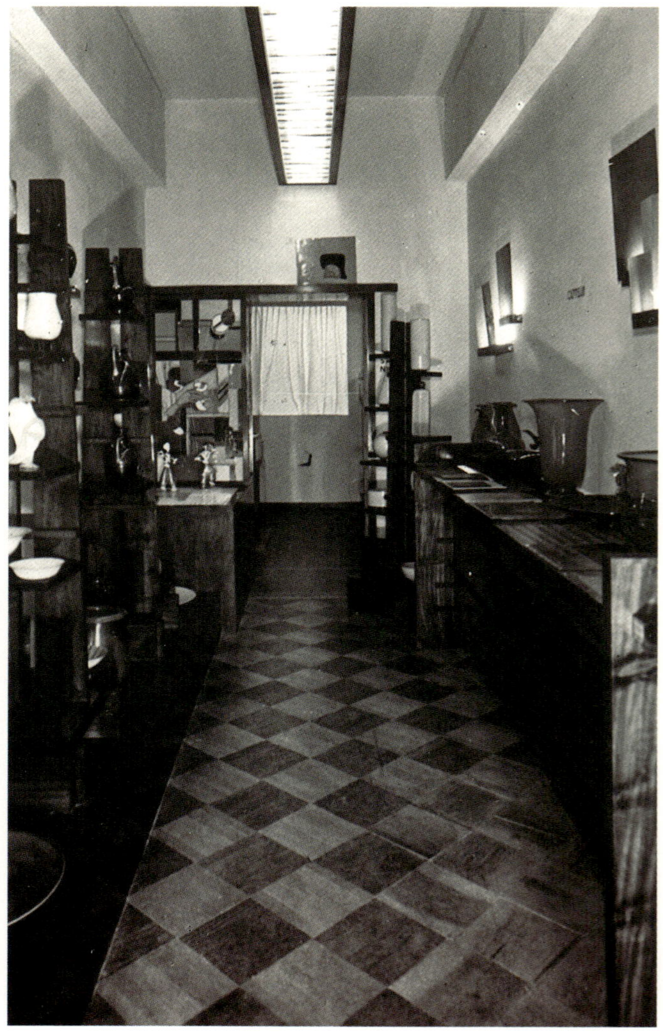

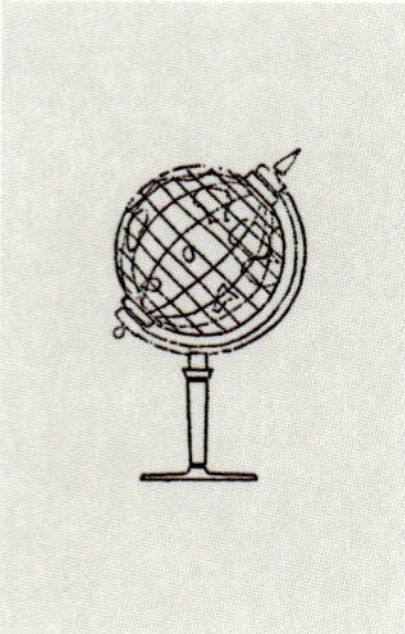

3022

Catalogue drawing for a table lamp in the form of a globe, AMuVe.
1929–30
55 cm high

>
Carlo Scarpa, studies for the support of a table lamp in the form of a globe, c. 1929, AMuVe

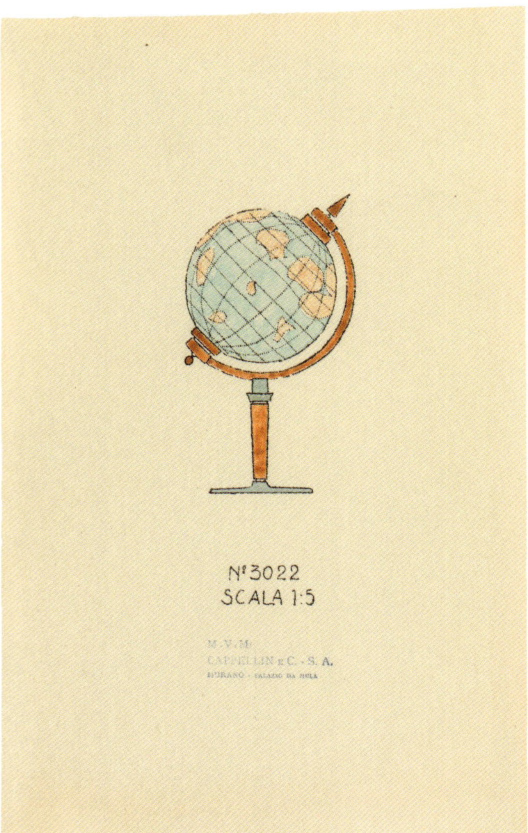

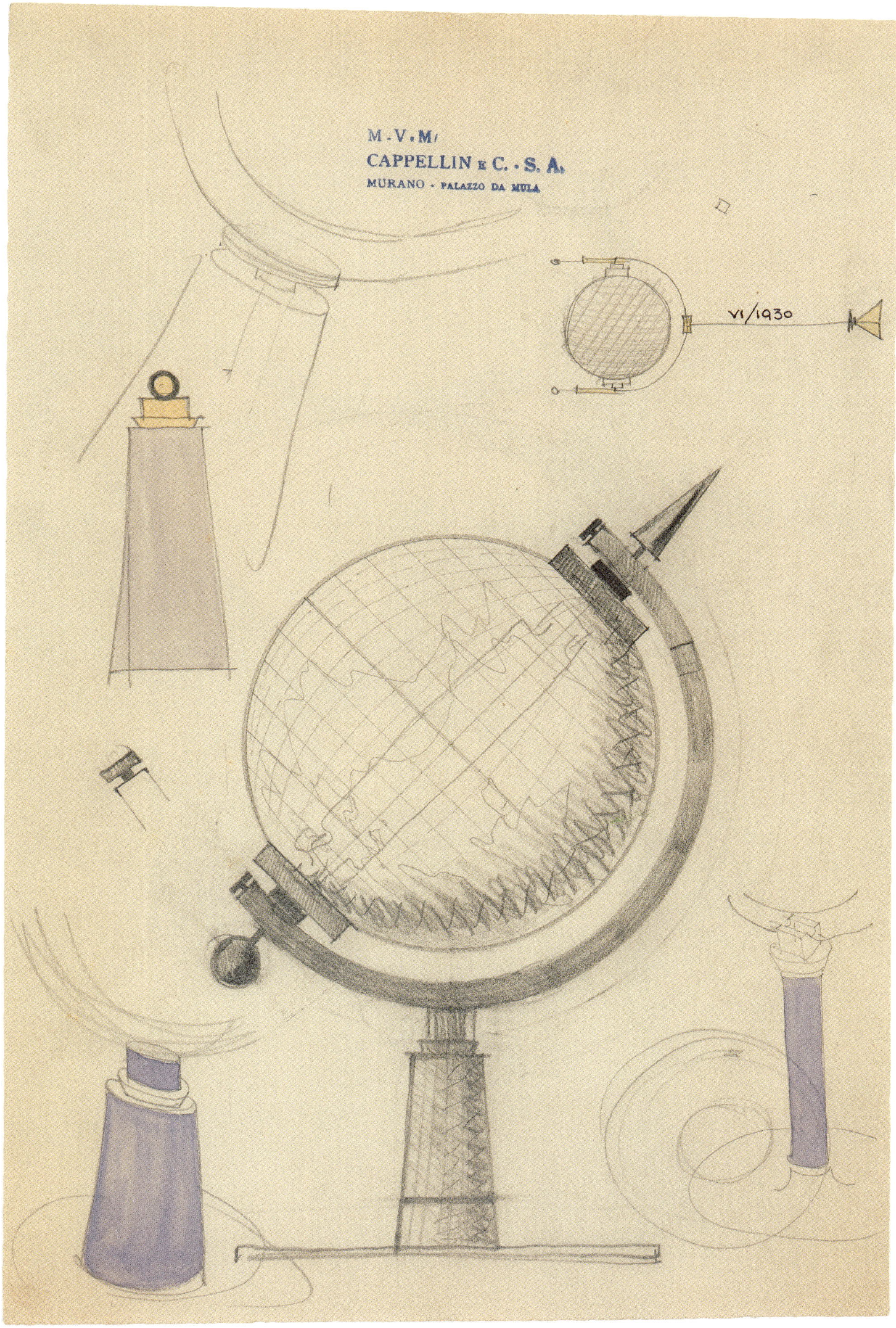

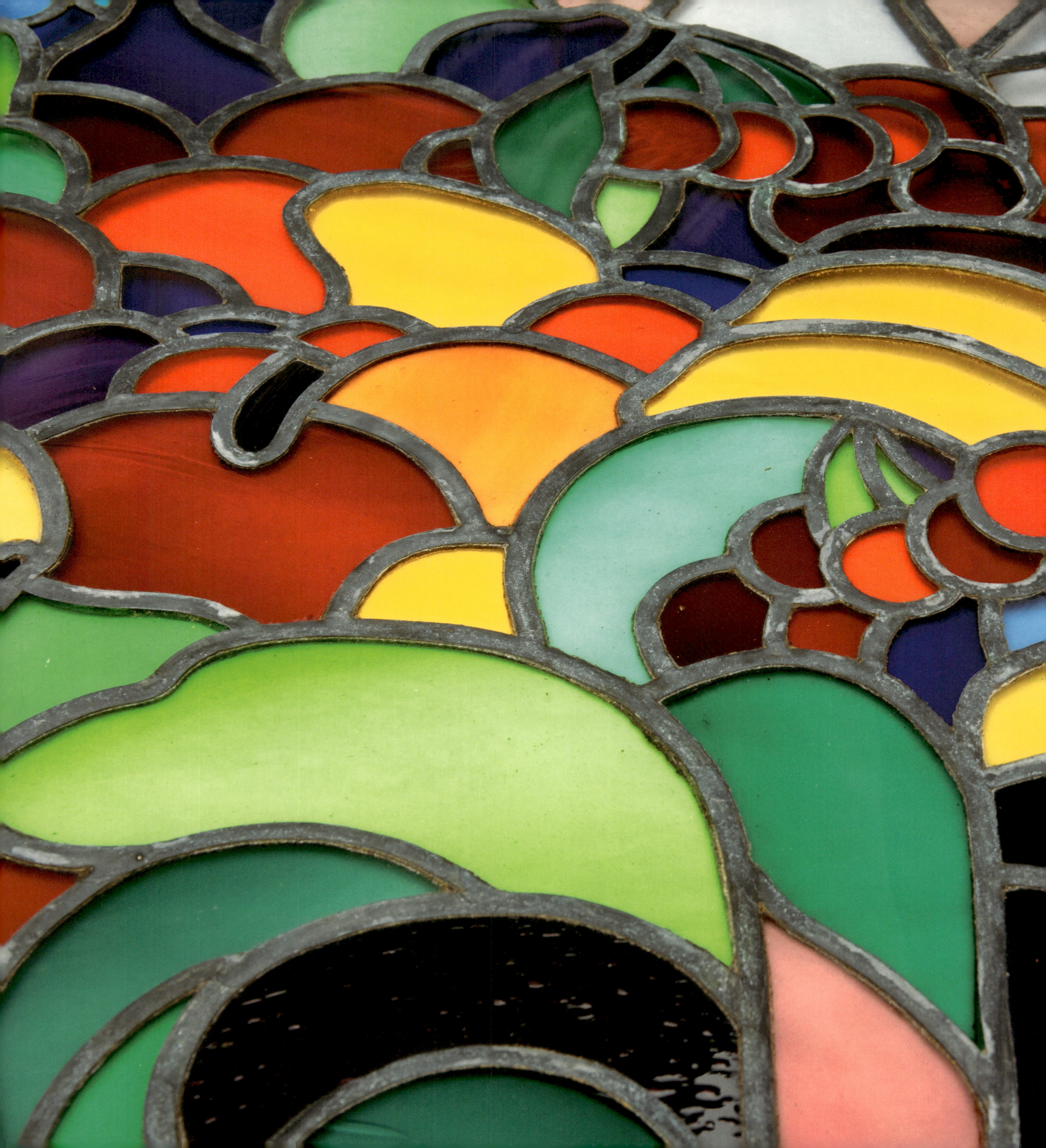

The production of decorative panels in Murano glass was one of the distinctive sectors, and an excellent one, of M.V.M. Cappellin & C. Beginning with the works of Vittorio Zecchin exhibited at the Biennale of Venice in 1926, this production was an important part of the activity of the glassworks up to its closure. In these years extraordinary stained-glass works were produced, after the cartoons of various authors, such as Casorati, Ortona, Sironi etc., many of which could be enjoyed on the occasion of the important decorative art exhibitions of the time. Carlo Scarpa also took up this theme when he designed stained-glass panels to be placed in the Cappellin shops in Florence and Turin, but also to be exhibited at the 4th Triennale of Monza in 1930.

STAINED GLASS 1927-31

Exhibitions

1927, Monza, 3rd *Mostra Internazionale delle Arti Decorative*
1927, Rome, 93rd *Esposizione di Belle Arti della Società amatori e cultori di belle arti*
1930, Monza, 4th *Esposizione Triennale Internazionale delle Arti Decorative ed Industriali Moderne*
1931, Amsterdam, *Tentoonstelling van glas, aardewerk en kant, van moderne italiaansche kunstnijverheid. Mostra di vetri, ceramiche e merletti d'arte moderna italiana*
1931, Rome, 1st Quadriennale d'arte nazionale

3RD MOSTRA INTERNAZIONALE DELLE ARTI DECORATIVE, MONZA, 1927

On the occasion of the 3rd Biennale of Monza (1927) the M.V.M. Cappellin firm also exhibited two stained glass pieces in its own room: one after a cartoon by Ugo Ortona, the other after a cartoon by Felice Casorati.

Ugo Ortona, *The Fawns*, polychrome stained glass in Murano glass with lead came, 1927
171 cm high; 143 cm wide
Carlo Moretti and Giovanni Moretti collection

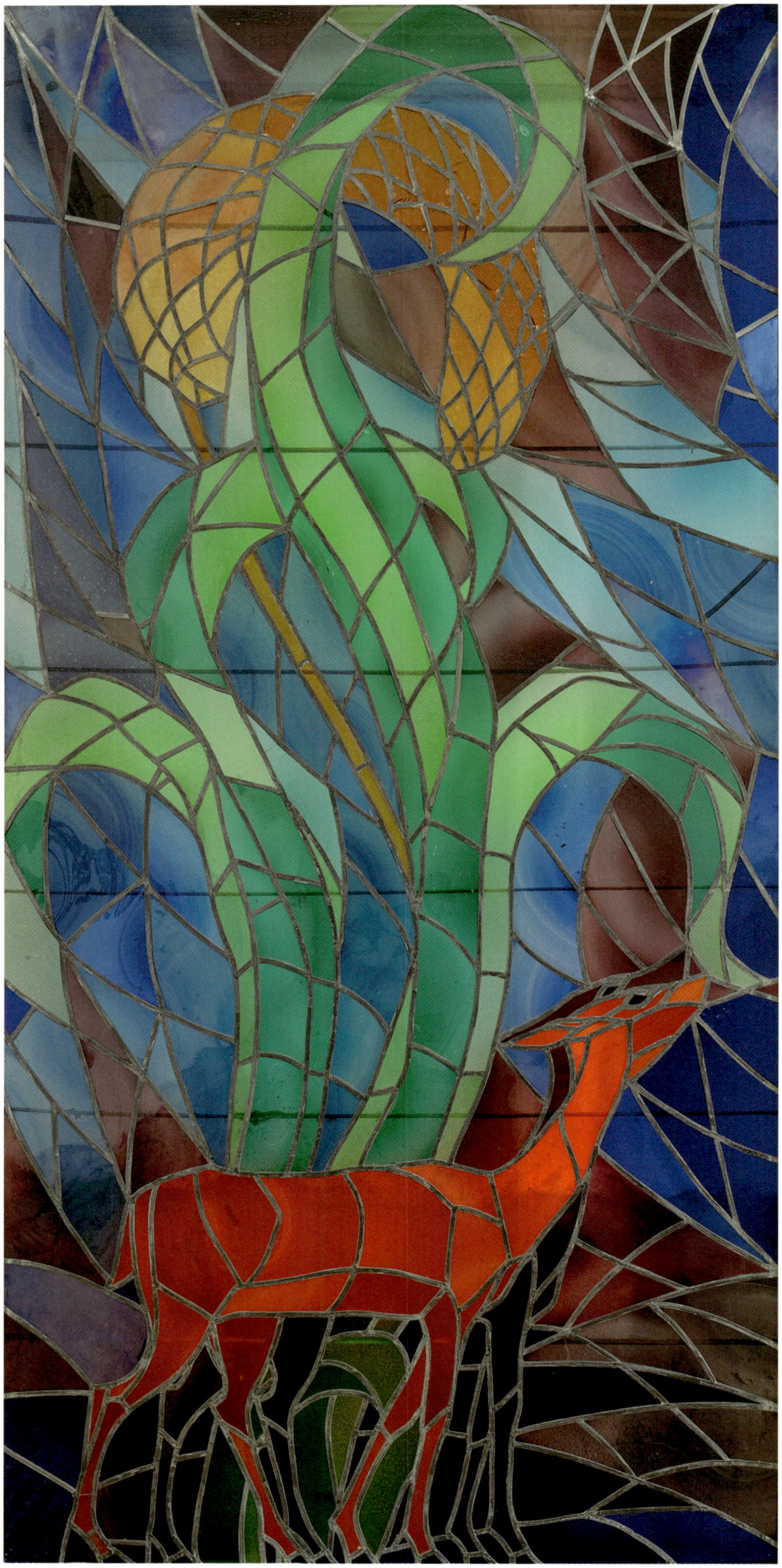

>
M.V.M. Cappellin room at the 3rd Biennale of Monza, 1927, ACSVr

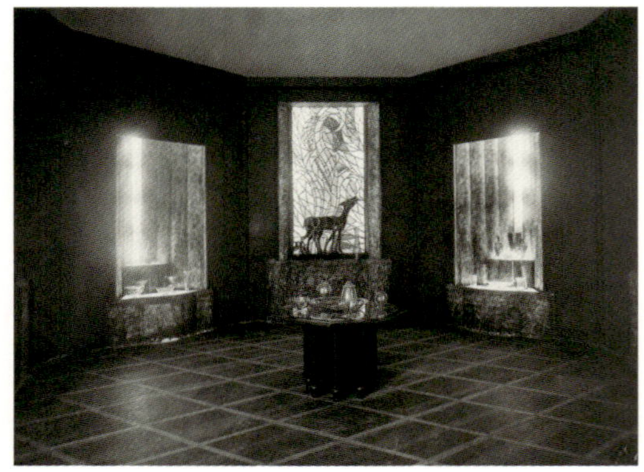

>
Sketch for the stained-glass piece The Fawns, *c. 1927, CSV*

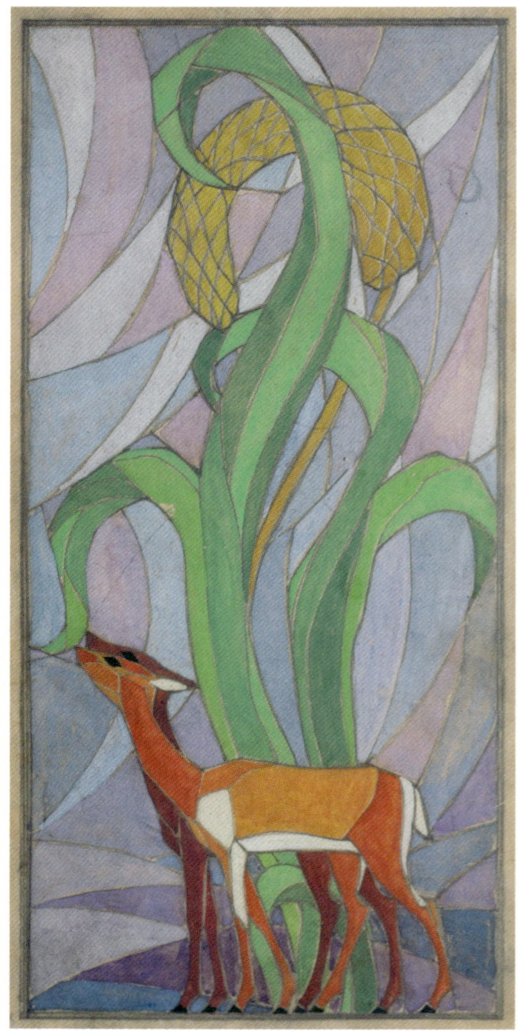

Felice Casorati, *Birch Trees and Starry Night*, polychrome stained glass in Murano glass with lead came, 1927, AFT.
The stained glass piece was exhibited in 1927, first at the 93rd Esposizione di belle arti of Rome and then at the 3rd Biennale of Monza.

>
Drawing for the stained glass Birch Trees and Starry Night, *c. 1930, G. De Carlo collection*

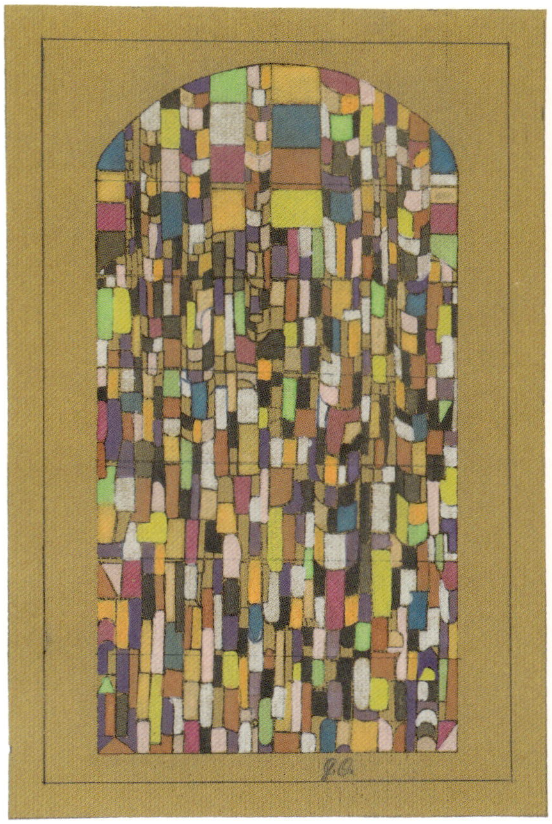

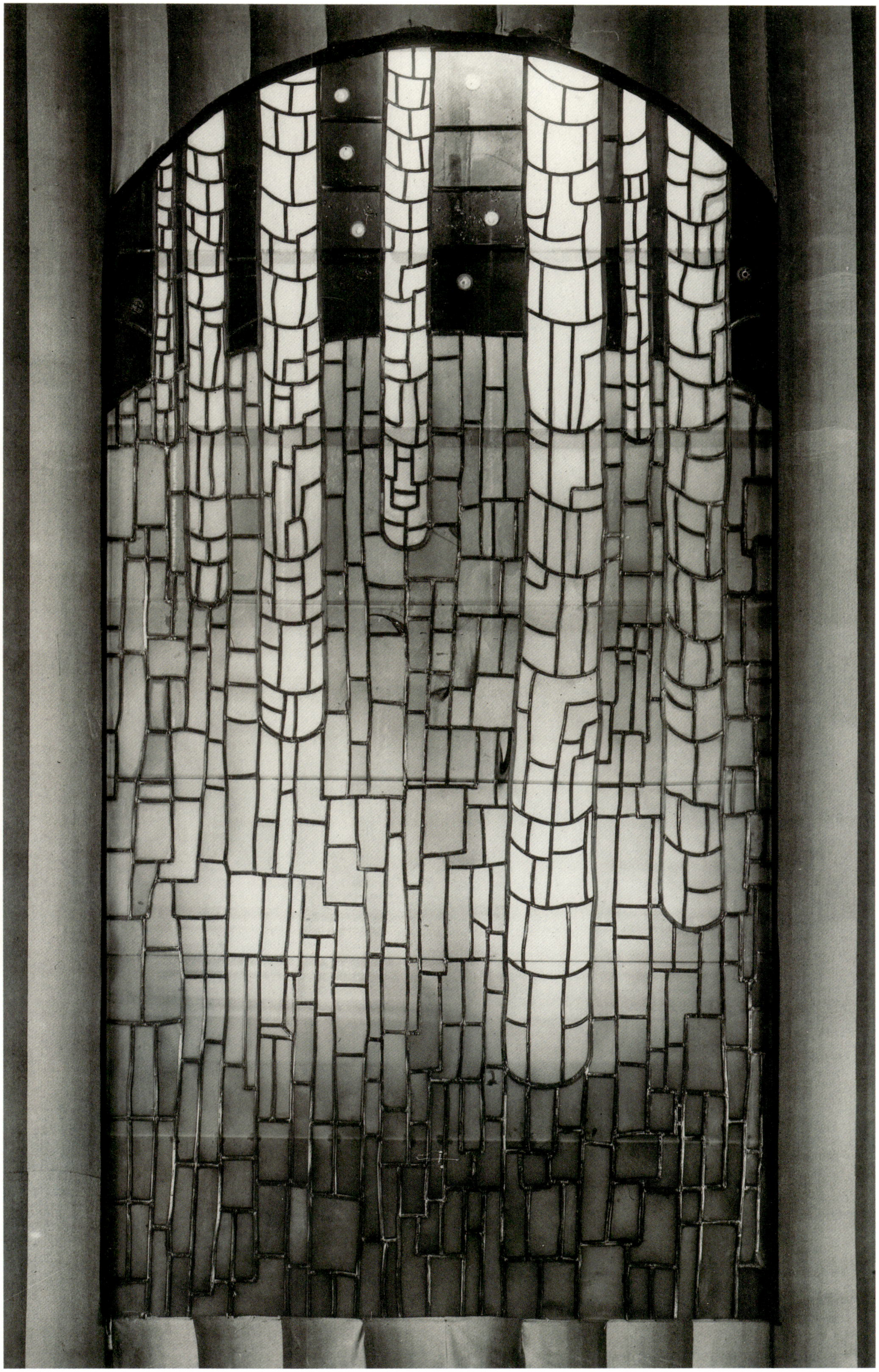

THE M.V.M. CAPPELLIN SHOP, FLORENCE, 1927-28

Thayat (E. Michahelles), polychrome stained glass piece in Murano glass with lead came, depicting a still life, 1927-28. The stained glass was originally located in the M.V.M. Cappellin shop in Florence and in 1969 was donated to the Museo del Vetro of Murano by the Baroness Lydia Quintili Cappa Bava. 115 cm high; 137 cm wide
Fondazione Musei Civici di Venezia collection, Museo del Vetro

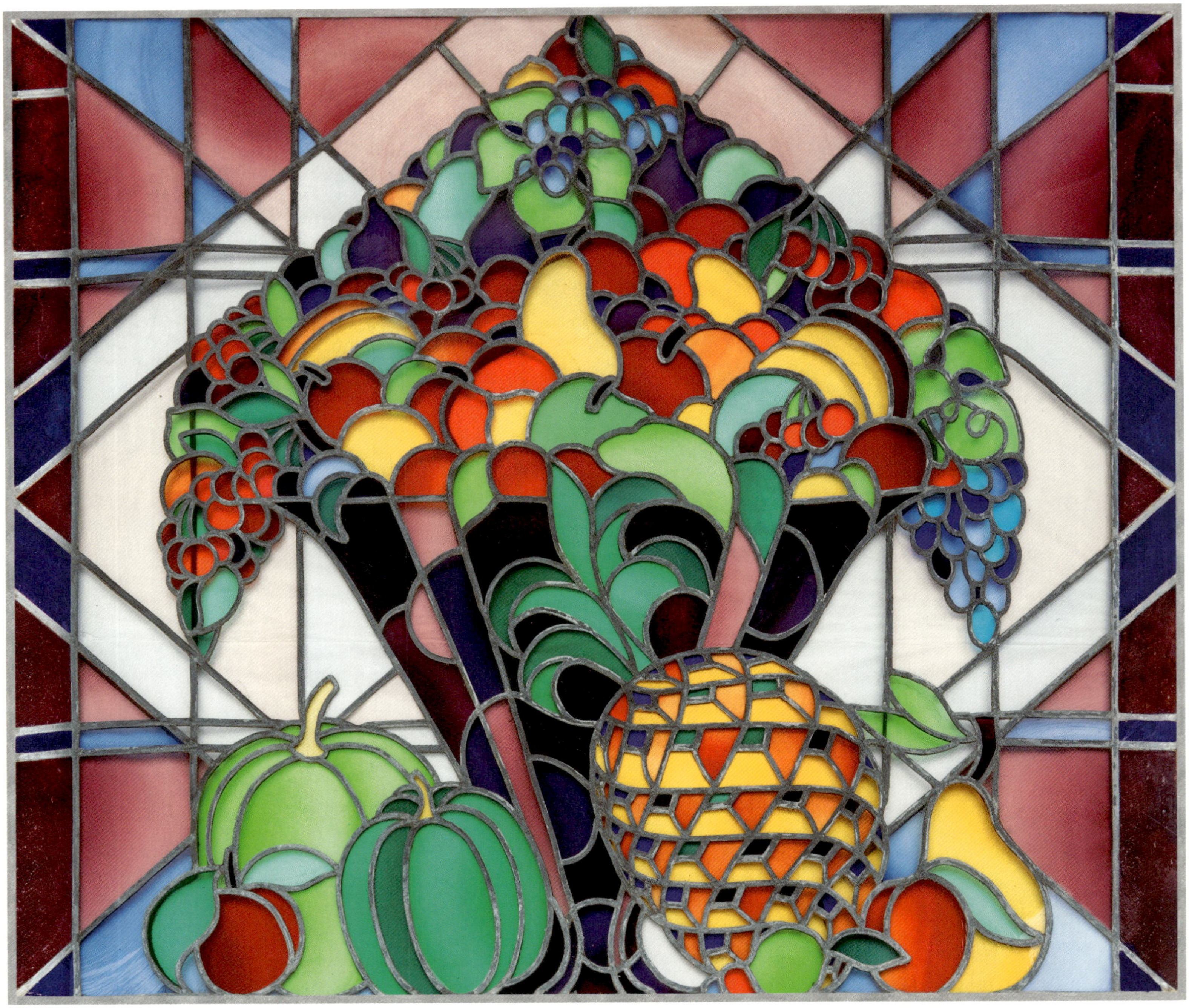

>
Study for stained glass with still life, c. 1928, CSV

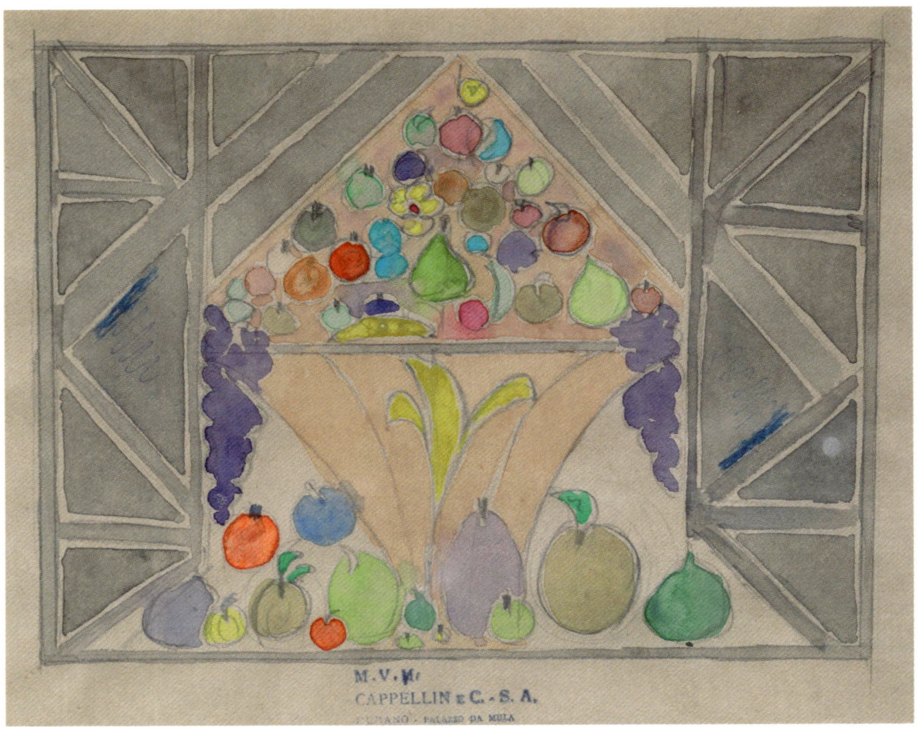

>
Interior of the M.V.M. Cappellin shop in Florence. Period photograph, 1928–29, AMB, detail

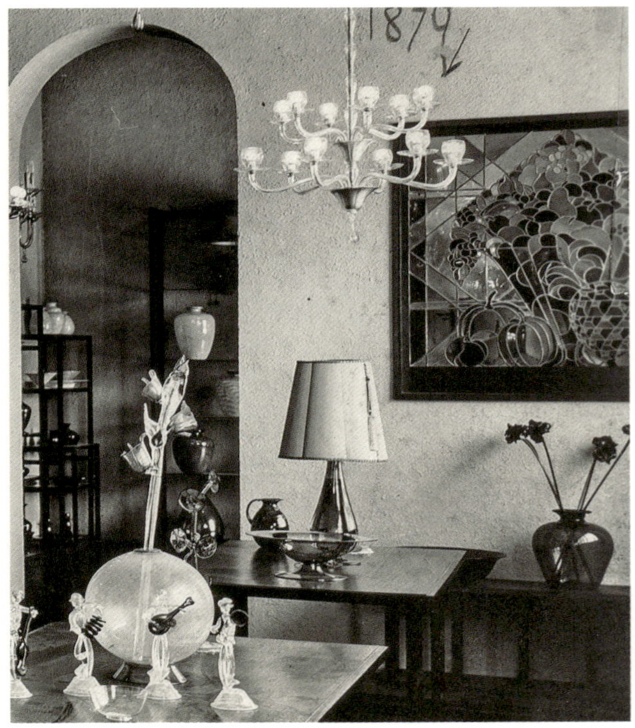

Thayat (E. Michahelles) (?), polychrome stained glass piece with floral subject in Murano glass with lead came (1927-1928), in the M.V.M. Cappellin shop in Florence. Period photograph 1928–29, AMB, detail. The stained glass was used to fill an arched window, so as to receive light from outside.

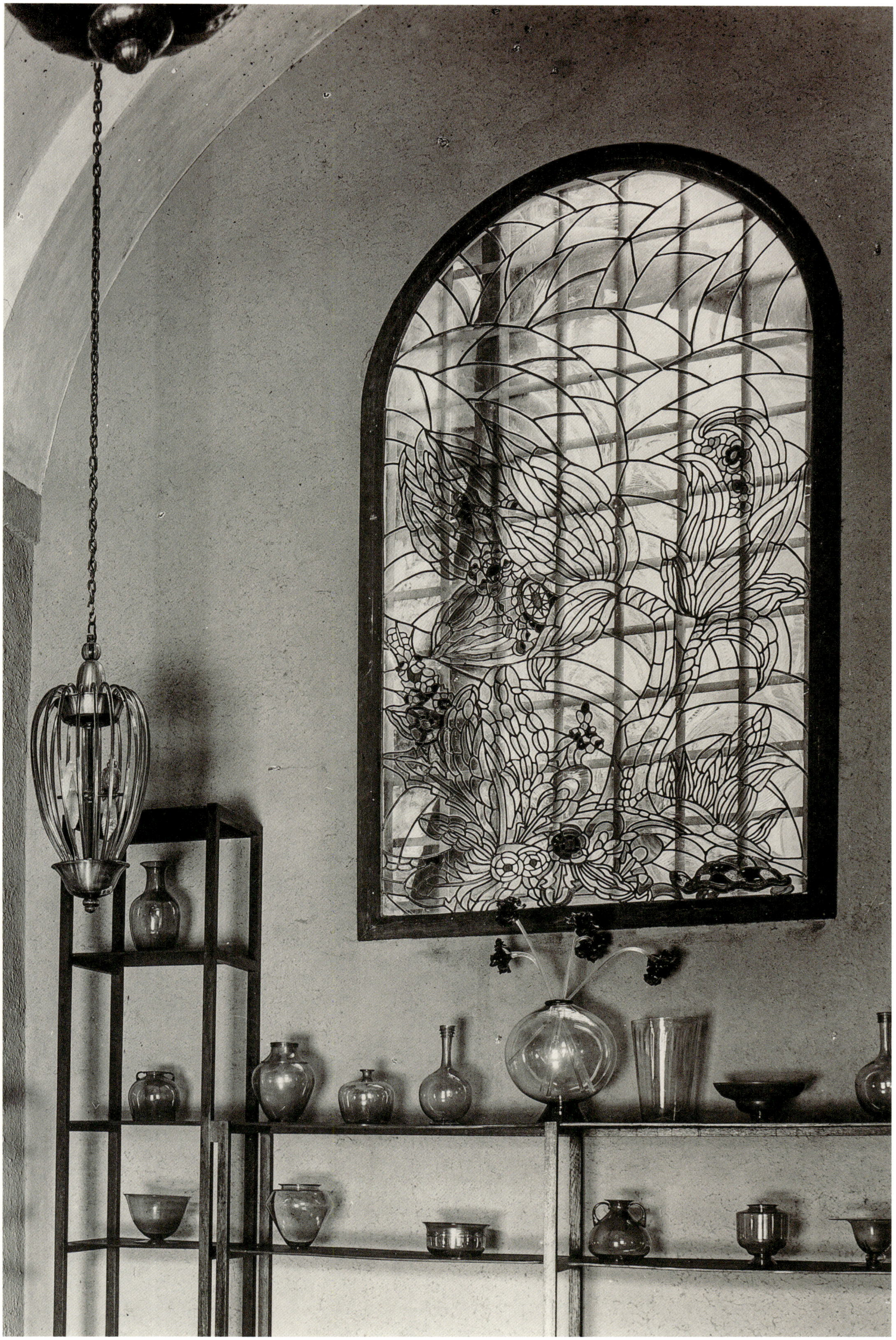

>
Sketch for stained glass with floral subject, c. 1927, ACSVr

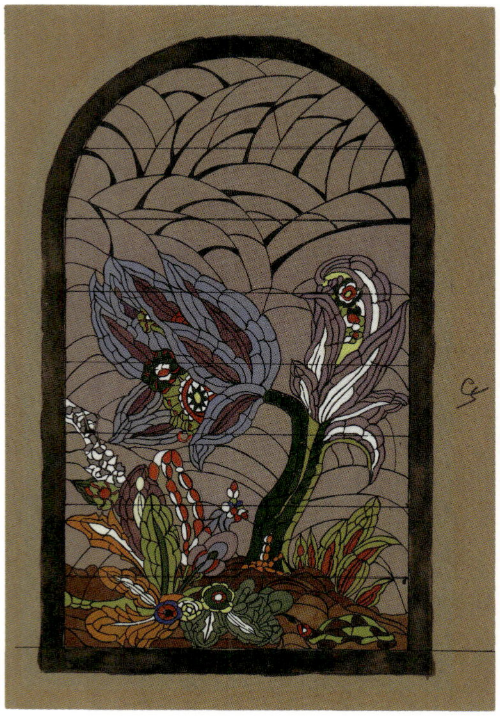

Carlo Scarpa, preparatory sketch for double-leaf stained glass window of the M.V.M. Cappellin shop in Florence, 1927-28, ACSVr. The theme of the stained glass took as its starting point the acronym M.V.M. in the name of the Murano company. Some preliminary studies document the various solutions envisaged by Scarpa before he arrived at the version produced. Most of these build upon the furnace work or the pieces produced there, but there is no lack of drawings of a more abstract kind.

>
Polychrome double-leaf stained glass window in Murano glass with lead came, in the M.V.M. Cappellin shop in Florence. Period photograph, 1928–29, AMB, detail

>>
Carlo Scarpa, studies for two-leaf stained glass window, 1927-28, CSV

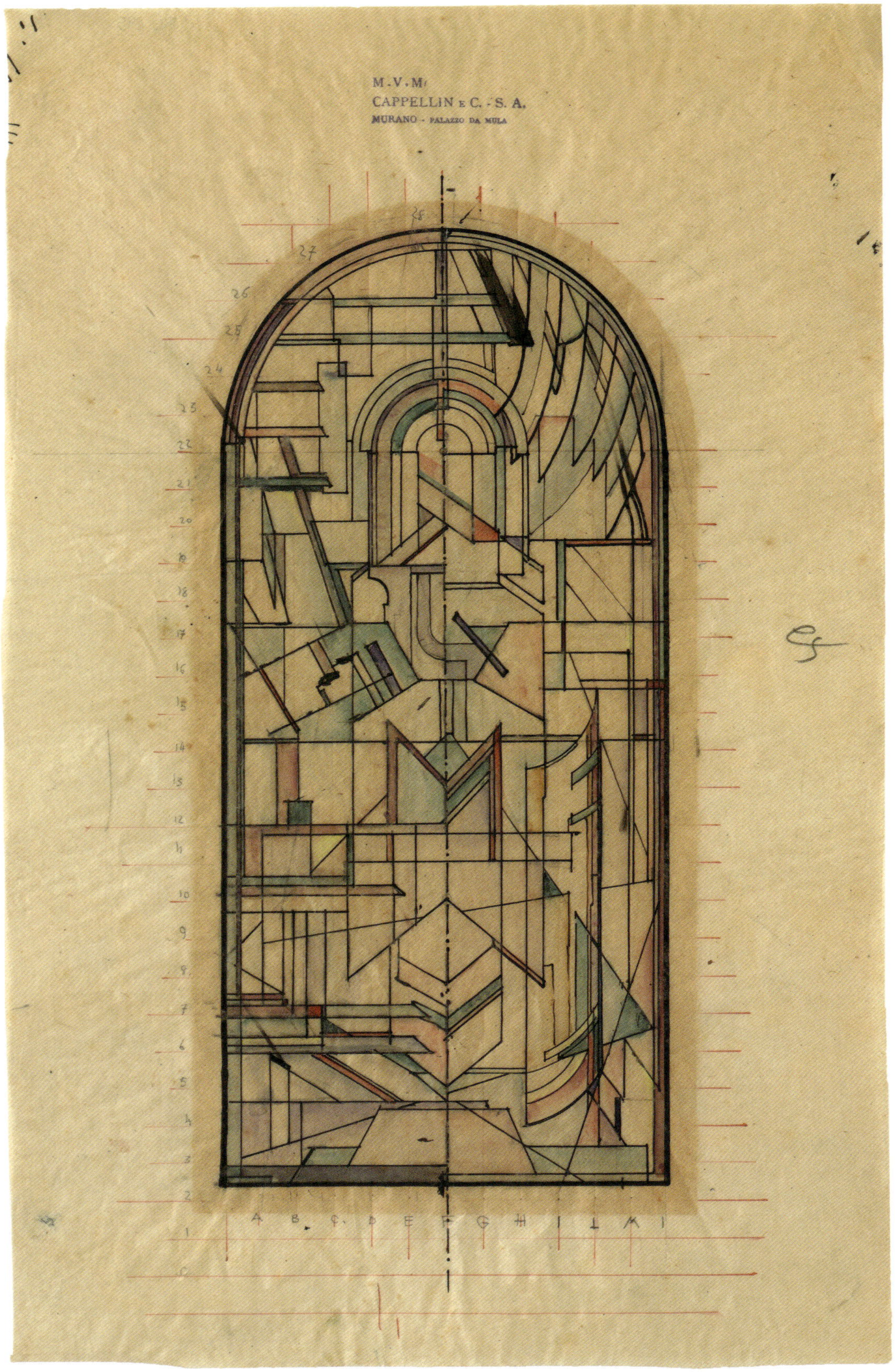

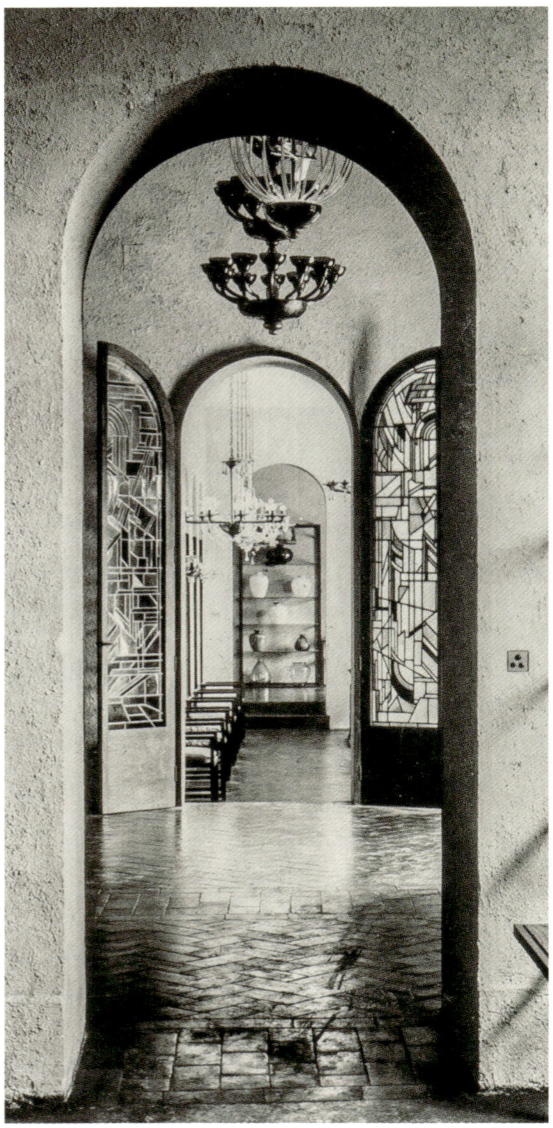

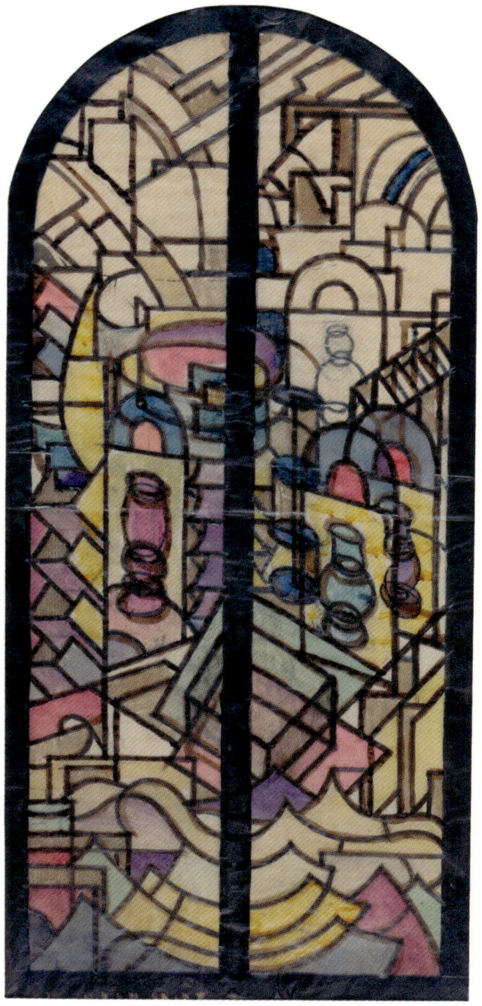

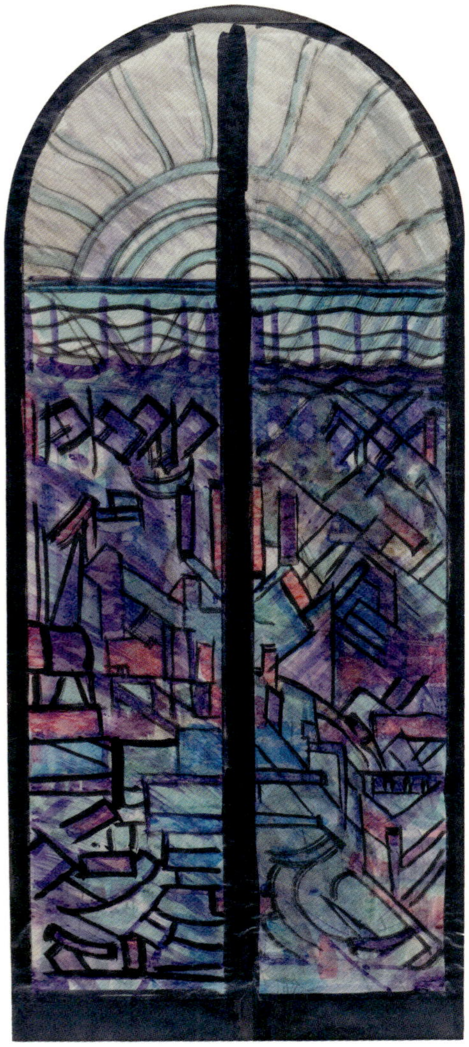

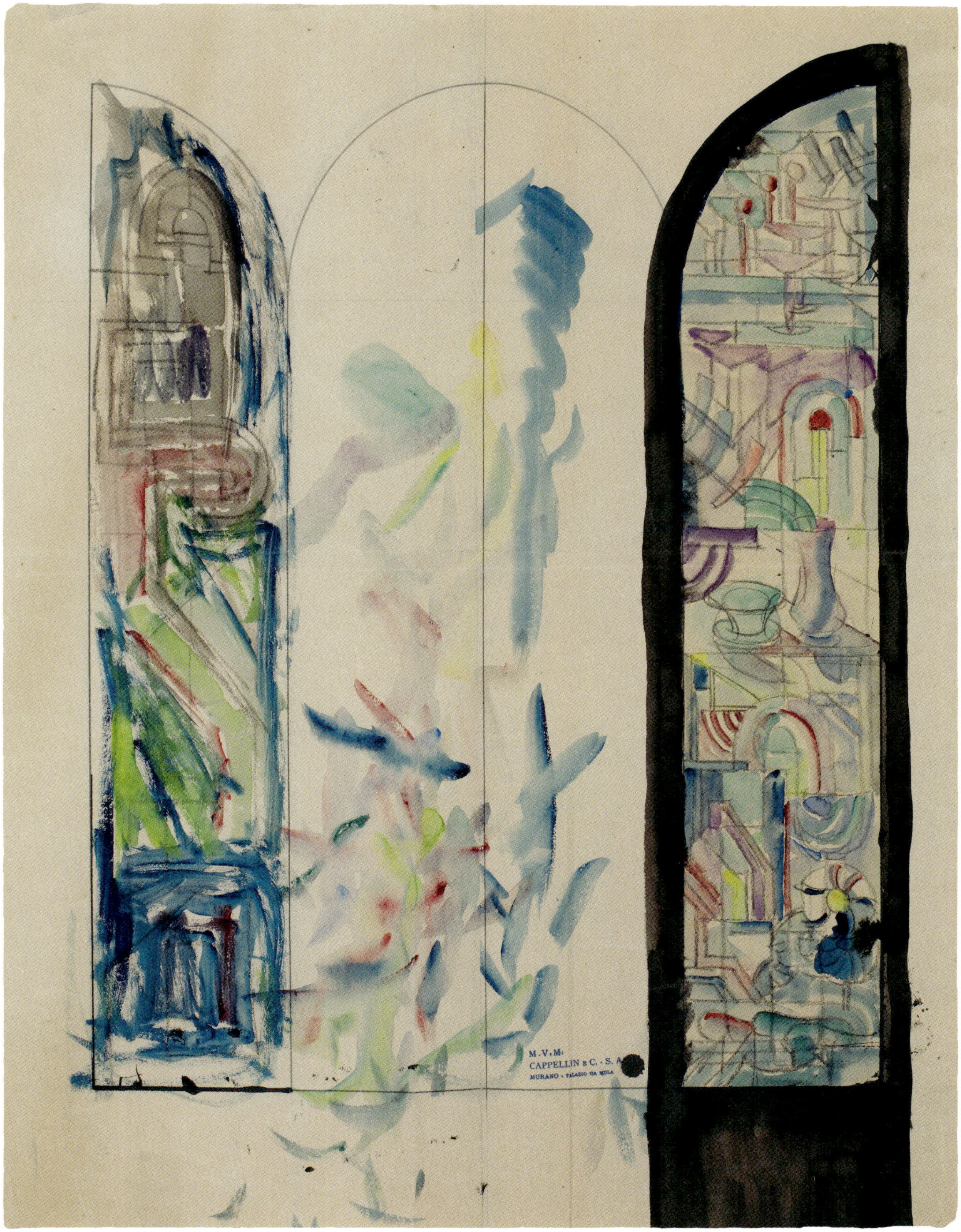
M.V.M.
CAPPELLIN & C. - S. A.
MURANO - PALAZZO DA MULA

THE M.V.M. CAPPELLIN SHOP, TURIN, 1927-28

Among the studies for stained glass windows attributable to the hand of Scarpa, one series has the common characteristic of the same form of the overall volume: a lower rectangular part and an upper lunette. Based on a factory drawing, it is possible to conjecture that they were studies for the decoration of one or more windows of the M.V.M. Cappellin shop in Turin. Among the subjects, that of the production of the glassworks is again present (vases, pages with designs for vases) together with an exotic theme in various interpretations (tigers and vases, vases with plants and snakes), with the association of more geometrical and abstract decorative motifs (vases with rods and weathervanes, polychrome lattices).

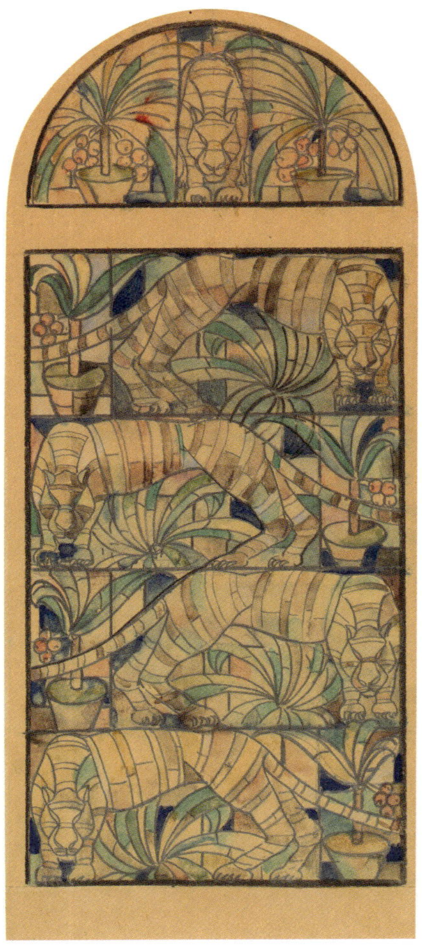
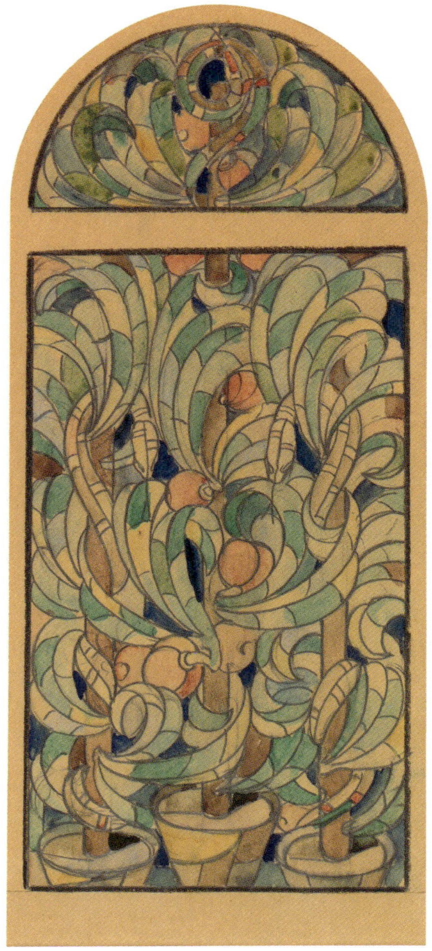
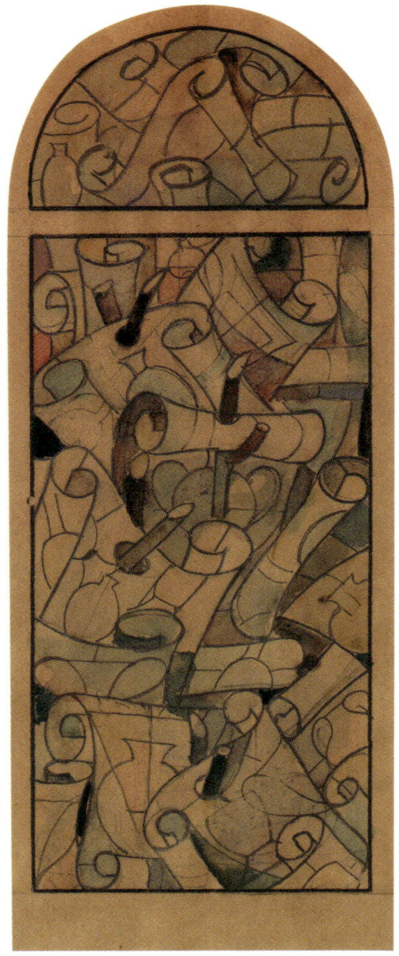
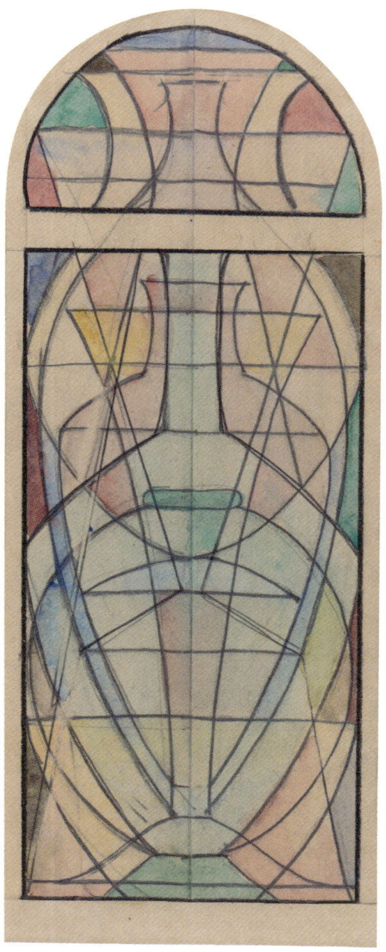
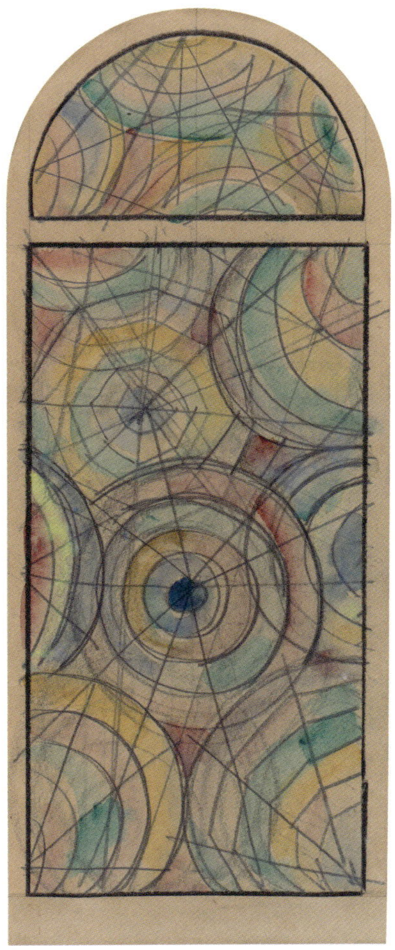
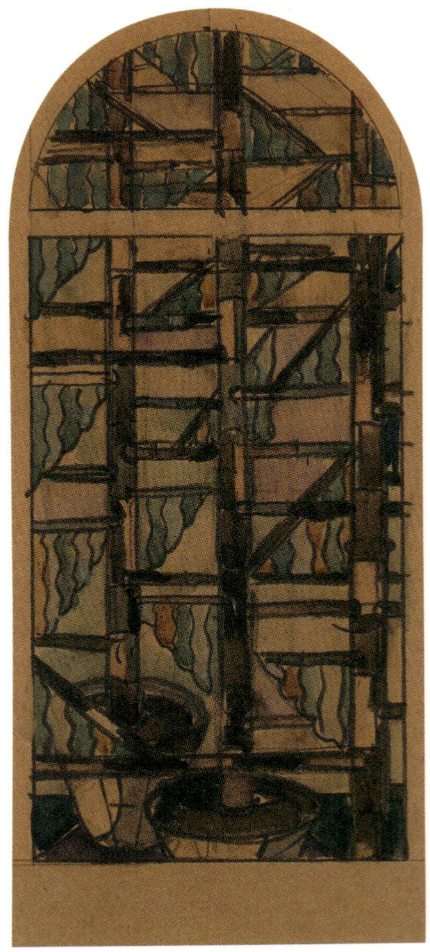

>
Carlo Scarpa, study for stained glass with tigers, 1927–28, G. De Carlo collection

>
Carlo Scarpa, study for stained glass with plants and snakes, 1927–28, G. De Carlo collection

>
Carlo Scarpa, study for stained glass with designs for vases, 1927–28, ACSVr

>
Carlo Scarpa, study for stained glass with silhouettes of vases, 1927–28, CSV

>
Carlo Scarpa, study for stained glass with polychrome lattice, 1927–28, CSV

>
Carlo Scarpa, study for stained glass with vases and rods with weathervanes, 1927–28, ACSVr

4TH ESPOSIZIONE TRIENNALE INTERNAZIONALE DELLE ARTI DECORATIVE ED INDUSTRIALI MODERNE, MONZA, 1930

On the occasion of 4th Triennale of Monza (1930) in the M.V.M. Cappellin exhibition room, four stained glass pieces after cartoons by Carlo Scarpa were on display. One of these, with a female nude, was located at the back of a long area dedicated to the glassworks, while the other three, situated at the opposite end, were arranged next to one another and placed at right angles, forming a variegated polychrome wing.

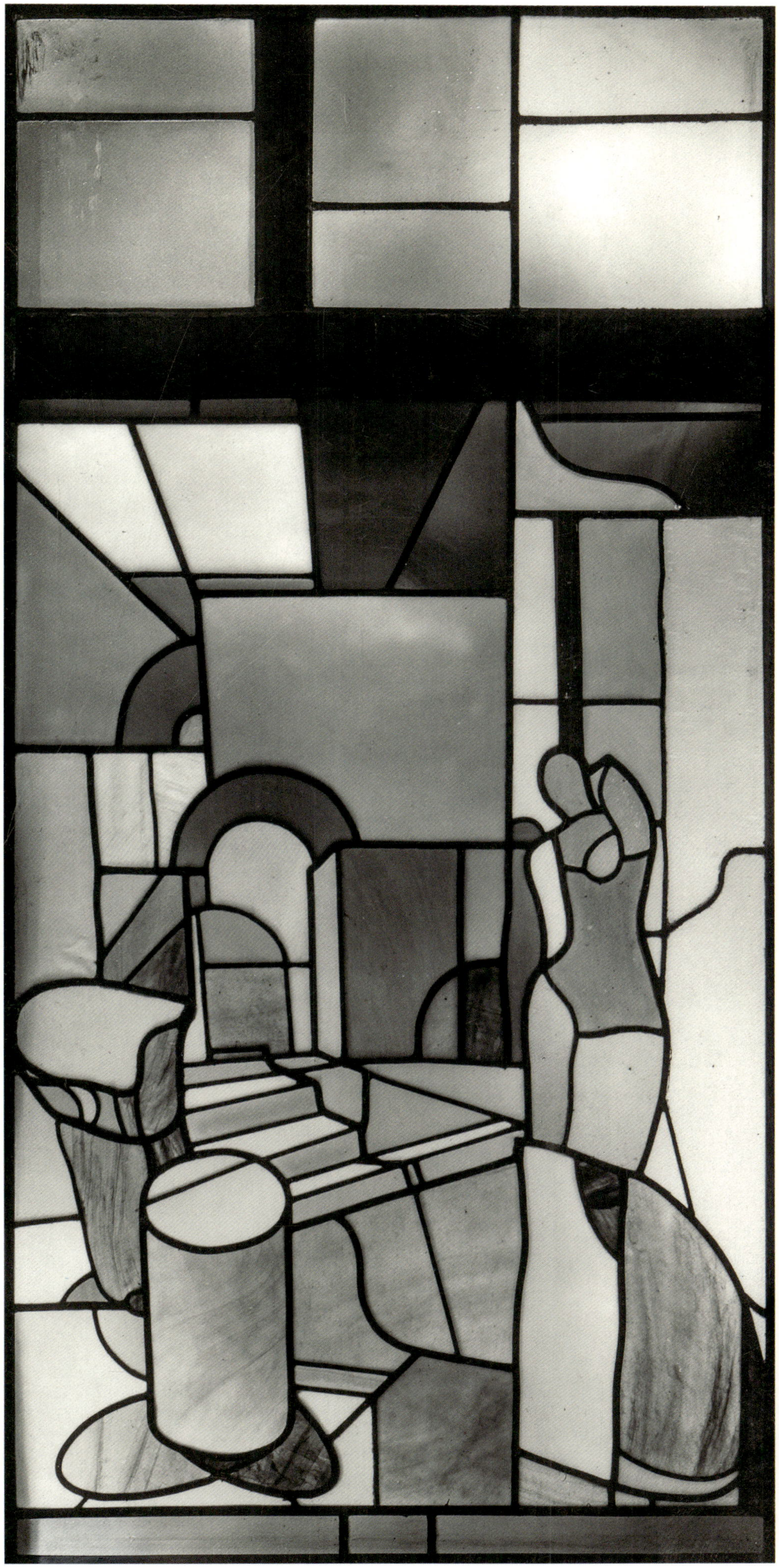

Carlo Scarpa, polychrome stained glass in Murano glass with lead came, with a female nude, 1930, AFT

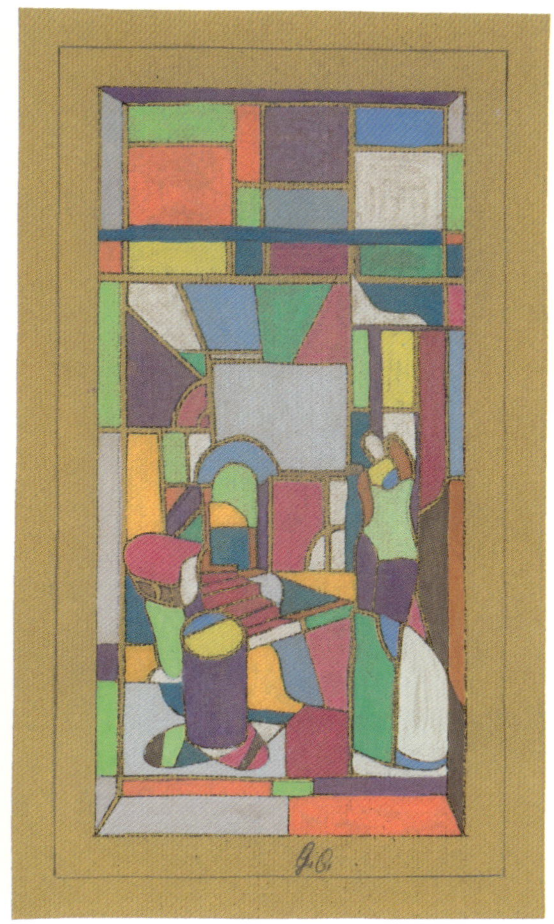

>
Drawing for the stained glass piece with a female nude, c. 1930, G. De Carlo collection

Carlo Scarpa, polychrome stained glass in Murano glass with lead came, 1930

Carlo Scarpa, polychrome stained glass in Murano glass with lead came, with pipe, 1930

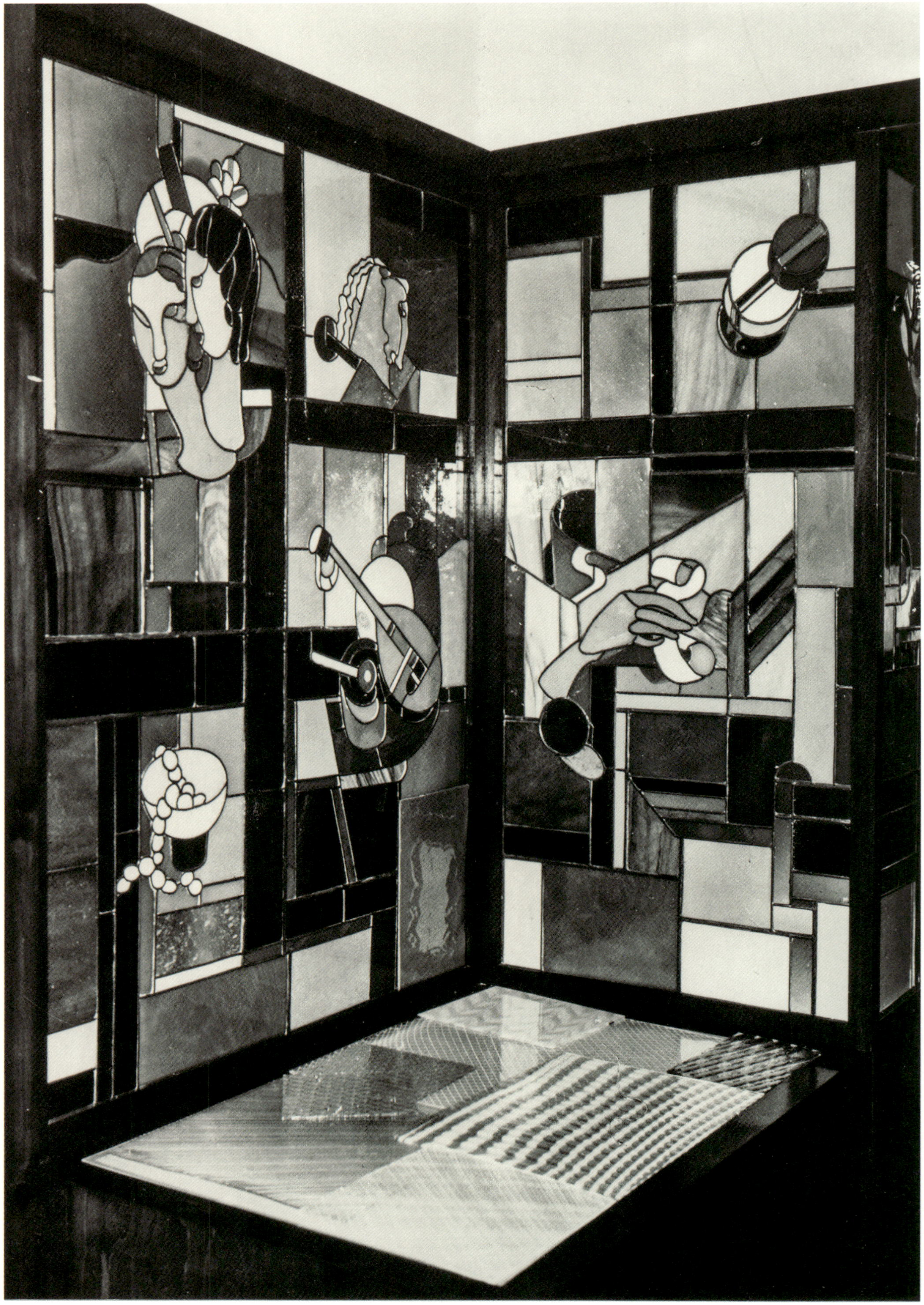

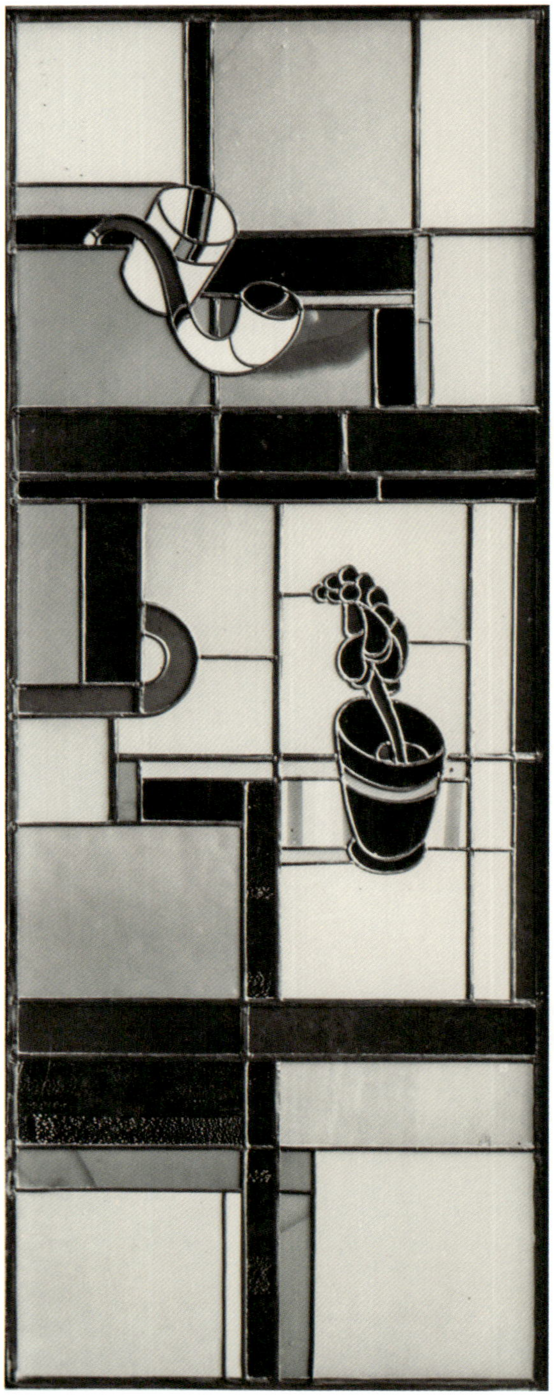

At the 4th Triennale of Monza in 1930, the glassworks were present with some stained glass pieces also in the Gallery of Graphic Arts. These were three panels produced after a cartoon by Mario Sironi. The largest one, with male figures, was placed in the last room. Another of smaller size and characterised by 'geometric figures [...] presented in bright red on a light background', was mounted on a high pedestal. Another, also with geometric figures, was inserted at the top of an arched recess (cf. Triennale archives).

Mario Sironi, polychrome stained glass in Murano glass with lead came, 1930. Certain panels are painted in *grisaille*. 212 cm high; 140 cm wide Raccolte d'Arte Applicata, Castello Sforzesco collection, Milano

Mario Sironi, polychrome stained glass in Murano glass with lead came in *La casa bella*, May 1930, p. 24

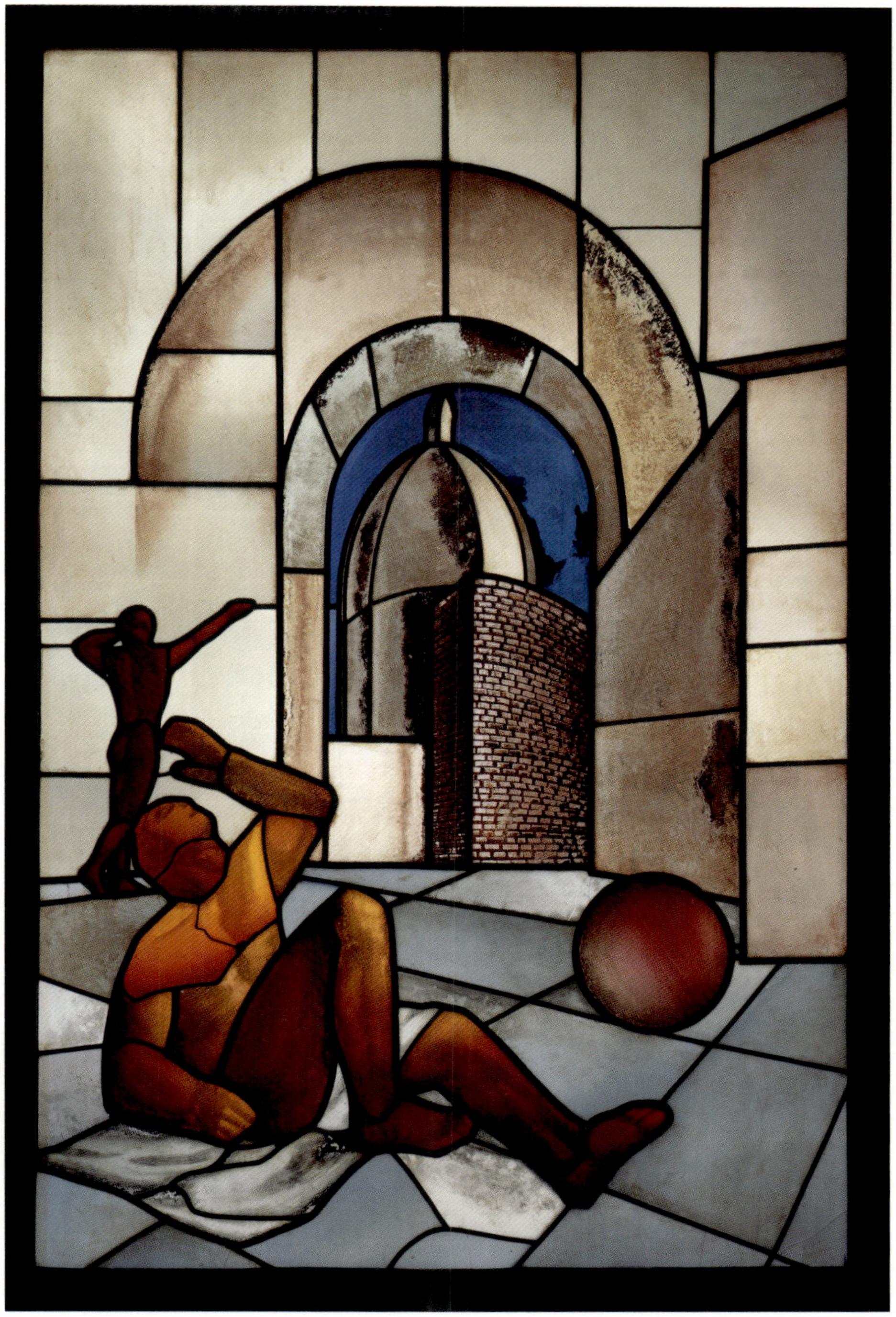

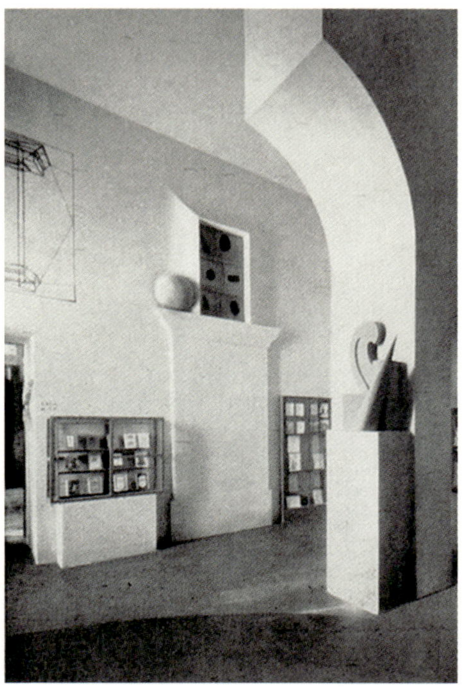

1ST QUADRIENNALE D'ARTE NAZIONALE, ROMA, 1931

On the occasion of the 1st Quadriennale d'arte of Rome the M.V.M. Cappellin firm was present with various stained glass pieces: three designed by Mario Romano and two by Enrico Del Debbio. The first of these, which closely followed the poster of the exhibition, was placed at the entrance.

Mario Romano, polychrome stained glass in Murano glass with lead came, 1930. Some of the panels were painted in *grisaille*. As certain commentators reported, the stained glass extended to about 10 square metres.

>
'Entrance to the Exhibition with the radiant Cappellin stained glass which reproduces the poster (architect Mario Romano)', in L'illustrazione italiana, 11 January 1931, p. 50

>
Drawing for stained glass at entrance to the 1st Quadriennale of Rome, c. 1931, G. De Carlo collection

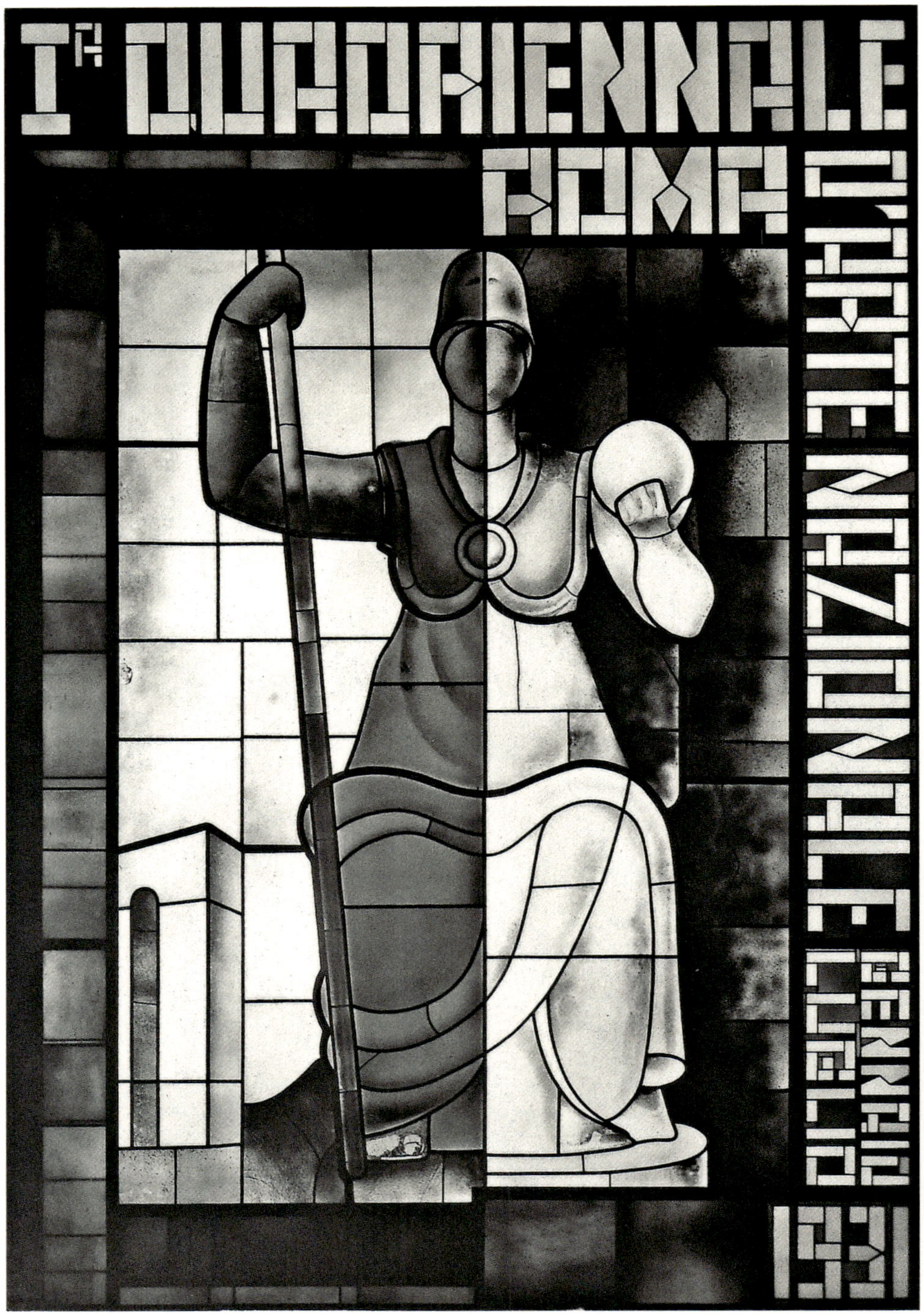

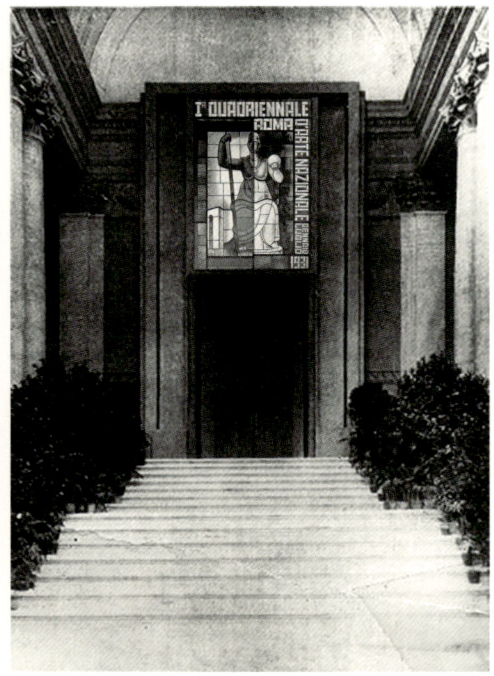

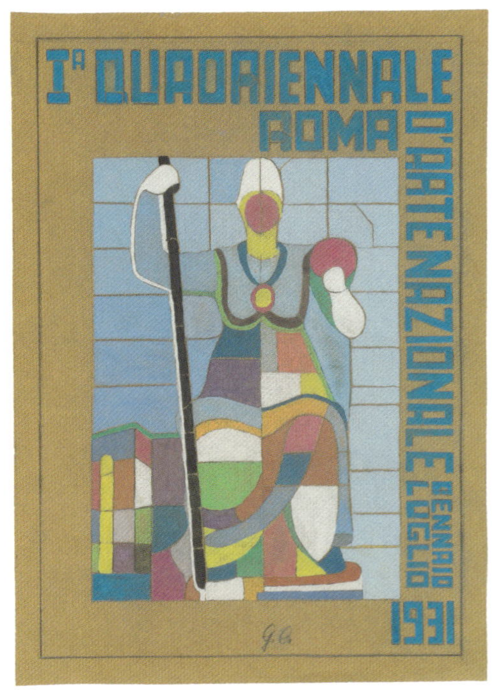

Mario Romano, *Sculpture*, polychrome stained glass in Murano glass with lead came, 1930

Mario Romano, *Painting*, polychrome stained glass in Murano glass with lead came, 1930

The stained glass panels (c. 245 x 610 cm) were placed in the area of the entrance, opposite the rotunda. Following the bankruptcy of the M.V.M. Cappellin glassworks, the panels were bought in 1933 by the Pauly glassworks and, mounted in a way that was not consistent with the original design, placed in the show-room of the company.

>
Show room of the Pauly glassworks. Period photograph of the 1930s. At the centre can be seen the stained glass panels presented at the 1st Quadriennale of Rome

>
Sketch for the stained glass piece, painting exhibited at the 1st Quadriennale of Rome, c. 1930, ACSVr

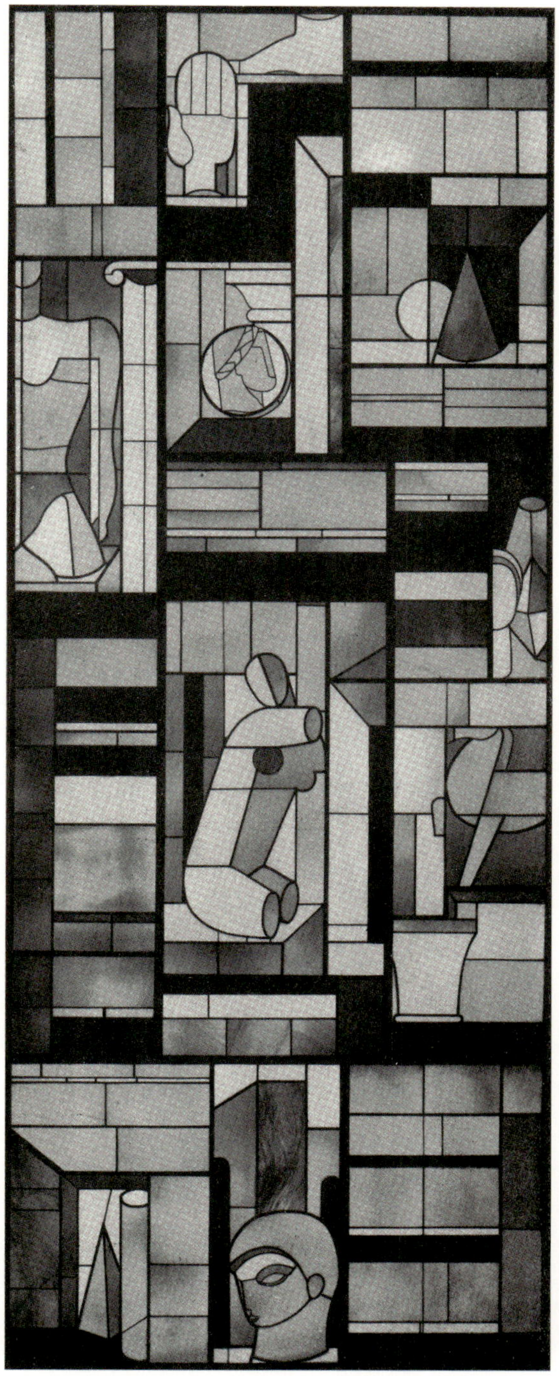

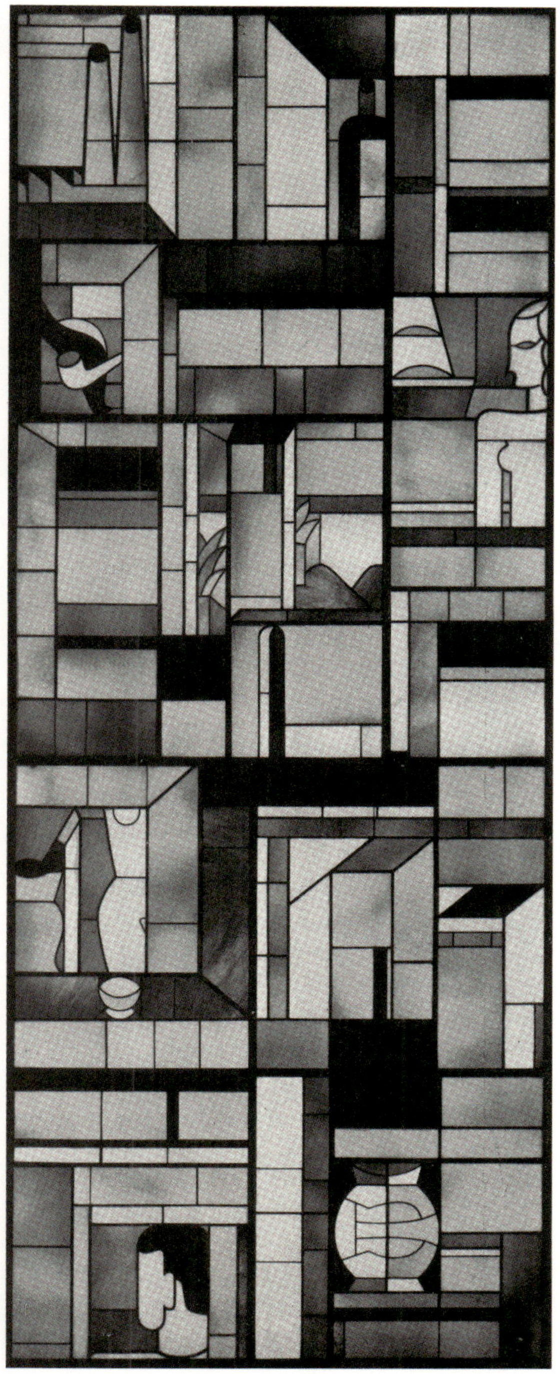

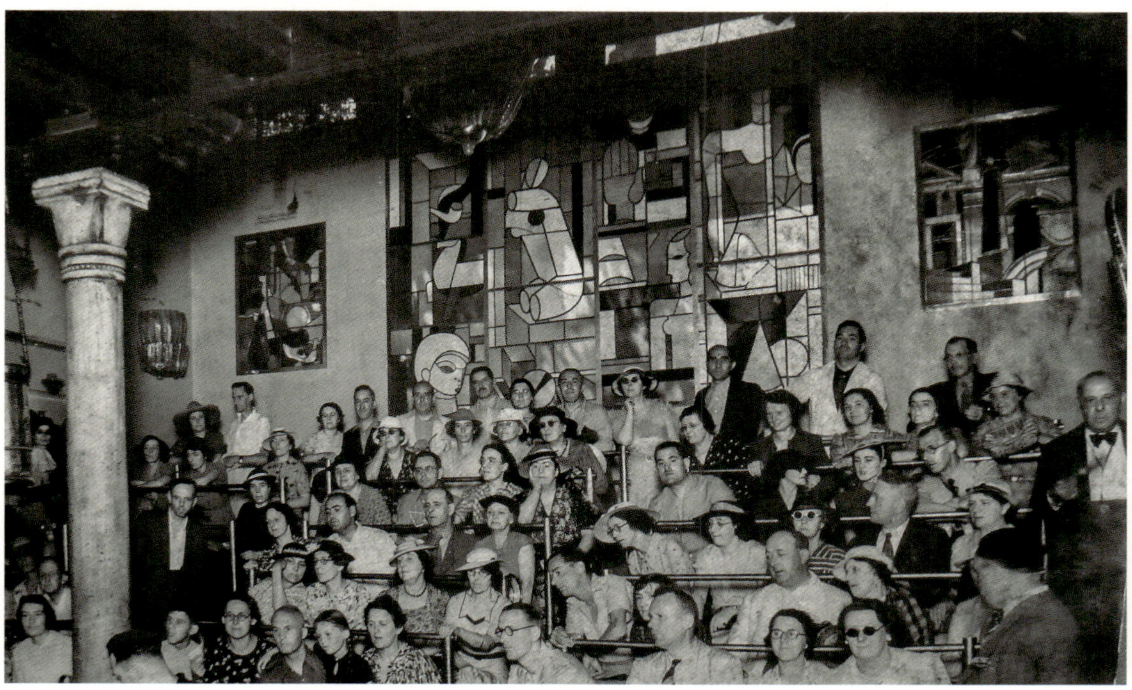

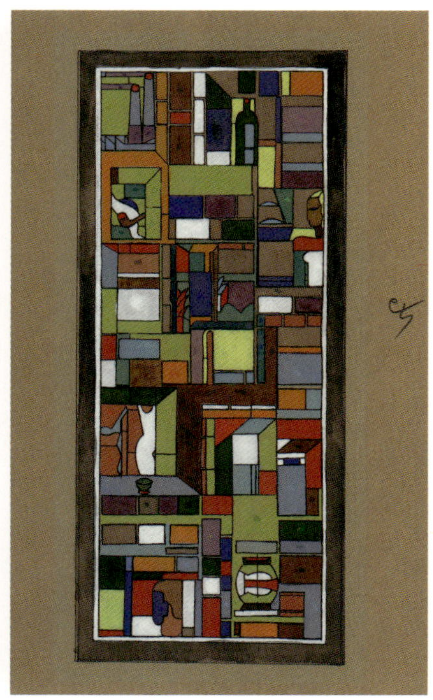

Enrico Del Debbio, *The Street* or *The Street in the Evening*, polychrome stained glass piece in Murano glass with lead came, 1930. The stained glass (c. 515 x 440 cm) was positioned in the area of the bookshop, placed between the gallery of the 'niches' and the winter garden.

Enrico Del Debbio, *The Arts*, polychrome stained glass in Murano glass with lead came, 1930. Certain panels were painted in *grisaille*.
The stained glass was positioned on the first floor in room 42.

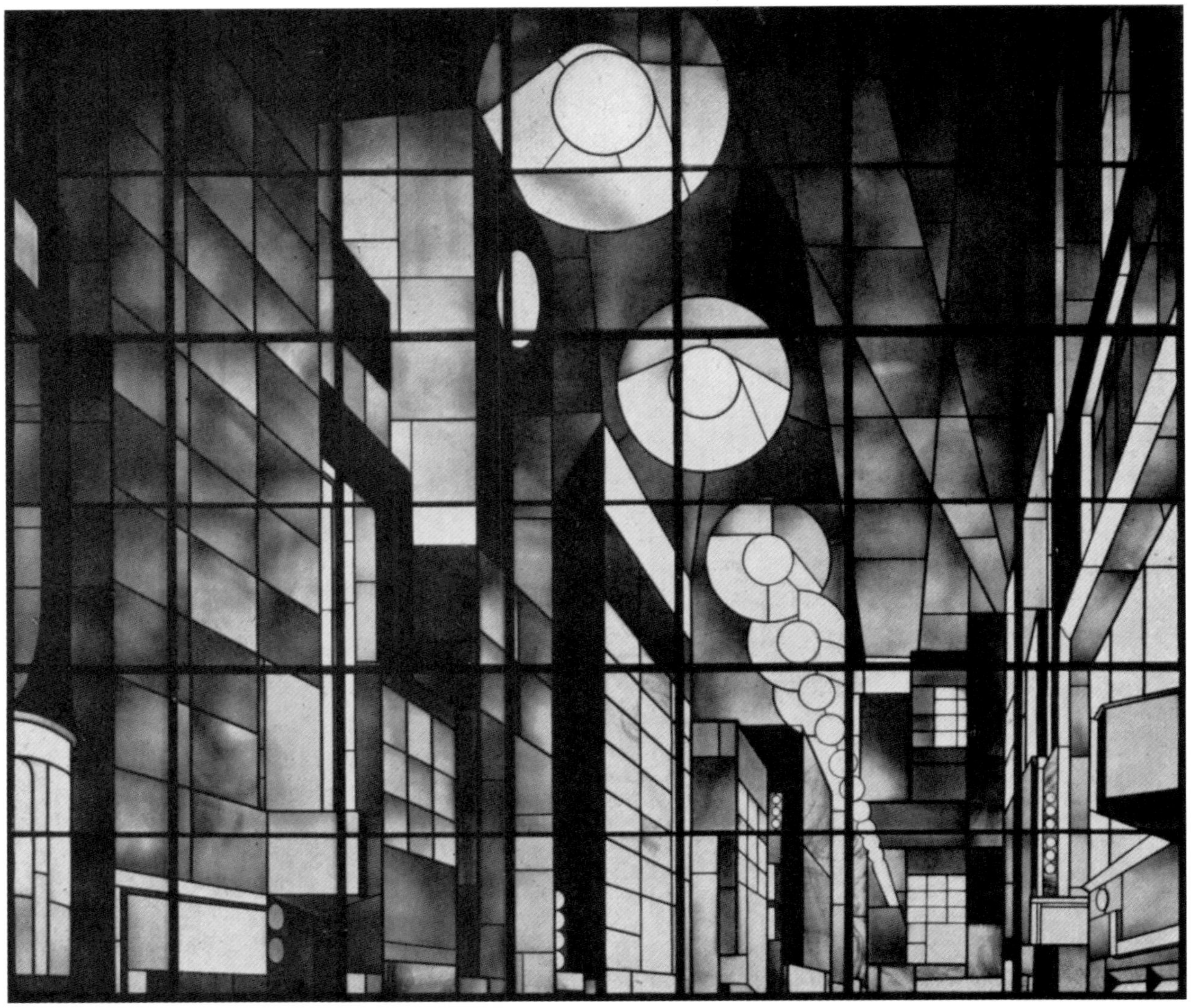

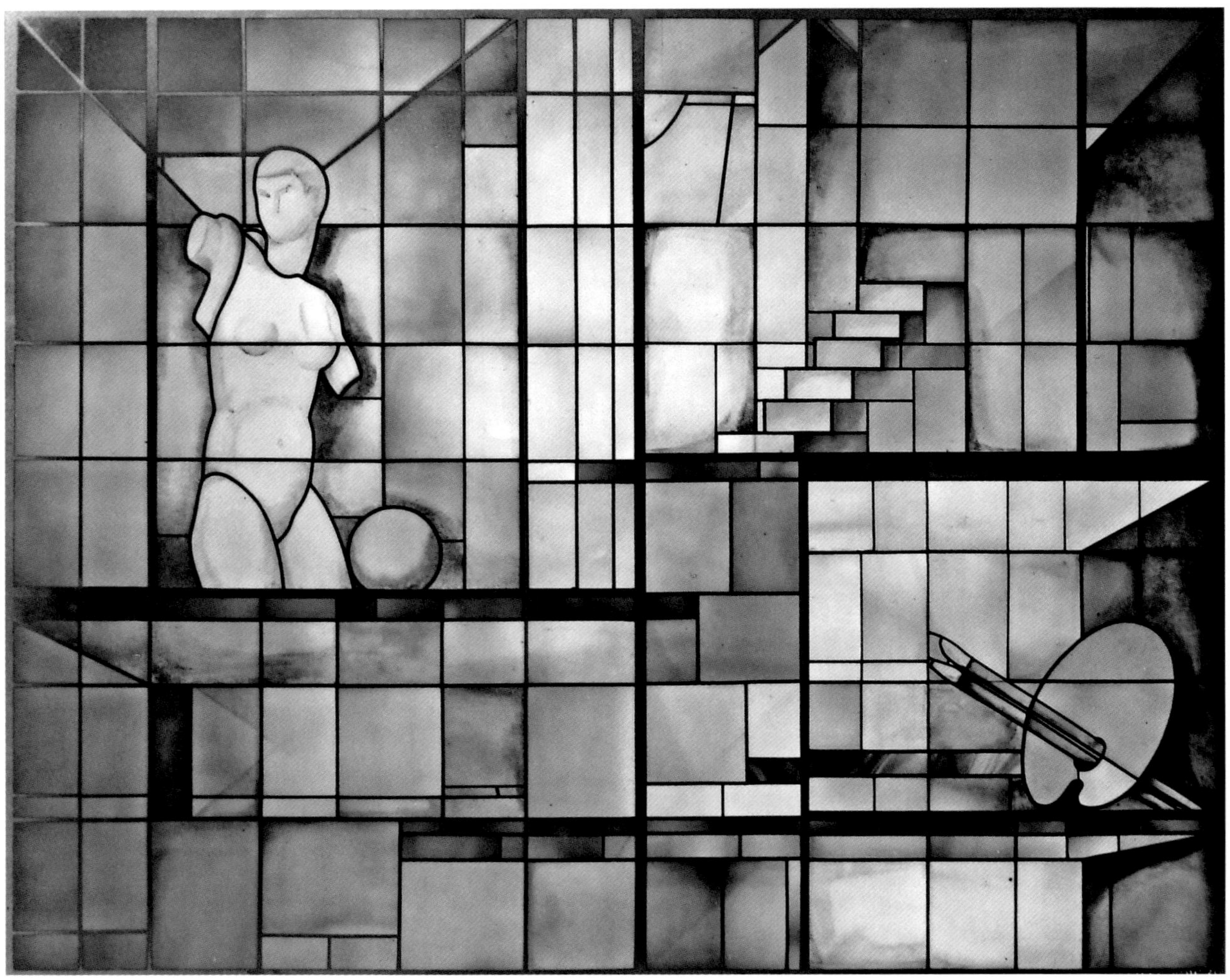

EXHIBITION OF MODERN ITALIAN ART GLASS, CERAMICS AND LACE, AMSTERDAM, 1931

Carlo Scarpa (?), polychrome stained glass piece in Murano glass with lead came, with female figure and palm, 1930–31.

The stained glass was published in *Tentoonstelling van glas, aardewerk en kant, van moderne italiaansche kunstnijverheid*, *Mostra di vetri, ceramiche e merletti d'arte moderna italiana* Venice 1931, p. 27, catalogue of the exhibition held in Amsterdam.

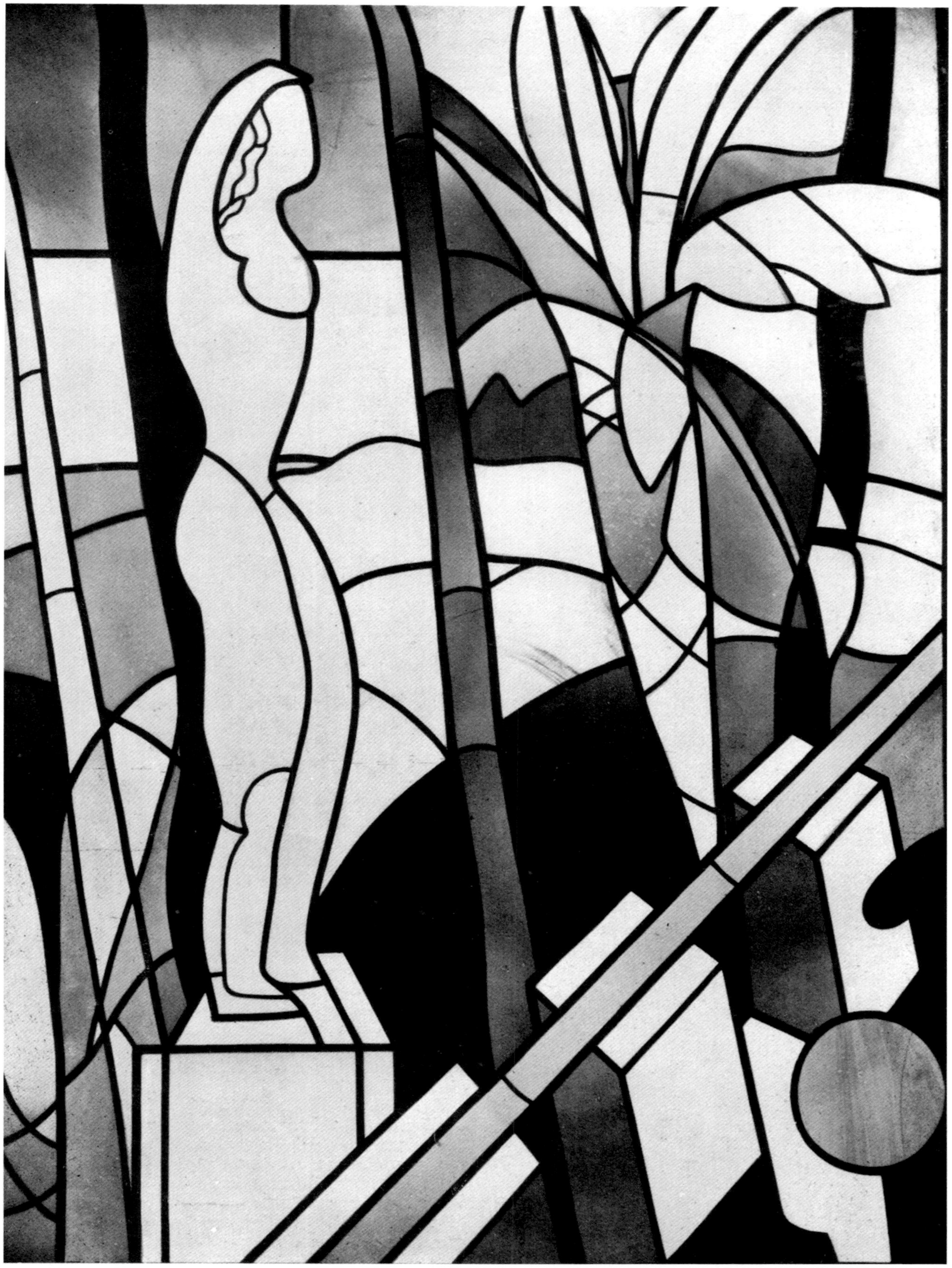

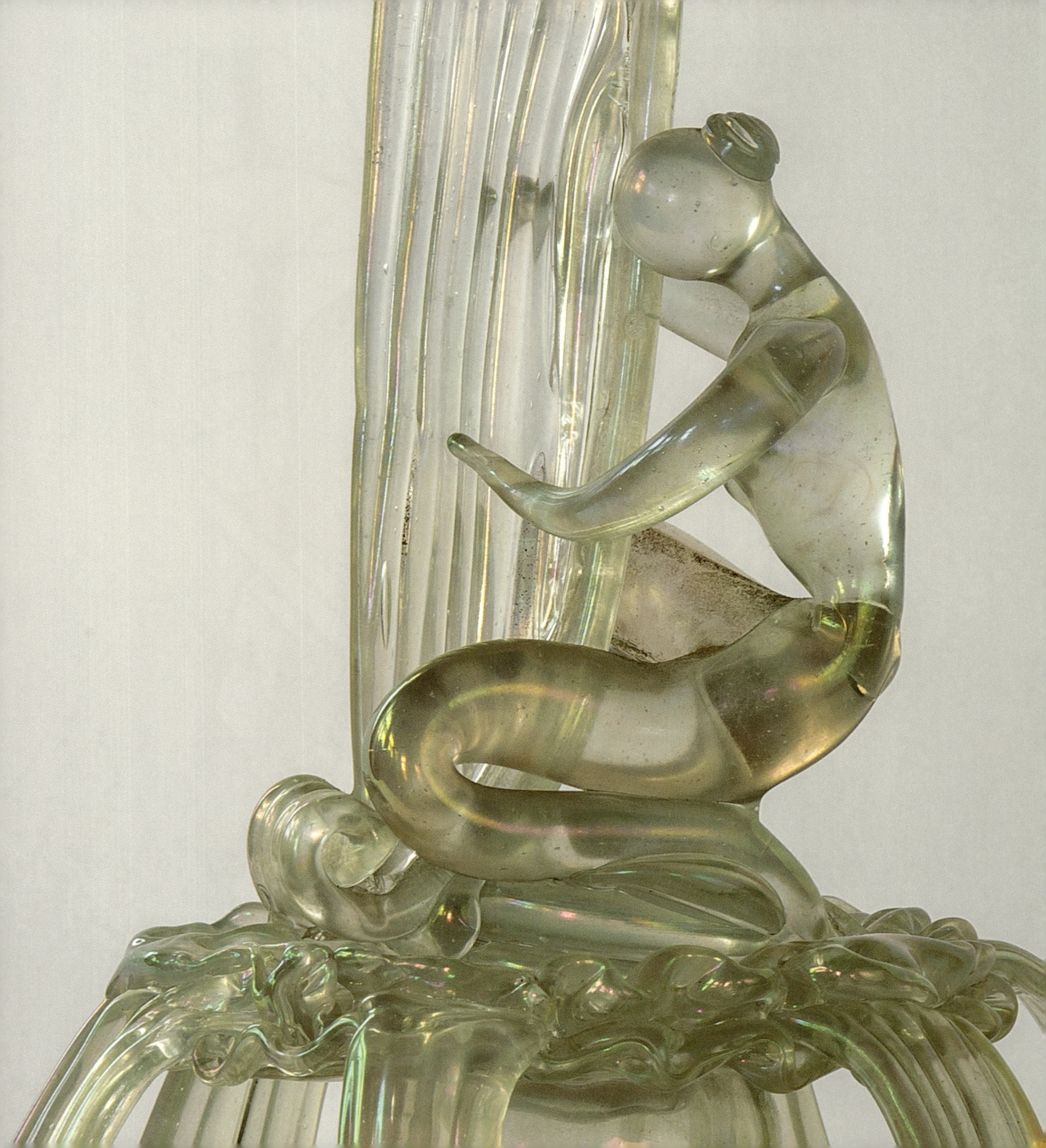

On the occasion of the Italian Garden exhibition at Palazzo Vecchio in Florence in 1931, the M.V.M. Cappellin firm presented, together with five small table centrepieces with flowers in glass, a centrepiece of monumental dimensions in iridised crystal glass with a base consisting of variously juxtaposed twisted canes . The composition was inspired by 18th-century centrepieces, known as 'deseri' (from the French *dessert*) which ornamented the Doges' banquet tables on solemn feast days. In particular, it seems to be inspired by the Italian glass centrepiece garden preserved at the Museo del Vetro of Murano.

TABLE CENTREPIECE c. 1931

Exhibitions

1931, Florence, Italian Garden Exhibition

>
'Monumental centrepiece presented by Cappellin', in Domus, July 1931, p. 52

>
Nude male at the top of a column, detail

>
Harpist, detail

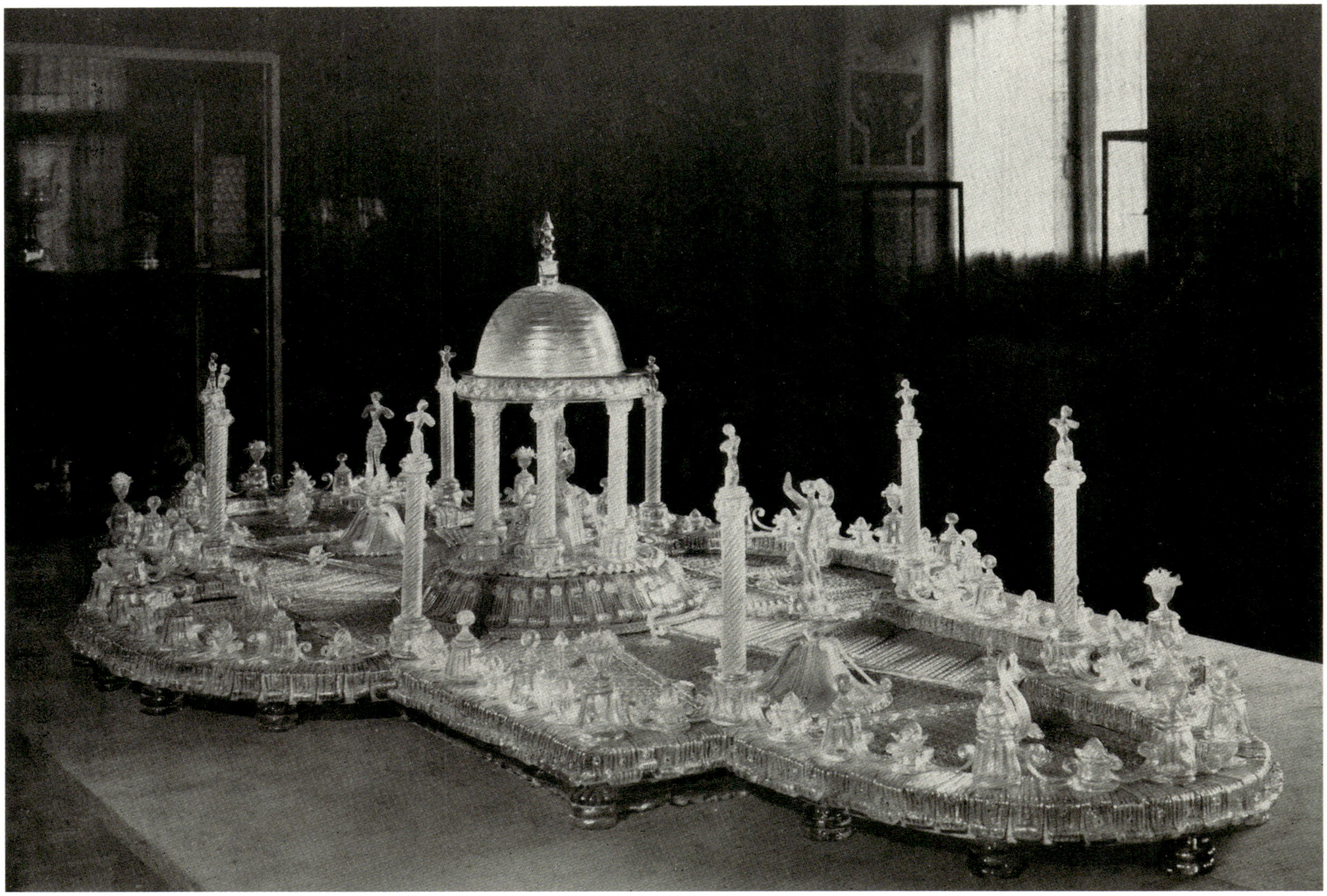

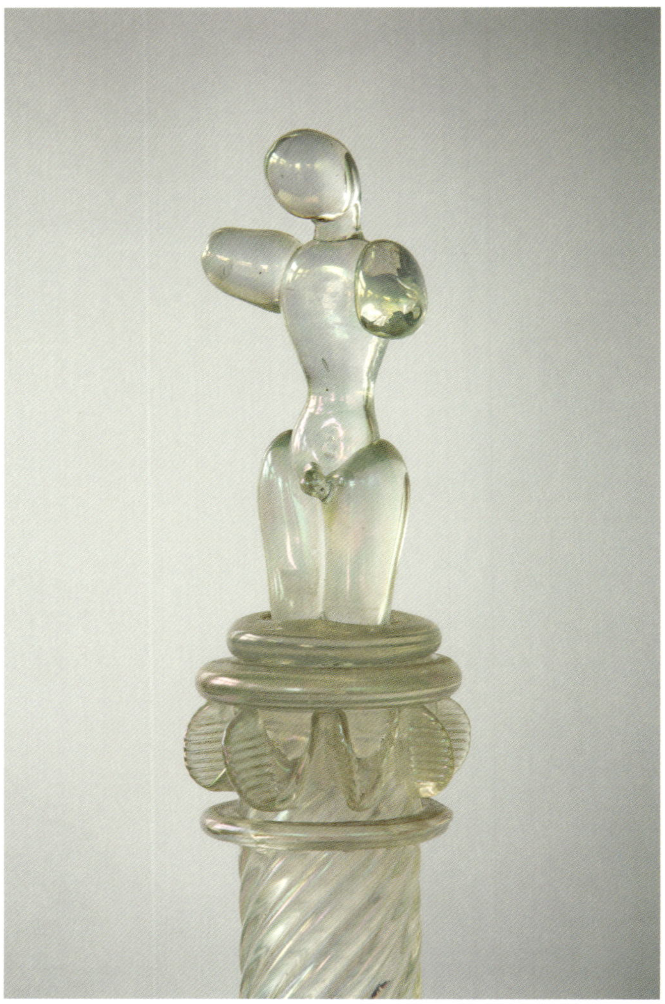

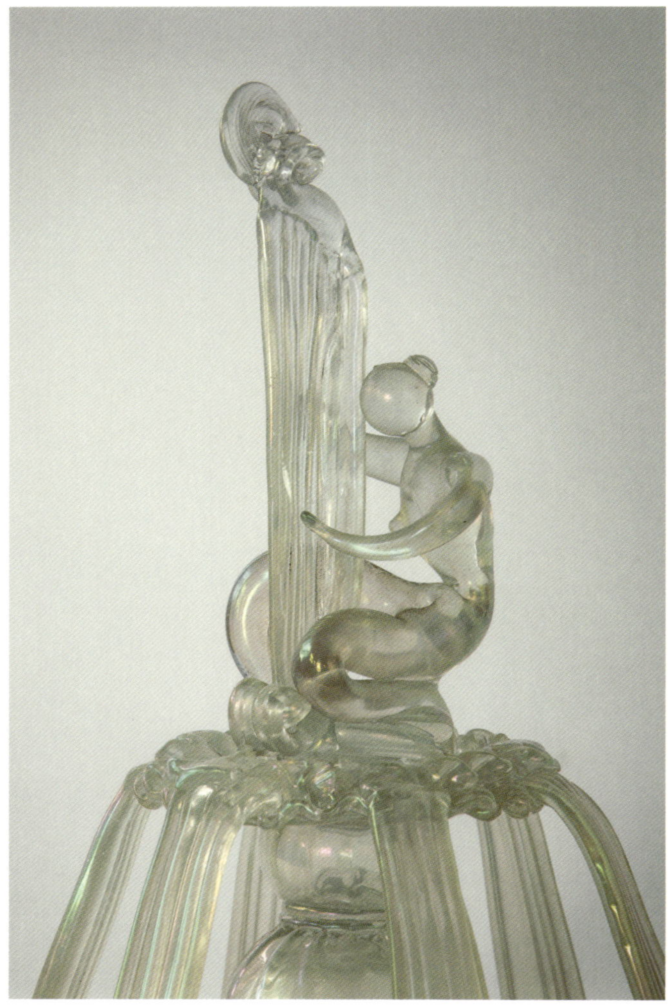

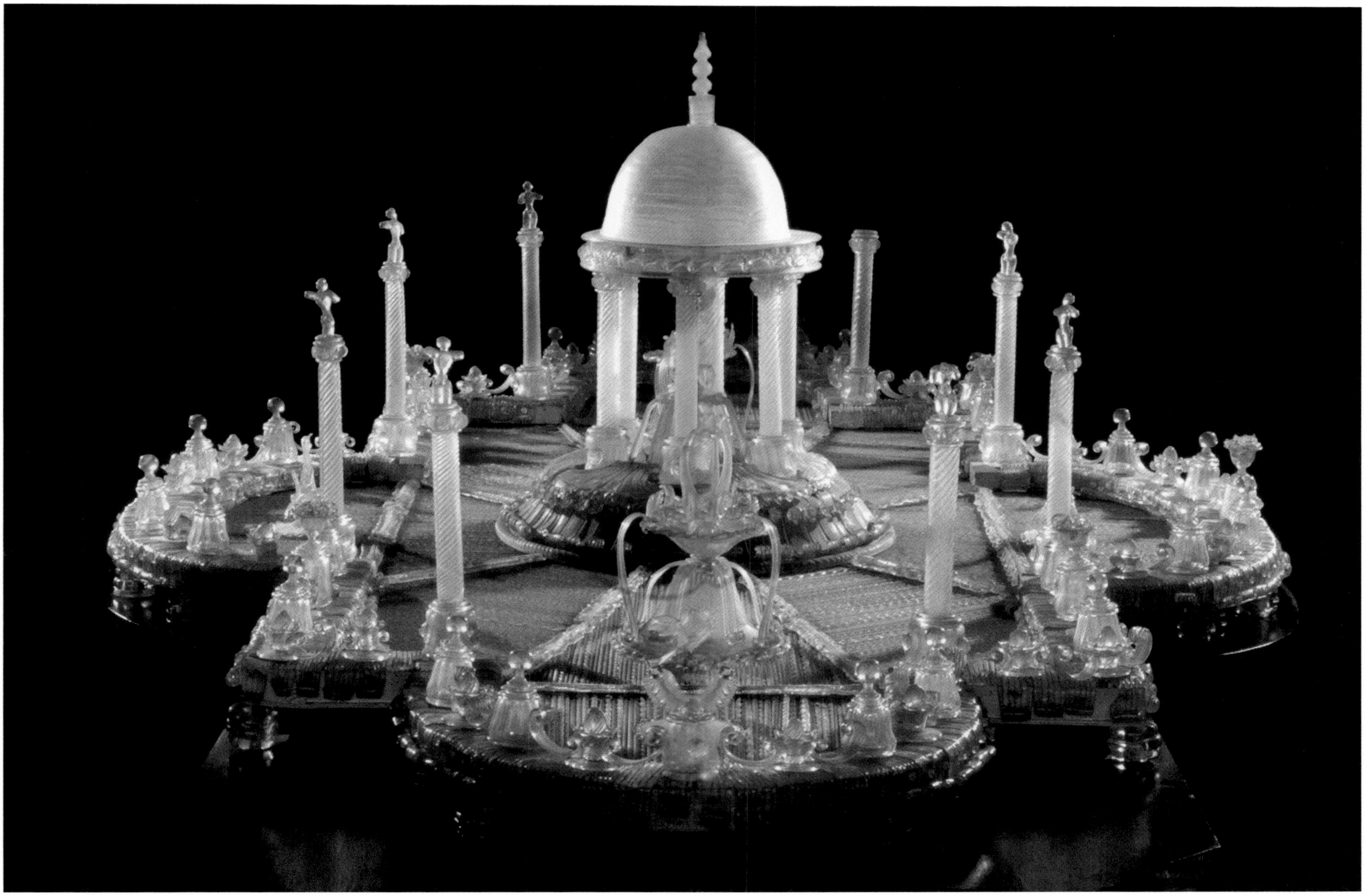

Table centrepiece in transparent glass. Surface with iridised finish. The structure of the composition is based on a circular temple covered with a dome with *decoro fenicio*, underneath which there is a female figure playing a harp made of solid glass. Around the perimetre eight twisted columns rise high, surmounted by nude males (now partially missing) alluding to antique statues. Among the elements of the composition there are also fish ('dolphins') which appear in the catalogue under no. 4758.
c. 1931
190 x 360 cm
Pauly collection

exhibitions
1931, Florence, Italian Garden Exhibition

ANNEXES

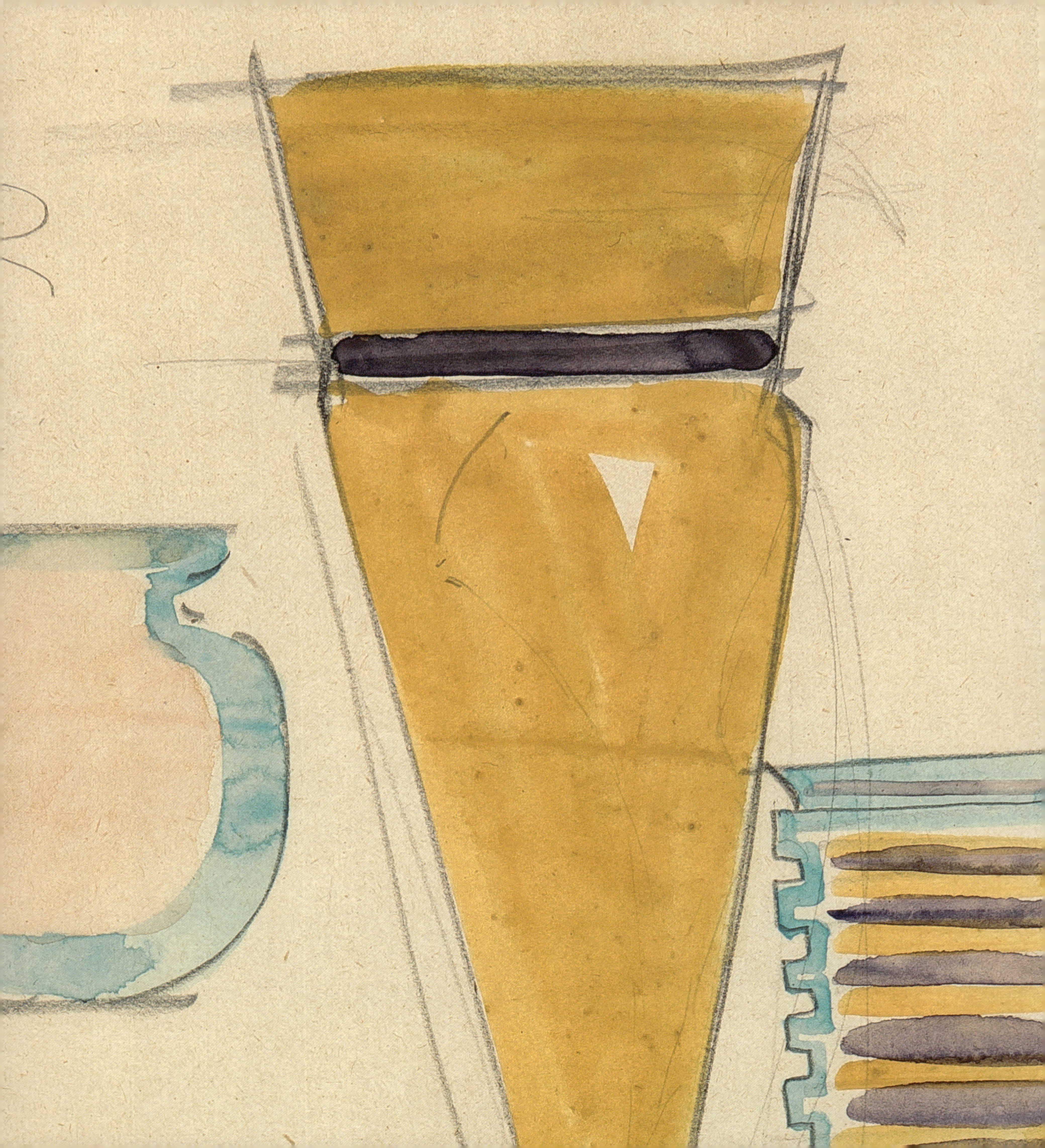

In the documentation we traced concerning the activity of M.V.M. Cappellin & C. and kept in different public and private archives, numerous drawings for glass pieces attributable to the hand of Carlo Scarpa have been identified. They constitute a heterogeneous collection of material documenting studies and proposals made during his artistic association with the glassworks. What emerges from an analysis of these sheets is a particular attention both to the shape of the objects and to their chromatic interpretation, in the variety of the working techniques used.

DRAWINGS

While not finding any great confirmation in the models produced, these studies show a particular attention to the juxtaposition of colours, associated to produce contrast or with the same chromatic shades.

>
Carlo Scarpa, studies for glass pieces, graphite and watercolour on paper, c. 1927, AMuVe

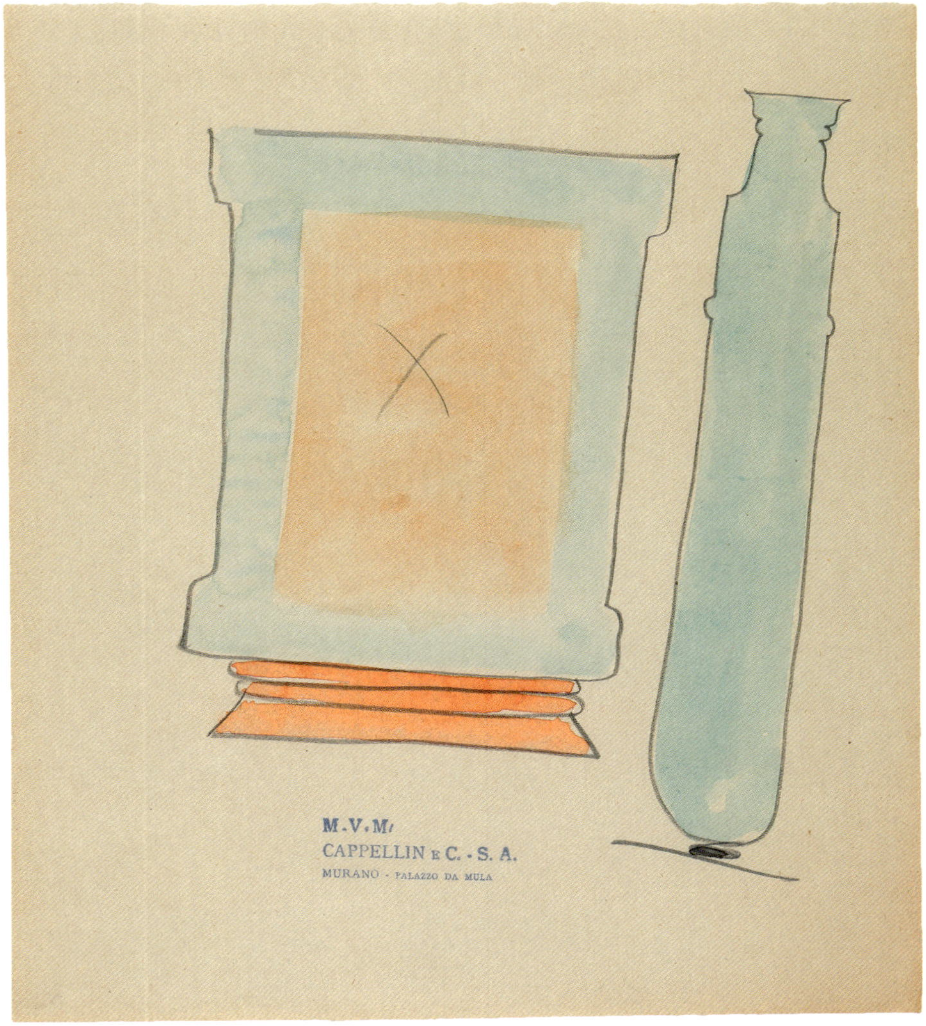

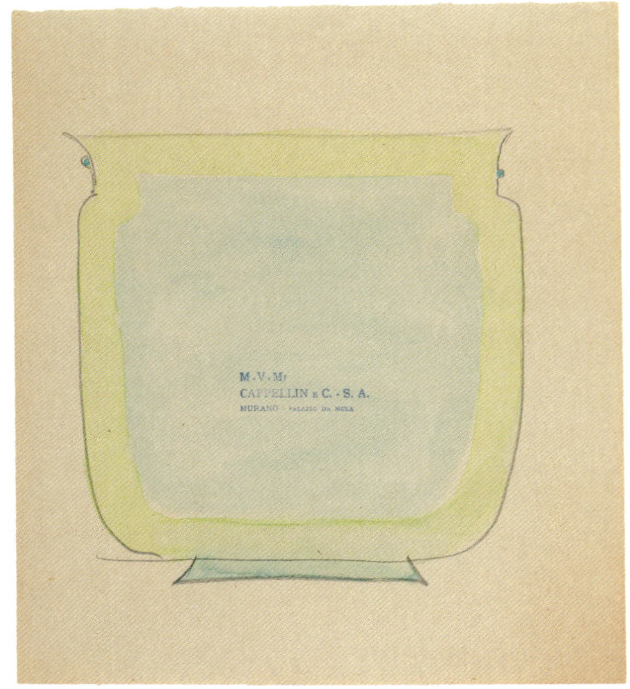

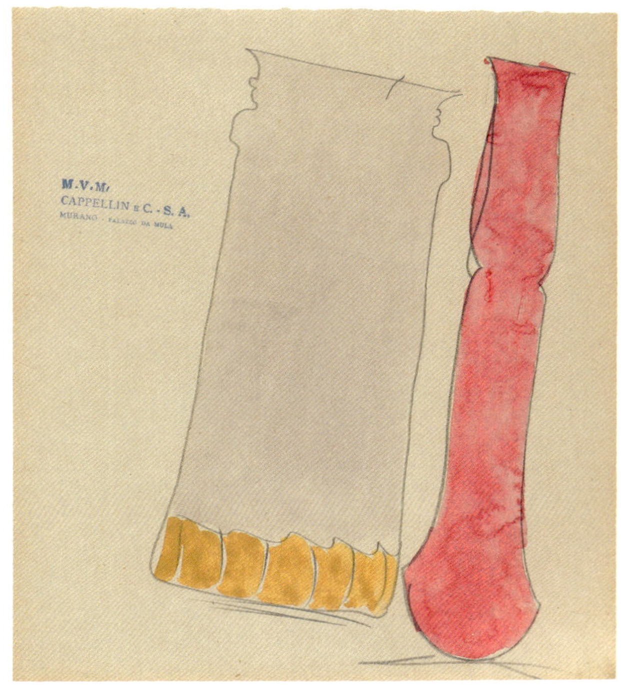

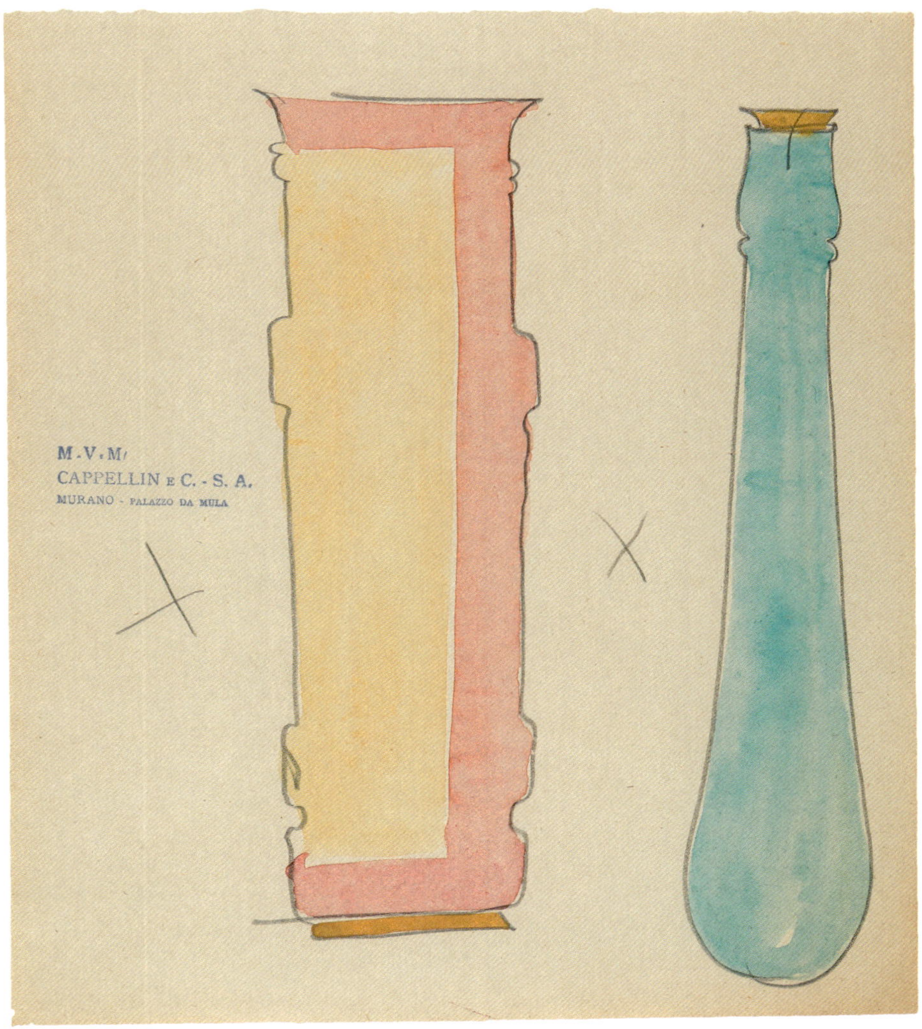

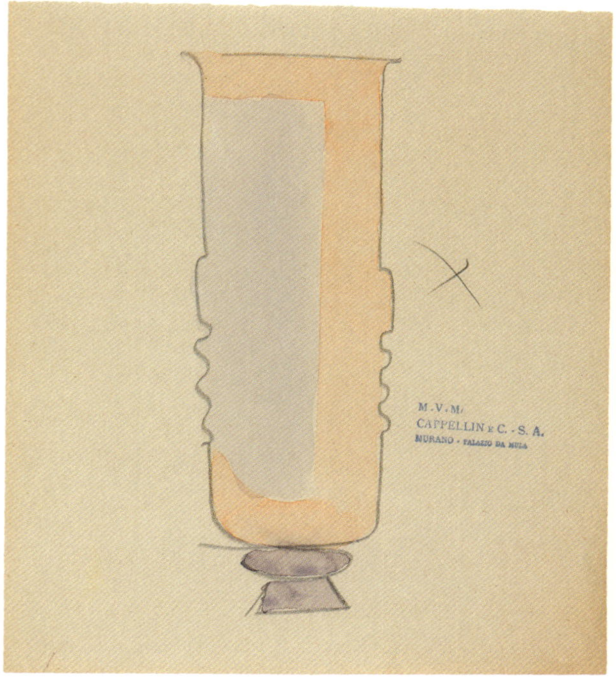
M.V.M.
CAPPELLIN E C. - S. A.
MURANO - PALAZZO DA MULA

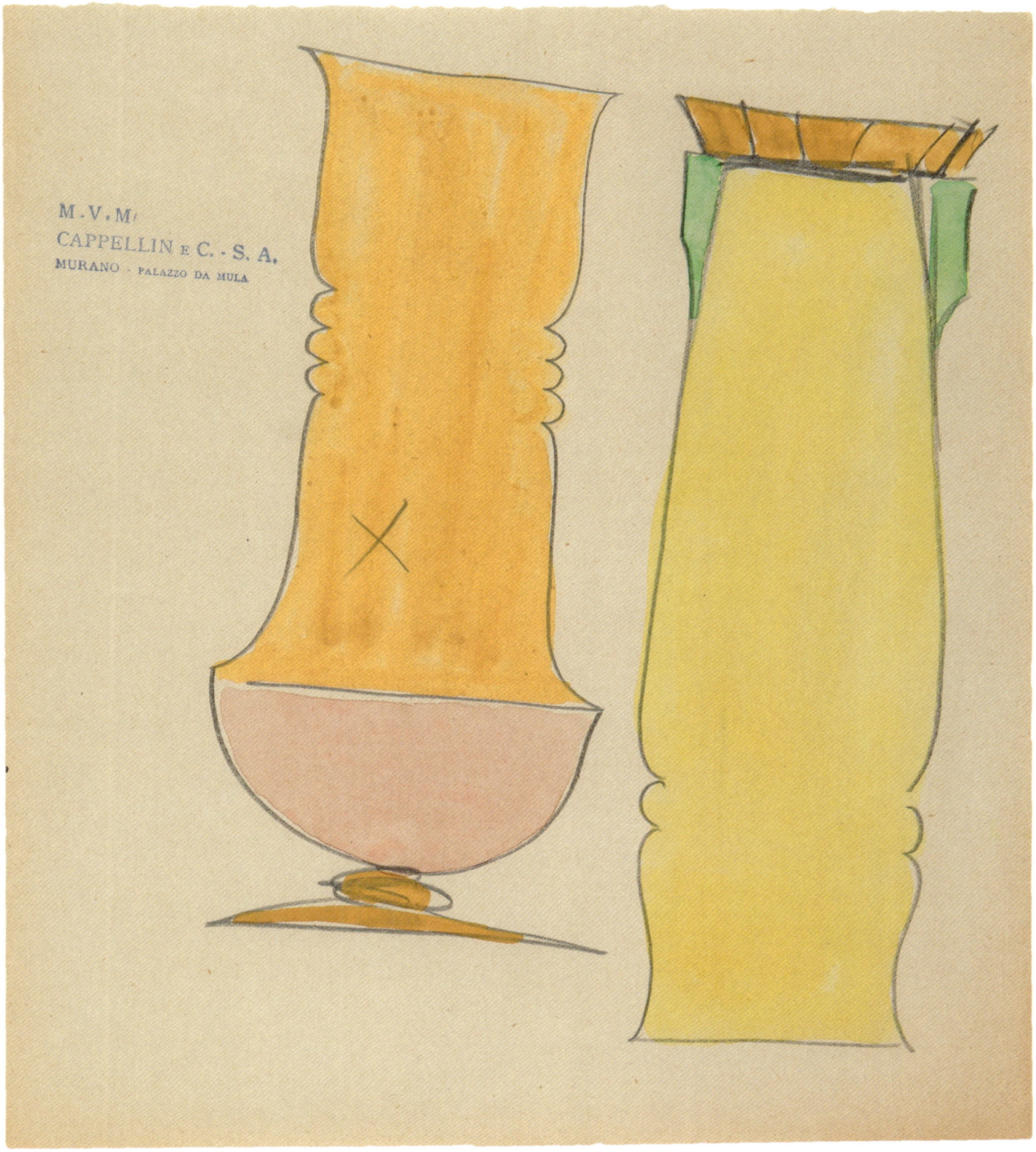
M.V.M.
CAPPELLIN E C. - S. A.
MURANO - PALAZZO DA MULA

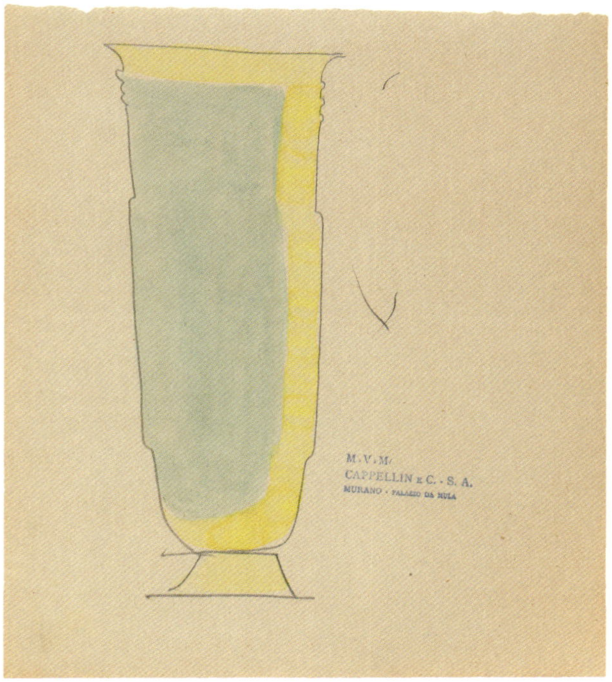
M.V.M.
CAPPELLIN E C. - S. A.
MURANO - PALAZZO DA MULA

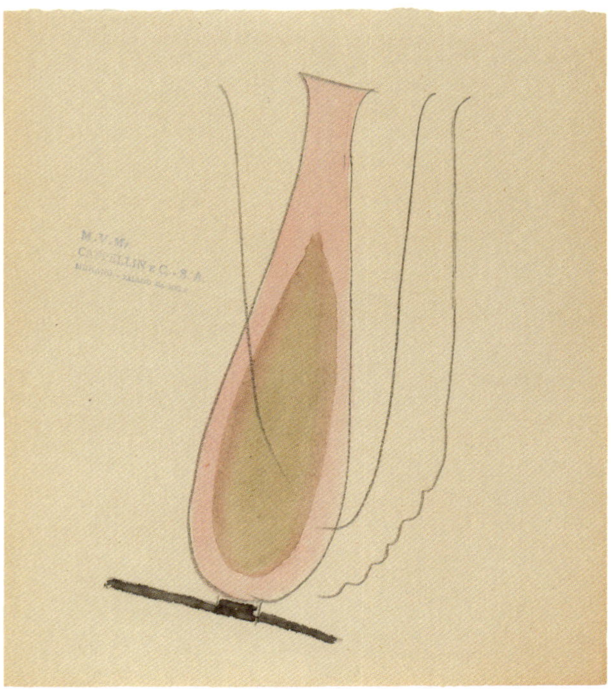
M.V.M.
CAPPELLIN E C. - S. A.

>
Carlo Scarpa, studies for glass pieces, graphite and watercolour on paper, c. 1927, AMuVe

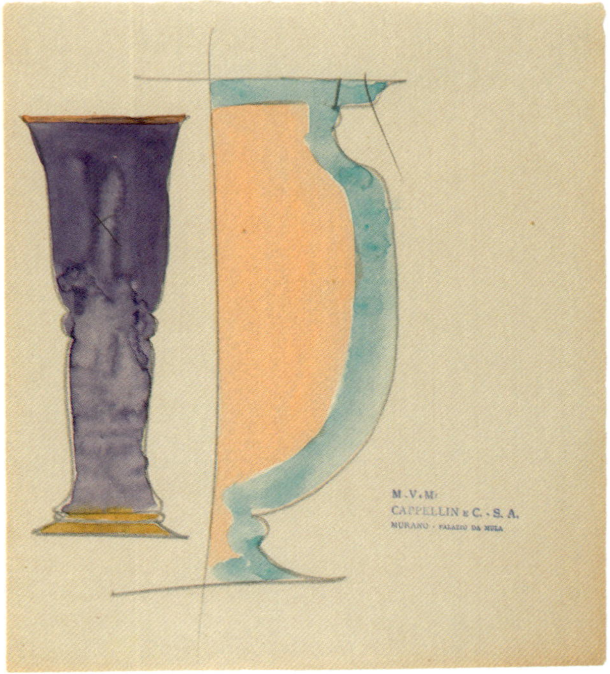

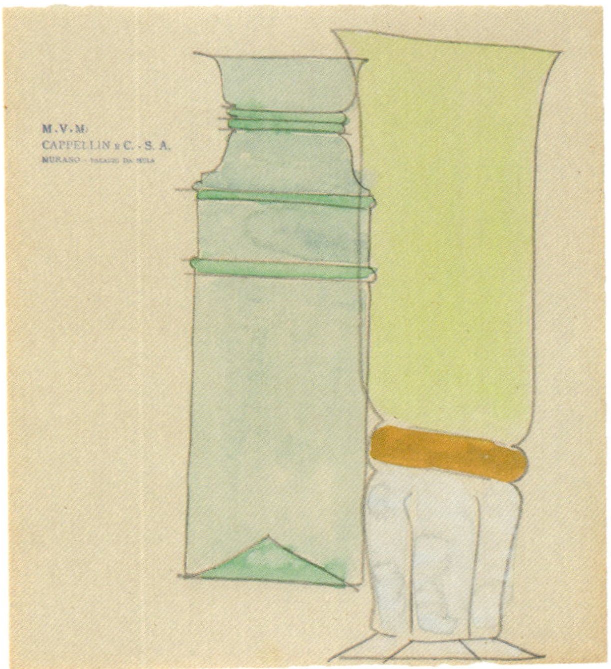

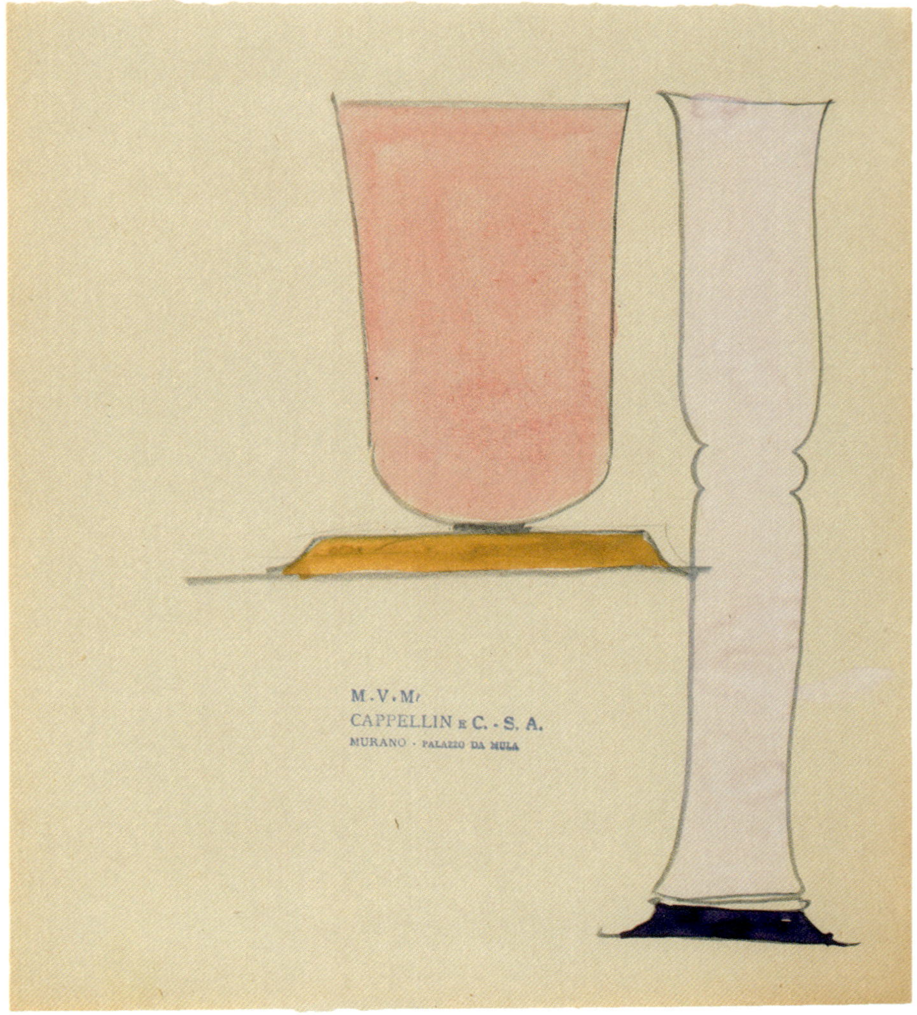

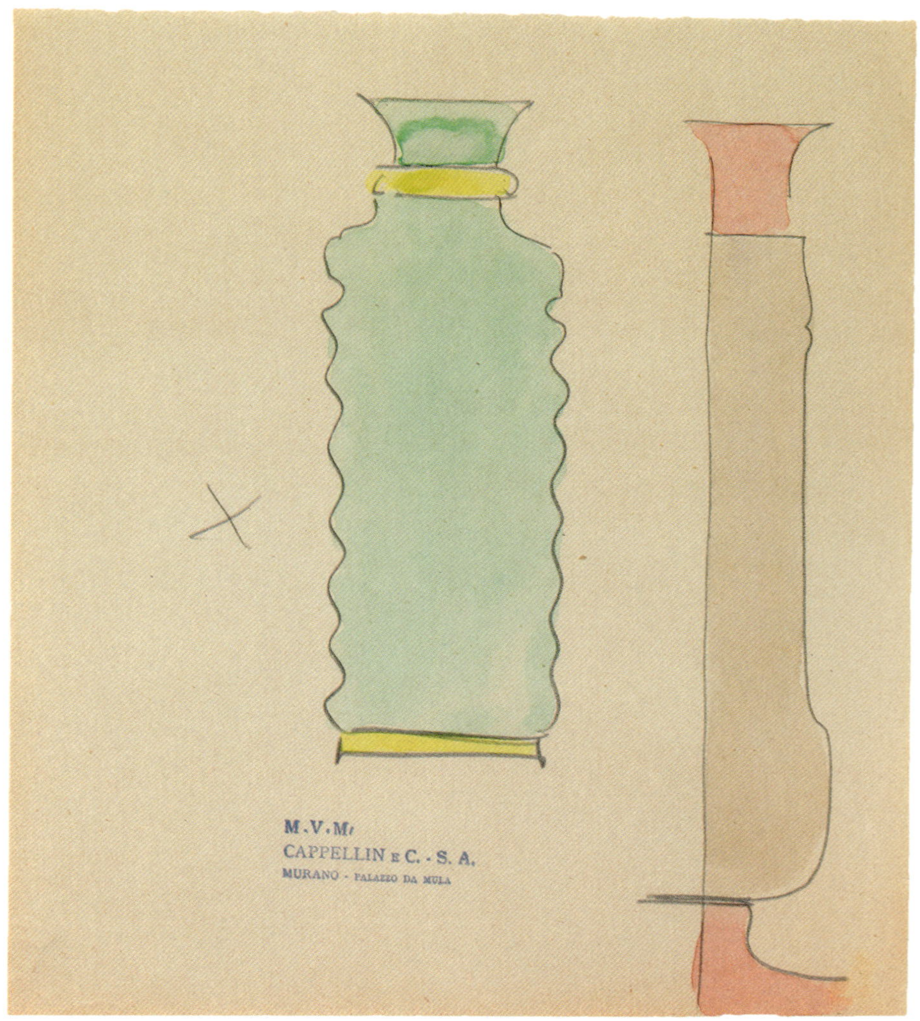

The examples of this series illustrate, in the characteristic drawing style of the author, proposals which would partially find concrete form in some glass pieces exhibited in 1930 at the 4th Triennale of Monza.

>
Carlo Scarpa, studies for glass pieces, charcoal and sanguine on paper, 1929-30, AMuVe

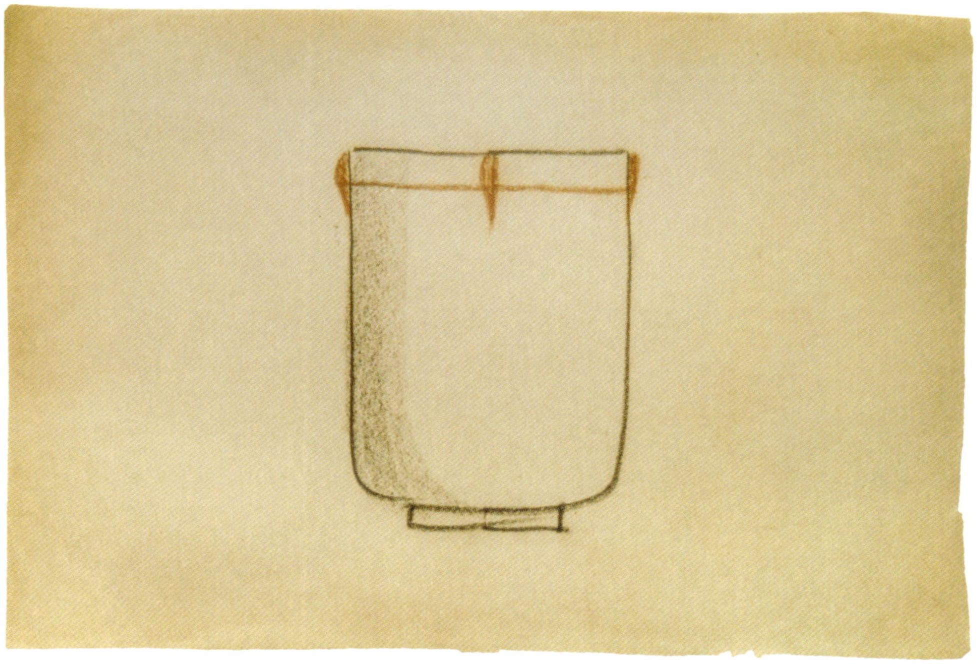

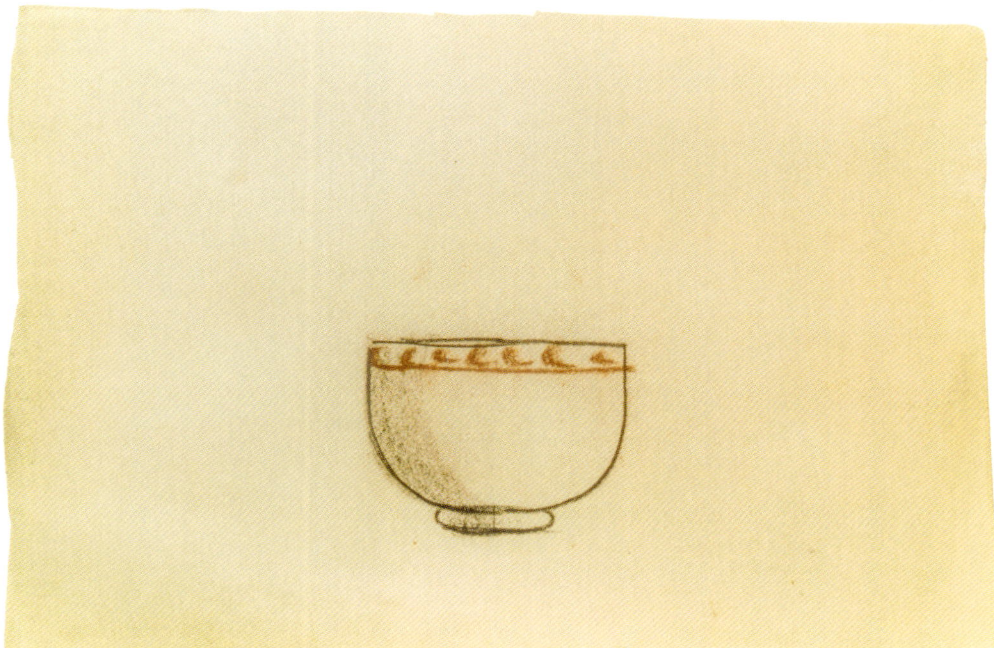

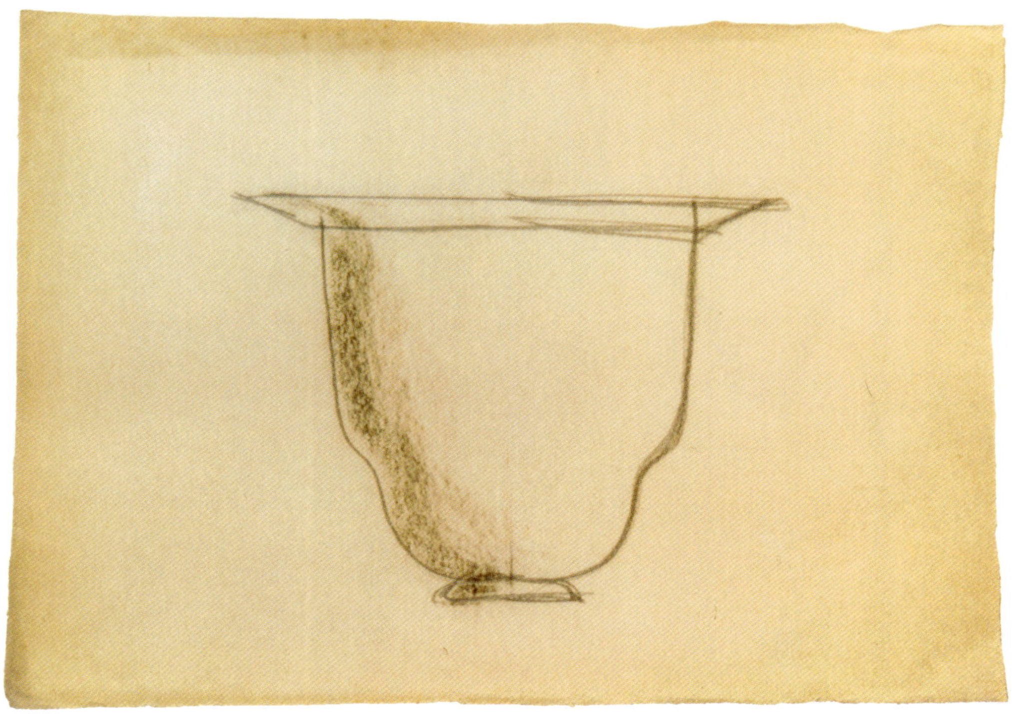

These compositions are a selection of drawings for new models for possible insertion in the catalogue. Evident in them is a search for form which, in the majority of the cases illustrated, is combined with a self-confident use of ancient techniques, such as *filigrana a retortoli*, *decoro fenicio* and *murrine*.

>
Carlo Scarpa, drawings for glass pieces, graphite on paper, c. 1930, AMuVe

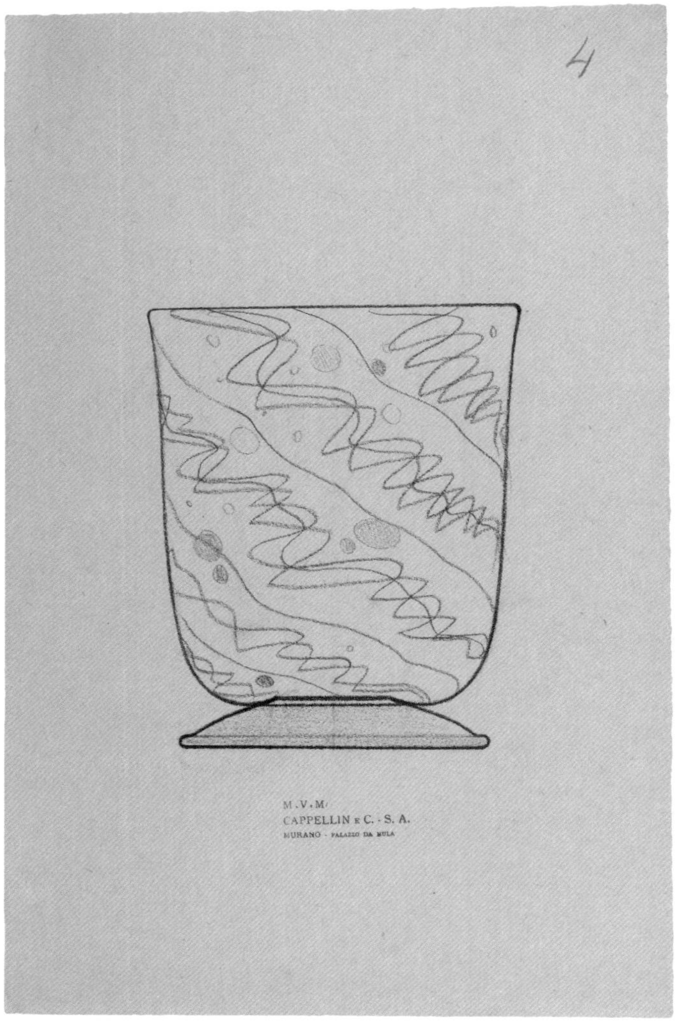

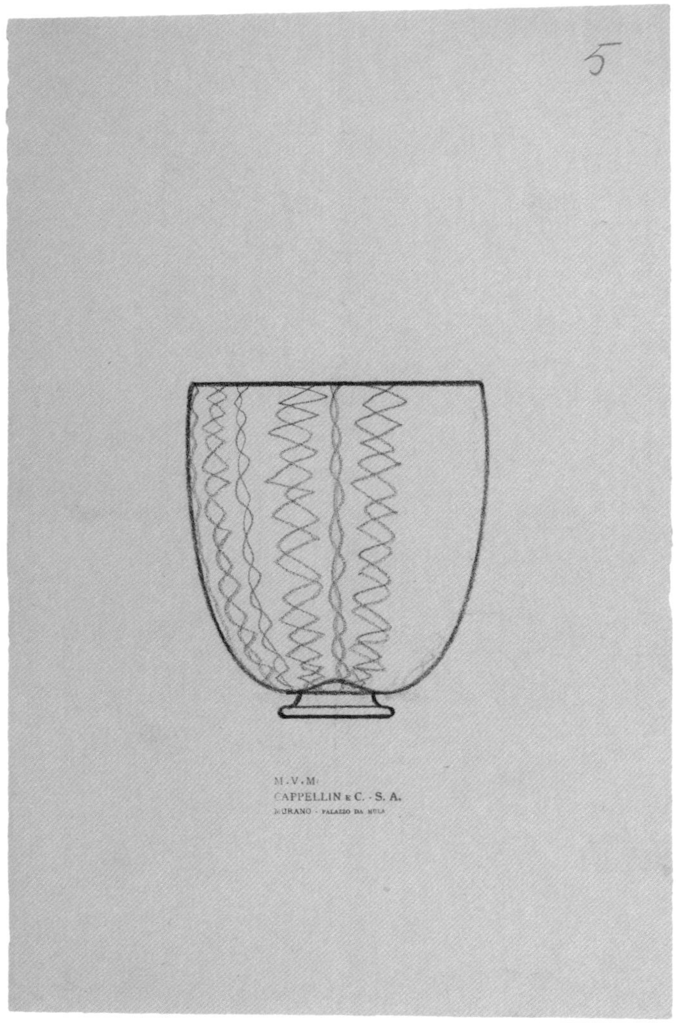

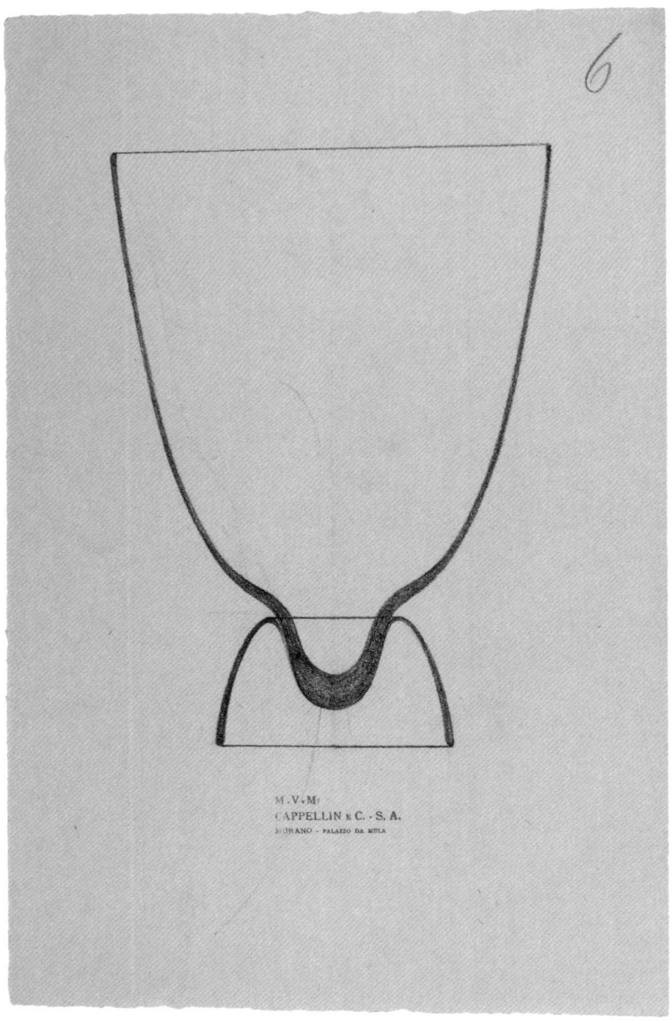

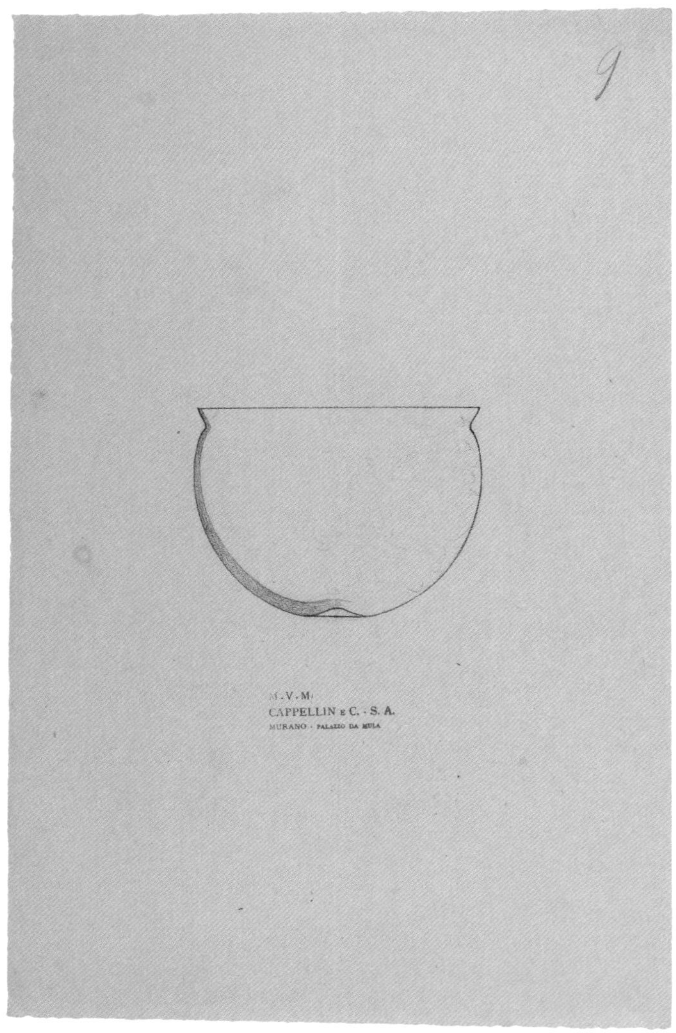

>
Carlo Scarpa, drawings for bowls, graphite on paper, c. 1930, AMuVe

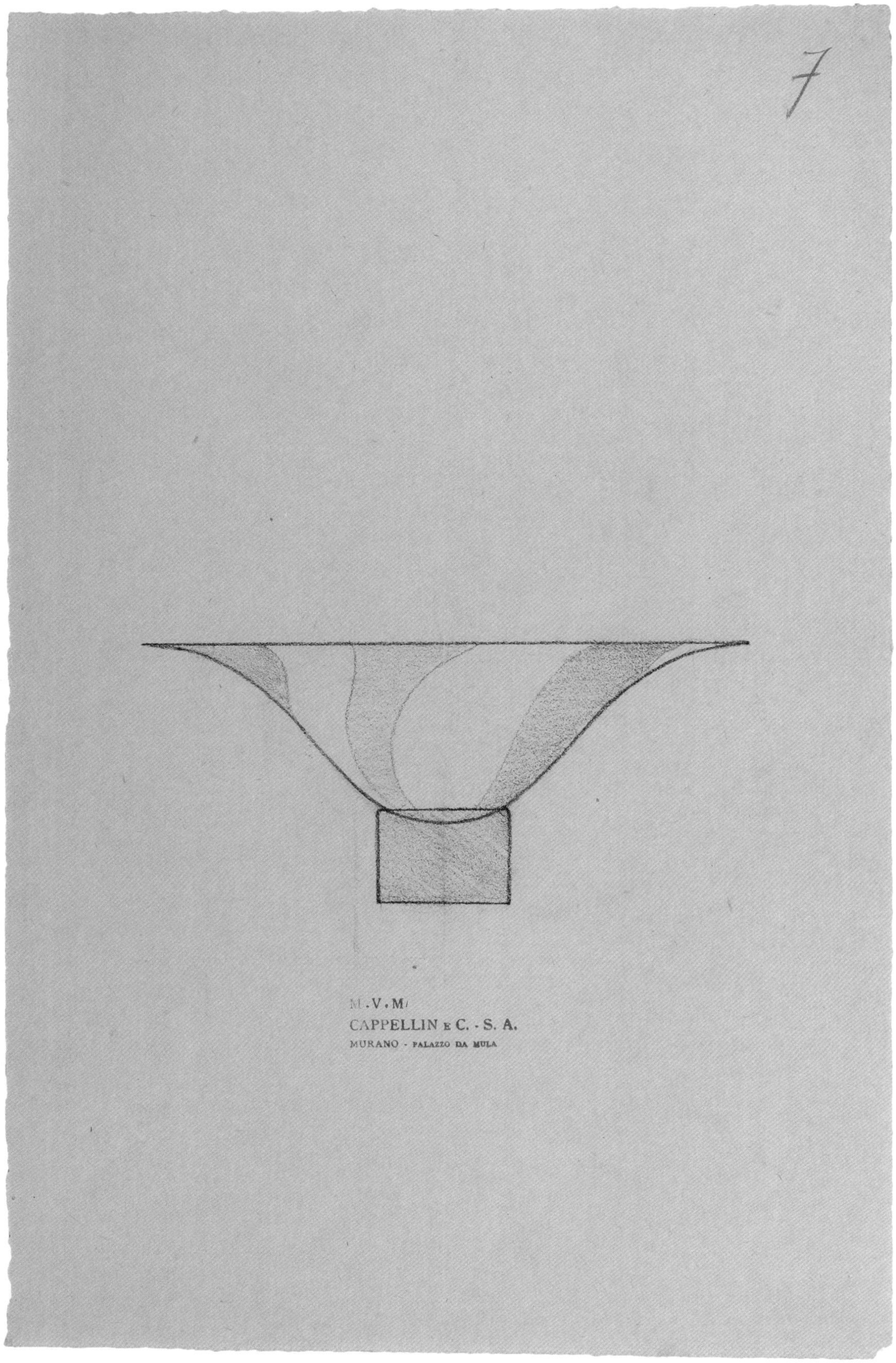

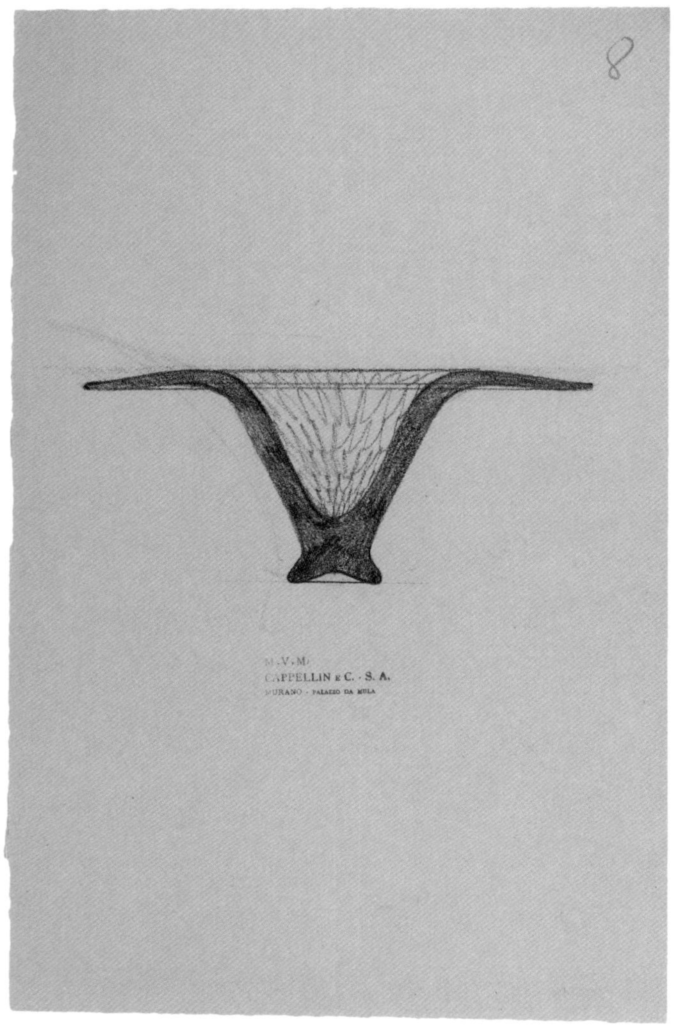

>
Carlo Scarpa, drawings for glass pieces with filigrana a retortoli *and* murrine*, graphite on paper, c. 1930, AMuVe*

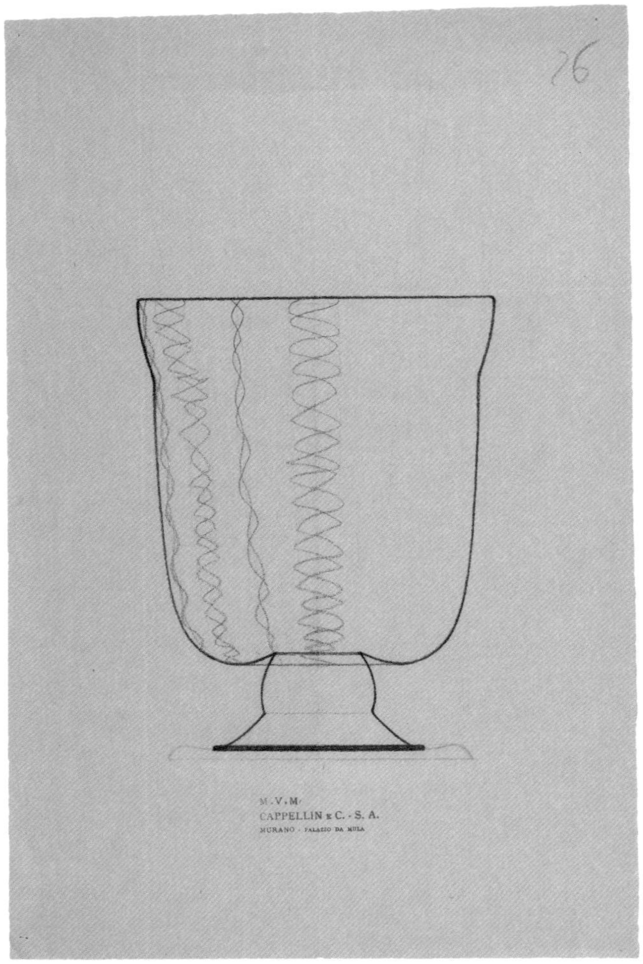

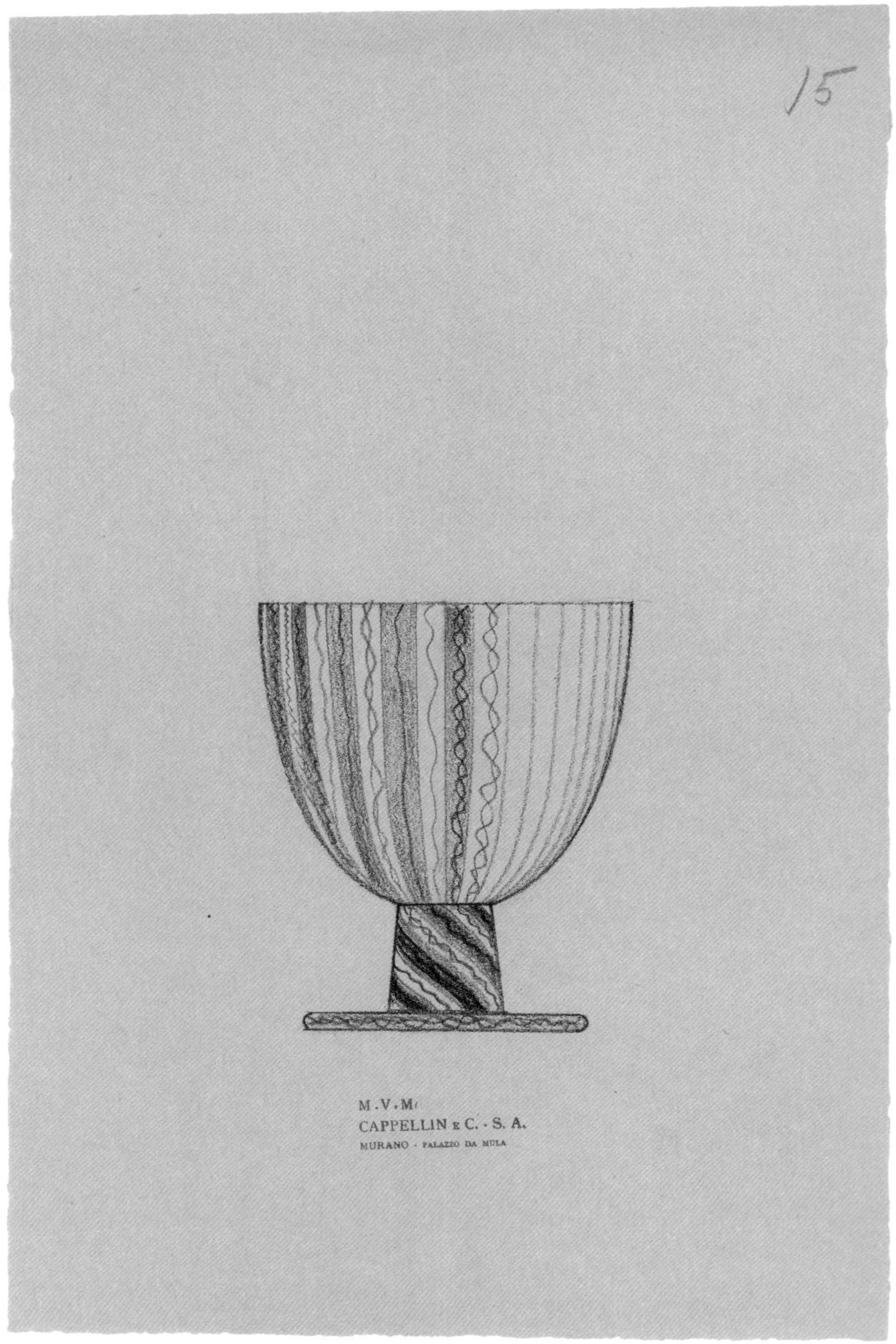

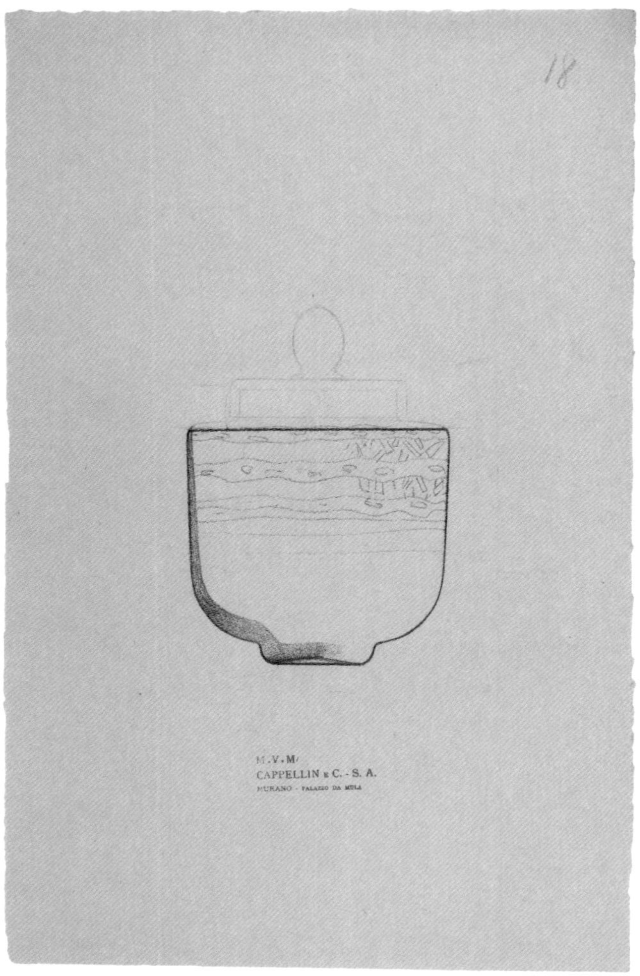

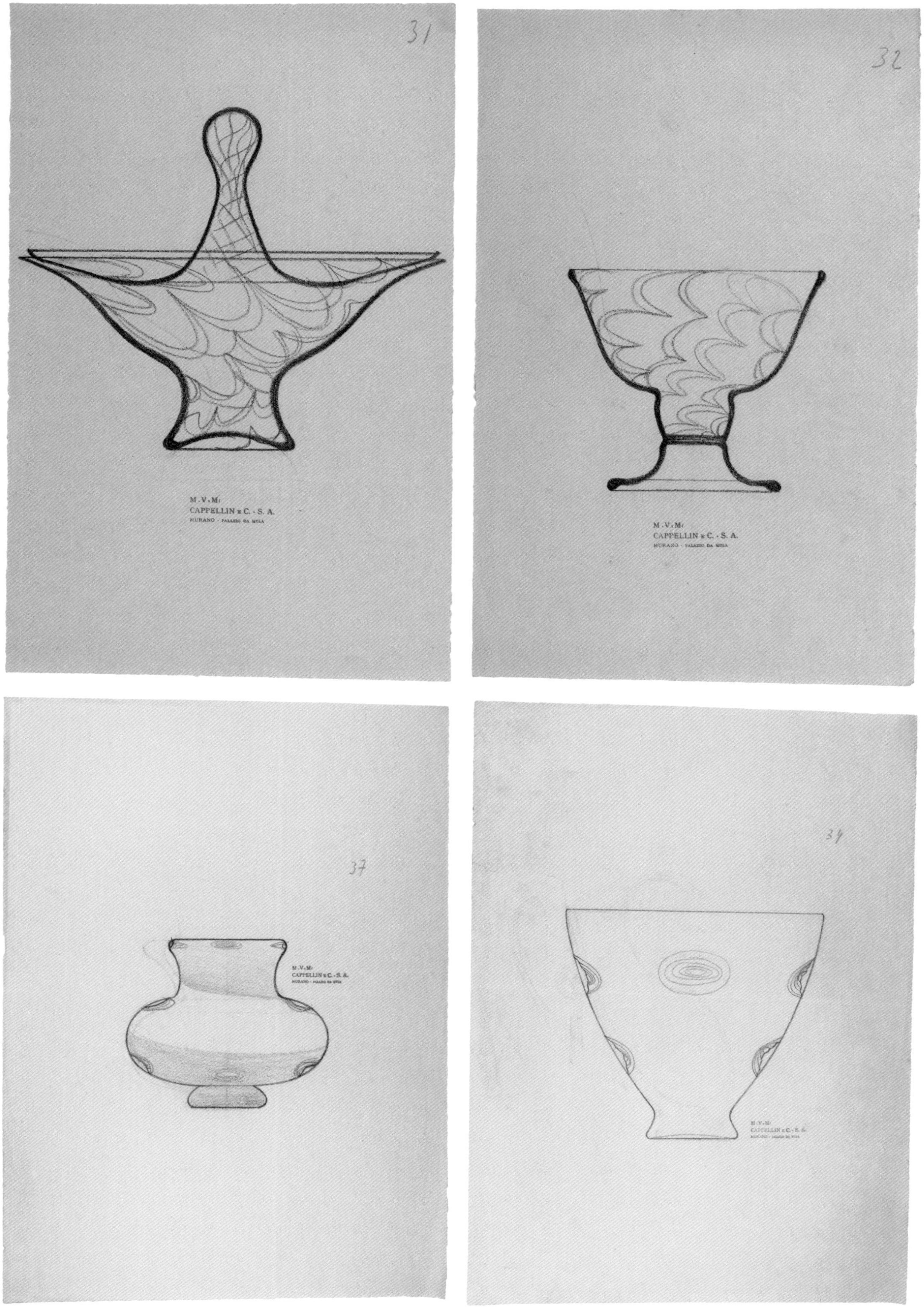

>
Carlo Scarpa, drawings for glass pieces with decoro fenicio *and decoration with* murrine*, graphite on paper, c. 1930, AMuVe*

These are some studies for new models of glass objects with *decoro fenicio*, *filigrana a reticello* and applied threads. It is legitimate to suppose in particular that those with *filigrana a reticello* are a prelude to those with a foot and *filigrana lattimo* exhibited in 1930 at the 4th Triennale of Monza.

>

Carlo Scarpa, studies for glass pieces, graphite and coloured pencil on paper, c. 1930, AMuVe

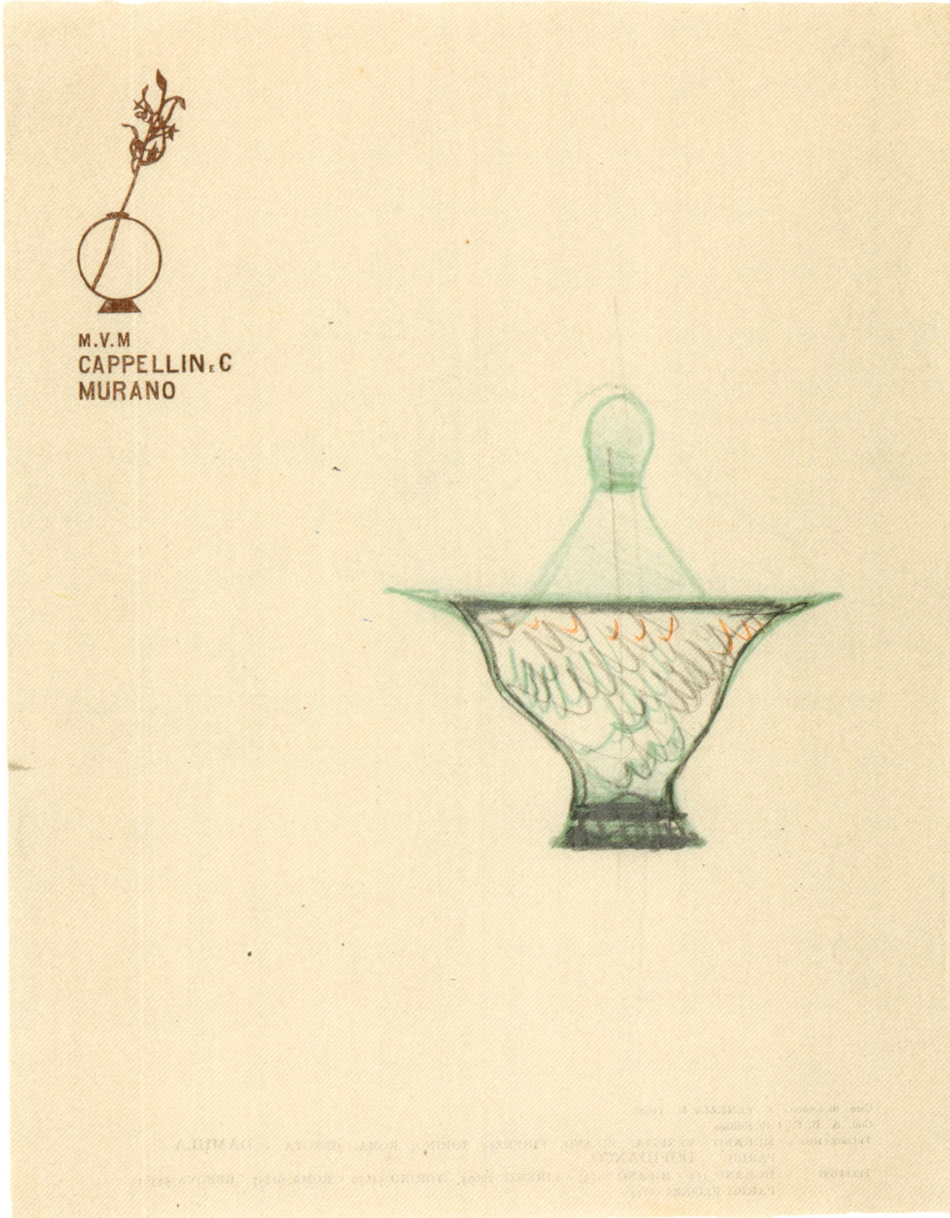

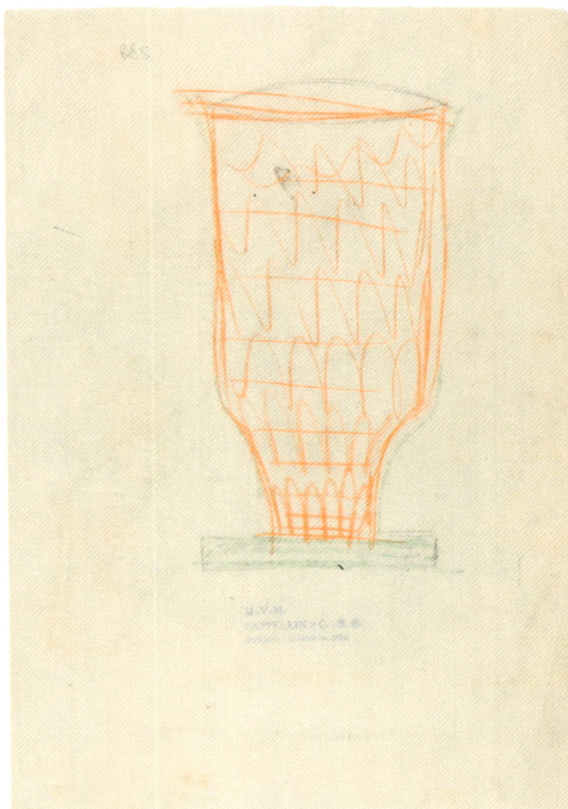

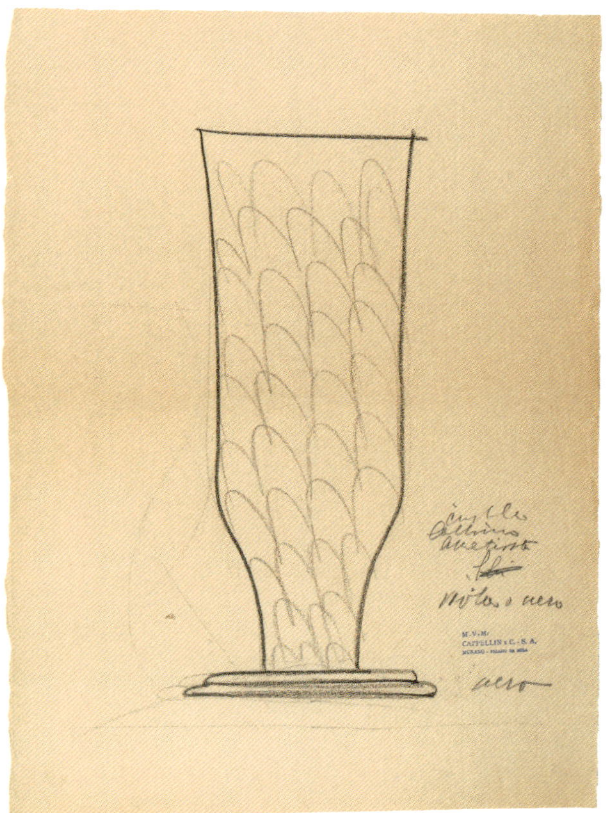

>
Carlo Scarpa, studies for glass pieces, graphite on paper, c. 1930, AMuVe

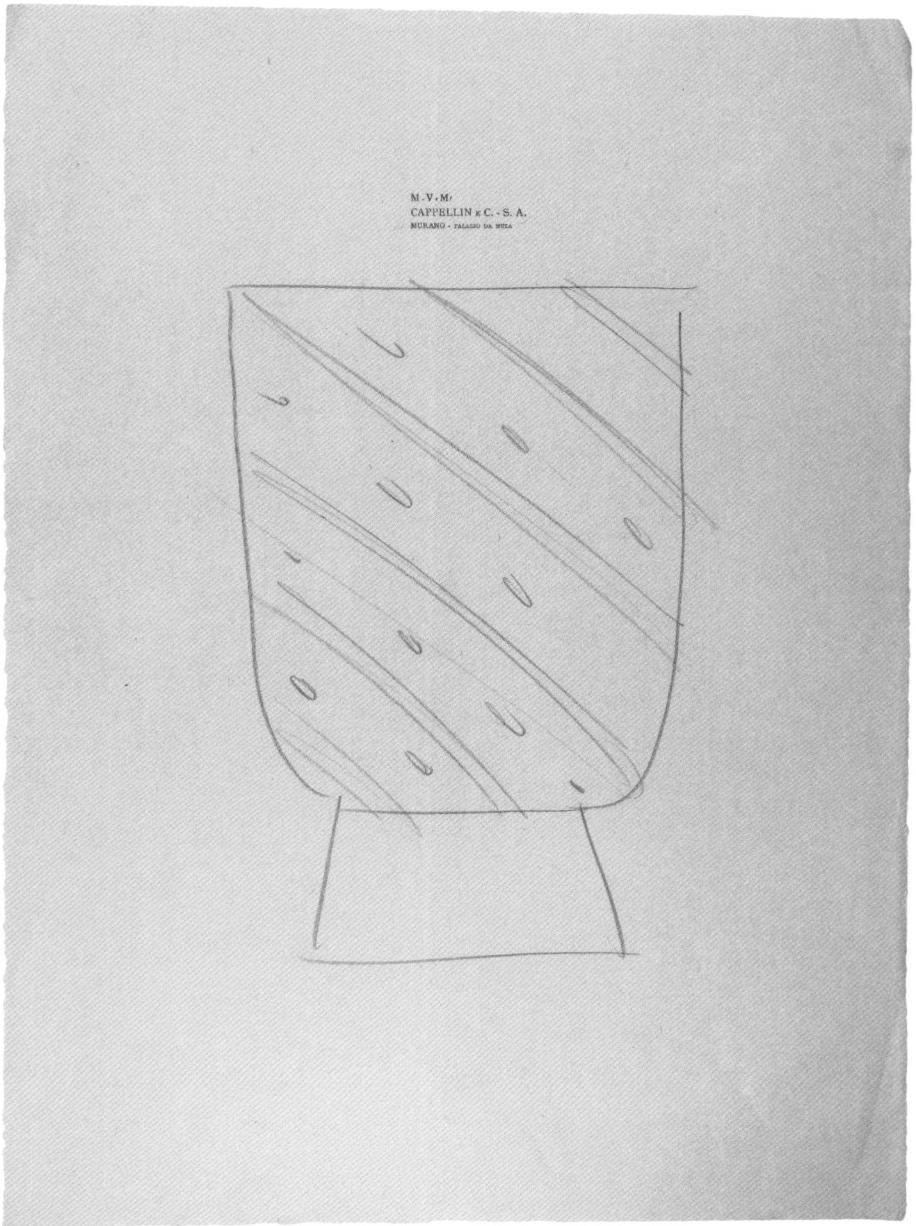

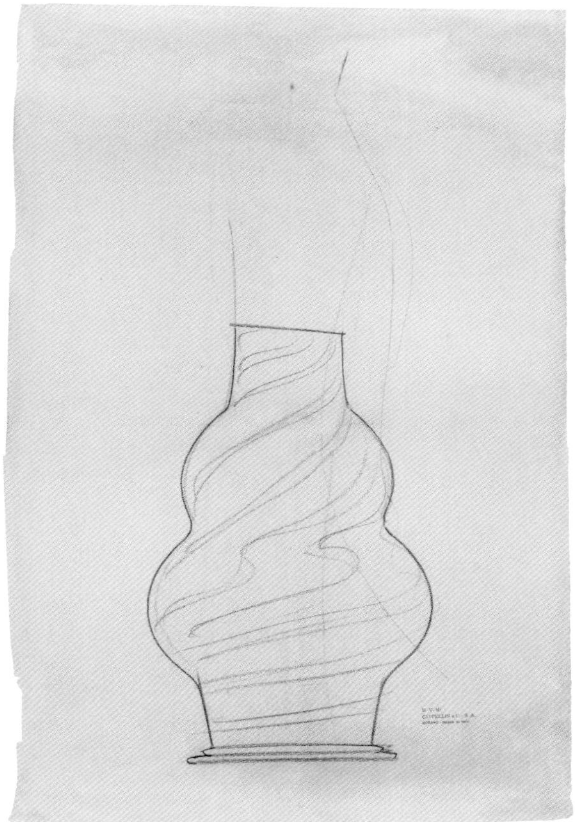

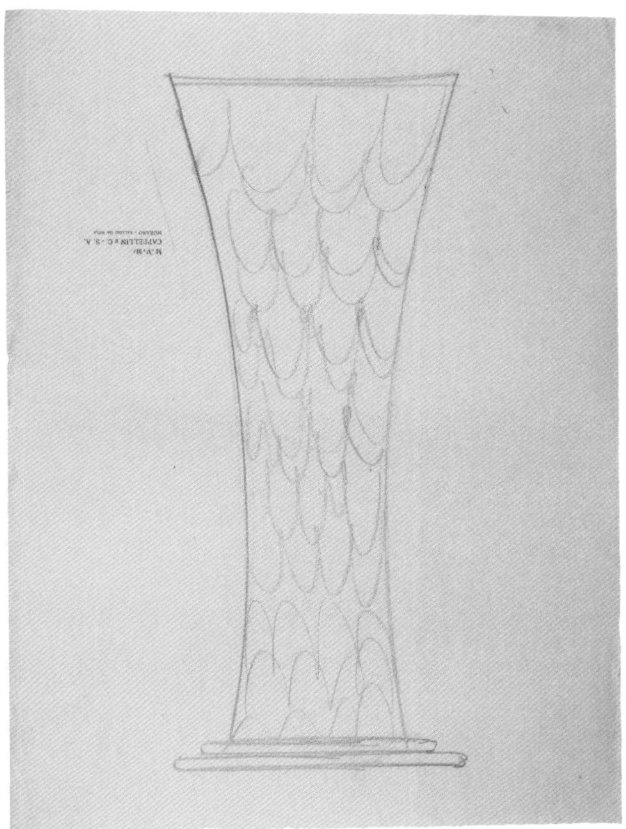

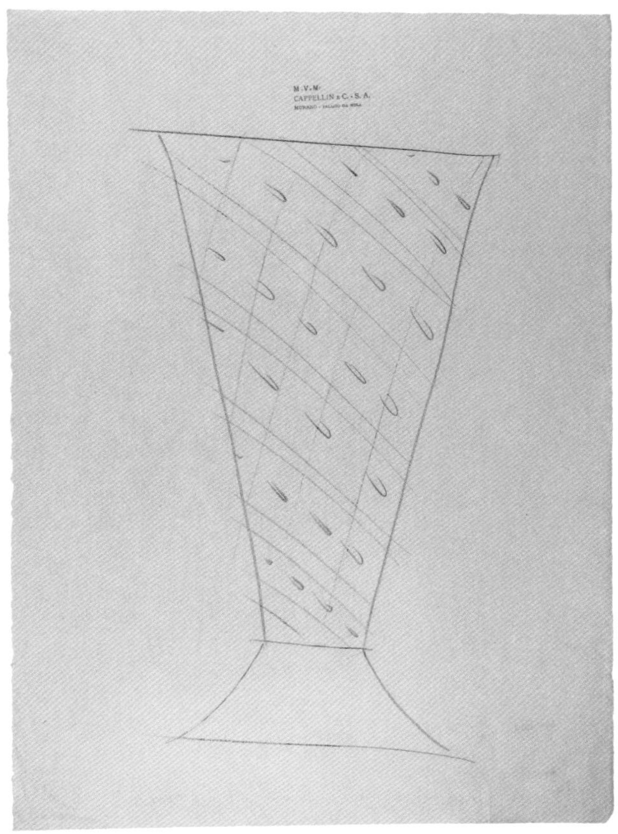

The studies and drawings for glass pieces and lighting illustrated here document research developed between 1929 and 1931, where the interest in colour in the definition of exceptional glass compositions is evident. The proposals for polychrome artefacts with inserted canes and murrine are highly original in the composing of original decorative motifs, sometimes of a figurative kind inspired by art déco, other times pronouncedly abstract.

>
Carlo Scarpa, studies for lamps and studies for vases, graphite and watercolours on paper, c. 1930, ED Gallery, Piacenza (above)

>
Carlo Scarpa, drawing and studies for glass pieces, graphite and watercolours on paper, 1930, C. Moretti and G. Moretti Fund, Centro Studi del Vetro, Giorgio Cini Foundation, Venice (below)

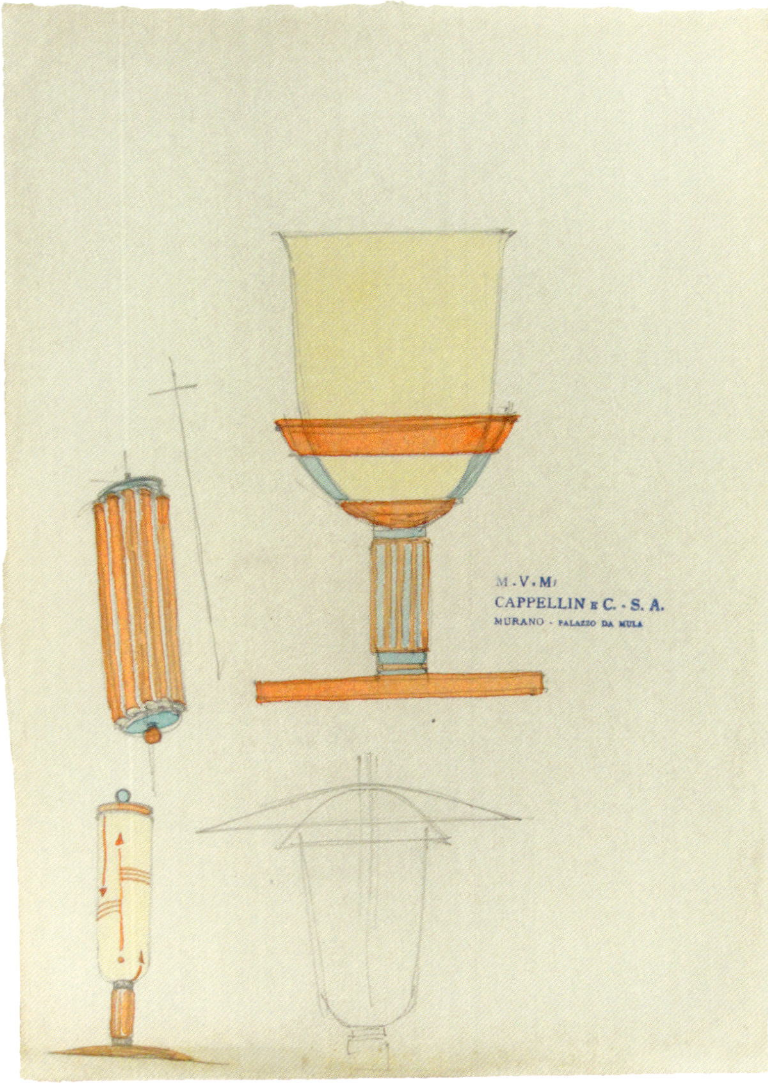

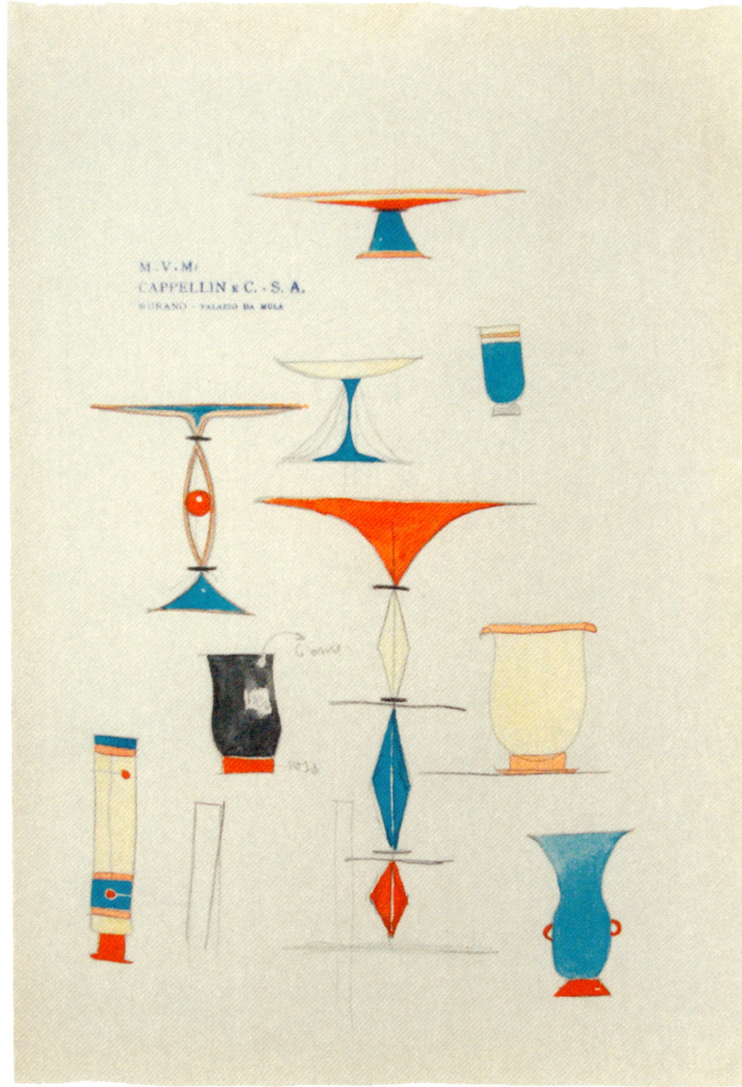

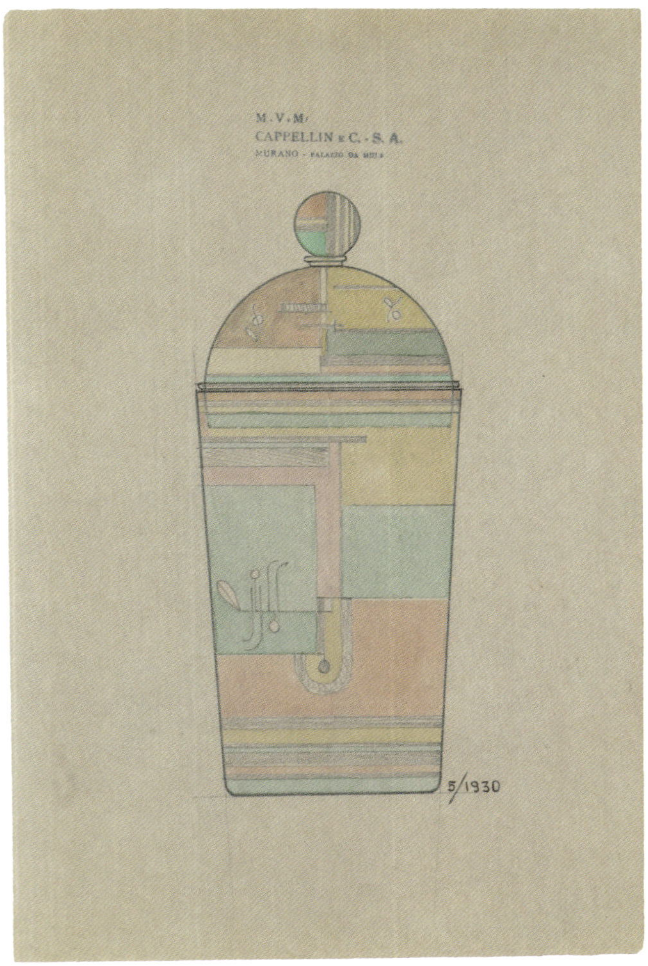

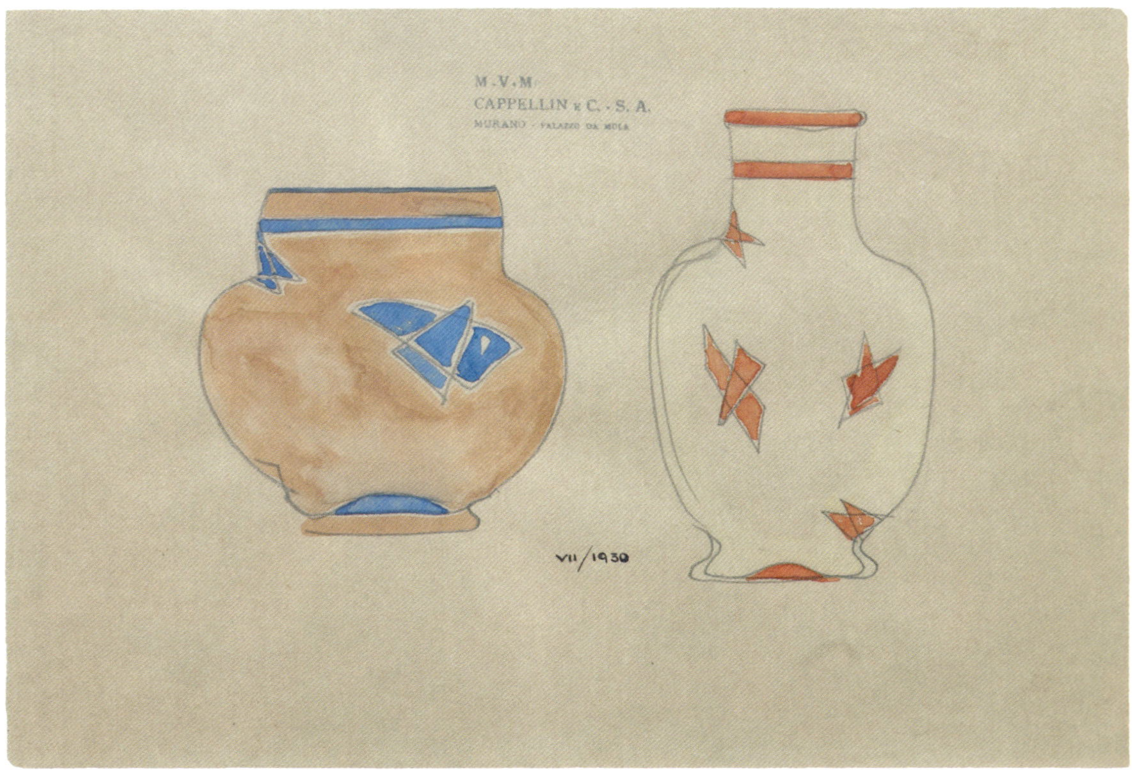

>
Carlo Scarpa, studies for glass pieces, graphite and water colours on paper, 1930-31, ED Gallery, Piacenza

>
Carlo Scarpa, studies for lamps, graphite and watercolours on paper, 1930-31, AMuVe

>
Carlo Scarpa, study for vase, graphite and watercolours on paper, 1930-31, AMuVe

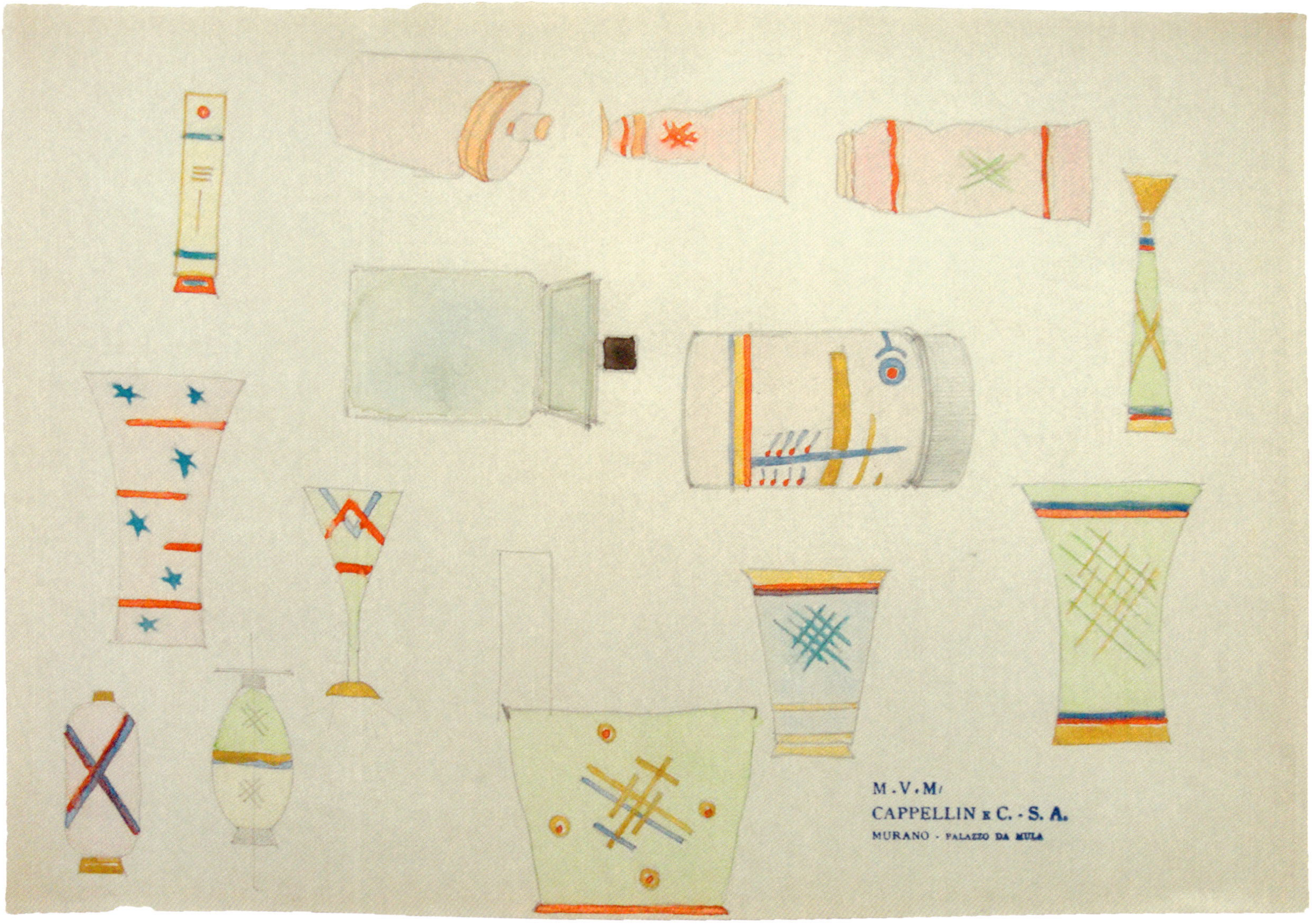

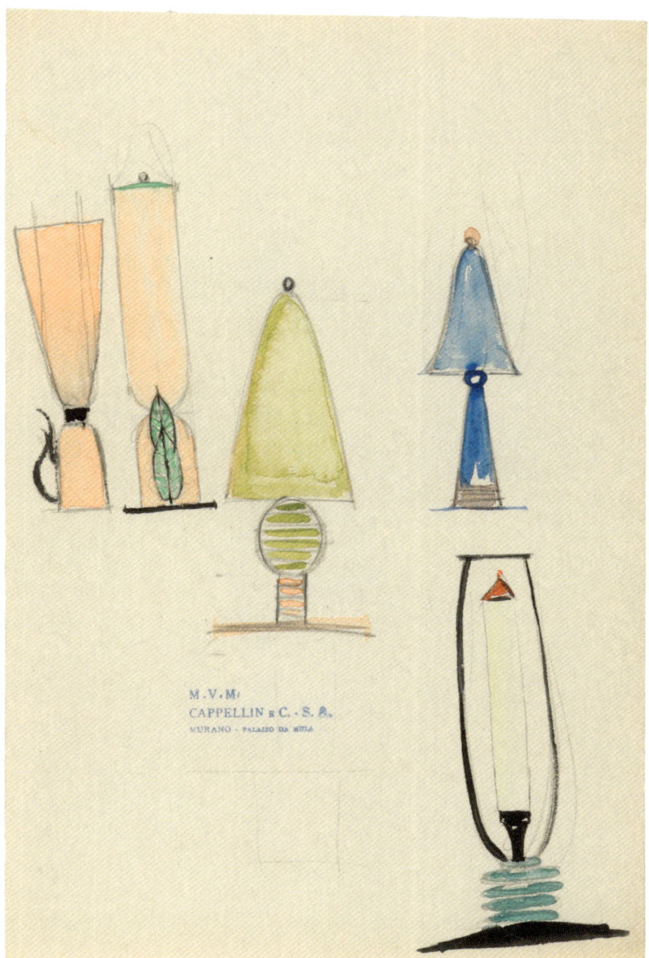

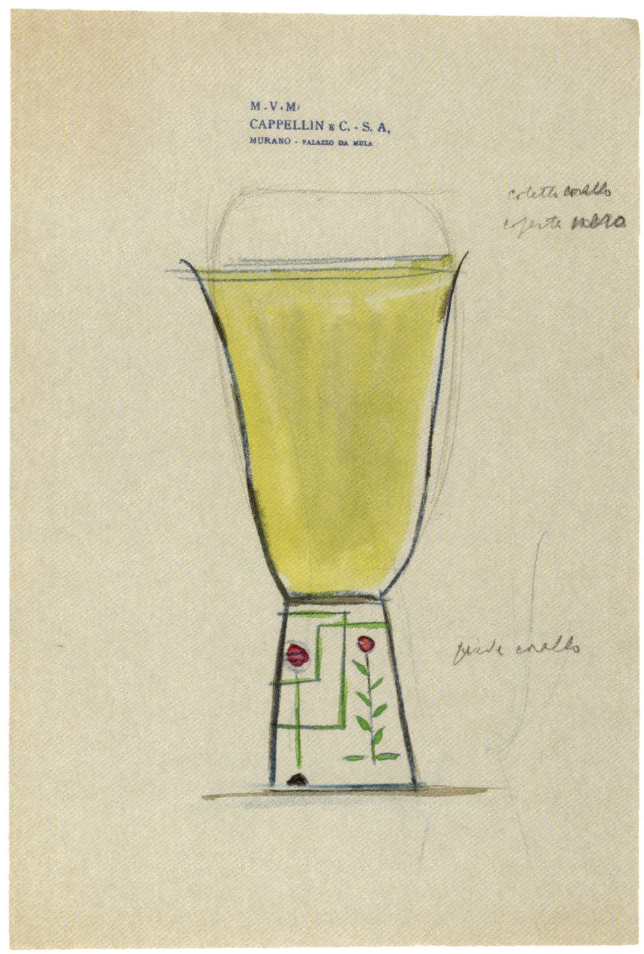

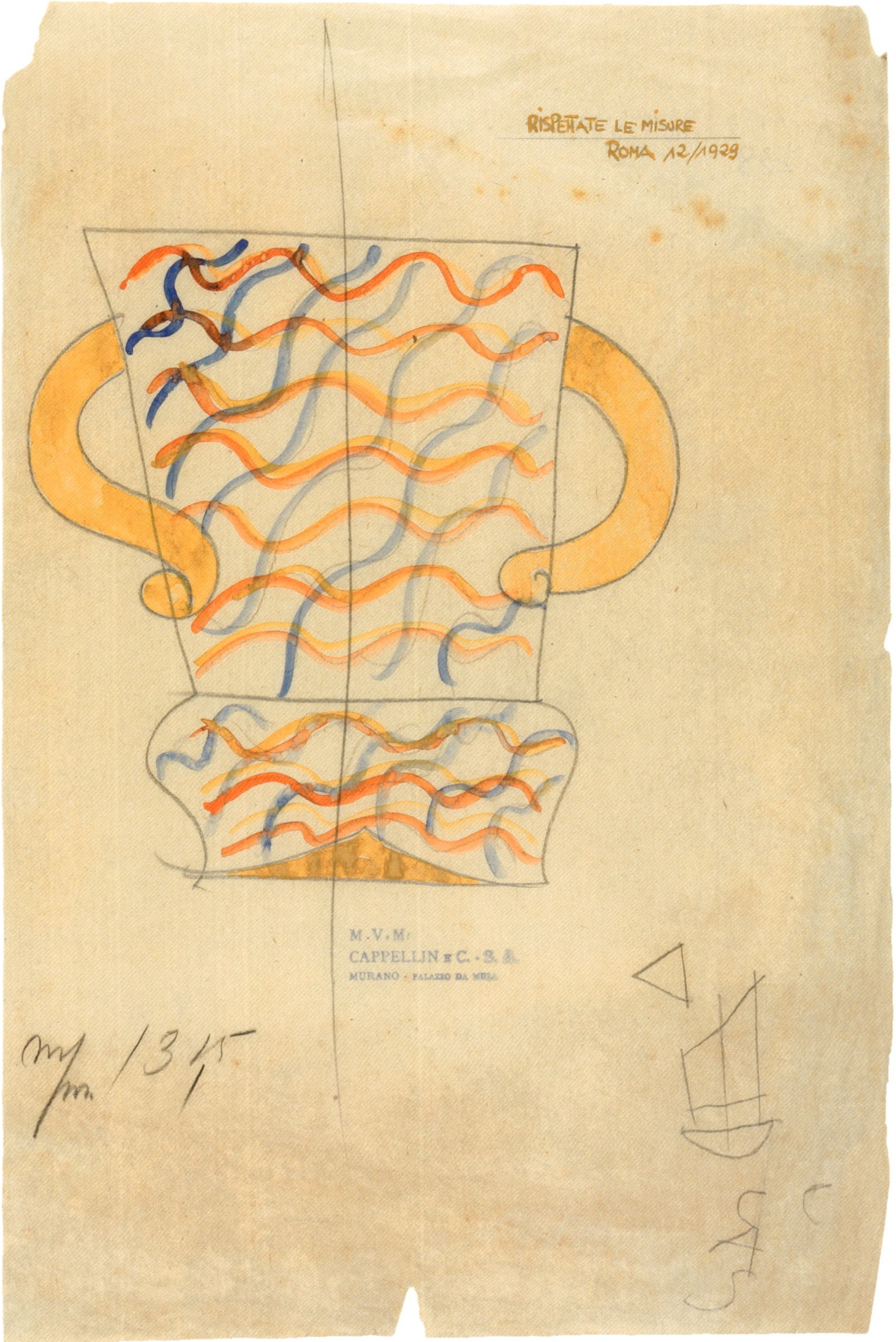

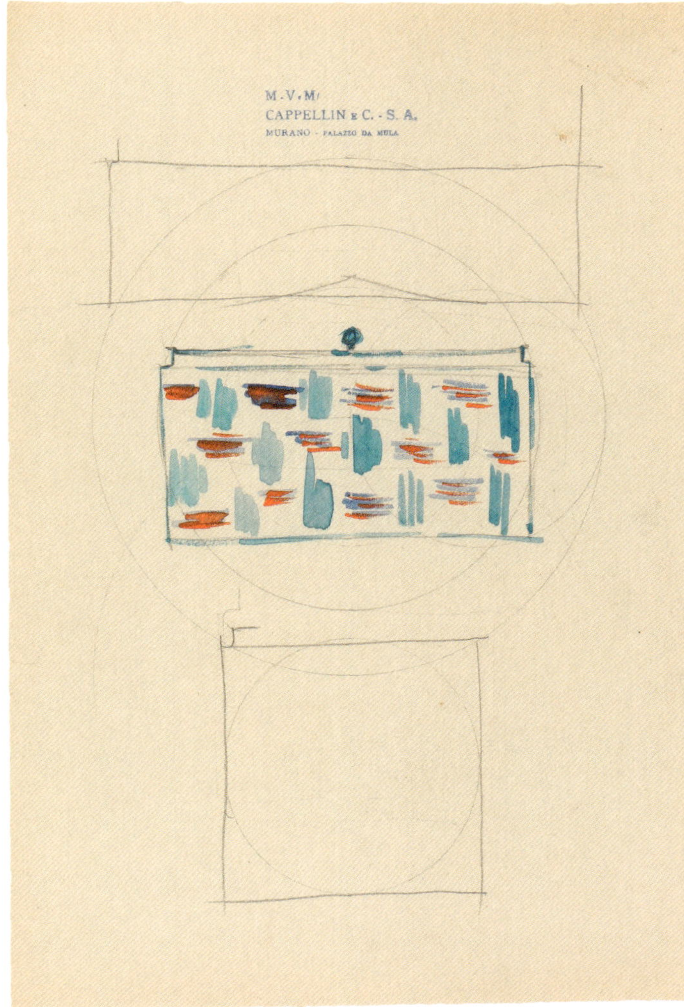

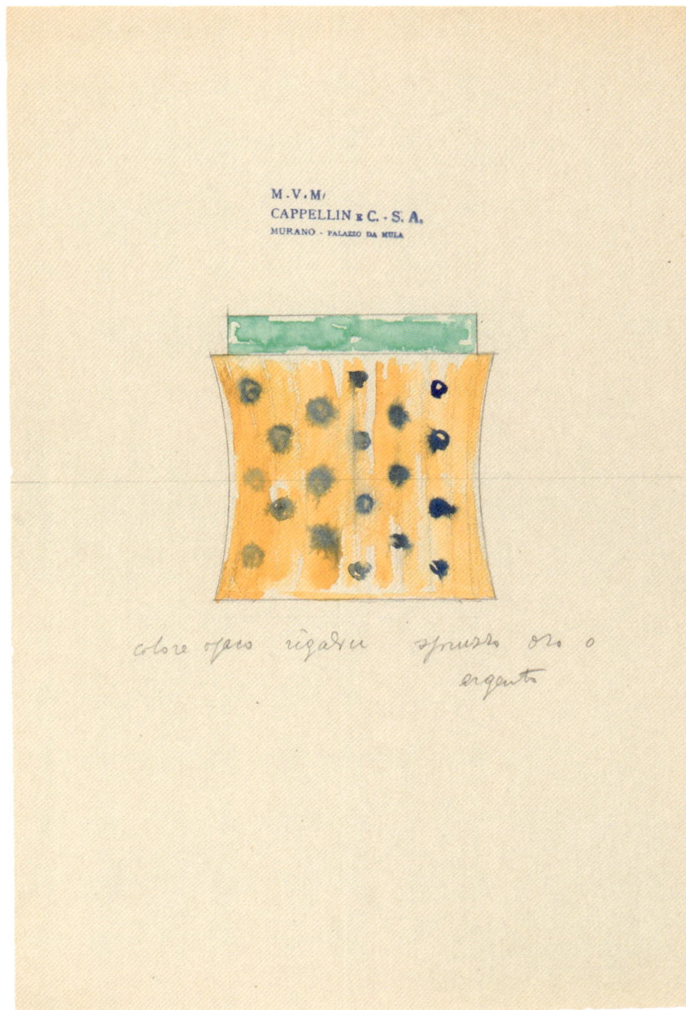

>
Carlo Scarpa, study for a vase, graphite and water colours on paper, 1929, AMuVe

>
Carlo Scarpa, study for compote bowl, graphite and watercolours on paper, 1930-31, AMuVe

>
Carlo Scarpa, study for glass piece, graphite and watercolours on paper, 1930-31, AMuVe

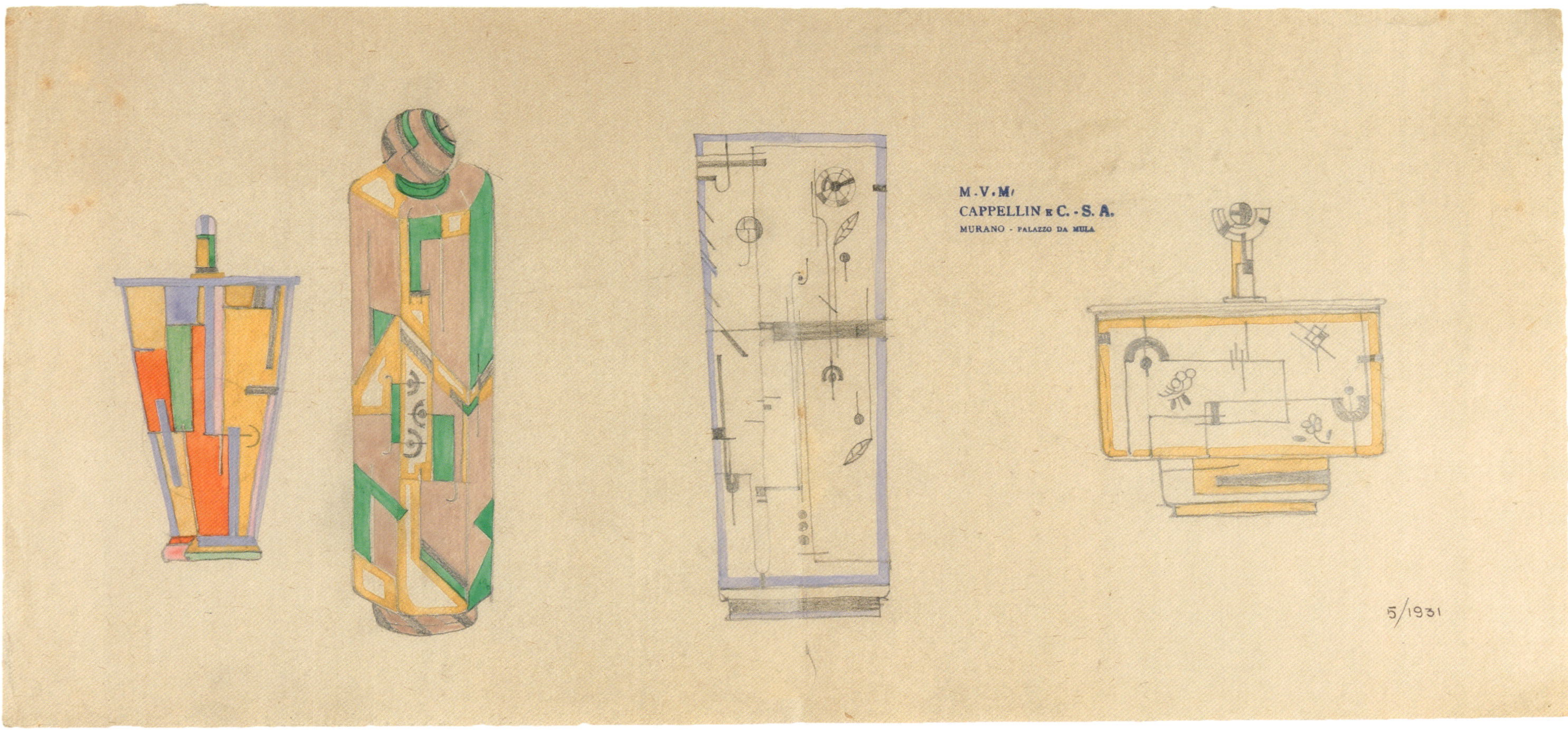

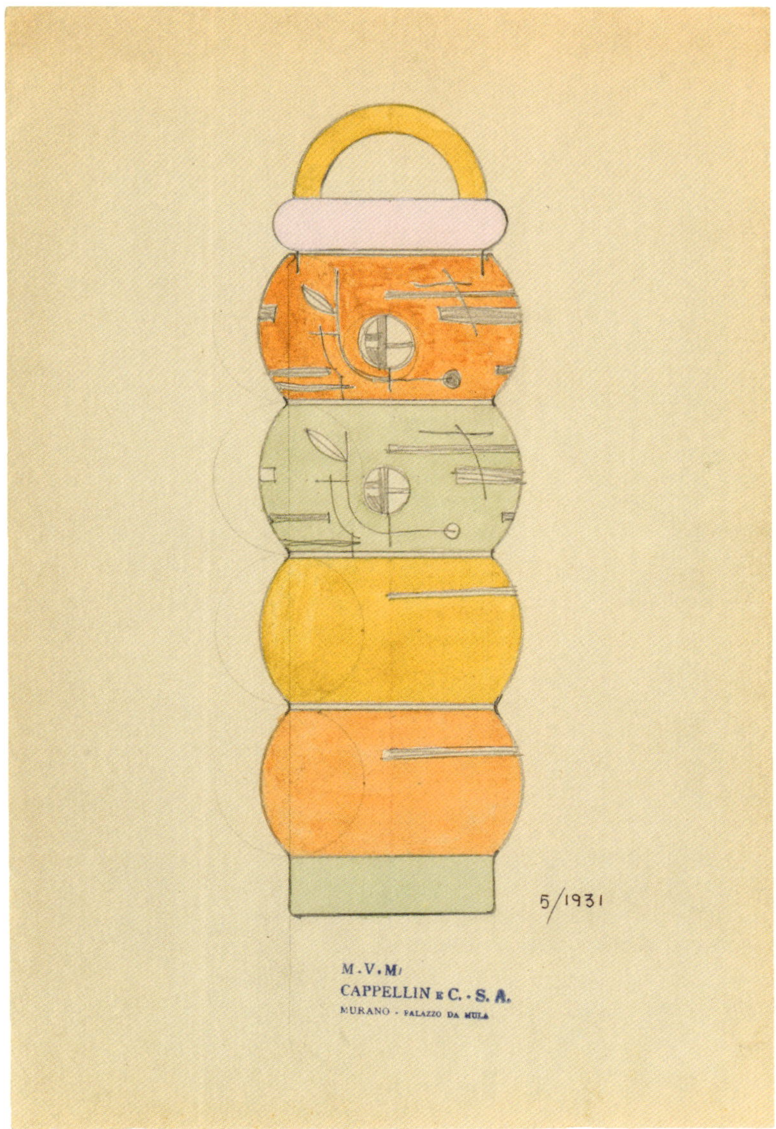

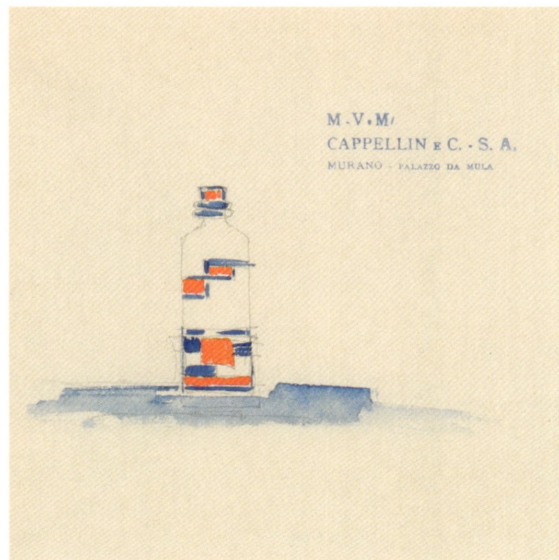

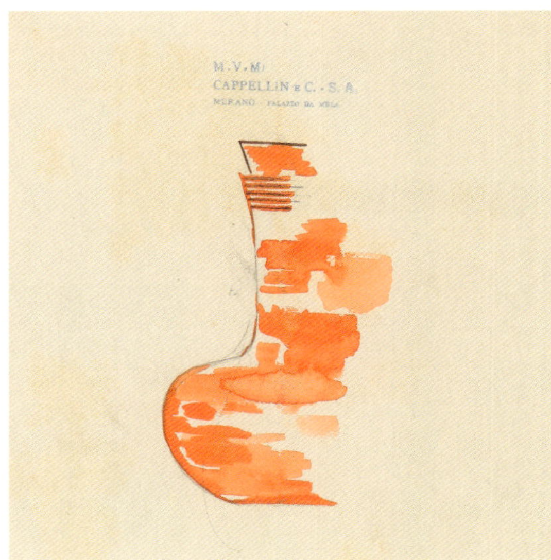

>
Carlo Scarpa, studies for glass pieces, graphite and watercolours on paper, 1931, AMuVe

>
Carlo Scarpa, study for vase, graphite and watercolours on paper, 1931, Marco Arosio collection

>
Carlo Scarpa, study for vase, graphite and watercolours on paper, 1930-31, detail, AMuVe

>
Carlo Scarpa, study for vase, graphite and watercolours on paper, 1930-31, detail, AMuVe

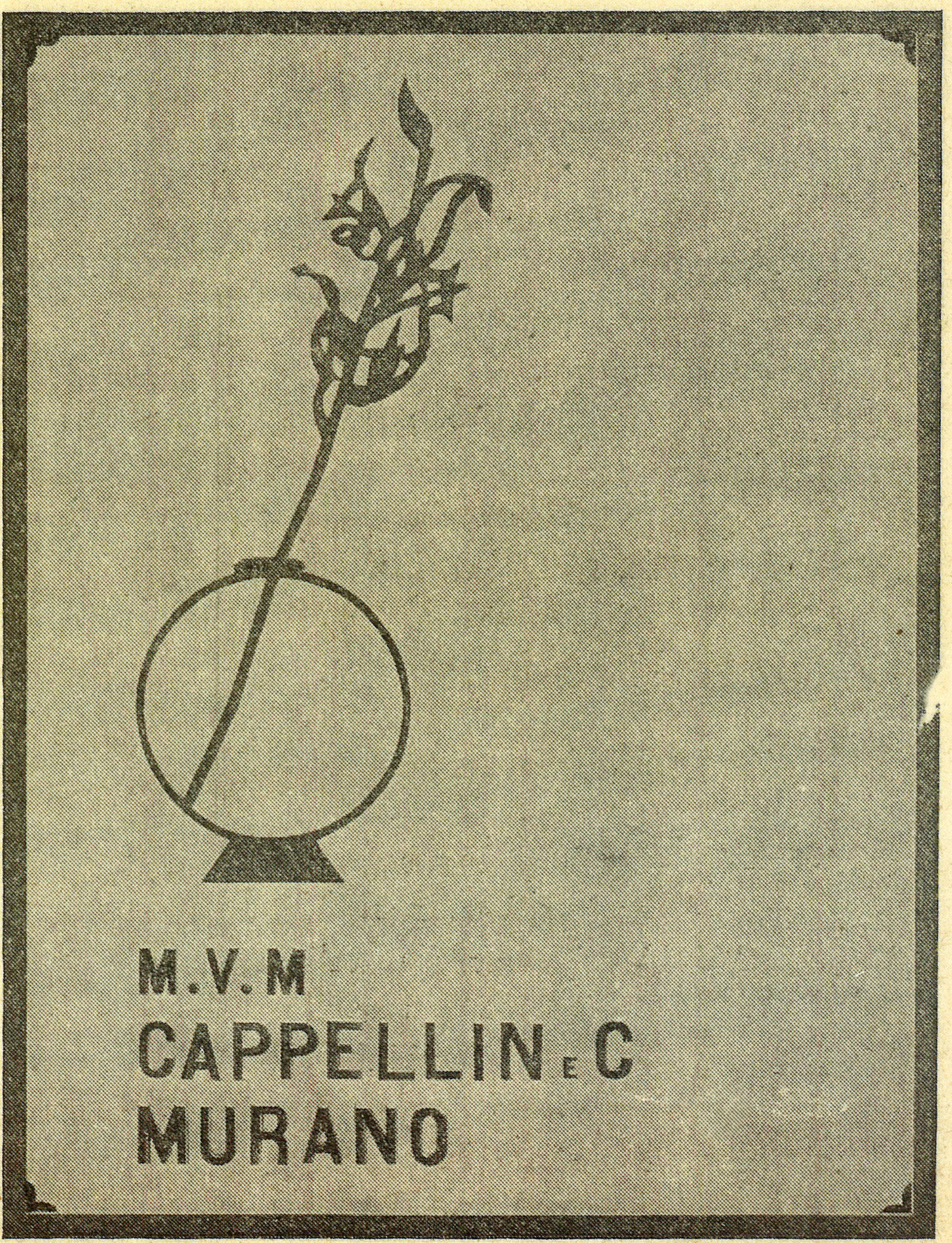
M.V.M
CAPPELLIN E C
MURANO

The M.V.M. Cappellin & C. catalogue documents the different models of vases produced by the glassworks over the course of its activities (1925-31). Hitherto unpublished, it has been reconstructed by putting together material coming from different archives, as after the bankruptcy of the company (1932) the furnace documentation was first acquired by Gaetano Ceschina, proprietor of the Pauly glassworks (1933), and later partly dispersed.

M.V.M. CAPPELLIN & C. CATALOGUE VASES 1925-31

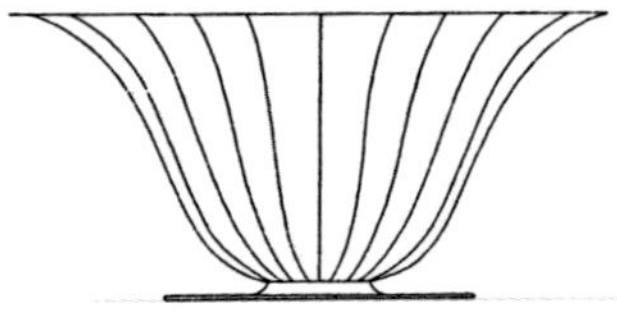
5001
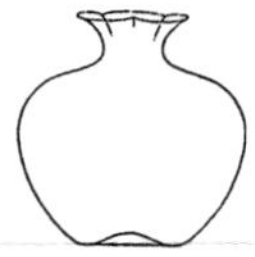
5002
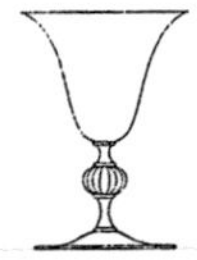
5003
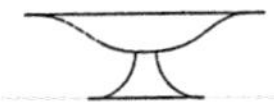
5004
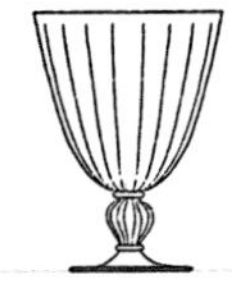
5005
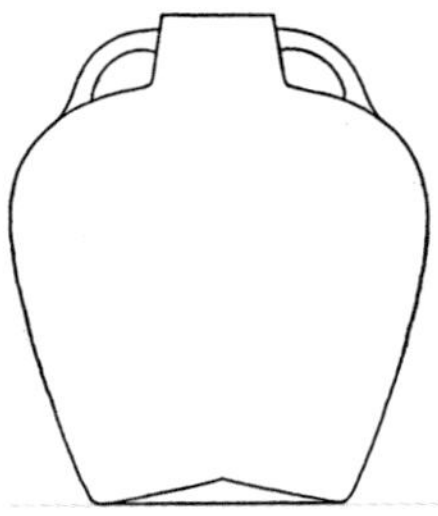
5006
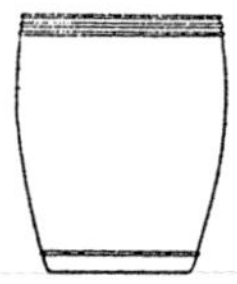
5007
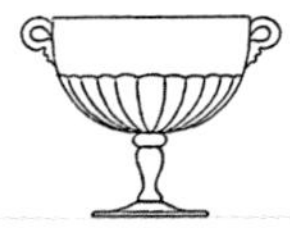
5008
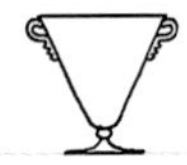
5009
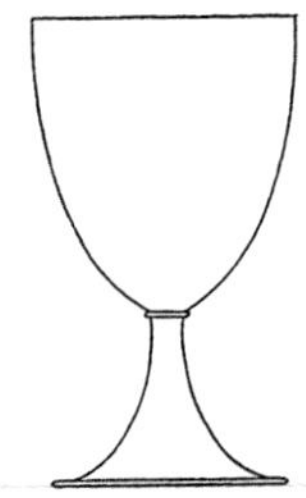
5010
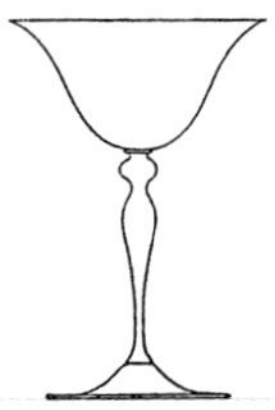
5011
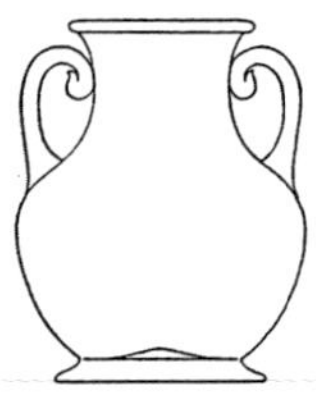
5012
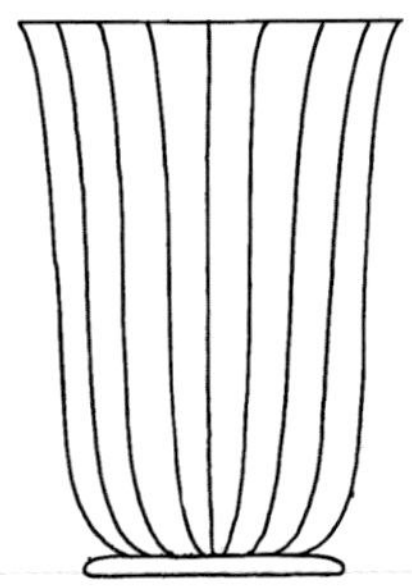
5013
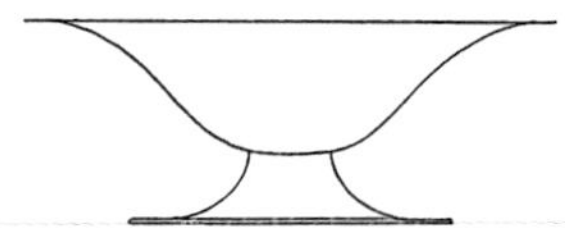
5014
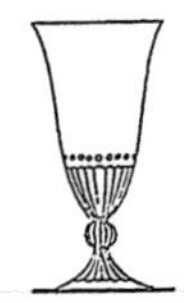
5015
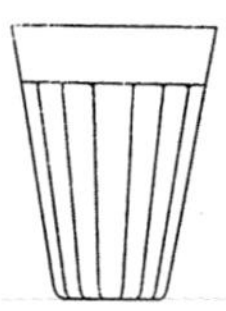
5016
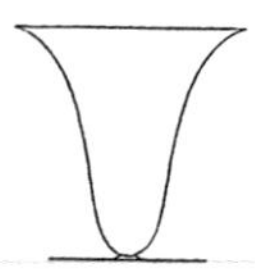
5017
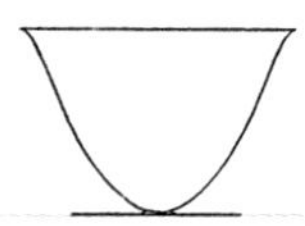
5018
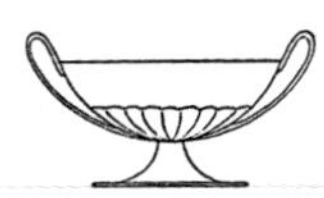
5019
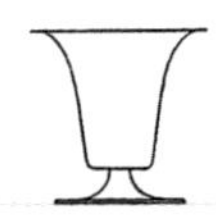
5020
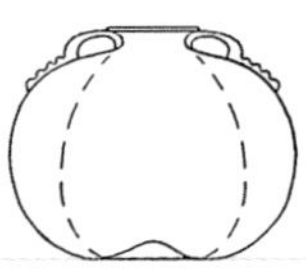
5021
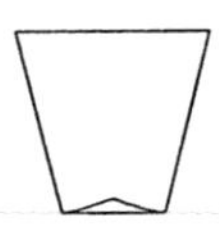
5022
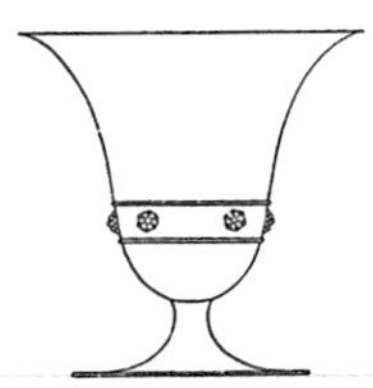
5023
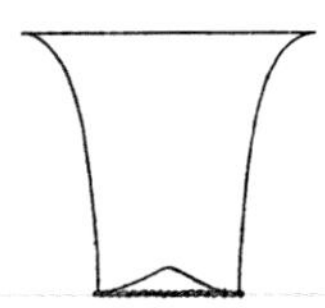
5024
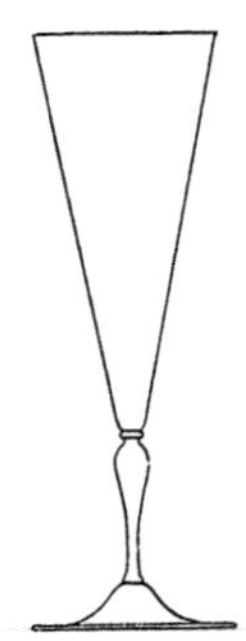
5025

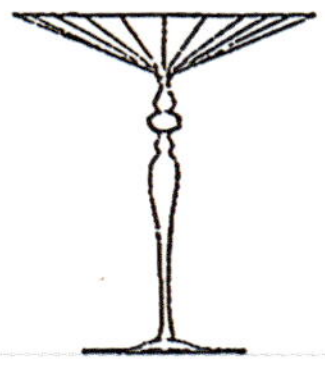
5026
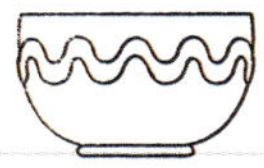
5027
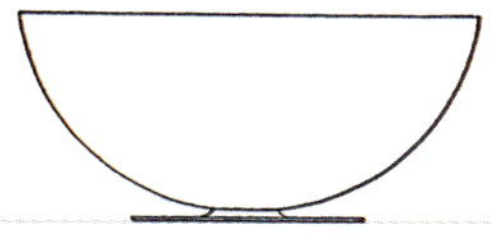
5028
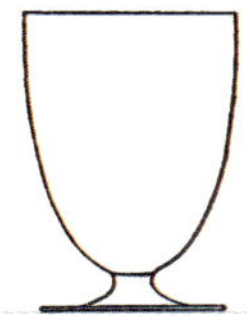
5029
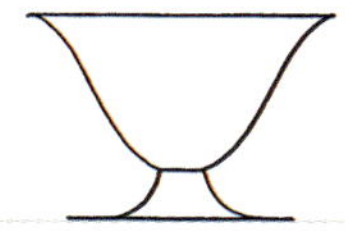
5030
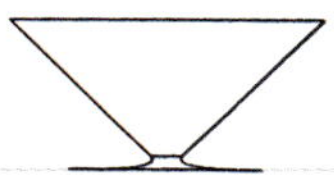
5031
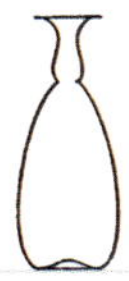
5032
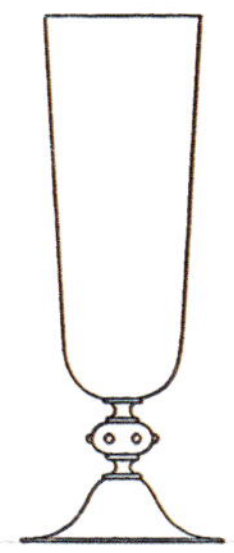
5033
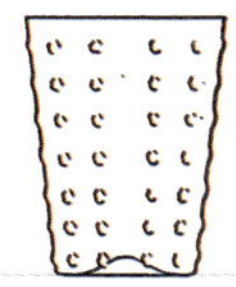
5034
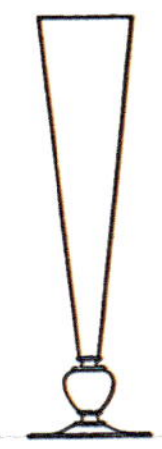
5035
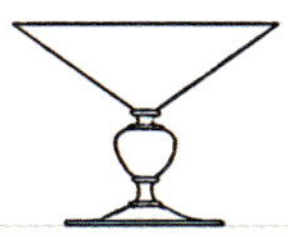
5036
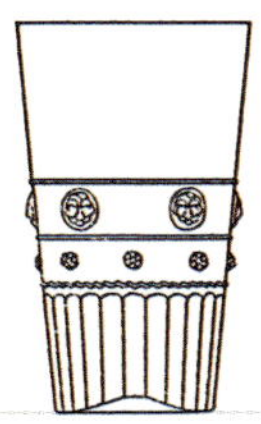
5037
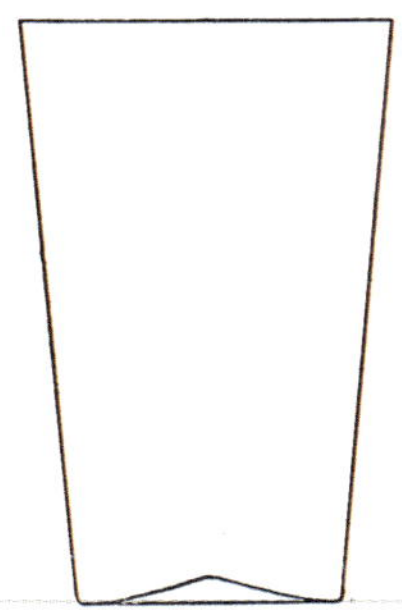
5038
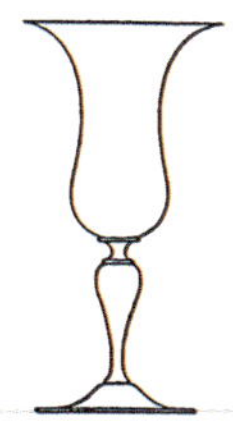
5039
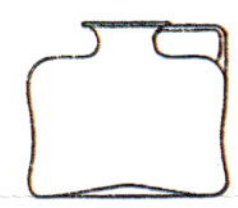
5040
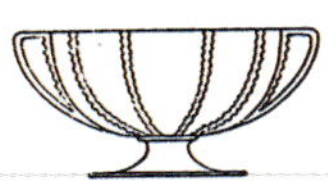
5041
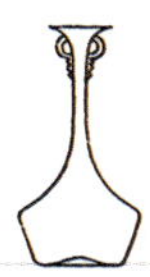
5042
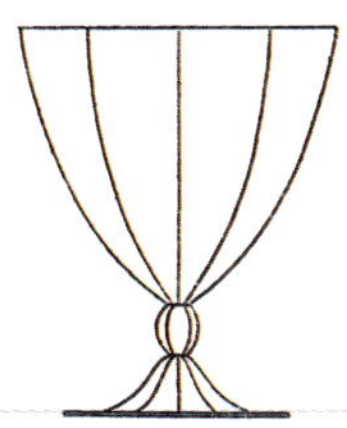
5043
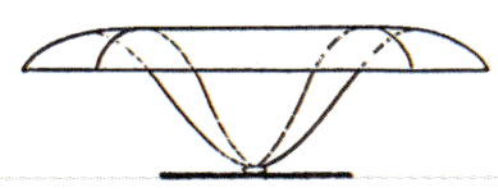
5044
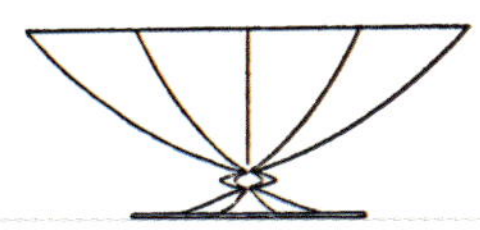
5045
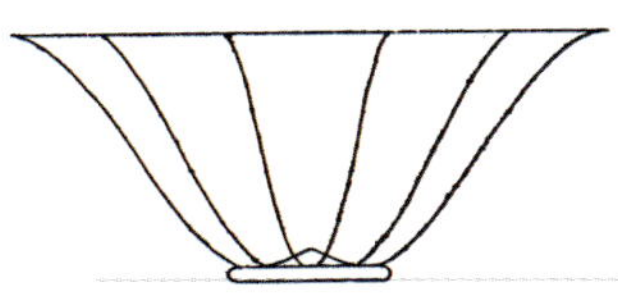
5046
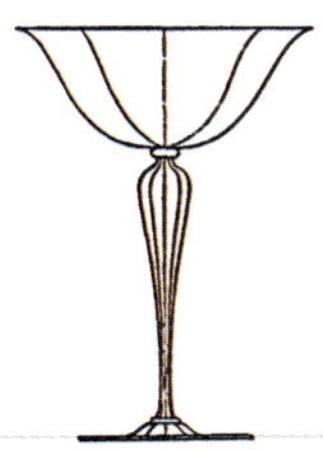
5047
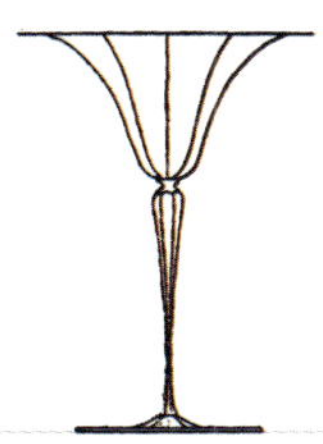
5048
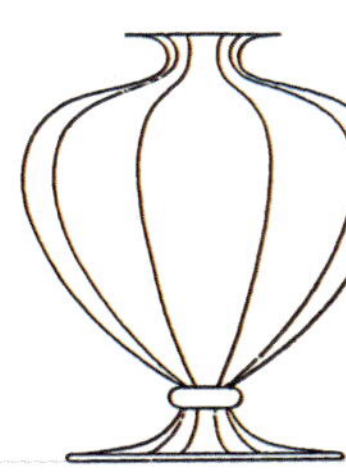
5049
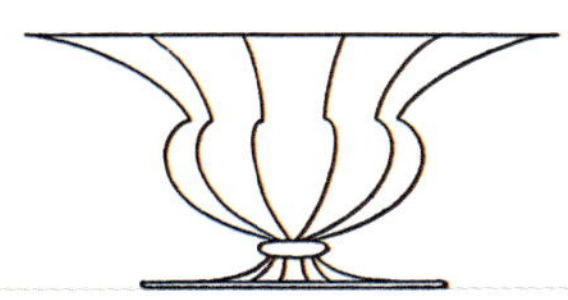
5050

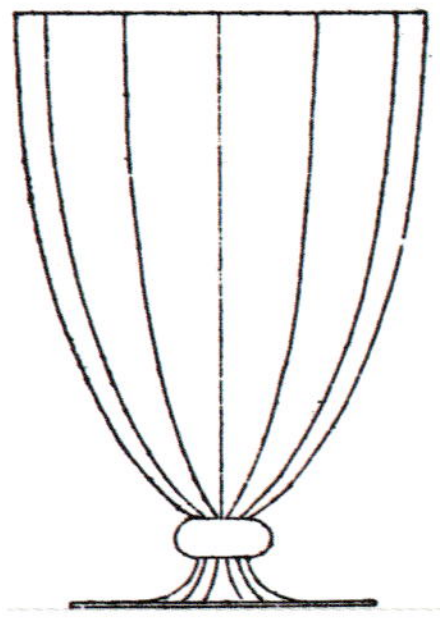
5051
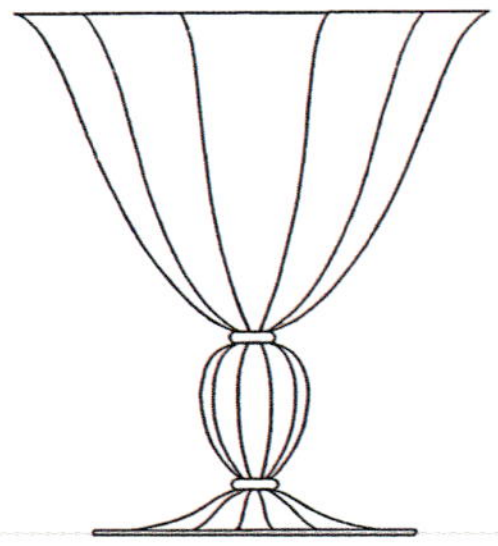
5052
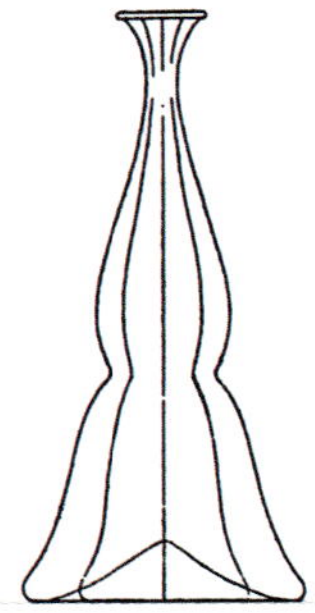
5053
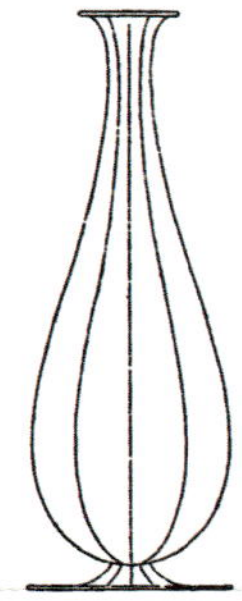
5054
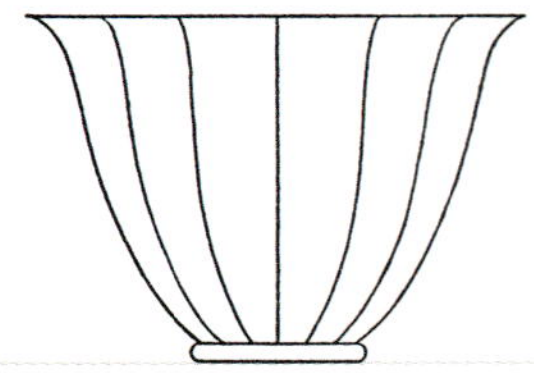
5055
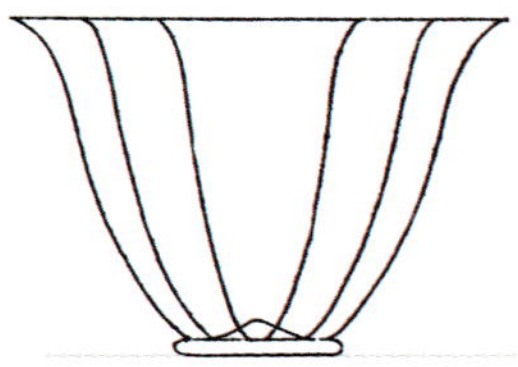
5056
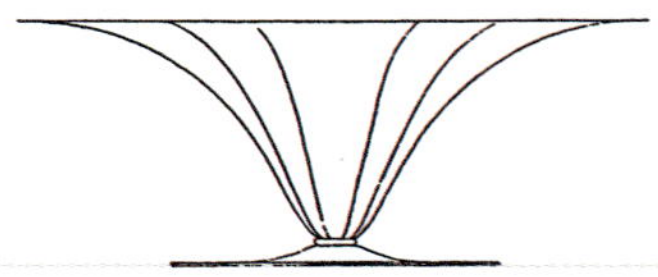
5057
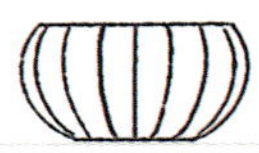
5058
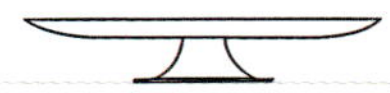
5059
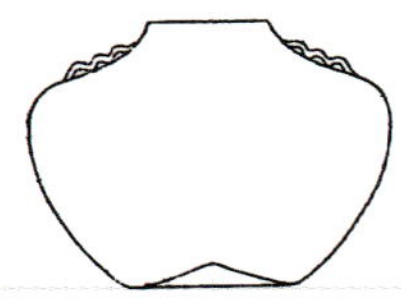
5060
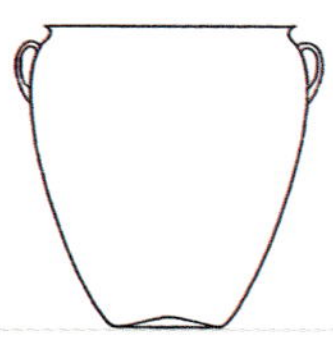
5061
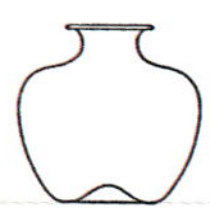
5062
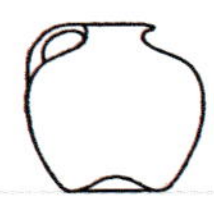
5063
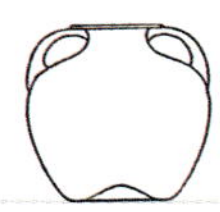
5064
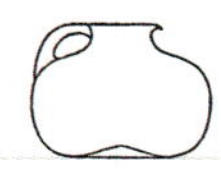
5065
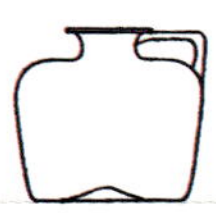
5066
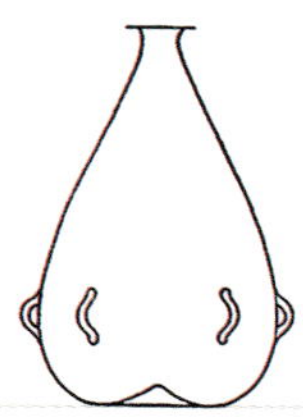
5067
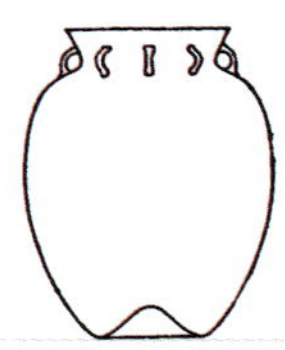
5068
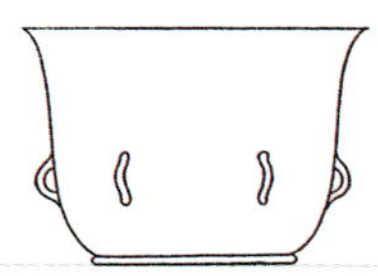
5069
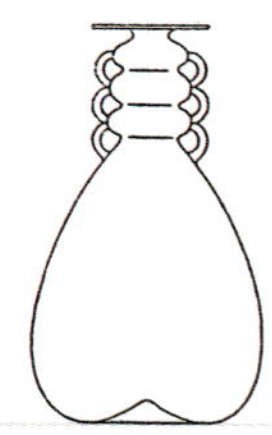
5070
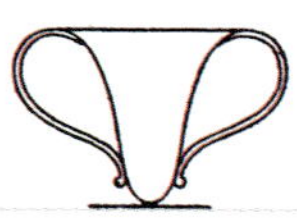
5071
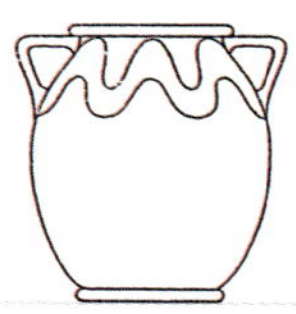
5072
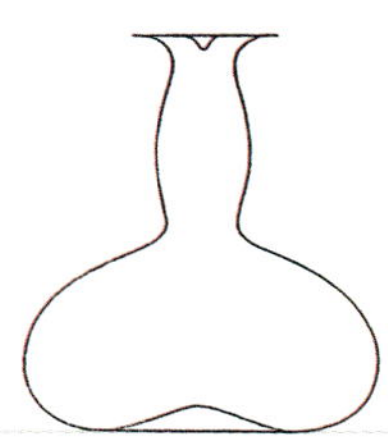
5073
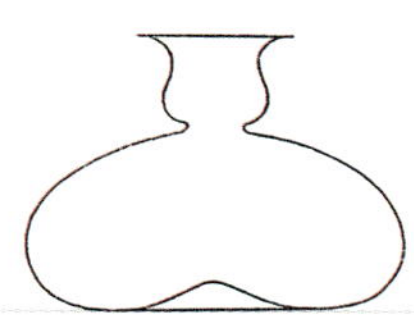
5074
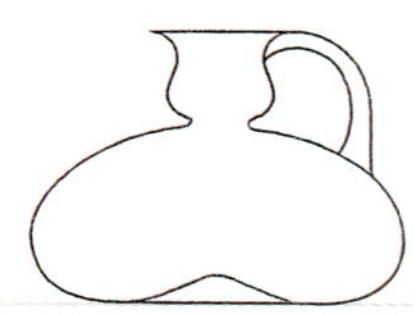
5075

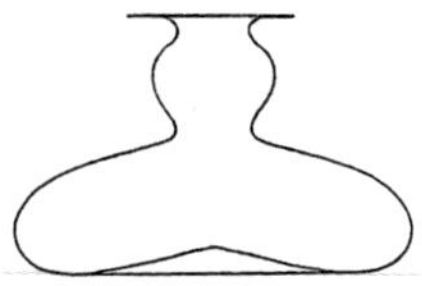
5076

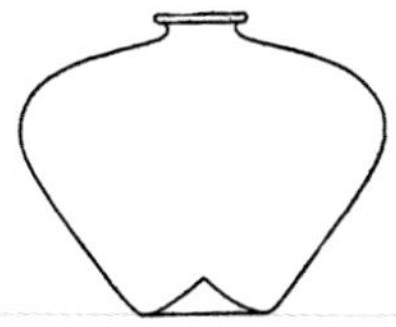
5077

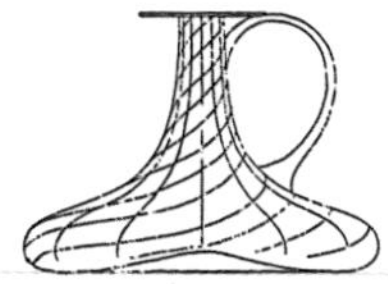
5078

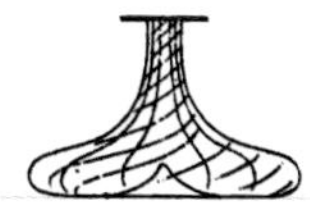
5079

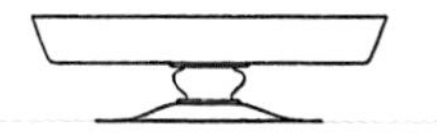
5080

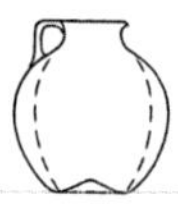
5081

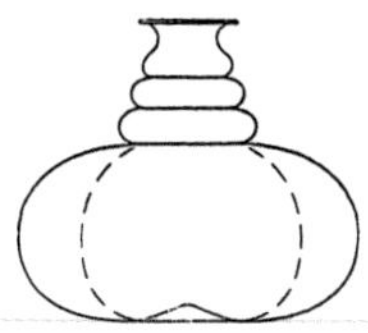
5082

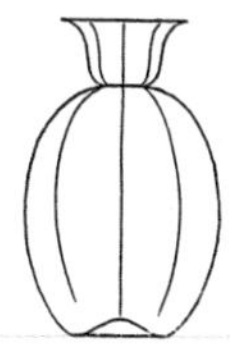
5083

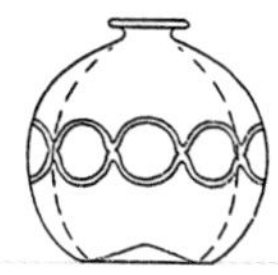
5084

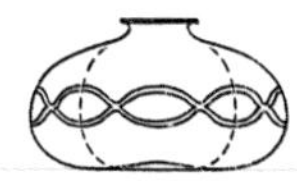
5085

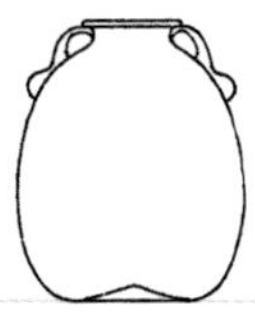
5086

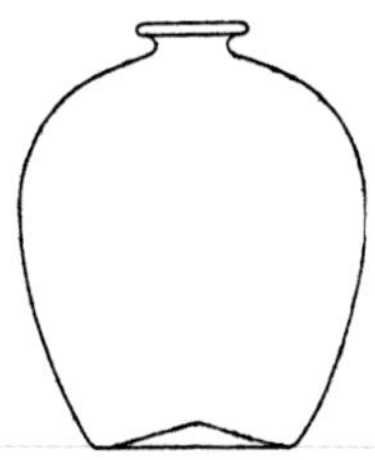
5087

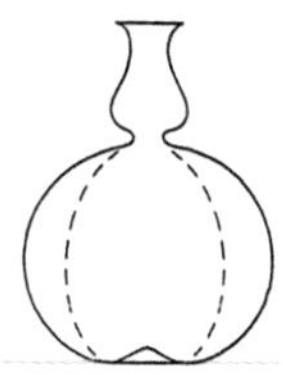
5088

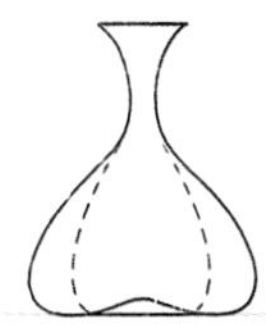
5089

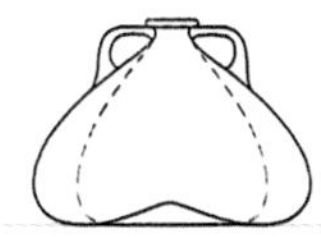
5090

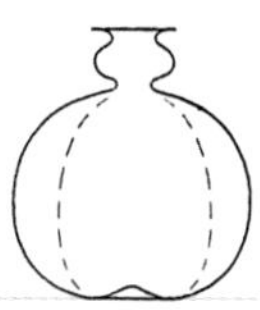
5091

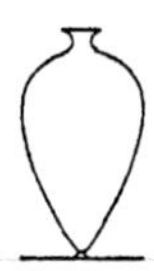
5092

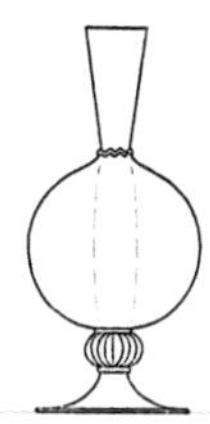
5093

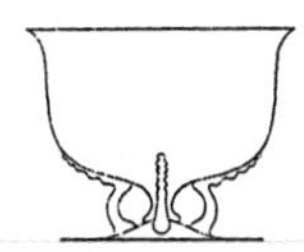
5094

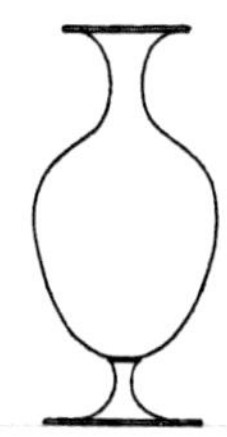
5095

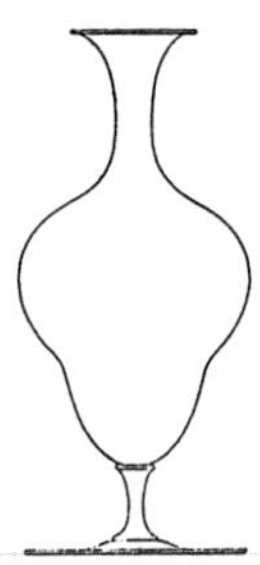
5096

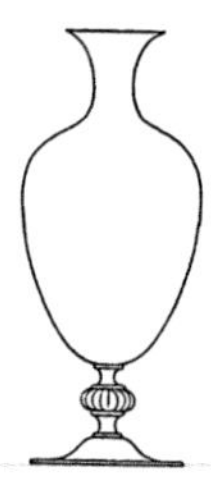
5097

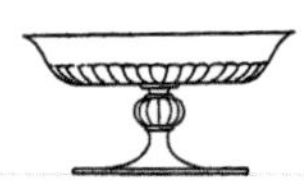
5098

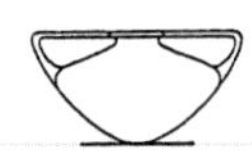
5099

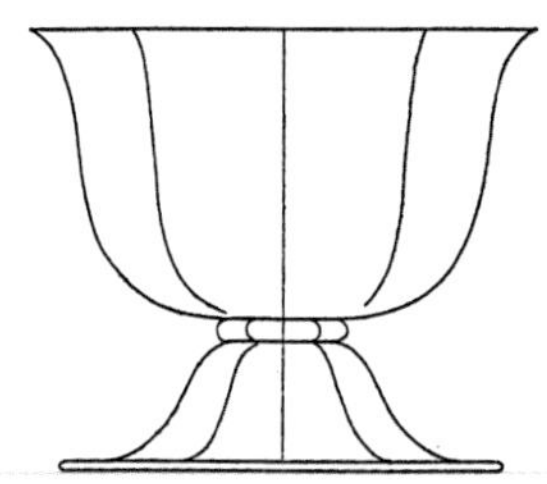
5100

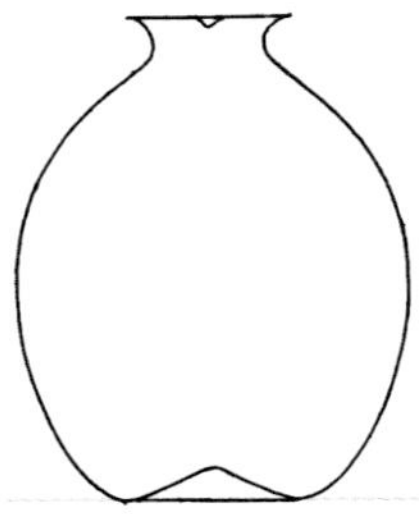
5101

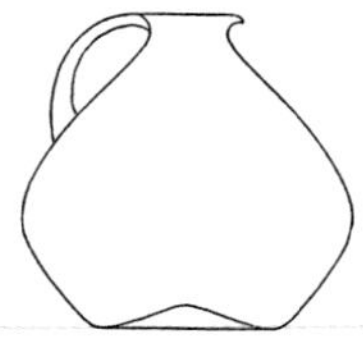
5102

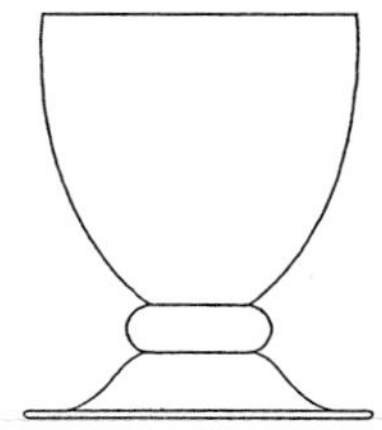
5103

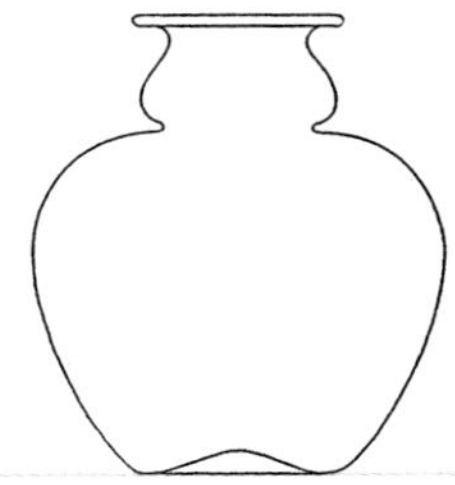
5104

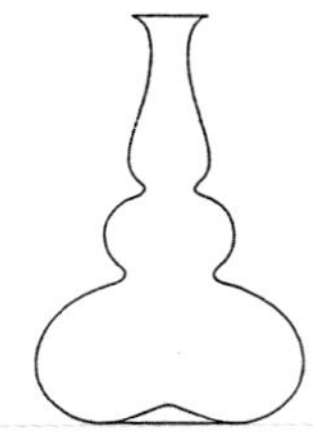
5105

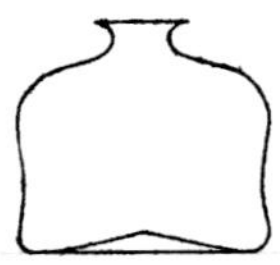
5106

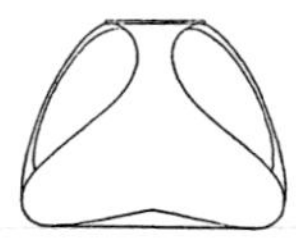
5107

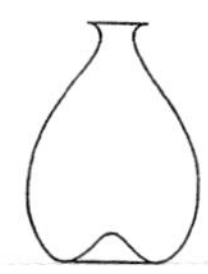
5108

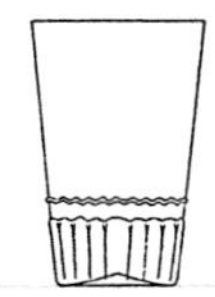
5109

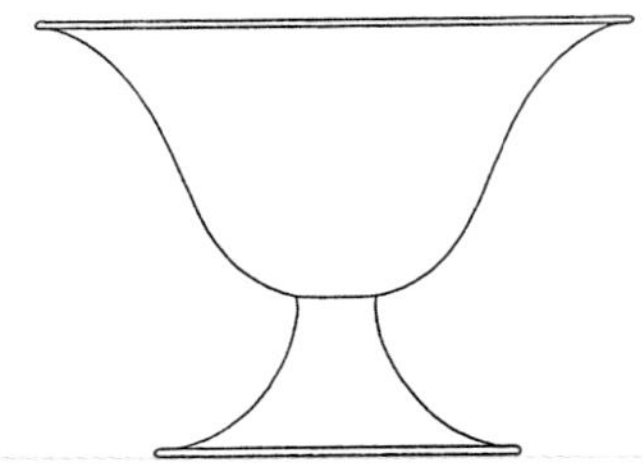
5110

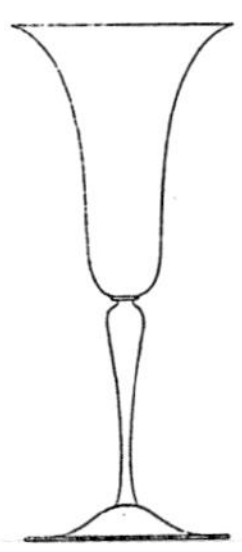
5111

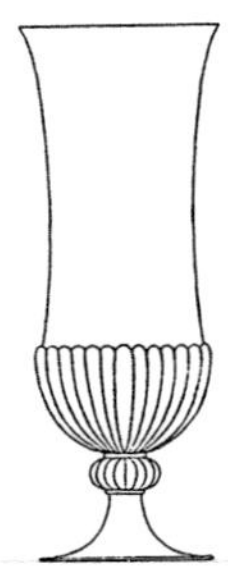
5112

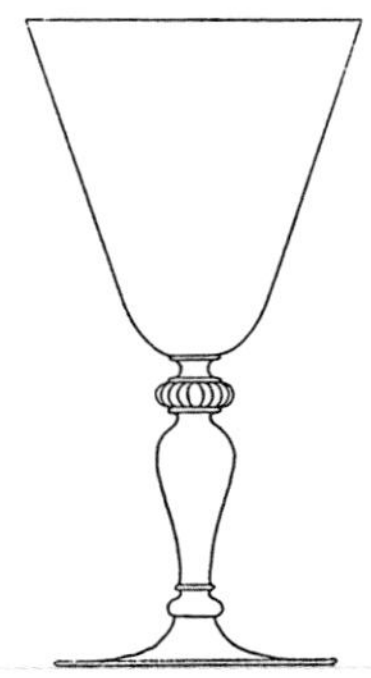
5113

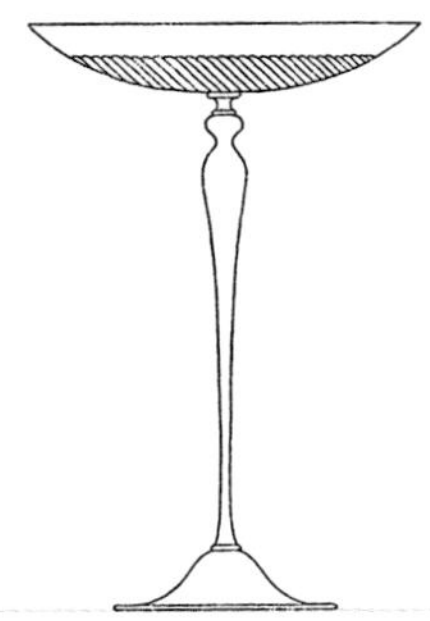
5114

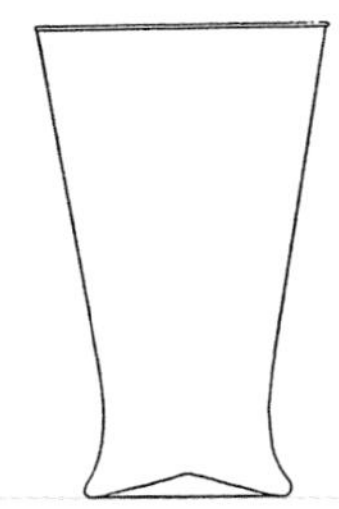
5115

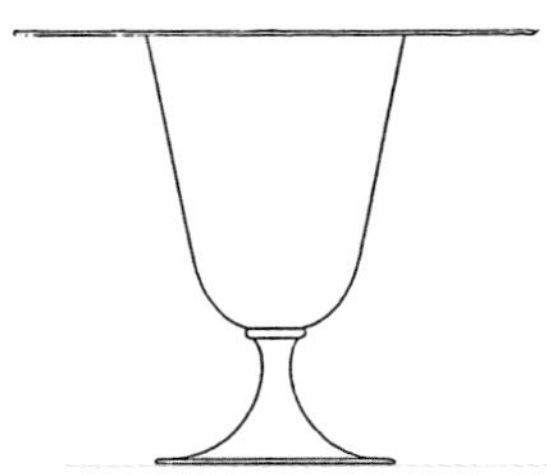
5116

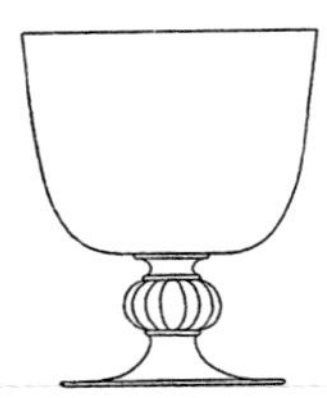
5117

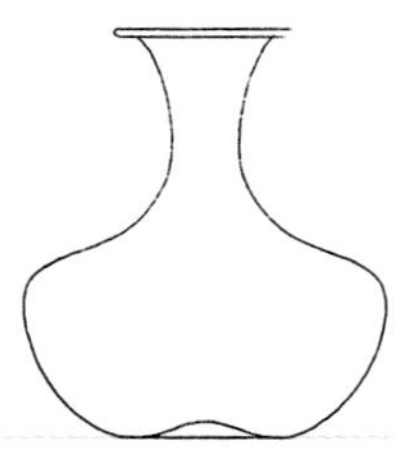
5118

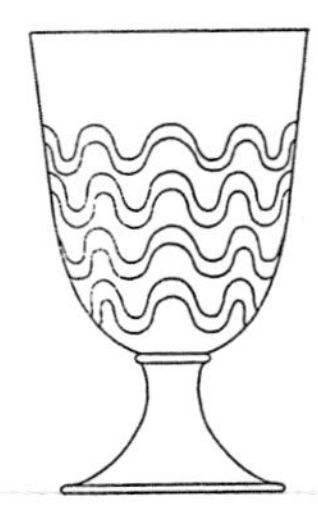
5119

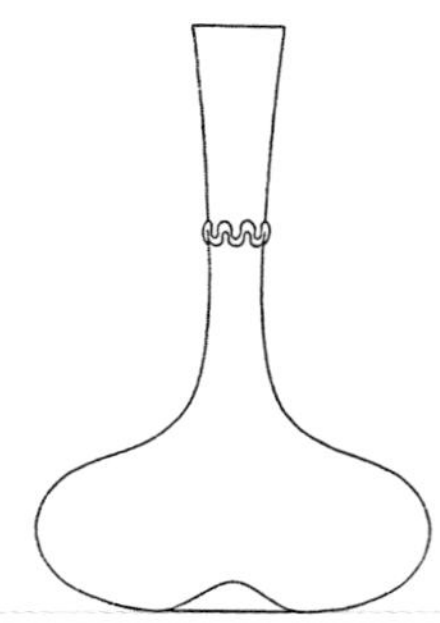
5120

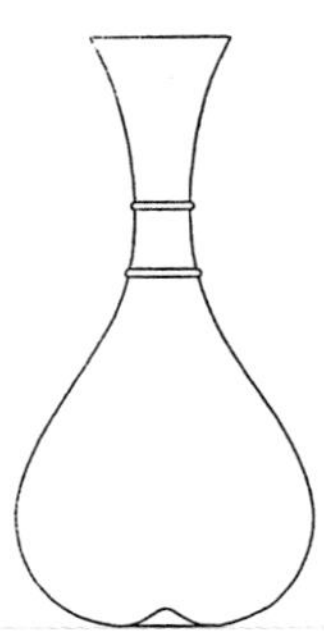
5121

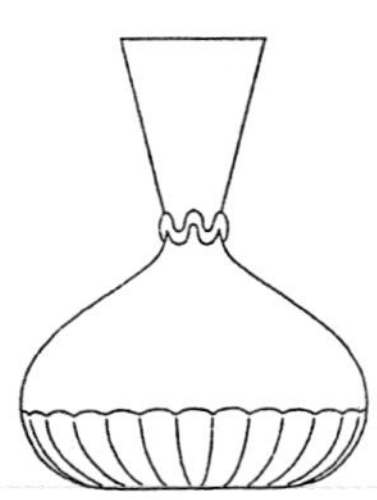
5122

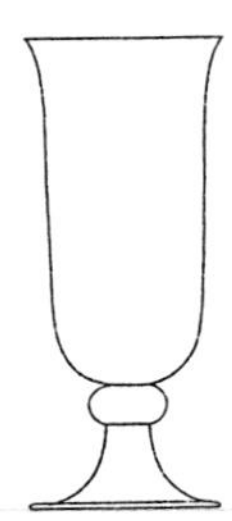
5123

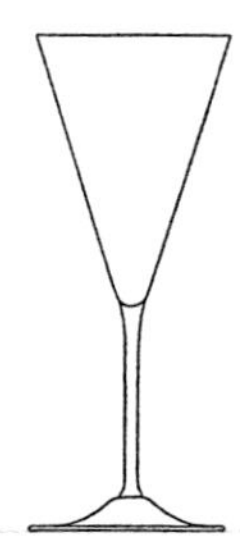
5124

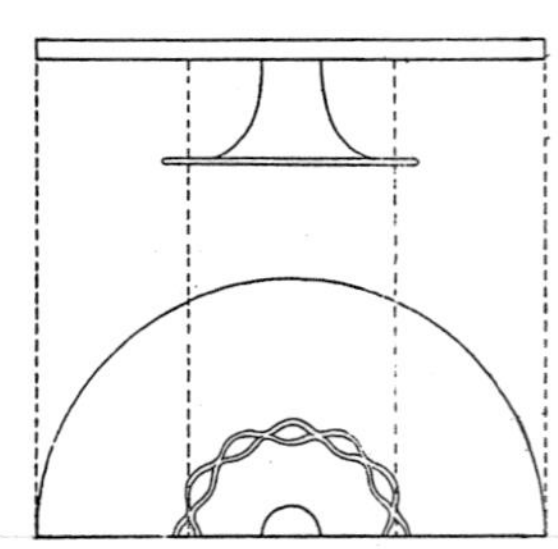
5125

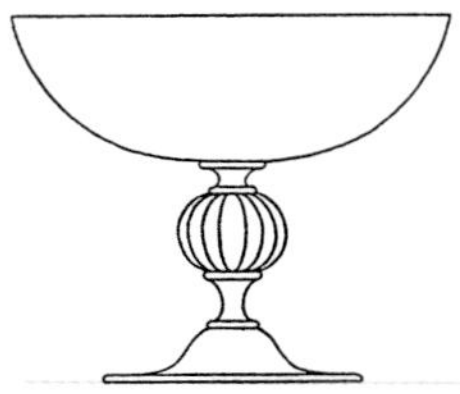
5126
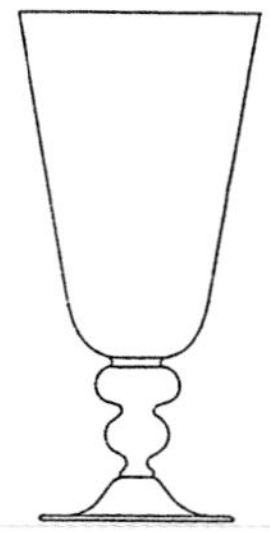
5127
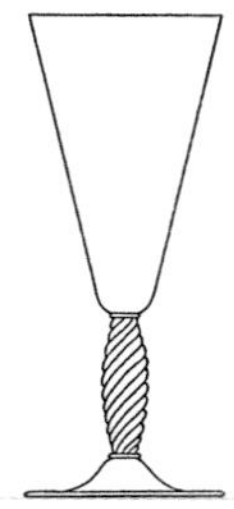
5128
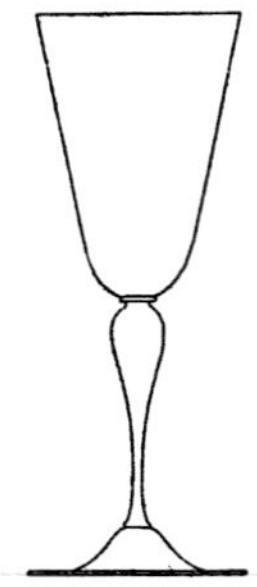
5129
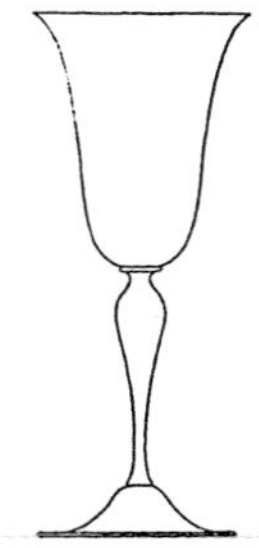
5130
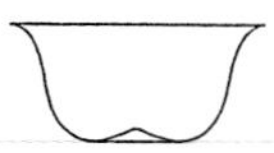
5131
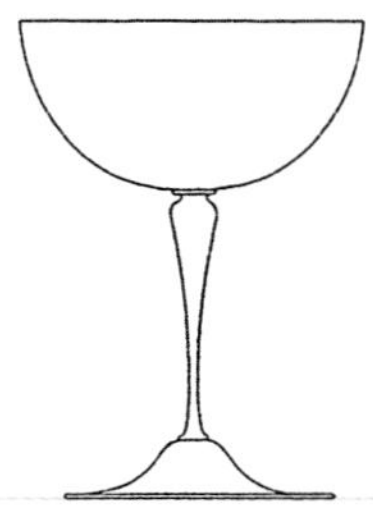
5132
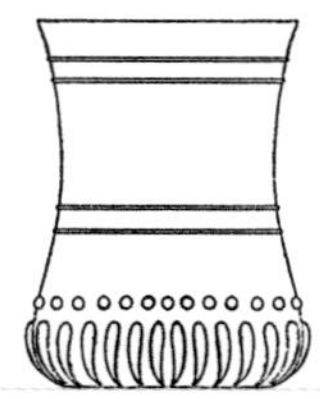
5133
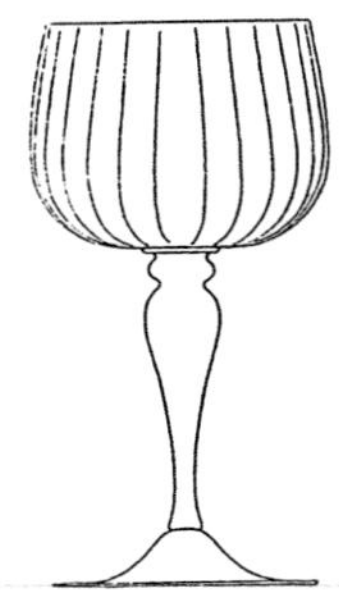
5134
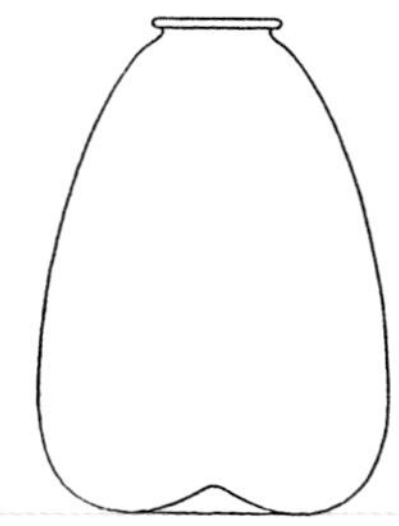
5135
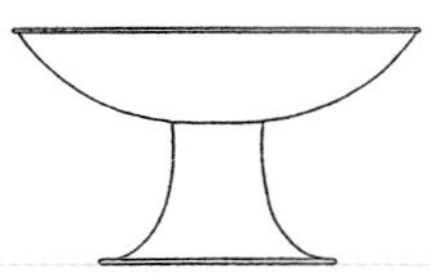
5136
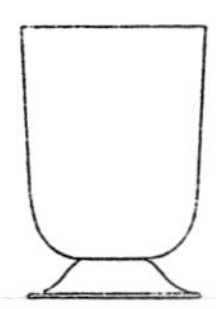
5137
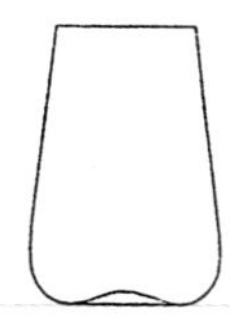
5138
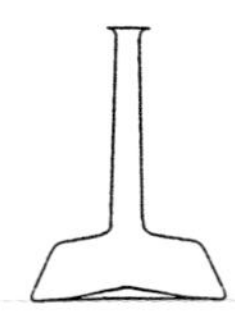
5139
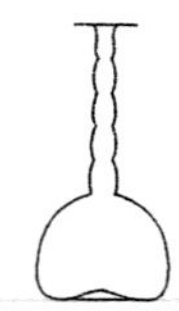
5140
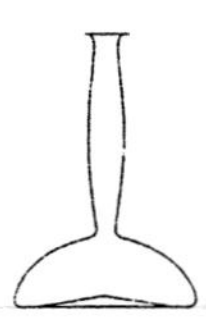
5141
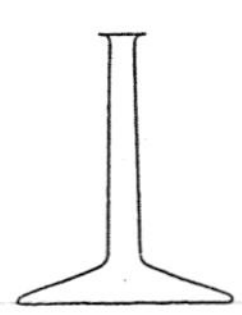
5142
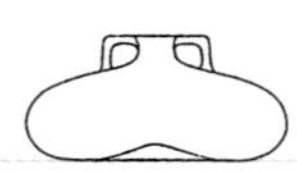
5143
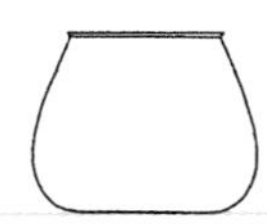
5144
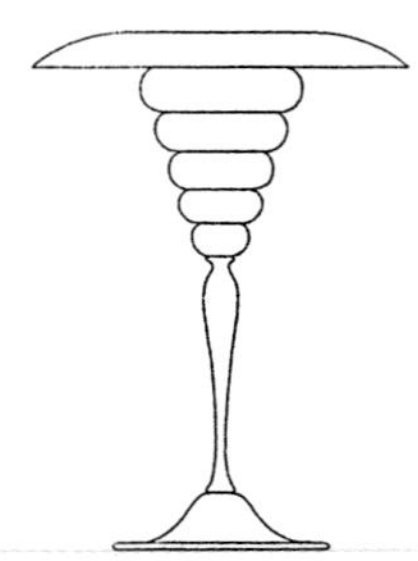
5145
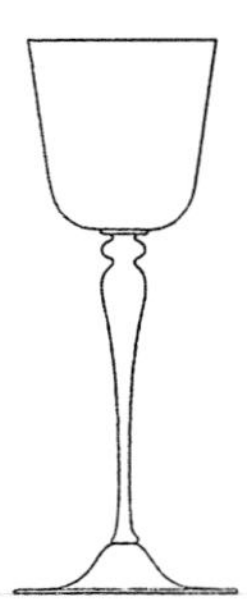
5146
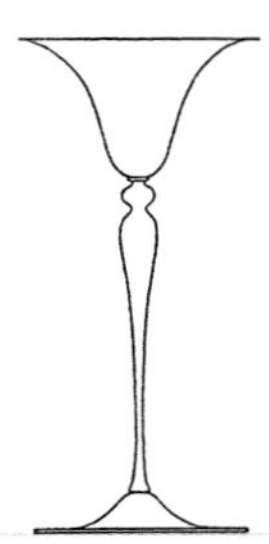
5147
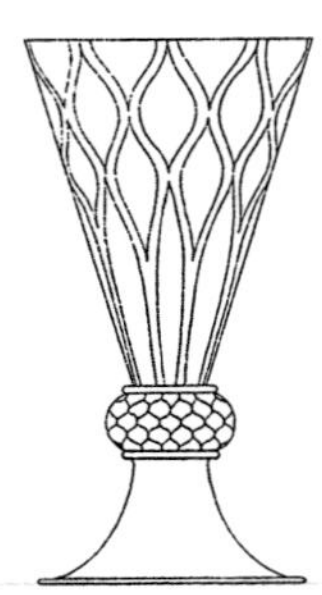
5148
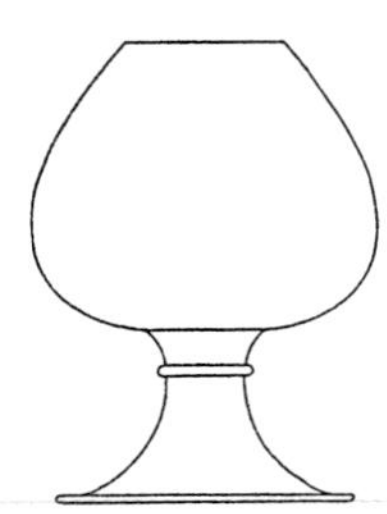
5149
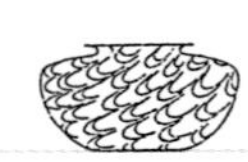
5150

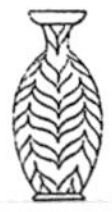
5151

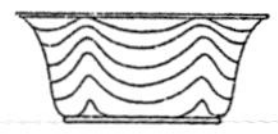
5152

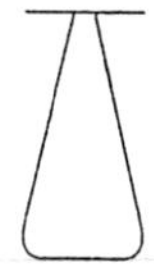
5153

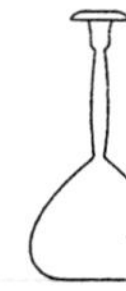
5154

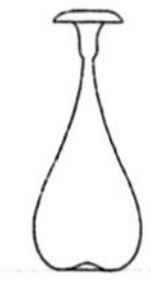
5155

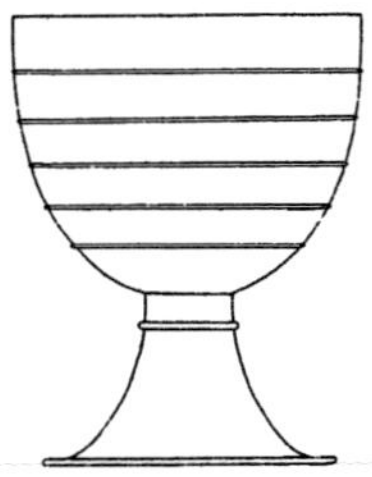
5156

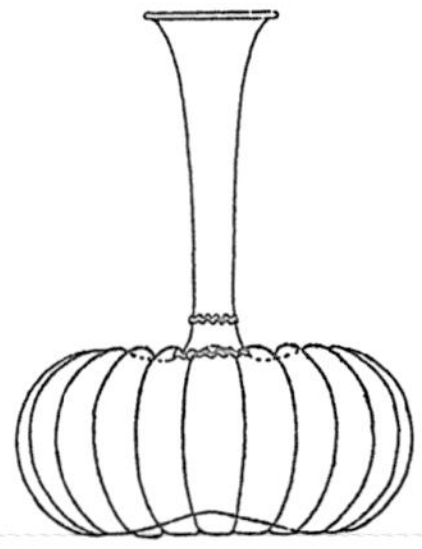
5157

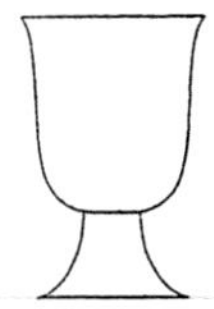
5158

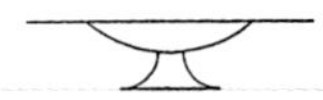
5159

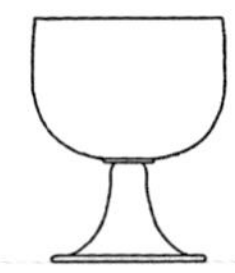
5160

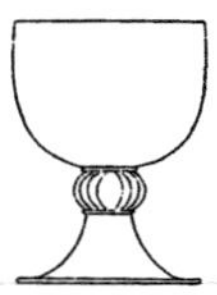
5161

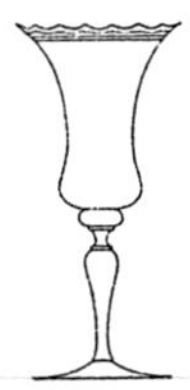
5162

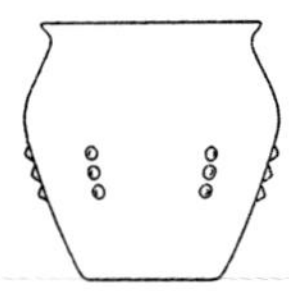
5163

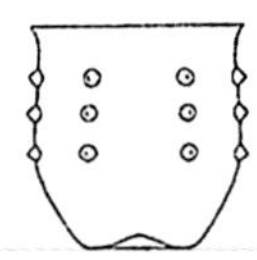
5164

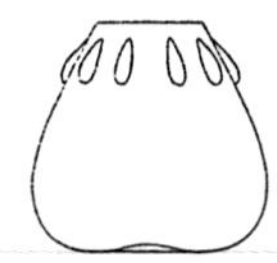
5165

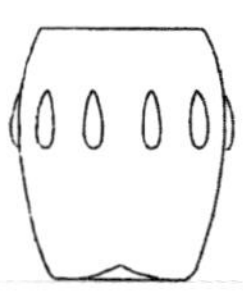
5166

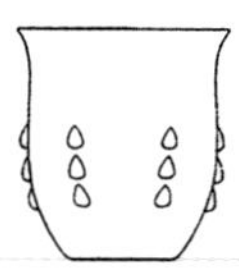
5167

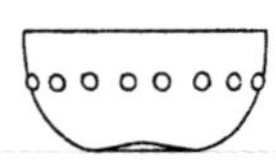
5168

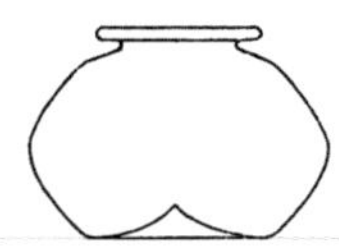
5169

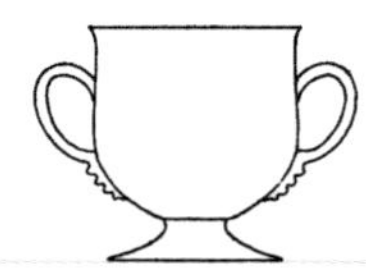
5170

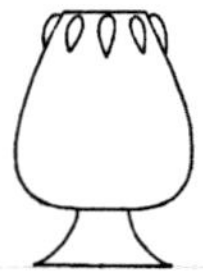
5171

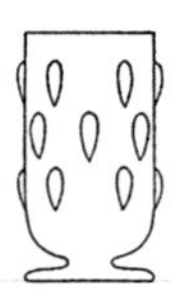
5172

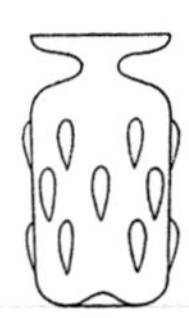
5173

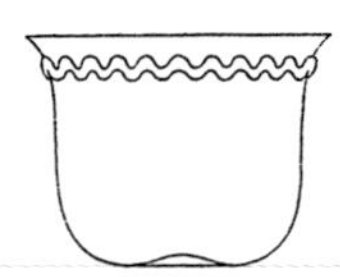
5174

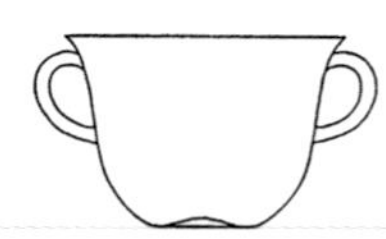
5175

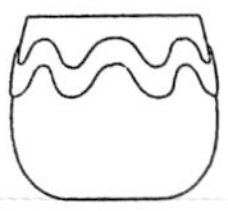
5176

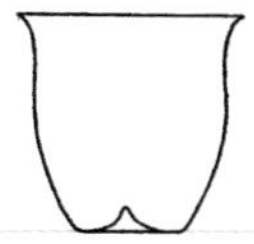
5177

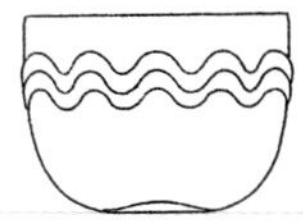
5178

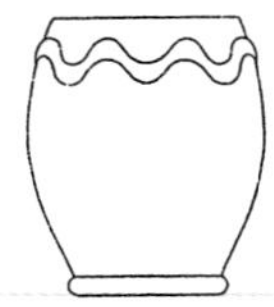
5179

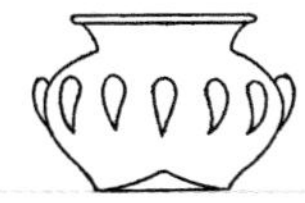
5180

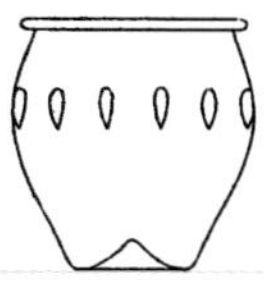
5181

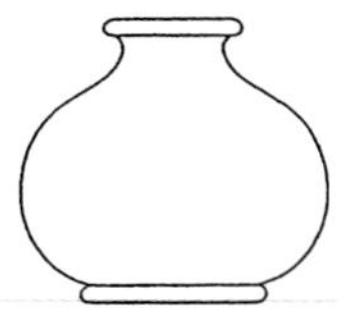
5182

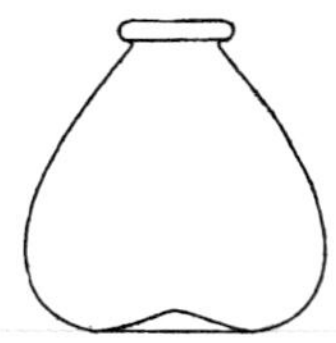
5183

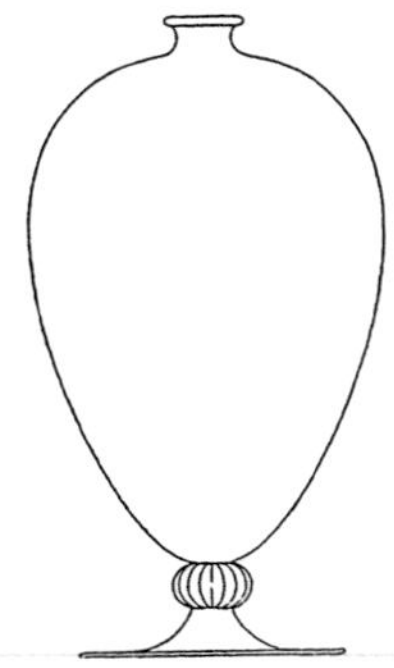
5184

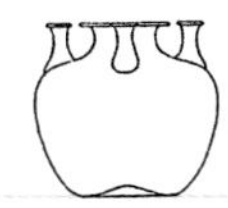
5185

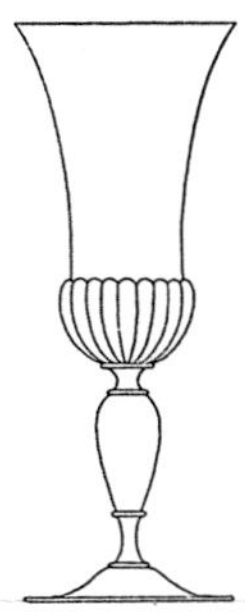
5186

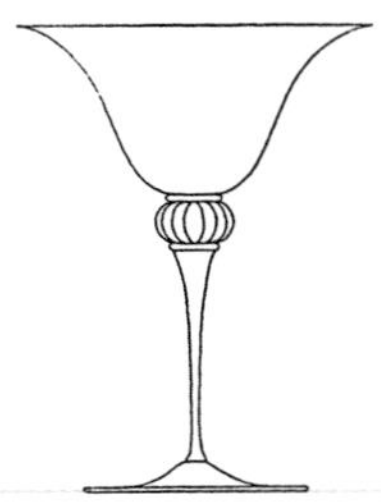
5187

5188

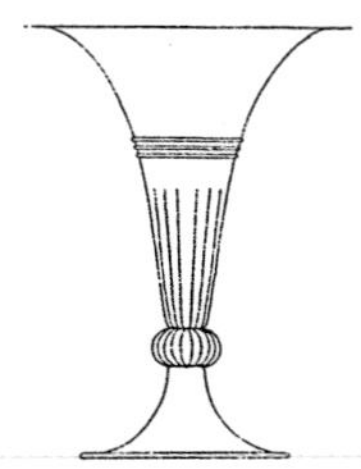
5189

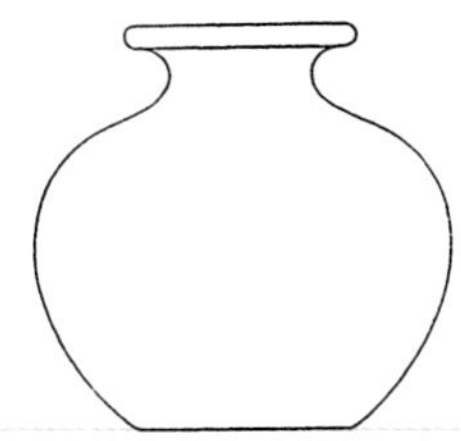
5190

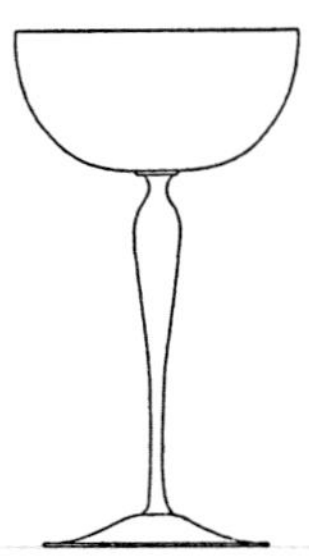
5191

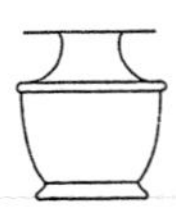
5192

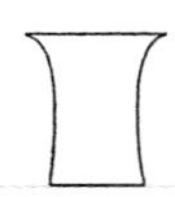
5193

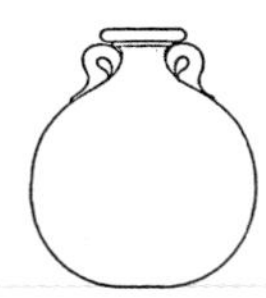
5194

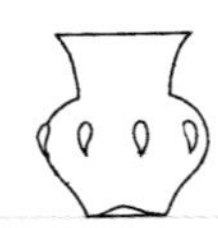
5195

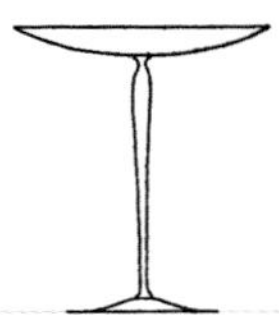
5196

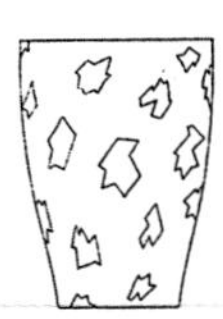
5197

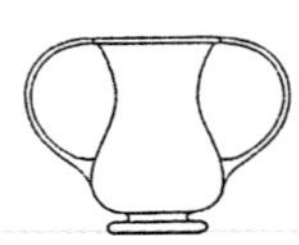
5198

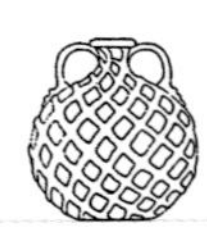
5199

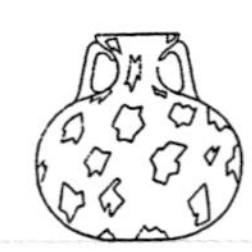
5200

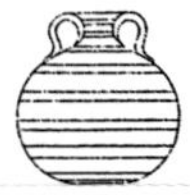
5201
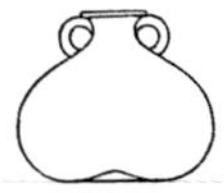
5202
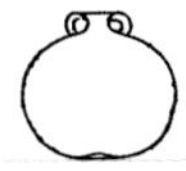
5203
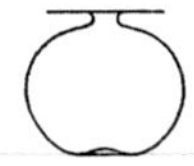
5204
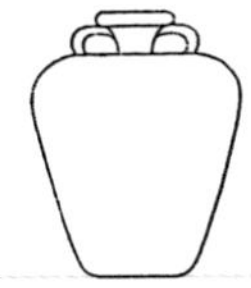
5205
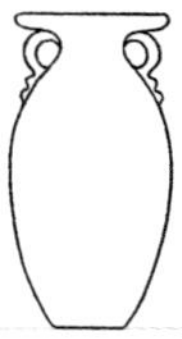
5206
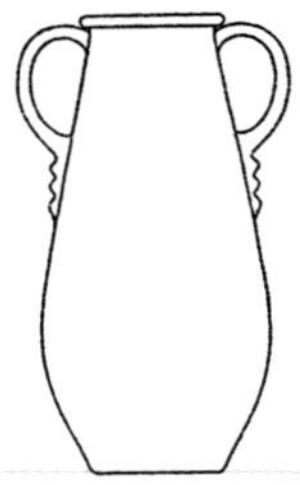
5207
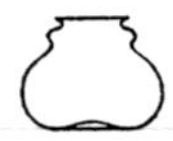
5208
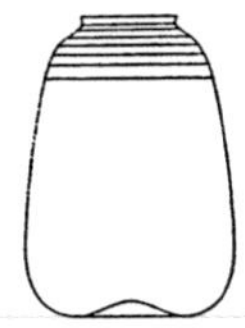
5209
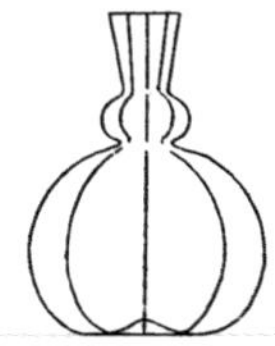
5210
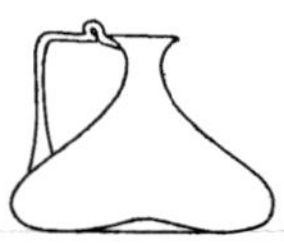
5211
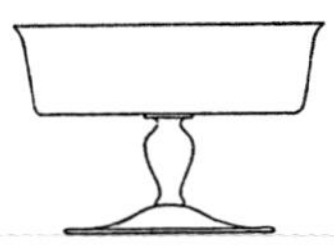
5212
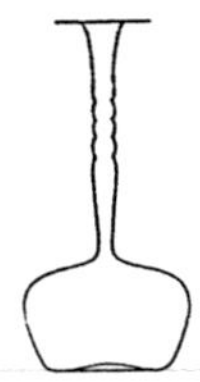
5213
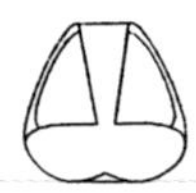
5214
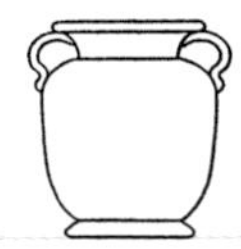
5215
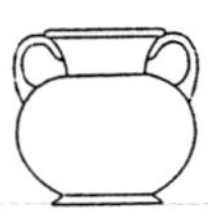
5216
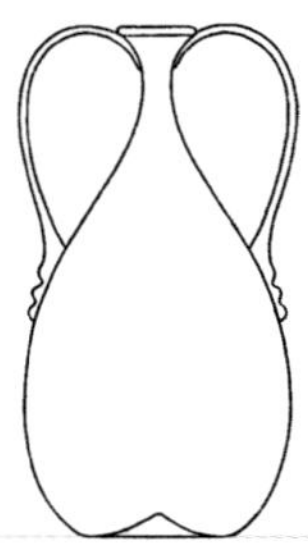
5217
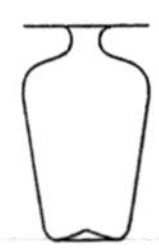
5218
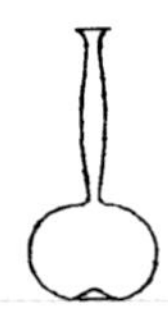
5219
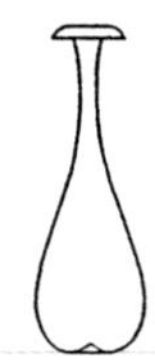
5220
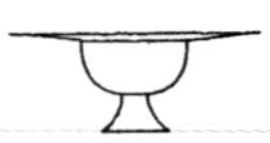
5221
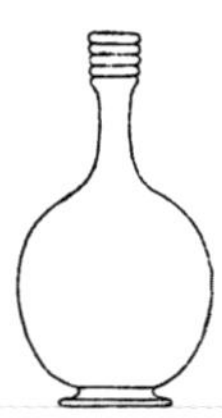
5222
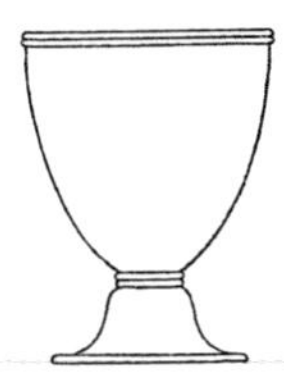
5223
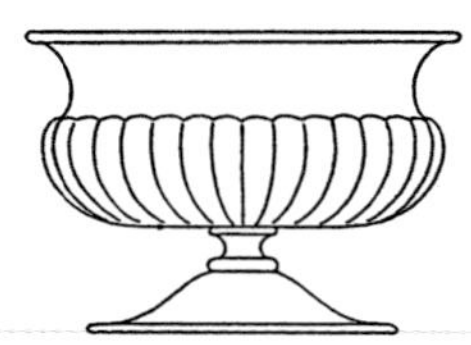
5224
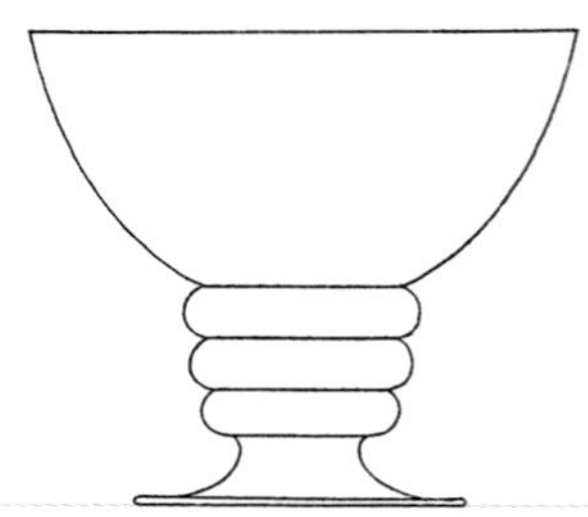
5225

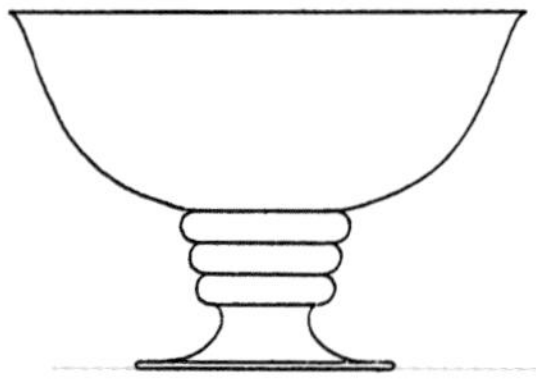
5226
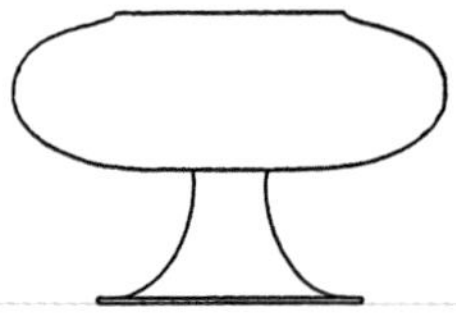
5227
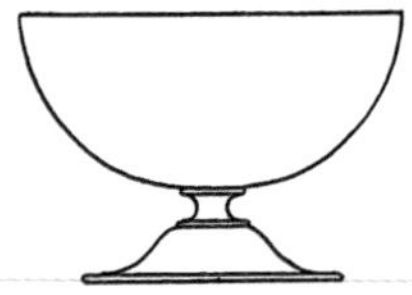
5228
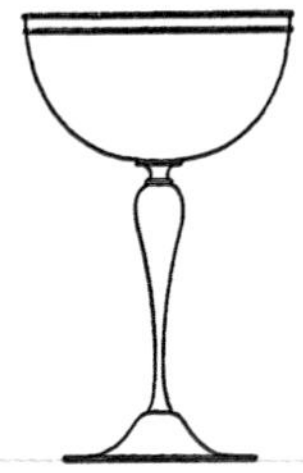
5229
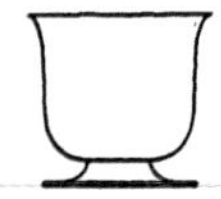
5230
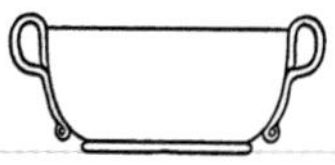
5231
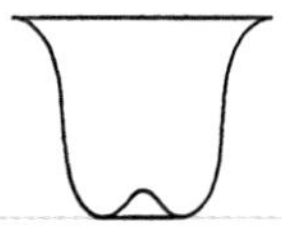
5232
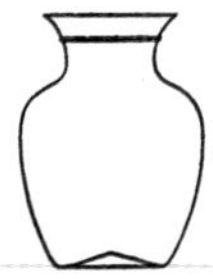
5233
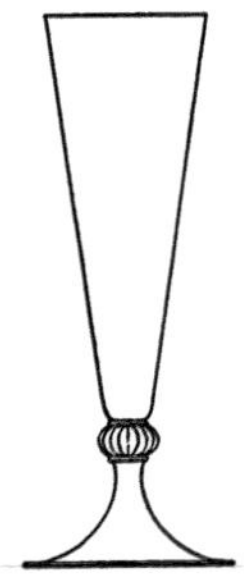
5234
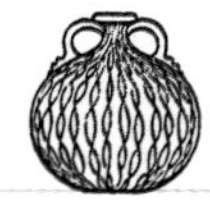
5235
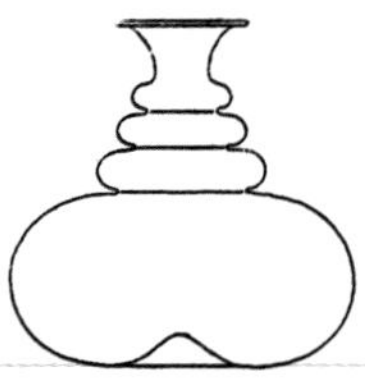
5236
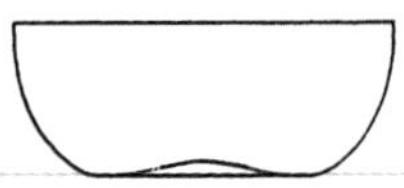
5237
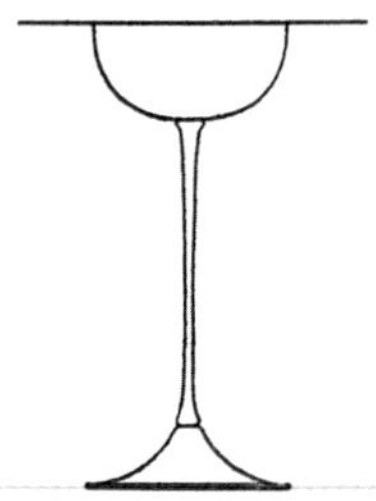
5238
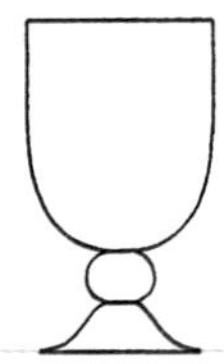
5239
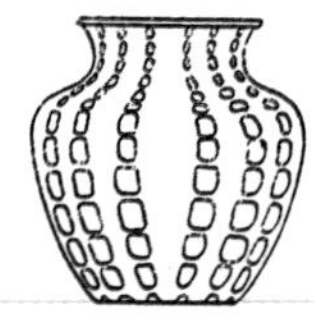
5240
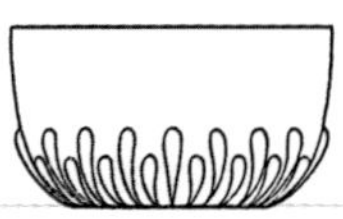
5241
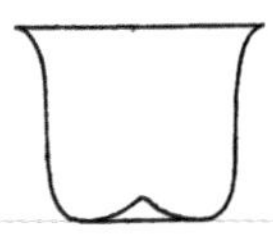
5242
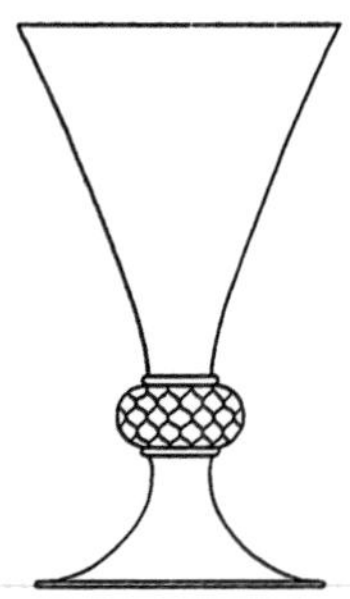
5243
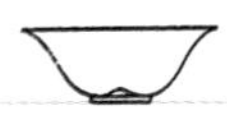
5244
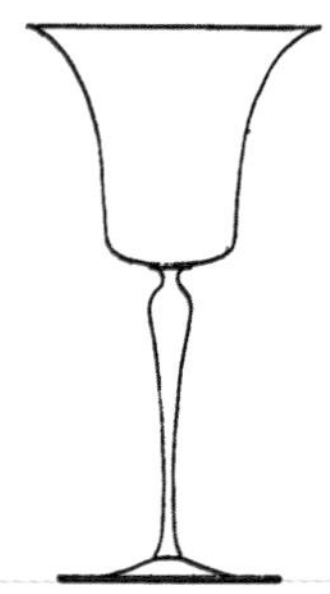
5245
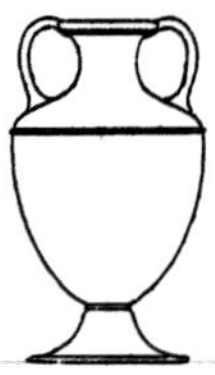
5246
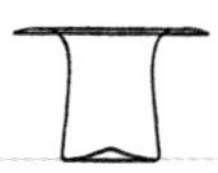
5247
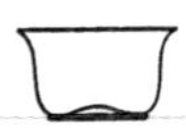
5248
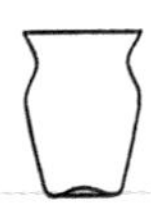
5249
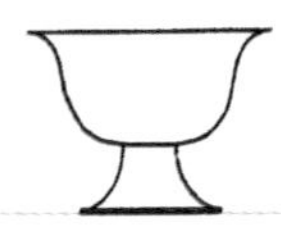
5250

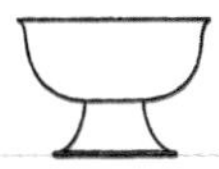
5251
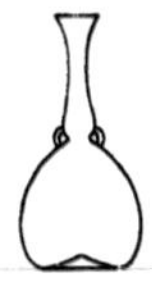
5252
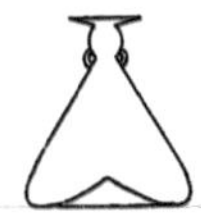
5253
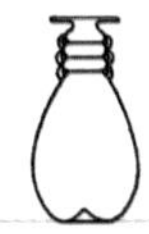
5254
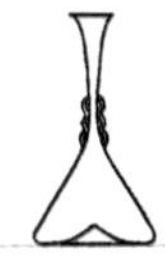
5255
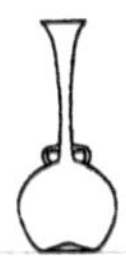
5256
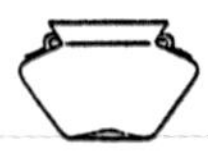
5257
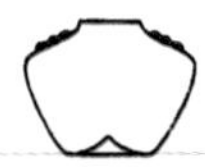
5258
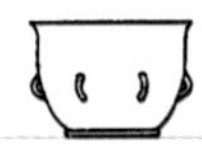
5259
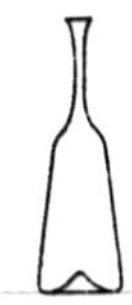
5260
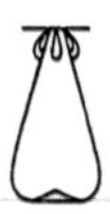
5261
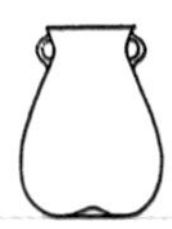
5262
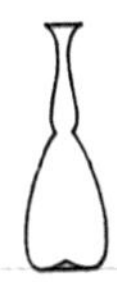
5263
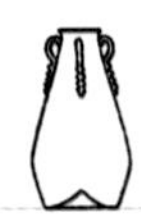
5264
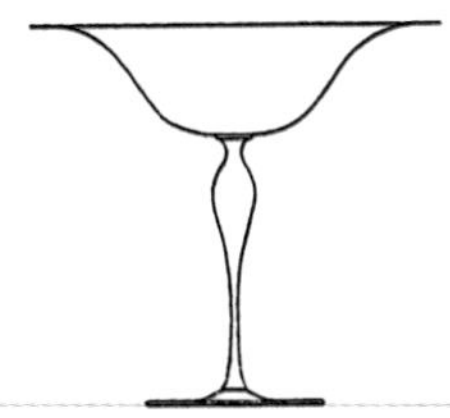
5265
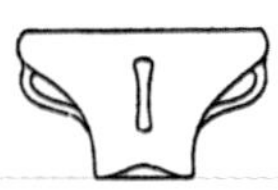
5266
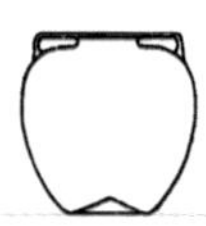
5267
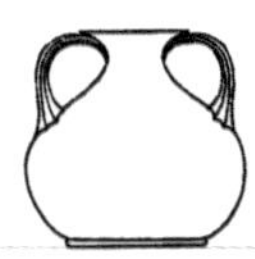
5268
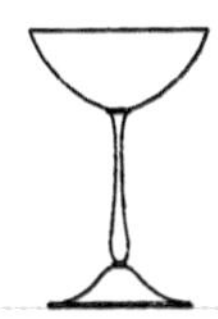
5269
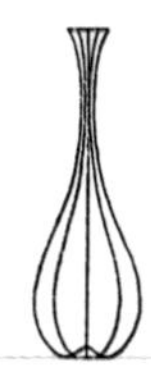
5270
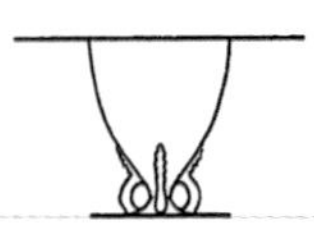
5271
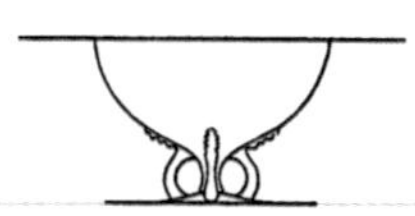
5272
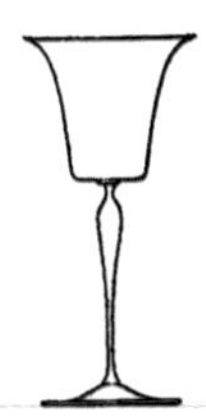
5273
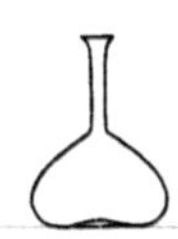
5274
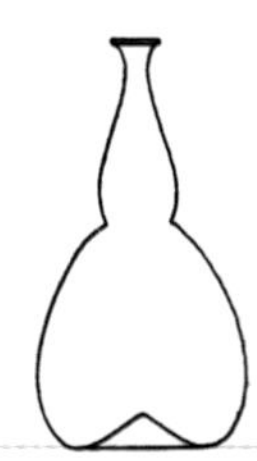
5275

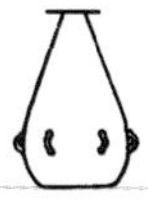
5276

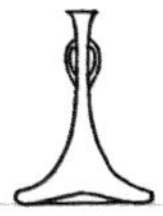
5277

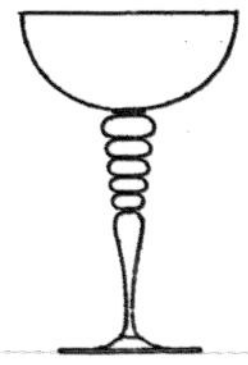
5278

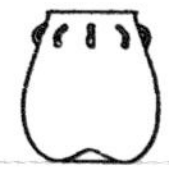
5279

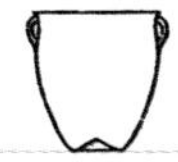
5280

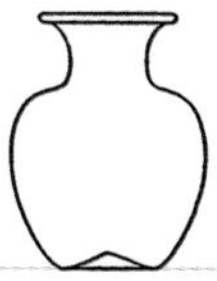
5281

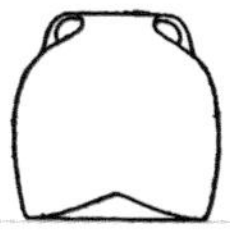
5282

5283

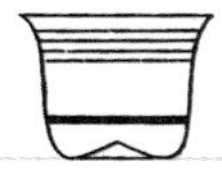
5284

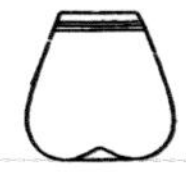
5285

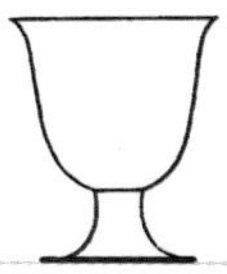
5286

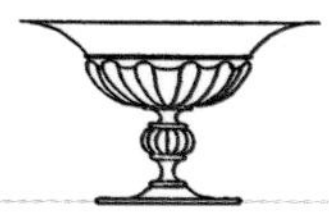
5287

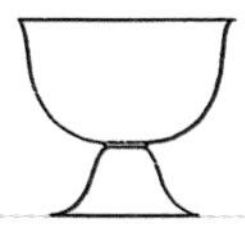
5288

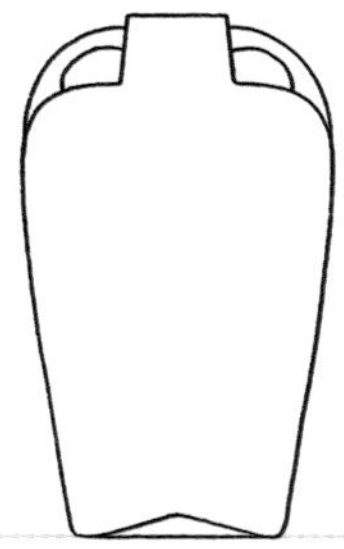
5289

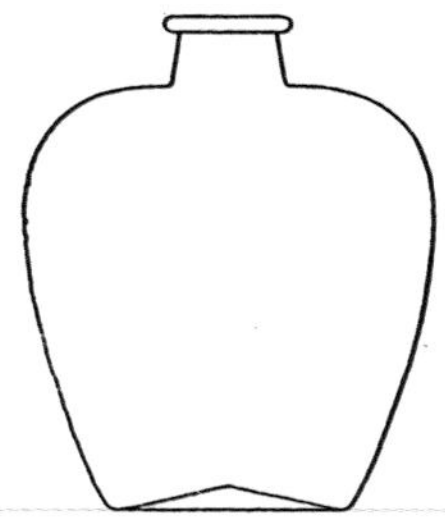
5290

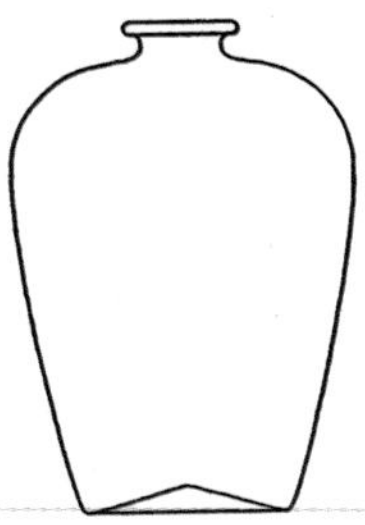
5291

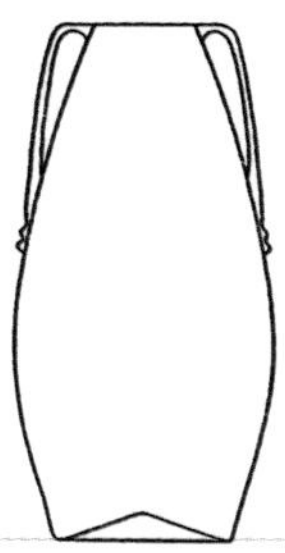
5292

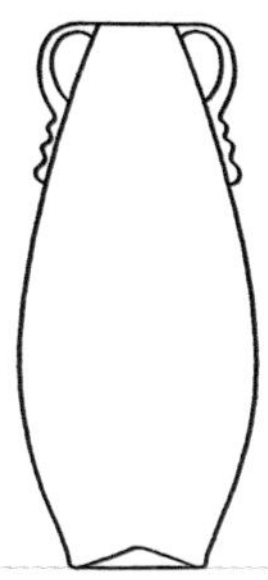
5293

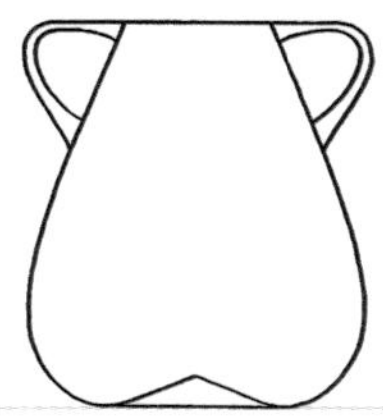
5294

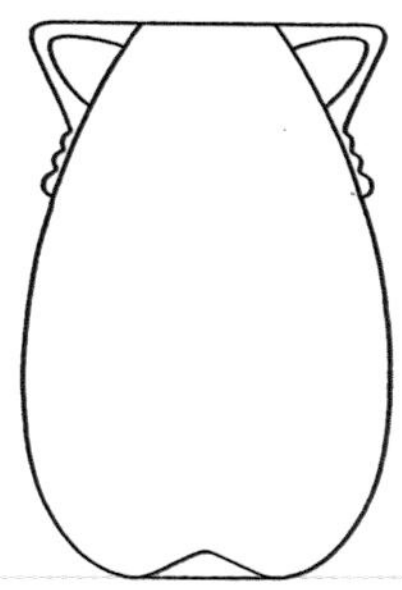
5295

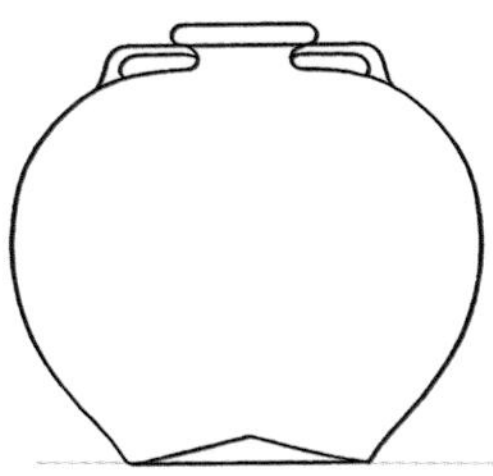
5296

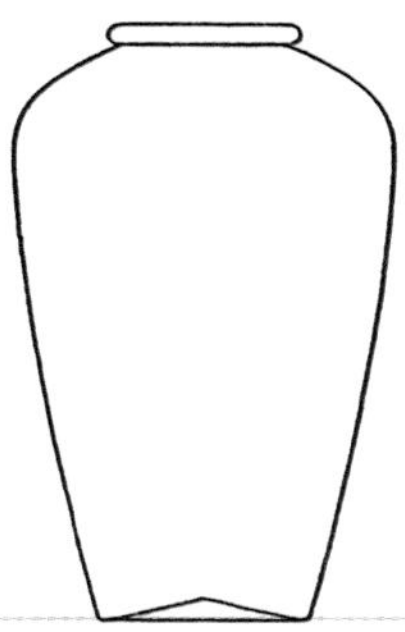
5297

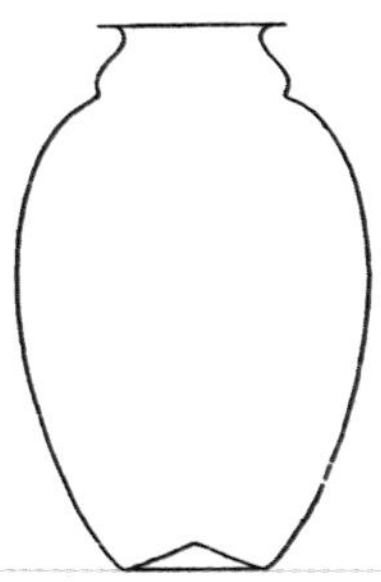
5298

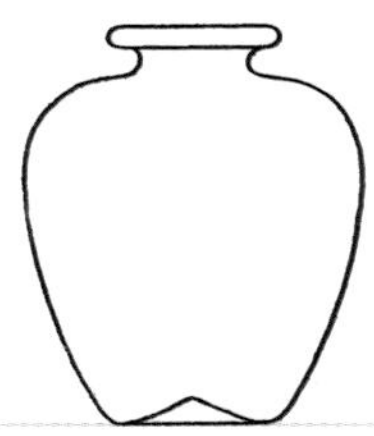
5299

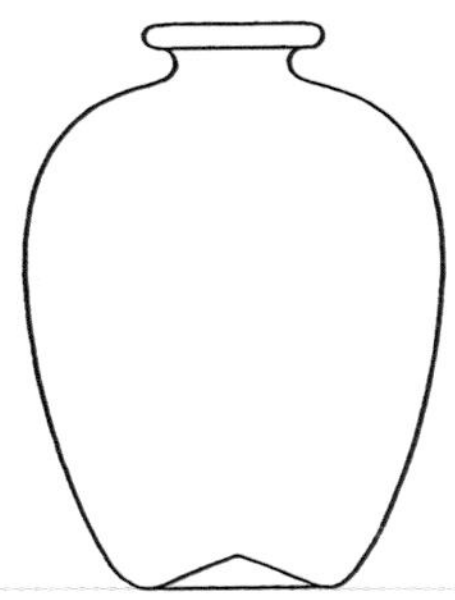
5300

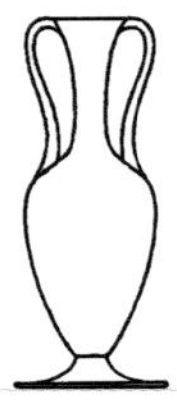

5301

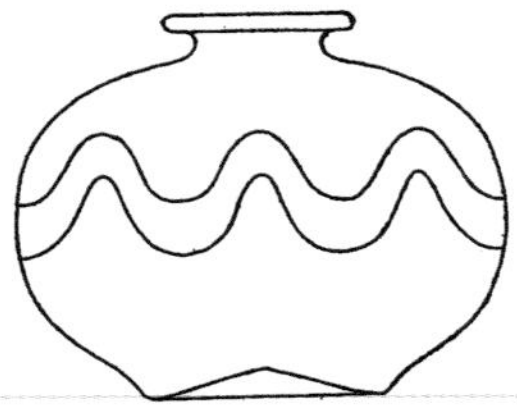

5302

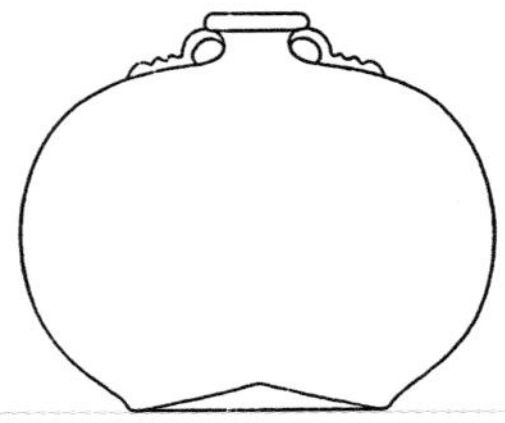

5303

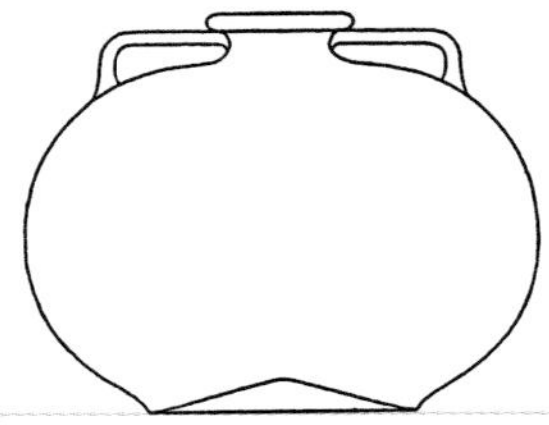

5304

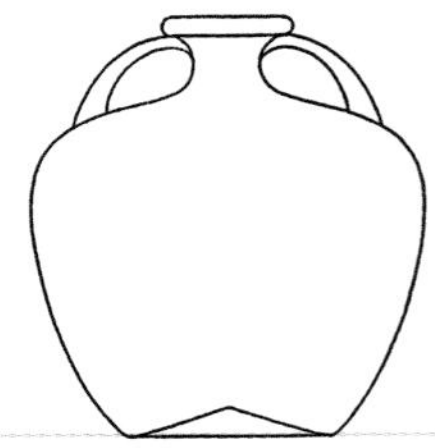

5305

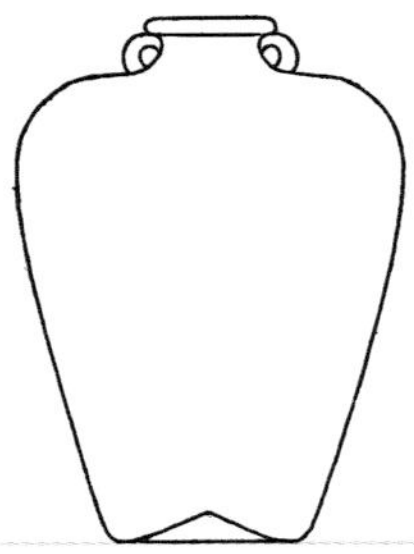

5306

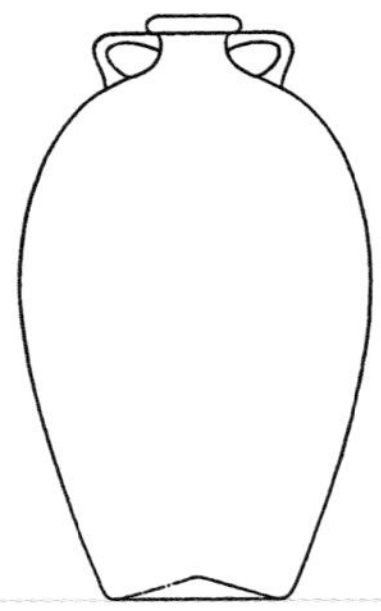

5307

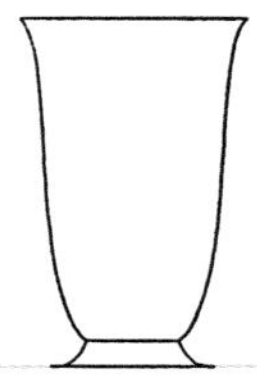

5308

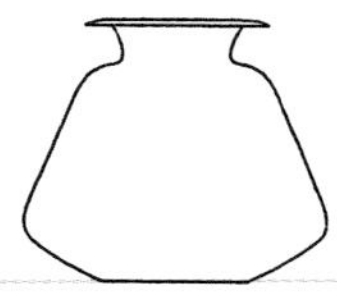

5309

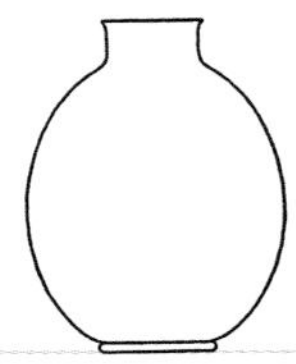

5310

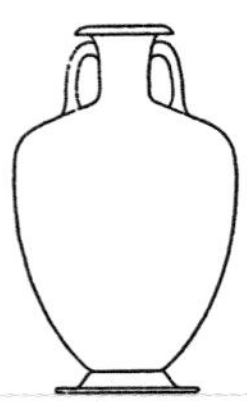

5311

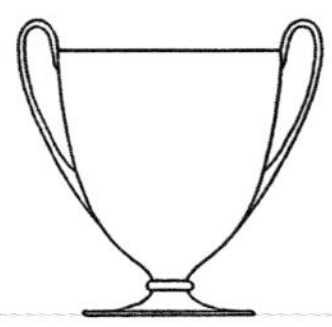

5312

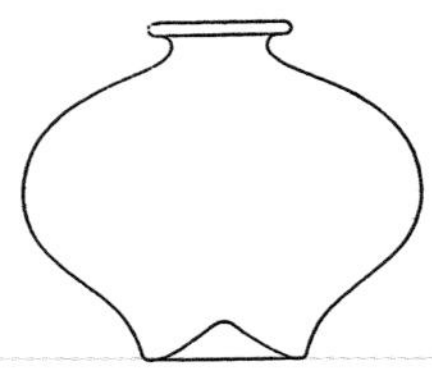

5313

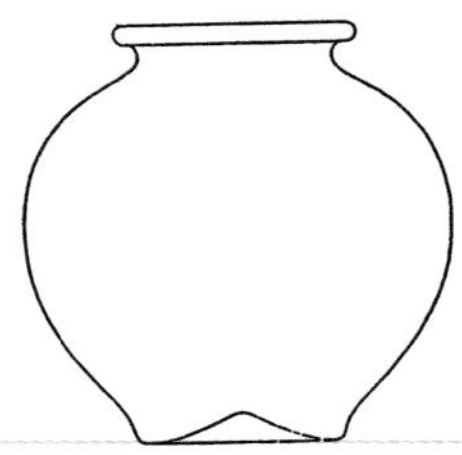

5314

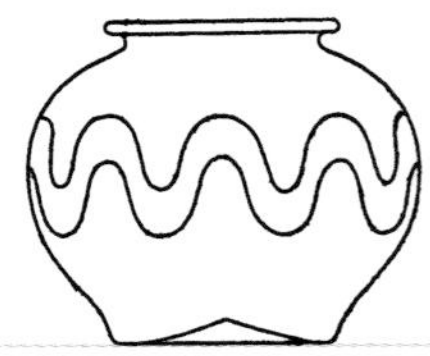

5315

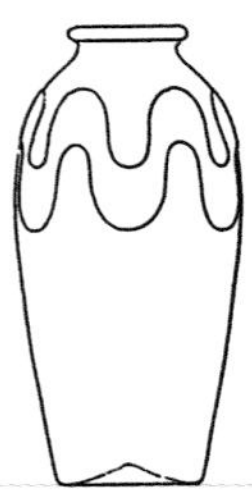

5316

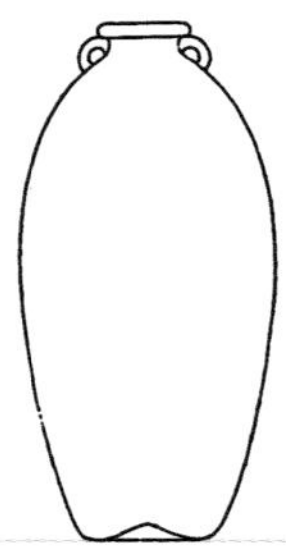

5317

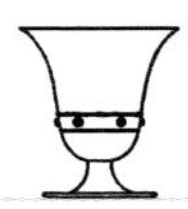

5318

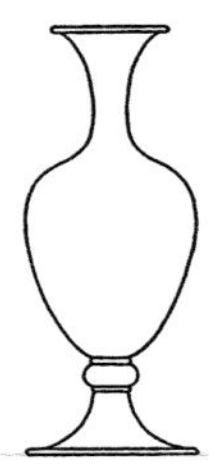

5319

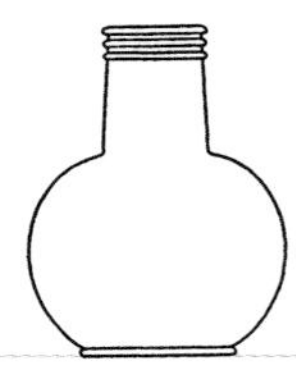

5320

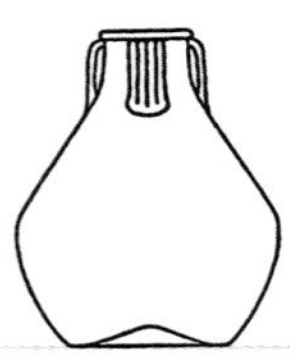

5321

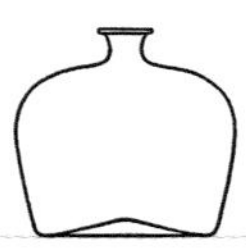

5322

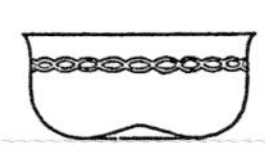

5323

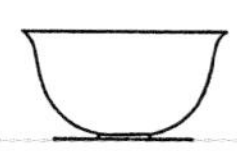

5324

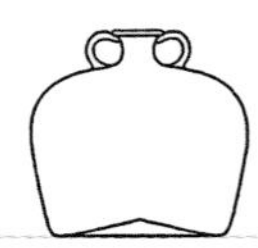

5325

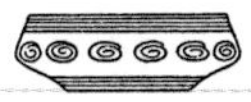
5326

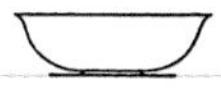
5327

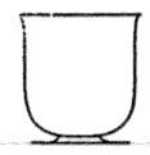
5328

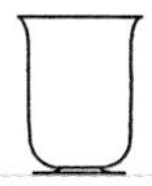
5329

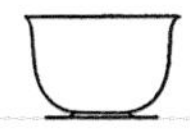
5330

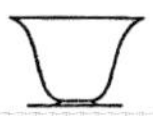
5331

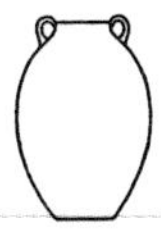
5332

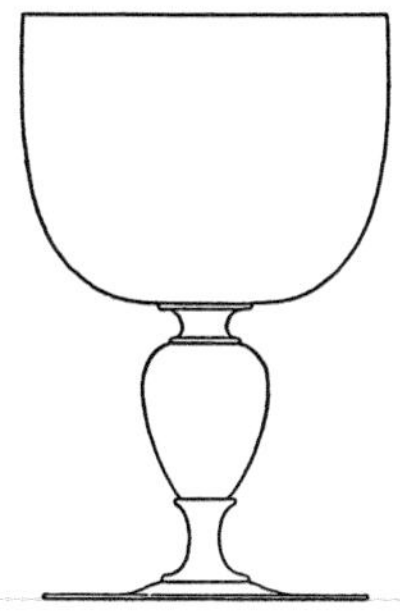
5333

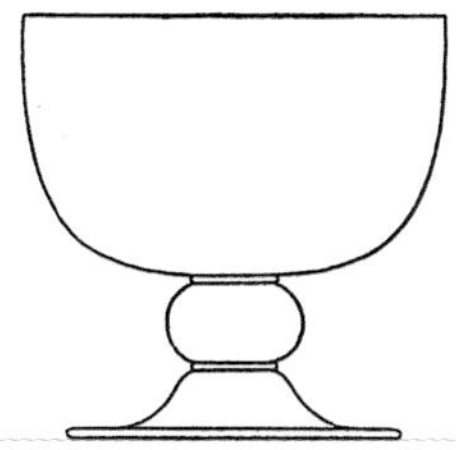
5334

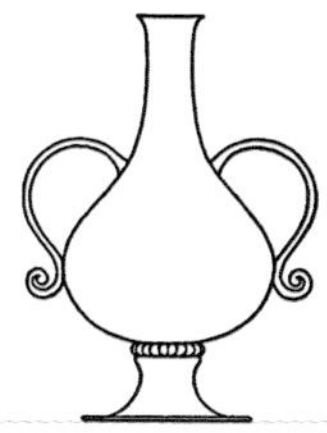
5335

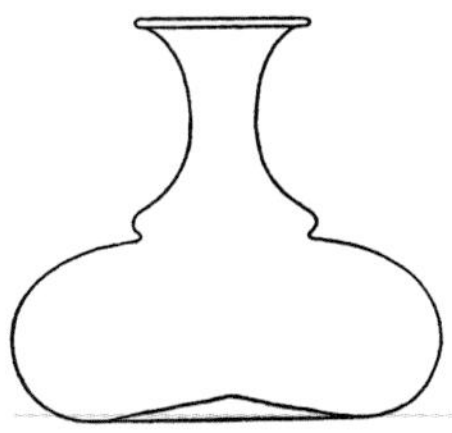
5336

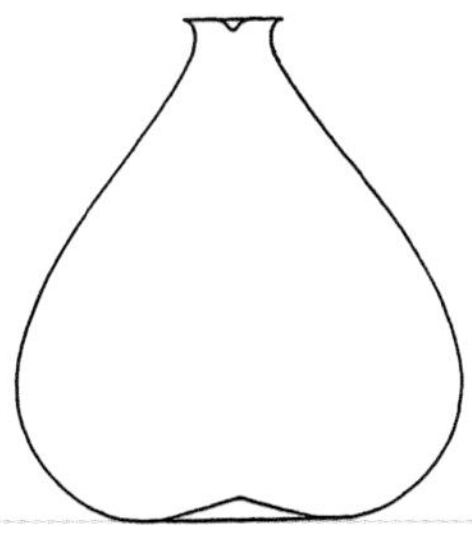
5337

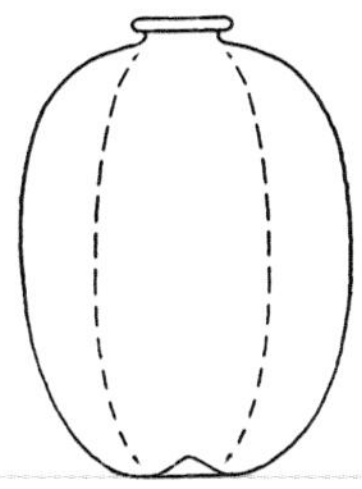
5338

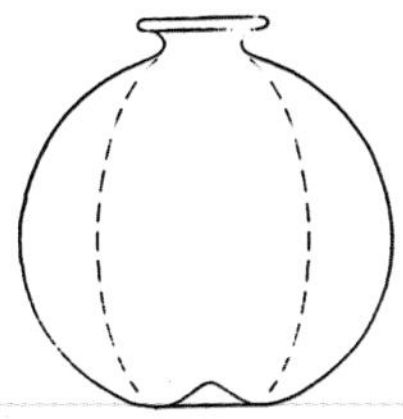
5339

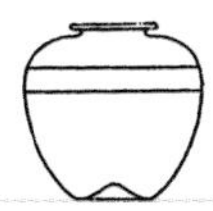
5340

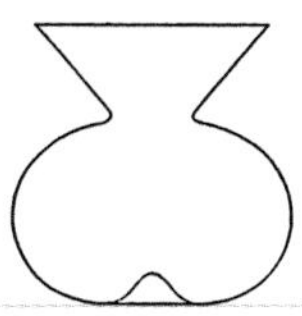
5341

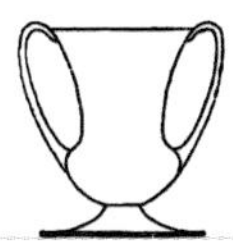
5342

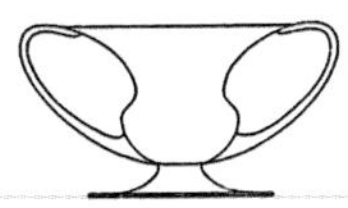
5343

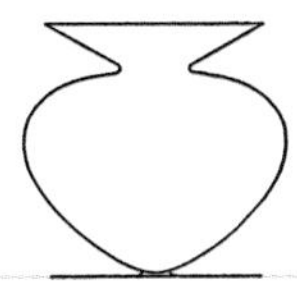
5344

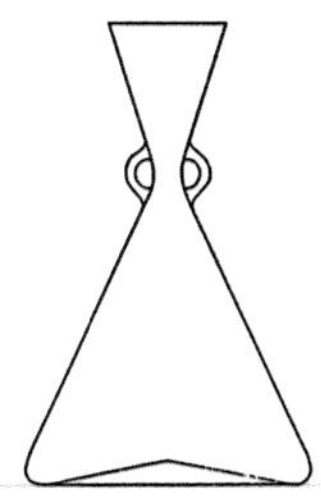
5345

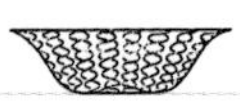
5346

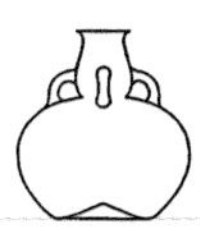
5347

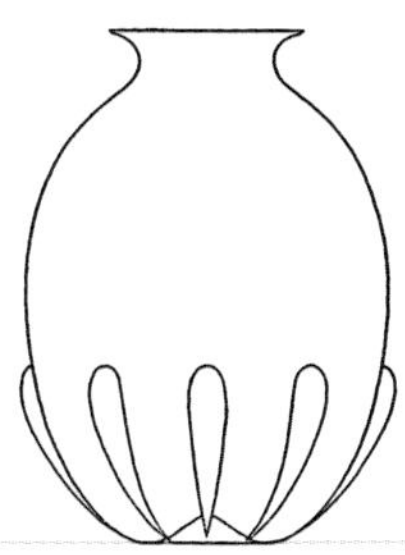
5348

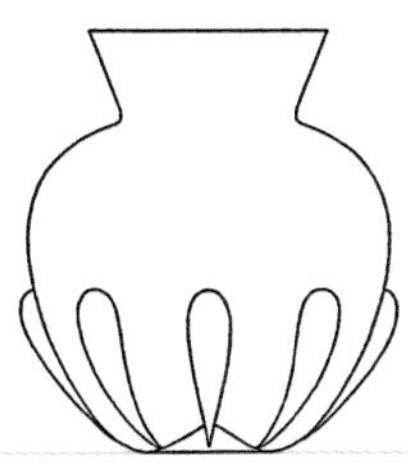
5349

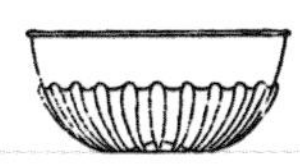
5350

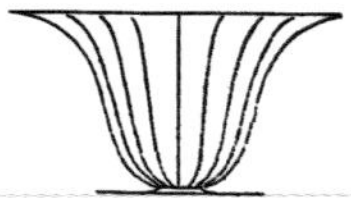
5351

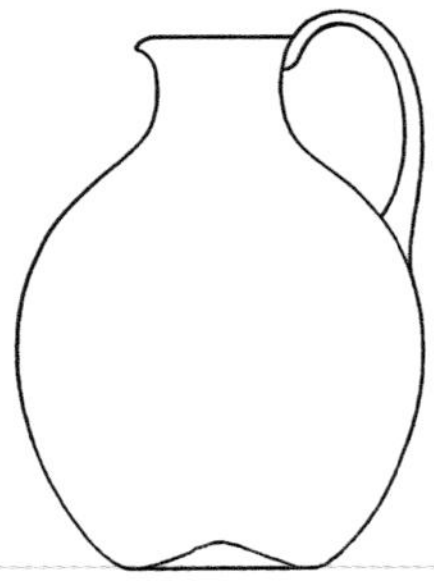
5352

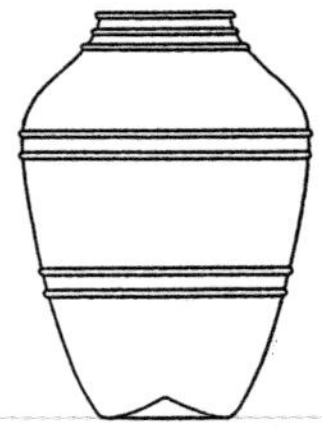
5353

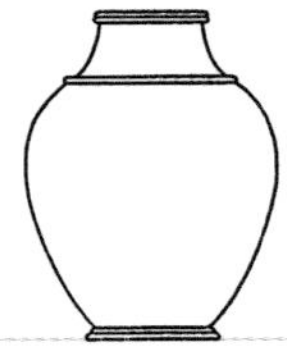
5354

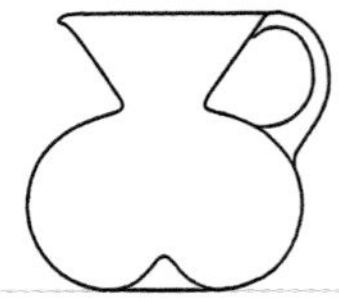
5355

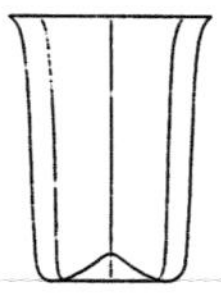
5356

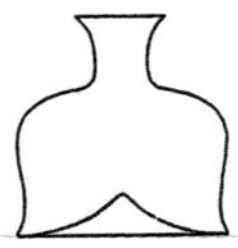
5357

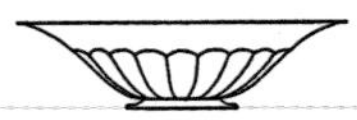
5358

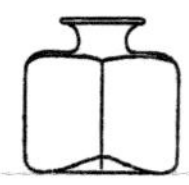
5359

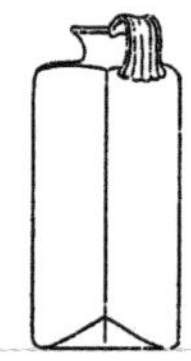
5360

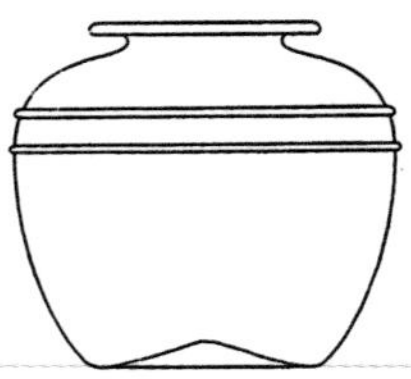
5361

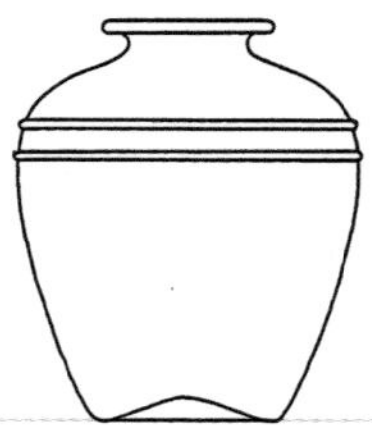
5362

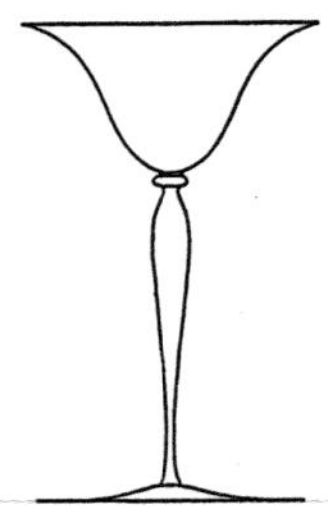
5363

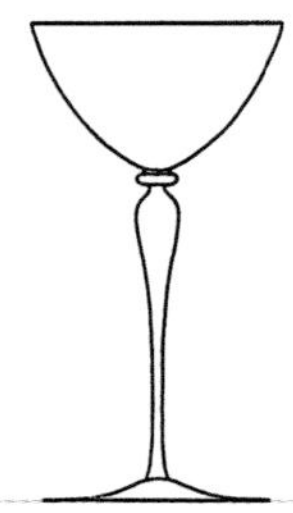
5364

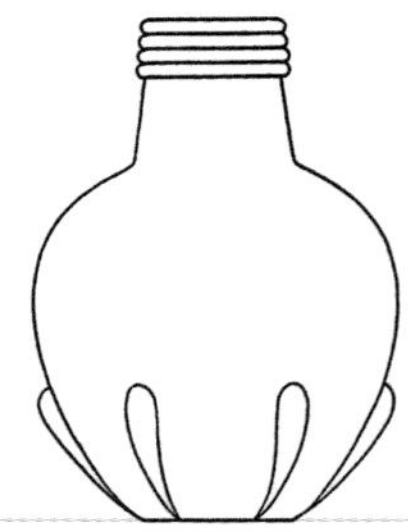
5365

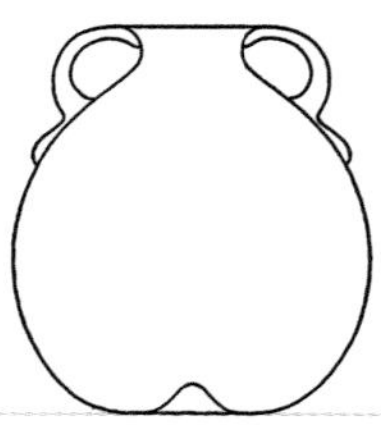
5366

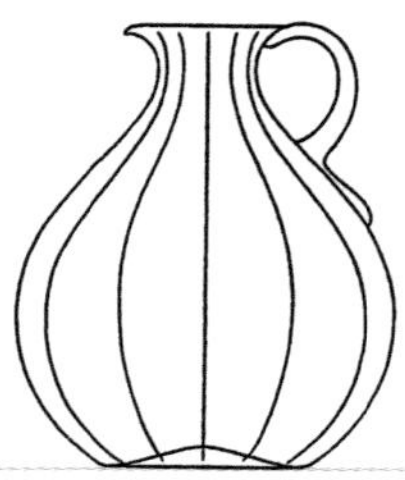
5367

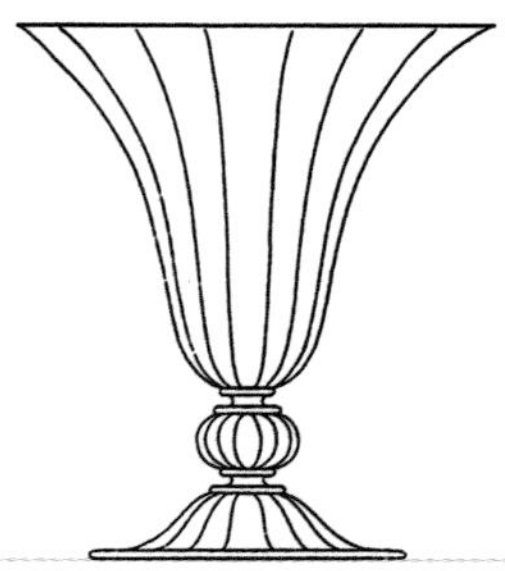
5368

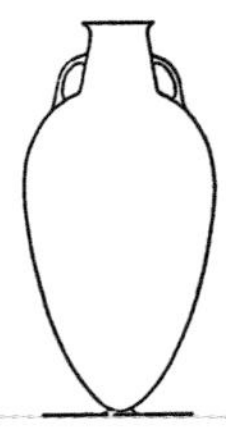
5369

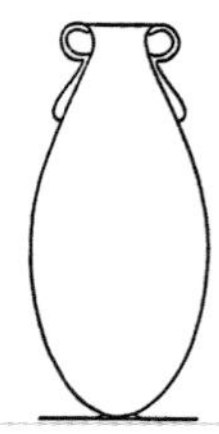
5370

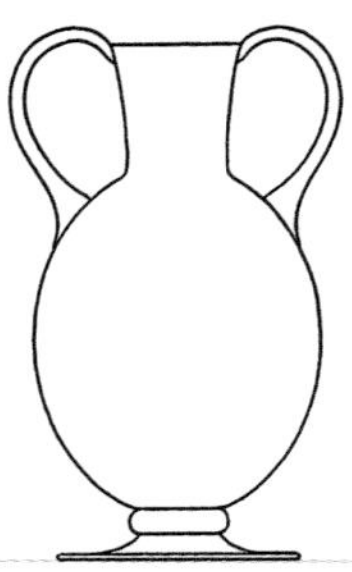
5371

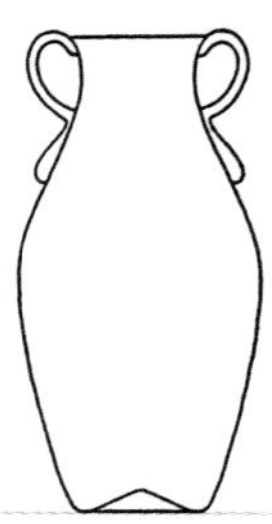
5372

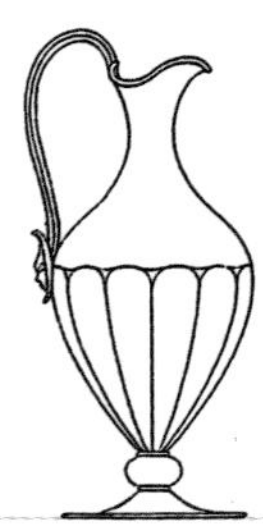
5373

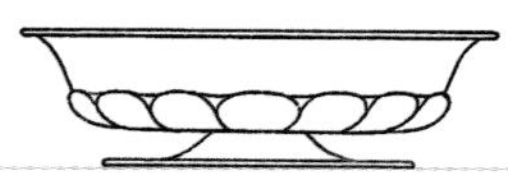
5374

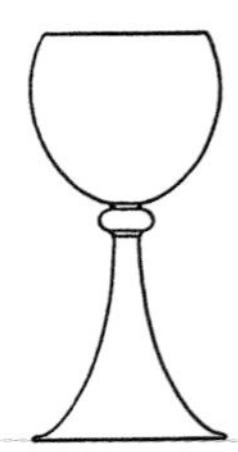
5375

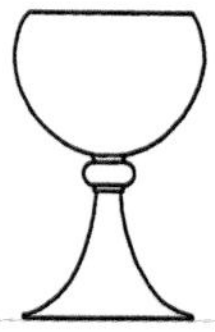
5376

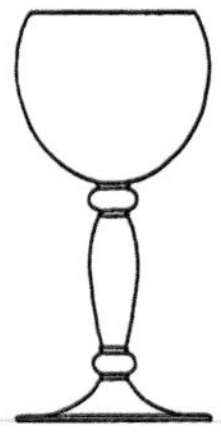
5377

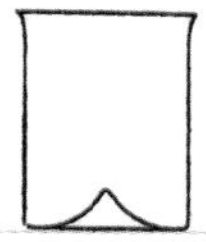
5378

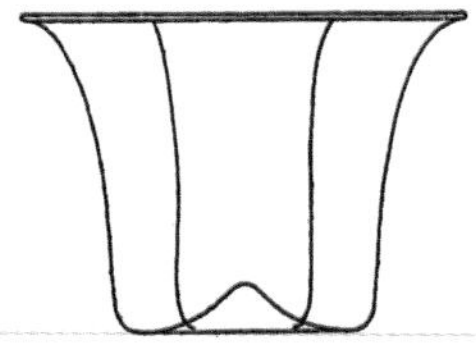
5379

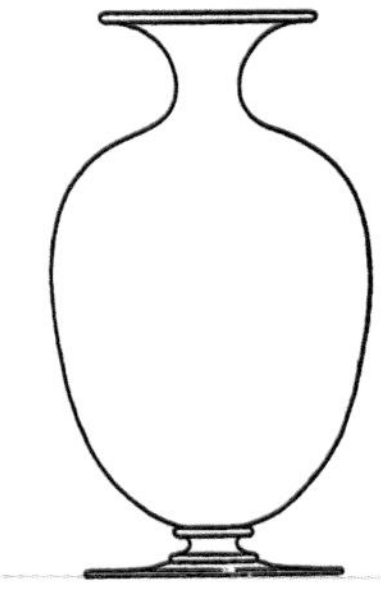
5380

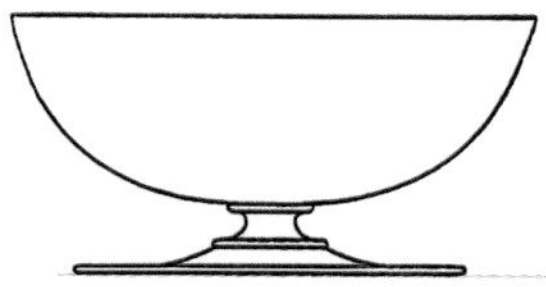
5381

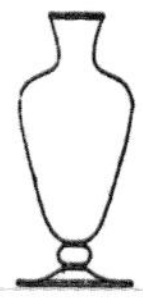
5382

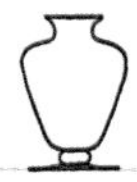
5383

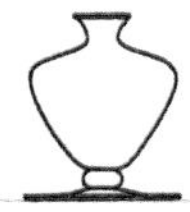
5384

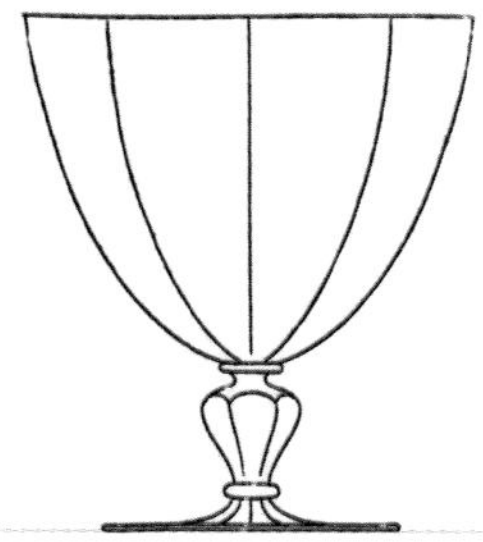
5385

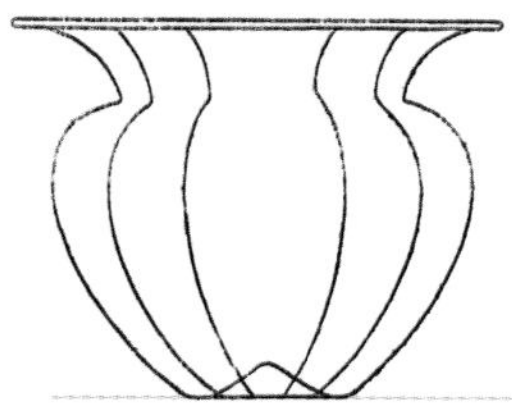
5386

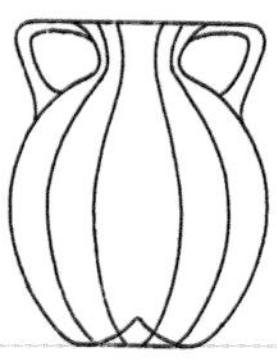
5387

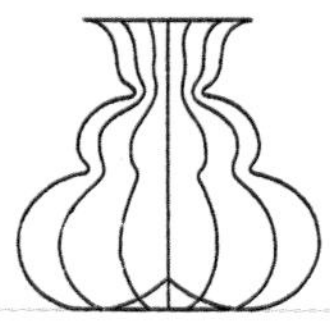
5388

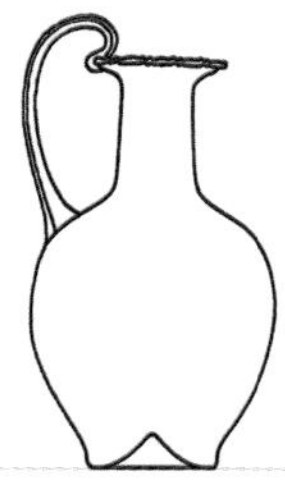
5389

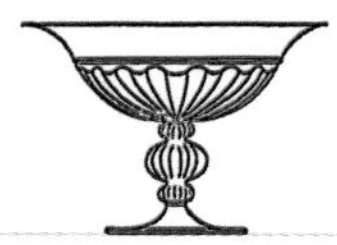
5390

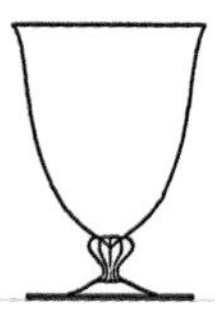
5391

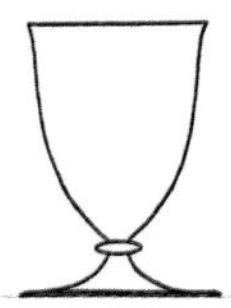
5392

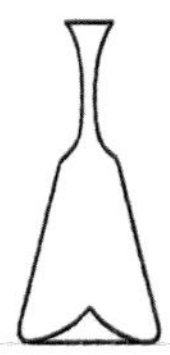
5393

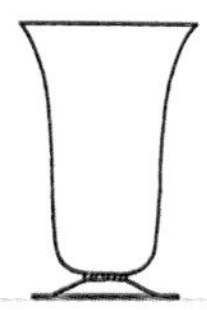
5394

5395

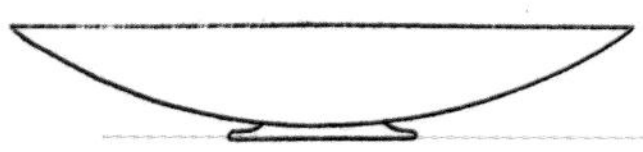
5396

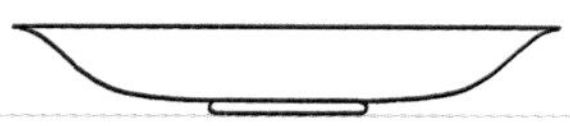
5397

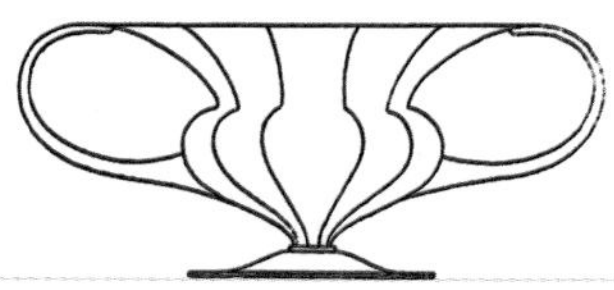
5398

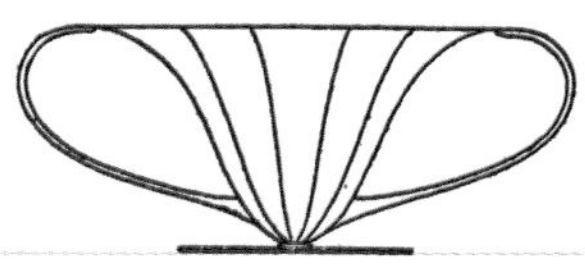
5399

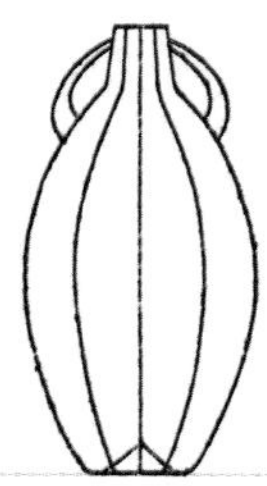
5400

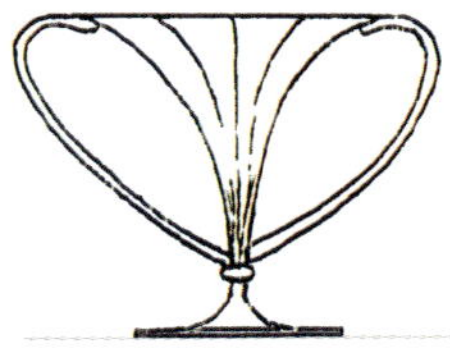
5401
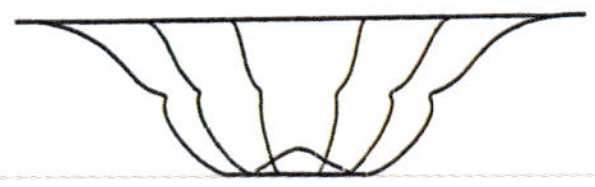
5402
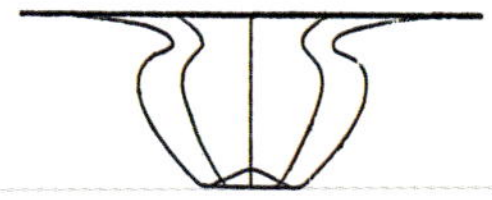
5403
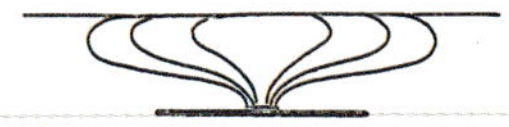
5404
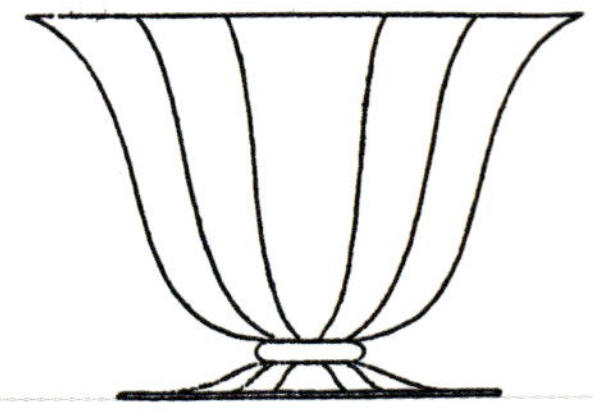
5405
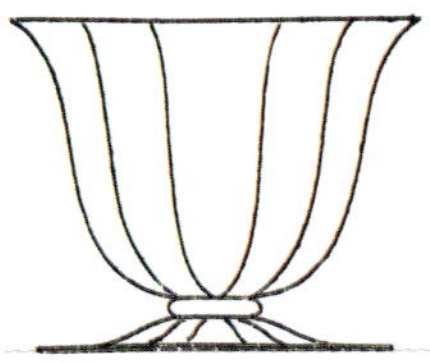
5406
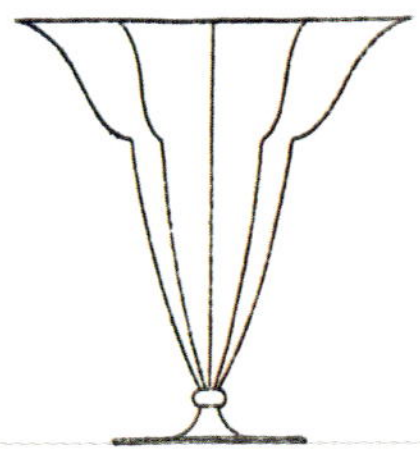
5407
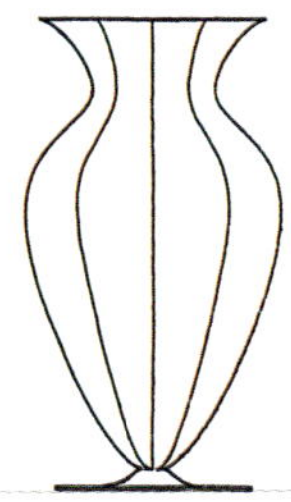
5408
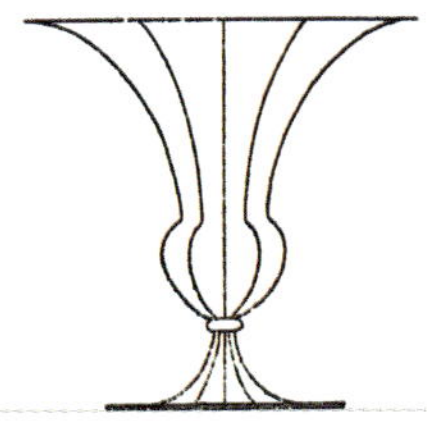
5409
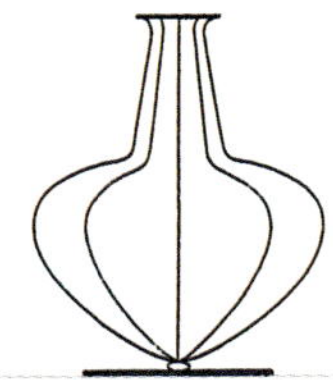
5410
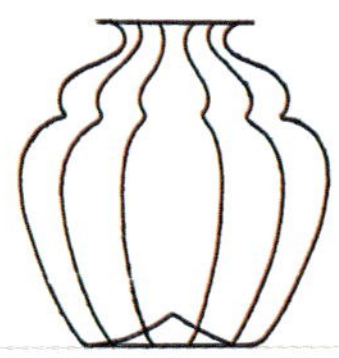
5411
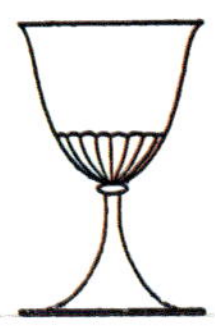
5412
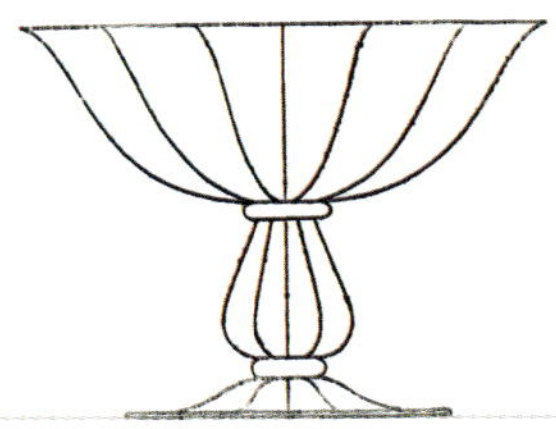
5413
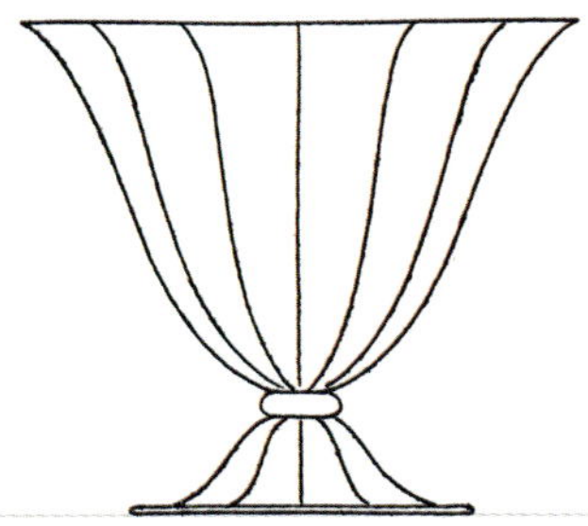
5414
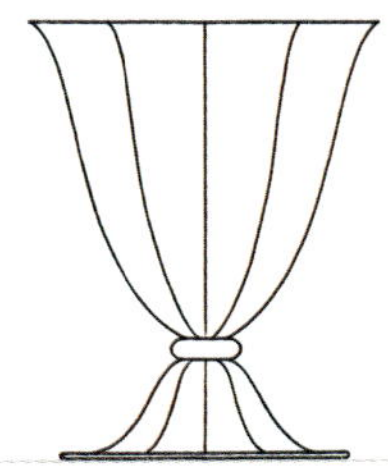
5415
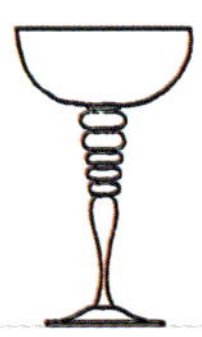
5416
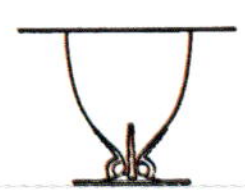
5417
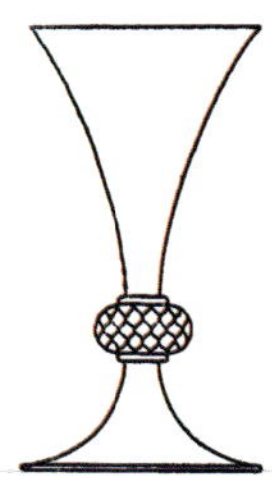
5418
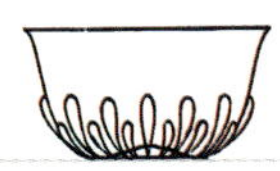
5419
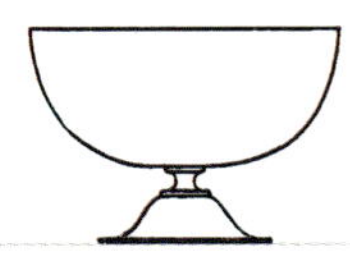
5420
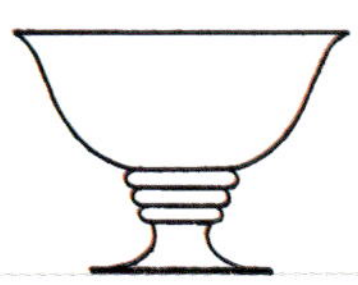
5421
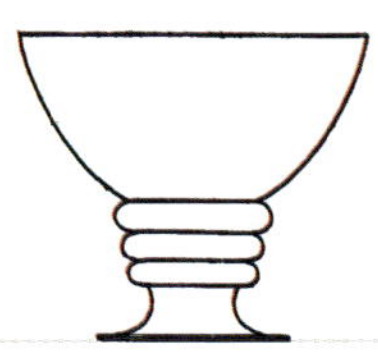
5422
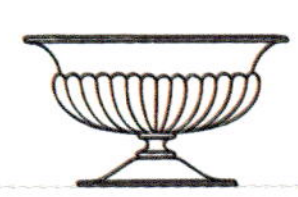
5423
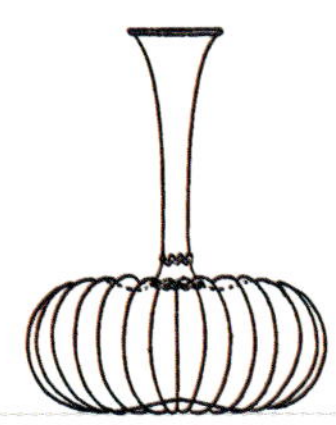
5424
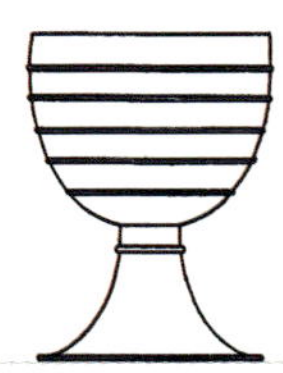
5425

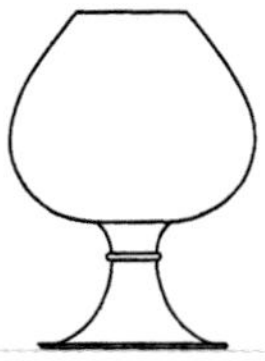
5426

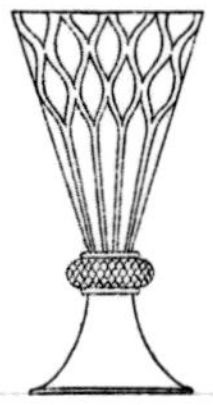
5427

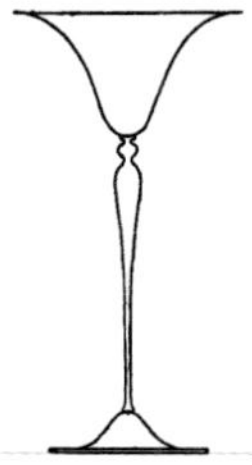
5428

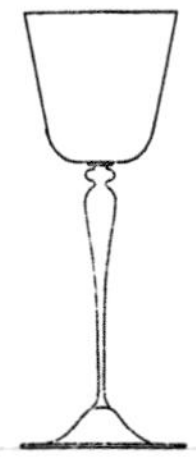
5429

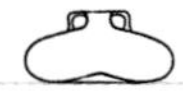
5430

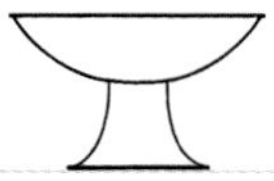
5431

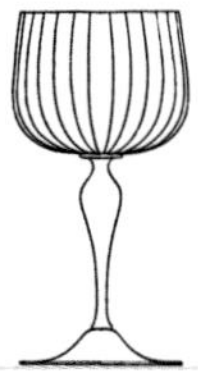
5432

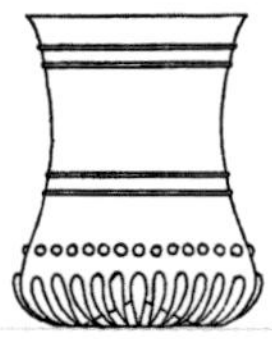
5433

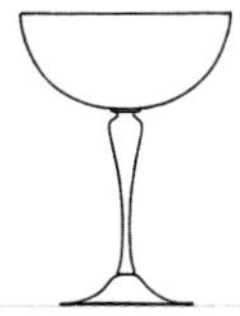
5434

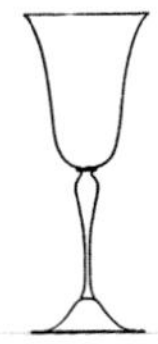
5435

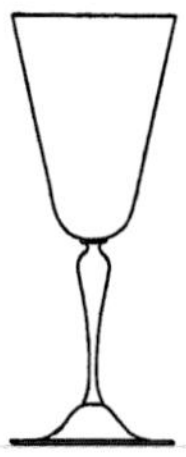
5436

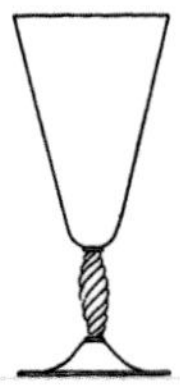
5437

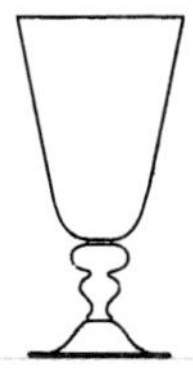
5438

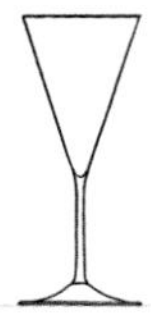
5439

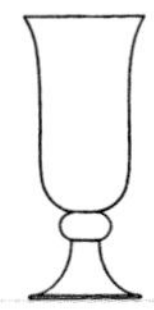
5440

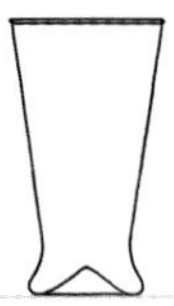
5441

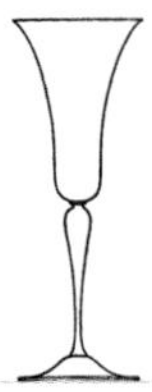
5442

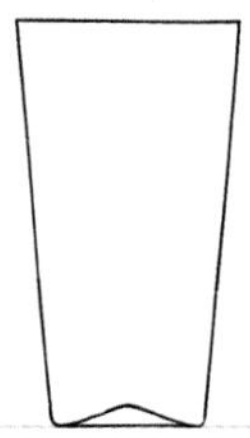
5443

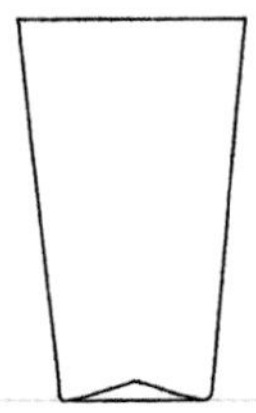
5444

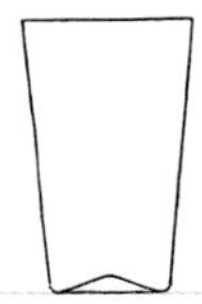
5445

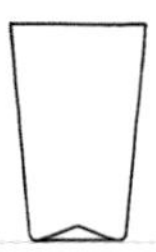
5446

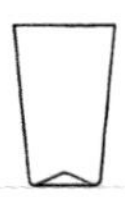
5447

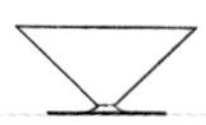
5448

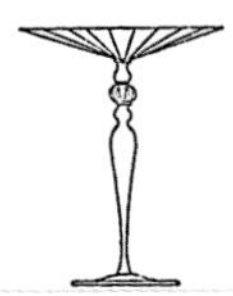
5449

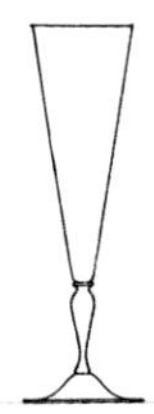
5450

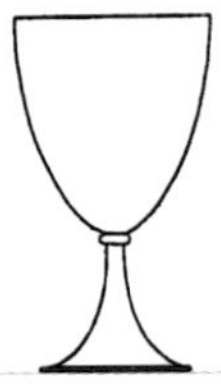
5451

5452
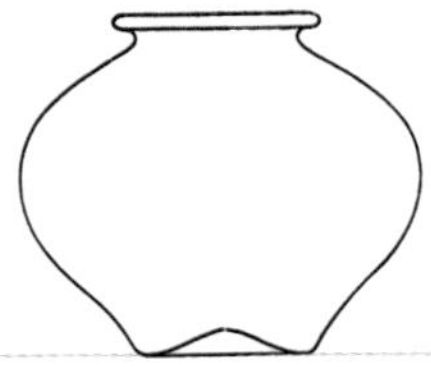
5453
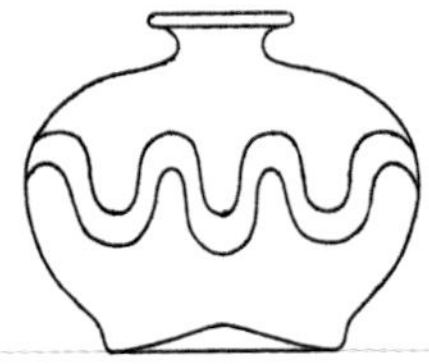
5454
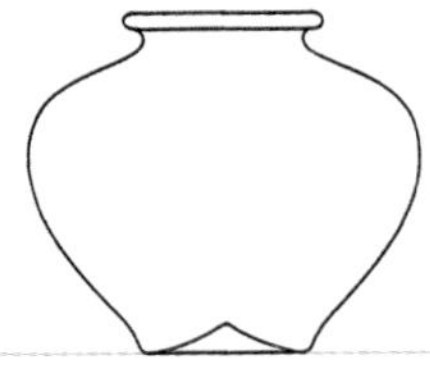
5455
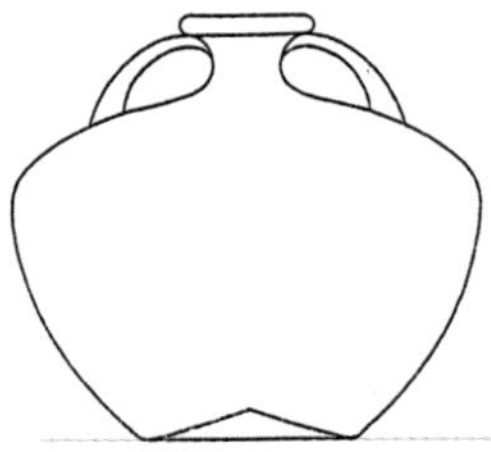
5456
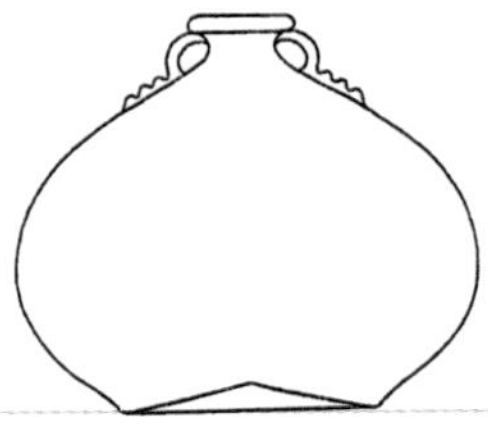
5457
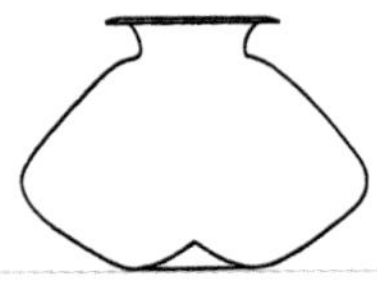
5458
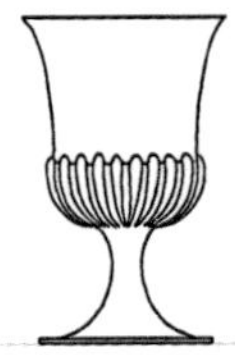
5459
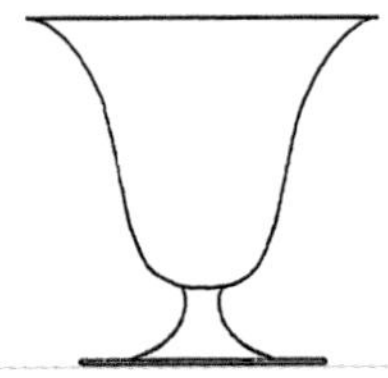
5460
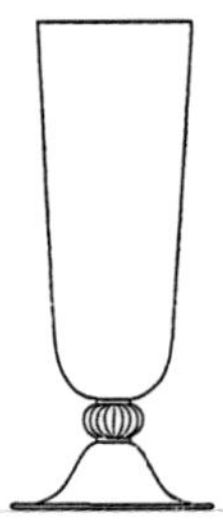
5461
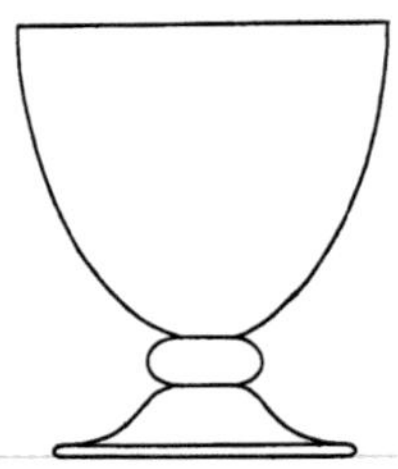
5462
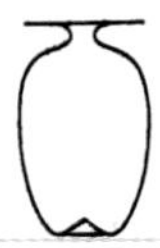
5463
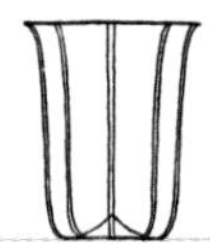
5464
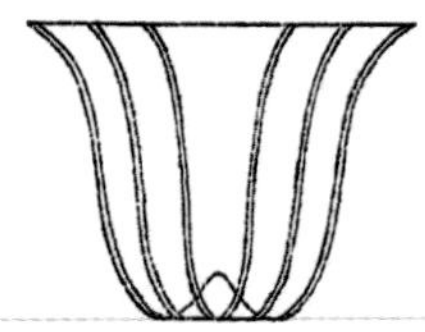
5465
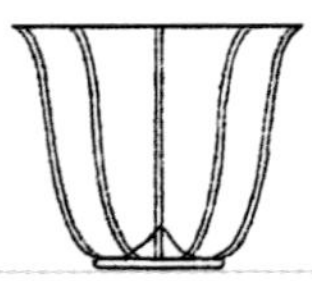
5466
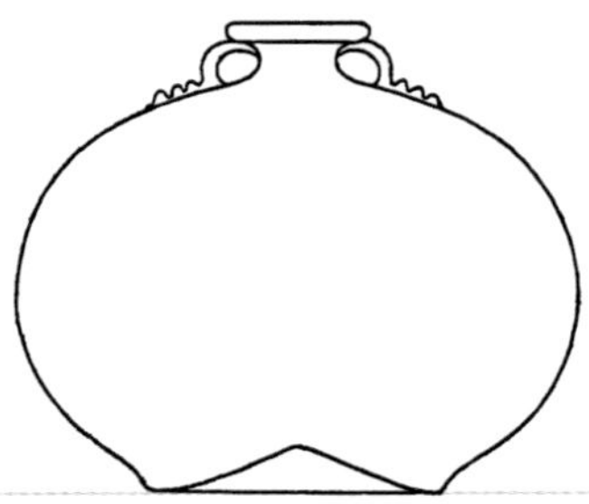
5467
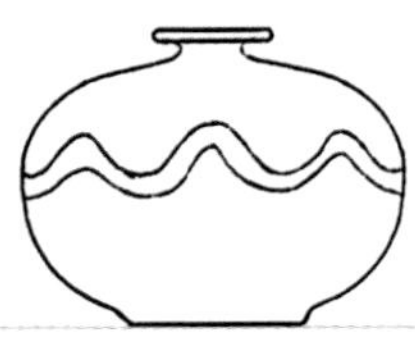
5468
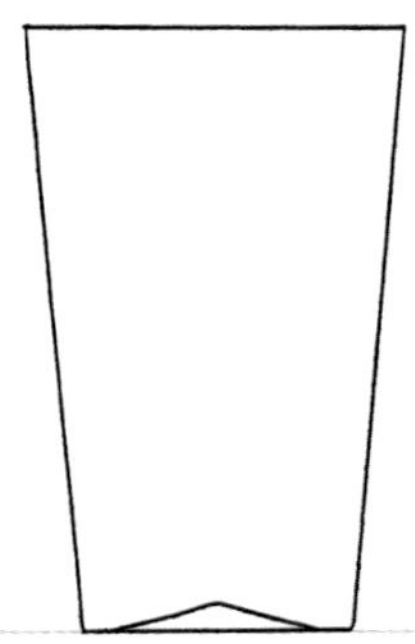
5469
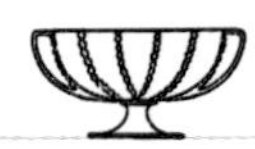
5470
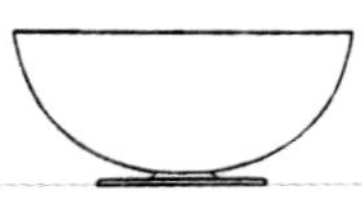
5471
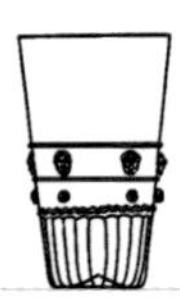
5472
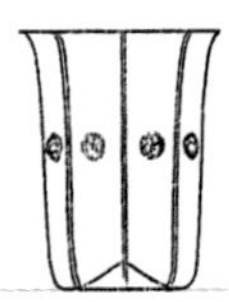
5473
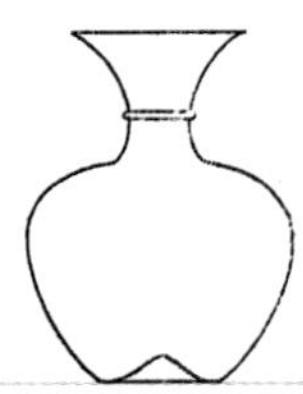
5474
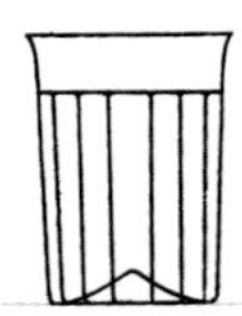
5475

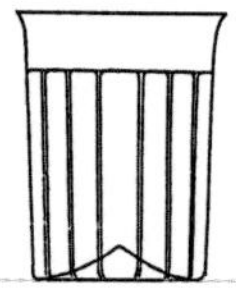
5476

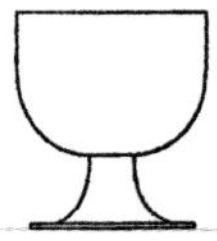
5477

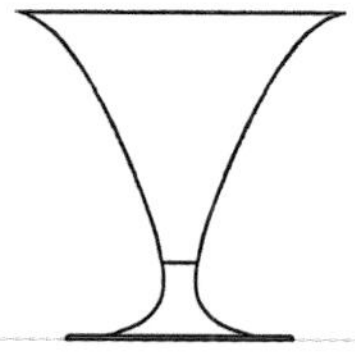
5478

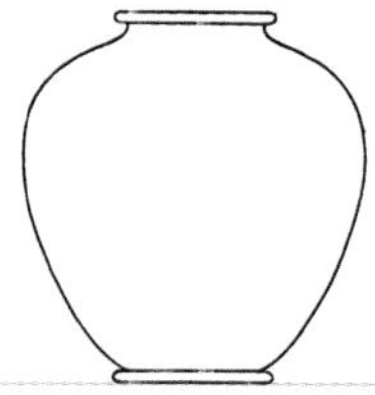
5479

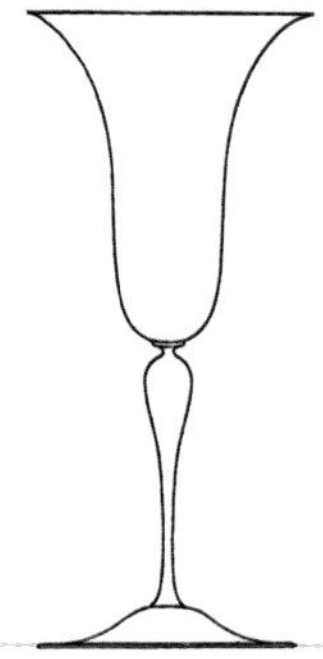
5480

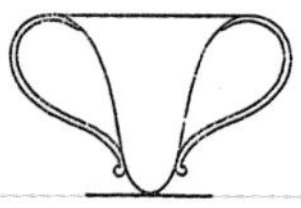
5481

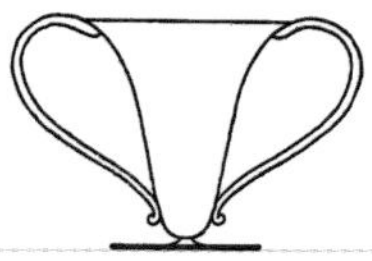
5482

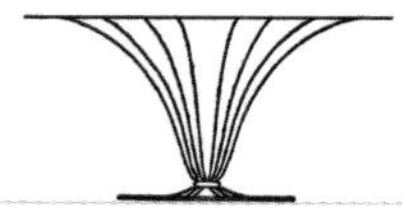
5483

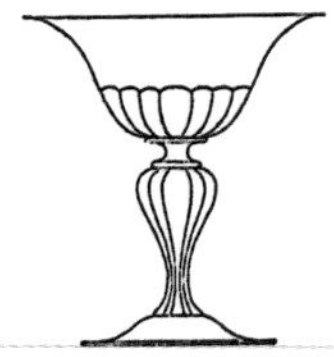
5484

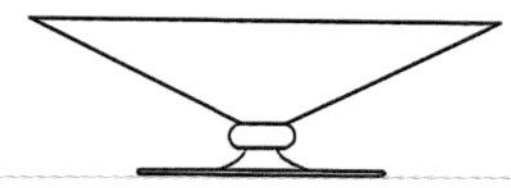
5485

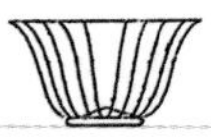
5486

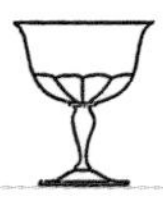
5487

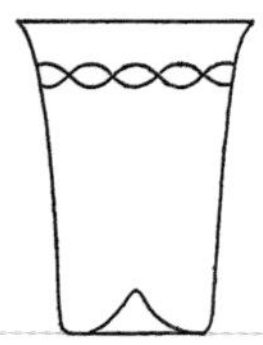
5488

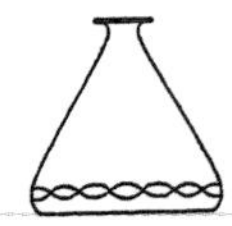
5489

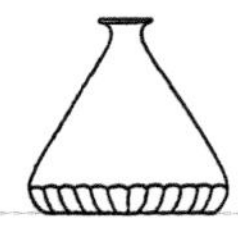
5490

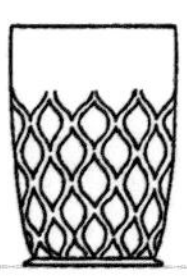
5491

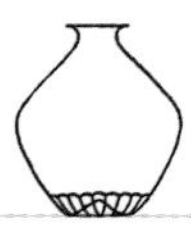
5492

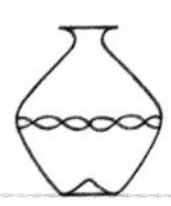
5493

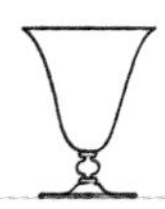
5494

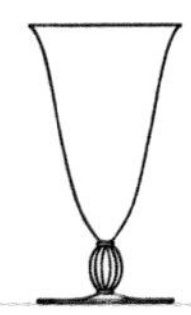
5495

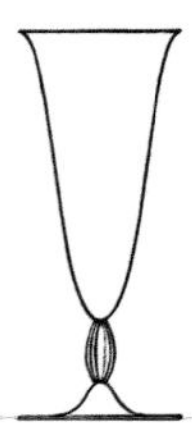
5496

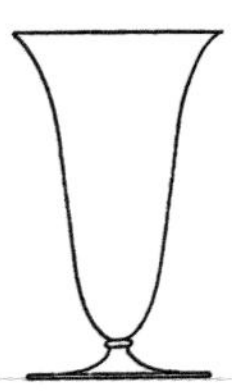
5497

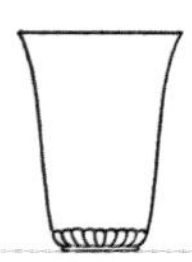
5498

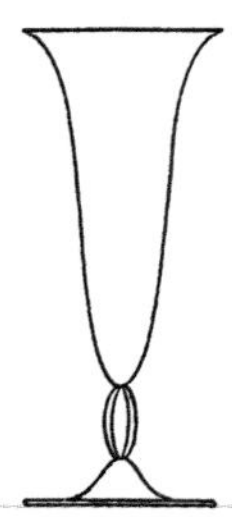
5499

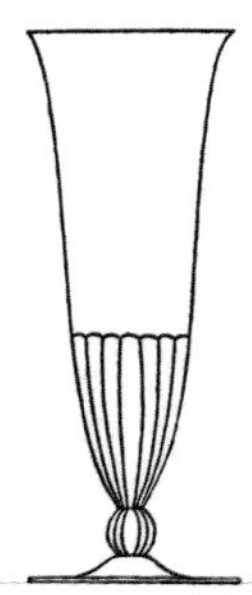
5500

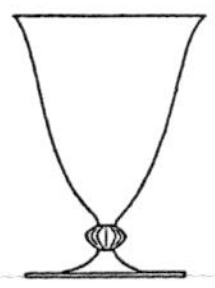
5501
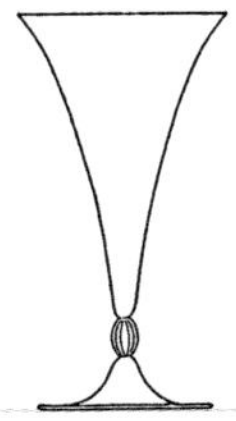
5502
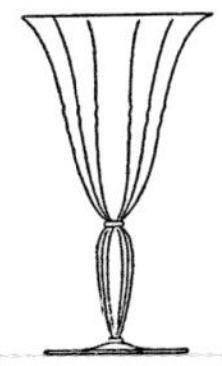
5503
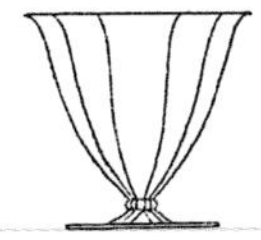
5504
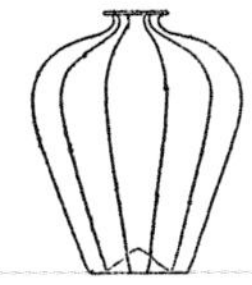
5505
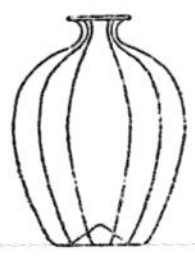
5506
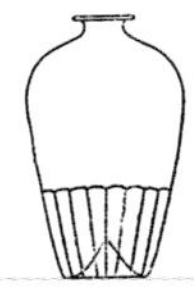
5507
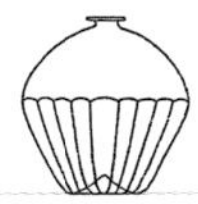
5508
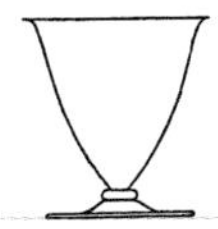
5509
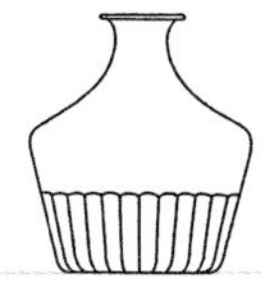
5510
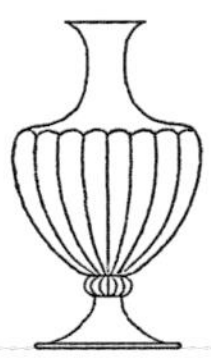
5511
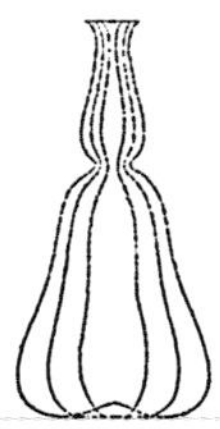
5512
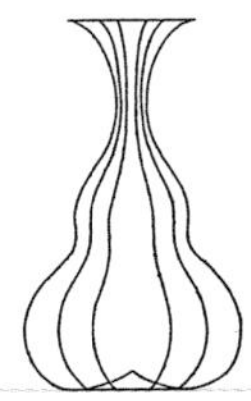
5513
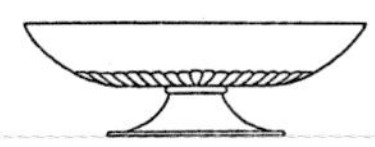
5514
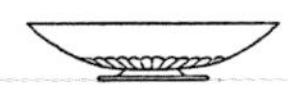
5515
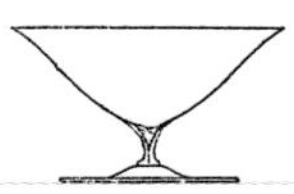
5516
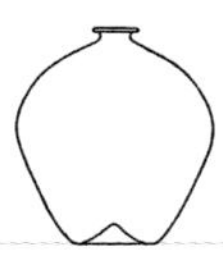
5517
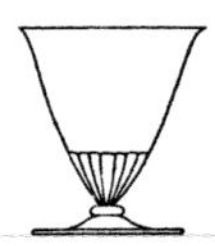
5518
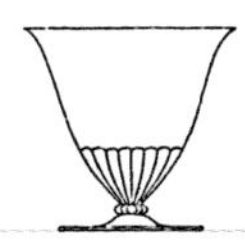
5519
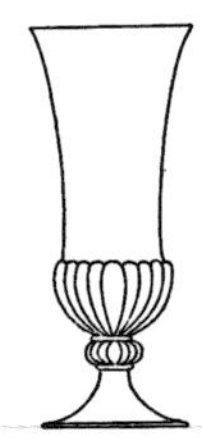
5520
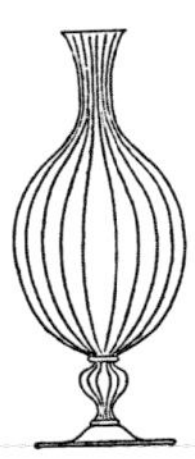
5521
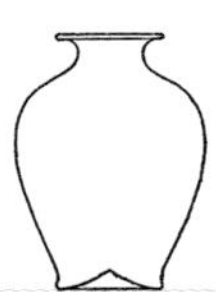
5522
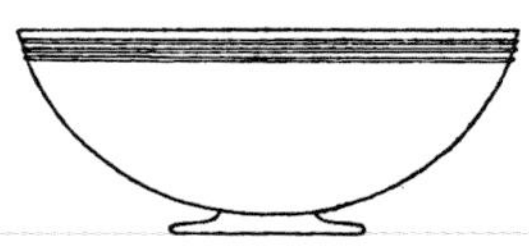
5523
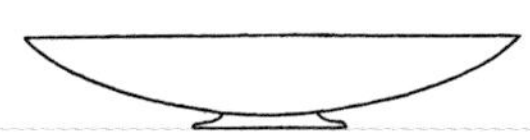
5524
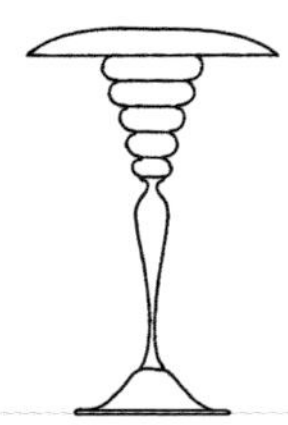
5525

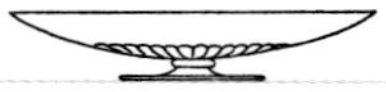
5526
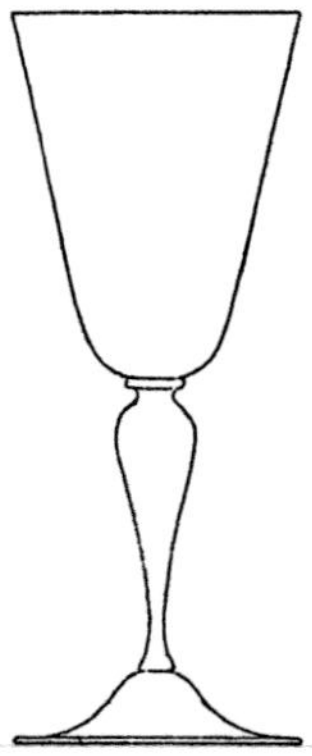
5527
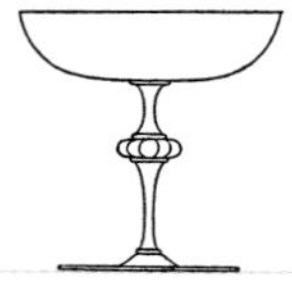
5528
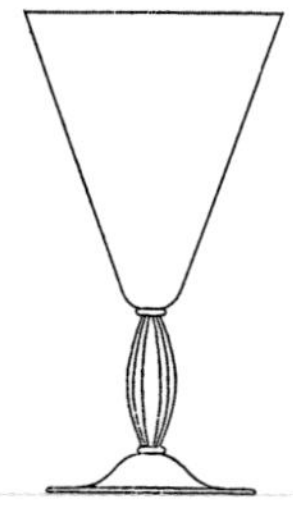
5529
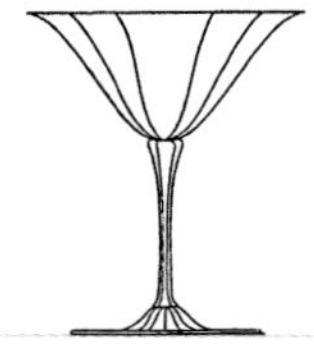
5530
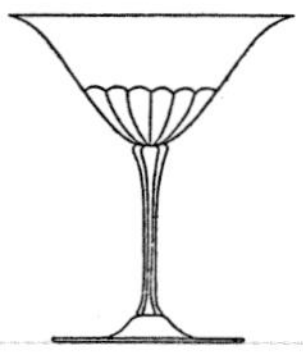
5531
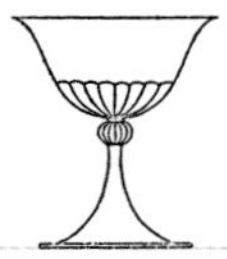
5532
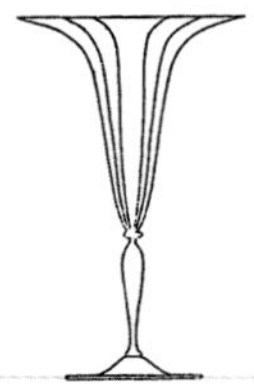
5533
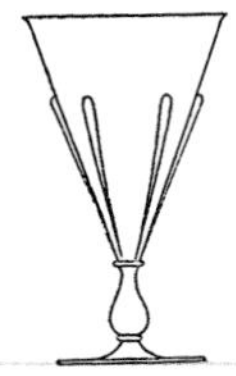
5534
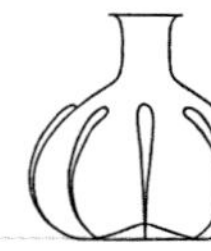
5535
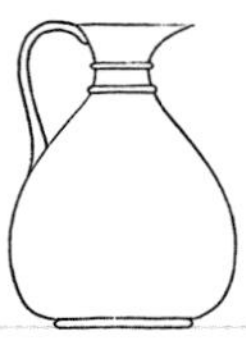
5536
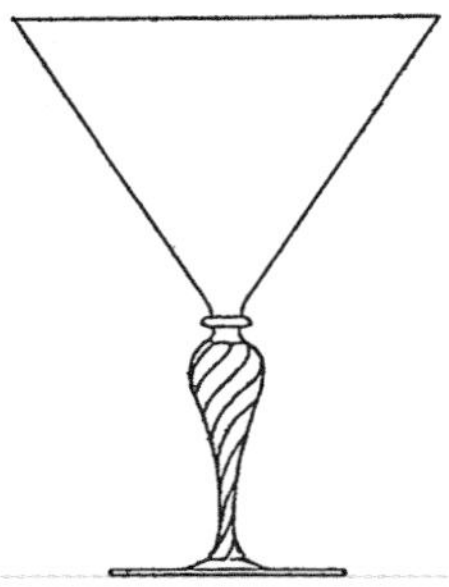
5537
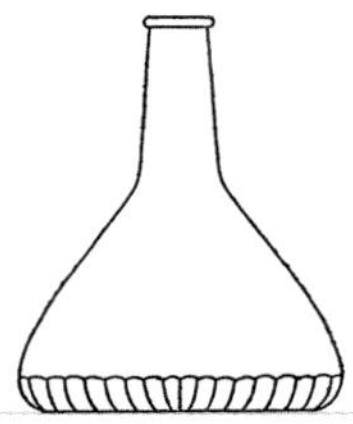
5538
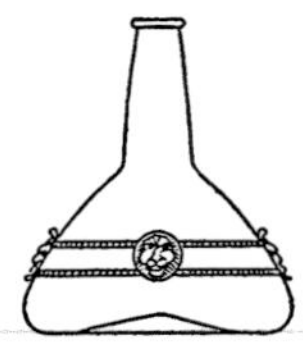
5539
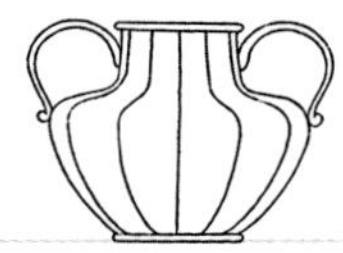
5540
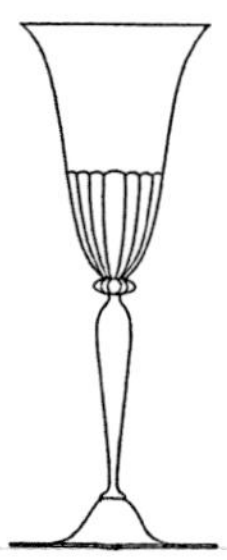
5541
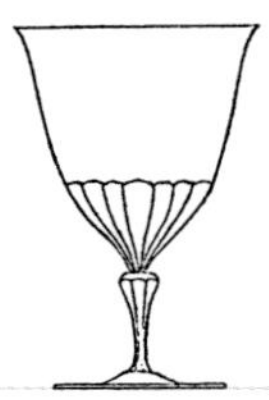
5542
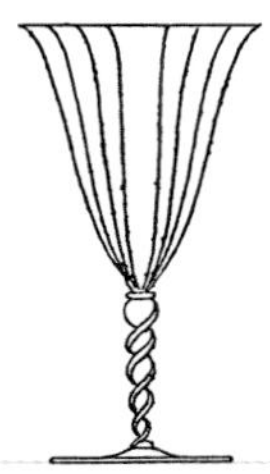
5543
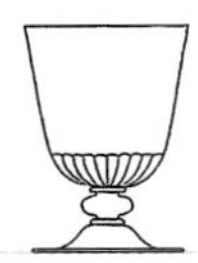
5544
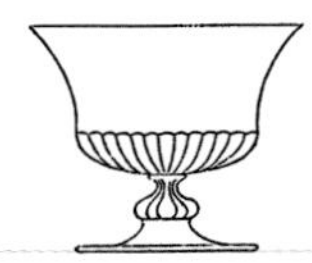
5545
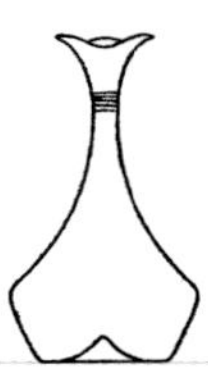
5546
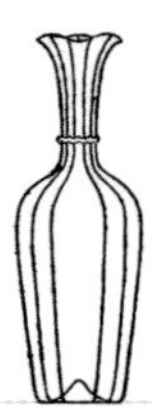
5547
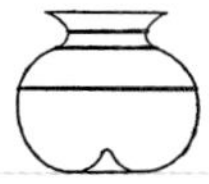
5548
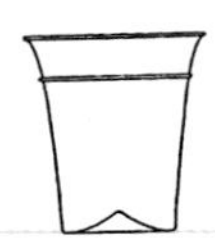
5549
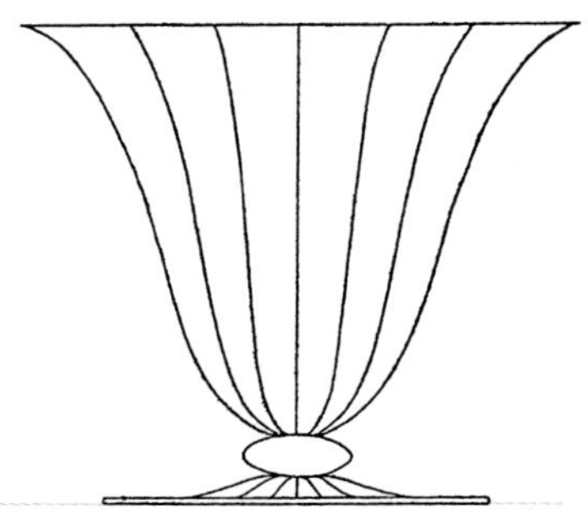
5550

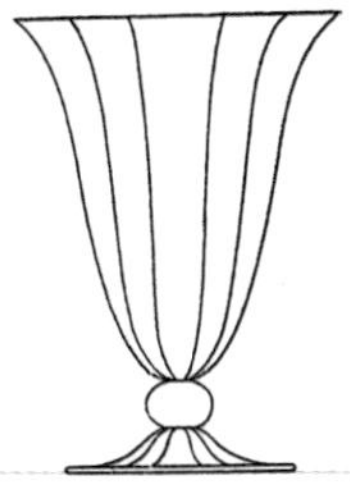
5551
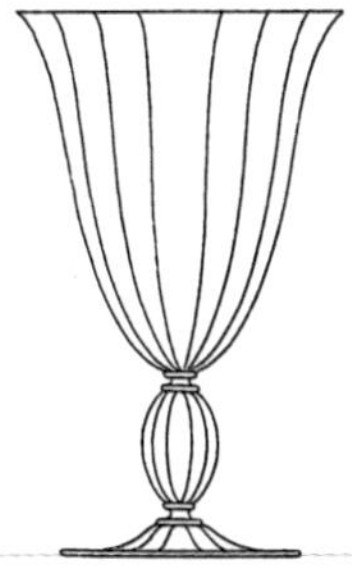
5552
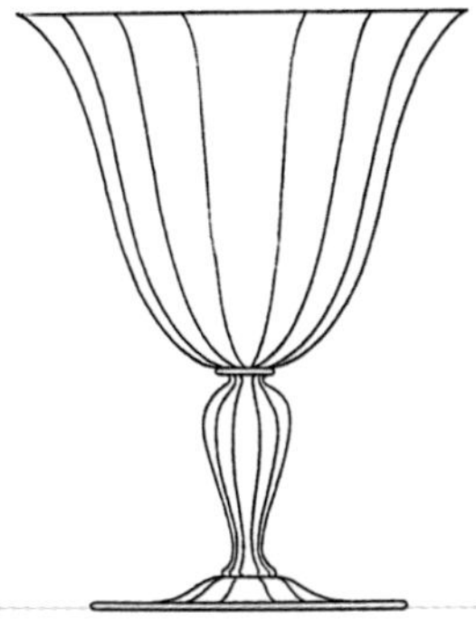
5553
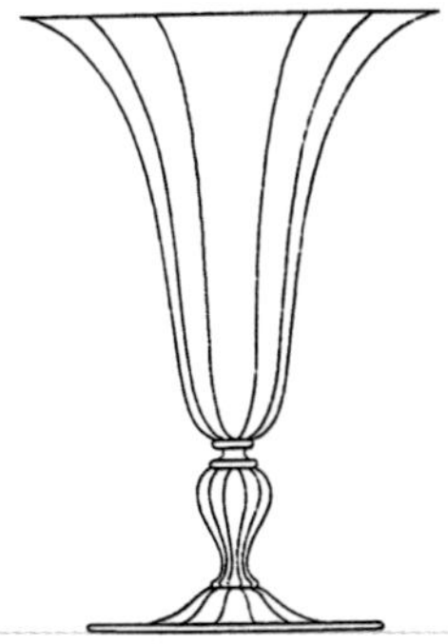
5554
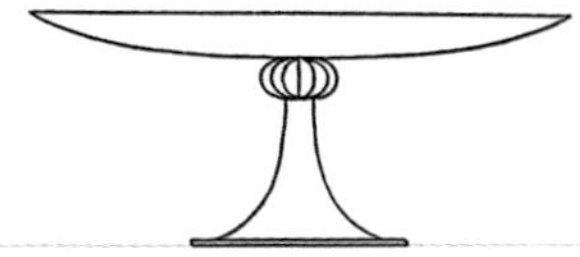
5555

5556
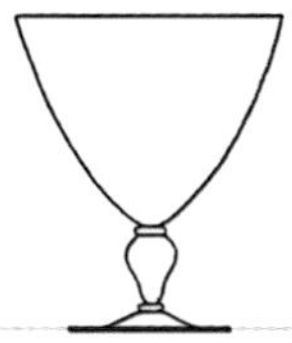
5557
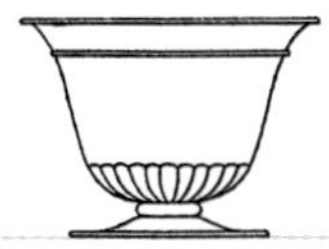
5558
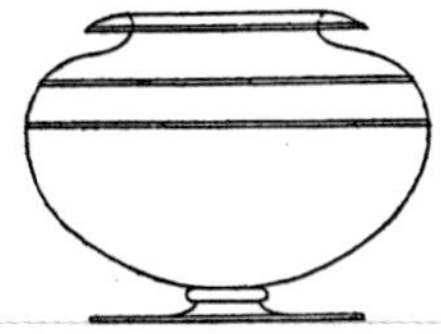
5559
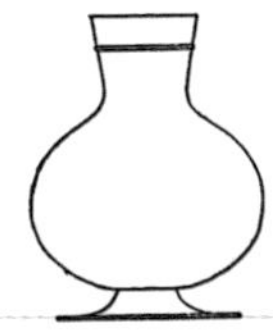
5560

5561
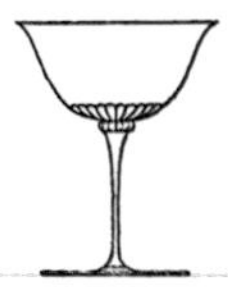
5562
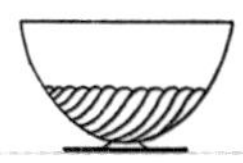
5563
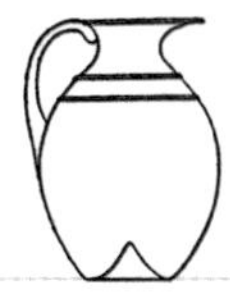
5564
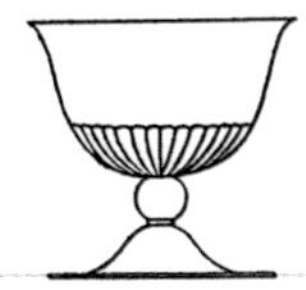
5565
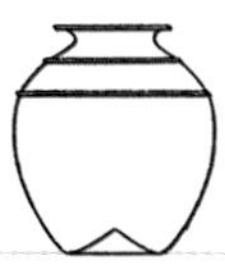
5566
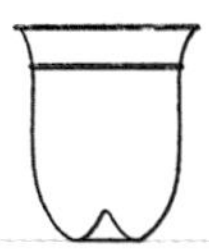
5567
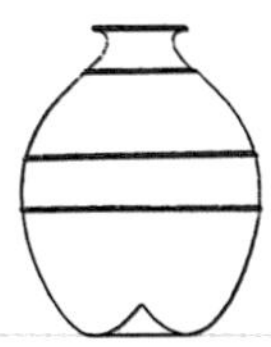
5568
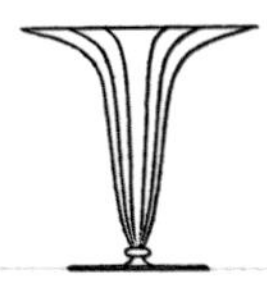
5569
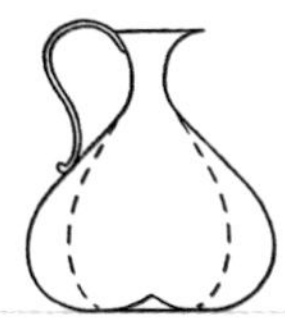
5570
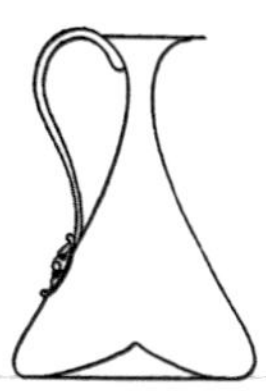
5571
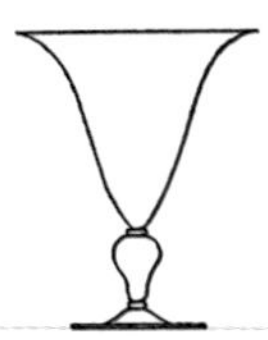
5572
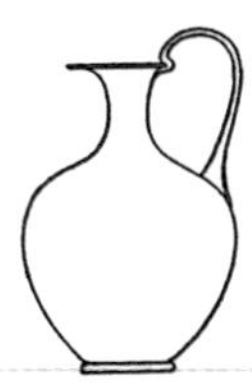
5573
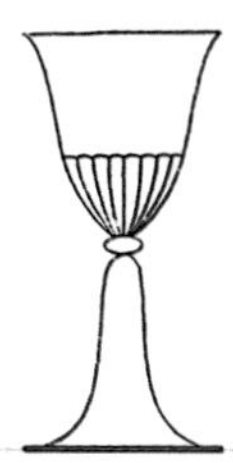
5574
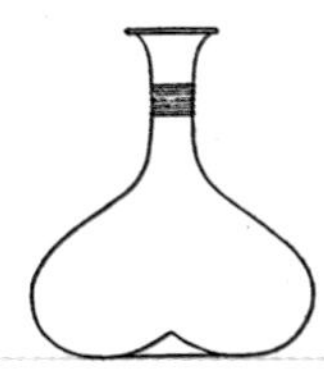
5575

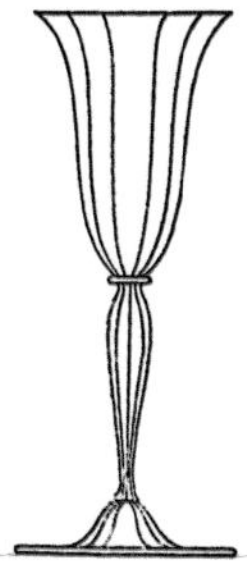
5576

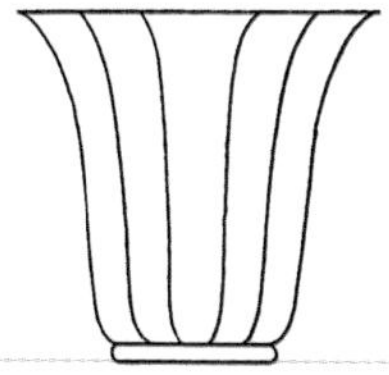
5577

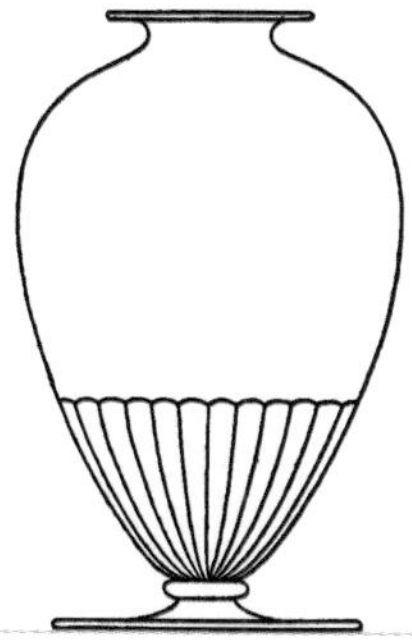
5578

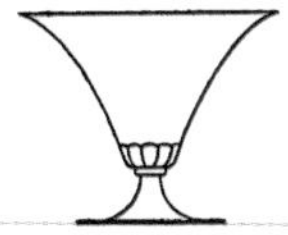
5579

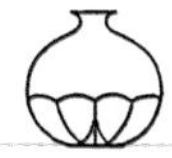
5580

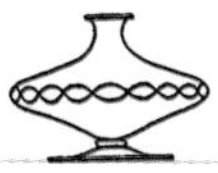
5581

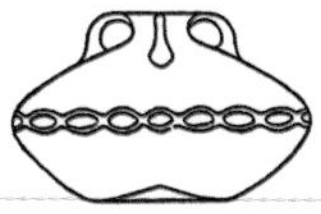
5582

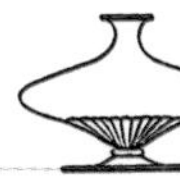
5583

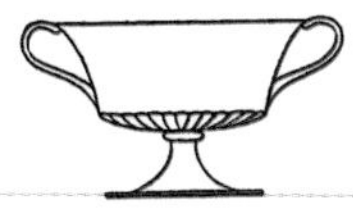
5584

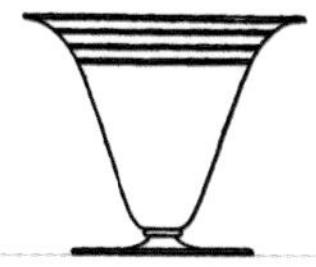
5585

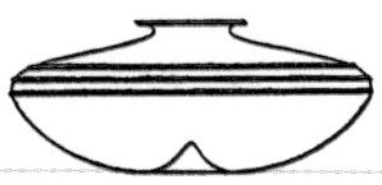
5586

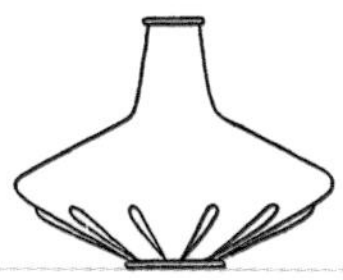
5587

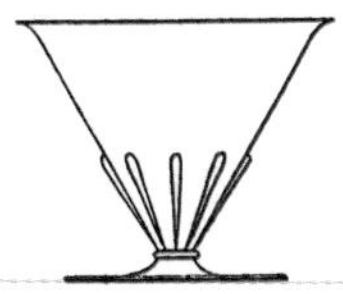
5588

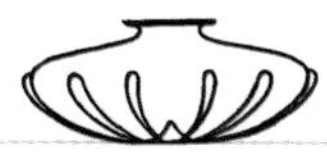
5589

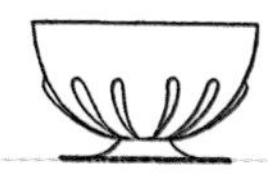
5590

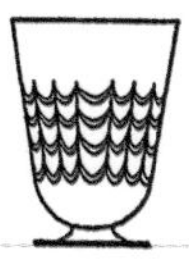
5591

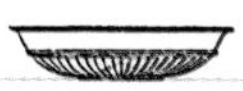
5592

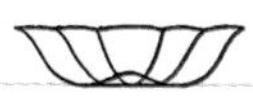
5593

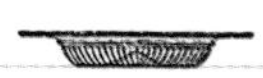
5594

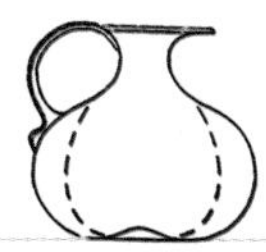
5595

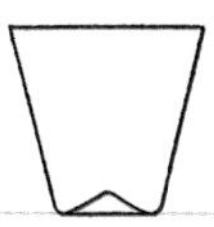
5596

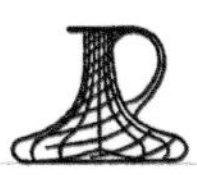
5597

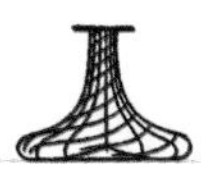
5598

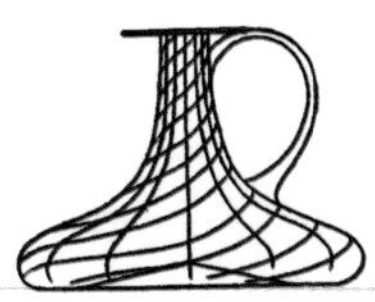
5599

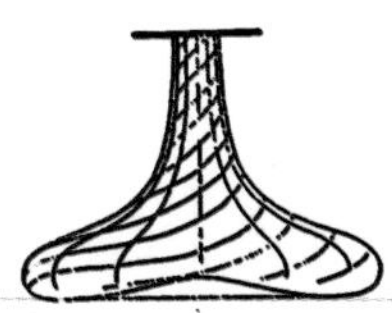
5600

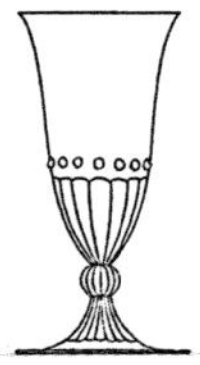

5601

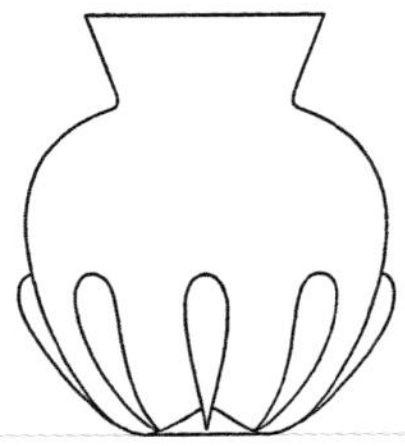

5602

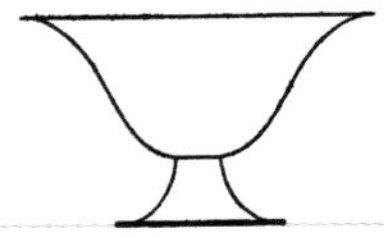

5603

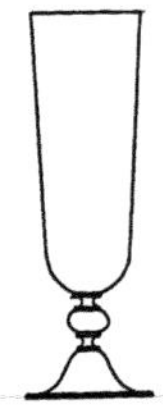

5604

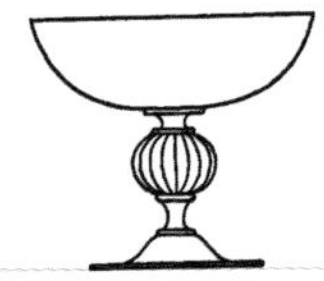

5605

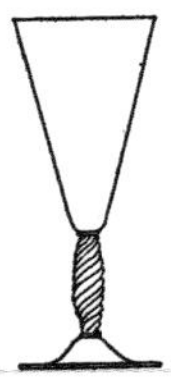

5606

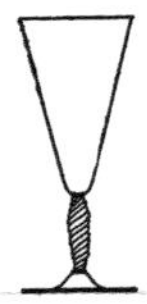

5607

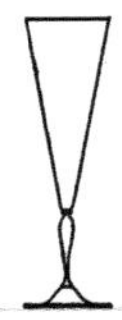

5608

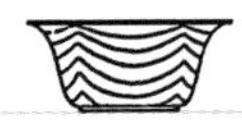

5609

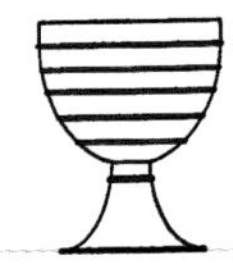

5610

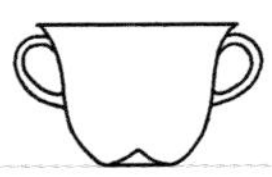

5611

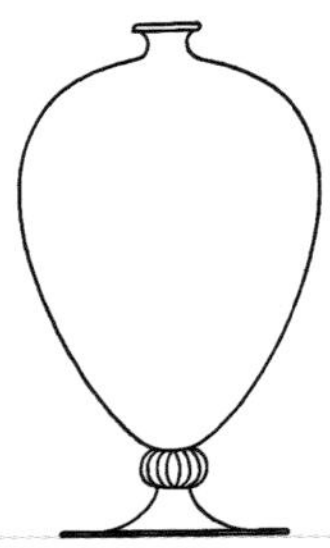

5612

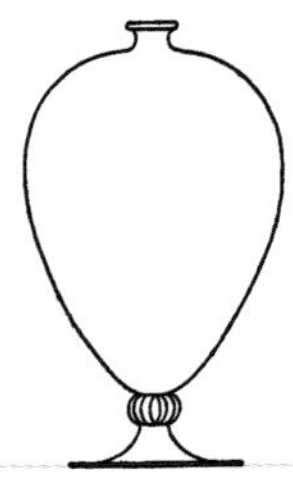

5613

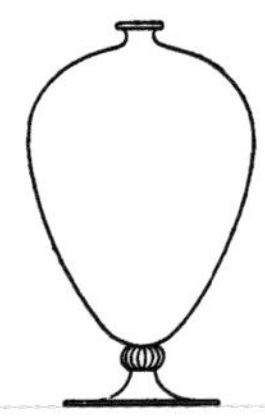

5614

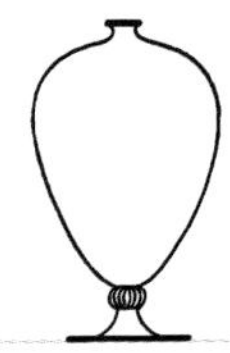

5615

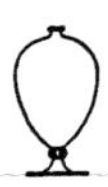

5616

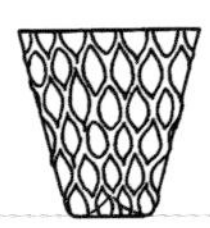

5617

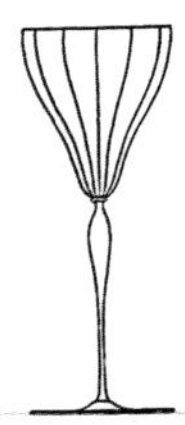

5618

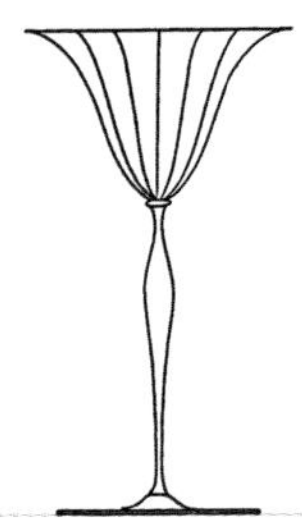

5619

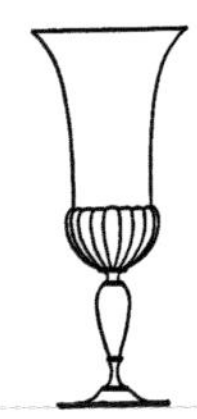

5620

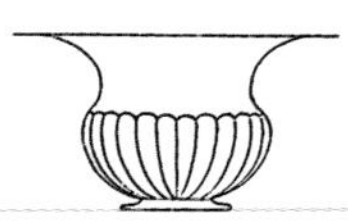

5621

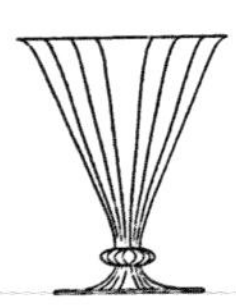

5622

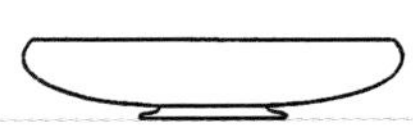

5623

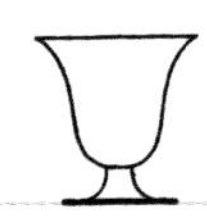

5624

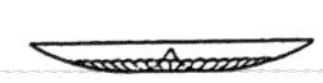

5625

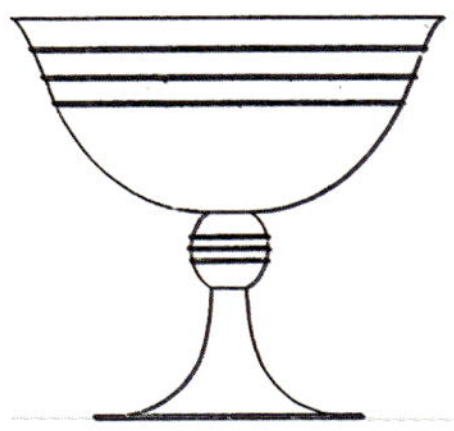
5626

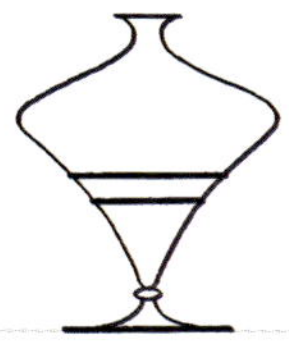
5627

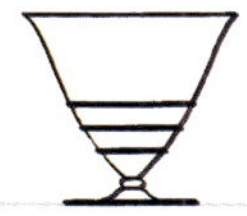
5628

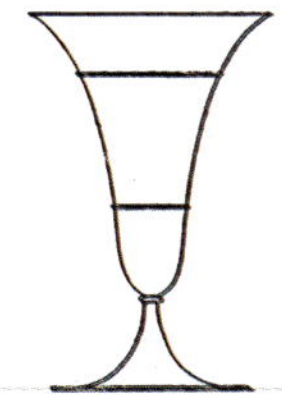
5629

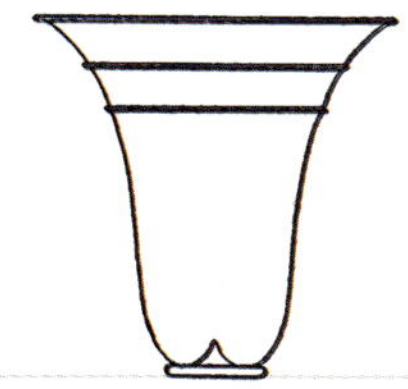
5630

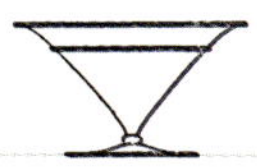
5631

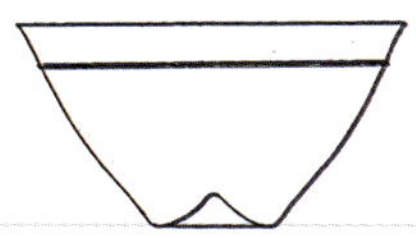
5632

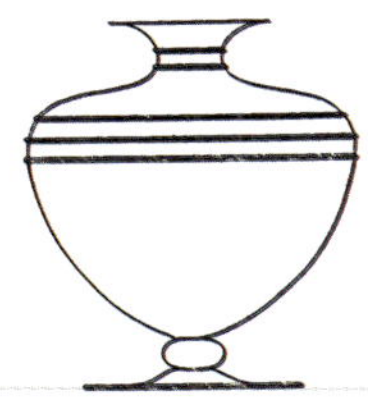
5633

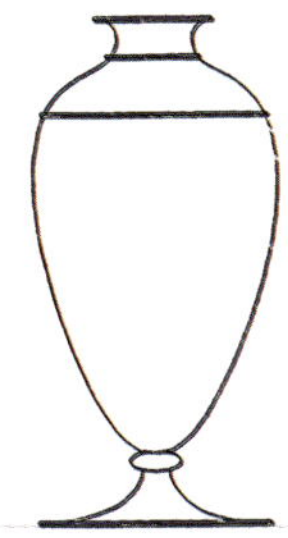
5634

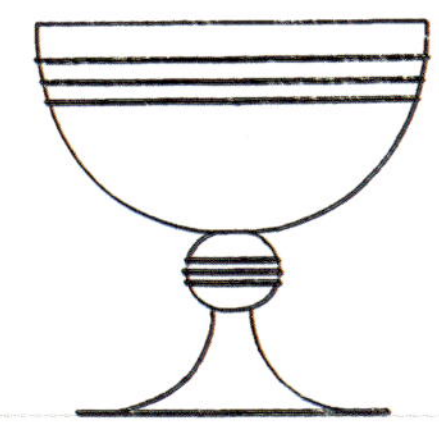
5635

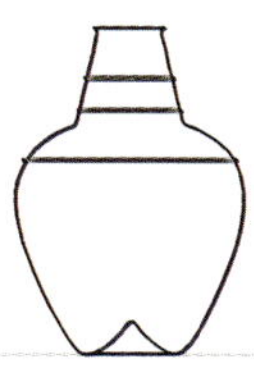
5636

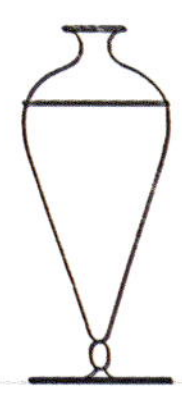
5637

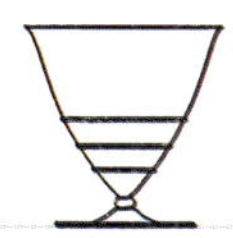
5638

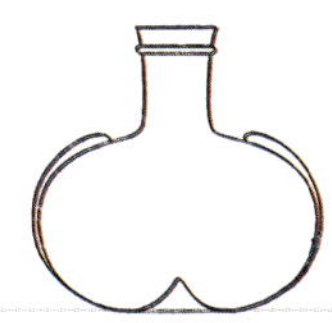
5639

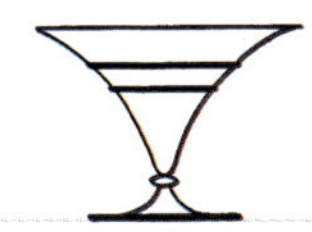
5640

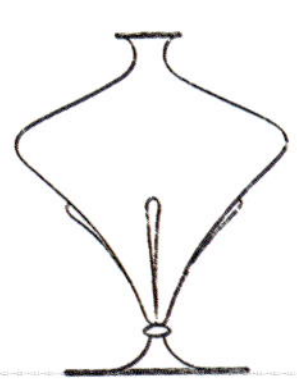
5641

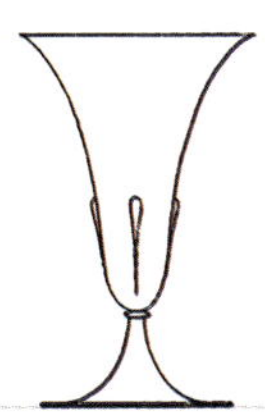
5642

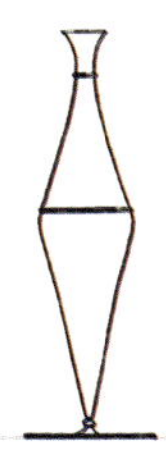
5643

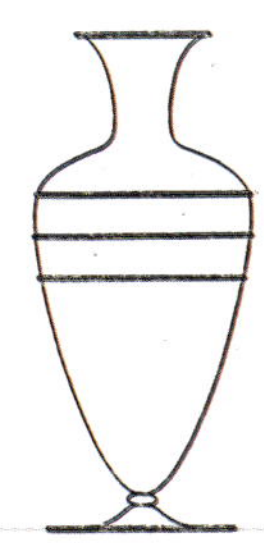
5644

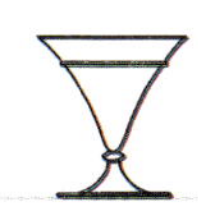
5645

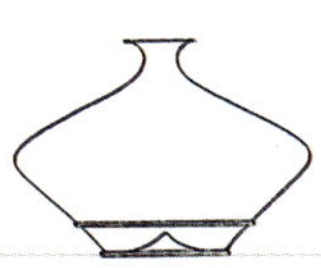
5646

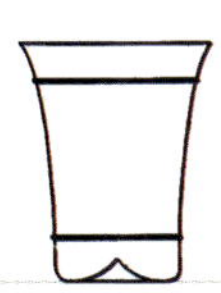
5647

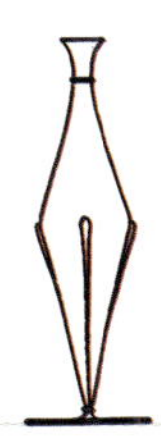
5648

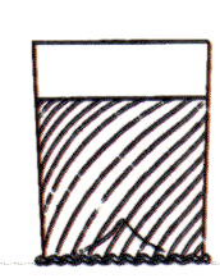
5649

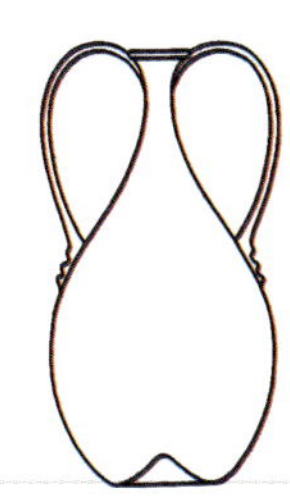
5650

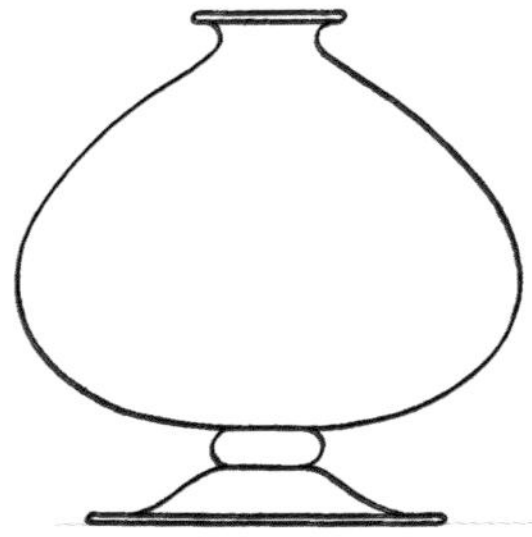
5651
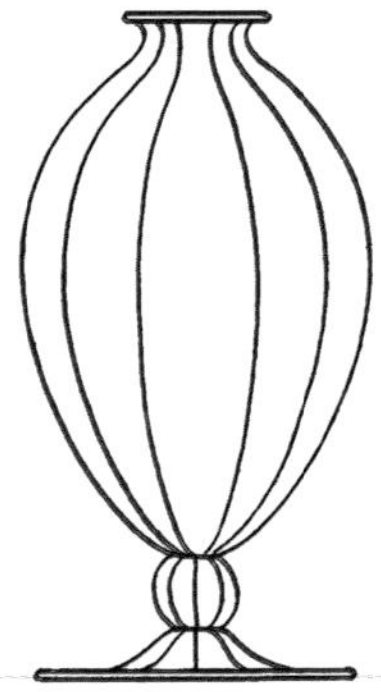
5652
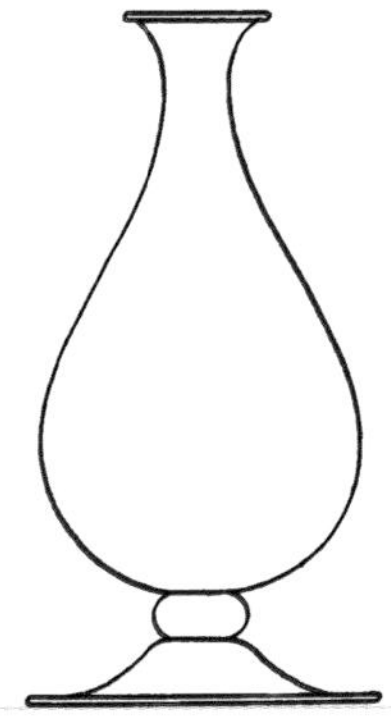
5653
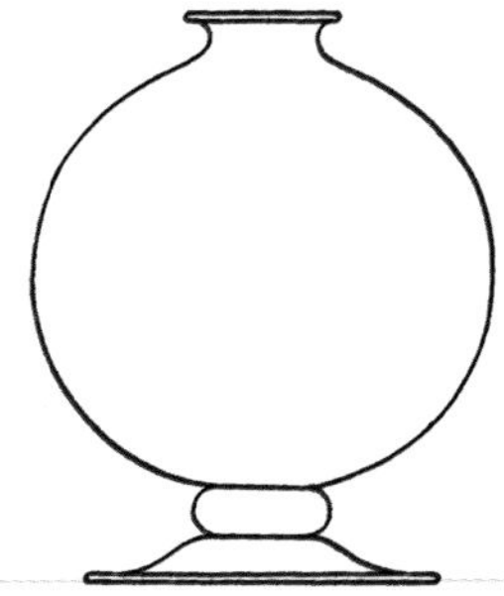
5654
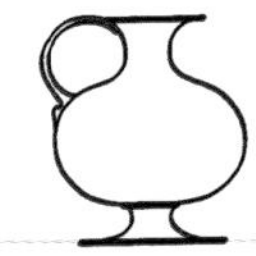
5655

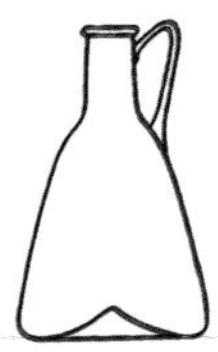
5656
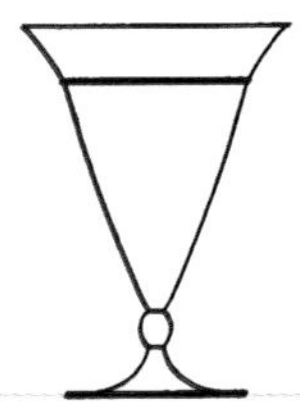
5657
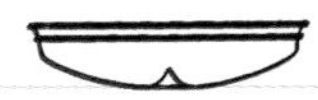
5658
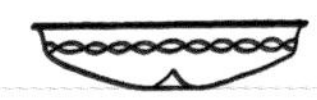
5659
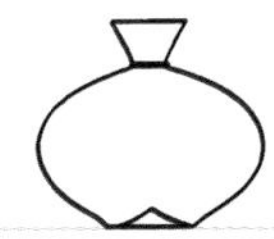
5660

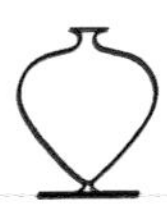
5661
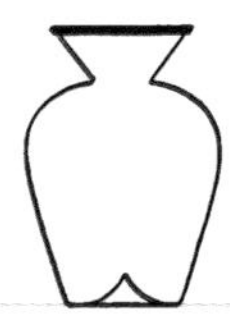
5662
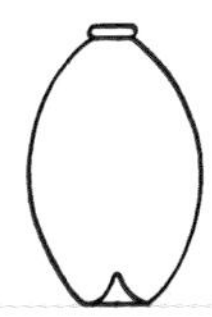
5663
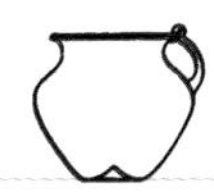
5664
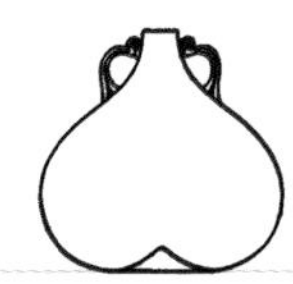
5665

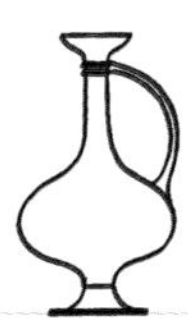
5666
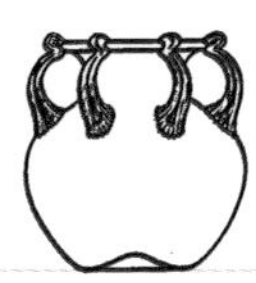
5667
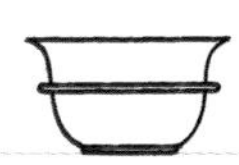
5668
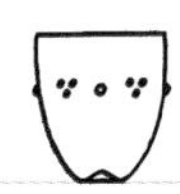
5669
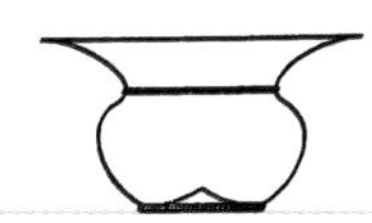
5670

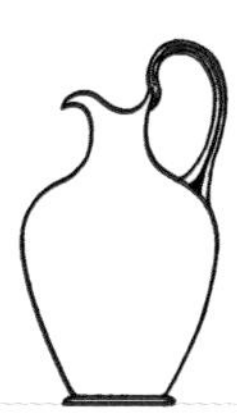
5671
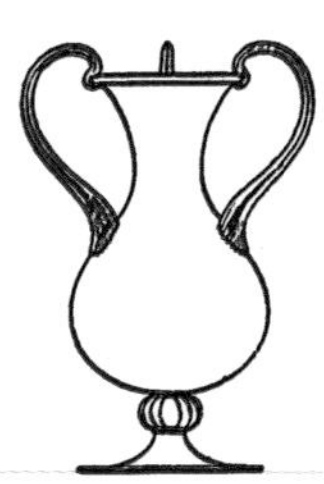
5672
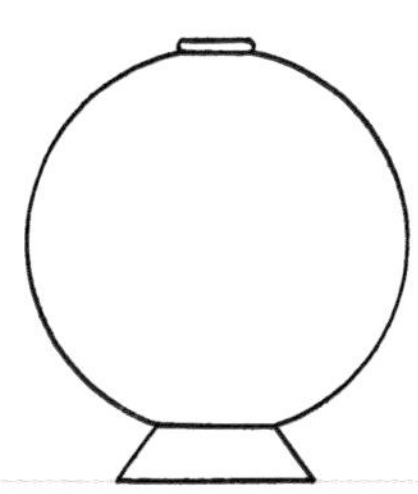
5673
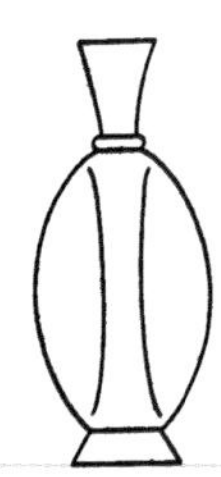
5674
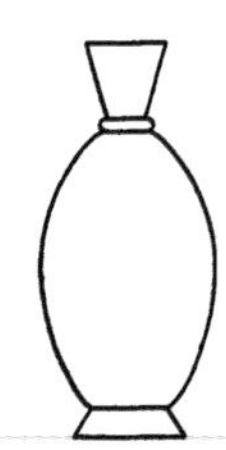
5675

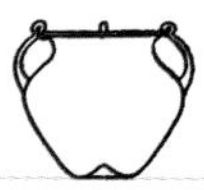
5676

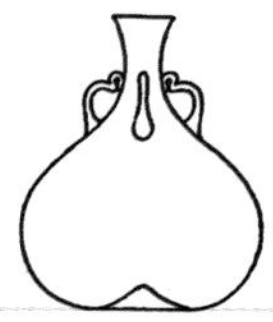
5677

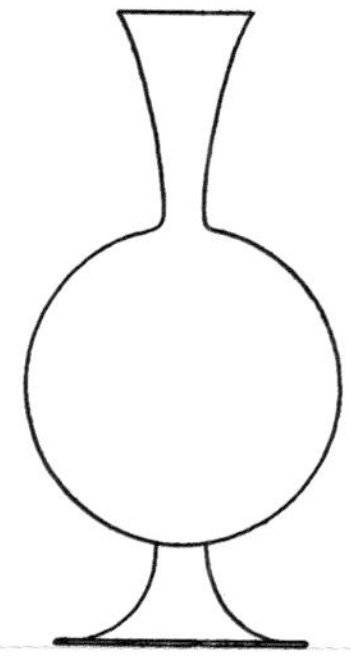
5678

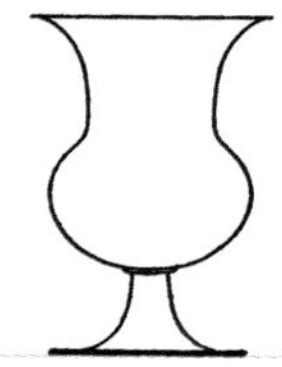
5679

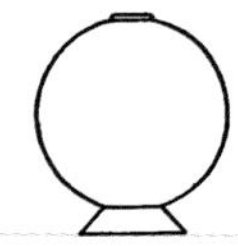
5680

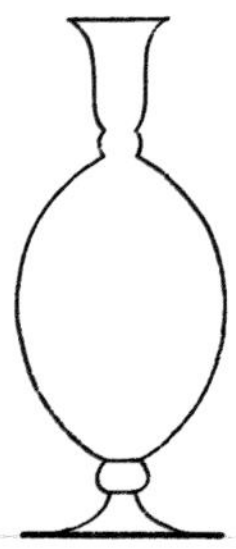
5681

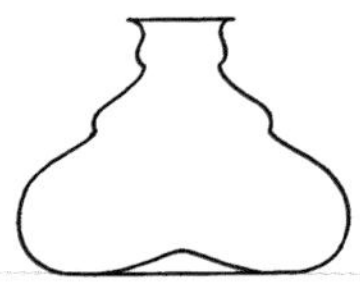
5682

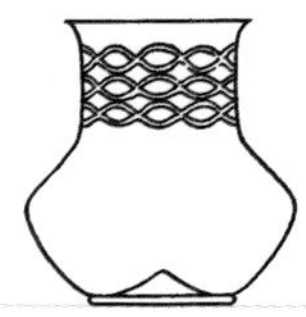
5683

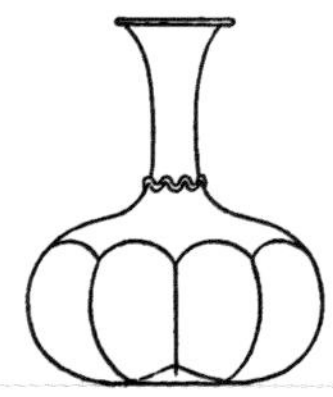
5684

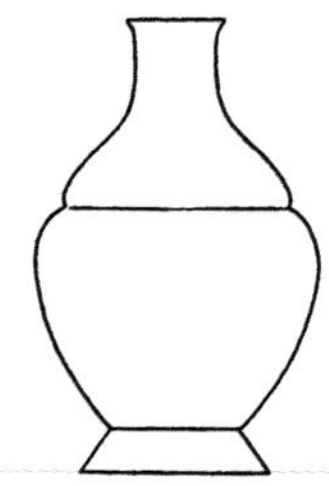
5685

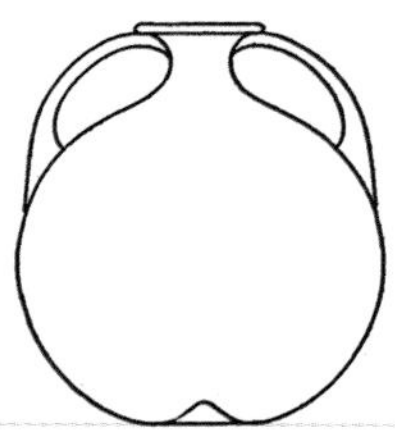
5686

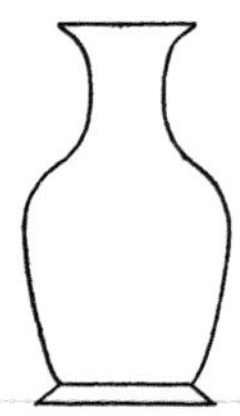
5687

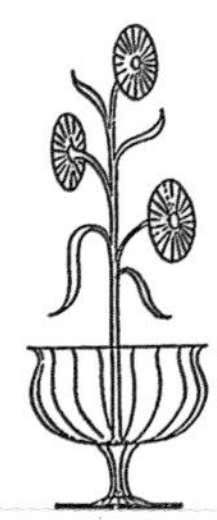
5688

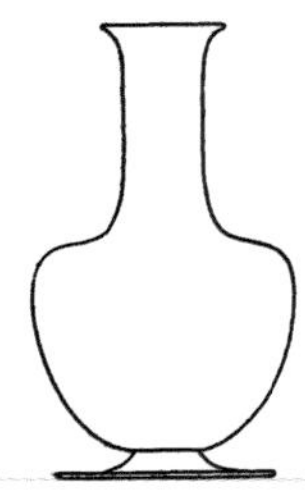
5689

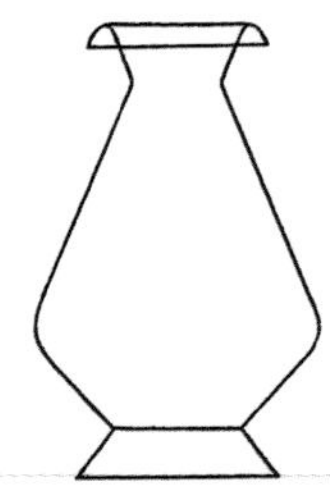
5690

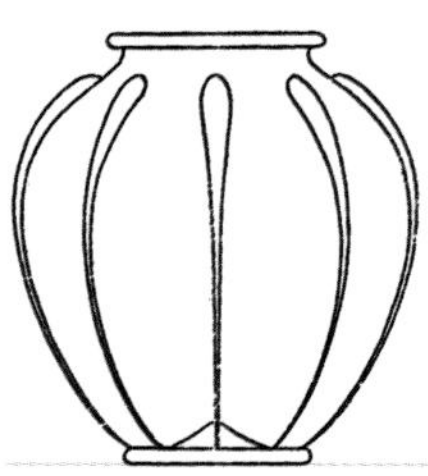
5691

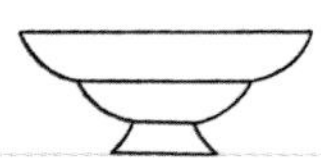
5692

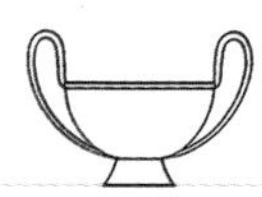
5693

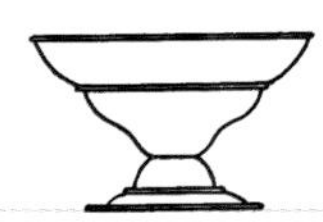
5694

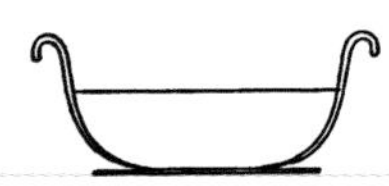
5695

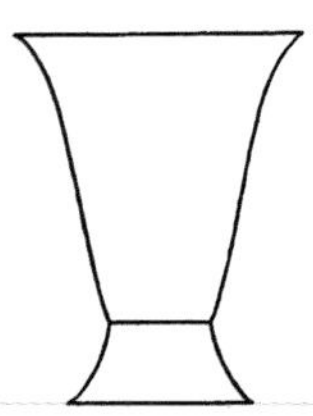
5696

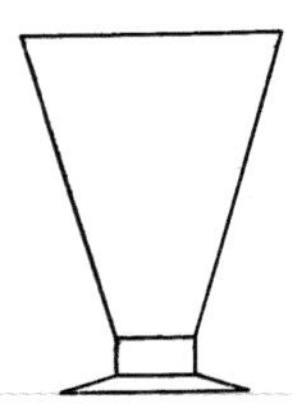
5697

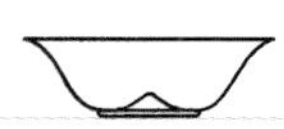
5698

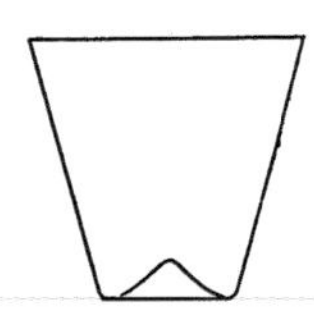
5699

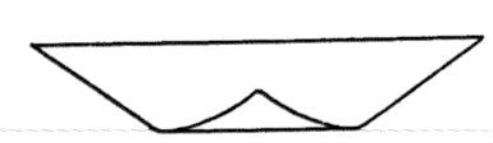
5700

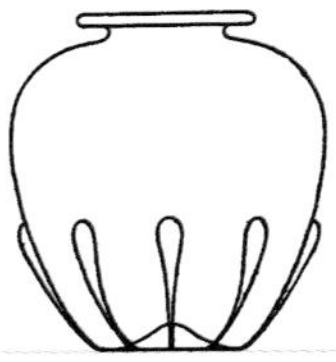
5701

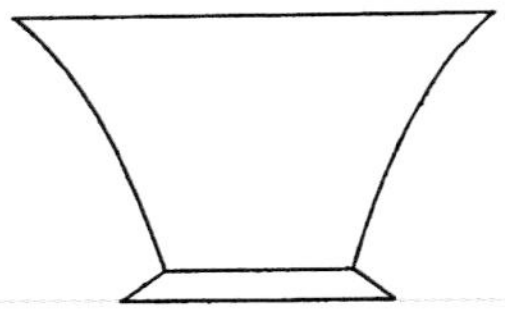
5702

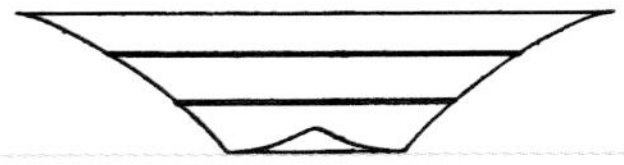
5703

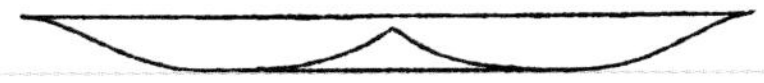
5704

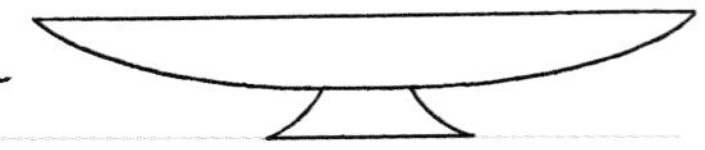
5705

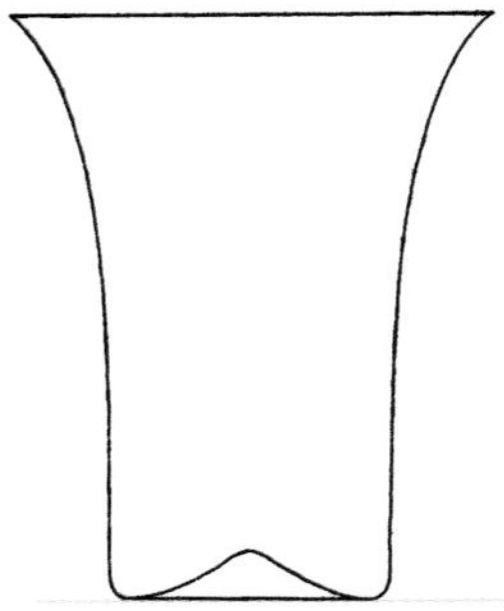
5706

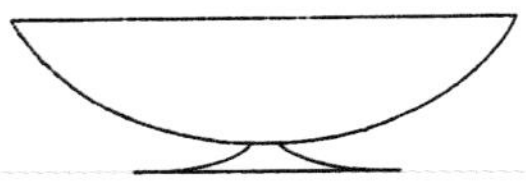
5707

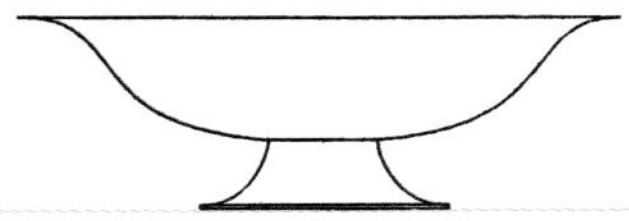
5708

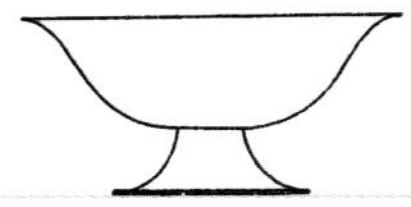
5709

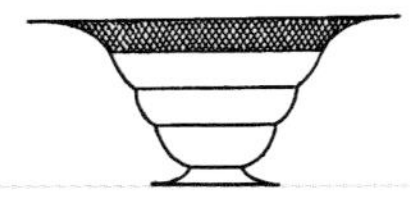
5710

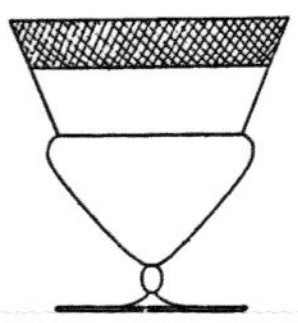
5711

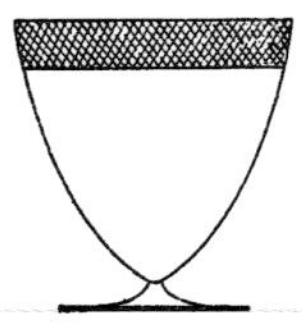
5712

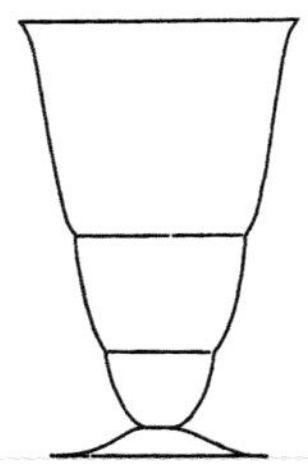
5713

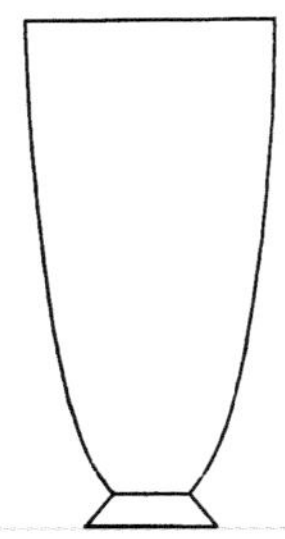
5714

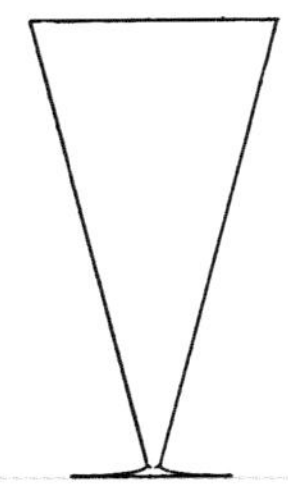
5715

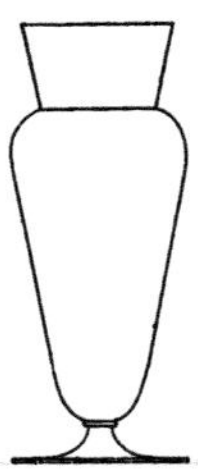
5716

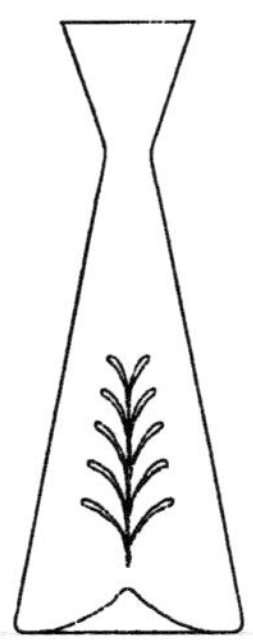
5717

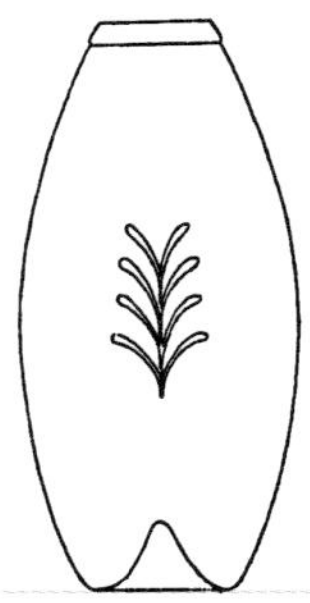
5718

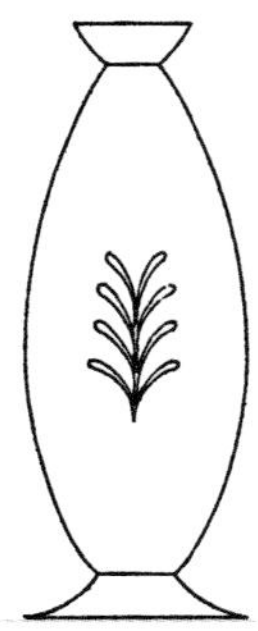
5719

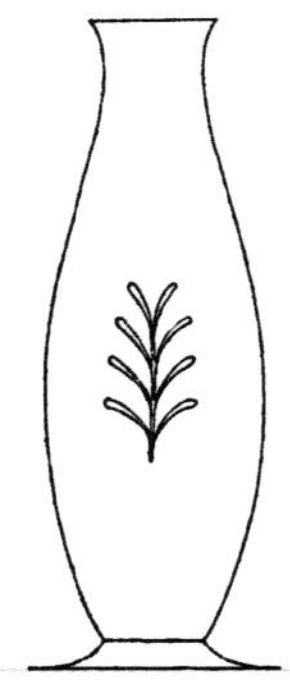
5720

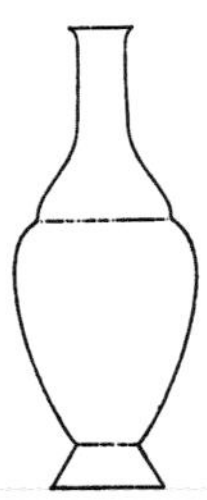
5721

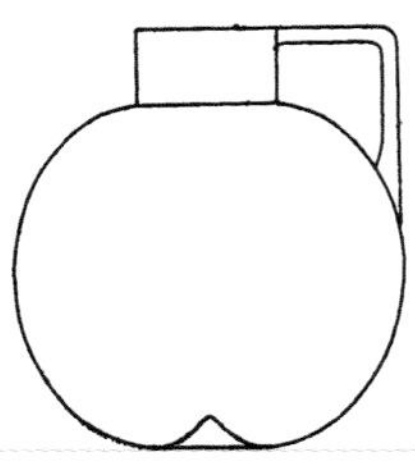
5722

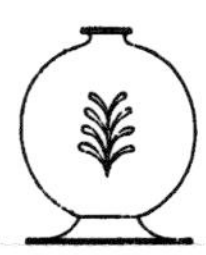
5723

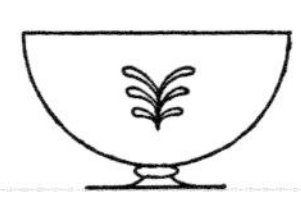
5724

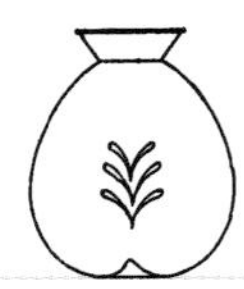
5725

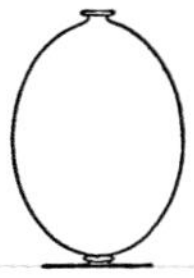
5726

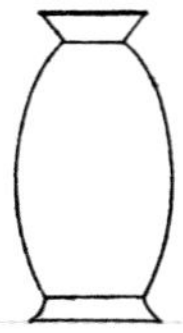
5727

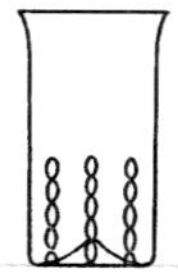
5728

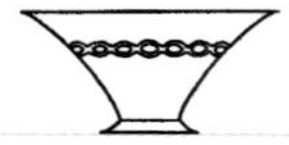
5729

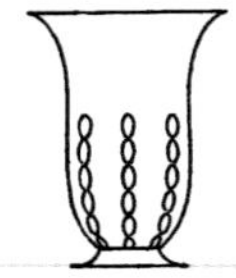
5730

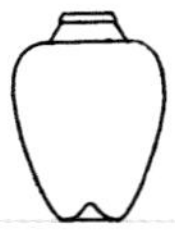
5731

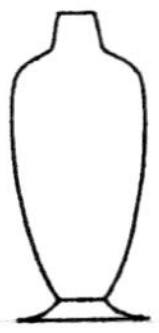
5732

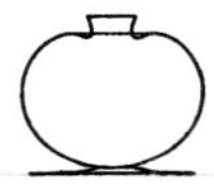
5733

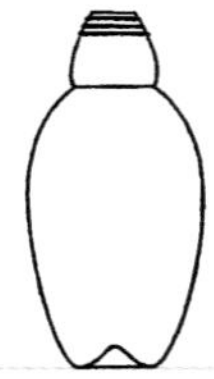
5734

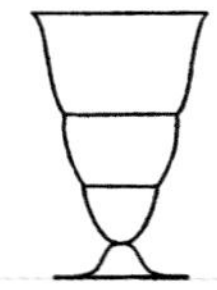
5735

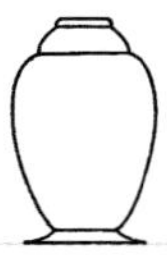
5736

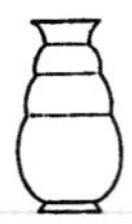
5737

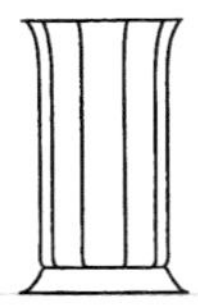
5738

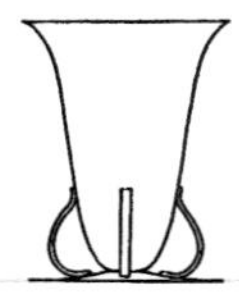
5739

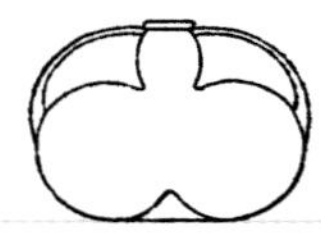
5740

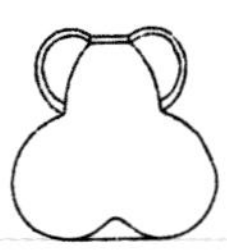
5741

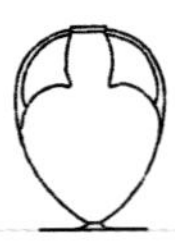
5742

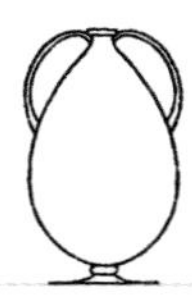
5743

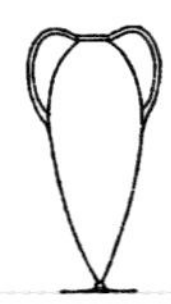
5744

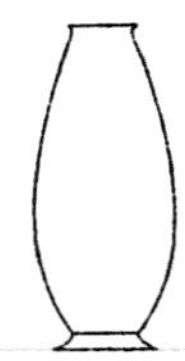
5745

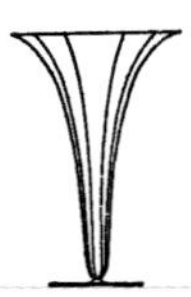
5746

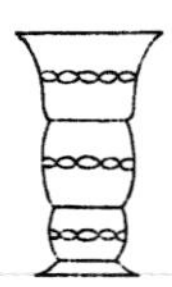
5747

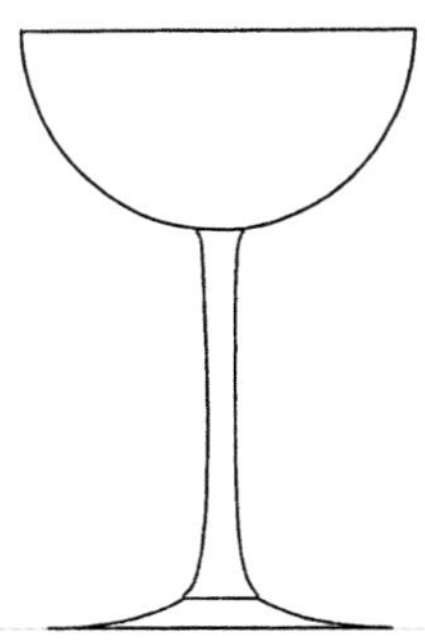
5748

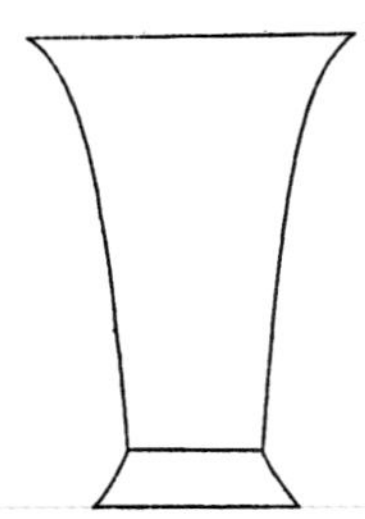
5749

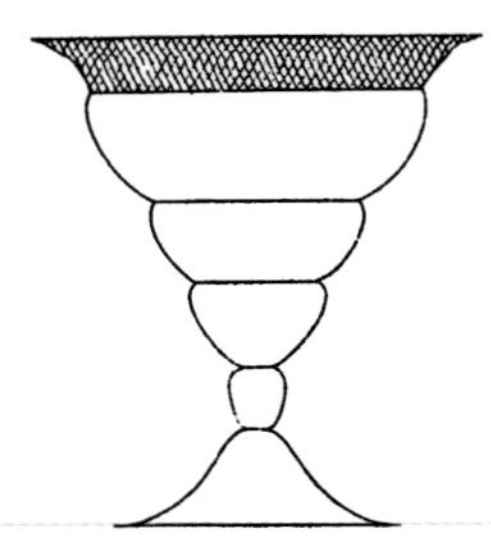
5750

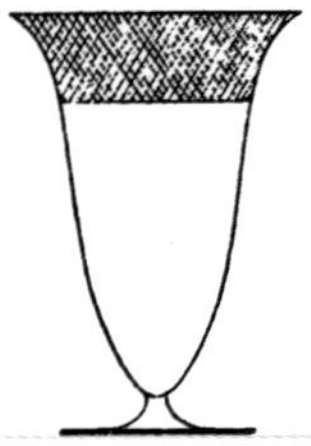
5751

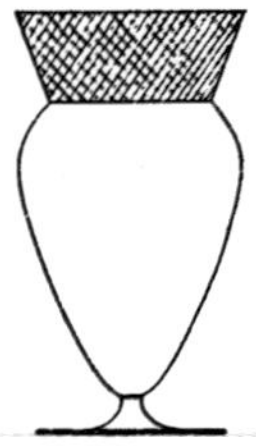
5752

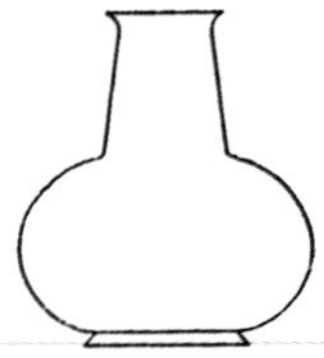
5753

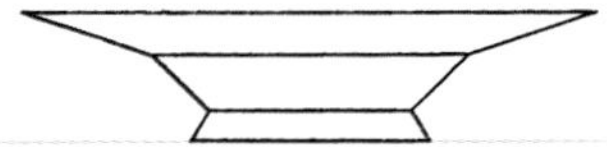
5754

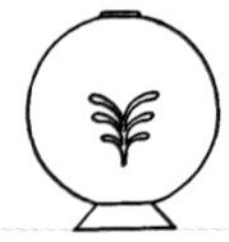
5755

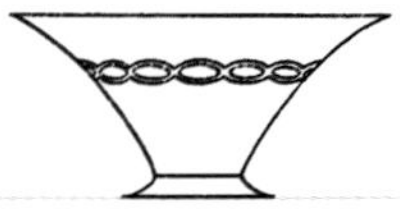
5756

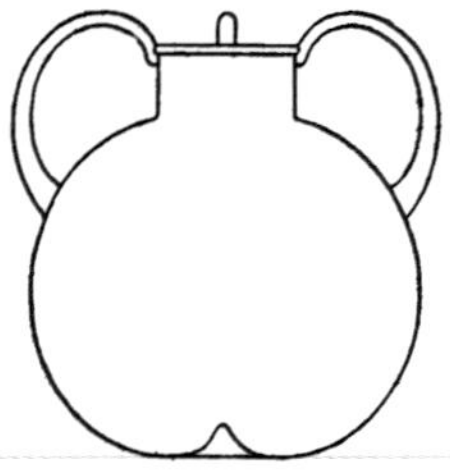
5757

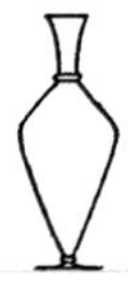
5758

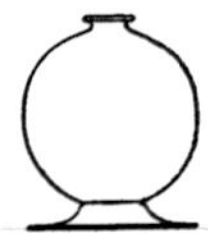
5759

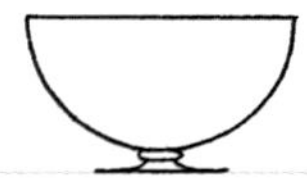
5760

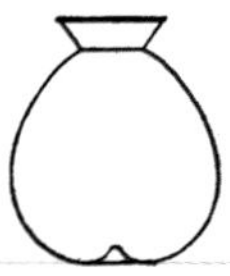
5761

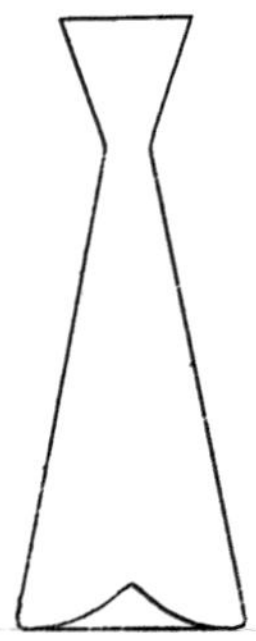
5762

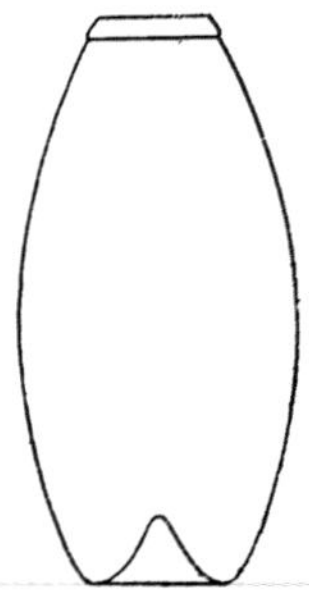
5763

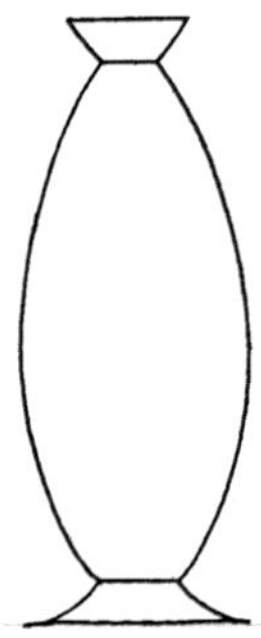
5764

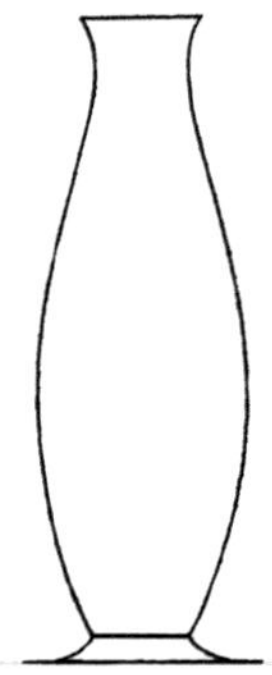
5765

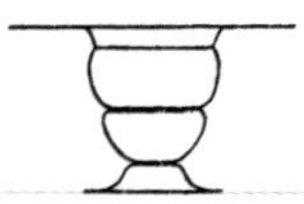
5766

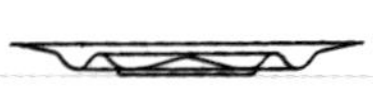
5767

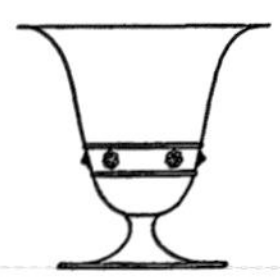
5768

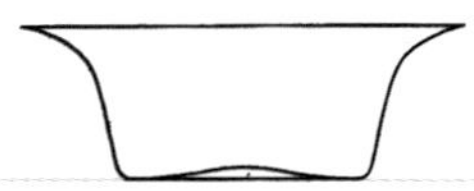
5769

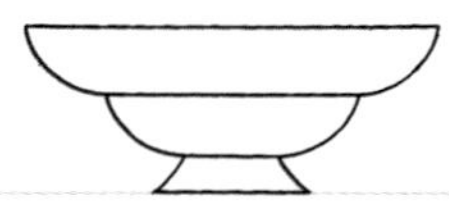
5770

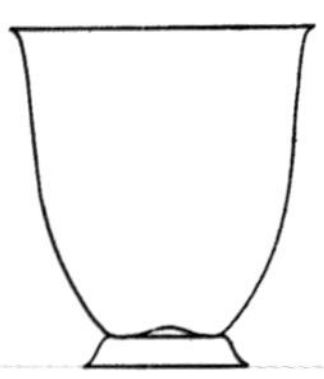
5771

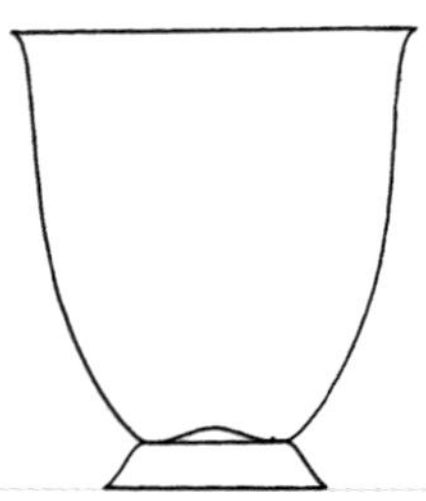
5772

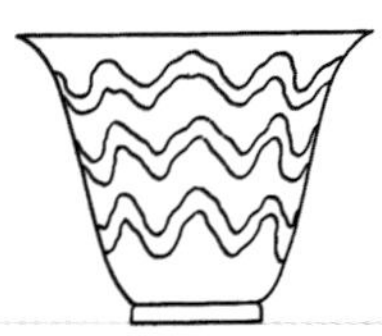
5773

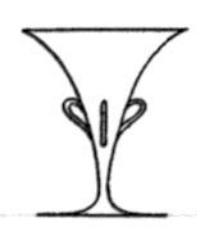
5774

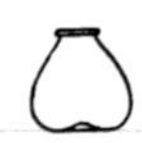
5775

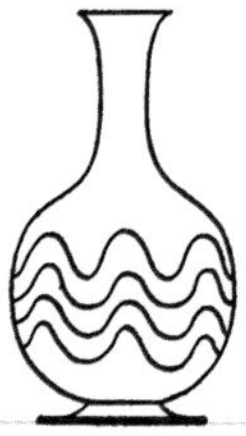
5776

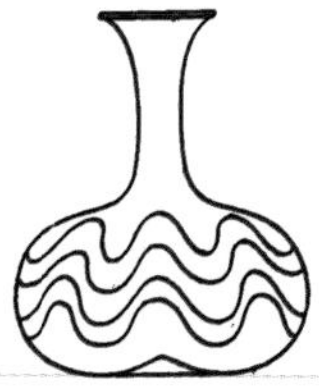
5777

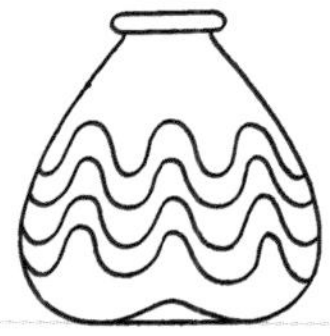
5778

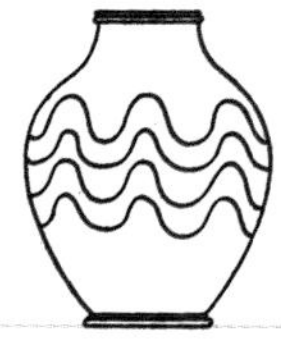
5779

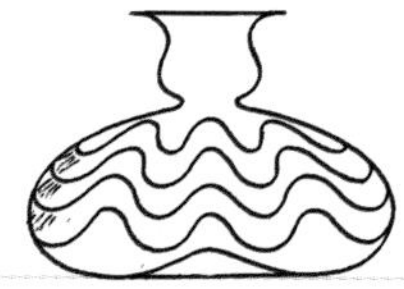
5780

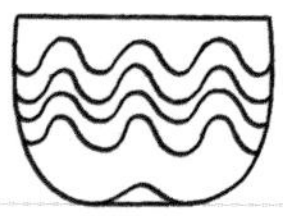
5781

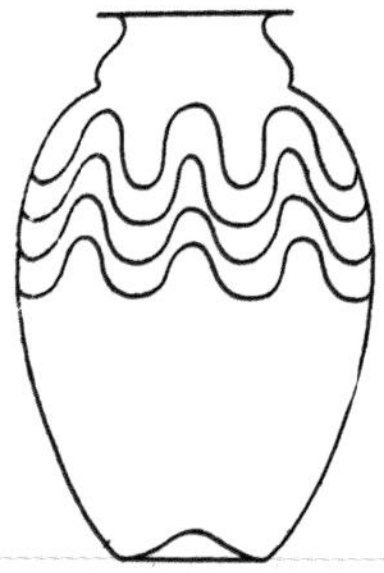
5782

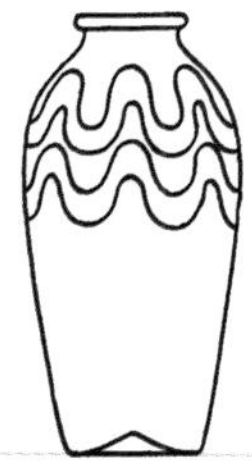
5783

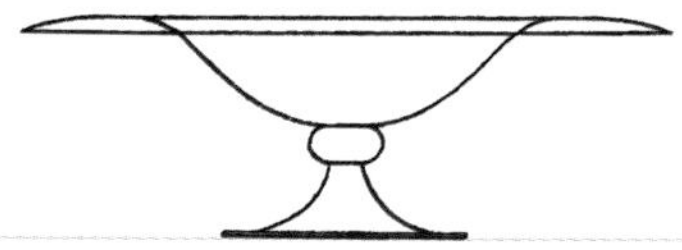
5784

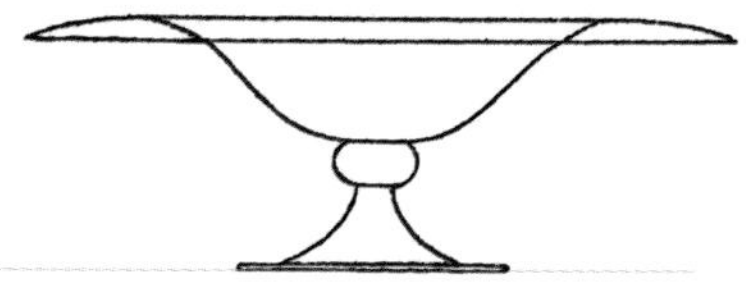
5785

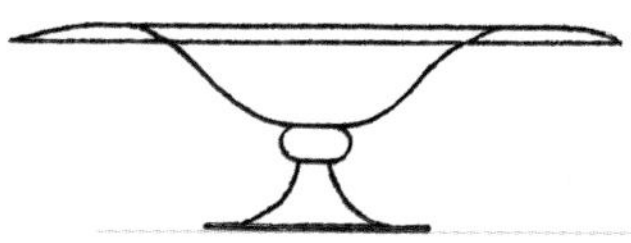
5786

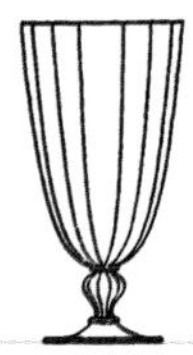
5787

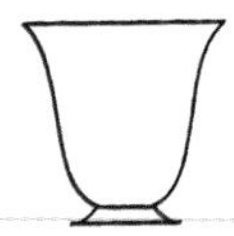
5788

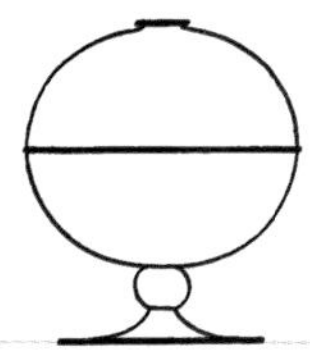
5789

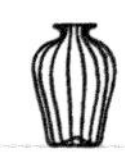
5790

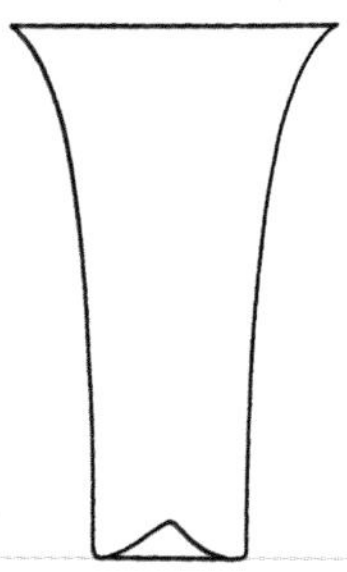
5791

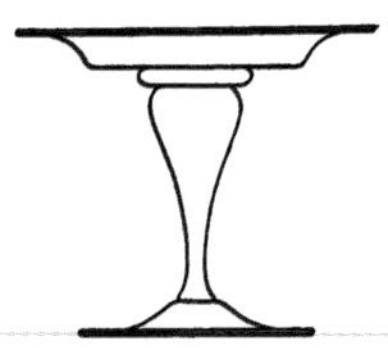
5792

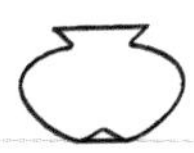
5793

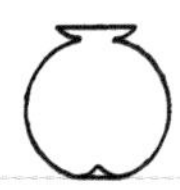
5794

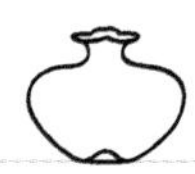
5795

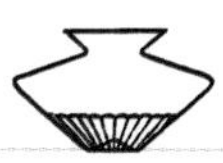
5796

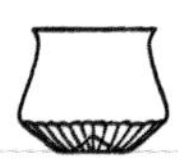
5797

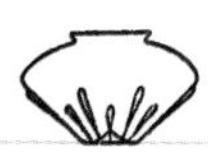
5798

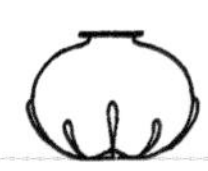
5799

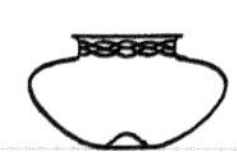
5800

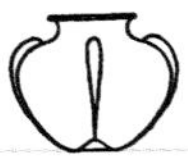
5801

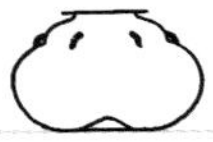
5802

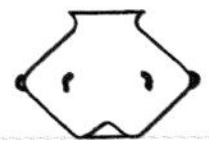
5803

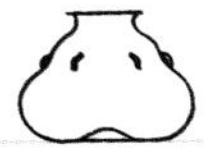
5804

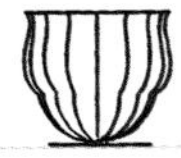
5805

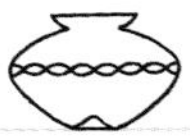
5806

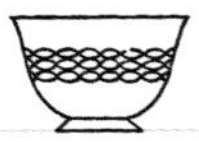
5807

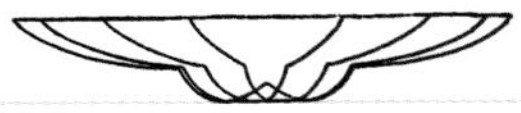
5808

5809

5810

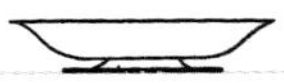
5811

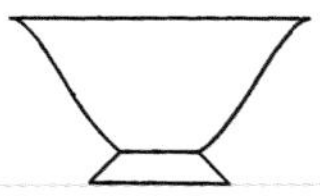
5812

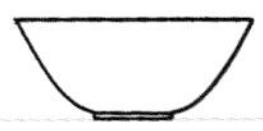
5813

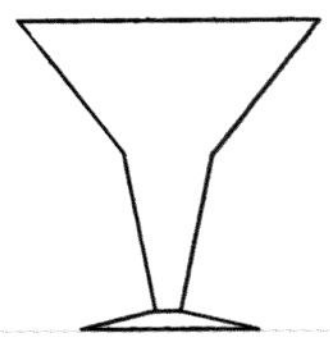
5814

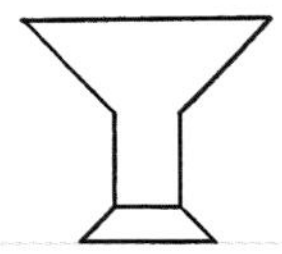
5815

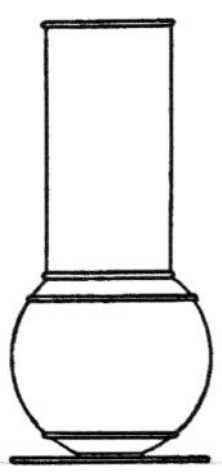
5816

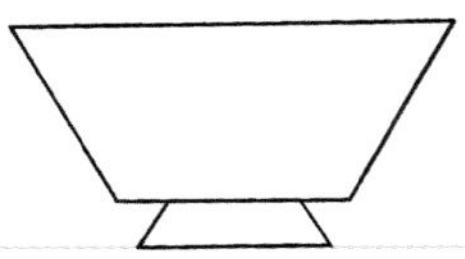
5817

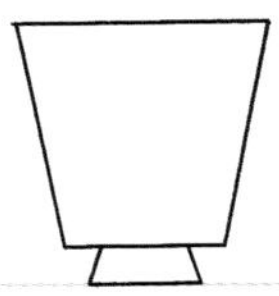
5818

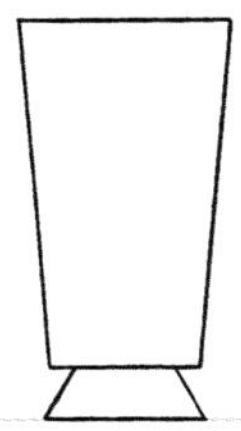
5819

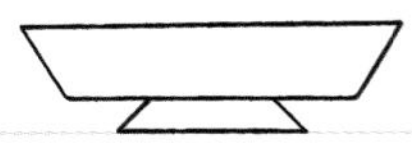
5820

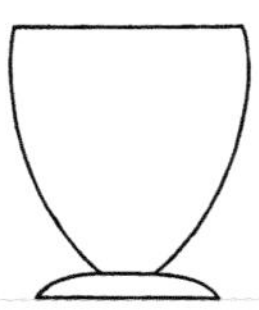
5821

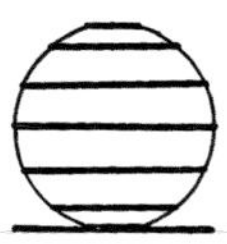
5822

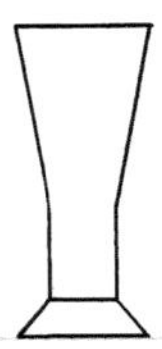
5823

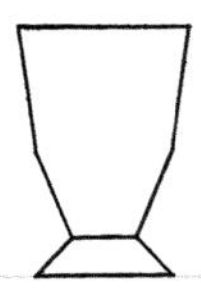
5824

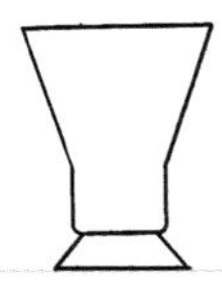
5825

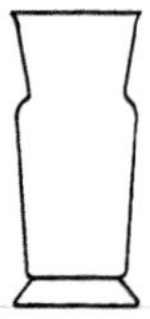
5826

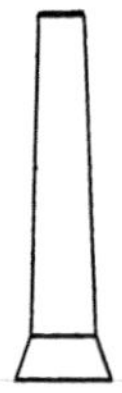
5827

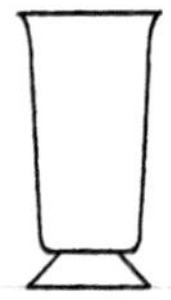
5828

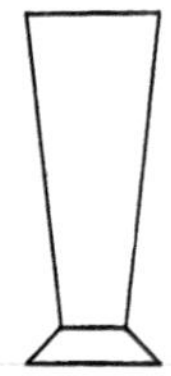
5829

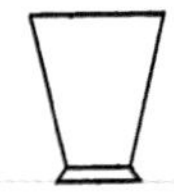
5830

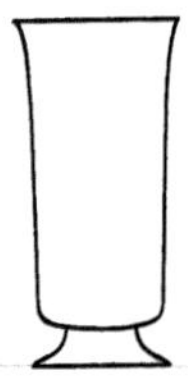
5831

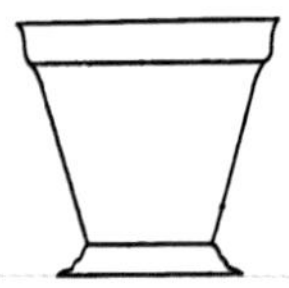
5832

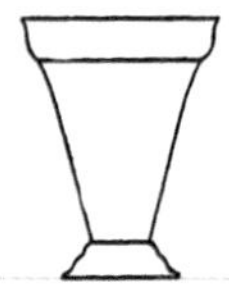
5833

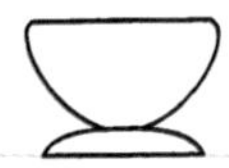
5834

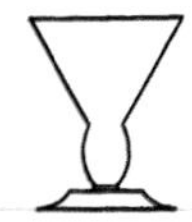
5835

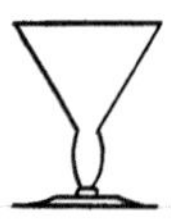
5836

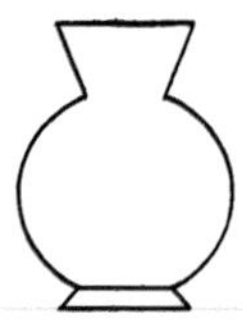
5837

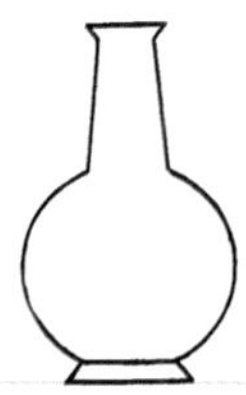
5838

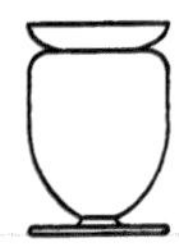
5839

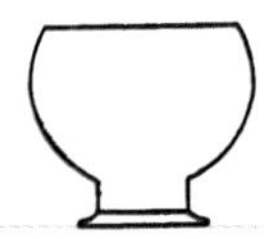
5840

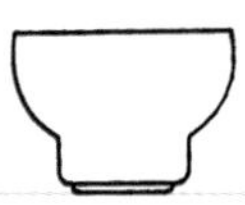
5841

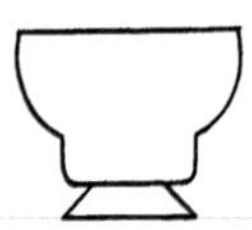
5842

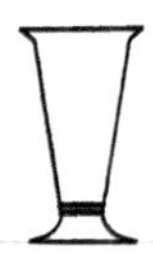
5843

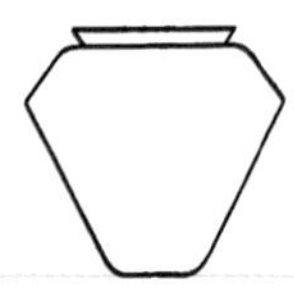
5844

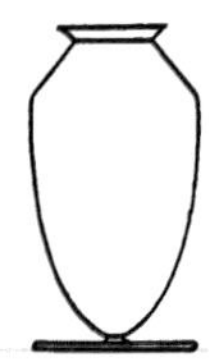
5845

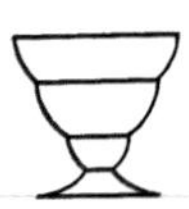
5846

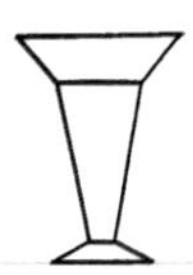
5847

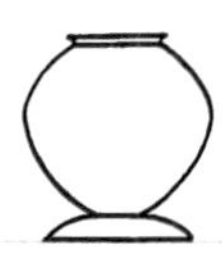
5848

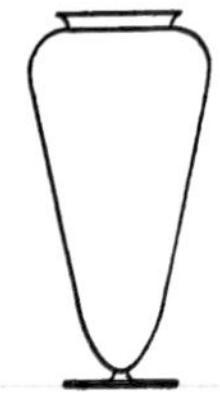
5849

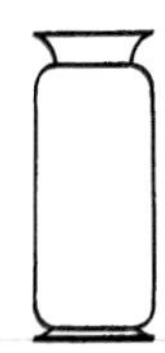
5850

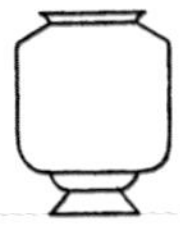

5851

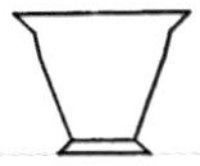

5852

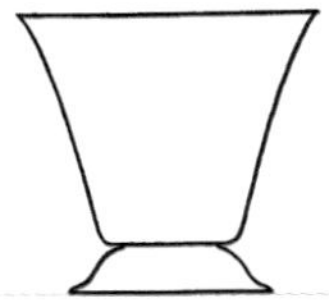

5853

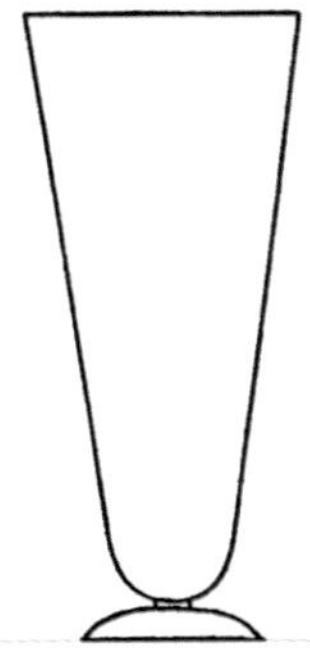

5854

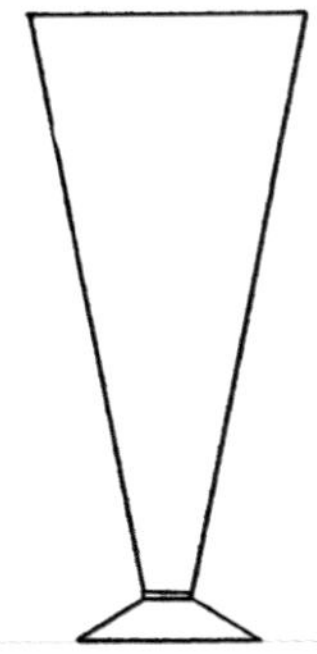

5855

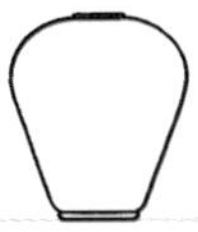

5856

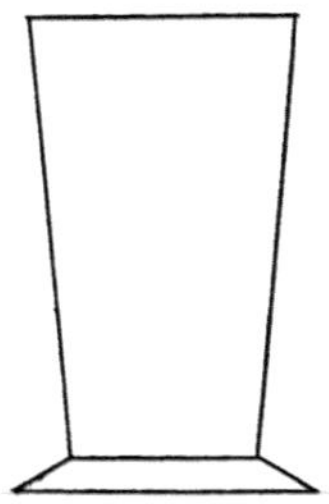

5857

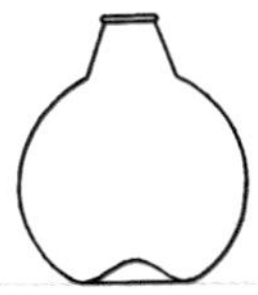

5858

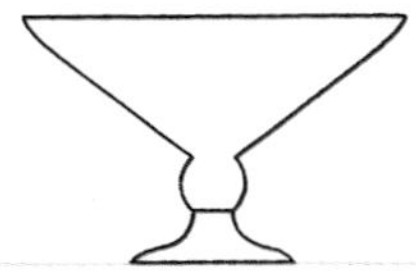

5859

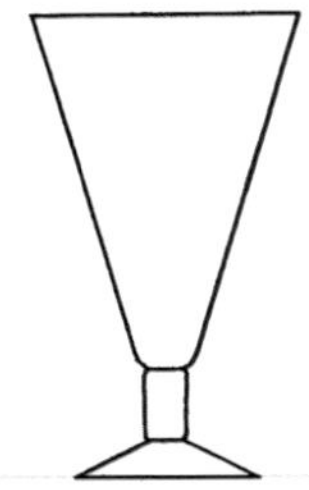

5860

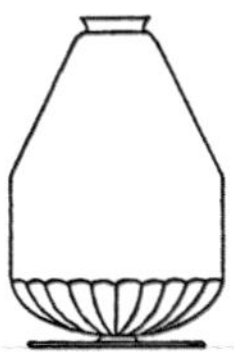

5861

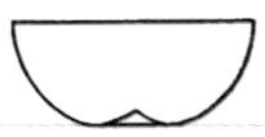

5862

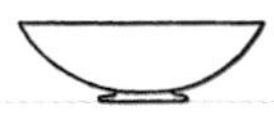

5863

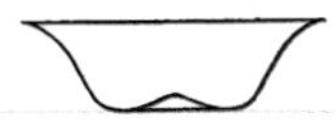

5864

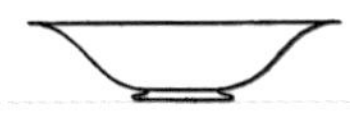

5865

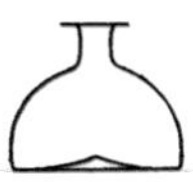

5866

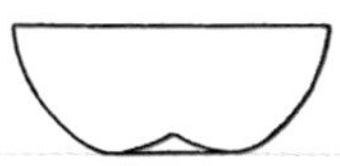

5867

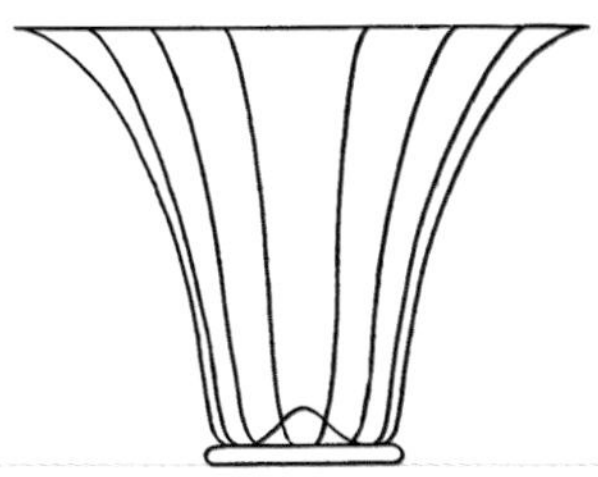

5868

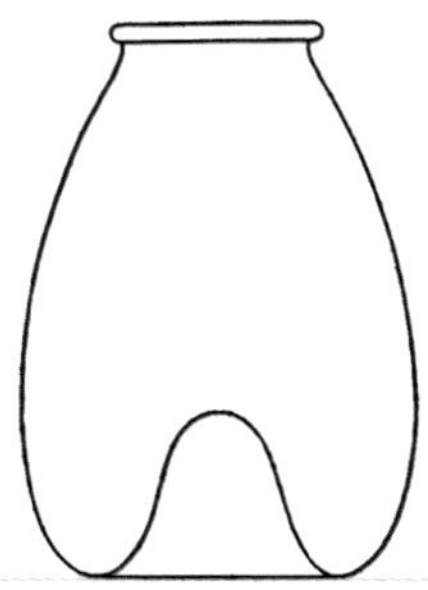

5869

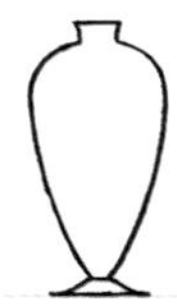

5870

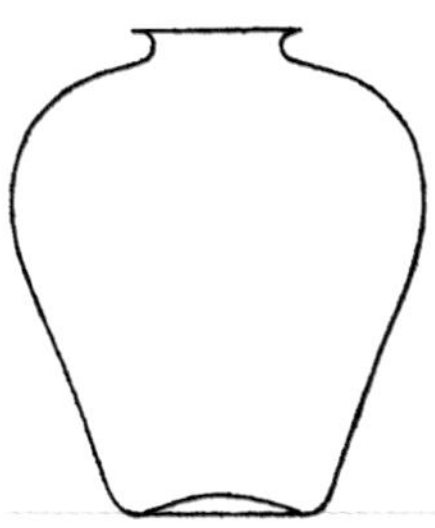

5871

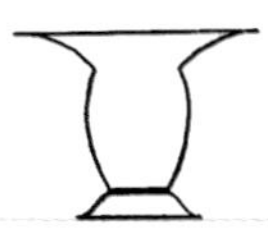

5872

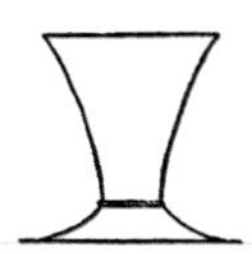

5873

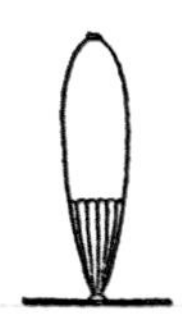

5874

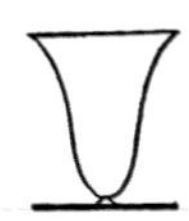

5875

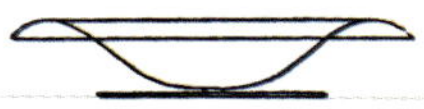
5876

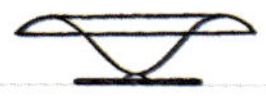
5877

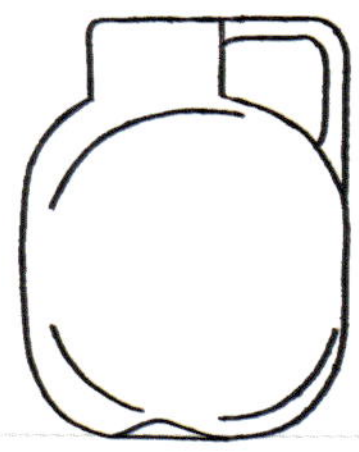
5878

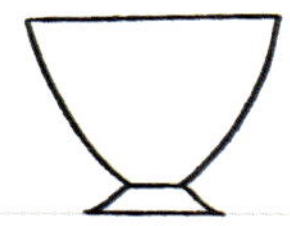
5879

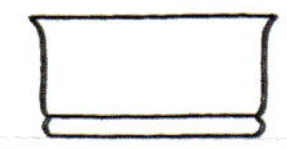
5880

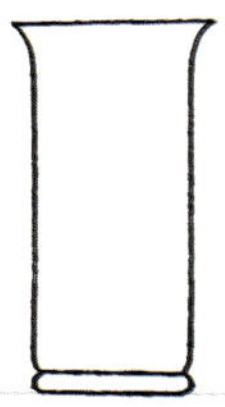
5881

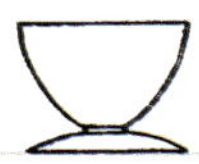
5882

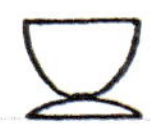
5883

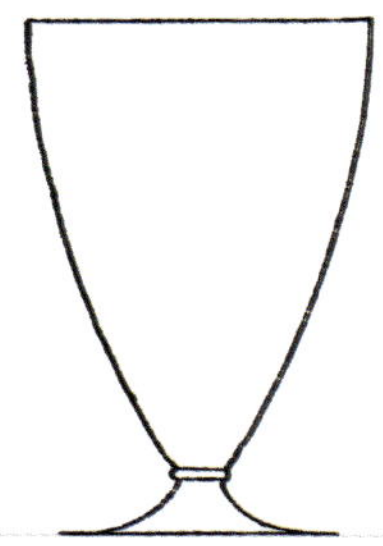
5884

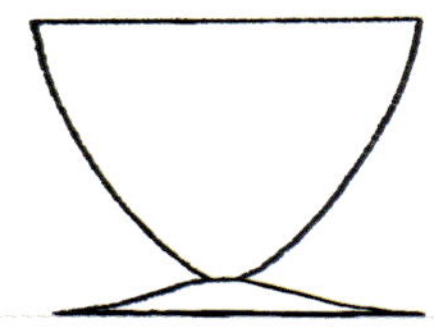
5885

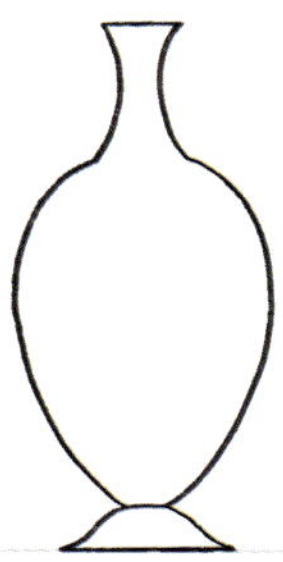
5886

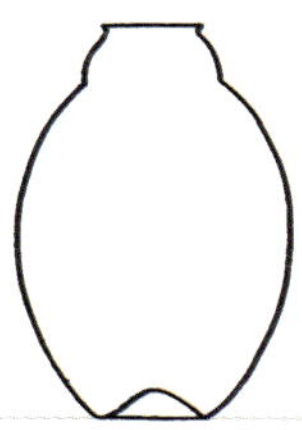
5887

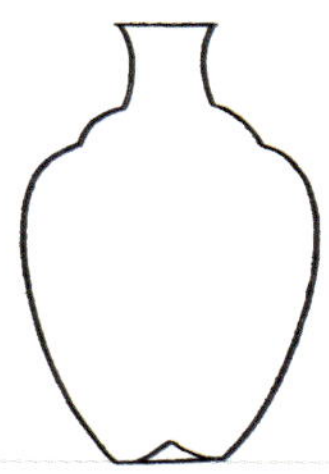
5888

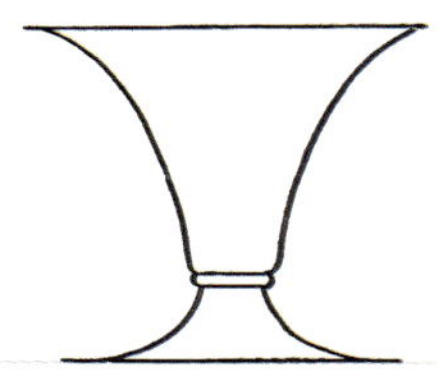
5889

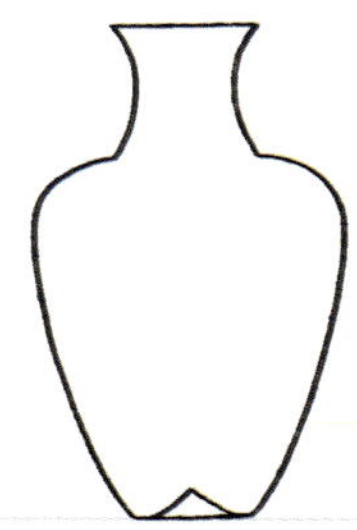
5890

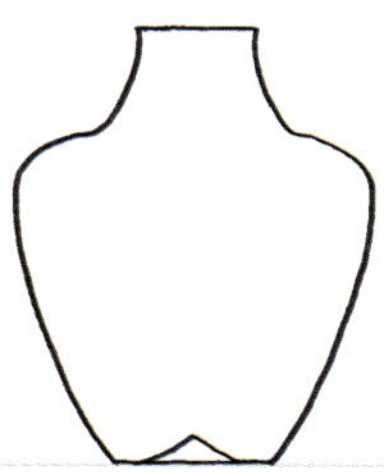
5891

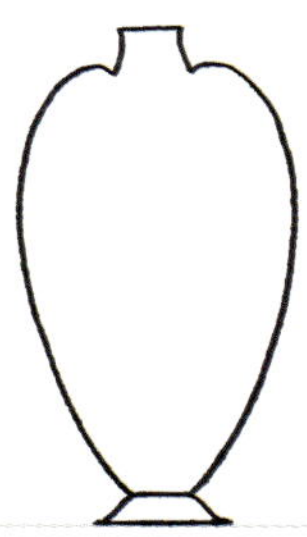
5892

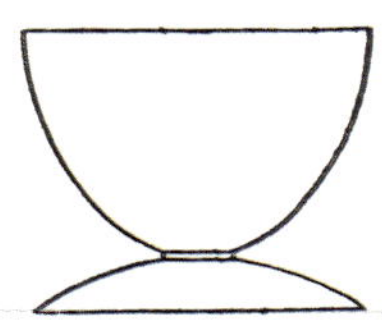
5893

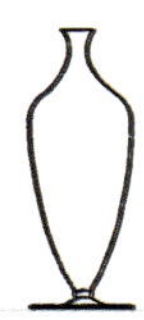
5894

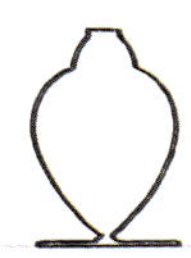
5895

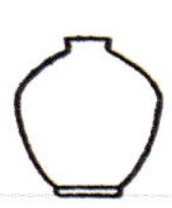
5896

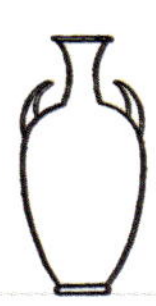
5897

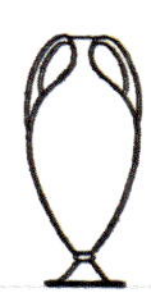
5898

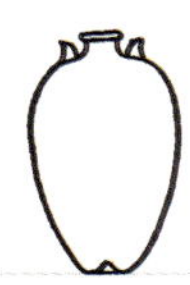
5899

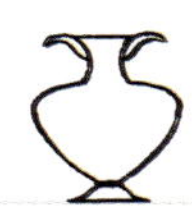
5900

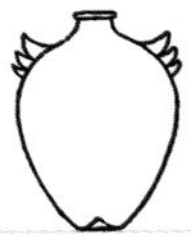
5901
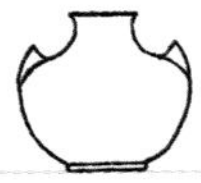
5902
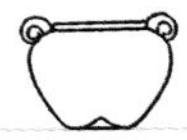
5903
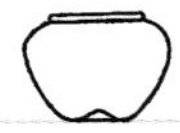
5904
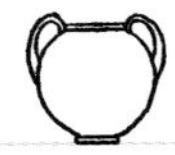
5905
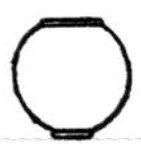
5906
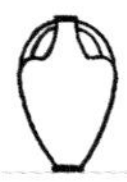
5907
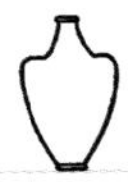
5908
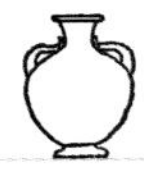
5909
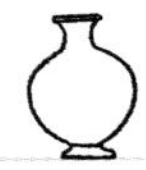
5910

5911
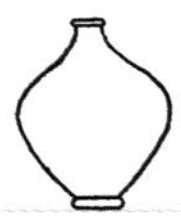
5912
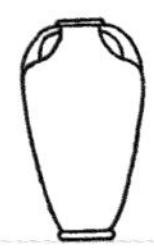
5913
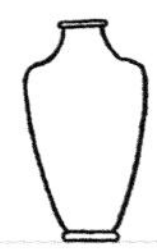
5914
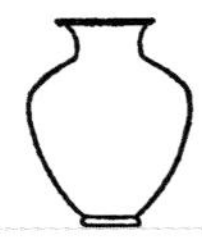
5915
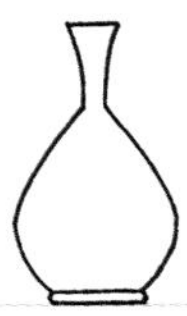
5916
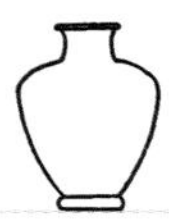
5917
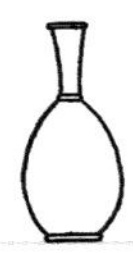
5918
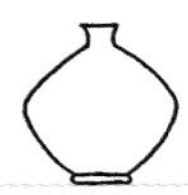
5919
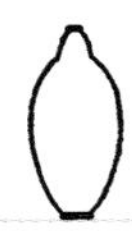
5920
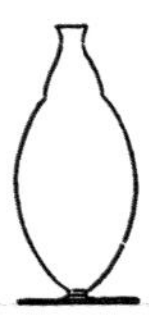
5921
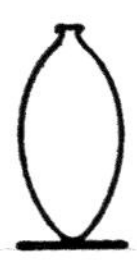
5922
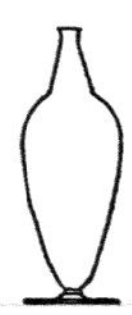
5923
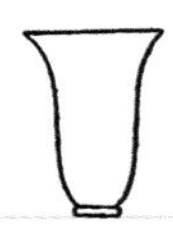
5924
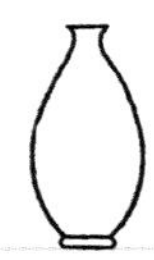
5925

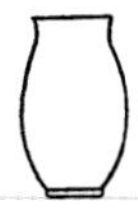
5926
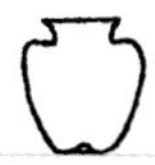
5927
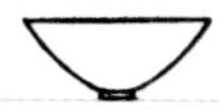
5928
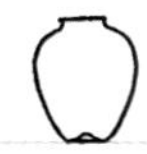
5929
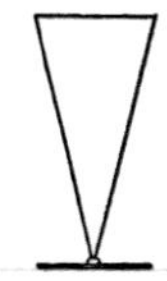
5930
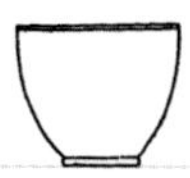
5931
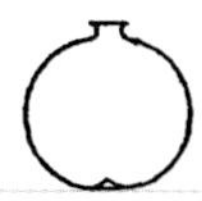
5932
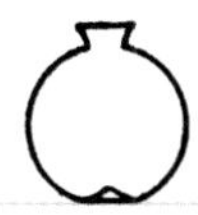
5933
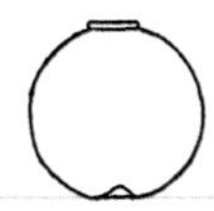
5934
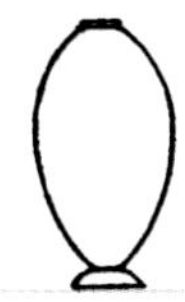
5935
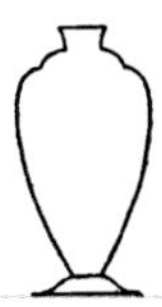
5936
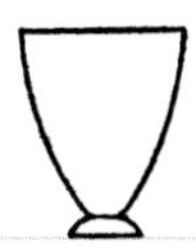
5937
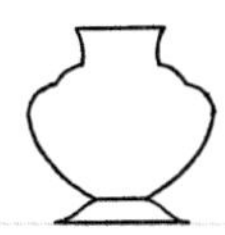
5938
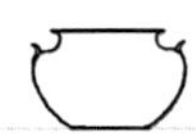
5939
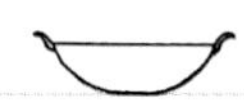
5940
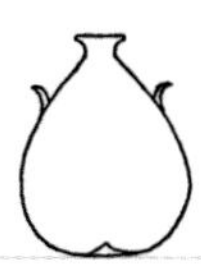
5941
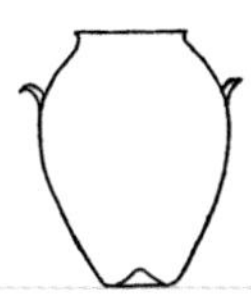
5942
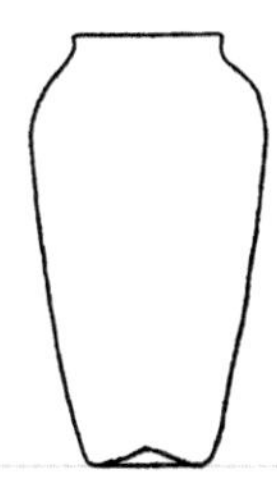
5943
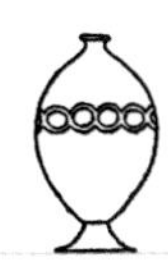
5944
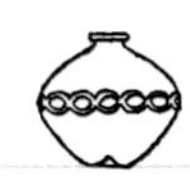
5945
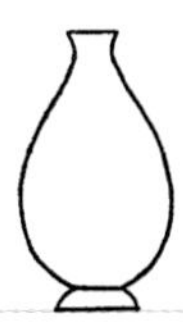
5946
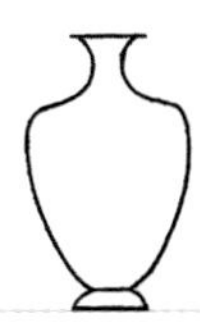
5947
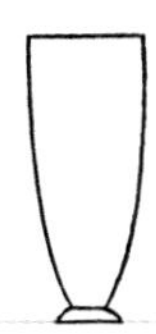
5948
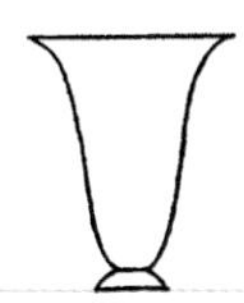
5949
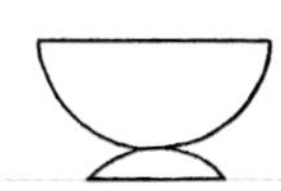
5950

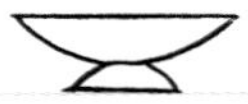
5951
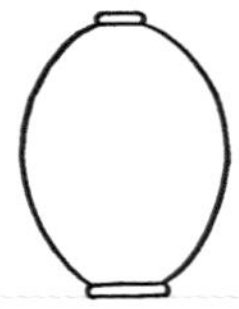
5952
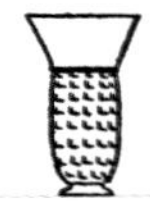
5953
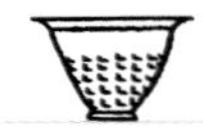
5954
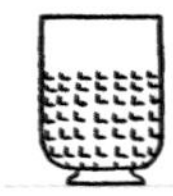
5955
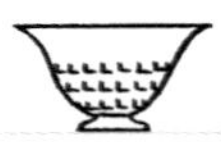
5956
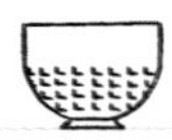
5957
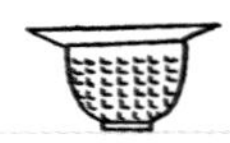
5958
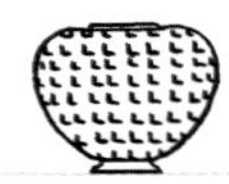
5959
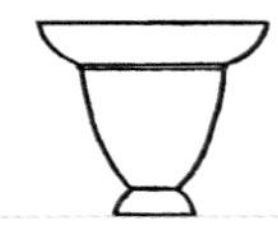
5960
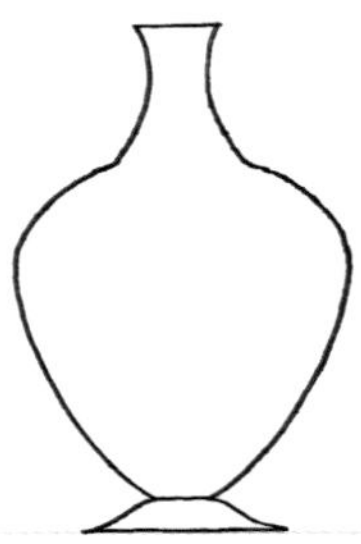
5961
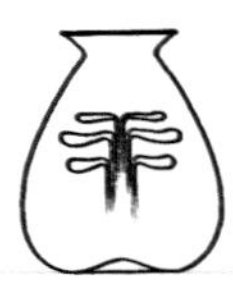
5962
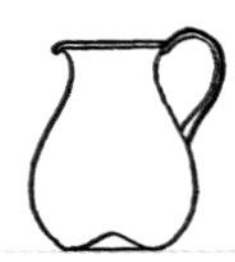
5963
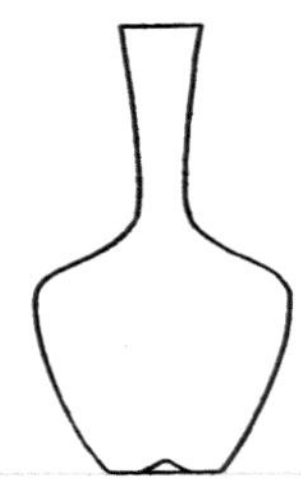
5964
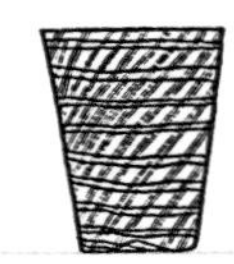
5965
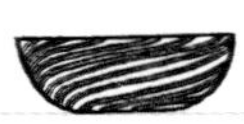
5966
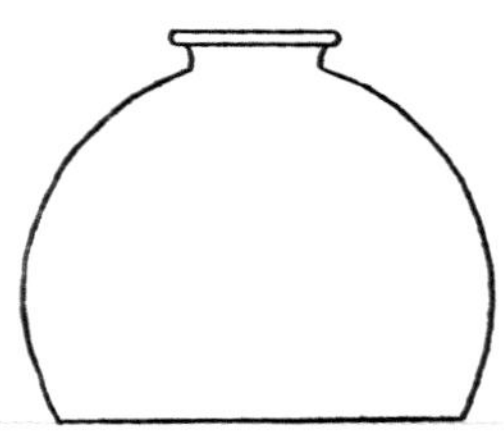
5967
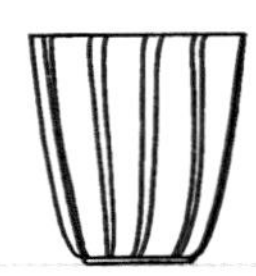
5968
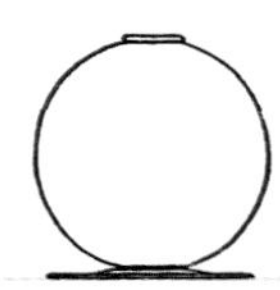
5969
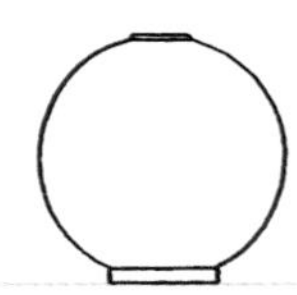
5970
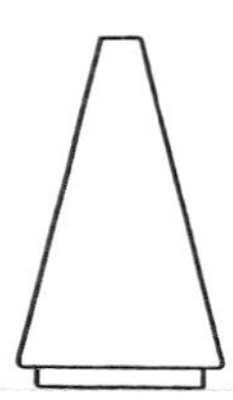
5971
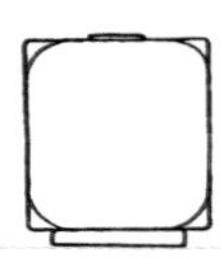
5972
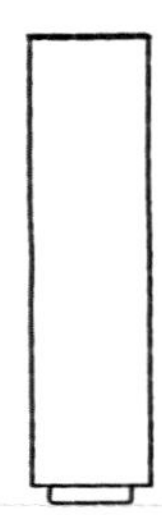
5973
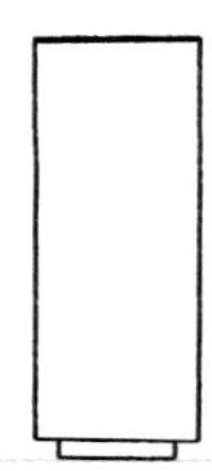
5974
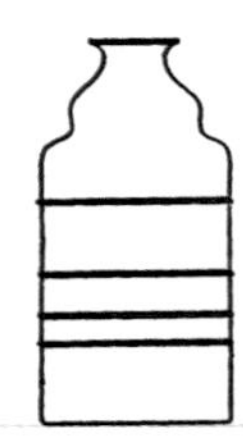
5975

(5976)

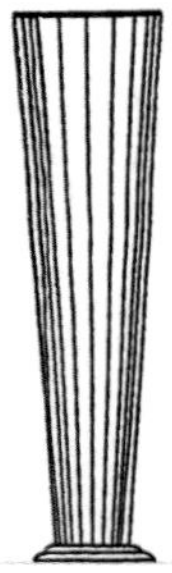

5977

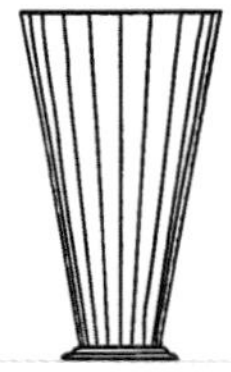

5978

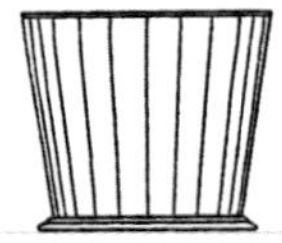

5979

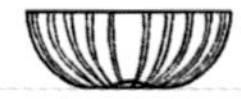

5980

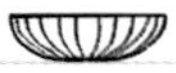

5981

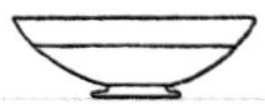

5982

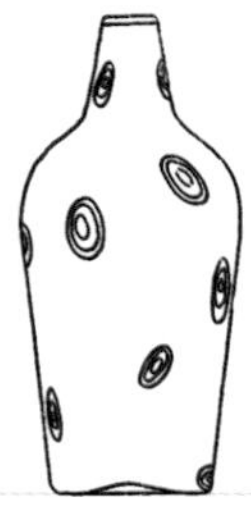

5983

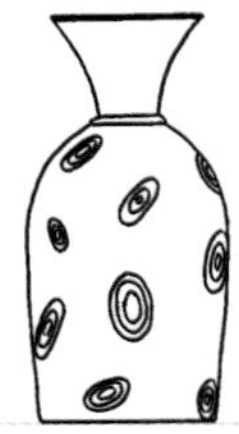

5984

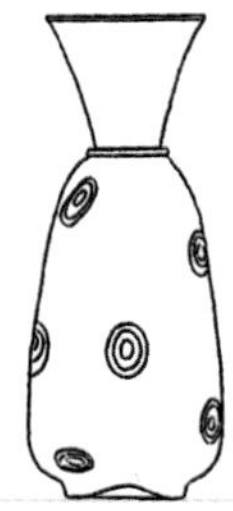

5985

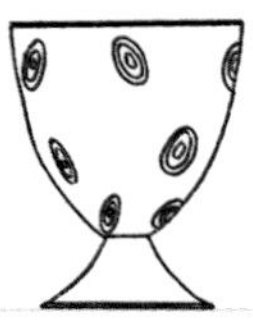

5986

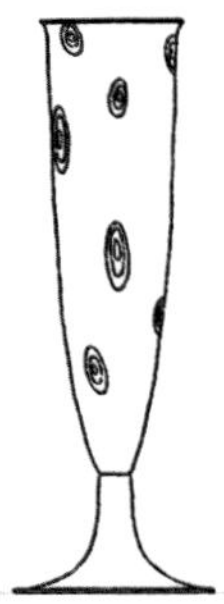

5987

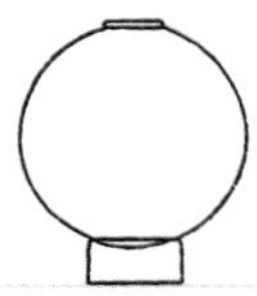

5988

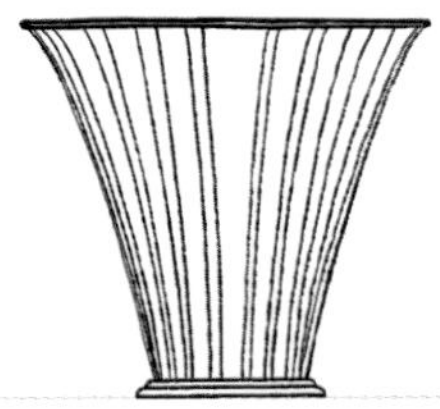

5989

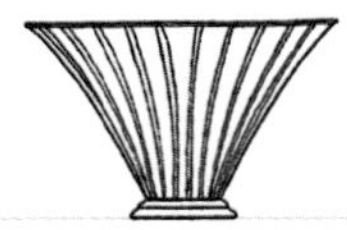

5990

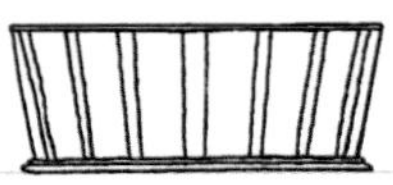

5991

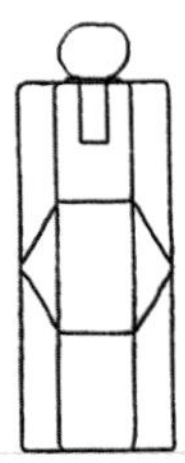

5992

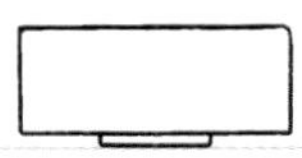

5993

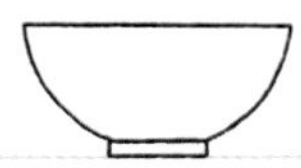

5994

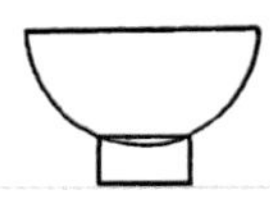

5995

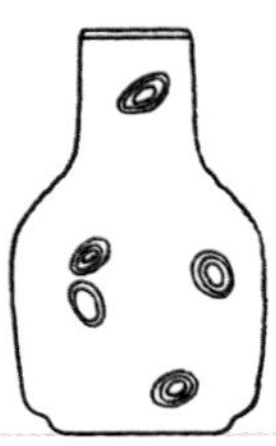

5996

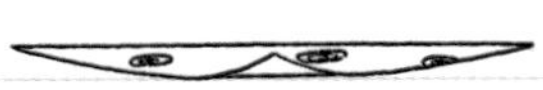

5997

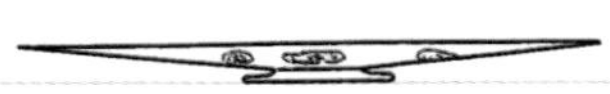

5998

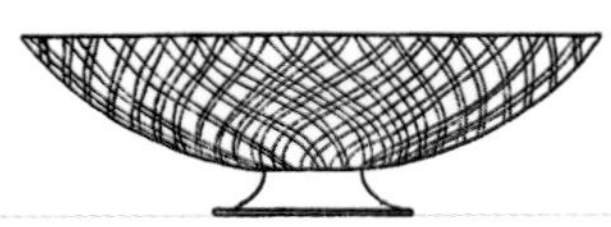

5999

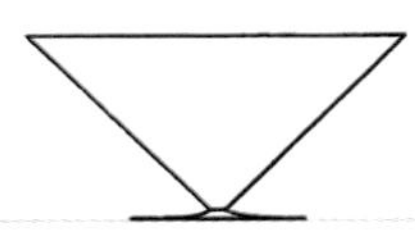

6000

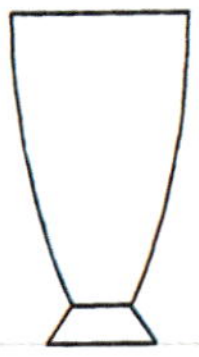
6001

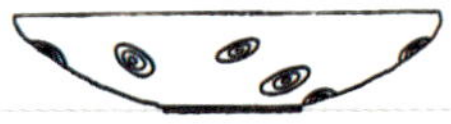
6002

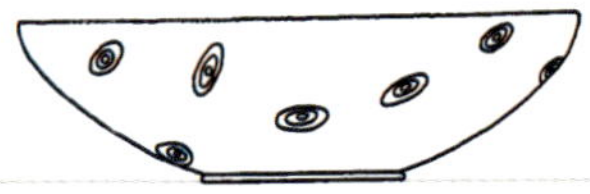
6003

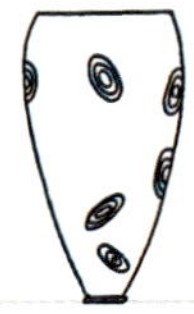
6004

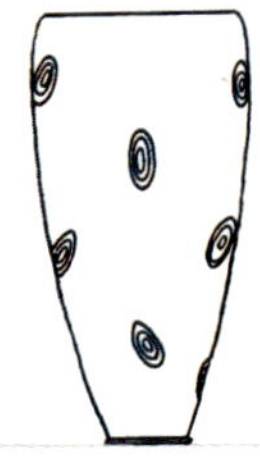
6005

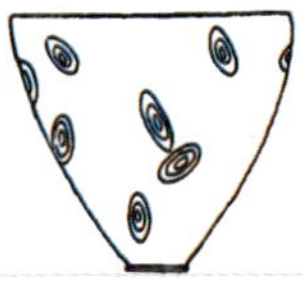
6006

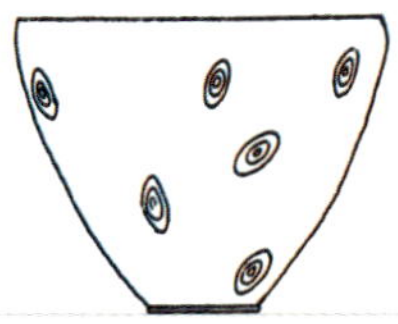
6007

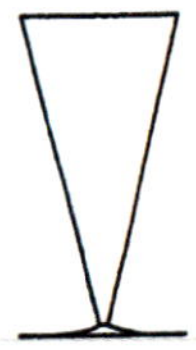
6008

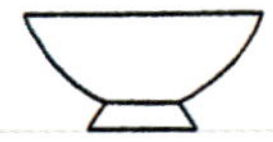
6009

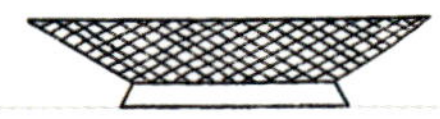
6010

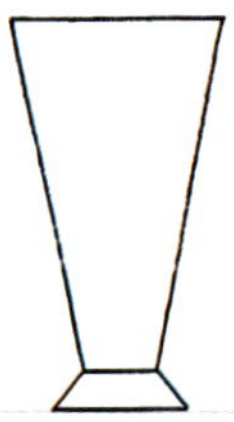
6011

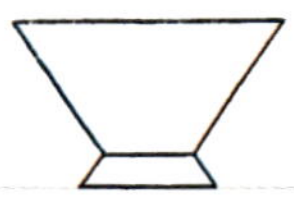
6012

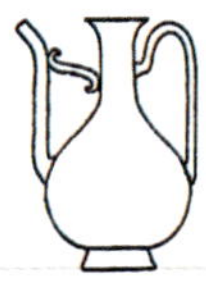
6013

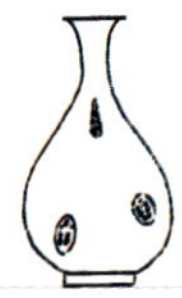
6014

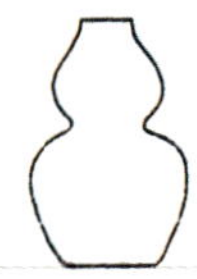
6015

(6016)

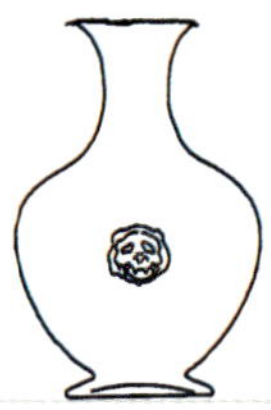
6017

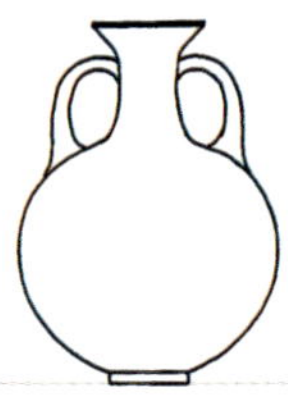
6018

(6019)

(6020)

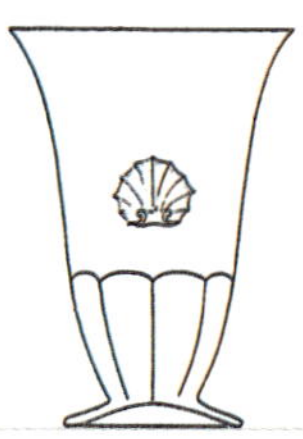
6021

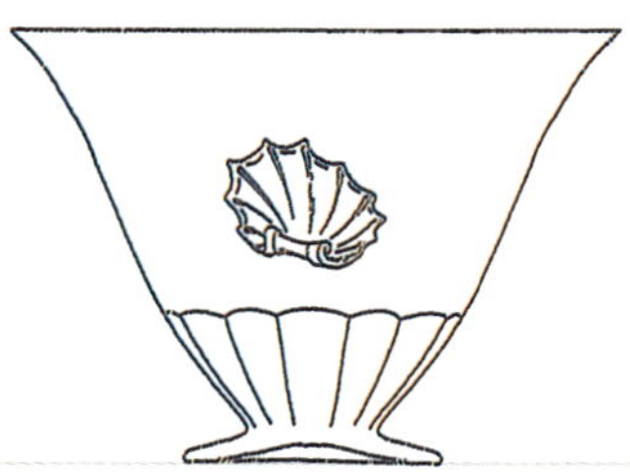
6022

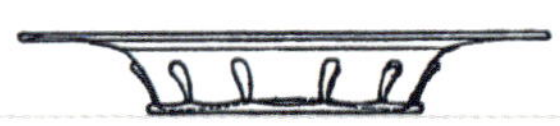
6023

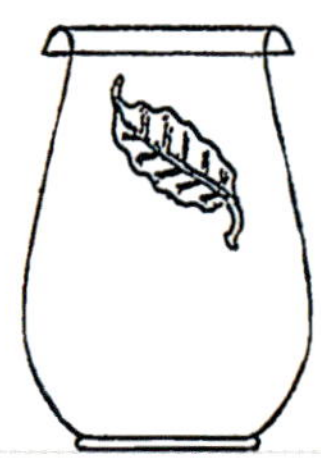
6024

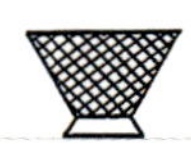
6025

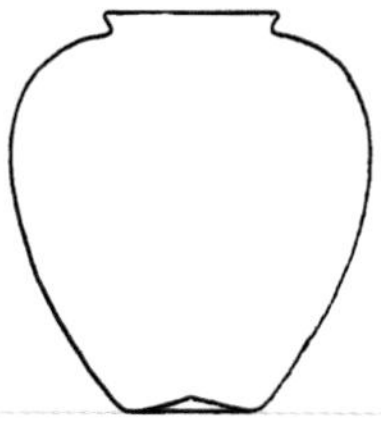
6026

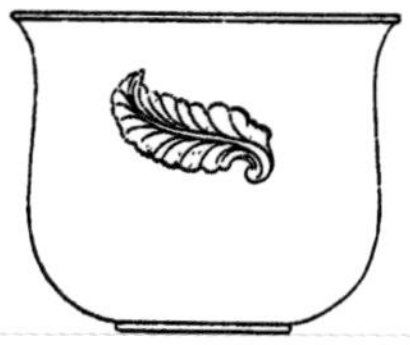
6027

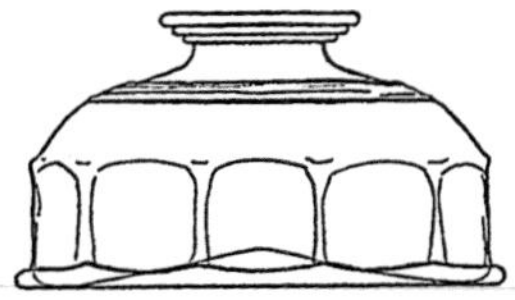
6028

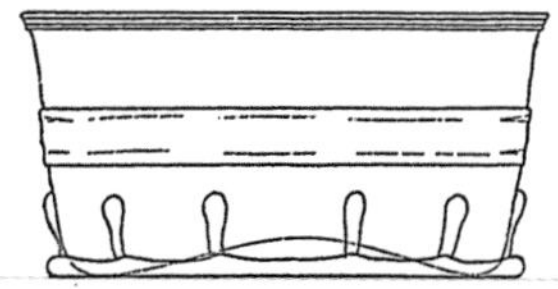
6029

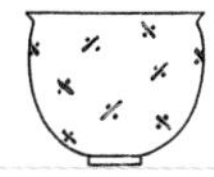
6030

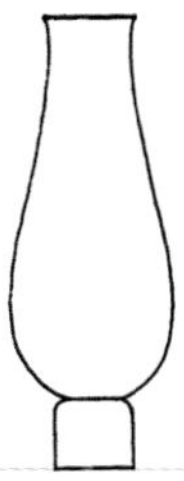
6031

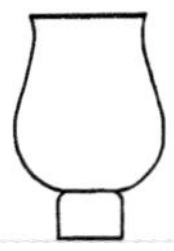
6032

(6033)

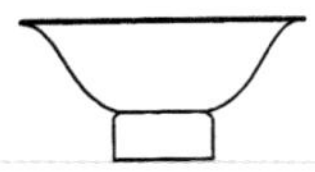
6034

(6035)

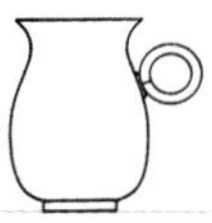
6036

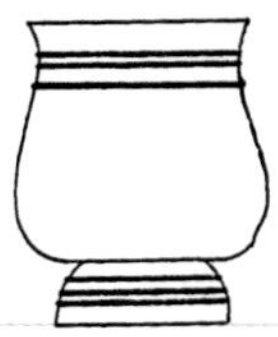
6037

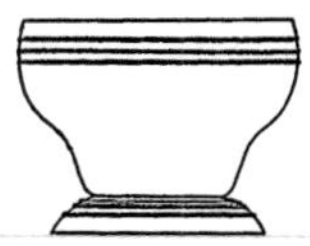
6038

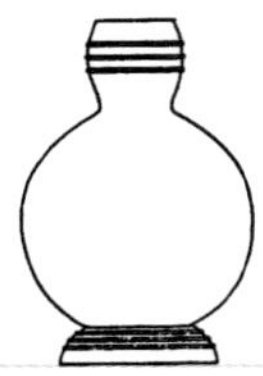
6039

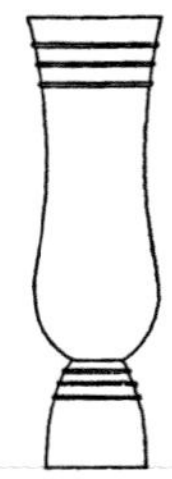
6040

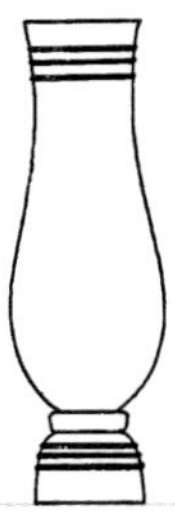
6041

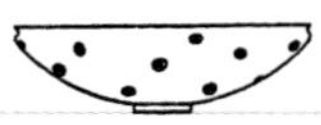
6042

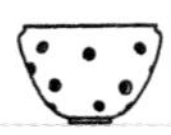
6043

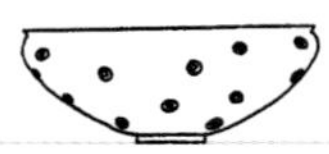
6044

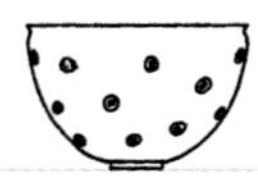
6045

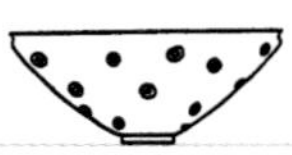
6046

(6047)

(6048)

(6049)

(6050)

EXHIBITIONS 1926-1931

edited by Carla Sonego

1926
Venice, 15th *Esposizione Internazionale d'Arte*
Venice, Lido, *Opera Bevilacqua La Masa*, 17th *Mostra degli Artisti di Ca' Pesaro*
Paris, *Salon d'automne*

1927
Milan, 8th Trade Fair
Monza, 3rd *Mostra Internazionale delle Arti Decorative*
Rome, 93rd *Esposizione di Belle Arti della Società amatori e cultori di belle arti*
Rome, 2nd *Mostra Nazionale d'Arte Marinara*
Venice, 18th *Mostra dell'Opera Bevilacqua La Masa*
Geneva, Rath Museum, *Exposition d'artistes italiens contemporains*
Leipzig, Grassimuseum, *Europäisches Kunstgewerbe 1927 (?)*
Paris, *Salon d'automne*
Tripoli, 1st International Fair

1928
Milan, 9th Trade Fair
Turin, *Torino 1928*
Venice, 16th *Esposizione Internazionale d'Arte*
Leipzig, Trade Fair
New York, Macy's, International Exposition of Art in Industry
Newark (NJ), Newark Museum, Five Hundred New Objects
Paris, *Art au Foyer*
Paris, *Salon d'automne*
Vienna, 15th International Fair, *Esposizione dei grandi porti della media Europa*

1929
Milan, Galleria Pesaro, *Mostra Artisti italiani contemporanei*
Milan, 10th Trade Fair
Barcelona, International Exhibition
Leipzig, Trade Fair
Paris, Musée de Galliera, *Le décor de la table*
Paris, *Salon d'automne*
Paris, *Salon des Artistes décorateurs*
Paris, Trade Fair

1929-30
USA, International Exhibition of Contemporary Glass and Rugs, travelling exhibition

1930
Milan, 11th Trade Fair
Monza, 4th *Esposizione Triennale Internazionale delle Arti Decorative ed Industriali Moderne*
Paris, Musée de Galliera, *Le décor de la table*

1931
Florence, Palazzo Vecchio, *Mostra del Giardino italiano, mostra del Fiore d'arte*
Rome, 1st *Quadriennale d'Arte Nazionale*
Venice, 20th *Esposizione dell'opera Bevilacqua La Masa*
Amsterdam, Stedelijk Museum, *Tentoonstelling van glas, aardewerk en kant, van moderne italiaansche kunstnijverheid. Mostra di vetri, ceramiche e merletti d'arte moderna italiana*
Athens, Italian Week in Athens
Paris, *Salon des artistes décorateurs*

BIBLIOGRAPHY

1912
Le industrie artistiche italiane. Vetri e vetrate, Rome

1922
Catalogo della XII esposizione d'arte della città di Venezia, Milan

1923
L. Linzler, 'Les verreries de Cappellin et Venini', in *Art et Décoration*, March 1923
R. Papini, 'La Mostra delle arti decorative a Monza III. Le arti della terra', in *Emporium*, 343, pp. 3-21

1924
XIV Esposizione internazionale d'arte della città di Venezia, 1924. Catalogo, Venice

1925
'La Missione argentina a Murano: Il ricevimento in Municipio', in *La Gazzetta di Venezia*, 9 July, p. 5
Seconda Mostra Internazionale delle Arti Decorative. Villa Reale di Monza. Catalogo, Milan

1926
'Cadeaux pour la Noël et le Jour de l'An', in *Les Echos des industries d'art*, November, p. 19
H. Clouzot, 'Les arts appliqués', in *La Renaissance de l'art française et des l'industries de luxe*, November, pp. 626-636
'Concours de Noël', in *Art et Industrie*, December, p. 43
C. [V. Costantini], 'Maestri vetrai muranesi "Cappellin"', in *Le arti plastiche*, 16 July
P. Du Colombier, 'Le Salon d'automne', in *Art et Décoration*, November, pp. 184-188
'Here and Everywhere', in *International Studio*, March, p. 92
Exhibition of Modern Italian Art, New York
'Il ripristino di Palazzo da Mula a Murano', in *La Gazzetta di Venezia*, 5 October
I Mostra Nazionale d'Arte Marinara. Promossa dalla Lega Navale Italiana. Roma MCMXXVI-MCMXXVII, Rome
R. L., 'La mostra degli artisti di Ca' Pesaro al Lido di Venezia', in *Le Tre Venezie*, July, pp. 51-52
L'Italia alla Esposizione internazionale delle arti decorative e industriali moderne. Parigi MCMXXV, Rome
G. Morazzoni (notes by), *Il palazzo da Mula di Murano*, limited edition, Milan
U. Nebbia, *La Quindicesima Esposizione d'Arte a Venezia - 1926*, Bergamo
Salon d'automne. Catalogue des Ouvrages de peinture, sculpture, dessin, gravure, architecture et art décoratif, Paris, p. 148
XV Esposizione internazionale d'arte della città di Venezia, Venice
E. Tisserand, 'L'Art Décoratif au salon d'automne', in *L'Art Vivant*, December, p. 894

1927
'Artisti italiani a Ginevra', in *Le arti plastiche*, 1 February
Austellung europäisches Kunstgewerbe 1927, Leipzig, p. 47
'Beauty Laboratory on Downtown Lease', in *New York Times*, 17 August, p. 40
M. Bernardi, 'Giacomo Cappellin (Maestri Vetrai Muranesi)', in *Exposition d'artistes italiens contemporains*, Turin, no page numbers
M. Bernardi, 'Le "botteghe" piemontesi e il mobile 1927 alla Mostra di Monza', in *La Stampa*, 11 June
Catalogo della XVIII Mostra dell'Opera Bevilacqua La Masa, Venice, p. 5
Catalogo ufficiale della III mostra Internazionale delle Arti Decorative, Milan
R. Chavance, 'L'Eclairage moderne', in *Les Echos des Industries d'art*, September, p. 32
'Civis, Alla III Biennale', in *Cittadino*, 18 August
'Cose veneziane', in *Le arti plastiche*, 16 August
V. Costantini, 'Murano e le sue celebri vetrerie' in *Le vie d'Italia e dell'America latina*, June, pp. 704-707
J. De la Tour, 'Le décor de la table', in *Les Echos de Industries d'art*, April, p. 17
G. Dell'Oro, 'I veneti alla III Biennale delle Arti Decorative di Monza', in *Le Tre Venezie*, August, pp. 41-43
A. De Stefani, 'Il trionfo del vetro a Monza', in *Il lavoro fascista*, 7 June
Exposition d'artistes italiens contemporains. Turin
C. A. Felice, 'La Biennale d'Arte Decorativa', in *La Fiera Letteraria*, 3 July
Fiera di Milano. Campionaria internazionale 12-27 aprile 1927. Catalogo ufficiale, Milan, p. 48
R. G., 'Il Padiglione Venezia alla I fiera di Tripoli', in *Rivista della Città di Venezia*, February, pp. 37-42
R.L. Hobson, *Chinese Art*, London
'L'art du verre', in *Les Echos de Industries d'art*, May, p. 32
'Le mostre di Ginevra', in *Le arti plastiche*, 1 March
G. Marangoni, *Arti del fuoco: Ceramiche-vetri-vetrate*, Milan
G. Marangoni, *La terza mostra internazionale delle arti decorative nella Villa Reale di Monza. Notizie, rilievi, risultati*, Bergamo
F. T. Marinetti, 'Gli avvisi luminosi. Lettera aperta a S.E. Mussolini', in *L'Impero*, 12 February
R. Papini, 'Le arti Monza nel 1927. I Gli italiani', in *Emporium*, 391, pp. 14-32
R. Papini, 'Le arti a Monza nel 1927. II Gli stranieri', in *Emporium*, 392, pp. 71-79.
Salon d'automne. Catalogue des Ouvrages de peinture, sculpture, dessin, gravure, architecture et art décoratif, Paris, p. 151
Seconda mostra d'arte marinara. Prima mostra d'arte fiumana. Catalogo, Rome, p. 83
G. Turati, 'Alla mostra di Monza. La bottega moderna', in *Domenica del Corriere*, 4 September
'Verrerie', in *Les Echos des Industries d'art*, April, p. 24
XCIII Esposizione di Belle Arti della Società amatori e cultori di belle arti. Catalogo, Rome, pp. 11, 25, 78, 81
X.X., 'Espositori a Monza. II', in *Le arti plastiche*, 16 September
E. Zorzi, 'Cronache d'arte', in *Le Tre Venezie*, August, p. 48
E. Zorzi, 'Cronache d'arte. Mostre in preparazione. Una mostra d'arte italiana a Ginevra', in *Le Tre Venezie*, February, p. 57
E. Zorzi, 'I vetrai a Lipsia', in *Le Tre Venezie*, March, p. 61

1928
'Art au Foyer', in *Art et industrie*, December, p. 35
'Bollettino delle istituzioni, La famiglia Artistica', in *Le Arti Plastiche*, 1 March
Catalogo ufficiale generale. Fiera esposizione di Milano 1928 A. VI., Milan, pp. 153, 512
Dir. [G. Ponti], 'Murano: oggi fatto d'arte', in *Domus*, December, pp. 59-62
Ente Nazionale Piccole Industrie [National Small Industries Body], *Esposizione, Torino, 1928. Catalogo*, Rome, pp. 21, 26
G. Dell'Oro, 'Piccole industrie venete all'esposizione di Torino', in *Le Tre Venezie*, July, pp. 45-47
Filigrane di Murano. 'Un'arte che riappare in forme nuove', in *Domus*, January, p. 29
'Il forno dei vetri dei Maestri Vetrai', in *Milano nel Mondo*, 12 April, p. 13
'Il porto di Venezia alla fiera di Vienna', in *Le Tre Venezie*, October, pp. 59-60
'Italiani e stranieri in una esposizione d'arte decorativa moderna a New York', in *Domus*, October, pp. 18-21
Italienische Musterschau im rahmen der Leipziger Messe, Rome, p. 3

G. Marangoni, 'La vetrata', in *La casa bella*, April, pp. 13-15
A. Marzocchi, 'Ambienti milanesi. La Famiglia Artistica', in *La casa bella*, December, pp. 26-29
U. Nebbia, *La XVI Esposizione internazionale d'arte Venezia – MCMXXVIII*, Milan-Rome, p. 42
E. Pansini (edited by), *XV Biennale internazionale d'arte della città di Venezia,1926,* Naples
A. Parini, 'La mostra piemontese dell'artigianato', in *La casa bella*, November, pp. 40-41
F. Reggiori, 'La terza Biennale delle arti decorative a Monza', in *Architettura e arti decorative*, 7th issue, pp. 300-318
F. Sollar, 'Le décor de la table', in *Les Echos des Industries d'art*, April, p. 17
Torino 1928. Catalogo ufficiale dell'esposizione, Turin, pp. 385, 392
P. Torriano, 'La III mostra d'arte decorativa a Monza. II', in *L'Illustrazione Italiana*, 9 October, pp. 289-292
'When Art goes to Work for Trade', in *Crockery and Glass Journal*, June, p. 29
XVI Esposizione internazionale d'arte della città di Venezia, 1928. Catalogo, Venice, 2nd ed. p. 108

1929
C. L. Avery, 'The International Exhibition of Contemporary Glass and Rugs', in *Bulletin of The Metropolitan Museum of Art*, November, pp. 288-294
S. Biazzi, 'Plastica vetraria', in *Poligono*, December, pp. 66-75
'Cappellin', advertising insert in *The New Yorker*, 19 October
E. Cary, 'Modernism in Glass and Rugs at Metropolitan', in *The New York Times*, 11 October
'Cappellin Glassware Inc.', advertising page in *The Glass and Art Shop*, November, p. 22
'Cappellin Glassware Inc.', advertising page in *Town & Country*, November, p. 34
'Cappellin Glassware Inc.', advertising page in *Town & Country*, December, p. 28
Catalogo ufficiale della IV Esposizione Triennale Internazionale delle Arti Decorative ed Industriali Moderne, Milan
Catalogue. International Exhibition. Contemporary Glass and Rugs, Portland, no page numbers
M. Croci, 'Gli italiani al Salon d'automne', in *La casa bella*, December, pp. 50-53
P. Bottoni, 'Ernesto Thayaht decoratore', in *Rassegna di architettura*, August, pp. 310-313
D. [M. Dazzi], 'Le industrie venete alla X Fiera Campionaria di Milano', in *Le Tre Venezie*, May, p. 37
Dir., 'Le piccole industrie italiane alla fiera di Lipsia', in *Domus*, June, pp. 36-37
Exposición Internacional de Barcelona 1929. Cátalogo de la sección italiana, Milan, pp. 80, 122
'Il vetro di Murano', in *La Fiera di Milano*, March, pp. 19-21
Italienische Musterschau im rahmen der Leipziger Messe, Rome, p. 4
'Le Salon des Artistes décorateurs', in *Art et industrie*, June, pp. 31-35
G. Levi-Montalcini, '(IV) La camera da letto', in *La casa bella*, May, pp. 20-26
L'Italie à la foire de Paris, Paris
Mostra Artisti italiani contemporanei. Galleria Pesaro, Milano, 8-24 novembre, Milan
'Nuovi vetri di Cappellin', in *Domus*, April, p. 35
'Nuovi vetri di Giacomo Cappellin', in *Domus*, October, pp. 30-31
A. Pica, 'Una villa moderna nella cittadella del "Rinascimento"', in *La casa bella*, December, pp. 18-23
LS [Santini], 'Au Salon des Artistes décorateurs', in *La Construction moderne*, 30 June, pp. 482-486
R.S., 'Le ultime settimane della mostra di Barcellona', in *L'Illustrazione Italiana*, 1 December, pp. 870-872
Salon d'automne. Catalogue des Ouvrages de peinture, sculpture, dessin, gravure, architecture et art décoratif, Paris, p. 134
X Fiera di Milano. Campionaria internazionale. 12-28 aprile 1929 VII. Catalogo ufficiale, Milan, pp. 319, 891
1930
'Alla Triennale di Monza', in *Domus*, July, pp. 34 and 41
'Alla Triennale di Monza. Cappellin nella galleria dei vetri d'arte', in *Domus*, September, p. 32
'Acquisti a Monza', in *Le arti plastiche*, 1 November
'Arte e modernità negli stands della S.a. Salone del mobilio', in *La casa bella*, April, pp. 68-69
'Cappellin Glassware Inc.', advertising page in *Arts and Decoration*, March, p. 9
'Cappellin Glassware Inc.', advertising page in *Arts and Decoration*, September, p. 103
'Cappellin Glassware Inc.', advertising page in *Arts and Decoration*, November, p. 1
'Cappellin Glassware Inc.', advertising page in *Home and Field*, November, p. 5
'Cappellin Glassware Inc.', advertising page in *The Gift and Art Shop*, August, p. 144
'Cappellin Glassware Inc.', advertising page in *The Gift and Art Shop*, March, p. 48
A. Cassi Ramelli, 'L'illuminazione alla triennale di Monza', in *L'illuminazione razionale*, 6-7, pp. 46-47
Catalogo ufficiale della IV esposizione triennale internazionale delle arti decorative ed industriali moderne, Milan
'City Museum Glass and Rug Exhibit for "Man in the Street"', in *The St. Louis Globe*, 4 May
'Come viene allestita la sede della prossima Quadriennale a Roma', in *Il Lavoro Fascista*, 12 November
J. Downs, 'An Exhibition of Glass and Rugs', in *The Bulletin of the Pennsylvania Museum*, January, p. 13
C.A. Felice, *Arti decorative 1930 all'esposizione di Monza*, Milan
'Guida all'arredamento', in *La casa bella*, August, pp. 9-10
Fiera di Milano 1930 VII. Catalogo ufficiale, Milan, pp. 710, 731
E. Jewett, 'Glass Exhibition at Art Institute Opens with Beautiful Colors', in *Chicago Daily Tribune*, 20 March, p. 35
'Il concorso per l'ornamento della tavola alla Triennale di Monza', in *Domus*, November, pp. 58-61
'I premiati alla Esposizione Internazionale di Barcellona alla mostra di Belle Arti', in *La casa bella*, February, p. 67
'La Biennale di Monza', in *Le arti plastiche*, 1 April
Le décor de la table, Paris, p. 3
'L'ornamento della tavola', in *Domus*, December, pp. 69-71
E. Lounsbery, 'From the Antique Galleries and Smart Shops', in *Arts and decoration*, October, p.18
'Maltecca & Taccani', advertising page in *Domus*, February, p. 8
Y. Maraini, 'The Modern Italians', in *Architectural Review*, February, pp. 101-106
'Monza II', in *Le arti plastiche*, 16 July
U. Nebbia, 'I veneti alle "arti decorative" di Monza', in *Le Tre Venezie*, June, pp. 13-14
'Nuovi vetri di Cappellin', in *Domus*, January, pp. 38-39
R. Pacini, 'La IV Triennale d'arti decorative a Monza. II - Le sezioni italiane', in *Emporium*, 431, pp. 259-276
R. Papini, 'Bilancio dell'esposizione di Monza 1930', in *L'Illustrazione Italiana*, 26 October, pp. 638-640
A. Pica, 'La mostra grafica a Monza', in *La casa bella*, May, pp. 21-24
'Premi a Monza', in *Le arti plastiche*, 16 October
C. Rager, 'Venetian Glass: Made in an Old Castle', in *Crockery & Glass Journal*, 1 March, p. 64
F. Reggiori, 'La Triennale di Monza. IV Mostra Internazionale delle arti decorative', in *Architettura e arti decorative*, 11th issue, pp. 481-526
'Seen in the Galleries', in *International Studio*, April, p. 10
E. Serretta, 'Una giornata alla Triennale di Monza', in *L'illustrazione italiana*, 22 June, p.1103-1106
R. Targetti (edited by), *La partecipazione Italiana all'Esposizione Internazionale di Barcellona a cura del Commissario generale del governo,* Milan
A. Wenzel, 'Alt Modern', in *Das schöne Heim*, November, pp. 48-51

1930-31
C. A. Felice, 'I vetri alla Triennale di Monza', in *Dedalo*, II, pp. 312-326
A. Maraini, 'La quadriennale di Roma', in *Dedalo*, III, pp. 682-707
1931
'Architetture alla Quadriennale', in *La casa bella*, March, pp. 16-17
'Arti applicate alla quadriennale romana', in *Domus*, April, pp. 68-69

'Beauty. Giving beauty anyway', in *Vogue*, November, p. 82
'Cappellin Glassware Inc.', advertising page in *Arts and Decoration*, October, p. 19
'Cappellin Glassware Inc.', advertising page in *Arts and Decoration*, November, p. 17
'Cappellin Glassware Inc.', advertising page in *Arts and Decoration*, December, p. 1
'Cappellin Glassware Inc.', advertising page in *Gift and Art Shop*, January, p. 173
Catalogo della XXII Esposizione dell'Opera Bevilacqua La Masa. II Regionale Veneta del Sindacato Fascista delle Arti. Arti decorative. Città di Venezia, Venice, p. 39
Catalogue 1931. Salon des artistes décorateurs, Paris, p. 104
G. D. O. [Dell'Oro], 'Le arti decorative alla mostra di Ca' Pesaro', in *Le Tre Venezie*, March, pp. 563-565
Gida [G. Dell'Oro], 'Giardini veneti e veneziani alla mostra di Palazzo Vecchio', in *Rivista di Venezia*, June, p. 212
G. Dottori, 'Prima Quadriennale d'arte Nazionale', in *Oggi e Domani*, 5 January
Erredi, 'Vetri, ceramiche e merletti alla mostra di Amsterdam, in *Le Tre Venezie*, March, pp. 166-167
F. Geraci, 'I veneti alla I Quadriennale romana d'arte', in *Le Tre Venezie*, February, pp. 77-81
'Il Duce assiste alla "vernice" della prima Quadriennale d'arte', in *Gazzetta di Venezia*, 4 January
'I muranesi alla mostra del fiore d'arte di Palazzo Vecchio a Firenze', in *Domus*, July, pp. 52-53
'Interni d'oggi in Italia', in *Domus*, February, pp. 45-48
A. Lancellotti, *La prima Quadriennale d'arte romana*, Rome, p. 236
'L'industria artistica italiana ad Amsterdam', in *Il Popolo d'Italia*, 17 March
G. Lorenzetti, *Vetri di Murano*, Venice, no. 23
Mostra del Giardino italiano. Catalogo, Florence, p. 153
Mostre veneziane, in *Le arti plastiche*, 16 August
U. Nebbia, 'Una mostra d'arte decorativa italiana ad Amsterdam', in *Emporium*, 483, pp. 54-58
U.N. [U. Nebbia], 'Nuova Arte Decorativa a Ca' Pesaro', in *Domus*, November, pp. 82-83
U. Nebbia, 'Quel che ho portato ad Amsterdam', in *Domus*, May, pp. 24-28
'Palazzina al Lungotevere Arnaldo da Brescia in Roma', in *Domus*, January, pp. 15-34
Prima Quadriennale d'arte nazionale. Catalogo, Rome, pp. 8, 28, 63
J. Rusconi, 'La mostra del giardino italiano a Firenze', in *Emporium*, 437, pp. 259-273
Settimana italiana in Atene, catalogo, Rome
Tentoonstelling van glas, aardewerk en kant, van moderne italiaansche kunstnijverheid. Mostra di vetri, ceramiche e merletti d'arte moderna italiana, Venice
P. Torriano, 'La prima Quadriennale nazionale d'arte a Roma', in *L'Illustrazione italiana*, 11 January, pp. 50-53

1932
'Bonwit Teller', advertising insert, in *New York Times*, 6 April
Ernesto Thayaht, Florence
'Marshall Field & Company', advertising insert, in *Chicago Daily Tribune*, 11 November
'Mobili d'oggi', in *Domus*, July 1932, pp. 424-425

1933
'Aspetti di una nuova dimora in Firenze', in *Domus*, October, pp. 527-529

1935
L. Arimattei, 'L'arte fenestriera veneta', in *L'industria del vetro e della ceramica*, pp. 13-15

1945
G. Samonà, 'Relazione per l'inaugurazione dell'anno accademico', in *Annuario I.U.A.V. 1945/46*, p. 6

1960
A. Gasparetto (edited by), *Vetri di Murano 1860-1960*, Verona, plate XVI

1964
Catalogo della XXXII Esposizione biennale internazionale d'arte Venezia, Venice, p. 314

1969
'Recenti doni al Museo Correr', in *Bollettino dei Musei Civici Veneziani*, 1, pp. 29-31

1979
B. Radice (edited by), 'Colloquio con Carlo Scarpa, un architetto a regola d'arte', in *Modo*, 16, pp. 19-21

1981
A. Dorigato, R. Barovier Mentasti, *Venezianisches Glas 19.bis 20 Jarhundert*, Berlin, no. 9
Vetri Murano oggi, Milano, no. 19

1982
R. Barovier Mentasti, *Il vetro veneziano*, Milan
Mille anni di arte del vetro a Venezia, Venezia, nos. 262-263

1984
F. Dal Co, G. Mazzariol (edited by), *Carlo Scarpa. Opera completa*, Milan
Mostra del vetro italiano 1920-1940, Turin, nos. 73, 75, 81, 85

1986
Les cahiers de la recherche architecturale, *Carlo Scarpa*, 19, single-subject number, pp. 48-55
R. Maggio Serra (edited by), *Arte moderna a Torino. 200 opere acquisite per la Galleria civica d'Arte Moderna*, Turin

1989
The Venetians. Modern glass 1919-1990, New York, no. 5

1990
Mario Sironi. Il mito dell'architettura, Milan

1991
M. Barovier, *Carlo Scarpa: I vetri di Murano 1927-1947*, Venice
Vetri di Murano del '900. 50 capolavori, Milan, nos. 20-21

1992
L. Baldin, 'Alcuni disegni inediti di Carlo Scarpa', in *Venezia Arti*, 6, pp. 138-140
R. Barovier Mentasti, *Vetro veneziano 1890-1990*, Venice, nos. 62, 66
L'arte del vetro: silice e fuoco; vetri del XIX e XX secolo, Venice, nos. 332-334

1993
M. Heiremans, *Art Glass from Murano 1910-1970*, Stuttgart, nos. 76, 78-87

1995
M. Barovier, R. Barovier Mentasti, A. Dorigato, 'Il vetro di Murano alle Biennali 1895-1972', Venice
'Eclettismo e innovazione. L'arte italiana negli intendimenti politici e nella pratica espositiva. 1931-1935: le prime Quadriennali romane', in G. Bonasegale (edited by), *Catalogo generale Galleria Comunale d'arte moderna di Roma*, Rome, pp. 57-78

1996
M. Barovier, A. Dorigato (edited by), *Il bestiario di Murano*, Venice, nos. 6-8
R. Barovier Mentasti, 'Venini a Murano e nel mondo', in A. Bettagno (edited by), *Gli artisti di Venini, per una storia del vetro d'arte veneziano*, Milan, pp. 13-28
F. Deboni, *Murano '900: Vetri e Vetrai*, Milan, pp. 32-34, nos. 61-65, 67-70
M. De Michelis, A. Kohlmeyer (edited by), *Bauhaus Bauhaus 1919 -1933*, Milan
M. Laudani, M. Romanelli, *Design: Nordest*, Milan, p. 19
M. Heiremans, *20th century Murano Glass: from craft to design*, Stuttgart, nos. 21, 25-26, 31- 32, 36

1997
C. Sonego, 'Carlo Scarpa alla Maestri Vetrai Muranesi Cappellin & C. 1925-31', in M. Barovier, *Carlo Scarpa: I vetri di un architetto*, Milan, pp. 43-46
M. Barovier, *Carlo Scarpa: I vetri di un architetto*, Milan, pp. 192-201, 225-231

1998
G. Mori, 'La collezione di vetri moderni del Castello Sforzesco acquisiti negli anni Venti e Trenta', in *Tra creatività e progettazione. Il vetro italiano a Milano, 1906-1968*, Milan, pp. 62- 67
Tra creatività e progettazione. Il vetro italiano a Milano, 1906-1968, Milan, nos. 36, 40-41, 51

1999
M. Barovier, *Il vetro a Venezia dal moderno al contemporaneo*, Milan, pp. 140-143
M. Carraro, *Venezia 1913-1947: architetti, istituzioni, città*, tesi di laurea, IUAV (Venice University Institute of Architecture), 1998-99
P. Nicoloso, *Gli architetti di Mussolini. Scuole e sindacato, architetti e massoni, professori e politici negli anni del regime*, Milan, pp. 70-74

2000
Venetian Glass: The Nancy Olnick and Giorgio Spanu Collection, Milan-New York, nos. 17-23, 25-32

2001
Murano. Vetri dalla collezione Olnick Spanu, Milan, nos. 30-37, 39-46, 48-50, 52-53

2002
A. Dorigato, *L'arte del vetro a Murano,* Venice, pp. 257-261, 269
M. Heiremans, *Murano Glass: themes and variations (1910-1970)*, Stuttgart, nos. 71, 92,147

2003
Murano. Glass from the Olnick Spanu Collection, New York, nos. 30-37, 39-46, 48-50, 52-53

2004
F. Antonacci, D. Lapiccirella, C. Cerutti (edited by), *Thayaht e Ram. Dal Futurismo al Novecento*, Rome
M. Barovier (edited by), *Venetian Art Glass: An American Collection 1840-1970*, Stuttgart, no. 152
M. Barovier, 'Il déco a Murano', in F. Benzi (edited by), *Il Déco in Italia*, Milan, pp. 204-221
F. Benzi (a cura di), 'Il Déco in Italia', in F. Benzi (edited by), *Il Déco in Italia*, Milan, p. 115
G. Ginex, 'La Gallerie delle Arti Grafiche alla IV Triennale di Monza', in A. Sironi (edited by), *Sironi. La grande decorazione*, Milan, pp. 192-197
M. Margozzi, 'La Carta del Lavoro', in A. Sironi (edited by), *Sironi. La grande decorazione*, Milan pp. 198-217
C. Salaris, *La Quadriennale. Storia della rassegna d'arte italiana dagli anni Trenta a oggi*, Rome-Venice
C. Sonego, 'Carlo Scarpa: Gli anni '20', in K. W. Forster and P. Marini (edited by), *Studi su Carlo Scarpa 2000-2002*, Venice, pp. 27-86

2005
F. Carli, 'Tra Déco e razionale. Allestimenti di qualità', in *XIV Quadriennale di Roma. Retrospettive 1931/1948*, Milan, pp. 49-58
M. Pratesi (edited by), *Futurismo e bon ton: i fratelli Thayaht e Ram*, Florence, pp. 84-85

2006
O. Lanzarini, 'Fermare la complessità La videobiografia. Un'ora con Carlo Scarpa, 1972', in J. Gudelj, P. Nicolin (edited by), *Costruire il dispositivo storico. Tra fonti e strumenti*, Milan, pp. 247-273

2007
A. Bassi, P. Marini, A. Di Lieto (edited by), *Vinicio Vianello. Il design del vetro*, Venice
C. Sonego, 'La cultura tecnica nella formazione di Scarpa', in W. Tegethoff, V. Zanchettin (edited by), *Carlo Scarpa. Struttura e forme*, Venice, pp. 83-97
S. Tosini Pizzetti (edited by), *Sironi metafisico. L'atelier della meraviglia*, Milan, pp. 167-169, 205-213

2009
P. Marini, 'L'Archivio di Carlo Scarpa a Castelvecchio', in M. Guccione (edited by), *Documentare il contemporaneo. Archivi e Musei di architettura, Atti della giornata di studio MAXXI Museo nazionale delle arti del XXI secolo*, Rome

2010
A. Bova (edited by), *L'avventura del vetro, dal rinascimento al Novecento tra Venezia e mondi lontani*, Milan, nos. IX.15, IX.19, IX.27

2011
I. Abbondandolo, *Carlo Scarpa e la forma delle parole*, Venice, p. 13

2012
M. Barovier (edited by), *Carlo Scarpa Venini 1932-47,* Milan
K.M. Pierconti, 'Appunti per uno scenario di cultura cinese nei vetri e nell'opera di Carlo Scarpa', in M. Barovier (edited by), *Carlo Scarpa Venini 1932-47*, Milan, pp. 85-98
C. Sonego, 'Carlo Scarpa nell'ufficio tecnico della Venini', in M. Barovier (edited by), *Carlo Scarpa Venini 1932-47,* pp. 45-65

2013
M. Barovier, *Vetri Déco: l'esperienza Muranese*, Milan, pp. 16-27
M.F. Giubilei, V. Terraroli (edited by), *La forza della modernità. Arti in Italia 1920-1950*, Lucca, nos. 195, 206
L. Mannini, 'La "colorata anima del Vittoriale". Aggiunte e precisazioni sulle vetrate di Pietro Chiesa', in *Quaderni del Vittoriale*, 9, pp. 105-119
P. Marini, A. Di Lieto, K. Bertolaso (edited by), *Castelvecchio sottotraccia,* Appendix to *Architettiverona*, 2, May-August
F. Rapisarda, 'Ospiti in museo', in *Castelvecchio Sottotraccia*, Appendix to *ArchitettiVerona*, 2, maggio-agosto
C. Sonego, 'I disegni per i vetri', in *Castelvecchio Sottotraccia*, Appendix to *ArchitettiVerona*, 2, May-August
G. Vigna, *Stati Naturali*, Milan

2015
M. Barovier, 'Le verre à Murano entre les deux guerres', in G. Cogeval, B. Avanzi (edited by), *Dolce vita? Du liberty au design italien 1900-1940*, Milan, p. 120
G. Cogeval, B. Avanzi (edited by), *Dolce vita? Du liberty au design italien 1900-1940*, Milan, no. 143
F. Mangone, G. Belli, M.G. Tampieri, *Architettura e paesaggi della villeggiatura in Italia tra Otto e Novecento*, Milan, p. 218

2016
L. Mannini, 'Italian Stained Glass in the 1920s and 1930s: Standing the Test of Modernity', in L. De Girolami Cheney (edited by), *Radiance and Symbolism in Modern Stained Glass: European and American Innovations and aesthetic interrelations in material culture*, Cambridge, pp. 315-342

2017
M. Barovier, C. Sonego (edited by), *Vittorio Zecchin: i vetri trasparenti di Cappellin e Venini*, Milan
G. Belli, E. Barisoni (edited by), *Uno sguardo sul '900. La collezione Chiara e Francesco Carraro*, Arezzo, pp. 90-91
M. Barovier, 'Le veneziane', in V. Terraroli (edited by), *Art Déco. Gli anni ruggenti in Italia*, Cinisello Balsamo, pp. 234-235

ABBREVIATIONS

AAAI	Archivi Arti Applicate Italiane del XX secolo, Rome
ACSVr	Archivio Carlo Scarpa, Museo di Castelvecchio, Verona
AFT	Archivio fotografico©La Triennale di Milano
AHL	Howard Lockwood Archive, Fort Lee (NJ-USA)
AMA	Marco Arosio Archive, Milan
AMB	Marino Barovier Archive, Venice
AMH	Marc Heiremans Archive, Brussels
AMuVe	Archivio Museo del Vetro, Murano, Venice
ASAC	Archivio Storico delle Arti Contemporanee, Venice
CSV	Fondo Carlo Moretti e Giovanni Moretti, Centro Studi del Vetro, Fondazione Giorgio Cini, Venice

DELIUS KLASING

BENGT STILLER

Von Rahmenbauern, Sammlern und Klassikern

RÄDER AUS STAHL

DELIUS KLASING VERLAG

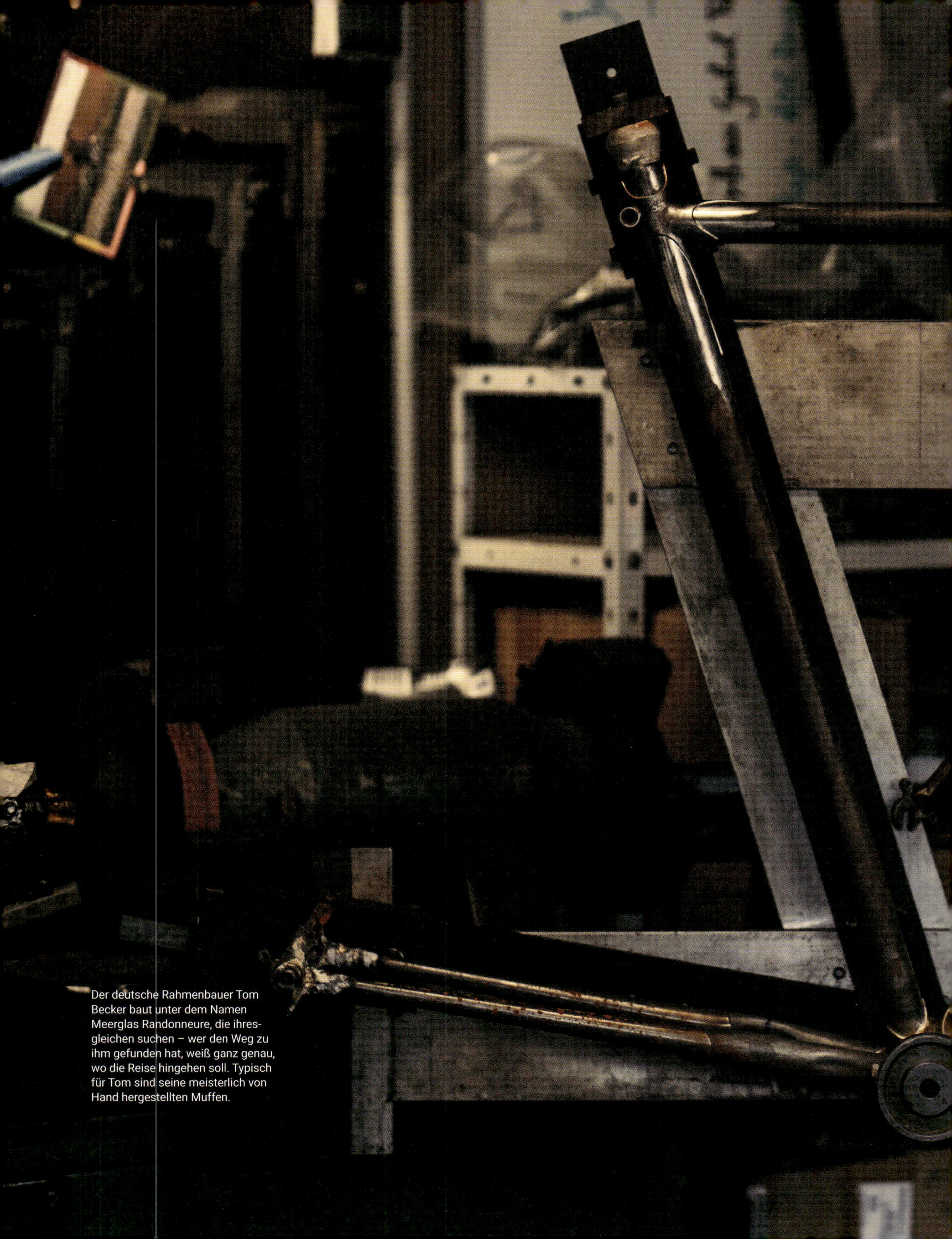

Der deutsche Rahmenbauer Tom Becker baut unter dem Namen Meerglas Randonneure, die ihresgleichen suchen – wer den Weg zu ihm gefunden hat, weiß ganz genau, wo die Reise hingehen soll. Typisch für Tom sind seine meisterlich von Hand hergestellten Muffen.

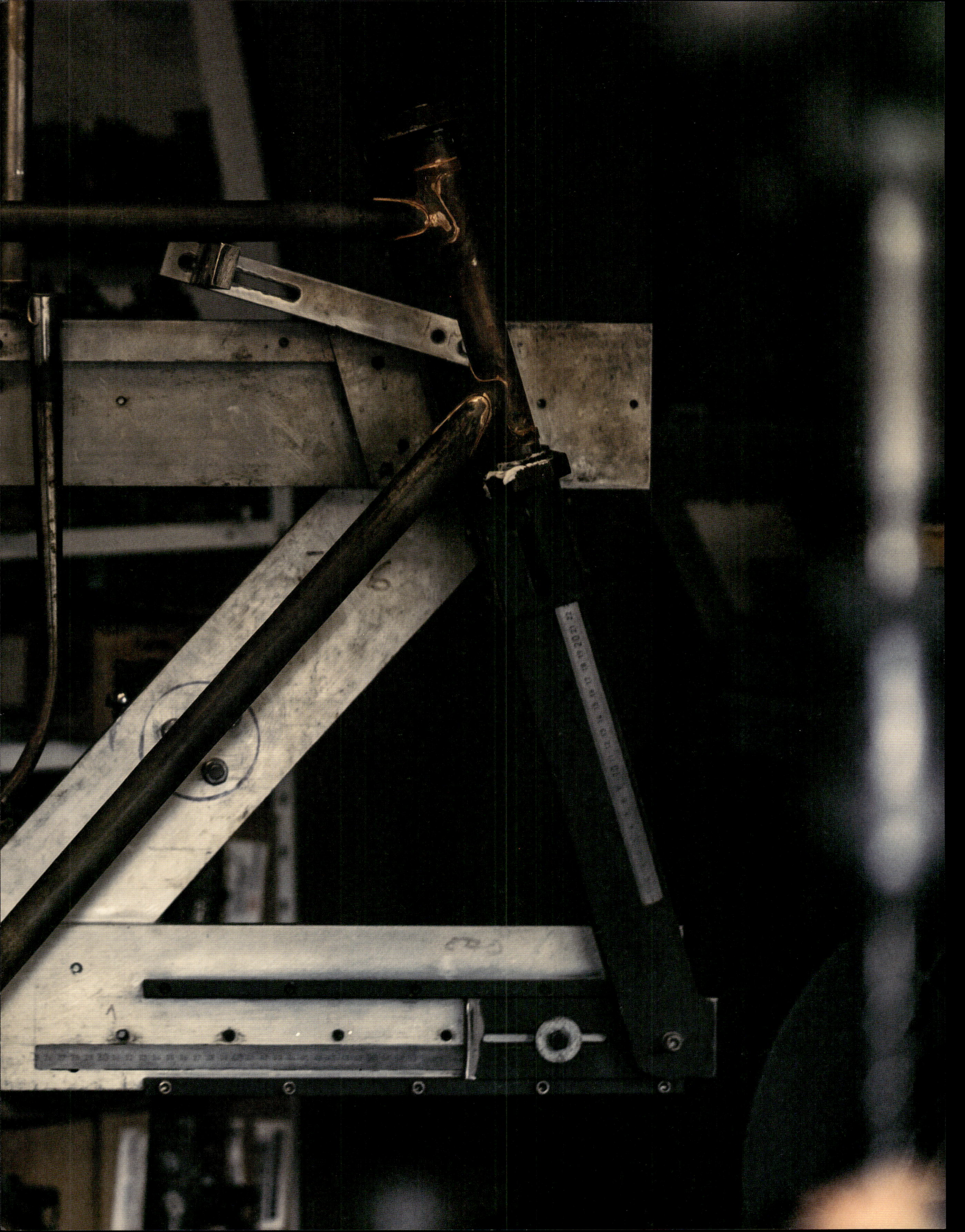

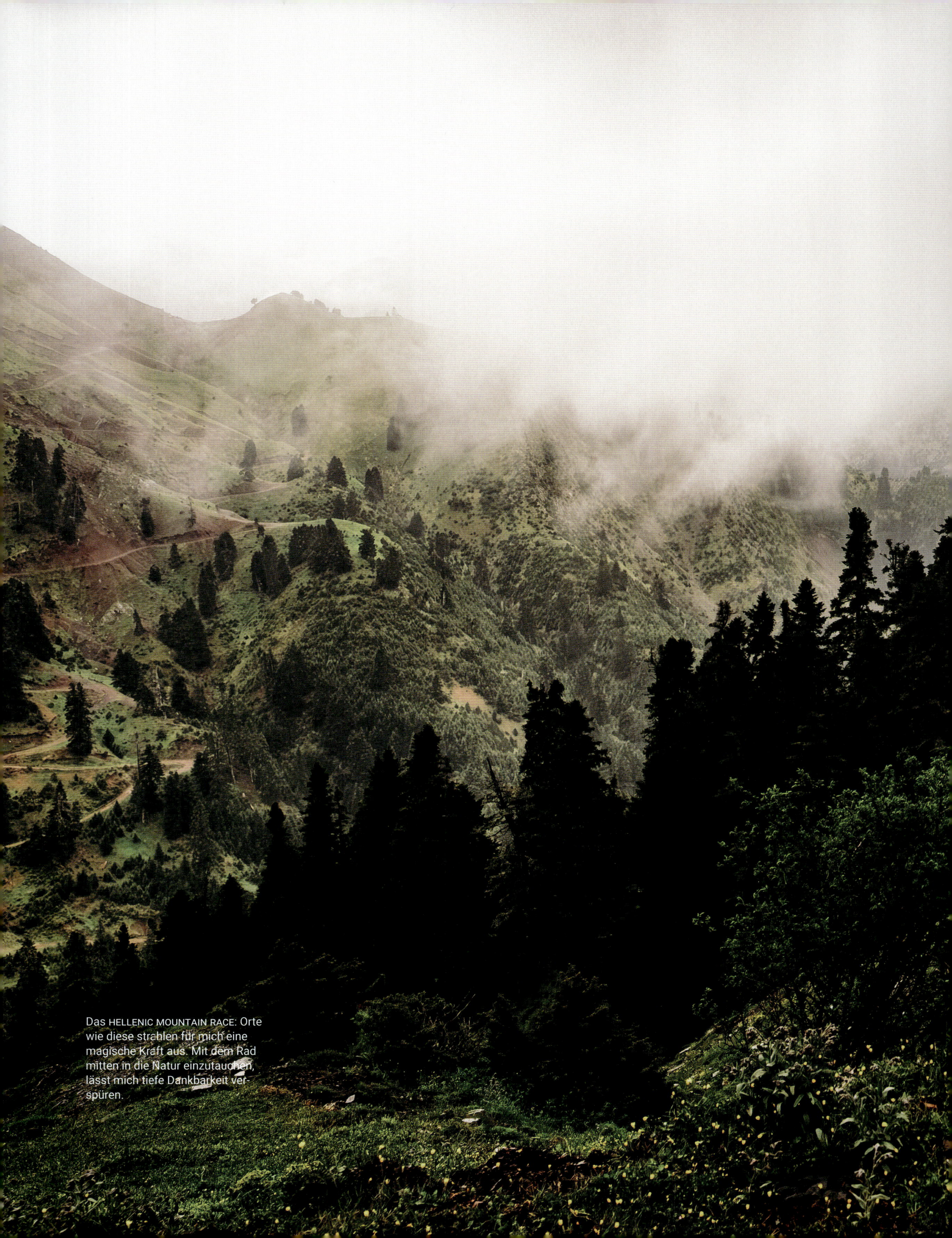

Das HELLENIC MOUNTAIN RACE: Orte wie diese strahlen für mich eine magische Kraft aus. Mit dem Rad mitten in die Natur einzutauchen, lässt mich tiefe Dankbarkeit verspüren.

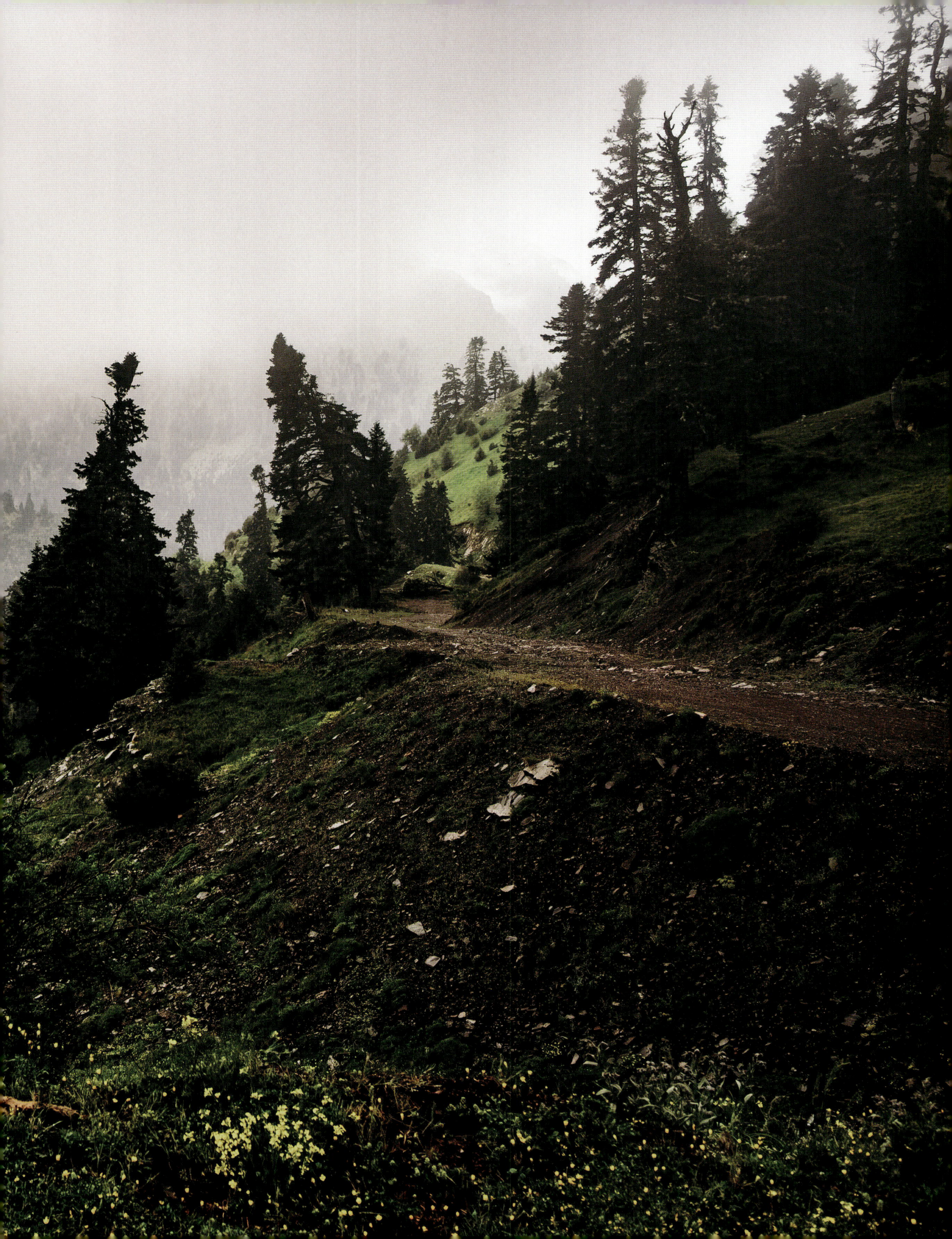

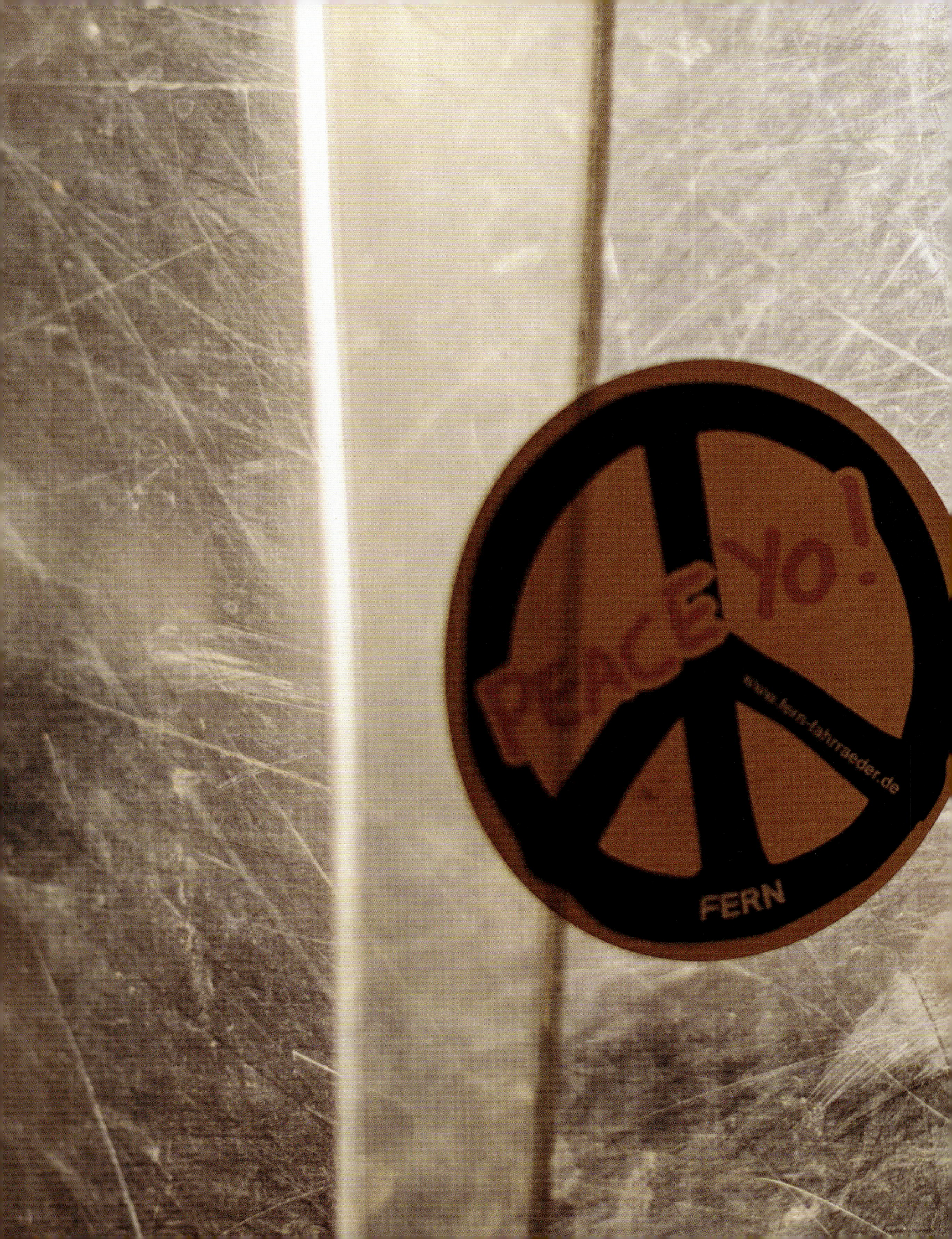
PEACE YO!
www.fern-fahrraeder.de
FERN

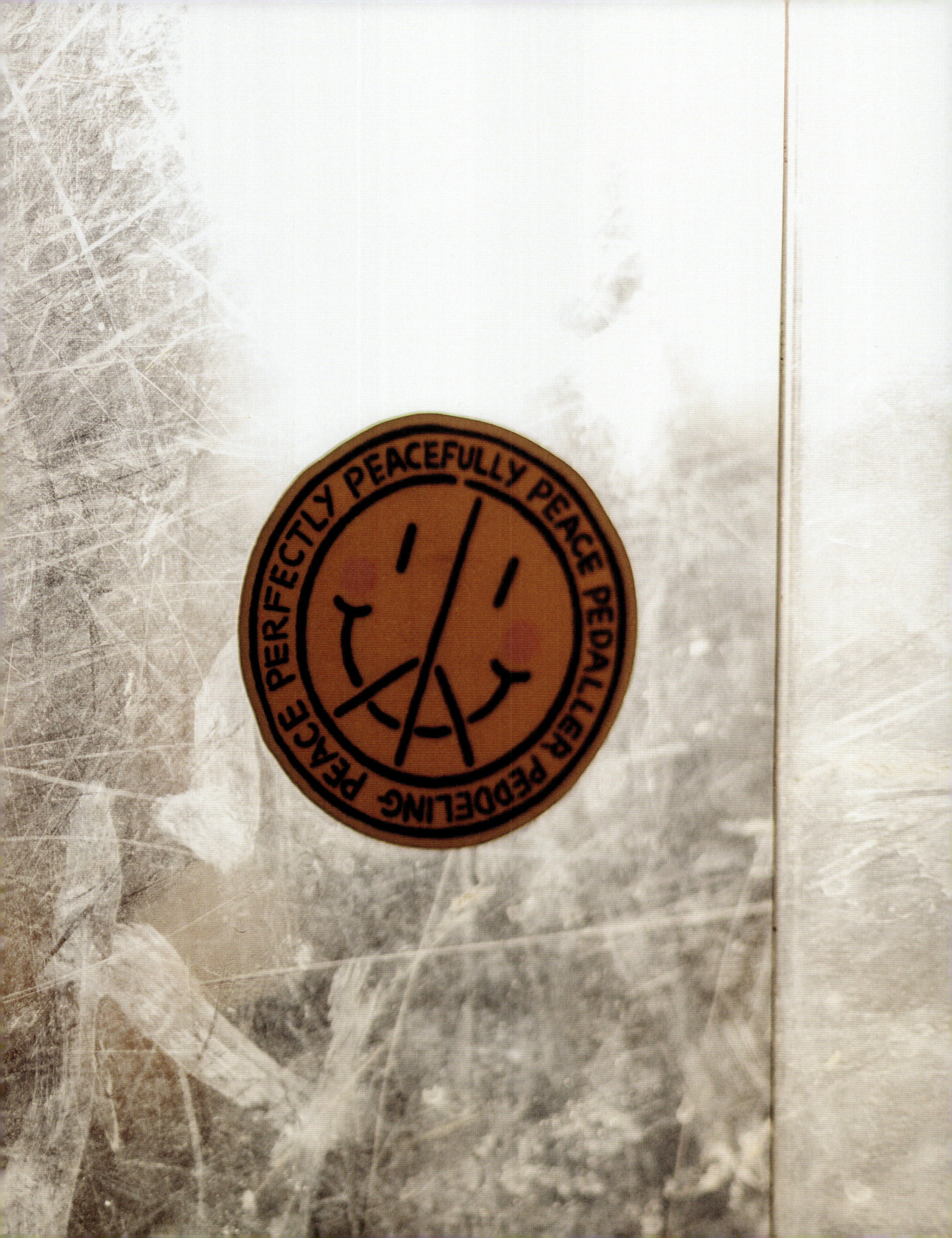
PEACE PERFECTLY PEACEFULLY PEACE PEDALLER PEDDELING-

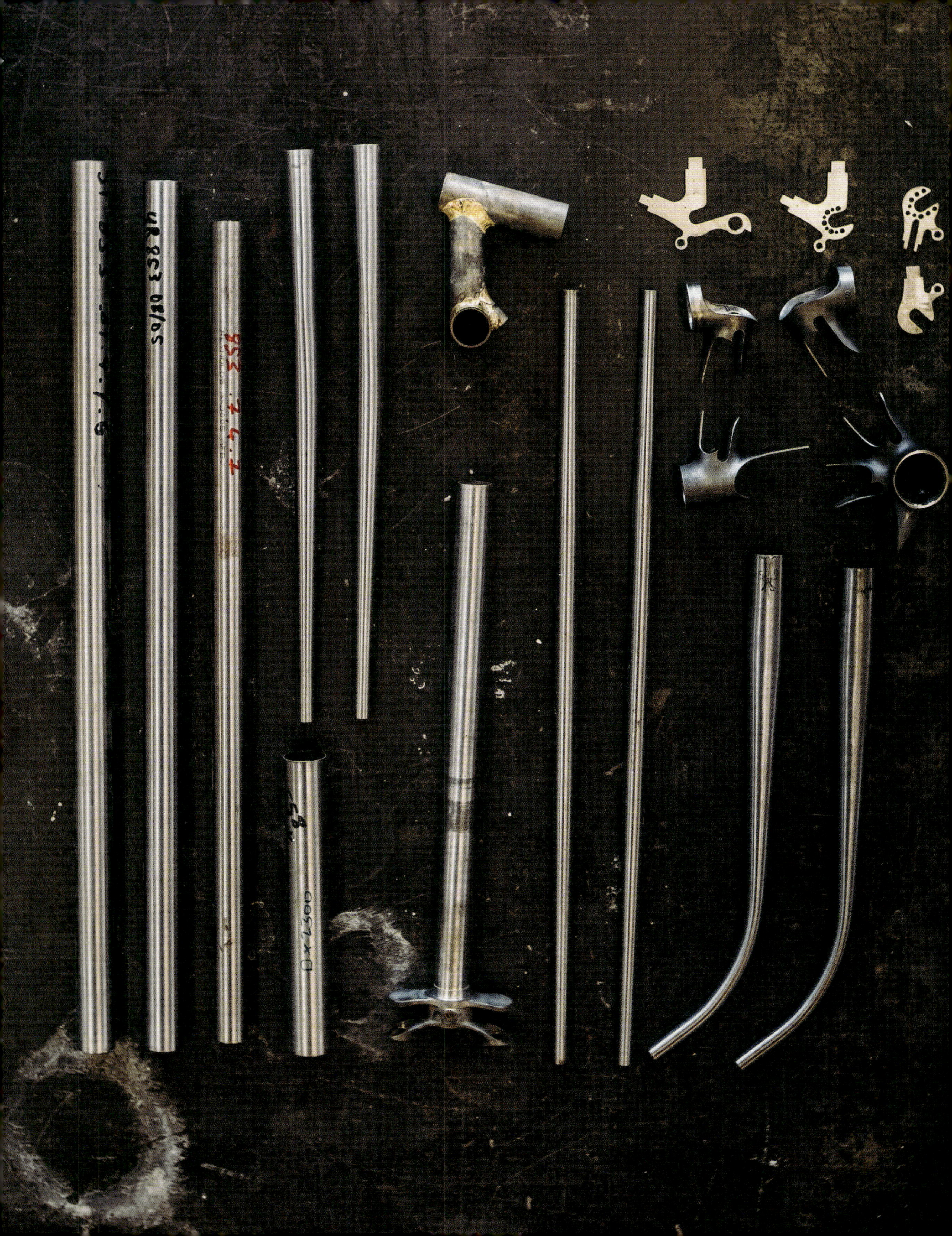

INHALT

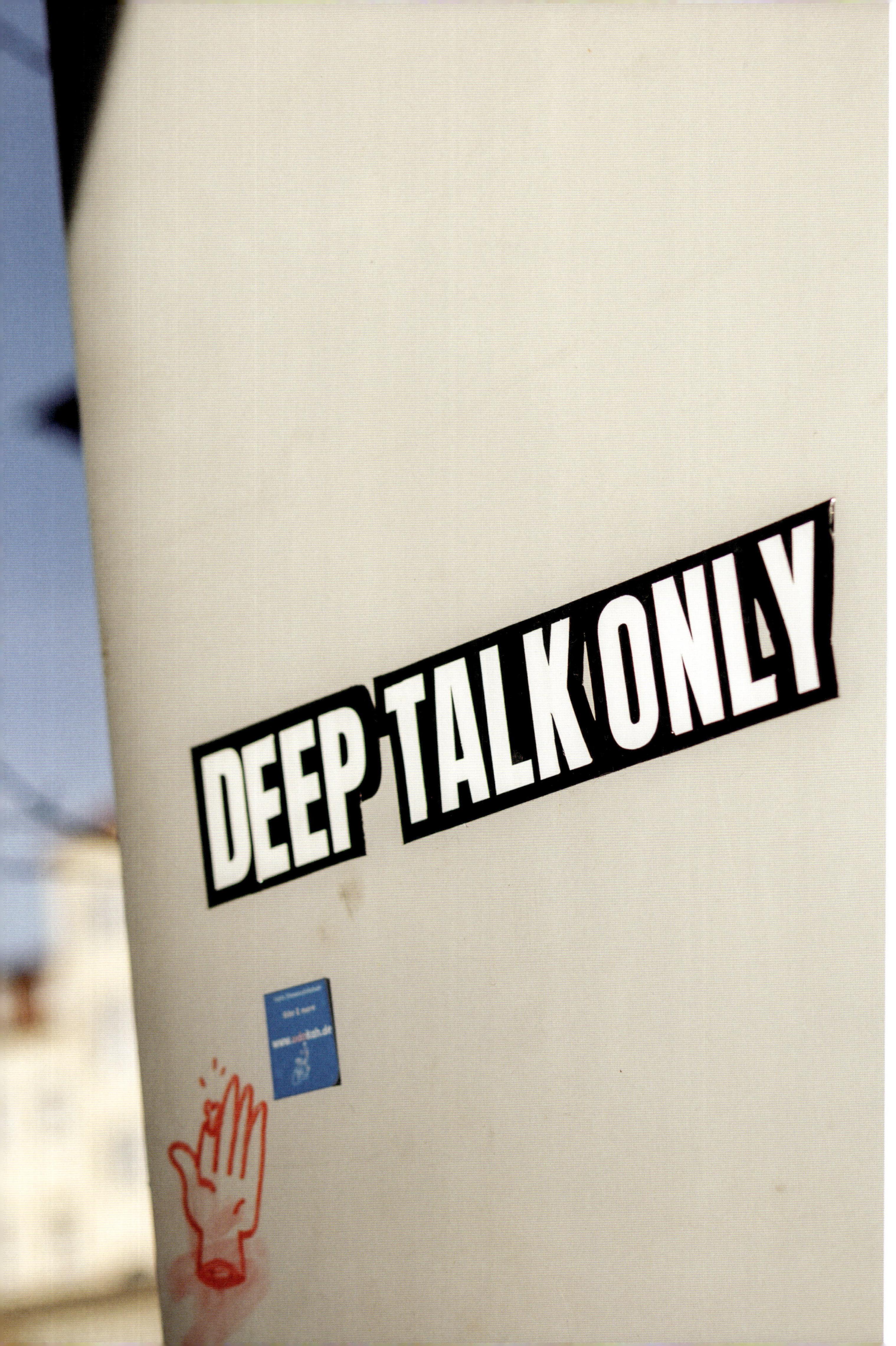
DEEP TALK ONLY

ES DREHT SICH,

UNAUFHALTSAM,

WEITER

BENGT STILLER

Als ich im Jahr 2013 die Jedermann-Ausgabe des Frühjahrsklassikers PARIS-ROUBAIX auf einem Stahlrad mit 28 mm breiten Reifen fuhr, hatte ich das Gefühl, die bestmögliche Wahl getroffen zu haben. Was für eine enorme Reifenbreite! Hatte ich doch vorher nur maximal 25 mm breite Reifen in die alten, klassischen Stahlrahmen montieren können. Die Entscheidung fiel also für maximalen Komfort – vermeintlich – als Antwort auf die 56 Kilometer Kopfsteinpflaster in der Hölle des Nordens. Welch ein Irrtum: Schon der erste Pavé-Abschnitt belehrte mich eines Besseren. Es war tatsächlich die Hölle ... 28 mm hin oder her!

Gut zehn Jahre später hat sich nicht nur meine persönliche Fahrrad-Welt extrem verändert, auch die Entwicklung in der Fahrradindustrie hat ungeahnte Veränderungen hervorgebracht. Der Siegeszug des Gravelbikes ist einer der unangefochten großen Treiber des letzten Jahrzehnts. Was einigen zu Beginn mehr als Marketing-Gag vorkam, hat sich flächendeckend durchgesetzt – mir fällt kein relevanter Hersteller ein, der nicht mindestens zwei Modelle im Programm hat. Durch diese neue Gattung erfolgte der Einzug der wirklich breiten Reifen und der damit verbundenen Möglichkeit, mit einem einzigen Rad ein sehr großes Spektrum an Untergründen abdecken zu können. Auf die neugewonnene Reifen-Freiheit folgte die Freiheit der Streckenwahl, wenn man so will. So findet man z.B. im Umland von Berlin auf Wald- und Feldwegen mehr Räder mit Rennlenker, als man auf Brandenburger Landstraßen jemals antreffen würde. Der Wunsch, dem Autoverkehr und der Hektik des Alltags entfliehen zu können, scheint einer der entscheidenden Treiber zu sein.

Was hat die Entwicklung des Gravelbikes nun mit dieser erweiterten Neuauflage zu tun? Ganz einfach: Es sind 54 Millimeter in Form eines seidenweichen Reifens eines japanischen Herstellers, die mein Interesse am Fahrradfahren komplett verändert haben.

Als im Jahr 2015 die Erstausgabe dieses Buches erschien, hatte ich bereits begonnen auf die amerikanische Szene und ihre MTB-Umbauten zu Rädern mit sehr breiten Rennlenkern und ultrabreiten Reifen zu schielen. Rahmentaschen und wilde Frontgepäckträger-Konstruktionen inkl. Taschen-Sonderanfertigungen sorgten bei mir für innere Ekstase. Es wirkte nicht nur extrem cool, nein, irgendwie schwang für mich dort die Vorstellung von Freiheit mit. Einer Freiheit, mit nur einem Rad alles machen zu können – von langer Tour bis zur schnellen Runde mit Freunden am Abend. Das Gefühl, als ich dann mein 26“-Stahl-MTB mit 54 mm breiten Slicks von René Herse bestückt hatte, werde ich nie vergessen. Da war es: das Rad, von dem ich immer geträumt hatte. Dieses alte Stahl-MTB wurde mit dem breitestmöglichen Rennlenker und den ultrabreiten Slicks zu meinem Traumobjekt – vorerst. Es war der Auftakt zu einer langen Reise.

Es folgte ein Abschnitt, auf dem mein persönlicher Weg alles andere als »glatt asphaltiert« war. Als mein Vater im Jahr 2018 durch einen ungeklärten Radunfall ins Koma fiel, verließ ich umgehend meine neue Wahlheimat Wien und zog nach Berlin. Es wurden zwei Jahre, in denen ich eher das Gefühl hatte, verkehrt herum auf dem Rad zu sitzen, blind steuern und dabei zuschauen zu müssen, wie mich das eigene Hinterrad bei voller Fahrt bergab überholt.

Und gleichzeitig war es in dieser schwierigen Situation das Fahrrad, das mir Kraft schenkte und zum Motor für Neues wurde. Es diente mir als Ventil, wenn die Schmerzen über den Verlust des geliebten Menschen zu groß wurden. Und als Triebwerk, wenn neue Energie notwendig schien. Es wurde zum Taktgeber meines Lebens.

Parallel hierzu veränderte sich mein sportliches Interesse in dieser Zeit maßgeblich: Seit 2018 hat mich das Fieber der Unsupported-Ultra-Endurance-Races gepackt. Ein Sport, in dem man in entlegenen Bergregionen Strecken über 1.000 Kilometer am Stück und mehr ohne Unterstützung bewerkstelligen muss. Wie sehr man sich bei diesen Rennformaten auf dem Rad jenseits der Komfortzone bewegt, beschreibe ich im Kapitel über das HELLENIC MOUNTAIN RACE. Gefahren auf einem maßgeschneiderten Stahlrahmen von FERN – mit 54 mm breiten Reifen. Die Reise auf Stahl hatte ihren Höhepunkt erreicht – vermeintlich.

So wie sich die Fahrradindustrie im letzten Jahrzehnt entwickelt hat, hat sich auch mein Blick auf die Stahlradszene über die Jahre grundlegend weiterentwickelt. Gleich geblieben ist meine Faszination für das Thema Rahmenbau und das damit verbundene Handwerk.

Einen Großmeister seines Fachs und Impulsgeber der aktuellen Rahmenbauer-Szene, Flo Haeussler, besuchen wir im Kapitel über FERN – maßgeschneiderte Stahl-Alleskönner Made in Berlin. Eine mir wichtige Ergänzung der bestehenden Kapitel dieses Bildbandes, die zeigt, wie schön und wie intensiv zugleich handgemachter Rahmenbau ist. Neben der Frage, was die Sammler, Rahmenbauer und Freaks, die ich porträtiere, antreibt und ausmacht.

Und nun, aufsatteln: Auf geht´s zu elf sehr unterschiedlichen Geschichten rund um das Thema: *Räder aus Stahl.*

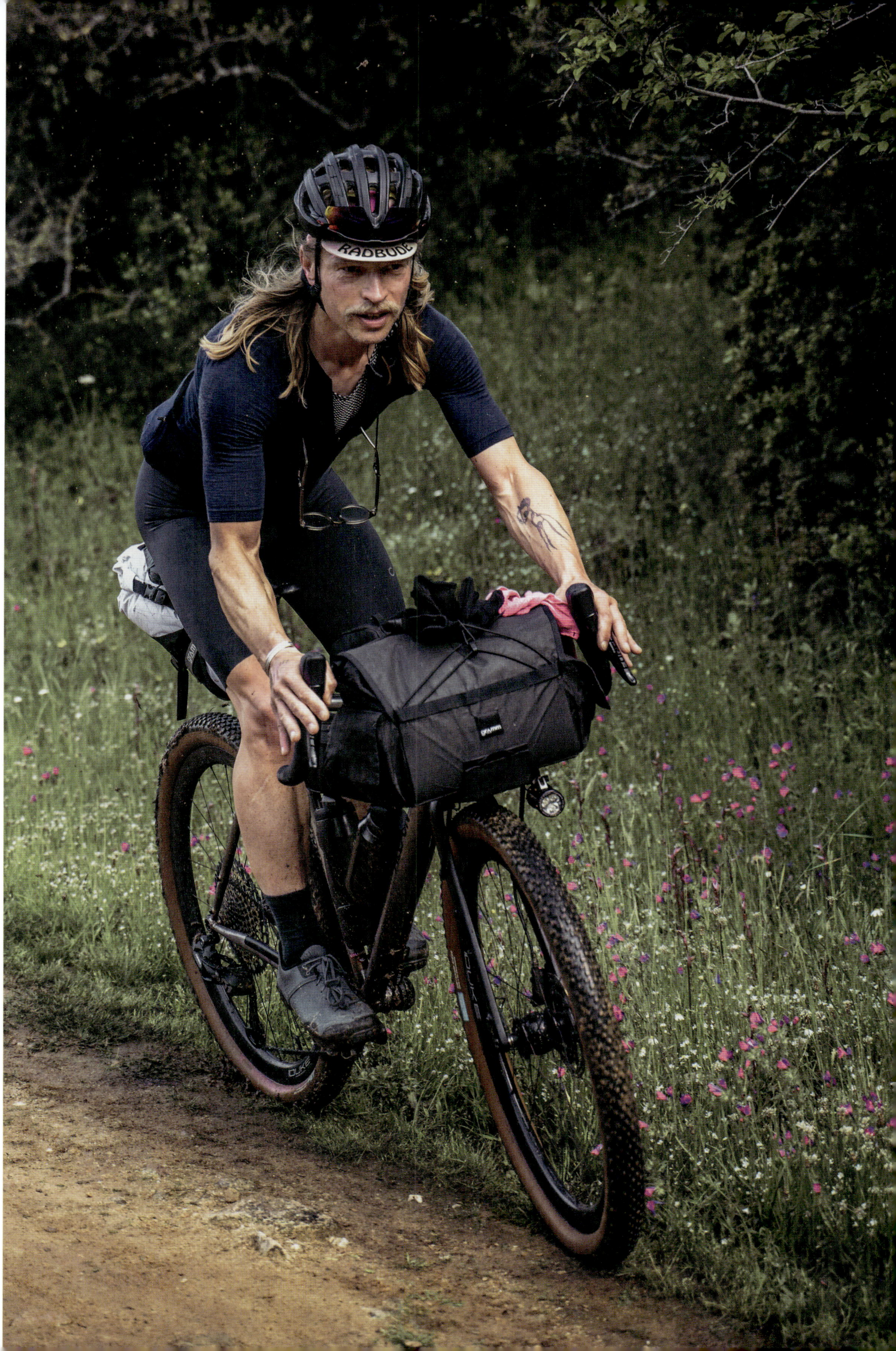
RADBUDE
DUKE

PELIZZOLI

Wo steckt der Typ nur?

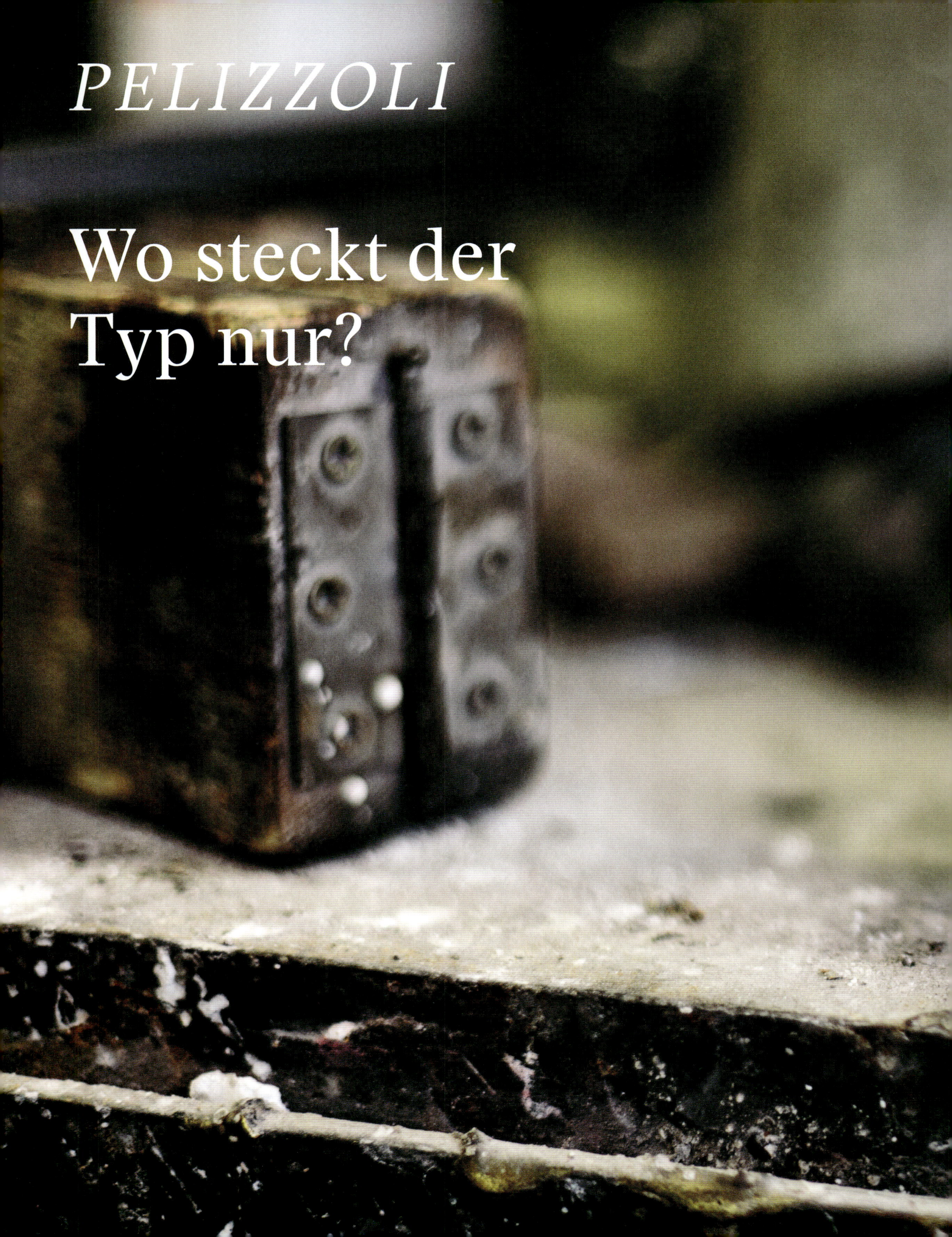

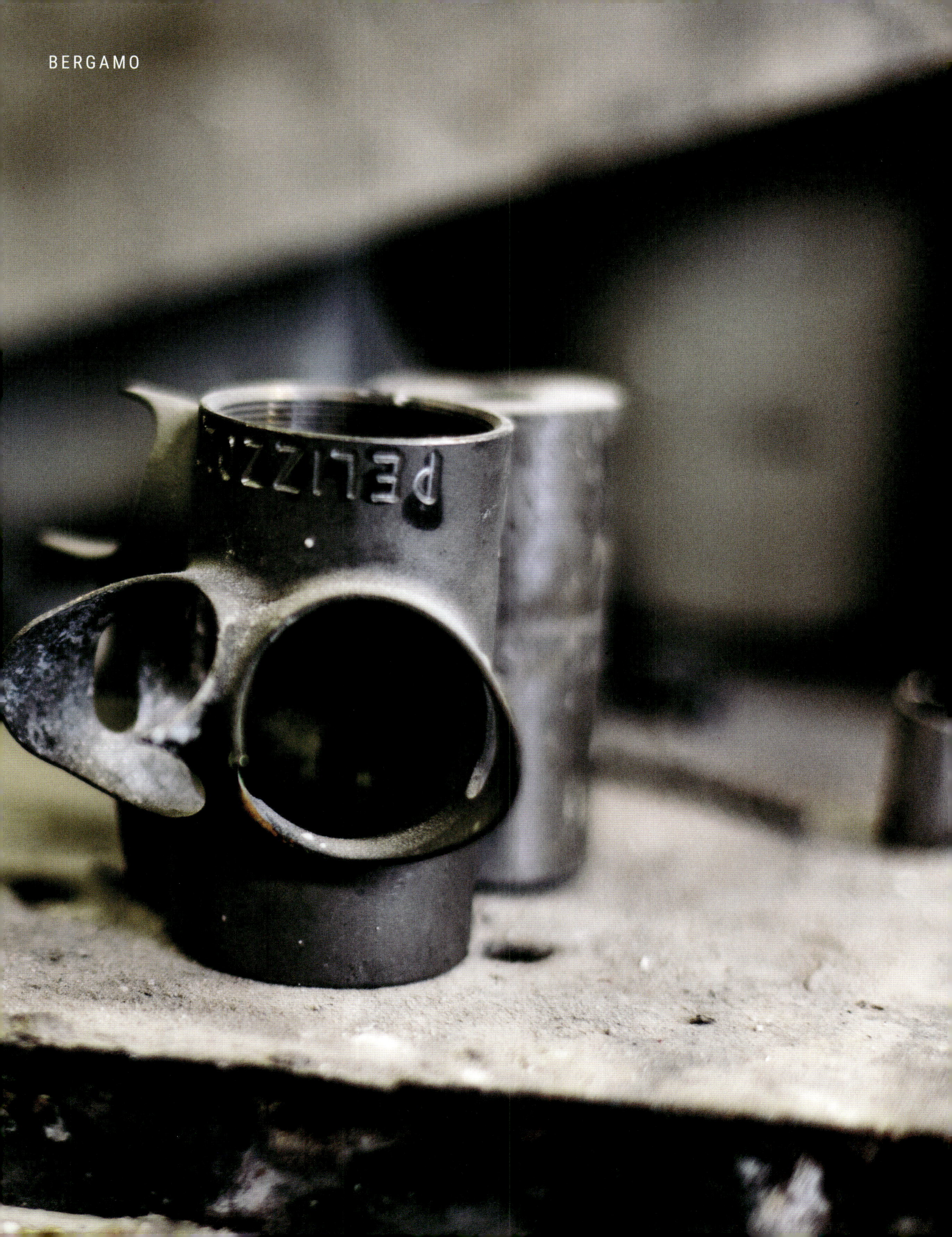

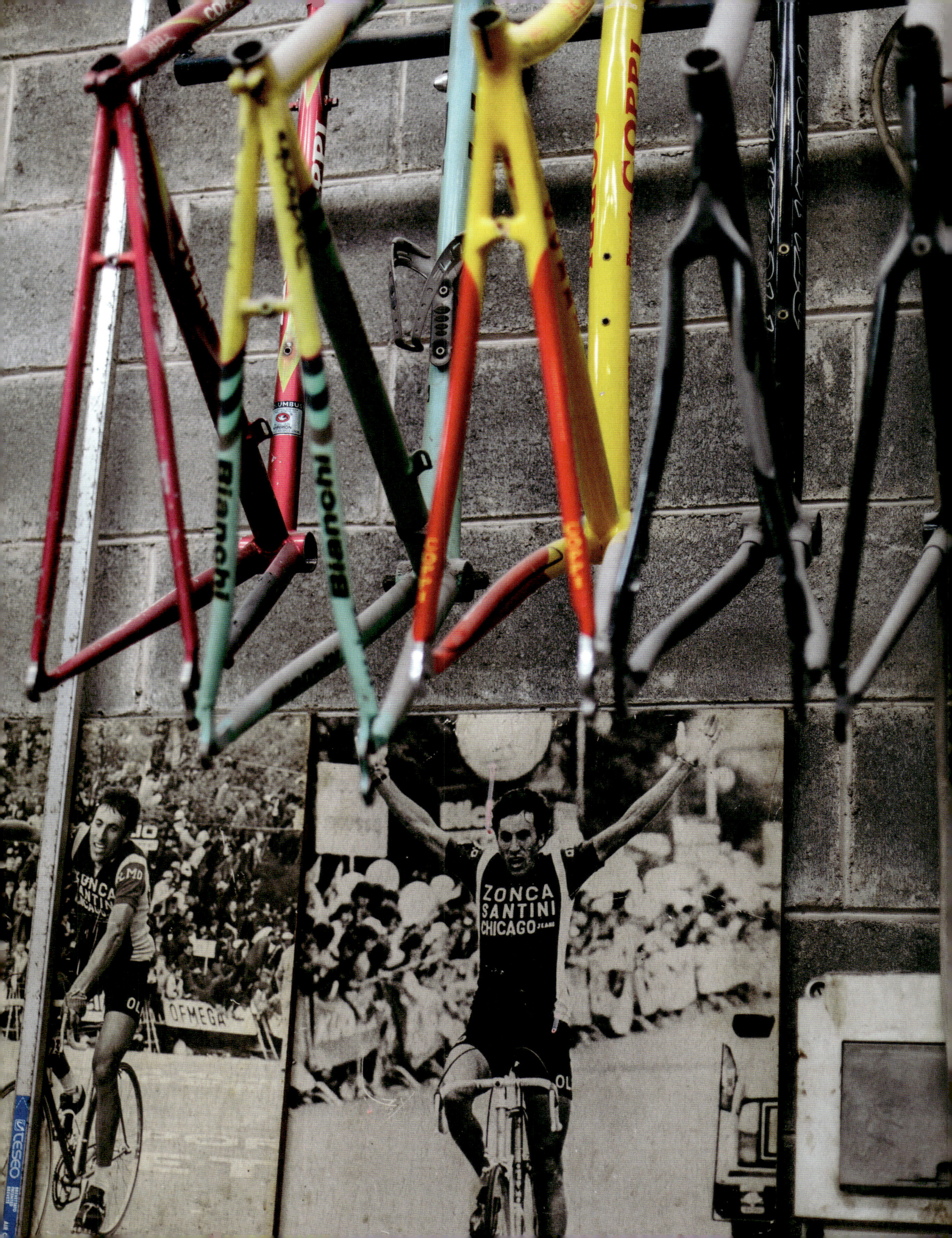
Bianchi
Bianchi
COPPI
ZONCA
SANTINI
CHICAGO

Müde und geschlaucht von der langen Fahrt im alten VW-Bus karriolen meine Freunde und ich nun schon ewig durch ein Industriegebiet in Bergamo und wissen nicht, wo wir überhaupt mit der Suche anfangen sollen.

Als ich einigen Fahrradenthusiasten vom Plan erzählte, auf dem Weg nach Frankreich bei PELIZZOLI in Bergamo vorbeizuschauen, war eine allgemeine Unruhe zu spüren, begleitet von eindringlichen Ratschlägen: »Schau dir vorher bloß dieses YouTube-Video an, sonst findest du den nie!« Gesagt, getan. Ich schaute mir das Filmchen an, bis ich die Straße blind hätte ablaufen und jede Strebe im Zaun hätte erkennen können – inklusive der individuellen Maserung. Das Bild der Halle war auch abgespeichert, und doch kam es genau so, wie befürchtet: Wo steckt nur dieser Pelizzoli?

Müde und geschlaucht von der langen Fahrt im alten VW-Bus karriolen meine Fahrrad-Freunde und ich nun schon ewig durch ein Industriegebiet in Bergamo und wissen nicht, wo wir überhaupt mit der Suche anfangen sollen. Nachdem wir jede Straße und jedes Gässchen abgesucht haben, zunächst ohne einen einzigen Moment des Wiedererkennens, stehen wir plötzlich doch vor der zuvor so sorgfältig eingeprägten Halle. Das Problem: Ein Tor zwischen uns und unserem Ziel. Keine Klingel – und kein Pelizzoli, der uns öffnet. Wo versteckt er sich nur?

Nach einer ganzen Weile fährt ein Punto vor. Darin sitzt er zwar auch nicht, doch wir sprinten trotzdem hinterher durch das sich bereits wieder schließende Tor. Die Halle steht offen, wir lassen uns selbst rein. Halle? Ach was, eine Schatzkammer voller Rohrsätze, Schaltaugen, Gabelenden, Ausfallenden, Kettenblätter und Rahmen, Rahmen, Rahmen. Überall! Egal, wohin man schaut. Vor einer Lackierkabine stehen Dutzende sauber aufgereiht und warten auf ihre Farbdusche. Die Produktion ist offensichtlich im vollen Gange.

Doch nicht nur dieses Ensemble bietet ein außergewöhnliches Bild. Zwischen all dem fällt mein Blick auf ein gerahmtes Foto. Es zeigt Pelizzoli, der gerade Papst Johannes Paul II. ein glänzendes Exemplar aus dieser Halle überreicht. Seine Heiligkeit sei begeisterter Rennradfahrer gewesen, erfahre ich von einem meiner Mitstreiter.

So, jetzt reicht's aber! Wir sind in Pelizzolis Werkstatt, vor mir sehe ich ihn auf einem Foto – von dem Mann selbst fehlt jedoch immer noch jede Spur. Plötzlich – wie aus heiterem Himmel – steht er da und wuselt sofort geschäftig herum. Herzlich, aber flink wie ein Wiesel begrüßt er uns. Im nächsten Moment steht er auch schon wieder an einem Montageständer und prüft einen Rahmen, lötet an einem

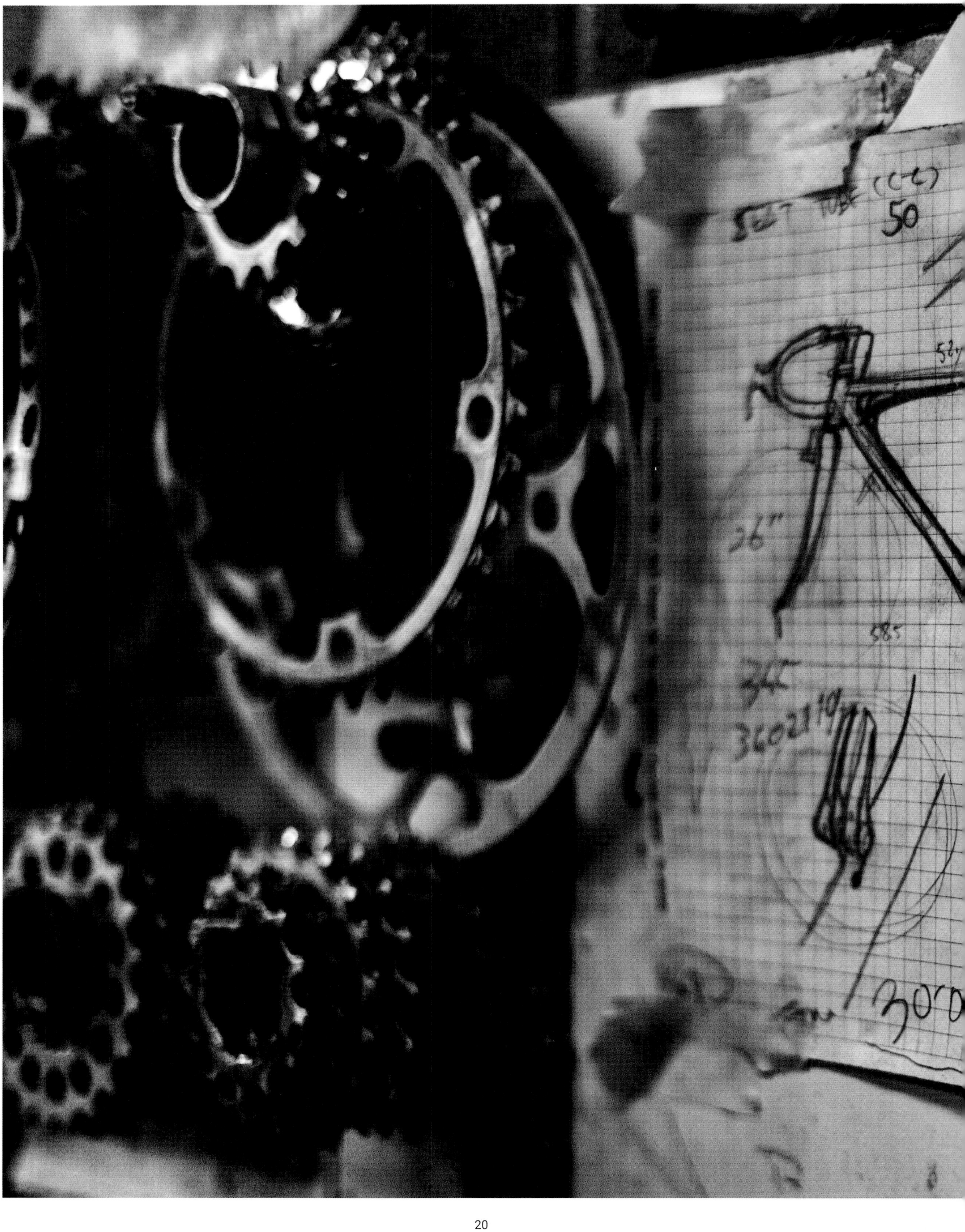
SEAT TUBE (C-C)
50
26"
585

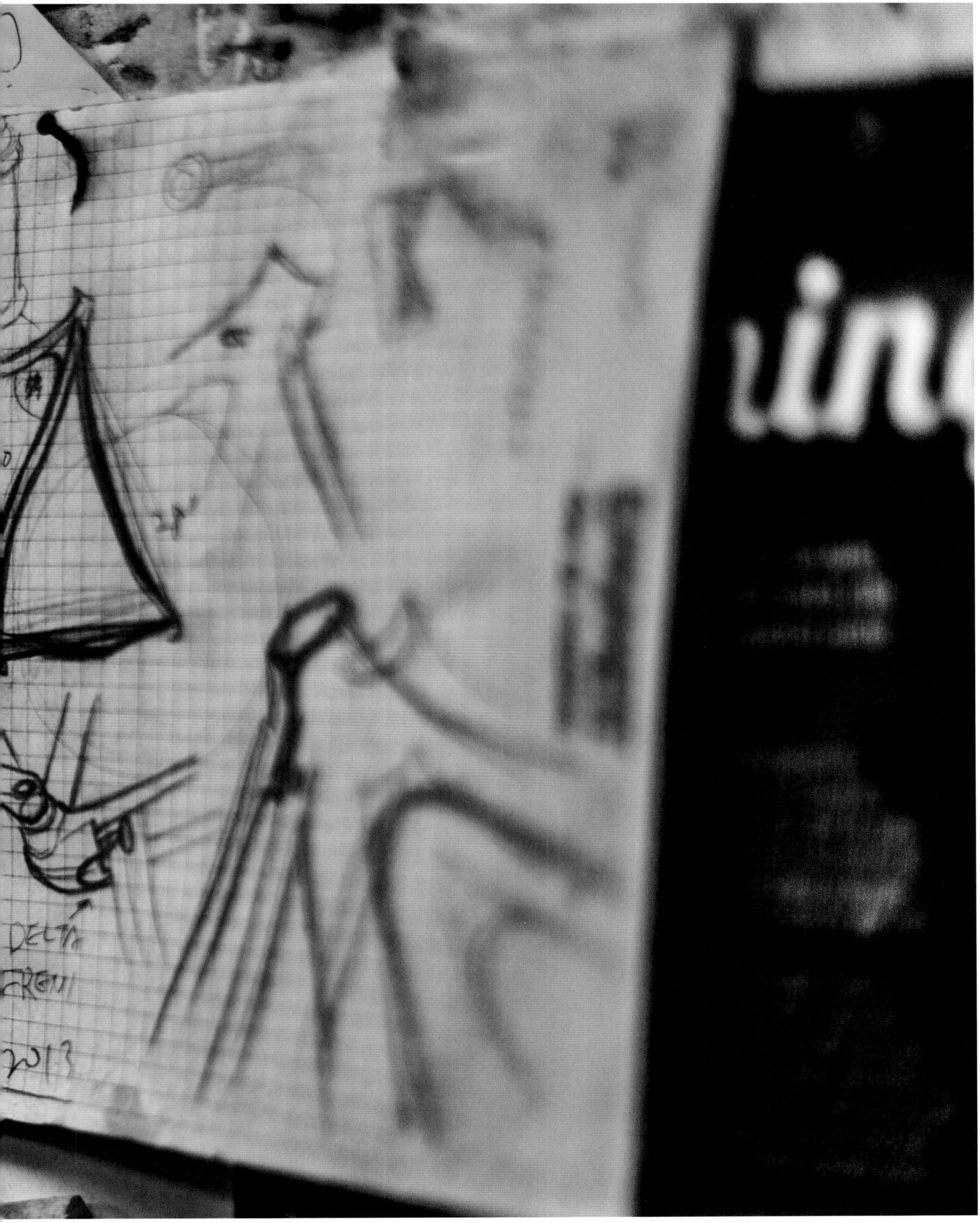

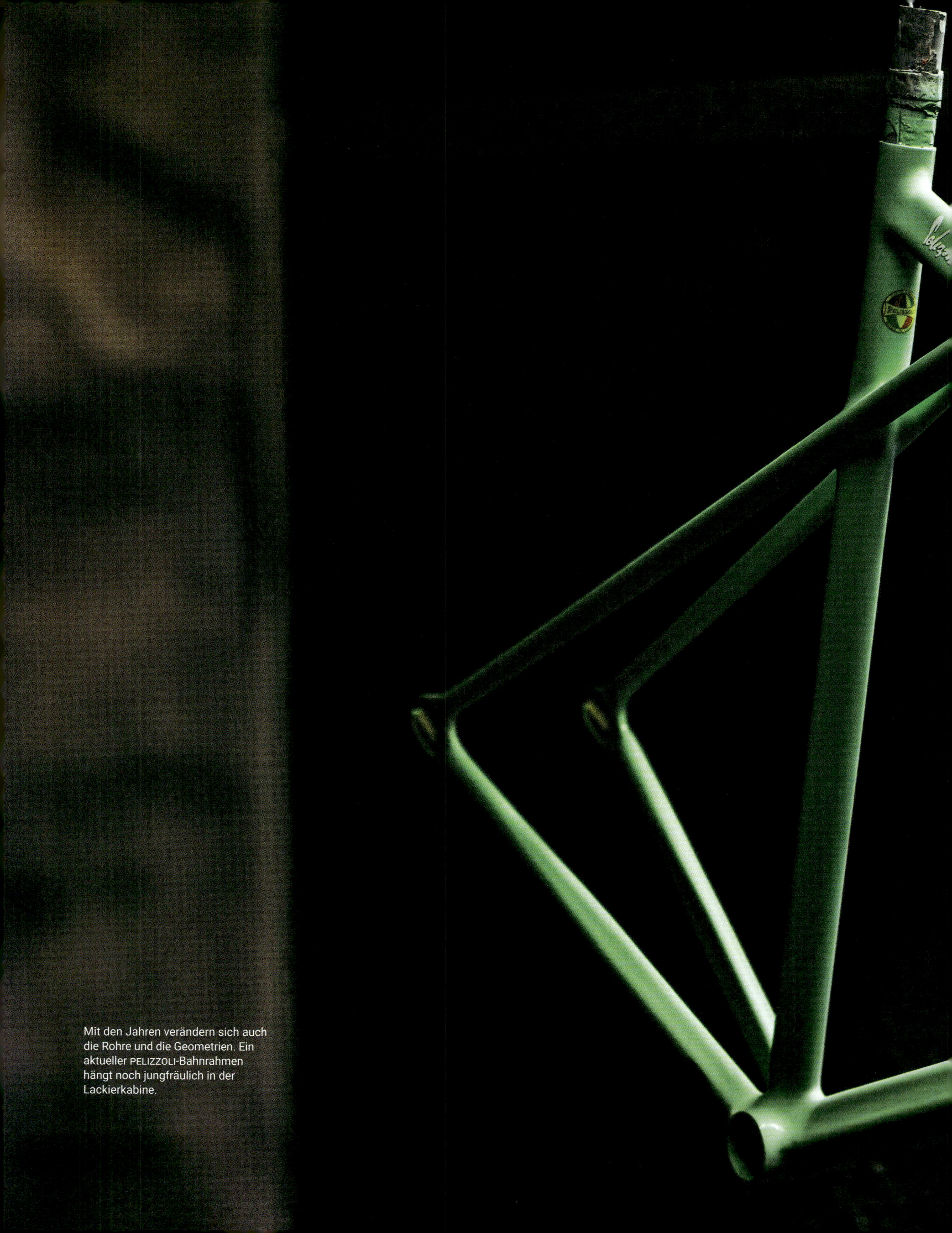

Mit den Jahren verändern sich auch die Rohre und die Geometrien. Ein aktueller PELIZZOLI-Bahnrahmen hängt noch jungfräulich in der Lackierkabine.

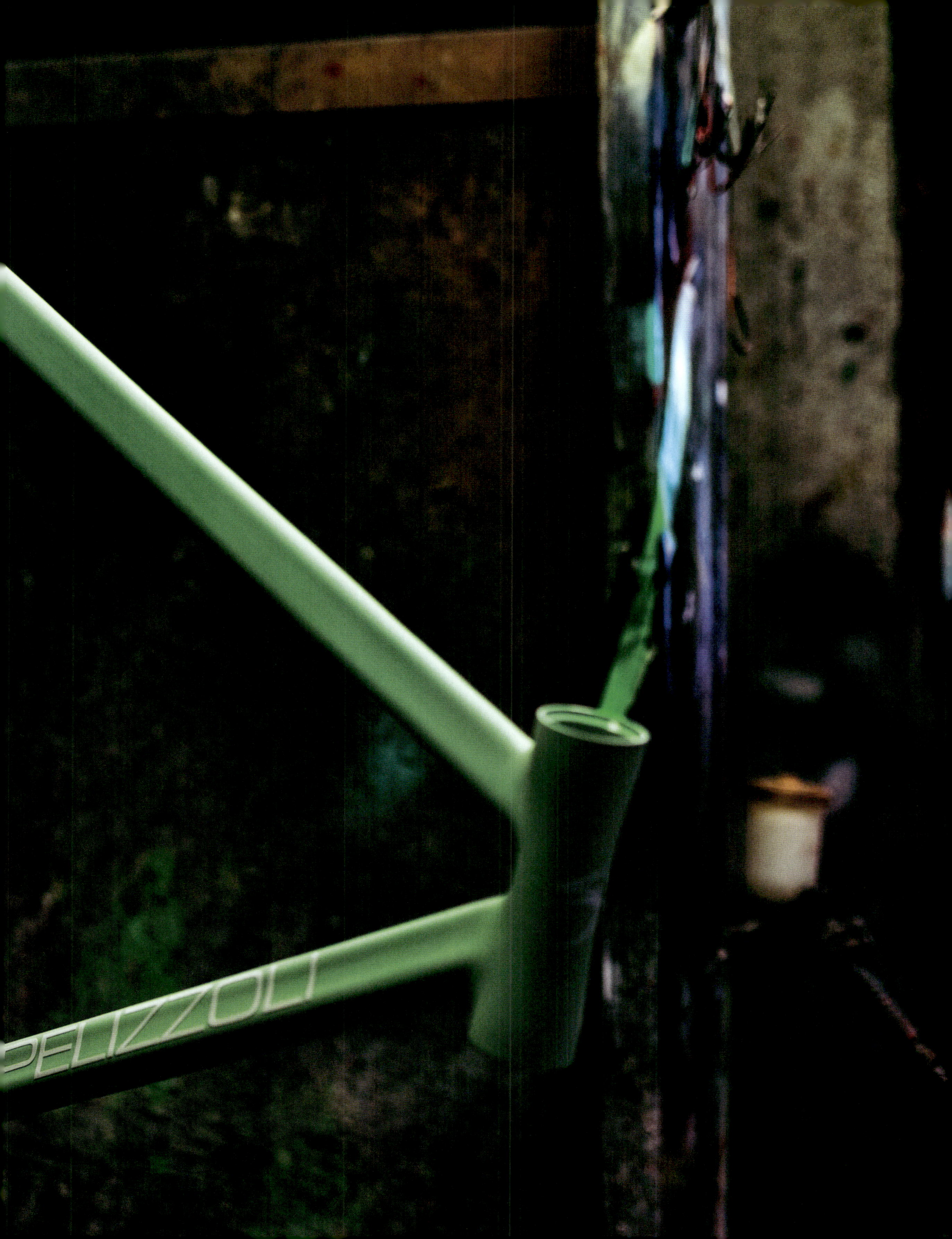
PELIZZOLI

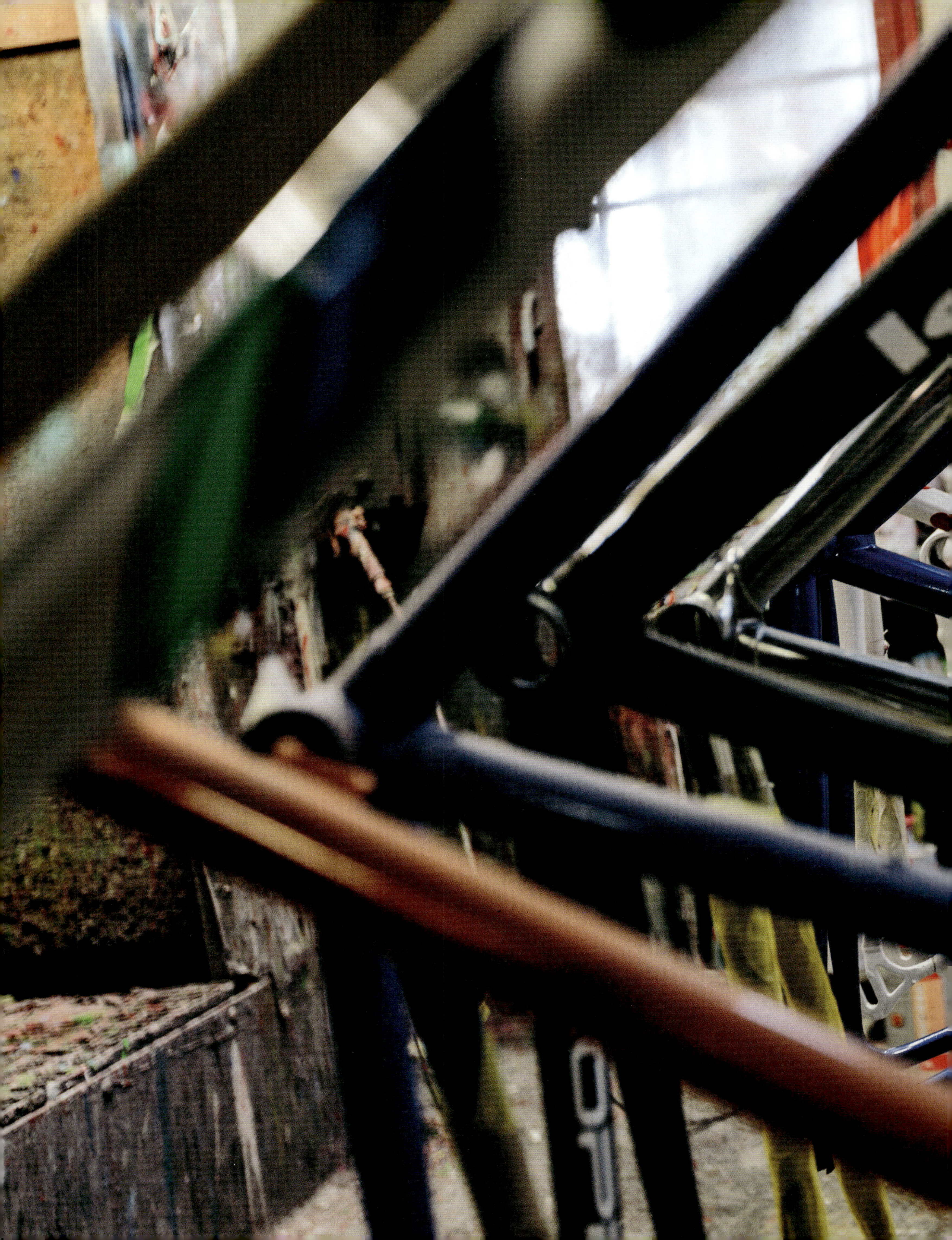

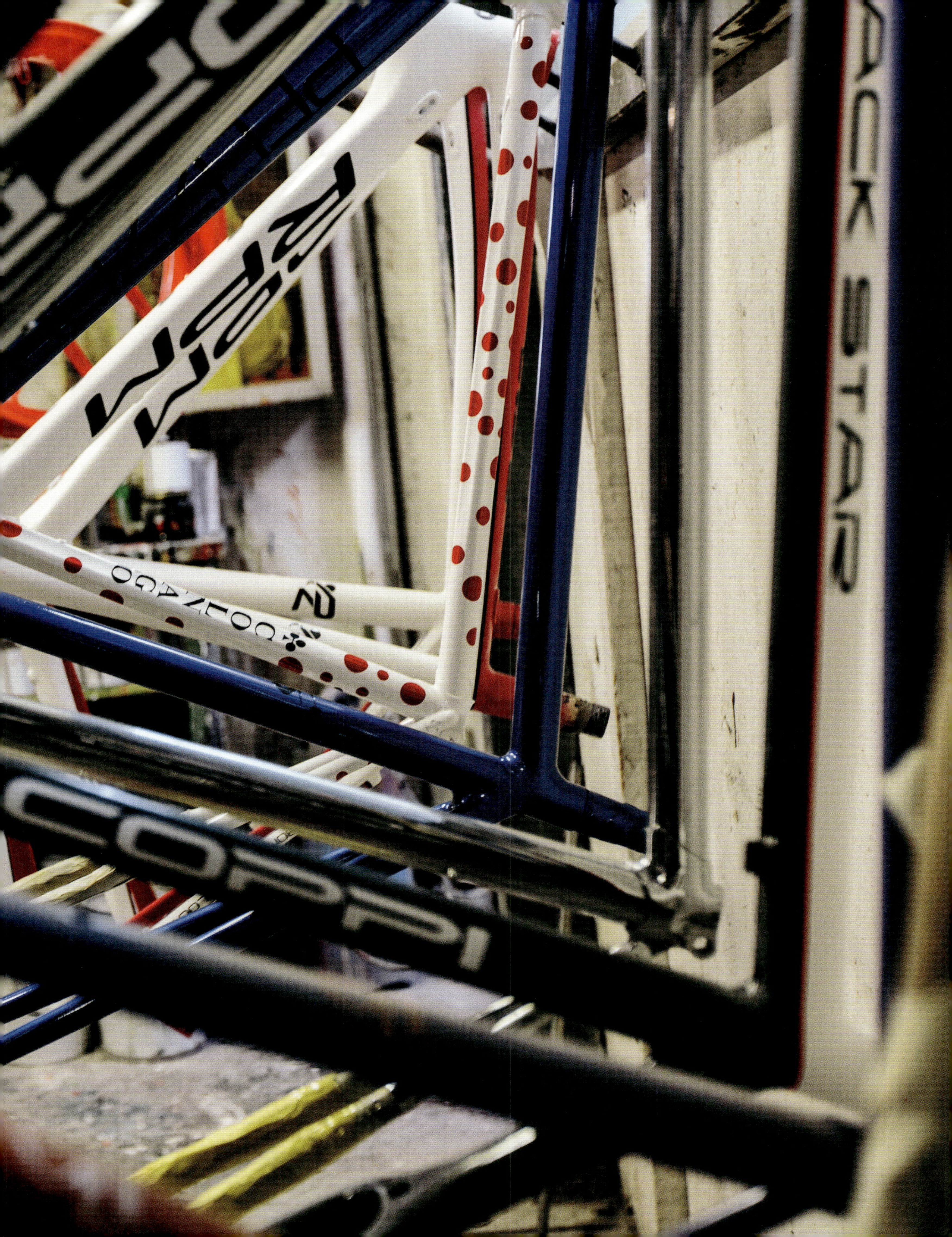
COPPI

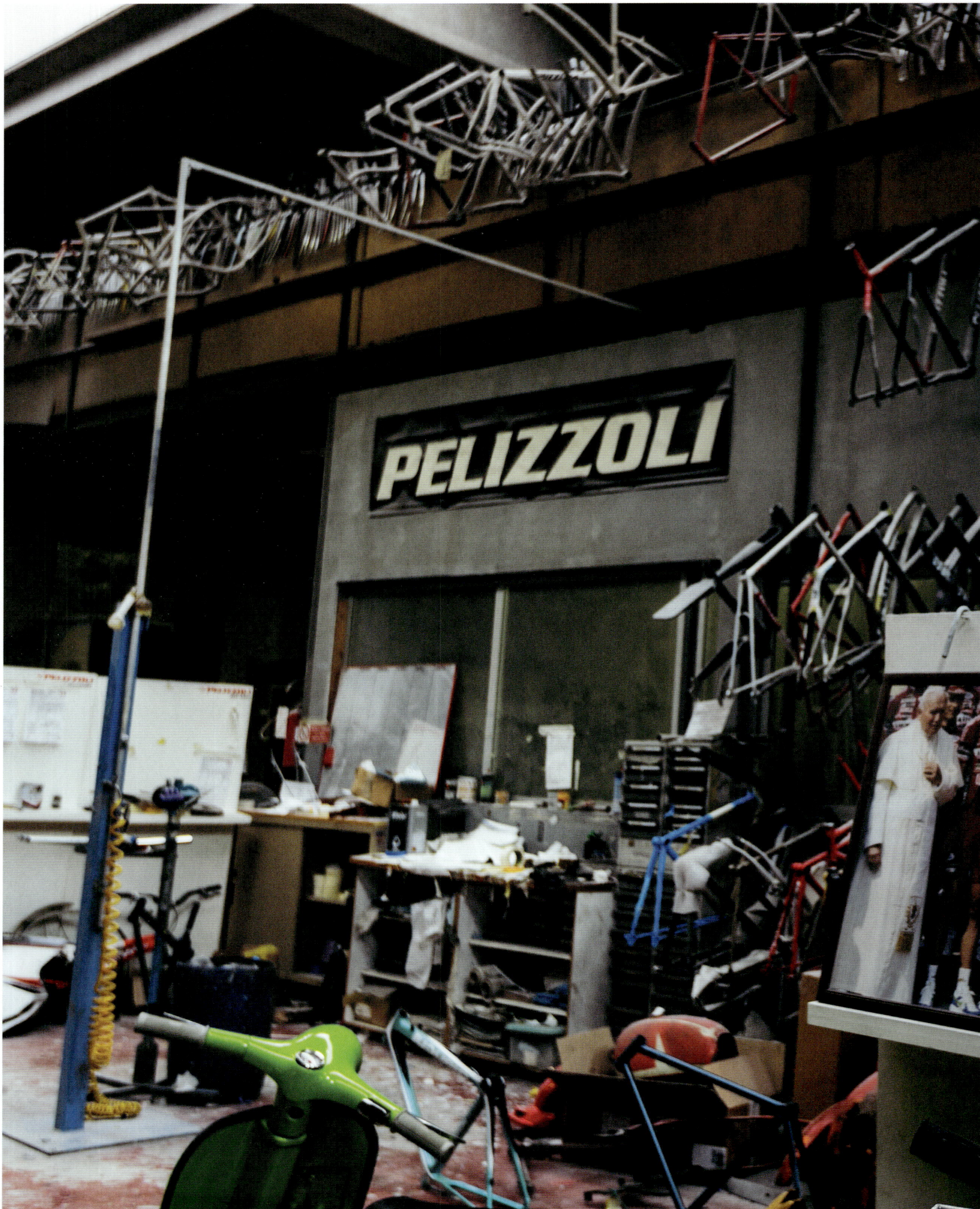
PELIZZOLI

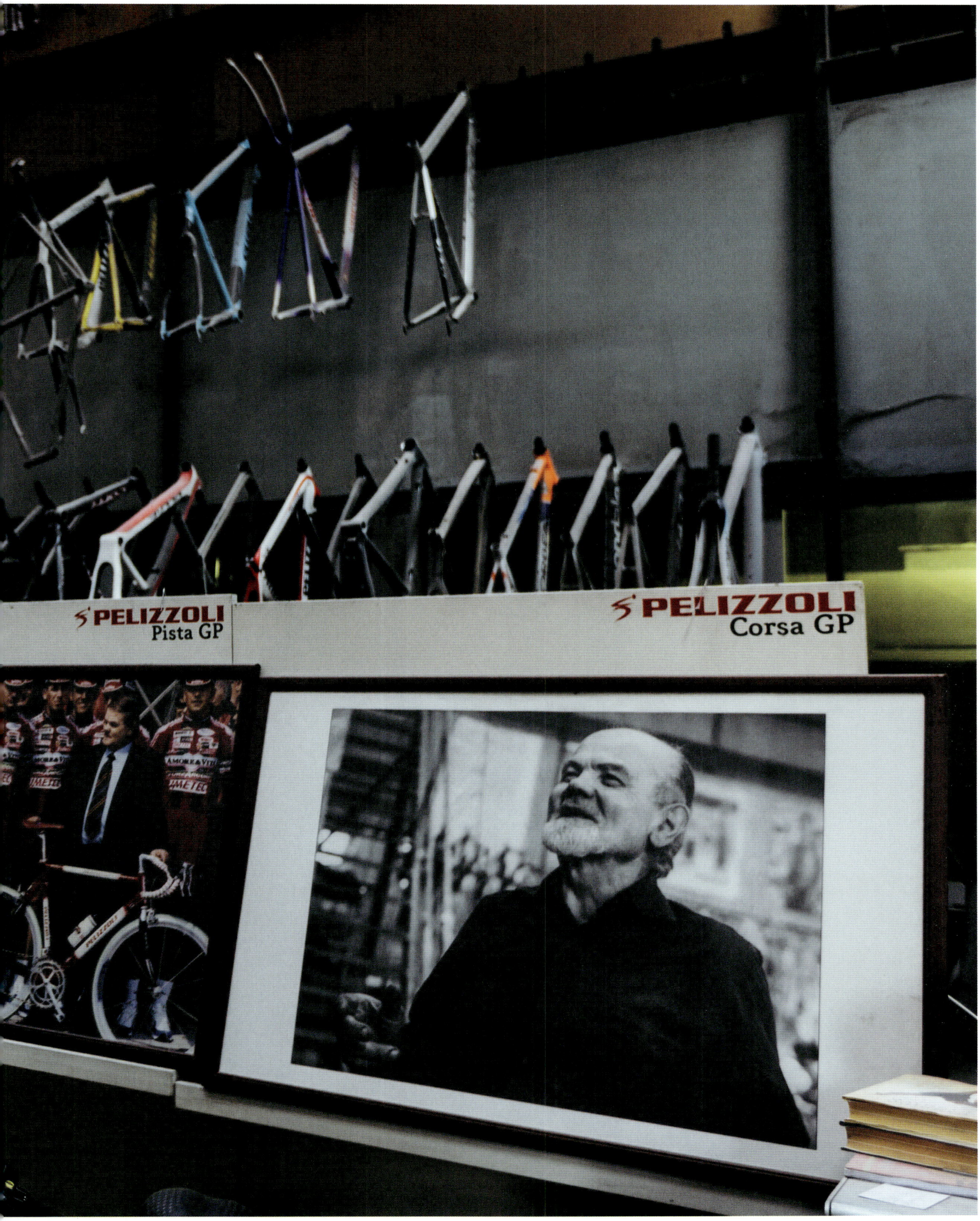
PELIZZOLI
Pista GP
PELIZZOLI
Corsa GP

Gabelkopf herum, und räumt Sekunden später ein paar Teile von rechts nach links.

Giovanni Pelizzoli steigt mit zwölf zum ersten Mal auf ein Rennrad; zwei Jahre später fährt er sein erstes Rennen. Mit 17 wird er Manager eines Juniorteams und hilft zeitgleich seinem Vater, einem Fahrradmechaniker, bei den täglichen Reparaturen. Doch Pelizzoli wünscht sich mehr. Er will eigene Rahmen entwerfen und geht zu einem Rahmenbauer in Bergamo in die Lehre. 1969 ruft er seine Marke CIÖCC ins Leben. Als Claudio Corti 1977 auf einem CIÖCC Weltmeister wird, widmet Pelizzoli diesem Erfolg seinen geliebten SAN-CRISTOBAL-Rahmen. 1980 verkauft er CIÖCC, drei Jahre später nimmt er die Produktion unter dem Namen PELIZZOLI auf. In den 1990er-Jahren arbeitet er außerdem für COPPI – als Verantwortlicher für Design und Produktion.

Doch PELIZZOLI ist und bleibt ein echtes Familienunternehmen. inzwischen führt sein Sohn Samuel, nach Jahrzehnten der Zusammenarbeit mit seinem Vater, die Produktion. Sie bauen und lackieren aber nicht nur für die eigene Herstellung. Schon immer haben sich italienische Rahmenmeister untereinander geholfen. Es war gar nicht so unüblich, dass auch einmal ein MASI oder COLNAGO aus dieser Halle kam. Masi und Pelizzoli beispielsweise verbindet nicht nur die Liebe zum perfekten Fahrrad, sondern auch eine alte Freundschaft. Und beide verstecken sich ganz erfolgreich – wenn auch nicht gut genug für mich.

Der ganze Stolz einer vergangenen Zeit: Pelizzoli präsentiert einen seiner Lieblinge. Ein vergoldetes CRISTOBAL-Jubiläumsmodell mit gravierte Leckereien.

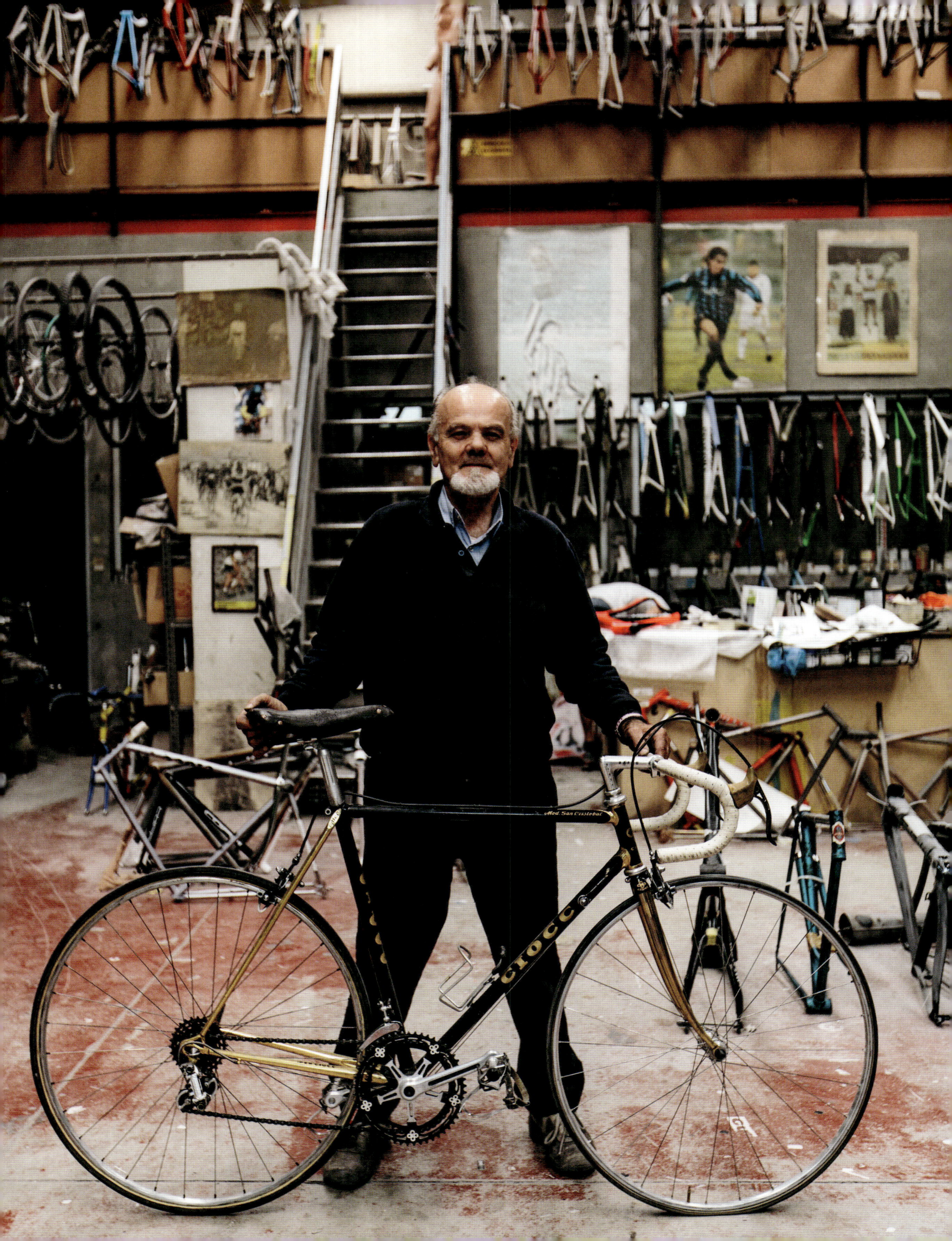
CIOCC

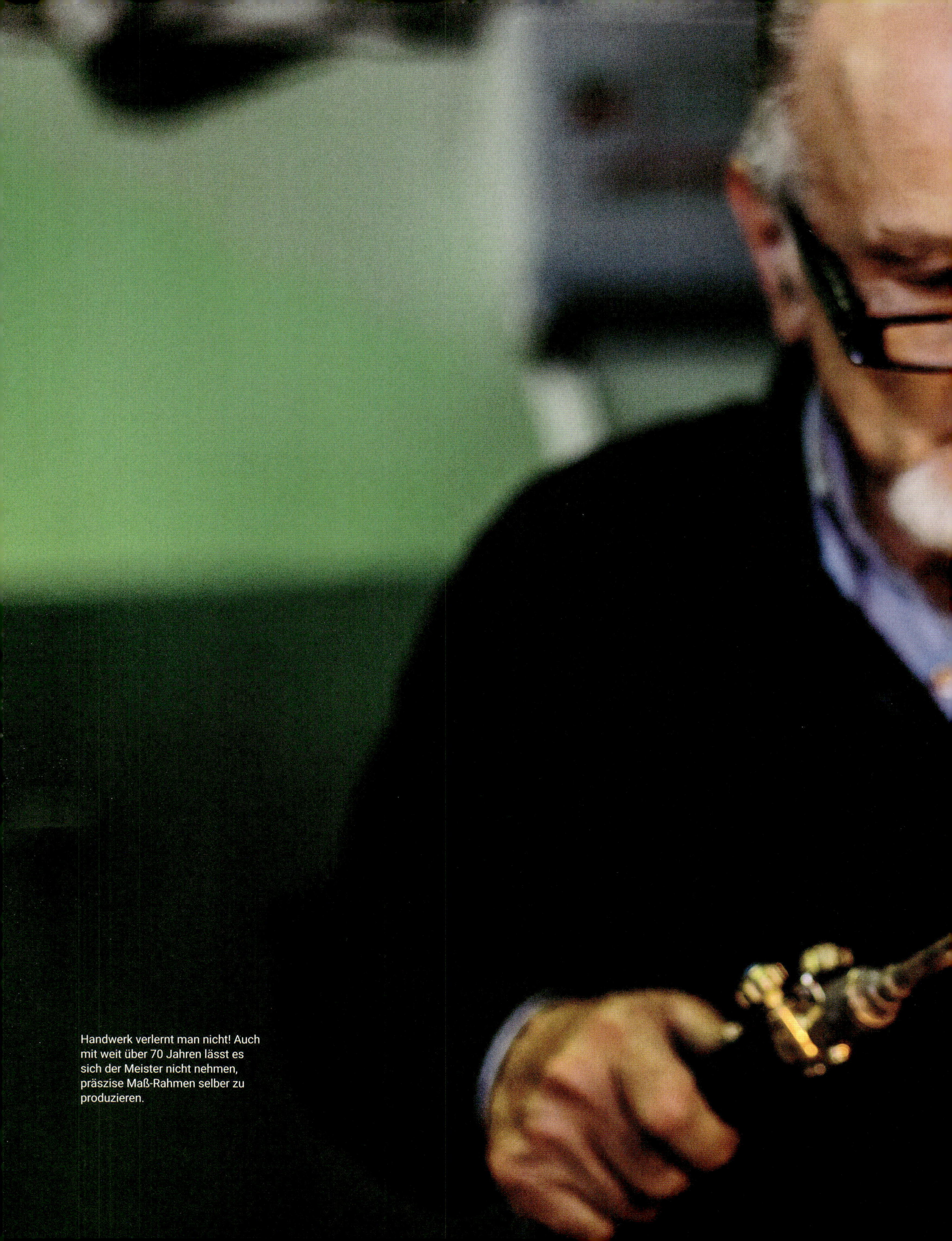

Handwerk verlernt man nicht! Auch mit weit über 70 Jahren lässt es sich der Meister nicht nehmen, präszise Maß-Rahmen selber zu produzieren.

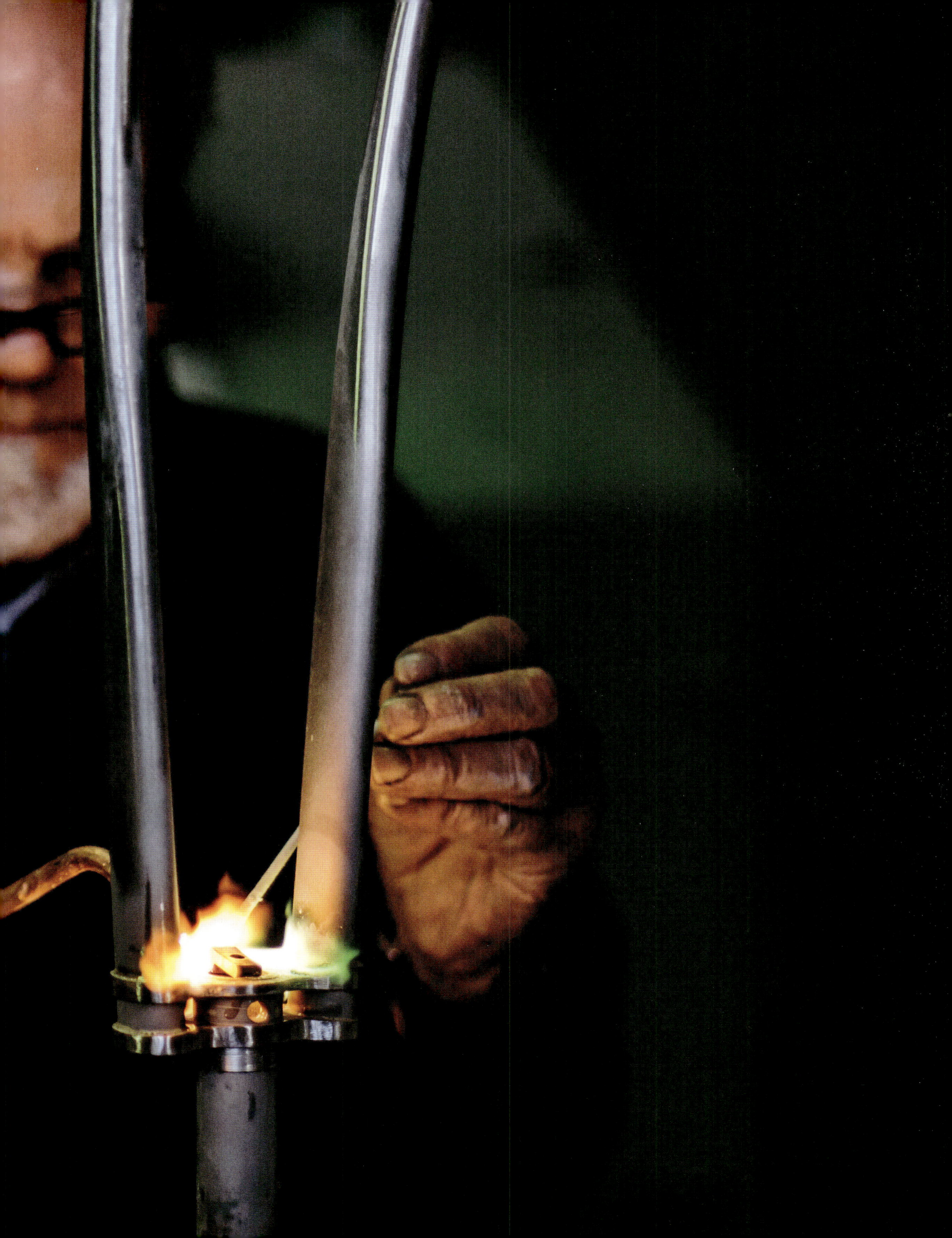

HERR S.

Wenn aus Platzangst Sammelwut wird

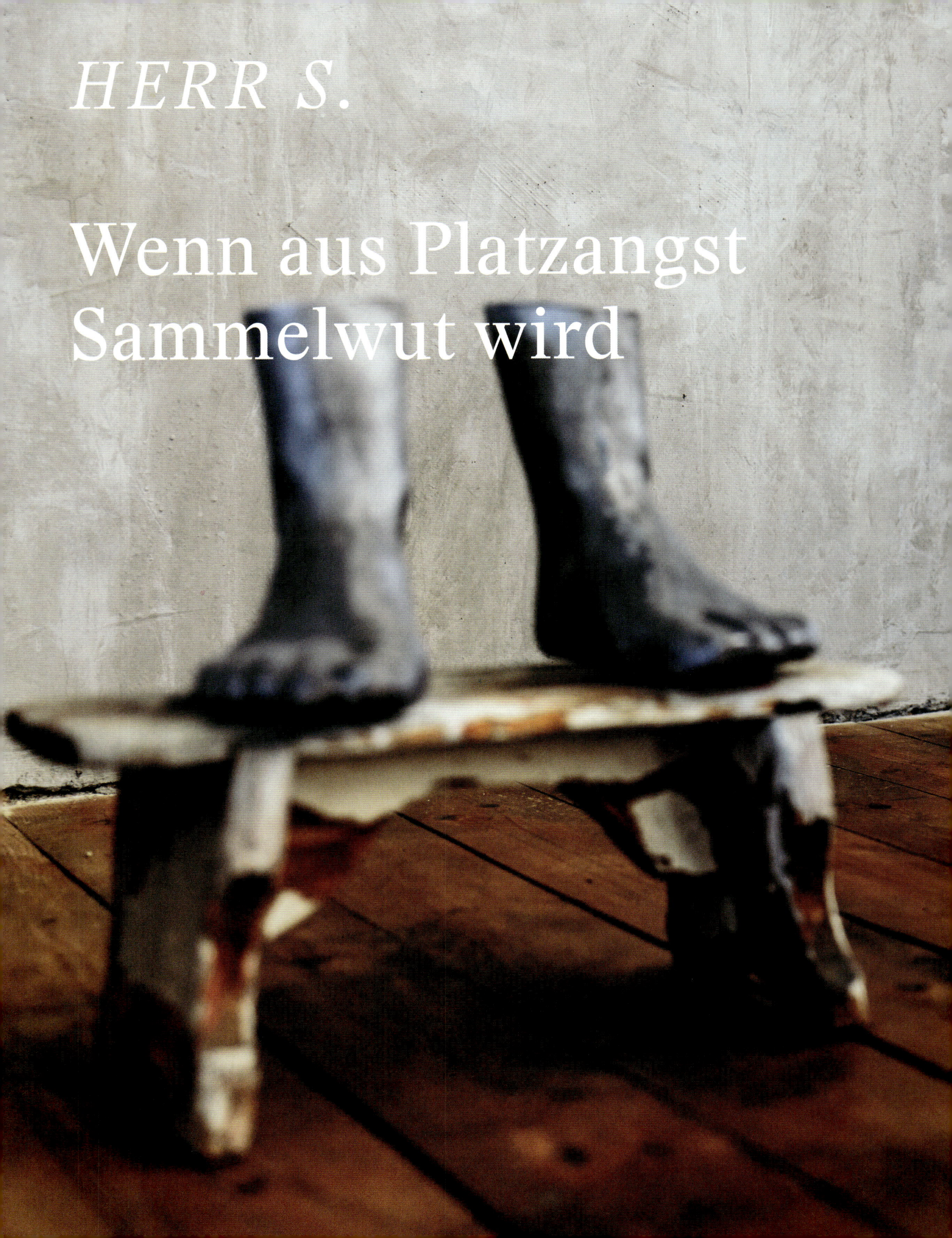

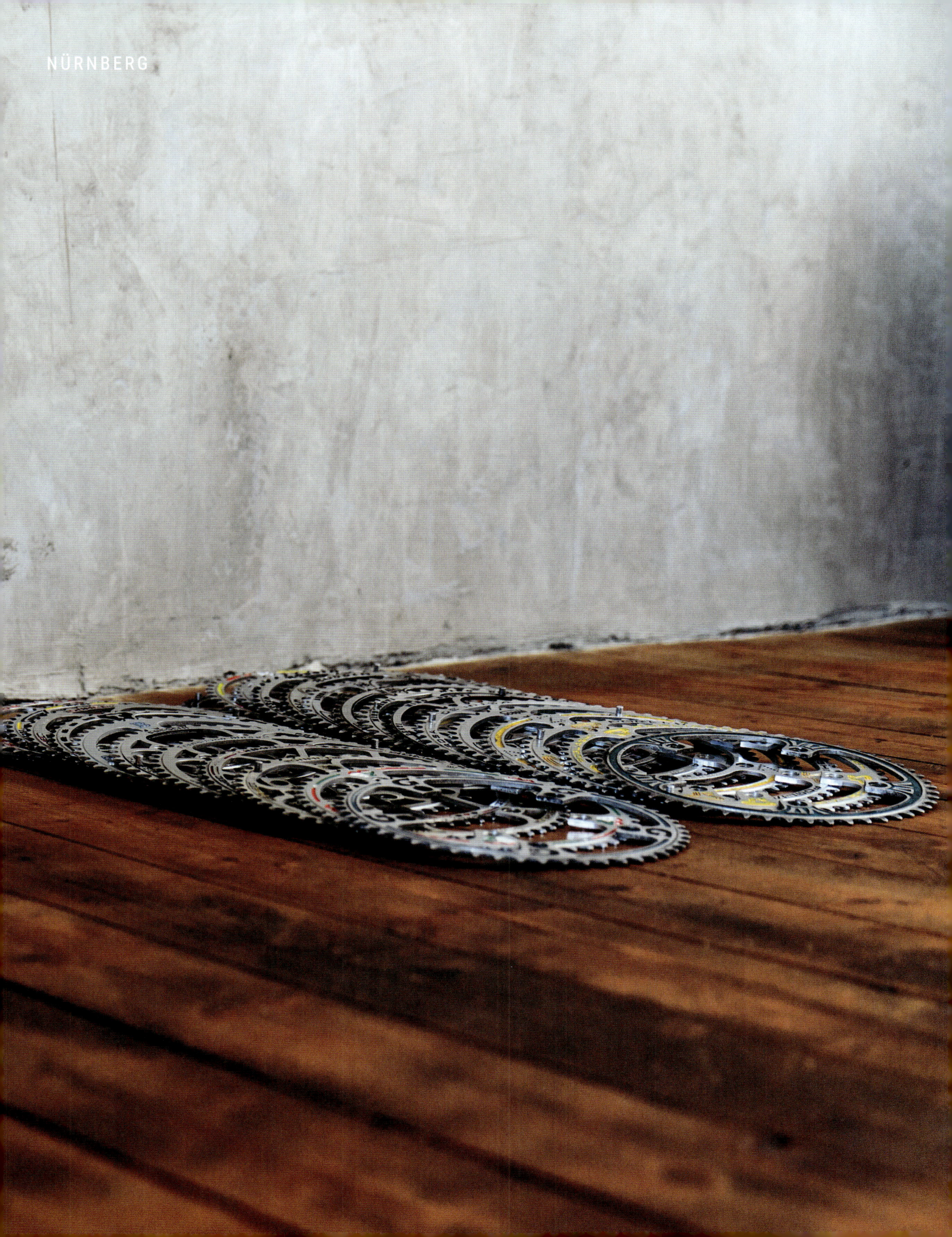

HERR S.

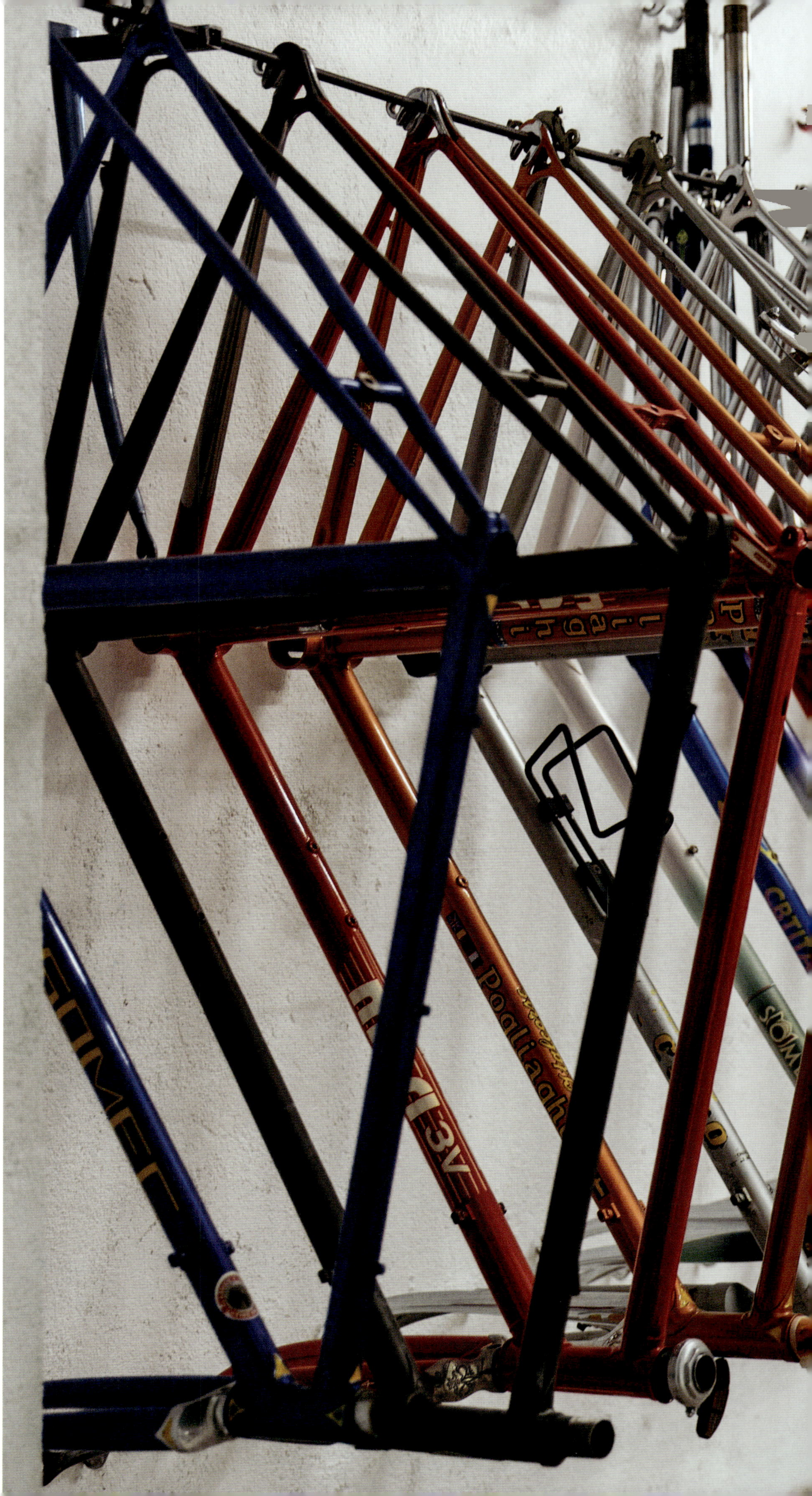

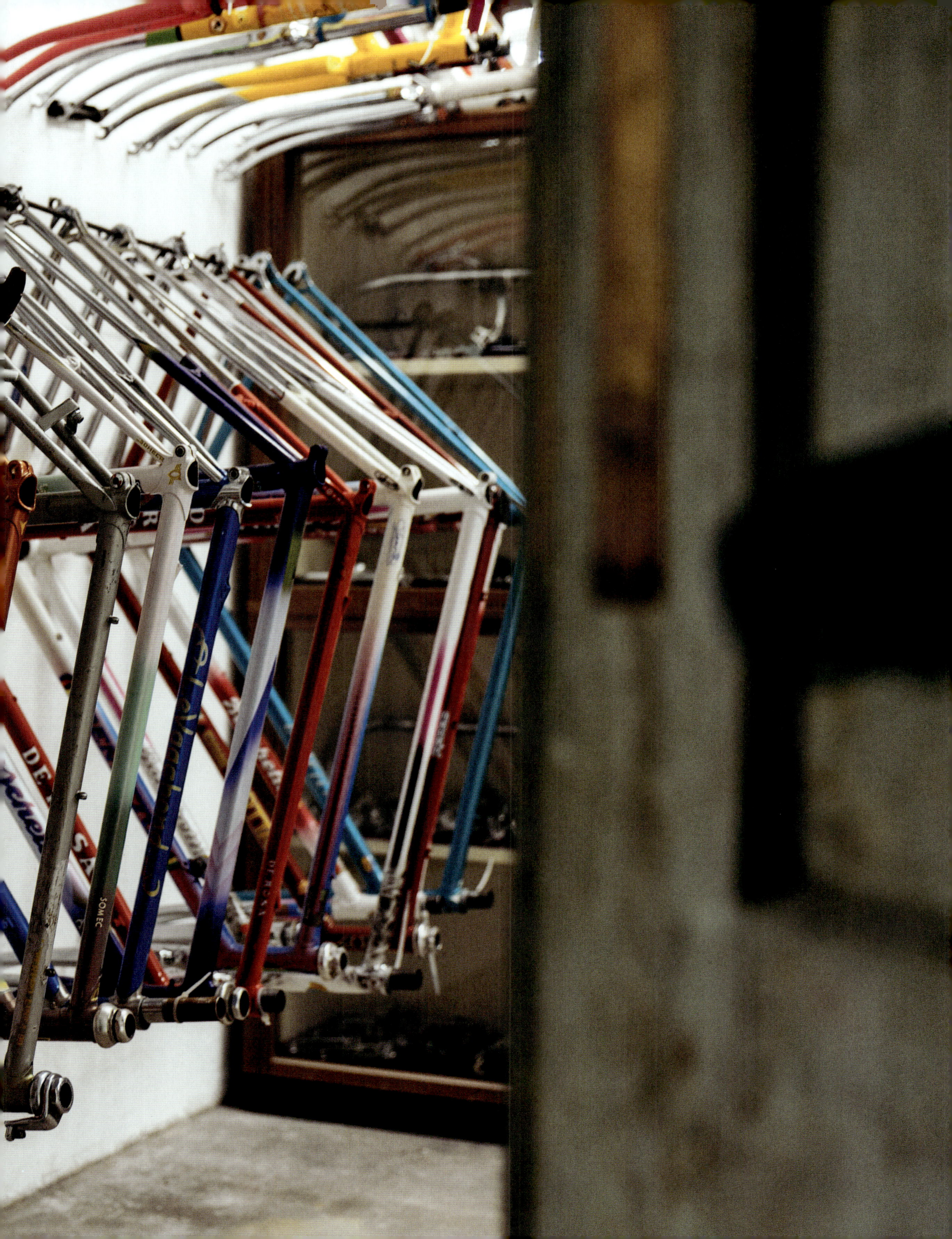

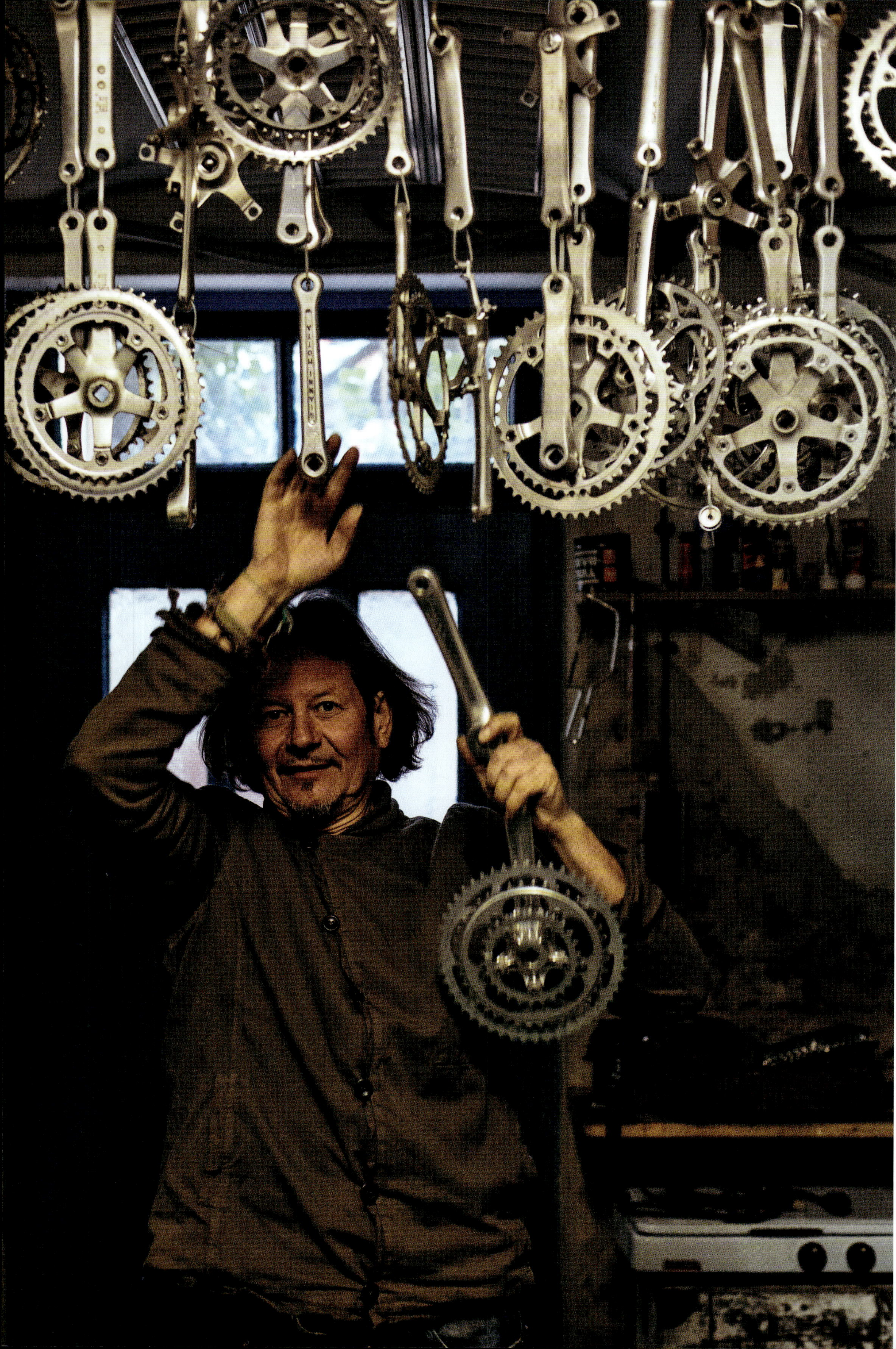

Kein Sexspielzeug, auch wenn's anmacht! Ralf in einem seiner vielen Räume. Nicht viele Sammler haben die Möglichkeit, so gut geordnet ihrer Sammlung Herr zu werden.

Ankunft am Hauptbahnhof Nürnberg. Mit dem Rad im Gepäck und gespannt auf das, was mich erwartete, wusste ich über Ralfs Sammlung bisher ja nur eins: Alle Räder sind mit CAMPAGNOLOS CROCE D'AUNE ausgestattet. Alle? Ich konnte es mir kaum vorstellen. Wer auch nur im Entferntesten etwas mit Sammeln zu tun hat, der weiß: Das Zeug ist ausgesprochen selten und meist sehr kostspielig. Aus einer Zeit stammend, in der bei CAMPA gerade der *Generationenwechsel* stattfand und der junge Campagnolo das Ruder übernahm, war die CROCE D'AUNE ein Versuch der Italiener, an SHIMANOS Vorherrschaft auf dem Weltmarkt heranzukommen. Das Design der Delta-Bremse entlockt Liebhabern auch heute noch »Ahhs« und »Ohhs«, auch wenn sich über deren Funktion lange diskutieren lässt.

Nürnberg also. Ralf kommt angerollt. Auf einem wunderschönen GIANNI MOTTA. Voll verchromt und funkelnd – mit DURA-ACE-Ausstattung. Wie jetzt, Ralf? Kein CAMPA? Er grinst. »Nee, nee, es gibt schon auch noch andere schöne Teile an Rädern!« Alles klar!

Als Ralf kurze Zeit später erzählt, wie er zum Rennradsammler wurde, stockt mir der Atem. In seinen besten Jahren als erfolgreicher Künstler, aber auch Café- und Clubbesitzer, beginnt für den 30-Jährigen auch die sicher dunkelste Zeit seines Lebens. Anstatt seinen Erfolg zu genießen, entwickelte er eine sogenannte Agoraphobie. Platzangst, Todesangst vor weiten Räumen; Ralf traut sich einfach nicht mehr aus dem Haus, so groß ist die Angst, den Weg zurück nicht mehr zu finden. Dieser Umstand macht dem ehemaligen Weltenbummler ein normales Leben unmöglich.

Da erinnert er sich an seine alte Leidenschaft aus der Jugend: Rennräder – und Rennradfahren. Ralf erwirbt also sein erstes neues *altes* Rennrad, ein COLNAGO SUPER in Rosa, und schafft es damit, um den Häuserblock zu radeln. Elektrifiziert von dem Erfolg will er mehr und kauft das nächste Rad. Mit dem schafft er es um die Innenstadt. Das dritte trägt ihn raus aus Nürnberg, etwa 30 weitere Modelle folgen und versammeln sich nun aufs Feinste restauriert in einem extra dafür eingerichteten Raum – und tatsächlich, fast alle mit der legendären CROCE D'AUNE. Von den anderen 40 Rahmen, die noch auf ihre Ausstattung warten, ganz zu schweigen. Wichtig ist, dass alle Teile pantografiert sind, die Namen ihrer Hersteller also auf so vielen Teilen am Rad wie nur irgendwie möglich eintätowiert wurden.

»Wenn Masi der Papst ist,
dann ist Pogliaghi Gott.«

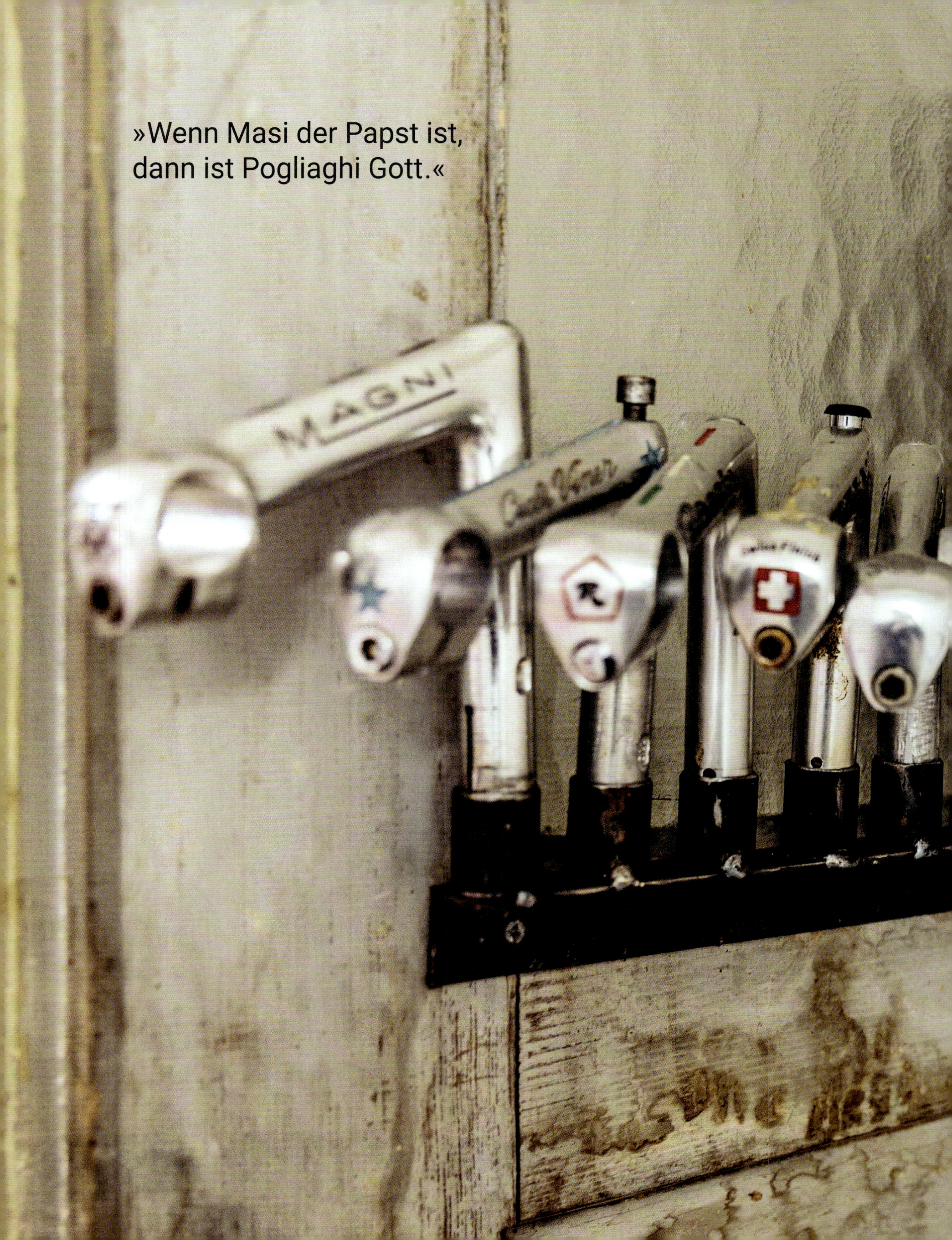

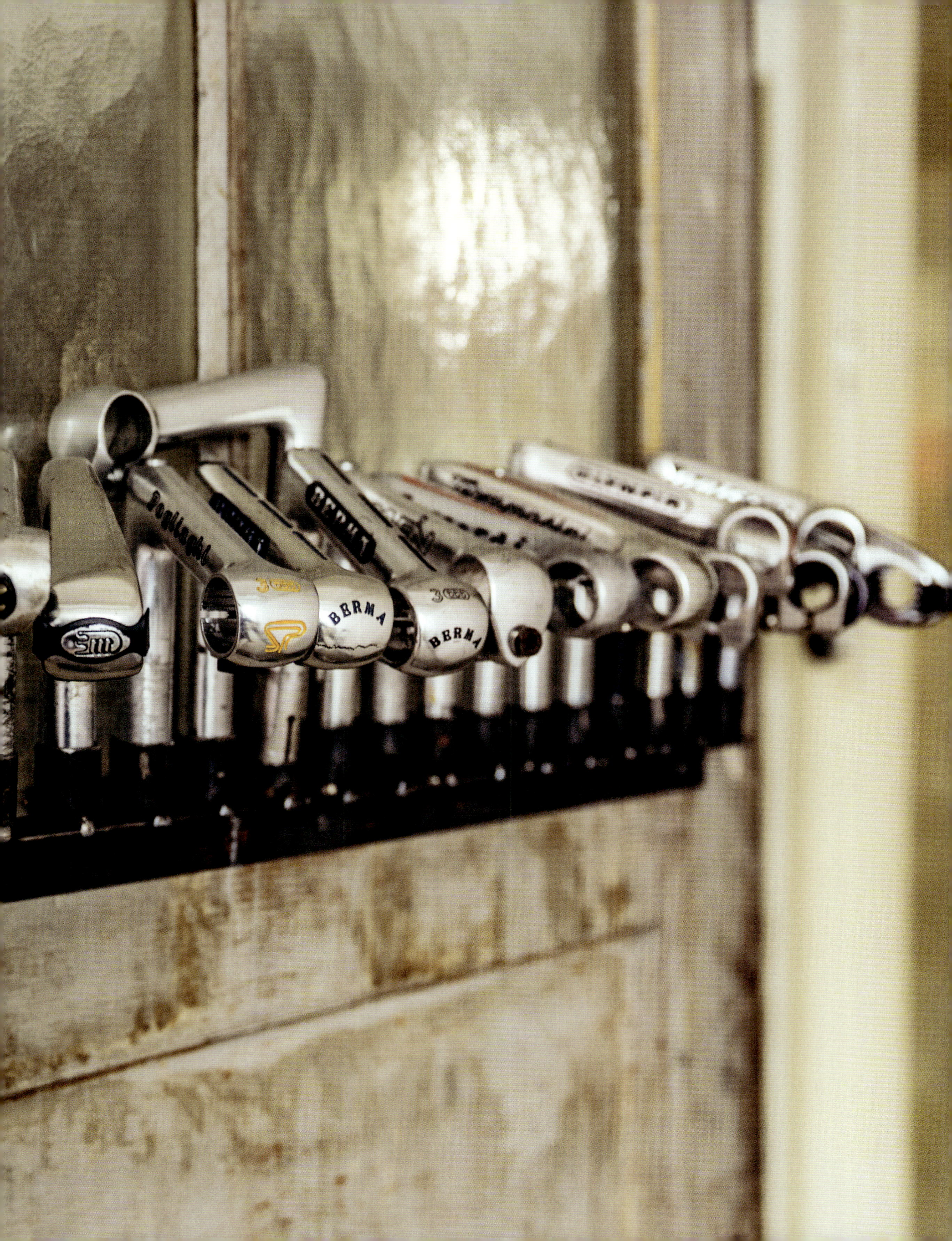
BERMA
BERMA

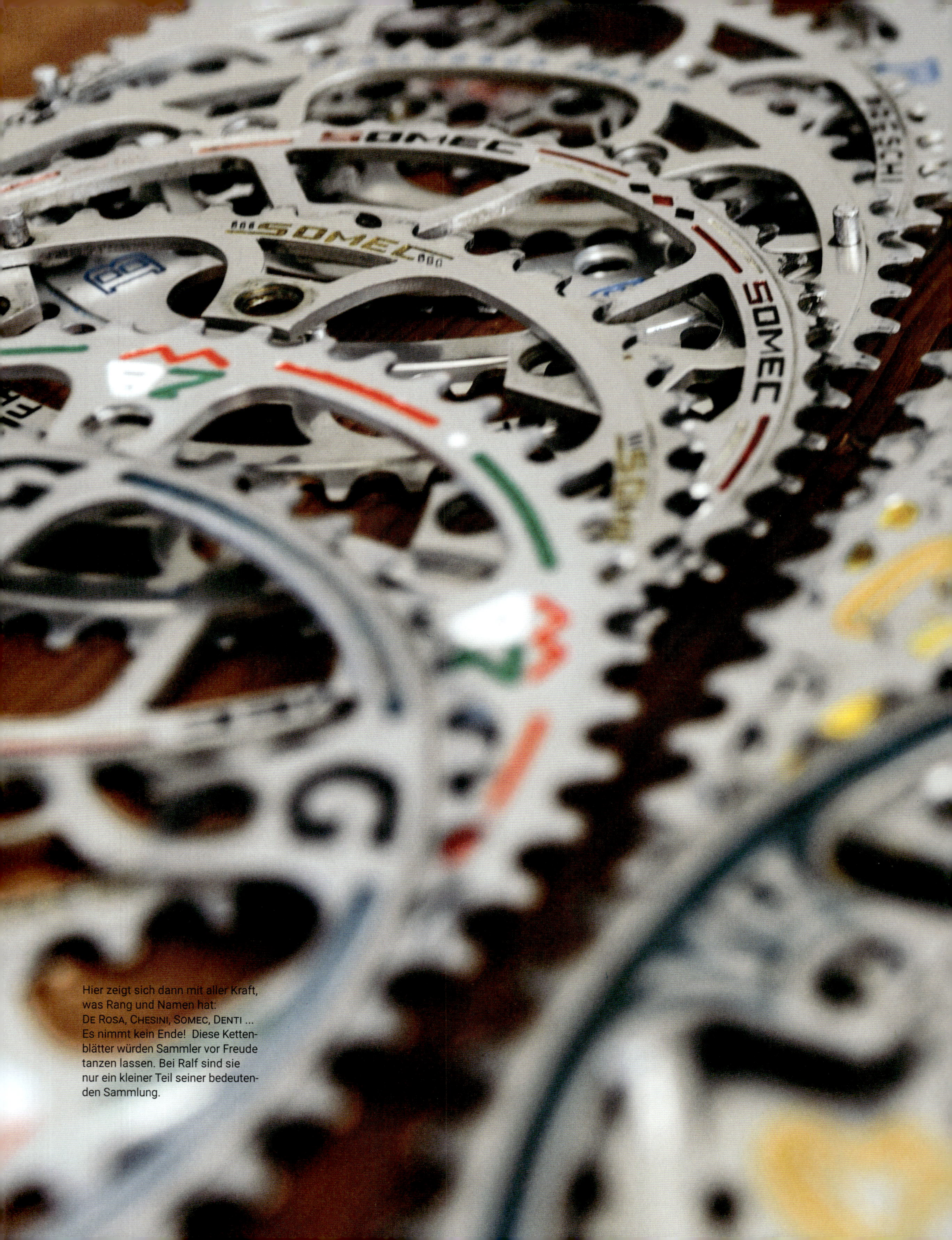

Hier zeigt sich dann mit aller Kraft, was Rang und Namen hat: De Rosa, Chesini, Somec, Denti ... Es nimmt kein Ende! Diese Kettenblätter würden Sammler vor Freude tanzen lassen. Bei Ralf sind sie nur ein kleiner Teil seiner bedeutenden Sammlung.

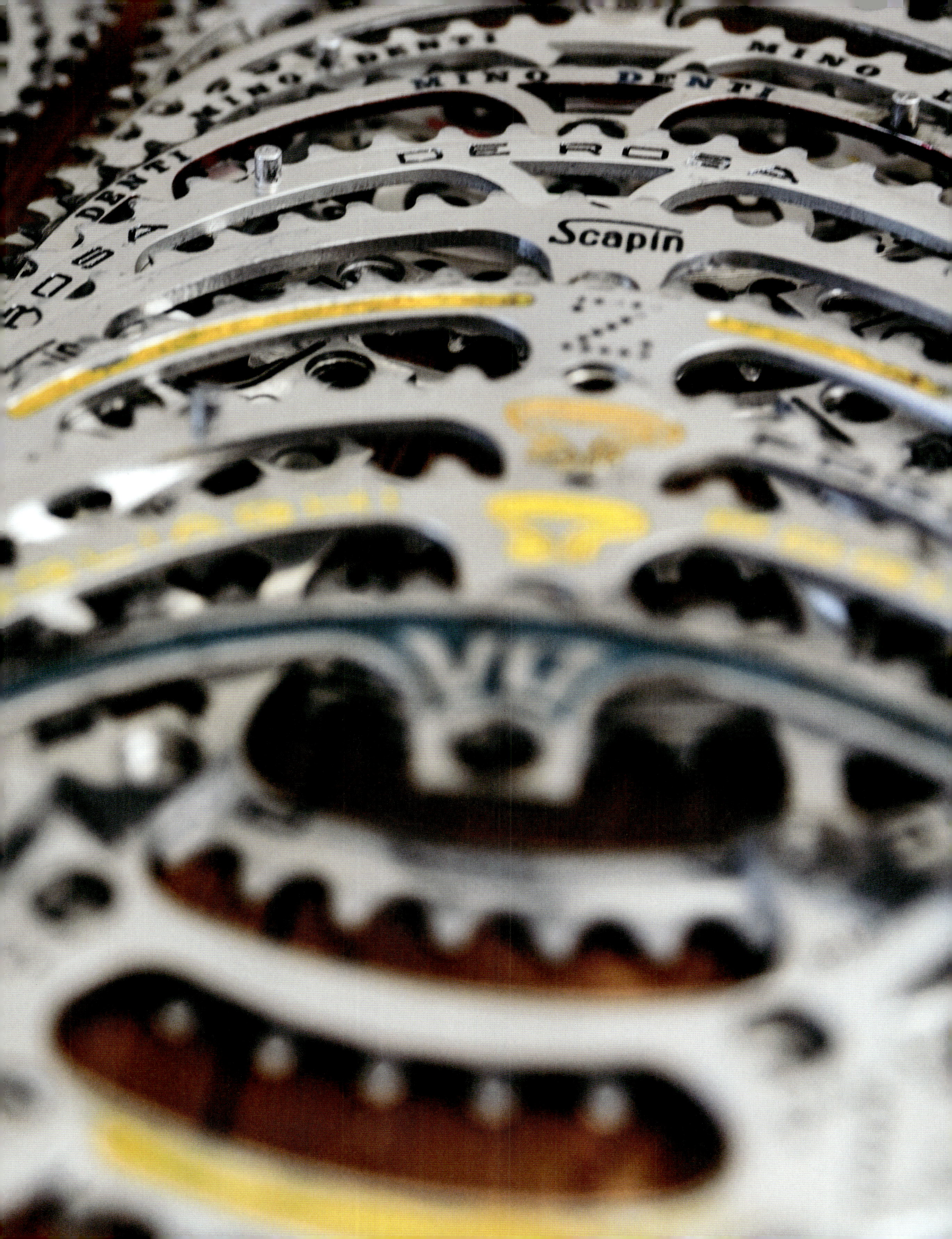
MINO DENTI
DE ROSA
Scapin

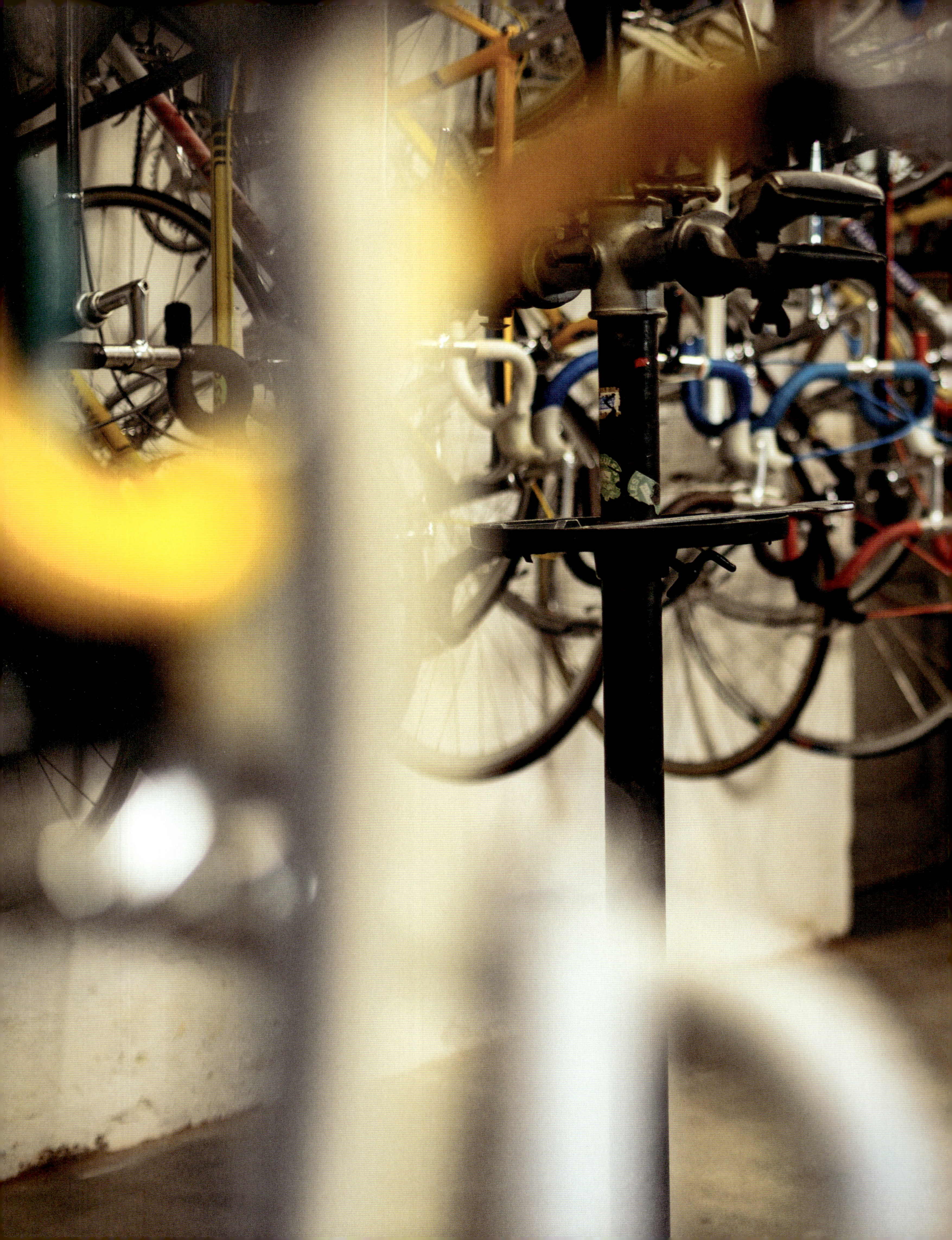

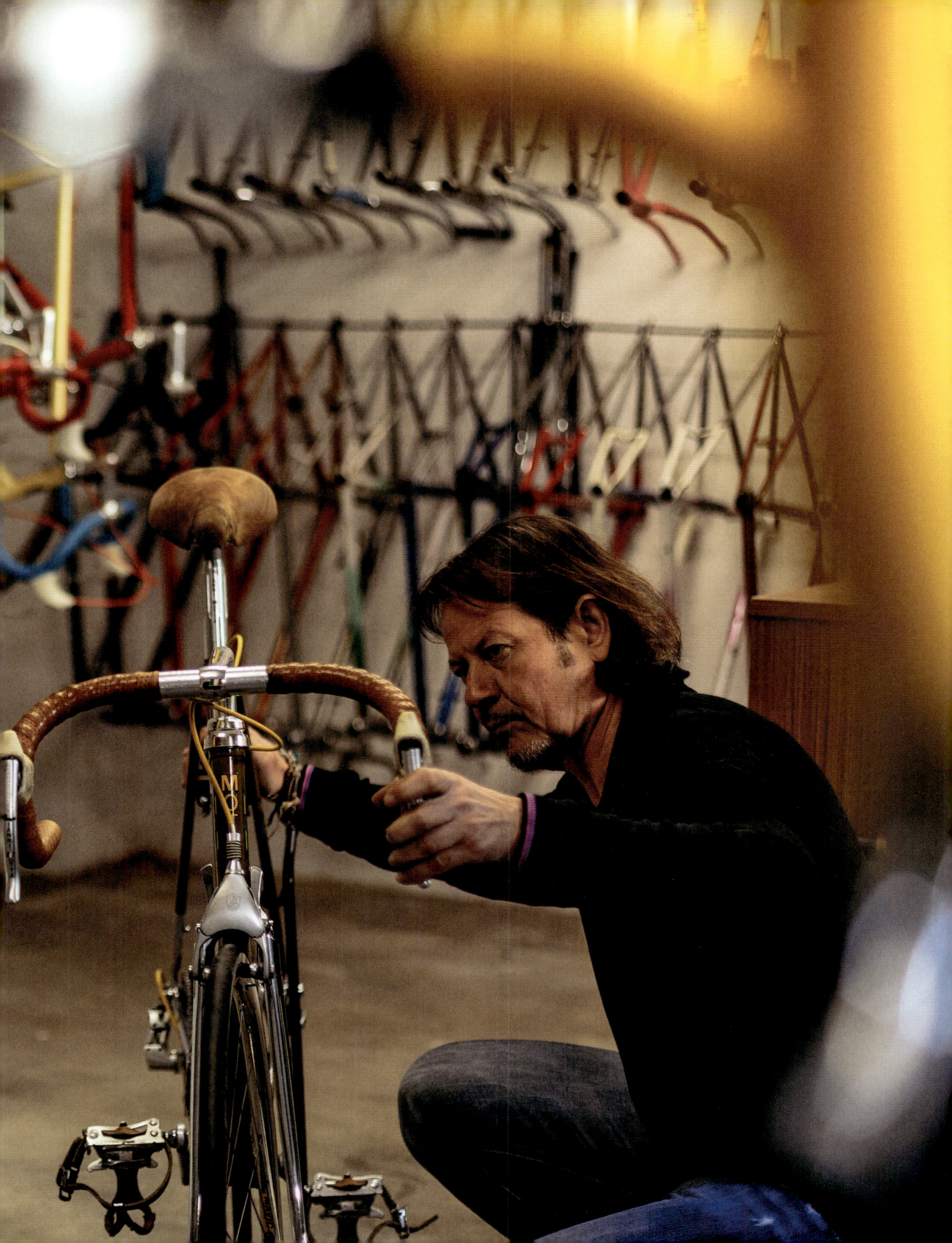

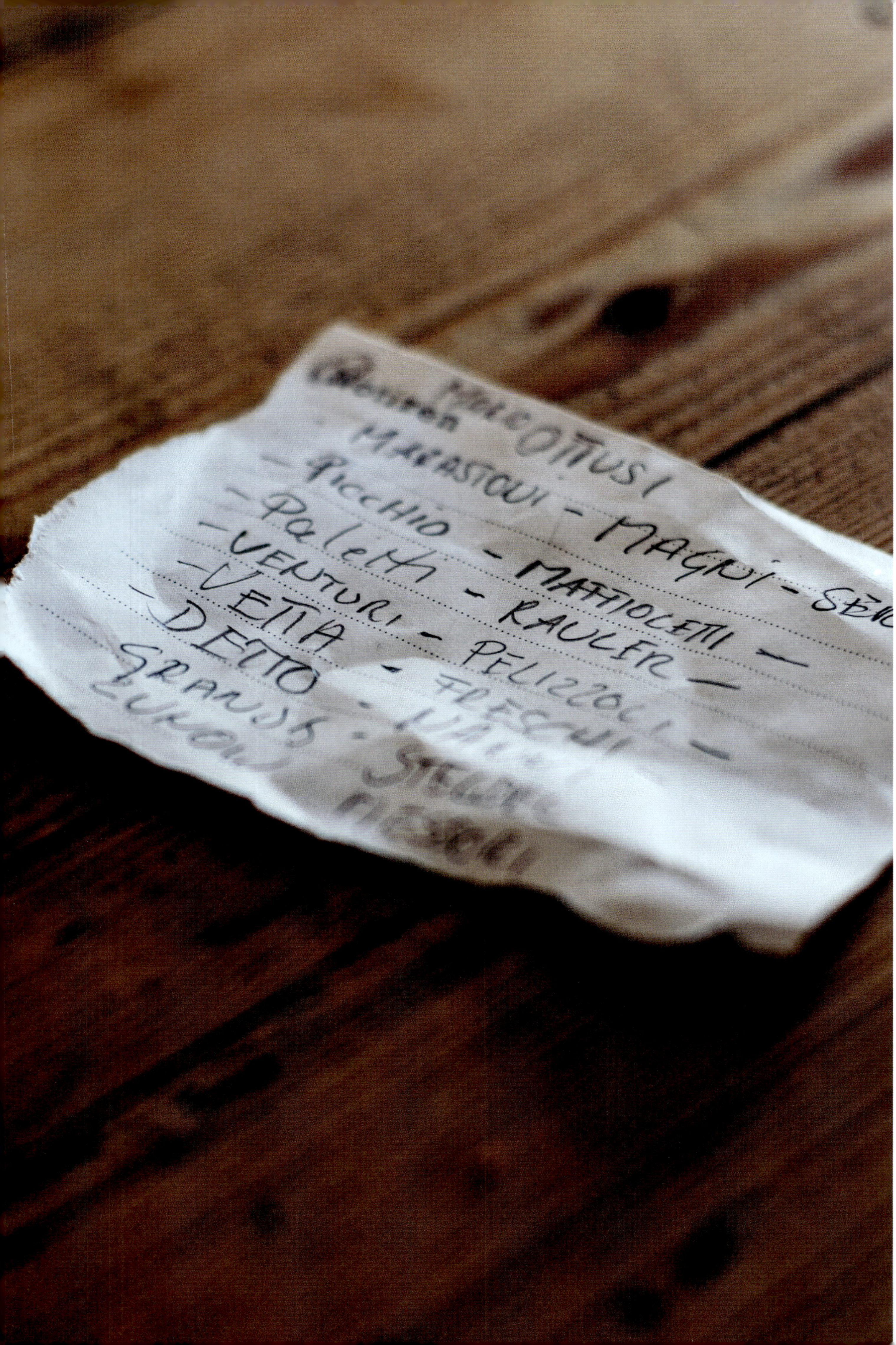

OTTUSI
MARASTONI - MAGNI
Picchio - MATTIOCETTI
Paletti - RAULER
VENTURI - PELIZZOLI
VETTA - FRESCHI
DETTO

Der geheime Spickzettel! Quasi Ralfs Heilige Schrift, mit der er jeden morgen eine Stunde im Netz nach fehlenden Teilen stöbert.

Dafür muss Ralf täglich mit seiner handgeschriebenen Suchliste das Internet durchforsten und schreitet erst zur Tat, wenn eine Gruppe komplett ist.

»Persönlich 'ne ganz dunkle Zeit« – die gesammelten Farben der 1980er-Jahre überstrahlen die Vergangenheit in leuchtendem Rot, Gelb, Grün, Türkis, Orange, Lila, Rosa, Hellblau oder Gold.

Keines der Räder ist ein Ausstellungsstück, jedes wird gefahren. Egal, ob es ein Rahmen von VITO ORTELLI aus Faenza ist, von WALTER DOSI aus Imola, RINO BOSCHETTI aus Marano Sul Panaro, dem Turiner ANGELO PICCHIO, EMILIANO FRESCHI aus Mailand oder von LUCIANO PALETTI aus Modena. Die gesammelte Schönheit Italiens!

»Lieblingsrad? Kannst du mich das bitte in einer Minute nochmal fragen?«

METALL AUF METALL

Ein Berliner Kraftwerk: FERN

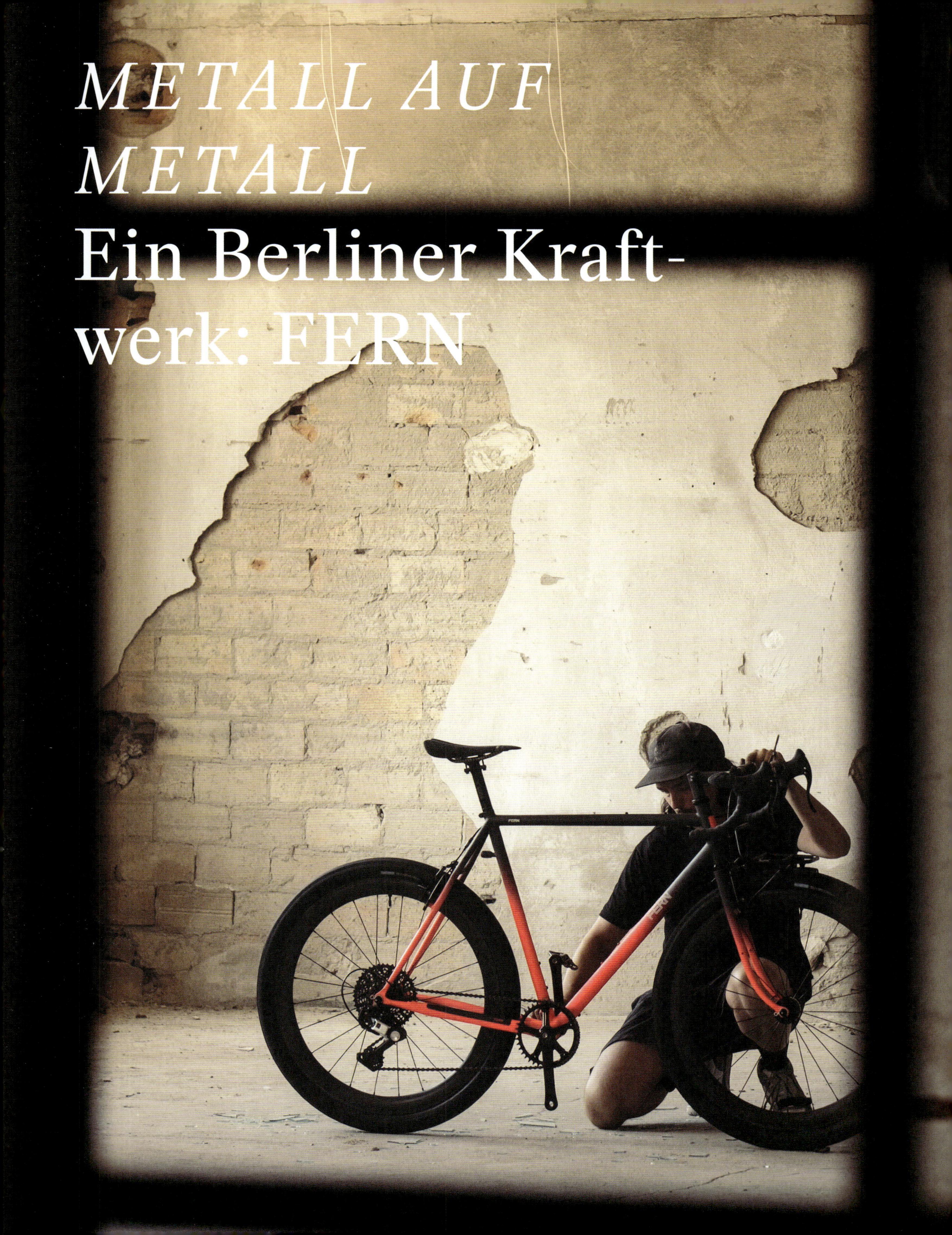

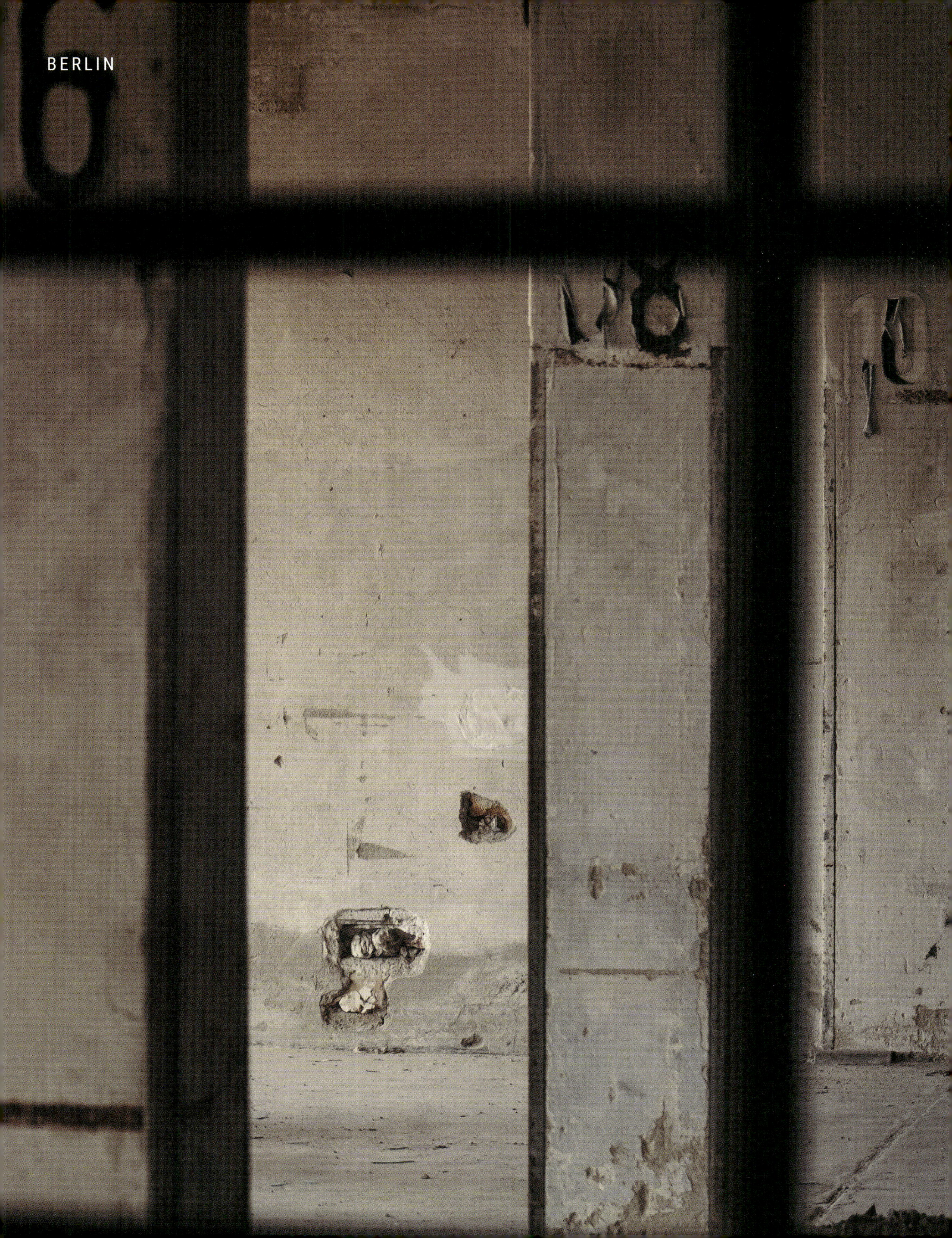

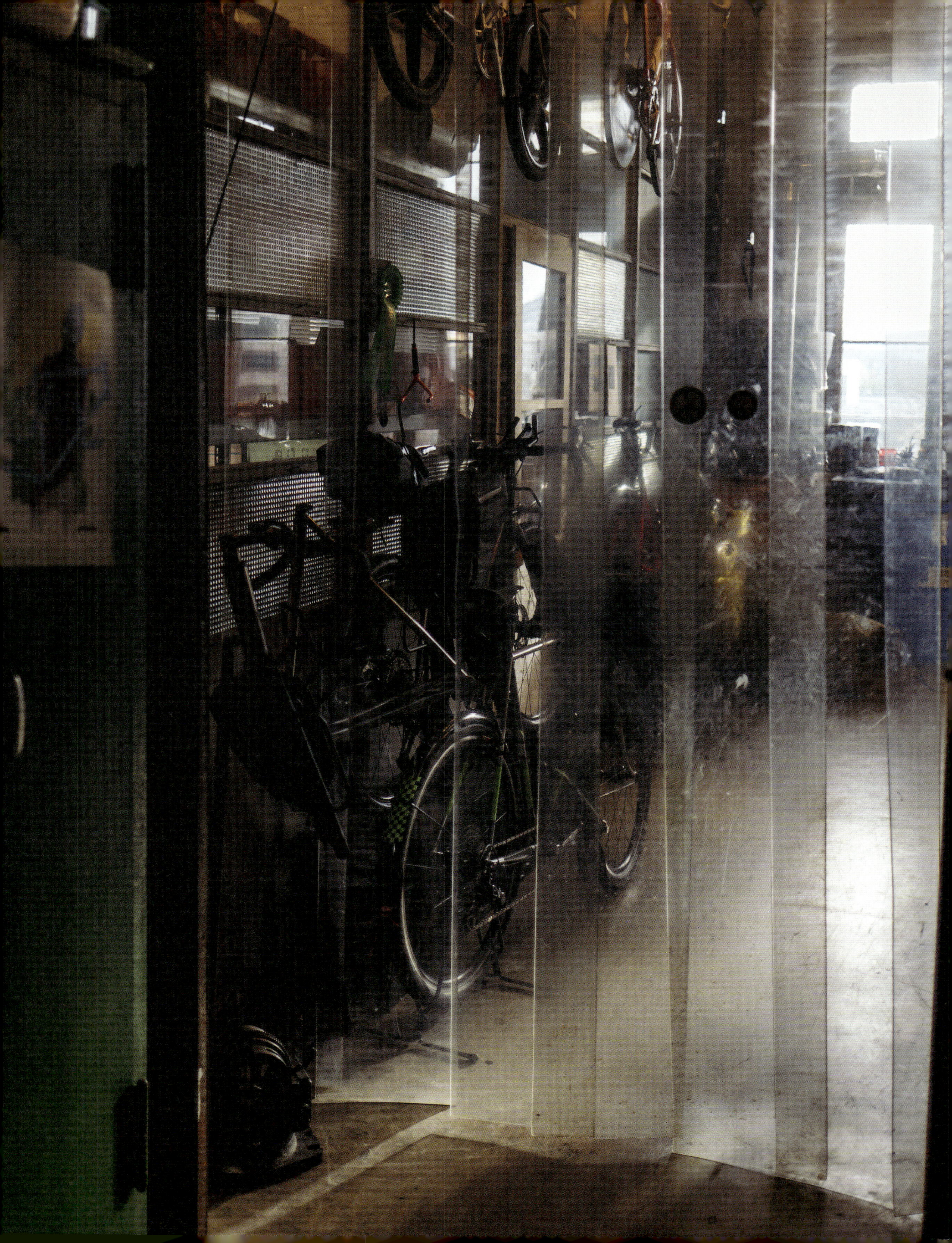

Aus dem Boxsport kommt die Redewendung, dass man nur K. o. gehen kann, wenn man den entscheidenden Schlag nicht kommen sieht. Den nun folgenden Schlag sehe ich nicht kommen, aber gehe auch nicht zu Boden. Ganz im Gegenteil, ich bin völlig euphorisiert: Auf der ersten und einzigen Wiener Fahrradschau im Jahr 2016, einem Ableger der damals tonangebenden Berliner Fahrradschau, treffe ich zum ersten Mal auf Florian Haeussler von FERN.

Mit meiner damals erschienenen Erstausgabe dieses Buches stelle ich zusammen mit vielen anderen Fahrrad-Enthusiasten in der Wiener MARX HALLE mein Buch und fotografische Arbeiten aus und feiere das Thema Fahrrad. In diesem Moment kann ich kaum glauben, was vor mir steht: Ein Gesamtkunstwerk aus klassischem Rahmendesign mit perfekt verschliffenen Lötstellen (*fillet brazed*), dazu Ton-in-Ton auf die sandfarbene und Neon-Orange Lackierung des Rades abgestimmte maßgeschneiderte Taschen von Flos kongenialer Partnerin Kristin Heil von Gramm-Tourpacking. Diese Liebe bis ins allerkleinste Detail hatte ich vorher noch nicht live gesehen. Es ist ein Wirkungstreffer von umfassender Reichweite!

Wenn man über den großen Innenhof des ehemaligen Industrie-Areals im Berliner Bezirk Lichtenberg die Räumlichkeiten von Flo in einem alten Fabrikloft betritt, meint man, eines der letzten »echten« Berliner Areale zu entdecken. Hinter einer schweren Stahltür öffnet sich ein riesengroßer Kreativraum: Meterhohe Metallregale, ein Künstleratelier, die Werkstatt von GRAMM-TOURPACKING, ein Vintage-Fahrrad-Teile-Handel mit einer offenen Küche in der Mitte.

Hier, hinter einem schweren Vinyl-Vorhang, befindet sich das Zentrum seines Schaffens: der Maschinenraum! Ein Raum, dessen Backsteinfassade durch deckenhohe Holzfenster gleichmäßig strukturiert ist und seine Drehbank, tonnenschwere Fräsen, einen Messtisch und seine Anvil-Rahmenbaulehre in wunderschönes natürliches Licht taucht.

Von den querliegenden Stahlbetonträgern unterhalb der Decke hängen Laufräder, Rahmenteile und Vintage-Räder. Dazwischen stolpert man über ein halbzerlegtes TRIUMPH-Motorrad vom Beginn des letzten Jahrhunderts, das Flo »nebenbei« Stück für Stück eigenhändig zerlegt und restauriert. An der Wand thront eine meterlange rote »Fahrrad«-Leuchtreklame eines alten Geschäfts, fast so,

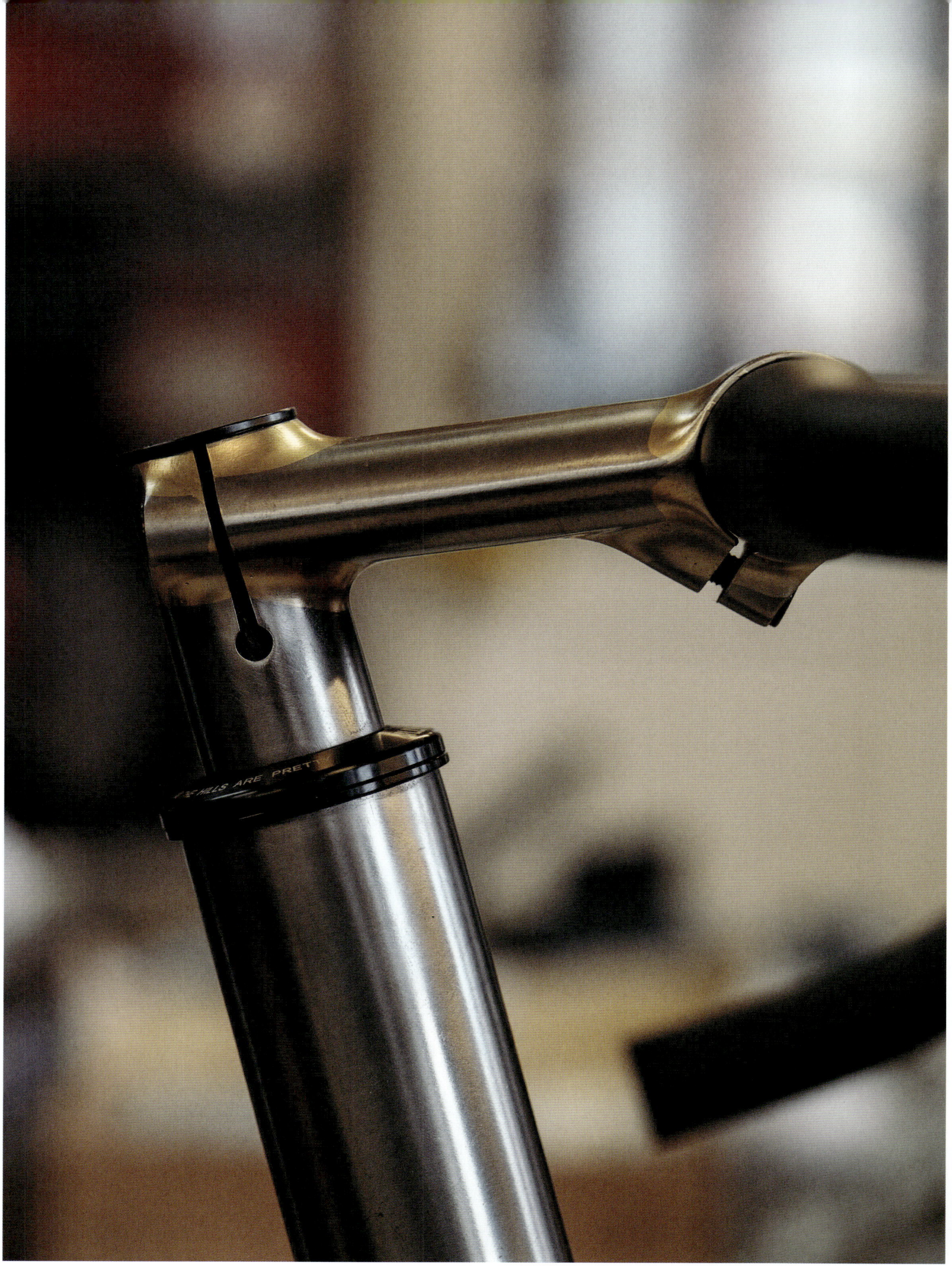
HILLS ARE

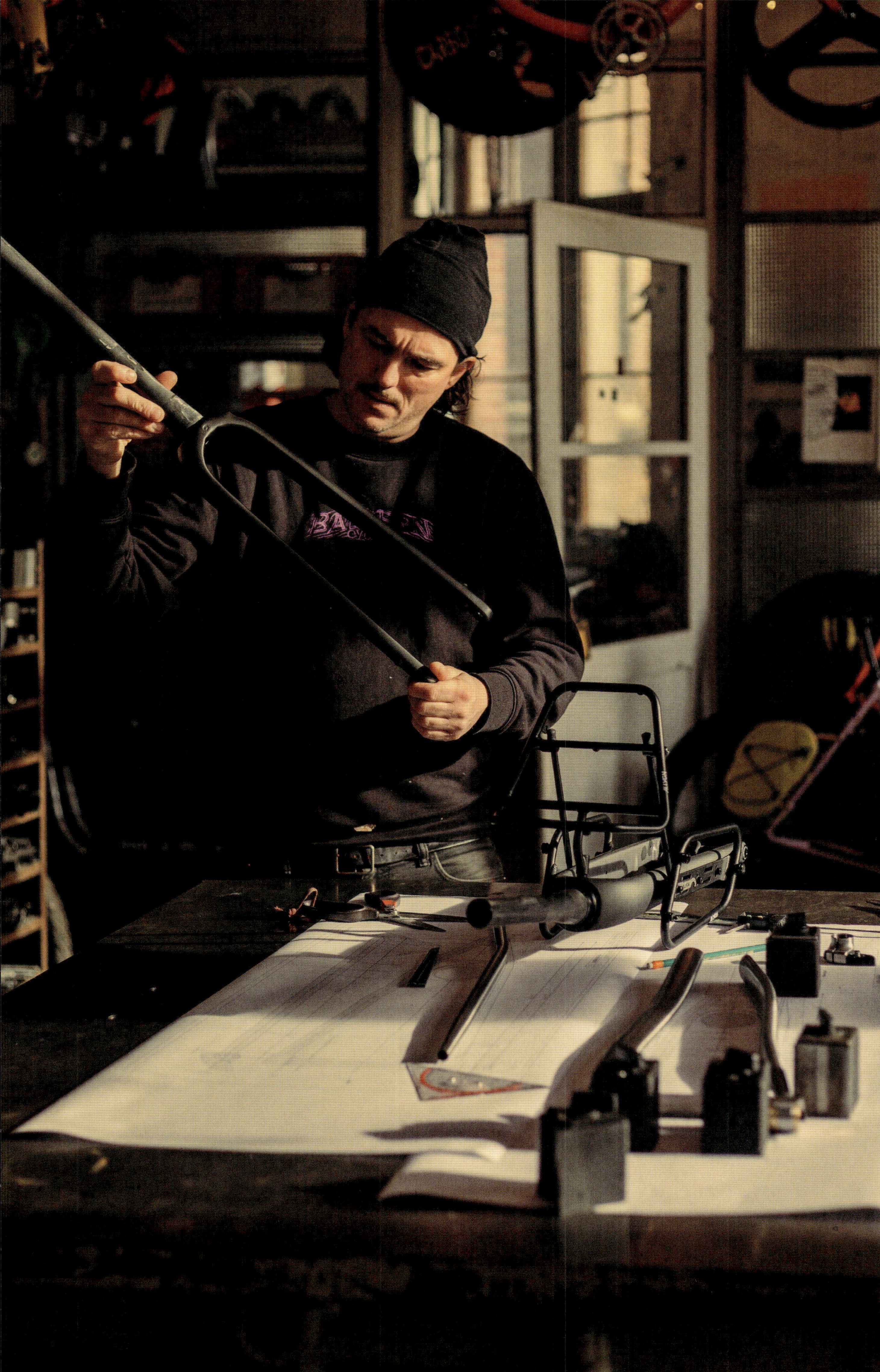

als müsse sich Flo immer noch in Erinnerung rufen, um was es hier eigentlich geht. Vielmehr entspricht der für den Raum etwas zu groß proportionierte Schriftzug dem tatsächlichen Motto seiner Ideenschmiede: Think Big! Denn »geht nicht«, gibt's für Flo einfach nicht.

»Dass es besser gehen muss«, denkt sich Flo nach seiner ersten großen Fahrradtour im Jahr 2006. Auf einem alten MTB, bei dem nur eine Bremse und schlappe sieben Gänge funktionieren, schafft es Flo auf seinem viel zu kleinen Fahrrad von Budapest schlussendlich bis nach Istanbul. Hier, so der Gründungsmythos, wird die Idee, Fahrräder selbst zu produzieren, geboren. Aus einem Fehlgriff heraus entsteht also die Notwendigkeit für Neues.

Ähnlich wie die legendäre deutsche Band *Kraftwerk* ihre wegweisenden Klangvorstellungen für ihr 1977 erschienenes Album Trans Europa Express nur durch den Bau eigener Musikinstrumente realisieren konnte, wächst auch bei Flo die Erkenntnis: Er muss das Objekt selbst bauen. Es folgen viele weitere Touren, bevor der studierte Industriedesigner mit zwei guten Freunden im Jahr 2012 die ersten selbstgefertigten Chacha-Prototypen quer durch Europa bis nach Istanbul fährt und sie seinen Erwartungen standhalten. Die Marke FERN wird, wenn man so will, aus einem »Trans Europa Express« heraus geboren – der Sound scheint gefunden.

Entgegen vielen schon damals scheinbar veralteten Vorstellungen folgt Flo seiner Idee und entwirft sein Ur-Modell CHACHA, mit 26"-Laufrädern und 54 mm breiten Reifen von René Herse bestückt – zu jenem Zeitpunkt ein absoluter Geheimtipp unter Kennern. Um diese Reifenbreite realisieren zu können, kombiniert Flo V-Brakes mit Rennradbremshebeln, auf denen er einen analogen Schalthebel montiert.

Es mussten noch viele Jahre vergehen, bis sich die Erkenntnis durchgesetzt hat, dass so voluminöse Reifen nicht langsamer als schlanke 28"-Laufräder rollen. Mit dem entscheidenden Vorteil, dass über das Volumen der Fahrkomfort extrem gesteigert wird, da mit ungewohnt niedrigem Reifendruck gefahren werden kann. Gerade dieser Pluspunkt setzt sich unter Langdistanzfahrern im Verlauf der Jahre durch. In das Zentrum des Antriebs setzt Flo eine SHIMANO-DURA-ACE-7400ER-Kurbel, die 1984 zum ersten Mal auf den Markt kam. Sie zeichnet sich durch

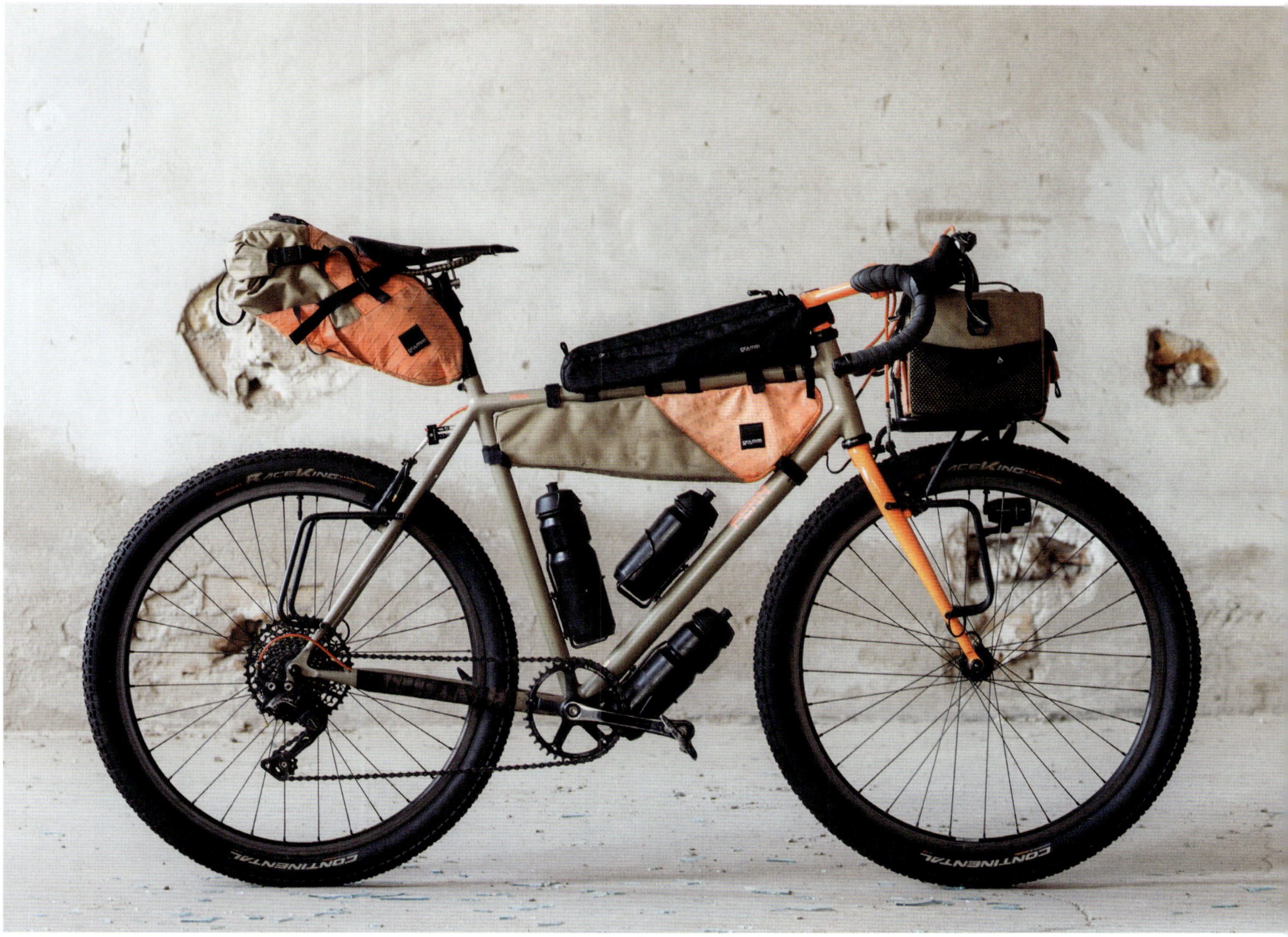

Modellname: Chuck. Ein Allroad-Traum in 26". Oder aber – das Problem, sich spontan zu verlieben. Als ich dieses Modell zum ersten Mal sehe, ist es um mich geschehen: Seit diesem Moment ist für mich klar, irgendwann möchte ich auch ein FERN besitzen.

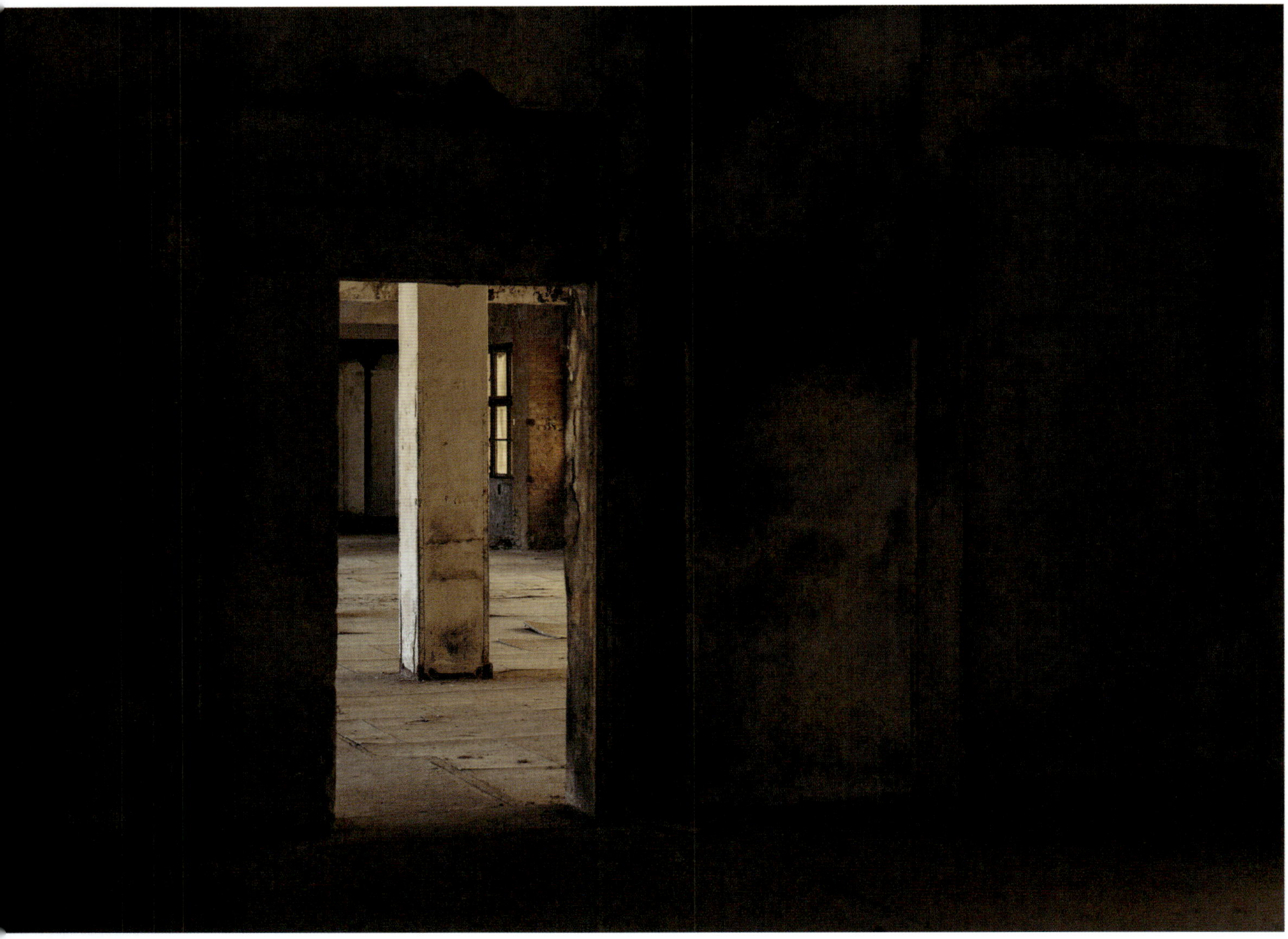

Das Motto seiner Ideenschmiede: Think Big! Denn »geht nicht«, gibt's für Flo einfach nicht.

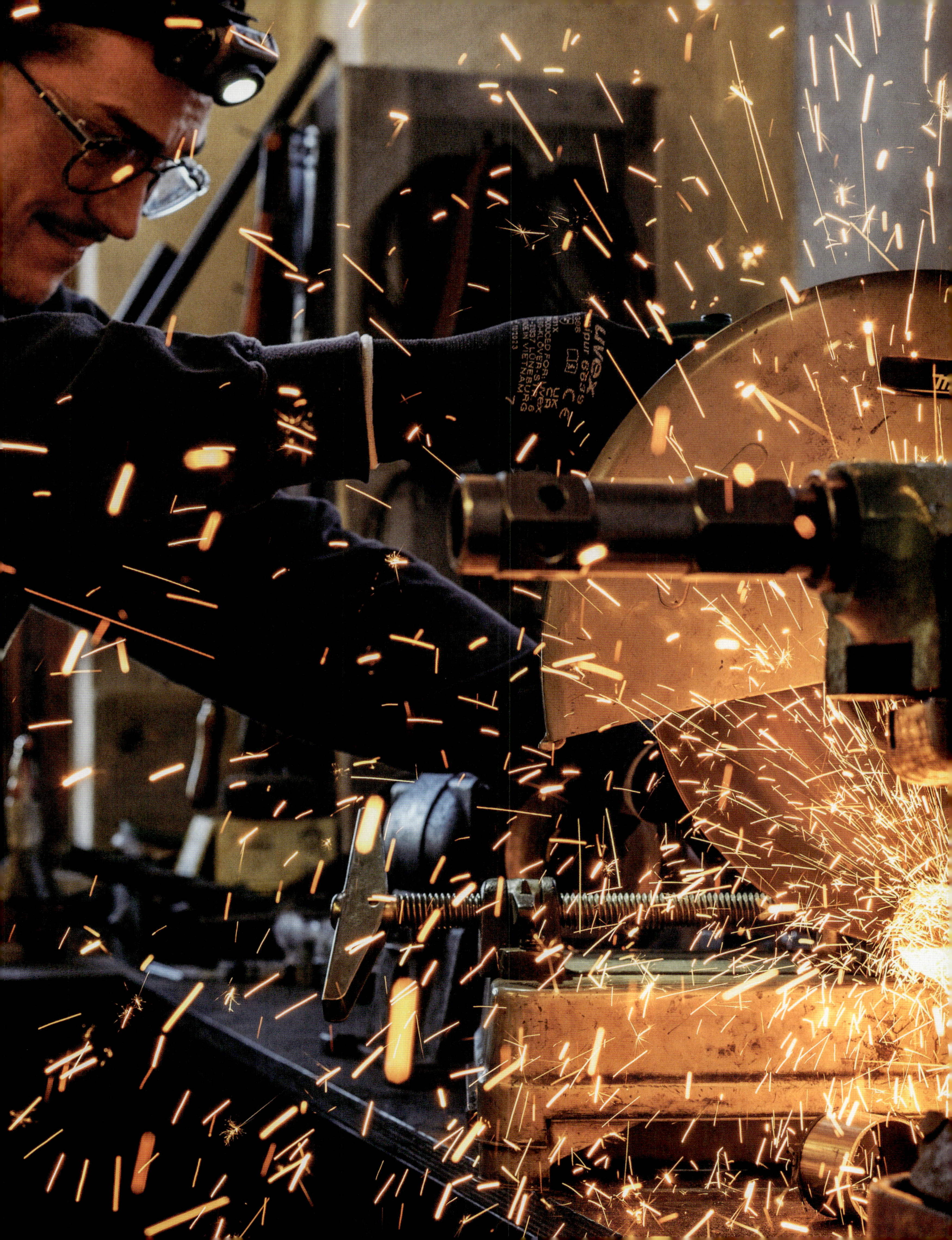
uvex

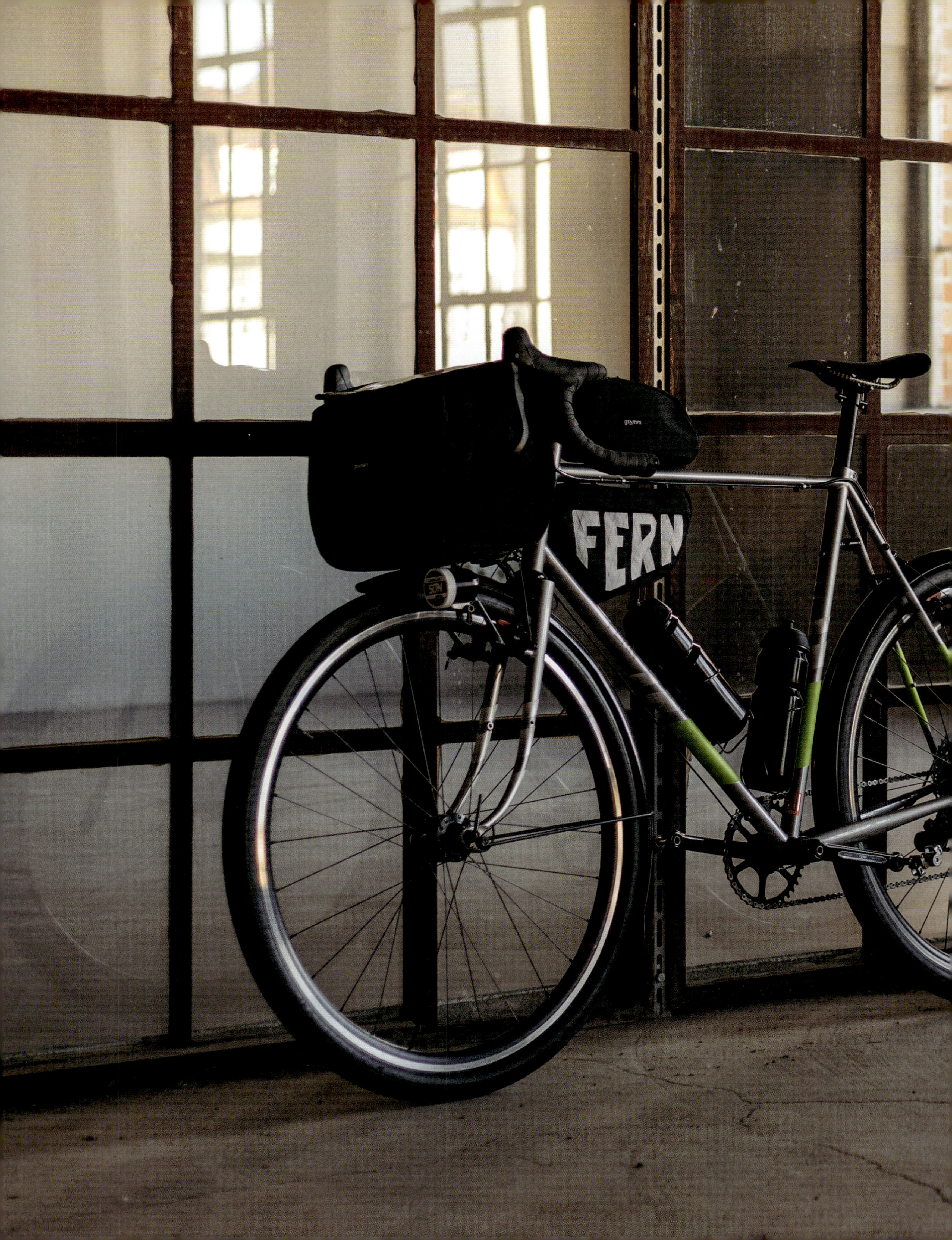
FERN

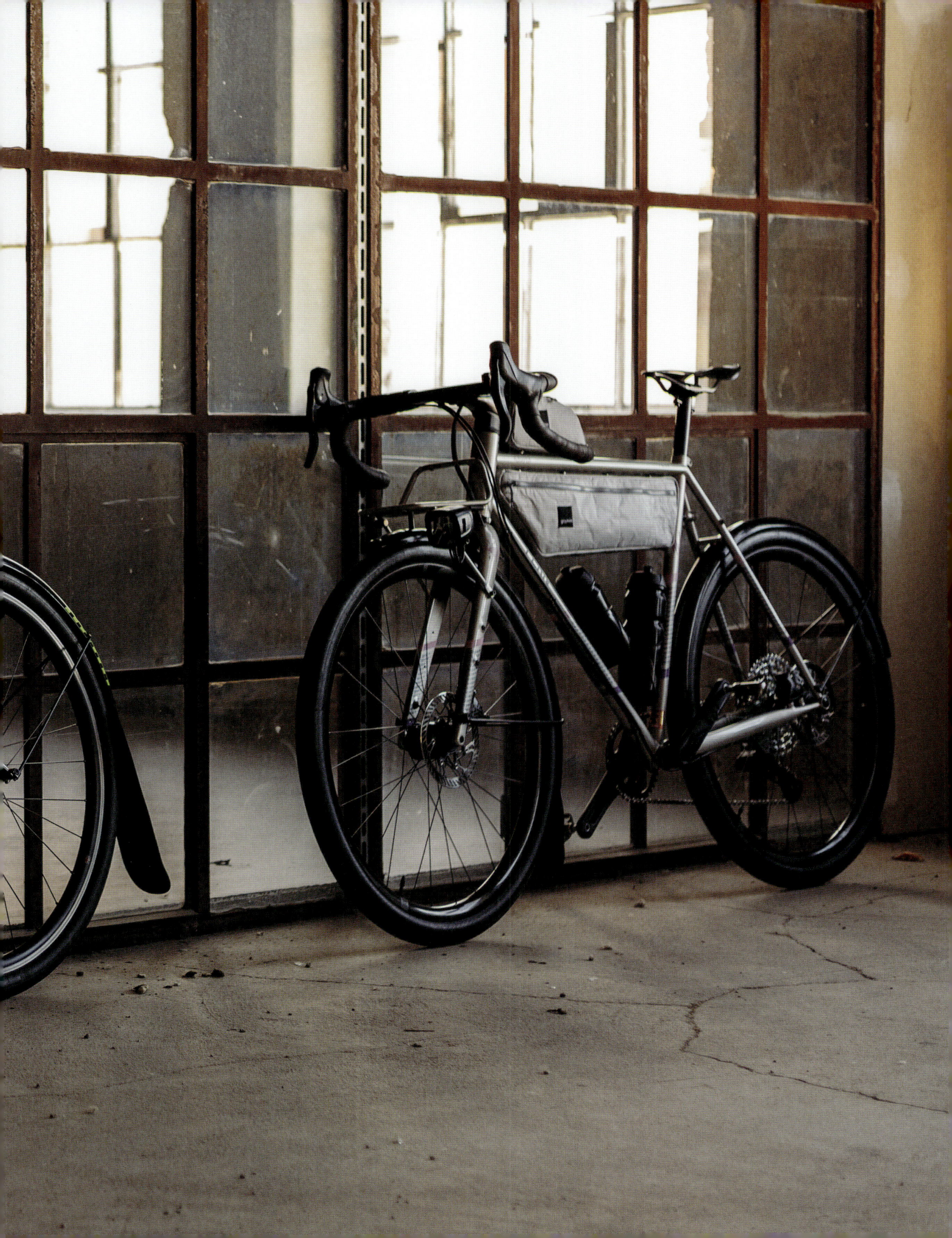

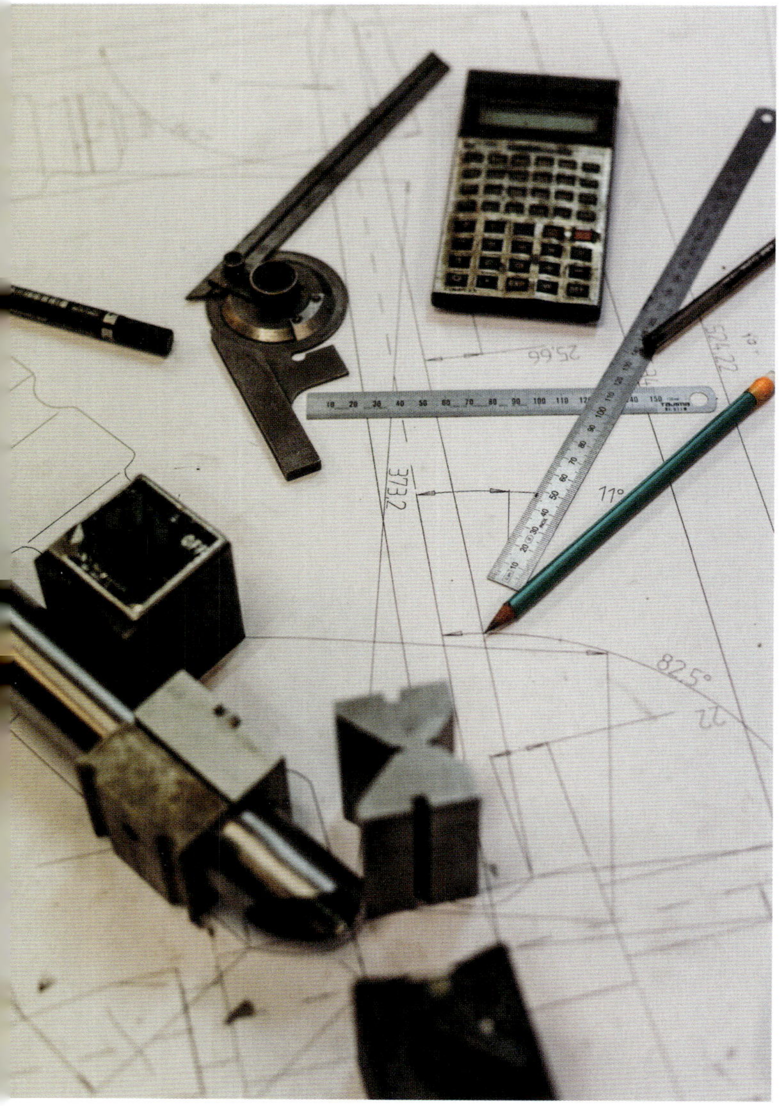

den Einsatz von hochfestem Aluminium aus, ermöglicht somit eine extrem schlanke Bauweise und damit einen sehr geringen Q-Faktor. Der ist für eine sehr sportliche Fahrweise und direkte Kraftübertragung entscheidend. Erneut re-interpretiert Flo etwas, das völlig »out-of-date« scheint. Es sind diese »Vintage-Kurbeln« die in der von ihm schwarz eloxierten Variante zu einem der »Signature«-Elemente eines echten FERN werden.

CHUCK heißt die Weiterentwicklung dieses Rades vor dem ich in Wien staunend stehe – Parallelen zum bekannten amerikanischen Action-Helden der 1970er sind rein zufällig. Die Farbkombination aus einem Sandton und Neon-Orange spiegelt exakt den Charakter des Rades wider: Eine Maschine, die jede Offroad-Strecke glattbügelt und mit der man trotzdem bei der Asphaltrunde zum Ortsschildsprint ansetzen kann.

Die Taschen, von GRAMM-TOURPACKING, in denselben Duo-Tönen gehalten, machen das Rad komplett. Am liebsten wäre ich mit dem Rad heimlich aus der Halle gefahren und nicht wieder zurückgekommen.

Flos Innovationsdrang hat über die letzten zehn Jahre glücklicherweise nicht bei einer Laufradgröße, einer Reifenbreite und dem Bestehen auf Felgenbremsen Halt gemacht. Zeigt doch die umfangreiche Galerie auf seiner Homepage, wie facettenreich sein Schaffen ist. Schnell wird deutlich, dass er spezielle Anforderungen, wie körperliche Besonderheiten oder Ausstattungswünsche, jeweils in genau das für die Person passende Rad übersetzen kann.

Eine besondere Stellung nimmt für mich ein Rad ein, das die Merkmale eines legendären Paris-Dakar-Motorrads des bayerischen Herstellers BMW in ein Fahrrad überführt: Darin wird Flos Stärke als Industriedesigner besonders deutlich, Dinge neu interpretieren zu können und in ein anderes Gewand zu packen. Wie viele Stunden Handarbeit in den jeweiligen Rädern steckt, lässt sich kaum ermessen.

Die Übergänge seiner Lötstellen sind von solch einer Perfektion, die sofort deutlich macht, warum Flo vor seiner Selbstständigkeit in der Automobilbranche bei namhaften deutschen Herstellern als »Surface-A-Class«-Designer tätig war: Wer in diesem Bereich für die entsprechend komplexen Freiformflächen zuständig ist, muss einer der Besten im CAD-Design sein.

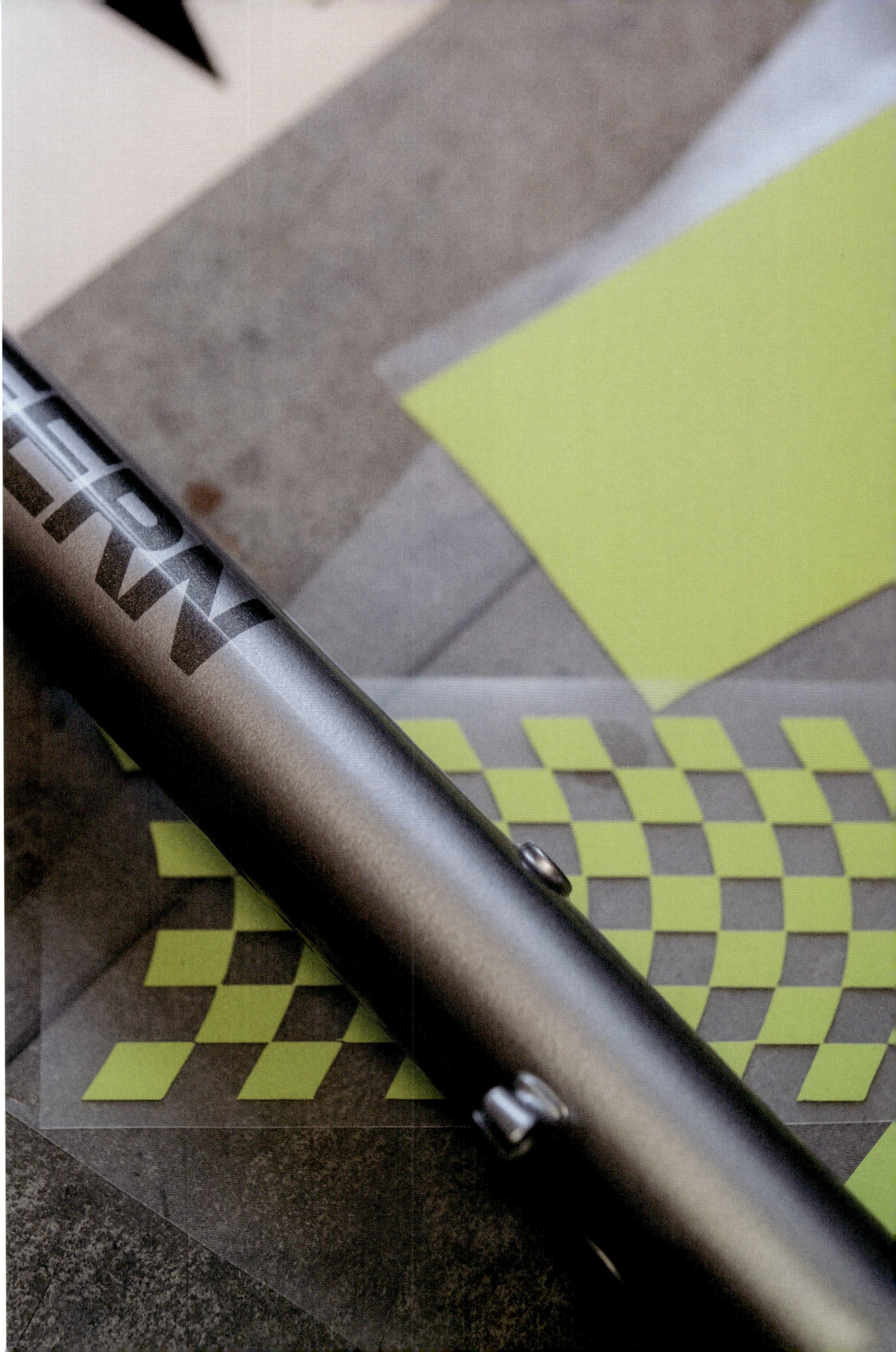

DIE MARKE FERN WIRD, WENN MAN SO WILL, AUS EINEM »TRANS EUROPA EXPRESS« HERAUS GEBOREN – DER SOUND SCHEINT GEFUNDEN.

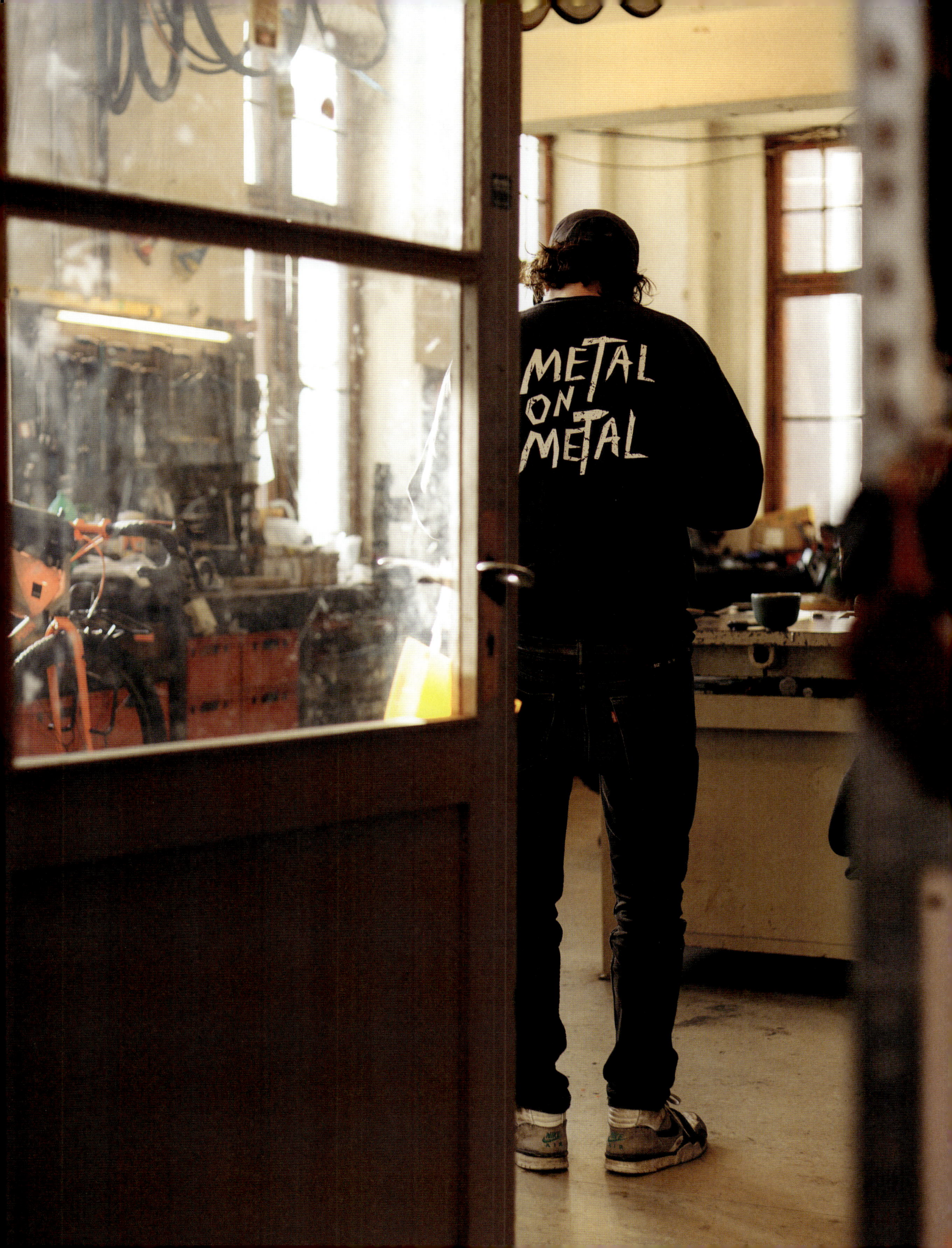
METAL
ON
METAL

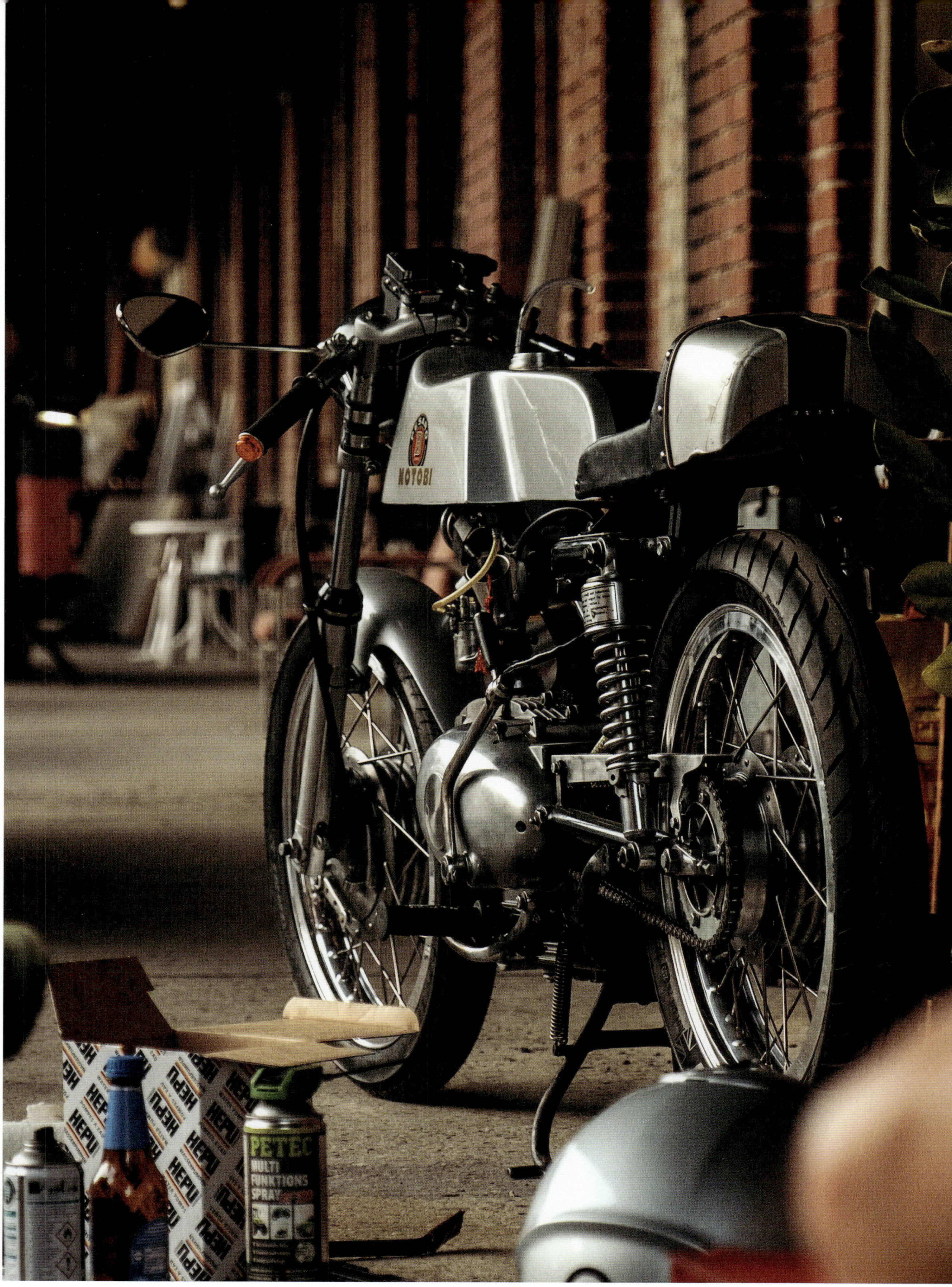
MOTOBI
HEPU
PETEC
MULTI
FUNKTIONS
SPRAY

Freizeitbeschäftigung eines Workaholic: Die MOTOBI, ein 70er-Jahre Coffee-Racer, ist frisch restauriert. Nur der Führerschein, der fehlt noch!

Aus dem Wunsch heraus, für seinen Maß-Rahmenbau die bestmöglichen Teile verwenden zu können, entwickelt Flo seit Jahren parallel zur Maßarbeit industriell gefertigte Anbauteile für Fahrräder, die unter dem Namen ALLYGN weltweit gefragt sind. Dazu gehören meiner Meinung nach auch einer der formschönsten Frontgepäckträger; eine Gravel-Carbongabel, die keine Wünsche offenlässt; extrem flexible Hinterradgepäckträger oder formvollendete Rücklichter.

Viele der Kernprinzipien des Bauhaus-Designs für industrielle Fertigung erkenne ich in seiner Arbeit: Jedes einzelne Objekt ist von funktionaler Klarheit geprägt, auf das Wesentliche reduziert, frei von Ornamentik und ganz im Stil der Serienfertigung, um seine wundervolle Gestaltung für jedermann zugänglich und erschwinglich zu machen. Die handwerkliche Romantik des Maß-Rahmenbaus weicht bei Flo einer Maschinenästhetik und betont vielmehr die klaren, oft abstrakten Linien und die Struktur.

In den letzten Jahren durfte ich Flo immer wieder beim Maß-Rahmenbau über die Schulter schauen. Besonders aufregend war die Produktions-Zeit meines eigenen FERN-Rahmens. Es sind zu viele einzelne Arbeitsschritte, die den Bau eines solchen Rahmens komplettieren, als dass ich sie alle beschreiben könnte.

Nachhaltig sind mir die Abläufe bei der Fertigung des Oberrohrs in Erinnerung geblieben: Ich konnte mir nicht vorstellen, wie viele Wiederholungen, erneutes Überprüfen der Vorbiegung, wiederholtes Einspannen und Quetschen notwendig sind, um ein einziges Rohr »querzuovalisieren« – ein Vorgang, bei dem ein Rohr von rund auf oval verformt wird. Natürlich kommt der besondere Design-Anspruch Flos »erschwerend« hinzu – bildet das Rohr am ovalisierten Ende doch zusammen mit den Sitzstreben und dem Sitzrohr den nach Flo benannten »Berliner Knoten« aus.

Er hört es nicht gern, aber er ist und bleibt einer der Impulsgeber der internationalen Rahmenbauerszene – bei ihm kommt, wie beim gleichnamigen Kraftwerk-Song, in allerbester Form: Metall auf Metall!

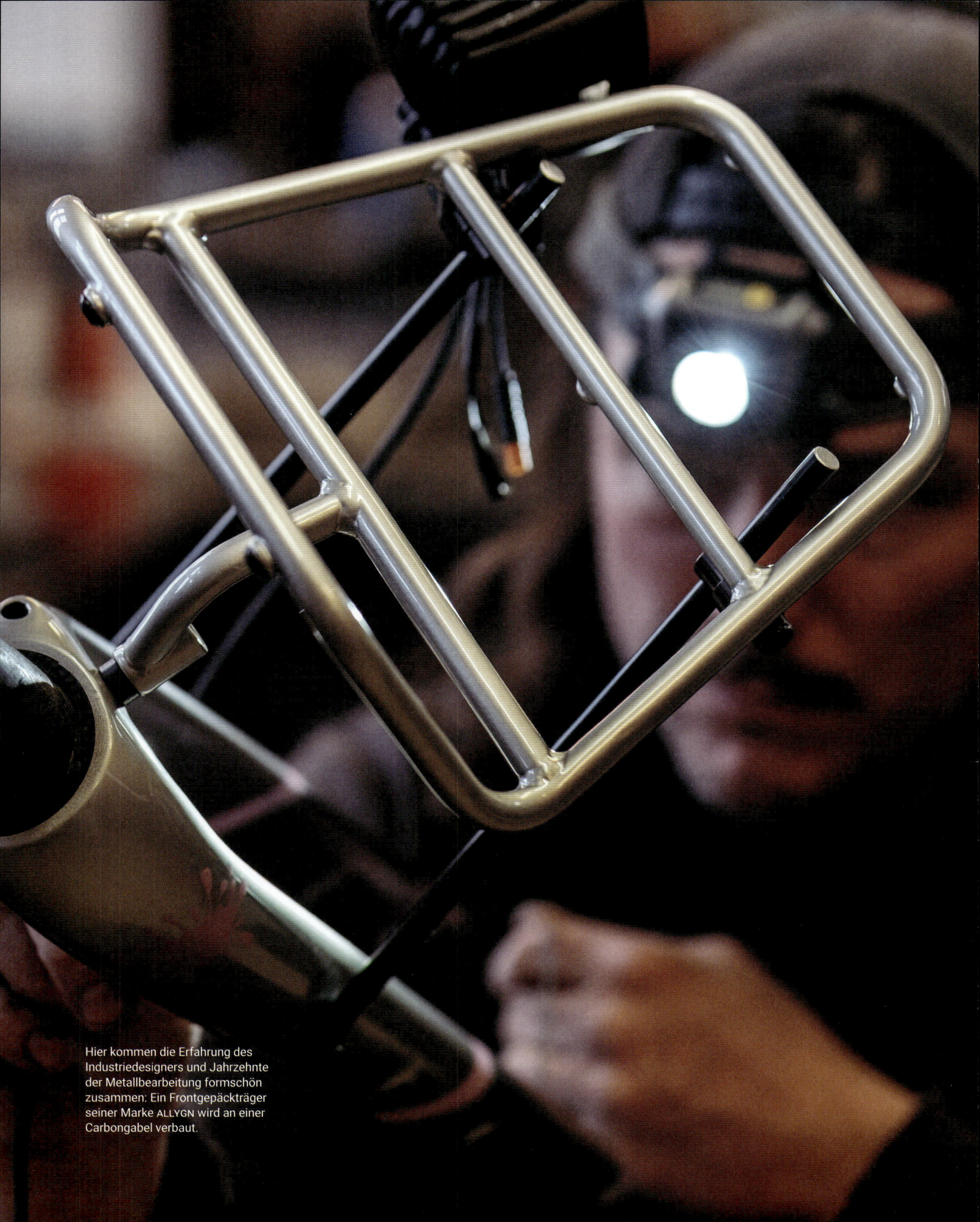

Hier kommen die Erfahrung des Industriedesigners und Jahrzehnte der Metallbearbeitung formschön zusammen: Ein Frontgepäckträger seiner Marke ALLYGN wird an einer Carbongabel verbaut.

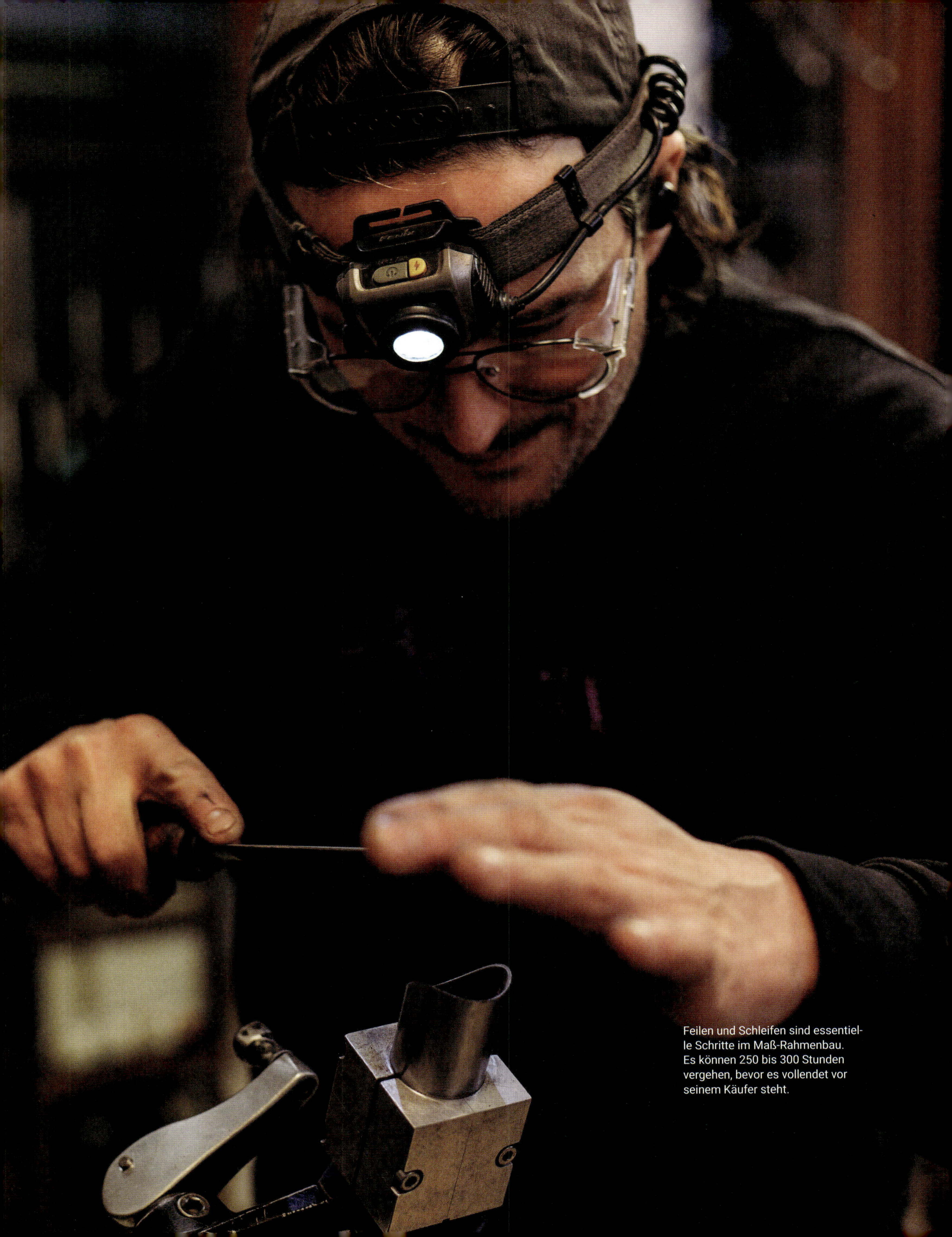

Feilen und Schleifen sind essentielle Schritte im Maß-Rahmenbau. Es können 250 bis 300 Stunden vergehen, bevor es vollendet vor seinem Käufer steht.

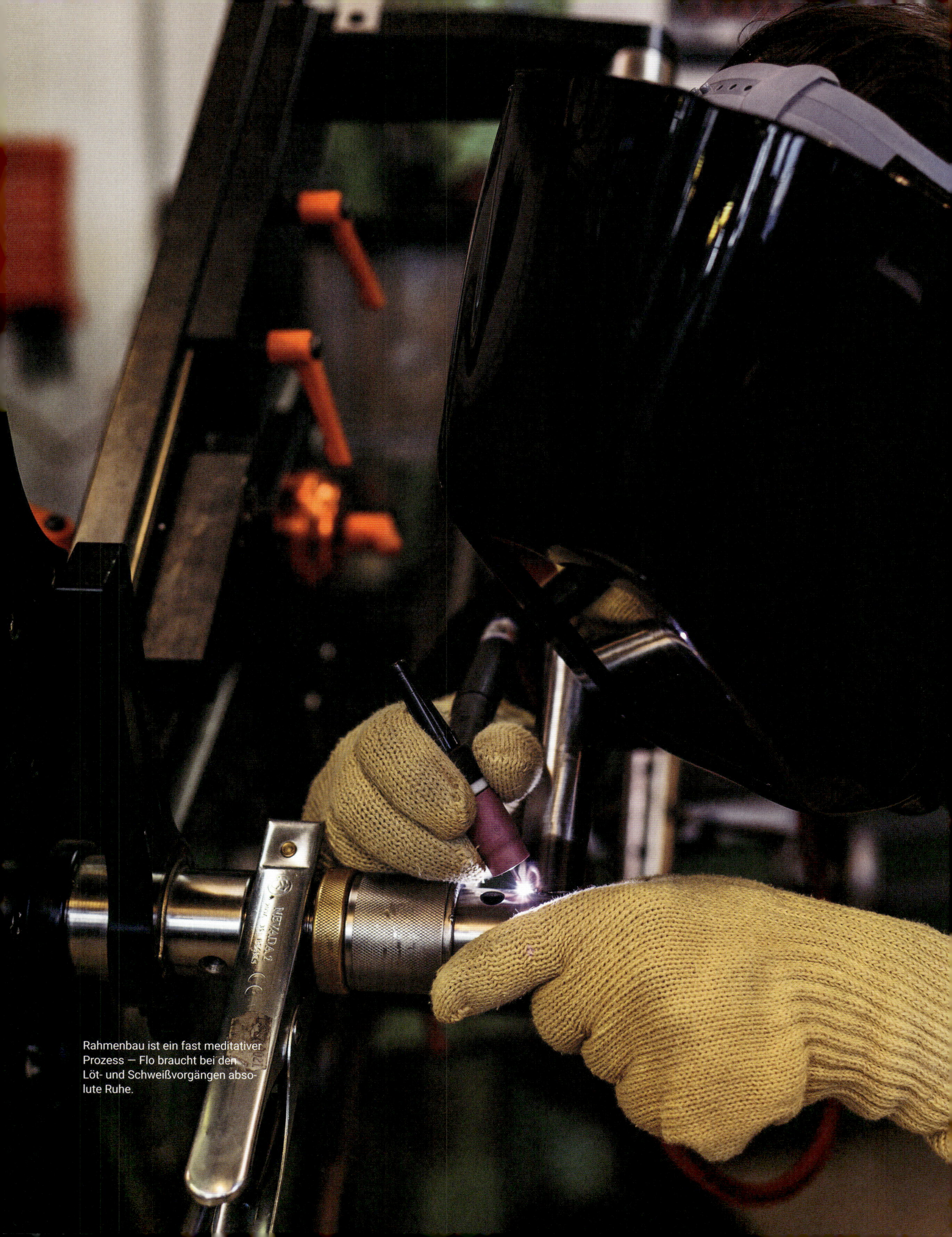

Rahmenbau ist ein fast meditativer Prozess — Flo braucht bei den Löt- und Schweißvorgängen absolute Ruhe.

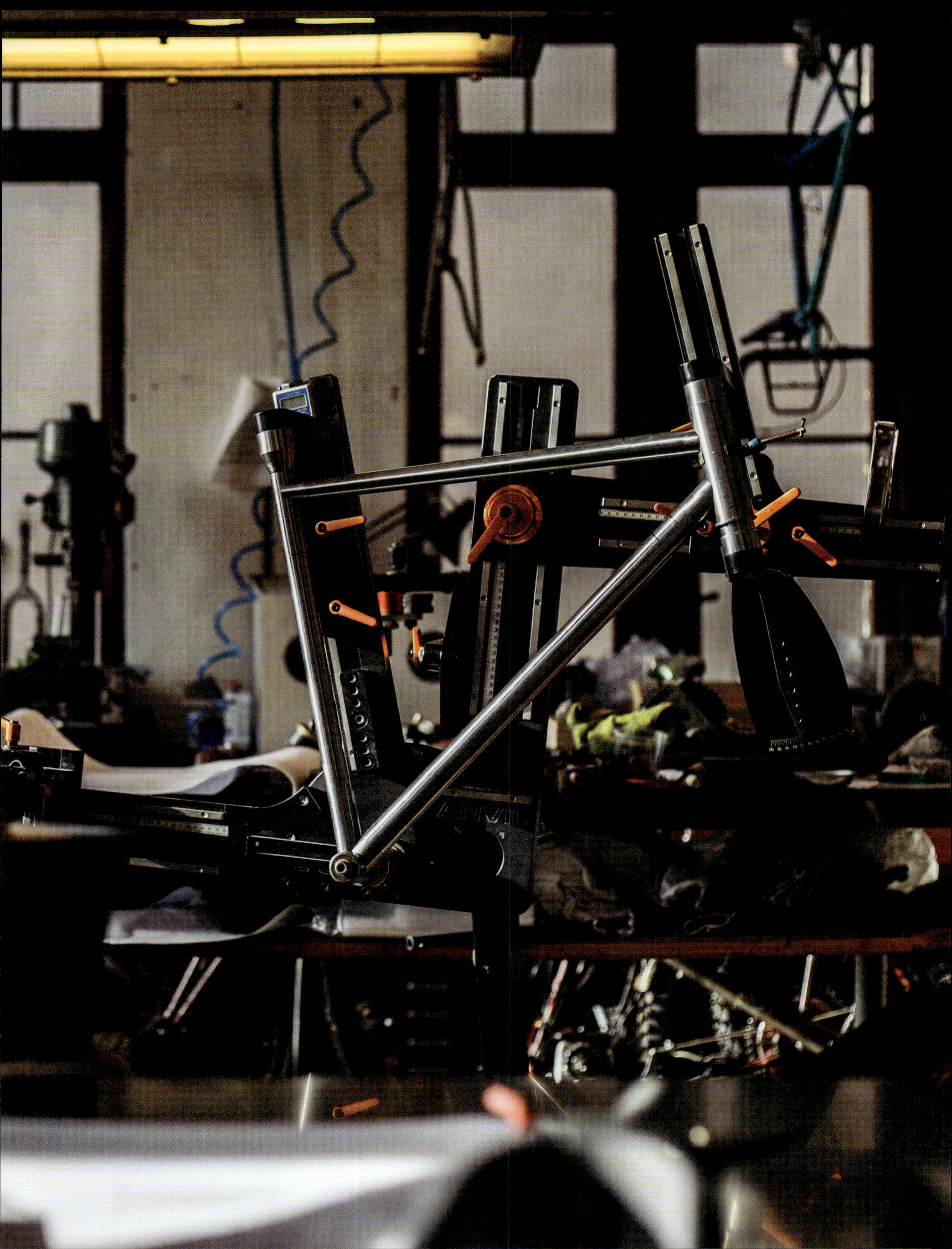
ANVIL

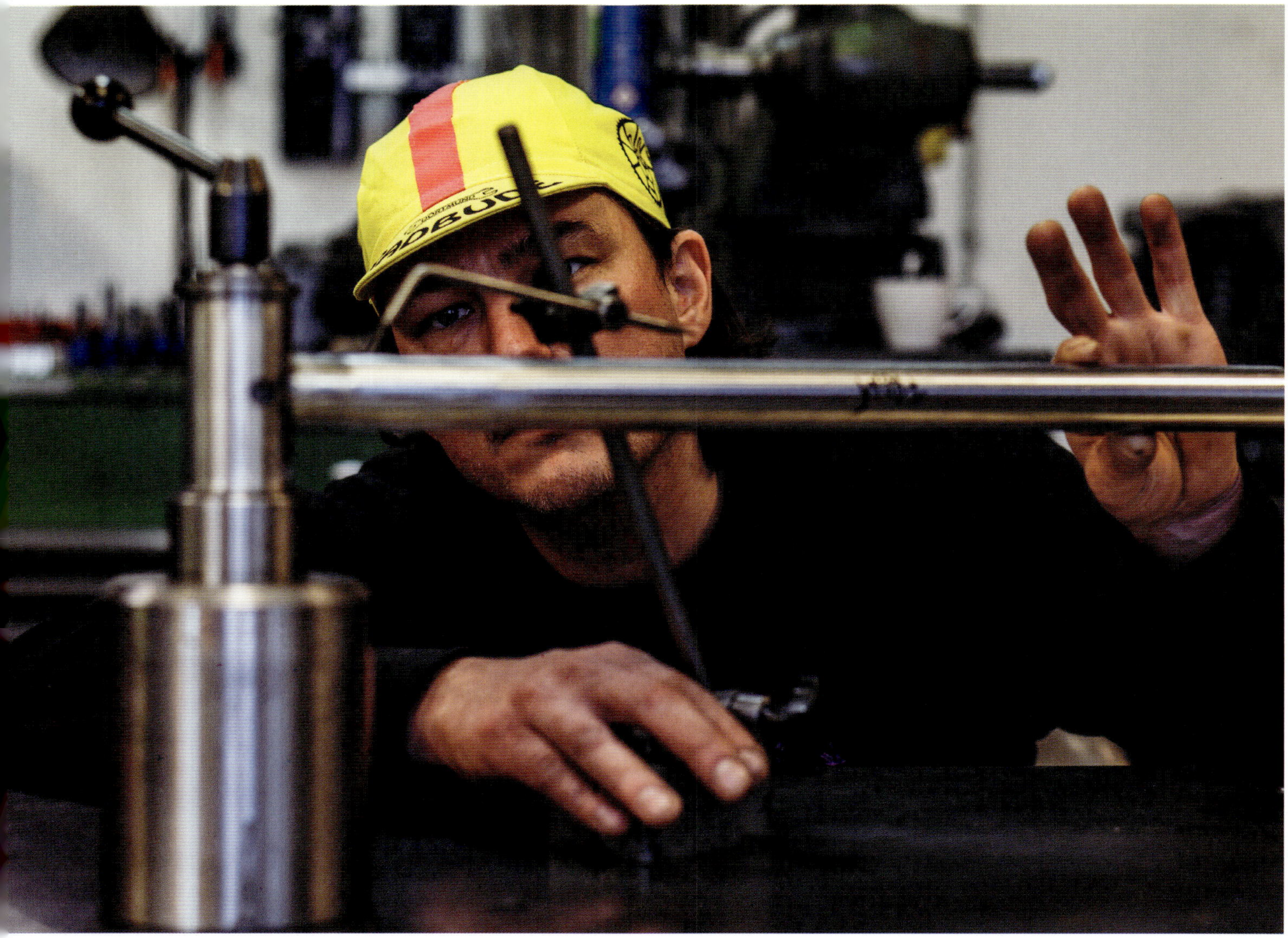

Eine Hommage an einen Freund: Eines der großen Vorbilder Flos ist der legendäre Rahmenbauer Sacha White, Begründer von Vanilla Workshop. Wir konnten es uns nicht nehmen lassen, sein Lieblingsbild von Sacha originalgetreu nachzustellen.

Rechts: Der Berliner Knoten – wie er unter Rahmenbauern voller Anerkennung genannt wird – ist eines von Flos besonderen Merkmalen. Hier kommen Oberrohr, Sitzrohr und Sitzstreben in Perfektion zusammen.

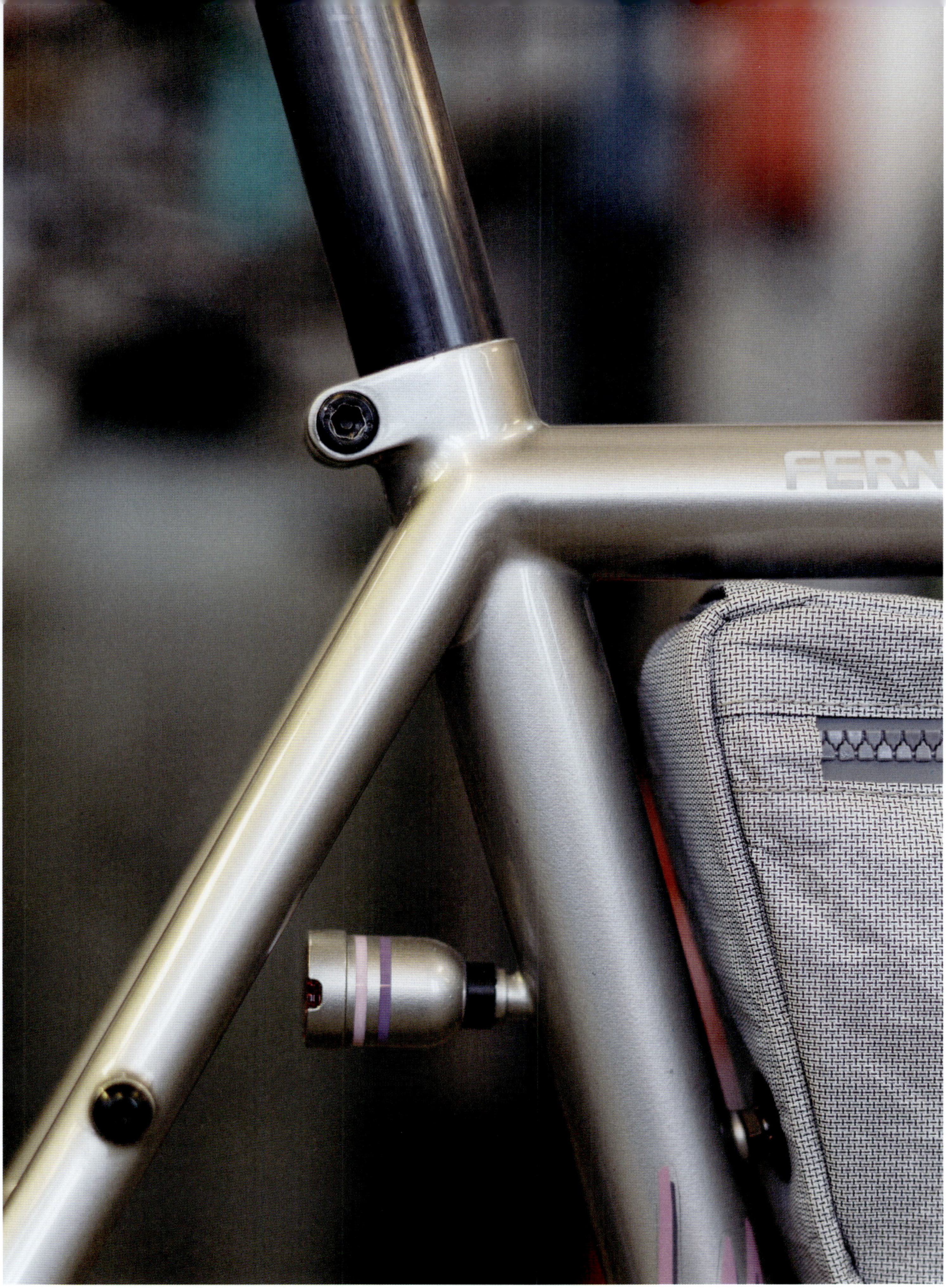
FERN

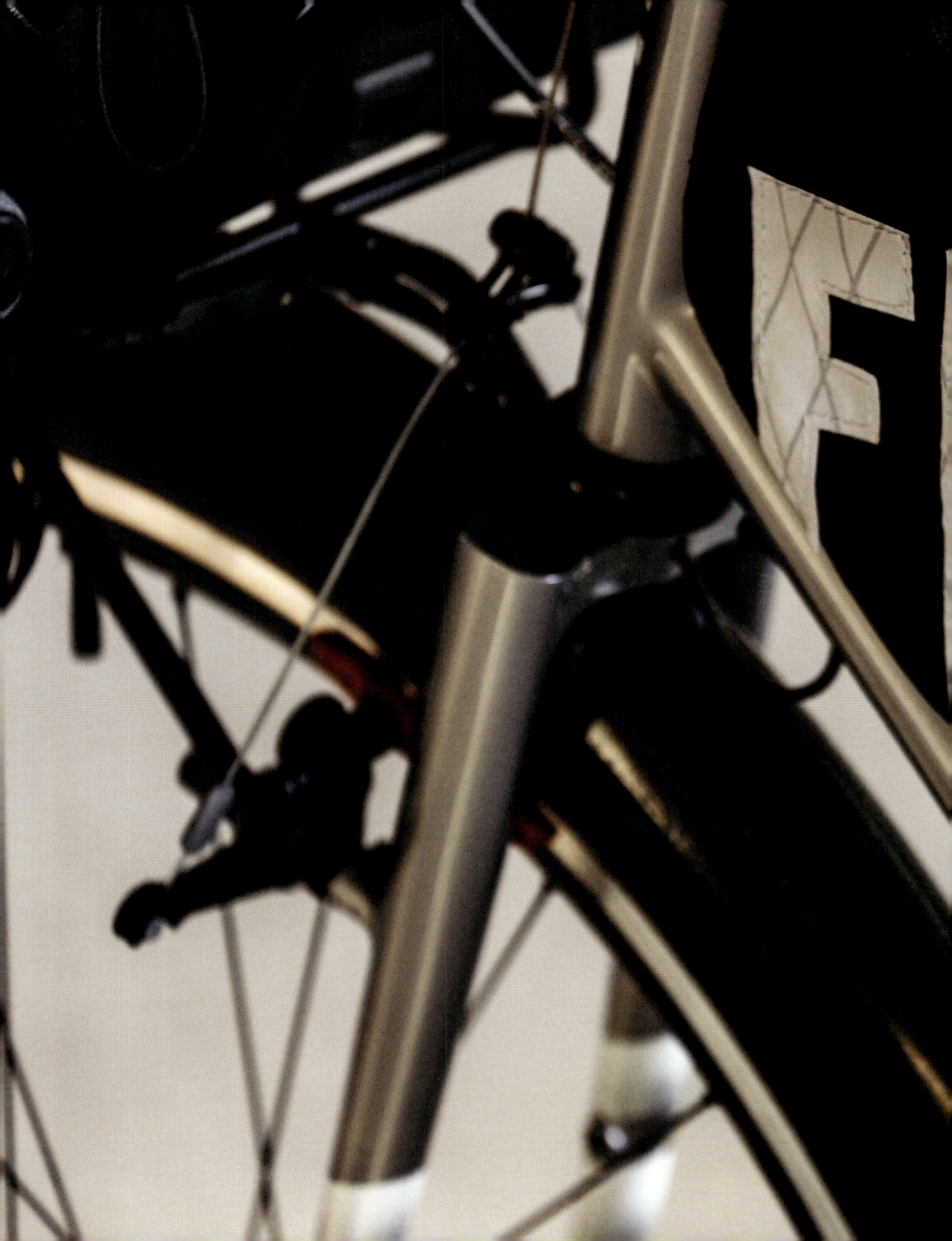

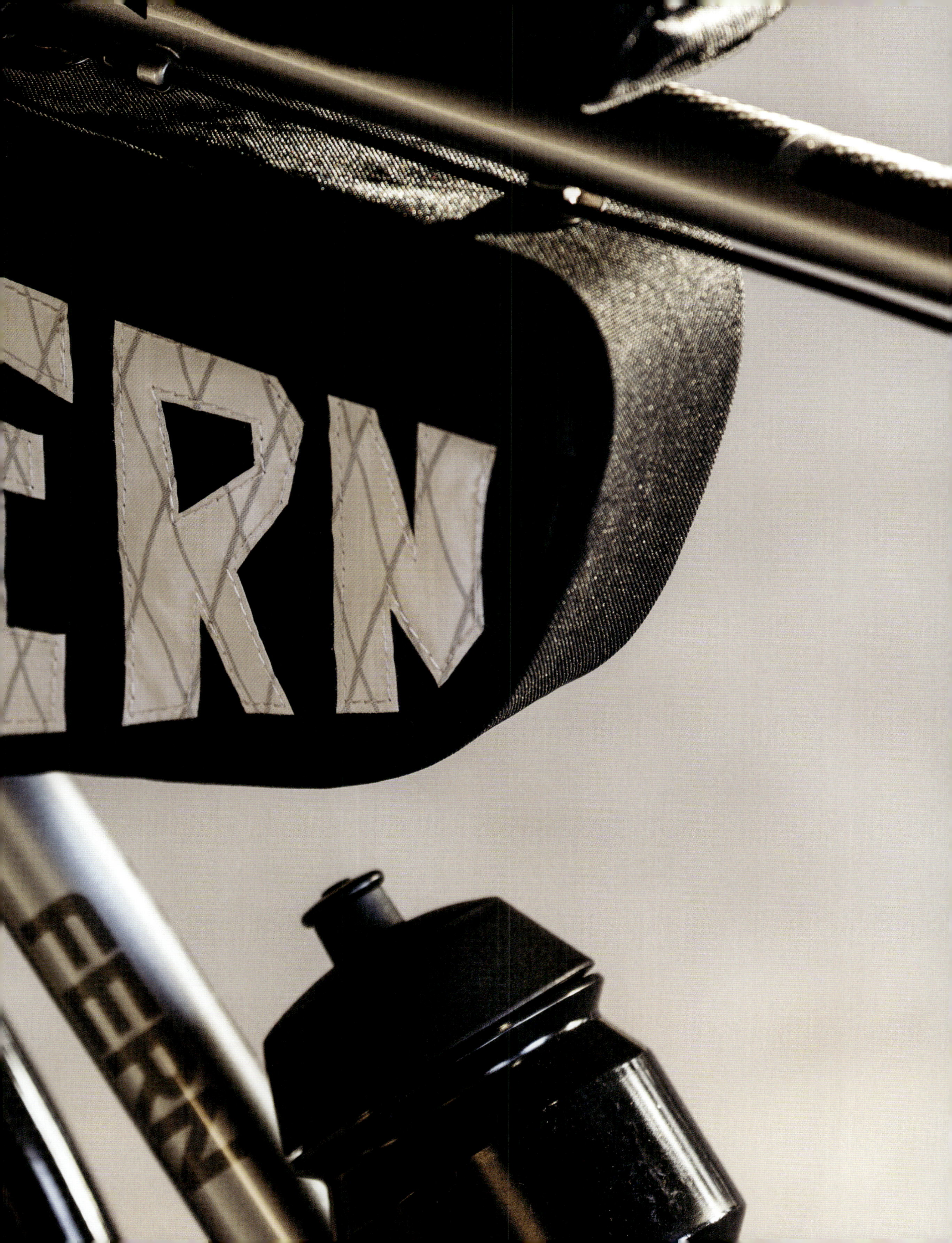

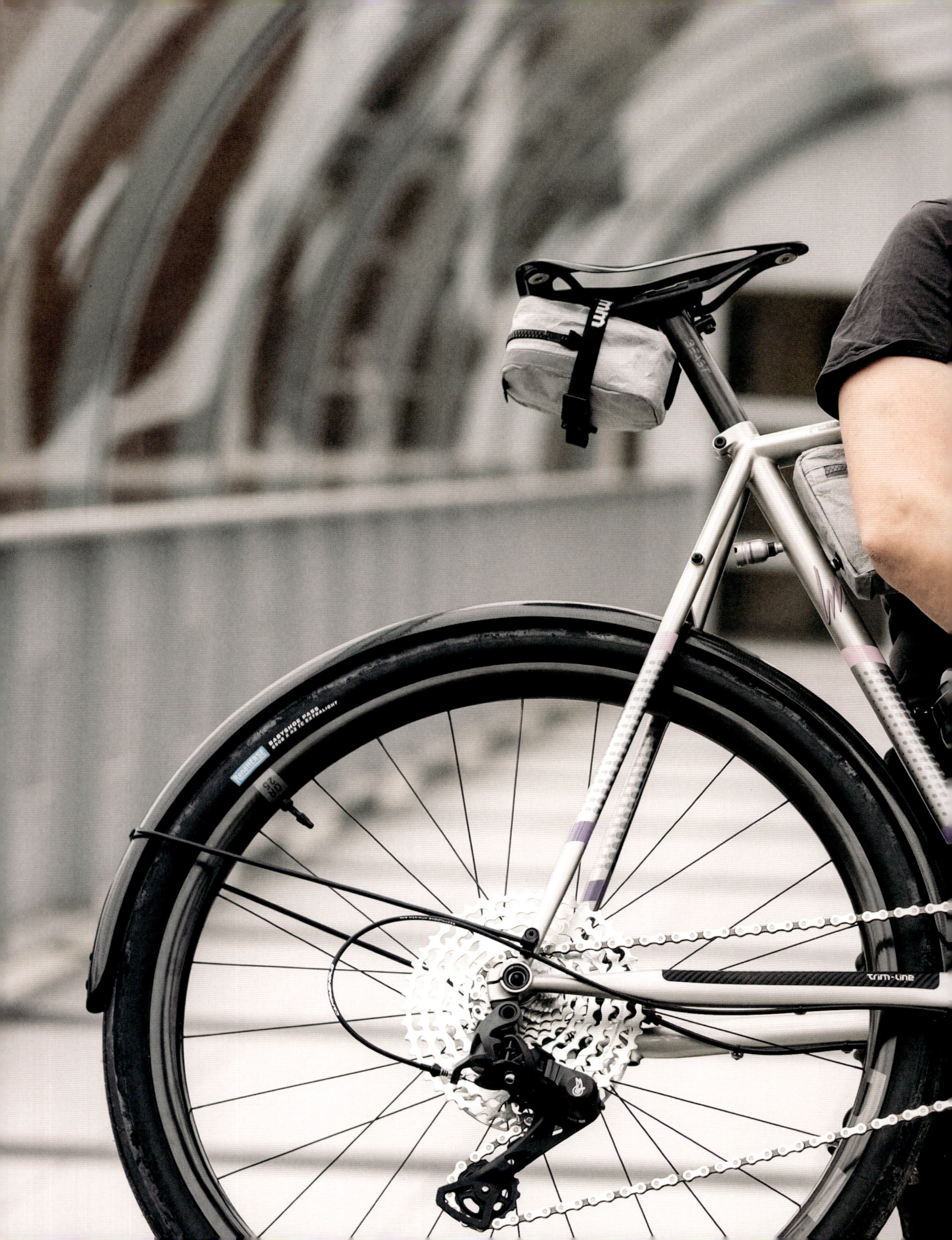
trim-line

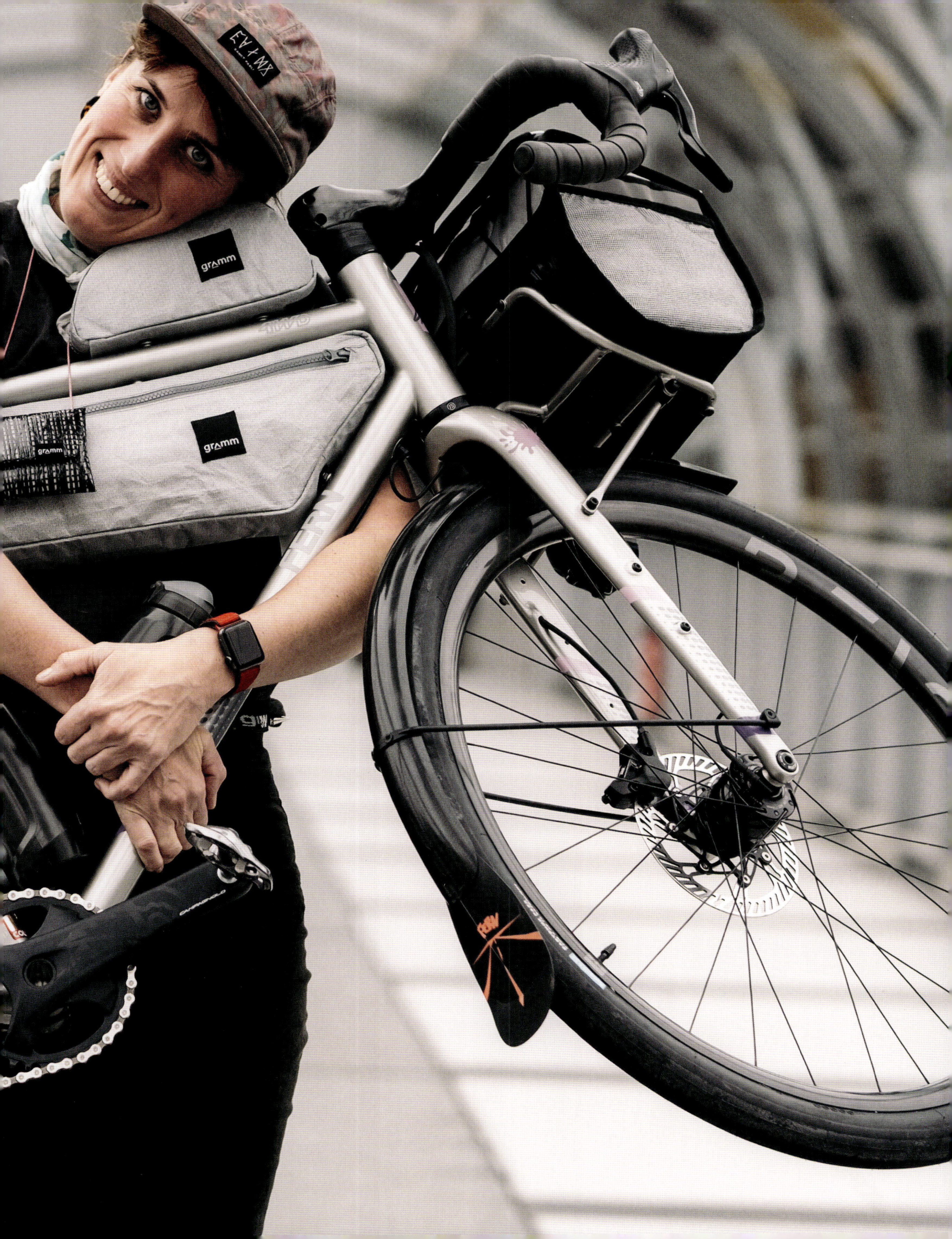
gramm
gramm
gramm

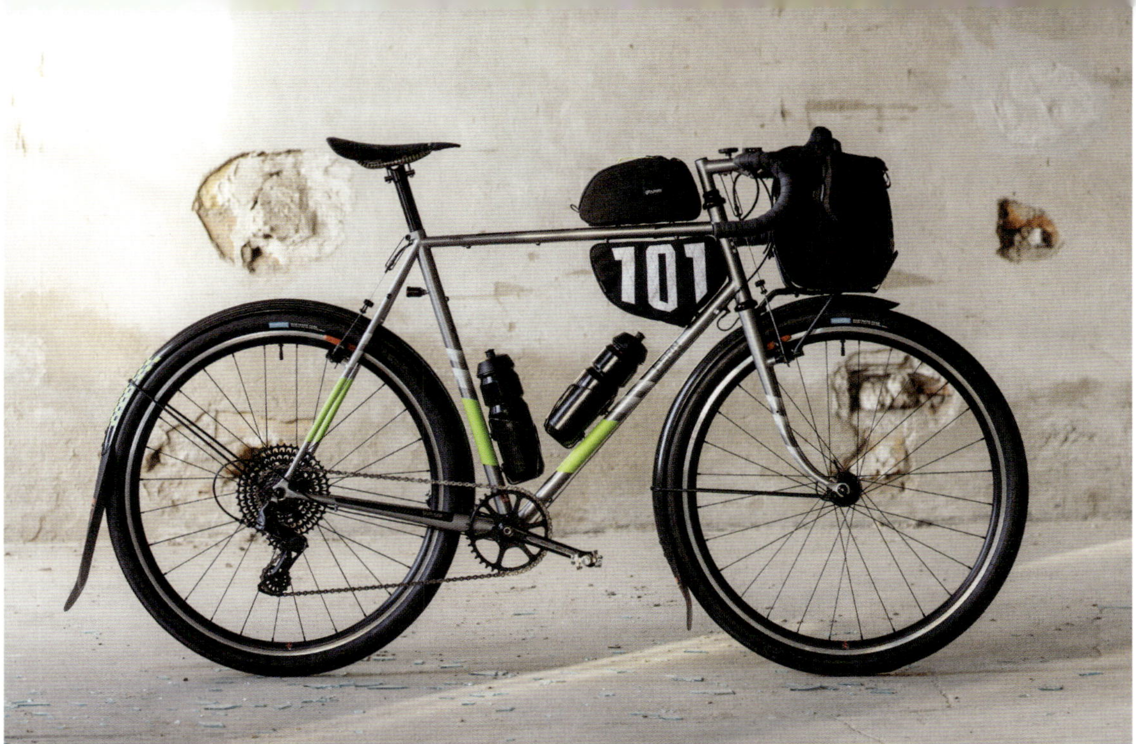
701

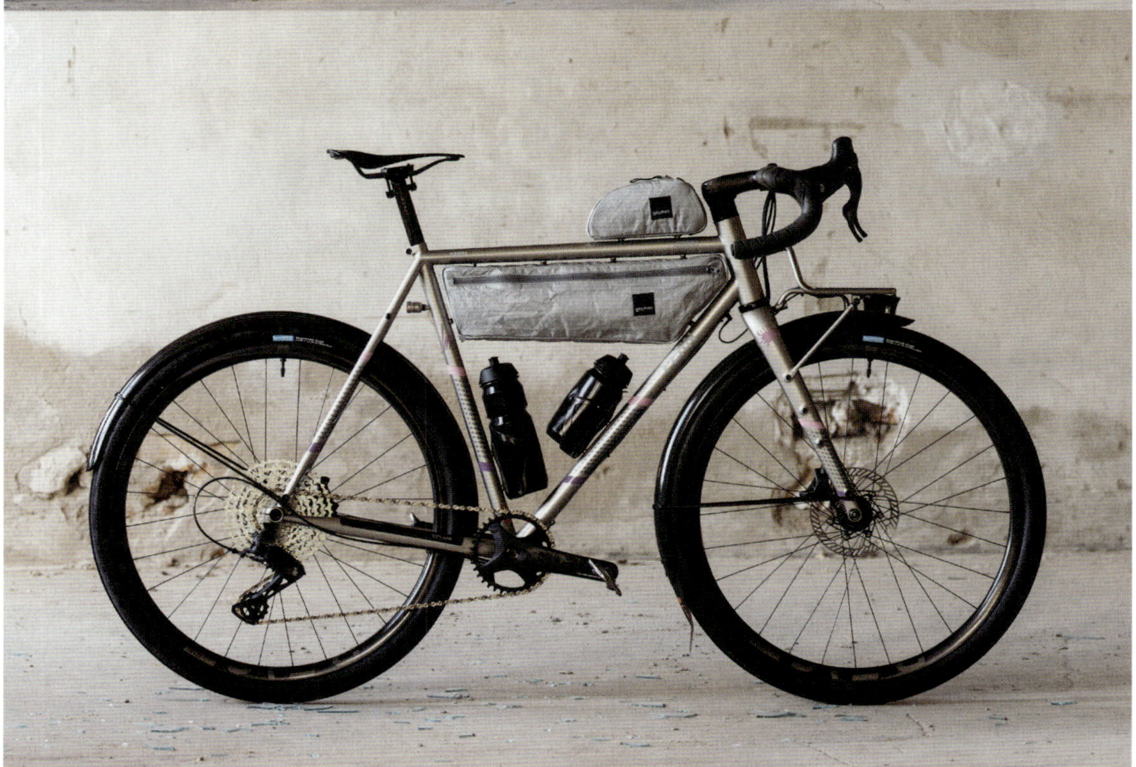

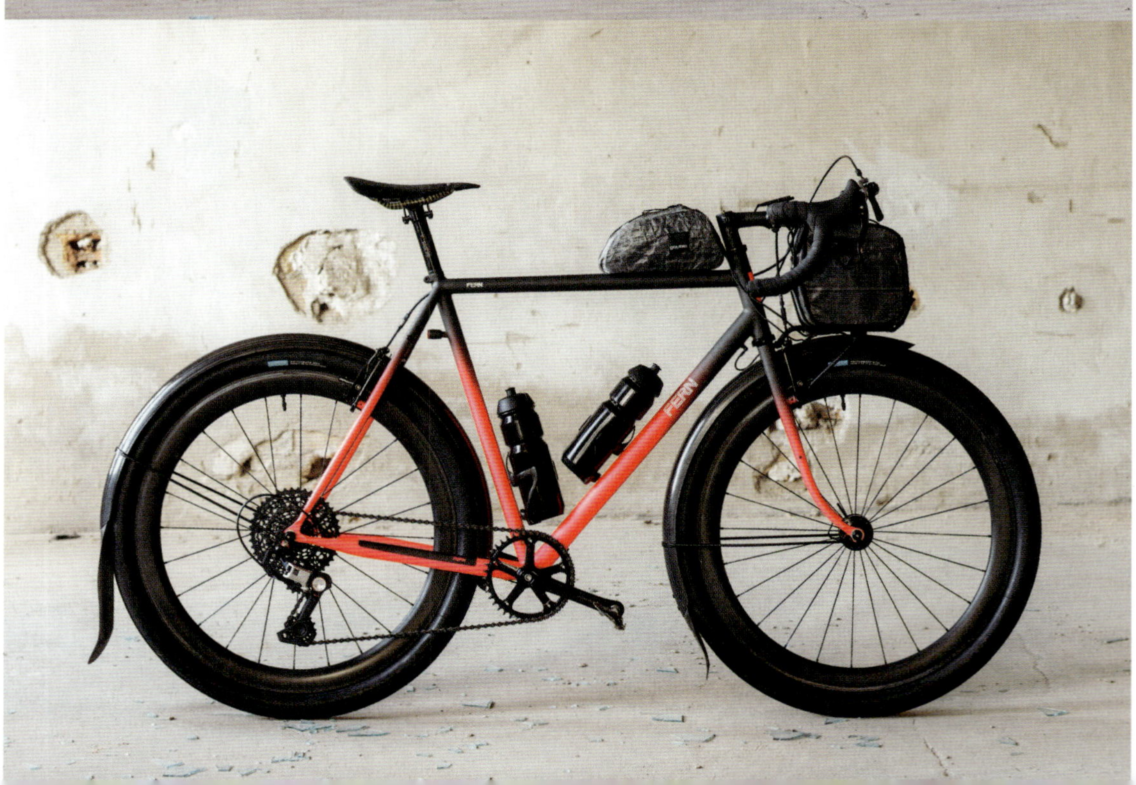
FERN

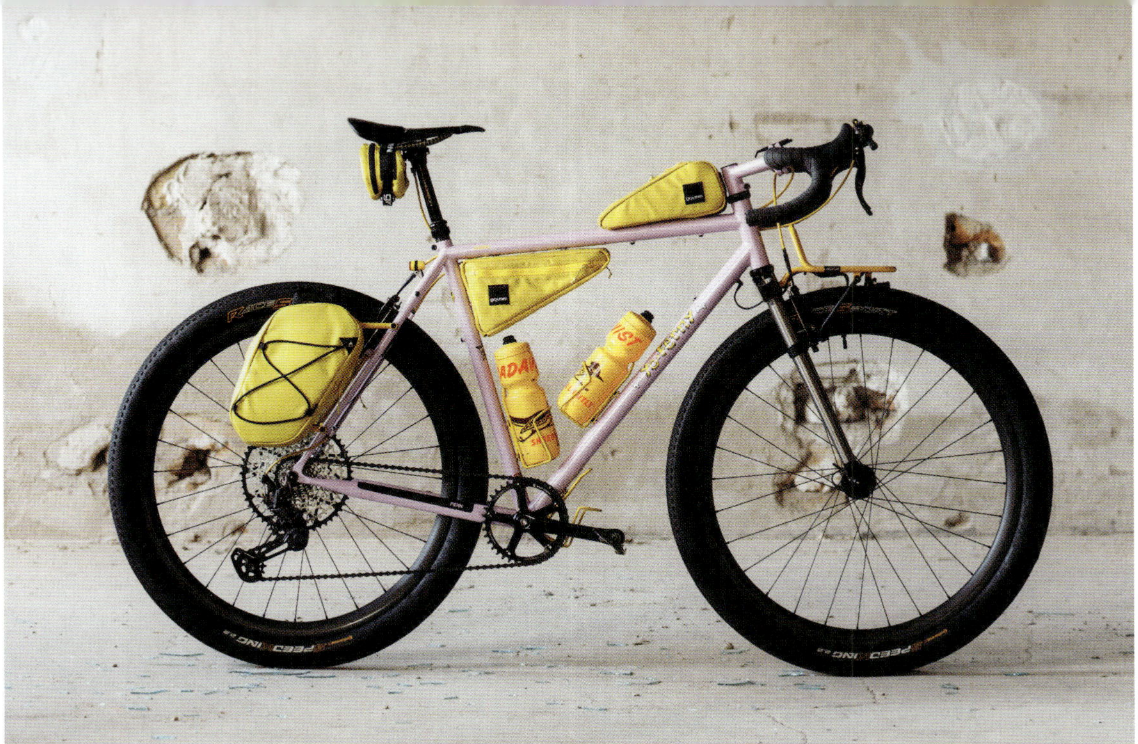

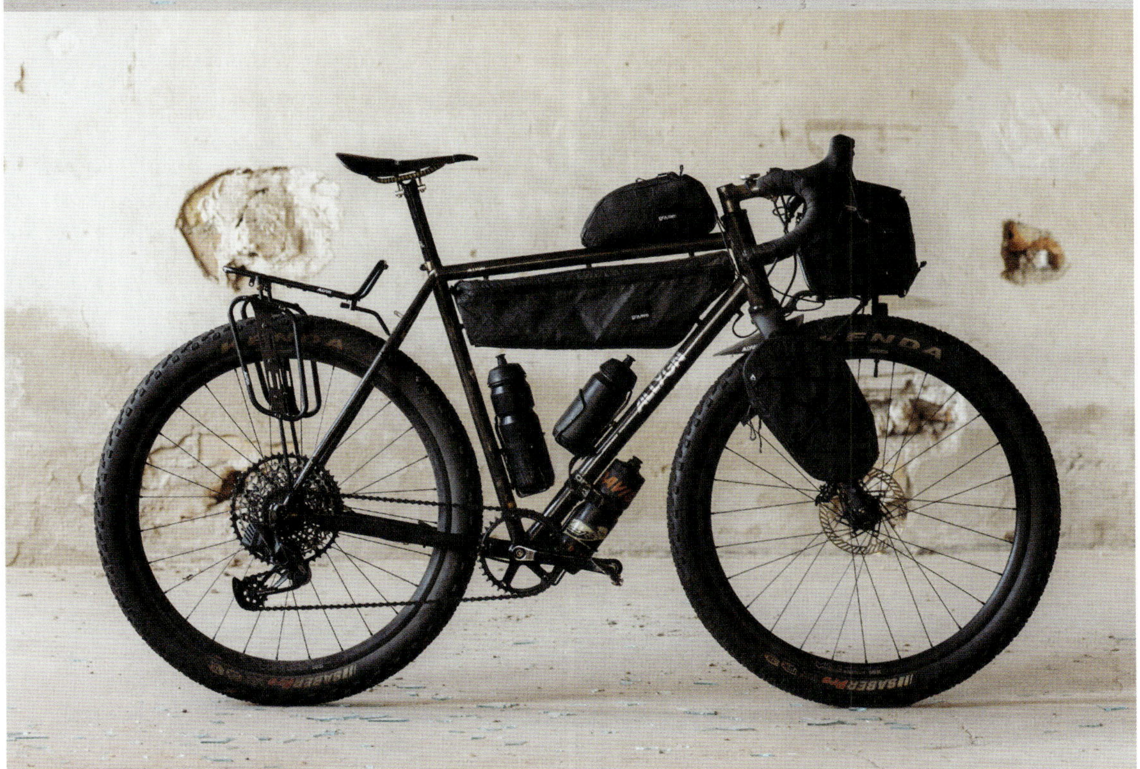

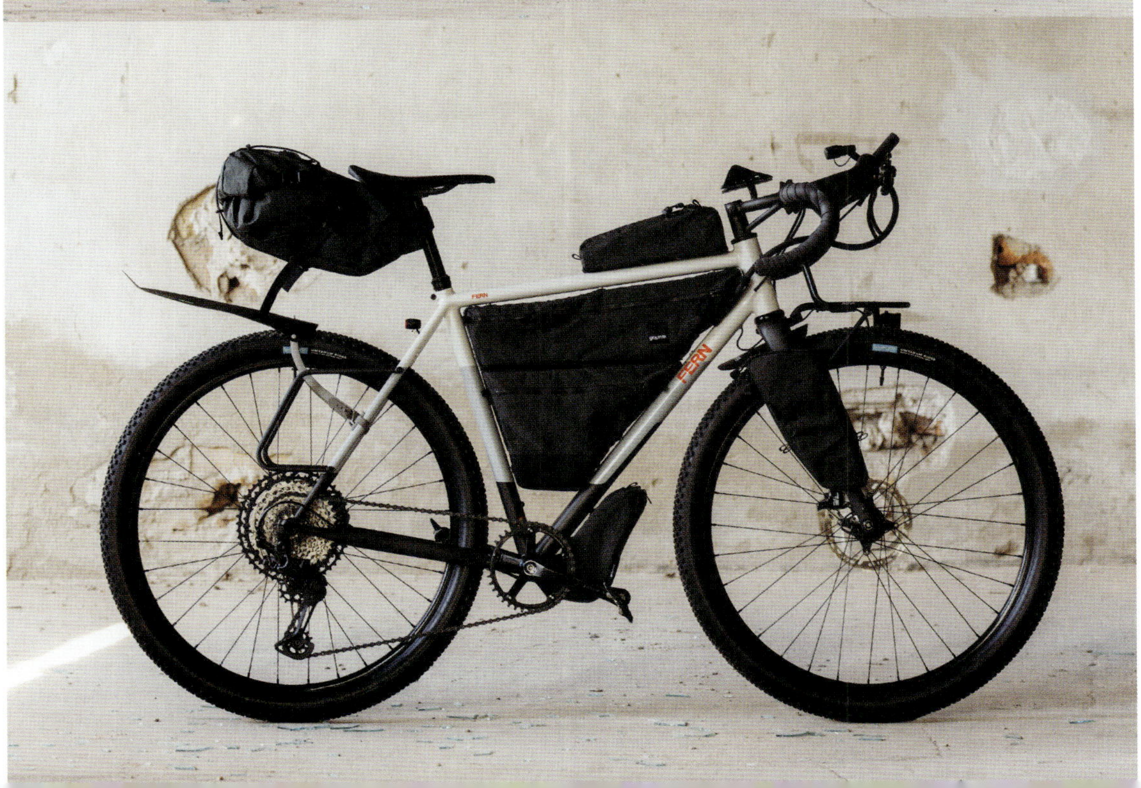

Das ist ein kleiner Auszug aus Flos Schaffen. Kombiniert mit den wunderbaren Taschen von Tine ergeben sich Rad-Gesamtkunstwerke. Fast allen Rädern ist gleich, dass sie für die lange Strecke gebaut sind – ohne die Möglichkeit, eine Gepäcktasche auf dem Frontgepäckträger mitführen zu können, entwirft Flo erst gar nicht. Deutlich ist zu erkennen, dass der Einsatzbereich auf der rechten Seite klar dem Geländeeinsatz zugeordnet werden kann.

Links oben sehen wir Flos aktuellen Ansatz, einen schlanken, klassisch anmutenden Fusion-Randonneur zu bauen: 1" Steuerrohr, geschraubter Klemmvorbau, eine klassische Rahmengeometrie – jedoch geschweißt, gespickt mit einer sehr alten T+A-Kurbel mit einem von ihm dafür entworfenen Kettenblatt, um nur einige Details aufzuzählen ... Wie immer steckt bei ihm der Teufel, oder besser gesagt, die Innovationskraft, im Detail!

Mein absoluter Traum: Nach zwei Jahren ist mein erstes FERN vollendet. Traumhafte Komponenten und eine irre Lackierung von WIDE-WAYS aus Leipzig runden das Beast-Of-The-East ab.

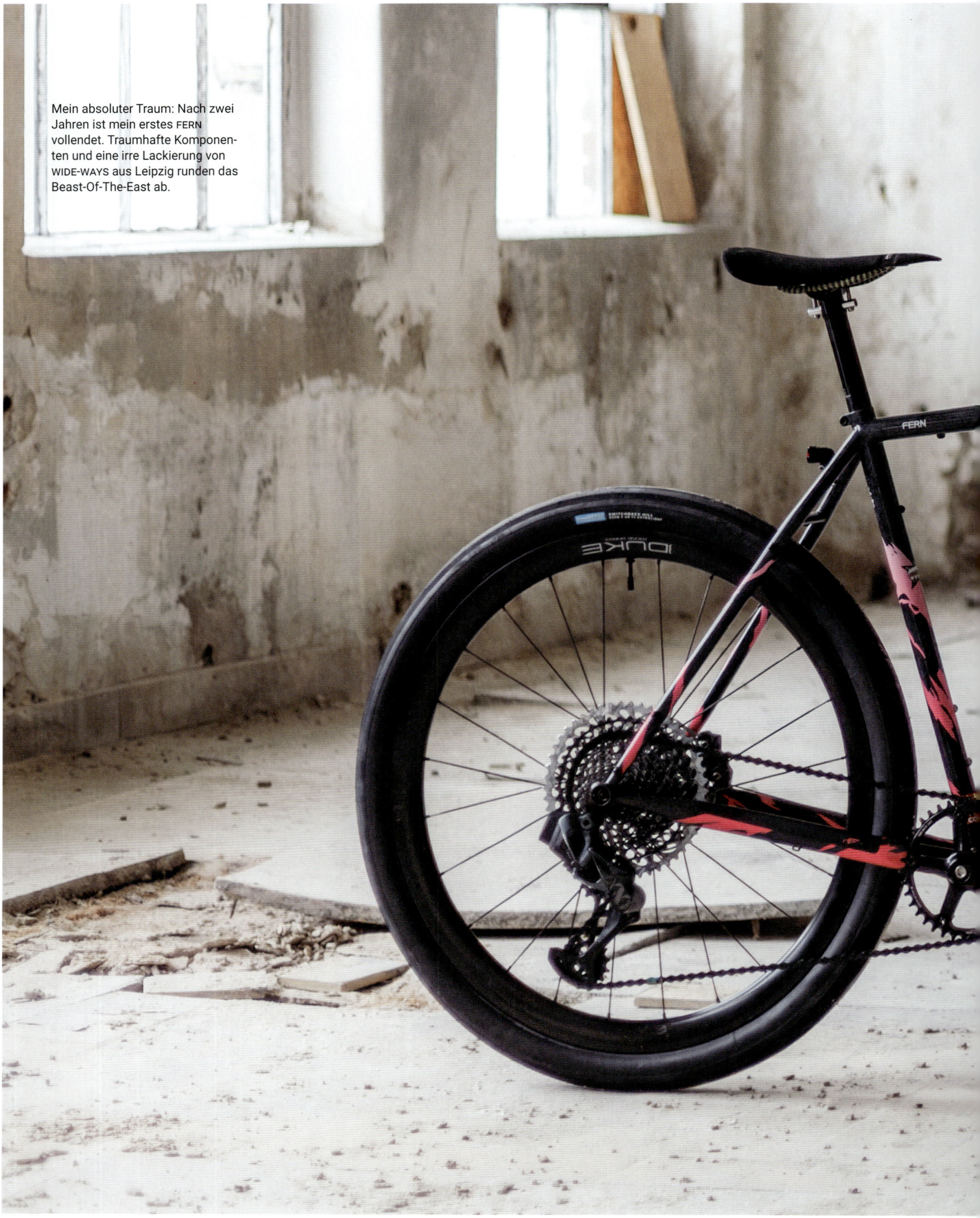

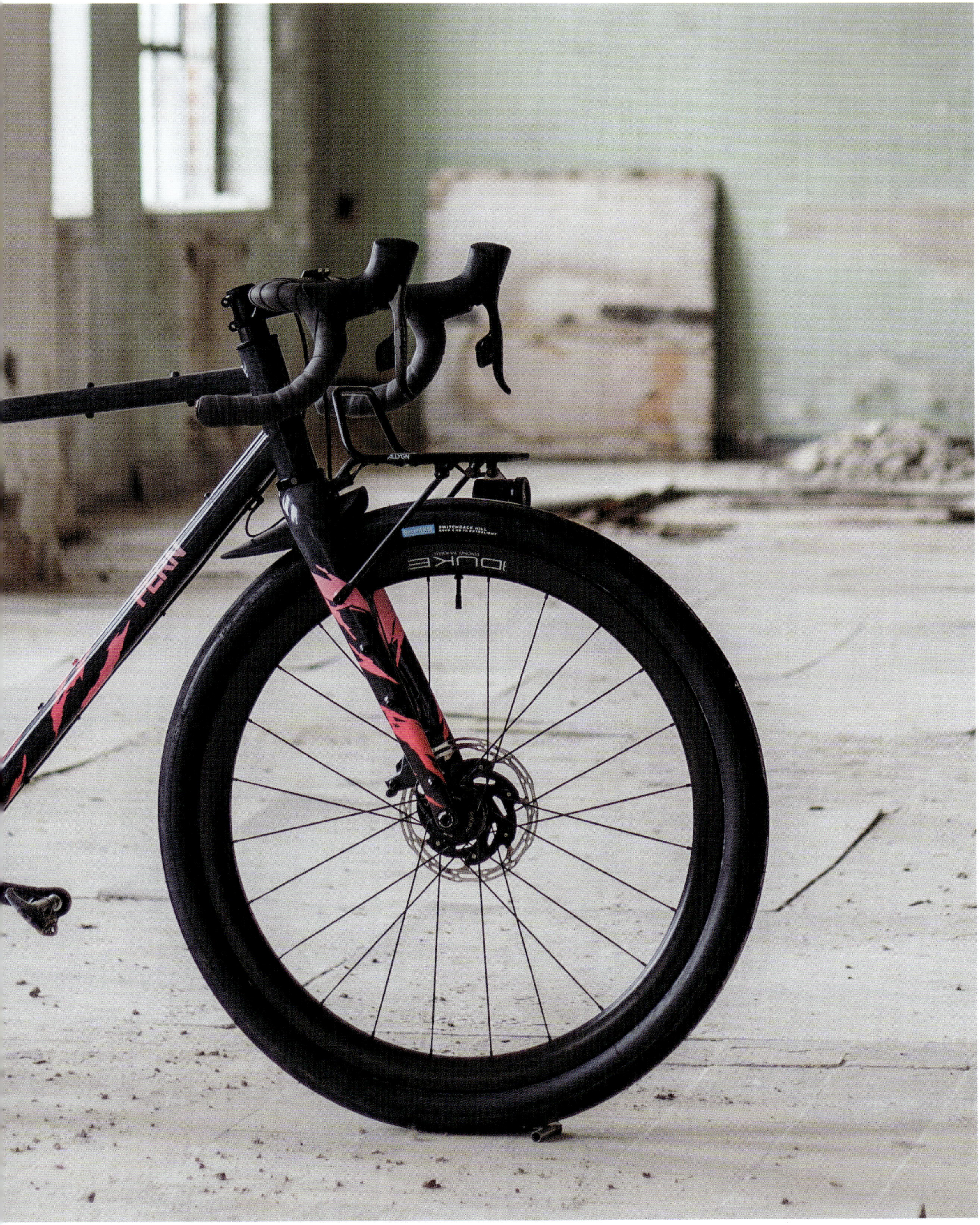

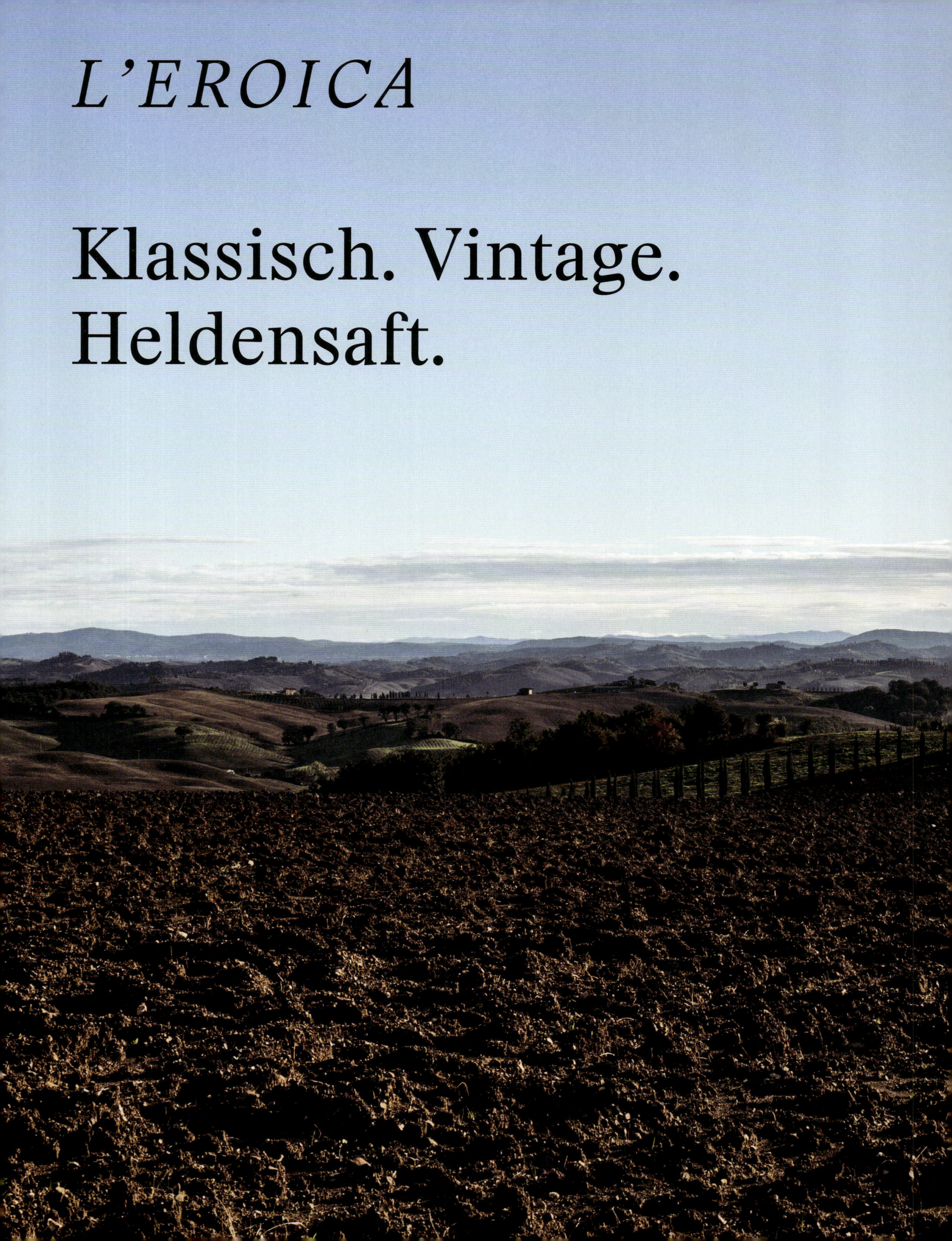
L'EROICA
Klassisch. Vintage.
Heldensaft.

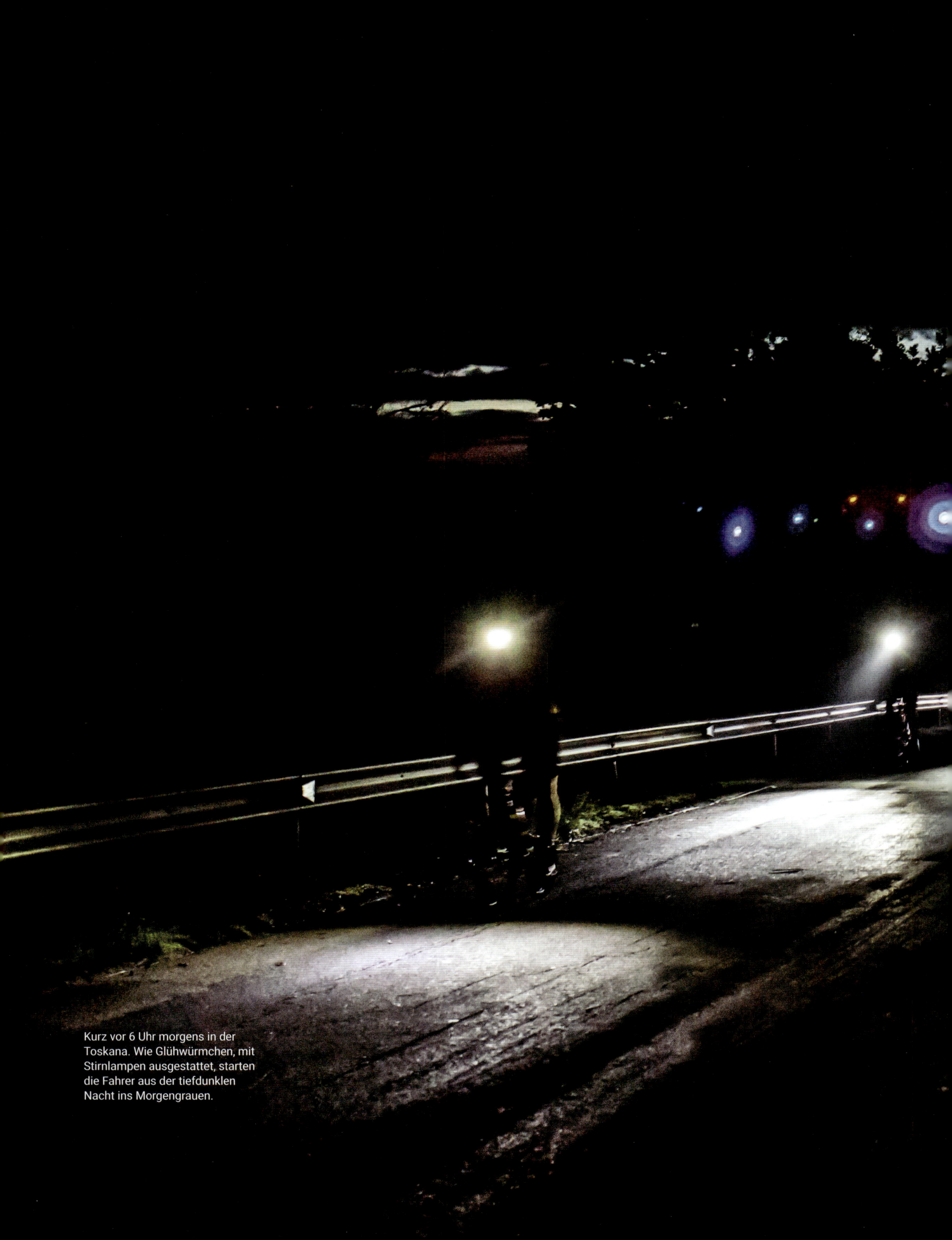

Kurz vor 6 Uhr morgens in der Toskana. Wie Glühwürmchen, mit Stirnlampen ausgestattet, starten die Fahrer aus der tiefdunklen Nacht ins Morgengrauen.

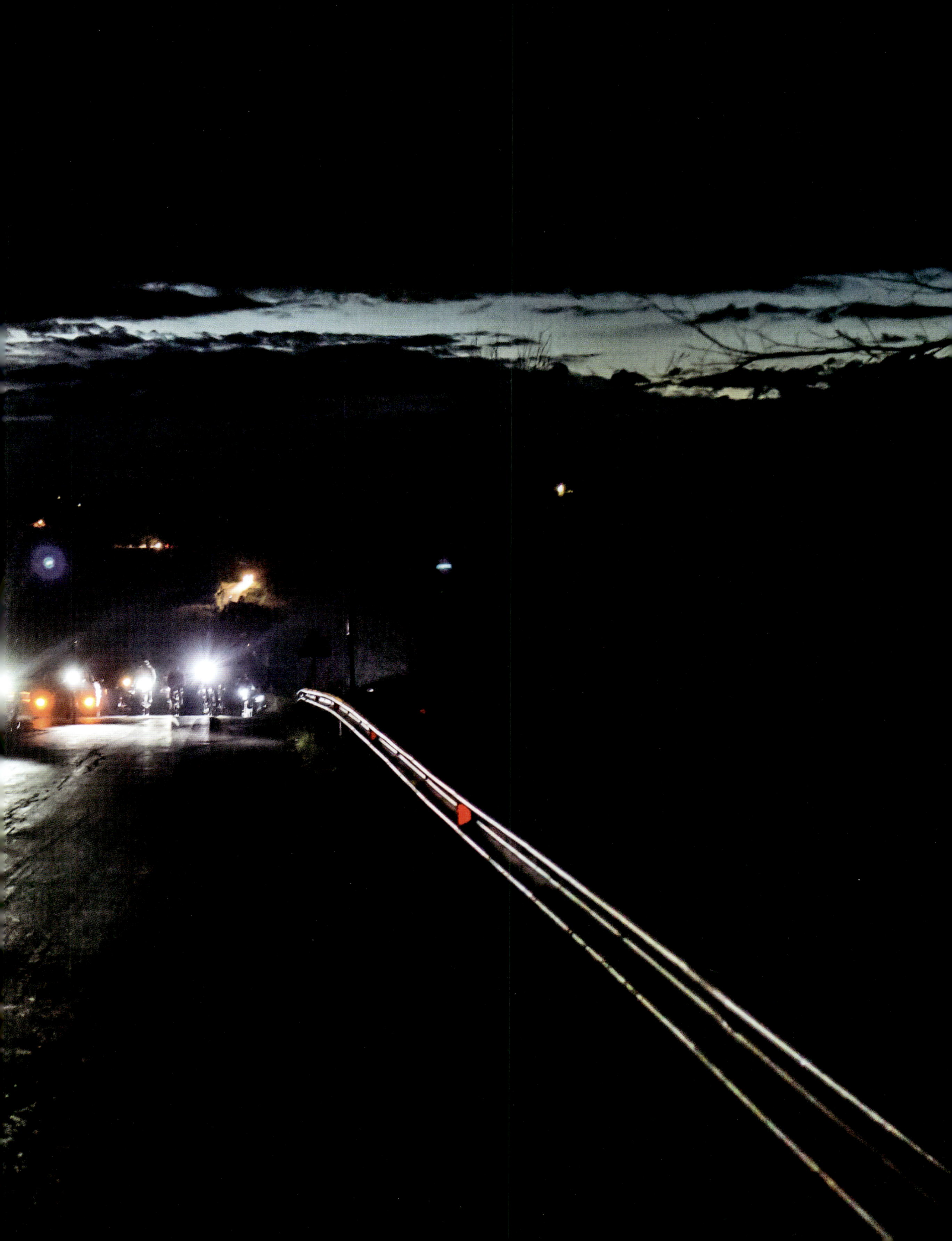

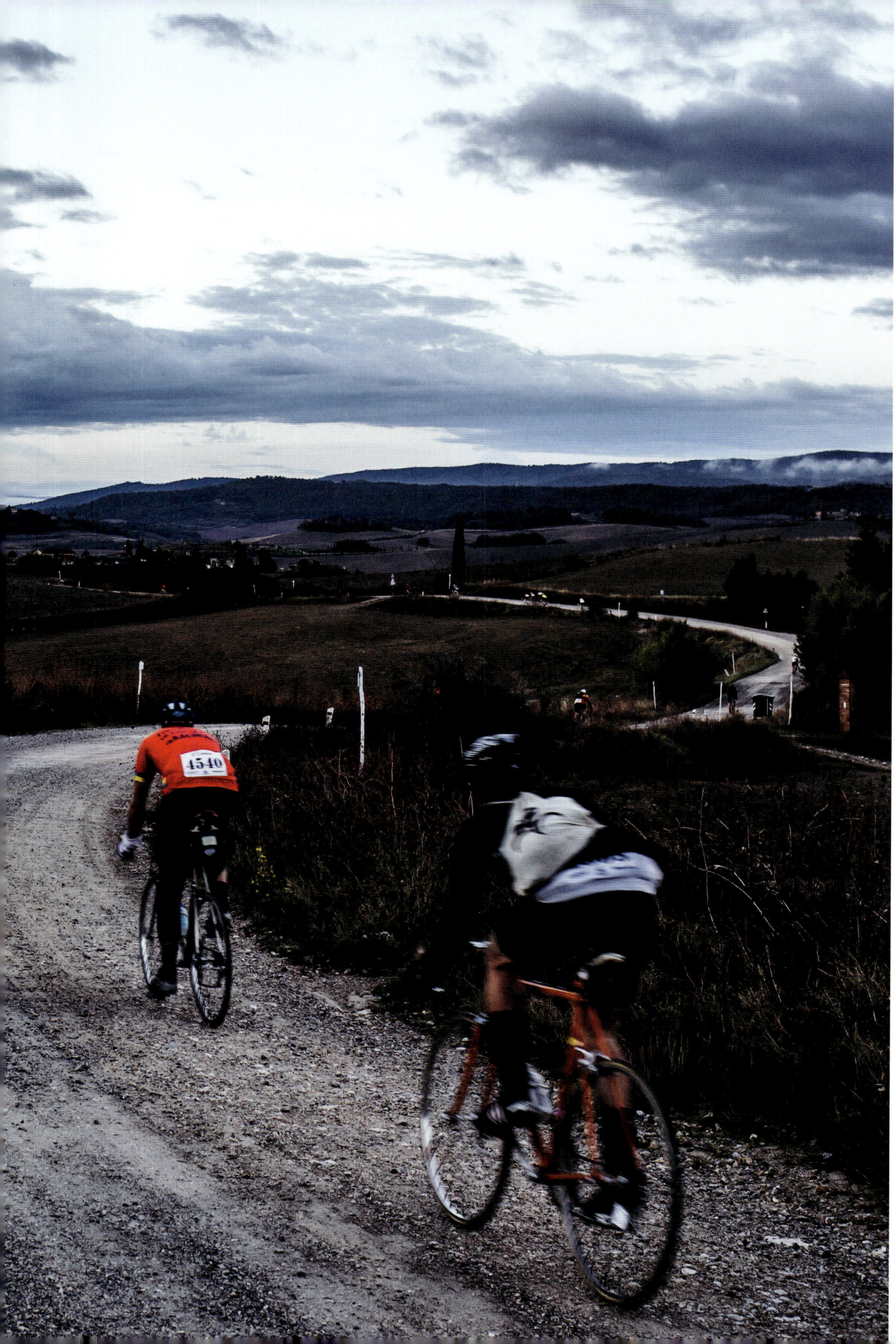
4540

Der Tag verdrängt die Nacht. Die letzten Wolken ziehen dahin. Es wird wieder ein warmer Tag werden im Oktober in der Toskana. Die Piste zieht sich wie ein Band durch die Hügel. Dank des Regens am Vortag wirbeln die Räder auf den STRADA BIANCA kaum Staub auf.

Es regnet seit 18 Stunden. Pausenlos. Ich liege in meinem Ein-Mann-Zelt auf einem kleinen Grünstreifen in einer Buskehre und schaue auf die Uhr: 1 Uhr. Wann soll das Rennen genau beginnen – wirklich um 5 Uhr in der Früh? Nicht im Ernst! Besserung scheint nicht in Sicht. Es schüttet wie aus Kübeln! Dann zieht ein Gewitter auf, und an Schlaf ist gar nicht mehr zu denken. Die Blitze peitschen so heftig, dass man bei geschlossenen Augen das Gefühl hat, an einem Schweiß-Kurs teilzunehmen. Das soll also dieses legendäre Rennen werden! Nee, is klar.

Ich bin bereits vor fünf Tagen aus Wien angereist. Hatte mein Rad zerlegt, in Frischhaltefolie gewickelt, um es als *Gepäckstück* im Zug zu transportieren, und war nach vier Mal Umsteigen endlich in der Toskana gelandet. Genauer gesagt in Gaiole, einem kleinen Ort, in dem 51 Wochen im Jahr wirklich überhaupt nichts los ist. Und dann kommt die L'EROICA. Platt bin ich, aber aufgeregt. In ein paar Tagen will ich mich auf die 209 Kilometer lange Schotterpisten-Strecke wagen. Mit Stahlrahmen, Körbchen statt Klickern und Wäscheleinen am Lenker.

Die frühe Anreise hat sich gelohnt. Die ersten Besucher trudeln langsam ein. Im Ort bauen die Verkäufer ihre Stände auf, begrüßen sich und schauen neugierig, was der Nachbar so im Angebot hat. Sammler können auch noch das eine oder andere Ersatzteil ergattern. Teile, Teile, Teile! Das ist die andere Seite der L'EROICA. Ob es nun Holzfelgen, Vintage-Wolltrikots, Lederschuhe aus allen Epochen, Karbidlampen oder spezielle Komponenten aus der ganzen Welt sind, es gibt scheinbar alles! Der Preis ist dann immer noch ein ganz eigenes Thema.

Umso näher das Rennen rückt, desto voller wird das Dorf. Eine Metamorphose. Der Sportplatz wird zum Campingplatz, der Hauptplatz zur Startrampe. In jedem Hauseingang öffnet plötzlich ein kleines Geschäft. Der Dorfschuster verpasst alten Lederschuhen neuen Halt im Pedal. Der hiesige Radladen macht wohl im ganzen Jahr nicht so viele Notfall-Reparaturen wie in diesen Tagen. Jeder will halt gut vorbereitet starten.

Schließlich fährt man den ganzen Tag lang immer wieder über die weißen Schotterpisten der Toskana. Egal, ob man sich für die 38, 75, 135 oder 209 Kilometer lange Strecke entscheidet. Und was ist für diese Region noch typisch? Richtig, die Berge. Rauf und runter geht es – da ist es schön, wenn die Bremsen zumindest ein bisschen ihren

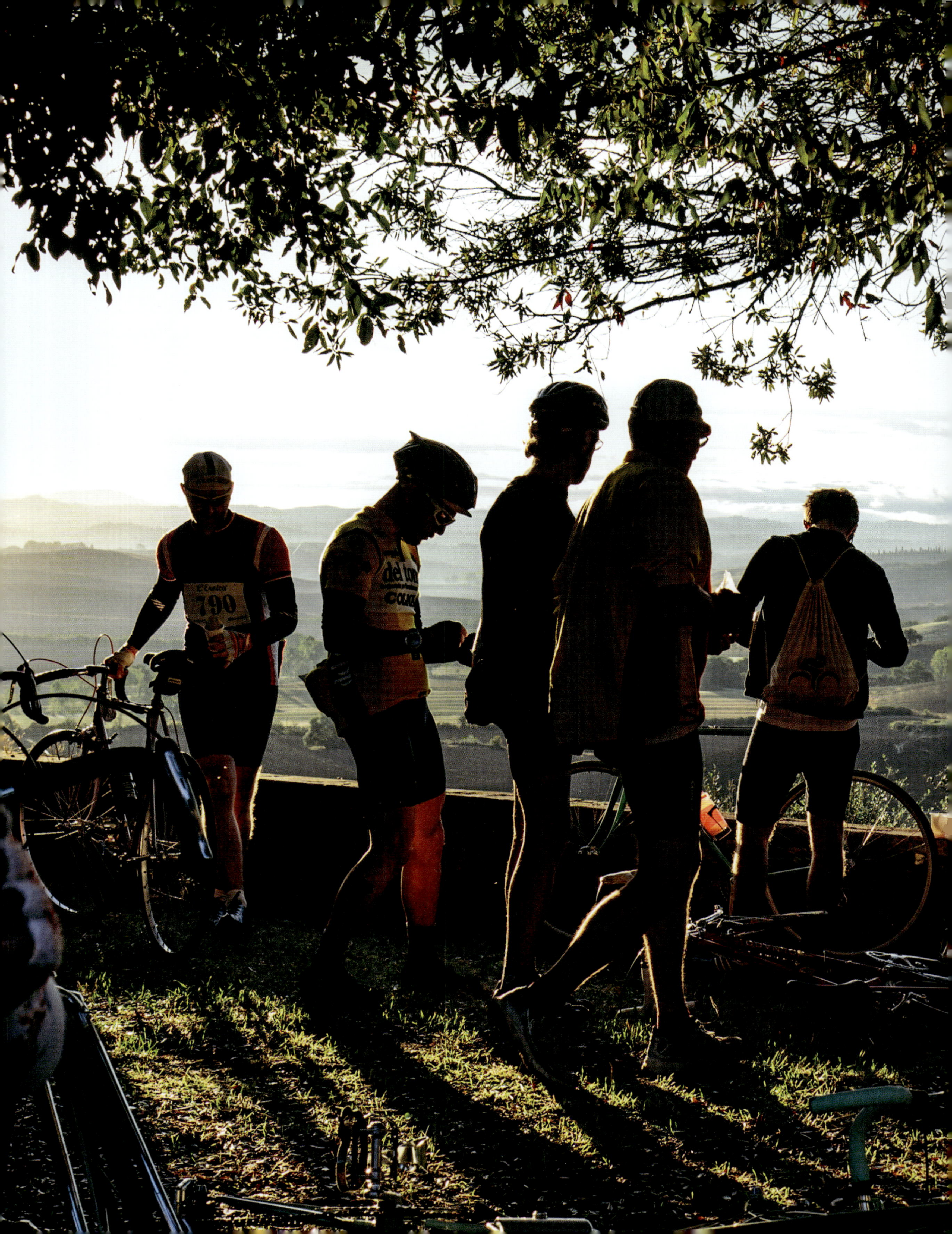
790

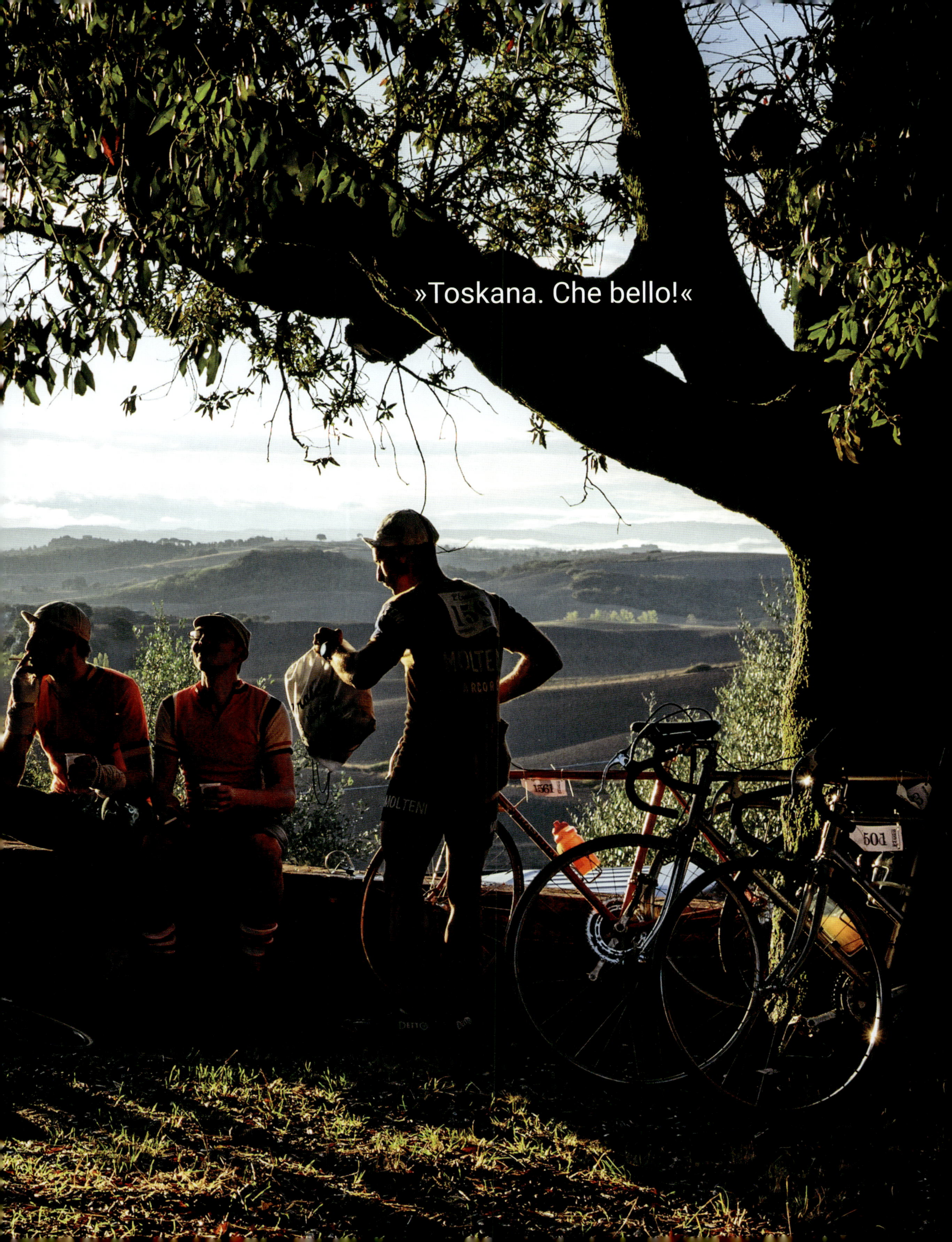

»Toskana. Che bello!«

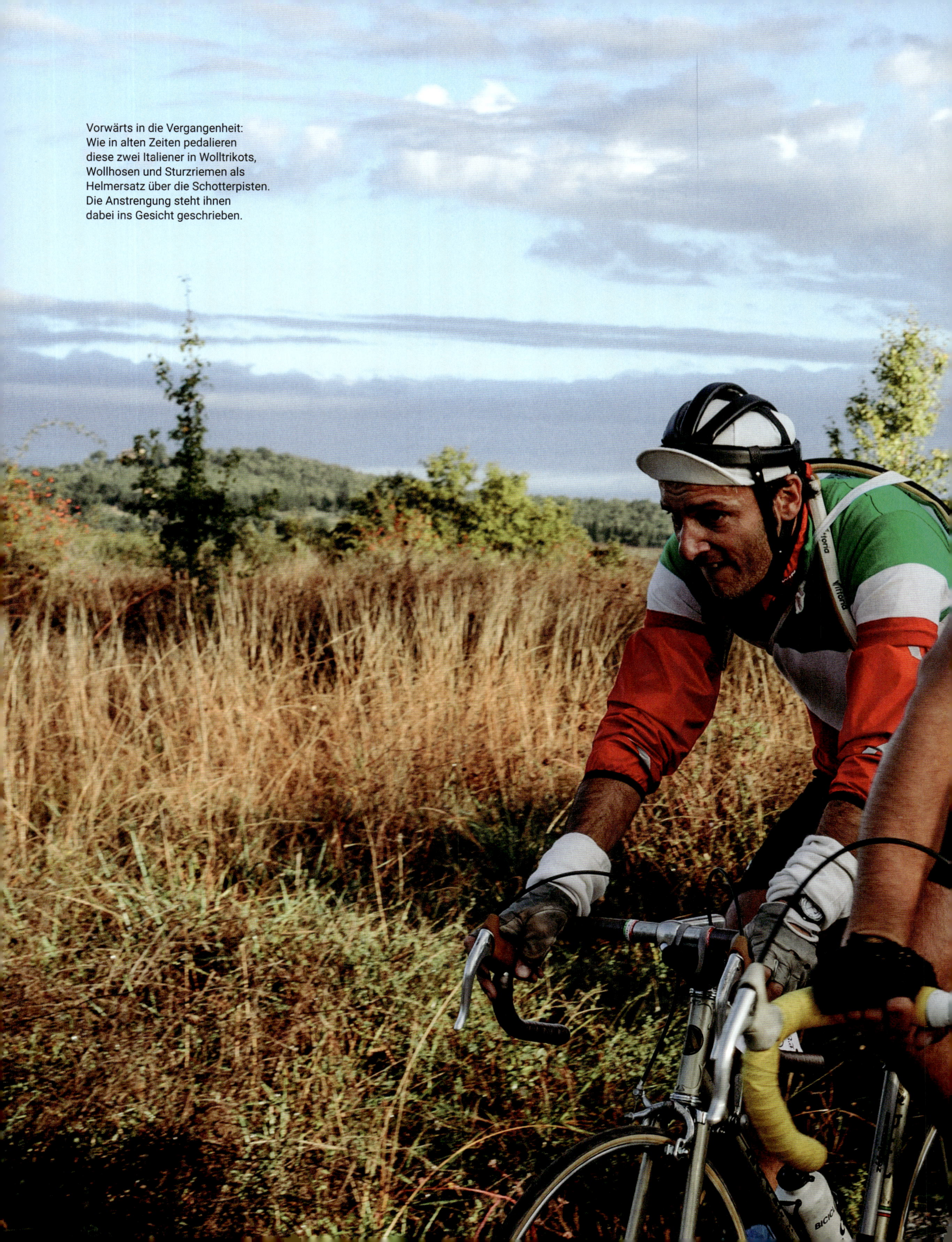

Vorwärts in die Vergangenheit:
Wie in alten Zeiten pedalieren
diese zwei Italiener in Wolltrikots,
Wollhosen und Sturzriemen als
Helmersatz über die Schotterpisten.
Die Anstrengung steht ihnen
dabei ins Gesicht geschrieben.

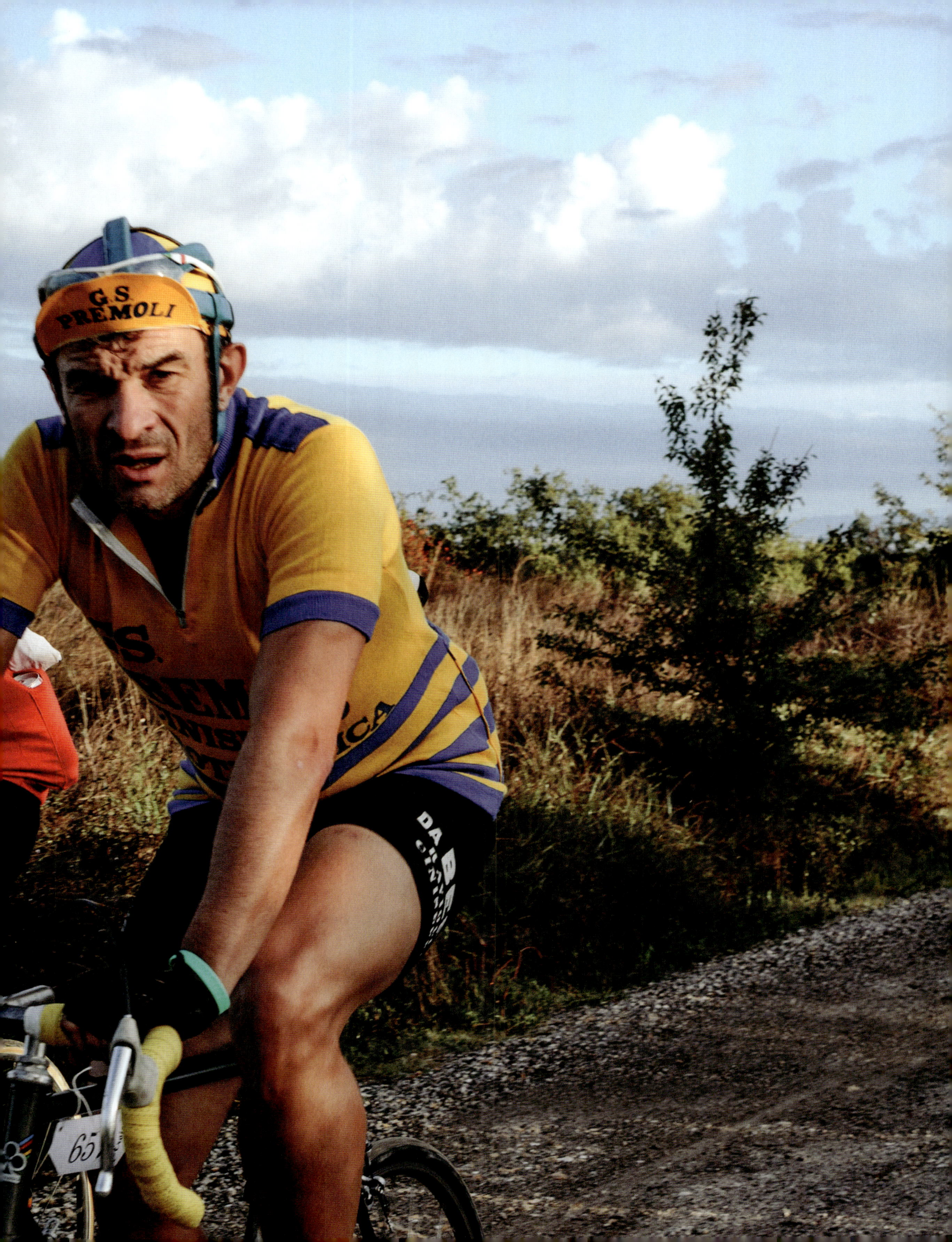
G.S
PREMOLI

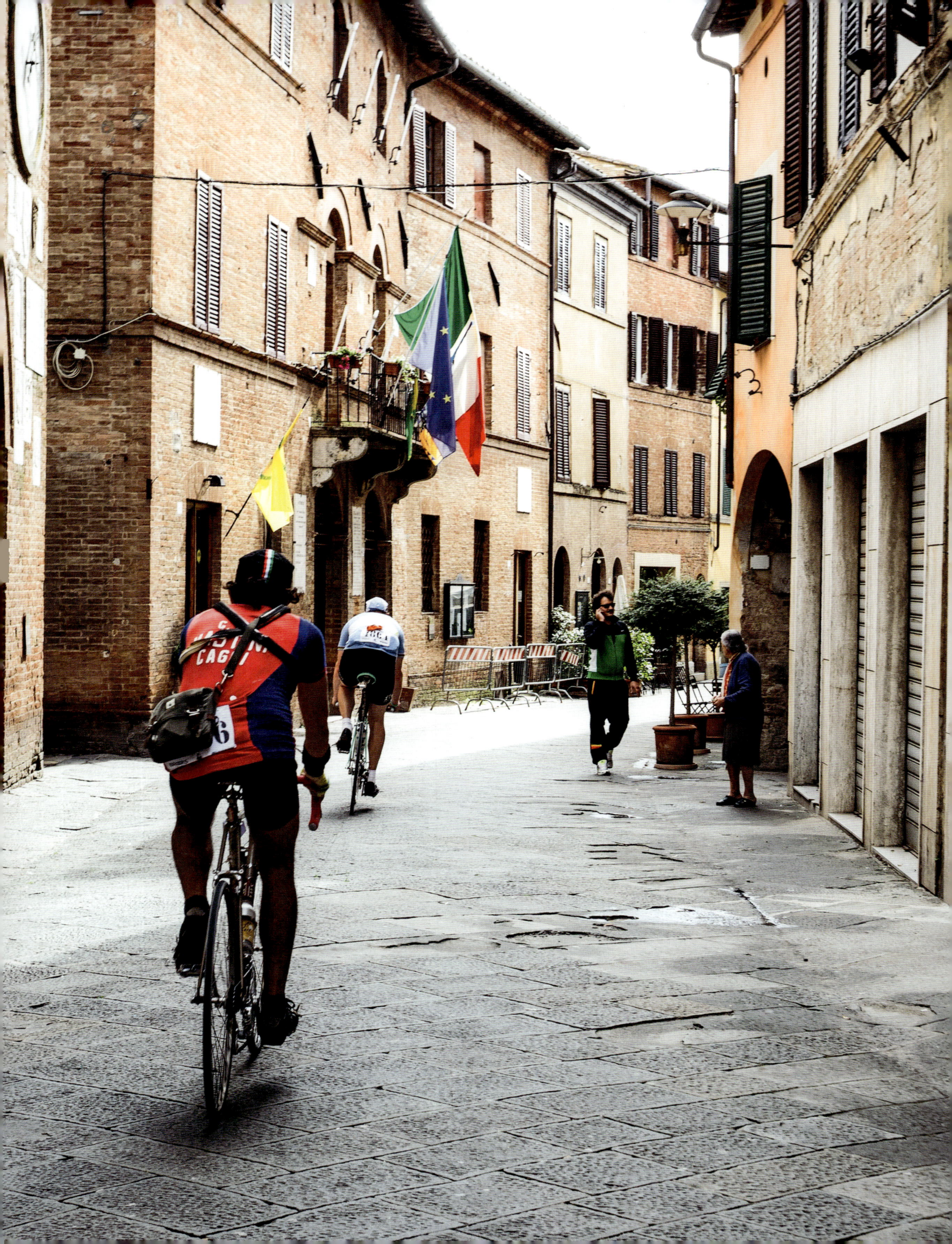

SCHNELLER, SCHNELLER, IMMER SCHNELLER! MIT VOLLGAS ÜBER STOCK UND STEIN, DURCH ABGESPERRTE ITALIENISCHE INNENSTÄDTE

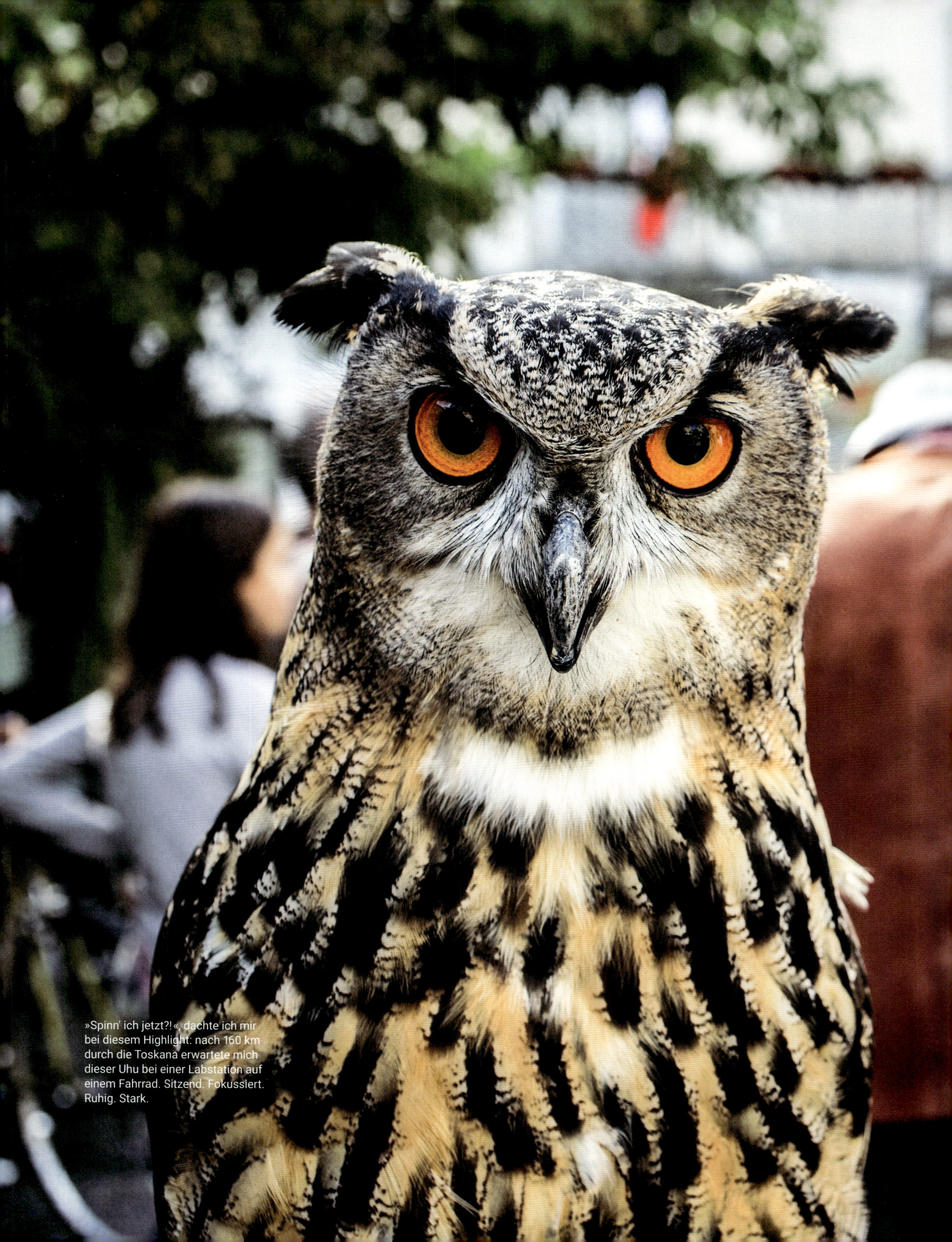

»Spinn' ich jetzt?!«, dachte ich mir bei diesem Highlight: nach 160 km durch die Toskana erwartete mich dieser Uhu bei einer Labstation auf einem Fahrrad. Sitzend. Fokussiert. Ruhig. Stark.

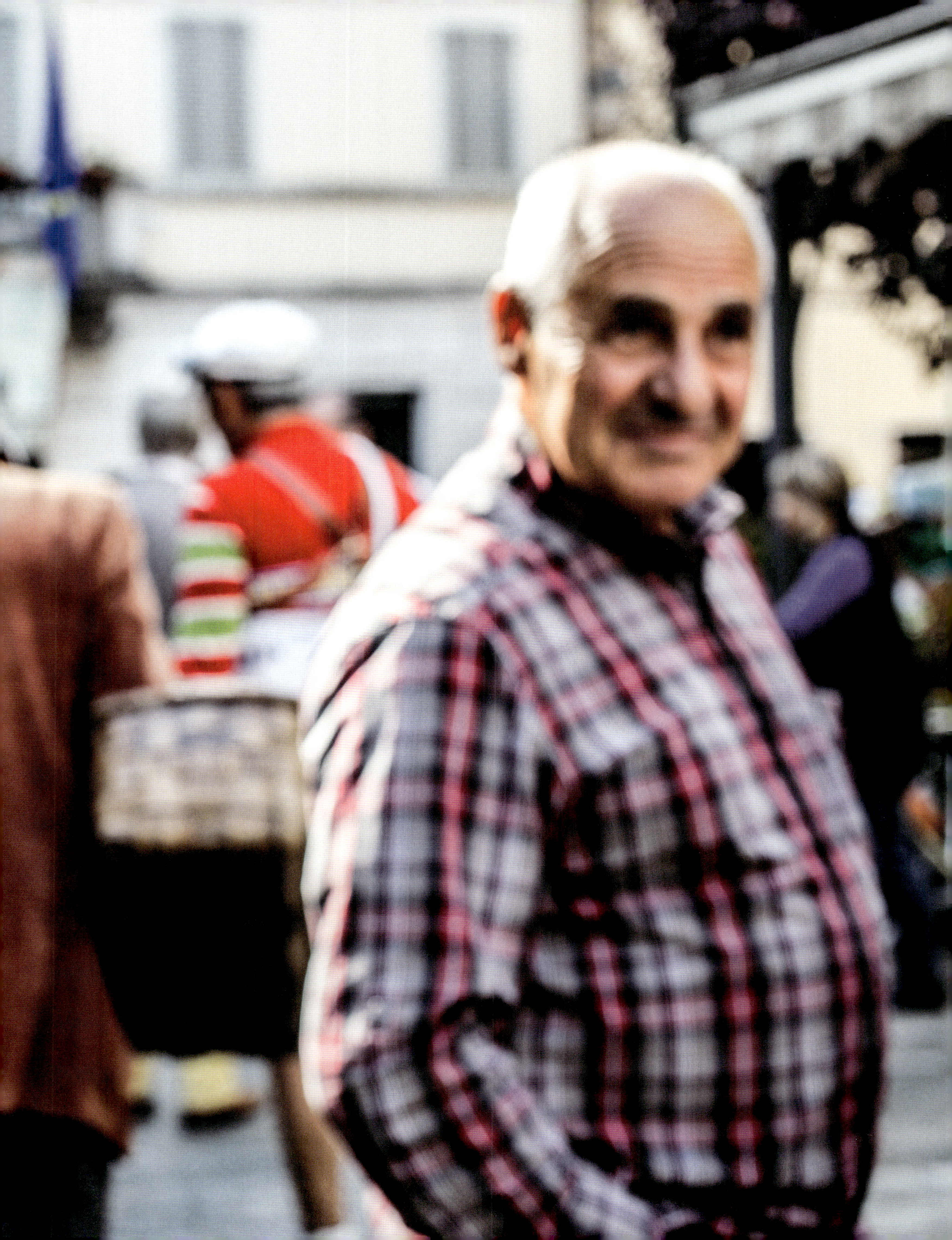

Ist das nicht ... ja, genau, er ist's! Erik Zabel, einer der größten Sprinter aller Zeiten und L'EROICA-Teilnehmer.

Job erledigen und das Vintage-Rad nicht auseinanderfällt. Spaß oder Tortur – die Idee der Veranstalter war es, ein Rennen ins Leben zu rufen, das an die ersten großen Radrennen vor fast einhundert Jahren erinnert. Als die Fahrer sich noch selbst halfen und der Schwächste im Notfall sogar sein Rad abgeben und laufen musste, wenn das der Mannschaft einen Vorteil versprach. Nicht umsonst heißt das Rennen übersetzt *Die Heldenhafte*. So fährt man auch heute mit Ersatz-Schlauchreifen um den Hals und der Trockenwurst am Rücken befestigt durch die malerische Weinregion.

Die L'EROICA ist aber weniger ein Rennen als eine Genuss- und Erlebnistour. Statt Energieriegel erwartet die Fahrer an den Stationen gezuckertes Brot, Käse, Schinken und Ribollita, ein traditioneller Kohleintopf, der über dem Feuer erhitzt und von freiwilligen Helfern in Trachten verteilt wird. Um das ganze italienische Ur-Raderlebnis komplett zu machen, trinkt man dazu Chianti statt isotonischer Drinks. Und auch die Tage vor dem Start sind dem Genuss verschrieben: Die Bäckerei des Ortes kommt mit ihren wundervollem Gebäck kaum mehr hinterher und auf die Pizza am Abend wartet man locker bis zu zwei Stunden. Kein Problem – Gespräche mit Engländern, Amerikanern und den vielen Italienern entwickeln sich überall.

Jetzt geht es los. Aus dem Zelt hinaus in die Nacht. Frisch ist es, doch zum Glück hat der Regen aufgehört. Es ist zwar alles klatschnass, aber das trocknet in der toskanischen Herbstsonne auch wieder. Aufgeregt, halb zitternd schälen sich die Teilnehmer aus ihren Zelten und zwängen sich in ihre Outfits. Der Marktplatz mitten in Gailoe gleicht um 4.40 Uhr einem bunten Ameisenhaufen. Hunderte Radfahrer von Anfang 20 bis Mitte 80 stehen dicht an dicht. Doppelte Espresso werden gekippt wie Schnäpse; anscheinend wird nicht nur mir erneut bewusst, dass es nun Realität wird – 209 Kilometer liegen vor uns. Meine neuen Radlerfreunde und ich sprechen uns Mut zu, während die letzten Vorbereitungen getroffen werden.

Sitzt die Mütze? Hängt das Startnummerschildchen aus Pappe korrekt am Rahmen? Ist die Bremse auch richtig festgezogen? Wo ist die Brevet-Karte? Nur wer am Ende trotz der wenigen Wegweiser immer richtig gefahren ist und jeden Stempel vorweisen kann, bekommt auch seinen Lohn. Einen Kuchen und eine Flasche Chianti. Salute!

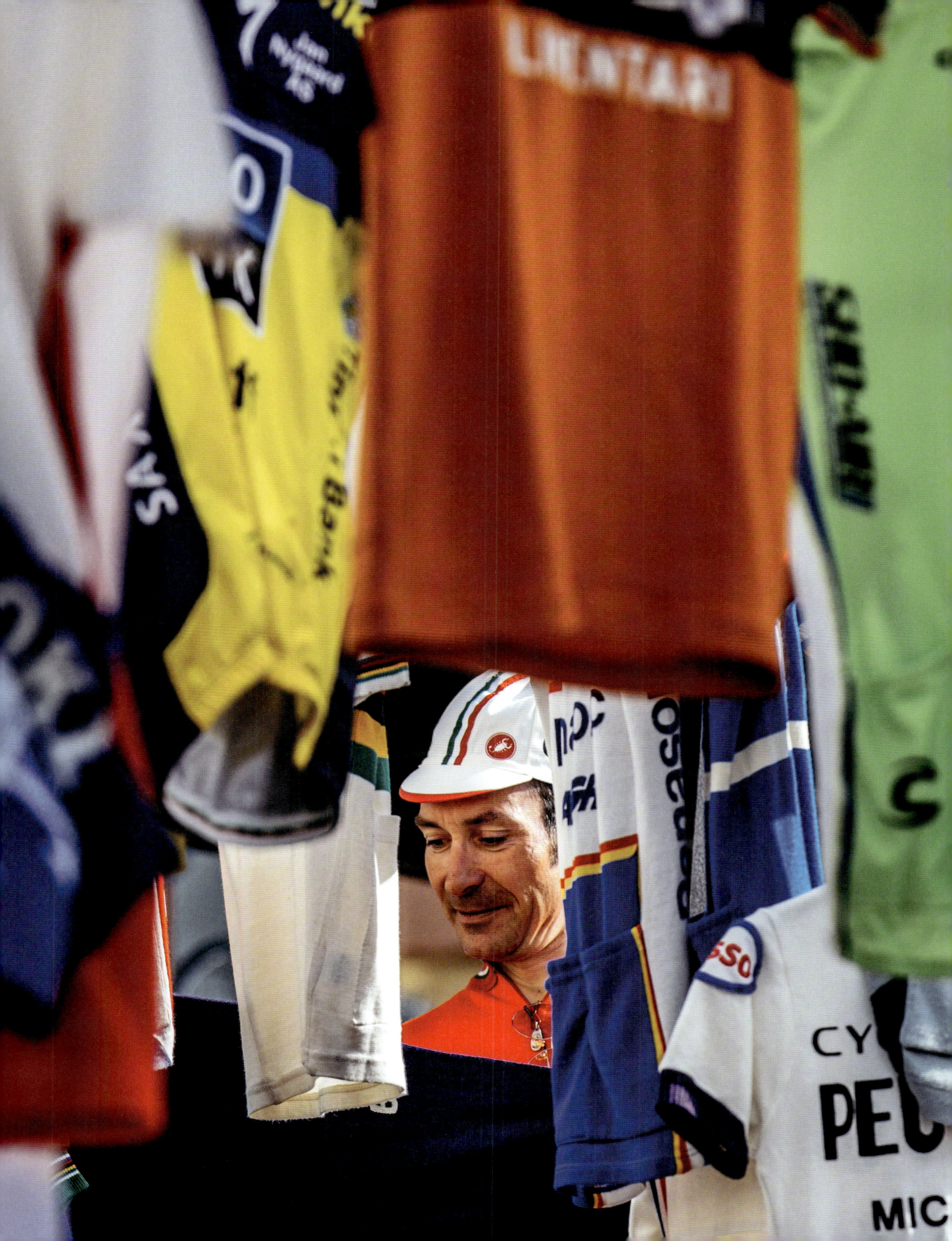

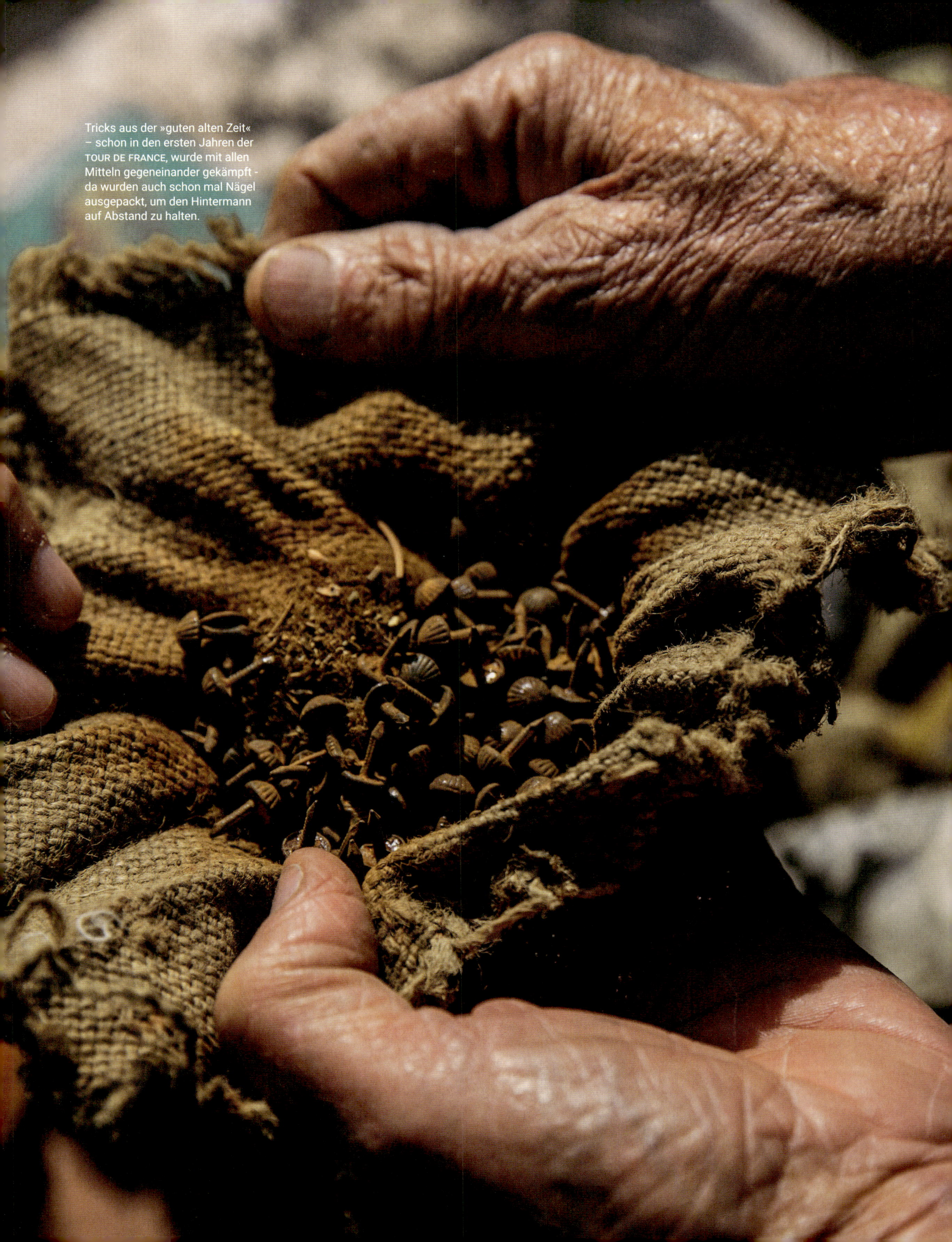

Tricks aus der »guten alten Zeit« – schon in den ersten Jahren der TOUR DE FRANCE, wurde mit allen Mitteln gegeneinander gekämpft - da wurden auch schon mal Nägel ausgepackt, um den Hintermann auf Abstand zu halten.

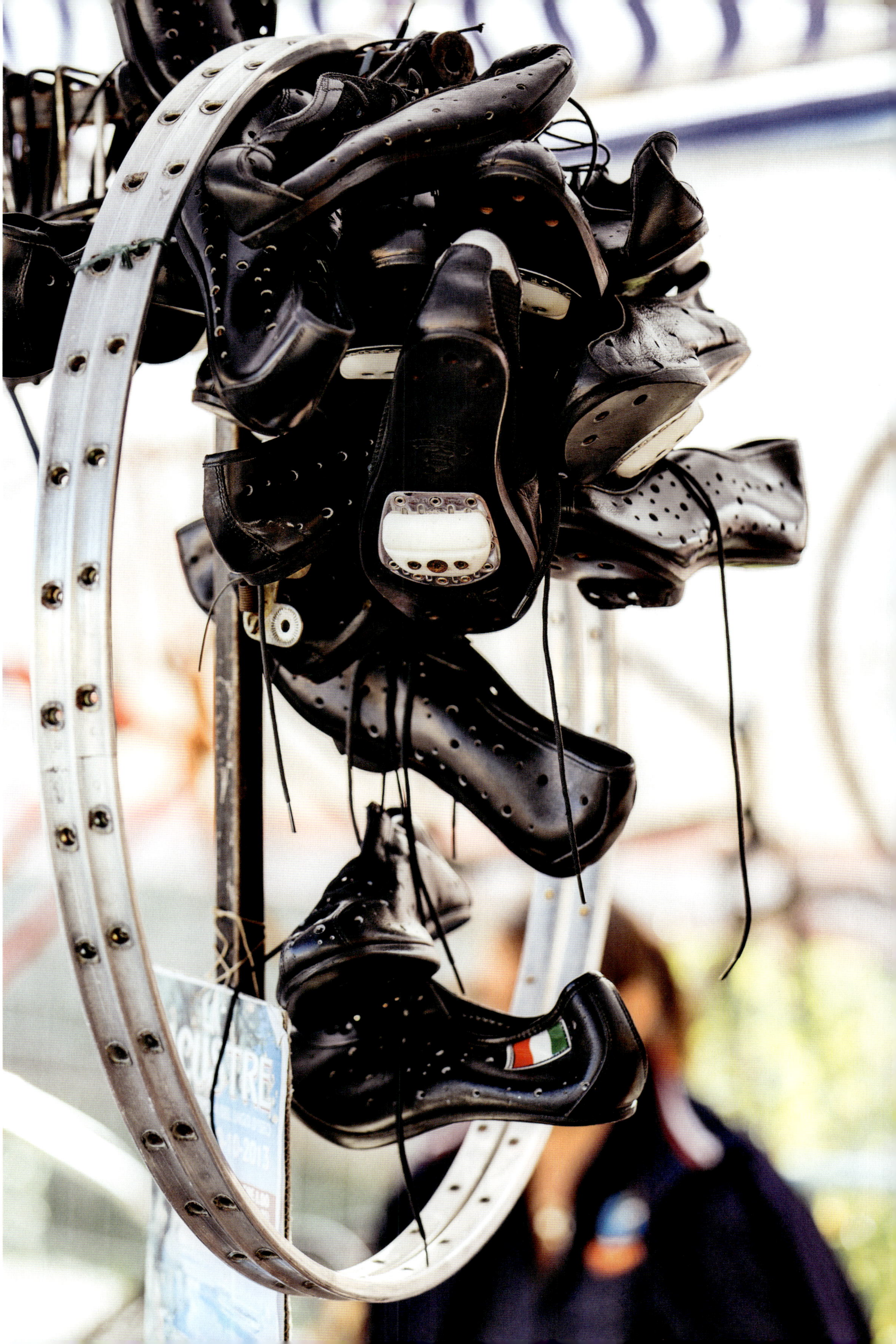

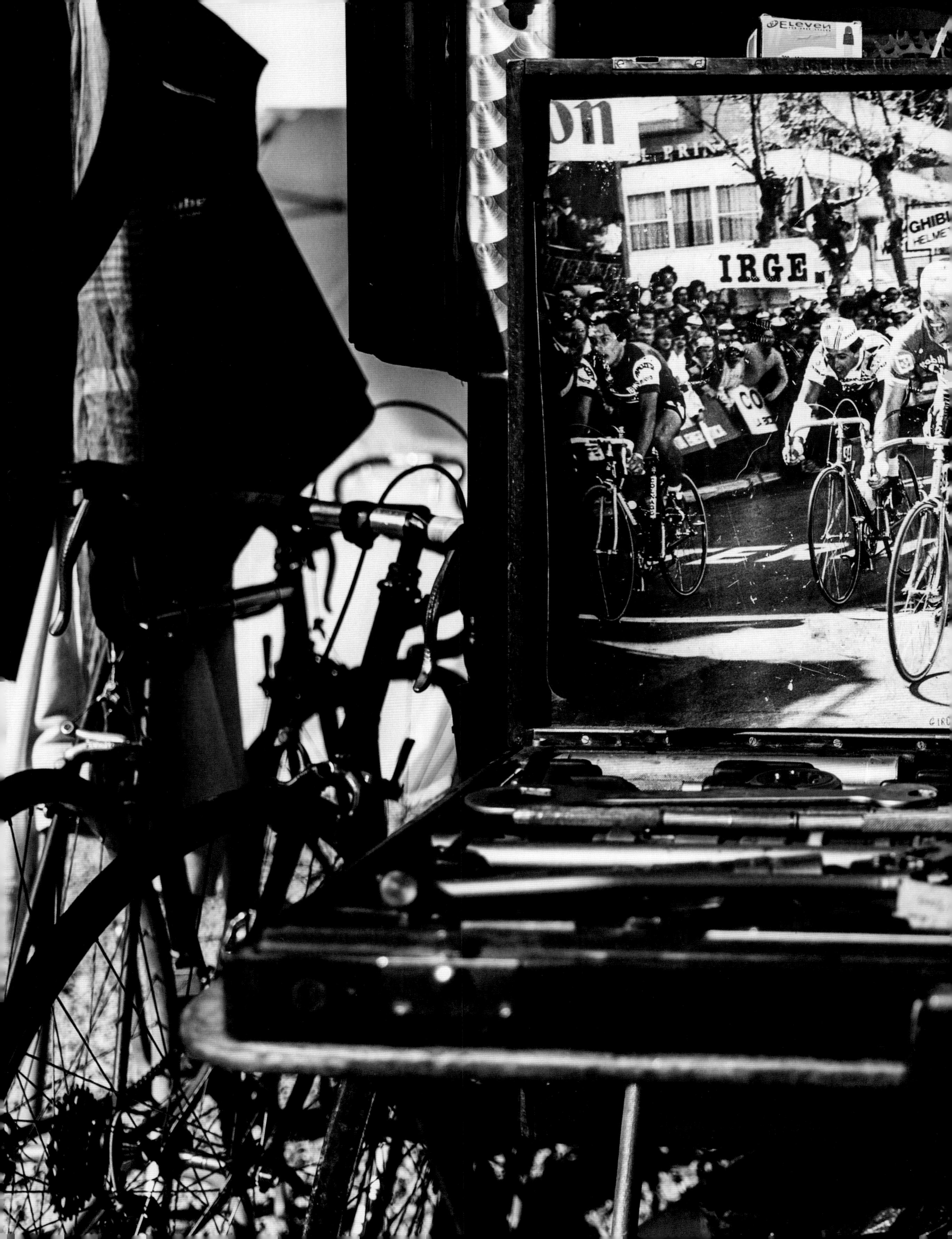
Eleven
IRGE

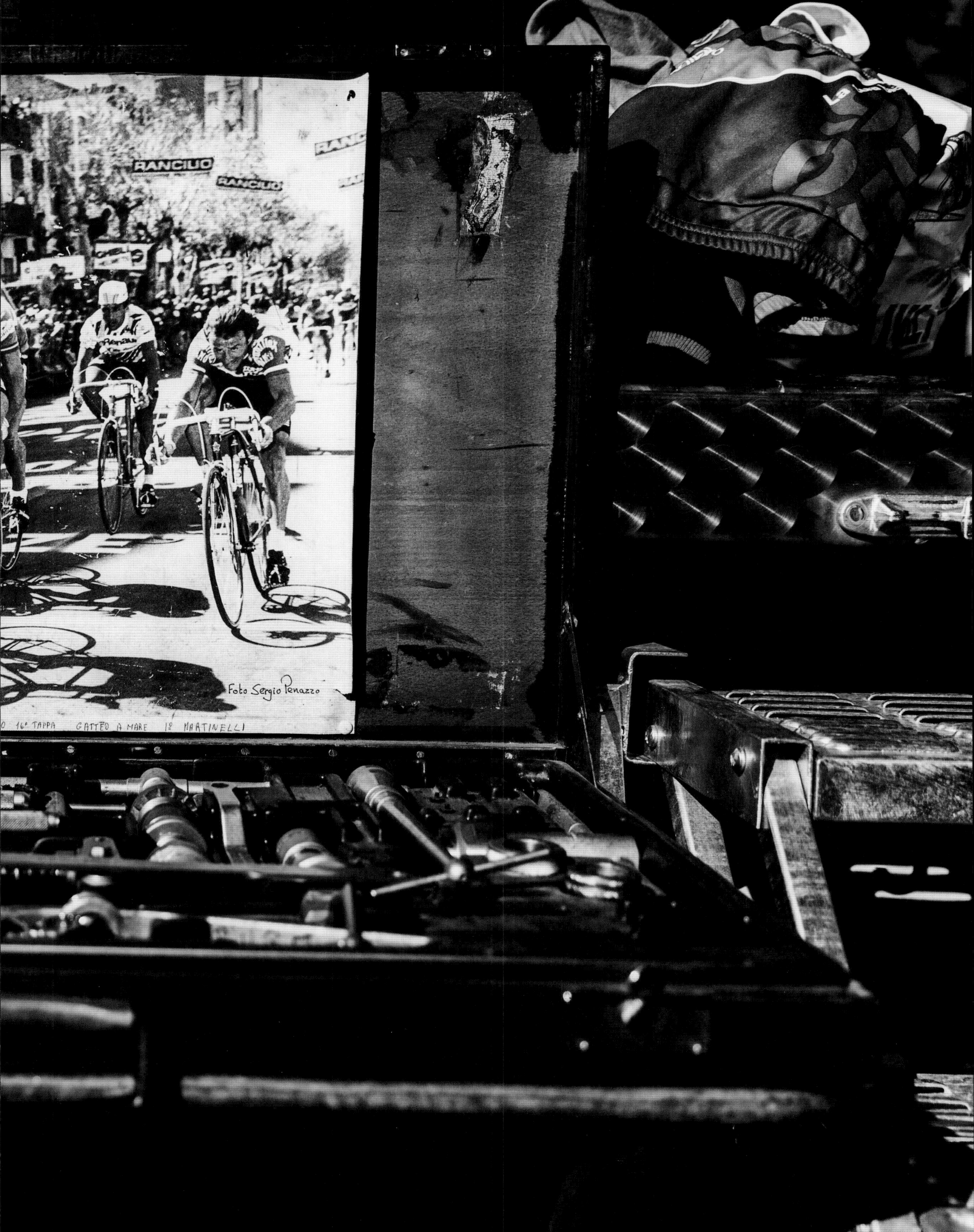
RANCIUO
RANCIUO
Foto Sergio Penazzo
O 16ª TAPPA GATTEO A MARE 1º MARTINELLI

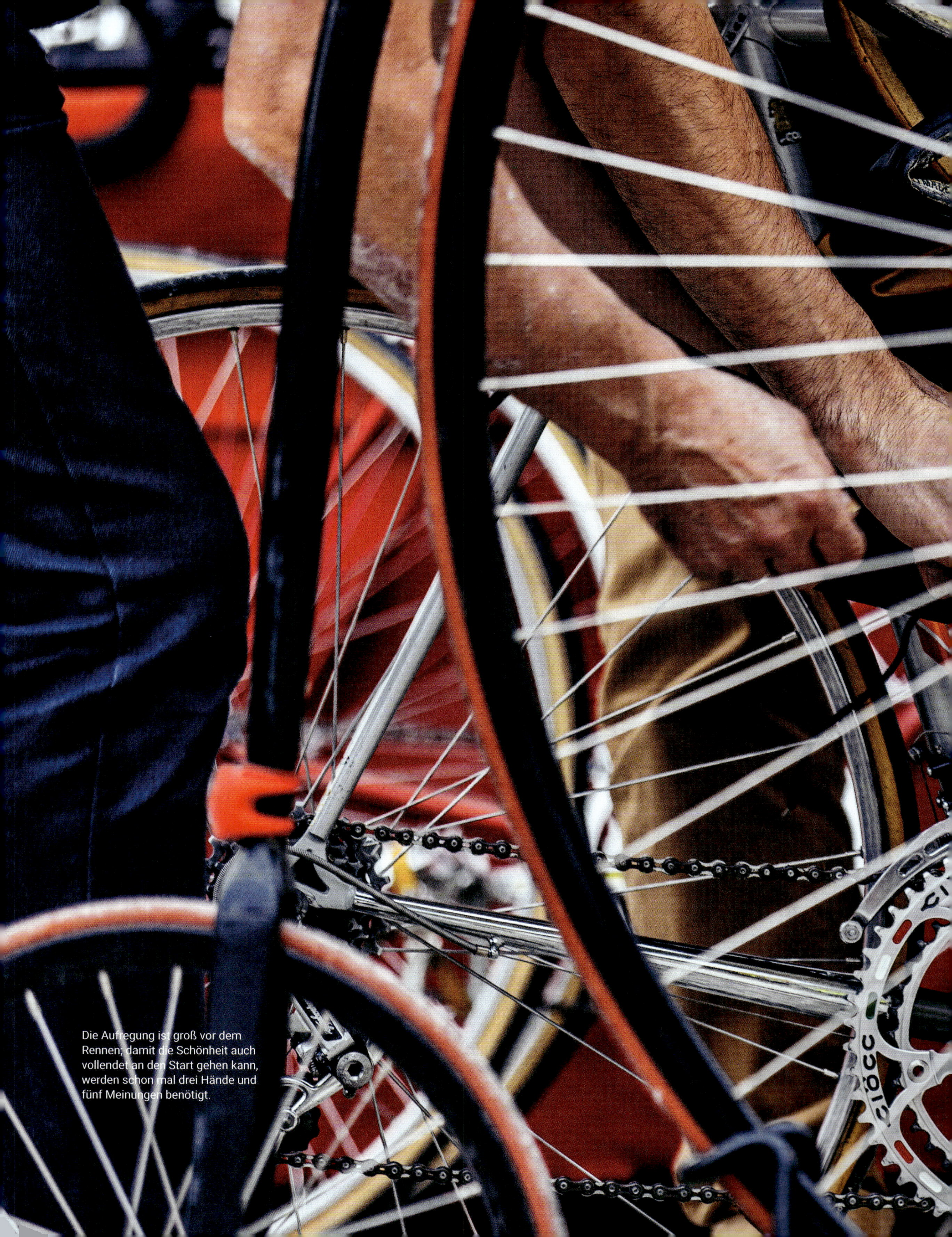

Die Aufregung ist groß vor dem Rennen; damit die Schönheit auch vollendet an den Start gehen kann, werden schon mal drei Hände und fünf Meinungen benötigt.

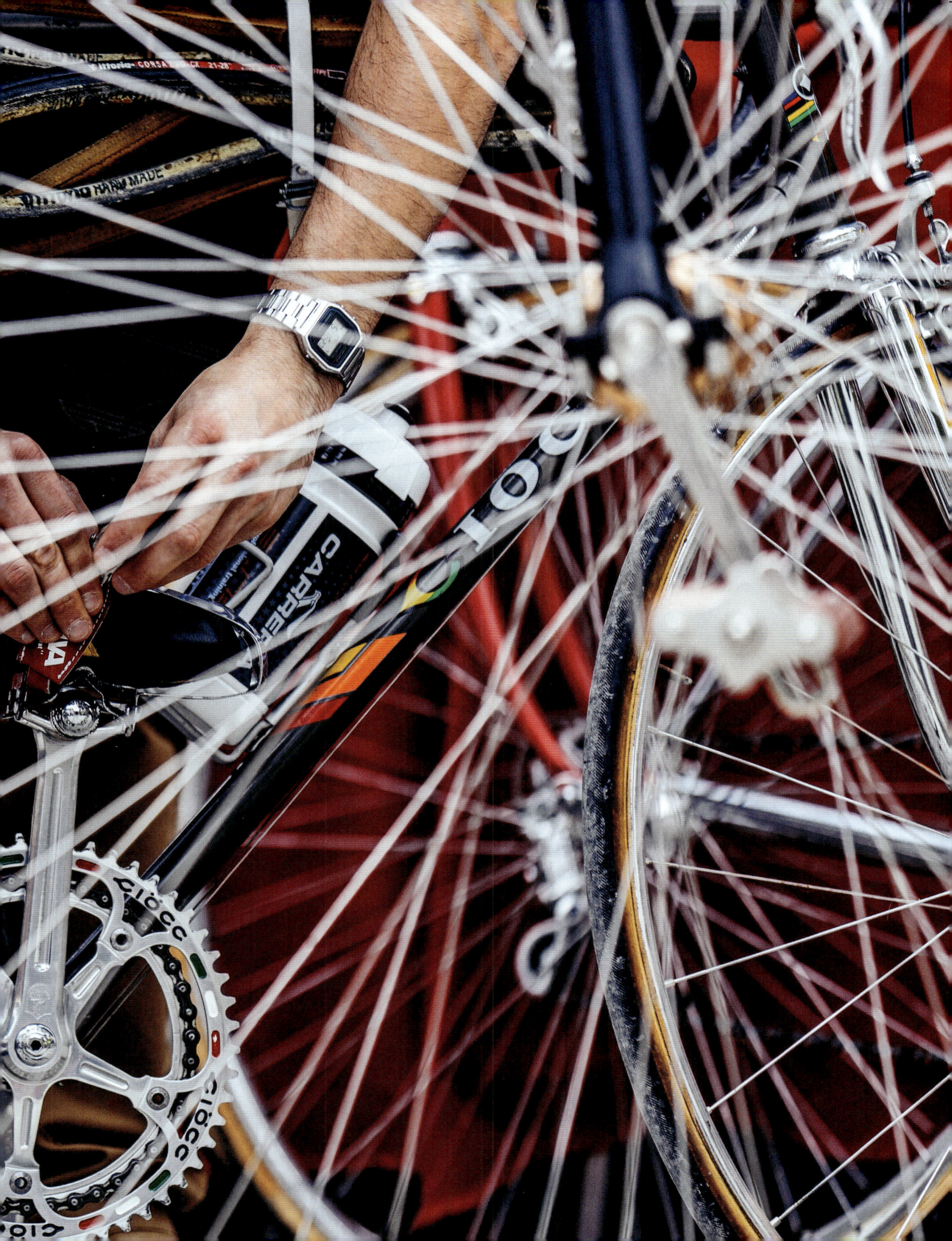

LINO

Der Profi und die Schildkröte.

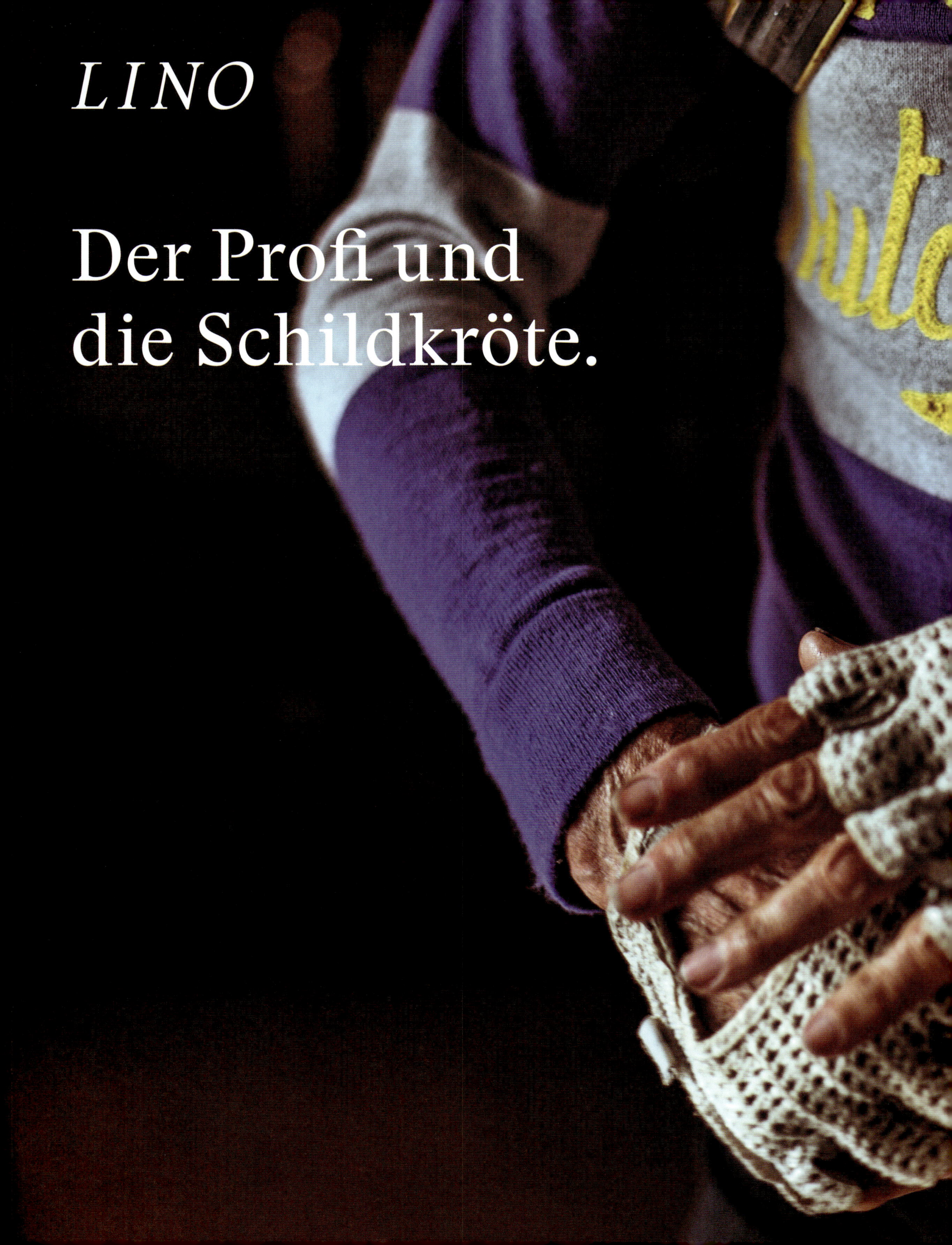

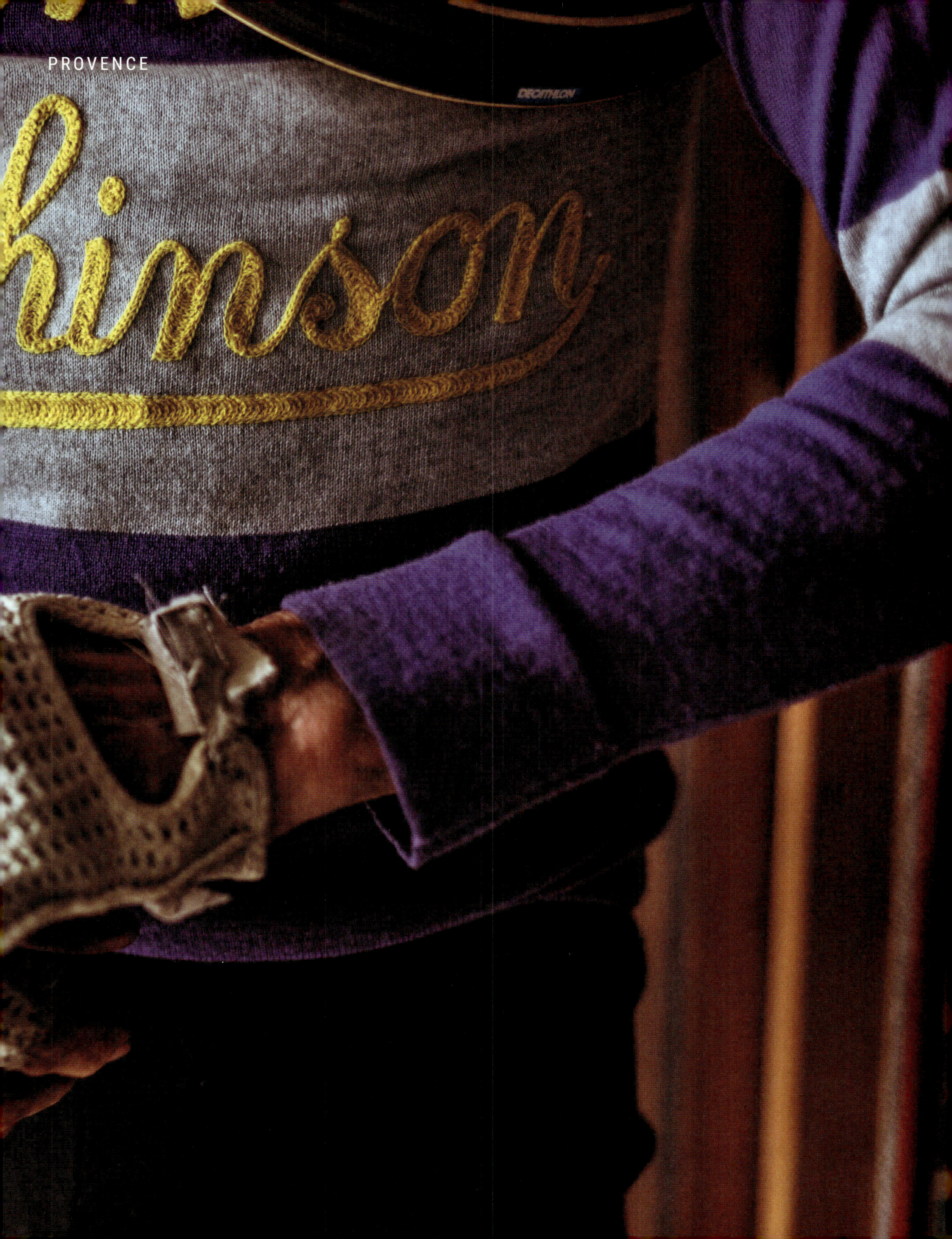

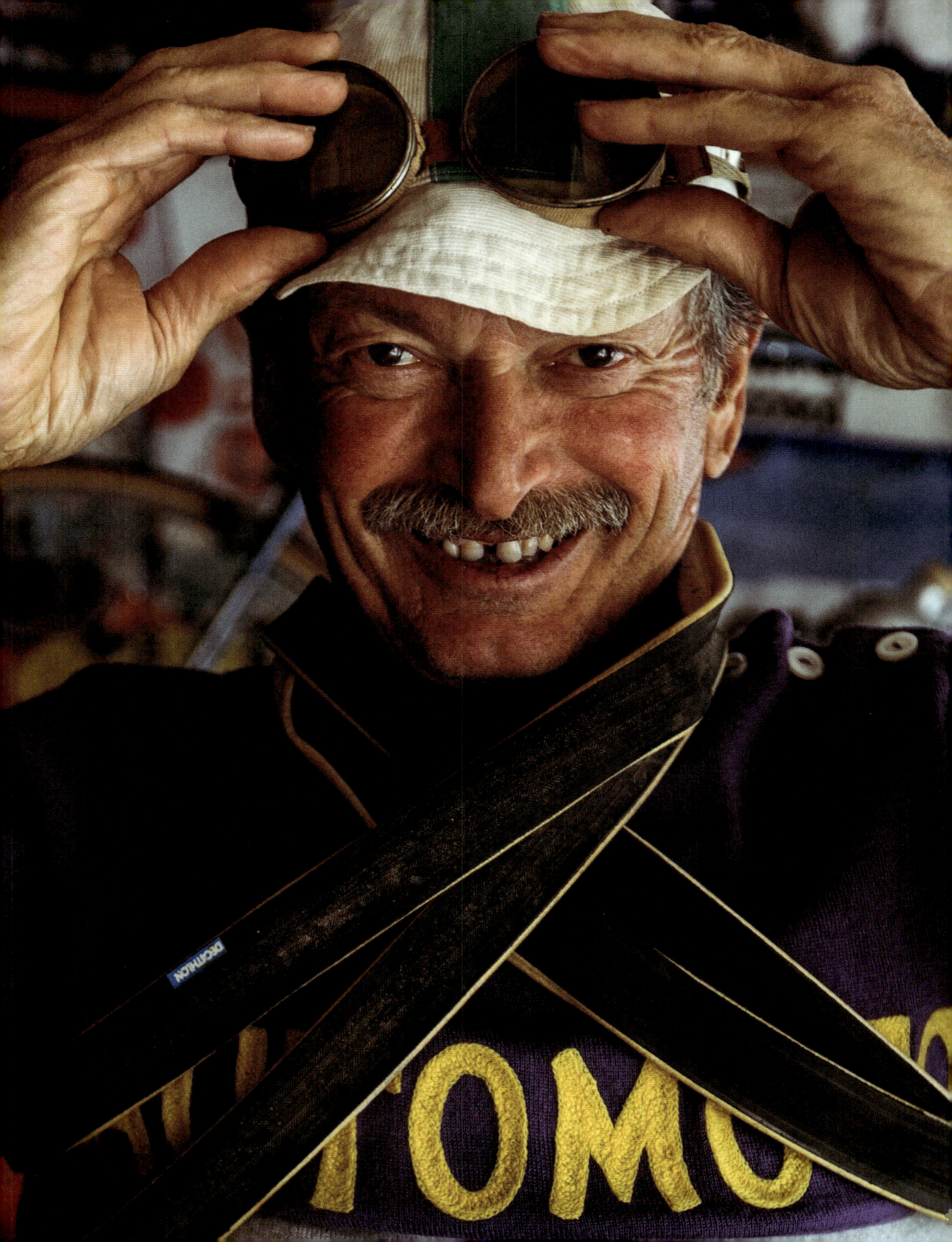
TOMO

Kann diese Zahnlücke lügen? Dieses Strahlen in den Augen – das Glück eines langen Radfahrerleben scheint hier zu uns zu sprechen.

Wenn Fahrradsammler ein bisschen verrückt sind, dann ist Lino Lazzerini vollkommen irre. Wunderbar durchgeknallt! Beginnen wir vorn. Traumhafte Landstraßen durch die Provence führen in die kleine Gemeinde Cavaillon. Von dieser Szenerie und dem Frühsommerlicht in die passende Stimmung versetzt mache ich mich also auf den Weg zu Lino.

Grauer Schnauzbart, Zahnlücke und ein Gesicht, das seine Jungenhaftigkeit auch im Alter nicht verliert. Nach einer herzlichen Begrüßung beginnt die Führung durch die Lino-Sammlung. Seit über 40 Jahren spürt der Mann die Fahrräder seiner Helden auf und fügt sie zu einem Gesamtkunstwerk zusammen. Rahmen, Sättel und komplette Räder, die Geschichten von 1900 bis heute erzählen könnten. Inzwischen hat sich der Schatz über das gesamte Grundstück ausgebreitet. Zwei Anbauten am Haus sorgen für ausreichend Platz.

Egal, in welche Richtung wir gehen oder schauen: Jeder Quadratzentimeter ist von Originaltrikots, signierten Fotos, Rennkappen, Streckenschildern vergangener Touren, Team-Trinkflaschen und Provianttaschen belegt. Und natürlich noch und nöcher Rennräder, so weit das Auge reicht. Nach dem ersten überwältigenden Eindruck wird mir bewusst, was genau diese Sammlung von anderen abhebt. Es ist schlicht und einfach die Menge! Trikots von Small bis Extra Large, von weiß über gelb, rosa, grün bis dunkelblau. Steuerkopfschilder, Schaltgruppen, Ledersättel. Alles en masse. Unter Plastikplanen abgedeckt verstecken sich die noch unbearbeiteten Schätze.

Wenn Fahrradsammler ein bisschen verrückt sind, dann ist Lino Lazzerini vollkommen irre.

Zwischen all dem taucht plötzlich eine Schildkröte, seine Tartaruga, auf. Die beiden passen gut zusammen und ich stelle mir vor, wie sie Lino schon sein ganzes Sammlerleben begleitet. Das Tier hat allerdings wenig für Radsport übrig, dafür aber viel für süße Leckerbissen. Die Himbeeren, die sie nun schmatzt, mag sie am liebsten, erklärt mir Lino.

Kurz mal weggeschaut, da steht der Gastgeber im kompletten Vintage-HUTCHINSON-Outfit vor mir. Natürlich eines von vielen Ensembles, die an ihm sitzen wie für ihn entworfen. Er grinst sein fröhliches Zahnlückengrinsen. Eine unglaublich ansteckende Lebensfreude! Auch mit über 70 hat er noch den Körper eines Radfahrers. Schlank und sehnig, Muskeln hauptsächlich an den Beinen. Eines ist ganz offensichtlich: Der Mann ist topfit!

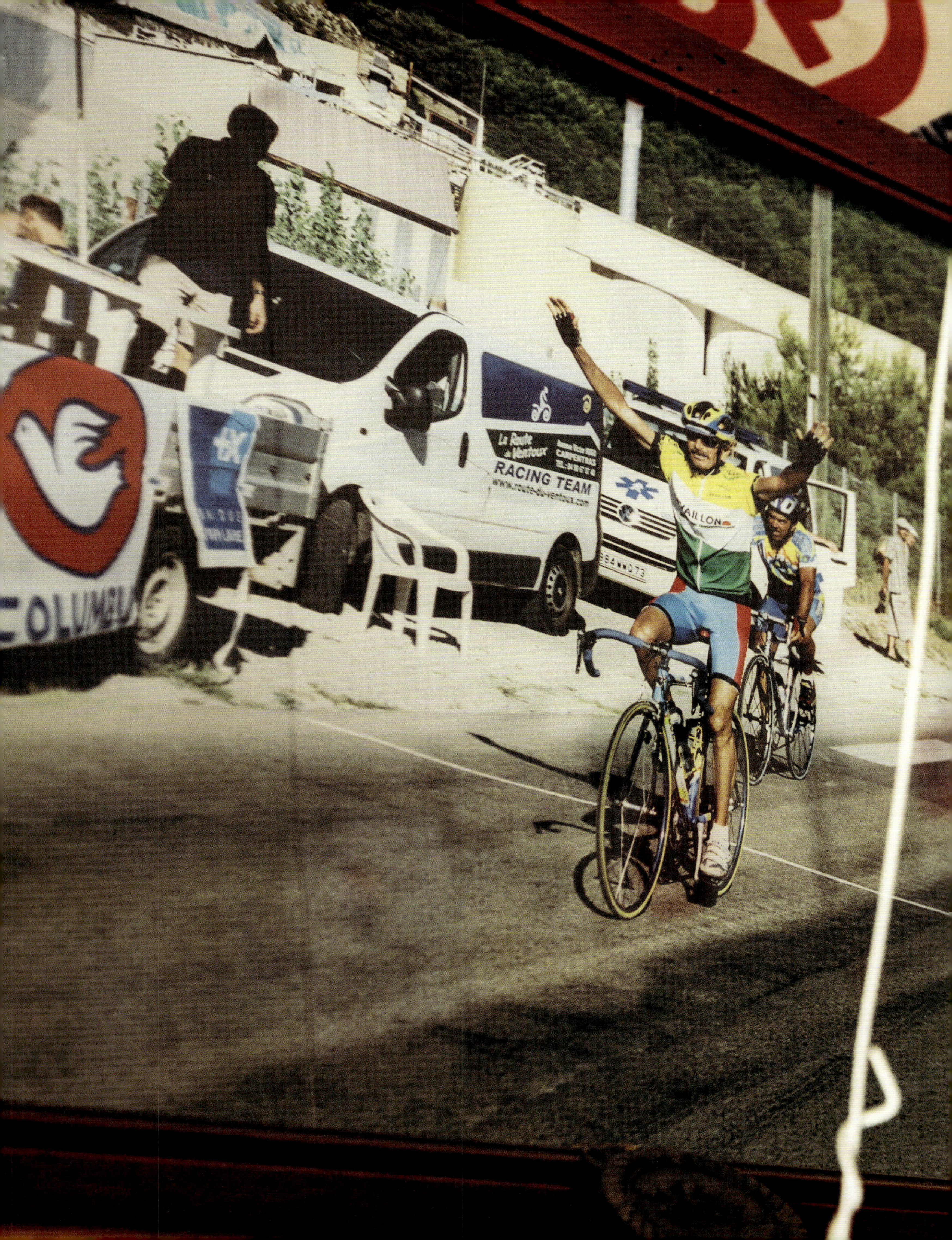

La Route du Ventoux
CARPENTRAS
RACING TEAM
www.route-du-ventoux.com

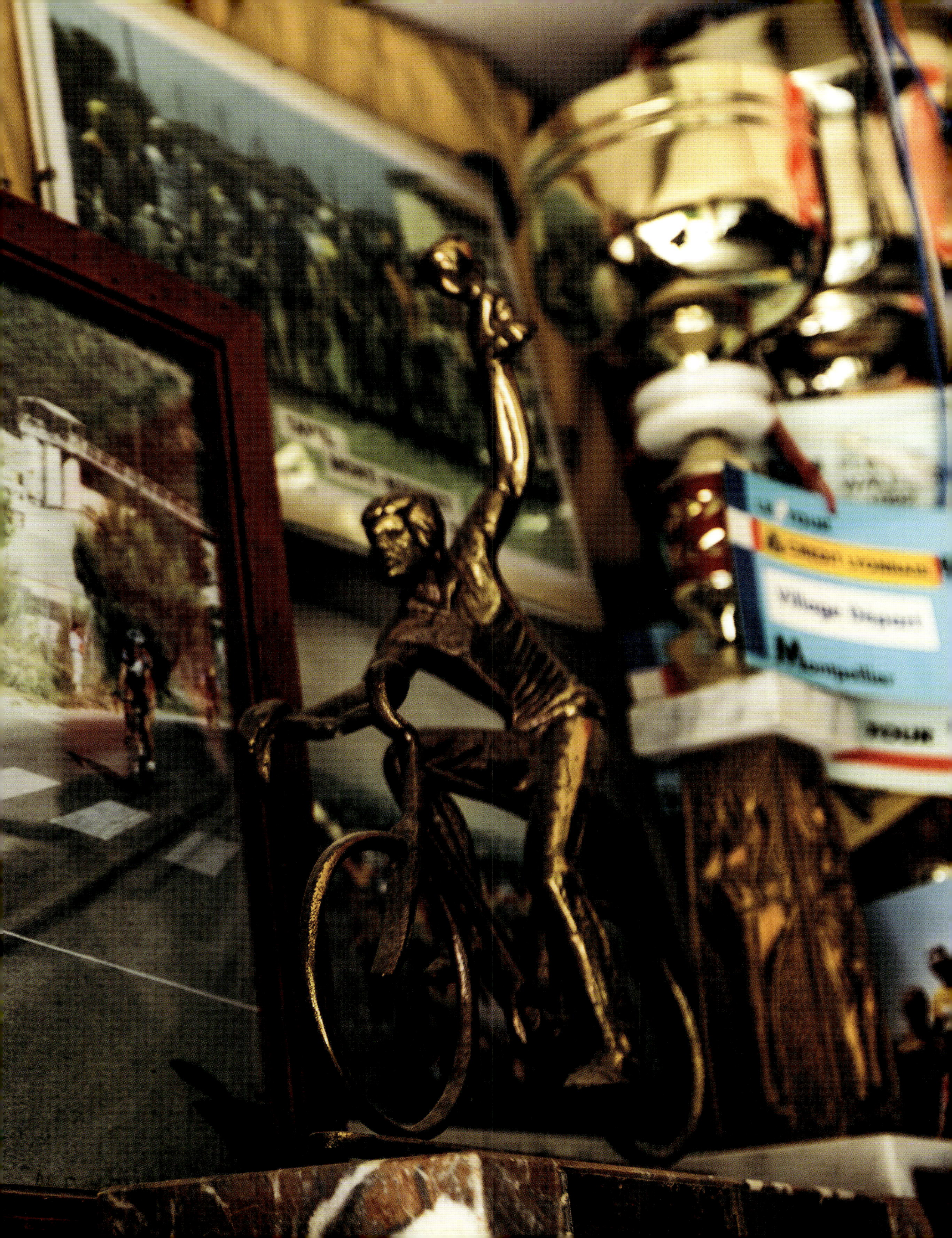

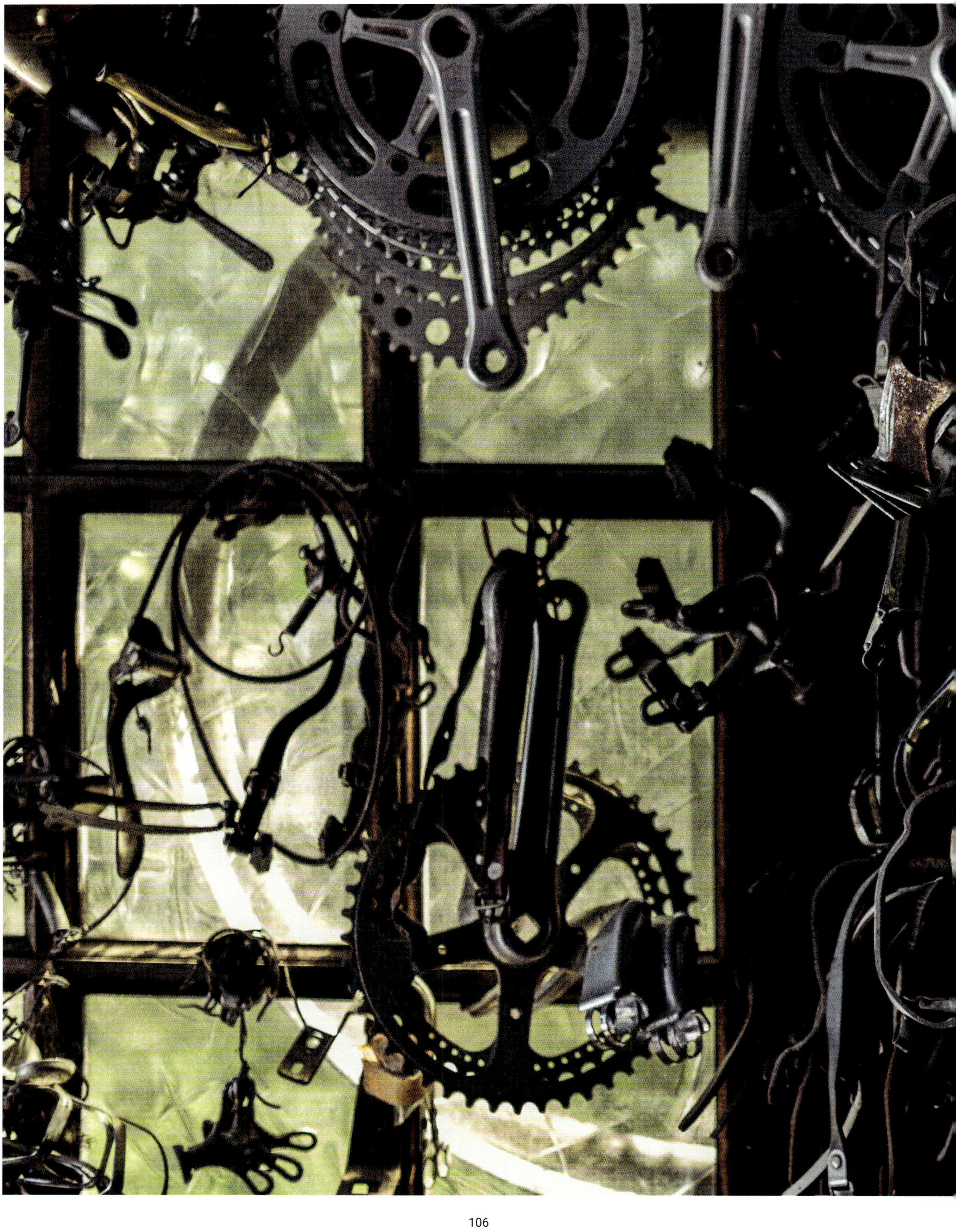

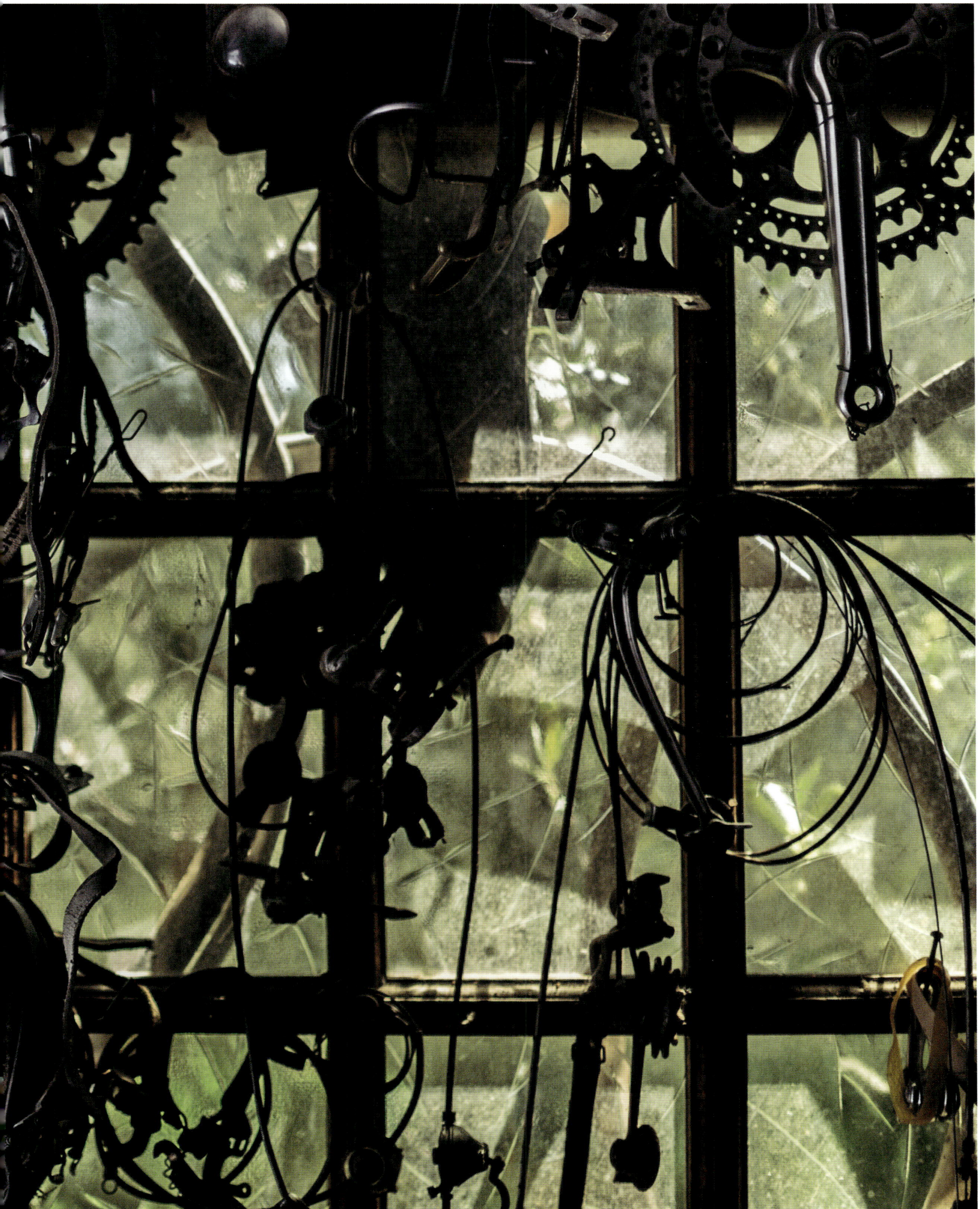

Lino hat nicht nur eine ganz bedeutende Sammlung, er hat auch sein Grundstück in ein Museum verwandelt. Hier versammeln sich hauptsächlich Räder der 1910er- bis 1930er-Jahre. In der hinteren Ecke befindet sich das Rad eines Helden der ersten TOUR DE FRANCE!

Einmal im Jahr organisiert er eine Klassikerausfahrt in Cavaillon. Am liebsten radelt der jedoch noch immer über den nahe gelegenen MONT VENTOUX. Als einer der *heiligen Berge* der TOUR DE FRANCE ist der Ventoux wohl auch einer der unnachgiebigsten Abschnitte der Region. Lino kennt die Herausforderung und ist an die Hitze und den Wind gewöhnt. Seine Fahrten unterbricht er nur hin und wieder, um heimlich Beeren und Kräuter zu sammeln, mit denen er seinen provenzalischen Schnaps brennt.

Den kredenzt uns jetzt seine Frau unter dem Olivenbaum auf einem kleinen Tisch voller Tartes, Oliven, Käse, selbstgebackenem Brot und Rotwein.

Doch auch Lino ist das Funkeln in meinen Augen nicht entgangen; grinsend taucht er noch ein weiteres Mal mit mir in die Tiefen seiner reichen Sammlung ein. Es gibt einfach noch massenhaft zu entdecken.

Trikots von Small bis Extra Large, von weiß über gelb, rosa, grün bis dunkelblau. Steuerkopfschilder, Schaltgruppen, Ledersättel. Alles en masse. Unter Plastikplanen abgedeckt verstecken sich die noch unbearbeiteten Schätze.

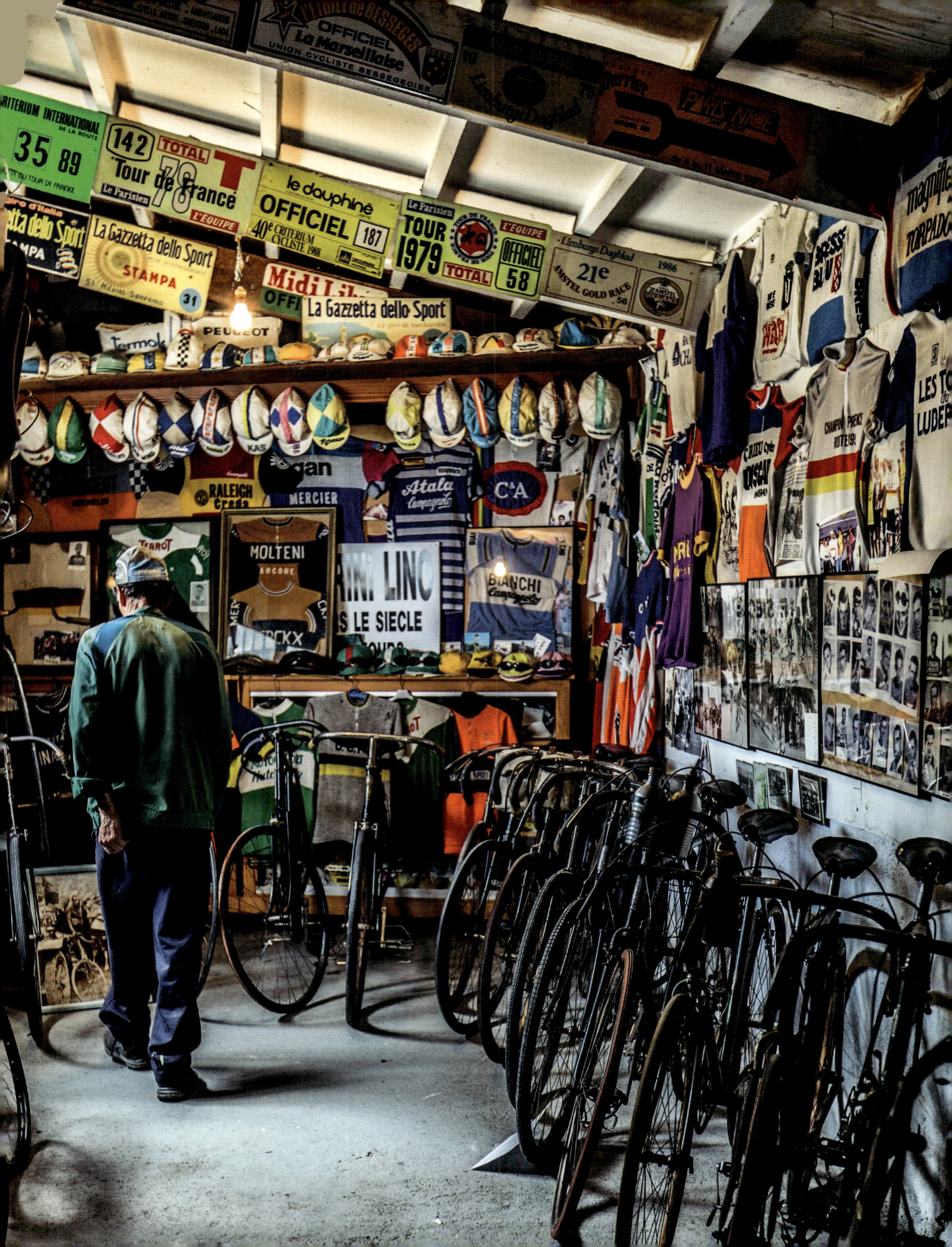
OFFICIEL
La Marseillaise
UNION CYCLISTE BESSEGEOISE
CRITERIUM INTERNATIONAL DE LA ROUTE
35 89
142
TOTAL
Tour de France
78
Le Parisien
L'EQUIPE
La Gazzetta dello Sport
STAMPA
31
le dauphiné
OFFICIEL
187
40e CRITERIUM CYCLISTE 1988
Le Parisien
TOUR 1979
L'EQUIPE
OFFICIEL
58
TOTAL
Limburgs Dagblad
1986
21e
AMSTEL GOLD RACE
PARIS NICE
Midi Libre
La Gazzetta dello Sport
PEUGEOT
RALEIGH
MERCIER
Atala
Campagnolo
C&A
MOLTENI
BIANCHI
Campagnolo
S LE SIECLE

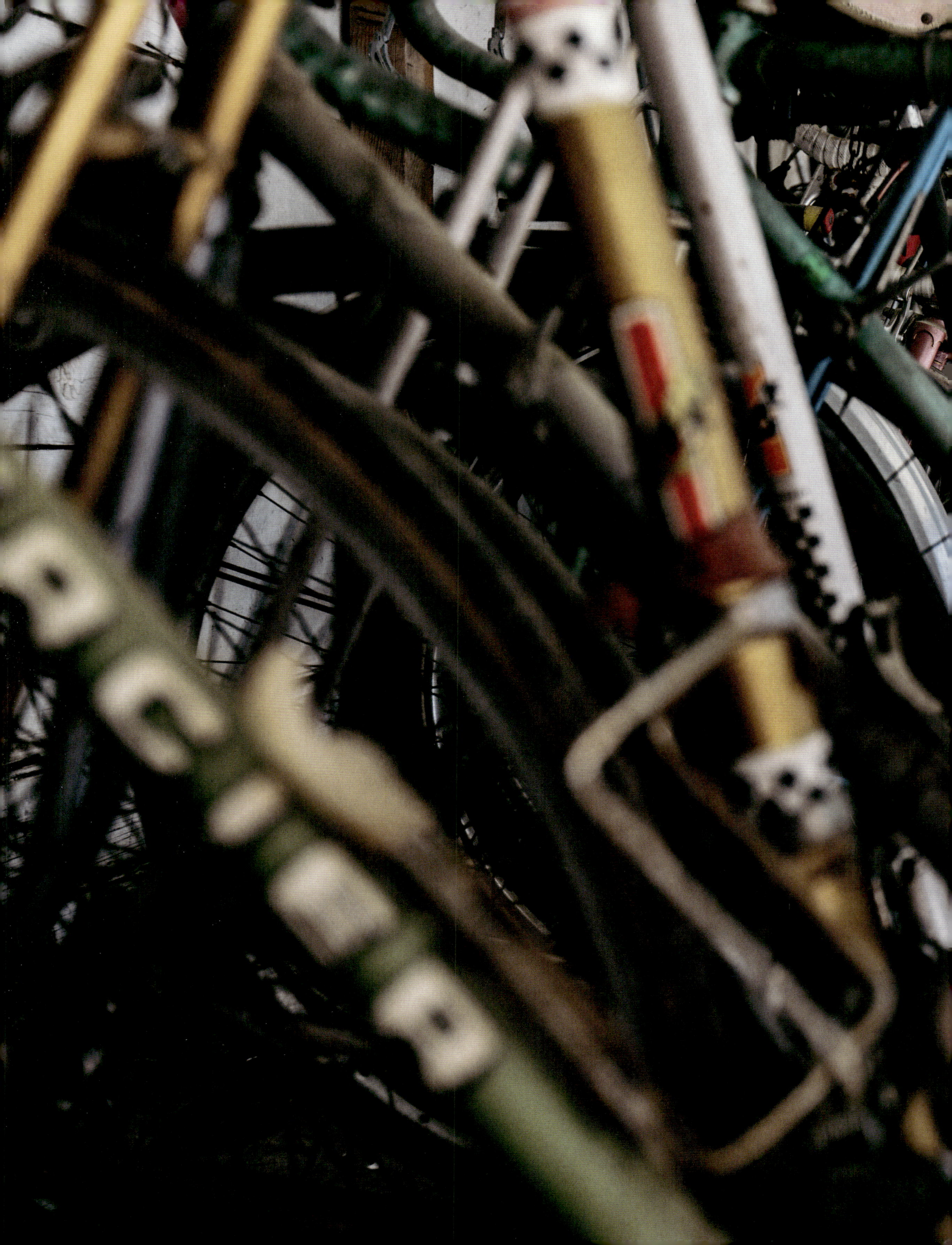

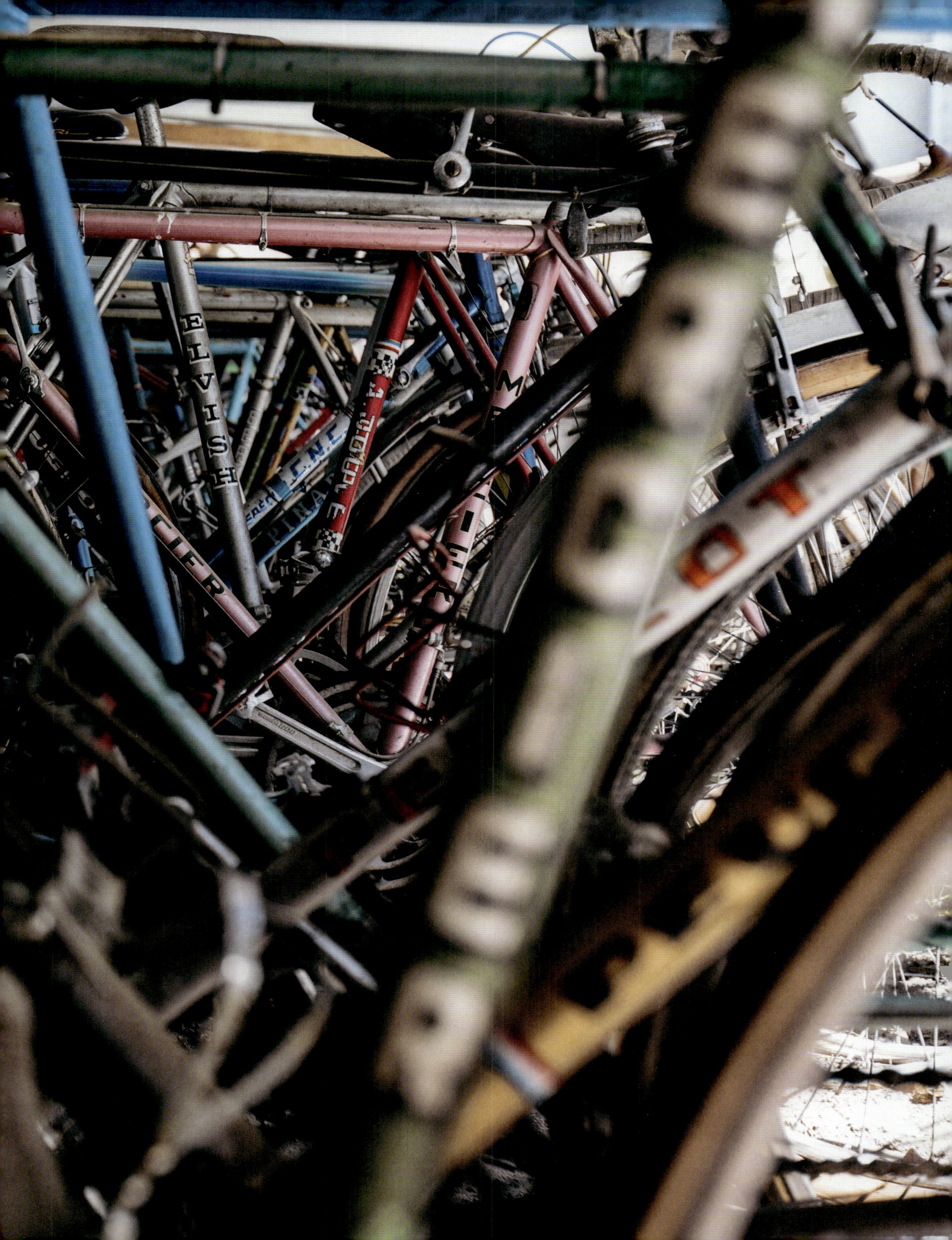
ELVISH

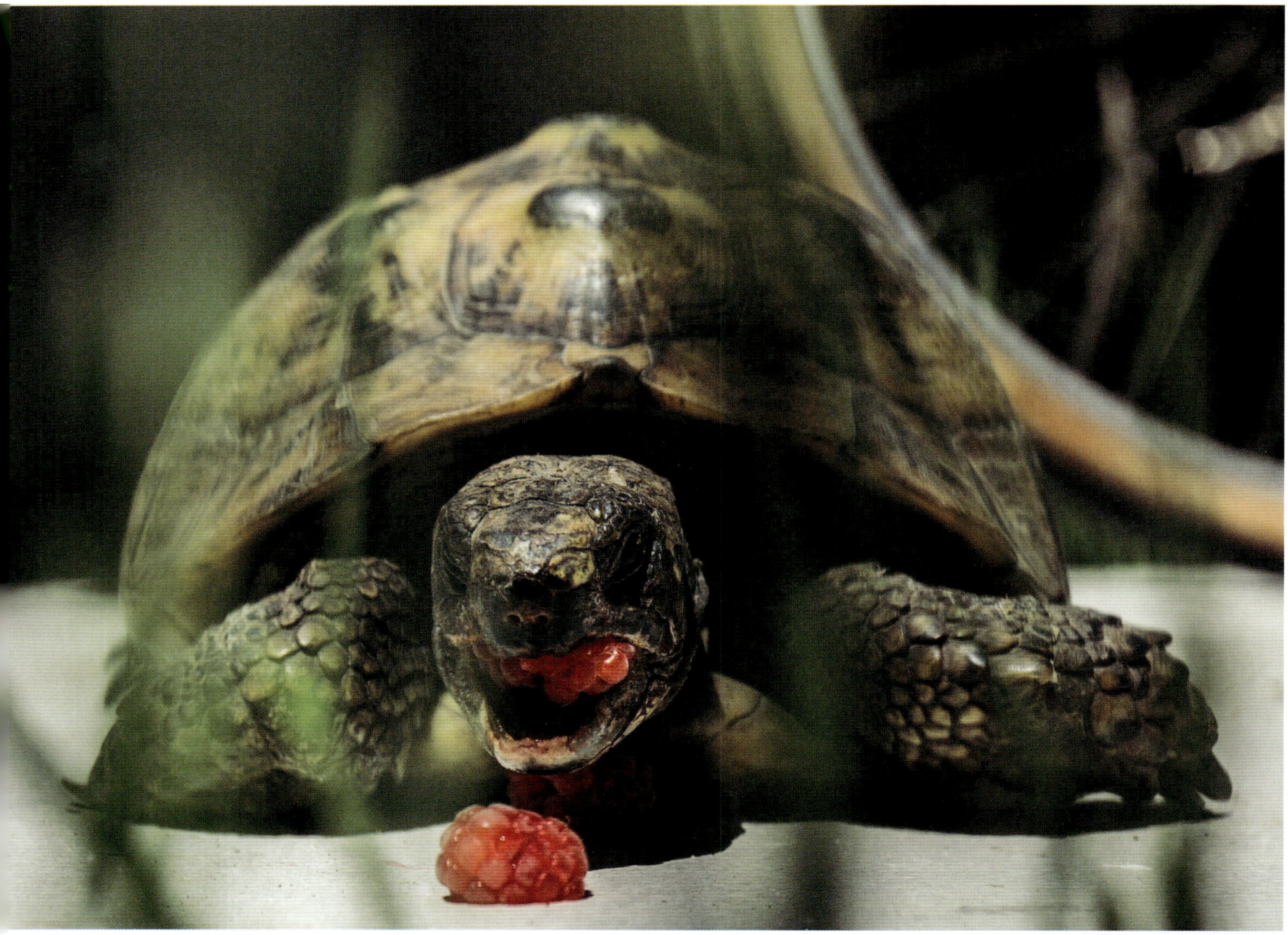

Was ist jetzt älter? Fahrrad oder Schildkröte? Man kann sich einfach nicht entscheiden. Beide sind gezeichnet von der Patina eines langen Lebens.

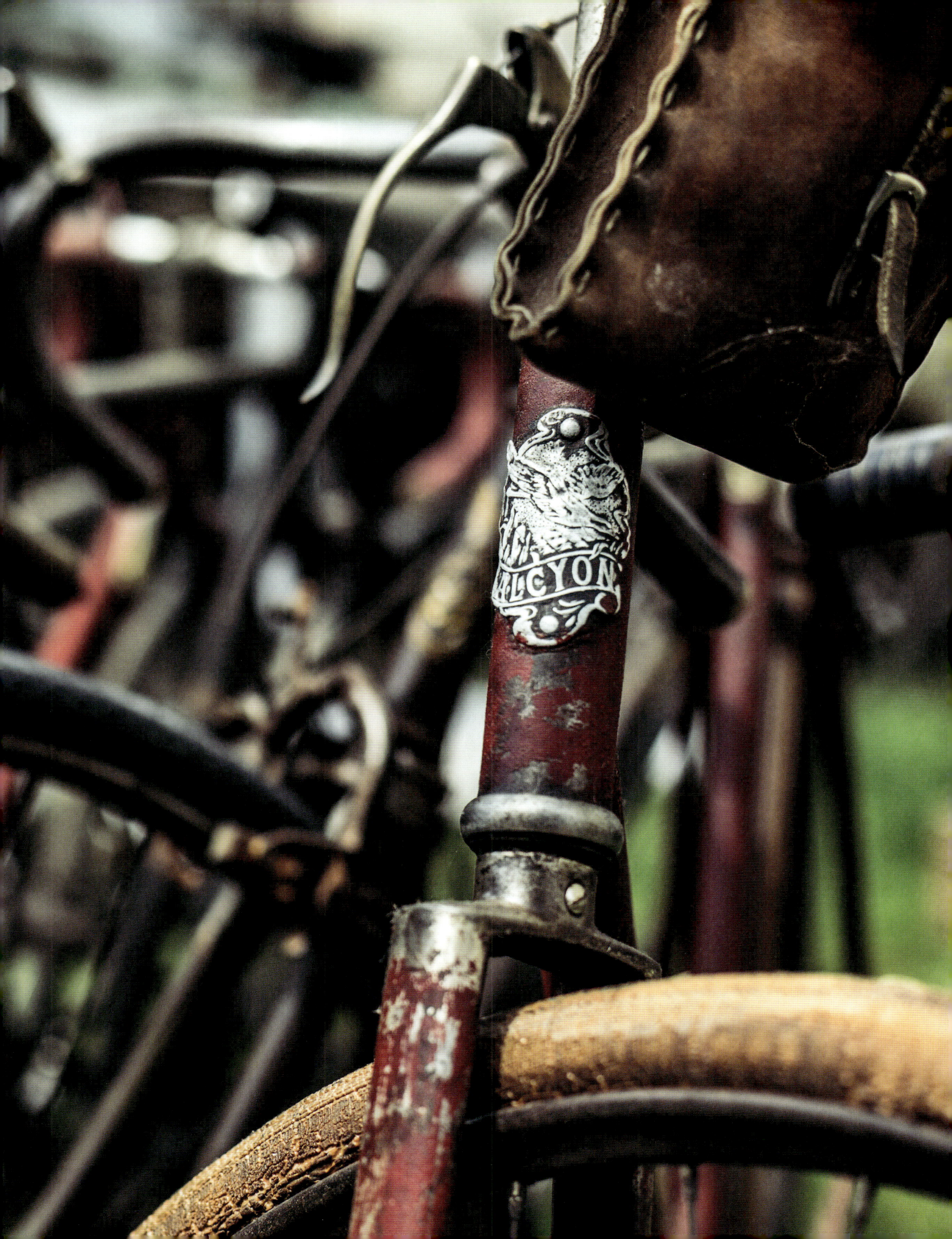
ALCYON

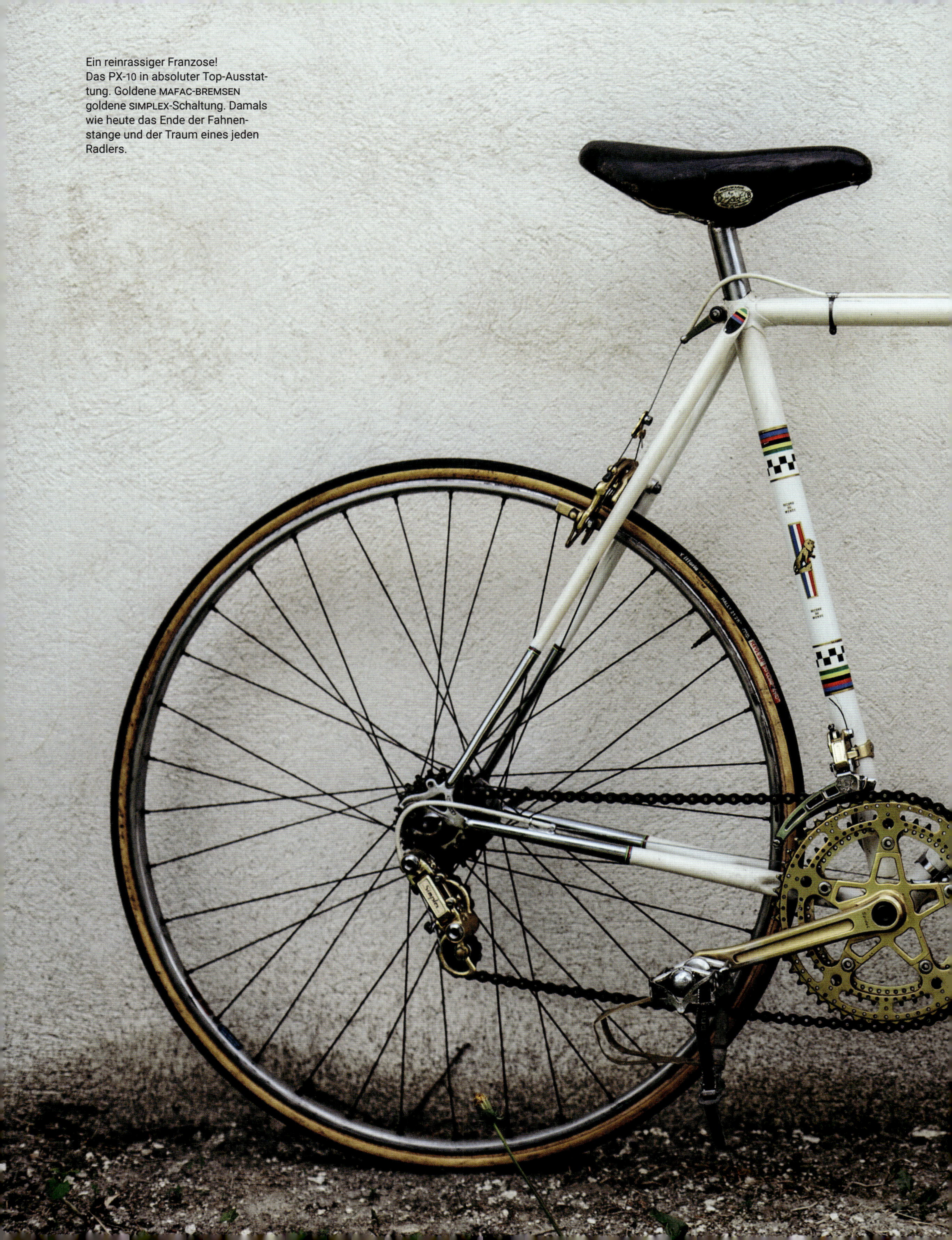

Ein reinrassiger Franzose!
Das PX-10 in absoluter Top-Ausstattung. Goldene MAFAC-BREMSEN goldene SIMPLEX-Schaltung. Damals wie heute das Ende der Fahnenstange und der Traum eines jeden Radlers.

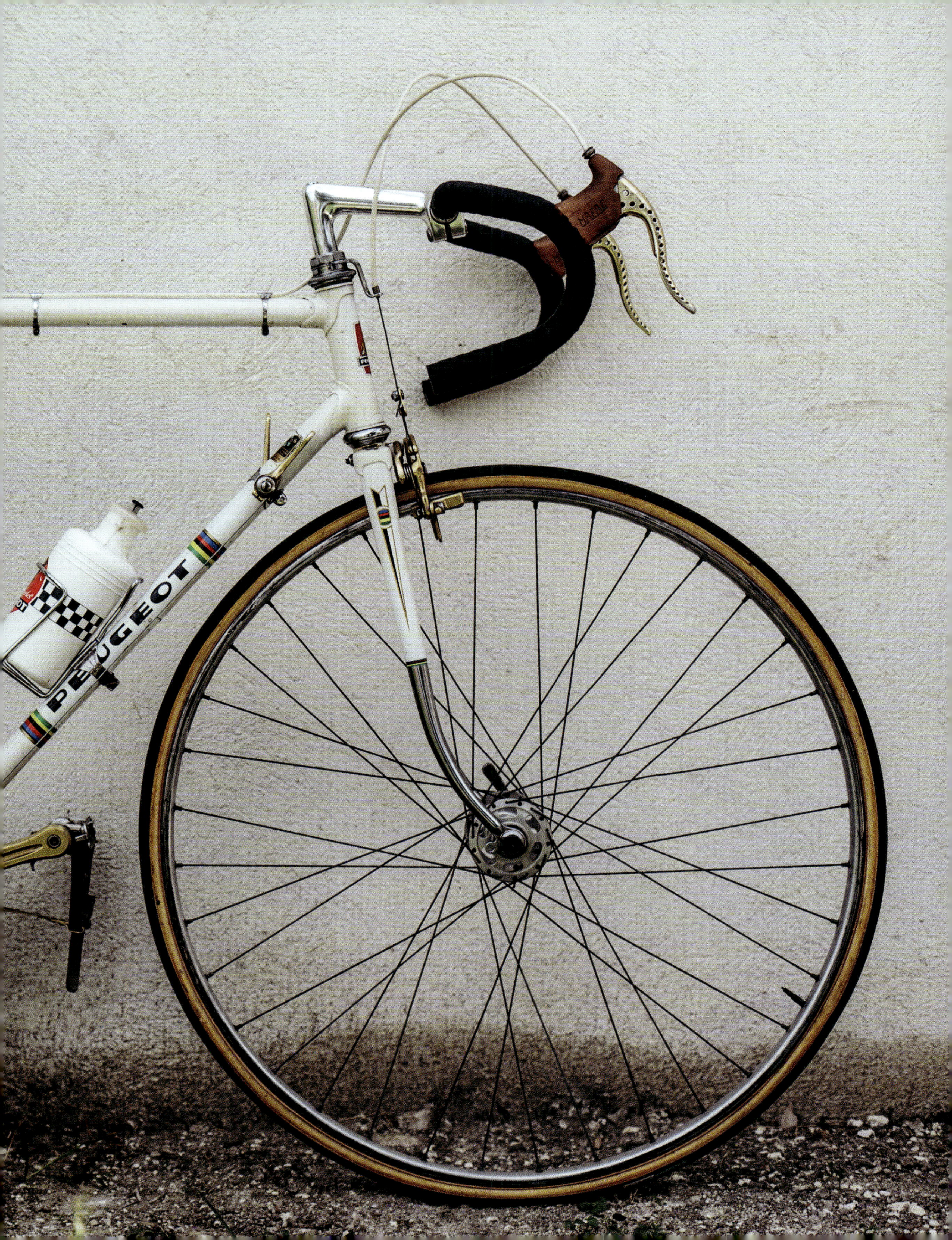
PEUGEOT

HELLENIC MOUNTAIN RACE

Hunde, die beißen, bellen nicht

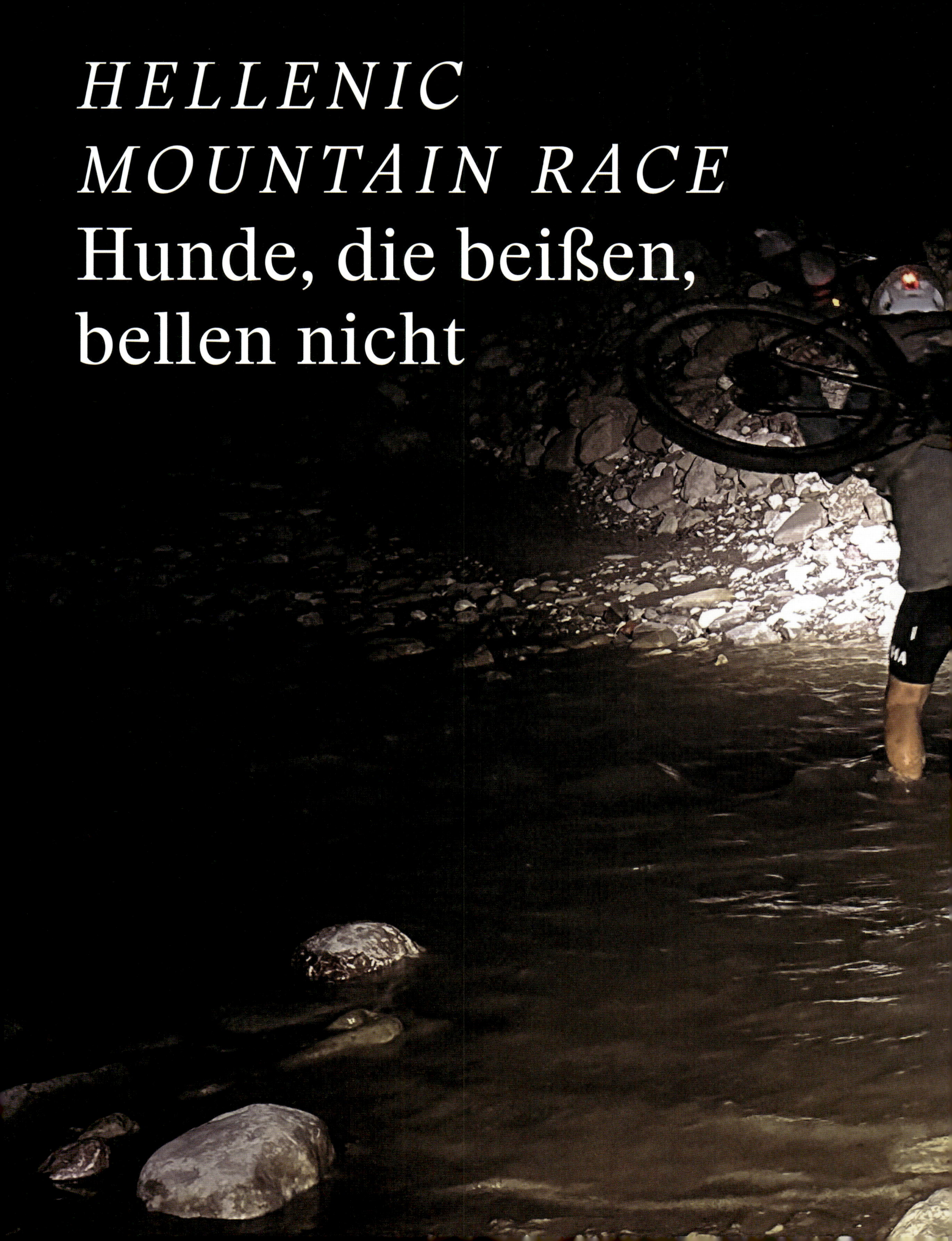

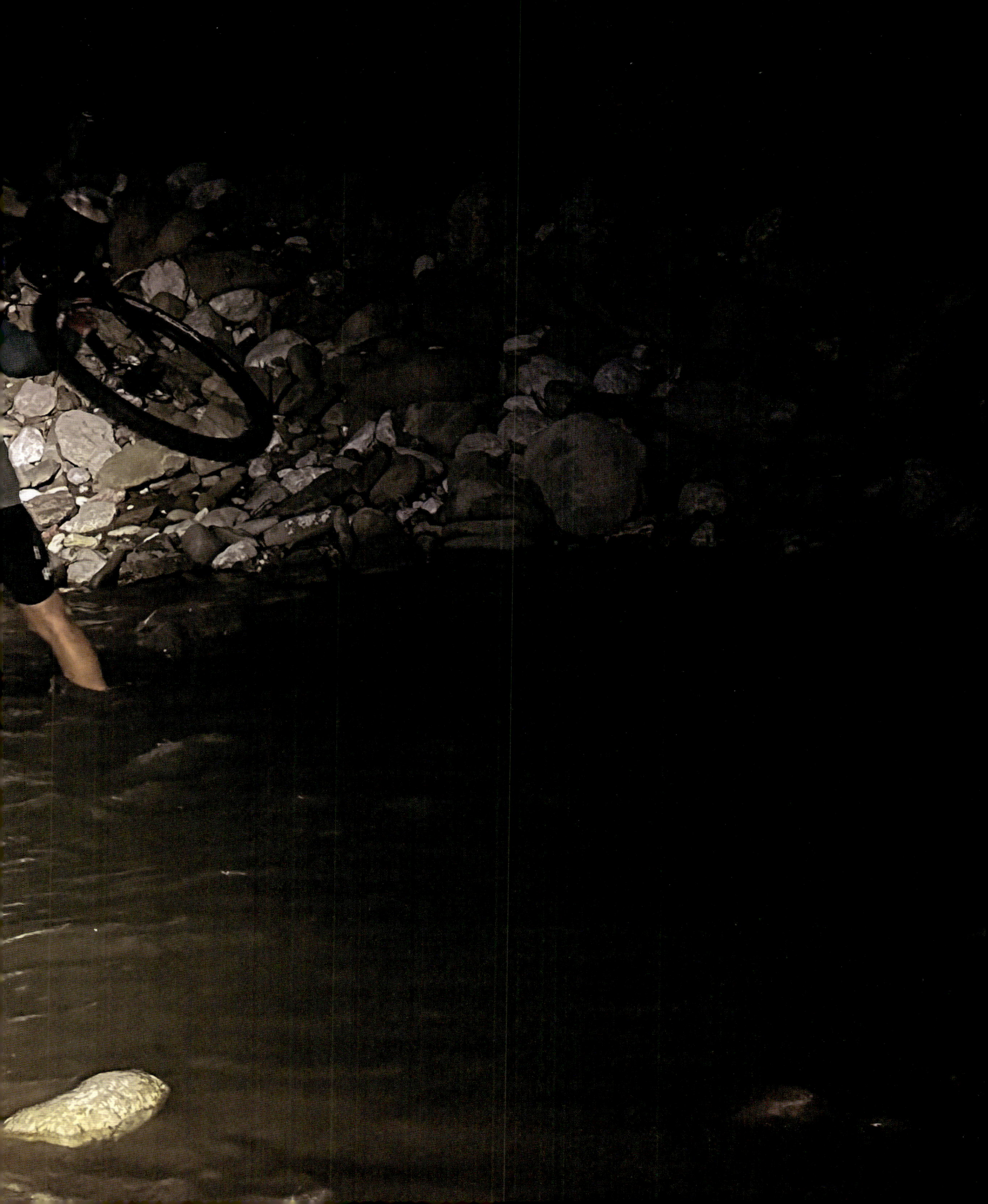

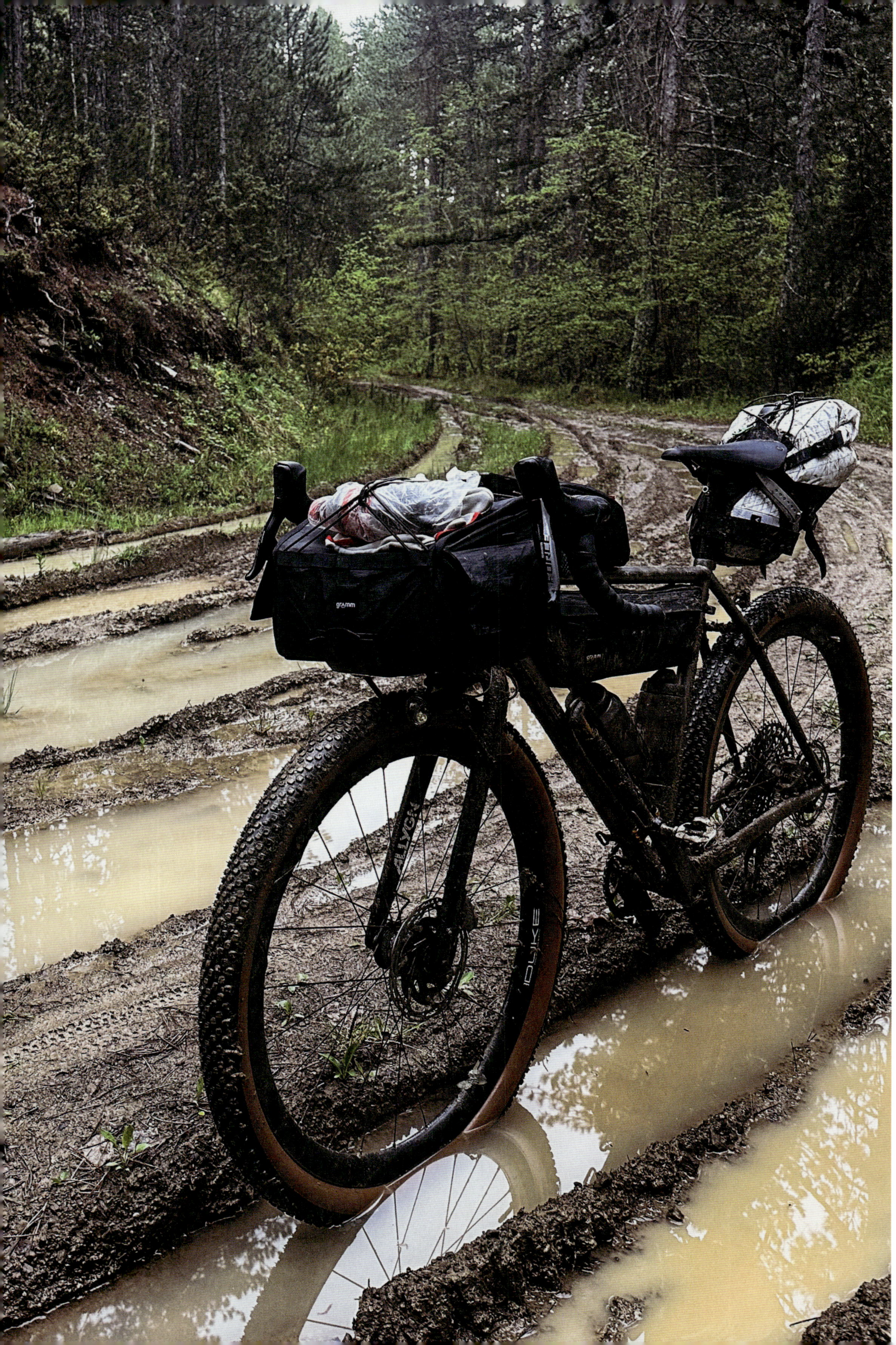

Da helfen auch 54 mm breite Stollenreifen nicht mehr weiter. Tagelanges fahren in tiefem Schlamm kann nicht nur körperlich, sondern auch psychisch eine extreme Herausforderung darstellen.

Als ich mir die Sprachnachricht erneut anhöre, die ich zwei meiner besten Freunde gerade geschickt habe, erschrecke ich ein wenig: war doch ein deutliches Wimmern in meiner Stimme zu hören.

Seit Stunden habe ich immer wieder leise vor mich hin geweint. Völlig verzweifelt, denn genau genommen ging nichts mehr, körperlich genauso wenig wie bei meinem Rad: der Untergrund hat sich zu einer Erdnussbutter artigen Masse geformt, wodurch sich zwischen Gabel und Reifen solche Mengen Schlamm festgesetzt haben, als wollte mir die Natur einfach nur den guten Rat geben, zurückzugehen, sich eine Möglichkeit raus aus diesem Schlamassel zu suchen und darüber nachzudenken, was genau ich hier mit einem Rad und ultraleichtem Gepäck zu suchen habe.

Zu diesem Zeitpunkt befinde ich mich seit gut drei Tagen zusammen mit 120 anderen Teilnehmern aus mehr als 20 Nationen in den tiefen Wäldern des Pindos-Gebirges im Nordwesten Griechenlands und bestreite die Erstausgabe des HELLENIC MOUNTAIN RACE (HMR). Ein Rennen, das sich auf 900 Kilometern Länge von seinem Startpunkt in Kalambaka, unweit der berühmten Klöster von Meteora gelegen, weit
in die Bergkette hineinzieht. Es führt uns an der atemberaubenden Vikos-Schlucht vorbei über antike Brücken und Berge, die aus weißem Kalkstein gestapelt zu sein scheinen. Am Ziel in der ursprünglich venezianischen Hafenstadt Nafpaktos werden dann knapp 30.000 Höhenmeter hinter uns liegen.

Das HMR ist Teil einer Serie, die der Abenteurer und Ausdauersportler Nelson Trees mit dem legendären SILK ROAD MOUNTAIN RACE (SRMR) 2018 in Kirgisistan ins Leben gerufen und mit dem ATLAS MOUNTAIN RACE (AMR) in Marokko 2020 fortgesetzt hat. Nun also die zivilere, weil nicht ganz so exponierte neue Variante dieses Rennformates in Griechenland im Mai 2023.

Die Regeln dieser Unsupported-Ultra-Endurance-Races sind vermeintlich einfach: Die Teilnehmer folgen einem vorgegebenen Track, müssen vor den Cut-offs drei Checkpoints passiert haben, dürfen keine Hilfe von außen annehmen, die nicht auch jedem anderen Teilnehmer zur Verfügung steht, also keinen persönlichen Support, und das Ganze als Einzel- oder Zweierteams fahrend. Die sogenannte Brevet-Karte nicht zu vergessen, auf der die Zeit der Durchfahrt

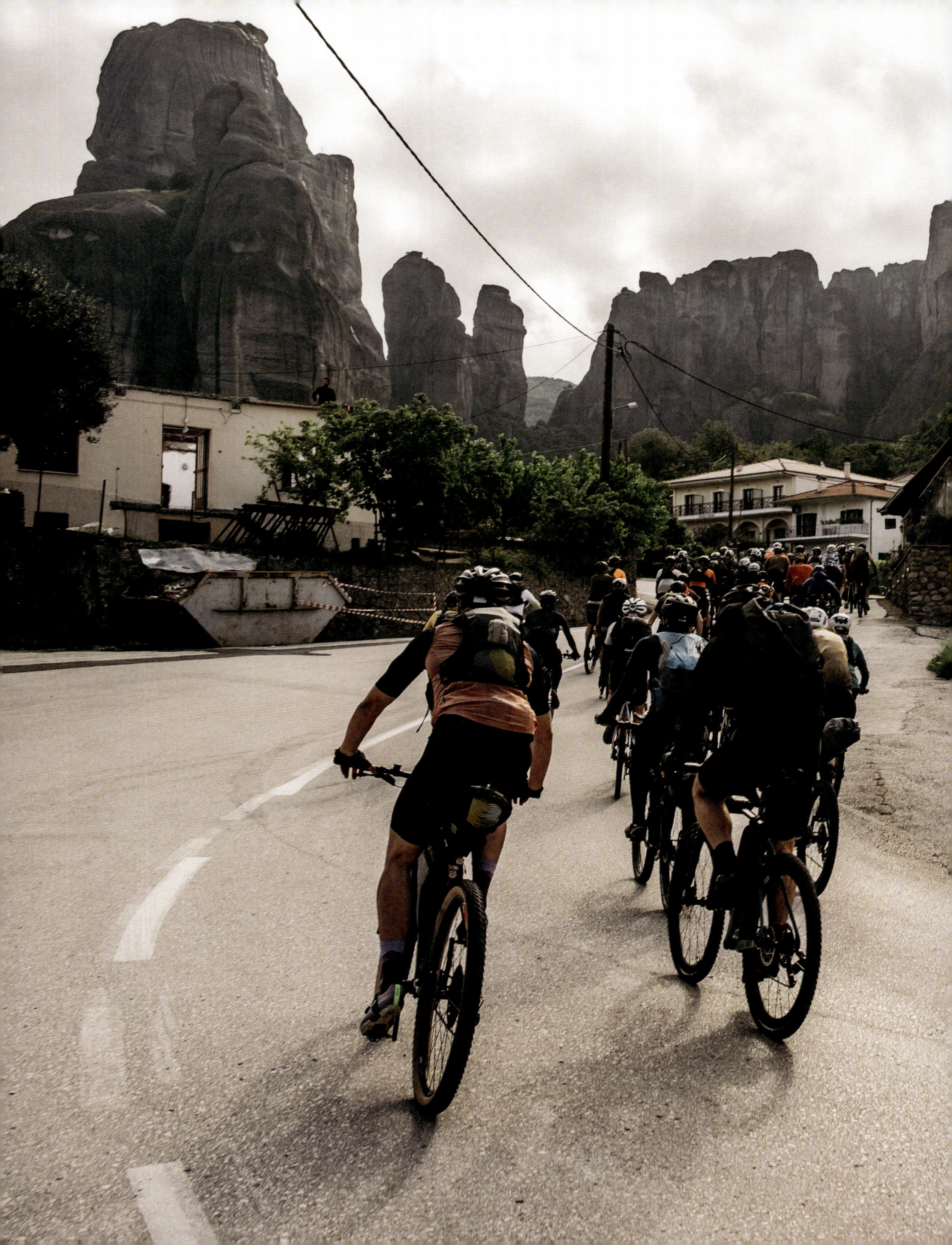

Was mache ich hier eigentlich? Wenn ich über den Grund nachdenke, der mich in diese Abenteuer hineinzieht, sind es vor allem die Geschichten und der Pioniergeist eines Reinhold Messner.

am jeweiligen Checkpoint eingetragen werden muss: keine Brevet-Karte im Ziel, kein Finish! So einfach.

»IT IS GOING TO BE A WET ONE!« – erscheint zwei Tage vor Rennstart die Nachricht auf meinem Handy. Nelson, in seiner Funktion als Rennleiter, kann uns über Sportify, eine App für Sportveranstaltungen, direkt kontaktieren. Die Großbuchstaben verdeutlichen die Ernsthaftigkeit der Nachricht: Jeder weiß, dass uns hiermit desaströse Verhältnisse bevorstehen. Zu allem Überfluss bahnt sich zeitgleich eine weitere Information ihren Weg zu den Startern: Einer der weltweit stärksten Fahrer, Sofiane Sehili, mehrfacher Gewinner zahlreicher Ultrarennen, war vor Ort bei einer Testfahrt von wilden Hunden attackiert worden und musste in der Notaufnahme eine zehn Zentimeter lange Fleischwunde verarzten lassen – der große Favorit war also schon raus.

Von da an konnte man in Kalambaka tagsüber hektisch umherschwirrende Ausdauersportler dabei beobachten, wie sie im lokalen Sportgeschäft verzweifelt nach Regenhosen oder Ponchos Ausschau hielten oder in Mini-Baumärkten wasserdichte Latexhandschuhe aus dem Reinigungssektor kauften. Ich suchte den lokalen Waffenhändler auf und fragte nach Pfefferspray – ich hätte Bären aufzuhalten!? Die Antwort kam prompt: »Nein, haben wir nicht!«

Was mache ich hier eigentlich? Wenn ich über den Grund nachdenke, der mich in diese Abenteuer hineinzieht, sind es vor allem die Geschichten und der Pioniergeist eines Reinhold Messner.

Als Kind habe ich seine Abenteuerberichte von Besteigungen der Achttausender verschlungen und wollte wie er, oder am besten mit ihm, auf die höchsten Gipfel dieser Welt steigen. Die Vorstellung davon, die eisigen Höhen auszuhalten und zu bezwingen, faszinierte mich. Allerdings muss die prägende Wirkung der Zigaretten-Werbung eines berühmten amerikanischen Konzerns im Kino der 1980er Jahre zusätzlich mit in Betracht gezogen werden: Die Bilder von endloser Weite, tagelangem Draußensein in wilder Natur, das Wetter über den Horizont hinweg ziehen zu sehen, Sonnenauf- und -untergänge, unzählige Sterne am Nachthimmel, Lagerfeuer und das ganze auf einem Pferd mit minimalem Gepäck – da war er, der Traum, der mich prägte. Ich würde lügen, wenn mich die Stiefel, die Leder-Chaps und die Flanell-Hemden nicht zusätzlich beeindruckt

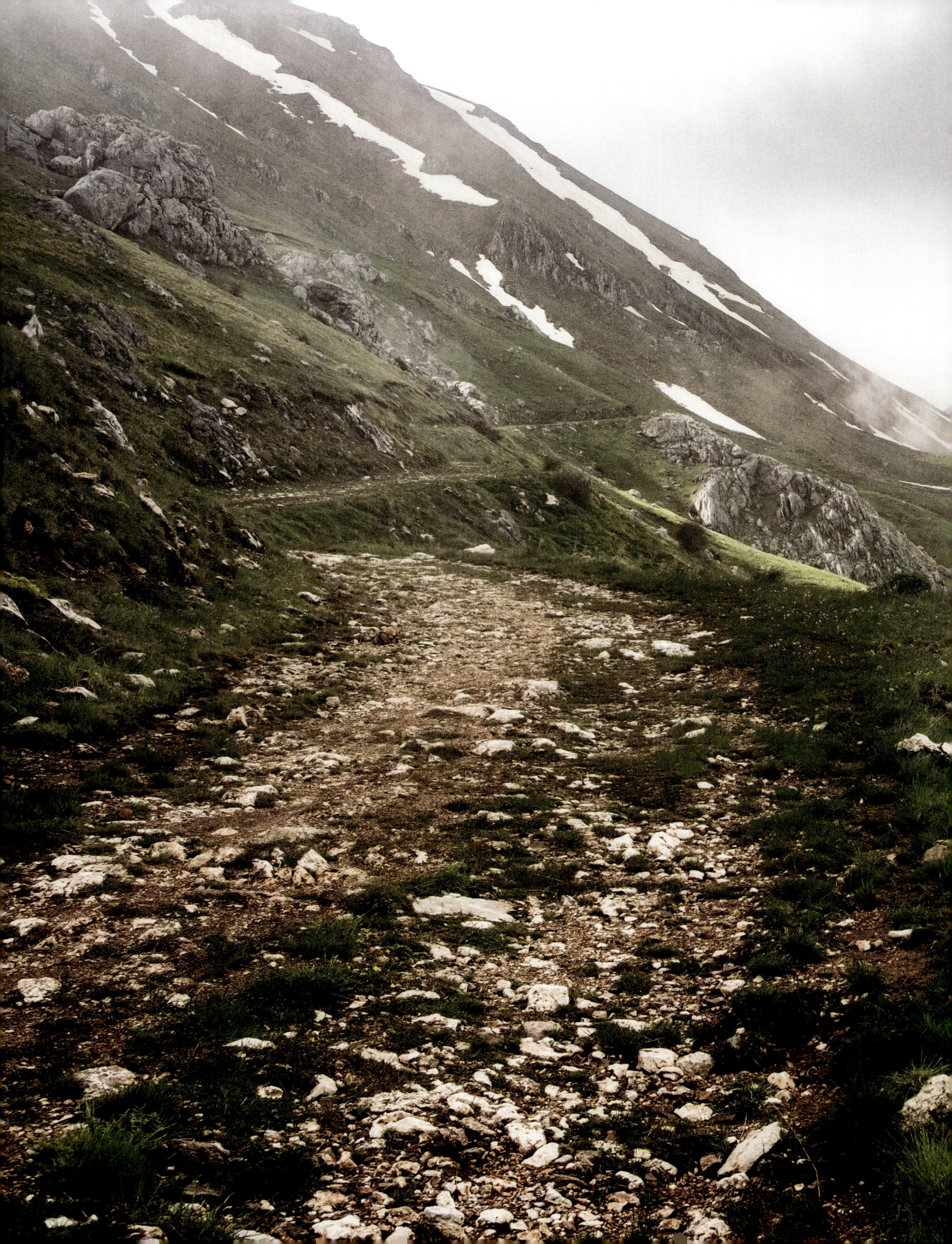

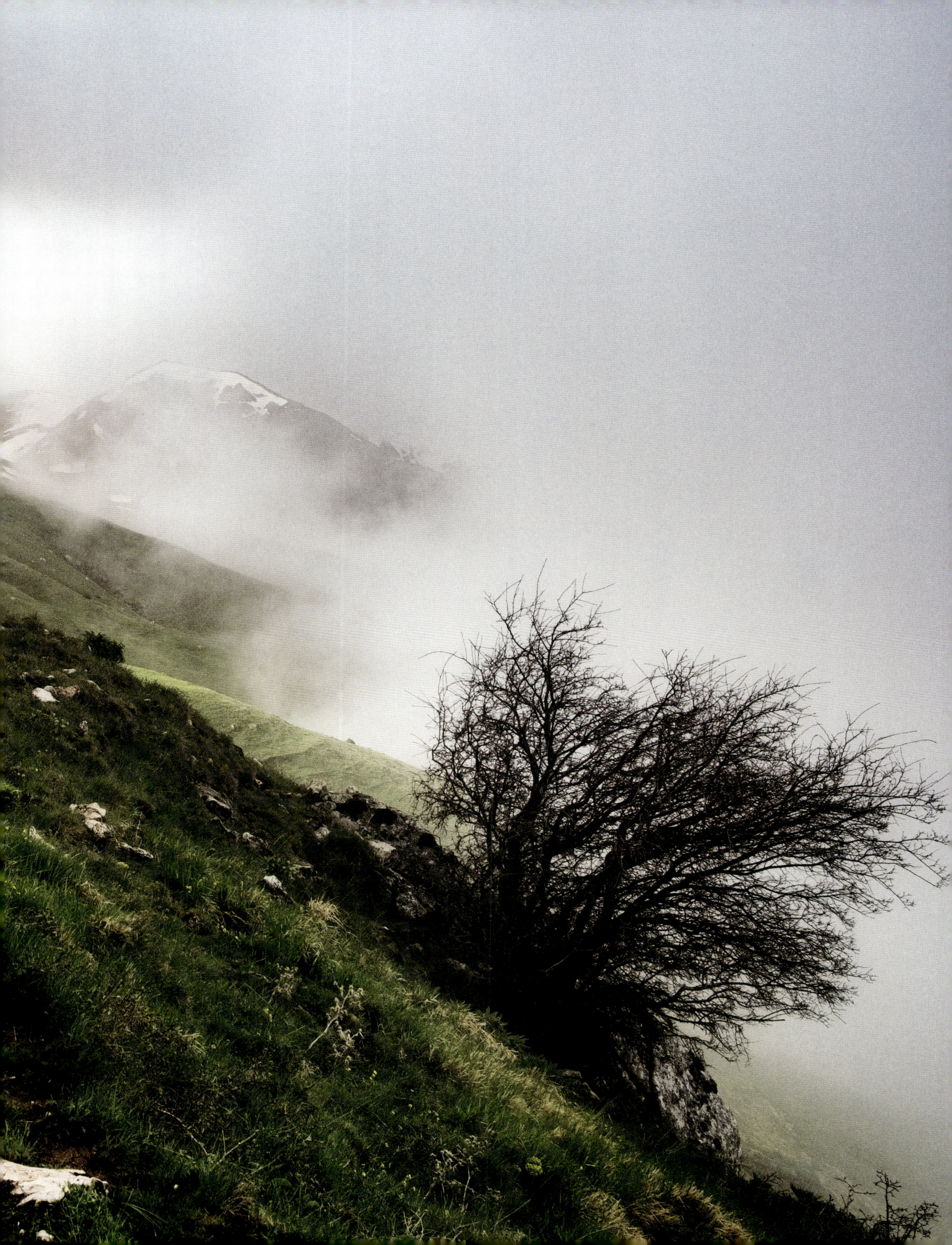

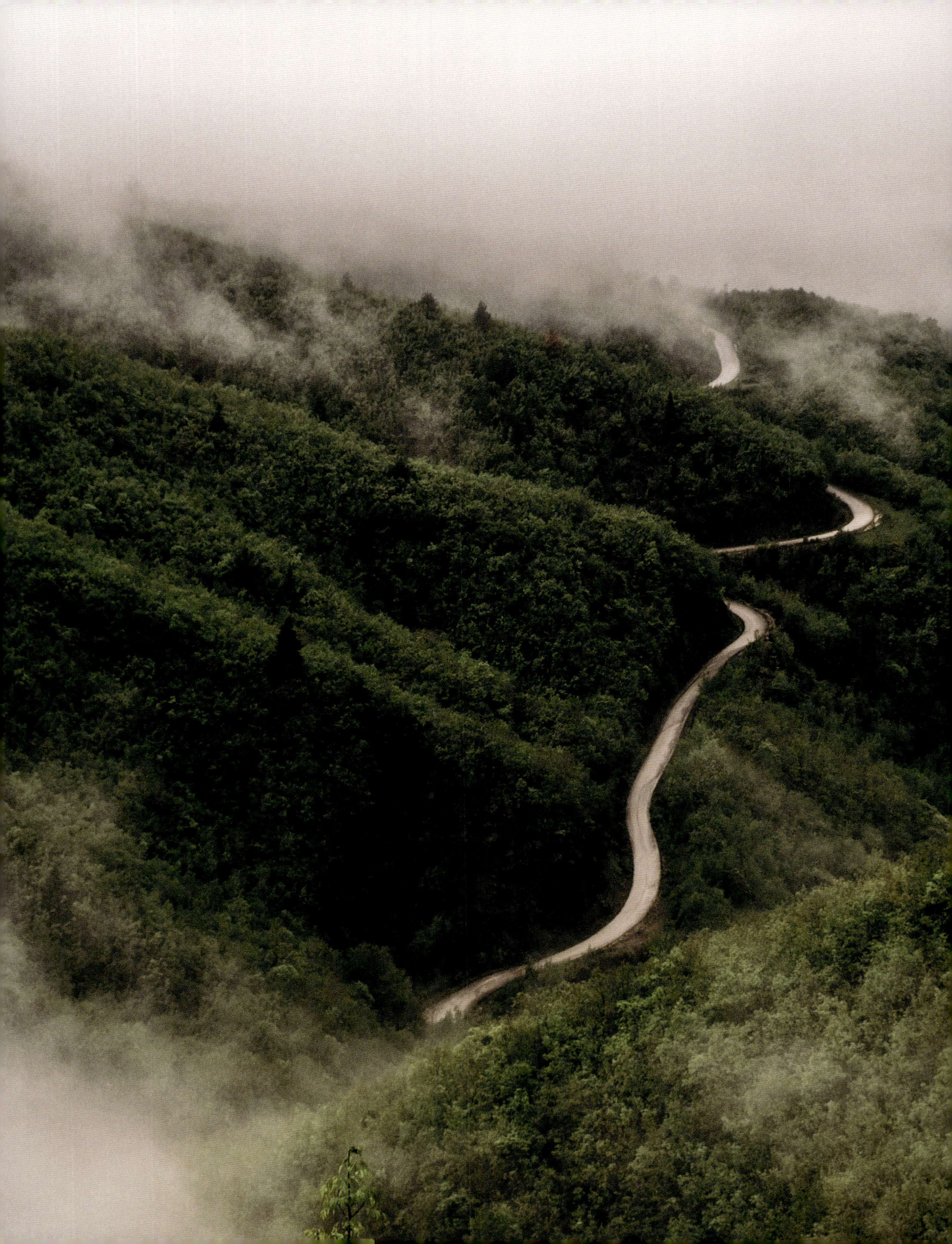

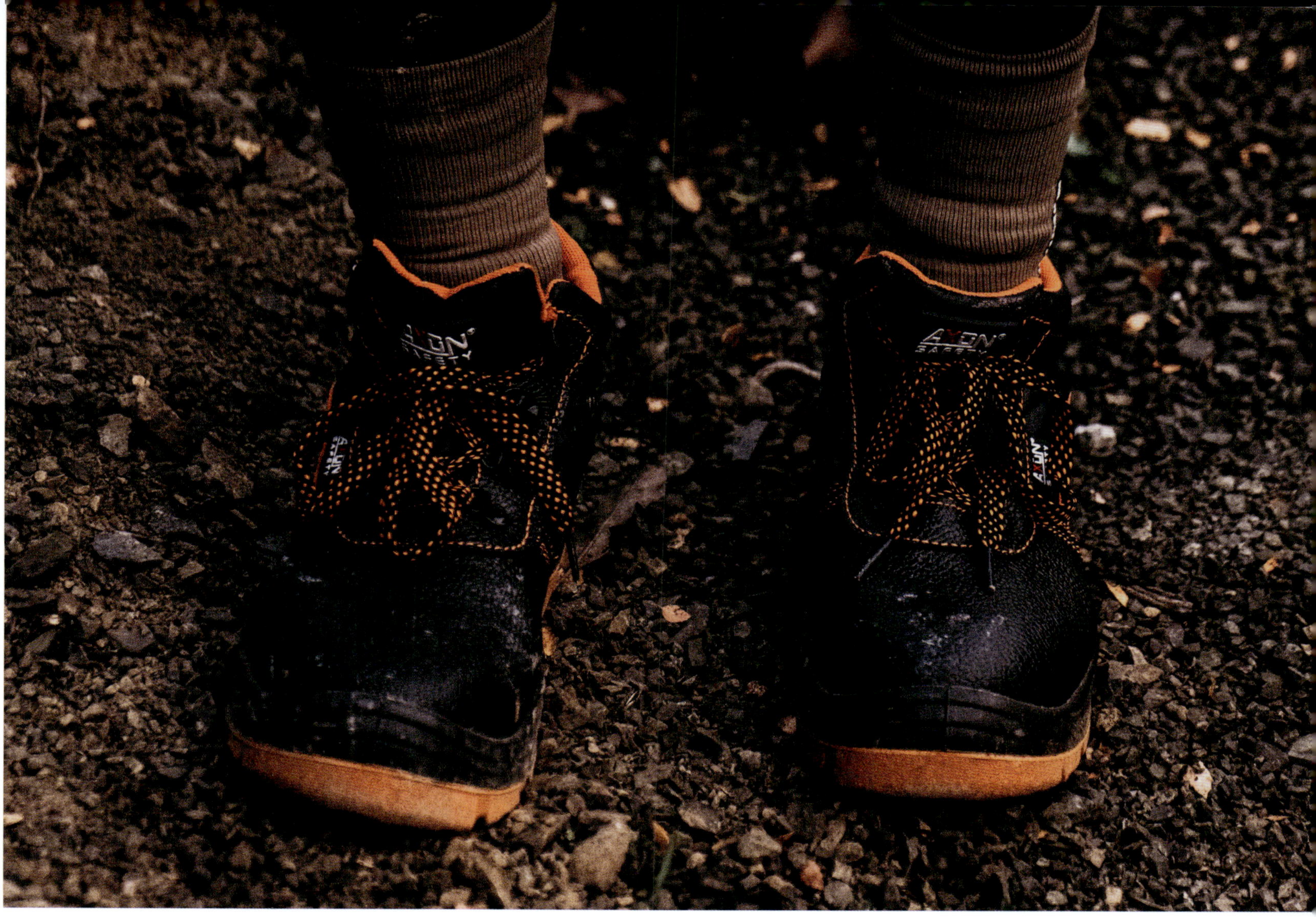

Sicherheitsschuhe mit Stahlkappen?! Ernsthaft? Ich verstehe es jedoch nur zu gut – passen doch meine geschwollenen Füße kaum mehr in die Schuhe.

In besonders gefährlichen Situationen wird auch bei diesem Rennformat die gegenseitige Achtsamkeit geduldet – die Unversehrtheit hat hier absoluten Vorrang.

hätten. Glücklicherweise habe ich mit dieser Art von Rennen meine Form der »Erstbesteigungen« gefunden.

Im Alpinismus werden Schlüsselstellen als die schwierigsten Stellen eines Aufstiegs beschrieben. Es mag verwunderlich klingen, dass gerade Ortschaften und Checkpoints während eines Ultra-Rennens für mich als geselligen Menschen sogenannte Schlüsselstellen darstellen: Beide strahlen mit ihren Menschen, Einkaufsmöglichkeiten und im besten Fall warmen und trockenen Schlafplätzen eine sehr reizvolle Wärme aus und laden zum (zu langen) Verweilen ein.

Der erste Checkpoint im malerischen Ort Papingo, am Fuße des Mount Tymfi gelegen, birgt zusätzlich das Problem, dass die Teilnehmer entweder den sehr schweren »Hike-A-Bike« auf 2.500 Meter Höhe noch vor sich haben und die Nacht zur Regeneration nutzen oder vom etwa acht Stunden langen Auf- und Abstieg zurückkehren und weiterrollen. Man trifft also zwangsläufig Teilnehmer die ein paar Stunden vor oder hinter einem liegen. Beide Situationen machen mich nicht glücklich und von erholsamer Ruhe kann ich in der Nacht am Checkpoint nicht sprechen.

Am nächsten Tag bin ich allerdings heilfroh, den schweren Auf- und Abstieg nicht bei Dunkelheit und weiterhin strömendem Regen angegangen zu sein. Kurze Zeit später sollten die ersten 20 Rennteilnehmer aus dem Rennen aussteigen ...

Im Vergleich zum SRMR FERN und dem AMR soll das Rennen in Griechenland weniger Tragepassagen, leichteres Terrain, kaum erwähnenswerte Flussquerungen und viele »gut rollbare« Abschnitte beinhalten – vom mutmaßlich guten Wetter Mitte Mai einmal abgesehen. Das klang im Vorfeld zumindest theoretisch nach einem Durchmarsch im Gegensatz zu den Strapazen, die ich bei den Erstausgaben der Rennen in Marokko und Kirgisistan auf mich genommen hatte: Extreme Kälte, Flussquerungen durch hüfthohes Wasser, Lebensmittelvergiftungen und alle Erscheinungsformen von Übelkeit und Durchfall ... Davon soll, so der Plan, hier nichts vorkommen.

Wenn man John Lennon glauben darf, »ist das Leben das, was passiert, während wir dabei sind, andere Pläne zu machen«. Ich hatte viele Pläne im Vorfeld gemacht und war

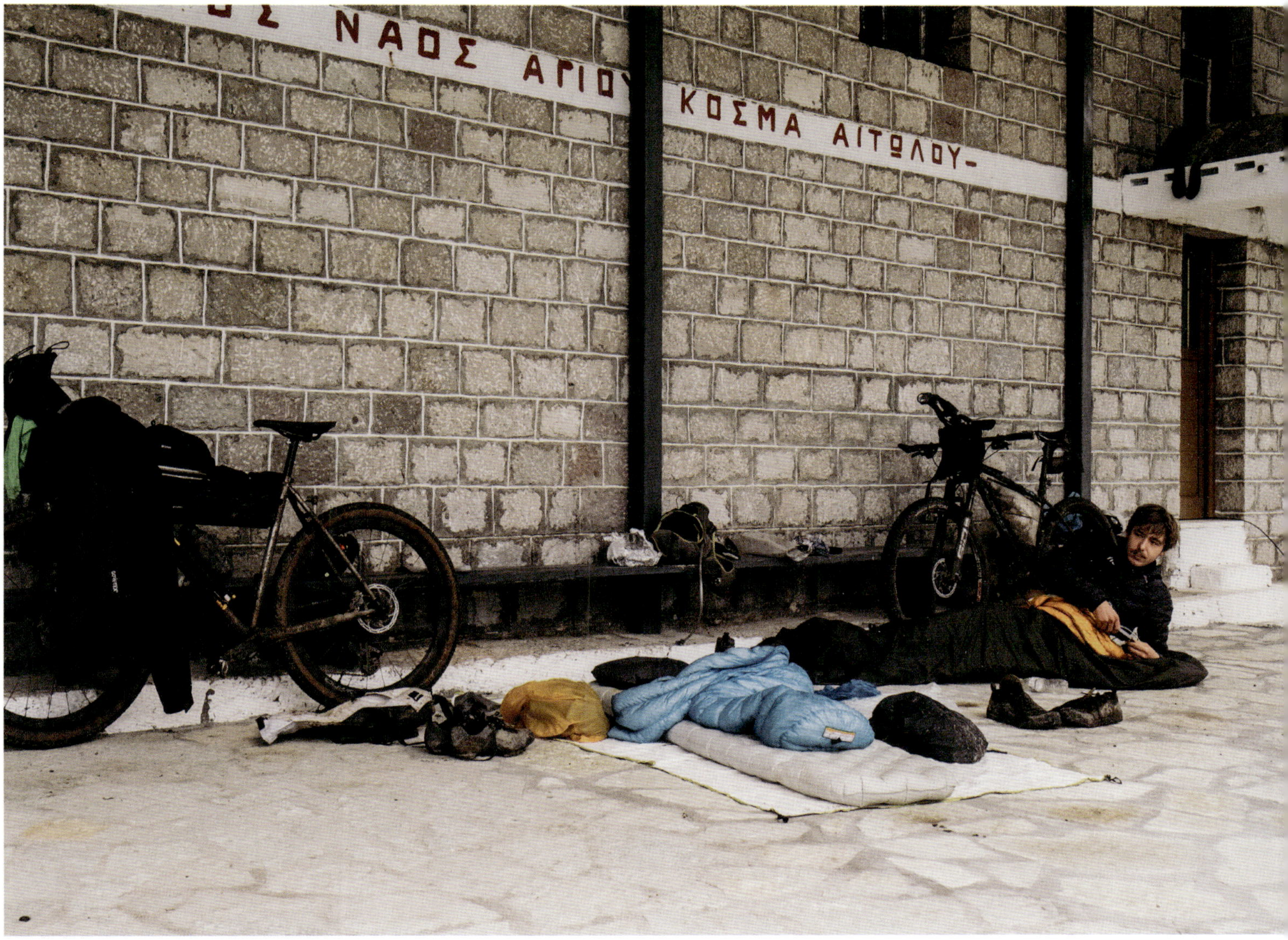

Ich war so unglaublich erleichtert, nach dem nächtlichen Gewaltmarsch über den zugeschneiten Pass, auf der anderen Seite einen Kirchenvorplatz als Schlafplatz nutzen zu können – ein paar Stunden durchatmen. Es war so nötig …

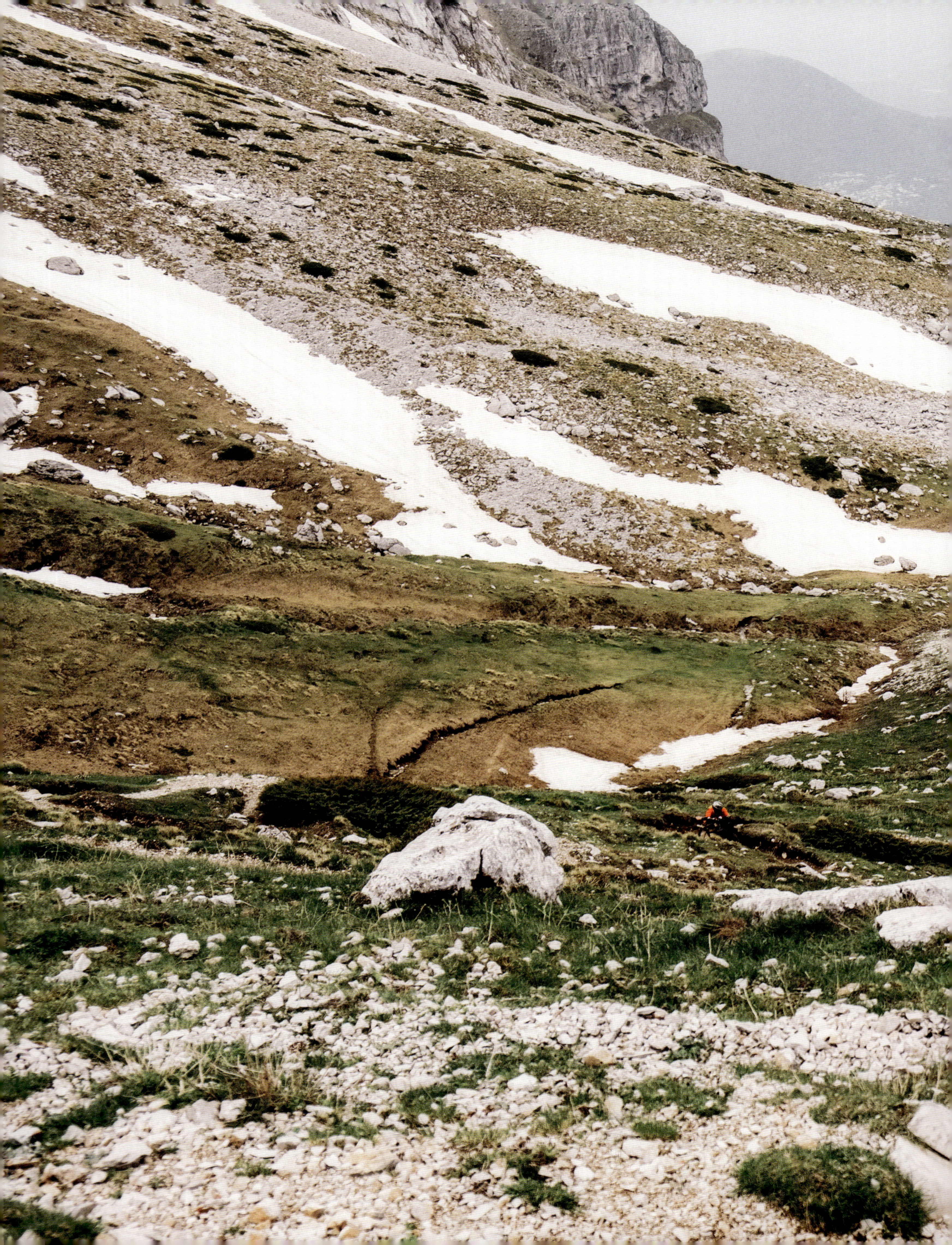

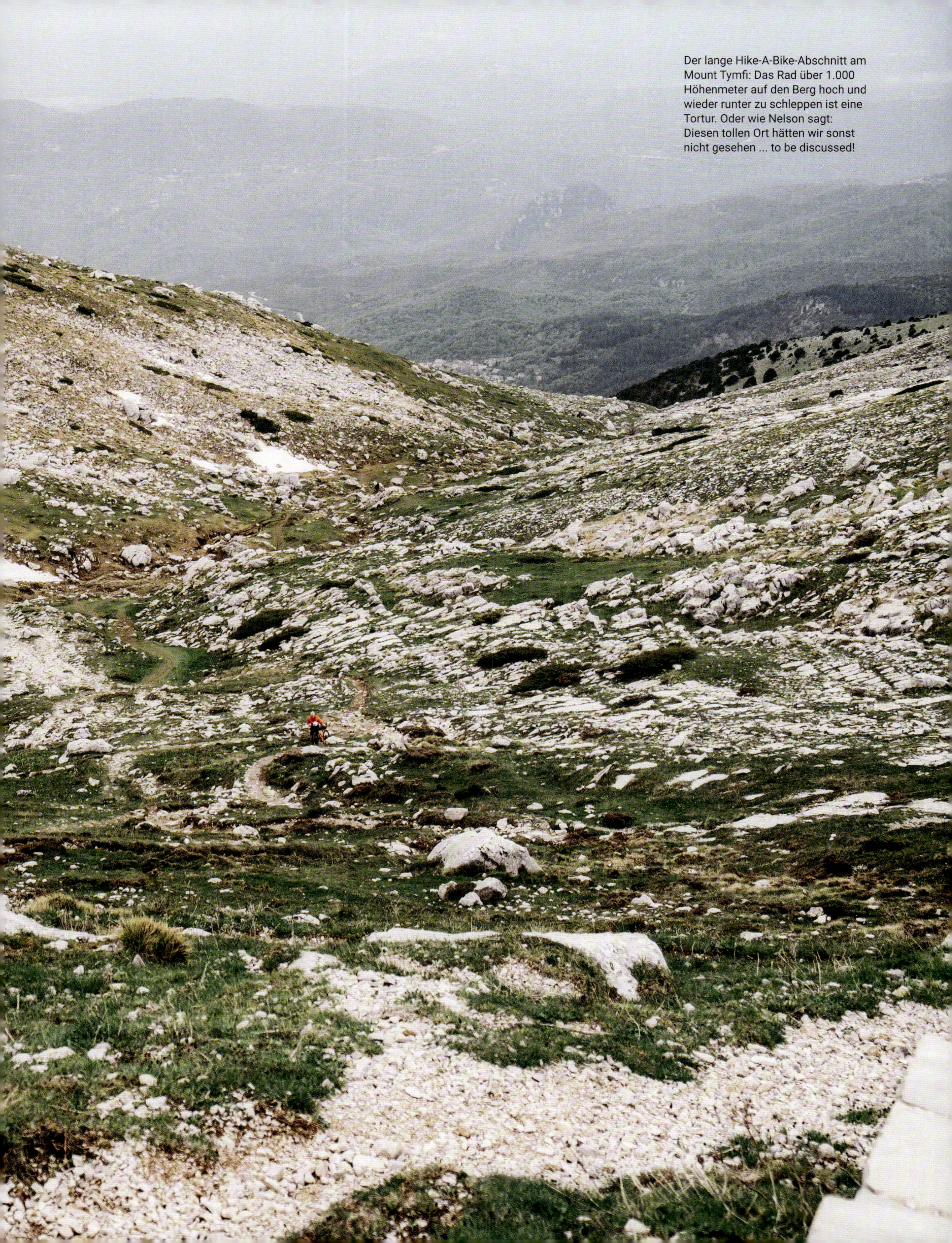

Der lange Hike-A-Bike-Abschnitt am Mount Tymfi: Das Rad über 1.000 Höhenmeter auf den Berg hoch und wieder runter zu schleppen ist eine Tortur. Oder wie Nelson sagt: Diesen tollen Ort hätten wir sonst nicht gesehen ... to be discussed!

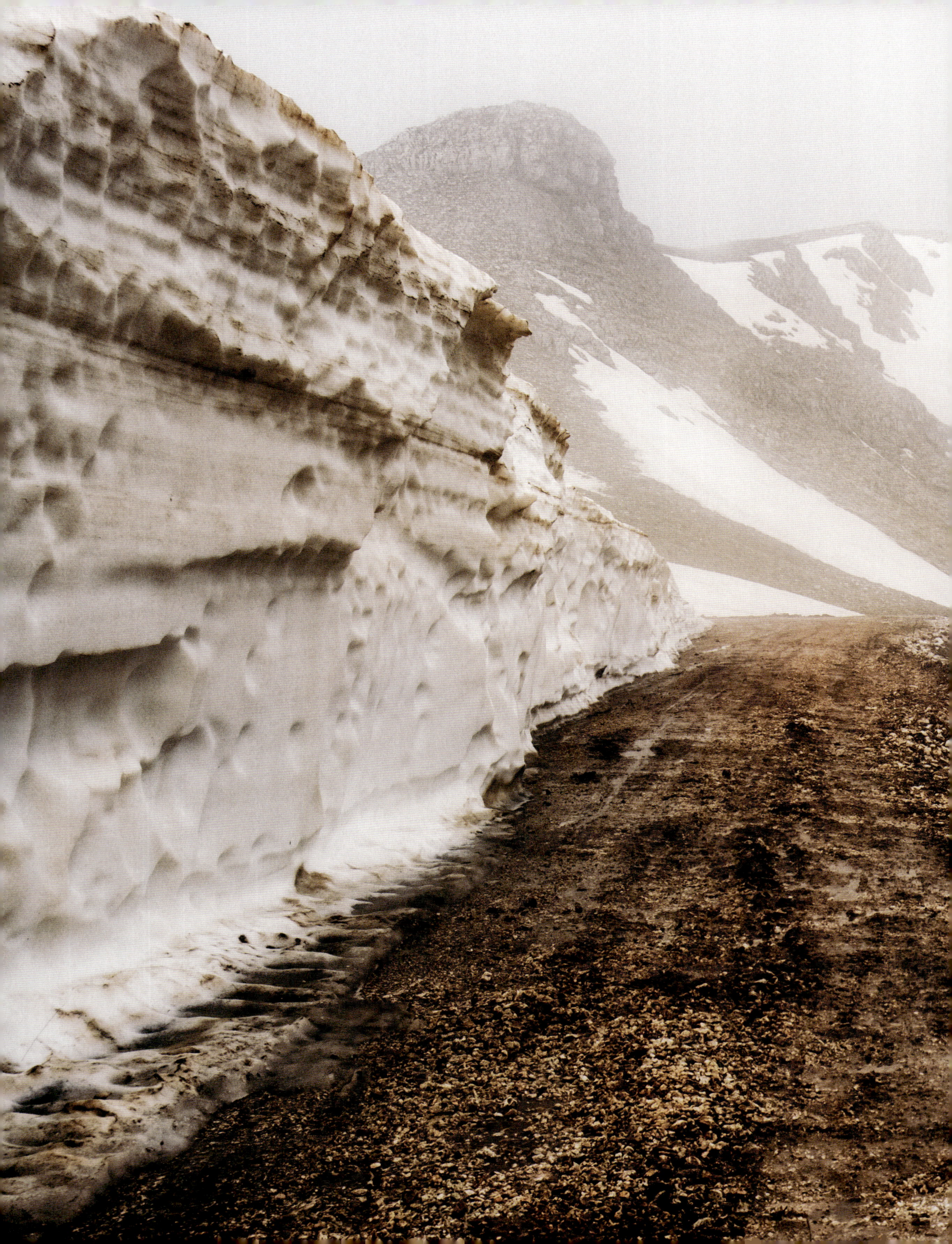

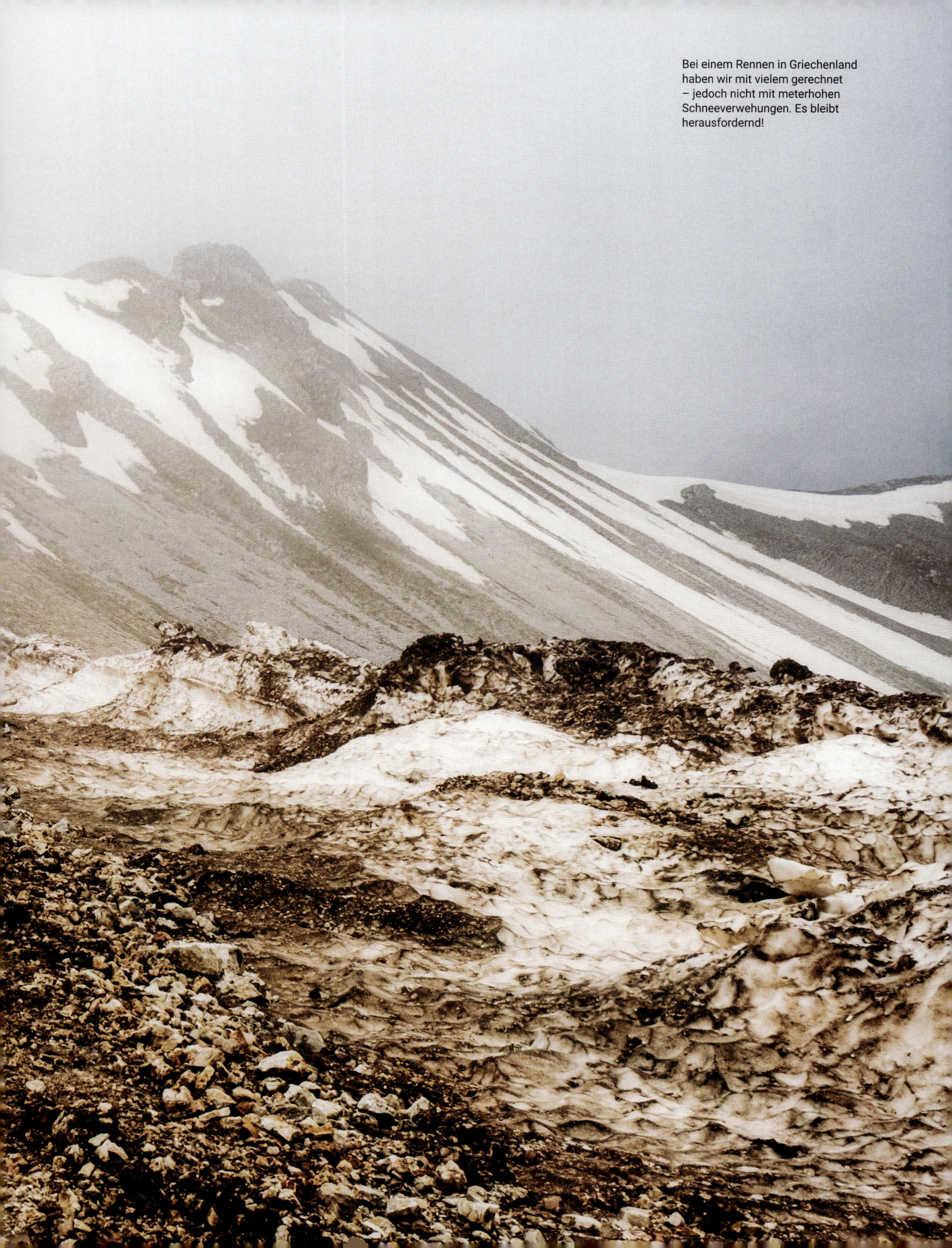

Bei einem Rennen in Griechenland haben wir mit vielem gerechnet – jedoch nicht mit meterhohen Schneeverwehungen. Es bleibt herausfordernd!

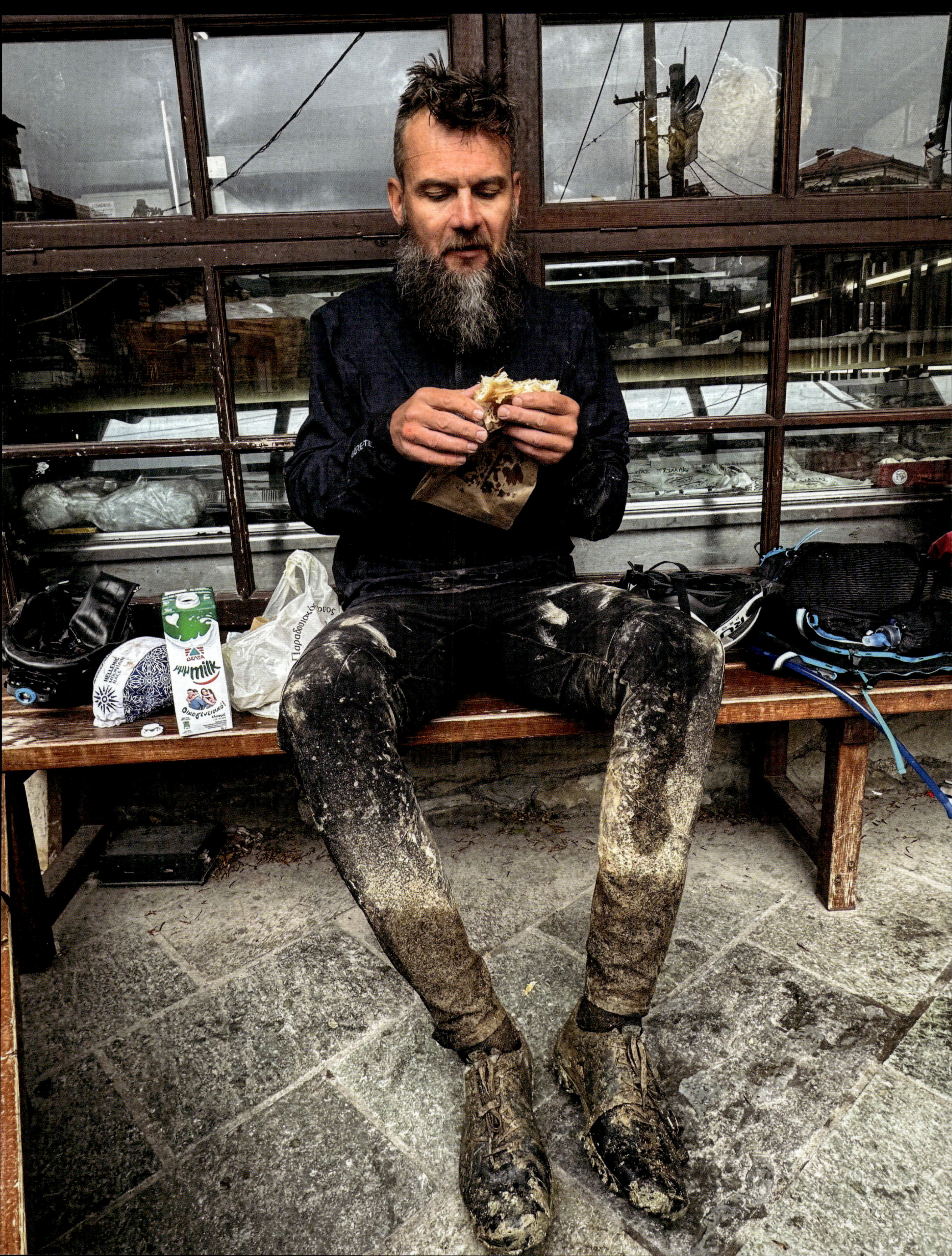
milk

Den Moment werde ich nie vergessen: Als ich nach vier Tagen in Metsovo auf Ondrey aus Tschechien treffe, zittern seine Hände minutenlang beim Versuch sein Börek zu essen. Brutale Anstiege, Regenmassen und Schlamm der letzten Tage haben ihre Spuren hinterlassen.

bestens präpariert: Mit Florian von FERN-Fahrräder und Tine, von GRAMM-TOURPACKING, beide in Berlin produzierend, entstand im Winter 2022 die Idee, mir mein Traum-Fahrrad inklusive Taschen für dieses Unterfangen zu bauen. Wir kennen uns schon lange, und doch konte ich mein Glück kaum fassen, bestand der Traum eines Rades nach Maß wiederum schon seit Jugendtagen. Und dann noch ein echtes FERN! Das top ausgestattete FERN-Rad bewahrte mich allerdings nicht vor der eingangs beschriebenen Verzweiflung.

Mit allem, was ich an Kraft aufbringen kann, schaffe ich am dritten Renntag durch den aufgeweichten Schlamm in acht Stunden doch kaum mehr als 60 Kilometer. Es ist eine reine Tortur. Über mir schließt sich das saftige Grün der Buchenwälder wie ein Torbogen zusammen, wirkt in seiner Endlosigkeit wie ein Tunnel ohne Ausgang. Ich bin im Loch. Alles Grau, kein Licht zu sehen. Kein guter Status, um ein Solorennen zu bestreiten. Ich will einfach nur raus.

Es sind auch hier, wie so oft im Leben, die kleinen Dinge, die eine Geschichte verändern. Als ich kurze Zeit später auf Conan Thai treffe, einen in New York lebenden Amerikaner, kann ich mein Glück kaum fassen. Ein Mensch, ein Bekannter, ein Freund, der überraschend aus dem vermeintlichen Nichts auftaucht. Wir kennen uns seit dem ersten SRMR 2018; schon da war mir Conan als besonderer Fahrer aufgefallen, fuhr er doch eines der härtesten Rennen der Welt auf einem Stahl-MTB mit 26“ und mit nur einem einzigen Gang: »Bengt, you just gotta do your math ...«, beschreibt er mir seine Herangehensweise an die brutalen Steigungen die in Griechenland auf uns warten. Er rechnet sich ganz einfach aus, wann er absteigen und wie lange er sein Rad wird schieben müssen. Ich bin erneut beeindruckt. Vielleicht war diese Begegnung der Grund, warum sich tief in mir neue Energie aufbaut.

Bei Erreichen der Berghütte am zweiten Checkpoint in Melissourgos ist meine Transformation zum aktiven Rennteilnehmer abgeschlossen. Endlich können wir die Tiefen der Wälder vollends hinter uns lassen, die Landschaft hat sich in Hochgebirgsterrain mit der ersehnten Weite verwandelt. Meine Stimmung ist kurz vor Ekstase – es fließen erneut Tränen, diesmal aber vor Glück. Zu einigen der vor mir liegenden Fahrer kann ich aufschließen und das geschäftige Treiben der anderen steckt mich an. Die Berg-

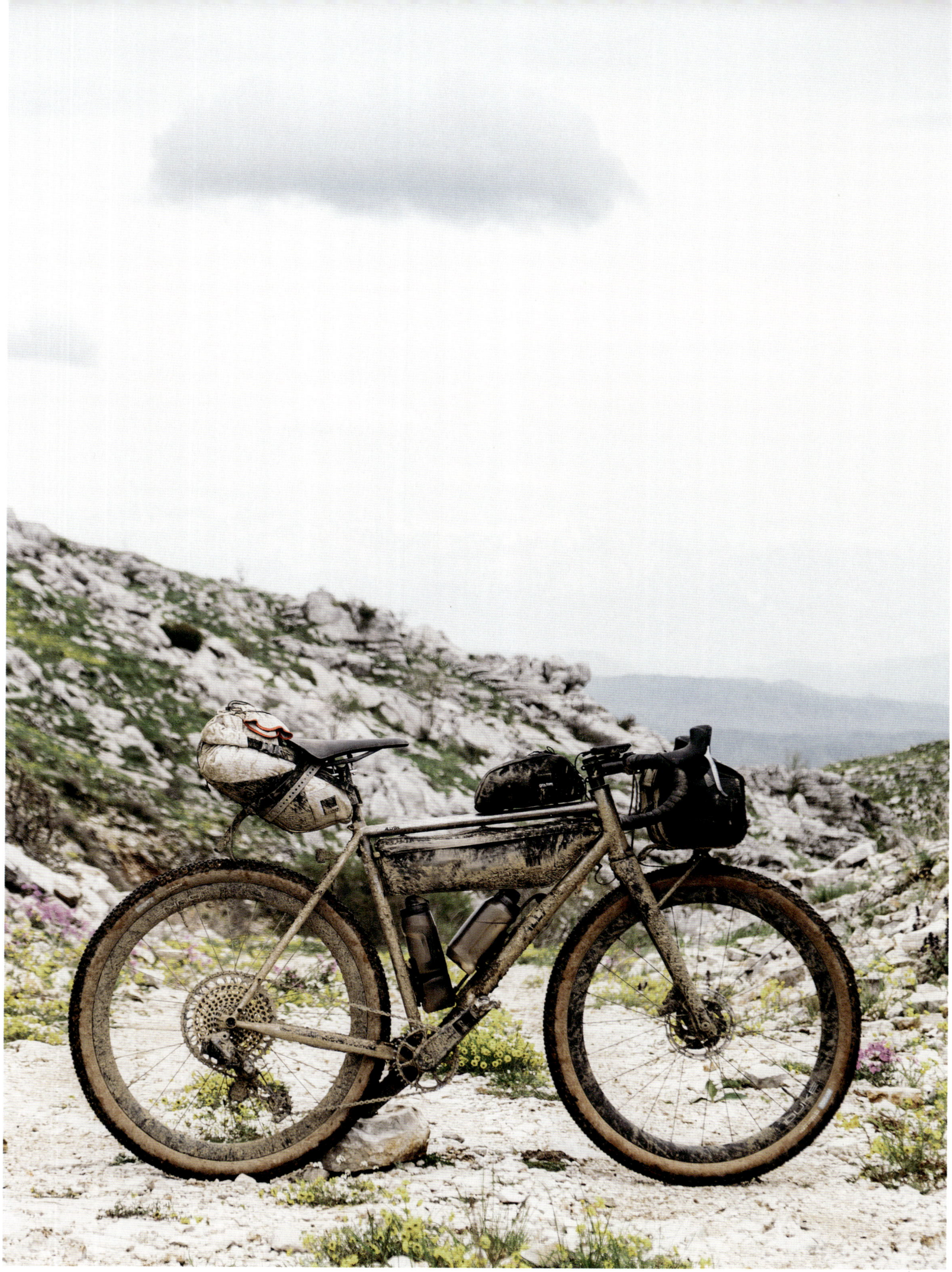

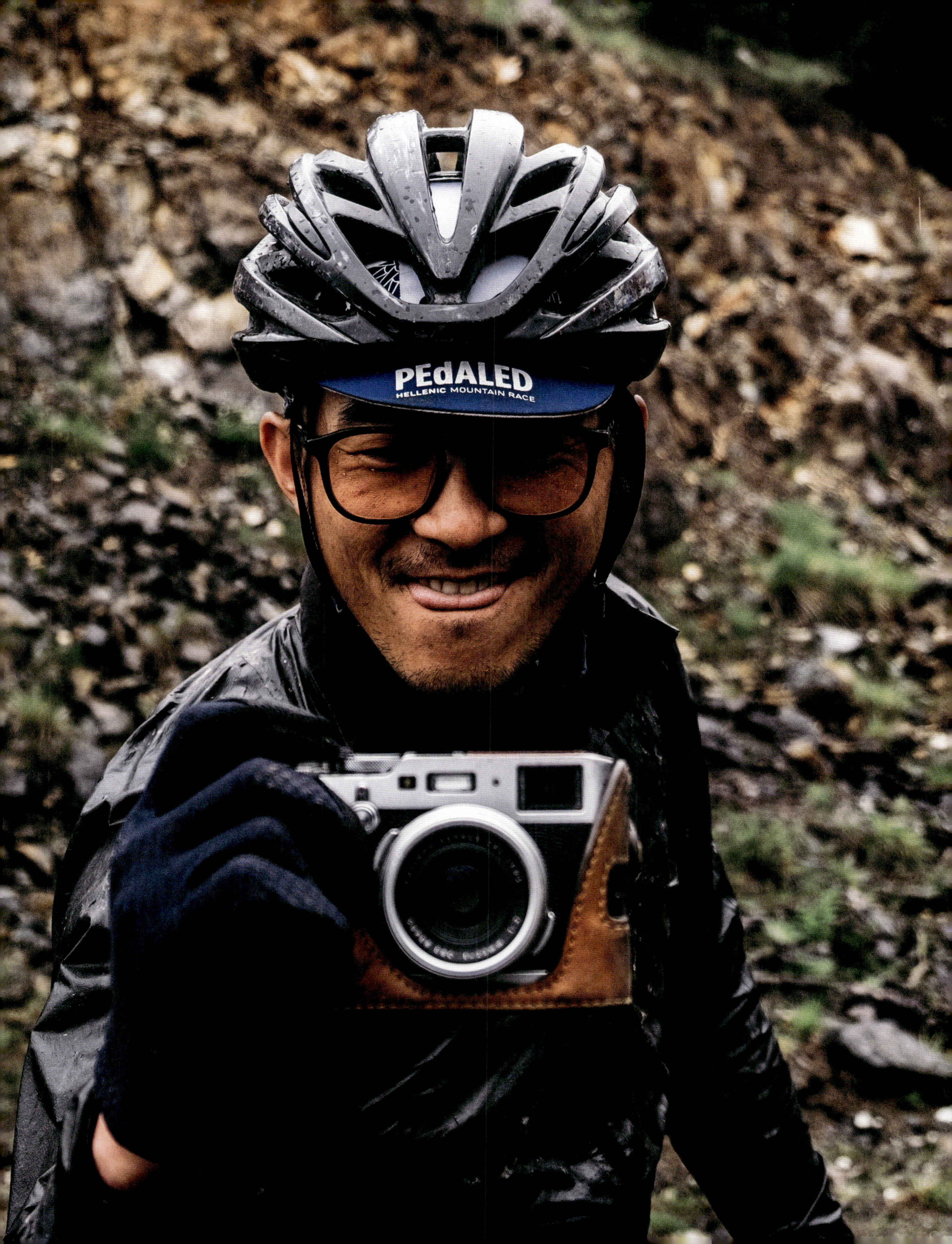
PEdALED
HELLENIC MOUNTAIN RACE

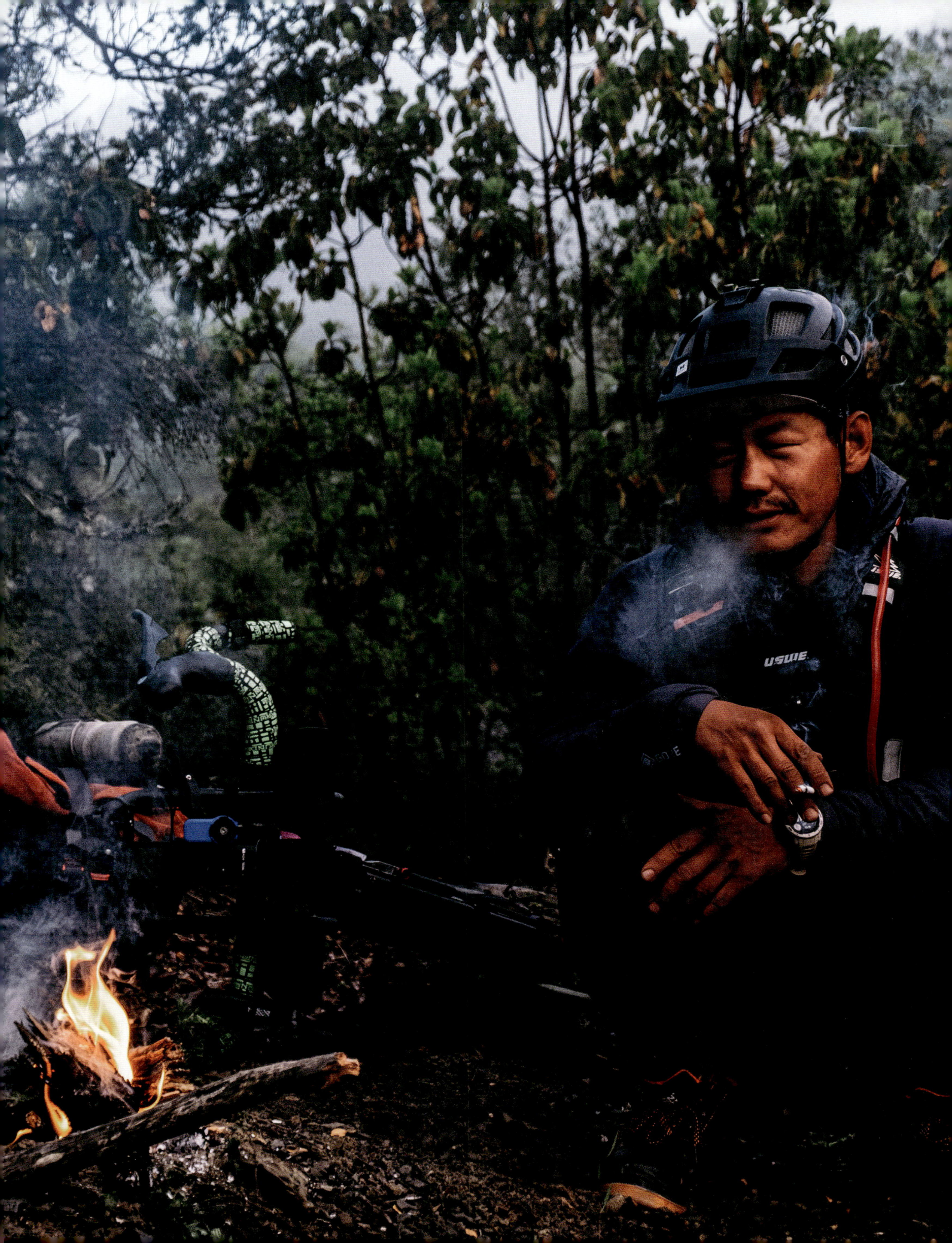

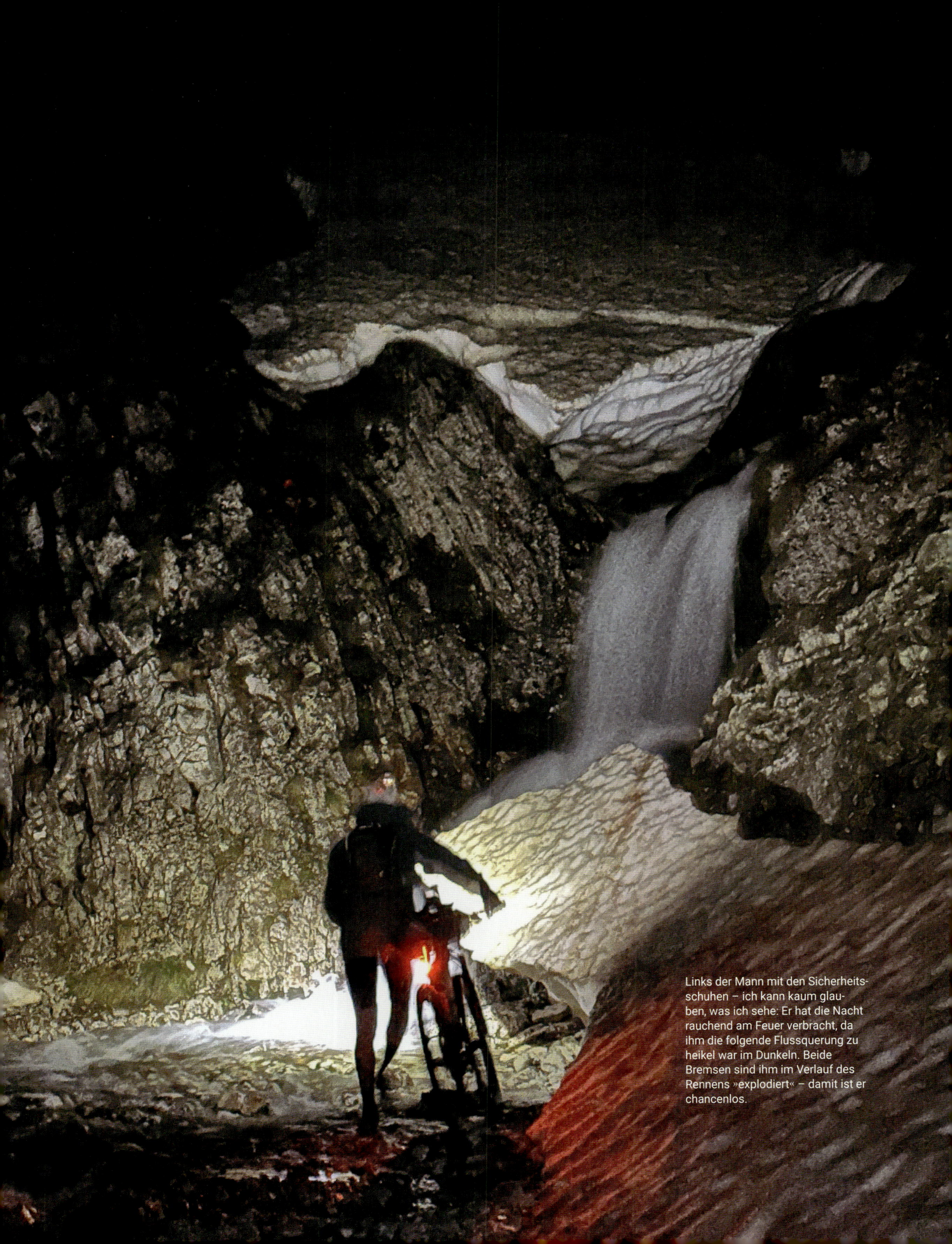

Links der Mann mit den Sicherheitsschuhen – ich kann kaum glauben, was ich sehe: Er hat die Nacht rauchend am Feuer verbracht, da ihm die folgende Flussquerung zu heikel war im Dunkeln. Beide Bremsen sind ihm im Verlauf des Rennens »explodiert« – damit ist er chancenlos.

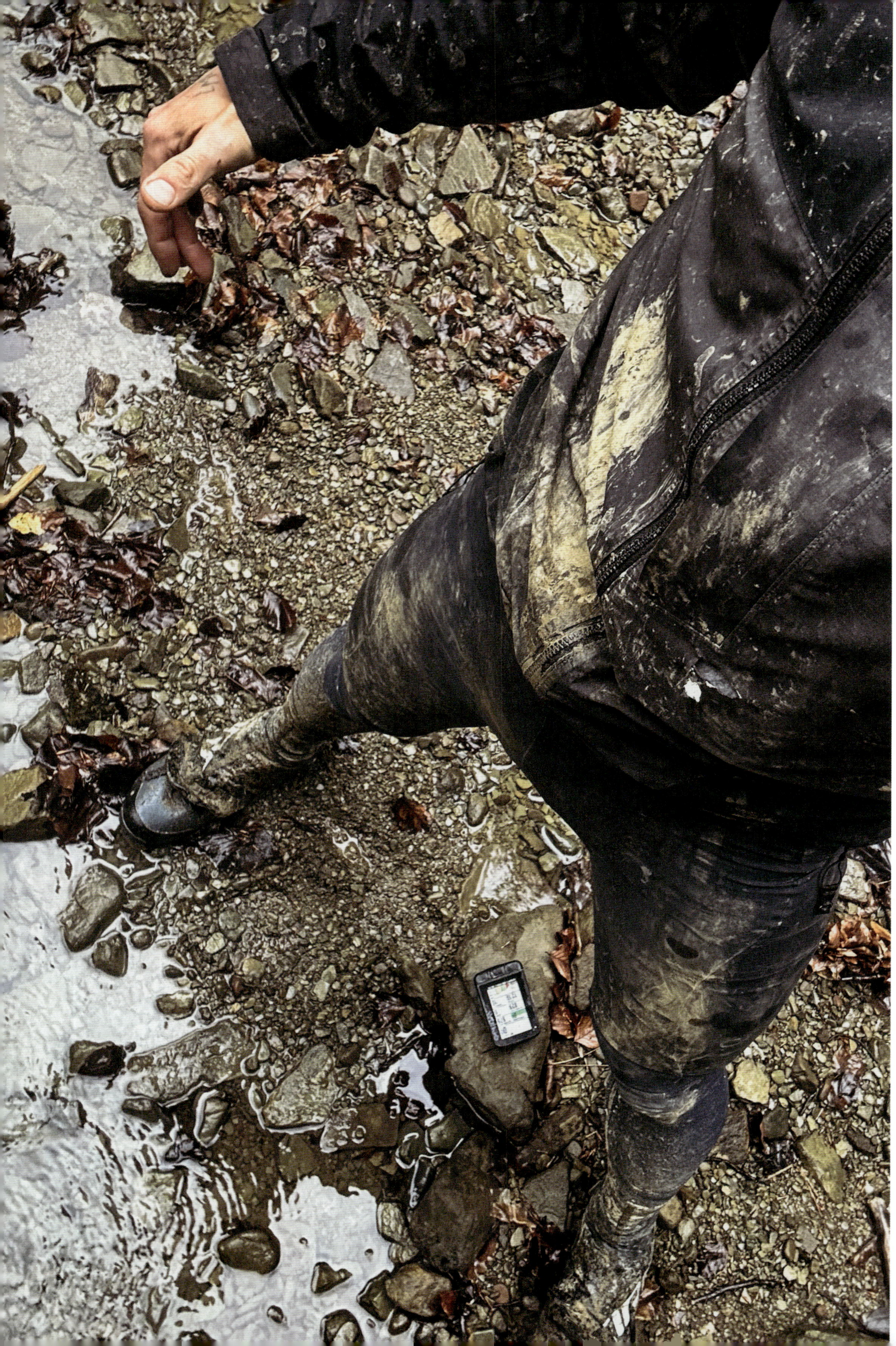

Bodenkontakt: Die Verhältnisse während des Rennens haben mehrfach dazu geführt, dass ich mir den Untergrund sehr genau angeschaut habe.

hütten-Crew versorgt uns mit Bergen von Pasta und an einem warmen Ofen kann alles Nasse kurz durchtrocknen. Es ist ein Traum. Dass ich überhaupt noch am Rennen teilnehme, ist der Tatsache geschuldet, dass am vorherigen Tag aufgrund des Wetters der Zielschluss um zwölf Stunden nach hinten verlegt wurde. Im kleinen Örtchen Metsovo wollte ich nach knapp der Hälfte des Rennens, so wie viele andere Teilnehmer auch, die Segel streichen. Es war der verlockende Reiz, sich mit öffentlichen Bussen an den Zielort fahren zu lassen, der sich hier bot.

Um Mitternacht überquere ich zusammen mit Jules, einem jungen Fahrer aus Belgien, den anspruchsvollsten Pass des Rennens. Tags vorher musste das Team um Nelson eine 30 Zentimeter breite Schneise in die meterhohe Schneewechte schaufeln – es herrschte Lebensgefahr. In besonders gefährlichen Situationen wird auch bei diesem Rennformat die gegenseitige Achtsamkeit geduldet – die Unversehrtheit hat hier absoluten Vorrang. In diesem Moment bin ich heilfroh, jemanden an meiner Seite zu haben.

Es folgen noch unzählige Situationen, in denen ich mich mit allem Mut, den ich aufbringen kann, durch die Horden von Hütehunden hindurcharbeite. Es hilft, ganz entgegen jedem Reflex nur das ruhige Absteigen vom Rad und mit voller Lautstärke zu brüllen, um sich Respekt zu verschaffen. Wir gehören als Radfahrer für sie hier nicht her – verständlich, dass da auf beiden Seiten Nervosität aufkommt.

Interessant ist, dass in der Rückschau diese Momente so viel Raum einnehmen, bilden sie doch nur einen Bruchteil der Erlebnisse während des großen Abenteuers ab. Wichtiger sind die Momente, die ich in dieser tollen Natur erleben darf. Ganz im Rhythmus des monotonen Tretens

Ich habe mein Rad noch nie so lang geschoben. Nach vier Tagen war mein Profil der Schuhe einfach weg. Komplett.

arbeite ich mich auf den wilden Pfaden und Wegen ab und sauge die Gerüche und wahnsinnigen Ausblicke förmlich auf.

Ich frage mich immer wieder, was eigentlich der Erkenntnisgewinn ist, den ich durch diese Form von Abenteuern erziele. Zum einen ist da das sehr tiefe Eintauchen in die jeweilige Umgebung mit ihren Gerüchen und Geräuschen, das direkte Spüren des Wetters, das In-der-Natur-sein, was unter der Belastung zu einer sehr intensiven Erfahrung werden kann. Die Sinne scheinen wie eine Membran aus zart gespannter Seide jede Regung der Natur aufzunehmen.

Durch das stunden- und tagelange gleichmäßige Treten transformieren sich mein Körper und Geist in einen tranceartigen Zustand – ich komme zur Ruhe. Zum anderen entstehen Situationen, die ich mit dem Verlassen der Komfortzone beschreiben könnte. Wenn man sich mit nicht mehr, als dem absolut Notwendigen hinauswagt, kann jede Situation zu einer Herausforderung werden und Antworten fordern, die wir im normalen Alltag nicht gewohnt sind, geben zu müssen.

Hier entsteht in einem Moment der extremen Fokussierung etwas, aus dem ich für meinen gewohnten Alltag ganz Entscheidendes mitnehme: Ich gehe Dinge ruhiger an und lerne dadurch Unsicherheiten zu vermeiden. Denn es kann ungemein helfen, gewohnte Abläufe immer wieder zu brechen oder andere Antworten finden zu müssen. Daraus resultieren eine Zuversicht in meine Kraft und ein Vertrauen in die Welt, die ich als wertvoll empfinde.

Als einer der 35 Finisher der ersten Ausgabe des HELLENIC MOUNTAIN RACE macht sich in mir ein kaum zu beschreibendes Gefühl der Zufriedenheit breit. Dass im Ziel die anderen Rennteilnehmer auf uns warten und uns beglückwünschen, zeichnet die Ultra-Race-Community aus. On to the next one!

Keine Kampfmaschine im Freizeitmodus, sondern lediglich ein ganz friedlicher Straßenhund auf der Suche nach seinen Buddies.

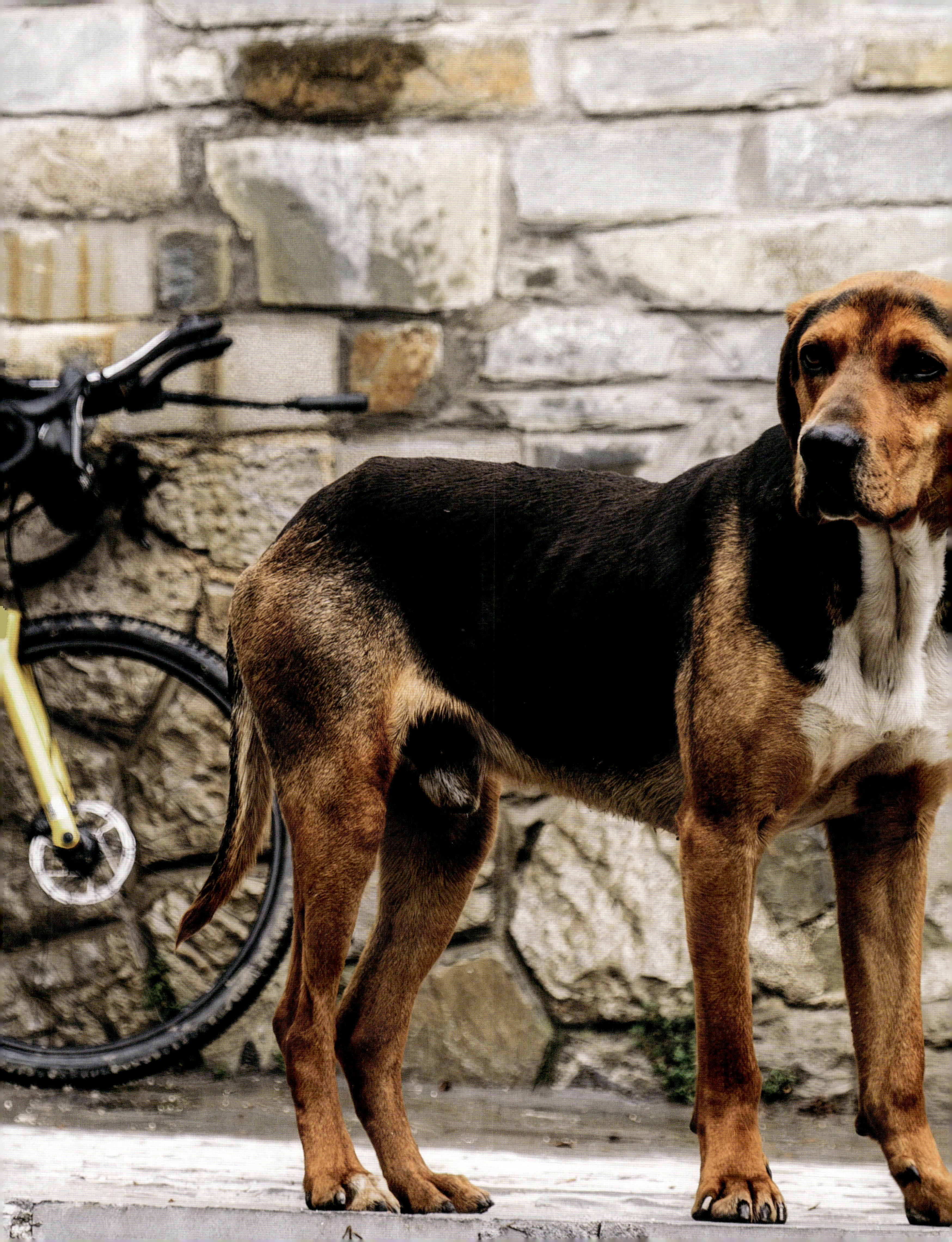

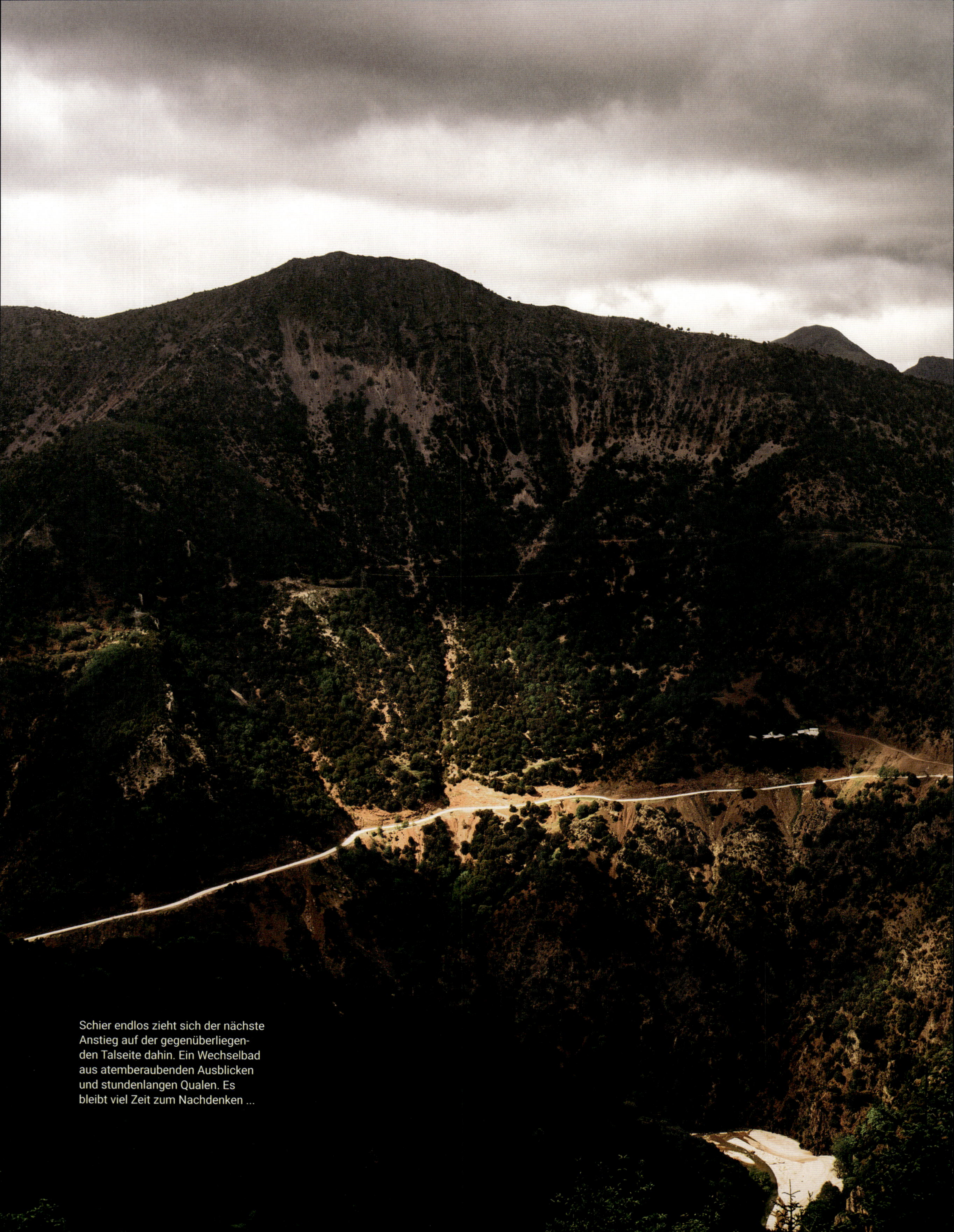

Schier endlos zieht sich der nächste Anstieg auf der gegenüberliegenden Talseite dahin. Ein Wechselbad aus atemberaubenden Ausblicken und stundenlangen Qualen. Es bleibt viel Zeit zum Nachdenken ...

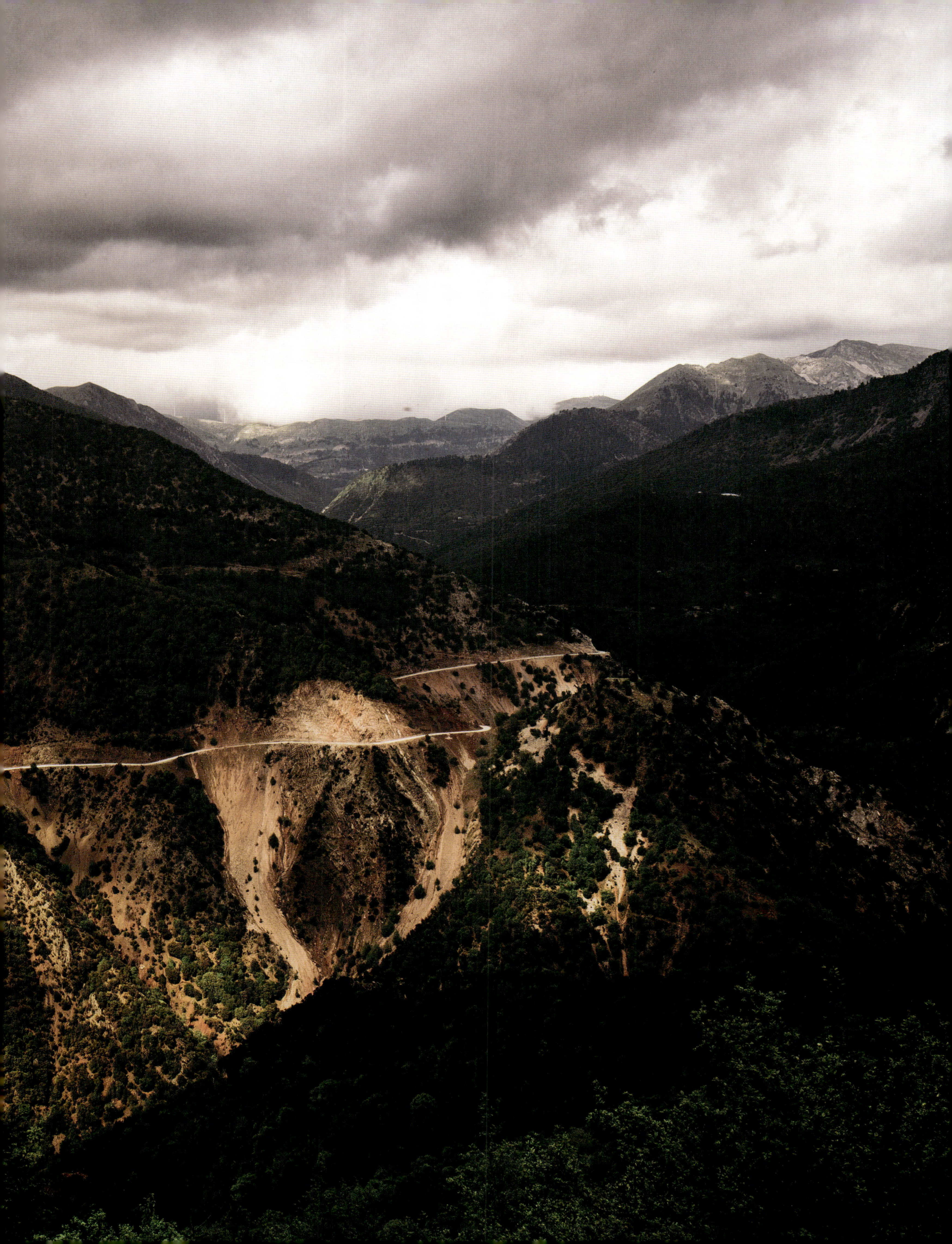

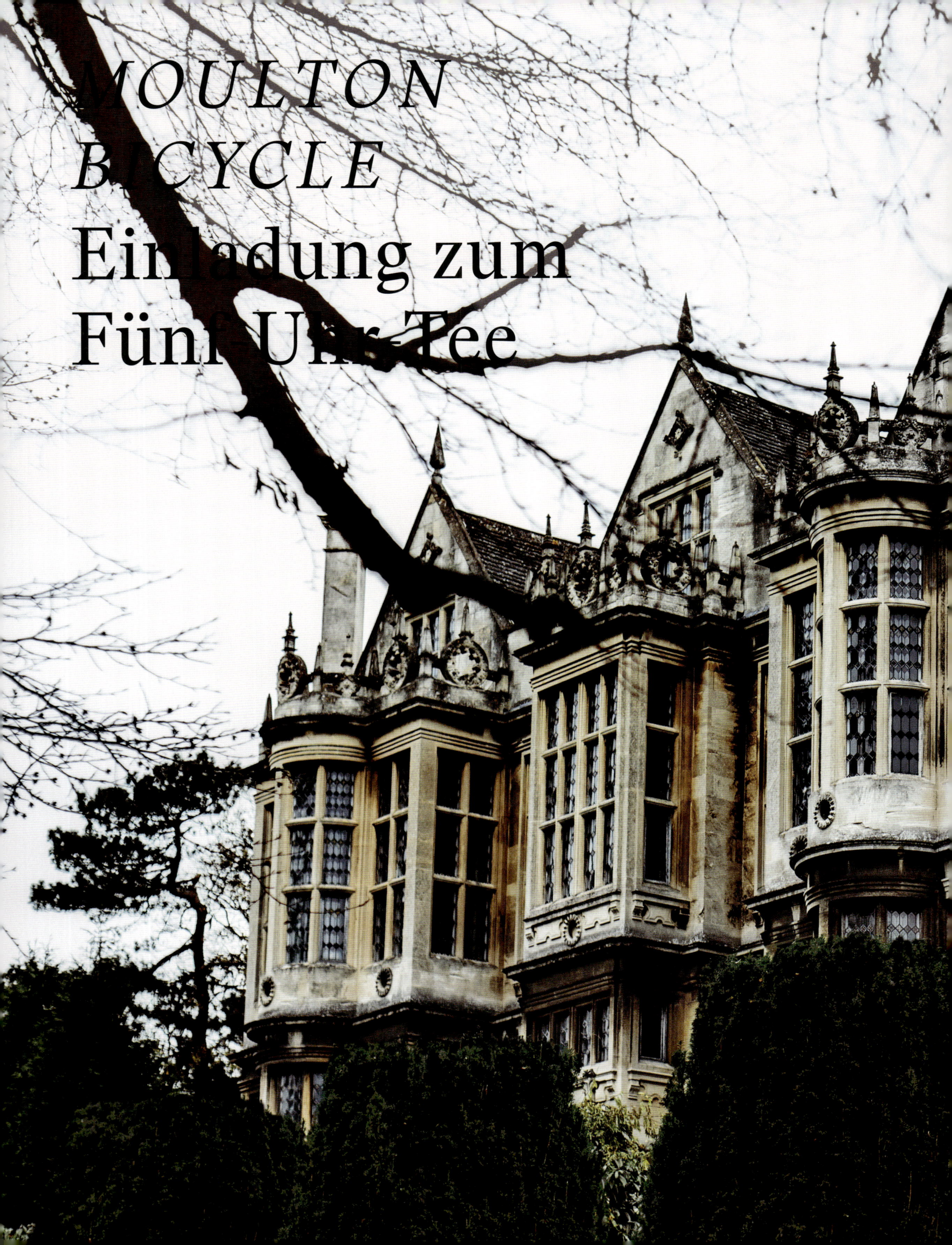

MOULTON BICYCLE

Einladung zum Fünf-Uhr-Tee

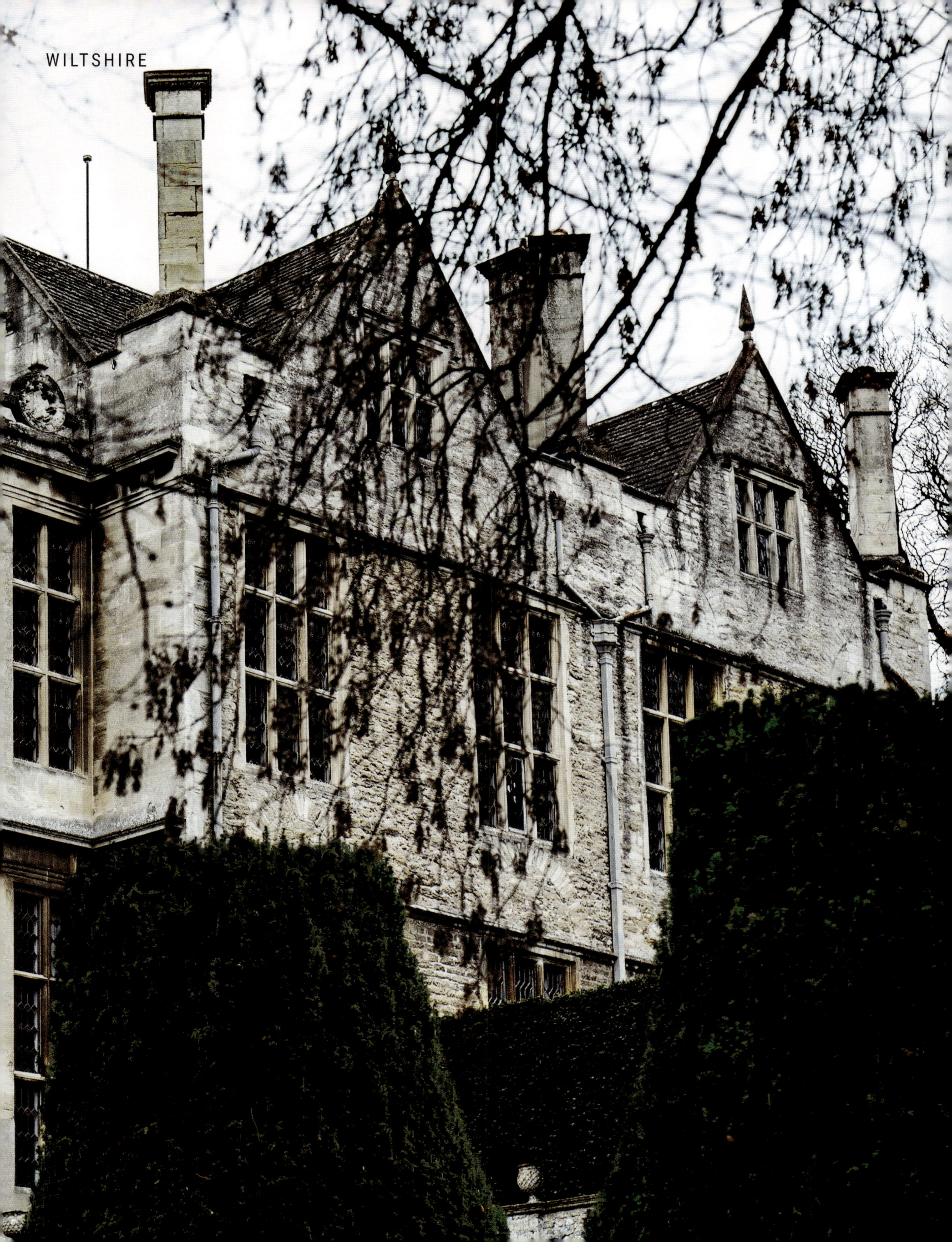

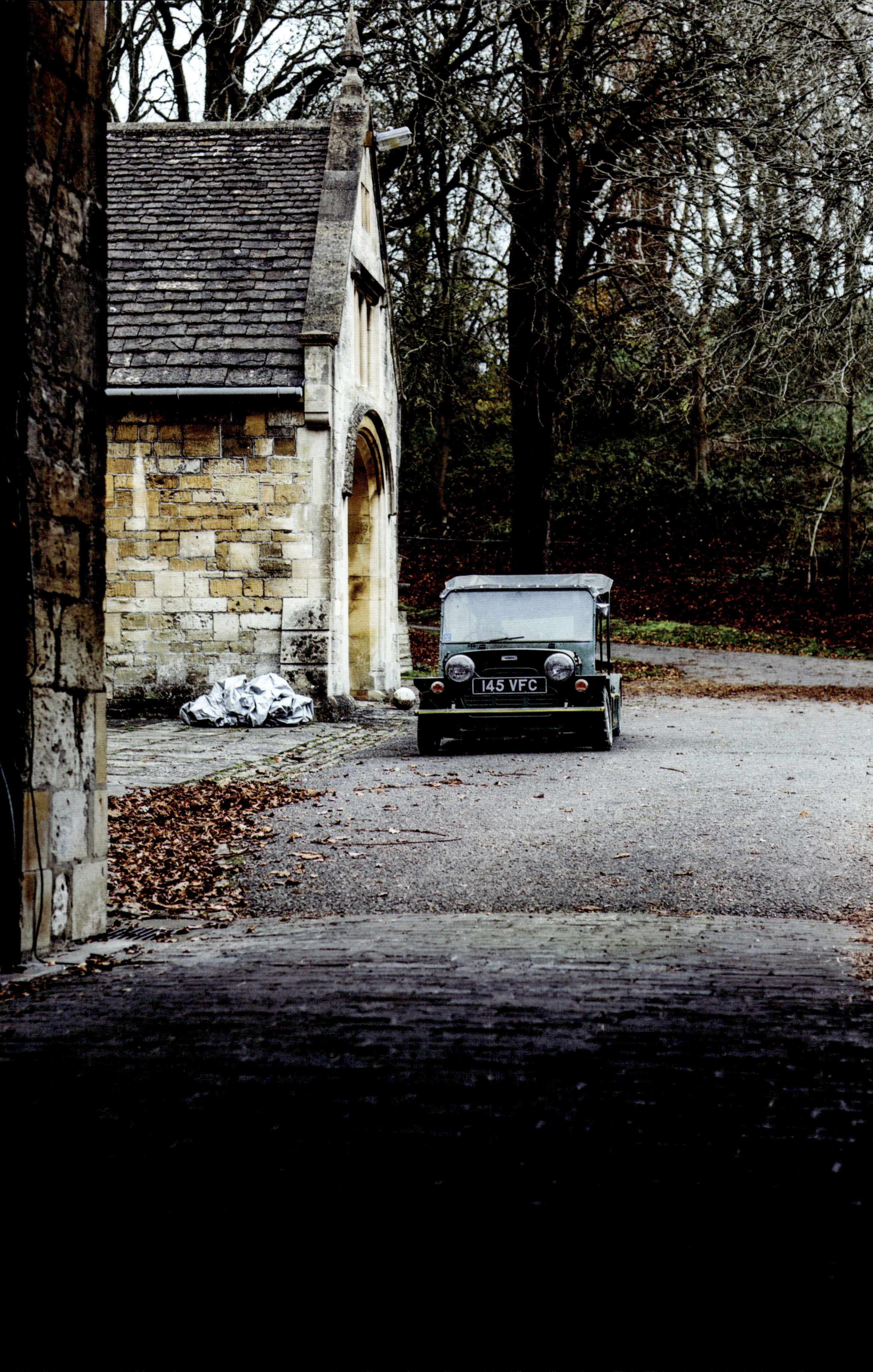
145 VFC

Im Haus hat sich anscheinend nichts verändert. Im Arbeitszimmer meint man plötzlich wirklich, seinen Atem noch spüren zu können.

Ein unvergesslicher Moment, als ich bei Moulton in Bradford on Avon, im südlichen England, auf das Grundstück fahre. Mein Kopf voller Klappräder, begrüßt mich als Erstes ein Auto. Es liegt da, platt wie eine Flunder. Bewegungsunfähig. Aufgebockt. Alle Räder von sich gestreckt. Es ist ein Morris, ein schickes Teil. Aber was macht der hier? Die Antwort gibt mir Shaun, der gerade auf mich zukommt und wunderbar typisch englisch ist. Er begrüßt mich und erklärt, dass die Wagen des verstorbenen Dr. Moulton nächste Woche bei SOTHEBY'S versteigert werden. Lauter kleine, aber feine Fahrzeuge. Diese Preziose benötige dafür noch neue Reifen.

Notiz am Rande: Bevor Dr. Alex Moulton sein inzwischen berühmtes Rad mit kleinen Rädern auf den Markt brachte, verkaufte er das familiäre Gummi-Unternehmen an DUNLOP und wirkte für eine Weile entscheidend bei der Entwicklung verschiedener Autotypen mit. Unter anderem war er zur Zeit der Suezkrise und des damit verbundenen Rohstoffmangels in das Projekt MINI involviert. Er war sich anscheinend sicher: Klein, stylishfein und mit der gleichen Leistung wie die großen Vorgänger muss möglich sein.

So entwickelte er in den 1960er-Jahren ein Fahrrad, das trotz seiner Minireifen gut fährt und das trotz seines Klapprahmens stabil und bequem ist. Der Prototyp wurde inzwischen zwar verbessert, wurde steifer, leichter und schneller gemacht – doch das Grundprinzip hat seinen Erfinder überdauert.

Im Alter von 92 Jahren starb Dr. Moulton. Das war 2012. Doch ein Rundgang durch seine Villa – oder ist das schon eher ein Schloss? – vermittelt den Eindruck, er wohne und arbeite noch immer hier. Im Haus hat sich anscheinend nichts verändert. Im Arbeitszimmer meint man plötzlich wirklich, seinen Atem noch spüren zu können. Auf seinem Schreibtisch reihen sich Notizen auf kleinen Zetteln aneinander, ein paar Stifte liegen verstreut herum, die Leselupe daneben. Sein voll verchromtes MOULTON steht auf der anderen Seite des Raumes. Es ist eine Sonderanfertigung, ein ganz persönliches Geburtstagsgeschenk seiner Mitarbeiter. Erinnerungen an über sechzig Jahre erfolgreiche Erfinderarbeit. Alles fügt sich zusammen und hat einen gewissen Charme, so auch die zwei EAMES-Sessel, die dem Ingenieur wohl zur Entspannung gedient haben. Reinsetzen und Geistesblitze à la Daniel Düsentrieb haben – das wäre es!

Da stand er noch, der kleine süße Wagen auf dem Anwesen Moulton. Stunden später war er weg. Moulton hätte sich sicher im Grabe umgedreht, wenn er gewusst hätte, dass seine feine Autosammlung bei SOTHEBY'S in London unter den Hammer kommen würde.

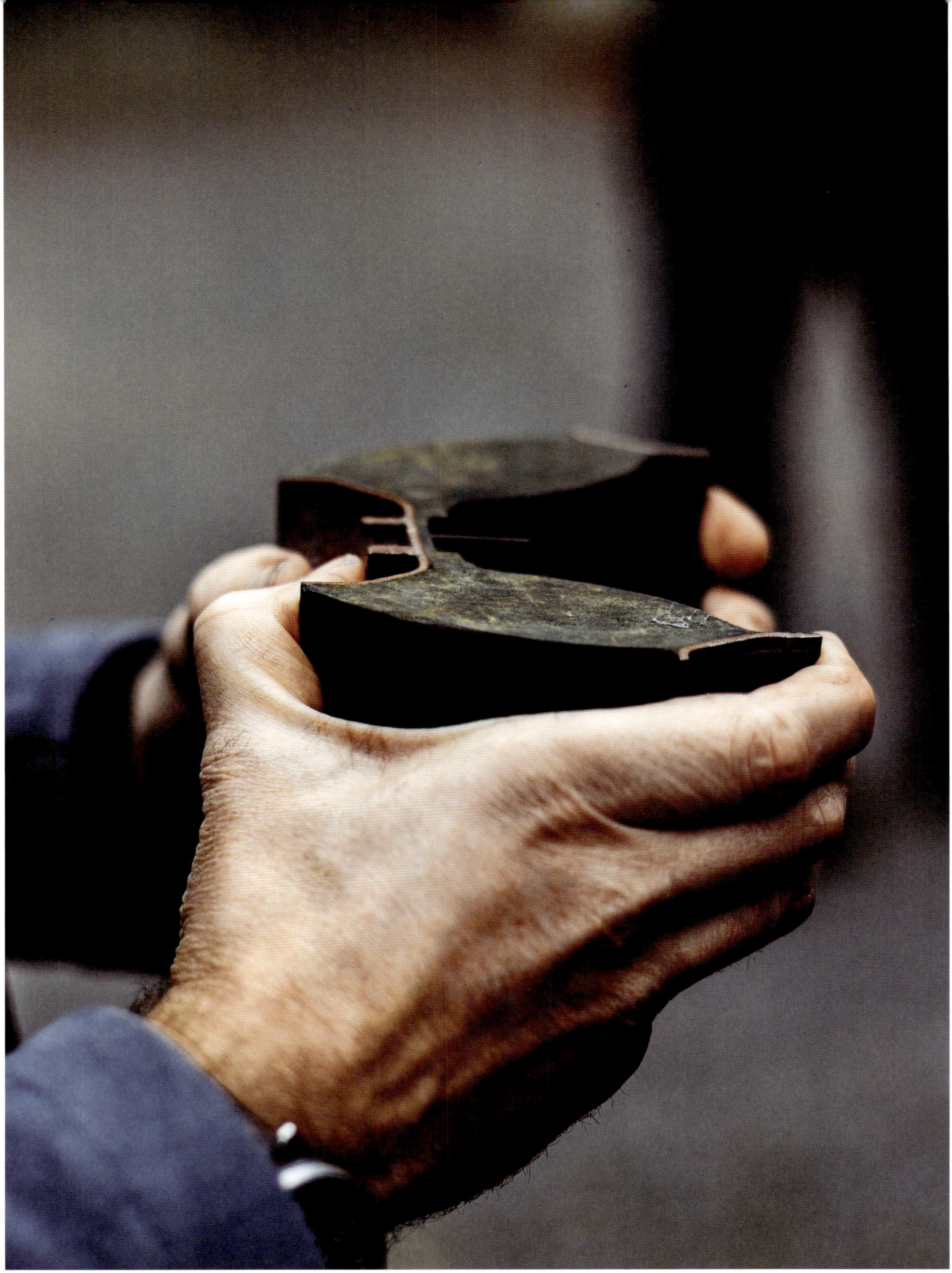

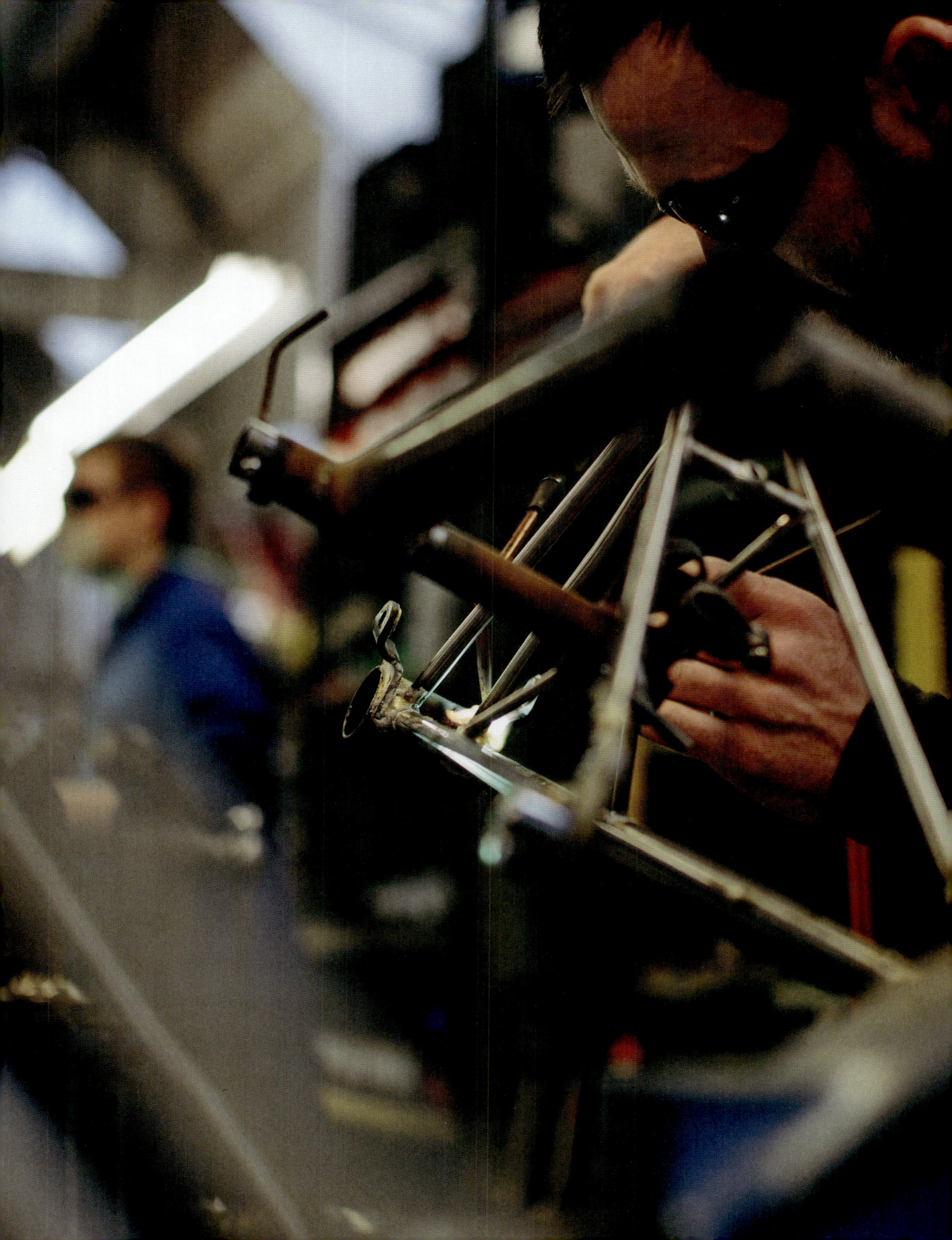

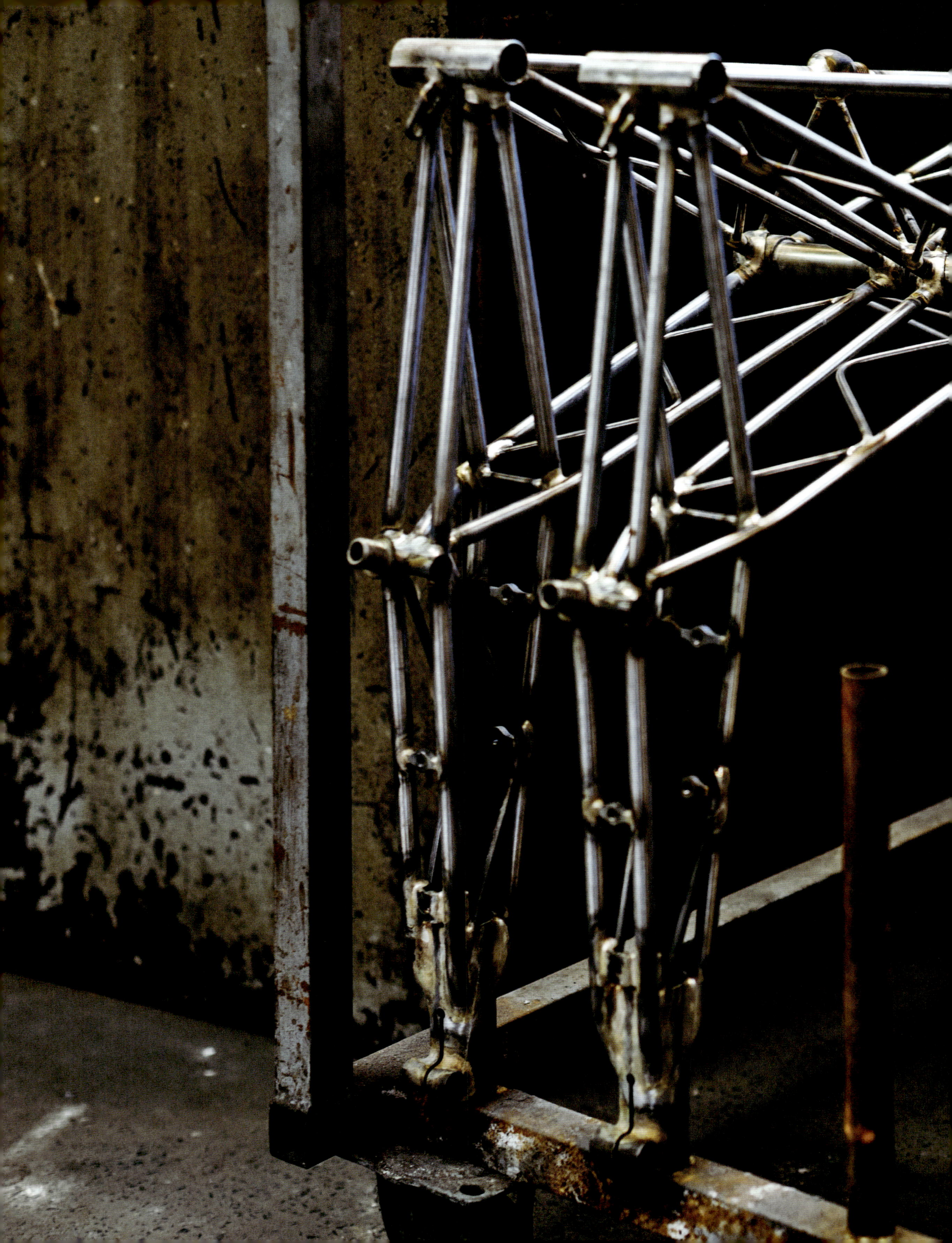

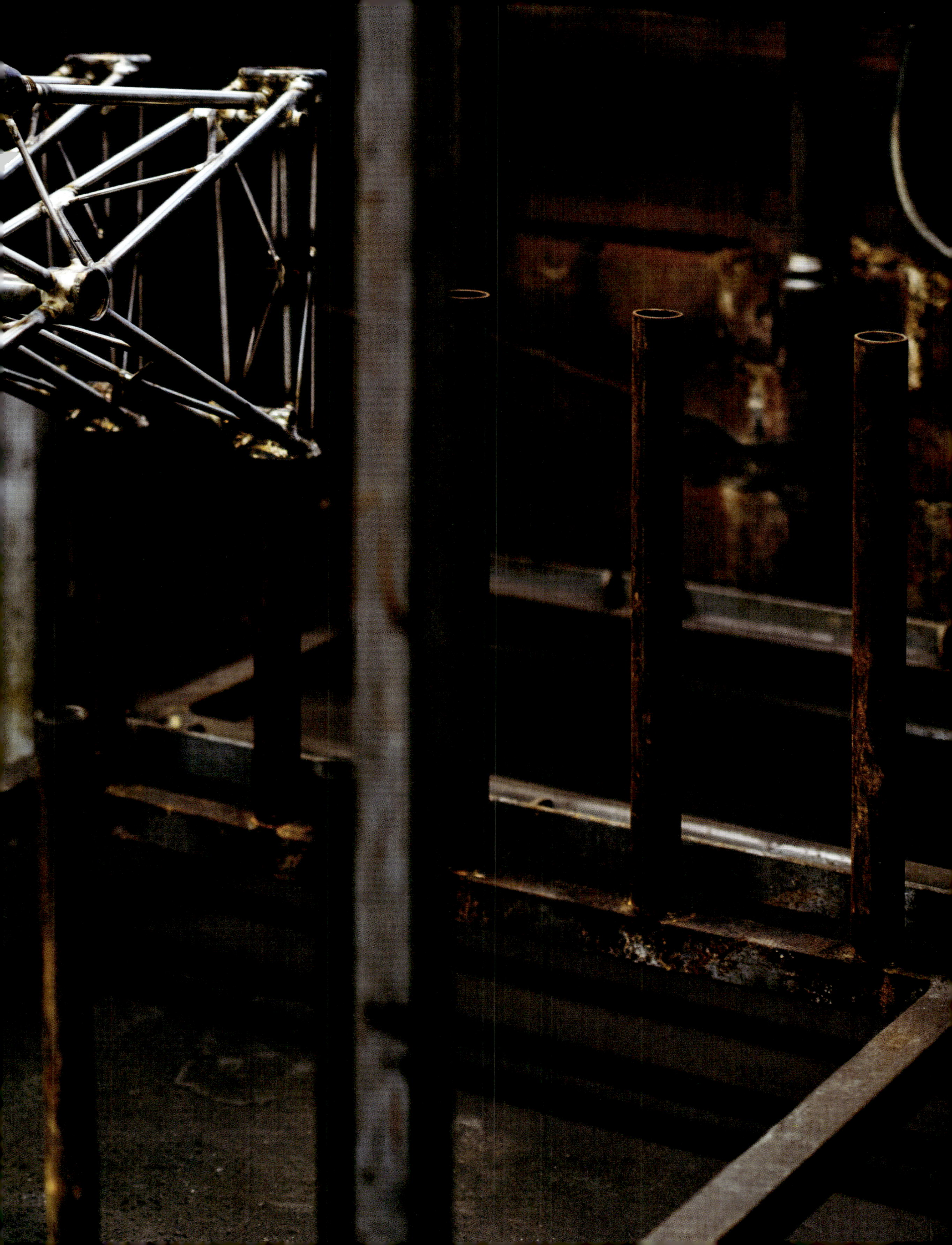

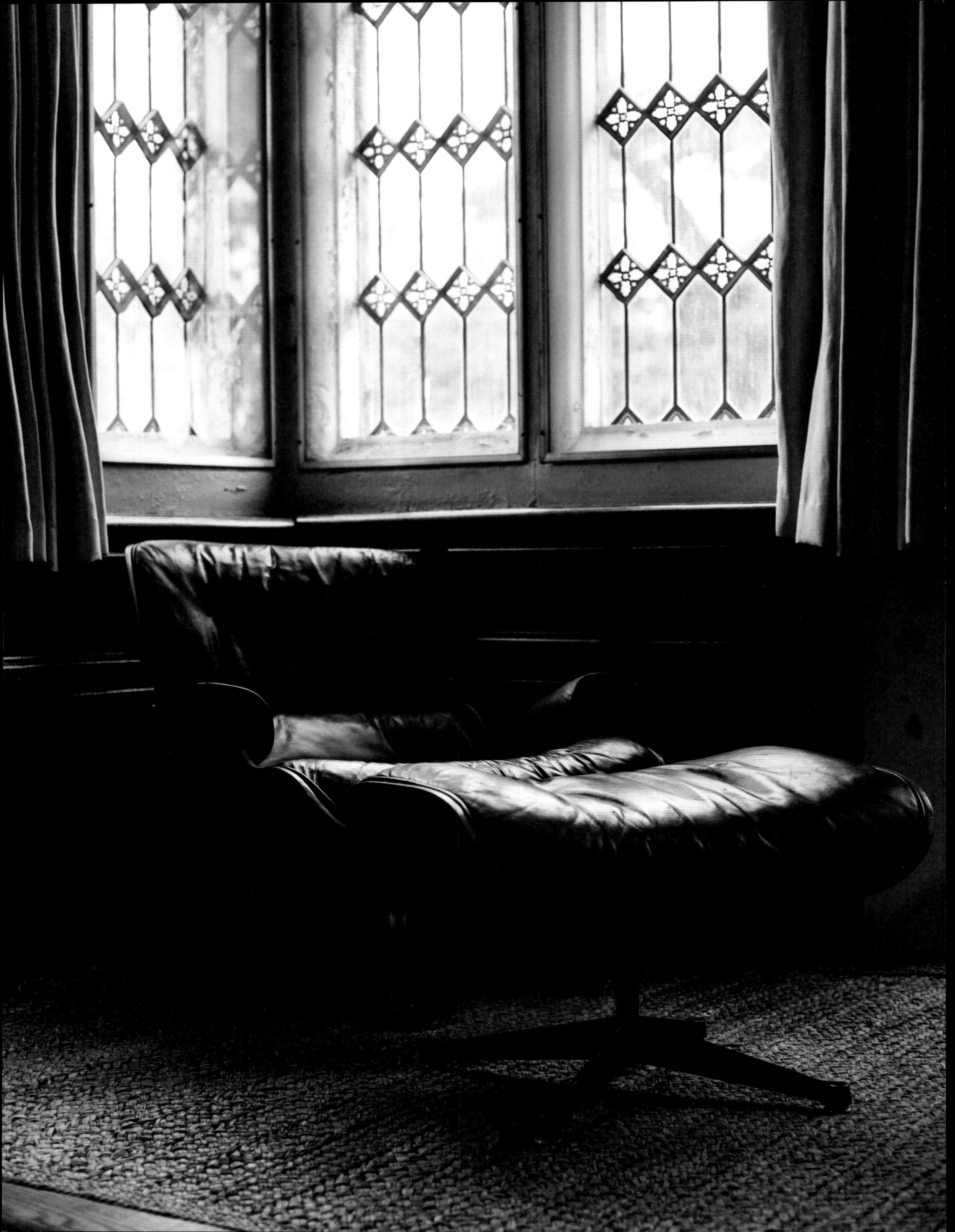

Dr. Alex Moulton. Eine Ikone. Sein Geist, so Steven, der Moulton Bicycles von seinem Onkel übernommen hat, wehe noch immer durch die Firma. Man darf also gespannt sein, was alles noch aus dieser Schmiede kommen wird.

Dann führt Shaun über das Gelände. Vorbei an brennenden Haufen, auf denen das immer noch fallende Herbstlaub langsam vor sich hin kokelt. Very british: Kurzgeschorene Rasenflächen, kleine Steinbrücken, sauber angelegte Wege, akribisch manikürte Hecken. Wir könnten auch durch einen Park laufen.

Kleine Steinhäuser, die sogenannten Sheds, ordnen sich um die riesige Villa. In ihnen produzieren die Mitarbeiter die meisterlichen Handwerksstücke. Eine Shed für die Metallarbeiten, das Löten, Ausrichten und Ablängen der Rohre. In der nächsten Shed lackieren die Experten in aufwendigen Prozessen. Dann kommen die fertigen Teile zur nächsten Station, die Hütte für die Endmontage. Und so trägt man ein Teil nach dem anderen von Häuschen zu Häuschen, bis nach endlosen Arbeitsschritten die Glanzstücke englischer Handwerkskunst den Weg zu ihren Liebhabern in der ganzen Welt finden. Wer einmal mit dem MOULTON-Virus infiziert ist, gibt sein Rad sowieso nicht mehr her, sondern fährt und fährt – und sammelt vielleicht auch irgendwann.

Shaun blickt auf die Uhr. Oha – es ist später Nachmittag und das kann in einem so ur-britischen Teil Englands ja nur eines bedeuten. Richtig vermutet, die Frage lässt nicht lange auf sich warten: »Would you like to join me for some tea?« »Well, sure.« Und so sitzen wir kurze Zeit später in der Herbstsonne vor der Villa, plaudern und schlürfen dabei genüsslich unseren Earl Grey. Mit Milch natürlich.

Und so sitzen wir kurze Zeit später in der Herbstsonne vor der Villa, plaudern und schlürfen dabei genüsslich unseren Earl Grey. Mit Milch natürlich.

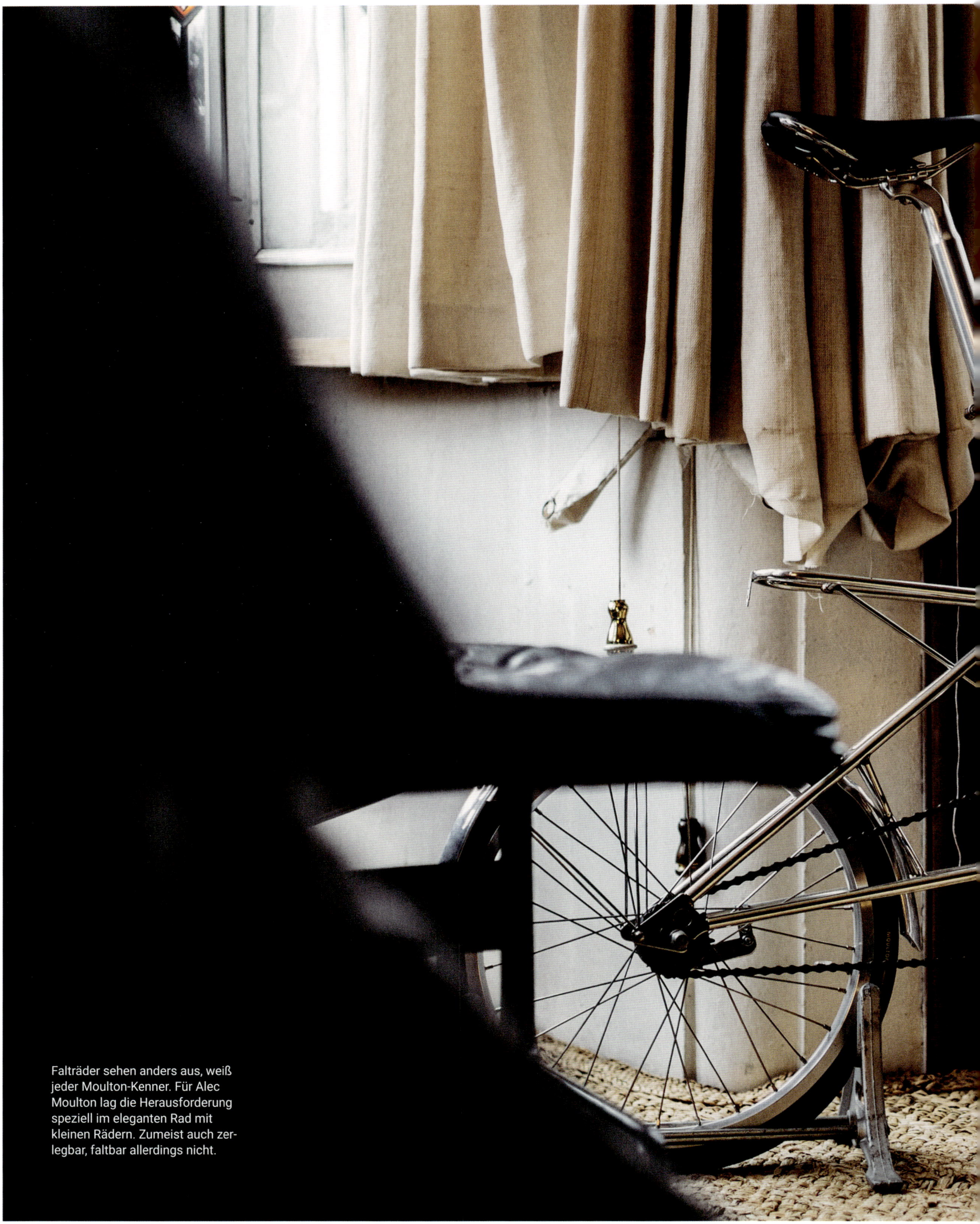

Falträder sehen anders aus, weiß jeder Moulton-Kenner. Für Alec Moulton lag die Herausforderung speziell im eleganten Rad mit kleinen Rädern. Zumeist auch zerlegbar, faltbar allerdings nicht.

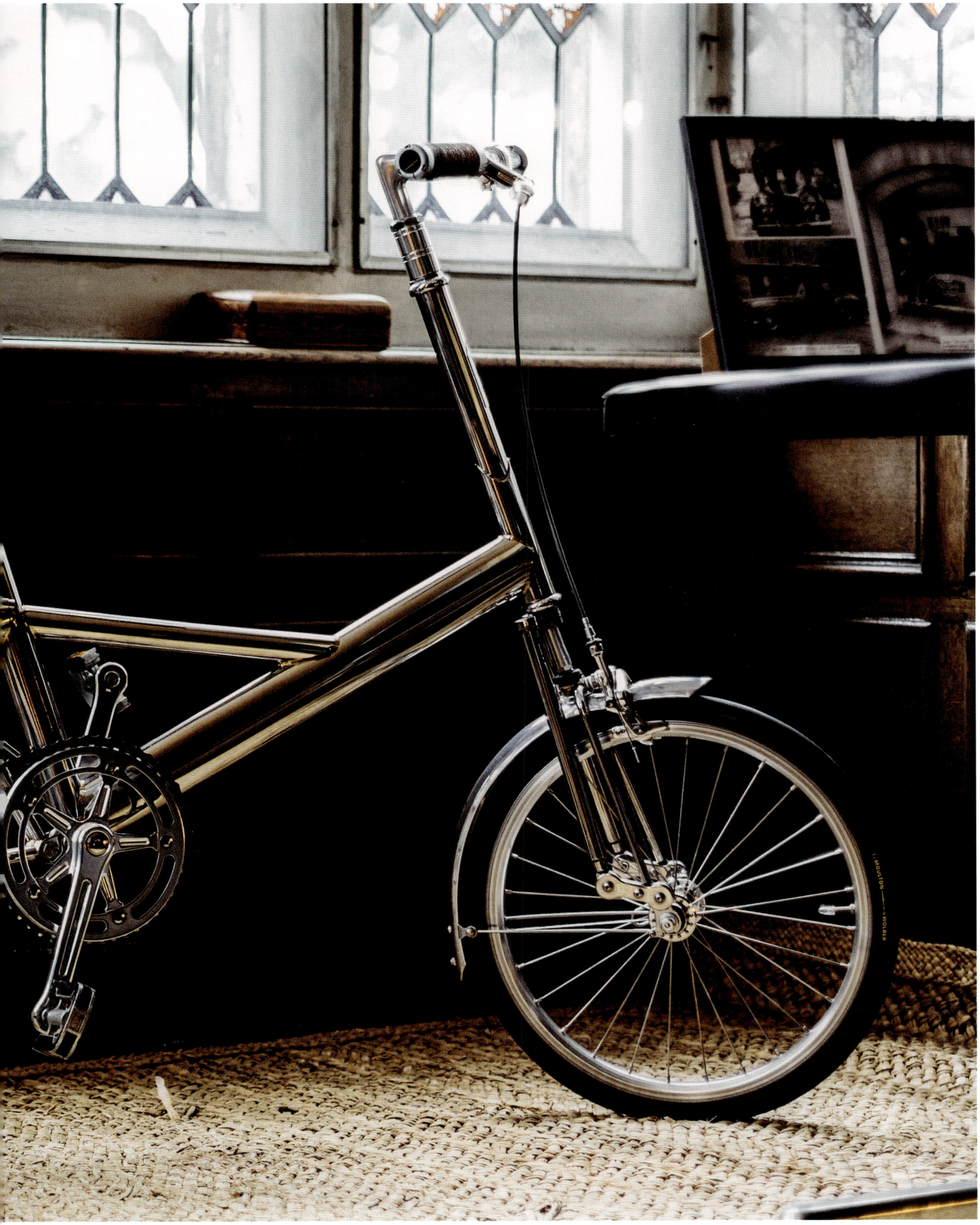

MICHAEL ZAPPE
Codename: Papst!

Der Papst gerahmt.
Der Rahmen im Rahmen, erwischt im Rahmen. Kein ALAN, sondern ein PUCH ALUTRON steckt da drin. Noch Fragen?!

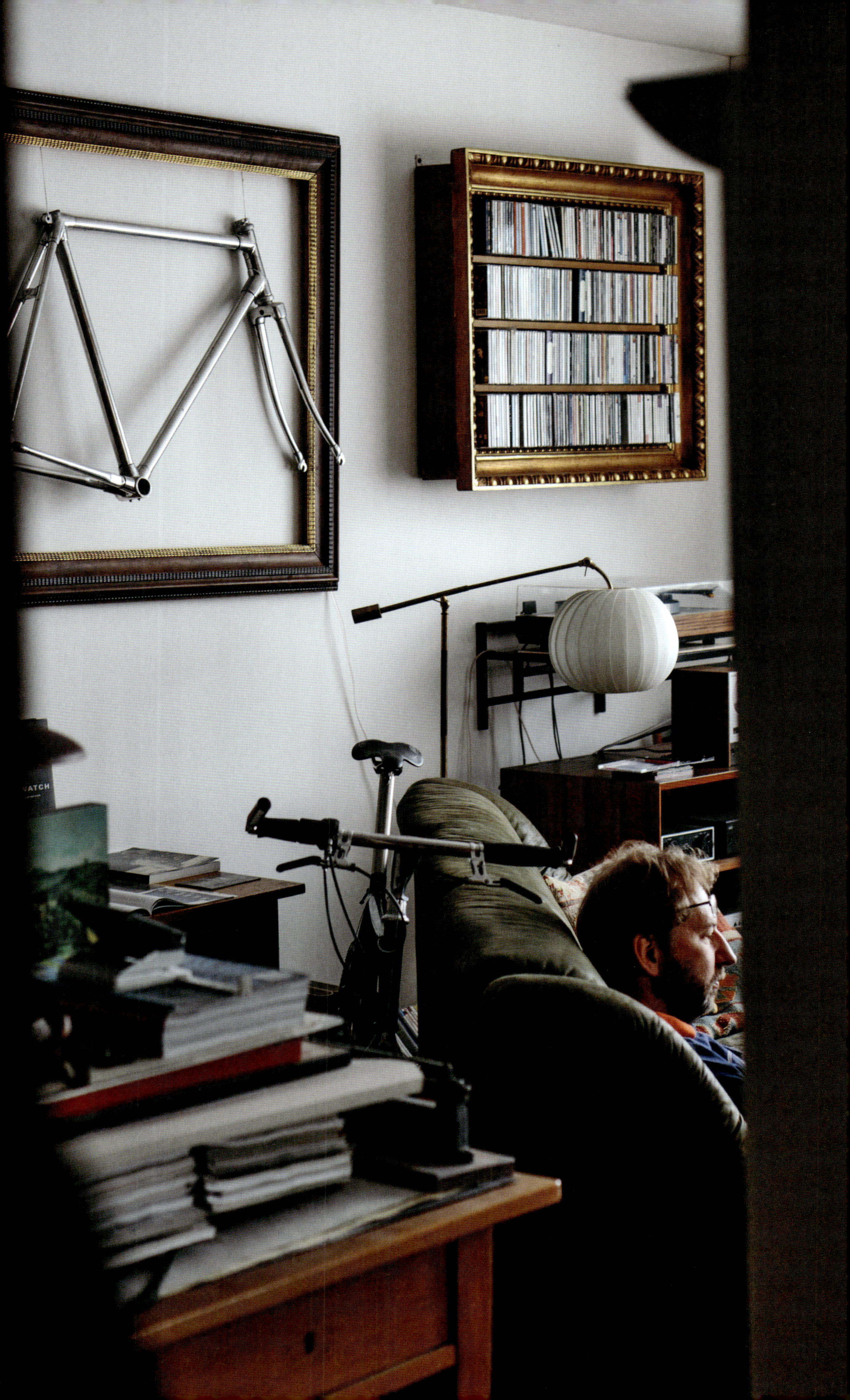

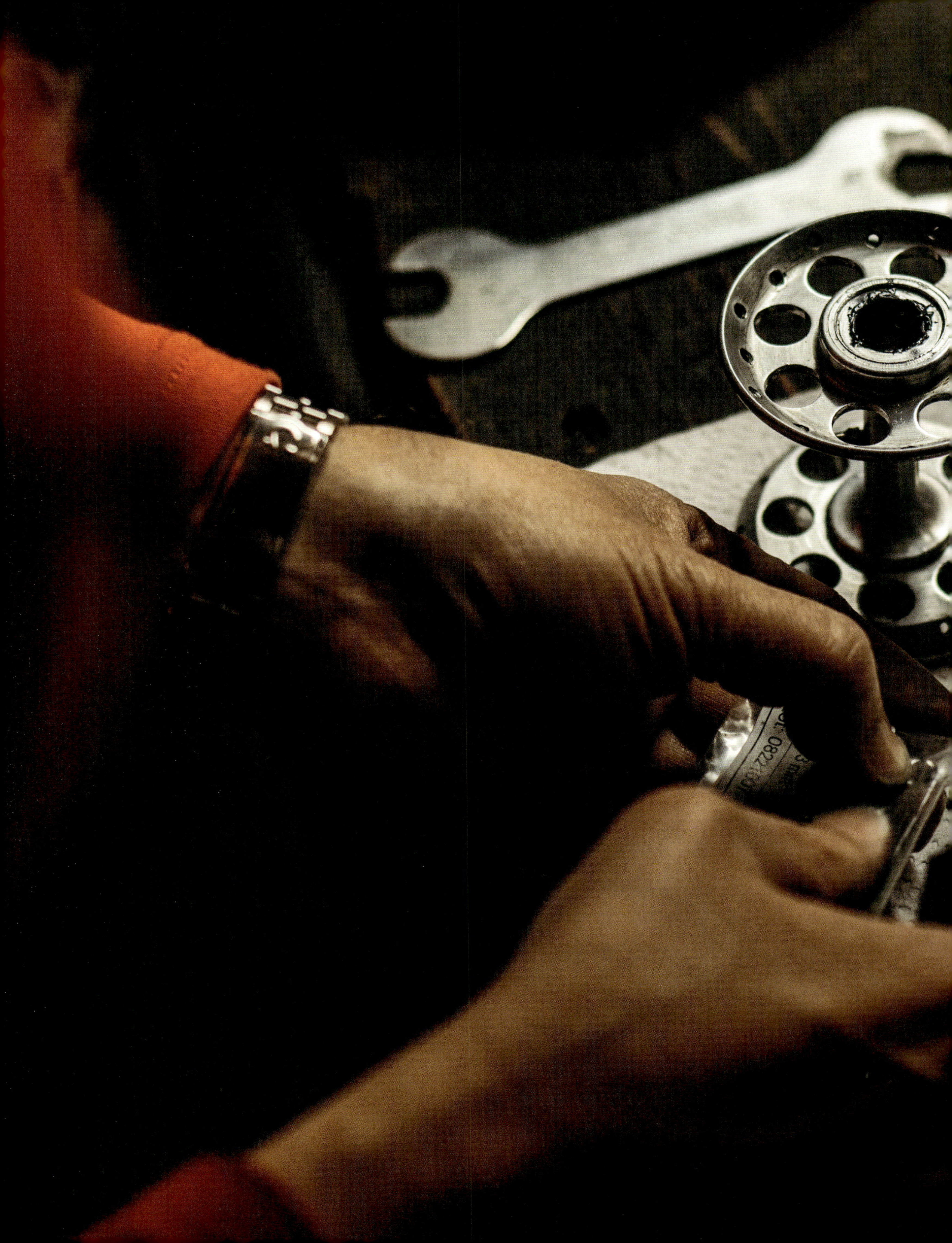

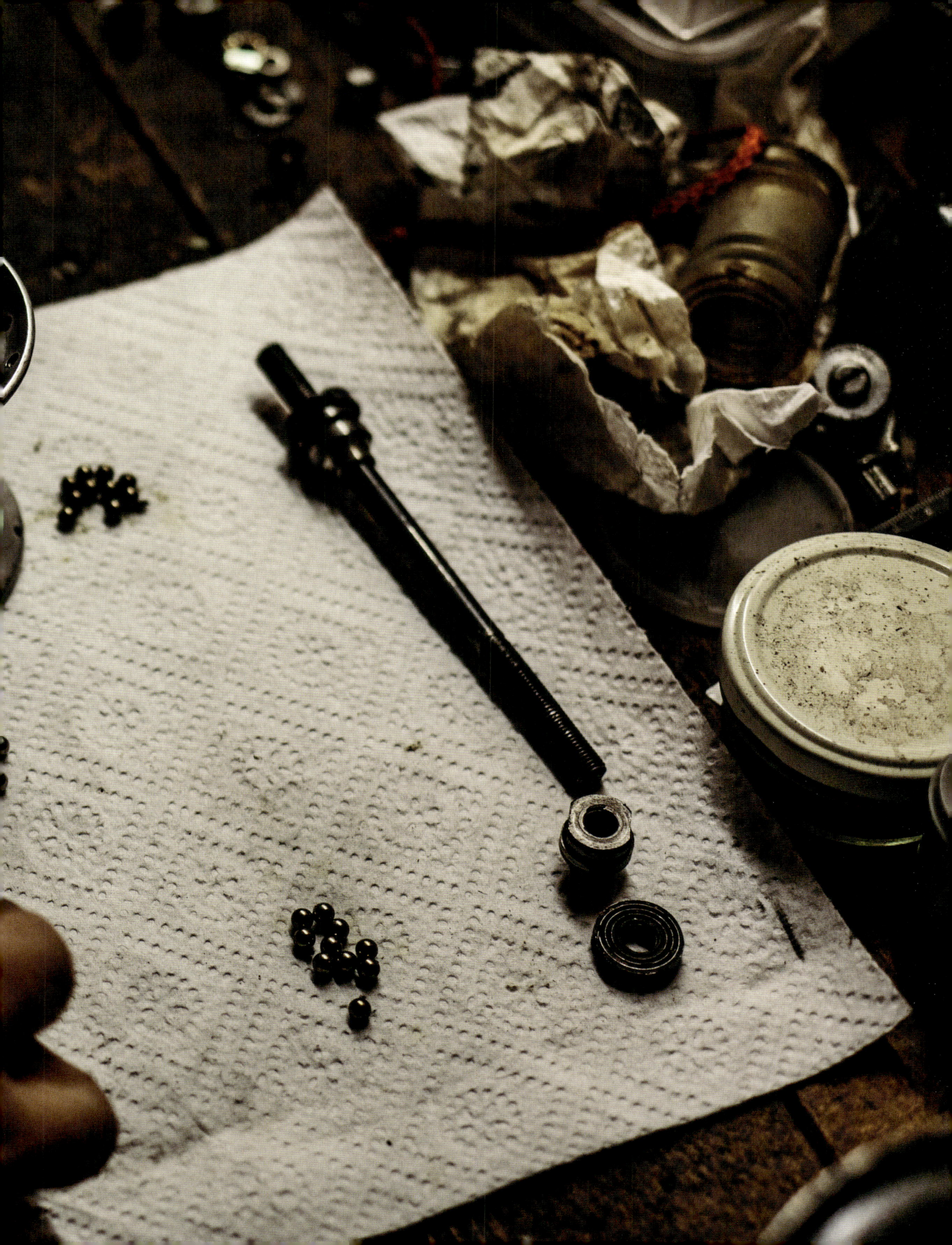

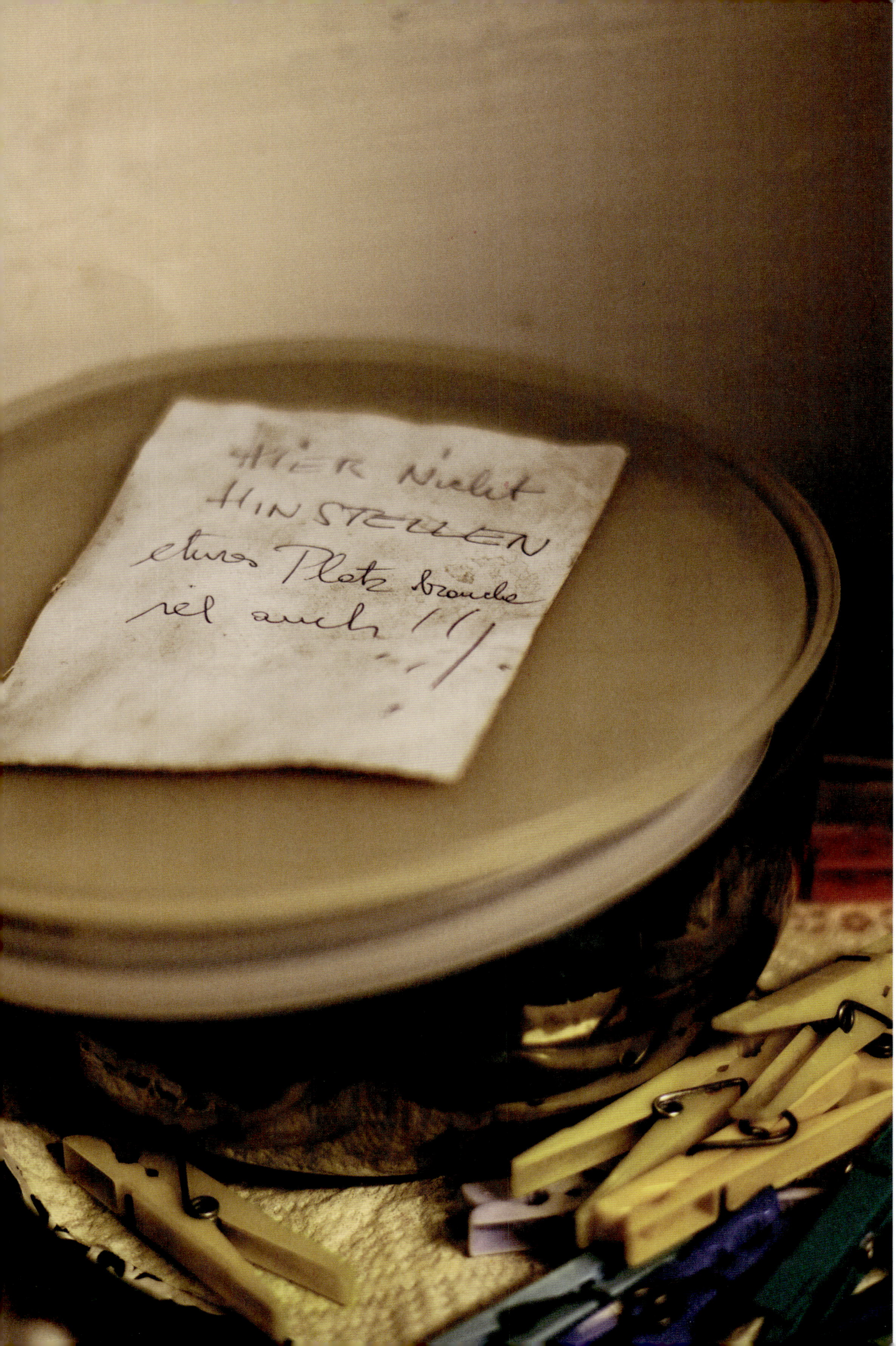
HIER Nicht
HINSTELLEN
etwas Platz brauche
ich auch !!

»Hier jetzt aber mal nichts hinstellen – etwas Platz brauche ich auch!«, versucht sich seine Mutter auch heute noch quadratzentimeter-weise Raum im einverleibten Keller zurück zu erkämpfen.

Gerade habe ich mich als Fotograf selbstständig gemacht, da stolpere ich sozusagen auf den Weg, der letztendlich die Geburtsstunde zu diesem Buch bildet. Naja, also erst einmal stolpere ich in das Wiener *Radlager*, einen der angesagtesten Läden für Vintage-Räder vor Ort. Gerade frisch eröffnet am neuen Standort braucht man auf jeden Fall neue Fotos – und eigentlich auch eine helfende Hand. Die Situation rüttelt bei mir sofort Erinnerungen an meine Zeit als Schrauber für einen großen Fahrradhändler in Hamburg wach. Ich mache also Fotos, ich biete meine Schrauberdienste an, und schwuppdiwupp bin ich mit dem Radlager und mit der Wiener Vintage-Radszene verzahnt.

Und immer wieder, wenn das eine oder andere besondere Stück – vielleicht ist es ein C.B.T ITALIA, vielleicht ein MASI oder doch ein Franzose – Fragen und scheinbar unlösbare Ersatzteilproblematiken aufwirft, heißt es: »Da fragst du am besten den Papst!«

Der Papst, wer soll das denn bitte sein? Doch es dauert nicht lange, da steht der größte aller Brückenbauer dann auch in Person vor mir. Mit Brille, Bart und einem richtigen Namen: Michael Zappe. Und es gibt wirklich nichts, worauf er keine Antwort weiß. Kein Rahmenbauer ist ihm unbekannt. Kein Teil, von dem er noch nichts gehört, das er nicht in einem Original-Katalog zu Hause hat. In seinem Archiv mit schier unabsehbaren Ausmaßen. Das Ganze wird begleitet von einer unglaublichen Portion an Humor und Lebensfreude. Zappe entspricht eben genau dem, was man Fahrradfahrern so oft nachsagt: Er ist glücklich! Und am glücklichsten ist er auf seinem Rad und in seinem Keller.

Die Sammelleidenschaft, die nimmt in den 1980er-Jahren ihren Anfang. Aus Speermüllcontainern und auf Schrottplätzen sammelt Michael mit seinen Kumpels die ersten Rahmen und Teile. So filigran, so schön – die kann man doch nicht einfach so vernichten. Einsammeln, Aufarbeiten und Anhäufen. So lässt sich die Leidenschaft in drei Worten zusammenfassen. Das Schraubenkrankenhaus – nichts kommt weg! Die Spezialität seiner Sammlung liegt in der Vielfalt.

»Hier jetzt aber mal nichts hinstellen – etwas Platz brauche ich auch!«, versucht sich seine Mutter auch heute noch quadratzentimeterweise Raum im einverleibten Keller zurück zu erkämpfen. Es ist ein harter Stellungskampf, in

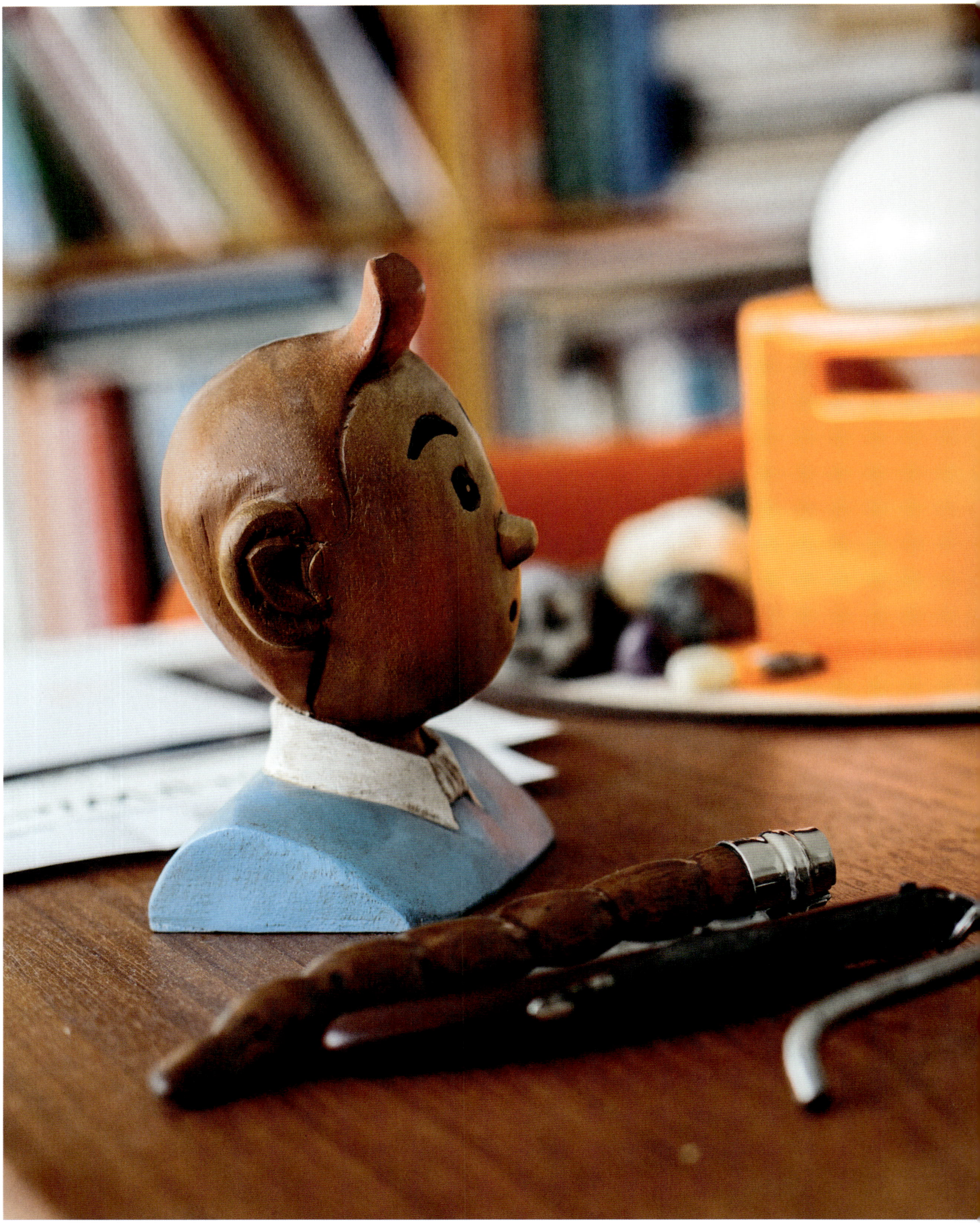

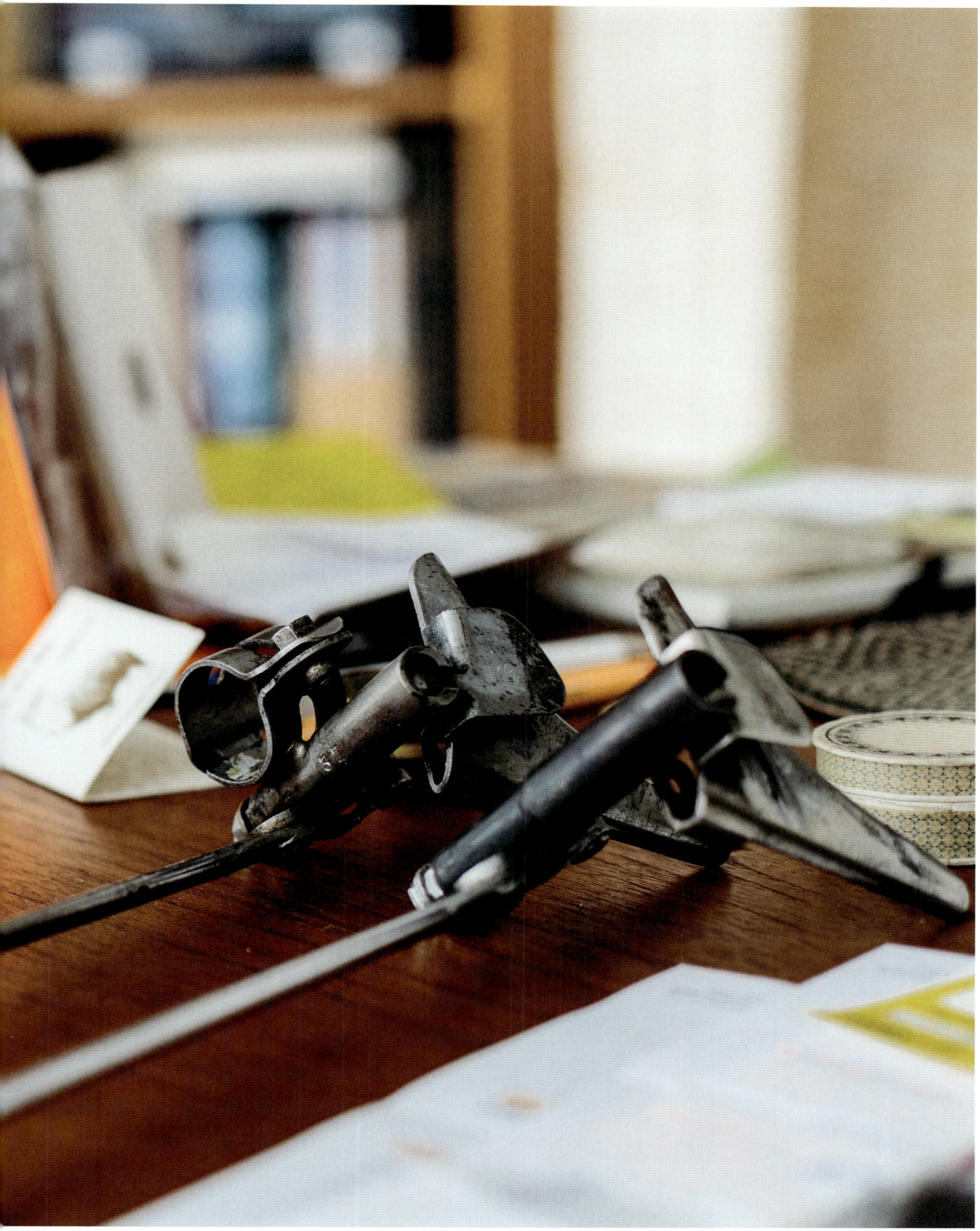

MICHAEL ZAPPE

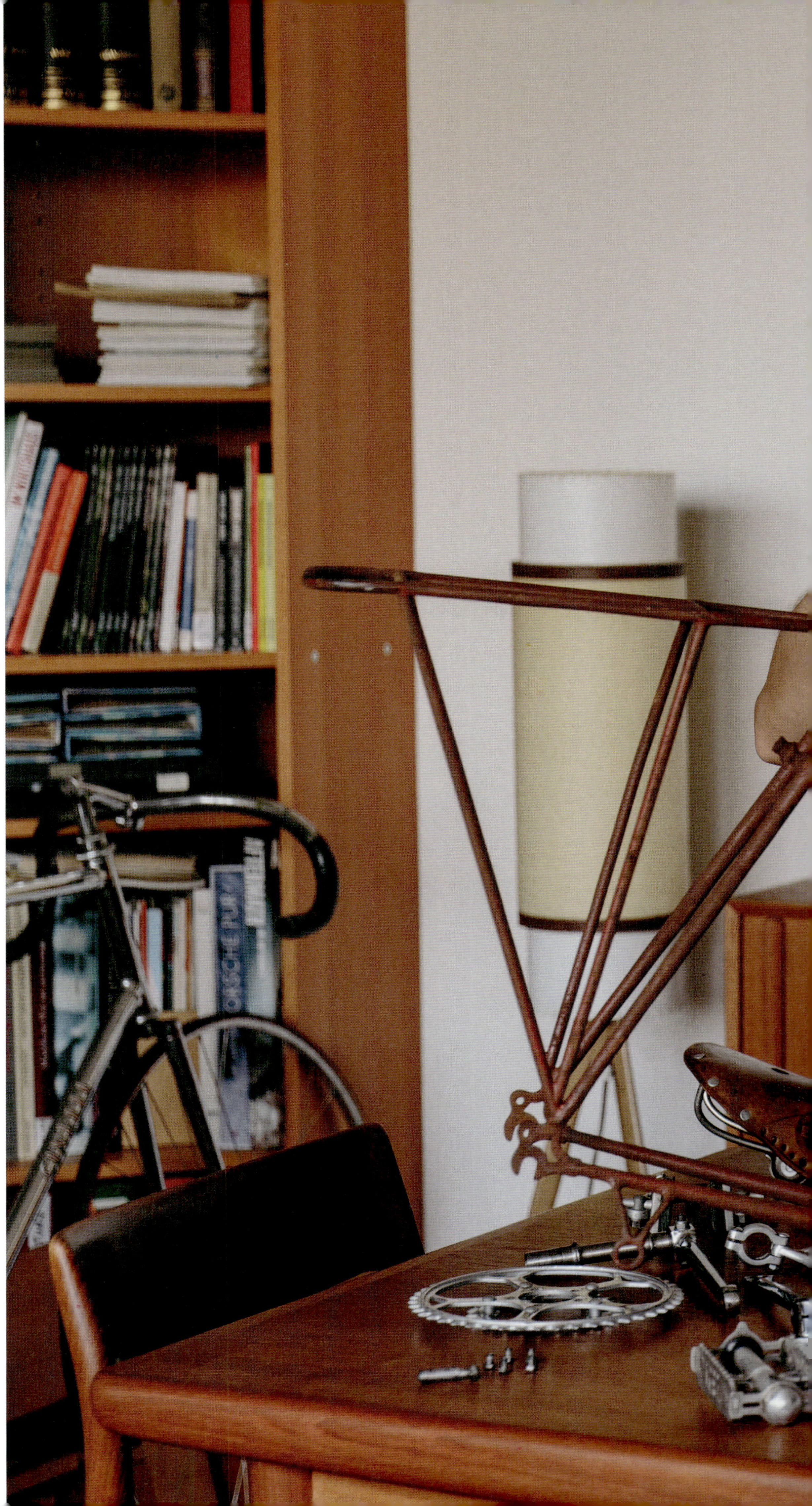

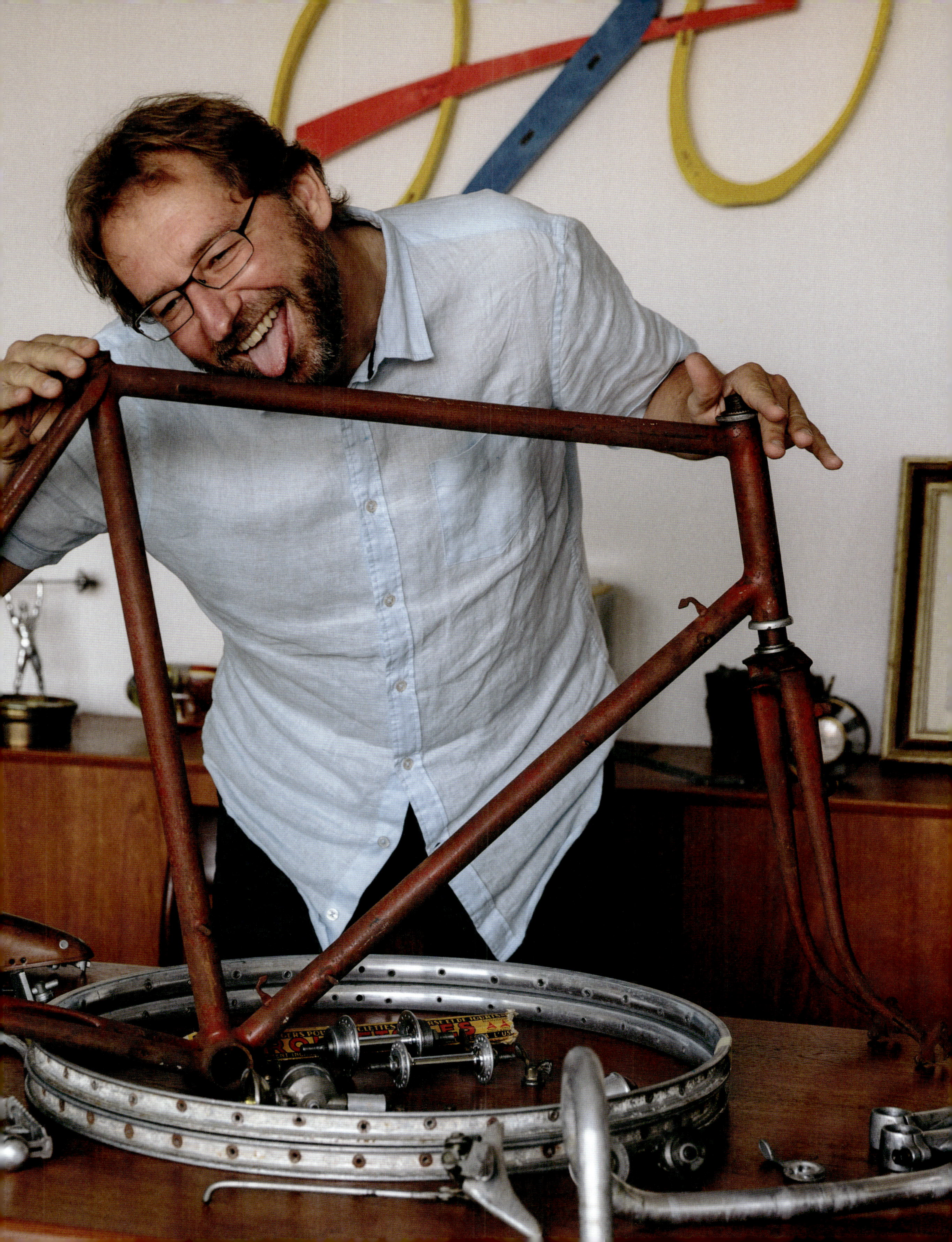

Akribie vom Feinsten:
Dieses Archiv kann immer wieder das entscheidende Detail hervorzaubern!

dem Wäscheklammern versuchen, gegen Speichennippel vorzurücken.

Besonders angetan haben es Zappe die kleinen Wiener Rahmenbauer. Aus dem Interesse entsteht das Buch *Wiener Mechanikerräder*, ein riesiges Fotoarchiv mit Originalaufnahmen, Prospekten, Katalogen und Beschreibungen der letzten acht Jahrzehnte. Wandkalender mit detaillierten Motiven folgen. Mitglied im englischen VETERAN-CYCLE-CLUB, Stammtische, Sammlerrunden, Reisen durch Europa zu anderen Rad-Verrückten – es gibt kaum jemanden in der weltweiten Szene, den der Papst nicht über irgendeine Ecke kennt. Und das auch schon, bevor das World Wide Web Querbeziehungen oder Recherchen in Spezialbereichen einfacher machte.

Ob technisch oder ästhetisch: Michael Zappe liebt Herausforderungen. Laufräder reizen ihn bis zum Exzess. Naben sind sein Traum. Spezielle Schaltungskombinationen lassen seine Augen erst so richtig leuchten. Richtig ansteckend ist seine Passion und ich bin mir sicher: Ohne ihn wäre dieses Buch niemals entstanden.

Mitglied im englischen VETERAN-CYCLE-CLUB, Stammtische und Sammlerrunden, Reisen durch Europa zu anderen Rad-Verrückten – es gibt kaum jemanden in der weltweiten Szene, den der Papst nicht über irgendeine Ecke kennt.

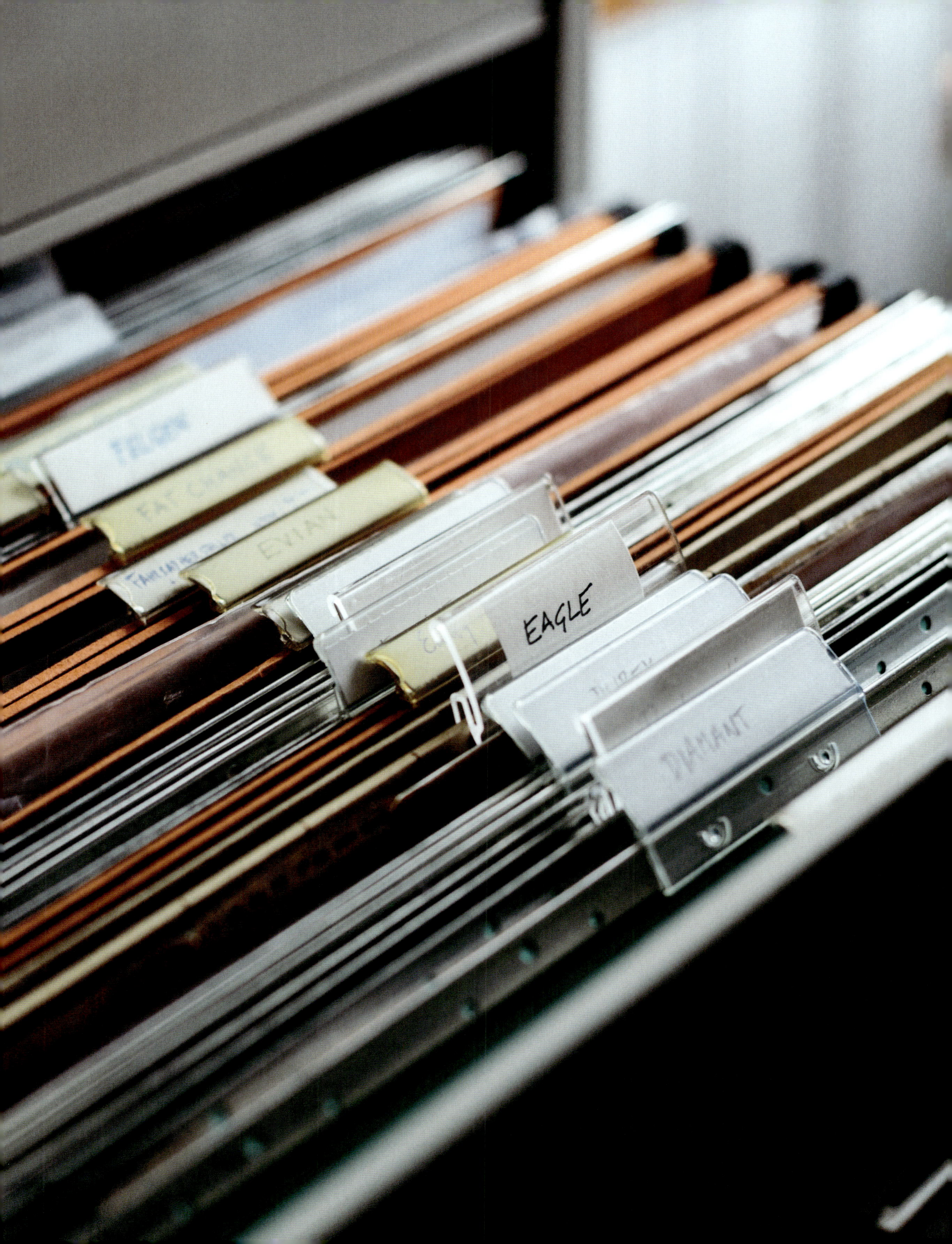
EVIAN
EAGLE

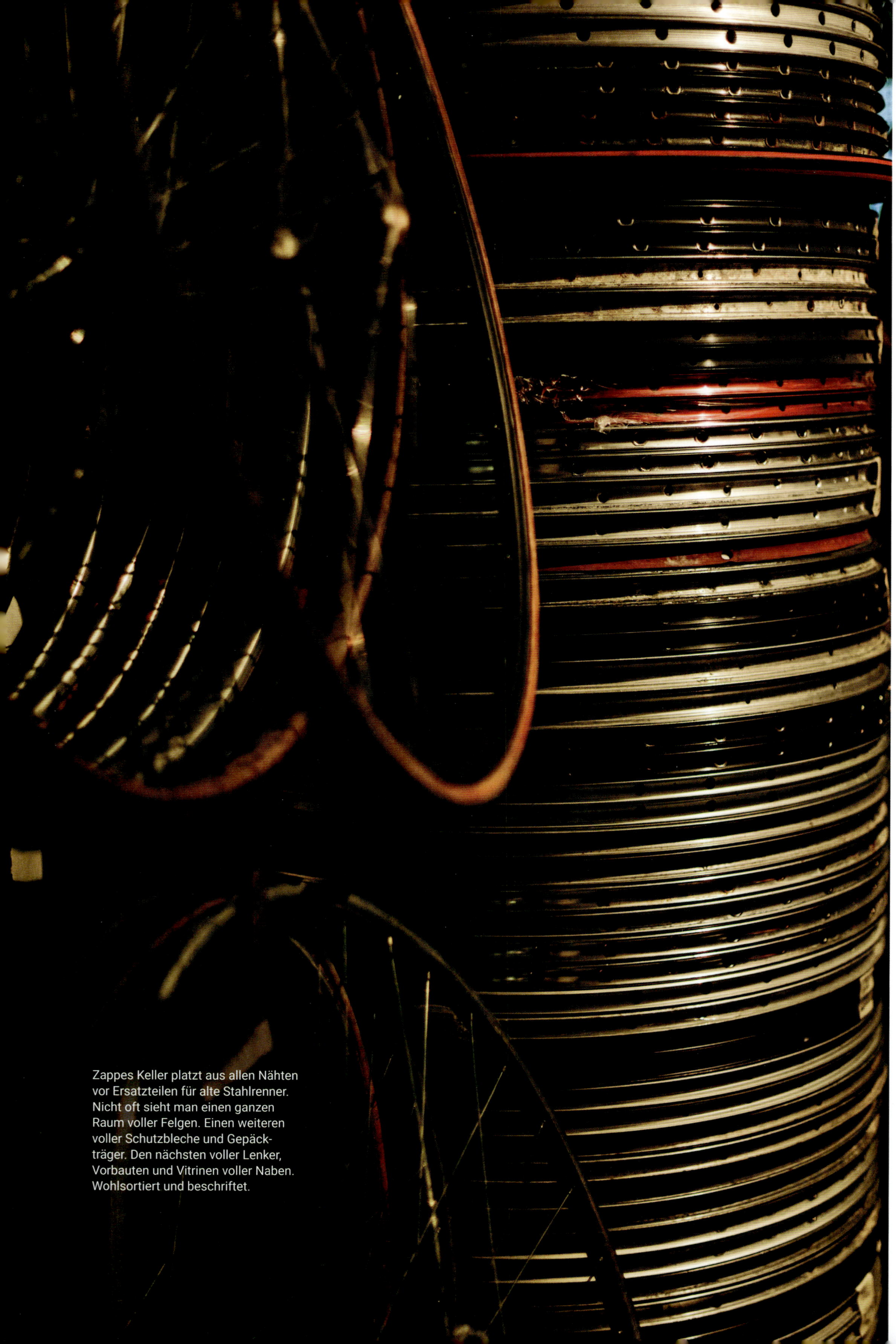

Zappes Keller platzt aus allen Nähten vor Ersatzteilen für alte Stahlrenner. Nicht oft sieht man einen ganzen Raum voller Felgen. Einen weiteren voller Schutzbleche und Gepäckträger. Den nächsten voller Lenker, Vorbauten und Vitrinen voller Naben. Wohlsortiert und beschriftet.

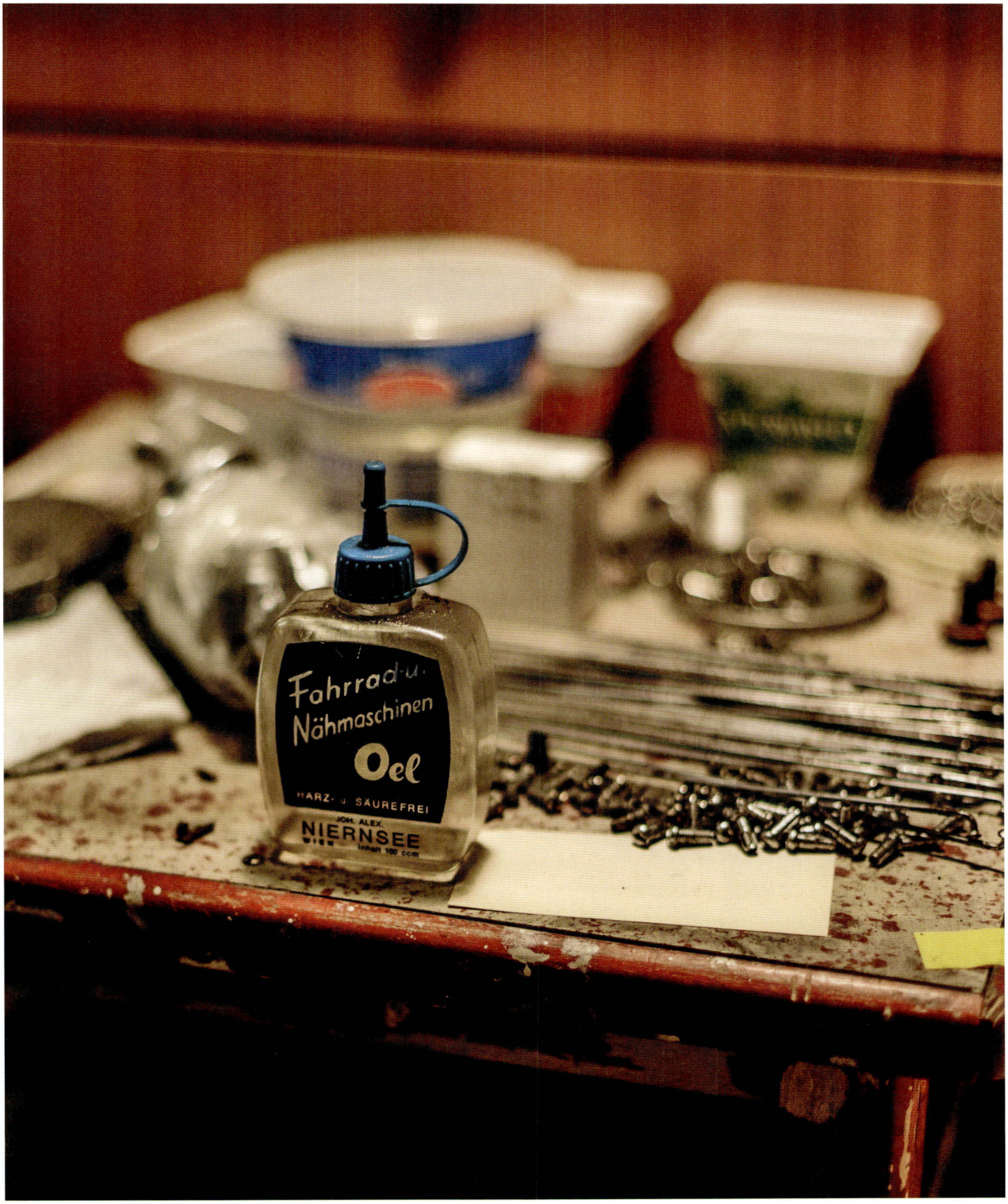
Fahrrad-u.
Nähmaschinen
Oel
HARZ- u. SÄUREFREI
JOH. ALEX.
NIERNSEE

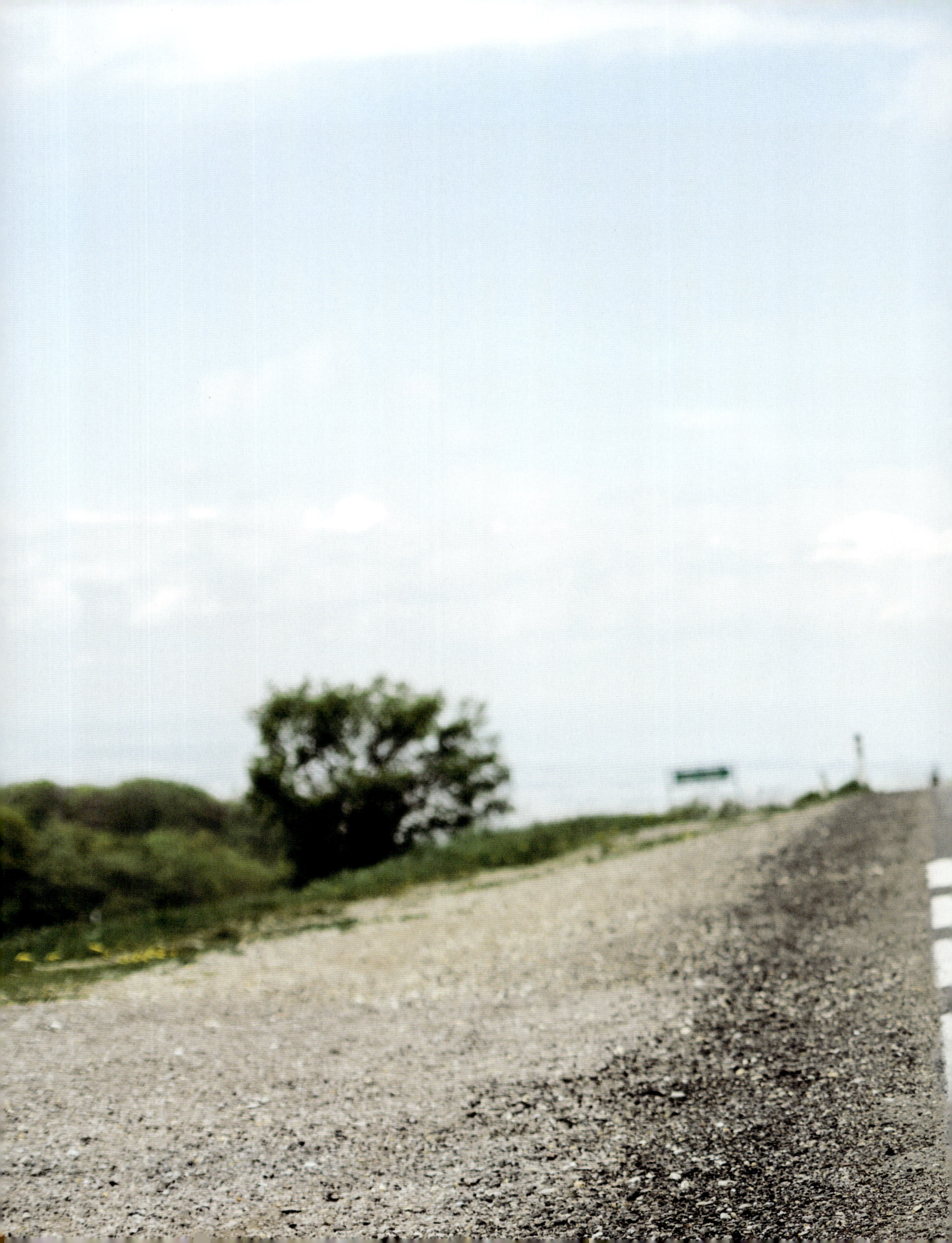

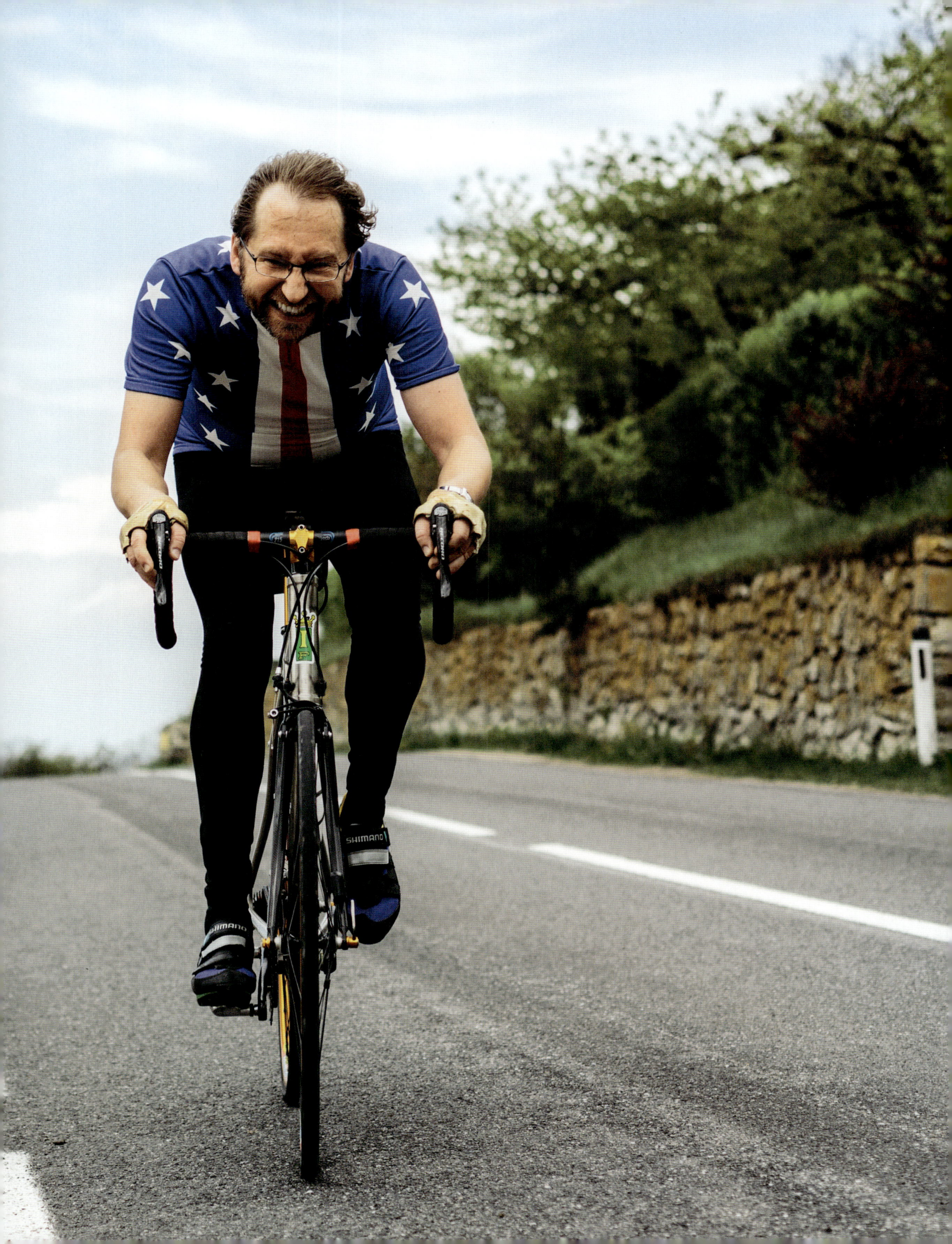
SHIMANO

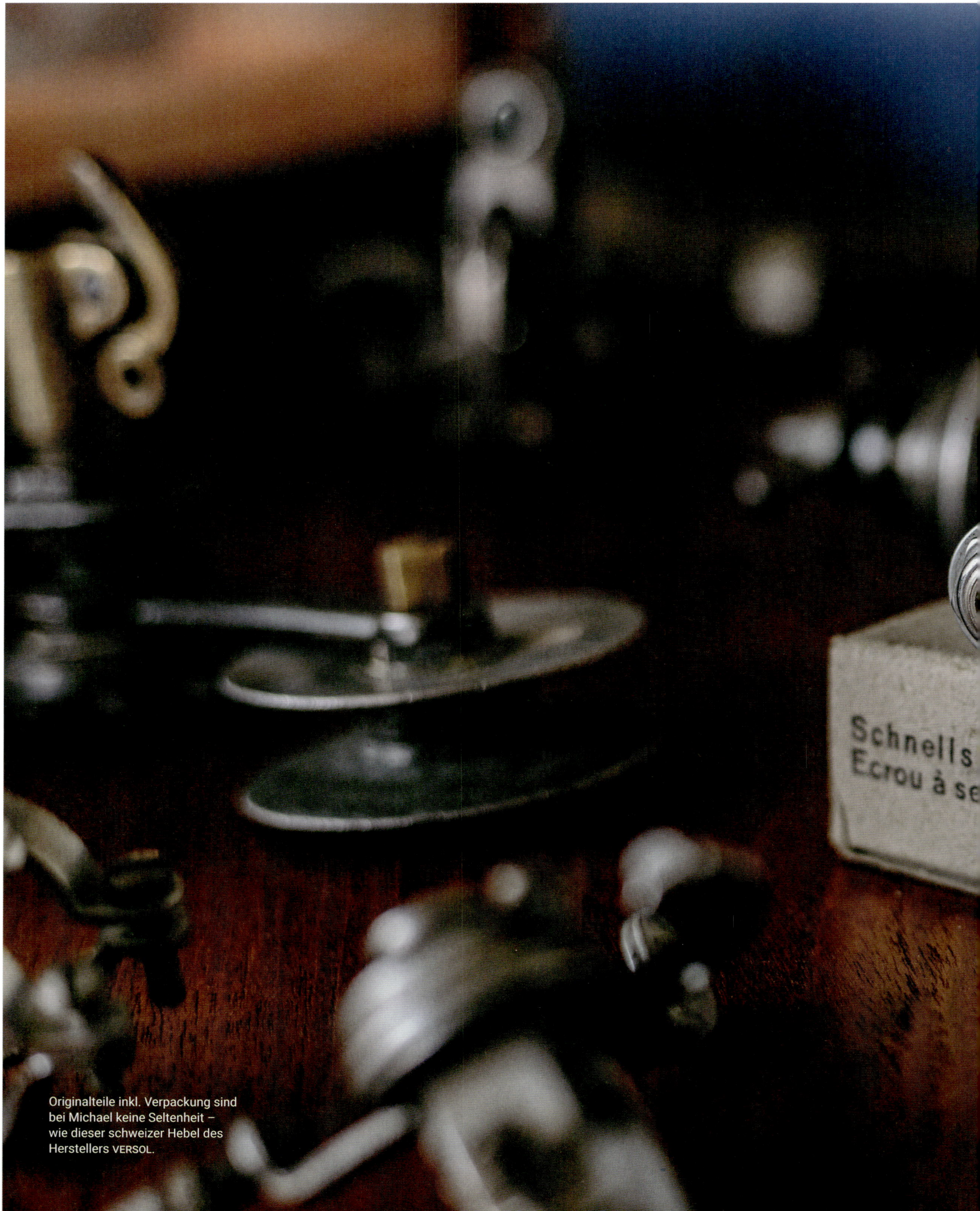

Originalteile inkl. Verpackung sind bei Michael keine Seltenheit – wie dieser schweizer Hebel des Herstellers VERSOL.

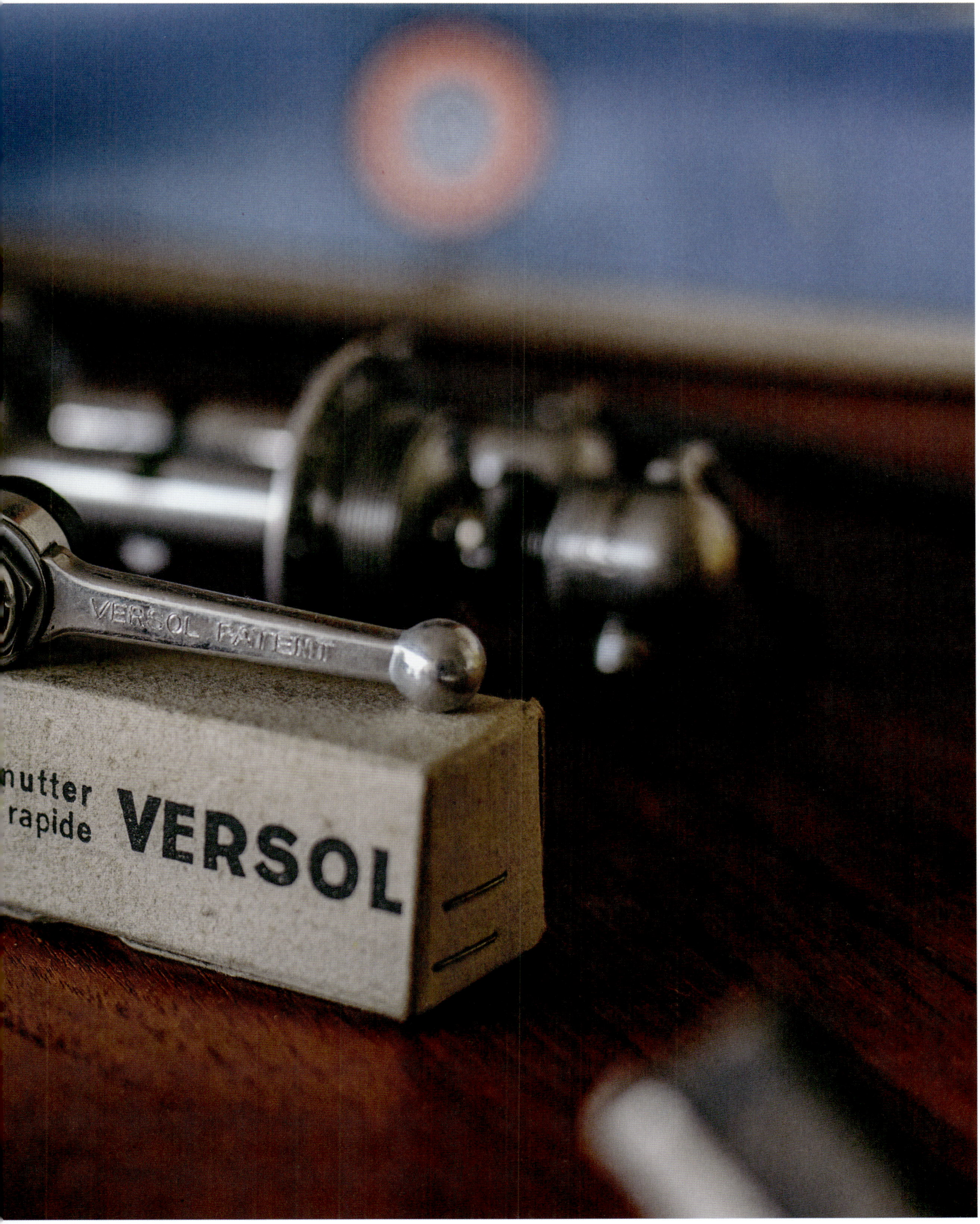
VERSOL PATENT
nutter
rapide
VERSOL

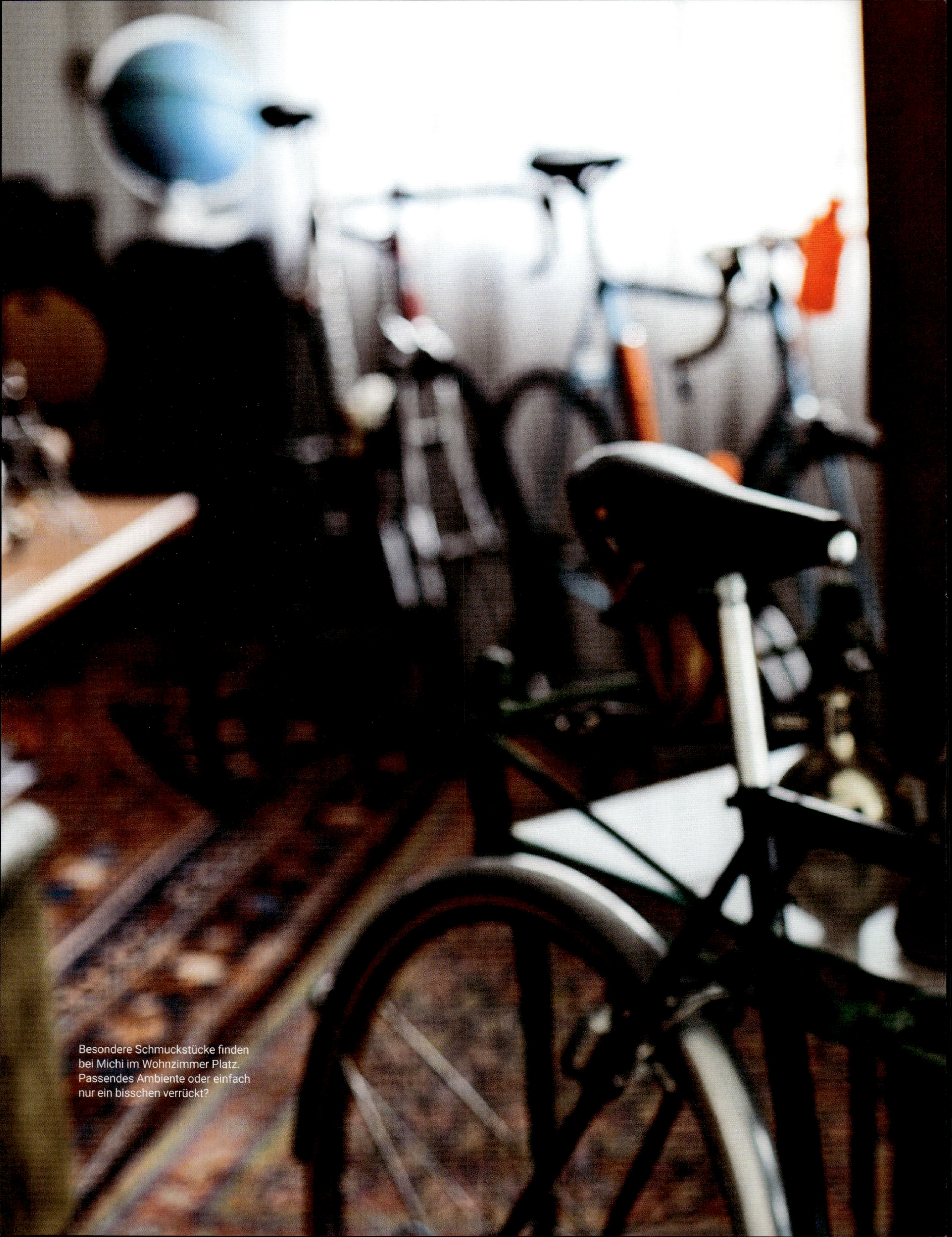

Besondere Schmuckstücke finden bei Michi im Wohnzimmer Platz. Passendes Ambiente oder einfach nur ein bisschen verrückt?

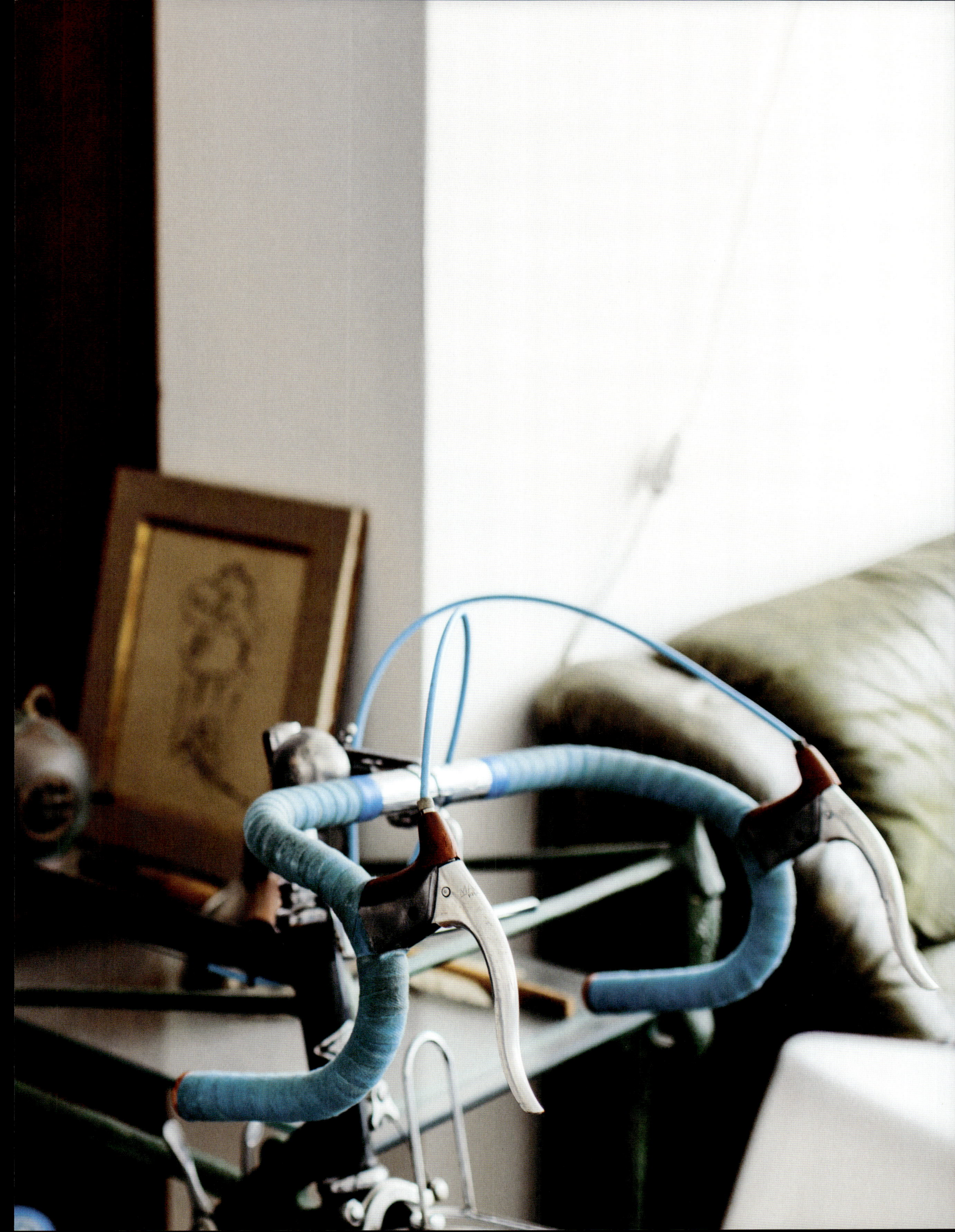

PARIS–ROUBAIX

Heldenwetter in der Hölle

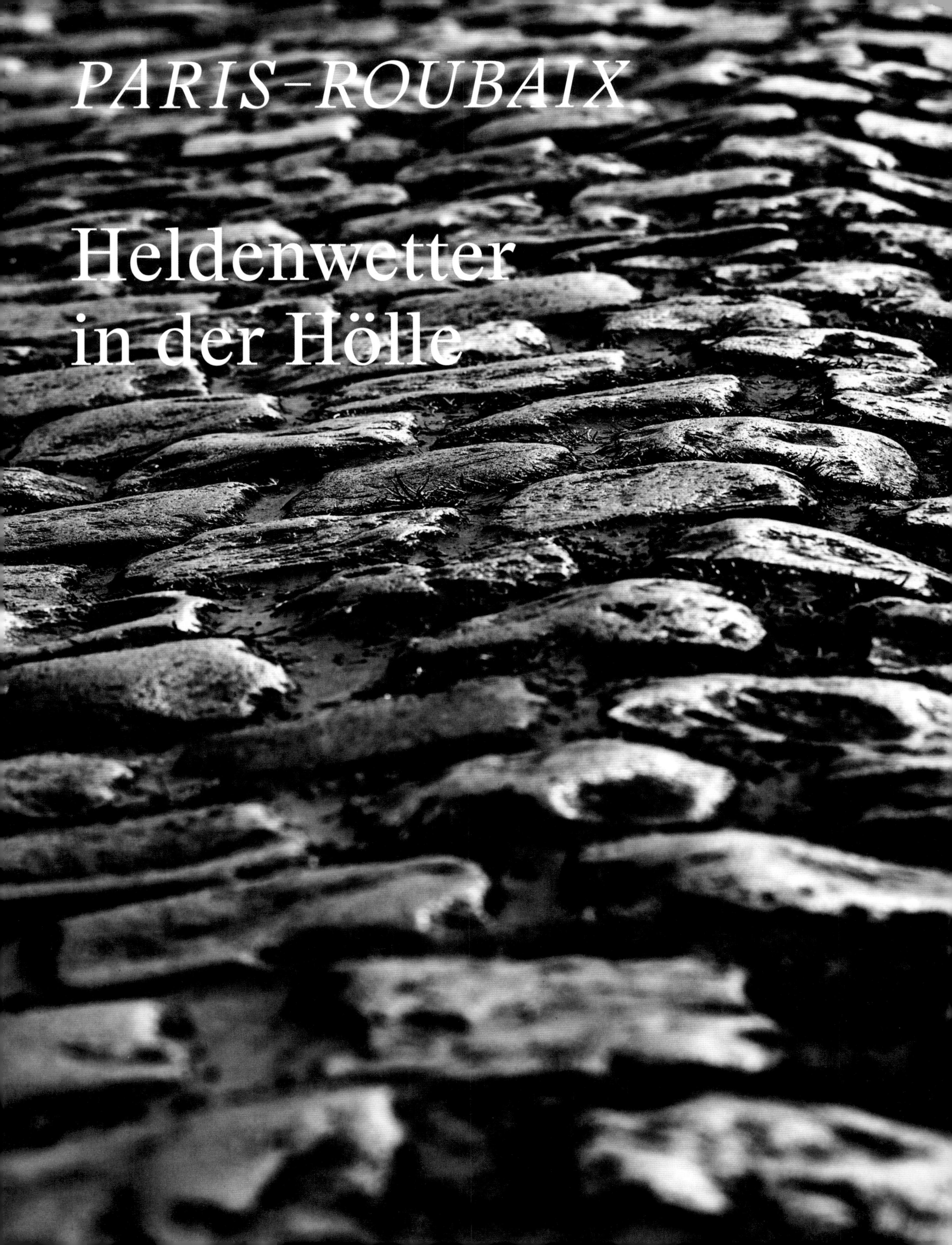

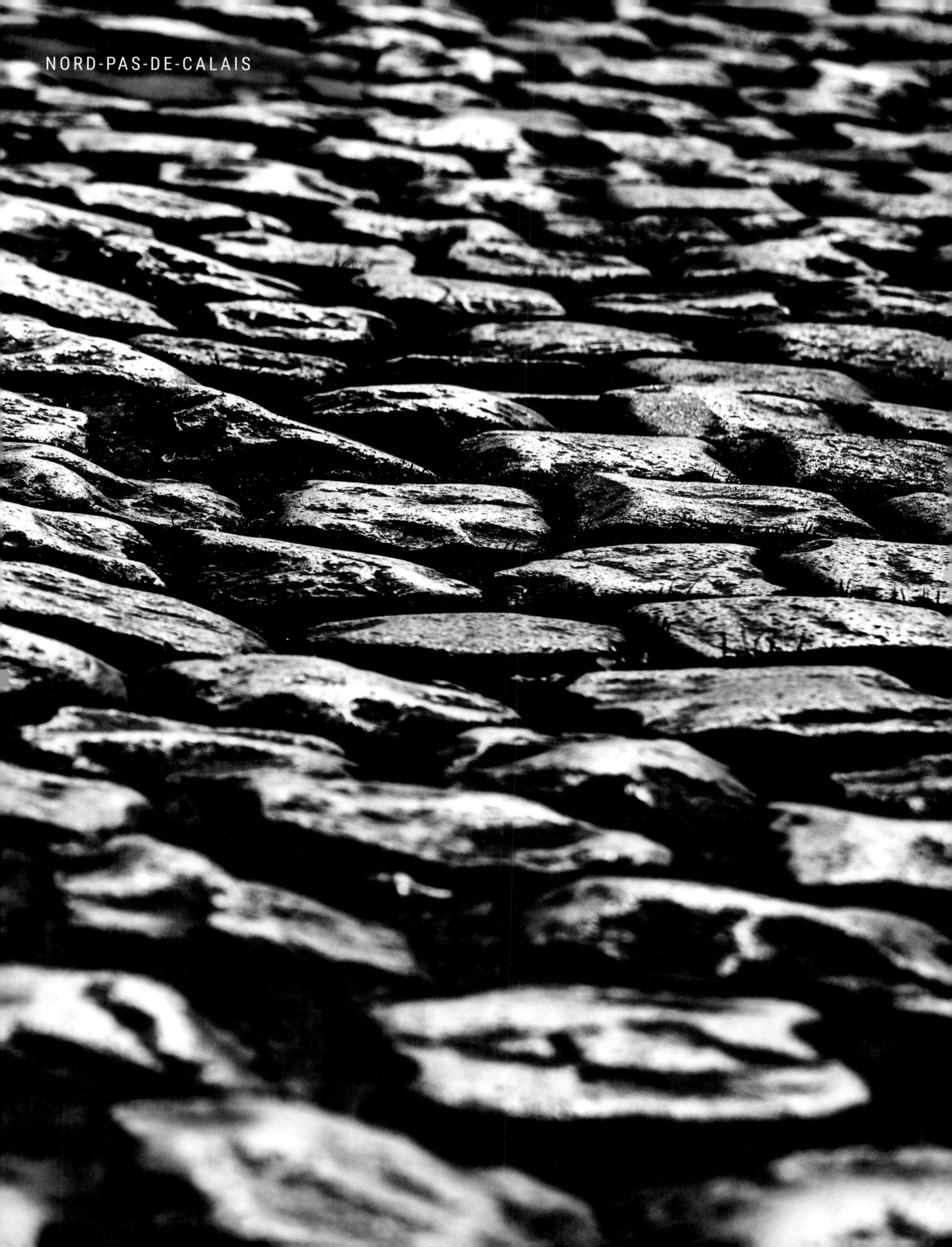
NORD-PAS-DE-CALAIS

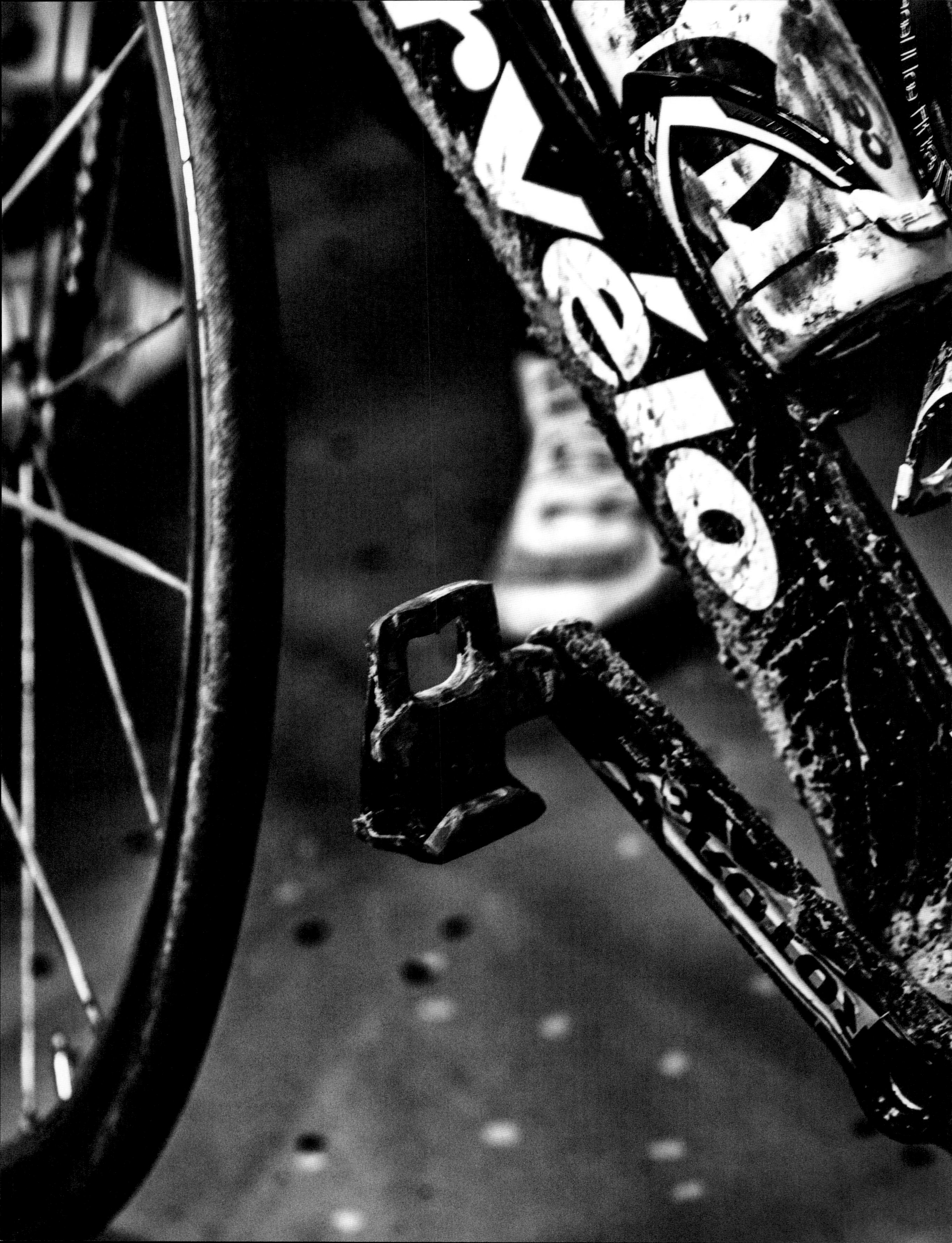

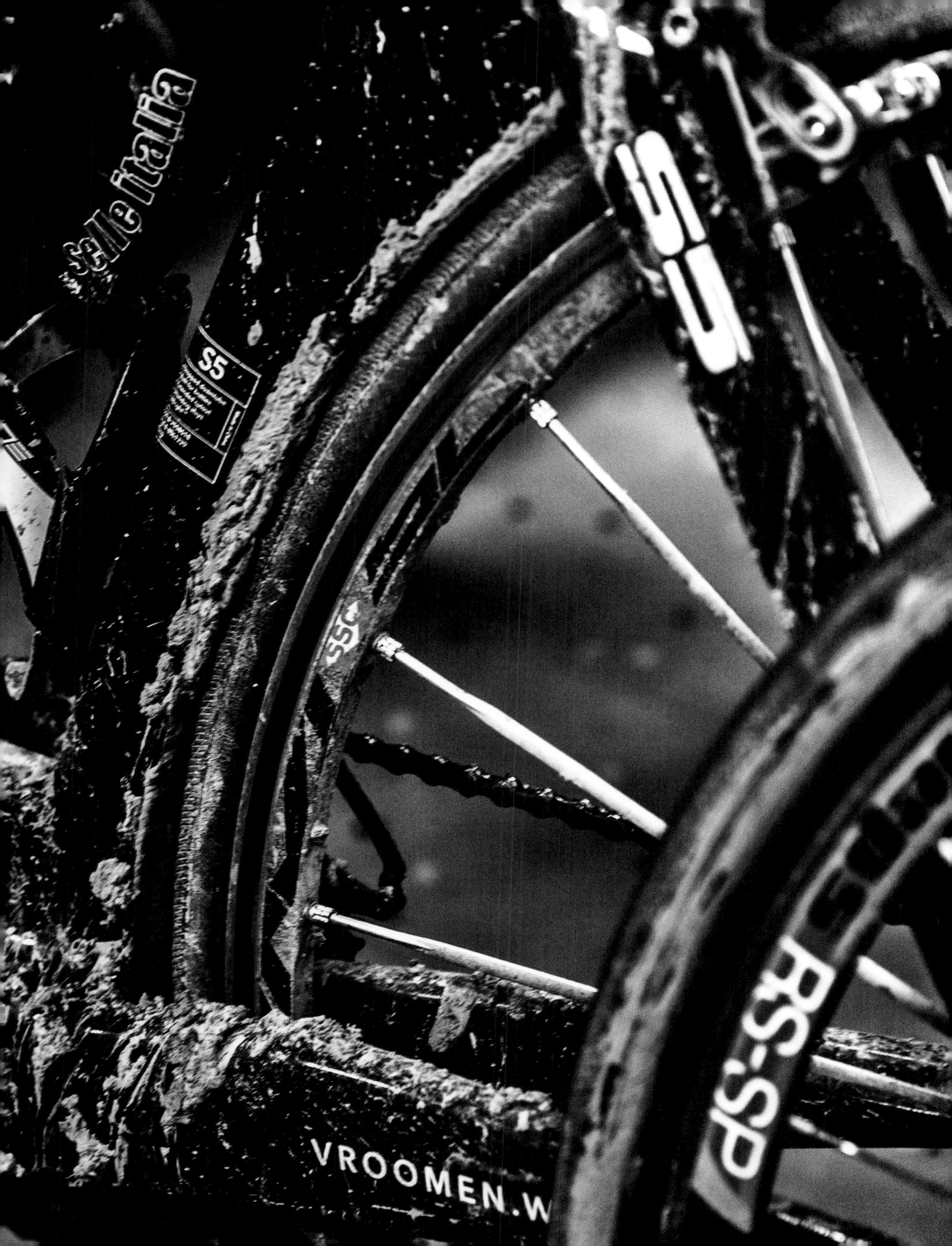
selle italia
S5
VROOMEN.W

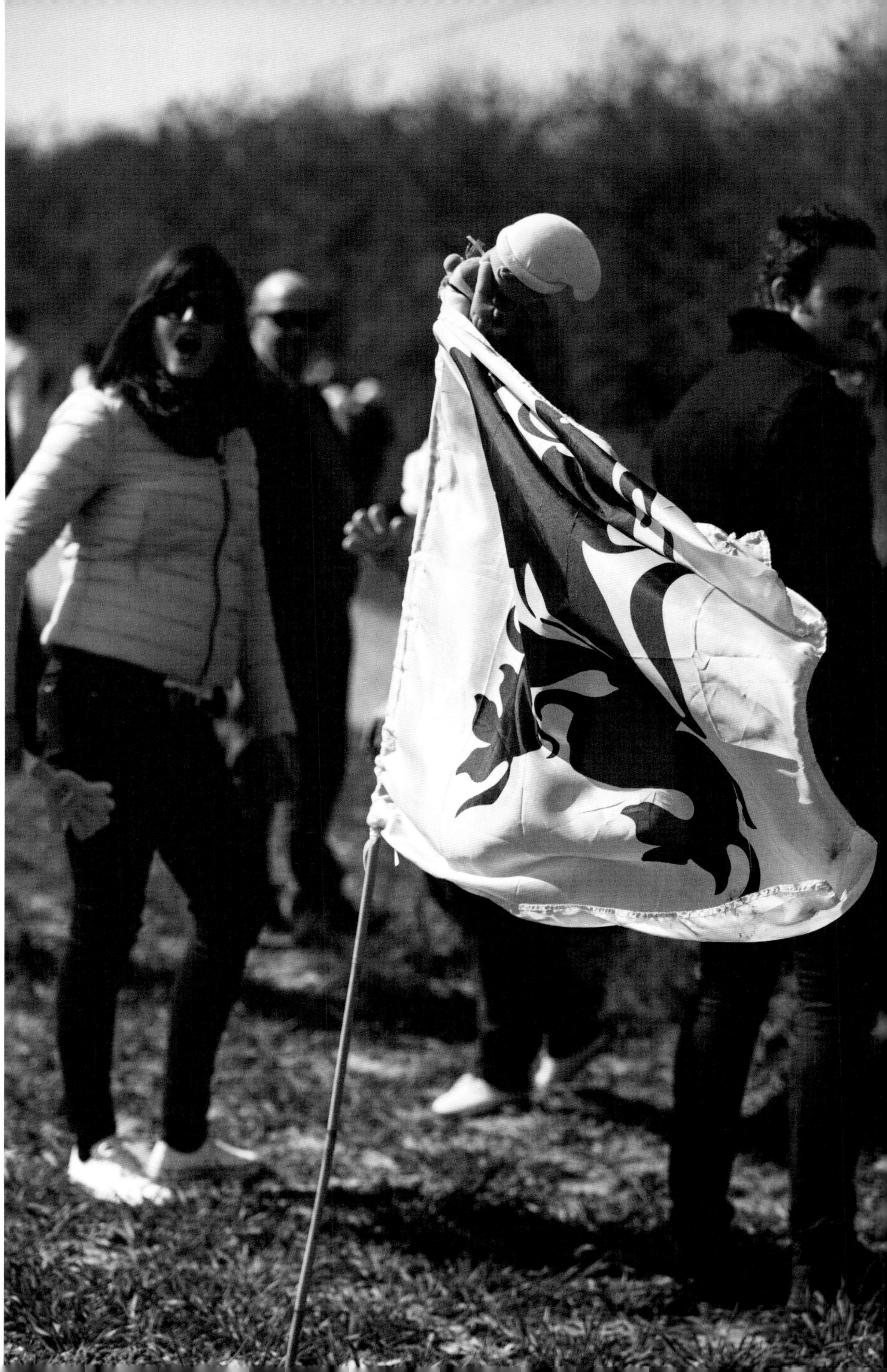

Schon beim zweiten Pavé denke ich: Noch einmal Kopfsteinpflaster überstehe ich nicht. Und das Rad erst. Ab dem dritten könnte die Bremse auch abmontiert sein.

Die Hölle ist mit Kopfstein gepflastert! Das wird mir irgendwo zwischen Paris und Roubaix bewusst. Keine Ahnung, wo genau. Wie ein Stier richte ich den Blick auf die 27 Millimeter vor mir. 170 Kilometer lang. Egal, es schüttet eh wie aus Kübeln und die Landschaft wird zur Nebensache. Wichtig ist nur: Hält das Zeug? Peng! Speichen gerissen? Nein! Und dann diese Angst vor einem Platten.

Schon beim zweiten Pavé-Abschnitt denke ich: Noch einmal Kopfsteinpflaster überstehe ich nicht. Und das Rad erst. Ab dem dritten könnte die Bremse auch abmontiert sein. Das Rad läuft sowieso von alleine geradeaus und in meiner Hand liegt nichts mehr. So etwas wie Gleichgültigkeit stellt sich ein. »Noch 23 Pavés«, schreit einer begeistert. Setzt so früh der Wahnsinn ein?!

Aus einer Laune heraus, von Übermut ganz besoffen, habe ich mich für PARIS-ROUBAIX angemeldet. Nun sitze ich hier auf dem Rad – mit 5000 anderen in der Amateur-Ausgabe. 50 Kilometer über das schlimmste Kopfsteinpflaster der Welt. Auf Stahl. So wie früher halt. Spannend und schnell und irgendwie spaßiger wirkt das im Film *Hell of the North*.

Es regnet echt pausenlos. Heldenwetter nennt man das, glaube ich. Wo wir gerade dabei sind: Was macht überhaupt diese Heldenkurbel an meinem Rad? Das 53er-Blatt sichert mir zwar den Respekt der Carbon-Crosser, dafür aber auch unglaubliche Qualen. Hilft nur der Wille, das Ding irgendwie durchzudrücken. Alle 50 Meter steht oder liegt wieder einer darnieder. Entweder ausgespuckt von der Strecke oder mit kaputtem Velo. »Sieh zu! Vorbei, der stürzt!«, schreit es um mich herum. Fünf Minuten später liege ich selbst in der Matsche. Das ist also PARIS-ROUBAIX! Schön, dich kennenzulernen.

Und dann kommt doch dieser ARENBERG. Der absolut schlimmste Streckenabschnitt, den man sich nur vorstellen kann. Wer genau hat diese losen Brocken in die Gegend gewürfelt, als hätte ein Riese Kniffel gespielt? Dir möchte ich begegnen.

Hintern? Nacken? Oberschenkel? Alles schmerzt. Die Hände spüre ich gar nicht mehr. Konzentration. Fokus. »Halt dein Maul!«, schreie ich dem inneren Schweinehund zu. Bis zum VÉLODROME DE ROUBAIX – ein Ziel, das sich anfühlt wie die Gnade.

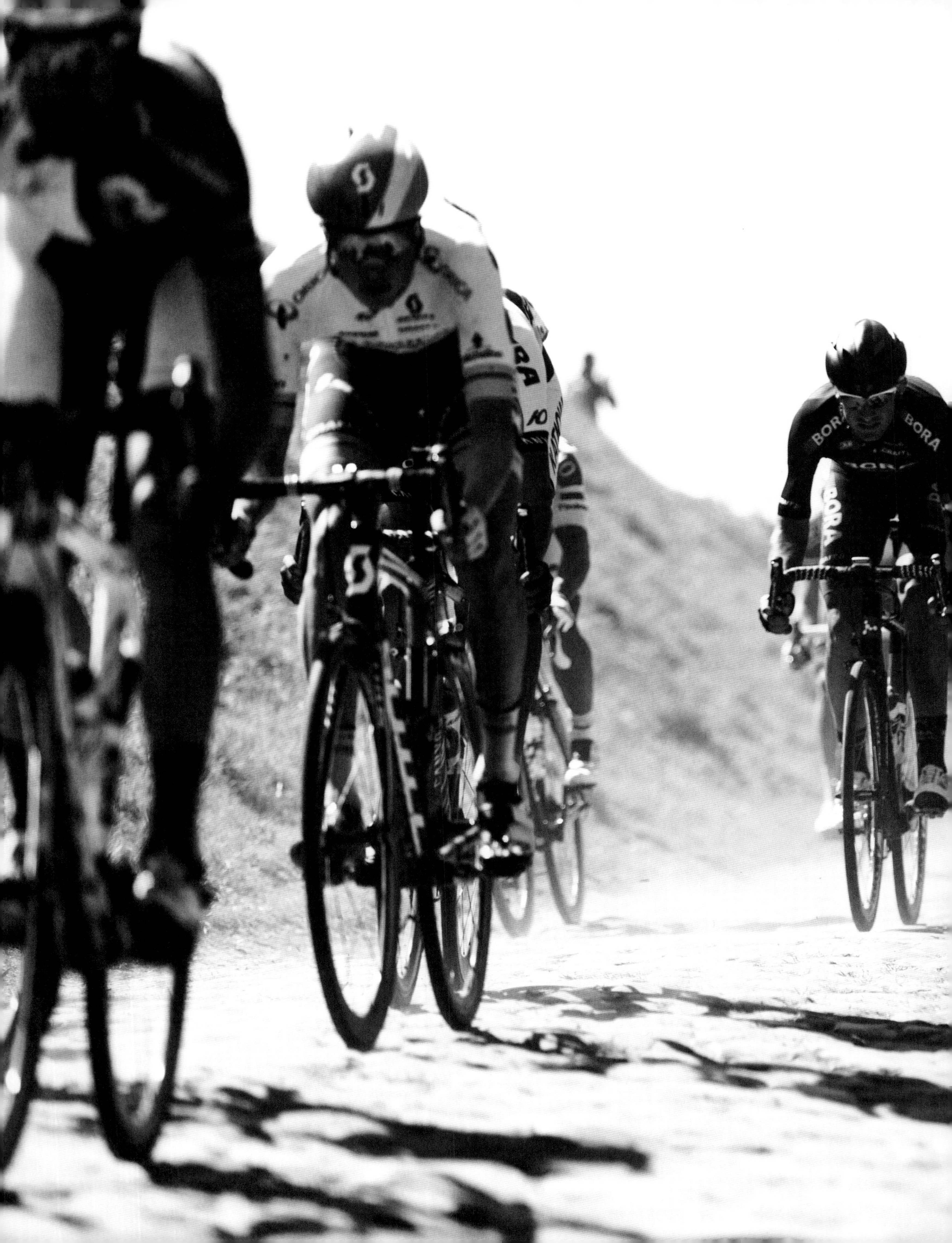
BORA

PARIS-ROUBAIX. Die Hölle des Nordens.
Blut, Schweiß und Tränen. Des Glücks.

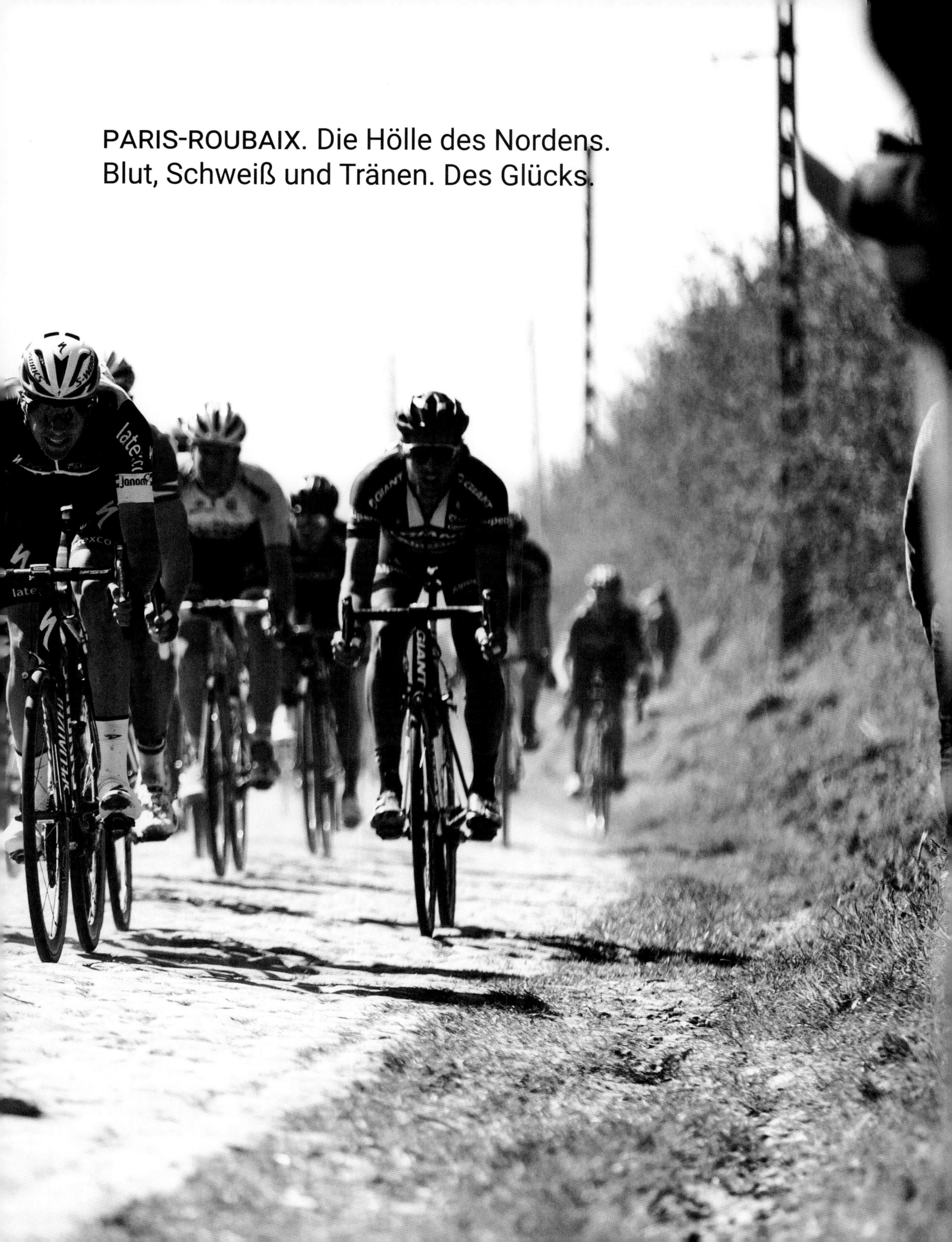

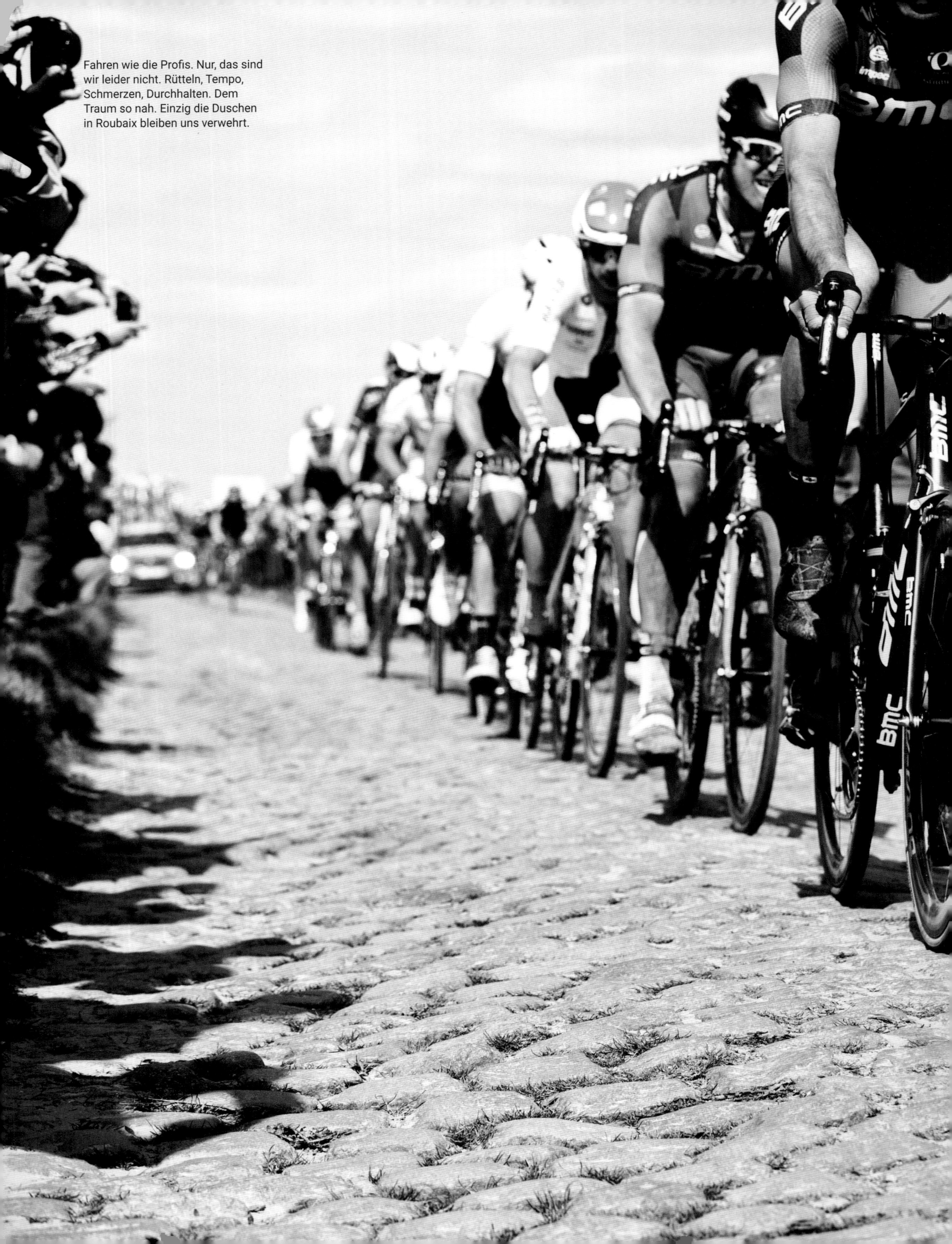

Fahren wie die Profis. Nur, das sind wir leider nicht. Rütteln, Tempo, Schmerzen, Durchhalten. Dem Traum so nah. Einzig die Duschen in Roubaix bleiben uns verwehrt.

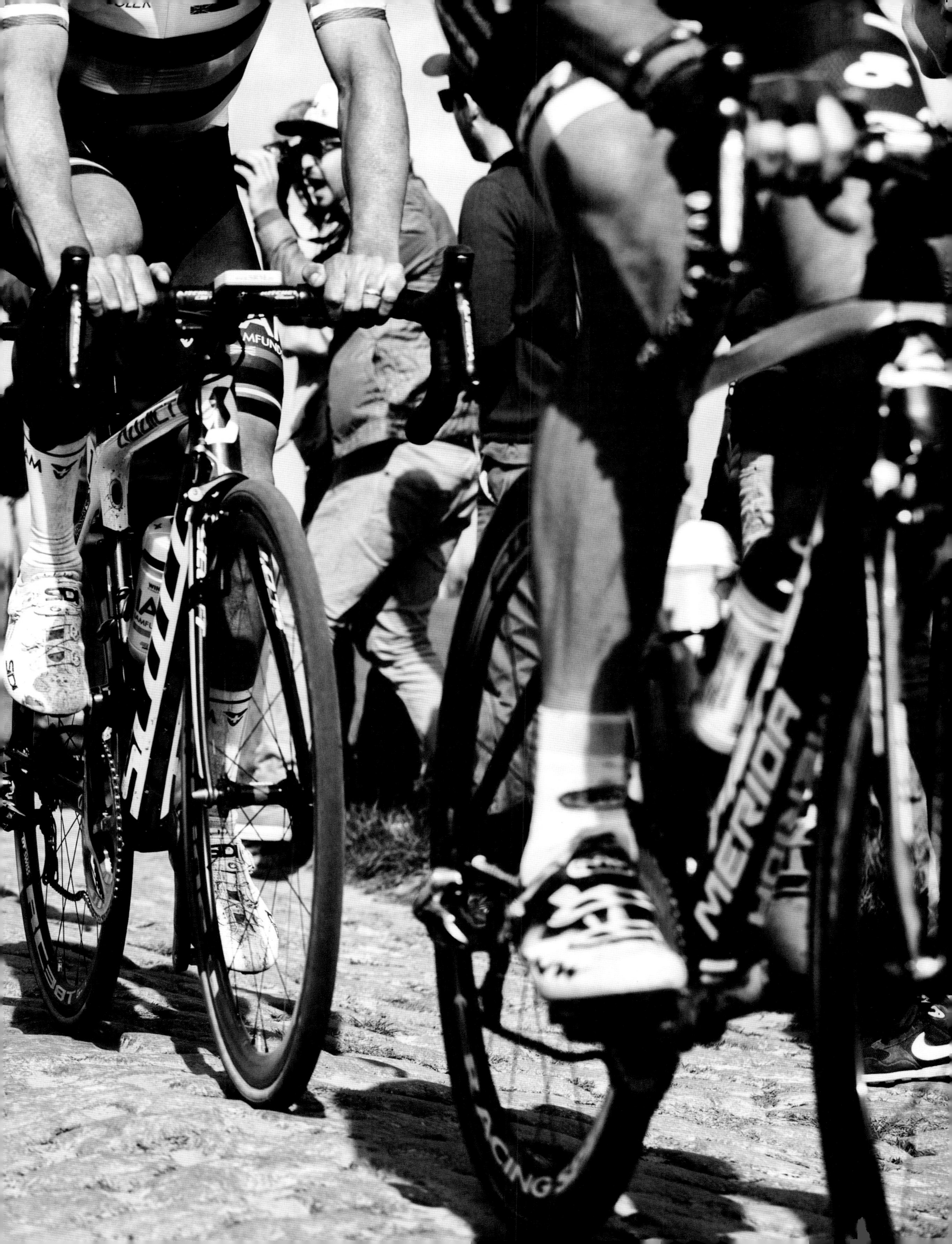
MERIDA

MASI

Hier kommt keine Schraube heimlich davon.

Das weiche Licht des Mailänder Spätsommers küsst Signore Masi. Seine Werkstatt im Velodrom ist eine besondere Perle des Rahmenbaus.

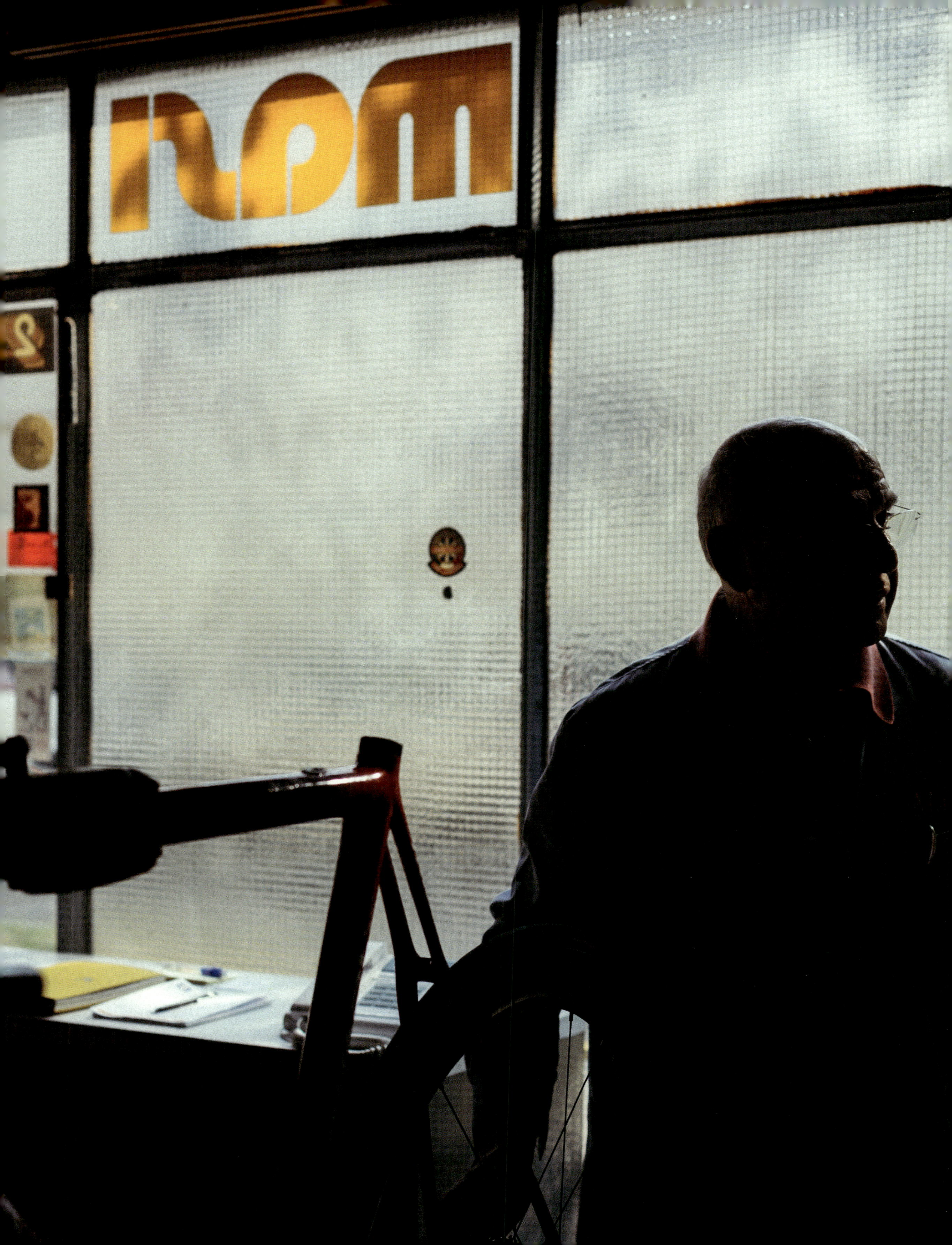

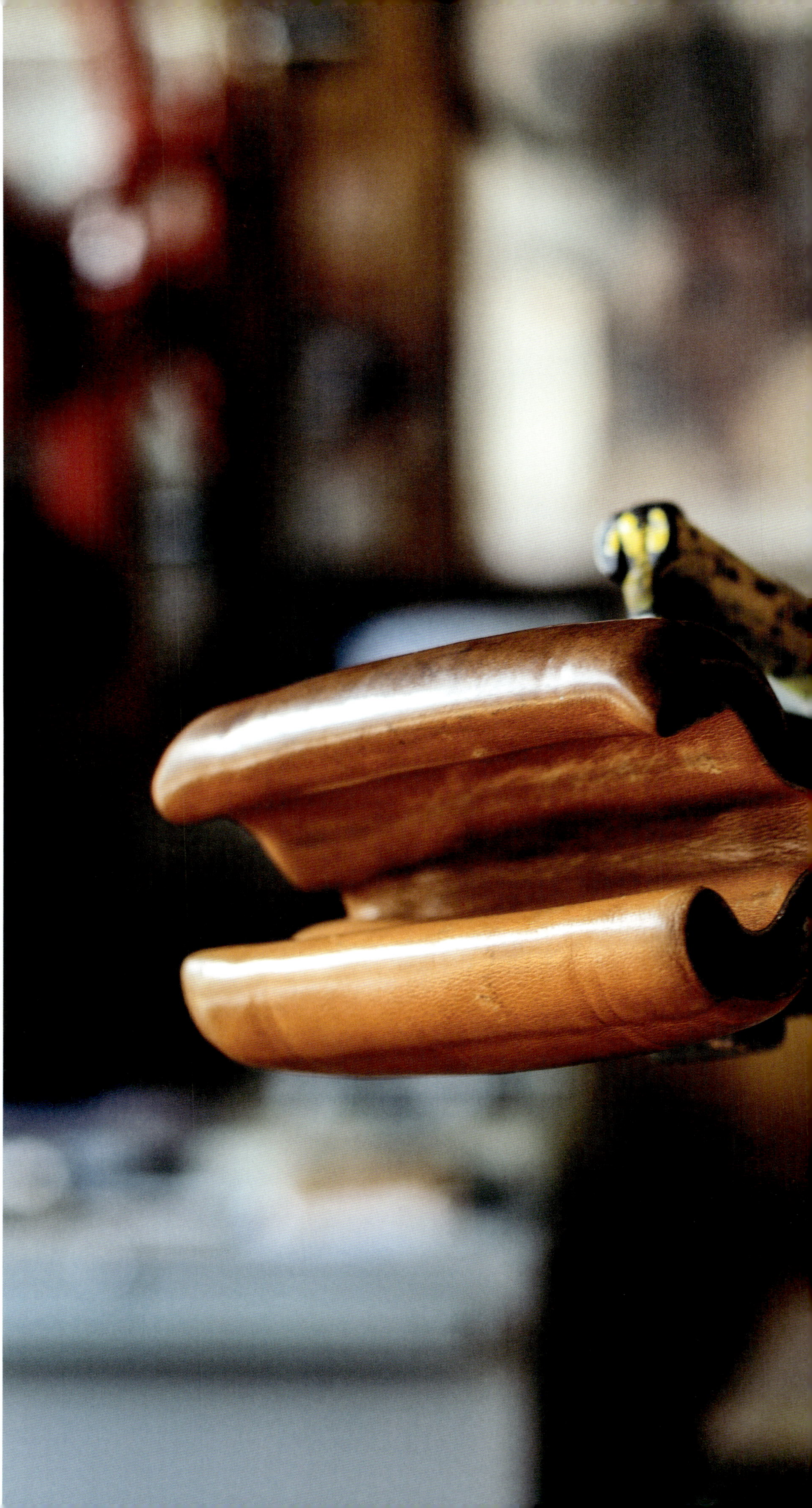

Nicht nur, dass man bei Masi vom Fußboden essen könnte, nein, auch dieser Montageständer ist ein wahres Juwel: An mit Leder bezogenen Klemmen kann der Meister an zwei Rädern gleichzeitig arbeiten.

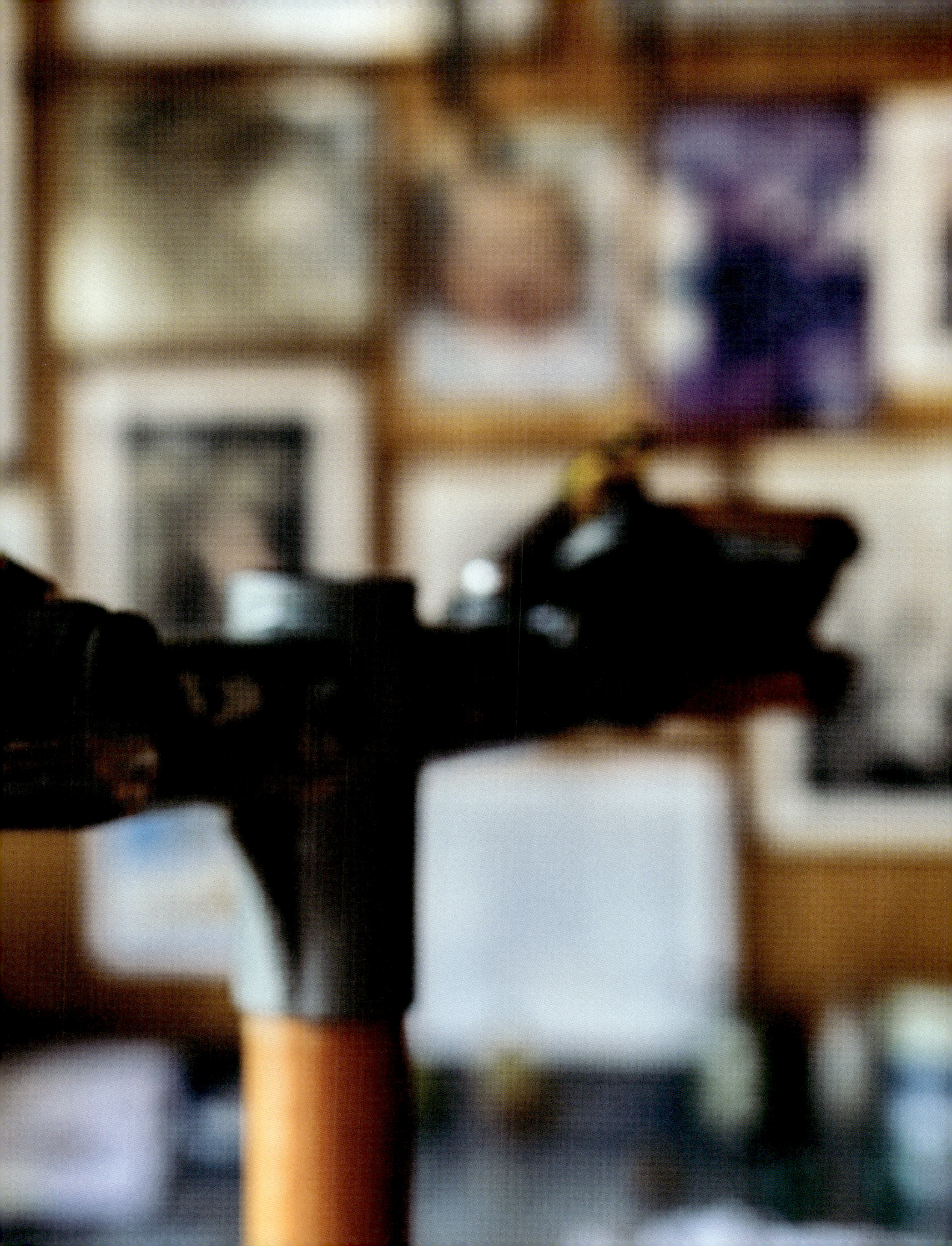

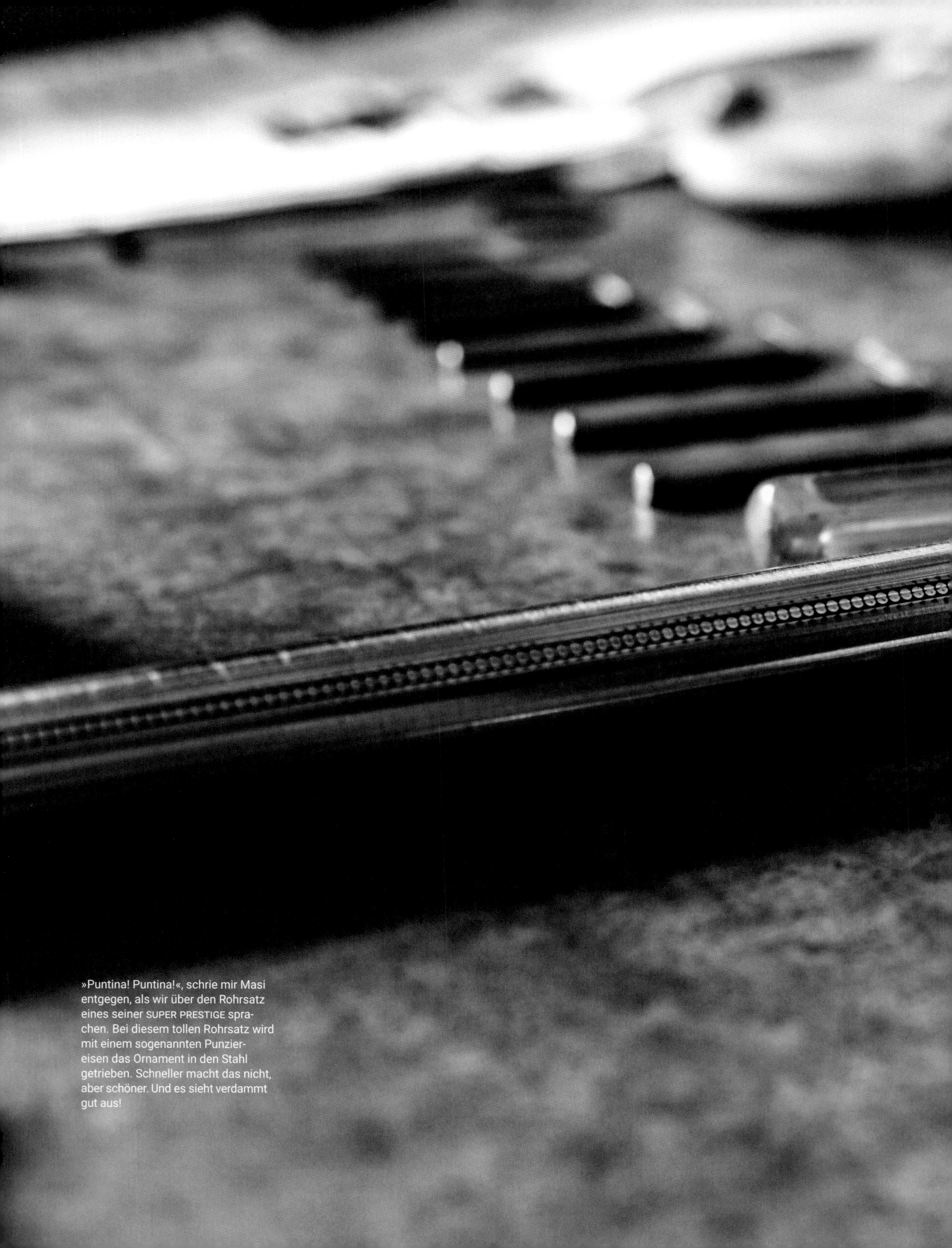

»Puntina! Puntina!«, schrie mir Masi entgegen, als wir über den Rohrsatz eines seiner SUPER PRESTIGE sprachen. Bei diesem tollen Rohrsatz wird mit einem sogenannten Punziereisen das Ornament in den Stahl getrieben. Schneller macht das nicht, aber schöner. Und es sieht verdammt gut aus!

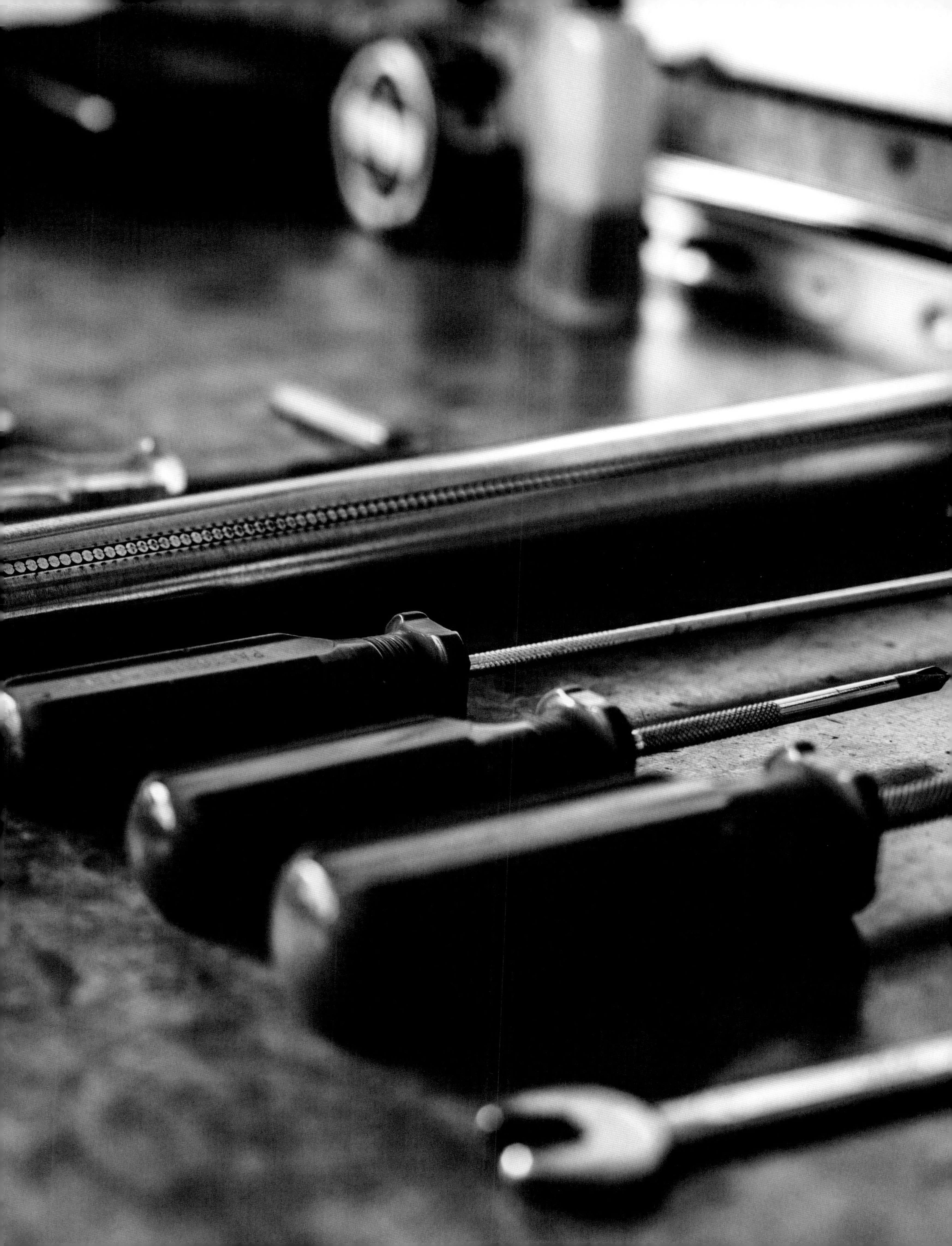

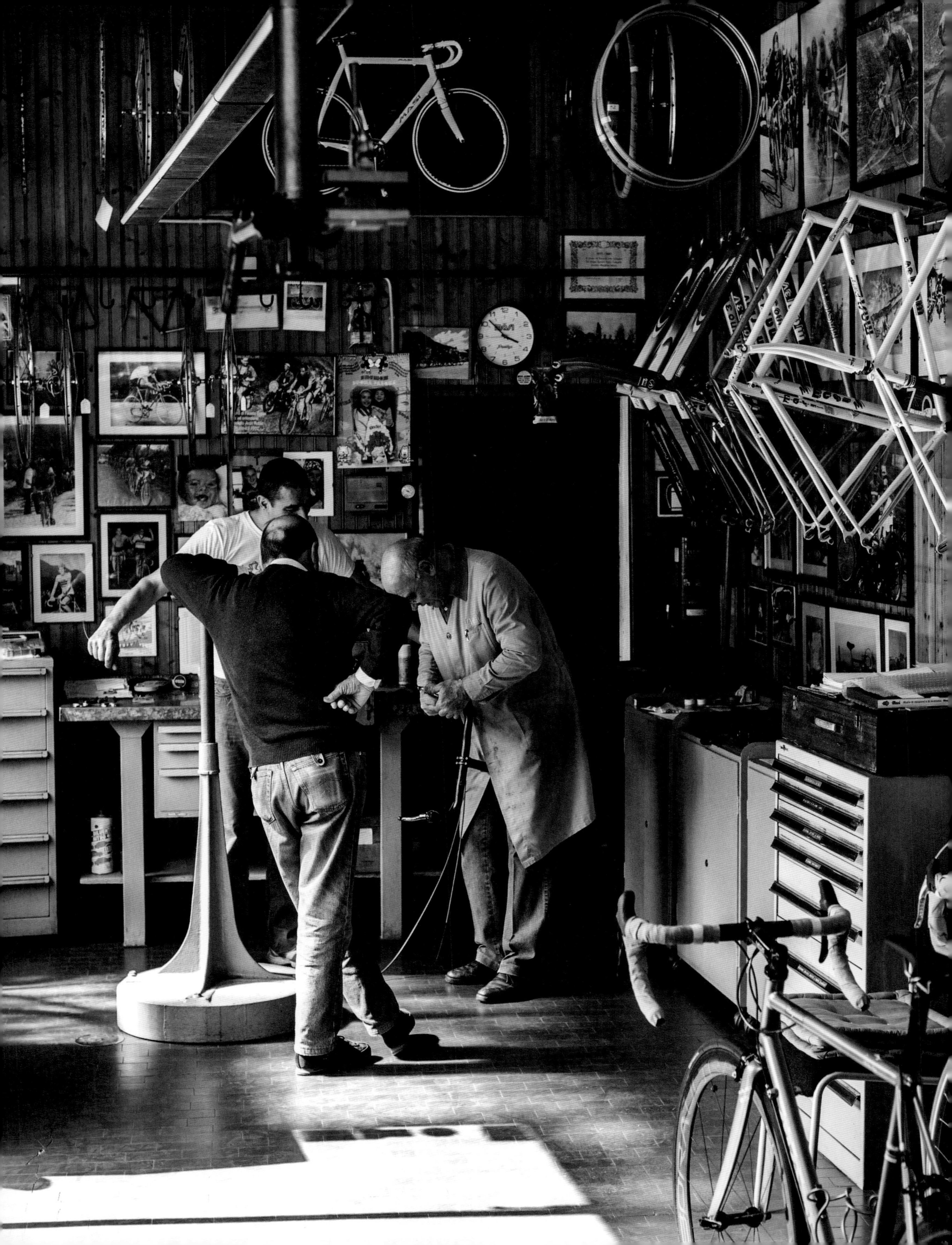

Und genau in diesem Moment fährt ein fitter, älterer Herr auf einem klapprigen PEUGEOT-Damenrad mit rostiger Kette vor. Signore Alberto Masi, der mich schnappt und in die hinter dem Rolltor gelegenen heiligen Hallen mitnimmt.

Schon allein die Werkstatt zu finden ist ein Akt! Aufs Rad gesetzt und raus aus der Mailänder Innenstadt, ein Park, eine Hauptstraße. Betonplatten aneinander gereiht, das Rad rattert – ab und zu entdecke ich ein kleines Schild, das mir Hoffnung gibt, nicht vollkommen verkehrt zu sein. Dann eine Wohnsiedlung; und plötzlich stehe ich vor dem VELODROMO MASPES-VIGORELLI.

Die Geschichte ist legendär: Zunächst in Rom erbaut, zog die sogenannte *Pista Magica* 1935 nach Mailand um, wurde bei einem Fliegerangriff fast vollständig zerstört und anschließend wieder aufgebaut. Neun Stundenweltrekorde wurden hier erzielt, und – das ist für mich das Relevante – irgendwo unter dem Stadion soll Masi noch immer seine Räder bauen. Das Velodromo selbst wird inzwischen nur noch hin und wieder für American-Football-Spiele genutzt. Was für eine Verschwendung!

So, und wo ist nun Masi? Ist er überhaupt noch hier? Ich entdecke ein mit Tusche bemaltes Pappschild, etwa DIN-A4-groß. MASI. Das Rolltor dahinter ist allerdings verschlossen. Verdammt! Warum habe ich nicht vorher angerufen? Ich setze meine Runde um das verlassene Stadion fort, komme an ein weiteres Tor und abermals lacht mich so ein handbemaltes Schild an. MASI. Mehr nicht.

Als ich einen Passanten nach der Werkstatt frage, sagt der nur »Komm um drei wieder!« Ich Idiot. Wie konnte ich nur die Tageszeit außer Acht lassen. Die sakrosankte italienische Mittagszeit. Der Mann war essen!

Und genau in diesem Moment fährt ein fitter, älterer Herr auf einem klapprigen PEUGEOT-Damenrad mit rostiger Kette vor. Signore Alberto Masi, der mich schnappt und in die hinter dem Rolltor gelegenen heiligen Hallen mitnimmt. Warmes Frühjahrslicht durchflutet den kleinen Raum. An den holzvertäfelten Wänden hängen Rahmen, Siegerräder von MERCKX und Co., dazwischen dicht an dicht Werkzeuge und Reliquien – und Fotos mit bedeutenden Momenten aus sechzig Jahren Radsportgeschichte. In der Mitte der Halle steht ein Montageständer. Mit Leder bezogen, damit die Schönheit der Räder vollkommen erhalten bleibt. Der Steinfußboden ist so sauber, dass man vom Boden essen könnte. Und auch sonst ist die Werkstatt ein Musterbeispiel an Ordnung und Reinlichkeit: Hier kommt keine Schraube heimlich davon!

Wenn Masi seinen blauen Kittel überstreift, beginnt die Transformation. Dann ist er nicht mehr der Signore, dann

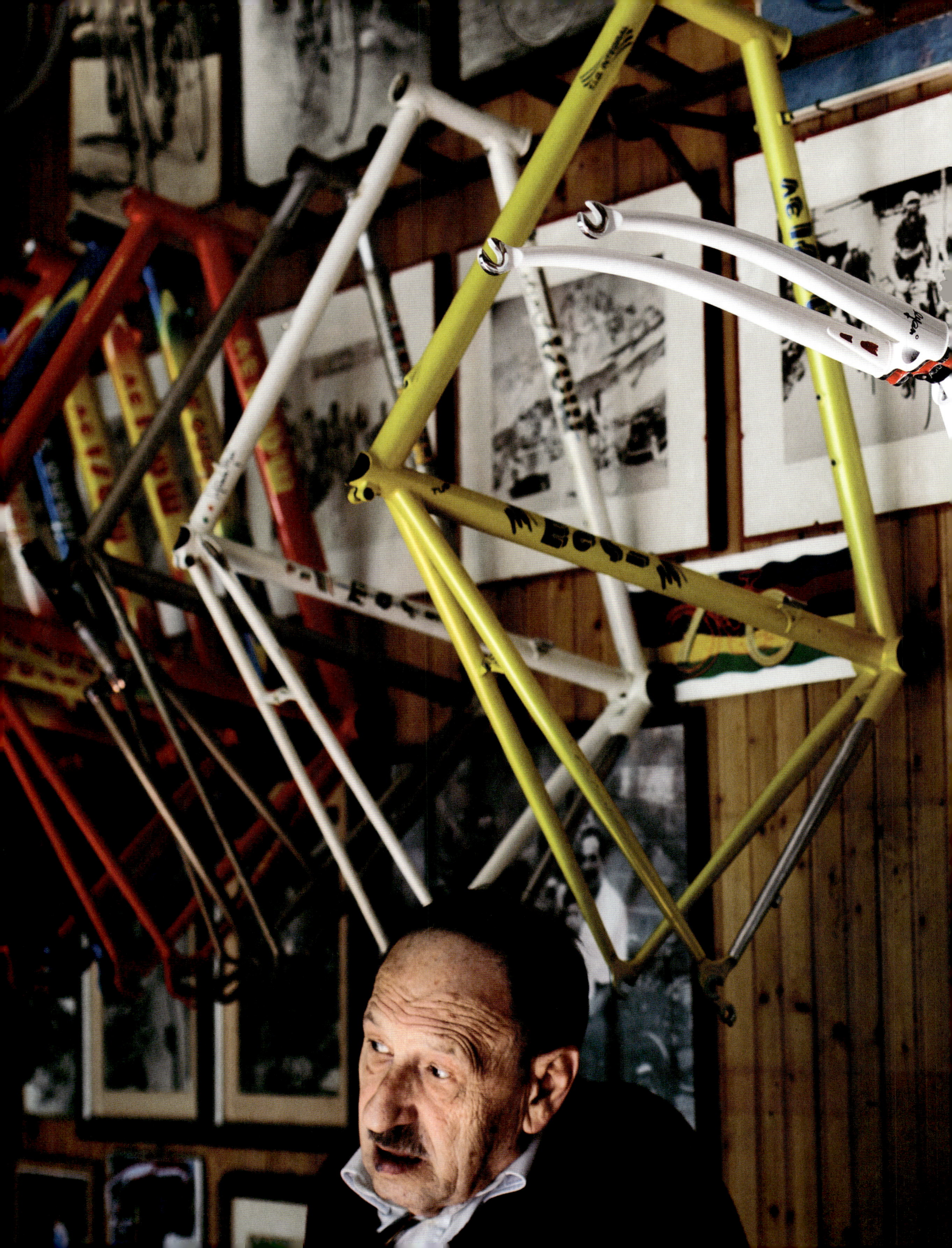

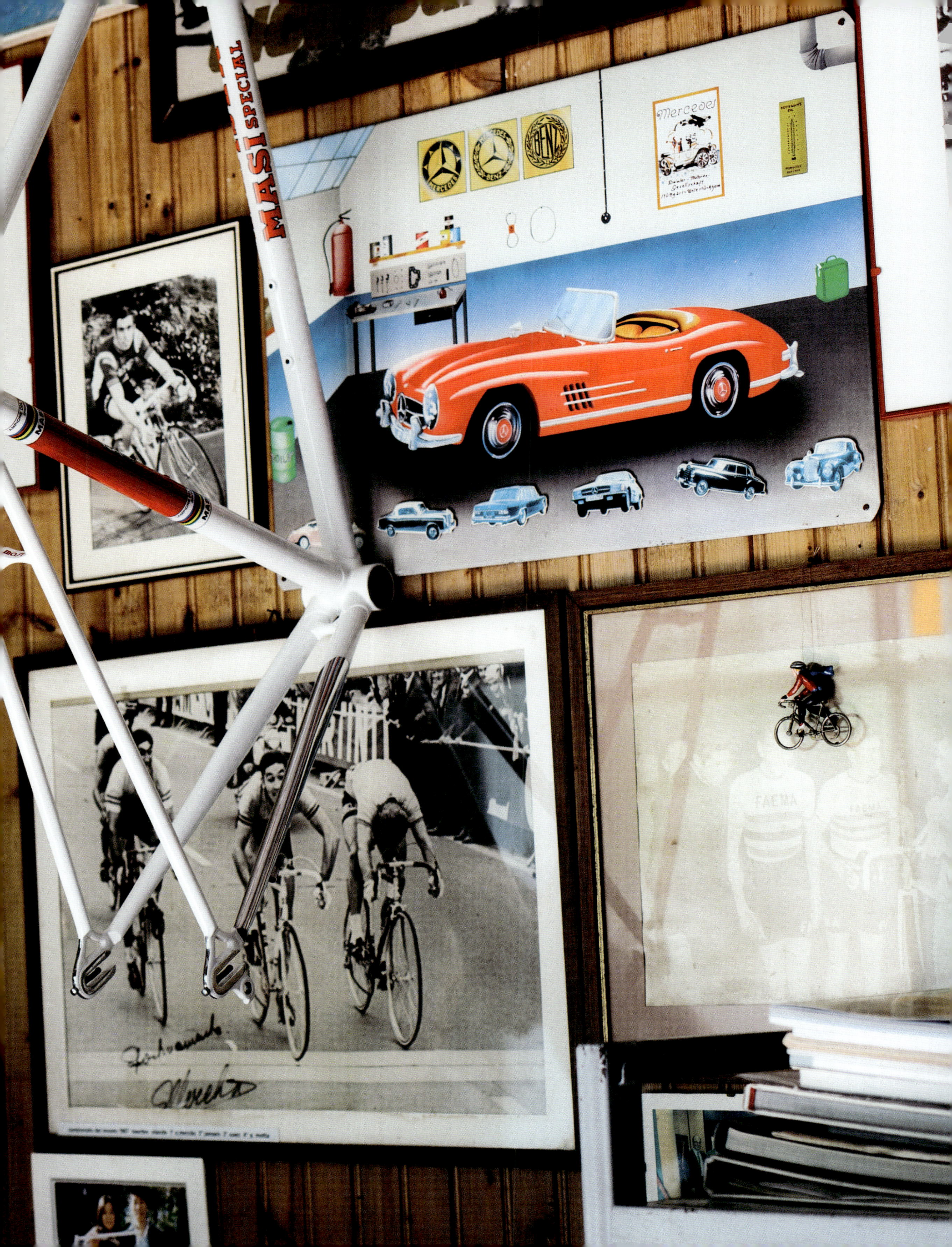
MASI SPECIAL
Mercedes
BENZ
FAEMA

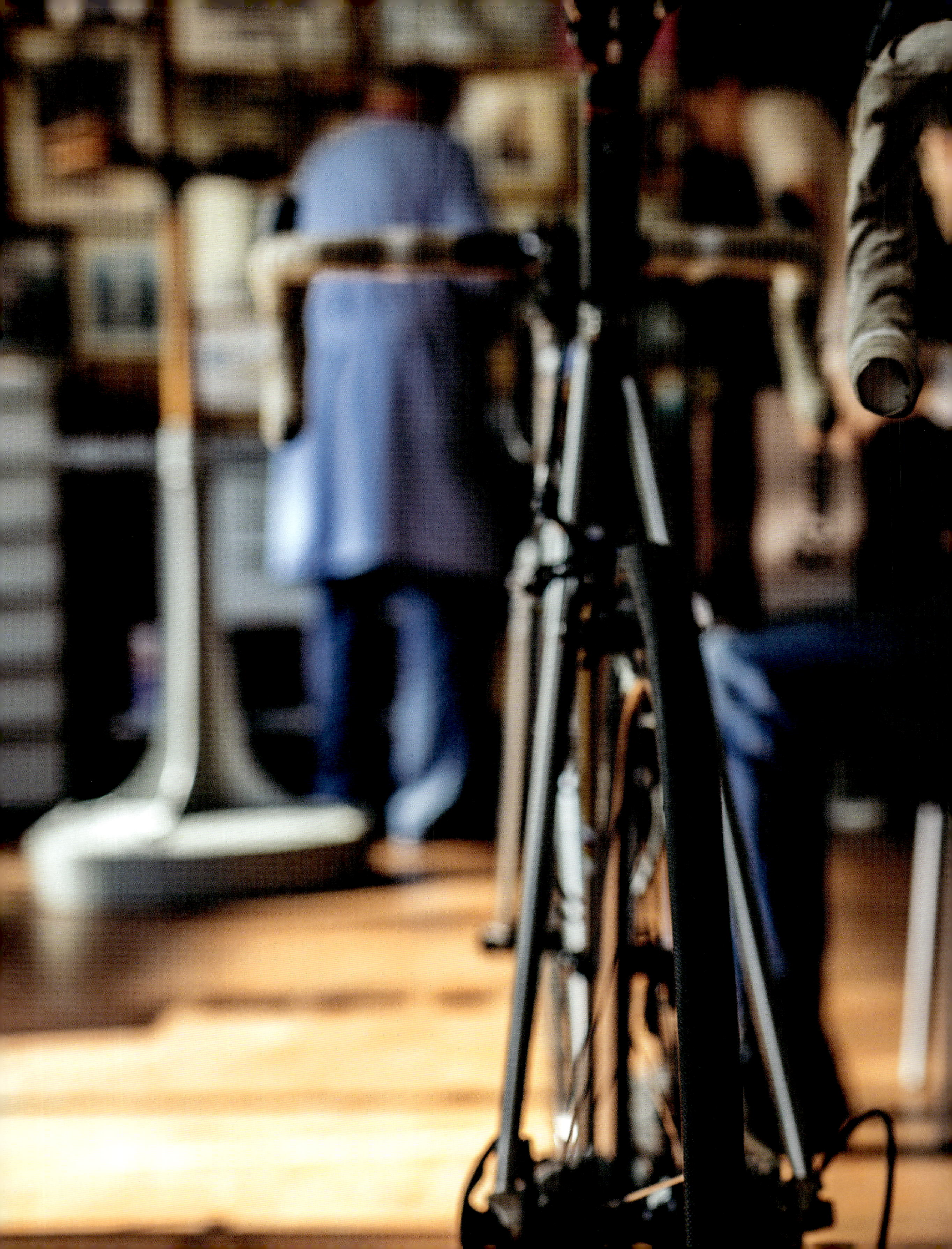

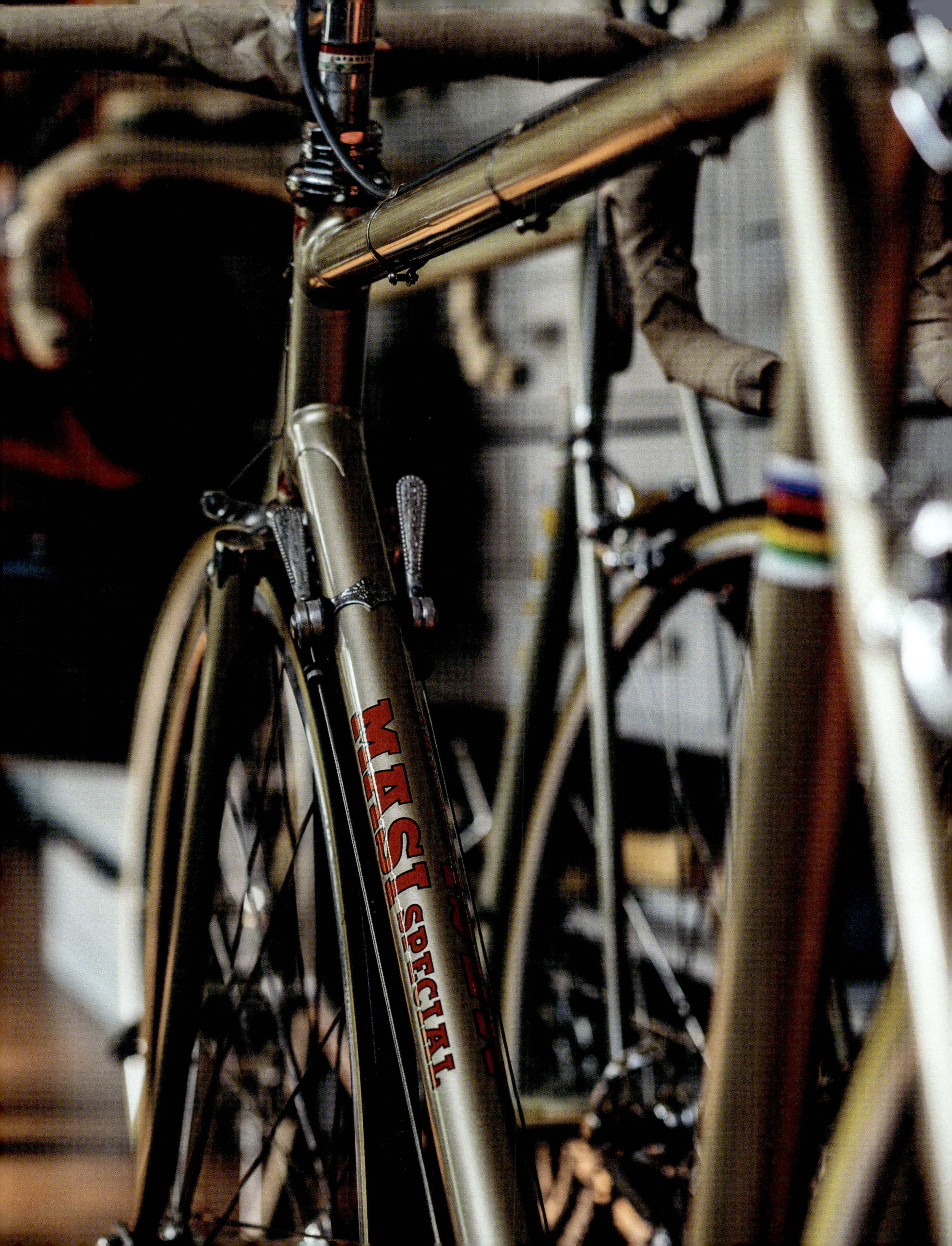
MASI SPECIAL

ist er einfach nur Masi. Sauber aufgereiht, wie beim Zahnarzt, liegt sein Werkzeug da – jeder Abstand zentimetergenau. Daneben Karteikarten mit den Bestellungen seiner Kunden. Jeder Wunsch, von Modell, Ausstattung, Einstellwinkel bis zur Sitzposition, ist in penibler Handschrift auf diesen Karten vermerkt. Maßgeschneiderte Einzelstücke!

Mithilfe von Assistent Angelo als Übersetzer unterhalten wir uns. Dafür lässt Masi die Arbeit für einen Moment liegen und nimmt seine Brille ab. Es ist ruhig in der Werkstatt, Kunden kommen nur mit Termin vorbei – die Auftragsbücher sind ohnehin voll. Masi greift gezielt einen Bildband von einem der vielen Bücherstapel und beginnt zu blättern. Dabei scheint er die alten Erfolge noch einmal zu durchleben. Still. Schweigend.

Plötzlich wird mir noch einmal richtig bewusst, wo ich mich hier befinde: In der Rahmenschmiede so vieler Fahrradlegenden wie Fausto Coppi, Fiorenzo Magni, Louison Bobet, Vittorio Adorni und, wie schon erwähnt, Eddy Merckx. Viele Sammler würden große Opfer bringen, um ein Rad dieses Mannes zu besitzen. Ein MASI, das ist so ungefähr der MAYBACH der Radrennszene.

Nun bin ich also hier, doch meine Hoffnung, einen Raum weiter zu gehen und den Bereich der Werkstatt zu sehen, um bestenfalls sogar den sagenhaften Rahmenbau in Aktion zu erleben, bleibt ein unerfüllter Wunschtraum. »Komm du auf deinem eigenen MASI vorgefahren, dann öffnen wir dir auch die hinteren Tore!«, wird mir mit einem Lächeln geantwortet. Diskussion beendet.

Ein bisschen erweichen lässt sich der Maestro dann aber doch. Er holt ein Oberrohr aus dem hinteren Raum und zeigt mit den Worten »Puntina! Puntina!« die für MASI so typischen Gravuren, die mit einem Punziereisen eingetrieben werden. Ein Foto darf ich machen, dann verschwindet das Teilchen aber auch gleich wieder.

So, nun muss er aber auch wieder los. Familiäre Verpflichtungen. Irgendein Kaffeetrinken zum Geburtstag, wenn ich das richtig verstehe. Ciao. Arrivederci. Doch auch nachdem der Großmeister mit einer kurzen Verabschiedung auf seinem klapprigen Damenrad davongerollt ist, der Kittel wieder sauber auf seinem Montageständer hängt und es noch stiller ist als zuvor – der Geist Masis ist noch immer spürbar. Ein kleines Stück Magie!

Doch auch nachdem der Großmeister mit einer kurzen Verabschiedung auf seinem klapprigen Damenrad davongerollt ist, der Kittel wieder sauber auf seinem Montageständer hängt – der Geist Masis ist noch immer spürbar. Ein kleines Stück Magie!

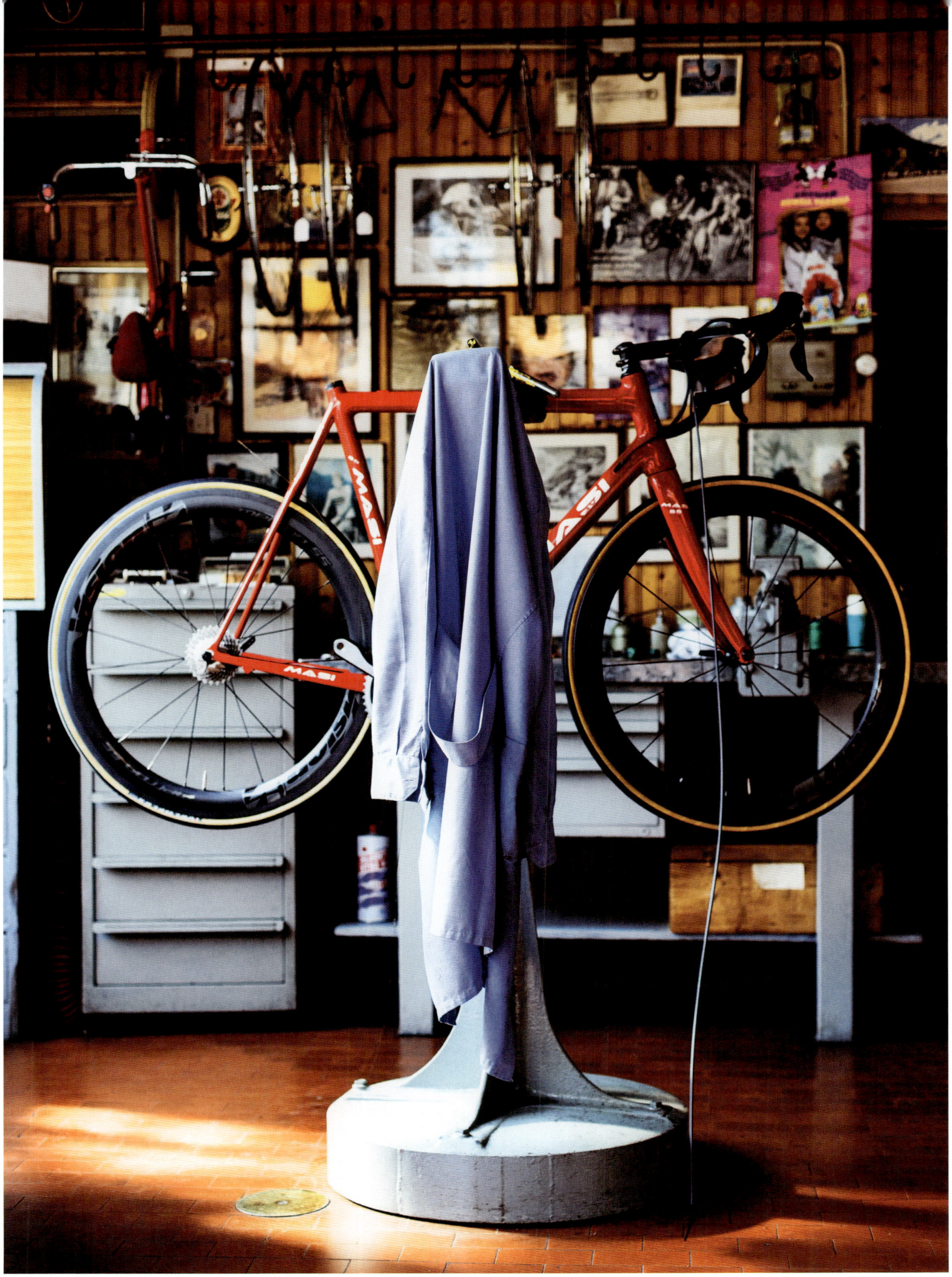
MASI
MASI
MASI

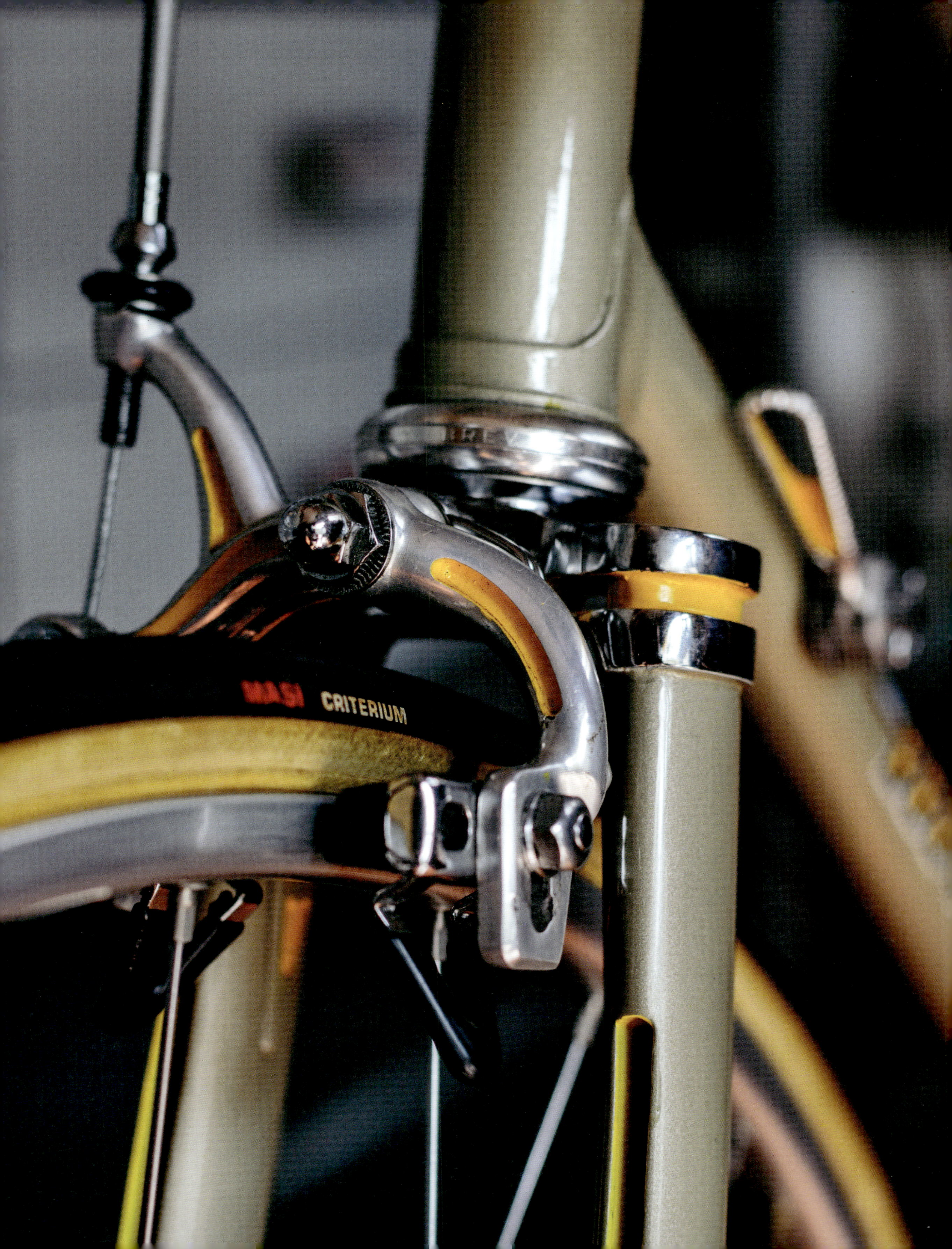
MASI CRITERIUM

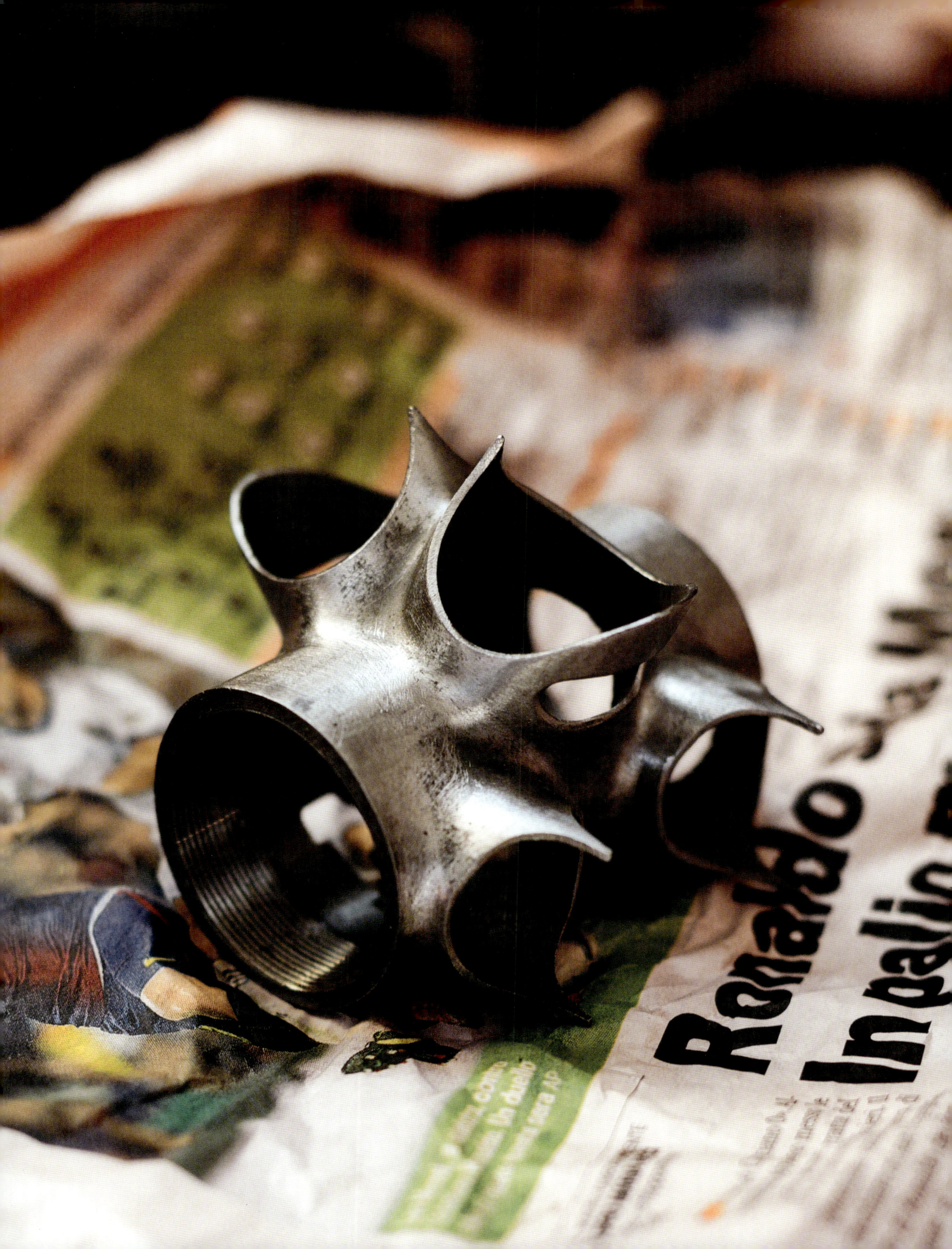
Ronaldo

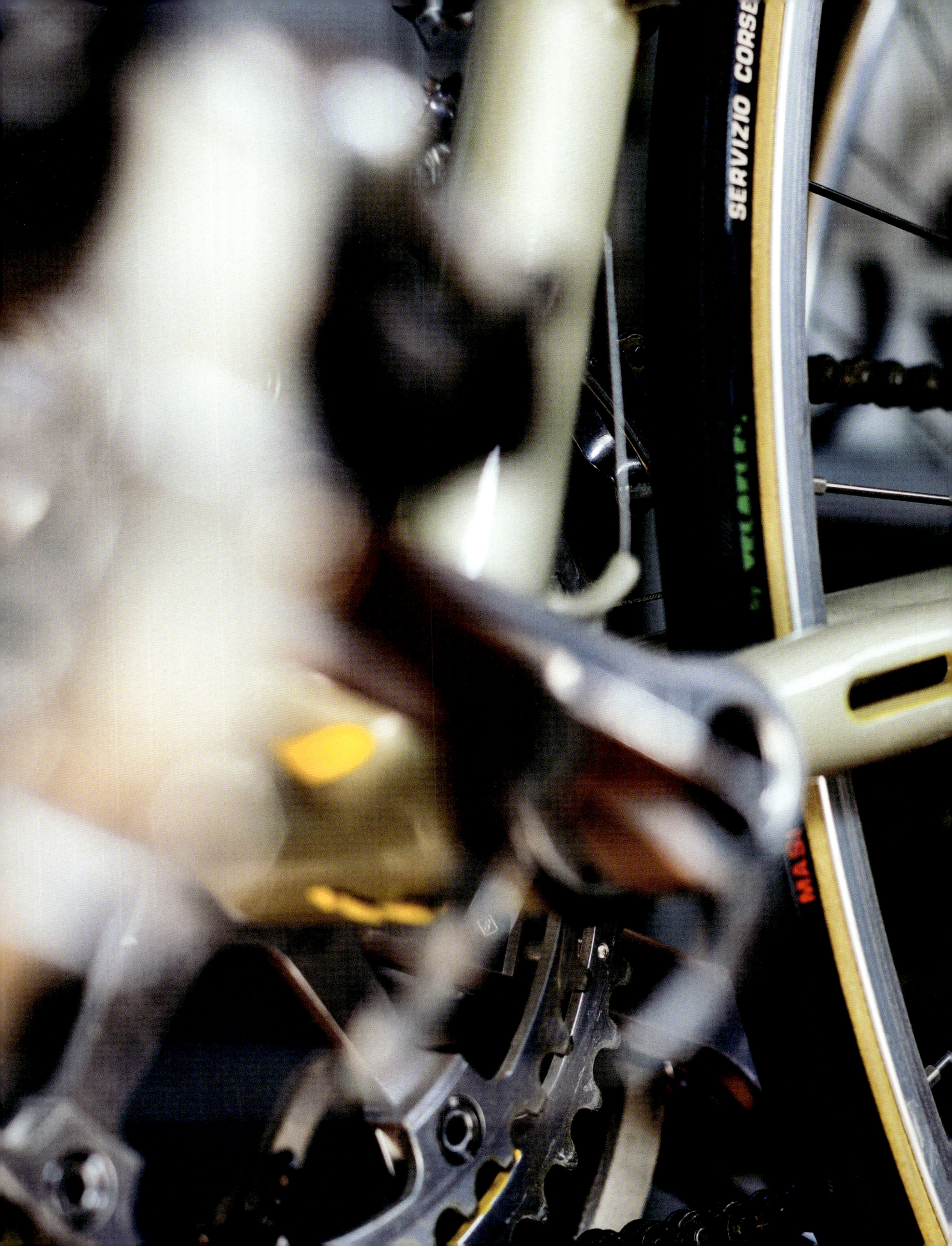
SERVIZIO CORSE
MASI

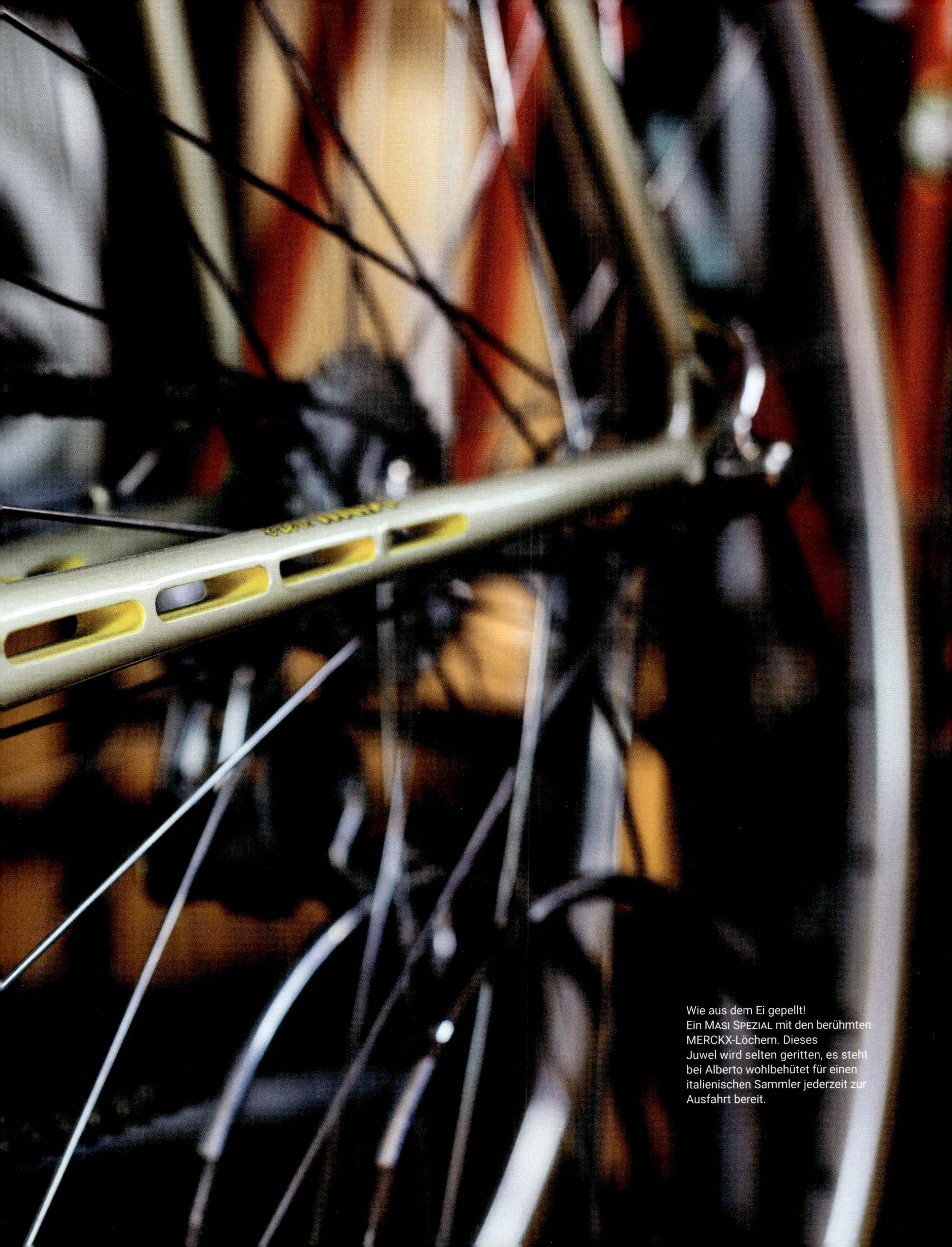

Wie aus dem Ei gepellt!
Ein Masi Spezial mit den berühmten MERCKX-Löchern. Dieses Juwel wird selten geritten, es steht bei Alberto wohlbehütet für einen italienischen Sammler jederzeit zur Ausfahrt bereit.

IN VELO VERITAS
Wadln, Wein und Wolle.

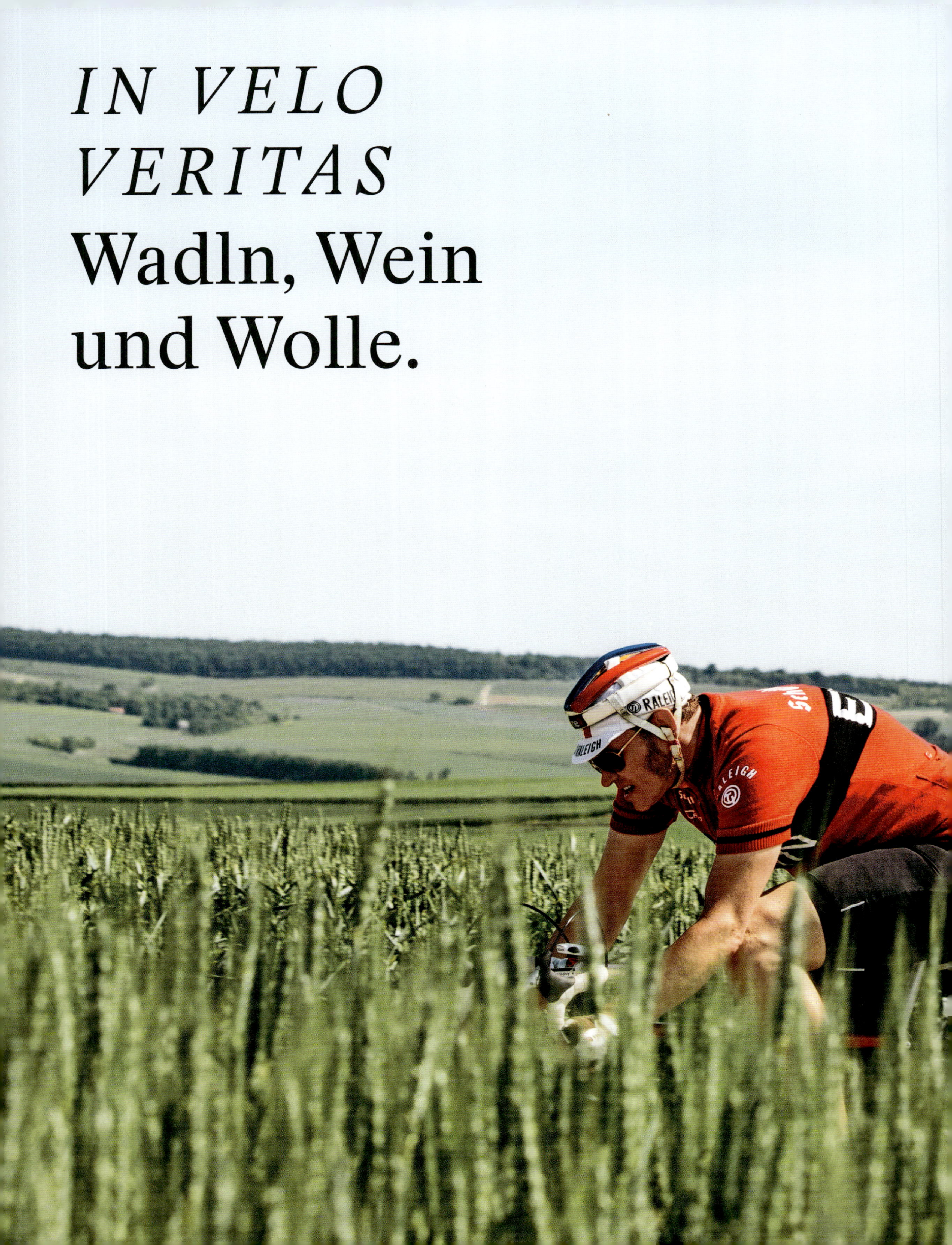

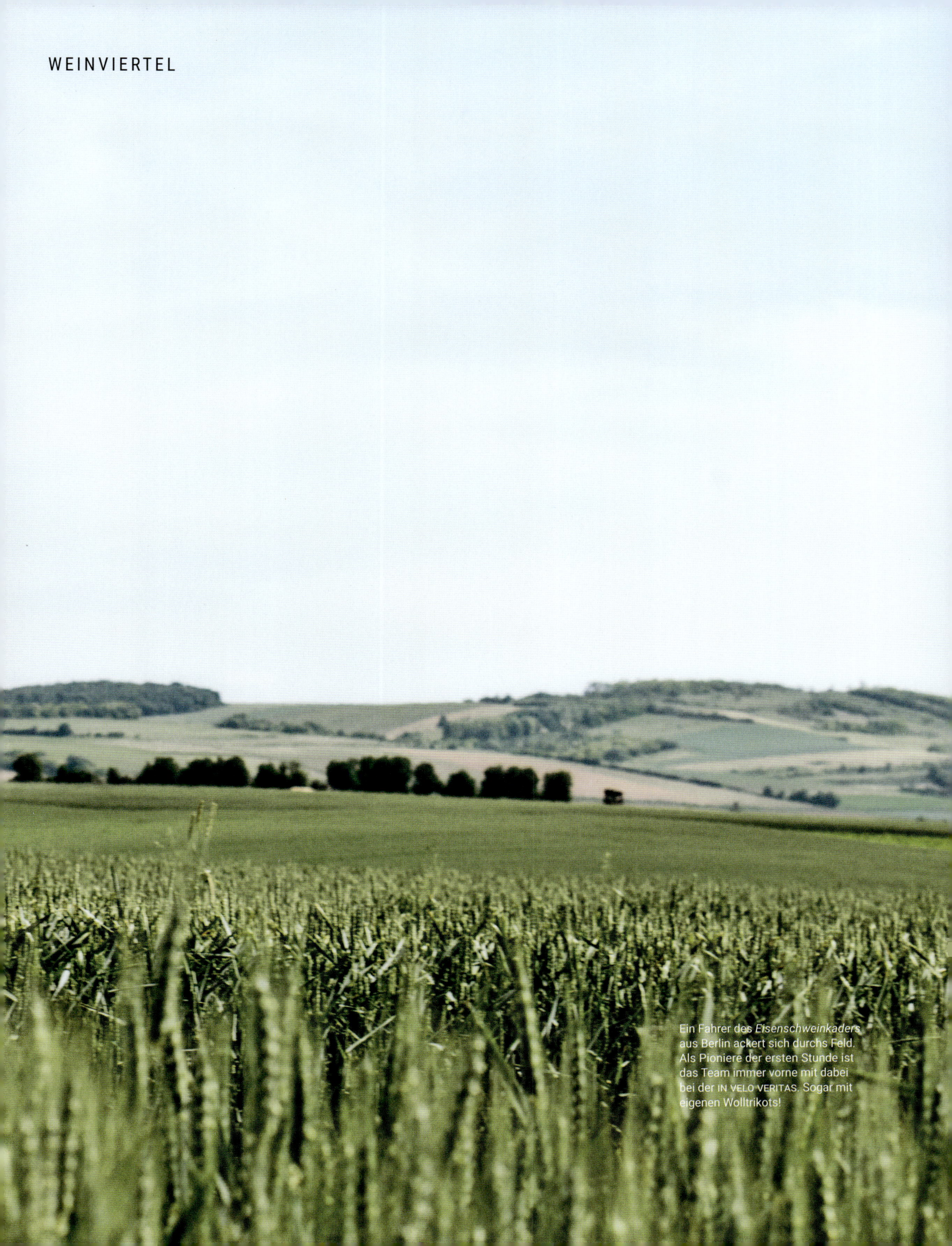

Ein Fahrer des *Eisenschweinkaders* aus Berlin ackert sich durchs Feld. Als Pioniere der ersten Stunde ist das Team immer vorne mit dabei bei der IN VELO VERITAS. Sogar mit eigenen Wolltrikots!

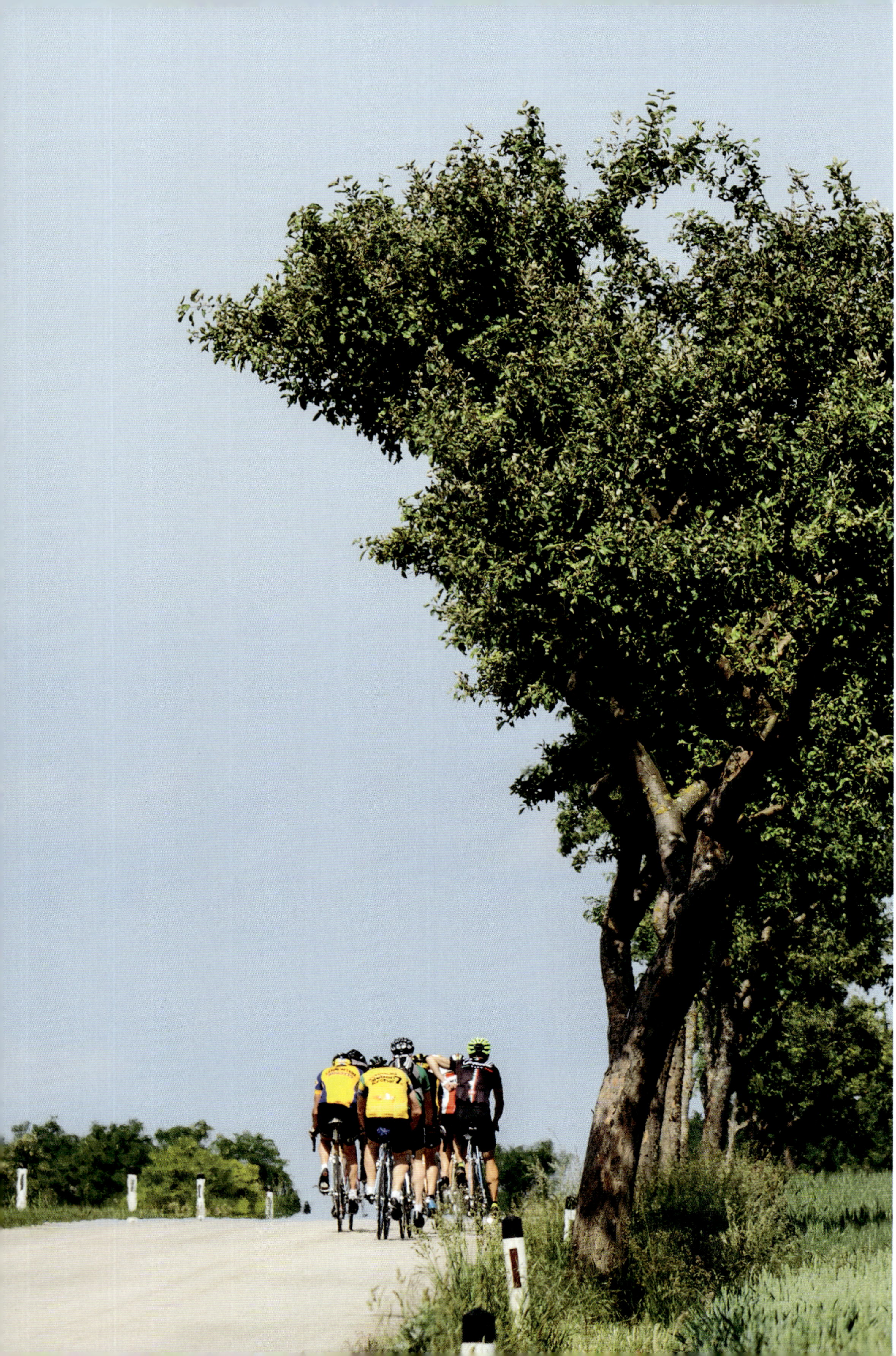

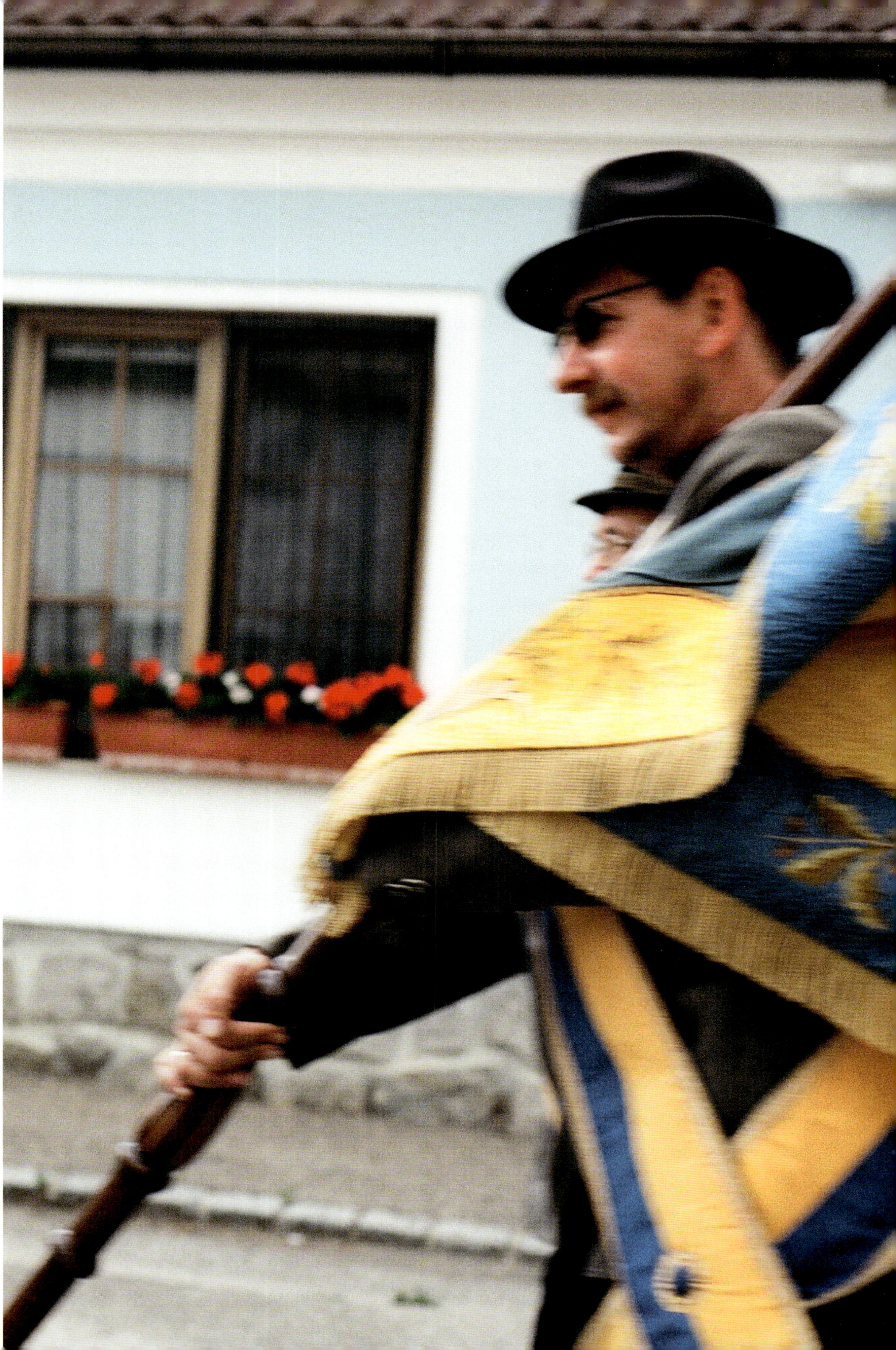

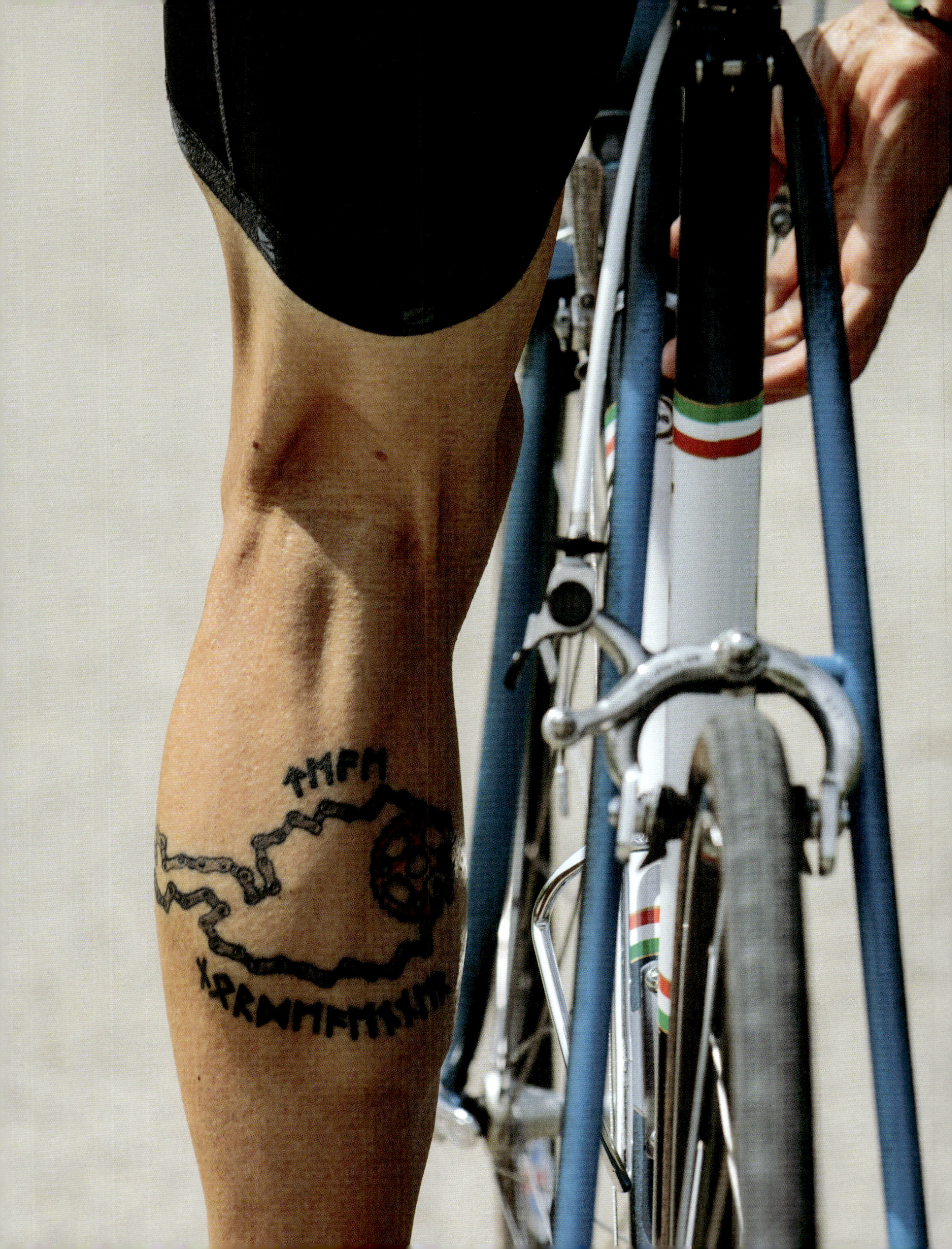

Nach Stahl, Alu und Carbon fährt die Welt wieder Stahl. Das ist in Wien nicht anders als in Stockholm, London und Kapstadt. Das gleichmäßige Surren der Ketten gibt den Takt vor.

Nach Stahl, Alu und Carbon fährt die Welt wieder Stahl. Das ist in Wien nicht anders als in Stockholm, London und Kapstadt. Und so wollen auch die Österreicher ihre eigene klassische Heldentour haben – angelehnt an die L'EROICA zeigt man der Welt: Auch jenseits der Toskana kann man genussvoll radeln. Kann man sich wie einer jener Rennhelden fühlen, die noch in Wolltrikots und mit eigenem Werkzeug in der Tasche ins Ziel geradelt kamen.

Die Radverrückten um Horst Watzl und Michl Mellauner erkundeten also die Region nahe Wien, um so etwas Ähnliches wie die legendäre L'EROICA auf die Beine zu stellen. Dafür ging es dann über Tausende von Kilometern durch das Weinviertel, um mögliche Routen zu erkunden. 2012 fiel dann zum ersten Mal ganz offiziell gemeinsam mit ein paar hundert Enthusiasten der Startschuss. IN VELO VERITAS nennt sich das Spektakel. Je nach Motivation, Fitnesslevel oder Mut können die Teilnehmer 70, 140 oder gar epische 210 Kilometer radeln. Die Routen, die zurückgelegt werden, sind anspruchsvoll und führen durch wunderschöne Weinberge, über knackige Anstiege in der tschechischen Grenzregion oder durch endlose Weizenfelder.

Beim Start morgens um sechs Uhr ist es frisch, bald wärmt aber die Sommersonne die Fahrer in ihren Wolltrikots. Wie Perlen an der Kette fädeln sich die Teilnehmer hintereinander auf, wenn es durch die schmalen Feldwege der Weinberge oder durch verschlafene Ortschaften geht. Das gleichmäßige Surren der Ketten gibt den Takt vor.

Die Labstellen, die zu jedem klassischen Rennen dazugehören, lassen es an nichts mangeln. Natürlich gibt es im Weinviertel Reben- und Traubensaft, dazu frisches Brot, selbst gemachte Aufstriche, Salate und Suppen. Hernach weiterzufahren wird da fast zur größten Herausforderung des Tages. Wie gut könnte man es sich doch noch ein wenig gemütlich machen beim Heurigen.

Jedes Jahr wächst die IN VELO VERITAS, und ist inzwischen ein fester Termin vieler Vintage-Fahrer und Sammler in ganz Europa. Die eine oder andere Gruppe reist schon auf klassischen Rennrädern aus der Schweiz oder Deutschland an, bevor es gilt, die Epische als letzte Etappe ihres Abenteuers zu bewältigen. Eine echte Radwanderung im Sinne der Radonneure.

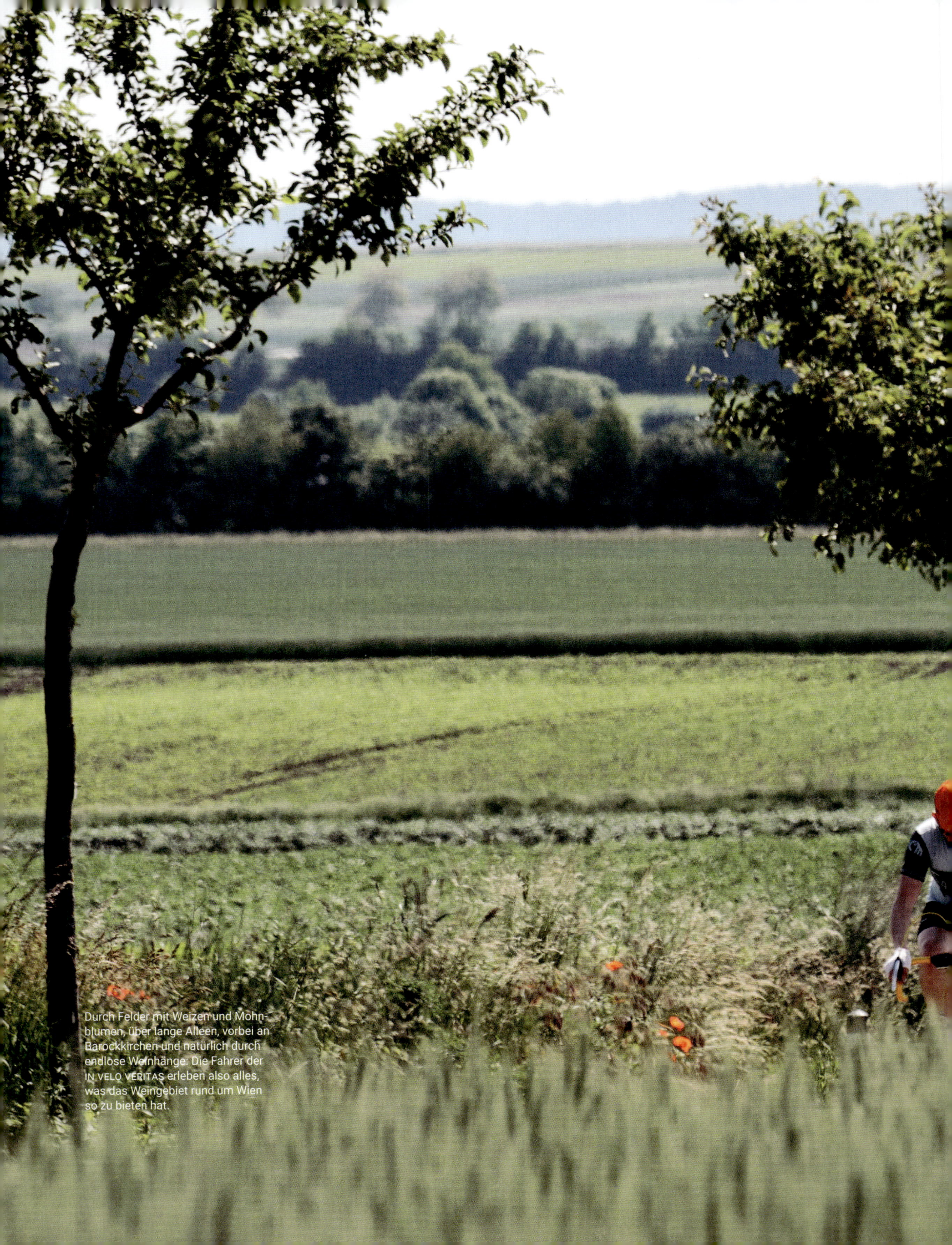

Durch Felder mit Weizen und Mohnblumen, über lange Alleen, vorbei an Barockkirchen und natürlich durch endlose Weinhänge: Die Fahrer der IN VELO VERITAS erleben also alles, was das Weingebiet rund um Wien so zu bieten hat.

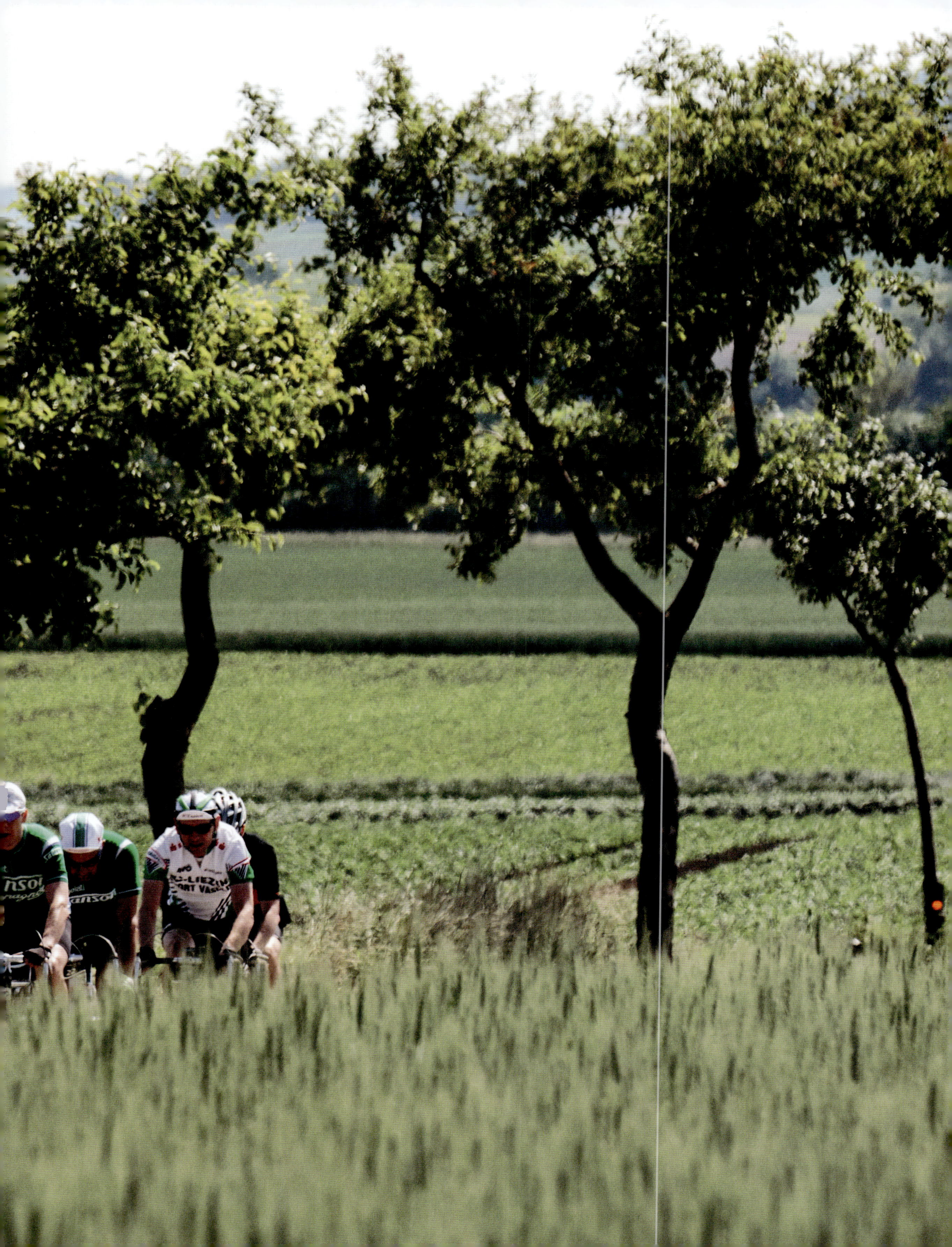

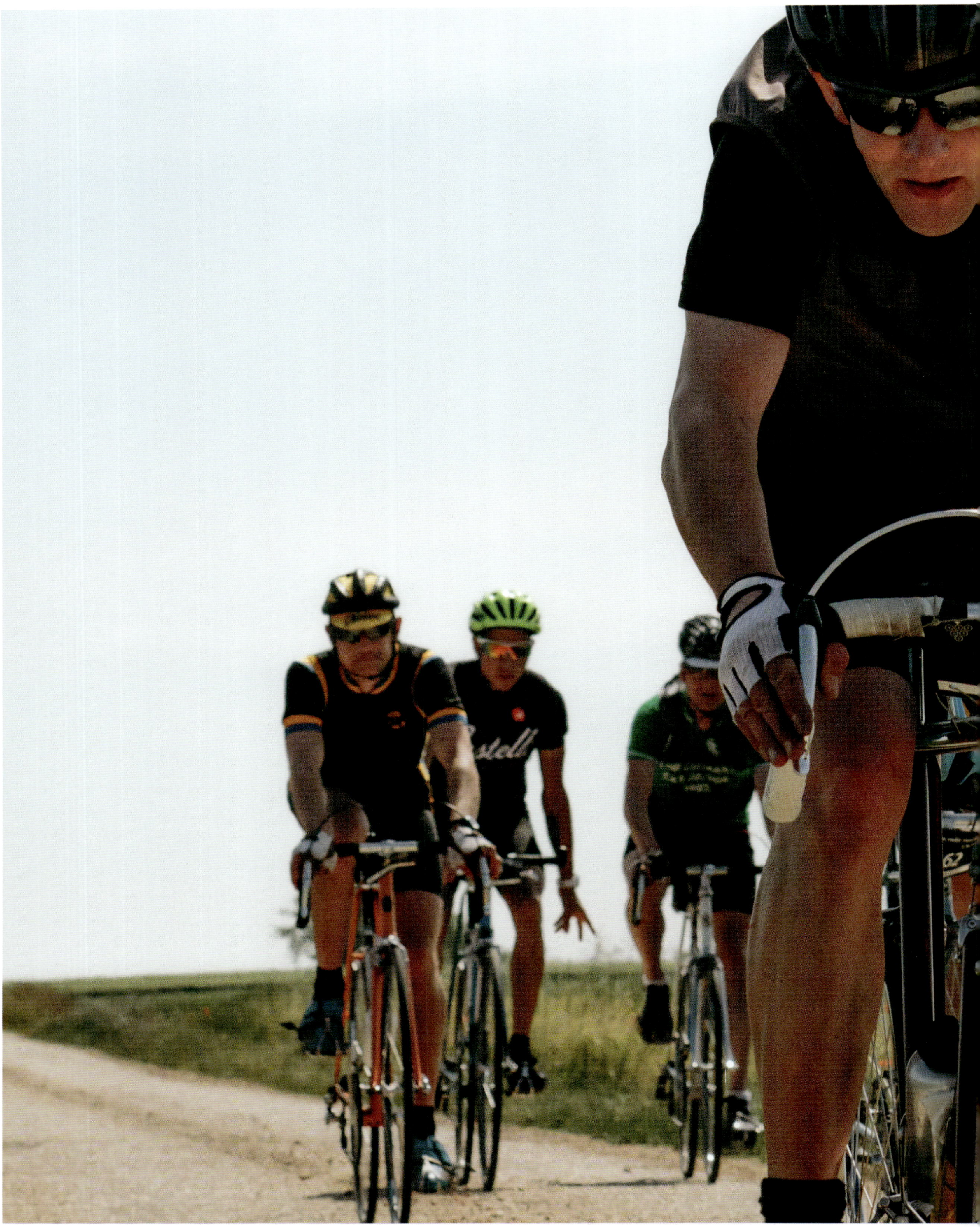

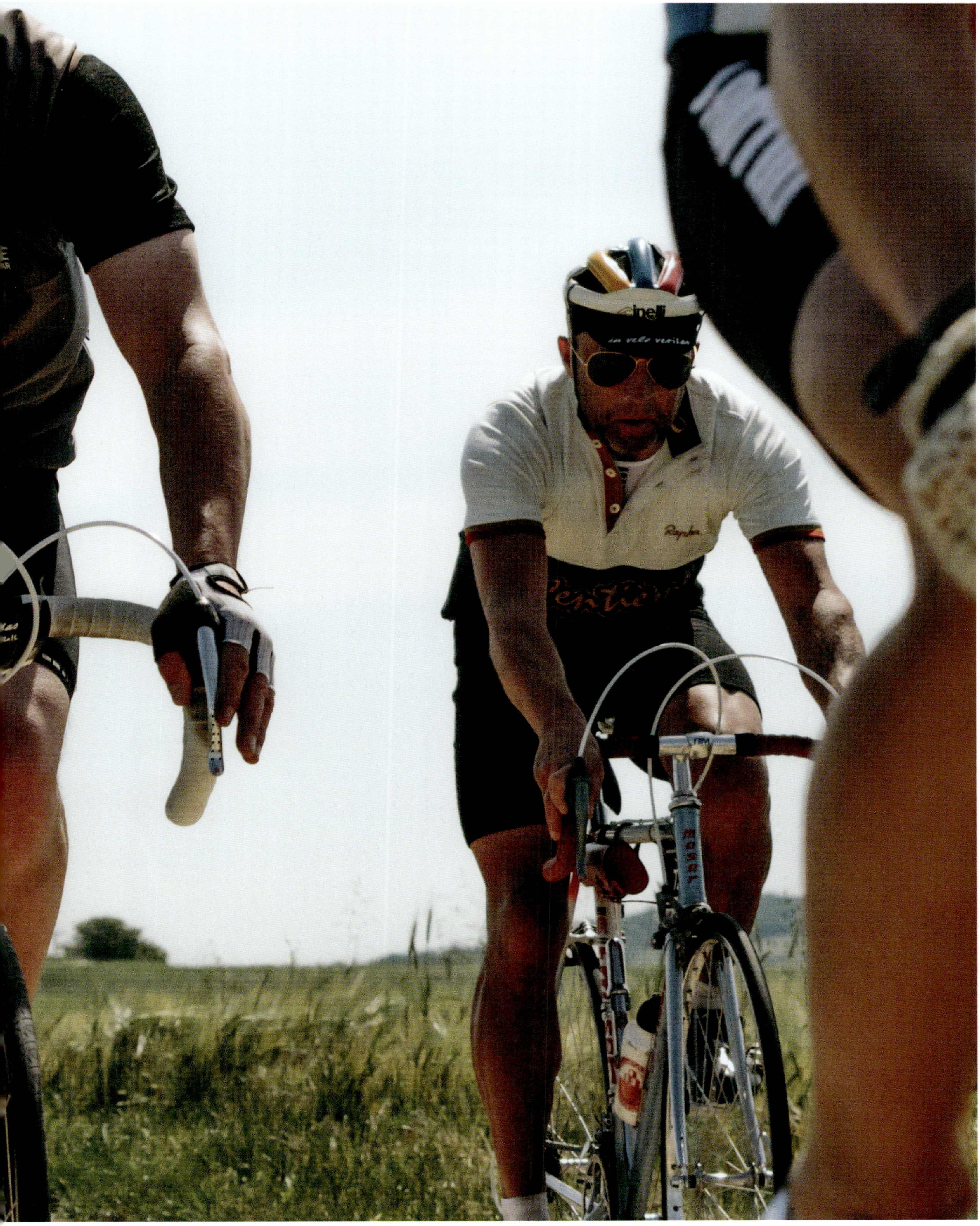
Rapha
moser

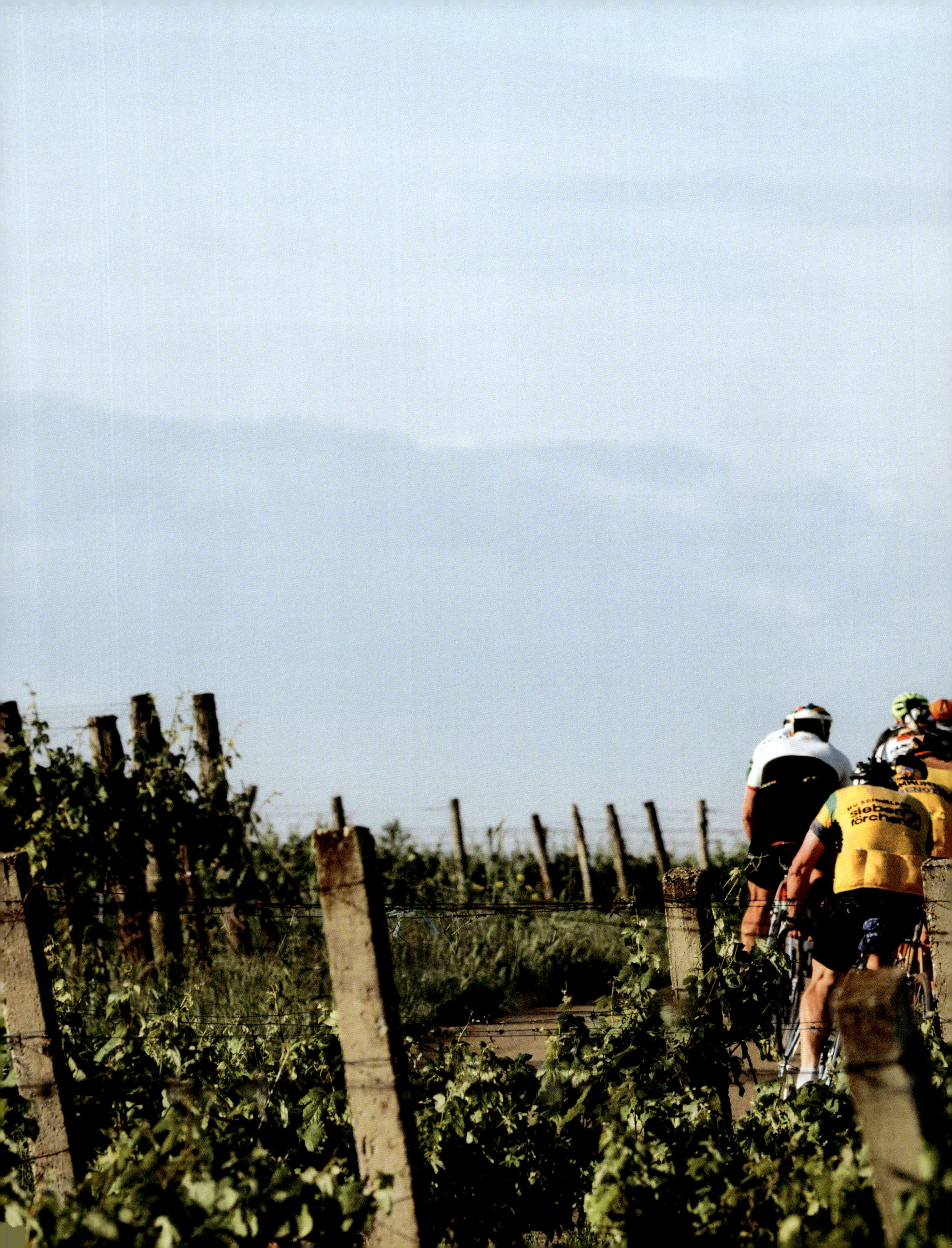

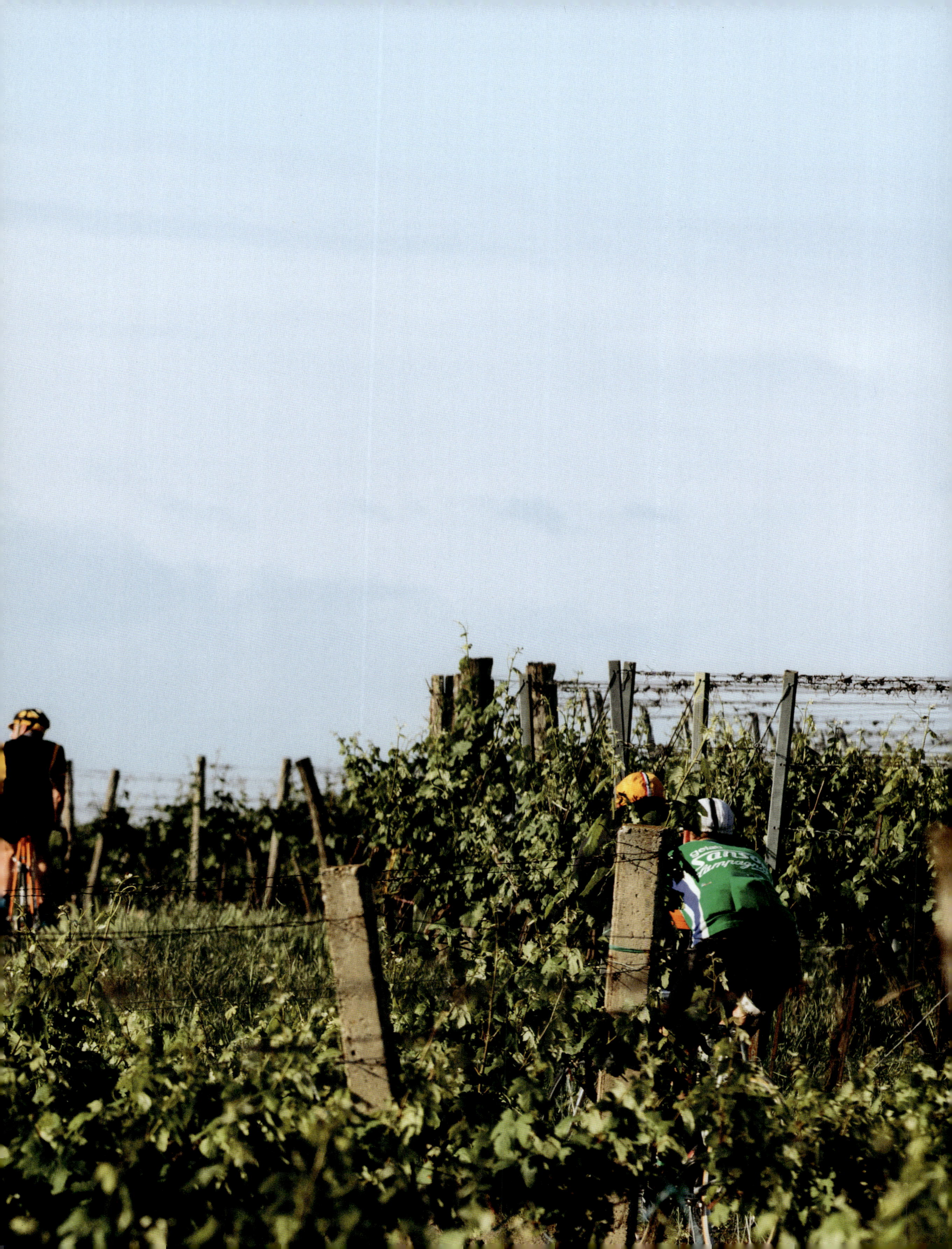

IMPRESSUM

Bibliografische Information der Deutschen Nationalbibliothek
Die Deutsche Nationalbibliothek verzeichnet diese Publikation in der Deutschen Nationalbibliografie; detaillierte bibliografische Daten sind im Internet über http://dnb.dnb.de abrufbar.

1. Auflage
ISBN 978-3-667-13209-3

Fotos und Text: Bengt Stiller
Portrait auf Seite 15 von Gavin Kaps

Textredaktion: Carina Wolfram, Stephanie Jaeschke, Erwin Puschkarsky

Layout und Gestaltung: Ute Schindler

Herstellung: Yvonne Bernardes

Druck: Print Consult GmbH, München

Printed in Czechia 2026

Delius Klasing Verlag GmbH
Siekerwall 21
D - 33602 Bielefeld
Tel.: 0521/559-0
E-Mail: info@delius-klasing.de
www.delius-klasing.de

Der Architekt, passionierter Schrauber und Extremsportler **Bengt Stiller,** Jahrgang 1978, lebt und arbeitet als Fotograf in Berlin. Seit jeher begeisterter Radfahrer, stehen unzählige Touren durch Europa und Vintage-Rennen in seinem Palmarès. Unter anderem begleitete er auf seinem Gravelbike Jonas Deichmann als Fotograf 2.000 Kilometer durch Schweden.

Danke an alle, die mich in den schwierigen Phasen meines Lebens unterstützt haben.

Besonderer Dank gilt meiner Mutter, die sich vehement weigert, nicht an mich und meine Fähigkeiten zu glauben. Und allen, die mit Material und Beratung so tatkräftig mithelfen. Ein Leben ohne Freunde ist möglich, aber sinnlos!

FSC www.fsc.org
MIX
Papier | Fördert gute Waldnutzung
FSC® C106855

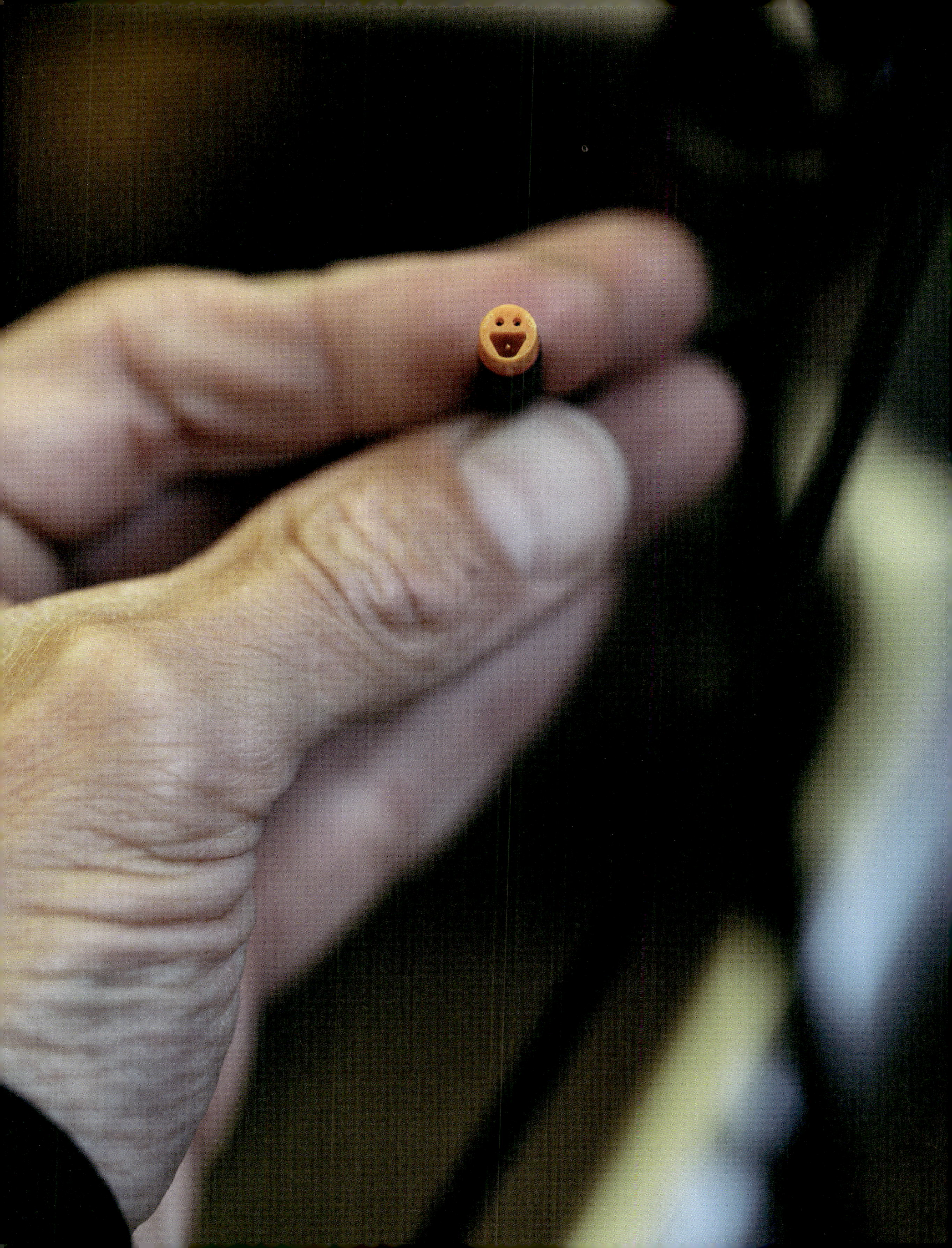

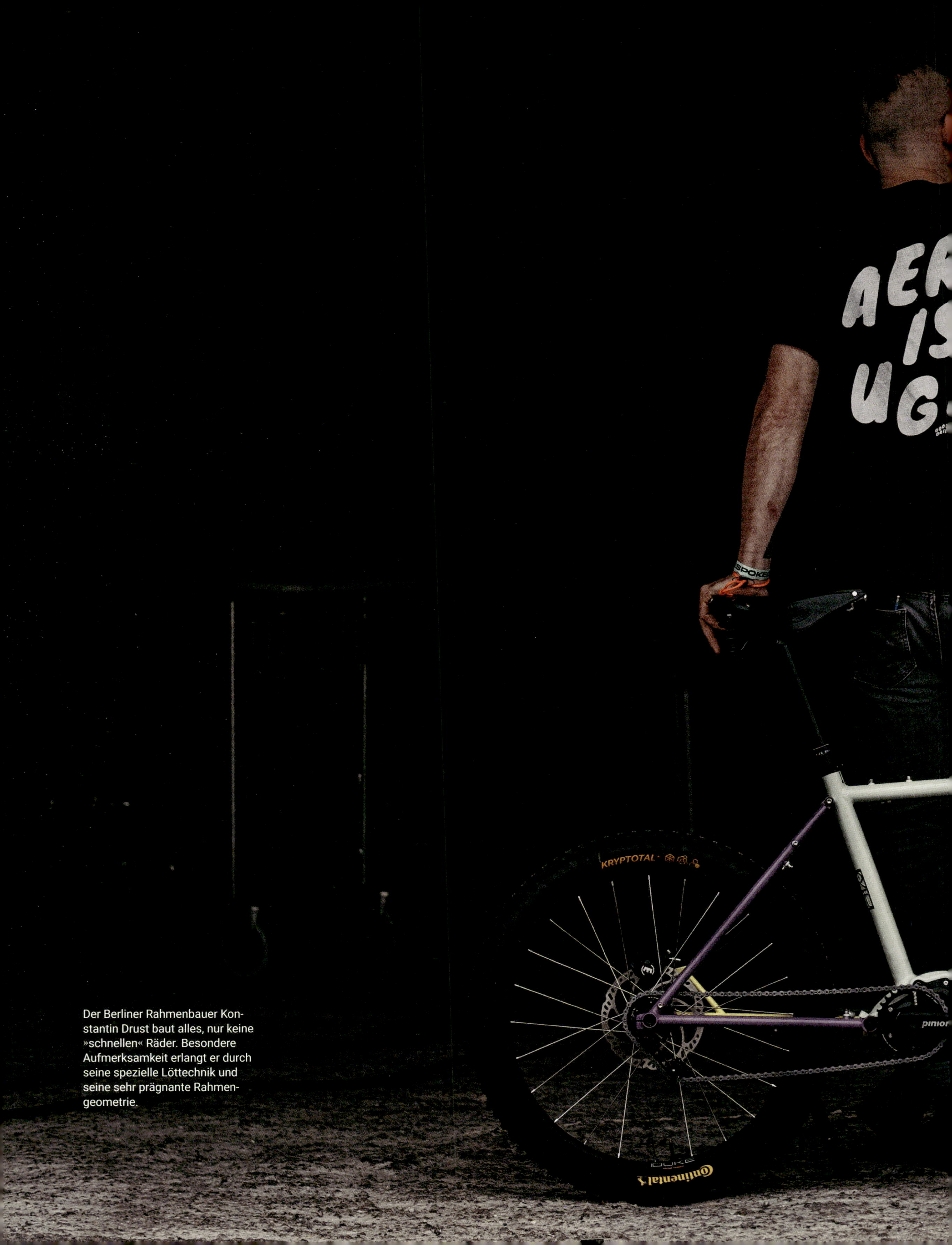

Der Berliner Rahmenbauer Konstantin Drust baut alles, nur keine »schnellen« Räder. Besondere Aufmerksamkeit erlangt er durch seine spezielle Löttechnik und seine sehr prägnante Rahmengeometrie.

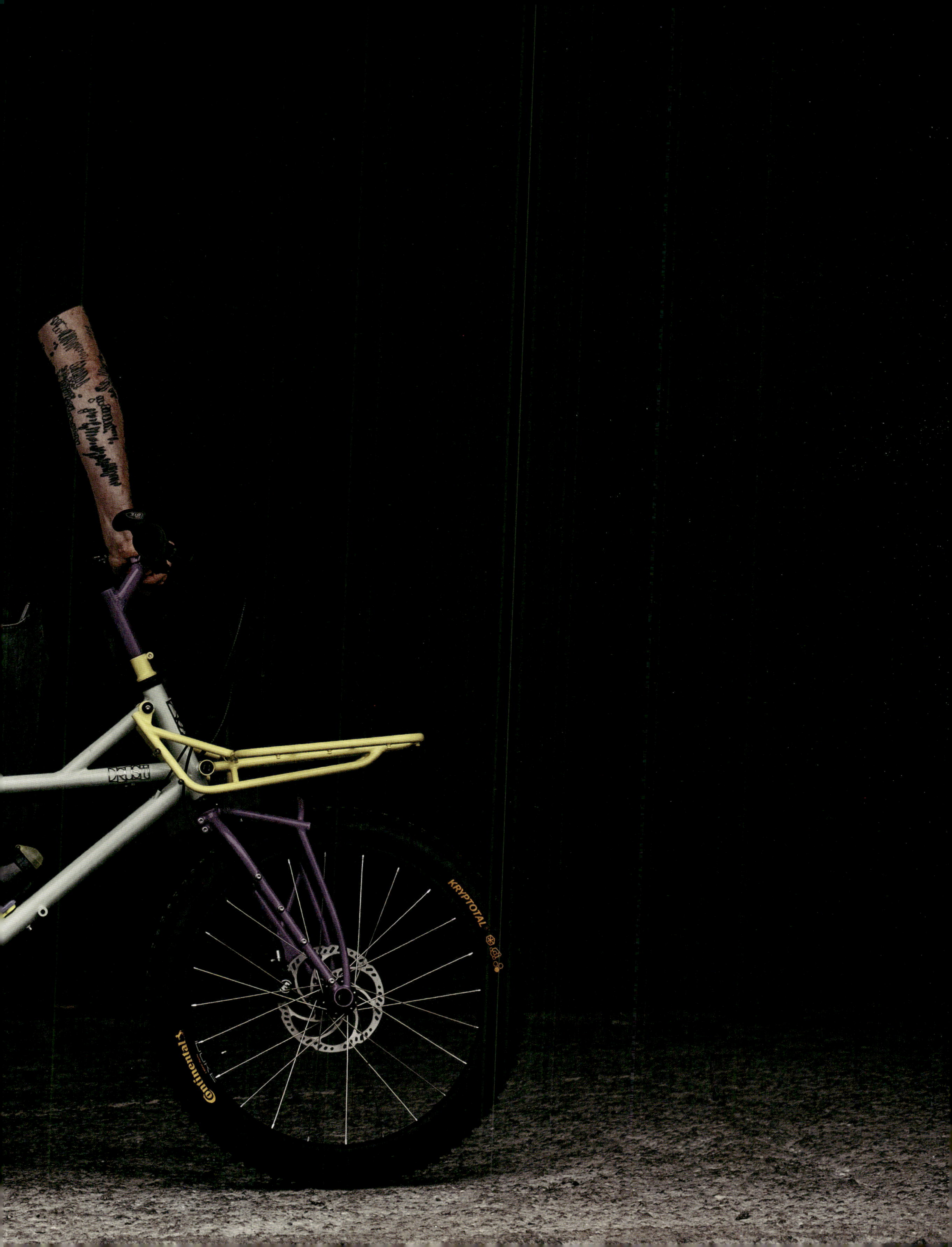
KRYPTOTAL
Continental

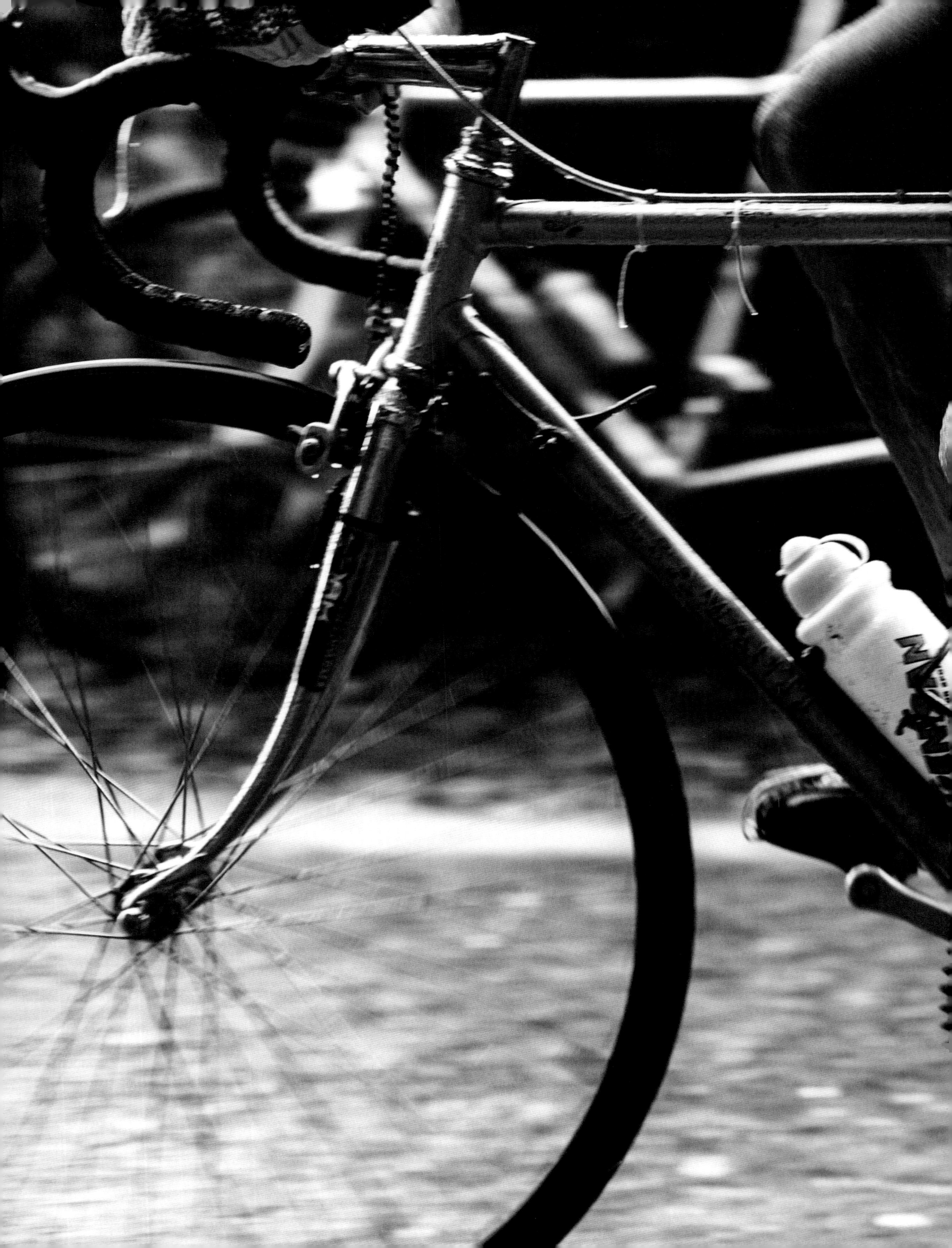

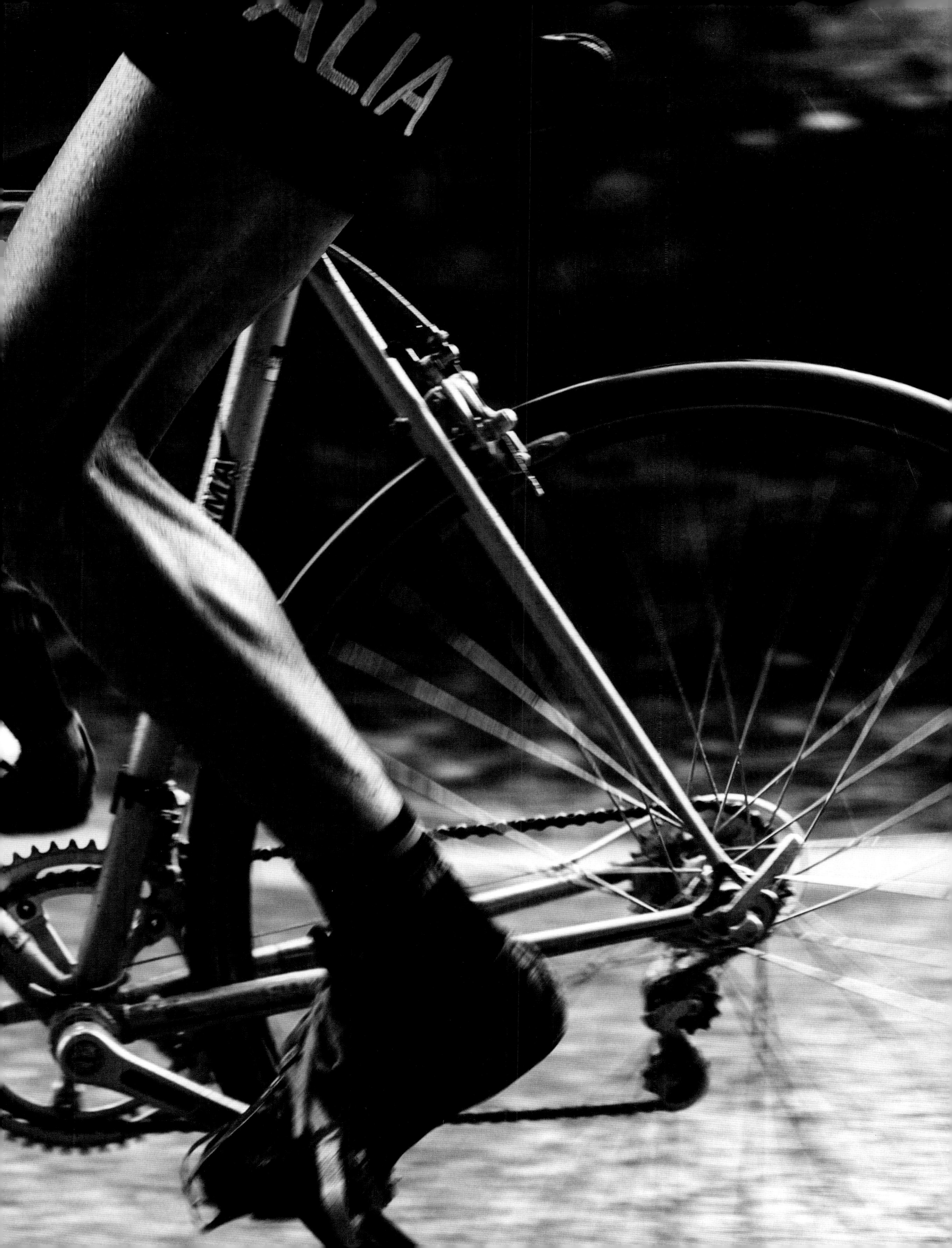
ALIA

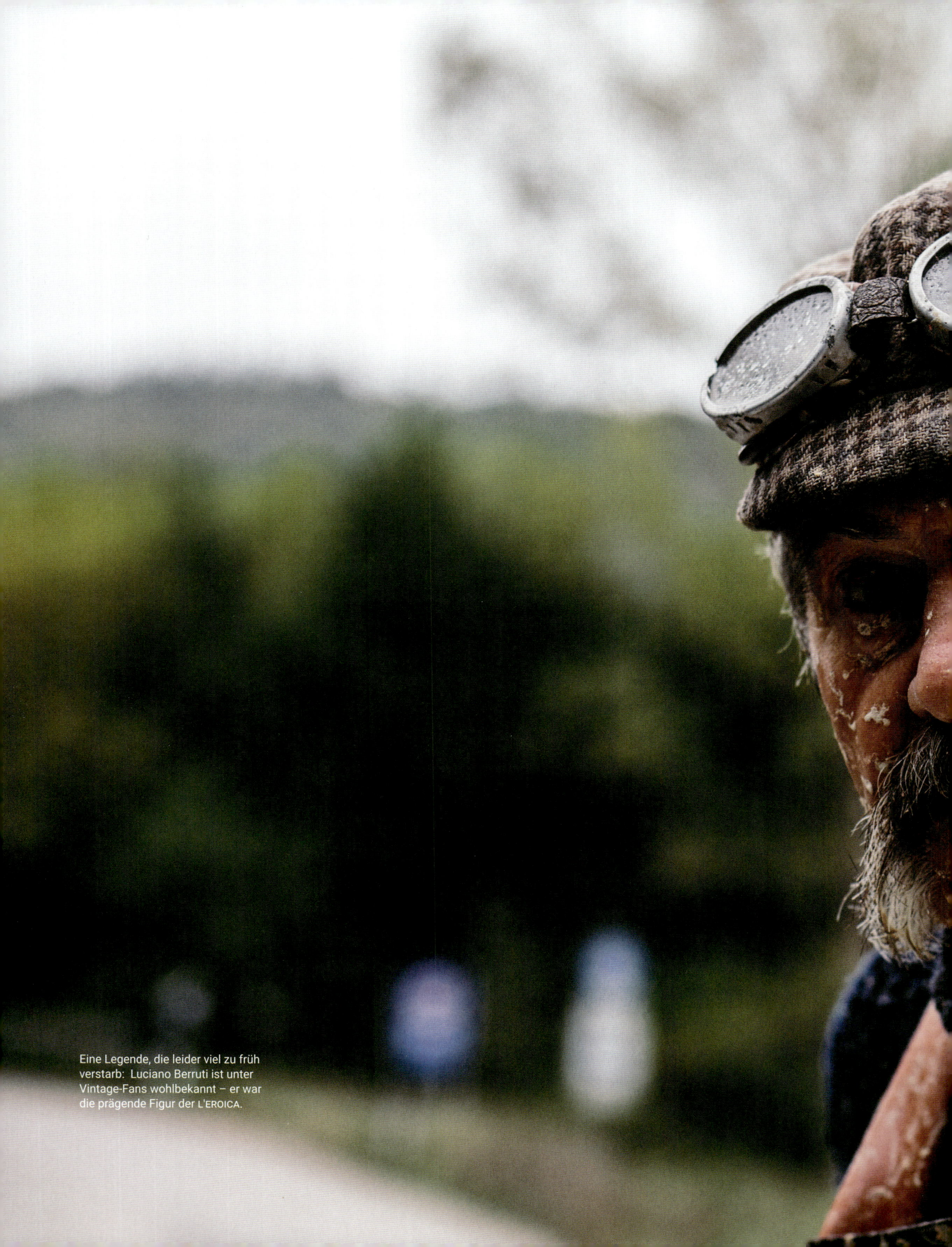

Eine Legende, die leider viel zu früh verstarb: Luciano Berruti ist unter Vintage-Fans wohlbekannt – er war die prägende Figur der L'EROICA.

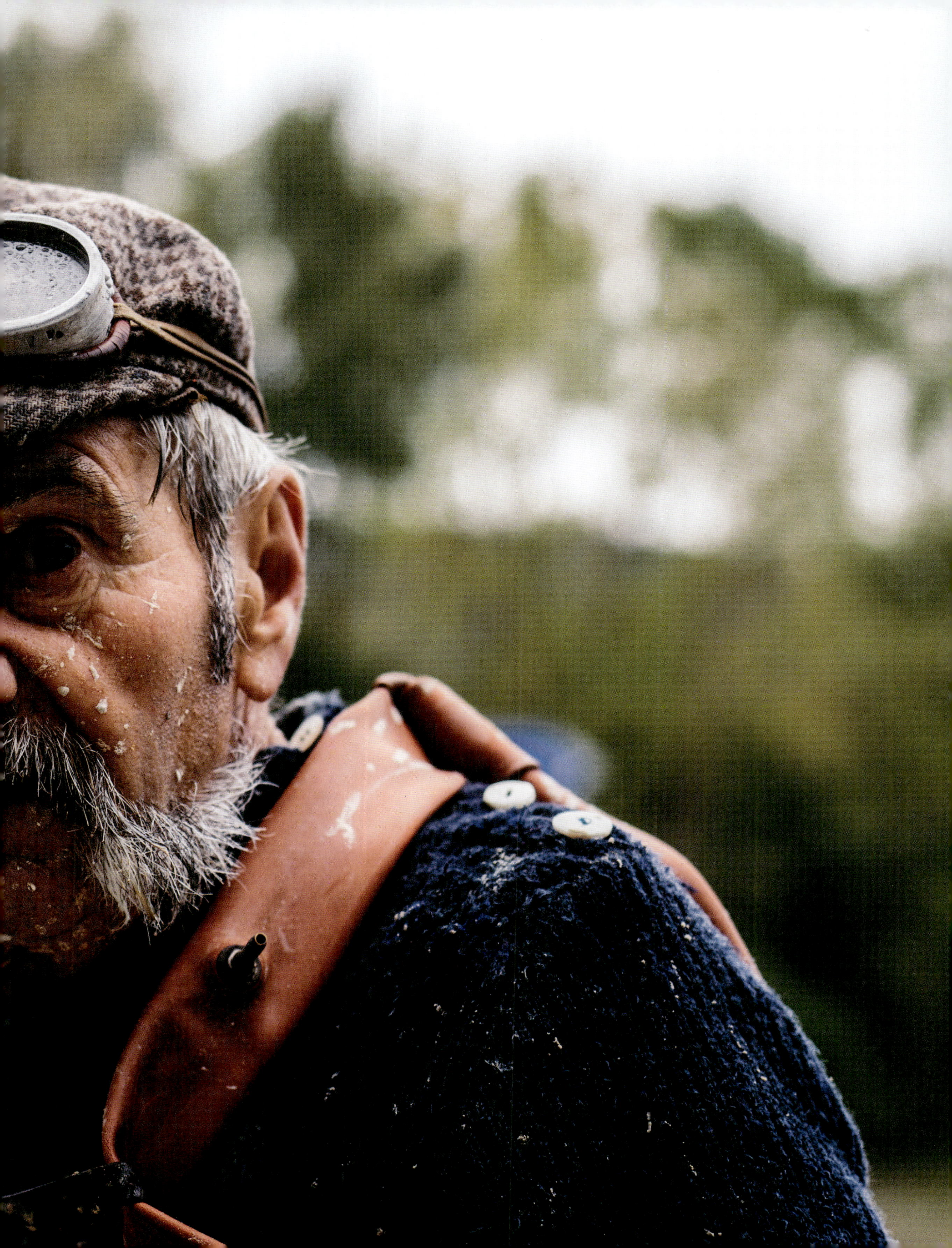